Biomag 96

Volume I

Springer Science+Business Media, LLC

Cheryl J. Aine Yoshio Okada
Gerhard Stroink Stephen J. Swithenby
Charles C. Wood

Editors

Biomag 96

Proceedings of the Tenth International Conference on Biomagnetism

Volume I

Springer

Cheryl J. Aine
Radiology Service Albuquerque
VA Medical Center
Building 49, 114M
1501 San Pedro SE
Albuquerque, NM 87108
USA

Yoshio Okada
Department of New Mexico School
of Medicine
Albuquerque, NM 87131
USA

Gerhard Stroink
Department of Physics
Dalhousie University
Halifax, Nova Scotia, Canada
B3H 3J5

Stephen J. Swithenby
Department of Physics and Astronomy
The Open University
Milton Keynes MK7 6AA, UK

Charles C. Wood
Biophysics Group (P-21), MS D454
Los Alamos National Laboratory
P.O. Box 1663
Los Alamos, NM 87545
USA

Library of Congress Cataloging-in-Publication Data
International conference on Biomagnetism (10th : 1999 : Sante Fe, N.M.)
Biomag 96 : proceedings of the Tenth International Conference on Biomagnetism / editors, Cheryl J. Aine . . . [et al.].
p. cm.
Includes bibliographical references and index.
ISBN 978-1-4612-7066-9 (hardcover : alk. paper)
1. Biomagnetism Congresses. I. Aine, Cheryl J. II. Title.
QP345.I56 1999
571.4'7—dc21 99-42829

Printed on acid-free paper.

Production managed by Jenny Wolkowicki; manufacturing supervised by Joseph Quatela.
Camera-ready copy provided by the editors.

9 8 7 6 5 4 3 2 1 SPIN 10742866

ISBN 978-1-4612-7066-9 ISBN 978-1-4612-1260-7 (eBook)
DOI 10.1007/978-1-4612-1260-7

Contents
Volume I

I Biomagnetic Instrumentation

1. **Biomagnetic Instrumentation** 3
Koch, H.

2. **Reference Noise Cancellation: Optimizations and Caveats** 7
Barnes, G.R., and Ioannides, A.A.

3. **Positioning of MCG Measurement Equipment** 11
Borchardt, M., Krause, E., and Lindner, G.

4. **A Low-Noise, Integrated DC SQUID Magnetometer for Applications in Biomagnetism** 15
Cantor, R., Vinetskiy, V. and Matlashov, A.

5. **Magnetocardiograms Recorded by First-Order Gradiometers using HTS rf SQUIDs** 20
Copetti, C.A., Soltner, H., Zhang, Y., Schuett, H., Rongen, H., Brandenburg, G., Clemens, U., Zimmermann, E., Borgmann, J., Tavrin, Y., Otto, R., Bousack, H., and Halling, H.

6. **A Realtime, High-Bandwidth, Programmable Digital Signal-Processing System for Monitoring Acquisition, Filtering, and Archival Storage of Biomagnetic Signals** 24
Corby, N.R., Hogle, R.A., Miller, P.D., and Pasquarelli, A.

7. **Magnetic Measurement Techniques for Locating Foreign Bodies in Humans** 28
Costa Monteiro, E., Kelber, C., Andrade Lima, E., Costa Ribeiro, P., Bruno, A.C., Hall Barbosa, C., Parente Ribeiro, E., and Boechat, P.

8. **Optimisation of Co-Registration Procedures for Biomagnetic Measurements using a Thermoplastic Mask** 32
Dammers, J., Barnes, G.R., Weise, F., Gross, J., Clewett, B.S., Ioannides, A.A., Müller-Gärtner, H.-W.

9. **Design of a Magnetically Shielded Room for Development of HTS SQUID Systems for Magnetocardiography** 35
Davidson, C.R., Griffin, R.H., Best, K.J., Maltin, L., Wolf, W., Zhang, Y., Bousack, H. and Soltner, H.

10. **A Digital-Signal-Processor (DSP) Flux-Locked-Loop (FLL) for LTC SQUIDs with Automatic Reset and Feedback Cancellation** 39
Flynn, E.R., Bracht, R., Kraus, R.H., Jr., Maas, P.M., Ruminer, P. and Stettler, M.

11. **A 32-Channel MCG System cooled by a GM/JT Cryocooler** 43
Fujimoto, S., Sata, K., Fukui, N., Haraguchi, E., Kido, T., Nishiguchi, K. and Kang, Y.M.

12. **A Cryocooled SQUID System with Minimal Clearance between the Pick-up Coil and the Specimen** 47
Fukui, N., Sata, K., Fujimoto, S., Kang, Y.M.

13. **A Comparison of Vector and Radial Magnetometer Arrays for Whole-Head Magnetoencephalography** 51
Hughett, P. and Miyauchi, S.

14. **32 Channel High Tc SQUID System for Biomagnetism** 55
Itozaki, H., Kugai, H., Nagaishi, T., Toyoda, H., Hirano, T., Haruta, Y., Tanaka, S. and Kado, H.

15. **Source Detectability for Multichannel Biomagnetic Sensors** 59
Johnson, R.T., Black, W.C. and Buchanan, D.S.

16. **A Digital Flux-Locked Loop for High Temperature SQUID Magnetometer and Gradiometer Systems with Field Cancellation** 63
Kraus, R.H., Jr., Bracht, R., Flynn, E.R., Jia, Q., Maas, P.M., Reagor, D. and Stettler, M.

17. **Characterisation of a Planar Second-Order Gradiometer in a Lightly Shielded Environment** 67
Lang, G., Hutson, D., Weston, R.G., Klein, U., Shahani, U., Weir, A.I., Maas, P.M., Pegrum C.M. and Donaldson, G.B.

18. **Magnetic Field Error of a Current Dipole in a Superconducting Shield** 71
Matsuba, H.

19. **High Tc Superconducting Shielded Biomagnetometer System** 75
Matsuba, H., Shintomi, K., Yahara, A., Imai, K., Irisawa, D. and Masuda, M.

20. Current Dipole Localization Errors as a Function of the System Noise and the Number of Sensors 79
Matsuba, H., Vrba, J. and Cheung, T.

21. Auditory Brainstem evoked Fields measured in a High Tc Superconducting Shield 83
Nogai, T., Yahara, A., Matsuba, H. and Kuriki, S.

22. High Performance Magnetically Shielded Room for Clinical Measurements 87
Paavola, M.A.J., Ilmoniemi, R.J., Sohlström, L., Meinander, T., Penttinen, A. and Katila, T.E.

23. An AC Susceptometer with a large Magnetised Volume for Biological Samples 91
Pasquarelli, A., Del Gratta, C., Della Penna, S., Di Luzio, S., Pizzella, V. and Romani, G.L.

24. A Six-Channel SQUID Biomagnetometer optimised for Fetal magnetocardiography 95
Rassi, D., Zhuravlev, Y.E., Josephs, O. and Swithenby, S.J.

25. High Temperature Superconducting Quantum Interference Devices in a Superconductor-Normal-Superconductor Geometry 99
Reagor, D., Jia, Q.X., Mombourquette, C., Foltyn, S., Houlton, R. and Wu, X.D.

26. A Digital SQUID Controller for Unshielded Biomagnetic Measurements 103
Robinson, S.E.

27. Investigation of the thin Film SQUIDs using Double Layer Model of the Secondary Sources 107
Romanovych, S.S.

28. DC-MEG Studies: Development of Techniques for resolving DC Fields 110
Saligram, U., Moran, J.E. and Tepley, N.

29. Development of a 61-Channel MEG System cooled by a GM/JT Cryocooler 114
Sata, K., Fujimoto, S., Fukui, N., Haraguchi, E., Kido, T., Nishiguchi, N. and Kang, Y.M.

30. Low-Noise, Single Layer $YBa_2Cu_3O_{7-\delta}$ DC SQUID Magnetometers for Shielded and Unshielded Operation 118
Schmidt, J.M., Lee, L.P., Matlashov, A., Teepe, M., Vinetskiy, V., Cantor, R. and DiIorio, M.S.

31. The Use of Software Magnetometers for Ambient Noise Cancellation in Unshielded Environment 122
Shabalin, A.V., Golyshev, N.V., Greenberg, Ya.S., Motorin, S.V., and Rogachevskij, B.M.

32. The use of a Current Feedback Loop for active Compensation of Magnetic Noise in Superconducting Magnetocardiograph 126
Shabalin, A.V., Golyshev, N.V., Motorin, S.V., and Rogachevskij, B.M.

33. A Skull-Based Multiple Dipole Phantom for EEG and MEG Studies 130
Spencer, M.E., Leahy, R.M. and Mosher, J.C.

34. High-Tc SQUID Gradiometers for Magnetocardiography in an Unshielded Environment 134
Tralshawala, N., Claycomb, J.R., Ying, Q., Miller, J.H., Jr., Nesteruk, K. and Jackson, D.R.

35. 143 Channel Whole-Cortex MEG System 138
Vrba, J., Angus, V., Betts, K., Burbank, M.B., Cheung, T., Fife, A.A., Haid, G., Kubik, P.R., Lee, S., Ludwig, W., McCubbin, J., McKay, J., McKenzie, D., Robinson, S.E., Smith, M., Spear, P., Taylor, B., Tillotson, M., Cheyne, D. and Weinberg, H.

36. SEFs from the Human Spinal Cord using a Novel Second Order Planar Gradiometer 142
Weir, A.I., Shahani, U., Hutson, D., Lang, G., Maas, P.M., Pegrum, C.M. and Donaldson, G.B.

37. High-Tc Superconducting rf Receiver Coils for High Resolution Magnetic Resonance Imaging of the Spine 146
Wosik, J., Nesteruk, K., L.M. Xie, Zhang, X.P., Jiao, C., Gierlowski, P. and Miller, J.H., Jr.

38. 3-D Magnetic Field Analysis of a Magnetically Shielded Room for Biomagnetic Measurements 150
Yamazaki, K., Uchikawa, Y. and Fujiwara, K.

39. A 129 channel Vector Neuromagnetic Imaging System 154
Yoshida, Y., Arakawa, A., Kondo, Y., Kajihara, S., Tomita, S., Tomita, T., Takanashi, Y. and Matsuda, N.

40. Magnetocardiography using SQUIDs cooled by Liquid Nitrogen 158
Zhang, Y., Schütt, H., Bousack, H., Wolf, W., Schirdewan, A., Burghoff, M. and Trahms, L.

II Models for the Biomagnetic Forward and Inverse Problems

41. Comparison of Different Approaches to the Biomagnetic inverse Problem—Workshop Report 163
Lutkenhöner, B., Greenblatt, R., Hämäläinen, M., Mosher, J., Scherg, M., Tesche, C., and Valdes-Sosa, P.

42. Distributed Source Reconstruction using a Non-Linear Spatiotemporal Regularization Method: An Alternative to LORETA 177
Baillet, S., Garnero, L. and Renault, B.

43. Localization of Extended Sources using a modified Linear Estimation Approach 181
Brauer, H., Ziolkowski, M., Tenner, U., Kosch, O., Wiechmann, H. and Arlt, A.

44. Inverse Magnetography for Reentry Currents 185
Brio, M., Marchesin, D., Costa Monteiro, E., Bruno, A.C. and Costa Ribeiro, P.

45. Frequency Domain Distributed Inverse Solutions 189
Casanova, R., Valdes-Sosa, P., Garcia, F.M., Aubert, E., Riera, J.J., Korin, W., and Lins, O.

46. Small Volume Resistivity Changes Observed using MEG 193
Clifton, Y.C., Moran, J.E., and Tepley, N.

47. A Weighting Matrix to remove Depth Bias in the Linear Biomagnetic Inverse Problem with Application to Cardiology 197
Crowley, T.A., Haupt C.D. and Kynor, D.B.

48. Imaging Properties of "Myocardial Current Imaging" 201
Dössel, O., Katscher, U., Köhler, Th., Fuchs, M., Wischmann, H.-A., Wagner, M., Drenckhahn, R. and Hoppe, W.

49. Practicability of Neuromagnetic Source Analysis using the Finite-Element Method 205
Drüen, K., Steinsträter, O. and Lütkenhöner, B.

50. A Numerically Stable Approximation for the Magnetic Field of the Conducting Spheroid close to the Symmetry Axis 209
Fieseler, T., Ioannides, A.A., Liu, M.J. and Nowak, H.

51. Source Reconstructions by Spatial Deviation Scans 213
Fuchs, M., Wischmann, H.-A., Wagner, M., Drenckhahn, R. and Köhler, Th.

52. Generalized Inverse Solution for Bimodal Electro-Magnetic Source Images 217
Gencer, N.G. and Williamson, S.J.

53. Can Compact Currents be uniquely determined? 221
Gorodnitsky, I.F.

54. Estimation of Distributed Sources with adjustable Contrast 225
Grunwald, A. and Lütkenhöner, B.

55. Aspects of Non-Uniqueness: How one Source affects the Reconstruction of Another 229
Hasson, R. and Swithenby, S.J.

56. The Theoretical Basis of Iterative Distributed Solutions to the Biomagnetic Inverse Problem 233
Hasson, R. and Swithenby, S.J.

57. Anatomically constrained vs unconstrained Dipole Fit in the Combination of MEG and MRI 237
Haueisen, J., Hajek, M., Huonker, R., Nowak, H. and Sauer, H.

58. One Effect of a Maximum possible Current Strength 241
Huang, M. and Heller, L.

59. Multi-Source Localization in MEG using Simulated Annealing: Model Order Determination and Parameter Accuracy 245
Huang, M., Supek, S., Aine, C.

60. A Three-Dimensional Rule assisted Interpolation Algorithm for Iso-Field Mapping of Neuromagnetic Signals 249
Hultin, L.

61. The Influence of Glial Buffering Currents on Sustained Neuromagnetic Fields 253
Hultin, L.

62. Magnetic Field Tomography: Theoretical basis, Performance Measures and Limitations 257
Ioannides, A.A., Bolton, J.P.R., Liu, M.J., Bamidis, P.D., Liu, L.C., Dammers, J. and Müller-Gärtner, H.W.

63. Algorithm for Magnetocardiography Analysis based on Multi Channel SQUID and Thoracic MR Images 261
Ishii, H., Niki, N., Nakaya, Y., Nishitani, H. and Kang, Y.M.

64. Estimation of Distributed Sources of Auditory evoked Magnetic fields in the Human Brain 265
Iwaki, S. and Ueno, S.

65. A New Approach to the Reconstructing Source Distribution 270
Kajihara, S., Tomita, T., Tomita, S., Kondo, Y., Arakawa, A., Yoshida, Y. and Matsuda, M.

66. A Comparative Study of Minimum Norm Inverse Methods for MEG Imaging 274
Leahy, R.M., Mosher, J.C. and Phillips, J.W.

67. Development, Implementation and Applications of Single Epoch Analysis of MEG Signals 278
Liu, L.C., Ioannides, A.A., Heer, H., Wagener, M., Delonge, F., Halling, H. and Müller-Gärtner, H.W.

68. Dipole Localization Accuracy as a Function of the Area Covered by the Magnetometer System 282
Lütkenhöner, B.

69. Towards a Standardized Representation of Neuromagnetic Data 286
Lütkenhöner, B., Grunwald, A. and Steinsträter, O.

70. Multiple Current-Dipole Distribution reconstructed by modified Selective Minimum-Norm Method 290
Matsuura, K. and Okabe, Y.

71. An Estimation Algorithm of Neuromagnetic Source using MRI Information 294
Mino, K., Niki, N., Nakasato, N. and Yoshimoto, T.

72. Realistic Modeling, Analysis and Animated 3D Visualization for MEG Studies 298
Morabito, M., Takeda, T., Kobayakawa, T. and Endo, H.

73. Subspace Angles: A Metric for Comparisons in EEG and MEG 302
Mosher, J.C.

74. Comparison of the Constant and Linear Boundary Element Method for EEG and MEG forward Modeling 306
Mosher, J.C., Chang, C.H. and Leahy, R.M.

75. MEG-based Imaging of Focal Neuronal Current Sources 310
Phillips, J.W., Leahy, R.M. and Mosher, J.C.

76. A Four-Sensor Skeleton Configuration for Single Dipole Localization 314
Prakash, N.M., Spelman, F.A. and Read, P.

77. The new Methods of Localization and Reconstruction of Biomagnetic Signal Sources 318
Primin, M.A. and Nedayvoda, I.V.

78. **Effects of Tissue Conductivity Variations on Cardiac Magnetic Fields** 322
Ramon, C., Czapski, P., Huntsman, L.L., Bardy, G.H. and Kim, Y.

79. **Discrete Spline Electric-Magnetic Tomography (DSPET) based on realistic Neuroanatomy** 326
Riera, J.J., Aubert, E., Valdes-Sosa, P., Casanova, R., and Lins, O.

80. **Strategy for objective Comparisons of Inverse Solutions** 330
Schlitt, H.A., Zhao, S.-R., Halling, H. and Müller-Gärtner, H.-W.

81. **A method for locating Regions containing Neural Activation at a given Confidence Level from MEG Data** 334
Schmidt, D.M. and George, J.S.

82. **Generalized Wiener Estimation of Three-Dimensional Current Distribution from Biomagnetic Measurements** 338
Sekihara, K. and Scholz, B.

83. **Localizing dipoles: The Correlation Coefficient and the Goodness of Fit in Source Localization of Filtered Noise** 342
Siedenberg, R., Goodin, D.S., Aminoff, M.J., Rowley, H.A. and Roberts, T.P.L.

84. **Monte-Carlo Analysis and Confidence Region Ellipsoids for Equivalent Current Dipole Solutions to EEG/MEG Data** 346
Singh, K.D. and Harding, G.F.A.

85. **Localization of Current Sources with Constraints of Solution Area** 350
Sugaya, H., Osozawa, S., Ishiyama, A. and Kasai, N.

86. **Source Localization by Genetic Algorithm** 354
Takeuchi, R., Ikeda, H., Ishiyama, A. and Kasai, N.

87. **Optimization of Conductivity and Permeability Parameters for Brain, CSF, Skull and Scalp using Implanted Sources in the Human Cranium** 358
Taylor, S.A., Rogers, R., McNay, D., Levesque, M., Akhtari, M. and Sutherling, W.

88. **Modeling of Spreading Cortical Depression using a realistic Head Model** 361
Tepley, N., Roth, B.J. and Wijesinghe, R.S.

89. **A Simulation Study on the Nature of Electrical Sources of the MCG** 365
Tilg, B., Wach, P., Lafer, G., Fleischmann, P. and Rucker, W.

90. **Global Optimization in the Localization of Brain Activity** 369
Uutela, K., Hämäläinen, M. and Salmelin, R.

91. **Variable Resolution Electric-Magnetic Tomography** 373
Valdes-Sosa, P., Marti, F., Garcia, F., and Casanova, R.

92. **Spatio Temporal Distributed Inverse Solutions** 377
Valdes-Sosa, P., Riera, J.J., and Casanova, R.

93. **Quantification of Deviations between Spherical and realistically Shaped Forward Models** 381
v. Rango, J., Schlitt, H.A., Halling, H., Terstegge, A. and Müller-Gärtner, H.-W.

94. **The Influence of Holes in a Layer with Low Conductivity on the Potential and Magnetic Field** 385
van den Broek, S.P., Reinders, K.F.J. and Peters, M.J.

95. **Estimation of Cardiac Sources on the Epicardium and Endocardium from Simulated and Measured MCG Data** 389
Wach, P., Tilg, B., Lafer, G., Kynor, D. and Rucker, W.

96. **Current Density Reconstructions using the L1 Norm** 393
Wagner, M., Wischmann, H.-A., Fuchs, M., Köhler, T. and Drenckhahn, R.

97. **Modelling Closed Surface Geometry for Skull and Torso using Snapping** 397
Yamrom, B.

98. **Locally Refined Head Geometries used to Evaluate Spherical Model Accuracy** 401
Yvert, B., Bertrand, O., Thévenet, M. and Pernier, J.

99. **Mesh Refinement of Boundary Element Head Models in MEG Source Reconstructions** 405
Zanow, F. and Peters, M.J.

III Magnetocardiography (MCG)

100. **New Trends in Clinical Magnetocardiography.** 410
Mäkijärvi, M., Brockmeier, K., Leder, U., Lesh, M., Moshage, W., Nakaya, Y., Nenonen, J., Nowak, H., Oeff, M., Stroink, G., Fenici, R.

101. **Revisiting the Fetal Magnetocardiogram and Fetal Magnetoencephalogram** 418
Wakai, R.T., Leuthold, A.C. and Martin, C.B.

102. **The Fetal Magnetocardiogram during Different Stages of Pregnancy** 422
Achenbach, S., Menéndez, T., Moshage, W., Flüg, M., Beinder, E., Bittl, A., Bachmann, K., and Lang, N.

103. **Clinical Experience with the Combination of Multichannel Magnetocardiography and Magnetic Resonance Imaging for the Localization of Idiopathic and Secondary Ventricular Cardiac Arrhythmias** 425
Achenbach, S., Moshage, W., and Bachmann, K.

104. **Localization of the Origin of Ventricular Arrhythmias: Comparison of ECG Morphology and Localization based on ECG Algorithms to Multichannel Magnetocardiography** 429
Achenbach, S., Moshage, W., and Bachmann, K.

105. **Comparison of the QRS duration and RMS40 in Signal-Averaged Recordings of the Multichannel MCG to the Signal-Averaged ECG** 433
Achenbach, S., Moshage, W., Flüg, M., Diem, B., Schibgilla, V., and Bachmann, K.

106. **Investigation of Fetal Cardiac Arrhythmias by Magnetocardiography** 437
Achenbach, S., Moshage, W., Menéndez, T., Beinder, E., Bittl, A., Hofbeck, M., Buheitl, G., Singer, H., and Bachmann, K.

107. **Contraventions of Homogeneity of Ventricular Repolarization according to Data of Magnetocardiography and their Connection with Ventricular Tachycardia** 440
Bobrov, V.A., Gapelyuk, A.V., Chaikovsky, I.A., Stadnyuk, L.A., Sosnitsky, V.N. and Kozlovsky, V.I.

108. **Ventricular Repolarization Disturbances' Diagnostics in Chronic Ischemia Patients Evidence Derived from MCG** 444
Chaikovsky, I.A., Lutay, M.I., Lomakovsky, A.N., Gapelyuk, A.V., Sosnitsky, V.N., Gushcha, B.G., and Minov, Yu.D.

109. **Induced Atrial Flutter in Rabbit Hearts Tissue and Inverse Magnetocardiography** 448
Costa Monteiro, E., Andrade Lima, E., Ornelas, P.H., Cruz, R.S., Picon, A.P., Kühner, G.S., Kelber, C., Parente Ribeiro, E., Bruno, A.C., Costa Ribeiro, P., Marchesin, D., Brio, M., and Leite, T.T.

110. **Cellular Automata Simulation in the Magnetic Study of Rotating Propagation Models Associated to Cardiac Arrhythmias** 452
Costa Monteiro, E., Correia, M.P., Hall Barbosa, C.R. and Costa Ribeiro, P.

111. **Comparison of Measured and Simulated MCGs with a Realistic Heart-Torso Model** 456
Czapski, P., Haueisen, J., Ramon, C., Nowak, H., Huntsman, L.L., Leder, U., Bardy, G.H., and Kim, Y.

112. **High Resolution Finite Element Simulations of Magnetic Fields resulting from Myocardial Infarction** 460
Czapski, P., Ramon, C., Huntsman, L.L., Bardy, G.H., and Kim, Y.

113. **Invasive Evaluation of MCG-Localization Results Importance of Measured Electrophysiological Time Intervals** 464
Gapelyuk, A., Copetti, C.A., Schirdewan, A., Schütt, H., Wiedemann, M., Meyerfeldt, U., Primin, M.A. and Maslenikov, Y.

114. **Changes in Magnetic Field Maps at QRS-Onset after Myocardial Infarction** 467
Hailer, B., Van Leeuwen, P., Donker, D., Rahn, N., Lange, S. and Wehr, M.

115. **The Automatic Selection of Decisive Parameters in the Magnetocardiographic Localization of the Accessory Pathway in Patients Suffering from the WPW Syndrome** 471
Jazbinsek, V., Trontelj, Z., Oeff, M. and Burghoff, M.

116. **Source Localization of Initial QRS Vector in Patients with Idiopathic Cardiomyopathy: Comparison with Normal Subjects and Patients with Anterior Myocardial Infarction** 475
Kishi, F., Nomura, M., Yukinaka, M., Saito, K., Nakaya, Y. and Ito, S.

117. **Linear and Nonlinear Analysis in the Parallel HR-MCG/HR-ECG Experiment in Arrhythmogenic Right Ventricular Cardiomyopathy** 479
Kowalik, Z.J., Mäkijärvi, M. Wichter, T., Fetsch, T., Elbert, T. Breithard, G. and Hoke, M.

118. **His-Purkinje Conduction System Maps: Fact or Fiction?** 483
Lamothe, M.J.R., Stroink, G. and Gardner, M.J.

119. **Spectral Analysis of Body Surface Potential Maps and Magnetic Field Maps of Patients with and without Ventricular Tachycardia** 487
Meeder, R.J.J., Stroink, G., Ritcey, S.P. and Gardner, M.J.

120. **Magnetocardiography—A New Approach for early Detection of Arrhythmogenic Right Ventricular Dysplasia** 491
Meyerfeldt, U., Schirdewan, A., Wiedemann, M., Schütt, H., Dietz, R., Thierfelder, L., Burghoff, M. and Trahms, L.

121. **Magnetocardiographic Functional Imaging Studies in the BioMag Laboratory** 494
Montonen, J., Ahonen, A., Hämäläinen, M., Ilmoniemi, R.J., Laine, P., Nenonen, J., Paavola, M., Simelius, K., Simola, J. and Katila, T.

122. **Spatial Aspects in Magnetocardiographic Time Domain and Spectrotemporal Post-Infarction Arrhythmia Risk Assessment** 498
Montonen, J., Korhonen, P., Paavola, M., Mäkijärvi, M., Toivonen, L. and Katila, T.

123. **Comparison of Different Imaging Modalities as Anatomical Reference Methods for Magnetocardiographic Localization** 502
Moshage, W., Achenbach, S. and Bachmann, K.

124. **Comparison of the Magnetocardiographic Localization Accuracy using Simultaneous and sequential recordings in a single and multiple planes** 506
Moshage, W., Achenbach, S., Flüg, M., Lohse, P. and Bachmann, K.

125. **Localization of Kent Bundles prior to HF Catheter Ablation: A Quantitative Comparison of the Localization Accuracy based on an Inverse Solution applied to Electrical and Magnetic Field Distributions** 509
Moshage, W., Achenbach, S., Killmann, R., Bruder, H. and Bachmann, K.

126. **Analysis of Electromotive Force by Magnetocardiogram— Is the Origin of Initial QRS Force (Septal Vector) of Electrocardiogram Intraventricular Septum?** 513
Nakaya, Y., Yukinaka, Y., Nomura, M., Kishi, F. and Saito, K.

127. **The Algorithm of Spatial-Temporary Analysis of Cardiomagnetic Signals** 517
Nedayvoda, I.V. and Primin, M.A.

128. **Measurement of Tangential Components of the Magnetic Field in the PR Segment in Man** 521
Noda, Y., Mitsui, T., Terada Y., Tsukada, K., Adachi, A., Yoshida, T. and Haruta, Y.

129. **Observation of a Difference Signal in MCG Measurements in Patients with Ventricular Extrasystoles** 525
Pedersen, J.R., Bloch Thomsen, P.E., Kofoed, B., Bak, C.K. and Saermark, K.

130. **The Analysis of Magnetocardiosignal at Switching of Heart from Rhythm to Rhythm: The Preliminary Researches** 529
Primin, M., Nedayvoda, I. and Gapelyuk, A.

131. **Vector Fetal Magnetocardiography** 533
Rassi, D., Zhuravlev, Y.E., Lewis, M.J. and Emery, S.J.

132. **Biomagnetic Evidence for Extreme Anisotropy—a Mechanism Limiting Conduction Capability in Accessory Pathways** 537
Schirdewan, A., Meyerfeldt, U., Wiedemann, M., Schütt, H., Burghoff, M., Gapelyuk, A., Copetti, C.A. and Dietz, R.

133. **Magnetocardiogram in 3 Patients after Aborted Sudden Cardiac Death without overt underlying Heart Disease** 540
Schütt, H., Stellbrink, C., Dammers, J., Boers, F., Wiedenau, G., Ioannides, A.A., Müller-Gärtner, H.W., Gapelyuk, A. and Schirdewan, A.

134. **Magnetocardiography and other Methods of Investigation in Diagnostics of Heart Remodelling after Myocardial Infarction and in Hypertension** 544
Shabalin, A.V., Nikitin, Y.P., Golyshev, N.V., Motorin, S.V., Rogachevsky, B.M., Kytmanov, A.V., Ermakova, E.N. and Tselmin, S.G.

135. **Comparative Data on Magnetocardiography and Clinical-Instrumental Methods of Investigation when some Hereditary Cardiovascular Diseases** 547
Shabalin, A.V., Nikitin, Y.P., Golyshev, N.V., Motorin, S.V., Rogachevsky, B.M., Lisithenko, O.V., Ermakova, E.N., Dolgich, M.M. and Kuroedov, A.Y.

136. **Atrial Depolarization Homogeneity Diagnostics on Evidence derived from MCG** 550
Stadnyuk, L.A., Chaikovsky, I.A., Sosnitsky, V.N., Gapelyuk, A.V., Minov, Yu.D. and Chervonopiskaya, E.M.

137. **Magnetocardiographic Localization of Focal Arrhythmias in Conjunction with Catheter Ablation** 554
Steiner, P.R., Kynor, D.B., Haupt, C.D., Rowley, H.A. and Lesh, M.D.

138. **Magnetocardiographic Assessment of Left Ventricular Hypertrophy: Correlation with Echocardiography** 558
Takala, P., Montonen, J., Aalto, T., Mäkijärvi, M., Nenonen, J., Yildirim, Y., Nieminen, M.S. and Katila, T.

139. **Localization of Atrial Depolarization Activity from Non-averaged MCG's in Congenital Heart Disease** 562
Thieme, R., Brockmeier, K., Brümmerhoff, A., Kissenbeck, S., Burghoff, M. and Bein, G.

140. Tangential (x- and y-) Component Measurement of Cardiac Magnetic Field and Comparison with Conventional z-Component Measurement 565
Tsukada, K., Kandori, A., Yoshida, T., Haruta, Y., Noda, Y., Horigome, H., Terada, Y. and Mitsui, T.

141. Visualization for Electric Activities in a Human Heart 569
Ueda, T., Kasai, N., Sagane, M., Chinone, K. and Awano, N.

142. Magnetocardiography in the Determination of Fetal Heart Rate Complexity 573
Van Leeuwen, P., Bettermann, H., Schüßler, M. and Lange, S.

143. Fetal Magnetocardiography: Spectral Analysis of Fetal Heart Rate Variability in the Course of Pregnancy 577
Van Leeuwen, P., Hartmann, M., Lange, S. and Schüßler, M.

144. Determination of Changes in Fetal Cardiac Time Intervals and T/QRS Ratio during Pregnancy using Magnetocardiography 581
Van Leeuwen, P., Lange, S., Schüßler, M. and Lajoie-Junge, L.

145. Fetal Magnetocardiography in Twin Pregnancy 585
Van Leeuwen, P. and Schüßler, M.

146. Magnetocardiographic Monitoring of the Fetal Autonomic Nervous System 589
Zhuravlev, Y.E., Lewis, M.J., Rassi, D. and Emery S.J.

Contents
Volume II

IV Biomagnetic Measures of the Gastro-Intestinal System, Susceptometry and Other Biomagnetic Measures

1. **Biomagnetic Research in Gastroenterology** 594
Baffa, O., Basile, M., Bradshaw, A., Forsman, M., Nakaya, Y. and Richards,W.O.

Biomagnetic Measures of the G-I System

2. **Magnetic Field Measurement of Rabbit Colonic Electrical Activity** 608
Bradshaw, L.A., Allos, S.H., Wikswo, J.P., Jr. and Richards, W.O.

3. **Esophageal Transit Time evaluated by means of the Biomagnetic and Scintigraphic Methods** 612
Daghastanli, N.A., Braga, F.J.H.N., Oliveira, R.B. and Baffa, O.

4. **Measurements of Colonic Motility using Ferrimagnetic Powder** 616
Forsman, M. and Abrahamsson, H.

5. **Detection of Stomach Contractions using a Biosusceptometer** 620
Miranda, J.R.A., Baffa, O., Oliveira, R.B., Sousa, P.L. and Braga F.J.H.

6. **Biomagnetic Measurement of Gastric Mechanical Motility using a High-Temperature SQUID Imaging.** 624
Nomura, M., Saijyo, T., Haruta, Y., Itozaki, H., Toyoda, H., Nakaya, Y., Ito, S. and Kado, H.

7. **Single and Multichannel Magnetic Field Measurements of Gastrointestinal Activity in the Pre- and Post-Prandial States** 628
Petrie, R.J., van Leeuwen, P., Brandts, B., Turnbull G., Veldhuyzen van Zanten, S.J.O. and Stroink, G.

8. Magnetogastrograms using 64-Channel SQUID System: Study on Clinical Usefulness 631
Saijyo, T., Nomura, M., Haruta, Y., Itozaki, H., Toyoda, H., Nakaya, Y., Ito, S. and Kado, H.

9. Noninvasive Measurement of the Vector Magnetic Field from Human Gastrointestinal Sources 635
Staton, D.J., Allos, S.H., Henry, V.K., Bradshaw, L.A., Ladipo, J.K., Richards, W.O. and Wikswo, J.P., Jr.

Susceptometry And Other Biomagnetic Measures

10. Human Alveolar Macrophage Activation in Healthy Subjects and in COPD-Patients Evaluated by Magnetopneumography 639
Barth, W., Möller, W., Meier-Sydow, J., Pohlit, W. and Stahlhofen, W.

11. Magnetic Markers linked to Monoclonal Antibodies allow *in vitro* Evaluation of Cell Population Size 643
Del Gratta, C., Della Penna, S., Battista, P., Di Donato, L., Vitullo, P., Romani, G.L. and Di Luzio, S.

12. Iron Quantification by Non-Invasive Biomagnetic Organ Susceptometry 647
Engelhardt, R., Fischer, R., Nielsen, P. and Gabbe, E.E.

13. The Use of Biomagnetic Liver Susceptometry in the Ferrara-Hamburg-Turin study on Thalassemia 651
Fischer, R., Piga, A., Tricta, F., Nielsen, P., Engelhardt, R., Garofalo, F., Di Palma, A. and Vullo, C.

14. Detection of Biodegradation of Ferrofluid by Relaxation Measurements 655
Kötitz, R., Trahms, L., Pfefferer, D., Kresse, M., Semmler, W. and Weitschies, W.

15. Distribution and Biotransformation Dynamics of Magnetic Contrast Agents in Organism on Evidence derived from SQUID-Magnetometry 659
Mikhailik, O.M., Pankratov, Yu.V., Dudchenko, A.K., Sosnitsky, V.N., Budnik, N.N., Minov, Yu.D., Sutkovoj, P.I. and Bakai, E.A.

16. Human Alveolar Long-Term Clearance in Healthy and Disease evaluated by Magnetopneumography 663
Möller, W., Barth, W., Meier-Sydow, J., Pohlit, W., Siekmeier, R. and Stahlhofen, W.

17. **A Model of Phagosome Motion within Cells incorporating Viscoelasticity based on Cytomagnetometric Measurements** 667
Nemoto, I.

18. **Magnetic Susceptibility Tomography with nonuniform Field** 671
Parente Ribeiro, E., Wikswo, Jr., J.P., Costa Ribeiro, P. and Szczupak, J.

19. **MCG studies in Rats using Ferrofluids** 675
Saligram, U., Moran, J.E. and Tepley, N.

20. **Magnetic Field and Electric Potential of Excited Plant Cell Chara Corallina: Calculation and Comparison with Experiment** 679
Slibar, M., Trontelj, Z., Jazbinsek, V., Thiel, G. and Müller, W.

21. **System for Magnetic Susceptibility Investigations of Human Blood and Liver** 683
Sosnitsky, V.N., Budnik, N.N., Minov, Yu.D., Sutkovoj, P.I. and Vojtovich, I.D.

22. **Registration of Iron Overload by Means of SQUID-Magnetometry when coupled with Low-Temperature ESR Spectroscopy** 687
Sosnitsky, V.N., Mikhailik, O.M., Lubyanova, I.P., Gapelyuk, A.V., Budnik, N.N., Minov, Yu.D., Sutkovoj, P.I., Pankratov Yu.V. and Bakai, E.A.

V Magnetoencephalography (MEG): Basic Research

23. **MEG Studies of Cognition** 692
Salmelin, R.

Basic Human MEG

24. **The Magnetic Field Components evoked by an Optical Illusion** 697
Adachi, Y., Kohda, M., Morita, T. and Takebayashi, M.

25. **Spatiotemporal Imaging of Human Cortical Areas sensitive to Visual Motion** 701
Ahlfors, S.P., Aronen, H.J., Belliveau, J.W., Dale, A.M., Huotilainen, M., Ilmoniemi, R.J., Korvenoja, A., Liu, A.K., Simpson, G.V., Tootell, R.B.H., and Virtanen, J.

26. **Task-Specific Magnetic Fields from the left Human Prefrontal Cortex** 705
Basile, L.F.H., Simos, P.G., Tarkka, I.M. and Papanicolaou, A.C.

27. Searching for the Oddball Response in the EEG and MEG 709
Berg, P., Kakigi, R., Koyama, S., Dobel, C., Zobel, E. and Scherg, M.

28. Magnetoencephalographic Measurements of a Developing Infant: A Pilot Study 713
Bowyer, S.M., Moran, J. and Tepley, N.

29. The Steady State Visual evoked Response and Estimates of Phase Velocity 717
Burkitt, G.R., Silberstein, R.B. and Wood, A.W.

30. Nonlinear Spatio-Temporal Interactions and Neural Connections in Human Vision using Transient and m-Sequence Stimuli 721
Chen, H.-W., Aine, C.J., Flynn, E.R. and Wood, C.C.

31. Spatial Frequency tuning Functions and Contrast Sensitivity at different Eccentricities in the Visual Field 725
Chen, H.-W., Aine, C.J., Flynn, E.R. and Wood, C.C.

32. Neuromagnetic Fields Evoked by Sine Tones, Sine Tone Complexes, Vowel Formants, and Two-Formant Vowels 729
Diesch, E., Luce, T.A. and Eyferth, K.

33. Source Location and Orientation, Dipole Moment, and Latency of Magnetic Mismatch Fields may mirror the Perceptual Salience of Phonetic Contrasts 733
Diesch, E., Luce, T.A. and Eyferth, K.

34. Auditory evoked Magnetic Fields after Stimulation with Test-Tones and Words-Studies on Experimentees with Good Hearing and with Long-Term Hearing Impediments 737
Emmerich, E., Meyer, R., Nowak, H., Huonker, R., Gießler, F., Meißner, W. and Beleites, E.

35. Analysis of Movement-Related Brain Activities elicited by External Instructions 741
Endo, H., Takeda, T., Kizuka, T., Masuda, T. and Kumagai, T.

36. Visual evoked Magnetic Fields: Bilateral Dipole Pattern for Full-Field Stimuli 745
Fujita, S., Nakasato, N., Seki, K., Kanno, A., Kawamura, T., Ohtomo, S., Fujiwara, S. Tamura, I. and Yoshimoto, T.

37. Evoked Magnetic Responses from Human V1: Evidence for a Chromatic-Sensitive Motion Mechanism 749
Fylan, F., Singh, K.D., Holliday, I.H., Anderson, S.J. and Harding, G.F.A.

38. **Evoked magnetic responses to Isoluminant Chromatic Stimuli from Human V1: Spatial Frequency Characteristics** 753
Fylan, F., Singh, K.D., Holliday, I.H., Anderson, S.J. and Harding, G.F.A.

39. **EEG and MEG Activity accompanying spontaneous Reversals of the Necker Cube** 757
Gaetz, M., Weinberg, H., Jantzen, K.J., Rzempoluck, E. and Millbank, A.

40. **MEG and EEG Responses to a Pitch Naming and Identification Task: Slow DC Shift predicts Performance** 761
Gordon, R., Weinberg, H., Cheyne, D. and Jantzen, K.

41. **Magnetic Detection of Activity of presumed Inhibitory Interneurons in the Human Somatosensory Cortex** 765
Hashimoto, I., Mashiko, T. and Imada, T.

42. **Effect of Color on Visual evoked Magnetic Fields with Pattern-Reversal Stimulation** 769
Hatanaka, K., Seki, K., Nakasato, N., Kanno, A., Ohtomo, S., Fujiwara, S. and Yoshimoto, T.

43. **Evoked Magnetic Fields at Semantic Judgment of Simple Japanese Sentences** 773
Hirata, Y., Sakai, T. and Kuriki, S.

44. **Neuromagnetic Correlates of Endogenous Auditory Sensations** 777
Hoke, E.S., Hoke, M. and Ross, B.

45. **Spatial and Temporal Properties of the slow Components of the Contingent Magnetic Variation in a warned Choice Reaction Time Task** 781
Hultin, L., Högstedt, P., Pizzella, V., Romani, G.L., Rossini, P. and Tecchio, F.

46. **Dependence of SEFs on the Interstimulus Interval: A Model for SEF Generation Mechanism at the Primary Sensorimotor Cortex** 785
Huttunen, J., Wikström, H., Korvenoja, A., Aronen, H. and Ilmoniemi, R.J.

47. **Somatosensory evoked Fields from SMI and SII During 'Interfering' Finger Movements and Tactile Stimulation** 789
Huttunen, J., Wikström, H., Korvenoja, A., Aronen, H. and Ilmoniemi, R.J.

48. Time Course of Neuromagnetic Signals related to Reading Japanese Phonograms 793
Imada, T., Kawakatsu, M., Ikeya, A., Kadota, R., Miyamoto, J., Nakayama, S. and Kotani, M.

49. Electric and Magnetic Brain Responses evoked by Deviant Tones in Melodies 797
Jürgens, R., Lopez, L., Diekmann, V., Becker, W., Ried, S., Grözinger, B. and Erné, S.N.

50. Multidipole Estimation of the Neuromagnetic Fields related to Auditory Discrimination 801
Kikuchi, Y., Yoshizawa, S., Kita, M., Nishimura, C., Tanaka, M., Endo, H., Kumagai, T. and Takeda, T.

51. Measurement of Gustatory-evoked Magnetic Fields with Sharp Stimulation 805
Kobayakawa, T.K., Endo, H.E., Saito, S.S., Ayabe-Kanamura, S.A., Kikuchi, Y.K., Yamaguchi, Y.Y., Ogawa, H.O. and Takeda, T.T.

52. Source Location Outside the Auditory Cortex: Auditory Event related Field to Tone-Duration in a Discrimination Task 809
Kofoed, B., Bak, C.K., Csepe, V., Pantev, C. and Saermark, K.

53. Measurement of MEG evoked by Visual Discrimination Task 813
Kumagai, T., Takeda, T. and Endo, H.

54. Cerebral Activities in the Occipital and Temporal Regions when Subjects perform a matching Task of Visual Stimuli 817
Kuriki, S. and Hirata, Y.

55. Magnetoencephalographic Study on Cerebral Cortical Activities related to Speech 821
Kyuhou, S., Sasaki, K., Nambu, A., Matsuzaki, R., Tsujimoto, T. and Gemba, H.

56. Magnetic Recording of the Propagation of Motor Unit Action Potentials in the Human Leg Muscles 825
Masuda, T., Endo, H., Kumagai, T. and Takeda, T.

57. Effect of Luminosity Change on Visual evoked Field 829
Matsuhashi, M., Oyanagi, H., Kumagai, T., Endo, H. and Takeda, T.

58. Evoked Magnetic Fields for Gustatory Stimuli on the Tongue 833
Murayama, N., Nakasato, N., Hatanaka, K., Fujita, S., Igasaki, T., Kanno, A. and Yoshimoto, T.

59. Different Effects of Stimulus Rate on the Somatosensory evoked Magnetic Fields and Potentials 837
Nagamine, T., Mäkelä, J.P., Mikuni, N., Mima, T., Nishitani, N., Sato, T., Ikeda, A., Kimura, J. and Shibasaki, H.

60. Binaural Fusion of Virtual Pitch in the Human Auditory Cortex 841
Pantev, C., Elbert, T., Ross, B., Eulitz, C. and Terhardt, E.

61. Left and Right Supratemporal Auditory Cortices are differentially modulated by Linguistic-Attentional Task Demands 845
Poeppel, D., Yellin, E., Phillips, C., Roberts, T.P.L., Rowley, H., Wexler, K. and Marantz, A.

62. Extensive Somatosensory Stimulation alters Somatosensory evoked Fields 848
Rockstroh, B., Vanni, S., Elbert, T. and Hari, R.

63. Laterality of Hippocampal Responses to Infrequent and unpredictable Omissions of Visual Stimuli 852
Rogers, R.L., Taylor, S.A., Akhtari, M. and Sutherling, W.W.

64. MEG Measurement of Auditory Sensory Memory Persistence via the M100 in Children and Adults 856
Rojas, D.C., Sheeder, J.L., Teale, P., Walker, J.R., Robertson, B.A., Selig, S. and Reite, M.

65. The Relation between Gustatory-Evoked Magnetic Fields and Reaction Times to different Taste Qualities 860
Saito, S.S., Kobayakawa, T.K., Ayabe-Kanamura, S.A., Endo, H.E., Yamaguchi, Y.Y., Kikuchi, Y.K., Ogawa, H.O. and Takeda, T.T.

66. SQUIDs and Stereopsis 864
Shahani, U., Weir, A.I., Lang, G., Mansfield, D.C., Halliday, D.M., Hutson, D., Maas, P.M. and Donaldson, G.B.

67. Source Localization of a N400-like Negativity in a Sentence-Reading Paradigm using Evoked Magnetic Fields and Magnetic Resonance Imaging 868
Simos, P.G., Basile, L.F.H., Zouridakis, G. and Papanicolaou, A.C.

68. Information Theory Predicts the Magnitude of an MEG Response related to the Operation of the Auditory Sensory Memory 872
Sinkkonen, J., Tiitinen, H., May, P., Ilmoniemi, R.J. and Näätänen, R.

69. Determination of the Nature of the Connection between the Auditory and Visual Cortices 874
Stephen, J.M., Broadhurst, J.H., Knuth, K.H. and Schwartz, B.J.

70. Visualization of Tonotopy of Auditory Cortex using a newly developed 129 Channel Vector Magnetoencephalography 877
Takanashi, Y., Iwamoto, K., Yoshikawa, K., Makino, M., Tomita, T., Tomita, S., Kajihara, S., Kondo, Y., Okamura, S., Yoshida, Y., Ueda, M. and Nakajima, K.

71. Measurement of Accommodation related MEG Responses 881
Takeda, T., Hashimoto, K., Kumagai, T. and Endo, H.

72. Electromagnetic source Localization of Cognitive Processes related to Attention to Single or Omitted tones 885
Tarkka, I.M. and Stokić, D.S.

73. A Cortical Substrate for Hand Skill: An MEG Evaluation 889
Thoma, R.J., Lewine, J.D., Davis, J.T. and Orrison, W.W., Jr.

74. Separation of Overlapping Sources induced by mixed Stimulations using 3-Dimensional MEG Mappings 892
Uchikawa, Y., Kobayashi, K., Sakawa, K. and Kotani, M.

75. Measurements of Visual evoked MEG Responses associated with Color Discrimination 896
Ueno, K., Ueno, S. and Weinberg, H.

76. Measurement and Source Analysis of MEG Activities associated with Short-Term Memory and Identification Processes 900
Yoshida, H., Ueno, S. and Weinberg, H.

MEG Measures Of Brain Oscillatory Responses

77. Oscillatory Brain Activities. 904
Lopes da Silva, F., Basar, E., van Dijk, B., Narici, L., Pantev, C., Ribary, U. and Salenius, S.

78. Time-Frequency Analysis of Oscillatory Gamma-Band Activity: Wavelet Approach and Phase-Locking Estimation 919
Bertrand, O., Tallon-Baudry, C. and Pernier, J.

79. Inverse Imaging of Distributed Oscillatory Activity 923
Bradshaw, L.A. and Wikswo, J.P., Jr.

80. MEG Measurements of 40Hz Auditory evoked Response in Human Brain 927
Ciulla, C., Takeda, T., Morabito, M., Kumagai, T. and Endo, H.

81. Oscillatory Neuromagnetic Activity induced by Verbal and non-Verbal Stimuli presented in Visual and Auditory Modalities 931
Eulitz, C., Maess, B., Pantev, C., Friederici, A., Feige, B. and Elbert, T.

82. Singular Spectrum Analysis applied to Brain Signals 935
Gaeta, R., Benzi, R., Peresson, M. and Narici, L.

83. Characterization of Photosensitive Discharges by Oscillatory responses to Driving Stimuli 939
Karhu, J. and Tesche, C.D.

84. Equivalent Current Sources of the Auditory evoked Gamma Band and Steady State Fields 943
Knuth, K.H., Broadhurst, J.H. and Schwartz, B.J.

85. Local Modulation of MEG Rhythmic Activities related to the alternation of Binocular Rivalry and Binocular Fusion 947
Kobayashi, T., Kato, K. and Kuriki, S.

86. Magnetic Oscillatory Components of Human Flash-VEP: Luminance and Eccentricity Functions and effect of Acute Scopolamine 951
Lopez, L., Narici, L., Conforto, S., Romani, G.L. and Sannita, W.G.

87. Increased Variability of the Period Duration for Waveforms in Epochs of MEG Data as a Function of Arousal State 955
Moran, J.E. and Tepley, N.

88. Suitability of FFT Spectra versus DPA Period Spectra for Identification of Arousal States 959
Moran, J.E. and Tepley, N.

89. Chaos in the Brain: A New Strategy to Discriminate Deterministic Low Dimensional Dynamics in the Spontaneous Activity of the Human Cortex 963
Narici, L., Boccaletti, S., Giaquinta, A. and Arecchi, T.

90. A New Procedure to discriminate and study different Rhythmical Cortical Activities on the basis of their responsiveness to Simple Sensory Stimulation 967
Narici, L. and Peresson, M.

91. Cognitive Temporal Binding and its Relation to 40Hz Activity in Humans: Alteration during Dyslexia 971
Ribary, U., Joliot, M., Miller, S.L., Kronberg, E., Cappell, J., Tallal, P. and Llinás, R.

92. A Technique for the Identification of Hippocampal Theta from MEG Data 975
Tesche, C.D.

VI Magnetoencephalography (MEG): Clinical Studies

93. Clinical Applications of MEG 980
Mäkelä, J.P., Elbert, T., Kakigi, R., Lewine, J., Lopez, L., Nagamine, T., Nakasato, N., Reite, M. and Sannita, W.

Clinical MEG

94. Changes of the Functional Organization of the Somatosensory Cortex in Chronic Pain Patients 991
Braun, C., Flor, H., Birbaumer, N. and Elbert, T.

95. Clinical MEG II: Development of a Normative Database 995
Davis, J.T., Lewine, J.D., Edgar, J.C., Hoesing, J. and Orrison, W.W., Jr.

96. Cortical Reorganization in Arm Amputees: Alterations of the Somatosensory Representation of the Intact Arm 999
Elbert, T., Sterr, A., Rockstroh, B., Charbonnier, D., Flor, H., Pantev, C., Wienbruch, C., Knecht, S. and Taub, E.

97. Comparison of Magnetoencephalographic and Intraoperative Electrocorticographic Localization of Sensory Response 1003
Greenblatt, R., Bucholz, R., Sobel, D., Schwartz, B. and Hirschkoff, E.

98. Origin of Slow Wave observed in Cerebrovascular Disease 1007
Hatanaka, K., Nakasato, N., Kanno, A., Ohtomo, S., Seki, K., Fujiwara, S. and Yoshimoto, T.

99. Magnetic Source Imaging in the Evaluation of Chronic Alcohol Abuse 1011
Hill, D., Waldorf, V.A., Lewine, J.D., Provencal, S.L., Moyers, T. and Yeo, R.

100. MEG and MCG in a Clinical Environment: BioMag Laboratory, Helsinki 1014
Ilmoniemi, R.J., Ahonen, A., Alho, K., Aronen, H.J., Huttunen, J., Karp, P., Mäkijärvi, M., Näätänen, R., Paetau, R., Pekkonen, E., Karhu, J., Standertskjöld-Nordenstam, C.-G., Tiihonen, J., Toivonen, L. and Katila, T.

101. MEG Traces of Painful Processes 1018
Kakigi, R., Kitamura, Y., Koyama, S., Hoshiyama, M., Shimojo, M., Watanabe, S., Takeshima, Y. and Nagata, O.

102. Normalized N100m Latency of the Auditory evoked Fields after Surgical Removal of Temporal Lobe Gliomas 1022
Kanno, A., Nakasato, N., Ohtomo, S., Seki, K., Kawamura, T., Fujita, S., Kumabe, T., Kayama, T. and Yoshimoto,T.

103. Reversed Hemispheric Asymmetry of Auditory N100m in Schizophrenics 1026
Karhu, J., Tiihonen, J., Pekkonen, E., Katila, H., Huotilainen, M., Virtanen, J. and Ilmoniemi, R.

104. Neuromagnetic Identification of the Somatosensory Cortex in cases with Arteriovenous Malformation Adjacent to the Central Sulcus 1030
Kawamura, T., Nakasato, N., Ohtomo, S., Seki, K., Kanno, A., Fujita, S., Fujiwara, S. and Yoshimoto, T.

105. Time Course of Abnormal MEG Activity associated with Transient Ischemic Attacks 1033
Kober, H., Vieth, J.B., Stippich, C., Kassubek, J.R. and Hopfengaertner, R.

106. Preoperative Localization of Sensorimotor Cortex: MRI versus MEG 1037
Lewine, J.D., Baldwin, N.G., Bucholz, R.D., Sanders, J.A., Halliday, A.L., Anson, J.A., Shih, J., Stearley, J., Davis, J.T., Astur, R.S., Paulson, K. and Orrison, W.W., Jr.

107. Clinical MEG I: Towards a Standardized Examination 1041
Lewine, J.D., Davis, J.T., Davis, L.E., Canive, J., Roberts, B., Graeber, D., Shih, J., Edgar, J.C., Provencal, S.L., Paulson, K., Meyers, J., Christner, R., Silveri, J., Rawcliffe, N., Tessman, C., Espinosa, M., Depper, M., Sanders, J.A. and Orrison, W.W., Jr.

108. Two Components of Giant Somatosensory evoked Magnetic Fields in Patients with Cortical Myoclonus 1045
Mima, T., Mikuni, N., Nishitani, N., Nagamine, T., Ikeda, A., Fukuyama, H., Takigawa, T., Kimura, J. and Shibasaki, H.

109. Deviations from the Tonotopic Map are correlated with Tinnitus Strength 1049
Mühlnickel, W., Flor, H. and Elbert, T.

110. The Determinism of MEG in Tinnitus Patients 1053
Mühlnickel, W., Hoke, E.S., Hoke, M. and Flor, H.

111. Auditory Signal Processing is impaired in Alzheimer's Disease: A Whole-Head MEG Study 1057
Pekkonen, E., Huotilainen, M., Virtanen, J., Näätänen, R., Ilmoniemi, R.J. and Erkinjuntti, T.

112. **SEFs to Median Nerve and Finger Stimulation: Analysis of Interhemispheric Asymmetries in Healthy Subjects and following Hemispheric Lesion** 1060
Pizzella, V., Rossini, P.M., Rossi, S., Tecchio, F., Lopez, L. and Romani, G.L.

113. **Effects of Acute, Oral Administration of Phenobarbital (1.6 mg/kg) to Healthy Subjects** 1064
Sannita, W.G., Lewine, J.D., Maclin, E.L., Orrison, W.W., Jr. and Robinson, S.

114. **Visual evoked Fields for Pattern Reversal Stimuli in Patients with Occipital Lobe Lesions** 1067
Seki, K., Nakasato, N., Fujita, S., Kanno, A., Hatanaka, K., Kawamura, T., Ohtomo, S., Fujiwara, S., Takahashi, A. and Yoshimoto, T.

115. **MEG and fMRI Localisation of Primary Sensory-Motor Cortex in Normals and a Patient with a Low-Grade Astrocytoma** 1071
Singh, K.D., Furlong, P.L., Weale, P., Harding, G.F.A. and Walsh, A.R.

116. **Comparison between the Processing of Static Images of Human Faces in High Functioning Autistic Subjects and Normal Controls** 1075
Swithenby, S.J., Bräutigam, S., Bailey, A.J., Jousmaki, V. and Tesche, C.

117. **A Comparison of MEG and EEG for the Assessment of Traumatic Brain Injury** 1079
Weinberg, H., Gaetz, M., Jantzen, K., Cheyne, D. and Deecke, L.

118. **Somatosensory evoked Magnetic Fields in Acute Cerebral Infarction** 1083
Wikström, H., Huttunen, J., Roine, R.O., Salonen, O., Aronen, H.J. and Ilmoniemi, R.J.

Clinical MEG: Epilepsy

119. **Magnetoencephalographic vs electrocorticographic Source Localization in Patients with Epilepsy** 1087
Aung, M., Sobel, D., Schwartz, B., Squires, K. and Hirschkoff, E.

120. Propagation of Epileptic Activity during Single (unaveraged) Interictal Spikes as studied with Magnetic Field Tomography (MFT) 1090
Bamidis, P.D., Hellstrand, E. and Ioannides, A.A.

121. Presurgical Evaluation of Methohexital-Induced Epileptiform Activity: Comparison of EEG-, MEG-, and Simultaneous EEG/MEG Models 1094
Diekmann, V., Becker, W., Jürgens, R., Grözinger, B., Kleiser, B., Wollinski, K.H. and Richter, H.P.

122. Intracortical Spread of Epileptic Spikes-Simultaneous Measurement of Magnetic Field and Electric Potential in Rabbits 1098
Eiselt, M., Zwiener, U., Gießler, F., Wagner, H., Haueisen, J., Huonker, R., Nowak, H. and Schack, B.

123. Effects of Volume Current on Localization of Focal Epileptic Activity 1102
Ikeda, H., Takeuchi, R., Osozawa, S., Ishiyama, A., Kasai, N. and Minami, T.

124. Evaluation of Abnormal Low Frequency Magnetic Activity (ALFMA) in Epilepsy 1106
Lewine, J.D., Davis, J.T. and Orrison, W.W., Jr.

125. Event-Related Magnetic Fields Associated with Auditory Oddball Paradigm-Postoperative Study in Patients with Temporal Lobe Epilepsy 1110
Nishitani, N., Nagamine, T., Mikuni, N., Ikeda, A., Kimura, J. and Shibasaki, H.

126. Use of Magnetic Source Imaging in Surgical Planning for Epilepsy Patients who have Undergone Prior Resection 1114
Sobel, D., Aung, M., Schwartz, B., Squires, K. and Hirschkoff, E.

127. Methohexital-Induced Changes in Spectral Power of Neuromagnetic Signals: Reduced β-Band Enhancement over the Hemisphere Ipsilateral to the Epileptogenic Focus 1117
Wienbruch, C., Eulitz, C., Lehnertz, K., Brockhaus, A., Elger, C.E., Elbert, T. and Hoke, M.

VII Multimodality Comparisons And Integration

128. Evaluation of Source Localization Models for Somatosensory evoked MEG with PET Verification 1122
Akhtari, M., McNay, D., Levesque, M., Rogers, R.L., Brown, C., Mandelkern, M., Babai, A. and Sutherling, W.W.

129. Combination of Point and Surface Matching Techniques for Accurate Registration of MEG and MRI 1126
Bamidis, P.D. and Ioannides, A.A.

130. EEG and MEG Source Analysis of Somatosensory evoked Responses to Mechanical Stimulation of the Fingers 1130
Cheyne, D., Roberts, L.E., Gaetz, W., Bosnyak, D., Weinberg, H., Johnson, B., Nahmias, C. and Deecke, L.

131. Dynamic Neuroimaging by MEG, constrained by MRI and fMRI 1134
George, J.S., Schmidt, D.M., Mosher, J.C., Aine, C.J., Ranken, D.M., Wood, C.C., Lewine, J.D., Sanders, J.A. and Belliveau, J.W.

132. Combining Modalities to investigate Cortical Retinotopy 1138
Hurdal, M.K. and McElwain, D.L.S.

133. fMRI constrained Linear Estimation of Cortical Activity from MEG Measurements: A Model Study 1142
Liu, A.K., Sereno, M.I., Rosen, B.R., Belliveau, J.W. and Dale, A.M.

134. Multiple Modality Biomagnetic Analysis System 1146
Martin, K., Erne, S., Law, C., Conforto, S., Mallick, J. and Tatar, B.

135. Measurement of Visually evoked EEG and MEG associated with Verbal Cognitive Processes 1150
Nakagawa, S., Ueno, S., Iramina, K. and Weinberg, H.

136. Superadditive Information from Simultaneous MEG/EEG Data 1154
Pflieger, M.E., Simpson, G.V., Ahlfors, S.P. and Ilmoniemi, R.J.

137. Multimodal Functional Imaging of the Sensorimotor Cortex with fMRI and Magnetic Source Imaging (MSI) 1158
Roberts, T.P.L. and Rowley, H.A.

138. Quantitative Comparison of Motor Control Studied with MEG and fMRI 1162
Russell, D.P., Schwartz, B.J. and Hirschkoff, E.C.

139. Generating Cortical Constraints for MEG Inverse Procedures using MR Volume Data 1166
Sandor, S.R., Leahy, R.M. and Timsari, B.

140. Correlation between Depth and MEG/EEG Surface recordings in Auditory Cortex. Contribution of Numerical Simulations 1170
Schwartz, D., Badier, J.M., Scarabin, J.M. and Liégeoix-Chauvel, C.

141. Brain Activities evoked by Missing Auditory Stimuli: A Combined Study using MEG and fMRI Data 1174
Takino, R., Miyauchi, S., Sasaki, Y., Pütz, B. and Sekihara, K.

142. Integration of Functional MRI and MEG of Median Nerve Stimulation 1178
Verbiesen, M.J.R., Peters, A.R., den Boer, J.A., Knösche, T.R. and Peters, M.J.

143. Comparison of Functional Localization in Human Visual Cortices using MEG and fMRI: A Preliminary Report 1182
Yoshikawa, K., Takanashi, Y., Iwamoto, K., Yoshida, Y., Ueda, M., Tanaka, C., Umeda, M., Ebisu, T., Fukunaga, M., Naruse, S., Sato, H. and Nakajima, K.

I. Biomagnetic Instrumentation

Workshop on Biomagnetic Instrumentation

Koch, H. (Organizer)
Physikalisch-Technische Bundesanstalt (PTB), Berlin, Germany

Introduction

The concept of the workshop was to initiate a guided discussion from start to end. Thus the sequence of topics to be covered have been announced to the panelists and the audience at the beginning of the workshop resulting in lively contributions from both sides. Members of the panel were S. Buchanan (San Diego, USA), O. Dössel (Hamburg, Germany), J. Knuutila (Helsinki, Finland), J. Vrba (P. Coquitlam, Canada), and Y. Yoshida (Kyoto, Japan), with the author being responsible for moderating the discussion. Although all panel members were representing their respective companies, it was strictly urged not to exploit this workshop for company promotion but to support unbiased technical discussions of general interest only.

Below, the main topics touched during the workshop are compiled. To stimulate the spirits and the discussion two recent success stories were presented as an introduction:

1. a graph on recent installations of biomagnetic multichannel LTS-SQUID-systems (LTS = Low Tc Superconductivity) (cf. Fig 1), and
2. results obtained with integrated HTS-SQUIDs (HTS = High Tc Superconductivity) with a field noise of <10 fT/√Hz at 1 kHz and <35 fT/√Hz at 1 Hz.

Although it was known that several other groups worldwide have obtained similar impressive achievements, it was discouraged to include these into this discussion in order to avoid a collection of "me too" contributions. Thus the presentation of Y. Q. Shen, Denmark (cf. Fig 2) acted as a representative introduction to the status of HTS-SQUIDs and their application in biomagnetic experiments. This opened the discussion on the potential impact of HTS on biomagnetism.

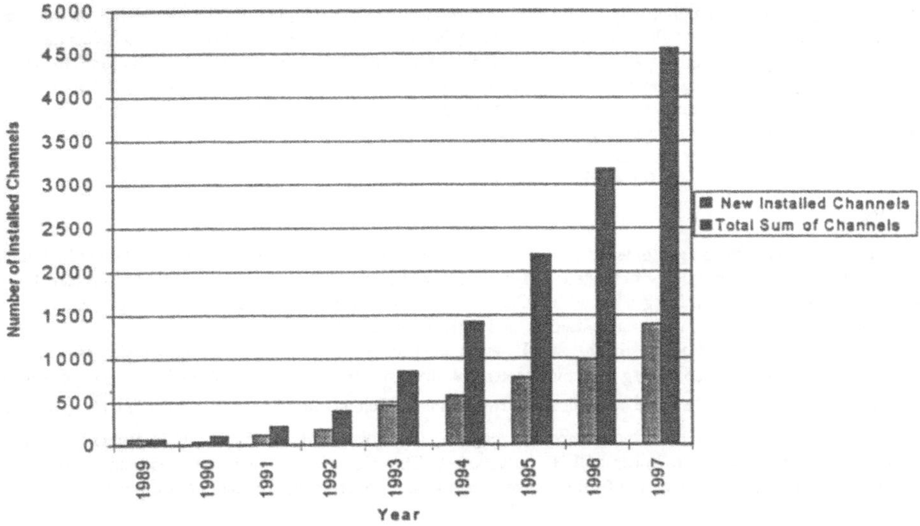

Fig. 1 Number of biomagnetic SQUID-channels installed worldwide (compiled by L. Trahms).

offer many of the same advantages realized by the direct-coupled readout electronics, in addition to the well-known advantages of ac measurement techniques: problems owing to dc signal drift are eliminated, and tuning the SQUIDs is greatly simplified and can be easily automated.

Methods

For a magnetometer, the relevant figure of merit for most applications is the rms magnetic flux density noise $S_B^{1/2}(f) = S_\Phi^{1/2}(f)/A_{eff}$. Here, $S_\Phi^{1/2}(f)$ is the rms flux noise and A_{eff} is the effective flux capture area of the magnetometer. For a dc SQUID that is free of resonances, it can be shown that the flux noise power $S_\Phi(f) \propto L^2/R$.[12] For optimum performance, the SQUID inductance is therefore usually made as low as possible. The effective flux capture area may be expressed in terms of the field-to-flux conversion efficiency B_Φ; that is, $A_{eff} = \Phi_0/B_\Phi$, where $B_\Phi = \Phi_0(L_p + L_{i,eff})/(M_i A_{p,eff})$. Here, L_p is the inductance of the pickup loop, $L_{i,eff}$ is the effective inductance of the input coil which includes the screening effect of the SQUID inductance, M_i is the mutual inductance of the input coil and SQUID, $A_{p,eff}$ is the effective area of the pickup loop including flux focussing effects, and Φ_0 is the flux quantum. Since $L_{i,eff} \approx L_p$ to satisfy inductance matching, $B_\Phi \propto (L_p/A_{p,eff})/M_i$. For a given chip size and reasonable values of L_p, the ratio $L_p/A_{p,eff}$ is a slowly varying function depending only on the linewidth of the pickup loop. Consequently, once the chip size has been fixed, B_Φ depends mostly on the mutual inductance M_i, which in turn depends on the SQUID inductance L and the design of the input circuit.[12]

Cantor *et al*[5] describe an integrated magnetometer that was designed using an optimization procedure that takes the adverse effects of parasitic capacitance and microwave resonances into account.[4] The magnetometer was designed for a nominally 4 mm × 4 mm chip size, and the design parameters were chosen by a minimization of the magnetic field noise.[5] The SQUID inductance L of the magnetometer was 51 pH, and the measured white flux noise at 1 kHz was 0.61 $\mu\Phi_0/\sqrt{\text{Hz}}$. The low SQUID inductance and optimal design resulted in very low flux noise, but the mutual inductance M_i was low, and consequently the field-to-flux conversion efficiency was 5.6 nT/Φ_0. Nevertheless, this resulted in a field noise at 1 kHz of 3.4 $\text{fT}/\sqrt{\text{Hz}}$, but the low flux noise placed stringent demands on the room-temperature electronics. In addition, the flux noise deteriorated at low frequencies, increasing to 1 - 2 $\mu\Phi_0/\sqrt{\text{Hz}}$ at 1 Hz. Interestingly, this value is frequently observed for many different SQUIDs, independent of the junction technology used.

The magnetometer described here is designed using a larger chip size which increases the flux capture area of the pickup loop and a much higher SQUID inductance in order to achieve a higher mutual inductance. In this way, a significantly improved field-to-flux conversion efficiency is obtained. The relatively high SQUID inductance, L = 190 pH, is transformer coupled via a multi-turn spiral input coil to a 7.75 mm × 7.75 mm pickup loop. The design of the SQUID has been chosen so that the microwave transmission line resonances of the input coil are well damped, and the resonant frequency of the gradiometric SQUID washers is well away from the intended operating frequency of the SQUID. An integrated *RC* shunt in the SQUID input circuit ensures that the SQUID characteristics are smooth over the entire intended operating range and that the noise minimum is very broad, both of which simplify operation in the flux-locked loop mode.

Results

The fabrication of the magnetometers is based on a simple, four-layer process[4] and a robust Nb/Al-AlO_x/Nb trilayer technology.[13] Only three metalization layers are required to fabricate the complete magnetometer. Using a series-configured double washer gradiometric SQUID design, the parasitic capacitance is kept to a tolerable level, even though only anodic niobium oxide is used as the interlayer dielectric.

Owing to the optimized designs of the SQUID and pickup loop, a high field-to-flux conversion efficiency of 0.83 nT/Φ_0 has been achieved. Using a picovoltmeter, the intrinsic rms white flux noise of the magnetometer is measured to be 1.6 $\mu\Phi_0/\sqrt{\text{Hz}}$. The corresponding rms magnetic field noise is better than 1.4 $\text{fT}/\sqrt{\text{Hz}}$ (see Fig. 1(a)). Because of the high SQUID inductance, the magnetic flux noise is higher than what is generally observed for magnetometers using much lower SQUID inductances. The high flux noise is compensated by the large effective area of the magnetometer and facilitates coupling to the room-temperature readout electronics.

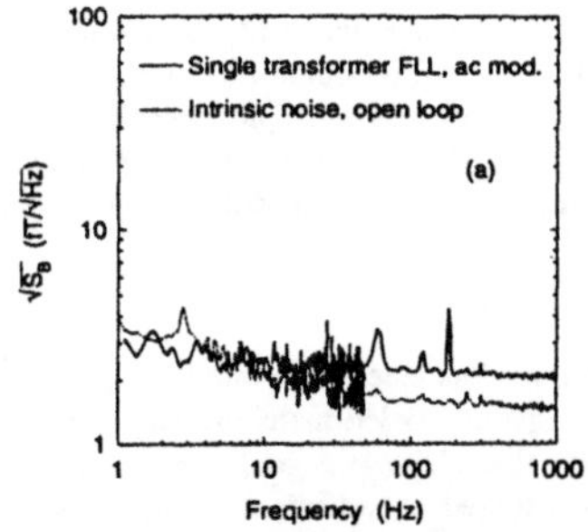

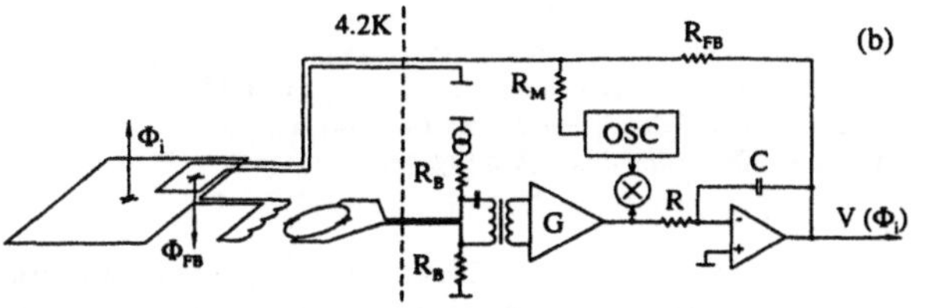

Fig. 1 (a) The rms magnetic field noise measured in flux-locked loop using flux modulation and a two-wire connection to the room-temperature preamplifier (solid line), and the intrinsic rms field noise measured open loop using a picovoltmeter (dotted line). (b) Schematic diagram of the SQUID and flux-locked loop electronics.

For multi-channel applications, we have developed flux-locked loop (FLL) electronics that do not require a cooled impedance matching network: a single transformer at room temperature is used to couple the SQUID output to the room-temperature preamplifier. Further, common bias current and signal leads are used so that each magnetometer requires only four wires to room temperature, ±Signal and ±Modulation, which simplifies multi-channel system design. A diagram of the SQUID and electronics is shown in Fig. 1(b). Although the two-wire configuration results in a slight noise increase, the rms magnetic field noise in the flux-locked loop mode is typically better than 3 $\mathrm{fT}/\sqrt{\mathrm{Hz}}$ at 1 kHz, increasing to less than 4 $\mathrm{fT}/\sqrt{\mathrm{Hz}}$ at 1 Hz (see Fig. 1(a)).

We have built a single-channel test probe for biomagnetic applications using two magnetometers vertically spaced by 53 mm. A first order electronic gradiometer with a baseline of 53 mm is formed by coupling the output signal from the first reference channel through a scaling amplifier to the feedback circuit of the primary magnetometer. In this way, uniform background noise may be significantly reduced by properly scaling the reference channel output.

We have used this probe to record spontaneous human brain activity in a clinical environment. The measurements were carried out in a magnetically shielded room (MSR) having two layers of μ-metal, a copper Faraday cage, and an aluminum eddy current shield. The MSR is in the Epilepsy and Brain Mapping Center in the Hospital of the Good Samaritan in downtown Los Angeles, CA. The single-channel gradiometer was adjusted by varying the scaling amplifier to minimize the 60 Hz background field. A 42 dB attenuation of the 60 Hz peak was achieved, while the attenuation of higher order harmonics was around 25 dB. A representative field noise spectrum is shown in Fig. 2 (a). The sharp peaks at low frequencies ($f < 50$ Hz) are believed to have been caused by microphonic effects. As can be seen from these data, the system noise level is just under 10 $\mathrm{fT}/\sqrt{\mathrm{Hz}}$ down to a few Hz.

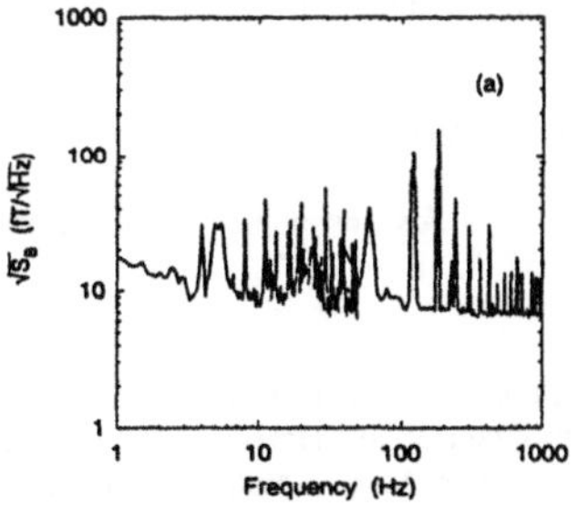

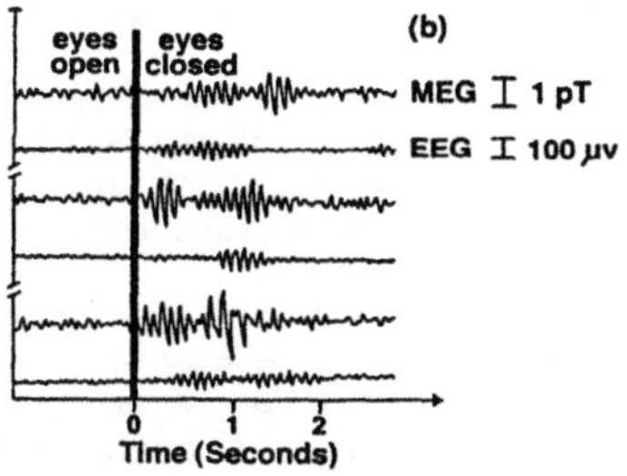

Fig. 2 (a) The rms magnetic field noise referred to the primary (lower) magnetometer in the electronic gradiometer mode. (b) Simultaneous MEG and EEG measurements performed by eye closure during alertness on multiple occasions in a normal 24 year old man.

The magnetoencephalogram was measured radially over the right occipital head region at O_2. In addition, an encephalogram was measured at O_2 referenced to the vertex. A 10-14 Hz, 4-pole Bessel bandpass filter was used for the MEG measurements. Alpha rhythm was produced by eye closure during alertness on multiple occasions. The MEG and EEG were recorded on a Grass polygraph in adjacent channels. The alpha rhythm appeared reproducibly in the MEG and EEG during eye closure as shown in Figure 2 (b). This demonstrates that spontaneous alpha rhythm can be identified using this SQUID system.

Discussion

We have designed and built a low-noise integrated magnetometer FLL readout electronics employing ac flux modulation. The magnetometer requires only four wires to room temperature and no cooled matching circuit which simplifies the design of multihannel systems. In addition, we have shown that electronic gradiometers may be formed using these magnetometers. The attenuation and system sensitivity achieved is sufficient for spontaneous human MEG measurements in magnetically shielded rooms in clinical environments.

References

[1] Jaycox, J. M. and Ketchen, M. B., "Planar coupling scheme for ultralow noise SQUIDs," IEEE Transactions on Magnetics, 1981, MAG-17: 400 - 403.

[2] Ryhänen, T., Seppä, H. and Cantor, R., "Effect of parasitic inductance and capacitance on the dynamics and noise of direct current superconducting quantum interference devices," Journal of Applied Physics, 1992, 71:6150 - 6166.

[3] Knuutila, J., Kajola, M., Seppä, H., Mutikainen, R., and Salmi, J. "Design, optimization, and construction of a dc SQUID with complete flux transformer circuits," Journal of Low Temperature Physics, 1988, 71: 369 - 391.

[4] Cantor, R., Ryhänen, T., Drung, D., Koch, H., and Seppä, H., "Design and optimization of DC SQUIDs fabricated using a simplified four-level process," IEEE Transactions on Magnetics, 1991, MAG-27: 2927 -2931.

[5] Cantor, R., Enpuku, K, Ryhänen, T., and Seppä, H., "A high performance integrated DC SQUID magnetometer," IEEE Transactions on Applied Superconductivity, 1993, 3: 1800 - 1803.

[6] Drung, D., Cantor, R., Peters, M., Scheer, H. J., and Koch, H., "Low-noise high-speed dc superconducting quantum interference device magnetometer with simplified feedback electronics," Applied Physics Letters, 1990, 57: 406 - 408.

[7] Kazami, K, Takada, Y., Uehara, G., Matsuda, N., and Kado, H., "Evaluation of Drung-type magnetometers for multi-channel systems," Superconductor Science and Technology, 1994, 7: 249-252.

[8] Drung, D., "The PTB 83-SQUID system for biomagnetic applications in a clinic," IEEE Transactions on Applied Superconductivity, 1995, 5: 2112 - 2117.

[9] Drung, D., and Koch, H., "An electronic second-order gradiometer for biomagnetic applications in clinical shielded rooms," IEEE Transactions on Applied Superconductivity, 1993, 3: 2594 - 2597.

[10] Seppä, H. "DC-SQUID electronics based on adaptive noise cancellation and a high open-loop gain controller," In: Koch, H., and Lübbig, H., Superconducting Devices and their Applications, Berlin, Springer-Verlag, 346.

[11] Koch, R H., Foglietti, V., Rozen, J. R., Stawiasz, K. G., Ketchen, M. B., Lathrop, D. K, Sun, J. Z., and Gallagher, W. J., "Effects of radio frequency radiation on the dc SQUID," Applied Physics Letters, 1994, 65: 100- 102.

[12] Ryhänen, T., Seppä, H., Ilmonierni, R., and Knuutila, J., "SQUID magnetometers for low-frequency applications," Journal of Low Temperature Physics 76: 287 - 386.

[13] Barfknecht, A. T., Ruby, R. C., and Ko, H., "A simple and robust niobium Josephson junction integrated circuit process," IEEE Transactions on Magnetics, 1991, MAG-27: 3125 - 3128.

Acknowledgments

The authors gratefully acknowledge the assistance of R.L. Rogers, Ph.D., S. Taylor, M.S., M. Akhtari, M.S., and W.W. Sutherling, M.D., of the Epilepsy and Brain Mapping Program at the Hospital of the Good Samaritan in Los Angeles. The authors also thank E. Flynn and R.H. Kraus of Los Alamos National Laboratory who supported the development of the magnetometer described here under a joint Cooperative Research and Development Agreement with Conductus.

Reference noise cancellation: Optimizations and Caveats

Barnes G.R. [1] **and Ioannides A.A.** [1,2]
[1] *Institute of Medicine, Forschungszentrum Jülich GmbH, Jülich D52425, Germany.*
[2] *Physics Department, The Open University, Milton Keynes MK7 6AA, U.K.*

Introduction

Reference noise cancellation algorithms theoretically allow the removal of external noise from biomagnetic measurements without attenuation of signal components in the same frequency band. The technique requires reference sensors that are insensitive to the biomagnetic signals of interest, yet subject to the same environmental noise conditions. By estimating the correlation between reference (e.g. distant magnetometers) and measurement (e.g. gradiometer) channels it is possible to form a prediction of the measurement channel based on reference channel information; this prediction is defined as noise and subtracted from the data. The filters considered in this paper predict the measurement channel from the superposition of linearly weighted reference channels; the weights are determined through a least squares minimization of the filter output power.

For time stationary white Gaussian data, adaptive filter theory provides good analytical estimates of filter performance [1,2]. Typical biomagnetic data sequences are, however, neither white nor stationary. This paper addresses two important implications of using linear least squares algorithms with real data:

1. The longer the recording session the more likely that the external noise environment, and hence the optimal filter weights, will change. The rate of weight update is defined by the filter time constant: filters with small time constants introduce steady state misadjustment or error, since the weight estimates are based on relatively few data and are therefore highly variant. Filters with large time constants incur little steady state error, since the weights update smoothly, yet may fail to track changes in the environment (non-stationary misadjustment). Optimizations of filter performance achieved by changing the filter time constant are examined.

2. Spectrally broad signals have a sharp temporal signature; it is therefore possible to make unequivocal correlation estimates between relatively short time sequences within well defined bounds [2]. However, the narrower the cross spectrum of two dominantly periodic sequences, the longer the time window required to make an accurate correlation estimate. In practice, reference noise cancellation is seen to be particularly effective for removing low frequency components from measured data. The question is whether the removed components were eliminated because of a true correlation with the reference data, or simply because of an insufficiently long time window.

Background

The acausal Wiener [3,5] and causal square-Root Recursive Least Squares (RRLS) [4,5] algorithms were used. In both cases, the filter output y is given by:

$$y[i] = x[i] - \mathbf{w}[i] \cdot \mathbf{r}[i] \qquad (1)$$

Where i is the time or sample index, x is the measured data channel, $\mathbf{r}$ is the vector of reference channels, and $\mathbf{w}$ is the weight vector. The dot product $\mathbf{w} \cdot \mathbf{r}$ is the filter's prediction of the external noise on the measured channel x. The RRLS algorithm updates the weight vector sample by sample; whereas in the Wiener case, stationarity is assumed, and $\mathbf{w}[i]$ can be replaced by the optimal weight vector for an N sample data block, $\mathbf{w}^*$. The weight vectors are calculated from analytical expressions which minimize:

$$< \sum_{i=1}^{N} y[i]^2 > \qquad and \qquad < \sum_{i=1}^{n} y[i]^2 \lambda^{n-i} > \qquad (2)$$

for the Wiener and RRLS algorithms respectively. Where N is the number of samples in the data block for the Wiener algorithm; and for the RRLS algorithm n is the current sample and λ is termed the 'forgetting factor'. The corresponding effective filter time constants (in samples) are $T \doteq N/4$ [2] and $T=1/(1-\lambda)$ [4].

Methods

The measure Misadjustment (M) [2] was used to quantify filter performance :

$$M = \frac{Error\ power}{Min.\ output\ power} = \frac{\sum_{i=1}^{N_t}(y[i]-t[i])^2}{\sum_{i=1}^{N_t} t[i]^2} \tag{3}$$

Where y is actual filter output, t the target or ideal filter output, and N_t the number of samples in the target. All data were recorded using the BTi Magnes II system (filters DC-400 Hz; sample rate 1042 Hz). A 24 s target pseudo-neuromagnetic response[1] was superimposed, at a time of high environmental noise, on the gradiometer output from a 400 s sequence of real recorded data. That is, the target represented true signal, and the gradiometer output represented noise. RRLS filter (using three magnetometers and a single DC reference) misadjustment was measured over a range of time constants. The misadjustment for the Wiener filter, applied to the DC corrected 400 s data block (3 magnetometer references), was also calculated.

In order to illustrate the limits of spectral resolution, the Wiener algorithm was appled to both simulated and real data sets.

The simulated data consisted of a single signal and a reference channel, both containing random phase sinusoids in white noise (SNR=10.0, 512 samples, sample rate 1042 Hz). The frequency of the signal sinusoid was fixed at 1.5 times that of the reference. As the signal and reference data were drawn from uncorrelated distributions, the target of the filter output was the input signal itself. The frequency of the reference sinusoid was varied from 0.1 to 20 Hz; at each frequency the average filter misadjustment was calculated from an ensemble of 10 signal realizations.

The real data consisted of 6 s of averaged gradiometer output from a CNV experiment (go condition, 30 trials). The data was DC corrected and Wiener filtered (3 averaged magnetometer references). The filter output was compared to that obtained when the magnetometer channels were replaced with those from recording sessions of other subjects. That is, the measurement and reference data were again drawn from uncorrelated distributions.

Results and Discussion

Figure 1 illustrates the trade off between steady state and non-stationary misadjustment as filter time constant is varied. Filters with large time constants (Figure 1C, T=100 s) fail to adapt to changes in the recording environment and excess external noise is incurred. In the limit, as RRLS filter time constant is increased, the results tend to those of the Wiener block solution (figure 1B, solid line; figure 1C, lower section) where the data in the 400 s recording period is assumed to be stationary. Filters with short time constants minimize filter output power (Figure 1B, dashed trace) yet also remove the lower frequency components of the target (Figure 1C, T=5 s). This high-pass filtering occurs because of the dominantly low frequency components in the reference channels: temporally, it can be thought of as the filter adapting so quickly that it tracks the useful signal; spectrally it is a leakage problem, considered in the next section.

Figure 2A shows the expected misadjustment when signal and reference are dominantly sinusoidal at a constant frequency ratio of 1.5. At lower frequencies, the algorithm clearly makes no distinction between signal and reference, and therefore there is cancellation of the true signal component. This misadjustment is given analytically by taking error power (equation 3) to be the expected value (i.e. half) of Fourier spectral leakage power (equation 4,[6]) between the corresponding frequency bins.

$$L(d) = \frac{(sin(\pi d))^2}{(N sin(\pi d/N))^2} \tag{4}$$

Where L is the spectral leakage power, d is the fractional difference between signal and reference frequencies in Fourier bins and N is the number of samples in the window. Figure 2B shows an averaged CNV response

[1] The target was constructed from filtered, averaged, Contingent Negative Variation (CNV) data.

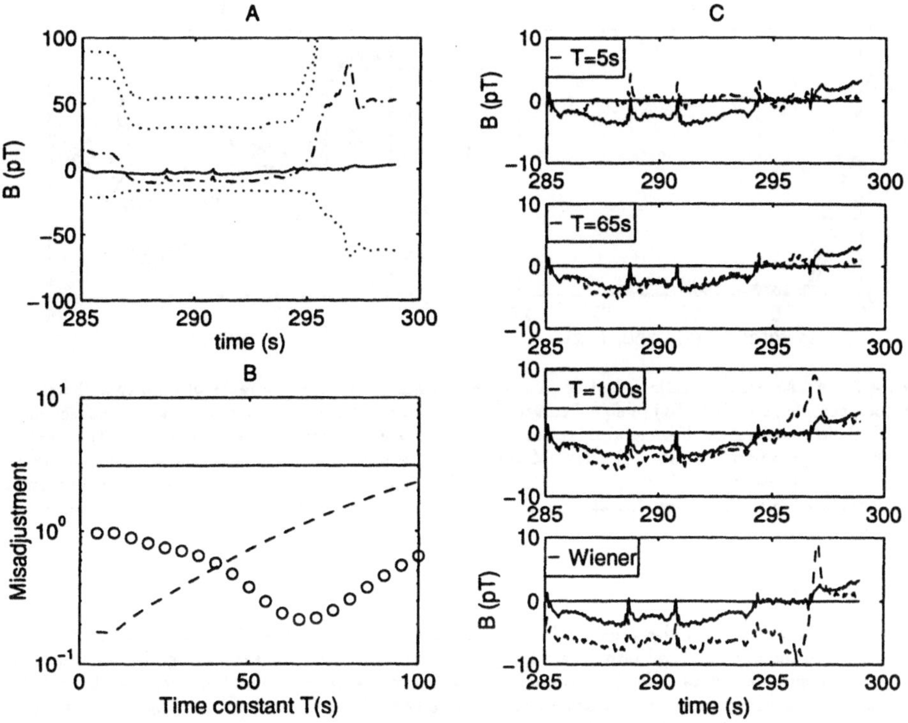

Figure 1: A. Segment of data from a 400 s recording: 3 magnetometer references (dotted trace; amplitudes scaled down by 10.0); the artificial target (solid); the target superimposed on gradiometer output (dash-dot). B. RRLS misadjustment to target as filter time constant is varied (circles); the solid line shows the level of misadjustment incurred by the Wiener solution; the dashed line is the fraction of filter output to target power. C. Top three sections show RRLS filter output (dotted) alongside target (solid) for selected time constants in B; the lower section shows Wiener filter output (dotted) alongside target (solid).

(circles) Wiener filtered (solid) using the corresponding averaged (3 magnetometer) reference channel data. The three dashed curves show filter output when the averaged reference channel data is taken from different experimental sessions (different subjects, days and dewar positions). The results of the filtering are strikingly similar indicating that, in these conditions, the filter action is independent of the true signal-noise correlation properties of the data.

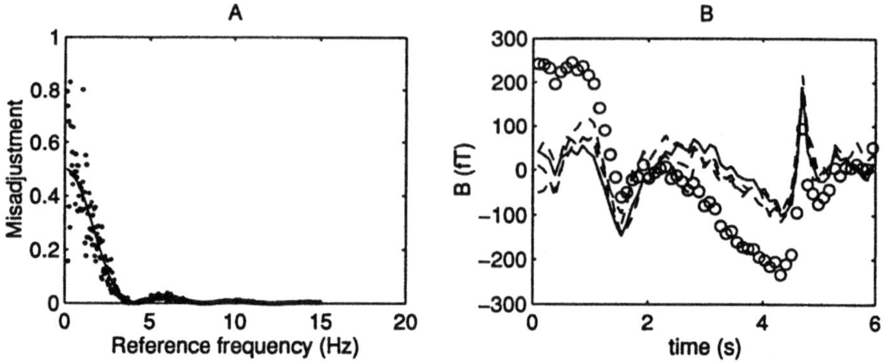

Figure 2: A. Average misadjustment of Wiener algorithm (dots) for ensemble of segments (length 512 samples, sample rate=1042 Hz) of signal sinusoids at 1.5 times reference frequency as reference frequency is varied. Solid curve shows theoretical misadjustment when replacing the filter error (equation 3) by half the spectral leakage power (equation 4). B. Real averaged CNV data before (circles), and after (solid) Wiener filtering (3 magnetometer references). The three dashed lines show filter output when the magnetometer data is replaced with that from separate experiments conducted with different subjects, dewar positions, and noise conditions.

Conclusions

Correlation estimates over finite time windows can be expressed as a discrete Fourier transform [6] and are constrained to the same spectral resolution. The goal of reference noise cancellation algorithms is to remove from the measurement data *only* those components which are correlated with the reference channel data; this cannot be achieved when both measurement and reference channels are dominated by low frequency (of period comparable to the window length) components.

However, for long data recordings, decreasing the time window length (i.e. decreasing the filter time constant) can significantly improve algorithm performance since such shorter segments are more likely to be statistically stationary.

In summary, reference noise cancellation filters with long time constants fail to cancel external noise as the environment changes; filters with short time constants are fast to track environmental changes, yet *introduce* noise in the form of signal removal.

References

[1] Widrow B., Glover J.R., McCool J.M., Kaunitz J., Williams C.S., Hearn R.H., Zeidler J.R., Dong E., Goodlin R.C. Proc IEEE, 1975, 6:1692-1716.

[2] Widrow B., McCool J., Larimore M.G., Johnson C.R. Proc IEEE 1976, 64:1151-1162.

[3] Robinson S.E. In: Williamson S.J., Hoke M., Stroink G., Kotani M. Advances in Biomagnetism, 1989, New York, Plenum Press.

[4] Porat B., Digital Processing of Random Signals, Prentice Hall, Engelwood Cliffs, NJ, 1993.

[5] Candy J.V., Signal Processing: the model based approach, McGraw-Hill 1986.

[6] Press H., Teukolsky S.A., Vetterling W.T., Flannery B.P., Numerical Recipes in Fortran, Cambridge University Press, 1992.

Positioning of MCG Measurement Equipment

Borchardt, M., Krause, E. and Lindner, G. [1]

Department of Mathematics, University of Greifswald, Germany
Physikalisch-Technische Bundesanstalt (PTB), Berlin, Germany

Introduction

The present study was designed to investigate data obtained from biomagnetic measurement of the human heart. In this paper we shall report on two issues. Firstly, we show how data may be visualized online in order to allow positioning of measurement device and secondly, we discuss the question of how to set up the algorithms needed for this online visualization on a transputer system.

The MCG-Data (magnetocardiogram) under investigation were recorded at the PTB-laboratories in Berlin at the Klinikum " Benjamin Franklin " and at the Berlin Magnetically Shielded Room at the PTB, respectively. The first measurement device consists of 83 SQUIDs (Superconducting Quantum Interference Device) in gradiometer mode [2], the second device consists of 37 SQUID magnetometers [1].

In contrast to the conventional ECG our data do not describe the electric potential generated by the excitation of the human heart but the time dependent changes of its magnetic field $\mathcal{B}(x, y, z)$. The magnetic field

$$\mathcal{B}(x,y,z) = \Big(\mathcal{B}_x(x,y,z), \mathcal{B}_y(x,y,z), \mathcal{B}_z(x,y,z)\Big)$$

is a vector field. For simplicity we introduce Cartesian coordinates. We consider the (x, y)–plane of the coordinate system to be a frontal plane of the patient where the positive x–axis denotes the lateral direction pointing to the left arm of the patient and the positive y–axis denotes the cranial direction. The positive z–axis is orthogonal on the (x, y)-plane and ventrally directed.

Both measurement devices are to be localized in the frontal plane $(x, y, z = c)$ directly over the chest of the patient and record the z–component $\mathcal{B}_z(x, y, z = c)$ measured in this plane. The gradiometer system allows to detect certain $\mathcal{B}_{x,y}$– field components as well, but we have not used these field components yet. Please note that until now only the z–component of the coordinate system's origin was fixed by the constant c, while the x– and y–components were not.

Methods

It is generally assumed that the magnetic field generated by the current flow in the human heart during its excitation can be approximated by the field generated by an equivalent electric current dipole [3, 5]. Clearly, this assumption is only reasonable whenever the signal is significantly unequal to zero. The projections of both extrema onto the plane $(x, y, z = c)$ allow to estimate the position of the equivalent electric current dipole in $\mathbb{R}^3$. Therefore, recording data in an appropriate way means to record both projections. Unfortunately it is not possible to meet this requirement over a whole heart cycle, and moving the device while recording would falsify the recorded data.

[1] We would like to thank L. Trahms, M. Burghoff and U. Steinhoff of the MCG-group of the PTB in Berlin for providing the data and many helpful discussions as well as C. Bandt and I. Bandlow of Greifswald University for their support and encouragement.

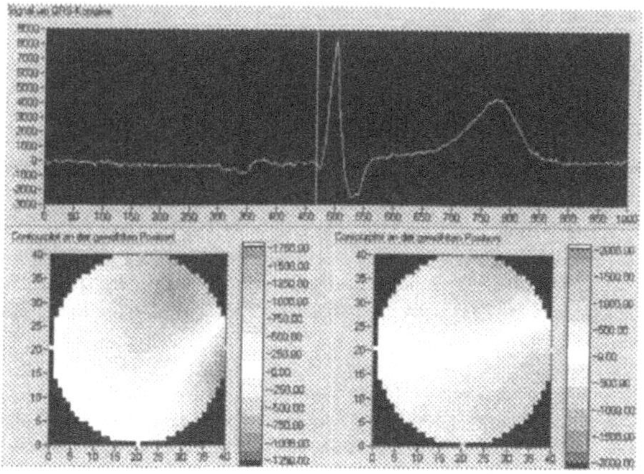

Figure 1: Left bottom: before positioning; Right bottom: after positioning

However, when looking for certain diseases such as WPW-syndrome one is especially interested in small segments of the signal which recur in every heart beat.

Our program detects a user defined point of time relative to the R-peak of the heart cycle and then online calculates a contour plot of all channels at this point of time. Clearly, the contour plot lets decide immediately whether both extrema are recorded or not and therefore it allows good positioning of the measurement device. Using this method it is possible to record data at the position of interest in an optimal way.

Two algorithms are the basis of our program,

1. finding a user defined point of time relative to the R-peak of the heart cycle in the following heart beats, and

2. calculating a contour plot at the found position.

1. When recording data the user first needs to select the channel he wants to use for defining a point of time relative to the R-peak. Let us denote this channel by $(M_t)_{t\in\mathbb{N}}$, the chosen position for the contour plot by $t = C$.

In order to find the first R-peak we take the measured data of the first 1.5 seconds into a buffer and calculate the position of the maximum of all absolute values in the buffer. This maximum position is to be taken as R_{Old}.

Suppose we have already found a maximum position R_{Old} and wish to detect the point of time R_{New} of the next R-peak.

Starting at this last position R_{Old} we take data of the next 1.5 seconds into the buffer and calculate the position of the maximum R_{New} of all absolute values.

Then we compare the RR-distance $R_{New} - R_{Old}$ with a user-defined minimal and maximal value. If it fits into the given interval, we take the position of the maximum to be R_{New} and calculate a contour plot at the point $C := R_{New} + \text{offset}$; otherwise we do not. In both cases we set $R_{Old} := R_{New}$.

Using all n previous positions of R-peaks up to R_{New}, we online estimate the heart rate variability (HRV). Denoting the time intervals between the positions of R-peaks by x_i and stressing the relation

$$s^2(n) = \frac{1}{n-1}\left[\sum_{i=1}^{n} x_i^2 - \left(\sum_{i=1}^{n} x_i\right)^2\right],$$

independent of the number of known R-peaks there are only five arithmetic operations needed to calculate the new estimation $s^2(n)$ for HRV from the old estimation $s^2(n-1)$.

2. Using the positions of SQUIDs P_i in the (x, y)–plane the area covered by the measurement device is triangulated. The measured data can now be converted into approximations of the magnetic field $\widetilde{M}(P)$ defined on an $(n \times n)$– array, where n may be defined by the user.

Solving the minimum problem

$$||P_i - P|| + ||P_j - P|| + ||P_k - P|| = \min!$$

for all points P covered by the measurement device we determine the closest three positions of SQUIDs P_i, P_j and P_k. Please note that P usually lies in the triangle $\Delta(P_i, P_j, P_k)$.

Using linear interpolation, we obtain as approximations $\widetilde{M}(P)$ at points P

$$\widetilde{M}(P) = \alpha \cdot [M(P_j) - M(P_i)] + \beta \cdot [M(P_k) - M(P_i)] + M(P_i).$$

The barycentric coordinates α and β are calculated for all points of the array using the equation

$$P = \alpha \cdot (P_j - P_i) + \beta \cdot (P_k - P_i) + P_i \text{ where } \alpha + \beta \leq 1.$$

The **transputer system** at the Klinikum Benjamin Franklin which is used to visualize the data is based on a MIMD distributed memory architecture. This transputer system works on the basis of the message passing programming model along synchronized channels. The data are read in continuous operation mode and monitored online on two monitors using two different graphic nodes [6, 7]. One monitor displays the overall view of all channels and the status window, the other displays the reference channel chosen by the user and the contour plot.

In order to compute the contour plot online with high accuracy for 49 channels and more, we split up the program and use different nodes of the transputer system. The *data input node* reads all data from the data acquisition system and then the *test node* checks underflow or overflow, respectively, and calculates the median of each channel. The median has turned out to be a reliable approximation of the zero line of a channel and therefore it is used for the calibration of measured data. From the test node data get to the processing node. The *processing node* is a PowerPC601 high performance node[2]. In the operation mode data of all channels are continuously read into an evaluation buffer. When the buffer is full, the user-defined point of time R of the R-peak and the contour plot at cursor position $C := R + \text{offset}$ are calculated and finally all data to be plotted are distributed to the two graphics nodes. The *host node* communicates both with the test node as well

[2]PowerPC601 is a trademark of International Business Machines

as the processing node and is used to compute the $(n \times n)$– array. Moreover, it is connected with the user's console and therefore makes interactive keyboard commands possible.

Discussion

The suggested program for online positioning of measurement device uses as trigger the R-peak of the signal, but does work for general triggers as well. The contour plot algorithm is based on triangulation of the measurement area and linear interpolation. The accuracy of these algorithms is apparently not the best but still sufficient for online visualization of a huge amount of data. Please note that the barycentric coordinates are to be calculated only once at the beginning of the program.

References

[1] D. Drung, R. Cantor, M. Peters, H.J. Scheer, H. Koch. Low-noise high-speed DC Superconducting Quantum Interference Device magnetometer with simplified feedback electronics. *Appl. Phys. Lett.* **57** (1990), 406-408.

[2] D. Drung. The PTB 83-SQUID system for biomagnetic application in a clinic. *IEEE Trans. Appl. Supercond.* (1994).

[3] R.S. Gonnelli and M. Agnello. Inverse problem solution in cardiomagnetism using a current multipole expansion of primary sources. *Phys. Med. Biol.*, Vol. **32** (1987), No. 1, 133-142.

[4] A. Link, L. Trahms, R. Zimmermann, M. Oeff. Complex binomial bandpass filters for analysing phase and envelope of high frequency components in cardiac signals. *IEEE Computers in Cardiology* (1994), 617-620.

[5] J. Sarvas. Basic mathematical and electromagnetic concepts of the biomagnetic inverse problem. *Phys. Med. Biol.*, Vol. **32** (1987), No. 1, 11-22.

[6] D. Stollfuss, G. Lindner. A transputer based real time visualization system for medical measurement data (in german). *Physikalisch-Technische Bundesanstalt, Jahresbericht* (1993), 281-282.

[7] G. Lindner. The use of formal description means as an useful thing for the program development of a real time visualization system (in german). *4. Fachtagung Entwurf komplexer Automatisierungssysteme, TU Braunschweig*, ISBN: 3-9803363-3-6, June 1995, 505-519.

A Low-Noise, Integrated DC SQUID Magnetometer for Applications in Biomagnetism

Cantor, R., Vinetskiy, V., Matlashov, A.
Conductus, Inc., Sunnyvale, CA, USA

Introduction

Applications such as magnetocardiography (MCG) and, in particular, magnetoencephalography (MEG) require ultra-low noise sensors to detect the extremely weak magnetic fields emanating from the human body. The dc Superconducting QUantum Interference Device (SQUID) is the most sensitive detector of magnetic *flux* available, but the low SQUID inductance, typically around 100 pH or less, results in a low effective flux capture area and consequently insufficient magnetic *field* resolution for such applications. To improve the field resolution, a separate pickup coil with a much larger flux capture area is used. The pickup coil is inductance matched to a multi-turn spiral input coil that is transformer coupled to the SQUID inductance. For best coupling, this is generally accomplished by patterning the SQUID inductance in the shape of a thin-film washer with the input coil integrated on top.[1] In this case, it is important to consider the potentially adverse effects of the parasitic capacitance and microwave resonances that are inherent to such thin-film structures.[2] By following a design optimization procedure that takes these parasitic effects into account,[3],[4] integrated dc SQUID magnetometers with very low field noise performance suitable for applications in biomagnetism have been demonstrated.[4],[5]

Alternatively, thin-film, fractional turn dc SQUID magnetometers have been developed whereby the SQUID inductance consists of several loops connected in parallel in order to achieve a high effective area while maintaining a tolerable total inductance that is nevertheless a few hundred picohenry.[6],[7] Because of the high SQUID inductance, the flux noise is also high. This is compensated by the large flux capture area of the sensor, resulting in sufficiently low magnetic field noise for biomagnetic applications.[8]

Another important consideration for multi-channel applications such as MCG and MEG is that the SQUID electronics be simple, easy to interface to the SQUID, robust against external radio frequency interference (RFI), and user friendly to operate. Conventional electronics employing ac flux modulation and phase sensitive detection to readout the SQUID signal satisfy most of these requirements, and presently available commercial SQUID electronics use this technique. Important advantages of the flux modulation technique include the elimination of certain types of $1/f$ noise and dc drifts in the room-temperature preamplifier, and ease of operation. Commercial dc SQUID systems include a simple autotune algorithm which enables the user to quickly and easily adjust all of the SQUID operating parameters. This is an important consideration for multi-channel SQUID systems. A disadvantage of the conventional readout electronics for multi-channel applications, however, is the cooled impedance matching circuit that is required to match the low source impedance of the SQUID to the much higher input impedance of the room temperature preamplifier. This complicates the assembly of multi-channel systems.

During the past few years, new types of direct-coupled readout electronics have been developed. It has been demonstrated that the SQUID output may be directly coupled to the room temperature preamplifier without any impedance matching circuitry if the noise of the preamplifier is either made much smaller than that of the SQUID using Additional Positive Feedback (APF) in conjunction with Bias Current Feedback (BCF),[9] or is cancelled using Adaptive Noise Cancellation (ANC).[10] Using common signal and bias current leads, the direct-coupled electronics with APF and BCF require as few as four wires per channel to room temperature. The direct-coupled electronics are very simple, and the reduced number of leads required per channel and elimination of the cooled matching circuit simplify the construction of multi-channel systems. Such dc measurement techniques, however, are susceptible to dc drifts, require careful adjustment of up to three three different dc feedback levels, and are more sensitive to external RFI than conventional electronics employing phase sensitive detection.[11]

We describe here the design of an advanced dc SQUID magnetometer that can be operated using conventional readout electronics with ac flux modulation but without the cooled impedance matching network typically used with this technique: a single transformer at room temperature is used to couple the SQUID output signal to the room-temperature preamplifier. This greatly simplifies and reduces the size of the SQUID magnetometer assembly. Further, common bias current and signal leads are used so that each SQUID requires only four wires to room temperature. Thus, the SQUID magnetometer and flux modulation electronics described here

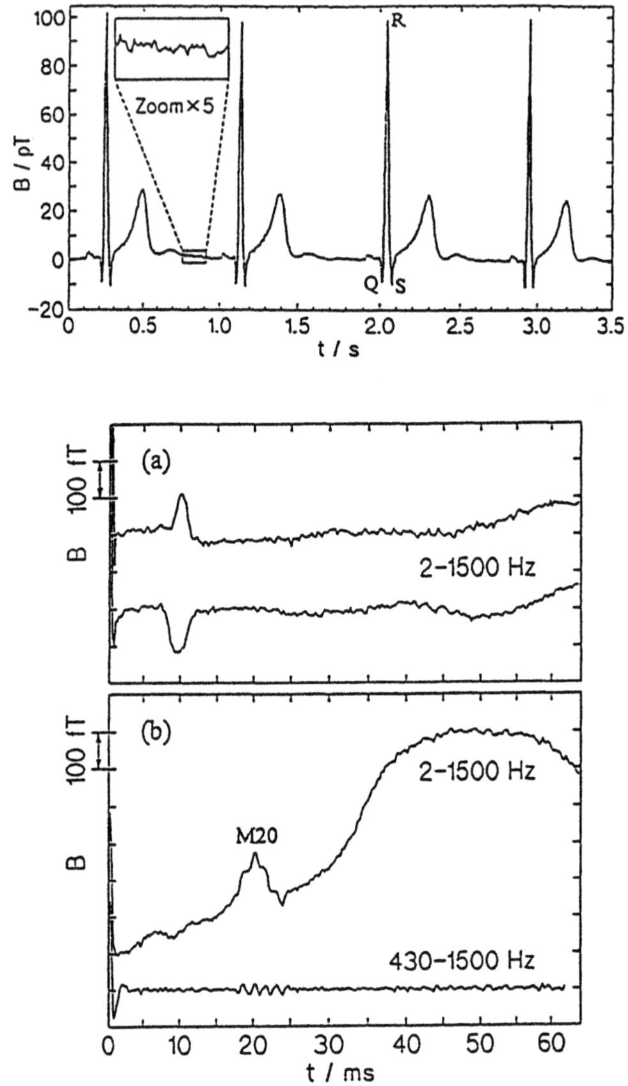

Fig. 2 Top: real-time MCG-signal obtained with a HTS-SQUID (NKT, Denmark) at 77 K inside the Berlin Magnetically Shielded Room. Center: evoked responses to electrostimulation of the right median nerve at the wrist measured with the same set-up above and below the clavicle. Bottom: responses at the left cerebral hemisphere. The lowest trace is obtained from the one above by using a 430 Hz highpass filter. Note the weak 600 Hz signal. Ref.: Appl. Phys. Lett., Vol. 68, No. 10 (1996), pp. 1421-1423.

HTS-SQUIDs

While the field sensitivity obtained with HTS-SQUIDs at 77 K (the temperature of liquid nitrogen) is already amazingly good, the big remaining problem is the performance in unshielded environments. For unshielded operation the performance is considerably deteriorated: necessary heavy filtering distorts the signal and handling in earth field leads to strong flux noise (strongly increased 1/f noise). G. Dantsker (Berkeley, USA) and R. Cantor (San Francisco, USA) reported on the difficulties of handling these delicate devices in order to avoid flux trapping. Primarily, field-free cooling and cycling lead to a greatly improved performance. Proposed measures like the "flux-dam" or "de-Gaussing" improve the performance of less sensitive HTS-SQUIDs but tend to fail in case of the very high sensitive devices. The quality of SQUID fabrication (grain boundaries) and SQUID design greatly influence the amount of flux noise. Bias-reversal electronics considerably reduces the 1/f noise and is essential at present. Approaches to inhibit flux creep by entering pinning centers into the material by radiation did improve properties of HTS-single crystals but not thin films.

In conclusion, the main effort necessary to achieve practical devices is to improve the HTS-SQUID performance for "field-proof" operation. Other problems pointed out by the panelists were the low fabrication yield of HTS-SQUIDs and the high costs of recent HTS-technology (e.g. price of bicrystal STO-substrates).

Two scenarios were seen for HTS-systems

1. large systems with >50 channels with <50 fT/4Hz field noise in shielded rooms, and/or
2. small portable systems for unshielded operation

While the first option seems to be several years away from completion and only moderate advantages or cost savings are seen (LN_2-cooling vs. LHe-cooling allows simpler dewars with longer holding time and reduced sensor-source distance), the second option seems to be already at hand as first companies appear to sell such systems.

Thus, LTS-systems with their advanced performance characteristics will remain the first choice for several years to come. It may even be foreseen that unshielded operation (and thus cheaper investments) could easier and earlier be possible with LTS- than with HTS-systems, which leads to the next point of discussion.

Towards Bed-Side Systems

The ultimate solution would be systems that are insensitive enough against signal interference from urban and/or clinical noise sources to be operated unshielded at the bed-side. This will be a dream for the next decade, but we are approaching this scenario. What is the status and what are the options?

From the beginning of the discussion it was obvious that the term "unshielded operation" is very ambiguous. First, it is often not specified in *what* environment unshielded operation is possible, i.e. what noise sources and what gradients are interfering. Second, it is often not made clear how the signal is affected by the interference *and* by the measures that enable the unshielded operation. In other words: to what extent is the signal distortion due to filtering and gradiometry tolerable from the clinical point of view?

A strong need was stated for a standardized quantitative specification of the parameters involved to help to create a generally accepted view. Another effect frequently not considered in this discussion is the body's susceptibility as pointed out by J. Wikswo (Nashville), who mentioned that up to 25% changes in the amplitude of the T-wave are due to this effect.

Each site must be thoroughly analyzed and evaluated in order to decide for an appropriate balance between the amount of shielding vs. filtering and gradiometry. Due to many imponderables one is always more secure when leaning to more shielding.

The arguments pro or contra *active* shielding were quite controversial as until now no really convincing practical solution has been demonstrated (this statement will not be shared by everyone). However, the recent efforts in understanding the effects of noise interference and the achievements made in developing hardware and software gradiometry methods to compensate this interference have very considerably improved. It was a generally accepted view that although shielding cannot be avoided in most practical cases in hospitals the necessary degree of shielding will be decreased more and more in the years to come.

There was a brief discussion of shields made of HTS (this conference). Although the shielding factors are impressive, the small bore size and the very high cost make these shields rather impracticable yet.

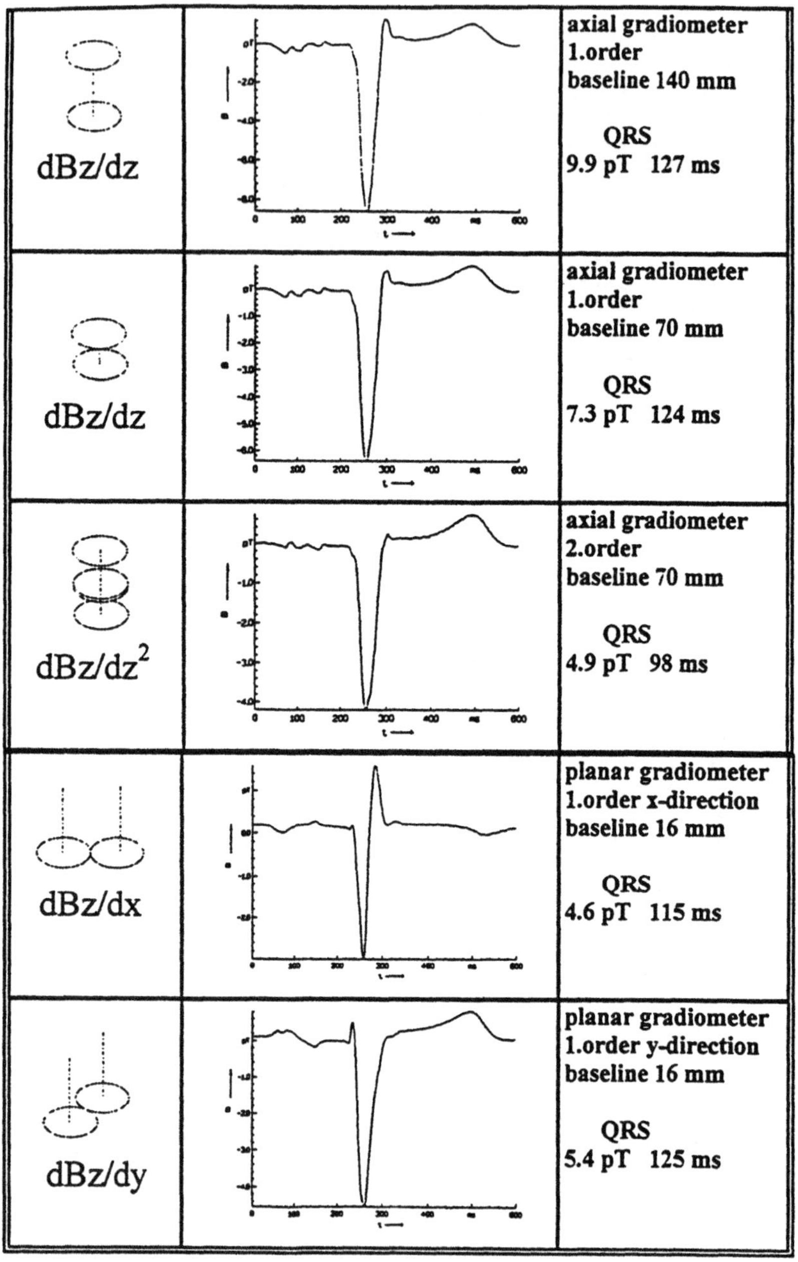

Fig. 3 Measured (top three signals) and calculated MCG-responses of different gradiometer types (from M. Burghoff, U. Steinhoff, and W. Haberkorn, Berlin).

Comparability of Biomagnetic Signals

All gradiometric methods do not only compensate unwanted external signals but distort the original source signals as well. This is not only the case for the derived spatial field distribution but also for the temporal signal evolution. Particularly the latter effect is often severely underestimated. M. Burghoff (Berlin) showed MCGs obtained simultaneously for first order gradiometers of different baseline and for a second order gradiometer (cf. Fig. 3). For an identical source the characteristic QRS-periods registered by different gradiometers differed by >25 % and the signal shape and amplitude were changed as well. Simulations for magnetometers and planar gradiometers showed an even larger effect.

This clearly demonstrates the dangers of utilizing gradiometers, particularly if they are adaptive electronic or software products. The effect done to the signals may be partly correctable by mathematical means, but one has to know what one has done to the signals all the time, i.e. one should critically consider the validity of the physical model underlying these methods.

M. Burghoff and J. Knuutila proposed two ways to convert the measurement data to a unified dataset that allows comparison of measurements obtained with different sensor configurations. It was agreed that whatever method is chosen, none of them can be perfect but will be very useful. The methods have to be worked out in more detail and tested for their practical merits.

It was felt by everyone in the audience that such a standardization is urgently needed in order to facilitate multi-center clinical studies with biomagnetic multichannel systems of different brands.

The newly available multichannel vector magnetometer systems (this conference) were discused. Y. Uchikawa (Tokyo) stressed the merits of the different type of isocontour map with its maximum at the center of a current dipole and the better dipole separation capabilities.

Conclusion

The progress observable in the development of biomagnetic instrumentation and its versatility is very impressive and led to a strong increase in sales and installed channels. However, several aspects need to be improved when a real breakthrough of the biomagnetic methods shall be achieved. Eliminating costly shields is a major goal and considerable progress in this direction is obvious but also the associated problems. Particularly several questions have to be solved concerning the standardization of data presentation to facilitate the exchange and comparability of results.

Other aspects, not touched in this workshop, are the need for better software solutions for data processing, real-time visualization and animation of results, improved gantry and dewar construction, etc.

It is felt that biomagnetism as a discipline and thus biomagnetic instrumentation experience a critical threshold situation: A great deal has been achieved, but a great deal is still to be done to achieve the breakthroughs needed to become a generally accepted clinical method.

Magnetocardiograms Recorded by First-Order Gradiometers Using HTS rf SQUIDs

Copetti, C.A.[1], Soltner, H.[1], Zhang, Y.[1], Schuett, H.[1], Rongen, H.[2], Brandenburg, G.[2], Clemens, U.[2], Zimmermann, E.[2], Borgmann, J.[3], Tavrin, Y.[1], Otto, R.[1], Bousack, H.[1], and Halling, H.[1]

Institute of Thin Film and Ion Technology (ISI)[1], Forschungszentrum Jülich GmbH (KFA), Jülich, Germany; Zentrallabor für Elektronik (ZEL)[2], Forschungszentrum Jülich (KFA), Jülich, Germany; Institut für Strahlen- und Kernphysik (ISKP)[3], Universität Bonn, Bonn, Germany,

Introduction

State of the art HTS SQUID sensors have already attained a sensitivity which makes them interesting for biomagnetical and especially cardiological applications. Many high quality measurements of magnetic signals emanating from the human heart have been performed with HTS sensors. With the improvement in sensitivity, the cancellation of environmental noise became one of the major problems for measuring high quality MCGs. Therefore, most of these measurements have been performed in expensive magnetically shielded rooms (MSRs) with high shielding factors (e.g. PTB, Berlin, Germany or SSL, Japan). A different, less expensive approach is the use of gradiometric arrangements of the sensors to cancel environmental disturbances. As a compromise, moderate shielding may be combined with gradiometry to keep the effort on both alternatives to a minimum.

As an initial step, first order gradiometers for recording MCGs can be assembled for use in more common moderate magnetic shielding. Tavrin et al. [1] have shown that with an electronic gradiometer of first or second order, comprising of an axial arrangement of SQUIDs, the recording of heart signals is possible even without any shielding. Although this system still was not sensitive enough to yield clinically relevant data especially because of its limited bandwidth, the concept can be adopted for use in standard MSRs. This contribution presents a HTS first order gradiometer which is able to yield clinically relevant data in MSRs.

The system

We assembled several types of axial gradiometers each comprising of a rigid arrangement of 8x8 mm^2 washer rf SQUIDs on 1 cm^2 substrates. The characteristics of these sensors have already been published elsewhere [2]. Briefly, the field resolution of these sensors in the white noise range is about 120 fT/√Hz and could be increased for use as a single magnetometer channel by a factor of 3 by the employment of a bulk flux concentrator with a diameter of 25 mm. The baseline of the gradiometers could be varied in the range from 2 cm to 11 cm to investigate the effect of signal reduction vs. noise cancellation. Different types of flux locked loop controllers, an analog and a digital one, were developed for use with these systems. Tab. 1 shows the performance data of the two approaches.

Tab. 1: Parameters of the analog and digital SQUID system used for the present investigations ([a] for a single channel in good magnetic shielding, [b] with a representative sensor)

	analog	digital
rf tank circuit frequency / MHz	200	320
base line / cm	2-10	5
bandwidth / kHz	20	10
slew rate / ϕ_0/sec	3.5×10^5	5×10^4
max. applicable flux / ϕ_0	10^4	4×10^6
dynamic range / dB	140	192
white flux noise / ϕ_0/√Hz [a]	65	70
magn. field res. B_N / fT/√Hz[a,b]	35	45

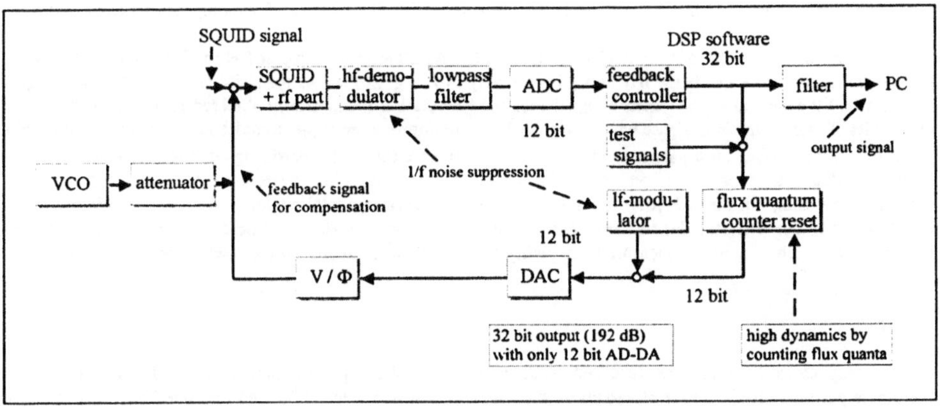

Fig. 1: Block diagram of the digital electronics for one channel.

Fig. 1 shows a block diagram of the digital control system for one channel [3]. In addition to flux compensation, the periodic SQUID signal is used for AD conversion. A flux quantum counter in the digital circuit confines the compensation signal to a range of 2 ϕ_0 by reducing the digital control signal (output signal) by an even number of flux quanta. Because this leads to changes by an integral number of quanta, the amplitutde of the SQUID signal remains unaffected. Thus the dynamic range may be increased to nearly any value and is -in principle- only limited by the precision of the DSP board (DSP 32). Contrary to analog controlling, the detected signal is available in digital form (32 bits). The analog part is connected to the DSP board via an optical link which galvanically decouples the SQUID from the DSP and the computer, thus avoiding electrical and magnetical disturbances. The control of the system is realized by a LabVIEW® program for adjusting the parameters of the electronics and allowing the storage, procession and visualization of the measured data.

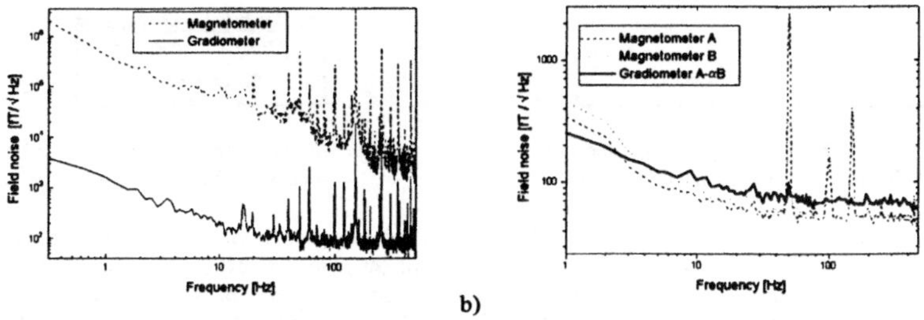

a) b)

Fig. 2: Noise spectra of the gradiometer: 2a) outside any shielding with analog electronics, 2b) with digital electronics inside the standard MSR at ISI.

The gradiometer was characterized by measuring the noise cancellation in unshielded laboratory environment (Fig.2a). For this purpose, we chose the analog control unit because of its higher slew rate. The baseline in this case was 3 cm. The reduction of environmental disturbances is about one order of magnitude in the white noise region and even about two orders of magnitude in the low frequency range. The performance of the system is worse in comparison to the system by Tavrin et al. [1] because of the smaller gradiometer balance (300) and its sensitivity to gradient fields. Due to the rigid arrangement of the SQUIDs, it is not possible to improve the balance by adjusting the orientation of the SQUIDs. Nevertheless, the noise reduction is sufficient for using the system in standard shielded rooms.

The shielding factor of the shielded room at ISI (KFA) is larger than 200 at 1 Hz and about 30.000 at 10 Hz [4]. The magnetic noise inside the laboratory is relatively low. Fig. 2b) shows the noise spectra of the two magnetometers and the resulting gradiometric signal measured inside the magnetically shielded room using the digital electronics. The baseline in this case was as large as 11 cm to improve the signal to noise ratio. We achieved a white noise level of 50 $fT/\sqrt{Hz}$ for the magnetometers and 70 $fT/\sqrt{Hz}$ for the gradiometric signal. The dominant noise contribution is due to the power line frequency at 50 Hz. The 50 Hz power line interference is strongly reduced in the gradiometric signal, but the reduction depends greatly on the environmental situation. At times, the 50 Hz signal could hardly be compensated due to large gradients of this component inside the shielded room. Nevertheless, this system worked stable enough to measure high quality MCG maps without any 50 Hz notch filter.

Results

Magnetocardiograms were recorded inside the MSR in ISI using the digital control with a subject lying in the supine position and the dewar placed above the anterior chest wall. One ECG lead was simultaneously recorded and served as a trigger for averaging over isomorphic heart actions. Figure 3 shows the magnetocardiogram of a healthy subject measured with this gradiometer during day-time. The number of averaged heartbeats is 100, the frequency bandwidth is 0.1 to 150 Hz. The signal to noise ratio, calculated as ratio of the R-peak and half of the peak to peak noise, is 120. Due to the small distance between HTS SQUID sensor and subject (distance dewar inside - outside 8 mm), a high signal to noise ratio is achieved.

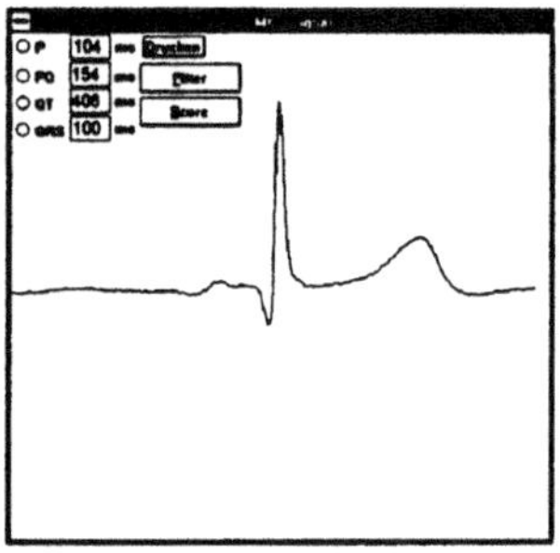

Fig. 3.: MCG of a healthy subject, recorded in the MSR in ISI by averaging of 100 heart beats.

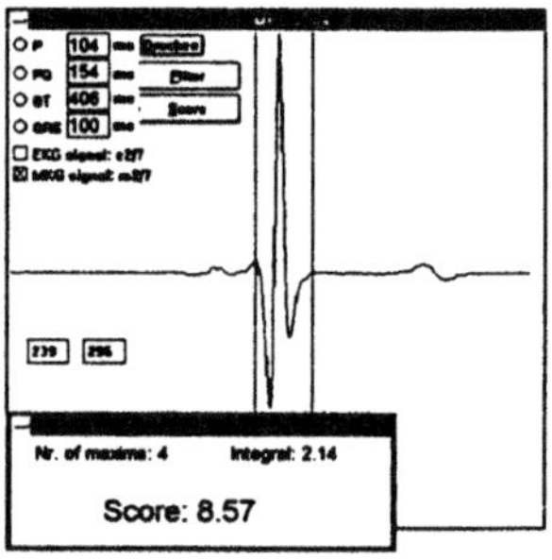

Fig.4: MCG signal processed with a binomial recursive filter algorithm developed by PTB

We plan to use this digital system for the investigation of patients with cardiological diseases. One issue will be the risk stratification of sudden cardiac death. For this purpose, a nonrecursive filter algorithm [5], which was developed in the Physikalisch-Technische Bundesanstalt, Berlin (PTB), for risk stratification in patients with coronary artery disease, was implemented. This filter uses the frequency range between 37 Hz and 90 Hz and evaluates the number and amplitude of extrema in the filtered data. As an example, Fig. 4 shows the filtered MCG signal of corresponding to fig. 3. The calculated score value is 8.6.

By moving the position of the subject under the SQUID-system, an MCG map can be obtained (Fig. 5), which served to reconstruct the magnetic field distribution above the subject's chest. This can be used for the noninvasive localization of the origin of cardiac arrhythmias, e.g. those responsible for atrial, atrio-ventricular, or ventricular tachycardias. The knowledge of the exact site of the arrhythmogenic tissue prior to catheder ablation can significantly shorten the time for this procedure. For this purpose, an algorithm for source localization is being implemented.

Outlook

A similar digital system will be installed at first in the PTB Berlin, where an evaluation of the signal quality in excellent shielding will occur. Then the system will be installed in the Franz-Volhard-Clinic (Berlin) in a standard magnetically shielded room, in order to carry out measurements with patients. While the cardiologists gain experience in using this system in clinical environment, a further optimization of this system in KFA will be carried out simultaneously. Thus, the interaction between users (cardiologists) and system developers will accelerate the development of HTS-SQUID systems for practical use in clinical routine.

Further improvement of the system will include a new generation of HTS SQUID sensors. By using multilayer technique we want to increase the sensor sensitivity at least by a factor of 2, thus significantly improving the signal to noise ratio.

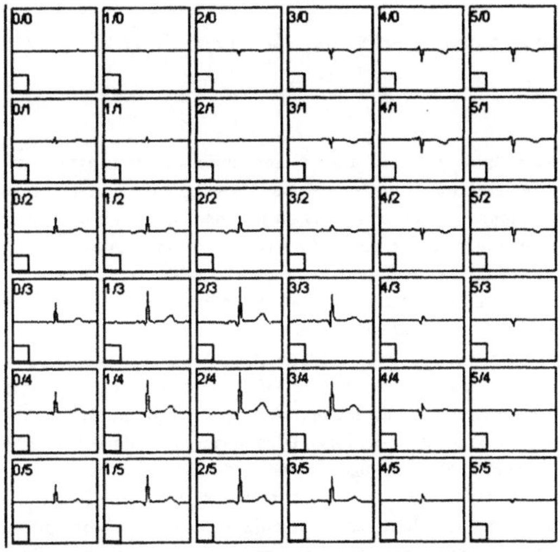

Fig. 5: An MCG map of a healthy subject recorded in the standard shielded room at ISI (KFA). The measuring points of neighbouring MCGs are separated by 4 cm.

Acknowledgements

We would like to acknowledge the support by W.Wolf.

References

[1] Tavrin, Y., Zhang, Y., Mück, M., Braginski, A.I., and Heiden, C., YBaCuO thin film SQUID gradiometer for biomagnetic measurements, in Appl. Phys. Lett. 62 (15), 12 April 1993

[2] Zhang, Y., Krüger, U., Kutzner, R., Wördenweber, R., Schubert, J., Zander, W., Sodke, E., Braginski, A.I., Strupp, M., Single layer YBaCuO radio frequency SQUID magnetometers with direct-coupled pickup coils and flip-chip flux transformers, in Appl. Phys. Lett. 65 (26), 26 December 1994

[3] Zimmermann, E., Brandenburg, G., Clemens, U., Halling, H., Kompensationsregelung für extrem nichtlinearen Sensor mit digitalem Signalprozessor, in Tagungsband DSP Deutschland 95, 133-142, 25.26 Sep. 1995, München

[4] Davidson, C.R., Griffin, R.H., Best, K.J., Maltin, L., Wolf, W., Zhang, Y., Bousack, H. Soltner, H. , Design of a magnetically shielded room for development of HTS SQUID systems for magnetocardiography, this conference

[5] Link, A., Trahms, L., Zimmermann, R., and Oeff, M., Complex binomial bandpass filters for analysing phase and envelope of high frequency components in cardiac signals. Computers in Cardiology, IEEE Computer society press 1994: 617-620

A Realtime, High-Bandwidth, Programmable Digital Signal-Processing System for Monitoring, Acquisition, Filtering, and Archival Storage of Biomagnetic Signals

Corby, N.R.[1], Hogle, R.A.[1], Miller, P.D.[1], and Pasquarelli, A.[2]

[1]GE Corporate Research and Development, Schenectady, N.Y., U.S.A.; [2]CNR, Rome, Italy

Introduction

As biomagnetism activities move from research to applications, there is greater demand for flexible, clinical-quality interfaces with ever-increasing performance levels. These operator interfaces require high-power graphical workstations to control the complex sensor apparatus during the acquisition session and to allow an operator or experimenter to comprehend and monitor the tremendous volume of data being acquired. The combined number of channels, sampling rates, and sampling resolutions for current applications demand extremely high digital bandwidths. Finally, the fast-paced development of sensors and the ever-expanding variety of applications have emphasized the need for highly flexible and expandable digital signal-processing power.

A new biomedical research center (the Istituto Tecnologie Avanzate Biomediche [ITAB]) is currently nearing completion in Chieti, Italy. This new institute, with a full suite of diagnostic imaging systems and bioelectric sensing systems, will provide researchers with the opportunity to study bioelectric phenomena at a new level of detail. A large team of private companies, national laboratories, and university researchers are cooperatively involved in developing equipment to allow acquisition and study of bioelectric signals. Researchers at the General Electric Company have become involved as team members in two areas. The first area is the development of high bandwidth realtime systems to acquire, filter, and archive the signals from biomagnetic or body surface potential sensors. The second area is the development of high-performance analysis systems for the postacquisition analysis of bioelectric signals. This paper describes activities in the first area.

Methods: Accommodating General System Requirements

The general design specification, determined by our ITAB collaborators, was for a sensing head (located within the shielded room) to deliver digitized signals over a single fiber link while receiving commands sent into the shielded room over a second fiber. It was desired that the acquisition and operator control systems for both instruments (an MCG system and a helmet-based MEG system) be modular and as similar as possible.

The following requirements were specified. Each system must accommodate up to 256 signal channels. A signal can be either magnetic (from a coil/SQUID detector) or electric (from a body electrode). Varying signal quantization of up to 24 bits have to be accommodated. Samples are formatted into 256-word packets, each of which represents a single time sample made simultaneously across all channels. The words not assigned to signal channels are free to be used to sample other processes (e.g., patient temperature) or to record housekeeping data (e.g., time stamps or status parameters). Packet composition must also be programmable.

The sensor provides digital samples taken at a fixed rate of 10,000 samples per second, per channel. The acquisition and control system must provide digital decimation so as to yield acquisition rates varying from 10,000 down to 100 samples per second, per channel. The processing system also is required to provide additional digital lowpass and highpass filtering with performance specified by the user.

The system must provide archival storage of up to 1 gigabyte per patient session, with a lifetime of at least 10 years. All acquisition parameters relating to the patient session (e.g., sampled signal waveforms, amplifier gains, physician's notes, sensing head geometry, software version numbers) are to be stored along with the sampled signal data on the archive media. During acquisition, single component hardware failures (e.g., disk or CPU) must not cause loss of recorded data, and all storage must be redundant until archiving has been accomplished.

The system must provide maximum clarity and fidelity when displaying data. For example, CRT stripchart-like displays must have accurate scrolling speeds and have vertical and horizontal gains that match paper strip chart devices. The operator console must provide for a range of operator skill levels, from automated operation by a less skilled operator, to full, on-line access of all system functions and parameters by a skilled physician or researcher. The operator console must also provide means to monitor over 200 signals while providing clinical-quality waveform displays on a smaller number of selected channels.

Results: Overall System Design

The systems that we have constructed meet the stringent design requirements. Fig. 1 is a block diagram of the three major components: the operator console, the acquisition and archive controller, and the realtime digital signal

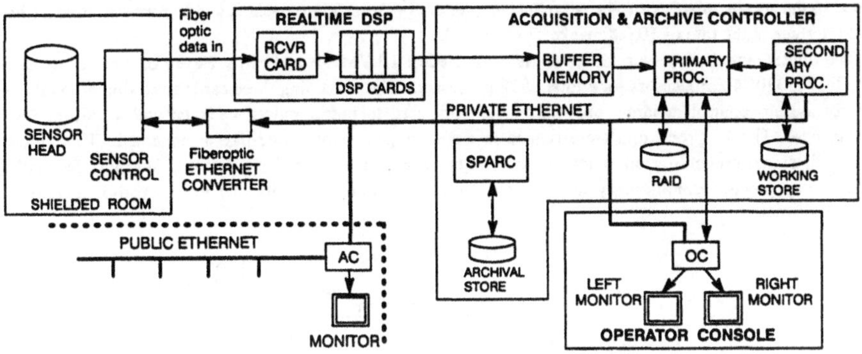

Fig. 1. The major components of an MSI system.

processor. Physically, the system consists of two units. The operator console is an HP 755 graphics workstation with two high-resolution displays. The left screen [see Fig. 2(a)] displays a large number of signal waveforms as small, low-resolution stripcharts. The right screen [see Fig. 2(b)] is primarily used for operator control and high-resolution display of selected channels.

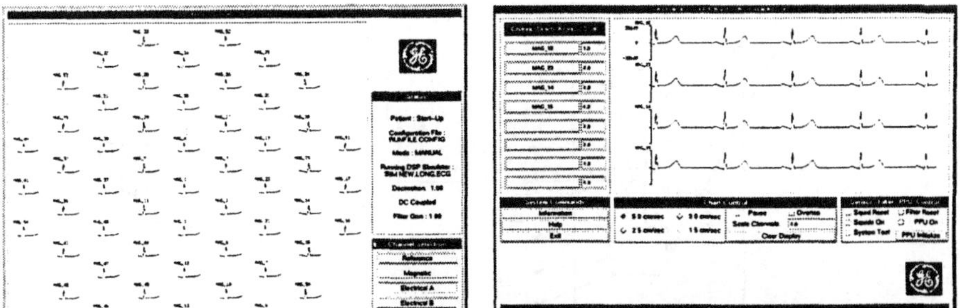

(a) (b)

Fig. 2. Left and right screens of the operator console.

The remainder of the system occupies a low-profile 19-inch equipment rack containing a VME chassis with embedded disk/tape storage units, and a 4-gigabyte RAID-3 disk array. Archiving is performed in the background to a magneto-optical disk. A dedicated Ethernet network is used to communicate status and control information among individual processors. This local Ethernet has a fiber optic extension to the sensor system located within the shielded room. Patents have either been granted or are under review for the primary features of the system described herein.

Discussion: The Realtime Digitial Signal-Processing System

The sensor head samples up to 256 signals at a fixed rate of 10,000 samples per second, per signal. This rate is high enough to capture high-frequency information residing within the signals of interest. The sampling is done synchronously across all channels. Fixed analog antialiasing filters are used prior to the analog-to-digital converters located in the sensor subsystem. The resultant samples are organized into packets and transmitted over a fiber-optic serial interface at the rate of 10,000 packets per second.

The digital signal processor subsystem is composed of 10 VME-based cards. Seven of these cards each carry four TI-TMS320C40 processors for a total of 28 processors. The remaining three cards are a single-board computer used for system control; a large, dual-ported memory card to buffer output packets; and a custom fiber-optic receiver card. The 28 processors are arranged in a four-stage, systolic, tree-like arrangement. This arrangement enables all signal-processing operations such as signal compensation, noise rejection, filtering, decimation, and formatting to proceed continuously in real time. Fig. 3 illustrates the units that make up the signal-processing subsystem.

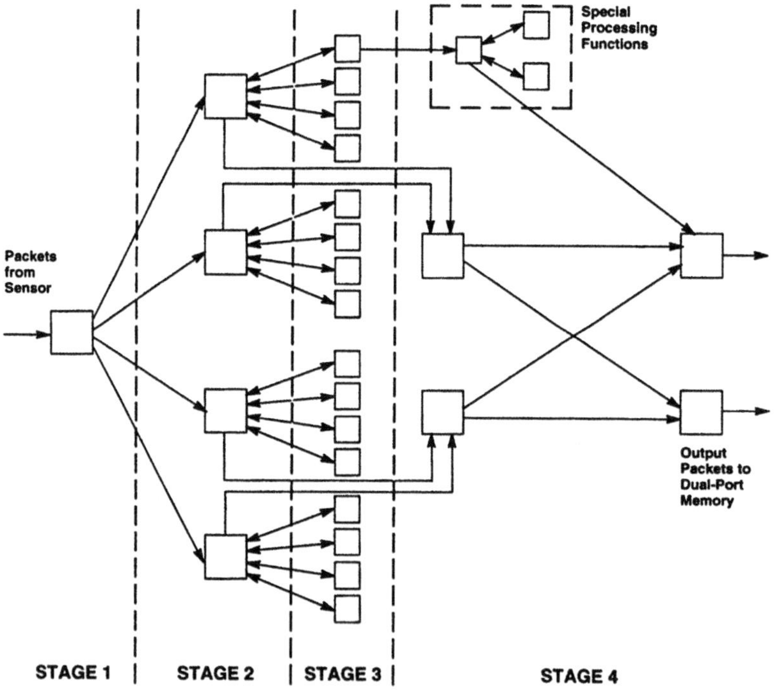

Fig. 3. Digital signal-processing array.

The data packets arriving via the fiber-optic link from the remotely located sensor head are received by a custom VME receiver card and flow into the first stage of the DSP processing array. The incoming sample rate is 2.56 megasamples per second. The format currently used specifies 3 bytes per sample (24 bits), giving an aggregate input rate of 7.68 megabytes per second.

The architecture of the DSP array is based on a pipelined tree structure. Each pipeline stage completes its work in one packet time (100 microseconds). The current implementation of the processing pipeline consists of four stages.

The first stage consists of a single input chip which performs data integrity checks on the incoming 256-word packet and maintains overall timing and control of the array processor. The input chip creates four new packets, each of which contains approximately 25% of the original data. Reference signal data, which are used for noise cancellation, are replicated into each of the four output packets.

The second stage consists of four chips connected to the input chip via commlinks. These second-stage processors perform input data conversions (all processing of signal data is done in floating point) and interchannel operations such as noise compensation. The compensation calculations allow a unique linear, weighted sum of specified samples to be added to each signal channel in real time.

The third stage consists of a 16-chip processing farm that performs channel-independent operations such as FIR and IIR filtering. High, low, and bandpass filtering are provided. Parameters, such as cutoff frequency and gain, are selectable by the operator. Filtering to allow variable digital decimation is also done in this stage. The operator can select from a menu of decimation values.

The processed data from the third stage are recombined by a fourth stage into output packets occurring at a rate of $10{,}000/D$ (where D is the user-selectable integer, decimation rate), and the packets are accumulated in an output buffer. The content, arrangement, and format of the processed data packets are completely programmable by the operator.

A second, parallel DSP pipeline processes operator-designated channels and performs two functions. First, the pipeline processes the data for the high-quality stripchart display. The data to be displayed must be lowpass filtered to minimize the aliasing artifacts on the CRT screen. Also, data resampling must occur to allow jitter-free scrolling of the stripchart-like displays on the operator console. Second, the pipeline can implement user-selected signal-processing routines such as computing signal power spectra or performing signal averaging functions. This second pipeline is shown as a small cluster of three processors in Fig. 3. The output of this pipeline is placed in the dual-port memory buffer for transmission to the HP 755 screen.

Status and Future Work

Two systems have been built and shipped to the Chieti site. The performance of the MEG system and the MCG system has been verified in our laboratory at rates exceeding the required data rates using a sensor simulator. The network of 28 processors performs well and has been experimentally expanded to 32 processors with little problem. The MCG acquisition and archiving system was shipped to Chieti in early 1995 and approved in March 1995. The MEG system is currently in Chieti awaiting assembly, with final acceptance testing scheduled for March 1996. It is anticipated that integration and final testing of the MCG system will occur this year as other team members complete their assigned subsystems and tasks. Implementation of the MEG is expected to follow soon after completion of the MCG system.

Magnetic Measurement Techniques for Locating Foreign Bodies in Humans

Costa Monteiro, E.[1], Kelber, C.[1], Andrade Lima, E.[1], Costa Ribeiro, P.[1], Bruno, A.C.[1], Hall Barbosa, C.[1], Parente Ribeiro, E.[1], and Boechat, P.[2]

[1]Departamento de Física, Pontifícia Universidade Católica, Rio de Janeiro, Brazil; [2]Hospital Fernandes Figueiras, Rio de Janeiro, Brazil

Introduction

There is a high incidence of needles accidentally introduced in the human body. The commonly used radiological techniques present several limitations. They usually do not give the appropriate informations for a surgical extraction procedure, like the depth relative to the skin and only the neddle projections are available. The development of a more accurate, non-invasive and innocuous method for a magnetic foreign body localization is of meaningful clinical importance. In this way, non-invasive magnetic field measurements, using a SQUID magnetometer, were performed to locate, for surgical purposes, needles lost in the bodies of 5 children. Also, another approach for detecting magnetic fields generated by foreign bodies, using a fluxgate magnetometer, is proposed.

Material and methods

SQUID measurements were performed to locate a hypodermic needle in a 3 year old child and sewing needles in four other children (4, 6, 8 and 11 years old). The present technique is an improvement of a procedure described elsewhere [1] and has been adapted for clinical use. It enables the computer acquisition of the entire magnetic field map in a couple of minutes and provides accurately the spatial position of the magnetic source. The reduction of the measuring time and the more systematic procedure make feasible the clinical use of the technique, since the patient, especially a child, is unable to remain at the same position for a long time. In order to track the needle position, a commercial graphics tablet was coupled to the X-Y bed over which the patient was laying and two laser beams were used to give information of the sensor position relative to the skin surface (Fig. 1).

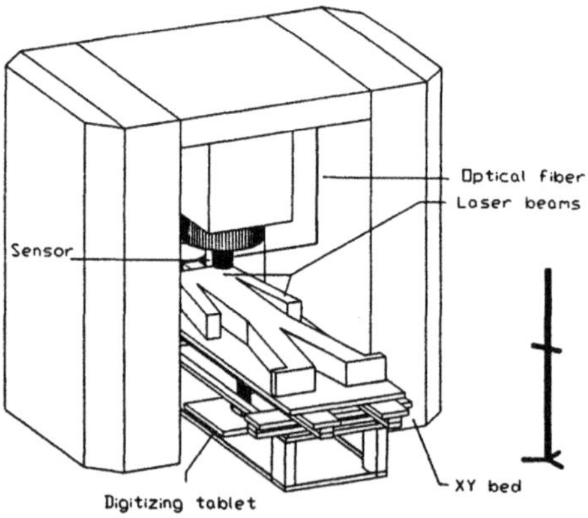

Fig. 1 - Arrangement of the magnetic sensing system. The patient is laid over a moving bed under the magnetometer and the laser beam gives information of the sensor position on the skin surface plane.

A computer program was developed, to acquire the SQUID output due to the magnetic field generated by the needle remanent magnetization, and displays it on-line on the computer screen using pseudocolor (see Fig. 2). Thus, while moving the X-Y bed, the different magnetic field intensities appear gradually on the computer screen making easy the identification of the two extreme field values. These points are marked on the patient skin with an uneraseble pen with the aid of one of the laser beams. Such references give information about the needle center and orientation. To monitor the distance between the Dewar tail and the patient skin, the second laser was used, horizontally fixed in a tripod.

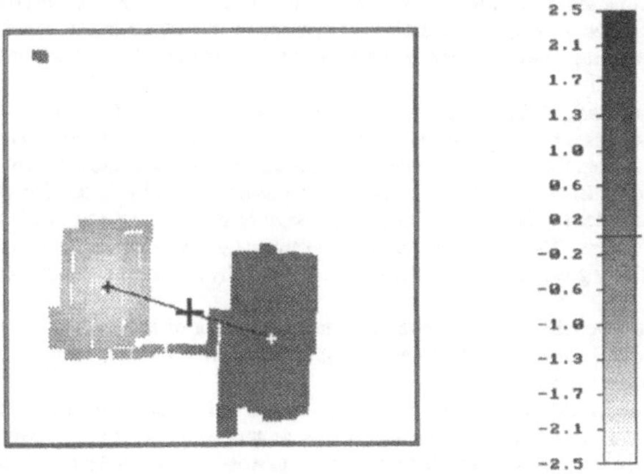

Fig. 2 - On-line map of the SQUID output as it appears on the computer screen.

The depth of the needle in relation to the skin is obtained from the distance between the points of magnetic field extremes, and from the ratio of their absolute values, using the graphic shown in figure 3. Two other graphics give the information about the inclination of the needle and the distance between its center and the center of the line joining the extrema.

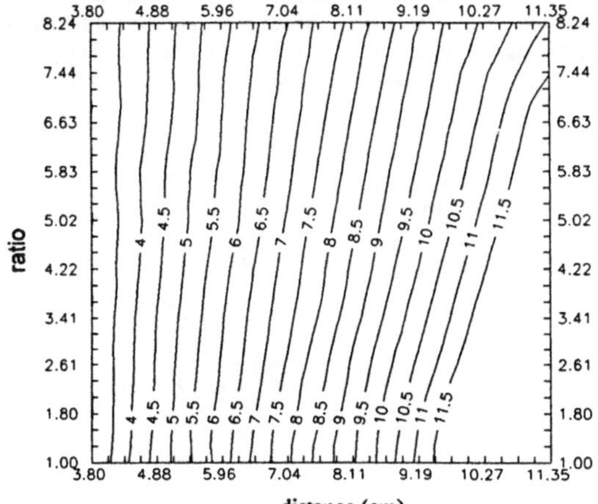

Fig.3 - Curves of ratio of amplitude at extrema vs. distance between field extrema for different depths of a 3 cm long needle.

These graphics were generated assuming the needle as a line of magnetic dipoles of finite length. The magnetic field produced by each dipole, given by the standard formula [2], is numerically integrated along the line of dipoles and the integration over the surface of each coil of the gradiometer is approximated using a technique described elsewhere [3].

Results and Discussion

In a patient with a sewing needle located close to the bladder, it was observed a systematic alteration of the needle depth and inclination although maintaining its orientation in relation to the plane of measurement, as a function of the bladder filling. Two different possible positions of the needle were given to the surgeon: depth, inclination and center position of the needle for the empty and full bladder. The depths were 2.3 cm (empty bladder) and 4.1 cm (full bladder). The needle was easily surgically removed within 15 minutes.

In the other four cases the magnetic mapping also allowed an easy surgical localization of the needles with a precision better than 2 mm. The depths of the center of the needles relative to the suface of the skin were found to be 0.8, 4.3, 5.0 and 6.5 cm. In one case, however, the needle was not surgically found. The child presented a strong difficulty of keeping quite during the magnetic mapping procedure and had a sewing needle located in a region of a very difficult surgical access. The conditions of the ambient magnetic noise in relation to the strength of the detected field were not very good in this case. Although magnetically localized it was not possible to extracted the needle surgically.

These results showed that this technique can reduce by a factor of at least 5 to 6 times the surgery time for foreign body extraction and diminishes its generally high risk failure. Without using this technique the surgery time can be easily a couple of hours.

Although the undeniable efficiency of the foreign body location performed using a SQUID sensor, there is a drawback when applying this technique in a hospital environment due to the high cost of maintenance of the SQUID system. The use of a simpler magnetic sensor, easier to work with, could be a considerable improvement. A fluxgate magnetometer has been considered as a good substitute for the SQUID. The remanent magnetization of the needle, due to its manufacturing process, varies among different samples. sometimes by more than an order of magnitude. The magnetic field generated by a typical sample of needle, distant 5 cm from the sensor, may be a fraction of nT. As the fluxgate sensitivity is of the order of nT, it would be necessary to increase the remanent magnetization. A typical needle remament magnetization due to the fabrication process is in general many orders of magnitude lower than its saturation value. The remanent magnetization can be increased by 2 to 3 orders of magnitude, by placing a 0.3 T permanent magnet on the surface of the patient skin even 5 to 10 cm over the place where needle is lodged. Due to the fabrication process, the magnetic domains are oriented in the needle axis. This assures that the final remanent magnetization will be always in the axis direction, so the exact orientation of the permanent magnet in relation to the needle axis is not critical. The approximate direction of the needle is given by an X-ray measurement.

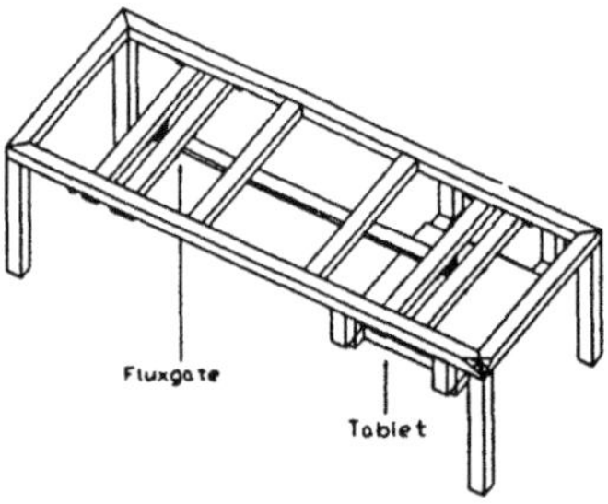

Fig. 4 - Movable Fluxgate assembly.

An assembly has been made with a movable fluxgate, the patient maintained in a fixed position (figure 4). This apparatus allows the magnetic field mapping and the needle location (figure 5). The maximum field obtained after the needle magnetization at a 5 cm distance was 300 nT.

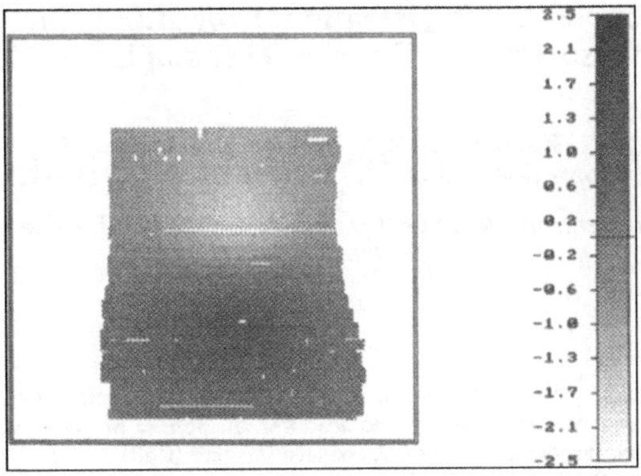

Fig. 5 - On-line map of the Fluxgate output as it appears on the computer screen.

Acknowledgements

This work was partially supported by CNPq grants (n.: 53.0484/93-5; 300665/93.8; 30.0959/81-8), by FINEP grant (n.: 6593056600), and by PADCT grant (n.:62.0708/94-8), Brazil.

References

[1] Ilmoniemi, R.J., Williamson, S.J., Kaufman, L., Weinberg, H.J. and Boyd, A.D., Method for Locating a Small Magnetic Object in the Human Body, IEEE Trans. Biomed.Eng., 1988, 35-7:561-564.

[2] Reitz, J.R. , Milford, F.J., Foundations of Electromagnetic Theory, Reading, Addison-Wesley, 2nd. ed, 1967.

[3] Roth, B. J., Sato, S., Accurate and efficient formulas for averaging the magnetic field over a circular coil, In: Hoke, M. et al., Biomagnetism: Clinical Aspects, Excerpta Medica, 1992, p.797-800.

Optimisation of Co-Registration Procedures for Biomagnetic Measurements Using a Thermoplastic Mask

Dammers, J.[1,2], Barnes, G.R.[1], Weise, F.[1], Gross, J.[1], Clewett, B.S.[2], Ioannides, A.A.[1,2], Müller-Gärtner, H.-W.[1]

[1]Forschungszentrum Jülich GmbH, Jülich, Germany; [2]The Open University, Milton Keynes, UK

Introduction

The BTi MAGNES II twin system was installed in June 1994 at the Research Centre in Jülich and will be exchanged soon for the new MAGNES WH whole head system. Besides Magnetoencephalography (MEG), the research centre also houses functional Magnetic Resonance Imaging (fMRI), Single Photon Emission Computed Tomography (SPECT) and Positron Emission Tomography (PET) neuro-imaging groups. The goal is to provide multi-modality imaging exploiting the strengths of each technique.

In any comparison between modalities, or indeed experimental repeats within one modality, it is necessary to establish an accurate and unequivocal subject head based coordinate system. For example, a traditional coordinate system is based around the nasion and the bilateral pre-auricular landmarks on the subject's head (see figure 1).

In this paper we set out to investigate a coordinate frame based upon a thermoplastic mask, moulded to the subject's head. The accuracy and reproducibility of the measurements is examined across operators.

Methods

Theory

For the BTi MAGNES II system LPA, RPA (left/right pre-auricular) and nasion are used to define the head coordinate system [1].

The origin of the head frame is defined as the midpoint between the LPA and RPA. The line extending through the origin and the nasion is the positive X axis of the head frame. The positive Z axis is a line extending through the origin and the top of the head, perpendicular to the line between the left and right PA. The positive Y axis is orthogonal to X and Z and extends from the origin through the left side of the head.

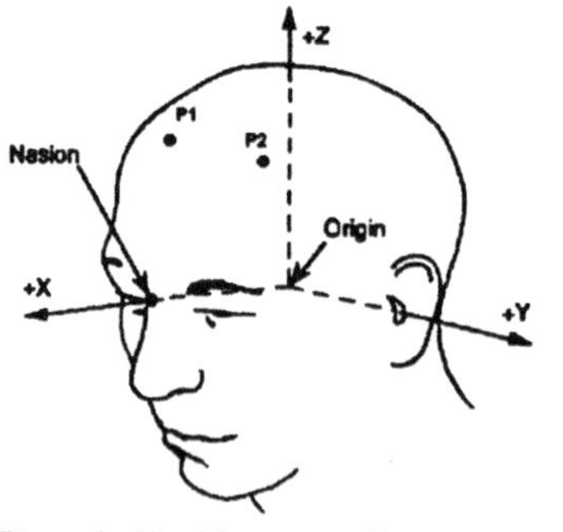

Figure 1. Head frame coordinate system
(from BTi Software Reference Manual, 1995)

Experimental procedure

A strip of thermoplastic material, with eye and mouth holes cut out, was heated to 60° C, stretched and moulded around the face of a volunteer. Two reference points (P1 and P2) were marked once on the subject's head and not erased during the experiment.

The subject lay on an adjustable bed supported by a vacuum cushion to avoid head movements during the experiment. All estimates of point positions were recorded with a 3D position indicator (SPI, 3-space tracker) and stored for further analysis.

The mask coordinate system was defined by three holes (diameter ≈ 2 mm) in the approximate positions of the nasion and the bilateral pre-auricular points. The head coordinate system was based around the nasion and pre-auriculars points identified visually on the subject's scalp.

Four operators were each asked to set the coordinate frame and locate the marked points P1 and P2 over a set of 32 trials performed over a four day period. Alternate trials were conducted with and without the use of the thermoplastic mask. The experiment compared the reproducibility of the points P1 and P2 in both the mask and head coordinate frames.

The accuracy of the SPI was tested by the four operators on an acrylic sphere phantom with permanently marked fiduciary points and an outer diameter of 20 cm. For each investigation the stylus was placed perpendicular to the outer surface of the subject/sphere.

Results and Discussion

Individual operator reproducibility

We calculated the individual operator means and the standard deviations (expressed as the length of the vector of standard deviations of the x,y,z components) for both fiduciary points P1 and P2 for the three conditions (no mask, mask and phantom). Since P1 and P2 remained fixed throughout the experiment, any variation in recorded position of P1 and P2 reflected a change of the coordinate system. Figure 2 shows the standard deviations of the point P1 for each of the four operators. The figure shows that by using the mask the individual operators did not improve their ability to consistently define a coordinate system. This could be due to the errors involved in fitting the mask to the subjects face, and due to the loss of visual information (the mask is opaque) when using the mask.

Reproducibility across operators

The all operator mean of the coordinates of P1 and P2 was calculated for both mask and no mask conditions. Figure 3 shows the inter-operator standard deviation from this mean for both conditions; the use of the mask has reduced the inter-operator deviation by a factor of 4.

Phantom measurement

The phantom measurements were used to estimate the accuracy of the SPI. This is illustrated by the standard deviations of P1 for each operator (see figure 2) and the standard deviations of the mean P1 and P2 across the four operators, as shown in figure 3. Both figures show that either the standard deviation for the individual operator and across the four operators is in a range of 0.2 mm and 0.4 mm.

Conclusion

The use of a thermoplastic mask provides a head based coordinate system which is robust across operators; it therefore paves the way to reliable multimodal imaging studies.

References

[1] BTi San Diego CA, Software Reference Manual, 1995, Volume I

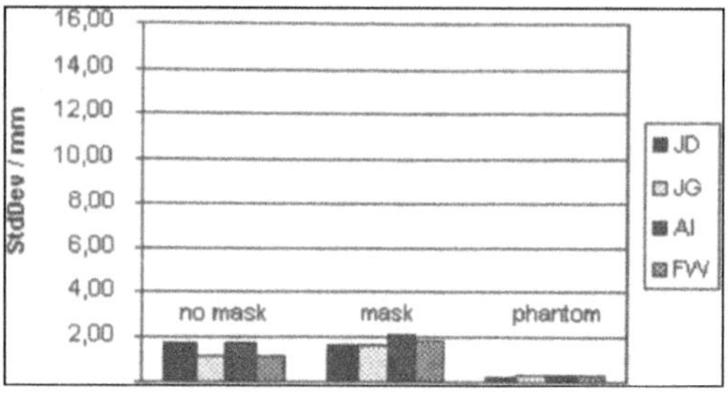

Figure 2. Standard deviation of one fiduciary point P_1 for three conditions (no mask, mask, phantom) measured by four operators.

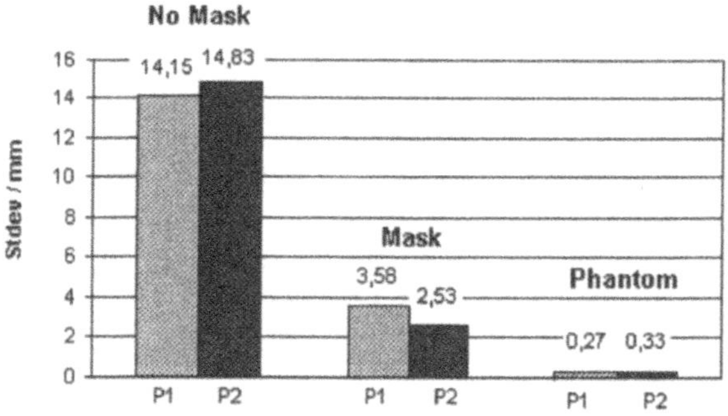

Figure. 3. Standard deviation of the mean fiduciary points P_1 , P_2 across four operators.

Design of a Magnetically Shielded Room for Development of HTS SQUID Systems for Magnetocardiography

Davidson, C.R.[1], Griffin, R.H.[1], Best, K.J.[2], Maltin, L.[1], Wolf, W.[3], Zhang, Y.[3], Bousack, H.[3] and Soltner, H.[3]

[1]AMUNEAL Manufacturing Corporation, Philadelphia, PA, 19124 USA; [2]Best Engineering, D-63452, Hanau, Germany; [3]Institute of Thin Films and Ion Technology (ISI), Forschungszentrum Jülich GmbH (KFA), D-52425 Jülich, Germany

Introduction

Using HTS rf washer SQUIDs together with bulk flux concentrators, members of ISI had recorded MCGs in the magnetically shielded room by Physikalisch-Technische Bundesanstalt (PTB), Berlin, already in 1994 [1]. The high signal to noise ratio of these MCGs make them relevant for clinical applications and offer the perspective that some of the clinical domains especially in cardiology with LTS SQUIDs will be taken on by HTS SQUIDs. A prerequisite for this goal is the applicability of HTS sensors in shielded rooms with a lower shielding factor than the one by PTB. For the development of HTS systems for cardiological research, AMUNEAL provided a moderately shielded room to ISI. Single channel and multichannel HTS SQUID systems will be developed in ISI in collaboration both for and with clinical partners in Germany.

One of the main characteristics of an MSR is the frequency dependent shielding factor for magnetic fields. Two different methods were suggested to us to measure this factor. To evaluate the performance of the MSR we applied both of the methods and will discuss them in detail in this contribution.

Methods

The MSR was set up by AMUNEAL in October 1995 and is located in the Institute of Thin Films and Ion Technology in an ordinary laboratory environment. It consists of two layers of μ-metal (thickness 1.27 mm each) and one layer of aluminum (thickness 12.7 mm) and has the following dimensions. (The x-direction is normal to the closed door of the MSR, y denotes the other horizontal direction, z is the height.) Outer dimensions (x = 290 cm, y = 290 cm, z = 345 cm), inner dimensions (x = 244 cm, y= 250 cm, z = 300 cm).

For the determination of the frequency dependent shielding factors for each of the three principal directions of the MSR a magnetic field of variable frequency is applied from the outside. The ratio of the theoretical value of the field at the center of the shielded room to the measured value, recorded with an appropriate sensor, is the shielding factor. In the most ideal case the applied magnetic field from the outside should immerse the shielded room as homogeneously as possible, i.e. it should be generated by a coil with much larger dimensions than the MSR. In practice, this is impossible and one has to make compromises, e.g. the employment of three coil pairs wound at the edges of the shielded room, with the individual coils of each pair facing each other. This configuration was also used for the evaluation of the shielding factor of our MSR and will be called the large coil method for simplicity and to distinguish it from the small coil method. The latter generates a magnetic field by a relatively small coil placed at a distance from one of the walls of the MSR but also uses a magnetic field sensor in its center.

Instrumentation

For the small coil method a hexagonal coil was placed at a distance of 247 cm to the outer wall of the MSR, and fed with a current of 14.142 A. Due to spatial restrictions only the measurements for the shielding along the x- and y- directions were carried out. A synchronous sine wave from 0.010 Hz to 1 kHz was generated using a FFT Network analyzer (Stanford Research System SR 770) to the input terminal of a power amplifier (Techron

Model 7571). A fluxgate magnetometer (Applied Physics System SR 770) was used to measure frequencies from 0.010 Hz to 100 Hz. The output signal from the fluxgate was sent back to the network analyzer. Current was monitored by measuring the voltage drop across 10 parallel 1Ω resistors, in series with a 10 turn 38.5 cm radius excitation coil at a distance of 247 cm from the outer wall. For frequencies from 300 Hz to 1 kHz the fluxgate was replaced by a ferromagnetic core (Meda MGCH 2) and a signal conditioner (Meda Sam-3).

For frequency generation and lock in detection of the response signal from the magnetic field sensor for the large coil method, we used a lock-in detector from Stanford Research Systems (Model SR830, DSP Lock-in Amplifier) providing frequencies in the range from 0.001 Hz to 10,000 Hz. The output voltage from the frequency generator of the lock-in detector was fed to the input of an amplifier (Model BOP50-8M, Bipolar Operational Power Supply, by KEPCO), which in turn fed the induction coils corresponding to the desired measurement direction. The output of the magnetic field sensor was analysed by the lock-in detector again. The current which was fed through the coils was determined by the voltage drop over an additional in line 1 Ω resistor using a digital oscilloscope for display (Hewlett Packard model nr. 54610 B) or at the very low frequencies by a multimeter in the current measuring mode. We used several magnetic field sensors such as a fluxgate, an rf-SQUID sensor and and a homemade induction coil with many turns [2]. The SQUID sensor and its electronics have already been described elsewhere [3]. The fluxgate was manufactured by Billingsley (TFM 100-LN). These sensors had the following characteristics:

low frequency coil	coil constant:	1.99 mV/(μT*Hz)	frequ. range: 0.05 - 20 Hz
fluxgate	flux to voltage conversion:	1000 mV/μT	frequency range: 0.001 - 200 Hz
rf SQUID	flux to voltage conversion:	10^6 mV /μT	frequency range: 0.05 - 20 Hz

The dc value of the shielding defined via the attenuation of the earth magnetic field was measured by the use of a fluxgate (Billingsley TFM 100-LN), which was oriented in the center of the MSR along the principle axes and comparing the results to the values far away from the MSR along the same directions.

Results

Fig. 1 and Fig. 2 show the shielding values for the x- direction and the y-direction obtained with the small coil method (square symbols). The shielding factor is about 30 to 300 in the range from 0.01 Hz to 1 Hz. A steep rise leads the shielding factor to values of more than 10,000 for frequencies higher than 10 Hz. For comparison the data of the large coil method obtained with the fluxgate for the corresponding directions have also been included (crosses). For comparison in the frequency range up to 1Hz the insets of both figures show the shielding factors on a linear scale.

The shielding factors for the principal directions of the MSR and for the magnetic field sensors are summarized in Figs. 3, 4, and 5 for the large coil method. The general behaviour of the shielding is the same for all the three directions, although the numeric values may be different by as much as a factor of 10. The data of the magne-

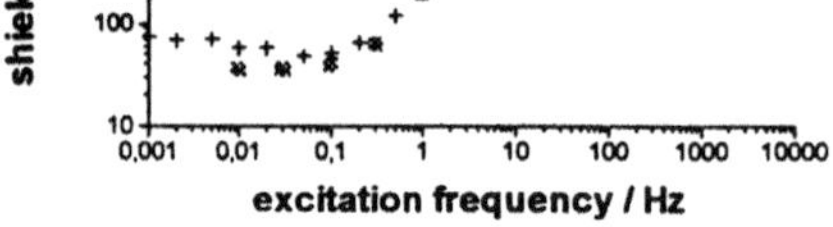

Fig. 1: Shielding along the x-direction determined by the small coil method (square symbols)

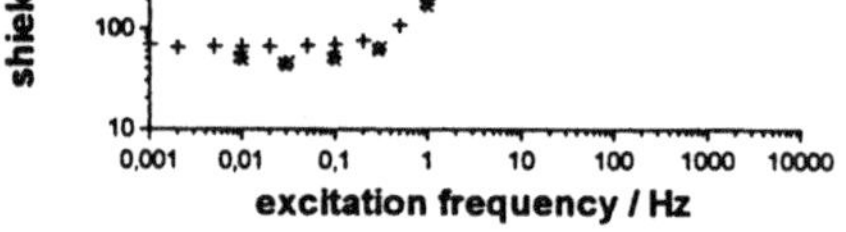

Fig. 2: Shielding along the y-direction determined by the small coil method (square symbols)

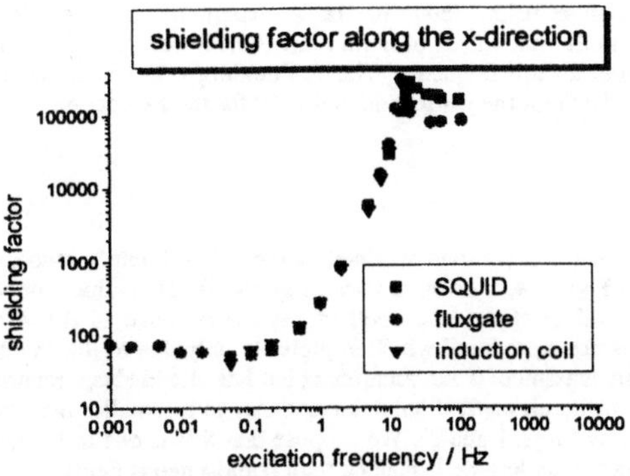

Fig. 3: Frequency dependence of the shielding factor along the x-direction (large coil method)

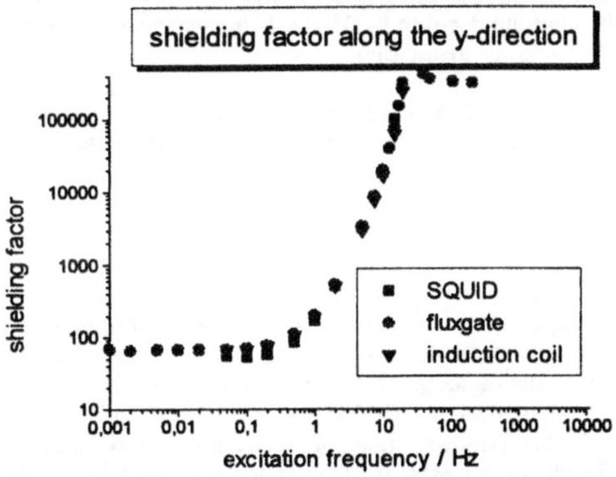

Fig. 4: Frequency dependence of the shielding factor along the y-direction (large coil method)

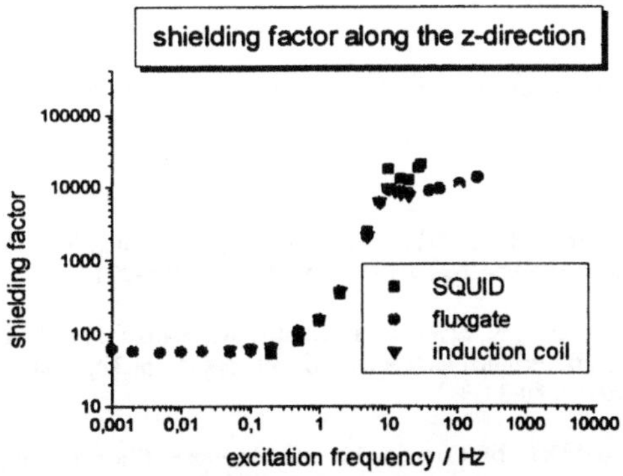

Fig. 5: Frequency dependence of the shielding factor along the z-direction (large coil method)

tic field sensors used are in very good agreement below about 10 Hz. A maximum of the shielding occurs at frequencies of about 12 Hz for all three directions with values of more than 100,000 for the x and y direction, the value for the z-direction being ten times smaller for this frequency. The dc shielding values obtained by the method given above are about 450 for the x-direction, 1000 for the y direction and 1300 for the z-direction.

Discussion

Fig. 1 and Fig. 2 show the correspondence of the data obtained by the two different methods (data for the flux gate measurements are reproduced from Figs 3, 4, 5 using crosses as symbols). The values for the small coil method are somewhat lower compared to the values of the other method, especially above 10 Hz. Also, the maximum of the shielding factor at about 12 Hz is not reproduced which is probably only due to the smaller density of data points for this method in this region. This maximum is not an artefact but has already been reported elsewhere [4]. The shielding factors obtained with the small coil method have the tendency to be smaller than those obtained with the other method (see especially the insets of figs. 1 and 2). We suppose that this is due to the fact that gradient fields (which the small coil method generates) are less well shielded than homogeneous fields.

To our knowledge the shielding of MSRs had not yet been measured below 0.01 Hz, and we had expected an upturn at these low frequencies extrapolating to the high dc value of several 100. Yet one result of this investigation is that the shielding remains practically constant down to 0.001 Hz. If the dc value should be continuously approached, this transition can only happen at even lower frequencies.

The plateau region of the shielding between 10 Hz and at least 100 Hz is lower for the z-direction values compared to those for the other directions. This is because the circumferential eddy currents responsible for the shielding are obstructed by the door of the MSR although care was taken to ensure good electrical contact.

Outlook

The shielding factor is maximum at a range between 10 Hz and 100 Hz and is more than 10,000, even for the z-direction. This frequency range is of particular interest for cardiological investigations as, e.g., the stratification of the risk of patients with malignant cardiac arrhythmia by PTB, Berlin, is performed by analysing the data in the range from 37 Hz to 90 Hz [5]. Thus, for these applications, the MSR should be well suited. For other applications, which more rely on the information in the low frequency range the shielding may be improved by the employment of an active compensation system. Results of this MSR in combination with the a system for active compensation will be reported after its implementation.

Ackowledgements

We thank Bork, J., Copetti, C.A., Görtz, A., and Krause, H.-J. for support and valuable discussions.

References

[1] Trahms, L., Burghoff, M., Schirdewan, A., Bousack, H., Schütt, H., Wolf, W., and Zhang, Y. Magnetokardiographie mit stickstoffgekühlten SQUIDs. Biomedizinische Technik 40, supp. 1, 1995: 199-200
[2] layout of the coils by K.J. Best
[3] Zhang, Y., Krüger, U., Kutzner, R., Wördenweber, R., Schubert, J., Zander, W., Sodtke E., and Braginski, A.I. Single layer $YBa_2Cu_3O_7$ radio frequency SQUID magnetometers with direct-coupled pickup coils and flip-chip flux transformers. Appl. Phys.Lett. 65 (26) 1994: 3380-3383
[4] Best, K.J., private communication
[5] Link, A., Trahms, L., Zimmermann, R., and Oeff, M. Complex binomial bandpass filters for analysing phase and envelope of high frequency components in cardiac signals. Computers in Cardiology, IEEE Computer society press 1994: 617-620

A Digital-Signal-Processor (DSP) Flux-Locked-Loop (FLL) for LTC SQUIDs with Automatic Reset and Feedback Cancellation

Flynn, E.R.[1], Bracht, R.[1], Kraus, R.H., Jr.[1], Maas, P.M.[2], Ruminer, P.[1] and Stettler, M.[1]

[1]Los Alamos National Laboratory, Box 1663, Los Alamos, NM, 87545, USA; [2]Strathclyde University, Dept. of Physics and Applied Physics, Glasgow G4ONG, Scotland

I. Introduction

Linear operation of SQUID sensors is normally provided by a flux-locked-loop, FLL, which provide a negative feedback mode of operation [1]. In the FLL, the signals are amplified, phase sensitively detected, filtered, integrated, and the resulting analog signal provides a measure of the magnetic field and noise in the SQUID output. The output is fed back as a flux to the modulation and feedback coil through a voltage-current converter to oppose the applied magnetic field. The resulting signal from the FLL is proportional to the change in the applied magnetic field. The FLL then provides the SQUID the ability to track changes in the applied field within a well defined linear range provided the slew rate or dynamic range of the SQUID and FLL system are not exceeded.

In the present application, a digital-signal-processing (DSP) chip has been used to replace the usual analogue (FLL) for a low temperature (LTC) SQUID. This method has a number of possible advantages: (1) software control of the SQUID operating characteristics including the addition of filtering and noise reduction, (2) automatic calibration of SQUID performance, (3) flux counting when a flux-quanta limit is reached and automatic reset with storage of flux counts, (4) a direct digital output with up to 32 bit accuracy, and (5) a direct channel into the feedback loop for other signals, such as a background sensor, permitting cancellation at the SQUID itself to further increase the dynamic range. This system has been described in detail elsewhere [2,3] with the present report describing the capability for rejecting external signals through the cancellation port. There are also several other references to the use of DSP in SQUID applications [4,5] and a report on their use for HTC SQUIDs elsewhere in these proceedings [6].

The DSP system described here is being used to provide FLL operation for a large 155 channel biomagnetometer for magnetoencephalography. The sensors are SQUID-magnetometer chips located on a head shaped surface below a superconducting imaging surface to provide gradiometer action while the superconducting surface yields background rejection of between 2200 and 3500.

II. Methods

Figure 1 is a block diagram of the DSP system illustrating the software flow. The dotted line contains the DSP tasks while outside of the dotted line are the ADCs and DACs necessary for digitizing and converting the SQUID output and feedback. The SQUID is also shown in a highly simplified sketch. Of importance here is the method for determining when the SQUID is to be reset and keeping track of resets for the reset integrator and the additional ADC used by the system for adding external signals. The DSP is a 80 ns AT&T WE DSP32C floating point DSP processor. A separate unit contained the ADCs and DACs used for input with these being 18 bit and 200 kHz speed.

The bandwidth of the DSP system has been measured in reference 2 and is displayed in Figure 2 in comparison to the analog circuit, both measured using a sine wave. The DSP is similar in bandwidth to the analog but has a rather constant amplitude as a function of frequency above 6×10^3 Hz. The same SQUID operating with an analog FFL has a dynamic range in a 1 Hz bandwidth of +/- 3.5×10^6 and a slew rate of 5.8×10^5 Φ_0/s, the latter number has not yet been measured for the DSP system. In the DSP results shown here, the temporary testing setup of the DSP and ADC/DAC units contributed to the noise.

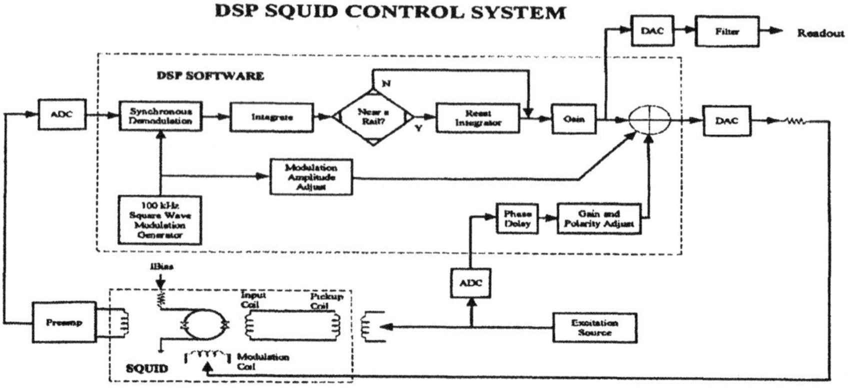

Figure 1

The majority of cancellation measurements were conducted in a shielded chamber shown in Figure 3. Here the static field is s and 100 nT for the horizontal and vertical components, respectively. The line frequency noise was also reduced substantially. This chamber thus permitted accurate measurements of the ability of the DSP FLL to compensate for external fields. The coil shown below the sensor was activated with various waveforms and this same signal was applied to the DSP through the ADC shown in the lower right hand corner of Figure 1. The phase delay and gain-polarity were then adjusted to achieve maximum cancellation of this applied signal. The DSP operated in the backplane of a PC computer and was connected to the SQUID preamp through 20 meter differentially driven cables to minimize noise pickup. However, the system is still not optimized against line frequency interference.

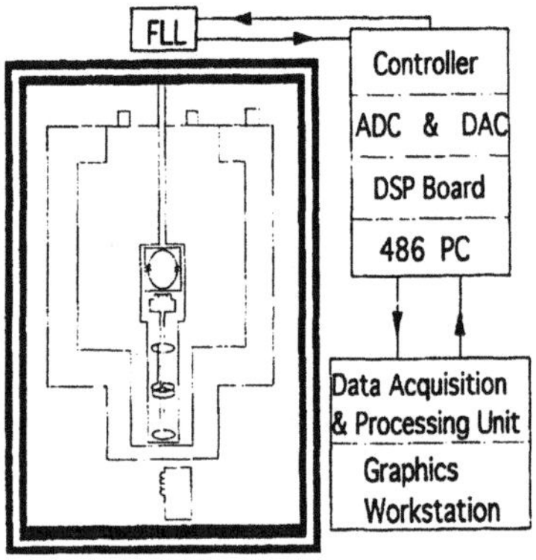

Figure 3

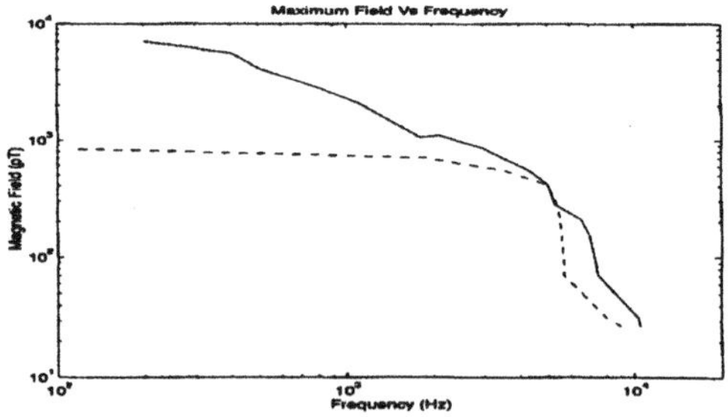

Figure 2

III. Results

A variety of waveforms were applied to the coil under the sensor in the shielded chamber and simultaneously applied to the DSP feedback. The phase and amplitude of the feedback could be adjusted through software commands (under the control of sliders using the data acquisition program Lab View). Figures 4 and 5 are examples of results of this study which was intended to simulate how well rejection of background or other signals could be accomplished by such feedback. In Figure 4, a sine wave is shown as applied to the lower coil, by the bottom trace. The sensors response (wired as a second-order gradiometer) is shown in the middle curve and closely follows the generating field, the phase difference simply being a matter of choice as to the output polarity. The root-mean-square of the observed sine wave was 254. When the same stimulating field was applied to the feedback loop but out-of-phase with the excitation coil, the upper trace was observed from the coil. This value is seen to be 1.9 in rms units or a reduction of 134. Actually, the result is seen to be in the noise (the analog output of the DSP was measured rather than the digital in these trials resulting in the digital bit pattern shown in the figure). The sine wave was of 20 Hz frequency.

A very similar pattern is observed for the triangle wave of Figure 5 where a reduction of 180 observed with the resulting signal again primarily sensor noise. It is thus shown that such digital cancellations performed within the 32 bit precision of the DSP can provide substantial cancellation of magnetic fields at the SQUID through the modulation coil. Such cancellations should be possible for other types of inputs such as reference sensors placed to observe primarily background fields.

When the SQUID sensor is removed from the shielded chamber, the pickup of fields from line frequency tends to dominate the device. This is shown in Figure 6 where the middle trace shows the output of the sensor with no external fields being generated other than the laboratory electromagnetic environment. The line frequency, in this case 60 Hz, dominates the spectrum with much slower deviation from unknown sources observed as well. A 60 Hz waveform, as shown on the bottom trace, was then applied to the DSP as a cancellation wave to the SQUID with the top waveform being the result. This waveform has been expanded to show that the line frequency is indeed gone and the remaining waveform is now dominated by slow changes that are present within the room due to changes in the magnetic field from machines, the earth, etc.

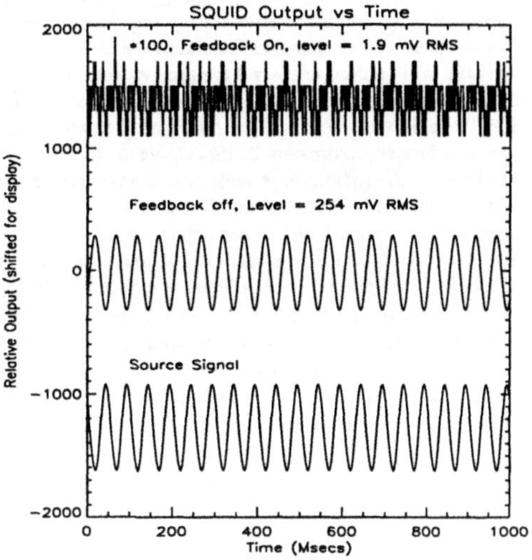

Figure 4

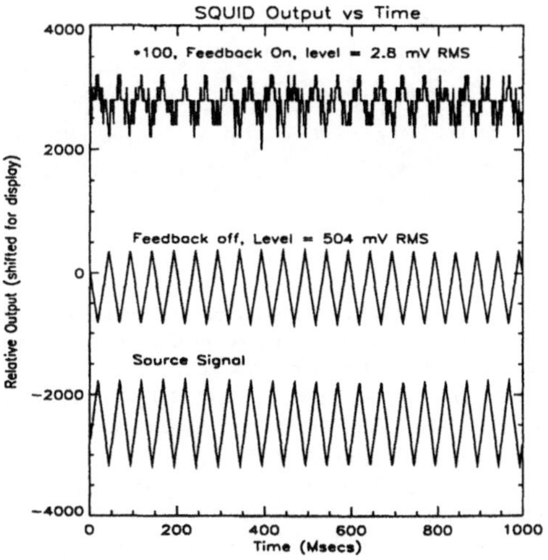

Figure 5

IV. Discussion

The use of a DSP as the FLL for linearizing the performance of a SQUID sensor has proven to be a useful concept with the added advantages of increasing the dynamic range through flux counting and providing a convenient mechanism for background cancellation through negative feedback. The DSP will be used to control the sensors associated with a large system illustrated in Figure 7 where some 155 SQUID-magnetometers will be distributed every 25 mm over a surface approximately 20 mm above the human head. As shown in the figure, these sensors will be placed as buttons on the ends of tubes, 20 mm below a superconducting surface of niobium which has been spun out of a flat sheet. A brim is added to this hemispherical shape to increase the rejection of background fields which are reduced by 3000.

To further increase the background rejection and permit this sensor system to operate in a minimally or unshielded room, additional sensors are provided above the shield. These will be used as negative feedback through the DSP to cancel residual fields as explained above.

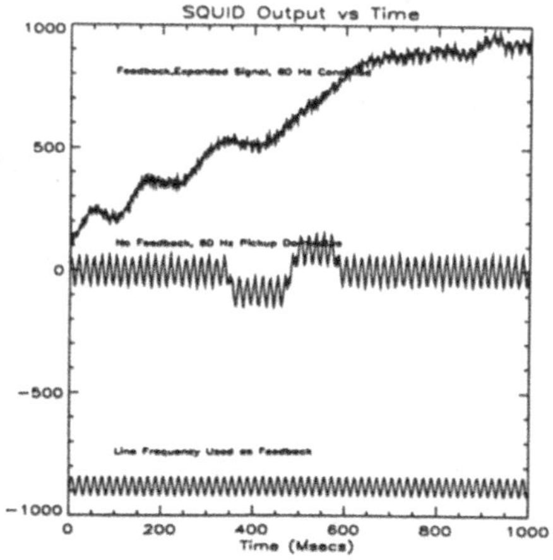

Figure 6

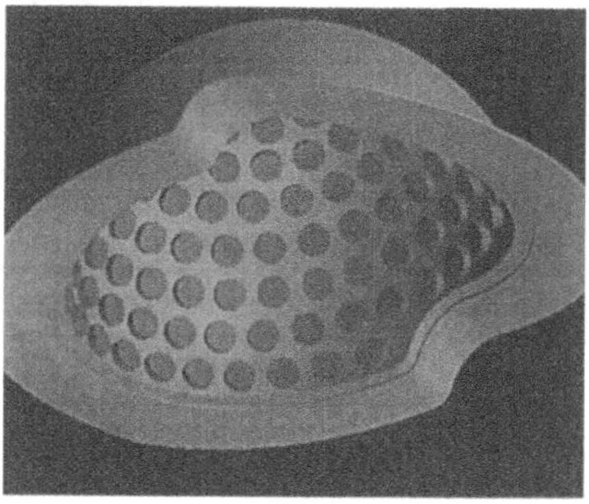

Figure 7

References:

[1] Wellstood, F., Heiden, C., and Clarke, J. "Integrated DC SQUID magnetometer with a high slew rate", Rev. Sci. Instrum. 1984, **55**, 952

[2] Bracht, R. R., Kung, P. J., Lewis, P. S., and . Flynn, E. R, "DSP control of superconducting quantum interference devices", in Procs. of the Int. Conf. on Signal Processing Applications and Technology, 1994, ICSPAT, 1146

[3] Kung, P. J., Bracht, R. R., Flynn, E. R., and Lewis, P. S., " A direct current superconducting quantum interference device gradiometer with a digital signal processor controlled flux-locked loop and comparison with a conventional analog feedback scheme", Rev. Sci. Instrum. 1996, **67**, 222

[4] Vrba, J., Betts, K., Burbank, M., Cheung, T., Cheyne,D., Fife, A. A., Haid, G., Kubik, P. R., Lee, S., McCubbin, J., McKay, Mckenzie, D., Mori, K., Spear, P., Taylor, B., Tillotson, M., and Xu, G., "Whole cortex 64 channel system for shielded and unshielded environments", Biomagnetism: Fundamental Research and Clinical Applications (Vienna, Austria), edited by Baumgartner, C., Deecke, L., Stroink, G., and Williamson, S. J., Elsevier-IOS, New York, 1995, 521

[5] Brandenburg G. Clements, U., Halling, H., and Zimmerman, F., Julich (KFA) internal report, (mentioned in Wikswo, J. P, IEEE Trans. on App. Superconductivity, 1995, **5**,, 74)

[6] Kraus Jr. R. H., Bracht, R. R., Flynn, E. R., Jia, Q., Maas, P., Reagor, D., Stettler, M., and Wu, X., "High-t emperature SQUID magnetometer and gradiometer systems with digital signal processor flux-locked loop for biomagnetic applications", Biomag96, Tenth International Conference on Biomagnetism, (Santa Fe, NM), edited by C. Aine, C. C. Wood, and Y. Okada (Springer-Verlag, New York, 1996)

Acknowledgment

This work was supported in part by a grant from NINDS and by the US DOE Contract W-7405-ENG-36

A 32-Channel MCG System Cooled by a GM/JT Cryocooler

Fujimoto, S., Sata, K., Fukui, N., Haraguchi, E., Kido, T., Nishiguchi, K. and Kang, Y.M.

MEC Laboratory, DAIKIN INDUSTRIES, LTD., 3 Miyukigaoka, Tsukuba 305, Japan

Introduction

Most SQUID systems for biomagnetic measurements use liquid helium as a cryogen. To reduce operating costs and to eliminate the need for handling liquid helium, SQUID systems cooled by small closed-cycle cryocoolers have been developed and have been used to successfully measure MCGs and MEGs [1-4]. Those systems were single-channel systems. We have developed a cryogen-free 32-channel MCG system cooled by a Gifford-McMahon/Joule-Thomson (GM/JT) cycle cryocooler. Other advantages for cryocooled systems include excellent S/N ratio because of the small clearance between the pickup coil and the subject [5], and the high degree of freedom in the tilt of the cryostat. On the other hand, two disadvantages are the magnetic noise produced by movement of the magnetic and metallic parts in the cryocooler and the mechanical vibration. To reduce this noise and the vibration, we have been improving the cryocooler in the system.

System

A block diagram of the 32-channel MCG system is shown in Fig.1. The system is cooled by a GM/JT cryocooler. It is suspended from a non-magnetic gantry, thus allowing it to be moved vertically and tilted in one direction. A non-magnetic bed is used for the subjects. The 32-channel SQUIDs are controlled by the FLL controller. All of these components are placed in a magnetically shielded room (VacuumschmelzeGmbH AK3a). Parallel analog signals from the 32-channel MCGs are sampled by the data acquisition unit through the low-pass filters (480 Hz) and the 16-bit A/D converters. The data sampling rate is 1 kHz. All data are stored on a computer (DAIKIN COMTEC Indigo), and continuously monitored on a real-time display. The data is also processed using digital filters and then averaged. Iso-contour maps and current dipoles are displayed with MR-images.

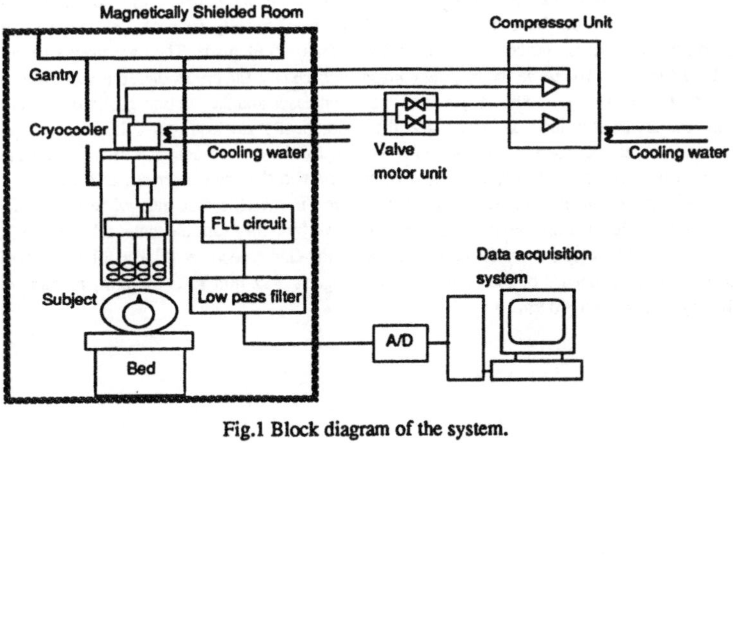

Fig.1 Block diagram of the system.

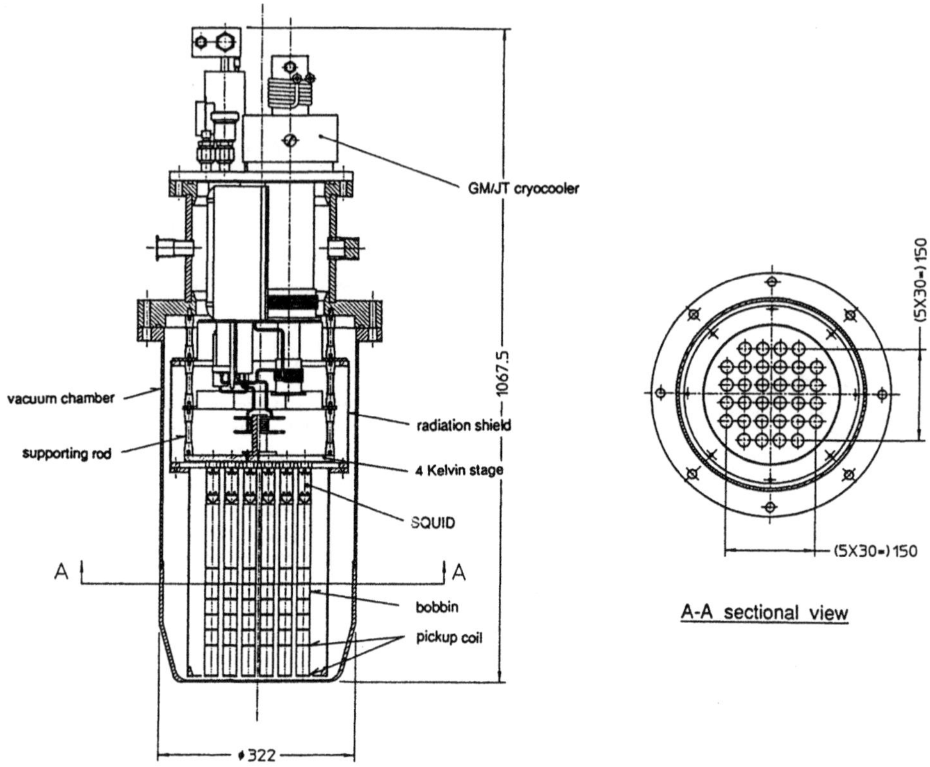

Fig.2 Sectional views of the cryostat.

Sectional views of the cryostat are shown in Fig.2. The pickup coils are arranged on a square matrix (6 by 6) in which the four corners coils are removed. The space of the matrix is 30 mm. The coils are first-order axial gradiometers, with 20 mm diameter and 37.5 mm baseline length. The space between the pickup coils and the outer surface of the cryostat is 14 mm before cooling and about 16 mm after cooling. The dc-SQUIDs are molded by the plastic material and each device is placed in its own individual superconducting shield cases. The pickup coils are wound on the bobbins. The bobbins are mechanically connected to the base on which the superconducting shield cases are mounted. Each sensor is thus a unit made. They are made of material that has good thermal conductivity. To reduce the magnetic noise produced by the cryocooler, the valve motor unit, which is used for switching between the exhaust gas from the compressor and the suction gas into the compressor, was separated from the cryocooler and was set outside the magnetically shielded room. Moreover, a portion of the magnetic or metallic materials used in the cryocooler were replaced by non-magnetic or non-metallic materials. The magnetic noise is produced by not only the magnetic materials in the cryocooler but also the metallic materials that have good electrical conductivity. Eddy current flows in the metallic materials when they reciprocate in the cryocooler. The cryocooler is connected to the valve motor unit by a 3-m flexible metal tube. The SQUIDs and the pickup coils are vibration-isolated from the cooling stage of the GM cryocooler [6].The total length of the cryostat is 1067.5 mm, and the diameters of the dome and the flange is 322 mm and 400 mm, respectively. The total weight of the cryostat is about 90 kg.

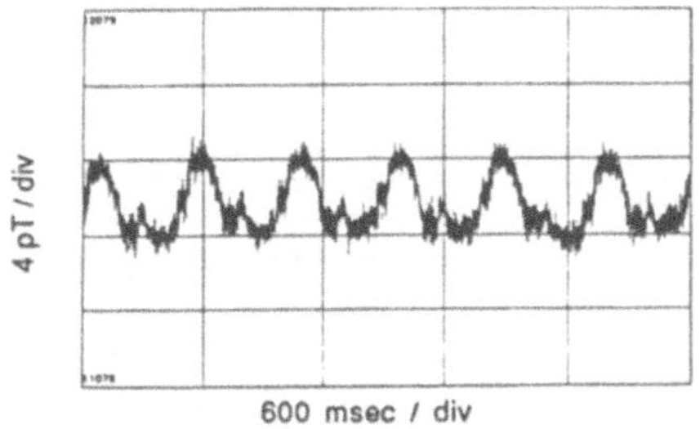

Fig.3 Typical system noise patterns.

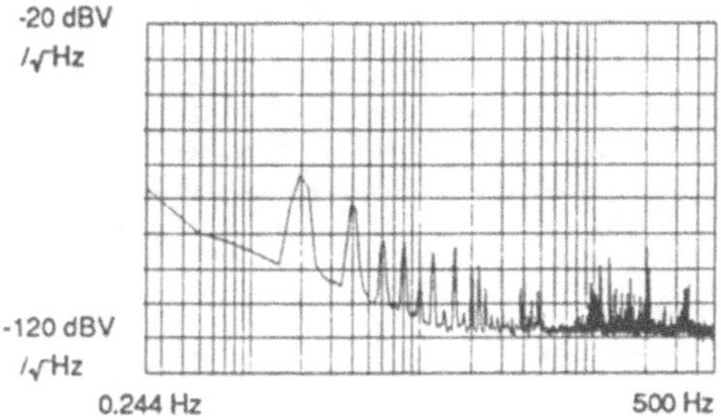

Fig.4 Typical raw noise spectrum.

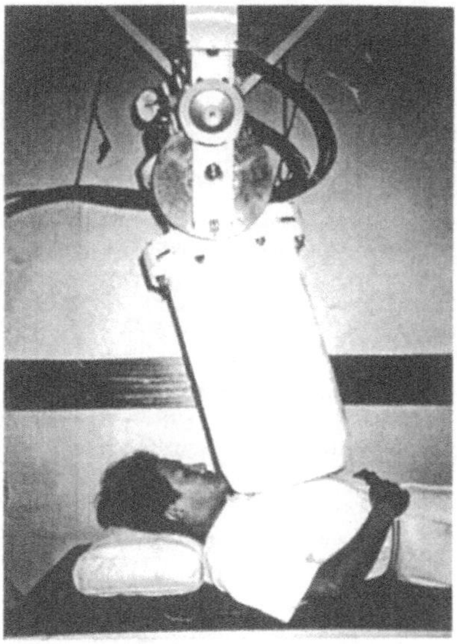

Fig.5 View of MCG measurement.

Results

Typical system noise patterns for the 32-channel MCG system are shown in Fig.3. The figure shows 2-Hz sinusoidal oscillations and sharp wave patterns are shown. The oscillations are possibly caused by the reciprocating motion of the metallic parts, and sharp wave patterns by the mechanical vibrations. Further improvements are needed to the cryocooler to reduce the 2-Hz periodic noise. Furthermore, improvements must be made so that the vibrations are minimized. Fig.4 is a typical raw noise spectrum, showing 2-Hz and its harmonics. The white noise level of the 32-channel signals ranged from 10 to 15 $fT/\sqrt{Hz}$. However, it is less than 7 $fT/\sqrt{Hz}$ when the SQUID sensors are cooled in liquid helium. By further optimizing the material of the sensor bobbins, it is expected to reduce the white noise level of all channels to less than 10 $fT/\sqrt{Hz}$.

The periodic noise can be processed by using a noise template. The MCGs and MEGs were then measured using this template. A view of the MCG system in operation is shown in Fig.5. The cryostat can be tilted to follow the surface of the subjects' breast. The raw signals of the 32-channel MCGs and the processed signals (the noise was canceled by the noise template and not averaged) are shown in Fig.6 and Fig.7, respectively. The iso-contour map of the MCGs is shown in Fig.8. The processed signals were averaged for about 180 beats.

Discussion

The cryocooled system can accurately measure MCGs and evoked fields from brain [7]. We can process and average the signals during the measurements of those fields. However, due to the necessity of the time for the data processing by using the noise template, this system does poorly when measuring spontaneous fields from brain or abrupt change of MCG. The sinusoidal noise is expected to be eliminated when the eddy current is prevented from flowing in the metallic parts. Because the magnetic noise produced by the mechanical vibrations will be residual, the speed of the noise template process must be increased by using devices such as a DSP (digital signal processor). With such improvements, spontaneous fields from brain or abrupt change of MCG can be accurately measured in real time.

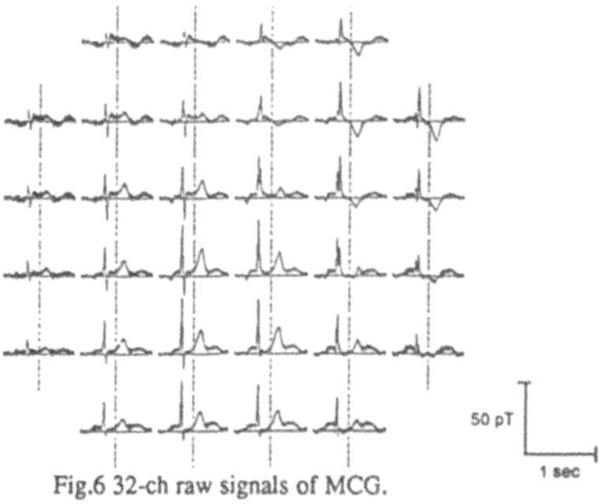

Fig.6 32-ch raw signals of MCG.

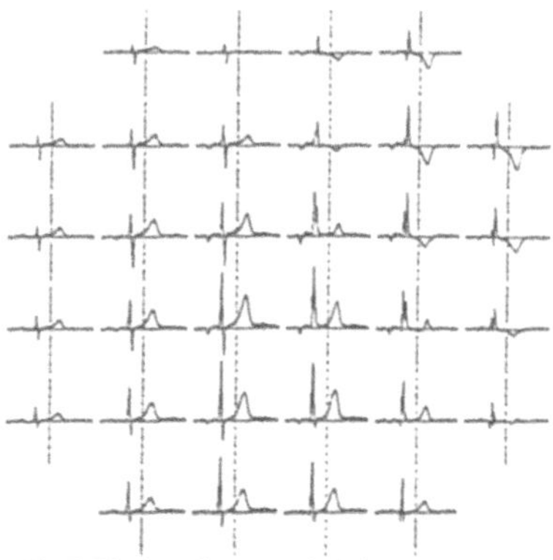

Fig.7 32-ch noise canceled signals of MCG.

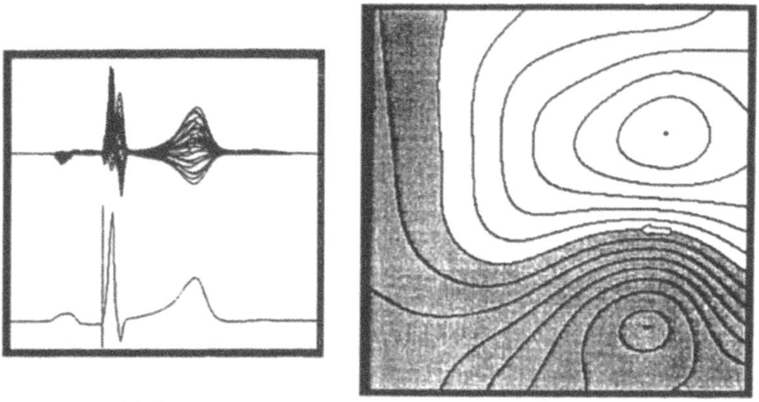

Fig.8 Iso-contour map of MCGs (iso-field step is 200 fT).

References:

[1] Sullivan, D.B., Zimmerman, J.E. and Ives, J.T. Operaion of a practical SQUID gradiometer in a low-power Stirling cryocooler, NBS Special Publication 607, 1981, 186-194

[2]Kang, Y.M., Ogura, T., Uosaki, S., Sakitani, K., Sochi, S., Fujimoto, S., Sata, K., Fukui, N. 4 K GM/JT cryocooler for cryogenic sensors, Proc. CRYOCOOLERS 6, 1990, Volume 1: 17-25

[3] Buchanan, D.S., Paulson, D.N., Klemic, G.A. and Williamson, S.J. Development of hybrid Gifford-McMahon Joule-Thomson based neuromagnetometer CryoSQUID, Proc. CRYOCOOLERS 5, 1988, 35-46

[4] Sata, K., Ueda, T.,Fukui, N., Yoshida, T., Kang, Y.M. and Nishiguchi, K. Biomagnetic measurements with a SQUID cooled by a GM/JT cryocooler, 8th International Conference on BIOMAGNETISM BOOK OF ABSTRACTS , 1991, 131-132

[5] Fukui,N., Sata, K., Fujimoto, S. and Kang, Y.M. Cryocooled SQUID system with minimal clearance between the pick-up coil and specimen, Proc. Biomag96,to be published

[6] Kang, Y.M., Sakitani, K., Taneya, S., Miura, K. Development of a 4 K GM/JT cryocooler, ICEC 12, Butterworths, U.K., 1988, 582-586

[7] Sata, K., Fujimoto, S., Fukui,N., Haraguchi, E., Kido,T., Nishiguchi,K. and Kang, Y.M. Development of a 61-channel MEG system cooled by a GM/JT cryocooler, Proc. Biomag96,to be published

A Cryocooled SQUID System with Minimal Clearance Between the Pick-up Coil and the Specimen

Fukui, N., Sata, K., Fujimoto, S., Kang, Y.M.

MEC Laboratory, DAIKIN INDUSTRIES, LTD., 3 Miyukigaoka, Tsukuba 305, Japan

INTRODUCTION

We have developed a SQUID system that have a clearance of about 3 mm between the pick-up coil and the specimen and is cooled by a Gifford-McMahon (GM)/Joule-Thomson (JT) cryocooler. Previous work showed that cryocooled SQUID systems, which do not require liquid helium, can successfully measure biomagnetic fields [1][2][3]. To eliminate the noise caused by the cryocooler, such systems either utilize digital filters or interrupt the cooler expansion operation during the measurements. In our new system, however, we reduced this noise by replacing the ferromagnetic materials of the cryocooler, and we measured magnetcardiogram (MCG) signals in real-time without the need for interrupting the cooler expansion operation.

In measurements of weak magnetic fields, the clearance between the specimen and the pick-up coil must be minimized because the measured field intensity attenuation depends on the distance from the field source. For measurements involving a shallow field source with high spatial resolution, minimal clearance combined with a small diameter pick-up coil increases the measurement accuracy. Previously, such systems were cooled by liquid helium. In some of those systems, minimal clearance is achieved not by cooling the SQUID and pick-up coil via immersion in a liquid helium bath, but by heat conduction from the bath just same as cryocooled system. Furthermore, in those liquid-helium-cooled systems, the effect of thermal contraction is minimized by using either a movable pick-up coil or a movable vacuum vessel. These systems achieve a clearance of about 1 mm with pick-up coils that have a few mm diameter [4][5][6]. We have developed a cryocooled system that achieves a clearance of about 3 mm, and is capable of measuring weak signals more strongly. Here, we present our system and show its performance characteristics.

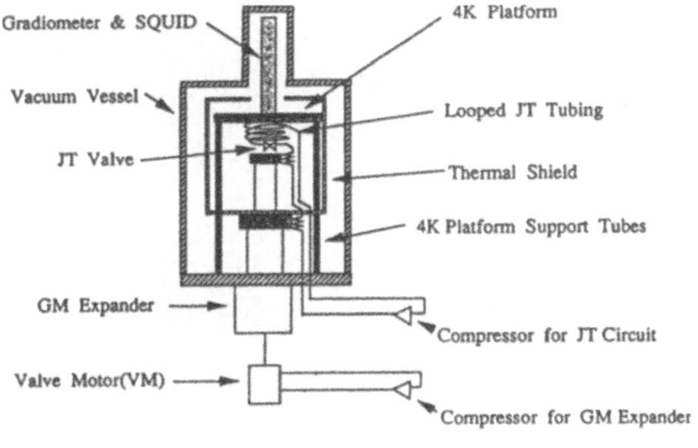

Fig. 1 Schematic of the SQUID and cryocooler used in our system.

SQUID SYSTEM

Figure 1 shows a schematic of the SQUID and cryocooler used in our system. A magnetic field is sensed by a thin film Nb/AlOx/Nb dc SQUID, which is attached to a superconducting pick-up coil and installed in a superconductive shield. This coil is a first-order gradiometer that has a 20-mm diameter and a 40-mm base length. The SQUID is cooled by a GM/JT cryocooler based on the commercial model DAIKIN CG304, which has a refrigeration capacity of 1.0 W at 4.2 K and a power consumption of 4.2 kW. To isolate the mechanical vibration from the expander, a platform at a temperature of 4 K where the SQUID has installed is supported by tubes extending up from the base flange, and to absorb the vibration, the platform is cooled by looped JT tubing [2].

To reduce the electromagnetic noise, we separated the expander and valve motor (VM), which is used to switch between high and low pressure helium gas. Another way to further reduce the noise from the cryocooler is to replace ferromagnetic materials in the cryocooler with non-ferromagnetic materials. This reduces the noise that are caused by cooler expansion operation. The most effective replacement is the back-up ring at the displacer seal that reciprocates in the expansion cylinder; the noise peak at the reciprocating frequency (2 Hz) is then decreased by about 40 %.

Excluding the compressor and VM, the total weight of the cryostat is about 50 kg and the length is about 1 m. By utilizing a main structure of a reliable commercially available cryocooler, our system can run continuously for 10,000 hours without maintenance.

MINIMAL CLEARANCE

To minimize the clearance between the pick-up coil and the specimen, we used a movable vacuum vessel. The fiber-reinforced plastic (FRP) vacuum vessel is divided into two sections that are connected by a vacuum-tight seal. One end of the vessel faces the specimen and can be moved toward the pick-up coil. Thermal contraction of the pick-up coil and bobbin causes the clearance between the pick-up coil and specimen to increase before cool-down. This thermal contraction is estimated to be about 2 mm. After cool-down, when we measure the biomagnetic field, we decrease the clearance between the pick-up coil and specimen by moving the vacuum vessel; the result is a stronger reading of the field. Figure 2 and 3 show a schematic and view, respectively, of this movable vacuum vessel.

The thickness of the vacuum vessel at the end nearest the specimen is 1 mm. The pick-up coil is 1 mm from the leading edge of the gradiometer bobbin. With the addition of the minimum space required for thermal insulation, the smallest clearance between the pick-up coil and specimen is about 3 mm.

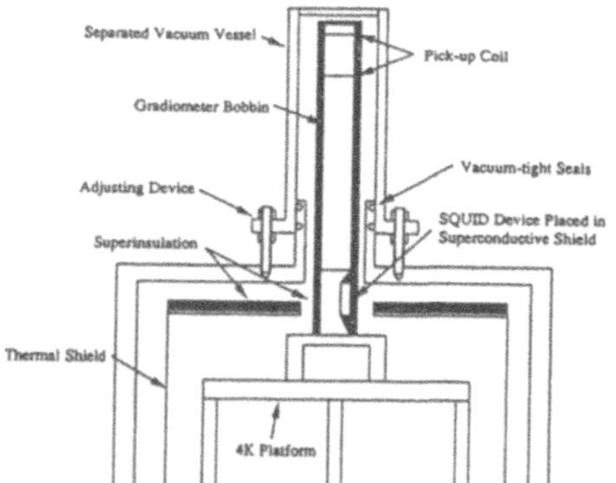

Fig. 2 Schematic of the movable vacuum vessel.

Fig. 3 Movable vacuum vessel.

MEASUREMENT SYSTEM

Figure 4 shows a schematic of our biomagnetic measurement system in operation. To reduce ambient magnetic noise, the SQUID system is located in a magnetically shielded room (MSR), and the compressor and VM are outside the MSR. The magnetic flux detected by the pick-up coil is converted to a voltage signal by the SQUID and the flux locked loop (FLL) controller, and then sent via a feedthrough connector to a recording device outside the MSR. A typical value of the shielding factor of the MSR (Vacuumschmelze AK3a) is 60 dB at 1 Hz.

The system is installed on a gantry that provides five degrees of freedom: rotation about the vertical and horizontal axes of the gantry, and movement along two horizontal directions and the vertical direction. By utilizing the mobility of the gantry, the pick-up coil can be turned to almost any desired direction, accommodating the measurement situation. The gantry is equipped with casters, and thus can be moved in and out of the MSR. Figure 5 shows a view of the cryostat and the gantry.

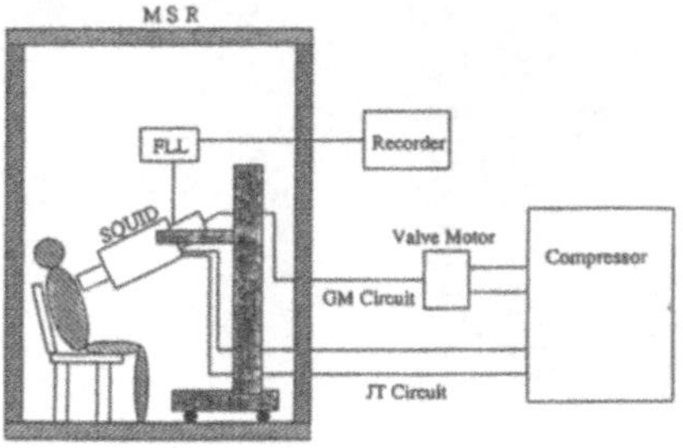

Fig. 4 Biomagnetic measurement system.

Fig. 5 Cryostat and gantry.

NOISE CHARACTERISTICS AND MCG MEASUREMENT

The white noise level of this system is about 20 $fT/\sqrt{Hz}$ and the noise peak at the expander driving frequency (2 Hz) is about 1 pT for an rms value. This value of the white noise level is slightly higher than the 10 $fT/\sqrt{Hz}$-to-15 $fT/\sqrt{Hz}$ range seen in our cryocooled multichannel system that employs the same SQUID devices [7]. This higher value is caused mainly from a mismatch in the inductance between the pick-up coil and the input coil. By examining other components between the system and multichannel system, the above consideration about noise source is valid. The noise peak at the driving frequency is caused by an eddy current in the reciprocating displacer. To increase the accuracy of the measurements, this noise peak must be reduced in future systems.

We applied the system to measurement of MCG signal and measured MCG signals in real-time without interrupting the cryocooler operation. To make certain the effect of small clearance, we compared measurements of MCG signals obtained for the smallest clearance with a clearance of 16 mm which is equal to the clearance of our cryocooled multichannel system. We used a high-pass filter at 0.1 Hz and a low-pass filter at 500 Hz. Figure 6 shows the measured raw MCG data for a clearance of 3 mm. Figure 7 shows raw MCG data for a clearance of 16 mm. A comparison of these two figures shows that the S/N ratio was improved by a factor of about two because of the small clearance. If we assume that a field source is located 30 mm beneath the surface of the skin of the subject, our calibration test reveals that the measured field intensity at a point 16 mm from the skin surface is less than half that at a point 3 mm from the skin surface. This tendency in the attenuation correlates with the distance at which the field source corresponds to the measured MCG signal.

Figure 8 is a view of the system during a measurement, illustrating the usefulness of having the cryostat being able to tilt to almost any desired direction. Such capability is achieved because no liquid helium is used in the system.

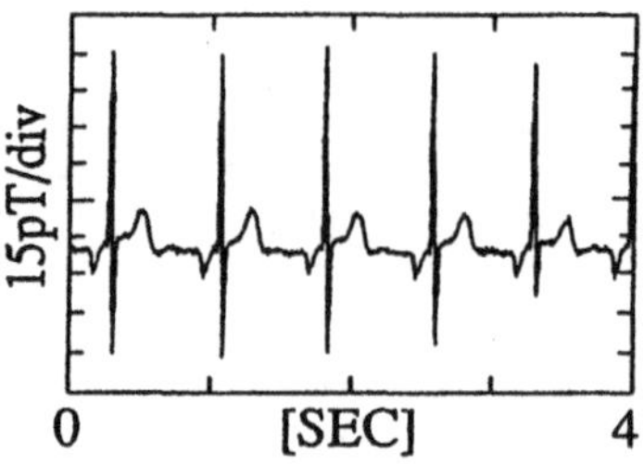

Fig. 6 Raw MCG data for a coil-to-specimen clearance of 3mm.

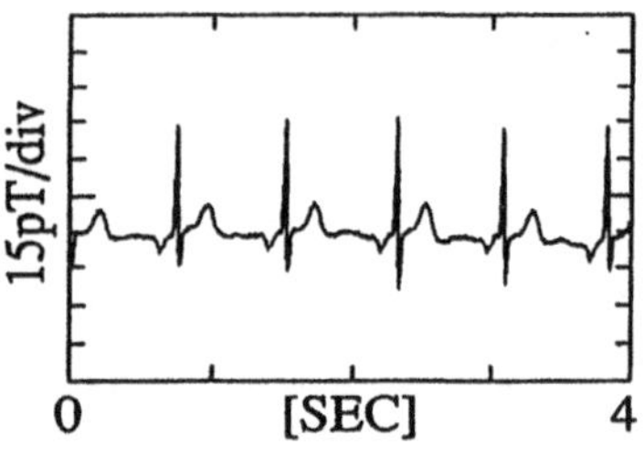

Fig. 7 Raw MCG data for a coil-to-specimen clearance of 16mm.

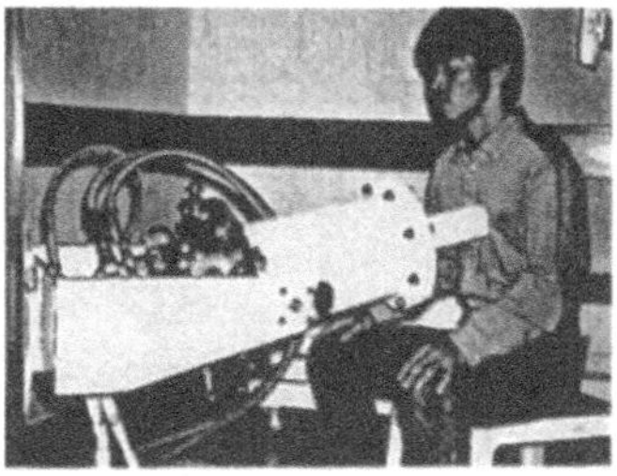

Fig. 8 Biomagnetic measurement system in operation.

CONCLUSION

We have developed a SQUID system that has a minimal clearance of 3 mm between the pick-up coil and the specimen and is cooled by a GM/JT cryocooler. We replaced the ferromagnetic materials of the cryocooler and measured MCG signals in real-time without having to interrupt the cooler expansion operation.

Future improvements to the system include (1) reducing the white-noise level to less than 10 $fT/\sqrt{Hz}$, (2) reducing the inherent noise peak, and (3) reducing the clearance between the specimen and the pick-up coil to less than 3 mm. The techniques used in this system can be applied to other systems, such as MEG or nondestructive testing (NDT) because of the small clearance achieved by our system.

REFERENCES

[1]Sullivan, D.B., Zimmerman, J.E., and Ives, J.T. NBS Special Publication 607, 1981, 186

[2]Kang, Y.M., Ogura, T., Uosaki, S., Sakitani, K., Sochi, S., Fujimoto, S., Sata, K. and Fukui, N. 4K GM/JT Cryocooler for Cryogenic Sensors, Proc. Cryocoolers 6, 1990, vol. 1: 17-25

[3]Buchanan, D.S., Paulson, D.N.,Klemic, G.A.,and Williamson, S.J. Development of a Hybrid Gifford-McMahon Joule-Thomson Based Neuromagnetometer CryoSQUID,Proc. Cryocoolers 5, 1988,35-46

[4]Buchanan, D.S.,Crum, D.B.,Cox, D., and Wikswo, J.P. MICROSQUID:A Close-Spaced Four Channel Magnetometer, In:Williamson, S.J., Hoke, M., Stroink, G., Kotani, M. Advances in Biomagnetism, 1989, 677-679

[5]Hibbs, A.D., Sager, R.E., Cox, D.W., Aukerman, T.H., Sage, T.A. and Landis, R.S. A High-resolution Magnetic Imaging System Based on a SQUID Magnetometer,Rev. Sci. Inseam., Vol. 63, No. 7, July, 1992, 3652-3658

[6]Cochran, A., Donaldson, G.B., Morgan, L.N.C., Bowman, R.M. and Kirk, K.J SQUIDs for NDT: The Technology and its Capabilities, British Journal of NDT, Vol. 35, No.4, April, 1993, 173-182

[7]Fujimoto, S., Sata, K., Fukui, N., Haraguti, E., Kido, T., Nishiguti, K., and Kang, Y.M. A 32-channel MCG System Cooled by a GM/JT Cryocooler, Proc. Biomag'96, 1996

A Comparison of Vector and Radial Magnetometer Arrays for Whole-Head Magnetoencephalography

Hughett, P.[1] and Miyauchi, S.[2]

[1]Lawrence Berkeley National Laboratory, Berkeley, CA, USA; [2]Communications Research Laboratory, Ministry of Post and Telecommunications, Tokyo, Japan

Introduction

The number of detectors in magnetometer arrays for magnetoencephalography (MEG) has been steadily increasing, with systems containing 200 or more detectors now possible. It is of considerable interest to know how best to configure such a large array. In particular, is it useful to measure all three components of the magnetic field, rather than just the radial component? This paper compares the information content provided by three different magnetometer arrays for whole-head measurements, using a definition of information content developed by Kemppainen and Ilmoniemi [2].

Magnetometer configurations

The three different magnetometer configurations shown below were compared. In all configurations, the magnetometer sites were approximately uniformly distributed over a spherical cap of radius 13 cm centered on the top of the head and extending 105° from the center to the rim, thus covering slightly more than a hemisphere. Each magnetometer channel was assumed to have independent zero-mean gaussian measurement noise with standard deviation 50 fT. The Name column of the table below gives the name used to refer to the magnetometer configuration in this paper. The Type column indicates whether radial or vector magnetomers were used. The Sites column gives the number of locations at which a magnetometer was placed, and the Channels column gives the total number of channels measured. The Information column gives the total information obtained, in bits, computed by the method described below.

Magnetometer configurations compared

Name	Type	Sites	Channels	Information
246R	Radial	246	246	150
246V	Vector	82	246	130
82R	Radial	82	82	98

Source model

The head was modelled as a homogeneous conducting sphere of radius 10.5 cm, and the volume currents and magnetic fields were computed using the equations derived by Sarvas [3]. The primary current sources in the brain were modelled as a grid of 1072 three-component current dipoles at a 1.0 cm spacing within a hemispherical volume of radius 9.0 cm. It is convenient to refer to the vertex of this source hemisphere as the north pole. Each dipole component was assumed to be a zero-mean independent gaussian random variable with standard deviation 1.0 nA-m; this was chosen to yield a peak signal amplitude of about 200 fT, which is typical for visual evoked field measurements in magnetoencephalography [1].

Figure 1 shows the RMS signal amplitude at each detector in the 246V configuration for this source model. The detectors are arranged so that detectors 1–82 measure the radial component of the field, detectors 83–164 the eastward component, and 165–246 the northward component. Within each block, the detectors are arranged from the north pole to the equator; the RMS amplitude decreases as the detectors move away from the north pole. The maximum signal amplitude is about 200 fT.

Information content

Suppose that $y = x + n$ is a measurement of a true signal x corrupted by independent additive measurement noise n, that the signal x is normally distributed with variance α^2, and that the noise n is normally distributed with variance σ^2.

We know from Shannon [4] that the information about the signal x contained in the measurement y is

$$R = H(y) - H(n) \quad ,$$

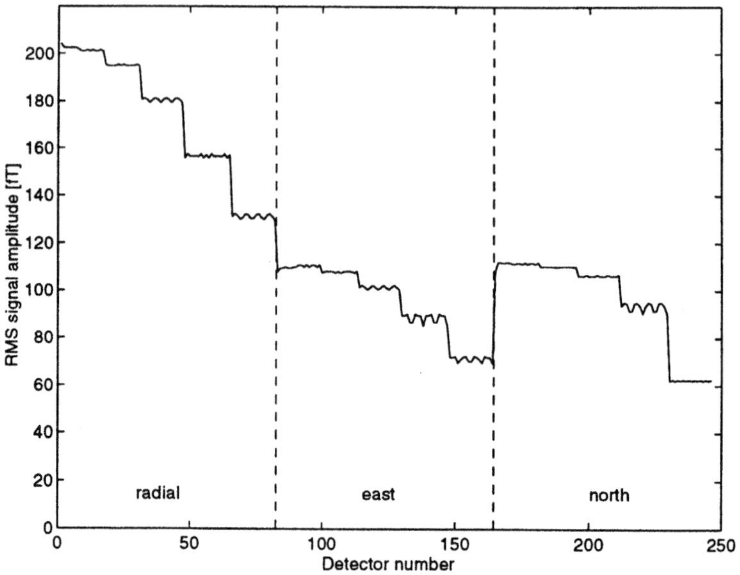

Figure 1. The RMS signal amplitude at each detector in the 246V configuration for this source model. The detectors are arranged so that detectors 1–82 measure the radial component of the field, detectors 83–164 the eastward component, and 165–246 the northward component. Within each block, the detectors are arranged from the north pole to the equator; the RMS amplitude decreases as the detectors move away from the north pole. The maximum signal amplitude is about 200 fT, which is typical for visual evoked field measurements.

which is the entropy of y minus the entropy contributed by the noise n. Furthermore, the entropy of a gaussian random variable with variance σ^2 is, in bits, $H = \frac{1}{2}\log_2(2\pi e\sigma^2)$. Then the measurement y is normally distributed with variance $\alpha^2 + \sigma^2$ and the received information is, in bits,

$$R = \frac{1}{2}\log_2\left(\frac{\alpha^2+\sigma^2}{\sigma^2}\right) = \frac{1}{2}\log_2\left(1+\frac{\alpha^2}{\sigma^2}\right) = \frac{1}{2}\log_2(1+\mathrm{SNR}) \quad ,$$

where SNR is the signal-to-noise power ratio α^2/σ^2.

Since the different channels of a magnetometer array are correlated, it is not legitimate to simply sum the information content of each channel. The singular value decomposition (SVD) or Karhunen-Loève transformation may used to define synthetic channels which are uncorrelated but contain the same information as the measured channels. The SNR for one of these synthetic channels is

$$\mathrm{SNR}_k = \frac{\alpha^2\lambda_k^2}{\sigma^2} \quad ,$$

where α^2 is the source variance, λ_k is the singular value for the kth synthetic channel, and σ^2 is the noise variance. Thus a semilog plot of the singular value spectrum is, apart from scale, a plot of the SNR (in dB) for these synthetic channels.

Furthermore, the total information content of all the channels is

$$R = \frac{1}{2}\sum_k \log_2\left(1+\frac{\alpha^2\lambda_k^2}{\sigma^2}\right) \quad ,$$

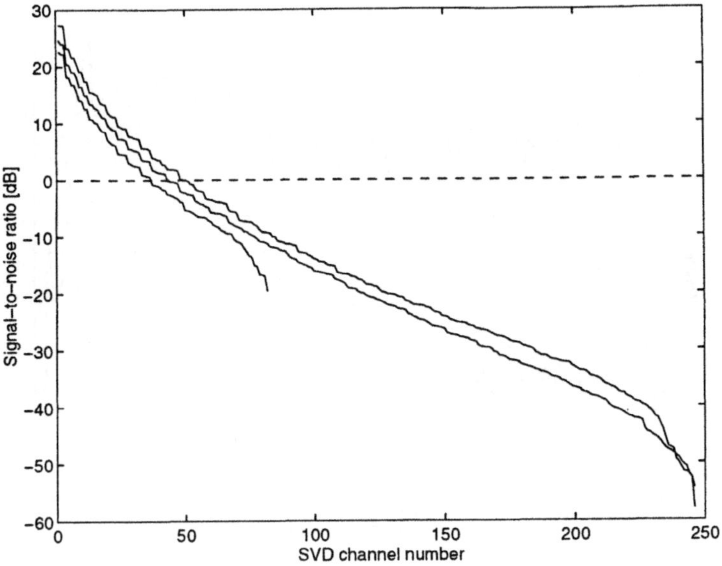

Figure 2. The three curves above correspond to the 246R (top), 246V, and 82R (bottom) magnetometer configurations. The information content of a configuration is roughly proportional to the area under the SVD spectrum and above the 0 dB line, so the 246R configuration provides the most information, followed by the 246V and 82R configurations. The curves do not diverge or cross until well below the 0 dB line, indicating that the magnetometer configurations differ in SNR but not in intrinsic resolution at the assumed source and noise levels.

which is roughly proportional to the area under the singular value curve and above the 0 dB line.

The number of synthetic channels with SNR greater than zero indicates the number of resolvable distinct measurements made by the system and may be taken as a very rough measure of the system resolution.

Results

The shape of the singular value spectrum, as shown in Figure 2, was almost identical for the three magnetometer configurations, but the 246R configuration was displaced upwards relative to the 246V, and the 82R displaced downward. Figure 3 shows the SNR difference in more detail and reveals that the 246R configuration was displaced upwards by about 2.3 dB, and the 82R displaced downward by about 2.5 dB. Thus the 246R configuration provided the most information, with the 246V and 82R configurations following in that order. The fact that the singular value spectrum is shifted but not changed in shape indicates that a 82R configuration with a 4.8 dB reduction in measurement noise, or a 246V configuration with a 2.3 dB reduction in noise, would provide the same information content as the 246R configuration.

Differences in the intrinsic resolution of the three configurations, as evidenced by the divergence or intersection of two SVD spectra, are not significant for the assumed source and noise amplitudes. The divergence between the 82R and 246V configurations appears at about −10 dB, so there would be no resolution difference until the SNR improves by 10 dB. Similarly, the intersection between the 246V and 246R curves would not be significant without a 40 dB improvement in SNR.

Thus, at the source and noise amplitudes assumed here, these configurations differ only in the SNR that they achieve but not in their resolution.

Moderate changes in the source configuration, in the radius and angular coverage of the detector array, or in the assumed signal and noise amplitudes did not affect the differences between the detector configurations.

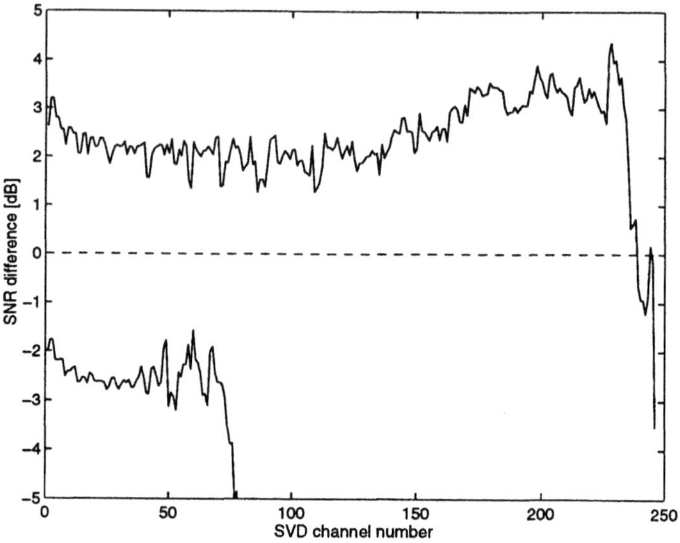

Figure 3. The top curve below shows the SNR improvement of the 246R configuration over the 246V configuration, for each SVD channel. The lower curve shows the 82R configuration relative to the 246V configuration.

Conclusions

For whole-head magnetoencephalography of distributed sources and assuming that the measurement error consists of uncorrelated sensor noise rather than correlated brain noise: (1) The spatial resolution is limited by the signal-to-noise ratio rather than by spatial sampling in the magnetometer array. (2) Each of the following changes will yield approximately the same improvement in information content: decreasing the RMS measurement noise by 3 dB; increasing the RMS signal amplitude by 3 dB (perhaps by moving the detectors closer to the head); doubling the number of detector sites; or replacing radial detectors by vector detectors at the same sites. (3) For a given number of sites, vector magnetometers provide more information; but for a given number of measurement channels, radial magnetometers provide more information.

Acknowledgements

The authors are grateful to Kensuke Sekihara for his advice and assistance. This work was supported in part by the National Science Foundation and the Center for Global Partnership of the Japan Foundation through the 1995 Summer Institute in Japan Program and in part by the Director, Office of Energy Research, Office of Health and Environmental Research, Medical Applications and Biophysical Research Division of the Department of Energy under Contract No. DE-AC03-76SF00098.

References

[1] Robert L. Fagaly. Neuromagnetic instrumentation. In Susumu Sato, editor, *Magnetoencephalography*, volume 54 of *Advances in Neurology*, pages 11–32. Raven Press, New York, 1990.

[2] P. K. Kemppainen and R. J. Ilmoniemi. Channel capacity of multichannel magnetometers. In S. J. Williamson, M. Hoke, G. Stroink, and M. Kotani, editors, *Advances in Biomagnetism: Proccedings of the Seventh International Conference on Biomagnetism*, pages 635–638. Plenum Press, 1989.

[3] Jukka Sarvas. Basic mathematical and electromagnetic concepts of the biomagnetic inverse problem. *Phys. Med. Biol.*, 32(1):11–22, 1987.

[4] Claude E. Shannon and Warren Weaver. *The Mathematical Theory of Communication*, pages 66–70 and 89–90. University of Illinois Press, 1963.

32 Channel High Tc SQUID System for Biomagnetism

Itozaki, H.[1], Kugai, H.[2], Nagaishi, T.[2], Toyoda, H.[1], Hirano, T.[1], Haruta, Y.[1], Tanaka, S.[1] and Kado, H.[3]

[1]Superconducting Sensor Laboratory and [2]Sumitomo Electric Ind., 1-1-1 Koyakita Itami Hyogo 664 Japan; [3]Electrotechnical Laboratory, Umezono 1-1-4, Tsukuba, Ibaraki 305 Japan

Introduction

Almost a decade has passed after the discovery of high Tc superconducting materials. One of the most promising applications is the SQUID. Speed of this progress is ahead of expectations. Magnetic field resolution became $8.5fT/Hz^{1/2}$[1] which is almost the same resolution of the Nb SQUID. The high Tc SQUID is expected to be used in various fields, because it can be used with liquid nitrogen. In the biomagnetism application, two-dimensional signals must be obtained simultaneously, because the biomagnetic signal sometimes does not repeat constantly. Multi-channel systems of the Nb SQUID have already been developed, such as a more than 100-channel system. In the high Tc SQUID, we reported a 4-channel system[2] and 16-channel system[3]. Here, we present a compact and easy handling 32-channel system. We also show its application, such as detecting heart activities with it.

High Tc SQUID chip

We used YBCO thin film SQUIDs with step edge Josephson junctions. The step height was 150nm and the film thickness was also 200nm. In order to develop the 32-channel system, a smaller SQUID must be developed with the large effective area, because we can make more number of the small size SQUID in a substrate. We adopted a slit center hole (5umX250um) for the 32-channel instead of a square hole (25umX25um) for the 16-channel. We also used a slit cover(1mmX2.4mm) of high Tc film to reduce open length of the slit to 100um. As this slit hole was more efficient than a square one, the 32-channel SQUID needs only a 5mm by 5mm washer size to have the same effective area as the 16-channel SQUID. So, it is possible to make 8 SQUIDs and a test element in a 20X20mm $SrTiO_3$substrate. Schematic figures of a SQUID pattern for the 32-channel system is shown in Fig.1. This SQUID had the effective area ($0.2mm^2$) and magnetic field resolution was ranged $300\text{-}800fT/Hz^{1/2}$ at 1Hz, $150\text{-}450fT/Hz^{1/2}$ at 10Hz and $70\text{-}250fT/Hz^{1/2}$ at 1kHz. This resolution was enough to catch the main peak of the magnetocardiogram, even though it was one order worse than that of the Nb SQUID ($5\text{-}10fT/Hz^{1/2}$). This chip was mounted on a chip carrier and was sealed with epoxy resin to protect it from moistureas shown in Fig. 2. This resin protection can keet its SQUID performance even after more than a 200 heat cycle test between room temperature and liquid nitrogen.

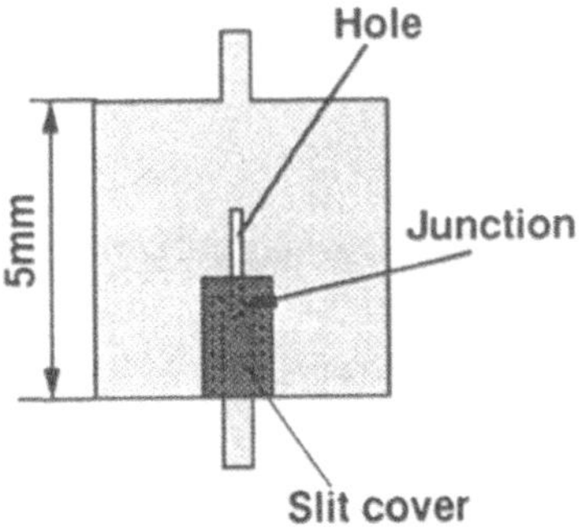

Fig.1 Schematic figure of SQUID for a 32-channel system. Outer size is 5mm. Hole size is 5um x 250 um.

Fig.2 High Tc SQUID device mounted on its chip carrier. It was sealed by a resin cap.

32-channel SQUID system

We developed a 32-channel high Tc SQUID system. The 32 SQUIDs were placed in a 6 by 6 array except for 4 corners at 40mm intervals (Fig.3). The 32-channel system could cover 7/3 larger area than the 16-channel one. One of the notable improvements is the flat liquid nitrogen dewar. Its diameter became 40cm which was slightly larger than that of the 16-channel system. However its height became about only 20cm (Fig.4), which was 2/3 of the 16-channel dewar. This system can work for more than 12 hours without more supply of liquid nitrogen, even though it contains only 7 liters of liquid nitrogen. This flat dewar makes a SQUID system more convenient than the Nb SQUID system, because our liquid helium dewar of a 16-channel Nb SQUID system has a diameter of 30cm and height of 120cm. We developed a 32-channel SQUID driving electronics unit which used a fluxed locked loop

Fig.3 32-channel high Tc SQUID array on the inner part of the flat dewar shown in Fig.3

Fig.4 Compact and flat liquid nitrogen dewar of 32-channel high Tc SQUID system.

method with the AC bias current. 1/f noise at low frequency is reduced by this electronics unit. Data were processed by a personal computer. Small magnetically shielded box (inner dimension:0.8mX0.8mX2m) was used for the measurement of a magnetocardiogram. Figure 5 shows a whole systemof the 32-channel high Tc SQUID.

Fig.5 32-channel high Tc SQUID system. The compact 32-channel probe is located in the magnetically shielded box. 32 channel SQUID electronics and its data aquisition personal computer are shown in the left side. All data of MCG are obtained with opening front door of the shielded box.

Biomagnetic application of high Tc SQUID

The multi-channel system was used for the investigation of biomagnetic applications of the multi-channel high Tc SQUID.

We have obtained clear signals from the human heart using the 32-channel system. Figure 6 shows one of the 32 magnetocardiograms near the center of the 32-channel array. It is similar to an electrocardiogram and shows typical signals such as R, S and T waves. Figure 6 also shows the magnetocardiac isofield contour map at these R, S and T waves. We can obtain clear contour maps. These maps shows healthy activities of the heart. If there is some disease which causes irregular current at the heart, this current makes an abnormal magnetocadiogram and we will be able to determine the existence of a disease of the heart. It is necessary for a medical doctor to make an in-depth examination of the magnetocardiogram of the patients and to prove the validity of the SQUID diagnosis.

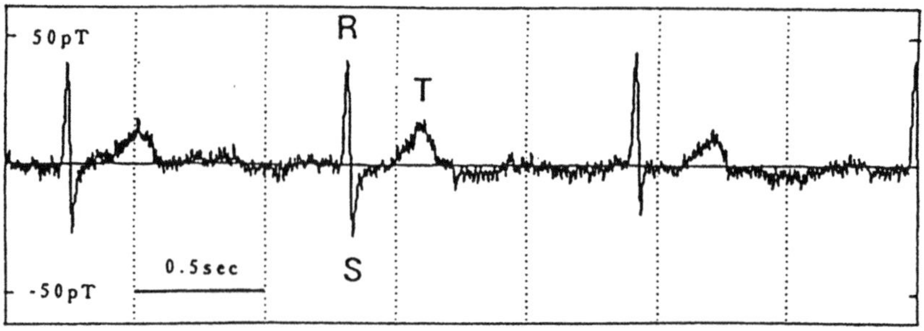

(a) (b) (c)

Fig.6 Upper figure: Magnetocardiogram near center channel of the 32-channel system without averaging. Lower figure: Magnetocardiac isofield contour map (20cm x 20cm). Interval of the contour line is 2pT. Solid line is positive and broken line is negative. (a)(b)(c) show contour maps at R, S, and T-wave, respectively.

Conclusion

A compact 32-channel high Tc SQUID system has been developed. It detects biomagnetic signals clearly. Since improvement of field resolution of the high Tc SQUID is expected to equal that of the Nb SQUID and is much easier to handle with lower operational costs, applications of the high Tc SQUID will extend to far broad fields than those of the Nb SQUID.

References

1. Dantsker, E., Ludwig,F., Kleiner,R., Clarke, J., Teepe, M., Lee, L. P. , Alford, N. McN., and Button, T., Low noise YBaCuO-SrTiO-YBaCuO multilayers for improved superconducting magnetometers Appl.Phy.Lett. 67, 725 (1995).
2. Itozaki, H., Tanaka, S., Harada, N., Adachi, T., Komuro, T., Toyoda, H., Okajima, K., Nagaishi, T., and Kado, H., Four channel YBCO DC-SQUID Extended Abstract of 4th Int. Superconductive Electronics Conf.,Boulder,1993, p138.
3. Itozaki, H., Tanaka, S., Nagaishi, T., and Kado H., Multi-channel high Tc SQUID, IEICE Trans. Electron., E77-C,1185 (1994).

Source Detectability for Multichannel Biomagnetic Sensors

Johnson, R.T., Black, W.C. and Buchanan, D.S.

Biomagnetic Technologies Inc., San Diego, USA

Introduction

It is useful to express the performance of multichannel biomagnetic sensors according to how well they are able to characterize sources of biomagnetic fields in the presence of measurement noise. Various authors have proposed performance measures based on the precision with which sources can be located [1-3] and on the minimum separation between two sources required to distinguish them from a single source [4]. We propose a new figure of merit for multichannel sensors, a measure of the weakest source that can be detected, which we call "detectability". This paper provides a simple definition of detectability and examines its implications for the performance of multichannel biomagnetic sensors designed to investigate magnetic signals generated by the human brain.

Methods

The objective is to estimate the strength of the weakest current source that can be detected at a specified location within a volume conductor by a multichannel sensor comprising an array of N magnetic detection coils. Since any source current distribution can be approximated by a superposition of current dipole elements, we consider the magnetic field signal S_i recorded by the ith detection coil from a current dipole of strength q located at position $\boldsymbol{r}$ with current direction described by the angle θ

$$S_i(\boldsymbol{r},\theta) = qG_i(\boldsymbol{r},\theta) + n_i, \tag{1}$$

where G_i is the field signal from coil i produced by a dipole source of unit strength at $(\boldsymbol{r},\theta)$ and n_i is the measurement noise in the ith sensor channel. The quantity G_i is the "forward field" commonly computed in source localization modeling. It takes into account the geometry of the detection coil (e.g., magnetometer, first-order axial or planar gradiometer, etc.) and its location and orientation relative to the source, and includes fields due to volume currents.

The least squares error estimate of the strength q of the dipolar source producing the recorded signals S_i is the value of q which minimizes the quantity $\Sigma(S_i - qG_i)^2$, where the sum extends over every detection coil in the measuring array. The result is

$$q = \Sigma(\omega_i S_i), \text{ where } \omega_i = G_i/(\Sigma G_j^2). \tag{2}$$

When the noise n_i in each channel is statistically independent and described by a standard deviation σ_i, the noise-induced uncertainty ε in measuring q is given by

$$\varepsilon = (\Sigma(\omega_i\sigma_i)^2)^{1/2}. \tag{3}$$

The quantity ε is a measure of the minimum strength source that can be detected at $(\boldsymbol{r},\theta)$ for the condition of unit signal-to-noise ratio. For the further special case that the σ_i are equal ($\sigma_i = \sigma$), we define detectability δ as the uncertainty in measuring q (or the minimum detectable q) per unit noise

$$\delta = \varepsilon/\sigma = 1/(\Sigma G_i^2)^{1/2}. \tag{4}$$

As defined by Eq. 4, detectability (units, Am/T) for an array of magnetic detection coils can be calculated directly from the array geometry, given a suitable model for the conducting environment in which the current source is embedded. For a given array, detectability is a function of source location and current direction; its value is dominated by detectors that are most sensitive to signals from the specified position $(\boldsymbol{r},\theta)$, i.e., those having the largest G_i values.

Detectability is a useful way to compare the performance of different multichannel sensor designs. As an example, we have calculated detectabilities for whole head arrays with magnetometer (M), axial gradiometer (AG) and planar gradiometer (PG) detection coils. The coils are uniformly distributed over the helmet-like surface shown in Fig. 1. For simplicity, the array models do not have facial cut-outs which results in a larger number of coils when their average spacing approximates that for commercially available sensors (see Table I). Detectability is calculated for dipoles in a homogeneous conducting spherical head model of 8.5 cm radius centered on the Z-axis and touching the inner surface of the helmet at the top (Fig. 1).

The assumption of equal noise in each channel is an appropriate approximation for sensor noise. However, in many studies of brain signals the dominant source of noise is so-called "brain noise", or background brain activity unrelated to the signals of interest. To investigate the minimum detectable dipole in the presence of brain noise, we simulated the generators of brain noise by a number of randomly activated dipoles (random in intensity and direction) distributed on a 1 cm grid inside a homogeneously conducting spherical brain model of 7.5 cm radius. The dipole

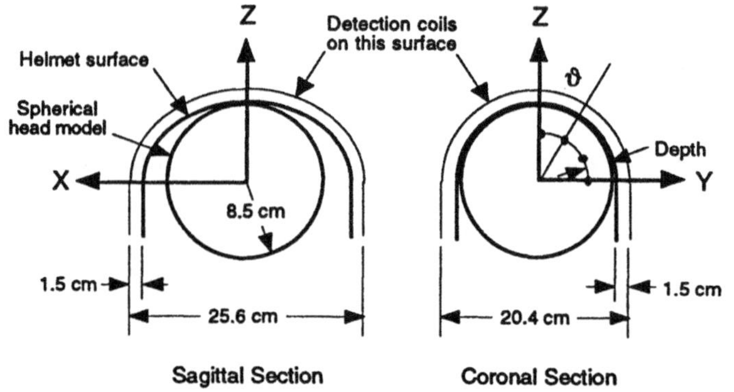

Fig. 1. Sensor geometry and spherical volume conductor used to model the head. The detection coils are distributed uniformly on the indicated surface.

Table I. Model Sensor Configurations

Detection Coil Type	Gradiometer Baseline(a)	Number of Coils	Average Spacing(a)
M	—	160	2.8 cm
AG	5 cm	160	2.8 cm
PG	1.65 cm	2 X 68(b)	4.3 cm

Notes: (a) Gradiometer baselines and average detection coil spacings are similar to those used in commercially available biomagnetic sensors. (b) The PG sensor has 68 measuring sites, each of which has two orthogonal planar gradiometer coils that measure both transverse derivatives of the field normal to the helmet surface.

intensity was scaled such that the rms field matches the value (30 fT/√Hz, including 5 fT/√Hz sensor noise) measured by Knuutila and Hämäläinen [5] with a 6 cm baseline axial gradiometer placed 1.6 cm from the head surface. This scaling simulates spontaneous brain noise; it could be reduced as appropriate to examine the effect of residual brain noise in signal averaged evoked response studies. The model allows us to estimate the value of σ_i in Eq. 3 for detection coils of any geometry and location with respect to the head, and thus to calculate the minimum detectable dipole strength when both sensor and brain noise are included. It should be noted that spatial correlation in the brain noise signals [5] has not been taken into account in these calculations.

Results

The results of the detectability calculations are given in Figs. 2 and 3. Fig. 2 shows how detectability varies with location at a depth of 1 cm below the head surface. Compared to the magnetometer array (M), detectability for the planar gradiometer array (PG) is both larger and exhibits much more variability with position; its smallest (best) values are in locations directly under one of the coils. This behavior at shallow depths is presumably a consequence of the spatially restricted PG lead field which causes the detectability to be dominated by the detection coil distribution in near proximity to the source. Variation for the M array is less because the M lead field is more extended, so the signal is distributed over a large number of coils.

As expected, the minimum detectable dipole intensity increases with depth for all arrays (Fig. 3). The average detectability for an array of magnetometers is best at all depths, followed closely by axial gradiometers. Detectability for the M array is approximately 2 to 4 times better than for the PG array over the range of depths where brain activity sources are commonly located (1 to 6 cm below the head surface).

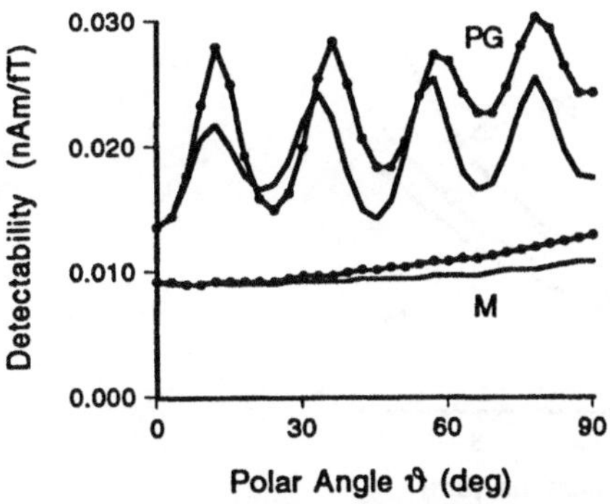

Fig. 2. Detectability for M and PG arrays at a depth of 1 cm as a function of polar angle ϑ defined in Fig. 1. To illustrate how detectability varies with proximity to the detection coils, values are plotted for two planes according to whether the plane passes mainly between (–) or through (•) the detectors in the upper portion of both arrays. The "between" plane is the mid-coronal (Y-Z) plane, and the "through" plane is rotated by 32.3 deg about the Z-axis. In both cases the source current is perpendicular to the source plane.

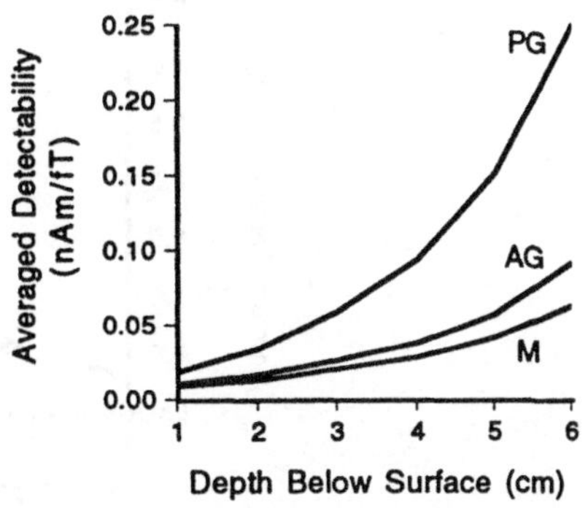

Fig. 3. Detectability for three arrays averaged over polar angles between 0 and 60 deg in the mid-coronal (Y-Z) plane as a function of depth below the surface of the head sphere. The source current is oriented along the X-axis (anterior direction).

The minimum detectable dipole strength when raw (unaveraged) brain noise is added to a typical sensor noise of 5 fT/√Hz is shown in Figs. 4 and 5. As for the case of sensor noise alone, the variation with polar angle at shallow depths is much larger for the PG array than for the M and AG arrays. At a depth of 1 cm, apart from regions located between the detection coils, the minimum detectable dipole for the PG array is smaller than either the M or AG arrays, and the M array value is the largest. This can be understood as a consequence of the relatively large reduction in brain noise picked up by the PG coils due to their more limited lead field distribution, and a similar but smaller effect for the AG coils. At a 2 cm depth the M array is slightly better overall than the others and at 6 cm depth the minimum detectable dipole for the PG and AG arrays are about 70% and 20% greater, respectively, than for the M array. When averaged over angles (Fig. 5), the M array is best when brain noise is taken into account for all depths below about 1.5 cm, though the advantage is reduced compared to the situation when only sensor noise is considered (Fig. 3).

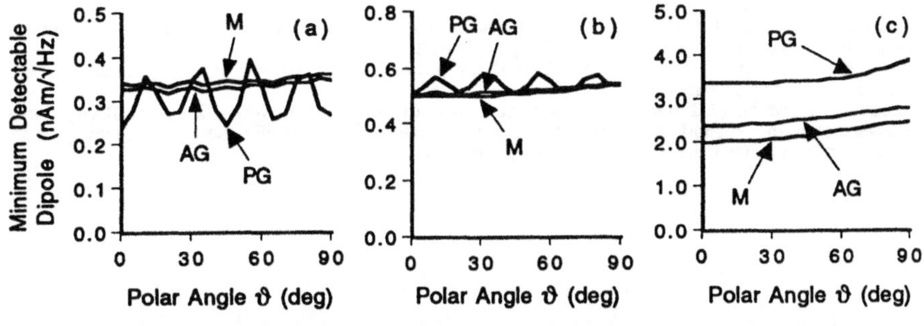

Fig. 4. Angular variation in strength of the minimum detectable dipole (1 Hz bandwidth, source current oriented along X-axis) at various depths in the mid-coronal plane in the presence of uncorrelated brain noise. Depths below the head surface are: (a) 1 cm, (b) 2 cm, and (c) 6 cm.

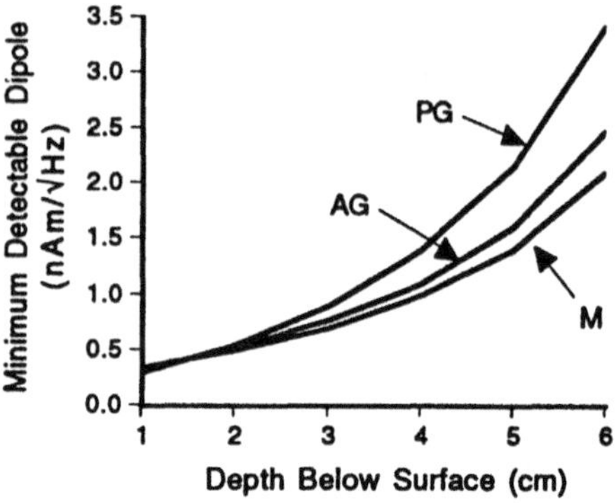

Fig. 5. Minimum detectable dipole averaged over polar angles ϑ between 0 and 60 deg as a function of depth in the mid-coronal plane in the presence of uncorrelated brain noise (source current oriented along X-axis).

Discussion

Detectability, a measure of minimum detectable dipole strength, is a useful way to characterize the "sensitivity" of an entire array of biomagnetic detection coils. Different sources of noise are readily accommodated by the formalism, e.g., sensor noise and brain noise. We have found detectability calculations to be valuable in the design of multichannel biomagnetic sensors when evaluating alternatives for detection coil design, the number and spacing of detection coils, and the shape and extent of the coil array.

The helmet-shaped array models investigated here for brain studies indicate that magnetometer coils provide better detectability than gradiometers. Short baseline planar gradiometers are found to have good detectability for shallow sources directly under each detection coil but compare less favorably for sources located elsewhere.

When brain noise is taken into account, ignoring correlations, the comparative advantage of magnetometer coils is reduced but is still significant for deeper sources. Short baseline planar gradiometer coils are found to be slightly better for sources located within about 1 cm of the head surface. The effect of spatial correlation in brain noise is yet to be investigated.

References

[1] Knuutila, J., and Hämäläinen, M.S. On the spatial locating accuracy of multichannel magnetometers, In: Williamson, S.J., Hoke, M., Stroink, G., and Kotani, M. Advances in Biomagnetism, Plenum Press, New York, 1989, pp. 713-716.

[2] Paulson, D.N., and Buchanan, D.S. Experimental accuracy of localization of current dipoles in a spherical phantom, In: Williamson, S.J., Hoke, M., Stroink, G., and Kotani, M. Advances in Biomagnetism, Plenum Press, New York, 1989, pp. 717-720.

[3] Buchanan, D.S., and Johnson, R.T., Sensor configuration study. Book of Abstracts, 8th International Conference on Biomagnetism, Münster, 1991, pp. 111-112.

[4] Flynn, E.R. Factors which affect spatial resolving power in large array biomagnetic sensors, Rev. Sci. Instrum., 1994, 65:922-935.

[5] Knuutila, J., and Hämäläinen, M.S. Characterization of brain noise using a high sensitivity 7-channel magnetometer, In: Atsumi, K., Kotani, M., Ueno, S., Katila, T., and Williamson, S.J. Biomagnetism '87, Tokyo Denki University Press, Tokyo, 1988, pp. 186-189.

A Digital Flux-Locked Loop for High Temperature SQUID Magnetometer and Gradiometer Systems with Field Cancellation

Kraus, R.H., Jr.[1], Bracht, R.[1], Flynn, E.R.[1], Jia, Q.[1], Maas, P.M.[2], Reagor, D.[1] and Stettler, M.[1]

[1]Los Alamos National Laboratory, Box 1663, Los Alamos, NM 87545, USA; [2]Strathclyde University, Dept. of Physics and Applied Physics, Glasgow G4ONG, Scotland

I. Introduction

Numerous challenges stand in the way of "fieldable" biomagnetic sensors that can be practically used unshielded in everyday ambient environments. A few of the possible applications of a fieldable biomagnetic sensor include bedside monitoring and emergency medical diagnosis of cardiac and brain function (i.e., trauma), and monitoring of cardiac or brain activity in specialized environments such as the NASA shuttle. One futuristic vision of a miniature portable biomagnetic sensor has been related by Wikswo[1]. To date, however, biomagnetic measurements of the heart and brain using Superconducting QUantum Interference Device (SQUID) sensors have been made almost exclusively inside shielded rooms or within systems of field cancellation coils, with only a few noted exceptions[2,3]. Shielding or active cancellation is usually necessitated by the debilitating affect of everpresent ambient background fields and radio-frequency (RF) interference on the performance of SQUIDs. Methods of shielding or compensating the ambient background fields use bulky hardware that is not practical for portable field applications and, in the case of shielded rooms, are extremely expensive. A further impediment to miniaturizing SQUID sensor systems has been the bulk associated with the cryogenic systems required to maintain low critical-temperature (LTC) devices at liquid helium temperatures (to maintain the superconducting state).

Recent advances in the noise characteristics of high critical-temperature (HTC) SQUIDs have opened the door for using these devices in both biomagnetic and non-biomagnetic applications. Because of the large heat content and low cost of liquid nitrogen (LN), HTC SQUID sensor systems are far more practical for portable field applications than LTC SQUID systems that use liquid helium (LHe). Not only is the cost of LN more than one hundred times less than for LHe per watt cooled, other factors including quantity of thermal shielding, volume of coolant, and boiling induced noise strongly favor HTC over LTC systems for portable field applications.

The SQUID sensor is typically operated in a null detector mode where an analogue flux-locked-loop, FLL, provides a negative feedback to maintain linear operation[4]. The modulated SQUID signal is amplified, filtered, demodulated, and integrated in the FLL. The resulting analog signal is a measure of the magnetic field and noise at the SQUID and is also fed back to the modulation and feedback (M&F) coil to null the flux at the SQUID to maintain the linear operating point. Thus, the FLL output signal is proportional to the change in magnetic field at the SQUID pickup coil, provided the slew rate and dynamic range of the SQUID and FLL system are not exceeded.

The goal of the work presented here is to advance technologies needed for a practical fieldable SQUID biomagnetic sensor. We used HTC SQUIDs to realize the benefits noted above. We also implemented the FLL algorithm on a digital-signal-processor (DSP) to realize a number of benefits including (1) software control of noise filtering and background rejection to enable unshielded use of SQUID sensors, (2) flux quanta countin and resetting SQUID operating point to increase system slew rate and dynamic range, (3) programmable FLL adaptable to numerous specific applications, (4) digital signal output (up to 32-bit precision), and (5) reduced FLL package cost. General aspects of this system concept have been described previously[5,6]. This paper presents results of external signal rejection for a sensor system using HTC SQUIDs, preamplifier circuit, and DSP FLL designed and built at our laboratory. We also note a companion paper in these proceedings[7] and other references to the use of DSP in SQUID applications [8,9].

II. Methods

Figure 1 is a block diagram illustrating the complete SQUID and DSP-FLL system and excitation source. All ADCs and DACs were 18-bit/10V, 200kHz devices. The dashed box labeled "SQUID" represents the components that are on the SQUID chip, consisting a superconducting-normal-superconducting (SNS) junction HTC SQUID with

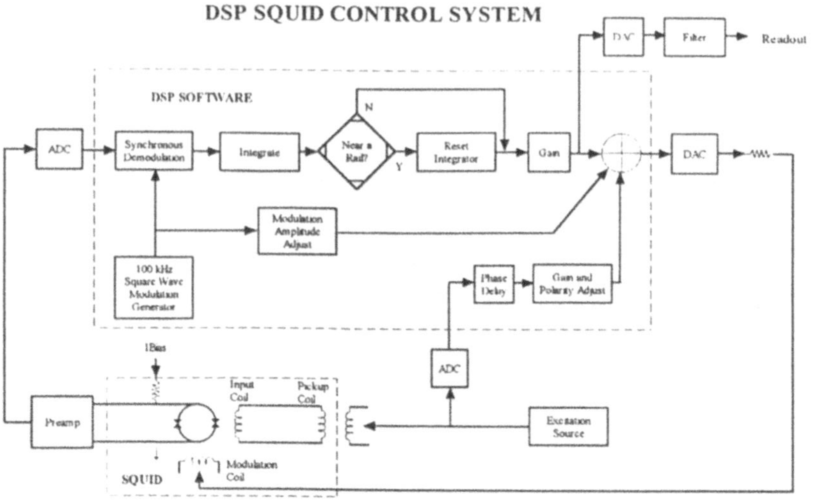

Figure 1

$0.04 mm^2$ area pickup coil and modulation coil. The SNS-junctoin SQUID was characterized open-loop with $30\mu V/\Phi_0$ modulation depth and $<20\mu V/\sqrt{Hz}$ noise after fabrication. The output of the preamplifier, which is directly coupled to the HTC SQUID, is digitized read by the DSP. The simplest algorithm implemented on the DSP controlled FLL is surrounded by the dashed box labeled "DSP Software." The DSP-FLL algorithm shown performs all of the functions of a traditional analog FLL and adds automatic SQUID reset and feedback cancellation capabilities. The summed digital M&F signals are converted to an analog signal by a DAC that drives the modulation coil. The demodulated digital SQUID output signal is also converted to analog, though this step is in general unnecessary and introduces additional noise, however was performed for compatibility with our existing data acquisition system.

Numerous preamplifier designs were tested including one- and two-transformer, and directly coupled designs. Although the two-transformer design was optimized with a narrow transmission peak at the modulation frequency to eliminate noise at other frequencies, it was impractical because the resonance peak would drift with time due to a variety of factors including transformer temperature. We chose a directly coupled low-noise broad-band preamplifier design, schematically shown in Figure 2. The modulation depth of the HTC SQUID mounted in our test apparatus and connected to the preamplifier and DSP-FLL was measured to be approximately $12\mu V/\Phi_0$. Trapped flux that could not be eliminated by cycling the SQUID is believed to cause the reduction in modulation depth. Substantial noise entered the system through the prototype preamplifier making SQUID noise measurements impossible.

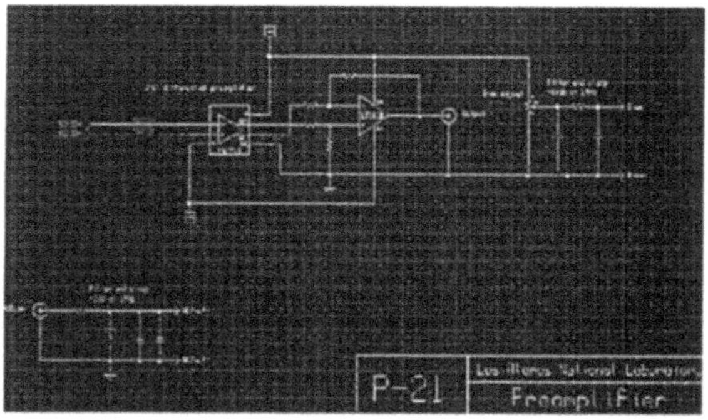

Figure 2

The SQUID sensor was mounted in a 6 inch diameter fiberglass dewar. A signal generator was used to drive a small coil placed inside a shielded chamber under the SQUID dewar. The SQUID signal leads were connected to the prototype preamplifier outside the shielded chamber. Ambient fields were reduced to approximately 100nT and other noise sources including 60Hz and harmonics were substantially reduced inside the shielded chamber. This allowed accurate measurement of the primary signal and feedback rejection for specific signals of interest. As shown in Figure 1, the same signal that was used to drive the excitation coil near the SQUID was also digitized and used by the DSP to sum into the M&F drive signal. In principle, proper adjustment of the phase delay and gain will cause the flux generated by the M&F coil to perfectly cancel the external flux carried to the SQUID by the pickup coil. A LABVIEW software interface was used to control the operational parameters of the DSP-FLL. An existing CAMAC data acquisition system was used to acquire the data, as noted above, because the PC-based LABVIEW system had insufficient bandwidth to perform this function.

III. Results

A variety of waveforms were used to drive the excitation coil placed in the shielded chamber with the HTC SQUID. The phase and gain (amplitude) of the digitized excitation coil drive signal were adjusted in the DSP algorithm by software command and optimized for maximal signal cancellation. These measurements were performed to simulate, in a controlled fashion how one might use such an algorithm to cancel a background or a primary excitation signal at the SQUID. The results of two sample measurements are shown in Figures 3 and 4. In each figure, the lower trace shows the SQUID sensor response to the excitation coil with the gain of the cancellation signal set to zero (no cancellation). This is analogous to a normal FLL control of the SQUID where the output is proportional to the rate of change of the flux in the pickup coil. The SQUID output observed for the unconcealed signal in both cases was between 400mV and 600mV. The upper trace in each figure shows the SQUID output when the feedback signal bias and gain are optimized for maximal cancellation. Note that the upper traces are multiplied by a factor of 100 to magnify the characteristics of the residual cancelled signal on the scale of the plot. The SQUID output as a function of magnetic flux was not calibrated for these measurements.

Figure 3 illustrates the measured signals resulting from a sine wave excitation. The rms amplitude of the unconcealed (feedback gain=0) SQUID output signal was 580mV. After optimizing the feedback gain (gain=281) and phase delay for maximal cancellation, the rms amplitude of the cancelled was observed to be 2±1mV. The feedback gain is

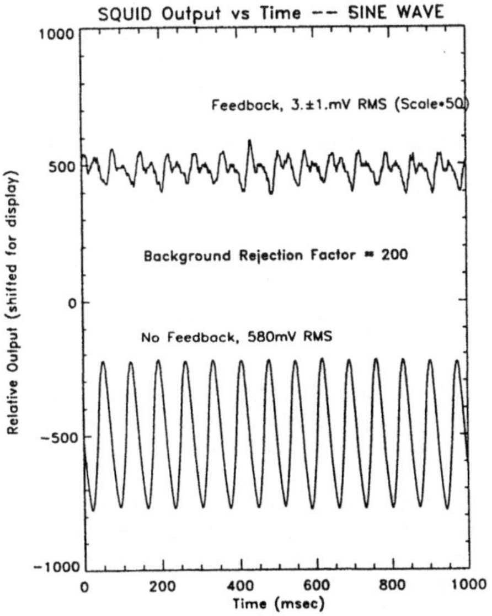

Figure 3

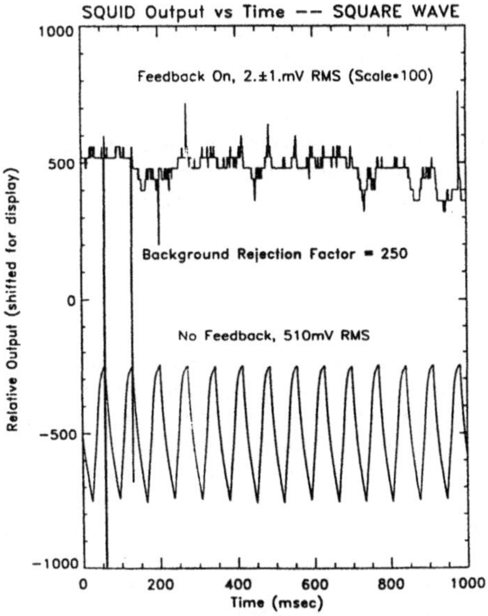

Figure 4

simply a binary value in the DSP between -32,768 and +32,768, however the relatively poor matching of the signal generator amplitude to the ADC dynamic range limited the precision of the feedback gain and consequently the precision with which the feedback could match the excitation signal at the SQUID.

Figure 4 shows the measured signals for a square wave excitation. The poor reproduction of the square waveform results from the long integration time required to optimize the system dynamic range. The reason for this is not understood at present. The cancelled signal residual shows a similar behavior to that noted in Figure 3. The large spikes shown in the cancelled signal are unique to the square wave data and were traced to least significant bit errors in the phase matching of the SQUID output and feedback signals. Only 8 bits were devoted to phase matching and this observation points to the need of greater timing precision in the cancellation loop.

IV. Discussion

A DSP controlled FLL system with feedback cancellation has been successfully demonstrated for a HTC SQUID sensor system that is easily portable and has a cryogen hold time of more than two weeks. More than 99% of simulated background or primary flux was cancelled by a feedback flux at the SQUID generated by the DSP SQUID control system. The primary limiting factor for the cancellation is the granularity in the feedback gain. More precisely matching the dynamic range of the feedback ADC to the input signal (from the signal generator) will substantially improve the cancellation factor. The cancellation can be further improved by increasing the gain of the preamplifier and reducing the noise introduced by the preamplifier. Although the dynamic range of the ADC is approximately 10^5 for input signals of 10V, input signals of 0.5V reduce the available dynamic range to approximately 10^4. Improving the performance of the preamplifer is in progress.

Finally, although powerful DSP units are relatively expensive, the per-channel cost can be dramatically reduced by controlling multiple SQUID channels with one DSP or by implementing the relatively simple algorithms in far less expensive and faster programmable logical gate arrays (PLGAs).

V. References

[1] Wikswo, J., "SQUID Magnetometers for Biomagnetism and Nondestructive Testing: Important Questions and Initial Answers," Plenary talk, 1994 Appl. Superconductivity Conf., IEEE Trans. App. Superconductivity, 1995, 5, 74

[2] Y. Tavrin, Y. Zhang, M. Mock, A.I. Braginski, and C. Heiden, "YBa2Cu3O7 Thin Film SQUID Gradiometer for Biomagnetic Measurements," Appl. Phys. Lett., 1993, 62, 1824

[3] Y. Tavrin, Y. Zhang, W. Wolf, and A.I. Braginski, "A Second-Order SQUID Gradiometer Operating at 77K," Supercond. Sci. Tech., 1994, 7, 265

[4] Wellstood, F., Heiden, C., and Clarke, J. "Integrated DC SQUID magnetometer with a high slew rate", Rev. Sci. Instrum. 1984, 55, 952

[5] Bracht, R. R., Kung, P. J., Lewis, P. S., and . Flynn, E. R, "DSP control of superconducting quantum interference devices", in Procs. of the Int. Conf. on Signal Processing Applications and Technology, 1994, ICSPAT, 1146

[6] Kung, P.J., Bracht, R.R., Flynn, E.R., and Lewis, P.S., "A Direct Current Superconducting Quantum Interference Device Gradiometer with a Digital Signal Processor Controlled Flux-Locked Loop and Comparison with a Conventional Analog Feedback Scheme," Rev. Sci. Instrum., 1996, 67, 222

[7] Flynn, E.R., Bracht, R., Kraus Jr., R. H., Maas, P., Ruminer, P., and Stettler, M., "A Digital- Signal-Processor Flux-Locked Loop for LTC SQUIDs with Automatic Reset and Feedback Cancellation", Biomag96, Tenth International Conference on Biomagnetism, (Santa Fe, NM), C. Aine, C. C. Wood, and Y. Okada, eds. (Springer-Verlag, New York, 1996, in press)

[8] Vrba, J., Betts, K., Burbank, M., Cheung, T., Cheyne, D., Fife, A.A., Haid, G., Kubik, P.R., Lee, S., McCubbin, J., McKay, R., Mckenzie, D., Mori, K., Spear, P., Taylor, B., Tilloston, M., and Xu. G., "Whole Cortex 64-channel System for Shielded and Unshielded Environments", Biomagnetism: Fundamental Research and Clinical Applications, Baumgartner, C., Deecke, L., Stroink, G., and Williamson, S. J., eds. (Elsevier, New York), 1995, 521

[9] Brandenburg G. Clements, U., Halling, H., and Zimmerman, F., Julich (KFA) internal report, (mentioned in Wikswo, J. P, IEEE Trans. App. Superconductivity, 1995, 5, 74)

Characterisation of a Planar Second-Order Gradiometer in a Lightly Shielded Environment

Lang, G.[1], Hutson, D.[1], Weston, R.G.[1], Klein, U.[1], Shahani, U.[2], Weir, A.I.[2], Maas, P.M.[1], Pegrum, C.M.[1] and Donaldson, G.B.[1]

[1]*Department of Physics and Applied Physics, University of Strathclyde, Glasgow, Scotland;*
[2]*Wellcome Biomagnetism Unit, Southern General Hospital, Glasgow, Scotland*

Introduction

In recent years biomagnetic studies using SQUID systems have mainly used axial or planar first-order gradiometers which measure the B_z component of the magnetic field. However, these systems tend to rely on large, expensive mu-metal shielding plus electronic channel-differencing to achieve reasonable signal-to-noise ratios. This is often inconvenient and in some situations can make measurements impossible. We have therefore chosen to concentrate on devices capable of operation with minimal magnetic shielding. We have developed a planar second-order gradiometer, integrated with a first-order gradiometric SQUID, for B_x/B_y measurements of magnetic field in lightly shielded environments[1]. They rely on a high degree of intrinsic geometric balance and no external balance mechanisms, making them appropriate for unshielded multi-channel applications. The devices are fabricated on two-inch silicon wafers using our fully-established niobium fabrication process at the University of Strathclyde[2]. Precise photolithographic patterning of the pick-up coils permits a high degree of intrinsic balance. We discuss the characterisation of our gradiometer by analysing its spatial response to magnetic sources and its immunity to background interference. The spatial response to a magnetic source, on-axis, is characteristic of the order and balance of the gradiometer and therefore indicates how well the device performs. We have modelled the response to an off-axis source, to better understand the spatial response of a tangential gradiometer. The noise performance of the device has been analysed and the sources of many discrete frequencies identified. This design of gradiometer has been successfully introduced into our biomagnetism programme at the Wellcome Biomagnetism Unit in the Southern General Hospital, Glasgow. Biomagnetic measurements using it are reported elsewhere[3,4].

Methods

We modelled the spatial response to a magnetic dipole for both first and second-order gradiometers. The model was confirmed experimentally using a small magnetic dipole, 4.5mm long and 2.5mm in diameter, in a three-dimensional grid around and below a first-order device. With the gradiometer stationary in the y-z plane the dipole was moved in the x-y and y-z plane and the response measured, see Fig.2(a). Data was obtained for three orientations of the dipole.

The balance of our second-order gradiometer was evaluated by analysing the response to a magnetic dipole

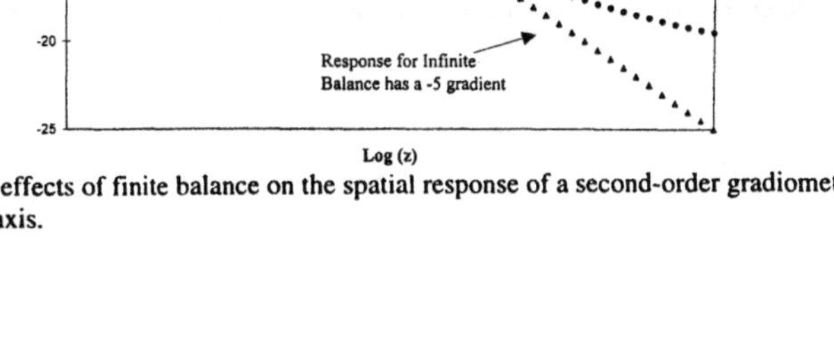

Fig.1. The effects of finite balance on the spatial response of a second-order gradiometer, to a magnetic dipole on-axis.

at a distance z along the main axis. The response A, for a second-order gradiometer, would asymptotically reach a gradient of -5 if the balance were infinite. Finite balance results in a deviation away from the asymptote to a gradient of -3 at far field, where the gradiometer assumes magnetometer behaviour to the dipole but with a much reduced sensitivity, Fig.1. The point of inflection where the deviation begins, can be determined experimentally. A degree of geometrical imbalance can be added to the spatial response model and the relationship between balance and distance to the point of inflection obtained. Using this relationship the amount of balance is estimated from the experimentally measured point of inflection.

The gradiometer is normally operated in a simple eddy-current shielded room with no additional shielding close to the device. The shielded room is a 4m cube of 2.5cm thick aluminium which attenuates the Earth's field by 20dB/decade above 1.4Hz. Tests have been carried out on the noise performance with and without magnetic shielding of the gradiometer. The discrete frequencies caused by our mechanical resonances of the shielded room have been identified using a commercial accelerometer.

Results and Discussion

The co-ordinate system used for the three-dimensional spatial response measurements is given in Fig.2(a). Fig.2(b) shows one half of the modelled off-axis response of a second-order gradiometer in the y-z plane, to a magnetic dipole in the x-direction. The response is modelled in the x-y plane 20mm under the gradiometer. This dipole orientation couples maximum flux to our tangential device. The response is a maximum directly under the device and decays with distance. However, as the dipole is moved along the x-axis the response goes through two zero points where the field coupled into the gradiometer by the dipole is cancelled in the gradiometer loops. There are two secondary peaks of alternating phase and decreasing amplitude. Where the zero points occur and the

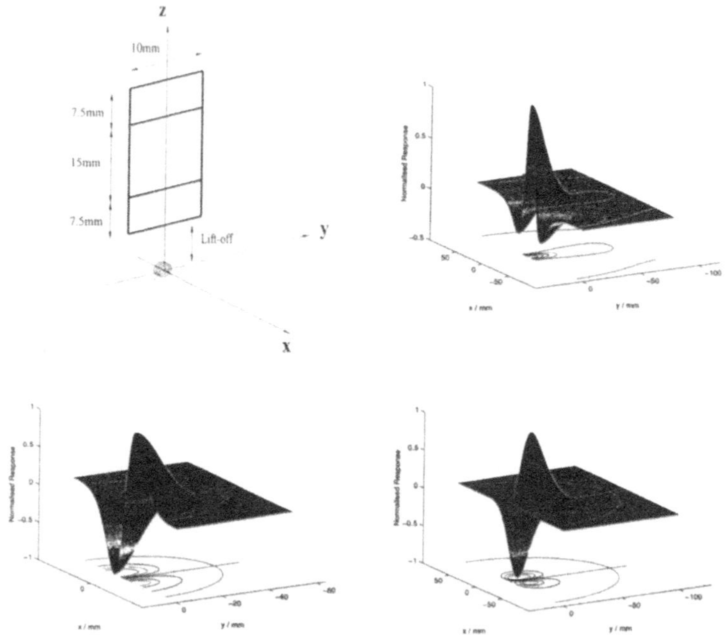

Fig.2(a)The co-ordinate system used in our experimental and modelled spatial response measurements. (b)One half, of a second-order gradiometer response to a magnetic dipole in the x-direction. Response in the x-y plane, 20mm below the gradiometer, (c)dipole in the z-direction, (d)dipole in the y-direction.

amplitude of the secondary peaks are dependent on the minimum gradiometer to source distance (i.e. lift-off). Fig.3 shows the relationship between lift-off and the distance from the main response peak to the zero point and to the second peak. Although not suitable for small amplitude, distant sources, this relationship could be a useful measurement of source depth. As the gradiometer order increases the number of phase changes and secondary peaks increases. The figure also shows the amplitude ratio of the second peak to main peak increasing with increasing lift-off. Therefore a small lift-off is not only better for obtaining a large main response, but also for a less significant secondary peak.

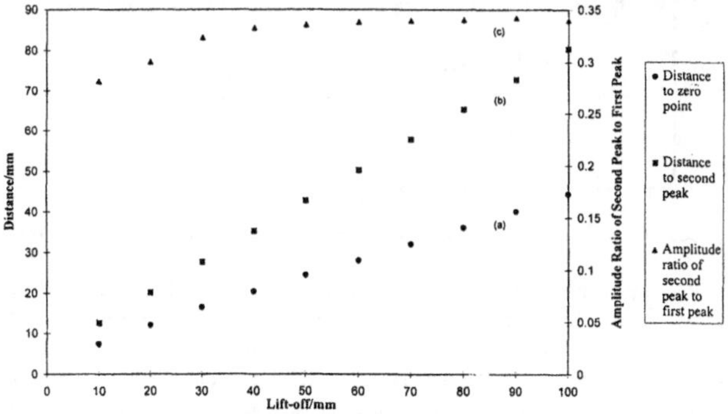

Fig.3. Relationship between lift-off and (a)distance to first zero point, (b)distance to second peak, (c)Ratio of the amplitude of the second peak to the amplitude of the main peak.

For a dipole orientated in the z-direction there is a zero response directly below the gradiometer and along the y-axis where no field lines cut through the gradiometer loops; (Fig.2(c)), but if the dipole is moved off in the ±x-direction, the response rises sharply to a maxima/minima. For a dipole orientated in the y-direction, the response is again zero directly below the gradiometer and along both the x and the y-axis, Fig.2(d). Along the y-axis no field lines enter the gradiometer loops and along the x-axis the field is cancelled due to the symmetric nature of the field of a dipole. Any deviation from the x or y-axis results in a response being detected. For these two orientations of dipole there is an outer ring of zero response and phase change, whose position is again dependent on the lift-off. The amplitude of these secondary peaks are negligible.

The modelled response in the same plane as the gradiometer, to a dipole in the x-direction, shows a broad circular shape of uniform field sensitivity. Therefore, for a lift-off of ~60mm the gradiometer could easily pick up cortical signals when recording the spine at the neck. With our fabrication limitations, a narrower response can only be obtained with higher-order gradiometers. The on-axis response of the second-order gradiometer to a magnetic dipole has an experimentally measured point of inflection 0.22m from the device, corresponding to a balance of 1.1 : 1000. It was initially thought that the balance was limited by the gradiometric SQUID acting as a magnetic anomaly very close to the gradiometer. The anomaly could be ignored by the use of asymmetric gradiometers where the SQUID is placed at a null point in the response[5]. We calculated the magnetic anomaly presented to the gradiometer by a SQUID in the form of a current loop of similar dimensions, and compared it to the SQUID intrinsic balance. This showed that it is the intrinsic balance of the SQUID that limits the overall balance of the device. The SQUID balance must be improved by a factor of 150 before the SQUID anomaly becomes a problem and asymmetric gradiometers beneficial. Imbalance in our devices is not purely a characteristic of design and fabrication. Trapped flux also causes imbalance and thermally cycling the device has on occasion improved the balance. This improvement has manifested itself in the noise spectra where many discrete frequencies have disappeared and the mains and overall white noise reduced. An improvement in the geometrical balance will therefore improve the noise performance of the device.

The noise is essentially white at $2\mu\phi_0/\sqrt{Hz}$ between 5Hz and 2kHz (Fig.4). Environmental noise dominates below 5Hz, but is outside our bandwidth of interest. We have found that while the gradiometric SQUIDs are resistant to rf interference, the integrated devices require careful suppression of both conducted and radiated rf

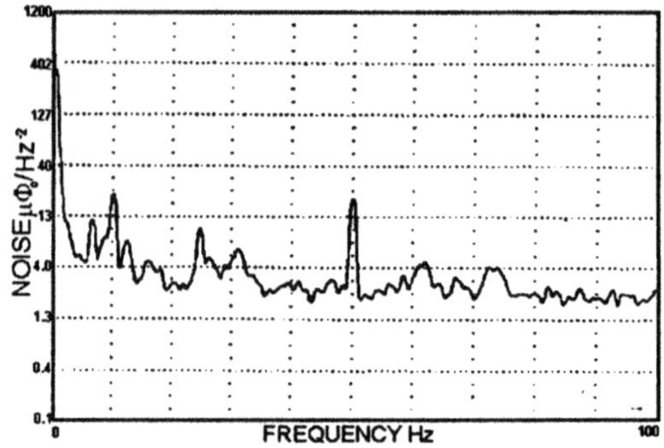

Fig.4. Noise spectra of a planar second-order gradiometer in eddy-current shielded room.

interference to obtain low white noise and instrumentational sensitivity. Tests have shown that enclosing the gradiometer in a superconducting shield increases the white noise level. By defining the effective area to be the bottom loop the calculated uniform field sensitivity is 5.3fT/√Hz. Most of the discrete frequencies observed were attributable to resonances in the cryostat (8Hz), gantry (15, 31Hz) and the shielded room (10, 12, 24.5, 61.5Hz). These resonances are caused predominantly by road traffic outside the building despite a moderate degree of vibration isolation.

Conclusions

We describe the characterisation of our design of tangential planar gradiometer, through measuring the spatial response, balance and noise performance. The off-axis spatial response of the gradiometer is non-trivial and the response to different orientations of dipole, other than that for maximum response, are not always negligible. This may complicate simple analysis but could help to identify sources, estimate source depths and locate the exact position of the gradiometer in the cryostat tail. With our present moderate degree of balance the device operates well in the Earth's field and rarely resets. The signal-to-noise ratio is currently large enough to measure biological signals from the heart, brain and peripheral nerves[1,3,4]. Future designs will concentrate on improving the balance of the integrated SQUID and gradiometer loops to reduce the noise and interference further. We plan to fabricate on three inch silicon wafers to allow an increase in the baseline and therefore improve sensitivity.

References

[1] Lang, G., Hutson, D., Weston, R.G., Bain, R.J.P., Maas, P., Pegrum C.M., Shahani, U., Weir, A.I. and Donaldson, G.B., Performance of a second-order planar gradiometer in the Earth's field, Applied Superconductivity 1995, Vol.2 :1573-6.

[2] Hutson, D., Weston, R.G., Bain, R.J.P., Lawrenson, M.J., Pegrum C.M., Popova K., Scarfone R., A robust fabrication process for a refractory integrated SQUID gradiometer, IEEE Trans. Appl. Superconductivity, 1995, Vol.5, No.2, 2295-8.

[3] Weir, A.I., Shahani, U., Hutson, D., Lang, G., Maas, P., Pegrum C.M., and Donaldson, G.B., SEF's from the spinal cord using a novel second order planar gradiometer. these proceedings.

[4] Shahani, U., Weir, A.I., Lang, G., Mansfield, D.C., Halliday, D.M., Hutson, D., Maas, P., and Donaldson, G.B., SQUIDs and stereopsis, these proceedings.

[5] Jones, A.E. and Bain, R.J.P., A generalisation of planar magnetic gradiometer design via orthogonal polynomials, J. Comp. Phys. 1995, 118:191-9.

Acknowledgements:

This work has been supported by the Wellcome Trust and EPSRC.

Magnetic field Error of a Current Dipole in a Superconducting Shield

Matsuba, H

Furukawa Electric Co., Ltd., Yokohama, Japan

Introduction

When we detect a current dipole in a brain in a superconducting shield, the image of the current dipole formed by the wall of the superconducting shield might give a considerable change to the magnetic field because the superconducting shield has a small dimension of 90cm in diameter. This also might result in localization errors of the current dipole. In this paper we evaluate the magnetic field of the image and estimate the contribution of the image field to the field of the current dipole.

Analytical approach

Magnetic field in a spherical superconducting shield

Magnetic fields of a current dipole in a spherical shaped brain are analytically obtained supposing that the superconducting shield is a sphere of radius R, Fig.1 shows a model for the analysis.

Magnetic fields of a current dipole of P_x in magnitude at z=a in a brain in an infinite space were introduced using a spherical-coordinate system whose origin is the brain center :[1]

$$B_r=\frac{P_x}{4\pi}\frac{a\sin\theta\sin\varphi}{r^3\gamma_a^{\frac{3}{2}}} \quad \text{--- (1)}$$

$$B_\theta=\frac{P_x}{4\pi}\frac{\sin\varphi}{r^2\gamma_a^{\frac{3}{2}}}\frac{r}{a\sin^2\theta}\left[\gamma_a^{\frac{3}{2}}\cos\theta+\left(\frac{a^2}{r^2}-1\right)\cos\theta\sin^2\theta+\left(\frac{a}{r}-\cos\theta\right)^3\right] \quad \text{--- (2)}$$

$$B_\varphi=\frac{P_x}{4\pi}\frac{\cos\varphi}{r^2\sin^2\theta}\frac{1}{\gamma_a^{\frac{1}{2}}}\left(\frac{r}{a}-\cos\theta-\frac{r}{a}\gamma_a^{\frac{1}{2}}\right) \quad \text{--- (3)}$$

where

$$\gamma_a=1-\frac{2a\cos\theta}{r}+\left(\frac{a}{r}\right)^2 \quad \text{--- (4)}$$

Then we introduce an image current dipole of P_{ix} in magnitude at z=A. The image dipole induces magnetic field :

$$B_{ir}=\frac{P_{ix}}{4\pi}\frac{A\sin\theta\sin\varphi}{r^3\gamma_A^{\frac{3}{2}}} \quad \text{--- (5)}$$

$$B_{i\theta}=\frac{P_{ix}}{4\pi}\frac{\sin\varphi}{r^2\gamma_A^{\frac{3}{2}}}\left[\cos\theta-\frac{A}{r}\right] \quad \text{--- (6)}$$

$$B_{i\varphi}=\frac{P_{ix}}{4\pi}\frac{\cos\varphi}{r^2\gamma_A^{\frac{3}{2}}}\left[\frac{A}{r}-\cos\theta\right] \quad \text{--- (7)}$$

where

$$\gamma_A=1-\frac{2A\cos\theta}{r}+\left(\frac{A}{r}\right)^2 \quad \text{--- (8)}$$

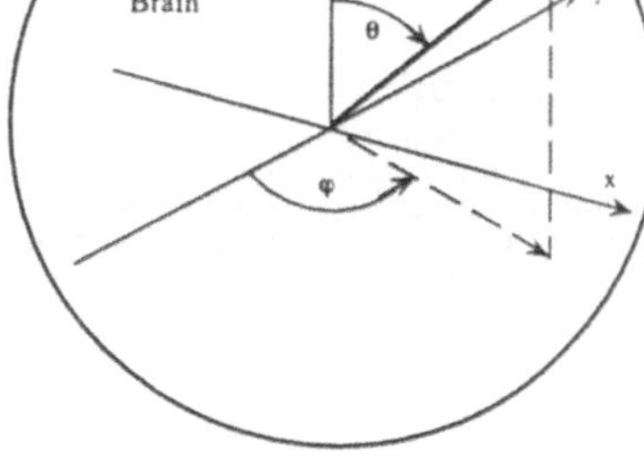

Fig.1 A model of a brain in a superconducting shield

The magnetic field in the superconducting shield is the sum of magnetic field of the two dipoles.
The boundary condition of the magnetic field at the surface of the superconducting wall is $B_r=0$ (at r=R). Hence

$$\frac{AP_{ix}}{\left[1-\frac{2A\cos\theta}{R}+\left(\frac{A}{R}\right)^2\right]^{\frac{3}{2}}}+\frac{aP_x}{\left[1-\frac{2a\cos\theta}{R}+\left(\frac{a}{R}\right)^2\right]^{\frac{3}{2}}}=0 \qquad \text{---}(9)$$

Solving the equation (9), we obtain relations between the two dipoles:

$$A=\frac{R^2}{a},\ P_{ix}=-\frac{R}{a}P_x \qquad \text{---}(10)$$

Magnetic field rrors due to the superconducting shield

The errors are defined as the ratio of magnetic field difference between a current dipole in a superconducting shield and the same current dipole in an infinite space, to the magnetic field of the current dipole in the infinite space.
From equations (1)~(8) and equation (10) we obtain the errors e_r, e_θ, e_φ corresponding to magnetic field components B_r, B_θ, B_φ ;

$$e_r=\frac{R^3}{a^3}\frac{\gamma_a^{\frac{3}{2}}}{\gamma_A^{\frac{3}{2}}}=\frac{R^3}{a^3}\left[\frac{1-\frac{2a\cos\theta}{r}+\left(\frac{a}{r}\right)^2}{1-\frac{2R^2\cos\theta}{ar}+\left(\frac{R^2}{ar}\right)^2}\right]^{\frac{3}{2}} \qquad \text{---}(11)$$

$$e_\theta=\frac{R}{a}\frac{\gamma_a^{\frac{3}{2}}}{\gamma_A^{\frac{3}{2}}}\frac{\cos\theta-\frac{R^2}{ar}}{\frac{r}{a\sin^2\theta}\left[\gamma_a^{\frac{3}{2}}\cos\theta+\left(\frac{a^2}{r^2}-1\right)\cos\theta\sin^2\theta+\left(\frac{a}{r}-\cos\theta\right)^3\right]} \qquad \text{---}(12)$$

$$e_\varphi=\frac{R}{a}\frac{\gamma_a^{\frac{1}{2}}}{\gamma_A^{\frac{3}{2}}}\frac{\sin^2\theta\left(\frac{R^2}{ar}-\cos\theta\right)}{\frac{r}{a}-\cos\theta-\frac{r}{a}\gamma_a^{\frac{1}{2}}} \qquad \text{---}(13)$$

When we consider the relation $R >> r \geq a$ in an actual superconducting shield dimensions, equations (11),(12) and (13) approxiamate

$$e_r^{R\to\infty}\approx\frac{a^3}{R^3},\ e_\theta^{R\to\infty}\approx\pm\frac{a^3}{R^3},\ e_\varphi^{R\to\infty}\approx\pm\frac{a^3}{R^3} \qquad \text{---}(14)$$

The equation (14) suggests that the errors are less than 1%, because R~45cm and a~10cm.

Deviation of the brain center from the center of the superconducting shield

Whenthe brain center deviates to a different direction than z direction, we may only consider the tangential components of the dipole because the radial component has no contribution to the magnetic field in the shield. Therefore, deviation to z direction gives the maximum errors. So, here, we evaluate the errors when the brain center deviates to z direction. Suppose that the deviation from the center of the shield is d, and that errors in case of the deviation d approximate the errors in case that a current dipole locates at a+d .

If we can neglect the volume current the approximation give the correct value. Hence the tangential field of the volume current brings unfitness to the approximation.

If the center of the brain does not locate at the center of the shield as shown in Fig.2, the tangential field $B_{\theta v}$ give a radial field B_{rv} at the superconducting wall. The radial field B_{rv} does not satisfy the boundary condition, so that it gives unfitness to the approximation. The fields $B_{\theta v}$ and B_{rv} are expressed as

$$B_{\theta v}=\frac{P_x}{4\pi}\frac{\sin\varphi}{r_1^2\gamma_a^{\frac{1}{2}}}\frac{r_1}{a\sin^2\theta}\left[\cos\theta\left(\frac{a}{r_1}\cos\theta-1+\gamma_a^{\frac{1}{2}}\right)+\frac{a}{r_1}\sin^2\theta\right] \qquad ---(15)$$

$$B_{rv}=B_{\theta v}\sin(\theta_1-\theta_2) \qquad ---(16)$$

B_{rv} take the maximum at $r_1=r_2=r_0$ as

$$B_{rv}^{max}=B_{\theta v}\frac{d}{r_0}\sqrt{1-\frac{d^2}{4r_0^2}} \qquad ---(17)$$

Therefore B_{rv}^{max} at the superconducting surface transverse to the orientation of the deviation gives major part of the unfitness. When the deviation is small compared with the radius of the superconducting shield, ($d<<r_0$), B_{rv}^{max} is so small that we may consider the approximation is reasonable.

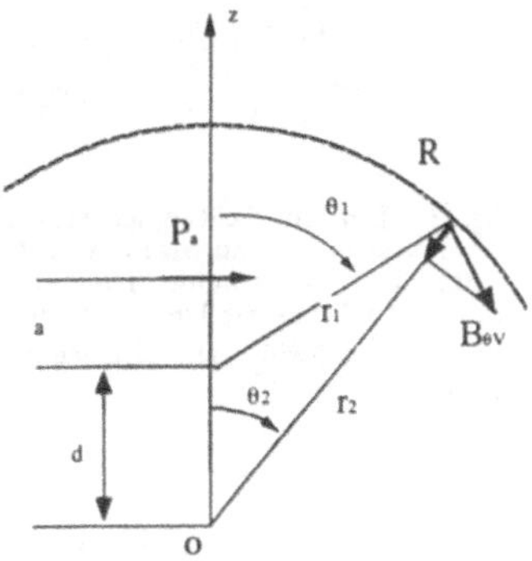

Fig.2, A model that the center of the brain does not locate at the center of the shield,

Calculation

According to the analytical method, Errors due to the superconducting shield are evaluated. The superconducting shield has a diameter of 90cm.

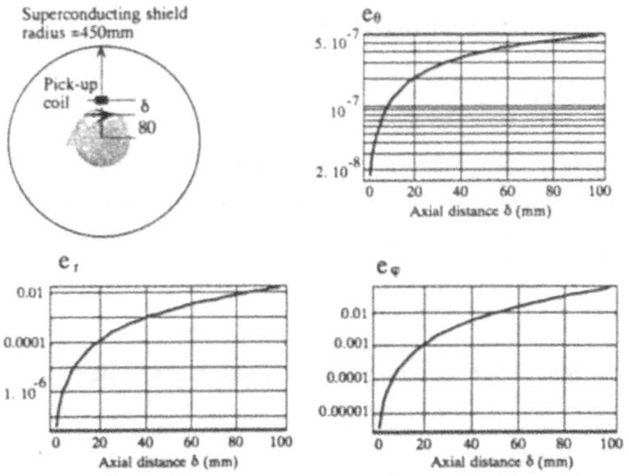

Fig.3 (a) Errors of field components as a function of the distance of a sensor from the current dipole. The brain is located at the center of the superconducting shield and the current dipole is located at a distance 80 mm from the brain center .

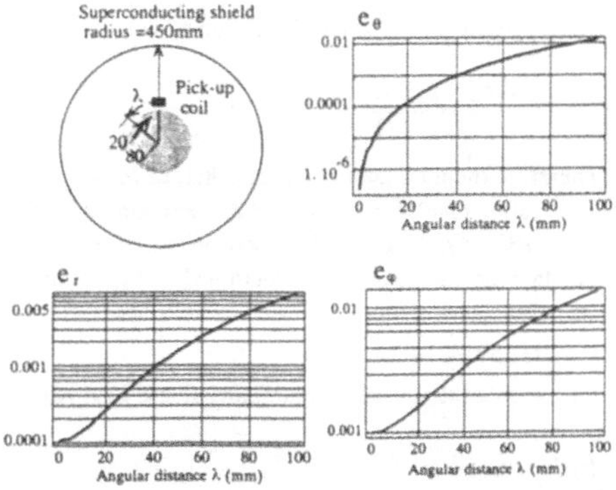

Fig.3 (b) Errors of field components as a function of the angular distance of a sensor from the current dipole. The brain is located at the center of the superconducting shield and the sensor is located at a distance 100 mm from the brain center.

Fig.3 The errors resulted from the image of a current dipole by a spherical superconducting shield . Errors are defined as the ratio of magnetic field difference between a current dipole in a superconducting shield and the same current dipole in an infinite space to the magnetic field of the current dipole in the infinite space.

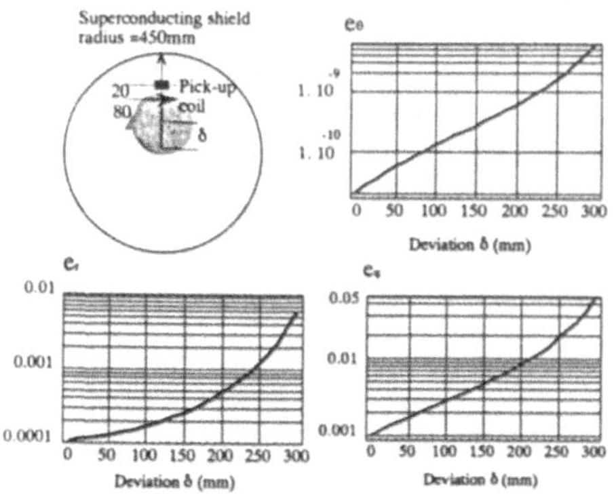

Fig.3 (c) Errors of field components as a function of the deviation of the brain from the center of superconducting shield. The current dipole is located at a distance 80 mm from the brain center.

The sensor is located at a distance 20mm from the dipole

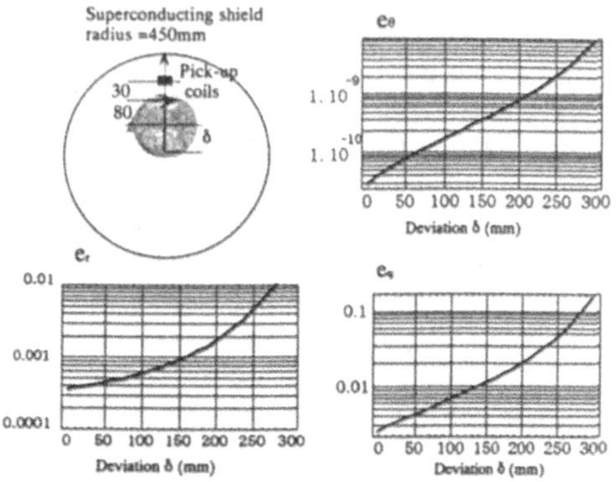

Fig.3 (d) Errors of field components as a function of the deviation of the brain from the center of superconducting shield. The current dipole is located at a distance 80 mm from the brain center.

The sensor is located at a distance 50mm from the dipole

Discussion

Localization errors due to the errors of magnetic field

A current dipole forms a magnetic field at a distance r as:

$$B \sim \frac{\vec{K}}{r^2} \qquad \text{---(18)}$$

We may write the change of location of the dipole Δr by a small change of magnetic field ΔB is

$$\frac{\Delta r}{r} = -\frac{1}{2}\frac{\Delta B}{B} \qquad \text{---(19)}$$

Therefore the localization errors is half of the errors of magnetic field ,when the errors are small.
Calculation results showed that a superconducting shield having diameter of 90cm gives small errors less than 1% in realistic cases. So, the localization errors of a current dipole is estimated to be less than 0.5 mm.
The fact means the superconducting shield gives no significant errors to the dipole localization.

Conclusion

It is revealed that a superconducting shield having diameter 90cm does not disturb biomagnetic measurement.

References

[1] Grynszpan, F. and Geselowitz, D. B. Model Studies of the Magnetocardiogram, Biophysical Journal, 1973, 13, 911-925.

High Tc Superconducting Shielded Biomagnetometer System

Matsuba, H., Shintomi, K., Yahara, A., Imai, K., Irisawa, D., and Masuda, M.

Furukawa Electric Co., Ltd., Yokohama, Japan

Introduction

Superconductors are expected to achieve perfect magnetic shielding, which will be useful for studies on neural activities of the brain especially in the lower frequency range down to DC. Operating in the environment of other noisy instruments and/or interference such as subways, cars, etc. will be available. Instead of gradiometers, higher sensitivity magnetometers are effective in the low noise space, which will allow detections of signals from deeper and/or weaker magnetic sources with a higher S/N ratio.

In this paper, we describe the advantages of the noise performance achieved in a high temperature superconducting shield, in comparison with that in a magnetically shielded room of ferromagnetic materials. Measurements and a current dipole source fitting of auditory evoked magnetic fields in a superconducting shielding system are demonstrated. Measurements with fewer averaging numbers using a magnetometer system are discussed.

System configuration

Fig 1 and 2 show the cross sectional schematic drawing and outside view of the system, respectively. The horizontal superconducting shielding cylinder of 90 cm in diameter and of 2 m in length encloses a human subject and a multichannel SQUID magnetometer system. A $Bi_2Sr_2CaCu_2O_x$ superconductor was developed to have a maximum shielding field of 2 mT, exceeding by 10 times the static earth field, at which field the flux relaxation noise was evaluated at less than 1 fT/s [1]. Ferromagnetic arrangements combined with the superconductor were designed to increase the field attenuation to shorten the length of the shielding cylinder [2]. A magnetic shielding factor exceeding 140 dB has been demonstrated in the frequency range below 10 Hz [3]. External ferromagnetic cylinders and a feedback coil system are included to reduce the dc field trapped during the cool down of the superconducting shield.

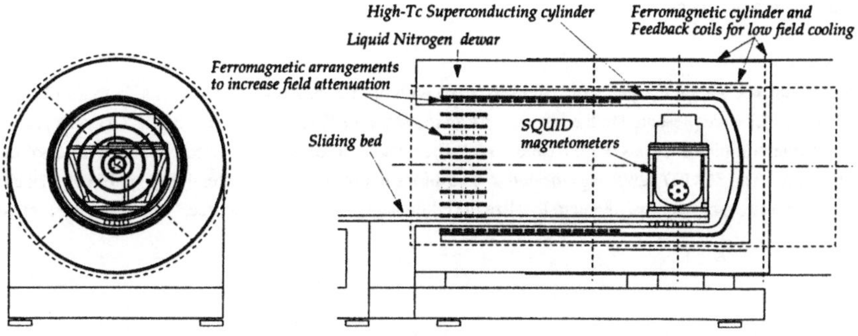

Figure 1. A cross sectional schematic drawing of the superconducting shielded MEG measurement system.

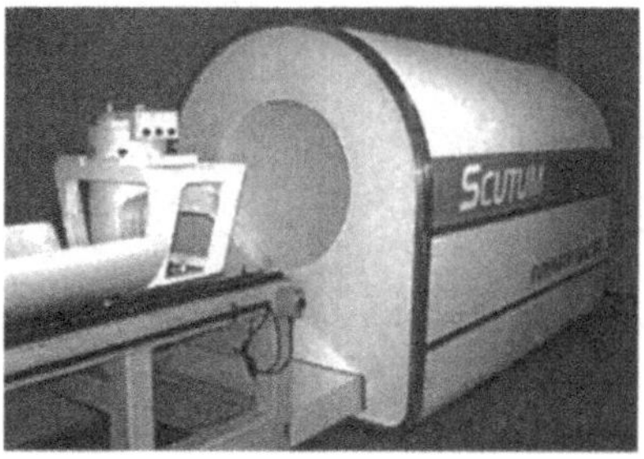

Figure 2. View of the superconducting shielded MEG system.

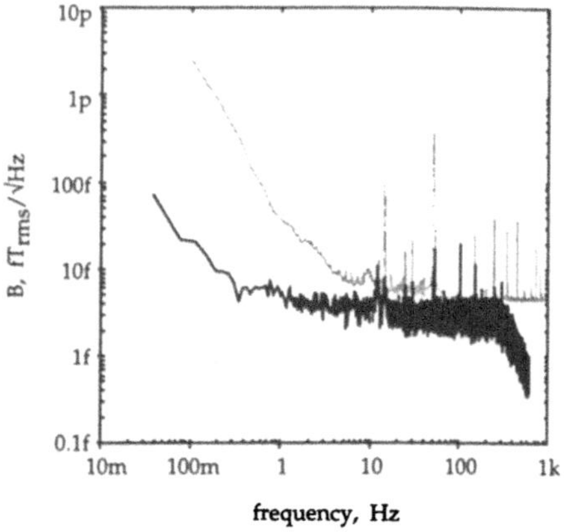

Figure 3. Magnetic noise spectrum in the superconducting shield (Solid), and an example of the noise spectrum measured in a magnetically shielded room[5] (Dots).

A 7 channel DC-SQUID magnetometer system (CTF Systems, Inc.) is installed for evaluation of the shield and for MEG (magnetoencephalography) measurements, as illustrated in Fig 1. The diameter and the separation of the magnetometer sensor coils are 15 mm and 55 mm, respectively.

Another both-end-open shielding system was constructed [4]. It is expected to improve the comfort of the subject in the shield and the accessibility from the outside.

Noise performance

The magnetic noise was measured at the bottom (as shown in Fig 1) of the superconducting shield. The spectrum is shown in Fig 3 (solid). An example of a noise spectrum measured in a ferromagnetic magnetically shielded room [5] is plotted as well (dots). In the superconducting shield, the white noise remains flat to less than 1 Hz, while the low frequency noise less than 10 Hz increases dramatically in the magnetically shielded room. This emphasizes the advantage of the superconducting shield in the low frequency region.

Performance of MEG measurements

MEG measurements were performed in the superconducting shield with the CTF 7 channel DC-SQUID magnetometer and acquisition system. In this section, long latency responses of auditory evoked fields are analyzed and averaging of fewer numbers are discussed.

Methods

The subject was a man (24 years old, right-handed) with normal hearing. Auditory 1 kHz tones of 200 msec duration (85 dB SPL) were presented to the left ear with an insert phone (Etymotic Research, ER-3A) via a plastic tube at randomized interstimulus intervals of 600-1000 msec. The evoked fields were collected with 625 Hz sampling after 0-150 Hz filtering. The SQUID dewar was located at 4 positions to cover the contralateral region. Finally, 28 channel data from the separate 4 sessions are joined to localize the equivalent current dipole source. The averaging number (N) was varied.

Results

In Fig 4, responses near the two extrema are illustrated with various averaging numbers N. In the raw data (Fig 4 (a)), spikes of more than 1 pT were observed periodically at approximately 0.9 s interval. With magnetometers, biomagnetic signals from the human heart were significant. In the case when 25 trials were averaged as detected (Fig 4 (b), N=25), the heart spikes were still observed, which might disturb the auditory responses. When the averaging number was increased to 200 trials (Fig 4 (d), N=200), the heart spikes seem to be reduced enough so not to disturb the auditory response of interest. In Fig 4 (e), a field map at the latency of 91.2 ms is drawn from the data of N=200. Because the sensor positions from the separate 4 sessions were not located on a simple surface, the sensors were projected on a model sphere in spherical coordinates and extended to a plane of polar coordinates in Fig 4 (e). A 3 Hz high pass filter was applied to the averaged data from 1 session out of 4 sessions. Single current dipole localization was calculated using a spherical model and Grynszpan-Geselowitz's equations [6]. The equivalent current dipole is shown in Fig 5. The dipole is confirmed to be localized at the proper position in the vicinity of the auditory cortex from MRI images.

Discussions

To evaluate the effectiveness of using less averaging (eg. N=25) for measurements in the low noise space affected by the heart signal disturbances, we tried to eliminate the heart signals by rejecting any trials containing spikes by a 1 pT criterion before averaging (N=25R). The averages of 25 trials after the rejection are shown in Fig 4 (c). Correlation coefficients between the averaged waves (b), (c) and (d) are calculated at the position A in Fig 4. The correlation coefficient between the averaged waves of N=25R (after the rejection) and N=200 was 0.85, while the correlation coefficient between N=25 (as detected) and N=200 were 0.71. Even with less averaging, comparable low noise responses can be obtained in the superconducting shield with heart signal elimination procedure.

References

[1] Imai, K., Matsuba, H., Spear, P., and Fife, A. A. Magnetic shields for HTc SQUIDs, IEICE Trans. Electron., 1993, E76-C: 1280.

[2] Matsuba, H., Irisawa, D., and Yahara, A. Superconducting shield for biomagnetic measurements coupled with ferromagnetic, ed: Hoke., M. et al, Biomagnetism clinical aspects, Amsterdam, Excerpta Medica, 1992, 863.

[3] Matsuba, H., Shintomi, K., Yahara, A., Irisawa, D., Imai, K., Yoshida, H., and Seike, S. Superconducting shield enclosing a human body for biomagnetism measurement, 9th Int'l Conf. Biomagnetism, Vienna, 1993, 147 (to be published).

[4] Shintomi, K., Yahara, A., Irisawa, D., Imai, K., and Matsuba, H. A large scale high-Tc superconducting shield and its cryostat, presented in CEC/ICME95, WE-PE1c-10, Columbus, OH, July 19, 1995 (to be published in Adv. Cryo. Eng., 1996, 41).

[5] Koch, H., Canter, R., Drung, D., Erné, S. N., Matthies, K. P., Peters, M., Ryhänen, T., Scheer, H. J., and Hahlbohm, H. D. A 37 channel dc SQUID magnetometer system, IEEE Trans. Magn., 1991, 27: 2793

[6] Grynszpan, F. and Geselowitz, D. B. Model Studies of the Magnetocardiogram, Biophysical Journal, 1973, 13: 911-925.

Acknowledgement

Authors appreciate Mr. Nogai, T., and Prof. Kuriki, S. of Hokkaido Univ. for the measurements of the auditory evoked fields and the discussions of the localization on the MRI images.

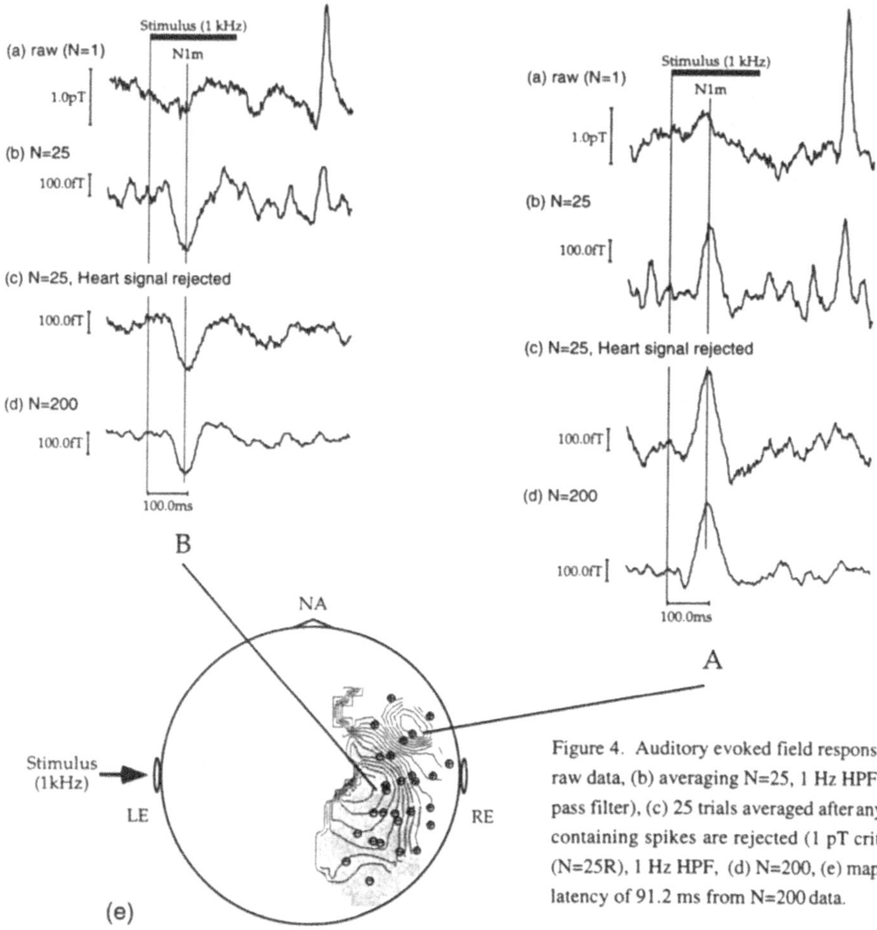

Figure 4. Auditory evoked field responses. (a) raw data, (b) averaging N=25, 1 Hz HPF (high-pass filter), (c) 25 trials averaged after any trials containing spikes are rejected (1 pT criterion) (N=25R), 1 Hz HPF, (d) N=200, (e) map at the latency of 91.2 ms from N=200 data.

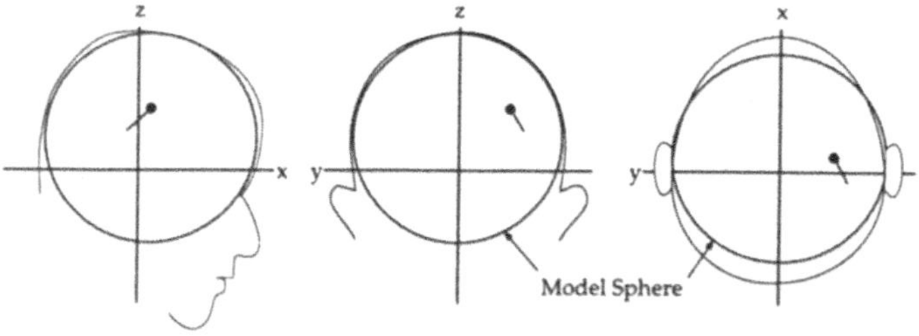

Figure 5. Equivalent single current dipole is localizedon (x, y, z) = (13, -52, 60) (mm) at the latency of 91.2 ms.

Current Dipole Localization Errors as a Function of the System Noise and the Number of Sensors

Matsuba, H.[1], Vrba, J.[2], and Cheung, T.[2]

[1]*Furukawa Electric Company, Ltd., Yokohama, Japan;*
[2]*CTF Systems Inc., Port Coquitlam, B.C., Canada*

Introduction

Multichannel MEG systems have been developed and operated by several groups, e.g. [1]. These systems have different sensor arrangements including radial or planar gradiometers, magnetometers, or even vector magnetometers. Even though the effectiveness of various sensor configurations has been actively investigated, e.g., [2, 3], there are still many questions remaining as to the best choice of the sensor configuration, and in the case of magnetometers, whether to use radial or vector sensors. In this paper the relative merits of radial and vector magnetometers are examined using a model of a current dipole in a spherical conducting medium. The dipole localization accuracy is evaluated analytically using a linear expansion of the dipole equations around its ideal position and compared with independent Monte Carlo simulations. It is found that the radial magnetometers yield better localization results than the vector magnetometers, the localization error is proportional to the sensor noise, and for a large number of channels the localization error scales as the inverse square root of the number of channels.

Analytical calculation

The dipole localization is usually performed by minimizing, in the least square sense, the difference between the signals detected by the sensors, S_i, and the forward solution, A_i, where i = 1, ..., N and N is the number of sensors. The error function is then given by

$$\varepsilon = \sum_{i=1}^{N}(A_i - S_i)^2 \tag{1}$$

It is assumed that a current dipole is embedded in a spherical conducting medium, as in Fig.1. The current dipole is characterized by 5 parameters for magnetic measurements since the radial dipole component is silent: 3 position parameters, the dipole magnitude, and the dipole orientation in the tangential plane, i.e. $(x_1, x_2, x_2, x_3, x_5) = (r, \varphi, \theta, q, \psi)$. In general, if more than one dipole is present, they are characterized by m parameters $(x_1, x_2, \ldots x_j, \ldots x_m)$, where m depends on the number of dipoles and their relationship. The forward solution is a function of the m parameters: $A_i = A_i(x_1, x_2, \ldots x_j, \ldots x_m)$. Further, the detected signal S_i is a combination of the detected magnetic field, B_i, and the sensor noise, R_i: $S_i = B_i + R_i$. The minimization condition for the error function ε in Eq. 1 is given by

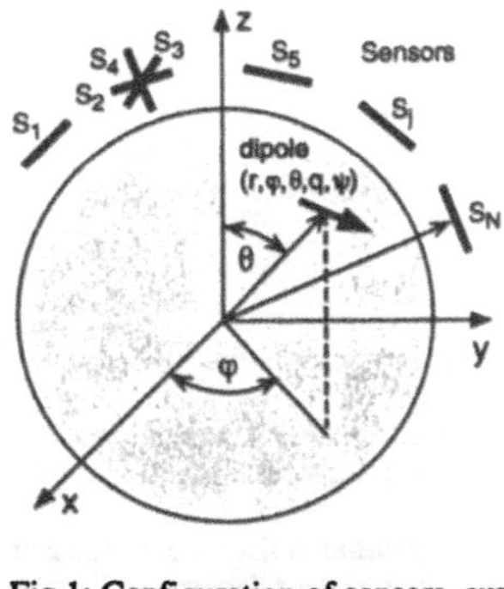

Fig.1: Configuration of sensors, current dipole, and conducting sphere.

$$\sum_{i=1}^{N}(A_i - B_i - R_i)D_{ij} = 0 \qquad \text{for } j = 1, \ldots. m \tag{2}$$

where

$$D_{ij} = \frac{\partial A_i}{\partial x_j} \tag{3}$$

In the absence of noise, the minimization of ε will result in $A_i - B_i = 0$, for all i's. When the noise is non-zero the difference $A_i - B_i \neq 0$ and the source parameters will exhibit deviations from their ideal values, $(\Delta x_1, \Delta x_2, \ldots \Delta x_j, \ldots \Delta x_m)$. For small noise, the difference $(A_i - B_i)$ can be expanded in terms of the source parameter deviations as

$$A_i - B_i = \sum_{k=1}^{m} D_{ik}\,\Delta x_k \tag{4}$$

Moving the term containing R_i to the right-hand side and substituting Eq.4 for A_i - B_i, Eq.2 may be rewritten as

$$\sum_{i=1}^{N} D_{ij}\left(\sum_{k=1}^{m} D_{ik}\,\Delta x_k\right) = \sum_{i=1}^{N} D_{ij}\,R_i \qquad \text{for } j = 1, \ldots m \tag{5}$$

Interchanging the summation order in Eq.5 results in a set of linear equations for Δx_k

$$\sum_{k=1}^{m}\left(\sum_{i=1}^{N} D_{ij}\,D_{ik}\right)\Delta x_k = \sum_{i=1}^{N} D_{ij}\,R_i \qquad \text{for } j = 1, \ldots m \tag{6}$$

The solution of Eq.6 for Δx_k's is then

$$\Delta x_k = \frac{\mathrm{Det}[k]}{\mathrm{Det}[0]} \qquad \text{for } k = 1, \ldots m \tag{7}$$

where

$$\mathrm{Det}[k] = \begin{vmatrix} \sum_{i=1}^{N} D_{i1}D_{i1} \cdots & \sum_{i=1}^{N} D_{i1}R_i \cdots & \sum_{i=1}^{N} D_{i1}D_{im} \\ \sum_{i=1}^{N} D_{i2}D_{i1} \cdots & \sum_{i=1}^{N} D_{i2}R_i \cdots & \sum_{i=1}^{N} D_{i2}D_{im} \\ \cdots & \cdots & \cdots \\ \sum_{i=1}^{N} D_{im}D_{i1} \cdots & \sum_{i=1}^{N} D_{im}R_i \cdots & \sum_{i=1}^{N} D_{im}D_{im} \end{vmatrix}, \quad \mathrm{Det}[0] = \begin{vmatrix} \sum_{i=1}^{N} D_{i1}D_{i1} \cdots & \sum_{i=1}^{N} D_{i1}D_{ik} \cdots & \sum_{i=1}^{N} D_{i1}D_{im} \\ \sum_{i=1}^{N} D_{i2}D_{i1} \cdots & \sum_{i=1}^{N} D_{i2}D_{ik} \cdots & \sum_{i=1}^{N} D_{i2}D_{im} \\ \cdots & \cdots & \cdots \\ \sum_{i=1}^{N} D_{im}D_{i1} \cdots & \sum_{i=1}^{N} D_{im}D_{ik} \cdots & \sum_{i=1}^{N} D_{im}D_{im} \end{vmatrix} \tag{8}$$

The noise R_i may be moved out of the determinant Det[k] and Eq.7 may be rewritten as a summation of the noise at each sensor weighted by coefficients Detd[u,k], as

$$\Delta x_k = \frac{\sum_{u=1}^{N} \mathrm{Detd}[u,k]\,R_u}{\mathrm{Det}[0]} \tag{9}$$

where

$$\mathrm{Detd}[u,k] = \begin{vmatrix} \sum_{i=1}^{N} D_{i1}D_{i1} & \cdots & D_{u1} & \cdots & \sum_{i=1}^{N} D_{i1}D_{im} \\ \sum_{i=1}^{N} D_{i2}D_{i1} & \cdots & D_{u2} & \cdots & \sum_{i=1}^{N} D_{i2}D_{im} \\ \cdots & \cdots & \cdots & \cdots & \cdots \\ \sum_{i=1}^{N} D_{im}D_{i1} & \cdots & D_{um} & \cdots & \sum_{i=1}^{N} D_{im}D_{im} \end{vmatrix} \tag{10}$$

Eq.9 represents a relationship between the source parameter errors, Δx_k, and the instantaneous value of the noise at each sensor, R_u. However, the quantity of interest is the standard deviation of Δx_k, which can be computed by squaring both sides of Eq.9 and taking its expectation value. The expectation value of the sensor noise R_u depends on the nature of the noise. The cases of random and correlated sensor noise will be discussed.

Random noise: Let the sensor noise be random with zero mean. The expectation value of its square is then equal to its standard deviation: $E[R_u^2] = \sigma_{R_u}^2$. If the noise at different sensors is considered independent, the expectation value of their product is zero, $E[R_i\,R_j] = 0$ for $i \neq j$. Further, assume that the noise is the same for all channels,

$\sigma_{Ru} = \sigma_{noise}$, u = 1, ... N. Denoting the expectation value of Δx_k^2 by $E[\Delta x_k^2] = \sigma_k^2$ (assuming that the mean value of Δx_k is zero), the square root of the expectation value of Eq.9 squared is

$$\sigma_k = \sigma_{noise} \frac{\sqrt{\sum_{u=1}^{N} (\text{Detd}[u,k])^2}}{\text{Det}[0]} \tag{11}$$

Further, assume that the sensors are uniformly distributed on some sensor surface S and the mean separation between the sensors is smaller than the inverse spatial frequency of the forward solution. Then the elements of the Det[0] can be approximated by an integral over the sensor surface, assuming that $\partial A/\partial x_j$ is a continuous function

$$\sum_{i=1}^{N} D_{ij} D_{ik} \approx n \int_S \frac{\partial A}{\partial x_j} \frac{\partial A}{\partial x_k} dS \qquad \text{for } j, k = 1, \dots m \tag{12}$$

where n is the sensor density, n = N/S. Hence, $\text{Det}[0] \propto n^m$. A similar approximation of $(\text{Detd}[u,k])^2$ in Eq.11 leads to the result $\sum_{u=1}^{N} (\text{Detd}[u,k])^2 \propto n^{2(m-1)}$. Applying this approximation to Eq.11 for large n yields

$$\sigma_k \propto \frac{\sigma_{noise}}{\sqrt{n}} \tag{13}$$

and it can be concluded that in the case of random noise, the standard deviation of the source parameters is proportional to the rms value of the noise and inversely proportional to the square root of the sensor density.

The calculations of the standard deviation of the dipole position according to Eq.11 were made using either Grynszpan-Geselowitz [4] or Sarvas [5] equations for a field generated by a current dipole in a homogeneous conducting sphere.

Correlated noise: (e.g. the brain noise). In this case $\langle R_i R_j \rangle \neq 0$ and Eq.11 is not valid. However, Eq.7 can still be used to estimate the dependence on the number of sensors. If the mean sensor spacing is less than the noise correlation distance, then the elements containing noise in Det[k] can be approximated, similar to Eq.12, as

$$\sum_{i=1}^{N} D_{ij} R_i \approx n \int_S \frac{\partial A}{\partial x_j} R \, dS \qquad \text{for } j = 1, \dots m \tag{14}$$

where *R* represents noise on the sensor surface (continuous function). The combination of Eqs.7, 12, and 14 then predicts that for correlated noise and a sufficiently high density of channels, the source parameter error, Δx_k, is independent of the channel density.

Monte-Carlo simulations of equivalent dipole fit

The dipole localization error was simulated using a homogeneous conducting spherical model. Radial or vector magnetometer point sensors were located on a hemispherical or spherical surfaces with a radius of 10.7 cm. The number of sensors was varied from 15 to about 600 and the sensor sites were uniformly distributed on the sensor surface (there were 3 orthogonal magnetometers located at each vector magnetometer site and 1 sensor at each radial magnetometer site). The current dipole was located on the z axis at a distance of a = 5 and 7 cm from the sphere origin and the dipole was oriented along the y axis with a dipole strength of 10 nA·m. The forward solution was computed first and then random noise with rms values of 15, 30, or 45 fT was added to each sensor output. The dipole was then localized using the ideal dipole position as a starting condition. The localization was repeated 200 times and standard deviations of the x, y, z components of the resulting positions were computed (σ_x, σ_y, σ_z). The mean standard deviation, $\sigma^2 = (\sigma_x^2 + \sigma_y^2 + \sigma_z^2)/3$, was then calculated and compared with the results of the direct computations according to Eq.11.

Results and discussion

The standard deviations of the dipole positions normalized by rms noise determined by simulations and by direct calculations using Eq.11 are in good agreement and are compared in Fig.2 for a hemispherical sensor surface. The simulations were computed for different rms noise values and the normalized standard deviations of the dipole position were found to be independent of the noise magnitude (for small values of the noise, the same type of sensor, and for larger number of channels), as expected on the basis of Eq.13. For more than about 80 channels, the normalized standard deviations depend on the number of channels as $1/\sqrt{N}$, again as expected from Eq.13. For small number of channels and especially for a deeper dipole the scatter of simulation results is large. In addition, for small number of channels, the results deviate from $1/\sqrt{N}$ because the dipole signals are undersampled and also 5 degrees of freedom are used for dipole parameters. Similarly, for a spherical sensor surface, the simulations and direct computations according to Eq.11 are also in good agreement. Simulations are also being carried out for correlated noise (randomly distributed random dipoles in a model sphere). The first results indicate that for a large number of channels, the dipole localization errors are indeed independent of the number of channels, as predicted by Eq.7 in the case of correlated noise.

Both simulation and direct calculations indicate that when the number of channels is equal and random noise is present, it is better to use radial magnetometers than the vector magnetometers because the radial magnetometers produce smaller localization error.

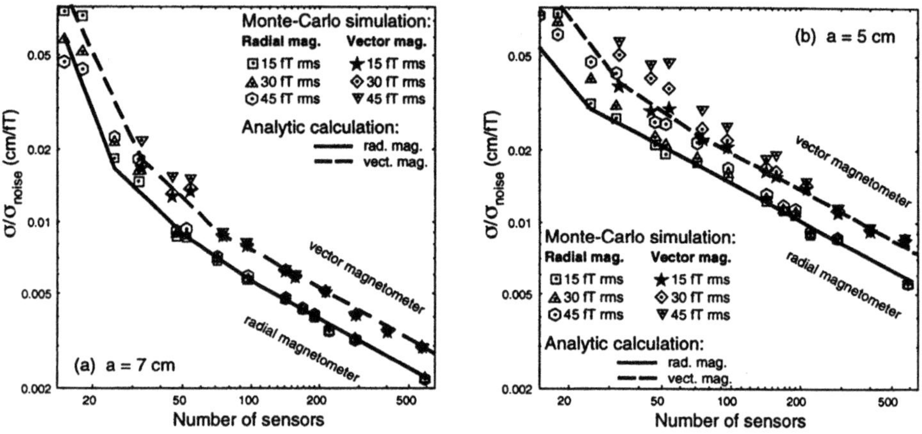

Fig.2: Comparison of normalized standard deviation of the dipole localization errors for hemispherical sensor surface determined by Monte-Carlo simulation (markers) and by evaluation of Eq.11 (curves). (a) shallow dipole, a = 7 cm; (b) deep dipole, a = 5 cm.

References

[1] Nakasato, N., Fujiwara, S., Yoshimoto, T., Fujita, S., Matani, A., Izuma, H., Tamura, I. The whole head magnetoencephalogram: localization of bilateral sources for unilaterally stimulated auditory evoked response, Electroencephalogr. Clin. Neurophysiol, 1993, **87**:S26.

[2] Mosher, J. C., Spencer, M. E., Leahy, R. M., and Lewis, P. S. Error bounds for EEG and MEG dipole source localization, Electroencephalogr. Clin. Neurophysiol., 1993, **86**: 303-321

[3] Sekihara, K., Scholz, B., Bruder, H., and Graumann, R. Reconstructing current distributions from biomagnetic measurements under large external disturbances, IEEE Trans. Med. Imag., 1994, **13**: 144-151

[4] Grynszpan, F. and Geselowitz, D. B. Model Studies of the Magnetocardiogram, Biophysical Journal, 1973, **13**, 911-925.

[5] Sarvas, J. Basic Mathematical and Electromagnetic Concepts of the Biomagnetic Inverse Problem, Phys. Med. Biol., 1987, **32**, 11-22.

Acknowledgment

The authors would like to thank Dr. Sekihara of Hitachi and Dr. Miyauchi of Communication Laboratory for their useful discussion and to Dr. S. E. Robinson for discussions regarding the current source analysis and to Mesrs. D. Hon and M. Laidlaw for carrying out the simulations.

Auditory Brainstem Evoked Fields Measured in a High Tc Superconducting Shield

Nogai, T.[1], Yahara, A.[2], Matsuba, H.[2] and Kuriki, S.[1]

[1]*Hokkaido University, Sapporo, Japan;* [2]*Furukawa Electric Co., Ltd., Yokohama, Japan*

Introduction

The early potential components at 1-8 ms elicited by a brief auditory stimulus are due to the responses through the auditory pathway in the brainstem. The auditory brainstem responses (ABR) have been useful methods both in clinic and in audiology. However, only a few studies describe the fields of the ABR [1][2], because of the weak signals. In this study, we report the auditory brainstem evoked fields (ABEFs) measured in an extremely low noise environment that was achieved in a superconducting magnetic shield. Because of the low field-noise, we could use magnetometers which are suited for the detection of deep sources in the brainstem.

Method

The measurements were made in a bithmus-based high critical temperature superconducting shielding enclosure. The magnetic shielding factor exceeds 140 dB in the frequency range lower than 10 Hz [3]. A 7-channel DC-SQUID magnetometer (CTF Systems, Inc.) was used with 2-turn sensing coils of 15 mm diameter. The total noise of the SQUID in the high-Tc shielding was typically 3 $fT/Hz^{1/2}$. Two female and one male right-handed subjects of normal hearing (25-35 years old) were measured. Auditory clicks of 0.7 ms width were presented monaurally to the left ear with an insertphone via a plastic tube at randomized interstimulus intervals (ISIs) of 100-140 ms. Multiple recordings were made at different locations over the lateral side of the head, contra-and ipsilateral to the click. Responses of 20000 epochs were averaged in a recording period of 90 ms including 36 ms prestimulus baseline. In another measurement for one subject (S1), extensive recordings were made using shorter ISIs of 50-100 ms in a 40 ms recording period with 14 ms baseline. The latency of the response was corrected with the measured transmission time of the sound through the plastic tube. The recorded fields were sampled at 3125 Hz after 1000 Hz low pass filtering. The averaged responses were finally filtered digitally between 10 and 600 HZ.

Results

Components of the response

In order to have a brief view of the components appearing in the evoked fields, contralateral responses obtained in the recordings of 100-140 ms ISIs at different locations are superposed in Fig. 1, where the responses were recorded at 45, 20, and 7 locations in S1, S2, and S3, respectively. In S1 main components were observed at about 10, 20, and 35 ms. These components are middle latency responses, and the 10 ms component corresponds to the earliest cortical response that was first observed in the previous work using first-order gradiometers in a conventional magnetically shielded room [4]. Note, that the 10 and 20 ms components have the same polarity of the field.

A response at about 15 ms was observed in S2 and S3. To our knowledge, this magnetic field response was newly observed in this study. The 10 ms component appeared at about 8 ms in S2, while the 20 ms component was delayed to 25 ms in S3. The wave of the ABEF was discernible at about 5 ms in S3. In S1 and S2 the ABEF was overlapped with the 10 ms wave in the superposed responses. In Table 1 the observed peak latencies of different components are summarized for the contra- and ipsilateral recordings.

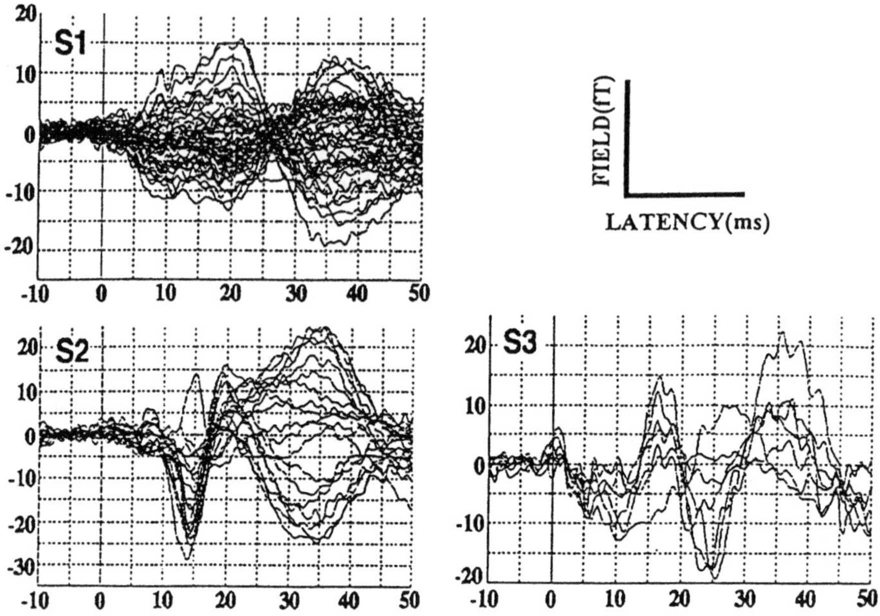

Fig. 1. Superposed responses of the auditory evoked fields to the click stimulus recorded at the contralateral side of the head in three subjects.

Table 1. Peak latencies in ms of different components of the evoked field responses.

SUBJECTS	M6	M10	M15	M20	M30
50-100ms ISI Recording					
S1 (contra)	6.4	-	-	-	-
(ipsi)	6.0	-	-	-	-
100-140ms ISI Recording					
S1 (contra)	5.5	9.9	-	19.3	36.2
(ipsi)	5.7	9.7	-	19.0	31.0
S2 (contra)	5.8	8.3	14.2	19.6	33.7
(ipsi)	5.5	-	13.7	21.4	32.5
S3 (contra)	5.3	10.8	16.8	24.8	33.6
(ipsi)	5.3	11.5	17.8	25.8	34.8

Auditory brainstem evoked field

Figure 2 shows the selected responses of the ABEF recorded over two hemispheres with 50-100 ms ISIs in S1, where the left ear was stimulated. At 6 ms after the click, distinct peak component (M6) of negative (inward to the skull) and positive (outward from the skull) field was detected, respectively, at the anterior-temporal and posterior-occipital location in the contralateral right hemisphere. The M6 peak was not clear at the posterior-occipital location in the ipsilateral left hemisphere because of high noise level of this channel sensor, but at the anterior-temporal location positive peak was observed. The M6 component was reproducibly recorded in repeated measurements, and therefore, may be the equivalent of wave V of the ABR [5]. The peak of M6 component was also observed clearly after fewer averages of 5000 epochs, suggesting the advantage of low environmental noise in the superconducting shield.

From the mapping recordings of the contralateral hemisphere of S1, isofield contour map (Fig. 3) was obtained at the peak latency of M6. Widely distributed fields in the entire region of the temporal side suggest subcortical deep sources, but tentative localization of a single current dipole source indicated a site deviated 2-3 cm from the brainstem. More extensive low-noise, i.e., high signal-to-noise ratio, recordings may be needed for the precise source localization.

Figure 4 shows the contra-and ipsilateral responses of M6 component in S2 and S3, selected from the recordings of the left ear stimulation with 100-140 ms ISIs. In S2 the polarity of the M6 and its change over the contra- and ipsilateral hemisphere were similar to those in S1. In S3, for whom recordings were limited at several locations, negative and positive peak of the M6 were observed in the contra-and ipsilateral side, respectively. These peak locations indicate different field distribution from those in S1 and S2, suggesting the variability of the ABEFs across subjects.

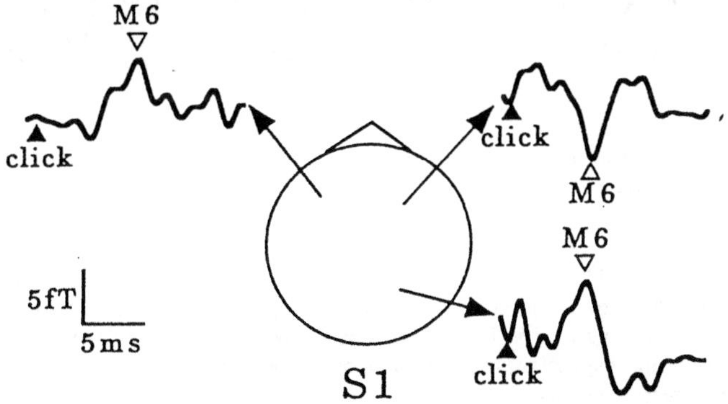

Fig. 2. Selected waveforms of the ABEFs recorded at the contra-and ipsilateral sides of the head to the left ear stimulation in S1.

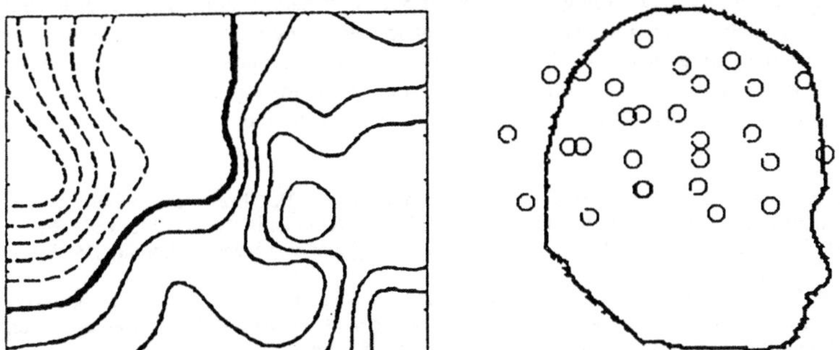

Fig. 3. Isofield map with 1 fT contours for the ABEF at 6.6 ms latency over the contralateral side of the head to the left ear stimulation in S1, where dotted and solid line indicate outward and inward field, respectively. Measurement locations are indicated by circles.

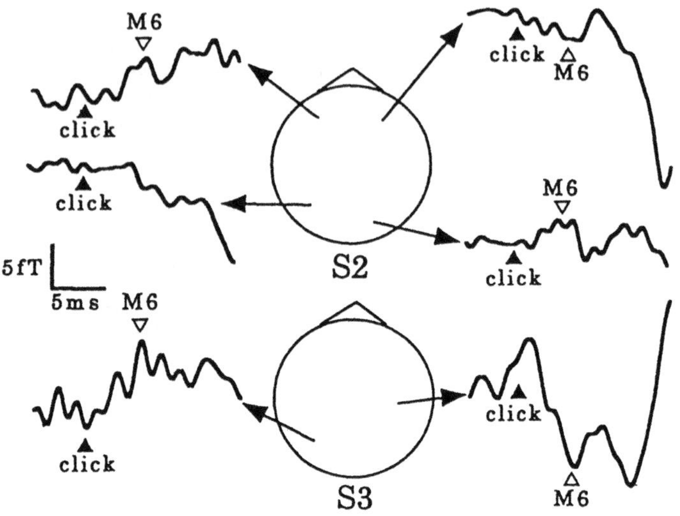

Fig. 4. Selected waveforms of the ABEFs recorded at the contra-and ipsilateral sides of the head to the left ear stimulation in S2 and S3.

Discussion

In addition to the middle latency responses at 10, 20, and 35 ms, a new response at 15 ms was observed in two of the three subjects. In the previous study of the middle latency evoked fields this 15 ms component was not observed in four subjects examined [4], where one subject (S1) is common in this study. These results suggest that this response originates in a source having mostly-radial currents, with tangential components which may vary among subjects. This field response would be the equivalent of the auditory evoked potentials occurring at 13-16 ms which were recorded with intracranial electrodes and have a cortical source in the primary auditory area [6].

High reproducibility in observing the ABEFs at 6 ms in different subjects indicates the advantage of using magnetometers in low noise superconductively-shielded environment. Positive maximum field of M6 at the ipsilateral anterior-temporal site is the new observation. Such widely distributed fields around the whole head is the consequence of the field generation from the deep brainstem. Computer simulation studies might help the understanding of the observed field distribution.

References

[1] Erne S.N., Hoke, M., Lukenhoner, B., Pantev, C., and Scheer, H.J. Brainstem auditory evoked magnetic fields. In: Atsumi., K. et. al. eds. Biomagnetism '87, Tokyo Denki Univ. Press, 1988, 158-161.

[2] Iramina, K., Uchizono, S., Yoshiura, T., and Ueno, S. (in Japanese) J. Japan Biomag. Bioelectromag. Soc., 1994, 7: 36-37.

[3] Matsuba, H., Shintomi, K., Yahara, A., Irisawa, D., Imai, K., Yoshida, H., and Seike, S. Superconducting shield enclosing a human body for biomagnetic measurement. In: Baumgartner et al. eds. Biomagnetism: Fundamental Research and Clinical Applications. 1995, 483-489.

[4] Kuriki, S., Nogai, T. and Hirata, Y. Cortical sources of middle latency responses of auditory evoked magnetic fields. Hearing Res. (in press).

[5] Jewett, D. L., and Williamson, J. S. Auditory-evoked far fields averaged from the scalp of human. Brain, 1971, 94: 681-696.

[6] Liegeois-Chauvel, C., Musolino, A., Badier, J.M., Marquis, P. and Chauvel, P. Evoked potentials recorded from the auditory cortex in man. Electroenceph. clin. Neurophyisol., 1994, 92:204-212.

High Performance Magnetically Shielded Room for Clinical Measurements

Paavola, M.A.J.[1], Ilmoniemi, R.J.[2], Sohlström, L.[3], Meinander, T.[4], Penttinen, A[4]. and Katila, T.E.[1]

[1]Helsinki University of Technology, Espoo; [2]BioMag Laboratory, Medical Engineering Centre, Helsinki Central Hospital, Helsinki; [3]Euroshield Ltd., Eura; [4]VTT Automation, Technical Research Centre of Finland, Espoo, Finland

Introduction

Developments in SQUID technology during the past decades have made it possible to measure extremely weak magnetic fields. This has enabled research in the area of biomagnetism, especially in magnetoencephalography (MEG) and magnetocardiography (MCG). When measuring weak fields, one must be able to reduce noise in measurements to get a high enough signal-to-noise-ratio. In our urban society, where electricity plays more and more important role, avoiding magnetic disturbances is impossible without specially designed shields. In 1994 Helsinki University Central Hospital (HUCH), University of Helsinki, and Helsinki University of Technology started a collaboration to evaluate clinical aspects of biomagnetism. For this purpose a biomagnetism laboratory called BioMag has been established within HUCH. For this laboratory, Euroshield Ltd. designed and built a magnetically shielded room (MSR) which is especially suitable for clinical measurements.

Magnetic shielding

In magnetic shielding one can distinguish four different methods. They are the use of soft ferromagnetic materials, conductive materials, superconductive materials, and active compensation coils. So far, superconducting technology has been much too expensive to be a real choice for large-volume magnetic shielding. There exists, however, some small volume superconductive shields (see for example [1]). Active shielding can be used for static field compensation and as additional shielding at low frequencies [2, 3, 4]. Currently, the most practical way to realize a large magnetically shielded enclosures is to combine ferromagnetic and eddy current shielding.

Typical materials used for ferromagnetic shielding are either iron-nickel or iron-cobolt based alloys. Ferromagnetic materials with a high magnetic permeability provide a pathway with a low reluctance for the magnetic flux. The interfering field stays within the shielding layer leaving the field inside the shield very small, almost zero as shown in Fig. 1. Conductors like aluminium provide good mechanical properties and can therefore be used for frame construction and to give additional eddy current shielding at higher frequencies.

The shielding factor is defined as

$$S = \frac{H_o}{H_i}, \tag{1}$$

where S = shielding factor, H_o = field in the volume without a shield and H_i = field in the same volume when the shield is installed. Usually the shielding factor is defined in desibels, given by $S_{dB} = 20 \log S$. The shielding factor can be analytically calculated only in a few special geometries. One example is a thin-walled sphere, for which the dc shielding factor is given by

$$S = 1 + \frac{2}{3}\frac{\mu d}{R}, \tag{2}$$

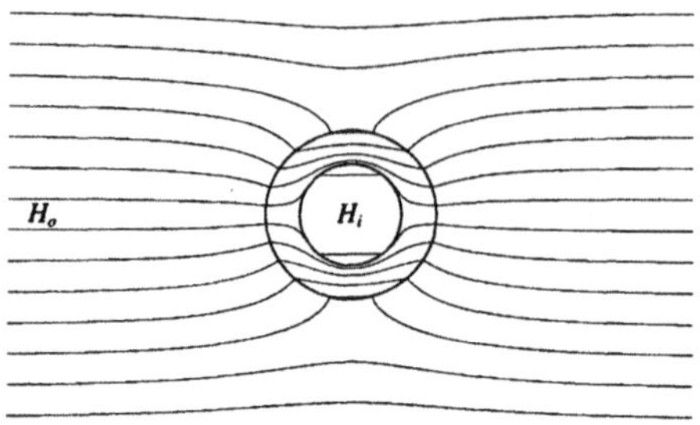

Fig. 1: *Ferromagnetic sphere in dc or low frequency magnetic field H_o. Because of the small reluctance (high permeability) in the ferromagnetic shielding layer, the interfering field goes within shield and the field inside the shield H_i remains very small.*

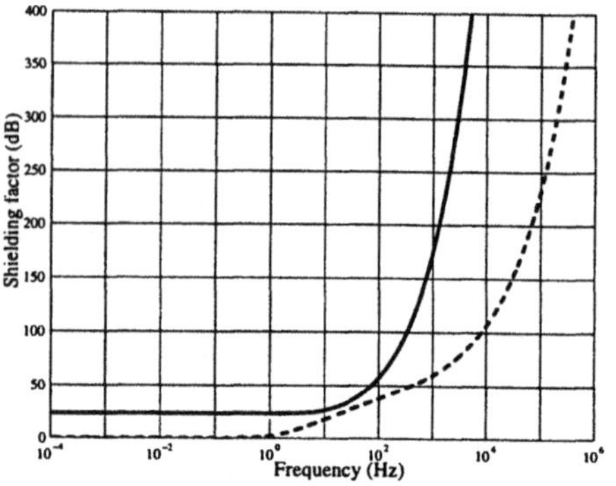

Fig. 2: *Theoretical shielding factor of a conductive (dashed line) and a ferromagnetic (solid line) sphere (radius = 1.5 m) as function of frequency. In the calculations parameter values equal to a 2 mm thick mumetal ($\mu = 15\,000$, $\sigma = 2 \times 10^6$ S/m) are used for ferromagnetic shield and those equal to a 5 mm thick aluminium plate ($\mu = 1$, $\sigma = 36 \times 10^6$ S/m) for conductive sphere.*

R = the radius of the sphere, d = the thickness of the wall and μ = the relative permeability of the material. The dc shielding factor for a cube is given by similar a equation, but the factor 2/3 must be changed. In literature one can find values from 0.6 to 0.8 [4] when the edge of the cube is given as $L = 2R$.

As seen from Eq. 2, the shielding improves if the wall thickness is icreased. Instead of one thick layer, it is generally more favorable to distribute the ferromagnetic material into several separate thinner layers. For a three-layer spherical shield the dc shielding factor is [4, 5]

$$S = 1 + S_1 + S_2 + S_3 + S_1S_2V_{12} + S_1S_3V_{13} + S_2S_3V_{23} + S_1S_2S_3V_{12}V_{23}\ , \tag{3}$$

where

$$S_i = \frac{2}{3}\frac{\mu d_i}{R_i} \tag{4}$$

$$V_{ij} = 1 - \left(\frac{R_i}{R_j}\right)^3,\ i,j = 1,2,3. \tag{5}$$

For a cubic threelayer shield, the factor 2/3 in Eq. 4 must again be replaced with a value between 0.6 and 0.8. From Eq. 3 one can see that additional layers in the shield make the overall shielding factor higher than the sum of the shielding factors for individual layers.

The information content of the biomagnetic signals is concentrated in the frequency range below 300 Hz, and naturally this is the frequency range where the shielding must be effective. Eddy current calculations [6] needs to be combined with the dc shielding calculations to obtain the shielding performance as a function of frequency. An example is shown in Fig. 2 for a sphere. Eddy current shielding increases with frequency, and it is also present in ferromagnetic layers. As the frequency increases, the penetration depth for electromagnetic fields gets smaller and becomes finally less than the wall thickness. Above this frequency, very high shielding values may be obtained. In practice, however, the shielding factor is limited by leaking through holes and imperfections in the structure [7, 8].

Construction

Calculation of shielding performance for practical shields must be carried out using element methods and iterative solutions. There are two main problems in the calculations: i) To know the field within a shielding layer, the permeability of the layer must be known, but the permeability depends on the local field and ii) precise geometries and material parameters, especially magnetic ones that are nonlinear and hysteretic, can not be known in detail. Fortunately, the equations introduced in the previous section give fairly reliable results, especially, if parameter values are adjusted to take into account the real situation.

The challenge of our MSR project was to combine the shielding requirements with the partially contradictory demands imposed by clinical measurements and patient handling. The biggest problems arose in connection with two fairly large monitor holes: one for visual stimulus and one for ECG-monitoring. When not in use these holes can be blocked by special covers providing both magnetic and acoustical isolation. Because of the patient measurements, the doorway had to be large enough to allow subjects to be brought in and out on a wheeled bed, if necessary. Illumination was specified to be noninterfering.

Based on shielding factor calculations by VTT Automation, it was decided that the MSR should be constructed as a three-layer enclosure with a self-supporting inner cabin. A combination of ferromagnetic and eddy current shielding is achieved by using several layers of mumetal and aluminium. With interlayer distance of 150 mm and free space of $4.0 \times 2.9 \times 2.7$ m^3 inside the cabin, the outer dimensions of the construction became $5.1 \times 3.9 \times 3.8$ m^3. The MSR weights about 12 000 kg. The walls are made of elements consisting of Al-mumetal-Al-plates (5 mm-2 mm-2 mm thick, respectively), which are attached onto an Al-frame. The measured relative permeability of the mumetal plates was about 30 000. During the construction work this value decreased to some extent, due to mechanical stresses introduced into the material.

The double sliding doors of the MSR are placed between the shielding layers, so that they are not exposed to the earth's magnetic field. The doorway is equipped with a nonmagnetic movable ramp allowing smooth access with wheeled beds. Both monitor holes are equipped with shields, which are specially designed to reduce the interference produced by the vertical deflection coils in crt-displays, which in experiments requiring optical stimuli are located close to the openings. The operation of the shields is based on field attenuation in a tube. Lighting is realized with the aid of optical fibers. The innermost shielding layer is mumetal, which minimizes the thermal noise produced by the innermost walls [9, 10]. Demagnetization coils integrated into the two outermost layers make it possible to demagnetize the MSR when necessary [4]. In order to minimize noise signals caused by mechanical vibrations, the self-supportive inner cabin can be either mechanically connected to or disconnected from the rest of the MSR.

Shielding measurements

Shielding factor measurements were conducted by VTT Automation. A known magnetic disturbance was generated outside the MSR by a large source coil (diameter 1.04 m, 28 turns, max. current 20 A) 10 m away from the centre of the room at 0.01, 0.03, 0.1, 0.3, 1, 3, 10, 20, 40, 60, 90, and 120 Hz. The attenuated signal components in three directions within the cabin were measured with a fluxgate and a SQUID magnetometer at the centre point of the MSR, and the residual static field was mapped with a fluxgate magnetometer. The measurements were performed in four phases: a reference measurement was done before the construction work had began, and the measurements were repeated every time a layer was completed. The layers were demagnetized before measurements. The curves in Fig. 3 represents the lowest shielding factors obtained in each direction at each frequency. The differences between directions were significant only with the first shielding layer. In later measurements with more layers the difference between shielding factors in different directions was of the same order as the measurement accuracy ($\pm$ 2 dB). The residual static field in the cabin remained less than 10 nT in the center and less than 20 nT near the walls.

Conclusions

A biomagnetic measurement site suitable for clinical work was constructed. The results clearly show that the specified shielding performance was exceeded. This is a result of carefull design, where the requirements imposed by the patient measurements were intoruduced already in the very beginning of the project. The shielding measurements show, also, that specifications were almost achieved with only two shielding layers. This suggests that good enough shielding for most of the clinical measurements is possible to realize with two-layer shields. In such a case, lower shielding could be

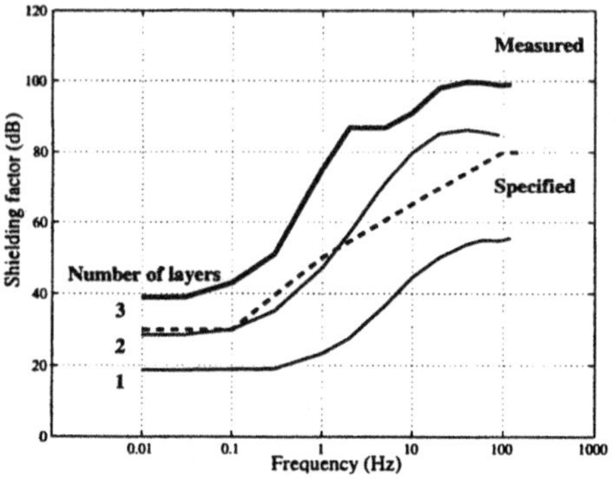

Fig. 3: *Measured and specified shielding performance of the MSR in BioMag Laboratory.*

partly compensated with a good site planning and by a proper instrumentation using gradiometric sensors. In addition to excellent shielding, the MSR in the BioMag Laboratory has proven to be easy to use and to provide a good clinical site for biomagnetic measurements.

References:

[1] Matsuba, H., Shintomi, K., Yahara, A., Irisawa, D., Imai, K., Yoshida, H., and Seike, S. *Superconducting shield enclosing a human body for biomagnetism measurement.* In: Baumgartner, C., Deecke, L., Stroink, G., and Williamson, J.S. Biomagnetims: Fundamental research and clinical applications, Amsterdam, Elsevier Science/IOS Press, 1995.

[2] Brake H. J. M., Wieringa H. J. and Rogalla H. *Improvement of the performance of a mumetal magnetically shielded room by means of active compensation*, Meas. Sci. Technol., 1991, 2: 596–601.

[3] Brake H. J. M., Huonker R., and Rogalla H. *New results in active noise compensation for magnetically shielded rooms*, Meas. Sci. Technol., 1993, 4: 1370–1375.

[4] Kelhä, V., Pukki, J., Peltonen, R., Penttinen, A., Ilmoniemi, R., and Heino, J. *Desing, Construction, and Performance of a Large-Volume Magnetic Shield*, IEEE Transactions on Magnetics, 1982, 18(1): 260–270.

[5] Schweizer, F. *Magnetic Shielding Factors of a System of Cocentric Spherical Shells*, Journal of Applied Physics, 1962, 33(3)nc.: 1001–1005.

[6] Kaden H. *Wirbelstrme und Schirmung in der Nachtrichtentechnik*, 2. Auflage, Berlin, Springer-Verlag, 1959.

[7] Ott, H. *Noise Reduction Techniques in Electronic Systems*, New York, John Wiley & Sons, Inc., 1976.

[8] Gluckstern R.L., and Diamond J.A. *Penetration of fields through a circular hole in a wall of finite thickness*, IEEE Transactions on Microwave Theory and Techniques, 1991, 39(2): 274–279.

[9] Katila, T., Leiniö, M., Montonen, J., and Nenonen, J. *Sensitivity limits in biomagnetic Measurements*, Acta Otolaryngol, 1991, Suppl. 491: 36-42.

[10] Nenonen, J., Montonen, J., and Katila, T. *Thermal Noise in Biomagnetic Measurements*, 1994, TKK Report-F-A677.

Acknowledgements

This research was partly supported by the Finnish Cultural Foundation.

An AC Susceptometer with a Large Magnetised Volume for Biological Samples

Pasquarelli, A.[1], Del Gratta, C.[2,3], Della Penna, S.[3], Di Luzio, S.[2,3], Pizzella, V.[1] and Romani, G.L.[1,2]

[1]*Institute of Solid State Electronics, CNR, Rome, Italy;* [2]*Institute of Medical Physics and* [3]*Institute of Advanced Biomedical Technologies, Gabriele D'Annunzio University, Chieti, Italy*

INTRODUCTION

Measurements of the magnetic susceptibility of human tissues has been applied to the study of patients affected by diseases like thalassemia or hemocromatosis that imply severe iron overload [1]. SQUID based instruments offer the sensitivity and bandwidth, extending to dc, needed to measure the low magnetic fields generated by magnetised biological tissues. The estimation of tissue susceptibility from field measurements require inverse calculations where several variables come into play. For example, the shape of body organs is needed to calculate the field. Moreover, susceptibility of organs that lie in the proximity of the investigated ones may affect the field value [2]. Generally, the quantity of interest is not the susceptibility itself but the concentration of a biochemical compound related to it. However, uncertainties in the susceptibility of these biochemical compounds may affect the estimated value of their concentration [3].

The motivation for the developement of a susceptometer for large samples lies in the need for a precise assessment of the susceptibilities of various substances. Such an instrument must feature a high sensitivity since usual susceptibility values are low - in the 10^{-5} SI units range. The distinctive features of our instrument are: 1) an ac applied field for sample magnetisation together with a lock-in detection of the response, 2) a feedback control for the rejection of the magnetising field sensed by the SQUID. In the following the technical solutions we adopted for our instrument and the results of preliminary calibration measurements are described.

METHODS

The susceptometer consists of a dcSQUID coupled to a first-order planar gradiometer. This configuration of the pick-up coil was chosen because it proved to be less sensitive to inhomogeneities in the applied field. The SQUID design is a variation of the Ketchen-Jaycox type with two washers connected in parallel in a gradiometric configuration, thus reducing by half the inductance and rejecting undesired external fields [4]. The detection coil consists of two circular coils, made of niobium wire, oppositely wound around a MACOR support. The coils are 0.01 m in diameter and the gradiometer baseline is 0.015 m. The accurate mechanical construction ensures an intrinsic balance of 5×10^{-4}. In order to effectively operate the active compensation system, to be described in the following, a proper flux coupling between the flux transformer and the feedback coil is used. Therefore, the coupling is almost ten times higher than for a typical biomagnetic sensor design.

In order to provide a spatially homogeneous applied field over a large volume, a pair of Helmoltz coils, 1.40 m in diameter, are used. The coils can generate a magnetic field of 5×10^{-5} T/A, uniform to 0.1% in a volume corresponding approximately to a cylinder, 0.16 m in radius and 0.24 m in height. The coils are connected in series and driven by a power amplifier able to supply up to $\pm$ 26 V to this load. This maximum value corresponds to a peak-to-peak field of 0.6 mT at 11 Hz.

The active compensation is achieved by sensing the current flowing in the magnetising coils which, after proper amplification and phase-shift, is injected into the feedback loop of the SQUID to cancel the flux induced by the magnetising field. After reaching stable operating conditions, the compensation is "frozen" during the measurements, which are performed by scanning the sample under the instrument. An overall compensation better than 1×10^{-7} is reached by utilising the active compensation. The residual signal contains higher harmonics of the driving frequency due to non linearities in the electronic chain of the compensation subsystem which give rise to beatings with the ambient power line field at 50 Hz. Therefore it is necessary to choose a value of the driving frequency such that the frequency range of these beatings does not overlap with the signal bandwidth.

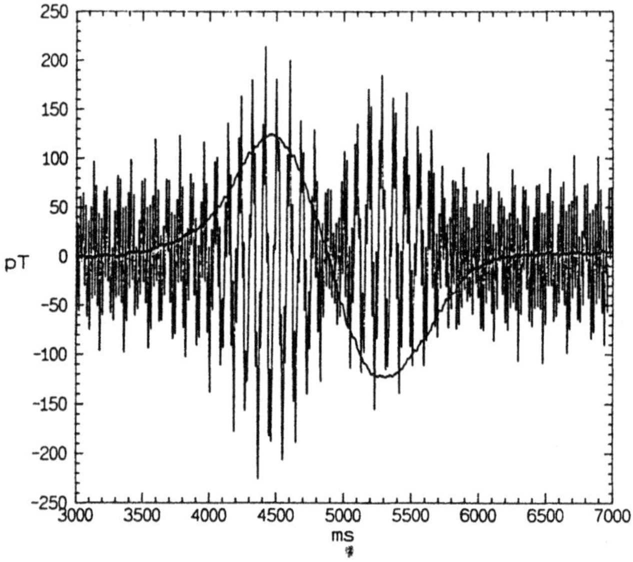

Figure 1:Experimental field recorded with a scan at a distance of 1.6×10^{-2} m over a capillary vessel filled with iron sulfate powder and magnetised by a field of 0.58 mT peak-to-peak, at a frequency of 11 Hz; the alternating signal containing the field generated by the sample at the driving field frequency and the instrument noise is displayed together with the related lock-in detection output obtained using an integration window of 364 ms.

The output signal from the SQUID electronics is the superimposition of an alternating signal at the same period of the magnetising field and the instrument noise. It is convenient to use a lock-in amplifier with the reference matched to the frequency and phase of the magnetising current. We first A/D convert the data and then use a digital lock-in which is more flexible and cheaper. The output of the lock-in is proportional to the magnetic susceptibility of the material under the probe. The selection of the length of the lock-in time window is related to the measurement procedure. Indeed the larger is this time the more accurate the time approximation; moreover a longer time window results in a smaller effective measurement bandwidth and consequently a smaller overall noise added to the signal. On the other hand, a consequence of the use of a lock-in detection technique is that the field variations along a single scan are spatially averaged over an interval equal to the lock-in integration time window multiplied by the scanning velocity. This results in an unwanted spatial low-pass filtering of the signal. Therefore the scanning velocity is carefully chosen to be low enough that the spatial variations of the signal are not filtered out.

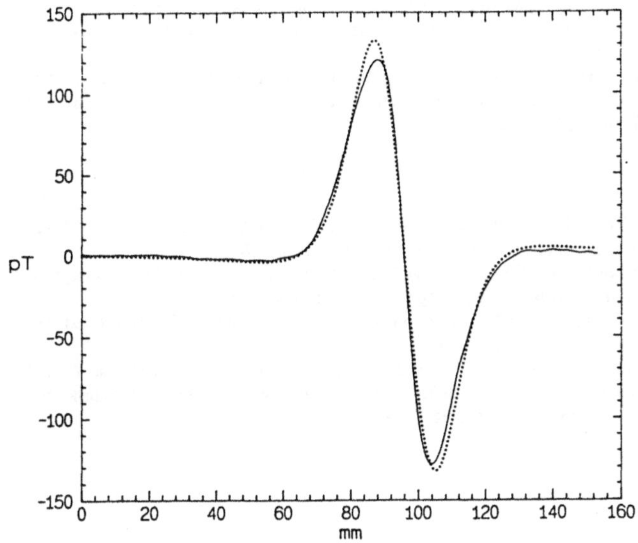

Figure 2: A comparison between measured and theoretical mean field threading the gradiometer coils: the continous line relates to the scan performed at a distance of 1.6×10^{-2} m over a capillary vessel filled with iron sulfate powder and immersed in an applied field of 0.58 mT peak to peak, at a frequency of 11 Hz; the lock-in integration window is 364 ms. The dotted line represents the theoretical field at the gradiometer generated by a thin cylinder uniformly magnetised; the susceptometer calibration value used in the calculations is the best fit value of $(6.5 \pm 0.3) \times 10^{-2}$ mV / pT.

RESULTS

We performed preliminary measurements with the aim at assessing the susceptometer calibration, the instrument behaviour as a function of the applied field frequency and the susceptometer sensitivity.

The susceptometer calibration factor has been calculated as the product of the coupling between the SQUID and the input coil and the SQUID output per flux quantum. The calibration factor was checked by measuring the magnetisation produced by a sample of known magnetic susceptibility. The sample consisted of a capillary vessel filled with a powder of hexahydrated iron sulfate; the measured field is displayed in Figure 1. The calibration was estimated by means of a least squares fit and was found in accordance with the calculated value; the theoretical field is shown in Figure 1. This shows on one hand that all the parameters of the detection system are carefully controlled, and on the other hand that AC susceptibility measurements with active compensation are feasible and consistent.

The measurements were repeated on the reference sample at various drive frequencies and were found to be consistent. Several operating frequencies were selected where the instrument was stable and not affected by beatings with the power line frequency.

The assessment of the instrument sensitivity has been performed by detecting the field produced by a sample consisting of a capillary vessel filled with pure water. the overall sensitivity of the instrument could be considered to be better than 7×10^{-11} Am^2.

REFERENCES

[1] D. E. Farrel, J. H. Tripp, P. E. Zanzucchi, J. W. Harris, G. M. Brittenham, and W. A. Muir, 1980, Magnetic measurement of human iron stores, IEEE Trans. Magn., 16:818-823.

[2] C. M. Bastuscheck and S. J. Williamson, 1985, Technique for measuring the ac susceptibility of portions of the human body or other large objects, J. Appl. Phys., 58:3896-3906.

[3] J. H. Bauman, J. W. Harris, 1967, Estimation of hepatic iron stores by in vivo measurements of magnetic susceptibility, J. Lab. Clin. Med., 70:246-257.

[4] V. Foglietti, A. Pasquarelli, V. Pizzella, G. Torrioli, G. L. Romani, S. Casciardi, W. J. Gallagher, M. B. Ketchen, A. W. Kleinsasser, R. L. Sandstrom, 1993, Operation of a hybryd 28-channel neuromagnetometer IEEE Trans. Appl. Supercond., 1993, 3:1890-1893.

A Six-Channel SQUID Biomagnetometer Optimised for Fetal Magnetocardiography

Rassi, D.[1], Zhuravlev, Y.E.[1,3], Josephs, O.[2], and Swithenby, S.J.[2]

University of Wales[1], Swansea, UK; Open University[2], Milton Keynes, UK; Institute of Radioengineering and Electronics[3], Moscow, Russia

Introduction

In recent years the trend in SQUID instrumentation for biomagnetic studies has been towards using large numbers of measurement channels. While this is necessary in applications requiring the mapping of the biomagnetic fields over extended areas, it has resulted in bulky equipment, often having large source-detector distances and requiring magnetic shielding. In addition to their high cost, shielded rooms have the disadvantage that many types of medical investigations cannot be carried out under the dual constraints of confined space and the demand for total "magnetic cleanliness" within the shielded room. There are, however, other applications for which it is highly desirable to have small, easily manoeuvrable SQUID biomagnetometers capable of operation in magnetically unshielded environments.

There have been very few attempts to explore the possibilities for developing such systems [1]. A pioneering multi-channel SQUID system for unshielded neuromagnetic measurements was the five-channel system introduced by Williamson et al [2]. It was shown in this work that the use of electronic noise cancellation by recording three vector components of the ambient magnetic noise makes it possible to achieve significant spatial filtering of distant noise sources. Recently, a more advanced design has been developed by Vrba et al [3] of a 64-channel whole-head system capable of unshielded operation. In this system additional reference SQUID magnetometers and first-order gradiometers are used for noise cancellation based on software formation of third-order gradiometers with large baselines. It was shown that this approach is very effective both in sensitivity and in noise suppression. The subtraction coefficients for this system were found to be very stable i.e. not needing adaptive adjustment. However, such a system may be too bulky for certain types of measurements such as, for example, recording of peripheral nerve signals.

One particular application which would benefit immensely from a compact SQUID system with a small number of channels and capable of unshielded operation, is fetal magnetocardiography (FMCG). It has been shown that measurements of the fetal heart's electromagnetic activity can provide clinically significant information [4,5] although by its nature, and for wide applicability, the measurement needs to be carried out in the hospital alongside other routine diagnostic techniques such as ultrasonography. In this paper we report on a six-channel SQUID gradiometer system, specially constructed for FMCG measurements, and suitable for hospital use.

Optimisation of the System Design

The design of this system was optimised by realistic modelling using a solution of the forward problem, and exploiting the specific features of the spatial distribution of the magnetic fields of the fetal heart. The fetal heart model was an adaptation of a previously described adult heart model [6]. In the case of the fetus, the distance between the source and the detection coil of the gradiometer is much bigger than the size of the heart and therefore the assumed isotropy of the myocardium is more valid than in the case of the adult. The magnetic field pattern predicted by the model can be described in terms of a rotating current dipole undergoing a near circular loop. Because of the random movements of the fetus it is important to have a system arrangement which gives maximal signal when it is placed at the nearest point to the fetal heart. The best achievable signal-to-noise ratio is essential for accurate real time triggering of the R-wave, while it is not practical to spend time searching for the location of maximum signal. Measurement in an unshielded environment is complicated by the fact that the system and the subject are exposed in the comparatively large geomagnetic field. This gives rise to a significant susceptibility signal (of the order of a few pT) from the maternal body surface, with the normal component of this 'noise' field being about an order of magnitude smaller than its tangential component. For this reason use of tangential magnetic measurements in an unshielded location needs compensation of the earth's magnetic field over a large volume (coil size of the order of 3 m) and with high levels (10 ppm) of homogeneity.

In this work we chose a regular array of 6 axial gradiometers lying on a circle of 6 cm diameter, for positioning directly and centrally above the fetal heart (which can be easily located by ultrasound). The separation of the measurement channels was the largest possible for our dewar. Use of a larger dewar with a curved bottom may be more efficient for the recording of FMCG signals above the central area of the maternal abdomen, but the lower lateral locations will become inaccessible.

A reference vector magnetometer, placed central to the six measurement channels is used for electronic noise cancellation of homogeneous noise. Symmetrical location and close proximity of the vector magnetometer to the measuring gradiometers allows optimal noise reduction.

In FMCG measurements the outputs of the diagonally opposite channels can be subtracted from each other. In this way, effectively third-order gradiometers are created with excellent ambient noise reduction effectively increasing the signal amplitude by spatial filtering. This is illustrated in Fig. 1 where signals are calculated for a fetal heart model of 2 cm size and 10 cm depth with 40x40x40 elements. As can be seen in Fig1(b) there is a twofold gain in signal amplitude after subtracting the outputs of opposite channels.

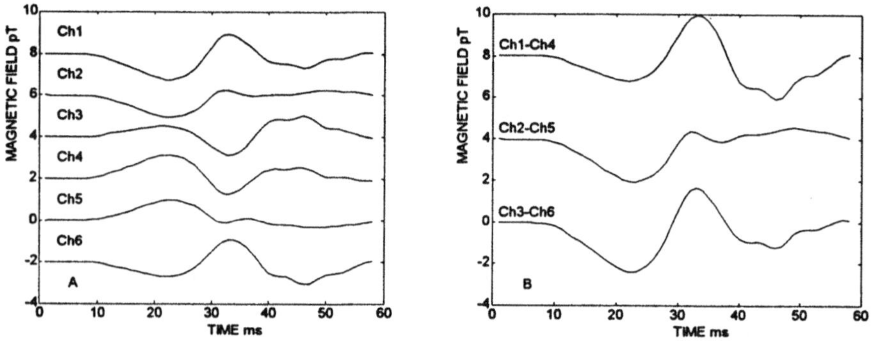

Fig. 1 The simulated magnetic fields due to a realistically-shaped fetal heart model within a cube of 2 cm side incorporating 40x40x40 elements, (A) coupled to the 6 second-order gradiometer measurement channels, and (B) obtained by subtracting the outputs of opposite pairs of channels

The Cryogenic Unit

Symmetrical wire-wound second-order axial gradiometers of 15 mm diameter and 60 mm baseline are made using 50 μm single-core niobium wire fixed in the sharp narrow grooves of a former made from graphite having a very small thermal expansion coefficient. The separate modular channels are rigidly fixed together by a fibreglass holder. The three-channel reference vector magnetometer (also made of wire-wound coils) is rigidly attached to the same holder. Special care is taken to minimise the noise which can arise from the vibration of the gradiometers in the earth's magnetic field.

The DC SQUID sensors are fabricated by the commonly used Nb/Al technology and have a coupled energy resolution of about 5×10^{-31} J/Hz or better. The noise of each signal channel was separately measured inside a superconducting lead shield and was found to be below 6 $fTHz^{-1/2}$.

The Electronics

Conventional flux-locked DC-SQUID electronics is used. The preamplifiers for all the channels are placed in small separate boxes at a distance of approximately 0.7 m from the top of the dewar, with the rest of the electronic units in a standard rack-mounted modular unit placed 5 m from the dewar. We have attempted to minimise the amount of electronics placed in the immediate vicinity of the system. Each module also incorporates scaling amplifier and 4th order Bessel antialiasing filter. The SQUID parameters (modulation and bias currents and DC offset) are digitally controlled by a micro-computer which is also used for data acquisition and processing.

The Dewar

The fibreglass dewar has the following dimensions: overall diameter 156 mm, overall length 730 mm, tail outer diameter 110 mm, neck/tail inner diameter 80 mm and the gap at the bottom of the dewar from outer to inner surfaces was measured (for the dewar filled with helium) to be 10 mm. Volume of the dewar is 4.5 litres and its helium boil-off rate was measured to be 1.3 litres per day, i.e. it is possible to keep the complete system cold over a week-end without the need for a refill. The weight of the dewar is 4 kg. This has been achieved by using very thin walls and avoiding any bulk metal parts in the neck region as had been used in previous systems for cooling the inner radiation shield. The intrinsic noise of the dewar in an unshielded space was measured by successively placing the lowest noise gradiometer centrally, near the edge of the tail and also raised by 5 cm above the bottom tail surface. The noise introduced by the dewar was thus measured to be about 5 $fTHz^{-1/2}$ at the centre, increasing to about 8 $fTHz^{-1/2}$ close to the inner tail wall.

Signal Evaluation

Preliminary testing of the system was carried out after assembly of the first three measurement channels. White noise performance of 10 $fTHz^{-1/2}$, measured in an unshielded normal physics laboratory, for one of the gradiometer channels is demonstrated in Fig. 2. Apart from 50 Hz mains noise, the two features at 25 and 70 Hz were due to laboratory sources. Typical measurements of the adult MCG measured in unshielded environment are shown in Fig. 3. The frequency bandwidth of 0.1 to 120 Hz was used and the signal-to-noise ratio was better than 60.Fig.4 demonstrates the performance of the system in recording the FMCG in the bandwidth from 0.1 to 12 Hz. These signals were used to reconstruct the heart vector loops [5].

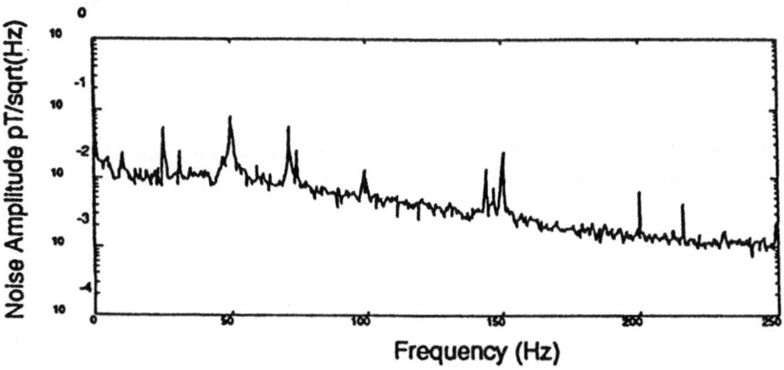

Fig. 2 The noise spectrum measured for one of the gradiometer channels in an unshielded laboratory

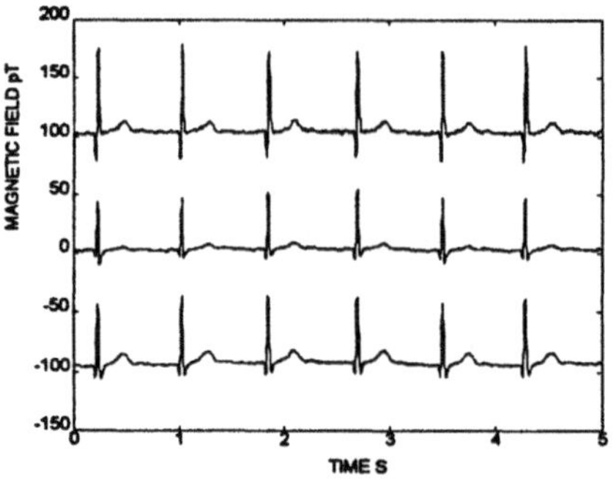

Fig. 3 Adult MCG measured in an unshielded environment in the frequency bandwidth 0.1-120 Hz

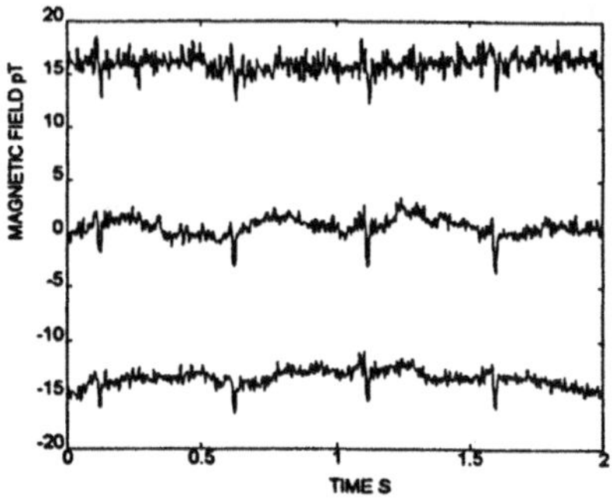

Fig. 4 Three-channel recording of FMCG signals in the bandwidth 0.1-120 Hz.

References

[1] Matlashov, A., Zhuravlev, Y., Slobodchikov, V., Bondarenko, N., Bakharev, A., and Rassi, D. Miniature DC SQUID magnetometers for clinical use, In: (eds. Baumgartner, C., Deecke, L., Stroink, G., and Williamson, S.J.) Biomagnetism: Fundamental Research and Clinical Applications, Amsterdam, Elsevier/IOS-Press, 1995, pp526-9.

[2] Williamson S.J., Pellizone M., Okada Y., Kaufman L., Crum D.D. and Marsden J.R. Five channel SQUID installation for unshielded neuromagnetic measurements, In: (eds. Wienberg H, Stroink G. and Katila T) Biomagnetism: Application and Theory, Pergamon Press, New York, 1985.

[3] Vrba J., Betts K., Burbank M. et al , Whole cortex 64 channel system for shielded and unshielded environments, In: (eds. Baumgartner, C., Deecke, L., Stroink, G., and Williamson, S.J.) Biomagnetism: Fundamental Research and Clinical Applications, Amsterdam, Elsevier/IOS-Press, 1995,pp521-5..

[4] Zhuravlev, Y., Rassi, D., and Emery, S. Clinical assessment of fetal magnetocardiography, In: (eds. Baumgartner, C., Deecke, L., Stroink, G., and Williamson, S.J.) Biomagnetism: Fundamental Research and Clinical Applications, Amsterdam, Elsevier/IOS-Press, 1995, pp700-3.

[5] Zhuravlev, Y.E., Lewis, M.J., Rassi, D., and Emery, S.J., Magnetocardiographic monitoring of the fetal autonomic nervous system (in this volume).

[6] Lipovitch, A.Y., Orlova, L.I., Zhuravlev, Y.E., and Baum O.V., Computer simulation of the heart magnetic field dynamics, Clin.Phys.Physiol.Meas., 1991, 12: 19-21.

ACKNOWLEDGEMENTS

This work is supported by a grant from EPSRC.(Grant GR/J72110) The authors wish to thank Mr A. Bakharev and his company, Krionit, for the design and production of the fibreglass dewar.

High Temperature Superconducting Quantum Interference Devices in a Superconductor-Normal-Superconductor Geometry

Reagor, D., Jia, Q.X., Mombourquette, C., Foltyn, S., Houlton, R., and Wu, X. D.

Materials Science and Technology Division
Los Alamos National Laboratory
Los Alamos, NM, USA

Introduction

The discovery of high temperature superconductors (HTS) has led to intense interest in the fabrication of thin films and Josephson Junctions from the new materials. Most of the Josephson junction work has focused on the bicrystal approach[1]. In this method bicrystal substrates are used to produce a single coherent large angle grain boundary in epitaxial films of HTS. This method has produced most of the best junction results to date [2], but requires an expensive substrate and limits the flexibility of the junction placement.

We have fabricated Josephson Junctions and Superconducting Quantum Interference Devices (SQUIDs) in an superconductor-normal-superconductor (SNS) configuration[3,4]. The SNS device geometry is an edge junction configuration, where the bottom layers of superconductor and capping insulator were patterned and etched to leave a sloped edge, and layers of normal material and superconductor were then deposited. The devices, either Josephson Junctions or SQUIDs, were formed by patterning and etching the N-layer and top S-layer. These devices are fully defined by photolithographic steps that allow great flexibility in layout and substrate choice.

In the following we report on SQUIDs made with this SNS technology. The devices typically have an IcRn product in excess of 100 μV and a flux-to-voltage transfer function ($\partial V/\partial \Phi$) of over 150 $\mu V/\Phi_0$ ($\Phi_0 = 2.07 \times 10^{-15}$ Wb is the flux quantum). These values are some of the highest reported to date and lead to very quiet devices. The flux noise is as low as 40 $\mu\Phi_0/\sqrt{Hz}$ at 1 Hz and 5 $\mu\Phi_0/\sqrt{Hz}$ at 1kHz.

Methods

The devices are intended for liquid nitrogen operation and, due to chemical compatibility considerations, we used $YBa_2Cu_3O_{7-x}$ related compounds for all the layers. The superconductor layers [1] are Ag doped $YBa_2Cu_3O_{7-x}$ approximately 3000 Å thick. The transitions temperature was over 90 K and the critical current was typically 4 MA/cm^2 at 77K, slightly improved over the best undoped $YBa_2Cu_3O_{7-x}$ films. The normal layer was $PrBa_2Cu_3O_{7-x}$ of various thicknesses, typically of order a few hundred angstroms.

Pulsed laser deposition (PLD) was used to deposit both top and bottom Ag:YBCO electrodes, the N-layer (PBCO), and the insulating layer (CeO_2). The Ag-doped YBCO target was a 5 wt.% Ag-doped $YBa_2Cu_3O_{7-x}$ pellet (atomic ratio of Ag in the film is Ag:Y = 1:10 based on Rutherford backscattering spectrometry analysis). The *in situ* deposition used a 308 nm XeCl excimer laser operating at a repetition rate of 20 Hz and producing 20 ns pulses with an energy density of 2 J/cm^2. Typically, a bilayer of superconductor (300 ± 20 nm) and insulator (300 ± 20 nm) was first deposited on a 1 x 1 cm^2 $LaAlO_3$ substrate. The Ag:YBCO was deposited at a substrate temperature of 775 °C and the CeO_2 at 650 °C, both at an oxygen pressure of 200 mTorr. The detailed experimental setup and deposition conditions have been previously described [3]. Conventional photolithography was used to pattern the first layer and a 170 °C postbake was used to lower the contact angle between the photoresist edge and the film surface. A 250 eV Ar ion beam tilted 60° from normal incidence and oriented into the photoresist edge was used to ion mill the first layer. The angle between the edge and the substrate surface, as measured by atomic force microscopy, was in the range of $15^\circ \pm 3^\circ$. The N-layer PBCO and the top Ag:YBCO electrode were deposited after stripping off the photoresist and cleaning the edge surface. The substrate temperature and oxygen pressure during both N-layer and top Ag:YBCO electrode deposition were the same as used for the deposition of the bottom Ag:YBCO electrode.

The devices were first characterized with standard I-V and V-H experiments. A discussion of the basic testing was presented earlier[5]. The noise measurements were performed in open loop mode using the apparatus shown in Fig. 1. The devices were biased with a dc current that was low pass filtered at the dewar connection. The magnetic field modulation was provided by a small 5 turn coil below the sample. A dc magnetic field was separately applied using a large coil wrapped around the sample space of the cryogenic probe. The SQUID

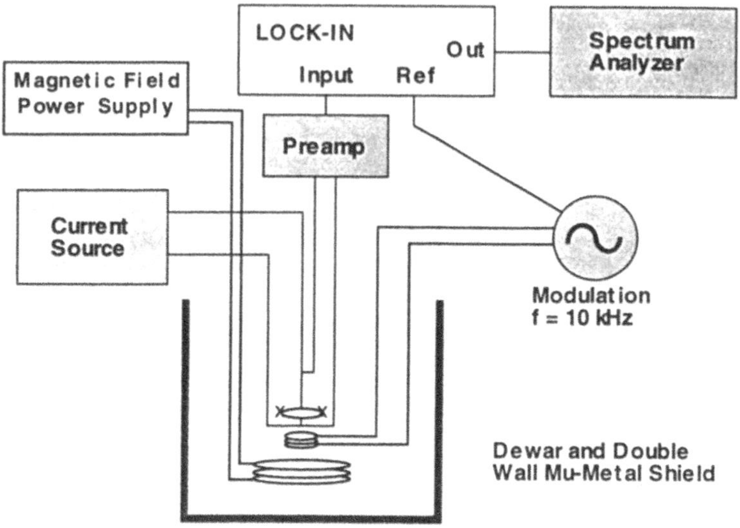

Fig. 1 Block Diagram of the experimental arrangement for the noise measurements.

voltage was input to the primary of a 5:1 cryogenic transformer and subsequently fed to the room temperature preamplifier with resistance wire. The preamplifier (Stanford Research model 552®) was modified by installing an 18:1 transformer to the differential inputs. The preamplifier signal was then fed to a lock-in amplifier referenced to the modulation frequency. In the configuration shown, with no bias current present, the voltage noise at 10 kHz was 90pV/√Hz. This led to a system noise floor below 1 $\mu\Phi_0/\sqrt{Hz}$ for the SQUIDs discussed here. The measurements were performed by first adjusting the magnetic field, bias current, and field modulation to maximize the lock-in voltage output. The magnetic field was then adjusted to zero the lock-in voltage (using the same operating point as a flux locked loop) and the voltage noise of the lock-in output was recorded with a FFT spectrum analyzer. Typically four 256 point spectra were recorded in partially overlapping frequency intervals. The voltage noise contained 60Hz and other kinds of interference that were removed for display. We then swept the magnetic field while recording the lock-in output and used this to measure the derivative of the lock-in voltage with respect to field (or flux) at the operating point. Dividing the voltage noise by the derivative yields the field noise (or flux noise). Care was taken that the integrated noise have an amplitude less than 0.1 Φ_0, maintaining the linearity assumed in the above analysis. This method determines the flux noise independent of the total voltage gain of the preamplifier system.

Results

The junctions and SQUIDs were initially characterized by voltage versus current and voltage versus field. The typical I_cR_n product was 120-140 µV with R_n over an ohm at 77 K. The voltage versus field measurements indicated voltage transfer functions as high as 150 µV per flux quanta. Many SQUIDs fabricated on the same chip or different chips from different runs show voltage modulation above 25 µV at 75 K. The yield of the devices, in terms of voltage modulation by magnetic field, is quite high: 11 of 12 SQUIDs on a typical 1 x 1 cm^2 chip exhibit voltage modulation at 75 K. All of the SQUIDs have a square center hole with dimensions of either 5 x 5 μm^2 or 10 x 10 μm^2. The bridge width was varied from 2 μm to 10 μm. It should be pointed out that the SQUID layout has not yet been optimized. The highest voltage modulation achieved was 49 μV at 75 K. Figure 2 shows the voltage modulation *vs* magnetic flux (Φ) with different bias currents at 75 K from another SQUID. At the optimum bias current the modulation depth is 45 μV. The curves are perfectly periodic and consist of positive and negative sweeps. No hysteresis was found.

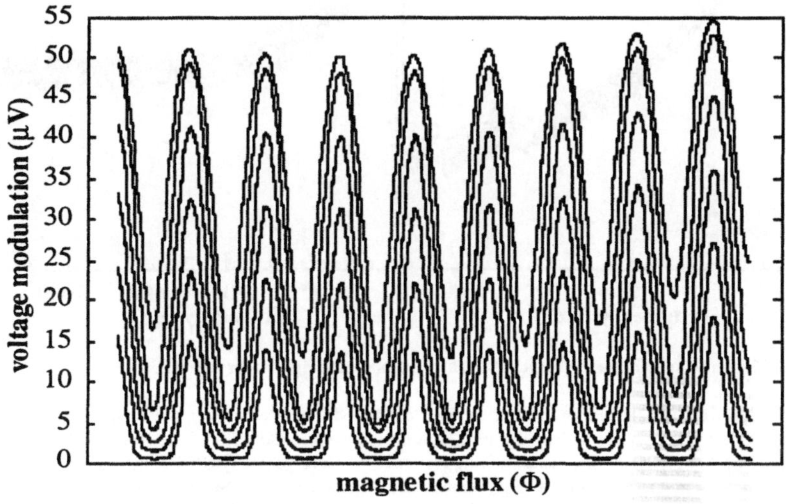

Fig.2 The voltage versus flux results for a small SQUID. The period is one flux quanta.

The relatively high flux-to-voltage transfer function ($\partial V/\partial\Phi$) of the SQUID, which is over 150 $\mu V/\Phi_0$ ($\Phi_0 = 2.07 \times 10^{-15}$ Wb is the flux quantum), leads to the possibility that the devices are fairly quiet. Noise spectra for the SQUIDs, performed with dc bias at 75 K, are shown in Fig.3. The left trace is the spectra for the SQUID with a 49 μV modulation depth. The right trace is for the device shown in Fig. 2 with 45 μV of modulation depth. The 49 μV device has a white noise floor of approximately 5 $\mu\Phi_0/\sqrt{Hz}$. This is one of the lowest values reported in the literature. In addition the I_c is 110 μA and the inductance is 20pH leading to an I_cL product very nearly equal to Φ_0. Under such a condition the white noise floor has been simulated [6] extensively and is very nearly given by $L(\alpha kT/R)^{1/2}$, where R is the measured resistance of the entire SQUID and α is a constant of order 15-20. For the device resistance of ~ 1.1 ohm we have a predicted white noise floor of 1-2 $\mu\Phi_0/\sqrt{Hz}$, in good agreement with the values found. The 45 μV device has a higher white noise floor along with smaller I_c and L. In this limit the ratio of LI_c to Φ_0 is 0.25, where the SQUID white noise is no longer optimized. The device on the right, however, has a much lower 1/f noise corner and a root mean square flux noise of 40 $\mu\Phi_0/\sqrt{Hz}$ at 1 Hz. Further studies of the variation of 1/f noise with dimensions and processing are underway.

Discussion

High-temperature superconductor Josephson junctions and dc SQUIDs have been fabricated using Ag:YBCO as electrodes with an edge-geometry SNS configuration. The devices routinely display I_cR_n products over 120 μV, $\partial V/\partial\Phi$ of over 100 $\mu V/\Phi_0$, and noise below 100 $\mu\Phi_0 Hz^{-1/2}$ at 1 Hz at 75 K. The improved device performance, based on our preliminary microstructural analysis of the material and evaluation of electrical

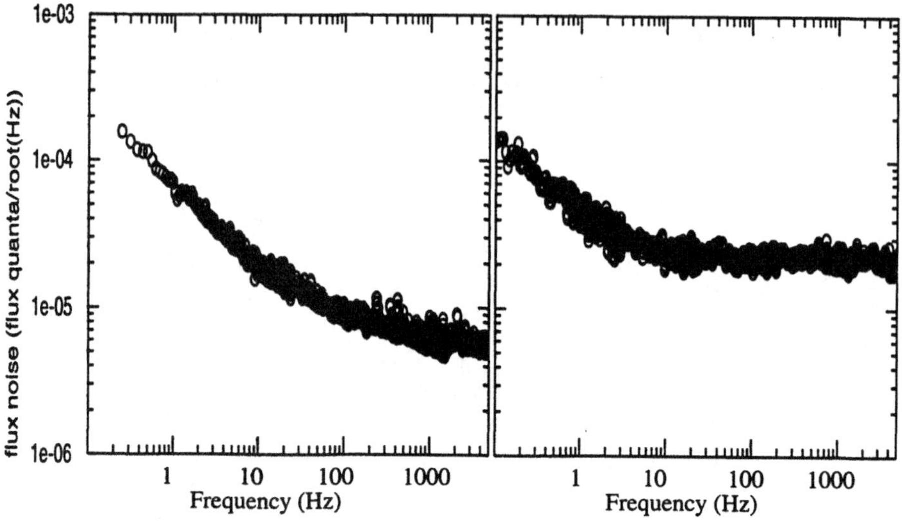

Fig.3. The noise spectra for two SQUIDs. The spectra on the right is for the SQUID in Fig.2.

properties of the thin film, is believed to be from the well controlled interface between the superconductor and N-layer material. Other factors could include the very high critical current densities of the Ag:YBCO material and high resistance of Ag:YBCO to degradation during processing. Edge-geometry SNS construction provides an alternative technology in the fabrication of high temperature superconductor dc SQUIDs and deserves further investigation.

Acknowledgements

The authors received assistance from Jonathan Sun of the IBM T.J. Watson Research center in the development of the noise measurements. We also acknowledge the help of Harriett Kung and Marilyn Hawley with microscopy services.

References

[1] D. Dimos, P. Chaudhari, J. Mannhart, and F. K. LeGoues, "Orientation dependence of grain-boundary critical currents in $YBa_2Cu_3O_{7-\delta}$ Bicrystals", Phys. Rev. Lett., 1988, 61:219-222.
[2] L. P. Lee, J. Longo, V. Vinetskiy, and R. Cantor, "Low-noise $YBa_2Cu_3O_{7-\delta}$ direct-current superconducting quantum interference device magnetometer with direct signal injection", Appl. Phys. Lett. 1995, 66: 1539-1541.
[3] Q. X. Jia, X. D. Wu, D. Reagor, S. R. Foltyn, C. Mombourquette, P. Tiwari, I. H. Campbell, R. J. Houlton, and D. E. Peterson, "High-temperature superconductor josephson junctions with a gradient Pr-doped Y1-xPrxBa$_2$Cu$_3$O$_{7-\delta}$ barrier", Appl. Phys. Lett., 1994, 65: 2866-2868.
[4] D. Reagor, R. Houlton, K. Springer, M. Hawley, Q. X. Jia, C. Mombourquette, F. Garzon, and X. D. Wu, "Development of high temperature superconducting quantum interference devices using low deposition temperature $YBa_2Cu_3O_{7-\delta}$ barriers", Appl. Phys. Lett. 1995, 66: 2280-2282.
[5] Q. X. Jia, X. D. Wu, D. Reagor, S. R. Foltyn, C. Mombourquette, and D. E. Peterson, "Edge-geometry SNS dc SQUIDs using Ag-doped YBa2Cu3O7-x electrodes", Electronics Lett., 1996, 32 :499-501.
[6] M.B. Ketchen, "Design considerations for dc SQUIDs fabricated in deep sub-micron technology", IEEE transactions on magnetics, 1991, 27: 2916-2919.

A Digital SQUID Controller for Unshielded Biomagnetic Measurements

Robinson, S. E.

Department of Physics and Astronomy, The University of New Mexico, Albuquerque, NM, USA[1]

Introduction

A biomagnetometer capable of achieving low noise performance without magnetic shielding would be attractive in a wide range of clinical and research applications. Magnetically shielded enclosures add significant initial cost to biomagnetometer systems and preclude their portability. The primary obstacle to unshielded operation is the immense dynamic range between the ambient magnetic noise and the SQUID noise floor. Ambient magnetic field noise found in a typical urban environment can be over eight orders of magnitude greater than the sensitivity of a modern thin film low-T_C superconducting (LTS) dc-SQUID magnetometer [1]. Gradiometric magnetic sensors and linear combinations of magnetic sensors, including references may be used to separate the low-amplitude biomagnetic signals from the much larger environmental magnetic noise [2], [3]. The accurate reproduction of fields by these sensors is a prerequisite to the effectiveness of linear noise reduction methods.

The limiting element for high-fidelity SQUID measurement is the analog feedback of the flux-locked loop (FLL), which is used to linearize the SQUID's transfer function [4]. The FLL must be capable of tracking the largest signals while resolving the SQUID noise floor. Although the dynamic range encountered by an unshielded magnetometer does not exceed the intrinsic capabilities of LTS SQUID devices, it may be too large to be accommodated by the FLL. The FLL applies negative flux feedback to the SQUID to maintain lock close to a single point on the V-Φ transfer function. Since the FLL loses lock when the measured signal exceeds the feedback range, the output signal from the FLL becomes discontinuous and severely distorted.

The SQUID's V-Φ transfer function is periodic in the flux quantum Φ_0. From this, it can be seen that by explicitly offsetting the FLL to a new lock point (some multiple of Φ_0) within the feedback range, whenever the required feedback will exceed that range, one extends the the dynamic range of the FLL. The output signal then becomes the sum of the feedback signal and the total flux quanta that have been "slipped." This sum must be represented digitally. Digital FLL electronics, designed for extending the dynamic range of LTS SQUIDs by slipping flux at Φ_0 intervals, have been commercially available for several years [5].

The flux slipping method extends the dynamic range of the FLL, but does not necessarily preserve linearity over the entire dynamic range of the SQUID device. When in lock, the FLL holds the flux in the SQUID constant. However, when a flux slip occurs, the flux in the SQUID is allowed to change. Since SQUID inductance is a function of flux, feedback flux induced in the SQUID by the modulation coil is no longer linearly dependent on feedback current. Although this effect is small for LTS SQUID magnetometers operating in the ambient magnetic field, it can be significant for high-T_C (HTS) SQUIDs operating in much lower fields. Additional distortion is introduced from the SQUID's nonlinear V-Φ curve when the FLL operates in the presence of high slew rate signals; the SQUID error voltage is nonlinear with respect to the instantaneous flux error. Finally, the flux slipping method requires careful adjustment of the flux slip threshold and gain so that measurement is "seamless."

A unique digital flux tracking loop (dFTL) has been developed for SQUID control, enabling wide dynamic range measurements and compensating for several sources of measurement nonlinearity [6]. The dFTL incorporates a four-phase modulation pattern, and feedback loops for tracking external flux and for regulating the modulation amplitude to precisely Φ_0 peak-to-peak. By measuring the feedback current per flux quantum, the dFTL is able to slip flux seamlessly at $\Phi_0/2$ intervals and compensate for variations in SQUID inductance with flux. In addition, the dFTL is self tuning, requiring no external adjustments for operation with a wide range of SQUIDs.

Methods

Experimental Hardware: Modulation and feedback current to the SQUID modulation coil is supplied by a single digital-to-analog converter (D/A). A second D/A converter is used to bias the dc-SQUID; this can also be used to generate bias modulation for reducing $1/f$ noise in HTS SQUIDs. The SQUID junction voltage is amplified by a low-noise preamplifier and digitized by an analog-to-digital (A/D) converter. A digital signal processor (DSP) implements all modulation, feedback, and detection functions, and represents the instantaneous SQUID flux by its

---

[1] Currently at CTF Systems Inc., 15-1750 McLean Ave., Port Coquitlam, BC, Canada

digital value. A block diagram of the dFTL is shown in Figure 1. The minimum D/A converter bit resolution for dFTL modulation and feedback can be approximated as follows:

$$N = ceil\left[\log_2\left(\frac{1.5}{S_\Phi \sqrt{f_{bw}}}\right)\right] + 1, \tag{1}$$

where S_Φ is the flux noise density of the SQUID and f_{bw} is the bandwidth. The dFTL requires a minimum D/A range of 1.5 Φ_0. Equation (1) assures that the D/A converter will resolve one bit more than the rms SQUID noise floor. The A/D converter used to read the SQUID voltage should have at least as good resolution. As an example, let us assume that the dFTL is used with a SQUID having a flux noise density of 10 $\mu\Phi_0$-$Hz^{-1/2}$ sampled in a bandwidth of 200 kHz. From (1) we find that the minimum converter resolution is 10 bits.

Modulation and Detection: An accurate measure of the feedback current per Φ_0 is needed in order to seamlessly extend the dynamic range by slipping flux. At the same time we must also measure the flux error signal from which feedback will be derived. A modulation signal for which the SQUID junction voltages convey information about both flux changes and Φ_0 amplitude can be created by selecting at least three points on the V-Φ curve such that two of the points are separated by an odd multiple of $\Phi_0/2$, and two are separated by an even multiple of $\Phi_0/2$. The lowest order set of points satisfying this are three points at $\Phi_0/2$ intervals (Φ_0 peak to peak). In the prototype dFTL, a four-phase modulation at the lock point and plus and minus $\Phi_0/2$ was used (Fig. 2).

When biased at the critical current, the dc-SQUID junction voltage V_j (neglecting SQUID noise) is periodic with respect to the magnetic flux Φ in the SQUID. This function is approximately sinusoidal:

$$V_j = \bar{V} - k\cos\left(\frac{2\pi\Phi}{\Phi_0}\right), \tag{2}$$

where $\bar{V}$ is the average junction voltage. Feedback current to the SQUID is adjusted such that at lock, the flux corresponding to each of the four modulation phases is:

$$\Phi(\phi_0) = \Phi(\phi_2) = \frac{n\,\Phi_0}{2} + \frac{\Phi_0}{4}, \tag{3}$$
$$\Phi(\phi_1) = \frac{[n+1]\,\Phi_0}{2} + \frac{\Phi_0}{4}, \text{ and}$$
$$\Phi(\phi_3) = \frac{[n-1]\,\Phi_0}{2} + \frac{\Phi_0}{4},$$

where n is an integer corresponding to the number of flux quanta in the SQUID. When the dFTL is operating, let the flux and modulation amplitude errors near lock, in radians, be denoted by Φ_ε and M_ε, respectively. The dc-SQUID junction voltages sampled during each of the four modulation phases, for small error values then are:

$$V_j(\phi_0) = V_j(\phi_2) = \bar{V} - k\sin(\Phi_\varepsilon) \cong \bar{V} - k\,\Phi_\varepsilon, \tag{4}$$
$$V_j(\phi_1) = \bar{V} - k\sin(\Phi_\varepsilon + [\pi + M_\varepsilon]) \cong \bar{V} + k(\Phi_\varepsilon + M_\varepsilon), \text{ and}$$
$$V_j(\phi_3) = \bar{V} - k\sin(\Phi_\varepsilon - [\pi + M_\varepsilon]) \cong \bar{V} + k(\Phi_\varepsilon - M_\varepsilon).$$

Two orthogonal detection channels — one for the flux error $V_{\varepsilon\Phi}$ and the other for the modulation amplitude error $V_{\varepsilon M}$ — are then derived by combining the measured SQUID junction voltages for each of the four phases:

$$V_{\varepsilon\Phi} = [V_j(\phi_1) - V_j(\phi_0)] + [V_j(\phi_3) - V_j(\phi_2)] \cong -4k\Phi_\varepsilon, \text{ and} \tag{5}$$
$$V_{\varepsilon M} = [V_j(\phi_1) - V_j(\phi_0)] - [V_j(\phi_3) - V_j(\phi_2)] \cong -2kM_\varepsilon.$$

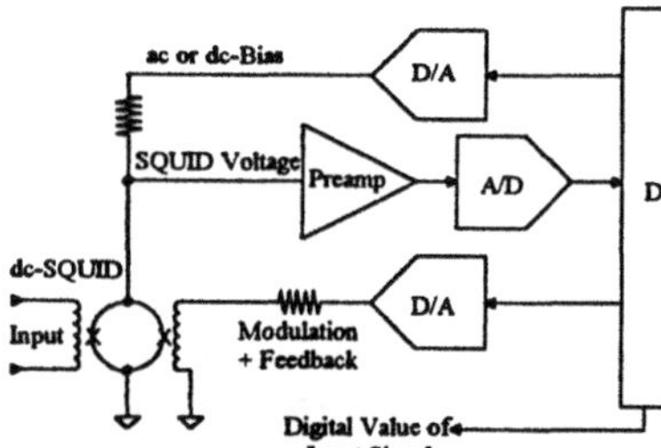

Fig. 1: Block diagram of the digital flux tracking loop electronics. Details of coupling the SQUID junction voltage to the low noise preamplifier are not shown.

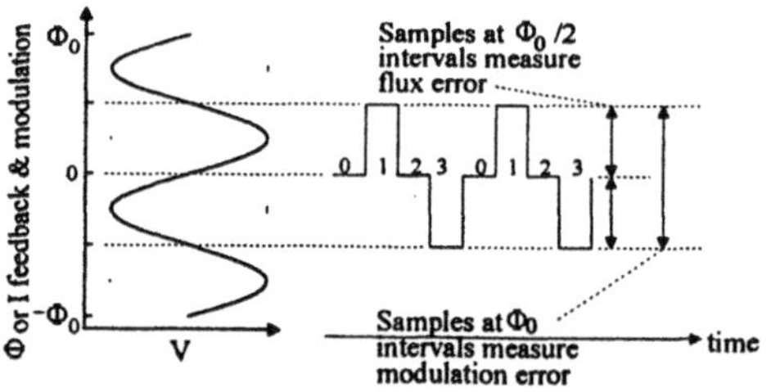

Fig. 2: dFTL four phase modulation pattern. Two orthogonal detection channels recover the flux error and modulation amplitude error.

Flux feedback for tracking the external signal is provided by integrating the flux error signal:

$$\Phi(t+1) = \Phi(t) + g_\Phi V_{e\Phi}(t). \tag{6}$$

The feedback gain function g_Φ can take the form of a simple integrator or higher-order filter. The feedback loop for controlling modulation amplitude is of similar form to (6), except that an integrator is sufficient, since the modulation amplitude (for the local flux quantum) does not change with time. The modulation integrator time constant must be long enough so that, in the presence of SQUID noise, the modulation amplitude for $\Phi_0/2$ can be determined to better than one LSB on the flux feedback D/A converter.

Seamless flux slipping: Since the dFTL modulation waveform is Φ_0 peak-to-peak, the minimum flux feedback range of the D/A converter must be greater than 1.5 Φ_0; this allows for two possible lock points (within the feedback range), $\Phi_0/2$ apart, at any time sample. The digital value for sum of feedback and modulation signals is computed prior to being sent to the D/A converter. Because the numerical range of the converter is known in advance, the required modulation value can be numerically tested against these limits prior to conversion. If modulation is within the D/A range, it is simply converted to its analog value. If the required modulation exceeds the D/A converter's range, the feedback value is offset by $\Phi_0/2$ so as to maintain the modulation pattern within the D/A range. The flux slip will occur without a discontinuity in the measurement because the numerical value corresponding to $\Phi_0/2$ is determined precisely by the modulation amplitude control loop. The dFTL represents the external signal measured by the SQUID as the digital value of the feedback plus the digital value for each $\Phi_0/2$ flux slip. Since flux is slipped in half quantum steps, the modulation phase is advanced by one cycle at each flux slip to maintain constant flux error feedback polarity. The sign of the modulation error reverses with each flux slip, and is corrected digitally. Without the advance in modulation phase, the flux error polarity would change at each flux slip and the modulation error polarity would remain constant.

Linearizing the dFTL for SQUID inductance changes and for transfer function nonlinearity: The modulation amplitude feedback loop regulates the D/A value of current per flux quantum in the vicinity of the current flux lock point. The dFTL builds and maintains a table of the digital values for $\Phi_0/2$ at each lock point, providing a piece-wise correction for inductive changes. The dFTL also linearizes the flux error signal. Let us assume that the SQUID V-Φ transfer function is approximately sinusoidal. The dFTL lock points are at the mean SQUID voltage (the zero crossings of the sinusoid at 0 and $\pm\pi$). The detected SQUID error signal for a flux error of Δ_Φ is:

$$\Delta_V = V_0 \sin\left(\frac{2\pi\Delta_\Phi}{\Phi_0}\right), \tag{7}$$

where V_0 is the peak to peak amplitude of the V-Φ transfer function (Fig. 2). Given V_0, we solve for the flux error, which will then be used for providing linear feedback:

$$\Delta_\Phi = \frac{\Phi_0}{2\pi} \sin^{-1}\left(\frac{\Delta_V}{V_0}\right). \tag{8}$$

Since the flux modulation amplitude is regulated by a feedback loop to precisely plus and minus $\Phi_0/2$ or Φ_{Mod},

$$\Delta_\Phi = \Phi_{Mod} \sin^{-1}\left(\frac{\Delta_V}{V_0}\right)/\pi. \tag{9}$$

However, the control loop for modulation actually determines the digital value of the current I_{Mod} corresponding to $\Phi_0/2$ for the local flux quantum that is locked. Furthermore, the peak-to-peak SQUID voltage V_0 is periodically read by the DSP whenever the flux slew rate is small. Consequently, the SQUID flux error voltage can be expressed as the amplitude ratio $R_V = \Delta_V/V_0$. It follows then that the linear feedback current must be:

$$I_\varepsilon = I_{Mod} \sin^{-1}(R_V)/\pi. \tag{9}$$

In practice, the V-Φ curve is seldom exactly sinusoidal. Instead, empirical measurement of the actual V-Φ transfer function is placed in a lookup table, making the feedback correction:

$$I_\varepsilon = I_{Mod} f(R_V). \tag{10}$$

Results

The dFTL algorithm was implemented on a general purpose standalone DSP board with built-in analog I/O (Sonitech DAQ-200). Modulation, feedback, and bias was generated by 18-bit D/A converters. SQUID junction voltage was measured by an 18-bit A/D converter. The analog I/O sample rate was 200 kHz. The DSP device (TI TMS320C31-33) implemented all communications, modulation, detection, and feedback processes in firmware, requiring less than 2.25μs per sample. Measurements were recorded on a PC. A shielded HTS integrated dc-SQUID magnetometer and low-noise preamplifier (Conductus iMAG) were used for testing. The sensitivity of this device was 11.1 nT/Φ_0. The SQUID noise floor, 45 fT-Hz$^{-1/2}$ above 100 Hz, was measured by the dFTL while a 100 nT rms test signal was applied to the SQUID. The spectrum contained no internally generated spurious responses. Harmonic and intermodulation distortion products were below the SQUID noise floor for the 100 nT rms test signal. The closed-loop bandwidth was 6.75 kHz, and the maximum slew rate 18 kΦ_0/second — smaller than the theoretical slew rate limit of $\Phi_0/4$ per sample period. This was attributed to sample delays in both A/D and D/A conversion, and to use of a simple feedback integrator in the firmware rather than a higher-order control loop. When unshielded, this dFTL SQUID magnetometer readily tracked the ambient field and power mains interference (at times, over 400 nT peak-to-peak) while resolving small signals near the SQUID noise floor (Fig. 3).

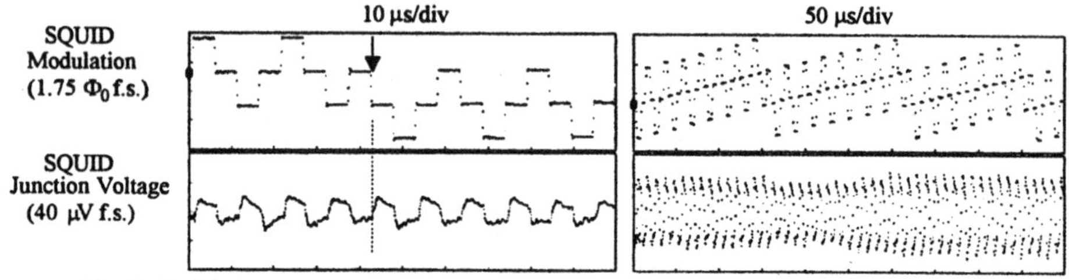

Fig. 3: Modulation and corresponding SQUID junction voltages for the prototype dFTL controlling an HTS integrated dc-SQUID magnetometer (unshielded). A flux slip, occurring at the arrow (left panel), is not associated with a change in error signal amplitude, implying that the flux slip was accurately regulated to one-half flux quantum. In the right hand panel, the dFTL is tracking ambient magnetic noise of about 100 nT peak-to-peak.

Conclusions

The digital flux tracking loop prototype successfully demonstrates a unique method for obtaining large dynamic range SQUID measurements with low noise and distortion. The multiple feedback loops of the dFTL also eliminate tuneup of SQUID modulation, feedback gain, and bias (and skew, for bias-modulated SQUIDs); this is particularly useful for maintaining systems having large numbers of SQUID channels. Slew rate could be improved by implementing higher-order feedback control, together with increasing the analog I/O sample rate. As a result, biomagnetometer performance should now be limited by the properties of the SQUID devices and not by their room-temperature electronics. The effectiveness of the dFTL electronics, when combined with advances in HTS SQUID performance, will enable development of biomagnetometer instruments for small clinics, including portable and hand-held instruments for rapid assessment of the magnetocardiogram and other vital functions.

References

[1] Ilmoniemi, R., Knuutila, J., Rhyänen, and Seppä, H. Multi-SQUID devices and their applications, In: Brewer, D.F. Progress in Low Temperature Physics, Vol. XII, North-Holland, Elsevier Science Publishers, B.V., 1989.

[2] Vrba, J. and McCubbin, J. First-gradient balancing of higher-order gradiometers, Il Nuovo Cimento, 1983, 2-D: 142-152.

[3] Williamson, S.J., Robinson, S.E., and Kaufman, L. Methods and instrumentation for biomagnetism, In: Atsumi, et al. Bio Magnetism '87, Tokyo, Tokyo Denki University Press, 1987.

[4] Forgacs, R.L. and Warnick, A. The flux locked loop, Rev. Sci. Instrum., 1967, 38: 214.

[5] Vrba, J., Fife, A.A., and Burbank, M.B. Digital SQUID electronics in geophysical applications, In: Weinstock, H. and Overton, W.C., Jr. SQUID Applications to Geophysics, Tulsa, The Society for Exploration Geophysicists, 1981.

[6] Robinson, S.E., Method and Apparatus for Extending the Dynamic Range of dc-SQUID Measurements Using a Flux Tracking Loop, U.S. Patent Number 5,469,057, November 21, 1995.

Acknowledgements

This work was supported, in part, by the Jet Propulsion Laboratory, Pasadena, California and Sandia National Laboratory, Albuquerque, New Mexico.

Investigation of the Thin Film SQUIDs Using Double Layer Model of the Secondary Sources

Romanovych, S. S.

Institute of Mathematical Machines and Systems Problems, Kiev, Ukraine

Introduction

In the most common formulation of a problem the computation of the SQUID contours magnetic flux-linkages is based on the solution of the boundary integral equations (IE), where the distribution of the secondary sources is the unknown function. So the 3D IE of the second kind respecting to density σ of the magnetic charges simple layer on the "volume" conductor surface S is formulated in the form [1]

$$\sigma(Q)+\frac{1}{2\pi}\int_S \sigma(M)\frac{\vec{r}_{MQ}\vec{n}_Q}{r_{MQ}^3}dS_M=-2\mu_0 H_n^0(Q)\,, \tag{1}$$

where μ_0 is the magnetic constant and H_n^0 is normal to S component of the external sources' intensity. For the decision of the plane-parallel (2D) problem, the regulated IE respecting to currents linear density σ on the contour L of the superconductor (also "volume") section by the plane of the progressive symmetry is written such a way [1]

$$\sigma(Q)-\oint_L \sigma(M)\left(\frac{\vec{r}_{MQ}\vec{n}_Q}{\pi r_{MQ}^2}-\frac{1}{L}\right)dL_M=-2H_n^0(Q)+\frac{I}{L}\,, \tag{2}$$

where L is the contour length, I is the current through the section. With axial symmetry of the field IE has the analogous form.

Considering the film superconductors as extremely thin (that is as "zero" volume) and transiting to the model of the secondary sources in the form of double layer of the magnetic charges, it is possible to simplify the machine analysis of the film superconducting devices and achieve the high decision accuracy of this in fact 3D problem.

Problem Statement and Solution method

Let's input double layer of the magnetic charges, allocated with the density τ at the thin superconducting films of common square S. Then the normal to S component of the magnetic intensity H_n at the observation point Q, $Q\in S$ will be expressed by the sum

$$H_n(Q)=H_n^0(Q)+H_n^\tau(Q)\,. \tag{3}$$

Here the first term of the sum is conditioned by the primary sources field and secondary field is defined by the integral

$$H_n^\tau(Q)=\int_S \tau(M)h(Q,M)\,dS_M\,, \tag{4}$$

where the kernel $h(Q,M)$ is the normal component of the intensity, formed at the point Q by unit magnetic moment, concentrated at the current point M.

Collecting (3) and (4) and taking into account condition of the magnetic field flowing around superconductor ($H_n=0$), we get the integral equation of the first kind

$$-H_n^0(Q)=\int_S \tau(M)h(Q,M)\,dS_M\,. \tag{5}$$

In 3D space the function

$$h(Q,M)=\frac{1}{4\pi}\left[\frac{3\left(\vec{e}_\tau(M)\vec{r}_{MQ}\right)\vec{r}_{MQ}}{r_{MQ}^5}-\frac{\vec{e}_\tau(M)}{r_{MQ}^3}\right]\vec{n}_Q\ , \tag{6}$$

it being known $\mathbf{e}_\tau$ is a unit vector of the magnetic moment $\tau(M)dS_M$, $\mathbf{n}_Q$ is a unit vector of the normal to S.

For the method illustration let's assume, that superconducting film is plane. Then kernel $h(Q,M)$ expressing normal to S intensity component because of S is the plane of moment symmetry, which is normal to its direction, will be satisfy equality (see (6))

$$h_p(Q,M)=-\frac{1}{4\pi\, r_{MQ}^3}\ , \tag{7}$$

where index "p" denotes the planar construction.

In order to transit from IE (5) to the set of linear algebraic equations (SLAE) it is needed to share the surface S onto elements areas ΔS_i, $1\leq i\leq n$ and find integrals $\int_{\Delta S_i}\tau(M)h(Q_i,M)\,dS_M$.

Let's present the double layer of the magnetic charges, allocated on ΔS as magnetic leaf, for which, as it is known, the density τ is constant. Then it may be carried out the sign of integral and SLAE matrix coefficient ($a(Q_i,M_j)$) may be defined, proceeding from expression:

$$a\left(Q_i,M_j\right)=\int_{\Delta S_j}h\left(Q_i,M\right)dS_M\ , \tag{8}$$

where Q_i, M_j are the centers of the elements ΔS_i, ΔS_j accordingly.

Taking into mind, that main details of the thin film SQUIDs such as receiving antennae, input contours, screens and so on have the rectangular outlines, share S onto element areas ΔS of the rectangular form, the sides of which are parallel to X, Y axes of the Cartesian system of coordinates (Fig. 1).

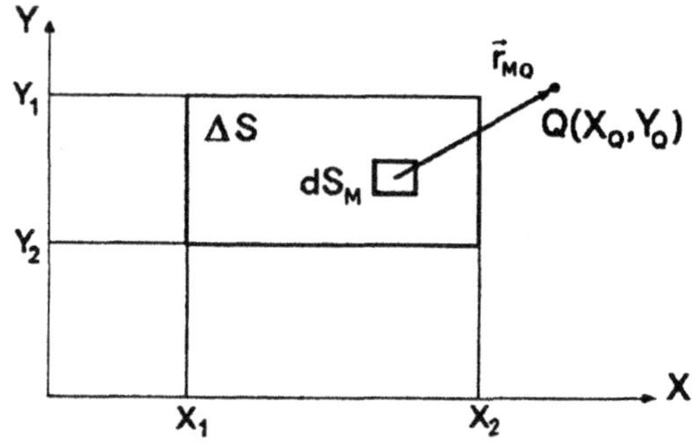

Fig. 1. Element area ΔS in coordinate system X, Y

Integral (8) on such element is taken in the finite form (see (7)), in the result of what the coefficient $a_p(Q_i, M_j)$ of SLAE matrix will be written so:

$$a_p\left(Q_i,M_j\right)=\frac{1}{4\pi}\left(\frac{R_{11}}{R_{x1}R_{y1}}+\frac{R_{22}}{R_{x2}R_{y2}}-\frac{R_{12}}{R_{x1}R_{y2}}-\frac{R_{21}}{R_{x2}R_{y1}}\right). \tag{9}$$

Here

$$\begin{gathered}R_{x1}=x_Q-x_1\ ,\ R_{x2}=x_Q-x_2\ ,\ R_{y1}=y_Q-y_1\ ,\ R_{y2}=y_Q-y_2\ ;\\ R_{11}=\sqrt{R_{x1}^2+R_{y1}^2}\ ,\ R_{12}=\sqrt{R_{x1}^2+R_{y2}^2}\ ,\ R_{21}=\sqrt{R_{x2}^2+R_{y1}^2}\ ,\ R_{22}=\sqrt{R_{x2}^2+R_{y2}^2}\ ,\end{gathered} \tag{10}$$

x_Q, y_Q - the observation coordinate points, x_1, y_1, x_2, y_2 define the location of the rectangle ΔS (see Fig.1). In the common case expressions for matrix coefficients are found by integrating on ΔS function (6).

Such a way, the density τ of the double layer on the superconducting plane film, presented by rectangle set, satisfy SLAE (see (9)).

$$\sum_{j=1}^{n} \tau\,(M_j)\,a_p(Q_i, M_j) = -H_n^0(Q_i), \quad 1 \le i \le n \tag{11}$$

The coefficients $a_p(Q_i, M_j)$ may be found analytically (but on more complex than (9) formulae) or calculated on any cubage if the form S is not rectangular.

The SLAE (11) right parts for planar film contours (films with holes) or screens (the entire films without holes) are defined by the field of the external known, or sought sources. In this case, it is calculated the mutual inductance of the given contour with any other counter streamlined by current, or the screen influence on proper or mutual inductance of other contours. For the superconducting contours, the proper inductance of which needs to be defined, the right part is defined by the strip with unit current that is located on little distance from being investigated contour.

After SLAE (11) decision and calculation of the distribution τ, the flux-linkages Ψ of superconducting contours are accounted as the vector magnetic potential **A** circulation: $\Psi = \oint_l \bar{A}d\bar{l}$, where l is contour internal boundary. Potential **A** founded by the double layer of density τ, allocated on the rectangular element ΔS we define by integrating on it of the corresponding expression. For vector **A** projections on coordinate axes (see (10)) we get

$$A_x^{\tau}(Q) = \frac{\mu_0 \tau}{8\pi} \ln \frac{(R_{x1}+R_{11})(R_{x2}+R_{22})(R_{x1}-R_{12})(R_{x2}-R_{21})}{(R_{x1}-R_{11})(R_{x2}-R_{22})(R_{x1}+R_{12})(R_{x2}+R_{21})},$$

$$A_y^{\tau}(Q) = -\frac{\mu_0 \tau}{8\pi} \ln \frac{(R_{y1}+R_{11})(R_{y2}+R_{22})(R_{y1}-R_{21})(R_{y2}-R_{12})}{(R_{y1}-R_{11})(R_{y2}-R_{22})(R_{y1}+R_{21})(R_{y2}+R_{12})}.$$

As an example of problem decision according to stated method take the result of the computation of the proper inductance of SQUID's input contour in the form of the square plate the size 500×500 μm^2 with central hole 200×200 μm^2. Because of such construction symmetry the beginning of coordinates was taken in the hole centre, and the account was performed for 1/4 of its part ($x \ge 0$, $y \ge 0$) with considering the symmetrical points influence. By found distribution in 450 nodes, corresponding to unit current through the normal section of the described superconducting contour, it is calculated its magnetic flux-linkage and proper inductance, the quantity of which is 339 pH and is agreement with physical experiment.

Summary

The proposed method allows to account the proper and mutual inductances of the thin film SQUID's contours of the complex form and the influence of the screening details on their parameters. Although the needed problem quantizing degree is rather high, it is possible to affirm that this method is available for accounting not only planar and coplanar superconducting films, but films of the arbitrary not plane configuration also. It concerns to thin film details of the super high frequency devices too.

References:

[1] Romanovych, S. S. Regularization of integral equations in problems of electro- and magnetostatics, Technique electrodynamics, 1986, 1: 44-49, (in Russian).

DC-MEG Studies: Development of Techniques for Resolving DC Fields

Saligram, U.[1,2], Moran, J.E.[1] and Tepley, N.[1,2]

[1]Henry Ford Hospital, Detroit, MI, USA; [2]Oakland University, Rochester, MI, USA

Introduction

Long duration magnetic field changes (DC) [1] in MEG may occur as a result of migraine [2], anoxia [3], epilepsy, closed head injury etc. These Field changes are easily masked by changes due to artifacts. It is therefore useful to separate the DC field changes of interest, i.e; arising from physiological phenomena, from those arising from external sources. The ability to measure these DC phenomena may lead to improvements in diagnosis and prognosis for treatment and recovery. With MEG we have successively measured DC field changes over periods of minutes to one hour. For these studies no absolute DC level can be established but shifts in (DC) levels over time can be recorded. However, we are presently developing new techniques to record changes of DC magnetic field amplitude which occur over a period of days or weeks. The problems that must be addressed to obtain meaningful DC measurements include arbitrary neuromagnetometer baseline, non-uniform residual magnetic field within the shielded room and the need to accurately reposition the subjects and control the distance between the cortex and sensor array. In this study, we describe results we have obtained for both experimental DC source models and actual DC field measurements recorded on human subjects on three consecutive days.

METHODS

A study using a glass dipole head model, was carried out. Twenty second epochs of continuos MEG data were collected from each of the seven placements of a seven channel BTi model 607 magnetometer. In each run, 2 seconds of baseline data were collected before slowly lowering the head away from the magnetometer probe over a period of 16 seconds. An additional 2 seconds of data were collected at this new level.

However in order to use this lowering technique in human studies it is necessary to separate the DC field due to the source of interest from the DC fields from other sources in the body. A second study using the dipole head model with an interfering DC source was carried out. The interfering source was a current loop wrapped around the glass head. Twenty second epochs of continuous MEG data were collected as described in the first study. In the first run both the dipole and the external wire were energized; the second run was carried out with only the dipole energized. The third run had only the external wire energized.

Based on the technique of lowering described above, a technique to measure the changes in DC fields from human subjects with changing distance from the sensor array, was developed in this study. The position of the magnetometer probe with respect to the shielded room was fixed, the orientation of the hexagonal array of sensors was also fixed. The wooden platform on which the subject lay was fixed with respect to the room and with respect to the magnetometer probe.

The subject lay on a wooden platform with his/her head in the head holder, which was fixed to the platform. The magnetometer probe was aligned such that it just touched the subject's head and the vertical axis was the Z-axis. This entire setup was kept the same throughout the study and through all three days of the study. For description of the apparatus refer to Fig. 1.

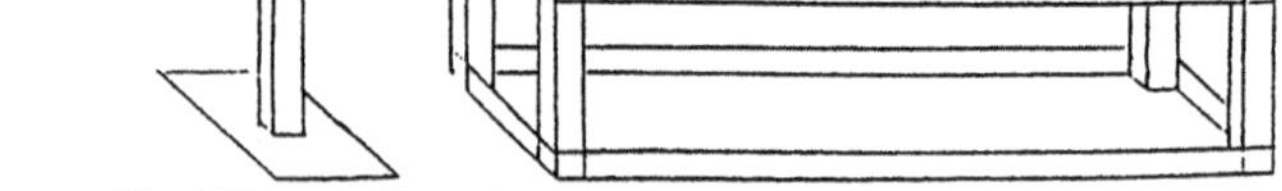

Fig. 1 The experimental apparatus.

Ten seconds of data were collected at this initial position, then the subject was slowly lowered away from the probe through a distance of 1.7 cm. Measurements were made after each step but not during the process of lowering. This lowering was done in seven equal steps. At each of these seven levels, ten seconds of data were collected. The subject was then raised all the up to the initial position, and the study repeated. The exact position of the subject was established using BTi's probe position indicator program A total of three successive runs of were carried out. This set of three runs was repeated thrice over three days, one set each day. Another study with no subject under the probe was carried out each day. Data from this study was used as the noise data for that day.

DATA ANALYSIS and RESULTS

The data were digitized using an HP 694A multiprogrammer. All the data were sampled at 130Hz and the high and low pass filters were set at 0 and 50 Hz respectively.

1) A dipole fitting algorithm was applied to the data from the first study with the glass dipole head at a latency point during or after the drop, and the model accepted only for dipole correlations greater than 0.91. The total drop was 3 cm. The shift in the DC level due to the drop was 15 pT(Fig. 2) at the field maximum. There was no discernible shift in the DC level when the head was lowered from the probe without the dipole being energized.

2) Data from the 3 runs of the second study with the glass dipole head were analyzed and it was seen that the DC shift due to the dipole could be extracted form the total shift (dipole+external wire) by a simple subtraction of the signal from the wire from the total shift (Fig. 3). This result validates our procedure for resolving the DC interference from the dipole source.

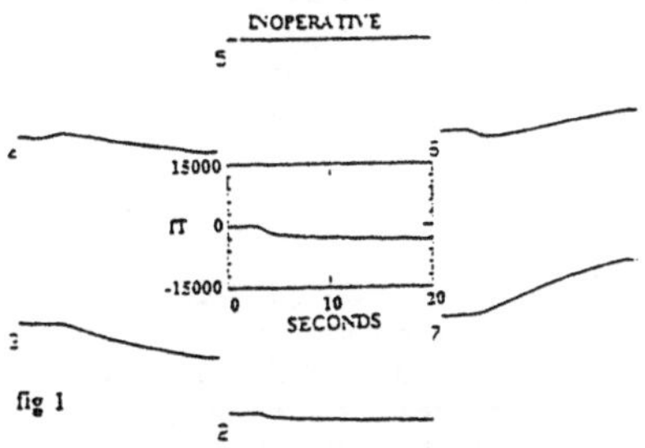

Fig. 2 The above figure shows the shift in the DC level (15pT), due to the lowering of the glass dipole head.

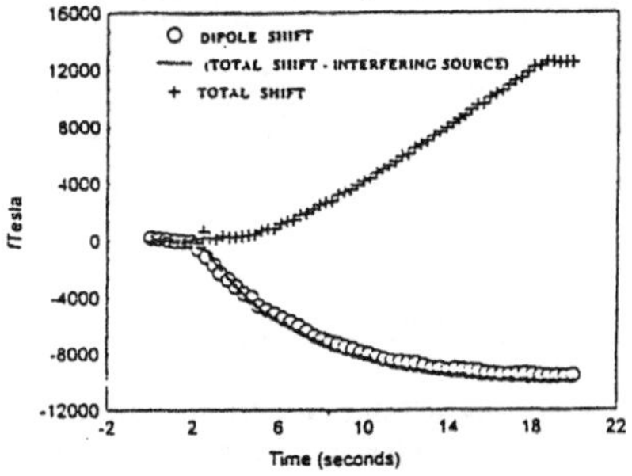

Fig. 3 The above figure shows the shift due to the dipole extracted from the total shift.

3) Data from the human studies were time averaged and decimated. The change in the DC magnetic field amplitude with distance moved is nearly linear. Therefore, we use the average change of the magnetic field amplitude with distance to quantify the DC field. The average change computed for one run was compared to the average change computed for each of the other runs of the same day, to asses run to run repeatability. Figs. 4,5,6 demonstrate run to run variation can be controlled by run averaging with our present positioning and movement control system. The DC field change per cm distance change for three successive days is plotted.(Fig. 7). The DC field amplitudes of the first and third day are nearly identical. Factors contributing to the different DC amplitudes recorded on the second day include alignment inaccuracies of the sensor array relative to the subject and the room plus physiological change.

4) The time averaged amplitude of the fast activity (AC) components of the spontaneous MEG data is also attenuated as the distance from the sensor array to cortex is increased. The amplitude of this activity is also expected to be altered by pathology. In addition, the amplitude of the fast activity and the DC amplitudes may be found to be correlated. Therefore, we have developed a mathematical model of the time averaged AC amplitude which assumes uniform cortical activity. Assuming uniform cortical activity change in amplitude with distance is found to be :

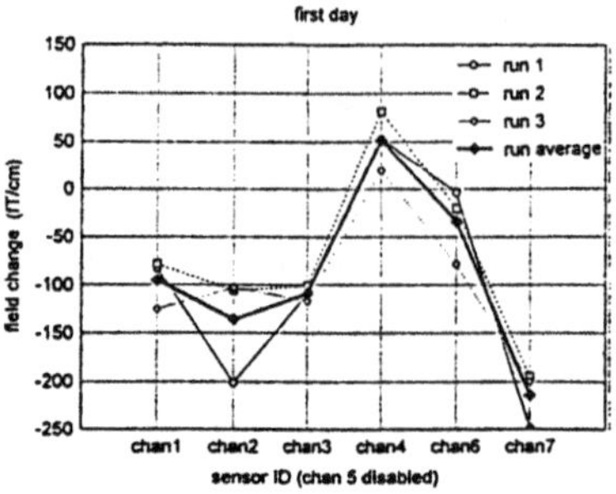

Fig. 4 First day's DC field amplitude measurements for each run and the average of these runs.

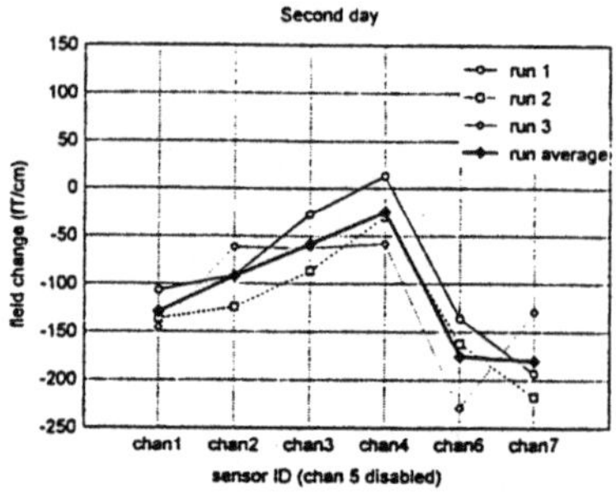

Fig. 5 Second day's DC field amplitude measurements for each run and the average of these runs.

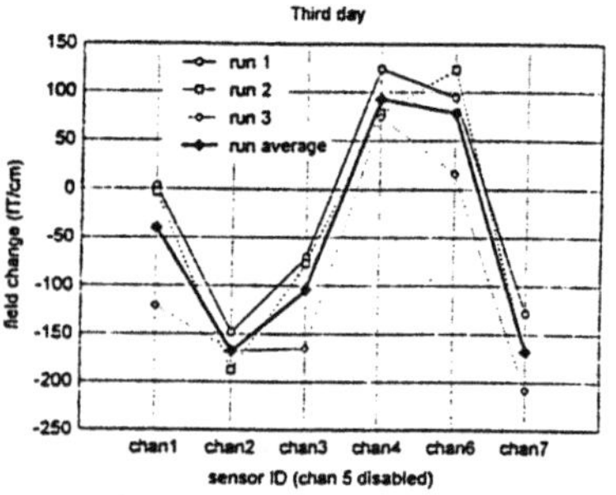

Fig. 6 Third day's DC field amplitude measurements for each run and the average of these runs.

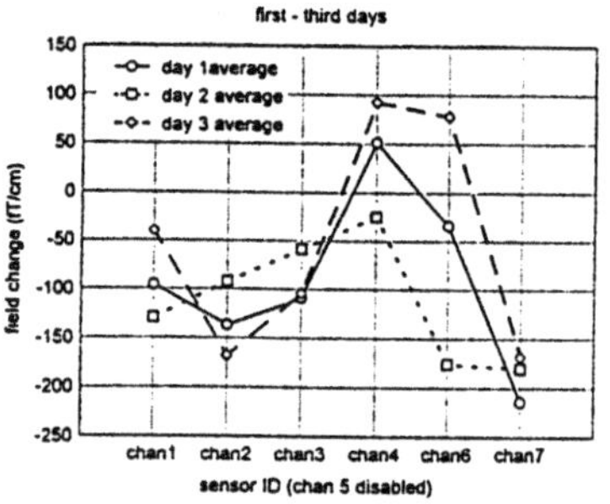

Fig. 7 The DC field change per cm distance change for three consecutive days is shown.

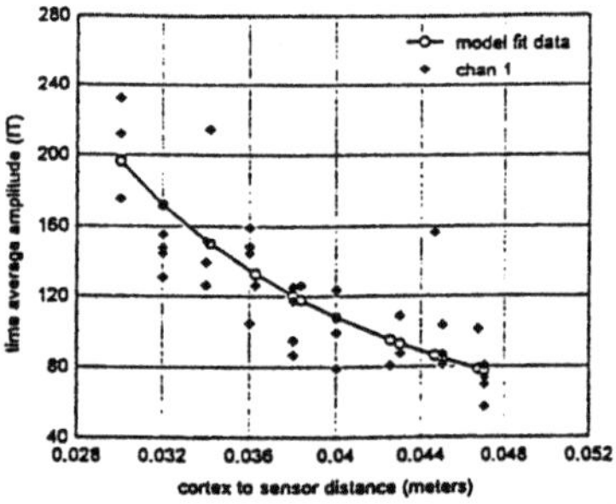

Fig. 8 The time averaged amplitude of the fast activity components of MEG data is attenuated as sensor - cortex distance increases. The solid line fits the equation used in the result and ρ was was found to be 2.1.

time averaged amplitude = constant / (distance to cortex)$^{\rho}$

For the data shown in Fig. 8 the solid line fits the above equation with ρ equal to 2.1. For the complete sensor array in this run, ρ was found to be 1.89 ± 0.15.

DISCUSSION

The technique developed in this study greatly reduces the time required to complete an entire study. Each run is completed in less than two minutes, hence subject fatigue is minimal. Also, since the total time that the subject is monitored is less than ten minutes, the subjects state of arousal remains nearly the same throughout the study. This eliminates the error introduced in the interpretation of the results due to comparison of data collected during two different states of arousal.

A range for variation in subjects with no pathology has to be established. Once this is established, further studies with subjects who have had strokes, closed head injury etc have to be performed. Data from these subjects could then be compared to data from the controls. Also the day to day variation in subjects with pathology could indicate changes in the physiology or function of the brain. It could also indicate change in function after surgery, indicating perhaps healing of the injured tissue.

The problems mentioned in the introduction have to be addressed in order to get any meaningful results with the technique we have developed. Reducing alignment problems of the sensor array is presently our greatest priority. The position of the subject with respect to the probe has to be kept the same in order to be able to compare any two sets of data and to be able to establish that the changes that we might observe have a physiological basis.

REFERENCES:

[1] Barkley, G.L., Moran, J.E., Yoshiaki Takanashi, and Norman Tepley. Techniques in DC-magnetoencephalography, Journal of Clinical Neurophysiology, 1991, 8(2):189-199.

[2] Barkley, G.L., Tepley, N., Nagel-Leiby, S., Moran, J.E., Simkins, R.T., and Welch, K.M.A. Magnetoencephalographic Studies of Migraine, Headache, 1990, 30(7): 428-434.

[3]Takanashi, Y., Chopp, M., Levine, S.R., Kim, J., Moran, J.E., Tepley, N., Chen, Q., Barkley, G.L., and Welch, K.M.A. Magnetic fields associated with anoxic depolarization in anesthetized rats, Brain Research, 1991, 562: 13-16.

This research supported by NIH/NINDs, grant no. 1R01 NS30914

Development of a 61-Channel MEG System Cooled by a GM/JT Cryocooler

Sata, K., Fujimoto, S., Fukui, N., Haraguchi, E., Kido, T., Nishiguchi, N. and Kang, Y.M.

MEC Laboratory, DAIKIN INDUSTRIES, LTD., 3 Miyukigaoka, Tsukuba 305, Japan

Introduction

Most SQUID-based systems used for biomagnetic measurements require liquid helium to achieve a low temperature environment. However, cryogenic temperatures can now be easily attained by using commercially available cryocoolers, thus eliminating the need for handling liquid helium. Sullivan et al. [1] and Kang et al. [2] successfully demonstrated that a SQUID system can be cooled using a cryocooler. We previously developed a single-channel system [3] and a 32-channel system [4] that are cooled by GM/JT (Gifford-McMahon / Joule-Thomson) cryocoolers and therefore do not need liquid helium. Here we designed a new 61-channel MEG system cooled by a GM/JT cryocooler. To demonstrate its performance, we measured the somatosensory-evoked magnetic field in a human subject using our 32-channel MCG system [4], which has a structure similar to that for our new MEG system.

Design of the MEG system

Fig.1 shows a block diagram of our MEG system. This diagram is the same as that for our single- and 32-channel systems. The valve motor unit for switching between high- and low-pressure helium gas, which is the circulating refrigerant in the GM cryocooler's circuitry, is set outside the magnetically shielded room. Thin-film dc SQUIDs and pickup coils are directly cooled by a GM/JT cryocooler without liquid helium. Output of the SQUIDs are linearized by FLL (Flux Locked Loop) units and then inputted into a graphics computer through an A/D converter unit.

Fig.2 shows a cross-section of the cryostat for our system. Not shown in this figure are the wiring, thermal anchors, and Super Insulator (SI, a kind of multiple-layered thermal insulator).

This cryostat has the same basic structure as that of our single- and 32-channel systems, but differs in bobbin arrangement of the pickup coils, in the base length of the gradiometers, and in the ability of the coils to face an optimal direction. For the pickup coils, we used first-order axial gradiometers, each with a diameter of 20 mm and a base length of 50 mm. We arranged 61 channels of these gradiometers on a spherical surface (125-mm radius) within an angle of 96 deg. Fig.3 shows the bobbin arrangement of the pickup coils. The vacuum chamber of the cryostat is made of GFRP (Glass-Fiber Reinforced Plastic) similar to other systems cooled by liquid helium.

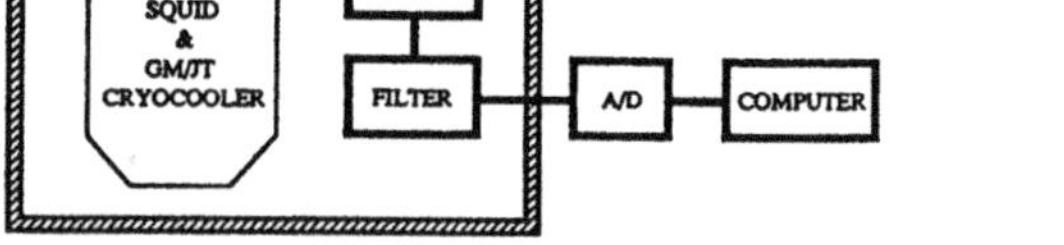

Fig.1 Block diagram of the 61-channel MEG system

The most significant difference in the construction of our systems compared with that for liquid-helium-cooled systems is that our systems require thermal conduction pathways from the corresponding cooling stage to the parts that must be cooled to predetermined temperatures, i.e., the thermal radiation shields at the first cooling stage of the GM cryocooler and the bobbins of the pickup coils at the final cooling stage. Therefore, the bobbins and the shields are made of hybrid material embedded with resin and copper wire. Although using such material for the bobbins induces a slight increase in the white noise level, this white noise can be controlled to a level low enough (about 10 $fT/Hz^{1/2}$) that the magnetic field from human brain can be accurately measured using an optimal quantity of copper wire in the bobbins. The bobbins of our MEG system are flexible so that they can face the pickup coil at any direction; thus allowing them to be arranged on a spherical surface that can fit a human head.

In general, heat inflow to the cryostat from outside is caused by thermal radiation and thermal conduction. In our system, SI is used to minimize the heat inflow from thermal radiation. Thermal conduction affects the cooling stages via two components of parts of the cryostat that are connected to areas at room temperature. One component is the eight rods supporting the final cooling stage and the thermal radiation shields. The other is the wiring between the connectors buried in the vacuum chamber and the SQUIDs on the final cooling stage. For reducing the noise caused by vibration of the GM cryocooler, the eight rods are attached around the base flange of the cryostat because it is the most stable position. The quantity of heat transferred through the rods can be ignored because the rods are made of GFRP, which has very low thermal conductivity. In our single-channel system, we use copper wire that has a diameter of 0.1 mm to 0.2 mm. If the same wire is used for our multi-channel systems (i.e., 32 and 61 channels), the heat quantity exceeds the cooling capacity of the cryocooler because of the wire's high thermal conductivity. To minimize such heat conduction, we used commercially available FPC (Flexible Printed Circuits) which has a minimum cross-section of $3.6X10^{-3}$ mm^2 per a line.

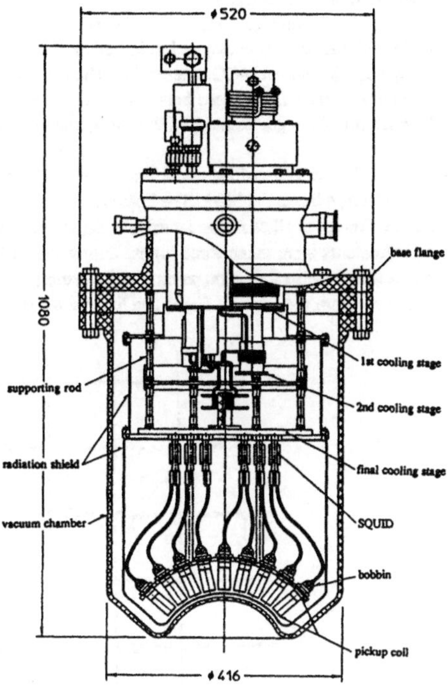

Fig.2 Cross-section of the cryostat

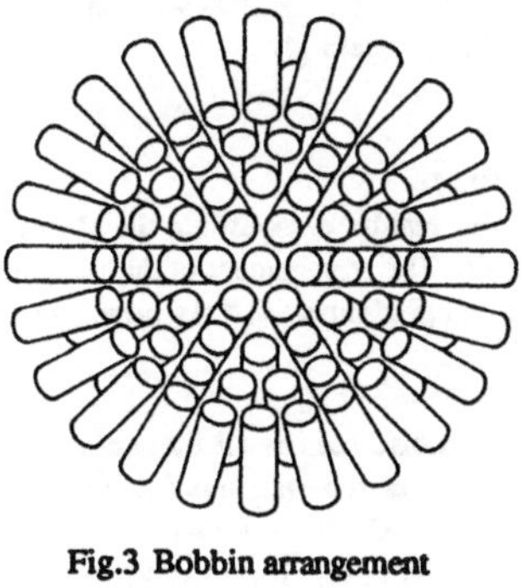

Fig.3 Bobbin arrangement

The expander moves in the cylinder of the GM cryocooler. A conventional expander thus generates significant noise because it contains ferro-magnetic and high-electric-conductivity material. To avoid such noise, we made the expander for the SQUID systems from non-magnetic and low-conductivity material.

Preliminary results

To demonstrate that our 61-channel MEG system is capable of measuring the magnetic field from a human brain with the same accuracy as liquid-helium cooled systems, we measured the somatosensory-evoked magnetic field using our 32-channel MCG system [4]. The reason for using the 32-channel system as a reference, is the similarity in their structure. Differences between the two systems are the arrangement of the bobbins of the pickup coils, the base length of the gradiometers (37.5 mm), and the flexibility of the bobbins mentioned above.

Random electric stimulation at a mean frequency of 1.7 Hz and a magnitude of 7 mA was applied to the left wrist of a human subject. The linearized output of the SQUIDs were inputted into a graphics computer through a 500-Hz low-pass filter for anti-aliasing and an A/D converter. The sampling frequency was 1000 Hz. Three-minute measurements were taken three times, during which 32 wave forms were monitored simultaneously. The data were processed for noise reduction using a noise-template [4] and a computer. Each data point was an average of about 900 responses from the stimuli. Fig.4 and 5 show the time-sequential wave forms and iso-contour maps, respectively.

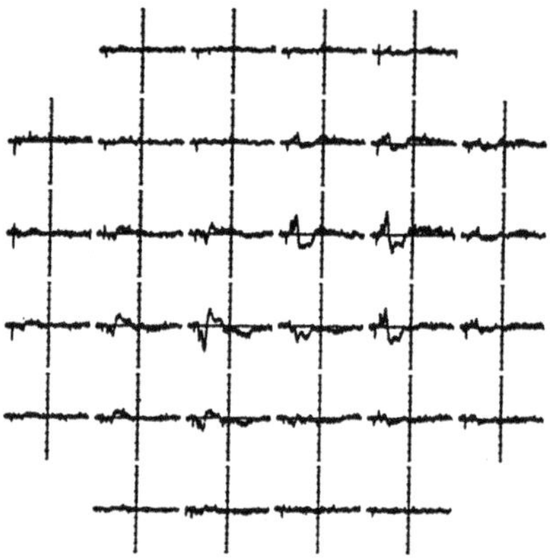

X axis : 50 fT/div
Y axis : 0 - 300 ms
Delay : 20 ms
Bandwidth : dc - 500 Hz
Average : ~ 900 times
Stimulation ...
mode : random
frequency : 1.7 Hz (mean)
magnitude : 7 mA
duration : 0.2 ms

Fig.4 Time-sequential waveforms of somatosensory-evoked response measured using a 32-channel MCG system [4].

Discussion

The results demonstrate the ability of the 61-channel MEG system to measure magnetic fields from a human brain. The arrangement of the pickup coils was designed to measure the MEG and the increment of the base length of the gradiometers. This arrangement improves the S/N ratio compared with the ratio for the 32-channel MCG system.

The noise seen in all three of our cryocooled systems consists of peaks at 2 Hz and its harmonics on the power spectrum of the system noise. This noise is caused by three factors. The first factor is that the bobbins swing in the magnetic field due to vibration of the GM cryocooler induced by the switching between high- and low-pressure refrigerant (helium gas). The second is the magnetization of materials in the displacer of the GM cryocooler. The last is an eddy current induced in the metallic material in the displacer. We achieved a 90 % reduction in such noise due to our design modifications of the cryostat. Further improvement in the components of the cryostat, especially in the expander and the bobbin of the pickup coil, will allow our 61-channel MEG system to control its own noise level at under 1 pT and to keep the white noise level at 10 $fT/Hz^{1/2}$. We believe that our MEG system will be a useful tool in biomagnetism research in the near future.

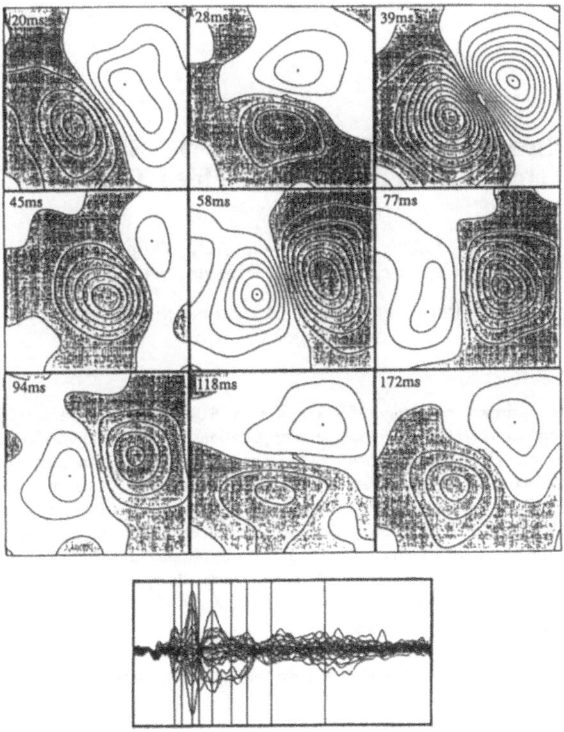

Fig.5 Iso-contour maps of somatosensory-evoked response measured using a 32-ch MCG system [4] at latency values of 20, 28, 39, 45, 58, 77, 94, 118 and 172 ms. The iso-field step was 16 fT. The data are from the data in Fig.4 processed using a 100-Hz low-pass digital filter. The lower graph shows all the data in one frame; the peaks (indicated by the vertical lines) represent the different latency values.

Reference

[1]Sullivan, D.B., Zimmerman, J.E., and Ives, J.T. Operation of a Practical SQUID Gradiometer in a Low-power Stirling Cryocooler, NBS Special Publication 607, 1981, 186-194

[2]Kang, Y.M., Ogura, T., Uosaki, S., Sakitani, K., Sochi, S., Fujimoto, S., Sata, K., Fukui, N. 4K GM/JT Cryocooler for Cryogenic Sensors, Proc. CRYOCOOLERS 6, 1990, Volume 1: 17-25

[3]Sata, K., Ueda, T., Fukui, N., Yoshida, T., Kang, Y.M., and Nishiguchi, K., Biomagnetic Measurements with a SQUID cooled by a GM/JT Cryocooler, Book of Abstracts, 8th International Conference on Biomagnetism, 1991, 131-132

[4]Fujimoto, S., Sata, K., Fukui, N., Haraguchi, E., Kido, T., Nishiguchi, K., and Kang, Y.M., to be published in Biomag 96.

Low-Noise, Single Layer $YBa_2Cu_3O_{7-\delta}$ DC SQUID Magnetometers for Shielded and Unshielded Operation

Schmidt, J.M.[1], Lee, L.P.[1], Matlashov, A.[1], Teepe, M.[1], Vinetskiy, V.[1], Cantor, R.[1] and DiIorio, M.S.[2]

[1]*Conductus, Inc., Sunnyvale, CA, USA;* [2]*Biomagnetic Technologies, Inc., San Diego, CA, USA*

Introduction

One of the primary advantages for the use of SQUID magnetometers fabricated from high temperature superconductors (HTS) in biomagnetic applications lies in the potential for greatly simplified and cheaper cryogenics. However, the costly and inconvenient magnetic shielding presently required also presents an obstacle to the widespread use of HTS magnetometers. One of the common problems associated with unshielded operation of HTS sensors is the adverse effect of the Earth's magnetic field, nominally 50 μT, on the sensitivity of these devices [1]. Although the noise at high frequencies, usually white in character, is not usually affected by fields of this magnitude, an increase in the low frequency noise is typically observed. This is generally believed to result from a hopping motion of magnetic flux vortices trapped in the film of the device. As the ambient field is increased, the density of trapped vortices also tends to increase, resulting in greater sensor noise.

Using magnetometers of various designs, we have performed a variety of tests which focus on characterizing the performance of these devices in shielded and unshielded operation.

Methods

In this study, we focus on three magnetometer designs, denoted as types A, B, and C. Several magnetometers of each design were fabricated using the Conductus single-layer laser deposition process: Films of c-axis oriented $YBa_2Cu_3O_{7-\delta}$ (YBCO), 150 to 300 nm thick, were deposited on 10×10 mm^2 $SrTiO_3$ bicrystal substrates with a 24° misorientation angle. While each of the three magnetometer designs consists of a large-area pickup loop directly coupled to a low-inductance SQUID [2], the designs vary in geometry and in effective area, that is, field-to-flux conversion ratio. In all the tests described here, the magnetometers were operated in feedback mode using Conductus flux modulation electronics employing current-bias reversal to reduce low-frequency noise.

Results

To illustrate the effects of the Earth's magnetic field on HTS SQUID magnetometers, we shall first compare the performance of commercially available HTS magnetometers, designated type A, operated first in a shielded laboratory environment and then unshielded in an open, outdoor environment. Because of the large geomagnetic signals present in the latter environment, we examined the differential output of two magnetometers mounted in a parallel configuration separated by a distance of 28.5 mm. In this way, an electronic gradiometer was formed which canceled external uniform signals common to both sensors. The output from this synthetic sensor was dominated by the intrinsic noise of the two magnetometers, which added incoherently. This was verified in the case of shielded operation by comparison with the noise of the sensors measured individually. In each measurement, the sensors were normalized, that is, heated above their transition temperature and cooled in the ambient field. The results of the two measurements are shown in Figure 1, where we plot the spectral density of the gradiometer output. While the white noise above a few Hertz remained unaffected by the Earth's field, the low frequency noise increased roughly by a factor of ten in power at 1 Hz.

The desire to improve the performance of HTS sensors in ambient fields motivated a modification to the geometry of the pickup loop in a second sensor design, type B. This sensor had a field-to-flux conversion ratio of 42 nT/Φ_0, compared to 11 nT/Φ_0 for type A. In a laboratory test, this sensor was

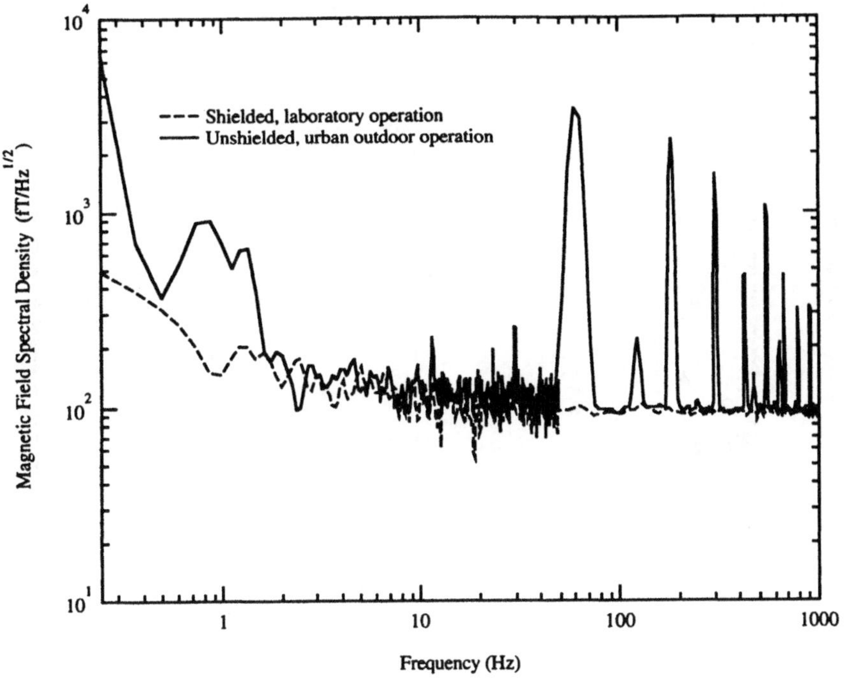

Figure 1. Spectral density of an electronic gradiometer formed by two HTS magnetometers of type A. Measurements performed in a magnetically shielded laboratory environment (dashed lines) and an unshielded outdoor environment (solid line) are shown.

magnetically shielded from external interference, and controlled static fields ranging from zero to 100 μT were applied. Measurements of the spectral density at 1 Hz were made to characterize the low frequency sensitivity as a function of the static background field. The results are shown in Figure 2. In one run, shown as open circles, the sensor was not normalized between measurements. In this case, the sensitivity dramatically decreased as the field increased. In a second run, shown as solid circles in Figure 2, the sensor was normalized at each value of applied field. As a result, greatly improved performance was realized. The noise level was not measurably affected up to fields of 15 μT, although at 50 μT the performance was about twice worse than in the nominally zero field.

A third sensor, denoted as type C, was of a design similar to type B, but with an improved field sensitivity of 11 nT/Φ_0. Similar tests to those performed on device B were performed on this device. Normalizing the sensor at each value of applied static field, we found that the sensitivity was not degraded over the full range of zero to 50 μT. Unfortunately, this particular sensor showed a flux noise of 12 to 14 $\mu\Phi_0$/ Hz$^{1/2}$, about twice the value we routinely obtain in our HTS sensors. This may be at a level which obscures effects of additional noise produced by the trapped flux which results from the sensor being cooled in a static ambient field.

We have used an HTS magnetometer of type C in conjunction with Conductus SQUID electronics to perform simple magnetocardiogram measurements. Representative data, obtained in real time using a single magnetometer, are shown in Figure 3. The R- and T-waves are highly pronounced in the unaveraged time trace. The measurements were made in a magnetically shielded room at Biomagnetic Technologies, Inc. (BTi), with the magnetometer mounted in a liquid nitrogen dewar specially designed (also by BTi) to

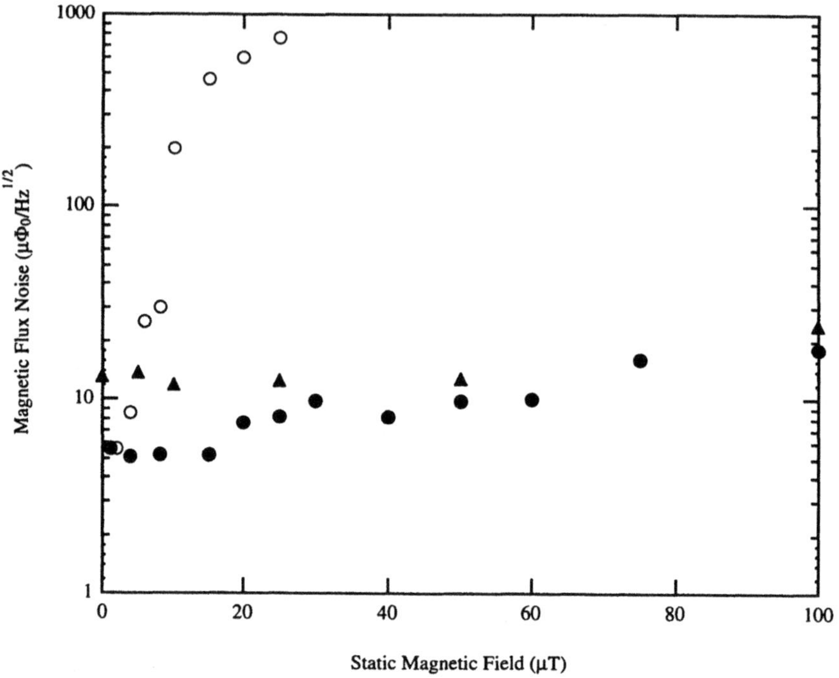

Figure 2. Flux noise of sensors B and C at 1 Hz versus static magnetic field. Shown are sensor B without normalization (open circle), sensor B with normalization (solid circle), and sensor C with normalization (solid triangle).

minimize the distance from the sensor to the subject [3]. The output signal from the readout electronics was connected to a standard data acquisition system supplied with commercial low-T_c systems [4]. No averaging or additional filtering beyond the 0.1 - 200 Hz bandpass was used to process the data shown in Figure 3. The primary noise source visible in the data is an environmental interference signal at 18.5 Hz.

Although these data were obtained in a carefully shielded environment, using this type of sensor to perform similar measurements in an unshielded environment one could clearly obtain useful results. We anticipate that the increased noise from operation in ambient fields would not adversely affect the results, however, interfering signals from other sources would necessitate suppression. This could be realized through the use of multiple or reference channels.

Discussion

We have found that while HTS magnetometers may be used in an unshielded environment, the presence of ambient magnetic fields, such as the Earth's field in particular, may degrade sensor performance at low frequencies. There are, however, steps that may be taken to improve upon the situation. First, if one normalizes the sensor in the ambient field, we have found that generally its performance improves. Secondly, although flux trapping may be an intrinsic shortcoming of present HTS film technology, design improvements may be made to the sensors which minimize the impact of ambient fields.

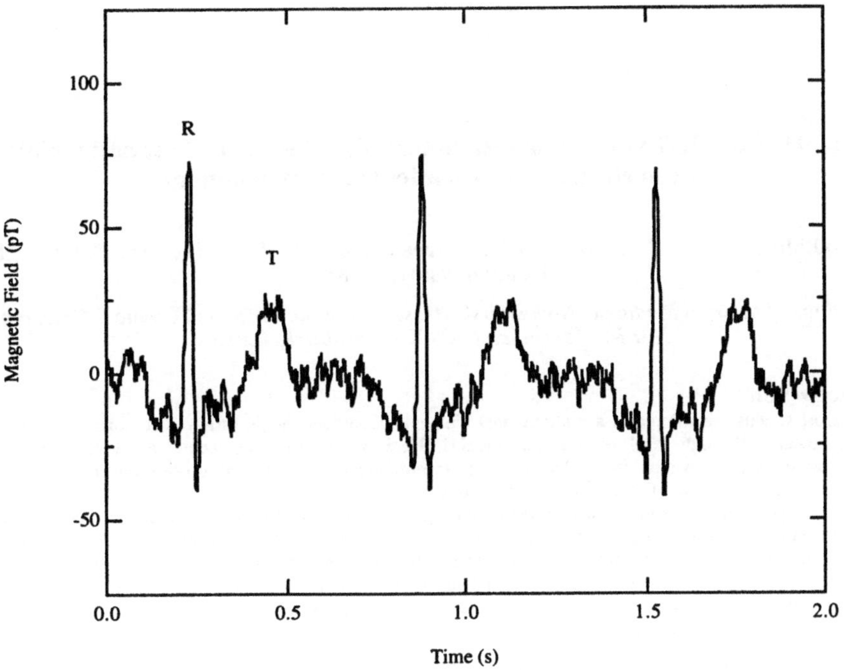

Figure 3. Magnetocardiogram measured using a HTS sensor of type C in a magnetically shielded environment. The R- and T-waves are particularly pronounced in these data, obtained in real time using a 0.1 - 200 Hz bandpass with no additional filtering. The primary noise source visible in the raw data shown here is an environmental interference signal at 18.5 Hz.

References:

[1] Miklich, A.H., Koelle, D., Shaw, T.J., Nemeth, D.T., Dantsker, E., Clarke, J., Alford, N.M., Button, T.W., and Colclough, M.S., Low-frequency excess noise in $YBa_2Cu_3O_{7-x}$ dc superconducting quantum interference devices cooled in static magnetic fields, Applied Physics Letters, 1994, 64:3494-3496.

[2] Lee, L.P., Longo, J., Vinetskiy, V., and Cantor, R., Low-noise $YBa_2Cu_3O_{7-\delta}$ direct-current superconducting quantum interference device magnetometer with direct signal injection, Applied Physics Letters, 1995, 66:1539-1541.

[3] DiIorio, M.S., Yang, K-Y., and Yoshizumi, S., Biomagnetic measurements using low-noise integrated SQUID magnetometers operating in liquid nitrogen, Applied Physics Letters, 1995, 67:1926-1928.

[4] Biomagnetic Technologies, San Diego, California 92121.

Acknowledgments

The work at Conductus, Inc. was partially supported by the US Navy's Office of Naval Research SBIR Project Office under contract N61331-94-C-0049, administered by the Coastal Systems Station.

The Use of Software Magnetometers for Ambient Magnetic Noise Cancellation in Unshielded Environment

Shabalin, A.V.[1], Golyshev, N.V.[2], Greenberg, Ya. S.[2], Motorin, S.V.[2] and Rogachevskij, B.M.[2]

[1]Institute of Internal Medicine, Novosibirsk, Russia; [2]Novosibirsk State Technical University, 20 K. Marx Ave., 630092 Novosibirsk, Russia

Introduction

The most serious requirements a multichannel biomagnetic system should satisfy if it works in unshielded environment with high level of ambient noise[1]. Here we describe the technique of electronic noise suppression we used in our three channel magnetocardiograph [2] which have been developed for use in unshielded environments of typical urban conditions.

The technique of noise suppression is based on the three additional reference magnetometer channels. Every signal channel has its own electronic compensation unit which is connected with the output of three reference magnetometer channels. The reference magnetometers feel homogeneous magnetic fields in three, in general not necessarily orthogonal, directions. The signals from output of the every reference channel are being electronically weighed, summed up and subtracted from the output of signal channel as shown below on schematic layout of electronic compensation unit.

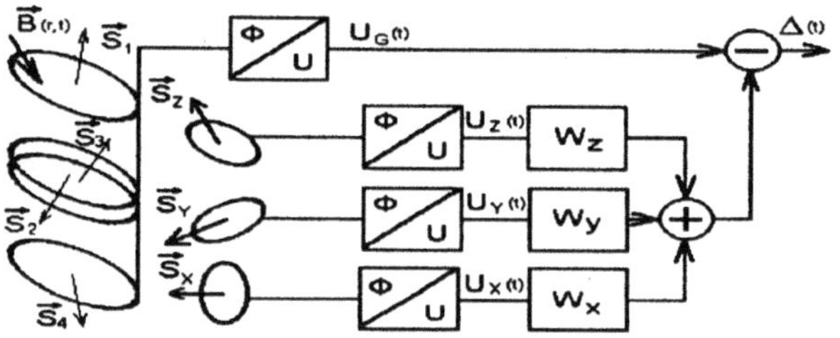

Therefore the output of a signal channel is the difference between gradiometer signal and the weighed sum of signals in reference magnetometer channels. Thus, our electronic system rejects only the " field imbalance" part of second order gradiometer. We describe here the algorithm for determination of the reference signal weights. In addition, we consider the noise contribution of reference channels and the dependence of a field imbalance on the frequency properties of reference and signal channels.

The noise suppression algorithm.

In our devices the sensing flux transformer loops in signal channels were the second order gradiometers. Due to technological imperfections the gradiometer pick up loops response to homogeneous magnetic field. The ratio of this response to that of a magnetometer with identical parameters is called a gradiometer technological imbalance factor δ_t. In practice the value of δ_t is of the order of 10^{-3} which is unsufficient for biomagnetic applications.

In homogeneous magnetic field the voltages U_i and U_G at the outputs of reference and signal channel could be written as follows:

$$U_i = k_i(\vec{B}\vec{S}) \quad i = 1,2,3 \quad ; \qquad U_G = k_G(\vec{B}\Delta\vec{S}_G) \qquad (1)$$

where k_i, k_G are the transfer coefficients which convert magnetic flux coupled with flux transformer sensing loops to the output voltages in reference and gradiometer channels; $\vec{S}$- directional vectors of sensing loops of reference magnetometer. These vectors are normal to the loop planes with the absolute value being equal to the plane area; $\Delta\vec{S}_G$ - the field imbalance vector of gradiometer flux transformer.
The signal at the output of compensation unit can be written as follows:

$$\Delta = W_1U_1 + W_2U_2 + W_3U_3 - U_G \tag{2}$$

where W_1, W_2, W_3 -are the weights of programmable voltage devider (PVD),which should be adjusted in such a way as to obtain the minimum signal amplitude Δ in homogeneous field.
The method of determination of W_i is based upon the fact that there always exist such values of W_i which cause the difference signal Δ at the output of compensation unit to be zero independently of the particular value and direction of the applied homogeneous magnetic field. In order to prove the statement one should put in (2) Δ=0 and use for U_i and U_G their expressions (1). We obtain the following equation which holds at any direction of magnetic field B:

$$\left[W_1k_1\bar{S}_1 + W_2k_2\bar{S}_2 + W_3k_3\bar{S}_3\right] = k_G\Delta\bar{S}_G \tag{3}$$

From the mathematical point of view the equation (3) is the decomposition of the vector ΔS_G on the set of three vectors S_1, S_2, S_3. As is known this decomposition in 3D space is unique unless the vectors S_1, S_2, S_3 lie in one plane. It follows from (3) that the weights W_i depend only on the loop directional vectors S_i, transfer coefficients k_i, k_G, but not on the magnetic field B. Therefore the compensation error does not depend on the degree of othogonality of the loops of reference magnetometers . In particular, for ideal orthogonal loops the solution of equation (3) is as follows:

$$W_i = \frac{k_{Gi}}{k_i}\frac{\Delta S_G}{S_i} \equiv \frac{k_{Gi}}{k_i}\delta_{t,i} \qquad i=x,y,z \tag{4}$$

where $\delta_{t,i}$ is the technological imbalance of *i*-th gradiometer.
Thus, we may say that the noise cancellation method we suggest makes it possible to completely eliminate interference from homogeneous field with magnetometer loops being arbitrary oriented in space. The only requrement is that three magnetometer loops should lie in the different not necessarily orthogonal planes.
For practical determination of the weights W_i it is necessary to measure the outputs of reference channels U_1, U_2, U_3 and the output of gradiometer channel U_G for three different not necessarily orthogonal directions of homogeneous magnetic field B, which we apply at the momets t_1, t_2, t_3. Therefore the weights W_i should satisfy the following set of three equations:

$$W_1U_1(t_i) + W_2U_2(t_i) + W_3U_3(t_i) = U_G(t_i) \quad (i=1,2,3) \tag{5}$$

In order to overcome the measurement errors in the output values U_1, U_2, U_3 and U_G we used the iteration scheme which is similar to that known from literature [3].
The main advantages of our algorithm are that it does not require the *a priory* knowledge of the direction of the test magnetic field or its value and it is robust against measurement errors.
The residual imbalance is determined mainly by three factors: the restricted number of bits of D/A convertors in compensation units; intrinsic noise of reference channels; frequency dependence of electronic circuit parameters. The first factor is not of principal importance, therefore below we consider in detail other ones.

The noise contribution of reference channels

Due to its statistical independence the intrinsic flux noise of reference magnetometer channels contribute coherently to the output noise in signal gradiometer channel.
The spectral densities of noise voltages at the outputs of magnetometer and gradiometer channel can be written down as follows:

$$S_{U,i} = W_i^2 k_{\Phi,i}^2 S_{\Phi,i}\,; \quad S_{U,G} = k_{\Phi,G}^2 S_{\Phi,G} \tag{6}$$

where $k_{\Phi,i}$, $k_{\Phi,G}$ are the transfer coefficients which convert the flux acting on the SQUID interferometer loop into the output voltage; $S_{\Phi,i}$ and $S_{\Phi,G}$ are the spectral densities of SQUID intrinsic flux noise in magnetometer and gradiometer channels respectively.

Obviously, the reference magnetometer channels increase the total output noise. It is convenient to express the noise increase in terms of a factor ν:

$$\nu = \left(1 + \frac{1}{S_{U,G}} \sum_{i=1}^{i=3} S_{U,i}\right)^{1/2} \tag{7}$$

With the help of (6) and (4) we can rewrite (7) as follows:

$$\nu = \left(1 + \sum_{i=1}^{i=3} \delta_i^2 \frac{k_{B,G}^2}{k_{B,i}^2} \frac{S_{\Phi,i}}{S_{\Phi,G}}\right)^{1/2} \tag{8}$$

where we have introduced the transfer coefficients $k_{B,i}$ and $k_{B,G}$ which convert the magnetic field applied to flux transformer loops into the magnetic flux applied to the interferometer loop. Obviously, the quantities k_i , k_G can be expressed in terms of $k_{\Phi,i}$, $k_{\Phi,G}$, $k_{B,i}$ and $k_{B,G}$ as follows:

$$k_i = k_{B,i}\, k_{\Phi,i}\,; \quad k_G = k_{B,G}\, k_{\Phi,G} \tag{9}$$

Therefore we see from (8) that the noise contribution from reference channels directly depends on the technological imbalance of gradiometer.

In order to estimate the increase of the overall noise due to noise contribution from reference magnetometer channels we take the following values: $\delta_i \cong 10^{-2}$-10^{-3}, the intrinsic flux noise of reference chanels (RF SQUIDs) $S^{1/2}_{\Phi,i} \cong 5\ 10^{-5}\Phi_0/\mathrm{Hz}^{1/2}$, the intrinsic flux noise of signal channel (DC SQUID) $S^{1/2}_{\Phi,G} \cong 1.5\ 10^{-6}\Phi_0/\mathrm{Hz}^{1/2}$, $k_{B,G} \cong k_{B,i}$. For these values the result for ν is: $\nu \cong 1.15$-1.02. Therefore, the increase of the noise can be considered as acceptable. However, in this case the dynamic range of the reference magnetometers should be no less than 130 dB, since the level of ambient noise is about 10^{-7} T.

It should be noted that the decrease of the reference magnetometers dynamic range at the expense of sensitivity (decrease of $k_{B,i}$ at the expense of decreasing of pick up loop areas S_i and the coupling of flux transformer to interferometer loop) is restricted by the acceptable rise of noise level. For example, if $k_{B,G} = 10\ k_{B,i}$ the dynamic range decreases to 110 dB, but the noise level is increased: $\nu \cong 5.85$-1.15.

The influence of frequency on the electronic imbalance.

As is known, in SQUID applications every channel has from the electronic point of view a feedback loop structure. It means that the coefficients $k_{\Phi,i}$, $k_{\Phi,G}$ depend on signal frequency:

$$k_{\Phi,i}(\omega) = \frac{\omega_{c,i}}{K_{fl}\sqrt{\omega^2 + \omega_{c,i}^2}} \qquad k_{\Phi,G}(\omega) = \frac{\omega_{G,i}}{K_{fl}^G\sqrt{\omega^2 + \omega_{G,i}^2}} \quad , \tag{10}$$

where ω_C , ω_G are the cutoff frequencies of the intergrators in magnetometer and gradiometer channel respectively; K_{fl}, K^G_{fl} are the feedback gain coefficients in magnetometer and gradiometer channels, respectively.

In what follows we show how the defference in cutoff frequencies ω_c, and ω_G influences the imbalance level. First we write U_i , U_G as follows:

$$U_i = k_{B,i}K_{\Phi,i}(\omega)(\vec{B}\vec{S}), \qquad i = 1,2,3 \tag{11a}$$

$$U_G = k_{B,G}k_{\Phi,G}(\omega)(\vec{B}\Delta\vec{S}_G) \tag{11b}$$

Assuming $\omega \ll \omega_{C,i}$, ω_G and using the exact solution for W_i (4) the imbalance factor could be written as $\delta(\omega)=\delta(0)+\delta_1(\omega)$ where the frequency dependent part is:

$$\delta_1(\omega) = \sum_{i=1}^{3}\delta_t\left(\frac{\omega}{\omega_{c,i}}\right)^2\left(1-\frac{\omega_{c,i}}{\omega_G}\right) \tag{12}$$

For the estimation of the frequency influence on imbalance due to the difference in cutoff frequencies we take $\delta(0)=10^{-5}$, $\omega_{C,i}/\omega_G=0.7\text{-}0.13$, and technological imbalance of gradiometer $\delta_{t,G}=10^{-2}$ .Then, at the frequency $\omega=0.1\omega_G$ the level of frequency imbalance is $\delta_1(\omega)=9\ 10^{-5}=9\delta(0)$, i. e., the total imbalance is being increased about an order of magnitude.

Experimental results

The effectiveness of our noise cancellation system has been investigated in unshielded laboratory of our University which is situated in a close vicinity of a main street traffic.

A test field has been generated by a pair of Helmholtz coils the field inhomogeneity being about 10^{-5} near the center of the coils.

The technological imbalance components of gradiometer were $\delta_{t,Z} \approx 3.5\ 10^{-3}$; $\delta_{t,X} \approx \delta_{t,Y} \approx 8\ 10^{-3}$. These values were frequency independent in the gradiometer bandwidth 5kHz. The electronic balancing has been performed by the method described above. At the first step (without iterations) we obtained the electronic imbalance about $3 10^{-2}$ for all three components.Then several (3-4) iterations were necessary to improve the imbalance by an order of magnitude. Using our algorithm we obtained for 10 Hz signal the following values of electronic imbalance along coordinate axes: $\delta_{e,Z} \approx 2\ 10^{-3}$; $\delta_{e,X} \approx \delta_{e,Y} \approx 4\ 10^{-3}$.Therefore the overall imbalance at 10 Hz is $\delta_Z \approx 7\ 10^{-6}$; $\delta_X \approx \delta_Y \approx 1.2\ 10^{-5}$.

Since the electronic imbalance is frequency dependent we performed measurements with signal frequencies up to 1 kHz. For frequencies less than 100 Hz the overall imbalance is: $\delta_Z < 3\ 10^{-5}$; $\delta_X \approx \delta_Y < 5\ 10^{-4}$. These values are more than adequate for biomagnetic measurements.

When traffic was not so intensive (as usual, at nights) we used for the test field a magnetic field generated by line frequency 50 Hz whose inhomogeneity is about 10^{-3}- 10^{-4}. Using this line frequency field instead of Helmholtz coils we obtained the imbalance level 10^{-4} which is sufficient for MCG recordings in a quiet noise environment.

In conclusion, we have describe here the method for effective electronic noise cancellation which is designed for application in multichannel biomagnetic systems working in noisy environment.

References

[1] Robinson, St. E. Environmental noise cancellation for biomagnetic measurements.- Proc. 7th Int.Conf. on Biomagnetism, August 13-18,1989, New York, Advances in Biomagnetism (eds. S.J.Williamson, M.Hoke, G. Stroink and M.Kotani) Plenum Press,1989, p.721-724.

[2] Golyshev, N. V.,. Rogatchevsky, B. M., and Motorin, S. V. Superconducting magnetocardiograph.-Proc. Europ. Conf. Appl. Supercond. (EUCAS'93) Gottingen, 1993, p. 1413-1414.

[3] Forsythe, G. E., Moler, C. B. Computer solution of linear algebraic systems, Englewood Cliffs., N. J. Prentice Hall, 1967.

The Use of a Current Feedback Loop for Active Compensation of Magnetic Noise in Superconducting Magnetocardiograph

Shabalin, A.V.[1], Golyshev, N.V.[2], Motorin, S.V.[2] and Rogatchevskij, B.M.[2]

[1]Institute of Internal Medicine, Siberian Branch of Russian Medical Academy, Russia;
[2]Novosibirsk State Technical University, Novosibirsk, Russia

1. Introduction

It is known that the active compensation of magnetic noise is widely used in superconducting magnetometry.

The method is based on a reference vector magnetometer which measures the noise field. The signal from the magnetometer output is being fed to the Helmholtz coils generating a compensation field. It results in the decrease of the current, generated by ambient magnetic noise in a flux transformer loop thus avoiding the overload of the signal channels at their inputs.

However the requirements for the homogeneity of compensation field is very high thus restricting the measuring volume. In addition, the large inductance of Helmholtz coils restricts the frequency range with acceptable level (of the order of 10^5) of the noise supression.

It should be noted that this problem can be solved by compensation at the outputs of signal channels. However, this approach has some drawbacks: it does not solve the problem of channels interference, and, in addition, the uncompensated field leads to the harmonic distortion of the signal under investigation [1].
Here we describe another technique of active compensation which is based on the use of the current feedback [2].

2. Analisis

The structure of superconducting magnetometer is shown on the following figure.

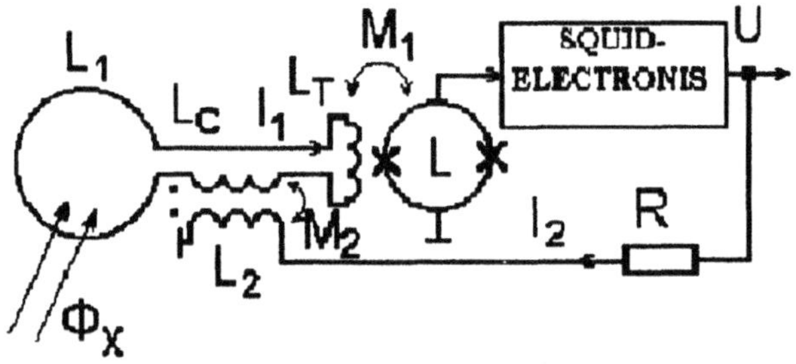

The flux transformer keeps the magnetic flux constant, which can be expresssed as:

$$\begin{cases} L_p \cdot I_1 - \Phi_x - M_1 \cdot I_s \mp M_2 \cdot I_2 = 0, \\ \Phi_e = M_1 \cdot I_1, \\ I_s = \Phi_e / L_s. \end{cases} \quad (1)$$

where $L_P=L_1+L_T+L_C$ is the total inductance of the flux transformer loop, L_C is the inductance of the flux transformer leads and coupling coil, I_S is the current in the interferometer loop, Φ_e -magnetic flux acting on the interferometer, L_S is the inductance of the interferometer loop.

The Laplace transform of (1) results in $\Phi_e=W_1\cdot\Phi_X(s)-W_2\cdot I_2$, where s -is the complex variable, $W_1=M_1/L_P\cdot(1-\alpha^2)$, $W_2=M_1\cdot M_2/L_P\cdot(1-\alpha^2)$, $\alpha^2=\beta\cdot L_T\cdot k^2/L_P$, $k^2=M_1/L_T\cdot L$, $\beta=L/L_S$. With the notations $\Phi_X'(s)=\Phi_X(s)\cdot W_1$, $\Phi_F(s)=-W_2\cdot I_2(s)$, $W_F=W_2/R$, $W_3=2\cdot\pi\cdot U_M/\Phi_0$, $W(s)=1/s\cdot T$, $\Delta\Phi(s)= \Phi_X'(s)-\Phi_F(s)$ the transfer function of magnetometer with unlocked feedback loop is $W=\omega_0/s$, where $\omega_0=2\cdot\pi\cdot U_M\cdot W_2/\Phi_0\cdot R$ -cutoff frequency of the amplitude-frequency charachteristic (AFC).

The expressions for $U_{OUT}(S)$, currents $I_2(s)$ and $I_1(s)$ take the following form:

$$\begin{aligned} &U_{out}(s) = H(s)\cdot\Phi_x(s)\cdot W_1 / W_{oc}, \\ &I_1(s) = \frac{\Phi_x(s) - M_2\cdot I_2(s)}{L_p\cdot\left(1-\alpha^2\right)}, \\ &I_2(s) = H(s)\cdot W_1\cdot\Phi_x(s)\cdot R / W_2 . \end{aligned} \quad (2)$$

where $H(s)=\omega_0/(s+\omega_0)$.

For transfer functions of AFC and phase-frequency charachteristic (PFC) H_U and H_I respectively, we obtain:

$$\begin{aligned} &\left|H_u(s)\right| = \frac{1}{\sqrt{1+\Omega^2}}\cdot\frac{M_1}{L_p\cdot\left(1-\alpha^2\right)}, \\ &\left|H_I(s)\right| = \frac{1}{\sqrt{1+\Omega^2}}\cdot\frac{1}{L_p\cdot\left(1-\alpha^2\right)}, \\ &\Psi_I = \frac{\pi}{2} - \mathrm{arctg}\,\Omega \end{aligned} \quad (3)$$

where $\Omega=\omega/\omega_0$.

One can see from (3) that the current $I_1(s)$ in flux transformer loop tends to zero with phase shift going to $\Omega \rightarrow 0$..It is important that $I_1(s)$ does not depend on coupling constant *k*.

In multichannel devices due to restricted volume occupied by sensing loops there exists a channel interference which is proportional to their mutual inductve copupling Mij (i,j-are the channels numbers).Taking into account the channel interference we rewrite the first equation in (1) as follows:

$$L_p \cdot I_{1i} - \Phi_{xi} + M_1 \cdot I_{si} - M_2 \cdot I_{2i} + \sum_{k=1}^{N} M_{ik} \cdot I_{ik} = n \cdot \Phi_0. \tag{4}$$

Other equations in (1) aquire the subscript I=1,......N, N being the number of channels.

We define the error which comes from channel interference as $\gamma = | U^{(N)} / U^{(1)} - 1 |$, where $U^{(N)}$ is the voltage at the output of any channel with all channels working simultaneously, $U^{(1)}$ is the voltage at the output of a single channel when other N-1 channels are switched off, the external magnetic field being the same in both cases.In order to simplify the problem we suppose all channels charachteristics to be identical. In addition, we introduce the channel effective number N_{eff} which defines the number of strongly coupled neibouring channels. This number depends on particular geometry of multichannel probe and as a rule N_{eff}=3-4.

Therefore if the interference from distant channels is neglected we can solve the set of equations (1) to obtain the interference coefficient in case of using a feedback current loop:

$$\gamma_I = \frac{N_{eff} \cdot M}{L_p^{(N)}} \cdot \left[\frac{\Omega^2}{p^2 \cdot \left(1-\alpha^2\right)^2 + \Omega^2 \cdot \left(1-\alpha^2 \cdot p^2\right)^2} \right]^{1/2}, \tag{5}$$

where $L_P^{(N)} = L_P + N_{eff} M$, $p = L_P / L_P^{(N)}$.

One can see from (5) that the interference scales as α^2 .If the channels AFC cutoff frequencies are chosen so that <0.01 the mutual intreference can be lowered to fractions of a per cent.

In analogy with known compensation methods [3] of magnetic noise at the output of magnetocardiograph we consider a compensation in flux transformer loop of a signal channel with the aid of additional vector magnetometer. In this case the expression for the current $I_1(s)$ is as follows:

$$\sum_{q=1}^{3} I_{1,q}(s) = \frac{\sum_{q=1}^{3} \Phi_x(s) \cdot \left(1 - \frac{R}{R_q}\right)}{L_p \cdot (1-\alpha^2)} \cdot (1 - H(s)), \tag{6}$$

where q denotes the corresponding component of magnetic flux noise $\Phi_X(s)$ (X,Y,Z), of current $I_1(s)$, and the resistance R_q which controls the value of compensation current.

Therefore, if we fed the compensation current into the inductance coil L_2 we can set the current $I_{1,q}(s)$ to zero either by changing R_q or by changing the transfer coefficient of a corresponding channel of a vector magnetometer. The appropriate values of the transfer coefficients can be found by the method described in [3].

3. Conclusion

As is seen from (3) the expression for $H_u(s)$ in case of a current feedback is similar to that for magnetometer transfer function in case of a flux feedback. However, in case of a current feedback the compensation is carried out at the input of a signal channel. It allows us to use our magnetocardiograph in unshielded invironment in the dynamic range of $1\Phi_0$.

The channels interference is of the order of 10^{-3} in the frequerncy range up to 200Hz. In addition, we solved the problems of signal distortions caused by harmonics of the line frequency.

One of the modification of our magnetocardiograph is being used in unshielded urban conditions for more than three years.

References

[1]Golyshev, N.V., Greenberg, Ya.S., Motorin, S.V., Rogachevsky, B.M. Harmonic Distortions in DC SQUID, Europ. Conf. on Appl.Superconductivity,-1993.-Gottingen (Germany), p. 1319-1321.

[2] Classen, I.H. Coupling considerations for SQUID devices, J. Appl. Phys., 1975, v.**46**, 5, pp.2268-2275.

[3]Golyshev, N.V. Superconducting magnetometers, Russian Journal Optoelectronics, Instrumentation and Data Processing.-1, 1995, p.1-21.

A Skull-Based Multiple Dipole Phantom for EEG and MEG Studies

Spencer, M.E.[1], Leahy, R.M.[1] and Mosher, J.C.[2]

[1]*University of Southern California, Los Angeles, California, USA;* [2]*Los Alamos National Laboratory, Los Alamos, New Mexico, USA*

Abstract

A versatile phantom for use in evaluating forward and inverse methods for MEG and EEG has been designed and is currently being constructed. The phantom consists of three major components: (i) a 32-element current dipole array, (ii) a PC-controlled dipole driver with 32 isolated channels allowing independent control of each dipole, (iii) spherical and human-skull mounts in which the dipole array is placed. Materials were selected throughout the phantom to produce minimal field distortions and artifacts to enable acquisition of high quality EEG and MEG data. The dipoles are made from a rigid narrow (.84mm) stainless steel coax cable. The dipole drivers can be configured as either current or voltage sources, are independently programmable and fully isolated, and are capable of producing arbitrary bipolar waveforms up to a 200Hz bandwidth. The spherical mount is a single shell sphere filled with conductive gelatin. The human skull mount has three shells: "brain" (conducting gelatin), "skull" (the skull is impregnated with a low conductivity conducting gelatin), and "scalp" (a thin layer of rubber latex mixed with NaCl to achieve a conductivity matched to the brain). The conductivities will be adjusted to achieve approximately an 80:1:80 ratio. Data collected to date from the spherical phantom shows excellent agreement between measured surface potentials and that predicted from theory (27 of the 32 dipoles give better than 99.9% rms fit) and negligible leakage between dipoles. We are currently completing construction of the skull mount.

I. Introduction

There has recently been great interest in developing accurate forward and inverse methods for mapping between neural current sources and the scalp potentials and external magnetic fields that they produce. Evaluation of these methods requires that data are available for which the "ground truth" is known. While the other functional modalities (fMRI and PET) offer the potential for providing ground truth for clinical and volunteer studies, these methods are currently not sufficiently well understood or mature to provide reliable validation. Evaluation can be performed using computer simulation. However, the majority of published results that use computer simulations assume simplified models for the head, instrumentation and noise. Typical simulations use a spherical head with point measurements of the scalp potential or magnetic field and additive white Gaussian noise. To establish realistic limits on the performance of forward and inverse methods, more realistic models are required. While it is possible to develop more elaborate simulations, there are advantages to performing evaluation using data collected directly from a physical system. Such studies have been performed using dipolar sources implanted in epilepsy patients [1], however the procedures required to implant these sources (e.g. temporary removal of part of the skull, and related tissue displacement and swelling) may result in severe distortion of volume currents. While such studies are important, they are not ideal for evaluation of general forward and inverse methods. Interesting studies have been performed with dipoles implanted in a cadaver head [2] and a gelatin filled skull [3], but these studies have been restricted to at most a few dipoles. A multiple dipole phantom was used in [4] for evaluation of several MEG imaging methods. This is a dry phantom that is based on the theoretical description in [5] in which it is shown that the resulting fields are identical to those produced by a current dipole in a uniformly conducting medium. The major limitation of this phantom is that is unable to generate the volume currents associated with realistic head geometries nor is it suitable for EEG.

Motivated by the desire to produce realistic data corresponding to complex spatio-temporal current sources and to include the effects of realistic head geometries, we have designed and fabricated a multiple dipole phantom consisting of 32 independently programmable and isolated dipoles which can be inserted in either a spherical or skull mount and used to collect both EEG and MEG data. In the following section we describe the phantom. We then report on the current status of the project and briefly summarize our preliminary results.

This work was supported by the National Institute of Mental Health, Grant No. R01-MH53213.

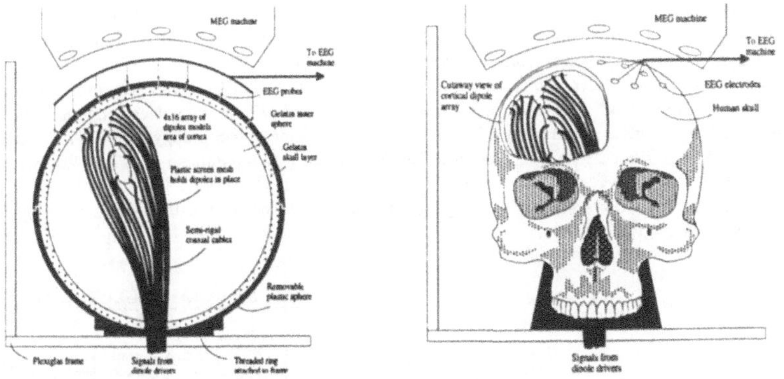

Fig. 1. Schematics for the spherical and skull phantoms.

II. Phantom Design

The phantom design consists of three components: (i) a 32-element current dipole array, (ii) a PC-controlled dipole driver with 32 isolated channels allowing independent control of each dipole, (iii) spherical and human-skull mounts in which the dipole array is placed. We describe each of these three components below.

Dipole Array

The dipoles were constructed from semi-rigid coaxial cable (UT-034-SS-SS, Micro-Coax, UTI, Collegeville, PA) consisting of a 0.84mm outside diameter shield made from stainless steel alloy (#304) tubing, a Teflon PTFE (Polytetrafluoroethylene) insulator and an inner stainless steel conductor. The semi-rigid coax was insulated using standard 3/64 inch heat shrinkable tubing and then striped back to expose 1 mm of the shield and 1 mm of the inner conductor to the volume conducting medium, Fig. 2. Thirty two dipoles were constructed and arranged to approximately follow a pair of sulcal folds: 16 were distributed along, and arranged normally to, a nominal central sulcus. Another 14 dipoles were arranged along a nominal calcerine fissure. The remaining dipoles were positioned to represent a pair of frontal sources. The semi-rigid nature of the cable allows us to bend the cable to individually position the dipoles for the desired locations and orientations. Once the dipoles are positioned, we can move the dipole array from the spherical to the skull mount with no significant change in the relative dipole locations.

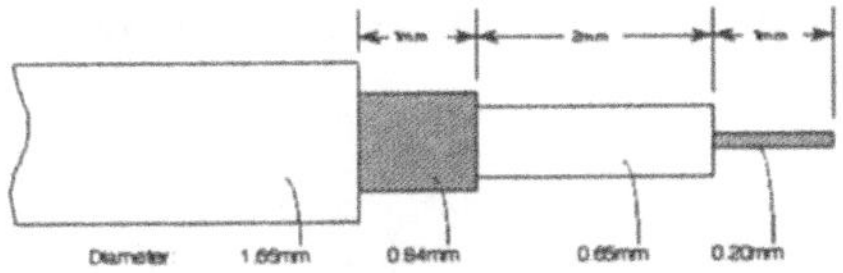

Fig. 2. Dimensions of the stainless steel coaxial dipole.

Driver Design

The dipoles are driven by isolated current sources which in turn are controlled through a special purpose interface to a personal computer. Arbitrary waveforms can be generated simultaneously at each of the 32 dipoles. These waveforms are generated using standard MATLAB code on the PC and directed to the dipole drivers. The disk file on our PC can contain over an hour of real time data with a sampling rate of 1000 samples per second per channel. Streamer software and the PDMA-16 Digital DMA Board (Keithley Metrabyte, Taunton, MA 02780) in the PC is used to clock the data at a constant rate in 16 bit words to the external dipole driver chassis.The electronics for the 32 isolated dipole drivers is housed in a 19 inch rack mount chassis and powered by the PC's 5V power supply. A six foot 37 conductor ribbon cable carries the 16 bit data words and clock signals from the PDMA-16 board in the PC to the dipole drivers, Fig. 3. A parallel interface and clock generator format the digital data for 32 electrically isolated Digital-to-Analog (D/A) converters. The sampling rate clock is divided by 1000 and is supplied to the EEG or MEG machine for synchronization when performing epoch-based acquisition.

The 32 channels are electrically isolated from ground and each other. Each channel has optical isolators for the serial data, clock and load signals and has a transformer isolated DC-to-DC power converter, Fig. 3. The digital data is converted to an analog signal using a 12bit D/A converter (Linear Technology, LTC1257). The signal from the D/A converter is then filtered with a second order bandpass filter with a passband of 1Hz to 200Hz and a transconductance amplifier is used to convert the voltage signal to a current in the dipole. The dipole is driven with a current source so that the resulting dipole current is independent of the impedance of the dipole. Alternatively, voltage output can be selected with a jumper change.

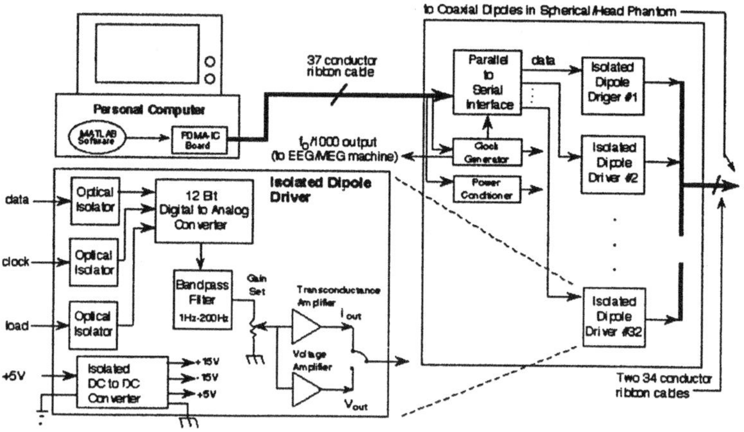

Fig. 3. The dipole driver chassis receives real-time digital signals from the PC and is formatted to drive 32 electrically isolated dipole driver circuits.

The Spherical and Skull Mounts

The single layer spherical phantom is constructed within a plastic sphere comprising two hemispheres that snap together to form a tight seal. The spherical single shell phantom will be used to validate the dipolar patterns of the coaxial sources and electrical isolation when running multiple dipoles.We can also compare results for identical source patterns for the spherical and skull mounts to study the effect of the non-spherical geometry on the accuracy of our inverse solutions. The plastic sphere (Mangelsen's, Omaha, NE) has an inside diameter of 136.0 mm +-0.2 mm and a wall thickness of 1.9 mm +-0.4 mm. The lower half is mounted to the base with a removable slip ring and the upper half is fitted with 34 silver silver-chloride electrodes (Neuromedical Supplies, Herndon, Va) including reference and a ground electrode. The electrodes are located in an irregular pattern over the sphere. The reference probe is positioned at the top dead center. Prior to data collection, the sphere was filled with a conductive gelatin made from water, sodium chloride (to control conductivity), Knox unflavored Gelatin (to solidify the solution), and dimethyl benzyl ammonium chloride (to prevent bacteria growth).

To model the irregular, eccentric morphology of the head, a human skull mount is currently under construction. The whole skull will be boiled in saline solution and gelatin to impregnate the bone to achieve an appropriate conductivity. The conductivity will be adjusted by changing concentrations of sodium chloride. Conductivity will be measured using samples of the prepared solution. Conductivity of the skull can be determined using calipers and resistance measurements at several places.The skull will then be filled with saline-gelatin mixture to represent the brain volume; the conductivity will be adjusted to provide an approximate 80:1 brain/skull ratio. To provide a scalp layer, the outside of the skull will be coated with rubber latex molding compound (GP 306, Gerisch Products, Inc.

20814 S. Normandy Ave. Torrance, CA90502). The latex will be mixed with sodium chloride to achieve the desired conductivity. X-ray CT images of the completed skull phantom will be collected to verify the dipole locations and to extract the skull and scalp surfaces.

The spherical and skull phantoms can be interchangeably attached to a plexi-glass base containing the dipole array. The base has a fill/drain plug for filling the brain volume of the phantom. A pressurized air volume in the neck pushes on a membrane that is in contact with the conducting solution to compensate for volume loss due to the small evaporation through the skull. Without volume compensation, evaporative losses through the skull and scalp layers can cause air gaps to form that will interrupt the volume currents in the skull. All materials in the base and phantoms are nonferrous. In Fig. 4 we show the various components of the phantom. We have collected 32 channel EEG data from the spherical phantom using a Neuroscan 32 channel EEG Synamps system. Of the 32 dipoles, 27 give better than a 99.9% rms fit between the measured and theoretical data based on a least squares single dipole fit to each coaxial dipole. The fit for the remaining 5 dipoles range from 86% to 94%. We believe these errors are primarily due to disruption of the volume currents by adjacent coaxial wires and can be reduced by repositioning the dipoles. We will then fill the skull mount and collect MEG and EEG data on the completed skull phantom.

Fig. 4. Photograph showing from left to right: skull mount on plexi-glass base with dipole array inside the skull; skull cap covered with latex scalp layer and EEG electrodes attached; spherical mount with Ag/AgCl electrodes attached; and the dipole driver chassis (DIPWIT32) below the head box for a 32 channel Synamps EEG system.

The authors thank Jeff Lewine of the New Mexico Institute of Neuroimaging in Albuquerque for his valuable advice on the design of the skull phantom.

III. References

[1] Cohen D, Cuffin BN, Yunokuchi K, Maniewski R, Purcell C. Cosgrove GR, Ives J, Kennedy JG, Schomer DL. 1990, "MEG versus EEG localization test using implanted sources in the human brain," *Ann Neurol,* 28:811-817.

[2] Barth D, Sutherling W., Broffman J., Beatty, J., "Magnetic localization of a dipolar current source implanted in a sphere and a human cranium," *Electroenceph Clin Neurophysiol.,* vol 63, pp 260-273, 1986.

[3] Lewine, JD, Edgar, JC, Repa, K, Paulson, K, Asture, RS, Orrison, WW Jr, "A Physical Phantom for Simulating the Impact of Pathology on Magnetic Source Imaging," Biomagnetism: Fundamental Research and Clinical Applications, eds C. Baumgartner et al. Elsevier,1995, pp. 368 - 372.

[4] Leahy, R.M., Mosher, J.C., and Phillips, J.W, "A comparative study of minimum norm inverse methods for MEG-imaging," (this volume)

[5] Ilmoniemi, R.J., Hamalainen, M.S., Knuutila, J, 1985, "The forward and inverse problems in the spherical model," in Biomagnetism: Applications and Theory, eds Weinberg, Stroink, and Katila, Pergamon: New York, pp.278-282.

High-T_c SQUID Gradiometers for Magnetocardiography in an Unshielded Environment

Tralshawala, N., Claycomb, J.R., Ying, Q., Miller, J.H., Jr., Nesteruk, K.[1], and Jackson, D.R.[2]

Texas Center for Superconductivity, University of Houston, Houston, Texas, USA;
[1]Department of Magnetic Materials, Institute of Physics, Polish Academy of Sciences, Warsaw, Poland;
[2]Department of Electrical and Computer Engineering, University of Houston, Houston, Texas, USA.

Introduction

We have designed a magnetocardiography (MCG) system that is capable of measuring magnetocardiograms in an unshielded environment. In order to carry out such a measurement, one has to contend with various ambient noise sources. These include power line and RF interference, microphonics pickup, fluctuations in the earth's magnetic field, and electrostatic pickup. Earlier solutions devised to overcome these problems have entailed the use of a second-order gradiometer inside a deep mine or inside a magnetically shielded enclosure [1]. Apart from the above problems which are common to both low-T_c and high-T_c SQUIDs, high-T_c SQUIDs exhibit additional 1/f noise [2]. This noise is the result of hopping of flux vortices that are trapped in the body of the field cooled high-T_c SQUID, and can only be eliminated by zero-field cooling the SQUID.

Methods

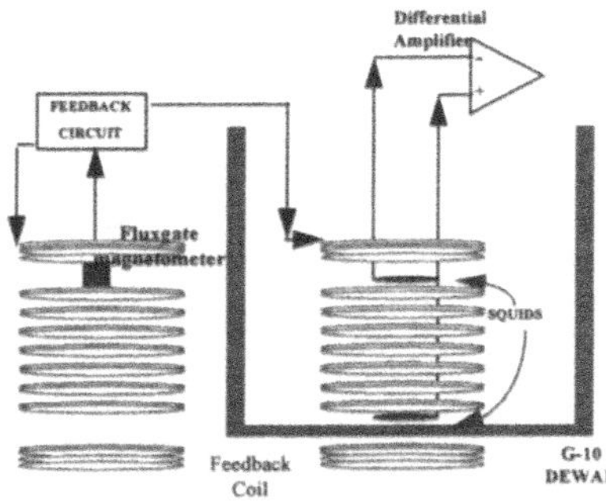

Fig. 1. Schematic of our MCG instrument. In this particular example, a room temperature fluxgate magnetometer is used as a reference sensor to primarily cancel earth's field and LF fluctuations.

We have implemented active noise compensation [3] techniques to eliminate some of the noise. Fig. 1 depicts a schematic of one of our systems. A fluxgate magnetometer is used to track earth's field and its low-frequency (LF) fluctuations as well as powerline noise. This also enables us to zero-field cool (to within 1% accuracy) our high-T_c SQUIDs [4] and passive shields. We have also incorporated small, localized high-T_c superconducting shields that selectively shield out uniform and magnetic dipole noise fields without affecting the current dipole source fields of interest [5]. The entire setup is supported by a simple non-metallic gantry to eliminate eddy current noise.

We have implemented a finite-impulse-response (FIR) comb filter to remove power line interference. This filter has stop bands at 0, ±60, ±120, ...etc., with stop-bandwidths of 1.4 Hz each. In addition, we have also implemented a nonlinear filter based on mathematical morphology operations to remove baseline wander. Namely, set-theoretical operations such as erosion and dilation are utilized (see [6] and references therein). Mathematical morphology has been extensively used in image processing, shape analysis and pattern recognition. Specifically, for our application, we can suppress baseline wander by choosing proper structuring elements. Fluctuations with widths less than that of the structuring element are suppressed. In an MCG signal, the P-QRS-T pulses are between about 75 msec to 250 msec in duration, so a simple flat structuring element is sufficient for our purposes.

These filters have been implemented on a dedicated DSP chip from Texas Instruments (TMS320C50). We have also implemented FIR filters and nonlinear median filters on a Windows NT workstation using the LabVIEW package. We have electronically compensated for the differences in the sensitivities between individual high-T_c SQUIDs. The individual SQUID outputs are then appropriately combined to make electronic gradiometers. Fig. 2 depicts system testing of our instrument, without active noise cancellation, for a model biomagnetic signal.

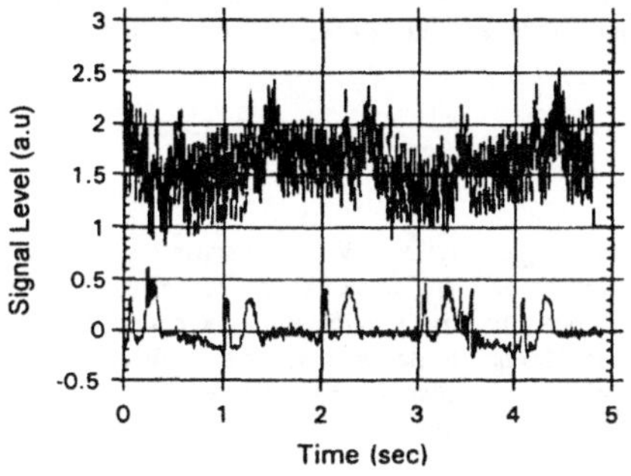

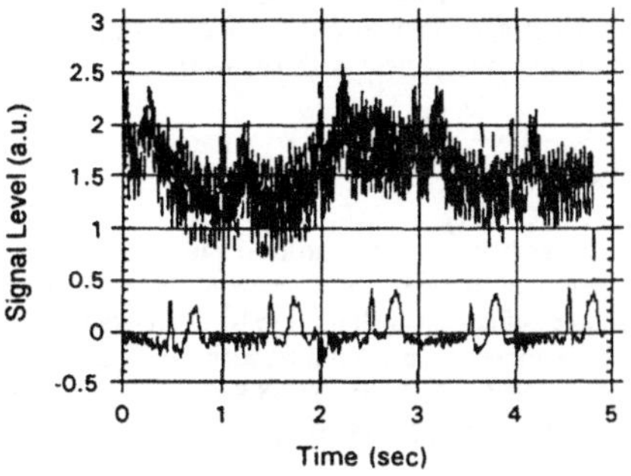

Fig. 2. System testing using a first-order planar gradiometer with a model biomagnetic signal. Graph at the left depicts outputs before (top) and after (bottom) the use of the FIR filter. The graph at the right depicts the outputs before (top) and after (bottom) the use of both the FIR and the nonlinear filter. The sampling rate is 300 Hz, and the time delay in the first case is about 0.8 seconds, while for the latter case is about 1.4 seconds.

Results

We have succeeded in obtaining an adult human MCG in an unshielded, clinical environment of a catheterization laboratory at Texas Children's Hospital. This data was obtained using a 3 cm baseline planar gradiometer and taking 100 ECG-gated averages.

Since then we have improved our instrument and have succeeded in obtaining real time, unshielded MCG data in our laboratory, which is next to a construction site and is located on the third floor. This data (Fig. 4) was obtained using a 9 cm baseline first-order axial gradiometer.

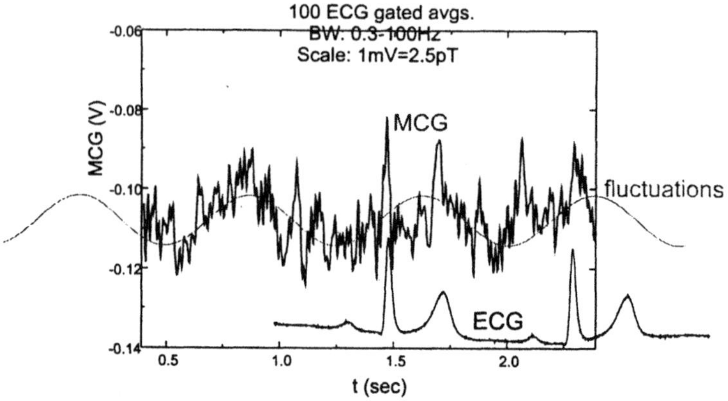

Fig. 3. ECG-gated MCG data obtained at Texas Children's Hospital. The graph also depicts LF fluctuations present. Due to variations in the heart rate over 100 averages, only the central QRS complex, which was at the chosen trigger point, is clearly visible, while one to the right is somewhat distorted (for comparison, the ECG record from the same patient is also shown).

Discussion

In summary, we have utilized a modular approach to designing our system. We have started out in a noisy environment and systematically removed the various contributions to ambient noise. At this stage, we have two major sources of noise which need to be eliminated. Namely, efforts are underway to decouple the entire setup from building vibrations (fundamental frequencies of about 1 Hz and 10 Hz) and to design a better RF shielding (one of the buildings next to ours has an 88.9 MHz FM transmitter). We are also planning to use three SQUID sensors, instead of two in use at present (Fig. 1), to further improve the noise cancellation.

References

[1] Zimmerman, J. E. SQUID instruments and shielding for low-level magnetic measurements, Jnl. of Appl. Phys., 1977, 48: 702-710.

[2] Miklich, A.H., Koelle, D., Shaw, T.J., Ludwig, F., Nemeth, D.T., Dantsker, E., Clarke, J., Alford, N.McN., Button, T.W., and Colclough, M.S. Low-frequency excess noise in YBCO dc superconducting devices cooled in static magnetic fields, Appl. Phys. Lett., 1994, 64: 3494-3496.

[3] Aarnink, W.A.M., van den Bosch, P.J., Roelofs, T.-M., Verbiesen, M., Holland, H.J., ter Brake, H.J.M., and Rogalla, H.. Active noise compensation for multichannel magnetocardiography in an unshielded environment, IEEE Trans. on Applied Superconductivity, 1995, 5: 2470-2473.

[4] Commercial high-T_c "iMAG" SQUID system made by Conductus, Inc., Sunnyvale, CA.

[5] Tralshawala, N., Miller, J. H., Jr., and Jackson, D. R. High-Tc superconducting image surface magnetometers, IEEE Trans. on Applied Superconductivity, 1995, 5: 2354-2357.

[6] Ying, Q., Tralshawala, N., Fan, N., Miller, J.H., Jr., and Jackson, D.R. Digital signal processing of magnetocardiography signals, Proceedings of the 1995 DSP^x Technical Program, 1995, pp. 520-527.

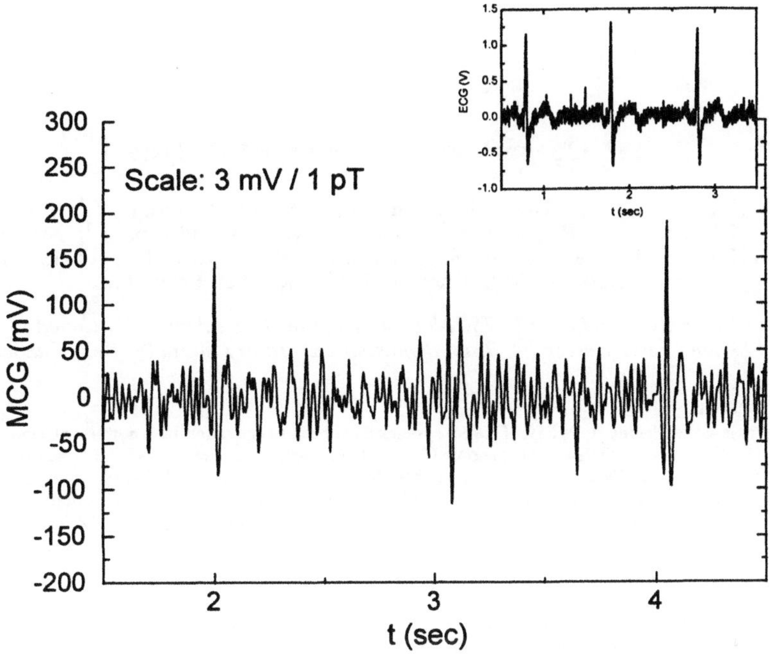

Fig. 4. Real time MCG obtained in our unshielded laboratory room. The QRS complex can be seen in the MCG. The simultaneously measured ECG signal is shown in the inset at the top right corner.

Acknowledgments

We gratefully acknowledge the assistance of Y. Bai, M. Boyd, A.K. Guzeldere, and L.M. Xie. This project was supported by the State of Texas through the Texas Center for Superconductivity and the Advanced Technology Program, and by the Robert A. Welch Foundation.

143 Channel Whole-Cortex MEG System

Vrba, J.[1], Angus, V.[1], Betts, K.[1], Burbank, M.B.[1], Cheung, T.[1], Fife, A.A.[1], Haid, G.[1], Kubik, P.R.[1], Lee, S.[1], Ludwig, W.[1], McCubbin, J.[1], McKay, J.[1], McKenzie, D.[1], Robinson, S. E.[1], Smith, M.[1], Spear, P.[1], Taylor, B.[1], Tillotson, M.[1], Cheyne, D.[1,2] and Weinberg, H.[2]

[1] *CTF Systems Inc., 15-1750 McLean Ave, Port Coquitlam, B.C., Canada;*
[2] *Simon Fraser University, Brain Behaviour Laboratory, Burnaby, B.C., Canada*

Introduction

Whole-cortex helmet-type SQUID MEG systems have been in operation in a number of laboratories since 1992, e.g. [1]. Recently, a whole-cortex integrated MEG/EEG system consisting of 143 MEG sensor channels, 40 EEG channels, 8 A/D and D/A channels and 8 logical channels was developed and operated in an unshielded environment and, more recently, in a shielded room. This paper presents the resulting noise performance and several examples of unshielded recordings of spontaneous and evoked brain activity with response to DC.

System description

The MEG/EEG system incorporates a SQUID sensing system, SQUID amplifiers and electronics, EEG amplifiers and electronics, a peripheral interface unit, a host computer, and a stimulus controller with associated stimulus presentation hardware. The SQUID sensing system consists of sensors (143) and references (29) enclosed in a helmet-shaped liquid Helium dewar. The SQUID sensors are 1st-order radial gradiometers with 2 cm coil diameters and 5 cm baselines, uniformly distributed on the helmet surface with mean spacing of 3.2 cm. The reference array contains magnetometers and 1st-order gradiometers and, together with the sensors, is used for noise cancellation by means of software formation of 2nd- and 3rd-order synthetic gradiometers (2nd-SG, 3rd-SG) [2]. The distance between the liquid Helium and the room temperature surface in the helmet region is 1.5 to 1.6 cm, and the boil-off rate is less than 11 liters/day. The SQUID sensors are operated by a digital feedback loop with resulting dynamic range of 32 bits and slew rates of up to 33 kΦ_o/sec. Data are further processed in a DSP processing bank and then collected by the host computer (HP 9000/700 Series or Apple Quadra). The Peripheral Interface interconnects auxiliary A/D's and D/A's with the external world, conditions various trigger inputs, and provides signals to 4 small coils used for head position determination. EEG electrodes are connected to a passive nonmagnetic head box near the subject. Signals from the head box are conditioned by the EEG pre-amplifiers and isolated EEG electronics where they are further amplified and digitized. The digitized EEG signals, control messages and synchronization are communicated to the control electronics via fiber optic links. The EEG channels have noise levels < 2 μV rms in a 1 kHz bandwidth and input impedance > 10 MΩ. The maximum MEG/EEG system sample rate for 228 channels is1,250 sam-

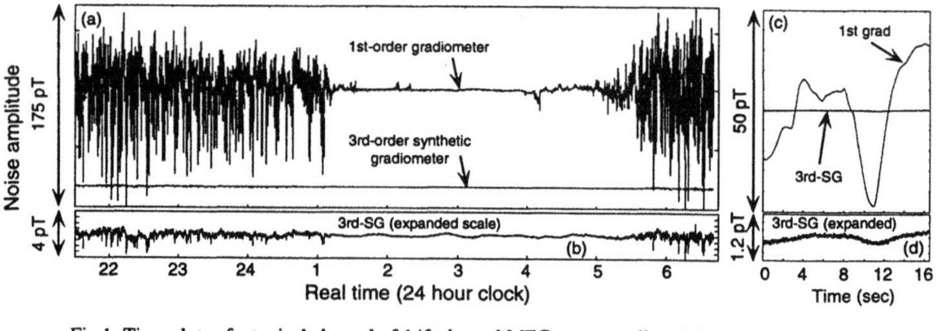

Fig.1: Time plots of a typical channel of 143 channel MEG system collected in a shielded room (Vacuumschmelze AK3b). (a) over-night recording, about 9 hours, fs = 1.25 samples/sec, low pass filter = 0.5 Hz, 1st-order gradiometers and 3rd-SGs on the same scale; (b) expanded scale of 3rd-SG over-night record; (c) short recording during the day, about 16 sec, fs = 1250 samples/sec, low pass filter = 300 Hz, 1st-order gradiometers and 3rd-SGs; (d) expanded plot of the 3rd-SG during the day.

ples/sec. With fewer selected channels, higher maximum sample rates are possible, e.g., 2,083 samples/sec with up to 143 channels, 2,500 samples/sec with up to 114 channels, 3,125 samples/sec with up to 76 channels and 4,167 samples/sec with up to 57 channels.

Noise performance

The 143 channel system has been operated in both shielded and unshielded environments. The noise levels of all channels were measured and a channel with average properties was selected to illustrate the noise behavior. Time plots of this channel operated in a shielded room are shown in Fig.1. Fig.1.a corresponds to a long, 9 hour overnight record and demonstrates, that even in a shielded room, the 1st-order gradiometers exhibit noise levels which can be meaningfully reduced by 3rd-SGs. An expanded plot of the 3rd-SG over the same time period is shown in Fig.1.b. In Fig.1.c and d the 1st-order gradiometer and 3rd-SG noise during the day-time is shown in a time interval of about 16 sec with behavior similar to the night-time data, i.e. the 3rd-SG significantly reduces the noise. Fig.1 thus illustrates how the software formation of high-order synthetic gradiometers provides effective noise cancellation even when the system is operated in a shielded room.

Noise spectra of a typical unshielded channel (from the vicinity of the histogram peaks in Fig.3, below), corresponding to field magnitude, 1st-order gradiometer sensor, and 3rd-SG, are shown in Fig.2.a in the frequency range from 0.1 to 400 Hz. The residual noise (largely environmental and system noise) for a 3rd-SG is nearly white and is lower than 10 fT rms/√Hz for frequencies above about 0.6 Hz. In addition to the white and low frequency noise backgrounds, the spectra also exhibit various peaks, mostly associated with power lines. In this example, the magnitude of the 60 Hz magnetic field is about 260 nT p-p and the residual 3rd-SG white noise level in a 1 Hz bandwidth is about 5 fT rms (≈ 25 fT p-p), corresponding to a minimum dynamic range requirement of $10^7 \approx 140$ dB. This large dynamic range requirement is representative of unshielded environments; the corresponding dynamic range requirement in shielded environments is considerably smaller. Noise spectra for a typical channel in a shielded room are shown in Fig.2.b in the frequency range from 0.002 to 300 Hz. Note that the 3rd-SG also completely eliminates the residual power line frequencies.

The residual (environmental plus system) noise floor for all channels in their 1st-order gradiometer and 3rd-SG modes and in the frequency range above 5 Hz for an unshielded and shielded 143 channel system were measured and are summarized as histograms with bin size of 1 fT rms/√Hz in Fig.3. (Various environmental lines were omitted and only the wide band noise floor was considered). In Fig.3.a, the noise histograms for 3rd-SGs are shown. Curve a_1 corresponds to the noise collected during a working day and has a mean and standard deviation of 6.6 ± 1.8 fT/√Hz and the highest observed unshielded noise was 13 fT/√Hz. Curve a_2 corresponds to a day-time recording in a shielded room and it has a mean value of 4.8 ± 0.8 fT/√Hz. Curve a_3 corresponds to shorted SQUID noise with a typical sensor flux transformer gain (3.4×10^{-6} Φ_0/fT) and has mean value of of 2.2 ± 0.5 fT/√Hz (the shorted

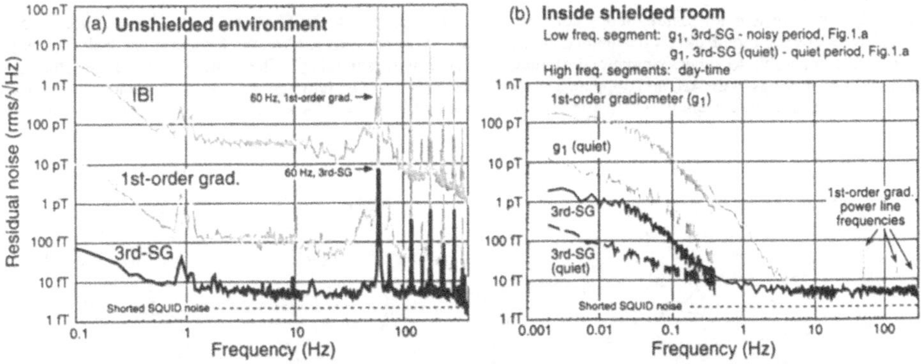

Fig.2: Noise spectra of a typical shielded and unshielded channel of 143 channel MEG system. Thin dashed line - a typical noise level of a shorted SQUID. (a) magnetic field magnitude, 1st-order gradiometers and 3rd-SGs, all unshielded. (b) 1st-order gradiometer and 3rd-SG in a shielded room in frequency range from 2 mHz to 300 Hz, the light gray lines correspond to the 1st-order gradiometer and black lines correspond to 3rd-SG, the low frequency segments were collected overnight and solid lines during noisy periods and dashed lines during the quiet periods (c.f. Fig.1.a), and the high frequency segments were collected during the day.

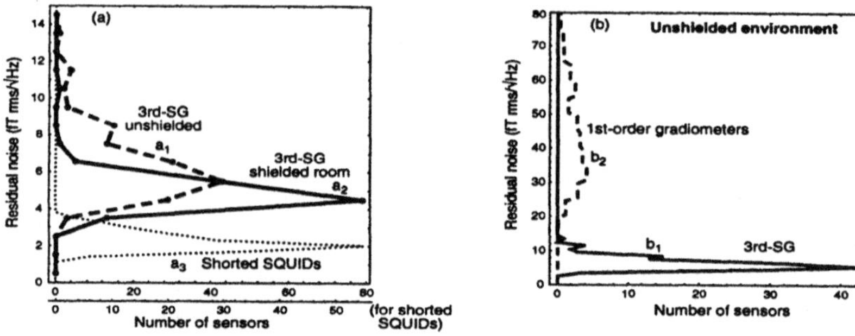

Fig.3: Noise histograms of 143 channel MEG system in shielded and an unshielded environments. Bin size = 1 fT rms/√Hz for 1st-order gradiometers and 3rd-SGs, 0.3 fT rms/Hz for shorted SQUIDs. Noise is sum of system and ambient noise in band of interest (excluding lines identified with known sources. (a) 3rd-SG, a_1: unshielded, a_2: shielded, a_3: shorted SQUIDs; (b) comparison of 1st-order gradiometers and 3rd-SGs in an unshielded environment, b_1: 3rd-SG, b_2: 1st-order gradiometer.

SQUID histogram has been computed with higher resolution and therefore it has different horizontal scale). Curve b_1 in Fig.3.b is the same as curve a_1 in Fig.3.a, and curve b_2 represents the corresponding 1st-order gradiometer noise (and is non-zero up to 150 fT/√Hz).

The MEG/EEG system has been used for numerous human brain experiments. Simultaneous MEG/EEG recordings in an unshielded environment are shown in Fig.4 for Auditory Evoked Fields (AEF) and in Fig.5 for interictal spikes in temporal lobe epilepsy. The epilepsy recordings are not averaged and are displayed in the bandwidth from DC to 40 Hz. Even though the measurement was unshielded, the baseline drift was less than 0.6 pT (with 0.7 pT standard deviation over all channels) over a 540 sec period. In Fig.5 only a short data segment in the vicinity of an epileptic interictal spike (arrow) is shown. In addition to the spikes, the frontal channels (RF and LF) also exhibited high amplitude slow variations while these slow variations were absent in the central channels (LC27 and RC27 in Fig.5), demonstrating that the source of the low frequency signals was the brain. Visual and tactile stimulations were also recorded in an unshielded environment [3].

Conclusions

A whole-cortex system for simultaneous collection of 143 MEG and 40 EEG channels has been constructed and operated for several months in an unshielded environment. In the ambient environment, the system had a noise performance comparable to other commercial biomagnetometers with shielding. The system demonstrated that good unshielded recordings of evoked and spontaneous (not averaged) human brain activity can be obtained down to DC. In a shielded room, the system provided effective cancellation of the residual low frequency and power line noise and

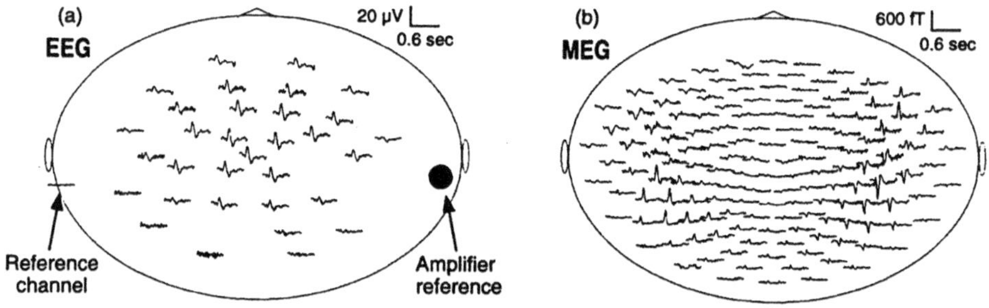

Fig.4: Simultaneous EEG and MEG recording, unshielded environment. AEF, 1 kHz tone into both ears, 50 msec duration, isi = 1.5 to 2 sec, f_s = 625 samples/sec, 50 averages. (a) 32 channels EEG, linked ears reference, BW = 0.16 to 40 Hz; (b) 143 channels MEG, 3rd-SG, BW = DC to 40 Hz.

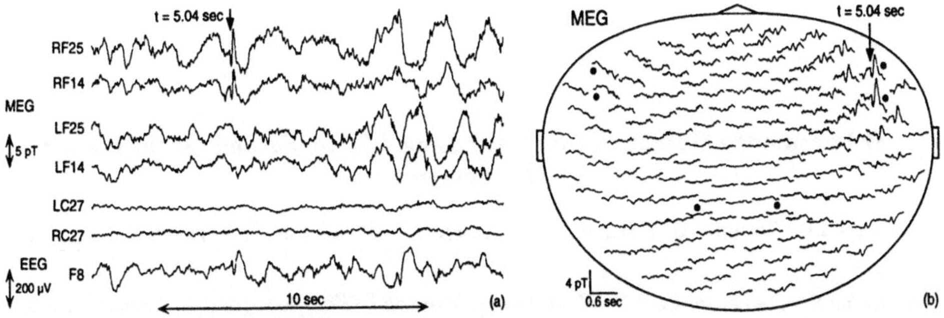

Fig.5: Temporal lobe epilepsy, complex partial seizures, measured in an unshielded environment. f_s = 250 samples/sec, no averaging, MEG BW = DC to 40 Hz, 3rd-SG, EEG BW = 0.16 to 40 Hz. This figure represents only a 15 sec segment of an 800 sec recording. (a) typical time traces (LF and RF - left and right frontal, LC and RC - left and right center), an interictal spike is shown at the arrow; (b) traces for all channels. Dots indicate channels shown as time traces in (a) and arrow indicates the interictal spike.

the combination of the shielded room attenuation and the gradiometer noise cancellation resulted in a very low noise background. The development of the 143 channel MEG system is an extension of the existing 64 channel system, with a significant increase in the channel density, as illustrated in Fig.6.

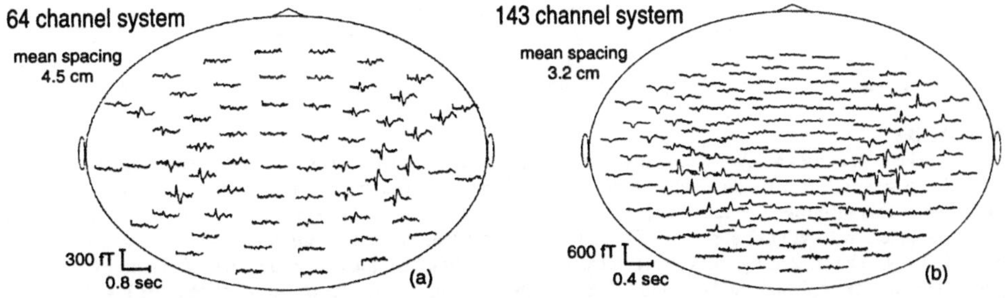

Fig.6: Comparison of 64 and 143 channel densities, unshielded environment, 3rd-SG, BW = DC to 40 Hz, f_s = 625 samples/sec, 100 averages. AEF, 1 kHz tone, 50 msec duration, isi = 1.5 to 2 sec. (a) 64 channel MEG system, right ear, (b) 143 channel system, both ears.

References

[1] Nakasato, N., Fujiwara, S., Yoshimoto, T., Fujita, S., Matani, A., Izuma, H., Tamura, I. The whole head magnetoencephalogram: localization of bilateral sources for unilaterally stimulated auditory evoked response, Electroencephalogr. Clin. Neurophysiol, 1993, **87**:S26.

[2] Vrba, J., Haid, G., Lee, S., Taylor, B., Fife, A.A., Kubik, P.R., McCubbin, J. and Burbank, M.B. Biomagnetometers for unshielded and well shielded environments, Clin. Phys. Meas., 1991, **12 B**: 81-86.

[3] Weinberg, H., Gaetz, M., Jantzen, K., Cheyne, D., and Deecke, L. Comparison of MEG and EEG for assessment of traumatic brain injury, These proceedings, 1996; Cheyne, D., Roberts, L.E., Gaetz, W., Bosnyak, D., Weinberg, H., Johnson, B., Nahmias, C. and Deecke, L. EEG and MEG source analysis of somatosensory evoked responses to mechanical stimulation of the fingers, These proceedings, 1996.

Acknowledgments

The work was supported in part by the Canadian Departments of National Defence, Supply and Services, Transport Canada, and the BC Science Council.

SEFs from the Human Spinal Cord using a Novel Second Order Planar Gradiometer

Weir, A.I.[1], Shahani, U.[1], Hutson, D.[2], Lang, G.[2], Maas, P.M.[2], Pegrum, C.M.[2], and Donaldson, G.B.[2]

[1]Wellcome Biomagnetism Unit, Southern General Hospital, Glasgow, Scotland; [2]Department of Physics and Applied Physics, University of Strathclyde, Glasgow, Scotland

INTRODUCTION

Somatosensory evoked fields (SEF) elicited by electrical and mechanical [1] stimuli have been recorded from the cortex using SQUID gradiometers of various designs, and the SI and SII sources localised [2]. In clinical practice the equivalent evoked potential (EP) study would always be combined with recordings of the afferent pathway in the peripheral nerve and spinal cord. These spinal evoked potentials are complex recordings of afferent volleys [5]. SEF recordings of these locations have been limited due to a) the lack of devices with appropriate geometry for all locations b) the depth and orientation of the sources c) the need for highly screened environments (particularly for spinal cord measurements) d) the need for clinically unacceptable numbers of stimulus averages to achieve acceptable signal/noise ratios. To date for recordings of evoked magnetic fields from the spinal cord the most successful has been the use of a magnetometer in a highly shielded room and up to 8000 averages [3].

This study reports preliminary results of the design and successful use of a novel thin film integrated second order SQUID gradiometer for spinal recordings in a simple eddy current shielded environment.

METHODS

The system uses a novel symmetric planar gradiometer oriented tangentially to the recording plane within the tail of a standard BTi single channel dewar. The full design is described in our accompanying paper [4] also presented at this meeting. The recordings were made in our eddy-current shielded room and on healthy normal adult volunteers with ethical approval. The left or right median nerve was stimulated to evoke somatosensory evoked fields. The recording sites were located over the brachial plexus, a number of locations over the posterolateral neck cephalad and lateral to C7 spinous process and over the contralateral cortex (see Fig. 1).

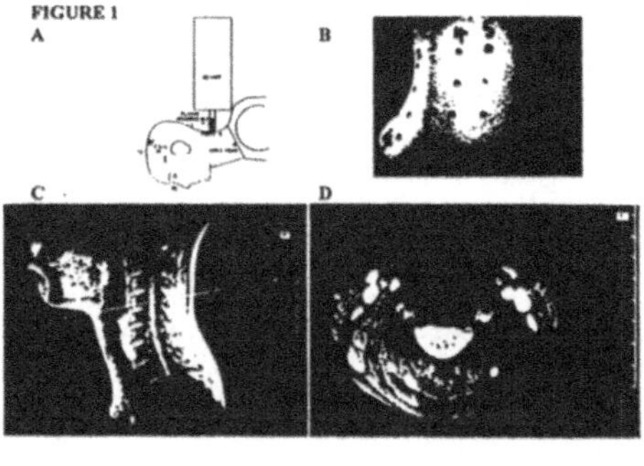

Fig. 1. (A) Schematic representation of recording sites. (B) recording sites marked over the posterolateral neck. Point 5[1] is seven centimeters cephalad to the C7 spinous process. Point 1[1] is eight centimetres to the left of 5[1] (Subject GL). (C) MRI view of the neck. (D) MRI tilted slice transversely at point 5[1].

The median nerve was stimulated electrically at the wrist above motor threshold with a 0.2ms constant current stimulus at 3 pps and responses 512 stimuli averaged for each run. Each run was reproduced at least twice. The raw signals were fed into a digital DC Squid controller (Quantum Design) and then into a single pole filter and preamplifier (PAR Model 113). Filter settings were at 1 Hz for high pass and 300Hz at low pass. The overall scale factor was 19 fT/division and gradiometer input loop referred SQUID noise was 5 $fT/Hz^{1/2}$. The SQUID signals were amplified both in the SQUID controller box and in the preamplifier. The signals were then fed into a Medelec Sapphire (2ME) EMG system for averaging. Consecutive EP recordings from comparable sites were made on the same Medelec Sapphire system to allow for latency comparisons. Bandpass settings for electrical measurements were set between 1Hz and 2KHz.

RESULTS

Consistent waveforms temporally related to the stationary spinal P/N13e, were recorded over the posterolateral neck in several adults. Here we present data from subject GL to stimulation of the median nerve, of magnetic and electrical recordings over Point 5^1 which is seven centimetres cephalad to the C7 spinous process. In Figure 2(A) traces 1 (512 averages) and 2 (1024 averages) are of SEFs recorded on separate occasions. The sum of the two is represented in trace three. Figure 2(B) shows the spinal evoked potential in trace one and the spinal evoked field in trace 2. The electrical and magnetic responses coincide at 13.5 ms.

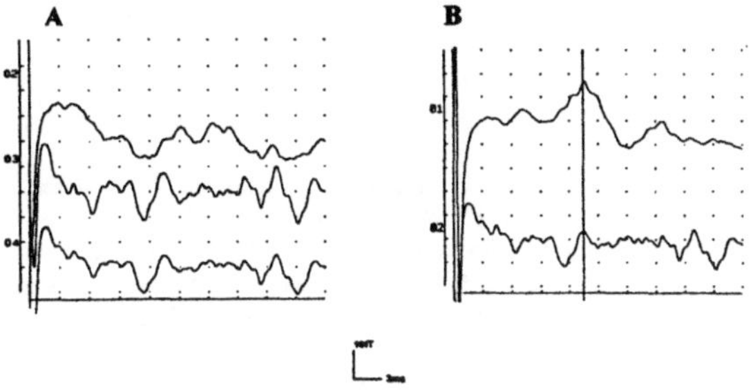

Fig. 2. (A) Composite records of SEFs in response to median nerve stimulation over Point 5^1. Top two traces are records taken on separate occasions. Bottom trace is the grand average of the first two traces. The amplitude of the P/N13m is 20-40 fT peak to peak, consistent with previous studies. (B) Comparisons between SEP (top trace) and SEF bottom trace. The N13e and N13m are coincident.

Figure 3 (A) shows spinal evoked fields from subject GL to ipsilateral median nerve stimulation with the planar gradiometer placed at Point 1^1, 8 centimetres to the left of Point 5^1. Traces one through to six are consecutive runs of 512 averages each. Trace seven is the addition of traces one to six. The N11m is apparent in these traces. (B) Comparison of SEP and SEF. Trace one is the SEP from position 5^1 (midline cephalad to C7 spinous process; see Figure 1(B)). Trace two is the SEF from Point 1^1. The N11e and the N11m are coincident at 11.8 ms.

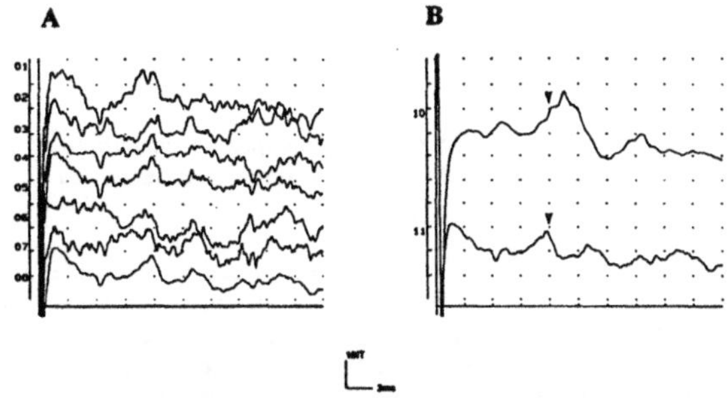

Fig. 3.(A) Composite records of SEFs in response to median nerve stimulation over Point 1^1. The N11m is apparent in these traces. (B) Comparisons between SEP (top trace) and SEF bottom trace. Trace one is the SEP from position 5^1 (midline cephalad to C7 spinous process; see Figure 1(B)). Trace two is the SEF from Point 1^1. The N11e and the N11m are coincident at 11.8 ms.

Direct recordings over the brachial plexus revealed a reliable N9m which was coincident with the N9e at 10.4 ms and these data are shown in Figure 4 (A and B). In (A) traces 1 and 2 are consecutive records of SEFs from Erb's point (512 averages) of responses to median nerve stimulation at the wrist. (B) Comparison of SEPs (trace one using non-cephalic and trace two using cephalic reference, 512 averages each) and SEFs (trace three).

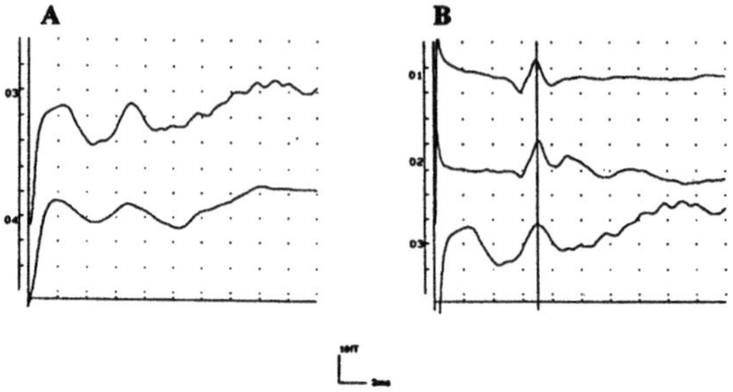

Fig.4. (A) Consecutive records of SEFs from Erb's point in response to ipsilateral median nerve stimulation. (B) Comparison of SEPs (non-cephalic and cephalic reference) and SEFs.

The cortical N20m was also located with the device close to the C3, C4 position on the head. Position C3/C4 of the 10-20 electrode placement system is thought to be over the Rolandic fissure and therefore considered to be close to the source of the N20m/e. Figure 5 (A) shows SEFs from Subject GL of the N20m recorded over C3/C4 position using our new planar SQUID gradiometer. Traces one through to 4 are consecutive records (100 averages each). Note the low noise level. Trace 5 is the summation of traces one to 4 (B) SEFs recorded from another subject in response to median nerve stimulation using a commercial single channel axial neuromagnetometer (BTi San Diego) at exteremata 5 and 12 (see Figure 1 (A)). The N20m is clearly seen and its amplitude is similar to recordings made with the conventional single channel axial neuromagnetometer.

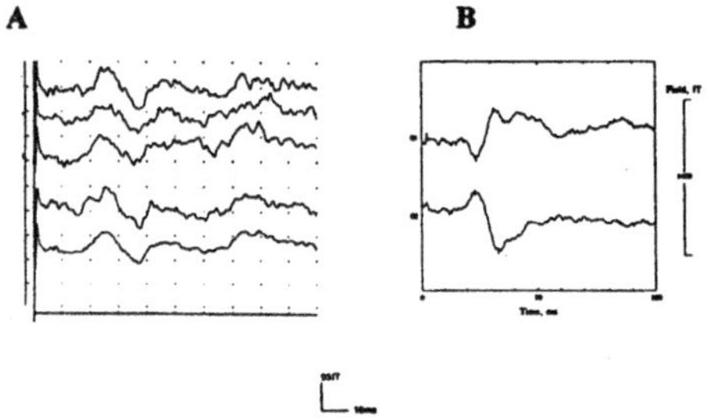

Fig. 5.(A) Cortical SEFs recorded over C3/4. Traces 1 to 4 are sequential repetitions of 100 averages of responses to contralateral median nerve stimulation. Trace 5 is the grand average of traces 1 to 4.

DISCUSSION

We believe that this is a major step forward in the development of a device suitable for clinical use in spinal recordings, with advantages both in signal/noise ratio within a limited shielded environment and the location and orientation of the gradiometer because of its tangential design. A significant advantage of our device for use in a clinical setting is that averaging of evoked magnetic fields from the spinal cord can be limited to a relatively small number with no diminution of the response. In practice this implies that less time is taken for tests and patient discomfort is thus minimised.

Our preliminary results indicate that the device is sensitive enough to record the P/N13m, the stationary post-synaptic dorsal horn potential over the posterior neck in similar orientation to the electrical evoked potential. In addition, coincident with the P/N11e, the latency of the incoming volley at entry into the spinal cord [5] we are able to record the P/N11m. Moreover, the ability of our device to record volleys through the brachial plexus suggests an immediate clinical use in a completely non-invasive manner.

REFERENCES:

[1] Forss, N., Salmelin, R., & Hari, R. Electroencephalography and Clinical Neurophysiology, 1994, 92:510-517.

[2] Hari, R., Karhu, J., Hamalainen, M., Knuttila, J., Salonen, O., Sams, M., & Vilkman, V. European Journal of Neuroscience, 1993, 5: 724-734.

[3] Curio, G., Mackert, B.-M., Burghoff, M., Koetitz, S., Drung, D., & Marx, P. In: Baumgartner, I., Deeke, L., Stroink, G., Williamson, S.J., Biomagnetism: Fundamental Research and Clinical Applications, 1995.

[4] Hutson, D., Lang, G., Shahani, U., Klein, U., Weston, R.G., Weir, A.I., Maas, P.M., Pegrum, C.M., and Donaldson, G.B. these proceedings.

[5] Desmedt, J.D. In: Maurer K., Topographic Brain Mapping of EEG and Evoked Potentials, 1989.

Acknowledgements : This work was supported by the Wellcome Trust and EPSERC.

High-T_c Superconducting rf Receiver Coils for High Resolution Magnetic Resonance Imaging of the Spine

Wosik, J., Nesteruk, K.,[1] L. M. Xie, Zhang, X. P., Jiao, C., Gierlowski, P.,[1] and Miller, J. H., Jr.

Texas Center for Superconductivity at the University of Houston, Houston, TX 77204, USA;
[1]Institute of Physics, Polish Academy of Sciences, Warsaw, Poland

ABSTRACT

A low-loss high temperature superconducting (HTS) *rf* receiver surface probe developed primarily for high-resolution magnetic resonance imaging imaging (MRI) of spinal cord injuries in small animals is described. This probe is used *in lieu* of an implanted copper coil being currently used in research on spinal cord injuries. The influence on the signal-to-noise ratio (SNR) of both an examined body and rf receiver coil was analyzed. It is shown that, for spine cord MRI, noise in the receiver coil limits the SNR. Numerical calculations were done using the complete Maxwell equations solution and the reciprocity principal for a coil of rectangular shape facing a finite loose dielectric cylinder. This approach can be extended to any shape of the coils and body. The effects of coil size, position, location of the NMR signal source as well as ultimate limit on the SNR were analyzed. The HTS probe was designed with a virtual ground plane, thus reducing the coil-to-ground losses and making its resonance frequency less sensitive to its proximity to the body. A resonant circuit for this probe is created by sandwiching a dielectric between two faces of the planar coils in order to achieve capacitive coupling. Each coil was fabricated by depositing superconducting $YBa_2Cu_3O_x$ onto 2" x 2" $LaAlO_3$ substrates using a large-area pulsed laser deposition system. The copper prototype of such probe was tested in a 2 Tesla (84.4) MRI scanner. Both the signal to noise ratio and field of view were measured.

INTRODUCTION

Magnetic resonance imaging has become one of the most widely used diagnostic tools in medicine. It is also a promising technique for studing spinal cord injury in clinical (humans) and experimental (animals) cases. Such studies, because of the object sizes, require high spatial resolution and signal-to-noise SNR ratios. Recently it has been demonstrated [1] that a substantial improvement in the image quality can be achieved by implanting a receiver coil near the structure of interest, such as heart, kidney, or spinal cord. The high SNR ratio obtainable with an implanted coil enables the acquisition of higher resolution NMR images and enhances possibility of *in vivo* localized NMR spectroscopy of the spinal cord and other structures. However, despite these advantages, there are a number of disadvantages with the use of such implanted coils. For example, the structure requires a large blunt dissection of a muscle from the spine, often resulting in inflammation of the adjacent musculature. In addition, coil failures due to broken solder joints are common in some animals, which result from additional stresses on the coil. Finally, this approach clearly can not be used in the case of human subjects.

We propose another approach to improve SNR ratios for the spine cord MRI. In the case of small-volume, high-resolution imaging, noise of the probe coil and/or noise of preamplifier set the system noise floor and hence the MIR performance; body noise no longer dominates the SNR of the system [2]. Thus, it is desirable to reduce the thermal coil noise to improve the image resolution and reduce image acquisition time. Since the Johnson noise is a function of the product of resistance and temperature, reduction of either or both of these parameters will enhance the SNR values. In addition, improved SNR can enable one to decrease voxel size and thereby increase the resolution of structural details. High-temperature superconductors are extremely attractive for such applications due to their very low losses. A superconductor is completely lossless (below the critical temperature) at zero frequency, provided that the *dc* current is less than the critical current. At higher frequencies, finite losses result when *rf* screening currents flowing near the surface within a penetration depth of about 0.3 μm. The loss is due to the presence of uncondensed normal carriers. Microwave measurements of $YBa_2Cu_3O_x$ films at 77 K yield a surface resistance R_s of 150 μΩ at 10 GHz [3], which is nearly two orders of magnitude lower than the surface resistance of OFHC copper at the same frequency and temperature. Scaling [4] both the values of R_s down to 100 MHz indicates that superconducting YBCO will have a surface resistance at least four orders of magnitude lower than that of copper at this frequency.

Indeed, preliminary studies have shown that for selected applications, HTS MRI receiver coils perform significantly better than comparable copper coils [5, 6, 7]. It was proven that in low-field MRI- and in high-resolution, small-volume MRI, thermal noise of the receiver coil and its associated electronics set the system noise floor and hence the performance. Even in high-field MRI, when the coil and the corresponding region of interest (ROI) are reduced in size, there is a crossover point beyond which the receiver noise becomes the dominant source of noise. Black et. al [6] discussed such scaling for the case when the coil and body sizes are the same. As an example to demonstrate the scaling laws for noise in MRI, a typical case was used where a human head is imaged. It was found that the coil noise becomes comparable to the sample noise when the field is reduced from 1.5 to 0.3 T, or when the diameter of the coil is reduced from 25 cm to 12 cm. In our design, the size of a typical dimension of the FOV is about 2-4 cm, while the operating field is 2.0 T.

PROBE DESIGN

Electric and rf magnetic fields can induce currents in the body tissue, which has a relatively high conductivity, causing significant losses in the resonant circuit of an MRI probe. Such losses reduce Q of the circuit and are more pronounced at the higher frequencies required in high-field MRI systems. Relatively little can be done to reduce currents generated by the rf magnetic field. Another method of reducing electric fields is to use a symmetrically balanced coil to introduce a virtual ground plane [8, 9]. The balanced coil, shown in Fig. 1, is of this type. In this case, the maximum voltage produced in the coil, with the respect to ground, is half of what it would be if only one end of the coil had been grounded. A balanced coil (with a virtual ground plane) not only reduces the coil-to-ground losses but, in addition, its resonance frequency is less sensitive to its proximity to the body. We have designed and fabricated a two-coil probe consisting of a pair of balanced planar HTS surface coils, consisting of patterned YBCO films on 2" x 2" $LaAlO_3$ substrates, suitable for imaging of vertebrae in the spine. In principle, this design follows that shown by R. Withers et. al. [5] A resonant circuit is created by sandwiching a dielectric (sapphire) between two faces of the planar coils in order to achieve capacitive coupling. Each coil was fabricated by depositing YBCO onto $LaAlO_3$ substrates using our large-area pulsed laser deposition system [10]. Standard positive resist photolithography and wet etching were used for patterning. The subject's spine would be parallel to the longer superconducting strips during imaging. Measurements of the Q of a copper two-coil probe at room temperature and 77 K showed $Q = 200$ and 640, respectively. By contrast, the Q of the YBCO two-coil probe shown in Fig. 1 was found to be more than an order of magnitude higher.

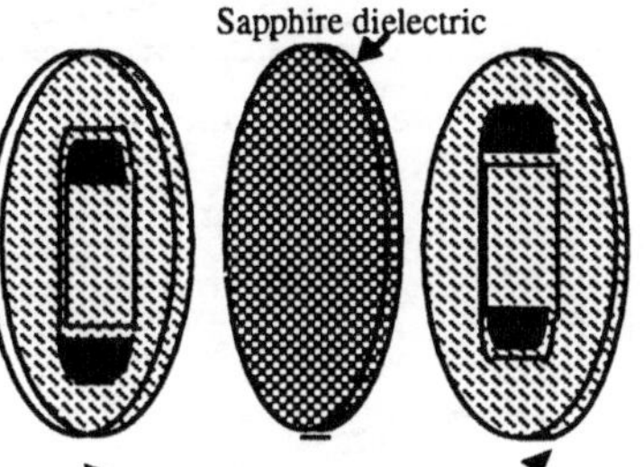

Fig. 1 A pair of planar balanced YCBO receiver coils, designed for imaging of the spine cord. A sapphire substrate, sandwiched between the two coils, capacitively couples the coils to create a resonant LC circuit. The wide strips are capacitive plates.

SIGNAL-TO NOISE RATIO CALCULATIONS

The design of surface coils optimized for maximum SNR requires knowledge of the basic physical effects determining the signal and noise generated in a coil by the body after the rf excitation. To achieve this goal the body was modeled as a 60 mm long dielectric cylinder of given conductivity σ. A rectangular (15 by 30 mm coil is facing the cylinder of radius a). A ΔV volume voxel was assumed to be 5 mm deep in the body. This configuration is adequate to the real situation for the spine cord imaging in rats using the 84.4 MHz MRI scanner. The body noise was simulated by losses in the coil due to the existence of the body and defined as a equivalent body resistance R_{body} for the coil.. Thus the rms thermal noise voltage of the coil is assumed to be caused by both the resistance of the coil R_{coil} and the body resistance R_{body} and can be expressed as: $V_n = \sqrt{4k\Delta f(R_{coil}T_{coil} + R_{body}T_{body})}$, where k is Boltzman constant and T and Δf are temperature and the bandwidth of a preamplifier attached to the coil, respectively. Signal and noise voltage calculations are carried out using the reciprocity principle and line integrals along the coil contour. The induced signal V_s in the coil can be expressed as: $S \propto V_s\omega^2(B_1)_{xy}$, where B_1 is the calculated sensitivity of the coil, and w and ΔV are the frequency and voxel volume, respectively. In this paper, SNR is define as the signal to noise voltage ratio and not the signal-to-noise power ratio. The SNR is calculated by dividing the magnitude of the signal voltage for the coil located at a distance z_p from the dipole source placed 5 mm deep in the body, by the thermal noise voltage of the coil. Thus the SNR as a function of the probe position z_p can be expressed as:

$$SNR(z_p) = \frac{V_s(z_p)}{V_n(z_p)} \tag{1}$$

Fig. 2 shows the variations of SNR with the respect to the coil position at 10 MHz and 84.4 MHz for 15 by 30 mm rectangular coil for a tissue of conductivity equal to 1 $(\Omega m)^{-1}$.

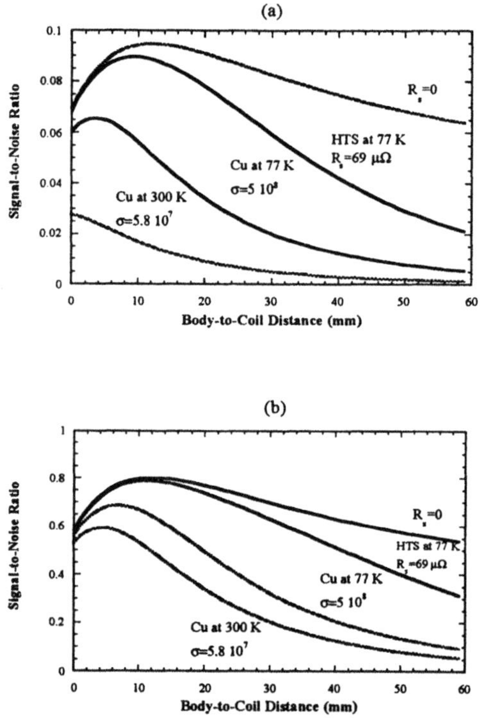

Fig. 2 The plots show the SNR dependence on the coil position away from the tissues for 10 MHz (a) and 84.4 MHz (b). The curves are calculated for the case of room temperature copper, liquid nitrogen temperature copper, an average HTS thin film and for the case when material does not have any Johnson noise.

The calculations were done for four cases: when the coil is made out of (1) copper and used at room temperature, (2) copper cooled to 77 Kelvin, (3) high temperature superconductor with R_s equal to 69 $\mu\Omega$, and (4) an idealized material with R_s equal to zero. Surface resistance value for the HTS thin film was chosen very conservatively. This is rather a high R_s value for superconductors at this frequency. The calculations were performed for two frequencies: 10 MHz and 84.4 MHz, which are shown in Fig. 2 as (a) and (b), respectively. It can be seen from this figure that there is a position away from the body which yields the highest SNR while imaging a voxel deep in the body. This is extremely important feature for the HTS application, where the cryostat implementation introduces a gap between the body and the coil. It proves that carefully a designed probe still can be placed at z_p equal to 10-15 mm away from the body at the maximum SNR point. The SNR of the HTS probe is about 5 times larger than that of the warm copper coil at 10 MHz. Such an improvement is smaller at 84.4 MHz, however still it is equal to about 2.

SUMMARY

We have designed a HTS rf receiver probe for the MRI of spine. This probe is used *in lieu* of an implanted copper coil being currently used in research on spinal cord injuries. The probe, due to the careful design which balances between the image uniformity and the need for a higher SNR, showed large field of view and very good field uniformity in the preliminary MRI tests of the probe prototype.

The SNR improvement due to the use of HTS thin films was theoretically analyzed for the case of the rat spine MRI. It was found that, for each coil size, there is a position away from the body which yields the highest SNR (while imaging the voxel deep in the body). This indicates that separation of the HTS probe from the body, resulting from the cryostat implementation, does not decrease the SNR improvement. The SNR improvement due to the use of HTS rf receiver coil, for the analyzed experimental MRI of a rat spine configuration, is at least five-fold. Even for the high field MRI at 84.4 MHz this improvement is about two.

REFERENCES

[1] J. C. Ford, D. B. Hackney, P. M. Joseph, M. Phelan, D. C. Alsop, S. L. Tabor, C. M. Hand, R. S. Markowitz, and P. Black, "A method for in vivo high resolution MRI of rat spinal cord injury," *Magn. Reson. Med.* 31, 218 (1994).

D. I. Hoult and P. C. Lauterbur, "The sensitivity of the zeugmatographic experiment involving human samples," *J. Magnetic Resonance* 34, 425 (1979).

[2] R. D. Black, T. A. Early, P. B. Roemer, O. M. Mueller, A. Mogro-Campero, L.G. Tuner, and G. A. Johnson, "A high-temperature superconducting receiver for nuclear magnetic microscopy," *Science* 259, 793 (1993).

[3] H. Piel, G. Muller, " The microwave surface impedance of high-T_c superconductors," *IEEE Trans. Magn.* 27, 854 (1991).

[4] J. H. Hinken, *Superconductor Electronics; Fundamentals and Microwave Application*, (Springer-Verlag, New York Berlin Heildelberg 1989).

[5] R. S. Withers, B. F. Cole, M. E. Johansson, G.-C. Liang, and G. Zaharczuk, *SPIE Proceedings Series, Vol. 2156, High-T_c Microwave Superconductors and Applications, 24-27 January 1994, Los Angeles, CA*, R. B. Hammond and R. S. Withers, editors, pp. 27-35.

[6] R. D. Black, P. B. Roemer, A. Mogro-Campero, L. G. Turner, and K. W. Rohling, "High temperature superconducting resonator for use in nuclear magnetic resonance microscopy," *Appl. Phys. Lett.* 62 771 (1993).

[7] J. G. van Heteren, T. W. James, and L. B. Bourne, "Thin film high temperature superconducting rf coils for low field MRI," *Magn. Reson. Med.* 32, 396 (1994).

[8] M. Decorps, P.Blondet, H. Reutenauer, and J. P. Albrand, "An Inductively Coupled, Series-Tuned NMR Probe," *J. Magn. Resonance* 65, 100 (1985).

[9] J. Murphy-Boesch, "An in Vivo NMR Probe Circuit for Improved Sensitivity," *J. Magn. Resonance* 54,526 (1983).

[10] M. F. Davis, J. Wosik, K. Forster, S. C Deshmukh, H. R. Rampersad, S. Shah, P. Siemsen, J. C. Wolfe and D. J. Economou, "Large area laser deposited superconducting YBCO thin films," *J. Appl. Phys.* 69, 7182, (1991).

ACKNOWLEDGMENTS

We are grateful to Dr. Ponnada Narayana for stimulating discussions and for making his MRI scanner available for measurements, and Dr. N. Tralshawala for very helpful suggestions. This work was supported by the Robert A. Welch Foundation (E-1221), by the State of Texas via Texas Center for Superconductivity at University of Houston and by the Texas Higher Education Coordinating Board (ARP grant #003652-238).

3-D Magnetic Field Analysis of a Magnetically Shielded Room for Biomagnetic Measurements

Yamazaki, K.[1], Uchikawa, Y.[2] and Fujiwara, K.[3]

[1]*Takenaka Corp., Chiba, Japan;* [2]*Tokyo Denki Univ., Saitama, Japan;* [3]*Okayama Univ., Okayama, Japan*

1. Introduction

As magnetic fields generated by the human brain are extremely weak, a magnetically shielded room (MSR) is required which can reduce magnetic noise near the brain in the range of 0.1Hz to radio frequency. As a sensor for such biomagnetic measurements, a SQUID magnetometer is widely used. When the SQUID magnetometer equipped with a differential type of sensing coil (gradiometer) is employed, resolution is affected by field gradients. In the multichannel SQUID magnetometer, the reliability of results measured also depends on the spatial distribution of magnetic noise, because the sensing coils are arranged three-dimensionally not only in the vertical direction but also in the horizontal directions.

Analytical expression of the shielding factor derived by Wills [1] and approximate formula by Mager [2] have been used to predict the shielding factor of MSR. These formulas, however, cannot calculate leakage magnetic fields from a ventilation outlet and a gap between a wall and door. There are a few examples illustrating the flux distributions in MSR by using only 2-D magnetic field analysis [3]. However, the 2-D analysis is not sufficient to predict the shielding factor of the practical MSR. In this paper, the 3-D magnetic field analysis of MSR is carried out by using the finite element method taking into account eddy currents.

2. Verification of 3-D Magnetic Field Analysis for MSR

Results computed are compared with those measured in a hexagonal type of MSR for biomagnetism [3] shown in Fig.1, which is installed at Tokyo Denki University. This MSR is constructed by using plywood panels of 24mm thick sandwiched by permalloy sheets of 0.5mm thick (two layers on each side). The magnetic field analysis is carried out in the cases when the door is open and closed. The magnetic noise of 1Hz is applied by a square coil (2,500mm×2,500mm) placed on top or in front of MSR. The current in the coil is determined so that the magnetic noise can be equal to 7.5mG at a point of 2,200mm far from the coil on the center line. Incremental permeability of permalloy sheet is set at $10^4\mu_0$ (μ_0: permeability of vacuum), which is measured by the vertical double-yoke type of single sheet tester. Conductivity is 1.63×10^6 S/m. The fundamental equations for analyzing 3-D magnetic field distribution are expressed as follows:

$$rot(\nu\, rot A) = J_0 - \sigma\left(\frac{\partial A}{\partial t} + grad\phi\right) \tag{1}$$

$$div\left\{-\sigma\left(\frac{\partial A}{\partial t} + grad\phi\right)\right\} = 0 \tag{2}$$

where A: magnetic vector potential, ϕ: electric scalar potential, J_0: magnetizing current density, ν: reluctivity, σ: conductivity.

The spatial distributions of magnetic noises in the MSR computed by the 3-D analysis show very similar tendency to those measured as shown in Figs.2 and 3. At the point more than 2,000mm far from the door (Fig.2) and 500mm (Fig.3), the magnetic field distribution in the case when the door is open does not differ from that when the door is closed. ThesOe figures demonstrate that the 3-D magnetic field analysis is applicable to evaluation of the shielding effect of MSR.

3. Relationship between the Magnetic Fields and Eddy Currents

The 3-D magnetic field analysis of a brick type of MSR shown in Fig.4 is carried out to grasp the relationship between the magnetic fields and eddy currents. This MSR is constructed by using plywood panels of 44mm thick sandwiched by permalloy sheet of 1mm thick (one layer on each side). It is assumed that there is a gap of 1mm between a door and wall and it isolates them magnetically and electrically. A doorframe made by permalloy sheet which covers around the doorway connects the first layer to the second layer of wall electrically.

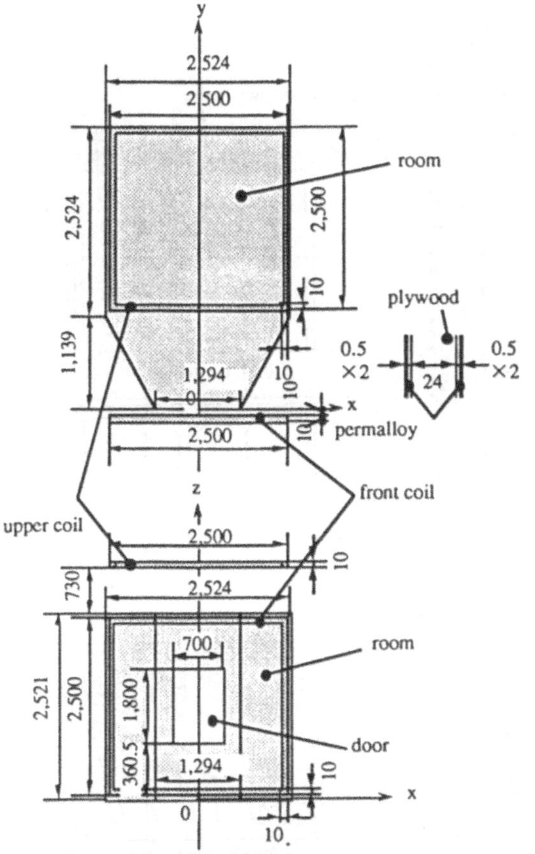

Fig.1 Magnetically shielded room (hexagonal type) and coil.

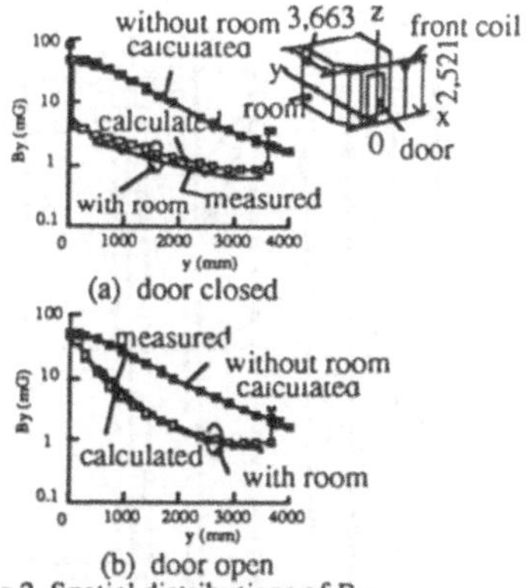

Fig.2 Spatial distributions of By (front coil, x=0, z=1,056mm).

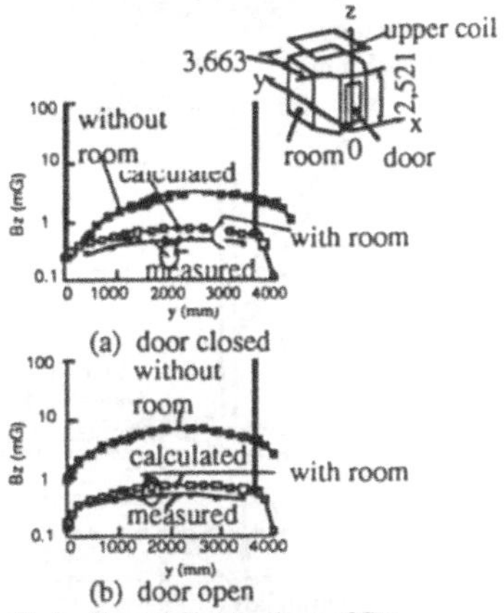

Fig.3 Spatial distributions of Bz (upper coil, x=0, z=1,056mm).

Analysis of a model without gap is carried out for comparison. The magnetic noises of 1Hz and 40Hz are applied by a pair of rectangular coils (2,300mm×1,800mm). The current of 4.0 AT is applied. The generated magnetic noises in the x- and z-directions are 25mG and 15mG in the case without MSR respectively. Fig.5 shows the spatial distribution of flux density vectors at 1Hz with a gap (a), at 40Hz with a gap (b) and without gap (c). Fig.6 shows spatial distributions of eddy current density vectors in the first layer of wall in the case of z-directional noise. At 40Hz with a gap, the direction of magnetic field especially near door is different from that of magnetic noise. This is due to the eddy current concentrating in the wall around the doorway as shown in Fig.6(b), which is much different from that in Fig.6(a). On the other hand, at 1Hz with a gap and 40Hz without gap, the magnetic field is uniform and its direction is in accordance with that of magnetic noise. In the case of x-directional noise, the magnetic field is almost uniform and its direction is in accordance with that of magnetic noise except near the door.

A gap between the door and wall, which isolates them magnetically and electrically, does not affect the magnetic field distribution at low frequency of less than about 1Hz. At frequency of more than several ten Hz, the magnetic field distribution becomes non-uniform, because of undesirable eddy currents due to the gap, especially in case of the z-directional noise.

4. Field Gradients in MSR with Entrance Tunnel without Door

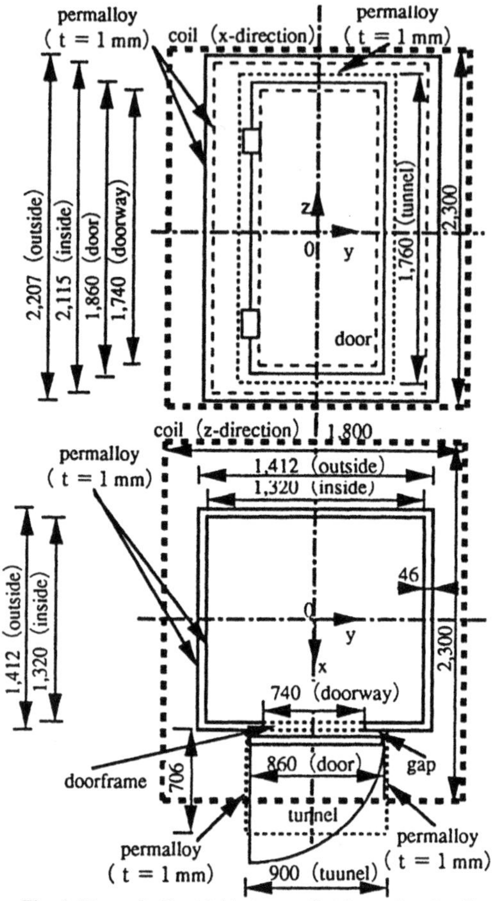

Fig. 4 Magnetically shielded room (brick type) and coil.

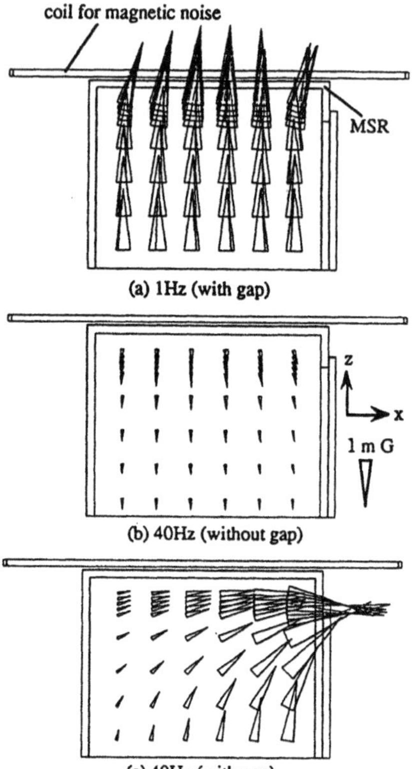

(a) 1Hz (with gap)

(b) 40Hz (without gap)

(c) 40Hz (with gap)

Fig.5. Spatial distributions of flux density vectors for z-direction noise (x-z plane, y=0).

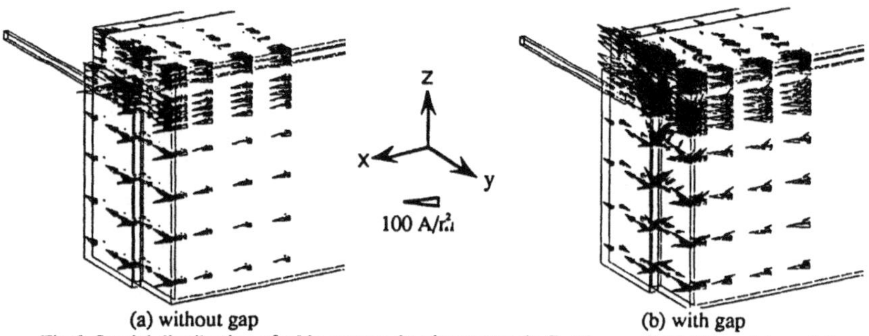

(a) without gap

(b) with gap

Fig.6 Spatial distribution of eddy current density vectors in first layer of shielding plate and door (40Hz, z-directional noise).

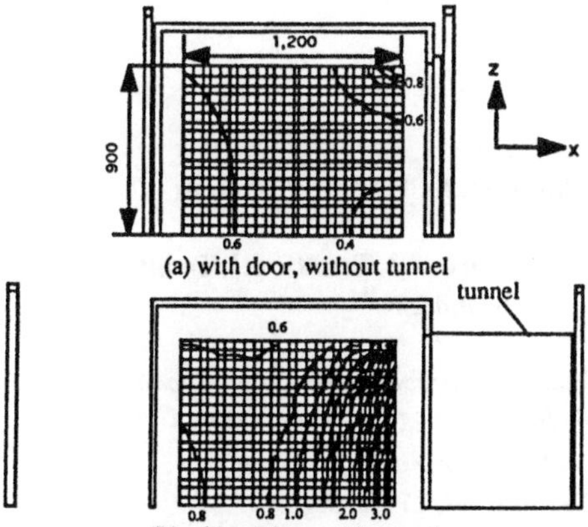

(a) with door, without tunnel

(b) without door, with tunnel

Fig.7 Spatial distributions of magnetic flux density Bx (40Hz, x-directional noise, x-z plane, y=0).

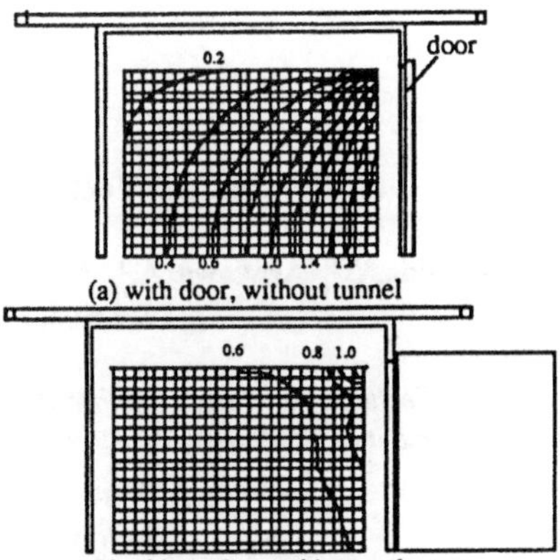

(a) with door, without tunnel

(b) without door, with tunnel

Fig.8 Spatial distributions of magnetic flux density Bz (40Hz, z-directional noise, x-z plane, y=0).

A doorless shielded room is desirable for magnetoencephalography (MEG), especially for a diagnostic study of normal subjects and studying of visual evoked brain magnetic fields [4]. Visual stimuli are presented by a large-screen video projector through the doorway easily, and a special air circulation equipment is not required. Therefore, the magnetic field distribution in doorless MSR with tunnel shown in Fig.4 is discussed. The condition of applied magnetic noise is the same as that described in the chapter 3. Fig.7 shows spatial distribution of magnetic flux density in the case of x-directional noise of 40Hz with a tunnel (a) and without tunnel (b). Fig.8 shows the spatial distribution of magnetic flux density in the case of z-directional noise. In the case of x-directional noise, the magnetic noise decays through the tunnel (Fig.7(a)). The x-component of flux density, Bx, of the center point of MSR is not less than that with a door without tunnel (Fig.7(b)). The tunnel needs to be lengthened to attenuate magnetic flux density as same as that with a door without tunnel. In the case of z-directional noise, spatial distribution of magnetic flux density with tunnel (Fig.8(a)) is more homogeneous than that with a door without tunnel (Fig.8 (b)).

5. Conclusions

The reliability of the 3-D magnetic field analysis to predict the shielding effect of MSR is confirmed. It is understood by the 3-D magnetic field analysis that the magnetic field distribution becomes non-uniform at frequency of more than several ten Hz, because of undesirable eddy currents due to the gap between a door and wall, especially when the magnetic noise is applied in the vertical direction. The shielding effect of MSR must be estimated not only in the direction of applied magnetic noise but also in the other directions. An entrance tunnel without a door contributes to the homogeneities of magnetic flux density in MSR, when the magnetic noise is applied in the vertical direction. In the future, the application of the 3-D magnetic field analysis to superconducting MSR will be reported.

References:

[1] Wills, A.P., On the Magnetic Shielding Effect of Tri-Lamellar Spherical and Cylindrical Shells, *Phys. Rev.*, Vol.IX, 1899: 193-213.

[2] Mager, A., The Berlin Magnetically Shielded Room, In: S.N. Erne, H.D. Hahlbohm, and H. Luddig, *Biomagnetism*, Berlin, Walter de Gruyer & Co., 1981: 51-78.

[3] Yamazaki, K., Okuda, K., Abe, T., Uchikawa, Y., Shimura, T., and Kotani, M., Development of a Prefabricated-Magnetically-Shielded Room of the Hexagonal Type for Biomagnetic Measurement (in Japanese), *Journal of the Magnetics Society of Japan*, Vol.16, No.6, 1992: 769-774.

[4] J. Edrich, P. Teale, M.R. Reite, and J.E. Zimmerman, Doorless and Economic Shield for Clinical Magnetoencephalography, *Proceedings of the Recent Advances Biomagnetism 9th International Conference on Biomagnetics*, 1993: 294-295.

A 129-channel Vector Neuromagnetic Imaging System

Yoshida,Y.[1], Arakawa, A.[1], Kondo,Y.[1], Kajihara, S.[1], Tomita, S.[1], Tomita, T.[1] Takanashi, Y.[2], and Matsuda, N.[3]

Technology Research Laboratory, Shimadzu Corporation[1], Kyoto, Japan
Department of Neurology, Research Institute for Neurological Diseases and Geriatrics, Kyoto Prefectural University of Medicine[2], Kyoto, Japan
Superconducting Sensor Laboratory[3], Chiba, Japan

Introduction

We developed a 129-channel vector neuromagnetic imaging system detecting magnetic fields as a vector value. It is expected that further information can be obtained from vector neuromagnetic field than a conventional system. An axial vector first-order gradiometer was developed as a detection coil, which consists of three first-order differential coils wound on the surface of the same bobbin. These three coils are perpendicular to each other. In addition to a radial component of magnetic fields, two tangential components can be calculated from the measured field at each measuring point. The measuring area with a diameter of nearly 20 cm has 43 measuring points. The DC-SQUID which we developed, adopted a double-washer superconducting ring and employed week link Josephson junctions instead of the conventional tunnel junction. Therefore magnetic flux penetrating the SQUID ring is canceled. In addition, the feedback coil is directly connected with the input coil magnetically so that the current in the detection coil can be canceled by the feedback flux. Therefore a crosstalk between adjacent detection coils is negligible. The read-out electronics is based on the Direct Offset Integration Technique. The system noise in a magnetically shielded room is less than $10fT/Hz^{1/2}$ above 10Hz. We confirmed that this system could measure various evoked responses. In this report the features of this system and an application will be shown.

SQUID and Flux Locked Loop

For a multi-channel use there are two problems; (1) Magnetic field which penetrates SQUID washer directly generates flux noise.
(2) Shielding current flowing in a detection-input loop makes magnetic field affecting other coils as a crosstalk. We have developed DC-SQUID in order to solve these problems[1]. Fig.1 shows a circuit diagram and a photograph of the DC-SQUID including a SQUID washer, an input coil and a feedback coil. The SQUID washer is made of two superconducting rings connected in parallel, so that the SQUID is insensitive to spatially uniform changes in background field. The feedback coil is directly connected with the detection-input loop magnetically instead of the SQUID ring, so that the currents in the detection-input loop are canceled. Therefore a crosstalk between adjacent detection coils is negligible. The characteristics

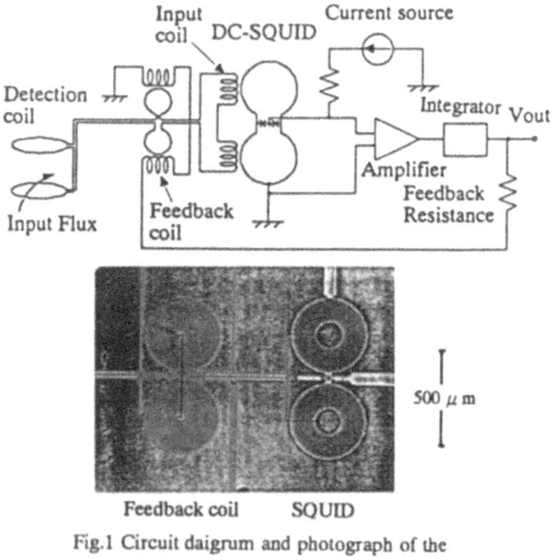

Fig.1 Circuit daigrum and photograph of the DC-SQUID and FLL

Table 1 Characteristics of the DC-SQUID

Critical current	30 μA
SQUID inductance	140pH
Input coil inductance	200nH
Mutual inductance between SQUID and input coil	5.1nH
Coupling cofficient	0.96

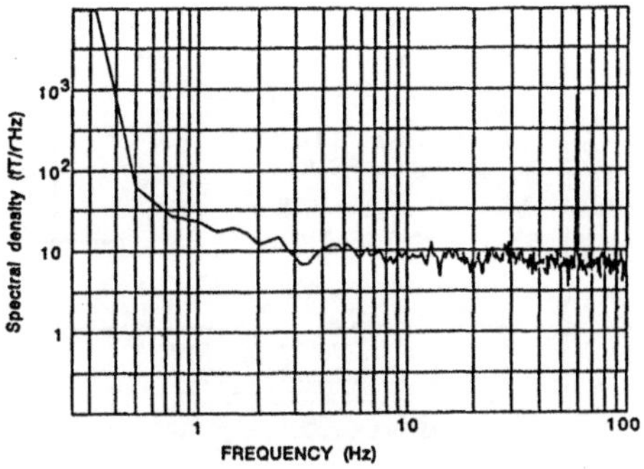

Fig.3 System noise

of the SQUID are shown in the Table 1. Also in Fig.1 shown is a circuit diagram of Flux Locked Loop, which has been developed, based on the Direct Offset Integration Technique[2]. Fig. 3 shows the system noise of the SQUID with the Flux Locked Loop. The noise level in a magnetically shielded room is lower than 10f/Hz$^{1/2}$ above 10Hz. Thus it was confirmed that the system had a sufficient sensitivity for biomagnetic measurement.

INSTRUMENTATION

Fig.4 shows a photograph of the instrument. MEG recordings are performed in a magnetically shielded room made of high-permeability materials. The sensor unit composed of a dewar, detection coils and SQUIDs is mounted on a counterbalanced gantry. So one can move the sensor unit up and down, rotate and tilt in order to detect the magnetic fields over an area of interest. The bed can also be moved toward the sensor unit and perpendicularly.

DC-SQUIDs and detection coils are arranged at the bottom of the dewar made of fiber reinforced plastic and filled with liquid helium. We have developed a vector gradiometer as a detection coil for biomagnetic field. As shown in Fig.5, in the same bobbin three axial first-order differential coils with a 50mm baseline are wound. Each coil of the vector gradiometer has the same elliptic shape; a large diameter of

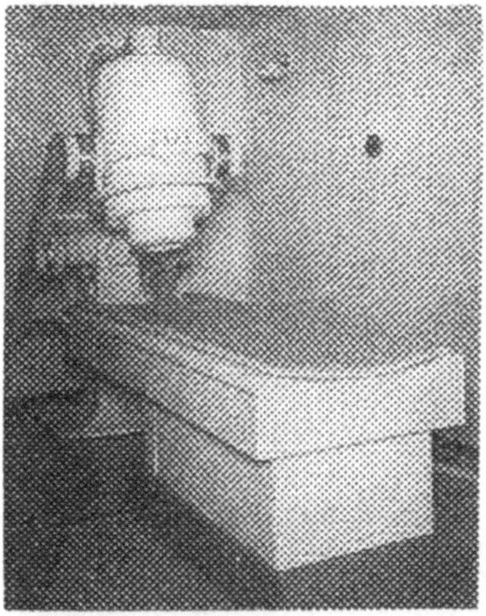

Fig.4 Potograph of the system
(sensor unit, gantory and bed in a magnetically shielded room)

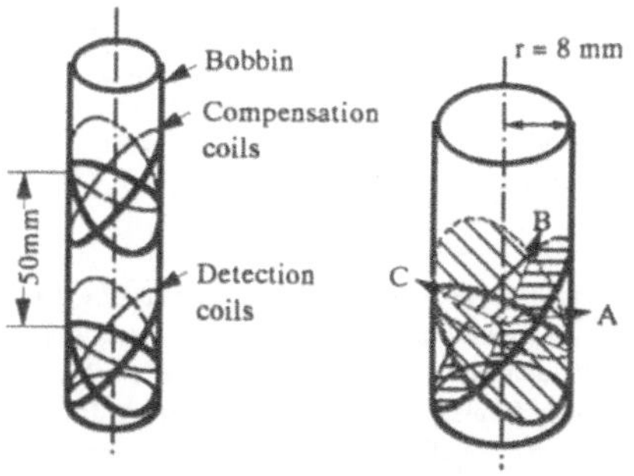

Fig.5 Configuration of the vector gradiometer

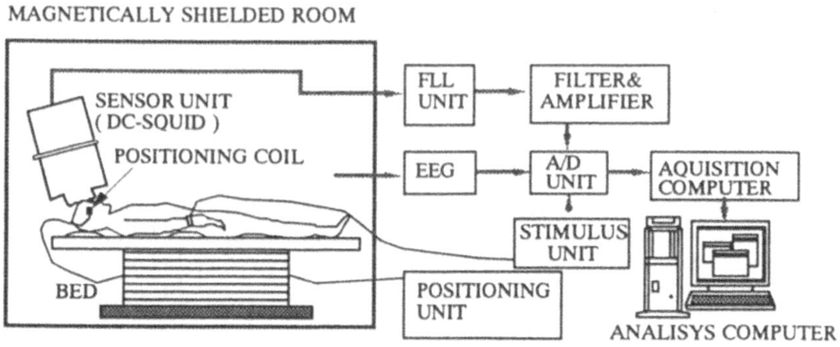

Fig.6 Block diagram of the neuromagnetic imaging system

27mm and a short diameter of 16mm. Three coils have the identical center, and planes being perpendicular to each other. Forty three vector gradiometers are arranged at intervals of 25mm over the area with a diameter of nearly 20cm. Thus there are 129 channels in the sensor unit. At each measuring point, magnetic fields of three directions can be measured. A radial magnetic field and two tangential magnetic fields can be calculated from the measured three values at each measuring point.

Fig.6 shows a block diagram of the neuromagnetic imaging system. To reject outside magnetic noise, the sensor unit is placed in the magnetically shielded room. The outputs of FLL are filtered, amplified digitized by AD converter and stored in a computer for analysis.

A spherical conductor model is used for localization of currents sources in the brain. The spherical conductor model is made by fitting the subject's scalp near the sensor unit to a sphere. Magnetic fields over the surface of the spherical conductor due to a current dipole in it can be expressed by a simple formula[3]. Therefore parameters of the dipoles are obtained by the least square fitting method. Obtained dipoles are superimposed on the MRI of the same subject.

Recording and analysis are performed as follows;

(1) Three positioning coils are set on the subject's head and fed tiny currents. The magnetic fields distribution from the coils are measured by the sensor unit and analyzed later by least square fitting method in order to obtain the position of the head on the coordinate of sensor unit.

(2) Evoked fields or spontaneous fields are measured.

(3) MR images of the subject are taken with three MRI markers made of NiCl set on the same position of the positioning coils.

(4) Positions of the three MRI markers are determined on the MR image manually. And a spherical model is determined from the best fitting sphere to the subject's scalp.

(5) The parameters of a current dipole are calculated by least square fitting on a basis of a sphere conductor model.

(6) The calculated current dipoles are displayed on the MR image of the subject.

Results

In Fig.7 shown are auditory evoked responses detected as a function of time over the right hemisphere. The right ear was stimulated by 1kHz tone bursts with 100 msec duration. The numbers of 1 to 43 indicate each position of vector gradiometer and A, B and C indicate magnetic fields detected by three gradiometers. The maximum peak occurs approximately 86 msec after the stimulus onset. This peak is called N100m, indicates that responses to stimuli reach the auditory area. Fig.8 shows a isofield contour map of the N100m component. The contour map of magnetic field being radial to scalp is shown in the left. The solid lines mean that the fields go out of the head, and the dashed lines mean that the fields come into the head. This is a typical field pattern from a current dipole. The dipole was estimated on the center of two streams. The contour map of magnetic field tangential to scalp is shown in the right. A peak appears just above the current source. One can recognize the tangential map more easily than the radial map.

Fig.9 shows a result of mapping a current dipole of N100m on MRI. Single current dipole fitting was done on the N100m component. The obtained dipole is superimposed on the MRI. The dipole is in the Sivilian fissure, thus the position is reasonable anatomically.

We are now estimating the performance of this system, applying to various evoked responses[4].

Conclusion

We have newly developed 129-channel vector neuromagnetic imaging system. It was confirmed that this system had a sufficient sensitivity for neuromagnetic measurements.

We detected auditory evoked responses successfully and a current dipole was located from the data by least square fitting using a sphere conductor model as the brain .

References

[1] Shinada K. Double Washer dc SQUIDs with Short Weak Link Junctions, ISCE95

[2] Drung D., Low-noise high-speed dc superconducting quantum interference device magnetometer with simplified feedback electronics, Appl. Phys. Lett., 1990, 57: 406-408

[3] Savas, J. Basic mathematical and electromagnetic concepts of the biomagnetic inverse problem, Phys. Med. Biol. 1987,32: 11-22

[4]Takanashi Y. , Visualization of Tonotopy of Auditory Cortex Using A Newly Developed 129 Channel Vector Magnetoencephalograpy, in this volume

Fig.7 Auditory evoked response

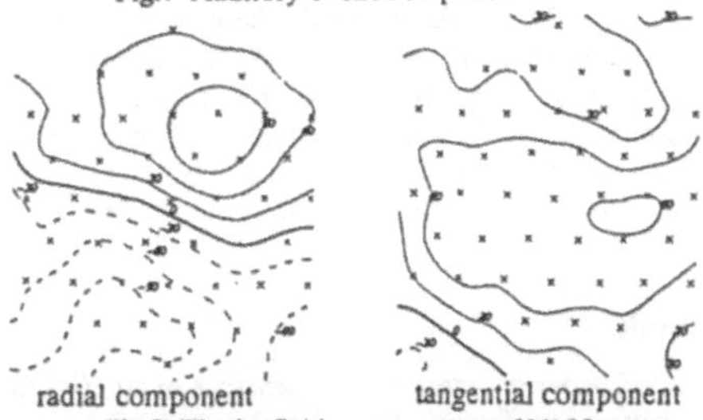

Fig.8 The isofield contour map of N100m

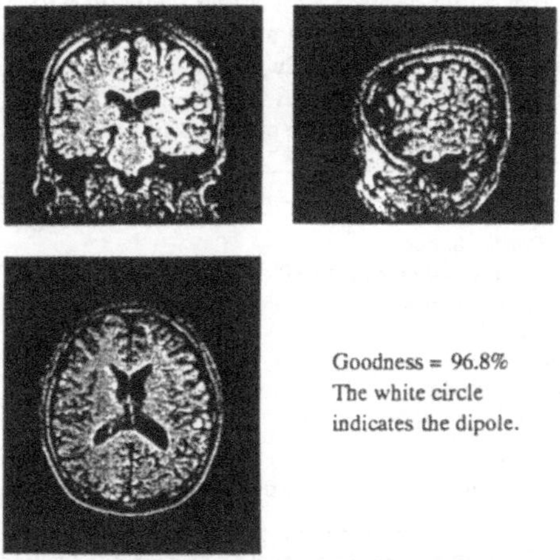

Fig.9 Current dipole mapping on MRI

Magnetocardiography Using SQUIDs Cooled by Liquid Nitrogen

Zhang, Y.[1], Schütt, H.[1], Bousack, H.[1], Wolf, W.[1], Schirdewan, A.[2], Burghoff, M.[3], and Trahms, L.[3]

Institute of Thin Film And Ion Technology, Forschungszentrum Jülich GmbH (KFA)[1], Jülich, Germany; Franz-Volhard-Klinik, Humboldt Universität[2], Berlin, Germany; Physikalisch-Technische Bundesanstalt[3], Berlin, Germany

Introduction

SQUIDs cooled by liquid helium have been used for more than 25 years for recording magnetocardiograms (MCGs). Nowadays, SQUIDs from high temperature superconductors (HTS) operating at the boiling point of liquid nitrogen, i.e. at 77 K, are available now, thus simplifying the handling and making systems more cost effective. So far, the relatively high noise level of these systems has prevented their use as a diagnostic tool. This paper gives evidence that state-of-the-art HTS SQUID sensors have a sufficiently low noise level to allow diagnostically relevant investigations of human heart activity by using the fragmentation of the QRS complex.

Methods

In order to assess the performance of an HTS SQUID system, we used a low temperature superconductor (LTS) system as reference. For optimal measuring conditions, we used both systems in magnetometric mode inside the magnetically shielded room of PTB Berlin (BMSR) [1]. The parameters of both systems are shown in table 1. The HTS SQUID system was a single channel rf SQUID system developed by KFA [2]. Together with a flux concentrator from bulk $YBa_2Cu_3O_7$ with a diameter of 25 mm, a white noise level of 35 - 40 fT/√Hz was obtained. The LTS dc SQUID system developed by the PTB Berlin [3] had a field resolution of 4.5 fT/√Hz. The distance of the sensors to the outer surface of the dewars was 6.5 mm (HTS) and 10.7 mm (LTS).

	HTS system (KFA)	LTS system (PTB)
type	HTS rf SQUID magnetometer	LTS dc SQUID magnetometer
number of channels	1	1
sensitivity at 200 Hz in BMSR	35 fT/√Hz	4.5 fT/√Hz
diameter of sensor	25 mm (concentrator)	7 mm (multiloop-SQUID)
band width	16 mHz - 250 Hz	16 mHz - 250 Hz
sample rate	1 kHz	1 kHz
transfer function	40 $\mu V/\Phi_0$	1 mV/Φ_0
field-flux-transfer coefficient	0.5 nT/Φ_0	0.47 nT/Φ_0
slew rate	$5 \cdot 10^4$ Φ_0/s	$1 \cdot 10^4$ Φ_0/s
distance sensor - outer surface of cryostat	6.5 mm	10.7 mm

Table 1: Parameters of the HTS and LTS system in the Berlin magnetically shielded room (BMSR)

Both sensors were positioned 125 mm caudal of manubrium sterni, with a distance of 11 mm between the SQUID and the chest, thus recording the normal component of the magnetic field of the heart. The distance of 11 mm was chosen to have the same measurement position for HTS and LTS SQUID. With the HTS system it is possible to reach a minimum distance of 7 mm, see table 1. After analog high-pass and low-pass filtering between 0.016 Hz and 250 Hz, the data were digitized with a sampling frequency of 1 kHz. After averaging over 100 beats the signal to noise ratio (S/N) was 1000 for the LTS and 150 for the HTS system. Here S/N is defined as ratio of base-to-R-peak to base-to-peak of the noise. The high values of S/N are due to the small distance between the source and the sensor and demonstrate the state of the art attained today for MCG measurements by LTS and HTS systems.

To extract diagnostically relevant information from these measurements, the high frequency components between 37 Hz and 90 Hz of the averaged MCGs were analyzed both for healthy subjects and patients prone to arrhythmias after myocardial infarction. The nonrecursive filter algorithm used for this purpose has already been described elsewhere [4]. After filtering, a score value S, which is used for quantifying the pattern of the filtered MCG, can be calculated according to Fig. 1:

$$S = \frac{N}{B_{f\,max}} \cdot \sum_{i=0,N-1} \left| B_f(t_{i+1}) - B_f(t_i) \right|$$

with

$$B_{f\,max} = Max_{i=0,N}\left[(B_f(t_i)\right] - Min_{i=1,N}\left[(B_f(t_i)\right]$$

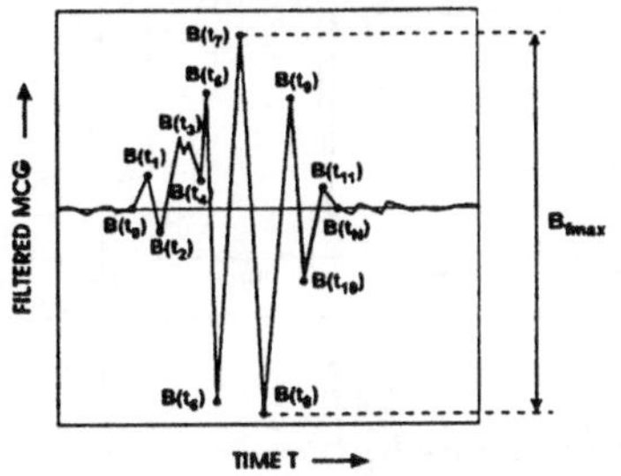

Fig. 1: Determenation of fragmentation score [4]

Results

The quality of the HTS MCG raw data is comparable to the quality of LTS measurements performed about ten years ago. Fig. 2 shows the MCGs of an healthy person after increasing the S/N values by averaging (100 beats). Only an enlargement of the fine pattern demonstrates the better noise level of the LTS SQUID system.

Fig. 3 shows the filtered HTS and LTS MCGs of an healthy person and an arrhythmia patient after signal processing with the nonrecursive filter algorithm. The high-frequency components for the healthy person obtained with the HTS and LTS SQUID system both show the well known regular morphologies of the QRS complex. An irregular fragmentation of the QRS complex of the arrhythmia patient was well resolved by both systems. This kind of fragmentation has been shown to be characteristic for patients with ventricular arrhythmias after myocardial infarction [5]. The comparison of HTS and LTS filtered data shows a good morphological resemblance. The score values derived from these data are indicated in fig. 3. In spite of the higher noise level of the HTS system compared to the LTS system, the score values of both systems differ only by 2 - 3%. This means that for this application the sensitivity of the HTS system is sufficient.

Fig. 4 shows a typical example of the dependence of the peak-to-peak QRS amplitude on the distance chest - HTS-SQUID. The increase of signal intensity with smaller distance is obvious. This makes evident, that HTS systems can access regions of high biomagnetic signal intensity that are more difficult to investigate with LTS systems.

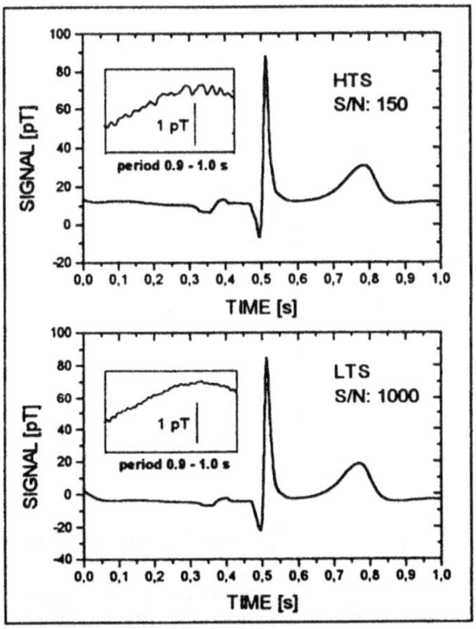

Fig. 2: MCGs of a healthy person taken by a HTS and LTS system in the Berlin magnetically shielded room after averaging over 100 beats

Discussion

This study shows that the resolution of today's HTS SQUID systems is an appropriate tool for risk stratification of life threatening ventricular arrhythmias. Score values received with HTS and LTS state - of - the - art systems in the good

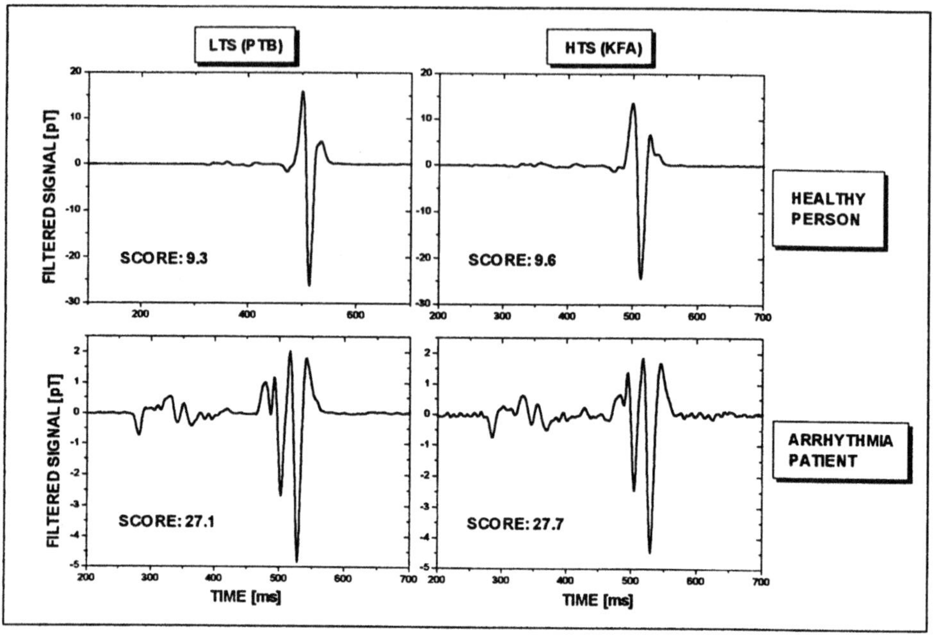

Fig. 3: Comparison of the filtered HTS and LTS MCGs of an healthy person and an arrhythmia patient. The calculated score of the different data sets are indicated separately.

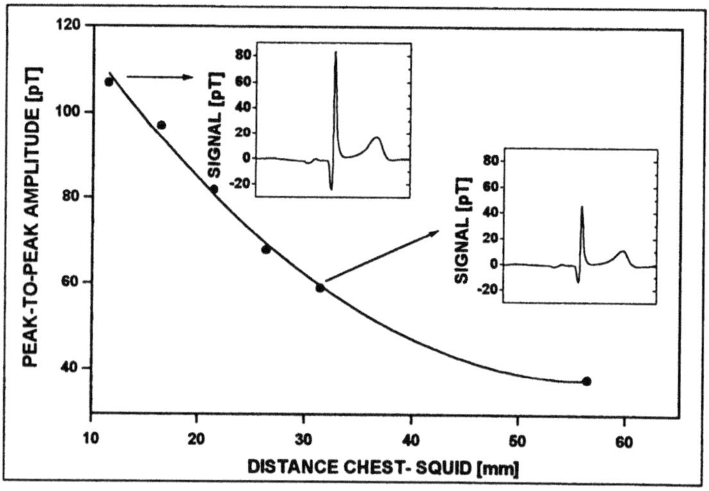

Fig. 4: Dependence of the peak-to-peak QRS amplitude on the distance chest - SQUID, recorded with the HTS system.

shielding conditions of the Berlin magnetically shielded room differ by only 2 - 3 %.

Because of the good shielding in this study, an HTS magnetometer SQUID system was used. For applications in medical clinics, only moderate standard shielding rooms are conceivable.

For this purpose, an HTS gradiometer system with sufficient field resolution has to be developed [6]. First results with a new designed first order gradiometer HTS SQUID system using a digitally controlled flux locked loop [8] in the new moderately shielded room [7] of KFA demonstrate that the new system is an excellent base for further developments. The system controlling, data acquisition and signal processing integrated with an easy-to-handle user interface of a personal computer is designed for the application in medical clinics.

A cost-effective HTS technology and an affordable magnetically shielded room can help to introduce magnetocardiography into medical clinics. Today, the system technology is at hand [9]. The cheapest solution would be an HTS system that operates with sufficient sensitivity in an unshielded environment. The experience with LTS systems for medical applications shows that this will not be reached as a general solution for all kinds of locations. An aluminum shielded room with active compensation could be a concept for minimum shielding.

To evaluate clinical applications of magnetocardiography, i. e. risk analysis of sudden cardiac death and localization of arrhythmogenic tissues with HTS one-channel and multi-channel systems, a group of medical clinics and research centres in Germany have joined. The task of this group is to establish magnetocardiography with HTS SQUID systems as an useful diagnostic tool in cardiology.

References:

[1] Erné, S.N., Hahlboom, H.D., Scheer, J., Trontelj, Z., The Berlin magnetically shielded room (BMSR) - Section B Performances, in: S. N. Erné, H.-D. Hahlbohm and H. Lübbig, Biomagnetism, Berlin, de Gruyter-Verlag, 1981.

[2] Zhang, Y., Krüger, U., Kutzner R., Wördenweber, R., Schubert, J., Zander, W., Sodtke E., Braginski, A.I., Strupp, M., Single-layer $YBa_2Cu_3O_7$-SQUID magnetometers with direct-coupled pick up coils and flip-chip flux transformers, Appl. Phys. Lett,1994, 65: 3380-3382.

[3] Drung, D., Koch, H., An electronic second-order gradiometer for biomagnetic applications in clinical shielded rooms, IEEE Trans. on Appl. Supercond., 1993, 3: 2594-2596.

[4] Link, A., Kittan, J., Burghoff, M., Trahms, L., Nichtrekursive Filter zur Selektion elektrophysiologischer Signale, Technisches Messen, 1994, 61: 23-32.

[5] Brockmeier, K., Schmidt, S., Kreiseler, D., Trahms, L., Schmitz, L., Oeff, M., Hahlboom, H.-D, Fragmented ventricular depolarization detectable in the QRS-complex of the magnetocardiogram after filtering with binominal non-recursive filtres, in: Computers in cardiology, IEEE Computer society press, 1994.

[6] Copetti, C. A., Borgmann, J., Soltner, H., Tavrin, Y., Zhang, Y., Otto, R., Brandenburg, G., Clemens, U., Zimmermann, E., Bousack, H., Halling, H., Magnetocardiogram recorded by first order gradiometers using HTS rf SQUIDs, this conference

[7] Davidson, C.R., Griffin, R.H., Best, K.J., Maltin, L., Soltner, H., Wolf, W., Zhang, Y., Bousack, H., Design of a magnetically shielded room for development of HTS SQUID systems for magnetocardiography, this conference

[8] Zimmermann, E., Clemens, U., Brandenburg, G., Halling, H., Kompensationsregelung für extrem nicht linearen Sensor mit digitalem Signalprozessor, in: Proceedings of the DSP Deutschland '95 conference, Munic, Germany, 1995

[9] Braginski, A.I., Borgmann, J., Bousack, H., Brandenburg, G., Clemens, G., Copetti, C.A., Halling, H., Otto, R., Soltner, H., Tavrin, Y., Zhang, Y., Zimmermann, E., Status and trends in liquid-nitrogen-cooled rf SQUID systems for biomagnetic applications, this conference

Acknowledgments

The authors gratefully acknowledge the support by A. I. Braginski, H. Koch and the SQUID system group of KFA. The work of KFA was supported by the German BMBF.

II. Models For The Biomagnetic Forward and Inverse Problems

Comparison between different approaches to the biomagnetic inverse problem — workshop report

[1]B. Lütkenhöner, [2]R. Greenblatt, [3]M. Hämäläinen, [4]J. Mosher, [5]M. Scherg, [3]C. Tesche, and [6]P. Valdes Sosa

[1]*Institute of Experimental Audiology, University of Münster, Münster, Germany*
[2]*Source/Signal Imaging, San Diego, USA*
[3]*Low Temperature Laboratory, Helsinki University of Technology, Espoo, Finland*
[4]*Physics Division, Biophysics Group, Los Alamos National Laboratory, Los Alamos, USA*
[5]*Department of Neurology, University of Heidelberg, Germany*
[6]*Cuban Neuroscience Center, Havana, Cuba*

Introduction

In view of the eminent importance models have for the interpretation of biomagnetic data, it is a matter of concern that a generally accepted strategy for the analysis of experimental data is not in sight. It seems evident that the future prospects of biomagnetic investigations as compared to other functional imaging techniques (e.g. PET, fMRI) will decisively depend on the success of new data analysis techniques. The necessity to develop new modeling strategies is underlined by the fact that the availability of whole-head magnetometer devices necessitates the development of better models: First, more realistic volume conductor models are required, because simple models like a homogeneous sphere are not capable of providing a good approximation if the measurement area covers the whole head rather than a relatively small area. Second, the number of sources to be considered generally increases with increasing measurement area so that simple source models like the equivalent current dipole become more and more obsolete.

These considerations gave rise to the idea to organize a workshop on modeling. In view of the given time constraints it was clear from the beginning that the workshop had to be confined to a relatively narrow topic. The topic chosen was *source modeling of the brain* with special emphasis to the so-called *inverse problem*, i.e., the estimation of the generators in the brain from magnetic field data measured outside the head. Compared to this problem, problems related to the volume conductor or other modeling aspects appear less serious since they are more of a technical rather than a principal nature. A second strategic decision was to organize the workshop so that experimenters and theoreticians could meet. A consequence of this decision was that complicated mathematical theories had to be banned, and a platform accessible to everybody in the audience had to be defined. Datasets fulfill this requirement. While experimental datasets have the indisputable advantage of being realistic (though any given data set can be representative only for a certain type of experiment), they have the disadvantage that results of different inverse procedures can be compared only among themselves, but not with actual facts. Thus, the final idea was that each panel member functions as the advocate of a certain modeling approach and had the opportunity to provide a dataset chosen so that the advantages of this approach become as clear as possible. The resulting datasets were to be analyzed by all panelists. The provider of a dataset kept the idea underlying his dataset secret so that all the other panelists had to do their analysis blindly. The intention of this kind of "game" was to give a fair chance to both the panelists (who were able to illustrate the advantages of the own approach with an optimally chosen data set) and the audience (which was confronted also with results obtained for non-optimally chosen data sets). It should be emphasized that this "game" was never intended to have a winner. It is in the very nature of the biomagnetic inverse problem that an algorithm superior under all conceivable conditions does not exist, i.e., an algorithm always pays a price for having advantages under a certain condition: disadvantages under other conditions. For this reason it would be counterproductive in the present stage to single out methods. The needs of the moment are understanding the advantages and disadvantages of the different approaches to the biomagnetic inverse problem and skillful design of modified approaches which adopt promising concepts from other methods.

To achieve the goals of the workshop, an intensive interactions between the panelists took place *before* the workshop — exploiting the new possibilities offered by the Internet. A detailed description of all data sets and all analysis results would evidently be beyond the scope of this report. Thus, in what follows, only a brief overview is given and a few details are spotlighted. Additional material (including the original data sets) are available on the Internet [11].

Model Environment for Comparison of Methods

The modeling environment as used in all simulations is displayed in Fig. 1. The head is represented by a homogeneous sphere. The source space (not visualized in the figure) was assumed to have a radius of 8 cm and to extend from the "north pole" to a plane 4 cm below the "equator" plane. The magnetometer system comprises 148 radially oriented coils in a distance of 12 cm from the center of sphere.

Data Sets

Simulation of Greenblatt

The dataset was chosen to illustrate in a simplified fashion some of the dynamics that might be encountered in an evoked response. Two discrete dipoles had relatively slow time courses, such as those produced by motor readiness fields. These sources were located approximately symmetrically with respect to the midsagittal plane. A third, lateral, dipole had a more rapid time course, similar to those encountered in a variety of sensory and motor evoked responses. The fourth dipole was located frontally and had a quasi-periodic time course characteristic of endogenous rhythmic activity.

Simulation of Hämäläinen

The simulation of Hämäläinen was designed to mimic actual evoked-response signals following the stimulation of the left median nerve at wrist. A total of five dipoles were inserted. The first two were at the right SI hand area, formed a small angle with respect to each other, and exhibited overlapping activity up to 110 ms. The next pair of sources was symmetrically located near the two Sylvian fissures, presumably at SII. These sources had almost identical time behavior. The fifth source was posterior to the right SI and closer to the midline, corresponding to the 'posterior parietal source' (PPC) recently found experimentally [3].

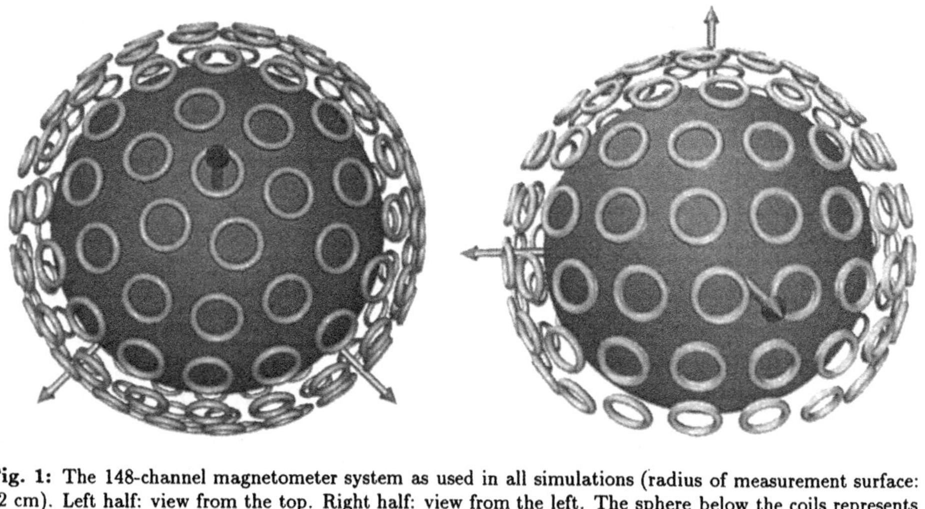

Fig. 1: The 148-channel magnetometer system as used in all simulations (radius of measurement surface: 12 cm). Left half: view from the top. Right half: view from the left. The sphere below the coils represents the scalp (radius 10 cm).

Simulation of Mosher

The simulation of Mosher was designed to test a few concepts in a simple manner. Three dipoles were placed in relatively close proximity near the top of the cortex, about 75 mm on the z-axis, with roughly a 3 cm spacing. Two of the dipoles were oriented mostly parallel and were each given a Hamming function time series (a single Gaussian-appearing "hump") that overlapped about 50%. A third dipole was oriented mostly orthogonal to the pair, but it's time series was a scaled copy of the first dipole; hence, this third dipole was synchronous with the first. White Gaussian noise was added to the data, such that noise equaled about 12% of the total variance of the data. Although the first two dipoles are asynchronous to the other, their parallel alignment allows the possibility of an intermediate single "rotating" dipole model, particularly in the presence of noise. Although the third dipole was orthogonally aligned, it's synchronous time series does not allow a standard MUSIC approach.

Simulation of Scherg

The simulation of Scherg consisted of a deep brainstem dipolar source, one deep and one shallow source dipole in the posterior bank of the right central sulcus (reafferent somatosensory activity), and bilateral synchronous sources in the cingulate gyri and prerolandic motor cortex with preceding slow activity patterns. All activities overlapped considerably in time. The simulation intended to demonstrate the usefulness of the equivalent dipole model in the case of multiple extended cortical activities. Hence, cingulate gyrus dipole sheets of about 40 × 25 mm were constructed to either side of the interhemispheric cleft with a separation of 12 mm (Fig. 2). Because the left sheet was more superficial, the left dipole field dominated in the sensors. The prerolandic motor cortex strip was modeled on each side with three dipoles expanding over a curved strip of about 40 × 10 mm. Activities were overlapped, and noise from 60 random dipole sources in the brain was added.

Methods

Signal-space Projection (Tesche)

The inverse problem in MEG consists of determining current flow in the brain from the magnetic field pattern measured by an array of detectors placed over the scalp. There are an infinite number of different current distributions which may have produced the measured fields. Any algorithm which selects out a single distribution must incorporate additional information which constrain the possible set of solutions. In many cases, it may be difficult to select appropriate constraints, even when guided by "physiological considerations". Thus, the question arises as to whether information can be extracted from MEG data without "solving" the inverse problem?

MEG provides excellent temporal resolution compared to many other imaging modalities. Thus it is useful to try to extract waveforms from the data for components of the signal which may represent neuronal populations or networks. In the formulation of signal-space projection (SSP) presented here, a component is identified by a pattern of signals in the array. In technical terms, a component of the signal is represented by a unit vector in a N-dimensional signal space, where N is the number of detectors in the array. The k-th element of the vector is the signal recorded by the k-th detector [8, 9, 10]. Each unit vector in signal-space which is used to identify a component may also be used to approximate a magnetic field topography over the scalp. The SSP waveforms determined from the data for each component describe how strongly each of these topographies is represented in the signal at every sampled instant. In particular, the sampling frequency of the SSP waveforms is identical to that of the original data.

SSP components may be selected directly from the data. However, it is also possible to calculate the magnetic field pattern for a specific current distribution using a spherical or realistic model for the brain. Anatomical information obtained from other functional imaging technologies may be included into the definition of such a current distribution. Moreover, ECDs or more complex current distributions determined by any of the techniques described elsewhere in this workshop may also be used to define SSP components from the corresponding forward-modeled magnetic field patterns. All of these components are described as vectors in a common signal-space, and thus may be used together in the analysis. Thus, SSP provides a very

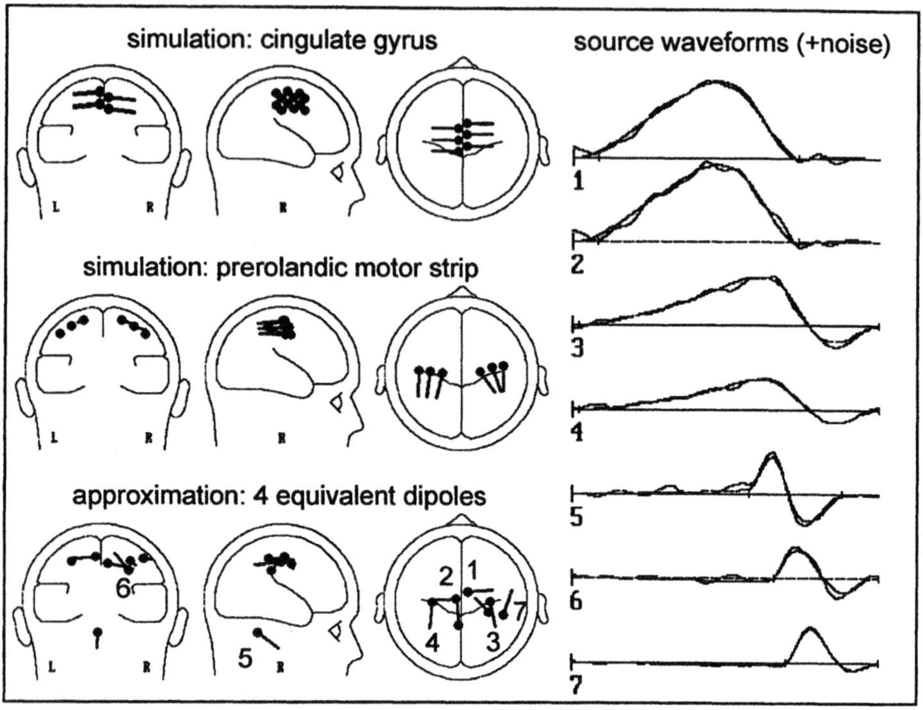

Fig. 2: Simulation of bilateral dipole sheets in cingulate gyrus (CG) and bilateral dipole strips in prerolandic motor cortex (PM). Source waveforms are identical within sheets and differ between cingulate gyrus (1-2) and motor cortex (3-4). Dipoles 1-4 are close to the center of the sheets they reflect. Brainstem and somatosensory cortex sources (5-7) are only shown in the equivalent solution with 7 dipoles (left bottom). On the right, source waveforms with and without added noise are displayed.

general framework for the description of the temporal behavior of neuronal populations, as long as they may be identified as SSP components.

SSP waveforms may be determined from the data using a simple algorithm (the pseudoinverse of the source matrix) [19]. Thus the SSP waveforms may be computed without determining a unique solution to the inverse problem, or even without identifying any specific current distributions which may have generated the SSP unit vectors. Moreover, the contribution of additive system noise (i.e. SQUID noise and ambient magnetic field fluctuations) to each SSP waveform can be estimated by applying the same algorithm to data recorded by the array without the subject [19, 20].

Spatio-temporal multiple source model (Scherg)

The spatio-temporal multiple source model [18] explains the surface recorded EEG/MEG (matrix **A**) as the linear superimposition of the compound activities arising from multiple equivalent sources. Sources are fixed in space and orientation, i.e. each source is associated with a stable spatial pattern (topography **C**). Changes over time are described by the time varying magnitudes of the source activity (source waveforms **S**). The number of sources is discrete, usually much smaller than the number of sensors, but for each active brain region at least one equivalent source is needed. The linear model is defined as: $\mathbf{A} = \mathbf{C} \cdot \mathbf{S}$ (matrix annotation) or $\mathbf{a}(t) = \sum \mathbf{c}_i \cdot \mathbf{s}_i(t)$ (vector annotation). Each topography $\mathbf{c}_i$ may be calculated from the forward model of an equivalent dipole, a dipole sheet or a more complex source distribution describing one brain activity with a particular temporal evolution. Linear combinations of measured topographies, for example from voluntary eye movements, may be used to separate overlapping artifacts or source activities for which a model has not been established. Components thus selected are equivalent to SSP components (Tesche). Each source or artifact activity must be described by at least one topography to be separable.

To find a consistent multiple source model various strategies have been proposed, for example sequential strategies using temporal information and regional source strategies [17]. A subset of sources can be fit to the data during one epoch and held fixed when fitting more sources in a (later) epoch when there is more complex overlap. Most of these strategies rely on a priori selection of the number of sources, operator interaction and use of temporal information. Recently, more robust automated approaches using genetic algorithms [22] or subspace scanning methods [12, 14] have been proposed to improve convergence to the global minimum.

Multi-dipole models optimized with a genetic algorithm (Hämäläinen)

The first source estimation approach taken in MEG and EEG was the use of a single moving current dipole. This model was subsequently extended to include multiple, fixed-location sources whose amplitudes are allowed to vary with time. The locations of the dipole sources have to be determined with a nonlinear least-squares search. Apart from simple situations with sources separated by several centimeters or forming a large angle with respect to each other, conventional search methods are not guaranteed to reach the global minimum of the least-squares criterion function without sufficiently good initial estimates for the dipole positions. To overcome this problem, a global genetic optimization algorithm was developed [22].

The idea of the genetic algorithm is to mimic the evolution of species. The parameters of the dipoles are coded as a binary sequence called a chromosome. The chromosomes are mutated, new offspring are produced by combining parts of existing chromosomes, and the chromosomes producing largest mismatch with the measured data are discarded. Both binary mutations (producing large changes to locate new viable source areas) and normally distributed mutations (to adjust the dipole parameters locally) were used. Only the dipole locations were coded into the chromosomes. At each set of candidate locations, the dipole amplitudes and orientations were optimized with linear optimization methods [14].

MUSIC Recursively Applied to the Generation of Source Models in EEG and MEG (Mosher)

The MEG-MUSIC algorithm [14] localizes multiple asynchronous dipolar sources by scanning a single dipole model across a signal subspace extracted from spatio-temporal MEG or EEG data. An extension of this method, Recursively APplied MUSIC (*RAP-MUSIC*), allows the use of the subspace scanning concept to build more complex source models [13]. RAP-MUSIC generalizes to combinations of single dipoles, synchronous dipole pairs, and quadrupoles. Furthermore, by recursively building the source model, RAP-MUSIC avoids the implementation problem encountered in MUSIC of searching for multiple peaks in the scanning metric.

Prior to processing the data, about 1000 dipole locations roughly evenly spaced over the 80 mm cortical surface were created. About another 1000 dipole locations were spaced at increasing separations as a function of depth, for a total of about 2000 dipole locations over the 80 mm radius spherical volume extending from z = -40 mm and upwards. At each of these dipole locations, the gain matrix was generated for the spherical forward model for radially oriented sensors; in Cartesian coordinates, this matrix was 148 by 3 for each dipole location. Each matrix was then decomposed with an SVD, and the two dominant left singular vectors were stored (i.e., the radial component was not).

Distributed source solutions using signal space covariance information (Greenblatt)

Distributed vs. discrete solutions. Distributed solutions generally assume a large number of possible sources, each one fixed to a lattice point. The lattice itself may be restricted to a representation of the cortical surface, if this information is available [2]. Using well understood mathematical and physical methods, a linear inverse operator may be constructed to estimate source magnitudes from the measured field data. Discrete solutions assume a relatively small number of dipoles whose positions may vary. Discrete solutions provide a simplified equivalent source model that is adequate for many purposes, but require significant human operator intervention, so that different starting condition may produce different models, all of which can account for the data. Distributed solutions are attractive because of their objectivity. Since the inverse solution depends only on explicit assumptions of the algorithm, all human operators using the same assumptions (and parameters) should obtain the same results.

Weighting and regularization. Conventional minimum norm inverse operator techniques [5] favor shallow sources, so weighting must be used to correct this. The most commonly used technique is lead field normalization [1], but signal space covariance [2, 4] may also be used. Unless suitable precautions are taken, generalized inverse solutions to the biomagnetic inverse problem are unstable in the presence of noise. Regularization refers to the class of methods employed to dampen unwanted noise sensitivity [6]. The most commonly used regularization methods in biomagnetism are Tikhonov regularization (which may include the use of a discrete Laplacian operator [15]) and truncated singular value decomposition.

Signal space covariance. The (zero lag) covariance between sensors i and j, $c_{i,j}$, for a signal sampled discretely with period Δt, is given by $c_{i,j} = \frac{1}{N}\sum_{n=1}^{N}(m_i(n\Delta t) - \mu_i)\cdot(m_j(n\Delta t) - \mu_j)$ where m_i is the measured signal at the i-th detector, and μ_i is its mean over the interval defined by N. The signal space covariance matrix is a square array whose elements are defined by this relationship. The covariance matrix is a statistical signature of the spatial patterns in the signal. If sources are not completely correlated, the sources are defined by their spatial patterns, and these may be used to obtain information about the activity of the sources. Statistical techniques [16] are available to estimate the number of components (or dimensions) needed to adequately reproduce the signal, given some noise level. These components span the signal subspace, and the remaining components span the noise subspace.

The inverse operator. Once the covariance matrix has been estimated for the interval of interest, the source space may be scanned iteratively (one lattice point at time) to determine the principal orientation, i.e. the orientation (at each location) associated with the smallest projection into the noise subspace. Each principal orientation may be assigned a weight based on the covariance matrix (the PRoMS algorithm as described in [4]). However, there are possible numerical instabilities using this approach with experimental data, and as an alternative, lead field normalization may be used, combined with regularization by singular value truncation. This approach (the principal orientation linear inverse operator) has been used for the analysis of the model data sets described below.

VARETA (Valdes Sosa)

Two types of solutions have been studied: *LORETA* (Low Resolution Electrical and Magnetic Tomography), a distributed solution that imposes a maximal spatial smoothness constraint (implemented according to [15] and detailed instructions from R. Pascual-Marqui), and a new technique, *VARETA* (Variable Resolution Electrical and Magnetic Tomography). Variable resolution allows VARETA to estimate simultaneously both discrete and distributed sources, thus breaking down the apparent distinction between dipole fitting and variants of the minimum norm approach. VARETA is based upon Tikhonov regularization and may be formulated in a Bayesian context. Specifically, it estimates the x-y-z components of the current density $\mathbf{J}$ at a grid of points inside the sphere for each time instant:

$$\hat{\mathbf{J}} = \arg\min_{J}\left\{\frac{\|\mathbf{B}-\mathbf{K}\cdot\mathbf{J}\|^2}{\sigma^2} + \frac{\|\mathbf{\Lambda}\cdot\mathbf{L}\cdot\mathbf{W}\cdot\mathbf{J}\|^2}{\mu^2} + \frac{\|\mathbf{L}\cdot\ln\left(\mathrm{diag}(\mathbf{\Lambda})\right) - \bar{\alpha}\|^2}{\tau^2}\right\} \tag{1}$$

where $\mathbf{B}$ denotes the magnetic measurements, $\mathbf{K}$ the lead field, and $\mathbf{L}$ the 3D Laplacian operator. $\mathbf{W}$ is a diagonal weight matrix introduced to avoid superficial concentration of solutions. $\mathbf{\Lambda}$ is a diagonal matrix of the weight given to regularization at each point in space. The hyperparameters are σ (measurement noise standard deviation), μ (standard deviation of the product $\mathbf{LWJ}$), τ (standard deviation of the logs of the diagonal elements of $\mathbf{\Lambda}$), and $\bar{\alpha}$ the mean of the logs of the $\mathbf{\Lambda}$). LORETA is a special type of VARETA that occurs when $\mathbf{\Lambda} = \mathbf{I}$ (the identitity matrix). Given the hyperparameters, there is an explicit estimate for LORETA and an iterative one for VARETA (the latter starts from an initial matrix ($\mathbf{\Lambda}$, usually $\mathbf{I}$). The hyperparameters are themselves taken to minimize the Generalized Cross Validation criterion GCV [23] common to all time instants. GCV measures the overall goodness of fit of the solution.

The λ_i on the diagonal of $\mathbf{\Lambda}$ determine the resolution of the given solution. When the λ_i are equal, VARETA has the same average (low) resolution as LORETA. At any point i, when $\lambda_i \to \infty$, the current density at that point is estimated to be zero, causing a data driven pruning of the variables to be estimated.

When $\lambda_i \to 0$, then the current at the point is not included in the regularization and is estimated as a discrete source, simultaneously with other possible discrete and distributed sources.

As all distributed solutions, both LORETA and VARETA generate waveshapes $J_x(t)$, $J_y(t)$, $J_z(t)$ for all 3 components of **J** at all grid points (either 3 × 1153 or 3 × 2048). This is an overwhelming mass of data, and new procedures are required for deciding which spatio-temporal regions of **J** are significant. As a first step, a principal component analysis of the three waveshapes for each point on the grid was carried out. Assuming a fixed direction for the source, this gives an estimate of the orientation of the source and its time waveshape $G_i(t)$. Volume maps of the standard deviation of $G_i(t)$ allowed the selection of either regions with point-like sources or with distributed regions of activation.

Results and discussion of aspects related to single approaches

Signal-space Projection (Tesche)

SSP components may be determined from the data in either the frequency or time domain [19, 20]. For example, an oscillatory component may be identified in the Greenblatt data by a feature in the imaginary part of the FFT of the waveform recorded by each sensor. The topography of this Fourier-transformed data at 40 Hz defines a SSP component. The vector space has the same structure for both frequency- and time- domain data. Thus, the SSP unit vector may also be used to define a component in the time domain. The waveform for this component may be determined along with those for other components identified in either the frequency or time domain or from forward modeling of specific current distributions. More general transformations are also possible. Averaged spectra computed subsequently from SSP waveforms may describe the frequency content of specific components with exquisite detail [20].

A SSP component may describe not only an interesting neuronal population but also an artifact such as a blink or heartbeat, or a source of ambient magnetic field noise. The 40 Hz component identified in the "Greenblatt data" may be an example of an artifact or noise component which may be projected out of the data [7]. Projection is also a useful technique to uncover weak signals which may be obscured by stronger components [21]. However, this projection may also remove salient parts of the magnetic field pattern for a neuronal population active at this latency.

Mirror waveforms often result from the inversion of data in multi-dipole models when there are two or more ECD sources at nearly the same location with nearly the same orientation. This problem also occurs in multi-component SSP analysis. The "difference" between two SSP components is characterized by the angle between the two unit vectors in signal space [9, 10, 21]. The angle is 0° for "identical" and 90° for "orthogonal" SSP components. The inclusion of a non-orthogonal component to a set of mutually orthogonal components may affect the SSP waveforms. In the "Mosher data", SSP components defined by the prominent features at 17, 30 and 47 ms are not all mutually orthogonal. The SSP waveform for the 17 ms component is not changed by the addition in the analysis of the 47 ms component (angle 89.9^o), but is altered by the inclusion of the 30 ms component (angle 46.3^o).

A set of orthogonal SSP components may be generated by choosing a component from the data, projecting this component out of the data, and then choosing the next component, etc. This gives interesting results in the "Scherg data" when applied sequentially backwards in time starting from the last time. However, this is not the only possible choice of SSP components. The brain may not generate population responses which are usefully described by a set of successive orthogonal SSP components, even for a very cleverly chosen experimental paradigm. Thus, although SSP does not require a solution to the inverse problem, it also does not eliminate the necessity for the inclusion of additional information. It is imperative, irrespective of the analysis method used, to include all available additional information at each stage of the analysis of MEG data.

Spatio-temporal multiple source model (Scherg)

Simulations of Greenblatt, Hämäläinen and Mosher. Using sequential temporal strategies, filtering techniques and subspace modeling within the BESA program (NeuroScan, version 2.2), three independent raters were able to solve the simulations of Greenblatt, Mosher and Hämäläinen within noise accuracy. In the simulation of Hämäläinen three frontal sources with overlapping activities were in close proximity. This constitutes a case close to a regional source with rotational invariance, i.e. linear combinations of the dipoles 1-3, and consequently of their source waveforms, yielded almost identical goodness of fit. This result illustrated that additional constraints on the timing of the source activities or on the orientation of the sources may have to be derived from independent physiological or anatomical evidence.

Simulation of Scherg. Seven equivalent dipole sources were fitted to the data. Residual variance was extremely low (0.93% with noise, 0.16% without). Source waveforms of the different processes were rendered very accurately (Fig. 2). These results indicated: a) equivalent dipoles are adequate models for extended cortical sheets provided that each active region is modeled by at least one source, b) source waveforms accurately depict the compound activities from these regions and separate the activities of the different regions, c) extended bilateral sources can be modeled by pairs of bilateral dipoles.

These results suggest that spatio-temporal discrete multiple source models are both valid and feasible if recently developed scanning and iterative techniques are employed to search for the global minimum (Hämäläinen, Mosher). Robustness of the solution can be improved by explicit use of the temporal information, even though this is implicitly used by the new techniques. For example, the use of digital filtering and projection into multiple subspaces defined over time or frequency bands may further improve stability of solutions.

Multi-dipole models optimized with a genetic algorithm (Hämäläinen)

In most simulations the method worked very satisfactorily. Fig. 3, showing the results obtained for the Greenblatt data, gives an example of good agreement between analysis results and original sources. The locations of the sources are virtually identical. There is a small discrepancy in the orientations.

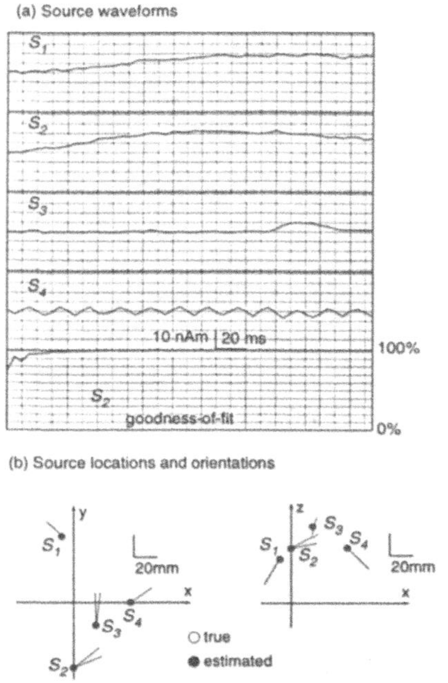

Fig. 3: Hämäläinen results, Greenblatt data. The source waveforms (a) display the time behavior of the four sources required to explain the data. The additional curve at the bottom shows the goodness-of-fit as a function of time.

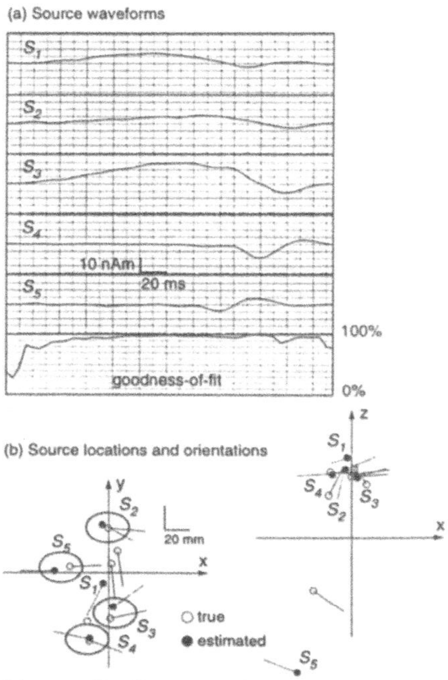

Fig. 4: Hämäläinen results, Scherg data. The format of the figure is identical to that in Fig. 3. The open and closed circles in (b) indicate the true and estimated source locations, respectively.

The most difficult data for this algorithm were those provided by Scherg, which were created on the basis of actual physiological experiments. This data set involved one very deep source and two symmetrically located source pairs. As shown in Fig. 4, the algorithm was able to discern the deep source (S5), one of the symmetrically located source pairs (S2 and S3), and the source near the somatosensory hand area (S4). It was not able to resolve the symmetrical source pair near the midline and the source between the locations of S3 and S4. The activities of all these *missing* sources are accounted for by source S1. Attempts to add more sources to the model were unsatisfactory. It thus seems that the completely automatic analysis of such a complex data set is quite difficult. One of the reasons for the algorithm to fail might be that no heuristic rules were implied to constrain the locations of the sources. Such a measure would clearly be helpful with the Scherg data set.

RAP-MUSIC (Mosher)

In the Greenblatt data, four asynchronous sources were present in relatively low noise. An SVD of the 148 x 50 matrix clearly revealed a rank 4 subspace, so we retained the four dominant singular vectors. Elsewhere in this volume [12], we introduce the metric of subspace angles as a multi-dimensional form of correlation analysis between model subspace and data subspace. We show how this correlation analysis can in turn be related to the more familiar measure of normalized residual variances. We also discuss how subspace angles are the core of the MEG-MUSIC algorithm. For the Greenblatt (and other) data, we retrieved each pair of dipole singular vectors and computed the two subspace angles between the two dipole singular vectors and the rank 4 subspace. We searched over the grid of 2000 dipoles for the dipole point with the highest

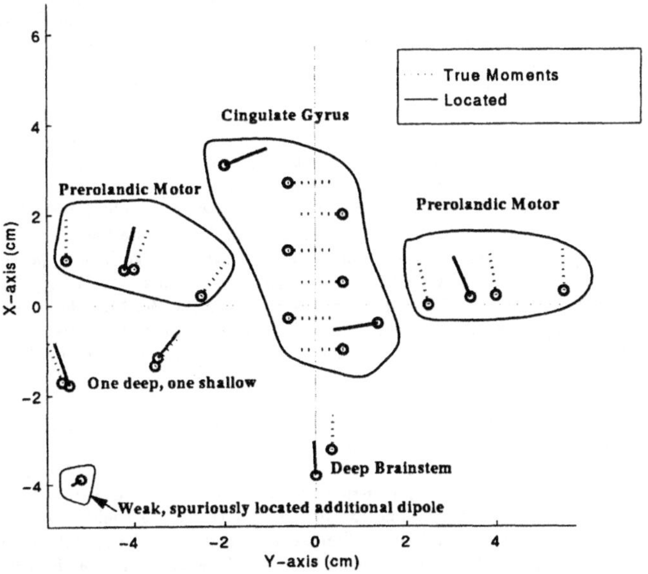

Fig. 5: RAP-MUSIC applied to the Scherg data. The true model was three asynchronous dipoles (marked "One deep, one shallow" and "Deep Brainstem"), six near synchronous dipoles ("Prerolandic Motor"), and twelve synchronous dipoles ("Cingulate Gyrus"). The true noise was colored, but not pre-whitened in the inverse procedure. The "Deep Brainstem" solution was mismodeled with an additional synchronous dipole, but this spurious dipole was given only a weak amplitude. The "Prerolandic Motor" sets were each modeled adequately by a single dipole. The "Cingulate Gyrus" was the final source recursively located, and the combined effects of colored noise and mismodeling appear to have perturbed one of the dipole equivalent solutions. Compare with Figures 2 and 4.

correlation (smallest subspace angle). This first dipole point is only as precise as the gridding, so we refined this point by using it as the initial point in a Nelder-Meade simplex directed search, with the cost function of minimizing the subspace angle. In the Greenblatt data, we thus found the first point with a correlation of 99.9%, indicating a near perfect fit between a dipole model and the data. We extracted from the best fitting dipole location the corresponding best fitting singular vector (see [12]).

In [14], we showed slices through the head volume which plotted the MUSIC metric as a function of location. We discussed how the other dipole locations are found as local peaks in the metric. Obviously, the largest peak is easy to find numerically, but the other peaks represent graphical and computational challenges in deciding which grid points are separate peaks or the same peaks. The RAP-MUSIC algorithm allows the second (and ultimately four) dipoles to be found numerically by computing the second through fourth subspace angles. To find the second dipole in the Greenblatt data, we returned to the grid of 2000 dipole points. At each point, we combined the best fitting singular vector from the first dipole with the two singular vectors for that point (or equivalently, the original three Cartesian-based vectors) to form a new model matrix, then again calculated the three subspace angles between the model and the rank 4 subspace. The first subspace angle was still equal to or better than the previously calculated single dipole angles, but we now minimize the second subspace angle. We searched over all 2000 grid points to find the point that yielded the best second subspace angle. We then refined this location again with a Nelder-Meade simplex directed search. Since the first singular vector from the first dipole was always held constant, the complexity of the search for the second dipole is nearly the same as for a single dipole. In this manner, we found a second dipole such that the second correlation was 99.9%. Thus two dimensions of the rank 4 subspace were now explained.

We repeated the process to find the third and fourth dipoles, looking for points that minimized the third and fourth subspace angles. The final result was four dipole locations that yielded four subspace angles corresponding to 100.0%, 99.9%, 99.8% and 99.5%. By recursively appending the next dipole model to the previous MUSIC results, we accurately located all four asynchronous dipoles in the Greenblatt data, without the need for graphically locating local peaks in the three dimensional MUSIC metric. We approached the Hämäläinen data in the same manner as the Greenblatt data, yielding five suitable dipoles, each recursively found as one of the five subspace angles. The RAP-MUSIC approach is not restricted to simply building up sets of asynchronous dipoles, as we demonstrate next with the Scherg and Mosher data.

An SVD of the Scherg data did not reveal a clear separation between signal and noise subspace, which is more typical of real data. As later revealed, the true data were actually formed from a combination of synchronous and asynchronous dipoles, to which spatially correlated noise had been added. We selected a rank 8 subspace that explained 99.75% of the data, since the MUSIC algorithm appears rather forgiving to over specifying the true subspace. As in the Greenblatt data, we first found two dipoles with a high correlation with the rank 8 subspace. However, when searching for the third dipole, no point in the search grid was found to correlate reasonably well with the data, i.e., no point in the dipole grid yielded a small third subspace angle. We therefore needed to generate some other type of source model to append to the first two dipoles.

We tested a dipole/quadrupole model at each point in a similarly sparse grid, but again no small third subspace angle was located. We then tested a two synchronous dipole model for this third source, and we located a pair of dipoles that yielded a small third angle. This pair of dipoles was found by first using an "alternating projections" or "projection onto convex sets (POCS)" approach to find a pair of dipole that minimized the third angle, then this two-dipole pair was used in a six parameter (the location parameters) Nelder-Meade simplex directed search to refine the fit. Alternatively, we could have searched over every combination of the 2000 grid points for four million pairs, which seemed impractical, or we could have confirmed our directed search results by repeatedly restarting the search with random starting points. In this particular data set, we found that the alternating projections approach produced a pair of dipoles with a small third subspace angle. We thus had an intermediate model comprising two asynchronous dipolar sources, and a third pair of synchronous dipoles.

We repeated this two dipole synchronous model to find two more pairs of dipoles with corresponding small fourth and fifth subspace angles. No suitable sixth angle was located, using even three synchronous dipoles and pairs of dipole/quadrupoles, but our five temporal sources (two asynchronous dipoles, three pair of synchronous dipoles) already explained 98.01% of the data, so we halted the model recursion. The final dipole locations and the corresponding time series qualitatively matched the true model quite well, which itself explained 98.63% of the variance in the data. Fig. 5 presents the dipolar solutions overlaid with the original set of dipoles. The Mosher data, with one asynchronous and two synchronous dipoles, achieved similar results with RAP-MUSIC.

To summarize, the RAP-MUSIC algorithm recursively builds a model that spans the signal subspace extracted from the data. In the Greenblatt data, each dimension of the signal subspace was the result of a single dipole, and the algorithm easily combined four single-dipole models to accurately locate these sources. In the more difficult Scherg data, the algorithm passed twice through a single dipole model, then three times through a two-dipole model to generate its final model. In the final recursion, we were effectively adding dipoles seven and eight to our model, yet we were still using only a relatively simple two-dipole search. A more traditional nonlinear least-squares approach would have been attempting an eight-dipole (24 parameter) directed search at a greater computational complexity and subject to trapping in the multiple local minima of the error surface.

Distributed source solutions using signal space covariance information (Greenblatt)

Covariance-based source estimation methods work best when the individual generators are not temporally correlated. This is true, for example, of the Hämäläinen dataset, which comprised 5 discrete dipoles. Using the principal orientation linear operator method, each of these dipoles could be identified. Fig. 6(A) illustrates the source map peaks corresponding to two of the dipoles which were simultaneously active. The Mosher

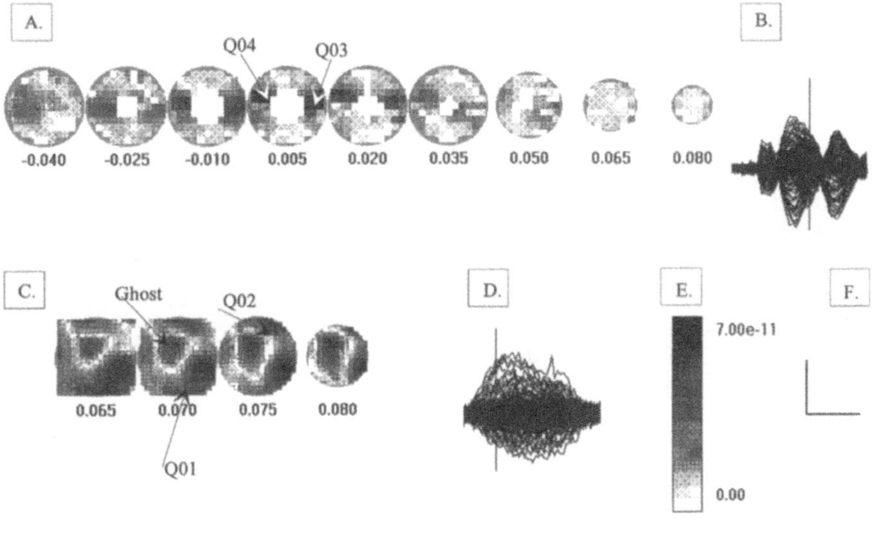

Fig. 6: Results obtained with PRoMS algorithm (Greenblatt). **A.** Source map at 1.5 cm resolution for Hämäläinen dataset. Voxels corresponding to locations of dipoles 3 and 4 are indicated. Source maps are shown as axial slices (z coordinate in meters under each slice). +X is to the right, and +Y upwards. **B.** Superimposed data time series display with latency used for source map in A shown by vertical line. **C.** Source map at 0.5 cm resolution for Mosher dataset. Axial orientation as in A. **D.** Superimposed data time series display with latency used for source map in C shown by vertical line. **E.** Source map scale (Am), maximum is 8×10^{-10} Am for A and 7×10^{-11} Am for C. **F.** Data time series scale; (3×10^{-13} T, 0.1 s) for B, (1.8×10^{-13} T, 0.02 s) for D.

dataset provides an example which included two perfectly correlated sources, along with a third uncorrelated source. As shown in figure Fig. 6(C), this configuration produced a ghost source in the inverse solution.

VARETA (Valdes Sosa)

VARETA and LORETA both dealt well with those examples where the underlying model was a simple set of dipoles (Greenblatt, Hämäläinen, and Mosher). A typical case is the solution for the Greenblatt data shown in Fig. 7. As can be seen, VARETA recovered the four sources, their orientations as well as the waveshapes of each source. Compared to LORETA, the solution is quite concentrated. This was reflected in a much lower value of GCV (0.003) than LORETA (3.4). A critical issue is the grid spacing: a sufficiently fine grid is necessary to avoid misallocation of source variance. This is evident in the LORETA solution where several of the sources are fused. However the source waves of LORETA were the same as those of VARETA at those grid points with maxima in the latter method (Fig. 8). Given that numerical methods may be found for increasing the number of grid points by several orders of magnitude and that criteria are established for selecting significant activated sites, distributed methods may prove to be competitive with the more established dipolar solutions.

The need for a fine grid and the consideration of all possible source regions is illustrated by the analysis of the Scherg data. An initial solution with VARETA using a coarse grid (758 points) and not including a search for sources in the inferior hemisphere, concentrated sources in the central region. The resulting waveforms were a composite of the real ones. A finer grid including the whole hemisphere (2048) correctly identified the deep source. The somatosensory cortical sources were also correctly identified. However, the grid size was apparently not sufficiently fine to distinguish the cingulate and motor cortex set of sources. Bilateral sources were found without the need of symmetry constraints, but the spatial extent of the sources was not estimated correctly, and the waveshapes of both sets were confounded. One of the techniques currently in exploration is to refine the grid adaptively. Such a multigrid strategy might provide the additional grid refinement without undue computational expense.

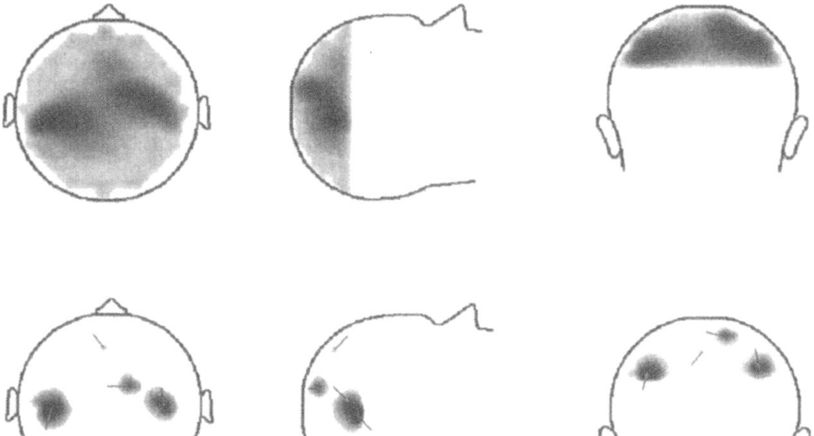

Fig. 7: Top: Maximal Intensity Projection Map (MIP) for LORETA solution of Greenblatt data. Head viewed from top, right side and back. Bottom: MIP for VARETA solution. Lines indicate source orientation at maximum.

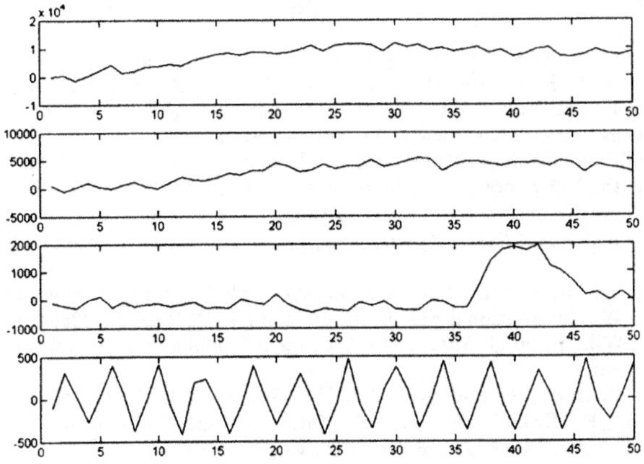

Fig. 8: Waveshapes of sources at local maxima in Fig. 7.

Conclusions

The study has shown that the interpretation of a dataset can be quite tricky. Nevertheless, when applying not just one single method, but exploring a great variety of approaches there is a good chance to arrive at an interpretation which comes relatively close to the truth, as can be seen by considering the results obtained for the Scherg data. The distributed source models failed to separate the bilateral synchronous and overlapping multiple source activities. In addition, centers of activity were mislocalized. The crucial test for the multiple discrete spatio-temporal source model appears to be the solution strategy rather than the validity of the model. The genetic algorithm solved most of the sources except for the bilateral cingulate gyrus sources. This may have been due to the limitation in the number of selected sources and may be overcome by incorporating the number of sources as an additional variable into the genetic algorithm. The subspace scanning method (RAP-MUSIC) was successful in modeling each of the simulation data as sums of asynchronous and synchronous dipoles. The Hämäläinen and Greenblatt data were adequately modeled as sums of asynchronous dipoles, while the Mosher and Scherg data required the extension of the search for synchronous dipole pairs. Each search procedure in the recursion was only as complex as a two-dipole search, even though the Scherg data were ultimatly modeled as two asynchronous dipoles and three pairs of synchronous dipoles.

These results are considered as encouraging, since they demonstrate that the biomagnetic inverse problem can be successfully attacked from many different sides. Comparisons like those carried out here certainly help to get deeper insight into the individual methods and to learn something about their strengths and weaknesses. Furthermore, they allow to assess in which situation to take what method. Thus, we are firmly convinced that the work we initiated should be continued. To stimulate further joint research, all the datasets, their descriptions and selected results are made publicly available [11]. The game should go on!

References

[1] Crowley, T., Haupt, C., and Kynor, D. A weighting matrix to remove depth bias in the linear biomagnetic inverse problem with application to cardiology (this volume).

[2] Dale, A. and Sereno, M. Improved localization of cortical activity by combining EEG and MEG with MRI cortical surface reconstruction: a linear approach. *J. Cognit. Neurosci.*, 1993, 5:162–176.

[3] Forss, N., Hari, R., Salmelin, R., Ahonen, A., Hämäläinen, M., Kajola, M., Knuutila, J., and Simola, J. Activation of the human posterior parietal cortex by median nerve stimulation. *Exp. Brain Res.*, 1994, 99:309–315.

[4] Greenblatt, R. Probabilistic reconstruction of multiple sources in the bioelectromagnetic inverse problem. *Inverse Problems*, 1993, 9:271–284.
[5] Hämäläinen, M. and Ilmoniemi, R. Interpreting measured magnetic fields of the brain: Estimates of current distributions. Technical Report TKK-F-A559, Helsinki University of Technology, Helsinki, 1984.
[6] Hansen, P. Regularization tools, a Matlab package for analysis and solution of discrete ill-posed problems (available in postscript from http://www.mathworks.com/), 1993.
[7] Huotilainen, M., Ilmoniemi, R., Tiitinen, H., Lavikainen, J., Alho, K., Kajola, M., and Näätänen, R. The projection method in removing eye-blink artefacts from multichannel MEG measurements at the site of pain. In Baumgartner, C., Deecke, L., Stroink, G., and Williamson, S., editors, *Biomagnetism: Fundamental Research and Clinical Applications*, pages 363–367, Amsterdam, 1995. Elsevier Science, IOS Press.
[8] Ilmoniemi, R. The projection method in magnetoencephalography. Finnish patent application No. 925461, Nov. 30, 1992. International patent application No. PCT/F193/00504, Nov. 30, 1993.
[9] Ilmoniemi, R. and Williamson, S. Analysis for the magnetic alpha rhythm in signal space. *Soc. Neurosci. Abstr.*, 1987, 13 (2):46.
[10] Ilmoniemi, R., Williamson, S., and Hostetler, W. New method for the study of spontaneous brain activity. In Atsumi, K., Kotani, M., Ueno, S., Katila, T., and Williamson, S., editors, *Biomagnetism '87*, pages 182–185, Tokyo, 1988. Tokyo Denki University Press.
[11] Lütkenhöner, B., Greenblatt, R., Hämäläinen, M., Mosher, J., Scherg, M., Tesche, C., and Valdes Sosa, P. *http://hobbes.uni-muenster.de/SantaFeWorkshop1996* (Homepage of the Santa Fe 1996 Modeling Workshop).
[12] Mosher, J. Subspace angles: A metric for comparisons in EEG and MEG (this volume).
[13] Mosher, J. and Leahy, R. Music recursively applied to the generation of source models in EEG and MEG (in preparation).
[14] Mosher, J., Lewis, P., and Leahy, R. Multiple dipole modeling and localization from spatio-temporal MEG data. *IEEE Trans. Biomed. Eng.*, 1992, 39:541–557.
[15] Pascual-Marqui, R., Michel, C., and Lehmann, D. Low resolution electromagnetic tomography: a new method for localizing electrical activity in the brain. *Int. J. Psychophysiol.*, 1994, 18:49–65.
[16] Sakamoto, Y., Ishiguro, M., and Kitagawa, G. *Akaike Information Criterion Statistics.* Tokyo, KTK Scientific Publishers, 1986.
[17] Scherg, M. Functional imaging and localization of electromagnetic brain activity. *Brain Topography*, 1992, 5:103–111.
[18] Scherg, M. and von Cramon, D. Two bilateral sources of the late AEP as identified by a spatio-temporal dipole model. *Electroenceph. clin. Neurophysiol.*, 1985, 62:32–44.
[19] Tesche, C., Uusitalo, M., Ilmoniemi, R., Huotilainen, M., Kajola, M., and Salonen, O. Signal-space projections of MEG data characterize both distributed and well-localized neuronal sources. *Electroenceph. Clin. Neurophysiol.*, 1995, 95:189–200.
[20] Tesche, C., Uusitalo, M., Ilmoniemi, R., and Kajola, M. Characterizing the local oscillatory content of spontaneous cortical activity during mental imagery. *Cog. Brain Res.*, 1995, 2:243–249.
[21] Uusitalo, M. The projection method in magnetoencephalography. Sc. thesis, Helsinki University of Technology (in Finnish), 1993.
[22] Uutela, K., Hämäläinen, M., and Salmelin, R. Global optimization in the localization of brain activity (this volume).
[23] Wahba, G. *Spline Models for Observational Data.* Philadelfia, Society for Industrial and Applied Mathematics, 1990.

Distributed Source Reconstruction Using a Non-Linear Spatio-Temporal Regularization Method: An Alternative to LORETA

Baillet, S.[1,2], Garnero, L.[1,2] and Renault, B.[1]

[1]LENA - CNRS URA 654 - Hôptital de La Salpêtrière, Paris, France; [2]Institut d'Optique Théorique et Appliquée - CNRS URA 14, Orsay, France

INTRODUCTION

In the EEG / MEG inverse problem solving, it appears that most of the time single dipole models fail to produce acceptable results - specially with the late components of evoked potentials - when space and time overlapping in the source pattern is very likely. At the same time, minimum-norm multi-source solutions exhibit over-smoothed magnitude patterns and, unless specified, do not respect brain anatomical constraints like sulcus borders. Moreover, these methods fail to recover deeper sources because dipoles located at the surface of the source space (the cortex surface, basically) would be privileged for the same data set. Recently Pascual-Marqui *et al.* have presented a generalized minimum-norm solution to the inverse problem that appears to overcome some of these limitations but still offers smoothed dipole images [1].

In this paper, LORETA is compared with an inversion method that explicitely makes use of heterogeneous *a priori* information in a Bayesian formalism with a maximum *a posteriori* (MAP) estimator of dipole activities, allowing some spatio-temporal regularization [2]. For instance, and when consistent with physiology, we show how it is possible to introduce discontinuities between magnitudes of some sources. We will particularly insist on the fact that adjoining temporal priors - like smoothness constraints on dipole moment time course - to spatial ones yields better and less time consuming results than with spatial regularization only. Results from an EEG simulated auditory experiment are shown and compared with those from LORETA.

METHODS

As in [3], dipoles are constrained perpendicularly to the cortex surface with changing magnitudes during a given experiment. Owing to previous PET or f-MRI images, a distributed source model is built on a patchwork of different regions that locally approximate some *a priori* active parts of the cortex surface. As an example, Fig. 1 shows active areas isolated from PET data in an auditory experiment [4]. Then, dipoles are distributed over this patchwork so as to create the source space.

Then we introduce two supplementary constraints: firstly, it is assumed that dipole magnitude evolution is smooth in time. Secondly, although we suppose that dipoles on the same surface patch are strongly connected, we assume that dipoles located on two different surface patches may have completely separated activities. For instance, one can assume that activities of dipoles located on two different walls of a sulcus, in some cases, could be really functionally non-correlated depending on the nature of physiological constraints.

It is possible to proceed with a *selective filtering* of the dipole image - preserving some intensity "edges" - owing to a MAP estimator of dipole magnitudes. Indeed, in a distributed source model using current dipoles with given positions and orientations, all the sources are gathered in a row vector so as to obtain the following linear system:

$$\mathbf{M}_n = \mathbf{G}\,\mathbf{J}_n + \mathbf{b}_n, \qquad (1)$$

where $\mathbf{M}_n$ is a N_M row vector containing (EEG or MEG) measurements at time n, and $\mathbf{G}$ is a N_M x N matrix containing global information about the head model, the source set pattern and measurement positions on the surface of skull. $\mathbf{J}_n$ is a N row vector containing the nth time sample of dipole magnitudes. Finally, data are corrupted by an additive measurement noise $\mathbf{b}_n$. This is a standard writing for many linear *forward problems* in image reconstruction.

A MAP estimator of dipole magnitudes taking all the previous *a priori* information into account consists in finding, for a given time sample n, an estimate of current dipole magnitudes $\hat{\mathbf{J}}_n$, that maximizes the posterior probability distribution of $\mathbf{J}_n$ given the measurements $\mathbf{M}_n$:

$$\hat{\mathbf{J}}_n = \max_{\mathbf{J}_n}(p(\mathbf{J}_n|\mathbf{M}_n)) = \max_{\mathbf{J}_n}(p(\mathbf{M}_n|\mathbf{J}_n).p(\mathbf{J}_n)). \quad (2)$$

An energy function in terms of $\mathbf{J}_n$ is associated to the previous probability distribution:

$$p(\mathbf{J}_n|\mathbf{M}_n) = \frac{1}{Z}\exp(-U(\mathbf{J}_n)), \quad (3)$$

Z is a normalization constant called partition function, and $U(\mathbf{J}_n) = U_1(\mathbf{J}_n) + U_2(\mathbf{J}_n)$ where $U_1(\mathbf{J}_n)$ and $U_2(\mathbf{J}_n)$ are associated with $p(\mathbf{M}_n|\mathbf{J}_n)$ and $p(\mathbf{J}_n)$ respectively. Then the MAP estimator of dipole magnitudes becomes :

$$\hat{\mathbf{J}}_n = \min_{\mathbf{J}_n}(U(\mathbf{J}_n)). \quad (4)$$

If for simplicity noise is assumed to be white, gaussian and zero-mean, one can write $U_1(\mathbf{J}_n)$ as:

$$U_1(\mathbf{J}_n) = \|\mathbf{M}_n - \mathbf{G}\mathbf{J}_n\|^2. \quad (5)$$

As a prior term, we can choose $U_2(\mathbf{J}_n) = U_s(\mathbf{J}_n) + U_t(\mathbf{J}_n)$ where $U_s(\mathbf{J}_n)$ introduces spatial priors and $U_t(\mathbf{J}_n)$ temporal ones. A widely used formalism for the prior spatial energy function $U_s(\mathbf{J}_n)$, consists in writing it as a sum of potential Φ-functions in terms of dipole magnitude gradients (noted as $u = \nabla\mathbf{J}_n$) [5]. For instance, Φ could be of the following Lorentzian form :

$$\Phi(u) = \frac{u^2}{1+\left(\frac{u}{K}\right)^2}, \quad (6)$$

where K is a scaling factor. Contrarely to the quadratic potentials used in minimum gradient norm methods, these Φ_K-functions possess limited cost for high intensity jumps which allows preservation of discontinuities when necessary. Anatomical priors can now be introduced by *locally* assigning different values to K ($K=K_1$ *or* $K=K_2$, *with* $K_1 >> K_2$) [6], which sets the local behavior of the algorithm. Spatial variations of dipole moments will be more smoothed in areas where K_1 is the scaling factor, than in regions scaled with K_2.

Temporal constraints can also be introduced by means of another energy function, $U_t(\mathbf{J}_n)$.We use $\hat{\mathbf{J}}_{n-1}$ as the unique temporal neighbor of $\hat{\mathbf{J}}_n$. Assuming that $\hat{\mathbf{J}}_{n-1}$ and $\hat{\mathbf{J}}_n$ may be really qualitatively close to each other means that the orthogonal projection of $\hat{\mathbf{J}}_n$ on the hyperplane $E^{\perp}_{\hat{\mathbf{J}}_{n-1}}$ perpendicular to $\hat{\mathbf{J}}_{n-1}$ is "small". Thus, $U_t(\mathbf{J}_n) = \beta\|\mathbf{P}^{\perp}_{n-1}\mathbf{J}_n\|^2$, where $\mathbf{P}^{\perp}_{n-1}$ is the projector onto $E^{\perp}_{\hat{\mathbf{J}}_{n-1}}$. The MAP estimator of dipole moments at time n is given by minimizing $U(\mathbf{J}_n)$, with :

$$U(\mathbf{J}_n) = \|\mathbf{M}_n - \mathbf{G}\mathbf{J}_n\|^2 + \lambda\sum_{k=1}^{N}\left(\Phi^x_k(\nabla_x\mathbf{J}_{n|_k}) + \Phi^y_k(\nabla_y\mathbf{J}_{n|_k})\right) + \beta\|\mathbf{P}^{\perp}_{n-1}\mathbf{J}_n\|^2, \quad (7)$$

where x (res. y) indexed values are for horizontal (res. vertical) gradients between cortical dipoles along the surface patch. λ and β are parameters allowing the weighting of the prior part of the global energy with regard to the data attachment term. A MAP estimate is then computed owing to a deterministic sub-optimal minimization procedure.

RESULTS

In this section, results from MAP estimation of dipole activities in an EEG experiment simulation are shown and compared with those of the LORETA method. 16 *a priori* active areas in a spherical 3 shell head model are isolated from an auditory experiment data (see Fig. 1). 128 dipoles (8 dipoles per patch) are distributed perpendicularly over these surfaces. The EEG recording is calculated on a 65 electrodes set. The temporal courses of active dipoles are given by a damped sine curve with different scale ratios and time delays. While there are 200 time samples, Fig. 2 shows a 3D meshing of this dipolar spatio-temporal mapping, from time sample 20 to 80. The patch numbers are on the x-axis. Noise level is set to 20dB (standard deviation scale).

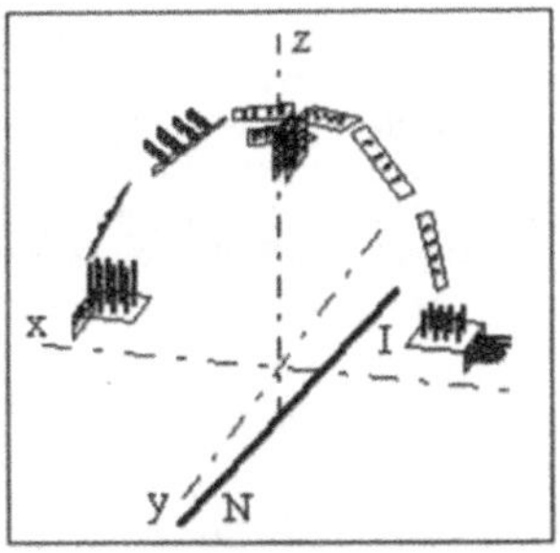

Fig.1- Original patch, (NI): Nasion-Inion line, gray segments are for active dipoles

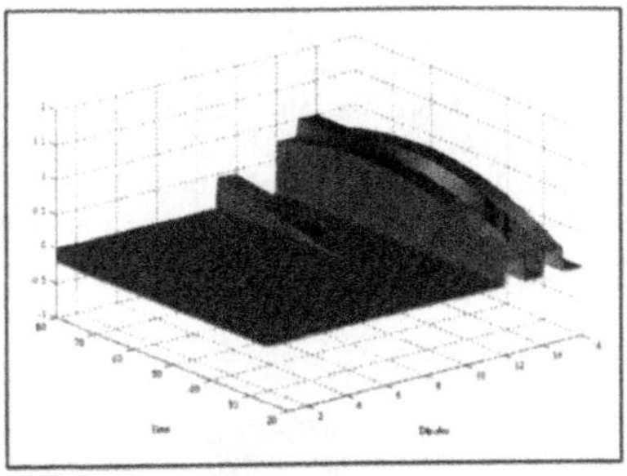

Fig. 2- Original dipole sequence

LORETA results are shown in Fig. 3 & 4. The global activity shape spreads over dipoles possessing roughly the same contributions to the data measurement set that the active ones. Nevertheless, most of the silent dipoles are pretty nicely recovered but spurious spikes appear, because of the high noise sensitivity of the Laplacian spatial constraints. Moreover, this noise sensitivity is enhanced because of the pseudoinverse method, which is a harsh way to regularize the inverse problem.

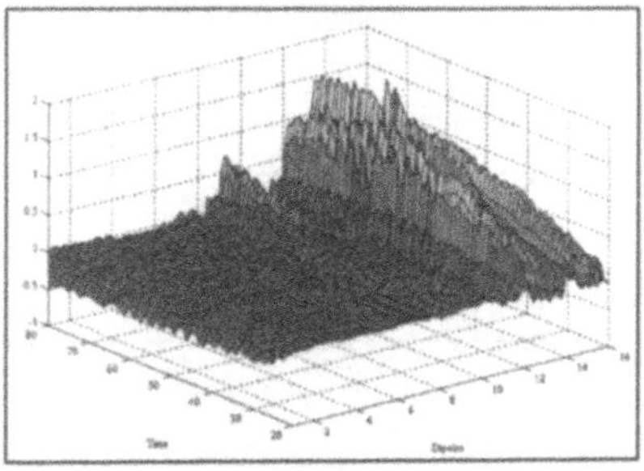

Fig. 3 - An active dipole time profile with LORETA

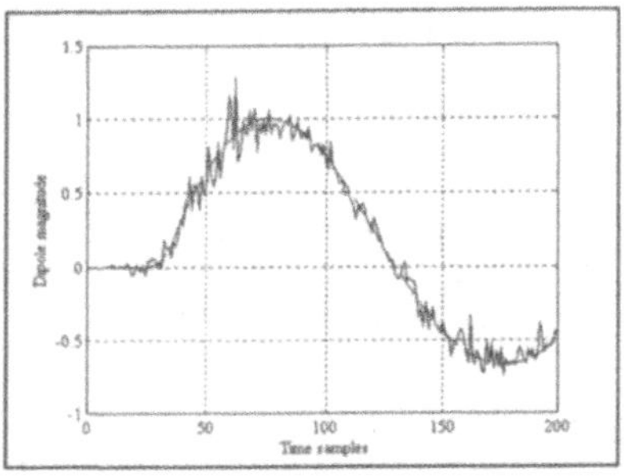

Fig. 4 - LORETA dipole sequence

MAP estimation results: Without considering any temporal constraints for reconstruction (S-MAP), results are shown on Fig. 5 & 6. Most edges are nicely recovered and there is very little spurious activity on silent dipoles. Fig. 5 shows an example of time profile for an active dipole. It is very clear that while the global temporal shape has been preserved, some sharp magnitude variations still occur (see for instance between time samples 50 and 100).

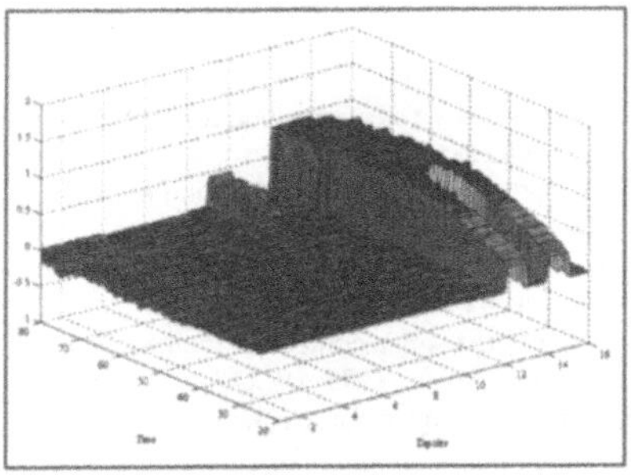

Fig. 5 - An active dipole time profile with S-MAP

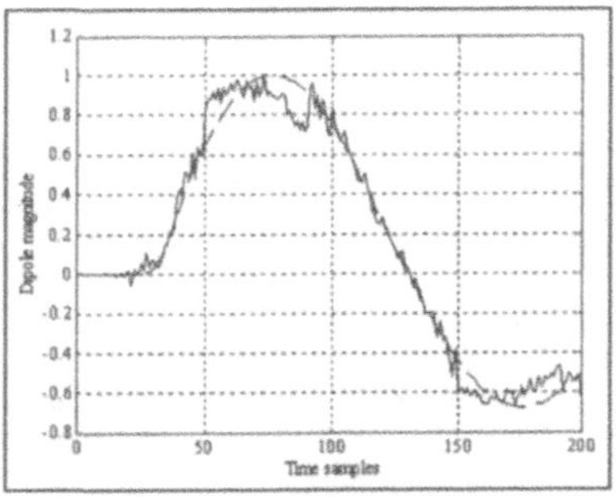

Fig. 6 - S-MAP dipole time sequence

Results obtained when adding temporal constraints (ST-MAP) may appear very similar to the previous ones (Fig. 8). But looking closer to the active dipole time profile (Fig. 7), it appears that the temporal evolution of dipoles is smoothed. Moreover, all original edges are recovered, and the total computation time is 22 % less than with spatial regularization only. This shows that well chosen priors, such as temporal constraints in addition to spatial ones, greatly improve the quality of the solution. And at the same time, they tend to help stabilizing the algorithm so that it seeks a solution in the vicinity of the ones found for previous time samples, and finally computing time is decreased.

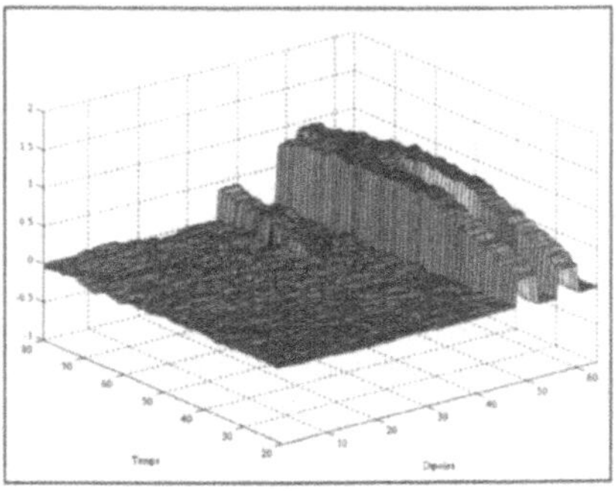

Fig. 7 - An active dipole time profile with ST-MAP

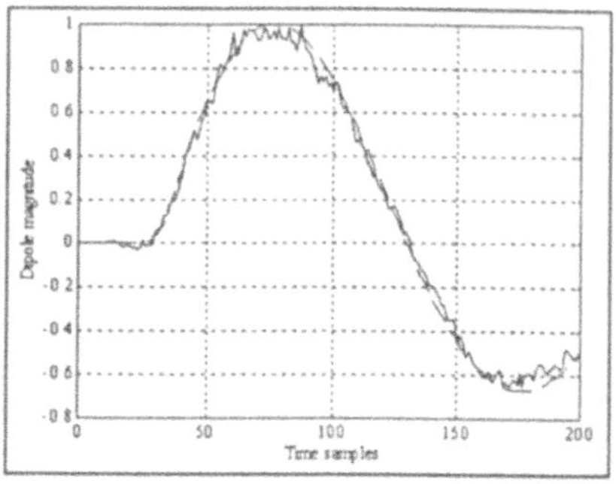

Fig. 8 - ST-MAP dipole sequence

The following table summaries the performance of these algorithms by means of the mean-square reconstruction error on dipole intensities.

	Data fit	Reconstruction error
LORETA	97.8 %	36.7 %
S-MAP - λ=3 10^5, β=0	97.3 %	11 %
ST-MAP - λ=3 10^4, β=1.5 10^4	97.8 %	6.3 %

DISCUSSION

Temporal constraints play a major role in both qualitative and quantitative enhancement of reconstruction results with regards to spatial priors only. Despite heavy blurring on potentials, it is now possible to recover sharp edges in the source space. In that sense, LORETA brings some improvement to classical quadratic regularization, and may offer satisfactory results as an initialization procedure for the MAP algorithm.

REFERENCES

[1] Pascual-Marqui, R.D., *et al.*, Low resolution electromagnetic tomography : a new method for localizing electrical activity of the brain, Int. Journal of Psychophysiology, 1994, 18:49-65.

[2] Baillet, S., and Garnero, L., A Bayesian approach to introducing anatomo-functional priors in the EEG / MEG inverse problem, submitted to IEEE Trans. on Biomed. Eng., february 1996.

[3] Dale, A.M., and Sereno, A.M., Improved localization of cortical activity by combining EEG and MEG with MRI cortical surface reconstruction: a linear approach, J. Cogni. Neurosci., 1993, 5:162-176.

[4] Tzourio, N., *et al.*, Mapping the auditory selective attention generators using a combined brain evoked potentials and PET activation study, Society for the Neuroscience Abstracts, 1993, 19: 1285.

[5] Geman, S., and Mac Clure, D.E., Statistical methods for tomographic image reconstruction, Inverse Problems, 1987, 21: 5-21.

[6] Gindi G., *et al.*, Bayesian reconstruction of functionnal images using anatomical informaton as priors, IEEE Trans. Med. Imaging, 1993, 12: 670-680.

Localization of Extended Sources Using a Modified Linear Estimation Approach

Brauer, H.[1], Ziolkowski, M.[2], Tenner, U.[1], Kosch, O.[1], Wiechmann, H.[1] and Arlt, A.[1]

[1]Department Fundamentals and Theory of Electrical Engineering, Technical University of Ilmenau, P.O. Box 0565, D-98684 Ilmenau, Germany; [2]Technical University of Szczecin, Al. Piastow 19, PL-70310 Szczecin, Poland

Introduction

Biomagnetic measurements using multichannel systems provide magnetic data sets for functional localization of heart or brain activity non-invasively. The interpretation of measured magnetic fields generated by the electrical active organs requires special algorithms to localize or reconstruct the sources. Numerous optimization algorithms to solve this inverse problem are under consideration, deterministic as well as stochastic one. In most cases non-linear, deterministic optimization procedures like Levenberg-Marquardt algorithm are used. We have found that Marquardt algorithm combined with Gauss-Seidel-iteration is the best inverse solution method in least-squares sense to localize up to three single dipoles [1]. But there appear problems with convergence of this method if deep lying or extended sources have to be localized. For these cases we developed a new version of the well-known minimum-norm solution based on the Linear Estimation Theory [2,3]. We introduced some modifications which enable us to reconstruct distributed sources too [4]. Therefore, the aim of this paper is to propose a new method for reconstruction of extended sources and estimating their extension.

Method

Biomagnetic signals arise from electric currents due to transmembrane ionic flow in excitable tissues. These magnetic fields are linked to bioelectric potentials, and it makes sense to consider both the magnetic field and the electric potential together. Since a great number of elementary sources is activated simultanously, it is convinient to represent them as a continuous macroscopic primary current distribution. The total current density incorporates this impressed current density, and the ohmic current density which is due to the effect of changes in conductivity in the volume conductor. Because the varaitions of varibles in time are relatively slow (below 1 kHz), treatment of sources and fields in quasi-static approximation is allowed. Then the electric potential obeys the Poisson-equation while the magnetic field due to the total current density is obtained from the Biot-Savart law.

If we assume an inhomogeneous volume conductor consisting of homogeneous subvolumes with constant conductivities the surface potential can be obtained from an integral equation. If the potential is known the magnetic field can be calculated in the above mentioned way. This calculation is called the solution of the forward problem, and it is necessary to solve the inverse problem. Because of the irregular shape of the body, calculations can be performed only numerically, using the finite element (FEM), or boundary element method (BEM). We applied the BEM because it is easier to perform on a workstation [4]. The volume conductor models were obtained from Magnetic Resonance Images (MRI) using specific mesh generation techniques [5]. The BEM approach was applied to a realistically shaped, 3-compartment thorax model (fig. 1). Single current dipoles as well as twin current dipoles were used to to solve the forward problem and calculate the magnetic flux density normal to that hypothetical measurement grid consisting of 6x6 equidistant sensor positions shown in fig. 1. The magnetic field was calculated using a linear BEM approach [4].

In the biomagnetic inverse problem, the objective is to estimate the source current density underlying the magnetic signals measured outside the body. Unfortunately, the primary current distribution can not be computed uniquely, even with a complete knowledge of the electric potential and the magnetic field everywhere at the surface and outside the body, respectively. However, it is often possible to use additional physiological information to restrict the problem and facilitate the solution. One often used approach to the restricted inverse problem is to replace the actual current sources by equivalent generators which are characterized by a few parameters only. A unique solution for the parameters may be obtainable from measured (or simulated) data, e.g., by a non-linear least-squares optimization method known as the Levenberg-Marquardt algorithm [6].

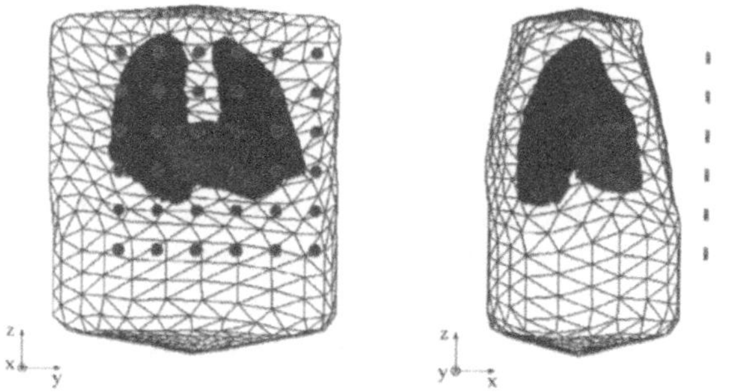

Fig. 1. Three-compartment BEM model of the human thorax with a hypothetical quadrilateral gradiometer array in front of the chest (6x6 equidistant sensor positions, distance 50 mm)

Because this method is only applicable to localization problems with less than three single dipoles great effort has been spent in developing procedures to reconstruct continuous source current pattern. Such methods are based on the minimum-norm approach [2], probabilistic entropy concepts [7] or general linear estimation considerations [3]. We performed a minimum-norm-solution technique based on the Linear Estimation Theory. In the common Linear Estimation Approach (LEA) the estimated region (heart or brain) is divided into voxels and only one unit current per voxel is assumed. These current dipoles are assumed to be fixed in their position and free in the moments. Usually this leads to a linear system of equations which is under-determined, i.e. the number of sensor points, N, is much smaller than the number of dipole moments, which is equal to 3M, if M is the number of assumed dipoles. The assumed sources can be evaluated by minimizing the fitness function

$$\mathrm{FIT} = | \underline{\mathbf{B}}_{meas} - \underline{\mathbf{B}}_{calc} | \, / \, | \underline{\mathbf{B}}_{meas} | .$$

The solution of the equation $\underline{\mathbf{A}} \cdot \underline{\mathbf{f}} = \underline{\mathbf{b}}$, where $\underline{\mathbf{A}}$ is the lead field matrix, $\underline{\mathbf{f}}$ are the unknown dipole moments, and $\underline{\mathbf{b}}$ is the measured flux density in the N sensors, can be found by taking the pseudo-inverse matrix (Moore-Penrose Inverse) and estimating the minimum-norm solution, i.e. minimizing the L_2 -norm of $\underline{\mathbf{f}}$. This solution can be found using the Singular Value Decomposition (SVD). But the result is the well-known so-called blurring effect, i.e. the reconstructed dipoles which are significant for the minimization are spread over the whole solution space. A reconstruction of distributed sources or a localization of source foci is not possible in this way. In the proposed Modified Linear Estimation Approach (MLEA) we applied some modifications.

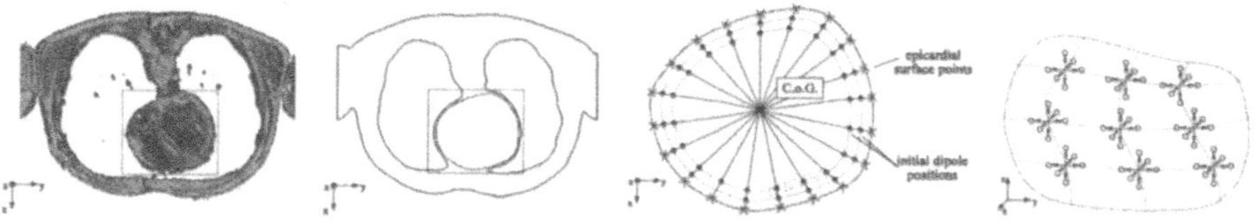

Fig. 2. MRI cross section of the human thorax and contour identification used for BEM triangularisation; Definition of initial dipole position just below the epicardial surface (similar distances to surface points) and visualization of possibilities for successive focussing of dipole positions

Starting from (equally distributed) points on the epicardial surface within the estimated region (myocard) some areas are defined where a set of current dipoles is assumed to be initially fixed in position. These positions are chosen to lie nearly in the same distance to the epicardial surface with respect to some anatomical and physiological constraints (fig. 2). To minimize the least-squares deviation between measured/simulated data and forward solutions (deviation scan) we realized a regularisation of the deviation scan. The best-possible dipole fitting was obtained if the lead field matrix was weighted with the dipole depths, i.e. the distances between sensor and source points, and with the field deviation. Furthermore, a successive focussing of dipole positions used for reconstruction was introduced. The originally fixed dipole positions were now randomly shifted by small distances (see fig. 2) to improve the deviation scan (or dipole fitting). A further focussing effect was achieved if the number of dipoles used for reconstruction was successively reduced according to given thresholds ignoring dipoles without significant electrical activity.

Results

All these modifications together with proper chosen starting points for dipole positions used for reconstruction enable us to reconstruct really 3D sources. To demonstrate the ability of the proposed technique we illustrate the simpliest case of reconstruction of a single current dipole located relatively deep inside the left ventricle of the BEM model (see fig. 1). But the intrinsic goal of the investigation is, of course, the reconstruction of extended sources. We are interested in finding (and separating) centers of electrical activity, estimating the extent of these centers, and calculating equivalent current dipoles for each center to estimate the source volume. The deviation of this volume from the sphere volume which is related to the calculated equivalent dipole could be a criterion for the extent of the source volume. To realize this we propose an algorithm for finding distributed sources (after application of the MLEA):

1. Find the smallest cube containing all reconstructed dipoles
2. Find the average distance between all dipoles
3. Use an average distance to calculate a partition of this cube constructing a finite element mesh consisting of cubes)
4. Find elements where dipoles exist
5. Calculate an average source volume current density for each finite element where dipoles exist using only dipoles which are in that cube
6. Show results: a) small cubes which show volume sources,
   b) euqivalent dipoles in the center of gravity of cube elements
7. Change partition of global cube, reconstruct the FE mesh and repeat steps 5 and 6
8. Change threshold of displaying dipoles and show results again (separate centers of activity)

Figure 3 gives an overview of the proposed algorithm for finding and separating centers of electrical activity.

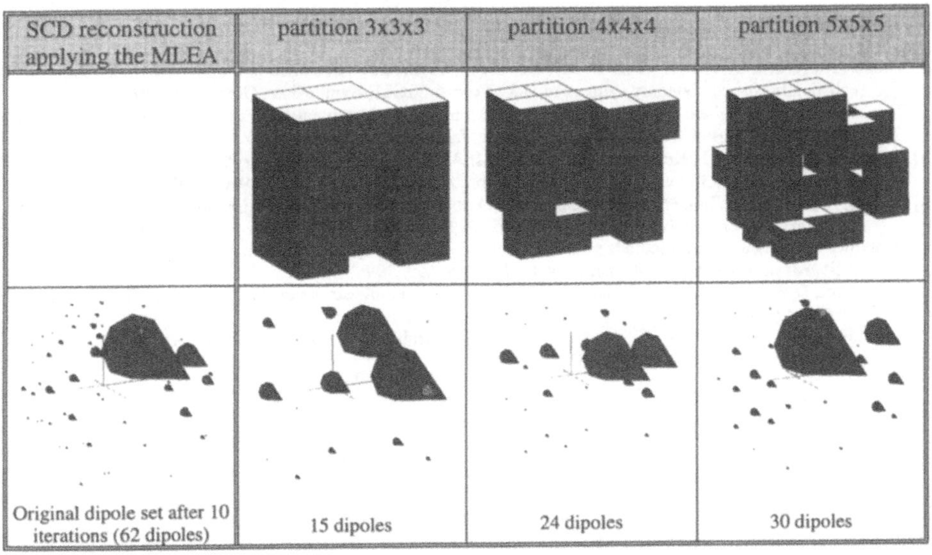

Fig. 3. Application of the proposed algorithm for estimating the extent of source volume; the finite element approximation using cube elements and the related equivalent dipoles located in the c.o.g. are shown assuming the MLEA was applied to reconstruction of a single current dipole and 62 dipoles were found

Summary

A modified Linear Estimation Approach was performed to reconstruct current sources within the heart. Based on MRI data sets the Boundary Element Method was used to create tailored multicompartment models of the human thorax which were used to solve the forward problem of magnetocardiography. The ability of the proposed method was demonstrated for the localization of a single current dipole as an example of a focal source.

By means of introducing small shiftings to all reconstruction dipoles during linear estimation solution as well as performing a successive focussing strategy ignoring places without significant electrical activity the method could easily be extended to the reconstruction of real 3D sources. Based on a special minimum-norm solution the source volume can be estimated applying a finite element approximation using cube elements. The size of an extended current source can be estimated by superimposing the reconstructed dipoles to an equivalent dipole and comparing the corresponding volume with the sphere which would be related to the equivalent dipole. The deviation of these volumes can be taken as a criterion for non-dipolarity of sources.

Conclusions

The proposed Modified Linear Estimation Approach can be used to reconstruct both depp focal sources and extended non-dipolar sources within the heart (or within the brain). To increase the ability of the proposed method the initial dipole positions must be better chosen with respect to anatomical and physiological constraints; an optimal number and location of surfaces for starting the reconstruction algorithm should be found. Before the method can really be applied to reconstruction of real 3D sources it should be validated whether the solutions of the forward problem could be further improved. Computer simulations have to establish whether the Finite Element Method (FEM) or the Multiple Multipole Method (MMP) can provide more precisely solutions of the biomagnetic forward problem. Furthermore, it should be studied how these methods could be applied in inverse solutions in an efficient way. The proposed algorithm for estimating the extent of the source volume could be further improved if better adapted finite element discretizations are used. Finally, it would be very helpful if these numerical results could be validated by experimental investigations. Hence, the development of physical thorax phantoms and performing corresponding magnetic measurements should be taken into account, we are going on to realize this.

References

[1] Brauer, H., Kosch, O., Arlt, A., Tenner, U., Michelsson, O. Inverse calculation of biomagnetic fields using realistically shaped boundary element models, Proc. 6th IGTE-Symposium, Graz 1994, 270-275

[2] Nenonen, J.T., Hämäläinen, M.S., Ilmoniemi, R.J., Minimum-norm estimation in a boundary-element torso model, Medical&Biol. Engineering&Computing, January 1994, 43-48

[3] Smith, W.E., Dallas, W.J., Kullmann, W.H., Schlitt, H.A., Linear estimation theory applied to the reconstruction of a 3-D vector current distribution, Applied Optics, vol. 29, 1990, 5: 658-667

[4] Brauer, H., Kosch, O., Tenner, U., Wiechmann, H., Arlt, A., A modified linear estimation approach for solving biomagnetic inverse problems, (to appear in: IEEE Trans. Magnetics, vol. 32, No. 5, May 1996)

[5] Ziolkowski, M., Brauer, H., Methods of mesh generation for biomagnetic problems, (to appear in: IEEE Trans. Magnetics, vol. 32, No. 5, May 1996)

[6] Marquardt, D.W., Analgorithm for least squares estimation of nonlinear parameters, Journal of SIAM, vol. 11, 1963, 2: 431-441

[7] Ioannides, A.A., Bolton, J.P.R., Clarke, C.J.S., Continuous probabilistic solutions of the biomagnetic inverse problem, Inverse Problems, vol. 6, 1990, 523-549

Inverse Magnetography for Reentry Currents

Brio, M.[1,2], Marchesin, D.[1], Costa Monteiro, E.[3], Bruno, A.C.[3] and Costa Ribeiro, P.[3]

[1]*Instituto de Matemática Pura e Aplicada, Rio de Janeiro, Brazil;* [2]*U. Arizona, Tucson, USA;* [3]*Pontifícia Universidade Católica, Rio de Janeiro, Brazil*

INTRODUCTION

We present a numerical study of an inverse biomagnetic problem – the recovery of atrial reentry current patterns. The reentry current is a circulating electrical wave pattern present in cardiac pathologies such as atrial or ventricular tachyarrythmias. The problem is to determine the center of rotation, the size, and the regions of slow conduction velocity of the reentry fronts. Mathematically, it shares important features with other biomagnetic inverse problems for locating stationary or moving sources [1]. It is also of interest in its own right, since the electrical activity of tachyarrythmias is not fully understood.

In this work, we extend and analyze a model introduced previously for treating magnetic data obtained in animal heart experimentation [2]. We propose a new method of solution based on a least squares fit of the information contained in the first few Fourier harmonics of the time series of magnetic data measured at several different spatial locations. This method combines spatial and temporal information contained in the magnetocardiogram.

Our starting point is the assumption that the electrical activity may be represented by a simple model consisting of a current dipole moving on a circular path, or circulating current dipole. In this paper we concentrate on the recovery of the equivalent circulating current dipole for synthetic data on points along a line parallel to the plane of rotation.

THE CIRCULATING CURRENT DIPOLE MODEL

Consider the magnetic field generated by the electrical activity of a uniformly rotating current dipole of constant strength p. Up to a phase shift, the normal component of the magnetic field on a plane parallel to the plane of rotation is given by Biot-Savart's law, [2],

$$B_n(x,y,t) = \frac{\mu_o}{4\pi} p \frac{r - s\cos(\omega t)}{(r^2 + s^2 + z_d^2 - 2rs\cos(\omega t))^{\frac{3}{2}}} \quad , \tag{1}$$

where r is the radius of the circulating current dipole and (x_d, y_d, z_d) is the position of the center of rotation, $z = 0$ and $z = z_d$ being the measurement and the dipole planes, respectively. The quantity $s = \sqrt{(x - x_d)^2 + (y - y_d)^2}$ is the distance between the measuring point $(x, y, 0)$ and the rotating dipole axis. The angular frequency of the dipole rotation is denoted by ω. This model may be viewed as a first approximation to more general models.

We will use the information contained in the amplitude of the first few Fourier harmonics of the measured magnetic field. For example, the first one is:

$$a_1(x,y) = \frac{\omega}{2\pi} \int_0^{2\pi/\omega} B_n(x,y,t)\cos(\omega t)dt \tag{2}$$

In numerical calculations, it is helpful to know that the Fourier coefficients with respect to time t in equation (1) may be expressed, using hypergeometric functions such as ${}_2F_1$, in terms of x, y, x_d, y_d, z_d. See [3].

Similarly, we can define the magnetic flux $A_{1model}(x,y)$ through a horizontal coil of a certain diameter centered at (x, y). In practice, a seventh order Gaussian quadrature formula is employed to approximate the magnetic flux [4].

THE INVERSE SOLUTION PROCEDURE

We will simulate the measurement of the vertical component of the magnetic field generated by flutter activity in horizontal atrial tissue. These measurements are taken sequentially at several points along a horizontal line approximately 1.5 cm above the tissue. Each measurement represents a 16 second-long time series of the magnetic flux through the magnetometer coil of 1.0 cm diameter. The amplitude $A_{1data}(i)$ of the first Fourier harmonic of the magnetic flux is then computed for each measuring point i; a detailed description of the actual experimental procedure utilizing a single SQUID is found in [5].

The next step is to find parameters x_e, y_e, z_e, r, p of the "equivalent circulating current dipole" described in the previous section, which reproduce as much as possible the measured data. Thus, we minimize the total deviation in amplitude between the first Fourier harmonic of the data $A_{1data}(i)$ and of the magnetic field generated by the circulating current dipole $A_{1model}(i)$ (see Equation(3)); phase lags in time are not used in the reconstruction because they cannot be accurately measured using a single SQUID at different positions and times in a real experiment.

The synthetic measurements were taken only along a line, rather than on a plane, because of two factors which limit the number of available measurements in the real experiment [5]. One is that only a single magnetometer is available. The other is that the induced flutter in the dissected atrial tissue lasts only a few minutes; during its lifetime its frequency decreases by 30%, indicating that the flutter pattern changes steadily. Thus it is important to attempt recovering the equivalent circulating dipole parameters with as few measurements as possible. Also, the measurements have to be taken as quickly as possible.

Only the amplitude of the AC component of the signals is considered because of current difficulty in measuring the DC component of the magnetic field generated by the reentry pattern in the actual experiment. This difficulty is temporary; the SQUID is very well suited for DC measurements, and we plan to incorporate the DC information into our method. Despite that Fourier harmonics higher than the first may carry useful information, they have been disregarded in the present work to eliminate noise. Figure 1 shows the power spectrum at a single position, using a Fast Fourier Transform on a 16 second long time series from the actual experiments in [5]. It corresponds to one of the spatial points in the experimental data taken at $x = 1.0$ in Figure 2.

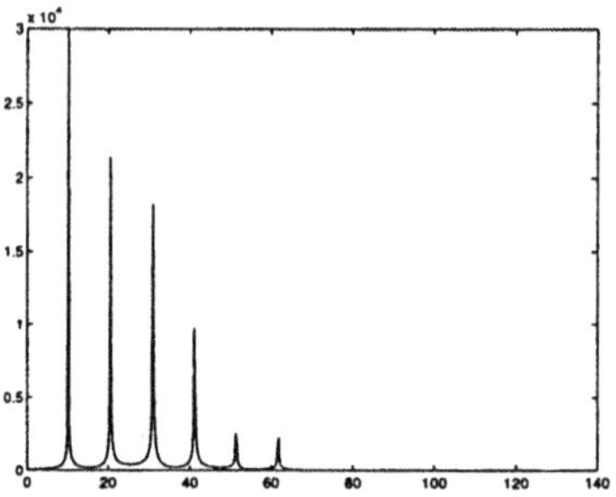

Figure 1:
Power spectrum of magnetic field

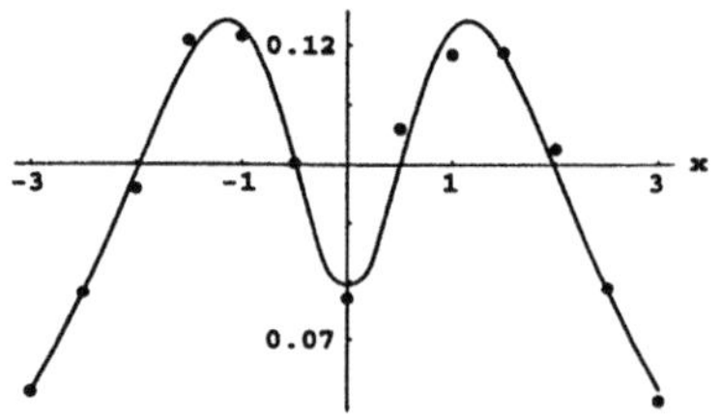

Figure 2:
Measured and recovered first harmonic

Using Figure 1, it is easy to obtain the amplitude of the first harmonic, as well as its frequency. Proceeding in a similar way, we obtain the amplitude of the first Fourier harmonic for the data at several positions along the measuring line, shown as dots in Figure 2. The measurements are spaced by 0.5 cm along the line.

The graph shows that useful information is obtained only if measurements are taken at enough locations to cover the extrema of the function. In the actual laboratory experiment, there were not enough accurate measurements, so the dots in Figure 2 represent simulated, rather than actual, data. They represent the first Fourier coefficient with 10% random perturbation from the correct values of $A_{1model}(i)$. This perturbation was introduced primarily to simulate the fact that the magnetic activity of the reentry current should be represented only approximately by the activity of a circulating dipole. We verified that SQUID and environmental noise is strongly attenuated by Fourier filtering, see [5], to a value much lower than 10%. The original circulating current dipole utilized to generate the synthetic data was positioned at x=0.0 cm, y=0.0 cm, and z=1.5 cm. The radius r of the circulating dipole was 0.5 cm. The measuring line is y=0.5 cm, and z=0.0 cm. The dipole strength was chosen as p=1.0 in arbitrary units (a.u.) for convenience. The radius of the magnetometer is 0.5 cm.

We now employ a least squares procedure to minimize the cost function $C(x_d, y_d, z_d, r, p)$ with respect to the unknown parameters (x_d, y_d, z_d), r, and p, representing the coordinates of the center, the radius, and the strength of the circulating dipole, respectively. The cost function for the amplitude of the first Fourier harmonic is given as

$$C(x_d, y_d, z_d, r, p) = \sum_i [A_{1model}(i) - A_{1data}(i)]^2 \tag{3}$$

Once the minimization procedure is complete, we obtain the parameters for the equivalent circulating dipole. They are x_e=0.006 cm, y_e=0.07 cm, z_e=1.47 cm, r_e=0.504 cm, p=0.93 a.u. We have a good agreement between the original and the recovered circulating dipole parameters. The graph of the reconstructed amplitude $A_{1model}(i)$ is shown in Figure 2. Notice that the agreement between the synthetic and the reconstructed values of the first harmonic is good.

In order to understand the significance of the parameters recovered, we show cuts of the graph of the cost function in Figure 3. Each cut was obtained by fixing all but one variable at a time for intervals of size 2 centered at (x_e, y_e, z_e, r_e). Figure 3a shows the cost function in terms of x, near x_e. It looks like a sharp parabola, and therefore it allows to locate the x position very accurately. Figure 3b shows the cost function in terms of y near y_e. It has two minima, reflecting the symmetry with respect to the measuring line y=0.5 cm. The curvature at the minima is not as pronounced as in the previous graph indicating the need for measurements at different y positions for accurate localization. Figure 3c shows the cost function in terms of z near z_e. It has a well defined minimum. Figure 3d shows the cost function in terms of r near r_e. The recovery of r (see Figure 3d) is delicate. Even though the negative radius does not make physical sense, equations (1) and (2) show that the first Fourier coefficient has radius multiplied by cosine, and therefore the sign can be compensated by the phase shift. The graph becomes flatter as the size of the magnetometer coil increases.

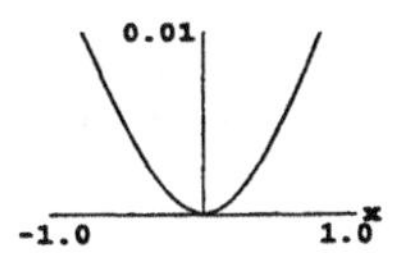

Figure 3a:
C as a function of x

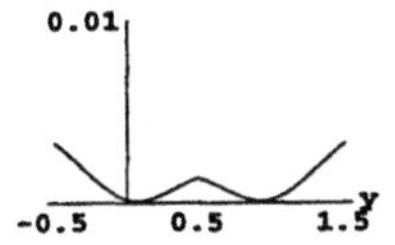

Figure 3b:
C as a function of y

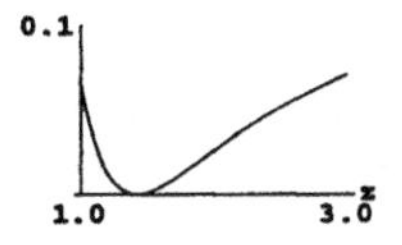

Figure 3c:
C as a function of z

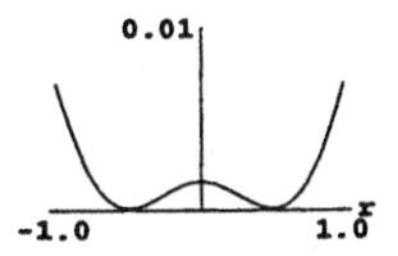

Figure 3d:
C as a function of r

We made experiments with cuts away from the minima. We saw that the cuts become flatter and that most information is contained in the measurements near the center of the activity. The flatness of the cost function also depends on the size of the detection coil; the recovery of r is especially sensitive to the magnetometer radius. If the radius of the magnetometer becomes comparable to that of the circulating dipole, the cost function becomes flatter, especially as a function of r. The convergence of the minimizing procedure will be ensured as long as one starts the search for the minimization procedure within the domain of attraction of the minima, the size of which depends dramatically on the quality of the data collected. Conversely, synthetic experiments show that the domain of attraction increases and the cost function becomes sharper for more accurate data and smaller magnetometer coil, respectively.

The cost function for the full magnetic field, including many harmonics, looks very similar to the one shown in Figure 3, as the contribution due to the cosine term in the denominator of equation (1) decreases for higher harmonics. Thus most of the information is contained in the first few Fourier harmonics. However, the full cost function is very susceptible to noise. The cost function becomes polluted by the ripples near the minima, and this could make the solution obtained by the least squares procedure unreliable.

CONCLUSION

In this paper we have proposed a solution for the inverse biomagnetic problem of detecting the reentry current in atrial flutter. It takes into account both spatial and temporal information. We have described the results for measurements taken along a single line, based on synthetic data simulating an actual experiment, but generalizations of the mathematical setup to two-dimensional, or to arbitrary geometry or synchronization of the measuring devices can be obtained easily.

Our method is based on utilizing the information contained in the first few Fourier harmonics of the measured time series for the magnetic data. The advantage of this procedure is that it automatically eliminates the high frequency noise. The version of the method presented here does not use phase information of the magnetic data, which may be unavailable if the experiments are done with a single SQUID magnetometer at different times and at different positions. The phase information of the reentry activity disappears if we use only information contained in the Fourier amplitude. To recover the phase information needed to determine the actual position of the regions with slow conduction velocity, a multichannel magnetometer and the use of several Fourier harmonics are probably needed.

REFERENCES

[1] Roth, B.J., Sepulveda, N.G., and Wikswo, J.P., "Using a Magnetometer to Image a Two-Dimensional Current Distribution", J. Appl. Phys. 65 (1), 1989, 361-372.

[2] Costa Ribeiro, P., Bruno, A.C., Saboia e Silva, P.L., Hall Barbosa, C.R., Parente Ribeiro, E., Costa Monteiro, E., and Fonseca Costa, A., "Detection of Reentry Currents in Atrial Flutter by Magnetography", IEEE Trans. Biomed. Eng. 39, 1992, 818-824.

[3] Abramowitz, M., and Stegun, I.A., Handbook of Mathematical Functions, National Bureau of Standards, Washington, 1964 (reprinted by Dover, New York, 1968).

[4] Roth B.J., and Sato, S., "Accurate and efficient formulas for averaging the magnetic field over a circular coil", in: "Biomagnetism: Clinical aspects", M. Hoke et al. (eds), 1992, Elsevier, Amsterdam.

[5] Costa Monteiro, E., Andrade, E., Ornelas, P.H.,Cruz, R.S., Picon, A.P., Kühner, G.S., Kelber, C., Bruno, A.C., Costa Ribeiro, P., Marchesin, D., Brio, M., Leite, T.T., "Induced Atrial Flutter in Rabbit Heart Tissue and Inverse Magnetocardiography", these proceedings.

ACKNOWLEDGEMENTS

M. Brio was partially supported by CNPq Grants nos. 300659/94-6 and 453497/95-0. D. Marchesin was partially supported by CNPq Grants nos. 300204/83-3 and 530054/93-0, and by FINEP Grant no. 65.92.03.11-00. E. Costa Monteiro was partially supported by CNPq Grant no. 300665/93-8. P. Costa Ribeiro was partially supported by CNPq Grants nos. 300959/81-8 and 530484/93-5, and by FINEP 65.930.566.00. Thanks are due to Monique O. Moura, who performed numerical experiments for a preliminary reconstructive procedure, to Beata F.B. Gundelach for many useful suggestions, to Tatiana T. Leite for processing the experimental data, and to Bradley J. Plohr for useful discussions.

Frequency Domain Distributed Inverse Solutions

Casanova, R.[1], Valdes-Sosa, P.[1], Garcia, F.M.[1], Aubert, E.[1], Riera, J.J.[1], Korin, W.[1] and Lins, O.[2]

[1]*Cuban Neurosciences Center, Havana, Cuba*
[2]*Rotman Research Institute, Toronto, Canada*

Abstract

Much interest has been generated in the past few years by Distributed Solutions that estimate the primary current distributions in the whole brain volume. This paper presents a LORETA based distributed method in the Frequency Domain. The method assumes a stochastic stationary time series model for generators that is masked by instrumental noise. Tikhonov regularization is used for dealing with the noise component. A common regularization parameter is computed by the generalized crossvalidation procedure for the complex valued FFT coefficients. It is shown by simulations that the method is effective for localising generators from noisy data and that a simple LORETA without regularization may fail to recover the generator spectra.

Introduction

Analysis in the frequency domain (FD) has a long history of revealing interesting structure in the MEG/EEG . A natural question that arises is the localisation of the sources of certain spectral components thus posing the inverse problem for the FD. Initial FD inverse solutions were of the dipole fitting variety [1],[2],[6]. Valdes[7] has also recently proposed a method for FD multiple dipole fitting which takes into account spectral and crosspectral information and the presence of spatially correlated "noise". However current interest has shifted from dipole solutions to Distributed Solutions that estimate the primary current distributions in the whole brain volume. LORETA[5] is a particularly successful variant of distributed solution that minimises a discrete laplacian roughness penalty. This paper presents a LORETA based distributed method in the FD. The method assumes a stationary time series model for generators masked by instrumental noise. Tikhonov regularization is used for dealing with the noise component. A common regularization parameter is computed by the generalised crossvalidation procedure for the complex valued FFT coefficients. It is shown by simulations that the method is effective for localising generators from noisy data and that a simple LORETA without regularization may fail to recover the generator spectra. In fact inadequate modelling of stochastic EEG source generators may originate "ghost solutions".

Methods

It is assumed that a multichannel EEG/MEG recording $\vec{V}_i(t)$ may be modelled as:

$$\vec{V}_i(t) = K \cdot \vec{J}_i(t) + \vec{E}_i(t)$$

or in the frequency domain, $\vec{V}_i(\omega) = K \cdot \vec{J}_i(\omega) + \vec{E}_i(\omega)$ (1)

where $\vec{V}_i(t)$ is the observed EEG/MEG time series, K is the lead field matrix, $\vec{J}_i(t)$ the primary current time series and $\vec{E}_i(t)$ spatially uncorrelated noise time series at the sensors. The suffix i = 1, ..., nw indexes the segments of collected data. All time series are considered to be second order stationary obeying some mixing condition [8]. Considering the complex matrices $\mathbf{V}(\omega) = \left[\vec{V}_1(\omega), \vec{V}_2(\omega), \dots, \vec{V}_{nw}(\omega)\right]$, $\mathbf{J}(\omega) = \left[\vec{J}_1(\omega), \vec{J}_2(\omega), \dots, \vec{J}_{nw}(\omega)\right]$ and $\mathbf{E}(\omega) = \left[\vec{E}_1(\omega), \vec{E}_2(\omega), \dots, \vec{E}_{nw}(\omega)\right]$, the problem is to estimate the complex Cross Covariance matrices $\mathbf{S}_{JJ}(\omega)$ of the generators from the data $\mathbf{V}(\omega)$ considered as a matrix of samples . For this purpose it is first necessary to estimate the $\mathbf{J}(\omega)$. The assumptions ensure that the FFT coefficients $\mathbf{V}(\omega)$ are approximately Complex multivariate Gaussian, each frequency independent of the other. To find $\mathbf{J}(\omega)$ the following problem is solved:

$$\min_{\mathbf{J}(\omega)}\left\{\left\|\mathbf{V}(\omega)-K\cdot\mathbf{J}(\omega)\right\|^2+\lambda\cdot\left\|L\cdot\mathbf{J}(\omega)\right\|^2\right\} \qquad (2)$$

where L=BW , B is the discrete laplacian operator , W is a diagonal weight matrix and ||X|| is the Frobenius norm of X. Note that (2) is a **matrix** regularization problem with complex data resulting from the FFT. In order to apply usual concepts developed for vector regularization problems (2) can be transformed to:

$$\min_{vec(\mathbf{J}(\omega))}\left\|vec(\mathbf{V}(\omega))-(I\otimes K)\cdot vec(\mathbf{J}(\omega))\right\|^2+\lambda\cdot\left\|(I\otimes L)\cdot vec(\mathbf{J}(\omega))\right\|^2 \qquad (3)$$

where $\otimes$ denotes the usual Kronecker product and *vec* is the vectorization operation. At this stage the usual vector-regression generalized crossvalidation procedure [9] can be applied, obtaining in this way a common regularization parameter for all samples of the frequency ω. Finally it can be shown that the solution to the problem (3) is :

$$\hat{\mathbf{J}}(\omega)=\left(K^T\cdot K+\lambda\cdot L^T\cdot L\right)^{-1}K^T\cdot\mathbf{V}(\omega) \qquad (4)$$

The complete crosspectral matrix estimator of the generators is $\hat{\mathbf{S}}_{JJ}=\frac{1}{nw}\cdot\sum_{i=1}^{nw}\hat{J}_i(\omega)\cdot\hat{J}_i^*(\omega)$ where $*$ denotes the conjugate transpose.

Results

All simulation studies were performed with the classical 3-shell head model, the brain in the upper hemisphere (R = 0.86), a regular cubic grid (1153 points) restricted to the brain volume and 120 sensors homogeneously distributed on the more external hemisphere. Two series of simulations were carried out. First, a dipole was simulated with a fixed position and orientation and with a gaussian stochastic amplitude. Second, simulations were carried out to study the methods performance in dependence of the noise level in what follows only results for spectrum estimates will be presented. Figure 1 shows how FD regularization recovers a single dipole , while failure to regularize produces a "ghost solution". Figure 2 shows how the quality of the reconstruction of the spectrum (as measured by the norm of the differences between the simulated and reconstructed spectrum current densities) worsens with the decrease of the signal to noise ratio.

Conclusions

Several authors (Pascual, personal communication), [3], [4] have suggested the use of regularization to avoid instability of the solution in the presence of noise. This has been confirmed by our results, when the simple LORETA without regularization is applied to noisy data a "ghost solution" concentrated at the top of the head is obtained, as can be seen from second row of figure 1. Meanwhile the use of Tikhonov regularization with a common smoothing parameter selected by generalized crossvalidation successfully recovers the location of generators. Matrix regularization with common parameter for all samples has a better performance that selection of individual parameters for each. Matrix Regularization with complex data is a feature of our approach.

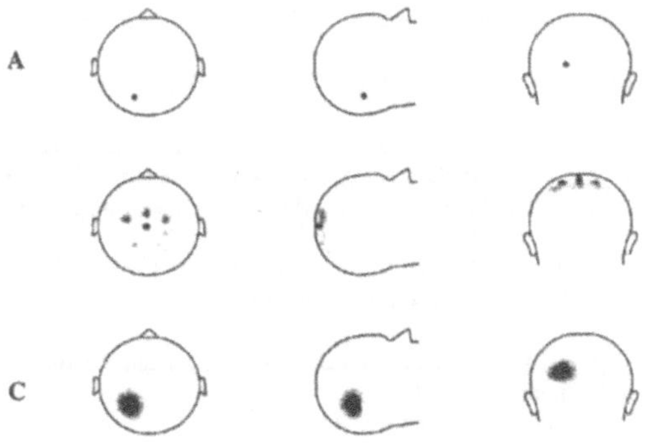

Fig 1. Each column of this figure corresponds with axial, sagital and coronal views formed by projecting maximal intensities values of FD solutions. Row A shows the position of one simulated dipole, B is LORETA estimate and C is the regularized LORETA in the FD.

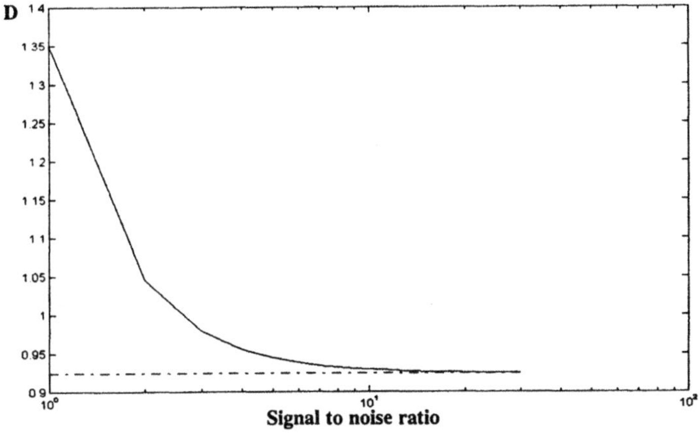

Fig 2 The dependence of the difference (D) between estimate solution and the original dipole on the signal to noise ratio is shown for unregularized LORETA (solid line) and Regularized LORETA (dashed line) in the FD.

References

[1] Lehmann, D., and Michel, C.M. Intracerebral dipole source localisation for FFT powers maps, Electroenceph. clin. Neurophysiol., 1990, 76:271-276

[2] Lutkenhoner, B., Frequency domain localisation of intracerebral dipole sources, Electroenceph. clin. Neurophysiol., 1992, 82:112-118
[3] Fuchs, M., Wagner, M., Wischmann, H-A, and Dosell, O. Cortical Current Imaging, Biomag93 Advances in Biomagnetism, 1993, 97
[4] Grave de Peralta, R., Oropesa, E., Gonzalez Andino, S.L. , Cabrera, D., and Aubert, E. An approach to meaningful distribute inverse solutions to the neuroelectromagnetic inverse problem, 1995, send to publish
[5] Pascual-Marqui, R.D., Michel, C.M. and Lehmann, D. Low Resolution electromagnetic tomography: A new approach for localising electrical activity in the brain, International Journal of Psychophysiology, 1994
[6] Raz, J., Biggins, C., Turetsky, B. and Fein, G. Frequency Domain Dipole Localisation: Extensions of the Method and Applications to Auditory and Visual Evoked Potentials. IEEE Transactions on Biomedical Engineering, 1993, vol 40, no. 9.
[7] Valdes, P., Bosch, J., Grave, R., Riera, J., Pascual, R. and Biscay, R. Frequency Domain Models of the EEG, Brain Topography, 1992, Volume 4, Number 4, 309-319.
[8] Brillinger, D.R., Time Series Data Analysis and Theory, 1975.
[9] Wahba, G. Spline Models for observational data, 1990.

Small Volume Resistivity Changes Observed Using MEG

Clifton, Y.C.[1], Moran, J.E.[2], and Tepley, N.[1,2]

[1]Oakland University Rochester, MI, USA; [2]Henry Ford Hospital, Detroit, MI, USA

Introduction

Small regions of increased electrical resistivity changes have been detected during spreading cortical depression (SCD) associated with seizures [1]. Localization of these regions may be useful for finding the epileptic foci since it locates the region of SCD. Boone, Levi, and Holder [1] used electrical impedance tomography (EIT) with sub-dural electrodes and a transverse applied DC current to identify epileptic foci in a rabbit model. Since EIT is notoriously low in resolution, a technique of measuring electric impedance which gives greater resolution is desirable. [2]

There has been a great deal of interest in finding methods to locate the focus of epileptic seizures with a less invasive method than cortical mapping because of the risks to patients created by extensive periods of this procedure [1] as well as the costs involved. This paper describes a method of identifying the region of spreading depression which is believed to be in vicinity of the epileptic focus using DC Magnetoencephalography (MEG) and provides a demonstration using a phantom, suggesting that this is a viable tool. MEG has already been used to localize epileptic foci with high spatial resolution [3], and the promise of additional information to confirm the localization could make MEG the preferred tool for preoperative diagnosis.

Method

Our study utilized a seven channel Neuromagnetometer (BTi model 607). Preliminary measurements were conducted using a cylindrically shaped phantom, 9 cm in diameter, containing 40 cc of saline. A current of 0.18 milliamps flowed through the saline between silver and silver chloride electrodes located diametrically opposite each other. A small plastic cylinder at the center of the phantom, hydraulically activated from outside the magnetically shielded room, was displaced approximately 1 mm giving rise to magnetic field changes of several hundred femto Tesla. The piston was triggered 50 seconds after the commencement of data acquisition, and 2.5 seconds were required for the piston to move the 1 mm distance. MEG data were collected at a rate of 130 Hz per channel, for ten 100 second epochs. A low pass filter was set to attenuate signal with frequencies of greater than 50 Hz. The measurements were made with the probe in several positions, and the data was mapped.

Theory

MEG measurements of DC field changes can be used to quantify small volume regions of electrical resistivity change, if the electric current in the region is known; The unit used to quantify the resistivity change is the magnitude of the current dipole (Q) which will be defined in a way permitting researchers to directly relate the volume in which the event takes place to the magnetic field measurement. The experiment described uses applied currents of known magnitude. Since such currents are already used in EIT, it is reasonable to use small applied currents in animal models or humans.

The shifts in magnetic fields which arise from shifts in current distributions can be described by the Biot-Savart equation:

$$\Delta\vec{B} = \frac{\mu_0}{4\pi}\int_{volume} \frac{\Delta\vec{j} \times \vec{R}dV'}{\left|\vec{R} - \vec{R}'\right|^3}$$

$\vec{R}$ is a radius vector from the center of the coordinate system to the point of integration. $\Delta\vec{j}$ is change in 'current density. $\vec{R}'$ is the distance from the center of the volume element dV' to the probe. μ_0 is the permeability of free space. dV' is the volume element.

If the resistivity change is confined to a small region of interest and $\vec{R} >> \vec{R}'$, then the equation simplifies further:

$$\Delta\vec{B} = \frac{\mu_{0\Delta}\vec{j} \times \vec{R}\Delta V}{4\pi\left|\vec{R}\right|^3}$$

ΔV is the volume element in which resistivity changes occur. For this experiment the current is set up by two electrodes in a cylindrical dish containing saline and the change in resistance is due to the motion of a piston which displaces a volume of saline. Approximating the electrodes as a capacitor, the electric field $\vec{E}$, can be approximated by $\vec{E} = \vec{U}(\Delta\Phi / D)\gamma$. U is an unit vector from the negative to the positive electrode, γ is a constant close to unity, D is the distance between the two electrodes. The magnitude of the current dipole is now: defined:

$$Q = \Delta\sigma(\Delta\Phi / D)\Delta V$$

$\Delta\Phi$ is the voltage between the two electrodes. $\Delta\sigma$ is the change in conductivity created by the motion of the piston. Since the conductivity of the piston is essentially zero, $\Delta\sigma$ = the conductivity of saline. The direction of the current element is defined by the unit vector U. (To a good approximation $\Delta\vec{j}$ is the current displaced by the piston.) It can be shown that the expression for magnetic field becomes:

$$\Delta\vec{B} = \frac{\mu_0 Q\vec{U} \times \vec{R}}{4\pi|\vec{R}|^3}.$$

The predicted value of Q for our experiment is 5 X 10^{-5} amp x meters which is consistent with our measured fields of several hundred femto Tesla.

Data Analysis

The data from each run were decimated 25 times, and a least squares fit to the following step function was performed on Statistica software [4]:

if t < 50 sec, step = 0 ; if 50 sec < t < 52.5 sec, step = (t-50 sec)/(2.5 sec) ; if t > 52.5 sec, step = 1.

The value of the correlation coefficients were averaged over ten runs. This process of averaging was repeated with the probe in each position where a measurement was made.

Results

Fig. 1 shows the relation of the initial data to the best correlation with the step function for one channel. The standard deviations of measurements determined using variance of ten measurements ranged from 30 to 85 femto Tesla.

Fig. 2 shows the relation of a computer simulated contour map of fields that would be detected by a second order gradiometer 4 cm above the top of the piston, to measured fields. The computer data, which is represented by contour lines was created using MathCad [5] assuming that parameters were identical to those discussed in the theory. The gray scale data is the mapped data obtained 4 cm above the top of the piston based on averaged correlation coefficient. The 7 channel probe made 10 acquisitions at each of the 4 positions (a total of 28 averaged points were used to create the map.) The data were smoothed using Spieglass Software [6]. Note the data is not symmetric about $\vec{B} = 0$, presumably due to the magnetic field created by the current in the rest of the phantom. Note the similarity of simulation to data. There is a local minimum not predicted by theory. This may be a result of a single very low data point which is erroneous (given the standard deviations as compared to measurements this is possible).

Discussion

If studies of rabbit brain performed using EIT are valid then [1] then the anticipated current dipole magnitudes could easily produce signals of 50 to 100 femto Tesla during spreading depression. Since Boone, Lewis, and Holder have proposed the use of sub-dural electrodes for Electrical impedance tomography (EIT) to find epileptic focus with minimal surgery, such an invasive method seems viable [1]. Considering the low spatial resolution of EIT [2], MEG may prove more useful for locating resistivity changes since MEG has already proven useful in finding epileptic foci. This may provide information in addition to the mapping of the epileptic foci. The value of Q can be estimated using values of ΔV and Δs obtained in earlier research [1,7]. If a current of less than 100 mamps were applied to the volume displaced by the piston with such values of ΔV and Δs, the theoretical estimate of the field measurements obtained 2 cm above the region of SCD would be close to one pico Tesla. This

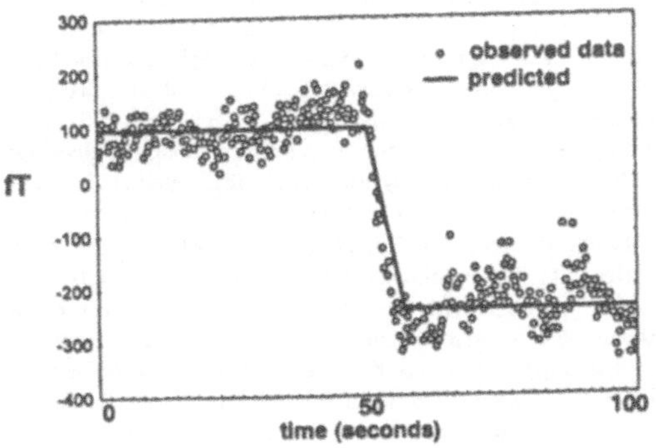

Fig. 1 A single channel of MEG data

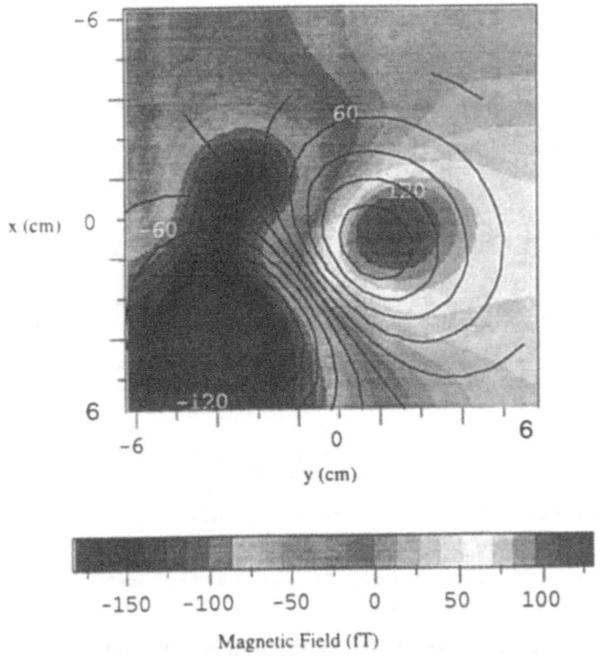

Fig. 2 Map of MEG data is described bu gray scale. Computer simulation of MEG is described by contour lines.

would involve much less current than Boone, Lewis, and Holder used to incite SCD [1]. Since it appears that such studies would give reasonably large signals and it is straight forward to locate the center of the region of SCD using this technique, it can be directly applied to clinical diagnosis. It might, for example, confirm the location of a focus which has been localized by mapping the epileptic spike. The advantage of this method is that the physical origin of the signal is well understood. The traditional method of locating an epileptic spike with MEG or EEG assumes that the signal is caused by a simple current dipole. There is evidence of a far more complex electrophysical process at the epileptic focus, for example the DC MEG data described by Vieth et al. (8), which could lead to an error caused by the model dependence of the measurement. Comparisons between *in vivo* measurements of detailed sources at various points in the cortical lamina in the animal model [9] failed to confirm MEG predications [9]. These animal studies used the penicillin model of epilepsy [10] to produce seizure like events in animals which can be studied with the head open. Neuroelectric fields measured with electrodes were compared to predications made with external magnetic fields measurements and the comparison suggested that the theory used to describe the relation of external magnetic fields to brain activity is imperfect [9]. Sutherling and Barth have pointed out that MEG data alone without confirmation by EGG often is not adequate to localize the focus and they have suggested EEG as a method of confirmation, but even EEG data assumes the same dipole model as MEG and measures the same electrical event [11]. The localization of the region of spreading depression could provide valuable confirmation of localizations of epileptic spikes.

There are several ongoing studies of spreading depression for which this may be relevant, including studies of anoxia in the animal model [12], and studies of migraine headaches in humans [13]. This form of mapping could be used to identify the location of spreading depression in animal models, but at present there is little to justify such invasive studies on human migraine patients.

Furthermore, the study of regions of electrical impedance has been applied to cardiac studies of blood flow and this may be a method to gain valuable cardiac information [14].

References:

[1] Boone K., Lewis A.M., and Holder D.S., Imaging of cortical spreading depression by EIT: implications of localization of epileptic foci, Physiological Measurements, 1994, 15 (Supple 2a): 189-198.

[2] Webster, G., Electrical impedance tomography and biomagnetisim, in Hoke M., Erne S.N., Okada Y.C., and Romani G.L., Biomagnetism: Clinical Aspects, Amsterdam, Excerpta Medica 1992, 675-685.

[3] Barth, J.S., The neurophysiological basis of epileptiform magnetic fields and localization of neocortical sources, Journal of Clinical Neurophysiology, 1993, 10: 99-107.

[4] Statistica Manual, Cambridge, Statistica inc., 1995.

[5] Mathcad User's Guide, Tulsa, Mathsoft inc., 1993.

[6] Spieglass Manual, Spieglass Inc. Savoy Illinois, 1995.

[7} Bures, J. Bursova, O., Krivanek, J.,: The Mechanism and Applications of Leao's Spreading Depression of Electroencephalographic Activity, New York, Academic Press, 1974. 23 -28

[8] Vieth, J., Schuller P., Sack, G., Ganslandt, O., Dratz, W., and Grimm, Extracranial slow magnetic field changes during epileptic activity, in U., in Haschki, W., Roitbak, A.I., and Speckmann, E.J., Slow Potential Changes. In The Brain, Birkhauser Boston, 1993, 243 - 249.

[9] Barth, D.S., and Sutherling W.S., Neuromagnetic studies of evoked and spontaneous activity in animals in Sato, S. ed Advances in Neurology Vol.. 54 : Magnetoencephalography, Raven Press, New York, 1990 119 - 131.

[10] Barth, D.S., Sutherling W.S., and Beatty J., Fast and slow magnetic phenomena in focal epileptic seizures, Science, 1986, 226: 855-857.

[11] Sutherling W.W. and Barth D.S., Magnetoencephalography in clinical epilepsy studies: the UCLA experience in Sato, S., Advances in Neurology Vol.. 54 : Magnetoencephalography, Raven Press, New York, 1990 231-245.

[12] Takanashi Y., Chopp, M., Levine, S.R., Kim, J., Moran, J.E., Tepley, N., Chen, Q., Barkley, G.L., and Welch, K.M.A. Magnetic fields associated with anoxic depolarization of anesthetized rats, Brain Research, 562 pp. 13-16 1991.

[13] Barkley G.L., Tepley N., Nagel-Leiby, S., Moran J.E., Simkins, R.T., and Welch, K.M.A., Magnetoencephalographic Studies of Migraine, Headache, 1993, 30: 7,428-434.

[14] Brown B.H., Barber D.C. and Leatherhard AD, Cardiac Respiratory Related Electrical Impedance Changes in the Human Thorax, IEEE Transactions of Biomedical Engineering, 1994, 41: 729-734.

Acknowledgment: This research supported by NIH/NINDS grant no. RO1 NS30914

A Weighting Matrix to Remove Depth Bias in the Linear Biomagnetic Inverse Problem with Application to Cardiology

Crowley, T.A., Haupt, C.D. and Kynor, D.B.

Biomagnetic Technologies Inc., San Diego, CA, USA

Introduction

The determination of unknown electrical sources within the body from external magnetic field measurements is known as the biomagnetic inverse problem. For electrical sources modeled by multiple current dipoles of fixed location and orientation and variable strength, the inverse problem is linear, and it is often ill-posed [1]. The ill-posedness can be addressed by specifying that the inverse solution be the minimum norm least squares estimate (MNLE) which exists and is unique [3]. The MNLE, however, has a major shortcoming in that it favors the sources that are closest to the magnetic field sensors [1][2]. This so called depth bias yields solutions that are physiologically implausible, especially for three dimensional source spaces. One method of altering the characteristics of the MNLE is to employ a weighting matrix in the formulation and to solve for the corresponding *weighted* minimum norm least squares estimate (WMNLE) [4]. The purpose of this paper is to present a specific weighting matrix that removes the bias to depth. Comparisons are made to both the unweighted case and the case where the weight is derived from column normalization of the forward matrix. Finally, the suggested weighting matrix is applied to magnetocardiographic data for a spherical multiple dipole model.

The Forward Solution

For a source space of n multiple discrete dipoles of fixed locations and directions, the forward solution maps a configuration of dipole strengths to an array of m magnetic field sensor values. This mapping is represented by the m x n forward matrix, **A**. The ij^{th} element of the forward matrix is equal to the ith sensor value produced by the jth dipole when it is of unit strength. The Euclidean norm of each column is referred to as the gain of the corresponding dipole. The gain of a dipole is inversely proportional to its distance from the sensors.

The Inverse Solution

Given a vector **b** of magnetic field measurements, the inverse solution is obtained by solving the matrix equation **Ax=b** for the vector **x** of dipole strengths. This system of equations is often ill-posed, meaning a unique solution does not exist. An alternate inverse solution is the minimum norm least squares estimate (MNLE) which always exists and is unique. The MNLE is the vector **x** of smallest norm that minimizes $\|\mathbf{Ax-b}\|$ [3]. All norms in this paper refer to the Euclidean norm. Since the solution vector consists of dipole strengths, its norm is minimized when the dipoles with the largest gain, namely those nearest the sensors, are favored.

The MNLE inverse solution can be modified by weighting the vector of source strengths. Let **W** be an invertible n x n matrix. The corresponding weighted minimum norm estimate (WMNLE) minimizes $\|\mathbf{Wx}\|$ subject to the constraint that $\|\mathbf{Ax-b}\|$ be minimized. Golub and Van Loan [4] refer to the introduction of **W** into the problem as column weighting.

Explicit Expressions for the MNLE and WMNLE

Explicit expressions for the MNLE and WMNLE are described as follows.
For an m x n forward matrix of rank k, the MNLE, denoted $\mathbf{x}^+$, is given by

$$\mathbf{x}^+ = \sum_{i=1}^{k}\left((\mathbf{b},\mathbf{u}_i)/\sigma_i\right)\mathbf{v}_i \qquad (1)$$

where $\mathbf{u}_1, \ldots, \mathbf{u}_m$ are the left singular vectors of the matrix $\mathbf{A}$, $\mathbf{v}_1, \ldots, \mathbf{v}_n$ are the right singular vectors, and $\sigma_1, \ldots, \sigma_k$ are the corresponding nonzero singular values listed in descending order. The notation $(\cdot,\cdot)$ denotes the inner product of two vectors.

A similar expression for the WMNLE is obtained by making the substitution $\mathbf{y}=\mathbf{Wx}$. The weighted problem then reduces to the unweighted one of selecting the vector $\mathbf{y}$ of smallest norm that minimizes $\|\mathbf{AW}^{-1}\mathbf{y}-\mathbf{b}\|$. When $\mathbf{W}$ is a diagonal matrix, the product $\mathbf{AW}^{-1}$ has the effect of dividing the columns of $\mathbf{A}$ by the corresponding diagonal element of $\mathbf{W}$. Hence the use of the term column weighting. By direct substitution into (1), an explicit expression for the WMNLE, denoted $\mathbf{x}_w^+$, is given by

$$\mathbf{x}_w^+ = \sum_{i=1}^{k}\left((\mathbf{b},\mathbf{u}_{wi})/\sigma_{wi}\right)\mathbf{W}^{-1}\mathbf{v}_{wi} \tag{2}$$

Here, $\mathbf{u}_{w1}, \ldots, \mathbf{u}_{wm}$ are the left singular vectors of the matrix $\mathbf{AW}^{-1}$, $\mathbf{v}_{w1}, \ldots, \mathbf{v}_{wn}$ are the right singular vectors of $\mathbf{AW}^{-1}$, and $\sigma_{w1}, \ldots, \sigma_{wk}$ are the corresponding nonzero singular values of $\mathbf{AW}^{-1}$ listed in descending order.

The Proposed Weight to Remove Depth Bias

Let $\mathbf{W}$ be a diagonal matrix with each diagonal entry equal to the square root of the gain of the corresponding dipole. That is, $\mathbf{W}=diag\{\sqrt{\|\mathbf{a}_i\|}\}$, where $\mathbf{a}_i$ is the ith column of the forward matrix $\mathbf{A}$. Then in the formulation of the weighted minimum norm problem, $\|\mathbf{Wx}\|^2 = \sum_{i=1}^{n}\|\mathbf{a}_i\|\cdot x_i^2$ where x_i is the ith element of the vector $\mathbf{x}$. Each dipole source is weighted exactly by its gain, penalizing the sources nearest the sensors, and eliminating the gain dependent depth bias. In Example 1, it is further illustrated how the proposed weight removes depth bias by examining the WMNLE expression (2) for a specific source space.

Example 1: The Proposed Weight

Suppose a source space consists of dipoles of different depths and that the gain of each dipole is inversely proportional to its distance from the sensors. Assume that the dipoles produce identical signal patterns that vary only in their respective gains. In this case, the columns of the forward matrix are multiples of one another. For any observation vector, $\mathbf{b}$, a depth unbiased solution would ignore the differences in depth or gain, and distribute the activity equally among the dipoles.

The WMNLE solution (2) requires computation of the right singular vectors of $\mathbf{AW}^{-1}$. These are the eigenvectors of the matrix $(\mathbf{AW}^{-1})^T(\mathbf{AW}^{-1})$ which, for the suggested weight based on the square root of the gain, has ij^{th} element $(\mathbf{a}_i,\mathbf{a}_j)/\sqrt{\|\mathbf{a}_i\|\|\mathbf{a}_j\|}$. By inspection, this matrix is of rank $k=1$ with nonzero eigenvalue $\lambda_{w1} = \sum_{i=1}^{n}\|\mathbf{a}_i\|$ and corresponding eigenvector $\mathbf{v}_{w1} = (\sqrt{\|\mathbf{a}_1\|}, \ldots, \sqrt{\|\mathbf{a}_n\|})^T$. From (2), the WMNLE is therefore a multiple of the vector $\mathbf{W}^{-1}\mathbf{v}_{w1} = (1, \ldots, 1)^T$. Thus, for the suggested weight, the WMNLE solution distributes the activity equally without bias to depth.

For comparison, Example 1 is reworked below for both the column normalization case and the unweighted case.

Example 2: Comparison to Column Normalization

The technique of column normalization has been used to address depth bias in the solution to the inverse problem [5]. This technique is equivalent to applying a diagonal weight $\mathbf{W} = diag\{\|\mathbf{a}_i\|\}$. Since $\|\mathbf{Wx}\|^2 = \sum_{i=1}^{n}\|\mathbf{a}_i\|^2\cdot x_i^2$, each dipole source is weighted by the square of its gain. For the forward matrix of Example 1, it can be shown that the WMNLE solution is a multiple of the vector $(\|\mathbf{a}_1\|^{-1}, \ldots, \|\mathbf{a}_n\|^{-1})^T$. This solution is biased toward dipoles with smaller gains, namely, the ones farthest from the sensors.

Example 3: Comparison to the Unweighted Case (MNLE)

In the unweighted case, W=I, the identity matrix. For the forward matrix of Example 1, it can be shown that the MNLE solution is a multiple of the vector $(\|\mathbf{a}_1\|, \ldots, \|\mathbf{a}_n\|)^T$. This solution, as reported earlier [2], is biased toward sources with the largest gains, the ones closest to the sensors.

The Ill-Conditioned Forward Matrix

In the above examples, for simplicity, the source space is chosen to produce a well-conditioned forward matrix of rank $k=1$. All of the dipoles produce identical signal patterns in the sensor array, and the columns of the forward matrix are exact multiples of one another. In practice, however, dipoles of varying depths produce signal patterns that deviate from one another and the columns of the forward matrix are not exact multiples. If the deviations are sufficiently small, then the angle between the columns is small, and the resulting forward matrix is ill-conditioned.

When the forward matrix is ill-conditioned, the results of the above examples still hold. For each weighting matrix, W, the first singular value of the matrix $\mathbf{AW}^{-1}$ is nonzero and the rest of the singular values are either nearly zero or identically zero (depending on the dimensions of the matrix). The matrix $\mathbf{AW}^{-1}$ can be conditioned through truncation of the smaller singular values [4]. Through truncation, the singular values that are nearly zero are set to zero and the rank $k=1$ in the expression (2) for the WMNLE. As in the well-conditioned case, the inverse solution is a multiple of the vector $\mathbf{W}^{-1}\mathbf{v}_{w1}$. In short, the proposed weight is effective in removing depth bias in an ill-conditioned setting.

Application to Magnetocardiography

The suggested weight based on the square root of the gain was tested on magnetocardiographic data, and compared to the performance of both the column normalization case and the unweighted case. The data were recorded using a BTi Magnes™ 37 channel biomagnetometer placed at several locations over the heart. A three dimensional source space corresponding to the myocardium was constructed using 8 coronal MR images of the patient's thorax, which is modeled as a sphere. The source space consists of 488 dipoles (two tangential components at 244 locations) lying in the 8 planes roughly parallel to the array of sensors. The planes are separated by 1cm. A representative image is shown in Figure 1a.

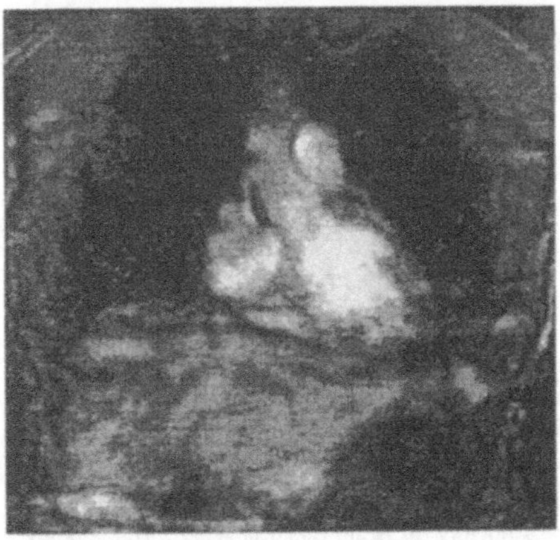

Figure 1a: Coronal MRI of Thorax

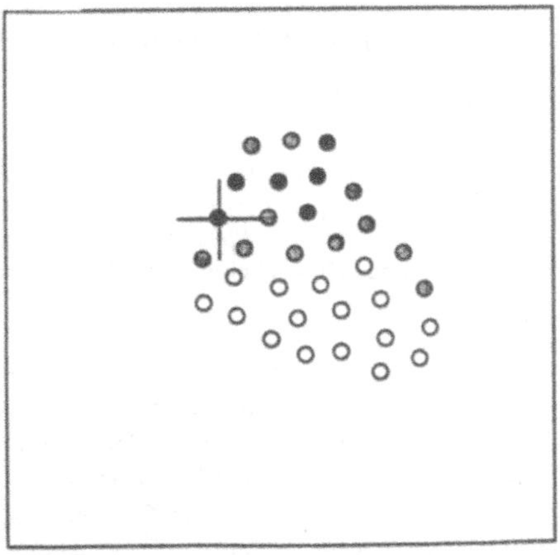

Figure 1b: Partial Display of Dipole Intensities

The images were registered with the biomagnetometer geometry using fiduciary markers common to both modalities. The multichannel MCG data were recorded at a latency corresponding to early atrial depolarization. WMNLE inverse solutions were calculated for the three cases. For each solution the relative distribution of the activity among the 8 planes was calculated by dividing the sum of squares of the dipole strengths in each plane by the total sum of squares of the strengths for all the planes. The relative distributions are given in Figure 2. It can be seen that the results are consistent with the earlier numerical examples: The proposed square root weighting distributes the activity more uniformly. The MNLE biases the activity closest to the sensors, while the column normalization weight biases the solution away from the sensors.

In Figure 1b, the inverse solution for the weight based on the square root of the gain is shown for just one plane of the complete three dimensional reconstruction. The displayed plane contains the greatest percentage of activity of the 8 planes, and corresponds to the MR image of Figure 1a. Dipole intensities are given in grey scale, darker colors corresponding to greater intensity, and a cross denoting the dipole of single greatest strength. As expected, the activity is concentrated in the atria.

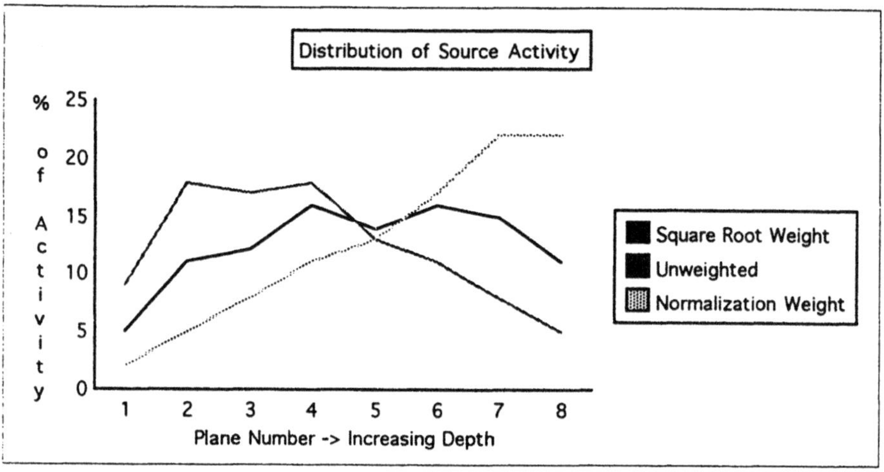

Figure 2: Distribution of source activity among 8 planes for each of three weights.

Conclusion

A weighted linear least squares solution to the biomagnetic inverse problem is proposed with a view toward eliminating any bias toward deep or shallow solutions. The proposed weight consists of the square root of the gains of the dipoles. When applied to numerical examples as well as physiological data, the proposed weighting appears to have eliminated depth bias from the solutions.

References

[1] Sarvas, J., Physics in Medicine and Biology, 1987, 32(1):11-22.

[2] Crowley, C., Greenblatt, R., Khalil, I., Proceedings of the Annual International Conference of the IEEE Engineering in Medicine and Biology Society, 1989, 11:1256-1258.

[3] Strang, G., Linear Algebra and Its Applications, Orlando, Harcourt Brace Jovanovich Inc.,1988.

[4] Golub, G., Van Loan, C., Matrix Computations, Baltimore, The John Hopkins University Press, 1983.

[5] Mosher, J., George, J., "Comments on Loreta," International Society for Brain Electromagnetic Tomography (ISBET) Newsletter No. 5, November 1994.

Imaging Properties of "Myocardial Current Imaging"

Dössel, O., Katscher, U., Köhler, Th., Fuchs, M., Wischmann, H.-A., Wagner, M., Drenckhahn, R. and Hoppe, W.

Philips Research Hamburg, Röntgenstr. 24-26, D-22335 Hamburg, Germany

Introduction

The imaging properties of the new technique "Myocardial Current Imaging" (MCI) [1] are investigated using computer simulations. MCI has the potential to depict complex current distributions on the human heart from multichannel Magnetocardiography. This might have important applications in diagnosis of complex arrhythmias or myocardial infarction. MCI basically is a minimum norm estimate using the morphological constraint that the impressed currents must be on the heart muscle.

The Point Spread Function (PSF) is an important property that is used to characterize medical imaging techniques. It delivers a quantitative measure of the smearing effects that show up if an image of a point source (single current dipole, δ-function of current density) is made. In biomagnetic imaging the PSF depends on the reconstruction algorithm and on location and orientation of the dipole. Line-sources and sources that are distributed over large areas can show different behaviour and cannot be extrapolated easily from the PSF because of the ill-posed nature of the problem. Noise will obviously increase the smearing effects and widen the PSF. So the PSF of biomagnetic imaging is a much more complex problem compared to other imaging modalities and we refer to a generalized PSF in the following.

The PSF of MCI was obtained from computer simulations using the software package CURRY™. First point sources (single current dipoles) were used for a forward calculation to determine the magnetic fields in a SQUID multichannel system. As an example 4*31 channels above the chest of a realistic torso were used. The "basic" imaging properties were investigated using noiseless data - for more realistic estimates synthetic noise was added. Many datasets were produced with various locations and different orientations of the dipoles. Using these synthetic data minimum-norm reconstructions were carried out and analyzed. The full-width-half-maximum was determined for different cases. In addition some examples of line-sources were investigated. Finally homogeneously distributed sources on the whole endocardium and with non-activated "holes" were analyzed.

Methods

The technique of "Myocardial Current Imaging" investigated here is based on minimum-norm estimates with morphological constraints and a Boundary Element Method (BEM) realistic volume conductor model [2,3]. The basic equations are:

$$\underline{j} = \underline{\underline{A}}^t * \left(\underline{\underline{A}} * \underline{\underline{A}}^t + \alpha^2 * \underline{\underline{1}}\right)^{-1} * \underline{\underline{M}} \qquad\qquad \underline{\underline{B}} = \underline{\underline{A}} * \underline{j}$$

with j: reconstructed current density, A: Biot-Savart-matrix, M: array of measured flux densities, α: noise regularization parameter, B: flux densities of the forward calculation. The minimum norm estimate includes a noise regularization and a depth normalization equalizing possible source locations with respect to their ability to generate flux density in the sensor matrix [4]. The morphological constraint is to use a set of roughly 400 points spread out with a mean distance of typically 7mm on the surface of the individual endocardium. With BEM the volume currents in the torso and lungs are modeled in a two compartment approach with isotropic conductivities. The surfaces of the boundaries are obtained from individual 3D-MR-data using semiautomatic segmentation techniques and fast triangulation algorithms. Fig. 1 depicts the scenario.

The computer simulations are started with the placement of single dipoles, lines of dipoles, or homogeneously spread out dipoles (all pointing in one direction) on the segmented endocardium. In this study a dipole strength of 5μAm is chosen because this is a typical value for the localization of the onset of an extrasystole. A "forward" calculation is carried out (BEM, two compartment approach) delivering the magnetic flux density gradients at 124 positions above the torso.

For the investigations on the influence of noise during a time window of 20 msec at every msec a random signal is added corresponding to a noise value of 5fT/√Hz, 10fT/√Hz, and 20fT/√Hz (measuring bandwidth 1Hz to 500Hz, 10 trials averaged). Another approach could have been the specification of the signal to noise ratio. We did not follow that approach, because the "mean" signal $(\sqrt{(\Sigma B_i^2)/N})$ depends strongly on the position of the magnetometers and the orientation of the dipoles. Radial and tangential dipoles e.g. would have lead to similar imaging properties hiding the fact, that radial dipoles must be much larger than tangential dipoles to produce the same "mean" signal. The selected noise values correspond to a mean SNR of 20, 10 and 5 respectively.

Then the inverse problem is solved using the minimum-norm estimate as described above. For the analysis of noisy signals several instants of time (with different values for the added noise) are selected and a typical reconstructed current density and PSF is taken. If the regularization parameter α is chosen properly, the reconstructed current density does not vary significantly with time.

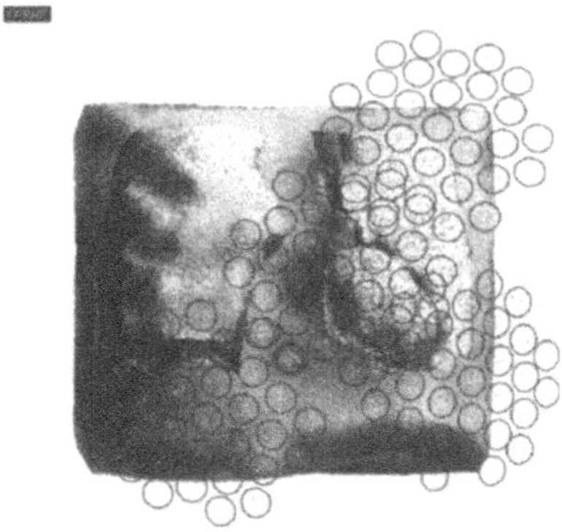

Fig.1: Segmented torso, lungs and left ventricle together with the sensor arrangement

The full-width-half-maximum (FWHM) of the image of a point source is determined in the following way: First the reconstruction point with the largest current density is looked up. Then the mean current density at the neighboring points (7mm distance) is determined. Finally a Gauss function is fitted to the maximum and the 7mm distance value. The FWHM of the Gaussian is taken as an estimate of the FWHM of the current density distribution. Fig. 2 shows a slice through a reconstructed current distribution together with the Gauss fit. We see that the approach is reasonable and that the PSF resembles a Gauss function. Other more elaborate ways of estimating the PSF can be thought of, but the basic results are certainly independent of the method.

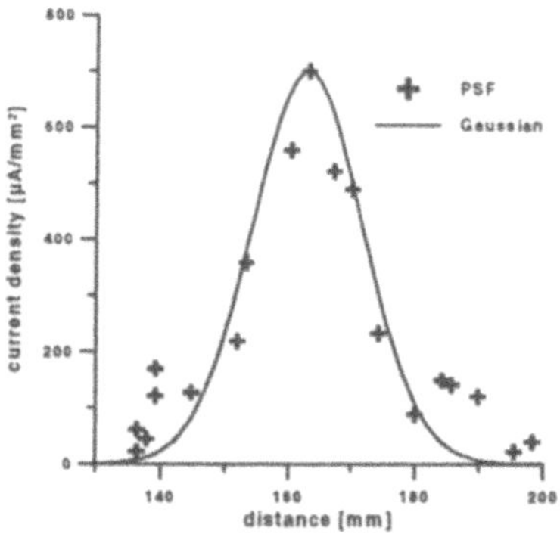

Fig. 2: Slice of a reconstructed minimum norm current distribution together with a fitted Gauss function. For the forward calculation a single current dipole was used (without noise).

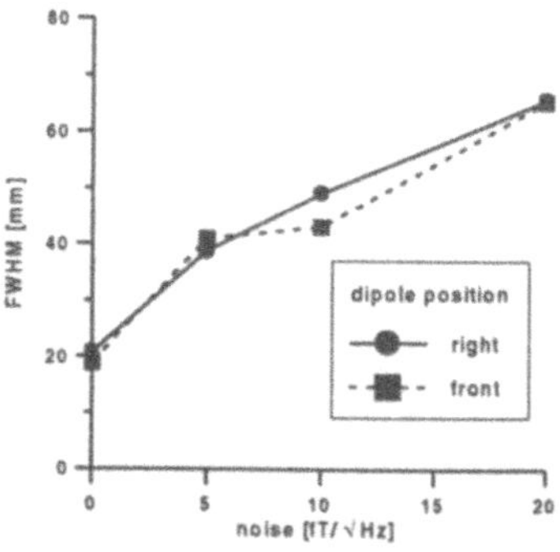

Fig 3: FWHM of the point spread function of single current dipoles with j=5μAm as a function of noise

Results

Ten single dipoles at various locations on the heart were simulated with and without noise. Fig. 4 depicts some reconstructed current densities. Fig. 3 shows the results for typical examples. The FWHM of the PSF is in the range of 20mm without noise and increases up to 60mm with 20 fT/√Hz. Radial dipoles (with respect to the body surface) can be found in the noiseless case only.

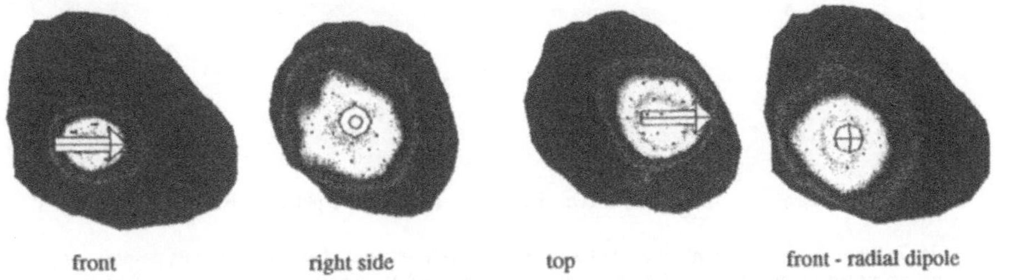

Fig. 4: Examples of reconstructed current densities of single current dipoles (without noise).

Two current dipoles can be clearly separated in the noiseless case provided their distance is larger than 30mm. Fig. 5 gives some examples.

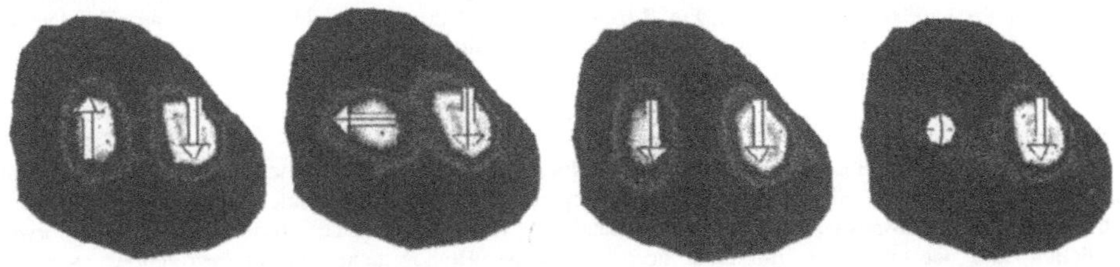

Fig. 5: Examples of reconstructed current densities of two current dipoles with a distance of 30mm

Activated rings around the heart can be reconstructed, but some artifacts show up. Obviously the simulated magnetic flux density can be reproduced with current distributions having a smaller integral current than the original one. (The minimum norm estimate picks the solution with the smallest currents!) Fig. 6 gives an example.

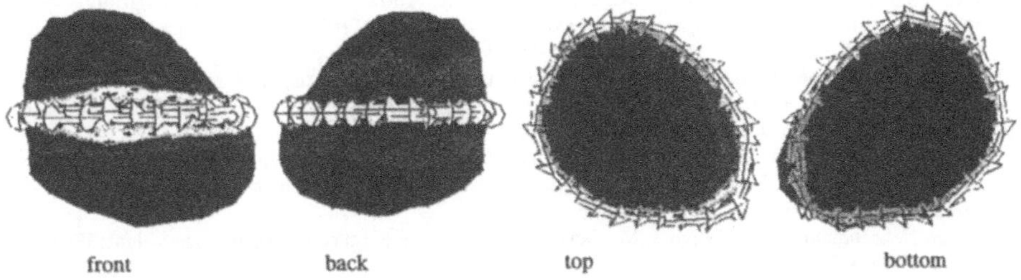

Fig. 6: Reconstructed current density of a ring of dipoles

A homogeneous activation of the heart was modeled with about 400 single dipoles distributed on the myocardium all pointing in the "upward" direction. Again small artifacts show up: Fluctuations of up to 20% of the current density are observed in the reconstructed image (Fig. 7 left).

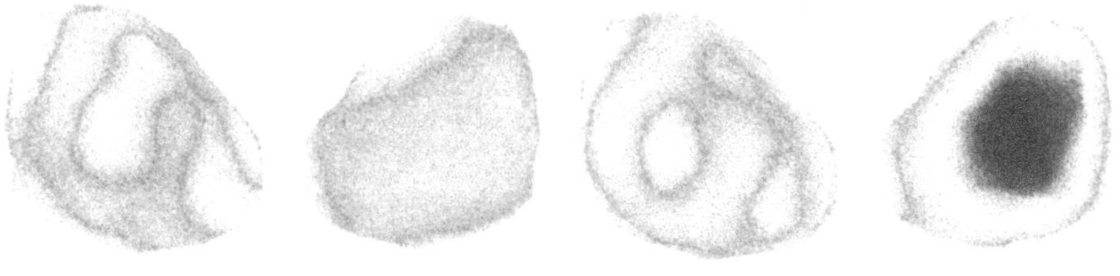

front - homogeneous back - homogeneous front - "inactive patch" back - "inactive patch"

Fig. 7: Reconstructed current density of a homogeneously activated heart (all currents pointing up) and reconstructed current density of a homogeneously activated heart with an "inactive patch" of 30mm diameter on the back of the heart

Again the artifacts can be explained by the minimum norm constraint. They can be ignored, if in pathological cases a large reduction of current density shows up that becomes clearly visible in these images. In fact this is the case as can be proven with simulations using homogeneously activated hearts with "black patches", where the dipoles have been switched off. Fig. 7 (right) shows that they can be clearly be seen provided they are larger than 30mm diameter. So if in pathological cases the current density on the myocardium is significantly reduced this may be imaged with MCI at the right position.

Discussion

The results show, that the overall spatial resolution of the method is in the range of 20mm depending mainly on the noiselevel. Single dipoles can be depicted reliably and accurately. More complex current distributions may not always be reconstructed accurately due to the fact that the "original" taken for the forward calculation does not fulfill the "minimum norm" condition. Homogeneous current distributions cannot be reconstructed exactly, but deliver an image with "fluctuating" current densities. So fluctuations of the current density in the order of 20% of the maximum cannot be interpreted as "reality". Holes in a mainly homogeneous current distribution can be seen provided they are larger than about 30mm diameter. They lead to "black patches" in the current density image that may in clinical cases be interpreted as a pathologically reduced activity in that specific region of the heart.

References

[1] Fuchs, M., Wagner, M., Wischmann, H.-A., Dössel, O., Cortical current imaging by morphologically constrained reconstructions, In: Baumgartner, C., Deecke, L., Stroink, G., Williamson, S.J., Biomagnetism: Fundamental Research and Clinical Applications, Amsterdam, Elsevier, 1995, 320-325

[2] Fuchs, M., Wischmann, H.-A., Wagner, M., Drenckhahn, R., Advanced biomagnetic and bioelectric reconstruction algorithms, In: Herrmann, H.J., Wolf, D.E., Pöppel, E., Supercomputing in Brain Research: From Topography to Neural Networks, London, World Scientific, 1995, 161-173

[3] Ferguson, A.S., Stroink, G., Localization of epicardial sources using magnetic and potential maps, In: Baumgartner, C., Deecke, L., Stroink, G., Williamson, S.J., Biomagnetism: Fundamental Research and Clinical Applications, Amsterdam, Elsevier, 1995, 641-646

[4] Fuchs, M., Wischmann, H.-A., Wagner, M., Generalized minimum norm reconstruction algorithms, ISBET Newsletter 5, 1994, 8-11

Practicability of Neuromagnetic Source Analysis Using the Finite Element Method

Drüen, K., Steinsträter, O. and Lütkenhöner, B.

Institute of Experimental Audiology, University of Münster, Germany

INTRODUCTION

It is still an open question as to what extent anatomical details have to be taken into account when constructing a volume conductor model for neuromagnetic applications. The approximation of the human head as a spherical model consisting of three or four concentric spheres with different conductivities (corresponding to brain tissue, liquor, skull and scalp) is poor when measurements cover a large area of the head. Investigations with more realistic models have shown the inappropriateness of spherical models especially for analysis of deep sources [1],[2]. Most of these numerical models are based on the boundary element method (BEM) resulting in a finite sum of bounded surface integrals. An alternative is the finite element method (FEM), which allows, in contrast to the BEM, not only a relatively easy handling of the complex geometry of the head, but also the consideration of anisotropic conductivities.

The purpose of this study is to explore the practicability of the FEM for neuromagnetic source analysis and to assess the performance of this method as a function of the discretisation quality.

METHODS

The mathematical formulation of the forward problem for a volume conductor leads to an elliptical differential equation of the Poisson type with additional boundary conditions. This representation of the problem can be transformed into a variational form, which is based on integrals over the volume of interest and its boundaries, respectively. The volume is discretized into small elements (tetrahedrons) which are supposed to be small enough to ensure that in each tetrahedron the conductivity tensor can be regarded as constant. The discretisation leads to a matrix system of N equations with N unknown, where N denotes the number of nodes.

A spherical volume conductor consisting of four concentric layers was discretized in terms of 8400 nodes and subdivided afterwards into about 47.000 tetrahedrons by means of a Delaunay Triangulation. The radii (in units of the outermost radius, having a value of 8 cm) were chosen to be 0.84, 0.87, 0.95 and 1. It is important that very flat tetrahedrons are avoided since serious numerical problems can arise otherwise. Local refinement was achieved by subdividing each tetrahedron within a certain distance from the actual position of the current source into eight new tetrahedrons. For this purpose each edge of the initial tetrahedron was split into two halves, resulting in six new nodes of the mesh. This method preserves the numerical quality of the discretisation since the new elements have the same angles as the initial one.

Current dipoles in the innermost sphere of a four layer model with the conductivities [3] of 0.45 (Brain), 1.5 (Liquor), 0.0063 (Bone) and 0.43 S/m (Scalp) were investigated. The resulting potential distribution V on the various surfaces of the model was obtained by solving the linear system of FEM equations with the method of Conjugate Gradients. The magnetic field vector of a location $\vec{r}$ outside the volume conductor was then determined by numerical integration over the four surfaces [4]:

$$\vec{B}(\vec{r}) = \vec{B}_0(\vec{r}) + \frac{\mu_0}{4\pi} \cdot \sum_{i=1}^{4} (\sigma_i^- - \sigma_i^+) \int_{S_i} \frac{V(\vec{r}')}{|\vec{r} - \vec{r}'|^3} \cdot (\vec{r} - \vec{r}') \times \vec{n}_i \cdot dS_i \tag{1}$$

where B_0 denotes the magnetic field due to the primary current source and $(\sigma_i^- - \sigma_i^+)$ reflects the change of conductivity at the adjacent boundaries. The magnetic field vector was calculated for the coil locations of a 37-channel gradiometer system similar to that of Biomagnetic Technologies (San Diego) with the center coil on the z-axis. These locations had a distance of 12 cm from the center of the spherical model.

The numerical results were compared with the analytical formulas for the magnetic field of a spherical conductor. It was essential to consider not only the radial field component (the component usually measured) because in a spherical model, volume currents do not contribute to this component. The relative difference measure (RDM) served as the error criterion. To quantify the accuracy of the numerical model, the determination of all three field components in every recording position resulted in a vector $\mathbf{B} = (B_x^1, B_y^1, B_z^1, ..., B_x^{37}, B_y^{37}, B_z^{37})$. Using this notation the RDM was defined as:

$$RDM = \sqrt{\sum_{i=1}^{3 \cdot K} \left(\frac{B_{ANA}^i}{|\mathbf{B}_{ANA}|} - \frac{B_{FEM}^i}{|\mathbf{B}_{FEM}|} \right)^2} \tag{2}$$

with K = 37. This measure quantifies the deviation between the field patterns obtained with the analytical formula and the FEM-model, but it does not detect a possible scaling error, owing to the special type of normalization used.

RESULTS

Since a primary current source located on an interface represents a singularity, the first question investigated was as to how close a dipole could approach the innermost boundary without causing unacceptable errors. For this purpose a dipole located on the z axis was shifted into the direction of the interface brain-liquor. Two different angles between dipole moment and z-axis were considered: 90° (x direction, representing a tangential source) and 45°. The results are shown in **Figure 1** (solid lines). At a relative dipole position of 0.74 the deviation from the analytical model was already about 5% and rises considerably for larger eccentricities.

This unacceptable effect was overcome by refining the existing mesh three times in small spheres (relative radii: 0.2, 0.14 and 0.07) around the actual dipole position, resulting in about 12.500 nodes (76.000 tetrahedrons). With this refinement the source could be placed at a relative position of 0.83 (first boundary at 0.84) without exceeding the 5% error limit (dashed lines in Fig. 1).

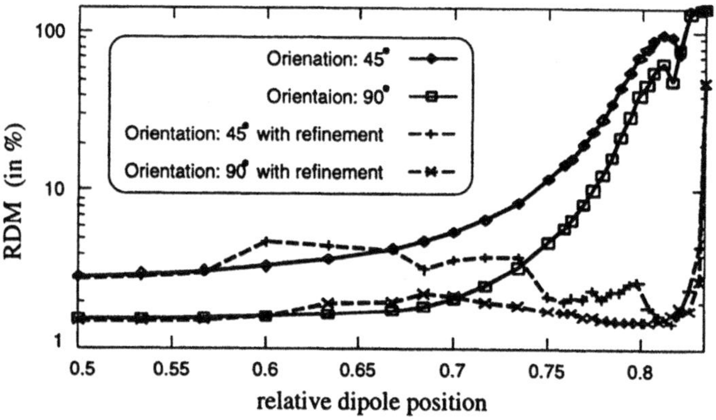

Figure 1: RDM versus relative dipole position in a four layer model with piecewise homogeneous layers.

The second question investigated was as to how good a thin layer could be approximated without losing accuracy of the numerical model. For this purpose the mesh was modified by reducing the thickness of the second layer (representing the liquor) without changing the thickness of the other layers. Three different positions on the z-axis of the primary source (position: 0.5, 0.72 and 0.81) with two orientations (angles between dipole moment and z-axis: 90° and 45°) respectively were considered. **Figure 2** shows the RDM-values for the different dipole positions and orientations as a function of the relative thickness of the second layer. Although the quality of the mesh decreased continuously when deforming the existing tetrahedrons the RDM-values remained very stable until a relative thickness of 0.003 (which corresponds to about 0.25 mm in practice). This value could be diminished further by increasing the maximum number of iteration steps when solving the linear system of equations.

The third point investigated was the influence of anisotropic conductivity of the third layer (corresponding

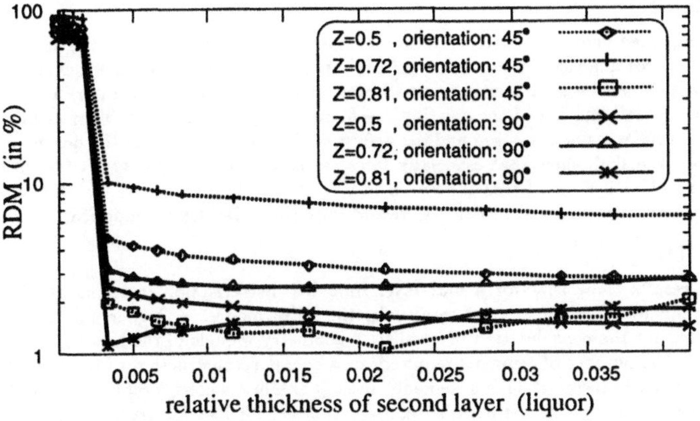

Figure 2: RDM versus relative thickness of the second layer (corresponding to the liquor) for different positions and orientations of a current dipole.

to the skull). The tangential conductivity σ_t of the third layer was systematically changed while keeping the radial conductivity σ_r constant. The conductivities in the other layers were assumed to be isotropic. The locations and orientations were the same as in the previous example. The numerical solution for the isotropic model was taken as the reference for the RDM.

Figure 3 shows that although the deviation from the isotropic model rises continuously when decreasing the ratio σ_r/σ_t the influence of anisotropy is rather small as long as the ratio is bigger than 0.2 - 0.3. For smaller ratios the deviation from the isotropic model rises up to 12%. The RDM-values obtained for a dipole orientation of 45° were consistently greater than those obtained for 90°. The influence of the dipole location was relatively small.

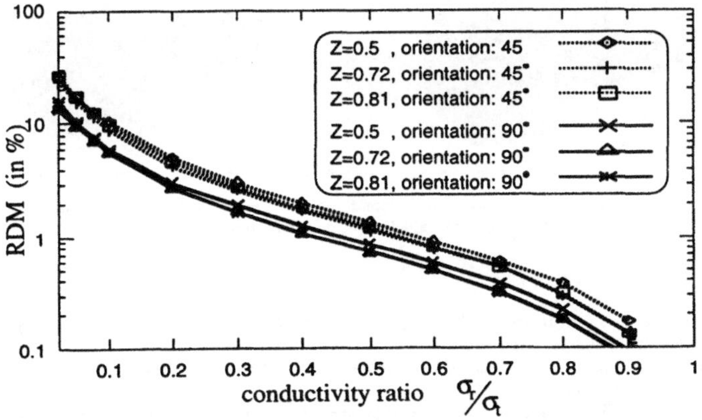

Figure 3: Influence of anisotropy of the third layer (representing the skull), the figure shows the RDM versus the conductivity ratio σ_r/σ_t (isotropic model as the reference).

DISCUSSION

All results reported here were obtained with a customary workstation. Thus, the use of FEM models for neuromagnetic source analysis represents no principle problem. With a suitable local mesh refinement the number of elements can be kept relatively small even in the case that the primary source is located near to an interface. A deformation of the existing mesh turned out to be no problem as long as the internal structure of the mesh was not destroyed and the deformed elements became not too degenerated. The studies of anisotropy for the skull show that especially for a realistic ratio of $\sigma_r/\sigma_t \approx 0.1$ the influence cannot be neglected. It shall be emphasized that the errors to be expected in practice are certainly smaller than those described here, since it is common practice to record only the radial field component.

PROSPECTS

The above results show that for a four layer model a discretisation in terms of 76.000 tetrahedrons (with a local refinement) leads to good results for the FEM-model. A higher number of elements is undoubtedly needed for the consideration of "realistic" geometry models of the human head. **Figure 4** gives an impression of the number of elements required for a coarse reconstruction of the brain. It demonstrates that a reconstruction in terms of 85.000 elements already exhibits a considerable amount of details. Further investigations have to show whether such a reconstruction is sufficient for neuromagnetic applications.

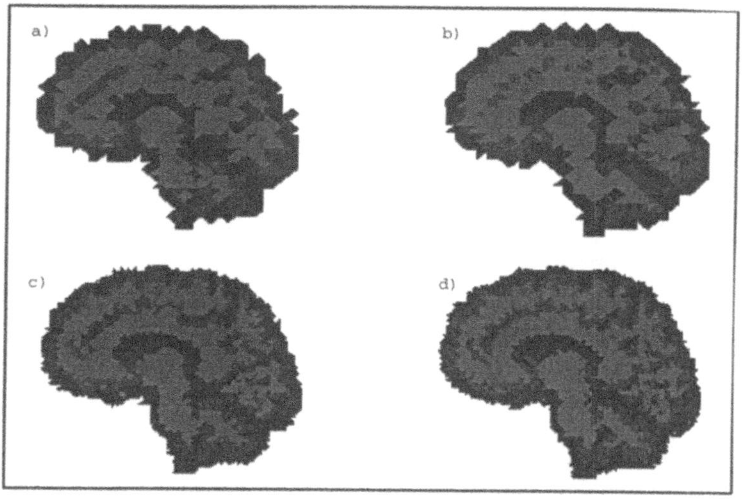

Figure 4: Cross sectional view of a realistic volume conductor discretized with a different number of elements: a) 8000, b) 37.000, c) 85.000 and d) 153.000 tetrahedrons , derived from MRI-data.

References

[1] Hämäläinen, M.,Sarvas,J., *Realistic conductivity geometry model of the human head for interpretation of neuromagnetic data,* IEEE 1989, Vol 36, page 165-171

[2] Menninghaus, E.,Lütkenhöner,B., Gonzalez,S., *Localization of a dipolar source in a skull phantom: realistic versus spherical model,*IEEE, 1994 ,Vol.41 page 986-989

[3] Geddes, L.A., Baker, L.E. *The specific resistance of biological material — a compendium of data for the biomedical engineer and physiologist. Med. & Biol. Eng., 1967*

[4] Hämäläinen, Matti, *Interpretation of neuromagnetic measurements,* PhD-thesis, Department of Technical Physics, Helsinki University of Technologies, Espoo 1987, Finland

A Numerically Stable Approximation for the Magnetic Field of the Conducting Spheroid Close to the Symmetry Axis

Fieseler, T.[1], Ioannides, A.A.[2,3], Liu, M.J.[3] and Nowak, H.[1]

[1]Biomagnetic Centre, Jena, Germany; [2]KFA Jülich, Germany; [3]The Open University, Milton Keynes, UK

Introduction

The homogeneous conducting spheroid is a convenient intermediate model between the widely used sphere model and complicated numerical models like boundary element method for an approximation of the conductivity profile of the human head. While a simple analytic expression exists for the magnetic field generated by a dipole in a sphere model [1], for the spheroid a series expansion [2] in terms of Legendre polynomials must be used. However, numerical problems arise in the vicinity of the symmetry axis of the spheroid. We present an expansion for this region that enables us to use the spheroid solution safely, even near the symmetry axis.

The break from spherical symmetry introduced by the spheroid makes the centre non-silent, and thus opens up the possibility of detecting generators, that are deep inside the head, e.g. the thalamus or brainstem, from measurements of the magnetic field they generate. As we will demonstrate the indications are encouraging, and further more an approximate description of the global geometry might be sufficient, provided sensors are placed all round the generator. Our tentative findings point to new and exciting applications of MEG with modern multichannel sensors.

Methods

If the Cuffin / Cohen series expansion is applied using the well known recurrence relations for the Legendre polynomials $P_n^m(x)$, the solution is numerically unstable over a critical region *near* the symmetry axis, and therefore not suitable for source points or field points very close to the symmetry axis.

As a criterion to distinguish a *critical region* of the spheroid where numerical instabilities are likely from the *safe region* where the general expressions can be safely applied we define the prolate coordinate parameters η_{crit} and ξ_{crit}. An example is shown on the left side of Fig. 1, the tiny spheroid corresponds to $\eta_{0_{crit}}$, the hyperboles correspond to ξ_{crit}.

For source points or field points within the critical region ($\eta_0 < \eta_{crit}$, $|\xi_0| < \xi_{crit}$ or $|\xi| < \xi_{crit}$) we use a series expansion of the standard expressions around the points $x = 1$ and $x = -1$, retaining terms up to second order in x; the critical expressions to be replaced are $P_n^m(x)$, $(x^2-1)^{-1/2}\ P_n^m(x)$ and $(x^2-1)^{1/2}\ P_n^{m\prime}(x)$ for x close to 1 or -1, where x can be η_0, ξ_0 or ξ. These expressions are partly given in the appendix.

The right side of Fig. 1 shows an example of instability in the general solution and the remedy provided by our approximation within the critical region. The display shows the modulus of the magnetic field due to a current dipole well away from the critical region (marked by the arrow inside the spheroid in the middle figure), computed on a line crossing the critical region (marked by the line on top of the spheroid in the middle figure).

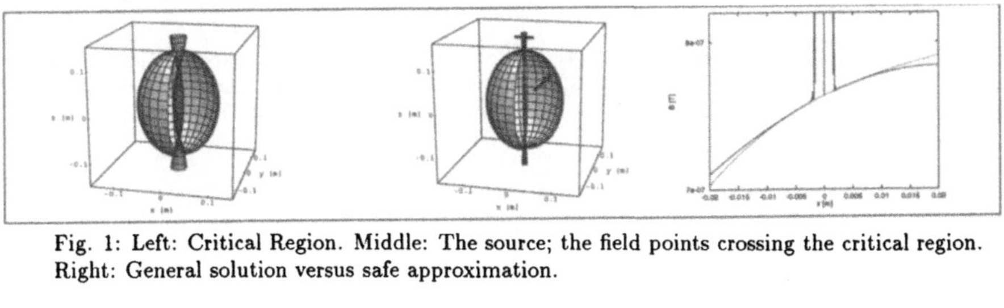

Fig. 1: Left: Critical Region. Middle: The source; the field points crossing the critical region. Right: General solution versus safe approximation.

The solid line shows the result of computation using the expressions in [2], while the dotted line shows our result. The example demonstrates that our approximation remedies the numerical instability (artificial peak around the symmetry axis). It is also worth noting that the choice of the values for η_{crit} and ξ_{crit} is not very critical, because the approximate expressions are accurate over a fairly large region. As a result the forward solution for the magnetic field of a current dipole inside the conducting spheroid becomes numerically stable regardless of the position of source point and field point. We are thus able to address the fundamental question of detectability of MEG signals generated deep inside the brain. We assume that the head *is* a homogeneous conducting spheroid (long half axis 12*cm*, short half axis 9*cm*) and use for simulations the twin dewar BTI MAGNES system, with each dewar placed on either side of the spheroid (Fig. 2).

The magnetic field is computed for different dipoles using our expressions for the spheroid, for all 74 channels. Random noise (20%) is added to the computer generated signal, which is then analysed using Magnetic Field Tomography (MFT) [3,4]. As found in previous MFT studies the resulting estimates for the primary current density, $\mathbf{J}(\mathbf{r})$ were robust to the level of noise we have used. A cylindrical source space was used placed in the centre of the spheroid and between the two sensor arrays as shown in Fig. 3. The radius of the cylinder is 6*cm*, the length 8*cm*. $\mathbf{J}(\mathbf{r})$ could be non-zero within this cylinder, but of course the full current was confined by the spheroidal or spherical shell. The whole setup of volume conductor, position of the dewars and the source space corresponds to measurements and analysis scenarios we are using routinely. Separate inversions were performed for different volume conductor choices: the *true* spheroid used to simulate the measurements, a slightly longer spheroid (half axis 13*cm* and 9*cm*), a slightly shorter spheroid (half axis 11*cm* and 9*cm*) and a sphere. The MFT analysis was done seperately for each volume conductor as described in detail elsewhere[4].

Fig. 3 shows the main result: the left part of the figure shows the general setup of volume conductor, source space and sensors used for the inverse calculations. On the right, the MFT estimates for $|\mathbf{J}(\mathbf{r})|^2$ are displayed in cross sections of the source space cylinders in different heights (x-values). Different columns correspond to the different conductivity profiles used.

Results

For most superficial sources there are only small differences in MFT solutions with the different volume conductors, which is in agreement with our previous results for measurements from one side of the spheroid [5]. With all three spheroids as well as with the sphere the data of the *correct* spheroid can be reconstructed very well. The effect of the different volume conductors used for the inversions are small variations of strength and location of the current density solutions.

For deep and central sources the results of inversions using spheroid and sphere are significantly different. The results are most obvious for inversions of forward data obtained with a dipole which lies exactly in the centre of the spheroid. We compare the two cases of a central dipole in x-direction that is radial orientated towards the dewar positions and a y-dipole which is tangentially orientated with respect of the detector positions (see in Fig. 2).

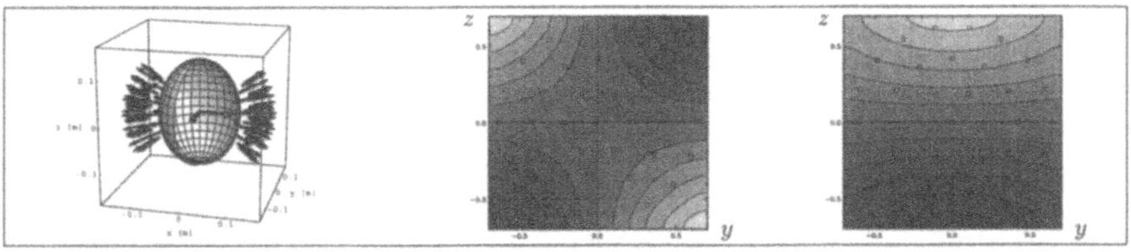

Fig. 2: Left: Spheroid used for the forward calculation, central x- and y-dipole, detectors. Middle: Magnetic field from the x-dipole (in right detector set). Right: Magnetic field from the y-dipole.

The magnetic field generated from these dipoles inside the *correct* spheroid in the (spherical) surface which is formed by the detectors is shown in Fig. 2. Due to the symmetry for each dipole the field pattern is identical for both dewars. In both cases the magnitude of the field is similar. The field is about 20 times weaker than the field of a superficial dipole ($2cm$ from the spheroid surface or $7cm$ from the centre) and about 10 times weaker than the field of a medium dipole ($4.5cm$ from the spheroid surface or $4.5cm$ from the centre) located directly under one of the dewars. Thus the magnetic fields for central dipoles could be strong enough to be detected in real measurements. The field patterns are more complicated than the familiar patterns generated by a single dipole inside a sphere. For the x-dipole there are four maxima in the corners of the field of view, for the y-dipole there are two maxima which are not entirely within the field view.

Since for the x-dipole the field pattern is much more structured than for the y-dipole we can expect more problems with an inversion of the latter case.

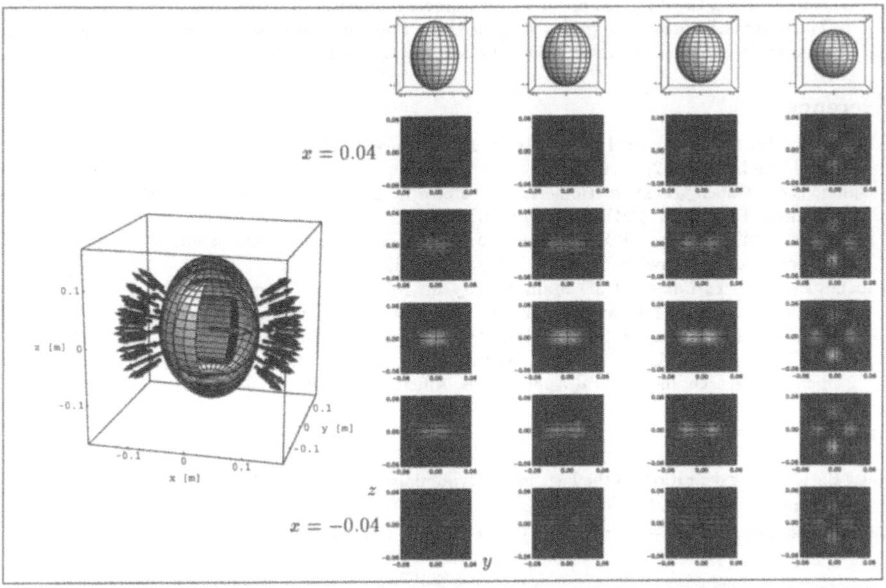

Fig. 3: Left: Detectors, volume conductors, source space cylinder; original source (x-dipole) Right: MFT inversions, plotted in yz-planes through the source space. The grey level represents the squared current density, the arrows correspond to the yz-projection of the current density. Columns from left to right: long spheroid, correct spheroid, short spheroid, sphere.

The results in Fig. 3 correspond to inversions for the fields of the x-dipole. As expected sensitivity in the depth is reduced, giving rise to a rather distributed, but clearly single source. The position of the maximum in x-, y- and z-direction is correct, i.e. in the centre of the volume conductor. The dependence on the spheroid details is weak: the solutions obtained with the correct spheroid and with spheroids that are slightly too long or too short are very similar. The transition away from the spherical configuration is sudden, but then stable, a result of considerable practical significance. Only the result for the sphere is clearly different. The complicated field pattern from Fig. 2 can not be approximated by a single generator inside a sphere, because no signal can be generated in this case by a source at the correct central position. This leads to the four maxima configuration away from the centre in the reconstruction.

The y-dipole can not be reconstructed correctly with just the two dewars in the arrangement of Figure 3. Instead of a distribution with one maximum at the volume conductor centre, the current density shows two maxima on either side of the spheroid symmetry axis. We expect that a wider sensor coverage (e.g. using a helmet type probe) will eliminate this defficiency.

Discussion

We have derived a series expansion for the spheroid which eliminates numerical instabilities in the forward solution, and applied it to the fundamental question of how accessible the centre of the head is with MEG. As expected, no striking differences are found between reconstruction of distributed sources using the spheroid and sphere for superficial sources. Also as expected, for very deep sources, corresponding to important structures like the thalamus in the brain, the sphere model is inadequate. But an even approximate spheroid model does remarkably well, when a central dipole is used, provided sensors are placed on either side of the head (spheroid) along the direction of the current dipole moment.

The optimistic message from this study is clear: there is good reason to believe that detectable, and (MFT) 'imageable' signal is generated from very deep regions in the brain.

References:

[1] Sarvas, J. Basic mathematical and electromagnetic concepts of the biomagnetic inverse problem. Phys. Med. Biol., 1987, 32(1):11–22.

[2] Cuffin, B.N., Cohen, D. Magnetic fields of a dipole in special volume conductor shapes. IEEE Trans. Biomed. Eng., 1977, BME-24(4):372-381.

[3] Ioannides, A.A., Bolton, J.P.R., Clarke, C.J.S. Continuous probabilistic solutions to the biomagnetic inverse problem. Inverse Problems, 1990, 5:523-542.

[4] Ioannides A.A., Estimates of 3D brain activity ms by ms from Biomagnetic signals: Method(MFT), results and their significance, In: Eiselt, M., Zwiener, U., Witte, H. Quantitative and Topological EEG and MEG analysis, Jena, Universitätsverlag Jena, 1995.

[5] Fieseler, T. Ioannides, A.A., Liu, M.J., Nowak, H. In: Baumgartner, C. et al. Biomagnetism: Fundamental Research and Clinical Applications, Amsterdam, Elsevier/IOS-Press, 1995, 445-449.

Appendix

We use the following definitions of the Legendre polynomials: $P_n = \frac{1}{2^n n!}\frac{d^n}{dx^n}(x^2-1)^n$, $P_n^m = (1-x^2)^{m/2}\frac{d^m}{dx^m}P_n$ for $|x| < 1$ and $P_n^m = (-1)^m(x^2-1)^{m/2}\frac{d^m}{dx^m}P_n$ for $|x| > 1$.
The series expansion of $(x^2-1)^{-1/2}\ P_n^m(x)$ for x close to 1 or -1 is

$$\frac{1}{(x^2-1)^{1/2}}P_n^0(x) = \frac{1}{\sqrt{2}}s_n^0(x)|x-1|^{-1/2} + \frac{(2n^2+2n-1)}{4\sqrt{2}}s_n^0(x)|x-1|^{1/2} + \frac{2n^4+4n^3-6n^2-8n+3}{32\sqrt{2}}s_n^0(x)|x-1|^{3/2} + O(|x-1|^{5/2})$$

$$\frac{1}{(x^2-1)^{1/2}}P_n^1(x) = -\frac{n(n+1)}{2}s_n^3(x) - \frac{(n-1)n(n+1)(n+2)}{8}s_n^1(x)|x-1| - \frac{(n-2)(n-1)n(n+1)(n+2)(n+3)}{96}s_n^3(x)|x-1|^2 + O(|x-1|^3)$$

$$\frac{1}{(x^2-1)^{1/2}}P_n^2(x) = \frac{(n-1)n(n+1)(n+2)}{4\sqrt{2}}s_n^0(x)|x-1|^{1/2} + \frac{(n-1)n(n+1)(n+2)(2n^2+2n-9)}{48\sqrt{2}}s_n^0(x)|x-1|^{3/2} + O(|x-1|^{5/2})$$

$$\begin{aligned}\frac{1}{(x^2-1)^{1/2}}P_n^3(x) &= -\frac{(n-2)(n-1)n(n+1)(n+2)(n+3)}{24}s_n^3(x)|x-1| - \\ &\quad \frac{(n-2)(n-1)n(n+1)(n+2)(n+3)(n^2+n-8)}{192}s_n^1(x)|x-1|^2 + O(|x-1|^3)\end{aligned}$$

$$\frac{1}{(x^2-1)^{1/2}}P_n^4(x) = \frac{(n-3)(n-2)(n-1)n(n+1)(n+2)(n+3)(n+4)}{96\sqrt{2}}s_n^0(x)|x-1|^{3/2} + O(|x-1|^{5/2})$$

$$\frac{1}{(x^2-1)^{1/2}}P_n^5(x) = -\frac{(n-4)(n-3)(n-2)(n-1)n(n+1)(n+2)(n+3)(n+4)(n+5)}{960}s_n^3(x)|x-1|^2 + O(|x-1|^3)$$

$$\frac{1}{(x^2-1)^{1/2}}P_n^m(x) = O(|x-1|^{5/2}) \quad \text{for} \quad m = 6, 8, 10, \ldots, \qquad \frac{1}{(x^2-1)^{1/2}}P_n^m(x) = O(|x-1|^3) \quad \text{for} \quad m = 7, 9, 11, \ldots$$

with

$$s_n^0(x) = \begin{cases} (-1)^n & x < 0 \\ 1 & x \geq 0 \end{cases}, \quad s_n^1(x) = \begin{cases} -(-1)^n & x < 0 \\ 1 & x \geq 0 \end{cases}, \quad s_n^3(x) = \begin{cases} (-1)^n & x \leq -1 \\ -(-1)^n & -1 < x < 0 \\ -1 & 0 \leq x < 1 \\ 1 & 1 \leq x \end{cases}, \quad s_n^4(x) = \begin{cases} -(-1)^n & x \leq -1 \\ (-1)^n & -1 < x < 0 \\ -1 & 0 \leq x < 1 \\ 1 & 1 \leq x \end{cases}$$

These expansions are substituted in the equations for the magnetic field of the spheroid with $x = \eta_0$, $x = \xi_0$ and $x = \xi$ for values of η_0, ξ_0 or ξ close to 1 or -1.

Source Reconstructions by Spatial Deviation Scans

Fuchs, M., Wischmann, H.-A., Wagner, M., Drenckhahn, R. and Köhler, Th.
Philips Research Hamburg, Röntgenstr. 24-26, D-22335 Hamburg, Germany

Introduction

Reconstructions of generators of biomagnetic and bioelectric activity are nowadays dominated by single or multiple equivalent current dipole models. These methods suffer from the dependence of their results from the start values of the nonlinear minimization algorithm that optimizes the dipole positions in order to achieve a minimum of the deviation Δ between the measured data $\underline{\mathbf{M}}$ and the forward calculated field distribution $\underline{\mathbf{L}}\,\underline{\mathbf{j}}$: $\Delta^2 = \| \underline{\mathbf{M}} - \underline{\mathbf{L}}\,\underline{\mathbf{j}} \|^2$. The lead field matrix $\underline{\mathbf{L}}$ depends on the sensor geometry, the dipole positions, and the volume conductor model. The optimum dipole components $\underline{\mathbf{j}}$ that minimize Δ^2 can be calculated by $\underline{\mathbf{j}} = (\underline{\mathbf{L}}^T \underline{\mathbf{L}})^{-1} \underline{\mathbf{L}}^T \underline{\mathbf{M}}$.

Thus, for given dipole positions, the (linear) problem of determining the best fit dipole components can easily be solved by linear algebra. The nonlinear part of the problem, namely of finding the dipole positions that have the minimum overall deviation, is handled by multidimensional (three coordinates per dipole) minimization procedures like the Nelder-Mead-Simplex algorithm. With noisy data and multiple dipole models, the solution found strongly depends on the start values for the nonlinear search method. That means that the minimizer can be trapped in local minima instead of finding the global minimum of the deviation hypersurface.

Methods

In order to overcome this problem one can scan the whole volume of interest by distributing support points e.g. on a regular threedimensional grid. Optimization of the computational performance of calculating the deviation or variance for given dipole positions $\Delta^2 = \| (\underline{\mathbf{1}} - \underline{\mathbf{L}} (\underline{\mathbf{L}}^T \underline{\mathbf{L}})^{-1} \underline{\mathbf{L}}^T) \underline{\mathbf{M}} \|^2$ allows several thousand support points to be tested per second on a standard workstation. So a complete noniterative search of the volume of interest is possible, that does not depend on any start values. The detection of the absolute minimum of the deviation hypersurface is guaranteed without the risk of being trapped in local minima, if the support point spacing is adequate (depends on the signal to noise ratio). By plotting $\mathbf{1} / \Delta^2$, the whole 3D-deviation hypersurface can be visualized with large values corresponding to good fit positions.

When more than one dipole is active simultaneously, the number of support point configurations **c** to be tested increases drastically : $\mathbf{c} = \mathbf{n}! / ((\mathbf{n} - \mathbf{d})!\, \mathbf{d}!)$ for **n** support points and **d** dipoles. E.g. for **n** = 1000 support points 10^3 deviations have to be calculated with a one dipole model, $5*10^5$ with two, $1.7*10^8$ with three, and $4*10^{10}$ with four dipoles. So without iterative refinement only one and two dipole models can be handled in reasonable time.

Results

The spatial resolution of the method is studied as a function of the signal to noise ratio with simulated data for one and two dipoles with different orientations. The results are compared to other reconstruction techniques such as equivalent dipoles and minimum norm current density approaches. Furthermore simulated EEG- and MEG-data evaluations are compared in view of their resolution power. Because of the uniqueness of the absolute minimum of the deviation hypersurface it is possible to figure out how many dipoles are reasonable to be fitted with a given signal to noise ratio and a given orientation without begining to fit the noise. Fig 1 displays the EEG and MEG sensor geometries used for the simulations, which were carried out with one, two, and three equally sized dipoles in the same depth with different orientations and synthetic noise levels. In addition different linear and planar arrays of dipoles have been studied without noise. For the reconstruction one, two, and three dipole deviation scans have been performed on regular 2D (horizontal and vertical planes) and 3D support points.

First different arrangements of dipoles were studied without additional noise, in order to find the potential / field differences of these set ups compared to the best fitting single equivalent dipoles. Table 1 shows the results of these examinations, the field deviations as well as the depth mislocalizations and magnitude errors are listed. Only small differences were found between the fields of the complex dipole sets and the single dipole fields. This means that the noise of real measurements and systematic modeling errors must be smaller than the field differences, in order to distinguish single from multiple dipoles - the larger the difference in the Table, the greater the chance to resolve more than just the center of activiy.

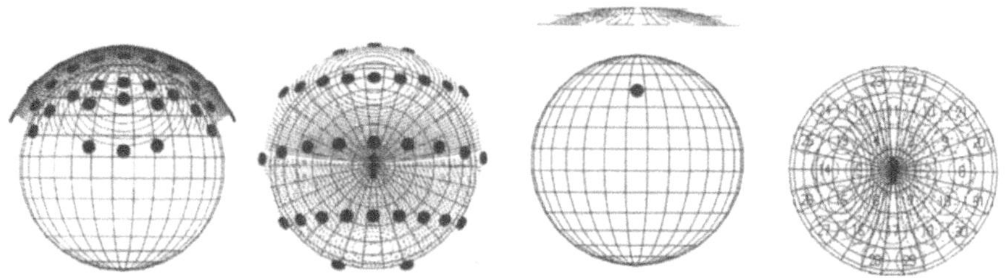

Fig. 1: Sensor geometries used for the 32 channel EEG- (left side, frontal and top views) and the 31 channel MEG-simulations (right side). The volume conductor was modeled by three spherical shells in the EEG case (radii: 76, 83, 90mm, conductivites: 0.33, 0.0042, 0.33 $1/\Omega m$) and a sphere in the MEG case (radius: 76mm).

Table: Field deviations (dev/%), depth mislocalizations (z/mm), and magnitude errors (q/%) of different dipole configurations fitted by single equivalent dipoles (EEG versus MEG). Horizontal dipole orientations are shown as arrows, vertical dipoles are depicted as circles.

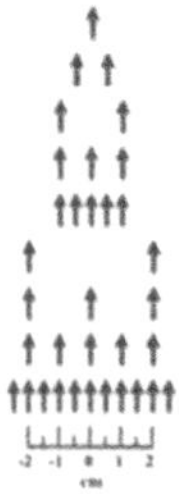

EEG 32 ch. dev/%	z/mm	q/%	MEG 31 ch. dev/%	z/mm	q/%
0,0	0,0	0,0	0,0	0,0	0,0
0,3	-0,2	0,2	0,5	-1,3	4,0
1,0	-0,7	0,7	1,8	-5,7	20,0
0,7	-0,5	-0,6	1,2	-3,6	10,6
0,5	-0,4	0,3	0,9	-2,6	8,4
3,9	-2,8	2,5	7,9	-38,2	625
2,6	-1,9	0,7	4,4	-17,7	88,8
2,0	-1,4	1,3	3,2	-12,0	50,4
2,4	-1,8	0,6	3,9	-15,6	71,5

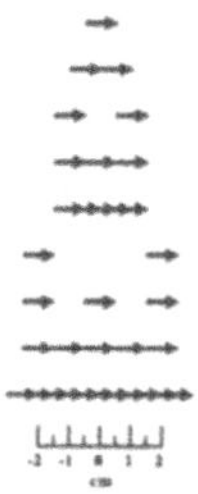

EEG 32 ch. dev/%	z/mm	q/%	MEG 31 ch. dev/%	z/mm	q/%
0,0	0,0	0,0	0,0	0,0	0,0
0,3	-0,4	0,2	0,5	-0,3	0,9
1,3	-1,4	0,9	1,9	-1,0	3,8
0,9	-0,9	-0,4	1,3	-0,7	1,5
0,6	-0,7	0,5	1,0	-0,5	1,8
5,1	-5,7	3,9	7,4	-4,2	17,1
3,4	-3,8	1,5	4,9	-2,8	9,9
2,5	-2,8	1,9	3,7	-2,1	7,9
3,1	-3,5	1,3	4,5	-2,5	8,7

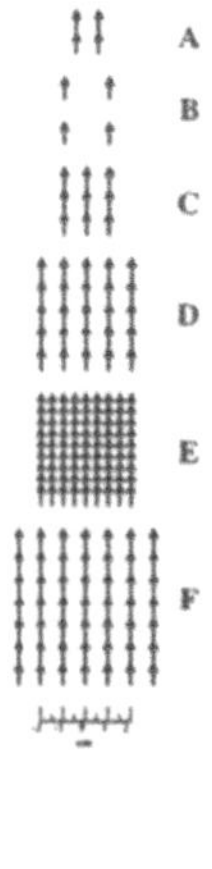

	EEG 32 ch. dev/%	z/mm	q/%	MEG 31 ch. dev/%	z/mm	q/%
A	0,2	-0,5	0,3	0,2	-1,6	5,0
B	0,8	-1,8	1,2	0,7	-6,8	25,2
C	0,6	-1,2	-0,2	0,4	-4,3	13,6
D	1,6	-3,6	2,4	1,0	-14,2	65,3
E	1,3	-3,0	-0,9	0,9	-11,4	43,9
F	3,0	-7,1	2,8	2,3	-35,3	425

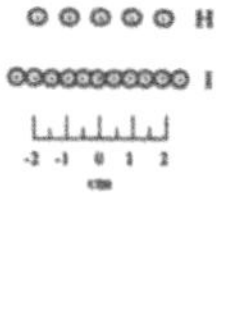

	EEG 32 ch. dev/%	z/mm	q/%
A	0,0	0,0	0,0
B	0,4	-0,4	0,4
C	1,7	-1,5	1,7
D	1,1	-1,0	0,1
E	0,9	-0,8	0,9
F	6,7	-5,5	6,5
G	4,4	-3,7	3,2
H	3,3	-2,8	3,3
I	4,1	-3,4	2,9
J	0,2	-0,6	0,8
K	0,8	-2,3	3,4
L	0,5	-1,5	1,2
M	1,5	-4,8	7,1
N	1,2	-4,0	2,8
O	2,8	-10,2	13,3

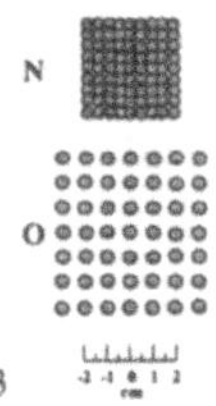

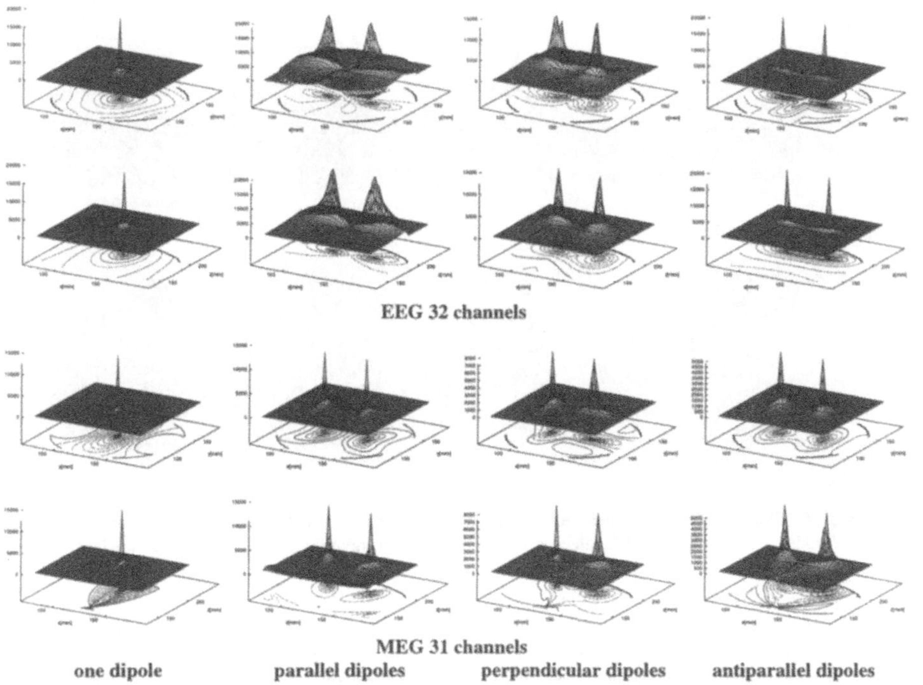

Fig. 2: 2D deviation scans with one and two dipoles (**$1/\Delta^2$** plotted in a.u.). Horizontal (first row) and vertical (second row) planes are scanned for the 32 channel EEG (upper rows) and the 31 channel MEG case (lower rows). From the left to the right one dipole, two parallel, perpendicular, and antiparallel dipoles (40mm distance) with an additional noise of 1% were used for the forward calculations.

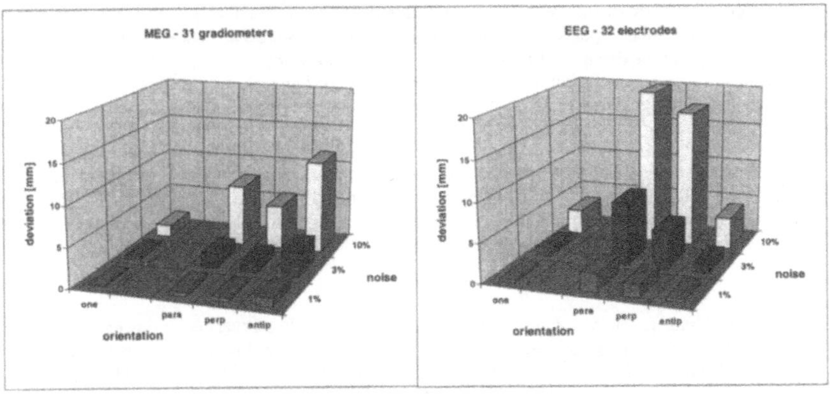

Fig. 3: Dipoles as in Fig. 2. Mean localization errors for the one dipole case (on the left) and the three different dipole pair orientations are plotted as a function of the noise added to the simulations.

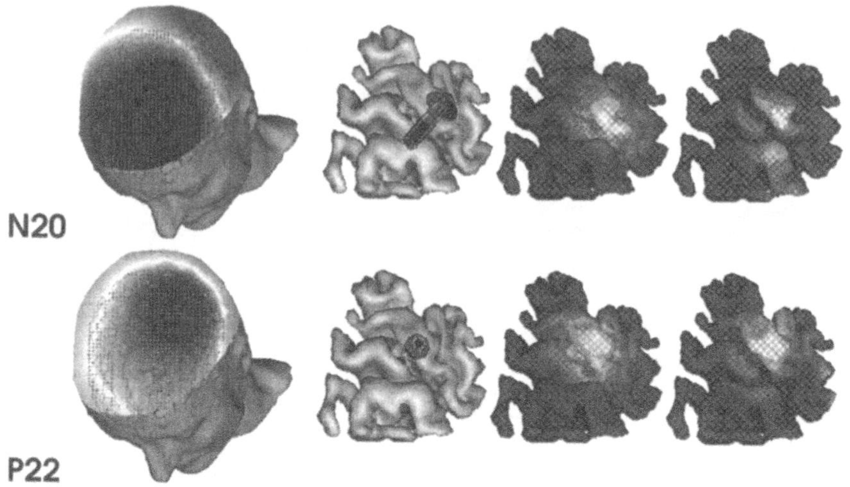

Fig. 4: Example for an application of cortical deviation scans and comparison to other reconstruction methods. SEP - N20 and P22 (electric Medianus nerve stimulation), from the left to the right: measured potential maps, single equivalent dipoles, cortical deviation scans, and cortical L_1-norm current density reconstruction.

Next two dipoles in parallel, perpendicular, and antiparallel orientations with different noise levels were fitted with two dipole deviation scans for both EEG and MEG sensor arrays. Fig. 2 shows typical results for a noise level of 1%, whereas Fig. 3 displays the localization errors as a function of different noise levels (signal to noise ratios). Finally Fig. 4 depicts an application example, where a 65 channel somatosensory evoked potential (SEP) measurement is analysed with different reconstruction methods at two different latencies (data kindly supplied by Prof. Dr. H. Buchner, RWTH Aachen).

Discussion

The calculated fields of different dipole configurations can be fitted very well with respect to the field differences by single equivalent dipoles. The centers of gravity of these setups are always found with good accuracy (in a radial projection), but the depth localization tends to too deep positions.

If the noise in the measurements is larger than these field differences, only one estimated center of activity (the single equivalent current dipole) can be determined uniquely. With good signal to noise ratios the multiple dipole deviation scans allow to find the unique global minimum of the error hypersurface without being trapped in local minima and without suffering from the start value problems of multidimensional nonlinear minimizers.

The implementation of more realistic volume conductor models (e.g. Boundary Element Method -BEM) can easily be done as well as constraining the support points to anatomically relevant structures (e.g. the gray matter layer of the cortex that can be segmented from individual MR-data)[1]. With more than two dipoles iterative variable scale support point distributions have to be used in order to achieve practicable computational efforts[2].

References

[1] Fuchs, M., Wischmann, H.-A., Wagner, M., and Drenckhahn, R. Advanced biomagnetic and bioelectric reconstruction algorithms, In: Herrmann, H.J., Wolf, D.E., and Pöppel, E. Supercomputing in Brain Research: From Tomography to Neural Networks, London, World Scientific, 1995: 161-173.

[2] George, J.S., Lewis, P.S, Schlitt, H.A., Kaplan, L., Gorodnitzky, I., and Wood, C.C. Strategies for source space limitations in tomographic inverse procedures, In: Baumgartner, C., Deecke, L., Stroink, G., and Williamson, S.J. Biomagnetism: Fundamental Research and Clinical Applications, Elsevier Science, IOS Press, 1995: 357-362.

Generalized Inverse Solution for Bimodal Electro-Magnetic Source Images

[1]Gencer, N.G., [2]Williamson, S.J.

[1]*Middle East Technical University, Ankara, Turkey;*
[2]*New York University, New York City, U.S.A.*

Introduction

It is commonly thought that an electric source image (ESI) and magnetic source image (MSI) provide at least partially complementary information about the spatial aspects of electrical activity of the cerebral cortex. This notion is supported by simulations performed using bimodal data for dipole fits [1-5] and by a linear approach based on computations of a generalized inverse based on various optimization methods [6,7]. Using both electric and magnetic data may provide enhanced reliability and accuracy in locating activity within the human brain [8]. We describe a method for exploiting this more complete information by computing an "electro-magnetic source image" (EMSI) obtained from simultaneous recordings of both EEG and MEG. We compare the quality of the image computed from bimodal data for low *versus* high signal-to-noise ratio, using a cortical geometry that represents the distribution of neural activity across a portion of the cerebral cortex.

Methods

In this assessment, the intracellular current is represented by current dipoles uniformly distributed on two parallel sulci joined by a gyrus, with each dipole perpendicular to the surface at its location (Fig. 1). This is a generalization of the geometry introduced by Wang et al. [1-3]. For our present purpose, the neural activity will be confined to the right-hand wall of the left sulcus. A spherical four-shell model represents the conducting geometry

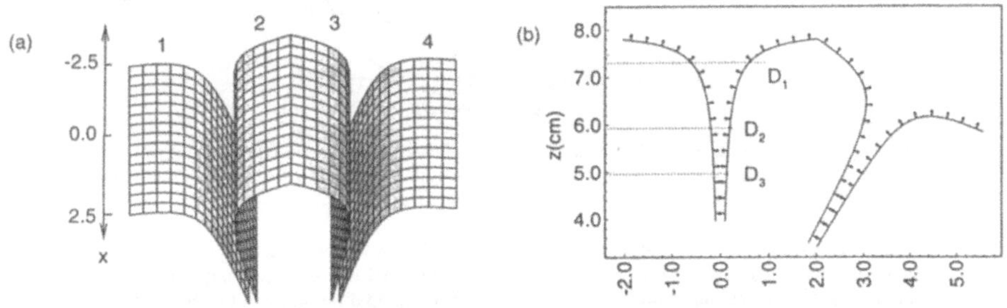

Fig. 1. (a) Representation of two parallel sulci of 5 cm length and walls denoted by 1, 2, 3, and 4; (b) Profile of the sulci viewed edge on, with dipoles designated by D_1, D_2, and D_3.

of the head, with inner radii of cerebrospinal fluid, skull, and scalp chosen as 7.9, 8.1, and 8.5 cm, respectively, and the surface of the scalp has a radius of 8.8 cm. The electrical conductivity of the brain volume is set to 3.3×10^{-3} S/m; whereas, the concentric shells have conductivities of 10×10^{-3}, 4.2×10^{-5}, and 3.3×10^{-3} S/m with increasing radius. A 15×15 grid divides each wall of a sulcus into 255 cells, providing a total of about 900 pixels across the model cortex, with each grid node having a current dipole oriented normal to the local cortical surface.

The locations of EEG electrodes on the scalp were chosen so that the distance between neighbors is about 2 cm, thus providing an array of 101 electrodes as depicted in Fig. 2. An additional electrode served as the reference. Each MEG sensor is radially oriented and positioned directly above each electrode, 9.5 cm from the center of the spherical head. The two equations that relate the set $\mathbf{b}$ of m = 101 measured magnetic field values and the set $\mathbf{v}$ of m = 101 measured voltages to the set $\mathbf{q}$ = 900 unknown current dipoles are expressed as:

(1) $$\mathbf{b} = \mathbf{L}_b\,\mathbf{q}, \qquad \mathbf{v} = \mathbf{L}_v\,\mathbf{q}.$$

where $\mathbf{L}_b$ is the $m \times 900$ magnetic lead field matrix for MSI and $\mathbf{L}_v$ is the $m \times 900$ electric lead field matrix for ESI.

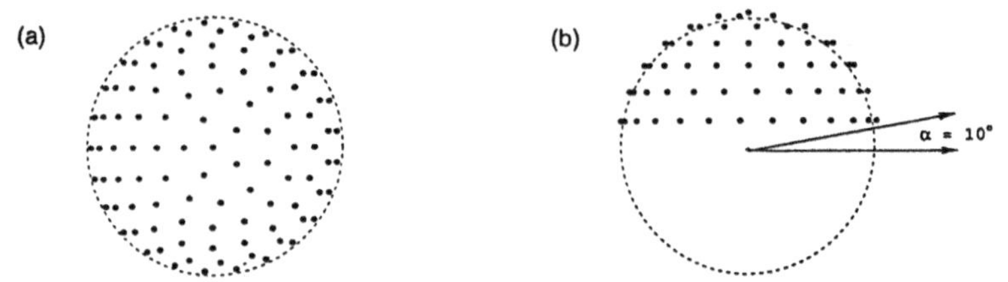

Fig. 2. Positions of 101 electrodes, plus a reference electrode, on a spherical head viewed from the top (A) and side (B). No electrodes are located within the angular range α above the transverse plane that passes through the center of the head.

The minimum-norm least-squares (MNLS) solution for a set of linear equations can be obtained with the Moore-Penrose Inverse ("pseudoinverse") of the lead-field matrix. Singular value decomposition of the individual lead-field matrices $\mathbf{L}_b$ and $\mathbf{L}_v$ yields two sets of orthonormal vectors $\mathbf{x}_j$ with solutions of the form:

(2)
$$\mathbf{q} = \sum_{i=1}^{n} c_i \mathbf{x}_i$$

The amplitudes c_i decrease with increasing i, as illustrated in Fig. 3, but there is no point to include contributions whose singular values are less than the noise level. An estimate for the optimal truncation index p_{opt} can be obtained by minimizing the error norm (9). Figure 3 gives an example of the singular value decay and a choice for the cutoff.

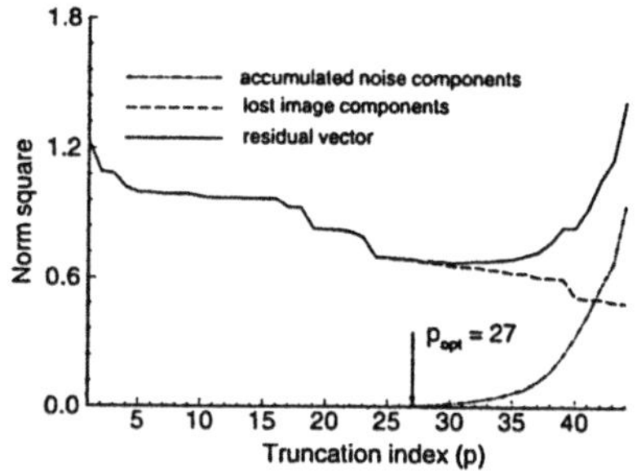

Fig. 3. Progressive decrease in the norm square of the basis functions representing features of higher spatial frequencies is compared with the accumulated noise components and the image components that are lost by truncating at the optimal index p_{opt}.

Because the number of measurements is small, the resolution and accuracy in the reconstructed image are limited. However, when EEG and MEG are used together, they provide complementary information which should help to stabilize the the image and improve localization [6]. The bimodal inverse can be obtained by combining Eqs. 1 and 2 into a single system of equations:

(4)
$$\mathbf{g}' = \begin{bmatrix} \mathbf{v} \\ \mathbf{b} \end{bmatrix} = \begin{bmatrix} \mathbf{L}_v \\ \mathbf{L}_b \end{bmatrix} \mathbf{q} = \mathbf{L}'_{vb} \mathbf{q}$$

where $\mathbf{L}'_{vb}$ is the bimodal lead-field matrix of dimension $2m \times n$, and $\mathbf{g}'$ is the $2m \times 1$ vector of EEG and MEG measurements. However, before solving these equations, both sides are multiplied from the left by a diagonal matrix $\mathbf{W}$ with elements $\mathbf{W}_v = \text{diag}(1/\varepsilon_1, 1/\varepsilon_2, ...)$ and $\mathbf{W}_b = \text{diag}(1/\beta_1, 1/\beta_2, ...)$. These are introduced to render the matrix elements dimensionless, so that electric and magnetic contributions can be combined. The scalar quantities ε_i and β_i are the standard deviations of EEG and MEG measurements, respectively. These weights emphasize the contribution of the modality having the higher signal-to-noise ratio.

Results

To provide an assessment of this method, the truncated pseudoinverse was used to obtain bimodal inverses for current-dipole sources at 3 different depths, as indicated by the locations of the crosses in Fig. 4a. For this example, the EMSIs were computed for highly favorable signal-to-noise ratios, with S/N_v = 145 db for electric measurements and S/N_b = 175 db for magnetic measurements. The quality of these images does not appreciably improve for data of higher quality. The center of each image lies within 3 mm of the dipole's location, even for the shallowest and the deepest sources. As illustrated in Fig. 4b, for data of lower quality or sources lying deep within the sulcus, the bimodal inverse imposes features on the sulcal wall facing the source, which we call a "shadow" of the image. If the source's current flows toward the cortical surface, the shadow's current on the opposite wall flows into the depth of the cortex. Note, however, that the shadow for high quality data (e.g., bottom panel of Fig. 4a) exhibits a shadow that is almost an order of magnitude weaker than the source. Consequently, it is possible to identify which wall of the sulcus has the source.

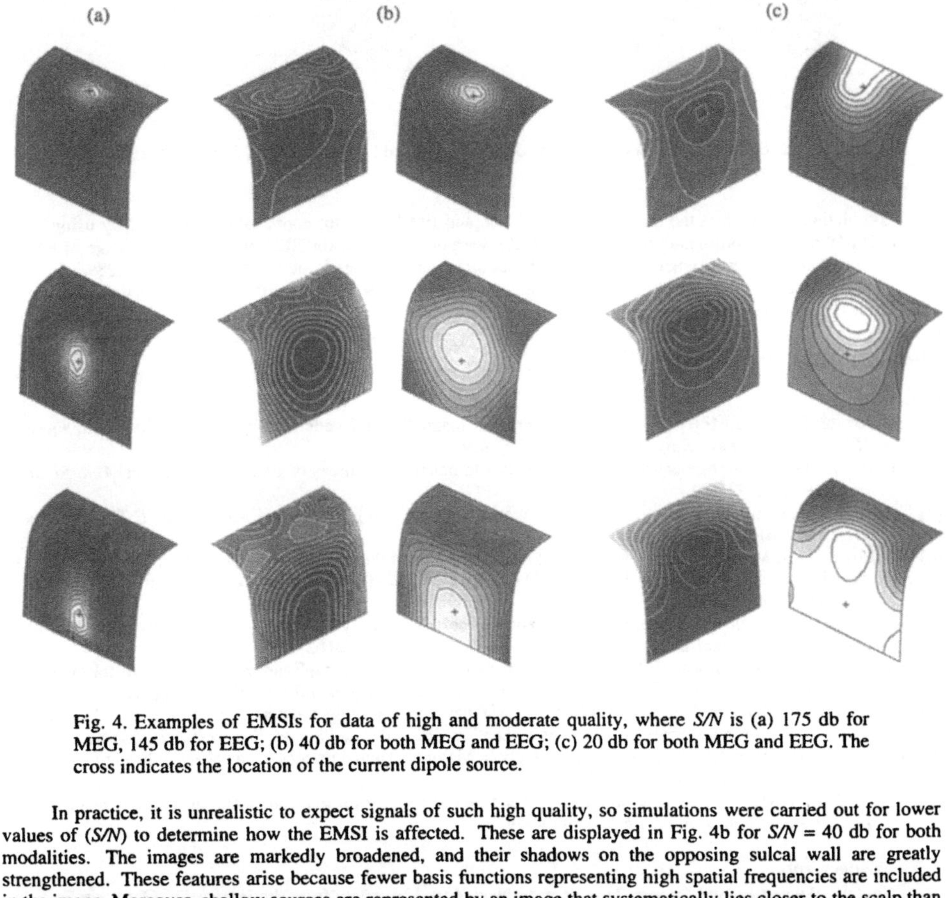

Fig. 4. Examples of EMSIs for data of high and moderate quality, where S/N is (a) 175 db for MEG, 145 db for EEG; (b) 40 db for both MEG and EEG; (c) 20 db for both MEG and EEG. The cross indicates the location of the current dipole source.

In practice, it is unrealistic to expect signals of such high quality, so simulations were carried out for lower values of (S/N) to determine how the EMSI is affected. These are displayed in Fig. 4b for S/N = 40 db for both modalities. The images are markedly broadened, and their shadows on the opposing sulcal wall are greatly strengthened. These features arise because fewer basis functions representing high spatial frequencies are included in the image. Moreover, shallow sources are represented by an image that systematically lies closer to the scalp than the actual neural source. This can be explained by the fact that the cost function of the MNLS inverse gains an advantage with the "minimum norm" criterion by depicting the image closer to the field sensors and voltage electrodes, thereby requiring less image strength to account for the measured fields and voltages. Moreover, when S/N is low, the penalty imposed on the least-squares fit by not accurately representing the field pattern is less sensitive to the position of the image. Consequently, upward shifts of 1 cm also appear in separate ESIs and MSIs for high noise levels, as illustrated in Fig. 5. Because of this, the total strength of the reconstructed image is weaker than that of the actual source. These deficiencies are further emphasized when *S/N* is reduced to 20 db for both recording modalities, as illustrated in Fig. 4c. This, as well as Fig. 4b, also reveal a significant shadow on the opposing (left) sulcal wall. The direction of the shadow's current being the same as that for the image on the opposing wall.

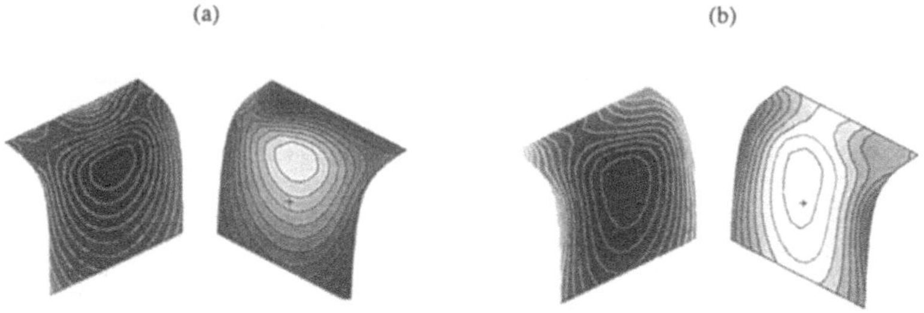

Fig. 5. Images for a dipole at intermediate depth as provided by: (a) ESI for *S/N* of 40 db and (b) MSI for *S/N* of 40 db.

Simulations reveal that the quality of MSIs, ESIs, and EMSIs are not appreciabily improved by using more sensors if *S/N* of the recordings is poor. For example, with *S/N* of 20 db for EEG and MEG, the image of a deep source is not improved even when the number of measurement positions is increased from 101 to 289. In such cases, it is not possible to accurately define localized cortical activity (9,10). Nevertheless, the bimodal inverse provides a means for interpreting complex potential and field patterns, and with increasing experience provided by computer simulations we may expect that these inherent biases in the MNLS inverse can be taken into account when interpreting EMSIs, as well as MSIs and ESIs.

References

[1] Geselowitz, D.B. and Miller III, W.T. Extracorporeal magnetic field generated by internal bioelectric sources. *IEEE Trans. Biomed. Eng.*, 1973, MAG-9:392-398.

[2] Cuffin, B.N. On the use of electric and magnetic data to determine sources in a volume conductor. *Ann. Biomed. Eng.*, 1978, 6:173-193.

[3] Cuffin, B.N. A comparison of moving dipole inverse solutions using EEG's and MEG's. *IEEE Trans. Biomed. Eng.*,1985, BME-32:905-910.

[4] Geselowitz, D.B. Model studies of electric and magnetic fields of the heart. *J. Franklin Inst.*, 1973, 296:379-391.

[5] Stok, C.J. *The Inverse Problem in EEG and MEG with Applications to Visual Evoked Responses*. Krisp. Repro. Meppel., Leiden, 1986.

[6] Dale, A.M. and Sereno, M.I. Improved localization of cortical activity by combining EEG and MEG with MRI cortical surface reconstruction, *J. Cognitive Neuroscience*,1993, 5:162-176.

[7] Gencer, N.G. and Williamson, S.J. Differential Characterization of Neural Sources with the Bimodal Truncated SVD Pseudoinverse for EEG and MEG Measurements. *IEEE Trans. Biomed. Engr.*, in press.

[8] Wang, J.-Z., Williamson, S.J., and Kaufman, L. Magnetic source images determined by a lead-field analysis: The unique minimum-norm least-squares estimation. *IEEE Trans. Biomed. Eng.*, 1992, 39:665-675.

[9] Wang, J.-Z. Minimum-norm least-squares estimation: Magnetic source images for a spherical head model. *IEEE Trans. Biomed. Eng.*, 1993, 40:387-396.

[10] Wang, J.-Z. MNLS inverse discriminates between neural activity on opposite walls of a sulcus of the brain. *IEEE Trans. Biomed. Eng.*, 1994, 41:470-479.

Acknowledgements

This work was supported in part by a grant to New York University from the Horace W. Goldstein Foundation. N.G. Gencer was supported in part by the Scientific Research Council of Turkey in the scope of the NATO Science Fellowships Program.

Can Compact Currents Be Uniquely Determined?

Gorodnitsky, I.F.

Cognitive Sciences Dept.-0515, University of California, San Diego, USA

Introduction

The EEG/MEG inverse problem is ill-posed and its solutions have two independent sources of non-uniqueness. The problem is ill-posed by the nature of the physics because infinitely many different current configurations can give rise to the same electro-magnetic fields. It is also mathematically underdetermined because the number of available mathematically independent data points is less than the dimension of the solution space.

Since electro-magnetic inverse solutions are non-unique, some criteria must be chosen by which to select a particular solution. Here we assume that neuronal generators of EEG/MEG data, and hence our desired solutions, are compact. Such an assumption, in one form or another, ranging from dipolar sources to sources being activations of small areas of arbitrary shape and location in the cortex, has been used in most neuro-magnetic inverse solutions. Here we take the latter case to be our target solutions. Intercortical recordings have shown a much greater range of activity in the cortex. We maintain, however, that our assumption is a likely one, since EEG/MEG data reflect only the synchronous activations on the order of at least 10,000-100,000 neurons. The spatial extent of the individual synchronous activations, at least in normal functional cases, is fairly small, as suggested by physiological data, by functional images from PET and fMRI, and by the EEG/MEG field waveforms. The degree of compactness of the currents, however, is not entirely known.

Although numerous algorithms have been developed to find localized solutions, the solutions from these algorithms do not necessarily agree. This suggests that the assumption of compactness by itself does not resolve the non-uniqueness of the electro-magnetic inverse problem. Since neural currents are not known to satisfy any particular optimization criterion, such as the minimum l_1 norm, we do not know which of the competing compact solutions is correct, as long such non-uniqueness prevails. This paper establishes the conditions under which the electro-magnetic inverse problem becomes unique.

Due to the space limitations here we were only able to provide a sparse summary of the results of our analysis. The full details of the analysis and the treatment of the equally important issue of methodologies for unambiguous identification of neural currents as suggested by the analysis, are presented in [1].

Uniqueness of electric field generators has also been studied in [2], where interpolation of finite measurements is suggested as a way to deal with the mathematical non-uniqueness part of the problem. Our results show that, in fact, interpolation cannot resolve this non-uniqueness.

Background

Forward model: The contributions to electric and magnetic fields from electric currents at different head locations add up linearly. The forward model describing this relationship is simply the equation

$$A\mathbf{x} = \mathbf{b}, \tag{1}$$

where $\mathbf{x}$ is a vector of n values of the current at n discrete points in the head, $\mathbf{b}$ is a vector containing m values of the electric/magnetic field measured at the m sensors and A is the $m \times n$ matrix that connects the two. To obtain a discrete representation of the current we subdivide the head into nodes spaced at some desired interval and represent the continuous current value at each node by a static dipole. The forward model A is found by discretizing a quasi-static approximation of Maxwell's equations. Expressions for forward models of virtually any complexity, as a function of the primary currents only, are now available, and we assume here that one such model has been selected. The analysis is independent of the model A that is chosen and applies to EEG as well as MEG data. By a generator we mean the set of all neurons that are firing simultaneously. By a source we mean a set of simultaneously firing neurons located in one continuous anatomical neighborhood.

We use the lead field formulation (1) is used in the analysis here, but another, namely a non-linear formulation of the inverse problem, is also possible. The results derived from the linear formulation, however, pertain to properties of the electric currents themselves, meaning that these results are true for any compact electro-magnetic inverse solution, irrespective of the forward model it is derived from.

[1]This work was supported by the ONR grant no. N00014-94-1-0856

Compact solutions: The compactness of a current distribution relates to the number of non-zero entries in $\mathbf{x}$. In solutions representing currents restricted to a few small regions in the brain, we expect most of the entries of $\mathbf{x}$ to be zero, indicating no current at those locations, and a few non-zero entries, indicating the presence of current at those few locations.

We define compact solutions (and currents) as ***solutions for which the number of nonzero entries in* $\mathbf{x}$ *is less than the number of data* m.**

The number m is the natural boundary for defining spatially compact solutions for a number of reasons. First, it provides mathematical tractability for the uniqueness results. Second, the mathematical properties of compact and non-compact solutions so defined are very distinct. For purely algebraic reasons, systems such as (1) always have an infinite number of solutions with m or more non-zero entries, regardless of the actual current that generated the given data. Alternatively, no compact solutions may exist for data generated by a non-compact source. The number of compact solutions, when they do exist, is very limited and can be one, as shown by the theory presented below. Finally, the assumption that sources don't extend over more than m grid nodes is already implicit in most popular temporal methods, such as adaptive filtering and MUSIC.

The dependence of the definition of compact currents on the choice of the grid spacing may appear to be a cause of concern, but such dependence cannot be avoided. The widely used temporal methods, for example, implicitly have this dependence. Here we assume that an 'optimal' grid has been chosen, that is, one that optimizes the trade-off between the smoothness of the sampled functions and the achievable resolution as determined by the level of noise. The topic of finding such an optimal grid is outside the scope of this paper. The whole issue should not be a concern however, as the results below show that the sizes of the currents of interest need to be comfortably below the $m/2$ threshold to assure their unique localization. Thus small changes in the grid spacing will not affect the definition of compactness for those sources.

For the analysis, it is also convenient to refer to the size of generators by their dimension, which we define as, ***the number of non-zero entries in* x.** Since each node is typically modeled by three components of x, the number of active nodes is 1/3 the dimension of x.

Results

Separation of physical and mathematical ill-posedness: The non-uniqueness of inverse solutions to (1) is embodied in the null space of A, denoted by $\mathcal{N}(\mathcal{A})$. It consists of all vectors $\{x_n\}$ that are mapped into the zero vector by the matrix A, i.e. $Ax_n = 0$. Thus non-uniqueness of solutions occurs because any current x_r can be replaced by a different current $x_r + x_n$ to produce the same data b. When $\mathcal{N}(\mathcal{A})$ exists, its set of vectors $\{x_n\}$ is infinite.

The null space exists under two conditions, when either $n > m$ for the matrix A in (1) or when the rank of A is less than m. The first condition, A having more columns than rows, relates to the mathematical non-uniqueness arising from the limited number of data, and the second, linear dependence within m columns of A, relates to the physical ill-posedness. This distinction between the two kinds of non-uniqueness should be obvious, but a more extensive discussion of this issue can be found in [1]. The distinction is also a key element that enables us to carry out the uniqueness analysis of the electro-magnetic inverse problem.

Non-uniqueness due to the physics of the electromagnetic problem: In a spherically symmetric, spherically homogeneous conductor, the radial component of the current produces no external magnetic field. In the general case of an arbitrarily shaped conductor, which includes the spherically symmetric case, a primary current $\mathbf{J}_0$ that is normal to a closed surface and has constant dipole moment per unit area is silent both electrically and magnetically. Any two primary currents differing by such $\mathbf{J}_0$ are indistinguishable magnetically and electrically. For two sources to differ by the same current J_0, they must be adjacent or overlapping. This means that non-unique generators of external electric/magnetic fields must be adjacent or overlapping as well. The same result for the electric field only was shown in [2].

We draw an important insight on the nature of physical non-uniqueness from the above discussion. For compact generators which consist of small spatially separated sources, the physics contributes only to local non-uniqueness within individual sources and does not affect the general locations of the sources. That is, the ambiguity is limited to the amplitude variations within each source and possibly to the shape of the boundaries of the sources only.

The problem of non-uniqueness is compounded significantly by the presence of noise and the finite precision of the instruments. We list some examples here. It is well known that the fields from two closely spaced dipoles with opposite orientations practically cancel out and cannot be detected. This effect can be observed on a relatively large scale, for example, in the lateral segment of the superior olivary nucleus of the cat, which consists of many neurons with variable orientations [3]. When oriented in the same direction, two closely spaced dipoles can appear, electrically and magnetically, as a quadrapole. Fields from two small, closely spaced, and not necessary point sources, can be fit within comparable error using a deeper and stronger single dipole [4]. We add that in fact the single dipole and the two (or one extended) sources, in the last example, are the lower and upper boundaries of a cone, within which an infinite number of current distributions can provide a comparable fit the same fields.

The columns of A describing non-unique solutions in the above examples are not linearly dependent, but they are nearly linearly dependent. These columns become effectively linearly dependent in the presence of noise, since the currents they describe are indistinguishable electrically and magnetically. Thus the ambiguity due to the physics extends to a greater number of currents in the presence of noise, but what is important is that it remains local.

Algebraic non-uniqueness of inverse solutions: To study the algebraic non-uniqueness we transform our problem into one that excludes the physical non-uniqueness, by assuming the following property of A.

Linear independence property: *Given a linear system (1), any m columns of A are linearly independent.*

The analysis in this section will refer to the transformed system only, and to simplify the presentation we will use the term (non-)uniqueness here to mean the algebraic (non-)uniqueness only. The key results are stated as theorems here. Their proofs are given in [1].

Theory in the noise free case

The following theorem provides the necessary and sufficient conditions for non-uniqueness of compact inverse solutions given EEG/MEG sampled fields.

Theorem 1 *Two generators $\mathbf{x}_1$ and $\mathbf{x}_2$ of EEG/MEG data are non-unique iff the vector $(\mathbf{x}_1 - \mathbf{x}_2) \in \mathcal{N}(\mathcal{A})$, where $\mathcal{N}(\mathcal{A})$ denotes the null space of A.*

Three main results come from Theorem 1. One is that compact non-unique inverse solutions can exist, because we can easily construct any two such compact sources satisfying the condition of Theorem 1. The second is that mathematical non-uniqueness is not dependent on location, that is non-unique generators of a certain size (see below) can occur anywhere in the cortex. What determines whether two or more generators become non-unique are the relative amplitudes of their currents, which must match exactly a vector in $\mathcal{N}(\mathcal{A})$. The likelihood of this happening depends on how many vectors are contained in $\mathcal{N}(\mathcal{A})$, i.e. the size of the null space of A. For EEG/MEG the size of $\mathcal{N}(\mathcal{A})$ is very large, on the order of $5,000 - 10,000$ dimensions. Thus the occurrence of non-unique generators is not that unlikely. The final result is that the combination of two non-unique generators must always extend over m voxels or more. There is however no restriction on the size of an individual non-unique generator, which can vary from a single dipole to a distribution of several sources extending jointly over $m - 1$ voxels.

The following example gives an illustration of the last result. Suppose the measurement system consists of 150 gradiometers and the activity is well approximated by a single dipole described by 3 components. The third result of Theorem 1 guarantees that no other distribution with less than 148 components can produce the same measurements. The gap in the size of the two solutions is quite large, so the smaller source is easily distinguishable. This is largely responsible for the success of many algorithms in finding sources of very small dimensions.

We now present the two conditions, besides the obvious utilization of priors from other modalities, under which non-uniqueness can be resolved.

Maximally sparse generators: The second result from Theorem 1 implies that there can never be more than one generator of dimension less than or equal to $m/2$. Thus, if we know that the true activation does not extend over $m/6$ nodes, we can always find it as the unique maximally sparse solution. The maximally sparse solution is the solution with the smallest number of nonzero terms. The following theorem formally states this result.

Theorem 2 *If the number of data samples is at least twice the dimension of the true generator, this generator can always be found as the unique maximally sparse solution.*

Note that the uniqueness of the generators of continuous fields shown in [2] follows directly from Theorem 2.

Temporal information: Because the vector made up of currents of two non-unique generators must match some particular null vector of A, the non-uniqueness of these two generators cannot be maintained if the amplitudes of the currents change. Since the intensity of neural activation tends to change rapidly, and provided that activations of the individual sources are at least partially uncorrelated, the data over several time points can consistently reflect only one compact generator. This also assumes that neural activations do not move rapidly in the cortex. Given that the assumptions are satisfied, consistency of a solution across several time points provides unambiguous means of resolving non-uniqueness.

Theory in the presence of Noise: The presence of noise makes non-uniqueness more prevalent, because candidate solutions need to provide only an approximate fit to the data. In the exact case, the two generators have to match a null vector of A exactly to be non-unique. With the addition of noise, this match doesn't have to be exact. Since the null space is orthogonal to the range space, even moderate perturbations in the current values of non-unique generators have a negligible effect on the data and thus their non-uniqueness properties. Hence the set of non-unique currents is increased.

The uniqueness result of Theorem 2 is also affected adversely by noise. If a true maximally sparse generator makes up a significant component of a vector that lies entirely or largely in $\mathcal{N}(\mathcal{A})$, there exist other sparse solutions with a comparable fit to the data. Some of these solutions are likely to be sparser than the true one. They also all share a common component in the range space of A. For these reasons, the algorithms for maximally sparse solutions are likely to find one of the 'phantom' solutions rather than the true one. For a true generator to make up a significant component of a vector in $\mathcal{N}(\mathcal{A})$, however, it has to be relatively large. Thus the described problem does not affect relatively small generators such as ones consisting of a few small sources, but becomes noticeable as the size of a generator approaches the $m/2$ threshold.

Note that the non-uniqueness described here should not be confused with the problem addressed by regularization, where the nature of incorrect solutions depends on the condition number of A.

Summary and Conclusions

The mathematical non-uniqueness manifests itself in the ambiguity in the general locations of the sources in the cortex. The physical ill-posedness manifests itself in ambiguity over current values within neighborhoods of individual sources. Thus estimation of a compact current is a two step process. We first need to find the general locations of the sources that make up the generator. We then estimate the optimal solution constrained to lie within these locations.

We can resolve the mathematical non-uniqueness in three ways: 1) use priors, such as fMRI or PET images; 2) when currents are known to be very small, solve for the maximally sparse solution with the understanding that this estimate is not always reliable; 3) use temporal information, provided the assumptions listed above are satisfied. Given the potential perils of the last two methods, they may be best combined into one algorithm. The optimal way of resolving physical non-uniqueness is to find the minimum norm estimate constrained by the approximate source locations from step 1 and by any other available information. Further discussion of these issues and the methodologies for finding compact sources motivated by the analysis are presented in [1].

References

1. I. F. Gorodnitsky. Uniqueness analysis of compact solutions of the electro-magnetic inverse problem: Methodological considerations for optimal reconstructions. *to be submitted*, May. 1996.

2. A. Amir. Uniqueness of the generators of brain evoked potential maps. *IEEE Trans. Biomed. Eng.*, 41:1–11, 1994.

3. P. L. Nunez. *Electric Fields of the Brain.* Oxford University Press, New York, 1981.

4. Y Okada. Discrimination of localized and distributed current dipole sources and localized single and multiple sources. In H. Weinberg, G. Stroink, and T. Katila, editors, *Biomagnetism: Applications and Theory*, pages 266–272. New York:Pergamon, 1985.

Estimation of Distributed Sources with Adjustable Contrast

Grunwald, A. and Lütkenhöner, B.

Institute of Experimental Audiology, University of Münster, Germany

Introduction

The aim of MEG source analysis is to characterize the sources of neuromagnetic fields in terms of location and moment. In the last years, two groups of models evolved: single-/multi-dipole models on the one hand and distributed-source models on the other hand. In this paper, a method is presented which allows a continuous change from distributed-source to single-/multi-dipole solutions.

Theory

It is assumed that the magnetic field is simultaneously recorded in K different channels (e.g. magnetometer, gradiometer). With b_k denoting the output of the k-th channel, $\mathbf{b}$ is a column vector $\in \mathbb{R}^K$. With N denoting the number of dipoles in the model, $\mathbf{b}$ can be calculated using

$$b_k = \sum_{n=1}^{N} \vec{q}_n \, \vec{l}_k(\vec{r}_n) \,, \tag{1}$$

where $\vec{l}_k(\vec{r}_n)$ is the leadfield function depending on the location $\vec{r}_n$ of the dipoles and on the volume conductor model as well as on properties of the k-th channel. $\vec{q}_n$ is the moment of the n-th dipole.

In the distributed-source model the source is discretized by a sufficient number of current dipoles ($3N \gg K$). Since the dipole locations are given a priori, the model can be described by the N vectors $\vec{q}_n$, assembled as a raw vector $\mathbf{j} \in \mathbb{R}^{3N}$. In this case, a linear leadfield operator can be defined such that

$$\mathbf{b} = \mathbf{L}\mathbf{j} \,. \tag{2}$$

The task now is to estimate a current distribution $\hat{\mathbf{j}}$ fulfilling

$$||\mathbf{L}\hat{\mathbf{j}} - \mathbf{b}|| \approx \sigma_{\text{noise}} \,, \tag{3}$$

where σ_{noise} is a measure for the standard deviation of the noise. The current distribution $\hat{\mathbf{j}}$ is typically calculated by the Tikhonov-Phillips method which provides a minimum-norm solution. The solution is unique in the sense that the contributions of all vectors which are not elements of the null space $\mathcal{N}$ of $\mathbf{L}$ are determined, whereas the contributions of the vectors lying in the null space are set to zero: since they do not provide a measurable field, their amplitude cannot be estimated.

Contrast algorithm

Particularly the lack of information about the null space $\mathcal{N}$ gives rise to the diffuse appearance of the distributions $\hat{\mathbf{j}}$, owing to the fact that the vectors spanning the null space are the high-spatial-frequency components [1]. By adding components of the null space $\mathcal{N}$ to the solution it is possible to fulfill constraint (3) and to produce a $\hat{\mathbf{j}}$ with higher contrast.

The central idea of the proposed algorithm is to add the required components by using a nonlinear contrast function $\kappa(\mathbf{j})$. Two different functions κ were considered which can be characterized by the following two transformations:

$$\text{A)} \quad \vec{q}_n \longrightarrow |\vec{q}_n|^{(\alpha-1)} \vec{q}_n \qquad \forall n \in [1, N] \,,\ \alpha \in [1, \infty) \,, \tag{4}$$

$$\text{B)} \quad \vec{q}_n \longrightarrow \beta \, \vec{q}_n \qquad \forall n \in \{ n \,|\, |\vec{q}_n| \text{ is local maximum} \} \,,\ \beta \in [1, \infty) \,. \tag{5}$$

The iterative algorithm starts with a standard solution $\hat{\mathbf{j}}_0 = \mathbf{R}\,\mathbf{b}$, where $\mathbf{R}$ is the Tikhonov-Phillips operator. Each iteration step is divided into three substeps:

1. In a first step the calculated moments $\hat{\mathbf{j}}_i$ are replaced by $\hat{\mathbf{j}}_i^c = \kappa(\hat{\mathbf{j}}_i)$.
2. In a second step $\hat{\mathbf{j}}_i^c$ is re-scaled so that the resulting $\hat{\mathbf{j}}_i^s$ has the property $||\mathbf{L}\hat{\mathbf{j}}_i^s|| = ||\mathbf{b}||$.

3. Finally, in a third step, a solution

$$\hat{\mathbf{j}}_{i+1} = \hat{\mathbf{j}}_i^s + \mathbf{R} \cdot (\mathbf{b} - \mathbf{L}\hat{\mathbf{j}}_i^s) \tag{6}$$

is calculated which serves as the starting point for the next iteration step.

Method

In order to illustrate the properties of the algorithm, a forward model consisting of four current dipoles is proposed. The dipoles were randomly distributed on a calotte with a radius of 8 cm and a span of 140°. Only dipole configurations where the minimal distance between a pair of dipoles was greater than 4 cm were considered. The orientations of the dipole moments were randomly selected, too. In each simulation the amplitudes of the dipoles were 10 nAm (two dipoles) and 5 nAm (the other two). The magnetic field arising from these dipoles was determined for 196 gradiometers on a calotte with a radius of 12 cm and a span of 126°.

The resulting data set was analyzed using the algorithm described above. A noiseless case ($\sigma_{\text{noise}}=0$) was considered to explain the principle of the method. To compare an estimated source distribution with the four-dipole configuration assumed in the forward calculation, the following criterion for the interpretation of a distributed source in terms of a few dominating sources ('dipole equivalents') was devised: A local maximum of the estimated current distribution represents a dipole equivalent if its strength is greater than 10% of the absolute maximum.

Results

Figure 1 shows the solutions for the two variants of the proposed contrast algorithm. A value of 1.2 was chosen for the parameters α and β. Already after the third step, the contrast of the distributed source solution is considerably enhanced for both variants of the contrast function. The activity centers of the contrasted solutions correspond to the locations of the true dipoles. This result suggests that a moderate enhancement of the contrast can be achieved without running the risk to deviate too much from the true solution. With an increasing iteration depth I the distributed-source solution becomes more and more dominated by a few 'foreground' dipoles, whereas all the other dipoles fade away and become 'background' dipoles. With contrast function A the solution finally turns into a solution dominated by a single dipole, despite the fact that four dipoles were assumed in the forward calculations. The remaining dipole (solid arrow) is located in the neighborhood of one of the true dipoles (outlined arrows). With function B, the asymptotic solution is comparable with a multi-dipole solution.

Figure 2 shows how the average number of estimated dipole equivalents depends on the iteration depth. To achieve a faster convergence, α and β were set to a value slightly larger than in the previous example (1.6). The average number of dipole equivalents was calculated on the basis of 100 independent replications of the forward model. In the case of contrast function A (□) the average number of dipole equivalents converges to the value 1, as in the example shown on the left of figure 1. In the case of contrast function B (◇), convergence is much slower, and a value between 4 and 5 is approached, i.e. the estimated number of dipole equivalents is close to the true number of dipoles. The dipole moments of the remaining background dipoles were about 5% (A) and 1% (B) of the moments of the foreground dipoles (omitted in the figure owing to their small amplitude).

Conclusions

The main conclusions of this study can be summarized as follows:

- Compared to related methods [2,3], the algorithm described in this paper is very efficient with regard to computation time since the Tikhonov-Phillips operator **R** has to be calculated only once.
- In the noiseless case ($\sigma_{\text{noise}} = 0$), the projection of $\hat{\mathbf{j}}_i$ into the space orthogonal to the null space is $\hat{\mathbf{j}}_0$, the original Tikhonov-Phillips solution.
- With only a few iteration steps, a moderate enhancement of the contrast can be obtained. With increasing iteration depth, the solution approaches a multi-dipole solution but still fulfills constraint (3).

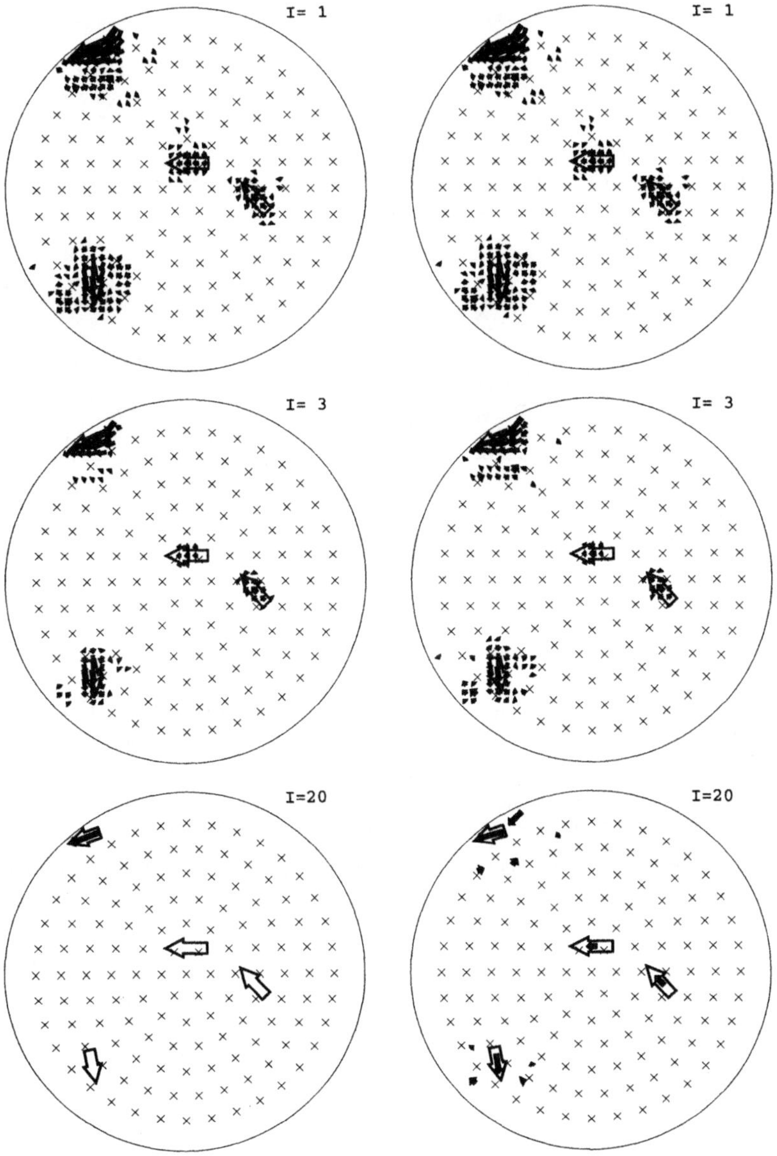

Fig. 1: Solutions for contrast functions A (left column) and B (right column). Original dipoles plotted as outlined arrows, estimated dipoles as filled arrows. Projections of gradiometer locations represented by the crosses.

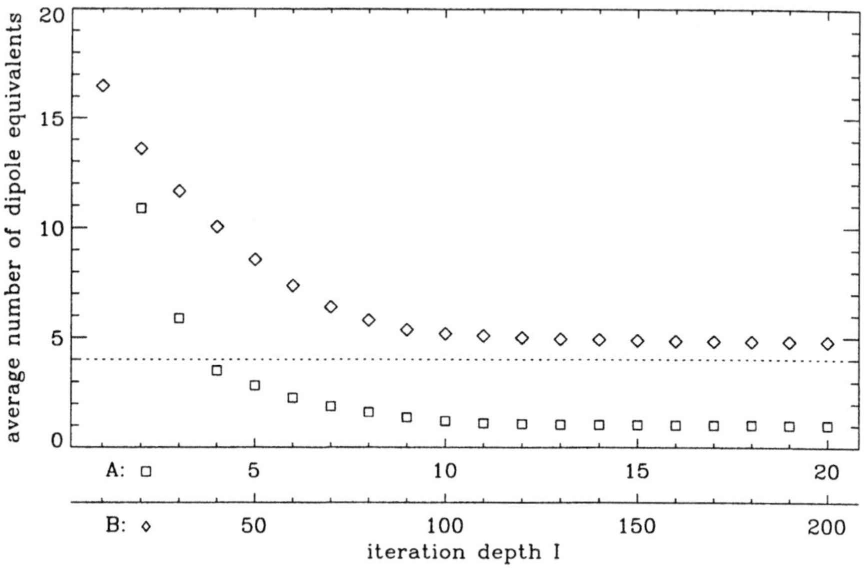

Fig. 2: Average number of estimated dipole equivalents as a function of the iteration depth I. The iteration depth for function B was scaled by a factor of 10.

- The properties of the solution critically depend on the choice of the contrast function. Only with contrast function B there is a good chance to obtain the true number of dipoles.
- The high contrast achieved by this method is only apparent. The effective resolution is limited by the resolution of the Tikhonov-Phillips method, which is used to get a start distribution for the iteration procedure. No mathematical trick can overcome this principal limitation.
- It is to be expected that more stable solutions can be achieved by a modified algorithm using temporal information.

References

[1] Louis, A.K. Inverse und schlecht gestellte Probleme, Stuttgart, Teubner, 1989.

[2] Gorodnitsky, I.F., George, J.S.,Rao, B.D. Neuromagnetic Source Imaging with FOCUSS: a Recursive Weighted MinimumNorm Algorithm, Electroenceph. clin. Neurophysiol., 1995, 95: 231–251.

[3] Matsuura, K., and Okabe, Y. Selective Minimum-Norm Solution of the Biomagnetic Inverse Problem, IEEE Trans. on Biomed. Eng., 1995, 42: 608-615.

Aspects of Non-Uniqueness: How One Source Affects the Reconstruction of Another

Hasson, R. and Swithenby, S.J.

The Open University, Milton Keynes, UK

Introduction

This paper discusses one of the issues that is at the heart of biomagnetism: the non-uniqueness of the inverse problem. Using a Bayesian statistical approach general reasoning is employed to place restrictions on the fidelity of ***any*** source reconstruction algorithm. This paper continues the theoretical development started in [1] and continued in [2] to extract probabilistic information about the biomagnetic inverse problem. The biomagnetic inverse problem is to reconstruct a current density, $\vec{j}(\vec{r})$, inside a volume Q (called the 'source space'). The geometry of the sensors and of the head are represented in the lead field distributions $\vec{L}_k(\vec{r})$ which relate current distributions to measurements by

$$m_k = \int_Q \vec{L}_k(\vec{r}) \cdot \vec{j}(\vec{r})\, \mathrm{d}\vec{r} \tag{1}$$

where m_k is the value of the kth sensor channel measuring the current distribution $\vec{j}(\vec{r})$. In this paper we are going to investigate the uncertainty of current density predictions that are intrinsic to the measurement geometry. Other types of uncertainty arise because of noise in the sensors and external noise sources, but this has been modelled before [3].

The limitations of the measurement geometry can be investigated in two stages. The first stage is to say, "Where are the sources that can't be reconstructed?" (i.e. the 'blind spots') and a method for investigating this was presented at the Vienna conference [2]. The question investigated in this paper is, "If this is the predicted current density, what other sources may exist that cannot be reconstructed?". A typical situation where this information might be useful is where two areas of localized activity are postulated and the output of a current density algorithm shows only one area of activity. We need to know whether the measurement instrument is capable of reconstructing the second source in the presence of the dominant source.

Method

The method relies heavily on the Bayesian formulation of the problem contained in [1]. In this paper there are two adjustable parameters, namely a weighting distribution $w(\vec{r})$ defined on the source space Q (which is a consistent way of inputing prior knowledge into the algorithm) and a positive number ζ (which reflects prior knowledge of the signal to noise ratio). In [1] a probability measure on the space of all possible current distributions is derived. In this paper we will use just one of the outcomes from this analysis, the covariance of two current density distributions. The formula for the covariance is:

$$\mathrm{cov}(\vec{j}, \vec{j}') = \zeta X^{\mathrm{T}}(P + \zeta I)^{-1} P^{-1} Y + \frac{1}{\zeta}\left[\gamma - X^{\mathrm{T}} P^{-1} Y\right] \tag{2}$$

where P_{ij} is the Gramm-Schmidt matrix and is defined by

$$P_{ij} = \int_Q \vec{L}_i(\vec{r}) \cdot \vec{L}_j(\vec{r})\, w(\vec{r})\, \mathrm{d}\vec{r} \tag{3}$$

and X_i, Y_i, γ are the overlap integrals defined by:

$$X_i = \int_Q \vec{j}(\vec{r}) \cdot \vec{L}_i(\vec{r})\, w(\vec{r})\, \mathrm{d}\vec{r}\,,\ Y_i = \int_Q \vec{j}'(\vec{r}) \cdot \vec{L}_i(\vec{r})\, w(\vec{r})\, \mathrm{d}\vec{r}\,,\ \gamma = \int_Q \vec{j}(\vec{r}) \cdot \vec{j}'(\vec{r})\, w(\vec{r})\, \mathrm{d}\vec{r} \tag{4}$$

The following method uses this formula to produce easily interpreted images.

1. Fix $\vec{j}(\vec{r})$ to be either a theoretical dipole source or a computed current density image from an experiment.
2. Pick a position $\vec{r'}$ in the source space Q.
3. Let $\vec{j}'(\vec{r}) = \hat{e}_x\delta(\vec{r} - \vec{r'})$ where $\hat{e}_x$ is a unit vector and $\delta(\ )$ is the Dirac delta function and compute $c_x = \mathrm{cov}(\vec{j}, \vec{j}')$
4. Repeat step 3 for the other two orthogonal unit vectors $\hat{e}_y$, $\hat{e}_z$ to get c_y and c_z.
5. Plot $\sqrt{c_x^2 + c_y^2 + c_z^2}$ at the point $\vec{r'}$.
6. Repeat steps 2–5 for different $\vec{r'}$ until the source space is adequately sampled.

The left hand diagram in Fig. 1 shows the measurement geometry for the theoretical studies. It shows the source space plane (z = 0.06m) below a flat hexagonal geometry 37 channel system (2nd order gradiometers, baseline 5 cm) in the plane $z = 0.11$ m. The head is modelled as a conducting sphere centered at the origin.

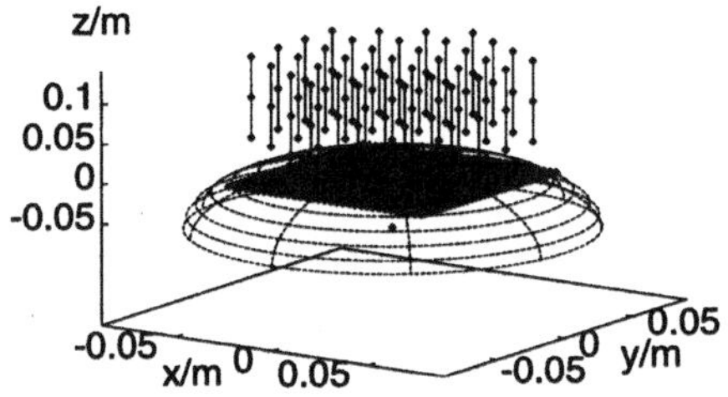

Fig. 1: (left) Measurement geometry for theoretical investigations.

Examples and Discussion

To find the 'blind spots' in the source reconstruction we must compute the overlap *in the space of all possible source distributions*, between the representation of the source present and the representations of all other possible sources. The Bayesian method described in [1] gives us a method of doing exactly this. In [1] a posterior probability distribution is derived on the infinite dimensional Hilbert space of all possible current density distributions. Using this probability measure we can derive the theoretical covariance in this Hilbert space of two current density distributions (equation 2). This is a measure of the overlap between two current density distributions. So, by fixing one current density distribution and mapping the covariance as outlined in this paper we can produce plots which indicate which sources are masked by the given source.

In Fig. 2 we see two such plots for dipolar sources (dipole are located at $(0, 0.01\,\mathrm{m}, 0)$ and $(0.01\,\mathrm{m}, 0, 0)$ resp., oriented along x and y resp.). This figure reproduces the pattern of insensitivity for a highly localised source which we might expect from previous experience with the biomagnetic inverse problem.

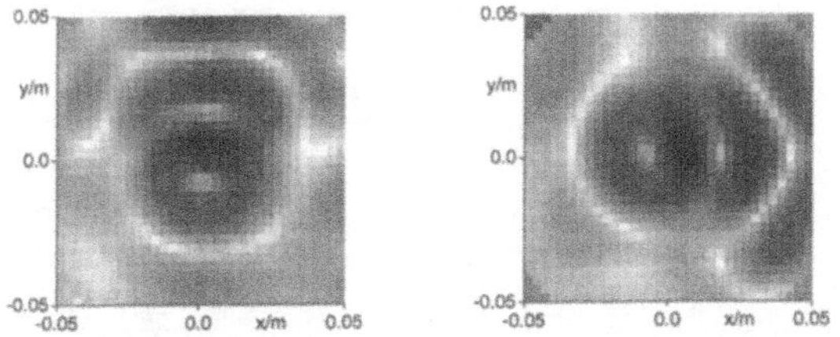

Fig. 2: (left) A dipole located at $(0.01\,\mathrm{m}, 0, 0)$ oriented along x. Magnitude of the covariance plotted so that black represents insensitivity to a second source.
(right) A dipole at the same position but oriented along y.

More interesting perhaps is the covariance maps produced for multiple sources. These are shown in Fig 3 for two dipoles located at $(-0.05\,\mathrm{m}, 0, 0)$ and $(0.05\,\mathrm{m}, 0, 0)$ respectively. The three plots correspond to different relative orientations. Notice that the patterns are not simply a superposition of the covariance maps for the individual dipoles. Note also that the hexagonal symmetry of the measurement system is just visible.

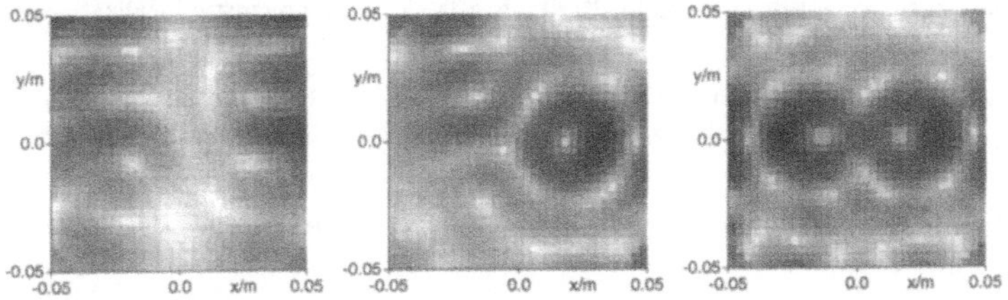

Fig. 3: (left) A dipole located at (-0.05 m,0,0) oriented along x and another dipole located at (0.05 m,0,0) also oriented along x. Magnitude of the covariance plotted so that black represents insensitivity to a second source.
(middle) The same dipole positions but dipoles oriented along x and y.
(right) The same dipole positions but dipoles oriented along y and y.

Now, for the one dipole example, imagine rotating the source space so that the reconstruction plane is perpendicular to the plane of the gradiometers (this means that the top edge of the reconstruction plane is only 1 cm from the bottom coils of the gradiometers). In Fig. 4 the covariance maps are displayed for such a reconstruction plane. Note that because the reconstruction plane is so close to the gradiometers, the lead fields oscillate rapidly at the top of the reconstruction plane.
The main features to note in Fig. 4 are:

- The degree of masking is greater for sources in the reconstruction plane (Fig 4 (right)) than for sources perpendicular to the reconstruction plane (Fig 4 (left)).
- The contours of equal insensitivity form ellipsoids. This agrees with experience of the biomagnetic inverse problem: the resolution in the transverse plane is greater than the depth resolution.
- Contrary to intuition, a superficial source does not mask sources directly underneath it, provided the sources are sufficiently well separated.

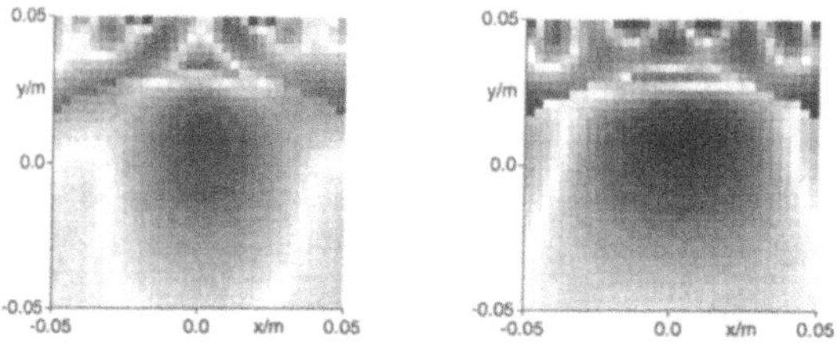

Fig. 4: (left) Depth profile of covariance for a dipole at the origin pointing perpendicular to the reconstruction plane.
(right) The same dipole position but with the dipole oriented in the plane of reconstruction.

We see this method as a means of evaluating whether a given instrument is capable of providing reliable information on a given postulated source distribution.

References

[1] Clarke, C.J.S., In: Nenonen, J., Rajala, H.-M., Katila, T. Biomagnetic localization and 3D modelling, Helsinki, TKK offset, 1991.

[2] Hasson, R., and Swithenby, S.J. In: Baumgartner, C., Deeke, L., Stroink, G., Williamson, S.J., BIOMAGNETISM: Fundamental research and clinical applications, 455–457, Amsterdam, Elsevier Science, 1995.

[3] Ioannides, A.A., Bolton, J.P.R., and Clarke, C.J.S. Inverse Problems, 1990, 6: 523–542.

The Theoretical Basis of Iterative Distributed Solutions to the Biomagnetic Inverse Problem

Hasson, R. and Swithenby, S.J.

The Open University, Milton Keynes, UK

Introduction

This paper aims to clarify the theoretical basis of iterative distributed solutions [1,2,3] to the biomagnetic inverse problem. A rationale for employing this technique is that it should produce more focal activity than the standard minimum norm algorithm [4]. We try to clarify the meaning of the images produced by deriving a precise functional that is minimised.

We will now review the algorithm for the generation of an iterated distributed current density image. There are five inputs to the algorithm. Firstly, we have the measurement data from a single time instant, collected into a vector m_k with $k = 1, 2, \ldots N$. Secondly, a volume Q in which the image current density, $\vec{j}(\vec{r})$, is to be reconstructed (called the 'source space'). Thirdly we need to know the geometry of the sensors and of the head which are represented in the lead field distributions $\vec{L}_k(\vec{r})$ defined by:

$$m_k = \int_Q \vec{L}_k(\vec{r}) \cdot \vec{j}(\vec{r})\, \mathrm{d}\vec{r} \tag{1}$$

Fourthly we can include a weighting distribution $w(\vec{r})$ defined on the source space Q which is a consistent way of inputing prior knowledge into the algorithm. Lastly we have a positive number ζ which reflects prior knowledge of the signal to noise ratio. The algorithm consists of carrying out the following steps:

1. The Gramm-Schmidt matrix is formed by computing the weighted overlap between the lead fields

$$P_{ij} = \int_Q \vec{L}_i(\vec{r}) \cdot \vec{L}_j(\vec{r})\, w(\vec{r})\, \mathrm{d}\vec{r} \tag{2}$$

2. The matrix equation $(P + \zeta I)a = m$ is solved for the coefficient vector a. The minimum norm solution is then given by

$$\vec{j}(\vec{r}) = \sum_{i=1}^{N} a_i w(\vec{r}) \vec{L}_i(\vec{r}) \tag{3}$$

3. A new weight distribution is defined by $w_1(\vec{r}) = w(\vec{r}) \times |\vec{j}(\vec{r})|^q$ where the parameter q is positive.
4. Steps 1 and 2 are repeated with the new weight distribution w_1 to find a new current density $\vec{j}_1(\vec{r})$.
5. A new weight distribution is defined by $w_2(\vec{r}) = w(\vec{r}) \times |\vec{j}_1(\vec{r})|^q$ (using the same value of q).
6. Steps 1 and 2 are repeated with the new weight distribution w_2 to find a new current density $\vec{j}_2(\vec{r})$ and the whole process repeats until the algorithm converges.

Derivation

Rather than derive the iterated algorithm alone we will show how the iterated algorithm arises as a natural extension of the more standard algorithm. We start by minimising the functional

$$\sum_{i=1}^{N} \left[m_i - \int_Q \vec{j}(\vec{r}) \cdot \vec{L}_i(\vec{r})\, \mathrm{d}\vec{r} \right]^2 + \zeta \int_Q \frac{\vec{j}(\vec{r}) \cdot \vec{j}(\vec{r})}{w(\vec{r})}\, \mathrm{d}\vec{r} \tag{4}$$

The Euler-Lagrange equations for this functional are:

$$-2\sum_{i=1}^{N} m_i \vec{L}_i(\vec{r}) + 2\sum_{i=1}^{N}\left(\vec{L}_i(\vec{r})\int_Q \vec{j}(\vec{r})\cdot\vec{L}_i(\vec{r})\,\mathrm{d}\vec{r}\right) + 2\zeta\frac{\vec{j}(\vec{r})}{w(\vec{r})} = \vec{0} \tag{5}$$

which holds pointwise for every vector $\vec{r}$ in the source space. This simplifies to:

$$\sum_{i=1}^{N}\vec{L}_i(\vec{r})\int_Q \vec{j}(\vec{r})\cdot\vec{L}_i(\vec{r})\,\mathrm{d}\vec{r} + \zeta\frac{\vec{j}(\vec{r})}{w(\vec{r})} = \sum_{i=1}^{N} m_i\vec{L}_i(\vec{r}) \tag{6}$$

This equation has been used directly to solve the biomagnetic inverse problem using an approach based on Tikhonov regularisation [5]. This equation has many solutions so we impose the very natural restriction that the solution current density $\vec{j}(\vec{r})$ lies in the subspace spanned by the sensitivities of the detectors (arguing that the measurement can contain no information about silent sources). To allow some flexibility in the method we include the weighting parameter $w(\vec{r})$ in the basis of the reconstruction space $\{w(\vec{r})\vec{L}_i(\vec{r}) : i = 1, 2, \ldots, N\}$. So we can write the reconstructed current density in terms of this basis as:

$$\vec{j}(\vec{r}) = \sum_{i=1}^{N} a_i\vec{L}_i(\vec{r})w(\vec{r}) \tag{7}$$

Substituting this into equation 6 gives:

$$\sum_{i=1}^{N}\left(\vec{L}_i(\vec{r})\int_Q\left[\sum_{k=1}^{N} a_k\vec{L}_k(\vec{r})w(\vec{r})\right]\cdot\vec{L}_i(\vec{r})\,\mathrm{d}\vec{r}\right) + \frac{\zeta}{w(\vec{r})}\sum_{i=1}^{N} a_i\vec{L}_i(\vec{r})w(\vec{r}) = \sum_{i=1}^{N} m_i\vec{L}_i(\vec{r}) \tag{8}$$

Now we can equate the coefficients of the basis element $\vec{L}_i(\vec{r})w(\vec{r})$ to get

$$\sum_{k=1}^{N} a_k\int_Q \vec{L}_k(\vec{r})\cdot\vec{L}_i(\vec{r})w(\vec{r})\,\mathrm{d}\vec{r} + \zeta a_i = m_i \tag{9}$$

which is usually written in matrix form as:

$$(P+\zeta I)a = m \text{ where } P_{ik} = \int_Q \vec{L}_k(\vec{r})\cdot\vec{L}_i(\vec{r})w(\vec{r})\,\mathrm{d}\vec{r} \tag{10}$$

This is a one step algorithm for solving the biomagnetic inverse problem [4,1]. Now we go on to derive the iterative method which has previously been used [1,2,3]. Start with the functional defined by:

$$\sum_{i=1}^{N}\left[m_i - \int_Q \vec{j}(\vec{r})\cdot\vec{L}_i(\vec{r})\,\mathrm{d}\vec{r}\right]^2 + \zeta\int_Q\frac{[\vec{j}(\vec{r})\cdot\vec{j}(\vec{r})]^{(q+2)/2}}{w(\vec{r})}\,\mathrm{d}\vec{r} \tag{11}$$

Before we derive an algorithm based on the above functional, we consider its physical significance. The integral in the first term is the forward problem for the current $\vec{j}(\vec{r})$, so the whole term is just the sum of the squared deviations of the predicted measurements from the actual measurements.

The second term is a regularization term to make the problem computationally tractible. It represents a cost associated with the magnitude of the $(q+2)$th moment of the field (and so $q = 0$ corresponds to the minimum power solution). In the denominator is the weighting factor $w(\vec{r})$ which means that there is less cost at points where the weight is large. So if $w(\vec{r})$ is normalized to one it can be interpreted as a probability weight. An obvious extension of the method as written here is to have a tensor weight $W(\vec{r})$ and this has been used to model anisotropies in the cortex. The other parameter in the functional is the scalar ζ which controls the compromise between fitting the data (first term) and smoothness (second term).

Now we follow the same steps as in the above derivation, first we write down the Euler-Lagrange equations and simplify to get:

$$\sum_{i=1}^{N}\vec{L}_i(\vec{r})\int_Q \vec{j}(\vec{r})\cdot\vec{L}_i(\vec{r})\,\mathrm{d}\vec{r} + \frac{\zeta(q+2)}{2w(\vec{r})}\vec{j}(\vec{r})\int_Q\frac{\left(\vec{j}(\vec{r})\cdot\vec{j}(\vec{r})\right)^{q/2}}{w(\vec{r})}\,\mathrm{d}\vec{r} = \sum_{i=1}^{N} m_i\vec{L}_i(\vec{r}) \tag{12}$$

This equation is not a linear equation unless $q = 0$, which was the case considered previously. So instead of just forming a matrix equation to solve this we must look for an iterative scheme. This scheme is the one outlined in the introduction. First compute the Gramm-Schmidt matrix:

$$P_{ik}^0 = \int_Q \vec{L}_k(\vec{r}) \cdot \vec{L}_i(\vec{r}) w(\vec{r}) \, d\vec{r} \tag{13}$$

and then use equation 10 to compute a current density $\vec{j}^0$. This starts the iteration process. Now at the nth step we use the current density estimate $\vec{j}^{n-1}$ to compute a new Gramm-Schmidt matrix:

$$P_{ik}^n = \int_Q \vec{L}_k(\vec{r}) \cdot \vec{L}_i(\vec{r}) w(\vec{r}) \left(\vec{j}^{n-1}(\vec{r}) \cdot \vec{j}^{n-1}(\vec{r})\right)^{q/2} \, d\vec{r} \tag{14}$$

and then we calculate a new current density, $\vec{j}^n$, by solving equation 10 with the new Gramm-Schmidt matrix P_{ik}^n. Clearly, if this iteration scheme converges then it converges to a solution of equation 12. We have not shown that the above scheme will converge in general, but in practice we find that for typical experimental arrangements it converges very rapidly (usually within three iterations).

Examples and Discussion

The general properties of the iterated current density reconstruction algorithm compared to the minimum norm reconstruction can be deduced from the form of the functional. In the introduction it was mentioned that one of the reasons for using this algorithm is that localised sources are reconstructed as more localised distributions. This may be seen for a 1-d example in Fig. 1. A localised source at the origin is reconstructed as a distributed source for $q = 0$ and $q = 2$ but the distribution is more sharply peaked for the latter. This is a direct consequence of the form of the minimized functional i.e. a quadratic polynomial ($q = 0$) falls off more sharply than a quartic polynomial ($q = 2$).

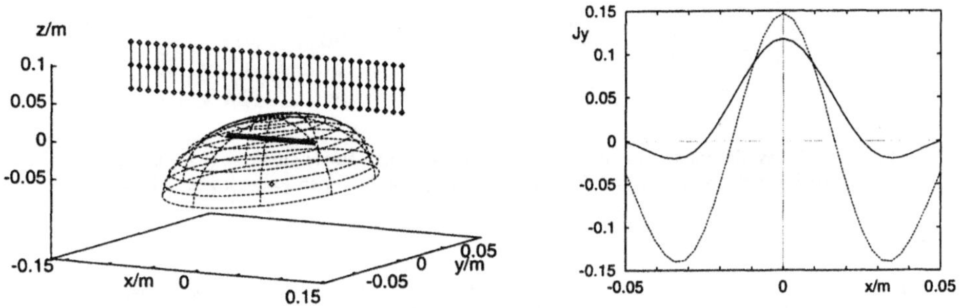

Fig. 1: (left) Geometry of apparatus which shows the positions of; the gradiometers (drawn as lines with dots at the position of the centre of each coil); the source space (a line); the conducting sphere. (right) One dimensional reconstruction. Solid line $q = 0$, dashed line $q = 2$.

The same localization effect can be seen in two dimension using data from a visually evoked response experiment. The data is the peak response (96ms) of the BT experiment described in [6]. The stimulus for the experiment was the pattern reversal of the bottom right quarter field. The left hand diagram in Fig. 2 shows the source space plane ($z = -7\,\text{cm}$) below a flat 6×7 grid of second order gradiometers, baseline 3.125 cm, center coil in the plane $z = 0\,\text{m}$. The head is modelled as a conducting sphere of radius 10 cm, with the centre at (0,4 cm,-14 cm). A localised source is thought to lie below (-0.02 m,-0.003 m).

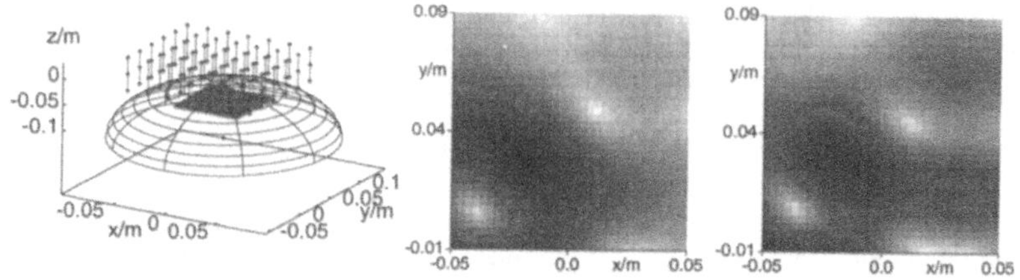

Fig. 2: (left) The arrangement for a visually evoked response experiment [6]. (middle) A real source reconstructed with a minimum norm algorithm ($\zeta = 1$). (right) The same source reconstructed using an interated distributed current algorithm ($\zeta = 1$, $q = 2$).

The iterated reconstructions become more and more localized for higher q, but as q increases the algorithm develops more and more instability with respect to multiple sources. Even if there is only one source, there can be competition between the true source and spurious sources (reconstructed by the aliasing of distant sources). This is the situation in Fig. 3 where the test source is a dipole located at $(0\,\mathrm{m}, 0.02\,\mathrm{m}, 0)$ oriented along x. A large amount of gaussian random noise (total noise power is twice the total signal power) has been added but the minimum norm algorithm reconstructs the source accurately (Fig. 3 (left)), as does the iterated algorithm with $q = 1$ (Fig. 3 (middle)). However, the reconstruction with $q = 3$ has magnified the spurious source and this source will dominate for all higher values of q. The point at which this phenomenon occurs depends on the measurement geometry, the measurement data and the value of the smoothing parameter ζ. In conclusion we should be wary of sources that occur in the iterated reconstructions that do not appear in the minimum norm reconstruction.

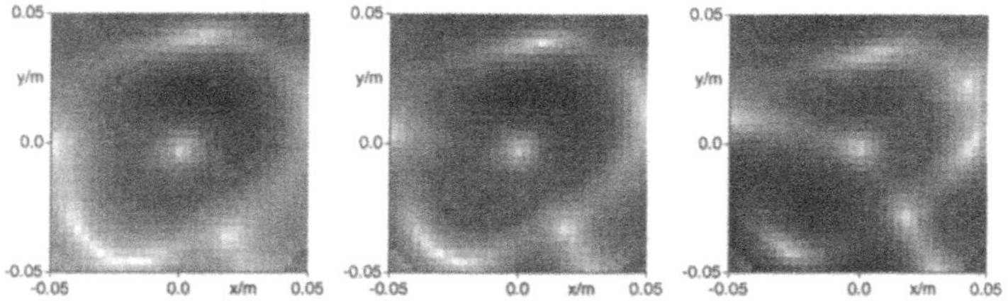

Fig. 3: (left) Minimum norm representation (i.e. $q = 0$) of a single dipole at $(0\,\mathrm{m}, 0.02\,\mathrm{m}, 0)$ with gaussian noise added ($\zeta = 0.5$). (middle) The same but with q=1. (right) The same but with q=3.

References

[1] Ioannides, A.A., Bolton, J.P.R., Hasson, R., and Clarke C.J.S In: Williamson, S.J., Hoke, M., Stroink, G. and Kotani, M. Advances in Biomagnetism, New York, Plenum Press,1989.

[2] Ioannides, A.A., Bolton, J.P.R., and Clarke, C.J.S. Inverse Problems, 1990, 6: 523–542.

[3] George, J.S., Lewis, P.S., Schlitt, H.A., Kaplan, L., Gorodnitsky, I. and Wood, C.C. In: Baumgartner, C., Deeke, L., Stroink, G., Williamson, S.J., BIOMAGNETISM: Fundamental research and clinical applications, 357–362 Amsterdam, Elsevier Science, 1995.

[4] Hämäläinen, M.S., and Ilmoniemi, R. Technical Report TKK-F-A559, Helsinki University of Technology, 1984.

[5] Nikulin, S.L., Zhuravlev, Y.E., Matlashov, A.N., Lipovitch A.Ya., Clinical Physics and Physiological Measurement, 1991, 12: 79–83.

[6] Janday, B.S., Swithenby, S.J., Thomas, I.M. In: Atsumi, K., Kotani, M., Ueno, S., Katila, T., Williamson, S.J., BIOMAGNETISM '87, 246–249, Tokyo, Denki University Press, 1987.

Anatomically Constrained vs Unconstrained Dipole Fit in the Combination of MEG and MRI

Haueisen, J.[1], Hajek, M.[2], Huonker, R.[1], Nowak, H.[1] and Sauer, H.[2]

[1]Biomagnetisches Zentrum; [2]Klinik für Psychiatrie, Friedrich-Schiller-Universität, Jena, Germany

1 Introduction

The most common solution approach to the inverse problem in neuromagnetism is the anatomically unconstrained current dipole fit with Simplex, Levenberg-Marquardt or a similar method. The result of this dipole fit is not correlated to the anatomy (anatomically unconstrained) and might yield unexpected dipole locations e.g. in the white matter or the liquor space. In contrast, the combination of MEG (magnetoencephalography) and MRI (magnetic resonance imaging) allows an anatomically constrained dipole fit. This was suggested e.g. by Wieringa et al. [6]. Recently, Lütkenhöner et al. [3] proposed a dipole patch model, where a number of dipoles (a patch) are riding on the cortical surface. A simpler model is used in this work. The model consists of a single current dipole riding on a surface 1 mm below the cortical surface and is called anatomically constrained dipole fit. For all anatomically constrained dipole fits is valid that only dipole positions in the cortex and dipole directions perpendicular to the cortex surface are considered as valid source space.

2 Methods

We compared the unconstrained dipole fit (Simplex method) with the anatomically constrained dipole fit for the N100m latency of an auditory evoked field. Ten male right handed schizophrenic patients and ten male right handed controls were recruited for this study. Both, controls and patients had normal hearing. All controls were without a history of neuropsychiatric illness. All patients met the DSM-III-R criteria for schizophrenia and received neuroleptic medication.

Tone burst at 1000 Hz were delivered to the right ear through a 6 m long plastic tube with an earpiece at the end. The duration of one burst was 50 ms with 10 ms rise and fall time. 256 trials were measured and averaged, where the stimuli was ramdomly distributed between 1000 and 2200 ms. The magnetic fields were recorded using a 31 channel Philips system. The magnetic signals were bandpass filtered from 0.1 to 40 Hz. Common mode rejection was applied in order to suppress background noise. The best fit time instance in the N100m latency (100 ms after the stimulus) for the unconstrained dipole fit was chosen for the comparison of constrained and unconstrained fit.

The 3-D MRI data sets were obtained from a 1.5 Tesla Philips Gyroscan ACS II system. The coordinate system transformation between MRI and MEG coordinate system was performed with the help of ten orthogonal coilsets in MEG and ten vitamin E capsules in the MRI scans at the same positions. From the 3-D MRI data sets one compartment Boundary Element Method models were constructed and used as volume conductor models. The surface of the posterior superior temporal gyrus (pSTG) was segmented as source space for the anatomically constrained dipole fit. The pSTG was hand segmented slice by slice and the results were also used in a volumetric study. Figure 1 illustrates the source space of the pSTG. The pSTG was chosen as source space because the N100m latency of the auditory evoked stimulus presented is expected to cause an activation in this region of the brain. Based on this source space an anatomically constrained deviation scan was calculated, where at each point of the source space the best fit dipole is calculated. From all dipoles computed the one with the best fit is chosen for the comparison with the unconstrained dipole fit. All data aquisition and analysis were performed using the CURRY^{TM} software.

The reconstruction results were transformed into a individual head-based cartesian coordinate system, where x indicates the medio-lateral axis, y indicates the posterior-anterior axis, and z the inferior-superior

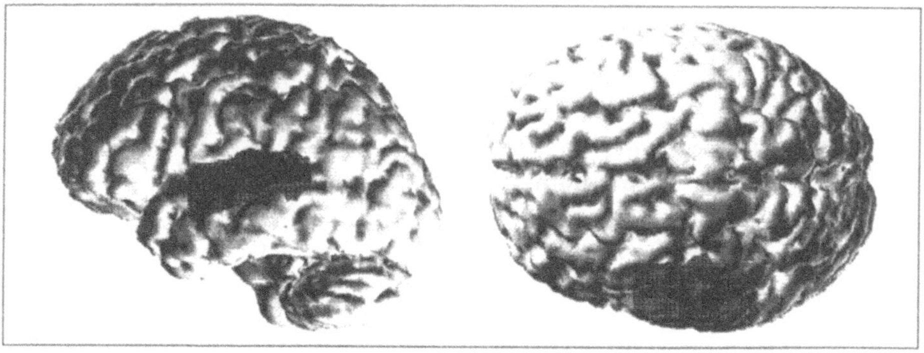

Figure 1: Human brain in left side view (left) and top view (right). The posterior superior temporal gyrus is indicated by the black area left temporal. The surface of the pSTG is used as source space.

axis (for details please refer to [1, 2]). Because the dipole locations and moments of the ten patients and ten controls, respectively were averaged this step is necessary.

3 Results

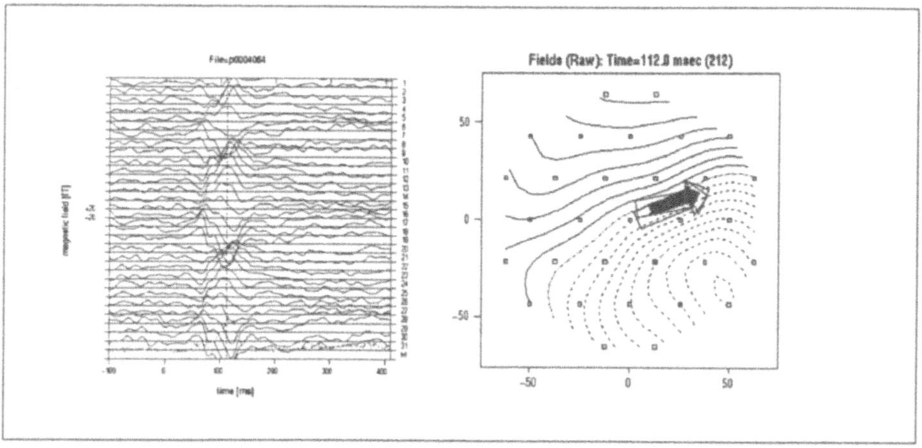

Figure 2: Time course of the measured magnetic signals (left) and magnetic field profile at the best fit time instance (right). The black arrow indicates the solution of the unconstrained dipole fit and the hollow arrow indicates the solution of the anatomically constrained dipole fit.

Figure 2 presents the measured magnetic field time course and profile at the best fit time instance. Figure 3 shows the result of constrained deviation scan and unconstrained dipole fit. The black arrow indicates the result of the unconstrained dipole fit and the black bars indicate the results of the anatomically constrained deviation scan. The largest bar is used as anatomically constrained dipole fit for comparison with the unconstrained dipole fit. Table 1 and 2 show the results for 10 patients and 10 controls, respectively.

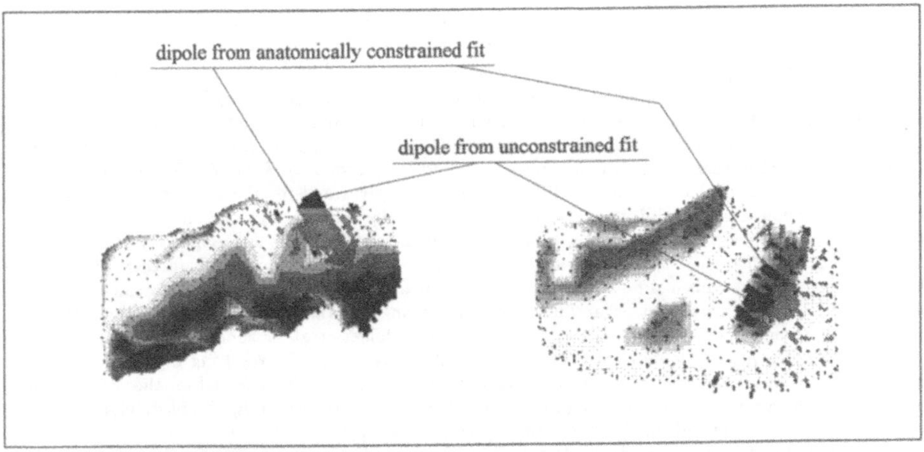

Figure 3: Source space of pSTG in left side view (left) and top view (right) with the result of the anatomically constrained deviation scan (black bars) and the unconstrained dipole fit (black arrow). For the sake of illustration the results are shown for one control person. The pSTG is the same as in figure 1 above.

parameter	unconstrained dipole fit	anatomically constrained dipole fit	difference
x in mm	53.8 (7.6)	50,6 (8.0)	3.2
y in mm	7.3 (8.1)	10.1 (6.4)	2.8
z in mm	44.7 (7.2)	42.7 (7.0)	2.0
q_x (normalized)	-0.07 (0.28)	-0.07 (0.43)	0
q_y (normalized)	0.69 (0.14)	0.62 (0.15)	0.07
q_z (normalized)	-0.62 (0.13)	-0.61 (0.16)	0.01
abs(q) in μ Amm	30.4 (12.5)	41.9 (19.0)	11.5
angle in degrees	42.3 (10.1)	44.3 (9.8)	2.0

Table 1: Comparison of anatomically constrained and unconstrained dipole fit for 10 patients. The mean value is given and the standard deviation in brackets. x, y, and z are the coordinates of the dipole position, q_x, q_y, and q_z are the dipole moments, and abs(q) is the absolute dipole strength. Angle indicates the $\arctan(\frac{q_z}{q_y})$ in degrees.

parameter	unconstrained dipole fit	anatomically constrained dipole fit	difference
x in mm	52 (6.2)	49.6 (8.2)	2.4
y in mm	6.2 (10)	8.1 (11.9)	1.9
z in mm	50.6 (7.3)	49.2 (7.3)	1.4
q_x (normalized)	0.02 (0.11)	-0.23 (0.48)	0.25
q_y (normalized)	0.51 (0.2)	0.36 (0.2)	.15
q_z (normalized)	-0.82 (0.12)	-0.71 (0.2)	0.11
abs(q) in μ Amm	33.2 (20.9)	42.6 (33.2)	9.4
angle in degrees	58.2 (13.6)	61.6 (14.4)	3.4

Table 2: Comparison of anatomically constrained and unconstrained dipole fit for 10 controls. Parameters as in table 1 above.

4 Discussion

The differences between the anatomically constrained dipole fit and the unconstrained dipole fit are in general not significant which indicates that the unconstrained dipole fit for the paradigm used and the latency considered is in good agreement with the anatomy. However one parameter is clearer in the case of the anatomically constrained dipole fit. This parameter is the angle $\arctan(\frac{q_x}{q_y})$.

The angle $\arctan(\frac{q_x}{q_y})$ is the main difference between controls and patients in the dipole localization of the N100m latency for investigated left hemispheric sources. This was first found by Reite [4]. A possible reason for these differences in the dipole orientation might be an altered structure of the gray matter in the pSTG. There is evidence for this altered structure of gray matter from a high number of volumetric MRI studies. Most of them found a reduction in the gray matter volume of the pSTG (see e.g. [5]).

For the unconstrained dipole fit the difference is 15.9 degrees and for the anatomically constrained dipole fit the difference is 17.3 degrees. Thus, the anatomically constrained dipole fit pronounces clearer than the unconstrained fit the difference between the patients and controls in the angle (last row in table 1 and 2). These differences are consistent with findings reported by Reite [4]. A further advantage of the anatomically constrained dipole fit is the direct connection to the anatomy, which can be used for investigations of correlation of structural and functional changes in the brain e.g. in schizophrenics.

In principle the dipole patch model suggested by Lütkenhöner et al. is superior to the anatomically constrained dipole fit. We would not expect significant differences in the resulting dipole locations and orientations for the anatomically constrained dipole fit and the dipole patch model for the measurements presented. If the activation area is small (2 mm extent) than there will be no significant difference between the anatomically constrained dipole fit and the dipole patch model.

Thus we conclude, that a single anatomically constrained dipole is sufficient for describing a source which has a small extent and no extended dipole patch model is necessary.

References

[1] R. Huonker, H. Nowak, M. Hajek, U. Leder, K. Wegner, R. Rzanny, and K. Rieke. 3D-Imaging biomagnetischer Quellen. *Biomedizinische Technik*, 40(Ergänzungsband 1):193 - 194, 1995.

[2] R. Huonker, H. Nowak, R. Rzanny, and K. Rieke. Combined 3D neuromagnetic source imaging and mri scans. *EEG Journal*, suppl.: Visualization of information processing in the human brain: Recent Advances in MEG and functional MRI, in press.

[3] B. Lütkenhöner, E. Menninghaus, O. Steinsträter, C. Wienbruch, H.M. Gißler, and T. Elbert. Neuromagnetic source analysis using magnetic resonance images for the construction of the source and the volume conductor model. *Brain Topography*, 7(4):291 - 299, 1995.

[4] M. Reite. Magnetoencephalography in the study of mental illness. *Adv. Neurol.*, 54:207 - 222, 1990.

[5] T.E. Schlaepfer, G.J. Harris, A.Y. Tien, L.W. Peng, S. Lee, E.B. Fereman, G.A. Chase, P.E. Barta, and G.D. Pearlson. Decreased regional cortical gray matter volume in schizophrenia. *Am. J. Psychiatry*, 151(6):842 - 848, 1994.

[6] H. J. Wieringa, M. J. Peters, and F.H. Lopes da Silva. The estimation of a realistic localization of dipole layers within the brain based on functional (EEG, MEG) and structural (MRI) data: A preliminary note. *Brain Topography*, 5(4):327 - 330, 1993.

One Effect of a Maximum Possible Current Strength

Huang, M. and Heller, L.

Biophysics Group, MS D454, Los Alamos National Laboratory, Los Alamos, NM 87545, USA

Introduction

One of the difficulties encountered when trying to interpret MEG/EEG experimental data is that the spatial extent of the neuronal (primary) current sources is often not known. For some classes of data it is thought that the sources are fairly well localized (*focal*); while for others they are expected to be distributed over a significant area or volume. In the former case, if the regions of activity are sufficiently small they can be well approximated by point current dipoles. If these are few in number there may be only one, or just a small set of acceptable solutions for the dipole parameters.

At the other extreme one assumes that current is present everywhere in the source space. Even with anatomical constraints taken into account, however, this leads to a highly underdetermined problem. For example, at 1 cm resolution on the cortex it would take ≈ 2,000 dipole moments to completely specify the current, and this is much larger than the number of measurements that can be made. Only by imposing physiological constraints can one hope to make this *distributed* source approach be of any use as a way to determine the actual current. This study examines the extent to which the ambiguity of a distributed source solution is reduced by imposing one specific constraint: A *maximum possible value for the dipole moment per unit volume, or per unit area of cortex.* There is experimental evidence in support of this limit [1].

We look at the consequences of such a limit on one particular issue, which is of considerable importance in some applications. Given a set of MEG data, can one be *certain* whether there is, or is not, a significant amount of current in a specified region of the source space? To study this question we choose a simple source space (shown in Fig.1) and subdivide it into two regions, I and II. If a set of magnetic field data can be fit in two different ways: (i) *All* the current is in region **I**; (ii) *None* of the current is in region **I**; then one cannot be certain whether or not region I contains current.

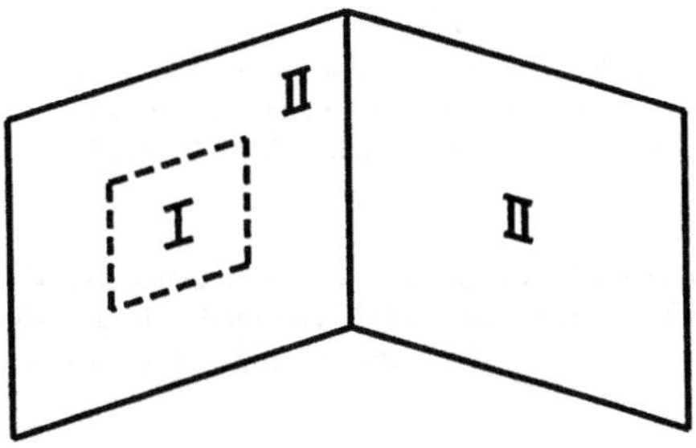

Fig.1. A source space consisting of two perpendicular planes. **I** is the region being tested. **II** is the rest of the source space.

If there are no constraints on the current, this situation can easily arise whenever the number of source parameters (dipole moments) is large compared to the number of measurements. Here we examine the extent to which a constraint on the maximum possible dipole moment per unit area can reduce this ambiguity.

To see why this might happen: Suppose there really is current in region **I**; a particular reconstruction algorithm may miss that current but still fit the data, by assigning *extra current* in region **II**. In this case, a constraint on the maximum possible dipole moment per unit area may be able to prevent this from happening, especially if region **I** is large.

Source Space and Sensor Array

To take a first look at this question, we have chosen the simple source space and sensor array shown in Fig.2.

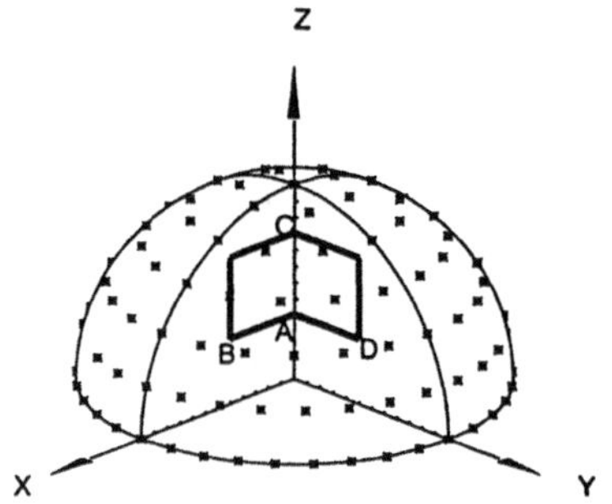

Fig.2. The two planes of the source space, ABC and ACD, and the array of magnetic field sensors.

The source space contains portions of two perpendicular planes, ABC and ACD. Each one is 5cm × 5cm and divided into 28 × 28 pixels. The dipoles are normal to the planes (tangential). The total number of dipole moments for this source space is 2 × 28 × 28 = 1,568. A spherical head model is assumed in our calculation. The sensor array forms a hemisphere 12 cm from the origin, and 127 radially oriented magnetic field sensors have been assigned to the array.

The number of parameters available to describe a current source (1,568) is very much larger than the number of measurements (127); consequently any given data set can be explained with a very large number (1,568 − 127 = 1,441) of different currents. This leads to a large degree of ambiguity for the inverse problem.

Procedure

We place a source in the center of the ABC plane as shown in Fig.3. The magnetic field produced by this source is then calculated at all the sensors. There is no noise added to the data. Note that the dipole moment per unit area of this source does not exceed 1 per pixel area. We take this to be the maximum allowed value for the dipole moment per unit area.

Here is the central question: Given this simulated magnetic field data, can we be *certain* that there is current in the central region of the ABC plane, if the only constraint is a maximum allowed value for the dipole moment per unit area? To answer this question, we perform the following steps. Force the current to be *zero* in the

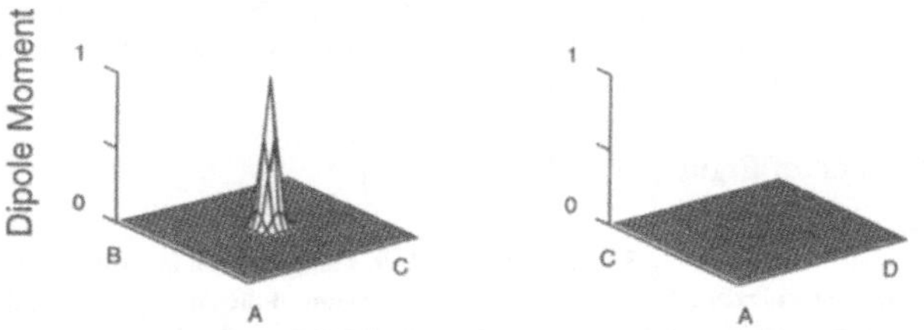

Fig.3. The original current source. It is located in the center of the ABC plane, has a Gaussian form with standard deviation of one pixel size (0.18cm), and a peak dipole moment of 1 (arbitrary units).

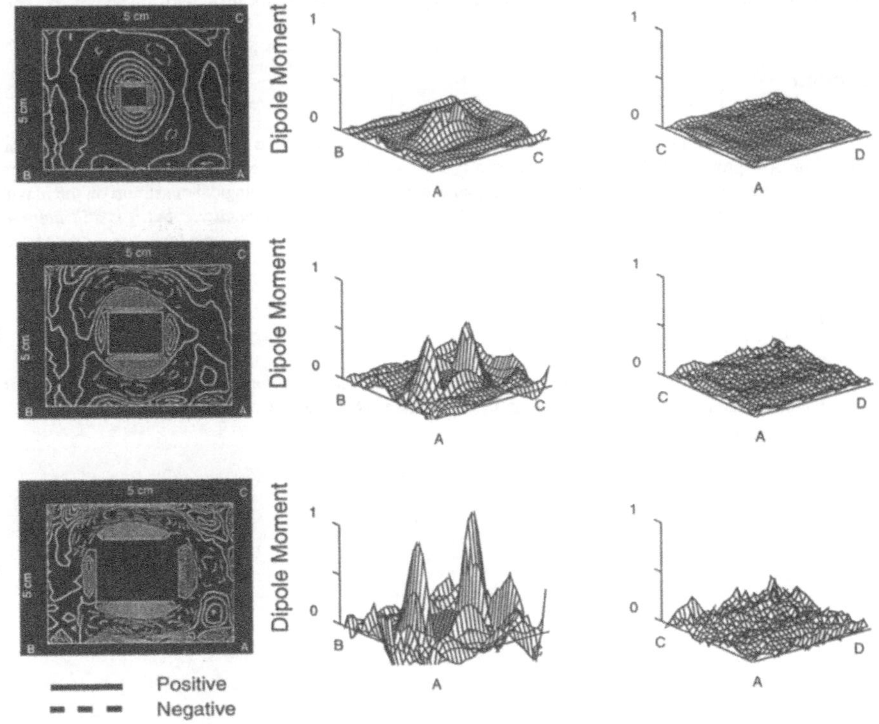

Fig.4. Three *distributed* current sources that produce exactly the same magnetic field (at 127 sensors) as the *focal* source shown in Fig.3. These sources have *zero current* in the central region of the ABC plane of size N × N pixels (pixel size=0.18cm).

(a) N=5 (0.9cm × 0.9 cm); (b) N=9 (1.6cm × 1.6cm); (c) N=13 (2.3cm × 2.3cm).

Left column: contour plot of the dipole moment strength in the ABC plane.

Center column: 3-D view of the same thing.

Right column: 3-D view of the dipole moment strength in the ACD plane.

central region of the ABC plane, of size N × N pixels. Then, construct a distributed current in the remainder of the source space that fits the magnetic field data. Finally, ascertain whether the dipole moment per unit area is everywhere less than the maximum allowed value.

Discussion of Figures 3 and 4

The current in Fig.3 is located *entirely* in a small region in the center of the ABC plane. The three sources in Fig.4 have *no current whatsoever* in the central region of the ABC plane. This region gets larger going from 4(a) to 4(c). All four sources produce exactly the same magnetic field at all the 127 sensors of the array.

In Fig. 4(a) and 4(b), the dipole moment per unit area is everywhere less than the maximum (1 per pixel area). But in Fig. 4(c), it exceeds the maximum. We find that in an area as large as 2.0cm × 2.0cm one could place *all* the current, or *none* of the current, and still fit the data *and* satisfy the constraint on the maximum dipole moment per unit area.

Conclusion

If the only constraint that is placed on the allowed current source is anatomical, the ambiguity of the MEG/EEG inverse problem for a *distributed* source is very great. Indeed, one cannot even know with certainty whether or not a large portion of the source space contains any current! A physiological constraint on the maximum allowed value of the dipole moment per unit area *reduces the extent of the ambiguity, but it is still quite large. Additional physiological constraints are needed if distributed source solutions, covering a large area, are to be useful in the electromagnetic inverse problem.*

Reference

[1] Okada Y.C., Papuashvili N., Xu C., *Maximum Current Dipole Moment Density as an Important Physiological Constraint in MEG inverse solutions*, Proceedings of the Biomag '96 Santa Fe Conference.

Multi-Source Localization in MEG Using Simulated Annealing: Model Order Determination and Parameter Accuracy

Huang, M.[1], Supek, S.[2], Aine, C.[1]

[1]Biophysics Group, MS D454, Los Alamos National Laboratory, Los Alamos, NM 87545, USA; [2]Department of Physics, Faculty of Science, Zagreb, Croatia

Introduction

Empirical neuromagnetic studies have reported that multiple brain regions are active at single instants in time as well as across time intervals of interest [1]. Determining the number of active regions, however, required a systematic search across increasing model orders using a reduced chi-square measure of goodness-of-fit and multiple starting points within each model order assumed [2]. Such a time-intensive strategy was necessary since initial source modeling assumptions for a given instantaneous or spatio-temporal data set are not available *a priori* and the minimization algorithms typically used rely on either a gradient-based search or simplex, both of which are reported to be sensitive to the initial starting points [2,3]. Simulated annealing was recently proposed for noiseless biomagnetic data [4,5] as an effective global minimizer. It does not require multiple starting points and with an appropriate cooling schedule, should converge to the global minimum. A modified cost function was also proposed to effectively deal with an unknown number of dipoles for noiseless, multi-source biomagnetic data [5].

Numerical simulation studies were conducted using simulated annealing as described in [6] to examine: 1) effects of a systematic increase in model order using both reduced chi-square [2] as a cost function as well as a modified cost function as proposed in [5]; and 2) effects of overmodeling on parameter estimation accuracy. Effects of different choices of weighting factors are also discussed. Simulated annealing was also applied to visually evoked neuromagnetic data and the effectiveness of both cost functions in determining the number of active regions was demonstrated.

Method

Instantaneous neuromagnetic field distributions were generated by 2- and 3-dipole configurations embedded in a spherical conductor. Field values were calculated at 135 measuring points distributed over a portion of a spherical surface (R=11.3 cm) as recorded by first-order gradiometers. Normally distributed noise of zero mean and a standard deviation of 20 fT was added. Simulated annealing was implemented as in [6]. The center of a spherical conductor was selected as the starting point in all cases. An iterative random search for dipole locations over the entire head volume was permitted. In order to approach the global minimum, the following annealing schedule was found as adequate: $T_{n+1} = (1-\varepsilon)T_n$, where ε was 0.05 for noisy and 0.15 for noiseless cases examined, and T_{n+1} and T_n are the temperature parameters for two consequent states. Best-fitting parameter estimates were obtained by minimizing: 1) chi-square (χ^2) as a cost function for every model order assumed [2]

$$\chi^2 = \sum_{i=1}^{N} \left(\frac{\mathbf{B}_i - \widehat{\mathbf{B}}_i(a_j)}{\sigma_i} \right)^2 \tag{1}$$

and 2) a modified cost function (MCF) as proposed by [5]

$$\mathrm{MCF} \equiv \chi^2 + w \sum_{p=1}^{P} \left| \widehat{\mathbf{B}}_p \right|^{1/2} \tag{2}$$

where $\mathbf{B}_i$ are noisy simulated field values at every sensor location i=1,...N, $\widehat{\mathbf{B}}_i(a_j)$ are field values for an assumed model described by parameters a_j; σ_i are standard deviations of gaussian noise added assumed to be equal for every sensor; and $\widehat{\mathbf{B}}_p$ are field values generated by p = 1,...P dipoles assumed in the model, typically greater than the actual or expected number of sources; w is a weighting factor. The second term in Eq.(2) is referred to as E_w. A range of weighting factors, used in the second term of the modified cost function, was examined.

Results

Both cost functions permitted the identification of the actual number of sources for 2- and 3-dipole configurations under noiseless and noisy conditions. While χ^2 required a search across model orders, a modified cost function allowed the proper identification of the actual number of sources when a model order greater than the actual number was assumed. However, different choices of weighting factors were necessary in order to assure: 1) adequate cost function values (at least for χ^2 for which the probability distribution is known) were obtained; and 2) the moments of extra dipoles added were close to zero. Table I. represents actual source parameters (panel A) and best-fitting parameter estimates (panels B and C) for noisy 2-dipole data for both cost function values when the model orders assumed were 3 and 4 (both of which represented overmodeling). In cases where adequate weighting factors

Table I: Noisy 2-Dipole Data (SD = 20 fT)

A. Actual Source Parameters

Dipole:	ρ (cm)	υ (^{o})	φ (^{o})	**P (nAm)**	ψ (^{o})
1	8.0	70.0	165.0	8.0	0
2	8.0	70.0	195.0	8.0	0

Dipole:	ρ (cm)	υ (^{o})	φ (^{o})	**P (nAm)**	ψ (^{o})	χ^2	$\chi^2 + E_w$
B1. Model order = 3; w = **0.025**							
1	**8.043**	**70.966**	**165.586**	**7.854**	**2.145**	117.490	166.209
2	**7.884**	**70.701**	**195.958**	**8.584**	**-0.814**		
3	9.999	33.458	178.669	0.204	0.714		
B2. Model order = 3; w = **0.01**							
1	**8.002**	**70.344**	**165.987**	**8.129**	**0.902**	132.176	309.098
2	**7.958**	**70.044**	**196.196**	**8.116**	**-0.147**		
3	0.008	143.205	309.478	0.115	311.746		
B3. Model order = 3; w = **0.02**							
1	**8.040**	**70.517**	**165.994**	**7.891**	**0.523**	133.753	484.928
2	**7.957**	**70.512**	**196.046**	**8.087**	**-0.039**		
3	0.000	174.128	107.094	0.061	219.199		
B4. Model order = 3; w = **0.03**							
1	6.416	67.896	180.000	19.998	0.000	397.362	729.431
2	0.009	95.623	223.362	0.000	333.031		
3	0.000	28.366	310.637	1.371	21.174		
C1. Model order = 4, w = **0.10**							
1	**8.052**	**71.079**	**165.657**	**7.837**	**-2.095**	131.266	307.815
2	**7.894**	**71.264**	**195.789**	**8.551**	**2.815**		
3	0.000	137.978	169.243	0.155	152.699		
4	0.000	138.887	223.289	0.6475	308.489		
C2. Model order = 4, w = **0.02**							
1	**8.099**	**71.462**	**165.729**	**7.548**	**-3.374**	134.262	483.349
2	**7.887**	**71.508**	**195.585**	**8.516**	**4.240**		
3	0.721	116.628	212.920	0.000	79.465		
4	1.356	117.025	116.144	0.000	167.918		

were selected, accurate parameter estimation of the actual sources was possible and the addition of extra dipoles resulted only in minor effects on parameter estimation accuracy. The moments of the extra dipoles were very close to zero (panels B2, B3, C1, C2) and their estimated locations were close to the center of the spherical conductor. However, selecting too large of a weighting factor (panel B4), strongly affected parameter estimates of all sources and resulted in too improbable χ^2 values indicating inadequacy of the model assumed. In this case, it was assumed that the particular choice of the weighting factor was inadequate, since overmodeling itself always lead to adequate χ^2 values when an optimal solution is reached. The optimal weighting factors for 3-dipole noisy data (not shown) were different than for the 2-dipole case (Table I.) (e.g., w = 0.075 for an assumed model order of 4 and w = 0.065 for an assumed model order of 5, respectively). While Haneishi and colleagues [5] suggested a weighing factor of about 0.5 for noiseless data, we found empirically that the ratio of 2:3 between first and second term in Eq.(2), resulted in an adequate choice of weighting factors for noisy simulated conditions.

Table II. shows best-fitting parameter estimates and cost-function values when different multi-dipole models were applied to visual neuromagnetic data evoked by a small Target stimulus placed 4° below the horizontal meridian and 3° to the right of the vertical meridian [8]. A 2-dipole model was identified in this case as an adequate model for an instantaneous field distribution at an early latency of 110 ms poststimulus by both cost functions. The identified brain regions were in agreement with solutions obtained by spatio-temporal modeling of the early evoked activity from 80-120 ms. Reduced chi-square values, rather than χ^2, are presented in panel A.

Table II. **Visually evoked neuromagnetic data (SD = 25.44 fT)**

A. Reduced chi-square - Increasing model order

Dipole:	x (cm)	y (cm)	z (cm)	P (nAm)	ψ (^{o})	χ_r^2	Q
1	-4.93	1.23	7.03	6.37	173.93	1.479	0.007
1	**-6.56**	**0.55**	**5.03**	**5.22**	**220.97**	*0.967*	*0.550*
2	**-4.39**	**3.09**	**5.23**	**8.48**	**115.52**		
1	-7.20	0.02	4.96	2.97	219.32	*0.867*	*0.749*
2	-4.91	2.44	5.57	7.18	140.05		
3	-7.55	4.86	4.02	0.20	17.71		

B. Modified cost function - Overmodeling: Model order = 4; w = 0.046

Dipole:	x (cm)	y (cm)	z (cm)	P (nAm)	ψ (^{o})	χ^2	$\chi^2 + E_w$
1	**-6.74**	**0.72**	**5.11**	**3.89**	**216.54**	54.36	88.03
2	**-4.68**	**3.60**	**5.16**	**5.21**	**117.99**		
3	-7.59	-4.26	8.33	0.08	54.51		
4	0.00	0.00	4.01	0.02	298.93		

Discussion

Both cost functions allowed for the identification of the actual number of sources in simulated cases (Table I) and the same adequate model order in empirical data (Table II). Best-fitting source parameters were either accurate estimates of the actual source parameters or represented physiologically plausible solutions which were similar to those obtained previously using other methods [7]. While χ^2 required a systematic increase in model order as demonstrated earlier [1], the use of a modified cost function allowed for more efficient model order determination while preserving the accuracy of parameter estimation. In essence, if one applies a model order which represents overmodeling, the correct number of dipoles can still be obtained since the "extra dipoles" would either have close to zero moments or they would be placed near the center of the head. Consequently, these "extra dipoles" are easy to identify. However, while a systematic search over different model orders might be avoided by using a modified cost function as proposed by [5], consideration of different weighting factors was necessary since both previous [5] and present simulations indicate that optimal values should be found empirically. Therefore, a search over model orders or over a range of weighting factors seems unavoidable. Since we found the optimal choice of the weighting factor to be related to the actual source configuration, which is unknown in empirical cases, it turns out that a search over model orders is likely to be more efficient. Previous simulations [5], as well as the present results, demonstrate the promise of simulated annealing as an efficient minimization approach for neuromagnetic applications. We are currently implementing simulated annealing for spatio-temporal analysis of neuromagnetic data [8] with χ^2 as a cost function of choice.

References:

[1] Aine, C.J., Supek, S., and George, J.S. Temporal dynamics of visual-evoked neuromagnetic sources: Effects of stimulus parameters and selective attention, Intern. J. Neuroscience, 1995, 80: 79-104.

[2] Supek, S., and Aine, C.J. Simulation studies of multiple dipole neuromagnetic source localization: Model order and limits of source resolution, IEEE Trans. Biomed. Eng., 1993, BME-40: 529-540.

[3] Press, W.H., Teukolsky, S.A., Vettering, W.T., and Flannery, B.P. Numerical Recipes in C, Cambridge University Press, 1992

[4] Sekihara, K., Haneishi, H., and Ohyama, N. Details of simulated annealing algorithm to estimate parameters of multiple current dipoles using biomagnetic data, IEEE Trans. Med. Eng., 1992, BME-11: 293-299

[5] Haneishi, H., Ohyama, N, Sekihara, K., and Honda, T. Multiple current dipole estimation using simulated annealing, IEEE Trans. Med. Eng., 1994, BME-41: 1004-1009

[6] Corana, A., Marchesi, M., Martini, C., and Ridella, S., Minimizing multimodal functions of continuous variables with the "simulated annealing" algorithm, ACM Trans. Mathem. Software, 1987, 13: 262-280

[7] Aine, C.J., Supek, S., George, J.S., Ranken, D., Lewine, J., Sanders, J., Best, E., Tiee, W., Flynn, E.R., and Wood, C.C. Retinotopic organization of human visual cortex: Departures from the classical model (in press, Cerebral Cortex, 1996)

[8] Huang, M., Supek, S., and Aine, C., Application of simulated annealing for multi-source spatio-temporal localization in MEG: Simulation and empirical study, 2nd International Conference on Functional Mapping of the Human Brain, June 17-21, 1996.

Acknowledgement

This research was supported in part by a grant from NEI (EY08610) and by the United States Department of Energy, contract W-7405-ENG-36.

A Three-Dimensional Rule-Assisted Interpolation Algorithm for Iso-Field Mapping of Neuromagnetic Signals

Hultin, L.

Department of Applied Electronics, Chalmers University of Technology, Gothenburg, Sweden

Department of Applied Electronics, Chalmers University of Technology, Gothenburg, Sweden

Introduction

Isofield maps are used as an important tool to evaluate the quality of recorded neuromagnetic signals and as a guide for choosing proper source models for source localizations. A commonly used method for calculating iso-fieldmaps is to project the recorded signal on a two-dimensional plane and calculate the map from the projected signals. This procedure implies that the information of the three-dimensional localization of the recorded signal is substantially reduced. The spatially warped information may introduce significant artifacts, especially close to the boundaries of the iso-fieldmap. The algorithm developed in the present project creates iso-fieldmaps using three-dimensional, rule-assisted interpolation on an ellipsoidal surface. Some of the benefits of the algorithm compared to the previously used algorithms are true three-dimensional interpolation, accurate handling of iso-map boundaries, and fast updating of iso-fieldmaps at time or view-angle changes. The algorithm is evaluated on recorded and simulated neuromagnetic fields, but is equally applicable to scalp-recorded EEG.

Methods

An ellipsoidal head model has been developed that is easy to relate to anatomical references of the skull [1]. The y-axis is parallel, and one centimeter superior, to a line from the inion to the nasion. The positive z-axis passes through the skull at the vertex, and the positive x-axis passes through the skull at the midpoint on a line from the inion to the nasion, 2-3 cm superior to the right pre-auricular point. The iso-fieldmaps are calculated on the surface of an ellipsoid, with the same origin as the head model, which best fits the recording sites. Field-maps are obtained in four separate steps:

1. A spatial grid, **G**, with equidistant Φ- and Θ-steps is calculated, defining the accurate map area and the grid points used (Fig. 1). Φ is the angle from the positive z-axis and Θ is the angle from the positive x-axis (Fig. 2).
2. The amplitude at each valid grid point in **G** is interpolated from the recorded signals.
3. A signal-space grid **W** (Fig. 1) is calculated, which contains the position, level and derivatives of iso-lines.
4. Iso-borders with equivalent level and derivative are connected by use of a rule-based algorithm.

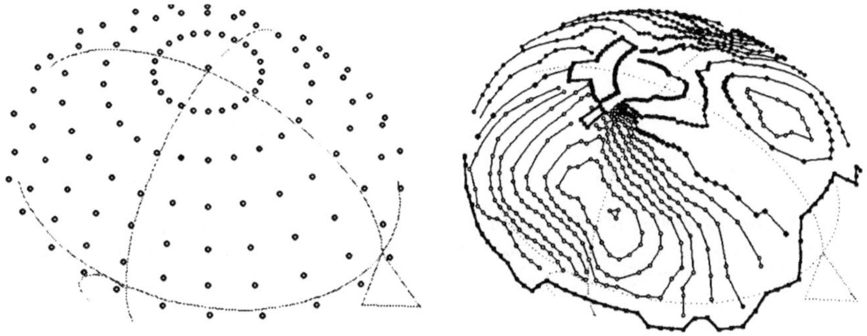

Fig. 1: An illustration of **G**-grid (left) and **W**-grid (right). In practical use, the number of Φ- and Θ-steps is about four times as high as in the illustration of **G**. Squares represent boundaries and '+' represent zero field level. Full circles represent negative, ingoing flux. Dashed lines represent the ellipsoidal head model.

1. Spatial grid

The viewing of iso-fieldmaps is an interactive process, where the progress of maps during a time interval often gives crucial information. The processing time needed for calculating a new iso-map, when the recorded signals change, is minimized by storing the time-invariant information in **G**. A grid point is declared *valid* if there exist more than N^r recording sites that are located within a distanc D^s from the gridpoint, and if the distance to the closest recording site is less than the distance D^M. For each grid point in **G** the positioning numbers, and distances to, the recorded signals that are within the distances D^s, as well as the Cartesian and spherical coordinates for the grid point, are stored for the interpolation algorithms.

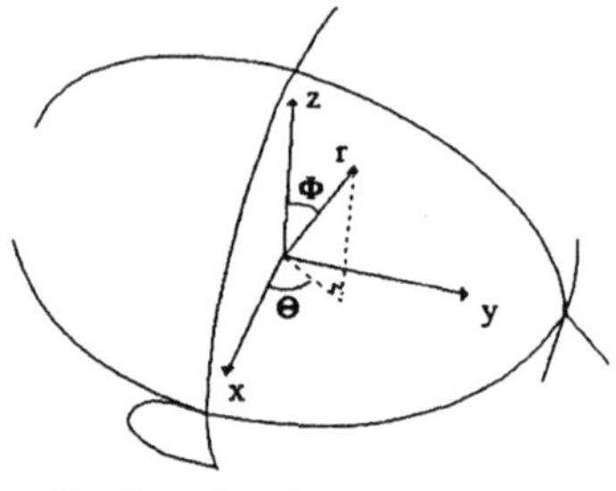

Fig. 2: Coordinates refer to either a Cartesian or a spherical coordinate system corresponding to the axes of the ellipsoidal model.

2. Interpolation and smoothing

At each grid point G_g the amplitude of the signal is interpolated from the recorded signals. The grid **G** has unequal Cartesian distances between the grid points. True spatial filtering of the interpolated signals therefore requires extensive, and time-consuming, data processing. A simplified smoothing algorithm is thus used, introducing a smoothing constant. The interpolated signal here is inverse-square dependent on the sum of the smoothing constant and the distance from recorded signals that are located within the distance D^s.

3. Signal-space grid

The grid **W** carries the information needed to draw iso-maps at a specific time period. The information in **W** is stored Θ-wise for monotonically increasing Φ-angles and indexed $w = [\theta, \phi]$. For each Θ-step the grid **G** is searched in the positive Φ-direction for consecutive grid points at separate iso-levels, and the Φ-angles for iso-lines are determined by linear interpolation. Each iso-level at each Θ-step becomes a grid point in **W**. At each grid point W_w the iso-level L_w, the derivative D_w in the positive Φ-direction and the Φ-angles Φ_w are stored.

To be able to use the same rules for boundaries as for iso-lines, iso-levels and derivatives of boundaries are set to numerical values that are non-existent for normal iso-levels but match the rule-based algorithm. If a **G**-grid point is *valid* and the next grid point in the positive Φ-direction is *nonvalid*, then L_w is set to a value defined as *boundary*, and D_w is set to a derivative defined as *boundary*. If a grid point is *nonvalid* and the next grid point is *valid*, then L_w is set as *boundary* and D_w is set as -1**boundary*.

4. Iso-line connections

Next gridpoint: Each grid point can be connected to two grid points at the previous Θ-step and two at the next Θ-step (Fig. 3). The connections to and from a grid point are determined by two pointers to grid points at the previous Θ-step and two pointers to grid points at the next Θ-step. For each Θ- and Φ-step, **W** is searched at the next Θ-step for the closest grid points in the upward Φ-direction N^U_w and for the closest grid points in the downward Φ-direction N^D_w. The pointers in the forward direction are set *nonvalid* if the next Φ is out of the specified range.

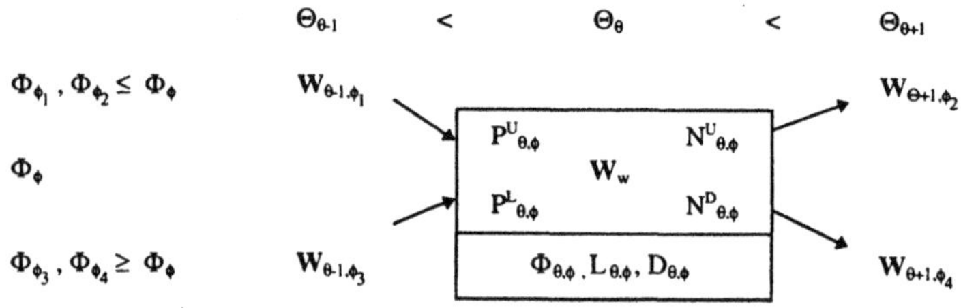

Fig. 3: The information in **W** is stored Θ-wise for monotonically increasing Φ-angles and indexed $w = [\theta,\phi]$. Each gridpoint W_w contains the Φ-angle, the iso-level, the derivative in the positive Φ-direction, two forward and two backward pointers.

The grid point at the previous Θ-step with the smallest Φ-angles that points to $\mathbf{W}_w$ is defined as P^U_w and the grid point with the largest Φ-angles pointing to $\mathbf{W}_w$ is defined as P^L_w. The pointers in the backward direction are set *nonvalid* if there is no previous grid point pointing to the current grid point.

Steep gradients: When the iso-map contains steep gradients in the Φ-direction, the forward pointers may not be set to the proper index, as the accurate connection might be more upward, or downward, than the grid point at closest Φ at the next Θ. Thus, in those cases where a *valid* forward pointer has not been found, the pointers to the next grid point are modified according to the following rules, where $\Phi' = N^U_w$ and $\Phi'' = N^D_w$: {A} If $(L_{\theta,\phi\text{-}1} = L_w - 1)$ then $(N^U_w = N^U_w - 1)$ as long as $(L_{\theta+1,\phi'} = L_w + 1)$. {B} If $(L_{\theta,\phi\text{-}1} = L_w + 1)$ then $(N^U_w = N^u_w - 1)$ as long as $(L_{\theta+1,\phi'} = L_w - 1)$. {C} If $(L_{\theta,\phi+1} = L_w - 1)$ then $(N^D_w = N^D_w + 1)$ as long as $(L_{\theta+1,\phi''} = L_w + 1)$. {D} If $(L_{\theta,\phi+1} = L_w + 1)$ then $(N^D_w = N^D_w + 1)$ as long as $(L_{\theta+1,\phi''} = L_w - 1)$.

Validity: When the pointers to the next Θ-step have been found, their validity is determined. N^U_w becomes a *valid* connection from ϕ_w to $\phi_{\theta+1,\phi'}$, where $\Phi' = N^U_w$, if $(L_w = L_{\theta+1,\phi'})$ and $(D_w = D_{\theta+1,\phi'})$ and $((L_{\theta,\phi\text{-}1} \neq L_w$ or $(\Phi_{\theta,\phi'} - \Phi_{\theta+1,\phi} < \Phi_{\theta,\phi\text{-}1} - \Phi_w))$. N^D_w becomes a *valid* connection from Φ_w to $\Phi_{\theta+1,\phi''}$, where $\Phi'' = N^D_w$, if $(L_w = L_{\theta+1,\phi''})$ and $(D_w = D_{\theta+1,\phi''})$ and $((L_{\theta,\Phi+1} = L_w$, or $(\Phi_{\theta,\phi''} - \Phi_{\theta+1,\phi} < \Phi_{\theta,\phi\text{-}1} - \Phi_w))$.

Endpoints: Endpoints of an iso-line are handled separately, as they follow specific rules. If two iso-lines at the same iso-level start, or end, at two consecutive Φ-angles at the same Θ-angle, there are five cases in which they normally are to be connected. When grid points at consecutive Φ-angles are connected, then the forward-pointer of the upper grid point and the backward-pointer of the lower grid point are marked as *vertical*.

If L_w is equal to $L_{\theta,\phi+1}$, and D_w is equal to $-1 * D_{\theta,\phi+1}$, then Φ_w and $\Phi_{\theta,\phi+1}$, are connected if one of the following statements is fulfilled: {A} P^U_w, P^L_w, $P^U_{\theta,\phi+1}$, $P^L_{\theta,\phi+1}$ are *nonvalid* and one, but not both, of N^U_w and N^D_w is *valid*, and one, but not both, of $N^U_{\theta,\phi+1}$ and $N^D_{\theta,\phi+1}$ is *valid*. {B} P^U_w, P^L_w, $P^U_{\theta,\phi+1}$, $P^L_{\theta,\phi+1}$ are *nonvalid* and N^D_w, but not N^U_w, is *valid*, and one of $P^U_{\theta,\phi+1}$ and $P^L_{\theta,\phi+1}$ is *nonvalid*. {C} P^U_w, P^L_w, $N^U_{\theta,\phi+1}$ and $N^D_{\theta,\phi+1}$ are *nonvalid*, and one, but not both, of $P^U_{\theta,\phi+1}$, $P^L_{\theta,\phi+1}$ is *valid*, and $N^D_{\theta,\phi+1}$ is *nonvalid*. {D} N^D_w, N^U_w and $N^U_{\theta,\phi+1}$ and $N^D_{\theta,\phi+1}$ are *nonvalid*, and one, but not both, of P^U_w and P^L_w is *valid*, and one, but not both, of $P^U_{\theta,\phi+1}$ and $P^L_{\theta,\phi+1}$ is *valid*.

Discrete Θ-steps: The grid **W** has a finite number of Θ-steps, all having the same step-length. The discrete Θ-step implies certain rules to handle end-lines of iso-levels within the Θ-grid. Start and end points of iso-lines are left unconnected if the endpoints are separated by single, or multiple, grid points at other iso-levels. In these cases, the Θ-*step* is divided into a number of semisteps and the endpoints are then connected (Fig. 4).

- If N^D_w is *nonvalid* and there exists a K such that: $L_{\theta,\phi+K} = L_w$, and $N^U_{\theta,\phi+K}$ is *nonvalid*, and all of $N^U_{\theta,\phi+1}$ to $N^U_{\theta,\phi+K-1}$ are *nonvalid*, and all of $N^D_{\theta,\phi+1}$ to $N^D_{\theta,\phi+K-1}$ are *nonvalid* or *vertical*, then {1} $M = 1 + \text{abs}(L_w - \text{maximum}(L_{\theta,\phi+1} \dots L_{\theta,\phi+K-1}))$, {2} N^D_w and $N^U_{\theta,\Phi+K} = \textit{semistep}_V$. , {3} M_w and K_w are stored to be used at the iso-line drawing, {4} $\Delta\Theta_w = \Theta\text{-steplength} * (M - 1) / M$, {5} $\Delta\Phi_w = (\Phi_{\theta,\phi+1} - \Phi_w) / 2$ and {6} $\Delta\Phi_{\theta,\phi+N} = (\Phi_{\theta,\phi+K-1} - \Phi_{\theta,\phi+K}) / 2$.
- If these conditions are not true but N^D_w is *nonvalid* and there exists a K such that: $L_{\theta,\phi+K}$ is a *boundary*, and all of $N^U_{\theta\phi+1}$ to $N^U_{\theta\phi+K-1}$ are *nonvalid*, and all of $N^D_{\theta\phi+1}$ to $N^D_{\theta\phi+K-1}$ are *nonvalid* or *vertical*, then: {1} $M = 1 + \text{abs}(L_w - \text{maximum}(L_{\theta,\phi+1} \dots L_{\theta,\phi+K-1}))$, {2} $N^D_w = \textit{semistep}_B$., {3} $M_{\theta,\phi}$ and $K_{\theta,\phi}$ are stored to be used at the iso-line drawing, {4} $\Delta\Theta_w = \Theta\text{-steplength} * (M - 1) / M$, {5} $\Delta\Phi_{\theta,\phi} = (\Phi_{\theta,\phi+1} - \Phi_{\theta,\phi}) / 2$ and {6} $\Delta\Phi_{\theta+K,\phi} = (\Phi_{\theta,\phi+1} - \Phi_{\theta,\phi+K}) / 2$.
- If P^L_w is *nonvalid*, P^U_w is not a *semistep*, and there exists a K such that: $L_{\theta,\phi+K}$ is equal to L_w, and $P^U_{\theta,\phi+K}$ is *nonvalid*, and all of $P^L_{\theta,\phi+1}$ to $P^L_{\theta,\phi+K-1}$ are *nonvalid*, and all of $P^U_{\theta,\phi+1}$ to $P^U_{\theta,\phi+K-1}$ are *nonvalid* or *vertical*, then: {1} $M = 1 + \text{abs}(L_w - \text{maximum}(L_{\theta+1,\phi} \dots L_{\theta+K-1,\phi}))$, {2} If P^U_w is *valid* then $P^L_w = \textit{semistep}_V$ else if P^U_w is *nonvalid* then $P^U_w = \textit{semistep}_V$. {3} $P^U_{\theta,\phi+K} = \textit{semistep}_E$, {4} M_w and K_w are stored to be used at the iso-line drawing, {5} $\Delta\Theta_w = \Theta\text{-steplength} * (M - 1) / M$, {6} $\Delta\Phi_w = (\Phi_{\theta,\phi+1} - \Phi_w) / 2$ and {7} $\Delta\Phi_{\theta,\phi+K} = (\Phi_{\theta,\phi+K-1} - \Phi_{\theta,\phi+K}) / 2$.
- If the above conditions are not true but P^L_w is *nonvalid*, and P^U_w is not a *semistep* pointer, and there exists a K such that: $L_{\theta,\phi+K}$ is a *boundary*, and all of $P^L_{\theta,\phi+1}$ to $P^L_{\theta,\phi+K-1}$ are *nonvalid*, and all of $P^U_{\theta,\phi+1}$ to $P^U_{\theta,\phi+K-1}$ are *nonvalid* or *vertical*, then: {1} $M = 1 + \text{abs}(L_w - \max(L_{\theta,\phi+1} \dots L_{\theta\phi+K-1}))$, {2} if P^U_w is *valid* then P^L_w is marked as $\textit{semistep}_V$, {3} if P^U_w is *nonvalid* then P^U_w is marked as $\textit{semistep}_V$, {4} $P^U_{\theta,\phi+K}$ is marked as $\textit{semistep}_E$, {5} M_w and K_w are stored to be used at the iso-line drawing, {6} $\Delta\Theta_w$ is set equal to the Θ-steplength * $(M - 1) / M$, {7} $\Delta\Phi_w$ is set equal to $(\Phi_{\theta,\phi+1} - \Phi_w) / 2$ and {8} $\Delta\Phi_{\theta,\phi+K}$ is set equal to $(\Phi_{\theta,\phi+K-1} - \Phi_{\theta,\phi+K}) / 2$.

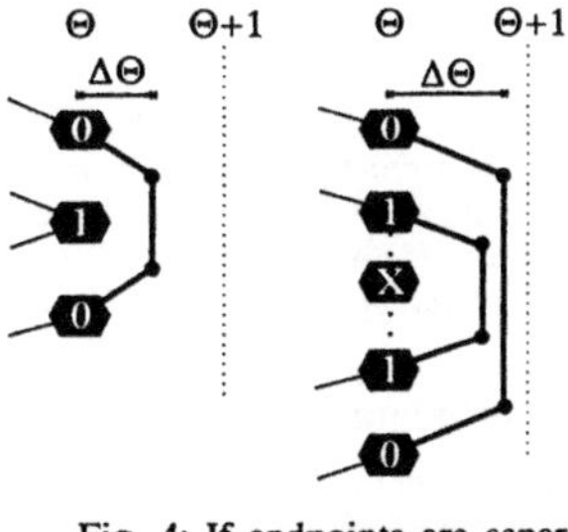

Fig. 4: If endpoints are separated by grid points at other iso-levels is the Θ-*step* is divided into a number of semi-steps and the grid points are then connected.

5. Isofield-map viewing

The three-dimensional surface can be rotated and viewed from any direction as a projection on a plane at the surface of the screen. The projection plane has its origin at the center of the head and is perpendicular to the normal given by the view-angle, specified by a Φ- and a Θ- angle. To give an impression of a 3D surface, only those coordinates that are located above the projection plane are plotted. To draw the iso-fieldmap, the three-dimensional iso-lines in **W** are projected and plotted on the screen. There are three different ways in which a grid point might be connected to surrounding points [1] : (A) if it is a *valid* pointer, then to a grid point at the next Θ-angles; (B) if it is a *vertical* pointer, to the following grid point at the next Φ-angles at the current Θ-angles; or (C) if it is a *semistep* pointer, to a grid point at a specified 'next Φ-angles' at the same Θ-angles. The iso-line from one grid point to another is segmented in steps of ΔΦ and ΔΘ before the iso-line is projected and plotted. ΔΦ and ΔΘ are set small enough to make the iso-line follow the surface of the head regardless of the distance. A segment is drawn if both the start and end points are located above the projection plane.

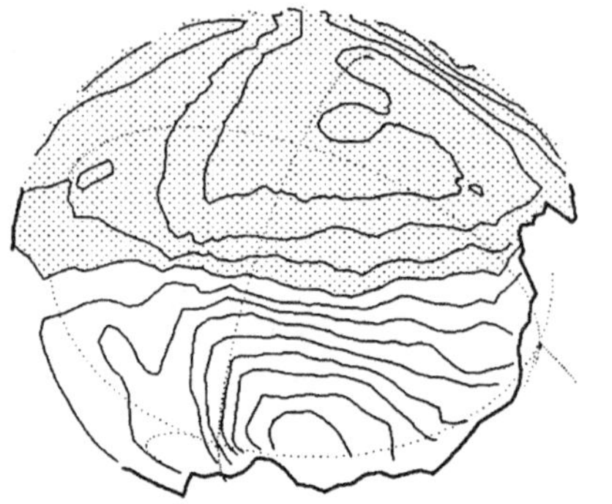

Fig. 5: An example of a recorded iso-fieldmap calculated on a grid containing 40 Φ-steps times 130 Θ-steps. Black lines represent positive (outgoing) flux, light gray lines represent zero-level, dark gray lines represent negative (outgoing) flux, and thick black lines represent the boundaries.

Results and discussion

The algorithms result in accurate iso-fieldmaps (Fig. 5), and the time needed for the calculations is short enough for interactive use. The possibility to interactively rotate the iso-fieldmap into any specified view-angles is a valuable feature in the analysis of complex neuromagnetic fields. The time needed to process an iso-fieldmap depends directly on the number of Φ and Θ grid points used. The number of grid points needed depends on the spatial frequency of the recorded field and the resolution required for the iso-fieldmap. Our experience from simulated and recorded neuromagnetic fields shows that the spatial resolution obtained by a limited number of Φ and Θ-steps is high enough for all practical purposes.

Iso-fieldmaps are used as a visualization tool and are a simplified presentation of complex signal-space information [2]. The iso-fieldmaps have to be easy to interpret visually, and easy to relate to the anatomical references of the head. The surface of the ellipsoidal model gives, under normal conditions, a good fit to the location of recording sites. The ellipsoidal model is easy to illustrate and hidden line removal is automatically achieved. For interactive use, more complex isofield-map surfaces are of limited value. The spatial smoothing method used gives fast and visually satisfying results, if the smoothing constant is carefully selected. The exact effects on the iso-fieldmaps are, however, difficult to evaluate and the actual spatial frequency is not known. True spatial filtering is therefore desirable.

References

[1] Hultin, L. 3D ISO-field mapping of neuromagnetic signals using a three-dimensional rule-assisted interpolation algorithm on an ellipsoidal model. Technical report 2:95, Research Laboratory of Medical Electronics, Chalmers University of Technology, Göteborg, Sweden. 1995

[2] Binnie, C. Macgillivaray, BB. Brain mapping--a useful tool or a dangerous toy? J Neurol Neurosurg Psychiatry, 1992, 55: 527-529

Acknowledgments

I want to thank Professor Gian Luca Romani, Dr. Vittorio Pizzela and Franca Tecchio for their support during the recording of the neuromagnetic data used for evaluation of the algorithms developed. I also want to express my gratitude to Per Högstedt and Professor Roland Kadefors for valuable comments on the manuscript.

The Influence of Glial Buffering Currents on Sustained Neuromagnetic Fields

Hultin, L.

Department of Applied Electronics, Chalmers University of Technology, Gothenburg, Sweden

Introduction

Neuromagnetic fields are commonly considered to originate from post-synaptic potentials, PSPs, and the relation between these potentials and neuromagnetic fields has been revealed by several authors [1, 2, 3]. Recordings of extracellular potassium and distribution of cortical potentials strongly suggest that transient event-related potentials of duration up to tens of ms are predominantly due to post-synaptic activity [4, 5]. The dominant sources of sustained potentials, SP, which are slow potential shifts lasting from hundreds of ms up to tens of seconds, are on the other hand proposed to be glial K^+ buffering current, GBC, [5, 6, 7]. Recordings of sustained magnetic fields, SF, such as contingent magnetic variation and readiness field, show a close relation to their electric counterparts [8, 9, 10]. Hence there are therefore reasons to assume that the SF are of glial origin, which has been suggested in the case of spreading cortical depression [11]. The aim of this paper was to discuss the relationship between GBC and neuromagnetic fields, and to evaluate consequences on source modeling and interpretation of neuromagnetic phenomena.

Astroglial cells in the central nervous system form an electric syncytium that surrounds cortical neurons. The syncytium extends from the cortical surface throughout the depth of the cortex. One important feature of the glial cells is to moderate the electrophysiological properties in the vicinity of active neurons and to ensure that the changes in extracellular ion concentration do not lead to neural hyperactivity. Neural activity is related to an equal exchange of potassium out of the cells and sodium cell influx. The most critical parameter for neural excitability is, however, the level of increased extracellular K^+ concentration, $[K^+]_e$, as the rise in $[K^+]_e$ depolarizes neighboring neurons and enhances their response to a given stimulus. The role of GBC is thus to limit the increase in $[K^+]_e$ related to enhanced neural activity, and to facilitate the recovery of ion buildup after the termination of activity. The GBC is estimated to be about five times more effective in redistributing the $[K^+]_e$ than passive diffusion [12]. The enhanced neural activity causes a local increase in $[K^+]_e$ in the narrow clefts that separate the neurons from the glial cells. Glial cells have a variety of different channels including active sodium and chlorine channels [13]. The glial cells are however preferentially permeable to K^+, and astrocytes in cell cultures have been shown to be ideal potassium detectors such that the glial membrane potential, E, follows the Nernst equation: $E \propto \ln([K^+]_e / [K^+]_g)$, where $[K^+]_g$ is the K^+ concentration in the glial cells. The exposed part of the glial membrane thus becomes depolarized as K^+ passes through the cell membrane, into the glial cell. Due to the initially high K^+ concentration in the glial cells, there will only be a relatively small increase in $[K^+]_g$. The potassium concentration gradient in the glial syncytium is thus a minor cause of K^+ redistribution. The local depolarization will however spread along the glial syncytium. Simultaneous recordings of glial cell potentials and $[K^+]_e$ in cat cortex reveal that the amplitude of the glial depolarization is relatively constant throughout the depth of cortex, in contrast to the laminar $[K^+]_e$ profile [14]. Glial cells at distal areas, with relatively lower $[K^+]_e$ than in the active area, will thus be depolarized with respect to the local K^+ diffusion potential. This induces a current that carries K^+ into glial cells at sites of maximal K^+ accumulation and transports K^+ out of glial cells at remote areas (Fig. 1). The corresponding extracellular current loop is predominantly carried by Na^+ and Cl^-. As only part of the extracellular potassium that is dispersed by the glial syncytium is replaced by sodium, this will result in a shrinkage of the extracellular space, ES, in areas with increased $[K^+]_e$ concentration and a corresponding expansion in those areas where K^+ is released. No significant intraneural current can be expected as a result of the changed intracellular ionconcentration, as the outflow of cations is balanced by an equal inflow of anions.

The restoration of $[K^+]$ after the end of the neural activity leads to a local hyperpolarization of the glial syncytium, at the site of those neurons that previously released the largest amount of K^+. This leads to a release of K^+ out of adjacent glial cells and a reversal of the glial buffer current. The K^+ thus slowly flows back to the cortical areas where the maximal activity previously occurred. An electrophysiological phenomenon related to this phase reversal is the post-imperative negative variation, PINV, which is manifested as a long-lasting positivity following the termination of the contingent negative variation.

Astrocytes are suggested to maintain their radial direction after their differentiation from fetal radial glial cells. The redistribution of K^+ is thus likely to be mainly perpendicular to the surface of the cortex [6]. The columnar organization of the glial syncytium is illustrated by recordings of glial membrane potentials in the visual cortex of cat. It is shown that the glial cells respond only to certain specific patterns of light [15]. The assumption of radial GBC is further supported by the low horizontal potential gradient seen in extracellular recordings of the cortex of cat during enhanced neural activity [16].

An asymmetrical K^+ flow is crucial to achieve an extracranial field of detectable strength at a sufficient distance from the cortical activity. The GBC is bi-directional from the site of maximal $[K^+]_e$, reported to occur in middle cortical layers (lamina III and IV) during enhanced activity in neocortex [15, 6]. Current density analysis shows a current sink at middle cortical layers and current sources at superficial and deeper layers during enhanced cortical activity [15, 17]. The superficial source is however found to be smaller than the deep source, and to be located closer to the current sink. The asymmetrical buffering current results in shrinkage of the ES which is largest in middle cortical layers, while the main expansion of the ES can be observed in deep cortical layers [17]. This asymmetry can be assumed to be due to neural activity in superficial layers and the non-symmetrical extension of the glial syncytium, where the surface of the cortex serves as an upper boundary.

The exact length constant, λ, for the glial syncytium in neocortex is not known. Single cell analysis has proposed a length constant of 300 μm for astroglial cells [18], 600 μm for Müller cells in retina [19] and up to 800 μm for glial cells in the Necturus optic nerve [20]. The glial membrane has high K^+ permeability due to passive K^+ channels. A depolarization of a glial cell in a cell culture thus leads to an instant outflow of potassium into the ES close to the site of depolarization. It is, however, shown that the potassium permeability is about ten times higher in the glial endfoots, which connect to the neurons, than for the rest of the glial membrane [21]. Furthermore, two types of K^+ inward rectifying channels have been demonstrated in astrocytes besides the passive K^+ channels [14, 22]. One is voltage-independent and one is a delayed rectifying channel, activated when the membrane becomes depolarized above -40 mV. A third type of active K^+ channel opens at hyperpolarization below -110 mV. Patch clamp recordings of ion channels in isolated glial cells from the retina of the turtle show that the predominant conductance in these cells was due to an inward rectifying potassium current [23]. This suggests that the glial syncytium has a buffering function, which increases the theoretical length constant for the glial syncytium, relatively glial cell in cell culture [24, 25].

Results

The asymmetrical GBC results in a net current flow that is directed perpendicular from the surface of the cortex. This represents an equivalent current dipole directed perpendicular from the surface of the cortex, the same as could be expected due to post-synaptic currents, PSCs. At a sufficient distance from the neural activity, the equivalent source model for the glial K^+ buffering current therefore does not differ from the model of PSCs. The current density, I(r), of GBC decreases exponentially with the distance, r , from the site of depolarization. The corresponding current dipole moment, Q, is thus:

$$Q = \int_0^\infty I(r)dr = I_0 \int_0^\infty e^{-\frac{r}{\lambda}} dr = \lambda I_0$$

The magnitude of the GBC has been estimated from extracellular potential gradients and shrinkage of ES. Dietzel et al. reported a current sink, corresponding to a removal of 0.1 - 0.5 mmol of monovalent cations per liter of brain per second from ES at middle cortical layers, during electrical stimulation of the ventrobasal thalamic nuclei of cats [16], the same as has been reported during epileptic activity [17]. This corresponds to a current flow of 10-50 $\mu A/mm^3$ brain tissue.

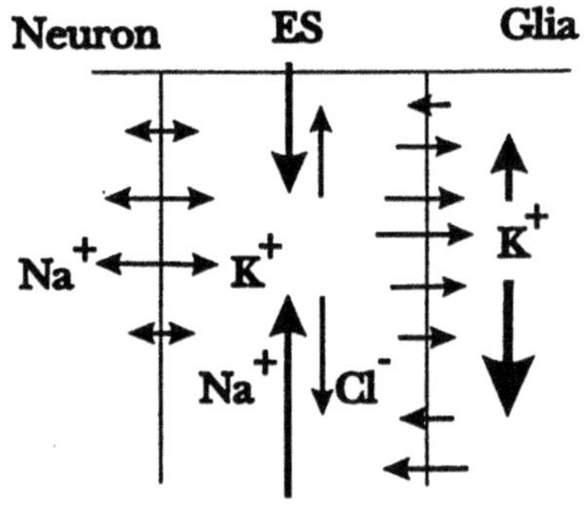

Fig. 1: Increased $[K^+]_e$ causes a local depolarization of the glial membrane. This induces a current that carries K^+ into glial cells at sites of maximal K^+ accumulation and transports K^+ out of glial cells at remote areas. The corresponding extracellular current loop is predominantly carried by Na^+ and Cl^-.

Discussion

A single cortical column has a diameter of 200 - 300 μm. The extension of the active cortical area thus depends on the number of cortical columns involved in the neurological process. A length constant of 400 μm, an active cylindrical region with a height of 0.6 mm and a diameter of 2 mm correspond to an equivalent current dipole moment of 2.5-13 nAm, if the dipole moment of the current flow directed towards the cortical surface is half the dipole moment of the current flow directed towards the depth of the cortex. At a cranial depth of 20 mm, this dipole moment generates a maximal extra cranial magnetic field (recorded radial to the head with a head-to-sensor distance of 15 mm) of 70-350 fT.

Neither the exact ratio between the K^+ flowing toward the cortical surface and the K^+ flowing from it, nor the equivalent lengths of the current dipoles, are known. The calculations show, however, that GBC is a potential source of the neuromagnetic fields. The possible contribution from PSCs in a corresponding area depends on the number of pyramidal neurons, the neural firing rate and the distribution of synaptic terminals on the neurons. PSCs contribution to scalp-recorded potentials could be due either to excitatory PSPs, EPSPs, in the distal part of the dendrite tree of a pyramidal cell, or to inhibitory PSPs, IPSPs, on or near the neural soma. It is commonly argued that IPSP's, due to their longer duration, have a stronger influence on the scalp-recorded potentials than do EPSPs. Neuromagnetic fields require an asymmetrical current flow. As the post synaptic potentials, PSPs, electrotonically spread bi-directionally from the site of activation, is it thus unlikely that PSPs close to the cell soma contribute to the extracranial magnetic fields. A post-synaptic volley with an amplitude of 10 mV, excited in the distal part of the dendrite tree of a pyramidal cell, have been estimated to gives rise to a current dipole of 30 fAm [3]. More than 250,000 simultaneously active EPSPs are thus required to evoke 200 fT, if the source is located at a cranial depth of 20 mm. In somatosensory cortex there are approximately 100,000 neurons/mm^2 and about 70% of these are pyramidal neurons (less than 5% of the pyramidal cells are giant Betz cells). If all pyramidal neuron in a cortical area with a diameter of 2 mm are active simultaneously the magnetic field evoked, under the conditions above, would be 180 fT. This is, due to spatial and temporal summation, not physiologically impossible but rather unlikely.

There is accumulated evidence that SP of long duration are of glial origin. The influence of GBC on evoked potentials with a duration of 100 ms or less is, however, not known. It is argued GBC is most effective during activity lasting seconds or more, and that it thus has a limited influence on transient evoked potentials [10]. Simultaneous recordings of glial cell potentials and $[K^+]_e$ show, nevertheless, that a single stimulus evokes a substantially increased $[K^+]_e$ and thus a related glial cell depolarization. Under physiological conditions, rises in $[K^+]_e$ have been shown to be limited to a "ceiling level" after only a few repetitive stimuli. The response by the first stimuli makes up one third of the total $[K^+]_e$ elevation [16]. Epileptic spike activity in cortex of cat gives rise to glial cell response with a rise time of about 70 ms and a fall time of 1200 ms [14]. Stimulation of peripheral nerve evokes transient extracellular K^+ signals with a rise time of 20-60 ms that last about one second or less [26]. The corresponding change in polarity of the SF could, however, be expected to be faster as it originates from the reversal of the K^+ current. Since the K^+ flowing into neurons is partially replaced by K^+ coming out of the glia, K^+ drops more slowly than the field potentials. Glial cells thus exhibit relatively fast transient responses and might consequently have influence on evoked potentials with relatively short duration. The relative contribution of GBC to transient neuromagnetic fields of short duration is, however, difficult to estimate as it is difficult to separate it from the contribution of PSCs.

The enhanced accuracy in localization and analysis of neural activity offered by neuromagnetic recordings stresses the current hypotheses of the origin of neuromagnetic fields and scalp recorded potentials. A better understanding of these complex relations is crucial for accurate electrophysiological source models to be used in source localization, and for accurate interpretation of the neural origin of neuromagnetic fields.

Some of the main features of GBC compared to PSCs are: its integrating properties, a phase reversal that occurs at the termination of neural activity, a close relation to the degree of neural activity, regardless of whether this is pre-synaptic, post-synaptic, excitatory or inhibitory activity, and finally GBC is not restricted to pyramidal neurons. It is thus essential to determine the origin of the neuromagnetic fields related to such electrophysiological phenomenon as: readiness fields, contingent magnetic fields, spreading cortical depression, phase reversal in evoked fields, interictal epileptic spikes and slow waves, neuromagnetic fields related to inhibitory neural activity and spontaneous rhythmic brain activity.

References

[1] Okada, Y. Neurogenesis of evoked magnetic fields, In: Williamson, S J. Romani, G L. Kaufman, L. Modena, I. Biomagnetism: An Interdisciplinary Approach, New York, Plenum Press, 1983.

[2] Okada, Y C. Recent Developments on the Physiological Basis of Magnetoencephalography (MEG), In: Williamson, S J. Hoke, M. Stroink, G. Kotani, M. Advances in Biomagnetism, New York, Plenum, 1989.

[3] Romani, G L. Fundamentals of neuromagnetism, In: Williamson, S J. Hoke, M. Stroink, G. Kotani, M. Advances in Biomagnetism, New York, Plenum, 1989.

[4] Speckmann, E J., and Walden, J. Mechanisms underlying the generation of cortical field potentials, Acta Otolaryngol Suppl Stockh, 1991, 491: 17-23.

[5] Roitbak, A.I., Machek, J., Pavlík, V., Bobrov, A.V., and Otcherashvili, I.V. The slow component of the direct cortical response and extracellulap potassium, In: Syková, E., Hník, P., and Vyklický, L. Ion-Selective microelectrodes and their use in excitable tissue, New York, Plenum, 1981.

[6] Laming, P R. Do glia contribute to behaviour? A neuromodulatory review, Comp Biochem Physiol A, 1989, 94: 555-568.

[7] Somjen, G G. Electrogenesis of sustained potentials, Progr. neurobiol., 1973, 1: 201-237

[8] Hultin, L., P. Rossini, G. L. Romani, P. Högstedt, Tecchio, F., and V. Pizzella. Neuromagnetic localization of the late component of the contingent negative variation, Electroencephalogr Clin Neurophysiol. Accepted for publication.

[9] Kristeva, R., Cheyne, D., and Deecke, L. Neuromagnetic fields accompanying unilateral and bilateral voluntary movements: topography and analysis of cortical sources, Electroencephalogr Clin Neurophysiol, 1991, 81: 284-298.

[10] Pantev, C. Eulitz, C. Elbert, T. Hoke, M. Origin of the auditory evoked sustained field, In: Baumgartner, C. Deecke, L. Stroink, G. Williamson, S J. Biomagnetism: Fundamental research and clinical applications, Amsterdam, Elseiver Ios Press, 1995.

[11] Okada, Y.C. Lauritzen, M. Nicholson, C. MEG source models and physiology, Med. Biol. Phys. 1987, 32:43-51

[12] Gardner Medwin, A R. A study of the mechanisms by which potassium moves through brain tissue in the rat, J Physiol Lond, 1983, 335: 353-374.

[13] Sontheimer, H., and Waxman, S G. Expression of voltage-activated ion channels by astrocytes and oligodendrocytes in the hippocampal slice, J Neurophysiol, 1993, 70: 1863-1873.

[14] Futamachi, K J., and Pedley, T A. Glial Cells and extracellular potassium: Their relationship in mammalian cortex, Brain research, 1976, 109: 311-322.

[15] Kelly, J. P., and Van Essen, D.C. Cell structures and functions in the visual cortex of the cat, J. Physiol, 1974, 238: 515-547.

[16] Dietzel, I., Heinman, U., and Lux, H. Relations between slow extracellular potential changes, glial potassium buffering, and electrolyte and cellular volume changes during neuronal hyperactivity in cat brain, Glia, 1989, 2: 25-44.

[17] Lux, H D. Heinemann, U., and Dietzel, I. Ionic changes and alterations in the size of the extracellular space during epileptic activity, Adv Neurol , 1986, 44: 619-639.

[18] Gardner-Medwin, A R. The amplitude and time course of extracellular potassium concentration changes during potassium flux through brain tissue, Journal of physiology, 1978, 234: 28P-39.

[19] Brew, H., and Attwell, D., Is the potassium channel distribution in glial cells optimal for spatial buffering of potassium? Biophys J, 1985, 48: 843-847.

[20] Orkand, R K., Nicholls, J G., and Kuffler, S W. Effect of nerve impulses on the membrane potential of glial cells in the central nervous system of amphibia, Journal of neurophysiology, 1966, 29: 789-806.

[21] Newman, E A. High potassium conductance in astrocyte endfeet, Science, 1986, 233: 453-454

[22] Korotzer, A R., and Cotman, C W. Voltage-gated currents expressed by rat microglia in culture, 1992, Glia, 6: 81-88

[23] Le Dain, A C., Anderton, P J., Martin, D K., and Millar, T J. A tetraethylammonium-insensitive inward rectifier K^+ channel in Muller cells of the turtle (Pseudemys scripta elegans) retina, J Membr Biol, 1994, 141: 239-245.

[24] Sontheimer, H. Voltage-dependent ion channels in glial cells, Glia, 1994, 11: 156-172.

[25] Tse, F W., Fraser, D D., Duffy, S., and MacVicar, B A. Voltage-activated K^+ currents in acutely isolated hippocampal astrocytes, J Neurosci, 1992, 12: 1781-1788.

[26] Keller, O., Ujec E., Pavlík, and Machek J. Potassium changes elicited in the feline somatosensory cortex by a single stimulus to the peripheral nerve, In: Syková, E., Hník, P., and Vyklický, L. Ion-selective microelectrodes and their use in excitable tissue, New York, Plenum, 1981.

Magnetic Field Tomography: Theoretical Basis, Performance Measures and Limitations

Ioannides, A.A.[1,2], Bolton, J.P.R.[1], Liu, M.J.[1], Bamidis, P.D.[1], Liu, L.C.[2], Dammers, J.[1,2] and Müller-Gärtner, H.W.[2]

[1]*Physics Department, The Open University, Milton Keynes MK7 6AA, UK;* [2]*Institute of Medicine, Research Centre Jülich, D-52425 Jülich, Germany*

1 Introduction

A general method for obtaining continuous solutions to the biomagnetic inverse problem was first presented during the 1989 Biomagnetism conference. From the outset, the method was developed for three-dimensional source spaces [1, 2] but, partly to reduce computational demands and partly for ease of presentation, the first major publication of the method [3] used two-dimensional reconstructions. Soon afterwards the algorithms were ported to a transputer, making possible the analysis of large sets of MEG data. The output (estimates of the primary current density in a three-dimensional source space) was displayed by taking slices (tomes) through the source space leading to a series of MFT (Magnetic Field Tomographic) images. Later forms of representation included time integrals and activation curves, describing the evolution of activity within specified Regions of Interest (ROI). Animations run on the transputer array and recorded on video, were by far the most effective presentation tool, showing in colour-coded form the changes in the activity against a background of coregistered MRI slices[4]. Early reconstructions used averaged data but the emphasis has recently shifted to analysis of single trial or continuous data [5, 6]. Each three-dimensional study was underpinned by thorough tests with computer generated data which, however, have been only superficially reported because of space limitations. We have previously described our philosophy and method [3], the transputer based implementation [7] and the logic in the various algorithmic steps [8]. However, the lack of published material on the background tests must be the cause of misrepresentations of MFT in some recent publications dealing with distributed source analysis, e.g. [9]. The present paper makes good this omission and uses computer-generated data to illustrate the power and limitations of MFT as a 3D imaging method, using realistic sensor configurations and signals that are neurophysiologically realistic in terms of location and temporal characteristics.

2 Methods

MFT algorithm: Very schematically, the expectation value of the primary current density is restricted within a source space S, and it is expanded in terms of lead fields $\phi^{(k)}$ associated with each of the n detectors:

$$\langle J(r)\rangle = \sum_{k=1}^{n} a_k w(r)\phi^{(k)}(r)$$

where $w(r)$ is an a-priori weight function chosen to enfold physiological knowledge or to compensate for known experimental insensitivities. The coefficients a_k are found by inverting the equations:

$$\sum_{k=1}^{n} P_{ik}m_k = \sum_{jk}(P_{ij}P_{jk} + \zeta P_{ik})\,a_k \qquad \text{with} \qquad P_{ik} = \int_S \phi^{(i)}(r)\cdot\phi^{(k)}(r)w(r)d^3r,$$

where the m_k are the detected signals and ζ is a regularisation parameter which strikes a compromise between improved spatial resolution and insensitivity to noise. This basic algorithm, or a very similar variant, is used in practically every distibruted source method proposed so far. In MFT three additional procedures are introduced. First, the weight function and the regularization parameter are optimised (trained) according to the given experimental arrangement; second, a one step iteration is applied to improve localisation; finally the computed current density may be integrated over space or time to construct whatever MFT display is required in the study.

MFT training: An MFT signal analysis is preceded by tuning of the inversion procedure to the experimental arrangement. First, the conducting volume is chosen. In the case of brain studies we generally use a sphere fitted to the curvature of the inner skull in the region closest to the sensors. The source space refers only to the primary current density so need not be the same as the conducting volume: we have used both cylinders and hemispheres. In order to compensate for experimental insensitivity to deep sources the weight function is generally taken to have the form $w(r) = e^{-(|r-r_0|/\lambda)^2}$, where r_0 is the centre of the conducting sphere and λ is a decay length. In order to select the decay length and regularisation parameter we generate a set of current dipoles distributed throughout the source space and solve the forward problem to provide a simulated data set. We then add random noise and invert for a range of λ and ζ, selecting the optimum values which lead to the best dipole reconstructions throughout the source space.

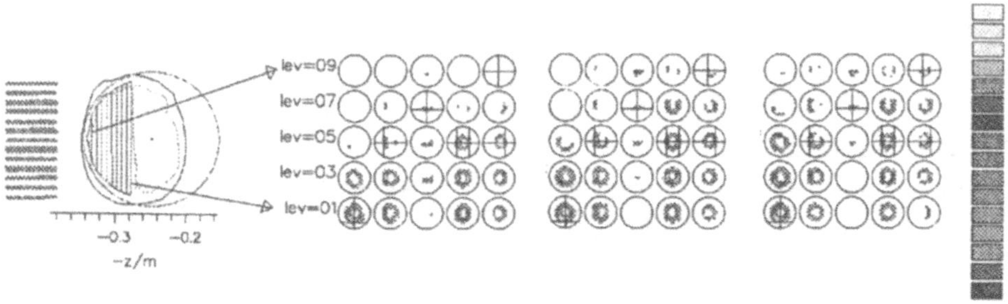

Figure 1: Choosing an appropriate value of λ. The left-hand diagram shows the conducting volume (circle), the hemispherical source space (6.1 cm thick) and the sensor positions in the KRENIKON system. 10 % random signal noise is added to signals. Inverse solutions are displayed as contour plots of $|J|^2$ for a range of dipole positions (crosses) and three different values of λ. For display purposes all tomes have been scaled to have the same radius.

Figure 1 shows one stage in the training process for the 37-channel Siemens KRENIKON system and a hemispherical source space. Inversions are shown for three values of λ. The left-hand diagram shows the effect of taking too small a value ($\lambda = 4.5$ cm). In this case deep sources are correctly recovered, but sources in middle and superficial layers extend over too great a depth. The right-hand diagram shows the effect of taking too large a value ($\lambda = 6.0$ cm): superficial sources are successfully recovered but sources in deeper levels (especially middle levels) are recovered at depths that are too small. A suitable compromise is given by the central diagram, for which $\lambda = 5.3$ cm. A similar process is carried out for the regularization parameter. The essential point is that having "trained" our inversion algorithms the subsequent analysis of experimental data is carried out blindly across all time slices with no adjustment of parameters.

3 Two examples chosen to challenge MFT

The first example simulates activity from the hippocampus in the presence of superficial activity in the parietal cortex (Figure 2). The smooth curve in the middle (bottom) panel on the right shows the activation profile of the "true" generator in the parietal (hippocampal) ROI. Random 10 % noise was added before the signal was subjected to MFT analysis. The left part of the diagram shows the MFT image at the parietal and hippocampal level, with $|J|^2$ integrated over the time interval between the dashed vertical lines in the corresponding panel on the right. The top panel on the right shows MFT estimates of the activity from each ROI, integrated over the ROI's marked on the left. In the lower two panels the MFT activation is compared with the "true" activation. As expected, MFT can recover each source when it is solely active. More impressively, when the sources are activated simultaneously, as between 55-80 ms, it is possible to recover the peak of the hippocampal activation between the two broader parietal activations. The limits of the technique, as currently implemented, are also apparent: for example, around 100 ms and 140 ms peaks in the parietal activity produce small blips in the hippocampal reconstruction. Moreover, not all the peak maxima are fully recovered - especially for the deep source.

The second example involves simultaneous bilateral measurements taken with the 2x37 channel BTi MAGNES system. This time, a cylindrical source space, shifted in depth is used. We often use this approach to home in

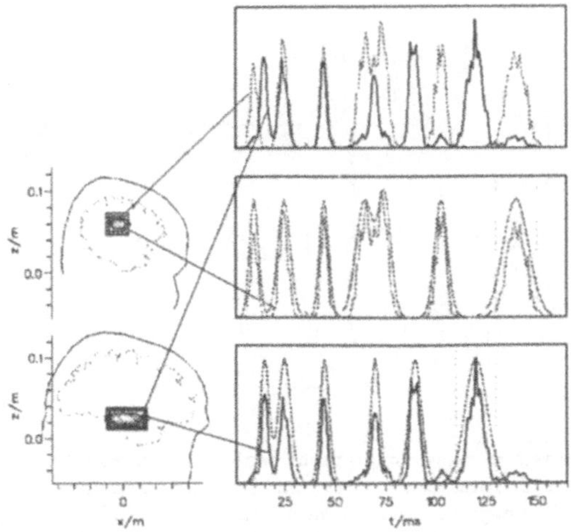

Figure 2: A simulated inversion using unilateral measurements. The bottom graph compares the source activation and $|J|^2$ integrated over the region indicated in the hippocampal region. The middle graph does the same for the parietal region and the upper graph gives a direct comparison between the two solutions. Vertical scales are in arbitrary units.

on deep activity, identified with a wider source space [6]. The sensor configuration is taken from a recent odd ball study, and the source modelling corresponds to activations in the left and right hippocampal regions. Each hippocampus is modelled by a dipole with activation profile displayed by the smooth light curves in the top (right hippocampus) and bottom (left hippocampus) panels of Figure 3. The MFT estimates for the activation profiles from the same areas are shown by the thicker and darker traces. The middle panel shows the activity integrated within each cylinder level and displayed as a contour plot of average intensity (across depth and time). Notice, the behaviour of the activation curves around 150 ms as compared with that around 260 ms. In the former case, a sharp (5 ms time constant) left hippocampal activation occurs together with the much broader (30 ms time constant) activation of the right hippocampus, while in the latter, both hippocampal regions undergo a broad activation (30 ms time constant).

4 Discussion and outlook

Our method is made for distributed sources and, of course, finds it easier to reconstruct superficial rather than deep activity, but our examples (based on localised sources with deep activity) show that the method can cope with a wide variety of sources, avoiding specific preconceptions about their number, position or localisation; unlike similar ideas recently introduced [9], our method does not rely on initial guesses. We stress again that training does not prejudice the inversion outcome; for example, (λ and ζ) are held constant throughout all time slices in Figures 2 and 3 while the activity shifts dramatically from one region to another. Another strength is the freedom for post-inversion processing, afforded because each timeslice is processed completely independently from each other, and in exactly the same way (determined by the prior training).

Our computer generated data correspond to real experiments we have analysed, generally involving 37 channel probes. The availability of helmet probes, covering the entire head, provides new opportunities and we are currently modifying our software to exploit them. We have emphasised the procedures we have used routinely during the last few years, but we note in passing that many refinements are now in progress. For

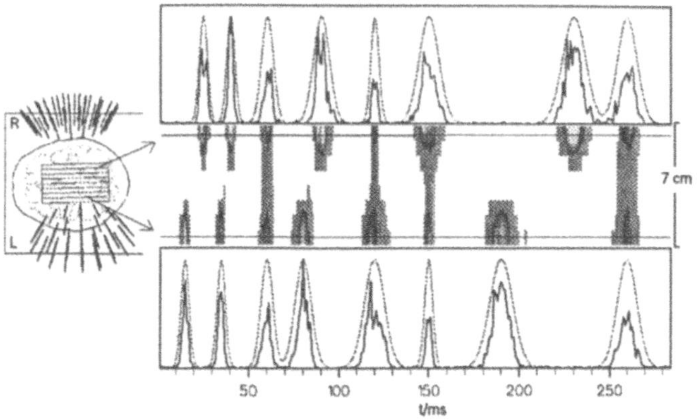

Figure 3: A simulated inversion using simultaneous bilateral measurements. The left-hand diagram shows sensor positions and the cylindrical source space. The bottom panel compares activation of a dipole in the left hippocampus with $|J|^2$ integrated over left-most layer of the cylinder. The top panel does the same for the right hippocampus. The central diagram further indicates the extent of localisation of activity. Random 10 % noise has been added prior MFT analysis.

example, the source space can be modelled closer to anatomy; time evolution can be constrained by (rather sparse) electrophysiological information; anisotropic or sensor-dependent weights can be used and the source space itself can be changed iteratively [6].

References

[1] Clarke CJS, Ioannides AA and Bolton JPR (1989), Localised and distributed source solutions for the biomagnetic inverse problem I. In: Advances in Biomagnetism, S.J. Williamson, M. Hoke, G. Stroink and M. Kotani (Eds.). Plenum Press, New York, pp 587-590.

[2] Ioannides AA, Bolton JPR, Hasson R and Clarke CJS (1989), Localised and distributed source solutions for the biomagnetic inverse problem II. In: Advances in Biomagnetism, S.J. Williamson, M. Hoke, G. Stroink and M. Kotani (Eds.). Plenum Press, New York, pp 591-594.

[3] Ioannides AA , Bolton JPR and Clarke CJS (1990) Continuous probabilistic solutions to the biomagnetic inverse problem. Inverse Problems, 6:523–542.

[4] Ioannides A.A. Searchlights into the brain. Open University video, August 1993.

[5] Liu LC and Ioannides AA. Single epoch analysis of MEG signals In *BIOMAGNETISM: Fundamental Research and Clinical Applications* (Elsevier, Amsterdam) 439–444.

[6] Bamidis PD, Hellstrand E, Lidholm H. ,Abraham-Fuchs K and Ioannides AA (1995) MFT in complex partial epilepsy: Spatio-temporal estimates of interictal activity. NeuroReport Neuroreport 7(1):17-23

[7] Liu MJ, Hasson R, and Ioannides AA (1993) "A transputer-based system for Magnetic Field Tomography". In: *Transputer Applications and Systems '93*, R. Grebe, J. Hektor, S.C. Hilton, M. Jane and P.H. Welch (eds.). IOS Press, Amsterdam, Vol. 2, pp. 1290-1297.

[8] Ioannides, A.A. Estimates of 3D Brain activity ms by ms from Biomagnetic Signals: Method (MFT), Results and their Significance *Quantitative and Topological EEG and MEG Analysis* Eiselt, M., Zwiener, U. and Witte, H. (Eds.) Universitätsverlag Jena. pp. 59–68 1995.

[9] George J.S., Lewiss P.S., Schlitt H.A., Kaplan L., Gorodnitsky I., and Wood C.C. Strategies for source space limitation in Tomographic Inverse Procedures. In *BIOMAGNETISM: Fundamental Research and Clinical Applications* (Elsevier, Amsterdam) 357–362.

Algorithm for Magnetocardiography Analysis Based on Multi Channel SQUID and Thoracic MR Images

Ishii, H.[1], Niki, N.[1], Nakaya, Y.[2], Nishitani, H.[2] and Kang, Y.M.[3]

[1]*Faculty of Engineering, University of Tokushima, Tokushima, Japan;* [2]*Medical School, University of Tokushima, Tokushima, Japan;* [3]*MEC Laboratory, DAIKIN Industries, Limited, Tsukuba, Japan*

INTRODUCTION

In recent years, the development of SQUID for biomedical measurement progresses rapidly. Since biomedical measurement is noninvasive measurement, it is considered that it reduces the patient's danger and burden remarkably, compared to the estimation of functional region using customary invasive measurement. It is considered that localization on time of extending the myocardial excitement is needed for heart disease diagnosis and to know their corresponding medical treatments. Now, magnetocardiogram is useful for myocardial infarction, bundle-branch block,cardiomegly, premature ventricular contraction and so on[1].

In this paper,we present a method for the localization of the heart magnetic sources, based on its magnetic field obtained by multi channel SQUID magnetometer. We estimate not only the localized excitement region but also the spread of the myocardial excitement at each time. We extract the myocardial regions from MR Images and we apply MUSIC algorithm[2][3] to the magnetic data obtained from multi channel SQUID magnetometer. Then, we investigate the distribution of magnetic sources based on anatomic information from MR Images.

METHODS

At first, we present the extraction of myocardial region from MR Images. As a first step, we make the binary images using Fuzzy C Means algorithm. However, these images contain the heart region and other regions like fat, bone and so on. So we make labels of the binary images to select the interested regions. As a second step, we search the body area in the labeled images by scanning the image from its four sides as shown in a result image of Fig. 1 (b). It is anatomically known that the heart region exists on a little upper and a little right from the center of this body area. So we make the square, which is centered there, and we consider that the biggest labeled region in this square is the region that contains the heart. Furthermore, we classify it into three classes as myocardial region, ventricle, atrium, and elsewhere based on the corresponding MR values using again Fuzzy C Means algorithm, and the result is shown in Fig. 2. Then we can extract the myocardial region as shown in Fig. 3.

It is known that the cardiac magnetic field is the induction of the heart voltage. If magnetic sources are defined as electric dipoles upon the half-infinite-plane model[3], magnetic field is shown by Biot Savart law as $\mathbf{B} = \mathbf{GQ}$. Then, we use MUSIC algorithm to estimate the electric dipole's location. Now, we assume that electric sources are n electric dipoles. If we measure the magnetic field which has k time samples using m sensors, Biot Savart law will be expressed as follows :

$$\begin{bmatrix} b_{1,1} & \cdots & b_{1,k} \\ \vdots & \ddots & \vdots \\ b_{m,1} & \cdots & b_{m,k} \end{bmatrix} = \begin{bmatrix} g_{1,1}^T & \cdots & g_{1,n}^T \\ \vdots & \ddots & \vdots \\ g_{m,1}^T & \cdots & g_{m,n}^T \end{bmatrix} \begin{bmatrix} q_1^T \\ \vdots \\ q_n^T \end{bmatrix} \tag{1}$$

When we estimate the electric dipoles' location and moment from measurement magnetic fields B_m, we use the following evaluation function :

$$\arg\min\|B_m - GQ\|^2 \tag{2}$$

If the electric dipoles are fixed dipoles which don't change both location and direction, formula (1) can be changed as follows :

$$\begin{bmatrix} b_{1,1} & \cdots & b_{1,k} \\ \vdots & \ddots & \vdots \\ b_{m,1} & \cdots & b_{m,k} \end{bmatrix} = \begin{bmatrix} g_{1,1}^T & \cdots & g_{1,n}^T \\ \vdots & \ddots & \vdots \\ g_{m,1}^T & \cdots & g_{m,n}^T \end{bmatrix} \begin{bmatrix} m_1 0 \cdots 0 \\ 0\ m_2 \cdots 0 \\ \vdots \ \ \ddots \ \vdots \\ 0\ 0 \cdots m_n \end{bmatrix} \begin{bmatrix} s_{1,1} & \cdots & s_{1,k} \\ \vdots & \ddots & \vdots \\ s_{n,1} & \cdots & s_{n,k} \end{bmatrix} = \begin{bmatrix} a_{1,1} & \cdots & a_{1,n} \\ \vdots & \ddots & \vdots \\ a_{m,1} & \cdots & a_{m,n} \end{bmatrix} \begin{bmatrix} s_{1,1} & \cdots & s_{1,k} \\ \vdots & \ddots & \vdots \\ s_{n,1} & \cdots & s_{n,k} \end{bmatrix} \tag{3}$$

Where, **G** means *m*3n* matrix of location, **M** means *3n*n* matrix of direction, **S** means *n*k* matrix of intensity and **A** means *m*n* matrix of electric dipole information. Formula (2) can be changed as follows using this connection :

$$\arg\min\|B_m - AS\|^2 \tag{4}$$

We use MUSIC algorithm to solve the formula (4). As a first step, we make sample covariance matrix R_B from measurement field B_m, and we make the eigen decomposition as follows :

$$R_B = \frac{1}{k} B_m B_m^T = [E_S\ E_N] \begin{bmatrix} \Lambda_S & \\ & \Lambda_N \end{bmatrix} \begin{bmatrix} E_S^T \\ E_N^T \end{bmatrix} \tag{5}$$

Second step, we divide the eigenvalues into the signal components and the noise components. Then, we minimize the following cost function using the eigenvectors E_N which correspond to the noise components.

$$J = \frac{|E_N^T A|^2}{|A|^2} = \frac{|E_N^T GM|^2}{|GM|^2} \tag{6}$$

At the above formula, if we recognize the matrix G, matrix M is the eigenvector which corresponds to the minimum eigenvalue of matrix $G^T E_N E_N^T G$.

We need to adjust the SQUID's coordinate and MR Image's coordinate to project the electric dipole, which is estimated using MUSIC, onto the MR Images. Therefore, the sensors on the SQUID system are placed as Fig. 4. Here, the sensor of a double circle in Fig. 4 is matched the subject's xiphisternum, and the localizational markers are attached on MR Images. Then, we transform the coordinate based on these markers.

RESULTS

In this paper, magnetic field data,which is used for estimating the electric dipole, is obtained from 32 channel SQUID system. We show the SQUID data's wave in Fig 5. Here, the upper left wave is the secondary inducing voltage of the heart. We show the SQUID data specifications at Table 1 and the photographing conditions of MR Images at Table 2.

Table 1. Specifications of SQUID

Pickup Coil	First Order Gradiometer
Sampling Rate	1,000 Hz
Data Length	180,000 Points (3min.)
Filter	500 Hz Low Pass Filter

Table.2 Photographing conditions of MR Images

Flip Angle	30 degrees	Scan Thickness	10.0 mm
TE	12 msec.	Interscan Space	0.3 mm
TR	21 msec.	Time Sequence	16
Field of View	32 cm square	Number of Slice	15

It took about 1 minute for one 2D image to extract the myocardial region using former manual tracing. So it took a long time to extract in all images. And manual tracing depends on observer's skill. On the other hand, it takes several seconds for 1 image using our auto extracting algorithm.

When we estimate the electric dipole using MUSIC algorithm, we need to divide the matrix $\mathbf{R}_B$ into signal components and noise components. So we determine the number of electric dipole using MDL criterion and decompose the matrix and we apply MUSIC algorithm. Then, we project the estimated electric dipoles distribution,which is shown in Fig. 6.

CONCLUSION

In this paper, we present an algorithm for extracting the myocardial region and localizing the electric dipoles. Our extracting algorithm, which extracts the myocardial region, is faster than former manual tracing and it reduces our labor. Then, we estimate the spread of the myocardial excitement at each beat. These estimation are based for the thoracic MR Images and MUSIC, and we measure the activity of the heart functional regions. And we visually grasp the active states of the heart by displaying the obtained images continuously.

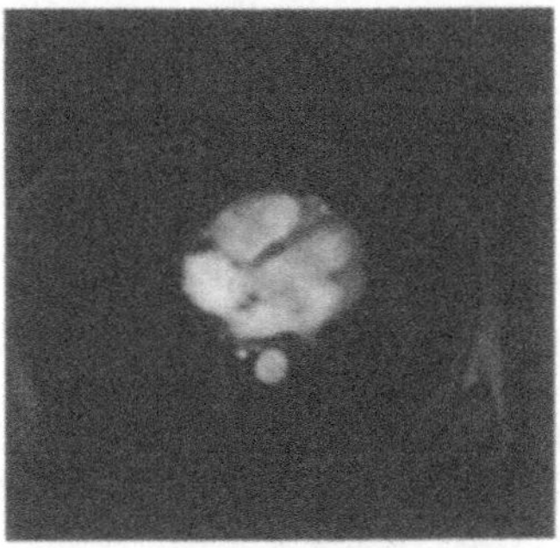

(a) 2D MR Image

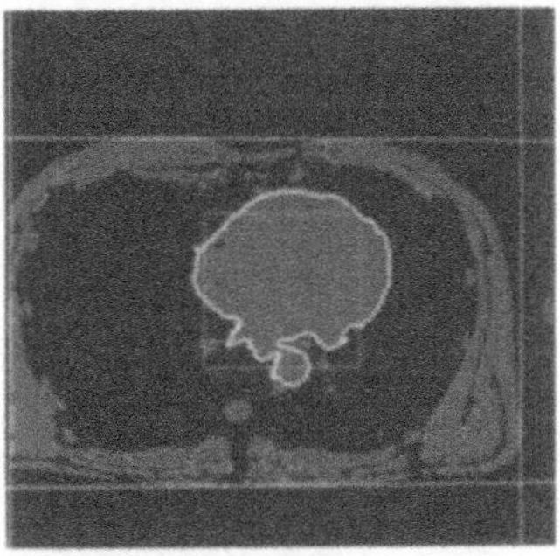

(b) Body area on binary image

Fig 1. MR Image and body area in labeled image

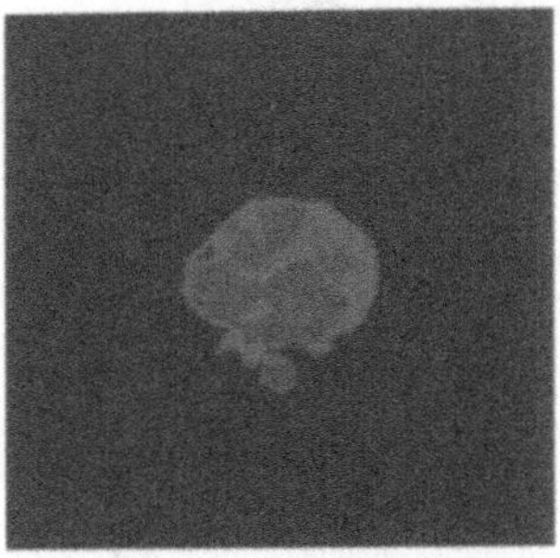

Fig. 2 Classify the heart region

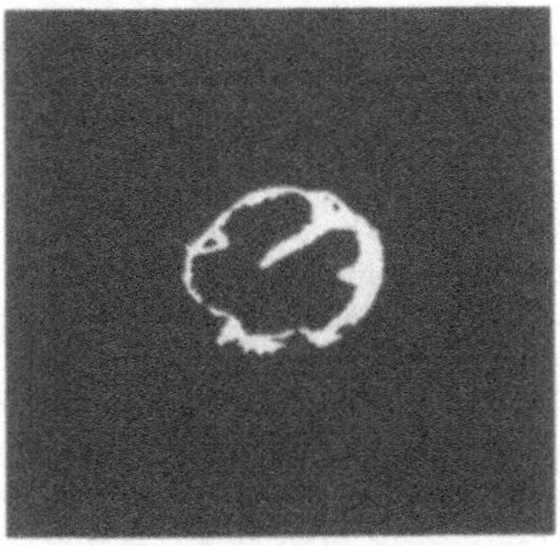

Fig. 3 Extracted myocardial region

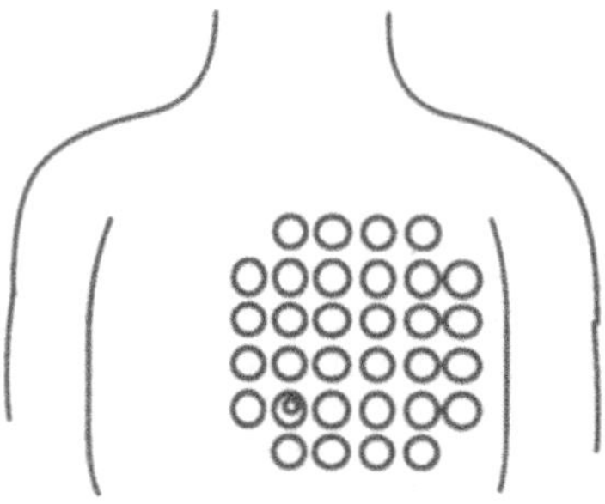

Fig 4. Placement of SQUID sensor

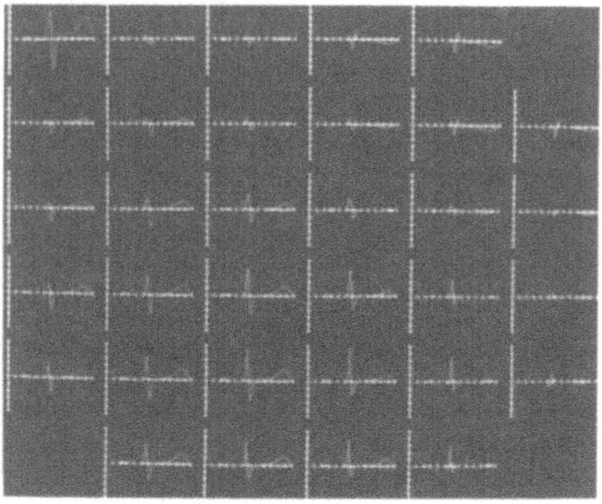

Fig 5. SQUID data from 32channel system

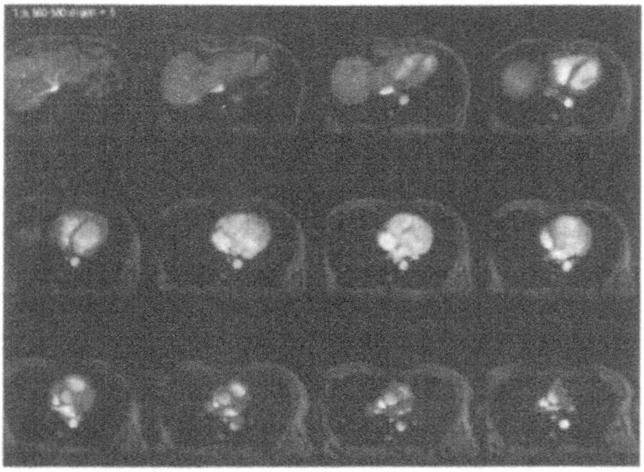

Fig 6. Estimated electric dipoles distribution at adjacent of the R wave

REFERENCES

[1] Nenonen, J., Katila, T., Leinio, M., et al. Magnetocardiographic Functional Localization Using Current Multiple Models, IEEE Trans. Biomed. Eng., 1991, 38, 7: 648-657

[2] Mosher, J. C., Lewis, P. S., Leahy, R. M., Multi Dipole Modeling and Localization from Spatio-Temporal MEG Data, IEEE Trans. Biomed. Eng., 1992, 39, 6:541-557

[3] Kudo, H., Maemura, T., Saito, T., Signal Source Localization from Spatio-Temporal Biomagnetric Data by Signal Subspace Method, Trans. IEICE, Japan, D-II, 1995, 78, 3:559-579

[4] Sarvas, J., Basic Mathematical and Electromagnetic Concepts of the Biomagnetic Inverse Problem, Phys. Med. Biol., 1987, 1:11-22

Estimation of Distributed Sources of Auditory Evoked Magnetic Fields in the Human Brain

Iwaki, S. and Ueno, S.
Institute of Medical Electronics, University of Tokyo, Tokyo 113, Japan.

Introduction

In recent years, many studies have been made on the techniques to obtain internal electrical source distributions in the human brain from surface measurements of potentials (electroencephalogram: EEG) or magnetic fields (magnetoencephalogram: MEG)[1][2]. However, in principle, because of the non-uniqueness of the inverse solutions, these bio-electric or bio-magnetic inverse problems cannot be solved in general. Furthermore, the existence of measurement noise makes it harder to obtain internal source distributions with precision. In neuromagnetic studies, we have to impose some constraints based on a priori information on physiological structures in order to estimate the internal electrical sources. We investigated the spatio-temporal properties of traveling electrical sources in the human brain using a sub-optimal least-squares subspace scanning technique[3].

In this study, we focused on the N1m and P2m components of auditory evoked magnetic fields which appear at the contralateral auditory cortex at about 100 msec for N1m and 170 msec for P2m after the onset of the auditory stimulus. Rogers et al. reported that the internal electrical sources of N100 propagate along the adjacent cortical columns within the primary projection area of the temporal lobe[4]. In this paper, we investigated the spatio-temporal distribution of the internal electrical sources associated with auditory N100 and P2 responses.

Methods

1. Forward Modeling

Normal components of the magnetic fields measured at *i*-th sensor position relating to a single current dipole can be written as:

$$B(i) = \vec{B}(i)\cdot\hat{n}(i) = \frac{\mu_0}{4\pi}\frac{\vec{Q}\times(\vec{R}(i)-\vec{L})}{|\vec{R}(i)-\vec{L}|^3}\cdot\hat{n}(i) = \frac{\mu_0}{4\pi}\frac{(\vec{R}(i)-\vec{L})\times\hat{n}(i)}{|\vec{R}(i)-\vec{L}|^3}\cdot\vec{Q} = \vec{g}(i)\cdot\vec{Q} \tag{1}$$

where $\vec{L}$ is the dipole position, $\vec{R}(i)$ is the *i*-th sensor location, $\hat{n}(i)$ is the unit orientation vector of *i*-th sensor, $\vec{Q}$ is the dipole moment, μ_0 is a constant ($4\pi\times10^{-7}$ T· m/A), $\vec{B}(i)$ is the magnetic flux density at *i*-th sensor position, and B(i) is the normal component of the magnetic flux density measured at *i*-th sensor. For the measurements at *m* measurement locations and *p* dipoles, we can extend equation(1) in a matrix form:

$$\begin{bmatrix} B(1) \\ \vdots \\ B(m) \end{bmatrix} = \frac{\mu_0}{4\pi}\begin{bmatrix} \frac{(\vec{R}(1)-\vec{L}_1)\times\hat{n}(1)}{|\vec{R}(1)-\vec{L}_1|^3} & \cdots & \frac{(\vec{R}(1)-\vec{L}_p)\times\hat{n}(1)}{|\vec{R}(1)-\vec{L}_p|^3} \\ \vdots & & \vdots \\ \frac{(\vec{R}(m)-\vec{L}_1)\times\hat{n}(m)}{|\vec{R}(m)-\vec{L}_1|^3} & \cdots & \frac{(\vec{R}(m)-\vec{L}_p)\times\hat{n}(m)}{|\vec{R}(m)-\vec{L}_p|^3} \end{bmatrix}\begin{bmatrix} \vec{Q}_1 \\ \vdots \\ \vec{Q}_p \end{bmatrix} = \begin{bmatrix} \mathbf{G}_1 & \cdots & \mathbf{G}_p \end{bmatrix}\begin{bmatrix} \vec{Q}_1 \\ \vdots \\ \vec{Q}_p \end{bmatrix} \tag{2}$$

where $\vec{L}_j$ is the j-th dipole position, $\vec{Q}_j$ is the j-th dipole moment, and $[\, G_1 \cdots G_p \,]$ is the gain matrix of the forward model.

2. Inverse Modeling using a Sub-optimal Least-squares Subspace Scanning Method

Next, to deal with spatially distributed source estimation, we used a linear parameter estimation framework with a sub-optimal least-squares subspace scanning method[3]. In this method, the following assumptions are required: (i) Gain matrix $[G_1 \ldots G_p]$ in equation (2) is of full column rank, (ii) Time series for different source components are asynchronous or linearly independent, (iii) Noise is considered temporally and spatially white. For m sensors and n time slices, we calculated the auto-correlation ($\mathbf{R}_F$) of $m \times n$ data matrix $\mathbf{F}$, performed eigen-decomposition of $\mathbf{R}_F$, and selected a separation point between signal and noise eigen-values.

$$\mathbf{R}_F = \frac{1}{n} \left(\mathbf{F}(i)\, \mathbf{F}(i)^T \right) \tag{3}$$

$$\mathbf{R}_F = \Phi \Lambda \Phi^T = \begin{bmatrix} \Phi_s & \Phi_n \end{bmatrix} \begin{bmatrix} \Lambda_s & \\ & \Lambda_n \end{bmatrix} \begin{bmatrix} \Phi_s & \Phi_n \end{bmatrix}^T \tag{4}$$

where Λ_s was a diagonal matrix containing larger (signal) eigenvalues, Φ_s was a set of eigenvectors corresponding to Λ_s (signal subspace), Λ_n was a diagonal matrix containing smaller (noise) eigenvalues which were, by assumption (iii), equal to the variance of noise, and Φ_n was a set of eigenvectors corresponding to Λ_n (noise subspace). Then, over fine scanning grids of three-dimensional locations in the head model, we calculated the gain matrix $\mathbf{G}_i$, obtaining principal left eigenvectors U_{G_i}, using singular-value decomposition (SVD) of $\mathbf{G}_i$:

$$\mathbf{G}_i = \mathbf{U}_{G_i} \Sigma_{G_i} \mathbf{V}_{G_i}^T \tag{5}$$

and evaluating cost function $J_h(i)$:

$$J_h(i) = \lambda_{min} \{ \mathbf{U}_{G_i}^T \Phi_n \Phi_n^T \mathbf{U}_{G_i} \} \tag{6}$$

which was a minimum eigenvalue of the matrix $\{U_{G_i}^T \Phi_n \Phi_n^T U_{G_i}\}$. This cost function $J_h(i)$ becomes 0 at the correct dipole location [5], and $1/J_h(i)$ can be used as the criterion of source existence.

Results and Discussion

1. Verifications of the Subspace Scanning Source Estimation Method

To verify the sub-optimal least-squares subspace scanning technique, we executed two computer simulations: a source estimation for spatially concentrated (single dipole) sources, and a source estimation for spatially distributed sources.

First, we executed a source estimation for spatially concentrated (single dipole) sources. We placed a single dipole at a depth of 0.60R (R:radius of the head model) as shown in Fig.1(a). The SNR (Signal-to-noise ratio) of the magnetic field due to this single dipole source was set to 12.5 dB (nearly equals to our SQUID system). The corresponding distribution of $1/J_h(i)$ (estimated source distribution) is shown in Fig.1(b).

Next, we executed a source estimation for spatially distributed sources. Fig.2(a) shows an example of a

Results

Ten single dipoles at various locations on the heart were simulated with and without noise. Fig. 4 depicts some reconstructed current densities. Fig. 3 shows the results for typical examples. The FWHM of the PSF is in the range of 20mm without noise and increases up to 60mm with 20 fT/√Hz. Radial dipoles (with respect to the body surface) can be found in the noiseless case only.

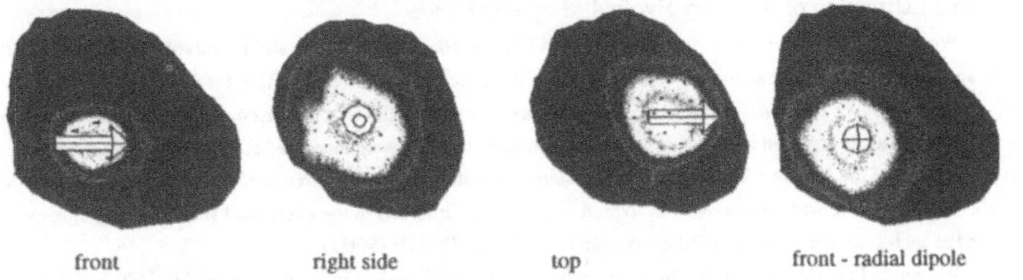

Fig. 4: Examples of reconstructed current densities of single current dipoles (without noise).

Two current dipoles can be clearly separated in the noiseless case provided their distance is larger than 30mm. Fig. 5 gives some examples.

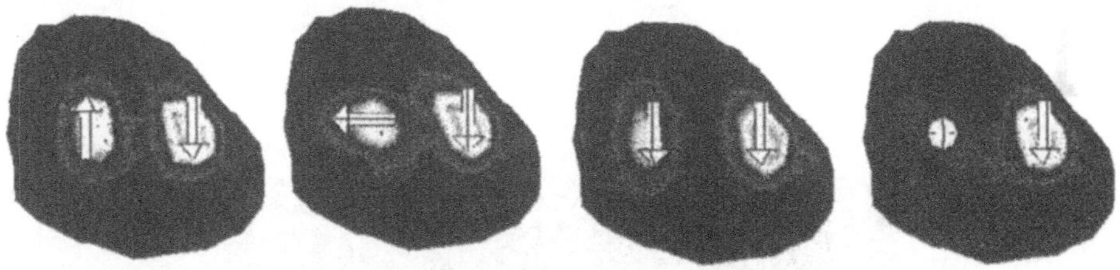

Fig. 5: Examples of reconstructed current densities of two current dipoles with a distance of 30mm

Activated rings around the heart can be reconstructed, but some artifacts show up. Obviously the simulated magnetic flux density can be reproduced with current distributions having a smaller integral current than the original one. (The minimum norm estimate picks the solution with the smallest currents!) Fig. 6 gives an example.

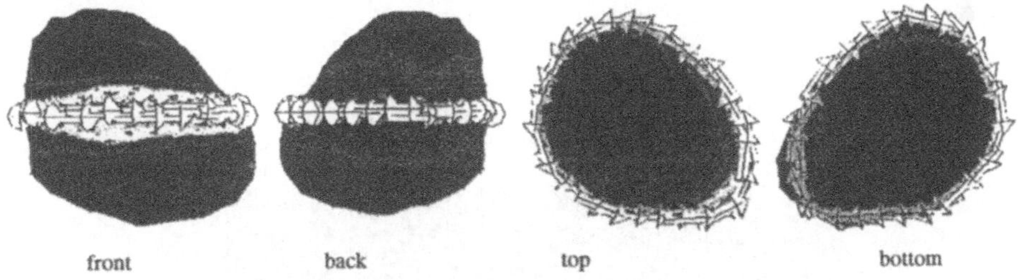

Fig. 6: Reconstructed current density of a ring of dipoles

distributed source model consisting of multiple dipoles distributed in the small area ($-0.05R < x < 0.05R$, $y = 0.00R$, $0.55R < z < 0.65R$). Again the SNR of the field due to this source was set to 12.5dB. The corresponding distribution of $1/J_h(i)$ is shown in Fig.2(b).

In each case, the spatial distribution of actual electrical sources was reasonably estimated by this method.

2. Source Estimation of Auditory Evoked Magnetic Fields

We focused on the source estimation of the N100 and P2 components of auditory evoked magnetic fields which appear at the contralateral auditory cortex at about 100msec for N100 and 170 msec for P2 after the onset of the auditory stimulus. We measured MEG activities with a seven channel DC-SQUIDsystem using first-order gradiometers in a magnetically shielded room. Auditory evoked magnetic fields were measured at 28 points over the contralateral hemisphere of the subject's brain, changing the sensing position repeatedly. We used 1.0 msec clicks as auditory stimuli. Masking noise was continuously delivered to the ear contralateral to the stimulated side. 500 responses were averaged at each sensor position after filtering with a band pass filter (1-150Hz)

We applied the sub-optimal least-squares subspace scanning method in the vicinity of the N100 and P2 peaks

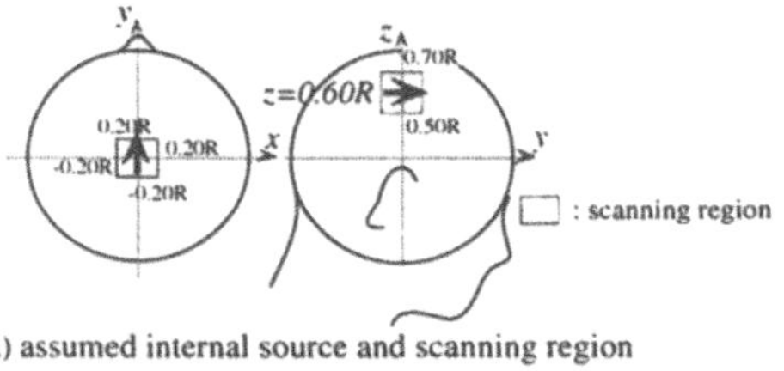

(a) assumed internal source and scanning region

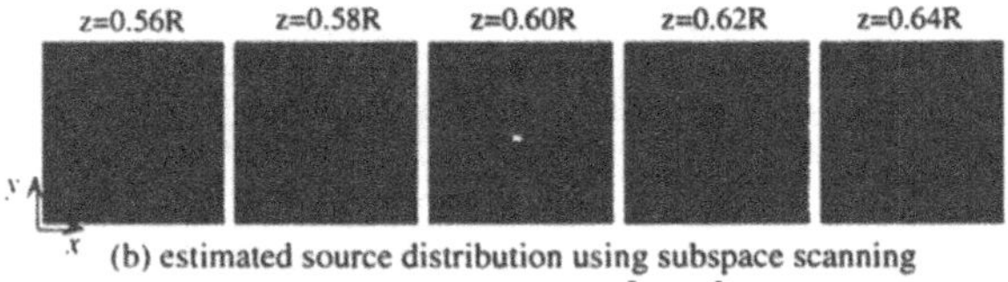

(b) estimated source distribution using subspace scanning

Fig.1 Simulation of spatially concentrated source estimation using subspace scanning

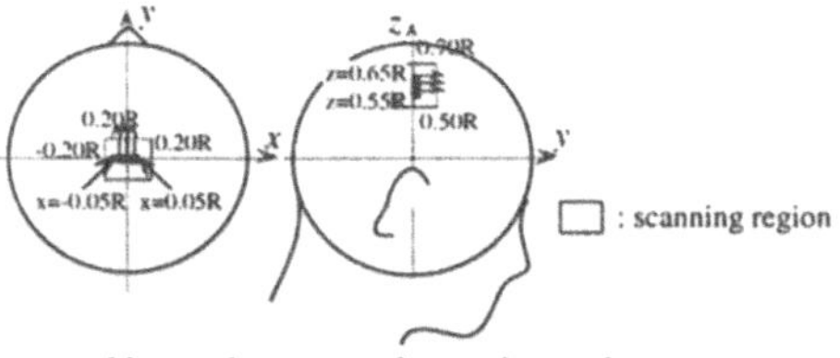

(a) assumed internal source and scanning region

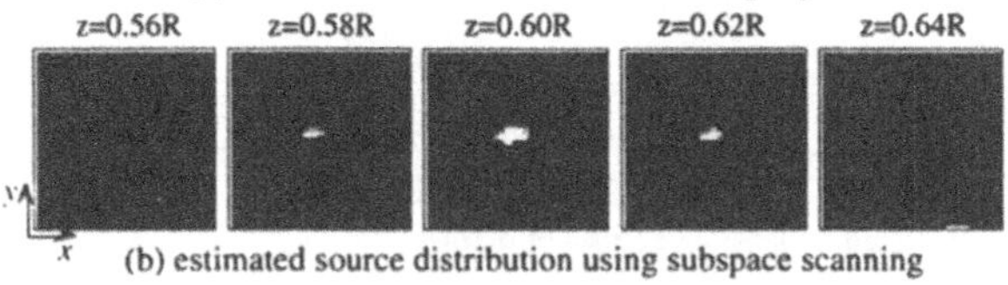

(b) estimated source distribution using subspace scanning

Fig.2 Simulation of spatially distributed source estimation using subspace scanning

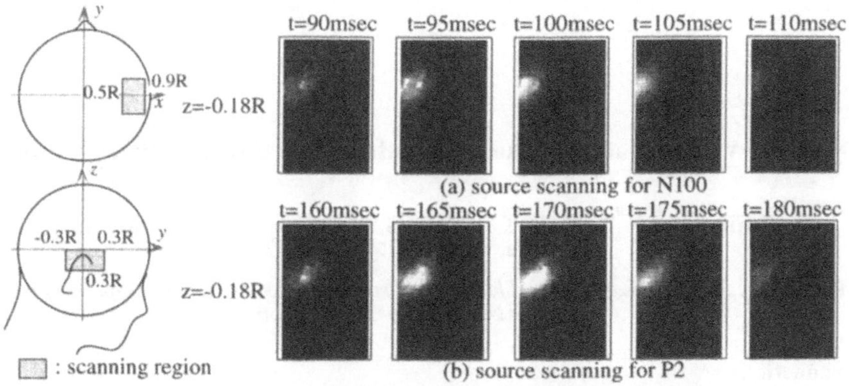

Fig.3 Results of source estimation using subspace scanning method

at intervals of 5msec. The results are shown in Fig.3(a) for N100 and (b) for P2. At the peak latencies (100msec for N100 and 170 msec for P2), estimated electrical sources showed similar distribution patterns. Furthermore, change of the estimated source distribution for N100 and P2 responses showed a similar pattern sequence with respect to the starting area and the direction of the propagation. These results indicate that the neural excitations which related to auditory N100 and P2 responses occur in the same or neighboring group of neuron in the human auditory cortex.

Conclusions

We estimated the traveling distributed electrical sources which represent the propagation of neural excitations in the human brain using a sub-optimal least-squares subspace scanning technique. The results show that the internal sources relating to auditory N100 and P2 responses have similar distribution patterns, and that the distributed sources of auditory N100 and P2 travel in the same direction within virtually the same region in the auditory cortex. It is suggested that two traveling auditory neural excitations with different latencies propagate along the same or neighboring group of neuron in the auditory cortex.

References:

[1] Ilmoniemi, R.J. Models of source currents in the brain. Brain Topography, 1993, 5: 331-336.

[2] Wang, J.Z., Williamson, S.J. and Kaufmann, L. Magnetic source images determined by a lead-field analysis: The unique minimum-norm least-squares estimation. IEEE Trans. on Biomedical Engineering, 1992, 39: 665-675.

[3] Mosher, J.C., Lewis, P.S. and Leahy, R.M. Multiple dipole modeling and localization from spatio-temporal MEG data. IEEE Trans. on Biomedical Engineering, 1992, 39: 541-557.

[4] Rogers, R.L., Papanicolaou, A.C., Baumann, S.B., Saydjari, C. and Eisenberg, H.M. Neuromagnetic evidence of a dynamic excitation pattern generating the N100 auditory response. Electroenceph. clin. Neurophysiol., 1990, 77: 237-240.

[5] Schmidt, R.O. Multiple emitter location and signal parameter estimation. IEEE Trans. on Antenn. Propagat., 1986, 34: 276-280.

A New Approach to the Reconstructing Source Distribution

Kajihara, S.[1], Tomita, T.[1], Tomita, S.[1], Kondo, Y.[1], Arakawa, A.[1], Yoshida, Y.[1] and Matsuda, M.[2]

[1]*Technology Research Laboratory, Shimadzu Corporation, Kyoto, Japan;* [2]*Superconducting Sensor Laboratory, Chiba, Japan*

Introduction

Recently, clinical applications of the source localization have been tried. In these application results, clinicians indicated some weak points of usual source localization methods and have hoped to develop a new localization method. In some problems, we discuss the problem of the distributed source. We think that the distributed source means the current groups with volume, not several dipoles. About the problem with the distributed source, it was indicated that the estimated source position was different from the actual source position [1].

We hope that the minimum norm technique or the least square technique is useful for the problem of the distributed source. Some researchers have proposed these methods, and obtained good results [2, 3, 4]. We referred to these results and proposed our own method. We confirmed the availability of our method by a computer simulation [5]. In this paper, we prepared the phantom with distributed sources and applied our method to the measured magnetic field.

Methods

Four steps in our method are described. The magnetic fields at M points were measured. In the first step, N grid points are set up within the analyzed space. The number of the grid points N is less than the number of M / 2. When source position (equal to grid point) is fixed, the relation between source moments and measured magnetic fields is shown as follows [6].

$$b = L\,p \qquad \text{Eqn. 1}$$

where the b is the M dimensional vector, and the elements are the magnetic fields. The L is the M × 2N matrix and is called lead field matrix. The p is the 2N dimensional vector, and the elements are the source moments.

In the second step, we solve Eqn. 1 as a linear inverse problem by "regularized g-inverse matrix" [7]. That is, we calculate the p in Eqn. 1 as follows.

$$p = (L^T L + r^2 W)^{-1} L^T b \qquad \text{Eqn. 2}$$

where the W is the 2N × 2N diagonal matrix (called penalty matrix), and the i-th diagonal element is the i-th column norm of matrix L in Eqn. 1. The r is the scalar, and it is computed as follows.

$$r = a\,d\,\| L \| / \| W \| \qquad \text{Eqn. 3}$$

where $\| \ \|$ shows the Euclidian norm and the a is the constant. The d is the distance of grid points, because we assume that sources are not scattered. If source moments P is computed as Eqn. 2, it will minimize the function as follows.

$$f = \| L\,p - b \|^2 + r^2 p^T W p \qquad \text{Eqn. 4}$$

The $\| L\,p - b \|$ in the first term means the error between the theoretical magnetic field and the measured magnetic field. The $p^T W p$ in the second term means the norm of the source moments that are corrected by the source depth [8]. So we will compute the source moments p under the condition that minimize the square error between magnetic fields and the norms of source moment.

In the third step, we judge whether the source moments p computed in Eqn. 2 are the best or not. By this judgment, the square error between the theoretical magnetic field and the measured magnetic field (that is the first term of Eqn. 4) is used. If this error is minimum, these source moments p and the source positions (equal to grid points) will reach the final result. Otherwise, we use the fourth step as follows.

In the fourth step, when the error is not minimum, some grid points having a small moment will be moved around the grid point having a large moment. But in this step, we compute the r as follows instead of Eqn. 2.

$$r = \| L \| / \| W \| \qquad \text{Eqn. 5}$$

After the grid points are moved, the second step is used again. These steps are repeated until the square error satisfy the condition in the third step.

Experiment

To confirm availability of our method in an experiment, we measured the magnetic fields of the phantom by our 129 channels vector neuromagnetic imaging system. For further details of this system, see the reference [9]. The prepared phantom is shown in Fig. 1. The phantom was a hollow sphere and the radius was 10 cm. It was made of acrylic resin and filled with physiological saline. The phantom had a cylindrical electrode. The figure of this electrode is shown in Fig. 2. It was made by two cylindrical conductors that were lined up closely. We connected the lead wire to each cylindrical conductor and twisted these wires together. And we covered this electrode with an insulator except one end side of each cylindrical conductor. We prepared two sizes of the electrode. The diameter of the large electrode was 20 mm and its length was 40 mm. The diameter of the small electrode was 10 mm and its length was 20 mm. The large electrode was at 30 mm deep from the phantom surface. The small electrode was 20 mm deep from the phantom surface. Each electrode was charged with electricity to generate the magnetic field that was as strong as biomagnetic fields. The current fed to a large electrode was about 250 nA. The current fed to a small electrode was about 500 nA. The measured magnetic field data was filtered within the bandwidth of 0.7 to 50 Hz, and averaged at 200 times. The maximum data of measured magnetic fields was about 300 fT.

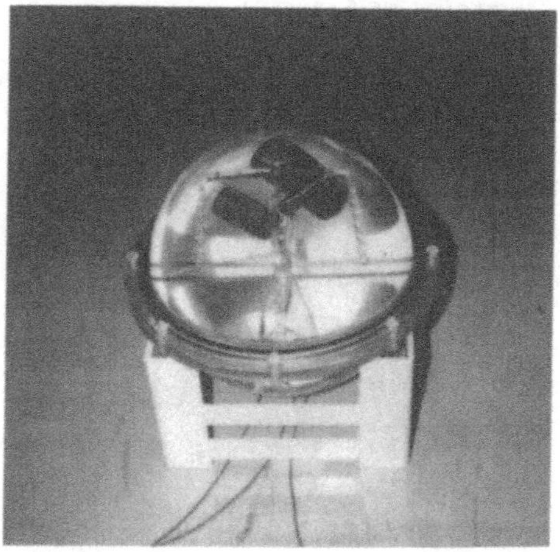

Fig. 1 The phantom with the distributed sources. In this figure, three distributed sources are shown at the same time. But each source is not charged at once in the experiment.

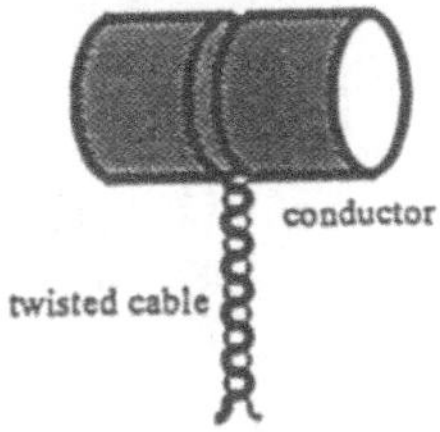

Fig. 2 The model of a distributed source. We assume that the electrode in this figure is the distributed source. The shade in this figure shows the insulator. The end of the twisted cable is connected to each cylinder.

Results

We applied our source localization method to the measured magnetic field data. In this study, the number of grid points was 32. Fig. 3 shows the reconstructed result for a large distributed source. Fig. 4 shows the result for a small distributed source. In these figures, the reconstructed source is projected on the X-Y, Y-Z and Z-X plains. The real source shape is shown by a solid line. The reconstructed source is shown by shade cubes. The darker the shade of the cube is, the stronger the source moment in this cube is. The center of the cube is the same as the grid point and a side of the cube is the same as the distance from the next grid point.

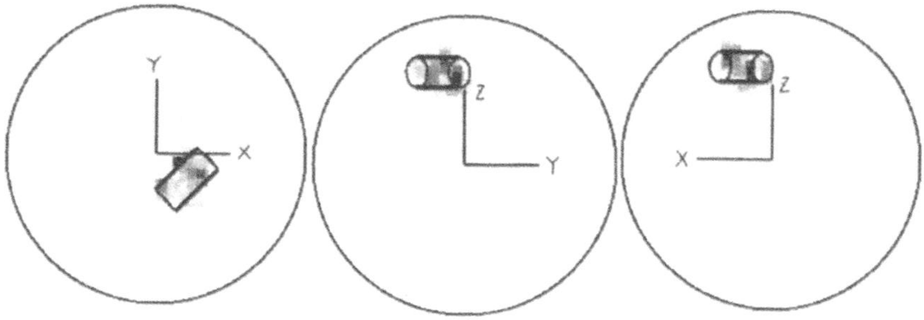

Fig. 3 Reconstructed source for a large distribution.
The cylinder in this figure shows the real source distribution. The reconstructed source is shown by the shading cubes, and is projected on the X-Y, Y-Z and Z-X plains. The circle in this figure shows the phantom shape. The arrow line in this figure shows the coordinate axes.

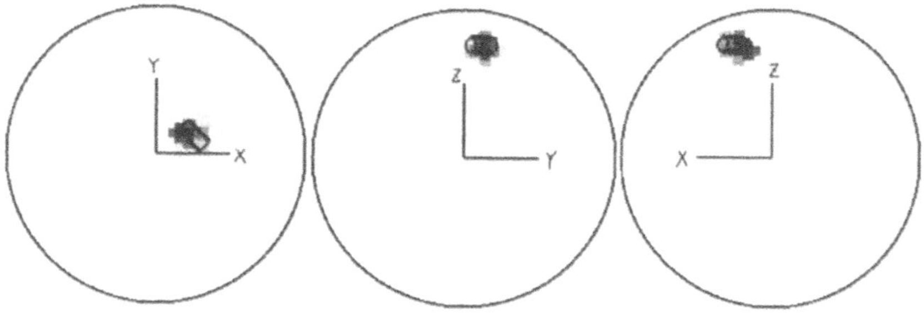

Fig. 4 Reconstructed source for a small distribution
The display in this figure is the same rule as in Fig. 3.

Discussion

As shown in these figures, the reconstructed sources were almost the same as the position of real sources and as the distribution of real sources. We can decide whether the real source is large or small by the reconstructed sources. But we can not tell by the reconstructed sources what is the size of real sources strictly. We could expect this result by our simulation studies [5]. Because only a little difference of the magnetic fields was appeared by the difference of the source's size. For example, the difference between the magnetic fields from 10×10×10mm cubic sources and the magnetic fields from 20×20×20mm was 0.2 %. We expected that it would be very difficult to reconstruct the size of sources strictly by the measured magnetic fields with the noise, because the details of the magnetic fields would be often covered by the noise.

In this study, the noise level was about 10 fT / √Hz. We assumed that this noise level was correspond to the level of the normal biomagnetic fields. We think that it is very difficult to reconstruct a strict source size under such noisy magnetic fields. But we confirmed by the phantom experiments that the reconstruct sources by our proposed method showed the relative size of the real source and showed an almost correct position of the real source. These results will be useful in the clinical. In next study, we will try to apply our methods to the biomagnetic measured fields.

References:

[1] Hari, R. On brain's magnetic responses to sensory stimuli, J Clin Neurophysiol, 1991, 8: 157-169.

[2] Okada, Y., Husang, J., and Xu, C. A hierarchical minimum norm estimation method for reconstructing current density in the brain form remotely measured magnetic fields, Biomagnetism, 1992, Clinical aspect: 729-734.

[3] Kuc, R., Szinger, J., Shimura, T., Goto, T., and Hayashi, T. Zoom-SVD biomagnetic imaging, The 8th International Conference on Biomagnetism, 1991, Book of Abstracts: 419-420.

[4] Wang, J.Z., Williamson, S.J., and Kaufman, L. Magnetic source images determined by a Lead-Field analysis : the unique Minimum-Norm Least-Squares estimation, IEEE Transactions on Biomedical Engineering, 1992, BME-39: 565-575.

[5] Kajihara, S., Tomita, T., Tomita, S., Kondo, Y., Arakawa, A., Yoshida, Y. and Ueda, M. The Journal of Japan Biomagnetism and Bioelectromagnetics Society, 1994, 1-1:150-153.

[6] Sarrvas, J. Basic mathematical and electromagnetic concepts of the biomagnetic inverse problem, Phys. Med. Biol., 1987, 32: 11-22

[7] Pratt, W.K. Digital imaging processing, John Wiley & Sons, New York, 1978

[8] Sekihara, K., Scholz, B., Bruder, H. Separating of correlated and uncorrelated source activities, Advances in Biomagnetism Abstracts, 1993, 233-234.

[9] Yoshida, Y., Arakawa, A., Kondo, Y., Kajihara, S., Tomita, S., Tomita, T., and Takanashi, Y. A 129 channel vector neuromagnetic imaging system, this volume

A Comparative Study of Minimum Norm Inverse Methods for MEG Imaging

Leahy, R.M.[1], Mosher, J.C.[2] and Phillips, J.W.[1]

[1]University of Southern California, Los Angeles, California, USA; [2]Los Alamos National Laboratory, Los Alamos, New Mexico, USA

The majority of MEG imaging techniques currently in use fall into the general class of (weighted) minimum norm methods. The minimization of a norm is used as the basis for choosing one from a generally infinite set of solutions that provide an equally good fit to the data. This ambiguity in the solution arises from the inherent non-uniqueness of the continuous inverse problem and is compounded by the imbalance between the relatively small number of measurements and the large number of source voxels. Here we present a unified view of the minimum norm methods and describe how we can use Tikhonov regularization to avoid instabilities in the solutions due to noise. We then compare the performance of regularized versions of three well known linear minimum norm methods [5][7] with the non-linear iteratively reweighted minimum norm method [1] and a Bayesian approach described in our companion paper ("MEG-based Imaging of Focal Neuronal Current Sources," Phillips J.W., Leahy R.M., Mosher J.C.).

The Minimum Norm Approach to MEG Imaging

Since the forward model in MEG is linear, we may relate the N sources $\mathbf{y}$ (Nx1) and the M MEG measurements $\mathbf{b}$ (Mx1) as $\mathbf{b} = \mathbf{G}\mathbf{y} + \mathbf{n}$, where the i'th row of the MxN system matrix $\mathbf{G}$ is a discrete representation of the lead field (sensitivity) of the i'th sensor. The j'th column of $\mathbf{G}$ specifies the gain vector for the j'th constrained dipole component. The Mx1 vector $\mathbf{n}$ represents noise generated within the sensor and by unwanted electromagnetic sources.

Since the inverse problem is inherently ill-posed, the search for an appropriate imaging method is concerned with finding a way to choose within a set of images that produce essentially the same fit to the data. The weighted minimum l_2-norm approach to linear inverse problems involves solving the constrained optimization problem:

$$\mathbf{y}_{\text{wmn}} = \arg\min_{\mathbf{y}} \; \mathbf{y}^{\mathrm{T}}\mathbf{C}_{\mathbf{y}}^{-1}\mathbf{y} \quad \text{such that} \quad |\mathbf{b} - \mathbf{G}\mathbf{y}|^2 = 0 \tag{1}$$

where $\mathbf{C}_{\mathbf{y}}$ is an arbitrary symmetric positive definite matrix. Writing $\mathbf{C}_{\mathbf{y}} = \mathbf{W}\mathbf{W}^{\mathrm{T}}$, we can form the solution as,

$$\mathbf{y}_{\text{wmn}} = \mathbf{W}\mathbf{W}^{\mathrm{T}}\mathbf{G}^{\mathrm{T}}(\mathbf{G}\mathbf{W}\mathbf{W}^{\mathrm{T}}\mathbf{G}^{\mathrm{T}})^{-1}\mathbf{b} = \mathbf{W}(\mathbf{G}\mathbf{W})^{+}\mathbf{b} \tag{2}$$

where $(\mathbf{G}\mathbf{W})^{+}$ denotes the pseudoinverse of $\mathbf{G}\mathbf{W}$. The weighted minimum norm solution can also be formed from the singular value decomposition (SVD) of $\mathbf{G}\mathbf{W}$:

$$\mathbf{y}_{\text{wmn}} = \mathbf{W}\sum_{i=1}^{M}\left(\frac{\mathbf{u}_i^{\mathrm{T}}\mathbf{b}}{s_i}\right)\mathbf{v}_i \tag{3}$$

where s_i, $\mathbf{v}_i$, and $\mathbf{u}_i$ are the i'th singular value and corresponding right and left singular vectors of $\mathbf{GW}$, respectively.

Several forms of $\mathbf{W}$ have been proposed for MEG imaging applications. In [6] and [7], the weight matrix is implicitly the identity matrix ($\mathbf{W}=\mathbf{I}$). In the normalized minimum norm method [4], $\mathbf{W}_{\text{norm}} = \text{diag}(1/\|\mathbf{g}_1\|, 1/\|\mathbf{g}_2\| \ldots, 1/\|\mathbf{g}_N\|)$, where $\|\mathbf{g}_i\|$ is the Euclidean norm of the i'th column of $\mathbf{G}$. This weighting is designed to compensate for the reduced sensitivity of MEG to deep sources resulting in a preference for superficial distributions when $\mathbf{W}=\mathbf{I}$. The NxN Laplacian operator $\mathbf{B}$ is commonly selected to smooth the reconstruction in minimum norm imaging, with $\mathbf{W} = \mathbf{B}^{-1}$. The LORETA technique [5] uses a weighting matrix $\mathbf{W}_{\mathrm{L}} = \mathbf{W}_{\text{norm}}\mathbf{B}^{-1}$. The Laplacian operator tends to select smoother reconstructions than either of the previous two weight functions.

[†]This work was supported by the National Institute of Mental Health Grant No. R01-MH53213, and National Eye Institute Grant No. R01-EY08610-04.

Exact matching of the data in (1) results in ill-conditioning and high sensitivity to noise. Regularized forms of weighted minimum norm methods lower this noise sensitivity. One popular technique is to simply truncate the summation in (3) at a threshold index. An alternative method, Tikhonov regularization [3], replaces the original problem (1) with the unconstrained minimization of a combination of the residual error norm and weighted l_2-norm of the solution vector:

$$y_{\lambda} = \underset{\mathbf{y}}{\arg\min} \quad |\mathbf{b} - \mathbf{G}\mathbf{y}|^2 + \lambda \mathbf{y}^T \mathbf{C}_{\mathbf{y}}^{-1} \mathbf{y}. \tag{4}$$

For any value of the regularization parameter λ, the solution can be found as:

$$\mathbf{y}_{\lambda} = \mathbf{W}\mathbf{W}^T\mathbf{G}^T(\mathbf{G}\mathbf{W}\mathbf{W}^T\mathbf{G}^T + \lambda\mathbf{I})^{-1}\mathbf{b} = \mathbf{W}\sum_{i=1}^{M} f_i\left(\frac{\mathbf{u}_i^T\mathbf{b}}{s_i}\right)\mathbf{v}_i \tag{5}$$

where the filter parameters $f_i = s_i^2/(s_i^2 + \lambda)$. Note that the filter coefficients decrease as s_i decreases. Therefore, the contributions of $(\mathbf{u}_i^T\mathbf{b}/s_i)\,\mathbf{v}_i$ to the solution from the smaller s_i are effectively filtered out. In our implementation of the regularized weighted minimum norm methods we use a *regularization toolbox* [3] which selects an optimal value for λ using the L-curve method.

The iteratively reweighted minimum norm approach, also known as focal underdetermined system solution (FOCUSS)[1], is a novel inverse method which iteratively updates the weight matrix using the result of the previous iteration. The algorithm first provides an initial estimate of $\mathbf{y}(0)$ using (2) with $\mathbf{W}_{\text{norm}}$ as the weighting matrix. At each step thereafter, a separate weight matrix $\mathbf{W}(k) = \text{diag}(\mathbf{y}(k-1))$ is formed. At each iteration, we update the solution using,

$$\mathbf{y}(k) = (\mathbf{W}_{\text{norm}}\mathbf{W}(k))\,(\mathbf{G}\mathbf{W}_{\text{norm}}\mathbf{W}(k))^{\dagger}\mathbf{b}. \tag{6}$$

For any starting point $\mathbf{y}(0)$ asymptotic convergence to a fixed point is guaranteed [1]. The fixed points are, however, unknown and highly dependent on the initial estimate. Also, the final error in fitting the data may be large, even for the noiseless case. In our implementation we use Tikhonov regularization and the L-curve method [3] at each iteration to avoid ill-conditioning.

The New Bayesian Approach

We have developed a new approach to MEG imaging. We use a Bayesian formulation of the inverse problem in which a Gibbs prior is constructed to reflect our specific expectations regarding the spatial distribution of sources. The reconstruction method uses this prior to resolve ambiguities that are inherent in the inverse problem. A thorough description of this Bayesian technique is given in our companion paper ("MEG-based Imaging of Focal Neuronal Current Sources," Phillips, J.W., Leahy, R.M., and Mosher, J.C.).

Basic studies of functional activation, such as somatotopic or retinotopic mapping using fMRI or PET, reveal the sparse and highly localized nature of activation in the cerebral cortex. Our Gibbs prior is therefore specifically designed to reflect the expectation that current sources tend to be sparse and focal. We combine this with a Gaussian distribution for active sources and an assumption of additive Gaussian noise to form the posterior distribution. An estimate of the primary current source distribution for a specific data set is formed by maximizing over the posterior probability.

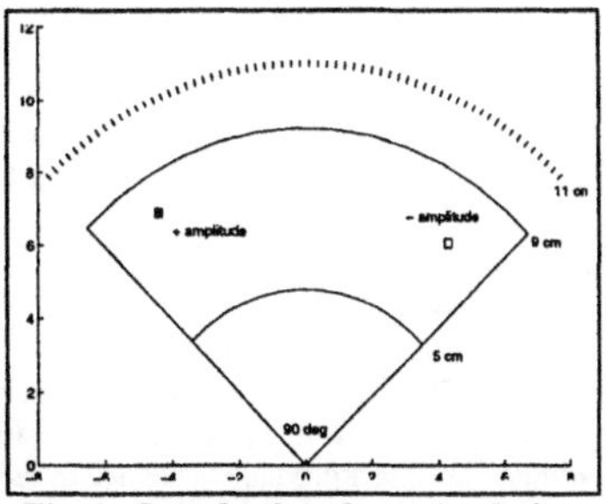

Fig. 1. Sample plot of one simulation solution showing representative positive and negative sources.

Simulations

We have conducted extensive simulations based on a simplified 2D source model with a 1D array of 64 sensors. All sources are constrained to the annular segment of the x-y plane shown in Fig. 1. The distance between source locations was set to 2.8 mm providing an image grid of 560 pixels. All dipoles were constrained in orientation perpendicular to the x-y plane. The source images were chosen using stochastic sampling from the prior distribution on the image. The results of one representative simulation are shown in Fig. 2 for two scenarios, one noiseless, and one with added white Gaussian noise. We compared the Bayesian technique to the methods previously discussed: the standard minimum l_2 norm technique, the normalized minimum norm technique, the LORETA method, and the iteratively reweighted minimum norm technique.

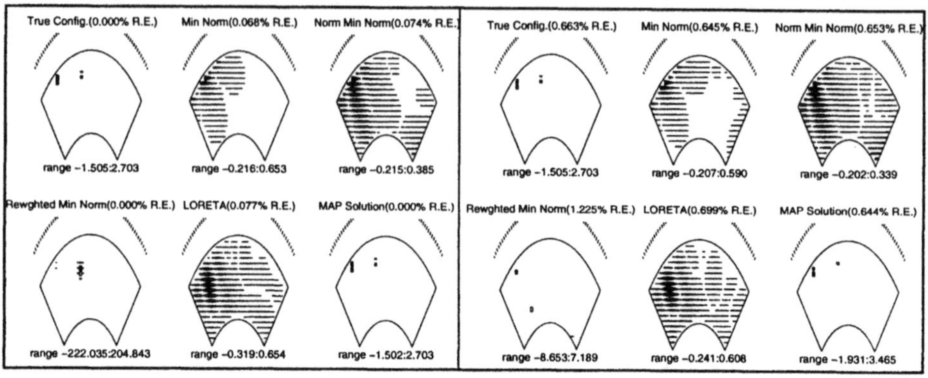

Fig. 2. Sample 2D simulation with a 1D sensor array. 64 point measurements, 560 point image grid. The percent residual error (% R.E.) is shown above and the range of source values (in nAm) is shown below each image. Two scenarios are presented: (left) No noise added to the computed data (right) White Gaussian noise added to the data to achieve a SNR of 20 dB

Phantom Experiments

We have applied the same techniques used in the simulations to experimental phantom data collected with a Neuromag-122 system [2] using the manufacturer's 26 dipole phantom. The Neuromag-122 system employs 61 dual-channel planar first-order gradiometer units in a helmet-shaped configuration at a radius of 10-11 cm, measuring the magnetic field gradient in two orthogonal tangential directions, for a total of 122 individual sensor measurements.

The phantom consists of two half circles with a 7 cm radius in the x-z plane and y-z plane, with dipoles in fixed positions in these planes oriented tangential to the outer edge. The image reconstruction grid consisted of 768 locations spaced 4 mm apart on two 180 degree annuli, with an inner radius of 3 cm and an outer radius of 7 cm. An example of the imaging surface with the surrounding sensor elements is shown in Fig. 3. The distance between sources and sensors creates a gain matrix which is more poorly conditioned than the one in the previous simulations, which increases noise sensitivity. In all cases, the gain matrix was constructed to include gradiometer effects and non-radial sensor orientations assuming a spherical source volume [6].

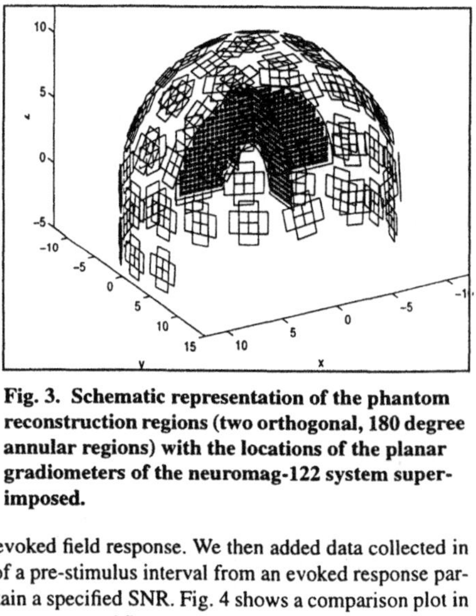

Fig. 3. Schematic representation of the phantom reconstruction regions (two orthogonal, 180 degree annular regions) with the locations of the planar gradiometers of the neuromag-122 system superimposed.

The phantom data was scaled to reflect a reasonable evoked field response. We then added data collected in the same system from a passive human subject (100 averages of a pre-stimulus interval from an evoked response paradigm). This background was added to the phantom data to obtain a specified SNR. Fig. 4 shows a comparison plot in which noise was added to obtain a SNR of 10 dB. The minimum norm and Bayesian techniques tested on phantom data performed comparably to the simulations.

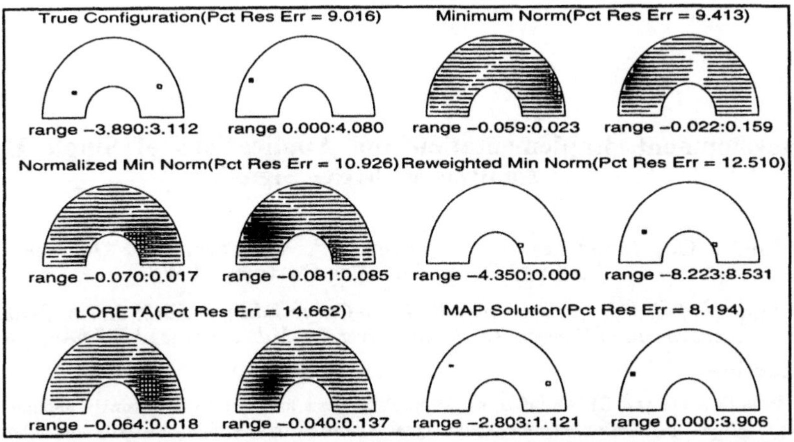

Fig. 4. Sample phantom results with three active dipoles on two orthogonal planes. 122 point measurements, 768 point image grid. Percent residual error (Pct Res Err) is shown above and the range of source values (in nAm) is shown below each image. Colored subject noise added to the data to achieve a SNR of 10 dB.

Conclusions

These results show a wide variation in the characteristics of the solution obtained using different weighting functions. All linear minimum norm methods produce results exhibiting a relatively large degree of smoothness. In comparison, both iteratively re-weighted minimum norm and the new Bayesian method produce very sparse solutions. For both simulations and the phantom experiment, all methods generally give similar residual errors in the fit to the data. In some instances, the iteratively reweighted minimum norm method gives larger errors since there is nothing inherent in the method to limit increases in the error from one iteration to the next. All results can be considered 'correct' in the sense that they are configurations that could have produced the observed data. This observation emphasizes the severely under-determined nature of this problem. Clearly, in order to select between these feasible solutions we must use additional information concerning the expected nature of the source. Our Bayesian approach specifically introduces the information that sources are sparse and focal into the reconstruction method which results in generally superior results for sources that exhibit these characteristics.

References

[1] I.F. Gorodnitsky, J.S. George, and B.D. Rao, "Neuromagnetic source imaging with FOCUSS: a recursive weighted minimum norm algorithm," *EEG and clinical Neurophysiol.*, 95, pp. 231-251, 1995.

[2] M. Hämäläinen, R. Hari, *et al.*, "Magnetoencephalography - theory, instrumentation, and applications to noninvasive studies of the working human brain," *Rev Mod Phys*, vol. 65(2), pp. 413-497, 1993.

[3] P.C. Hansen, *Regularization Tools, A Matlab Package for Analysis and Solution of Discrete Ill-Posed Problems*, available in Postscript from http://www.mathworks.com/, March 1993

[4] B. Jeffs, R.M. Leahy, and M. Singh, "An evaluation of methods for neuromagnetic image reconstruction," *IEEE Trans on Bio. Eng*, vol. 34, pp. 713-723, 1987.

[5] R.D. Pascual-Marqui, C.M. Michel, D. Lehmann, "Low resolution electromagnetic tomography: A new method for localizing electrical activity in the brain," *Int. Journ. of Psyc.*, vol. 18, pp. 49-65, 1994

[6] J. Sarvas, "Basic mathematical and electromagnetic concepts of the biomagnetic inverse problem," *Physics in Medicine and Biology*, vol. 32, pp. 11-22, 1987.

[7] J.Z. Wang, S.J. Williamson, and L. Kaufman, "Magnetic Source Images Determined by Lead-Field Analysis: The Unique Minimum-Norm Least-Squares Estimation," *IEEE Trans Bio. Eng.*, vol. 39(7), pp. 665-675, 1992.

Development, Implementation and Applications of Single Epoch Analysis of MEG Signals

Liu, L.C.[1], Ioannides, A.A.[1,3], Heer, H.[2], Wagener, M.[2], Delonge, F.[2], Halling, H.[2] and Müller-Gärtner, H.W.[1]

[1]Institute of Medicine; [2]Electronics, Research Centre Jülich, D-52425 Jülich, Germany; [3]Department of Physics, The Open University, Milton Keynes MK7 6AA, UK

Introduction

In both MEG and EEG studies, single epoch analysis has become increasingly popular [1, 2]. In EEG the relationship between generators and signal is blurred by the high resistivity and inhomogeneity of the skull, making single trial analysis problematic because of low signal-to-noise ratio. For MEG, the signal-to-noise ratio in single trials is high, in the sense that, contributions from small cortical regions are easily detectable, although not easily distinguished from other contributions. In addition, single trial analysis of MEG signals incurs a high computational burden and produces huge amount of output. In our earlier work [3, 4], we have addressed these problems by introducing a vector signal transformation called $\mathbf{V}_3$, which is the MEG analogue of the well-known Laplacian method of EEG. We have tested the accuracy of the $\mathbf{V}_3$ based estimates with computer generated data and compared the results with the ones obtained from Magnetic Field Tomography (MFT) analysis of the same data. These tests have shown that the $\mathbf{V}_3$ based estimates are remarkably accurate when superficial generators are involved [4]. Estimators constructed from correlational measures based on the $\mathbf{V}_3$ and integrated over a period of time and/or a region, can be used to scan quickly and efficiently through all single trials. The $\mathbf{V}_3$ analysis has already been applied to the identification and quantification of auditory cortex activation and the investigation of gamma-band activity in single trials [4, 5]. Further correlation analysis of spatio-temporal $\mathbf{V}_3$ templates, has enabled us to describe qualitatively and quantitatively how patterns observed in the average signal are rèpresented in single trials [6]: we showed conclusively that the average is a superposition of histories each reflected in different subsets of the epochs. In this paper, we will show another related application of the $\mathbf{V}_3$: it can be used as the basis for easy and fast inspection of data, and for identification of correlations in activity, in space and time, within the same hemisphere or across the hemispheres.

Methods

The radial component of the magnetic field is zero when a superficial generator is directly below a sensor. In contrast, the $\mathbf{V}_3$ highlights the position and vector properties of nearby generators. It was originally designed as an unbiased estimator for focal as well as extended sources. For measurements of the z-component of the magnetic field S_z, it is defined by $\mathbf{V}_3(x,y) = \nabla(S_z)\times\hat{\mathbf{e}}_z$. The sensitivity of derivatives to noise is ameliorated by the use of a spatial smoothing operation [3, 4]. Finally, a correlation is computed between $\mathbf{V}_3$ templates extracted from the average and single trials, which makes it possible to scan through each and every single trial for specific signatures quickly [6].

The original method was implemented for planar sensor arrays, and it has since been extended to the geometry of the BTi twin MAGNES probe (2×37 channels). A software package based on our earlier published work, has been implemented by the Institute of Electronics (ZEL); it is routinely used to interrogate and visualize the MEG data from the twin MAGNES system at the Institute of Medicine (IME). Fig. 1 shows the graphical user interface from the package. The display shows the typical N1m response at about 80 ms after the onset of the stimulus, clearly identified in the average signal from a single channel (Fig. 1b), and in the familiar dipolar pattern when the actual average signal is displayed (Fig. 1c). The display of the $\mathbf{V}_3$ transformed signal (Fig. 1d) identifies a strong source below the central channel (A1), which fits in location and direction the expected auditory cortex activation.

When a single focal generator is present, the advantage of the $\mathbf{V}_3$ is marginal; the trained eye can easily pick up the same information from the original contour plot of the signal. The advantage becomes decisive when the generators are geometrically complex and animations are used to show changes in the signal. Animations of the signal show some systematic behaviour, but it is difficult for the eye to place the significant elements in space and time. In contrast, animations of the $\mathbf{V}_3$ show spatio-temporally systematic

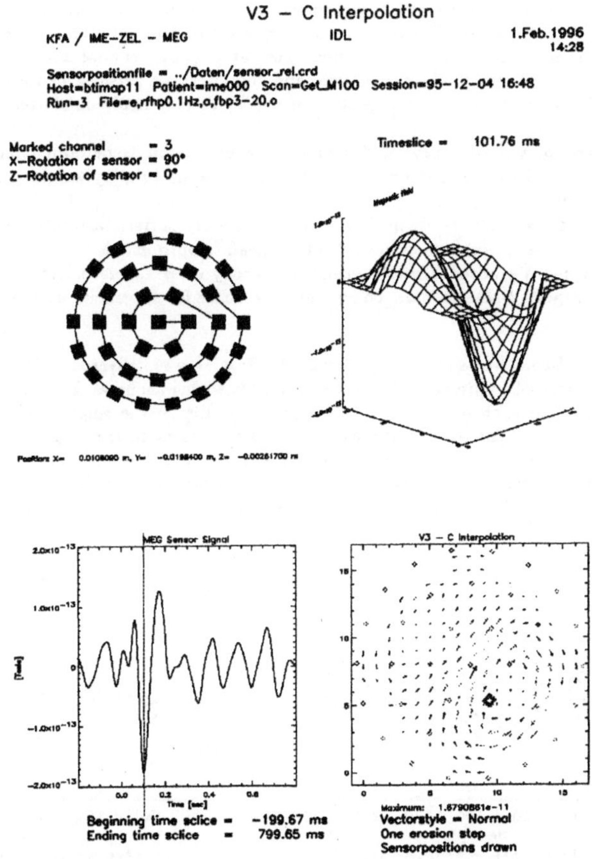

Figure 1: The user interface for the V_3 package. The example shows the MEG average signal elicited by a simple 1 kHz tone delivered to the left ear, after band pass filtering in the 3-20 Hz range: (a) the top-left figure displays the layout of the channels from one of the BTi twin MAGNES probe – probe A, which was placed over the right hemisphere of the subject; (b) the bottom-left figure shows the average signal from one marked channel in (a) – channel A3; (c) the top-right figure shows the distribution of magnetic field across the sensor array at the timeslice marked by the straight line in (b). Finally, the bottom-right figure (d), shows the same signal as (c) after the V_3 is applied.

behaviour which is directly related to the properties of the generators. Interactions between superficial regions become evident, provided they are separated by 2 *cm* or more and they all within the range of the sensor array. Animations from both probes simultaneously help to visualize and compare the responses in the left and right hemispheres at the same time, and the interplay between hemispheres as a function of time. For superficial generators, visual inspection of the V_3 display is often sufficient to identify candidate patterns.

The concept of virtual sensor and signal space projection is becoming increasingly popular in MEG [2, 7]. The V_3 and the derived linear combinations of appropriately weighted signals can be viewed as a virtual sensor, which are designed to improve sensitivity to nearby superficial sources. In fact a family of V_3 based virtual sensors can be defined with wide range of spatial sensitivity. We have used the twin MAGNES system with each probe on either side of the head to optimize the contribution from deep sources. We are currently using V_3 based virtual sensors to identify deep generators in single trials, but we will restrict the results reported here to more conventional searches for superficial generators.

Results

In the experiment we report here, a simple 1 kHz tone was delivered to the left ear of the subject. Fig. 2 shows estimates of auditory cortex responses in both hemispheres from the average for all trials, average for 5 trials, and each of these 5 trials: the variability in the single trials is not evident in the averages; the N1m component seen in the averages is also present in the single trials, albeit at different timeslices in relation to the onset of the stimulus. This is in agreement with one of our earlier results, i.e., the sequence of events observed in the average is not representative of what happens in any single trials, but it consists of contributions from different sets of trials, with each set probably reflecting the activation of a different pathway.

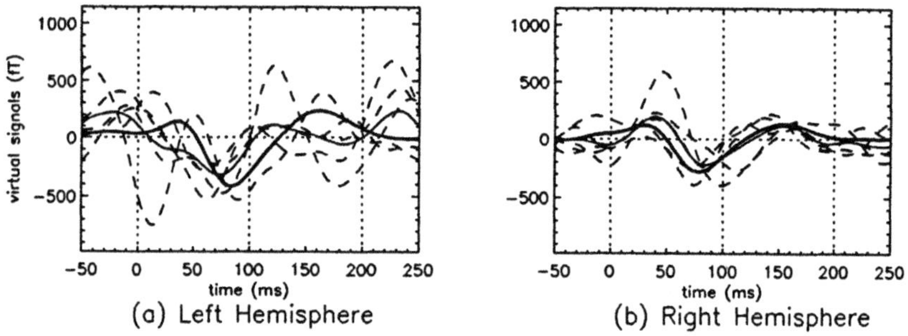

Figure 2: Estimates of auditory cortex responses in left and right hemispheres using virtual sensor constructs for average for all single trials (heavy solid line), average for 5 single trials (light solid line) and each of these 5 single trials (long dashed lines). The raw signals were filtered at 3–20 Hz.

The V_3 correlation analysis provides a quantitative measure of how representative is the description of events extracted from the analysis of the average signal. Fig. 3 shows the result for a typical auditory evoked response. A target template was first constructed from the V_3 of the average signal around the N1m peak. We then computed the correlation between this spatio-temporal pattern with a similar template from each single trial. A high correlation is marked by a dot in Fig. 3a. The total number of single trials having high correlation with the N1m template in the average is plotted as a function of time in Fig. 3b. The figure shows clearly the dramatic rise in the number of epochs with high correlation around 80 ms; in addition it shows how widespread the N1m pattern is in the single trials.

Discussion

Using the V_3 based correlation and virtual sensor constructs, we have applied the analysis to experiments using a variety of protocols, including auditory response to a simple tone and movement control and inhibition for both the average and single trial signals. Results from these studies support our early

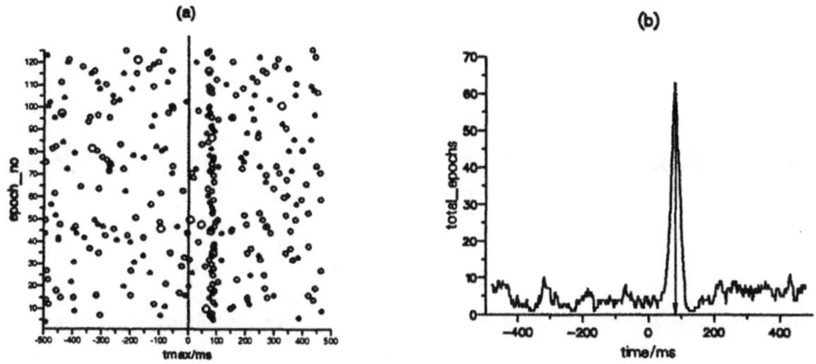

Figure 3: Correlation as a function of time (0 – onset of the stimulus) of the N100m pattern in the average with those in 125 single trial signals without any digital filtering: (a) the correlation for the individual single trials, filled and big dots have higher correlation than those empty and small dots; (b) the total number of high-correlated single trials: the arrowed line indicates that at 81 ms, there are 64 single trials (51.2% of all the trials) highly correlated with the average.

findings and extend the range of cortical events we can study to both hemispheres as well as deep brain regions. In this paper, we have shown a software package developed and implemented recently for easy visualization and fast inspection of MEG data. Our single trial analysis of the N1m pattern demonstrates the feasibility of tackling unaveraged MEG data, and how much information retains compared with what is lost by averaging. We believe that single trial analysis is necessary if the dynamics of brain function are to be properly understood.

References

[1] A.R. Haig, E. Gordon, G. Rogers, and J. Anderson. 'Classification of Single-trial ERP Sub-types: Application of Globally Optimal Vector Quantization using Simulated Annealing'. *Electroenceph. Clin. Neurophysiol.*, 94:288–297, 1995.

[2] B. Lütkenhöner, C. Pantev, A. Grunwald, and E. Menninghaus. 'Source-space Projection of Single Trials of the Auditory Evoked Field'. In C. Baumgartner, L. Deeke, G. Stroink, and S.J. Williamson, editors, *'Biomagnetism: Fundamental Research and Clinical Applications'*, pages 347–351. Elsevier Science, Amsterdam, 1995.

[3] A.A. Ioannides, R. Hasson, and G.J. Miseldine. 'Model-dependent Noise Elimination and Distributed Source Solutions for the Biomagnetic Inverse Problem'. In A.F. Gmitro and et al., editors, *'Digital Image Synthesis and Inverse Optics'*, pages 471–481. *Proc. SPIE 1351*, 1990.

[4] L.C. Liu. 'Single Epoch Analysis and Bi-hemispheric Study of Magnetoencephalographic (MEG) Signals using Vector Signal Transformation V_3 and Magnetic Field Tomography (MFT)'. PhD thesis, Department of Physics, The Open University, U.K., February 1995.

[5] L.C. Liu and A.A. Ioannides. 'Single Epoch Analysis of MEG Signals'. In C. Baumgartner, L. Deeke, G. Stroink, and S.J. Williamson, editors, *'Biomagnetism: Fundamental Research and Clinical Applications'*, pages 439–444. Elsevier Science, Amsterdam, 1995.

[6] L.C. Liu and A.A. Ioannides. 'A Correlation Study of Averaged and Single Trial MEG Signals: The Average Describes Multiple Histories Each in a Different Set of Single Trials'. *Brain Topography*, 1996. Accepted for publication.

[7] M. Huotilainen, R.J. Ilmoniemi, H. Tiitinen, J. Lavikainen, K. Alho, M. Kajola and R. Näätänen. 'The Projection Method in Removing Eye-blink Artefacts from Multichannel MEG Measurements'. In C. Baumgartner, L. Deeke, G. Stroink, and S.J. Williamson, editors, *'Biomagnetism: Fundamental Research and Clinical Applications'*, pages 363–367. Elsevier Science, Amsterdam, 1995.

Dipole Localization Accuracy as a Function of the Area Covered by the Magnetometer System

Lütkenhöner, B.

Institute of Experimental Audiology, University of Münster, Münster, Germany

Introduction

Dipole localization accuracy is dependent on numerous factors. Most important is certainly the signal-to-noise ratio. Since the amplitude of the magnetic field produced by a current dipole decreases with increasing dipole depth, the question arises as to what extent deep sources are accessible to neuromagnetic investigations [7]. To answer this question, not only the ***amplitude*** of the magnetic field has to be considered, but also its *spatial pattern*: The fact that the magnetic field distribution generated by deep dipoles is broader than that generated by more superficial ones improves the effective signal-to-noise ratio, provided that the measurement area is large enough to cover the most essential parts of the magnetic field distribution. This is illustrated in Fig. 1 for a spherical model. The left half of the figure shows the magnetic field pattern arising from a superficially located dipole (distance from center of sphere: $r = 7$ cm), whereas the right half shows the magnetic field pattern arising from a deep dipole ($r = 2$ cm). Customary 37 channel systems (on the right) are obviously well-suited for relatively superficial dipole locations, but in the case of deep dipoles they provide only part of the information principally available. In the present example, they clearly miss the highest signal amplitudes and, even worse, the field pattern in the area covered (characterized by a polarity reversal) is very similar for a large range of dipole depths. This explains why customary 37 channel systems provide only a relatively poor estimate for the depth of a deep source. It is obvious that the availability of whole-head magnetometer systems will lead to a considerable improvement of the accessibility of deep sources. But it is hard to predict to what extent the localization of superficial dipoles can profit as well from an increased measurement area. In the example shown on the left of Fig. 1, most of the coils are placed over regions with a relatively low field amplitude so that it could be speculated that source localization accuracy would improve if some of the coils were discarded. Questions of this type gave rise to the present study.

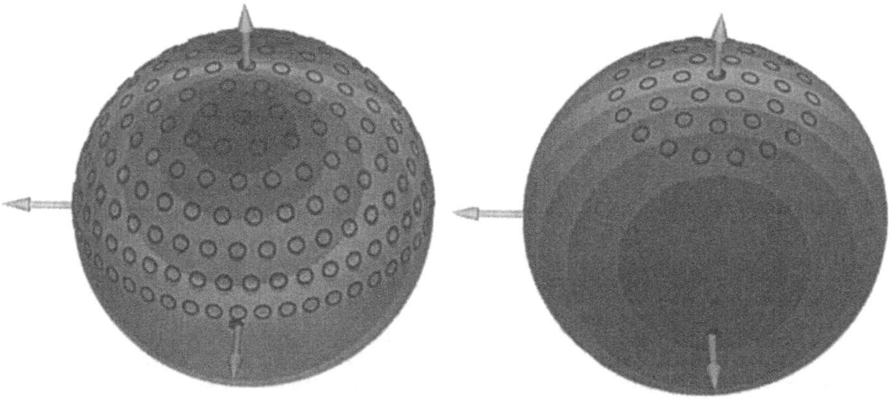

Fig. 1: Maps of radial component of magnetic field arising from a current dipole in a homogeneous sphere. The dipole was located on the z-axis (superior direction), and its moment was oriented in x direction (pointing to the left). The distance between dipole and center of sphere was 7 cm (left half) and 2 cm (right half), respectively. The gray-scale corresponds to a step size of 25 fT on the left and 3 fT on the right. The figure also shows two of the magnetometer systems simulated: A 37-channel system on the right and a 217 channel system on the left. The distance between coils and center of sphere was 12 cm.

Methods

Eight different magnetometer systems were investigated by means of Monte Carlo simulations. Starting point for the construction of the eight systems was the 37-channel system illustrated on the right side of Fig. 1, corresponding to the 37-channel system of Biomagnetic Technologies, Inc. [1]. Omitting outer rings of coils resulted in systems with 7 and 19 channels, respectively. Adding rings of coils by generalizing the idea underlying the design of the 37 channel system ($6i$ coils on the i-th circle, $i \geq 1$; span of i-th circle equal to $i \times 21^o$), resulted in systems with 61, 91, 127, 169, and 217 coils, respectively. The left half of Fig. 1 gives an impression of the 217-channel system (8 rings of coils, span of the measurement area corresponding to 168°).

The magnetic field was assumed to arise from a single current dipole located on the z axis, with a moment parallel to the x axis. It was assumed that the magnetic field measured by the magnetometer is composed of the radial component of the magnetic field produced by the dipole and Gaussian noise (with no correlation between channels). The superimposed noise was derived from pseudo random numbers (Gaussian deviates which were obtained from uniformly distributed random numbers provided by the routine ran3 of Press et al. [8]).

To estimate the dipole parameters from a simulated magnetic field pattern, the least-squares error was minimized. In a first step (rough optimization), the dipole moment was optimized for all points on a regular grid (± 3 cm around the true value of each dipole coordinate, with 5 mm spacing between adjacent grid lines). The grid point resulting in the smallest least-squares error served as the starting point for the second step: an iteration procedure where dipole location and dipole moment were optimized simultaneously. To solve this mixed linear-nonlinear optimization problem, the original least-squares estimation problem was transformed into a minimization problem for the non-linear parameters (dipole coordinates) by replacing the linear parameters (tangential components of the dipole moment) with the algebraic solution available for their least-squares estimates [4, 6]. The resulting minimization problem was solved iteratively by means of Powell's method (e.g. [8]). In general, this procedure resulted in dipole locations close to the true dipole locations. Exceptional cases where the Euclidean distance between estimated and true dipole location was greater than 5 cm (this happened especially in the case of deep dipoles and gradiometer systems covering a relatively small area) were discarded. The differences between estimated and true parameter values were finally used to calculate the covariance matrix of the parameter estimation errors (on the basis of 1000 trials).

Results

Fig. 2 shows exemplary results for two different measurement systems (37 coils on the left, 217 coils on the right). The symbols show the dipole locations obtained in 100 independent Monte Carlo simulations: as a projection into the x-y plane (top) and the x-z plane (bottom). Two different distances between dipole and center of sphere were considered: z =2 cm and z =7 cm (represented by the circles and the symbols

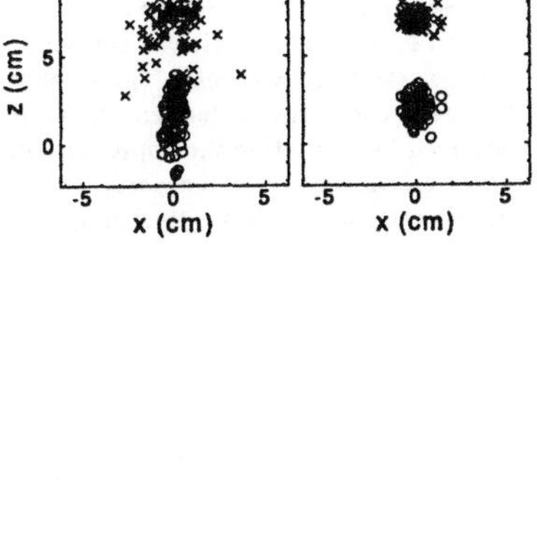

Fig. 2: Dipole locations obtained in 100 independent Monte Carlo simulations projected into the x-y plane (top) and the x-z plane (bottom). Two different measurement systems (37 coils on the left, 217 coils on the right) and two different distances between dipole and center of sphere (z = 2 cm and z = 7 cm, represented by the circles and the symbols "x", respectively) are considered. The superimposed white noise had a standard deviation of 10 fT in all cases. For the superficial dipole location the dipole moment was 2 nAm, and for the deep dipole location it was either 20 nAm (for the 217 coils) or 200 nAm (for the 37 coils).

"x", respectively). The figure demonstrates that the estimates for the three Cartesian dipole coordinates have quite different properties. The estimation of the coordinate perpendicular to the dipole moment (y) is obviously unproblematic: For all four conditions, the variance of this coordinate is significantly smaller than the variances of the other two coordinates. With the 217-channel system, a relatively precise estimation is possible also for the x coordinate (direction of dipole moment) and the z coordinate (corresponding to the radial direction in the present case). While the variances of these two coordinates are comparable in the case of the dipole at z=7 cm (the cluster of symbols has a spherical appearance), in the case of the dipole at z=2 cm the variance of the z coordinate is clearly greater than that of the x coordinate. With only 37 coils available, the localization accuracy deteriorates, especially for the deep dipole, where even a tenfold increase of the dipole amplitude could not prevent that the z estimates on the left half of Fig. 2 are more scattered than those on the right half.

A more systematic study of the influence of the measurement area on dipole localization accuracy is summarized in Fig. 3, which shows the standard deviations of the estimates for x, y, and z as functions of the distance between dipole and center of sphere (corresponding to the true value of z in the present case). Eight different instruments, with the number of coils ranging from 7 to 217, are considered in this figure. The results of Monte Carlo simulations, performed as described above, are represented by the dots. In addition, theoretical functions are plotted which were derived analytically [3] by considering a simulated MEG measurement with an ideal magnetometer system (providing the radial field component as a continuous function). For systems with 37 channels or more, the agreement between theoretical predictions (curves) and simulations (dots) is obviously very good, and for the 19-channel system the agreement is still reasonable.

Fig. 3 substantiates the preliminary conclusions derived from Fig. 2. The advantage of an increased measurement area can be assessed now in a quantitative way. A comparison of the curves obtained for 37 channels with those obtained for 217 channels shows, e.g., that, for distances between dipole and center of sphere of 1 cm, the estimation errors for the radial dipole coordinate, z, differ by one order of magnitude. Even in the case of more superficial source locations (e.g. 5 cm distance between dipole and center of sphere), a system with 217 coils is highly superior to a system with only 37 coils. The differences obtained for the x

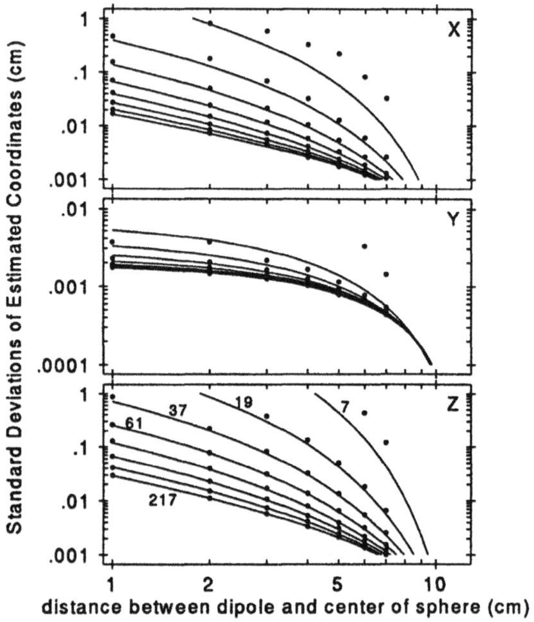

Fig. 3: Comparison of magnetometer systems with 7, 19, 37, 61, 91, 127, 169, and 217 coils (span of the measurement area equal to 21°, 42°, ···, 168 °). The system with 37 coils roughly corresponds to the 37-channel system of Biomagnetic Technologies (San Diego), whereas the system with 217 coils can be considered as a whole-head magnetometer system. The localization errors for the dipole coordinates x (longitudinal direction), y (transverse direction), and z (radial direction) are presented in three separate panels. The curves show the theoretical predictions, whereas the dots show the results of Monte Carlo simulations. The dipole moment was 1000 nAm, and the superimposed noise had a standard deviation of 10 fT. The latter values were chosen for the sake of convenience. The localization errors for other values can be easily obtained by exploiting the approximately linear relationship between parameter estimation errors and noise-to-signal ratio.

coordinate are smaller, but they are still quite significant. The differences obtained for the y coordinate are irrelevant in practice, since the estimates for this coordinate are extremely precise in any case.

Fig. 3 can be used to derive conclusions about the asymptotic behavior of the estimation errors for the three coordinates: Whereas the standard deviations of x and z are proportional to $1/r$, with r denoting the distance between dipole and center of sphere, the standard deviation of y approaches a constant value.

Discussion

The results of this study suggest that the availability of whole-head magnetometer systems will improve the precision of dipole source localization considerably (up to an order of magnitude for deep sources). The explanation of this effect is complex, which is expressed e.g. by the fact that the properties of the three dipole coordinates show fundamental differences. An important inference from the above results is that an increased measurement area is beneficial even in cases where the magnetic field recorded by the additional measurement coils has a relatively small amplitude. Consequently, it is in general not recommendable to confine a dipole source analysis to the coils with the highest amplitudes. This conclusion is remarkable insofar as the signal-to-noise ratio (defined as the variance of the signal divided by the variance of the noise) is obviously improved by considering such a subset of coils.

A simple consideration explains why including a channel with zero field amplitude (i.e., a channel contributing noise only) has, in general, no detrimental effect on the precision of a dipole source analysis. In the case of a homogeneous spherical volume conductor the parameter optimization algorithm has exactly 5 degrees of freedom (the three Cartesian coordinates and the amplitudes of the two tangential components of the dipole moments). Since the number of channels is usually much greater than 5, the fit algorithm must concentrate on those regions of the field distribution where the amplitude of the field is high, i.e., regions dominated by noise have little influence on the fit result. Technically, this is accomplished by associating channels having a bad signal-to-noise ratios with a low weighting factor. Since this is a property of the algorithm itself, selection of a subset of channels is in general superfluous. It shall be mentioned, finally, that a recording location where the field amplitude is zero can convey a significant amount of information: In the region where the magnetic field changes its sign, the gradient of the magnetic field has a maximum, and the line where the field amplitude is zero defines the dipole coordinate parallel to the moment.

In practice, the superimposed noise exhibits spatial correlation, which was not taken into account in the present study. Such noise could be modeled e.g. by assuming a large number of current dipoles randomly distributed in the volume conductor [2, 5]. In that case it is advantageous to replace the least-squares estimation procedure by a maximum likelihood estimation [9]. It is to be expected that the essential conclusions regarding the influence of the measurement area will be effected very little by the properties of the superimposed noise.

References

[1] Benzel, E. C., Lewine, J. D., Bucholz, R. D., and Orrison, W. W. Magnetic source imaging: a review of the Magnes system of Biomagnetic Technologies Incorporated. *Neurosurgery*, 1993, 33:252–259.

[2] de Munck, J., Vijn, P. C., and Lopes da Silva, F. A random dipole model for spontaneous brain activity. *IEEE. Trans. Biomed. Eng.*, 1992, 39:791–804.

[3] Lütkenhöner, B. Current dipole localization with an ideal magnetometer system. [submitted].

[4] Lütkenhöner, B. *Möglichkeiten und Grenzen der neuromagnetischen Quellenanalyse.* Münster, Lit, 1992.

[5] Lütkenhöner, B. Magnetic field arising from current dipoles randomly distributed in a homogeneous spherical volume conductor. *J. Appl. Phys.*, 1994, 75:7204–7210.

[6] Lütkenhöner, B., Lehnertz, K., Hoke, M., and Pantev, C. On the biomagnetic inverse problem in the case of multiple dipoles. *Acta Oto-Laryngol. (Stockh.)*, 1991, suppl. 491:94–105.

[7] Menninghaus, E. and Lütkenhöner, B. How silent are deep and radial sources in neuromagnetic measurements? In Baumgartner, C., Deecke, L., Stroink, G., and Williamson, S., editors, *Biomagnetism: Fundamental Research and Clinical Applications*, pages 352–356, Amsterdam, 1995. Elsevier Science, IOS Press.

[8] Press, W. H., Teukolsky, S. A., Vetterling, W. T., and Flannery, B. P. *Numerical recipes in C. The art of scientific computing.* Cambridge, Cambridge University Press, second edition, 1992.

[9] Sekihara, K., Ogura, Y., and Hotta, M. Maximum-likelihood estimation of current-dipole parameters for data obtained using multichannel magnetometer. *IEEE Trans. Biomed. Eng.*, 1992, 39:558–562.

Towards a Standardized Representation of Neuromagnetic Data

Lütkenhöner, B., Grunwald, A. and Steinsträter, O.

Institute of Experimental Audiology, University of Münster, Münster, Germany

Introduction

In contrast to electroencephalography, where the scalp provides a natural reference surface for the visualization of raw data, there is no such well-defined surface in the case of magnetoencephalography. Apart from the diversity of measurement devices being installed, elaborate measures (which are not practicable in routine applications) are required to reproduce given measurement locations and directions in subsequent sessions. This does not represent a problem if the measured data can be interpreted in terms of simple models (e.g. equivalent current dipole) since model optimization procedures can take into account the special conditions of a measurement. However, owing to the non-uniqueness of the solution of the inverse problem to be solved, such an interpretation is often not available. As a consequence, a direct comparison of results from different sessions (e.g. clinical follow-up study) is not feasible. The problem becomes even worse if different measurement devices are involved. For this reason it would be useful to have a standardized representation of biomagnetic data. It is obvious that such a standardization requires a thorough discussion of the various alternatives available. The purpose of this contribution is to initiate such a discussion by outlining some general concepts and making a concrete suggestion.

General concepts

A standardized representation should have the following properties: First, it should be related as closely as possible with the data actually measured, i.e., it should be affected as little as possible by modeling assumptions. Second, it should be intuitively appealing, i.e., it should not operate in abstract mathematical spaces. Third, the effort for obtaining the standardized representation should be reasonable. In practice, a compromise has to be found which reconciles these three requirements. Especially the latter requirement will probably be a subject of controversies, owing to the fact that it depends on equipment and experience whether the effort required for a procedure will be characterized as "low" or "high". To be more concrete, it has to be decided, for example, whether magnetic resonance images should be used, if available.

Methods

These relatively general ideas were explored by considering a demanding application: visualization of data obtained with the virtual 296-channel system shown in Fig. 1. A problem with this system is that the measurement coils are located on eight different surfaces (spheres with the same radius, but generally different centers) so that isocontour mapping in the standard way (using an interpolation formula for a surface) is not feasible. To overcome this problem, first, a reference surface has to be defined and, second, the measured data have to be "projected" onto this surface.

With regard to the reference surface, the first two alternatives considered were a sphere and a surface obtained by re-scaling a geometrical reconstruction of the scalp. Both surfaces were fitted to the 296 measurement locations by adjusting the respective parameters: radius and center in the case of the sphere, geometrical scaling factor and origin in the case of the re-scaled scalp. The maximal distance between measurement locations and reference surface was typically of the order of one centimeter (RMS error of a few millimeters). In what follows, only the results obtained with the re-scaled scalp as the reference surface will be shown. In addition, the scalp itself (without re-scaling) and the surface of the cortex (derived from magnetic resonance images [3, 4]) were used as the reference surface.

The projection of the measured data onto the reference surface was achieved following an idea of Hämäläinen and Ilmoniemi [1] who suggested to estimate, in a first step, a distributed source using the minimum norm technique and to calculate then, in a second step, the magnetic field at the locations of interest. The distributed source was represented by 152 current dipoles on a spherical surface with 4 cm radius. Regularization was implemented by using a singular value decomposition and adjusting the number

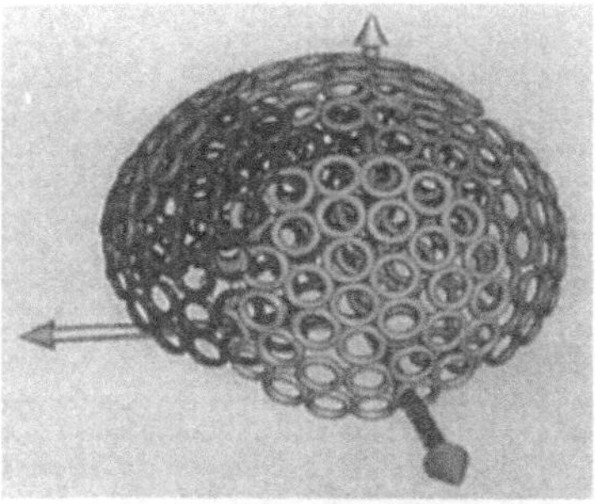

Fig. 1: A virtual 296-channel whole-head magnetometer system realized by combining eight subsequent measurements with a 37-channel system (Biomagnetic Technologies, San Diego): 3 measurements on each hemisphere, 2 measurements above central regions. The arrow pointing to the left (anterior direction) represents the x axis, whereas the arrow through the vertex (superior direction) corresponds to the z axis.

of singular values so that the deviation between the data and the field resulting from the estimated source corresponded to the standard deviation of the superimposed noise. After having estimated the distributed source, the radial component of the magnetic field was calculated for each discretization point on the reference surface, provided that the distance to the nearest coil did not exceed a certain limit (4 cm for the inflated scalp, 8 cm for scalp and cortex). Points outside this range were labeled with the color code "neutral" (light gray in the plots given here).

To test the above procedure, the magnetic field generated by a well-defined source configuration, consisting of 4 dipoles per hemisphere, was considered. The dipole parameters are given in table I. A spherical volume conductor model with the center corresponding to the origin of the coordinate system was assumed. The noise in the 296 channels was assumed to be uncorrelated and to have a Gaussian distribution with a standard deviation of 10 fT. It is worth mentioning that dipole #1 in the right hemisphere has a distance of only 2 cm from its pendant in the left hemisphere and that the moments of the two dipoles have an anti-parallel orientation. Thus, the magnetic fields arising from these two sources partially extinguish each other so that the remaining field has a quadrupolar appearance.

Results

Fig. 2 shows a comparison between the theoretical field map (left half), calculated directly from the source model specified in Table I, and the field map extrapolated on the basis of a simulated measurement with the virtual 296-channel system (right half). The plots in the upper row were obtained for an inflated scalp (inflation factor 1.29), whereas those in the bottom row were obtained for the scalp itself. The figure demonstrates that the proposed extrapolation technique works excellently if the inflated scalp serves as the reference surface: at least for the areas covered by the magnetometer coils (upper half of the head) the differences between theoretical and extrapolated field pattern are quite insignificant. These differences can be put down to the fact that the simulated field measurement was affected by superimposed noise and that a regularization was required for the current density estimation. The considerable differences found below the line nasion—ear are not very surprising, since the extrapolation can work, of course, only for areas adequately covered by measurement points. The bottom row of Fig. 2 demonstrates that the proposed extrapolation technique can be applied also if the scalp itself (rather than the inflated scalp) is used as the reference surface, though the differences between theoretical and extrapolated field pattern are much larger than those found for the inflated scalp. The corresponding results for the cortical surface are presented in Fig. 3. Theoretical and extrapolated field pattern show considerable differences now, but these difference typically correspond to only one or two isocontour levels.

	location (cm)			moment (nAm)		
#	x	y	z	Q_x	Q_y	Q_z
1	0	±1	7	0	±60	0
2	6	±3	5	5	±10	0
3	-5	±3	6	10	0	0
4	3	±4	5	10	0	10

Table I: Parameters of the source configuration used in the simulations (8 dipoles). The source configuration in the right hemisphere (y and Q_y with negative sign) is a mirror image of the configuration in the left hemisphere.

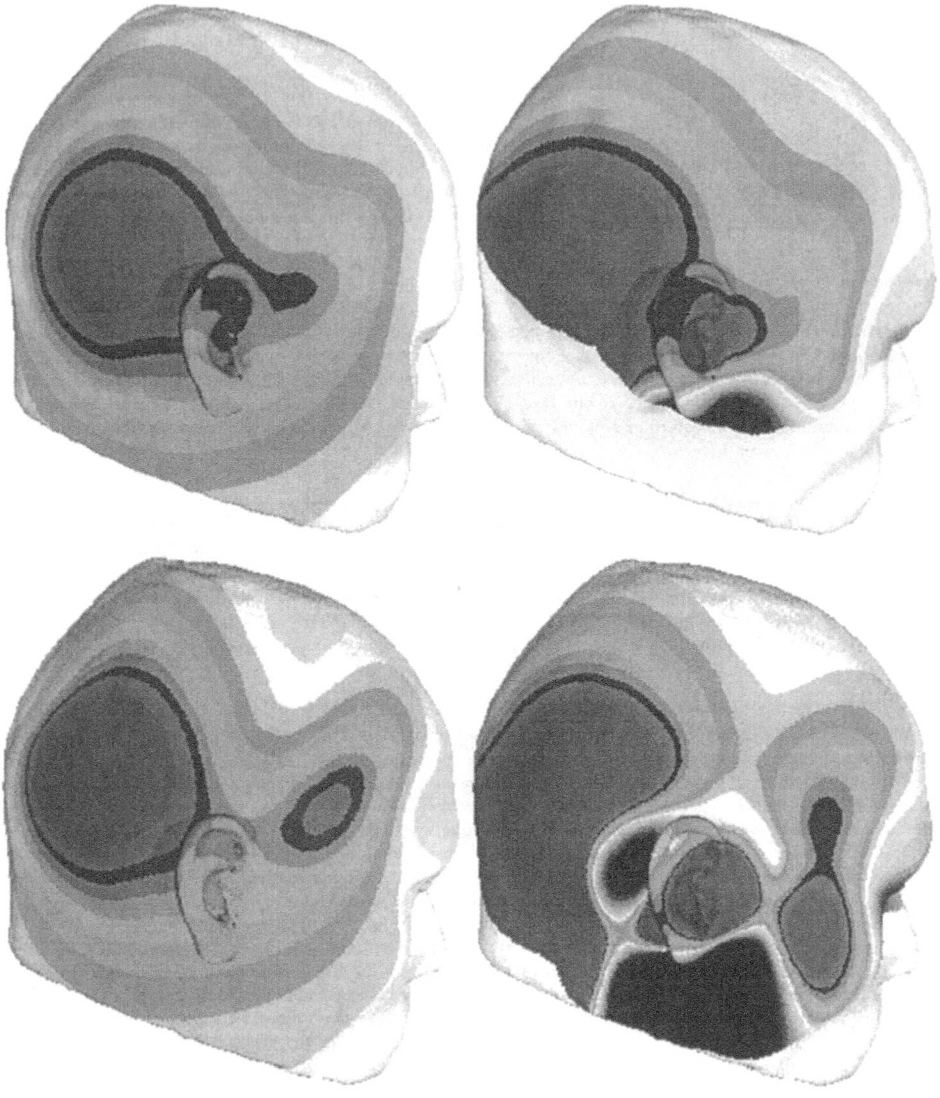

Fig. 2: Map of radial component of magnetic field on a surface corresponding to the scalp inflated by a factor of 1.29 (upper row) and on the scalp itself (bottom row). Left: Theoretical field distribution. Right: Field distribution extrapolated on the basis of a simulated 296-channel measurement.

Discussion

The results of this study suggest that a properly inflated scalp is a very appropriate reference surface. The choice of the sphere carrying the distributed source is probably relatively unimportant, provided that the

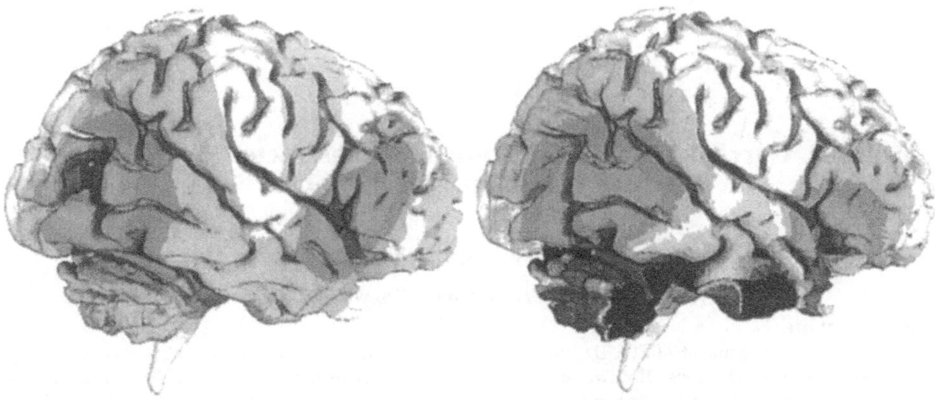

Fig. 3: Map of radial component of magnetic field on the cortical surface. Left: Theoretical field distribution. Right: Field distribution extrapolated on the basis of a simulated 296-channel measurement.

distances between reference surface and actual measurement locations are relatively small. Instead of the radial component of the magnetic field other quantities could be mapped as well, e.g. the magnetic scalar potential [5], which would have the advantage that the definition of a center of sphere is not required.

A comparison between the theoretical field patterns on the inflated scalp (Fig. 2, upper row on the left), the scalp itself (Fig. 2, bottom row on the left), and the cortical surface (Fig. 3, left half) illustrates that increasing the distance between sources and measurement locations results in a relative attenuation of high spatial frequencies. A projection of a measurement (carried out in a certain distance from the scalp) onto the scalp or the cortical surface can be interpreted as the attempt to overcome this attenuation. In the ideal case (no noise, no discretization errors, measurement e.g. on a complete spherical surface) this attempt would be indeed successful (the problem can be considered as the magnetic analogue of the inward harmonic continuation problem described for the electroencephalogram [2, 6]). However, in practice (i.e., in the presence of noise) the attenuation of high spatial frequencies corresponds to an irreversible loss of information, and a high price has to be paid for the attempt to compensate for this attenuation: since there is, of course, no way to recover information irreversibly lost, an increased contrast is associated with an increased uncertainty.

References

[1] Hämäläinen, M. S. and Ilmoniemi, R. J. Interpreting magnetic fields of the brain: minimum norm estimates. *Med. Biol. Eng. Comput.*, 1994, 32:35–42.

[2] Kearfott, R., Sidman, R., Major, D., and Hill, C. Numerical tests of a method for simulating electrical potentials on the cortical surface. *IEEE Trans. Biomed. Eng.*, 1991, 38:294–299.

[3] Lütkenhöner, B., Elbert, T., Menninghaus, E., Steinsträter, O., and Wienbruch, C. Electro- and magnetoencephalographic source analysis: current status and future requirements. In Herrmann, H., Wolf, D., and Pöppel, W., editors, *Workshop on Supercomputing in Brain Research: From Tomography to Neural Networks*, pages 175–192, Singapore, 1995. World Scientific.

[4] Lütkenhöner, B., Menninghaus, E., Steinsträter, O., Wienbruch, C., Gißler, H., and Elbert, T. Neuromagnetic source analysis using magnetic resonance images for the construction of source and volume conductor model. *Brain Topography*, 1995, 7:291–299.

[5] Sarvas, J. Basic mathematical and electromagnetic concepts of the biomagnetic inverse problem. *Phys. Med. Biol.*, 1987, 32:11–22.

[6] Sidman, R., Vincent, D., Smith, D., and Lee, L. Experimental tests of the cortical imaging technique — applications to the response to median nerve stimulation and the localization of epileptiform discharges. *IEEE Trans. Biomed. Eng.*, 1992, 39:437–444.

Multiple Current-Dipole Distribution Reconstructed by Modified Selective Minimum-Norm Method

Matsuura, K. and Okabe, Y.

RCAST, the University of Tokyo, Tokyo 153, Japan

Introduction

Recent development of SQUID systems enables us to measure biomagnetic fields simultaneously with more and more channels. How to relate the measured fields with multiple-dipole distribution is one of the most important software issues in biomagnetism. We proposed Selective Minimum-Norm (SMN) method to estimate multiple but sparse dipole distribution [1],[2] and modified it for robust reconstruction [3]. This paper discusses a noise-tolerance parameter introduced to the modified SMN method and presents inverse-analysis examples of event-related fields.

The ordinary SMN method uses a spatially-discrete model where two tangential components $q_j^{(\theta)}$ and $q_j^{(\phi)}$ of each dipole moment are regarded as independent variables. The modification of the SMN method consists of two features: First, each moment is represented by using the weighted magnitude q_j and orientation ψ_j. This representation is derived from

$$\left\|\boldsymbol{l}_j^{(\theta)}\right\| q_j^{(\theta)} = q_j \cos\psi_j \quad , \quad \left\|\boldsymbol{l}_j^{(\phi)}\right\| q_j^{(\phi)} = q_j \sin\psi_j \ , \tag{1}$$

where $\left\|\boldsymbol{l}_j^{(\theta)}\right\|$ and $\left\|\boldsymbol{l}_j^{(\phi)}\right\|$ are the Euclidean norms of lead-field vectors. Second, in order to reduce the effect of noise, we consider inequality constraints $\boldsymbol{b} - \boldsymbol{\varepsilon} \leq L(\boldsymbol{\psi})\boldsymbol{q} \leq \boldsymbol{b} + \boldsymbol{\varepsilon}$ instead of equality constraints $L(\boldsymbol{\psi})\boldsymbol{q} = \boldsymbol{b}$, where $\boldsymbol{b}$ is a measured magnetic-field vector and $L(\boldsymbol{\psi})$ is a parametric system matrix. $\boldsymbol{\varepsilon} = {}^t(\varepsilon, \varepsilon, \cdots, \varepsilon)$ is a noise-tolerance vector.

Imposing these inequality constraints, we use a parametric simplex algorithm to minimize the L_1-norm of $\boldsymbol{q}$. This minimization problem is implemented by using slack variables $s_1, s_2, \cdots, s_M$ as

$$\sum_{j=1}^{N} q_j \rightarrow \min . \quad \text{subject to} \quad \begin{cases} \sum_{j=1}^{N} l_{i,j}(\psi_j) q_j + s_i = b_i + \varepsilon & (1 \leq i \leq M) \\ \sum_{j=1}^{N} l_{i,j}(\psi_j) q_j - s_{i+M} = b_i - \varepsilon & (1 \leq i \leq M) \\ {}^\forall q_j, {}^\forall s_i \geq 0 & \end{cases} \tag{2}$$

where M is the number of the measurement points and N is the number of the grid points. Details of the simplex algorithm are reported elsewhere [4].

The tolerance vector employs the estimated value of the standard deviation of noise as each element ε. In clinical-data analysis, however, estimation error of the standard deviation of noise might disturb solution-dipole distribution. For example, overestimate can cause too sparse distribution to cover all the real source-dipoles. The worst case is a completely-silent solution occupied by slack variables. This occurs when $\varepsilon \geq \max\{|b_i|\}$, since $\boldsymbol{q}$=**0** satisfies the inequality constraints with such a large tolerance.

Methods

First, rates of overestimate were investigated based on Monte-Carlo simulation: we assigned different values to the noise-tolerance and examined whether solution-dipoles clustered around each source-dipole. The number of trials for each tolerance value was one hundred. This simulation used a spherical model described in Fig. 1. Signal sources consisting of two dipoles with the same magnitude were randomly located and oriented in the reconstruction region. Simulated random noise was generated according to uniform distribution whose mean value was zero.

Next, the modified SMN method was applied to the inverse analysis of right-head MEG responses to sequential visual stimuli of Japanese letters [5]: three "sample" letters followed by a "probe" letter. The sample letters were Hiragana letters presented simultaneously and the probe letter was a Katakana letter[1].

[1] Hiragana and Katakana letters have common readings but different shapes like lower and upper cases.

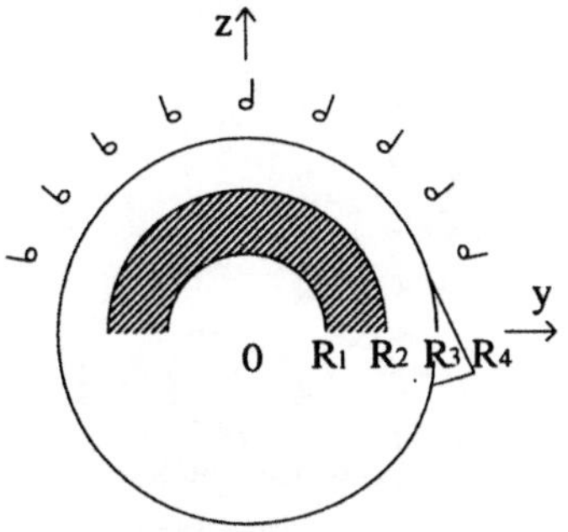

Fig. 1: Simulation model whose radius is R_3. M=57 sensors are on the upper hemisphere ($z > 0$) whose radius is R_4. The radial components of the magnetic fields are measured. N=2220 grid points are set up in the reconstruction region ($R_1 < r < R_2$, $z > 0$). Model parameters used in the simulation are: R_1=5.4cm, R_2=8.5cm, R_3=9.0cm, R_4=11.7cm.

Both of the stimuli had 0.2s duration and they were separated by randomized interval of 1.8–2.8s. The subject was instructed to discriminate whether the probe letter had the same reading as one of the sample-letter readings. The MEG responses measured with a 37-channel system (BTi) were averaged over 100 trials. The standard deviation $\sigma^{(est)}$ of the measured fields before the stimuli was assigned to the noise-tolerance. Solutions resulting from larger tolerances were also examined.

Results

The Monte-Carlo simulation gave the rates of overestimate shown in Fig. 2. This curve suggests that roughly 100%-error is allowed in the estimation of the standard deviation of noise as far as the overestimate is concerned: at least 95% of modified SMN solutions cover both of the source-dipoles even if the estimated standard deviation is twice as large as the real value. Figure 3 describes a simulation example with the deviation-estimation error of 60%: although the noise-tolerance ε is 1.6 times as large as the real standard deviation of noise σ, the significant solution-dipoles cluster around both of the source-dipole locations. When gradually larger tolerances were used, insignificantly-silent solutions appeared first: solution-dipole distribution not surrounding one of the real source-dipoles but clustering around the other. Completely-silent solutions were finally obtained when far more tolerant constraints with $\varepsilon > 5\sigma$ were used.

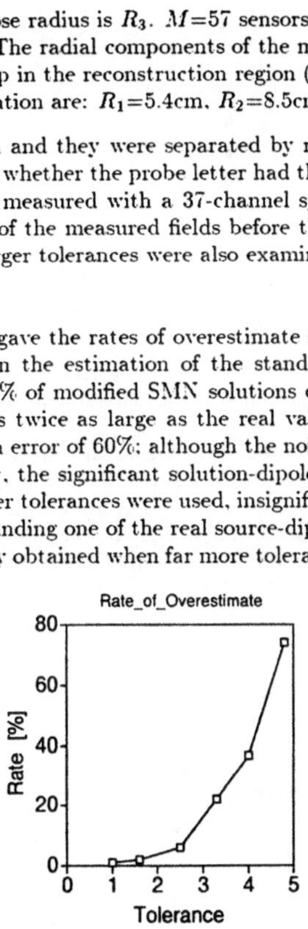

Fig. 2: Rates of overestimate investigated by Monte-Carlo simulation with 100 trials for each noise-tolerance. The horizontal axis is normalized by the real standard deviation of noise. Since source-dipoles and noise were randomly generated in each trial, Signal to Noise (S/N) ratios slightly differ from one trial to another: 8.3±0.5 [dB].

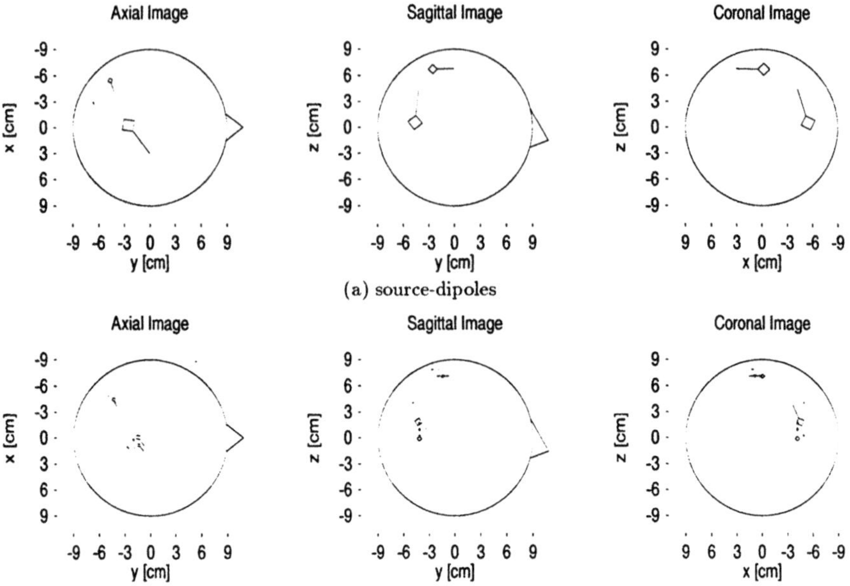

Fig. 3 panels: (a) source-dipoles; (b) solution-dipoles

Fig. 3: A simulation example when the tolerance is 1.6 times as large as the standard deviation of noise. S/N ratio is 7.7dB, which corresponds to 17%-noise in power. The location of each dipole is the center of the corresponding tadpole-shaped mark.

At a latency with the MEG contour map similar to a dipole pattern, tolerance parameter $\varepsilon = \sigma^{(est)}$ gave a solution composed of concentrated dipoles described in Fig. 4. This concentrated region was reconstructed even with larger tolerances up to $5\sigma^{(est)}$. A more complicated field gave a solution composed of multiple active regions shown in Fig. 5. These regions were obtained even with larger tolerances up to $2\sigma^{(est)}$, which is consistent with the result of the Monte-Carlo simulation.

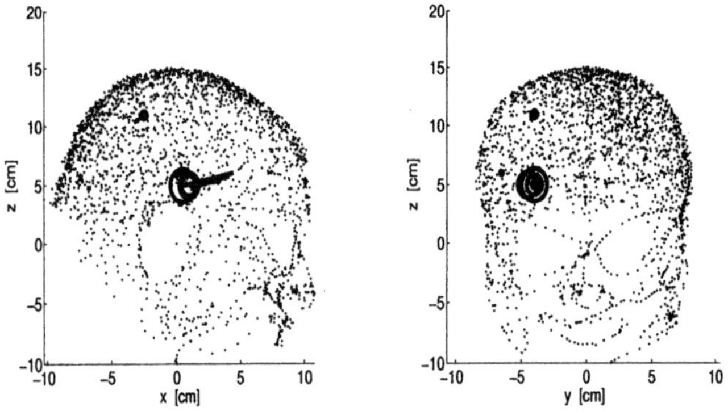

Fig. 4: Modified SMN solution for MEG response at 272msec after the probe onset.

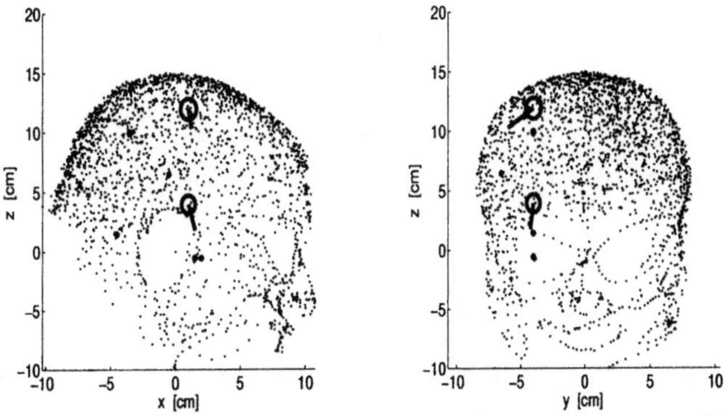

Fig. 5: Modified SMN solution for MEG response at 366msec after the probe onset.

Discussion

The robustness of a modified SMN method for multi-dipole reconstruction was examined by using Monte-Carlo simulation. Not only this simulation but also a real MEG study suggested that roughly up to 100%-overestimate is allowed when we estimate the standard deviation of noise to assign a moderate value to the noise-tolerance used in the modified SMN method. Even when noise is overestimated, survivor dipoles reconstruct some of the source-dipoles with no artifacts if the overestimate is insignificant, roughly less than 400%.

The MEG analysis might show dynamic activation of different cortical areas and should go on to further studies combined with anatomical information. Future simulation studies will include an examination of underestimate.

References

[1] Matsuura, K., and Okabe, Y. Selective minimum-norm solution of the biomagnetic inverse problem, IEEE Trans. Biomed. Eng., 1995, 42: 608–615.

[2] Matsuura, K., and Okabe, Y. Multi-dipole estimation by Selective-Minimum-Norm method, In: Baumgartner, C., Deecke, L., Stroink, G., and Williamson, S. J. Biomagnetism: Fundamental Research and Clinical Applications, Amsterdam, Elsevier, 1995.

[3] Matsuura, K., and Okabe, Y. A parametric simplex algorithm for multi-dipole estimation, Human Brain Mapping, 1995, Suppl. 1: 104.

[4] Matsuura, K., and Okabe, Y. Reconstruction of sparse dipole source by a modified SMN method, Proceedings of 10th Tokyo Institute of Psychiatry International Symposium (in press)

[5] Kuriki, S., Hirata, Y., Fujimaki, N., Tsuchiya, N., and Kobayashi, T. Neuromagnetic study of human cortical areas related to visual/auditory information processing, Human Brain Mapping, 1995, Suppl. 1: 237.

Acknowledgements

The authors wish to thank Prof. S. Kuriki and his colleagues in Hokkaido University for providing the MEG data for us.

This work was partly supported by Grant-in-Aid for Scientific Research (No. 00074035) from the Ministry of Education, Science, Sports and Culture of Japan.

An Estimation Algorithm of Neuromagnetic Source Using MRI Information

Mino, K.[1], Niki, N.[1], Nakasato, N.[2], Yoshimoto, T.[2]

[1]*Department of Information Science University of Tokushima, Tokushima, JAPAN;*
[2]*Department of Neurosurgery, Tohoku University School of Medicine, Sendai, JAPAN*

INTRODUCTION

Recently, the imaging technique of the brain function by using the SQUID technology has been developed since technical advancements of the SQUID make it possible to measure feeble magnetic fields of the brain with multichannel [1]-[4]. In this work we estimate the magnetic sources in the cortical region. The estimation region is limited to the surface of the cortex. For this limitation, the cortex is extracted from MR images. We assume that the points on the extracted cortex are current dipoles. All the obtained points are considered as candidate sources. We solve the inverse problem by means of the Multiple Signal Classification (MUSIC) method [2],[3] which is applied to the limited region, i.e. all the points on the surface of the cortex. In the MUSIC method, the measured magnetic fields data is divided into two subspaces, the signal subspace and the noise subspace. The Minimum Description Length (MDL) criterion is used for this division [5]. We describe here the estimation algorithm, and we present the computer simulations with the application to the real human brain data.

METHODS

Electric phenomena due to the neural activities in the human brain produces a magnetic field outside the head. We assume that this generated field is due to multi-current dipoles, and that the head is a spherically symmetric conductor. Then the relationship between the magnetic field and current dipole is described [1] by

$$\boldsymbol{b}(\boldsymbol{r}) = \frac{\mu_0}{4\pi F^2}\left(F\boldsymbol{q}\times\boldsymbol{r}' - (\boldsymbol{q}\times\boldsymbol{r}'\cdot\boldsymbol{r})\nabla F\right), \quad \begin{array}{c} \boldsymbol{a}=\boldsymbol{r}-\boldsymbol{r}',\ a=|\boldsymbol{a}|,\ r=|\boldsymbol{r}|,\ F=a\left(ra+r^2-\boldsymbol{r}'\cdot\boldsymbol{r}\right), \\ \nabla F=\left(r^{-1}a^2+a^{-1}(\boldsymbol{a}\cdot\boldsymbol{r})+2a+2r\right)\boldsymbol{r}-\left(a+2r+a^{-1}(\boldsymbol{a}\cdot\boldsymbol{r})\right)\boldsymbol{r}' \end{array} \tag{1}$$

where $\boldsymbol{b}(\boldsymbol{r})$ is a magnetic field at $\boldsymbol{r}$ and $\boldsymbol{q}$ is a dipole moment at $\boldsymbol{r}'$. As this equation includes the effect of the volume current, we can calculate the magnetic field more effectively. A radial component of the dipole moment makes no external magnetic field, so that only two tangential components contribute to the field. The measured field is represented by using the sensor direction $\boldsymbol{s}$, $b(\boldsymbol{r}) = \boldsymbol{s}^T\boldsymbol{b}(\boldsymbol{r})$.

Here, if a dipole has a fixed location for all measurement time, i.e. don't change the site, we can separate the parameter $\boldsymbol{q}$ from (1) [2],[3]. This separation indicates the linear relationship between the magnetic field and the dipole moment. Moreover we assume that the number of sensors is m, the number of dipoles is n, and the number of time samples is k, then the above equation becomes $\boldsymbol{B} = \boldsymbol{G}\boldsymbol{Q}$. $\boldsymbol{B}$ is an $m\times k$ matrix. $\boldsymbol{G}$ is an $m\times 2n$ matrix which stands for the relationship between the sensors and the dipole locations. $\boldsymbol{Q}$ is a $2n\times k$ matrix of dipole moments. From the linear relationship, we have only to obtain $\boldsymbol{G}$. $\boldsymbol{Q}$ is then calculated by using the Moore-Penrose pseudoinverse [1]-[4]. If the dipole has also a fixed orientation, the dipole moment is further partitioned into a unit orientation and a magnitude, $\boldsymbol{Q} = \boldsymbol{M}\boldsymbol{S}$. $\boldsymbol{M}$ is a $2n\times n$ matrix and $\boldsymbol{S}$ is an $n\times k$ matrix. In this case, the dipole is called a fixed dipole.

The estimation is generally performed by the parameter fitting in the least squares sense. But it becomes difficult when many dipoles exist. So we apply the MUSIC method [2],[3] to the spatio-temporal data $\boldsymbol{B}$. At first, the eigendecomposition of $\boldsymbol{R_B}$ which is the sample covariance matrix of the measured magnetic fields $\boldsymbol{B}$ is performed. R_B is decomposed as

$$R_B = \frac{1}{k} B B^T = [U_s \;\; U_n] \begin{bmatrix} \Sigma_s & \\ & \Sigma_n \end{bmatrix} [U_s \;\; U_n]^T , \tag{2}$$

where U_s and Σ_s are the matrix containing eigenvectors and the diagonal matrix whose diagonal elements are eigenvalues, of the signal subspace, and U_n and Σ_n are those of the noise subspace, respectively. If the noise is the white Gaussian noise, the eigenvalues of the noise subspace are equal to the variance of the noise σ^2. For division of the signal and the noise subspace, we use the MDL criterion [5]. This criterion selects the separation point between the signal and the noise subspace from the obtained eigenvalues automatically. If the magnetic fields B is due to the fixed dipoles, the number of eigenvalues corresponding to the signal subspace is equal to the number of dipoles n.

Next, we limit the search region to the cortex. The brain has surface layer, the cortex, and inner layer, the medulla. The cortex is called the gray matter because it looks gray on which a lot of neurons exist. Therefore neural activities are concentrated in this layer, we may regard it as the area where there are the magnetic sources. So we extract this gray matter from MR images and we fix the extracted area as source positions. We remove the skull and the scalp from MR images by fuzzy clustering, and we extract only the brain image, then an outline is extracted from this brain image. This outline is the surface of the cortex. We assume that the magnetic sources exist on the points of the outline. A coordinate transformation of the extracted points between MRI and SQUID system is performed [5]. The obtained points are the dipoles positions to be used for the calculation of the matrix G_i, as will be seen later.

The estimation is performed by calculating the following equation in the search region.

$$J = \lambda_{\min}\left(U_{G_i}^T U_n U_n^T U_{G_i}\right) \tag{3}$$

where $\lambda_{\min}(\cdot)$ is the minimum eigenvalue of the bracketed term, U_{G_i} is obtained from the singular value decomposition (SVD) of matrix G_i which is an $m \times 2$ matrix for one dipole, and U_n is the matrix containing eigenvectors of the noise subspace. The MUSIC requires the one dipole matrix G_i, regardless of the true number of dipoles, i.e. we have only to calculate (3) on each point in the search region. We adopt the cortex obtained from MR images for this search region. Each point on the cortex is used for calculation of G_i. On the cortex, we find the area where (3) becomes small.

RESULTS

We performed computer simulations to evaluate our algorithm. Simulated cortical shape was obtained from the practical MR Image as shown in Fig. 1. The edge line represents the surface of the cortex. We make a simulation of magnetic data by putting dipoles on the edge line, and we numbered the points on the cortex in order. Simulated dipoles were assumed to be on arbitrary points. Fig. 1 (a) is one dipole, (b) is the distributed source. They were oriented normal to the surface of the cortex. We had dipole magnitude fluctuate with time. If some dipoles were assumed, they had the same maximum magnitudes. It was assumed that the magnetic fields were measured with 66 sensors, and 300 time samples were generated on each sensor. We performed the estimation for the data simulated by the above conditions.

At first, one dipole and two dipoles simulations were performed. In the case of two dipoles, no correlation existed between dipoles. The cost function was calculated on each point. The result of the two dipoles is shown in Fig. 2. The index of horizontal axis indicates the numbered point on the cortical shape. Fig. 2 shows that the cost function value becomes small on the position (displayed by dash lines) where simulated dipole exists. Thus, we can see that the estimation result is satisfactory.

Next, we assumed a distributed source. This model was defined as a sequence of dipoles on a section represented by bold line, like Fig. 1 (b). When we think about the distributed source, there exist two cases.

These are about time variation on dipoles magnitude. One is that all of dipoles are active synchronously (t1), and another is that they are asynchronously active (t2, t3), as shown in Fig. 3. The former represents synchronous activation over large area, the latter represents propagation activation. In this simulation, we arranged 10 dipoles. Estimation results are shown in Fig. 4. The indexes corresponding to from 3 to 12 (within dash lines) indicate the positions where we arranged 10 dipoles. In the case of t1, the minimum value is the biggest. This indicates that it is difficult to estimate the sources if they are active synchronously. But, in each case, the cost function values become small on this section. In the case of many dipoles exist densely like this simulation, the estimation with the conventional method, the parameter fitting, becomes difficult. But the limitation to the surface of the cortex can remove this difficulty.

We also applied this algorithm to the AEF (auditory evoked magnetic field) data of normal person. Auditory stimulus was given to the subject's left ear. The magnetic fields were measured with 66 sensors. A number of time samples was 313 on each sensor. By using this data, we performed the eigendecomposition. The MDL criterion selected the 18 largest eigenvalues as the signal subspace. The rest corresponded to the noise subspace eigenvalues. In computer simulations, the separation point was clear from eigenvalues. In the application to the AEF data, it was not clear. This indicates that the measured magnetic fields include several noise such as the brain noise caused by other sources. The noise subspace divided by the MDL criterion has eigenvalues corresponding to the white Gaussian noise elements. We suppose that the signal subspace has eigenvalues corresponding to the signal elements and the brain noise elements, and the latter elements make the selection of the separation point between two subspaces difficult in the real data. We regarded the separation point as 18th eigenvalue, and performed the estimation. Fig. 5 shows the results synthesized with MR images. The cost function values are small around right temporal lobe. Thus it became possible to calculate (3) in the cortex region, and to judge the result visually. The cost function values is shown in Fig. 6. Each line indicates the values calculated on each MR image. In the second line, the value became minimum. From this result, the source of the AEF was considered as one dipole.

DISCUSSION

We have presented an estimation algorithm of the magnetic sources. We have assumed that the points on the surface of the cortex were the current dipoles, and have performed the estimation using the MUSIC method. The separation point between the signal subspace and the noise subspace eigenvalues was selected by the MDL criterion. In the application to the real human brain data, we found that the measured magnetic fields included several noise such as the brain noise. From the result of simulations, we consider that the magnetic source of the AEF data can be approximated to one dipole. But small cost function values like simulations was not obtained. We suppose that the brain noise effects on the estimation.

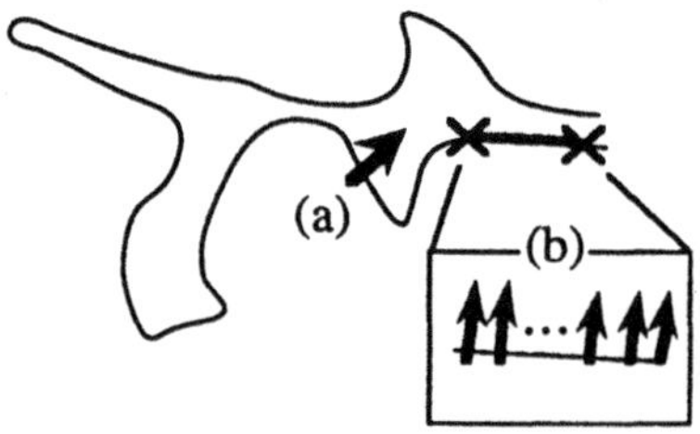

Fig. 1. Simulated cortex and dipoles.

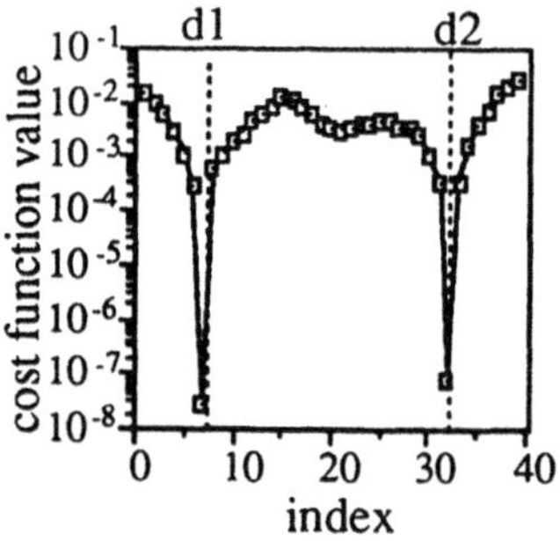

Fig. 2. Estimation result of two dipoles.

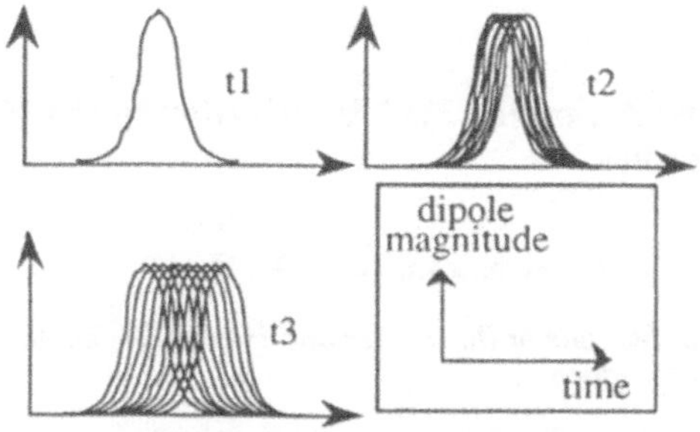

Fig. 3. Three time variations on dipole magnitude.

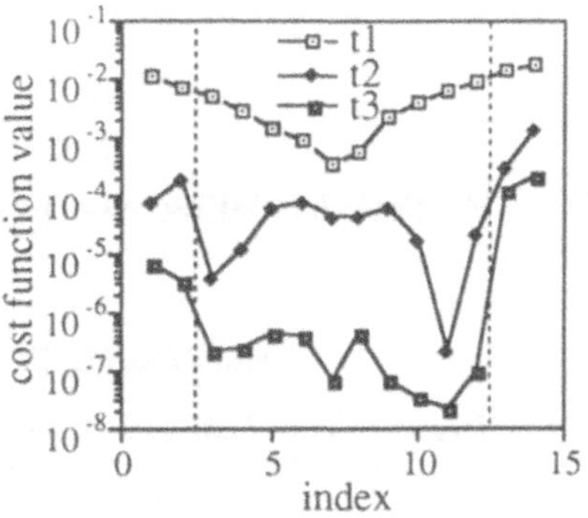

Fig. 4. Estimation results of the distributed source.

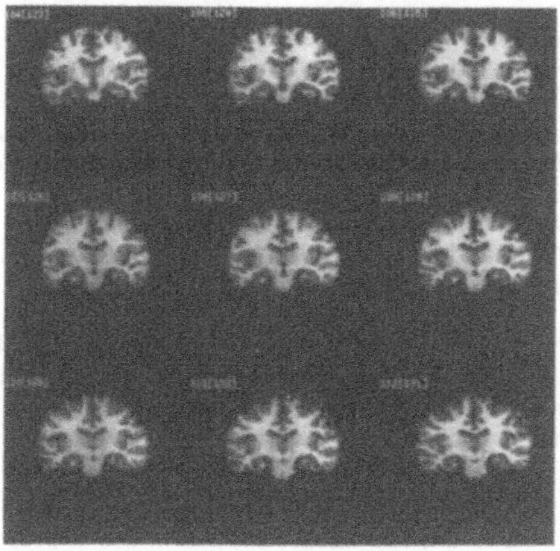

Fig. 5. Results synthesized with MR images.

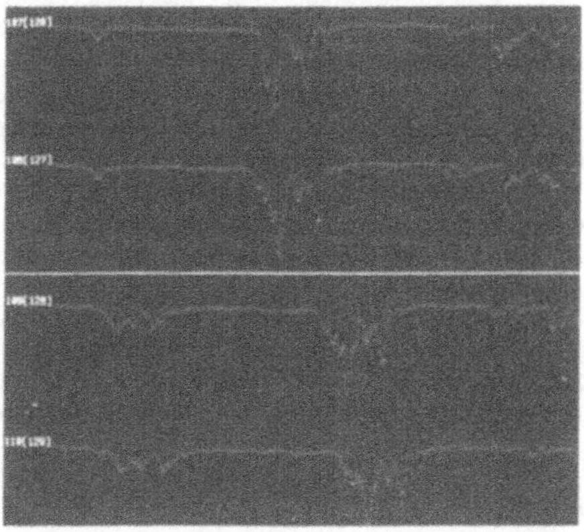

Fig. 6. Cost function values of the AEF data.

REFERENCES

[1] Sarvas, J. Basic mathematical and electromagnetic concepts of the biomagnetic inverse problem, Phys. Med. Biol., 1987, 32: 11-22.

[2] Mosher, J. C., Lewis, P. S., Leahy, R. M. Multiple dipole modeling and localization from spatio-temporal MEG data, IEEE Trans. Biomed. Eng., 1992, 39: 541-557.

[3] Mosher, J. C., Leahy, R. M., Lewis, P. S. Matrix kernels for MEG and EEG source localization and imaging, Proc. ICASSP-95, 1995, Detroit, MI.

[4] Wang, J.-W., Williamson, S. J., Kaufman, L. Magnetic source images determined by a lead-field analysis: the unique minimum-norm least-squares estimation, IEEE Trans. Biomed. Eng., 1992, 39: 665-675.

[5] Mino, K., Niki, N., Tamura, I. Neuromagnetic source analysis with a 64-Channel SQUID system and MR imaging, Proc. of the IEEE Workshop on Biomed. Image Analysis, 1994, Seattle: 255-260.

Realistic Modeling, Analysis and Animated 3D Visualization for MEG Studies

Morabito, M., Takeda, T., Kobayakawa, T., Endo, H.

Behavior Control Laboratory, National Institute of Bioscience and Human-Technology, Tsukuba, Japan

Introduction

With the availability of new analysis techniques and visualization capabilities for biomagnetic studies, it is possible to improve the significance and reliability of the results. For example, it is nowadays common to solve the biomagnetic forward problem modeling the head as a sphere: with such a model the computations are easy and fast. Per contra, its simplicity brings some unavoidable inaccuracies, that can be avoided using a more realistic head model. A physically plausible shape can be obtained from anatomical MRI data, and the corresponding computations simplified using the Boundary Element Method (BEM).

Nevertheless, it is at the same time essential to facilitate the judgement of the final user, developing an easy-to-use, comprehensive interface for the data [1]. To achieve this goal, we are using the Application Visualization System (AVS), implementing a comprehensive, tridimensional view and analysis of MEG and MRI data, on a common workstation.

A realistic head model has been (virtually automatically) built from MRI data; the forward problem has been solved using the BEM, and the inverse problem with the Simplex Method. Preliminary results using actual experimental data seem very promising. Anatomical structures and (animated) measured and computed activities are visualized in 3D and the system is ready for future upgrading.

Methods

A. Realistic head modeling

The inversion of magnetoencephalographic data is basically influenced by three choices: the inversion technique, the source model and the head model (used to solve the forward problem). For the latter, the first and easiest choice is a simple sphere; an analytical solution of the forward problem exists, and the spherical symmetry introduces some simplifications, therefore making the computations much easy [2]. This comes at the cost of unavoidable inaccuracies, like for example in the temporal area, where there is no meaning in defining a common radius of curvature for the local head structures. In general, the physical plausibility of a sphere as a model for the human head is highly questionable [3], and a realistic shape is needed to improve the forward problem solution. It can be built using anatomical data, such as those obtained with Magnetic Resonance Imaging (MRI). Without loss of accuracy, the corresponding volumic problem can be solved via surface computations utilizing the Boundary Element Method (BEM), that is via the analysis of the (triangulated) boundaries ("layers") of regions with different conductivities [4]. In the case of the human brain, the BEM is usually applied using either a single-layer or a three-layer model, respectively considering only the outer brain surface or the three discontinuities brain-skull, skull-skin and skin-air [5].

B. Shape triangulation

We have developed an almost completely automatic process for the single-layer model. The procedure is started by the user with a simple clicking of the mouse to select the brain on a single slice of MRI data (semi-automatic segmentation). All the subsequent steps are automatic. First of all, the outer contour is generated using all the slices. The resulting points are "decimated": for example, a row of points is simply represented by the 2 extrema. A triangulation algorithm is then applied on subsampled data; the amount of subsampling determines the accuracy of the resulting model and the number of triangles used. Finally, a simple algorithm computes the externally-oriented triangle normals. On a Sun Sparc10 workstation, a few minutes are enough to complete this entire procedure. The result is currently being used to realize a 2350-triangle plastic model of the brain of one subject.

Another, completely automatic segmentation algorithm, firstly developed by H.J. Wieringa [5], has been applied to extract the outer head surface (skin). It presently takes 10 minutes to perform. We are working to extend it in order to triangulate the skull and the skin and incorporate them in a 3-layer realistic head model.

C. Animated 3D visualization and the AVS environment

The work presented here belongs to a wider project aimed to the development of increasingly more sophisticated data-analysis and data-visualization tools. Such a project is based on the use of programs able to process all the kinds of data used in the analysis, and to make them available to the end user in an easy-to-understand way. For this task we have chosen to work with the Application Visualization System (AVS) [6]. It is a very flexible data-flow oriented environment, especially suited for (animated) 2D and 3D scientific visualization. The user interface and even the data structure can be customized. AVS runs on a common workstation (Unix/X-Window). The software developer defines (graphically) the structure of the processing pipeline, connecting a series of basic processing modules. More than 60 modules have been developed in our laboratory till now, ranging from the animated 3D representation of MEG to their superposition to MRI data, to the solution of the biomagnetic forward and inverse problems. The result is a comprehensive, tridimensional view and analysis of the data.

D. Data analysis with BEM

In the present implementation the sources are modeled as "moving Equivalent Current Dipoles" (ECDs) and the inverse problem is solved using the Simplex Method. We are planning to implement other source models and inversion techniques [7] as well. As a preliminary step, in order to assess the characteristics of the BEM in the case of 1-layer modeling, we analyzed the inversions of biomagnetic data obtained with visual stimulation (checkerboard). We solved the inverse problem at the latency of the N100m peak, for a single ECD. For the forward problem, we compared the solutions obtained using as head model: a) a sphere (analytical formulas, "analytical sphere"); b) a triangulated sphere (BEM, 480 triangles); c) a realistic shape of the outer brain surface (BEM, 1413 triangles). For the cases a) and b), the sphere has been fitted to the occipital area. To triangulate the sphere for the case b), we employed a resolution of 0.5cm along the up-bottom (Z) direction, sampling 16 points (angular distance = 22.5deg) at each height. For the case c), the original data have been subsampled to a resolution of 0.45cm along the Z and front-to-back (X) directions, and of 0.8cm along the left-to-right (Y) direction.

Results

All the pictures have been obtained from AVS views of 3D "geometric" objects whose views and positions are in complete user control.

Fig.1 shows the segmented brain surface (left) and its corresponding triangulation (right, 1413 triangles). With a subsampling of 0.38mm along X and Z, and of 0.2mm along Y, we have computed up to 6411 triangles. In Fig.2 the measured MEG data (upper left) and 3 solutions of the biomagnetic inversions are shown (tridimensional view from the back): respectively using the analytical sphere model (upper right), the triangulated sphere BEM model (lower left) and the realistic shape BEM model (lower right).

Finally, Table 1 shows the respective numerical results. The goodness-of-fits are very similar, with a maximum difference of 2.40%. The dipole positions differ by 9.2mm (between the realistic shape BEM and the analytical sphere), and by 2mm between the latter and the triangulated sphere BEM. The dipole strength in the case of realistic shape BEM is 40% smaller than in the case of the analytical sphere, and 35% smaller than in the case of triangulated sphere. In all the three cases, the Y components of the dipole directions are almost identical, while there are differences in the X and Z components.

Discussion

In the course of developing our analysis-visualization project, we have realized a realistic model of the human brain, starting from MRI data and with a minimal intervention by a human operator. Using AVS, the 2D and 3D visualizations, the temporal animation and the analysis of the data have been implemented, thus realizing the integration of MEG and MRI data. The user is able to study both anatomical and functional brain imaging data in a great variety of ways, consistently using the same interface. We believe this is an important step: "study of the temporal dynamics...requires animation...there might be new opportunities for insight into how the mind's processes are constrained not only in the space of the brain but in the time course of its neurophysiological operations" [1].

The solution of the forward problem by using the BEM has shown good performances in the case of a spherical model; the errors on noise-free simulated data have been negligible even using only a few hundreds triangles (data not shown here). In the preliminary results presented here, the realistic shape BEM has given promising results when applied to the inversion of actual experimental data (visual evoked fields): the goodness-of-fit is comparable in magnitude to that obtained using the analytical sphere model; there is a 9mm difference in position, and 5nAm difference in strength. The reliability of the realistic-shape BEM will be further assessed, also using a plastic head model. As next computational step, the single-layer BEM will be extended to 3 layers, representing the effects of brain, skull and skin.

References

[1] Tucker, D.M., Liotti, M., Potts, G.F., Russell, G.S., and Posner, M.I. Spatiotemporal analysis of brain electrical fields, Human Brain Mapping, 1994, 1:134-152

[2] Hamalainen, M.S., Hari, R., Ilmoniemi R. J., Knuutila, J., and Lounasmaa, O. V. Magnetoencephalography: theory, instrumentation, and applications to noninvasive studies of the working human brain, Review of Modern Physics 1993, 65: 413-497.

[3] Hamalainen, M.S., and Sarvas, J. Feasibility of the homogeneous head model in the interpretation of neuromagnetic fields, Physics in Medicine and Biology, 1987, 32:91-97.

[4] Hamalainen, M.S., and Sarvas, J. Realistic conductivity geometry model of the human head for interpretation of neuromagnetic data, IEEE Transactions on Biomedical Engineering, 1989, 36:165-171.

[5] Wieringa, H.J. MEG, EEG and the Integration with Magnetic Resonance Images (PhD. Thesis), Twente, University of Twente, 1993.

[6] Upson, C., Faulhaber T.jr., Kamis, D., Laidlaw, D., Schlegel, D., Voam, J., Gurwitz, R., and van Dam, A. The Application Visualization System: a computational environment for scientific visualization, IEEE Computer Graphics Application, 1989, 9:30-42.

[7] Lewis, P.S., Mosher, J.C., and Lehay, R.M. Neuromagnetic source reconstruction, Proceedings of IEEE International Conference on Acoustics, Speech and Signal Processing (ICASSP), Detroit, 1995.

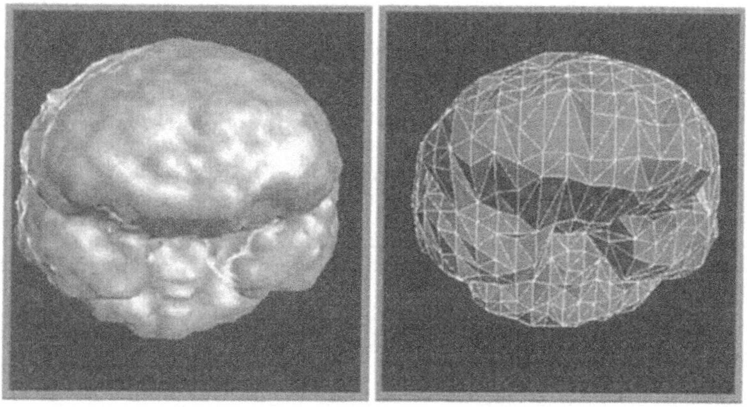

Fig1. Outer brain surface (left) and its triangulation (right) (1413 triangles).

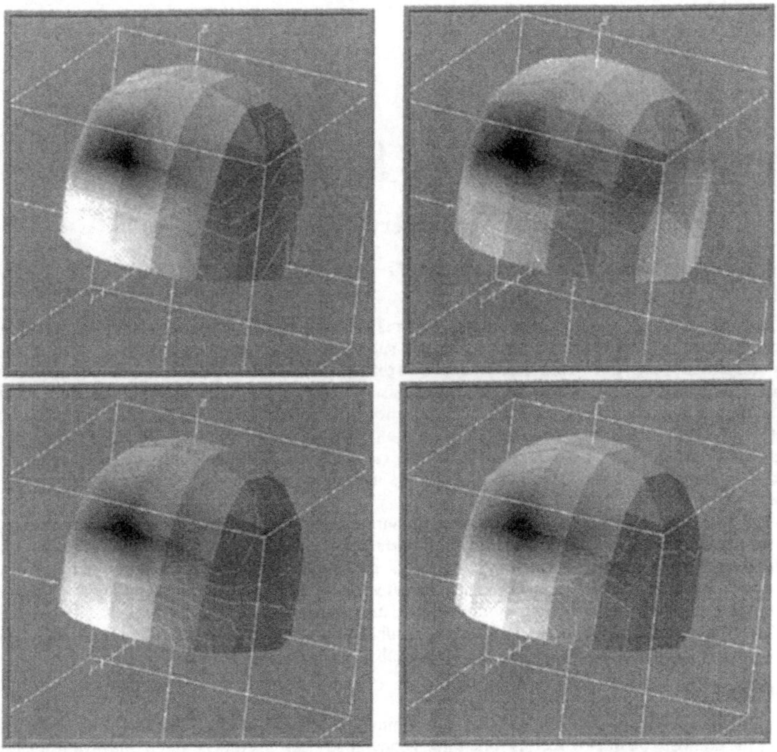

Fig.2 Magnetic fields after inversion of visual evoked field data: (upper left) measured data; (upper right) analytical sphere; (lower left) spherical shape BEM; (lower right) realistic shape BEM. Black is used for negative fields, white for positive fields.

	Analytical	BEM spherical shape	BEM realistic shape
fit %	89.38	87.63	86.98
x (mm)	-56.60	-57.59	-62.39
y (mm)	18.40	18.57	11.67
z (mm)	52.30	51.36	57.37
direction x	-0.09	0.25	-0.03
direction y	0.95	0.97	0.97
direction z	0.3	-0.07	-0.23
strength (nAm)	14.36	12.90	8.42

Table 1. Goodness of fit and dipole positions, directions and strengths in the 3 cases

Subspace Angles: A Metric for Comparisons in EEG and MEG

Mosher, J.C.

Los Alamos National Laboratory, Los Alamos, New Mexico, USA

Introduction

In forward head modeling, various approximations are made in order to keep the problem tractable. Simplifications can yield models ranging from simple spherical models to multi-tessellated arbitrary surfaces in a boundary element model (BEM). Spherical head models differ in the number of shells and the assumed conductivities. Other assumptions in the BEM include the choice of basis sets, such as constant, linear, or quadratic variations of the voltages across the individual areal elements, or the selection of error-weighting method, such as collocation, Galerkin, or "direct" methods. Numerical versus analytic integration can also yield numerical differences. These differences in parameters and approximations can yield models whose external fields (EEG potentials or MEG magnetic fields) differ for the same internal source configuration. Quantitative measures are needed to determine if these differences are significant.

The conventional approach is to define the source to be an arbitrarily placed single current dipole, then compare the field pattern generated. If two forward models generate similar patterns for many different placements of the dipole, then the presumption is that the two models are comparable. Several different methods of comparison between models have been proposed, such as the residual variance (RV), which we briefly review. We then propose a new method that effectively performs a multidimensional correlation between the two models, yielding their *subspace angles*. The method is specifically presented for head model comparisons, but we will also briefly show that the MEG-MUSIC algorithm [1] can be easily explained by these subspace angles.

Methodology

The concept of subspace angles may be explained by some examples. Let us assume first a very simple example of the comparison of two vectors (one-dimensional matrices). Vectors $\boldsymbol{a}$ and $\boldsymbol{b}$ may for instance be the gain vectors for a constrained dipole orientation model, each generated by a different method. The goal is to quantitatively describe how the two vectors are alike or different. For the measure of *residual variance*, we compare the difference between the vectors, scaled by one of the norms, $\text{RV}(\boldsymbol{a}, \boldsymbol{b}) = (\boldsymbol{a}-\boldsymbol{b})^T(\boldsymbol{a}-\boldsymbol{b})/a^2$ (alternatively, one may take the square root of this value for the *relative difference measure* RDM). This gives us a measure of how different $\boldsymbol{b}$ is from $\boldsymbol{a}$ in an average sense.

A problem with RV may arise when $\boldsymbol{b}$ is simply a scaled version of $\boldsymbol{a}$, for instance $\boldsymbol{b} = c\boldsymbol{a}$. Then the RV is $(1-c)^2$, indicating a significant difference between the two vectors. However, a scalar difference between two vectors may not be significant in some instances. For instance, in the EEG scalp potentials of a single shell spherical model, a scalar difference between two models may simply be due to a difference in assumed conductivities. The shape of the two vectors are otherwise identical, and a localization algorithm would perform identically for the two cases. The difference between the two models would arise in the amplitude of the assumed source, not in its localization. Similarly, one vector may simply be the negative of the other due to sign conventions, in other words with a scale factor of $c = -1$.

We would therefore like a measure that quantifies how similar the two patterns are, i.e., the correlation between the two. The correlation is found as the inner product between $\boldsymbol{a}$ and $\boldsymbol{b}$, divided by their norms, $\text{CORR}(\boldsymbol{a}, \boldsymbol{b}) = (\boldsymbol{a}^T\boldsymbol{b})/(ab)$. A correlation of unity indicates perfect agreement of the two waveforms, within a scale factor. The scale factor may easily be determined by the ratios of the norms, b/a, if so desired. A perfect negative correlation simply indicates a negative scale factor. The correlation and scale factor may be combined to generate the RV,

$$\text{RV}(\boldsymbol{a}, \boldsymbol{b}) = \left(1 - \frac{2b}{a}\text{CORR}(\boldsymbol{a}, \boldsymbol{b}) + \frac{b^2}{a^2}\right) \tag{1}$$

Thus we might view (1) as a method of partitioning the RV into two constituent components, the correlation and the scale factor. We will make one other adjustment to (1) by separating out the sign of the correlation, $\text{SGN}(\boldsymbol{a}, \boldsymbol{b})$, such that we will use a definition of the absolute correlation ACORR and its sign separately,

$$\text{RV}(\boldsymbol{a}, \boldsymbol{b}) = \left(1 - \frac{2\text{SGN}(\boldsymbol{a}, \boldsymbol{b})b}{a}\text{ACORR}(\boldsymbol{a}, \boldsymbol{b}) + \frac{b^2}{a^2}\right). \tag{2}$$

The (absolute) correlation between $\boldsymbol{a}$ and $\boldsymbol{b}$ has a geometric interpretation. Consider $\boldsymbol{a}$ and $\boldsymbol{b}$ to be two vectors in a multidimensional space. The correlation between $\boldsymbol{a}$ and $\boldsymbol{b}$ is the cosine of the angle between them, as

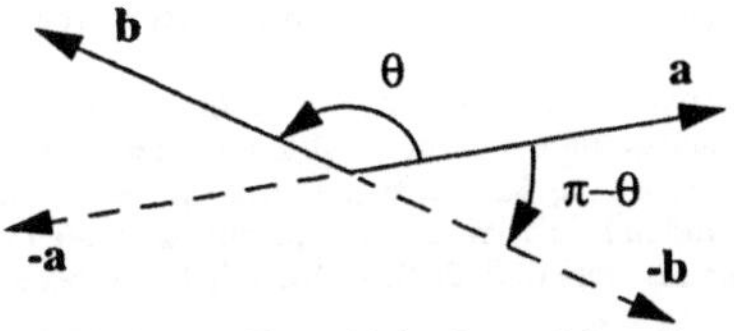

Figure 1. The correlation between vectors $\boldsymbol{a}$ and $\boldsymbol{b}$ is the cosine of the angle between them, θ. The absolute correlation between $\boldsymbol{a}$ and $\boldsymbol{b}$ is the cosine of the acute angle formed by the pair of intersecting lines. We can refer to the acute angle (θ or $\pi - \theta$) in a more general sense as the *subspace angle* between $\boldsymbol{a}$ and $\boldsymbol{b}$.

shown in Fig. 1. If we think of $\boldsymbol{a}$ and $\boldsymbol{b}$ as lines in space (i.e., allowing both positive and negative directions), then by convention we consider the angle between the two lines to be the lesser of $\pi - \theta$ or θ. The absolute correlation is the cosine of this acute angle between these two lines. This acute angle is also known in a more general sense as the subspace angle between $\boldsymbol{a}$ and $\boldsymbol{b}$.

Now we consider a more general case of two matrices, $\boldsymbol{A}$ and $\boldsymbol{B}$, each containing two columns and two or more rows. These matrices could represent the gain matrix relating a tangentially oriented dipole to the MEG sensor measurements, with each column representing one of the two tangential dipole components. The two matrices may represent two methods of generating this *forward* model. We would like to compare these two matrices to understand how they are similar, again using correlation analysis. We form the vector $\boldsymbol{a} = \boldsymbol{A}\boldsymbol{x}$, the linear combination of the columns of $\boldsymbol{A}$, where $\boldsymbol{x} = x[\cos\alpha, \sin\alpha]^T$ with magnitude x and a unit-norm two-dimensional orientation represented by $\alpha, 0 \le \alpha < \pi$. Similarly, we form $\boldsymbol{b} = \boldsymbol{B}\boldsymbol{y}$, $\boldsymbol{y} = y[\cos\beta, \sin\beta]^T$. We then calculate the absolute correlation between these two vectors, $\mathrm{ACORR}(\boldsymbol{a}, \boldsymbol{b}) = \mathrm{ACORR}(\boldsymbol{A}, \alpha, \boldsymbol{B}, \beta)$; considerations of the magnitudes x and y can be disregarded in the correlation, as can α and β greater than π. We generate a grid of values of α over the range 0 to π, and for each α we calculate the absolute correlation using β over the same grid. A coarse grid might yield the results shown in Table 1.

Table 1: Absolute correlations between $\boldsymbol{a}$ and $\boldsymbol{b}$ as a function of α and β.

	$\beta = 0$	36	72	108	144	Row Max
$\alpha = 0$ (degrees)	0.6735	0.1848	0.3722	0.7862	0.9027	0.9027
36	0.9406	0.6802	0.1610	0.4174	0.8388	0.9406
72	0.8484	0.9158	0.6327	0.1108	0.4545	0.9158
108	0.4322	0.8016	0.8628	0.5968	0.1034	0.8628
144	0.1492	0.3812	0.7633	0.8548	0.6218	0.8548
Column Max	0.9406	0.9158	0.8628	0.8548	0.9027	

We extract from this matrix of data the two vectors of data labeled "Column Max" and "Row Max." We are interested in two quantities. The first is the overall maximum correlation. We find this quantity as the maximum of either the Column Max or the Row Max. In the table presented, we see this occurs for $\alpha = 36$, $\beta = 0$ degrees, for a maximum correlation of 0.9406. Thus, based on this coarse table, the best way to correlate matrix $\boldsymbol{A}$ with matrix $\boldsymbol{B}$ is to form $\boldsymbol{a} = \boldsymbol{A}[\cos 36°, \sin 36°]$ and $\boldsymbol{b} = \boldsymbol{B}[\cos 0°, \sin 0°]$. The second quantity of interest is the minimum of the Column Max or Row Max, which occurs here for $\alpha = 144$, $\beta = 108$ degrees, for a correlation of 0.8548. This *minimax* correlation is the worst way to linearly combine $\boldsymbol{A}$, since it yields a maximum correlation with $\boldsymbol{B}$ of only 85%, the *minimum* of the *maximum* correlations.

Of course, the grid in Table 1 is too coarse to draw such conclusions, and we should grid more finely. However, a far more elegant method bypasses the need to search over these grids and is easily extended to arbitrary dimensional matrices. The steps are [2] (p. 585),

1) Perform a singular value decomposition (SVD) of $\boldsymbol{A}$, such that $\boldsymbol{A} = \boldsymbol{U}_A \boldsymbol{S}_A \boldsymbol{V}_A^T$. Similarly decompose $\boldsymbol{B}$. Retain only those components of $\boldsymbol{U}_A$ and $\boldsymbol{U}_B$ that correspond to non-zero singular values, i.e., the number of columns in $\boldsymbol{U}_A$ and $\boldsymbol{U}_B$ correspond to their ranks, and the other matrices are square, with dimension equal to the ranks.

2) From $\boldsymbol{C} = \boldsymbol{U}_A^T \boldsymbol{U}_B$, and decompose as $\boldsymbol{C} = \boldsymbol{U}_C \boldsymbol{S}_C \boldsymbol{V}_C^T$. Let r be the minimum of the ranks of $\boldsymbol{A}$ and $\boldsymbol{B}$. Let $s_1, \ldots, s_r$ be the r singular values extracted from the diagonal of $\boldsymbol{S}_C$, $s_1 \ge \ldots \ge s_r$.

3) Optionally form $\boldsymbol{U}_a = \boldsymbol{U}_A \boldsymbol{U}_C$, $\boldsymbol{U}_b = \boldsymbol{U}_B \boldsymbol{V}_C$, $\boldsymbol{Q}_a = \boldsymbol{V}_A \boldsymbol{S}_A^{-1} \boldsymbol{U}_C$, and $\boldsymbol{Q}_b = \boldsymbol{V}_B \boldsymbol{S}_B^{-1} \boldsymbol{V}_C$.

The largest singular value s_1 is the maximum absolute correlation between $\boldsymbol{A}$ and $\boldsymbol{B}$, and the smallest singular value, s_r, is the minimax absolute correlation. The best way to arrange $\boldsymbol{A}$ is found in the first column of $\boldsymbol{Q}_a$, $\boldsymbol{a}_1 = \boldsymbol{A}\boldsymbol{q}_{a1}$, which is best correlated with $\boldsymbol{B}$ when it is arranged as $\boldsymbol{b}_1 = \boldsymbol{B}\boldsymbol{q}_{b1}$. The first columns of $\boldsymbol{U}_a$ and $\boldsymbol{U}_b$ are

a_1 and b_1 normalized to unit-norm. Similarly, the worst way to arrange A is $a_r = Aq_{ar}$, which is best correlated with $b_r = Bq_{br}$ with a correlation of only s_r.

We designate θ_1 as the *principal subspace angle*, such that $\cos\theta_1 = s_1$ and similarly designate the other subspace angles. The matrices U_a and U_b are each orthogonal, i.e., each column is orthogonal to all other columns within the matrix, and each column is unit-norm.The columns of U_a and U_b are known as the *principal vectors*. In the example presented, we find that the maximum correlation is indeed 95.23% and the minimax is 86.1%, corresponding to subspace angles of 18 and 31 degrees. Since our matrices were rank 2, there are only two subspace angles.

Results

In a companion paper [5], we generate the forward EEG or MEG model for a dipole in a sphere, but using different approaches of the boundary element method (BEM). The differences in our approaches focus on the polynomial approximation assumed across each tessellated triangle, the error weighting used to control the fit, and the inclusion of an isolated skull approach. The subspace angle was one of a couple metrics used to quantify the differences between models. In this paper, we present another use of subspace angles, for the comparison of different EEG models. In Table 2, we present the subspace angles for four EEG head models, as compared to a fifth EEG head model.

Table 2: Comparison of Different Spherical Head Models Using Subspace Angles.

Head Model	Radii (fraction of 88 mm)	Conductivities (mho/meter)	Location (mm) (% True Location)	Minimax	
				Correlation (%)	Scalar Ratio
de Munck ([7])	[0.63, 0.88, 0.91, 0.95, 1]	[5, 2, 3, 0.0040, 1]	[0, 0, 54]	Base Model	1
Stok ([6], [7])	[0.84, 0.8667, 0.9467, 1]	[0.33, 1, 0.0042, 0.33]	[0.12, 0.16, 51.4] (95.3%)	100.00	12.7
Rush/Driscoll ([6]) Ary ([7])	[0.87, 0.92, 1]	[0.33, 0.0042, 0.33]	[0.22, 0.25, 48.8] (90.3%)	100.00	13.9
Cuffin & Cohen 1979[[6]]	[0.8977, 0.9205, 0.9659, 1]	[0.33, 1, 0.0042, 0.33]	[0.3, 0.2, 43.7] (80.9%)	99.96	16.9
Single Shell	1	1	[0.61, 0.44, 31.7] (58.8%)	99.94	6.96

The head model parameters were extracted from [6] and [7], as indicated, and all were applied to a 88 mm outer spherical radius model. The number of shells ranged from 1 to 5, and 528 EEG sensors were spaced about 1 cm apart on the outer hemisphere. The de Munck head model was selected as the base model, and a dipole was placed just 1.4 mm from its innermost shell. The other models were then used in a three-dimensional Nelder-Meade Simplex minimization algorithm, with the goal of minimizing the minimax (third) subspace angle with the de Munck model, by changing the location of the dipole. The results confirm that each single dipole model has a corresponding dipole location in the other head models, and in particular in the single shell model. The minimax moment orientation was in every case z-directed, but the minimax correlation was nonetheless 99.9% or better over all five models. Although the results of Table 2 are presented for a single dipole location, other locations yielded similar results. For example, we used the Nelder-Meade Simplex algorithm to minimize the minimax subspace angle between the Single Shell model and the Cuffin & Cohen model, for 1000 dipoles spread mostly uniformly with depth. This minimization function converged well in all cases to locations 99% or better correlated.

The percentages in column 4 indicate the percentage of the radial distance to the dipole versus the true distance. The locations are almost, but not exactly on the z-axis, which is also the radial line through the true location. If the search were restricted to the radial line through the true location, then these percentages would be equivalent to the "Ary" correction factor. In general, the single dipole correction factor is a function of eccentricity and reveals how different head models may be interchanged by adjusting the dipole's depth between models. Two synchronous dipole models will in general not use the same correction factors as one dipole; see [6], [7] for further discussions on improvements for the multiple dipole cases. In these prior works, empirical results were often generated by fixing not only the base models location, but its orientation as well. Using subspace angles, we do not have to fix the orientation; indeed, we let the subspace angles select the worst orientations.

Discussion

The measure presented is based on the subspace angles between two matrices [2], a multi-dimensional extension of correlation analysis. A singular value decomposition is used to extract the *principal angles* and *principal vectors* between two subspaces, such as the subspaces spanned by two competing head models. If two head models are mostly similar, then we find that the largest principal angles are still small (the minimax correlations are still high), which can be interpreted to mean that no dipoles exist in one head model that cannot be adequately modeled by another dipole in the other head model, irrespective of dipole orientation. By not performing Step 3, we can rapidly calculate just the singular values in Step 2 for literally thousands of comparison points between two models. By examining the minimax correlations, we can identify those comparison points where the two models yield the poorest comparisons, then investigate just these points further to understand the differences between the two models.

"Small" and "adequate" must be interpreted relative to a given noise level. Our experiences in dipole localization errors [3] [4] indicate that a minimax correlation angle of less than 18 degrees (correlation above 0.95) should indicate that two models are mostly indistinguishable, given the noise conditions we see in EEG and MEG. These principal angles emphasize the *pattern* of the measurements each model can make at the sensors, regardless of the actual dipole intensity or orientation. If model A requires a dipole ten times stronger and at a moment angle 45 degrees rotated from that of model B, then we can still obtain a principal angle of zero degrees (perfect correlation) if the *shape* of the measurements is identical between the two models. In other words, correlation is scale independent, and using subspace angles, we achieve a performance measure independent of dipole orientation and intensity.

In inverse modeling, with the uncertainties of the true orientation of the equivalent current dipole, and with the approximation of the gains by uncertain conductivities, the emphasis in model comparison is properly placed on this shape, which leads directly to localization. We can, however, also easily extract orientation and gain differences from this same procedure, as shown in Step 3. The principal vectors of each subspace can be directly interpreted to indicate orientation differences between head models, and the gain difference between two models is simply found as a power ratio.

The subspace angle is also useful in the inverse problem. The MEG-MUSIC algorithm [1] simply generates the subspace angle between the signal subspace extracted from the data and the subspace spanned by the desired model. As discussed in [1], if each dipole has a time series linearly independent of the other dipolar time series, then we can decompose the spatio-temporal data matrix to extract a signal subspace. Each dipole's spatial model will be correlated with this signal subspace. We therefore form the gain matrix (forward model) for a dipole and calculate the subspace angles between the gain matrix and the signal subspace. In MEG with a spherical head model, the gain matrix is rank 2, so there are at most only two subspace angles. A MUSIC metric of 1.0 is the same as a minimum subspace angle of zero degrees, indicating that a dipole of some fixed orientation (found from Step 3) matches the data. If the second angle, corresponding to the minimax correlation, is also near 1.0, then we have indications of a "rotating" dipole, as discussed in [1].

We therefore scan over a grid of candidate dipole locations, looking for correlations near unity with the signal subspace. Peaks near unity can be refined by further gridding, or the subspace angle metric may simply be used in a directed search, such as a Nelder-Meade Simplex algorithm. If the rank of the signal subspace is underestimated, we may destroy the correlation of the dipolar model with the data. On the other hand, if we somewhat overestimate the rank, then it is unlikely that we will falsely find a correlation between a dipole model and the extraneous subspace components, since those components are drawn from a randomly oriented noise subspace. Thus we generally recommend that the signal subspace be somewhat overestimated.

Although in [1] the source model was limited to dipoles, the MUSIC method is general for any desired parametric source model, such as pairs of dipoles or combined dipole and quadrupoles. We simply generate the subspace angles between the signal subspace and the source model.

References

[1] Mosher JC, Lewis PS, and Leahy RM, "Multiple dipole modeling and localization from spatio-temporal MEG data," *IEEE Trans. Biomedical Eng*, 1992, 39:541 – 557.

[2] Golub GH, Van Loan CF, *Matrix Computations*, second edition, Johns Hopkins University Press, 1984.

[3] Mosher JC, Lewis PS, Leahy RM, *Biomagnetism: Fundamental Research and Clinical Applications*, Proceedings of the Biomag '93 Vienna Conference, Elsevier, 1995, 330-334.

[4] Mosher JC, Spencer ME, Leahy RM, Lewis PS, *Electroenceph. and clin. Neurophys*. 1993, 86:303 – 320.

[5] Mosher JC, Chang, CH, Leahy RM, "Comparison of the constant and linear boundary element method for EEG and MEG forward modeling" this volume.

[6] Berg P, Scherg M, "A fast method for forward computation of multiple-shell spherical head models," *EEG Journal*, 90 (1994) 58-64.

[7] Zhang Z, "A fast method to compute surface potentials generated by dipoles within multilayer anisotropic spheres," *Phys. Med. Biol*, (1995) 335-349.

Acknowledgment

This work was supported by the National Institute of Mental Health grant no. R01-MH53213 and the National Eye Institute grant no. R01-EY08610-04.

Comparison of the Constant and Linear Boundary Element Method for EEG and MEG Forward Modeling

Mosher, J.C.[1], Chang, C.H.[2] and Leahy, R.M.[2]

[1]*Los Alamos National Laboratory, Los Alamos, New Mexico, USA;* [2]*University of Southern California, Los Angeles, California, USA*

Abstract

We present a comparison of boundary element methods for solving the forward problem in EEG and MEG. We use the method of weighted residuals and focus on the *collocation* and *Galerkin* forms for *constant* and *linear* basis functions. We also examine the effect of the *isolated skull approach* for reducing numerical errors due to the low conductivity of the skull. We demonstrate the improvement that a linear Galerkin approach may yield in solving the forward problem.

Introduction

The forward problem in EEG and MEG is to determine the potential or magnetic field generated by neural current sources. The standard model for the head is a set of connected volumes, typically representing the brain, skull and scalp. If the conductivities within each of these regions are isotropic and constant, the volume integral equations relating the neural current sources and the external fields can be reduced to surface integrals. If these regions of constant conductivity can also be modeled as concentric spheres, then analytic solutions are known for EEG and MEG [1], [2]. If more accurate shapes of regions can be obtained from MR or CT images, a more realistic head model can be found. For an arbitrarily shaped head model, however, no analytic solution exists, and the forward problem must be solved numerically. Here we use the boundary element method (BEM) which involves tessellating the surfaces into small areal elements (planar triangles) and approximating the surface potential with a set of basis function.

We assume the head is modeled by a piecewise homogeneous volume conductor G. Denote the surfaces between compartments with different conductivity as $S_1, S_2, \ldots, S_m$. Under the quasi-static approximation, the potential $v(\boldsymbol{r})$ on the jth surface is governed by the Fredholm integral of the second kind, $\boldsymbol{r} \in S_j$, (c.f. [5]),

$$\frac{2}{\sigma_j^- + \sigma_j^+} v_\infty(\boldsymbol{r}) = v(\boldsymbol{r}) - \frac{1}{2\pi} \sum_{i=1}^{m} \frac{\sigma_i^- - \sigma_i^+}{\sigma_j^- + \sigma_j^+} \int_{S_i} v(\boldsymbol{r}') \frac{(\boldsymbol{r}' - \boldsymbol{r})}{\|\boldsymbol{r} - \boldsymbol{r}'\|^3} \cdot d\boldsymbol{s}_i', \; v_\infty(\boldsymbol{r}) = \frac{1}{4\pi} \int_G \boldsymbol{j}^P(\boldsymbol{r}') \cdot \frac{\boldsymbol{r} - \boldsymbol{r}'}{\|\boldsymbol{r} - \boldsymbol{r}'\|^3} dv', \tag{1}$$

where σ_j^- and σ_j^+ are the conductivity inside and outside the jth surface, $d\boldsymbol{s}_i'$ is a vector element of surface S_i oriented along the outward unit normal of S_i, and $\boldsymbol{j}^P(\boldsymbol{r})$ is the primary current. Once the potential $v(\boldsymbol{r})$ is obtained on all surfaces, the magnetic field can be computed by the Geselowitz' formula (c.f. [2]),

$$\boldsymbol{b}(\boldsymbol{r}) = \boldsymbol{b}_\infty(\boldsymbol{r}) + \frac{\mu_0}{4\pi} \sum_{i=1}^{m} (\sigma_i^- - \sigma_i^+) \int_{S_i} v(\boldsymbol{r}') \frac{(\boldsymbol{r} - \boldsymbol{r}')}{\|\boldsymbol{r} - \boldsymbol{r}'\|^3} \times d\boldsymbol{s}_i', \; \boldsymbol{b}_\infty(\boldsymbol{r}) = \frac{\mu_0}{4\pi} \int_G \boldsymbol{j}^P(\boldsymbol{r}') \times \frac{\boldsymbol{r} - \boldsymbol{r}'}{\|\boldsymbol{r} - \boldsymbol{r}'\|^3} dv', \tag{2}$$

where $v_\infty(\boldsymbol{r})$ and $\boldsymbol{b}_\infty(\boldsymbol{r})$ are the potential or magnetic field due to $\boldsymbol{j}^P(\boldsymbol{r})$ in an infinite homogeneous medium.

Both $v_\infty(\boldsymbol{r})$ and $\boldsymbol{b}_\infty$ are related to the primary current and therefore directly depend on the specific source model. The most widely used source model is a current dipole with strength concentrated at a point, $\boldsymbol{j}^P(\boldsymbol{r}) = \boldsymbol{q}\delta(\boldsymbol{r} - \boldsymbol{r}_q)$, where $\boldsymbol{r}_q$ and $\boldsymbol{q}$ denote the dipole location and moment, respectively. The second terms in (1) and (2) are contributions from the volume or return currents, and they depend on the head geometry and conductivity profile. Only for some special cases (e.g., a spherical volume) can (1) and (2) be obtained analytically [1], [2]; in most cases, they need to be treated numerically. The high resistivity of the skull has been shown to lead to numerical inadequacies in calculating the potentials on the scalp, and the *isolated skull approach* (ISA) was devised to slightly alter (1) by assuming a perfectly insulating skull, then folding its calculation back into (1) [3]. In the following, we examine use of the method of weighted residuals [4] to numerically solve (1) and (2), with and without the ISA alteration.

Method

We can express (1) as a linear operator acting on $v(\boldsymbol{r})$, $L(v(\boldsymbol{r})) = g(\boldsymbol{r})$, where $g(\boldsymbol{r})$ is proportional to the infinite homogeneous potential [5]. In the forward problem, the source and hence $g(\boldsymbol{r})$ is known, and the task is to determine $v(\boldsymbol{r})$, such that residual $L(v(\boldsymbol{r})) - g(\boldsymbol{r})$ is as small as possible. We distribute the residual or error function in the domain by weighting it by $w(\boldsymbol{r})$, i.e.,

$$\int (L(v(\boldsymbol{r}')) - g(\boldsymbol{r}'))w(\boldsymbol{r}')ds' = 0 \text{ or equivalently, } (w(\boldsymbol{r}), g(\boldsymbol{r})) = (w(\boldsymbol{r}), L(v(\boldsymbol{r}))) \tag{3}$$

where (,) denotes the inner product of the two functions. We represent both the weighting function and the unknown potentials as combinations of linearly independent basis functions $\psi_n(\boldsymbol{r})$ and $\varphi_n(\boldsymbol{r})$ respectively,

$$w(\boldsymbol{r}) = \sum_{n=1}^{N} \beta_n \psi_n(\boldsymbol{r}), \text{ and } v^h(\boldsymbol{r}) \approx \sum_{n=1}^{N} v_n \varphi_n(\boldsymbol{r}) \tag{4}$$

where β_n are arbitrary coefficients. The solution to (3) should hold for all $w(\boldsymbol{r})$ in the space spanned by the basis functions $\psi_n(\boldsymbol{r})$, i.e., the β_i coefficients are arbitrary, such that (3) leads to N equations,

$$(\psi_i(\boldsymbol{r}), g(\boldsymbol{r})) = (\psi_i(\boldsymbol{r}), L(v(\boldsymbol{r}))), i = 1, 2, \ldots, N . \tag{5}$$

The coefficients for $v^h(\boldsymbol{r})$ are the *nodal parameters*, which are functions of the *nodes* or *nodal points* $\boldsymbol{r}_n, v_n = v^h(\boldsymbol{r}_n)$. The basis function φ_n is chosen with the property that $\varphi_n(\boldsymbol{r}_n) = 1$ and zero at all other nodes, i.e., $\varphi_m(\boldsymbol{r}_n) = \delta_{mn}$. Substituting $v^h(\boldsymbol{r})$ into (5) yields a system of equations representable in matrix form as

$$\begin{bmatrix} (\psi_1(\boldsymbol{r}), g(\boldsymbol{r})) \\ \ldots \\ (\psi_N(\boldsymbol{r}), g(\boldsymbol{r})) \end{bmatrix} = \begin{bmatrix} (\psi_1(\boldsymbol{r}), L(\varphi_1(\boldsymbol{r}))) & \ldots & (\psi_1(\boldsymbol{r}), L(\varphi_N(\boldsymbol{r}))) \\ \ldots & \ldots & \ldots \\ (\psi_N(\boldsymbol{r}), L(\varphi_1(\boldsymbol{r}))) & \ldots & (\psi_N(\boldsymbol{r}), L(\varphi_N(\boldsymbol{r}))) \end{bmatrix} \begin{bmatrix} v_1 \\ \ldots \\ v_N \end{bmatrix} \tag{6}$$

Many different methods for selecting the basis functions in (4) exist (c.f. [4]). Among these methods, we will focus on the collocation form and Galerkin form. In the *collocation* form, $\psi_i(\boldsymbol{r})$ is chosen as the Dirac delta function $\delta(\boldsymbol{r} - \boldsymbol{r}_i)$, where $\boldsymbol{r}_i$ is the nodal point, usually at the centroids or vertices of triangles. In other words, (5) is satisfied at certain collocation points instead of in an average sense. The obvious advantage is the inner product reduces to a function evaluation at the nodal point. The *Galerkin* form is a particular weighted residual method for which the weighting functions belong to the same set as the approximating functions, i.e. $\psi_i(\boldsymbol{r}) = \varphi_i(\boldsymbol{r})$.

We examine two widely used basis functions, *constant* and *linear*. Let $\boldsymbol{r}_i$, $\boldsymbol{r}_j$, and $\boldsymbol{r}_k$ be three vertices of the nth triangle, and they are ordered in such a way that the permutation $\boldsymbol{r}_i \to \boldsymbol{r}_j \to \boldsymbol{r}_k$ corresponds by the right-hand rule to the outward normal of the surface. The basis functions are defined as [3], [6]:

$$\varphi_n^c(\boldsymbol{r}) = 1 , \quad \varphi_i^l(\boldsymbol{r}) = \frac{\boldsymbol{r} \cdot (\boldsymbol{r}_j \times \boldsymbol{r}_k)}{\boldsymbol{r}_i \cdot (\boldsymbol{r}_j \times \boldsymbol{r}_k)}, \quad \varphi_j^l(\boldsymbol{r}) = \frac{\boldsymbol{r} \cdot (\boldsymbol{r}_k \times \boldsymbol{r}_i)}{\boldsymbol{r}_i \cdot (\boldsymbol{r}_j \times \boldsymbol{r}_k)}, \quad \varphi_k^l(\boldsymbol{r}) = \frac{\boldsymbol{r} \cdot (\boldsymbol{r}_i \times \boldsymbol{r}_j)}{\boldsymbol{r}_i \cdot (\boldsymbol{r}_j \times \boldsymbol{r}_k)} \tag{7}$$

The formulation originally derived by Lynn and Timlake [5] is effectively the Galerkin form using constant basis functions. In many other papers, the "geometry" matrix in the Galerkin form of (6) is approximated by the values at the triangular centroids (c.f.[3]), effectively a mix between collocation and Galerkin. The methods due to de Munck [6], Schlitt et al. [7] and Ferguson et al. [8] are collocation forms using linear basis functions. We introduce in this paper the comparison of these prior results with the linear Galerkin BEM.

Results and Discussion

A simple but illustrative example will serve to explain the differences these error weightings have on EEG and MEG. Let $f(x) = x^2$ be a simple quadratic function to be approximated by a constant or a linear function. The weighted residual expression is therefore $\int (f(x) - x^2) w(x) dx$, which we will evaluate on the interval $[0, 1]$. For constant collocation, we approximate $f(x)$ as a constant function, $\varphi_0(x) = 1$, with a nodal point at the midpoint, $x = 0.5$. Substitution into the weighted residual expression and minimization yields $v_0 = 1/4$. Similarly, for linear collocation, the two basis functions are $\varphi_0(x) = 1 - x$ and $\varphi_1(x) = x$ with corresponding nodal points 0 and 1. Minimizing this weighted residual yields $v_0 = 0$ and $v_1 = 1$.

Both the constant and linear approximations to $f(x)$ are overlaid in Fig. 1, and we see that both collocations effectively yield the correct values of $f(x)$ at their nodal points. Next we consider Galerkin weightings over the same interval,

$$\int_0^1 \left(\sum v_i \varphi_i(x) - x^2\right) \varphi_j(x) dx \tag{8}$$

yielding $v_0 = 1/3$ for constant Galerkin, and $v_0 = -1/6$, $v_1 = 5/6$ for linear Galerkin. These solutions are overlaid on Fig. 1.

Two features distinguish the Galerkin solution from the collocation: 1) the Galerkin approximations to $f(x)$ have larger errors at the nodal points than the collocation results (which are perfect at the nodal points in this example), and 2) both Galerkin approximations integrate to the same value as the true function over this interval, $\int f(x) dx = 1/3$. By comparison, the constant and linear collocation forms integrate to $1/4$ and $1/2$ respectively, where we note that linear collocation error is twice as great as the constant collocation.

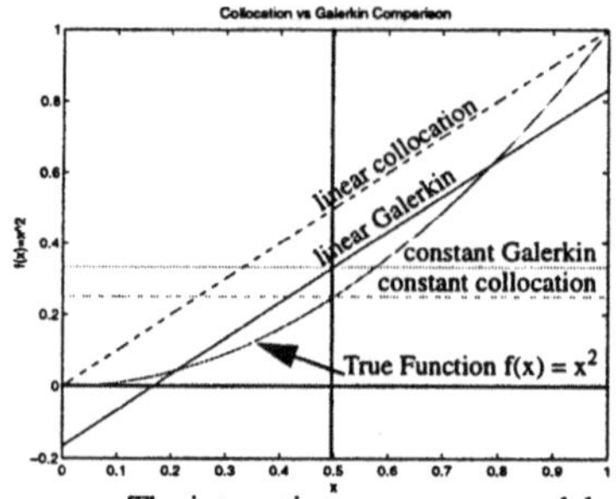

Figure 1. Simple example suggested by Ferguson [11] to show the effects of approximation and error weighting. The true function over the interval is $f(x) = x^2$, and its true integral over the interval is 1/3. The collocation forms precisely match the true function at their respective nodal points, x = 0.5, and x = 0,1. The linear collocation has twice the integration error over that of constant collocation for this interval. The Galerkin forms are in error at the nodal points, but both forms precisely integrate to the correct value, 1/3.

The integration error versus nodal evaluation error in this simple example has impacts on the evaluation of EEG and MEG. The potentials are embedded in a Fredholm integral of the second kind, such that their calculation is sensitive to integration issues. In EEG, the potential is both inside and outside the integral, while in MEG, the potential is completely contained in the integral, suggesting that MEG may be more sensitive to integration. In EEG, we are primarily interested in the evaluation of the potential at discrete sensor sites about the scalp. Thus the collocation approach implies we might actually achieve better error performance than Galerkin if we assign nodal points to the sensor sites. In MEG, however, the sensor sites are a measurement of the integral of the potential over all surfaces (the fictitious currents), and the Galerkin approach implies that we might achieve better error control over the continuum of potentials and hence better MEG approximations.

With this simple insight, we address the results of simulations with a known spherical model. We applied the constant and linear basis functions, with collocation and Galerkin methods, with and without the isolated skull approach, to a three shell spherical model. The radii of the model were 88, 85 and 81 mm, and the conductivities were 0.33, 0.0042, 0.33 $\Omega^{-1}m^{-1}$ for the scalp, skull and brain respectively. The test dipoles were arranged tangentially in the x-direction, and 104 MEG sensors were oriented in this same direction over the upper hemisphere, 120 mm from the center. This non-radial orientation was chosen to suppress the contribution from $b_\infty(r)$ in (2), which is analytically known, and therefore highlights the BEM calculation of the contribution from the volume currents.

The EEG electrodes were placed at the nodal points (centroids or vertices) of the triangles on the upper hemisphere. We used 492 and 1016 triangles per surface for the constant and linear BEM respectively, so that the degrees of freedom for the linear and constant cases were approximately equal. The matrices for the forward model were properly "deflated" [5] and efficiently computed into forward gain matrices, [9], [10]. A tangential dipole, $q = [1, 0, 0]$, was moved along the z-axis from [0,0,3] mm to [0,0,78]mm, i.e, to within 3 mm of the vertex of a triangle tessellated on the inner most shell; z represents the vertical axis in a right-handed Cartesian coordinate system. The comparison metric used was the relative difference measure [7] and defined as:

$$RDM = \sqrt{(b_{th} - b)^t(b_{th} - b)/(b_{th}^t b_{th})} \quad (9)$$

where b_{th} and b denote the analytic and numeric EEG or MEG sensor values, and "t"denotes the vector transpose.

In Fig. 2, the RDM for EEG and MEG are shown. For the EEG results without ISA, we observe that constant Galerkin (effectively the original Lynn and Timlake [5] approach) and linear collocation do not generally improve the RDM over that of the relatively simpler constant collocation. This result is consistent with linear collocation yielding a worse integral than constant collocation, and with constant Galerkin yielding a worse nodal evaluation. Linear Galerkin achieves a better RDM over constant collocation, but we note in all EEG cases without ISA that the RDM is on the order of unity (100%) error as the dipole approaches the innermost surface. The EEG results with ISA show a remarkable improvement in the RDM, and we see that both the constant collocation and the linear Galerkin results are about 8% error as the dipole approaches the inner surface.

In the MEG plots, two curves are shown. The dashed line indicates the MEG solution if the potentials at the nodal points were precisely known, therefore highlighting the error we make in MEG if we apply constant and linear approximations to (2). The solid line is the MEG RDM using the BEM calculated potentials. We observe that in general the RDM is below 10%, a significant improvement over the EEG RDM, suggesting that MEG is in general less sensitive to the BEM errors. The true potentials (dashed) curves reveal that the constant approximation is better than the linear approximation, if we fix the nodal points at the true values, which is consistent with our simple one-dimensional example where linear collocation had twice the integration error of constant collocation. We note that ISA does not improve the MEG results, which suggests that although ISA improves the scalp potentials for EEG, the cost is a perturbation of the stronger innermost surface potentials that are also detected by MEG. Near the innermost surface, the constant model results show errors of only a few percent, but we see that the linear Galerkin results are dramatically superior to all others.

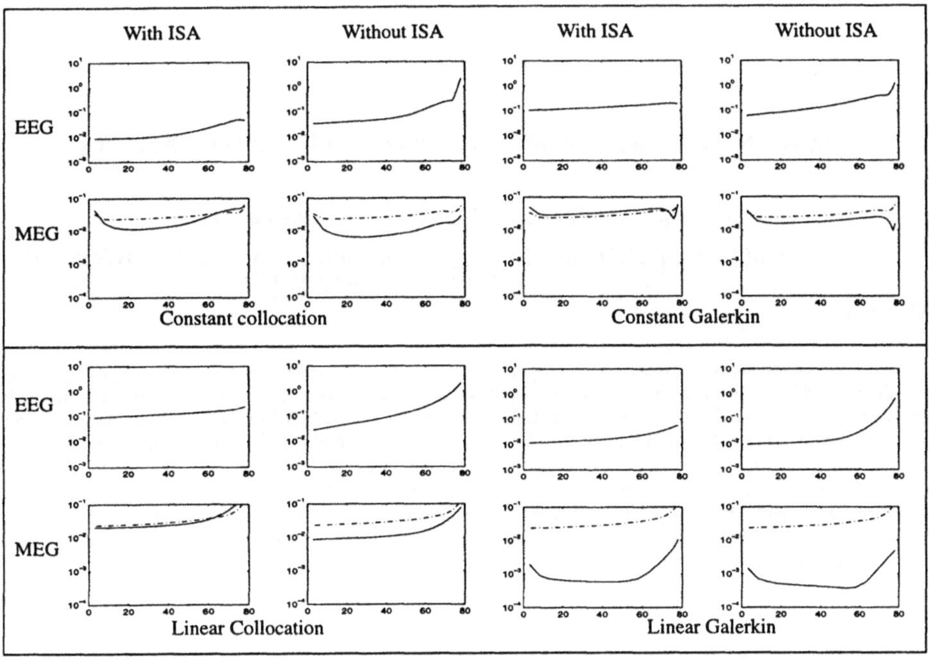

Figure 2. The RDM (in log scale) versus a tangential dipole along the z-axis from 3 mm to 78 mm. The radius of the innermost of the three spherical shells is 81 mm. In the MEG case, the dash-dot line denotes the RDM using the exact potentials at the nodal points for the constant or linear approximation. The EEG RDM is calculated at the nodal points on the upper hemisphere (scalp). The MEG RDM is calculated at 104 sensors spaced 120 mm from the origin about the upper hemisphere.

References

[1] Zhang, Z, "A fast method to compute surface potentials generated by dipoles within multilayer anisotropic spheres," *Phys. Med. Biol.*,1995,40:335-349.

[2] Sarvas, J., "Basic mathematical and electromagnetic concepts of the biomagnetic inverse problem," *Phys. Med. Biol.*,1987,32:11-22.

[3] Hämäläinen, M.S., Sarvas, J., "Realistic conductor geometry model of the human head for interpretation of neuromagnetic data," *IEEE BME,* 1989, 36: 165-171.

[4] Brebbia, C.A., Telles, J.C.F., Wrobel, L. C., *Boundary element techniques,* Springer Verlag, Berlin 1984.

[5] Lynn,M.S., Timlake,W.P., "On the numerical solution of the singular integral equations of potential theory," *SIAM, J. Numer. Anal.*, 1968,11: 77-98.

[6] de Munck, J.C., "A linear discretization of the volume conductor integral equations using analytically integrated elements," *IEEE BME, 1992*, 39: 986-990.

[7] Schlitt, H.A., Heller, L. , Aaron, R., Best, E., Ranken, D. M., "Evaluation of boundary element method for the EEG forward problem: Effect of linear interpolation," *IEEE BME,* 1995, 42: 52-57.

[8] Ferguson, A.S., Zhang, X., Stroink, G., "A complete linear discretization for calculating the magnetic field using the boundary element method," *IEEE BME,* 1994, 41: 455-459.

[9] Mosher, J.C., Lewis, P., Leahy,R.M., "Multiple dipole modeling and localization from spatio-temporal MEG data," *IEEE BME*,1992, 39: 541-557.

[10] Mosher, J.C.,Leahy,R.M.,Lewis,P.S., "Matrix kernels for MEG and EEG source localization and imaging," *IEEE Proceedings of ICASSP-1995*, pp. 2943 – 2946.

[11] Ferguson, S., 1996, private communication, this conference.

Acknowledgments: We thank Stuart Ferguson for suggesting the simple example. This work was supported by the National Institute of Mental Health Grant No. R01-MH53213 and the National Eye Institute Grant No. R01-EY08610-04.

MEG-Based Imaging of Focal Neuronal Current Sources†

Phillips, J.W.[1], Leahy, R.M.[1] and Mosher, J.C.[2]

[1]*University of Southern California, Los Angeles, California, USA;* [2]*Los Alamos National Laboratory, Los Alamos, New Mexico, USA*

Abstract

We describe a new approach to imaging neuronal current sources from measurements of the magnetoencephalogram (MEG) associated with sensory, motor, or cognitive brain activation. Previous approaches to this problem have concentrated on the use of weighted minimum norm inverse methods. While these methods ensure a unique solution, they do not introduce information specific to the MEG inverse problem, often producing overly smoothed solutions and exhibiting severe sensitivity to noise. We describe a Bayesian formulation of the inverse problem in which a Gibbs prior is constructed to reflect the sparse focal nature of neuronal current sources associated with evoked response data. The prior involves a binary process indicating active sources and a continuous Gaussian process designating associated amplitudes. An estimate of the primary current source distribution for a specific data set is formed by maximizing over the posterior probability with respect to the binary and continuous variables.

Introduction

An array of SQUID biomagnetometers may be used to measure the spatio-temporal MEG signal produced by the brain. We wish to construct an image of neural activity which produced this signal. Physiological models for the MEG assume primary sources are constrained to the cortex with current flow oriented normal to the local surface [7]. The image can therefore be constrained to the cortical surface which can be extracted from a registered volume MR image of the subject's head. By tessellating the cortex with N disjoint regions and representing the sources in each region by an equivalent constrained current dipole oriented normal to the surface with amplitude y_i, the MEG inverse problem can be expressed in terms of a linear model. The linear forward model relating the N sources $\mathbf{y}$ (Nx1) and the M MEG measurements $\mathbf{b}$ (Mx1) can be written,

$$\mathbf{b} = \mathbf{G}\mathbf{y} + \mathbf{n} \tag{1}$$

where the i'th row of the MxN system matrix $\mathbf{G}$ specifies a discrete representation of the lead field (sensitivity) of the i'th sensor. The j'th column of $\mathbf{G}$ specifies the gain vector for the j'th constrained dipole component. The Mx1 vector $\mathbf{n}$ represents noise generated within the sensor and by unwanted electromagnetic sources (power lines, the heart, background brain activity, etc.).

The search for an appropriate imaging method is primarily concerned with finding a way to choose within a set of images that produces essentially the same fit to the data. Weighted minimum norm inverse methods [8,9] typically find solutions which match the data while minimizing a weighted l_2-norm on the solution vector. These techniques tend to smear sources over the entire reconstruction region and are generally unstable due to ill-conditioning of the system matrix. The instability can be overcome using Tikhonov regularization [6] but the reconstructions remain smeared. The iteratively reweighted minimum norm method [5] is a nonlinear approach to overcoming the problem of smeared sources in which the norm weighting is updated at each iteration based on the result of the previous iteration. This method uses a weighting matrix which, as the iterations proceed, reinforces strong sources and reduces weak ones. This results in very sparse solutions, but again the method is extremely sensitive to noise and highly dependent on the initial estimate.

Here we propose an alternative approach to the inverse problem based on a Bayesian formulation. Rather than use an arbitrary weighted l_2-norm to select the solution, we introduce a prior distribution on the source which is used to resolve the ambiguities inherent in the inverse problem. This prior is constructed to favor the reconstruction of physiologically plausible solutions. Basic studies of functional activation, such as somatotopic or retinotopic map-

†This work was supported by the National Institute of Mental Health Grant No. R01-MH53213, and National Eye Institute Grant No. R01-EY08610-04.

ping using fMRI or PET, reveal the sparse and highly localized nature of activation in the cerebral cortex [1]. Our prior is therefore specifically designed to reflect the expectation that current sources tend to be sparse and focal. This prior is combined with a Gaussian likelihood model for the data which is based on the linear model (1) and an assumption of additive white Gaussian noise. Maximization over the resulting posterior probability results in a maximum *a posteriori* (MAP) estimate of the primary current sources.

Bayesian Technique Description

Since the primary sources of the MEG are widely accepted to be sparse and focal, we suggest that this information be used in reconstructing the image. This information can be naturally introduced into the problem using the Bayesian paradigm in which the source is modeled as a random field. Since we assume that sources are sparse, the large majority of source pixels will have zero amplitude. We therefore use a binary indicator process **x** to model whether each source dipole is *on* ($x_i = 1$) or *off* ($x_i = 0$). Those sites that are active are assumed to have a Gaussian amplitude, z_i. We can then write the source image vector **y** as $\mathbf{y} = \mathbf{x}.*\mathbf{z} = \mathbf{Xz}$, where ".*" signifies the Schur product (element by element pair-wise multiplication) and $\mathbf{X} = \mathrm{diag}(\mathbf{x})$. Assuming independence of the indicator and amplitude processes, we can write the posterior probability for **x** and **z** given the MEG data **b** as,

$$p(\mathbf{x}, \mathbf{z}|\mathbf{b}) = \frac{p(\mathbf{b}|\mathbf{x}, \mathbf{z})p(\mathbf{x})p(\mathbf{z})}{p(\mathbf{b})}. \tag{2}$$

We find a MAP estimate $\mathbf{y}^*$ of **y** as $\mathbf{y}^* = \mathbf{x}^*.*\mathbf{z}^*$ where,

$$\mathbf{x}^*, \mathbf{z}^* = \arg\max_{\mathbf{x}, \mathbf{z}} \; p(\mathbf{x}, \mathbf{z}|\mathbf{b}). \tag{3}$$

The joint probability $p(\mathbf{x})$ is chosen to reflect the expectation that the sources are sparse and focal. To achieve this goal, we use a Markov Random Field (MRF) [4] model for which sparse focal sources have a higher probability of occurring than more distributed sources. We define $p(\mathbf{x})$ to be a Gibbs distribution,

$$p(\mathbf{x}) = (1/K)\exp\{-V(\mathbf{x})\} \tag{4}$$

where K is a proportionality constant and the energy function $V(\mathbf{x})$ is given by,

$$V(\mathbf{x}) = \sum_i \Big[\underbrace{\alpha_i x_i}_{\text{Sparseness Term}} + \underbrace{\beta_i C_i\{x_i, x_j \;\; j \in \xi_i\}}_{\text{Clustering Term}} \Big] \tag{5}$$

where the parameters $\alpha_i > 0$ and $\beta_i > 0$ determine the relative weights of the sparseness and clustering terms. The potential function $C_i\{x_i, x_j \;\; j \in \xi_i\}$ is defined in terms of each pixel and its neighbors ξ_i as,

$$C_i\{x_i, x_j \;\; j \in \xi_i\} = \Big[\sum_{j \in \xi_i} (x_i - x_j)^2\Big]^Q. \tag{6}$$

This clustering term is small if neighboring pixels are of the same magnitude. The exponential parameter Q determines the strength of the clustering. As Q increases, the size of the clusters tends to increase. In the results presented below we constrain sources to a plane and use a four nearest-neighbor interaction. Some examples of binary images, produced by sampling from this prior using a Gibbs sampler [4], are shown in Fig. 1. The source amplitude process, **z**, is assumed to be a set of independent zero mean Gaussian random variables with covariance $\mathbf{C_z}$.

Using the definitions above and assuming the noise process in (1) is zero mean Gaussian with covariance $\mathbf{C_n}$, we can write,

$$p(\mathbf{x}, \mathbf{z}|\mathbf{b}) = \frac{1}{Z}\exp\{-U(\mathbf{x}, \mathbf{z}|\mathbf{b})\} \tag{7}$$

where Z is the posterior partition function. The posterior energy function is given by:

$$U(\mathbf{x}, \mathbf{z}|\mathbf{b}) = \frac{1}{2}[\mathbf{b} - \mathbf{GXz}]^T \mathbf{C_n}^{-1} [\mathbf{b} - \mathbf{GXz}] + \frac{1}{2}\mathbf{z}^T \mathbf{C_z}^{-1}\mathbf{z} + V(\mathbf{x}). \tag{8}$$

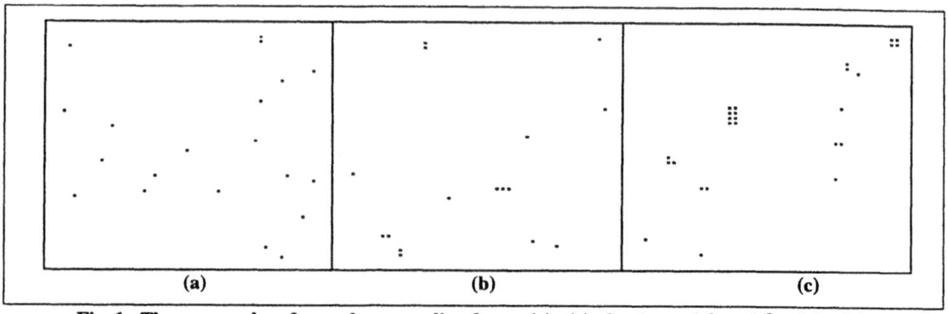

Fig. 1. Three examples of a random sampling from $p(\mathbf{x})$. (a) $Q = 1, \alpha = 1.0$, and $\beta = 0.4$ (b) $Q = 2$, $\alpha = 1.0$, and $\beta = 0.2$ (c) $Q = 3, \alpha = 1.0$, and $\beta = 0.1$.

The MAP estimate is found by maximizing over the log-posterior, or equivalently,

$$\mathbf{x}^*, \mathbf{z}^* = \underset{\mathbf{x}, \mathbf{z}}{\arg\min} \quad U(\mathbf{x}, \mathbf{z}|\mathbf{b}) . \tag{9}$$

The solution to the optimization problem (9) provides our estimate of the neural current sources. The solution is clearly a function of the parameters of the likelihood function and the prior probabilities. Methods for selecting these parameters will be addressed in a future publication. However, we note that since the parameters α_i represent the relative probabilities that each source pixel is active, it is straightforward to include pixel-wise probability weightings determined from fMRI or PET activation studies to influence the formation of these sparse images.

Mean Field Annealing

Minimization of $U(\mathbf{x}, \mathbf{z}|\mathbf{b})$ is difficult since the optimization must be performed over a mixture of discrete and continuous variables. Since this function is quadratic in the continuous variables, $\mathbf{z}$, we can derive a closed form expression for the optimal $\mathbf{z}$ as a function of any particular indicator process $\mathbf{x}$, i.e. the vector $\mathbf{z}^*(\mathbf{x})$ which minimizes (8) given the binary vector $\mathbf{x}$ is given by,

$$\mathbf{z}^*(\mathbf{x}) = \mathbf{C}_z\mathbf{X}\mathbf{G}^T(\mathbf{G}\mathbf{X}\mathbf{C}_z\mathbf{X}\mathbf{G}^T + \mathbf{C}_n)^{-1}\mathbf{b} . \tag{10}$$

Substituting $\mathbf{z}^*(\mathbf{x})$ into $U(\mathbf{x}, \mathbf{z}|\mathbf{b})$ results in a new energy function $U(\mathbf{x})$, a function of $\mathbf{x}$ only. We can therefore first find the optimal indicator process by minimizing $U(\mathbf{x})$, then substitute this result in (10) to obtain the optimal amplitude process.

Coordinate-wise optimization with respect to a collection of binary variables using, for example, iterated conditional modes (ICM) [2], tends to produce rapid convergence to an undesirable local minimum. Here we use, instead, a continuation method based on mean field annealing(MFA) [3]. We visit each pixel in turn using the following update strategy:

$$x_i^{(n+1)} = E_T\{x_i | x_j^{(n)} \quad \forall\ j \neq i\} \tag{11}$$

where the conditional expectation is computed with respect to the modified joint probability,

$$p_T(\mathbf{x}) = \frac{1}{Z_T}\exp\{-\frac{1}{T}U(\mathbf{x})\} . \tag{12}$$

The temperature parameter T is slowly reduced as the iterations proceed. As $T \rightarrow 0$ the iteration will converge to a binary solution which is a local minimum of $U(\mathbf{x})$. MFA typically results in a better local minimum than is achieved using ICM. We will address the convergence behavior and the basis for the mean field approach in a future extended publication.

Simulations and Phantom Experiments

We have conducted extensive comparisons of this technique with minimum norm based methods using simulated and experimental phantom data. Some example images of representative comparisons are shown in our sister publication in this conference proceedings ("A comparative study of minimum norm inverse methods for MEG imaging," R.M. Leahy, J.C. Mosher, and J.W. Phillips). We found that all of the linear minimum norm methods produced results exhibiting a relatively large degree of smoothness. For all additive noise scenarios tested, all methods generally gave similar residual errors in the fit to the data. All results can be considered 'correct' in the sense that they are configurations that could have produced the observed data. This observation emphasizes the severely under-determined nature of this problem. Clearly, in order to select between these feasible solutions we must use additional information concerning the expected nature of the source.

Conclusions

In the simulations and phantom studies we have conducted, our MAP solution was generally superior to those obtained using minimum norm methods. However, this is true only when the sources exhibit the sparse focal characteristics on which our method is based. We stress that all of the methods provide good fits to the data, and hence are physically (if not physiologically) plausible. This ambiguity is inherent in attempting to infer spatial information from on the order of 100 external sensor measurements. Specific prior information is essential if useful spatial information is to be extracted from the data.

In this work we have developed a Bayesian framework for image estimation from MEG data. This approach can be extended to include information from other modalities (fMRI or PET) as well as using anatomical MR images to constrain sources to the cortex. The method can also used for combined MEG/EEG data. We can also directly extend the model for dynamic imaging by simply replacing each of the amplitude processes, z_i, in our model with a time series model, $z_i(t)$.

The results that we have presented assume sources are constrained to 2D planes. Future research will focus on sources constrained to a realistic cortical surface. Only then can be we begin to establish realistic limits on the ability of MEG to usefully image neural activity.

References

[1] C.J. Aine, "A Conceptual Overview and Critique of Functional Neuroimaging Techniques in Humans: 1. MRI/fMRI and PET," *Critical Reviews in Neurobiology*, vol. 9, nos. 2 and 3, pp. 229-309, 1995.

[2] J. Besag, "On the statistical analysis of dirty pictures (with discussion)," *J. of the Royal Statist. Soc., Series B*, vol. 48, pp. 259-302, 1986.

[3] G. Bilbro., *et al.* "Optimization by Mean Field Annealing," in: Touretzky D.S., *Advances in Neural Information Processing Systems*, San Mateo, CA: Morgan-Kauffman, pp. 91-98, 1989.

[4] S. Geman and D. Geman, "Stochastic Relaxation, Gibbs Distributions, and the Bayesian Restoration of Images," *IEEE Trans on Patt. Anal. and Mach. Int.*, vol. 6, no. 6, pp. 721-741, 1984.

[5] I.F. Gorodnitsky, J.S. George, and B.D. Rao, "Neuromagnetic source imaging with FOCUSS: a recursive weighted minimum norm algorithm," *EEG and clinical Neurophysiol.*, 95, pp. 231-251, 1995.

[6] P.C. Hansen, *Regularization Tools, A Matlab Package for Analysis and Solution of Discrete Ill-Posed Problems*, available in Postscript from http://www.mathworks.com/, March 1993.

[7] Y. Okada, "Neurogenesis of Evoked Magnetic Fields," in: S.J. Williamson, G.L. Romani, L. Kaufman, and I. Midena, *Biomagnetism, an Interdisciplinary Approach*, Plenum Press, N.Y., pp. 399-408, 1983.

[8] J. Sarvas, "Basic mathematical and electromagnetic concepts of the biomagnetic inverse problem," *Physics in Medicine and Biology*, vol. 32, pp. 11-22, 1987.

[9] J.Z. Wang, S.J. Williamson, and L. Kaufman, "Magnetic Source Images Determined by a Lead-Field Analysis: The Unique Minimum-Norm Least-Squares Estimation," *IEEE Trans. on Bio. Eng.*, vol. 39, no. 7, pp. 665-675, 1992.

A Four-Sensor Skeleton Configuration for Single Dipole Localization

Prakash, N.M., Spelman, F.A. and Read, P.

Center for Bioengineering, University of Washington, Seattle, USA

Introduction

Magnetic dipole localization is used in the study of epileptic foci, cardiac arrythmias, lung contaminants, and gastrointestinal motility. The source of the magnetic field is described in terms of a limited number of parameters, which are found by the use of an appropriate least squares search method [1]. The number of parameters for a magnetic dipole in three-dimensional space is six (three position parameters, two orientation parameters and one dipole moment). We have devised a method to localize a single magnetic dipole that reduces the number of parameters to three.

Theory

If a magnetic dipole with a dipole moment p is positioned at the origin (see Fig. 1), the z-component of the magnetic field (B_z) at any position (x, y, z) is given by [2]

$$B_z = \mu_o/4\pi\,[3\,x\,z\,r^{-5}p_x + 3\,y\,z\,r^{-5}p_y + (3\,z^2 - r^2)\,r^{-5}p_z] \tag{1}$$

where, p_x, p_y, and p_z are the orthogonal components of the dipole moment (see Fig. 1)

The variables in Equation (1) are the position parameters (x, y, z), the orientation parameters (α, β), and the magnitude of the dipole moment (p).

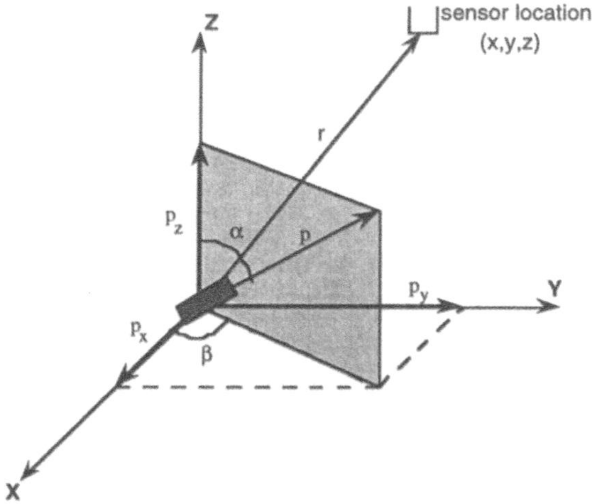

Fig. 1. Components of the dipole moment of the magnet. The magnet is at the origin of the coordinate space. The sensor is located at (x, y, z) in rectangular coordinates. The distance of the magnet from the sensor is r. The dipole moment p is oriented at an angle α to the z-axis and the projection of p on the x-y plane is at an angle β to the x-axis. The components of the magnetic dipole moment p are p_x, p_y, and p_z.

The Configuration

In our four-sensor skeleton configuration, the sensors are at the corners of a rectangle (see Fig. 2). Let the distances of each of the sensors from the dipole be r_1, r_2, r_3, and r_4, respectively, and let B_z at each of the four positions be B_1, B_2, B_3, and B_4, respectively. The relation between the magnetic field values and the distances from the dipole is given by the following equation:

$$B_1 r_1^5 - B_2 r_2^5 - B_3 r_3^5 + B_4 r_4^5 = 0 \quad (2)$$

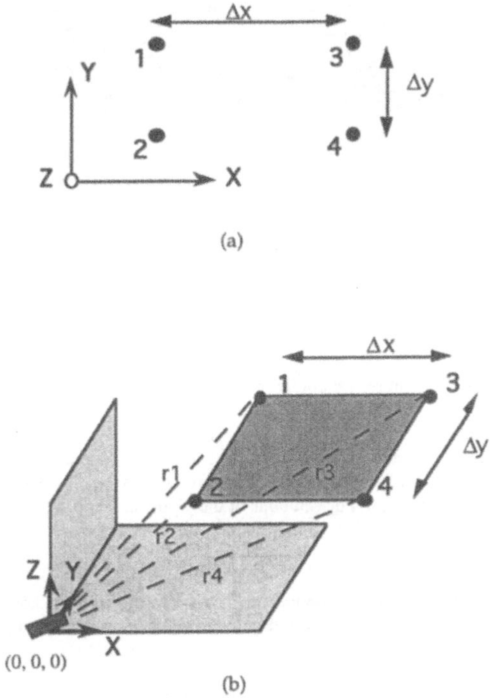

Fig. 2. Four-sensor skeleton configuration (a) two dimensional view, (b) three-dimensional view. The axes are set such that the y-axis is parallel to the line connecting sensors 2 and 1, and the x-axis is parallel to the line connecting sensors 2 and 4. Δx and Δy represent the distances along the x- and y- axes respectively. The distances of sensors 1, 2, 3, and 4 from the magnetic dipole (origin) are represented by r1, r2, r3, and r4.

In equation 2, the only variables, in addition to the magnetic field values at the four sensor positions, are the position coordinates of the sensors with respect to the dipole. The orientation parameters and the magnitude of the dipole moment are no longer a part of the equation for B_z. The number of degrees of freedom is now three.

If we set up three four-sensor skeleton configurations, the three resulting equations (similar to Equation 2) can be solved numerically to obtain the exact position of the dipole. The orientation parameters and the dipole moment can then be calculated using Equation (1).

Simulation

The minimum number of sensors required for setting up three independent equations to localize a single magnetic dipole is eight. An eight-sensor configuration for dipole localization, shown in Fig. 3, was modeled for a dipole with a dipole moment of 0.004 Am^2 [3]. The magnetic field values at the sensor locations were generated using Equation (1). The eight-sensor configuration and localization algorithm was implemented using the software MATHEMATICA.

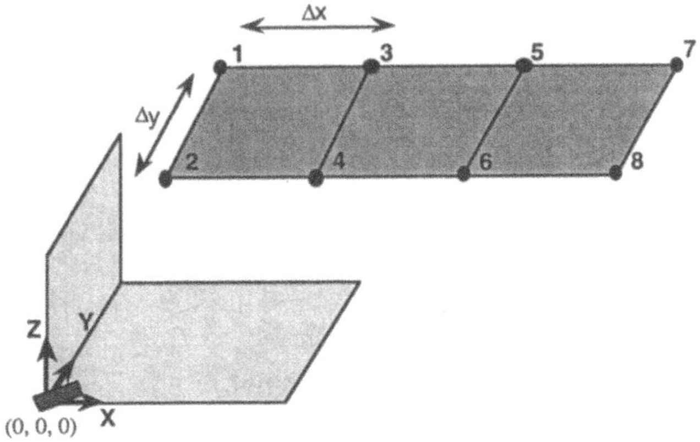

Fig. 3. Eight sensor configuration for single dipole localization.

The spacing between adjacent sensors was set at 5 cm, and an accuracy of 0.01 mG was used for the magnetic field values. The position of sensor number 4 with respect to the dipole was (x, y, z) = (2.0, 3.5, 4.5) cm. The algorithm was implemented for several orientations of the dipole at that position.

Results

The results (see Table) of the simulation indicate that the localization method is very effective. Localization is achieved for all orientations to an accuracy of within 1 mm. When oriented in a planar configuration using the

Table: The exact and calculated position parameters for a magnetic dipole obtained by using the eight sensor configuration shown in Fig. 3.

Exact			Calculated			Error
Position	Orientation		Position (cm)			Δ r
(x, y, z)	α	β	x	y	z	(cm)
x = 2.0 cm y = 3.5 cm z = 4.5 cm	90°	0°	1.99	3.53	4.54	0.051
	90°	45°	1.98	3.53	4.50	0.037
	90°	90°	2.00	3.50	4.55	0.050
	90°	135°	2.01	3.51	4.50	0.014

four-sensor skeleton configuration, a minimum of eight magnetic field sensors yield information about the position orientation, and moment of the magnetic dipole.

Discussion

The four sensor skeleton configuration reduces the number of parameters required for localization of a single magnetic dipole from six to three. The number of degrees of freedom in any reconstruction methodology is consequently reduced. This localization technique can however, be used only for a single magnetic dipole.

The use of the four sensor skeleton configuration, outlined in this paper, offers a simple and accurate method for localizing a magnetic dipole. The use of this skeleton configuration will reduce the number of sensors employed for localization of a dipole. As a consequence, it will reduce the size, complexity, and cost of multi-sensor localization instruments used for localization of single dipoles.

References

[1] Sarvas, J., "Basic mathematical and electromagnetic concepts of the biomagnetic inverse problem," Physics in Medicine and Biology, 1987, 32:11-22.

[2] Brauer, F., and Stroink, G., "Multipole analysis in magnetopneumography—a model study," IEEE Transactions on Biomedical Engineering, 1985, 32:385-391.

[3] Spelman, F.A., Nelson, J.A., Heitkemper, M., Read, P., Prakash, N.M., Seibold, C.J., and Pope II, C.E. "Magnetometer monitoring of human gut transport," Gastroenterology, 1994, 107:1237.

Acknowledgments

This work was supported by the Washington Technology Center, grant RR00166 of the National Institutes of Health, and the Royalty Research Fund of the University of Washington.

The New Methods of Localization and Reconstruction of Biomagnetic Signal Sources

Primin, M.A. and Nedayvoda, I.V.

Institute for Cybernetics, Kiev, Ukraine

ABSTRACT. The algorithms of biomagnetic signal localization and reconstruction are developed on the basis of magnetostatic inverse problem analytical solution. The algorithms of data processing can be used in magnetometric systems with "on-line" gradiometers or - planar ones (including those based on HTc).

1. INTRODUCTION.

The already known methods of magnetic fields spatial analysis permit to locate and to determine the integrated parameters of "effective" source, which magnetic field at the supervision points corresponds to the measured distribution. This is used for the localization of activity zones in miocard. For the field source (magnetic dipole) "point model" of and for field source treated as a planar layer of current this problem is solved analytically in [1-4]. In this report we take into consideration the algorithm of initial data determination for the field sources localization and reconstruction. This very technique of biomagnetic signals preliminary processing is also applicable for the majority of known algorithms of measurements and it explicitly takes into account the design of squid-sensor's magnetic flux transformer.

2. STATEMENT OF THE PROBLEM. METHOD OF SOLUTION.

It is supposed, that the Cartesian system of coordinates XYZ origin is located within the volume V of biomagnetic signal source (in our case - the human heart), and the OZ axis direction coincides with the direction of normal to surface of the patient chest. Then the majority of existing designs of magnetic flux transformers may be divided on two groups:

(1) Squid-gradiometers that have their each pick-up coils located in $z=z_0$ plane (z_0 - is the coordinate of the plane of measurements) and they may have planar technology design, including high-temperature superconductors (HTc) basis. In this case the flux transformers are used after the one of the following schemes $d_{1x} = B_z (x-l, y, z_0) - B_z (x + l, y, z_0)$ or $d_{1y} = B_z (x, y-l, z_0) - B_z (x, y + l, z_0)$. Here d_{1x}, d_{1y} signify the squid-gradiometer measured signals , which are proportional to the appropriate difference of magnetic field component, created by the source at the point of observation; l is the distance ("base") between the coils of flux transformer along the axis X and Y correspondingly.

(2) Gradiometer's pick-up coils are placed along the OZ axis in planes $z=z_0$, $z=z_0 + l$, $z=z_0 + 2\,l$ and they are connected in such a manner that the registered signal is proportional to the values of $d_{1z} = B_z (x, y, z_0) - B_z (x, y, z_0 + l)$ or $d_{2z} = B_z (x, y, z_0) - 2\,B_z (x, y, z_0 + l) + B_z (x, y, z_0 + 2\,l)$.

In recent years the squid-gradiometers, that have the output signal correlated to the direction orthogonal to the normal to $z=z_0$ plane and it has the form of $d_{1zx} = B_x (x, y, z_0) - B_x (x, y, z_0 + l)$ and $d_{1xx} = B_x (x-l, y, z_0) - B_x (x + l, y, z_0)$, have been developed. Hereinafter we'll show, that in this case also the data processing after the suggested algorithm has no special features when compared to the pair of allocated groups of squid-gradiometers.

The magnetic field of arbitrary system of currents, limited by surfaces S, has no breakpoints within the whole space outside the volume of source. As far as the powers lines of magnetic fields are always closed, the integral from any of listed parameters of biomagnetic signal on $z=z_0$ plane is equal to zero (it is assumed, that the size of plane is not limited). Therefore it is possible to find the Fourier-image F [s] and its values at the set points of the wave plane with the help of two-dimensional Fourier transformation. It is assumed here that the s-parameter is determined throughout the whole plane of $z=z_0$ as follows: it is measured at the regular grid nodes, that permits to apply the F[s]-calculations standard techniques[5]; it is equal to zero at each point, located away from the measurement area on distances greater then 3 L (L is the characteristic size of the area of observation), this assumption is proved in [3]; and it is the solution for the bicubic polinom, which factors are determined after the values of s at the boundaries with respect to the condition of s(3 L, y, z0) = s(-3 L, y, z0) = s(x, -3 L, z0) = s(x , 3 L, z0) = 0.

From the definition of the Fourier transformation it follows, that the d1x, d1y, d1xx -parameters' Fourier-image is related to the Fourier-image of magnetic induction vector's appropriate component by simple equations

$$F[B_z]=2iF[d_{lx}]/\sin(k_x l)=2iF[d_{ly}]/\sin(k_y l)\quad k_x,k_y\neq 0 \qquad (1a)$$
$$F[B_x]=2iF[d_{lxx}]/\sin(k_x l) \qquad (1b)$$

The expressions (1) are fair for the arbitrary closed system of currents. The similar formulas of transition may be obtained for "on-line" gradiometers too, however, it is necessary to use the existing physical restrictions on measurements plane's z0-coordinate and biomagnetic signal excitation area characteristic sizes.

As researches show [1,4], in randomly-taken electrocardiocycle moment it is possible to find the most remote (in electromagnetic meaning) point of the grid, where the biomagnetic signal is described using the dipole approach

$$\mathbf{B}(\mathbf{r})=g(\mathbf{r})\mathbf{M},\ g_{ij}=\mu_0/4\pi r^5(3r_ir_j-3r^2\delta_{ij}),\ i,j=x,y,z, \qquad (2)$$

Here **r**, **M** - radius-vector and magnetic moment of dipole, $\mu_0 = 4\pi\, 10^{-7}$ Hn/m - magnetic constant, δ_{ij} - Cronecer symbol. When "on-line" gradiometers are in use, these are the points, where the measured parameters d_{1z},d_{2z} at the moment of examination differ mostly (when investigating the magnetocardiosignal - up to 20%) from the values of B_z (usually, these are the extreme series of points of observation on spatial grid or the extreme probes of the multichannel measuring system). In the case of planar gradiometers the values of d_{1x}, d_{1y} practically coincide with the spatial derivatives $\partial B_z/\partial x$, $\partial B_z/\partial y$ in these points.

The next step in cardiomagnetic signal spatial analysis becomes the field source presentation in a form of a current with its density of **j** (j_x, j_y, 0), that is distributed as the thin layer within the plane, parallel to the measurement plane $z=z_0$. Meantime z-coordinate of this plane corresponds to effective dipole's r_z -coordinate (occurrence depth). Then it is possible to describe the magnetic field at the point of observation (measurement) by the following equation

$$\mathbf{B}(\mathbf{r})=\mu_0/4\pi\int_V[\mathbf{j}(\mathbf{r}')\times(\mathbf{r}-\mathbf{r}')]\,dV/|\mathbf{r}-\mathbf{r}'|^3 \qquad (3)$$

Applying (2)-(3) to the results of "on-line" gradiometers measurements at each point of wave plane one can obtain the following expressions of magnetic induction component's transition to Fourier-image

$$F[B_z] = F[d_{1z}]/\{1-\exp(-kl)\} = F[d_{2z}]/\{1-2\exp(-kl)+\exp(-2kl)\} \qquad k=(k_x^2+k_y^2)^{1/2} \qquad (4a)$$
$$F[B_x] = F[d_{1zx}]/\{1-\exp(-kl)\} \qquad (4b)$$

Having performed the return Fourier transformation, one can obtain in explicit form the value of B_z or B_x at the same points of measurements plane with respect to the contribution of all flux transformer's coils to the registered signal. Besides this, the described scheme of "restoring" the parameters B_z (B_x) permits to allocate automatically (for the moment of time investigated) the observation points, where it is admissible to use the biomagnetic signal source dipole model on the first stage of problem solution (2).

Let's prove that the received information is quite enough for determining the parameters B_x, B_y, B_z, $\partial B_i/\partial r_j$, $\partial^2 B_z/\partial r_i\partial r_j$, which are necessary for the inverse problem solution. As far as the measurements are executed in uniform, non-magnetic, non-conducting environment, the matrix of spatial derivatives of magnetic induction D_1, D_{2x}, D_{2y} and D_{2z} are symmetric and magnetic induction vector satisfies the Maxwell equations Curl**B**=0, div**B**=0. It follows, that the Fourier-images of this vector component are linearly dependent and they are determined through the values of F [B_z] (or F [B_x] in the case of (1 b), (4 b)) after the plain algebraic expressions

$$F[dB_z/dx]= ik_xF[B_z],F[dB_z/dy]= ik_yF[B_z],F[dB_z/dz]= -kF[B_z] \qquad (5)$$
$$ik_xF[B_x] + ik_yF[B_y] - kF[B_z] = 0, \qquad (6a)$$
$$-ik_xF[B_z]-kF[B_x]=0,\ -ik_yF[B_z]-kF[B_y]=0,\ ik_yF[B_x]-ik_xF[B_y]=0 \qquad (6b)$$

And in wave plane one can obtain

$$F[\partial B_x/\partial x] = k_x^2\, F[B_z]/k\, ,\ F[\partial B_y/\partial y] = k_y^2\, F[B_z]/k\, ,\ F[\partial B_x/\partial y] = k_xk_y\, F[B_z]/k\, ,$$
$$F[\partial B_x/\partial z] = i\, k_x\, F[B_z]\, ,\ F[\partial B_y/\partial z] = i\, k_y\, F[B_z]\, ,\ F[\partial^2 B_z/\partial x^2] = -\, k_x^2\, F[B_z]\, , \qquad (7)$$
$$F[\partial^2 B_z/\partial y^2] = -\, k_y^2\, F[B_z]\, ,\ F[\partial^2 B_z/\partial x\partial y] = -\, k_xk_y\, F[B_z]\, ,$$
$$F[\partial^2 B_z/\partial x\partial z] = -\, ik_xk\, F[B_z]\, ,\ F[\partial^2 B_z/\partial y\partial z] = -\, ik_yk\, F[B_z]$$

After executing the return Fourier transformation we obtain at each point of measurement plane: the measured values of biomagnetic signal, the components of magnetic induction vector B_x, B_y, B_z and their first and second spatial derivatives, as well as the values of Fourier-images of these parameters at the points of wave plane (k_x, k_y, 0). It enables the analytical solution for the inverse problem [1,4].

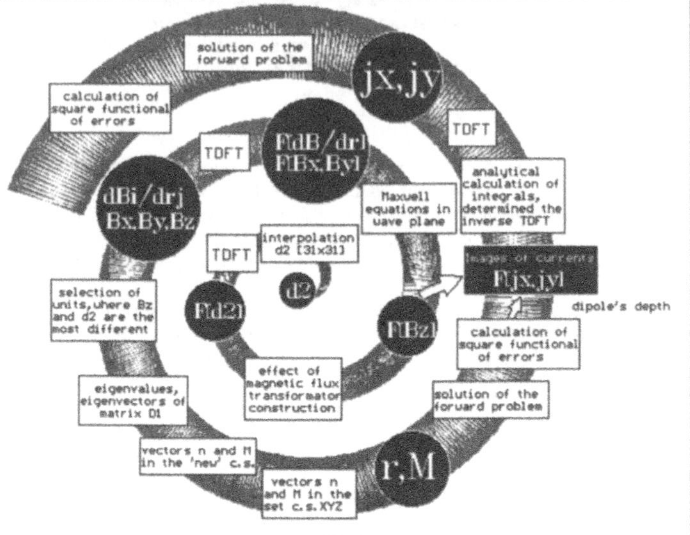

Fig.1

The block diagram of the inverse problem solution algorithm is submitted on fig.1 and it shows the mutual self-consistency of the two branches of algorithm (for effective dipole and system of currents) when processing the biomagnetic data.

3. RESULTS

Figs. 2-3 present the results of processing the biomagnetic signal caused by the activity of normal person heart. The initial data for fig.2 comprise the results of measuring the parameters of the human heart magnetic field with the help of single-channel, the second-order squid-gradiometer d_{2z} within the non-shielded room, at 36 nodes of rectangular grid (6 x 6 nodes of 4 cm- interval).

For the given moments of QRS-complex the fig.2 shows: the spatial configuration of the effective system of currents as the elements of current (j_x, j_y); the projection of observation points upon the source plane and contours of hearts, obtained by the x-ray investigations; an averaged electrocardiocycle; three series of parameter d_{2z} distributions; the measured field (series of pictures at the top) with display of located dipole source; the magnetic field distribution within the plane of measurements that is simulated under assumptions, that the source of field is the "effective" dipole (the series of pictures in the middle); the magnetic field in the plane of measurements that is simulated under assumptions, that the field source is the planar currents system (bottom series of pictures); the values of functional

$$G = \Sigma \{ d_{2meas}-d_{2calc} \}^2 / \{ N \Sigma [d_{2meas}]^2 \}, \qquad (8)$$

which gives (in certain degree) the quantitative estimation of error caused by the description of biomagnetic signal distribution when the chosen model of the field source is used. The results of processing the normal person magnetocadiosignal, that is measured simultaneously at 30 nodes of regular planar grid using the multichannel, first-order squid-gradiometer "Siemens" [6] are submitted on fig.3. For the moment of time corresponding to the beginning of QRS-complex, there are shown: the two pictures of MRI-tomograms of patient chest with allocated heart, location of the reference points and position of located dipole source; the averaged electrocardiocycle; the fragment of tomograms of area of the patient heart and the spatial configuration of the planar current system (j_x, j_y, 0), reconstructed after the developed algorithm (fig.1); the distribution of magnetic field within the measurement plane that is measured and simulated after accepted models.

This way obtained parameters of dipole and spatial configurations of effective system of currents are in qualitative agreement with the distribution of excitation wave in heart of the normal person and with the results of other authors.

The research described in this publication was made possible in part by Grant N K1 U100 from the Joint Fund of the Government of Ukraine and International Science Foundation.

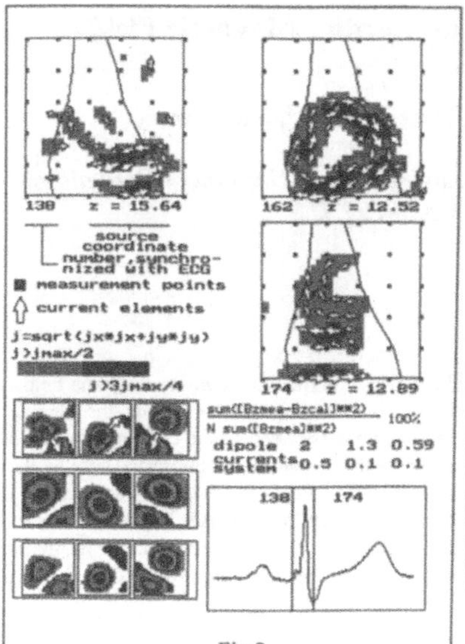

Fig.2

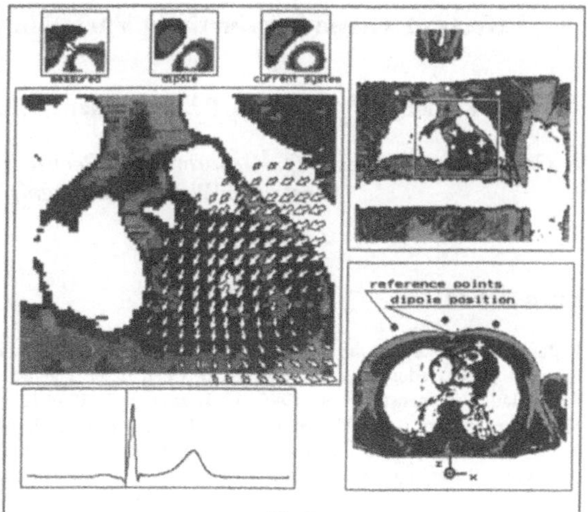

Fig.3

4. REFERENCES.

1. Primin M.A., Gumeniuk-Sychevskij V.I.,Nedayvoda I.V. Methods and algorithms of localization of magnetic field source (in Russian), Kiev, "Naukova dumka",1992.

2. Primin M., Gumeniuk-Sychevskij V., Nedayvoda I. Algorithms of data processing for distributed multichannel squid-gradiometer system in the case of magnetic field source localization, Cryogenics (1992), v.32, p.529-532.

3. Primin M., Gumeniuk-Sychevskij V., Nedayvoda I., Mathematical models and algorithms of information conversion in spatial analysis of weak magnetic fields, Int.J.Applied Electromagn.in Materials 5 (1994),311-319.

4. Primin M., Nedayvoda I. Mathematical model and measurement algorithms for the location of a dipole source, Int.J.Applied Electromagn. In Materials (in press).

5. Oppenheim A., Schafer R., Digital signal processing, Prentice Hall, Inc., Englewood Cliffs, New Jersey, 1975,480 p.

6. Biomagnetic multichannel system with integrated SQUIDs and first order gradiometers operating in a shielded room, H.Hoenig, G.Daalmans, W.Folberth, H.Reichenberger, S.Schneider, Seifert H., Cryogenics, (1989) 809-813 pp.

Effects of Tissue Conductivity Variations on Cardiac Magnetic Fields

Ramon, C.[1,2], Czapski, P.[1], Huntsman, L.L.[2], Bardy, G.H.[3], and Kim, Y.[1]

[1]*Center for Bioengineering,* [2]*Department of Electrical Engineering, ,* [3]*Division of Cardiology*
University of Washington, Seattle, WA 98195, U.S.A.

Introduction

In this study we examine the dependence between thoracic tissue conductivities and resulting cardiac magnetic fields. Previous studies addressed the relations between tissue conductivities and voltage gradients during the cardiac defibrillation (Jorgenson et al., 1995 [1]; Karlon et al., 1994 [2]), and relations between tissue conductivities and electrocardiograms (Lepeshkin et al., 1970 [3]; Rudy et al., 1979 [4]). There are, however, no published data on the dependence of the magnetic fields on tissue conductivities. We simulated magnetic fields from the cardiac excitation process while varying the conductivity values of major tissues in the torso. We examined changes in conductivity of blood, myocardium, lungs, muscle, fat and bone on the simulated cardiac magnetic fields. We found that the magnetic fields are most sensitive to variations of the conductivity of blood and myocardium. These findings are reported below.

Methods

A high resolution finite element model of a heart and torso was used in this study. The model was constructed from segmented CT images. A semi-automatic process was used to classify the images into 19 different tissue types. Each tissue was assigned a resistivity value based on data reported in the literature (Geddes and Baker, 1967 [5]; Foster and Schwan, 1989 [6]). The resistivity values of the major tissue types used in the model are given in Table 1. This procedure produced a model with 2,554,610 elements and 2,634,573 nodes, and resolution of 1.5625 × 1.5625 mm in the horizontal plane and 3.0 mm in the vertical direction.

Table 1. Resistivity values of the major tissues types used in the simulation.

Tissue type	lungs	muscle	fat	blood	bone	myocardium
Normal Resistivity Ohm cm	1500	700	2500	154	16000	420
Resistivity Range Ohm cm	2500 - 500	1500 - 250	3000 - 1500	420 - 100	20000 - 7000	750 - 100

The normal cardiac excitation was modelled with multiple dipolar sources in the heart wall using methods similar to those used by other researchers (Gulrajani, 1988 [7]). The positions and magnitudes of the sources were determined from experimental data of Durrer et al. (1970) [8]. The entire ventricular depolarization process was assumed to last for 55 ms. Dipole magnitudes were determined by a typical shaped action potential with a resting potential of -90 mV and a peak value of +10 mV. The ventricular gradient of the action potential duration was included in the model. A primary gradient of the action potential, from the epicardium to the endocardium with a typical value of 20 ms/mm, and a secondary gradient from the base to apex with the value of 7.5 mm/ms, were used to simulate the T wave and the S-T segment of the magnetocardiogram. Volume currents were computed from the potentials distribution from the finite element model. Biot-Savart's law was then used to calculate the magnetic fields on a sampling plane located in front of the chest. The total current distribution inside the heart and torso and magnetic fields outside of the chest were computed on a Cray YMP/C-90 computer. Changes in one tissue conductivity were introduced while all remaining tissues were kept at their normal values. The effects of variations in myocardial and blood resistivities were evaluated by means of magnetic field contour maps.

Results and Discussions

Results for five conductivity values of blood are shown in Figure 1. Because of the close proximity of the blood to the sources in the myocardium and the high value of its conductivity, the field profiles change both in values and in spatial location of their peaks and valleys. Lowering blood resistivity by 35% increases peak values of the magnetic field by a factor of 5. Locations of peaks are also shifted in space. Increasing the blood resistivity by approximately 30% decreases the negative peak value by a factor of 1.6 but increases the positive peak value by a factor of 1.3 (Figure 1; 200 Ω·cm). Peak locations are also altered. Additional increases in blood resistivity decrease both the negative and positive peak values, but the relative change is not as high as when decreasing the resistivity. This might be explained by the electrical shunting effect of the blood. Lowering the resistivity value causes larger currents to flow through the blood in the chambers of the heart offsetting the source currents and thus producing much larger changes in the simulated magnetic fields. Increasing the resistivity decreases the shunting currents, allowing the sources to dominate the total field profiles.

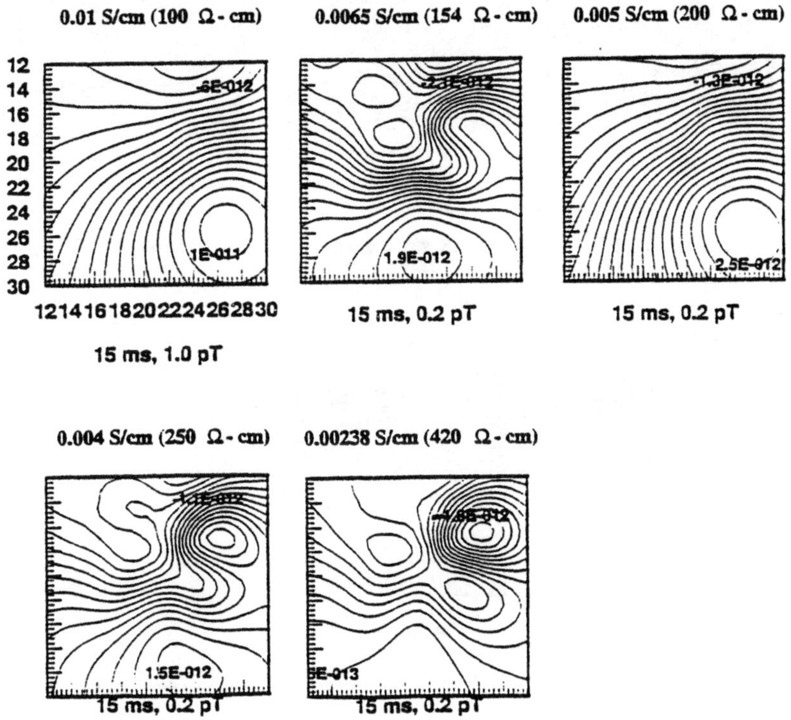

Fig. 1. Magnetic field profiles at different blood conductivities. Notice the shift in position and changes in the peak intensities as one moves from one profile to another.

Data for the changes in the myocardial resistivity are shown in Figure 2. Both decreasing and increasing of the resistivity value increases the peak values of the magnetic field and changes locations of the peaks. This might be explained by a complex interaction between the source and volume currents in the myocardial tissue and blood chambers. In our model each dipolar source is represented by a pair of voltages. Therefore, a change in the resistivity of the myocardium alters not only the volume currents but the source currents as well. Decreasing the resistivity increases the primary currents and increases the volume currents in the heart, and thus the observed increase in the peak values of the magnetic fields. When the resistivity of the myocardium increases, the source currents are lowered, as are the volume currents in the myocardium. However, the low blood resistivity causes the volume currents in the blood chambers to increase, thus increasing the peak values of the magnetic fields. In addition, increasing the resistivity increases the distance between the peaks of the field, and decreasing the resistivity brings these closer to each other. This is consistent with our analysis of the volume versus primary currents. Decreasing the resistivity causes the source region to dominate, thus the field is produced mostly by a confined source area in the heart wall. When the resistivity increases, there is more contribution from volume currents in the ventricles corresponding to the larger spatial region.

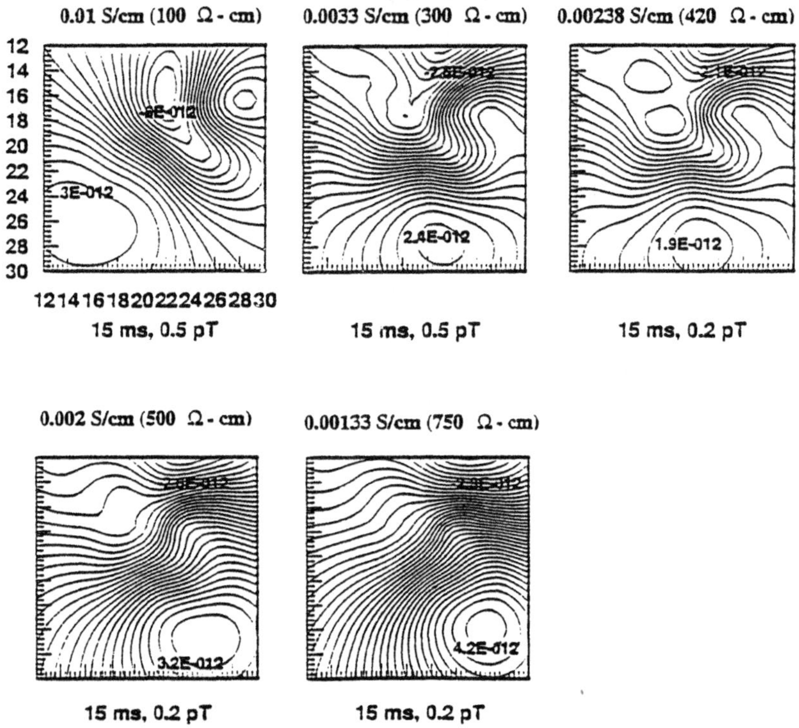

Fig. 2. Magnetic field profiles at different values of myocardial conductivity.

The computed results show the sensitivity of the magnetic fields to changes in the conductivity of blood and myocardium. These results are relevant to future modeling of magnetocardiograms and solving the inverse problem. They also emphasize the importance of careful modeling of the blood and heart regions.

References

[1] Jorgenson, D. B., Haynor, D. R., Bardy, G. H., and Kim, Y. Computational studies of transthoracic and transvenous defibrillation in a detailed 3-D human thorax model, IEEE Trans. on Biomed. Engr., 1995, 42: 172-184.
[2] Karlon, J., Lehr, J. L., and Eisenberg, S. R. Finite element models of thoracic conductive anatomy: sensitivity to changes in inhomogeneity and anisotropy, IEEE Trans. Biomed. Eng., 1994, 41: 1010-1017.
[3] Lepeshkin, E., Rush, S., Zao, Z. Z., Zakopoulos, K. S., and Mashima, S. Effects of thoracic conductivities on cardiac vectors, Vectorcardiog., 1970, 2: 18-29.
[4] Rudy, Y., Plonsey, R., and Liebman, J. The effects of variations in conductivity and geometrical parameters on the electrocardiogram using eccentric spheres model, Circ. Res., 1979, 44: 104-111.
[5] Geddes, L. A., and Baker, L. E. The specific resistance of biological material - a compendium of data for the biomedical engineer and physiologist, Med. Biol. Engr., 1967, 5: 271-293.
[6] Foster, K. R., and Schwan, H. P. Dielectric properties of tissues: a critical review, CRC Handbook of Biological Effects of Electromagnetic Fields, 1989, 17: 25-104.
[7] Gulrajani, R. M. Models of the electrical activity of the heart and computer simulation of the electrocardiogram CRC Critical Reviews in Biomedical Engineering, 1988, 16: 1-66.
[8] Durrer, D., van Damm, R. T., Freud, G. E., Janse, M. J., Meijler, F. L., and Arzbacher, R. C. Total excitation of the isolated human heart, Circulation, 1970, 41: 899-912.

Acknowledgements

This work was supported by the National Science Foundation grant #BCS-9209938, an equipment grant from the Seattle Foundation and the Graduate School Research Fund of the University of Washington. We would like to express our gratitude to Cray Research, Inc. and especially to Mr. Steve Baumann, Mr. James Harcourt and Mr. Thomas Ashbrook for allowing us to use their Cray YMP/C-90 computer.

Discrete Spline Electric-Magnetic Tomography (DSPECT) Based on Realistic Neuroanatomy

Riera, J.J.[1], Aubert, E.[1], Valdes-Sosa, P.[1], Casanova, R.[1] and Lins, O.[2]

[1]Cuban Neurosciences Center, Havana, Cuba
[2]Rotman Research Institute, Toronto, Canada

Introduction

Electric Magnetic Tomography (EMT) may be defined as the estimation of the primary current density $\vec{J}(\vec{r},t)$ in the brain from time series of EEG/MEG measurements [1]. Estimation is, therefore, the solution of a Fredholm Integral equation of the first type (FIE) that relates $\vec{J}(\vec{r},t)$ to the actual measurements via the lead field. This is a problem which does not have a unique solution and that is highly ill-conditioned. One general Bayesian solution strategy for solving FIE has been proposed by Wahba ([2] pp. 101-102): the use of spline models which select that solution that minimises a roughness penalty functional. The approach, applied to EMT leads to **Spline EMT (SEMT). Low Resolution EMT (LORETA)** is a particular discrete version of SEMT [3] based upon a vector laplacian penalty. LORETA has become popular due to its ease of implementation as well as the excellent results obtained in estimating $\vec{J}(\vec{r},t)$. An improvement is **Variable Resolution EMT (VARETA)** which allows a better reconstruction of both distributed and localised $\vec{J}(\vec{r},t)$ by specifying a spatially adaptive amount of smoothing [4].

In spite of the usefulness of these methods, they share a major drawback with many other types of "volumetric" distributed solutions: anatomical information available from MRIs is not used. In the first place, information contained in the measurements is "wasted" in estimating $\vec{J}(\vec{r},t)$ placed in white matter and CSF. Furthermore, the requirement of spatial smoothness as defined for LORETA and VARETA can only be accepted as a first approximation [5] since jumps in the strength and direction of $\vec{J}(\vec{r},t)$ will be probable on opposite sides of sulci in spite of proximity in 3D space. These problems were avoided by Dale and Sereno [6] who specified that the primary sources be placed on the cortical surface and oriented perpendicular to it. However, their selection of a spatial a priori covariance matrix for the generators used in Bayesian inference is somewhat ad hoc. Additionally, [6] does not explicitly address the issue of compact subcortical generators.

This paper presents a new type of SEMT (DSPET) based on neuroanatomy. It combines the use of a triangulated surface representation for the cortex with compact voxel sets for subcortical structures. The spline formulation allows a unified treatment that introduces a simple and natural a priori covariance matrix which is proportional to the squared inverse of the surface/volume laplacian. Essential new features of the method are: a) the use of a discrete version of the laplacian that preserves the intrinsic properties of the cortical surface defining smoothness as a function of geodesic distance [7], b) the use of a constraint which avoids inter hemispheric a priori covariances, and c) the use of a VARETA like technique for both surface and volume currents $\vec{J}(\vec{r},t)$. The basic theory of the method is presented as well as simulation results for a discretized spherical cortex with two separated hemispheres. Finally, an illustration with a high resolution EEG recording of the α rhythm is included. A preliminary version of this paper was presented in [8] [1].

---

[1] During the preparation of this manuscript we have become aware of other attempts at cortical surface restricted SEMT. Volumetric LORETA [3] has been restricted to the cortex [9], a method not based on an accurate triangulation of the cortex. Wagner et al. [10] propose a laplacian operator based upon a triangulated cortex similar to that proposed in this paper. However the laplacian operator used [11] does not take into account an irregular triangulation as discussed in [7].

Method

DSPET estimates the primary current density on a discrete mesh of points in the brain, defined for a given subject from his MRI. The current density at the *j-th* node $\vec{r}_j$ for time t is denoted as $J_{jt} = \vec{J}(\vec{r}_j, t)$ The estimation mesh comprises two distinct subsets of nodes:

a) Ω_{SC}: those nodes obtained from a 3-D regular grid of voxels in the grey matter of subcortical nuclei. The volumetric current densities for these nodes may be arranged into the matrix $J^{SC} = \{J_{jt} | j \in \Omega_{SC}; 1 \le t \le T\}$.

b) Ω_C: those nodes obtained from a triangulation of the cortical surface. The superficial density currents, will be assumed to have a fixed orientation $\vec{\mu}_j$ perpendicular to the local tangential plane and a time varying amplitude $\gamma_j = \{\gamma_{jt}\}$. That is, $J^C = M\gamma$, where M is a certain patterned matrix [1] and $\gamma = (\gamma_1' \ \cdots \ \gamma_s')'$. Each $\vec{\mu}_j$ is a unit length orientation vector obtained as the average of the normals of the neighbouring triangles (see fig. 1).

The forward problem of EMT can then be formulated as:

$$O = K^C M\gamma + K^{SC} J^{SC} + E$$

where O is the data matrix and E the matrix of sensor noise. As discussed in [1], [4] the forward problem leads to a regularization problem in which the following functional must be minimised:

$$f(\gamma, J^{SC} / O) = \left|H\left(O - K^C M\gamma - K^{SC} J^{SC}\right)\right|^2 + \left|\Lambda^C L^C W^C \gamma + \Lambda^{SC} L^{SC} W^{SC} J^{SC}\right|^2 + g\left(\Lambda^C, \Lambda^{SC}\right)$$

where H is a centring matrix that eliminates the reference for electrical data, W^C and W^{SC} are weight matrices and L^C, L^{SC} are a scalar surface laplacian operator for the cortex and a volume vector laplacian for subcortical structures, respectively, as will be defined bellow. Λ^C and Λ^{SC} are the spatially adaptive regularization parameters and $g(\Lambda^C, \Lambda^{SC})$ a functional that ensure their spatial smoothness [4].

The surface laplacian L^C may be defined as follows [7]. Let the cortical surface point situated at the j-th vertex of the triangulated surface be surrounded by the set ψ_j of N_j neighbours. Figure 1 shows a schematic representation of the triangulated surface with the following definitions: 1) the distance r_{jk} between the point and the k-th neighbour; 2) the mean distance $\bar{r}_j$ between the point and all the neighbours; and 3) the angles ϕ_{jk}^+ and ϕ_{jk}^- for right and left parts of the k-th neighbour, respectively. Define the following symbols: $\Phi_{jk} = \frac{1-\cos(\phi_{jk}^-)}{\sin(\phi_{jk}^-)} + \frac{1-\cos(\phi_{jk}^+)}{\sin(\phi_{jk}^+)}$ and

$$\theta_{jk} = \frac{4\,\Phi_{jk}}{\bar{r}_j \left(\sum_{l=1}^{N_j} \Phi_{jl}\right) r_{jk}}$$

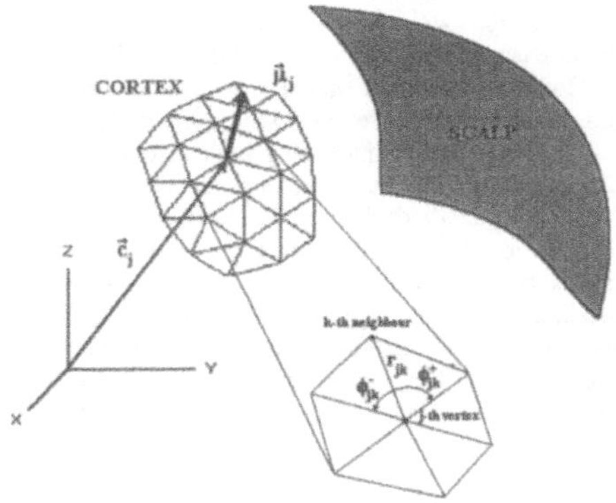

Figure 1

From this, the sparse surface laplacian and diagonal lead field normalisation matrices can be defined as $L^C = \{L_{jk}^C\}$ and $W^C = \{\omega_{jj}^C\}$, respectively, where $L_{jj}^{C} = -\sum_{k \in \Psi_j} \theta_{jk}$ $\quad L_{jk}^{C} = \begin{cases} \theta_{jk} & k \in \psi_j \\ 0 & otherwise \end{cases}$ $\quad \omega_{jj}^{C} = \left\| \left\{ HK^{C} M \right\}_{j} \right\|$

L^{SC} and W^{SC} are defined as in [3] with slight modifications.

The solution of the regularization problem was implemented using sparse matrix operations in MATLAB. An iterative procedure was used that minimised the Generalised Cross Validation Criterion [2].

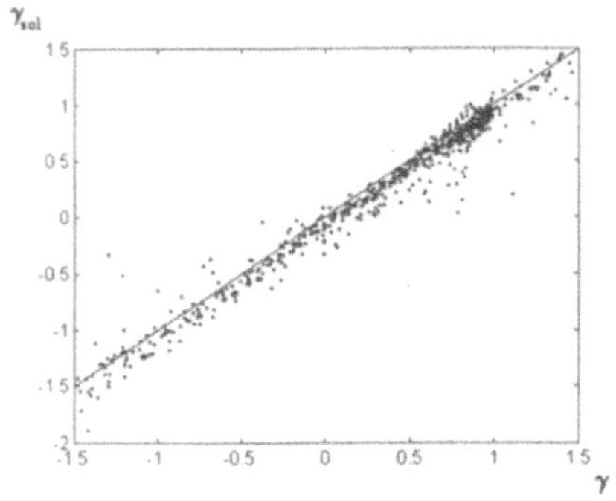

Figure 2

Simulation on a spherical surface

A single sphere head model was use to produce an external magnetic field at 148 magnetometers. The "head" contained an additional eccentric sphere modelling the "cortex". On the right hemisphere the current source density was taken to be proportional to the spherical harmonic $Y_{11}(\vartheta,\varphi)$ and on the left proportional to the function $\cos(\vartheta)\cos(\vartheta_o)+\sin(\vartheta)\sin(\vartheta_o)\cos(\varphi-\vartheta_o)$, with $\vartheta_o = 1.41$ and $\varphi_o = 1.63$. Instrumental noise was added at the sensors with a SNR of 10. DSPET was applied to this data using a common regularization parameter. Figure 2 shows the simulated versus estimated cortical current density. The results show an adequate reconstruction, well within the margin of experimental error.

Cortical sources of the α rhythm

The scalp and cortical surfaces of a subject were segmented and triangulated by special purpose software. A 3-D digitizer was used in conjunction with external landmarks to co-register electrode positions and the MRI. The lead field for the three shell model was obtained by fitting the external sphere to the scalp surface. The spectra of cortical sources of a high resolution (120 channels) EEG recording was obtained by DSPET using methods described in [12]. Figure 3 shows the cortical distribution of the α spectral peak. Note that alpha activity seems to be quite concentrated in the occipital region (arrow) in good agreement with both frequency domain dipole fitting [13] and frequency domain volumetric VARETA [4].

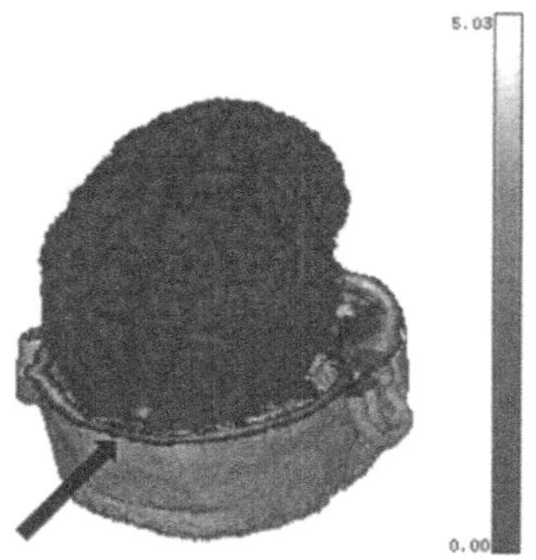

Figure 3

Extensions and further developments

As discussed in [1] the formulation presented in this paper does not consider the temporal dynamics of neural sources, an extension that will depend on the implementation of efficient algorithms that take into account the complex patterns of spatio-temporal correlations. Also additional improvements in accuracy will accrue from introducing improved roughness penalties. Instead of using the discrete laplacian, work is in progress that will evaluate the performance of the Thin Plate Spline functional:

$$h(\gamma) = \iint_S \left[\left(\frac{\partial^2 \gamma}{\partial x^2} \right)^2 + \left(\frac{\partial^2 \gamma}{\partial y^2} \right)^2 + 2\left(\frac{\partial^2 \gamma}{\partial x \partial y} \right)^2 \right] dS$$

numerically calculated in such a way as to take into consideration the geodesic curvature of the cortical surface [7].

References

[1] Valdes, P., Riera, J., Casanova, R. and Lins, O., Spatio Temporal Distributed Inverse Solutions, This Volume.

[2] Wahba, G., Spline Models for Observational Data, SIAM, Philadelphia, 1990.

[3] Pascual-Marqui, R.D., Low resolution electromagnetic tomography: a new method for localising electrical activity in the brain, Int. J. Psychophysiol., 1994, vol. 18, pp. 49-65.

[4] Valdes, P., Marti, F., Garcia, F. and Casanova, R., Variable Resolution Electric Magnetic Tomography, This volume.

[5] Mosher, J., and George, J.S., Comments on Loreta, ISBET Newsletter, 1994, no. 5, ISSN 0947-5133.

[6] Dale, A.M. and Sereno, M.I., Improved Localization of Cortical Activity by Combining EEG and MEG with MRI Cortical Surface Reconstruction: A Linear Approach, Journal of Cognitive Neuroscience, 1993, 5:2, pp. 162-176.

[7] Huiskamp, G., Difference Formulas for the Surface Laplacian on a Triangulated Surface, Journal of Computational Physics, August 1991, vol. 95, no. 2, pp. 477-496.

[8] Riera, J., Valdes, P., Jimenez, J.C., Casanova R. and Bosch J., Mixed Distributed and Discrete Electromagnetic Brain Tomography. Special Lecture given by Valdes P. at the 6th International Congress of The International Society for Brain Electric and Magnetic Tomography, October 1995, Tokushima, Japan.

[9] Pascual-Marqui, P. ISBET Newsletter, 1995, no 6. ISSN 0947-5133.

[10] Wagner, M., Fuchs, M., Wischmann, A.A., Drenckhahn, R. and Köhler, T., Smooth Reconstruction of Cortical Sources from EEG or MEG recordings, Abstract presented at the Human Brain Mapping Meeting, July 1996, Boston.

[11] Huiskamp, G. and van Oosterom, A., The Depolarization Sequence of the Human Heart Surface Computed from Measured Body Surface Potentials, IEEE Trans. Biomed. Eng., 1988, vol. 35, no. 12, pp. 1047- 1058.

[12] Casanova, R, Valdes, P, Garcia, F, Aubert, E, Riera, J, Korin, W. and Lins, O. Frequency Domain Distributed Inverse Solutions, This Volume.

[13] Valdes, P., Bosch, J., Grave de Peralta, R., Hernandez, J., Riera, J., Pascual, R. and Biscay, R., Frequency Domain Models for the EEG. Brain Topography, 1992, vol 4, (4), pp. 309-319.

Strategy for Objective Comparisons of Inverse Solutions

Schlitt, H.A., Zhao, S.-R., Halling, H. and Müller-Gärtner, H.-W.

Forschungszentrum Jülich GmbH, Jülich, Germany

Introduction

There is much discussion in the Biomagnetism community about the differences in and usefulness of 'dipole solutions' and 'minimum norm solutions'. By 'dipole solutions' we mean the wide variety of procedures that assume the current in the brain is well modeled by a small number of current dipoles, and determine the locations and moment vectors of those dipoles. By 'minimum norm solutions' we mean the equally diverse group of procedures that do not assume the source is dipolar, and instead minimize the norm of the current, $|\mathbf{J}|$, or a weighted version of the current. We do not wish to discuss the validity of the assumptions made in order to obtain a dipole or minimum norm solution. Our goal is a quantitative measure for comparing dipole solutions to minimum norm solutions when the true current source is known. We hope that such an objective measure will lead to a more fruitful discussion of the usefulness of dipole solutions and minimum norm solutions, although we acknowledge it will not add anything to the discussion about the validity of the assumptions used to obtain the solutions.

The idea of creating an extensive collection of magnetic fields to be used to test reconstruction methods has been informally discussed for some time. The nature of the magnetic fields in the test-bed has been the primary focus of the discussion; what is missing from this concept is a way of objectively determining how well a particular method has reconstructed the original source. If the original source consists of a known distribution of a small number of dipoles, then the reconstruction error for a dipole solution can be quantified in terms of the correct number of dipoles, and the distance between fit and true dipoles. Quantifying reconstruction error for minimum norm solutions is more difficult. The peaks of the minimum norm reconstruction can be compared to the dipole locations, but that doesn't account for the 'extra' current that was reconstructed at other locations.

Quantitative Description of Sources

We propose describing sources in terms of the location of the mean of the distribution, and the spread of the source (the standard deviation). Because the peak of the minimum norm reconstruction is often used, we also consider the peak of the distribution and the standard deviation with respect to the peak.

For any source distribution, the mean and standard deviation can be calculated. If the source is a single dipole, then the mean is the dipole location and the standard deviation is zero. For a distributed source (more than one dipole), the weighted mean and standard deviation are calculated. Consider a distributed source consisting of N_d dipoles at locations $(x_i, y_i, z_i), i = 1, ..., N_d$ and with orientations (q_{xi}, q_{yi}, q_{zi}). Then the mean weighted x location is:

$$< x >= \frac{\sum_{i=1}^{N_d} |\mathbf{q}_i| x_i}{q_{tot}} \qquad (1)$$

Similarly for $< y >$ and $< z >$. The standard deviation in the x direction is:

$$\sigma_x = \sqrt{\frac{\sum_{i=1}^{N_d} |\mathbf{q}_i| (x_i)^2}{q_{tot}} - < x >^2} \qquad (2)$$

And similarly for σ_y and σ_z. Both equations have been normalized to the total magnitude of the current:

$$q_{tot} = \sum_{i=1}^{N_d} |\mathbf{q}_i| \qquad (3)$$

To calculate the standard deviation with repect to the peak, the location of the peak is used in eq. 2 instead of the location of the mean.

Methods

We have developed a limited test-bed of magnetic fields, and are applying this quantitative method to the results of the dipole and minimum norm reconstruction methods available to us.

To demonstrate the usefulness of this approach, we quantatively compare the results from only two different reconstructions methods: a current dipole procedure and a weighted minimim norm procedure. We have chosen 4 source distributions (Table 1, Fig. 1), ranging from a single current dipole to a distribution of current described in terms of 317 dipoles. Source B is confined to the plane $z = 8$ cm, and Source D is confined to the plane $x = 0$ cm. We choose these sources in part because Sources A and B produce exactly the same magnetic fields, and Sources C and D produce exactly the same magnetic fields. We expect the dipole methods to reconstruct the dipolar sources (A and C) better than the distributed sources (B and D), and vice versa for the minimum norm reconstruction methods.

Figure 1: Quantitative Description and Plots of Original Current Sources. The plots for Sources A, B, and C are all in the plane $z = 8$ cm, while the plot of Source D is in the plane $x = 0$ cm. The scale of the arrows is arbitrary and not consistent across the plots.

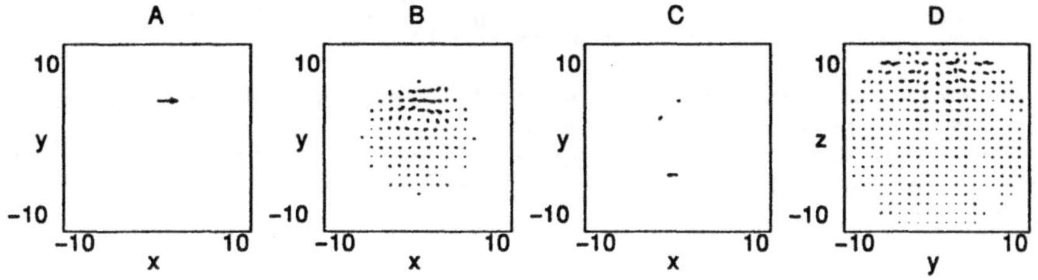

Source	Number of Dipoles in Source	Location of Mean [cm]			Std. Deviation w.r.t. Mean [cm]			Location of Peak [cm]			Std. Deviation w.r.t. Peak [cm]		
		$<x>$	$<y>$	$<z>$	σ_x	σ_y	σ_z	x_p	y_p	z_p	σ_x	σ_y	σ_z
A	1	0.0	4.0	8.0	0.0	0.0	0.0	0.0	4.0	8.0	0.0	0.0	0.0
B	113	0.0	2.3	8.0	2.2	2.6	0.0	0.0	4.0	8.0	2.2	0.0	0.0
C	3	-0.4	-1.8	6.7	0.9	3.3	0.9	0.0	-4.0	6.0	0.9	0.0	3.0
D	317	0.0	0.4	4.1	0.0	4.4	4.0	0.0	-4.0	8.0	0.0	1.9	0.0

Table 1: Quantitative Description of Original Current Sources.

The conductor geometry is a single homogeneous, isotropic sphere centered at the origin of the coordinate system. The sensor array is an idealized BTi 37-channel array (first-order axial gradiometer), with the middle sensor centered on the z-axis 12 cm from the origin. For this paper, we present only results with no added noise.

Dipole Solution

We use a non-linear least squares fitting routine (with the Nelder-Mead simplex algorithm, [1]) to find the location, strength and orientation of the dipole. We use a variety of starting locations and numbers of dipoles to fit the data. The quantity that is minimized is the Relative Difference Measure (RDM) [2]:

$$RDM = \sqrt{\frac{\sum_{i=1}^{N_{sensors}} (B_i^{measured} - B_i^{calculated})^2}{\sum_{i=1}^{N_{sensors}} (B_i^{measured})^2}} \tag{4}$$

Minimum Norm Solution

We use a non-iterative, weighted minimum norm solution (inspired by [3], [4], and described in more detail in [5]) to solve the underdetermined set of equations:

$$\mathbf{LWJ} = \mathbf{B} \tag{5}$$

where **L** is the Lead Field, **J** is the current distribution, and **B** is the measured magnetic field [6]. The estimate of the current distribution is:

$$\mathbf{J} = \mathbf{W}(\mathbf{LW})^T[\mathbf{LW}(\mathbf{LW})^T + \zeta\mathbf{I}]^{-1}\mathbf{B} \tag{6}$$

The regularization parameter ζ is defined as:

$$\zeta = k\frac{tr[(\mathbf{LW})(\mathbf{LW})^T]}{N_{sensors}} \tag{7}$$

where k is a user defined parameter, $tr\{\cdot\}$ denotes the trace of the matrix, and $N_{sensors}$ is the number of sensors. The diagonal elements of the weights matrix, W, are defined as

$$W = exp(\frac{-|\mathbf{r}_i|^2}{\lambda^2 R^2}) \tag{8}$$

where $\mathbf{r}_i$ is the location of the ith grid point, λ is a user-defined parameter which controls the depth of the reconstructed source, and R is the radius of the source space.

There are 1453 source locations in a hemispherical source space ($R = 11$ cm). The constant k in Eq. 7 is set to 0.1, and λ is set to 0.7; these values were chosen arbitrarily and seemed to have little affect on the final reconstructions.

Results

As expected, the dipolar sources were adequately reconstructed using the Dipole Method (Fig. 2, I and II). Source A (the single dipole) was easy to reconstruct; most starting locations converged to the true locations, and when more than one dipole was used the final magnitudes were zero for all dipoles except the one in the correct location. Source C (3 dipoles) was more sensitive to the starting locations and number of dipoles used, but the algorithm was usually able to correctly reproduce the original source. Sources B and D (the distributed currents) could not be adequately reconstructed even when 7 dipoles were used. The largest number of dipoles that could be fit was 7 because each dipole requires 5 degreees of freedom and there are only 37 degrees of freedom available because we used only 37 sensors.

The weighted minimum norm solutions (Fig. 2, III and IV) were able to find either the mean location of the single dipole or the peak location, but not both at the same time.

Discussion

We won't be able to tell if the original simulated source adequately describes true sources in the brain, but we can tell if a particular reconstruction method adequately reconstructed the original source. By quantifying the source in terms of the location of its mean and its spread we can objectively compare reconstruction methods.

For these particular sources, the dipole solutions matched the mean and peak locations better than the weighted minimum norm solutions did. Surprisingly, the dipole solutions also matched the standard deviations slightly better. What is important to note is that we have can objectively compare the reconstruction methods.

Acknowledgement:

We are indebted to D. Ranken (at Los Alamos National Laboratory) for the use of his MRIview software, and to H. Heer and M. Wagener (at Forschungszentrum Jülich) for their programming assistance.

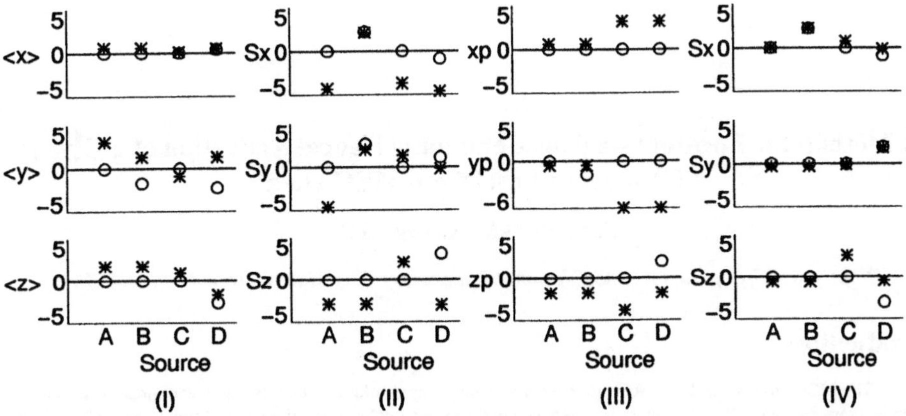

Figure 2: Errors in the location of the Mean (I) and Standard Deviation with respect to the Mean(II), and in the location of the Peak (III) and Standard Deviation with respect to the Peak (IV). The open circles represent the errors in the dipole solutions and the stars represent the errors in the weighted minimum norm solutions.

References:

[1] Press, W.H., Teukolsky, S.A., Vetterling, W.T., and Flannery, B.P. Numerical Recipes, 2nd Edition, Cambridge University Press, 1992

[2] Meijs, J.W.H., Peters, M.J., Boom, H.B.K., and Lopes da Silva, F.H., Relative Influence of Model Assumptions and Measurement Procedures in the Analysis of the MEG, Med. & Biol. Eng. & Comput., 1988, 26:136-142.

[3] Clarke, C.J.S. Probabilistic methods in a biomagnetic inverse problem, Inverse Problems, 1989, 5:999-1012

[4] Ioannides, A.A., Bolton, J.P.R., and Clarke, C.J.S. Continuous probabilistic solutions to the biomagnetic inverse problem, Inverse Problems, 1990, 6:523-542

[5] Zhao, S.-R. and Halling, H. Evaluation of Minimum Norm Methods, to be written

[6] Jeffs, B., Leahy, R., and Singh, M. An Evaluation of Methods for Neuromagnetic Image Reconstruction, IEEE Trans. BME, 1987, BME-34(9):713-723

A Method for Locating Regions Containing Neural Activation at a Given Confidence Level from MEG Data

Schmidt, D.M., George, J.S.

Biophysics Group, Los Alamos National Laboratory, Los Alamos, New Mexico, USA

Introduction

The MEG inverse problem does not have a general, unique solution. Unless restrictive model assumptions are made, there are generally many more free parameters than measurements and there exist "silent sources"—current distributions which produce no external magnetic field. By weighting solutions according to how well each fits our prior notion about what properties good solutions should have, it may be possible to obtain a single current distribution that "best fits" the data and our expectations. However, in general there will still exist a number of different current distributions which fit both the data and our prior expectations sufficiently well. For example, a simulated data set based on a single or several dipoles can generally be fit equally well by a distributed current minimum-norm reconstruction. In experimental data it is often possible to find a relatively small number of dipoles which both fit the data and have a norm not much larger than that of the minimum-norm solution. Moreover, the few-dipole solutions often have currents in different regions than the corresponding minimum-norm solution. Because there exist well-fitting current distributions which may have current in significantly different locations, it can be misleading to infer locations of stimulus-correlated neural activity based on a single, best-fitting current distribution. We demonstrate here a method for inferring the location and number of regions containing neural activation by considering *all possible current distributions* within a given model (not just the most likely one) weighted according to how well each fits both the data and our prior expectations.

Methods

We employ a Bayesian approach which considers a *posterior* probability distribution, $P(j \mid data)$, that assigns each current distribution, j, a relative likelihood based on how well each distribution fits the data and meets one's prior expectation—explicitly defined by a *prior* probability distribution, $P(j)$. Specifically, Bayes theorem[1] states that

$$P(j \mid data) \propto P(data \mid j)P(j)$$

where $P(data \mid j)$ is the likelihood of the data given the current distribution, j. In some over-determined problems the information from the data dominates, and the posterior probability distribution is highly peaked. In such cases inferences from the most likely solution may be reliable. However, with MEG data the posterior distribution is generally quite broad because of the ill-posed nature of the problem and because there are often more free parameters than measurements. Most methods widely used for analysis of MEG data infer neural activity based on a single best-fitting current distribution within the context of a given source model. Implicit model assumptions strongly influence the nature of the reconstructed current distribution but the consequences of model assumptions are seldom considered. In our Bayesian approach, model assumptions are made explicit in the form of the prior probability distribution. In addition, we *infer activation from the full posterior distribution*, which we believe will yield different and more reliable estimates than inferences made from a single "most likely" current distribution.

In our model, we assume that there are a small number of compact regions that contain the stimulus-correlated neural activity. This is appropriate, for example, for stimuli designed to generate a limited, focal evoked response. What we want to determine is how many of these regions are needed, n, and where they are located, w. Following Bayes theorem,

$$P(n, w \mid data) \propto P(data \mid n, w)P(n, w)$$

where $P(n, w \mid data)$ is the posterior probability of the number and location of regions containing activity given the data, $P(n, w)$ is the prior probability for the number and location of these regions, and $P(data \mid n, w)$ is the likelihood of the data given a particular number and location of regions. The data, however, do not depend directly on the number and location of regions; rather they depend on the neural current distribution. Thus, we need to sum over all possible current distributions consistent with a given number and location of activation regions:

$$P(data \mid n, w) = \sum_j P(data \mid j) P(j \mid n, w)$$

where $P(j \mid n, w)$ is the prior probability for a current distribution, j, given the number and location of the regions of activation, n and w. Putting this together we have:

$$P(n, w \mid data) \propto \sum_j P(data \mid j) P(j \mid n, w) P(n, w).$$

What we have essentially done is to change the problem from finding the best current distribution, which generally has a large number of parameters, to finding the best set of small regions, which has a much smaller number of parameters yet retains a rich set of possibilities within current space. To do this, we sum over all possible current distributions—a process known as marginalization.

In principle, determining confidence intervals within the Bayesian formalism is straight forward because all relevant information is contained in the posterior distribution. The 90% confidence interval for $\{n, w\}$, for example, is simply the set, $\{n, w\}_{90}$, for which the posterior, $P(n, w \mid data)$, is greater than the value of the posterior for any other set of $\{n, w\}$ *and* the integral of the posterior over $\{n, w\}_{90}$ is 90% that of the integral over the full range of $\{n, w\}$. In practice we determined the confidence region by sampling the posterior using a Monte Carlo integration technique and constructing a histogram similar to that shown in Fig. 1. The integral of the posterior over the full range of $\{n, w\}$ is approximated by summing the product of the average value of the posterior at each histogram bin and the frequency of occurrence for that bin. To determine the 90% confidence interval we simply found the cut-off value of the posterior probability along the horizontal axis, α, such that the integral of the posterior from this value to the maximum value is 90% of the value of the integral over the whole range. The 90% confidence interval region consists of all values of $\{n, w\}$ for which $P(n, w \mid data)$ is greater than or equal to this cut-off value.

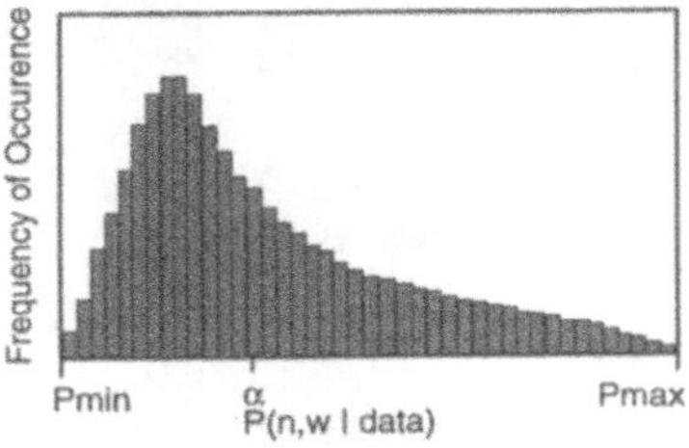

Fig. 1, Illustration of a histogram of values sampled from the posterior distribution, $P(n, w \mid data)$.

Results

The use of this algorithm is demonstrated with simulated data consisting of a range of numbers and locations of current dipole sources. In order to be as realistic as possible we used a sensor configuration from a real experiment in which a visual evoked response was studied. In this experiment there were two placements of a 37 channel first order gradiometer array (BTI) around the back of the subject's head. The Sarvas model[2] was used to calculate the output from the 74 sensors for a given set of current dipoles. Gaussian noise with standard deviation 1 fT was added to the simulated data which resulted in a noise level that was between 1% and 10% that of average signal strength, depending on the locations of the sources. The source volume consisted of 17,342 cortical volume elements (voxels) from the subject's anatomical MRI scan with the cortex segmented using tools developed in our laboratory and described elsewhere.[3, 4]

For the compact regions of activation which make up our model we used spheres of 1 cm radius such that any voxel within a sphere is considered to be part of the "active" region. We allowed no more than 4 regions with relative prior probability of 1., 1., 1., 1., .7 for 0 through 4 regions respectively. This specifies $P(n, w)$. For the prior probability of the current we used

$$P(j \mid n, w) \propto Exp\left(-\frac{1}{2}\sum_i \frac{\vec{j}_i^2}{V}\right)$$

where i enumerates all voxels within the regions specified by $\{n, w\}$, and $\vec{j}_i$ is the current dipole for voxel i. The parameter V represents the value above which the squared current dipole moment for each voxel starts to become significantly improbable. This prior tends to exclude large currents. We used $V = 500\, nA^2\, m^2$, which is fairly large since the sources used in our simulation all had a strength of $10\, nA\, m$. Because Gaussian noise was added to the data, the likelihood of the data given a current distribution, j, is itself a Gaussian distribution:

$$P(data \mid j) \propto Exp - \frac{1}{2} \sum_k \left(\frac{d_k - \sum_i \vec{A}_{k,i} \cdot \vec{j}_i}{\sigma_k} \right)^2$$

where d_k is the data value for channel k, σ_k is the noise standard deviation for channel k, and $\vec{A}_{k,i}$ is the forward field value for channel k with a unit dipole at voxel i.

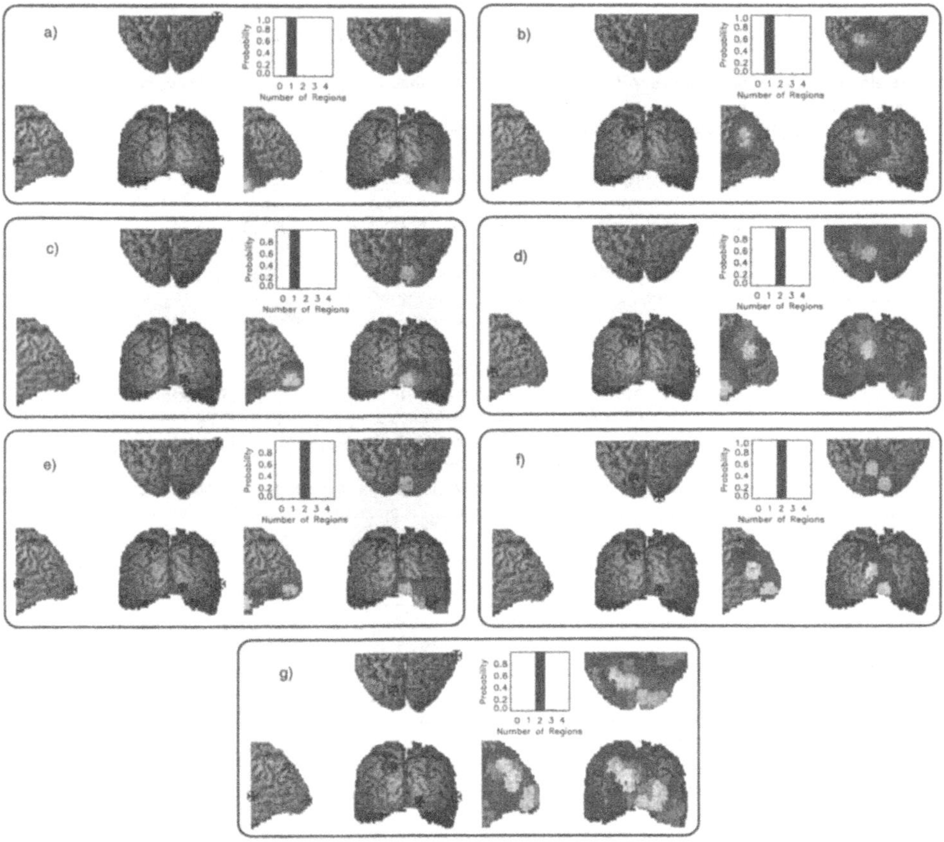

Fig. 2, Results of our method with simulated data. The location and number of true sources are shown in the left 3 cortical views within each panel. The 99.9% confidence intervals are shown as smooth grey patches on the right 3 views within each panel. Also shown is the posterior probability for the number of regions needed.

We evaluated the posterior by first randomly selecting values for the number and the locations of regions, $\{n, w\}$. For each set of $\{n, w\}$ we evaluated the posterior, $P(n, w \mid data)$, by integrating over all current values, $\vec{j}_i$, for voxels within the selected regions of activation. This integral was evaluated semi-analytically, using the fact that the current is quadratic in the exponential; the computation was therefore relatively fast. The value of the posterior was recorded in a histogram. In addition, the maximum posterior value in which a voxel is active was recorded for each voxel. After sufficiently sampling $\{n, w\}$, the histogram of posterior values was examined to determine the cut-off value for the desired confidence level. All voxels whose maximum posterior value was greater than or equal to the cut-off value were assigned to the confidence level region. In addition, we integrated (or marginalized) over location, w, in order to determine the posterior probability for n, the number of regions needed.

The results are shown in Fig. 2. Each panel represents a different source configuration, with the total number of sources ranging from one to three. Within each panel the left three cortical orthogonal renderings show the projection of the location and number of the true sources, marked with a black symbol. A maximum intensity projection of the 99.9% confidence region needed to contain the activity is superimposed on the right three cortical renderings within each panel. The confidence regions appear as smooth grey-scale patches on top of the textured cortical surface. The shade of grey indicates relative likelihood within the confidence region (white being most likely). In addition, the marginal posterior probability for the number of regions, n, is shown in the bar chart within each panel.

In most cases the true number of regions is reconstructed and the true location of activity is within the confidence region. This region is not guaranteed to contain activity, but rather to probably contain activity at the given confidence level. Nevertheless, this is much better than the level of accuracy and confidence obtained with just the most likely solution—given by the lightest patches in the images.

Discussion

We have demonstrated a Bayesian approach for inferring the number and location of regions of neural activation from MEG data. Our model consists of a small number of small size regions and is not parameterized by a given current distribution. Rather, we marginalize over the rich set of current distributions consistent with our model in order to determine the relative likelihood for the location and number of regions of activation. This is different from an equivalent current dipole approach, for example, which severely restricts the set of possible current distributions in order to obtain a more well-behaved, over-determined problem. The equivalent current dipole solution is merely representative of the actual neural current and it is left to the experimenter to guess at where the actual locations of activation are. Furthermore, we use the Bayesian formalism to solve the "model order" problem by constructing a prior probability consisting of all possible numbers of regions we care to examine, with relative weights for each. Marginalization of the posterior probability then gives the relative likelihood of the number of regions needed from the information contained in the data. This is a promising and flexible approach. Future work will examine the effects of different sizes and numbers of regions, and will incorporate information about temporal as well as spatial aspects of the data for assessment of conditional probabilities.

References

[1] Bayes, T., An essay towards solving a problem in the doctrine of chances, Biometrika, 1958 (Reprint of 1763), 45: 293-315.

[2] Sarvas, J., Basic mathematical and electromagnetic concepts of the biomagnetic inverse problem, Phys. Med. Biol., 1987, 32: 11-22.

[3] Ranken, D.M., and George, J.S., MRIVIEW: An interactive computational tool for investigation of brain structure and function, Visualization '93, IEEE Computer Society, 1993: 324-331.

[4] George, J.S., Aine, C.J., Mosher, J.C., Schmidt, D.M., Ranken, D.M., Schlitt, H.A., Wood, C.C., Lewine, J.A., Belliveau, J.W., Mapping function in the human brain with MEG, anatomical MRI and functional MRI, J. Clin. Neurophysiology, 1995, 12: 406-431.

Acknowledgments

This work was supported in part by the National Institutes of Health, grant number DA/MH09972-01.

Generalized Wiener Estimation of Three-Dimensional Current Distribution from Biomagnetic Measurements

Sekihara, K.[1] **and Scholz, B.**[2]

[1]*Hitachi Central Research Laboratory, Tokyo, Japan;* [2]*Siemens AG, Erlangen, Germany*

Introduction

The inherent difficulty in the biomagnetic inverse problem is that a three-dimensional current distribution must be estimated from a biomagnetic field measured on a two-dimensional surface close to a human head or body. In such cases, the estimation is ill-posed. To reduce this ill-posedness, the current distribution to be estimated is often modeled by using the equivalent current dipole (ECD), which assumes a highly localized current source. However, when the source current distribution is not localized, or when no information regarding the spatial extent of the source distribution is obtained, we cannot rely on ECD modeling. The least-squares minimum-norm method [1] has been developed that does not use a particular source model and, therefore, can estimate non-localized sources. This method has been successfully applied in two-dimensional cases where the biocurrent distribution is assumed to be confined to a single or a few planes. Its three-dimensional application, however, poses a serious problem in that the signal current is estimated to be closer to the detector coils than its actual location [2]. This paper presents a method that can provide three-dimensional reconstruction where the conventional minimum-norm method fails. It utilizes the principle of the generalized Wiener estimation with the assumption that the activities of the biocurrent sources are uncorrelated.

Method

Let us define the magnetic field measured by the mth detector coil at time t_k as $b_m(t_k)$, and a vector $\boldsymbol{b}(t_k) = (b_1(t_k), b_2(t_k), \ldots, b_M(t_k))^{\mathrm{T}}$ as a set of measured data at time t_k, $k = 1, 2, \ldots, K$. Here, M is the total number of detector coils, K is the total number of time points, and the superscript T indicates the matrix transpose. Let us also define the signal primary current distribution at time t_k as a vector $\boldsymbol{f}(t_k) = (f_1(t_k), f_2(t_k), \ldots, f_{3N}(t_k))^{\mathrm{T}}$. Here, N is the total number of pixels and the x, y, and z components of the primary current located at the ith pixel are assigned to $f_{3(i-1)+1}(t_k)$, $f_{3(i-1)+2}(t_k)$, and $f_{3(i-1)+3}(t_k)$, respectively. We define the lead field matrix $\boldsymbol{L}$, which is a $M \times (3N)$ matrix. Its elements, $L_{m,\,3(i-1)+1}$, $L_{m,\,3(i-1)+2}$, and $L_{m,\,3(i-1)+3}$, represent the sensitivity of the mth detector to the x, y, and z components of the primary current at the ith pixel. The column vector of the lead field matrix defined by $\boldsymbol{l}_p = (L_{1p}, L_{2p}, \ldots, L_{Mp})^{\mathrm{T}}$ is introduced. The lead field matrix is expressed using its column vectors as, $\boldsymbol{L} = (\boldsymbol{l}_1, \boldsymbol{l}_2, \ldots, \boldsymbol{l}_{3N})$. The relationship between $\boldsymbol{b}(t_k)$ and $\boldsymbol{f}(t_k)$ is given by $\boldsymbol{b}(t_k) = \boldsymbol{L}\,\boldsymbol{f}(t_k) + \boldsymbol{n}(t_k)$, where $\boldsymbol{n}(t_k)$ is the noise vector. The mth element of this noise vector is the noise contained in the mth detector measurement at time t_k. The problem that this paper deals with is how to find the optimum estimate of the signal primary current source distribution $\hat{\boldsymbol{f}}(t_k)$ at all time instants $k = 1, 2, \ldots, K$, from a given spatio-temporal data set of measured data $\boldsymbol{b}(t_k)$, $k = 1, 2, \ldots, K$.

The method proposed in this paper utilizes the minimization of the following least-squares-error expressed in the source current space. That is, the cost function $\mathcal{F} = \sum_{k=1}^{K} ||\boldsymbol{f}(t_k) - \hat{\boldsymbol{f}}(t_k)||^2$ is minimized to obtain the optimum estimate $\hat{\boldsymbol{f}}(t_k)$. Here, $\boldsymbol{f}(t_k)$ is the true current distribution at time t_k, and the measured data $\boldsymbol{b}(t_k)$ is expressed as $\boldsymbol{b}(t_k) = \boldsymbol{L}\,\boldsymbol{f}(t_k)$. The solution which minimizes the above cost function is known to be the generalized Wiener estimate or the minimum mean square error estimate [3]. This is expressed as $\hat{\boldsymbol{f}}(t_k) = \boldsymbol{S}\,\boldsymbol{L}^{\mathrm{T}}\boldsymbol{D}^{-1}\boldsymbol{b}(t_k)$. Here, defining the time average of A as $\langle A \rangle$, $\boldsymbol{D}$ is the covariance matrix of the measured data defined as $\boldsymbol{D} = \langle \boldsymbol{b}(t_k)\boldsymbol{b}^{\mathrm{T}}(t_k) \rangle$, and $\boldsymbol{S}$ is the covariance matrix of the signal current sources defined as $\boldsymbol{S} = \langle \boldsymbol{f}(t_k)\boldsymbol{f}^{\mathrm{T}}(t_k) \rangle$

When applying the above equation to an actual problem, it is usually difficult to obtain an accurate estimate of the signal-source covariance matrix $\boldsymbol{S}$. When all the non-diagonal elements of $\boldsymbol{S}$ can be assumed to be zero, estimating the signal covariance matrix is performed in the following manner. Using the column vectors of the lead field matrix, let us express Wiener estimate as

$$\hat{f}_p(t_k) = S_{pp} l_p^{\mathrm{T}} D^{-1} b(t_k) \,, \tag{1}$$

where S_{pp} is the pth diagonal element of the signal covariance matrix. Defining the estimate of S_{pp} as $\hat{S}_{pp}$, we have

$$\hat{S}_{pp} = \langle \hat{f}_p(t_k) \hat{f}_p(t_k) \rangle = \langle (S_{pp} l_p^{\mathrm{T}} D^{-1} b(t_k))(S_{pp} l_p^{\mathrm{T}} D^{-1} b(t_k))^{\mathrm{T}} \rangle = S_{pp}^2 l_p^{\mathrm{T}} D^{-1} l_p \,.$$

Assuming that $\hat{S}_{pp} = S_{pp}$, one can obtain the relationship $\hat{S}_{pp} = 1/(l_p^{\mathrm{T}} D^{-1} l_p)$. Thus, the proposed Wiener estimation becomes

$$\hat{f}_p(t_k) = \frac{l_p^{\mathrm{T}} D^{-1} b(t_k)}{l_p^{\mathrm{T}} D^{-1} l_p} \,. \tag{2}$$

One problem with this estimation method is that it gives erroneous results when the orientations of current sources are completely fixed during the measurement. The extension of the methods to accommodate such cases can be performed in the following manner. Let us define the normal vector representing the orientation of the source current located at the pth pixel as e_p, and the $M \times 3$ matrix L_p as $L_p = [l_{3(p-1)+1}, l_{3(p-1)+2}, l_{3(p-1)+3}]$. This e_p can be obtained by solving the following eigenvalue problem at each pixel,

$$L_p^{\mathrm{T}} D^{-1} L_p e_p = \lambda_{min} e_p, \tag{3}$$

where λ_{min} is the minimum eigenvalue. Once e_p is obtained, the amplitude of the current element at the pth pixel is obtained using

$$\hat{f}_p(t_k) = \frac{(L_p e_p)^{\mathrm{T}} D^{-1} b(t_k)}{(L_p e_p)^{\mathrm{T}} D^{-1} (L_p e_p)} . \tag{4}$$

Computer Simulation and Results

Computer simulations were performed to show the effectiveness of the proposed method. A magnetometer with thirty-seven channels that simulated the BTi 37-channel biomagnetic measurement system was assumed. The detector frame with its origin at the center of the coil alignment was used. The values of the coordinate (x, y, z) were expressed in centimeters. Since the homogeneous spherical conductor was assumed, only the two tangential components, ϕ and θ components, of the primary current vectors were considered.

Two time-varying magnetic-field sources were assumed to exist: the first source at $(-4, 1, -4)$ and the second source at $(3, 1, -6)$. This source configuration is schematically shown in Fig. 1. The ϕ and θ components of those sources randomly fluctuated, and the activities of the two sources were assumed to be uncorrelated with each other. The time instant at which the relative value of the (ϕ, θ) components of the first and second sources were equal to $(0, 1)$ and $(1, 1)$ was chosen, and the field map at this time instant was calculated. The noise was added to set the signal-to-noise ratio (SNR) of this simulated data equal to 25.

First, conventional minimum-norm reconstruction was performed using $\hat{f}(t_k) = L^{-} b(t_k)$, where $L^{-} = L^{\mathrm{T}}(L L^{\mathrm{T}})^{-1}$. The field map obtained above was used and the reconstruction volume covers $-8 \le x \le 8$, $-8 \le y \le 8$, and $-2 \le z \le -8$. It consists of $17 \times 17 \times 7$ pixels. The plane, y=1, in the three-dimensional reconstruction is shown in Fig. 2. Here, the ϕ and θ components are separately displayed. These results clearly show that the minimum-norm method provides severely distorted three-dimensional reconstruction in this case. The covariance matrix of the measured data D was estimated from 500 data generations; This simulates data acquisition for 0.5 s with a sampling interval of 1 ms. The proposed Wiener reconstruction using Eq. (2) was applied with this covariance matrix. The results are shown in Fig. 3. Here, both sources are reconstructed, clearly demonstrating this method's three-dimensional reconstruction capability.

Next, sources having fixed orientations was reconstructed. In this experiment, the data used for obtaining $\boldsymbol{D}$ was generated assuming that the (ϕ, θ) components of the first and the second sources were fixed at $(0,1)$ and $(1,1)$, respectively. The results obtained using Eq. (2) are shown in Fig. 4. The ϕ component is correctly reconstructed. However, the reconstruction for the θ component contains severe distortion caused by non-diagonal terms that cannot be ignored. The results obtained using Eqs. (3) and (4) are shown in Fig. 5. Here, the two components are reconstructed. These results indicate that the effectiveness of Eqs. (3) and (4) for reconstructing sources with fixed orientations.

Conclusion

This paper proposes a method based on the principle of generalized Wiener estimation for obtaining a three-dimensional biocurrent distribution from spatio-temporal biomagnetic data. The method is formulated under the assumption that current sources are uncorrelated, and the method is extended to reconstruct current sources with fixed orientations. Computer simulation shows that the proposed method can reconstruct three-dimensional current distribution where the conventional least-squares minimum-norm method fails.

Acknowledgment

The authors wish to acknowledge Dr. A. Oppelt for the helpful discussion and for his encouragement to proceed this work.

References:

[1] Hämäläinen, M.S. and Ilmoniemi, R.J., "Interpreting measured magnetic fields of the brain: Estimates of current distributions," Helsinki University of Technology, Report TKK-F-A559, 1984.

[2] Jeffs, B., Leahy, R., and Singh, M., "An evaluation of methods for neuromagnetic image reconstruction", IEEE Trans. Biomed. Eng., 1987, 34: 713-723.

[3] Kay, S.M., Fundamentals of Statistical Signal Processing: Estimation Theory, Englewood Cliffes, New Jersey, Prentice-Hall, Inc., 1993.

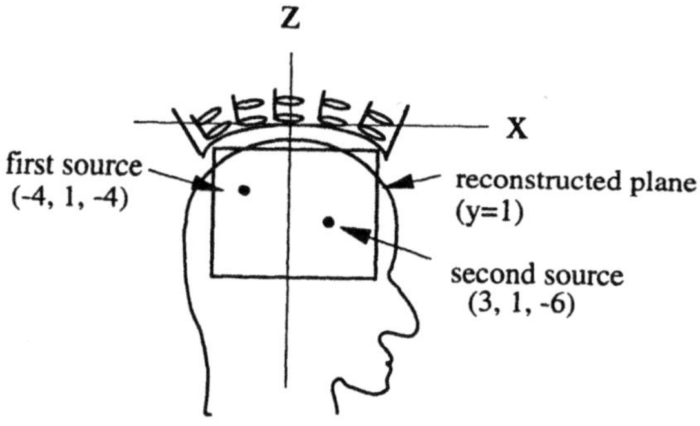

Fig. 1 Source current configuration assumed in computer simulation.

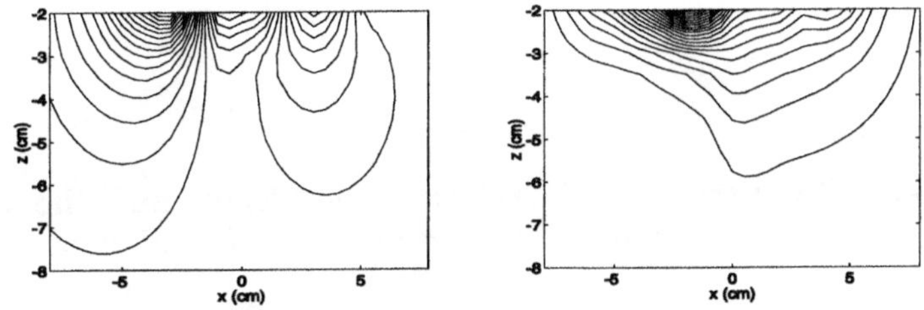

Fig. 2 Minimum norm reconstruction results.
Dstribution of θ component (left) and that of ϕ components (right).

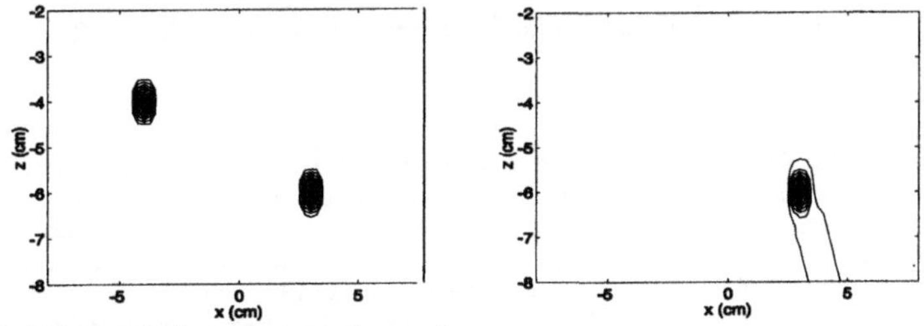

Fig. 3 Proposed Wiener reconstruction results.
Dstribution of θ component (left) and that of ϕ components (right).

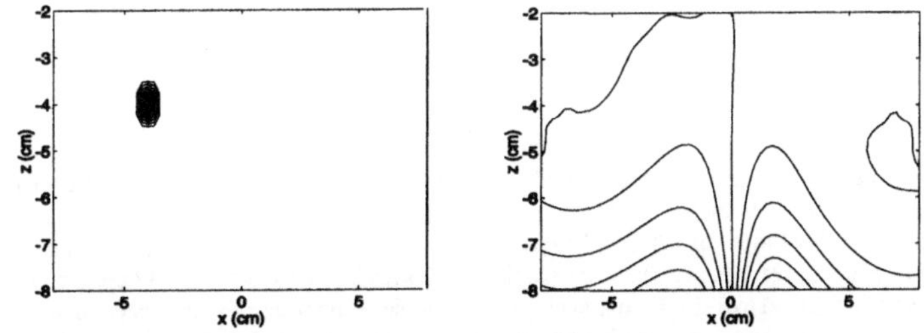

Fig. 4 Results of reconstruction with Eq. (2) for fixed orientation sources.
Dstribution of θ component (left) and that of ϕ components (right).

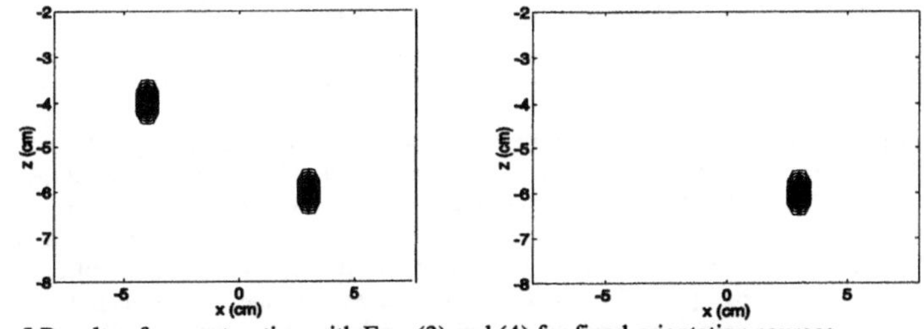

Fig. 5 Results of reconstruction with Eqs. (3) and (4) for fixed orientation sources.
Dstribution of θ component (left) and that of ϕ components (right).

Localizing Dipoles: The Correlation Coefficient and the Goodness of Fit in Source Localization of Filtered Noise

Siedenberg, R., Goodin, D.S., Aminoff, M.J., Rowley, H.A. and Roberts, T.P.L.

University of California, San Francisco, San Francisco, USA

Introduction

In source localization using magnetoencephalography (MEG) the correlation coefficient and the goodness of fit are commonly computed as a means of estimating how well one or more dipoles account for the observed data. However, the magnitude of the correlation coefficient or the goodness of fit necessary for confidence that a computed dipole corresponds to an actual generator is not known. It cannot be calculated mathematically and must therefore be determined empirically. Different authors have defined minimum correlation coefficients or minimum goodnesses of fit, ranging from 0.70 to 0.99 [e.g. 1, 2, 3, 4], as evidence of source localization but the basis for selecting these values is unclear. We have therefore studied the parameters of correlation coefficient and goodness of fit in two different recordings of noise in order to define the specificity of these parameters in source localization of electromagnetic brain activity in evoked magnetic field research.

Methods

Two experiments were undertaken. In the first experiment the magnetic field was recorded using a 37-channel magnetometer (Magnes™, Biomagnetic Technologies, Inc., San Diego, Ca) in a magnetically shielded room without any subject present. Two blocks of 420 epochs were recorded with a time window of 1.5 seconds, a sampling rate of 520.8 Hz, and a bandpass of 1 to 200 Hz. Each block was divided into 10 averages of 42 epochs. These averaged original data were subjected to a single equivalent current dipole fit algorithm for each data point . After further filtering (bandpass 1-20 Hz) of the data a second dipole fit was performed. For each data point the correlation coefficient and the goodness of fit were computed. The mean correlation coefficient and the mean goodness of fit were calculated for original data and filtered data. Peaks of correlation and goodness of fit were defined independently as the local maxima of the measure within a time window of at least 40 ms. The length of the time window was chosen because preliminary analysis showed that generally both the correlation and the goodness of fit had peaks occuring within a time window of about 40 ms. In each averaged segment the number of peaks of correlation or goodness of fit above 0.90 was counted.

In the second experiment the MEG was recorded from each of four subjects (aged 28 to 46 years, 1 woman and 3 men). The subjects were seated comfortably in a magnetically shielded room, in a quiet awake state, eyes closed, without external stimulation. The 37-channel magnetometer was centered over the vertex at a distance of about 0.5 cm. Recording parameters were the same as in the first experiment. For two subjects the methods of analysis were the same as in the first experiment, resulting in 20 averages for each subject. In order to estimate the random replicability for each peak of correlation exceeding 0.99 found in an averaged segment, we measured the correlation at the same latency for the remaining 19 averages of the same subject and calculated the mean correlation coefficient, which we defined as the latency replication value. In addition, for each dipole fit with a correlation in excess of 0.99 the local maxima of correlation within a 40 ms time window were also measured in the remaining 19 averages, defining a peak replication value. The location of the corresponding dipoles was determined and the mean distance between the dipoles with a correlation exceeding 0.99 and their replications was calculated. The orientation of these dipoles was classified by the octant of a sphere into which the dipole pointed and orientation was judged to be the same if the dipole from both trials pointed into the same octant.

In addition to the averages containing 42 epochs, an analysis of averages containing 360 epochs was undertaken to ensure that the findings did not spuriously reflect the small number of epochs. For each of the four subjects, two averages of 360 epochs each were computed and the resulting eight averages were subjected to a single dipole fit algorithm. After filtering of the data with a bandpass of 1-20 Hz, a second dipole fit was performed for each data point. A third dipole fit was performed after filtering of the data with a bandpass of 8-40 Hz. The analysis of the correlation coefficient and the goodness of fit was as above. Statistical analysis was performed using Student's t-test.

Results

In the first experiment the original data had a mean correlation of 0.43 and a mean goodness of fit of 0.34 (Table I). The original data never achieved a correlation of higher than 0.70 or a goodness of fit of higher than 0.85. After filtering with a bandpass of 1-20 Hz, however, the mean correlation and the mean goodness of fit increased significantly (Table I) and the maximum correlation and the maximum goodness of fit reached 0.93 and 0.98 respectively.

Table I. Mean and maximum correlation (R) and goodness of fit (GOF) for MEG recordings of noise

MEG recording of noise without a subject present

	Bandpass		Significance
	1-200 Hz	1-20 Hz	
mean R (SD)	0.43 (0.08)	0.58 (0.21)	$p<0.05$
mean GOF (SD)	0.34 (0.17)	0.74 (0.12)	$p<0.0001$
maximum R	0.70	0.93	
maximum GOF	0.85	0.98	

MEG recording of noise with a subject present

	Bandpass		Significance
	1-200 Hz	1-20 Hz	
mean R (SD)	0.66 (0.16)	0.86 (0.09)	$p<0.0001$
mean GOF (SD)	0.66 (0.19)	0.88 (0.10)	$p<0.0001$
maximum R	0.96	0.99	
maximum GOF	0.97	0.99	

In the second experiment (with subjects present during the MEG recording), the mean correlation and the mean goodness of fit of the original data were consistently higher for the averages of 42 epochs than in the recordings of environmental noise. Filtering again caused a significant increase of both measures (Table I). The peak correlation for the original data never exceeded 0.96 and the peak goodness of fit never exceeded 0.97. After filtering, however, there was a substantial increase in the number of peaks of high correlation and high goodness of fit (Table II). Thus, 13 out of 40 averages contained peaks of correlation in excess of 0.99, and 14 of 40 averages contained peaks of goodness of fit in excess of 0.99. The mean latency replication value was 0.88 (SD 0.11), which was significantly less than 0.99 ($p<0.001$). Similarly the peak replication value was 0.94 (SD 0.05), which was also significantly below 0.99 ($p< 0.001$). The mean distance between the dipole location and the location of its random replication was 4.5 cm (SD 2.7 cm), ranging from 0.2 to 15.9 cm. Fifteen percent of the replications were oriented in the same direction as the reference dipole and this was not significantly different from random chance.

Averages of 360 epochs resulted in a markedly lower correlation and goodness of fit in the original data than that found in the original data of averages of 42 epochs. There was no peak correlation exceeding 0.76 and no peak goodness of fit exceeding 0.75. However, filtering of the data again caused a substantial increase in the number of peaks of high correlation and high goodness of fit. Thus, after filtering with a bandpass of 1-20 Hz, 6 of 8 averages contained peak correlations in excess of 0.99, and 6 of 8 averages contained peaks of goodness of fit in excess of 0.99 (Table II). Filtering of the data with a bandpass of 8-40 Hz did not substantially alter these findings.

Table II. Mean number of peaks of correlation (R) and goodness of fit (GOF) per average

Average of 42 epochs (bandpass 1-20 Hz)

R	Mean	(SD)	GOF	Mean	(SD)
≥ 0.99	0.9	(1.4)	≥ 0.99	0.9	(1.2)
≥ 0.98	3.7	(4.6)	≥ 0.98	4.2	(3.9)
≥ 0.97	6.4	(5.4)	≥ 0.97	8.3	(5.3)
≥ 0.96	9.6	(6.2)	≥ 0.96	11.5	(5.5)
≥ 0.95	12.3	(6.8)	≥ 0.95	14.7	(5.3)
≥ 0.94	14.8	(7.1)	≥ 0.94	17.1	(5.4)
≥ 0.93	17.3	(7.1)	≥ 0.93	19.5	(5.0)
≥ 0.92	19.4	(6.5)	≥ 0.92	21.4	(4.8)
≥ 0.91	21.2	(6.2)	≥ 0.91	22.6	(4.4)
≥ 0.90	22.7	(5.6)	≥ 0.90	23.9	(4.1)

Average of 360 epochs (bandpass 1-20 Hz)

R	Mean	(SD)	GOF	Mean	(SD)
≥ 0.99	1.4	(1.1)	≥ 0.99	1.8	(1.4)
≥ 0.98	5.0	(3.5)	≥ 0.98	6.5	(3.7)
≥ 0.97	9.5	(6.1)	≥ 0.97	10.1	(5.2)
≥ 0.96	13.6	(8.4)	≥ 0.96	13.6	(5.8)
≥ 0.95	15.3	(8.3)	≥ 0.95	16.3	(6.0)
≥ 0.94	17.1	(7.3)	≥ 0.94	18.6	(5.8)
≥ 0.93	18.6	(6.9)	≥ 0.93	19.6	(5.9)
≥ 0.92	20.4	(7.0)	≥ 0.92	21.6	(5.0)
≥ 0.91	21.9	(7.1)	≥ 0.91	23.6	(4.3)
≥ 0.90	22.3	(6.7)	≥ 0.90	24.6	(4.1)

Conclusion

1) This study demonstrates that single equivalent current dipoles with high correlation and high goodness of fit can be computed for background brain activity or even environmental noise, i.e., magnetic activity that in neither case is thought to have a single generator. Therefore it is not justified to conclude that a high correlation or goodness of fit for an equivalent current dipole computed for stimulus-locked brain activity supports in itself the assumption of a single source.

2) The values of correlation and goodness of fit, both for MEG recordings with and without a subject present, are well in excess of those cited in the literature as criteria of good dipole fits. This can be interpreted as a failure of both the correlation coefficient and the goodness of fit in distinguishing stimulus-related brain activity from background noise.

References

[1] Huttunen, J., Hari, R., and Leionen. L. Cerebral magnetic responses to stimulation of ulnar and median nerves. Electroenceph. clin. Neurophysiol., 1987, 66: 391-400

[2] Pantev, C., Hoke, M. Lehnertz, K., Lütkenhöner, B. , Anógianakis, G., and Wittkowski, W. Tonotopic organization of the human auditory cortex revealed by transient auditory evoked magnetic fields. Electroenceph. clin. Neurophysiol., 1988, 69: 160-170

[3] Rogers, R.L., Baumannn, S.B., Papanicolaou, A.C., Bourbon, T.W., Alagarsamy, S., and Eisenberg, H.M. Localization of the P3 sources using magnetoencephalography and magnetic resonance imaging. Electroenceph. clin. Neurophysiol., 1991, 79: 308-321

[4] Gallen, C.C., Schwartz, B., Rieke, K., Pantev, C., Sobel, D., Hirschkoff, E. and Bloom, F.E. Intrasubject reliability and validity of somatosensory source localization using a large biomagnetometer. Electroenceph. clin. Neurophysiol., 1994, 90: 145-156

Acknowledgments

Dr. Siedenberg was supported by the Deutsche Forschungsgemeinschaft.

Monte-Carlo Analysis and Confidence Region Ellipsoids for Equivalent Current Dipole Solutions to EEG/MEG data

Singh, K.D., Harding G.F.A.
Clinical Neurophysiology Group, Dept. of Vision Sciences, Aston University, Birmingham, U.K.

Introduction

Although equivalent current dipoles (ECDs) are point source representations of the underlying neuronal currents they are in reality approximations to the true focus or foci of activity. The actual neuronal distribution is not point-source in nature, and may be distributed over a significant portion of the cortex. If an equivalent current dipole is a successful model for this distribution, either the neuronal current is localised or the spatial distribution is such that it is amenable to representation as an ECD. However, noise of all kinds will lead to an indeterminacy in the localisation of the ECD. Noise in the case of Magnetoencephalography (MEG) or Electroencephalography (EEG) refers to brain noise, signal noise, and experimental errors such as mis-registration of the detectors with the head. The estimation and display of positional uncertainties due to noise is the subject of this paper.

Despite the fact that a consideration of the accuracy of source localisation is essential in the assessment of whether source localisations are valid and are similar or different to other ECDs, it is still common to see dipole solutions presented with only a statistical justification of the validity of the localisation - e.g. whether the ECD meets a particular χ^2 criteria. We believe that ECDs should never be co-registered with MRI and then displayed as point source representations, but that some representation of the confidence volume should always be plotted.

Methods

Monte-Carlo Analysis

The first stage in the depiction of confidence volumes for ECDs is to quantify the relationship between noise in the data and positional accuracy of the ECD. ECD fitting is a mixture of linear and non-linear algorithms [1] involving a least-squares fit of the dipole parameters, and the relationship between the noise and the positional parameters is complex. An elegant approach is to use Monte Carlo Simulation (MCS), as proposed by Medvick et al. [2]. MCS has two keys advantages in that it is not dependent on any particular ECD localisation algorithm and it is valid for any type (signal, positional) or model (e.g. non-normal) of noise. In our laboratory we use MCS together with an ellipsoid fitting algorithm to generate a representation of the dipole confidence volume.

The first step in the MCS is to estimate the signal noise in the data. In every evoked response experiment that is carried out we calculate the average response in channel c at time t, $A(c,t)$, using expression (1.0), where $M(e,c,t)$ is the signal in channel c at time t in epoch e, and may be MEG or EEG data. By analogy, we can calculate the anti-average, $\bar{A}(c,t)$ using expression (1.1).

$$(1.0) \quad A(c,t) = \sum_{e=1}^{Epochs} M(e,c,t)$$

$$(1.1) \quad \bar{A}(c,t) = \sum_{e=1}^{Epochs} M(e,c,t)\cdot\delta \qquad \delta = \begin{cases} 1 & \text{if e is odd} \\ -1 & \text{if e is even} \end{cases}$$

For a perfect evoked response with no noise, the anti-average is zero and can thus be considered as an estimator of the noise in each channel as a function of time [3]. To calculate an estimate of the noise performance of each channel in the detector array, the standard deviation, σ_C, of the anti-average is calculated.

The dipole is then fitted at appropriate latencies using a standard weighted least-squares fitting algorithm minimising expression (2.0) with respect to the dipole parameters:-

$$(2.0)\quad \chi^2 = \sum_{c=1}^{chans} \frac{\left(M_c(t) - \hat{M}_c(t)\right)^2}{\sigma_c^{\ 2}}$$

Where M_c is the measured signal and $\hat{M}_c$ is the estimated signal from the dipole. Note how the anti-average estimated σ_c is used as a weight for each channel. Once a valid fit has been obtained, the MCS procedure is carried out as follows: The field from the final dipole solution is calculated in each measurement channel and random gaussian noise with magnitude σ_c is added to each channel. The dipole is then refitted to yield a new dipole position. This procedure is carried out many times (30-100) to yield a set of dipole solutions which are spread about the true location [2].

Ellipsoid Confidence Volumes

It is possible to simply plot these dipoles on a graph or an appropriate MR slice. It is, however, advantageous to model the spread of dipole solutions as a volume, both for quantification purposes and simplicity of graphical presentation. Of course, one could simply plot the standard deviation of each of the positional parameters x, y and z. However, the maximum positional error may be along a direction axis which is some linear combination of x, y or z i.e. the confidence volume is a vector quantity with a maximum in a particular direction. The simplest volume that can be used to simulate this distribution is an ellipsoid, the size and orientation of which must be fit to the spread of dipole positions.

The ellipsoid is defined in terms of its centre, $\bar{r}_{mean}$ and its three radii vectors ($\vec{S}_{major}, \vec{S}_{minor}, \vec{S}_{int}$) which define the major, minor and intermediate dimensions of the ellipsoid, respectively. The algorithm we use iteratively finds the direction axes using the Powell algorithm [4]. Firstly, the major direction axis is found by maximising expression (3.0) with respect to the direction components of S_{major}. Then, the minimum axis of the ellipsoid is found by *minimising* expression (3.1) with respect to the direction components of S_{minor}.

$$(3.0)\quad \sum_{j=1}^{trials} \left((\bar{r}_j - \bar{r}_{mean}) \cdot \hat{S}_{major}\right)^2 \qquad (3.1)\quad \sum_{j=1}^{trials} \left((\bar{r}_j - \bar{r}_{mean}) \cdot \hat{S}_{minor}\right)^2$$

The final ellipsoid axis is found by defining it to be orthogonal to the major and minor axes. Once the direction axes have been defined, an iterative search algorithm increases the size of the ellipse along its axes until 95% of all dipoles are contained within the ellipsoid. The ellipsoid is thus a 95% confidence volume for the fit.

Co-registration with MRI

MEG/EGG source localisations are most useful when co-registered with anatomical MR images and we believe it is essential that any graphical representation of these solutions includes a depiction of the confidence volume. Several techniques could be used to plot the ellipsoid volume (such as a 3D wireframe model), but we have found the clearest representation is simply to plot the projection of the ellipsoid perimeter onto whichever MR orientation view is selected.

Simulation of a sensory evoked response

To demonstrate the use of MCS and ellipsoid confidence volumes, we simulated an MEG evoked response experiment by placing a trial dipole in a region approximating human sensory cortex. In the computer simulation the dipole moment was set to be 25nAm, which is in the physiological range. A simulated 19 channel magnetometer was used with the detector positioned above the C4 position. When the forward problem was calculated using a homogenous sphere model a clear dipolar pattern was elicited in the detector array. In the absence of signal noise, the MEG data was re-inverted using our dipole fitting algorithm and the position was found to have zero error with respect to the true dipole position.

Monte-Carlo simulations were then carried out by adding random gaussian noise of a particular magnitude to the test dipole field and re-fitting the data. The noise magnitudes chosen were 5, 10, 20 and 40% of the peak signal value. For each level of noise, 30 iterations were performed and then ellipsoids were fitted to the MCS distribution.

Results

Table 1 summarises the resultant ellipsoids for each of the simulations. The results show that as the noise increases, the positional errors increase, as does the ellipsoid volume and the major ellipsoid radius. This table shows that for the 5% case the positional errors are distributed from 0.7 to 4.8 mm. At the high noise level of 40% the variation is much larger. This is because at high noise levels, it is possible that for any one particular trial the additive noise may be large enough to defeat the minimisation algorithm completely.

In Fig. 1, the ellipsoid for the 10% noise case is shown on several MR views for the test subject used in the simulation. Note that the ellipsoid is biggest in one particular dimension, reflecting the maximum positional uncertainty.

Table 1

Noise Level (Percentage of Peak Signal)	Range of Positional Errors / *cm*	Ellipsoid Volume / cm^3	Major ellipse Radius S_{major} / *cm*
+/- 5%	0.07 to 0.48 $\bar{x} = 0.27$	0.060	0.46
+/- 10%	0.13 to 1.08 $\bar{x} = 0.60$	0.525	1.08
+/- 20%	0.23 to 2.67 $\bar{x} = 1.44$	11.53	2.50
+/- 40%	0.66 to 3.49 $\bar{x} = 2.07$	74.75	3.20

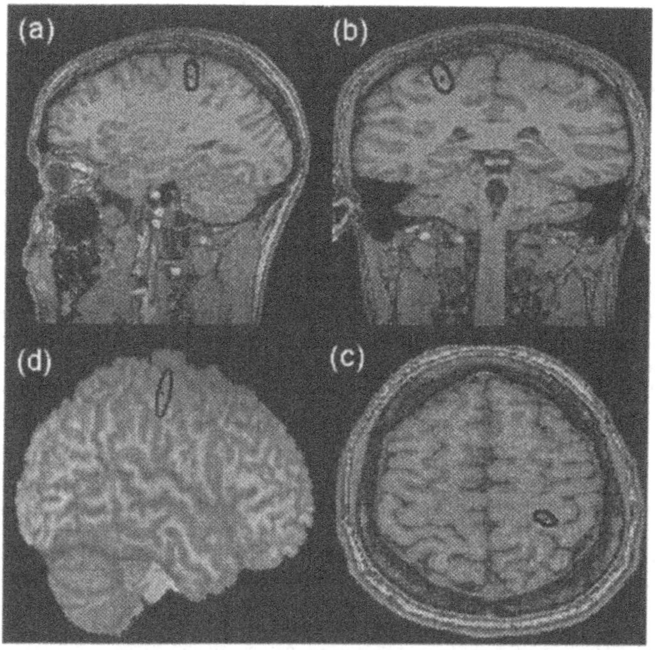

Fig. 1. Several MR views of the simulated 95% confidence ellipsoid for 10% Noise. (a) shows a sagittal slice viewed from the left, (b) a coronal slice viewed from the front. (c) shows an axial slice viewed from above with the nasion at the top of the image. (d) shows the ellipsoid projected onto a cortical surface image of the extracted brain. The true simulated dipole was located at the centre of the ellipses shown above

Discussion

We have demonstrated how a consideration of signal noise can be used to estimate the stability of localisation using a Monte-Carlo simulation, and the modelling of the resultant error distribution using a 3D ellipsoid volume.

The ellipsoid is a more informative representation of the confidence volume than a plot of the standard deviations of the dipole parameters as it identifies the maximum error direction and the error volume. The maximum error direction and ellipsoid volume are functions of the experimental parameters (number of detectors, detector orientation etc.), the source configuration (dipole position and moment, whether a dipole is a correct model) and noise in the data. The procedure outlined here should therefore be carried out for each individual EEG/MEG measurement.

Although in this paper we have only referred to the utility of MCS in single ECD fitting, we note that it can be used in any source localisation algorithm such as multiple dipoles or distributed current models. A further advantage not demonstrated here is that any other noise or errors in the experiment, such as misalignment of the detectors, may also be included in the MCS.

There are two potential drawbacks to the MCS procedure outlined in this paper. Firstly, in this simulation we have used additive noise distributed with a symmetrical gaussian probability and if this were not the true noise model then the MCS procedure would be invalid. However, this is true for all confidence estimate approaches, which often have the further constraint that the noise *must* have a normal distribution - MCS does not require any particular noise model, only that the model should be known. The second disadvantage of MCS is that instead of fitting one dipole, typically 30-100 dipoles must be fit. This can be reduced by only fitting dipoles at latencies of high signal to noise and only carrying out the MCS procedure on latencies which have been successfully fit using an ECD model. This disadvantage is further reduced by the use of increasingly powerful workstations.

The plotting of the ellipsoids representing MCS confidence volumes is a clear and unambiguous way of identifying which region of cortex is producing a particular MEG/EEG response. Unlike the more common point-source representations, it allows the uncertainty of the localisation to be easily seen. This is especially important in clinical applications such as the pre-surgical localisation of sensory/motor cortex, where the accuracy of identifying which sulcus is the central sulcus is paramount.

These techniques may also be used in the comparison of localisations obtained with different imaging modalities such as functional MRI (fMRI). Using confidence regions for the EEG and MEG data would allow a decision to be made as to whether the modalities are imaging the same source.

A further use of MCS and ellipsoid volumes is in the comparison of the results from different experiments. If the stimulus parameters are varied, the dipole localisation may change. The procedures outlined in this paper allow an assessment of whether this change in localisation is statistically significant. This is achieved by pasting the two ellipsoids onto 3 orthogonal views and observing whether they overlap. If there is still some uncertainty it is a relatively simple procedure to determine mathematically whether the two ellipsoids overlap.

It is clear from the advantages outlined here that it is essential that when the results of any source localisation are presented, the confidence region should be displayed, and that simply displaying the point of localisation is an incomplete representation. We propose that the use of MCS and ellipsoid volumes is an excellent way of achieving this goal.

References

[1] Mosher, J.C., Lewis, P.S., and Leaky, R.M. Multiple dipole modelling and localization from spatio-temporal MEG data. IEEE Trans Biomed Eng 1992; 39: 541-557.

[2] Medvick, P., Lewis, P.S., Aine, C.J., and Flynn, E.R.. Monte Carlo analysis of localization errors in magnetoencephalography. In: Williamson, S.J., Hoke, M., Stroink, and G., Kotani, M. Advances in Biomagnetism, New York, Plenum Press, 1989.

[3] Regan, D., Human brain electrophysiology: evoked potentials and evoked magnetic fields in science and medicine, New York, Elsevier, 1989.

[4] Press, W.H., Flannery, B.P., Teukolsky, S.A., and Vetterling, W.T. Numerical recipes in C, Cambridge, Cambridge University Press, 1989.

Localization of Current Sources with Constraints of Solution Area

Sugaya, H. [1] **,Osozawa, S.**[1]**, Ishiyama, A.** [1]**, and Kasai, N.** [2]

Department of Electrical Engineering, Waseda University[1]*, Tokyo, Japan*
Electrotechnical Laboratory[2]*, Tsukuba, Japan*

Introduction

A localized current source is well estimated using a single current dipole model. In contrast, it is difficult to estimate distributed current sources using multi-dipole model without any constraints. In the MCG study, a single current dipole model is applied to the localization of small activated region; for example, localizing the ventricular pre-excitation site in a Wolff-Parkinson-White (WPW) syndrome. While, at the beginning of QRS complex the cardiac depolarization wave-front propagates spreading in the ventricle. Various multi-dipole models have been studied to estimate the excited area. Several localization methods using multi-dipole model with anatomical constraints using MRI information were reported [1]. We propose a localization method of distributed current sources using physiological information with constraints of the solution area. In this paper, to obtain solutions of wave-front which proceeds more smoothly in spatio-temporal space, the information of the direction and the velocity of excitation is utilized.

Methods

In the proposed method, dipoles at a time, t_2, could be estimated using the condition of the position and the moment of dipoles at the previous time, t_1, ($t_2 = t_1 + \Delta t$), ; that is, the solution area can be confined in the sector for 2-D model (spherical cone for 3-D model) shown in Fig. 1. The center of the sector is the position of current dipoles at t_1. The radius r of the sector is determined with regard to the spreading velocity of excitation. The φ is given by experiences. This method has two merits. One is shortening the time of calculation due to diminishing the search area. Another is that the dipoles can be estimated to propagate smoothly, as described below.

The cost function E to be minimized is given by

$$E = \frac{\sum_{i}^{coil} (Bm_i - Bc_i)^2}{\sum_{i}^{coil} Bm_i^2} + \alpha n + \beta D(\theta), \qquad (1)$$

where Bm_i is the measured flux density and Bc_i is the flux density calculated from current dipoles. In this paper, we used multi-dipole model and homogeneous conducting half-space model. The n is the total number of current dipoles. $D(\theta)$ is the restrict condition of the dipole moment, where the θ is an angle between the direction of dipoles at t_2 and t_1. This condition is that the current dipoles need to move smoothly and continuously. The α and β are the constant giving weights. In this work, Boltzmann machine is used for the minimization of the cost function[2]. The solution area is presented by a 2 or 3-dimensional lattice.

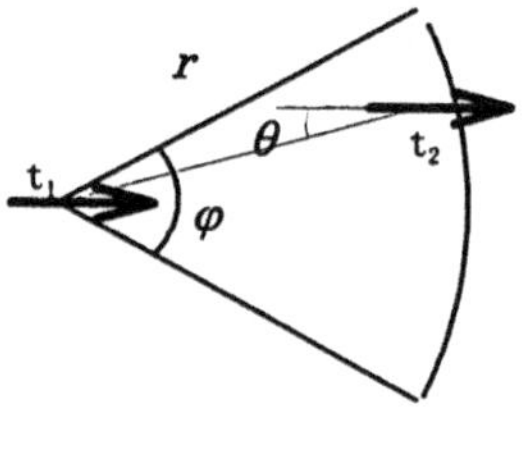

Fig. 1 Constraints of solution area

Simulation

Preliminarily, a computer simulation was carried out with a 3 dimensional model. The model is shown in Fig. 2. The solution area of the model includes three layers. One layer consists of 245 grids. 121 pick up coils are positioned on a measuring plane (100mm × 100mm) parallel to the layers. We computed the flux density normal to the measuring plane. The distance between the upper layer in the model and the measuring plane is 10mm, and that of each layer is 5mm. Each grid is 5mm apart. The source current dipoles model used in this simulation is shown in Fig. 3. The dipoles move continuously from the position at t_1 to the position at t_6. The dipoles at each time (t_1~ t_6) are shown together in Fig. 3. Using this model, the flux density at each time were computed. The initial condition in solving inverse problem is presented by the current dipoles at t_1 in Fig. 3. By means of this method, the position and the moment of the following current dipoles were estimated sequentially. The value of each parameter used in this simulation is as follows; $\varphi = 80°$, $r = 0.03$[m], $\theta = 40°$, $\alpha = 0.004$, $\beta = 0.004$.

The results are shown in Fig. 4. The estimated results are well in agreement with source model. It seems that there is a possibility for distributed current source estimation by means of this method.

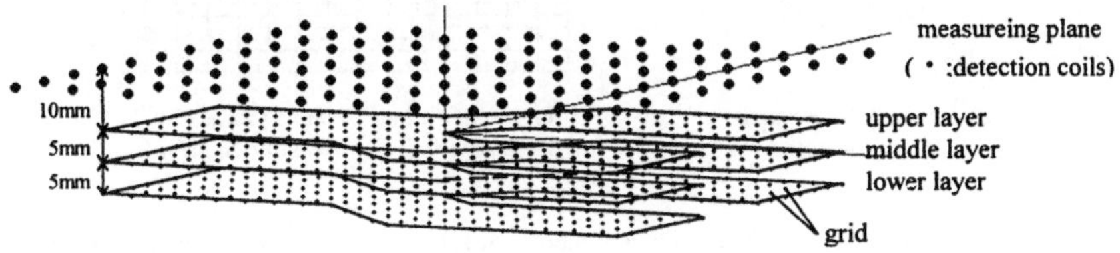

Fig. 2 The position of pickup coil above the solution area

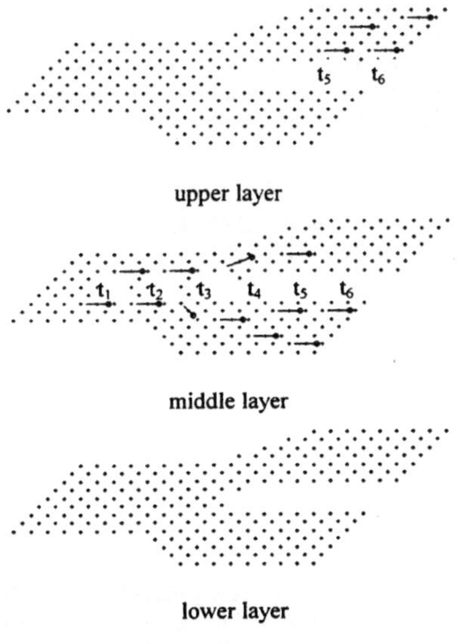

Fig. 3 Source dipole model

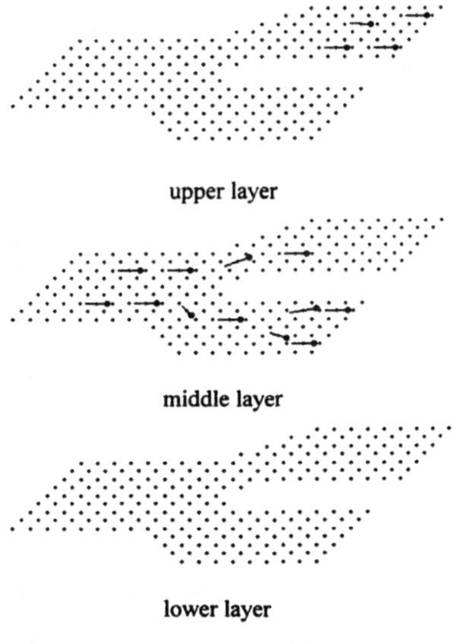

Fig. 4 Estimated results

MCG localization

The result of the simulation seems to present the possibility of the presented method in distributed source localization. Therefore, we applied this method to a real MCG. We measured MCG data of a normal male subject, aged 34 years old , by using a 7-channel dc-SQUID system in the magnetically shielded room. The sensitivity of flux density was 19fT/$\sqrt{}$Hz. The detection coil of the SQUID is the first order gradiometer, that has a radius of 6.5mm, the baseline of 80mm and the turn of two. Schematic position of the detection coils is shown in Fig. 5. The number of measurement points is 126 (7×18) in total. The MCG data is shown in Fig. 6.

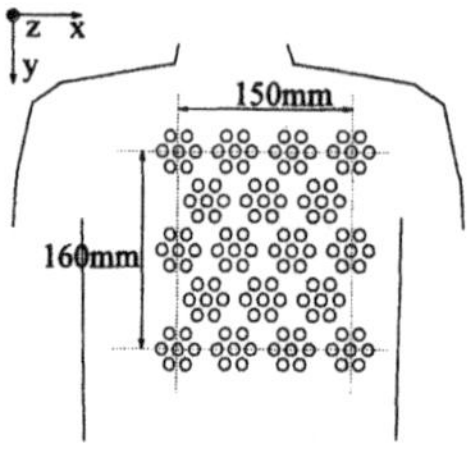

Fig. 5 Schematic location of detection coils

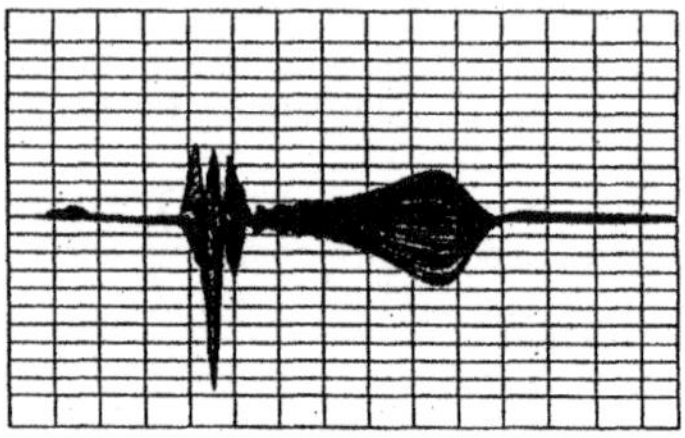

Fig. 6 MCG data (Time [50ms./div.] MCG data[0.2volts/div.6400 digits/volt])

This method needs a initial condition of dipoles. It is known that the excited area is localized in the ventricle at the beginning of QRS complex. We estimated the location and the moment of a dipole at 192ms as the initial dipole using a single current dipole model. To minimize the residual between measured and computed MCG maps, we used the non-linear least squares optimization method known as the quasi-Newton method. To estimate the following dipoles we applied our source localization method with constraints of solution area. The total solution area consists of 9261 (21×21×21) grids that formed around the initial dipole location as shown in Fig. 7. The distance between the grids is 2mm. The 3-dimensional solution area is a 40mm cube. In this solution area, MCG localization is performed by this method using multi-dipole model. We used the MCG data from 192ms to 200ms at 1ms intervals.

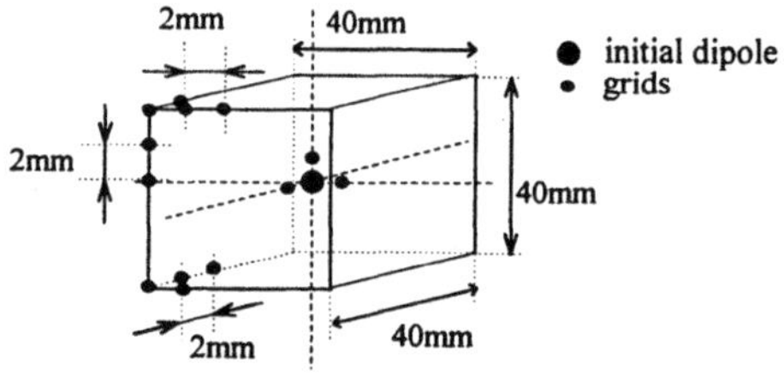

Fig. 7 Solution area (9261 grids) for MCG localization

Results

In Fig. 8, the estimated dipoles at each time are shown together on the x-y plane by arrows. The center of the each arrow represents the estimated location. The direction and the length of each arrow represent the direction and the magnitude of the dipole moment respectively. We tried to construct the excited wave-front form using the information of the estimated dipoles at each time as follows. First of all, the wave-fronts perpendicular to the estimated dipoles are formed. The width of the wave-front is in proportion to the magnitude of the dipole moment. Next, the ends of the neibouring wave-front are connected as disappeared wave-fronts. The constructed wave-fronts at 1ms intervals are shown in Fig. 9. The propagation of the wave-front is presented by gradation. The wave-fronts on MRI of the subject are shown in Fig. 10. The wave-front looks to propagate to the right-, up- and downward from the middle point on the left surface of the interventricular septum.

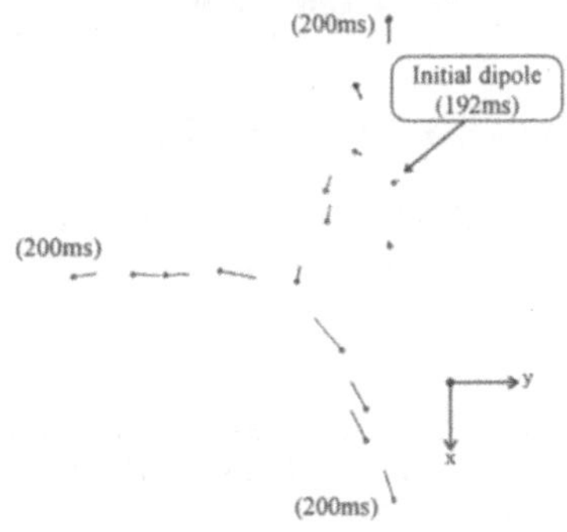

Fig. 8 Estimated results

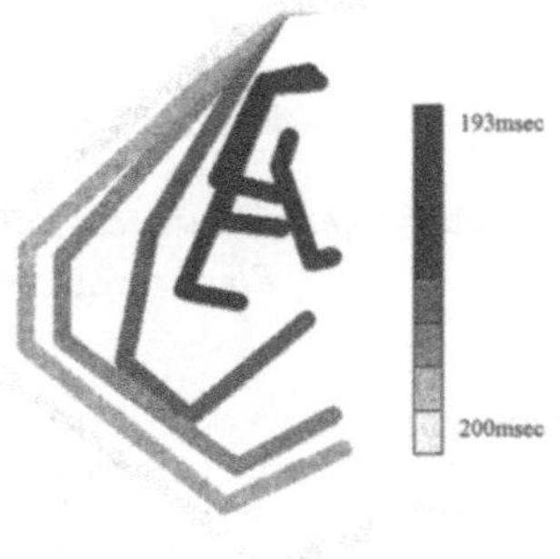

Fig. 9 Reconstruction of wave-front

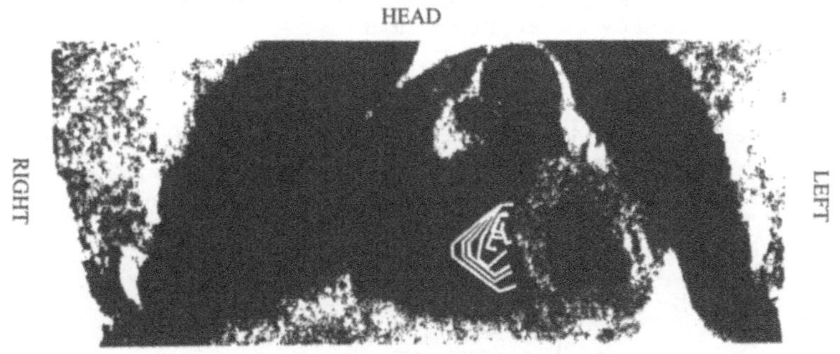

Fig. 10 Estimated results on MRI

Summery

We proposed a new method for the MCG localization with the constraints of the solution area. The results of 3-D model simulation suggest that this method has a possibility to estimate distributed current sources. We tried to apply this method to real MCG data of a normal subject. The results of MCG localization are consistent for the physiological information.

References:

[1] Moshage, W., Seese, B., Achenbach, S., Killmann, R., Bascmann, k. Abstracts of Biomag '93, 1993, 376-377

[2]Osozawa, S., Miyashita, S., Ishiyama, A., Kasai, N. The Journal of Japan Biomagnetism and Bioelectromagnetics Society (in Japanese), 1995, 8: 286-287

Source Localization by Genetic Algorithm

Takeuchi, R.[1], Ikeda, H.[1], Ishiyama, A.[1], and Kasai, N.[2],

Department of Electrical Engineering, Waseda University[1], Tokyo, Japan
Electrotechnical Laboratory[2], Tsukuba, Japan

Introduction

Recently, it is expected to estimate widely distributed current sources or multiple-dipole sources. Various optimization algorithms are proposed for these estimations, while there must be some difficulties such as local minimization or time consuming problems in those methods. For example, simulated annealing (SA) takes a lot of time to search optimal solutions because this kind of algorithm is not allowed to narrow the search space. Therefore, we consider that genetic algorithm (GA), which is an optimization algorithm based on a mechanics of natural evolution, is effective in rapidly searching the global solution[1][2]. However, specific techniques are necessary to apply GA to biomagnetic source localization. In this paper, we try to improve an application of GA to be more simple, and to assess what effect it will have on these problems.

Methods

Source localization was carried out with simple genetic algorithm (SGA)[1], which is a standard algorithm of GAs. The flow chart of SGA is represented in Fig. 1.

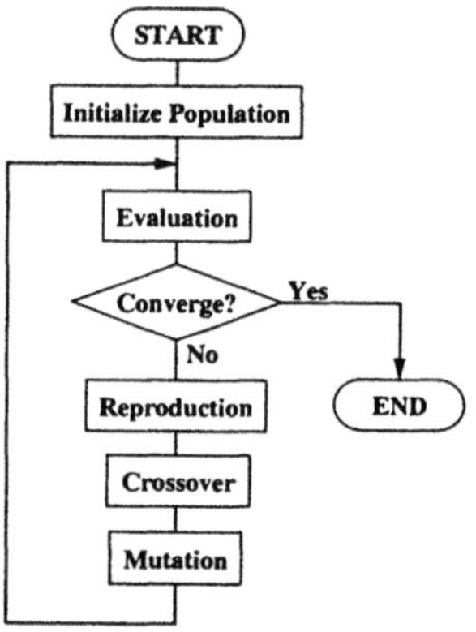

Fig. 1 Flow chart of simple genetic algorithm.

The model of a search space is shown in Fig. 2. The space is a cubic lattice. 512 (= 8x8x8) grids were set up in this cube with 1cm distance. The current dipoles would exist on these grids. 36 (= 6x6) pick up coils were 1cm distant from the search cube, in x-y plane with 2cm distance each other and faced to z-coordinate. The black circles show the pick up coils in this figure.

In this paper, we suppose that the biomagnetic conductor is a semi-infinite homogeneous medium. So the magnetic field $\boldsymbol{B}$ at the point P due to the impressed current $\boldsymbol{J}^i$ is given by Biot-Savart law;

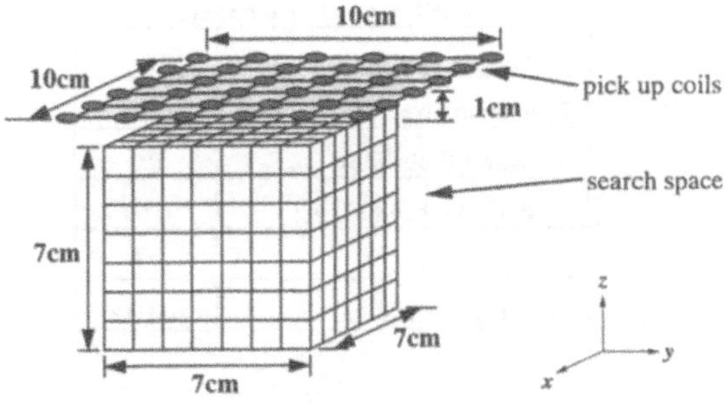

Fig. 2 Model of simulation.

$$B(r_P) = \frac{\mu_0}{4\pi}\int_v J^i(r) \times \nabla \frac{1}{|r - r_P|} dv \, . \tag{1}$$

According to SGA, we estimated the position and the moment of dipoles on the grids. Therefore, each genotype s_i was expressed by a string as described in eq. (2). In this string, 1 gene means 1 dipole. The strength of the moment was fixed in 1 in this case. Variable θ means the angular displacement of the moment.

$$s_i = \langle \underbrace{\underbrace{001}_{x}\ \underbrace{100}_{y}\ \underbrace{101}_{z}}_{\text{position}}\ \underbrace{00100110}_{\text{moment } \theta} \rangle , \tag{2}$$

where 17bit means 9 bit ($512 = 2^9$ grids) plus 8 bit (θ is divided into $256 = 2^8$ directions on x-y plain, because the coils were faced to z-axis). The directions of current dipoles were tangential to pick up coils. These genes were coded by gray code.

The objective function f has the following form:

$$f = \sum_{i=1}^{36} \left(\frac{B_i^m}{B_1^m} - \frac{B_i^c}{B_1^c} \right)^2 , \tag{3}$$

where B_i^m is measured magnetic field at ith coil, and B_i^c is calculated magnetic field at ith coil. Both measured and calculated data are divided by each data at 1st coil, because the calculated pattern is to be compared with the measured pattern.

In the routine of SGA, the fitness function *fit* has the following form using the objective function f:

$$fit = \frac{1}{1+f} , \quad 0 \le fit \le 1. \tag{4}$$

An individual is evaluated higher if *fit* comes close to 1. We maximized this function *fit* in the algorithm.

GA parameters are shown in Table. I. GA operators are described as follows:

selection: roulette wheel and elitist preserving selection,
crossover: 2-point crossover.

TABLE. I GA parameters.

Bit Length l	17
Population Size N	20
Crossover Probability p_c	0.6
Mutation Probability p_m	0.04
Generation Gap G	0.8
Maximum Generation t_{max}	100,000

Simulation

The source localization is carried out in the following two cases: (case 1) single dipole, and (case 2) two dipoles. In case 2, an individual consists of two genes.

case 1. single dipole

(1A) 1 dipole is located on a grid.
(1B) 1 dipole is located between adjacent grids.

case 2. two dipoles (all on the grids)

(2A) 2 dipoles are located in series with the same direction.
(2B) 2 dipoles are located in parallel with the same direction.
(2C) 2 dipoles are located in parallel with the opposite direction.
(2D) 2 dipoles are located in layers with the same direction.

Results

In the simulation (1A), the solution was converged within several 10 generations, and the true value of the single dipole was obtained. In (1B), the solution was not converged although the dipole was estimated just close to the true source.

In case 2, the optimal solutions were obtained respectively. However, the solution was not fully converged in the simulation (2B) and (2C) sometimes. The results of (2A) and (2B) are shown in Fig. 3-5. Fig. 5 shows the result of (2B) that is not converged within the specified generation t_{max}. In this figure, dipoles were estimated far from the optimum although the contour map by estimated sources was similar to that of true sources.

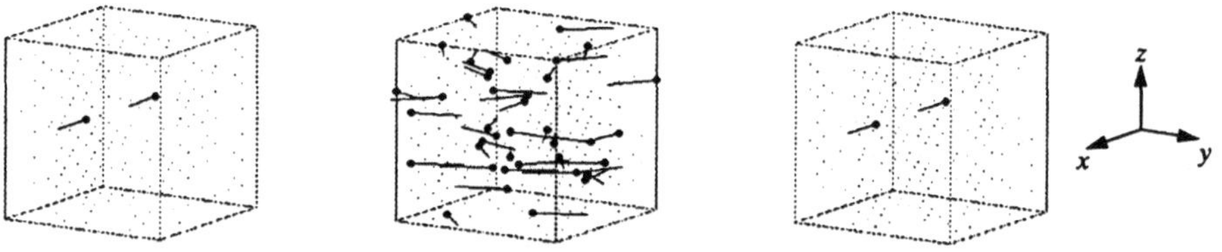

(a) (b) (c)

Fig. 3 Source localization (2A): (a) true sources, (b) initial sources, and (c) estimated sources.

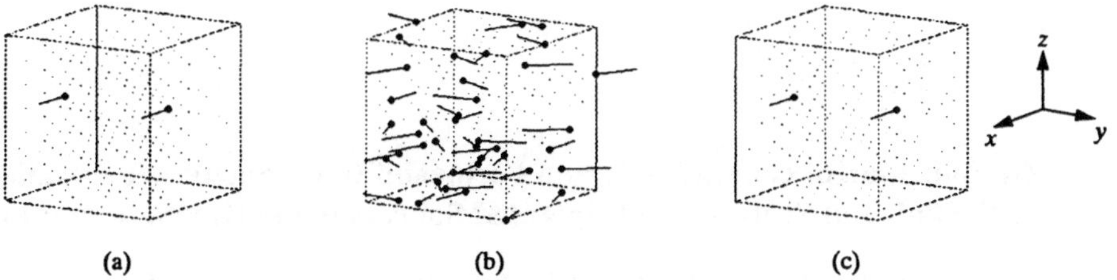

Fig. 4 Source localization (2B): (a) true sources, (b) initial sources, and (c) estimated sources.

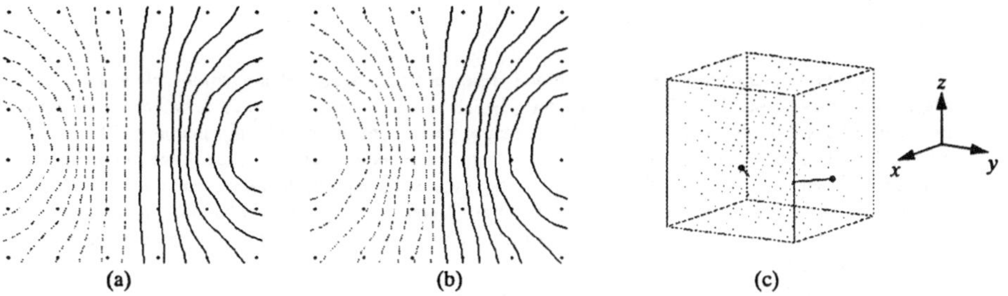

Fig. 5 Source localization (2B, not converged):
(a) contour map by true sources, (b) contour map by estimated sources, and (c) estimated sources.

Summary

We have examined the possibility of multi-dipoles estimation by GA. The current sources were well estimated on these examples. However, the optimal solutions were neither obtained nor converged under the max generation in some cases. It is considered to take some measures as described below against these problems.

In case that the dipoles are not on the grids like (1B), it is necessary to lattice the search space with finer mesh step by step. On the other hand, if magnetic field has a small gradient like (2B), it is thought that a lot of individuals would have high fitness values. Therefore, adjustment of GA parameters would be essential to keep the diversity of the population. (*i.e.* to extend population size, generation gap, etc.)

We will try to apply the above described algorithm to localization of epileptic foci after we investigate its possibility of further improvement.

References:

[1] Goldberg, D. E. Genetic Algorithms in Search, Optimization, and Machine Learning, Addison-Wesley Publishing Company, 1989.

[2] Hatano H. Journal of Japan Artificial Intelligence Conference, 1993, 8-3: 312-319. (in Japanese).

Optimization of Conductivity and Permeability Parameters for Brain, CSF, Skull and Scalp Using Implanted Sources in the Human Cranium

Taylor, S.A., Rogers, R., McNay, D., Levesque, M., Akhtari, M., Sutherling, W.

Neuromagnetism Laboratory, Epilepsy & Brain Mapping Center, Good Samaritan Hospital, Los Angeles, CA 90017

Introduction

By invoking the dipole in a sphere model, researchers elude the volume current effects in MagnetoEncephaloGraphic (MEG) localizations [1]. The drive for better accuracy, however, has led to the Boundary Element Model (BEM) [2] which more closely represents head and brain anatomy. The drawback, of course, is that volume currents are not suppressed in a BEM as they are in the dipole in a sphere model. Proper account of the volume current in the BEM requires the macroscopic conductivities associated with the four major head tissue layers: brain, CerebroSpinal Fluid (CSF), skull, and scalp. Unfortunately, prior studies to determine these parameters [3-5] have never been confirmed in the context of MEG localization studies.

The key to accuracy lies not only in choosing the most realistic model possible, but in tuning the model parameters. Clearly, researchers can tune the BEM geometry parameters to each patient's head anatomy using the patient's MR and CT scans. The next step is to tune the BEM conductivity parameters.

Methods

The only way to tune conductivities in any model is to hold all other model parameters fixed to known values and perform a least squares search over the conductivities until the model's forward solutions match the measured data. Among the model parameters held fixed are the dipole parameters. In order to hold dipole parameters fixed, we need a reproducible and known dipole source. The artificial dipole generated using Stereotactic ElectroEncephaloGraphic (SEEG) electrodes implanted in the patient's head [6] has all these qualities. Since epileptic patients have SEEG electrodes implanted as part of their clinical evaluation, they are good candidates for this study. Preliminary data indicating that this process produces a dipole source can be seen fron the EEG data in Figure 1. What follows is a proposed procedure for tuning the BEM conductivities using MEG data.

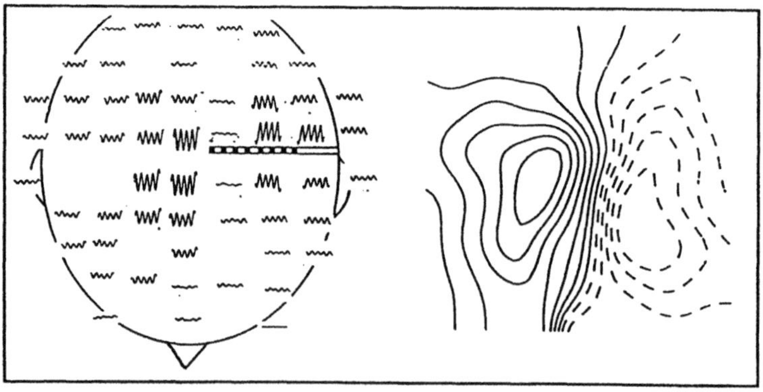

Figure 1: Artificial dipole - (Left) Averaged waveforms showing phase revesal and attenuation with distance from the extrema. (Right) Countour map generated by interpolation of a peak in the waveforms to help visualize the dipole pattern.

A patient with implanted SEEG electrodes is placed in a shielded room. One of the patient's SEEG electrodes is connected to an isolated current supply. A 13 Hertz symmetric (no DC offset) bipolar sinusoid generates the dipole between two contacts from the SEEG electrode which are spaced about five millimeters apart. We choose the sine wave amplitude such that the peak of the sine wave sets up 10 or 15 microAmps of current from one contact to the other, making the peak to peak current 20 or 30 microAmps.

MEG sensors coordinated into an array covering the entire scalp monitor the sinusoid. The MEG signals are tightly filtered over a five Hertz band centerd at 13 Hertz. Center frequency gain is set to an appropriate, properly calibrated known value. Analog to digital converters digitize the filtered MEG signals at 4095 samples per second. Although the sinusoid can be digitized continually, it is best to break up the data collection into 20 second epochs where the patient can suppress all movement, including breathing, for the duration. The current source can be turned off between epochs to give the patient's brain a rest and the digitized data can be saved. Three or four epochs should be enough for each dipole. Most implanted patients have several SEEG electrodes placed all about the head, thus affording the opportunity to collect data from five or six dipoles throughout the patient's brain.

Simultaneous with the digitization of the MEG signals, a clean input signal template should also be digitized. Optimization techniques can be used to accurately estimate the digital frequency of the signal template. This estimate may then be used with fourier series techniques to extract extremely low noise estimates of the amplitudes and phases (either phase synchronized or phase reversed) relative to the signal template of the MEG signals. At this point, the data can match up against forward solutions of the BEM in a search algorithm to find the best fit conductivities.

Discussion

This proposed conductivity study can have three possible outcomes. They all depend on the consistency of the conductivity values among different dipole locations in a single patient and also from patient to patient. With good data, the three possible outcomes imply the following interpretations in descending order of desirability:

1. consistent conductivities within and among patients - implies that the BEM is very good and the conductivities are realistic
2. consistent conductivities within, but not among patients - implies BEM is a good model, but that tuning conductivities for each patient is required to achieve best possible results
3. inconsistent conductivities within and among patients - implies BEM is imperfect and its conductivity parameters have very little physical relevance

Under the best circumstances where conductivities are fully consistent, further research into measuring them for the sake of MEG is unnecessary, except for confirmation. If tuning is required for every patient, we must develop a noninvasive means of doing so. One possible approach is to employ functional MRI coupled with evoked stimuli to generate dipoles in well mapped anatomical regions of the patient's brain. These same dipoles could be used as the reproducible and known dipoles for a conductivity study on the patient. The drawback is that while dipole location is known, we may not find a similarly reliable and accurate procedure for determining dipole moment. Perhaps a mathematical work around can be derived for the BEM that allows analysis of the data that depends only on dipole location. Such has been derived already for analyses with other MEG models [7].

This line of research affords further opportunities. Since the BEM used for EEG studies is a subset of the BEM used in MEG [2], simultaneous EEG and MEG data can confirm the validity of the BEM independently. The conductivities should be consistent between the two modalities. The same data can be used to directly compare the capabilities of EEG and MEG, thus extending comparative research previously done using artificial dipoles [6].

References

[1] Williamson, S., Kaufmann, L. Analysis of neuromagnetic signals. In: Gevins, A.S., Remond, A. Methods of Analysis of Brain Electrical and Magnetic Signals, London, Elsevier, 1987.
[2] Hamalainen, M.S., Sarvas, J. Realistic conductivity geometry model of the human head for interpretation of neuromagnetic data. IEEE Transaction on Biomedical Engineering, 1989, 36: 165-171.
[3] Rayport, M., Sandler, B. Six-electrode method for the measurement of the passive electrical properties of brain envelopes. Medical and Biological Engineering, 1969, 7: 321-324.
[4] Rush, S. and Driscoll, D. EEG electrode sensitivity: an application of reciprocity, IEEE Trans, 1969, BME-16: 15-22.
[5] Geddes. L.A. and Baker, L.E. The specific resistance of biological materials - a compendium of data for the biomedical engineer and physiologist, Medical and Biological Engineering, 1967, 5: 271-293.
[6] Cohen, D., Cuffin, B.N., Yunokuchi, K., Maniewski, R., Purcell, C., Cosgrove, R., Ives, J., Kennedy, J.G., Schomer, D.L. MEG versus EEG localization test using implanted sources in the human brain, Annal of Neurology, 1990, 28: 811-817.
[7] Mosher, J.C., Lewis, P.S., Leahy, R., Singh, M. Multiple dipole modeling of spatio-temporal MEG data, Proceedings of the International Society for Optical Engineering, 1990, Vol. 1351: 364-375.

Modeling of Spreading Cortical Depression Using a Realistic Head Model

Norman Tepley[1,3], Bradley J. Roth[2] and Ranjith S. Wijesinghe[3]

[1]Department of Physics, Oakland University, Rochester, MI, U.S.A.; [2]Department of Physics, Vanderbilt University, Nashville, TN, U.S.A.; [3]Department of Neurology, Henry Ford Hospital, Detroit, MI, U.S.A.

Introduction

Spreading Cortical Depression (SCD) has been an interesting yet physiologically complex question among researchers during the past half century. SCD is a slowly moving hyper-excitation followed by suppression of spontaneous cortical electrical activity in the cortex. Leao and Morrison [1] suggested that SCD might be related to migraine and epilepsy. Later SCD was studied extensively [2, 3, 4] both with EEG and MEG using animal models. SCD in humans has also been observed using cortical electrodes during neurosurgery [5] and in a single event, signals suggesting SCD were observed in positron emission tomography (PET) [6]. Magnetic signals consistent with what would be expected from SCD in humans have been observed by Barkley and colleagues [7] in migraine patients. We have previously shown that SCD can be modeled by using dipoles [13].

Methods

In this calculation, we have used a realistically shaped head model to investigate SCD. This model includes three different compartments for the scalp, skull, and brain. We have represented the sources, which are in the excited area of the cortex due to the propagating SCD, by dipoles which are located perpendicular to the surface of the cortex. During this investigation, we assumed that the shape of the sulcus is parabolic. At the macrocolumn scale, the dipole moment per unit volume and the current density are roughly equal. If $\vec{p}\,(p_x, p_y, p_z)$ represents the dipole moment of a unit volume of the cortex, then the magnetic field produced by current dipoles in a volume dV of the sulcus can be evaluated using the following equations;

$$B(\vec{r}) = B_0(\vec{r}) + \frac{\mu_o}{4\pi}\sum_{j=1}^{3}(\sigma_j^- - \sigma_j^+) \bullet \int_S V_j(\vec{r}\,')\frac{\vec{r}-\vec{r}\,'}{|\vec{r}-\vec{r}\,'|^3} \times d\vec{S}_j \qquad (1)$$

where μ_o is the permeability, J^s is the current dipole moment per unit volume of the sources, σ_j^- and σ_j^+ are the conductivities on the inside and outside of surface ds_j, V_j is the potential on that surface.

To evaluate the magnetic field given by Eq. (1), we divided the scalp, skull, and brain surfaces into triangles [9] as shown in Fig. 1. The magnetic field corresponding to the dipole $\vec{p}\,(p_x, p_y, p_z)$ in an unbounded volume conductor is given by the first term;

$$B_0(\vec{r}) = \frac{\mu_o}{4\pi}\frac{\vec{p}\times(\vec{r}-\vec{r}\,')}{|\vec{r}-\vec{r}\,'|^3} \qquad (2)$$

The second term represents the contribution of the volume current on the scalp, skull, and brain surfaces, where the integral is over the jth surface, $\vec{S}_j$, and $V_j(\vec{r}\,')$ is the potential on the jth surface. In the second term, if the potential is taken as constant over each triangle on the surface, the integral becomes [9]

$$\int_{S_j} V_j(\bar{r}') \frac{\bar{r}-\bar{r}'}{|\bar{r}-\bar{r}'|^3} \times dS_j = \sum_{k=1}^{N} V_{jk} \int_{\Delta jk} \frac{\bar{r}-\bar{r}'}{|\bar{r}-\bar{r}'|^3} \times d\bar{S}_{jk} \quad (3)$$

where the integrals are over the kth triangle, Δ_{jk}, on the jth surface. These integrals depend only on the head and coil geometries, and not on the dipole position.

The potential of a dipole in a homogeneous, unbounded volume conductor, V_o, is calculated at the center of each triangle using the equation

$$V_0(\bar{r}) = \frac{1}{4\pi\sigma} \frac{\bar{p} \bullet (\bar{r}-\bar{r}')}{|\bar{r}-\bar{r}'|^3} \quad (4)$$

where $\bar{r}$ is the position where the potential is calculated, $\bar{r}'$ is the position of the dipole, $\bar{p}$ represents the dipole moment of a unit volume of the cortex, and σ is the conductivity. The electrical potential is calculated using the algorithm developed by Hammalainen and Sarvas (1989). The integral equation for the potential is approximated as a set of linear equations [9];

$$V_{ik} = \sum_{j=1}^{3} \sum_{l=1}^{L(j)} B_{ijkl} V_{jl} + g_{ik} \quad (5)$$

where

$$V_{ik} = \frac{1}{\mu_{ik}} \int V(\bar{r}) dS_{ik}, \quad g_{ik} = \frac{1}{\mu_{ik}} \frac{2}{\sigma_i^- + \sigma_i^+} \int V_o(\bar{r}) dS_{ik}$$

$$B_{ijkl} = \Gamma_{ij} \frac{1}{\mu_{ik}} \frac{1}{2\pi} \int \Omega_{il}(\bar{r}) dS_{ik}, \text{ and } \Gamma_{ij} = \frac{\sigma_j^- - \sigma_j^+}{\sigma_i^- - \sigma_i^+}. \quad (6)$$

Here μ_{ik} is the area of the kth triangle (k=1,...,L(i)) on the ith surface (i = 1,2,3), V_{ik} is the potential averaged over that triangle, $\Omega_{jl}(r)$ is the solid angle subtended from $\bar{r}$ by the lth triangle (l = 1,...L(j)) on the jth surface (j = 1,2,3) σ_i^- and σ_i^+ are the conductivities inside and outside of the ith surface, L(i) is the number of triangles in the ith surface, and $V_o(\bar{r})$ is the potential produced at $\bar{r}$ by a dipole in an unbounded conductor with a conductivity 1 S/m. We modified Hamalainen and Sarvas's algorithm slightly by using a low-order integration formula to solve the integrals in Eq. 6 instead of evaluating the integrals at the center of each triangle . Solid angles are calculated using the analytical expressions presented by van Oosterom and Strackee [10].

Equation (5) has no unique solution because the potential is defined only up to an arbitrary constant. Therefore, there is a resulting ambiguity in the solution [11]. By means of deflation we can remove the ambiguity with the following alteration

$$C_{ijkl} = B_{ijkl} - 1/N \quad (7)$$

where N is the total number of triangles. Our deflated potential obeys the equation

$$V_{ik} = \sum_{j=1}^{3} \sum_{l=1}^{L(j)} C_{ijkl} V_{jl} + g_{ik} \quad (8)$$

This equation can be rearranged to give

$$\sum_{j=1}^{3}\sum_{l=1}^{L(j)}(I_{ijkl}-C_{ijkl})V_{jl}=g_{ik} \quad (9)$$

where

I_{ijkl} = {1 if i=j and k=l) or (0 otherwise)}

is the identity matrix. We use a direct solution of matrix equation using Gaussian elimination [12]. An accurate calculation of the potential using Eqs. (5) and (6) is difficult if the skull conductivity is much less than the conductivity of the scalp and brain [11]. Therefore, we have implemented the scheme presented by Hamalainen and Sarvas [11] to first determine the potential on the brain in air, and then use this result to get the potential on all three surfaces.

When SCD propagates over the cortex, both the directions and positions of dipole moments change. Hence, it is difficult to formulate a general mathematical formula for the magnetic field created by a dipole moment located at any arbitrary location in a volume conductor. Therefore, we divide dipole moments into their x, y, and z components (p_x, p_y, p_z). Then we consider these components individually and calculate the corresponding magnetic fields. By summing these magnetic field components over the active region of the cortex, we can obtain the total magnetic signal arising from the SCD which is detected by the SQUID coil.

For calculating the magnetic field arising from SCD as it propagates across a sulcus, it is necessary to represent the curvature of the sulcus by a mathematical expression. Without knowing the shape of sulcus, the path of the traveling disturbance associated with SCD can not be determined. For this investigation, the used cross-sectional shape of the sulcus is given by Eq. (10);

$$x^2=\frac{z-d_s}{(a-z)}, \quad \text{for} \quad a > z \geq d_s \quad (10)$$

where d_s is the depth of sulcus measured from the origin of the spherical coordinate system and a is the distance to the skull-brain boundary from the origin of the coordinate system.

Evaluating the first derivative of z with respect to x, we obtain an expression for the gradient of this curve at any point on it. Using the value of the gradient, we can determine the orientation of the dipole at the same point where the gradient was obtained. Since, in our model, we assume that the sources associated with SCD are oriented perpendicular to the surface of the sulcus, we can obtain the orientations of these sources using these gradients. Thus we can evaluate the magnetic field created by this dipole at the point where the magnetometer pickup coil is located. As described earlier, in order to deal with the rapidly changing orientations of dipoles encountered by SCD as it travels across a sulcus, we have divided the sulcus region into very small surface elements and represented them by tiny individual dipoles.

Results and Discussion

Figure 2a shows the recorded LAW signals from a patient during migraine aura. We simulated the signals shown in Fig. 2b for comparison with the signals shown in Fig. 2a. These two figures are sufficiently similar to suggest that reasonable selection of model parameters can lead to the modeling of real data.

Additional simulations, not shown for brevity, suggest that amplitudes and durations of the simulated signals increase with the increase of active sulcus area. This is because when active sulcus area increases, the total number of dipoles involved in the simulation increases. Similarly when the active region is wider, the time required for the SCD region to propagate across the sulcus increases leading to the longer duration signal.

Thus, we show that the large amplitude waves (LAWs), reported by Barkley and coworkers [7] in time series magnetoencephalography (MEG) recordings from migraine patients but not controls, may arise from propagation of Spreading Cortical Depression (SCD) across a sulcus. SCD propagates very slowly across the cortex in all species in which it has been observed. In our model, current dipoles represent the excitable neurons in the cortex. Using this model, we are able to simulate signals very similar to the LAWs. The simulated waveforms suggest that shapes, amplitudes, and durations of the SCD signals depend on the size of the active area of cortex involved in SCD, as

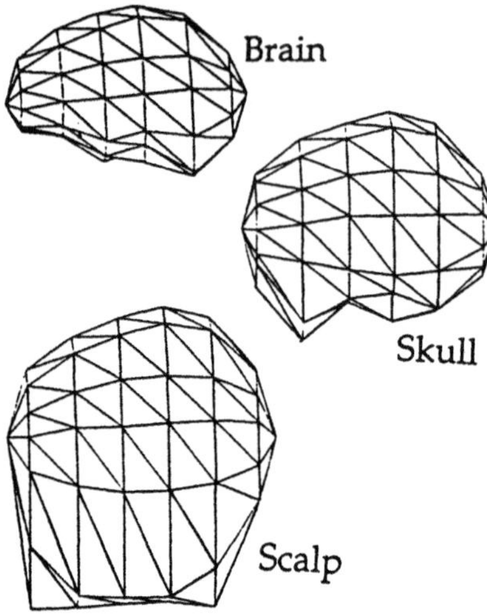

Fig. 1 Digitized scalp, skull, and brain surfaces.

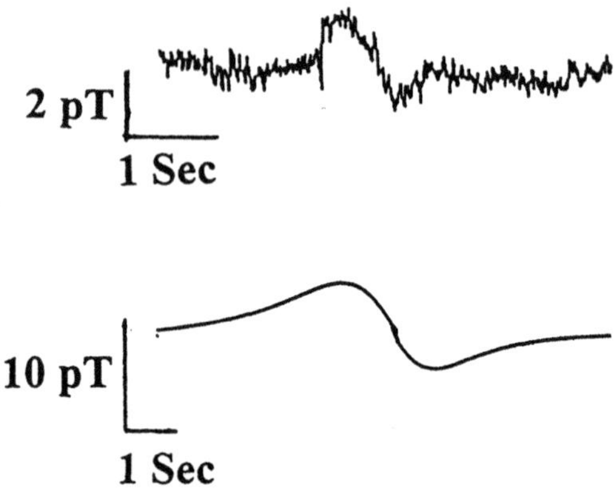

Fig. 2. a) MEG signal from reference 7 detected from migraine patient. b) Simulated signal for SCD using our model.

well as the relative orientation of the pickup coils. The shape of the simulated LAW waveform is strongly influenced by the relationships between the detector location and orientation, the propagation direction of the SCD wave, and the orientation of the sulcus.

References

[1] Leao, A.A.P., and Morrison, R.S. Propagation of Spreading Depression. J. Neurophysiol., 1945, 8: 33-45.

[2] Bures, J., Buresova, O., and Krivanek, J. The mechanism and applications of Leao's spreading depression of electroencephalographic activity. New York: Academic Press; 1974.

[3] Okada, Y., Lauritzen, M., and Nicholson, C. Magnetic field associated with spreading cortical depression: a model for detection of migraine. Brain Research, 1988, 442:185-190.

[4] Gardner-Medwin, A.R.; Tepley, N.; Barkley, G.L.; Moran, J.; Nagel, L.S.; Simkins, R.T.; Welch,W., Magnetic fields associated with spreading depression in anaesthetized rabbits. Brain Research. 540: 153-158; 1991.

[5] Sramka, M., Broack, G., Bures, J., and Nadvornik, P. Functional ablation by spreading depression. Possible use in human stereotactic neurosurgery. Appl. Neurophysiol., 1977, 40: 48-61.

[6] Woods, R.P., Iacoboni, M., and Mazziotta, J.C. Bilateral Spreading cerebral Hypoperfusion During Spontaneous Migraine Headache. N. Engl. J. Med., 1994, 331: 1689-1692.

[7] Barkley, G.L., Tepley, N., Nagel-Leiby, S., Moran, J.E., Simkins, R.T., and Welch, K.M.A. Magnetoencephalographic Studies of Migraine. Headache, 1990, 30: 428-434.

[8] Cuffin, N.B. Eccentric sphere models of the head. IEEE Trans. on Biomedical Eng., 1991, 58: 871-878; 1991.

[9] Roth,B.J., Balish,M., Gorback,A., and Sato,S., How well does a three-sphere model predict positions of dipoles in a realistically shaped head? Electroenceph.Clin.Neurophysiol., 1993, 87: 175-184.

[10] Van Oosterrom,A., and Strackee, J., The solid angle of a plane triangle, IEEE Trans. on Biomedical Eng., 1983, 30: 125-126.

[11] Hamalainen,M.S., and Sarvas,J., Realistic conductivity geometry model of the human head for interpretation of neuromagnetic data, IEEE Trans. on Biomedical Eng., 1989, 36: 165-171.

[12] Purcell,C.J., and Stroink,G., Moving dipole inverse solutions using realistic torso models, IEEE Trans. on Biomedical Eng., 1991, 38: 82-84.

[13] Tepley,N., and Wijesinghe,R.S., A dipole model for spreading cortical depression, Brain Topography, In press.

Acknowledgment

This research supported by NIH/NINDS grant Nos. R01 NS30914 and P01 NS32399.

A Simulation Study on the Nature of Electrical Sources of the MCG

Tilg, B.[1], Wach, P.[1], Lafer, G.[1], Fleischmann, P.[1], and Rucker, W.[2]

Department of Biophysics, Institute of Biomedical Engineering[1], Graz, Austria
Institute of Theory in Electrical Engineering[2], Stuttgart, Germany

Rationale

The purpose of this study is the investigation of the relationship between the electrical sources (distribution of the primary current densities or the transmembrane potential) within a model of the entire human heart (cellular automaton approach) and the simulated magnetocardiogram (MCG) including an inhomogeneous boundary-element (BE) thorax model for solving the forward problem.

Method

Cardiac modelling on a macroscopic level is performed since several years in order to investigate the source structure and its relation to the observable electric and magnetic fields and macroscopic reentry phenomena like supraventricluar or ventricular tachycardias [1, 4, 5, 6, 7, 8, 10, 11].

In a voxel of the heart tissue, equivalent current dipoles located in the intracellular and extracellular space represent an equivalent electrical activity [1]. In case of electrical isotropy (uniform dipole layer (UDL) theory) a antiparallel dipole configuration occurs during the process of depolarization and repolarization. No net current exists as long as the depolarization or repolarization wave does not impinge on a discontinuity and as long as the tissue conductivity is isotropic. For this reason no magnetic field or body surface potential can be detected anywhere outside the myocardium. On the other hand if a wave front impinges on a discontinuity (e.g. endocardium, epicardium, border zone of an infarcted area) the dipoles can not compensate each other any longer. This effect causes a detectable extracorporal magnetic field and body surface potential. Following the bidomain theory [1] the myocardial tissue forms a bisyncytium, consisting of two interpenetrating domains, the intracellular space and the extracellular space. Across the depolarization and repolarization front there exists a macroscopic gradient of the transmembrane potential.

The impressed primary current density is defined by the negative macroscopic gradient of the transmembrane potential multiplied by the intracellular effective conductivity. In case of electrical isotropy these impressed primary current densities are directed perpendicular to the wave front, whereas in case of electrical anisotropy the current densities are oblique to the wave front because of the local conductivity tensor [1]. The impressed primary current density is given by [1]

$$\bar{J}_o = -\tilde{\kappa}_{in} \nabla \Phi_m \qquad (1)$$

with the intracellular effective conductivity $\tilde{\kappa}_{in}$ and the macroscopic gradient of the transmembrane potential Φ_m. Considering a voxel of heart tissue ΔV_{Heart} one can form a primary current dipole (PCD) vector

$$\bar{J} = \int_{\Delta V_{Heart}} \bar{J}_o dV_{Heart} \qquad (2)$$

describing the equivalent (and averaged) electrical activity in this voxel of heart tissue. After spatial discretization of the cardiac muscle (in the heart model volume elements with a grid spacing of $2.5 \cdot 10^{-3}$ m or $1.25 \cdot 10^{-3}$ m are used, respectively) the electrical activity of the cardiac muscle can be represented by a time-variant PCD distribution during depolarization and repolarization. In principle these PCD distributions are the origin of the MCG and body surface potential but one has to consider the fact that sometimes they are electrically and magnetically silent because of cancellation effects. This means that as long as the wave front propagates within the (homogeneous and isotropic) cardiac muscle, no external magnetic or electric signal can be detected. This can be shown theoretically by the application of the Gaussian integral theorem [1]. The volume integral containing the primary current density can be substituted by a surface integral of the transmembrane potential distribution on the heart's surface in case of electrical isotropy [1]:

$$\int_{V_{Heart}} \vec{J}_e(\vec{r}') \times \nabla G(\vec{r},\vec{r}')dV_{Heart} = \tilde{\kappa}_m \oint_{A_{Heart}} \Phi_m(\vec{r}')(\nabla G(\vec{r},\vec{r}') \times \vec{n}(\vec{r}'))dA_{Heart} \qquad (3)$$

A cellular automaton model of the entire human heart considering 15 different tissue types and the dependence of the refractory period and conduction velocity on the cycle length is used for studying the nature of sources [4, 8, 10]. Anisotropy of the conduction velocities and optionally of the electrical conductivities in the ventricles (UDL or oblique dipole layer (ODL) theory) is assumed. Each voxel is described by a transmembrane potential determined by twelve parameters in dependence on the cycle length. The Huygen's propagation principle is applied for calculating the isochrones [4]. The heart model is positioned inside a thorax model (286 isoparametric quadrilateral 8-node BEs [9] are used for modelling the outer thorax surface and the surfaces of the lungs from magnetic resonance imaging cross-sections) in a similar position as the real subject's heart. In this study the MCG is calculated above the thorax surface at 37 sensor positions on the anterior and at 37 positions on the posterior side of the thorax (first-order axial gradiometers). Besides the normal case the cellular automaton allows the calculation of pathological MCGs (e.g. acute phase of infarction, bundle branch block, Wolff-Parkinson-White-syndrome) by changing the electrophysiological parameters (conduction velocitiy, shape of the action potential and refractory period dependent on the cycle length) of the respective volume elements, as well.

Results

The Fig. 1. depicts a transversal cross-section of the heart model geometry (z=35, ventricles) with the epicardial and endocardial surfaces. The dashed line inside the septum and the myocardium depicts the isochrone at 0.175s after sinus node depolarization (SND) for the normal case (onset of ventricular depolarization: 0.144s after SND). In z=35 the 'breakthrough' occurs at 0.174s on the epicardial surface of the right ventricle (RV). In case of electrical isotropy the PCDs are directed perpendicular to this isochrone. Fig. 2. shows the activation time map on the ventricular epicardium for an anteroseptal acute phase of infarction in a grey level pattern from an anterior view. The ischemic area includes 9.8% of the ventricular mass. The ischemic area is activated later than in the normal case because of the reduced conduction velocities in the ischemic central and border zone.

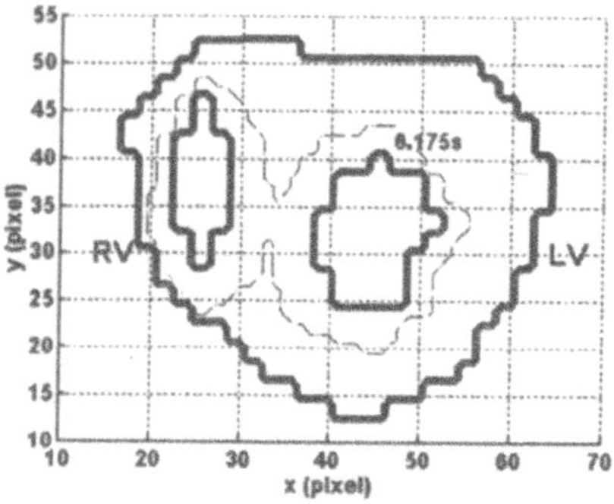

Fig. 1. Vertical cross-section (z=35) of the heart model geometry with the isochrone at 0.175s after sinus node depolarization (RV ... right ventricle, LV ... left ventricle) for the normal case (1 pixel = $2.5 \cdot 10^{-3}$ m).

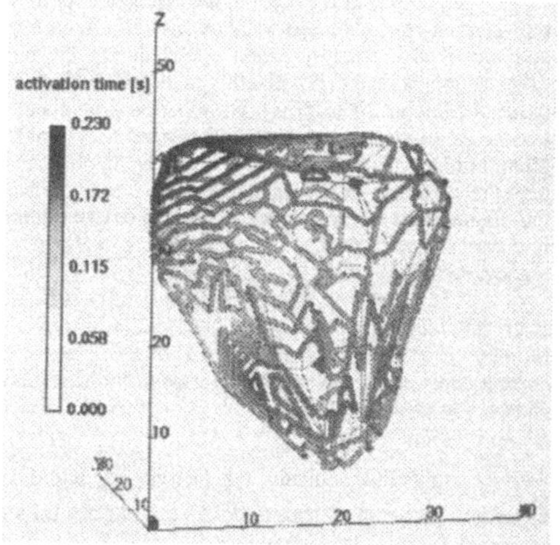

Fig. 2. Isochrone map (activation times) on the ventricular epicardium of the heart model geometry for an anterospetal acute phase of infarction (ischemia) in a grey level pattern.

The Fig. 3. shows the simulated PCD distribution in a grey level pattern at 0.186s after SND within the heart model when electrical isotropy is assumed. The PCDs are calculated from the distribution of the transmembrane potential following equation (1) and (2). At this instant most of the PCDs are located near the left heart base. Only the conduction system and a ring of the atrial mass are depicted. The dipoles in the atrium and in the remaining ventricular mass are caused by the repolarization process and are termed 'background activity' in the approach presented. Fig. 4. shows the calculated MCG for this normal case during one heart cycle following the UDL theory. As mentioned above in this case no MCG can be detected above the thorax as long as the excitation wave does not impinge on the heart's surface (epicardium and endocardium for the normal case). In case of the ODL theory (the PCDs are directed oblique to the wavefront), an additional contribution to the MCG is generated. Assuming electrical isotropy the real signal generators are located on the heart's surface only. Solving the inverse problem from the MCG (or from the body surface potential) one can estimate the transmembrane potential surface distribution at a given time instant or the activation time map during the P-, or QRS-wave on the atrial and ventricular endocardium and epicardium [2, 3].

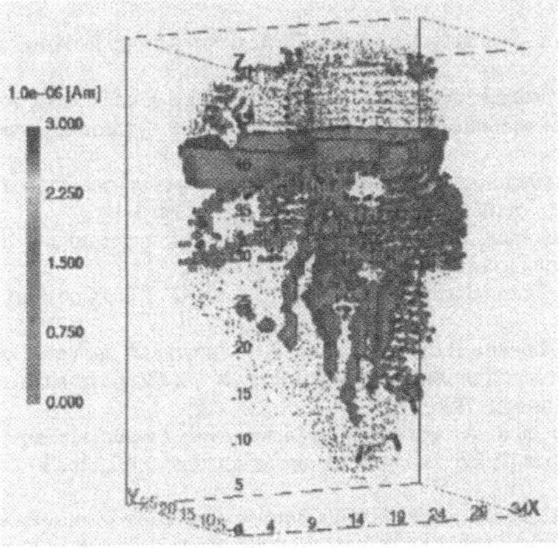

Fig. 3. Primary current dipole distribution for the normal case at 0.186s after sinus node depolarization. Only the conduction system and a ring of the atrial mass are visualized. Most of the depolarizing dipoles are located near the left heart base.

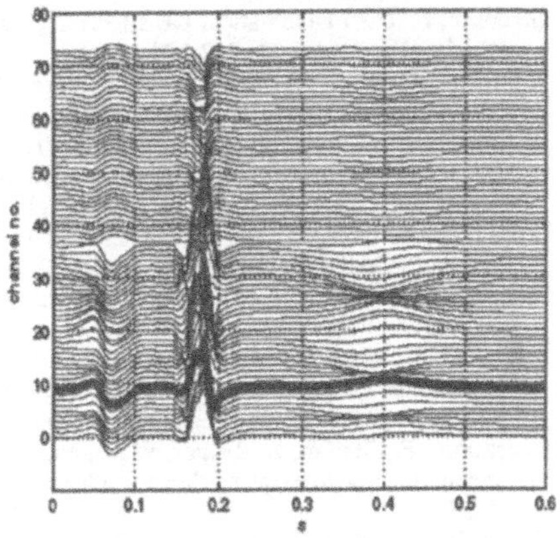

Fig. 4. Simulated magnetocardiogram (MCG) at the 37 anterior (channel no. 1 to 37) and 37 posterior sensor positions for the normal case. The bold line represents the MCG time course for sensor no. 9 (anterior probe).

The numerical evaluation of both sides of equation (3) is implemented in the heart model. In case of electrical isotropy one can show that these different numerical approaches result in the same MCG time courses at the different sensor positions. In case of electrical anisotropy an additional contribution to the MCG signal is produced in dependence on the conductivity ratio in the different myocardial areas. As shown in [8] the modelling of the electrical anisotropy requires the consideration of a changing fibre direction in the voxels of the heart model during the contraction in order to achieve a correct T-wave.

Discussion

By varying the conduction velocities, the refractory periods and the action potential shapes in dependence on the cycle length the model allows the calculation of normal and pathological excitation patterns [4] and the according body surface potential and magnetic field data. In particular the modelled background activity seems to be an important feature, especially for developing reconstruction algorithms. Neglecting this 'background activity' in the foward problem results in a too less realistic modelling approach, because this 'background activity' has a remarkable smearing effect on the estimated inverse solution. In simulations with the heart model it was found that electrical anisotropy influences much more the time course of the T-wave than that of the QRS-wave. This justifies the assumption of electrical isotropy when reconstruction is carried out during ventricular depolarization only. It was shown that in case of electrical isotropy the real signal generators are located on the heart's surface only. This enables the possibility of determination of the transmembrane potential or of the activation times on the endocardium and epicardium in the inverse problem. Some details of the underlying theory for estimating the activation times are discussed in [2, 3]. The method seems to be a very promising approach for investigating the sequence of depolarization by a fully noninvasive functional imaging method.

References

[1] Geselowitz, D.B. A bidomain model for anisotropic cardiac muscle. Annuals of Biomedical Engineering, 1985, 11: 191-206.

[2] Greensite, F., Huiskamp, G., and Oosterom van A. New quantitative and qualitative approaches to the inverse problem of electrocardiology: Their theoretical relationship and experimental consistency. Medical Physics, 1990, 17(5): 569-579.

[3] Greensite, F. The mathematical basis for imaging cardiac electrical function. Critical Reviews in Biomedical Engineering, 1994, 22: 347-399.

[4] Killmann, R., Wach, P., and Dienstl, F. Three-dimensional computer model of the entire human heart for simulation of reentry and tachycardia: gap phenomenon and Wolff-Parkinson-White syndrome. Basic Research in Cardiology, 1991, 86: 485-501.

[5] Leon, L.J., and Horacek, B.M. Computer model of excitation and recovery in the anisotropic myocardium: I. Rectangular and cubic arrays of excitable elements. Journal of Electrocardiology, 1991, 24: 1-15.

[6] Leon, L.J., and Horacek, B.M. Computer model of excitation and recovery in the anisotropic myocardium: II. Excitation in the simplified left ventricle. Journal of Electrocardiology, 1991, 24: 17-31.

[7] Plonsey, R., and Barr, C.C. Mathematic modelling of electrical activity of the heart. Journal of Electrocardiology, 1987, 40: 219-226.

[8] Renhardt, M., Wach, P., Moshage, W., Dienstl, F., Killmann, R., Tilg, B., and Fleischmann, P. Influence of contraction caused changes in the cardiac fiber direction on the magnetocardiogram - a computer simulation study. Advances in Biomagnetism, Book of Abstracts, 1993, 370-371.

[9] Schlemmer, E., Steffan, J., Rucker, W.M., and Richter, K.R. Accuracy improvement using a modified Gauss-quadrature for integral methods in electromagnetics. IEEE Transactions on Magnetics, 1992, 28: 1755-1758.

[10] Wach, P., Killmann, R., Dienstl, F., and Eichtinger, C. A computer model of human ventricular myocardium for simulation of ECG, MCG, and activation sequence including reentry rhythms. Basic Research in Cardiology, 1989, 84: 404-413.

[11] Wei, D., Yamada, G., Musha, T., Tsunakawa, H., Tsutsumi, T., and Harumi, K. Computer simulation of supraventricular tachycardia with the Wolff-Parkinson-White syndrome using three-dimensional heart models. Journal of Electrocardiology, 1990, 23: 261-273.

Acknowledgement

This work was supported by the Austrian Science Foundation, grant P10749 and grant P9995.

Global Optimization in the Localization of Brain Activity

Uutela, K., Hämäläinen, M. and Salmelin, R.

Low Temperature Laboratory, Helsinki University of Technology, Finland

Introduction

The parameters of a multidipole model are usually estimated by finding the values that minimize the sum-of-squares error. Conventional dipole localization usually works well if the number of sources and their approximate locations are known. When data from the whole cortex are analyzed, however, several interacting sources are seen; conventional methods are then often inadequate because the optimization becomes easily trapped in local minima. We studied three different global optimization methods that localize several current dipoles simultaneously without need for any initial guesses.

Methods

Clustering Method

Global minimum of the error function could be found by selecting a large number of initial locations and starting a local search from each. This is, however, inefficient; many of the starting points lead to same the local minima and, therefore, the same areas are searched several times. We used a method [1] that starts local optimizations from random locations (Fig. 1 a). After a few iterations, the candidate locations concentrate around the local minima (Fig. 1 b). We separated the clusters using a *c-means* algorithm [2]. The optimization was then completed for each cluster (Fig. 1 c).

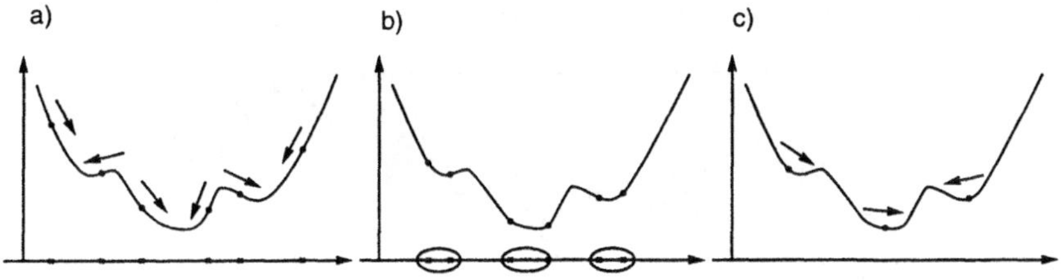

Fig. 1. The clustering method

Simulated Annealing

Minimization with simulated annealing [3] is based on the Metropolis simulation [4] used in statistical physics. The value of the error function of a random location is calculated at each iteration. If the new value is smaller, the new location is always accepted. If the value is larger, it is accepted with a probability depending on the change of the error function Δf and parameter T, analogous to temperature:

$$P(\Delta f, T) = e^{-\Delta f / T}.$$

In the beginning, T is large and almost all the new locations are accepted. When T decreases, the search concentrates near some minimum. If T is decreased slowly enough, the global optimum is found [5].

The version we used [6] selected the new locations from a normal distribution centered on the previous location. After a few evaluations the variance of the distribution (step length) was adjusted according to the percentage of accepted changes. After a few iterations T was decreased by a constant factor.

Dipole fitting with simulated annealing has been previously tried in simulated measurements [7, 8].

Genetic Algorithms

The idea of genetic algorithms [9] is to mimic the evolution of species. Each candidate location is described as a binary sequence called a chromosome. The chromosomes are mutated, new offspring are produced by combining parts of existing chromosomes, and the worst chromosomes are discarded (Fig. 2).

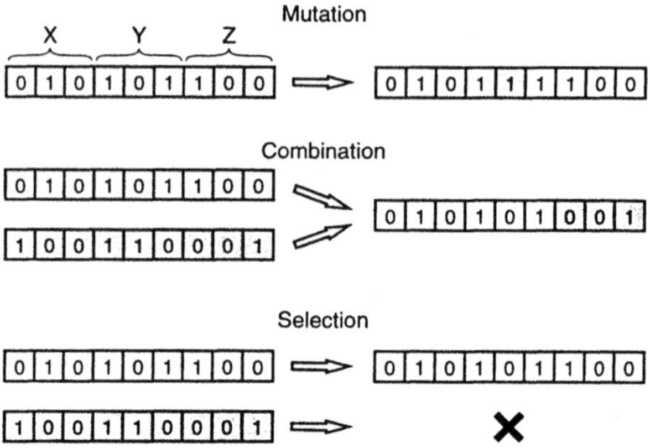

Fig. 2 The basic operations of genetic algorithms

We coded each dipole coordinate as a 16-bit integer: source space width of 20 cm led to a 0.003 mm accuracy in representation of the coordinates. We selected the surviving chromosomes in small tournaments of four chromosomes. The sum-of-squares error of the corresponding dipole model was calculated, the two chromosomes with largest errors were discarded, and new ones were created by selecting the dipole coordinates randomly from the two best chromosomes.

Two different mutation types were used: a binary mutation, which randomly changed one bit of the chromosome, and a normally distributed mutation. The former could produce large changes and find new source areas, whereas the latter was more efficient in adjusting the coordinates locally.

The idea of the algorithm was to search many different areas in the beginning and later concentrate on the good ones. Therefore, the number of chromosomes was decreased during the optimization.

Our genetic algorithm made use of some properties of the other algorithms. The Metropolis-criterion and the adaptive step of the simulated annealing method were used to discard the worst mutations and to select the variance of the normally distributed mutations. The *c-means* algorithm was used to select chromosomes representing different clusters when the number of chromosomes was decreased.

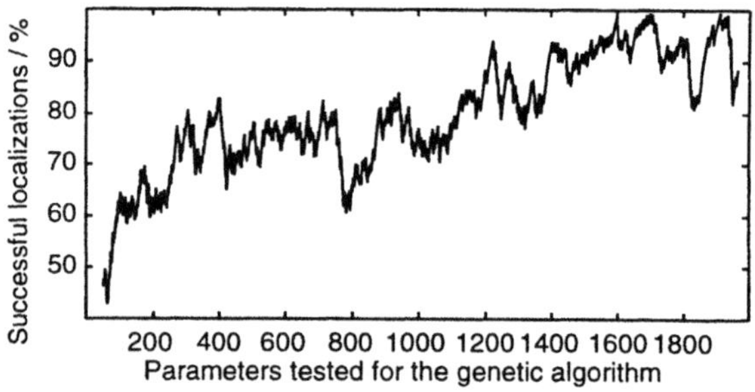

Fig. 3 The percentage of successful localizations using the genetic algorithm during optimization of its parameters.

Parameters of the Algorithms

The algorithms have many parameters that affect their efficiency and reliability. We used a separate genetic algorithm to select the optimal parameters. The parameter combinations were evaluated by fitting a two dipole model into a simulated low-noise measurement. In each trial the number of tested candidate locations was constrained to 10000. During the optimization of the parameters the percentage of successful localizations increased (Fig. 3). For each algorithm, we selected the parameters that were most common in the end of the optimization.

Source Model

The locations of the current dipoles were determined by the global optimization methods. The orientation and amplitude were selected separately for each candidate location to reduce the number of free parameters in the actual optimization problem. The orientations were identical at all time points. An appropriate number of dipoles can be determined by the saturation point of tne goodness of fit curve (Fig. 4 a).

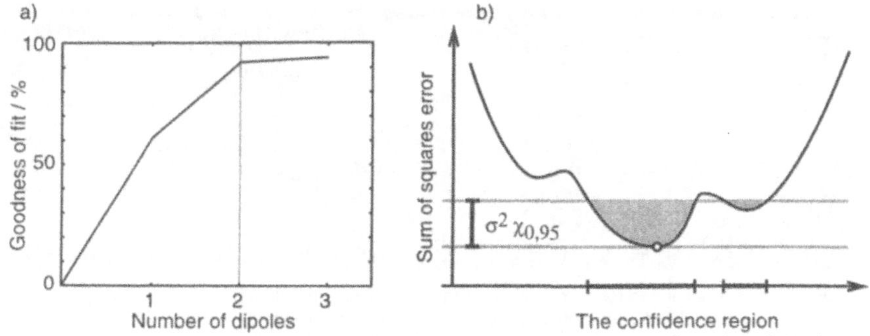

Fig. 4. An appropriate multidipole model explains the data well with a small number of dipoles (a). The confidence region can be estimated from the evaluated locations (b).

Confidence Regions

Confidence regions of the source localization problem consist of the locations with smallest sum-of-squares error (Fig. 4 b). The distribution of the values of the error function can be approximated with a χ^2-distribution [10]. The threshold level for 95 % confidence regions is $\sigma\, \chi_{0.95}$, where σ is the standard deviation of the channel noise and $\chi_{0.95}$ is the 95 % point of the χ^2-distribution with as many free parameters as there are in the multidipole model. Confidence regions of multidipole fitting are irregular and may consist of several disjoint areas. The method resembles the Bayesian error estimation method [11] which would require, however, evaluated sample points from the whole source area.

Results

Comparison of the Methods

The methods were compared by locating sources in simulated measurements. We used both spherical and realistic [12] head models. The genetic algorithm always found the correct sources (Fig. 5).

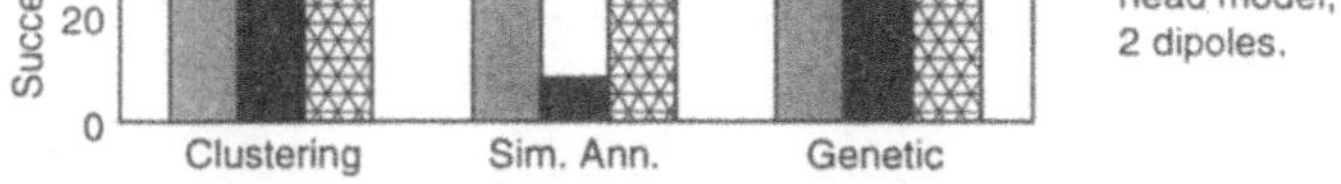

Fig. 5. Reliability of different methods. 10000 function evaluations were used for localization of two sources and 20000 evaluations for four sources. There were 100 trials in the 2 dipole sphere case and 40 in the others.

Measured data

The methods were validated with measured data. We used simple stimuli that could be reliably analyzed with conventional methods. Sources of somatosensory evoked fields in the contralateral S1 area and both S2 areas were localized with a three dipole model with a time span 20–120 ms (Fig. 6 a,b). Sources of auditory evoked fields in both auditory cortices were localized with a two dipole model with a time span 70–130 ms (Fig. 6 c).

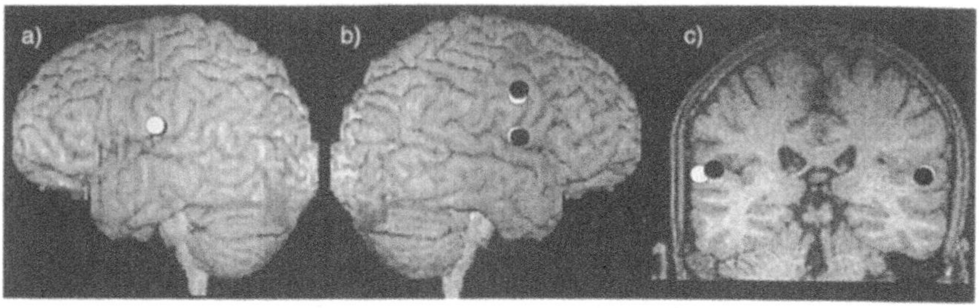

Fig. 6. Source locations for responses to left median nerve stimulation (a,b) and tones (c). The sources were localized with global optimization (black circles) and conventional methods (white circles).

Discussion

Three global optimization algorithms were applied to multidipole model parameter estimation. The genetic algorithm was most reliable. Global optimization was successful in the analysis of both simulated and measured data.

References

[1] Törn, A., and Zhilinskas, A. Global Optimization, Springer-Verlag, 1989.

[2] Schalkoff, R. Pattern Recognition: Statistical, Structural and Neural Approaches,, John Wiley & Sons, 1992.

[3] Kirkpatrick, K., Jr., C.D.G., and Vecchi, M.P. Optimization by simulated annealing, 1983, 220: 671-680.

[4] Metropolis, N., Rosenbluth, A.W., Rosenbluth, M.N., Teller, A.H., and Teller, E. Equation of state calculation by fast computing machines, J. Chem. Phys., 1953, 21: 1087-1092.

[5] Geman, S., and Geman, D. Stochastic relaxation, Gibbs distribution and the Bayesian restoration of images, IEEE Trans. Pattern Anal. Mach. Intell., 1984, 6: 721-741.

[6] Corona, A., Marchesi, M., Martini, C., and Ridella, S. Minimizing multimodal functions of continuous variables with the simulated annealing algorithm, 1987, 13: 262-280.

[7] Haneishi, H., Ohyama, N., Sekihara, K., and Honda, T. Multiple current dipole estimation using simulated annealing, IEEE Trans. Biomed. Eng., 1994, 41: 1004-1009.

[8] Gerson, J., Vardenas, V.A., and Fein, G. Equivalent dipole parameter estimation using simulated annealing, Electroenceph. Clin. Neurophysiol., 1994, 92: 161-168.

[9] Holland, J.H. Adaptation in Natural and Artificial Systems, University of Michigan, 1975.

[10] Cox, D.R., and Hinkley, D.V. Theoretical statistics, Chapman and Hall, 1974.

[11] Hämäläinen, M., Hari, R., Ilmoniemi, R.J., Knuutila, J., and Lounasmaa, O.V. Magnetoencephalography — theory, instrumentation, and applications to noninvasive studies of the working human brain, Rev. Mod. Phys., 1993, 65: 413-497.

[12] Hämäläinen, M.S., and Sarvas, J. Realistic conductivity geometry model of the human head for interpretation of neuromagnetic data, IEEE Trans. Biomed. Eng., 1989, 36: 165-171.

Variable Resolution Electric-Magnetic Tomography

Valdes-Sosa, P., Marti, F., Garcia, F. and Casanova, R.
Cuban Neuroscience Center, Havana, Cuba

Abstract

A new type of Distributed Inverse Solution (DIS) is proposed: Variable Resolution Electric and Magnetic Tomography (VARETA). This method is a weighted minimum norm solution in which a regularization parameter is used that varies spatially at each point of the solution grid. Usual DIS (including LORETA) are particular case of VARETA in which the regularization parameter is a constant. At points in which the regularization parameter is small the source is treated as concentrated. When the regularization parameter is large the source is estimated to be zero. Thus Vareta incorporates automatically both concentrated and distributed sources as well as dealing with the pruning of points at which there is no activity.

Introduction

Initial attempts to solve the Brain Electric and Magnetic Inverse Problem consisted in fitting specific parametric models of concentrated sources such as dipoles to the data [4], [9]. A more recent approach is represented by the Distributed Inverse Solutions (DIS). DIS attempt to solve for the current density on a grid of points in the brain, the indeterminacy of the problem being eliminated by imposing some type of constraint. One well known example of DIS is LORETA [6], obtained by minimizing a functional that reflects a trade off between fidelity to the data and spatial smoothness.

To date, concentrated solutions and DIS are considered to be radically different. This paper introduces a new technique, VARETA (VAriable Resolution Electrical and magnetic TomogrAphy) that reconciles this difference, incorporating both concentrated and distributed solutions into a single framework. This is achieved by using a spatially variable constraint in conjunction with any type of DIS. Applied to LORETA this means that a data driven procedure selects the amount of spatial smoothness required at each grid point in the brain. In the limit extremely high smoothing amounts to pruning these points from the solution grid (estimating that the current density is zero). Also in the limit, zero smoothing is equivalent to fitting that grid point as a point source. This paper presents the theoretical foundations of VARETA, some comments on numerical implementation as well as examples of its use. Further examples are described in the report on the workshop on inverse solutions in this same volume.

Classical DIS

Classical DIS of the matrix of current densities **J** are given by :

$$\hat{\mathbf{J}}_{DIS} = \arg\min_{\mathbf{J}} \left(\|\mathbf{O} - \mathbf{K}\cdot\mathbf{J}\|^2 + \lambda^2 \|\mathbf{P}_k \cdot \mathbf{J}\|^2 \right) \tag{1}$$

where **O** denotes the observations (electric and/or magnetic), **K** is the lead field matrix and $\mathbf{P}_k$ is the stabilizer or regularization operator. Expression (1) originates an explicit linear estimator that depend on λ, the regularization parameter which may be determined by procedures such as Generalized Cross-validation (GCV), AIC, etc. Particular choices of $\mathbf{P}_k$ determine the type of inverse solution. $\mathbf{P}_0 = \mathbf{I}$ leads to the (unweighted) Minimum Norm Solution (MNS) [11], $\mathbf{P}_1 = \mathbf{W}$, where **W** is a certain weight matrix [2], defines the weighted MNS (WMNS). LORETA [6] obtains when $\mathbf{P}_2 = \mathbf{L}_3\cdot\mathbf{W}$, where $\mathbf{L}_3 = \mathbf{L}\otimes\mathbf{I}_{(3\times 3)}$ is a discrete vector Laplacian formed from the Kronecker product of a nonsingular univariate discrete Laplacian **L** with the identity matrix. LORETA is, in effect, a discrete spline solution to the Fredholm integral equation of the Electric/Magnetic forward equations. We are currently exploring other possible spline version DIS (See companion paper by Riera J.J. et al. " Discrete Spline Electric-Magnetic Tomography (DSPET) based upon realistic neuroanatomy").

Variable Resolution Extensions of DIS

A problem with classical DIS is that the same amount of regularization is imposed at all points in the brain grid. For example LORETA assumes that the same degree of spatial smoothness is required everywhere as reflected by the use of a single regularization hyperparameter λ. As a consequence smooth solutions are recovered correctly, but concentrated sources are blurred. A way to improve the resolution of classical DIS is to allow the regularization parameter to vary with the position in the brain grid $\mathbf{x}$. Use of a variable resolution function $\lambda(\mathbf{x})$ would allow spatially adaptive solutions. When $\lambda(\mathbf{x}) = \lambda_{const}$ the variable resolution method reduces to the classical DIS. If $\lambda(\mathbf{x}) \to \infty$, then the Laplacian of the current density at $\mathbf{x}$ is estimated to be zero, causing a data driven pruning of the variables to be estimated. When $\lambda(\mathbf{x}) \to 0$, then the Laplacian at $\mathbf{x}$ is not included in the regularization and this point is estimated as a discrete source (simultaneously with other possible discrete and distributed sources).

The problem of spatially adaptive regularization has only been addressed recently. In [10], [5], [8] $\lambda(\mathbf{x})$ is expanded as $\lambda(\mathbf{x}) = \sum \lambda_k \phi_k(\mathbf{x})$ on a set of spatial basis functions $\phi_k(\mathbf{x})$. The λ_k are then estimated by GCV [6], the evidence procedure [5] or by a plug-in approach [8]. These methods are only practical for a small set of spatial basis functions which, in turn, imposes excessive spatial smoothness on $\lambda(\mathbf{x})$. We have therefore chosen a different approach for implementing VARETA, which additionally allows more a priori information (such as that obtained from other imaging modalities) to be incorporated into the model.

VARETA

Instead of enforcing a spatially smooth $\lambda(\mathbf{x})$ by means of a truncated spatial basis function, this smoothness may be built into the model as a prior distribution in a Bayesian context. In effect, one may require that the Laplacian (or any other regularizing operator) of $\ln \lambda(\mathbf{x})$ be small. VARETA is a DIS specified by:

$$\widehat{\mathbf{J}}_{VARETA} = \arg\min_{\mathbf{J},\Lambda} \left(\|\mathbf{O} - \mathbf{K}\cdot\mathbf{J}\|^2 + \|\Lambda\cdot\mathbf{L}_3\cdot\mathbf{W}\cdot\mathbf{J}\|^2 + \tau^2 \|\mathbf{L}\cdot\ln(\Lambda) - \boldsymbol{\alpha}\|^2 \right)$$

where all the symbols are defined as in (1), Λ is a diagonal matrix of regularizing parameters, and the additional hyperparameters τ and $\boldsymbol{\alpha}$ are introduced. We have assumed a gaussian prior for $\ln \lambda(\mathbf{x})$ in which τ is the inverse variance and controls the amount of smoothness required . $\boldsymbol{\alpha}$ is the mean vector of the prior and controls the relative importance given to each point in space. When no a priori information is available $\boldsymbol{\alpha}$ may be assumed to be a constant vector (as in what follows). However information from other image modalities (such as PET or fMRI) may be incorporated into $\boldsymbol{\alpha}$. LORETA is a special case of VARETA that occurs when $\Lambda = \lambda\cdot\mathbf{I}$.

The price paid for this more general model is that the estimator is no longer explicit but must be calculated iteratively as follows. Starting from a given initial value for $\mathbf{J}^{[0]}$, Λ^i is estimated from $\mathbf{J}^{i-1}$, then $\mathbf{J}^i$ is estimated from Λ^i until either $\mathbf{J}^i$ or Λ^i converge. The hyperparameters are adjusted to minimize the Generalized Cross Validation criterion GCV [6] common to all times or frequencies as described in [1]. An efficient MATLAB version has been implemented that is only an order of magnitude more time consuming than a classical DIS. Different choices of initial values have been experimented with, but good results are obtained by taking $\mathbf{J}^0 = \widehat{\mathbf{J}}_{LORETA}$. This corresponds to a noninformative estimate of $\lambda\cdot\mathbf{I}$ for Λ^0.

A one dimensional example will illustrate the differences between VARETA and classical DIS. The actual current density is a square wave, a spike and a gaussian function which is projected onto a signal space to which noise is added. Fig.1 (top) shows the true current density (solid line) as well as the LORETA solution. As is to be expected the gaussian is recovered with an acceptable error, the square wave is smoothed and the spike (one dimensional dipole) is grossly underestimated as a small bump. This is due to the use of a common regularization parameter $1/\lambda$ shown in Fig.1 (bottom). Fig.2 (top) shows the VARETA solution which has a higher spatial adaptation. $1/\lambda(x)$ Fig.2 (bottom) reflects the fact that there are regions without activity and captures nicely the presence of a concentrated source and sharp edges in the pulse.

A critical issue found in simulations with VARETA is the need for a very fine grid spacing. Experience from the examples constructed in the inverse problem workshop indicate that a small enough spacing between grid points is necessary to avoid misallocation of source variance.

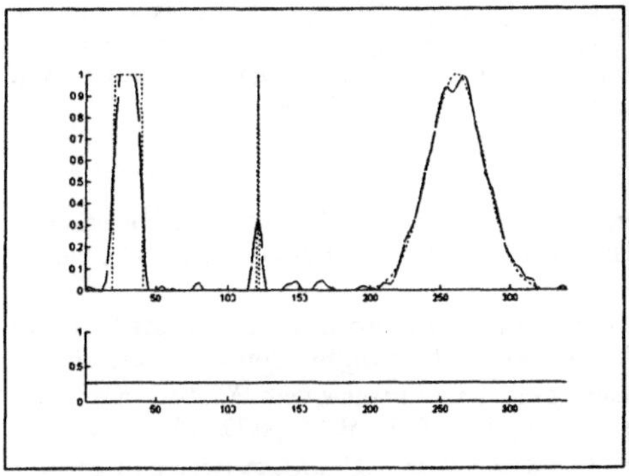

Figure 1

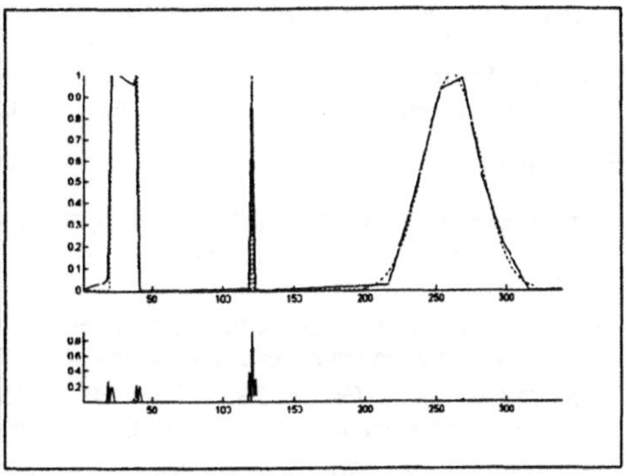

Figure 2

Results with α Rhythm

VARETA was applied to 120 channel EEG recordings of the α rhythm. A frequency domain version of both LORETA and VARETA has been implemented that estimates regularization parameter that are common to several (complex valued) FFT coefficients [1]. Analyses with LORETA reveal a predominantly occipital locus for the spectral peak of the α rhythm (Fig. 3). However activity was also present at other sites and the question arises as wether these are "ghost solutions". The VARETA solution Fig. 4 reveals that this may be the case, since this method indicates a strict localization that includes asymmetries not present in the LORETA solution. Compared to LORETA, the solution is a more parsimonious model as reflected in a much lower value of GCV (GCV_{VARETA} =0.003) than LORETA (GCV_{LORETA} =3.4). This analysis was in good correspondence with previous frequency domain analysis of dipolar sources [9] .

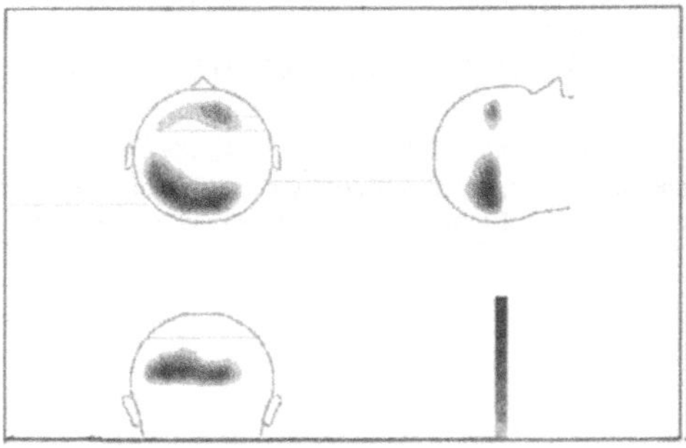

Figure 3

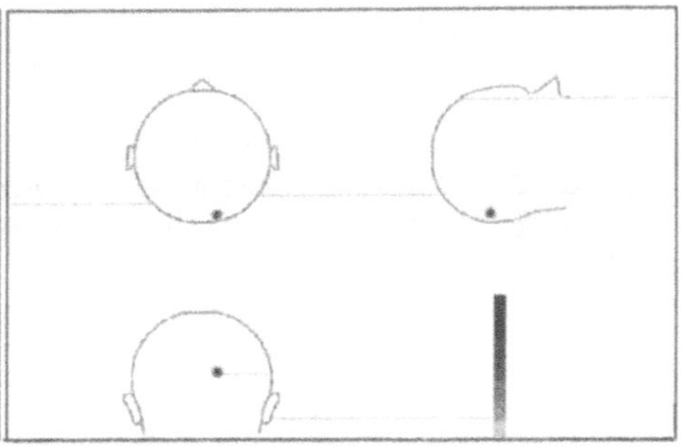

Figure 4

Conclusions and Discussion

Classical DIS provide true tomographic solutions but tend to smear out actual current densities. Variable Resolution DIS offer a bridge between concentrated solutions and classical DIS that may provide a competitive alternative to the highly successful multiple dipole fitting methods. VARETA seems to be a natural complement to LORETA since it "sharpens" initial LORETA solutions.

There have been several recent attempts to achieve less diffuse DIS. A proposal by Phillips et al. [7] uses a Gibbs prior to introduce the a priori requirement of sparse focal activation. VARETA seems to achieve this same objective automatically, without the need to exclude possibly extensive sources a priori. VARETA is also related to another recently proposed method: FOCUSS [3]. FOCUSS iteratively changes the weights in a WNS according to the strength of the solution obtained in the previous step. If continued long enough FOCUSS converges to a set of concentrated sources equal in number to that of the sensor array. VARETA is based upon Bayesian considerations and does not necessarily converge to a set of dipoles if the original sources are distributed.

References

[1]Casanova, R., Valdes, P., Garcia, F., Aubert, E., Riera, J., Korin, W., and Lins, O., , Frequency Domain Distributed Inverse Solutions, Proceedings of the 10th International Conference on Biomagnetism, 1996, Biomag '96, Santa Fe, New Mexico, USA.

[2]George, J.S., Lewis, P.S., Schlitt, H.A., Kaplan, L., Gorodnitsky, I., and Wood, C.C., Strategies for source space limitation in tomographic inverse procedures .In: Advances in Biomagnetism, Abstracts book, 1993.

[3]Gorodnitsky, L.F., George, J.S., and Rao, B.D., Neuromagnetic source imaging with FOCUSS: a recursive weighted minimum norm algorithm, EEG and clinical Neurophysiol., 1995, vol. 95, pp.231-251.

[4]Hämäläinen, M., Hari, R., Ilmoniemi, R.J., Knuutila, J., and Lounasmaa, O.V., Magnetoencephalography - theory, instrumentation and applications to noninvasive studies of the working human brain, Rev. Mod. Phys., 1993, vol. 65(2), pp. 413-497.

[5]MacKay, D., and Takeuchi, R., Interpolation models with multiple Hyperparameters. Tech. report IP (13.111.48.24) cd pub/www/mackay/newint.ps..z.alpha.ps.z., 1996.

[6]Pascual-Marqui, R.D., Michel, C.M., and Lehmann, D., Low resolution electromagnetic tomography: A new method for localizing electrical activity in the brain, Int. Journ. of Psyc., 1994, vol. 18, pp. 49-65.

[7]Phillips, J.W., Leahy, R.M., and Mosher, J.C.,MEG-Based Imaging of Focal Neuronal Current Sources, Proceedings of the 10th International Conference on Biomagnetism, 1996, Biomag '96 Santa Fe, New Mexico, USA.

[8]Sano, A., Optimally regularized inverse of singular value decomposition and application to signal extrapolation, Signal Processing, 1993, 30.

[9]Valdes, P., Bosch, J., Grave, R., Hernandez, J., Riera, J., Pascual, R., and Biscay, R., Frequency Domain Models for the EEG, Brain Topog., 1992, vol 4, (4), pp. 309-319.

[6]Wahba, G., Spline Models for Observational Data, Society for Industrial and Applied Mathematics, 1990, Philadelphia.

[10]Wahba, G., Discussion of the paper by Donoho et al., J. R. Statist. Soc. ser B., 1995, 57, (2), pp. 360-361.

[11]Wang, J.Z., Williamson, S.J., and Kaufman, L., Magnetic Source Images Determined by a Lead-Field Analysis: The Unique Minimum-Norm Least Squares Estimation, IEEE Trans. Biomed. Eng., 1992, vol. 39, (7), pp. 665-667.

Spatio Temporal Distributed Inverse Solutions

Valdes-Sosa, P., Riera, J.J. and Casanova, R.

Cuban Neuroscience Center, Havana, Cuba

Introduction

Initial efforts for the solution of the Brain Electrical/Magnetic Inverse Problem were centered on the use of single or multipole dipole models [3]. At first, these were fitted at each time instant separately. A step forward was the introduction by Scherg and von Cramon [10] of the class of *spatio temporal models* (STM). STM assume that the current generators are fixed in position and orientation and that the strength of each generator changes as a smooth function of time. Besides being physiologically plausible, these assumptions underlie the development of interactive algorithms that have been quite successful in constructing possible inverse solutions (B. Lückenhöner et al., "Comparison between different approaches to the biomagnetic inverse problem- workshop report", this volume). Recent interest has been turned to the class of Distributed Inverse Solutions (DIS). Rather than fitting several elemental generator configurations to the data, DIS attempt to estimate the primary current density $\vec{J}(\vec{r},t)$ in the whole brain volume from the time series of EEG and/or MEG data [17][8]. The hope is that DIS can provide a type of Electric/Magnetic Tomography that will not require the interactive input that current dipole fitting programs seem to need, thus bringing this technique closer to other, more established types of functional imaging. However, to date there has been little work on spatio temporal formulations of DIS [6][1]. What has been done is to add additional simple temporal smoothness constraints to existing solutions. Very little has been done in terms of modeling sources as statistical time series. In view of the fact that the strengths of EEG/MEG are precisely for exploring temporal dynamic, more work on time series analysis of DIS seems warranted.

A note on notation. Hereafter matrices will be denoted by boldface letters. The i-th row vector of $\mathbf{X}$ will be denoted by $\mathbf{X}_{i,\circ}$, the j-th column vector by $\mathbf{X}_{\circ,j}$, the transpose by $\mathbf{X}^t$, the inverse by $\mathbf{X}^{-1}$, the Frobenius norm by $\|\mathbf{X}\|^2$.Also, $\mathbf{0}$ will denote the vector or matrix of zeros, $\mathbf{I}$ the identity matrix. Vectors will also be denoted with lowercase boldface letters excepting 3-D vectors identified as $\vec{x}$. The vectorization of a matrix will be indicated by the function vec operator. The symbol $\sim$ reads "is distributed as", i.i.d means "independent and identically distributed" and $\mathfrak{N}^p(\mu,\Sigma)$ denotes the p-dimensional multivariate normal distribution with mean μ and covariance matrix Σ.

Classical types of DIS

A discrete approximation is defined for a mesh of points $\{\vec{r_j} \mid j \in \Omega\}$ and a set of time instants $\{t_k \mid 1 \leq k \leq T\}$. Let $\mathbf{J}_{j,\circ} = \left[\vec{J}(\vec{r_j}, t_k)\right]_{\{1\leq k\leq T\}}$. The Forward problem may then be expressed as:

$$\mathbf{O} = \mathbf{K} \cdot \mathbf{J} + \mathbf{E} \tag{1}$$

where the $\mathbf{O}$ is the data matrix (electric and/or Magnetic) and $\mathbf{K}$ is the lead field matrix as discussed in [9]. It is assumed that $\mathbf{E}_{\circ,t} \sim \mathfrak{N}^{N_\bullet}\left(\vec{0}, \Sigma_{\mathbf{E}}\right)$ i.i.d., where $\Sigma_{\mathbf{E}}$ is the sensor noise covariance. The probability density (p.d.) of data at time t is then $\mathrm{p}\left(\mathbf{O}_{\circ,t} \mid \mathbf{J}_{\circ,t}, \Sigma_{\mathbf{E}}\right) \propto \exp\left(-\frac{1}{2}\|\mathbf{O}_{\circ,t} - \mathbf{K}\cdot\mathbf{J}_{\circ,t}\|^2_{\Sigma_{\mathbf{E}}}\right)$,where $\|x\|^2_\Sigma = x^t\Sigma^{-1}x$. Inverse theory [12] also poses an a priori probability density (p.d.) for the $\mathbf{J}_{\circ,t} \sim \mathfrak{N}^{N_\bullet}(0, \Sigma_{\mathbf{J}})$ i.i.d. Thus $\mathrm{p}(\mathbf{J}_{\circ,t} \mid \Sigma_J) \propto \exp\left(-\frac{1}{2}\|\mathbf{J}_{\circ,t}\|^2_{\Sigma_{\mathbf{J}}}\right)$. The Maximum aposteriori estimate (MAP) for $\mathbf{J}$ is then obtained by maximizing:

$$\mathrm{p}(\mathbf{J} \mid \mathbf{O}, \Sigma_{\mathbf{E}}, \Sigma_{\mathbf{J}}) \propto \prod_t \left[\mathrm{p}\left(\mathbf{O} \mid \mathbf{J}_{\circ,t}, \Sigma_{\mathbf{E}}\right) \cdot \mathrm{p}(\mathbf{J}_{\circ,t} \mid \Sigma_{\mathbf{J}})\right] \tag{2}$$

where the independence of the $\mathbf{J}_{o,t}$ at different times is crucial for obtaining all joint densities. All DIS that do not consider temporal correlations (such as imposing temporal smoothness on source waveforms) are of this type. As an example take the family of regularized solutions which result from taking $\mathbf{\Sigma_E} = \sigma^2\mathbf{I}$ and separately minimizing $-\frac{1}{2}\ln\left[\mathrm{p}\left(\mathbf{J} \mid \mathbf{O}, \sigma^2\mathbf{I}, \mathbf{\Sigma_J}\right)\right]$ for each time instant .

$$\widehat{\mathbf{J}}_{o,t} = \underset{\mathbf{J}_{o,t}}{\arg\min} \left(\frac{1}{\sigma^2} \left\| \mathbf{O}_{o,t} - \mathbf{K} \cdot \mathbf{J}_{o,t} \right\|^2 + \left\| \mathbf{\Sigma}_\mathbf{J}^{-\frac{1}{2}} \cdot \mathbf{J}_{o,t} \right\|^2 \right) \tag{3}$$

The parameters σ^2 and $\mathbf{\Sigma_J}$ are then obtained by Bayesian methods, generalized maximum likelihood or generalized cross validation. Specific examples are: the Minimum Norm [17] and LORETA [8] which result when $\mathbf{\Sigma_J} = \tau^2\mathbf{I}$ and $\mathbf{\Sigma_J} = \tau^2\left(\mathbf{L}^{-1} \cdot \mathbf{L}^{-t}\right)$ respectively, where $\mathbf{L}$ is a discrete vector Laplacian operator. Allowing a more general sensor noise covariance improves estimation performance [4][11]. There have been recent proposals for $\mathbf{\Sigma_J}$ designed to require spatially adaptive laplacian regularization [15] as well as to deal with realistic neuroanatomy based upon a Thin Plate Spline measure of smoothness [9] .

Spatio temporal DIS (STDIS)

The classical DIS approach does not use the *fixed position and direction* constraint that has been so useful in dipole modeling. As seen below, this is easily remedied and estimation is possible using an extension of a technique developed in [5]. In addition these methods ignore *the temporal dependencies among the* $\mathbf{J}_{o,t}$. This is very unrealistic since single neurons not only exhibit complex temporal dynamics, but are also highly interconnected. Introducing temporal constraints is not simple. Spline theory has been recently applied by Wahba to space-time series by using an ANOVA type decomposition for the smoothness operator [16]. Rather more simple approaches [6][1] simply add an additional temporal smoothness functional to the regularization problem which does not deal with space-time interactions. This paper proposes a specification of time dependencies by use of an extended state-space model within the framework of Spline theory. The model can be stated as follows:

1. **Observation Equation**. The generators at each grid point j have a fixed orientations $\vec{\mu}_j$ and a fluctuating amplitude $\gamma\left(\vec{r}_j, t\right)$. Thus $\vec{J}\left(\vec{r_j}, t\right) = \vec{\mu}_j \cdot \gamma\left(\vec{r_j}, t\right)$. Defining $\gamma_{j,o} = \left[\vec{\gamma}\left(\vec{r_j}, t_k\right)\right]_{\{1 \le k \le T\}}$:

$$\mathbf{O} = \mathbf{K} \cdot \mathbf{M} \cdot \gamma + \mathbf{E} \tag{4}$$

where $\mathbf{M}$ is the patterned matrix: $\begin{bmatrix} \vec{\mu}_1 & \vec{0} & \cdots & \vec{0} \\ \vec{0} & \vec{\mu}_2 & \cdots & \vec{0} \\ \cdots & \cdots & \cdots & \cdots \\ \vec{0} & \vec{0} & \cdots & \vec{\mu}_{N_g} \end{bmatrix}$ and N_g denotes the number of generators. As in the classical DIS, $\mathbf{E}_{o,t} \cdot \sim \mathfrak{N}^{N_e}\left(\vec{0}, \mathbf{\Sigma_E}\right)$i.i.d. and

$$\mathrm{p}\left(\mathbf{O} \mid \mathbf{M}, \gamma, \Sigma_\mathbf{E}\right) \propto \prod_t \exp\left(-\frac{1}{2} \left\| \mathbf{O}_{o,t} - \mathbf{K} \cdot \mathbf{M} \cdot \gamma_{o,t} \right\|_{\Sigma_\mathbf{E}}^2\right)$$

Some elements of $\mathbf{M}$ may be known, e.g. in realistic cortical triangulations.

2. **System Dynamics Equation**. Consider the $\gamma_{o,t}$ as generated by a general p-th order Multivariate Auto-Regressive (MAR) time series model: $\gamma_{o,t} = \mathbf{\Pi}_{o,t} + \mathrm{f}\left(\gamma_{o,t-1}, \dots, \gamma_{o,t-p}; \mathbf{\Theta}_{o,t}\right) + \mathbf{\Xi}_{o,t}$

where $\mathbf{\Pi}_{o,t}$ is a constant, $\mathrm{f}(\circ)$ is a vector function [13] of the auto-regressive process, $\mathbf{\Theta}$ is the matrix of parameters of the process , and $\mathbf{\Xi}_{o,t}$ is the innovation process which is i.i.d.$\sim \mathfrak{N}^{N_g}\left(\vec{0}, \Sigma_\Xi\right)$,where Σ_Ξ is the innovation covariance. Note that this model is quite general. With a sufficiently large number p of pasts most stochastic processes may be approximated to all practical purposes [18]. In addition several particular cases can be derived. When $\mathbf{\Pi} = 0$, equation (5) may model spontaneous source activity and event driven source activity in otherwise. $\mathrm{f}(\circ)$ may be specified to be linear and will serve as a basis for linear statistical time series analysis [14][2] or may taken as nonlinear and lead to nonlinear dynamical analysis [7] of sources. While Θ_t may in general depend on time, as a first approximation the "background source activity"will be considered stationary and therefore Θ a constant. Model (5) specifies the following p.d. for γ :

$$p(\gamma \mid f, \Theta, \Sigma_{\Xi}) \propto \prod_t \exp\left[-\frac{1}{2}\|\gamma_{o,t} - \Pi_{o,t} - f(\gamma_{o,t-1}, \ldots, \gamma_{o,t-p}; \Theta_{o,t})\|^2_{\Sigma_{\Xi}}\right]$$

3. **Apriori probability densities for Θ and Σ_{Ξ}.:**

$$\mathrm{vec}(\Theta) \sim \mathfrak{N}(0, \Sigma_\Gamma), \mathrm{vec}(\mathrm{LN}(\Sigma_{\Xi})) \sim \mathfrak{N}(0, \Sigma_\Delta)$$

where LN is the matrix logarithm and Σ_Γ and Σ_Δ are spatial covariances the specify the following a priori p.d. $p(\Theta \mid \Sigma_\Gamma)$and $p(\Sigma_{\Xi} \mid \Sigma_\Delta)$.

It must be noted that the STDIS introduced includes classical DIS as a special case when $\Pi = f(\circ) = 0$, $\Sigma_\Delta = 0$.

Estimating STDIS models

A possible approach to estimate γ, M, and Θ is to first use the classical DIS procedure to estimate the $J_{o,t}$ and then:

* estimate $\overrightarrow{\mu}_j$ as the first (normalized) eigenvector of the matrices $S = J_{j,o} \cdot J^t_{j,o}$ and $\widehat{\gamma}_{j,o}$ as $\overrightarrow{\mu}^t_j \cdot J_{j,o}$.

* apply conventional linear or nonlinear time series analyses to γ in order to estimate Θ.

This procedure gives acceptable results when the $J_{j,o}$ are weakly correlated and may serve as initial estimates for the iterative procedures described below. However from the previous section it is evident that the statistically efficient procedure would be to maximize the full a posterior p.d.:

$$p(M, \gamma, \Theta \mid O, \Sigma_E, \Sigma_\Gamma, \Sigma_\Delta) \propto p(O \mid M, \gamma, \Sigma_E) \cdot p(\gamma \mid f, \Theta, \Sigma_{\Xi}) \cdot p(\Theta \mid \Sigma_\Gamma) \cdot p(\Sigma_{\Xi} \mid \Sigma_\Delta)$$

or minimize:

$$\sum_t \left[\|O_{o,t} - K \cdot M \cdot \gamma_{o,t}\|^2_{\Sigma_E} + \|\gamma_{o,t} - \Pi_{o,t} - f(\gamma_{o,t-1}, \ldots, \gamma_{o,t-p}; \Theta_{o,t})\|^2_{\Sigma_{\Xi}}\right] + \ln\left[p(\Theta \mid \Sigma_\Gamma)\right] + \ln\left[p(\Sigma_{\Xi} \mid \Sigma_\Delta)\right]$$

This is in general not feasible except for small data sets. Instead a strategy is here proposed that consists of alternatively maximizing two sets of p.d.s, which at the $k-th$ step will be:

1. maximize $p(O \mid M, \gamma, \Sigma_E) \cdot p(\gamma \mid f, \Theta, \Sigma_{\Xi})$. This is equivalent to considering the dynamical parameters Θ and Σ_{Ξ} as known and then using Kalman Filtering (Extended Kalman Filtering for nonlinear models)[18] to estimate $M^{[k]}$ and $\gamma^{[k]}$. This stage will also require iterating between $M^{[k]}$ and $\gamma^{[k]}$ in a generalization of a computational strategy proposed in [5]

2. maximize $p(\gamma \mid f, \Theta, \Sigma_{\Xi}) \cdot p(\Theta \mid \Sigma_\Gamma) \cdot p(\Sigma_{\Xi} \mid \Sigma_\Delta)$. This is equivalent to consider the multivariate time series γ as known and then performing classical linear or nonlinear analysis to estimate the dynamical parameters with the proviso that they be spatially smooth.

This approach has currently been implemented for linear and additive bilinear MAR models for the generators. A set of efficient procedures based upon a sparse representation of all data structures has been implemented in MATLAB. This toolkit has served to demonstrate the feasibility of this approach which shall be reported in future publications.

References

[1]Baillet , S., A Bayesian approach to introducing anatomo-functional priors in the EEG-MEG inverse problems. submitted to IEEE trans. on biomedical computing, Feb. 1996.

[2]Casanova, R., Valdes, P., Garcia, F., Aubert, E., Riera, J., Korin, W., and Lins, O., Frequency Domain Distributed Inverse Solutions, This Volume.

[3]Cabrera, D., Grave de Peralta, R., and Gonzalez, S., Some limitations of spatio temporal source models, Brain Topography. In press.

[4]Dale, A.M., and Sereno, M.I., Improved Localization of Cortical Activity by Combining EEG and MEG with MRI Cortical Surface Reconstruction: A Linear Approach, Journal of Cognitive Neuroscience, 1993, 5:2, pp. 162-176.

[5]de Munck, J. C., The estimation of time-varying dipoles on the basis of evoked potentials, Electroenceph. Clin. Neurophysiol., 1990, 77, pp., 156-160.

[6]Grave de Peralta, R., Oropesa, E., Gonzalez, S.L., Cabrera, D., and Aubert, E., An approach to physiologically meaningful distributed inverse solutions to the neuro-electromagnetic, Submitted for publication.

[7]Hernandez, J.L., and Valdes, P., EEG spike and wave modeled by a stochastic limit cycle, Neuroreport, 1996, In press .

[8]Pascual-Marqui, R.D., Michel, C.M., and Lehmann, D., Low resolution electromagnetic tomography: A new method for localizing electrical activity in the brain, Int. Journ. of Psych., 1994, vol. 18, pp. 49-65.

[9]Riera, J., Aubert, E., Valdes, P., and Casanova, R., Discrete Spline Electric-Magnetic Tomography (DSPET) based on realistic Neuroanatomy. This volume.

[10]Scherg, M., and von Cramon, D. A., A new interpretation of the generators of BAEP waves I-V: Results of spatio-temporal dipole model, Electroenceph. Clin. Neurophysiol., 1985, 62, pp. 290-299.

[11]Sekihara, K., Scholz, B., Generalized Wiener estimation of three dimensional current distribution from biomagnetic measurements. This volume.

[12]Tarantola, A., Inverse problem theory . Methods for fitting data and model parameter estimation, Elsevier Science Publishers B.V., 1987.

[13]Tong H., Non-linear time series., Clarendon press Oxford, 1990.

[14]Valdes, P., Bosch, J., Grave, R., Hernandez, J., Riera, J., Pascual, R., and Biscay, R., Frequency Domain Models for the EEG, Brain Topog., 1992, vol. 4, (4), pp. 309-319.

[15]Valdes, P., Marti, F., Garcia, F., and Casanova, R., Variable Resolution Electrical Tomography. These volume.

[16]Wahba, G., Spline Models for Observational Data, Society for Industrial and Applied Mathematics, 1990, Philadelphia.

[17]Wang, J.Z., Williamson, S.J., and Kaufaman, L., Magnetic Source Images Determined by a Lead-Field Analysis: The Unique Minimum-Norm Least Squares Estimation, IEEE Trans Biomed Eng, 1992, vol. 39, (7), pp. 665-667.

[18]Wei, W-S., Time Series Analysis, Addison-Wesley Publishing Co. Inc., 1990, pp. 364-402.

Quantification of Deviations Between Spherical and Realistically Shaped Forward Models

v. Rango, J., Schlitt, H.A., Halling, H., Terstegge, A. and Müller-Gärtner, H.-W.

Forschungszentrum Jülich GmbH, Jülich, Germany

Introduction

Knowledge of the forward operator and its inverse are essential for assessing the quality of reconstructions. In biomagnetism, the current sources in the brain are reconstructed from the measured magnetic field $\vec{b}$. The relationship between $\vec{b}$ and the source distribution is expressed as $\mathbf{A}\vec{q}(\vec{r}) = \vec{b}$, where $\vec{q}(\vec{r})$, is the magnitude and orientation of the reconstructed source at fixed location $\vec{r}$. The forward operator $\mathbf{A}$ is defined by the conductivity distribution and the sensor geometry. In this paper, we use the well-developed tools of operator theory to investigate two properties of the forward operator. First, we use forward operators to find the best spherical approximation to a realistic conductor geometry. Then, we investigate the effect of the conductor geometry on the reconstructed magnitude and orientation of the source.

For realistic conductor geometries the forward operator, $\mathbf{A}_r$, is calculated using Boundary Element Methods (BEM, e.g. LAHEM [1]), Finite Integration Techniques (FIT, e.g. MAFIA [2]), or other Finite Element Techniques (FEM). The conductor geometry is often approximated with a sphere because the forward operator for a sphere, $\mathbf{A}_s$, is defined analytically [3] and is therefore much faster to compute. The location of the sphere center is usually determined by geometrical constraints: The surface of the sphere is fit to the local curvature of the inner skull surface in the region near the source. Unfortunately, this geometrical matching of the conductor surfaces does not guarantee that the corresponding forward operators will match. There is actually no need to find a proper approximation of the surface, because the analytic formula of the magnetic field of a dipole in a sphere does not depend on the diameter of the sphere. Therefore, any sphere with fixed center and arbitrary diameter fits the realistic forward operator with the same deviation. In this paper, we present a method for determining the sphere center based on matching the forward operators instead of the geometry.

We calculate the deviation, $\Delta\vec{q}$, of the reconstructed source magnitude and orientation due to the use of a spherical conductor as an approximation to the realistic conductor:

$$\Delta\vec{q} = (\mathbf{A}_r^+ - \mathbf{A}_s^+)\vec{b} + \mathbf{A}_s^+\Delta\vec{b}, \qquad (1)$$

where $\mathbf{A}^+$ denotes the pseudo-inverse of the operator $\mathbf{A}$. The first term describes the model dependent deviation caused by the spherical approximation and is the subject of this paper. The second term describes the error due to differences in the measured data (e.g. noise) and is not discussed here. The influence of the operator difference on the inverse must be calculated for each source location. In this paper we consider only one source location at a time, although in general any number of source locations can be used. Two different measures are presented that estimate deviations arising from the applied spherical approximation for a source at a given location.

Methods

We use a homogeneous inner skull shaped conductor model and a sensor array consisting of 148 point magnetometers spaced approximately 2.5 *cm* apart (fig. 1). To date we have studied 7 dipole locations, and present the results of 4 representive dipoles in this paper. The coordinate system used is the standard head-centered one, where the origin is defined as the midpoint of the line connecting the periauricula, so that the x-axis is through the nasion, the y-axis is through the left ear and the z-axis is through the top of the head.

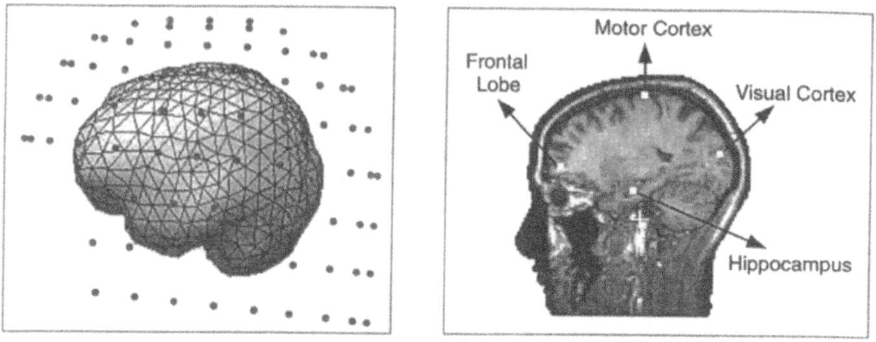

Figure 1: Left: Volume within the inner surface of a skull which is used for realistic forward calculations and the location of the 148-magnetometer sensor-array. Right: Location of the presented dipoles.

Within a chosen of region of interest (e.g. hippocampus) we fix a location (fig. 1) and perform three independent calculations for dipoles pointing in the $\hat{x}$, $\hat{y}$ and $\hat{z}$ directions. The corresponding (148×3)-lead field matrix is a representation of the realistic forward operator for an arbitrarily oriented dipole with fixed location. The spherical forward operator is calculated using the Sarvas formula and is also represented by a (148×3)-matrix.

Determination of the best fitting sphere

The general spherical forward operator for an arbitrary sphere containing a fixed dipole can be written as a function of the sphere center:

$$\mathbf{A}_s = \mathbf{A}_s(c_x, c_y, c_z), \tag{2}$$

where (c_x, c_y, c_z) denotes the coordinates of the center of the sphere. By iteratively varying (c_x, c_y, c_z), we minimize the error between the two lead field matrices with a least square estimation:

$$\begin{aligned} LSE &= \frac{1}{\|\mathbf{A}_r\|} \cdot \|\mathbf{A}_r - \mathbf{A}_s(c_x, c_y, c_z)\| \\ &= \sqrt{\frac{\sum_{i=1}^{148} \sum_{j=1}^{3} \left(\mathbf{a}_r^{ij} - \mathbf{a}_s^{ij}(c_x, c_y, c_z)\right)^2}{\sum_{i=1}^{148} \sum_{j=1}^{3} \left(\mathbf{a}_r^{ij}\right)^2}}. \end{aligned} \tag{3}$$

Note that we minimize the error for all three columns of the lead field at once. This means that we find the best fitting sphere for a dipole at a fixed location but *with any orientation.* Plots of the error, *LSE*, for dipoles in the hippocampus and the visual cortex as a function of the sphere center are shown in fig. 2. Coordinates of the sphere centers for dipoles in various regions of the brain are listed in table 1.

	Hippocampus			Frontal Lobe			Visual Cortex			Motor Cortex		
	c_x	c_y	c_z	c_x	c_y	c_z	c_x	c_y	c_z	c_x	c_y	c_z
	cm	cm	cm	cm	cm	cm	cm	cm	cm	cm	cm	cm
LAHEM	0.5	0.3	5.7	3.6	0.2	5.2	-0.9	0.3	4.4	-0.1	0.2	4.1
MAFIA	0.6	0.4	5.8	3.8	0.7	5.3	-1.0	0.4	4.6	0.1	0.2	4.3

Table 1: Locations of the center (c_x, c_y, c_z) of the best fitting sphere for dipoles in different regions in the brain for realistic forward calculations with LAHEM and MAFIA.

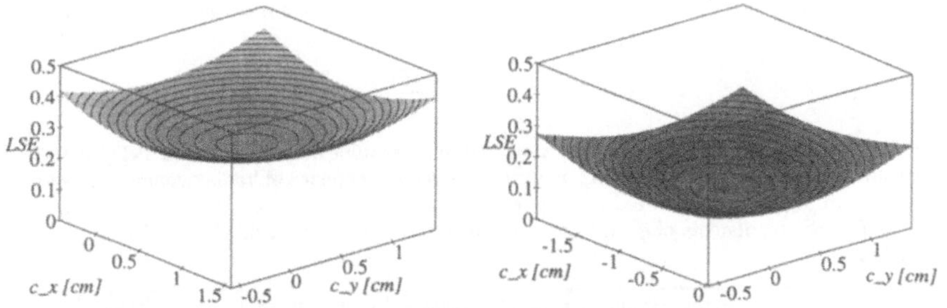

Figure 2: *LSE* values as a function of x- and y-locations of the center of a fitting sphere for a dipole in the hippocampus (left) and in the visual cortex (right). The z-components are each fixed at the global minima of the *LSE* distribution. The plots are each centered in the global minima.

Error estimation

Once the best fitting sphere has been found, it is important to determine the effect of the spherical approximation on the reconstructed magnitude and orientation of the source. We consider two quantities that measure the effect of the spherical approximation:

- **Singular value decomposition**
  We apply a singular value decomposition (SVD) on the (148×3) matrix representation of the realistic forward operator, and get three singular values (SV) and the related (right-handed) singular vectors. For a spherical model one SV is always 0, because radial components of currents in spherical conducters produce no external magnetic field. For a realistically shaped conductor a dipole pointing in the direction of the singular vector that corresponds to the smallest SV produces the smallest magnetic field. This provides us with useful definitions of 'radial' and 'tangential' for dipoles in realistically shaped conductors. The ratio:
  $$r_{SV} = \frac{SV_{min}}{SV_{max}},$$
  provides information about the 'sphereness' of the realistic forward operator. If the ratio is small, there is one dipole orientation which yields a small magnetic field outside compared to perpendicular dipoles with the same magnitude and the same location. Values of r_{SV} are shown in table 2.

- **Deviations of the forward operator**
  In addition we determine the error of the components of the reconstructed source for a fixed location. The forward problem with a realistic conductor geometry is:
  $$\mathbf{A}_r \vec{q}_r = \vec{b}_r, \qquad (4)$$
  and for the fully ranked matrix $\mathbf{A}_r$ we get the inverse solution:
  $$\vec{q}_r = \mathbf{A}_r^+ \vec{b}_r. \qquad (5)$$
  When we apply an approximation $\mathbf{A}_s^+$ (e.g. spherical model) for the realistic inverse operator we get the model-dependent error:
  $$\Delta\vec{q} = (\mathbf{A}_r^+ - \mathbf{A}_s^+)\vec{b}_r. \qquad (6)$$
  Inserting the forward solution for $\vec{b}_r$ (eq. 4) into eq. 6 yields:
  $$\begin{aligned} \Delta\vec{q} &= (\mathbf{A}_r^+ - \mathbf{A}_s^+)\mathbf{A}_r \cdot \vec{q}_r \\ &= \mathbf{E} \cdot \vec{q}_r. \end{aligned} \qquad (7)$$

The maximum eigenvalue, $\lambda^{\mathbf{E}}_{max}$, of the error matrix $\mathbf{E}$ corresponds to the largest relative error that appears in the inverse calculation. Therefore we get:

$$\lambda^{\mathbf{E}}_{max} = \left(\frac{\|\Delta\vec{q}\|}{\|\vec{q}\|}\right)_{max} \tag{8}$$

For the calculation of $\lambda^{\mathbf{E}}_{max}$ we apply a truncated singular value decomposition (TSVD) on the numerical pseudo-inverse matrices $\mathbf{A}_r^+$ and $\mathbf{A}_s^+$ in order to avoid an influence of 'radial' components of the source, which can not be reconstructed by a spherical approximation. We can consider $\lambda^{\mathbf{E}}_{max}$ as the relative error for the components of $\vec{q}$ which produce an external magnetic field. Values for $\lambda^{\mathbf{E}}_{max}$ are shown in table 2.

	Hippocampus		Frontal Lobe		Visual Cortex		Motor Cortex	
	r_{SV}	$\lambda^{\mathbf{E}}_{max}$	r_{SV}	$\lambda^{\mathbf{E}}_{max}$	r_{SV}	$\lambda^{\mathbf{E}}_{max}$	r_{SV}	$\lambda^{\mathbf{E}}_{max}$
LAHEM	0.088	0.22	0.032	0.30	0.022	0.09	0.033	0.04
MAFIA	0.075	0.22	0.026	0.26	0.025	0.09	0.020	0.06

Table 2: Deviation parameters for spherical approximations of the forward model.

Results and Discussion

We think it is more appropriate to choose a spherical approximation based on the forward operators instead of geometrical constraints. From the minimization of *LSE* we find larger differences between the spherical forward operator and the realistic forward operator for deep dipoles and dipoles in the frontal lobe than for regions with shallow dipoles (e.g. motor cortex). The values for the x-coordinate of the center of the best fitting sphere vary by about 4.5 *cm*, while the variations in the z-coordinate are less than 2 *cm*. Note that the error surface is reasonably flat near the minima and the choice of the sphere center is not too critical.

From the analysis of the SVD results we get useful definitions of 'radial' and 'tangential' for dipoles in realistically shaped conductors by using the orientations of the singular vectors of $\mathbf{A}_r$. We find a good agreement between the orientation of a vector which starts in the center of the best fitting sphere and points to the location of the dipole and the singular vector which belongs to the smallest singular value of $\mathbf{A}_r$. That property yields a graphical understanding of the calculated singular vector which belongs to SV_{min} and might be helpful for determining the best fitting sphere, because it restricts the center of the sphere close to a line passing through the dipole location in the direction of that singular vector.

The r_{SV} values for shallow dipoles are in general smaller than for deep dipoles. This implies that for shallow dipoles there is one dipole orientation which yields a negligible field relative to the magnetic field produced by a 'tangential' dipole with the same magnitude and location. A deep source, on the other hand, has no such orientation that produces a negligible magnetic field relative to a 'tangential' dipole. A small value for r_{SV}, however, does not mean that the largest relative error in the reconstructed orientation, $\lambda^{\mathbf{E}}_{max}$, will be small. For example, for a dipole in the frontal lobe r_{SV} is small but $\lambda^{\mathbf{E}}_{max}$ is large, while the reverse is true for a dipole in the hippocampus.

Acknowledgements

We would like to thank the Biophysics Group at Los Alamos National Laboratory for the generous use their mriview software.

References

[1] Schlitt, H., Heller, L., Aaron, R., Best, E., and Ranken, D., IEEE Trans. Biomed. Eng. 1995, 42:(1):52-58
[2] Weiland, T. Proc. of URSI International Symposium on Electromagnetic Theory, 1986, Part B:537-542
[3] Sarvas, J. Phys. Med. Biol., 1987, 32:11-22

The Influence of Holes in a Layer with Low Conductivity on the Potential and Magnetic Field

van den Broek, S.P., Reinders, K.F.J. and Peters, M.J.

University of Twente, Enschede, The Netherlands

Introduction

The volume conductor in bioelectric and biomagnetic research is usually described by a compartment model, each compartment of which has a homogeneous and isotropic conductivity. Models used to localize sources within the brain from measured MEG and/or EEG data, often consist of three compartments representing the brain, the skull and the scalp. A model used to explain fetal MCG's and ECG's after the 28th week of gestation may also consist of three compartments, representing the fetus, the layer of vernix caseosa and the maternal abdomen [1]. The skull conductivity is about 80 times smaller than the conductivity of the brain and the scalp. The ratio between the conductivity of the vernix and the conductivity of the surrounding tissue is in the order of 10^{-6}. Thus a poor conducting layer is present in both cases. Such layers give rise to large errors when using the boundary element method (BEM). Moreover, both the skull and the layer of vernix may contain holes. In the head for example, there are openings in the skull, such as the occipital hole. As the conductivity of the tissues in this hole is much higher than that of the skull, a large influence on the EEG can be expected, as has already been shown by Bertrand *et al.* [2]. Fetal ECG's, measured after the layer of vernix caseosa is formed, show that the amplitude of the potential diminishes. After some weeks it increases again, however, then the distribution of the potential on the maternal abdomen hardly varies during a heart cycle. This can be explained by the assumption that open patches appear in the layer [1]. The BEM is not capable of handling volume conductors with holes in a layer. Therefore the finite-element method (FEM) is used to calculate the potential. Subsequently, the magnetic field is calculated from the potential distribution, by applying the Biot-Savart law [3]. Calculations are performed to assess the accuracy using this procedure. In order to examine the influence of holes, forward calculations are carried out using a model consisting of a sphere and two concentric spherical layers, the middle compartment having a hole. Subsequently, using the calculated potential or field distribution, inverse calculations are carried out for a three-sphere model, with no hole present. It turns out that for the skull the influence of a hole on the potential distribution is substantial and on the magnetic field negligible. The influence of holes in the layer of vernix caseosa on both the potential and magnetic field is large.

Method

The volume conductor problem is posed as a Poisson equation for the potential Φ. The boundary condition is that the normal component of the current density is continuous at interfaces and zero at the outer surface. This problem is solved by means of the FEM. The singularity of the dipolar current source is removed by subtracting from Φ the potential Φ_∞, which is the potential in a homogeneous medium of infinite extent [4]. The Poisson equation now reduces to a Laplace equation with solution $\Psi = \Phi - \Phi_\infty$. This procedure leads to a different boundary condition. As a consequence, an extra integral over all interfaces between elements with differing conductivity has to be computed [3]. The infinite medium potential, for which a simple analytical solution is known, is added again after the FEM calculations. From the potential, the current density in each element is calculated. The magnetic field can then be obtained using the Biot-Savart law, summing the contributions of the current densities in all elements. The volume is divided into tetrahedral elements, the so-called mesh. The same mesh can be used for the potential calculation as well as for the calculation of the magnetic field. In order to limit the number of nodes needed to obtain a required accuracy, locally refined meshes are used. Several ways of refinement are tested. A cross section of a refined mesh that required a low number of elements to obtain a certain accuracy, both for the potential and the magnetic field, is shown in Fig. 1. This refinement is based on two arguments. Firstly, the mesh is refined in regions where the function Ψ is expected to change rapidly, in order to allow that within each element Ψ is approximated by a linear interpolation of the values in the vertices. Secondly, because the magnetic field is influenced most by current elements nearby the point of observation and by current elements where the current density is relatively high, the density of nodes is largest in those regions.

The accuracy of the method is assessed for a model consisting of three concentric spheres. In order to do so, the results obtained with the method described above are compared with those obtained with analytical expressions. The radii of the spheres are respectively 87, 92 and 100 mm, the normalized conductivities 1, 0.0125 and 1S/m. Calculations are carried out for radial and tangential dipoles located at different eccentricities. The potential is calculated in all nodes on the outer surface. Three orthogonal components of the magnetic field are computed in 59 points on a concentric hemisphere with a radius of 120 mm. Meshes with regularly distributed nodes are used, as well as one which is locally refined. As a measure of the error the relative difference measure (RDM) is used, which is defined as:

$$RDM = \sqrt{\sum_i \left(V_i - V_{an,i}\right)^2 \Big/ \sum_i V_{an,i}^2}$$

where V_i is the potential or a component of the magnetic field, computed using the FEM and the Biot-Savart law, and $V_{an,i}$ the analytically calculated value for the same location.

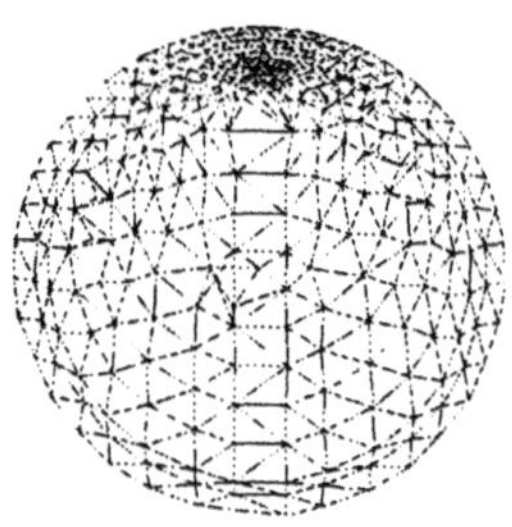

Fig. 1 A cross section of a locally-refined mesh used to describe a volume conductor consisting of three concentric spheres. The dipole is on the vertical axis.

Using a regular mesh, the error in the potential increases when the dipole is closer to the surface of the inner sphere. This is true both for tangential and radial dipoles. The results for the potential shown in Fig. 2a are for a dipole at an eccentricity of 80 mm. In order to obtain an error of less than 10 % a regular mesh consisting of at least 8000 nodes is needed. Using the locally-refined mesh with only 3073 nodes leads to an error that is much smaller, as can be seen in Fig. 2b, where the RDM for this mesh is given as a function of the dipole eccentricity.

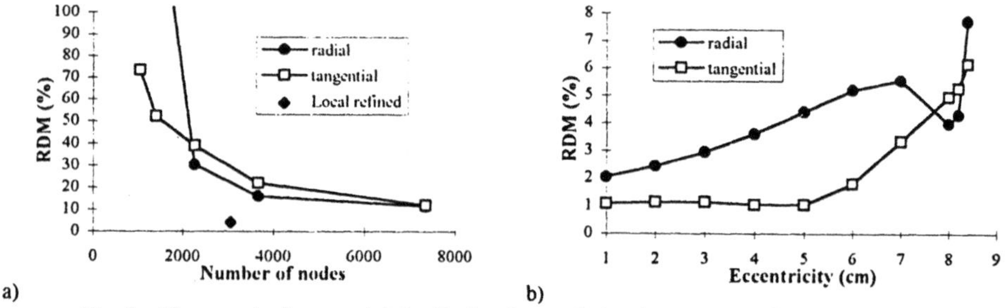

Fig. 2 a. The error in the potential distribution for a radial and a tangential dipole as a function of the number of nodes. The dipole has an eccentricity of 80 mm; b. The error in the potential distribution for a locally-refined mesh. The error is given as a function of the eccentricity of a tangential or radial dipole.

The magnetic field is calculated from the potentials generated by tangential dipoles, since for radial dipoles the magnetic field is zero outside a spherical symmetric volume conductor. As for the potential, using a regular mesh, the error in the magnetic field increases when the dipole is closer to the surface of the inner sphere. In this case an increase in the RDM value is also observed for dipoles closer to the center, as the magnetic field is smaller for such dipoles, and consequently the relative errors are larger. The results for the magnetic field are shown in Fig. 3a for a dipole at an eccentricity of 80 mm. Using the locally-refined mesh errors less than 3% are obtained for dipoles up to an eccentricity of 82 mm, which is 5 mm under the surface of the inner sphere. This can be seen in Fig. 3b, where the RDM for this mesh is given as a function of the dipole eccentricity.

A hole is modeled by giving some elements of the poor conducting layer the same conductivity as that of the surrounding media. Extra points are added to the mesh on the border of the holes. An impression of such a layer with a hole is given in Fig. 4.

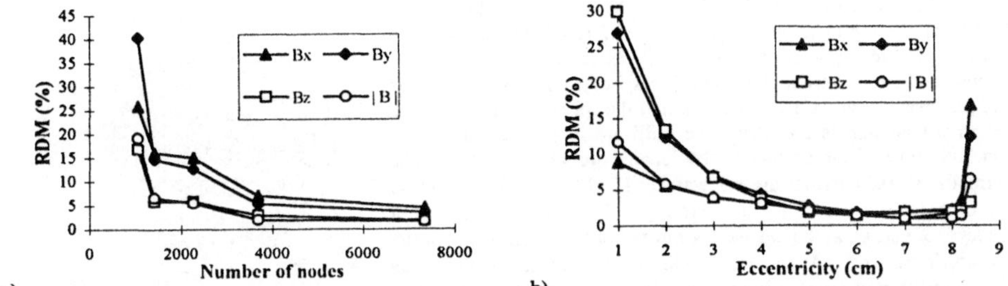

a) b)

Fig. 3 a. The error in three orthogonal components and the magnitude of the magnetic field for a tangential dipole as a function of the number of nodes. The dipole has an eccentricity of 80 mm; b. The error in the magnetic field for a locally-refined mesh. The error is given as a function of the eccentricity of the tangential dipole.

Results

We have used the method described above to estimate the influence of holes in the skull on the localization of dipoles from EEG's and MEG's. The head is modeled by three concentric spheres, describing the brain, the skull, and the scalp. The radii of the spheres are respectively 87, 92 and 100 mm, the normalized conductivities 1, 0.0125 and 1 S/m. A circular hole with a diameter of 30 mm is created in the skull layer. Forward calculations are carried out using tangential and radial dipoles on the z-axis situated at several eccentricities. These dipoles are shown in Fig. 5a., where the size of the arrows is a measure of their strengths. The azimuthal angle between the z-axis and the vector from the center of the spheres to the center of the hole is varied. Using this model, the radial component of the magnetic fields is computed in 59 points on a hemisphere with a radius of 120 mm at a constant distance to the outer surface. Subsequently, the results are used in an inverse calculation. The inverse calculations are based on a three-sphere model, where all spheres are closed, using an analytical expression. For the case that the azimuthal angle of the hole is 15 degrees, the results are depicted in Fig. 5b. As shown, the dipoles are practically the same as the original ones. This result is typical for all simulations carried out. In other words, the influence of the hole on the radial component of the magnetic field is negligible. The same procedure is followed for the potential distribution at the scalp. Results are shown in Fig. 5c, again for an azimuthal angle of the hole of 15 degrees. Not only the location, but also the orientation and the strength of a dipole are greatly influenced by the presence of the hole. It looks as if the dipoles are drawn towards the hole. These results are representative for all simulations.

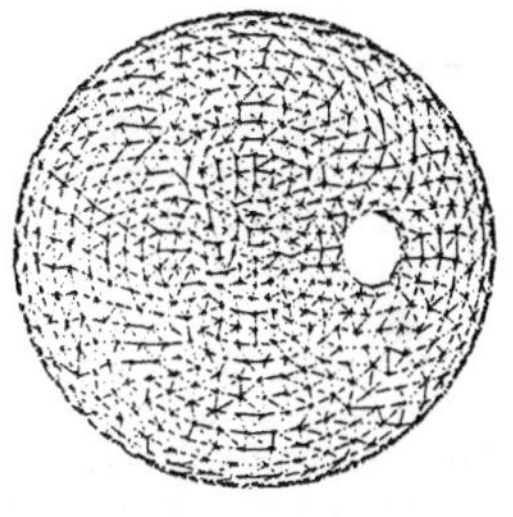

Fig. 4 The skull compartment with a hole at 30 degrees from the z-axis

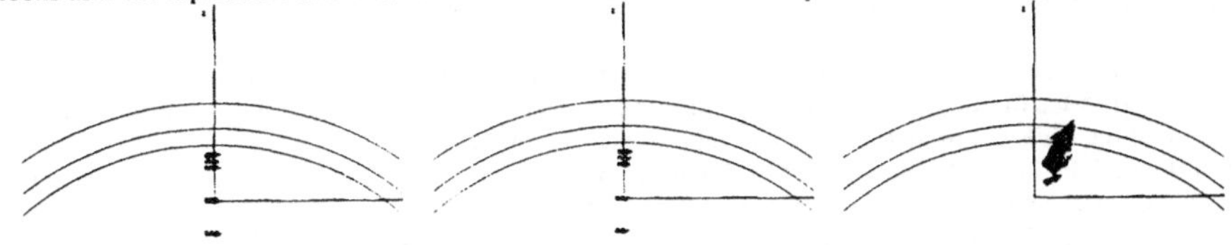

a) b) c)

Fig. 5 a. The dipoles used in the forward calculations; b. The dipoles found in the inverse solution based on the magnetic field; c. The dipoles found in the inverse solution based on the potential.

Simulations of fetal MCG's near the maternal abdomen and fetal ECG's at the maternal abdomen are performed. The model used consists of three spheres, describing the fetus, the layer of vernix caseosa and the maternal abdomen. The spheres describing the fetus and the layer of vernix are concentric and the sphere representing the maternal abdomen is eccentric. The radii are respectively 150, 160 and 300 mm and the centers of the spheres are 120 mm apart. The normalized conductivities are respectively 1, 10^{-6} and 1 S/m. A cross section of the this model is pictured in Fig. 6. Two holes are created in the layer of vernix, at the top and bottom end of the vertical axis of the concentric spheres. A current dipole at the center of the fetus compartment, and rotating in the x-z plane, represents a cycle of fetal heart activity. The mesh that is used consists of 3382 nodes. The magnetic field is calculated in 76 locations on a hemisphere, as indicated in Fig. 6. It is found that the potential on the maternal abdomen as a function of the dipole orientation, only varies in strength. The more the dipole is pointing towards a hole, the stronger the signal. The spatial distribution however, is almost constant during a heart cycle. The magnetic field also diminishes in strength when the dipole is not pointing towards a hole, but a direction dependence can still be seen. However, a clear influence of the holes is visible.

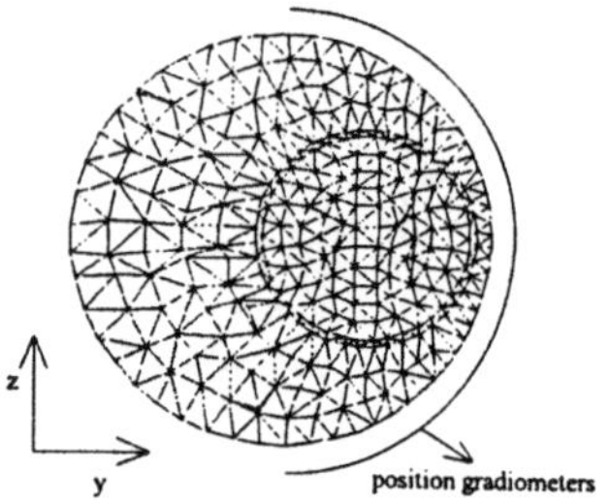

Fig. 6 A cross section of the mesh used to simulate the fetus

Discussion

Comparison with analytical solutions show that the combination of the FEM and the Biot-Savart law yields accurate results in computing the potential and magnetic field. Using a locally refined mesh, only 3073 nodes are needed to obtain an RDM of less than 5% in the potential and 3% in the magnetic field for dipoles 5 mm under the surface of the inner sphere.

Simulations of EEG's and MEG's show that the influence of a hole in the skull on the potential is large, and on the magnetic field small. Simulations of fetal MCG's and fetal ECG's show that both the magnetic field and the potential distribution are influenced by the presence of open patches in the poorly conducting layer of vernix. The spatial distribution of the potential on the maternal abdomen is almost independent of the orientation of the dipole describing the fetal heart activity. The magnetic field patterns, although clearly distorted, still show some source orientation dependence. In both cases the amplitude of the signal is much larger if the dipole is pointing towards a hole.

Values found in the literature for the electric properties of the vernix caseosa and for fat make it plausible that capacitive effects cannot be neglected. Therefore, the boundary condition used to solve the forward problem may not be valid anymore. In this case, the layer will function as a high-pass filter. This may further distort the signal.

References

[1] Oostendorp, T.F. Modelling the fetal ECG. Ph.D. Thesis, Catholic University of Nijmegen. The Netherlands, 1989

[2] Bertrand, O., Thevenet, M., Perrin, F. and Pernier, J. Effects of skull holes on the scalp potential distribution evaluated with a finite element model. Satellite Symp. on Neuroscience and Technology, 14th An. Int. Conf. of the IEEE Engin. in Medicine and Biology Soc., Lyon, 1992: 42-45

[3] van den Broek, S.P., Zhou, H., and Peters, M.J. Computation of neuromagnetic fields using the finite element method and the Biot-Savart law, Med. & Biol. Eng. & Comput., *in press*

[4] Zhou, H. Anisotropic Volume Conduction in Biological Media. Ph.D. Thesis, Catholic University of Nijmegen, The Netherlands. 1993:47-51

Estimation of Cardiac Sources on the Epicardium and Endocardium from Simulated and Measured MCG Data

Wach, P.[1], Tilg, B.[1], Lafer, G.[1], Kynor, D.[2] and Rucker, W.[3]

[1]Department of Biophysics, Institute of Biomedical Engineering, Graz, Austria; [2]Biomagnetic Technologies, Inc., San Diego, California; [3]Institute of Theory in Electrical Engineering, Stuttgart, Germany

Rationale

The estimation of the transmembrane potential (TMP) on the epicardium and endocardium from simulated and measured magnetic heart field data using the Tikhonov's regularization approach is investigated including an inhomogeneous boundary-element (BE) thorax model. For the generation of synthetic normal and pathological magnetic field data a model of the entire human heart is applied. The model is positioned inside the subject's thorax. Additionally from this normal subject the magnetic field data were collected by two Magnes® 37 channel gradiometer probes on the anterior and posterior side of the thorax. Synthetic magnetic field data for the normal case, for an anteroseptal acute phase of infarction and for a Wolff-Parkinson-White (WPW) syndrome are simulated on the same sensor positions.

Methods

The volume integral containing the primary current density can be substituted by a surface integral of the TMP distribution on the heart's surface in case of electrical isotropy [2]:

$$\int_{V_{Heart}} \vec{J}_o(\vec{r}') \times \nabla G(\vec{r},\vec{r}')\,dV_{Heart} = \tilde{\kappa}_{in} \oint_{A_{Heart}} \Phi_m(\vec{r}')\left(\nabla G(\vec{r},\vec{r}') \times \vec{n}(\vec{r}')\right)dA_{Heart} \quad (1)$$

$\vec{J}_o$	impressed primary current density	$\vec{r}$	observation point
$\vec{r}'$	source point	$\vec{n}$	normal vector
Φ_m	transmembrane potential	$\tilde{\kappa}_{in}$	effective intracellular conductivity

Following the uniform double layer (UDL) theory (for electrical isotropy the primary current dipoles are always oriented perpendicularly to the wavefront) no magnetocardiogram (MCG) and electrocardiogram (ECG) can be detected above the thorax as long as the excitation wave does not impinge on the heart's surface. In case of the oblique dipole layer (ODL) theory (in this case the impressed dipoles are directed obliquely to the wavefront) an additional small contribution to the MCG and ECG is generated. According to the UDL theory the signal generators are located on the heart's surface only [2, 3, 4]. Consequently one can try to estimate the transmembrane potential distribution on the endo- and epicardium for reconstruction. Single- and multiple dipole structures used in inverse solutions can be derived from the distribution of the transmembrane potential on the heart's surface at least qualitatively.

The main advantage of this TMP approach is the fact that the number of unknowns is reduced by a factor of 3 in comparison to pseudo primary current dipole estimation if the direction of the dipoles is unrestricted. The minimum-norm solution of the estimated transmembrane potential is given by:

$$\hat{\Gamma} = L^T(LL^T + \varepsilon^2 E)^{-1} B \quad (2)$$

B is the magnetic induction data vector, **L** the magnetic lead field matrix, and **E** the unity matrix. The regularization parameter ε is determined by the L-curve method [5]. The algorithm is applied in a first step to synthetic MCG data computed by a cellular automaton model of the entire human heart [6]. The heart model is positioned inside the thorax model with a position similar to the subject's heart. Synthetic MCG data are simulated for a normal case, for an anteroseptal ischemia, and for a WPW-syndrome at the locations of the 74 detectors. In the second step the inverse algorithm is applied to the real world MCG data. The MCG for a normal subject was simultaneously measured by two 37 first-order axial gradiometer probes (Magnes®, BTi, Inc.) on 74 observation points on the anterior and posterior side of the thorax. An inhomogeneous thorax model [1, 7, 8] of the investigated subject is built up from magnetic resonance imaging (MRI) cross-sections by 278 isoparametric quadrilateral 8-node BEs for modeling the outer thorax surface and the surfaces of the two lungs [8]. The influence of the blood mass is neglected so far. The epicardium and endocardium are extracted from cine MRI cross-sections.

Results

Fig. 1 depicts the MCG time course for the normal case (solid line), acute phase of infarction (dash-dotted line) and for the WPW-syndrome (dashed line) at channel no. 1 of the anterior probe. In case of the WPW-syndrome the accessory pathway is located on the right ventricular free wall. As can be seen the characteristical δ-wave can be simulated. In this case the depolarization wave spreads through the pathway and activates thereby the right ventricle much earlier than in the normal case. For the acute phase of infarction the injury currents are flowing from the normal to the ischemic tissue and produce a deviation in the MCG and ECG time course during the ST-segment. Simulating myocardial ischemia the conduction velocities are reduced in the ischemic central zone by 40% and in the ischemic border zone by 20%. The intracellular effective conductivity is not changed from its normal value. In comparison to the normal case the action potential is modelled by a more positive resting membrane potential and a smaller action potential duration. After having calculated all the synthetic MCG maps TMP estimation is performed on the epicardium and endocardium of the heart model geometry. Fig. 2. depicts the TMP distribution at 72ms after the onset of VD (ventricular depolarization) for the normal case (left panel) and for the acute phase of infarction (right panel) from an anterior view. The TMP distribution on the epicardium is shown in a grey level pattern and the darkest grey level represents the maximal TMP. The distribution on the endocardium is not shown. In the normal case at 72ms the depolarization wave impinges on the right anterior heart base. In case of the anteroseptal ischemia (right panel) at 72ms an area with a positive TMP can be reconstructed simultaneously in the pathological area because of its delayed activation and near the right anterior heart base.

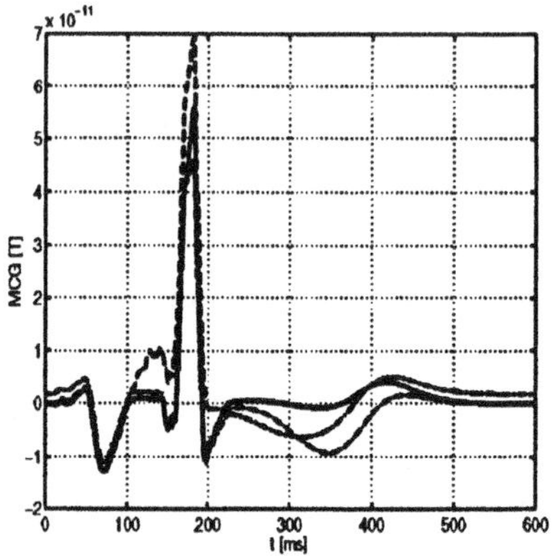

Fig. 1. Simulated MCG time course for the normal case (-), WPW-syndrome (--) and for the acute phase of infarction (· -).

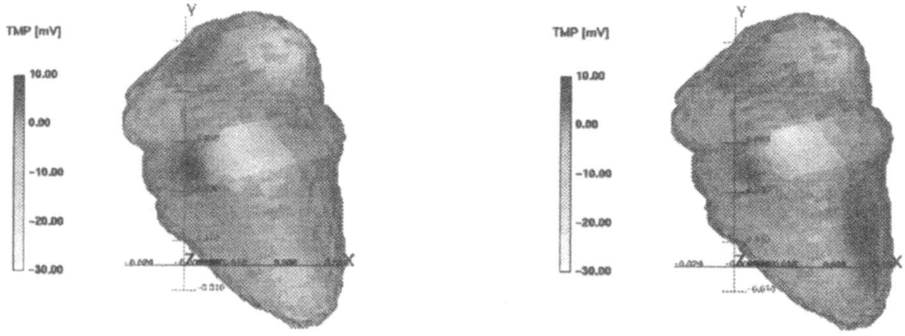

Fig. 2. TMP distribution for the normal case (left panel) and for the anteroseptal acute phase of infarction (right panel) at 72ms after the onset of ventricular depolarization.

For the WPW-syndrome an area with a positive TMP can be located on the endocardium and epicardium in the right ventricular free wall, when the excitation spreads through the accessory pathway.

In a next step reconstructions during ventricular depolarization are applied to measured (50 heart cycles are used for averaging) normal MCG data. The atrial and ventricular epicardium and endocardium of the normal subject's heart are modelled again by 232 and 368 BEs, respectively. The resulting 1806 nodes of the BE structure are used as reconstruction points. Fig. 3 depicts the TMP distribution 4.8ms after the onset of VD. As can be seen signal generators can be located on the apical endocardium of the right ventricle.

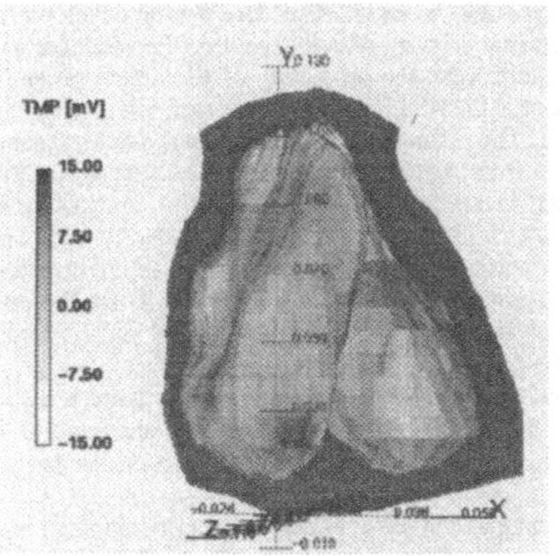

Fig. 3. TMP distribution at 4.8ms after ventricular depolarization on the endocardium of the normal subject's heart.

The epicardium is depicted only schematically. At this instant no signal sources can be reconstructed on the epicardial heart surface. By performing reconstructions for later time instants breakthroughs on the right and left epicardium can be observed. On the left panel of Fig. 4 the TMP distribution on the epicardium is depicted at 67ms after the onset of VD when the depolarization wave activates the myocardial mass near the left heart base. The corresponding signal generators can be located on the endocardium and epicardium of the left ventricle. Later when the depolarization wave spreads towards the middle anterior heart base and the right heart base the corresponding areas with the positive TMPs can be located there. On the right panel of Fig. 4 the TMP at 88ms after the onset of VD is depicted. At this instant the activated region is near the right heart base. In Fig. 4 the TMP distribution on the endocardium is not shown.

A seperation of the activity on the endocardium from the opposite epicardium was possible in reconstructions with synthetic TMP distributions. Regarding measured MCG data a more detailed investigation must be done in further studies to clarify this important question. Because of the close distance of the endocardium and epicardium (sometimes only a few millimeters) such a seperation needs the inclusion of specific routines in the optimization procedure. In Fig. 4. the reconstructed activity on the atrial epicardium results from the atrial repolarization during the QRS complex.

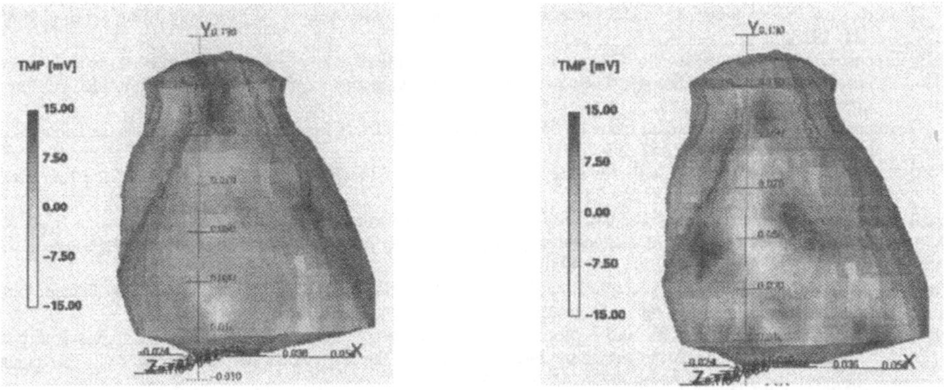

Fig. 4. TMP distribution at 67ms (left panel) and at 88ms (right panel) on the epicardium of the normal subject's heart.

In simulation studies with the heart model it was found that the reconstructed TMP solutions are in good agreement with the propagation sequence and the TMP distribution modelled in the forward problem. The same inverse methods as used for analyzing synthetic MCG data are applied to measured MCG data. From the results achieved one can conclude that the method presented is well suited for the imaging of the real signal generators on the epicardium and endocardium during the process of depolarization and repolarization.

Discussion

The heart model used in forward simulation studies turned out to be an excellent tool for the development of inverse algorithms because of its capability to model a rather realistic primary source structure. The model allows the computation of the activation sequence [6] and thereby (together with the modelling of the action potential shape in dependence on the cycle length) the transmembrane potential distribution on the endocardium and epicardium at any time instant during depolarization and repolarization. The estimation of the transmembrane potential turned out to be a very effective alternative to pseudo primary current dipole estimation. It was found that this transmembrane potential approach enables a much higher contrast in the reconstructed solution in comparison to the non-iterative minimum-norm pseudo primary current dipole search [8]. In the approach presented no temporal regularization was included so far. Furthermore the solution vector was not restricted to the interval -80mV (resting membrane potential) to +10mV (overshoot potential). Temporal regularization as treated in e.g. [3, 4] for the estimation of the activation times should be included in future studies in order to consider the physiological interdependence of the solutions during subsequent time instants of reconstruction.

Results obtained from measured MCG data are qualitatively in good agreement with the pattern of the spread of excitation in the ventricular mass of a normal subject. In future research work these results have to be validated by invasive techniques, e.g. measurements of the potential or of the activation time map on the heart's surface.

Although up to now biomagnetic cardiac functional imaging is not used in clinical routine applications, some excellent research work has been done in basic research. From the authors point of view further basic research has to be performed during the next years in order to create a new powerful, fully noninvasive functional diagnostic tool for cardiologists.

References

[1] Ferguson, A.S., Zhang, X., and Stroink, G. A complete linear discretization for calculating the magnetic field using the boundary element method. IEEE Transactions on Biomedical Engineering, 1994, 41(5): 455-460.

[2] Geselowitz, D.B. A bidomain model for anisotropic cardiac muscle. Annuals of Biomedical Engineering, 1985, 11: 191-206.

[3] Greensite, F., Huiskamp, G., and Oosterom van A. New quantitative and qualitative approaches to the inverse problem of electrocardiology: Their theoretical relationship and experimental consistency. Medical Physics, 1990, 17(5): 569-579.

[4] Greensite, F. The mathematical basis for imaging cardiac electrical function. Critical Reviews in Biomedical Engineering, 1994, 22: 347-399.

[5] Hansen, per Ch. Numerical tools for analysis and solution of Fredholm integral equations of the first kind. Inverse Problems, 1992, 8: 849-872.

[6] Killmann, R., Wach, P., and Dienstl, F. Three-dimensional computer model of the entire human heart for simulation of reentry and tachycardia: gap phenomenon and Wolff-Parkinson White syndrome. Basic Research in Cardiology, 1991, 86: 485-501.

[7] Nenonen, J.T., Hämäläinen, M.S., and Ilmoniemi, R.J. Minimum-norm estimation in a boundary-element torso model. Medical and Biological Engineering and Computing, 1994, 52: 45-48.

[8] Tilg, B., Wach, P., Rucker, W., and Kynor, D. Biomagnetic functional localisation: iterative approach to estimation of electrical sources within the human heart from the magnetocardiogram. Medical and Biological Engineering and Computing, 1995, 55: 258-240.

Acknowledgement

This work was supported by the Austrian Science Foundation, grant P10749. Real world MCG and MRI data were made available by Biomagnetic Technologies, Inc., 9727 Pacific Heights Blvd., San Diego, CA 92121-5719.

Current Density Reconstructions Using the L1 Norm

Wagner, M., Wischmann, H.-A., Fuchs, M., Köhler, T. and Drenckhahn, R.

Philips Research Hamburg, Röntgenstr. 24-26, D-22335 Hamburg, Germany

Introduction

A major goal in biomagnetism is the reconstruction of current distributions without making preliminary assumptions about number and temporal properties of the sources to be reconstructed. We propose a new current density reconstruction method that computes source images with high resolution and small blurring. A nonlinear functional, the L1-norm, is used as a source constraint. The L1-norm is the sum of absolute current densities. A simple simulation shows, that the L1-norm does neither impose artificial smoothness nor sharpness upon the reconstructed sources. We have implemented three different minimization schemes for the L1-norm, which we compare against each other and against the MNLS (minimum norm least squares, L2-norm) method [1]. Unlike the L1-norm method described earlier [2], our approaches tolerate noisy data and arbitrary source orientations.

Methods

A current distribution may be discretized into a large number of dipolar sources covering the space of possible activity. To account for noisy data, regularization techniques must be used. A χ^2 data term and a model term are minimized simultaneously. In matrix formulation, we have to minimize

$$\Delta^2(\underline{J}) = \left\| \underline{\underline{C}}\left(\underline{M} - \underline{\underline{L}}\,\underline{J}\right)\right\|^2 + \lambda f_p\left(\underline{\underline{W}}\,\underline{J}\right) \tag{1}$$

The first term is the data term, the second is the model term. For m sensors, n a priori possible source locations, and c source components per location, $\underline{M}$ is the m-dimensional vector of measured EEG and MEG data. $\underline{J}$ is the nc-dimensional vector of unknown currents. $\underline{\underline{L}}$ is the m x nc lead field matrix modeling the Biot-Savart law and the conducting properties of the human head. $\underline{\underline{C}}$ is the m x m sensor equalization and EEG/MEG combination matrix given by the Cholesky decomposition of the sensor covariance matrix $\underline{\underline{S}}$:

$$\underline{\underline{C}}^T \underline{\underline{C}} = \underline{\underline{S}}^{-1} \tag{2}$$

The regularization parameter λ is adjusted so that the (gaussian distributed) χ^2 data term matches its expectation value of m within its standard deviation $(2m)^{1/2}$. $\underline{\underline{W}}$ is the diagonal nc x nc lead field normalization matrix as in [3], p denotes the used norm, and $f_p(\underline{X})$ is the constraint model itself:

$$f_p(\underline{X}) = \left(\sum_{i=1}^{n} h_p\left(\left|\underline{X}_i\right|\right)\right)^{1/p} \tag{3}$$

where $\underline{X}_i$ is the c-dimensional subvector of weighted source components at location i. For the p-Norm,

$$h_p(x) = |x|^p \tag{4}$$

For the minimization of Δ^2 using the L1-norm (p=1), we have implemented three different approaches. The first two methods use iteratively modified model terms to overcome the discontinuity in $h_1(x)$ at x=0. This discontinuity can trap optimizers evaluating gradient information, such as the conjugate gradient method. In the first approach, we start with p=2, which is equivalent to MNLS. Then p is slowly reduced to 1, while the result of each minimization serves as the initialization of the next one.

In our second approach, we use a modified term $h_1(x,\xi)$ instead of $h_1(x)$. In $h_1(x,\xi)$, the portion of $h_1(x)$ between $-\xi$ and ξ is replaced by a quadratic term. In an iterative minimization, ξ starts at an arbitrary value higher than the largest MNLS current and is reduced to some small ε. Again, the result of the preceding iteration is used to initialize the next one:

$$h_1(x,\xi) = \begin{cases} \dfrac{1}{2\xi}\left(x^2 + \xi^2\right) & \text{for } |x| < \xi \\ |x| & \text{else} \end{cases} \tag{5}$$

The third approach introduces a modified data term and a modified $\underline{J}$: Instead of minimizing eq. 1, we

minimize $f_1\left(\underline{\underline{W}}\,\underline{J}^+\right)$ under the constraint $\left\| \underline{\underline{C}}\left(\underline{M} - \underline{\underline{L}}\,\underline{J}^+\right)\right\|_\infty \le \sigma_{\max}$ (6)

where σ_{max} is the largest permissible normalized deviation. σ_{max} is chosen so that $\chi^2=m$ (see eqs. 1,2). All elements of $\underline{J}^+$ are positive:

$$j_k^+ \geq 0 \text{ for } 1 \leq k \leq nc \tag{7}$$

The number of components c must now be chosen to fit the positivity constraint on $\underline{J}^+$. For given source orientations, e.g. cortical constraints [4], we need two opposed components of $\underline{J}^+$ per location:

$$c{=}2:\ \underline{J}_i^+ = \begin{bmatrix} j_{i,1} & j_{i,2} \end{bmatrix}^T \text{ with } j_{i,1} \geq 0 \text{ and } j_{i,2} \geq 0 \tag{8}$$

where $\underline{J}_i^+$ is the c-dimensional subvector of source components at location i. For the resulting source vector at location i, $\underline{J}_i$, and a normalized source orientation $\underline{N}$, we obtain the relations

$$\underline{J}_i = \underline{N}(j_{i,1} - j_{i,2}) \text{ and } \|\underline{J}_i\|_1 = j_{i,1} + j_{i,2}\,. \tag{9}$$

If the source orientations are not known a priori, the source space must still be reached by the components of $\underline{J}^+$. We need a planar arrangement of components for MEG reconstructions in a spherical volume conductor model, and a three-dimensional arrangement if EEG data or realistical volume conductor models [5] are used. In analogy to eq. 9,

$$\|\underline{J}_i\|_1 \overset{!}{=} \sum_{k=1}^{c} j_{i,k} \tag{10}$$

which can only be achieved asymptotically for large c. For a planar arrangement, c=8 may be used, which means that the elementary sources point to the corners of an octagon. For 3D arrangements, sources may span an icosahedron (c=12) or a dodecahedron (c=20). For components $j_{i,k}$ and normalized orientations $\underline{N}_k$, we get

$$\underline{J}_i = \sum_{k=1}^{c} \underline{N}_k j_{i,k} \tag{11}$$

Now that the nonlinearity in $h_l(x)$ has been transformed into the positivity constraints in eq. 8, a linear optimization problem with inequality constraints remains. The simplex technique of linear programming [6] solves the corresponding combinatorial problem. Eq. 6 yields the optimization function plus 2 inequality constraints per sensor: ($[\underline{X}]_i$ here denotes the i-th element of $\underline{X}$)

$$-\sigma_{max} \leq \left[\underline{\underline{C}}\left(\underline{M} - \underline{\underline{L}}\,\underline{J}^+\right)\right]_i \text{ and } \left[\underline{\underline{C}}\left(\underline{M} - \underline{\underline{L}}\,\underline{J}^+\right)\right]_i \leq \sigma_{max} \text{ for } 1 \leq i \leq m \tag{12}$$

There will be at most m currents that are not zero, because only one constraint per sensor can be critical. From eqs. 1 and 6 follows that, for a data term of m, σ_{max} must be ≥1, depending upon the number of critical constraints.

Simulations and Experiments

In order to compare different norms, we have computed data and model terms for a simple model ($m=3$, $n=7$, $c=1$, $\lambda=1$, $\underline{\underline{C}}=\underline{\underline{I}}$, $\underline{\underline{W}}=\underline{\underline{I}}$) and five source configurations. The goal was to determine, which of three values for p (0.5 or 1 or 2) lead to the minimum Δ^2 for the correct solution. Two simulations were carried out: In case 1, a point source was simulated, and in case 2, a distributed source was simulated. The left columns of table 1 show the forward solutions and data terms for five source configurations per case: a) a point source, b) a distributed source, c) an even more distributed source, d) the simulated source, but at a different location, and e) a large point source and two nearby sources with opposite sign. The simulated sources (1.a and 2.b) and their forward terms are printed in boldface. Of course, their data terms are zero.

In order to test the three different minimization schemes, we performed a second set of simulations. Synthetic measurements for four different MEG systems were generated: a 31-channel system with axial gradiometers and coplanar coils (Philips), a 37-channel system with axial gradiometers and cospherical coils (BTI), a 109-channel helmet system with axial gradiometers, and a 2x61-channel helmet system with planar gradiometers. Two orthogonal dipoles were located in a horizontal plane 40mm above the volume conductor center with a distance of 40mm. Noise was added to obtain a signal-to-noise ratio of 30. A spherical volume conductor was used. For the reconstructions, source locations were distributed on an 11x11x11 grid centered around the dipole positions with dx=dy=10mm and dz=5mm. This resulted in 1159 locations inside the sphere. Fig. 1 a-d) shows the geometries used.

Finally, we compared the linear programming method to MNLS and the single equivalent dipole model in the reconstruction of the M20 peak of the SEF in an MEG measurement performed with the Philips system. The signal-to-noise ratio was 18, a realistical volume conductor was used, and 9776 locations on the segmented cortex, together with their cortical surface normals [4], were used for a cortically constrained reconstruction. Fig. 1 e) shows the measurement setup.

Table 1: data, model, and Δ^2 terms (eq. 1) of a point source (case 1) and a distributed source (case 2), each compared to four other source configurations.

case	sources	forward solution		p=0.5		p=1		p=2	
	J^T	$(LJ)^T$	data	model	Δ^2	model	Δ^2	model	Δ^2
1. a)	**[0,0,0,3,0,0,0]**	**[7.68,12.00,7.68]**	**0.00**	3.00	**3.00**	3.00	**3.00**	3.00	3.00
b)	[0,0,1,1,1,0,0]	[7.72,11.57,7.72]	0.19	9.00	9.19	3.00	3.19	1.73	**1.92**
c)	[0,½,½,1,½,½,0]	[7.73,11.03,7.73]	0.95	14.66	15.61	3.00	3.95	1.41	2.36
d)	[0,0,0,0,3,0,0]	[5.75,11.36,9.72]	8.30	3.00	11.30	3.00	11.30	3.00	11.30
e)	[0,0,-1,5,-1,0,0]	[7.64,12.43,7.64]	0.19	17.94	18.13	7.00	7.19	5.20	5.39
2. a)	[0,0,0,3,0,0,0]	[7.68,12.00,7.68]	0.19	3.00	**3.19**	3.00	3.19	3.00	3.19
b)	**[0,0,1,1,1,0,0]**	**[7.72,11.57,7.72]**	**0.00**	9.00	9.00	3.00	**3.00**	1.73	1.73
c)	[0,½,½,1,½,½,0]	[7.73,11.03,7.73]	0.19	14.66	14.85	3.00	3.19	1.41	**1.60**
d)	[0,0,0,1,1,1,0]	[5.87,11.03,9.59]	7.21	9.00	16.21	3.00	10.21	1.73	8.94
e)	[0,0,-1,5,-1,0,0]	[7.64,12.43,7.64]	0.74	17.94	18.68	7.00	7.74	5.20	5.94

Results

The right columns of table 1 show the results of the comparisons of norms: For each case, the model and the Δ^2 terms of eq. 1 were evaluated for p=0.5, p=1, and p=2. The aim was to find the value of p, that minimizes Δ^2 for the correct source configuration. In case 1, p=0.5 and p=1 both gave the correct solution. In case 2, there was a bias towards more focal sources for p=0.5, and towards more distributed sources for p=2. Only p=1 showed no bias.

Fig. 2 shows reconstruction results for two orthogonal dipoles and four different reconstruction techniques for the Philips system. Computation times on a sparcstation 10/51 were 3s for MNLS [3], 160s for variation of p, 140s for variation of ξ, and 60s for the linear programming approach. MNLS reconstructed a much more smeared distribution of activity than the L1-norm, and linear programming gave the most focal results within the shortest computation times. Fig. 3 shows reconstruction results for all simulated MEG systems using linear programming.

Fig. 4. shows cortically constrained SEF reconstructions. The L1-norm result clearly confirms that the single dipole model is appropriate. MNLS again reconstructed a smeared source image.

Discussion

The first set of simulations shows, that the L1-norm is the best source model in the absence of a priori knowledge. It reflects the additive superposition of electromagnetic fields and neither imposes artificial smoothness nor sharpness. Silent and large compensating sources are nevertheless rejected completely.

The second set of simulations reveals, that helmet systems provide a slightly better localization compared to quasiplanar sensor arrays. For the Philips system, one of the simulated sources was spread over three distinct locations. Linear programming is especially suited for the reconstruction of cortical sources, where source

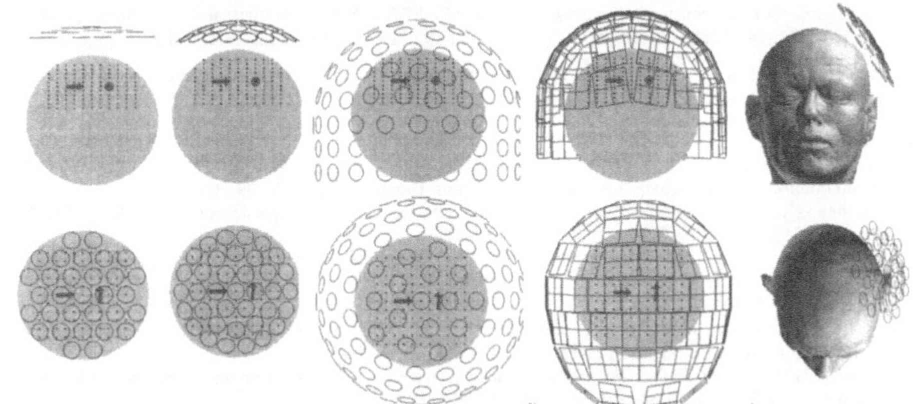

a) b) c) d) e)

Fig. 1: sensors, volume conductors, sources, and reconstruction grids, for a) Philips, b) BTI, c) helmet system, axial gradiometers, d) helmet system, planar gradiometers, top: frontal view, bottom: top view

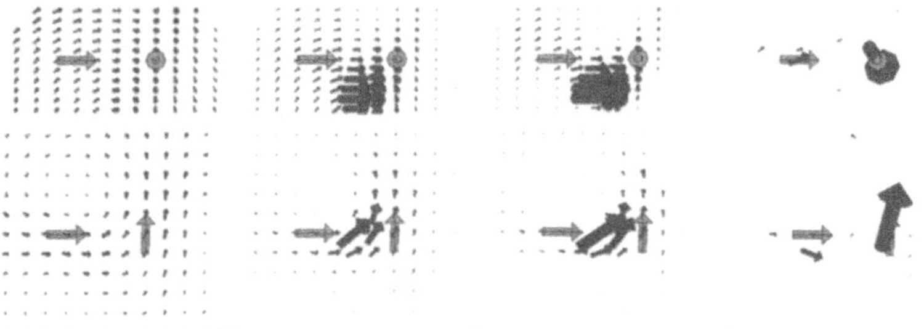

a) b) c) d)

Fig. 2: reconstruction results (dark gray, not to scale) for orthogonal dipoles (light gray), Philips system, a) MNLS, b) L1, lim $p \to 1$, c) L1, lim $\xi \to 0$, d) L1, linear programming, top: frontal, bottom: top view

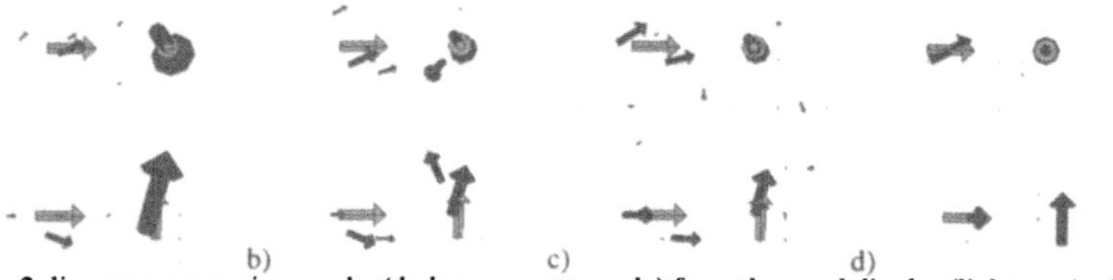

a) b) c) d)

Fig. 3: linear programming results (dark gray, not to scale) for orthogonal dipoles (light gray), a) Philips, b) BTI, c) helmet, axial gradiometers, d) helmet, planar gradiometers, top: frontal view, bottom: top view

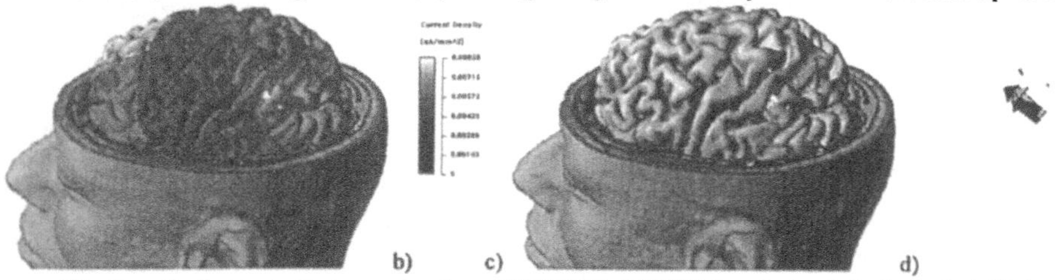

a) b) c) d)

Fig. 4: cortically constrained reconstructions of SEP M20 a) MNLS (overlay) and single dipole (white), b) MNLS colorscale, c) L1-norm (white), d) L1-norm (dark gray, not to scale) and single dipole (light gray)

orientations are fixed to the normals of the segmented cortical layer. The linear programming method replaces the a priori unknown parameter λ by σ_{max}, which is known to have a value between 1 and 5.

References

[1] Wang, J.Z., Williamson, S.J., Kaufmann, L., Magnetic source images determined by a lead-field analysis: the unique minimum-norm least-squares estimation, IEEE Trans. Biomed. Eng., 1992, 39:665-675.

[2] Matsuura, K., Okabe, Y., Selective minimum-norm solution of the biomagnetic inverse problem, IEEE Trans. Biomed. Eng., 1995, 42:608-615.

[3] Fuchs, M., Wagner, M., Wischmann, H.-A., Generalized minimum norm least squares reconstruction algorithms, ISBET Newsletter (ISSN 0947-5133), 1994, (5):8-11.

[4] Wagner, M., Fuchs, M., Wischmann, H:-A., Ottenberg, K., Dössel, O., Cortex segmentation from 3D MR images for MEG reconstructions, In: C. Baumgartner et al, Biomagnetism: fundamental research and clinical applications, Amsterdam, Elsevier/IOS Press, 1995, 433-438.

[5] Meijs, J.H.W., Weier, O.W., Peters, M.J., van Oosterom, A., On the numerical accuracy of the boundary element method, IEEE Trans. Biomed. Eng., 1989, 36:1038-1049.

[6] Kuenzi, H.P., Tzschach, H.G., Zehnder, C.A., Numerical methods of mathematical optimization, New York, Academic Press, 1971.

Modelling Closed Surface Geometry for Skull and Torso Using Snapping

Yamrom, B.

GE Research and Development Center, Schenectady, New York, USA

Introduction

Many branches of biomedical science and engineering require geometric modelling of human body parts. The success of magnetic resonance imaging (MRI) and other imaging modalities of computer-aided tomography has lead to many achievements in the scientific visualization of parts of the human body [1],[2]. Typical methods based on slice data generate geometries containing hundreds of thousands triangles that provide excellent details for 3D visualizations, but cannot be used directly for scientific analysis with numerical methods because of to their huge size and lack of "smoothness."

Certain methods exist that reduce substantially (more than 90%) the amount of triangles in the model [3], but the geometries that these methods produce are still unsatisfactory for the purpose of analysis. These methods were designed specifically to preserve the topology of original geometry produced from slice data and, therefore are subject to possible noise influence that results in holes and other irregularities of geometry. On the other hand, to reduce the amount of triangles in the model, these methods force constraints on the aspect ratio of the triangles in the model that leads to the meshes with narrow triangles, which is unsatisfactory for the application of numerical methods.

Finally, these methods still do not reduce the amount of triangles to the desired low level. To alleviate these problems, scientists use drastically simplified models for their analysis. These models may either be spheres or cylinders of appropriate size and position [4]. We have created a method that generates models to fit a niche between these two extremes. The models generated with this method satisfy the requirement of computational simplicity and at the same time can approximate the analyzed organs with much better fit than spheres and cylinders. Another approach to geometry construction can be found in [5].

Methods

The novel idea in the method is to use a closed regular mesh, a prototype, with a relatively small number of triangles that is placed inside or outside of "raw" complex model containing many thousands of triangles and "snap," or project nodes of small regular mesh onto the surface representing the "raw" data. By using this method we satisfy two conditions: (1) the model contains a small number of triangles; (2) the model is closed and fits the original surface. Because of details described below, any holes or irregularities in the original model are sealed and smoothed by the process of snapping of closed predefined mesh. There are two modifications of the basic idea specialized for two qualitatively different organ geometries. One type of geometry is represented by a torso that is an open-ended generalized cylinder and is drawn in Fig. 1 (3059 triangles, which is greatly reduced by decimation algorithm [3] from more than 200,000 triangles in the original data set produced by Cyberware [6] laser scanner) and another type of geometry is a skull and is drawn in Fig. 1 (4310 triangles in the decimated model). In the first case, a cylinder with ellipsoidal caps on top and bottom can be used for prototype (see Fig. 2 right). In the second case, a sphere triangulated by recursive subdivision of octahedron (see Fig. 1 left) is used. Because of differences in geometry, two slightly different algorithms are used.

The description and implementation of both algorithms are facilitated by a special data structure called compact cell structure [7]. The cell structure is a compact general data structure for representing n-dimensional topological constructs such as unstructured grids, polygonal, or triangle strip representations. The cell structure provides access methods for a wide variety of visualization algorithms. For purposes of this paper it is enough to say that the compact cell structure stores a list of 3D *points* representing nodes of triangular mesh of a surface together with lists of *edges* and triangles, or *elements*, of the triangulation. Several access methods are used in the algorithms: *get_cell_edge_neighbors, get_node_cells, get_cell_nodes, etc.* For example, *get_cell_edge_neighbors* returns an element neighboring specified element at specified edge and *get_node_cells* returns elements that use (contain) specified node. The full list of access methods is presented in [7].

Spherical snapping algorithm

In spherical snapping algorithm the user can select the position of the center of a sphere inside the surface that has to be approximated. Projecting a node of sphere triangulation onto the surface amounts to finding an intersection point between a ray from the center of the sphere passing through the node and the surface. The routine that computes intersection returns intersection point and the distance from it to the center of projection. In case of a hole in the corresponding direction the distance is set to 0. The intersection routine can be substantially accelerated by introducing additional hierarchy on top of the compact cell structure. Since both algorithms described below do not change the topology of the prototype (a sphere or a capped cylinder), its compact cell structure provides the final closed geometry with the original list of 3D nodes replaced by new nodes resulted from projection. The following steps constitute the spherical snapping algorithm:

1. Find a point P of the start node projected onto the surface.
2. If the start node does not project to the surface, find the first node in the sphere's node list that does by looping through all nodes.
3. Let *intersection* be the distance from the center of projection to the point P found in 1 or 2.
4. Find a list of triangles in the sphere triangulation containing the starting node or the node found in 2. Let *element* be the first triangle in that list.
5. Call routine process_element(*element, intersection).*

Six steps constitute the recursive subroutine process_element(e, t):

1. Mark e as visited element.
2. Let p_1, p_2 and p_3 be element's e nodes.
3. For each node p_i find distances d_i from the center of projection to the projected point on the surface. If for one of the nodes, distance does not exist (projection to a hole in the surface), or is such that $d_i \notin [s, 1/s]$, where s is a "smoothness" control parameter specified by the user, then set $d_i = t$.
4. Let e_1 be e's neighbor element at the edge (p_1, p_2). If e_1 is not visited process_element $(e_1, (d_1 + d_2)/2)$.
5. Let e_2 be e's neighbor element at the edge (p_1, p_3). If e_2 is not visited process_element $(e_2, (d_1 + d_3)/2)$.
6. Let e_3 be e's neighbor element at the edge (p_2, p_3). If e_3 is not visited process_element $(e_3, (d_2 + d_3)/2)$.

Marking of visited elements and closeness of the sphere triangulation guarantees that the call to recursive subroutine process_element in the spherical snapping algorithm will be terminated.

Cylindrical snapping algorithm

A similar, though slightly different, approach is used to generate torso model. Original torso geometry has vertical axial symmetry and two wide openings at the top and at the bottom. Spherical prototype with one point center of projection will create nonuniform distortion of triangles when they are projected onto the torso. A cylinder aligned with the torso axis and scaled to its height would be more useful in this case. Furthermore, instead of having one center of projection, we extend it along the cylinder axis. More specifically, we first add elliptical caps at the top and at the bottom of the cylinder to make it a closed surface; then for each point of a finite set of points along the axis of the cylinder we project a set of rays from this point in horizontal plane and snap cylinder nodes along these rays onto the torso surface. As in spherical snapping algorithm, we have to devise special processing of points that are projected to holes in the original geometry.

To simplify the description of the algorithm, we show how to build the top half of the mesh in Fig. 3 based on the top half of the torso (Fig. 1) and the top half of the capped cylinder. The closed surface is produced by repeating the application of this algorithm to the bottom half of the torso and merging two surfaces together.

Let *points* define a profile of capped cylinder in X,Y-plane and a cylinder mesh is generated by rotating the profile around the Y-axis. Profile points are ordered in increasing values of Y. Let the *resolution* be defined as the number of equidistant meridians in the cylinder mesh. In the algorithm description *point.x* and *point.y* are coordinates of profile points. The following steps constitute the cylindrical snapping algorithm:

1. $\Delta = 360 / resolution$.
2. *coeff = 1, old_radius = 0.*
3. For each *angle* in the set {Δ, 2*Δ, ...,*360*} do:
4. For each *point* in profile do:
5. origin = *(0,point.y,0), direction = (cos(angle),0 sin(angle)).*
6. Find the cylinder node corresponding to *point* and *angle.*

7. Calculate intersection P between the ray and the torso and set *d* to be the distance from P to Y-axis.
8. *radius = old_radius.*
9. If we have an intersection, snap the cylinder node to point P; *radius = d; coeff = point.x or 1, if point.x is 0; old_radius = d.*
10. If we do not have intersection, move the cylinder node to *direction * radius / coeff + origin.*
11. End of 4.
12. End of 3.

Results

We have implemented both algorithms and have provided a graphics-based user interface to them in the scope of Biomagnetic Analysis System developed at the GE Research and Development Center [8]. User interface allows for selection of various level of detail in sphere and capped cylinder triangulation. Choices of the number of triangles in the sphere are 8, 32, 128, and 512. The capped cylinder triangulation can be specified by axial and angular resolutions. The closed surface geometries derived from surfaces in Fig. 1 are presented in Fig. 3. For a skull original surface we are able to snap to it from inside and outside generating approximations to internal and external surfaces of the skull. The difference between internal and external snapping is in the choice of the closest and the farthest from the center of projection intersection points between a ray and the original surface.

Discussion

The algorithms we have described generate skull and torso closed surfaces used for solving inverse problems of biomagnetism. User controls the approximation by varying coarseness of prototype surface, smoothness parameter, and selection of the center of projection. Because of the nature of the algorithm, it is nonsensitive to the holes and topological irregularities of original geometries. One of the algorithms completely relies on the compact cell structure and uses recursive subroutine to accomplish construction; another uses special symmetry of prototype surface. Our approach in building closed surfaces is not limited to the choice of prototypes selected here for illustration: any simple closed surface may be used. With some experience the user can create a library of prototypes and can use one that is a better fit to approximated geometry.

The evaluation of quality of fit as a function of prototype triangulation and its positioning relative to approximated surface is beyond the scope of this project; however, this evaluation can be performed with the Biomagnetic Analyses System.

We have not discussed other aspects of model building related to original data collection. For our experiments we used CT scan and laser scan data. These are not always available, and a system of transformations can be developed that transforms one of several canonical models created and stored in advance to fit anthropomorphic measurements of a particular patient. The choice of those transformations and measurements can be a subject of separate study. Finally, though the algorithms were developed for the Biomagnetism Analysis System, they can be used to build geometrical models of human organs for other biomedical applications.

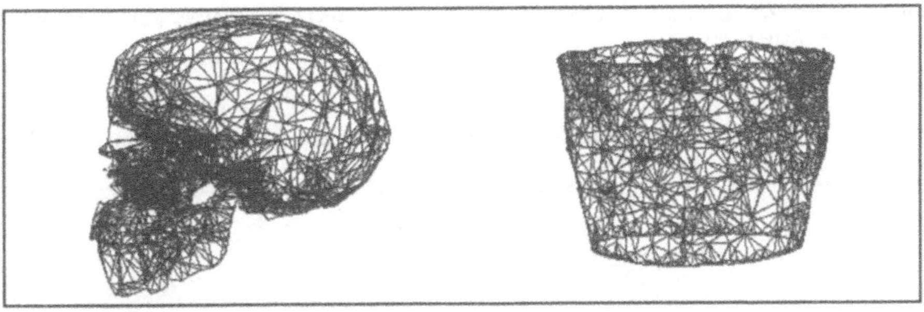

Fig. 1. Preprocessed skull (4310 triangles) and torso (3059 triangles).

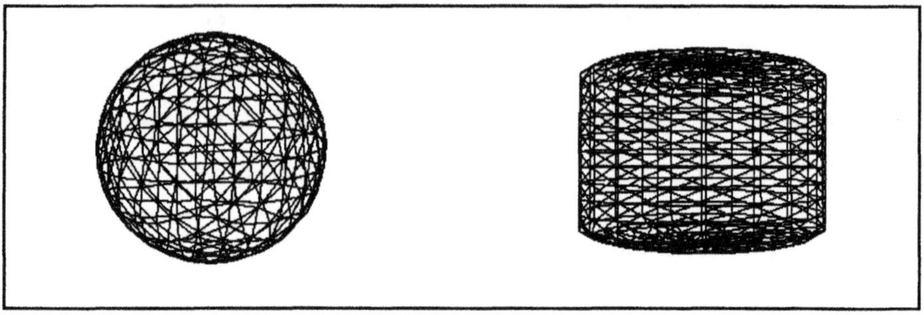

Fig. 2. Sphere and capped cylinder (512 and 880 triangles).

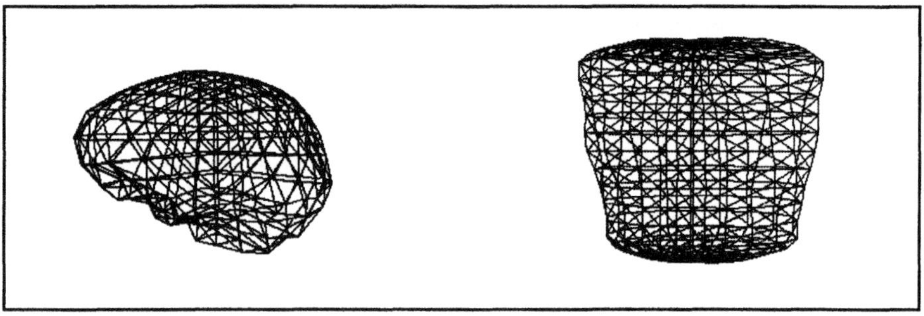

Fig. 3. Closed models of skull (512 triangles) and torso (880 triangles).

References

[1] Kerster, E. Imaging Systems for Medical Diagnostics, Berlin, Siemens, 1990.

[2] Lorensen, W.E., Cline, H.E., "Marching cubes: a high resolution 3D surface construction algorithm," Computer Graphics, 1987, 21 (3): 163-169.

[3] Schroeder, W.J., Zarge, J.A., Lorensen, W.E., "Decimation of triangles meshes," Computer Graphics, 1992, 26, (2): 65-70.

[4] Lutkenhoner, B., Pantev, C., Hoke, M., "Comparison between different methods to approximate an area of the human head by a sphere," In: Grandory, F., Hoke, M., Romani, G.L., Auditory Evoked Magnetic Fields and Electric Potentials, Basel, Karger, 1990.

[5] Hren, R., Vardy, D., Miller, R., Ferguson, S. Stroik, G., "Approximation of torso geometries by surface harmonic expansion," Biomag'93, 1993, 360-361.

[6] Cyberware, 2110 Del Monte Avenue, Monterey, CA 93940.

[7] Schroeder, W.J., Yamrom, B. "A compact cell structure for scientific visualization," Tutorial Notes, SIGGRAPH 95, 1995.

[8] Martin, K. et al., "Multiple modality biomagnetic analysis system," Biomag'96, Santa Fe, February 16-21, 1996.

Locally Refined Head Geometries Used to Evaluate Spherical Model Accuracy

Yvert, B., Bertrand, O., Thévenet, M. and Pernier, J.

"Brain Signals and Processes Laboratory", INSERM U280, Lyon, France

INTRODUCTION

An increasing number of studies on realistic head modelling for solving the inverse EEG and MEG problems have been emerging since the late 1980s. Recent evaluations of realistically meshed models showed that it was possible to correctly compute forward problems for any dipole depth and orientation, by using locally refined meshes [1, 2]. This being achieved, it is now possible to evaluate what error is made using various spherical models. A few previous works already adressed this question either using EEG [1,3] or MEG [4] data. These studies tested a small number of dipole locations, namely of the order of 10 and found localization errors of about 1cm for MEG and 1-2cm for EEG, assuming negligible errors on the forward solution. However, the numerical precision turns out to significantly depend upon the dipole orientation and depth within the brain.

The following study aims at systematically evaluating the localization accuracy that can be expected from the spherical model. About 2000 dipoles were randomly chosen on the segmented surface of a particular subject, obtained from MR images. For each dipole, the EEG forward solution was computed using a realistic head model being locally refined near the source location in order to ensure a good numerical accuracy. Then, inverse calculations were perfomed for each dipole using 4 different spherical models. Similar evaluations were investigated for 2 dipoles located in the vicinity of both auditory cortices. The influence of the number of electrodes and the effect of additional real EEG noise were particularly studied.

METHODS

The inner surface of the cortical gray matter of a particular subject was segmented using a region-growing algorithm applied to a stack of 1mm-spaced horizontal MR images. The surface resulted in a collection of 224,000 points (Fig 1). 2003 dipoles were randomly chosen on this surface such that their orientations were perpendicular to the local cortical surface. They were selected on 10 different 10mm-spaced horizontal slices, and in each slice, the average distance between 2 dipoles was 5mm.

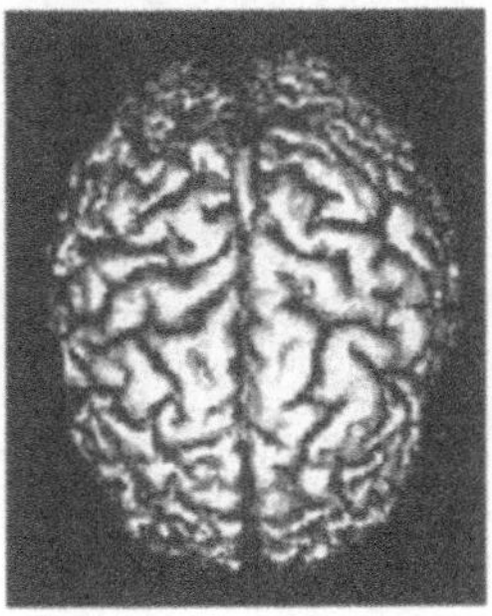

Fig 1 : segmented cortical surface (top view).

Realistic 3-shell Boundary Element models were used to perform forward calculations. For better computation accuracy, locally refined meshes had to be considered when dipoles were located at depths lower than 20mm below the model brain surface [2]. Hence, 2 types of model were considered : on one hand, one uniformly meshed model was used for deep dipoles, having a global element density of 1.3 triangles/cm2 (Fig 2a); on the other hand, 51 different locally refined models were used to cover all shallow dipole sites, all having a global density of 0.8 triangles/cm2 and a local density of 10 triangles/cm2 (Fig 2b). The size of the refined area was such that it covered all traingles at a distance less than 30mm from the dipole location. Forward solutions were computed at 63 electrode locations of the extended 10-20 system, that were positionned on each model.

For each dipole, inverse calculations were performed using 4 different analytic spherical models, with normalized radii : 1, 0.92, 0.87 (The conductivities of forward and inverse models were respectively 0.45, 0.05625, and 0.45 Siemens/m for the brain, skull and scalp shells).

Model 1 : Spherical model having its center roughly at the intersection of FPz-OZ and T3-T4, and whose external radius was the mean of all electrode radial coordinates.

Model 2 : Spherical model with external radius and center chosen to best fit all electrode locations.

Model 3 : "Multi-spherical" model having the same center as Model 2, and whose external radius was, for each electrode, the radial coordinate of the electrode.

Model 4 : Same model as Model 2, except that dipole coordinates were radially "stretched" toward a spherical spline passing through all electrodes.

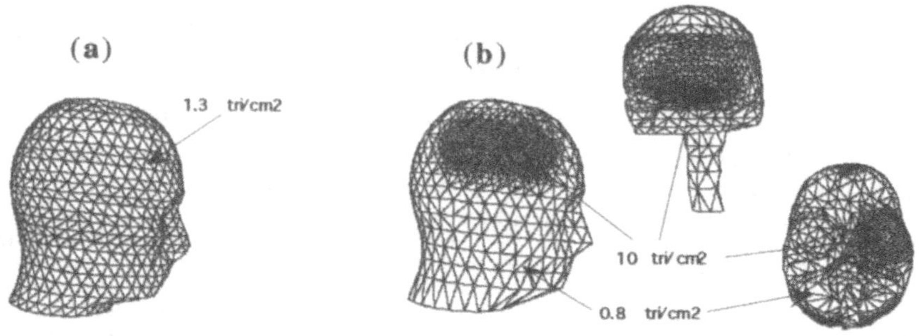

Fig 2 : (a) The uniformly meshed model used for all deep dipoles; (b) : 3 examples of locally refined models used for shallow dipoles.

ONE DIPOLE RESULTS

Localization and orientation errors obtained with each model in the absence of noise are given in Fig 3. 63 electrodes were considered. It turns out that errors are lower in the upper part of the head (5-6mm, 4-7 degrees) vs the lower part (15-25 mm, 10-20 degrees). Furthermore, in the lower part of the head, localization is better at the back than it is at the front (7-15mm vs 10-20mm).

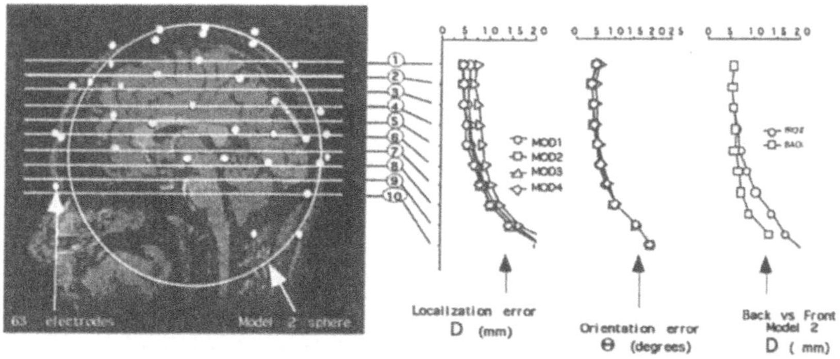

Fig 3 : localization (D) and orientation (Θ) errors for one dipole fit in the absence of noise. Each symbol represents the error averaged over all dipoles of the slice. Slices are shown by horizontal straight lines on the sagittal image. The graph on the far right reports the behavior of D with respect to the location of the dipole in each slice (back vs front of the head).

All models give significantly different errors. Using a paired wilcoxon test, it appears that Models 2 and 4 give the lowest localization errors for all slices ($p<0.0005$). Localization errors are of different types depending upon the dipole orientation : Dipoles oriented vertically are found too much anterior (3mm), whereas dipoles oriented horizontally are found too much up (5mm especially in the lower part) and anterior (3mm).

In a separate study, real EEG was recorded using 32 electrodes. 10 sets of 200 single trials were selected, and trials were averaged in each set. Each set was next added to each forward solution with a Signal-to-Noise ratio S/N = 5. Inverse solutions were then performed with Model 2. Results are reported in Fig 4. It appears that localization accuracy is particularly affected in the upper part of the head (2-3mm worse), and so is the orientation accuracy (3-4 degrees worse).

Fig 4 : localization (D) and orientation (Θ) errors for one dipole fit in presence of noise (S/N=5). Each symbol represents the error averaged over all dipoles of the slice. The graph on the far right reports the behavior of D with respect to the location of the dipole in each slice (back vs front of the head).

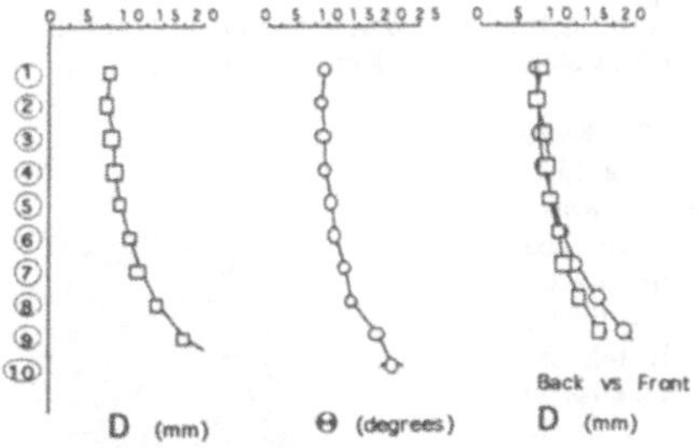

INFLUENCE OF THE NUMBER OF ELECTRODES

In a separate study, 3 different set-up were considered, respectively having 19, 32 and 63 electrodes. Fig 5 shows the electrode locations with respect to the anatomy of the subject on sagittal MRIs.

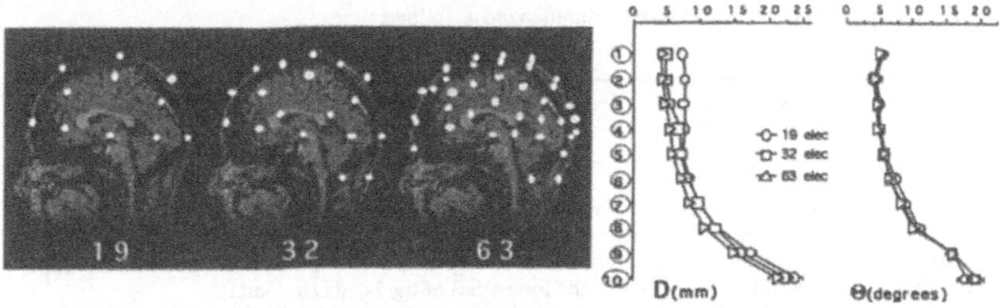

Fig 5: Influence of the number of electrodes used for inverse calculations.

It appears that using 32 electrodes rather than 19 improves the accuracy of the localization in the upper part of the head (3mm better). Improvement in the lower part can be achieved using 63 electrodes (1-2).

TWO DIPOLE RESULTS

100 pairs of dipoles located in the vicinity of the auditory cortex were randomly chosen. Inverse calculations were performed using 32 electrodes and model 2. Fig 6 shows the average localization and orientation errors that were obtained in absence of noise. Dipoles are found to much eccentric, anterior and up. The average localization error is found to be about 12mm, and the average orientation error 12 degrees.

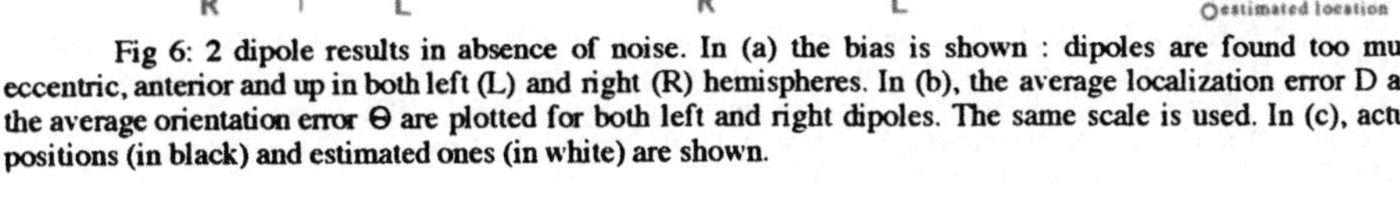

Fig 6: 2 dipole results in absence of noise. In (a) the bias is shown : dipoles are found too much eccentric, anterior and up in both left (L) and right (R) hemispheres. In (b), the average localization error D and the average orientation error Θ are plotted for both left and right dipoles. The same scale is used. In (c), actual positions (in black) and estimated ones (in white) are shown.

Next, for each pair of dipoles, real EEG noise was added to the forward problem solution, and then inverse calculation were performed. 10 different averaged EEG trials were taken for each dipole location. Fig 7 reports the corresponding results, together with the location of actual and estimated dipoles in the case of one particular noise situation. It appears that errors greatly increase with respect to the no-noise situation, and may attain values as high as 20mm and 22 degrees on average.

FIG : 2 dipole results in presence of EEG noise. In (a), actual positions (in black) and estimated ones (in white) are shown for one particular noise situation. In (a), the average localization error D and orientation error Θ are plotted for both left and right dipoles. The same scale is used.

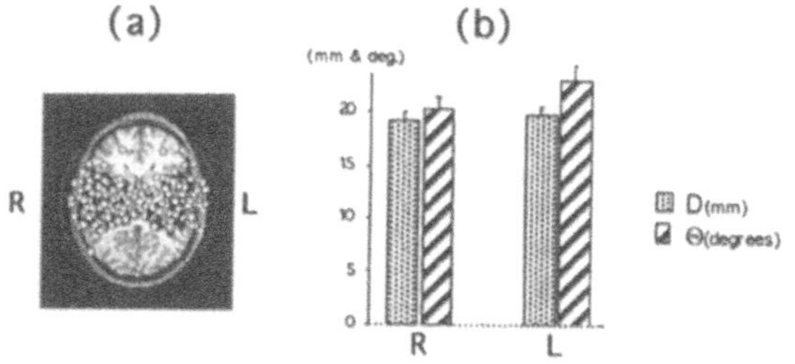

DISCUSSION

The spherical model accuracy can be summarized as follows :

		No Noise	Noise S/N=5
1 dipole	upper part	5-6mm	8-9mm
	lower part	15-25mm	15-25mm
2 dipoles		12mm	20mm

Our results in absence of noise are in accordance with those found by other authors. Indeed, localization error for one dipole is on average of the order of 1cm. However it clearly appears that the localization accuracy greatly depends upon the dipole location within the brain. Especially, it is particularly good in the upper part of the head (3-5mm), whereas it is much worse in the lower part of the head (15-25mm).

In order to appreciate what accuracy threshold should be considered as "acceptable" or "good", these results have to be compared with inverse calculation results in realistic geometries. A few studies [2,5] converge to the conclusion that the intrinsic error of realistic models is of the order of 2-3mm. Hence, errors as low as 3-5mm obtained using a spherical model should be considered as very good. By contrast, errors of more than 15mm should be considered as very poor by comparison with what can be expected from realistic models.

Our results using noisy data can be put in parallel with those of Zanow et al. [1] and Cuffin et al.[6]. The first group considered dipoles in the upper part of the head with 52 electrodes in this region. Using Gaussian noise, and S/N of 10 and 20, they found localization errors ranging from 5 to 13mm, which is in accordance with our results in the upper part of the head. Cuffin et al. evaluated the precision of the spherical model by generating 28 dipoles at different locations in the brain of epileptic patients. Inverse calculations used 16 scalp electrodes in the vicinity of the dipole locations, and led to average and maximum localization errors of 11 and 25.7mm, which again is in accordance with our results.

High errors in the lower part of the head may be explained by both : the lack of electrodes below T3, T4, FPz, Oz, and the non-spherical shape of the head structures in this region. For these reasons, the choice of the electrode locations (especially in the lower part of the head) may be critical. However, 32 electrodes appears to be sufficient for spherical inverse modelling, provided that they sufficiently cover the lower part of the head.

REFERENCES

[1] Zanow, F., and Peters, M.J., Med. & Biol. Eng. & Comput., 1995, 33: 582-588.
[2] Yvert, B., Bertrand, O., Echallier, J.F., and Pernier, J. Electroenceph. clin. Neurophysiol., 1995, 95: 382-391.
[3] Roth, B.J., Balish, M., Gorbach, A. and Sato, S. Electroenceph. clin. Neurophysiol., 1993, 87: 175-184.
[4] Schlitt, H., Heller, L., Best, E., and Ranken, D. Human Brain Mapping, 1995, Suppl 1:96 (abstract).
[5] Yvert, B., Bertrand, O., and Pernier, J. Electroenceph. clin. Neurophysiol., 1996, in press.
[6] Cuffin, B.N, Cohen, D., Yunokuchi, K., Maniewski, R., Purcell, C., Cosgrove, G.R., Ives, J., Kennedy, J., and Schomer, D. Ann. Neurol., 1991, 29: 132-138.

Mesh Refinement of Boundary Element Head Models in MEG Source Reconstructions

Zanow, F. and Peters, M.J.

Low Temperature Division, University of Twente, Enschede, The Netherlands

Introduction

In MEG source reconstruction, the use of realistically shaped volume conductor models for the head is becoming more and more common. Among several approaches to solve the volume conduction problem numerically, the boundary element method (BEM) is usually preferred. The main reason for this is that most of the blurring effects of the head on MEG arise from the skull inner boundary, which divides regions of very different conductivities. Thus, a homogeneous conductor with the shape of the inner skull surface is an adequate model. The skull inner boundary can be discretized by a two-dimensional triangular mesh, and the so-called fast forward solution of BEM [1] can be applied to compute the MEG caused by dipolar sources.

The effort of modelling the realistic shape of the subject's inner skull surface is only useful if computations performed with this model provide a higher accuracy than those with spherical models. For the reconstruction of sources close to the skull inner boundary, however, BEM gives rise to substantial numerical errors if the triangular mesh is too coarse to describe the potential gradients on the boundary. Overall refinement of the triangulation seems an easy solution, but is generally limited by the computer internal memory and slows down all computations. Therefore, we applied the idea of multiresolution analysis to our boundary description and investigated how the contribution of surface elements to the numerical error of the model can systematically be described. We developed a local refinement strategy, which relies on the measurement information and adaptively refines only those regions which actually determine the numerical accuracy of the forward solution.

The performance of this kind of mesh refinement was tested for spherical models and superficial dipoles with varying eccentricity. Furthermore, a realistically shaped inside of the skull model with adaptive mesh refinement was generated for a somatosensory evoked field experiment measured with a set of 19 + 31 gradiometer channels, and was applied to the source reconstruction.

The description of head boundaries by means of triangle subdivision surfaces

The application of adaptive mesh refinement to BEM requires a mesh generation scheme which is able to adapt *locally* the size of the surface elements within a wide range. Strategies for mesh refinement are usually based on the definition of a refinement region [2][3]. Seeking a more flexible mesh generation we found that the multiresolution analysis approach, which has found application in computer graphics surface rendering [4], also fits to adaptive mesh refinement of boundary element models. Assume that the actual geometry of the head compartment is described by a surface interpolation, e.g. two-dimensional spline functions [5]. A basis triangulation can be performed for a small number of points on the surface, which represents the lowest level of spatial resolution of the model. Each of the triangles is then subdivided into four smaller ones by splitting the edges and creating new nodes. It is very important that the new nodes are not just set at the middle of the edges, but lie on the interpolated surface and thus add geometry information to the model. The subdivision procedure is repeated successively until a certain geometry approximation level is reached, setting the maximum spatial resolution possible with this surface representation. The principle of subdivision is illustrated in figure 1.

The flexibility of mesh generation based upon triangle subdivision surfaces ranges from coarse basis triangulation up to the maximum subdivision level. The local element size which is actually needed for accurate magnetic field estimation must be derived from the measurement configuration and a predicted origin of the measured activity.

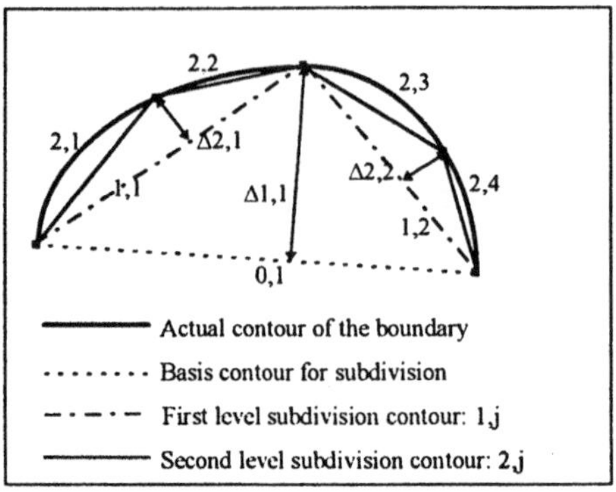

Figure 1: The general principle of subdivision is shown here for the one-dimensional case. The curved boundary contour is approximated by a set of piecewise linear functions (left figure). The subdivision contours are constructed using edge information of the preceding level and additional detail information.

Criteria for adaptive mesh refinement

Our approach to adaptive mesh refinement is based upon the assumption that we are able to predict the brain activity with moderate accuracy, e.g. by means of a spherical model. This preliminary source distribution and the position and orientation of the gradiometer determine how the resolution level on the subdivision surface has to be adapted in order to minimize the numerical error of the boundary element model. Thus, the solution to the inverse problem becomes a two-step process: first, the brain activity is estimated with a simple volume conductor model, and the boundary element model with adaptive refinement is constructed using this kind of information. Second, the adaptively refined mesh is used as a numerically accurate head model for the source reconstruction.

The criterion for the generation of the adaptive mesh refinement must be derived from the numerical approach which is applied to estimate the magnetic field. For a homogeneous volume conductor with a closed boundary of arbitrary shape, the magnetic field $\mathbf{B(r)}$ is [6]

$$\mathbf{B(r)} = \mathbf{B}^{\infty}(\mathbf{r}) + \frac{\mu\sigma}{4\pi}\int_{S}\phi(\mathbf{r}')\nabla'\left(\frac{1}{|\mathbf{r}-\mathbf{r}'|}\right)\times d\mathbf{s}'. \quad (1)$$

The linear discretization is applied for the surface potentials $\phi(\mathbf{r}')$ on boundary S as well as for the surface integral over the triangular elements in eqn. 1, using the analytical expressions described in [7] and [6]. Hence, only those surface elements, whose integrals significantly deviate from linear behavior, contribute to the numerical error in the estimated magnetic field. We developed an estimator for the deviance-from-linearity error based on the multiresolution analysis (eqn. 2). It relies on the integral of eqn. 1 and is computed at each node of the subdivision surface. The maximum of G over all channels i (position $\mathbf{r}_i$, orientation $\mathbf{n}_i$) is taken to obtain the maximum variation in the integral.

$$G(\mathbf{r}') = \max_{c}\left(\left[\phi(\mathbf{r}')\nabla'\left(\frac{1}{|\mathbf{r}_c-\mathbf{r}'|}\right)\times \mathbf{n}_{\mathbf{r}'}\right]\cdot \mathbf{n}_c\right),\ c=1,\ldots,N_{channels} \quad (2)$$

$$D(\mathbf{r}'_{i,j}) = \left|G(\mathbf{r}'_{i,j}) - \frac{\left(G(\mathbf{r}'_{i-1,k}) + G(\mathbf{r}'_{i-1,k+1})\right)}{2}\right|,\ \text{i: subdivision level; j, k: node index.} \quad (3)$$

The deviance-from-linearity criterion D can be computed for all triangle edges except those of highest refinement level as the difference between predicted integral value $G(\mathbf{r}'_{i,j})$ at the refinement point of the respective edge and the mean value of the two edge nodes (eqn. 3). Non-linear behavior of the integral in eqn. 1, e.g. at regions near the sensors, results in subsequent subdivision of the triangle edges in that region.

Simulations studies

We compared sphere-shaped models with adaptive mesh refinement to unrefined models of a comparable number of triangles. MEG forward solutions were estimated in a 19-channel planar gradiometer for superficial tangential dipoles with varying eccentricities (dipole distance from center of the sphere divided by sphere radius). The refined meshes were generated from a subdivision surface based on a spline interpolated sphere and ranged from 48 (basis triangulation) to 12288 (4th subdivision level) triangles; the refinement was estimated for the dipole with smallest eccentricity. The three different models were obtained by setting different tolerance levels for the deviance-from-linearity criterion. It can be shown that most of the gain in numerical accuracy can be reached with less than 1000 triangles, and that further refinement only increases the computational burden.

The numerical error in our simulated fields could exactly be measured because an analytical forward solution is available for spherical volume conductors. The root mean square (RMS) error of the fields predicted by the unrefined and adaptively refined models is depicted in figure 2. Except improved accuracy, the refined models perform better with increasing dipole eccentricity, due to the locally high spatial resolution of the model close to the dipole. Interestingly, the refined model with largest number of elements (2196 triangles) has a slightly increased RMS for the dipole with highest eccentricity, reflecting the limitation in spatial resolution by the highest level of triangle subdivision.

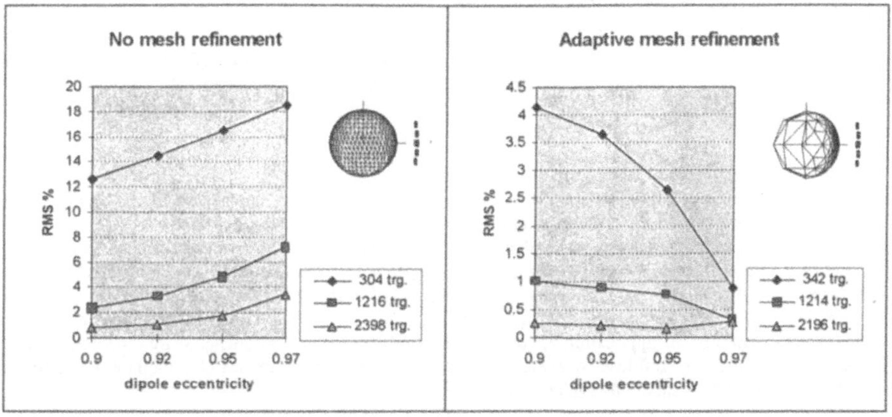

Figure 2: The RMS (root mean square error) of magnetic fields caused by a tangential dipole with varying eccentricity, compared to the analytical reference solution. Without mesh refinement, the error of the numerically computed magnetic field increases when the dipole moves towards the spherical boundary. The numerical error is generally much smaller for adaptively refined meshes of comparable number of elements, and decreases with increasing dipole eccentricity.

Dipole fit with an adaptively refined head model

The MEG of a left hand median nerve stimulation experiment with a 29 years old male right-handed volunteer was analyzed by means of a dipole fit. For the first cortical response at 20 milliseconds after stimulus, a superficial dipolar source was expected in the hand area of the contralateral somatosensory cortex. We first performed a dipole fit for this latency with a spherical head model. Based on the resulting source parameters, we estimated surface potentials on the subdivision surface of the subject's inside of the skull representation and applied the deviance-from-linearity criterion (eqn. 3). The resulting locally refined model (816 triangles) is shown in figure 3 together with the 19+31 channels gradiometer configuration. The dipole fit was repeated with this model, and, for comparison, with a triangular model of the same number of elements without local refinement. We here only considered the found dipole locations, which are shown in table 1. The differences in location due to different head models are small, but we found that the solution with locally refined model results in the smallest residual variance.

Table 1: Reconstructed dipole positions and residual variance, obtained with three different head models

Model	$(x,y,z)_{Dipole}$ mm	Residual Variance %
Sphere	(22.8,40.5,106.6)	2.7
BEM, no refinement	(24.8, 42.6,108.5)	2.3
BEM with refinement	(21.6,41.9,105.4)	2.2

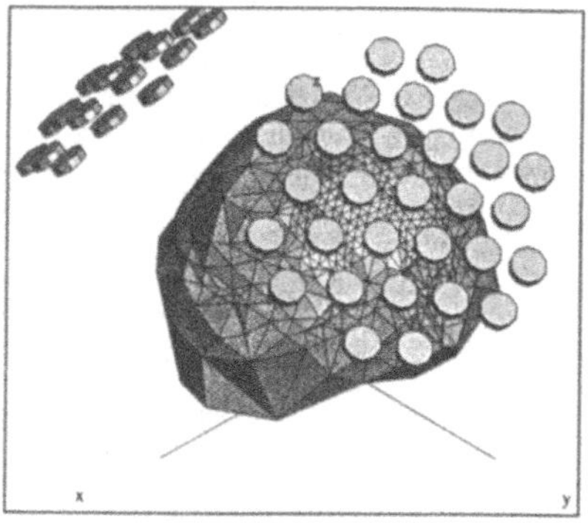

Figure 3: The locally refined inside of the skull model and gradiometers (31 channels contralaterally, 19 channels ipsilaterally)

Conclusions

Adaptive mesh refinement based on multiresolution analysis enables us to estimate BEM solutions to the MEG forward problem with a high numerical accuracy and a feasible computational effort. The meshing strategy relies on a prediction of the local numerical error at different discretization levels.

The influence of the volume conductor on MEG is generally small for superficial sources. Hence, significant improvements can only be expected for deeper sources or within head regions which notably deviate from spherical shape. Considering the mesh generation method described in this paper, it should be investigated to what extent the predicted error information correlates with the actual numerical error due to the model. For the application of adaptive mesh refinement to a parameter search method such as non-linear dipole fit, the influence of source parameter variations within the search region on the accuracy of the forward solution should be studied.

References

[1] Nenonen, J. et al Magnetocardiographic functional localization using a current dipole in a realistic torso, IEEE Trans. Biomed. Eng.,1991, BME-38:658-664.

[2] Meijs, J. W. H. et al On the numerical accuracy of the boundary element method, IEEE Trans. Biomed. Eng.,1989, BME-36:1038-1049.

[3] Yvert, B. et al Improved forward EEG calculations using local mesh refinement of realistic head geometries, Electroenceph. Clin. Neurophysiol., 1995, 89: 381-392.

[4] Lounsbery, M., DeRose, T., and Warren, J. Multiresolution analysis for surfaces of arbitrary type, Tech. Report 93-10-05B, Univ. of Washington, Dept. Comp. Science and Eng., Oct. 1993.

[5] Zanow, F., and Peters, M. J. Individually shaped volume conductor models of the head in EEG source localisation, Med. & Biol. Eng. & Comput., 1995, 33: 582-588.

[6] Ferguson, A. S., Zhang, X., and Stroink, G. A complete linear discretization for calculating the magnetic field using the boundary element method, IEEE Tans. Biomed. Eng., 1994, BME-41:455-459.

[7] de Munck, J. A linear discretization of the volume conductor boundary integral equation using analytically integrated elements. IEEE Trans. Biomed. Eng.,1992, BME-39:986-990.

III. Magnetocardiography (MCG)

New Trends in Clinical Magnetocardiography

Mäkijärvi, M.[1] (Co-Organizer), Brockmeier, K.[2], Leder, U.[3], Lesh, M.[4], Moshage, W.[5], Nakaya, Y.[6], Nenonen, J.[1], Nowak, H.[7], Oeff, M.[8], Stroink, G.[9], Fenici, R.[10] (Co-Organizer)

[1]Department of Internal Medicine, Division of Cardiology, and BioMag Laboratory, Helsinki University Central Hospital, Helsinki, Finland; [2]Department of Pediatric Cardiology, Virchow Medical Center, Humboldt University, 13533 Berlin, Germany; [3]Klinikum der Friedrich-Schiller-Universität, Klinik für Innere Medizin III, and Biomagnetisches Zentrum, and Institut für Diagnostische und Interventionelle Radiologie, Jena, Germany; [4]Department of Cardiac Electrophysiology, University of California, San Francisco, USA; [5]Medical Clinic II and Policlinic, University of Erlangen-Nürnberg, Erlangen, Germany; [6]Department of Nutrition, University of Tokushima, Tokushima, Japan; [7]Klinikum der Friedrich-Schiller-Universität, Klinik für Neurologie, Biomagnetisches Zentrum, Jena, Germany; [8]Internal Medicine, Cardiology and Pulmology, Free University of Berlin, Berlin, Germany; [9]Department of Physics, Dalhousie University, Halifax, Canada; [10]Centro di Biomagnetismo-Fisiologia Clinica, Catholic University, Rome, Italy

Introduction (M. Mäkijärvi)

Magnetocardiography was introduced to clinical cardiologists in the early 1970's. The measurement devices were technically unreliable, signal-to-noise ratio was poor and measurements were time consuming. Most measurements were performed in ambient environment without any shielding. Patient material studied included at first normals and myocardial infarction patients. Electromagnetic activity produced by the His-Purkinje conduction system also drew a lot of attention. Later on, pregnant women (fetal MCG), patients with ventricular hypertrophy, Wolff-Parkinson-White (WPW) syndrome patients as well as patients with tachycardias were studied. The analysis of MCG signals in those days was rather primitive: mostly descriptive, qualitative, and morphological data was published. In order to be useful in the clinic there were many things to be improved: low signal-to-noise ratio, limited mapping and analysis possibilities, and the measurement systems were mostly located outside hospital.

In the 1980's the methods developed a lot. First multichannel (7 - 37 channels) systems were introduced and they could be operated in good quality magnetically shielded rooms with highly improved acceptable signal-to-noise ratio. Also patient measurements in a catheterization laboratory took place validating for the first time the localization accuracy of the source localization. Patient series were not much larger but the amount of topics under study increased markedly: localization of myocardial infarction, risk identification after infarction and in long QT syndrome, ventricular hypertrophy, tissue rejection after heart transplantation, normal and abnormal His-Purkinje activity, localization of accessory pathways in WPW, and mapping of tachycardias. The analysis software took several important steps forward implementing signal-averaging, high-pass filtering based analysis, cardio-thoracic modeling, and introducing new concepts like late fields and risk score for clinical use. Still some problems remained to be solved: analysis was mostly done off-line and too limited and slow for clinical purposes, modeling was too primitive to give accurate results, patient populations were still too small to convince larger clinical audience. Lack of adequate clinical reference data and the non-clinical measurement environment decelerated the development of the method towards a real clinical tool.

Many of these problems have been solved when arriving into the 1990's. Most importantly, the installation of the advanced multichannel (up to 67 channels) MCG measurement systems in hospital environment has been decisive in the process of turning the MCG mapping from a research method to a clinical tool. In addition, signal-to-noise ratio exceeding the one reached with conventional ECG enlargens the spectrum of the possible clinical applications of the method. Larger scale clinical series are currently being collected including localization studies (WPW, tachycardias), risk analysis (infarction, long QT syndrome), as well as the fetal MCG. The analysis software includes improved modeling achieved with the anatomic reference data provided by magnetic resonance imaging as well as new signal-processing techniques like spectral turbulence analysis. Refined methods for detection of ST-segment changes and injury currents have been developed. Despite these great improvements there are still some obstacles to be overcome before wider clinical acceptance of the MCG will take place. The on-line analysis possibilities are too limited, comfortable user interfaces do not exist, lack of experimental data makes the interpretation of pathological findings somewhat uncertain, MCG is still a rather expensive technique and is currently available only for limited number of patients, the method is not yet flexibly applicable in cardiac catheterization laboratory, and the need for specially trained personnel to operate the current systems increases the running costs markedly.

The Biomag96 meeting presents with altogether 53 new MCG studies, which clearly demonstrate the vast amount of active research in this field. The methods used in these studies include several interesting features ranging from real time multichannel and multiplane recordings performed in clinical environment to advanced theoretical modeling and basic research. The topics under study are more numerous than ever: tachycardias, localization of arrhythmogenic substrates, fetal MCG, infarction, ischemia, arrhythmogenic right ventricle, hypertension, antiarrhythmic drugs etc. The analysis methods presented include excellent modeling, advanced signal processing, repolarisation analysis, remodeling, injury currents, new algorithms complemented with adequate reference methods like ECG, body surface potential mapping, intracardiac ECG, magnetic resonance imaging, and echocardiography.

From these presentations it can be concluded that in 1996 clinical MCG means multichannel MCG systems in clinical environment enabling fast measurements of large sample sizes and widening the spectra of potential clinical applications as well as stimulating basic and experimental research. The technical development in the near future will probably provide high-Tc SQUIDs, light shielding/no shielding, on-line analysis, and some measurement standardization, making the MCG method everyday routine in clinical cardiology.

This manuscript summarizes the invited oral presentations and the panel discussions held in the Workshop of Clinical MCG in the Biomag96 Meeting. The latest developments and clinical applications of the MCG are exciting and promise interesting new results for the clinical community in the near future.

Multimodal Imaging of Cardiac Electric Sources (J. Nenonen)

Cardiac studies in the BioMag Laboratory

The BioMag Laboratory at the Helsinki University Central Hospital is a new world-wide unique research center for non-invasive functional studies of the human heart and brain. The core of the facility, the BioMag Laboratory, was taken in use at the Helsinki University Central Hospital at the beginning of 1995. The BioMag Laboratory is owned together by three large educational institutions: Helsinki University of Technology (HUT), Helsinki University Central Hospital (HUCH), and University of Helsinki (UH). The BioMag facilities offer excellent possibilities for multidisciplinary cardiac research, including multichannel magnetocardiography (MCG) and electrocardiography (ECG) in the BioMag Laboratory, and the medical imaging modalities available at HUCH (anatomical and functional magnetic resonance imaging, ultrasound, X-ray, and clinical data). Most of these facilities, including the Laboratory of Biomedical Engineering at HUT, are connected to the BioMag laboratory via a fast digital ATM (Asynchronous Transfer Mode) network.

A 67-channel dc-SQUID system for cardiac measurements was designed and constructed by Neuromag Ltd. (Helsinki, Finland). The system was installed in the BioMag Laboratory in April 1995 [5]. The Laboratory of Biomedical Engineering was involved in the design of the system, which has 7 axial-type and 60 planar-type first-order gradiometers on a slightly curved surface with a diameter of 30 cm. The sensor array covers the whole anterior chest in most of the patients. The system is feasible in patient measurements and has proved very reliable performance. During the first year of operation, about 160 cardiac patients have been studied with this system.

Body surface potential mapping

Besides the cardiomagnetometer, the BioMag Laboratory is equipped with two ECG measurement systems that can be applied simultaneously with MCG. One of them is designed both for 64-channel ECG studies and for electroencephalographic brain recordings; this system is sharing the data acquisition system with corresponding biomagnetic measurements. The other system has a portable 128-channel front-end and utilizes a PC-based data acquisition. The front-end electronics was designed and manufactured in Amsterdam [4] while the electronics and software for data acquisition were developed in the Laboratory of Biomedical Engineering.

The electrodes used in the system are small Ag/AgCl pellets with the diameter of 1 mm to achieve radiolucency. The 128-channel system was taken in use at HUT in October 1995. During the first four months of operation, the system has been used at HUCH in over 40 patient studies. Besides normal patient measurements, the system has been applied during invasive catheterization and X-ray fluoroscopy. It has already been successfully applied in guiding the catheter ablation of the ventricular pre-excitation in the Wolff-Parkinson-White syndrome as well as in ablation of ventricular tachycardia, using the method reported by SippensGroenewegen et al. [9].

Mathematical modelling and medical image processing

The Laboratory of Biomedical Engineering was among the first to apply boundary element techniques in the inverse problem of MCG [6.7]. We have applied the technique in over 50 patients in the localization of the ventricular pre-excitation site. The average localization accuracy in these studies has been about 2 [8]. Even though this accuracy is satisfactory in pinpointing the anatomical region of the arrhythmia, significant improvements can be expected by applying more individualized patient models from medical images, and by developing better registration methods for combining the MCG results with the heart anatomy.

Magnetic resonance (MR) images of the heart have been acquired with a Siemens Magnetom Vision 1.5 T system at the Department of Radiology of HUCH. After testing several imaging sequences, a T1-weighted turbo-flash sequence was selected for cardiac imaging. MR images have been taken at various phases of the heartbeat, starting at the R-peak of the QRS complex and continuing at 40 ms steps. The MR images of the heart have to be segmented, i.e., anatomic objects are defined from the images for the purposes of boundary element modelling and three-dimensional visualization. Segmentation is performed with software developed in the Laboratory of Biomedical Engineering on the basis of interactive region growing and active contour models. The segmentation process is semi-automatic: the user can correct the result when the software has difficulties in distinguishing boundaries of the desired object.

In the Laboratory of Biomedical Engineering methods have been developed for constructing individualized torso models consisting of triangulated surfaces. Typically, a few hundred triangles are used in the boundary element calculations. Several triangulation methods have been tested on the basis of contours extracted from segmented MR images. Because the selection of the nodes is usually limited to parallelling contours, a new Delaunay based algorithm was developed to apply arbitrary numbers and placements of the nodes.To display MCG and ECG localization results accurately in respect to the anatomy of the heart, the results need to be visualized in medical images (MRI, X-ray, or ultrasound). Special software have been developed for presenting the localization results overlaid on a two-dimensional MR-slice, or in a three-dimensional rendered volume model.

Recently a pilot study has been started, in co-operation with the Italian National Research Council (CNR), to locate the site of a non-magnetic pacing catheter [2] from multichannel MCG recordings performed in the BioMag Laboratory. The catheter was invented and tested by prof. R Fenici at the Catholic University of Rome, and he is now responsible, in the framework of the BIRCH Cardiomagnetism Programme, to provide a further validation of his method and catheter for MCG-guided EPS and catheter ablation of arrhythmias [2]. Five patients, investigated so far, underwent first standard electrophysiological studies in the Cardiovascular Laboratory of HUCH, and then they were transferred to the BioMag laboratory for MCG recordings. Preliminary analysis showed that the MCG mapping was able to find the correct anatomical location of the amagnetic catheter's tip in all five cases. This study therefore verified that the MCG can provide accurate localization of arrhythmogenic sources, to drive catheter in their proximity. However, for clinical application, more accurate integration of the MCG results with cardiac anatomy and real time presentation of localization figures need still to be developed [3].

References

[1] Fenici R, Consiglio Nazionale delle Ricerche, United States Patent Documents (19), patent N. 5 056 517, Oct. 15, 1991

[2] Fenici R, Melillo M, Biomagnetically localizable multipurpose catheter and method for MCG guided intracardiac electrophysiology, biopsy and ablation of cardiac arrhythmias. International J of Cardiac Imaging 7: 207-215, 1991.

[3] Grönros K, Fenici R, Nenonen J, Mäkijärvi M, Korhonen P, Katila T. Magnetocardiographic Localization of a Non-Magnetic Pacing Catheter. Proc. XVIII Annual Internat. Conf. of IEEE Eng. Med. Biol. Soc. (Amsterdam, November 1996).

[4] MettingVanRijn AC, Kuiper AP, Linnenbank AC, Grimbergen CA. Patient isolation in multichannel bioelectric recordings by digital transmission through a single optical fiber. IEEE Trans. Biomed. Eng. 40(3), 302-308, 1993.

[5] Montonen J, Ahonen A, Hämäläinen M, Ilmoniemi RJ, Laine P, Nenonen J, Paavola M, Simelius K, Simola J, Katila T. Magnetocardiographic functional imaging studies in the BioMag Laboratory. This volume.

[6] Nenonen J, Purcell C, Horacek M, Stroink G, Katila T. Magnetocardiographic functional localization using a current dipole in a realistic torso. IEEE Trans. Biomed. Eng. 38(7), 658-664, 1991.

[7] Nenonen J, Mäkijärvi M, Toivonen L, Forsman K, Leiniö M, Montonen J, Järvinen A, Keto P, Hekali P, Katila T, Siltanen P. Non-invasive magnetocardiographic localization of ventricular pre-excitation in the Wolff-Parkinson-White syndrome using a realistic torso model. Eur. Heart J. 14, 86-93, 1993.
[8] Nenonen J. Solving the inverse problem in magnetocardiography. IEEE Eng. Med. Biol. Magazine 13, 487-496, 1994.
[9] Sippens Groenewegen A, Spekhorst H, van Hemel N, et al. Localization of the site of origin of postinfarction ventricular tachycardia by endocardial pace mapping. Circulation 88, 2290-2306, 1993.

Single-Plane versus Multiplane Magnetocardiography (W. Moshage)

The development of biomagnetic multichannel systems was the basic requirement for clinical applications of the magnetocardiogram (MCG) in cardiology [4]. Multichannel systems are necessary to coherently detect the rapidly changing magnetic fields caused by the beating and moving heart. Especially the magnetic fields of singular, short events such as single ectopic beats or haemodynamically effective runs of tachycardia have to be captured in a "one-shot"-recording.

The single-plane multichannel systems provided by several manufactures have permitted to develop a number of cardiac applications of possible clinical relevance during the last years: For example, several indicators for risk stratification concerning sudden cardiac death and the detection of graft rejection after heart transplantation have been developed, such as time-domain and frequency-domain analyses of late ventricular activity and other MCG data or the determination of the homogeneity of cardiac de- and repolarization [1]. The biomagnetic recording of normal heart beats and also arrhythmias of the fetus (fetal MCG) are promising applications in the important field of prenatal monitoring and diagnostics. For these applications, single-plane recordings will probably be sufficient, and in fetal MCG even systems comprised of a single or a limited number of channels, possibly in high-temperature technology, will be sufficient.

The situation is different concerning the clinically very important aspect of non-invasive 3-dimensional localization based on the magnetocardiogram. This problem not only concerns focal sources such as cardiac arrhythmias, for example ventricular and supraventricular ectopy and tachycardias as well as accessory pathways, but also the analysis of distributed sources using current density reconruction s. The latter are mainly used for the visualization of ischemic areas during stress ("stress-MCG") and the detection of viable myocardium after myocardial infarction.

The localization accuracy of biomagnetic recordings depends on many factors including data acquisition, signal processing, source and volume conductor modeling, and correlation to anatomical refernce methods. It is necessary to conduct careful experimental studies to exactly determine the influence of each of these factors, and, prior to patient investigations, phantom measurements have to be generally demanded in order to verify the localization accuracy in the complex interaction of biomagnetic hard- and software.

Consistent observations in phantom studies, invasive pacing studies in patients and comparisons of MCG data to reference methods such as radionuclide ventriculography, invasive electrophysiological mapping, and successful catheter ablation indicate that the accuracy of MCG localizations using simplified volume conductor models strongly decreases with growing distance from the source to the sensor plane, the limit for clinically acceptable localizations is probably a depth of around 10 to 15 cm. If the distance from the source to the sensor plane is larger, the localization accuracy mainly concerning the depth of the localization result decreases rapidly impairing thus the overall localization accuracy [2].

Recent investigations showed that the three-dimensional localization accuracy of MCG recordings can be increased by combining sequential or simultaneous measurements of one or several dewars in different planes (multiplane recordings). Modern systems permit the simultaneous recording from the ventral and dorsal side of the patient. Phantom and patient studies demonstrated an increased localization accuracy, mainly concerning the depth of the localization result, with this approach [3]. However, there is room and need for improvement in the hard- and software necessary for these recordings. This can be achieved only in close cooperation of manufacturers and applicants of the biomagnetic systems. The optimal solution for MCG recording would probably be a dedicated thorax system.

References
[1] Montonen J.: Magnetocardiography in identification of patients prone to malignant arrhythmias. In: Biomagnetismus: Fundamental research and clinical applications, Baumgartner Ch., Deecke L., Stroink G., Williamson eds, Elsevier press, 606-611 (1995)
[2] Moshage W., Achenbach S., Göhl K. et al.: Evaluation of the non-invasive localization accuracy of cardiac arrhythmias attainable ba multichannel magnetocardiography (MCG). International Journal of Cardiac Imaging, in press.
[3] Moshage W., Achenbach S., Flüg M. et al.: Comparison of the magnetocardiographic localization accuracy using simultaneous and sequential recordings in a single and multiple planes. This Conference, Santa Fe (1996)
[4] Schneider S., Hoenig E., Reichenberger H. et al.: Multichannel biomagnetic system for study of electrical activity in the brain and heart. Radiology 176, 823-839 (1990)

Biomagnetic Analysis of Distributed Cardiac Sources with Myocardial Current Imaging (U. Leder)

The ill-posed inverse problem of biomagnetism can only be solved using constraints in modelling the generator distribution. The most common, simple, and widely used model in cardiomagnetic research in the past was the single equivalent current dipole which obviousely fails in the explanation of more complex source structures like distributed myocardial activation wavefront. Therefore, we work on new reconstruction algorithms, which are able to visualize localization of non-focal cardiac activity. The calculation of distributed sources can only be solved approximately with multichannel MCG data by constraining the generator structure. We call this Myocardial Current Imaging (MCI) by anatomical constraints.

A first attempt to this problem was the use of depth normalized minimum norm estimation in patients with myocardial infarction and known regions of large myocardial damage. In patients with coronary artery disease and impaired left ventricular function, the distinction between ventricular dysfunction arising from myocardial fibrosis and ischemic, but viable myocardium has important clinical implications. The identification of hibernating myocardium is valuable in selecting patients, who will benefit from coronary revascularization. The purpose of this study was to determine whether magnetocardiographic reconstruction algorithms could identify regional myocardial fibrosis.

Methods

Patients/Normals

We investigated altogether 40 patients with a history of myocardial infarction and normals. All patients underwent 12-lead-electrocardiogram, stress-test, echocardiography, coronary angiography, and most of them Tl-SPECT. Only patients with well defined and clearly identified zones of localized myocardial fibrosis were included and compared to normals. Normals were defined by history, ECG, stress-test and Holter monitoring.

Magnetocardiographic Mapping

The magnetic fields were detected with a 50(31-19) channel SQUID system (Philips) with first order gradiometers (20 mm diameter, 70 mm baselength). The two turn gradiometer coils are hexagonally arranged with a center to center distance of 25 mm. The system is installed in a magnetically shielded chamber, and the overall noise is below 10 $fT/Hz^{-1/2}$. Nonsimultaneously measured MCG mapping positions were combined using averaging of a simultaneously recorded ECG channel. His way about 100 sensor signals could be used for data input.

Coordinate Transformation

The determination of a subject´s position relative to the multichannel magnetometer system was performed by the localization of the center coordinates of coilsets fixed to certain points on the subject´s anterior chest wall. A sinusoidal current was fed into each coil of the coilsets. At least 3 coilsets within the field of view of the sensor array are neccessary for stable localization. The localization of 10 coilsets took about 30 seconds of data acquisition time and was performed before and after each MCG measurement to recognize artefacts caused by patient movement. Coilsets were changed for vitamin E capsules prior to MRI investigation.

3D reconstruction of MRI

Individual volume conductor properties and physiologically meaningful constraints for current generator reconstructions were extracted from Tl-weighted 3D-MR-images. A series of transversal MR-slices (256 x 256 pixels, 12 bit resolution, 53 slices, 5 mm spacing) formed the input data for the 3D-graphics. The original sclices were reduced to 8 bits resolution and gaussean interpolated to get an isotropic threedimensional data set, which is essential for 3D-volume rendering techniques. The segmentation of the lung surface and the outer surface of the torso was realized by an automated procedure. A threedimensional region growing algorithm with a single threshold value started into the lung volume, ensuring that a closed surface was generated. Closing and opening procedures avoided bubbles within the volume. With a mouse controlled plane adjustment tool cut planes were determined (top and bottom of the torso to avoid exit of the region growing algorithm from internal compartments). The aim of the myocard segmentation was to provide a priori information about the possible locations of generators in the individual anatomy. This information is organized as a list of points. Each point, located on the left ventricular myocardial surface, stands for a small patch of the heart muscle.

Current Density Reconstruction

Current density was calculated for myocardial support points using depth normalized minimum norm least square estimates and overlayed with 3D-MR-images as color coded surfaces (consistant logarihmic scale for all timepoints). Characteristic torso compartments with isotropic conductivities were used for the generation of individually shaped Boundary Element Models for modelling the volume conductor geometry (conductivities torso:lung=5:1). Current density was calculated in steps of 5 ms from the onset of QRS complex to the end of T wave.

Results

Current density is increasing from QRS onset to QRS maximum. The majority of healthy myocardial regions has high current densities. Corresponding to the absence of electrical active myocardial tissue, MCI shows reduced current density for regions of infarcted myocardium. Infarcted areas have current densities below one $\mu A/mm^2$. The current density distribution correlates with other cardiological investigations (eg. left heart catheterization, scintigraphy, echocardiography). These are promising results for further application of current density reconstruction algorithms. In 55% of 40 patients/normals investigated, MCI had complete correspondence with the other cardiological findings, in 27% we found partly correspondence and 18% MCI failed to provide correct diagnosis.

Discussion

Myocardial current imaging (MCI) is a new promising method for assessing of myocardial viability. The assessment of myocardial viability by using standard imaging techniques is often imprecise. No criteria have been identified to accurately distinquish hibernating myocardium from myocardial fibrosis on the basis of regional ventricular function measurements obtained by contrast ventriculography, radionuclide angiography, or echocardiography. In contrast to these relatively poor results PET technology gave better results. MCI should be compared to all these methods in further investigations. Up to now the diagnostic sensivity and specifity of this reconstruction method in compare with other diagnostic performances are still unclear. After further checking of the reliability we will apply MCI to the detection of hibernating myocardium in patients with coronary artery disease.

References

[1] Dössel, O., David, B., Fuchs, M., Krüger, J., Lüdeke, K.-M., Wischmann, H.-A., 1993, A 31-channel SQUID system for biomagnetic imaging, Applied Superconductivity 1:1813-1825

[2] Fuchs, M, Wischmann, H.-A., Dössel, O., 1992, Overlay of neuromagnetic current density images morphological MR images, Proc. of Visualization in Biomedical Computing. Chapel Hill, SPIE 1808:676-684

Acknowledgement

This work was supported by the German BMBF grant 13N6736.

Stress Magnetocardiography (K. Brockmeier)

There are indications that the repolarization of the ventricular myocardium (the ST-T segment of the cardiac signal) shows a different behaviour in the electrocardiogram (ECG) and the magnetocardiogram (MCG) under particular conditions [2]. The standard to detect possible repolarization disturbances is stress testing. In cases of myocardial ischemia ST-T segment changes occur or deteriorate with increasing stress. Recently we observed significant ST-segment changes in the MCG of healthy individuals under physical exertion concomitantly with a normal ECG [1]. In order to further investigate this presumed discrepancy we designed a multichannel setting of both the MCG and the ECG.

Three healthy individuals were investigated (and two re-investigations) after verification of cardiovascular normality (bicycle stress test, echo-doppler, Holter, and, in two subjects, radionuclide perfusion scintigraphy). The 37-channel magnetometer SQUID system of the Physikalisch-Technische Bundesanstalt was used for rest and stress MCG [3]. Simultaneously, a 32-lead rest and stress ECG was recorded from the anterior and posterior thorax of the subjects. In order to guarantee identical positions during rest and stress, we used pharmacological induction of stress.

The ECG showed normal findings during rest and stress in all investigations. In contrast, comparison of rest and stress MCG showed distinct ST-T changes, mainly T-wave inversions, in all subjects at heart rates above 115 - 120 beats per minute. Isofield map examinations revealed T-wave onset as the instant with the strongest field changes, when rest and stress MCG were compared. Therefore, a quantification of the components of equivalent current dipole and magnetic dipole moments was performed and averaged over 10 ms of T-wave onset.

	subject 1	subject 2	subject 3a	subject 3b	subject 3c
m_z rest [nAm^2]	23.9 (1.4)	43.2 (1.4)	29.1 (2.5)	26.2 (1.9)	
m_z stress [nAm^2]	-54.3 (1.5)	-5.3 (0.6)	-51.3 (3.8)	-75.5. (1.0)	-137.0 (4.5)

The table shows the results from calculations of the moments of the *z* component of the magnetic dipole at T onset (mean and (SD)). The displayed changes of the the *z* component of the magnetic dipole were partly explanatory for changes of magnetic field distributions under stress. The different findings between ECG and MCG could be sufficiently explained by the sensitivity of the MCG vortex components of the current field, to which the ECG is insensitive. The physiology of the underlying effect is unclear. It is concluded that under defined conditions clear differences between ECG and MCG exist and that the MCG findings were of physiologic origin.

References
[1] Brockmeier K, Comani S, Erné SN, Di Luzio S, Pasquarelli A, Romani GL. J Electrocardiol 1994;27:137-142.
[2] Cohen D, Norman JC, Molokhia F, Hood W Jr. Science 1971;172:1329-1333.
[3] Drung D, Zimmermann R, Cantor R, Erné SN, Koch H, Matthies KP, Scheer HJ, Stollfuss D. Clin Phys Physiol Meas 1991;12 Suppl. B:21-29.

Panel Discussion

The panelists had to comment on 4 different questions raised by the chairmen. In the following there is a brief summary of the discussions.

1. Name 3 major steps forward in clinical MCG since 1993.

The majority of the panelists were naming the progress achieved in myocardial current imaging as the most important step forward during the last 3 years. Other steps forward were: improvement in signal-to-noise ratio, multichannel and multiplane systems with invasively verified localization accuracy, fetal MCG as a clinical application, more realistic torso models, increased amount of clinical MCG results, and development of MCG methods for the evaluation of cardiac risk.

2. Which will be the 3 most important clinical applications of MCG in the year 2000?

According to most panelists the detection of viable or ischemic myocardium will be the most important clinical application of the MCG in the near future. Also portable high-Tc multichannel MCG systems, fetal MCG, tachycardia localization in the catheterization laboratory, and noninvasive risk stratification were mentioned.

3. What are the 3 biggest obstacles of MCG?

The need for an expensive, space taking and operations limiting shielded room was clearly the biggest technical obstacle named by the panelists. Other discussed obstacles were: inefficient and undeveloped measurement and analysis software, lack of MCG data with clinical reference data, no clearcut important clinical application, high costs of MCG sensors and methods providing anatomical reference data.

4. What should be changed in order to optimize MCG systems for clinical use?

Almost every panelist wanted to have a high-Tc MCG recording system working with good signal-to-noise ratio outside the shielded room e.g. in the catherization laboratory. Additional improvements suggested by the panel were: improvements in analysis and modelling software, standardization of measurement devices and procedures, demonstration of the clinical validity by clinical multicenter studies, automated systems relating position of the patient and measuring probes.

Concluding Remarks (R.R. Fenici)

Magnetocardiography, although 25 years old, is still not considered an "adult" as a method. In fact clinical cardiologists find it hard to believe that with a single instrumentation, although still expensive and complicated, it is possible to investigate very different aspects of cardiac pathology, which usually require different equipment and special skills. Clinicians are still reluctant to believe that with the same MCG mapping it is possible, for instance, to localize an arrhythmogenic substrate and at the same time to identify areas of stunned myocardium, in patients with ischemic cardiomyopathy. The lack of user friendly instrumentation at reasonable cost impede large scale investigation at clinical level. However, during this very interesting workshop it has become evident that there is a new trend towards the solution of those basic problems. There are companies investing in the development of small, low-cost, almost portable equipments, even multichannel devices, some of them based on high-Tc SQUIDS, which should favour a wider involvement of clinician in biomagnetic measurements. On the other hand, fundamental investigations carried out with state-of-the-art large scale instrumentation provide us with updated validation of MCG accuracy for source localization, and insight into new potential clinical applications of biomagnetism for clinical cardiology. Both approaches are necessary and will contribute to a faster progress towards the final goal, which certainly is to have biomagnetic technology operating in unshielded clinical laboratories and interventional rooms, possibly with new reliable non cryogenic sensors. Let this be the challenge for the year 2000!.

Revisiting the Fetal Magnetocardiogram and Fetal Magnetoencephalogram

Wakai, R.T., Leuthold, A.C. and Martin, C.B.

University of Wisconsin-Madison, Madison, Wisconsin, USA

Introduction

The first studies of fetal neuromagnetism and fetal cardiomagnetism were published more than ten and twenty years ago, respectively [1,2]. Despite considerable interest generated by these reports, relatively few follow-up studies appeared. In retrospect, this is somewhat surprising given that 1) in several respects, magnetic detection is almost uniquely suited to intrauterine evaluation, and 2) there is a paucity of alternate fetal evaluation techniques that provide functional information. During the last few years, interest in fetal biomagnetism has reemerged, as a number of groups have reported findings at meetings and in the literature. This paper presents fetal MCG and fetal MEG studies performed recently by our group at the University of Wisconsin-Madison.

Fetal MCG

Fetal MCG recordings were made in more than fifty subjects at gestational ages ranging from 21-41 weeks, using a seven-channel, second-order gradiometer (Biomagnetic Technologies, 607 Neuromagnetometer) in an eddy-current shielded room.

The major advantages of fetal MCG, compared with fetal ECG, are manifest in the quality of the recordings (Fig. 1). Fetal ECGs recorded from the maternal surface typically show strong maternal interference and low signal-to-noise from 25-36 weeks gestation. These problems are considerably less severe for the fetal MCG, resulting in a qualitative improvement in signal quality. In addition, mapping studies show that, in contrast to fetal ECG, fetal MCG has signal transmission properties consistent with volume conduction, providing a basis for waveform interpretation [3].

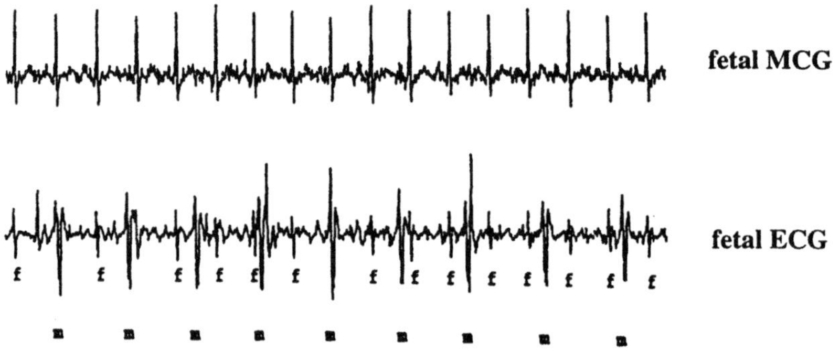

Fig. 1. Simultaneous recording of the fetal MCG and abdominal fetal ECG from a subject at 38 weeks. The fetal ECG contains both fetal (f) and maternal (m) complexes, while the fetal MCG is remarkably free of maternal interference. From [3].

Clinically, one of the most important potential applications of fetal MCG is assessment of fetal well-being. A number of obstetrical conditions, e.g. growth retardation, preeclamsia, diabetes, and postdatism, place the fetus at increased risk of hypoxic insult due to uteroplacental insufficieny. These high-risk pregnancies typically undergo twice weekly monitoring starting at 25-30 weeks gestation. Presently, no one monitoring technique has high predictive value. The primary clinical method involves assessment of fetal heart rate (FHR) tracings, obtained by Doppler ultrasound. Another widely used procedure is biophysical profile testing-- a composite test, consisting of FHR monitoring, estimation of amniotic fluid volume, and assessment of fetal behavior, using real-time ultrasound.

One shortcoming of present technology is the inability to accurately assess beat-to-beat FHR variability-- a major diagnostic parameter. FHR variability is seen as small fluctuations superimposed on baseline FHR, produced by asynchronous activity of the autonomic nervous system. The presence of normal FHR variability implies intact feedback between the central nervous system and SA node. Clinically, a reduction in variability is associated with hypoxic or pharmacological depression. However, short-term, or beat-to-beat, FHR variability is technically difficult to assess during pregnancy. It is generally believed that Doppler ultrasound does not accurately measure beat-to-beat FHR variability; only fetal ECG and MCG have adequate precision. One demonstration of the precision of fetal MCG is the common observation in our FHR tracings of small, rapid FHR oscillations associated with fetal breathing movements, known as respiratory sinus arrhythmia (RSA; Fig. 2). In contrast, FHR tracings obtained with Doppler ultrasound typically show no increase in beat-to-beat FHR variability during periods of fetal breathing movements.

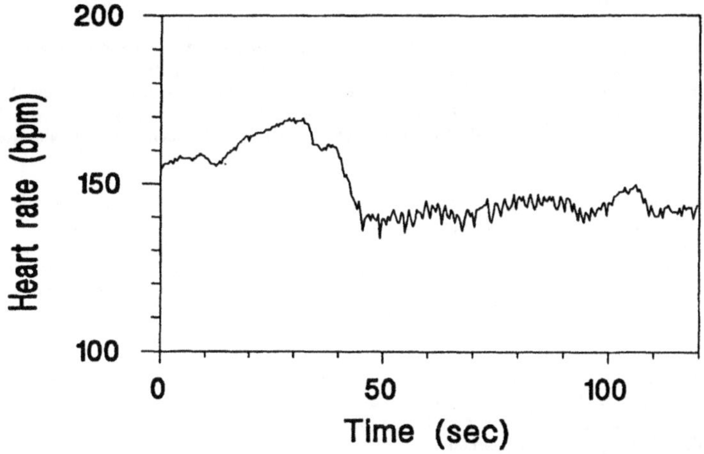

Fig. 2. Fetal heart rate tracing showing onset of respiratory sinus arrhythmia (RSA). These fetal heart rate oscillations are associated with fetal breathing movements and have frequencies in the range 0.4-1.0 Hz. A distinctive characteristic of RSA is greatly diminished amplitude during fetal heart rate accelerations.

A further advantage of fetal MCG is that the signal quality is adequate to provide meaningful waveform information. By averaging consecutive fetal complexes, the P-wave and QRS complex generally can be detected with good signal-to-noise ratio; however, the T-wave typically has low amplitude [6]. For reasons not completely understood, the T/QRS ratio is small compared to fetal ECG recordings and even neonatal MCG recordings taken from the same subject. Measurement of waveform intervals is not difficult, but requires some care due to spatial variation of the signal. The configuration of the P-wave, in particular, often shows spatial variation strong enough to confound measurements of PR interval or P-wave width based on inspection of a single channel.

There is evidence from animal studies [4] and from human studies during labor [5] that measurement of waveform parameters such as T/QRS ratio and PR-RR correlation can provide an early and sensitive indicator of hypoxia. Potentially this information can be used to improve the predictive value of fetal monitoring by resolving uncertainties that arise from the use of fetal heart rate monitoring alone. Fetal MCG may be the only technique that can assess beat-to-beat FHR variability and waveform parameters with good success during pregnancy.

Waveform parameters should also be useful for diagnosis of fetal heart disorders, such as fetal arrhythmias. In one subject with a complicated heart malformation, MCG was used to detect an unusual fetal arrhythmia, where normal sinus rhythm predominated when RSA was absent from the fetal heart rate tracings and a junctional rhythm predominated when RSA was present [7]. The junctional rhythm was manifested by a P-wave of reduced amplitude and altered configuration and a greatly shortened PR interval (Fig. 3). It was hypothesized that changes in autonomic activity associated with fetal breathing movements were largely responsible for causing the rhythm alternations. This case was notable because it provided an example of a rhythm disturbance that was not detected by ultrasound, but could be analyzed in considerable detail using fetal MCG.

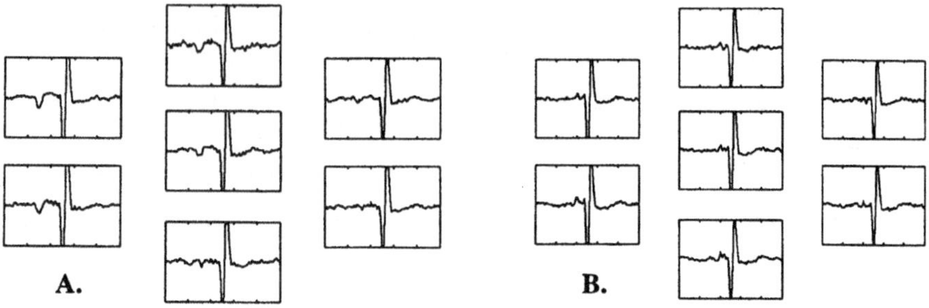

Fig. 3. Averaged seven-channel fetal MCG waveforms recorded at the same probe position from a subject with a complicated heart malformation at 34 weeks gestation. The amplitude scale ranges from -1.0 to +1.0 pT. The time axis has duration 0.5 sec. A) Fetal MCG consistent with normal sinus rhythm. B) Fetal MCG consistent with a junctional rhythm. The junctional rhythm was largely associated with the presence of respiratory sinus arrhythmia (RSA). Sinus rhythm was largely associated with an absence of RSA.

Fetal MEG

Perhaps the most intriguing application of fetal biomagnetism is intrauterine assessment of fetal brain function, which has been a longstanding objective of basic and clinical perinatology. In the last few decades many advances have occurred in the development of functional brain imaging technologies- e.g. PET, SPECT, fMRI. However, these techniques pose unacceptable or potential hazard to the fetus and are not considered suitable for routine fetal evaluation. Magnetoencephalography (MEG) is an ideal technology for fetal investigation in that it directly records neuronal activity, has high temporal resolution, and is completely safe and noninvasive.

Recently, we undertook to record fetal AERs from fetuses at 36-40 weeks gestation and neonates at 1-2 weeks [7]. Ultrasound was used to guide the placement of the probe. Typically the fetal scalp was 3-5 cm below the maternal surface. The stimulus was a 20 msec tone burst of frequency 1.5 kHz and intensity 100 dB, delivered from speakers placed 1.5 m from the subject. One hundred trials were collected, using a random intertrial interval of mean duration 2-sec. The analog passband was 1-50 Hz and the signal was digitized at 128 Hz. The AERs were averaged and digitally filtered, using a 2-20 Hz passband.

Four of fourteen fetal subjects showed distinct fetal AERs with latencies near 200 msec (Fig. 4A). The largest of these signals was approximately 100 fT. Smaller deflections with latency 100-300 msec were also observed. Two of these subjects and four others were studied after birth and all showed neonatal AERs similar to the fetal responses (Fig. 4B).

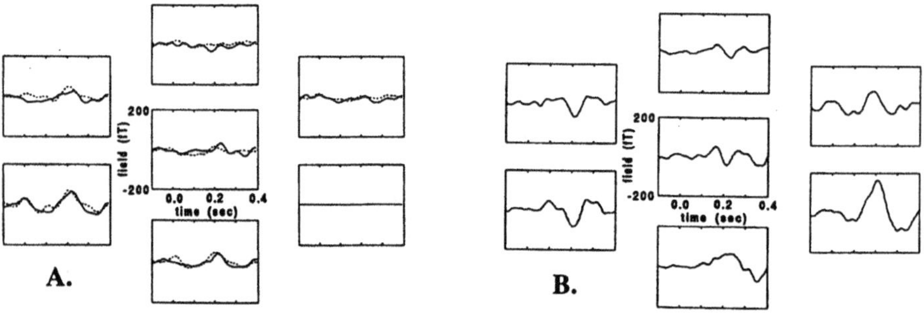

Fig. 4. A) Six-channel fetal auditory evoked responses from a subject at 36 weeks gestation. Shown are two separate measurements taken from similar probe positions. One of the channels was not working during the measurement. B) Seven-channel neonatal AER at age 10 days from the same subject as in A. A phase reversal of the signal is seen between the left and right channels. From [8].

The relatively low success rate can be attributed to a number of factors, including immaturity of the fetal brain, movement or unfavorable position or orientation of the fetal head, and low sound intensity in the womb. In addition, it should be noted that both the fetal heart and brain recordings reported here were made using detectors designed for adult brain studies. Major improvements in signal-to-noise can be expected from instruments designed specifically for fetal studies.

Discussion

As evidenced by the papers in this volume, the quantity and quality of work in fetal biomagnetism has progressed significantly since the time of the initial studies. While it is still early to evaluate the potential for clinical applications, it can be argued that fetal MCG provides more clinically-relevant information than other methods of fetal heart rate monitoring and, undoubtedly, is an excellent tool for detection and investigation of fetal rhythm disturbances. Among existing functional brain imaging modalities, fetal MEG may be the only viable technology for detection of human brain activity *in utero*. Despite formidible practical and technical challenges that lie ahead, the advantages of fetal MCG and fetal MEG for these applications appear compelling.

References

[1] Kariniemi, V., Ahopelto, J., Karp, P.J. and Katila, T.E. The fetal magnetocardiogram. J Perinat Med., 1974, 2: 214.
[2] Blum, T., Saling, E. and Bauer, R. First magnetoencephalographic recordings of the brain activity of a human fetus. Br J Obstet Gynaecol., 1985, 92: 1224-1229.
[3] Wakai, R.T., Wang, M., and Martin, C.B. Spatiotemporal properties of the fetal magnetocardiogram, Am. J. Obstet. Gynecol., 1994, 170, 770-6.
[4] Rosen, K.G. and Kjellmer, I. Changes in the fetal heart rate and EGG during hypoxia, Acta Physiol Scand., 1975, 93: 59-66.
[5] Murray, H.G. The fetal electrocardiogram: current clinical developments in Nottingham, J. Perinat. Med., 1986, 14: 399-404.
[6] Wakai, R.T., Wang, M., Leuthold, A., and Martin, C.B., Waveform analysis of the fetal magnetocardiogram, Society for the Study of Fetal and Neonatal Physiology, July, 1994, Melbourne, Australia.
[7] Wakai, R.T., Leuthold, A.C., and Martin, C.B., Association of fetal junctional rhythm and respiratory sinus arrhythmia detected by magnetocardiography, Ped. Cardiol., in press.
[8] Wakai, R.T., Leuthold, A.C., and Martin, C.B., Fetal auditory evoked responses detected by magnetoencephalography, Am J Obstet Gynecol, in press.

Acknowledgements

This work was supported by grant R01-HL54579 from the National Institutes of Health

The Fetal Magnetocardiogram During Different Stages of Pregnancy

Achenbach, S.[1], Menéndez, T.[1], Moshage, W.[1], Flüg, M.[1], Beinder, E.[2], Bittl, A.[2], Bachmann, K.[1], and Lang, N.[2]

[1]Medical Clinic II (Cardiology), [2]Department of Obstetrics and Gynecology, University of Erlangen-Nürnberg, Germany

INTRODUCTION

It has been demonstrated that the fetal magnetocardiogram can be recorded as early as the 13th week of pregnancy [2, 3]. In contrast to the fetal ECG recorded from the mother's abdominal surface [1, 4], it remains detectable throughout late pregnancy and thus constitutes a non-invasive tool for the monitoring of intrauterine development. We studied the development of the fetal magnetocardiogram during pregnancy.

PATIENTS AND METHODS

We investigated 33 women in 45 recordings (1 to 5 recordings per individual). The gestational age ranged from the 16th to 40th week. The recordings were performed inside a magnetically shielded room using a Krenikon® or MagnesII® 37-channel SQUID system in a hospital environment. In 19 recordings, ultrasound was used to detect the fetal heart position and place the dewar as close as possible to the fetal heart. In the other recordings, the dewar was placed in different positions over the mother's

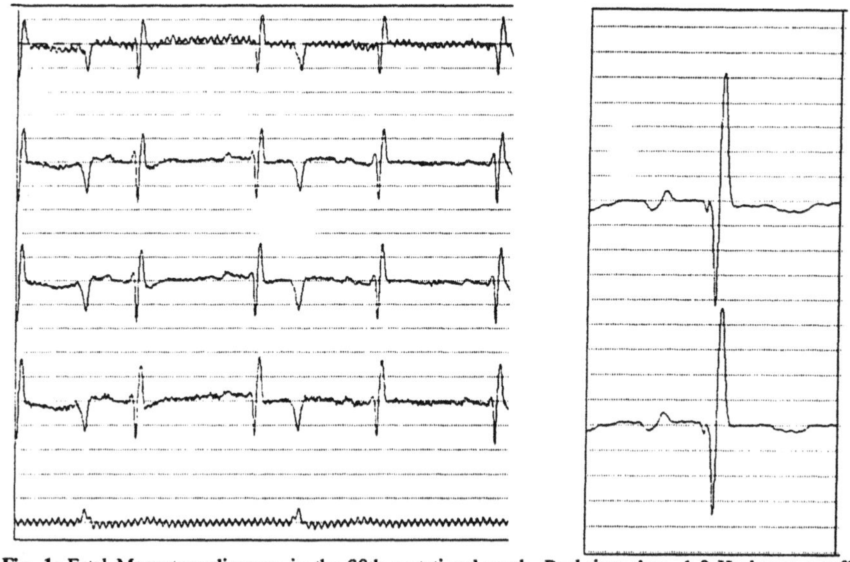

Fig. 1: Fetal Magnetocardiogram in the 38th gestational week. Real-time data, 1.0 Hz lower cutoff frequency. 4 magnetic channels (top) and one maternal ECG lead are shown. Fetal QRS complexes and P-waves can be detected in the magnetic traces. The fetal heart rate is 143/min.
Right: Two averaged MCG leads for determination of time intervals and amplitudes

abdomen until optimal discrimination between maternal and fetal heart beats was achieved in the on-line raw data. 37 magnetic channels and 6 maternal ECG leads were recorded with a sampling frequency of 1000 Hz and a bandwidth of 0.016 to 200 Hz. The duration of the recordings was 300 seconds.

For data analysis, the raw data was visually screened for artifacts and arrhythmic events. Only if necessary, suitable filters were applied to remove artifacts. The fetal heart rate was determined based on the frequency of fetal QRS complexes. After averaging of fetal heart beats, the PQ-interval and the duration and maximum amplitudes of the P-wave, QRS complex and T-wave were measured. In order to document the behaviour of these parameters during pregnancy and partly compensate for the high inter-individual variability, the average values for 5-week periods of gestational age were calculated.

RESULTS

In 4 out of 45 recordings, the data was heavily distorted due to background noise and fetal heart actions could not be reliably detected (3 recordings in the 15th to 20th week of pregnancy, one recording in the 27th gestational week). Of the remaining 41 recordings, 5 had to be excluded from further evaluation due to dominant arrhythmias such as AV-block or supraventricular tachycardia. Table 1 summarizes the mean values obtained for the amplitides and intervals as well as the percentage of recordings that could be evaluated during different stages of pregnancy.

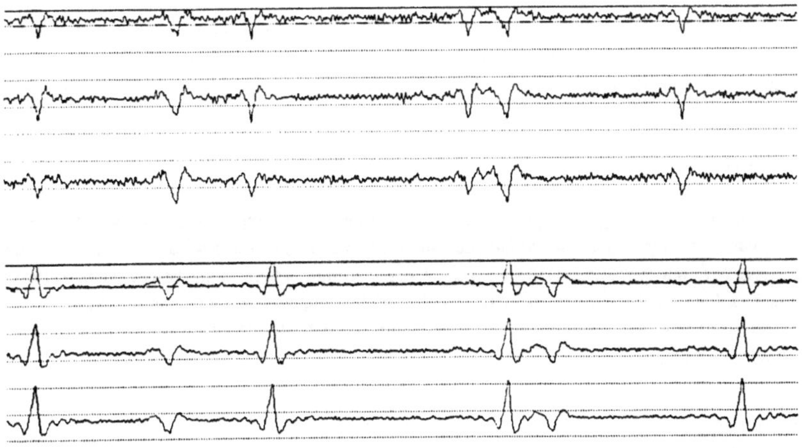

Fig. 2: fMCG recorded in the 26th week (top), 3 magnetic leads and 33rd week (bottom, 3 magnetic leads) of a normal pregnancy. Fetal heart rate 143/min (26th week) and 132/min (33rd week)

Table 1: Mean values for fetal heart rate as well as the amplitudes and time intervals of the fetal P-wave and QRS-complex during different gestational ages.

gestational week	n	evaluation possible (%)	P-wave detectable (%)	heart rate	amplitudes (pT)		intervals/durations (ms)		
					P	QRS	P	P-Q	QRS
15-20	4	25	0	157 ±0	-	0.30 ±0	-		43 ±0
21-25	3	100	0	138 ±8.5	-	0.35 ±0.1	-	-	27 ±7.7
26-30	15	93	47	139 ±10.1	0.33 ±0.18	1.07 ±1.1	45 ±8.1	109 ±11	40 ±10
31-35	11	100	73	136 ±8.0	0.46 ±0.42	1.59 ±0.97	41 ±5.3	103 ±9.3	46 ±6.5
36-40	7	100	85	135 ±4.9	0.46 ±0.10	1.88 ±0.94	51 ±12	121 ±12	51 ±8.5

Depending on the gestational age, the QRS complexes were detectable in 25% to 100% of recordings and the P-wave could be documented in 0 to 85%. The T-wave was detectable only in 3 recordings, the reason for its rare detectabilty most probably being its low frequency content.

DISCUSSION

Fetal magnetocardiography permits the detection and monitoring of fetal heart actions even in early pregnancy. Average values for amplitudes and time measures of the fetal MCG tend to increase during pregnancy, but the interindividual variation remains high, which might be caused by different recording positions and background activity. Larger studies, improvement of recording systems, and standardization of recording positions would be necessary to establish reference values for the fMCG in different stages of pregnancy.

REFERENCES

1) Carter, M.C. and Beard, R.W. Fetal heart rate monitoring using the abdominal fetal electrocardiogram, Br J Obstet Gynecol 1980, 87: 393-401
2) Dunajski, Z. and Peters, M. Development of the fetal magnetocardiogram from the 13th week of gestation onwards. In: Baumgartner, C., Deecke, L., Stroink, G., and Williamson., S. Biomagnetism: Fundamental research and clinical applications, Amsterdam, Elsevier/IOS, 1995, 704-708
3) Kariniemi, V., Ahopelto, J., Karp, P.J., and Katila, T.E. The fetal magnetocardiogram, J Perinat Med 1974, 2: 214-216
4) Roche, J.B., and Hon, E.H. The fetal electrocardiogram. Comparison of lead systems, Am J Obstet Gynecol 1965, 92: 1149-1159

Clinical Experience with the Combination of Multichannel Magnetocardiography and Magnetic Resonance Imaging for the Localization of Idiopathic and Secondary Ventricular Cardiac Arrhythmias

Achenbach, S., Moshage, W., and Bachmann, K.
Medical Clinic II, University of Erlangen-Nürnberg

INTRODUCTION

Multichannel magnetocardiography permits the localization of ventricular cardiac arrhythmias with a spatial accuracy of about 10 to 15 mm [3]. The combination with magnetic resonance imaging permits the correlation of MCG localization results to cardiac anatomy, and, in case of structural heart disease, to the pathological substrate. The following study gives an overview of MCG localization results in patients with and without structural heart disease who presented with ventricular arrhythmias.

PATIENTS AND METHODS

A total of 50 patients with premature ventricular complexes or ventricular tachycardia of idiopathic nature or due to underlying heart disease were investigated. 12 patients presented with ventricular tachycardia (VT), 3 of these had coronary artery disease, one cardiomyopathy, and in 8 VT patients structural heart disease had been ruled out. 38 patients had premature ventricular complexes, in 15 patients due to coronary artery disease, in 2 patients associated to right ventricular dysplasia, in 1 patient after chest trauma, one patient due to a large postoperative hematoma compressing the left ventricular wall, and in 19 patients of idiopathic nature. In addition to magnetocardiography, the patients were investigated by echocardiography, stress testing, magnetic resonance imaging, and, if clinically indicated, by coronary angiography and laevocardiography, in 19 cases also by invasive electrophysiologiocal mapping.

The patients were investigated using biomagnetic multichannel equipment (Krenikon® or MagnesII®) inside a magnetically shielded rrom. The MCG recordings lasted for 2 to 10 minutes, depending on the frequency of the arrhythmia. After signal-averaging and correction for baseline-shift and superimposed fields, the localizations of electric activity were determined based on the field distribution at the onset of the ectopic beats using the model of a current dipole in a homogeneous volume conductor. The localization results were transferred to the patients' MR images by use of fiducial markers.

MR images were acquired immediately after the MCG investigation using a Magnetom® 1.5T MR imaging system. Cross-sectional images in axial and coronal, sometimes also sagittal planes were obtained with spin-echo sequences (TR 20ms, slice thickness 8 mm, gap 1 - 2 mm).

RESULTS

In all patients with structural heart disease, the localization results could be assigned to the site of myocardial damage (margin of ischemic area or aneurysm in patients with coronary artery disease, right ventricular myocardium in patients with arrhythmogenic right ventricular dysplasia, right ventricluar anterior wall in a patient after chest trauma, and to the site of left ventricular impression by a large postoperative hematoma in one patient after cardiac surgery, see figg. 1- 3). In the majority of patients with idiopathic ventricular arrhytmias, the origin of ectopic beats was found to be in the right ventricular outflow tract (see table 1).

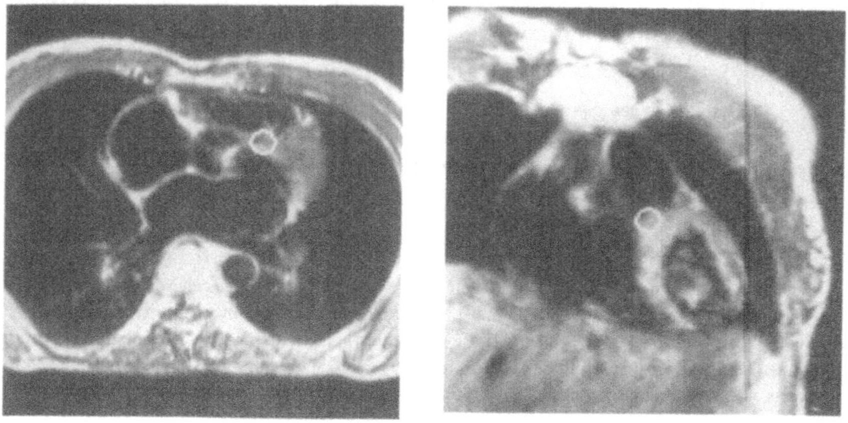

Fig. 1: Localization of idiopathic PVC in the right ventricular outflow tract (left, axial and right: frontal MR image)

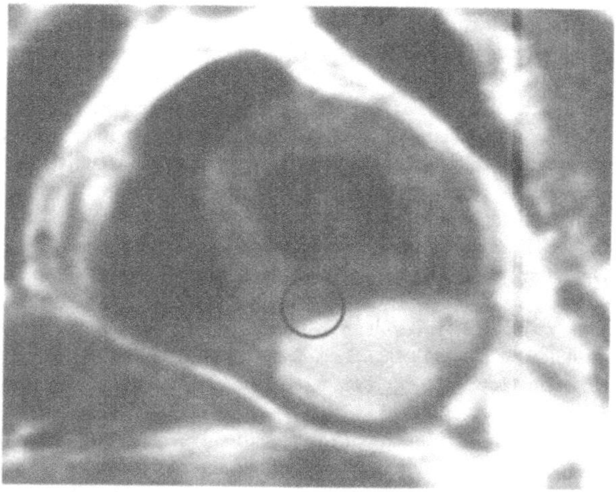

Fig. 2: MCG localization result of PVC in a patient with a large intrapericardial hematoma after cardiac surgery (aortic valve replacement). The localization result is found at the diaphragmal left ventricular wall, the site of impression of the left ventricle by the intrapericardial mass (coronal MR image).

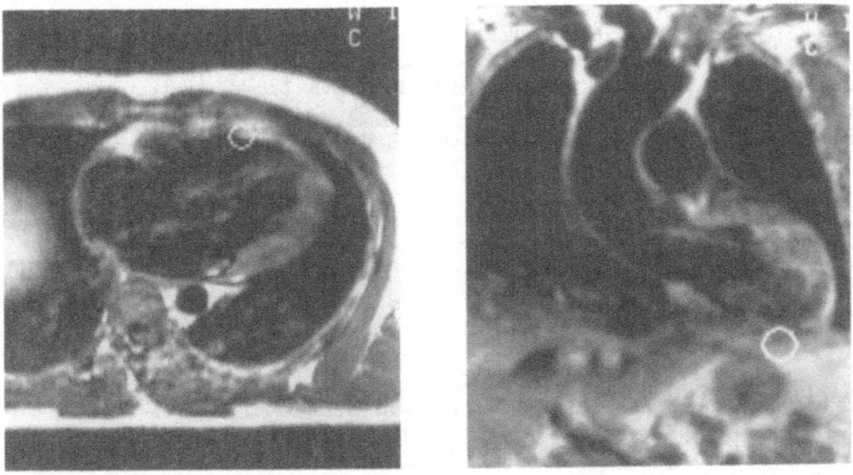

Fig. 3: Right: MCG localization result in a patient after anterior myocardial infarction. The origin of PVC is found in the left ventricular apex, where the coronal MR image shows a reduced thickness of the myocardial wall.
Left: MCG localization result of PVC in a patient with right ventricular dysplasia, the origin is found in the subtricuspid region of the right ventricular anterior wall, a predelection site for arrhythmias in right ventricular dysplasia [1, 6].

Table 1: Summary of MCG localization results in 50 patients with idiopathic or secondary cardiac arrhythmias.

heart disease	n	MCG localization
coronary artery disease	18 (15 PVC, 3 VT)	left ventricle, at the margin of ischemic area (18)
right ventricular dysplasia	2 (PVC)	right ventricular anterior wall (2)
chest trauma	1 (PVC)	right ventricular anterior wall (1)
dilatative cardiomyopathy	1 (VT)	interventricular septum (1)
intrapericardial mass (hematoma)	1 (PVC)	diaphragmal wall of LV (site of impression by hematoma)
idiopathic PVC/VT, heart disease ruled out	27 (19 PVC, 8 VT)	right ventricular outflow tract (21) proximal interventricular septum (1) right ventricular anterior wall (1) left ventricular lateral wall (1) left ventricular posterior wall (3)

DISCUSSION

Magnetocardiography in combination with magnetic resonance imaging permits to establish the correlation between functional cardiac disorder and structural myocardial damage. While in 23 patients with underlying heart disease, the localization results of PVC and VT could be assigned to the site of structural damage, the majority of idiopathic PVC and VT were found to originate in the right ventricular outflow tract, as has been observed before [2, 4].

REFERENCES

1) Breithardt, G., Borggrefe, M., and Wichter, T. Catheter ablation of idiopathic right ventricular tachycardia, Circulation 1990, 82: 2273-2276

2) Buxton, A.E., Marchlinski, F.E., Simson, M.B., Cassidy, D., and Josephson M.E. Repetitive, monomorphic ventricular tachycardia: Clinical and electrophysiological characteristics in patients with and patients without organic heart disease, Am J Cardiol 1984, 54: 997-1002

3) Moshage, W., Achenbach, S., Göhl, K., and Bachmann, K. Evaluation of the non-invasive localization accuracy of cardiac arrhythmias attainable by multichannel magnetocardiography (MCG), Int J Cardiac Imaging 1996, 12 (in press)

4) Ritchie, A.H., Kerr, C.R., Qi, A., and Yeung-Lai-Wah, J.A. Nonsustained ventricular tachycardia arising from the right ventricular outflow tract, Am J Cardiol 1989, 64: 594-598

5) Wichter, T., Borggrefe, M., and Breithardt, G. Die arrhytmogene rechtsventrikuläre Erkrankung, Z Kardiol 1991, 80: 107-125

Localization of the Origin of Ventricular Arrhtyhmias: Comparison of ECG Morphology and Localization Based on ECG Algorithms to Multichannel Magnetocardiography

Achenbach, S., Moshage, W., and Bachmann, K.
Medical Clinic II, University of Erlangen-Nürnberg, Germany

INTRODUCTION

Multichannel magnetocardiography in combination with magnetic resonance imaging permits the reliable localization of ventricular arrhythmias and assignment to anatomical structures [5]. Especially prior to interventional treatment of ventricular arrhythmias, the localization of the arrhythmogenic substrate is important. Some parameters of the surface electrocardiogram can suggest the site of origin of arrhythmias. Additionally, algorithms have been published that permit the assignment of the origin of ventricular arrhythmias to segments of the left or right ventricle based on the morphology of the beats in the 12-lead surface ECG [2, 3, 4].

The aim of this study was to compare the ECG morphology and the results of 3 published algorithms for the localization of ventricular arrhythmias based on the 12-lead ECG to the non-invasive multichannel magnetocardiographic localization in 21 patients.

PATIENTS AND METHODS

21 patients with premature ventricular complexes (PVC) or ventricular tachycardia (VT) due to coronary artery disease, other heart disease, or of idiopathic nature were investigated. The patient group consisted of 18 men and 3 women. 15 patients presented with premature ventricular complexes (PVC), 6 had ventricular tachycardia (VT). The arrhythmias were due to coronary artery disease in 11 patients (6 patients with myocardial infarction, 5 patients CAD without myocardial infarction). One patient had arrhythmogenic right ventricular dysplasia, one other had developed arrhythmias after chest trauma, 8 patients had idiopathic ventricular arrhythmias.

The magnetocardiogram was recorded with a multichannel system (Krenikon® or Magnes II®) in a magnetically shielded room for 2 to 10 minutes (1000 Hz sampling rate, bandwidth 0.1 to 200 Hz). The 12-lead ECG was recorded inside the shielded room, simultaneous to the magnetic recordings. MCG data were evaluated by signal-averaging, correction for superimposed fields and baseline offset, and localization using the model of a single moving current dipole in a homogeneous volume conductor (halfspace or sphere). Using fiducial markers, the MCG localization results were transferred to the patients' magnetic resonance images in axial, frontal and in some cases sagittal cross-sections for anatomical correlation.

Algorithms published by Josephson [2], Miller [4], and Kuchar [3], were applied to the patients' 12-lead surface electrocardiograms. Following the algorithms, the VT and PVC were classified as to determine their origin. The algorithms published by Miller and Kuchar were applied only to arrhythmias arising from the left ventricle.

Localization results obtained by magnetocardiography and based on the ECG algorithms were compared.

RESULTS

All 8 PVC/VT with left ventricular origin as determined in the magnetocardiogram displayed right bundle branch block (RBBB) morphology in the surface ECG, all 8 PVC/VT with right ventricular origin had left bundle branch block (LBBB) morphology. In 5 ectopies with a septal origin according to the MCG, RBBB was found in 3 and LBBB in 2 cases. All 7 PVC/VT with an origin in the right ventricular outflow tract (all of idiopathic nature) displayed left bundle branch block morphology and an inferior axis in the ECG (see fig. 1).

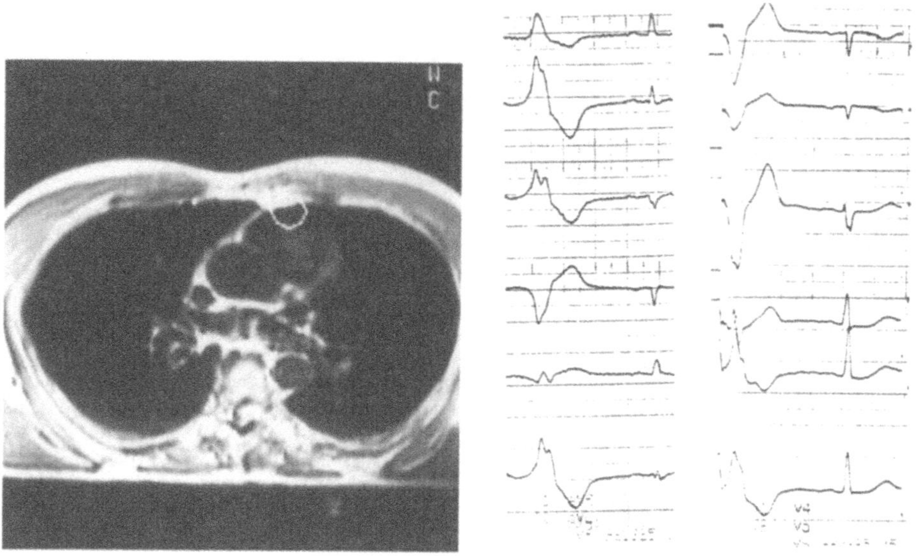

Fig. 1: 32 year-old female patient with idiopathic PVC and VT. The MCG localization result is in the right ventricular outflow tract (left, transferred to the axial MR image). The ECG (right) shows an inferior axis and left bundle branch block morphology.

When the algorithm according to Josephson [2] was applied to the 21 ventricular arrhythmias, 13 cases could not be classified. Out of the remaining 8 cases, 7 were localized at the same or adjacent to the segment which was identified as the origin by MCG.

The algorithms published by Miller [4] and Kuchar [3] are limited to arrhythmias arising from the left ventricle, therefore, a comparison to MCG was attempted only in 13 patients with a left ventricular or septal origin of the ectopic beats according to the MCG. The algorithm given by Miller yielded the same localization result as the MCG in one case, in 3 cases, an adjacent segment was determined as the respective site of origin, one localization result was more than on segment distant from the MCG localization, and the algorithm yielded no localization in 8 of the 13 patients. In 4 patients, the algorithm published by Kucher yielded the same localization as MCG, adjacent segments were determined in 5 cases, and in 4 patients, the localization results were more than one segment apart.

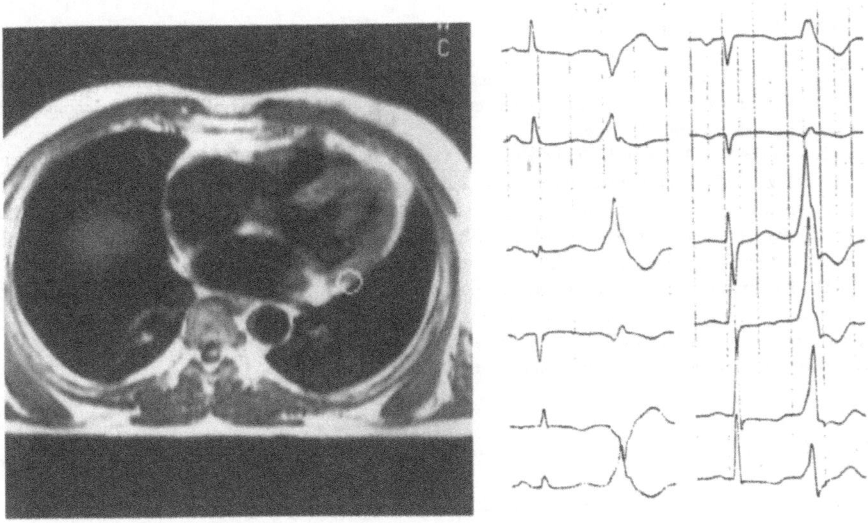

Fig. 2: 56 year old male patient with coronary artery disease. MCG localization result in the left lateral wall (left, axial MR image). Surface ECG (right): No classification possible using the algorithms published by Josephson and Miller, localization result accoriding to Kuchar: "mid-ventricular (long axis)"; "anterior, central (short axis)".

Table 1: Summary of the correlation between MCG and ECG localization results of ventricular tachycardia and PVC in 21 patients (algorithms given by Miller and Kuchar were applied only to 13 cases with left ventricular origin of the arrhythmia)

author	classification impossible	same segment as MCG	adjacent segment to MCG localization	more than 1 segment distant to MCG localization
Josephson [2]	13/21	3/21	4/21	1/21
Miller [4]	8/13	1/13	3/13	1/13
Kuchar [3]	0/13	4/13	5/13	4/13

DISCUSSION

In 21 patients with ventricular arrhythmias, the morphology of the surface ECG and the results of published localization algorithms based on characteristics of the surface electrocardiogram were comperd to magnetocardiographic localization results.

The comparison between MCG localization and axis in the surface ECG as well as bunlde branch block morphology confirmed the expected behaviour: left ventricular arrhythmias had RBBB morphology, while all arrhythmias arising in the right ventricle displayed LBBB morphology. Arrhythmias with an origin in the septum according to the MCG could have either LBBB or RBBB morphology. All arrhythmias arising from the right ventricular outflow tract had LBBB pattern and an inferior axis, as priorily reported [1, 6].

The localization results obtained with the published ECG algorithms did not correspond well to the MCG localization results. However, these algorithms were developed exclusively for ventricular tachycardia in patients after myocardial infarction which might have influenced the results in our patient group, which included only 6 patients after myocardial infarction.

REFERENCES

1) Buxton, A.E., Marchlinski, F.E., Simson, M.B., Cassidy, D., Josephson M.E. Repetitive, monomorphic ventricular tachycardia: Clinical and electrophysiological characteristics in pations with and patients without organic heart disease, Am J Cardiol 1984, 54: 997-1002

2) Josephson, M.E., Horowitz, L.N., Waxmann, H.L., Cain, M.E., Speilman, S.R., Greenspan A.M., Marchlinski, F.E., and Ezri, M.D. Sustained ventricular tachycardia: Role of the 12-lead electrocardiogram in localizing site of origin, Circulation 1981, 64: 257-272

3) Kuchar, D.L., Ruskin, J.N, and Garan, H. Electrocardiographic localization of the site of origin of ventricular tachycardia in patients with prior myocardial infarction, J Am Coll Cardiol 1989, 13: 893-900

4) Miller, J.A., Machlinski, F.E., Buxton, A.E., and Josephson, M.E. Relationship between the 12-lead electrocardiogram during ventricular tachycardia and endocardial site of origin in patients with coronary artery disease, Circulation 1988, 77: 759-766

5) Moshage W., Achenbach, S., Göhl, K., Weikl, A., Bachmann, K., Wegener, P., Schneider, S., and Härer, W. Biomagnetic localization of ventricular arrhythmias, Radiology, 1991, 180: 759-766.

6) Ritchie, A.H., Kerr, C.R., Qi, A., Yeung-Lai-Wah, J.A. Nonsustained ventricular tachycardia arising from the right ventricular outflow tract, Am J Cardiol 1989, 64: 594-598

Comparison of the QRS Duration and RMS40 in Signal-Averaged Recordings of the Multichannel MCG to the Signal-Averaged ECG

Achenbach, S., Moshage, W., Flüg, M., Diem, B., Schibgilla, V., and Bachmann,K.

Medical Clinic II, University of Erlangen-Nürnberg, Germany

INTRODUCTION

The signal-averaged ECG is an accepted tool for the identification of patients prone to malignant arrhythmias. There have been several reports of risk assessment using the magnetocardiogram (SA-MCG) [2, 3, 5]. Reports of time-domain analysis of the signal-averaged QRS complex in the MCG using the Simson criteria [4], especially the filtered QRS duration, have been inconsistent as to the correlation of MCG values and predictive accuracy to the SA-ECG [2, 3, 5]. We investigated the correlation of the time-domain analysis of signal-averaged multichannel MCG recordings to the signal-averaged ECG.

PATIENTS AND METHODS

38 male patients were investigated (age: 40 to 75 yeras). 32 patients had coronary artery disease, 25 of these presented after myocardial infarction. 2 patients had cardiomyopathy, 1 aortic stenosis. Cardiac disease had been ruled out in 2 patients.

The magnetocardiogram was recorded for 240 seconds at a sampling rate of 1024 Hz using a MagnesII® multichannel system inside a magnetically shielded room.

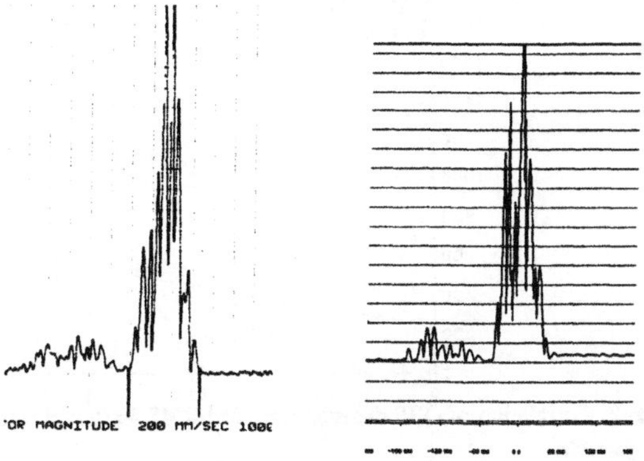

Fig. 1: SA-ECG (left) and SA-MCG (right) in a 51-year old patient after myocardial infarction. RMS channels of the 3 ECG leads (x/y/z) and the 37 magnetic channels are shown.

SA-ECG:
QRS duration 86 ms,
RMS40 37 μV;

SA-MCG:
QRS duration 85 ms,
RMS40 290 fT.

37 magnetic channels were recorded, the dewar with the sensors was placed in a standardized position over the patient's chest: the midpoint of the sensor plane was positioned as close as possible over the chest at the border between the lower 1/3 and upper 2/3 of a connecting line between the left mamilla and jugulum. Immediately after the MCG recording, the signal-averaged ECG (SA-ECG) was recorded inside the magnetically shielded room using a LP Plus® SA-ECG system; the x-, y-, and z-leads (Frank leads) were recorded with online-averaging of the QRS complexes until a noise level below 0.7 μV was reached. MCG data was averaged off-line. Subsequent evaluation followed the ACC guidelines [1], including 40Hz high-pass filtering. For the RMS channel of the 37 magnetic leads and 6 ECG leads, respectively, the noise level, total QRS duration, and RMS40 (amplitude of the terminal 40 ms of the QRS complex) were determined, the corresponding MCG and ECG values were compared.

RESULTS

The mean QRS duration in MCG was 90.7±20.0 ms vs. 99.6±16.5 ms in the SA-ECG. The correlation between SA-ECG and SA-MCG was poor (r = 0.51). RMS 40 also correlated poorly between SA-MCG (mean: 283.8±131 fT) and SA-ECG (mean: 20.93±8.14 μV) with a correlation coefficient of r = 0.22. If only recordings with a noise level below 10 fT were taken into account (n = 23), a close correlation between ECG and MCG QRS duration was found (r = 0.94, mean values: SA-MCG 96.1 ms vs. ECG 99.0 ms), while the correlation for RMS40 remained low (r = 0.67), probably due to high interindividual variation of the amplitude in the MCG recordings.

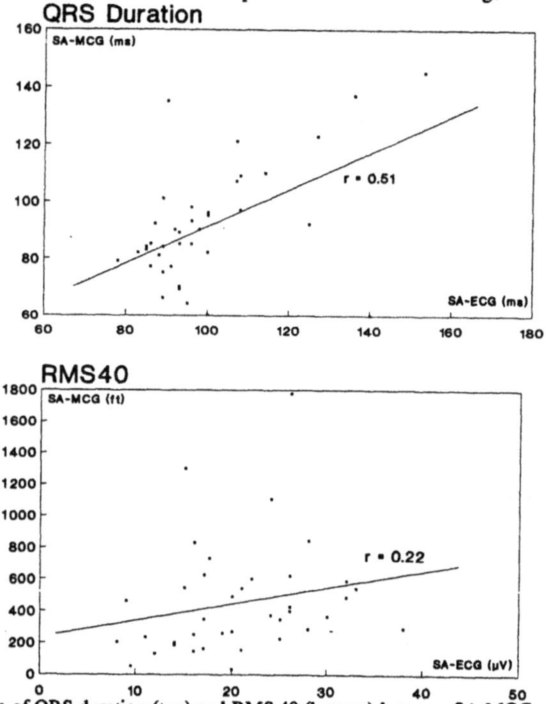

Fig. 2: Correlation of QRS duration (top) and RMS 40 (bottom) between SA-MCG and SA-ECG for all 38 patients.

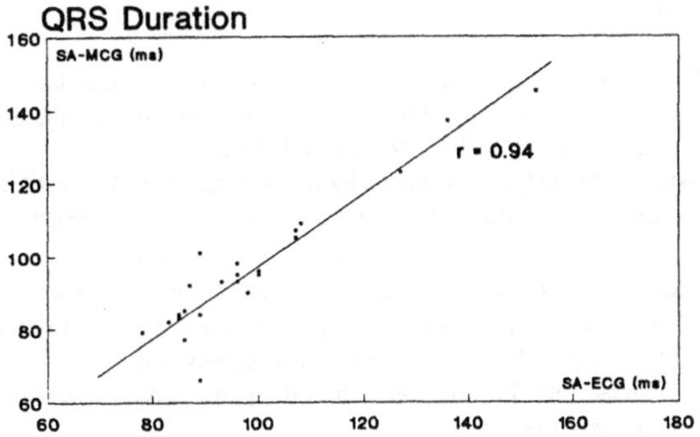

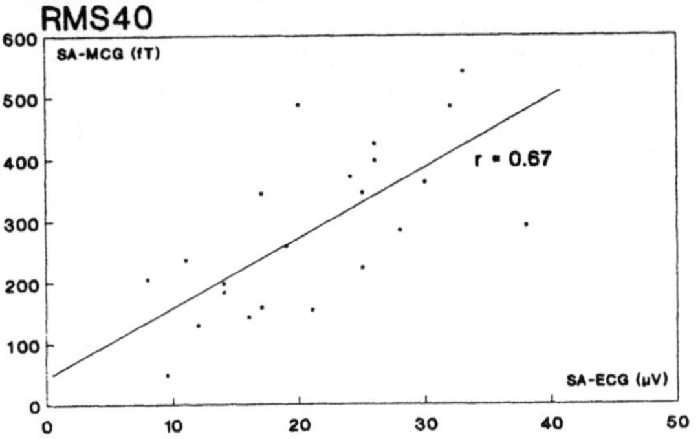

Fig. 3: Correlation of QRS duration (top) and RMS 40 (bottom) in 23 patients with a noise level below 10 fT in the MCG recordings. While a close correlation is observed for QRS duration, the correlation is still rather low for RMS40.

DISCUSSION

QRS duration and RMS40 were found to correlate poorly between SA-MCG and SA-ECG. The reason for the low correlation probably was a higher noise level in the MCG recordings. Since the end-point of the QRS complex is defined according to the baseline noise, higher noise in the MCG recordings causes early definition of the end of the QRS complex and thus shorter QRS duration and higher RMS 40 than in ECG recordings. An improvement in correlation was found if only MCG recordings with a low noise level (<10 fT) were included in the evaluation.

While the QRS duration seems to correlate well between SA-MCG and SA-ECG if and only if a satisfactory noise level is reached, the correlation of RMS40 remains low. Further studies will have to collect more data on the predictive accuracy of the SA-MCG in comparison to the SA-ECG.

REFERENCES

1) ACC policy statement: Standards for analysis of ventricular late potentials using high-resolution or signal-averaged electrocardiography. JACC 1991, 17: 999-1006

2) Mäkijärvi, M. Magnetocardiography and risk analysis, In: Hoke, M., Erné, S.N., Okada, Y.C., and Romani, G.I. Biomagnetism: Clinical Aspects, Amsterdam, Excerpta Medica, 1992, 523-529

3) Montonen, J. Magnetocardiography in identification of patients prone to malignant arrhythmias, In: Baumgartner, C., Deecke, L., Stroink, G., and Williamson S.J. Biomagnetism: Fundamental Research and Clinical Applications, Amsterdam, Elsevier, 1995: 606-611

4) Simson, MB. Use of signals in the terminal QRS complex to identify patients with ventricular tachycardia after myocardial infarction, Circulation 1981, 64 (2): 235-242

5) Weismüller, P., Richter, P., Abraham-Fuchs, K., Härer, W., Schneider, S., Höher, M., Kochs, M., Edrich, J., and Hombach, V. Spatial differences of the duration of ventricular late fields in the signal-averaged magnetocardiogram in patients with ventricular late potentials, PACE 1993, 16: 70-79

Investigation of Fetal Cardiac Arrhythmias by Magnetocardiography

Achenbach, S.[1], Moshage, W.[1], Menéndez, T.[1], Beinder, E.[2], Bittl, A.[2], Hofbeck, M.[3], Buheitl, G.[3], Singer, H.[3], and Bachmann, K.[1]

[1]Medical Clinic II, [2]Department of Obstetrics and Gynecology, [3]Department of Pediatric Cardiology, University of Erlangen-Nürnberg, Germany

INTRODUCTION

While the recording of the fetal ECG is difficult or impossible especially during late pregnancy [1, 4], fetal magnetocardiography (fMCG) permits the registration of fetal cardiac activity during the second half of pregnancy, in some cases even much earlier [2, 3, 5]. We studied the ability of the fetal magnetocardiogram to non-invasively document and analyze fetal cardiac arrhtyhmias.

PATIENTS AND METHODS

33 women in their 16th to 40th week of pregnancy were investigated. In 5 of these pregnancies, fetal cardiac arrhythmias had been priorily detected in doppler ultrasound investigations, while no fetal arrhythmias were known before the MCG investigation in the other 28 pregnant women. All women gave informed consent to the investigation.

The fetal magnetocardiogram (fMCG) was recorded over the mother's abdomen for 3 to 5 minutes using biomagnetic multichannel systems (Krenikon® or MagnesII®) inside a magnetically shielded room. If possible, the position of the fetus was determined immediately before the investigation by ultrasound in order to place the dewar as close as possible to the fetal heart. The recordings were performed with a sampling rate of 1000 Hz, the bandwidth was 0.016 to 250 Hz. 37 magnetic channels were recorded along with 6 maternal ECG leads.

In postprocessing, filters were applied only if necessary to suppress background noise or artifacts due to maternal respiration and pulse. In selected cases, maternal heart beats were removed from the magnetic channels by cross-correlation, aligment and subtraction from the raw data. The recordings were analyzed for fetal heart rate and arrhythmias, averaging of the fetal heart beats yielded noise-reduced data for further analysis.

RESULTS

In 12 cases, one or several kinds of fetal arrhythmias were detected by fMCG. The most common arrhythmia were supraventricular ectopic beats which were recorded 7 cases, only one of which had been priorily detected by doppler ultrasound). Since atrial activity could not in all cases be reliably detected in the fMCG recordings, supraventricular arrhythmias could appear as premature beats with narrow QRS complexes or, in case of blocked conduction to the ventricles, pauses in the normal heart rhythm.

Fig. 1: fMCG showing ectopic beats (arrows) with narrow (right) and wide QRS complexes (left) in the 30th week of an otherwise normal pregnancy.
Real-time data, 3 magnetic channels and maternal ECG lead shown.

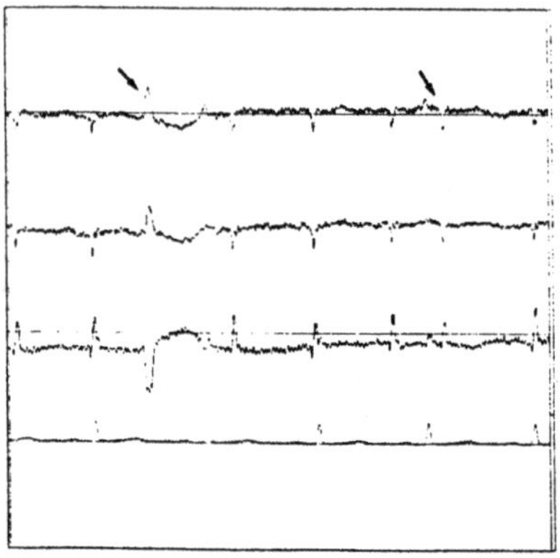

In 2 cases, premature beats with wide ventricular complexes were observed, corresponding to supraventricular beats with aberrant conduction or bundle branch block or ventricular ectopic beats (see figg. 1, 2). In one case, atrial flutter with 2:1 to 4:1 conduction to the ventricles was documented in a fetus with hydrops due to cardiac failure (see fig. 3). In another fetus with cardiac failure, torsade de pointes could be documented which were confirmed in the postpartal ECG. Fetal arrhythmias with a slow heart rate were detected in two cases, one fetus had complete AV-block with a ventricular rate of 78/min, the other AV-Block II° with 2:1 conduction to the ventricles and a heart rate of 70/min.

DISCUSSION

Fetal magnetocrdiography permits the recording and analysis of fetal cardiac activity, first reports of the fMCG date back to 1974 [1]. This, to our knowledge, is the first systematic investigation about the detection and analysis of fetal cardiac arrhythmias by magnetocardiography. We could demonstrate that fetal magnetocardiography permits the reliable detection and investigation of fetal cardiac arrhytmias. However, the analysis of single premature beats can be complicated due to

Fig. 2: Signal-averaging of normal (top) and premature QRS complexes permits further analysis due to improved signal-to-noise ratio.
The lack of atrial activity in the ectopic beats (bottom) suggests ventricular origin of the arrhythmia.

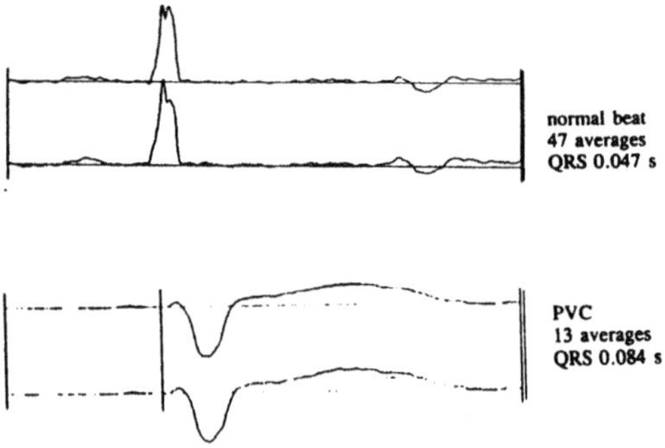

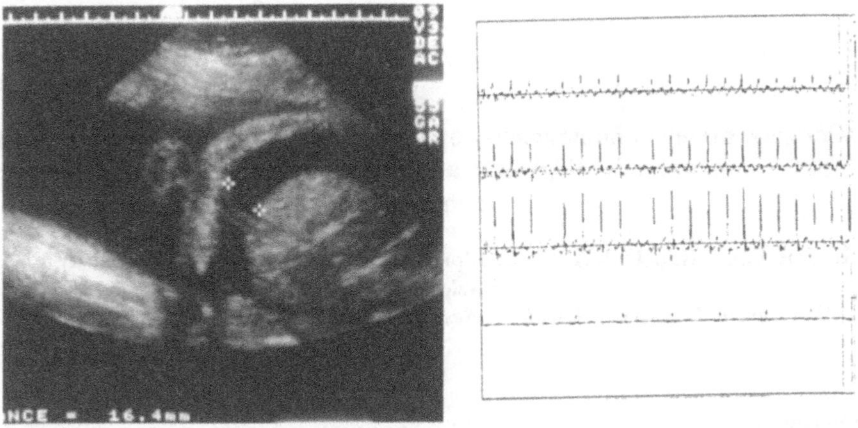

Fig. 3: Left: Fetal hydrops (ascites) due to heart failure seen in ultrasound, 36th week of pregnancy. Right: fMCG reveals atrial flutter with 2:1 and 4:1 conduction to the ventricles, which was confirmed after preterm delivery, when transoesophageal overdrive stimulation to sinus rhythm was performed.

the fact that atrial activity cannot be reliably detected in all recordings. The overall occurence of fetal cardiac arrhythmias is frequent even in otherweise normal pregnancies, and the detection of arrhythmias is more frequent in MCG recordings than in doppler sonographic studies.

REFERENCES

1) Carter, M.C. and Beard, R.W. Fetal heart rate monitoring using the abdominal fetal electrocardiogram, Br J Obstet Gynecol 1980, 87: 393-401

2) Dunajski, Z. and Peters, M. Development of the fetal magnetocardiogram from the 13th week of gestation onwards. In: Baumgartner, C., Deecke, L., Stroink, G., and Williamson., S. Biomagnetism: Fundamental research and clinical applications, Amsterdam, Elsevier/IOS, 1995, 704-708

3) Kariniemi, V., Ahopelto, J., Karp, P.J., and Katila, T.E. The fetal magnetocardiogram, J Perinat Med 1974, 2: 214-216

4) Roche, J.B., and Hon, E.H. The fetal electrocardiogram. Comparison of lead systems, Am J Obstet Gynecol 1965, 92: 1149-1159

5) Zhuralev, Y.E., Rassi, D., and Emery, S.J.: Clinical assessment of fetal magnetocardiography. In: Baumgartner, C., Deecke, L., Stroink, G., and Williamson., S. Biomagnetism: Fundamental research and clinical applications, Amsterdam, Elsevier/IOS, 1995, 700-703

Contraventions of homogeneity of ventricular repolarization according to data of magnetocardiography and their connection with ventricular tachycardia

Bobrov V.A., Gapelyuk A.V., Chaikovsky I.A., Stadnyuk L.A., Sosnitsky V.N., Kozlovsky V.I.

Biomagnetic Laboratory, Ukrainian Research Institute of Cardiology, Kiev, Ukraine

Introduction

Paroxysmal cardiac rhythm disturbance (PCRD) is one of the most urgent problems of modern cardiology. According to data from the American Heart Association sudden death of 300-600 thousand of people from PCRD occurs annually. After the first arrhythmia attack there generally follows a decrease in physical tolerance and often invalidity. It is known that inhomogeneity of myocardial excitation is the main factor in generation of arrhythmia. Cardiomagnetic field mapping is an efficient way of evaluation of the homogeneity of the underlying electrophysiological processes. The advantages of this approach are high spatial resolution and sensitivity to local tangential currents which appears on the borders of myocardial zones with different electrical properties. Several methods for estimation of homogeneity have been proposed [1,2]. Such methods provide stable results for measurements in shielded environments yet are less robust in unshielded conditions [3].

The objective of this study is to examine new criteria of ventricular repolarization disturbances suitable for analysis of magnetocardiograms obtained in unshielded environments.

Methods

25 patients (19 male and 6 female, mean age 45 ± 3.5) suffering from ventricular tachycardia (VT) were examined. This group was divided into subgroups depending on VT etiology: 17 of the patients had Ischemic Heart Disease (IHD) and VT; 8 of the patients, mainly with non-ischemic etiology, had myocarditic cardiosclerosis and VT. For comparison, two control groups without arrhythmia were examined: 14 male patients with IHD and 18 healthy persons.

MCG mapping was performed by means of the single-channel SQUID-magnetometer system coupled with an axial second order gradiometer at the Biomagnetic Laboratory of the Ukrainian Cardiology Center. Magnetic field registration was carried out using a 6*6 point rectangular grid of 4 cm step. A selective averaging algorithm was used to synchronise data from each measurement position and improve signal to noise ratio. During the ST-T interval, 32 maps of the magnetic field distribution were constructed over evenly spaced time intervals.

Two criteria for ventricular repolarization homogeneity were used:

1. The number of local extrema (positive and negative) was determined in the first and seconds halves of the examined interval.
2. The homogeneity coefficient (HC) was calculated in the following way:
   - isomagnetic maps were constructed for the 32 slices of the ST-T interval (first point J, last point peak of T wave)
   - for each examination the field values, over the whole interval, were normalized relative to the maximum;
   - We defined the measure Coefficient of Extremality (CE) as the sum of all differences between magnetic field strengths in the local extrema positions to neighbouring measurement points.
   - for the set of maps of ST-T interval dependence between CE and number of maps is well approximated by a second order polynomial function (Fig.1);
   - for estimation of deflection of real CE values from square dependence we used the cross - correlation between the CE values and calculated values for fitting square function;
   - this criterion was named Homogeneity Coefficient (HC);

- HC was approximately equal to 1.0 for normal subjects (Fig. 1), and demonstrated substantially fewer values for patients with ventricular rhythm abnormalities (Fig. 2).

Fig 1.

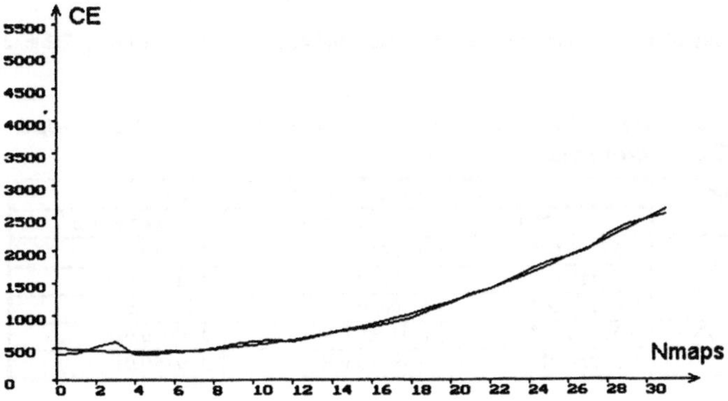

Fig 2.

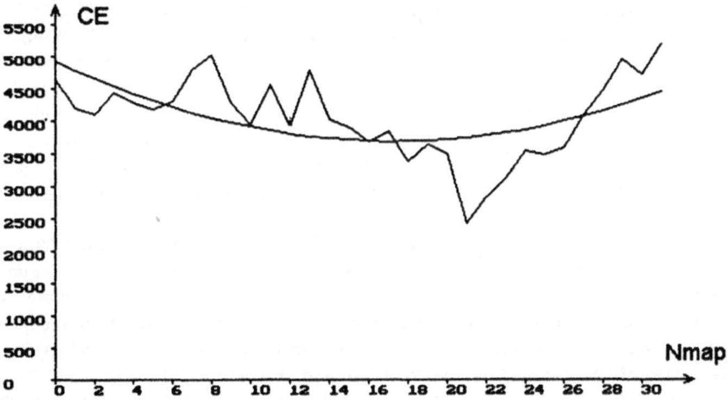

Results and discussions

The estimation results of ventricular repolarization homogeneity are presented in Tables 1,2.

Table. 1. The extrema number on magnetic maps of ventricular repolarization in patients groups with VT, IHD without VT and controls.

Examined groups	Extrema number in ST-T interval		
	1st half	2nd half	as a whole
VT pts. as a whole	44.4±2.6*	34.4±1.6*	78.7±3.6*
Pts. with recurring VT	60.1±4.6*	41.1±3.0*	100.2±6.0*
Pts. with IHD and VT	42.3±2.5*	34.3±2.0*	76.7±3.7*
Pts. with IHD and no VT	50.5±4.9*	30.1±1.9	80.7±5.5*
Control	26.5±0.8	24±0.4	50.5±0.8

Note: * - distinction of data compared to control is significant ($P<0.05$).

As seen from table 1, the total number of local extrema on magnetic field distribution maps during ventricular repolarization (VT) in patients is significantly higher than in normals, particularly those with recurring VT. A tendency to higher number of extrema in the second half of ventricular repolarization is detected in patients with IHD complicated by VT, than in IHD pts with no rhythm disturbances.

Calculation results of homogeneity coefficient in the studied groups are presented in Table. 2.

Table. 2. The HC values of ventricular repolarization in patients groups with VT, IHD without VT and controls.

Examined groups	HC values in ST-T interval		
	1st half	2nd half	as a whole
VT pts. as a whole	0.579±0.03*	0.754±0.05*	0.811±0.07*
Pts. with recurring VT	0.454±0.07*	0.688±0.07*	0.621±0.05*
Pts. with IHD and VT	0.611±0.06*	0.806±0.04*	0.844±0.03*
Pts. with IHD and no VT	0.551±0.07*	0.946±0.02	0.906±0.02*
Control	0.939±0.01	0.995±0.006	0.991±0.01

Note: * - distinction of data compared to control is significant ($P<0.05$).

First of all the high, near to 1.0 HC data, should be marked in healthy persons (Fig.1).

Then lower HC values were observed in VT pts, especially at recurring paroxysms (Fig.2). The evident difference between HC values in the whole VT pts. group and the group with recurring VT should be stressed. A considerable difference is also marked between HC values for IHD pts. groups with VT and without it. This difference was significant in the second part of the analyzed segment. However, at individual analysis of the results obtained in group IHD with VT pts, it was shown that repolarization homogeneity by both used criteria are normal in a considerable part of pts. (41%), but direction of equivalent current dipole (ECD) is different from normals. This subgroup included pts., who suffered from singular spontaneous or induced VT paroxysm at the acute period of myocardial infarction. This proves once more the idea of arrhythmogenic substrate instability in such pts [4].

Thus connection of the magnetic field fragmentation of ventricular repolarization and ventricular paroxysms severity was observed. A rise of extrema number and lowering of HC were found on the magnetic field maps of pts., suffering from arrhythmia. These changes were the more expressed the more severe was arrhythmic syndrome. As know additional extrema on the magnetic field maps originate from local tangential current, appearing on the borders of myocardial zones with different excitation conduction rate [1, 5]. Such electrophysiological state of myocardium favors to appearing of re-entry.

Remarkable are the considerable differences between ventricular repolarization homogeneity in different VT pts. groups. The group of pts. with recurrent VT is significantly differs from control, and from the IHD without VT group both by extrema number and by HC in the whole interval studied. This suggests evident inhomogeneity, fragmentation of ventricular repolarization and hence electrical myocardial instability in such pts. The group of pts. with IHD complicated by VT also considerably differs from control by both used criteria. However the difference between this group and the group of IHD pts. without VT is significant only in the final part of ST-T segment.

It is known, that disturbance of ventricular repolarization takes place in IHD pts. due to myocardial ischemia and cardiosclerosis [6]. In the group of IHD without VT such disturbances are observed only in the initial part of ventricular repolarization, but in group of IHD complicated by VT, ventricular repolarization inhomogeneity remains along the whole ST-T segment.

Thus it may be supposed that just repolarization inhomogeneity stability and just its presence in the final part of ST-T segment is a marked of high VT risk.

References:

[1] Lant, J., Stroink, G., Vardy, D., et al. Discrimination between patients with miocardial infarction and those with ventricular tachycardia using magnetic and potential map extrema and trajectory plot, Advances in Biomagnetism, 1989,

[2] Erne, S.N., Fenici, R.R., Hahlbohm, H.D., et al. High resolution izofield mapping in magnetocardiography. Nuovo Cimento, 1983, 2D, 291-300.

[3] Fenici, R.R., Melillo, G., 1993. Magnetocardiography:ventricular arrhythmias, European Heart J., 1993, 14(E), 53-60.

[4] Kushakovsky, M.S., Cardiac arrhythmias, St.Peterburg, Hyppocrat, 1992 (in Russian).

[5] Nomura, M., Fujino, K., Katayama, M., et al. Analysis of the T-wave of the Magnetocardiogram in Patients With Essential Hypertension By Means of Isomagnetic and Vector Arrow Maps, J. of Electrocardiology, 1988, 21: 174-182.

[6] Fu, L., Takanashi, N., Iinuma, H. et al. Experimental study on ischemic T inversion, Jap. Circulation J., 1971, 35: 1233.

Ventricular Repolarization Disturbances' Diagnostics in Chronic Ischemia Patients Evidence Derived from MCG

Chaikovsky, I.A., Lutay, M.I., Lomakovsky, A.N., Gapelyuk, A.V., Sosnitsky, V.N., Gushcha, B.G., and Minov, Yu.D.

Biomagnetic Laboratory, Ukrainian Research Institute of Cardiology, Kiev, Ukraine

Introduction

Analysis of electrophysiological processes occurring in ischemic heart disease (IHD) pts myocardium is still one of important tasks of electrocardiology. It is well known that the rest ECG abilities are very limited in this sense because 70-85% of ECG patterns in chronical IHD pts do not differ from normal [1]. During latest years MCG has more often been used in addition to ECG methods.

Compared to ECG this method has a series of principle advantages, in particular it has a high spatial resolution. A lot of investigations are devoted to analysis of MCG in IHD pts. with postinfarctional cardiosclerosis [2-5]. However there is known just a limited number of works, treating repolarization as a continuos process in IHD with no postinfarctional cardiosclerosis.

The aim of our work was to study disturbances features of ventricular repolarization in chronic IHD pts with help of MCG-mapping.

Materials and methods

16 male and 2 female pts (mean age 55 ± 3.6) suffering from IHD were observed, excertional angina pectoris of I-II functional classes (FC) -- 4 pts, and III-IV FC -- 14 pts. IHD was diagnosed by clinical (angina pectoris) bicycle ergometria positive data and coronary angiography (stenosing coronary atherosclerosis) data. Pts with concomitant hypertension, compound rhythm disturbances, cardiac failure of a higher than IInd stage by NIHA were not considered Control group -- 18 healthy persons (mean age 28.2 ± 2.9). MCG was carried out in all pts at admission and in 10 pts of them repeated after 10 days of antianginal therapy.

MCG mapping was performed by means of the one-channel SQUID-magnetometer at Biomagnetic Laboratory installed in the Ukrainian Cardiology Center. Magnetic field registration has been carried out at a rectangle grid 6×6 with 4 cm pitch. Measurements in each point during 30 seconds have been provided with simultaneous recording of the reference ECG-channel. On the base of 36 synchronous averaged MCG curves for a selected time interval of a cardiocycle (ST-T interval as the most sensitivity to ischemic alterations), isomagnetic maps in 32 sequential points at equal time intervals of the ST-T interval were built with help of two-dimensional interpolation algorithms. The state of main (greatest by magnetic induction value) positive and negative extrema was evaluated in every MCG map. The direction of equivalent current dipole (ECD) vector in the measuring plane was defined by these data. The ordinal map number was determined, starting from which the ventricular repolarization ECD direction becomes steadily normal, with the ECD being directed right-to-left, up-to-down.

Results and discussions

Steady normal ECD direction was detected in the control group members already from 1st-2nd maps of magnetic field of ST-T segment (see Table 1).

In IHD pts a different ECD direction in the initial ST-T segment was observed, more often left-to-right down-to-up and right-to-left down-to-up. ECD direction normalization took place much later than in control group, at average normalization being observed in map 24.6 ± 2 in pts of III-IV FC and in I-II FC pts in map 18.8 ± 1.7. Normalization of ECD direction in ST-T interval occurred earlier under antianginal therapy (Fig.1).

Table. 1. Alterations of ventricular repolarization ECD direction in IHD pts. before and after antianginal treatment course compared to control

Examined groups	Ordinal number of map, starting normalization of ECD direction	
	before treatment	after treatment
IHD pts. with exertional angina pectoris I-II FC	18.8±1.7	12.0±1.8*
IHD pts. with exertional angina pectoris III-IY FC	24.6±2.0	18.2±2.3*
Control	2.4±0.1	

Note: * - distinction of data compared to control is significant (P<0.05).

Fig 1 Isomagnetic maps for ST-T interval, (first map 50 msec after point J, step between maps is 50 msec, the arrow is ECD direction) for patient with IHD, excertional angina pectoris 3 FC.

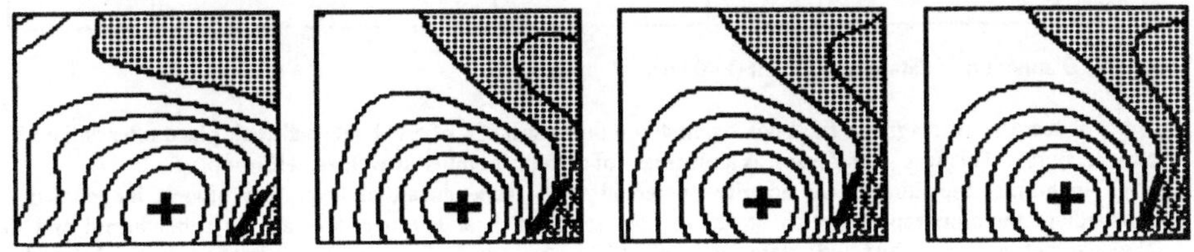

a) before antianginal treatment

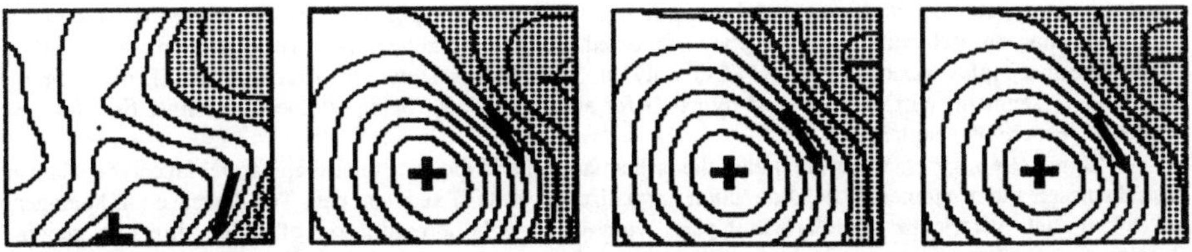

b) after antianginal treatment

To compare sensitivity of electro- and magnetocardiographic methods the ventricular repolarization analysis of the rest ECG of IHD pts. and control group members has been carried out (see Table 2).

Table. 2. Alterations of T-wave and ST-T segment by rest ECG data in IHD pts. and controls.

Examined groups	flattening T-wave	negative T-wave	high T-wave	ST -T segment depression
IHD pts. with exertional angina pectoris I-IV FC	13 (72%)	3(17%)	-	2(11%)
Control	-	-	2(11%)	-

Inversion of the terminal part of ventricular complex was registered only on ECG of 5 (27%) pts and merely little-flattening of T-wave was observed in others.

Besides ECD direction we analyzed homogeneity of ventricular repolarization of magnetic field distribution maps before and after courses of antianginal treatment. For that purpose the number of magnetic field extrema, in the first and second halves of studied intervals and within the whole interval were determined.

Homogeneity coefficient (HC) was calculated additionally that allows to perform integral estimation of the number and values of extrema [6]. Results are presented in tabel.3.

Table 3. Ventricular repolarization homogeneity alterations in IHD pts. before and after antianginal treatment course.

Examined groups	Extrema number as a whole	Extrema number in 1st half of ST-T segment	Extrema number in 2nd half of ST-T segment	HC as a whole	HC in 1st half of ST-T segment	HC in 2nd half of ST-T segment
before treatment	67.3±3.1	37.6±2.4	29.7±2.1	0.882±0.07	0.701±0.05	0.912±0.04
after treatment	81.7±3.9*	51.5±2.1*	30.2±1.4	0.734±0.07*	0.564±0.03*	0.881±0.05

Note: * - Distinction of data is significant ($P<0.05$).

A decrease of magnetic field homogeneity on the first half of ST-T interval took place after antianginal treatment course This was manifested as an increase of extrema number as well as HC drop.

In normal the duration of activity potential in subendocardial layers is always longer than in subepicardial, therefore repolarization starts in the subepicard and moves towards the subendocard and the total electromotive force is directed right - to - left and up - to - down. Such an ECD direction was observed in the control group during the whole ventricular repolarization process.

Transmembrane action potential of ischemic region is longer then normal and therefore the subendocardial layers are repolarized earlier for subepicardial ischemia. In this case the repolarization process moves from subendocardium to subepicardium [7,8]. Various types of T-wave alterations are found on ECG depending on ischemia level: these are flattened, isoelectric and negative. The inversion of terminal part of ventricular complex was observed by ECG only in 5 (27%) pts. However pathological ECD direction in the initial ST-T segment part was detected by MCG for all the patients, being retained for longer time for more expressed excertional angina.

It may be suggested that the higher the ischemia level and the later the subepicardial layers depolarization end, the nearer the occurence of ECD direction normalization to ST-T segment end. The decrease of homogeneity of the initial ventricular repolarization part, combined with normalization of ECD direction, may in our opinion, be of the same nature as rhythm disturbances for reperfusion syndrome. However this changes are not steady, and do not spread to the final part of ventricular repolarization. Thus the rest MCG mapping demonstrates high sensitivity towards repolarization direction changes in pts with myocardial ischemia, and may be used for evaluation of electrophysiological processes changes in ventricular myocardium.

References:

[1] Orlov, V.N., Electrocardiography manual, Moscow, Meditzina, 1984, (in Russian).

[2] Murakami, M., Nakaya, Y. Studies on the QRS wave of the magnetocardiogram in myocardial infarction, Jpn. J. Appl. Physiology, 1984, 14: 14-25.

[3] Nomura, M., Nakaya, Y., Fujino, K. et al. Magnetocardiographic studies of ventricular repolarization in old inferior myocardial infarction, Europ. Heart Journal, 1989, 10: 8-15.

[4] Oja, O.S., Nousiainen, J., Malmivuo, J., Uusitalo, A. Magnetocardiogram in anteroseptal and inferior infarctions, Biomagnetism: Clinical Aspects. Proc.of the 8th Intern. Conf. on Biomagnetism, Munster, 1991, 487-490.

[5] Lant, J., Stroink, G., Voorde, B. et al. Complementary nature of electrocardiographic and magnetocardiographic data in patients with ischemic heart disease, J. Electrocardiology, 1990, 23: 315-322.

[6] Bobrov, V.A., Frolov, A.I., Simorot, V.N. et al The estimate of the effect of propaphenon on electrophysiological indices of heart in patients with supraventricular tachycardia from data of electrophysiological and magnetocardiographic examination, Superconductive electronics and biomagnetism, Kiev V. M. Glushkov Institute of Cybernetics NAS of Ukraine, 1994.
[7] Spear, J. F., ,Horowitz, L. N., Hodess, A.B., Cellular electrophysiology of human myocardial infarction. 1. Abnormalities of cellular activation, Circulation, 1979, 59: 247-256.
[8] Chou, T.C. Electrocardiography in Clinical Practice, 2nd ed., New York, Grune&Stratton, 1986.

Induced Atrial Flutter in Rabbit Hearts Tissue and Inverse Magnetocardiography

Costa Monteiro, E[1], Andrade Lima, E[1], Ornelas, P.H.[1], Cruz, R.S.[1], Picon, A.P.[1], Kühner G.S.[1], Kelber, C.[1], Parente Ribeiro, E[1], Bruno, A.C.[1], Costa Ribeiro, P.[1], Marchesin, D.[2], Brio, M.[2,3], Leite ,T.T.[2]

Pontifícia Universidade Católica [1], Rio de Janeiro, Brazil; Instituto de Matemática Pura e Aplicada [2], Rio de Janeiro, Brazil; U. Arizona [3], Tucson, USA

Introduction

The possibility of non-invasive magnetic identification of electrical tissue activation during atrial flutter in humans is of great clinical interest. Most of the known evidence through electrical [1,2,3] and magnetic [4,5] measurements suggests the existence of a circular motion of the electrical activation (reentry) pattern as the underlying mechanism of this arrhythmia. A new animal experimentation protocol has been developed here in order to allow controlled studies of the reentry circuit in isolated pieces of rabbit atrium during experimentally induced flutter arrhythmia. In this preparation, the size of the tissue, its anatomical characteristics and the distance from the plane of the sample to the magnetic sensor can be taken into account. The underlying basis for the present work is the "equivalent circular dipole model" presented elsewhere [4]. Therefore, both the animal experimentation and signal processing protocols were based on the rotationally invariable structure of this model. Such structure led to good results, overcoming experimental difficulties due to small number of measurement positions and low sensor sensitivity.

Material and Methods

Isolated rabbit hearts were connected to a Langendorff perfusion system with a Tyrode solution (NaCl, glucose, $NaHCO_3$, KCl, NaH_2PO_4, $MgCl_2$ and $CaCl_2$), bubbled with carbogen (95% O_2 and 5% CO_2). Then, as described previously [6], dissected atrial tissue was isolated and maintained in a planar configuration between two netlike supports on the surface of a Tyrode solution bath. The sample obtained measured about 2 x 3 cm and, since the sinus node was maintained, it showed spontaneous activity. The superfusion fluid was saturated with a mixture of carbogen and kept at 37°C. Tachycardias were induced by adding acetylcholine (26.4 x 10^{-7} mM) to the volume conductor bath and by pacing (22 V, 16 Hz, 4 ms). Electrical measurements were performed to accompany the flutter firing, using extracellular glass electrodes immersed in the solution. One of the electrodes was positioned close to the tissue surface generating a signal very similar to an electrogram record. Detection of magnetic signals was carried out inside a shielded chamber consisting of two concentric cylindrical μ-metal layers. The one-channel MCG system used consists of a symmetric second order gradiometer, 1.5 cm diameter and 4 cm baseline coupled to an rf-SQUID. The pick-up coil was positioned between 1.5 and 2 cm above the sample, in different experiments. Successive magnetic recordings as a function of time were performed at different locations of a horizontal straight line above the isolated tissue by means of a non-magnetic sample positioner. During each recording the carbogen injection and the temperature control system were turned off in order to avoid magnetic noise. The shape and surface area (254 cm^2) of the volume conductor surrounding the tissue sample reduced to a negligible value its contribution to the measured magnetic signal [7], contributing to a more accurate solution of the inverse problem.

Magnetic and electric signals were analog filtered with a passband from 0.1 Hz to 50 Hz, and then acquired with a sampling frequency of 550 Hz for 11 positions on the straight line, ranging from x = -4 cm to x = 4 cm. Only the positions that seemed to have low noise in the fundamental frequency were used, while the other measurements were discarded. Before the computation of the Fast Fourier Transform, 8192-point data sequences were downsampled by a factor of 4, windowed and then padded with 30720 zeros in order to achieve a better frequency discrimination. The use of the Blackman Window led to wider and smoother peaks, allowing better determination of the amplitude of the harmonics. Noise-free time series were obtained by using the magnitude and phase of the first six harmonics and then calculating the Fourier Series.

Then the position of the center of the equivalent circulating dipole, its radius and the dipole strength were calculated using a least square procedure [8], attempting to reproduce the amplitude of the fundamental frequency at the points used.

Results and Discussion

As compared to data presented elsewhere [4,5], the signal-to-noise ratio was greatly improved by performing the magnetic measurements inside the magnetically shielded chamber. Also, positioning the isolated tissue in a planar configuration at the bath surface simplifies considerably the numerical solution of the inverse problem. First, such configuration reduces the number of geometrical parameters of the reentry path to be recovered. Second, since the atrial flutter frequency is a multiple of the ventricular frequency, the absence of ventricular electrical contribution greatly facilitates the processing of the magnetic signal (which is purely atrial and has an approximate rotational symmetry) by means of Fourier transforms.

Trying to minimize the time spent in the measurement, the inverse problem was solved using the data obtained along a single line passing over the tissue. Shortening this time also avoids having to process data with large changes in the tachycardia rate, which is not completely steady in time [5]. A simple reentry path model developed previously [5], where an equivalent current dipole describes a circular path, leads to a specific pattern for the isoamplitude map. The amplitude of the AC component of the magnetic field has a minimum over the center of the circular path. The amplitude of the AC component along a straight line passing near the center of this map also shows a minimum. In the present study, Fast Fourier Transforms (FFT) of the magnetic signals were computed, clearly showing amplitude peaks at the flutter fundamental frequency and at its harmonics (Fig. 1). The amplitude of the first harmonic was used to identify the propagation path of the equivalent current dipole during the tachyarrhythmia. Differently from the previous methodology [5], the amplitude of the first harmonic was used to construct the amplitude graph along the line of measurements.

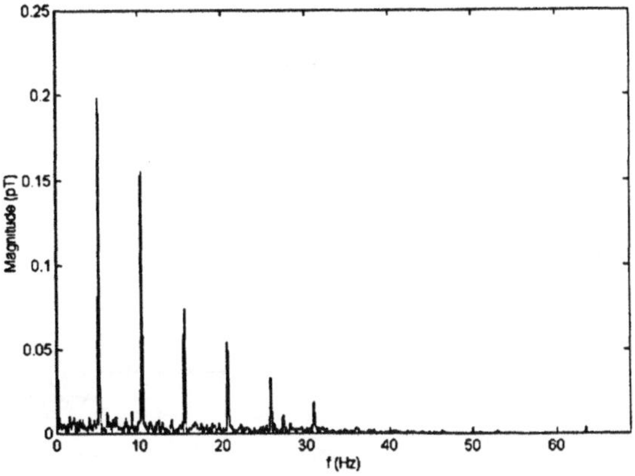

Fig. 1 - Fast Fourier Transform plot of the magnetic field oscillation generated by an isolated rabbit atrium in which flutter arrhythmia was experimentally induced. The magnetic signal was detected at a particular position.

Fig. 2a shows a tracing of the detected magnetic field during a flutter episode induced in an isolated atrial tissue and fig. 2b presents the reconstruction of the flutter oscillation wave from the harmonic data. Such processed signal shows a significant improvement in the signal-to-noise ratio. Figure 3 shows the fundamental frequency amplitude of the atrial flutter measured for different positions along the line where measurements were taken, parallel to the tissue plane. Fig. 4 shows the peak-to-peak value of the reconstructed signal as a function of these positions. As mentioned previously in this paper and also shown in isoamplitude maps [4,5], a point of minimum amplitude is present in both graphs. The similarity of Figs. 3 and 4 indicates the overall predominance of the first harmonic in the magnetic signal.

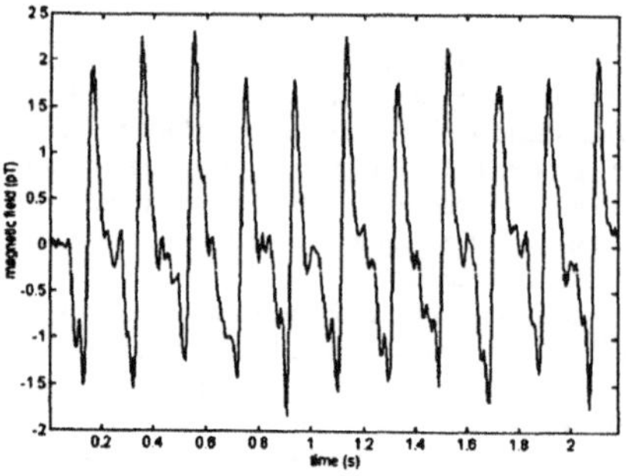

Fig. 2a

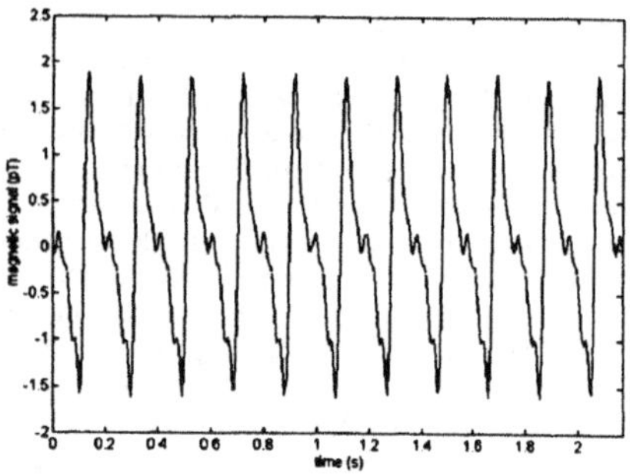

Fig. 2b

Fig. 2 - a) Magnetic field of the flutter oscillation wave, as a function of time, measured at a particular position. b) Magnetic field as a function of time reconstructed using the first six harmonics from the FFT data based on the magnetic signal shown in Fig. 2a.

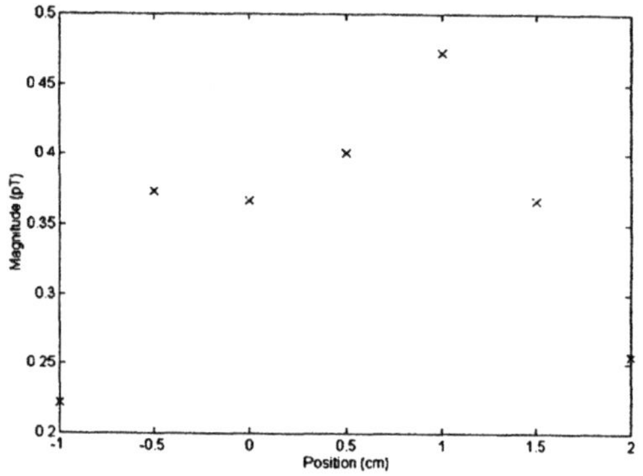

Fig. 3 - Amplitude of the first harmonic of the magnetic field as a function of position along the measuring line. The point zero indicates a position close to the sample center.

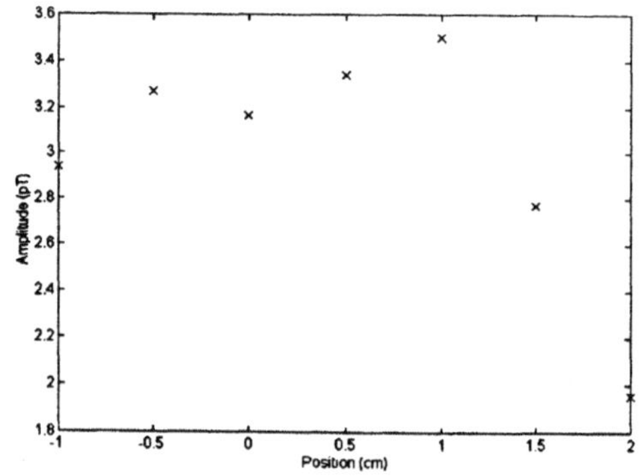

Fig. 4 - Peak-to-peak value of the reconstructed magnetic field as a function of position along the measuring line.

Ideally the graphs of amplitude as a function of position (figures 3,4) should be symmetric double hump graphs. However, we used only 7 data points, a number probably insufficient to resolve such a shape. These data points have errors due to changes in the reentry current pattern; these changes are reflected by changes in the tachycardia rate, which decreased by 20% from the first to the seventh measurement. As a result, the vertical distance z was recovered accurately, since we actually know the distance between the tissue and the magnetometer coil. An error analysis indicates that the distance across the x-axis was also accurate. However, the horizontal distance y and the radius of the circulating dipole could not be ascertained. The calculation was compatible both with a zero radius and with a 0.6 cm radius. Synthetic experiments reported in a companion paper [8] indicate that the difficulty in recovering the radius may be due to the fact that the measuring coil radius (0.75 cm) is larger than the radius of the circulating dipole, which has been estimated as about 0.5 cm in the experiments. Therefore, the electrical activation of the tissue indeed appeared to move along a circular path, characterizing the rotating pattern of propagation as the underlying mechanism of the experimentally induced arrhythmias. Amplitude graphs validate results previously obtained in animal experimentation [4,5] and in humans [9].

References

[1] Klein G.J., Guiradon G.M., Sharma A.D., Milstein S. Demonstration of macroreentry and feasibility of operative therapy in the common type of atrial flutter, Am. J. Cardiol. 1986; 57:587-591.

[2] Allessie, M.A., Lammers, W.J.E.P., Bonke, I.M. and Jan Hollen, E.E. Intra-atrial reentry as a mechanism for atrial flutter induced by acetylcholine and rapid pacing in the dog, Circulation, 1984, 70-1: 123-135.

[3] Isber, N., Restivo, M., Gough, W.B., Yang, H., El Sherif, N. Circus movement atrial futter in the canine sterile pericarditis model, Circulation, 1993; 87: 1649-1660.

[4] Costa Ribeiro, P., Bruno, A.C., do Carmo, J.S., Parente Ribeiro, E., Costa Monteiro, E. and Fonseca Costa, A. Detection of magnetic atrial flutter waves in isolated rabbits hearts, In: Willianson, S.J., Hoke, M., Stroink, G., Kotani, M. Advances on Biomagnetism, Pergamon Press, 1989, p.389-393.

[5] Costa Ribeiro, P., Bruno, A.C., Saboia e Silva, P.L., Hall Barbosa, C.R., Parente Ribeiro, E., Costa Monteiro, E. and Fonseca Costa, A. Detection of reentry currents in atrial flutter by magnetocardiography, IEEE Trans. on Biomed Engineering, 1992, 39-8: 818-824.

[6] Costa Monteiro, E., Louro, S.R.W., Bruno, A.C., Costa Ribeiro, P. and Fonseca Costa A. Detection of oblique wave-front depolarization by means of magnetic field measurements, In:, Atsumi, K., Kotani, M., Uemo, S., Katila, T. and Williamson, S.J. Biomagnetism '87, Tokyo Denki University Press, 1988: 310-314.

[7] Eghrari, I.R., Costa Monteiro, E., Costa Ribeiro, P., von der Weid, J.P., Oliveira, P.M. and Paes de Carvalho, A On the influence of the volume conductor in magnetocardiography, Il Nuovo Cimento, 1983, 2: 346-356.

[8] Brio, M., Marchesin, D., Costa Monteiro, E., Bruno, A.C. and Costa Ribeiro, P. Inverse magnetocardiography for reentry currents, these proceedings.

[9] Costa Ribeiro, P., Parente Ribeiro, E., Andrade Lima, E., Bruno, A.C., Correia, M.P., Costa Monteiro, E. and Thiesen S. Detection of reentry currents in atrial flutter by magnetocardiography in man, In:. Baumgartner, C., Deecke, L., Stroink, G. and Williamson S.J. Biomagnetism: Fundamental research and clinical applications, 1995: 562-565.

Acknowledgments

We thank Prof. Yoshio Okada (Magnetophysiology Laboratory, V. A. Medical Center, Albuquerque, USA) for providing the shielded chamber. This work was partially supported by CNPq grants (n.: 53.0484/93-5; 300665/93.8; 30.0959/81-8; 453497/95-0; 530054/93-0; 300659/94-6) and by FINEP grants (n.: 6593056600; 65.92.03.11-00), Brazil.

Cellular Automata Simulation in the Magnetic Study of Rotating Propagation Models Associated to Cardiac Arrhythmias

Costa Monteiro, E., Correia, M.P., Hall Barbosa, C.R. and Costa Ribeiro, P.

Departamento de Física, Pontifícia Universidade Católica, Rio de Janeiro, Brazil

Introduction

Periodic tachyarrhythmias like atrial flutter have generally been associated to a circular path of electrical activation. Realistic two-dimensional finite-element models have already been used to simulate the electrical [1,2] and magnetic [3] signals generated by the propagation of these electric tissue activations. Characterization of the tachycardia mechanism is difficult to carry out by means of electrical measurements, even by performing electrograms directly over the cardiac tissue surface. Beyond the common circular reentry moving along a one-dimensional path around a conduction obstacle in its center ("ring" model), there is another possible propagation model of circular activation. The latter do not involve an anatomical obstacle in its center but simply a functional heterogeneity. It propagates in two dimensions as a single or double spiral, corresponding respectively to the "leading circle" [4] and "figure eight" [5] concepts. In the present study, using a two dimensional cellular automata computer program, we calculate the magnetic field produced by these two different models for the propagation of rotating waves, generated in a plane of simulated cardiac atrial cells.

Material and Methods

The computer program we used was adapted from another program previously developed [3]. It consists of a two-dimensional square grid of 50x50 hexagonal cells (figure 1). Each cell, conductively linked to its six neighbors, may be in one of 3 different states: excited (white hexagons), refractory (black hexagons) or quiescent (gray hexagons). The activation spreads from the cells in the excited state to the cells in the quiescent state. Once excited, the cell goes in the next iteration to the refractory state, remaining there for the period of its recovery time, and then returning to the quiescent state, which allows the cell to be excited again. The recovery time τ is randomly ascribed according to a gaussian distribution with mean value $\overline{\tau}$ and standard deviation σ, which leads to a spatial dispersion on the recovery time.

In order to calculate the magnetic field generated, both depolarization and repolarization are considered, by means of their equivalent dipoles. The depolarization dipole is due to the cells in the excited state. Wherever there is one cell in the excited state adjacent to one or more cells in the quiescent state, a unit dipole is assigned to each interface. The equivalent depolarization dipole P_d of this excited cell is then obtained by vectorially summing the unit dipoles and by normalizing the result. In the next iteration, this cell passes to the refractory state, and a repolarization dipole will be associated with this cell for the duration of its refractory time τ. This equivalent dipole is calculated by: $P_r = -P_d/\tau$.

Thus, the magnetic field has two factors: the field due to depolarization dipoles of the cells in the excited state, B_d, and the field due to repolarization dipoles of the cells in the refractory state, B_r.

These fields are calculated by using the following formulas:

$$B_d = \sum \frac{P_d \times r}{|r^3|} \quad \text{and} \quad B_r = \sum \frac{P_r \times r}{|r^3|}$$

where r is a vector from each cell to the chosen measurement point. The total magnetic field is then calculated by $B_{total} = B_d + B_r$. Only the component perpendicular to the plane of cells was taken into account.

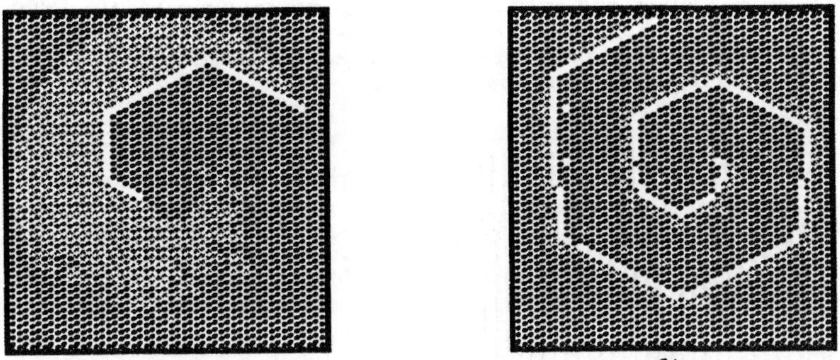

(a) (b)

Figure 1: Grid of 50x50 hexagonal cells. The white cells have just been activated and can activate the quiescent cells (in gray). After been activated, the cells remain in a refractory state (black). (a) "ring" model of propagation of activation with a conduction obstacle present in the center of the grid. (b) "Spiral" model.

In our synthetic experiments, we included heterogeneities of electrophysiological properties such as conduction obstacles and blocks in one direction. For the "ring" model (figure 1a) both anomalies are present. The spiral models were easily induced by a unidirectional block in a line segment suitably positioned in the plane of cells at the beginning of the process. Depending on its location in the grid either a single or a double spiral is started. The magnitude of the mean recovery time $\overline{\tau}$ and its spatial dispersion σ determines the stability and persistence of the rotating propagations. The magnetic field generated by the different configurations of spreading activation was calculated as a function of time (number of iterations) at various points on a plane parallel to the grid of cells, and the isofield and isoamplitude maps were computer processed.

Results

The magnetic signal calculated for the different configurations of simulated electric propagations have been analyzed by means of various signal processing techniques. By recording the signal as a funcion of time (number of iterations), it was observed that the morphology of the magnetic field oscillations is clearly preserved for all the records obtained with the conduction obstacle model. On the other hand, when no obstacle was present in the center of the simulated rotating activity, depending on the parameters considered a morphology disturbance was observed (figures 2a and 2b). In this case, the morphology from one cycle to the other is altered when a spatial dispersion of the refractory period is introduced in the simulation.

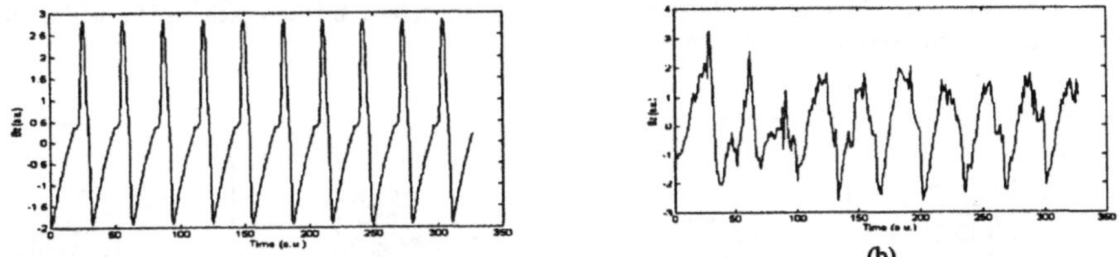

(a) (b)

Figure 2: Calculated magnetic signal as a function of iterations recorded for a "spiral" model activation with $\overline{\tau} = 30$ iterations and for two different values of spatial dispertion of this recovery time: (a) no spatial dispersion and (b) 16% spatial dispersion.

An interesting feature observed in the results is the behavior of the frequency. In the case of the "ring" model, independently of the eletrophysiologic parameters, the frequency of the rotating wave is the same. For the single spiral model, however, the frequency depends on the refractory period duration introduced. As smaller is the mean recovery time, as higher is the rotating frequency (figure 3a). Also with this latter condition (spiral model with short refractory period), the incorporation of a spatial dispersion of the mean recovery time leads to a chaotic record (figure 3b). The pattern of activation for the simulated cells that generate the record of figure 3b is the one shown at figure 1b.

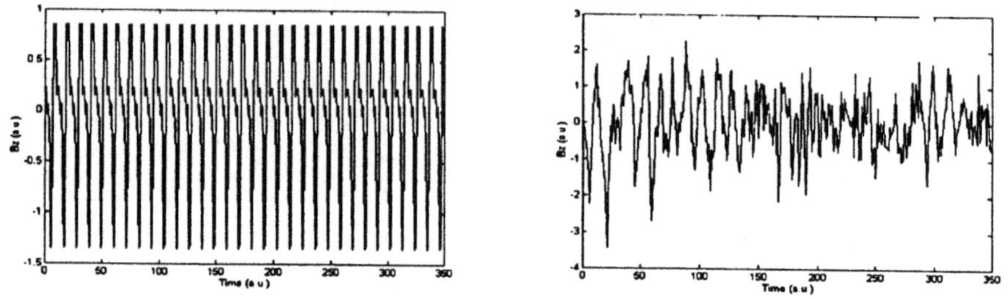

(a) (b)

Figure 3: Calculated magnetic signal as a function of iterations recorded for a "spiral" model activation with $\bar{\tau}$ = 10 iterations and for two different values of spatial dispertion of this recovery time: (a) no spatial dispersion and (b) 20% spatial dispersion.

The isofield maps calculated with the data of the "ring" model show a dipolar pattern for any instant of the cycle. The equivalent dipole moment describes a circular path, independent of the magnitude and spatial dispersion of the mean recovery time ($\bar{\tau}$) (figure 4). For the "spiral" model propagations, however, multipolar configurations are frequent in the isofield maps. Dipolar patterns are seldom observed and are particularly scarce for shorter refractory period (figure 5). The isoamplitude maps of the ac component of the magnetic field show a point of minimum amplitude in the center of the rotating wave for both the "ring" and single spiral models of propagation.

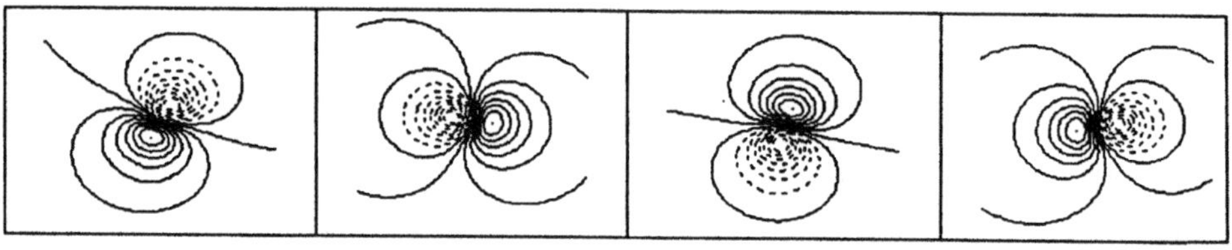

Figure 4: Isofield maps obtained at 4 different instants with 9 iterations between each one for a simulated "ring" pattern of activation with $\bar{\tau}$ = 30 and σ = 16%.

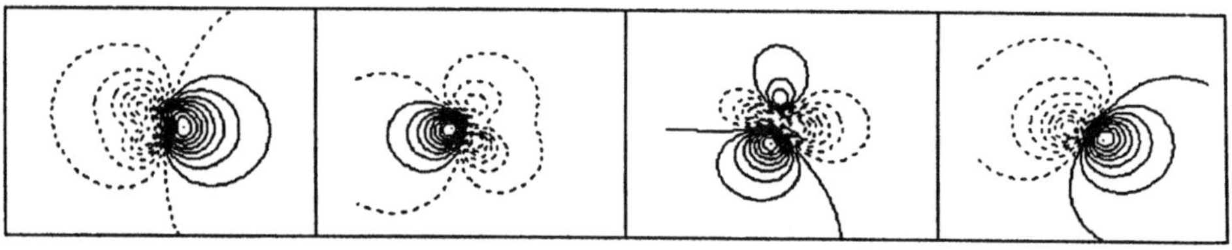

Figure 5: Isofield maps obtained using a simulated "spiral" pattern of activation with $\bar{\tau}$ = 30 and σ = 16%, at the same 4 instants considered for figure 4.

Discussion

Two different types of underlying mechanism supposedly responsible for human atrial flutter rhythms were studied by means of computer simulation. Atrial flutter is clinically classified in two types primarily based on differences in atrial rate. Clinical and experimental data show that the human type I atrial flutter is due to a reentry mechanism with an excitable gap (region of quiescent cells between the crest of activated cells and the tail of refractory cells), the here simulated "ring" model. There is a common believe that the "spiral" model of propagation may be responsible for the rapid type of atrial flutter. Since in this type of circus movement the length of the circuit is considered to be equal to the distance between the crest of activation of the circulating impulse and its tail, there exists no gap of full excitability. The rate of such a reentrant rhythm is primarily governed by the refractory period of the tissue in which the impulse circulates. As observed in the present study, the rotating frequency obtained using data generated by the "spiral" model simulation was as fast as the highest possible rate that can be followed by all parts of the circuit. With a longer refractory period, the rotating frequency of the simulated "spiral" was similar to the one observed using the "ring" model.

Incorporating a spatial dispersion of the refractory period in the simulated "spiral" model of propagation with a short mean refractory period, the obtained record approaches to a chaotic state reminding cardiac fibrillation (figure 3b) although the clearly organized underlying pattern of activation such as a single spiral wave (figure 1b).

Isofield maps configuration showed that the current distribution in the "ring" model can be well represented by an equivalent current dipole circulating along a round path (figure 4). Such was not the case for the "spiral" model (figure 5). A more complex current distribution is necessary in order to solve the inverse problem in this case.

References

[1] Moe, G.K., Rheinbolt, W.C. & Abildskov, J.A., A computer model of atrial fibrillation, Am. Heart J., 1964, 67, 200-220

[2] Smith and Cohen. Simple finite-element model accounts for wide range of cardiac dysrhythmias, Proc. Natl. Acad. Sci. USA, 1984, 81: 233-237.

[3] Hall Barbosa, C.R.; Costa Ribeiro, P.; Costa Monteiro, E.; Bruno, A. C.; Parente Ribeiro, E. and. Fonseca Costa, A., A finite-element model to study the magnetic detection of flutter waves, In: M. Hoke S. N. Erne, Y. C. Okada, G. L. Romani. Biomagnetism: Clinical aspects., Elsevier Science publishers B. V., 1992

[4] Alessie, M. A; Bonke, F.I.M.; Schopman, F.J.G., Circus movement in rabbit atrial muscle as a mechanism of tachycardia: III. The "leading circle" concept. Circulation Research, 1977, 41: 9-18.

[5] Schoels, W, Restivo, M, Caref, E.B., Gough, W. B. and El-Sherif, N., Circus movement atrial flutter in canine sterile pericarditis. Circulation, 1991, 83:1716-1730.

Acknowledgments

This work was partially supported by CNPq grants (n.: 53.0484/93-5; 300665/93.8; 30.0959/81-8) and by FINEP grant (n.: 6593056600), Brazil.

Comparison Of Measured And Simulated MCGs With A Realistic Heart-Torso Model

Czapski, P.[1], Haueisen, J.[2], Ramon, C.[1,3], Nowak, H.[2], Huntsman, L. L.[3] Leder, U.[2], Bardy, G. H.[4], and Kim, Y.[1]

[1]Department of Electrical Engineering, [3]Center for Bioengineering, [4]Division of Cardiology, University of Washington, Seattle, WA 98195, U.S.A.; [2]Biomagnetic Center, Friedrich Schiller University, Jena, Germany

Introduction

We have compared simulated magnetocardiograms (MCGs), calculated using a high resolution heart-torso model of an adult subject, with measured MCGs acquired from the same individual. Most of the channels showed good qualitative agreement between measured and simulated waveforms, suggesting that the model accounts for the most important mechanisms underlying the MCGs and may be useful for cardiac magnetic field modeling under normal and diseased states.

Methods

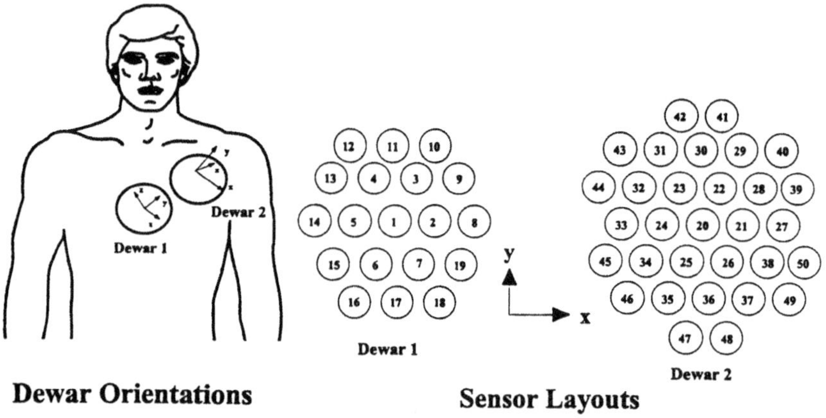

Fig. 1. Orientation of Dewars and sensor layouts for MCG data collection. Dewars were tilted at an angle of 30 degrees from the normal to the chest. Sensor layout is given in local coordinates for each Dewar.

MCGs were measured with a Phillips twin Dewar first order gradiometer (diameter of the pick up coil: 20 mm; base-length: 70 mm) SQUID-system consisting of 31 channels in one tank and 19 channels in the other. Both Dewars were pointing at an angle away from the normal to the chest surface. Orientation of the Dewars and the sensor layouts in each Dewar is shown in Fig. 1. This arrangement allowed for simultaneous measurement of the MCGs using both Dewars.

Our heart-torso model was constructed from segmented MRI images from a single subject. An operator-assisted process classified each area of each image into one of 19 different tissues. Each tissue was assigned a conductivity value based on data reported in the literature (Geddes and Baker, 1967 [1]; Foster and Schwan, 1989 [2]). A finite element mesh was generated with brick-shaped hexahedral elements and with one-to-one correspondence between elements and pixels in the classified images. The model consist of 2,578,630 brick-shaped elements and 2,506,844 nodes and has a voxel resolution of 1.56×1.56×3 mm. The small grid size of the model allowed polarization and depolarization processes in the heart wall to be represented as multiple dipole sources, with the total number of dipoles exceeding several hundred. Modeling of the cardiac excitation process was based on human potential map data (Durrer et al., 1970 [3]). The electrical activity was represented by the distributed dipolar sources in the heart wall similar to methods used by other researchers (Gulrajani, 1988 [4]). The three-dimensional propagation isochrones at 5 ms time steps were interpolated from the data of Durrer et al. (1970 [3]). These isochrones were overlaid on the finite element mesh. Each dipolar source was then specified by placing voltages at two nodes lying on opposite sides of the isochrones. These nodes were selected in such a way that every dipolar source was perpendicular to the isochrones. The number of dipoles at each time step was proportional to the surface of the corresponding isochrone. The entire ventricular depolarization process was assumed to last for 55 ms. Note that the typical excitation process lasts for 70 - 80 ms. We were unable to model the last 15 ms of the QRS complex because of the finite discretization of the model . Simulation was carried out in 5 ms time steps. A minimum of 2 dipoles was used at the beginning of the depolarization process and a maximum of 640 dipoles was used at 40 ms when the largest portion of the myocardium was depolarized (Czapski et al., 1996 [5]). Dipole magnitudes were determined by a typical shaped action potential (Miller and Geselowitz, 1978 [6]) with a resting potential of -90 mV and a peak value of +10 mV. The ventricular gradient of the action potential duration was included in the model (Sekiya et al., 1983 [7]). A primary gradient of the action potential, from the epicardium to the endocardium with a typical value of 20 ms/mm, and a secondary gradient from the base to apex with the value of 7.5 mm/ms, were used to simulate the T wave and the S-T segment of the magnetocardiogram. The resulting action potentials had durations between 220 and 265 ms. These dipolar sources were included in the model as boundary conditions for solving the Laplace's equation. Current distributions inside the torso and the magnetic field at the gradiometer coil locations were computed at each time step. Computations were done on a Cray YMP/C-90 machine.

Results and Discussions

Averaged measured and simulated MCGs are shown in Fig. 2. The measured raw MCGs were averaged over 28 beats to obtain averaged measured MCG for each sensor channel. Overall, general features of the measured MCGs are reasonably represented by the simulated data for most, but not all the channels. Several discrepancies in the progression of the QRS complex and polarity of the T wave can be seen. In most cases the simulated T waves have higher magnitudes than the measured ones. Moreover, T wave deflection is negative in the measured data from sensors 1-19, but the simulated data shows positive deflections for the same sensors. The T wave deflections agree for the remaining 31 sensors. Differences in the T wave segments are most likely due to the changes in the heart and blood volume during the repolarization process. Volume of the ventricles can change by as much as 60% and this is not included in our model. Because of this change in shape and volume, the isochrones of the repolarization have different shapes than during the depolarization. Our model assumes that the repolarization isochrones are the same as the depolarization isochrones. Therefore, both the error in the source positions and the difference in the volume current contribution are caused by the assumption of the constant heart shape during the ventricular repolarization. Another source of errors is the subject-to-subject difference in the excitation process. In our simulation we are using isochrones measured for one subject (Durrer et al., 1970 [3]) and superimposing them on the heart of another. The differences in the depolarization patterns and heart shapes introduce an error to the simulation. At this point there is no better depolarization data, and it is not possible to measure the isochrones for a modeled subject.

To quantify the accuracy of simulations, the mean absolute differences, standard deviations, and root mean square (RMS) error were all computed for the differences between the simulated and measured waveforms. These measurements were done for the QRS complex only and also for the whole heart beat. The results were very similar. The mean and standard deviations for the QRS complex are shown in Fig. 3. For the first Dewar, channels 1-19, error seems to be very uniform. However, for the second Dewar, channels 20-50, mean and standard both exhibit a periodic pattern. Channels with very high errors , i.e., 23, 31, and 43, are all located in the upper left corner of the sensor layout for the Dewar 2. This suggests that there might have been slight error in alignment of the Dewar 2 positions with the MRI data set.

For a comparative analysis of all the channels, we calculated the mean and standard deviations of the RMS errors. This was done for all 50 channels. For 50 sensors and all the time steps (whole heart beat) the mean error was 0.85 pT with standard deviation of 1.06 pT. The average RMS error value was 0.172 pT. These values are relatively small compared to the magnitude of the QRS complex (approximately 7 pT). The larger standard deviation indicates that the errors could be relatively high for some channels, but the low mean and RMS values indicate a good overall performance of the simulations. This suggests that the model accounts for the most important determinants of the MCGs and could be a useful tool for the study of cardiac magnetic fields. A strength of the model is its relatively small voxel size which allows for an accurate modeling of the electrical activity of the heart and analysis of volume currents in the whole torso. Further analysis needs to be done to reduce the errors between measured and simulated MCGs for the Dewar 2 sensors.

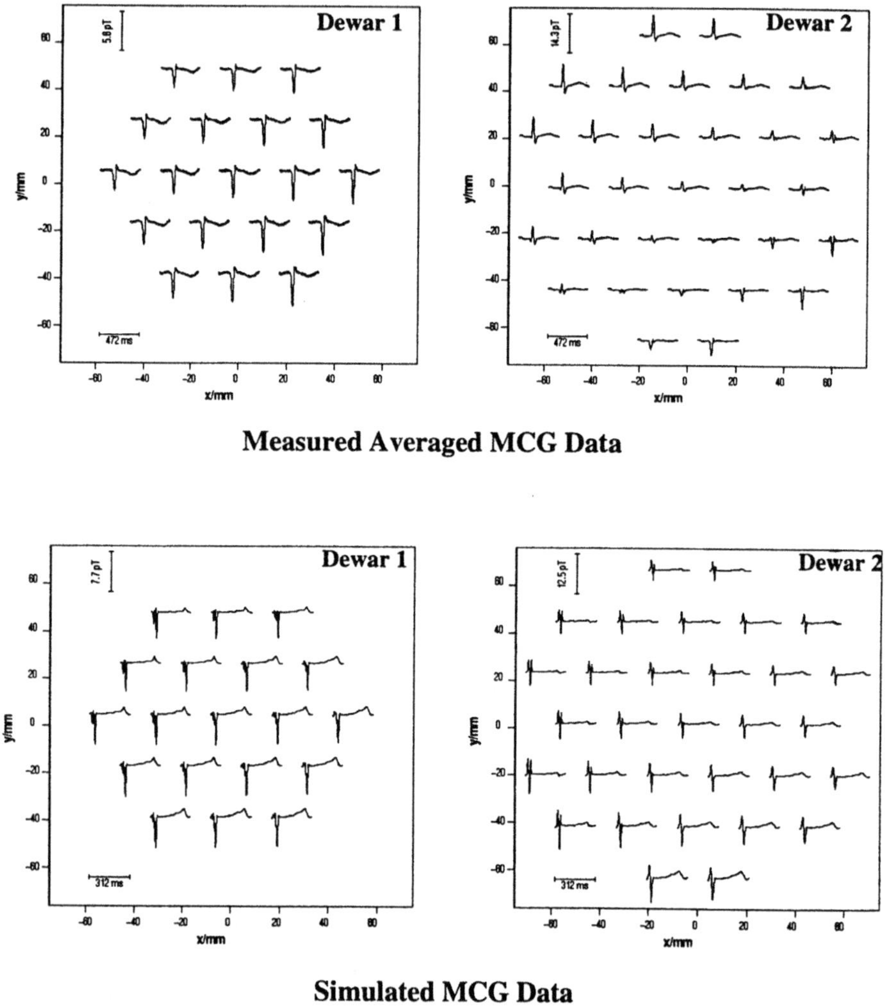

Fig. 2. Measured averaged MCGs and simulated MCGs of the same subject.

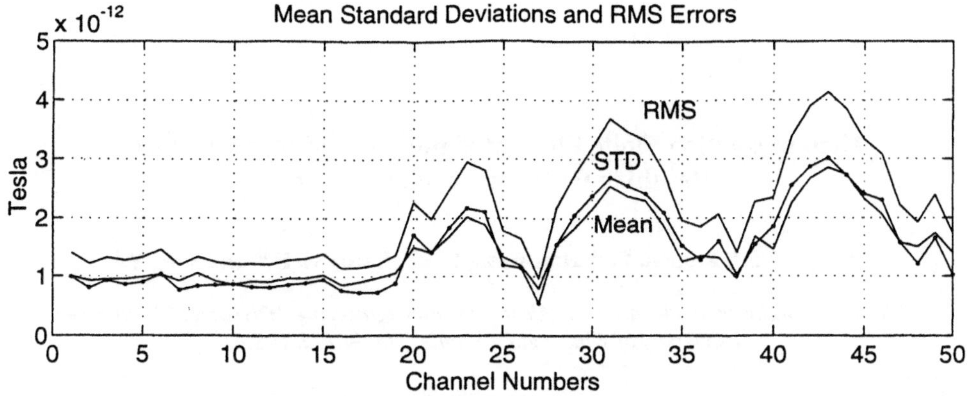

Fig. 3. Mean, standard deviations (STD) and root mean square (RMS) of the errors for all channels. Here error is the absolute difference between the measured and simulated MCGs.

References

[1] Geddes, L. A., and Baker, L. E. The specific resistance of biological material - a compendium of data for the biomedical engineer and physiologist, Med. Biol. Engr., 1967, 5: 271-293.

[2] Foster, K. R., and Schwan, H. P. Dielectric properties of tissues: a critical review CRC Handbook of Biological Effects of Electromagnetic Fields, 1989, 17: 25-104.

[3] Durrer, D., van Damm, R. T., Freud, G. E., Janse, M. J., Meijler, F. L., and Arzbacher, R. C. Total excitation of the isolated human heart, Circulation, 1970, 41: 899-912.

[4] Gulrajani, R. M. Models of the electrical activity of the heart and computer simulation of the electrocardiogram, CRC Critical Reviews in Biomedical Engineering, 1988, 16: 1-66.

[5] Czapski, P., Ramon, C., Huntsman, L. L., Bardy, G. H., and Kim, Y. On the contributions of volume currents to the total magnetic field resulting from the heart excitation process: A simulation study, IEEE Trans. Biomed. Engr., 1966, 43: 95-104.

[6] Miller, W. T., and Geselowitz, D. B. Simulation studies of the electrocardiogram: I. the normal heart, Circ. Res., 1978, 43: 301-315

[7] Sekiya, S., Ichikawa, S., Tsutsumin, T., and Harumi, K. Distribution of action potential durations in the canine left ventricle Jpn. Heart J., 1983, 25: 181-194.

Acknowledgements

This work was supported by the National Science Foundation grant #BCS-9209938, an equipment grant from the Seattle Foundation and the Graduate School Research Fund of the University of Washington. We would like to express our gratitude to Cray Research, Inc. and especially to Mr. Steve Baumann, Mr. James Harcourt and Mr. Thomas Ashbrook for allowing us to use their Cray YMP/C-90 computer.

High Resolution Finite Element Simulations of Magnetic Fields Resulting from Myocardial Infarction

Czapski, P.[1], Ramon, C.[1,2], Huntsman, L. L.[2], Bardy G. H.[3], and Kim, Y.[1]

[1]Electrical Engineering Department, [2]Center for Bioengineering, [3]Division of Cardiology University of Washington, Seattle, WA 98195, U.S.A.

Introduction

Data from simulations of the magnetic fields of a normal heart, and a heart with anterior myocardial infarction (AMI) and inferior myocardial infarction (IMI) are presented. Simulated fields and magentocardiograms (MCGs) show noticeable differences between the normal and infarcted heart. There are also differences in the fields of the two types of infarcts modeled here suggest that magnetic field profiles have a potential to detect and localize the infarcts in a heart.

Methods

A high resolution finite element model of a heart and torso was used in this study. The model was constructed from segmented CT images. A semi-automatic process was used to classify the images into 19 different tissue types. Each tissue type was assigned a conductivity value. A finite element mesh was generated with brick-shaped elements with one-to-one correspondence between elements and pixels in the classified images. This procedure produced a model with 2,554,610 elements, 2,634,573 nodes, and resolutions of 1.5625×1.5625 mm in the horizontal plane and 3.0 mm in the vertical direction (Czapski et al., 1996 [1]). The normal cardiac excitation was modeled with multiple dipolar sources. Their positions and magnitudes were determined from experimental data (Durrer et al., 1970 [2]). The propagation of the electrical excitation in the heart wall was modeled by 11 isochrones with a 5 ms interval. This allowed us to model the QRS activity from 0 to 55 ms..

One infarct located in the anterior section of the left ventricle and a second one in the inferior wall of the left ventricle were modeled. Myocardial infarcts were introduced by assigning constant shapes of the action potential in the necrotic regions of the myocardium. These regions were assumed to be surrounded by two ischemic areas of decreasing degrees of ischemia with changes in the action potential shapes based on the experimental data. This procedure allowed comparing of the magnetic fields from the model of the same subject with and without the myocardial infarction The total current distribution inside the heart and torso and magnetic fields outside of the chest were computed on a Cray YMP/C-90 computer. The magnetic field in a planar sampling surface in front of the torso was computed. The size of the sampling surface was 18×18 cm, large enough to cover the whole heart. The whole sampling surface was divided in to 300×300 pixels. The magnetic field profiles with a 5 ms interval were simulated for the whole heart from 0 to 330 ms.

Results and Discussion

Contour maps of the magnetic fields of a normal heart, AMI and IMI are shown in Figure 1. Contour maps of the difference between the normal and AMI, and normal and IMI magnetic fields are shown Figure 2. The contour maps in Figure 1 show that there are noticeable differences between the normal and infarcted magnetic field profiles. The features are more pronounced if one looks at the difference magnetic field profiles shown in Figure 2. These contour maps show almost constant shapes throughout the entire duration of the QRS complex. Contour lines are close to dipolar and only magnitudes of field values change during the QRS complex. Only three contour plots at different time frames in a QRS complex are shown, but the features described here were also observed at other time frames in the QRS complex.

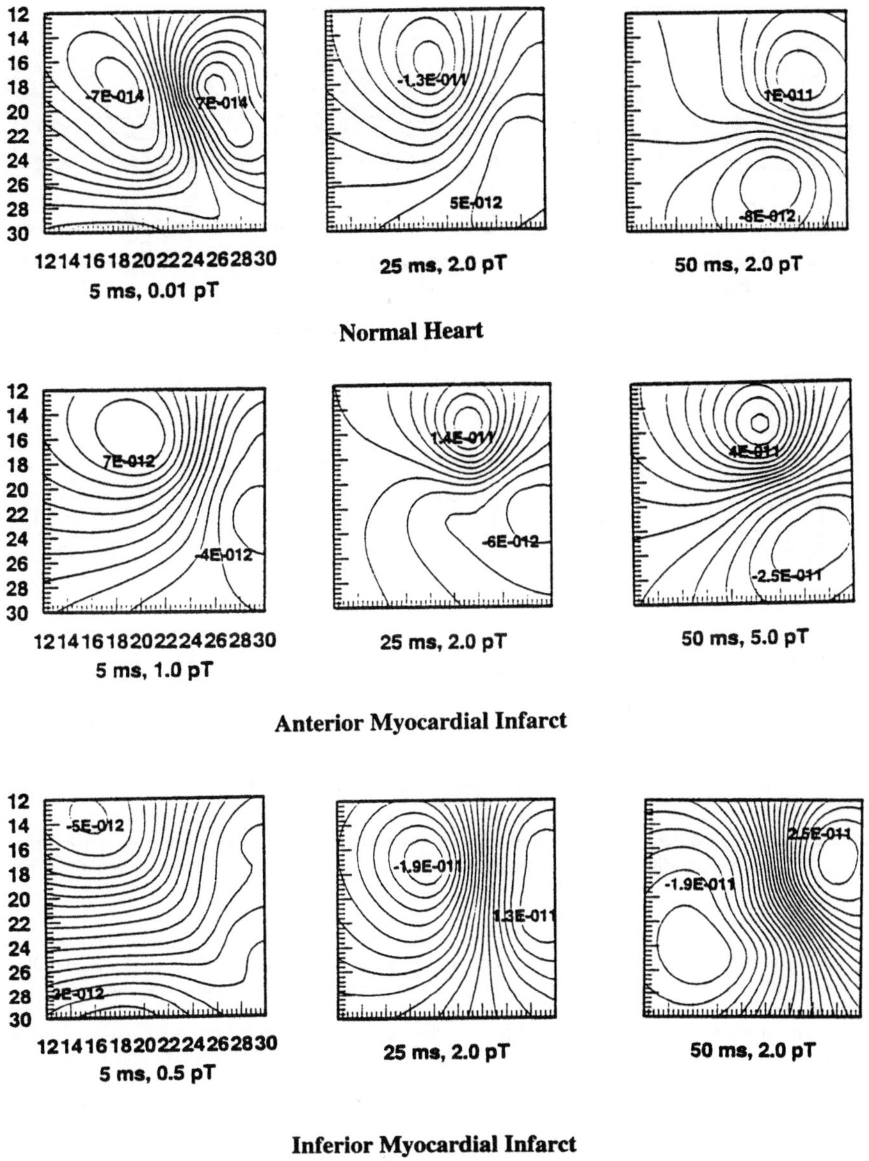

Figure 1. Magnetic field profiles of normal and infarcted hearts; top row of profiles for a normal heart, middle row for the anterior myocardial infarct, and bottom row for the inferior myocardial heart.

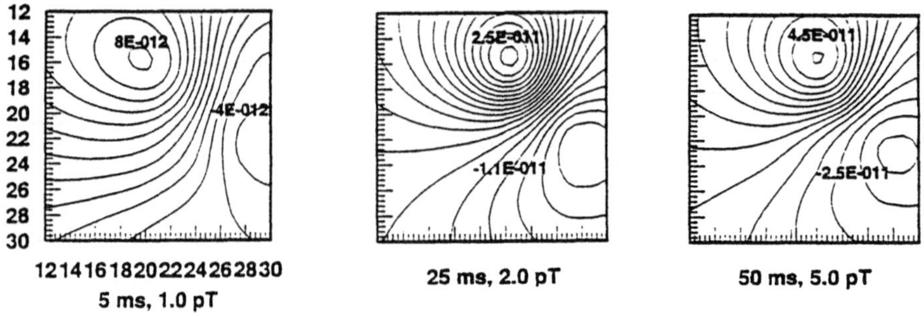

Anterior Myocardial Infarct

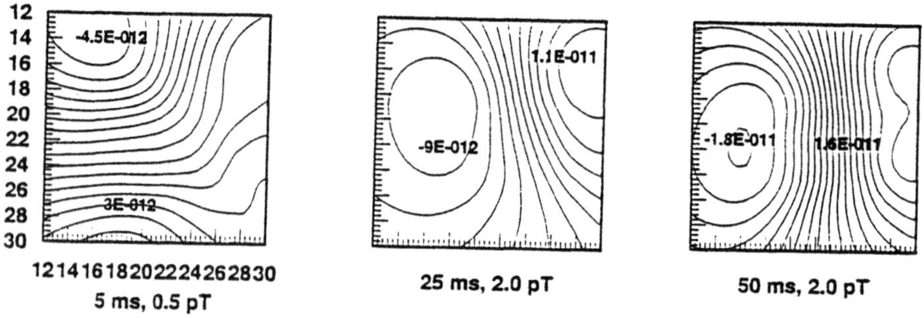

Inferior Myocardial Infarct

Figure 2 Top: difference of the magnetic field of a normal heart and a heart with an anterior myocardial infarct; bottom: difference of the magnetic field of a normal heart and a heart with an inferior myocardial infarct.

The magnetic field profiles shown in Figure 2 suggest that the contribution of the infarcted region to the total magnetic field can be accounted for by an equivalent current dipole. The orientation of the dipolar field is different for the AMI and IMI cases. This suggests that it might also be possible to detect infarct from these difference fields constructed for different cases of myocardial infarction.

The simulated magnetocardiograms (MCGs) at three different locations in the sampling surface directly above the heart are shown in Figure 3. The MCGs are shown for a normal heart, and AMI and IMI infarct. The T waves are elevated for the infarcted hearts as compared to a normal heart.

These simulations show that our models are capable of simulating infarcts, and that our simulated results are similar to what people have seen in MCG measurements (Siltanen, 1989 [3]).

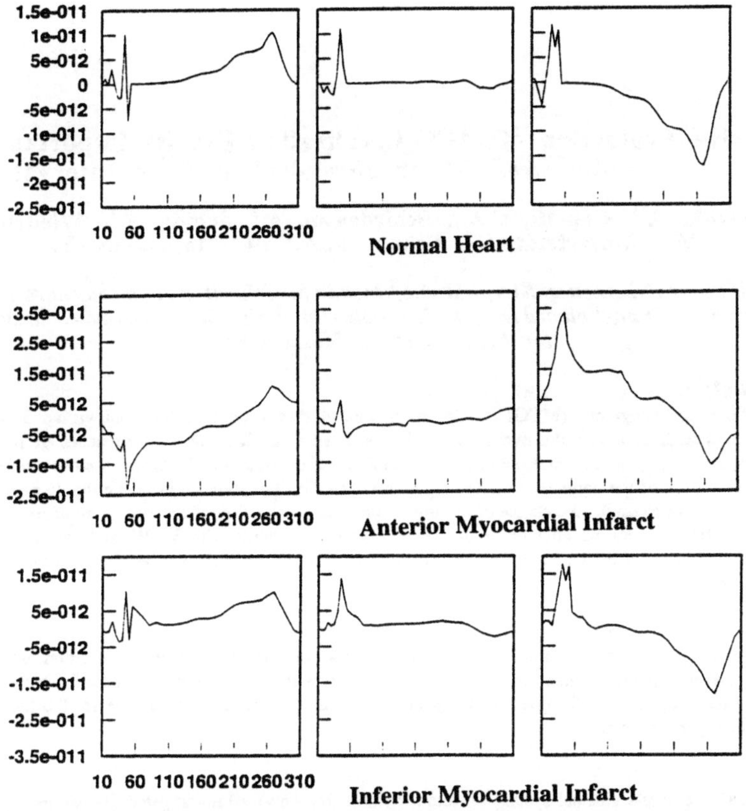

Figure 3 Simulated magnetocardiograms. Top: for a normal heart; middle: for a heart with an anterior myocardial infarct; bottom: for a heart with an inferior myocardial infarct.

References

[1] Czapski, P., Ramon, C., Huntsman, L. L., Bardy, G. H., and Kim, Y. On the contributions of volume currents to the total magnetic field resulting from the heart excitation process: A simulation study, IEEE Trans. Biomed. Engr., 1966, 43: 95-104.

[2] Durrer, D., van Damm, R. T., Freud, G. E., Janse, M. J., Meijler, F. L., and Arzbacher, R. C. Total excitation of the isolated human heart, Circulation, 1970, 41: 899-912.

[3] Siltanen, P. Magnetocardiography. In: MacFarlane, P. W., Veitch Lawrie, T. D. Comprehensive Electrocardiology: Theory and Practice in Health and Diseases, New York, Pergamon Press, pp. 1405-1438, 1989.

Acknowledgments

This work was supported by the National Science Foundation grant #BCS-9209938 and an equipment grant from the Seattle Foundation. We would like to express our gratitude to Mr. Steve Baumann, Mr James Harcourt and Mr. Thomas Ashbrook of Cray Research, Inc., for allowing us to use their Cray YMP/C-90 computer.

Invasive Evaluation of MCG-Localization Results Importance of Measured Electrophysiological Time Intervals

Gapelyuk, A.[1], Copetti, C.A.[2], Schirdewan, A.[3], Schütt, H.[2], Wiedemann, M.[3], Meyerfeldt, U.[3], Primin, M.A.[1] and Maslenikov, Y.[4]

[1]Institute for Cybernetics, Kiev, Ukraine;[2] Institut für Schicht- und lonentechnik, Jülich, Germany; [3]Franz-Volhard Hospital Humboldt-University, Berlin, Germany; [4]Institute for Radioelectronics, Moscow, Russia

Introduction

Magnetocardiography (MCG) is used as a clinical diagnostic tool for noninvasive 3-dimensional localization of cardiac sources. For evaluation of the localization results, different approaches are published by using various topographic reference methods. In the majority of cases, surface ECG was used as a time reference and the calculation of the equivalent current dipole was based on the presence of stable dipole patterns of the magnetic field. For example, the choice of the time points for localization of the site of an accessory pathway ranged from 10 up to 50 ms after onset of the delta-wave in patients with WPW-syndrome in the studies. Usually the accuracy of biomagnetic localization results, obtained by the described "conventional mode", is approximately 1-3 cm.

Purpose of the study

In the present study, patients with various cardiac arrhythmias (reentry-tachycardias due to overt, concealed and multiple atrioventricular accessory pathways; repetitive monomorphic ventricular tachycardias and ventricular reentrant tachycardias) were investigated with the aim to find the correct time for the localization of the arrhythmogenic substrate.

Methods

Using a single channel system (Krioton, Krionit, Moscow) we investigated 10 patients (pts) in the clinical environment without shielding prior to catheter ablation. The algorithm of the source localization, which is based on the analytical methods of inverse problem solution, was applied.

The system consists of LWTc DC SQUID coupled with axial wire second-order gradiometer. The baseline is 60 mm, the diameter of pickup coil 15 mm. The noise level of the system was 40 fT/Hz-1. Three or more MCG investigations were performed for all patients: two before the ablation procedure and one or more after the ablation procedure. We used a rectangular mapping grid of 6x6 gridpoints with a distance of 4 cm both in x and y directions. The bandwidth was 0, 1-120 Hz and the sample frequency was 500 or 1000 Hz.

The algorithm of the source localization, which is based on the analytical methods of inverse problem solution, was applied. First the localization procedure was performed in "conventional mode", i.e., the time of localization was selected by analysis of the surface ECG and the appearance of spatial stability of the current dipole sources and the magnetic field. Second, the registration of endocardial electrical potentials, which were related to the arrhythmogenic substrate, served as a reference for time analysis of the endocardial electrogram to estimate the correct time for localization of arrhythmogenic substrate.

A comparison between both sites of biomagnetic localization with the site of catheter tip were performed and documented by x-ray imaging in posterior-anterior direction in a representative phase of heart movement using a ring with 1 cm diameter on the anterior chest wall, presenting the magnetocardiographic localization result.

Results

We report the first results in 10 patients. Seven pts suffering from atrioventricular reentry tachycardias due to accessory pathways (1 pt with Mahaim-bundle; 2pts with concealed left-sided APs, 1 pt with 2 APs: right posteroseptal overt and left lateral concealed, 1 pt with left-sided overt AP, 1 pt with right lateral overt AP and 1 pt with posteroseptal epicardial AP), 2 pts suffering from repetitive monomorphic ventricular tachycardia and 1 pt had an incessant ventricular tachycardia.

In pts with accessory pathways, the difference between the site of the arrhythmogenic substrate and the site calculated by "conventional" MCG localization was approximately 10-30 mm in relation to both the x- and the y-coordinates. A higher spatial correspondence of the site of the catheter tip with the site of MCG localization was found when the intracardiac time reference describing the excitation of the arrhythmogenic substrate or the earliest pre-excitation of the ventricular muscle was used. The correspondence of the MCG localization result with the invasive localized structure was more closed, approximately 5 mm.

This high correlation of the MCG locations to the arrhythmogenic substrate was observed only for a few milliseconds, up to a maximum of 6 ms. This findings could be demonstrated for all sites of accessory pathways.

In pts with ventricular tachycardias caused by focal mechanism, similiar results were obtained. The origin of the ventricular tachycardias were endocardial identified by the earliest local excitation, 20-40 ms before onset of QRS-complex in surface ECG.

The best correspondence between magnetocardiographic localization and VT-origin, determined by RF-catheter ablation, was found up to 10 ms before the onset of the QRS complex of premature ventricular beat or during tachycardia.

Case report

Magnetocardiographic mapping was performed on a 40 year old female with atrioventricular reentry tachycardias due to WPW-syndrome, in order to localize the accessory pathway during sinusrhythm:

1. 5 days before ablation session with a multichannel SQUID-LTS-system in a shielded room with standard localization procedure.
2. 2 hours before ablation session with a single channel SQUID-system (nonshielded environment in the hospital).

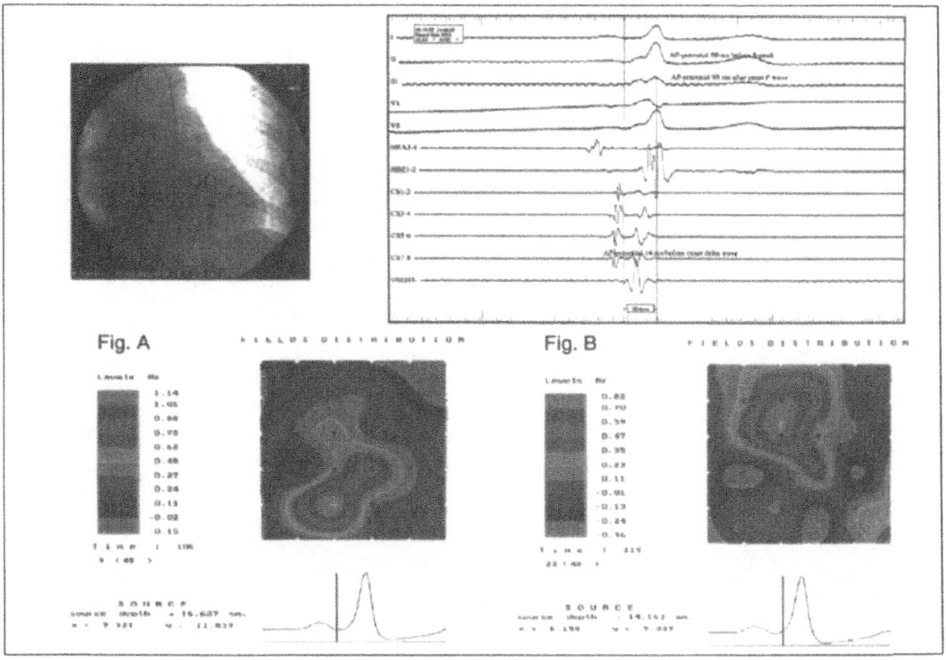

The x-ray shows the heart silhouette in a posterior-anterior view: Three rings are representing the localization results by MCG. The right lateral ring shows the localization result of the multichannel system investigation. The criterion of AP localization was development of stable magnetic field pattern (25 ms after onset delta wave). The ring in the middle represents the location of the accessory pathway determined by measurement with the single channel system using the standard localization program. The AP localization was performed 10 ms after onset delta wave (Fig. B). The left sided ring represents the localization result of AP fitted by using intracardiac electrogram 5 ms before onset of delta wave (Fig. A).

Discussion and Conclusions

We conclude from our results, that an improvement of the accuracy of MCG-localization of cardiac sources can be achieved by using the correct time for localization. A detailed analysis of the relationship between dipole trajectory and the corresponding structures within the heart is the basis for successful analysis of more complex arrhythmias and for increasing diagnostic power of MCG.

In more than the half of our investigated patients, where the time of localization corresponded strongly with the site of arrhythmogenic substrate, we did not find stable dipole patterns. These findings suggest, that stable dipole patterns appear at a time where the excitation of the arrhythmogenic substrate and the direct adjacent tissue is over. In fact, the time for "conventional localization mode" includes a distance to the correct site of the studied cardiac source caused by a to late time of localization.

More direct comparison between prospective MCG-localization and invasive determined site of arrhythmogenic substrate and fitting of the MCG-results during the invasive procedure are needed to further insights.

Changes in Magnetic Field Maps at QRS-Onset After Myocardial Infarction

Hailer, B.[1], Van Leeuwen, P.[2], Donker, D.[1], Rahn, N.[3], Lange, S.[2] and Wehr, M.[1]

[1]Department of Cardiology, Augusta Kranken Anstalt, Bochum, Germany; [2]Department of Biomagnetism, Research and Development Center for Microtherapy (EFMT), Bochum, Germany; [3]Medical Engineering, Siemens, Erlangen, Germany

Introduction

Several magnetocardiographic studies have been performed which investigated changes in cardiac magnetic field distribution for a number of diseases involving structural and functional changes in the myocardium, such as in hypertrophy or myocardial infarction. In such studies, different approaches have been used to examine and quantify the characteristics of the fields maps [1,2]. However, the continuous change of the map features over time during depolarisation has not been investigated. Here we introduce parameters which permit the discrimination of the effects of recent myocardial infarction on changes in conduction after Q onset as reflected in changes of the magnetic field pattern over time. Goal was to acquire more information on the underlying disease by identifying changes in the magnetic field distribution caused by ischemia or structural heart disease.

Method

Subjects: We examined 13 healthy subjects (9 males) between 20 and 51 years of age as well as 17 patients (16 males) between 37 and 79 years of age who had suffered recent (≤6 weeks) myocardial infarction (MI). Of these patients, 7 had had inferior infarction, 9 anterior and 1 both; 7 had single vessel disease, 10 multiple; 8 had an ejection fraction >60%, 9 ≤60%; 3 had slight, 5 had slight to moderate and 8 had moderate wall motion disturbances; 5 had left ventricular hypertrophy.

Data Acquisition: MCG was performed using a 37 channel biomagnetometer (Siemens Krenikon). The diameter of the probe is 19 cm. Each sensor consists of a first order gradiometer with a baseline of 7 cm. System noise is less than 30 fT/Hz^{-1} for frequencies > 2 Hz and 10 fT/Hz^{-1} for frequencies > 10 Hz. To reduce the influence of electromagnetic artifacts, the measurements were performed in a shielded room (Vakuumschmelze). Data were recorded for a minimum of sixty seconds with a bandpass of 1-200 Hz at a sampling rate of 1 kHz.

Data Analysis: Single beats in the raw data were averaged to reduce noise. The magnetic field map of the averaged beat was examined from Q-wave onset to R-wave maximum using the ECG (limb lead II) as the reference. We characterized the magnetic field distribution using 3 parameters: strength, orientation and width. Field strength was defined as the sum of the absolute values of the maximum and minimum magnetic field strength values of the recorded channels. By determining the center of gravity of the values in the positive and negative channels, respectively, the orientation of the complete field was given by the angle formed by the line joining the centers of gravity and the x-axis of the reference system (right-left across the torso). Field width corresponded to the distance between the centers of gravity.

The course of each of these parameters was time normalized to take differences in duration of the QR intervals between subjects into account. Because the distance between the sensors and the heart could not be standardized due to anatomical constraints, absolute signal intensity could not be compared directly between subjects. Therefore, field strength values for each subject were calculated relative to that subject's maximum value. The normal course of each parameter was defined as the course of the means ± 2 standard deviations (SD) of each data point of the data of the healthy subjects. On the basis of observed differences between the courses of the healthy subjects and those of the MI patients, the following criterion was taken as a marker for disease: ≥ 10% of the data points between Q and R_{max} lay beyond the ±2 SD limit of the healthy subjects.

Results

The time normalized course of each parameter was very similar for all healthy subjects (see Fig. 1). Relative field strength increased throughout the QR-interval, more rapidly in the second half, peaking at or shortly before R_{max}. Field orientation rotated by circa 70° clockwise from shortly before to shortly after Q onset, remained roughly stationary to about halfway through the QR-interval and then rotated back 120° up to R_{max}. Field width varied considerably before Q, lay between 6-9 cm in the first half of the QR-interval and, after a brief decrease, increased up to values between 9-9.5 cm at R_{max}.

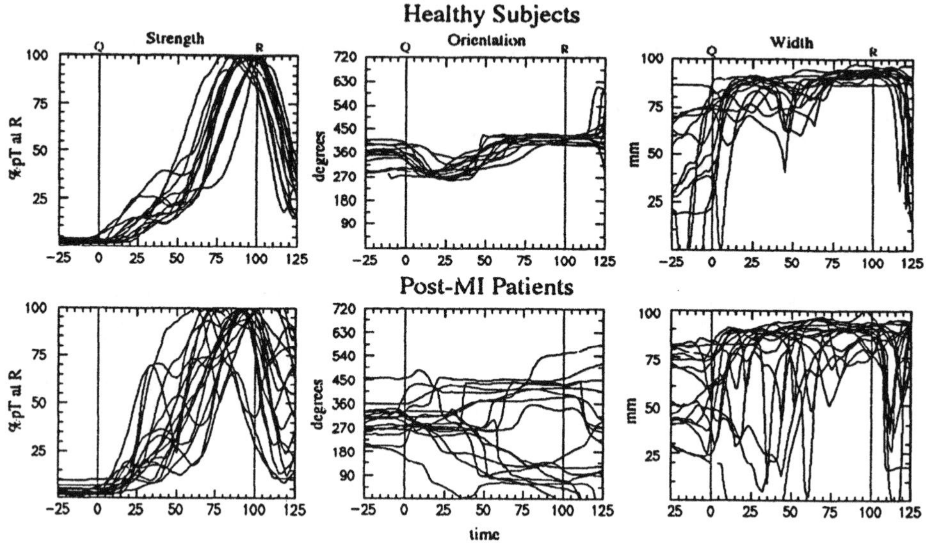

Fig. 1: Time normalized QR course of field strength, orientation and width for healthy subjects and post MI patients.

For a majority of the MI patients the course of the parameters deviated from those of the healthy subjects, either in terms of the values or in terms of the direction of change, or both (Fig. 1). Strength values tended to diverge most in the mid course and at R_{max}, orientation values deviated increasingly as QR progressed, and width values were altered constantly throughout except for an increase at R_{max} (Fig. 2). On the basis of the 10% limit on the number of deviations from the course, strength identified 10 patients (59%), orientation 17 patients (100%) and width 11 patients (65%).

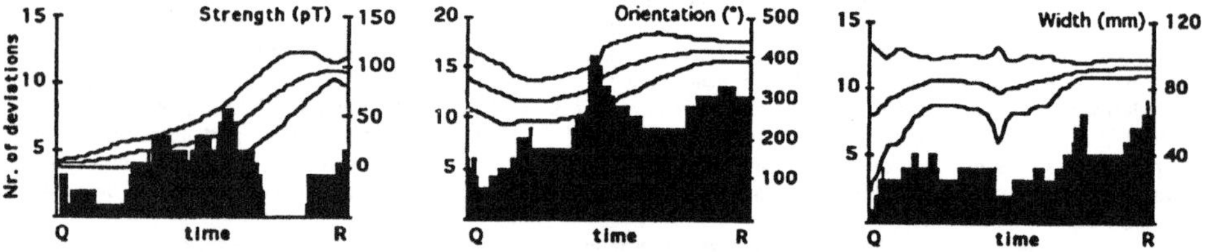

Fig. 2: Mean ±2 SD course of strength, orientation and width values of the healthy subjects (right scale) and the number of deviations of the post MI data from the normal data (left scale).

Consideration of clinical characteristics showed tendencies for certain patient groups toward a higher number of deviations, in particular for patients with more severe disease. Patients with hypertrophy tended to deviate more from the normal values for strength and orientation compared to those without hypertrophy. Patients with more severe wall motion disturbances tended to deviate more from the normal values for width. Furthermore, patients with multiple vessel disease tended to deviate more from the normal values for orientation than those with single vessel disease. Finally, patients with an ejection fraction ≤60% tended to deviate more from the normal values for orientation and width.

Discussion

The analysis of the magnetic field distribution has been used to characterize patients with structural or functional heart disease. Several routes have been taken to this end and have examined depolarisation as well as repolarisation. The construction of iso-integral maps permits the differentiation not only between healthy subjects and patients post MI but also the classification of anterior, inferior and non Q wave infarction [3]. In that study, Q wave and Q zone intervals identified patients with non Q wave and/or anterior MI whereas patients with inferior MI were identified using T wave intervals. The detection of patients after myocardial infarction with high risk for ventricular tachycardias was possible using so-called trajectory plots and iso-integral maps in the QRS and ST-T intervals [2]. It was suggested that multipolar isomagnetic fields, in particular during ST-segment, may indicate inhomogenous repolarisation. Furthermore magnetocardiography allows the visualization of pathological injury currents during stress induced ischemia or after acute infarction by applying current density reconstruction to the MCG data [4]. The identification of disturbed myocardial areas was also possible using so-called residual maps [5]. Certain components of magnetic field distribution during time intervals such as Q-wave, late QRS, ST and T-wave were determined for a group of healthy subjects and these values were subtracted from the respective field maps of patients. The results thus acquired may give certain reference points for myocardial areas with changed activity after myocardial infarction.

Here we examined the effects of pathological changes in the cardiac muscle resulting from myocardial infarction. In contrast to integrating field values over a certain cardiac time interval, as it is done in the studies described above, the parameters we present here to characterize magnetic field distribution, observe aspects of the magnetic field as they change during the time interval that is being analyzed. The data show that the courses of the three parameters during the QR-interval display a high degree of similarity for the healthy subjects. The results for field strength agree well with those obtained by others using vector MCG in normal subjects [6]. The coherence within our normal data suggests that these parameters describe underlying cardiac processes consistently. Furthermore, the parameters allow a good differentiation between healthy subjects and post MI patients. The changes of field strength in MI patients suggest that myocardial ventricular activation is changed, in particular during the R upstroke. This interpretation is further supported by the fact that absolute field strength values tended to be lower in the patient group. These results are also similar to those reported by Nousiainen et al. [6]

The results for field orientation discriminated optimally between both groups, identifying all patients. This is due in part to the minimal variation between the courses of the healthy subjects, a finding which is surprising as one would expect a certain amount of variability due to the differences in position and orientation of the individual heart relative to the sensor. Our results seem to indicate that this parameter may not be that sensitive to probe placement. Be that as it may, in the patient group, we may assume that the deviation from the normal field orientation was due to the changes in conduction patterns which result from myocardial damage. It remains to be determined which factors influence particular deviations from the normal course.

The majority of MI patients could be identified on the basis of their deviation from normal field width values with most of the discriminatory power at R_{max}, suggesting inhomogeneity around the maximum of depolarisation. Overall, the results show that deviations from the normal course in all three parameters occurred repeatedly shortly before or at R_{max}. As this time interval reflects the greatest change in activation of muscle mass, the detection of a loss of activation due to necrosis would be more probable at this time.

Furthermore, the results showed a moderate positive relationship between the degree of deviation from the normal course and the severity of the disease, as reflected in clinical parameters such as coronary status, wall motion disturbances, left ventricular hypertrophy and ejection fraction. It is worthy of note that, in contrast to the studies on magnetic field distribution mentioned above, the data presented here were not able to discriminate between different infarct locations. It therefore seems appropriate to view the results obtained to date with caution for several reasons: the patient group consisted of cases with only slight to moderate symptoms and disturbance of left ventricular function; as the number of subjects studied was small, the reliable examination of subgroups is questionable; also the study groups were neither age nor sex matched.

This study served to heuristically develop new parameters derived from the magnetic field distribution which reflect pathological changes in cardiac conduction. It could be shown that such parameters can discriminate well between healthy subjects and post MI patients and that the results may be interpreted in terms of the underlying disease. Given this and the facile acquisition of data using multichannel systems, it seems justified to further explore and develop this method on a wider range of pathological conditions and for other time intervals of the cardiac cycle. The long term goal would be to determine whether the magnetic field distribution permits the identification of myocardial ischemia. Thus, magnetocardiography would not only be a useful non invasive method in the diagnosis, but also in the further therapeutic management of patients with coronary heart disease, for example, in the verification of the hemodynamic relevance of coronary stenosis.

References

[1] Nakaya, Y., Takeuchi, A., Nii, H., et al. Isomagnetic maps in right ventricular overloading, J Electrocardiol, 1988, 21: 168-173.

[2] Stroink, G., Lant, J., Elliott, P., Charlebois, P., Gardner, M.J. Discrimination between myocardial infarct and ventricular tachycardia patients using magnetocardiographic trajectory plots and iso-integral maps, J Electrocardiol, 1992, 25: 129-142.

[3] Lant, J., Stroink, G., Montague, T.J., Gardner, M.J., Mieszkowski, M. Discrimination between myocardial infarct groups through the use of iso-integral magnetic field maps, Am J Noninvas Cardiol, 1991, 5: 215-222.

[4] Seese, B., Moshage, W., Achenbach, S., Killmann, R., Bachmann, K. Magnetocardiographic (MCG) analysis of myocardial injury currents, In: Baumgartner, G., Deecke , L., Stroink, G., Williamson, S.J. Biomagnetism: Fundamental Research and Clinical Applications, Amsterdam, Elsevier/IOS Press, 1995: 628-632.

[5] Stroink, G., MacAulay, C., Montague, T.J., Horacek, B.M. Normal and abnormal components in magnetocardiographic maps of a subject with myocardial infarction, Med Biol Eng Comp, 1985, 23: 61-62.

[6] Nousiainen, J., Oja, O.S., Malmivuo, J. Comparative analysis of the spatial sensitivities of VMVG an VECG, In: Hoke, M., Erné, S.N., Okada, Y.C., Romani, G.L. Biomagnetism: Clinical Aspects, Amsterdam, Excerpta Mediac, 1992: 497-501.

The Automatic Selection of Decisive Parameters in the Magnetocardiographic Localization of the Accessory Pathway in Patients Suffering from the WPW Syndrome

Jazbinšek[1], V., Trontelj[1], Z., Oeff[2], M., Burghoff[3], M.

[1]*Institute of Mathematics, Physics and Mechanics, University of Ljubljana, Slovenia;*
[2]*Klinikum Benjamin Franklin, Freie Universität Berlin, Germany;*
[3]*Physikalisch-Technische Bundesanstalt, Institut Berlin, Germany*

Introduction

The Wolff–Parkinson–White (WPW) syndrome is characterized by the accessory conduction pathway (AP) of excitation between atria and ventricles. To find a functional localization [1, 2] of the AP, noninvasive magnetocardiographic (MCG) mapping data were used with the current dipole source in different volume conductor models.

The AP is usually clearly manifested by a small deflection called Δ wave in a time evolution of the electric potential (ECG) or the MCG signal. In our study we have introduced an automatic selection of the AP onset as well as an automatic choice of the best reference signal interval for the magnetic data baseline correction. The stability of MCG inverse solutions has also been studied. For all cases the calculated AP spatial coordinates were compared to the corresponding results obtained by the invasive electrophysiological study (EPS).

Methods

We have studied 22 patients suffering from the WPW syndrome. All MCG measurements were performed in the Berlin Magnetically Shielded Room [3] by means of 37-SQUID multichannel magnetometer system [4]. Up to 5 ECG channels were simultaneously recorded. To reduce the effect of baseline shifts between different magnetic channels a reference time interval of 10 *ms* length has been chosen. The best reference interval has been found automatically by finding the minimum of root mean square value of standard deviations (SD) of signals in all magnetic channels

$$\mathrm{SD}(j) = \sqrt{\frac{1}{N - N_{chan}} \sum_{i=1}^{N_{chan}} \sum_{k=j-5}^{j+5} (B_{ik} - \overline{B}_i)^2}, \quad \overline{B}_i = \frac{1}{11} \sum_{k=j-5}^{j+5} B_{ik}, \tag{1}$$

where $N = 11 \times N_{chan}$, N_{chan} is the number of channels, B_{ik} is the magnetic field of the i-th channel in the k-th time, and $\overline{B}_i$ is the mean magnetic field in the i-th channel in the time interval $[j-5, j+5]\,ms$. A reference signal, which is the mean magnetic field on the best reference interval, is then subtracted from magnetic data in a whole time sequence. The effect of different reference intervals is shown on Fig. 1. The SD(j) reflects how much fluctuating activity around the j-th time is present. In addition to baseline corrections a subtraction of the reference signal, determined before the Δ wave, reduces the influence of the P-wave repolarization [1]. For all studied cases the very beginning of the AP was automatically selected by finding the time when the level of the SD in Eq.(1) had been four times the value of the SD on the best reference interval in the region before the Δ wave. Automatically determined onsets were confirmed by the visual inspection of magnetic as well as electric data.

As a model of a current source in a localization procedure a moving equivalent current dipole has been used. We have tested several volume conductor models like a half-space model, an elliptic cylinder and three different realistic torso models (see Fig. 2). For the volume conductor of an arbitrary shape the Boundary Element Method (BEM) has been used [5, 6].

For all studied cases the calculated MCG localization results were compared to the corresponding results obtained by the EPS.

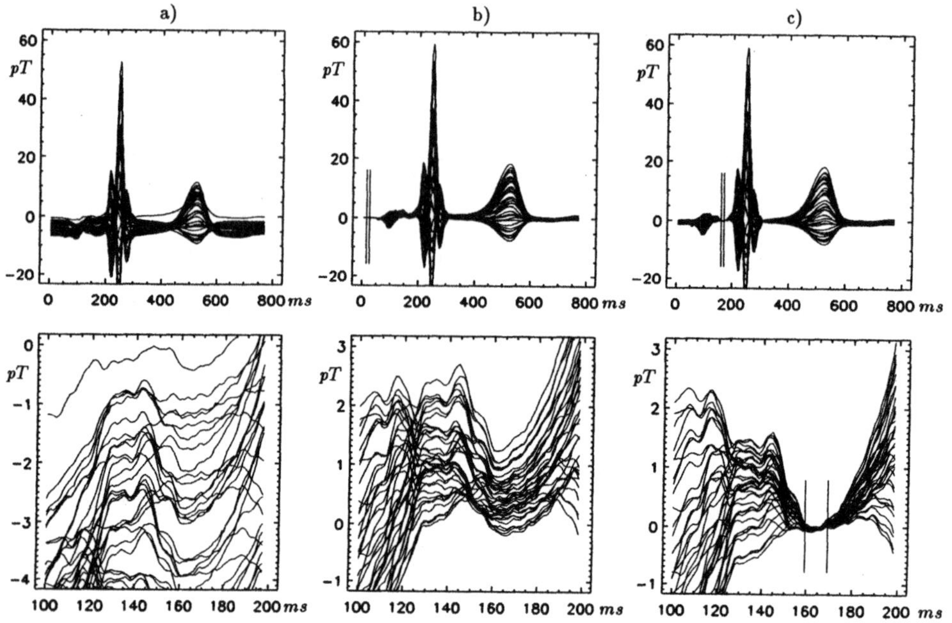

Figure 1: The effect of the reference interval: a) no reference, b) reference before the P wave and c) reference before the Δ wave. Superposition of all magnetic channels for one patient is displayed. The whole heart beat and zoomed signals around the Δ wave are shown in the first and the second raw, respectively. Two vertical lines denote reference intervals. The automatically selected AP onset in this particular case is at $182\,ms$.

Results

The locations of equivalent current dipole in different volume conductor models were determined for each ms in the $40\,ms$ time interval after the AP onset. For each model and time point the stability of the calculated localization site was estimated by averaging results of five neighboring time points, two before and two after the current time point. It was shown in [7], where 12 patients with reference intervals in the region before the Δ wave were studied, that the MCG inverse solution became more stable 10 to $20\,ms$ after the AP onset. The reason for that are very low magnetic signals in the very beginning of AP which are slightly above the noise level. This is even more evident in the present study where MCG inverse solutions with reference intervals before the P wave are also considered.

To compare the noninvasive MCG localization with the invasive EPS localization it is necessary to match a coordinate system during the MCG measurement with a coordinate system during the EPS. The patient positioning during all MCG measurements was done manually. The estimated error of this procedure is around $0.5\,cm$ for the frontal (xy) plane and about $1\,cm$ for the z coordinate (depth). The positioning of patients during the EPS was done with markers on the skin. The ablation catheter was recorded by X-rays in the frontal and sagittal planes. The X-ray film was afterwards projected on a screen and the position of the catheter ablation tip was determined manually. This procedure introduces some additional errors. However, they can be reduced considerably provided that the EPS recordings are digitized and the complete geometry of the X-ray optics is known. EPS data for the group of patients in the present research were not obtained by such an ideal treatment. Therefore the comparison between both modalities gives only a rough estimation of the MCG localization accuracy. These comparisons, displayed in Tables 1 and 2, indicate that

Table 1: The comparison between MCG and EPS localization results for the patient data shown on Fig. 1. Negative Δx means that result of the MCG localization is more to the right, negative Δy more cranial and negative Δz more anterior than the corresponding EPS localization and $\Delta r = \sqrt{\Delta x^2 + \Delta y^2 + \Delta z^2}$. P and Δ denote results for the reference interval before the P and the Δ wave, respectively. Different rows for each item correspond to the AP onset , AP onset + 10 *ms* and AP onset + 20 *ms*, respectively. All data are in *cm*.

	half-space		Elliptic cylinder		Torso 1		Torso 2		Torso 3	
	$(\Delta x, \Delta y, \Delta z)$	Δr	$(\Delta x, \Delta y, \Delta z)$	Δr	$(\Delta x, \Delta y, \Delta z)$	Δr	$(\Delta x, \Delta y, \Delta z)$	Δr	$(\Delta x, \Delta y, \Delta z)$	Δr
	(-12.8,-4.4,-5.9)	14.7	(4.4, 1.6,-2.6)	5.3	(3.5, 0.7,-2.0)	4.1	(2.9,-7.2,-3.4)	8.5	(3.9,-2.6,-2.3)	5.2
P	(2.6,-6.6,-7.9)	10.6	(3.0, 0.7,-1.9)	3.6	(2.5,-0.5,-1.8)	3.1	(3.5,-3.8,-0.9)	5.2	(3.7,-3.6,-2.0)	5.5
	(4.4, 1.6,-1.3)	4.9	(1.7, 1.3,-2.7)	3.4	(1.3, 0.5,-2.5)	2.8	(1.3, 1.3,-1.8)	2.6	(1.2, 2.2,-2.2)	3.3
	(6.0,-6.5,-3.7)	9.6	(3.2, 0.4,-1.1)	3.4	(3.0,-2.5,-1.0)	4.0	(3.4,-3.5,-0.6)	4.9	(4.0,-4.3,-1.7)	6.1
Δ	(6.5,-3.0,-1.5)	7.3	(2.0,-0.9,-0.9)	2.4	(2.0,-1.9,-0.9)	2.9	(2.5,-1.2,-1.0)	3.0	(2.6,-1.3,-1.5)	3.3
	(3.7, 1.6,-1.1)	4.2	(1.2, 1.1,-2.5)	2.9	(0.8, 0.4,-2.4)	2.6	(1.1, 1.0,-1.9)	2.4	(0.9, 1.8,-2.4)	3.1

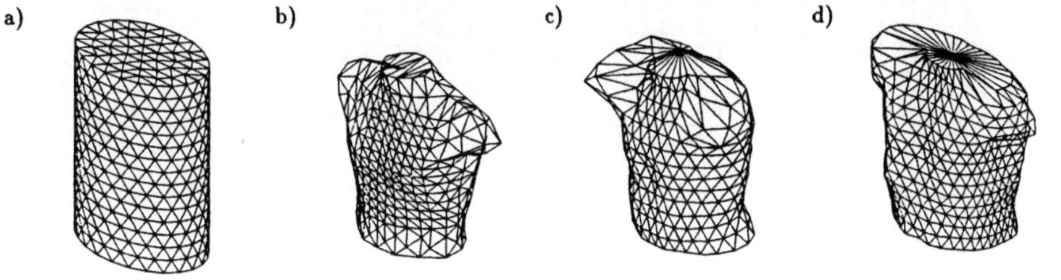

Figure 2: Volume conductor models: a) elliptic cylinder (axes of elliptic base 24&36 *cm*, height 50 *cm*), b) Torso 1 – standard torso developed by van Oosterom [8], c) Torso 2 – standard torso developed by Horaček [6] and d) Torso 3 – torso constructed by us from Magnetic Resonance Images [9].

better results are obtained for data corrected by the reference interval before the Δ wave, especially for the AP early phases.

It is interesting to compare SD values Eqn. (1) for the best reference interval before the P and the Δ wave. For data shown in Fig. 1 these values are 15 *fT* and 45 *fT*, respectively. Whereas the mean values and their standard deviations for all 22 studied patients are $(23 \pm 14)\,fT$ and $(41 \pm 26)\,fT$, respectively. The SD before the P wave reflects a data noise, which is the summation of noise due to the measuring system and noise caused by patients themselves (breathing, moving, etc.), while the SD before the Δ wave can in addition to the noise also contain some heart signals like the P wave repolarization and the His–Purkinje system depolarization.

Conclusions

Our study has shown that the automatic determination of the best reference interval and the automatic selection of the AP onset are both essential since the visual selection from the measured signals is very subjective and strongly dependent on the person in charge to evaluate data.

The MCG inverse solution is more stable if magnetic data is corrected by the reference interval before the Δ wave, especially in the first 20 *ms* after the AP onset.

We have also found that for the simplified torso model in the form of an elliptic cylinder the average differences between noninvasive MCG localizations and invasively obtained EPS results were approximately the same as those for the three different realistically shaped torso models, while the half-space model gave only a rough estimation of the AP site.

Table 2: The comparison between MCG and EPS localization results for 16 typical patients. $|\Sigma|$ is the average of absolute values of $(\Delta x, \Delta y, \Delta z)$ and Δr with standard deviation $\sigma_{|\Sigma|}$. The meaning of $(\Delta x, \Delta y, \Delta z)$ and Δr is described in Table 1 legend. Three rows for each item correspond to the AP onset , AP onset + 10 ms and AP onset + 20 ms, respectively. P and Δ denote results for the reference interval before the P and the Δ wave, respectively. All data are in cm.

		half–space $(\Delta x,\Delta y,\Delta z)$ Δr	Elliptic cylinder $(\Delta x,\Delta y,\Delta z)$ Δr	Torso 1 $(\Delta x,\Delta y,\Delta z)$ Δr	Torso 2 $(\Delta x,\Delta y,\Delta z)$ Δr	Torso 3 $(\Delta x,\Delta y,\Delta z)$ Δr
P	$\lvert\Sigma\rvert$	(5.7,3.2,4.0) 8.2	(3.3,3.2,2.7) 5.9	(3.5,3.0,3.1) 6.4	(4.2,2.9,3.5) 6.8	(4.7,2.8,3.2) 6.9
		(3.6,2.6,4.3) 7.0	(3.1,1.8,2.3) 4.8	(3.6,1.7,1.6) 4.6	(3.3,2.2,2.5) 5.3	(3.6,2.2,2.1) 5.1
		(4.0,2.5,3.3) 6.3	(3.2,2.0,1.9) 4.7	(2.3,1.7,1.5) 3.6	(2.9,1.8,3.0) 5.2	(3.3,2.1,3.1) 5.5
	$\sigma_{\lvert\Sigma\rvert}$	(3.1,2.0,2.4) 3.1	(2.1,2.1,2.4) 2.8	(2.3,2.6,2.0) 2.4	(2.2,1.9,2.1) 1.9	(2.4,1.4,1.8) 1.7
		(3.1,2.6,2.9) 3.4	(2.5,1.4,1.4) 2.3	(2.7,1.4,1.2) 2.7	(2.4,1.5,1.9) 2.3	(2.6,1.3,1.3) 2.3
		(1.6,2.2,1.9) 2.1	(2.5,1.6,1.5) 2.4	(1.5,1.1,1.2) 1.5	(2.5,1.4,2.2) 2.4	(1.8,1.6,2.7) 2.8
Δ	$\lvert\Sigma\rvert$	(4.2,2.9,3.5) 6.9	(3.0,2.4,2.3) 5.0	(2.9,1.9,3.1) 5.3	(2.1,2.4,2.7) 4.6	(2.9,2.5,2.3) 5.0
		(4.1,2.8,3.6) 6.9	(3.0,2.1,2.1) 4.8	(3.1,2.2,1.6) 4.6	(3.0,2.3,2.7) 5.3	(3.7,2.3,2.7) 5.7
		(4.1,2.3,3.4) 6.2	(2.2,2.3,1.8) 4.1	(1.6,2.0,1.5) 3.6	(2.4,1.6,2.6) 4.5	(2.5,2.3,2.4) 4.9
	$\sigma_{\lvert\Sigma\rvert}$	(3.8,2.0,2.2) 3.6	(1.9,1.6,1.7) 2.0	(2.2,1.3,2.2) 2.1	(1.8,1.1,1.8) 1.9	(2.4,1.7,2.0) 2.6
		(3.0,2.3,2.6) 3.1	(2.2,1.9,1.5) 2.3	(2.8,1.4,1.2) 2.7	(2.3,1.7,1.7) 2.1	(2.6,1.8,1.8) 2.5
		(2.4,1.9,1.5) 2.5	(1.8,2.1,1.1) 2.2	(1.6,2.4,1.2) 2.3	(2.4,1.4,1.8) 2.4	(2.2,1.8,1.9) 2.0

References

[1] Nenonen, J., Katila T., Leiniö, M., Montonen, J., Mäkijärvi, M., Siltanen, P. Magnetocardiographic Functional Localization Using a Current Multipole Models. IEEE Trans. Biomed. Eng. 1991, BME-38:648–57.

[2] Nenonen, J., Purcell, C.J., Horaček, B.M., Stroink, G., Katila, T. Magnetocardiographic Functional Localization Using a Current Dipole in a Realistic Torso. IEEE Trans. Biomed. Eng. 1991, BME-38:658–64.

[3] Erné, S.N., Hahlbohm, H-D., Scheer, H., Trontelj, Z. The Berlin Magnetically Shielded Room (BMSR). In: Erné, S.N., Hahlbohm, H-D., Lübbig, H. Biomagnetism, Berlin, Walter de Gruyter, 1981, 79–87.

[4] Koch, H., Cantor, R., Erné, S.N., Matthies, K.P., Peters, M., Ryhänen, T. Scheer H.J., Hahlbohm, H-D. A 37 channel dc SQUID magnetometer system. IEEE Transaction on Magnetics 1991, 27:2793–6.

[5] Barnard, A.C., Duck, I.M., Lynn, M.S., Timlake, W.P. The application of electromagnetic theory to electrocardiography II: The numerical solution to the integral equation. Biophys. J. 1967;7:463–91.

[6] Horaček, B.M. Digital model for studies in magnetocardiography. IEEE Trans. Magn. 1973, MAG-9:440–4.

[7] Jazbinšek, V., Burghoff, M., Oeff, M., Trontelj, Z. Magnetocardiographic localization of accessory conduction pathway in patients suffering from WPW syndrome, In: Murray, A., Arzbaecher, R. Computers in Cardiology 1995, Piscataway, IEEE Press, 1995, 417–20.

[8] van Oosterom A. Triangulating the human torso. Comput. J. 1978;21:253-258.

[9] Jazbinšek V, M Burghoff, M Oeff, R Zimmermann, Z Trontelj. Improved torso modeling and its influence on the localization of arrhythmogenic center(s) in heart, In: 10th International Symposium on Biomedical Engineering 1994. Zagreb: CROMBES & KOREMA Proceedings, 1994, 183–6.

Source Localization of Initial QRS Vector in Patients with Idiopathic Cardiomyopathy: Comparison with Normal Subjects and Patients with Anterior Myocardial Infarction

Kishi, F.[1], Nomura, M.[1], Yukinaka, M.[1], Saito, K.[1], Nakaya, Y.[2] and Ito, S.[1]

[1]Second Department of Internal Medicine and [2]Department of Nutrition, The University of Tokushima, Tokushima 770, Japan

INTRODUCTION

The magnetocardiographic three-dimensional (3-D) source localization using superconducting quantum interference device (SQUID) gradiometer has an excellent spatial and temporal resolution. Abnormal Q wave is frequently seen in the electrocardiogram (ECG) in patients with idiopathic cardiomyopathy, even in those without myocardial infarction. In the present study, to assess the origin of abnormal Q wave in the idiopathic cardiomyopathy, current sources were analyzed by the magnetocardiograms (MCGs), and the results were to that of compared with abnormal Q wave in patients with anterior myocardial infarction.

SUBJECTS AND METHODS

MCGs were recorded in 18 patients with idiopathic cardiomyopathy (ICM group), 14 normal subjects (control group) and 16 patients with anterior myocardial infarction (MI group). The ICM group were subdivided into 2 subgroups; the patients with abnormal Q wave in anterior chest leads (ICM Q group; 7 cases) and those without Q wave (ICM non-Q group; 11 cases).

The MCG was recorded with a second derivative SQUID gradiometer (Model BMP, SHE Corp.) [1]. The patient laid on the wooden bed in the supine position, the detector was placed close to the front of the chest and the magnetic shield. Fig. 1 shows a recording positions of the MCG. A 2.5 cm $\times$2.5 cm rectangular grid system consisting of six horizontal (A-F) and six vertical (1-6) lines was used with the xyphoid serving as a reference point (E-2) for the grid. As a reference, the lead II ECG was recorded at a paper speed of 200 mm/sec simultaneously with the MCG. A 50 Hz high cut filter was used to record the MCG. From the beginning of the QRS waves of the ECG II, the amplitude of the MCG was measured every 10 msec. The isofield lines were drawn at each time (5.2×10^{-9} gauss/line).

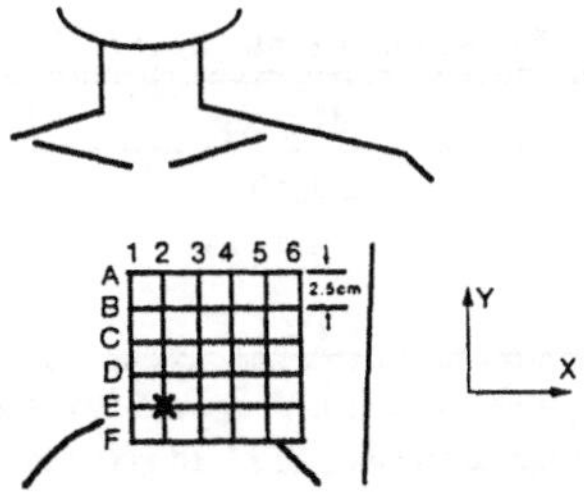

Fig. 1 Recording positions of the MCG

For source localization analysis, a single dipole model was used and the 3-D location of the equivalent current dipole was computed [2,3]. The 3-D dipole sources obtained from the MCGs were superimposed onto the magnetic resonance image (MRI) of the subjects providing anatomical localization.

RESULTS

Fig. 2A shows the isofield map of anterior chest of a normal subject. In the initial QRS pattern (20 msec) of the isopotential map, the minimum was located in the lower part of the map and the maximum was located superiorly. Fig. 2B shows the isofield map of a patient of anterior myocardial infarction. In this map, the positive area was located in the lower portion and the negative area was in the superior portion. Fig. 2C shows the isofield map of a patient of idiopathic cardiomyopathy. This map is similar to Fig. 2B. The isofield maps at 20 msec from the onset of QRS wave of the control group showed negative area in the lower anterior chest wall and positive area in the upper anterior chest wall, predicted rightward equivalent current dipole directed rightward. On the other hand, the isofield maps of the anterior myocardial infarction and ICM Q groups showed the reversed pattern from the control group.

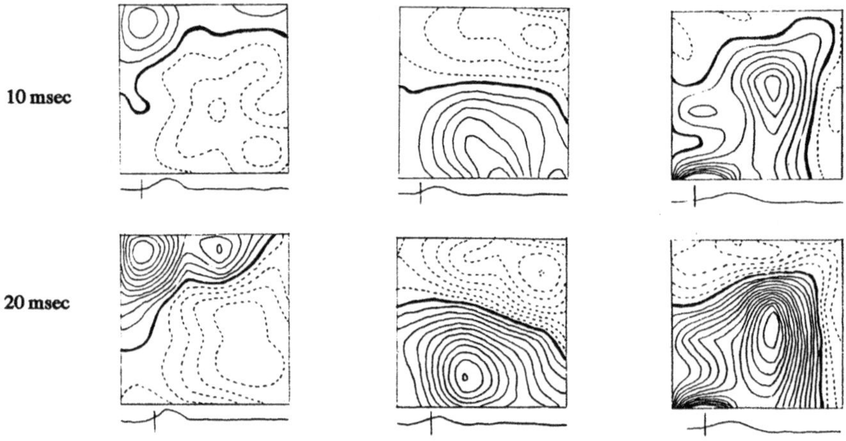

Fig. 2A (normal) Fig. 2B (MI) Fig. 2C (ICM)

Fig.2 Isofield maps at 10 msec and 20 msec

The dipoles at the early QRS of the anterior MI group and ICM Q group were displaced posteriorly, compared with those of control group and ICM non-Q group (Table 1). The deduced dipoles superimposed onto MRI in the control group were on the ventricular septum but those in MI or ICM Q groups were not on the ventricular septum. (Fig.3).

Table 1 Three-dimensional Source Localization in Each Group

	Control group	MI group	ICM Q group	ICM non-Q group
X (cm)	4.5±0.6	2.9±0.6*	2.9±2.3	3.6±0.6
Y (cm)	3.2±0.4	5.0±0.6*	3.9±0.8	3.6±0.4
Depth (cm)	5.6±0.2	7.1±1.7*	7.0±0.5*	6.1±0.6

X: left side from the xyphoid process (origin); Y: upward form the origin: Depth: distance between detector coil and dipole source. *$p<0.05$ Values are mean±SEM. (dipole sources were at 20 msec from the onset of the QRS wave)

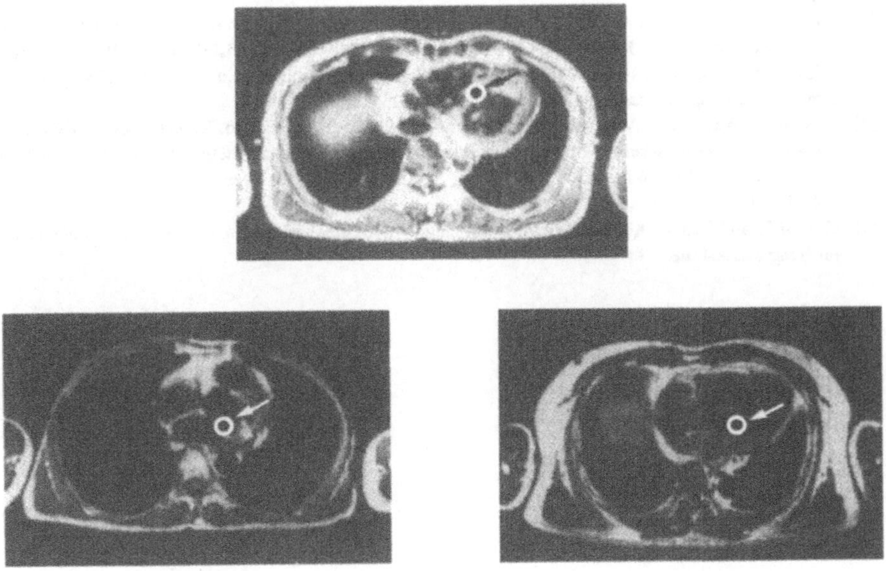

MI **ICM**

Fig.3 The deduced dipoles superimposed onto MR imaging

DISCUSSION

The biomagnetic recording method using a second derivative gradiometer provides an excellent spatial resolution that is capable of source localization of the equivalent current dipole with a high degree of accuracy and excellent temporal resolution (<1 ms). In our previous studies, the dipole location of patients with WPW syndrome was computed with a biomagnetometer onto MRI plans in order to study the positional relationship vis-a-vis the heart [3].

In the present study, we investigated the origin of abnormal Q wave in the idiopathic cardiomyopathy by isomagnetic mapping. Abnormal Q waves resembling myocardial infarction are particularly common in the obstructive type of cardiomyopathy. It is reported that primary myocardial disease of the nonobstructive type may also cause an infarction pattern containing abnormal Q wave in the ECG, and that this finding is the result of localized area of myocardial destruction. Microscopic lesions were thought to be responsible for conduction disturbances resulting in aberration of the initial portion of the QRS forces [4]. In the present study, however, the amplitude of the dipole was increased and the location of the abnormal Q vector was not in the ventricular septum and displaced posteriorly in the idiopathic cardiomyopathic patients with abnormal Q waves, suggesting the increased electromotive force of the posterior wall of the left ventricle. This study indicated that abnormal Q wave in the ECG in cardiomyopathic patients was not due to conduction disturbances in the ventricular septum but to the increased electromotive force of left ventricular free wall in cardiomyopathy.

REFERENCES:

[1] Mori, H. and Nakaya, Y. Present status of clinical magnetocardiolgraphy. CV World Report, 1988, 1: 78-86.

[2] Nomura, M., Nakaya, Y., Ken, S., Kishi, F., Wakatsuki, T., Miyoshi, H., Nishikado, A., Bando, S., Ito, S., Nishitani, H., Wada, M., Fujita, S. and Tamura, I. Noninvasive localization of accessory pathways by magnetocardiographic imaging. Clin Cardiol, 1994, 17:239-244.

[3] Nomura, M., Nakaya, Y., Watanabe, K., Katayama, M., Takeuchi, A., Fujino, K. and Mori, H. Detection of accessory pathway in patients with WPW syndrome by means of the isomagnetic map and MRI. In advances in Biomagnetism (Eds. Williamson SJ, Hoke M, Stroink G, Kotani M). Plenum Press, New York, London, 1989, 373-376.

[4] Tabel, M.E. and Fish, C. Abnormal Q waves simulating myocardial infarction in diffuse myocardial diseases. Am Heart J, 1964, 68: 534.

Linear and Nonlinear Analysis in the Parallel HR-MCG/HR-ECG Experiment in Arrhythmogenic Right Ventricular Cardiomyopathy

Kowalik, Z.J.[1,2], Mäkijärvi, M.[3], Wichter, T.[4], Fetsch, T.[4], Elbert, T.[2], Breithard, G.[4] and Hoke, M.[1]

[1]*Institute of Experimental Audiology, Biomagnetism Center, University of Münster, Germany;* [2]*Department of Psychology, University of Konstanz, Germany;* [3]*Cardiovascular Laboratory, First Department of Medicine, Helsinki University Central Hospital, Finland;* [4]*Department of Cardiology and Angiology University of Münster, Germany*

Introduction

The Uhl's disease, known also as ***arrythmogenic right ventricular cardiomyopathy*** (ARVC) may be a cause of sudden cardiac death in young people [3]. Application of the standard linear and new nonlinear methods to investigation of macroscopic physical aspects of this pathology, especially those which provide information for risk analysis, and predict the development of this disease in its early stage, seems to be a very promising task for the biomagnetic studies especially when combined with multichannel high-resolution ECG measurement. The aim of our study was to compare the information content of simultaneously measured ECGs and MCGs and the detailed spatio-temporal structure of multichannel patterns, with standard linear and modern nonlinear algorithms.

Methods

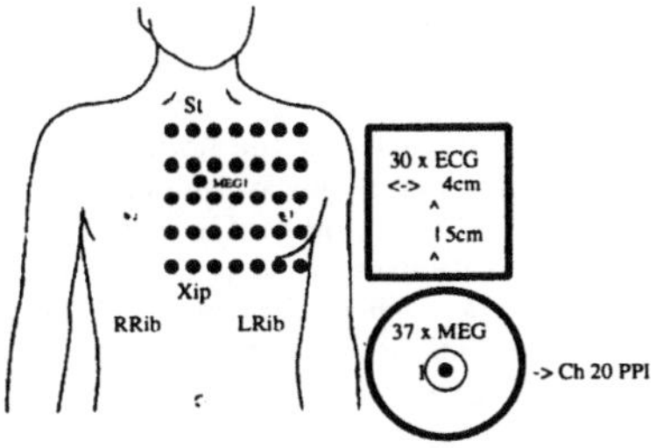

Figure 1: Experimental setup. The SQUID magnetometer was positioned in the central part of the chest. ECG's were mapped by use of the same pattern with a grid 4x5cm.

We used a 37-channel magnetometer (BTi/San Diego, USA) for the MCG and 32-channel EEG/ECG (NeuroScan/Houston, USA) for the ECG measurements, respectively. Four ARVC patients (males, age 46, 47, 61, 62 ys.) and four healthy controls (males, age 34, 35, 35, 36 ys.) were tested. The experimental setup is shown in Fig. 1. The experiment was performed in a magnetically shielded room to avoid environmental magnetic and electric noise. The data from 37 magnetic and 30 electric channels was first analysed with software included in the BTi-system. The results presented below were obtained with the BTi-MCG-Software-Tool (linear analysis: data clusterizing, filtering, FFT analysis), and our own nonlinear software packet (Correlation Dimension D2, and local Lyapunov Exponent LE).

Results

Linear Analysis

Linear procedures were based on BTi-MCG "preproc" batch program for both MCG and ECG. Briefly, the following procedures were performed: (1) calculation of spatial RMS-average for all magnetic channels at all points: $< s >= \sqrt{a_1{}^2 + a_2{}^2 + ... + a_{37}{}^2}$, (2) detection of the chosen characteristic events (in our case R-peak), (3) calculation of the average, minimum and maximum time-distances between events, (4) generating of epoch-files containing extracted cardiac beats, (5) clusterizing of beats, (6) averaging of beat clusters. Figure 2 presents time- and frequency domains of measured cardiograms.

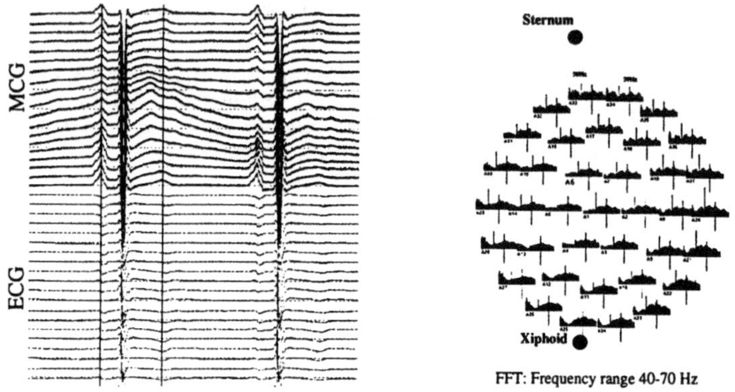

Figure 2: Most interesting interval occurs after R-peak. Note change of phase in this region (left). Close to the parasite 50 Hz line frequency there is a region with very distinct maximum in frequency band 40-70 Hz (right).

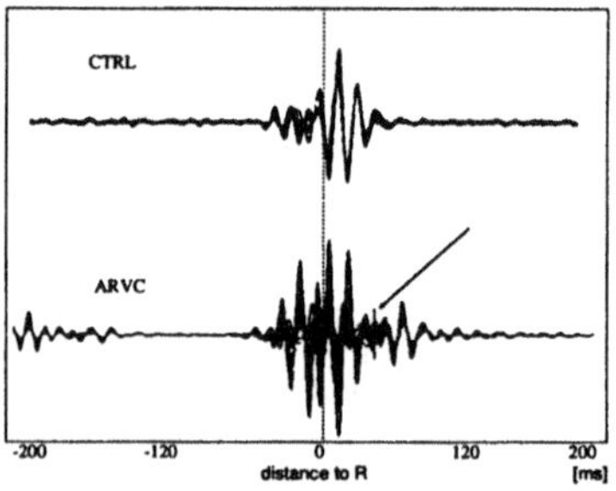

Figure 3: Overplots of all MCG-channels in normal and in ARVC patients.

Dephasing effect. These analyses revealed a "dephasing" effect. The clusterized patterns (correlation threshold 75%) were filtered (bandpass - 6dB/oct, cut-off frequencies of 52 and 99 Hz). The overplot of all measured ECG channels exhibited an almost fully overlapping patterns for both, the ARVC and healthy groups. However, MCG overplots represented in Fig. 3 exhibited a distinct difference between the normal (upper) and the pathological (lower) subjects. A "dephasing effect" was evident, i.e., different channels are not-in-phase, so that a "blurred" image appears. This effect observed for all ARVC-patients occurs in the S-T interval, about 30-80 ms after the R-peak.

Nonlinear analysis

Most of the available nonlinear analysis methods are based on the reconstruction of a phase space by the embedding technique. This can be done even when the real dimensions, i.e. the independent variables whose interaction constitutes an empirical system, are not known. The calculation of the correlation dimension D2 [1] and Lyapunov Exponents (LE) are based on this quasi attractor representation of a time series.

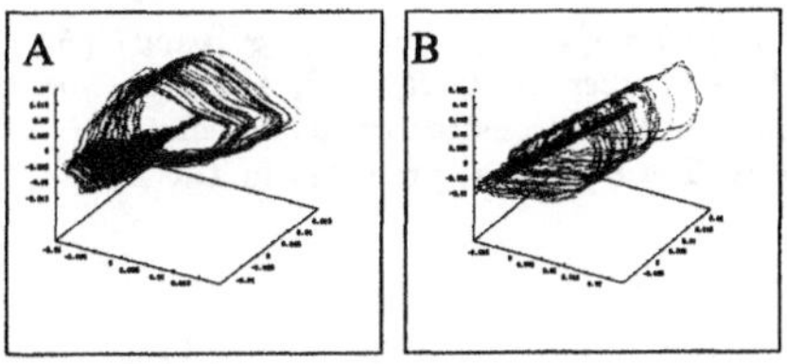

Figure 4: Comparison of the topology of 5-dimensional MCG-attractors (3D-projection), for one normal (A) and one ARVC (B) subject.

Practical reconstruction is performed by constructing an $N \times \mathcal{E}$–dimensional (embedding) matrix from N-elements length scalar time series, wherein the successive (in $\mathcal{E}$) matrix elements are delayed about time τ. τ is there determined with help of the characteristic frequency ω_0 in the power spectrum (the largest maximum in the spectrum lying far from $\omega = 0$) whereby: $\tau_0 = \frac{2\pi}{\omega_0}$ and $\tau = \frac{\tau_0}{4}$. Examples of reconstructed attractors for healthy subjects and ARVC patients are shown in Fig. 4. The volume of ARVC attractors shown there is distinctly smaller than the volume of CTRL-attractors.

Local Lyapunov exponent (LE). An essential characteristic of chaotic processes is their so-called sensitive dependence on initial condition. This means that even minuscule disturbances may completely alter the subsequent course of a trajectory, even though the generating system remains the same. Even in non-stationary systems (for instance, physiological time series) the rate of the divergence of close trajectories can be measured in the phase space by means of the Lyapunov exponents [2]. The Lyapunov exponent is defined as the average exponential divergence (if positive) or convergence (if negative) of nearby trajectories in the phase space of a given system. Wolff [4] proposed to estimate this measure through averaging of the following quantities:

$$\lambda_{i,m} = \frac{1}{m}\frac{1}{n_i}\sum_{j \in S_j} \log \left| \frac{X_{i+m} - X_{j+m}}{X_i - X_j} \right| \tag{1}$$

i=1,......,n-m; $S_i = \{j : 0 < |X_i - X_j| \leq h\}$, $n_i = \#(S_i); m \in N$, $h > 0$. Representative spatial maps of local LE are presented in Fig. 5. Two characteristic maxima (peaks) are visible for all cases occur at both ends of heart's axis. Noteworthy are (i) a more ordered spatial (!) pattern in healthy subjects and (ii) the specific shape of the pattern in these subjects. Performing group statistics for the estimated D2-

PID/Measure	D2(ecg)	SD2(ecg)	D2(mcg)	SD2(mcg)
ctrl-z10	1.812602	0.340892	2.024888	0.339570
ctrl-z11a	1.834800	0.885758	2.219154	0.389377
ctrl-t44	1.825118	0.349549	1.910880	0.324748
ctrl-z15a	1.850036	0.350655	2.161658	0.371567
arvc-z16	1.601862	0.294741	1.687413	0.294592
arvc-z17a	1.470085	0.481024	1.996400	0.337596
arvc-z18a	1.969467	0.384368	2.041851	0.359005
arvc-z14	1.795002	0.390125	1.621669	0.278256

Table 1: D2-dimension ($\mathcal{E}$ =5, τ =20) and its variability S(D2) averaged over time and all channels.

dimensions get following values: D2(ECG,CTRL)=1.83; SD2(ECG,CTRL)=0.48; D2(MCG,CTRL)=2.08; SD2(MCG,CTRL)= 0.36 and D2(ECG,ARVC)=1.71; SD2(ECG,ARVC)=0.37; D2(MCG,ARVC)=1.83; SD2(MCG,ARVC)=0.31. Briefly, these results suggest: D2(MCG)>D2(ECG) $\Rightarrow$ MCG supports more information. D2(CTRL)>D2(ARVC) $\Rightarrow$ "Normal" temporal pattern is more complex than arrhythmogenic. SD2(ARVC)<SD2(CTRL) $\Rightarrow$ The larger variability the more "healthy state" is observed.

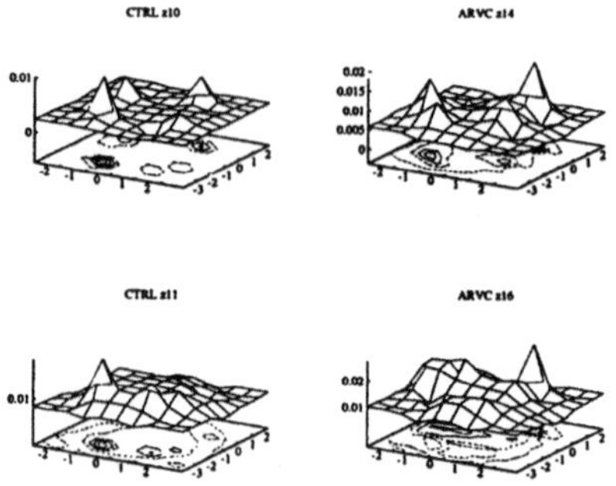

Figure 5: LE-distribution of MCGs.

Spatial distribution of correlation dimension D2 of MCG's and ECG's.

The correlation dimension was estimated after Grassberger and Procaccia algorithm [1] from following equations:

$$D2 = \lim_{R \to 0} \frac{\log C(R)}{\log R}; \tag{2}$$

with $C(R) = \frac{1}{N^2} \sum_{i \neq j}^{N} \Theta(R - |x_i - x_j|)$ and $\Theta(R) = 1$ if $R > 0$ otherwise $\Theta(R) = 0$, and x_i, x_j are the points in the reconstructed phase space (τ =10, $\mathcal{E}$ =5). The results are presented in Tab. 1.

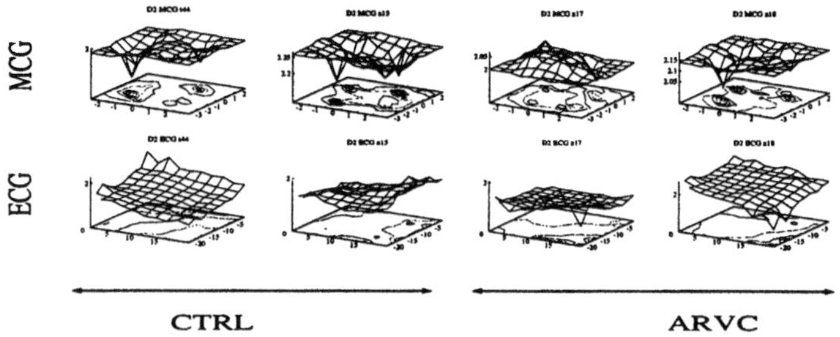

Figure 6: Distribution of D2-dimension measured in MCG and ECG channels.

Conclusions

Linear analysis

- Differences between groups appear in frequency band 30-100Hz, 30-80 ms after R-peak.
- The phase of a high frequency components is variable in both the spatial and temporal domain in the case of the MCG. This effect was not found in ECG.

Nonlinear analysis

- D2 differentiates the groups although the significance of this effect has to be proved for larger groups.
- MCG provides more information about heart dynamics than does ECG.
- The spatial distribution of D2 and of the local LE shows precisely the heart activity and especially its temporal changes. Unlike temporal structure, the spatial order observed via LE-maps is characteristic for healthy subjects but not the patients group.

References

[1] Grassberger, P. and Procaccia, I. Measuring the strangeness of strange attractors, Physica D9, 189-208, 1983.

[2] Kowalik, Z.J., and Elbert, T. A practical method for the measurements of the chaoticity of electric and magnetic brain activity, Int.J. Bifurcation and Chaos, 5, 475-490, 1995.

[3] Thiene, G., Nava, A., Corrado, D., Rossi, L., and Penelli, N. Right ventricular cardiomyopathy and sudden death in young people, N.Engl.J.Med., 318, 129-133, 1988.

[4] Wolff, R.C.L. Local Lyapunov exponents: Looking closely at chaos, J.R.Stat.Soc. B54, 353-371, 1992.

His-Purkinje Conduction System Maps: Fact or Fiction?

Lamothe, M.J.R.[1], Stroink, G.[1,2] and Gardner, M.J.[2,3]

[1]Departments of Physics, [2]Physiology & Biophysics, and [3]Medicine; Dalhousie University, Halifax, NS, Canada

Introduction

Electrical conduction between the heart's atria and ventricles occurs via the His-Purkinje conduction system (HPS) formed by the His bundle, bundle branches and Purkinje fibre network. The first attempts to record HPS electrocardiograms and magnetocardiograms were made by Berbari et al.[1] and Farrell et al.[2], respectively. When high resolution body surface mapping became available, efforts in HPS recordings turned toward the detection of spatial features of the HPS[3,4]. One particular interest in mapping the HPS stems from a belief that the small, localized structure of the His bundle could be reasonably modelled with a single-current-dipole source. Measurements of the spatial features of the HPS could then be used to test and validate current dipole inverse solutions. The results of these studies were not immediately accepted.

One particular issue, addressed in this work, is the degree to which atrial repolarization (Ta) also contributes to the maps recorded during the PR segment. At one end of the spectrum, there is published work[5] which supports the claim that the ramp pattern could be associated entirely with Ta. However, more recent work[6] provides convincing evidence that at least the bundle branch signal is detectable at the body surface. Believing that a combination of Ta and HPS contributes to the signal measured in the PR-segment, several signal-extraction techniques have been used by different authors[3,4,8] to separate these concurrently active signals.

To determine the origin of these extracted signals, we first undertook an extensive analysis of the nature of atrial activation and recovery as expressed in cardiac magnetic field (MFM) and body surface potential (BSPM) maps. Following this, the different signal-extraction techniques were implemented and applied to the data. These results were then evaluated with particular attention given to the electrophysiology of the sources in question and mathematics behind the various extraction techniques. We note that this work will not address whether the spike, bump or step-like signals reported by others[7] are due to HPS.

Methods and Theory

In the present work, 56-point MFM and 117-lead BSPM were recorded in 30 subjects, 6 with long PR intervals. The noise levels of these systems are $30\mathrm{fT}/\sqrt{\mathrm{Hz}}$ and $70\mathrm{nV}/\sqrt{\mathrm{Hz}}$, respectively. A detailed description of this recording system can be found elsewhere[9]. After signal averaging of 30 second recordings, maps of the PR-interval were produced every 2ms. Atrial activation and recovery was studied particularly from the perspective of the temporal evolution of the spatial features of isocontour maps.

To isolate the HPS signal from the concurrently active atrial repolarization signal, four techniques have been suggested. These are: (1) a correlation technique[3]; (2) a temporal averaging technique[4]; (3) Karhunen-Loève orthogonal expansion[8]; and (4) singular value decomposition[10]. We will only discuss the details of the two first techniques in this paper. To begin, definitions which apply to both theories are given. First, we will represent the MFM or BSPM data symbolically by the matrix $\mathbf{X}_{p\times t}$. The columns of $\mathbf{X}$ have p entries, and form either MF or BSP maps. The rows of $\mathbf{X}$ represent the time axis, t, and form ECG or MCG traces. Subsets of $\mathbf{X}$ represent segments of an MCG or ECG trace. The focus of the theory is on the separation of concurrent events. In this context, we will define a matrix $\mathbf{A}$ to be associated with the Ta source, and $\mathbf{H}$ will represent the HPS signal of interest. The sub-matrix $\mathbf{A}_e$ refers to the early PR segment where only Ta sources are active. HPS begins to depolarize at ~50ms before ventricular activation. During this concurrent activation phase, a combined signal $\mathbf{X}_l = \mathbf{A}_l + \mathbf{H}$ of the late atrial repolarization signal, $\mathbf{A}_l$, and the HPS signal, $\mathbf{H}$, will be measured.

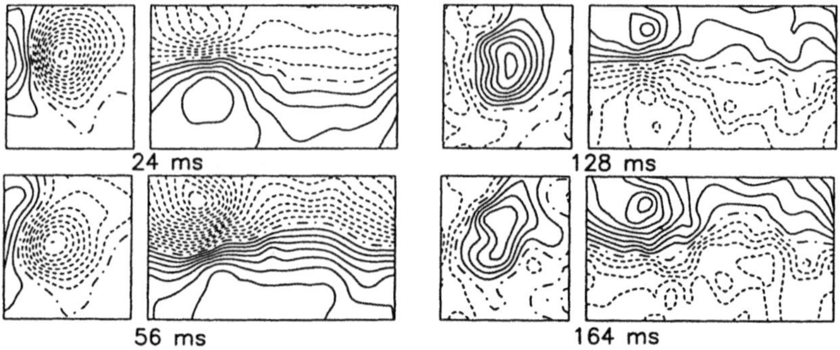

Fig. 1: Typical MF and BSP PR-interval maps (normal male, 59 yrs, P-wave 100ms, PR-interval 182ms). Fifty-six-point MFM are shown on the left and 117-lead BSPM on the right of each panel. Times, relative to P-wave onset are displayed below each pair of maps. Positive potentials and magnetic fields are plotted with solid lines, negative values with dashed lines, and the zero levels with dash-dot contours. Contour levels are 200fT (MFM) and 10μV (BSPM) at 24 and 56ms, and 100fT and 5μV at 128 and 164ms.

Temporal averaging technique

The temporal averaging technique (TAT) is the simplest and most commonly used technique investigated. It relies on the assumption that the background Ta source pattern and its amplitude do not change during the time interval of interest. Having selected t consecutive maps to represent an early part of the PR segment during which only atrial activation is present, an average Ta map is computed. Symbolically, $\overline{\mathbf{X}_e} = \overline{\mathbf{A}_e} = \frac{1}{t}\sum_{i=1}^{t}\mathbf{A}_i$ is the average Ta map. The difference, $\mathbf{R}^{TAT}$, between this early average Ta map and the late PR-segment map where both the Ta and HPS sources are active becomes,

$$\mathbf{R}^{TAT} = \mathbf{X}_l - \overline{\mathbf{A}_e} = \mathbf{H} + \mathbf{A}_l - \overline{\mathbf{A}_e}. \tag{1}$$

If assumptions stated earlier are valid, then $\mathbf{A}_l = \overline{\mathbf{A}_e}$, and the residual maps represented by the matrix $\mathbf{R}^{TAT}$ should reveal the HPS source, $\mathbf{H}$.

Correlation technique

The correlation technique (CT), proposed by Horan et al.[3], removes the assumption that the atrial repolarization source amplitude remains constant in time. After obtaining the average early Ta map, $\overline{\mathbf{A}_e}$ a correlation is made between this average and the late PR-segment maps of combined activity, $\mathbf{X}_l$. This scaling factor accounts for changes in the amplitude of the Ta source. The correlated fraction of $\overline{\mathbf{A}_e}$, defined by $\mathbf{C}_{X_l} = \frac{\overline{\mathbf{A}_e}\cdot\mathbf{X}_l}{\overline{\mathbf{A}_e}\cdot\overline{\mathbf{A}_e}}$ is subtracted from the combination map to give the residual matrix,

$$\mathbf{R}^{CT} = \mathbf{X}_l - \mathbf{C}_{X_l}\overline{\mathbf{A}_e} = \mathbf{H} + \mathbf{A}_l - \mathbf{C}_{X_l}\overline{\mathbf{A}_e}. \tag{2}$$

In this case, if the assumption of a stable pattern is correct, the residual map $\mathbf{R}^{CT}$ is expected to reveal an HPS map.

Results and Discussion

Typical PR-interval MF and BSP maps are shown in Fig. 1. The onset of atrial depolarization (24ms of Fig. 1) is marked by MFM and BSPM patterns which are consistent with a pacemaker complex located in the right atrium. As atrial activation progresses into the left atrium (56ms BSP map), the potential maximum is seen to migrate left parasternally (centre of BSP map), with the minimum remaining relatively stationary. During the second half of the P wave, in most subjects, this minimum continues to move down, as the positive peak moves further to the back (right side of BSP map). This reflects an increasingly large anterior to posterior component of atrial current flow. As the depolarization wave spreads into the left atrium, the magnetic field maps show an

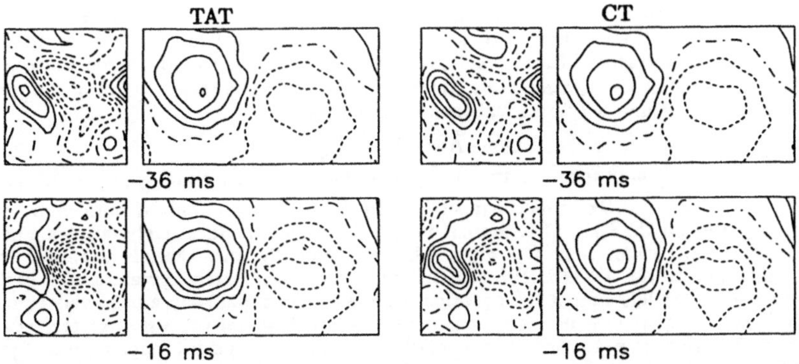

Fig. 2: TAT and CT residual distributions for the same subject as shown in Fig. 1. Times below maps are relative to Q-wave onset. A segment from -70 to -50ms relative to P-wave onset was used to characterize early Ta activity. Plotting conventions are same as Fig. 1. Contour levels are 50fT (MFM) and 5μV (BSPM).

initial (clockwise or counter-clockwise) rotation of the pattern (56ms map). When all depolarization ends, and only repolarization remains, the usual pattern (128ms map) which emerges is similar, but opposite in polarity, to P-wave onset in both MFM and BSPM plots. It generally remains stable until Q-onset (ventricular depolarization). In a few cases (164 ms map), where the PR segment was longer and the S/N was sufficiently high, the minimum in the late PR segment of the BSPM plots migrates onto the back (right of map). It is similar in distribution, though opposite in polarity, to late P-wave activity (56ms map).

Experimental epicardial measurements have demonstrated that the atrial regions which depolarize first will be the first to repolarize[11]. The similarity of early repolarization maps with early P wave, along with the epicardial results support the conclusion that the early PR-segment maps reflect primarily right atrial repolarization (RR). Previous descriptions of the repolarization process in body surface measurements[12,13] report that the repolarization phase is characterized by a stable pattern, in which the minimum is sometimes seen to migrate slightly to the left of the body. A similar leftward migration of the maximum during the P wave, resulting from a growth in left depolarization (LD) currents and a decrease in right depolarization, suggests that the BSPM map at 164ms of Fig. 1 represents left atrial repolarization (LR). The MFM distributions at these same times, do not correlate with LD. They continue to resemble either RR, some intermediate pattern between LR and RR, or, at times, loose any dipolar features as S/N levels decrease significantly. The explanation for this observation may be that since the LR sources are relatively far from the magnetic sensor and small in amplitude, only residual RR signals will be detected.

In the BPSM and MFM data of all patients, all HPS extraction techniques studied gave very similar, and at times identical, residual spatial distributions for a given subject. Typical results for the TAT and CT techniques are shown in Fig. 2. The residual MFM patterns are similar to those measured by others[4,8]. The extrema locations and orientation are consistent with a single current dipole representing the HPS source. In the past, this pattern has been used to conclude that HPS activity can be detected in magnetic field maps. The BSPM residuals suggest a posteriorly to anteriorly directed source, which is not consistent with HPS. It was also observed that the amplitudes of the residual maps reach minimum values during times corresponding to the segment used to represent Ta, and grow in time moving away from this segment. The dipolar maps before and after this time segment are similar but of opposite polarity on either side of this segment. This behavior can be modelled by assuming a single moving source, representing Ta activity only[10].

With these observations at hand, a theoretical comparison of the different extraction techniques was made. Careful consideration of the extraction techniques suggests that if the HPS contributes a significant fraction to the PR-segment signal, the TAT and CT results should not be the same. The following proves that the residual spatial distributions are not due to a significant contribution of an HPS signal. The similarity in the residual maps of Fig. 2 suggests an equality between Eqs. 1 and 2. If we assume that the residual maps represent the HPS source, then in both equations, all other terms besides **H** must (a) sum to remove most of the atrial source, and (b) sum in such a way that the two residuals, $\mathbf{R}^{TAT}$ and $\mathbf{R}^{CT}$ are the same. Ignoring the term **H** in both equations, the remaining terms will then require that

$$\mathbf{A}_l - \overline{\mathbf{A}_e} = \mathbf{A}_l - \mathbf{C}_{X_l}\overline{\mathbf{A}_e}. \tag{3}$$

This equality can only hold if $\mathbf{C}_{X_l} = \mathbf{C}_{A_l} + \mathbf{C}_H \simeq 1$. From the definition of $\mathbf{C}_{X_l}$, this means that $\mathbf{X}_l$ is about equal to $\overline{\mathbf{A}_e}$. One would expect $\mathbf{X}_l$ and $\overline{\mathbf{A}_e}$ to be equal only if both maps represent atrial activity and no HPS signal is present in the PR segment. Intuitively, we can understand that if the HPS source is present, $\mathbf{C}_H$ will not be zero generally since there is no reason to expect that the maps due to the Ta and HPS sources will not at least partly correlate with each other. Since the term on the left side of the equation represents the atrial source only, this equality can hold only if $\mathbf{H} = 0$, *that is if no significant amount of HPS signal is present in the PR-segment maps.* From the evidence presented, it is concluded that the PR-segment maps represent primarily Ta activity and the extracted maps actually reflect the continual changes in atrial repolarization current sources as a function of time, as reflected at the body surface.

References

[1] Berbari, E.J. et al. Noninvasive technique for detection of electrical activity during the P-R segment. Circulation, 1973, 48:1005-1013.

[2] Farrell, D.E., Tripp, J., and Norgren, R. Non-invasive information of the P-R segment of the cardiac cycle: An assessment of the clinical potential of the electric and magnetic methods. Proc. SPIE, 167, 1978.

[3] Horan, L.G., Flowers, N.C., and Sohi, G.S. A dynamic electrical record of the pathway of human His bundle activation from surface mapping. Circ. Res., 1982, 50:47-54.

[4] Erne, S.N. et al. Magnetocardiographic study of the PR segment of normals. In Weinberg, H., Stroink, G., and Katila, T. (Eds), Biomagnetism: Applications and Theory, Pergamon Press, New York, pages 132-136, 1984.

[5] Lorenzana, H.E. et al. A Study of the P-R Segment in the Human Magnetocardiogram, In Weinberg, H., Stroink, G., and Katila, T. (Eds), Biomagnetism: Applications and Theory, Pergamon Press, New York, pages 142-147, 1984.

[6] Leiniö, M. et al. ECG and MCG studies on the His-Purkinje activity in a case of congenital heart block. In Williamson, S.J. et al. (Eds), Advances in Biomagnetism, Plenum Press, New York, pages 377-380, 1990.

[7] Berbari, E.J. Recording from ventricular conduction system. In Webster, J.G. (Ed.), Encyclopedia of Medical Devices and Instrumentation, John Wiley & Sons, New York, pages 291-314, 1988 (and references therein).

[8] MacAulay, C. et al. Analysis of MCG spatial maps during the PR-interval. Med. Biol. Eng. Comp., 1985, 23(SUPPL.2):1479-1480.

[9] Stroink, G. et al. Body surface potential and magnetic mapping. In Atsumi, K. et al. (Eds.), Biomagnetism '87, Tokyo Denki Univ. Press, pages 74-81, 1988.

[10] Lamothe, R., and Stroink, G. Orthogonal expansions: Their applicability to signal extraction in electrophysiological mapping data. Med. Biol. Eng. Comp., 1991, 29:522-528.

[11] Spach, M.S. et al. Electrical potential distribution surrounding the atria during depolarization and repolarization in the dog. Circ. Res., 1969, 24:857-873.

[12] Stilli, D. et al. Newer data on the configuration and variability ranges of body surface maps in a sample of normal subjects. J. Electrocardiology, 1988, 21:1-14.

[13] Mirvis, D.M. Body surface distribution of electrical potential during atrial depolarization and repolarization. Circulation, 1980, 62:167-173.

Acknowledgements

Financial support for this work was obtained from the Heart and Stroke Foundation of Nova Scotia.

Spectral Analysis of Body Surface Potential Maps and Magnetic Field Maps of Patients With and Without Ventricular Tachycardia

Meeder, R.J.J.[1], Stroink, G.[1,2], Ritcey, S.P.[2] and Gardner, M.J.[3]

[1]Department of Physiology and Biophysics, [2]Physics, and [3]Medicine, Division of Cardiology, Dalhousie University, Halifax, Canada

Introduction

Non-invasive prediction of future arrhythmic events in patients who have suffered a myocardial infarction (MI) is extremely important in adequately treating these patients. It has been shown that patients who are convalescing from MI often have abnormal electrocardiograms (ECG). In the past years, much effort has been devoted to signals present in the terminal portion of the QRS complex of the signal-averaged ECG [1]. Such signals are believed to arise from semi-viable tissue in the infarcted area. This region is presumed to represent the origin of arrhythmic events. However, low predictive accuracy (17% to 29%) of techniques involving analysis of the terminal QRS has prompted investigators to investigate ECG signals in new ways. Recently a number of groups have investigated whether abnormal signals may be present during the entire QRS complex or in other non-QRS regions of the cardiac cycle. Although signals may be stronger in the terminal QRS region, it is physiologically possible (indeed, not unlikely) for aberrant electrical activity to be present during other regions of the cardiac cycle. In the following study we investigate a technique of spectral analysis [2,3] of spatially optimized ECG and MCG signals using electrical information obtained from a large area of the torso.

In a recent article Kavesh, *et al* [4] describe a technique for spatially optimizing ECG leads using a standard orthogonal lead system to determine the cardiac dipole parameters. Using a matrix of transfer functions, artificial ECG leads were generated on a standard torso model which were then subjected to spectral analysis. The spectral analysis consisted of examining the total power in certain frequency bands of the power spectrum over the entire body surface. Spatial maps of this spectral power, known as isoharmonic maps, were then plotted. From these maps, leads could be identified which contained the most power in a selected frequency range of the power spectrum. The optimum spectra, contained in these leads, could then be compared for different patient groups. We propose to extend this technique using real patient data rather than simulated data. This can be accomplished with the information contained in body surface potential maps (BSPM) and magnetic field maps (MFM) where a large number of leads (117 and 56, respectively) are used to measure the potential and magnetic fields. Isoharmonic maps of the entire torso (using BSPM) or the anterior torso (using MFM) can be generated from directly measured averaged data. Regions of the torso displaying relative dominance over certain frequency bands can be identified and these leads can be selected for further analysis.

Methods

Patient groups: Three patient groups were recruited for BSPM and MFM measurement. The first group (n=14) consisted of MI patients who showed no signs of VT (MI/VT-). The second group (n=10) consisted of MI patients with documented sustained VT episodes as confirmed by Holter monitoring and/or electrophysiological studies (MI/VT+). All patients were measured within 6 months following release from hospital. Two patients in the VT group were taking anti-arrhythmic medication. As a control group, 15 healthy volunteers were measured (NORM). Volunteers were considered healthy on the basis of their history of no heart disease, normal 12-lead ECG, normal echocardiogram and physical examination. Further details are given in Table 1.

Table 1: Patient groups.

Patient group	number	age (range)	male/female
NORM	15	54 ± 8 (43–66)	9/6
MI/VT-	14	57 ± 12 (44–67)	12/2
MI/VT+	10	58 ± 8 (34–76)	8/2

BSPM recordings: The 117 BSPM unipolar leads referenced to WCT were recorded simultaneously during normal sinus rhythm for a period of 30 seconds with a sampling rate of 500 Hz. The leads cover the entire anterior and posterior torso (Fig 1). Patients were supine. The recording bandwidth was 0.05 to 125 Hz. Averaging was performed off-line and the averaged beats were baseline corrected such that the first and last datapoints were isopotential. Each averaged beat (1 second) represented the entire cardiac cycle. Datasets for each lead (500 samples/lead) were zero-padded to yield 512 points per lead giving spectral line separation (bin width) of 0.977 Hz. Bad leads were identified visually and were corrected by averaging surrounding leads or by rejecting them from the data sets (if they were on the map boundary).

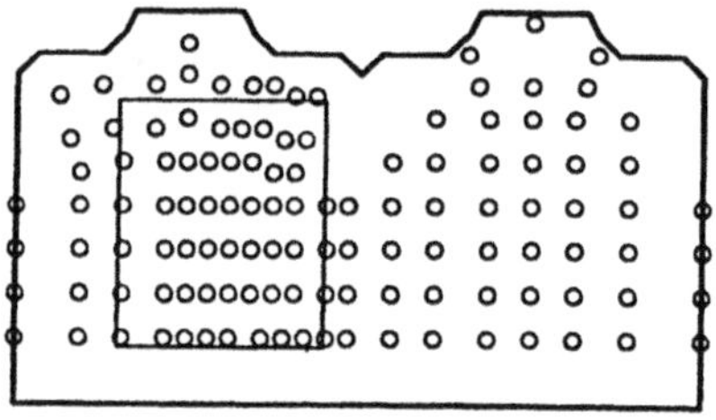

Fig 1: Anterior and posterior torso showing the 117 BSPM lead locations and relative location of the 7x8 MFM grid.

MCG Recordings: The 56 MFM sites were recorded sequentially using a single-channel 2nd-order gradiometer. Measurements were done in a aluminum shielded room with patients in the supine position. The sites cover most of the anterior torso and are centered over the cardiac region (Fig. 1). Recordings at each site were made for 30 seconds with a sampling rate of 500 Hz and a recording bandwidth of DC-125 Hz. Averaging and zero-padding was performed off-line using the same method as used for the BSPM recordings.

Spectral Analysis: To obtain isoharmonic maps, the power spectrum for each BSPM lead or MFM site was identically calculated for the entire cardiac cycle. Spatial isoharmonic maps were then generated using values obtained by integrating the power spectra over specified frequency bands (2-7 Hz, 7-25 Hz, 25-40 Hz, 40-60 Hz, 60-80 Hz, 1-100 Hz, 13-56 Hz and 68-125 Hz). Power spectra of leads located at these isoharmonic map maxima were then identified for each patient and averaged for all patients in the group to obtain the group mean power spectra ($PS_{2\text{-}7}$ (fig. 2), $PS_{7\text{-}25}$, $PS_{25\text{-}40}$, $PS_{40\text{-}60}$, $PS_{60\text{-}80}$, $PS_{1\text{-}100}$, $PS_{13\text{-}56}$, $PS_{68\text{-}125}$).

The average power spectrum (PS_{avg}) of a patient was calculated by averaging all the 56 or 117 calculated power spectra (MFM and BSPM, respectively). The group mean average spectrum is then obtained by averaging these spectra for all patients within the group.

The maximum power spectrum (PS_{max}) of a patient is composed of the maximum power in each frequency bin of the 56 or 117 power spectra. Again, the group mean maximum power spectrum is obtained by averaging these spectra for all patients within the group. Inter-group comparisons were made for these various group mean power spectra and the significance of any differences between MI/VT+ and MI/VT- groups was assessed using the Student's t-test. Furthermore, the diagnostic ability of spectral differences to differentiate between MI patient groups is expressed by the sensitivity, specificity and predictive accuracy.

Results

Location of isoharmonic map maxima: The location of the maxima on the isoharmonic maps showed much variability from patient to patient for both MFM and BSPM. In all cases, the maxima locations were centered over the cardiac region of the anterior torso.

Group differences in power spectra at isoharmonic map maxima: For BSPM data, $PS_{2\text{-}7}$ (fig.2), $PS_{7\text{-}25}$, $PS_{60\text{-}80}$, $PS_{1\text{-}100}$, and $PS_{68\text{-}125}$ all differed significantly in the low-frequency range (4-13 Hz). Using the total power in this range as the criterion to identify MI patients at risk for VT, the best sensitivity (70%), specificity (86%) and predictive accuracy (79%) was obtained using total low-frequency power (4-13 Hz) in $PS_{68\text{-}125}$. Such significant low-frequency spectral differences were also noted for the MFM recordings ($PS_{2\text{-}7}$ (fig.2), $PS_{7\text{-}25}$, $PS_{40\text{-}60}$, $PS_{1\text{-}100}$, and $PS_{14\text{-}56}$) however sensitivity, specificity and predictive accuracy was, at best, lower than for BSPM (70%, 79% and 75%, respectively). Significant differences in the high-frequency portions of these spectra occurred more often for spectra obtained from MFM where significant differences in the 39-41 Hz and 42-43 Hz ranges were found in $PS_{25\text{-}40}$ and $PS_{40\text{-}60}$, respectively.

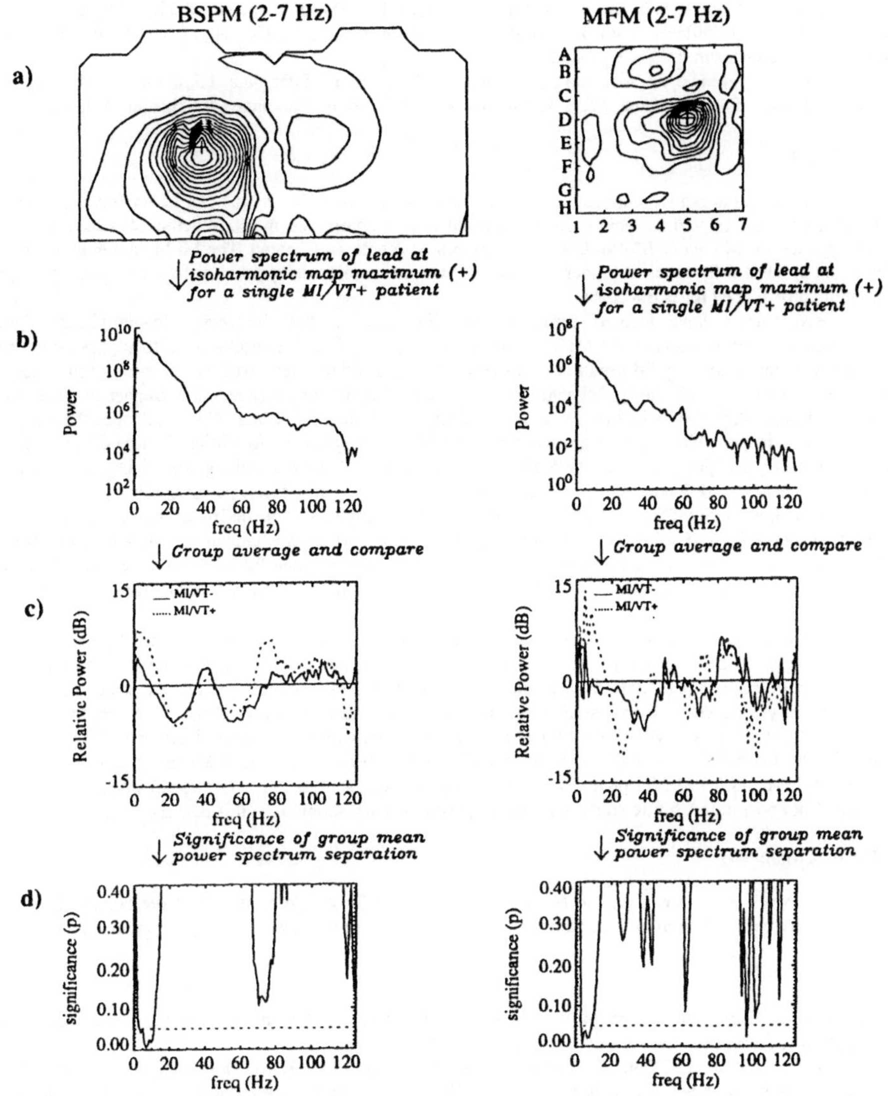

Fig 2: Method of calculating group mean power spectra. First, the isoharmonic maps (BSPM and MFM) of a subject are generated for the band 2-7 Hz (a). The power spectrum ($PS_{2\text{-}7}$) of the lead at the map maximum is then determined (b) and averaged for all patients in the group. The resulting group mean power spectra are then plotted relative to the normal group (straight line at 0 dB) (c). The significance of differences (p) in these group mean spectra can then be plotted as a function of frequency (d). A p-value of <0.05 (dotted line) indicates significant spectral separation.

Group differences in average power spectra (PS_{avg}): For BSPM data, again the low-frequency range (4-13 Hz) was significantly different with best separation occurring in the 5-11 Hz range ($p<0.01$). For MFM data, no significant differences in PS_{avg} were found.

Group differences in maximum power spectra (PS_{max}): For BSPM data, PS_{max} was significantly different for 6-11 Hz range. For MFM data, PS_{max} was significantly different for the ranges 4-8 Hz and 41-43 Hz.

Discussion

The results presented here suggest that the MI/VT- and MI/VT+ groups can be separated on the basis of spectral differences of various power spectra derived from isoharmonic maps. Whether this technique is more powerful when MFM data or BSPM data is used cannot be conclusively stated. The MFM data more often claimed significant spectral differences at higher frequencies, whereas significant spectral differences at lower frequencies were more pronounced for BSPM data.

Past studies have focused primarily on differences in high-frequency, low-amplitude signals to differentiate between patients of the MI/VT- and MI/VT+ groups [1]. Our results seem to suggest, however, that the amount of low-frequency information (4-13 Hz) best characterizes these groups. For every group mean power spectrum calculated in this study, MI patients with VT had significantly greater low-frequency power than the other two groups. A physiological basis for this finding is yet to be established. This challenges the notion that a VT substrate is responsible only for significant amounts of high-frequency information. It should be noted however that the currently accepted theory of the high-frequency nature of late potential signals arising from the potential VT substrate is not at all jeopardized by these results. In the three group mean power spectra ($PS_{25\text{-}40}$, $PS_{40\text{-}60}$ and PS_{max}) which showed significant differences in the 39-43 Hz range (part of the frequency range often associated with late potentials) the power spectra in this frequency range were always greater for MI/VT+ patients than for MI/VT- patients. The relative abundance of signals in this frequency range for MI/VT+ patients could be used as an argument for high-pass filtering at 25 Hz rather than 40 Hz when doing standard Frank lead late potential analysis.

The classification power of this technique is not nearly as strong as that reported by Kavesh *et al.* for Frank-lead derived BSPM data [4]. This may be due to the inadequacy of their technique in that they essentially extrapolate body surface potentials from a relatively limited lead set. Spatially optimized power spectra in such cases mean very little since little spatial information existed in their measured data. Moreover, their results are confounded by comparing a MI/VT+ group which was measured within 8-10 days of suffering MI with a MI/VT- group with healed MI. In our study we have looked only at patients with healed MI using actual spatio-spectral data from BSPM and MFM. Yet predictive tests, such as spectral analysis, are most useful when conducted before discharge from hospital. A future study, with larger patient groups, should take this into account.

Acknowledgments

This work is supported by the Nova Scotia Heart and Stroke Foundation and the Natural Sciences and Engineering Research Council of Canada. Discussions with B.M. Horacek were greatly appreciated.

References

[1] Breithard, G., Borggrefe, M., Pathophysiological Mechanisms and Clinical Significance of Ventricular Late Potentials, European Heart Journal, 1986, 7:364-385

[2] Hosoya, Y., Kubota, I., Shibata, T., Yamaki, M., Ikeda, K., Tomoike, H. Spectral Analysis of 87-lead Body Surface Signal-Averaged ECGs in Patients With Previous Anterior Myocardial Infarction as a Marker of Ventricular Tachycardia, Circulation, 1992, 85:2060-2064.

[3] Montonen, J, Makijarvi, M., Paavola, M., Leinio, M., Siltanen, P., Katila, T. Comparison of Time Domain and Spectrotemporal Domain Analysis of High Resolution Magneto- and Electrocardiographic Recordings of MI Patients With and Without Ventricular Tachycardia, In: Baumgartner, C., Deeke, L., Stroink, G., Williamson, S.J. Biomagnetism: Fundamental Research and Clinical Applications, Elsevier Science, IOS Press, 1995

[4] Kavesh, N.G., Cain, M.E., Ambos, H.D., Arthur, R.M. Enhanced Detection of Distinguishing Features in Signal-Averaged Electrocardiograms From Patients With Ventricular Tachycardia by Combined Spatial and Spectral Analyses of Entire Cardiac Cycle, Circulation, 1994, 90:254-263.

Magnetocardiography - A New Approach for Early Detection of Arrhythmogenic Right Ventricular Dysplasia

Meyerfeldt, U.[1], Schirdewan, A.[1], Wiedemann, M.[1], Schütt, H.[1], Dietz, R.[1], Thierfelder, L.[2], Burghoff, M.[3] and Trahms, L.[3]

[1]Franz-Volhard-Clinic, Berlin, Germany; [2]Max-Delbrück-Centrum, Berlin, Germany; [3]Physikalisch-Technische-Bundesanstalt (PTB), Berlin, Germany

Introduction

The arrhythmogenic right ventricular dysplasia or cardiomyopathy (ARVD) is one of the major causes of juvenile sudden cardiac death and severe ventricular arrhythmias. ARVD is a primary myocardial disease, pathologically characterized by fibrofatty replacement of the right ventricular myocardium, occurs in families, and it is inherited as an autosomal dominant trait. The phenotypical expression in any particular families can be very variable. The conventional diagnosis of ARVD based on the presence of major and minor criteria encompassing structural, histological, electrocardiographic and genetic factors, detected by echocardiography, right ventricular angiography, right ventricular myocardial biopsy, magnetic resonance imaging (MRI) and ECG. According to the clinical, angiographic and electrophysiological characteristics a severe form of ARVD with full manifestation of the disease and a "concealed" form with only mild morphological and ECG abnormalities can be distinguished.

Purpose of the study

The aim of our investigation is the early identification and phenotypical characterization of the donors of ARVD-gene in a family consisted of 25 persons by biomagnetic measurements. Based on a precise phenotypical characterization a localization of the ARVD-gene by genetic methods in this family is planned. Now we want to present the study design and first results.

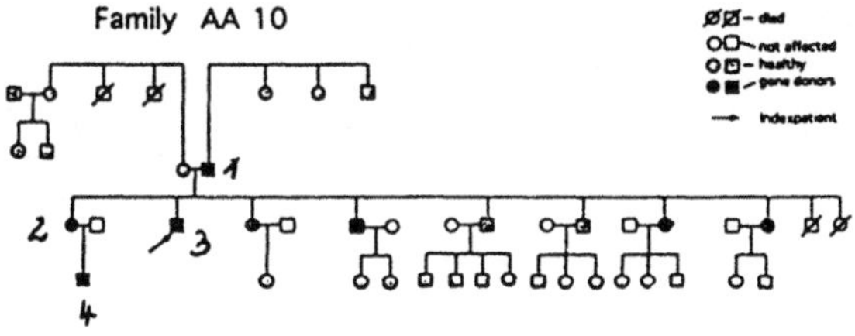

Figure 1 Pedigree of the ARVD family

Methods

We started a long-term follow up investigation in the family members over three generations (14 females and 11 males, age 12-72 y.). 2 males of them are survivors of sudden cardiac death (SCD) and had recurrent sustained ventricular tachycardias of left bundle branch block morphology. Because of ineffective pharmacological therapy in both patients a cardioverterdefibrillator was implanted. In the first step all persons were conventionally examined by ECG, Holter-ECG, signal-averaged ECG, exercise ECG test, echocardiography, Thallium-Myocardial SPECT and MRI. The magnetocardiographic investigations were performed by multichannel LTS-SQUID system in the magnetic shielded room of the PTB, Berlin-Steglitz. At first the magnetocardiograms (MCG) and the magnetic field distribution were analyzed by nonrecursive binominal highpass filter, 37-90 Hz (1). Additional signal processing methods of the MCG are prepared.

First results

On the conventional diagnostic methods the two survivors of SCD (pt 3 and 4) fulfilled all criteria of the severe form of ARVD with right ventricular dilatation, localized right ventricular aneurysms, fibrofatty replacement of right ventricular myocardium in endomyocardial biopsy and inducible ventricular tachycardias in the electrophysiological study. The mother and grand father of pt 4 (pt (2 and 1) had only a small dilatation of the right ventricle in echocardiography and MRI. The other patients did not have any changes on conventional diagnostic methods.

Patient	12 lead ECG	Echocardiogr.	SA-ECG	MRI	RV-angiogr.	MCG
1		(+)		(+)		+
2	-	-	-	(+)	(+)	+
3	+	+	+	+	+	+
4	+	+	+	+	+	+
5	-	-	-	-	0	0
6	-	-	-	-	0	-
7	-	-	-	-	0	-
8	-	-	-	-	0	-
9	-	-	-	-	0	-
10	-	-	-	-	0	-

Table 1 Results of conventional diagnostics

The filtered magnetocardiograms of pt 3 and 4 (survivors of SCD) showed an impressive fragmentation in the range of the late QRS complex (fragmentation score 96 and 105). Pt. 1 and 2 had similar magnetocardiographic phenomena with strong fragmentation of QRS (score 67 and 80). The other persons of the family had not such a fragmentation of the QRS complex (score 32-56)

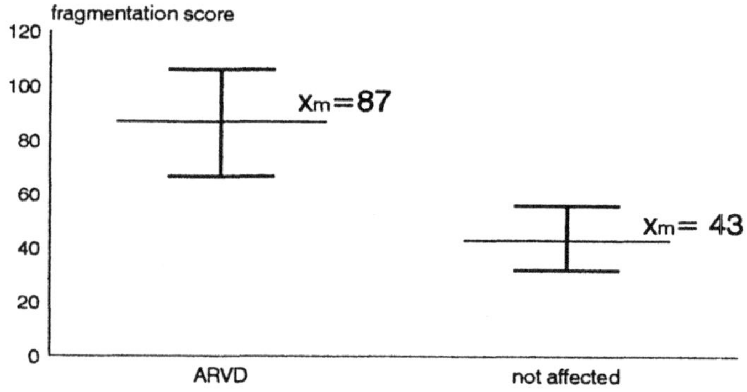

Figure 2 Fragmentation score

In knowledge of the autosomal dominant trait of ARVD we concluded that person 1 and 2 with the magnetocardiographic fragmentation are donors of the ARVD-gene with a "concealed" form.

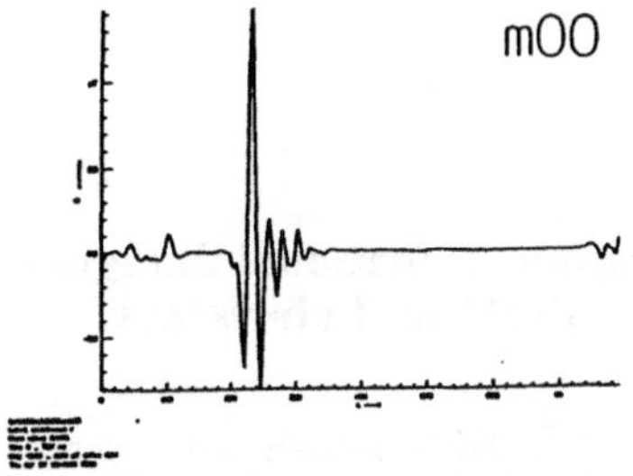

Figure 3 MCG of a ARVD patient

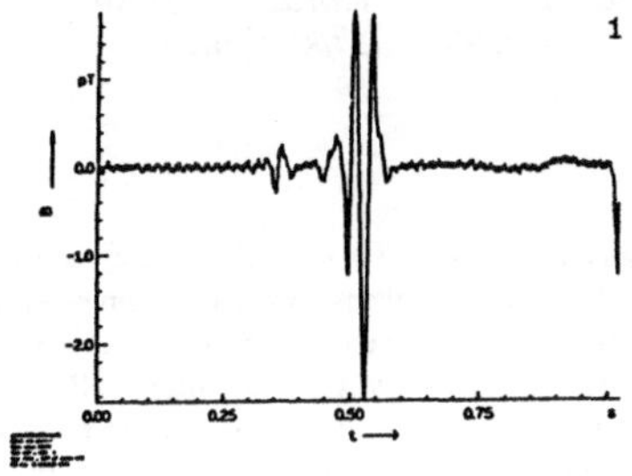

Figure 3 b MCG of a normal person

Conclusions

The first results of our study support the hypothesis that it is possible to detect donors of ARVD-gene early and to characterize the phenotype in "concealed" forms of ARVD by magnetocardiographic investigation before the development of macroscopic and electrocardiographic changes of the right ventricle by conventional methods can be proved. Even small fibrolipomatous infiltrations of the right ventricular myocardium seem to change the functional electrical properties of the myocardium especially the patterns of electrical current distribution determining the arrhythmogenic risk, which can be detected more sensitive by magnetocardiographic as by electrocardiographic approaches. The magnetocardiography as a sensitive method for identification of ARVD gene-donors can furthermore make an important contribution to localize the gene-locus and to elucidate the arrhythmogenesis of life-threatening ventricular tachyarrhythmias.

References

(1) Trahms, L., Hahlbohm, H.-D., Kreiseler, D. and Zitzmann, H. In: Trahms, L. Diagnostic methods to investigate cardiac arrhythmias, Lectures on the 114th PTB-Seminar, Braunschweig and Berlin, Wirtschaftsverlag NW (Bremerhaven), 1995

Magnetocardiographic Functional Imaging Studies in the BioMag Laboratory

Montonen, J.[1,3], Ahonen, A.[2], Hämäläinen, M.[2], Ilmoniemi, R.J.[3], Laine, P.[2], Nenonen, J.[1,3], Paavola, M.[1,3], Simelius, K.[1,3], Simola, J.[2] and Katila, T.[1,3]

[1]Laboratory of Biomedical Engineering, Helsinki University of Technology, 02150 Espoo; [2]Neuromag Ltd, 00510 Helsinki; [3]BioMag Laboratory, Helsinki University Central Hospital, 00290 Helsinki, Finland

Introduction

The BioMag Laboratory was established in 1994-1995 in the Helsinki University Central Hospital (HUCH) to provide a facility for versatile biomagnetic studies in a clinical environment. A three-layer magnetically shielded room (MSR) [7] together with a 68-channel cardiomagnetometer and a 122-channel neuromagnetometer [2, 3] form the core of the laboratory, owned together by HUCH, Helsinki University of Technology (HUT) and the University of Helsinki. In this paper, we focus on magnetocardiographic (MCG) functional imaging, started in May 1995 in the BioMag Laboratory.

Arrhythmias, and especially the non-invasive localization of arrhythmogenic sources and arrhythmia vulnerability assessment form the main content of the present cardiac studies at the BioMag Laboratory. However, the facilities have been designed also to fulfill the varying needs of other clinical and basic cardiac research. For example, fetal MCG and measurements of neonates and children are possible, and equipment for exercise MCG is under construction. A 128-channel body surface potential mapping (BSPM) device, modern imaging facilities in HUCH, and a fast asynchronous transfer mode (ATM) network in the hospital and between the Finnish universities give a unique framework for the studies. In the following, the hardware and software resources of the BioMag Laboratory are first described, then the connection of the BioMag Laboratory with some other facilities available at HUCH is presented, and finally an overview of the ongoing patient studies with some examples is given.

Installation

The 68-channel cardiomagnetometer was designed in collaboration between Neuromag Ltd. and HUT. It was constructed by Neuromag Ltd. and installed in the MSR in April 1995. The sensors consist of six axial and 62 planar dc-SQUID gradiometers in a hexagonal array (see Fig. 1). The wire-wound axial first-order gradiometers have a baseline of 80 mm and the pick-up coil radius is 37.5 mm. Each of the 31 planar thin-film gradiometer units consists of two perpendicular figure-of-eight loops detecting $\partial B_z/\partial x$ and $\partial B_z/\partial y$, where B_z is the magnetic field component perpendicular to the plane of the unit [3]. The baseline of the planar gradiometers is 16.5 mm, and the imbalance between the sensing loops is below 0.1%. Four of the planar units located at the edge of the array are positioned perpendicularly to the dewar bottom while the other 27 units lie tangentially. The system is modular in the sense that the axial and planar gradiometers are interchangeable, except for the four vertical planar units. The sensitivity of the axial and planar gradiometers is about 3 $\mathrm{fT}/\sqrt{\mathrm{Hz}}$ and 5 $\mathrm{fT/cm}\sqrt{\mathrm{Hz}}$, respectively, in the frequency range of 5 to 100 Hz.

The sensors are immersed in a 130-cm-long cylindrical dewar with inner and outer diameters of 30 cm and 42 cm, respectively (see Fig. 2). The dewar has a cylindrically curved bottom with a radius of curvature of 79 cm. The distance from the liquid helium space to the outer surface of the dewar is about 25 mm and the helium boil-off rate in normal operation is about 6 l/day. The dewar is supported by a gantry which is easy to move on an air cushion. The gantry system allows both vertical adjustment and tilting by 20 degrees in any direction. During the recording the patient is lying on a wheeled non-magnetic bed designed for biomagnetic measurements. The plane the patient is lying on can easily be moved horisontally along the long axis of the bed allowing a fine-scale positioning without moving the whole bed.

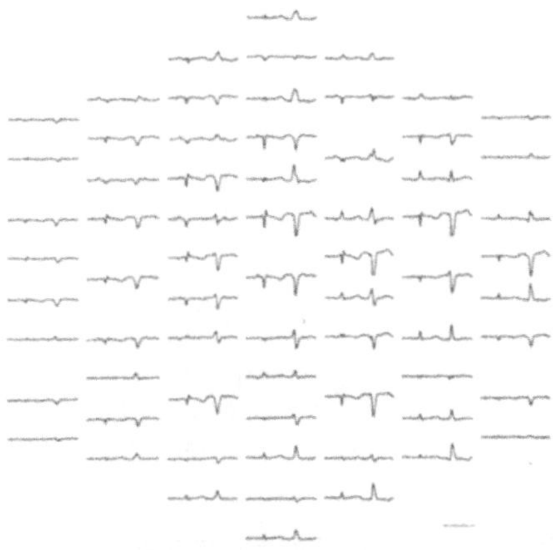

Fig. 1: *Morphological cardiac signal recorded by the 68-channel cardiomagnetometer from a patient with idiopathic ventricular tachycardia. A sinus beat is followed by a spontaneous ventricular premature beat. In the lower right corner, the time tick is 500 ms and the amplitude tick is 5 pT/cm for the planar sensors (two at one location) and 20 pT for the six axial sensors.*

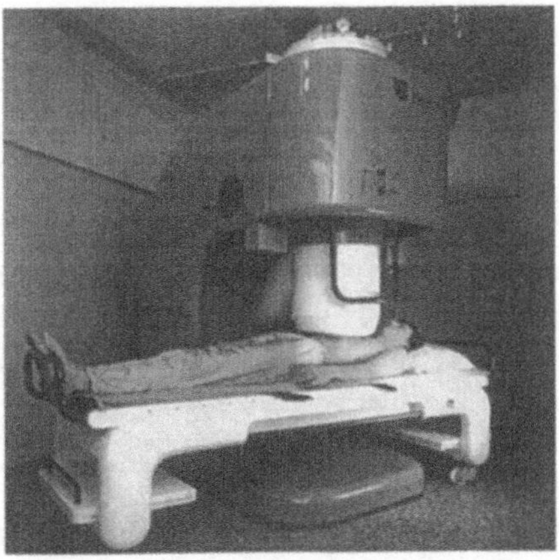

Fig. 2: *Cardiomagnetic measurement setup in the BioMag Laboratory.*

Simultaneously with the 68 magnetocardiographic channels, up to eight bipolar and 64 unipolar high-resolution electrocardiographic (ECG) leads can be applied. This includes options for conventional 12-lead ECG and bipolar orthogonal *x*-, *y*-, and *z*-leads. Also, the respiratory phase, based on detecting air pressure changes in a tube fixed over the diaphragm, can be continuously recorded during MCG. Further, a 128-channel portable high-resolution electric body surface potential mapping (BSPM) device, designed in collaboration with the University of Amsterdam, has been operational in the BioMag Laboratory since December 1995. Simultaneous BSPM recordings with MCG are possible with this battery powered device.

To maintain a contact from outside into the MSR during measurements, an audio connection and a video monitoring are always used. In addition, continuous independent ECG monitoring is possible. In a case of emergency, resuscitation equipment including a portable defibrillator are available outside the MSR.

Data acquisition and analysis

All measured data are collected by an HP data acquisition unit connected to an HP real time processor which is controlled by an HP725 workstation. Anti-aliasing filter and the sampling rate are adjustable by the software, the maximum sampling rate depending on the total number of channels recorded. In the beginning of the recording session, the position of the dewar and the sensors with respect to the patient is measured with a method analogous to Ref [1], utilizing three small, field-generating coils attached to the patient's skin. During the recording, user selected MCG and/or ECG channels can be monitored on-line. The raw data is stored off-line on CD-ROM disks and a back-up is made on DAT tapes.

After the recording all or selected channels of the stored data can be viewed with ajustable time and amplitude scales. Displays like in Fig. 1 are also available. When needed, noise from external local sources can be reduced using a signal subspace projection method [8]. Two-phase template-based averaging can be performed automatically or with user-specified parameters. In the averaging process, the signal quality on all channels is taken into account. For the analysis of the data, user-adjustable digital filters and fast Fourier transform are available and isofield and isointegral

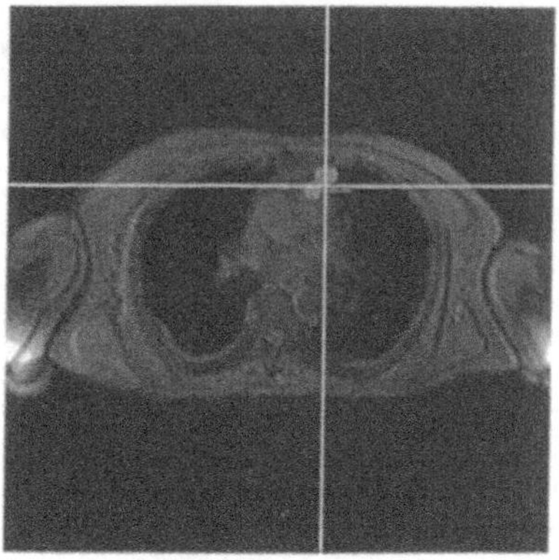

Fig. 3: *Magnetocardiographic localization of the initial activity of a spontaneous ventricular premature beat (gray dots) and the location of the successful ablation site (the cross-hair cursor) shown on a magnetic resonance image. The patient had idiopathic ventricular tachycardia.*

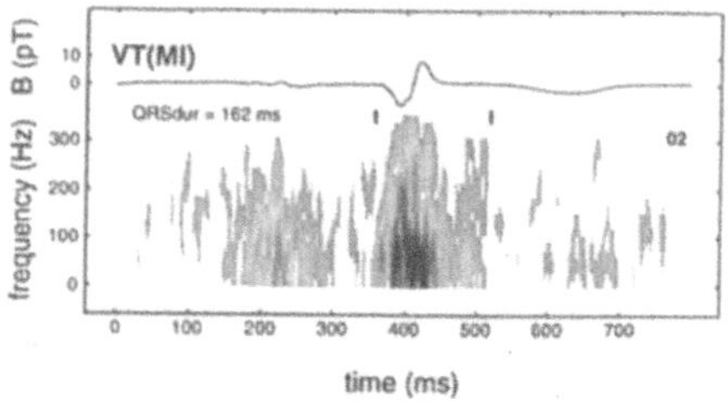

Fig. 4: *A magnetocardiographic spectrogram of a long QRS-complex with abnormally high spectral variability. The patient has had paroxysmal sustained ventricular tachycardias after myocardial infarction.*

maps can be formed using minimum-norm estimation [5].

Cardiac source localization is performed using a standard or an individual realistically shaped boundary-element torso model with a single or multiple current dipole source model [6]. The source locations can be presented, e.g., on magnetic resonance (MR) images; also three-dimensional visualization is possible. For arrhythmia risk stratification both time domain and spectrotemporal analyses of morphological waveforms are possible [4] and eigenvalue decompositions of isofield and isointegral maps can be formed. Also, an algorithm for automatic calculation of the QT_{max} and QT_{end} times has been developed. A fast ATM network between the Finnish universities allows to follow the patient measurements on-line from more distant locations, such as HUT, where off-line analysis can be continued.

Related hospital modalities

Both 1-T and functional 1,5-T MR-imaging devices and a digital X-ray at HUCH have been used to support the analysis of the MCG data and visualize the results. Sometimes also a digital bi-plane cine-X-ray from a pre-ablative electrophysiological study can be used in the MCG analysis. However, the cine-X-ray is more often used to document the ablation site for validation of the MCG localization result. Also, it has been demonstrated to be possible to insert in the cardiovascular laboratory a non-magnetic catheter into a chosen location in the heart and to transfer the patient to the BioMag Laboratory for an MCG measurement with catheter stimulation. The ATM network in HUCH connecting the Department of Radiology with the BioMag Laboratory enables a quick transfer of the image data.

Patient studies

A typical MCG measurement in the BioMag Laboratory starts by typing in the patient data and selecting the measurement parameters for an initialization file on an X-terminal. When the patient is lying on the measurement bed in the preparation room, ECG leads and the three coils, used in indicating the patient's position with respect to the dewar, are attached and the coordinates of the coils and other landmarks on the chest are stored using a three-dimensional digitizer. Next, the patient on the bed is wheeled into the MSR and after positioning the dewar over the chest the recording is started. Typically, data are collected for 5 minutes while the whole measurement procedure takes 15 to 30 minutes per patient. Thus, the maximal recording capacity of the laboratory is around 20 patients during an eight-hour working day.

Patients with malignant ventricular arrhythmias have been studied so far most extensively. These include post-myocardial infarction patients, patients with different cardiomyopathies and patients with idiopathic ventricular tachycardia (VT). When ventricular extrasystoles or arrhythmia have occurred spontaneously during the MCG recording, the source active in the beginning of the abnormal activity has been localized. The results have been compared to results from successful catheter ablations presented over X-ray and MR-images, when available. For an example of such a study see Fig. 3. Arrhythmia risk analysis has also been applied to the patients with the same underlying heart diseases with and without VT. An example of the arrhythmia vulnerability assessment studies is shown in Fig. 4. Several other patient groups have been studied as well: Patients with Wolff–Parkinson–White syndrome have been measured to validate the non-invasive localization procedure of accessory pathways; hypertensive patients have been studied to develop a method to identify left ventricular hypertrophy; in patients with long-QT syndrome, symptom-specific MCG characters are searched; and patients with different atrial arrhythmias have been measured. In the average, 30 patients a month have been measured so far; as of January 1996 the total number of cardiac patients amounts to 120.

Conclusion

The BioMag Laboratory opens a variety of new opportunities for basic and clinical cardiac studies, especially in comparing and combining magneto- and electrocardiographic recordings, in interventive studies, and in verifying biomagnetic methods in larger patient populations. The present experience already shows that practical patient work can be run fluently, and patients are highly compliant in participating cardiomagnetic studies. Many visiting researchers have already contributed to the work in the laboratory and this trend will be strongly favored in the future as well.

References:

[1] Ahlfors, S., Ilmoniemi, R.J.: Magnetometer position indicator for multichannel MEG, In: Williamson, S., Hoke, M., Stroink, G., Kotani, M. *Advances in Biomagnetism*, New York, Plenum Press, 1989.

[2] Ilmoniemi, R.J., Ahonen, A., Alho, K., Aronen, H., Huttunen, J., Karhu, J., Karp, P., Mäkijärvi, M., Näätänen, R., Paetau, R., Pekkonen, E., Standertskjöld-Nordenstam, C.-G., Tiihonen, J., Toivonen, L., Katila, T: MEG and MCG in a clinical environment, *this book.*

[3] Knuutila, J., Ahonen, A., Hämäläinen, M., Kajola, M. Laine, P., Lounasmaa, O., Parkkonen, L., Simola, J., Tesche, C.: A 122-channel whole-cortex SQUID system for measuring the brain's magnetic fields, *IEEE Trans. Magn.*, 1993, **29**: 3315-3320.

[4] Montonen, J., Mäkijärvi, M., Paavola, M., Leiniö, M., Siltanen, P., Katila, T.: Comparison of time domain and spectrotemporal domain analysis of high resolution magneto- and electrocardiographic recordings of myocardial infarction patients with ventricular tachycardia, In: Baumgartner, C., Deecke, L., Stroink, G., Williamson, S.J. *Biomagnetism: Fundamental Research and Clinical Applications*, Elsevier Science, 1995.

[5] Nenonen, J.T.: Solving the inverse problem in magnetocardiography, *IEEE Eng. Med. Biol.*, 1994, **13**: 487-496.

[6] Nenonen, J.T., Hämäläinen, M.S., Ilmoniemi, R.J.: Minimum-norm estimation in a boundary-element torso model, *Med. & Biol.. Eng. & Comput.*, 1994, **32**: 43-48.

[7] Paavola, M., Ilmoniemi, R.J., Sohlström, L., Meinander, T., Penttinen, A., Katila, T: High performance magnetically shielded room for clinical measurements, *this book.*

[8] Tesche, C.D., Uusitalo, M.A., Ilmoniemi, R.J., Huotilainen, M., Kajola, M., Salonen, O: Signal-space projections of MEG data characterize both distributed and well-localized neuronal sources, *Electroenceph. Clin. Neurophysiol.*, 1995, **95**: 189-200.

Spatial Aspects in Magnetocardiographic Time Domain and Spectrotemporal Post-Infarction Arrhythmia Risk Assessment

Montonen, J.[1,3], Korhonen, P.[2,3], Paavola, M.[1,3], Mäkijärvi, M.[2,3], Toivonen, L.[2,3] and Katila, T.[1,3]

[1]Laboratory of Biomedical Engineering, Helsinki University of Technology, 02150 Espoo; [2]Cardiovascular Laboratory and [3]BioMag Laboratory, Helsinki University Central Hospital, 00290 Helsinki, Finland

Introduction

Means to evaluate the risk of malignant ventricular arrhythmias after myocardial infarction are of considerable clinical importance. Magnetocardiography (MCG) has been suggested for this purpose e.g. by analyzing spatiotemporal features of MCG mappings [4]. We have applied time domain (TD) and spectrotemporal (ST) analysis on high-resolution (HR) MCG recordings with results comparable to signal-averaged (SA) ECG [2]. Recently, an observation of one TD-parameter to show a spatial maximum in MCG-mappings was reported [5]. As a preliminary study for clinical multichannel MCG recordings of post-infarction patients in the Biomag Laboratory [3], we have now extended our earlier work I) to study the spatial distribution of TD and ST parameters in MCG mappings and II) to obtain criteria for these parameters in risk stratification.

Material and methods

We studied 31 patients with old myocardial infarction, 20 of whom suffered from documented paroxysmal sustained ventricular tachycardias (VT). None of the patients had bundle branch block.

MCG mappings were performed in the Otaniemi magnetically shielded room with a single channel axial gradiometer at instrumental noise level below $4\text{fT}/\sqrt{\text{Hz}}$. The standard MCG measurement grid with an extra column on the patient's left was used (see Fig. 1), except for five patients with VT from whom only the minigrid was measured. Consecutively, HR-ECG recording of the orthogonal x-, y- and z-leads was performed. MCG and ECG recordings were selectively averaged by a partitional clustering based algorithm to an approximately constant noise level. The noise level, as measured from the TD 25 Hz high-pass filtered data was below 30 fT for the MCG and below 1 μV for the ECG. Averaged MCG and ECG data were analysed by a TD high-pass filtering based method (see Fig. 2) and by the spectral turbulence analysis [1] (see Fig. 3), as described in Ref. [2]. For the MCG data, spatial distributions of all the TD and ST parameters over the whole measurement grid were plotted for each patient as isovalue contours (for an example, see Fig. 4). An average of each MCG parameter was calculated in the 3×3 *minigrid*, with upper right corner at grid location c3. To find out if the lower and left part of the MCG grid would give additional information to improve the identification of patients with VT, four other indexes were formed from the area of a 5×4 *reduced grid* as follows: Designating the four largest or, correspondingly, the smallest values of a parameter over the reduced grid as p_1, p_2, p_3, and p_4, in order, the indexes are $I_1 = p_1$, $I_2 = p_2$, $I_3 = (\sum_{n=1}^{3} p_n)/3$, $I_4 = (\sum_{n=2}^{4} p_n)/3$.

In the statistical analysis, Student's t-test was used to produce two-tailed p-values and the level of significance was set at $p<0.05$. The cut-point values for significant parameters were determined by maximizing the product of sensitivity and specificity. In the statistical analysis BMDP biostatistic software was used.

Results

In the minigrid, TD and ST QRS durations (QRSd) were the most significant parameters ($p<0.00005$) in identification of the patients with VT. Both the TD parameters of low amlitude signal (LAS) duration and RMS amplitude at the end

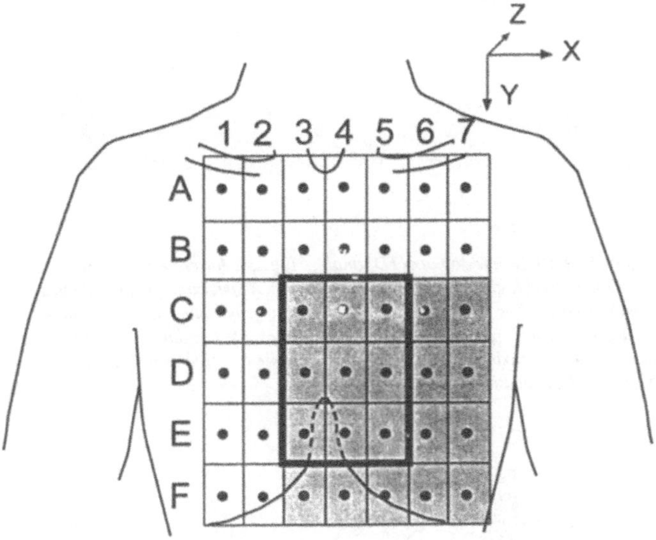

Fig. 1: *The measurement grids used in the study: The* standard *magnetocardiographic measurement grid with an extra column on the patient's left, containing 42 locations marked with black dots. The 20 location* reduced grid *is shown by gray shading. The* minigrid *with 9 locations is outlined by a thick line.*

of the QRS, and the ST analysis parameters of low slice correlation ratio (LSCR) and spectral entropy (SE) were also highly significant in the minigrid (see Table 1). These results were comparable to those obtained by SA-ECG.

The MCG TD and ST parameters performing best in the minigrid were, in all patients, observed to have their extrema frequently outside the minigrid. In addition, the spatial distributions of the TD parameters were continuous, showing often extrema on the left or at the bottom of the reduced grid. Distributions of the ST parameters were found more random. Both the TD and ST indexes (I_n) obtained from the reduced grid showed high levels of significance for VT identification. However, no clear improvement in comparison to the minigrid was obtained. Also, no obvious spatial differences were observed between patients with and without VT or between patients with anterior and inferior infarct locations.

Finally, using the most significant parameters and cutpoints optimized for them, we obtained the following citeria for VT identification, based on the averages over the minigrid. For the TD analysis: QRSd>110 ms and either LAS_1pT>40 ms or RMS40ms<750 fT, giving sensitivity of 90% and specificity of 100% in the present material. For ST analysis: QRSd>130 ms and either LSCR>74.0 or SE>17.5, giving sensitivity of 85% and spesificity of 100% in the present material.

Discussion

Averaging to a constant noise level was found more adequate than averaging a fixed number of beats. The lack of improvement in the MCG performance when using the 5×4 grid may be due to the high performances already reached with the minigrid, in this study with limited patient material. The qualitative results of the study suggest multichannel MCG mappings covering both the front and also to some extent the patient's left side, close to skin, to be optimal in TD and ST arrhythmia risk studies of infarction patients.

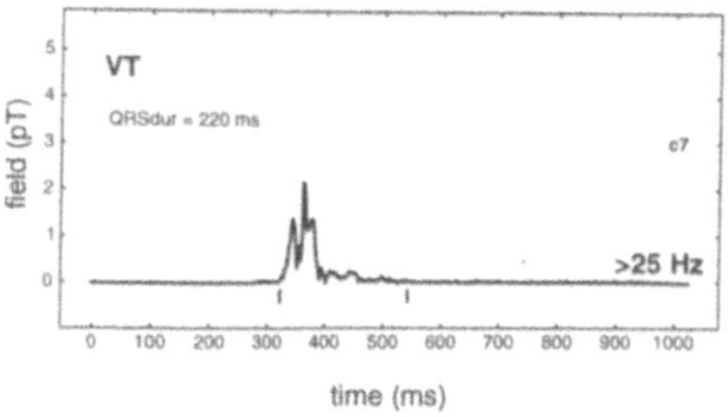

Fig. 2: *An example of the MCG time domain (TD) analysis where bidirectional 25 Hz high-pass filtering (HPF) of the QRS has enhanced the low amplitude field of the late ventricular depolarization. This patient with ventricular tachycardia had the maximum QRS duration at the left side of the grid at location C7, shown in the figure.*

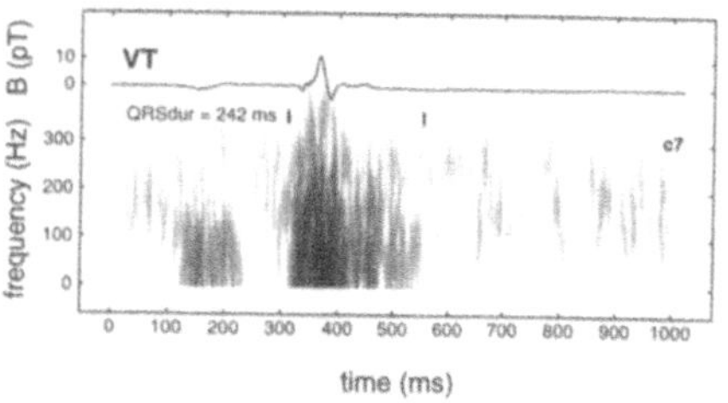

Fig. 3: *An example of the spectral turbulence analysis: A spectromagnetocardiogram is shown in logarithmic gray scale with the averaged morphological signal above it. The data is from the same patient and from the same location as in Fig. 2.*

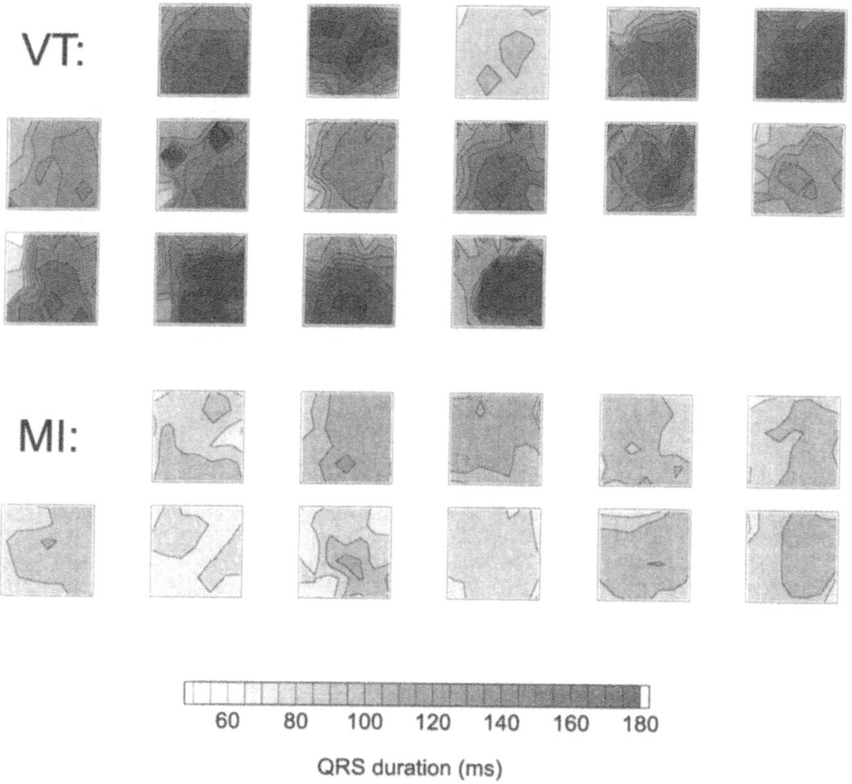

Fig. 4: *Spatial distributions of time domain QRS durations after 25 Hz HPF in the 42 location grid.*

	Parameter	Index	p	s./sp.	
TD	QRSdur	minigrid	< 0.00005	93	100
		I_1, I_2, I_3, I_4	< 0.00005	93	100
	LAS_1pT	minigrid	< 0.00005	93	100
		I_1, I_2, I_3, I_4	< 0.0015	87	64
	RMS40	minigrid	0.0014	87	91
		I_1, I_2, I_3, I_4	< 0.02	87	82
ST	QRSdur	minigrid	< 0.00005	93	100
		I_1, I_2, I_3, I_4	< 0.00005	93	100
	LSCR	minigrid	0.0006	80	77
		I_2, I_3, I_4	0.035	67	91
	SE	minigrid	0.0027	73	77
		I_1, I_2, I_3, I_4	NS		

Table 1: *The performance of TD and ST parameters in the minigrid and in the reduced grid. For TD parameters 25 Hz HPF was used. s. = sensitivity, sp. = spesificity, QRSdur = QRS duration, LAS_1pT = duration of low amplitude signal below 1 pT at the terminal QRS, RMS40 = root mean square value of the the last 40 ms of the filtered QRS, LSCR = low slice correlation ratio, SE = spectral entropy.*

References:

[1] Kelen, G.J., Henkin, R., Starr, A.-M., Caref, E., Bloomfield, D.A., El-Sherif, N.: Spectral turbulence analysis of the signal averaged electrocardiogram and its predictive accuracy for inducible sustained monomorphic ventricular tachycardia, *Am. J. Cardiol.*, 1991, **67**: 965-975.

[2] Montonen, J., Mäkijärvi, M., Paavola, M., Leiniö, M., Siltanen, P., Katila, T.E.: Comparison of time domain and spectrotemporal domain analysis of high resolution magneto- and electrocardiographic recordings of myocardial infarction patients with and without ventricular tachycardia, In: Baumgartner C., Deecke, L., Stroink, G., Williamson, S.J. *Biomagnetism: Fundamental Research and Clinical Applications*, Elsevier, 1995.

[3] Montonen, J., Ahonen, A., Hämäläinen, M.S., Ilmoniemi, R.J., Laine, P., Nenonen, J.T., Paavola, M., Simelius, K., Simola, J., Katila, T.E.: Magnetocardiographic functional imaging studies in the BioMag Laboratory, *this book*.

[4] Stroink, G., Lant, J., Elliot, P., Charlebois P., Gardner, M.J.: Discrimination between myocardial infarct and ventricular tachycardia patients using magnetocardiographic trajectory plots and iso-integral maps, *J. Electrocardiol*, 1992, **25**(2): 129-142.

[5] Weismüller, P., Richter, P., Abraham-Fuchs, K., Härer, W., Schneider, S., Höher, M., Kochs, M., Edrich, J., Hombach, V.: Spatial differencies of the duration of ventricular late fields in the signal-averaged magnetocardiogram in patients with ventricular late potentials, *PACE*, 1993, **16**(I): 70-79.

Comparison of Different Imaging Modalities as Anatomical Reference Methods for Magnetocardiographic Localization

Moshage, W., Achenbach, S., and Bachmann, K.
Medical Clinic II (Cardiology), University of Erlangen-Nürnberg, Germany

INTRODUCTION

Magnetocardiographic localization results need to be combined with anatomical imaging in order to be of clinical usefulness. Usually, magnetic resonance imaging is preferred in clinical MCG applications. We have compared different imaging modalities as to their applicability in combination with magnetocardiographic localization.

METHODS

More than 150 patients have been investigated with multichannel magnetocardiography inside a magnetically shielded room with the aim to localize the origin of cardiac arrhythmias. The MCG recordings were performed using a Krenikon® and MagnesII® biomagnetic system. In a varying number of patients, we have applied the following imaging modalities for anatomical correlation of the localization results:

1) Magnetic resonance imaging. In all patients, magnetic resonance (MR) images were obtained in axial, frontal, sometimes also sagittal planes with a Magnetom® 1.5T MR system (T1-weighted spin-echo-sequences, TE 20ms, TR according to heart rate, slice thickness 8 mm, gap 1mm). Special imaging sequences such as gradient-echo-techniques, doupble-echo sequences or oblique planes were applied if they were clinically necessary. Coordinate transfer to the MCG was achieved by the use of common reference points [3].

2) Chest X-ray. Chest X-ray images were obtained with the patient in supine position and in a neutral breathing position of the patient. Metal marks on the chest served as reference points for coordinate transfer to MCG.

3) Echocardiography. With the patient remaining in the magnetically shielded room after MCG recording, cardiac ultrasound was performed. 40 to 60 ultrasound images were obtained to cover the complete heart. The ultrasound transducer's position and orientation was recorded by a mechanical tracking system, computerized coordinate transfer permitted for assignment of the MCG localizations to the ultrasound images [2].

4) Electron Beam Tomography (EBT, "cine CT", "ultrafast CT"). EBT is a new X-ray tomographic imaging modality with extremely high spatial and temporal resolution [3]. Axial images of the heart were acquired during breathhold and triggered to the ECG after i.v. application of contrast agent (slice thickness 3mm, aquisition time 100ms/image). Metallic markers on the chest provided for coordinate transfer.

RESULTS

In all patients in which several imaging modalities were used for the assignment of localization results, the different imaging procedures yielded equivalent results. However, the spatial accuracy and resolution of the methods is not equal. Table 1 summarizes the observed advantages and disadvantages of the different imaging methods applied in combination with mutlichannel magnetocardiography.
While MR imaging permits ideal cardiac imaging without exposure to radiation, the procedure cost is high and the investigation time is high. Chest w-ray is a fast and widely available technique, but the image transfer is very inexact. Echocardiography could theoretically be performed within short time at the very site of the MCG investigation, but it is very difficult to cover all regions of the heart, especially with the patient in supine position. Electron Beam Tomography, finally, also yields extremely exact anatomical information, but the method is not widely available, requires intravenous injection of contrast agent, and exposure to radiation.

Table 1: Advantages and disadvantages of Magnetic resonance imaging, chest x-ray, echocardiography. and electron beam tomography ("ultrafast CT") for anatomical assignment of MCG localization results.

	advantages	disadvantages
Magnetic Resonance Imaging	good visualization of all cardiac structures high availability no radiation	high cost long investigation time
Chest X-ray	fast high availability inexpensive	imaging in only one plane no ECG trigger radiation
Echocardiography	investigation at MCG site inexpensive no radiation	difficult interpretation (lack of standardized planes) some areas of the heart impossible to document tracking of the transducer and computerized coordinate transfer necessary
Electron Beam Tomography	high resolution good visualization of all cardiac structures	radiation and contrast agent high cost limited availability

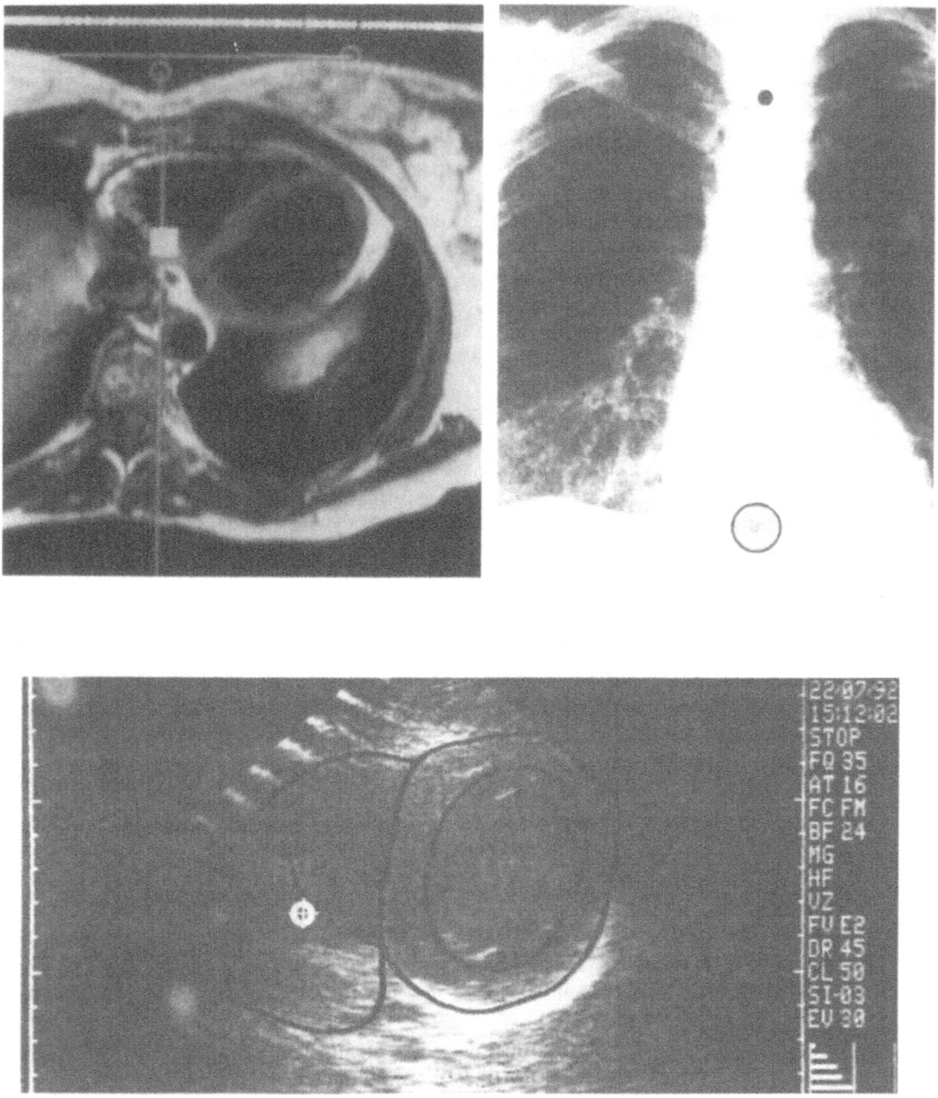

Fig. 1: Magnetocardiographic localization results in a patient with WPW-syndrome and accessory pathway in right posterior position (confirmed by invasive mapping and ablation). Identical result of MCG localization in all imaging modalities. Top left: magnetic resonance imaging (square), top right: chest x-ray (circle), and bottom: echocardiography: the result is in right posterior position between right atrium (RA) and right ventricle (RV).

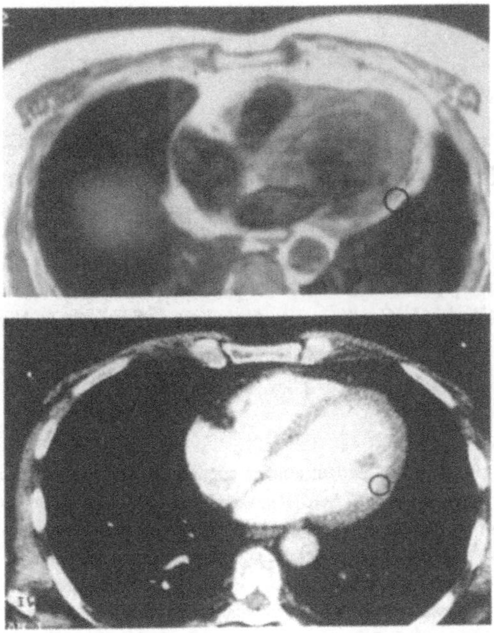

Fig.2: Identical localization result in MRI (top) and electron beam tomography (bottom). Origin of PVC localized in the left lateral wall.

DISCUSSION

While at the moment, MRI seems to be the ideal imaging method for magnetocardiography, EBT is also a very interesting alternative. Both methods can be triggered to the ECG and yield information on the complete chest anatomy, which is important for segmentation in realistic torso models. However, the use of EBT is restricted due to its very limited availability. Chest X-ray and - at present - echocardiography seem to be of limited applicability for anatomical imaging in combination with MCG.

REFERENCES

1) Boyd, D., Gould. R.G., Quinn, J., Sparks, R., Stanley, R., and Hermannsfeldt, W. A proposed cardiac 3-d densitometer for easy detection and evaluation of heart disease. IEEE Trans Nucl Sci 1979, 26: 2724-2727

3) Killmann, R., Moshage, W., Achenbach, S., Hönig, D., Konestabo, S., Reichenberger, H.: Combination of Biomagnetic source localization with ultrasonic images ofthe heart. Technology and Health care, Amsterdam, Elsevier, 1993, 214-215

3) Moshage, W., Achenbach, S., Schneider, S., Weber, U., Horn, S., Gebhardt, M., and Bachmann, K. Optimierung der Patientenpositionierung zum Koordinatentransfer zwischen Vielkanal-MKG-Messung und Kernspintomographischer Bildgebung, Biomed Technik 1992, 37 (Suppl): 168-169

Comparison of the Magnetocardiographic Localization Accuracy Using Simultaneous and Sequential Recordings in a Single and Multiple Planes

Moshage, W., Achenbach, S., Flüg, M., Lohse, P., and Bachmann, K.
Medical Clinic II (Cardiology), University of Erlangen-Nürnberg, Germany

INTRODUCTION

Thorough phantom and patient studies conducted during stimulation at a known site using a biomagnetic multichannel system (37 SQUID system Krenikon®) have demonstrated a localization accuracy of about 15 mm for recordings in a single plane. [1]. With a growing distance from the source to the sensors, an increasing spatial error, mainly concerning the depth of the localization result was observed, which could impair especially cardiac investigations [2]. In the course of technical progress, multiplane systems (e.g., 2x37 channels) have become available for cardiac applications. In the MagnesII® system (BTI, San Diego), the second set of sensors can be approached to the patient's back from a dorsal position.

METHODS

We investigated the biomagnetic localization accuracy of recordings in a single plane (dewar diameter 144 mm), simultaneous recordings in two planes (2x37 channels, dewar diameter 144 mm) and sequential recordings in multiple planes (horizontal plane over the chest, inclination of 20° and 40° to the left lateral side). 8 recordings during stimulation with a specially designed non-magnetic pacing catheter [3] in a torso-shaped phantom and 13 patient recordings (12 pt. with ventricular and 1 pt. with supraventricular ectopic beats) were evaluated. Based on the recorded field distributions, the electrical activity (stimulus at catheter tip or origin of the ectopic beat) was localized using the model of a single moving current dipole in a simplified homogeneous volume conductor (halfspace or sphere) and transferred to the patient's or phantom's magnetic resonance images. In phantom studies, the distance of the localization result to the postion of the pacing catheter tip permitted exact quantification of the localization error.

RESULTS

For typical bipolar field distributions, the localization results confirmed the known spatial errors of about 15 mm. No substantial advantages of multiplane recordings as compared to recordings with a single dewar position were observed. For dominantly unipolar field distributions, sequential recordings in 3 planes showed slightly better localization results in phantom studies than recordings in a single plane (see fig. 1). In patient investigations, the advantage of multiplane recordings were seen as a reduced scattering of dipole localizations during the first 10 to 20 milliseconds of ectopic activity as compared to single plane recordings. For very deep sources (distance of the source to the sensor plane above 15 cm), the simultaneous recordings in two planes showed slightly better localization results than recordings in a single or several sequential planes (see table 1).

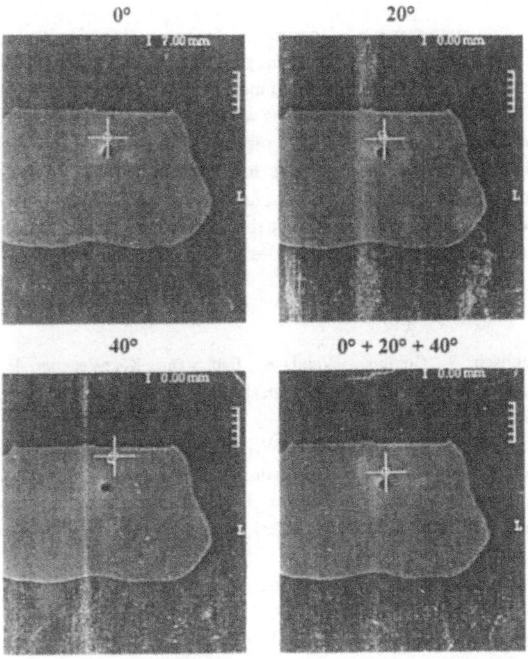

Fig. 1: Localization of the stimulus (circle/cross) applied at the catheter tip which can be seen as a dark mark in the axial MR images of the torso phantom. A) recording in one horizontal plane, B) dewar tilted 20° to the left, C) dewar tilted 40° to the left, and D) localization based on the recordings in all 3 planes.

Table 1: Localization error (mm) for "deep sources depending on the sensor position (n = 25)

Localization error (mm)	X	Y	Z
A 0°	3	17	65
A 0° + 15° + 30°	4	12	94
B 0°	2	6	17
B 30 °	?	5	6

A = ventral dewar (dipole-dewar distance 18 cm)
B = dorsal dewar (dipole-dewar distance 10 cm)

DISCUSSION

The observed slight improvements of localization accuracy using simultaneous recordings in two planes for deep sources within the heart (e.g., diaphragmal wall of left ventricle) have to be verified in further studies. Up to a distance of 12 to 15 cm from the dewar to the source, recordings with one dewar position seem to be sufficient for multichannel systems with a dewar diameter of 144 mm if the sensors are placed in a central position above the source. If the recorded field distribution is not bipolar, the combination of mutliple dewar positions permits improved localization accuracy of biomagnetic investigations in cardiology. However, this requires either simultaneous recordings with several sensors which have to be specially designed in order to permit simultaneous positioning close to the patient's chest or the events must be repetitive in order to be recorded in sequential measurements.

REFERENCES

1) Moshage, W., Achenbach, S., Göhl, K., Weikl, A., Bachmann, K., Wegener, P., Schneider, S., and Härer W. Biomagnetic localization of ventricular arrhythmias. Radiology 1991, 180: 685-692

2) Moshage, W., Achenbach, S., Göhl, K., Härer, W., Schneider, S., and Bachmann, K.: Magnetocardiography in Combination with MRI: Verification of localization accuracy with a nonmagnetic pacing catheter. In: Hoke., M., Erné, S., Okada, Y., and Romani, G.L. Biomagnetism: Clinical Aspects, Amsterdam, Ecrepta Medica, 1992, 447-451

3) Moshage, W., Achenbach, S., Bolz, A., Weikl, A., Wegener, P., Bachmann K., and Schaldach, M. A non-magnetic, MR-compatible pacing catheter for clinical application in magnetocardiography. Biomed. Technik, 1990, 35 (Suppl. 3): 162-163

Localization of Kent Bundles prior to HF Catheter Ablation: A Quantitative Comparison of the Localization Accuracy Based on an Inverse Solution Applied to Electrical and Magnetic Field Distributions

Moshage, W.[1], Achenbach, S.[1], Killmann, R.[2], Bruder, H.[2], and Bachmann, K.[1]
[1]Medical Clinic II, University of Erlangen-Nürnberg, [2]Siemens AG, Erlangen, Germany

INTRODUCTION

ECG mapping for the localization of ventricular arrhythmias or accessory pathways has up to now been performed based on pattern recognition of spontaneous or paced signals. The aim of our investigation was the direct comparison of the localization obtained using an inverse solution for both electrical and magnetic field distributions in patients, using successful ablation of accessory pathways as reference method.

PATIENTS AND METHODS

We investigated two patients with symptomatic Wolff-Parkinson-White syndrome due to a single accessory pathway. Prior to invasive catheter ablation, the Kent bundles were localized based on the body surface potential maps (BSPM) and magnetocardiographic (MCG) maps recorded during the delta-wave.

1) Body surface potential mapping (BSPM):
Body surface potential maps were recorded with 42 electrodes distributed over the patient's chest (see figg. 1 and 2). The electrode positions were recorded using a Polhemus Isotrak® system. Localization of the accessory pathway was performed in an inverse solution using the model of a single moving dipole whithin the patient's realistic outer torso boundaries as obtained from the MR images.

2) Multichannel magnetocardiography (MCG):
The patients' magnetocardiograms were recorded with a 37-SQUID multichannel system (Krenikon®) in a magnetically shielded room [3]. The sampling rate was 1000 Hz, the frequency range 0.016 - 250 Hz. Localization was performed using the model of an equivalent moving current dipole both in a simplified homogeneous volume conductor model (sphere or halfspace) [1] and in a realistic torso model.

The localization results of both methods were assigned to the patients' magnetic resonance images (frontal and axial cross-sections) and the localization error was quantified by comparison to the catheter ablation sites (see fig. 3)

RESULTS

Table 1 summarizes the localization errors obtained for both patients as well as the mean localization error for body surface potential mapping and multichannel magnetocardiography using a torso or a simplified volume conductor model.

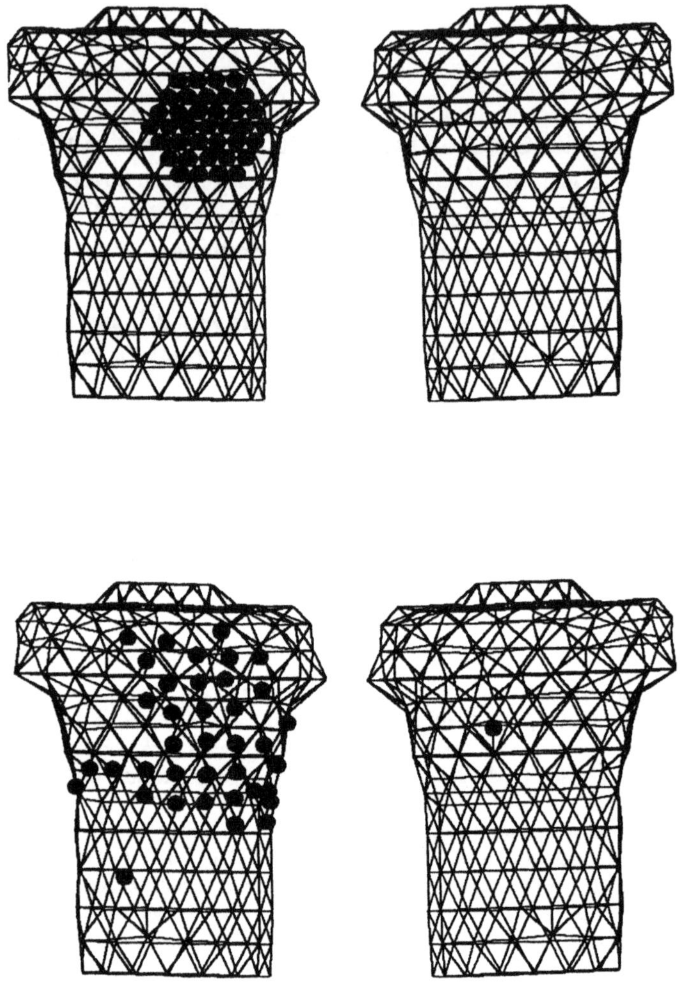

Fig. 1: Position of the senors (left: ventral view, right: dorsal view) with respect to the torso boundaries in patient 1 with right-posterior accessory pathway.
Top: Positions of the biomagnetic sensors (37-channel SQUID system). Bottom: ECG electrodes for BSPM.

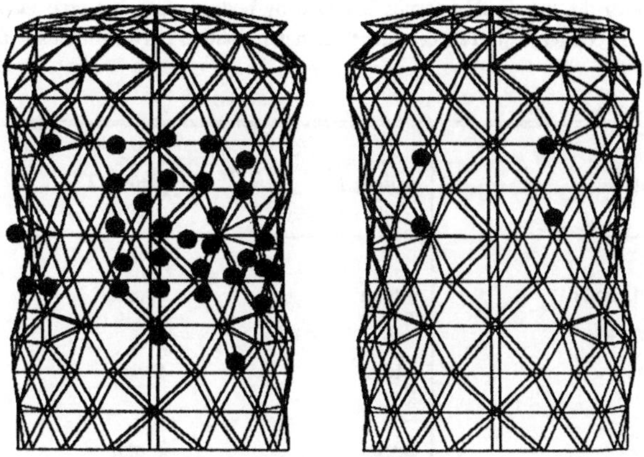

Fig. 2: Position of the ECG electrodes in patient 2 with left lateral accessory pathway

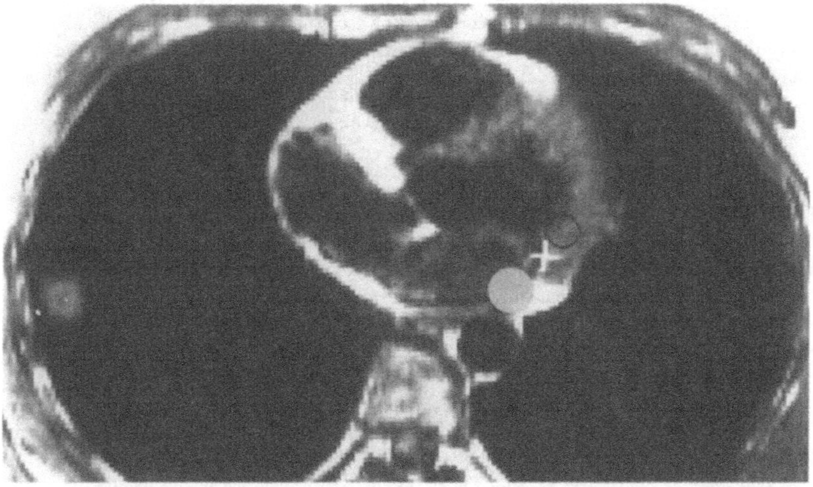

Fig. 3: Localization results for BSPM (solid circle), MCG (circle) in comparison to the site of successful HF catheter ablation (cross) in patient 2 with left lateral accessory pathway (axial magnetic resonance image).

Table 1: Summary of the localization results obtained by body surface potential mapping (BSPM) and multichannel MCG in comparison to successful HF catheter ablation in two patients with WPW-syndrome

LOCALIZATION ERROR	patient 1 (right posterior)	patient 2 (left lateral)	mean
MCG (simplified model)	12mm	14mm	13mm
MCG (torso)	14mm	18mm	16mm
BSPM (torso)	20mm	25mm	23mm

DISCUSSION

In conclusion, the localization error of simplified modeling (13mm) was within the known range of 10-20mm. The reason why complex modeling did not improve MCG localization accuracy is not yet clear. For MCG and BSPM, the quantitative error of accessory pathway localization using realistic modeling was in the same range. While MCG requires expensive recording equipment and magnetic shielding, data registration and evaluation are rather quick and simple. On the other hand, BSPM requires a very time-consuming registration and evaluation process. Larger investigations are necessary to further evaluate these early results.

REFERENCES

1) Achenbach, S., Moshage, W., Graumann, R., Horn, S., and Bachmann, K. Klinische Anwendung einfacher geomatrischer Volumenleitermodelle in der Magnetokardiographie. Biomed Technik 1992, 37 (Suppl): 170-171 (1992)

2) Ferguson, A. S. and Stroink, G. Localization of epicardial sources using magnetic and potential maps. In: Baumgartner, C., Deecke, L., Stroink, G., and Williamson S.J.: Biomagnetism: Fundamental research and clinical applications, Amsterdam, Elsevier, 1995, 641-646

3) Moshage, W., Achenbach, S., Göhl, K., Weikl, A., Bachmann, K., Wegener, P., Schneider, S., and Härer W. Biomagnetic localization of ventricular arrhythmias. Radiology 1991, 180: 685-692

Analysis of Electromotive Force by Magnetocardiogram - Is the Origin of Initial QRS Force (Septal Vector) of Electrocardiogram Intraventricular Septum?

Nakaya, Y.[1], Yukinaka, Y.[2], Nomura, M.[2], Kishi, F.[2] and Saito, K.[2]

[1]Department of Nutrition and [2]The Second Department of Internal Medicine, University of Tokushima, Tokushima 770, Japan

Introduction

Since the original investigations of Lewis and Rothsenchild (1915) [1], septal activation has been thought to occur first on the left septal surface and thereafter to spread form the left side to the right side of the intraventricular septum. Thus, the initial QRS vector in electrocardiogram is considered to be due to activation of intraventricular septum and called as septal vector. At this time (within 20 msec from the onset of the QRS wave), posterobasal area is not activated. However, normal septal vector disappears in cases of inferior infarction as well as anteroseptat infarction. Then arises the question whether the initial QRS vector is really composed of intraventricular septum alone. Therefore, we studied the origin of initial QRS force by magnetocardiogram. The magnetocardiographic three-dimensional (3-D) source localization using superconducting quantum interference device (SQUID) gradiometer has an excellent spatial and temporal resolution. To assess electrocardiographic origin of the infarction vector (abnormal Q vector), the magnetocardiograms of normal subjects and the patients with inferior myocardial infarction (MI) were recorded in an rf-shielded room using biomagnetic measuring system. We also studied the change of the initial QRS vectors in left ventricular hypertrophy. The results of the study provide a new interpretation of ECG wave forms.

Methods

We studied 20 normal subjects (N group) and 20 patients with old inferior myocardial infarction (MI group). We also studied 28 patients with left ventricular overload (16 hypertension, 8 aortic valvular disease and 4 mitral regurgitation), in which initial QRS vector is also frequently deviated [2]. The diagnosis of myocardial infarction was made by history increase in CPK, the ECG and echocardiogram.

The MCG was recorded over the anterior chest wall with a second-derivative superconducting quantum interference device (SQUID) gradiometer, and the magnetic field perpendicular to the anterior chest wall was measured. In order to localize the current dipole, the 3-dimentional dipole locations (single moving dipole model) were computed every 4 msec during the initial Q wave by the least square method, and were superimposed onto the Magnetic Resonance Image (MRI) of the individual subjects providing anatomical localization in the ventricle. We also studied vector arrow maps of both groups to study the activation of other parts of the ventricles during first 40 msec [3].

Results

Initial QRS vector of normal subjects:

Fig. 1a shows the isofield map at 10 msec from the onset of the QRS wave of a normal subject. The location of the early excitation was 47 mm left of the median line, 28 mm up from the xyphoid process and 70 mm below the detector coil of the SQUID sensor. Fig. 1b shows the MRI of this subject in which the dipole deduced by isofield mapping were superimposed. The deduced dipole was on the ventricular septum. Mean values of induced dipole of the initial 20 msec vector were 48$\pm$18 mm left ward, 58$\pm$24 mm superior to the xyphoid process and 41$\pm$13 mm deep from the surface of the gradiometer.

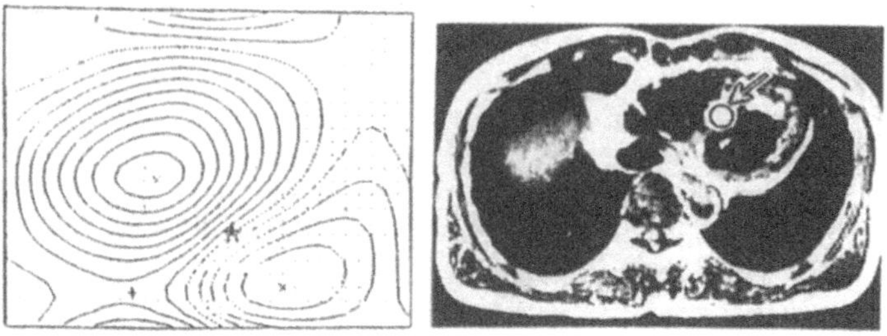

Fig. 1 (a) Isofield map at 10 msec from the onset of the QRS wave of a normal subject. (b) MRI of this patient. The deduced current dipole was superinposed on MRI and indicated by an arrow.

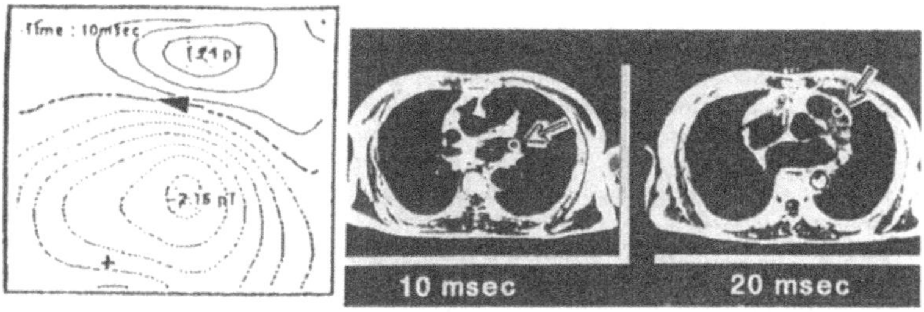

Fig. 2 (a) Isofield map at the 10 msec from the onset of the QRS wave in a patient with inferior MI. (b) MRI of this patient in which the dipole deduced by isofield mapping were superimposed.

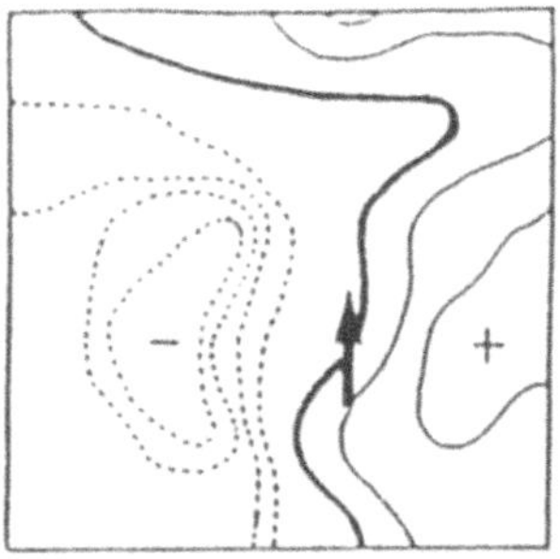

Fig.3 Departure map between normal subject and patients with inferior infarction

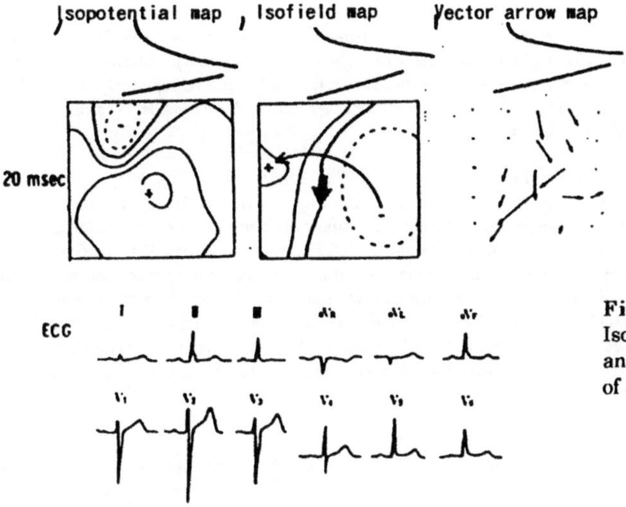

Fig. 4
Isopotential, isofield and vector arrow maps of a normal subject

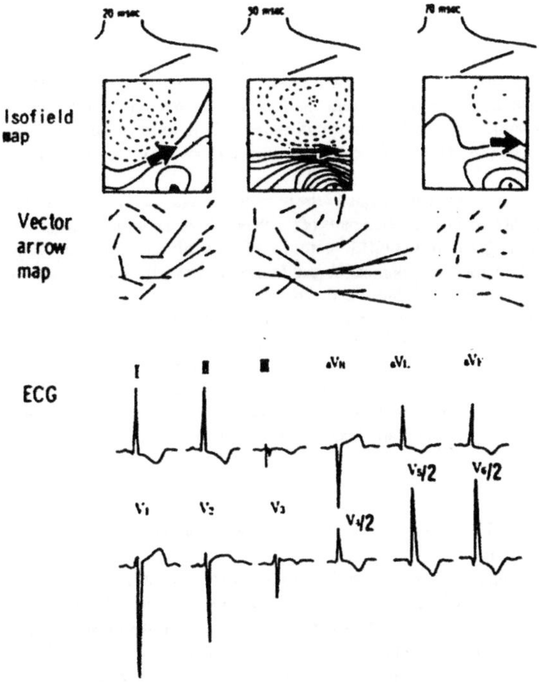

Fig. 5
Isofield and vector arrow maps of a patient with left ventricular overload

Initial QRS vector of patients with MI:

Fig. 2a shows the isofield map at the 10 and 20 msec from the onset of the QRS wave in a patient with inferior MI. Fig. 2b shows the MRI of this patient in which the dipole deduced by isofield mapping were superimposed. The deduced dipole was not on the ventricular septum and displaced superiorly. Fig. 3 shows the departure map between the mean values of N and MI groups. The subtracted QRS vector roughly corresponded inferior wall (infarcted area).

Initial QRS vector in LVO group:

Fig.4 shows ECG, isofield map and vector arrow map of a normal subject. The pattern of the isofield map at 20 msec indicated the presence of the current dipole directing to the right. Vector arrow map at 20msec aslo shows vector arrows directing to the right and inferiorly. Fig. 5 shows ECG, isofield map and vector arrow map of a patient with left ventricular overload without q waves in left precordial leads (V_5 and V_6). Both maps show that in this patient vector arrows directing to the left and located in the left part of the map are increased in magnitude compared to that of the normal subject. The direction of deduced single dipole in LVO group was also deviated to the left in all patients without q wave in the left precordial leads. Similar diviation of initial QRS force was also observed even in most of the LVO group with a q wave in the left precordial leads.

Discussion

The present study shows that the location of infarction vector was displaced superiorly from the infarct area and not in the necrotic infarct zone. Moreover, the subtracted QRS vector in the departure map between normal subjects and patients with infarction was located in the infarct area, suggesting that this hypothetical vector (infarction vector) is derived from necrotic myocardium. The loss of voltage due to inferior infarction results in a new balance of electrical forces, which oriented away from the regions of tissue loss. These results strongly support the idea that the initial QRS vector is not originated from ventricular septum alone, and that many areas of ventricle contribute the formation of initial QRS vector.

In left ventricular hypertrophy, the initial QRS vector is frequently deviated, which was explained by septal fibrosis due to hypertrophy. However, in the present study, vector arrow map showed that loss of normal rightward vector was due to the increased electromotive forces of the left ventricle and not due to the loss of septal activation. Thus, normal rightward vector was displaced both in location and its direction in left ventricular overload, because the balance of the electromotive force between the left and right ventricles was altered.

The initial QRS vector has been considered to be due to the intraventricular activation. However, the present study shows that the posterobasal area and other parts of the ventricle also contribute to generate initial QRS vectors, as evidenced by a changes in initial QRS vector in inferior infarction and left ventricular overloading. Thus the initial QRS vector is a hypothetical resultant or mean vector representing the average direction and magnitude of all electric forces produced at this time.

We used single dipole estimation and vector arrow map [3] to deduce electromotive force of the heart, as we did not have the methods which can deduce multiple dipoles (such as 100 or more) by magnetocardiograms. These methods still can estimate the current sources at each time. In agreement with our studies, three endocardial areas are synchronously depolarized from 0-5 msec after the onset of ventricular activation In the studies of human heart [4]. These studies also support the results of the present studies that the initial QRS vector is summation of electromotive forces of many areas of the both ventricles.

References

[1] Lewis, T., and Rothschild, M.A. The excitatory process in the dog's heart, Philos. Trans Royal Society of London, 1915, 206: 181.

[2] Gaum,W.E., Chou,T.C., and Kaplan, S. The vectorcardiogram and elelctrocardigram in supravelvular aortic stenosis and coarctation of the aorta. American Heart Journal, 1972, 84: 620.

[3] Nakaya, Y., Sumi, M., Saito, K., Fujino, K., Murakami, M., and Mori, H. Analysis of current sources of the heart using isomagnetic and vector arrow maps. Japanese Heart Journal, 1984, 25: 701.

[4] Durrer,D., van Dam,R.T., Freud, G.E., Janse,M.J., Meijler,F.L., and Arzbaecher, R.C. Total excitation of the human heart, Circulation, 1970, 41: 899.

The Algorithm of Spatial-Temporary Analysis of Cardiomagnetic Signals

Nedayvoda, I.V. and Primin, M.A.

Institute for Cybernetics, Kiev, Ukraine

ABSTRACT. Report considers the new algorithms on spatial analyses and visualization of biomagnetic data, which permit to establish the availability of "norm" or degree of deviation from norm for the patient under study with respect to the conventional methods of preliminary data processing.

1. INTRODUCTION. STATEMENT OF THE PROBLEM.

It is known that the parameters of heart's magnetic fields - magnetocardiosignal - are registered in the air (above the chest of patient) at the intersection points of a regular grid, that correspond to the anatomic points. The number of points of observation (measurement) at standard technique of measurement by single-channel squid-gradiometer, usually makes 36 (6 x 6 points in the plane, 4 cm offset). Having been filtered and temporally averaged the measurement data for the given moment of time that is synchronized to the appropriate moment of averaged electrocardiocycle, the results are treated by software means. The values of heart magnetic field parameters at the points that do not belong to nodes grid are determined with the help of smoothing and data interpolation.

The spatial analysis of magnetocardiographic data with the help of magnetostatic inverse problem methods makes it possible to locate a zone of miocard activity at the investigated moment of time. For the cases of "point" model of the field source and when the source of field is treated as a planar layer of current this task has its analytical solution [1,2].

The effective dipole trajectory and the spatial configurations of currents system display the temporal change in biomagnetic field structure during the studied interval of time. But when analyzing these changes and interpreting the measured results (in medical meaning) it is possible to use another approach, that does not need the difficult inverse problem to be solved:

basing upon the values of magnetic field already known for the nodes of regular grid at the regular moments of time t_0, $t_0 + dt$, $t_0 + 2\,dt$,..., t_N, one can reveal the field's spatial characteristics, that are changed from one interval to another, and find their representative form, which might have provided the quantitative estimation of these changes.

2. METHOD OF SOLUTION.

At present time the visual analysis on magnetic field "instant" maps sequence (dynamic mapping method [3]) or standard methods of data applied analysis is generally used to solve the delivered problem. Both in the first case, and in the second, the analysis of magnetic field parameters spatial distribution does not make use of the known information about these parameters dependence upon the model of the field source ("effective" magnetic dipole or, for example, planar layer of current), that enables the measured signal simulation at the point of observation. It is known, that for the most remote points of observation (extreme rows of points of measurement grid or extreme probes of multichannel system) the measured signal corresponds with high accuracy to the dipole signal

$$\mathbf{B}(\mathbf{r}) = g(\mathbf{r})\,\mathbf{M},\quad g_{ij} = \mu_0/4\pi r^5\,(3r_ir_j - 3r^2\delta_{ij}),\ i,j=x,y,z, \tag{1}$$

And at other points the signal is dependent upon the layer of current that is distributed within the plane parallel to the measurement plane [1,2]

$$\mathbf{B}(\mathbf{r}) = \mu_0/4\pi \int_V [\mathbf{j}(\mathbf{r}') \times (\mathbf{r}-\mathbf{r}')]\,dV/2\,|\mathbf{r}-\mathbf{r}'|^3 \tag{2}$$

In (1)- (2) **B** is the vector of magnetic induction at the point of observation; **r**, **r'** - radius-vectors of observation point and arbitrary point of volume V in the coordinates system, aligned with the source; μ_0 is the magnetic constant; **M** is the vector of field source's magnetic moment.

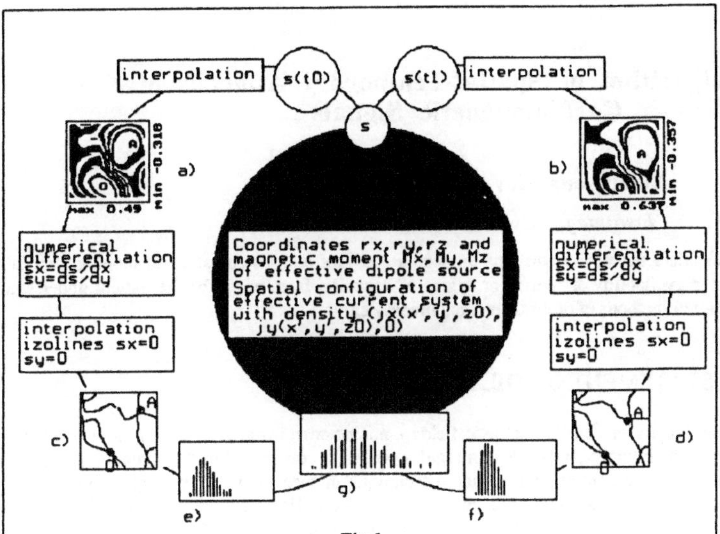

Fig.1

The block diagram of the algorithm is submitted on fig.1. As far as the algorithm does not assume the solution of magnetostatic inverse problem, the integrated parameters of cardiomagnetic field's effective source on fig.1 "are contained" within the circle of radius a (the parameters are unknown). The data of measurements s= { B_z, or $d_1=B_z$ ($z=z_0$)-B_z ($z=z_0$ + 1), or $d_2=B_z$ ($z=z_0$)-2 B_z ($z=z_0$ + 1) - B_z ($z=z_0$ + 2l)} are located at this circle. They are at the focus of epicycloide, which right and left-hand "branch" correspond the two different moments of time within the studied interval [t_0, t_N]. In this case the moments t_0 and t_N coincide with the beginning of segment ST and the maximum T of averaged electrocardiocycle. In the second focus of epicycloide we place the final results of analysis, which includes the following sequence of data processing operations:

(1) Interpolation of measured signal values s at the given nodes of planar uniform rectangular grid (in our case 31 x31 of nodes).

(2) Displaying and visualization the sequence of s-parameter maps in the form of isolines with the observation area borders shown, there are shown also: the maximal and minimum values of s for the studied moment of cardiocycle, as well as the location of plane points, where the measured s-parameter has its maximal and minimum values i.e. points O and A (fig.1a,b).

(3) Calculation of spatial derivatives $s_x=\partial s/\partial x$ and $s_y=\partial s/\partial y$ (x, y are the measuring system coordinates) at each node of rectangular grid.

(4) Interpolation of isolines $s_x=0$, $s_y=0$. The points of these isolines intersection give (by definition) the position of all the extremes of measured s-parameter, including points O and A (fig.1c,d) within the plane of measurements.

(5) Spatial derivatives of s-parameters displaying in the form of areas ($s_x>0$, $s_y<0$), ($s_x<0$, $s_y>0$), ($s_x>0$, $s_y < 0$), ($s_x < 0$, $s_y < 0$). The physical meaning of these areas may be interpreted basing upon the Maxwell's equations and Bio-Savar-Laplas expressions (2). Really, if the element of current with density (s_y, $-s_x$, 0) is located in each point of plane $z=z_0-d$ ($d \Rightarrow 0$), then the magnetic field of such a system practically coincides with the measured distribution of s. The area of "the most probable location of field source" is between the points of global extremes of measured distribution of the field and is limited by isolines $s_x=0$, $s_y=0$ [2].

(6) Analysis on the values of parameter g=sqrt $\{s_x(i,j)^2 + s_y(i,j)^2\}$ for the discrete set of points within the fragment OA with coordinates (r_{xi}, r_{yi}) = [r_{xO} + i dx, r_{xi} ($r_{yA}-r_{yO}$) / ($r_{xA}-r_{xO}$) + (r_{yA} - $r_{xO}r_{yO}$ - r_{xA}) / ($r_{xA}-r_{xO}$)], i=0, I. We suppose that each point within the fragment of OA (r_{xi}, r_{yi}) corresponds to the point with coordinates (r_{xi}, 0), and both of them are assigned to the same value of g. When determined so, the quantity I of spectrum points depends upon both z_0 -coordinate of plane of measurements and the direction of excitation wave at the studied moment of time (vector of dipole magnetic moment **M** of the source). In frameworks of described "effective" system of currents the g-parameter corresponds to the density "amplitude" of the current at the given points of the plane.

As a result we receive the spatial spectrum of magnetocardiosignal (fig.1e,f) for each moment of time $t_0,..., t_N$ within the chosen interval.

4. SPATIAL-TEMPORARY SPECTRUM OF MAGNETOCARDIOSIGNAL.

To provide the analysis on changes of magnetocardiosignal spatial structure the results of measurements are normalized and we into consideration the spectra for the "current" moment of time t_0 + k dt (j=0, N-1) and the following moment t_0 + (k + 1)dt. We assume, that during dt interval between the studied moments of time the values of integrated parameters of biomagnetic field source have changed: the z_0 coordinate ("occurrence" depth of miocard tissue active part with respect to the plane of measurements) of the field source, its dipole moment orientation or spatial distribution of sources intensity within the areas of signal excitation etc. The signal description technique we use now may drive at the appearance of spectrum points, where at the moment of tk "the spectral line" is equal to zero, and at the moment of t_k + 1 it is not equal to zero and vice versa (changes in spectrum length). The mutual correspondence between the spectrum lines can change also, i.e. $g (t_k, i) > g(t_k + 1, i)$ or $g(t_k, i) < g(t_k + 1, i)$. The qualitative map of magnetocardiosignal changes during dt interval can be presented by displaying the spectra g (t_k) and g (t_k + 1) on the same picture (PC monitor screen). Fig.1g shows the spatial-temporary spectrum of magnetocardiosignal for real measurements results.

It is known, that within the limits of temporary interval, chosen by us, the amplitude of biomagnetic signal is "normally" increased from previous moment of time to the next one. That's why it is possible to suppose, that at each spectrum point the "second" line should be greater then "the first" one. The total quantity of "inversions" at all the spectrum points, where at least one line is not equal to zero, can be defined as follows

$$v (t_k) = \Sigma \{ a_i \}, i = 0,1,.... M (g (t_M + 1)=g (t_M + 1)=0) \tag{3}$$
$$a_i = 0 \quad \text{if} \quad g (t_k, i) \leq g (t_k + 1, i), \quad a_i = 1 \quad \text{if} \quad g (t_k, i) > g (t_k + 1, i)$$

The generalization of described technique upon the whole temporary interval [t_0, t_N] produces the spatial-temporary spectrum of magnetocardiosignal and its quantitative estimation after the inversions methods in the following form

$$w (i) = \Sigma \{ a_j \}, i=0, N; j=0, M \tag{4}$$

where the summation is provided in accord with the number of temporal moment within the limits of current spectrum point numbered as I. The averaged factor of inversions makes

$$W = \Sigma \{ w (i) \} \tag{5}$$

It is possible to show, that the received estimation does not depend on both patient's anatomic features and initial values of magnetocardiosignal source integrated parameters and it is determined by the temporal changes in magnetocardiosignal spatial structure as well as by the number of spectrum lines. This number is determined by the length of temporal interval [t_0, t_N], which is the patient's individual characteristic. Therefore this step-type behavior of readout $dt=t_i- t_{i-1}$ (i=1, N) is taken as a constant for all the patients. In our studies the number of readouts (the number of investigated maps of heart magnetic fields parameters distribution) was not identical, but not less then 10 for each patient. And the maximum of QRS-complex of averaged cardiocycle corresponded to the same number of readout for each patient.

4. RESULTS.

Fig.2 illustrates the preliminary results of using the developed algorithms for spatial analysis of data, dealing with the activity of normal person heart and the heart of a patient with ishaemia (diagnosis is confirmed by other methods). The normal person data are logged as 00013201, while the ishaemia patient data have the log of 000055001.

There are submitted for each patient: the averaged electrocardiocycle with time moment marked, for which the magnetic map has been investigated; the spatial-temporary spectrum of biomagnetic signal (at the top - for normal person, at the bottom - for patient with ishaemia); the values of factor of inversions for each point of spectrum and using the measurement data. Thus for all the patients without anomalies in heart function the factor of inversions does not exceed 2, while for the patients with ishaemia this factor exceeds 10.

The research described in this publication was completed using the partial financial support by Grant N K1U100 from the Joint Fund of the Government of Ukraine and International Science Foundation.

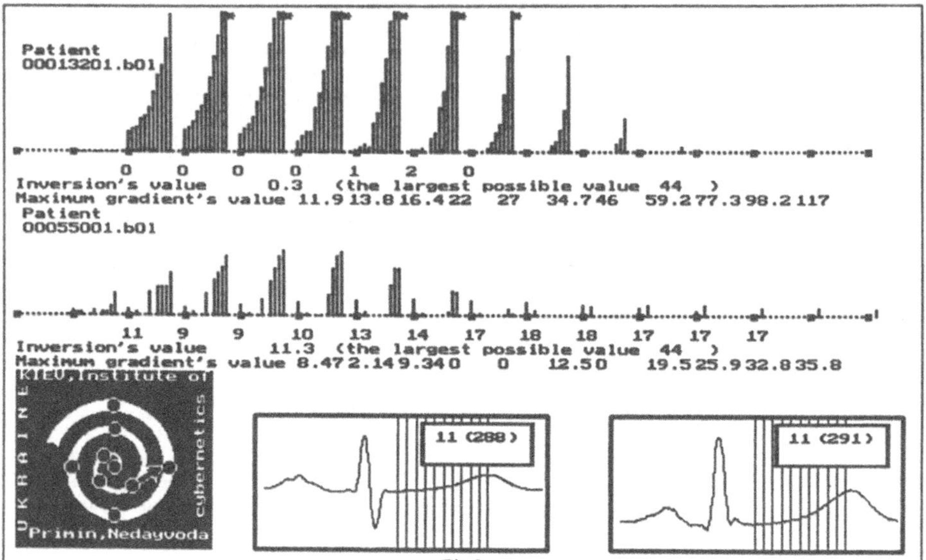

Fig.2

REFERENCES.

1. Primin M.A.,Gumeniuk-Sychevskij V.I.,Nedayvoda I.V., The methods and algorithms of localization of magnetic field source (in Russian), " Naukova dumka",Kiev,1992.

2. Primin M., Gumeniuk-Sychevskij V., Nedayvoda I., Mathematical models and algorithms of information conversion in spatial analysis of weak magnetic fields, Int.J.Applied Electromagn.in Materials 1994 5: 311-319.

3. Erne S., Fenici R., Hahlbohm H., Jaszczuk W., Lehmann H., Masseli M. High-resolution isofield mapping in Magnetocardiography, Il Nuovo Cimento 2 D, Editrice Compositori, Bologna 1983: 291-300 pp.

Measurement of Tangential the Magnetic Field in the PR Segment in Man

Noda, Y.[1], Mitsui, T.[1], Terada, Y.[1], Tsukada, K.[2], Adachi, A.[2], Yoshida, T.[2] and Haruta, Y.[2]

[1]*Department of Surgery, Institute of Clinical Medicine, University of Tsukuba, Tsukuba, Japan;* [2]*Superconducting Sensor Laboratory, Inzai, Japan*

Introduction

Noninvasive recording of the electrical activity of the His-Purkinje system has been reported by Berbari [1], Flower [2] and Hishimoto [3]. Their data are characterized by ramp-like patterns starting approximately 40-50 ms before the onset of QRS. A few years later MCG measurement of a ramp pattern attributed to the His-Purkinje system was reported by Farrell [4]. Fenici [5] has reported that the magnetic PR waveform is characterized by a "spike" followed by a "ramp" pattern. But the origin of the ramp pattern is still controversial. Moreover, the conventional method of measuring MCG was measurement of normal component of MCG only. In this paper, we report our measurements of the tangential components of the magnetic field in the PR segment using the 36 ch SQUID system.

Methods

We used the SQUID system with first-order gradiometers set in the Superconducting Sensor Laboratory. Pick-up coils facing tangential axes were placed 6 x 6 on a plane in this system, so tangential components of the magnetic field at the 36 points could be measured simultaneously. The sensor array of our measurements is shown in Fig. 1. Magnetocardiograms and electrocardiograms were measured in a magnetically shielded room, and the white noise level of this system was less than 7 fT/$\sqrt{}$Hz. We studied eight male volunteers, all of whom were normal according to their clinical histories, standard ECGs, chest X-rays, and physical examinations. The bottom of the dewar was set as closely as possible the surface of the anterior chest wall. The sampling rate was 2 KHz and the magnetic field and electrocardiograms data were averaged 300-1200 times. We also took an electrocardiogram (Ⅱ lead and bipolar lead between V1 and V4) simultaneously. Bipolar ECG between V1 and V4 was also averaged and obtained signal averaged ECG (SAE) data to compare with MCG data.

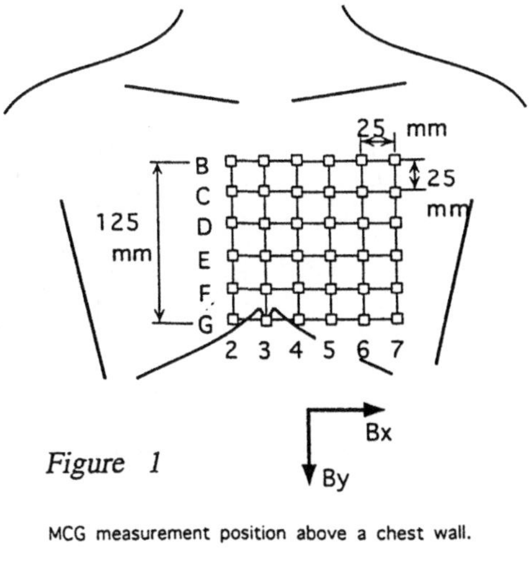

Figure 1

MCG measurement position above a chest wall.

Results

Figure 2 shows ***B*x** and ***B*y** (0.1-300 Hz) of one subject (Case 1: 32 year-old male). Spikes were observed approximately 60 ms before QRS onset in ***B*x** at the position almost above the AV node and His bundle which were B4~F5. These spiky waveforms were not clear at upper right and lower left regions of the measurement area. Sharp, spiky waveforms as seen in ***B*x** were not observed in ***B*y**. Figure 3 shows the ***B*x** and ***B*y** of two subjects (Case 1 and Case 2: 33 year-old male) at E4 which was considered near the AV node and His bundle. The V1/V4 surface electrocardiogram (80-300 Hz) taken simultaneously is shown at the bottom of the figure. In both cases, negative "spikes" were observed soon after the P wave in ***B*x**. In Case 1, the maximum amplitude of the magnetic signal was approximately 150 fT and its duration was 25 ms. The time position relative to the onset of QRS was 60 ms. The SAE recorded simultaneously also revealed a spiky signal 60 ms before QRS. After the spike there was a ramp pattern which continued to the beginning of QRS. In Case 2, the maximum amplitude of the magnetic spike was 200 fT and its duration was 20 ms in ***B*x**. The time interval from the beginning of the magnetic spike to QRS onset was 65 ms, which is equal to the interval observed in the SAE. These characteristic MCG patterns were observed in all other subjects, and these time intervals are shown in Table 1. The duration of "spike" was 20-30 ms, and it was easily observed in the ***B*x** component. The time interval between onset of spike and QRS was 60-90 ms, almost equal to the interval in SAE.

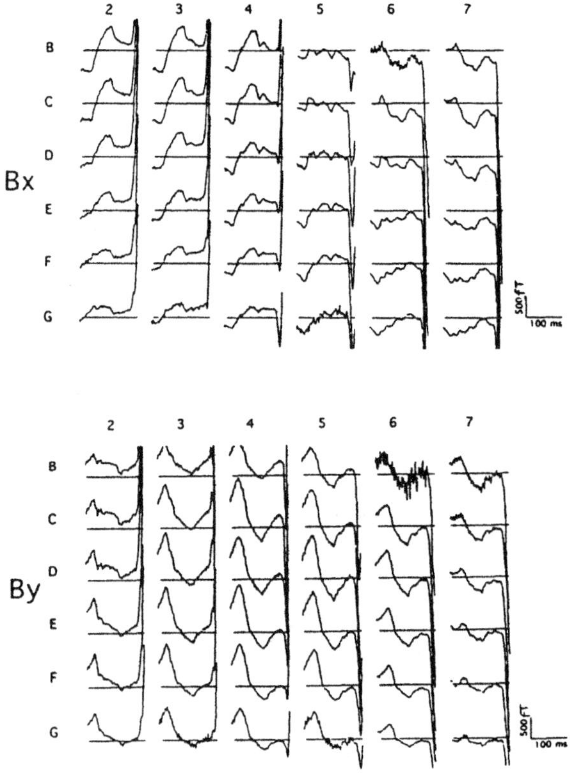

Figure 2

By & By waveforms in PR segment in Case 1

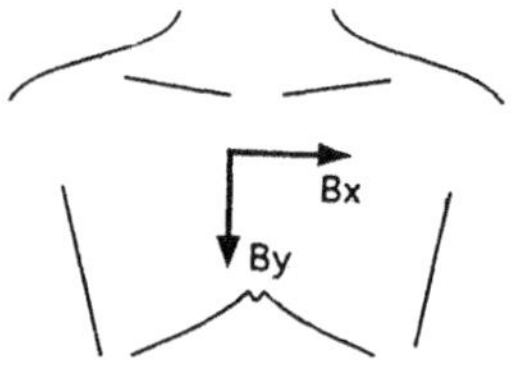

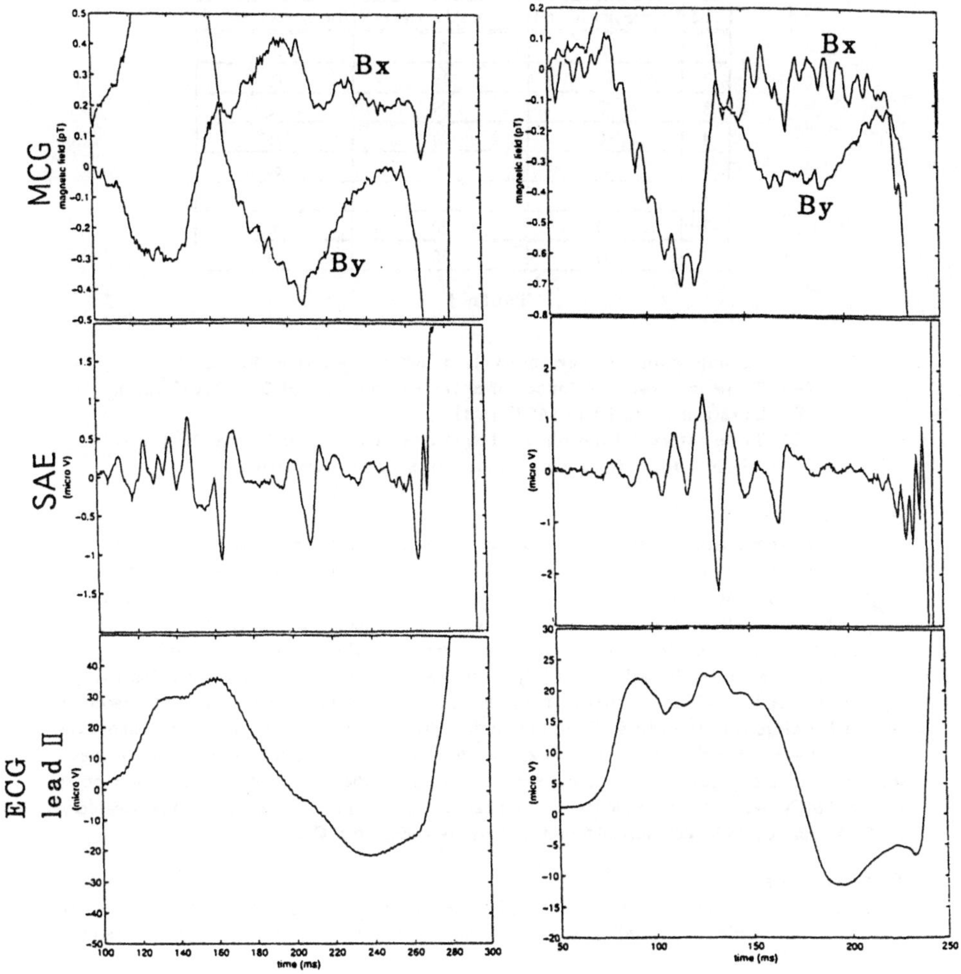

Figure 3

Comparison of MCG and ECG in PR segment

Case	Time interval in MCG (A)	Magnetic spike duration (B)	Time interval in SAE (C)
1	60	20	60
2	65	15	65
3	70	25	70
4	70	20	75
5	85	25	80
6	90	25	NI
7	90	20	80
8	85	20	NI

Table 1

Comparison of time intervals of MCG signals and SAE
(A): Time interval from onset of spike to onset of QRS in MCG (ms).
(B): Duration of spike in MCG (ms).
(C): Time interval from onset of spike to onset of QRS in ECG (ms).
NI (Not Identified) means that spike is not identified.

Discussion

In our experiment, "spike" and "ramp" patterns were recorded in all cases in averaged MCG, and especially clearly in the *B*x component. The time correlation between the magnetic signals and SAE signals was good, but the time interval from onset of spike to onset of QRS was a little longer than the HV time (time interval between His bundle depolarization and ventricular depolarization) that was already well known by electrophysiological studies. In this study, we didn't examined electrophysiological study using catheter, but Vencent [6] and Berbari [1] reported a strong correlation between the HV time measured by electrode catheter and duration of the His-Purkinje waveform in SAE. In our study, spikes were most clearly identified above the AV node and His-Purkinje system. It is known that the tangential components of MCG reveals maximum intensity just above the current. According to these evidences, spike observed in averaged MCG was also associated with electrical activation of His bundle, and it was considered that HV time could be determined noninvasively by averaged MCG.

References

[1] Berbari, E.J., Scherlag, B.J., El-Sherif, N., Befeler, B., Aranda, J.M., Ralph, L. The His-Purkinje electrocardiogram in man: An initial assessment of its uses and limitations. *Circulation*. 1976, 54: 219-224.

[2] Flower, L.G. Surface recording of electrical actvity from the region of the Bundle of His. *Am J Cardiol*. 1974; 33: 384-389.

[3] Hishimoto, Y., Sawayama, T. Non invasive recording of His bundle potential in man. *Br Heart J*. 1975, 37: 635-639.

[4] Farrell, D.E., Tripp, J.H., Norgren, R. Magnetic study of the His-Purkinje conduction system in man. *IEEE Trans Biomed Eng*. 1980, 27: 345-350.

[5] Fenici, R., Romani, G.L., Barbanera, S., Zeppilli, P., Carelli, P., Modena, I. High resolution magnetocardiography: Non-invasive investigation of the His-Purkinje system activity in man. *G Ital Cardiol*. 1980, 10: 1366-1370.

[6] Vencent, R., Stroud, N.P., Jenner, R., English, M.J., Woollons, D.J., Chamberlain, D.A. Noninvasive recording of electrical activity in the PR segment in man. *Br Heart J*. 1978, 40: 124-130.

Observation of a Difference Signal in MCG Measurements in Patients with Ventricular Extrasystoles

Pedersen, J.R.[1], Bloch Thomsen, P.E.[2], Kofoed, B.[1], Bak, C.K.[1] and Saermark, K.[1]

[1]Department for Physics, The Technical University of Denmark, DK-2800 Lyngby Denmark; [2]Gentofte University Hospital, DK-2800 Hellerup, Denmark

Introduction

Four patients, all suffering from pronounced ventricular extrasystoles (VES), were examined by means of magnetocardiography (MCG), with the aim to perform a source localization for the pathological site(s). Here we in particular wish to report an observation on a difference in behaviour of the signal as recorded in a normal-beat/abnormal-beat time segment relative to the signal as observed in a normal-beat/normal-beat time segment, both segments beginning at the position of an R-peak. The difference signal, itself, begins at the end of the S-wave, i.e. more than 200 msec prior to the earliest start of VES as seen in the EEG.

Methods

The measurements were performed by means of a 7-channel second order SQUID-system equipped with a probe-position-indicator (PPI) system. All measurements were taken in an unshielded room well separated from external magnetic noise sources. For each patient the MCG was recorded in a total of 100 or more channel positions, each recording consisting of 90 sec of data with a sampling time of 1 msec and a hard-ware high pass frequency of 0.3 Hz. The spatial location of the channel positions were determined by means of the PPI-system in a coordinate system with the y-axis along the breast bone (positive direction: inferior-superior) and the x-axis pointing to the left side of the patient. The z-axis, oriented outwardly, is normal to the chest and (x,y,z)=(0,0,0) is at the midpoint of the breast bone. In addition, two electrocardiographic leads, V1 and V6, were simultaneously recorded. We used a template technique to perform an averaging of the abnormal beats and the immediately preceding normal beat. The template had a width of 250 msec and was centred on one abnormal beat as observed in the lead V1 in the very first recording.

Results

For all four patients the recordings showed a very regular appearance of VES'. For two patients (YL and UA), in particular, the regularity was so pronounced that the recordings had the character of an alternating series of normal/abnormal beats. For YL we illustrate this in Fig. 1 (top panel) where we show simultaneous recordings for the V1-lead (uppermost curve, arbitrary units) and a magnetic channel (units fT). For this patient 2-3 consecutive normal beats could be found only once in each of about half of the total number of recordings, i.e. in less than approx. 5% of the total measuring time. We illustrate one such case in Fig. 1 (lower panel) where we show superposed simultaneous recordings for the V1-lead (arbitrary units) and a magnetic channel (units fT). For the patient UA the occurrence of 2-3 consecutive normal beats were even more rare, in fact they only appeared once or twice in each of 21 recordings out of a total of 140 recordings.

From Fig. 1 it is visible that there is a clear tendency to a time-locking of the abnormal beat to the immediately preceding normal beat. A closer examination of the template-averaging of the recordings for the V1 and V6 leads, not shown here, reveals that for the patient YL the time-locking is almost perfect - in the sense, that there is a vanishingly small scattering in the time separation between the abnormal peak and the preceding normal peak - whereas for the patient UA the time-locking is still present, but not to quite the same degree as for YL. This time-locking motivates a comparison of the signal observed in a normal-beat/normal-beat interval with the one observed in a normal-beat/abnormal-beat interval. In Fig. 2 (top panel) we show for a time segment of 500 msec beginning at the R-peak (t=0) the behaviour for the signal observed, respectively, in a normal-normal and a normal-abnormal interval. One notes that

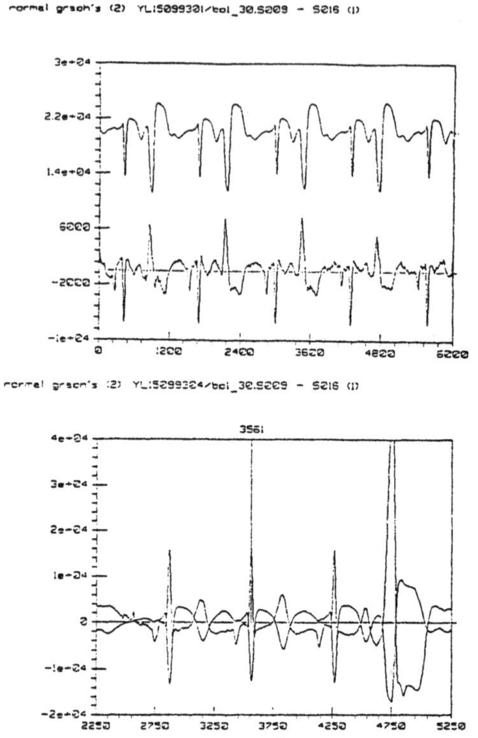

Fig.1 Top panel: Upper trace, recording from the V1 lead (arbitrary units). Lower trace, recording from a magnetic channel (units fT).
Bottom panel: Superposition's from a V1 lead (arbitrary units) and a magnetic channel (units fT).Both time axes in msec.

beginning at a latency of 30-40 msec there is a rather abrupt and well-defined deviation of the two signals from each other until - at a latency between 300 and 400 msec - the large abnormal peak of opposite sign set in. Clearly, the signal in the normal-abnormal time interval is suppressed relative to the signal in the normal-normal time interval. In Fig. 2 (lower panel) we show the difference between the above-mentioned two signals for 7 magnetic channels for each of the two patients YL and UA. The anatomical position for the 7 magnetic channels is nearly the same for the two patients and close to the midpoint of the breast bone. For both patients one observes an almost identical time course of the difference signal except, possibly, for a slightly different onset time for the large abnormal peak. The precise value of the latency for the onset of the abnormal peak is of course of interest, but is difficult to estimate from Fig. 2. To elucidate this point we show in Fig. 3 the spatial variation of the experimentally measured fields for the normal-normal time interval (left column) and for the normal-abnormal time interval (right column) and for the two latencies 280 msec (upper row) and 310 msec (lower row). For the latency 280 msec (and, in fact, for all lower latencies) the spatial field distribution is topologically almost the same in the two cases (Fig. 3 upper row) whereas for the latency 310 msec (and higher latencies) the field distribution is vastly different in the two cases (Fig. 3, lower row). This could indicate a beginning of the abnormal beat at a latency slightly larger than 300 msec.

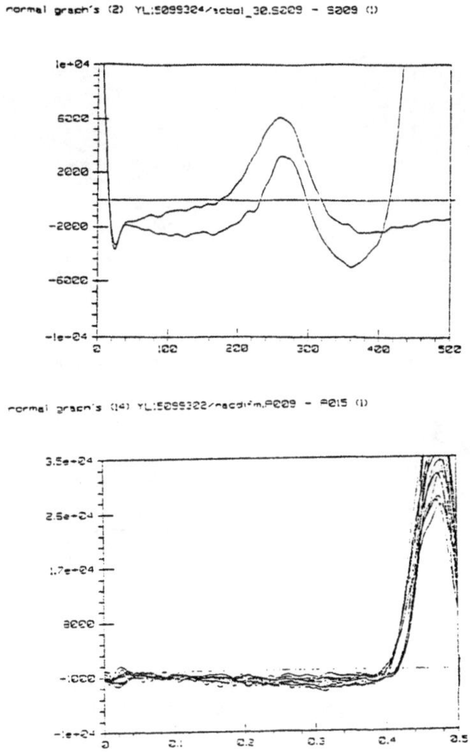

Fig.2 Top panel: Experimental recording from a magnetic channel in a normal-beat/normal-beat time interval (upper curve) and in a normal-beat/abnormal-beat time interval (lower curve). Time axis in msec, t=0 indicating the time position of the preceding R-peak. Units fT.
Lower Panel: Superposition of 7 difference signals (magnetic fields) for each of the two patients YL and UA. Time axis in sec, field values in fT.

We have applied the MUSIC-algorithm [1] in an attempt to localize the source(s) for the abnormal signal, but have - as yet - only a preliminary result to quote. Firstly, we note that for the patient UA the difference signal showed a clear polarity reversal between two groups of measuring channels (each group consisting of 7 magnetic channels), which allows for a rough estimate of a source location in the x,y-plane, but not as regards the position in the z-direction. This "source position" was 2-3 cm to the left of the breast bone and 1-2 cm below the midpoint of the breast bone. Next, the Music-algorithm was applied to a time interval of 40 msec centred at the peak of the abnormal signal for the patient YL. A contour plot of the cost-function showed well-defined peaks of the cost-function all lying close to a plane z=-4, i.e. 4 cm below the surface of the chest. Thus, two source regions were found at the positions: (x,y)=(9.5, 1.5) and (x,y)=(2,-0.5); these two source regions were not found in a MUSIC analysis of the normal beat for YL.

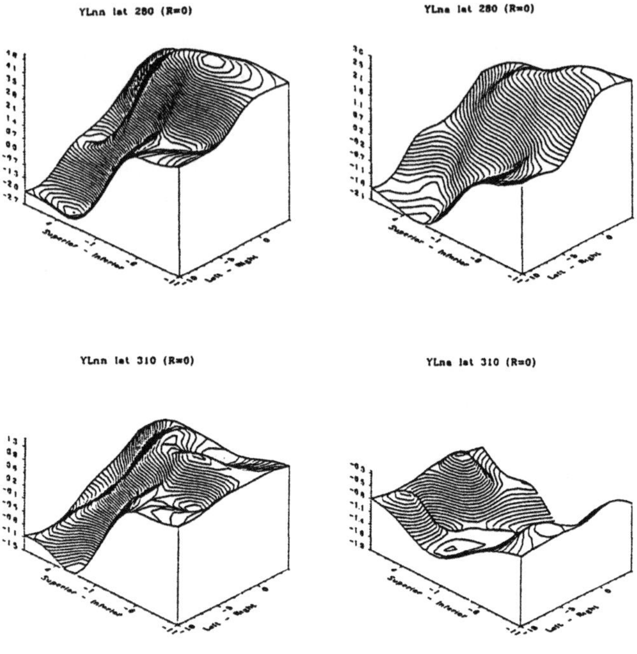

Fig.3 Spatial field distribution at two latencies in the normal-beat/normal-beat interval (left column) and in the normal-beat/abnormal-beat interval (right column). Vertical scale in arbitrary, but fixed units. Left-right indicates the patients left/right side.

Conclusions

On the basis of the above results we conclude that of the four patients examined, all suffering from pronounced VES', two showed a marked time-locking of the abnormal beat to the preceding normal beat; the remaining two patients did not show this time-locking ; one of them showed parasystoli, the other several morphologically different types of abnormal signals. Due to the time-locking it was possible in a consistent way to deduce a difference signal which begins 30-40 msec after the R-peak of the preceding normal beat and which at a latency slightly larger than 300 msec (see Fig. 2) end up with a transition to the large anomalous peak of the VES of opposite sign indicating a current reversal. We regard this as being in agreement with the "usual" explanation of VES in terms of a re-entry mechanism with a block occurring 30-40 msec after the preceding R-peak, thus causing a changed current path, followed by a release of the block 300 -325 msec later allowing for a reversed current direction. Finally the MUSIC algorithm was applied to a preliminary source localization.

References

[1] Mosher,J.C.,Lewis,P.S.,Leahy,R.M. IEEE Trans.Biomed.Eng. 1992,39; 541-557

Acknowledgements

We are grateful to the Danish Natural Science Research Council for financial support.

The Analysis of Magnetocardiosignal at Switching of Heart from Rhythm to Rhythm: The Preliminary Researches

Primin, M., Nedayvoda, I. and Gapelyuk, A.

The Institute of Cybernetics, Kiev, Ukraine

1. INTRODUCTION.

At present the level of knowledge in areas of biophysics permits to identify the parameters of electrodynamic models of heart (dipole, quadrupole, double layer of charges and etc.) on magnetic measurements in the air. After it the group of parameters sensitive to changes, previous to switching from one rithm to another,is marked and hypothesis about possible gears of this change is built.And in some cases (WPW, extrasystole with ektopic focus, some types of tachicardia) - localization of arithmogenic substrates in miocard is carried out.The known methods of cardiomagnetic researches assume the "work" with one, as a rule, averaged cardiocycle irrespective of the fact, which cardiocycle preceded to it and which followed it. After it the group of parameters sensitive to changes, previous to switching from one rithm to another,is marked and hypothesis about possible gears of this change is built.And in some cases (WPW, extrasystole with ektopic focus, some types of tachicardia) - localization of arithmogenic substrates in miocard is carried out.The known methods of cardiomagnetic researches assume the "work" with one, as a rule, averaged cardiocycle irrespective of the fact, which cardiocycle preceded to it and which followed it. The information about temporary structure of magnetocardiographic data is then lost and the complexes are separate investigated or all sinus, or all extrasystolic, but not interrelation of them. In the report we present some results of research of this interrelation with the purpose of investigation of changes in the electromagnetic activity of the heart, which precede the moments of rithm switching.

2. DATA REGISTRATION AND PROCESSING.

For research of magnetic field of the human heart the system,based on the single-channel gradiometer, was used.This permits:

- to execute noncontact measurements of magnetic field parameters by the second order squid-gradiometer $\partial^2 B_z/\partial z^2$ in the points of standard grid, located above the patient chest,under occurance of industrial noises (without application of shielding);
- to enter synchronously magnetocardiosignal (MCS) and electrocardiosignal (ECS) to computer;
- to adaptive filtering,to average and to allocate in arbitrary order and combination (in pairs) magnetocardiosignal;
- to interpolate the averaged data and to construct the "instant" maps of the field distribution above surface of the patient chest in the measurement plane ;
- to locate the source ofbiomagnetic signal by analytical methods (to solve the inverse problem) and to construct the trajectory of its displacement [1];
- in the plane, parallel plane of measurements and located on "occurance" depth of "effective" source (is found on previous step),to find the distribution of density of current, simulated the measured magnetic field [2];
- to execute the spatial analysis of magnetic field of heart, using the algorithm [3] without solution of inverse problem.

Having the opportunity to allocate from measurements data magnetocardiosignal, appropriated to two cardiocycles, and after making their analyses it is possible to be revealed some features,which precede switching of heart from rithm to rithm.On fig. 1-4 are adduced the preliminary results of the spatial analyses of magnetocardiosignal for patient with extrasystole.The analysis of measurements data is executed for moment of time $t_0,\ldots, t_N$, appropriate with ST-segment of averaged electrocardiocycle. On the fig. 1 we shown the measured distribution of parameters of magnetic field of heart in air for two normal cardiocycle before (top series of staff) and afterwards (third series of staff on fig. 1) extrasystole.Data processing on fig. 1 are executed for the same moment of time by algorithm [3] .

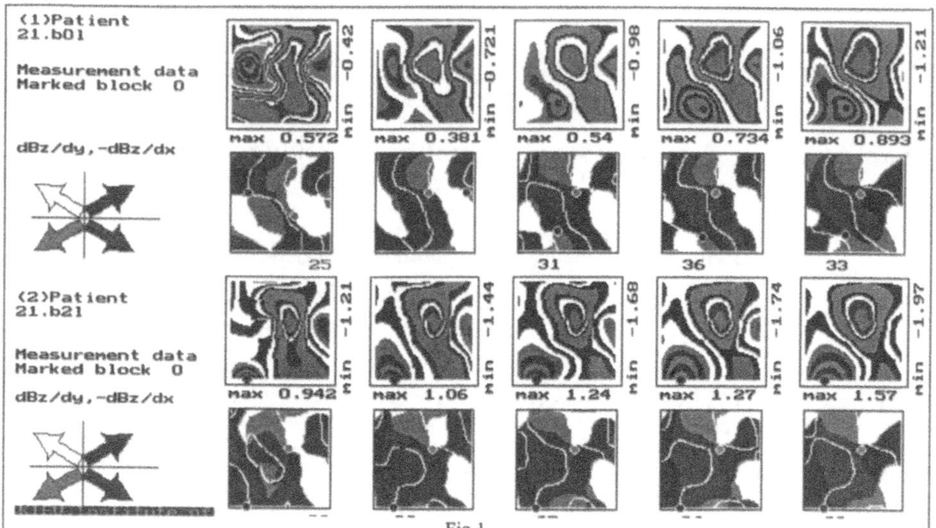

Fig.1

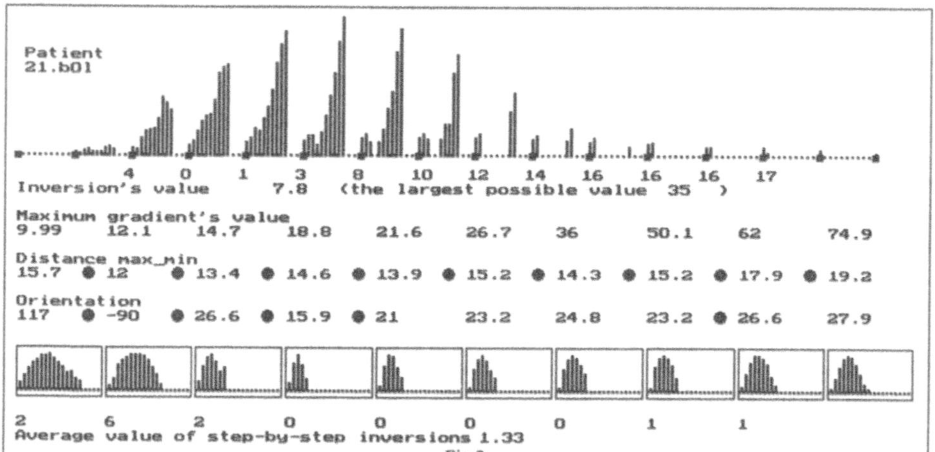

Fig.2

For the same moment of time on fig. 2 and fig. 3 we constructed the spatial-temporary spectrum of magnetocardiosignal and dispayed the factor of inversions for each point of spectrum and the averaged factor of inversions; the maximal significance of gradient's of field along section between point A and O (point of intersection of isolines $sx=\partial s/\partial x=0$, $sy=\partial s/\partial y=0$, where the measured size s accepts the maximal and minimum significance); the significance of angle Teta between normal vector to the section OA and axis OX of measuring lattice; the spatial spectra $g(t_0)$, $g(t_1)$,..., $g(t_N)$ with account of step-by-step factors of inversions. On fig. 4 the results of magnetostatic inverse problem solution are shown in the view of spatial configuration of effective system of currents with density $\mathbf{j}=(j_x, j_y, 0)$. On this figure three measured distributions of parameter d_2 [1] are adduced in the view of "map" of magnetic field.

The left and right maps (distribution) correspond to the normal cardiocycles before and afterwards extrasystolic one, the central map (distribution) correspond to extrasystole. The location of the investigated time moment for these figures was selected as the beginning of Q-wave for normal cardiocycle and the corresponded time moment for extrasystole. The magnetocardiosignal data processing on fig. 4 is executed on algorithm [2].

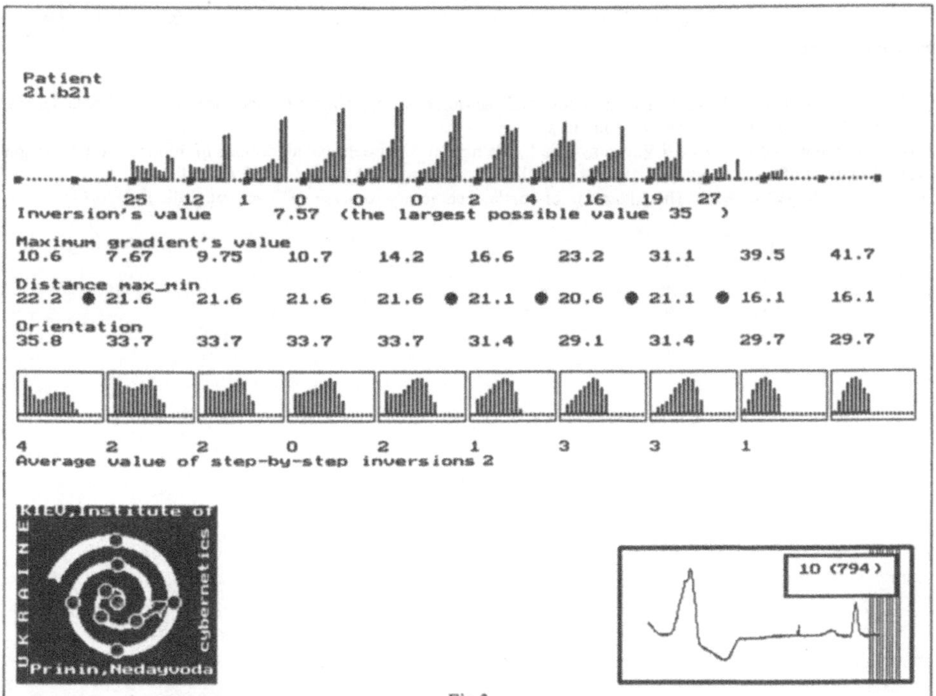

Fig.3

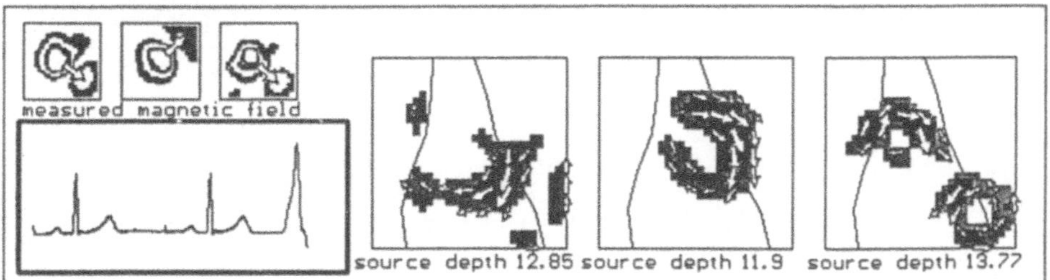

Fig.4

3. DISCUSSION.

We understand the complexities connected with interpretation of data processing of magnetocardiosignal. But even the visual analysis of results, submitted on fig. 1-4, shows, that magnetometric research would be more informative, if the data analysis is executed for time appropriate, for example,with two cardiocycles.

The set of parameters of biomagnetic field, which was used on this stage of researches for spatial analysis of magnetocardiosignal, is result of preliminary approbation of described technique in conditions of clinic in cooperation with Biomagnetism Laboratory (Kiev,Sosnitsky V.) and requires the further refinements.

The research described in this publication was made possible in part by Grant No K1 U100 from the Joint Fund of the Government of Ukraine and International Science Foundation.

REFERENCES.

1. Primin M., Nedayvoda I. Mathematical model and measurement algorithms for the location of a dipole source, Int.J.Applied Electromagn. In Materials (in press).

2. Primin M.,Gumeniuk-Sychevskij V.,Nedayvoda I., Mathematical models and algorithms of information conversion in spatial analysis of weak magnetic fields,Int.J.Applied Electromagn.in Materials 5 (1994),311-319.

3. Nedayvoda I.V.,Primin M.A. The Algorithm of Spatial-Temporary Analysis of Cardiomagnetic Signals (this issue).

Vector Fetal Magnetocardiography

Rassi, D.[1], Zhuravlev, Y.E.[1,3], Lewis, M.J.[1], and Emery, S.J.[2]

University of Wales[1], Swansea, UK; Singleton Hospital[2], Swansea, UK; Institute of Radioengineering and Electronics[3], Moscow, Russia

Introduction

Conventionally, the electrical activity of the human heart has been investigated using electrocardiography (ECG). An electric current dipole or 'double-layer' model has been widely used, and a plot of the dynamic variation of the magnitude and orientation of this 'equivalent dipole' throughout the cardiac cycle is known as a vectorcardiogram or 'dipole loop'. The vectorcardiogram may also be obtained from measurement of the cardiac magnetic field, for which both current dipole and magnetic dipole models have been considered. There have been very few reports in the literature regarding the equivalent cardiac current generator in the fetal heart. Those fetal vectorcardiographic investigations which have been described have involved measurement of the abdominal fetal electrocardiogram (FECG), with the analysis being mainly carried out in terms of the current dipole [1].

A major problem in abdominal fetal electrocardiography during the second half of gestation is the distortion of the signal caused by the electrically insulating layer of *vernix caseosa* which is formed around the fetal body. Because of this, the shape of the FECG waveform changes unpredictably with the position and orientation of the fetus, and it has been shown that conductive pathways have an increasingly important effect on the signals detected by surface electrodes [2]. Magnetic fields, on the other hand, are much less affected by conductivity variations and high-quality fetal magnetocardiogram (FMCG) recordings can be obtained from the 23rd week of gestation onwards [3]. The problem, however, remains of the variable fetal position and orientation with respect to the detector. Vector magnetocardiography, in which the rotation of the cardiac vector is reconstructed from independent measurements of its components, can provide a position-independent waveform of the heart's electromagnetic activity. Several lead systems /detector coil configurations have been suggested for vector cardiographic recording, e.g. the unipositional lead system [4]. The fetal heart is much smaller than its distance from the maternal abdominal surface and, therefore, a single current dipole model holds to a good approximation. As a result, a simple three-channel measurement system is sufficient to reconstruct the fetal heart vector.

Methods

In this work we report vector FMCG measurements using a SQUID system [5] equipped with three axial second-order gradiometers of 1.5 cm diameter and 6 cm baseline. The gradiometers were positioned at the vertices of an equilateral triangle of 3 cm (or 6 cm) side although it was not possible to change between these two alternative arrangements during measurements. The field sensitivity of the system was better than 15 $fT/Hz^{1/2}$ and all measurements were performed in a normal magnetically unshielded laboratory. Six normal subjects with gestational ages of between 29 and 38 weeks were investigated. Prior to FMCG measurements, the location of the fetal heart was determined using ultrasonography and this allowed the system to be positioned at minimum source-to-detector distance.

Computer software was developed to calculate the cardiac electrical generators (i.e. dipole vectors) consistent with the measured magnetic field data. The measurement system configuration was accurately modelled and algorithmic implementation of the reciprocity theorem [6] and lead field analysis [7,8] was carried out. Consecutive calculations of the current and magnetic dipole vectors were made at 2 ms intervals. We attempted to minimise the dipole reconstruction error associated with the inherent uncertainty in source-detector distance (which was adjusted within the range of anatomically realistic values), and applying a set of 'compatibility' criteria to the amplitude of the vector and the structure of the loops.

Results

Fetal magnetocardiographic recordings were obtained, with a typical signal-to-noise ratio of 5 (R-wave), using the above three-channel SQUID system. Lead field theory involves analysis of the sensitivity of a detector coil to the position and orientation of a current density distribution. In the present case, this constrains the reconstructed current dipole vector loop to a single plane (perpendicular to the axis of the gradiometers). A different approach, based on the reciprocity theorem, allows the reconstruction of the magnetic dipole vector loop in three-dimensional space.

Fetal vectorcardiographic measurements carried out external to the maternal abdomen yield data in the laboratory (or maternal) reference frame. No quantitative information can be derived from vector loops in the maternal reference frame as the orientation of the fetus *in utero* is variable. In the absence of accurate information (e.g. from ultrasound or MRI scans) regarding fetal position and orientation within the uterus, the 'maximum area reference frame' has been used to provide a standard reference frame for inter-individual vector loop comparison [1]. This approach is also adopted in the present work.

Fig. 1 presents typical results for a 37-week old fetus, showing the 100-beat averaged QRS complexes detected by the three gradiometers (a), and the reconstructed current (b) and magnetic (c) dipole vector loops. The current dipole vector loop is presented in the x-y plane of the maternal reference frame (which is parallel to the measurement plane i.e. perpendicular to the gradiometer axis), with the y-axis being the long axis of the body and the x-axis being the left-to-right direction. The three projections of the magnetic dipole vector loop (which was reconstructed in 3D) shown in Fig. 1(c) correspond to the 'maximum area' reference frame (in which the x'-y' plane contains the maximum area projection of the 3-dimensional vector loop). The 'x' and 'o' symbols in Fig.1(b) and Fig.1(c), respectively, denote the starting location of the vector and the arrows indicate its sense of rotation.

Discussion

The reconstructed fetal cardiac vector loops based on electric potential measurements (abdominal FECG) during late pregnancy are severely distorted due to the preferred conduction pathways (e.g. the mouth and the umbilical cord) between the fetus and the maternal abdomen. The open structure of the current dipole and magnetic dipole loops observed in this study for subjects between the 29th and 38th week of gestation demonstrates that the FMCG technique overcomes this difficulty and can therefore be used to obtain the cardiac signals in the fetal reference frame. It remains to be determined whether the current dipole, the magnetic dipole or a combination of these two yields the most clinically useful information.

Although a magnetic dipole located at the centre of the heart is consistent with the observed magnetic flux measurements, the physiological interpretation of such a model is not clear since a physical description of the distributed volume current within the myocardium in terms of magnetic dipoles is not straightforward. The current dipole, on the other hand, would appear to be a more realistic model for the propagating wavefront of electrical activity during the cardiac cycle. Previous clinical studies have analysed the relationship between the electric and magnetic heart vectors for the adult [9]. It has been demonstrated that the two dipole vector loops for the adult are approximately orthogonal to each other [10], and this was reported to be consistent with the double-layer model of the cardiac excitation wavefront. Our results show a similar orthogonal relationship between the electric and magnetic dipole loops of the fetal heart during the ventricular depolarisation phase. It is known that for the adult this relationship is severely distorted by ventricular conduction defects [11] and it may therefore be possible to perform similar diagnostics on the electrical conduction system of the fetal heart.

In general, the quantification of a dipole (of known location) requires three independent magnetic field measurements. In our present SQUID system the three detection coils are in the same plane which is sufficient for the reconstruction of a magnetic dipole. For the three-dimensional reconstruction of the current dipole loop, however, at least one of the three detection coils must be in a plane orthogonal to the others. Our present detector geometry produces a two-dimensional projection of the current dipole loop. By repeating the measurement at two different angles we were able to derive some information about the three-dimensional behaviour of the current dipole loop although it is recognised that more accurate results can be achieved by using a more suitable detector geometry.

For all the subjects investigated, both current and magnetic dipole vector loops exhibited open (i.e. near circular) structure. This result is in marked contrast to vector FECG results in late (i.e. after 28 weeks) gestation

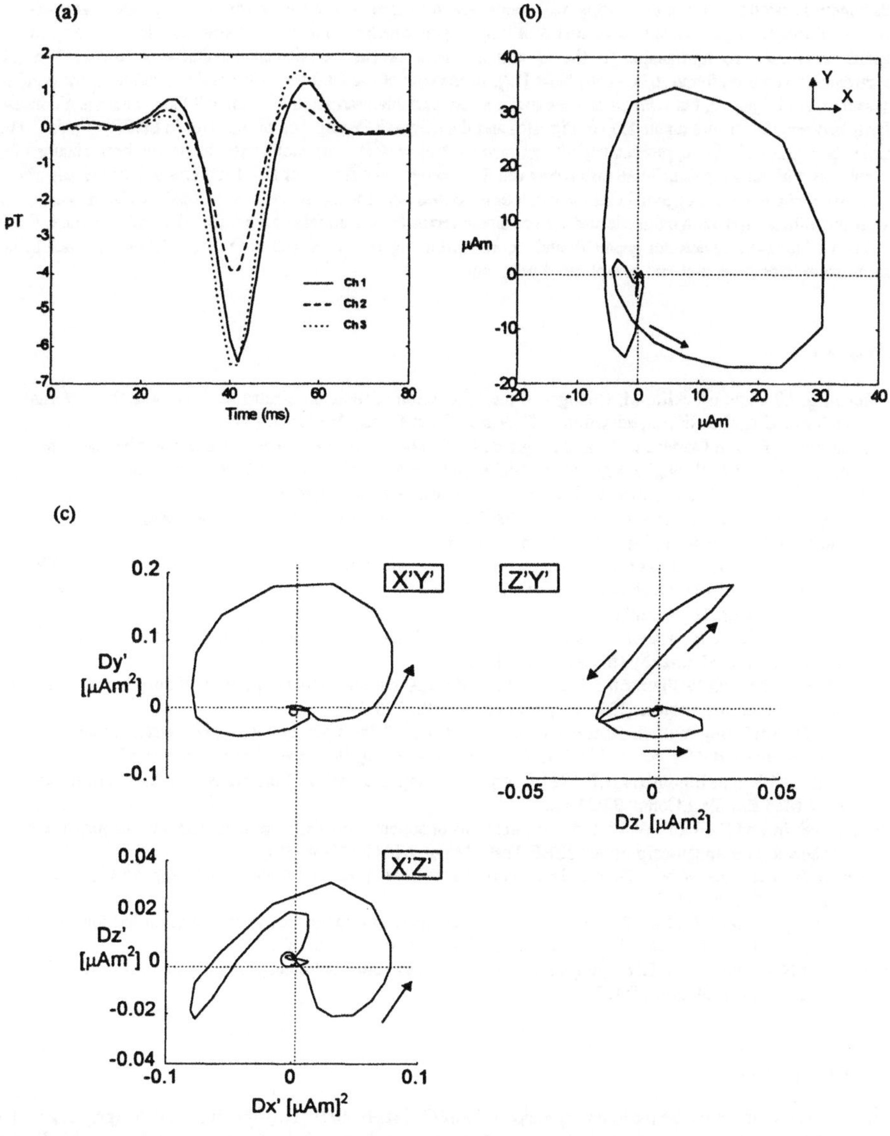

Fig. 1 (a) Averaged FMCG curves (80 ms segment centred on R-wave) measured from a 37-week old fetus; (b) the current dipole vector loop in the maternal reference frame reconstructed from these data; (c) the magnetic dipole vector loop reconstructed using the same data, presented in the three orthogonal planes of the maximum area reference frame.

which generally produce closed (i.e. elongated) loops which cannot be used for position-independent analysis. Our reconstructed dipole loops generally consisted of a larger open structure, and a secondary 'tail' loop which was often elongated. This may be attributable to the Q and S waves of the ventricular depolarisation phase. The adult vectorcardiogam has been found to be near-planar [12]. In the case of the fetal vector loops discussed here, the dominant part was found to be planar, but with an out-of-plane portion possibly associated with the tail loop. There is a noticeable similarity between the current dipole loop of Fig. 1(b) and the magnetic dipole loop of Fig. 1(c) in the X'-Y' plane. There appears to be a phase shift of approximately 90° between the two vectors, with similar results having been obtained from all the subjects, although the exact relationship between the two reference frames of Fig. 1 requires further investigation.

In conclusion, it is suggested that the methods described here for the analysis of the fetal cardiac magnetic fields allow the quantification of both magnetic and current dipole vector loops throughout the second-half of pregnancy. Further work is needed to optimise detector geometry and reconstruction algorithm, as well as to assess the clinical usefulness of the information in both normal and complicated pregnancy.

References

[1] Oldenburg, J.T., and Macklin, M. Changes in the conduction of the fetal electrocardiogram to the maternal abdominal surface during gestation 1977 Am.J.Obstet.Gynecol. 129:425-433.

[2] Oostendorp, T.F., van Oosterom, A. and Jongsma, H.W. The effect of changes in the conductive medium on the fetal ECG throughout gestation 1989 Clin.Phys.Physiol.Meas. 10(Suppl. B):11-20.

[3] Zhuravlev, Y., Rassi, D., and Emery, S. Clinical assessment of fetal magnetocardiography In (eds. C.Baumgartner *et al.*) Biomagnetism: Fundamental Research and Clinical Applications, Amsterdam, Elsevier/IOS-press, 1995. pp700-703.

[4] Malmivuo J. On the detection of the magnetic heart vector- an application of the reciprocity theorem 1976 PhD Thesis, Acta Polytechnica Scandinavica, Electrical Engineering Series 39, Finnish Academy of Technical Sciences, Helsinki.

[5] Rassi D., Zhuravlev Y.E., Josephs O. and Swithenby S.J. A six-channel SQUID biomagnetometer optimised for fetal magnetocardiography (in this volume).

[6] Bastuscheck C.M. and Williamson S.J. Technique for measuring the ac susceptibility of portions of the human body or other large objects 1985 J.Appl.Phys. 58(10):3896-3906.

[7] Tesche C.D. Analyzing multichannel magnetometer data 1992 In Superconducting Devices and their Applications (ed. H. Koch and H. Lübbig) (Springer-Verlag, Berlin Heidelberg) pp551-556

[8] Hämäläinen M.S. and Ilmoniemi R.J. 1994 Interpreting magnetic fields of the brain: minimum norm estimates 1994 Med.Biol.Eng.Comp. 32:35-42.

[9] Wikswo J.P.Jr. and Fairbank W.M. 1977 Applications of superconducting magnetometers to the measurement of the vector magnetocardiogram IEEE Trans.Magn. MAG-13:354-357.

[10] Barry W.H., Fairbank W.M., Harrison D.C., *et al.* Measurement of the human magnetic heart vector 1977 Science 198:1159-1162.

[11] Leifer M. C., Griffin J. C., Iufer E. J., *et al.* An integrated system for magnetic assessment of cardiac function 1981 In: Erné S. N., Hahlbohm H-D., Lübbig H., eds. Biomagnetism. Berlin: DeGruyter, pp123-37.

[12] Pipberger H.V. and Carter T.N. Analysis of the normal and abnormal electrocardiogram in its own reference frame 1962 Circulation 25:827.

Acknowledgments

The authors wish to acknowledge the following support: EPSRC (Grant GR/J72110), the Royal Society (project grant), the University of Wales, Swansea (studentship MJL) and the School of Postgraduate Studies in Medical and Health Care, Swansea (provision of nursing support).

Biomagnetic Evidence for Extreme Anisotropy - A Mechanism Limiting Conduction Capability in Accessory Pathways

Schirdewan, A.[1], Meyerfeldt, U.[1], Wiedemann, M.[1], Schütt, H.[3], Burghoff, M.[2], Gapelyuk, A.[4], Copetti, C.A.[3] and Dietz, R.[1]

[1]*Franz-Volhard Hospital Humboldt-University, Berlin, Germany;* [2]*Physikalisch-Technische-Bundesanstalt, Berlin, Germany;* [3]*Institut für Schicht- und lonentechnik, Jülich, Germany;* [4]*Institute of Cybernetics, Kiev, Ukraine*

Introduction

The conduction capability of accessory atrioventricular pathways (AP) varies widely among patients. In 1971 de la Fuenta, Sasyniuk and Moe postulated the existence of accessory pathways that are capable only of retrograde conduction. The authors concluded that the unidirectional conduction in concealed AP is caused by an "impedance mismatch" at the interface between the AP and the ventricular myocardium. Kuck et al. demonstrated in 1990 by direct registration of the AP-potentials, that the anterograde block in concealed pathways occurs at or near the ventricular interface. However, impedance mismatch alone does not seem to account for all patterns of conduction in APs. Additional factors which could explain all limitations in AP conduction are dynamic interactions between interconnecting fibers in left sided AP's and anisotropy.

The aim of this study was to investigate the conduction patterns in concealed pathways using the biomagnetic method in order to test the ability of MCG to localize the site of ventricular insertion and to get further information about the mechanism of limiting anterograde conduction in AP's.

Methods

Sixteen patients with typical atrioventricular tachycardias without signs of ventricular preexcitation during sinus rhythm (SR) in the 12 lead ECG underwent noninvasive biomagnetic mapping during SR prior to electrophysiological investigation and catheter ablation by radiofrequency. The site of successful catheter ablation using the retrograde left ventricular approach was by definition the site of ventricular insertion of the APs.

The biomagnetic measurements were performed with a 37 multichannel SQUID system with a white noise level 5-8 ft/ Hz-1. Signals were recorded with sampling rate 1000 Hz with a band pass from 16 mHz-250Hz in 10 patients. These measurements were performed in the Berlin magnetic shielded room of the Physikalisch-Technische-Bundesanstalt. The biomagnetic mapping in the other 6 patients were performed in nonshielded environment in our hospital, using the single channel system Krioton, Krionit (Moscov).

We analyzed the magnetic fields in the time between magnitude of P-wave and magnitude of R-wave by calculating the electrical or magnetic current dipoles, generating the fields. A concealed accessory pathway was diagnosed by the criterion: a spatial jump in dipole distribution, corresponding with the site of the successful catheter ablation of the AP.

Results

Electrophysiological results

In the electrophysiological studies 17 AP's were found in 16 patients (pts). The site of the AP's were in 1/16 pts right posterolateral, in 15/16 pts left sided. The sites of ventricular insertion of AP's were determined in the 15/16 pts:

left posterior (2x), left posterolateral (3x), left lateral (10x), left midseptal (1x).
1 patient had 2 concealed accessory pathways: 1 left posterolateral and 1 left lateral AP.

All patients underwent successful ablation procedure by radiofrequency energy. There was no change in parameters of the surface ECG before and after the ablation therapy in 14/16 pts (PR-interval and QRS-duration). 2 pts developed a temporary bundle branch block after ablation.

Biomagnetic results

Fifteen AP's in 14 pts were detected by MCG. The MCG localized the site of the AP's correctly in 9 cases and in 5 cases near to the successful ablation site. The right-sided AP and the left midseptal AP were not detected by MCG. The characteristic magnetic pattern of concealed AP were:

1. The dipole distribution in the time of ventricular excitation was different from normal controls.

2. There was a jump directed to the location of the AP.

3. This AP related pattern was only existent for 4-12 ms.

4. After successful ablation procedure, the dipole distribution was normalized in 11/14 pts.

5. In the 14 pts, the localization of concealed AP was possible within the Q-wave but not at the time of onset of the ventricular excitation.

Using the biomagnetic localization approach in 14/15 patients with left-sided APs, magnetic fields were recorded at the end of the Q-wave in ECG. The current generators of these fields could be localized near or close to the successful ablation site. These magnetic fields exist only for a period of a few milliseconds (6-12 ms). The time of the earliest appearance of such fields, in relation to the intracardiac local electrogram, was in most cases directly before the onset of the local ventricular spike, suggesting that there was no preexcitation. In the follow up MCG measurements after successful ablation therapy, these magnetic fields disappeared, confirming the connection between concealed accessory pathways and recorded magnetic fields. Such fields were not recorded only in 2 patients (1 right sided and 1 left midseptal AP).

MCG prior to RF-ablation

At the end of the Q-wave, we recorded a distorted magnetic field pattern with a duration of 8-10 ms. The magnetic dipole sources correspond to the location of the concealed accessory bypass, estimated by invasive procedure.

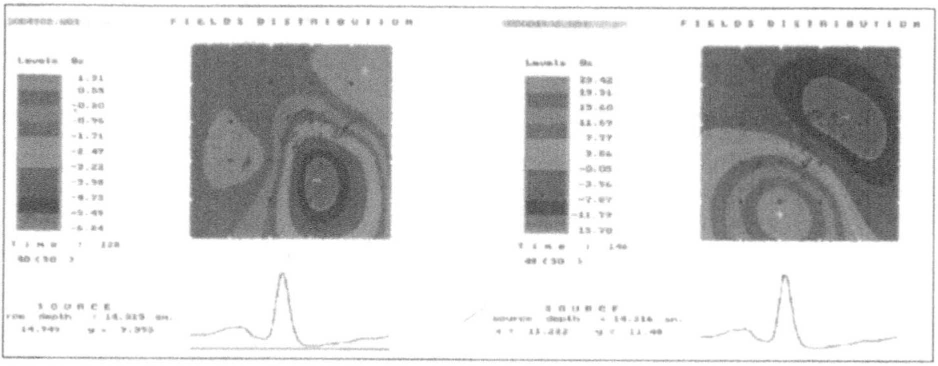

MCG after successful RF-ablation procedure

The magnetic field shows a normalized field pattern related to the physiological ventricular excitation. The magnetic field disappeared, which was corresponding to the concealed accessory pathway. The ECG (PR-interval, QRS-complex) and the invasive recorded intracardiac electrogram were not altered by the ablation procedure, suggesting there wasn't any ventricular preexcitation.

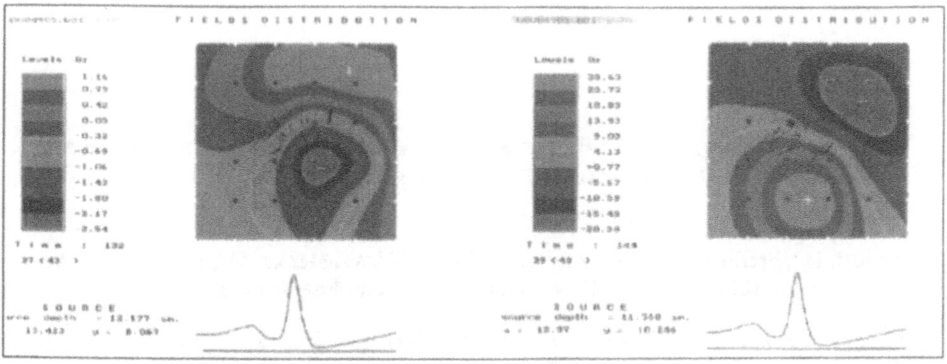

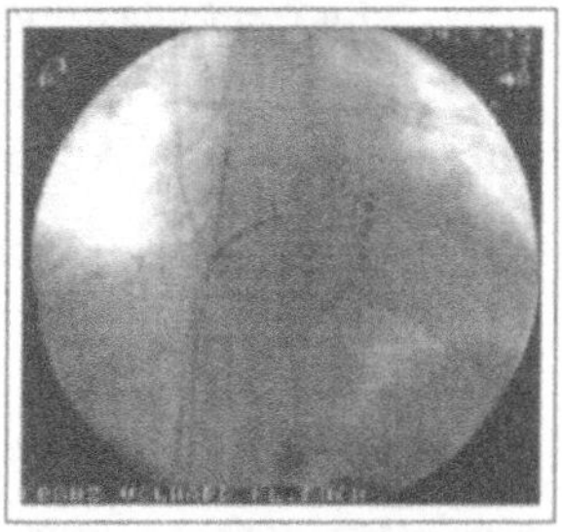

The X-ray shows the site of ventricular insertion of the concealed AP determined by successful RF-ablation. The ring with a diameter of 1 cm represents the magnetocardio-graphic localization of the same AP. The measured magnetic field related to the concealed AP dominates the field pattern.

Discussion and Conclusions

In the present study, we demonstrated that concealed accessory pathway can be successfully localized during SR by the biomagnetic method.

Left sided accessory pathways are composed of a network of myocardial fibers. At the ventricular sides, the bundles divided into smaller branches which continuously fused with the ventricular myocardium. In most cases, described in the literature, the smaller bundles were connected with the myocardium like the roots of a tree.

In contrast to left-sided accessory pathways, right-sided accessory pathways can be composed of cells typical of those found in specialized conduction tissue. The magnetocardiographic findings suggest that the site of the anterograde block of the accessory pathways is located within the region including roots of the Kent bundle and parts of the adjacent ventricular myocardium.

Because the AP related magnetic fields come into prominence in competition with the field produced by normal activation of ventricles in absence of preexcitation, the magnetic fields could not be generated by anterograde and/or retrograde conduction in the accessory pathway. It seems, that the magnetic fields are generated due to a more complex mechanism based on anterograde excitation of AP and additional influence of electrotonic changes during the onset of ventricular excitation. On the assumption of the anisotropic bidomain model, we can conclude that the measured magnetic fields are produced by extended current loops due to extreme unequal anisotropies. These loops can not be measured electrically. Only the mechanism of anisotropy can explain that we measure magnetic fields without electrical equivalent in surface and intracardiac ECG.

In the case with a right-sided accessory pathway, we could not record magnetic fields related to the bundle. We assume the Kent bundle is composed of specialized tissue and its functional behavior is like an electrical wire.

The precondition for magnetic fields generated by closed current loops due to extreme anisotropy is a myocardial network like it is realized in the left-sided accessory pathways itself and the junction to the surrounding ventricular myocardium.

Magnetocardiogram in 3 Patients After Aborted Sudden Cardiac Death Without Overt Underlying Heart Disease

Schütt, H.[1], Stellbrink, C.[2], Dammers, J.[3], Boers, F.[3], Wiedenau, G.[3], Ioannides, A.A.[3], Müller-Gärtner, H.W.[3], Gapelyuk, A.[4], and Schirdewan, A.[5]

Institute of Thin Film and Ion Technology (ISI) [1], Forschungszentrum Jülich GmbH (KFA), Jülich, Germany; Medical Clinic I [2], RWTH Aachen, Aachen, Germany; Institute of Medicine (IME) [3], Forschungszentrum Jülich GmbH (KFA), Jülich, Germany; Institute of Cybernetics [4], Kiev, Ukraine; Franz-Volhard-Clinic [5], Humboldt-University, Berlin, Germany

Introduction:

Sudden Cardiac Death (SCD) is an important public health problem. It is estimated that 350.000 people die suddenly every year in the USA [1]. Main cause of SCD are ventricular tachyarrhythmias [2]. Most patients with SCD suffer from coronary artery disease, but in some patients no organic heart disease can be detected using standard clinical methods (idiopathic ventricular fibrillation, IVF). In contrast to earlier observations [3], more recent studies have shown a significant recurrence rate of 25 % concerning major arrhythmic events and a 1-year mortality of 11 % in this patient group [4]. The development of therapeutic strategies for these patients is greatly dependent on the understanding of involved pathophysiological mechanisms.

Methods:

We analyzed the magnetocardiogram (MCG) of 3 patients after aborted SCD without overt underlying heart disease. All patients had normal coronary arteries, normal left and right ventricular cineangiograms and normal left ventricular ejection fractions. The results of the electrophysiological studies (EPS) and further patients characteristics are listed in Table 1.

Table 1: Patients characteristics

	patient 1	**patient 2**	**patient 3**
sex	male	male	male
age	31	43	55
aborted SCD	1993	1989	1990
result of EPS	VF inducible	VF inducible	no arrhythmia inducible
medication at time of MCG recording	none	amiodarone	metoprolol

Two different methods were used for magnetocardiographic analysis of the depolarization and repolarization process.

1. Depolarization process

MCGs were recorded using the MAGNES II system (BTI) of the Institute of Medicine, Forschungszentrum Jülich (shielded room: VAC AK3B). The upper dewar was located above the thorax of the patient in the supine position, the location of the central sensor was 12.5 cm below the manubrium sterni in the midline of the thorax with a distance thorax-dewar as small as possible (1-2 cm). MCG was recorded for 2 minutes and the signal was averaged ($\geq$ 100 beats).

For evaluation of the depolarization process, we used a nonrecursive binomial filter (37 - 90 Hz) to analyze the high frequency component of the central sensor's signal and calculated a fragmentation score of the QRS complex [5]. This procedure has been used previously in investigations of patients with coronary artery disease and normal subjects. Higher fragmentation scores (> 30) indicated disturbances in the depolarization process [6, 7]. We compared the results of our patients and one male healthy subject to this historical control group.

2. Repolarization process

For evaluation of the repolarization process, MCGs were recorded using a single-channel second order low-temperature gradiometer operating without shielding (Krioton / Krionit Moscow) at 36 measurement points (grid of 6 x 6 points with 4 cm distance) above the thorax of the investigated person in the supine position. Measurement point 1/3 (line/point) of this grid was located above the manubrium sterni. At each position, 30 heart cycles during sinus rhythm were recorded and averaged. Then the magnetic field (MF) distribution was computed.

The homogeneity of the MF distribution was analyzed between the end of the QRS complex and the T wave maximum. The duration of this ST-T segment was normalized to 32 time intervals, and the number of the MF extrema in the first and second part of this period was calculated.

We compared the results of our patients to 8 male healthy controls (age: 32.5 ± 6.1 years).

Results:

1. Depolarization process

Because of the nonplanar dewar construction of the MAGNES II system, there was an unavoidable distance between central sensor and thorax. Therefore, the amplitudes of the signals are relatively small (5-20 pT). The MCG signals of one healthy subject and our patients and their morphological differences are shown in Fig. 1.
However, the fragmentation scores of the patients and the healthy control were within a normal range of < 30 (25,4 - 28,0). This is in agreement with the findings of other authors in healthy volunteers [6, 7].

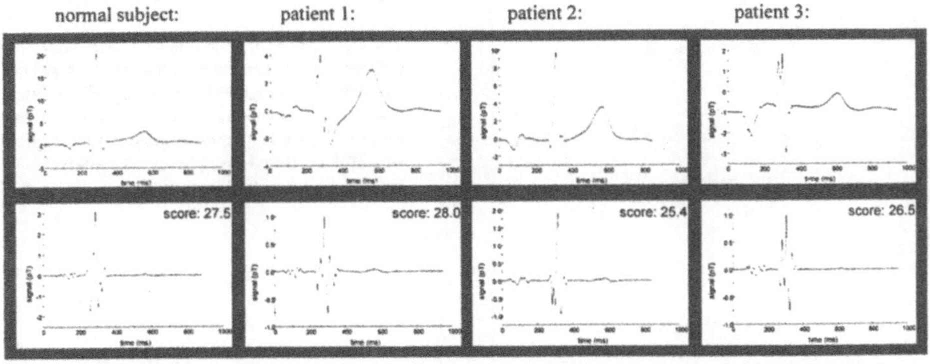

Fig. 1: MCG of a healthy subject (age: 29) and patients 1-3: averaged signal (upper panel) and filtered signal (lower panel) with fragmentation score. Note that fragmentation score is within normal range in all 4 subjects.

2. Repolarization process

MF distribution during repolarization was more inhomogeneous in patients with SCD compared to controls as shown by a larger number of extrema. This difference was statistically significant in the second part of repolarization, but should be interpreted with caution because of the small number of investigated subjects (see Table 2). Fig. 2 compares ECG and MCG signals of a healthy subject and patient 2, Fig. 3 and 4 compare the homogeneity of the repolarization process in this healthy subject and in patient 2.

Table 2: Number of MF extrema during repolarization period

	patient 1	patient 2	patient 3	patients mean (± variance)	8 controls mean (± variance)
number of extrema during 1. part of REP	2.66	3.08	3.82	3.19 (± 0.17)	2.69 (± 0.15)
number of extrema during 2. part of REP	2	3	3.16	2.72 * (± 0.2)	2.14 (± 0.02)

REP = repolarization period; * $p< 0.05$

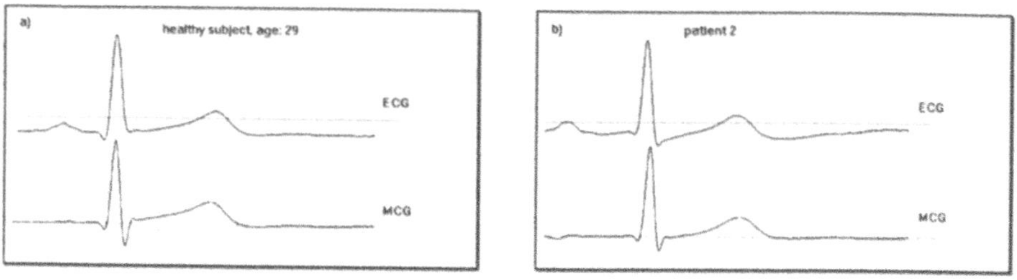

Fig. 2: ECG lead II (upper panel) and one selected MCG signal (lower panel) of a healthy subject (a) and patient 2 (b)

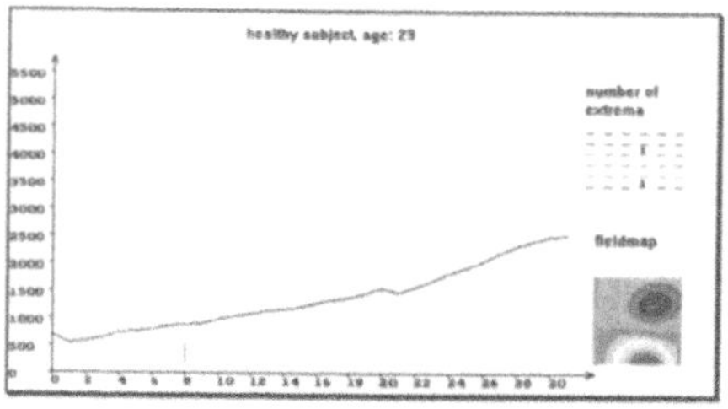

Fig. 3: MF distribution in a healthy subject

The curve represents the sum of all MF strength differences between the extrema positions and the corresponding neighboring points (y-axis) plotted against the time interval of repolarization (x-axis). The fieldmap and extrema displayed at the right side of the picture belong to the time interval indicated by the small vertical line.

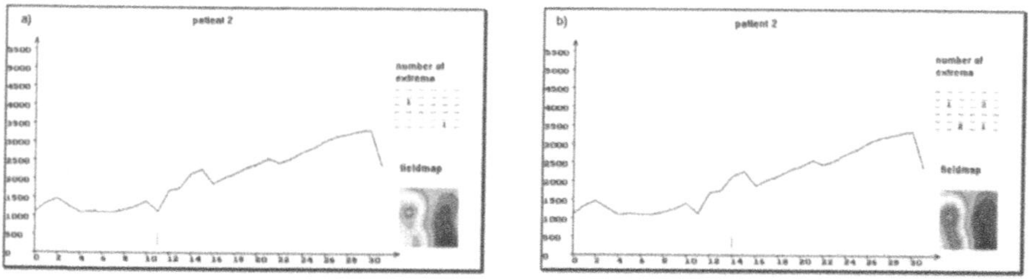

Fig. 4: MF distribution in patient 2. Parts a) and b) represent fieldmaps taken at 2 different time intervals. Note that in patient 2 there is 1) a higher sum of MF strength differences between the extrema po sitions and the corresponding neighboring points at each time interval, 2) a greater inhomogeneity of MF distribution during repolarization as indicated by the fluctuations of the curve and 3) a varying number of MF extrema at different time intervals [2 extrema in a) and 4 extrema in b)].

Discussion:

Besides the possibility of identification and localization of arrhythmogenic tissues as for example accessory pathways in patients with Wolff-Parkinson-White syndrome [8], magnetocardiography is a noninvasive method for evaluation of the de- and repolarization process in patients with malignant ventricular arrhythmias. In patients with inducible ventricular tachycardias for example, ventricular late fields could be demonstrated [9].

However, not much is known about the mechanisms of SCD especially in patients without underlying heart disease. Josephson et al. [10] suggested that inhomogeneities in ventricular activation and recovery as well as shortened ventricular refractoriness could provide the unstable substrate needed for development of ventricular fibrillation.

We reported the first results of an ongoing study concerning possible mechanisms of SCD and have investigated 3 male patients after aborted SCD without organic heart disease.

In these patients with IVF, no abnormalities of depolarization using a QRS complex fragmentation score could be found. However, an inhomogeneous distribution of MF extrema during repolarization could be demonstrated in our patients. The higher number of MF extrema in patients provides some evidence for differences between healthy subjects and patients with IVF during repolarization.

This indicates that repolarization rather than depolarization abnormalities are possible mechanisms for SCD in this specific patient population. Further prospective studies in larger patient groups are needed to confirm these results and to define the role of MCG as a diagnostic tool in patients with SCD.

Acknowledgment:

We want to thank U. Steinhoff (Physikalisch Technische Bundesanstalt Berlin) for calculation of fragmentation scores.

References:

[1] Manolio, T.A., Furberg, C.D. Epidemiology of sudden cardiac death. In: Akhtar, M., Myerburg, R.J., and Ruskin, J.N. Sudden cardiac death: prevalence, mechanisms, and approaches to diagnosis and management. Philadelphia, Williams & Wilkins, 1994

[2] Kempf, F.C., Josephson, M.E. Cardiac arrest recorded on ambulatory electrocardiograms. American Journal of Cardiology, 1984, 53: 1577-1582

[3] Belhassen, B., Shapira, I., Shoshani, D., Paredes, A., Miller, H., Laniado, S. Idiopathic ventricular fibrillation: inducibility and beneficial effects of class I antiarrhythmic agents. Circulation, 1987, 75: 809-816

[4] Viskin, S., Belhassen, B. Idiopathic ventricular fibrillation. American Heart Journal, 1990, 120: 661-671

[5] Link, A., Trahms, L., Zimmermann, R., and Oeff, M. Complex binomial bandpass filters for analysing phase and envelope of high frequency components in cardiac signals. Computers in Cardiology, IEEE Computer society press 1994: 617-620

[6] Trahms, L., Hahlbohm, H.-D., Kreiseler, D., Zitzmann, H. Non-recursive filters for ECG and MCG. In: Trahms, L. Diagnostic Methods to Investigate Cardiac Arrhythmias: Lectures on the 114th PTB-Seminar. Braunschweig and Berlin, Wirtschaftsverlag NW Bremerhaven, 1995

[7] Endt, P., Hahlbohm, H.-D., Kreiseler, D., Link, A., Oeff, M., Steinhoff, U., Trahms, L., and Zitzmann, H. Ein Algorithmus zur Quantifizierung der Fragmentation des MKSs im QRS-Komplex. Biomedizinische Technik (Biomedical engineering), 1995, 40 suppl. 1: 279-280

[8] Mäkijärvi, M., Neonen, J., Leiniö, M. et al. Localization of accessory pathways in Wolff-Parkinson-White-syndrome by high resolution magnetocardiographic mapping. Journal of Electrocardiology, 1992, 25: 143-155

[9] Weismüller, P., Abraham-Fuchs, K., Killmann, R. et al. Magnetocardiography: three-dimensional localization of the origin of ventricular late fields in the signal averaged magnetocardiogram in patients with ventricular late potentials. European Heart Journal, 1993,14: 61-68

[10] Josephson, M.E., Spielman, S.R., Greenspan, A.M., and Horowitz, L.N. Mechanism of ventricular fibrillation in man: observation based on electrode catheter recordings. American Journal of Cardiology, 1979, 44:623-631

Magnetocardiography and Other Methods of Investigation in Diagnostics of Heart Remodelling After Myocardial Infarction and in Hypertension

Shabalin, A.V.[1], Nikitin, Y.P.[1], Golyshev, N.V.[2], Motorin, S.V.[2], Rogachevsky, B.M.[2], Kytmanov, A.V.[1], Ermakova, E.N.[1] and Tselmin, S.G.[1]

[1]*Institute of Internal Medicine, Siberian Branch of Russian Medical Academy, Russia;*
[2]*Novosibirsk State Technical University, Novosibirsk, Russia*

1. Introduction

Magnetocardiogram (MCG) and electrocardiogram (ECG) both originate from the current source produced by electromotive forces of the heart. But the MCG is expected to provide more detailed and accurate information. The aim of our work was to compare diagnostic value of non-invasive methods of investigation of the heart in patients with Ischemic Heart Disease (IHD) and arterial hypertension (AH):MCG, ECG and echocardiography (EchoCG).

2.Methods

44 patients with IHD (group 1) and 27 patients with AH (group 2) at the age 46-70 have been investigated.In the 1st group 22 patients were after myocardial infarction (MI).The control group consisted of 37 healthy males of the same age. Biomagnetic apparatus: the second order gradiometer with superconducting quantum interference device (SQUID) was designed and constructed at Novosibirsk State Engineering University without any shielding. [1] Calibration signal was 25 pT. 12 standard,36 precordial ECG leads and EchoCG were recorded.

3.Results

MCG patterns of healthy subjects were obtained. Amplitude characteristics of P,R,S,T waves of MCG at each recording position was studied in the control group. Detailed characteristics of these MCG elements were given earlier [2].

In the 1st group (IHD) the prevalence of pathological complexes QRS of qR type, Qr, QS was more often revealed by ECG in 36 leads (12%,P<0,02) and by MCG (9%,P<0,05) than by ECG in 12 standard leads (2%).Repolarization disturbances were more often revealed by MCG (6%,P<0.05) and by ECG in 36 leads (6%,P<0.05) than by ECG in 12 standard leads (3%).The signs of left ventricle hypertrophy (LVH) were found by standard ECG in 10 patients, by EchoCG-in 15 patients, and by MCG-in 14 patients.Left auricle hypertrophy (LAH) was revealed only by EchoCG in 15 patients.

Diagnostic value of ECG and MCG criteria in revealing LVH was calculated:sensitivity-47% and 87% correspondingly, specificity-90% and 96%, prognostic value in negative tests-76% and 93%, in positive tests-70% and 93% (EchoCG-standard).We also tried to reveal the relations between morphological changes of MCG wave (not found by MCG in control group) and the data of ultrasonic investigation. We revealed that according to EchoCG patients having ventricular septum hypertrophy (VSH) > 1.1 cm (32% of patients (14 persons) had also MI in anamnesis (in 13 patients of 14), 5 of them had transmural MI and in 7 patients MCG showed signs of LVH wich repolarization disturbances in low leads. (Fig. 1) . With increasing thickness of the back wall of left ventricle (more then 1.1 cm), which was revealed by EchoCG in 4 patients, 3 of them had MI in anamnesis, 2 of them had transmural MI, MCG stated the signs of LVH. When hypokinesis of the back wall of LV in 10 patients (23%), (5 of them had transmural MI, and all the rest had MI without Q of the back wall) MCG showed 2 variants of changes. The first variant: 7 patients had segment ST depression and negative T-wave in the left low leads (E, F - 2~4). The second variant: 3 patients had low-amplitude ventricle complexes. For hypokinesis of the front wall of LV, MCG revealed low-amplitude complexes QRS (3 of them had transmural MI). In 9 patients (20%) EchoCG stated hyper- and/or akinesis of VSH for patients with MI in anamnesis. MCG showed negative T wave in left low leads in 5 patients (E,F - 3.4), and in 2 patients signes of LVH combined with repolarization disturbances were found.

In the 2d group (AH) the signs of LVH were revealed by EchoCG in 18 patients, by MCG-in 20 patients, that is more often (P<0,05) than by standard ECG (in 8 patients). Ventricle repolarization disturbances were more often revealed by MCG-in 14 patients, mainly with signs of LVH in MCG and EchoCG. LAH was revealed by EchoCG in 14 patients by ECG in 2 patientes and by MCG - in 6 patients.

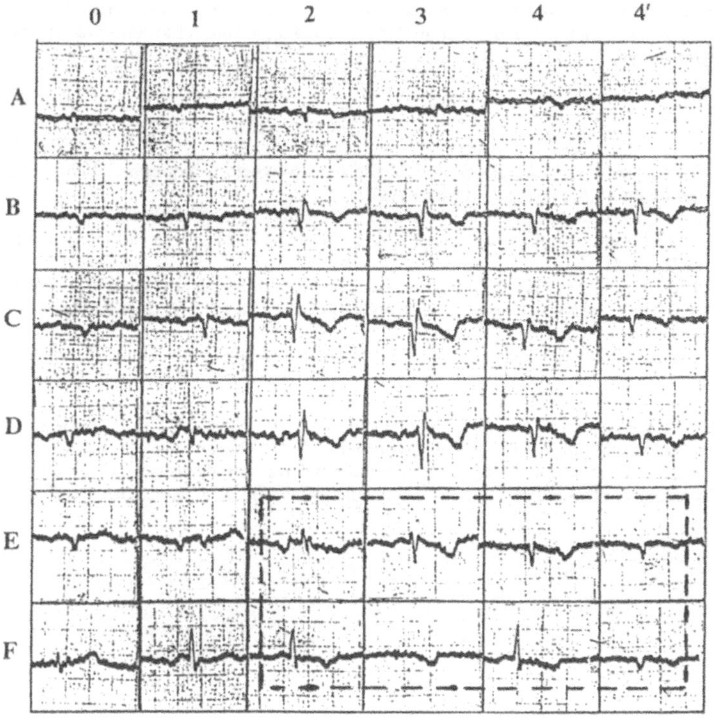

Fig. 1 MCG of patient with ischemic heart disease (55 years, male)

Diagnostic value of ECG and MCG criteria in revealing LVH was calculated: sensitivity - 39% and 94%, correspondingly, specificity - 89% and 75%, prognostic value in negative tests - 42% and 86%, in positive tests - 87.5% and 89%.

Diagnostic value of ECG and MCG criteria in revealing LAH was calculated: sensitivity - 7% and 43%, correspondingly, specificity - 92% and 100%, prognostic value in negative tests - 46% and 62%, in positive tests - 52% and 100% (EchoCG - standard).

4. Discussion

In patients after MI the prevalence of pathological QRS in MCG and ECG in 36 leads is the sign of post-infarction heart affection.We have reported about it in [3]. LVH is one of the components of post-infarction ventricular remodelling [4].We compared diagnostic value of ECG and MCG in revealing LVH in patients with IHD, MCG turned out to be more accurate. Repolarization disturbances are better revealed by MCG and ECG in 36 precordial leads. LVH and LAH are the early signs of "hypertensive heart" [5] in patients with AH. LVH and LAH were more precisely diagnozed by EchoCG and MCG. Repolarizational myocardial disturbances when LVH are more often registered by MCG, which is in accordance with data [6] about early diagnostics of LVH. We consider that the changes in repolarization complex in patients with AH is the reflection of differently expressed hypertrophy of some parts of LV myocard and ventricular septum. This was confirmed by EchoCG.

Thus, we may conclude that MCG is important for the revealing of signs of hypertrophic and post-infarction remodelling of myocardium.

References:

[1]Golyshev,N.V.,Motorin,S.V.,Rogachevsky,B.M.,Shabalin,A.V. Magnetocardiographic studies in the clinical diagnosis of heart diseases . Medicinskaya technika ,1995, 4:3-5.

[2]Nikitin,Yu.P.,Shabalin,A.V.,Kytmanov,A.V.,Tselmin,S.G.,Ermakova,E.N.,Golyshev,N.V.,Rogachevsky,B.M., Motorin, S.V. Magnetocardiography in complex clinical investigation of patients with ischemic heart disease . Kardiologiya, 1995, 4:63.

[3]Nikitin,Yu.P.,Shabalin,A.V.,Ermakova,E.N.,Kytmanov,A.V.,Tselmin,S.G.,Golyshev,N.V.,Rogachevsky,B.M., Motorin, S.V. Magneto- electro- and echocardiography in evaluation of changes of the heart in patients after myocardial infarction. Kardiologiya, 1995, 9: 45-48.

[4] Mitchell,G.F.,Lamas,G.A.,Vanghan,D.E. et al. Left ventricular remodelling in the year after first anterior myocardial infarction:A quantitative analisis of contricle segment lengths and ventricular Sharple. Am.J.Coll.Cardiol.,1992,19:1136-1144.

[5] Frohlich.E.D. Pathophysiology of essential hypertension, the Next Decade Verapamil in Focus/Eds A.Fleckenstein,J.H.Laragh, Edinburg: Churchil Livingstone, 1987,6-15.

[6] Nomura,M.,Fujino,K.,Watanabe,K. et al. Studies on the repolarization abnormality in systolic overloading of the ventricle by means of magnetocardiography. Biomagnetism.6th Int.Conf.Biomagnetism, Tokyo, 1987,152.

Comparative Data on Magnetocardiography and Clinical-Instrumental Methods of Investigation When Some Hereditary Cardiovascular Diseases

Shabalin, A.V.[1], Nikitin, Y.P.[1], Golyshev, N.V.[2], Motorin, S.V.[2], Rogachevsky, B.M.[2], Lisithenko, O.V.[1], Ermakova, E.N.[1], Dolgich, M.M.[1] and Kuroedov, A.Y.[1]

[1]*Institute of Internal Medicine, Siberian Branch of Russian Medical Academy, Russia;*
[2]*Novosibirsk State Technical University, Novosibirsk, Russia*

1. Introduction

During the last decade the interest of researchers to hereditary diseases is constantly growing, and first of all - to cardiovascular inherited diseases. The aim of our work was to study the data of magnetocardiography (MCG) in comparison with electrocardiography (ECG), Echocardiography (EchoCG) and other methods of investigation in patients with hereditary cardiovascular diseases.

2.Methods

13 males and 2 females at the age of 16 - 50 years were investigated. The patients were divided into 3 groups: the 1-st group- 9 patients with different types of connective tissue displasia (4 patients had not less than 3 signs of displasia, 5 patients - Marfan disease complex). The 2d group - 4 patients with hereditary - supraventricle paroxysmal disturbances of cardiac rhythm. The 3d group - 2 patients with carpal tunnel syndrome. The control group included 37 healthy males.

ECG was recorded in 12 standard and 36 multiple precordial leads [1], calibration signal was 1 mV = 10 mm. Registration of magnetic field was done in the same 36 points as in ECG recording in accordance with standard rules for MCG [2,3]. Regulation of the system was done automatically by IBM - PC computer, calibration signal was 25 pT.. Technical parameters are described earlier [4]. Methods of electrophysiological investigation, Holter monitoring, ECG and invasive probing of heart cavities (1 patient) were used. Scanners 'Accuson - 128' and 'Hewlett Packard'(USA) with electronic probes 2.5 MGz were used for EchoCG.

3.Results

In most patients heart conduction impairments were not found.There were no statistically significant differences in the length of P wawe,PQ and QT intervals in patients and controls.Only 5 patients had QT>0.44 sec : 3 of them were from the 1 group (Marfan disease) and 2 patients were from 2 group.LVH according to Sokolov-Lion was revealed in 1 patient from 1 group and 1 patient from 3 group.And only in one patient from 2 group LAH was revealed by ECG and confirmed by ultrasonic investigation.

Multiple additional chords of LV were found by EchoCG in 4 patients and mitral valve prolapse in 3 patients of 1 group.The signs of aortic insufficiency were revealed by EchoCG in 2 patients with Marfan disease.In one patient with Marfan disease the coarctation of aorta in usual place (confirmed by invasive methods of investigation).The other patient with similar pathology had implanted (8 years ago) spherical mitral valve because of the expressed mitral insufficiency.When ultrasonic investigation he had expressed LV dilatation whith the signs of stagnant heart insufficiency.In 2 patients from 2 group ECG revealed 3 signs at the moment: LAH and hypertrophy of ventricular septum, as well as mitral valve prolapse, 1 stage.In all the rest patients of the 3 group (a family) only one EchoCG sign was stated: hemodinamically insignificant prolapse of mitral valve, 1 stage.And in 2 patients from 3 group mitral valve prolapse was also found, in one of them multiple additional chords of LV were revealed.Direct or indirect signs of LVH according to EchoCG criteria were revealed in 4 patients from group 1 and 2 patients from group 2 ,MCG revealed signs of LVH in 7 patients (5 patients in 1 group ,1 patient in 2 group, and 1 patient in 3 group).LAH was not revealed by MCG.

Left ventricle hypertrophy (LVH) was revealed by ECG, EchoCG and MCG in 2, 6 and 7 patients correspondingly. EchoCG was used as standard method for calculating diagnostic value of ECG and MCG in revealing LVH: sensitivity - 17 % and 83 %, specificity - 89 % and 78 %, prognostic value in negative tests - 68 % and 87 %, prognostic value in positive tests - 50 % and 71 % correspondingly.

We analysed the interrelations between clinical symptoms, morphology of MCG-curve and ECG, EchoCG data. We found that ECG registered QTc >0,44 sec. in 5 patients, 3 of them were from the 1st group. MCG showed repolarization changes in 4 patients - negative T-wave in left low (E, F - 3,4) leads.

In the 2d group layer QTc >0,44 sec. was found in 2 patients. Electrophysiological investigation did not confirm sick sinus syndrome. MCG showed similar repolarization changes as we mentioned above.In the 3d group QTc >0,44 sec. was not found. The data showed that among 9 patients with mitral valve prolapse interval QTc > 0,44 sec. was found only in 2 patients, but repolarization disturbances were revealed by MCG in 7 patients. EchoCG revealed additional chords of left ventricle in 5 patients. MCG showed in 4 patients high-voltage R waves in middle (C, E - 2,3) leads.

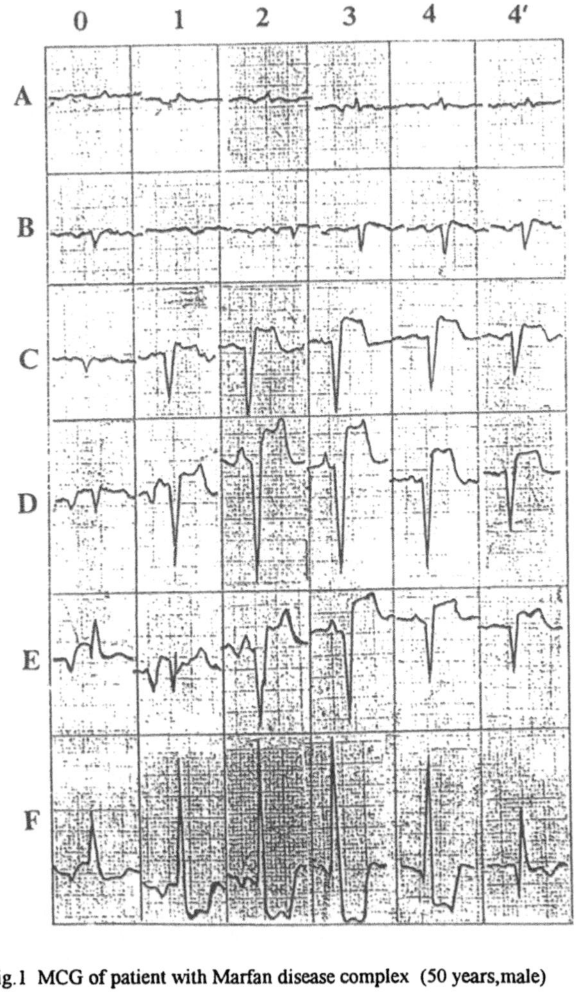

Fig.1 MCG of patient with Marfan disease complex (50 years,male)

Fig.1 shows MCG data of the patient group 1 (Marfan disease): implanted mitral valve, aortic insufficiency.MCG shows ventricle complexes as QS in middle right and left precordial leads (C-E-1-4') and discordant high-amplitude R in low right and left leads (F-1-4) with expressed repolarisation disturbances: ST segment depression and negative T waves.Catamnesis stated the patient's death in 1,5 years offer the investigation due to the growing heart insufficiency.

4.Discussion

In our work we studied 3 groups of hereditary diseases: system pathology of connective tissue, cardiac arrhythmias in families and one of the types of progressing neuromuscular diseases. ECG revealed possible arrhytmogenic factor - lenghthening of electric systole of ventricles. In 1975 P. Schwartz et al. [5] tried to explain mechanizm of arrhythmia when longer QT interval syndrome by "dispersion of refractóriness" - non-homogenic process of repolarization in myocardium. Taking into account literary data [6] we used MCG method in our study.The received data showed that ventricles repolarization disturbances (not registered by ECG and MCG in control group) were found when QTc > 0,44 sec. and when hemodinamically non-significant mitral valve prolapse. Attention should be paid to the interrelation between the presense of additional multiple chords in left ventricle (EchoCG) and high-amplitude R in MCG. Small number of investigated cases gives no ground for exact explanation of these interrelations.

References

[1] Mori,H.,Nakaya,J. Present Status of clinical magnetocardiography C.V.World Report. -1988 -Vol.16., N 2. - P.78-86.

[2] Biomagnetism:Application and theory / Ed. by H. Weinbery e.a.New-York etc:Pergamon Press.1985.

[3] Lepesckin. E., Progress in magnetocardiography J.Electrocardiol.,1976,9:295-296.

[4] Golyshev.N.V., Motorin.S.V., Rogachevsky,B.M. Superconducting magnetic cardiographic systems Medicinskaya technika , 1995, 3:14-18.

[5] Schwartz.P.,Pelliti,M.,Malliani,A. The long QT syndrome. Am.Heart J.,1975,89:378-390.

[6] Nomura.M.,Fujino,K.,Watanabe,K. et al. Studies on the repolarization abnormality in systolic overloading of the ventricle by means of magnetocardiography. 6th Int.Conf. Biomag., Tokyo,1987,152.

Atrial Depolarization Homogeneity Diagnostics on Evidence Derived from MCG

Stadnyuk, L.A., Chaikovsky, I.A., Sosnitsky, V.N., Gapelyuk, A.V., Minov, Yu.D. and Chervonopiskaya, E.M.

Biomagnetic Laboratory, Ukrainian Research Institute of Cardiology, Kiev, Ukraine

Introduction

A lot of works, devoted to studies of patients suffering from cardiac rhythm disturbances by magnetic mapping appeared during the last years. The majority of this works is concerned with ventricular rhythm disturbances. We are aware of only several papers, dealing with experimental and clinical studies of supra-ventricular arrhythmia (SVA) by magnetocardiography (MCG) [1-3]. At the same time, paroxysmal SVA, particularly atrial fibrillation, is the most prevalent variety arrhythmia occurring after extrasystolia, and influences considerably the clinical course of diseases and prognosis. As the result of SVA paroxysms lowering of physical tolerance and progress of heart failure and coronary pathology take place in these pts. This arrhythmia variant may also cause an increase of peripheral embolization risk and, as shown on big population, its correlated with mortality elevation.

The aim of this research is development of sufficiently simple technique of evaluation of expression degree of atrial depolarization contravention on the base of MCG-mapping, determination of connections of these contraventions with development of SVA and efficiency of its treatment.

Material and methods

Eighteen patients (13 men and 5 women in the age from 26 to 57 years, average age 44 ± 7.1) with paroxysms of supraventricular tachicardia and atrial fibrillation on the background of myocarditic cardiosclerosis have been examined. Paroxysm frequency has fluctuated between 2 and 5 per week (3.1 ± 1.6). The patients suffering from arterial hypertension, endocrine diseases, alterations in the mitral valve apparatus of the heart have not been involved into research. All the patients have been subject to the transesophagial cardiac pacing (TECP) of atrial according to the usual procedure to verify rhythm contraventions.

As a control group, 15 persons (14 men, 1 woman, average age 28 ± 3.0 years) without clinical symptoms of cardiac diseases with normal results of an objective examination, ECG, echocardiography, stress-testing have been examined.

MCG mapping was performed by means of the one-channel SQUID-magnetometer at Biomagnetic Laboratory installed in the Ukrainian Cardiology Center. Magnetic field registration has been carried out at a rectangle grid 6 × 6 with 4 cm pitch. Measurements in each point during 30 seconds have been conducted with simultaneous recording of the reference ECG-channel. On basis of these data for each selected time interval of a cardiocycle, instantaneous equiinductional maps of heart's magnetic field distribution in the plane of measurements have been plotted by means of two-dimensional interpolation algorithms. In order to estimate the fragmentation degree of the picture of magnetic field of the atrial depolarization wave, the time interval corresponding to the P wave, has been conventionally divided in four equal parts. Six instantaneous maps have been plotted in them. In order to analyze their fragmentation degree, the number of extreme points (positive and negative for each group of six instantaneous maps) during the first and second halves as well as during the entire atrial depolarization wave has been determined.

Recording of an initial MCG map has been carried out for all the persons under examination prior to prescribing preparations and for 10 pts repeated after a two-week treatment with propaphenonum, dizopiramidum or athenololum, The final effect of the treatment has been evaluated according to changes in the frequency and degree of expression of spontaneous SVA paroxysms, and to possibility to provoke them during the TECP and stress-testing.

Results and discussion

The obtained results of research with initial recording of the MCG mapping and evaluation of degree of their fragmentation are presented in the Table 1.

Table 1. Number of extrema in magnetic P-wave maps of the patients with SVA and of the control group (M+m).

	Number of extreme points							
Extreme point polarity	-				+			
Moment of determination	1	2	3	4	1	2	3	4
Healthy group n=15	10.0 ±0.1	6.4 ±0.4	9.8 ±0.7	13.5 ±1.0	9.0 ±0.8	8.3 ±0.5	8.0 ±+0.7	7.2 ±0.5
Patients n=18	7.5 ±0.5	6.2 ±0.5	11.3 ±1.1	15.8 ±0.8	13.9 ±0.9	10.9 ±+0.9	14.6 ±0.8	13.3 ±1.1
P	>0.05	>0.05	>0.05	>0.05	<0.01	<0.05	<0.01	<0.01

Note: 1; 2; 3; 4 are the numbers of time intervals, into which the P-wave has been divided.

As it can be seen from the table, an essential difference between the groups has been found out in the quantity of positive extreme points (Fig. 1).

Fig. 1. presents isomagnetic maps for second part of P wave, time interval between maps is 4 msec.

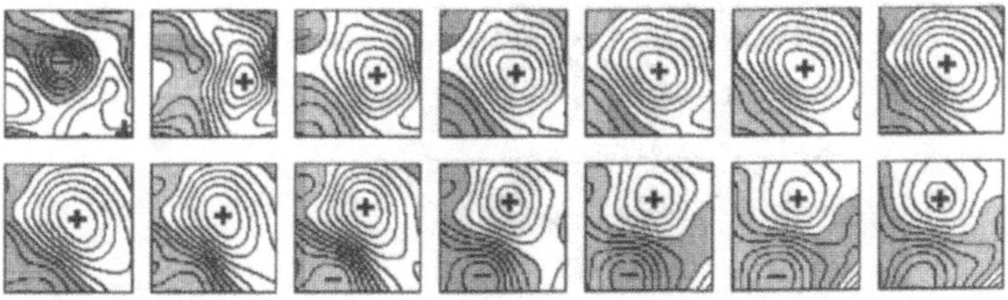

a) for normal subject

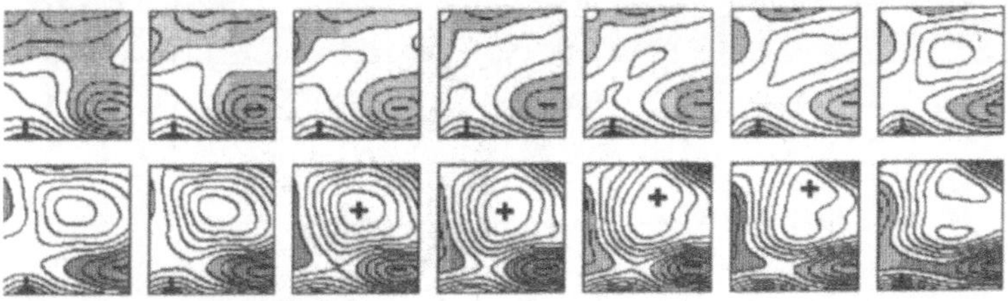

b) for patient with atrial fibrillation.

It should be admitted that the average number of positive extreme points per 1 map for the entire time interval at patients with arrhythmias is essentially (P< 0.01) higher compared to that at healthy persons (2.2 ± 0.2 against 1.3 ± 0.1).

Than analysis of correlation the atrial depolarization magnetic field homogeneity with the antiarrhythmic therapy (AAT) efficiency has been carried out. Four patients of 10 suffering from SVA have experienced remarkable improvement during the two-week treatment, no expressed effect has been observed at 6 patients or the antiarrhythmic preparation has had arrhythmogenic influence on them. The changes of the number of positive extreme points influenced by the treatment is displayed in the Table 2.

Table 2. Changes of the number of extrema during atrial depolarization under influence of treatment (M+m)

Subgroups of patients	Number of extreme points				
	1	2	3	4	Average
With positive treatment effect n=4	-5.8 ±2.2	-4 ±1	-4.3 ±2.1	-4.3 ±1.7	-4.6 ±0.8
Without positive treatment effect n=6	2.3 ±2.6	1 ±1.1	4.5 ±1.2	4 ±1.7	3 ±0.9
P	>0.05	<0.05	<0.01	<0.05	<0.01

Note: - number of extrema are decreased.

As it can be seen from the Table 2, for patients with positive effect of antiarrhythmic therapy the number of positive extreme points decreases (especially at the second half of the P-wave) (Fig.2).

Fig. 2. Isomagnetic maps for second part of P wave for patient with atrial fibrillation, time interval between maps is equal to 4 msec.

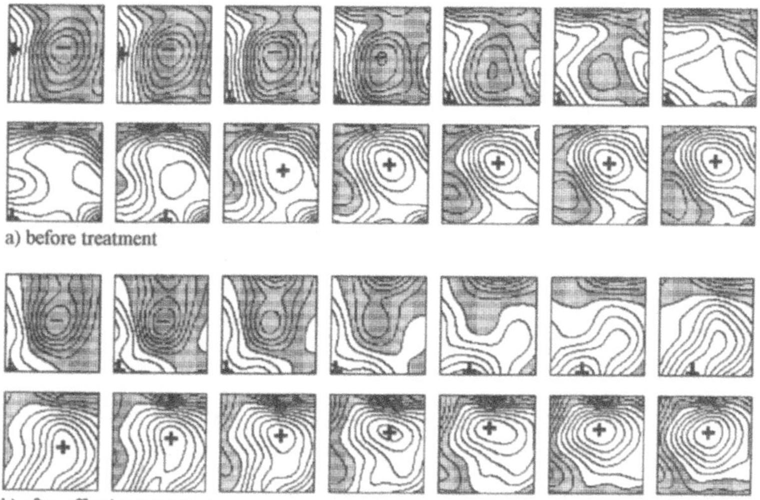

a) before treatment

b) after effective treatment.

To define connection between the homogeneity of the atrial depolarization and arrhythmias more precisely MCG was registered before and after transesophagial cardiac pacing induced and spontaneously ceased atrial fibrillation paroxysm in 6 pts. Immediately after the paroxysm the number of positive extrema in maps of P-wave's second part was significantly higher than before paroxysm (28.0 + 1.4 against 21.6 +1.0)

These investigations demonstrated a connection of the magnetic field fragmentation of atrial depolarization and supraventricular paroxysms severity. An increase of positive extrema number in the second part of P-wave was detected in pts., suffering from paroxysmal SVA. This may be evidence of the fact, that the arrhythmogenic substrate in the group of pts. examined by us has been developed, on the whole, in the myocard of the left atrial. At antiarrhythmic therapy, efficient according to clinical criteria, the number of extrema on magnetic field maps was decreased clinical investigations have shown that asynchrony increase of excitation conduction in atria is combined with a rise of atrial fibrillation paroxysms frequency [4]. As it has been already mentioned above, the most efficient technique used at the present time at the clinic for evaluation of efficiency and individual selection of antiarrhythmia preparations is the electrophysiological examination. However, the results obtained by means of this method do not coincide in 30-70% of cases with the clinical effect [5].

In our opinion magnetic mapping may be a valuable supplement of electrophysiological testing at individual drug chose and evaluation of antiarrhythmic therapy efficiency in pts., suffering from SVA paroxysms.

References

[1] Ribeiro, R. C., Bruno, A.C., Ribeiro, E.P., et al. Detection of magnetic atrial flutter waves in isolated rabbit heart, 1989, Proc. of 7th Int. Conf. on Biomagnetism, New York, 389-392.

[2] Fenici, R., Melillo, G., Capelli, C. et al. Atrial and ventricular tachycardias: Invasive validation and reproducibility of magnetocardiographic imaging, Ibid., 441-444.

[3] Nomura, M., Nakaya, Y., Satio, K. et al. Atrial excitation onto the MRI using magnetocardiogram, 1995, Proc. of 9th Int. Conf. on Biomagnetism, Amsterdam, Elsevier, 652-654.

[4] Campbell, P. W., Janse, M.I., Cardiac arrhythmias: The Management of atrial fibrillation, Berlin et al., Springer-Verlag, 1992.

[5]. Steinbeck, G., Luderitz, B., Andersen, D. Control of antiarrhymic drug therapy by electrophysiologic testing, European Heart J., 1989, 10(Suppl.E), 19-30.

Magnetocardiographic Localization of Focal Arrhythmias in Conjunction with Catheter Ablation

Steiner, P.R.[1], Kynor, D.B.[2], Haupt, C.D.[2], Rowley, H.A.[1], and Lesh, M.D.[1]

University of California, San Francisco[1], San Francisco, CA, USA; Biomagnetic Technologies Inc.[2], San Diego, CA, USA

Introduction

Catheter ablation has become the treatment of choice for many types of arrhythmias.[1] Ablative treatment of arrhythmias is frequently curative and is less expensive than surgery or a lifetime regimen of medications.[2],[3],[4],[5],[6],[7],[8],[9],[10]

Catheter ablation requires that the electrophysiologist precisely locate the region of the myocardium that is responsible for the arrhythmia. In current practice, target ablation sites are determined by the careful analysis of a series of multiple simultaneous intracardiac electrograms. The time required to locate the appropriate site can be lengthy, especially in patients with arrhythmias that either are difficult to induce or are poorly tolerated. Because catheter positioning must be performed under the guidance of fluoroscopic imaging, many of these cases involve significant radiation exposure for both patient and staff.[11]

The goal of this study was to investigate magnetocardiography (MCG) as a non-invasive technique for localization of sites of arrhythmogenesis and to determine if MCG data is useful for guiding subsequent electrophysiologic (EP) evaluations and catheter ablation procedures.

MCG was used to localize the onset of arrhythmias in patients scheduled to undergo an invasive EP study and catheter ablation procedure as part of their normal course of care. Localization of the arrhythmogenic regions as determined by MCG were superimposed on magnetic resonance images (MRI) and compared with fluoroscopic images depicting a catheter at the site of arrhythmogenesis as determined during the EP study.

Methods

MCGs were obtained on patients prone to spontaneous and well-tolerated arrhythmias who were scheduled for invasive EP mapping and radio frequency (RF) ablation. Fifteen patients had their typical arrhythmic episodes or related ectopic beats during MCG acquisition and completed all aspects of the study. These patients were selected for participation in the study based on the following screening criteria:

- Clinical need for electrophysiological evaluation and subsequent catheter ablation of their arrhythmia
- One of two arrhythmia types: idiopathic ventricular tachycardia or automatic atrial tachycardia
- No contraindications to MRI
- Willingness to participate in the study.

Simultaneous MCG and electrocardiogram (ECG) recordings were obtained from each subject using a 37 channel biomagnetometer (Biomagnetic Technologies, Inc.) in a magnetically-shielded room. Signals in the 0.1-520 Hz bandwidth were acquired over a period of three to ten minutes with the biomagnetometer probe placed at 1-4 locations over the precordium. Patients who had spontaneous arrhythmic episodes during the measurement session were included in the subsequent data analysis.

Low frequency measurement noise was removed from the MCG by high-pass filtering or subtraction of an interpolated polynomial curve that had been fit to latencies associated with isoelectric intervals of the cardiac cycle.

Correlation analysis was performed to extract cardiac complexes having similar morphologies. Complexes having similar morphologies were averaged together prior to source localization. In patients exhibiting only single ectopic beats, localization was performed on unaveraged data.

Source localization was performed by modeling the cardiac currents with an equivalent current dipole located inside a sphere with homogeneous conductivity. A modeling sphere was positioned and sized for a 'best-fit' to the region of the thorax immediately under the biomagnetometer probe during data acquisition.

Anatomical imaging was performed by obtaining axial and coronal volumetric MR images at multiple phases of the cardiac cycle. Oil-filled capsules were placed on several fiduciary points (suprasternal notch, xiphoid, and locations at the lateral thoracic locations) to aid in coregistration of MCG and MR data. The image containing superimposed MCG and MR data is called a magnetic source image (MSI) (Fig. 1, upper panels).

Invasive electrophysiologic mapping was performed under biplane fluoroscopic guidance using transvascular tip-deflectable multipole catheters. The sites of origin for automatic atrial and ventricular tachycardias were determined utilizing pace mapping and localization of the earliest onset of electrical activation relative to a P wave or QRS complex recorded on the body surface. Once the site was identified by its electrophysiologic features, the location was noted by fluoroscopic imaging and subsequently compared to MCG localizations (Fig. 1, lower panels). The morphologies of recordings obtained from surface ECG leads taken during the MCG measurement and during the electrophysiological mapping procedure were compared to ensure that the patient exhibited the same arrhythmia in both cases.

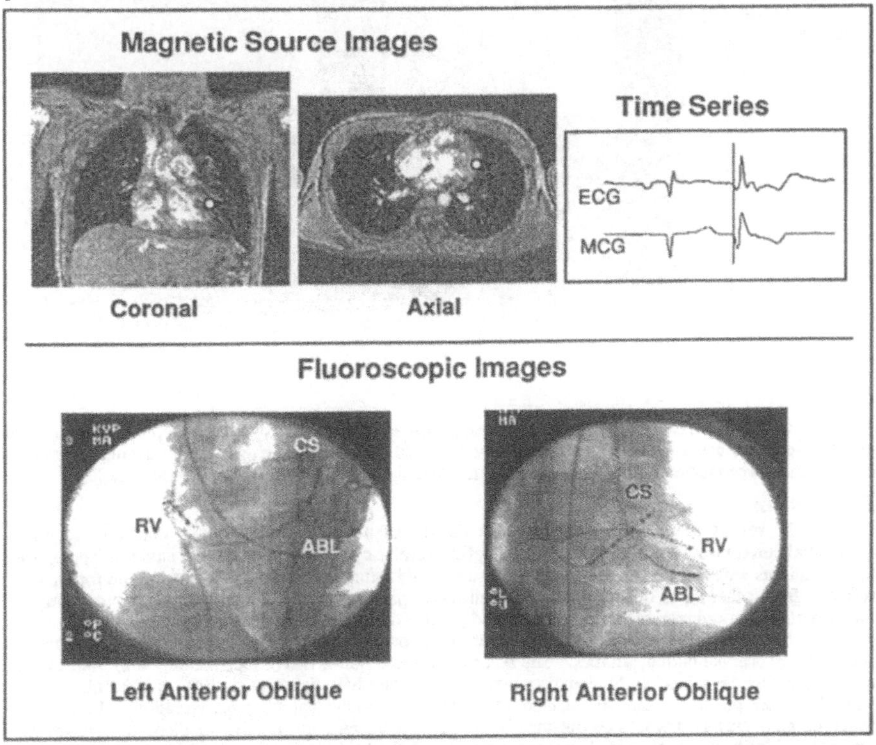

Fig 1. Magnetic source and fluoroscopic images from a patient with idiopathic ventricular tachycardia. The white dots on the MSI show the site of arrhythmogenesis as determined non-invasively during the MCG measurement. The horizontal line on the time series shows the latency used for MSI localization. The fluoroscopic images show three catheters: a decapolar catheter located in the coronary sinus (CS), a multipole catheter located in the right ventricular (RV), and the ablation catheter (ABL) at the site of successful ablation (the base of the posterolateral papillary muscle in the left ventricle).

Results

Comparisons of MCG localizations and the localization results obtained in the cardiac electrophysiology laboratory were made in fifteen patients, including seven with atrial tachycardia, and eight with ventricular tachycardia. The localization estimates obtained using MCG were within two centimeters of the localizations obtained using standard catheter mapping techniques in six patients and in the two to four centimeter range in another six patients (Fig. 2). In three additional patients, a qualitative comparison was performed due to insufficient information for three-dimensional image coregistration, limiting analysis to two dimensions. In each of these patients, the MCG localization was consistent with results obtained in the electrophysiology laboratory.

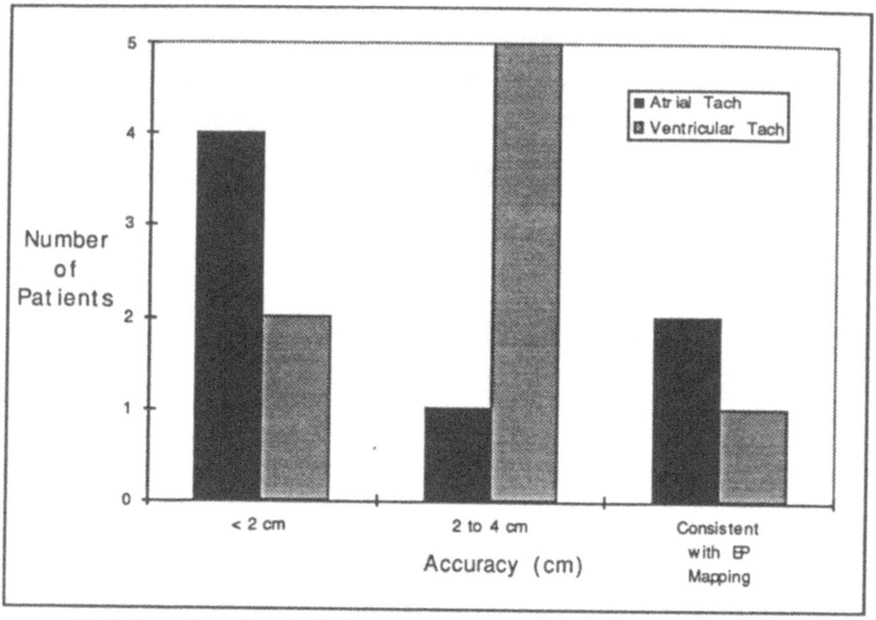

Fig 2. Comparison of results obtained using MSI and electrophysiological mapping. Discrepancy measurements are based on visual comparison of MSI and fluoroscopic images. Bar shading is used to distinguish between patients with automatic atrial tachycardia and idiopathic ventricular tachycardia.

Discussion

The results of this study are based on the use of a single current dipole model to represent myocardial current flow during a specific instant of the cardiac cycle. We feel that this model is appropriate for the patients with automatic arrhythmias examined in this study, but may not be appropriate for patients suffering from other types of arrhythmias. For instance, patients with reentrant arrhythmias may require source models that can represent widely distributed patterns of cardiac electrical activity.

This study demonstrates the ability of magnetic source imaging to provide regional localization of automatic myocardial current sources. The accuracy of these non-invasive localizations may improve as more experience is gained with: coregistration of MCG and MR data, integration of data obtained at multiple sensor positions, and new techniques for addressing individual anatomic variations. In addition, better accuracy estimates would be obtained by utilizing automated quantitative methods to compare the successful ablation site as seen in fluoroscopic views with the MSI localization.

Magnetic source imaging of cardiac arrhythmias can be readily applied in a clinical environment, and may provide a powerful tool for guiding subsequent electrophysiologic evaluations and catheter ablation procedures.

References

[1] Calkins, H., Sousa, J., el Atassi, R., Rosenheck, S. M., Kou, W.H., Kadish, A.H., Langberg, J.J., Morady, F. Diagnosis and cure of the Wolff-Parkinson-White syndrome or paroxysmal supraventricular tachycardias during a single electrophysiologic test, New England Journal of Medicine, 1991, 324(23):1612-8.

[2] Lesh, M.D., Van Hare, G.F., Epstein, L.M., Fitzpatrick, A.P., Scheinman, M.M., Lee, R.J., Kwasman, M.A., Grogin, H.R., Griffen, J.C., Radiofrequency catheter ablation of atrial arrhythmias - Results and mechanisms, Circulation, 1994. 89(4):1074-1089.

[3] Jackman, W.M., Wang, X.Z., Friday, K.J., Roman, C.A., Moulton, K.P., Beckman, K.J., McClelland, J.H., Twidale, N., Hazlitt, H.A., Prior, M.I., Catheter ablation of accessory atrioventricular pathways (Wolff-Parkinson-White syndrome) by radiofrequency current. New England Journal of Medicine, 1991. 324(23):1605-11.

[4] Lesh, M.D., Van Hare, G.F., Schamp, D.J., Chien, W., Lee, M.A., Griffin, J.C., Langberg, J.J., Cohen, T.J., Lurie, K.G., Scheinman, M.M., Curative percutaneous catheter ablation using radiofrequency energy for accessory pathways in all locations: results in 100 consecutive patients. Journal of the American College of Cardiology, 1992. 19(6):1303-9.

[5] Van Hare, G.F., Lesh, M.D., Scheinman, M.M., Langberg, J.J., Percutaneous radiofrequency catheter ablation for supraventricular arrhythmias in children. Journal of the American College of Cardiology, 1991. 17(7):1613-20.

[6] Gillette, P.C., Wampler, D.G., Garson, A.J., Zinner, A., Ott, D., Cooley, D. Treatment of atrial automatic tachycardia by ablation procedures. Journal of the American College of Cardiology, 1985. 6:405-9.

[7] O'Nunain, S., Linker, N.J., Sneddon, J.F., Debbas, N.M., Camm, A.J., Ward, D.E., Catheter ablation by low energy DC shocks for successful management of atrial flutter. British Heart Journal, 1992. 67(1):67-71.

[8] Saoudi, N., Atallah, G., Kirkorian, G., Touboul, P., Catheter ablation of the atrial myocardium in human type I atrial flutter. Circulation, 1990. 81(3):762-71.

[9] Feld, G.K., Fleck, R.P., Chen, P.S., Boyce, K., Bahnson, T.D., Stein, J.B., Calisi, C.M., Ibarra, M., Radiofrequency catheter ablation for the treatment of human type 1 atrial flutter. Identification of a critical zone in the reentrant circuit by endocardial mapping techniques. Circulation, 1992. 86(4):1233-40.

[10] Walsh, E.P., Saul, J.P., Hulse, J.E., Rhodes, L.A., Hordof, A.J., Mayer, J.E., Lock, J.E., Transcatheter ablation of ectopic atrial tachycardia in young using radiofrequency current. Circulation, 1992. 86(4):1138-46.

[11] Lindsay, B., Eichling, J., Ambos, H., and Cain, M., Radiation Exposure to Patients and Medical Personnel During Radiofrequency Catheter Ablation for Supraventricular Tachycardia, The American Journal of Cardiology, 1992 70: 218-223.

Magnetocardiographic Assessment of Left Ventricular Hypertrophy: Correlation with Echocardiography

Takala, P.[1,3], Montonen, J.[1,3], Aalto, T.[2], Mäkijärvi, M.[2,3], Nenonen, J.[1,3], Yildirim, Y.[2], Nieminen, M.S.[2] and Katila, T.[1,3]

[1]*Laboratory of Biomedical Engineering, Helsinki University of Technology, 02150 Espoo;*
[2]*Cardiovascular Laboratory and* [3]*BioMag Laboratory, Helsinki University Central Hospital, 00290 Helsinki, Finland*

Introduction

The presence of left ventricular hypertrophy (LVH) has been shown to be a powerful risk factor for cardiovascular morbidity and mortality. The 12-lead electrocardiogram (ECG) is the easiest and the most widely used noninvasive method for the diagnosis of LVH. Unfortunately, though the specifity of the conventional ECG criteria is high, the sensitivity is low [3]. The use of sophisticated multivariate ECG methods have resulted in only a modest improvement in the sensitivity [1]. Echocardiography, on the other hand, is an accurate but more demanding method for determining the increased left ventricular (LV) wall thickness and LV mass. However, the value of echocardiography is dependent on the experience of the physician performing the investigation.

In magnetocardiography (MCG), magnetic fields originating from the heart are recorded outside the body. Biomagnetic measurements are fully noninvasive and with a multichannel device a magnetocardiographic mapping is fast and simple to perform. In a magnetocardiogram, left ventricular overload has been shown to increase the S wave amplitude in the upper anterior part of the thorax and the R wave in the lower left part of the thorax [2].

In this study, we investigated methods for the assessment of LVH using magnetocardiography (MCG). Magnetocardiographic indexes for the assessment of LVH were developed and the correlation of the indexes with the echocardiographic LV mass was calculated. Also, comparisons of MCG indexes and electrocardiographic Sokolow–Lyon voltage were made.

Our hypothesis was that an increace in LV mass in left ventricular hypertrophy causes an increase in the amplitude and area of the Q, R and S deflections corresponding the ventricular depolarization. Also, it was hypothesised that the T wave area diminishes in LVH and that the T wave area has a negative correlation with the LV mass.

Methods

Patient material consisted of 13 patients with hypertension. MCG, trans thoracic echocardiogram and 12-lead electrocardiogram were recorded and analysed for the assessment of left ventricular mass of the patients. In addition, 11 of the patients had a standard biplane x-ray imaging of the thorax.

Echocardiographic LV mass was determined by a cardiologist. The echocardiographic mass index was calculated by dividing the LV mass by the body surface area. Patients having echocardiographic LV mass index $> 130\ g/m^2$ were considered having LVH. Magnetocardiograms were recorded in 13 hypertensive patients. The recordings were performed in the magnetically shielded room at the Helsinki University of Technology. The standard MCG measurement grid [4] with an extra column on the patient's left was used. At each measurement location 60 seconds of MCG signal was recorded with the sampling rate of 2 kHz and using bandpass filtering from 0.03 to 300 Hz. Standard 12-lead electrocardiograms were recorded and the Sokolow-Lyon voltage SV_1+RV_{5-6} [5] was calculated.

Each recording location has a different distance from the heart. Also, since the recording is made outside the thorax, the average distance of the recording locations from the heart is dependent on the size of the patient. In order to make the signals recorded from different locations and different patients comparable with each other, they were scaled. The approximation of the magnetic field being proportional to the inverse of the squared distance from the source was made. For the scaling, the distance from the heart to each point in the recording grid was measured. Here the biplane x-ray images of the patient's thorax were used. The x-ray images were available for only 11 patients, which limited the patient population. The signals were scaled to correspond a recording distance of 15 cm by multiplying the signal by a factor $(d/15)^2$, where d is the distance from the recording location to the center of the heart in cm. Also, before calculation of magnetic indexes, the baseline field was subtracted from the signal. The baseline field was determined as an average of the field during a 10 ms time interval in the P–Q segment.

The area of the QRS complex was determined by integrating the absolute value of the averaged and scaled MCG over the QRS complex at each of the 42 recording locations. The sum of the QRS areas in the MCG grid was calculated and used as a magnetocardiographic index of LVH. The method is illustrated in Fig. 1 (B) for one grid location.

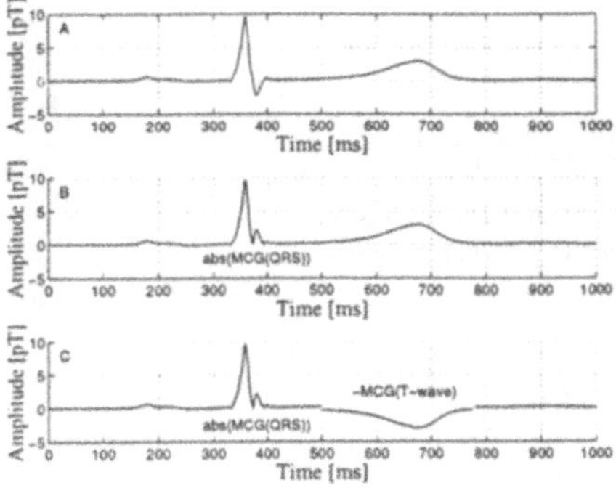

Fig. 1: *Magnetocardiogram of one cardiac cycle (A), QRS area of the magnetocardiogram (B) and the T wave area subtracted from the QRS area (C).*

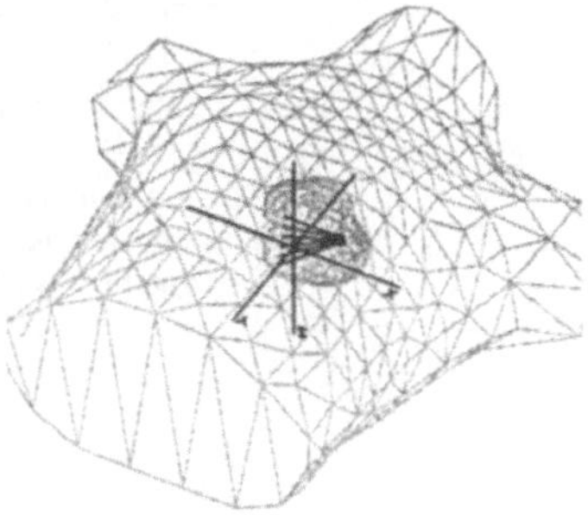

Fig. 2: *The boundary element method model of homogeneous human torso and current dipoles in the heart.*

The T wave areas were determined from the averaged and scaled MCG data using a similar procedure as in the determination of the QRS areas. However, in order to take into account the polarity of the T wave, an absolute value was not formed. Three patients had low frequency noise in their MCG and they were left out from the patient population because the T wave areas could not be determined accurately.

According to the hypothesis, LVH should increase the QRS area and diminish the area of the T wave. In order to take into account both of these features, a combination of the areas was formed by subtracting the T wave area from the area of the QRS complex. The combination of the areas is illustrated in Fig. 1 (C).

To estimate the electrical energy of the ventricular depolarization, a current dipole fit to the measured field distribution was made with 0.5 ms time intervals over the QRS complex. For the localization of a freely moving current dipole, a standard homogeneous boundary element torso model was used (see Fig. 2). A magnetocardiographic index for the assessment of LVH was formed as the time integral of the squared dipole amplitude over the QRS complex.

Results

The indexes calculated from the MCG showed a close correlation with the echocardiographic LV mass. The index calculated using dipole fitting showed correlation 0.70 with the echocardiogaphic mass index ($p = 0.007$, $n = 13$, Fig. 3). The correlation of the MCG index based on the QRS area with the echocardiographic LV mass index was 0.71 ($p = 0.015$, $n = 11$, Fig. 4). Without scaling with the recording distance, the correlation was 0.65 ($p = 0.031$, $n = 11$). The T wave based index showed correlation -0.80 ($p = 0.018$, $n = 8$, Fig. 5) with echocardiography. Index derived by subtracting the T wave integral from the QRS–integral, showed a correlation of 0.86 ($p = 0.006$, $n = 8$, Fig. 6) with echocardiography. Correlation of 12–lead ECG index ($SV_1 + RV_{5-6}$) with the echocardiography was poor, 0.17 ($p = 0.59$, $n = 13$, Fig. 7).

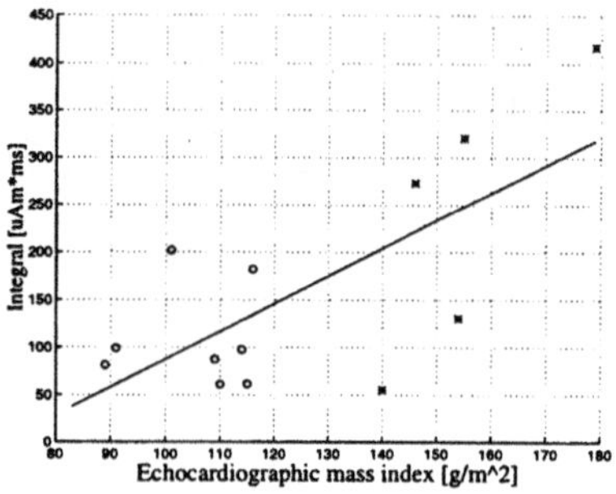

Fig. 3: *Dipole moment QRS–integral as a function of echocardiographic mass index (r = 0.70, p = 0.007).* o = *hypertension,* * = *LVH.*

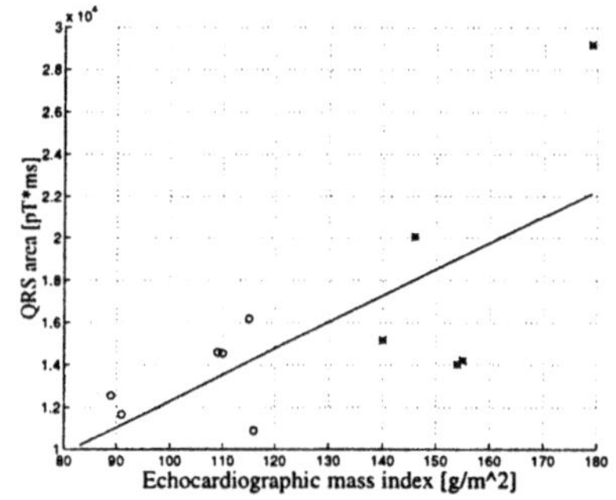

Fig. 4: *QRS area of the MCG grid as a function of the echocardiographic mass index (r = 0.71, p = 0.015).*

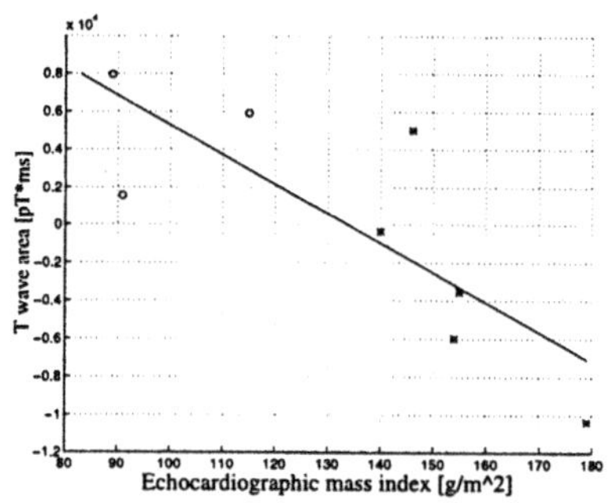

Fig. 5: *T wave area of the MCG grid as a function of the echocardiographic mass index (r = -0.80, p = 0.018).*

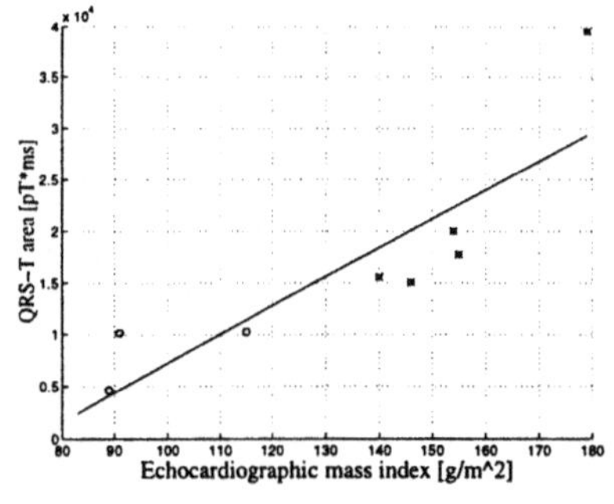

Fig. 6: *T wave area of the MCG grid subtracted from the total QRS area of the grid as a function of the echocardiographic mass index (r = 0.86, p = 0.006).*

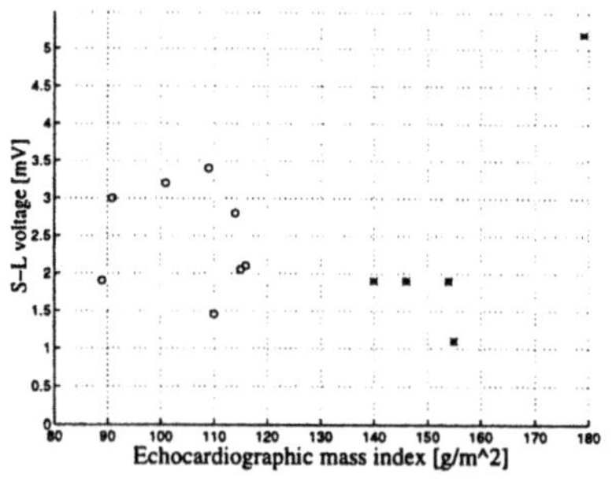

Fig. 7: *Electrocardiographic Sokolow–Lyon voltage as a function of the echocardiographic mass index (r = 0.17, p = 0.59).*

Discussion

The results support the hypothesis that LVH increases the area of the QRS complex and diminishes the T wave area in MCG. When echocardiographic mass index was used as a reference standard, MCG performed well. All MCG indexes showed significant correlation with the echocardiographic mass index. Compared with ECG, MCG performed clearly better. The scaling with the recording distance improved the correlation of the area of the QRS complex with the echocardiographic mass index.

In this study the patient population was small. Thus the results are not statistically comprehensive and they should be taken only as directive. Also, echocardiography which was used as a reference, is not the "golden" standard in the assessment of the LV mass. More patients with enlargened left ventricle need to be measured in order to assess the value of this method in the assessment of LVH.

MCG covers a limited area of the torso surface. This complicates the approximation of the activation strength with the area parameters. Also, individual torso models might improve the results of the dipole localization method. Outer dimensions of the patient or magnetic resonance images could be used as the parameters determining the shape of the thorax.

A study is in progress in the BioMag Laboratory to measure patients with multichannel MCG and to determine the LV mass from magnetic resonance (MR) images. In this study, MR images will be used in the forming of individual torso models. If an MCG index determinable without x–ray or MRI images turns out to correlate with LV mass, it will provide a fast method for the assessment of LVH. With MCG the assessmet of LV mass could be done without the presence of a doctor and with a better sensitivity than using 12–lead electrocardiography.

References:

[1] Casale, P.N., Devereux, R.B., Alonso, D.R., Campo, E., Kligfield, P. Improved sex–specific criteria of left ventricular hypertrophy for clical and computer interpretation of electrocardiograms: validation with autopsy findings, Circulation 75, No. 3, 565–572, 1987.

[2] Fujino, K., Sumi, M., Saito, K., Murakami, M., Higuchi, T. Nakaya, Y. Mori, H. Magnetocardiograms of patients with left ventricular overloading recorded with a second-derivative SQUID gradiometer, J Electrocardiology 17:219, 1984.

[3] Levy, D., Labib, S.B., Anderson, K.M., Christiansen, J.C., Kannel, W.B., Castelli, W.P. Determinants of sensitivity and specificity of electrocardiographic criteria for left ventricular hypertrophy, Circulation 1990;81:815–820.

[4] Saarinen, M., Siltanen, P., Karp, P.J., Katila, T.E. The normal magnetocardiogram: I morphology, Annals of clinical recearch, Vol. 10 Sulppl. 21 1978.

[5] Sokolow, M., Lyon, T.P. The ventricular complex in left ventricular hypertrophy as obtained by unipolar precordial and limb leads, Am Heart J. 1949;37:161–86.

Localization of Atrial Depolarization Activity from Non-Averaged MCG's in Congenital Heart Disease

Thieme, R.[1], Brockmeier, K.[1], Brümmerhoff, A.[1], Kissenbeck, S.[1], Burghoff, M.[2] and Bein, G.[1]

[1]Pediatric Cardiology, Virchow Medical Center, Humboldt University, D-14059 Berlin, Germany; [2]Physikalisch-Technische Bundesanstalt, D-10587 Berlin, Germany

Patients with complete transposition of the great arteries (TGA) have been surgically corrected by an intra atrial conversion of the venous blood, i. e. directing the pulmonary venous blood to the tricuspid valve and systemic venous blood to the mitral orifice (physiologic correction or Mustard procedure). The long term results in these patients with the right ventricle responsible for the systemic blood demand, are complicated by loss of sinus rhythm, atrial flutter, right ventricular dysfunction, and sudden death [1]. Most of the problems are probably related to impaired atrial depolarization activity. This study was performed to localize early atrial depolarization and to find out specific patterns of activation compared to normal activity in healthy subjects. For this purpose non-averaged MCG-signals were processed in order to evaluate beat-to-beat changes.

Methods & Materials: 9 patients with Mustard corrected TGA, mean age 10.2 ± 4 y, and 10 controls, mean age 15 ± 4 y, were included in the study. The measurements were performed in the facilities of the Berlin magnetically shielded room [2]. The system has 37 independent SQUID magnetometer units [3]. Notch filtering for suppression of power line disturbances in the MCG recordings was avoided by using a signal triggered compensation unit [4]. Data acquisition: 1 kHz A/D conversion rate, filtering 0.016 - 250 Hz, and a recording time of 100 seconds. The P wave begin was interactively determined in the y lead of the ECG.

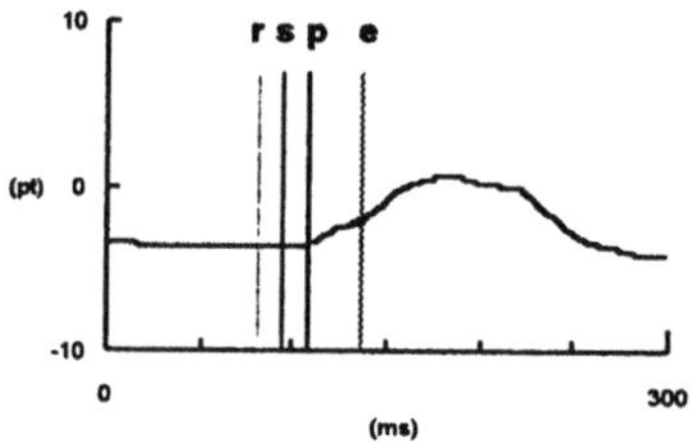

Fig. 1: Showing the P wave begin (p), the localization start (s), the end point of localization (e), and the reference point (r).

Localization procedure for each beat: Baseline reference (r) at -15 ms, localization start (s) at -10 ms, and end of the fitting calculations at +20 ms (e) of the interactively determined P onset (p) (**Fig.1**). Localizations were computed in 1 ms intervals (Gauss-Seidel algorithm) (equivalent current dipole, semi-infinite half space model). Digital low pass filtering (FIR cut-off at 29 Hz) was applied to enhence S/N ratio of the MCG rawdata. The coordinates, moments, and the error of estimation (σ) for dipole fitting were computed starting from the P wave onset. A $\sigma < 50$ was defined as sufficiently accurate. Then values of 3 consecutive coordinates of each P wave were compared for all beats in patients and controls respectively. Mean values and SD of the spatial distribution of early P wave activity were calculated for all cardiac beats. The peak-to-peak signal/noise ratio (S/N ratio) was calculated for the P wave at the $\sigma < 50$ instant, and close to the baseline reference in a 150 ms window, respectively.

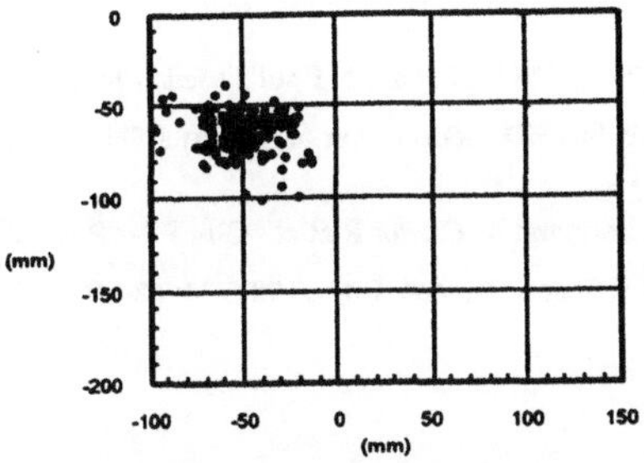

Fig.2: Example for the distribution of early atrial activity in a healthy subject. Displayed on a scatterplot which represents a plane parallel to the anterior chest wall. 0 in the horizontal axis was above the sternal line. 0 in the vertical axis was at the level of the jugulum sterni.

Results: S/N ratio in patients and controls showed no significant difference: 3.5 ± 0.8 in the TGA group and 4.5 ± 2.4 in controls. Sensor to source distances showed no significant differences between patients and controls and did not correlate with the quality of the localization results. After low pass filtering the mean range of P wave localization was x: ±4.2 cm, y: ±4.1 cm, z: ±5.2 cm (**Fig.2**). All TGA patients were in sinus rhythm. 4/9 patients had supraventricular premature activity (PAB). Clusters of PAB in TGA patients were localized at different sites compared to normal atrial activation, which typically described an array of 5*5 cm x-y diameter (**Fig.3**). 2/9 patients had a pattern of beat-to-beat P wave localization undistinguable from controls. The center of mean early P wave activity was localized right and cranial in controls, in contrast, in the TGA group the center was shifted to the left (n.s.). The spatial distribution of the PABs described a downward slope, projected from the upper right to the lower left side of the anterior chest wall.

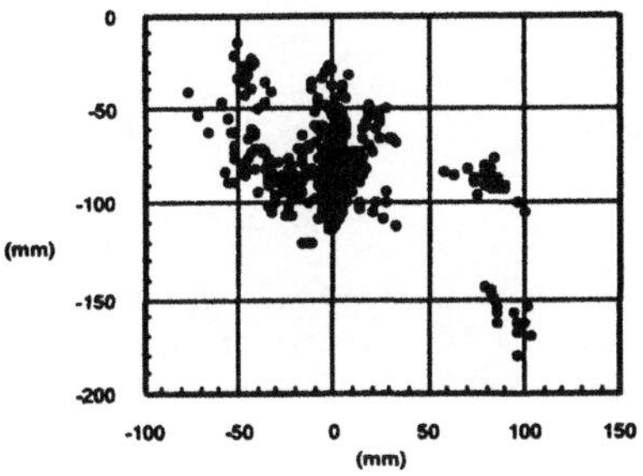

Fig.3: Localization result of the TGA group, showing clusters of normal atrial, supraventricular premature and ventricular premature activity.

Conclusions:

1. Beat-to-beat localization of early P wave depolarization, using data from a 37 channel MCG is possible
2. Different pattern of localization sites of P wave activation were found in patients and controls
3. The presented approach of localization of cardiac activity from multi-channel data opens new ways of automated signal determination in the time domain

References:

[1] Flinn CJ, Wolff GS, Dick M et al.: N Engl J Med, 1984, 310:1635-1638

[2] Erné SN, Hahlbohm HD, Scheer H et al: In: Erné SN et al. (eds): Biomagnetism. Berlin, De Gruyter, 1981

[3] Drung D, Zimmermann R, Cantor R et al.: Clin Phys Physiol Meas 12 Suppl. B: 21-29, 1991

[4] Scheer HJ: IEEE Trans. Biomed. Eng., 1987, 34: 68-69.

Tangential (x- and y-) Component Measurement of Cardiac Magnetic Field and Comparison with Conventional z- Component Measurement

Tsukada. K,[1] Kandori. A,[1] Yoshida. T,[1] Haruta. Y,[1]
Noda. Y,[2] Horigome. H,[2] Terada. Y,[2] and Mitsui. T[2]

Superconducting Sensor Laboratory,[1] Inzai, Chiba, Japan
Institute of Clinical Medicine,[2] University of Tsukuba, Tsukuba, Japan

INTRODUCTION

Magnetocardiograms (MCG) have better spatial resolution than ECGs and reveal abnormalities that cannot be seen in ECGs, but the MCG is not yet a standard diagnostic tool. One reason for this is that the inverse magnetic problem cannot be solved easily, and this creates the difficulties especially when the myocardium has a widespread active area. Even though the magnetic field is a vector, only the normal (or z-component) has usually been measured because the volume current contribution is thought to be small. Some reports, however, related to the measurement of MCG vectors have appeared[1]. A 32-channel SQUID system for measuring tangential (x- and y-) components by using planar-type gradiometers[2] and a conventional-type of 16-channel SQUID system for measuring the z-component by using coaxial-type gradiometers were fabricated. Using these instruments, we investigated the vector components of the MCG of normal subjects and the relation between the MCG and myocardium activity.

SQUID SYSTEM

To measure the three-dimensional components of the cardiac magnetic field, we constructed two types of SQUID systems: one for measuring tangential (x- and y-) components, and one for measuring the normal component. The planar-type gradiometer used to measure the tangential components had counter-wound 10-mm square-shaped coils connected in series, and its base line was 40 mm. Formed on a planar glass substrate, this configuration made the gradiometer sensitive to the tangential components when it was set vertical to the x-y plane. A coaxial type of pick-up coil was used for measuring the z-component. It too used square-shaped coils and its baseline was 40 mm. Both SQUID systems were used in the Tsukuba University Hospital (Fig. 1, 2), where the magnetic field originating from the heart was measured in a double-layer magnetically shielded room (MSR). The signals from the gradiometers were linearied by flux-locked-loop circuits and were amplified and filtered by the amplifier-filter units. They were then digitized by an A/D converter, and the data was stored and analyzed in a personal computer. The flux noise of the these gradiometers were less than 10 fT/√Hz.

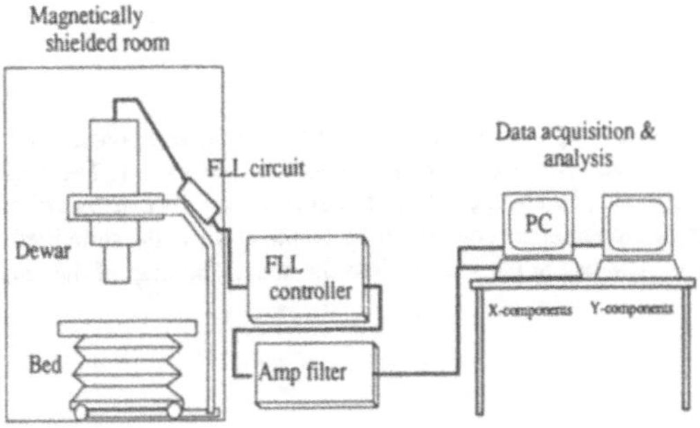

Fig. 1. Schematic diagram of the SQUID system.

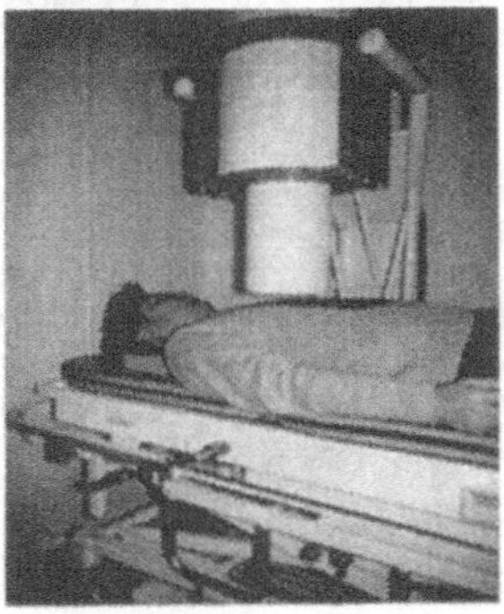

Fig. 2. Photograph of measuring equipment.

SIMULATION

To simplify the simulation of the biomagnetic field, we used a spherical homogeneous medium model. The magnetic field distributions calculated using the formula derived by Saravas [3] are shown in Fig. 3. The observation plane parallel to the chest was considered the x-y plane. There are three orthogonal components of the magnetic field generated by a single dipole parallel to the y axis. The z-component of the magnetic field has two field extrema, and their positions and strength determine the location and magnitude of the dipole. Tangential components, on the other hand, show different patterns. The magnetic field perpendicular to the dipole shows a single pole pattern just above the dipole, and the field parallel to the dipole is almost null. This means that the peak magnetic field of two-dimensional vector $\boldsymbol{B}xy=\boldsymbol{B}x+\boldsymbol{B}y$ coincide with the position of dipole.

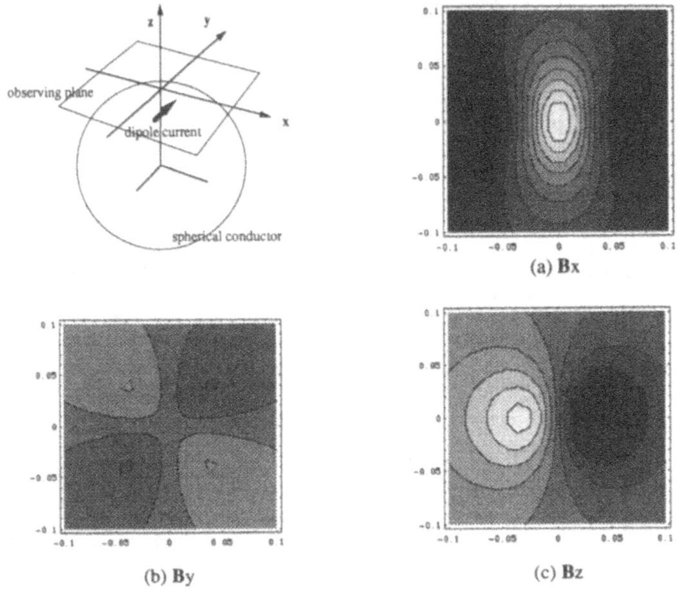

Fig. 3. Simulated magnetic field distributions generated in a spherical conductor.

MEASUREMENT

To measure tangential components and the z-component, we prepared two dewars equipped with the two types of gradiometer. The xiphisternum of the midline was the datum point to determine the position of the measurement on the chest wall, and the MCG and lead II ECG were measured at the same time. The 175 x 175-mm detection area (Fig. 5) on the chest was made up of individual detection areas (Fig. 4). The signal from the gradiometers were filtered by 0.1-100Hz bandwidth and sampled at 1kHz. To adjust the position between measuring plane and heart, a magnetic resonance image was taken for each subjects.

RESULTS

Figure 6 shows the sequence of the typical isofield map of the QRS wave of tangential components and the z-component. A tangential map was obtained by combination of x- and y-components: $|\boldsymbol{B}xy|=|\boldsymbol{B}x+\boldsymbol{B}y|$. The map of the z-components showed the typical pattern widely reported [4]. The location of electroactive area is usually estimated to be between the positive and the nagative peaks in the z-component countour map. At the Q wave, the simplified dipole directed from the left to the right of ventricular muscle cell at the septum near the apex. The map of the tangential

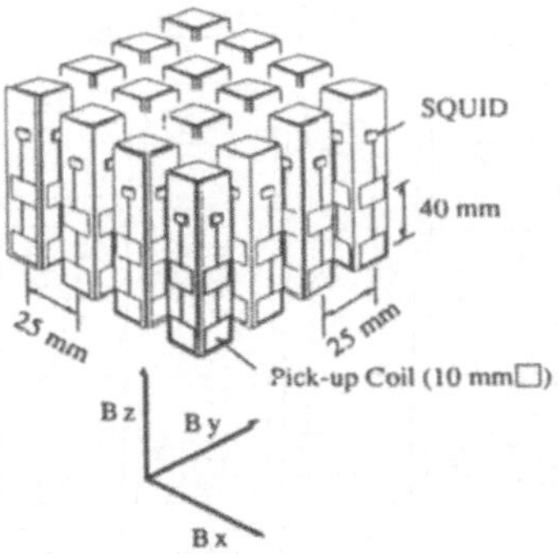

Fig. 4. Gradiometer arry.

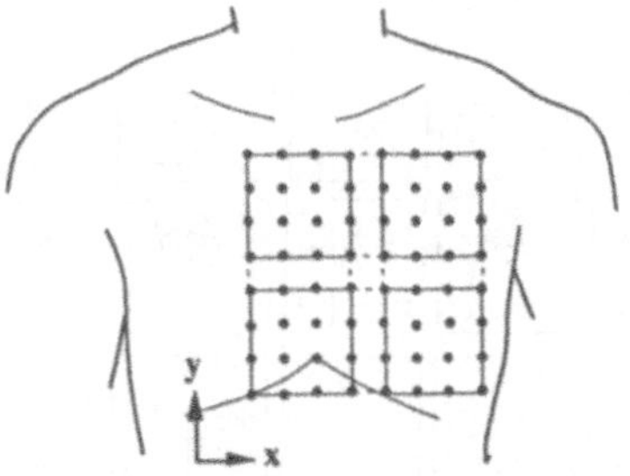

Fig. 5. MCG recording sites.

components, on the other hand, showed a different pattern. The peak was located at the same point of the estimated dipole in the z-component map. After the Q wave a new leftwardly directed simplified dipole appered, and the largest dipole directed leftward and inferior appeared gradually. At the maximum point of the R wave, the dipole moved leftward and was located in the left ventricle. The tangential map expressed a more complicated cardiac generator. More than two different directed dipoles, leftward and inferior, are easily recognized. At the S wave the direction of the dipole changed rightward and superior, and the location of the active area was thought to correspond to the movement from apex to base in the tangential map.

To image the propagating direction of the current source, we calculated a current arrow map. In the case of z-component, an arrow is defined as $a = \partial Bz/\partial y x - \partial Bz/\partial x y$ [5]. The arrow in the Bxy map is expressed by rotated 90 degrees counterclockwise of magnetic vector because the tangential components has linear relationship with x- and y-components of current density. These arrow maps, on the countour maps, are shown in Fig. 7. There is no distinctive difference between the two kinds of maps.

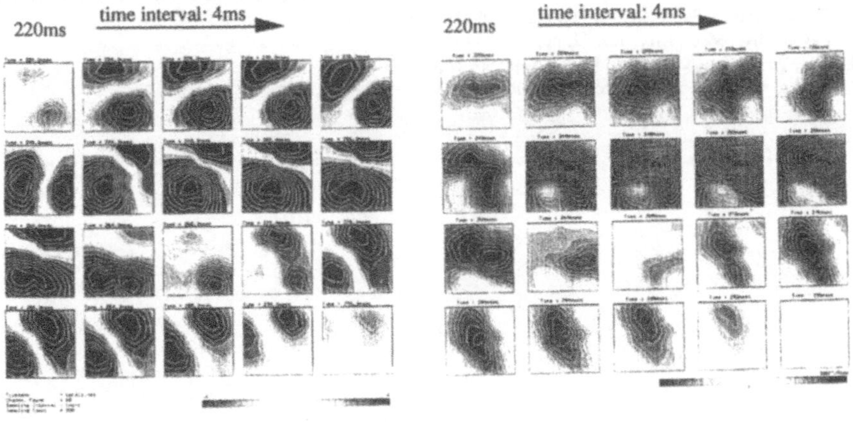

Fig. 6. Sequential isofield maps of the QRS wave.
(Q: 232ms, R: 252ms, S:280ms)

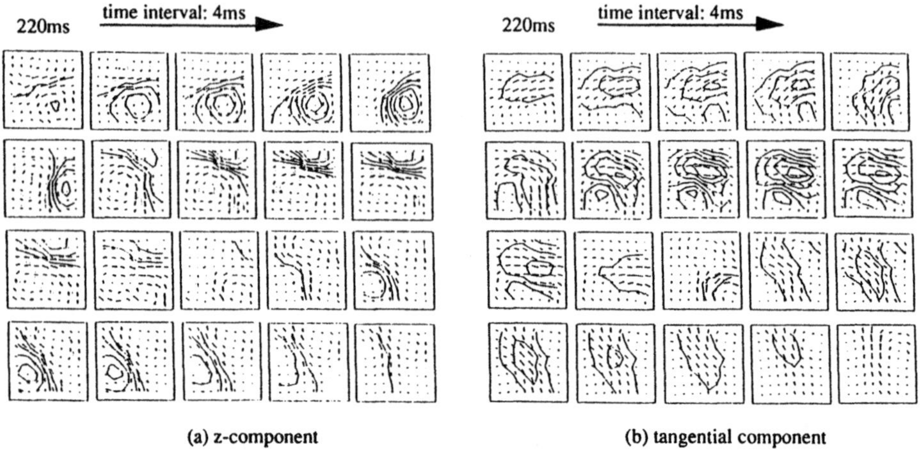

Fig. 7. Arrow maps.
(Q: 232ms, R: 252ms, S:280ms)

CONCLUSION

Tangential component mapping can be recognized as a projection of the cardiac current distribution in the x-y plane. Thus it is easy to visualize electrophysiological behavior without solving an inverse problem. On the other hand, tangential and z-components make similar arrow maps. The combination of the z-component and tangential components gives us more detailed information and will be useful for diagnosing heart disease.

REFERENCES

[1] J. O. Nousiainen, J. O. Lekkala, and J. A. V. Malmivuo, J. Electrocardiology, vol. 19, pp. 275-290 (1986).

[2] K. Tsukada, Y. Haruta, A. Adachi, H. Ogata, T. Komuro, T. Ito, Y. Takada, A. Kandori, Y. Noda, Y. Terada, and T. Mitsui, Rev. Sci. Instrum., vol. 66, pp. 5085-5091 (1995).

[3] J. Saravas, Phys. Med., Biol. vol. 32, pp. 11-22 (1987)

[4] H. Mori, Y. Nomura, CV World Report 1, pp. 78-86 (1988).

[5] H. Hosaka, and D. Cohen, J. Electrocardiology, vol. 9, pp 426-432 (1976).

Visualization for Electric Activities in a Human Heart

Ueda, T.[1], Kasai, N.[2], Sagane, M.[1], Chinone, K.[3], and Awano, N.[4]

DAIKIN Industries, LTD.[1], Kusatsu, Japan; Electrotechnical Laboratory[2], Tsukuba, Japan; SEIKO Instruments Inc.[3], Matsudo, Japan; OHARA Medical Centor[4], Fukushima, Japan

Introduction

The reconstruction of distributed current sources in a human heart seems to be an interesting aspect in clinical diagnosis like myocardial ischemia and infarction. There are several studies to reconstruct pseudo-continuous current density distributions using lead field with anatomical constraints [1]. In this paper, we propose a novel method to estimate distributions of source currents from magnetocardiographic data without anatomical constraints. The proposed method have been investigated by using real MCG data of a normal subject and a subject with right bundle branch block.

Methods

In this method, we use a micro current source with a single unknown parameter (u_i). As shown in Fig.1, let us assume that a MCG source is a group of micro current sources. In the case that we measure normal component (z component) of MCG, two sources are fixed on every grid point in a search space. One source is assigned to a x-component of current vector and the other is assigned to a y-component.

The estimated value O_j for the output of sensor j is given by

$$O_j=\sum_{i=1}^{m}\alpha_{ij}u_i \qquad (1)$$

where m is the number of sources, α_{ij} is the sensitivity coefficient of sensor j for a unit source located at the position i.

The evaluation function E is given by

$$E=\sum_{j=1}^{n}(S_j-O_j)^2 \qquad (2)$$

where n is the number of sensors, S_j is the value measured by sensor j.

The E can be expanded concerning u_i as follows:

$$\begin{aligned}E&=\sum_{j=1}^{n}\{(S_j-O_j+\alpha_{ij}u_i)-\alpha_{ij}u_i\}^2\\&=\sum_{j=1}^{n}(S_j-O_j+\alpha_{ij}u_i)^2-2u_i\sum_{j=1}^{n}\{\alpha_{ij}(S_j-O_j+\alpha_{ij}u_i)\}+u_i^2\sum_{j=1}^{n}\alpha_{ij}^2\\&=(A_iu_i-B_i)^2+C_i-B_i^2\end{aligned} \qquad (3)$$

where

$$A_i^2=\sum_{j=1}^{n}\alpha_{ij}^2, \qquad (4)$$

$$A_iB_i=\sum_{j=1}^{n}\{\alpha_{ij}(S_j-O_j+\alpha_{ij}u_i)\}, \qquad (5)$$

$$C_i=\sum_{j=1}^{n}(S_j-O_j+\alpha_{ij}u_i)^2. \qquad (6)$$

The equation (3) shows that the E is minimized at $u_i=B_i/A_i$ and the curvature of the E depends on the distance

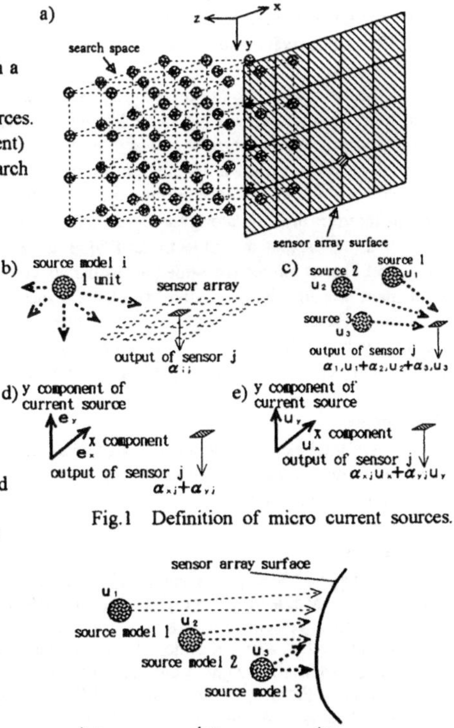

Fig.1 Definition of micro current sources.

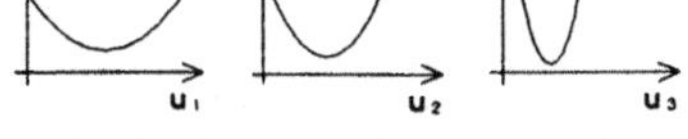

Fig.2 Concern with E, u_i and length.

between the sensor and the current source as shown in Fig.2.

But in the case that m is greater than n, we propose the base transition law to search the most appropriate micro source by the normalized evaluation function E_i as follows:

$$\Delta u_i = \frac{1}{\sum_{j=1}^{n} \alpha_{ij}^2} \sum_{j=1}^{n} \{\alpha_{ij}(S_j - O_j)\} \quad , \tag{7}$$

$$u_i \leftarrow u_i + \varepsilon \Delta u_i \quad , \tag{8}$$

$$E_i = \frac{1}{\sum_{j=1}^{n} \alpha_{ij}^2} \sum_{j=1}^{n} (S_j - O_j)^2 \quad , \tag{9}$$

where ε is a infinitesimal value.

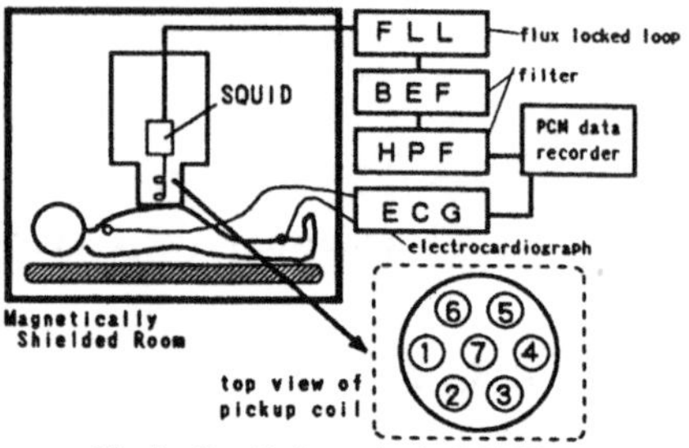

Fig.3 Our 7-channels dc-SQUID system.

The sensitivity constant α_{ij} is determined only by the geometric relationship between grid point i and measuring point j. In this paper, it was calculated by using the Biot-Savart law.

Measurements and Results

The MCGs were measured over the chests of the subjects by using a 7 channels dc-SQUID system with 1st order gradiometer in a magnetically shielded room [2] as shown in Fig.3. The measurement points were 126 points. The sensitivity of the system was 19fT/√Hz. Fig.4 shows the positional relationship between the base point and anatomical features. Fig.5 shows each position of measuring points.

The distributions of source currents at the beginning of QRS complex were estimated by the above method using the MCGs. Some examples of the results with the normal subject (male, 34yrs.) and the subject (male, 57yrs.) with right bundle branch block (RBBB) are shown in Fig.6 and in Fig.7 superimposed on their own MRI respectively. The circles show the positions of current sources and its radius shows the strength of the current source.

In the case of the normal subject, cardiac excitements seem to begin at about the superior middle-interventricular septum and proceed to the right part and the left part in the ventricular septum. Afterwards, the right excitement proceeds to right inferior direction and the left one spread gradually toward anterior inferior direction along the interventricular septum.

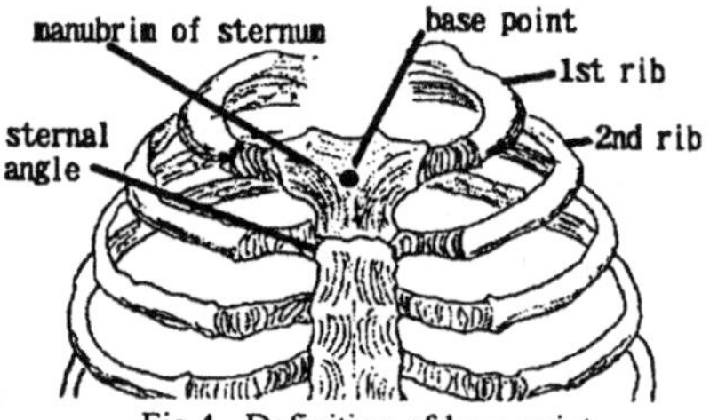

Fig.4 Definition of base point.

In the case of the subject with RBBB, the excitement seems to begin at almost the same part as the above case. The excitement branches into four and two of the branches proceed to the anterior interventricular septum. The another proceeds to the superior wall of left ventricle and the other proceeds to the right inferior and posterior wall through the interventricular septum.

Discussion

On the basis of experimental MCG data, it has been demonstrated that this method is useful to reconstruct distributed current sources.

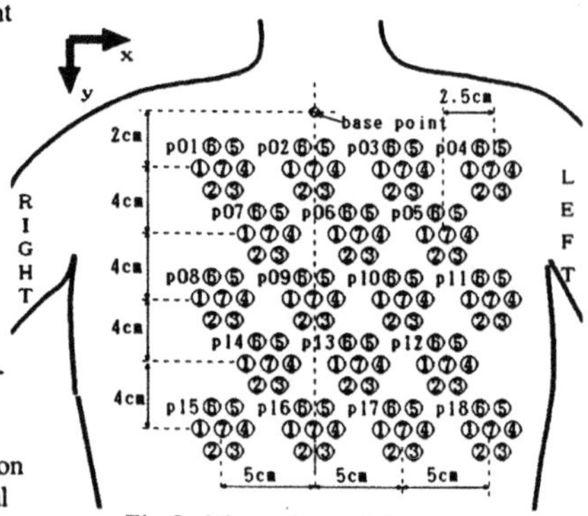

Fig.5 Measuring points.

References

[1] W.Moshage, B.Seese, S.Achenbach, R.Killmann, K. Bachmann, "Magnetocardiographic (MCG) Analysis of Myocardital Injury currents", Abstracts of Biomag'93: 376-377.

[2] T.Kokubun, N.Awano, K.Kido, M.Ono, Y.Maruyama, R. Horiuchi, N.Kasai, "Magnetocardiographic Source Localization of Accessory Pathway: Comparison with Electrophysiological Study", Abstracts of Biomag'93: 319-320.

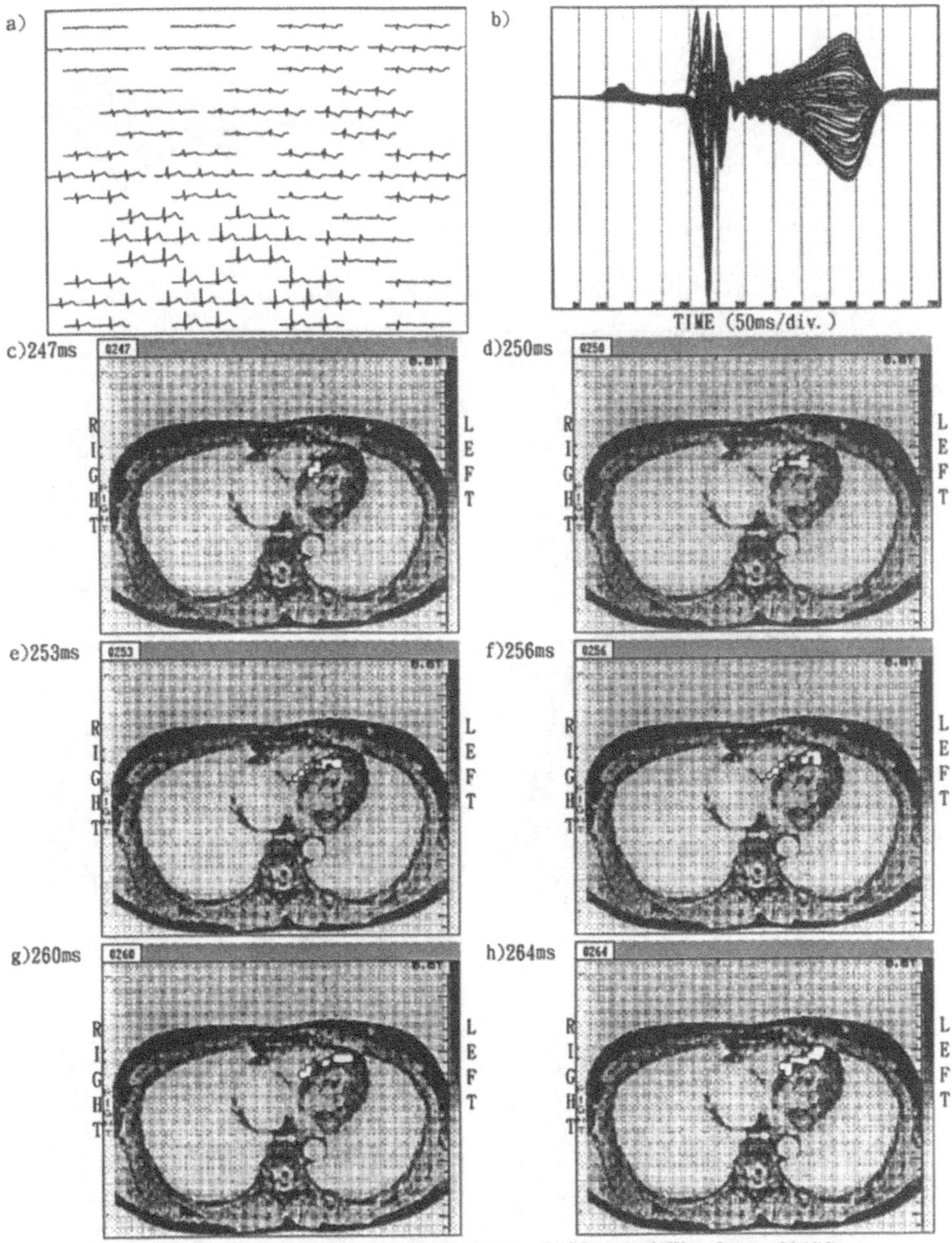

Fig. 6 Results of a normal subject (sex:male age:34). a)MCG maps. b)Waveforms of MCGs. c)-h)Current maps on his MRI sample. The depth of those currents is ignored in these figures.

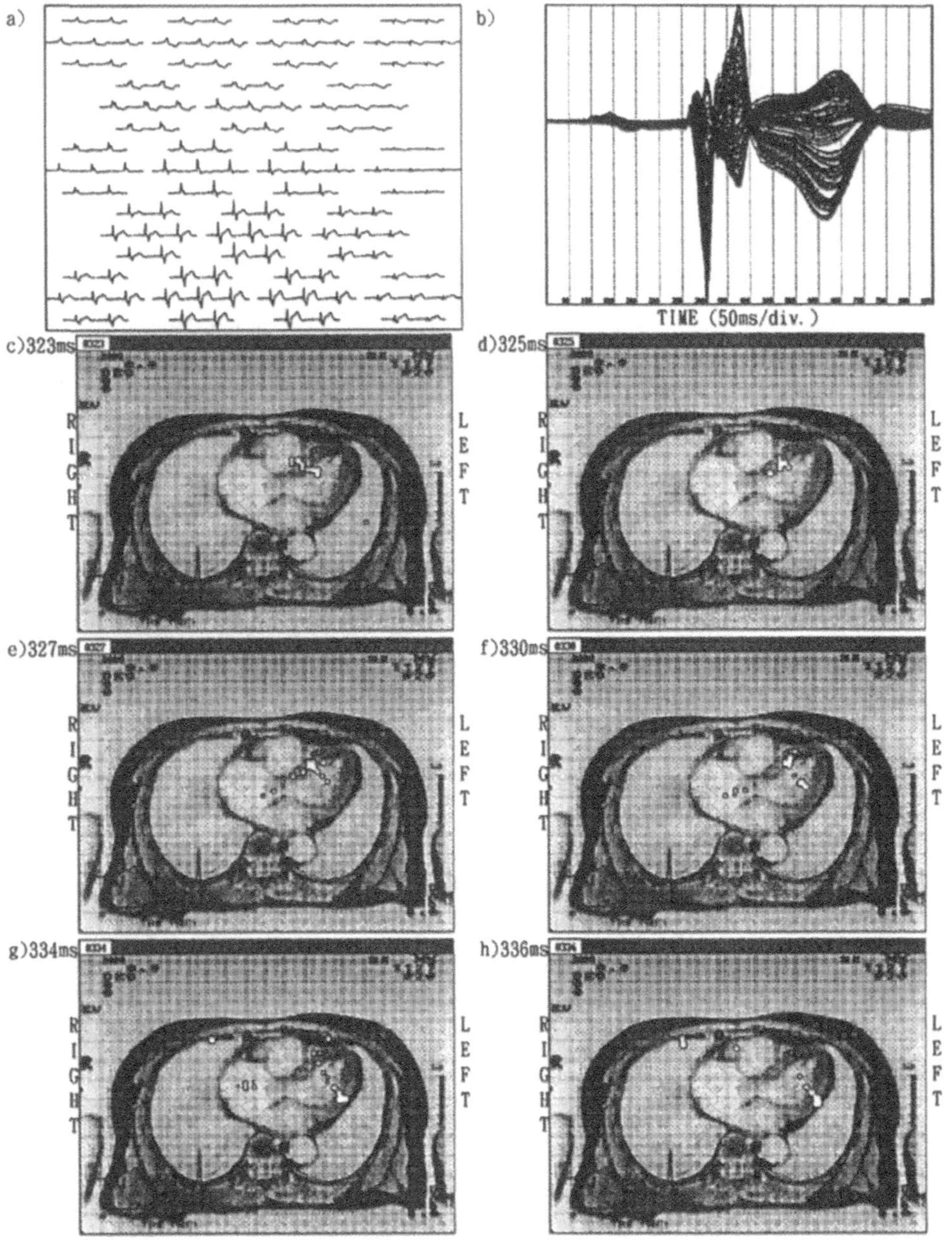

Fig.7 Results of a subject with RBBB (sex:male age:57). a)MCG maps. b)Waveforms of MCGs. c)-h)Current maps on his MRI sample. The depth of those currents is ignored in these figures.

Magnetocardiography in the Determination of Fetal Heart Rate Complexity

Van Leeuwen, P.[1], Bettermann, H.[2], Schüßler, M.[3] and Lange, S.[1]

[1]Department of Biomagnetism, Research and Development Center for Microtherapy (EFMT), Bochum, Germany; [2]Institute for Mathematics, University of Witten/Herdecke, Witten, Germany; [3]Department of Obstetrics, Marienhospital, Witten, Germany

Introduction

Heart rate variability (HRV) is an important marker for fetal well-being. Various methods have been used to analyse HRV, in particular parameters from the time and frequency domain [1]. Recently mathematical methods have been introduced which go beyond the examination of overall variance or rhythmic changes in heart rate [2,3]. The goal is to quantify temporal complexity by examining the chaoticity of time series in a reconstructed state space. These methods can be applied to fetal heart period time series to examine whether irregular but typical nonlinear dynamical heart rate patterns can be identified. These patterns might then be linked to fetal well-being.

Fetal heart rate monitoring is generally performed with the cardiotocogram (CTG) or the fetal electrocardiogram (FECG). Both methods have certain drawbacks. The former delivers solely the momentary heart rate with a limited accuracy. The latter produces the electric waveform but, due to artifacts and low signal to noise ratio, can only be applied later in pregnancy. As such these methods are not suited for the acquisition of data earlier during gestation and as such do not allow for an analysis of heart period complexity during the course of pregnancy. On the other hand, it has been shown that magnetocardiography can be used to register fetal cardiac activity as early as the beginning of the second trimester with sufficient precision to identify single QRS complexes and that, in comparison to FECG, it is not influenced by the insulating effect of the vernix caseosa between the 28th and 30th week of gestation [4,5]. It is therefore possible to quantify fetal heart period with high accuracy, allowing for the analysis of time series of fetal heart beats starting before the 20th week of gestation.

Goal of this study was to combine these new methods in the registration and analysis of fetal cardiac activity in order to examine changes in fetal heart period complexity in the course of pregnancy. If recognizable patterns can be established for normal gestational development then these methods will open up new diagnostic possibilities which have the added advantage of being completely non invasive.

Methods

Subjects: We investigated eleven women who presented with healthy singleton pregnancies in the second trimester. Goal was to perform magnetocardiographic registration of fetal cardiac activity (FMCG) at regular intervals of roughly four weeks up to the time of birth. In order to investigate the time course at least three measurements with data amenable to evaluation were required for each fetus. On that basis three fetuses were excluded from the study. One further data set was excluded because gestosis developed. Of the remaining seven fetuses, 36 recordings were acquired between the 16th and 40th week of gestation. The quality of five recordings, all before the 20th week of gestation, was too poor for complexity examination, leaving 31 for evaluation. The mean number of recordings per fetus was 4.4 ± 1.1 ($\pm$ standard deviation, range 3-6), mean time between recordings was 4.1 ± 2.2 weeks (range 1-11). Due to calculation restrictions, the parameter ApD1 could only be calculated for 24 recordings.

Data Acquisition: FMCG recordings were performed with a 37 channel biomagnetometer (Siemens Krenikon). The diameter of the probe is 19 cm. Each sensor consists of a first order gradiometer with a 7 cm baseline. System noise is less than 30 fT/Hz^{-1} for frequencies > 2 Hz and 10 fT/Hz^{-1} for frequencies > 10 Hz. To reduce the influence of electromagnetic artifacts, the measurements were performed in a shielded room (Vakuumschmelze). The position of the probe was placed over the maternal abdomen such that the fetal signal was optimized. By recording a limb lead maternal ECG the fetal QRS complexes could be clearly identified. Data was recorded for five minutes with a bandpass of 1-200 Hz at a sampling rate of 1 kHz.

Signal analysis: In order to extract the fetal heart period, fetal QRS complexes were marked using an appropriate template in a channel of raw data with strong fetal signal and minimal interference of maternal components. Complexes not automatically recognized were visually identified and added manually; similarly, falsely marked artifacts were removed. Heart period was determined with an accuracy of ±0.5 ms as the time between consecutive QRS complexes. The time series of a complete set of fetal heart periods served as the basis for further analysis.

Complexity analysis: Two parameters were used to describe the heart dynamics of each fetus at the time of recording: the approximate information dimension (ApD1) and the approximate entropy (ApEn). In contrast to moment statistics, these measures take the order of the data into consideration. Both are based on the correlation integral which reflects the number of data points which are within a defined distance embedded in a *k*-dimensional state space. The complexity measures ApD1 and ApEn characterize the scaling (static) properties of the distribution as well as the expansion behavior (dynamical properties) of the generating dynamics. Methods used in the determination of the approximate dimension and approximate entropy have been described elsewhere [2,6-8]. Apart from the original definitions, we used a strictly local approach for the approximation of the information dimension and the Kolmogorov entropy. By strictly local we mean the computation of local or pointwise dimensions and entropies for reference points generally with their symmetric neighborhood in time. The local approximate dimensions and the local approximate entropies were averaged over the entire data set and provided the values for ApD1 and ApEn (see Appendix for details).

ApD1 and ApEn were determined for each time series. Furthermore, time domain statistics in the form of the mean beat duration, the standard deviation (SD) and the root mean square of successive differences (RMSSD) were also calculated. For the calculation of the time domain statistics, outliers were visually identified in the heart period plot and eliminated from the time series.

In order to examine the dependency of the parameters on gestational age (GA), simple linear regression analysis was performed for each parameter for each fetus over week of gestation. The slope of the regression was taken as a measure of the degree of change of each parameter over time. All values are expressed as mean ± SD. Linear regression analysis was performed according to the method of least squares. r^2 was determined to control the quality of the linear model.

Results

The number of beats registered per five-minute recording ranged from 629 to 964 (mean: 744 ± 65). The average beat duration was 412 ± 24 ms (heart rate: 146 ± 9 bpm). The values of beat period, SD, RMSSD, ApEn and ApD1 tended to increase for most fetuses during pregnancy (see Fig. 1).

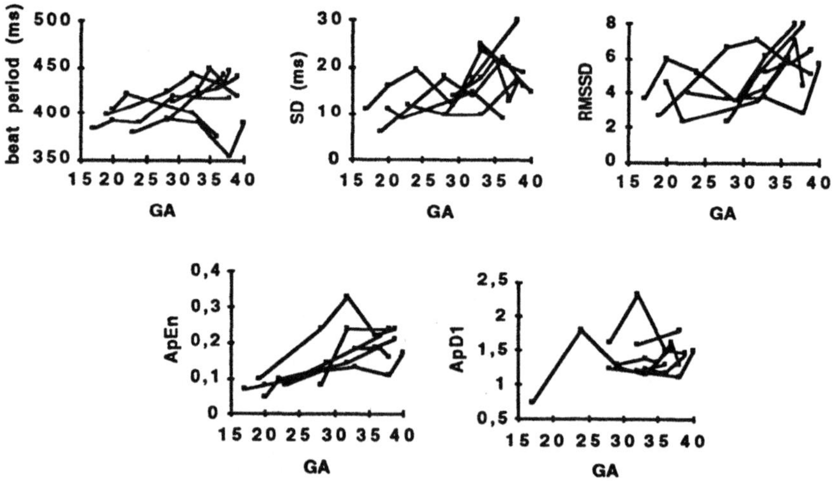

Fig. 1: Course of beat period, SD, RMSSD, ApEn and ApD1 for each fetus with respect to gestational age (GA).

Regression analysis indicated that there was a clear increase in beat period over the gestational period observed for five of the fetuses (Table 1), reflecting a decrease in heart rate of approximately 25 bpm. Increases in the time domain measures of HRV, SD and RMSSD, were measurable on the basis of positive slopes in six subjects. However, the r^2 values were generally low. In all seven fetuses the slopes for ApEn were positive, for ApD1 in six, indicating an increase in complexity in the course of pregnancy. In particular, the results for ApEn demonstrated high r^2 values; the lower correlation results for ApD1 may be due to the lower number of observations.

Table 1: Slope and r^2 of the linear regression of beat period, SD, RMSSD, ApD1 and ApEn with respect to gestational age for each subject.

ID	n	beat period / r^2	SD / r^2	RMSSD / r^2	ApEn / r^2	ApD1 / r^2
Subj. 1	4	-2.09 / 0.66	0.07 / 0.03	0.11 / 0.31	0.007 / 0.90	0.03 / *
Subj. 2	5	-0.64 / 0.08	0.32 / 0.43	0.04 / 0.07	0.004 / 0.58	0.01 / 0.13
Subj. 3	5	3.52 / 0.73	0.14 / 0.02	0.24 / 0.39	0.003 / 0.32	0.02 / 0.18
Subj. 4	6	2.84 / 0.91	0.41 / 0.40	0.13 / 0.39	0.008 / 0.97	0.02 / 0.20
Subj. 5	3	2.49 / 0.64	2.00 / 1.00	0.57 / 0.98	0.015 / 0.64	0.04 / *
Subj. 6	3	2.83 / 0.91	-1.00 / 0.96	-0.01 / 0.02	0.005 / 0.96	0.04 / 0.59
Subj. 7	5	1.41 / 0.43	0.58 / 0.58	0.17 / 0.57	0.007 / 0.42	-0.03 / 0.15
mean ± SD		1.48 ± 2.09	0.36 ± 0.89	0.18 ± 0.19	0.007 ± 0.01	0.02 ± 0.02
Min / Max		-2.09 / 3.52	-1.00 / 2.00	-0.01 / 0.57	0.003 / 0.015	-0.03 / 0.04

SD, standard deviation; RMSSD, root mean square of successive differences; ApD1, approximate dimension; ApEn, approximate entropy.

*No calculation of r^2 because <3 values for estimation of regression model.

Discussion

Analogous to time and frequency domain analysis, complexity analysis can be understood as a time series analysis in the state space domain. The generating dynamics in state space possesses folding and stretching properties which may be analysed by appropriate dynamical measures. ApD1 and ApEn characterize both the scaling (static) properties and the expansion behavior (dynamical properties) of the generating dynamics. Application of these measures on physiological parameters have demonstrated their ability to discriminate between states. For example, the examination of heart rate and blood pressure variability in adults have shown decreasing complexity with age [7].

Several studies have been done on fetuses during labor, applying these analysis techniques on data acquired by scalp ECG. The feasibility of dimensional analysis was thus demonstrated by Chaffin et al [8]. Also, on the basis of ApEn, a reduction in heart rate complexity could be shown in fetuses with acidosis and distress [3]. In contrast, time domain variables such as near-term HRV and mean heart rate did not distinguish between the groups. Furthermore, a decrease in ApEn after drug administration could be documented.

These studies suggest that complexity measures may be a sensitive analytic tool in quantifying fetal heart rate complexity. Their application during pregnancy depends on the possibility of acquiring data with high temporal resolution. At present cardiotocography is the clinical method of choice in the monitoring of fetal heart rate. However it is not suited for analysis of complexity due to limitations in accuracy. The precision in the determination of fetal heart beat duration can in principle be greatly improved by electrocardiography but, among other things, practical problems in signal acquisition and processing have hampered implementation during pregnancy. Our results and other studies, e.g.[4,5], show that FMCG enables the simple and effective acquisition of fetal data. The high quality of the magnetocardiographic signal permits the construction of a time series with the necessary temporal accuracy as early as the second trimester.

The analysis of the data collected here showed that the complexity of fetal heart rate increased throughout the second and third trimester of normal pregnancy. In particular, ApEn was very consistent and its reliability was reflected in high correlation values. ApD1 increased for all but one fetus over time, although the results must be viewed with caution on the basis of the r^2 values. However, these low correlations may well be due to the lower number of values that could be calculated. This disadvantage will be overcome in future by a different approach in the analysis. As both parameters are independent of the mean and variance of the signal, they have advantages over time domain statistics: they are unaffected by noise, robust to outliers and model independent [6]. In contrast, in order to obtain reasonable values for beat period, SD and RMSSD from the data recorded here, the elimination of outliers was necessary. In spite of this correction of the data, the results for these statistical measures were not as consistent as the complexity measures. Nonetheless, it must be noted that, in the analysis of HRV, such measures are complementary [6] and reflect different aspects of the structure of the time series.

In summary, FMCG may be used as a method to obtain data amenable to the analysis of the temporal complexity of the fetal cardiac system. The observed increase in complexity during the course of pregnancy may be interpreted in terms of the development of the cardiac system. The degree to which this reflects the maturation of the autonomic nervous system and the growth of the fetus as a whole is still to be determined. Further studies with more observations are required to define norm values and to examine the effect of pathological conditions.

References

[1] Cerutti, S.S., Baselli, G., Civardi, S., Ferrazzi, E., Marconi, A.M., Pagani, M., Pardi, G. Variability analysis of fetal heart rate signals as obtained from abdominal ECG recordings, J Perinat Med, 1986, 14: 445-452.

[2] Kaplan, D.T., Furman, M.I., Pincus, S.M. Techniques for analyzing complexity in heart rate and beat-to-beat blood pressure signals, In: IEEE Comput Cardiol, Los Alamitos, CA, IEEE Computer Society Press, 1990: 243-246.

[3] Pincus, S.M. , Viscarello, R.R. Approximate Entropy: A Regularity Measure for Fetal Heart Rate Analysis, Obstet Gynecol, 1992; 79: 249-255.

[4] Dunajski, Z., Peters, M. Development of the magnetocardiograms from the 13th week of gestation onward. In: Baumgartner, C., Deecke, L., Stroink, G., Williamson, S.J. Biomagnetism: Fundamental Research and Clinical Applications, Amsterdam, Elsevier/IOS Press, 1995: 340-341.

[5] Quinn, A., Weir, A., Shahani, U., Bain, R., Maas, P., Donaldson, G. Antenatal fetal magnetocardiography: a new method for fetal surveillance? Br J Obstet Gynaecol, 1994, 101: 866-870.

[6] Pincus, S.M. Quantification of evolution from order to randomness in practical time series analysis, Methods in Enzymology, 1994, 240: 68-89.

[7] Kaplan, D.T., Furman, M.I., Pincus, S.M., Ryan, S.M., Lipsitz, L.A., Goldberger, A.L. Aging and the complexity of cardiovascular dynamics, Biophys J ,1991, 59: 945-9.

[8] Chaffin, D.G., Goldberg, C.C., Reed, L.R. The dimension of chaos in the fetal heart rate, Am J Obstet Gynecol, 1991,165: 1425-1429.

Appendix

Without loss of generalisation, we interpret the RR-tachogram as a 'time' series with an one beat delay. k successive RR-intervals form a vector in the reconstructed state space indexed by the beat number of the leading RR-interval:

$$\vec{RR}_i = (RR_i, RR_{i+1}, \ldots, RR_{i+k-1}) \in \text{reconstructed state space.}$$

Both ApEn and ApD1 are based on the correlation integral $C_i^k(l)$ which is defined as the relative number of k-dimensional embedded data points within a l-sphere around a reference point with index i, as a function of the distance l:

$$C_i^k(l) = \frac{1}{N-k+1}\{\text{number of vectors } \vec{RR}_j \text{ with } \left|\vec{RR}_i - \vec{RR}_j\right| < l\},\ j \in \left[i - \text{trunc}\frac{N}{2}, i-k+1+\text{trunc}\frac{N}{2}\right]$$

For the determination of the pointwise approximation of fetal heart beat period complexity the following parameter values were set:

- N = 501 RR-intervals. This allowed the calculation of roughly 150-350 local complexity values per recording.
- Embedding dimension: $k = 3$.
- Fixed tolerance for ApEn: $l = 10$ ms.
- The local approximate dimension is defined as the slope of the mean of the logarithmic local correlation integrals between two length scale values l_a and l_b which are chosen such that

$$\log_{10} C_i^k(l_a) = -2.5 \text{ and } \log_{10} C_i^k(l_b) = -0.5.$$

Fetal Magnetocardiography: Spectral Analysis of Fetal Heart Rate Variability in the Course of Pregnancy

Van Leeuwen, P.[1], Hartmann, M.[1], Lange, S.[1] and Schüßler, M.[2]

[1]*Department of Biomagnetism, Research and Development Center for Microtherapy (EFMT), Bochum, Germany;* [2]*Department of Obstetrics, Marienhospital, Witten, Germany*

Introduction

Spectral analysis of heart rate has shown that changes in the power of frequency bands below 1 Hz reflect physiological changes associated with the autonomic nervous system (ANS) [1]. In particular the low frequency band (LF) between 0.04 and 0.15 Hz is often associated with sympathetic as well as parasympathetic activity, whereas the high frequency band (HF) in the range from 0.15 to 0.4 Hz is mediated by parasympathetic activity. An overall index of sympathovagal balance in the control of the heart rate is given by the ratio LF/HF. Furthermore, in adults the HF peak reflects the modulation of heart rate due to respiration known as respiratory sinus arrhythmia (RSA).

Using various forms of data acquisition, spectral analysis of fetal heart rate (FHR) has found similar results for the unborn child: LF and HF peaks can be identified and can be associated with fetal state and respiration [2-7]. However, little work has been done to examine progressive changes in FHR spectra during the course of gestation. One of the factors hindering such work is likely the fact that common methods of fetal heart rate monitoring such as cardiotocography and fetal electrocardiography do not offer the possibility of high quality data acquisition from an early stage of pregnancy onwards. Magnetocardiography, on the other hand, offers the possibility to record fetal cardiac signals with high spatial and temporal resolution as early as the 13th week of gestation and data acquisition is not affected by the electrical insulatory effects of the vernix caseosa [8]. Thus, our goal was to examine fetal heart rate variability on the basis of spectral analysis as gestation progresses, using fetal magnetocardiography (FMCG) registered regularly during pregnancy.

Material and Methods

Subjects: We investigated nine women who presented with healthy singleton pregnancies as of the second trimester. Magnetocardiographic registration of fetal cardiac activity was performed at regular intervals of roughly four weeks up to the time of birth. On that basis 39 recordings were acquired between the 16th and 40th week of gestation. The quality of four recordings, all before the 20th week of gestation, was poor, leaving 35 for evaluation. The mean number of recording per fetus was 3.9 ± 1.8 (± standard deviation, range 1-7), the mean time between recordings was 4.3 ± 2.4 weeks (range 1-11 weeks).

Data Acquisition: FMCG recordings were performed with a 37 channel biomagnetometer (Siemens Krenikon). The diameter of the probe is 19 cm. Each sensor consists of a first order gradiometer with a 7 cm baseline. System noise is less than 30 fT/Hz^{-1} for frequencies > 2 Hz and 10 fT/Hz^{-1} for frequencies > 10 Hz. To reduce the influence of electromagnetic artifacts, the measurements were performed in a shielded room (Vakuumschmelze). The position of the probe was placed over the maternal abdomen such that the fetal signal was optimized. By recording a limb lead maternal ECG, the fetal QRS complexes could be clearly identified. Data was recorded for five minutes with a bandpass of 1-200 Hz at a sampling rate of 1 kHz.

Signal Analysis: Signal analysis was performed in three steps: 1) determination of a time series of beat-to-beat intervals, 2) transformation of the time series into a time vector with length equal to a power of two and 3) implementation of the fast Fourier transformation (FFT). In order to extract fetal heart periods, the fetal QRS-complexes were marked using an appropriated template in a channel of the raw data with a strong fetal signal, low noise and minimal interference of the maternal components. Complexes not automatically recognized were visually identified and added manually; similarly, falsely marked artifacts were removed. Heart period was determined with an accuracy of ±0.5 ms as the time between consecutive QRS complexes. The time series of a complete set of fetal heart periods over five minutes, thus extracted from each recording, served as the basis for further analysis.

In order to change the event based ordinal scale (beats) to a time based scale, a step function was defined in which step height and duration were determined by beat period length. The use of the FFT on data in the time domain, i.e. avoiding the use of event steps, is more appropriate and easier to interpret. In order to avoid distortions resulting from zero padding, the step function was ranged in a vector with a length equal to a power of two. Subsequently, the vector underwent an FFT using the Danielson Lanczos-algorithm.

The resulting frequency spectra were normalized with respect to the total power and were band integrated with an increment of 0.1 Hz in the range between 0.0-1.0 Hz. Furthermore sympathovagal balance was calculated as LF/HF with LF being the integral between 0.04 and 0.15 Hz and HF the integral between 0.15 and 0.4 Hz. The results were grouped and averaged in 5 week intervals starting with the 16-20 week of gestation and ending with the 36-40 week (phase 1-5; see Table 1).

Results

Several bands showed distinct tendencies to change progressively in the course of gestation. The low frequency band from 0.0-0.1 Hz increased by 20% whereas the bands between 0.2 and 0.6 Hz decreased by 20-42%. Between 0.7 and 1.0 Hz the changes were minimal and not ordered (Fig.1, Table 1). LF/HF as an expression for sympathovagal balance displayed an increasing trend with the course of gestation (Fig. 2).

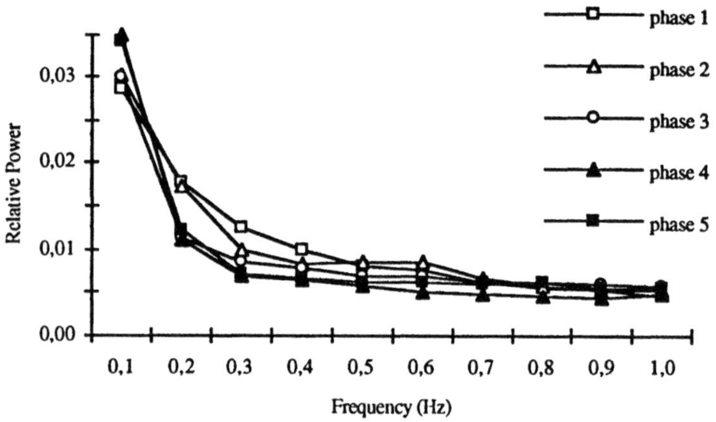

Fig. 1: Power in 0.1 Hz bands between 0 and 1 Hz at various stages in pregnancy.

Table 1: Power in 0.1 Hz bands between 0 and 1 Hz (mean ± SD) at various stages in pregnancy.

	phase 1	phase 2	phase 3	phase 4	phase 4
week of gestation	16-20	21-25	26-30	31-35	36-40
n	5	5	6	8	11
0.0-0.1 Hz	.029 ± .015	.030 ± .012	.030 ± .010	.035 ± .009	.034 ± .007
0.1-0.2 Hz	.018 ± .011	.017 ± .009	.011 ± .005	.011 ± .003	.012 ± .005
0.2-0.3 Hz	.012 ± .006	.010 ± .003	.009 ± .004	.007 ± .002	.007 ± .001
0.3-0.4 Hz	.010 ± .005	.008 ± .002	.008 ± .003	.006 ± .002	.007 ± .001
0.4-0.5 Hz	.008 ± .004	.009 ± .004	.007 ± .001	.006 ± .001	.006 ± .001
0.5-0.6 Hz	.008 ± .003	.008 ± .004	.007 ± .001	.005 ± .002	.006 ± .001
0.6-0.7 Hz	.006 ± .002	.007 ± .001	.006 ± .000	.005 ± .002	.006 ± .001
0.7-0.8 Hz	.006 ± .002	.006 ± .001	.006 ± .000	.005 ± .002	.006 ± .001
0.8-0.9 Hz	.005 ± .001	.005 ± .001	.006 ± .001	.004 ± .002	.006 ± .001
0.9-1.0 Hz	.005 ± .001	.005 ± .001	.006 ± .001	.005 ± .002	.006 ± .001

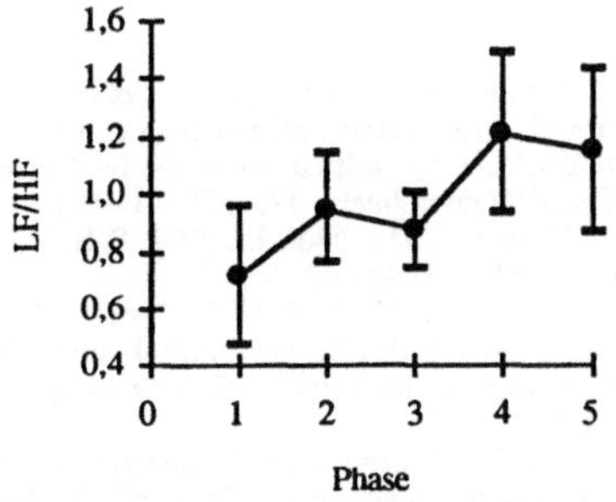

Fig. 2: Change in LF/HF balance (mean ± SD) in the course of pregnancy.

Discussion

In adults and newborns, physiological function as mediated by the ANS may be identified by examining the frequency spectrum of heart rate changes. Sympathetic and parasympathetic tone are reflected in selected spectral bands at frequencies below 1 Hz. In particular, a LF and a HF peak, between 0.04-0.15 Hz and 0.15-0.4 Hz respectively, can be associated with the inputs of the ANS to the sinoatrial node which modulate heart rate. One may accordingly expect that the analysis of the spectrum of fetal heart rate fluctuations will give insight into the development and state of the ANS during gestation. This in turn may well reveal information on the development and well-being of the unborn child as a whole.

To date, the examination of fetal heart rate traces acquired using cardiotocography, fetal ECG as well as FMCG have allowed investigators to identify a LF peak around 0.1 Hz [2,4,7]. One of these groups was able to associate changes in peak height to growth retardation at term [2]. Our data show a large proportion of the power in a band up to 0.1 Hz, which increases as pregnancy progresses. To the degree that this frequency band reflects sympathetic activity, these results may be interpreted as an increase in sympathetic tone which accompanies increased wakefulness and fetal activity with greater gestational age. This interpretation is further supported by the reported examination of the LF peak in preterm infants (< 33 weeks) which demonstrated that this peak increased with gestational age [9].

Karin et al. [4] have described increased power in the 0.2-0.7 Hz band around the 24th week of gestation, which was reduced in mature fetuses (38th week). They suggest that the higher power early in pregnancy is a result of the immature ANS which generates unarrayed changes in heart rate. Thus, reduced power in this band later in pregnancy would be indicative of a coordination of function which accompanies maturation of the ANS. Our results confirm those findings and support them further by demonstrating that the decrease in power changes continuously and consistently in this band throughout gestation.

A further index describing sympathovagal balance is the ratio of LF to HF. Activity in the HF band is mediated by solely vagal system whereas both the sympathetic and parasympathetic systems influence the LF band. Examining the LF/HF ratio in the course of pregnancy of the fetuses studied here, indicate an increase in values from prior to the 20th week to the period between the 20th and 30th week and a further increase after the 30th week. On the one hand, this doubling of LF/HF over the observed period further supports the notion mentioned above that sympathetic tone increases with gestational age. On the other, by taking various bands into consideration, this increase may also reflect the status of the maturing ANS and, as such, serve as a measure for the development of neural coordination. It remains to be seen to what degree various pathological conditions can lead to deviations from the patterns described here.

Our data did not show evidence of respiratory activity which is usually characterized by a clear peak centered between 0.4 and 1 Hz [3,5-7]. This may, among other things, be due to the length of the data set examined. As breathing episodes tend to be between 30-60 seconds, such episodes, should they be present, would be smeared out. It is expected that examination of shorter traces will reveal respiratory peaks. Detailed spectral examination of heart beat time series, in particular in combination with other measures of variability, may reveal more about behavioral state in the fetus.

References

[1] Akselrod, S., Gordon, D., Ubel, F.A., Shannon, D.C., Barger, A.C., Cohen, R.J. Power spectrum analysis of heart rate fluctuation: a quantitative probe of beat-to-beat cardiovascular control, Science, 1981, 213: 220-222.

[2] Breborowicz, G., Moczko, J., Gadzinowski, J. Quantification of the fetal heart rate variability by spectral analysis in growth-retarded fetuses, Gynecol Obstet Invest, 1988, 25: 186-191.

[3] Divon, M.Y., Yeh, S.Y., Zimmer, E.Z., Platt, L.D., Paldi, E., Paul, R.H. Respiratory sinus arrhythmia in the human fetus, Am J Obstet Gynecol, 1985, 151: 425-428.

[4] Cerutti, S.S., Baselli, G., Civardi, S., Ferrazzi, E., Marconi, A.M., Pagani, M., Pardi, G. Variability analysis of fetal heart rate signals as obtained from abdominal ECG recordings, J Perinat Med, 1986, 14: 445-452.

[5] Karin, J., Hirsch, M., Akselrod, S. An estimate of fetal autonomic state by spectral analysis of fetal heart rate fluctuations, Ped Res, 1993, 34: 134-138.

[6] Wakai, R.T., Wang, M., Pedorn, S.L., Reid, D.L., Martin, C.B. Spectral analysis of antepartum fetal heart rate variability from fetal magnetocardiogram recordings, Early Hum Dev, 1993, 35: 15-24.

[7] Rassi, D., Lewis, M.J. Power spectral analysis of the fetal magnetocardiogram, Physiol Meas, 1995, 16: 111-120.

[8] Dunajski, Z., Peters, M. Development of the magnetocardiograms from the 13th week of gestation onward. In: Baumgartner, C., Deecke, L., Stroink, G., Williamson, S.J. Biomagnetism: Fundamental Research and Clinical Applications, Amsterdam, Elsevier/IOS Press, 1995: 340-341.

[9] Van Ravenswaaij-Arts, C., Hopmann, J., Kollée, L., Stoelinga, G., Van Geijn, H. Spectral analysis of heart rate variability in spontaneously breathing very preterm infants, Acta Paediatr 1994, 83: 473-80.

Determination of Changes in Fetal Cardiac Time Intervals and T/QRS Ratio During Pregnancy Using Magnetocardiography

Van Leeuwen, P.[1], Lange, S.[1], Schüßler, M.[2] and Lajoie-Junge, L.[1]

[1]Department of Biomagnetism, Research and Development Center for Microtherapy (EFMT), Bochum, Germany; [2]Department of Obstetrics, Marienhospital, Witten, Germany

Introduction

Several studies have indicated that fetal cardiac time intervals (CTI) derived from the electric or magnetic cardiac signal increase in the course of pregnancy [1-3]. Such changes can be related to the size, development and maturity of the fetal heart and may thus serve as a measure reflecting not only normal cardiac function, but also the intrauterine growth state of the fetus as a whole. Another parameter which can be determined from the fetal cardiac signal is the amplitude of the signal at depolarisation and repolarisation times. In particular, the ratio of the T wave amplitude to the QRS complex amplitude (T/QRS) has been shown to increase in response to hypoxia during labor [4]. Little is known about changes in such amplitudes in the course of pregnancy. Among other things, this is due to the difficulty in obtaining viable electrocardiographic (ECG) signals consistently at different gestational ages (GA). ECG examination of the fetal cardiac signal is limited by noise, maternal artifacts and the insulating effects of the vernix caseosa. In this respect, fetal magnetocardiography (FMCG) has been shown to be a simple, non-invasive, and effective method in fetal monitoring, registering high quality fetal cardiac activity anytime during the pregnancy from the second trimester onward. The acquisition of the fetal cardiac magnetic signal enables the analysis of signal morphology for the complete beat with great accuracy [2]. The purpose of this study was to register fetal cardiac activity at regular intervals in the course of healthy pregnancies with magnetocardiography in order to determine fetal CTI and signal amplitudes accurately and to investigate the changes of these parameters over time.

Material

Subjects: We investigated eleven women who presented with healthy singleton pregnancies in the second trimester. Goal was to perform magnetocardiographic registration of fetal cardiac activity at regular intervals of roughly four weeks up to the time of birth. In order to investigate the individual time course of single fetuses, at least three measurements with data amenable to evaluation were required for each fetus. On that basis three fetuses were excluded from the study. One further data set was excluded because gestosis developed. Of the remaining seven fetuses, 36 recordings were acquired between the 16th and 40th week of gestation. The mean age of the seven remaining women was 30 ± 2 years (± standard deviation, range 27-34 years). The parity of three women was zero, the others were Para 1. The quality of two recordings, both acquired prior to the 20th week of gestation, was poor, leaving 34 for evaluation. The mean number of recordings per fetus was 4.9 ± 1.1 (range 3-6), the mean time between recordings was 4.5 ± 2.7 weeks (range 1-12 weeks).

Data Acquisition: FMCG recordings were performed with a 37 channel biomagnetometer (Siemens Krenikon). The diameter of the probe is 19 cm. Each sensor consists of a first order gradiometer with a 7 cm baseline. System noise is less than 30 fT/Hz^{-1} for frequencies > 2 Hz and 10 fT/Hz^{-1} for frequencies > 10 Hz. To reduce the influence of electromagnetic artifacts, the measurements were performed in a shielded room (Vakuumschmelze). The position of the probe was placed over the maternal abdomen such that the fetal signal was optimized. By recording a limb lead maternal ECG the fetal QRS complexes could be clearly identified. Data was recorded for five minutes with a bandpass of 1-200 Hz at a sampling rate of 1 kHz.

Signal Analysis: In order to extract fetal heart periods, the fetal QRS-complexes were marked using an appropriate template in a channel of the raw data with a strong fetal signal, low noise and minimal interference of the maternal components. Complexes not automatically recognized were visually identified and added manually; similarly, falsely marked artifacts were removed. Each recording consisted of ca. 600 - 800 fetal beats. These were averaged and DC-offsetted in an interval prior to P wave begin. In order to assure maximal precision in the determination of CTI, the deflections of the FMCG signal in the averaged beat were examined in all channels, thereby taking differences in signal polarity and intensity in individual channels into account. In addition, an integral signal was formed by summating the absolute values of each data point in each channel over all channels. This was used to complement the analysis of the single channels in the identification of cardiac events.

The following events were identified visually for all recordings: P begin and end, Q begin, S, T begin and end. On that basis, the following CTI were calculated to an accuracy of 1 ms: P wave (from P begin to P end), PQ interval (P end to Q begin), QRS complex (from Q begin to S), ST interval (S to T begin), T wave (T begin to T end), and QT interval (Q begin to T end). Simple linear regression analysis was performed for all intervals for each fetus over week of gestation. The slope of the regression was taken as a measure of the degree of change of each interval over time. In addition, the field integral was used to quantify the amplitudes of the P and T wave as well as of the QRS complex, from which, in turn, the T/QRS ratio was calculated. In 5 cases the maximum of the T wave could not be established, leaving 29 cases for T wave and T/QRS determination. Changes in the amplitudes of the P, T and QRS as well as the T/QRS ratio were then similarly examined over time.

Results

Although all signal defllections could be identified for all traces, their detection became increasingly easier as pregnancy progressed due to the increase in signal amplitude. The recognition of T begin and T end was particularly difficult in the single channels but was greatly facilitated by the use of the integrated signal. Of the intervals calculated, the duration of the P wave and the QRS complex increased consistently during pregnancy (see Fig. 1). This was confirmed in the regression analysis: the data for all seven fetuses showed positive slopes and high correlation values (r^2) (Table 1). Neither T wave duration nor the PQ, QT and ST intervals displayed a clear tendency to increase or decrease in the course of pregnancy.

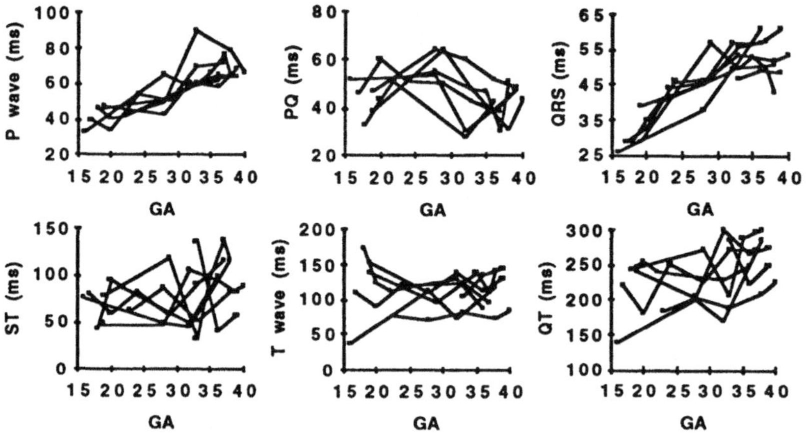

Fig. 1: Course of the CTI with respect to gestational age (GA) for the individual fetuses.

Table 1: Slope and r^2 of the linear regression of the CTI with respect to gestational age for each subject.

ID	P / r^2	PQ / r^2	QRS / r^2	ST / r^2	T / r^2	QT / r^2
Subj. 1	1.16 / 0.88	-0.22 / 0.03	1.54 / 0.95	0.72 / 0.05	-2.58 / 0.46	-0.32 / 0.07
Subj. 2	1.89 / 0.47	-1.03 / 0.58	0.60 / 0.80	1.03 / 0.18	0,27 / 0.13	1.91 / 0.60
Subj. 3	1.35 / 0.75	-0.03 / 0.00	0.87 / 0.38	1.67 / 0.10	-0.13 / 0.00	2.41 / 0.22
Subj. 4	1.85 / 0.86	-0.64 / 0.45	1.48 / 0.94	1.74 / 0.37	0.22 / 0.01	3.45 / 0.60
Subj. 5	1.46 / 0.79	-0.56 / 0.23	1.22 / 0.93	1.09 / 0.10	3.42 / 0.68	5.73 / 0.70
Subj. 6	0.50 / 0.36	0.83 / 0.31	0.33 / 0.75	-13.17 / 0.60	6.33 / 0.90	-6.51 / 0.38
Subj. 7	1.12 / 0.67	0.04 / 0.00	1.15 / 0.92	2.50 / 0.52	-1.22 / 0.19	2.43 / 0.27
mean ± SD	1.33 ± 0.48	-0.23 ± 0.60	1.03 ± 0.45	-0.63 ± 5.56	0.90 ± 3.01	1.3 ± 3.89
Min / Max	0.50 / 1.89	-1.03 / 0.83	0.33 / 1.54	-13.17 / 2.5	-2.58 / 6.33	-6.51 / 5.73

SD, standard deviation.

The amplitudes of P, QRS and T increased in all fetuses in the course of pregnancy (Fig. 2). Accordingly, regression analysis of P, T and QRS with respect to gestational age resulted in solely positive slopes with r^2 values between 0.25 and 0.99 (Table 2). As both T and QRS amplitudes increased at a similar rate over time, T/QRS ratio remained constant. Regression analysis showed no slopes greater than 0.003. Mean T/QRS over all measurements was 0.06 ± 0.03.

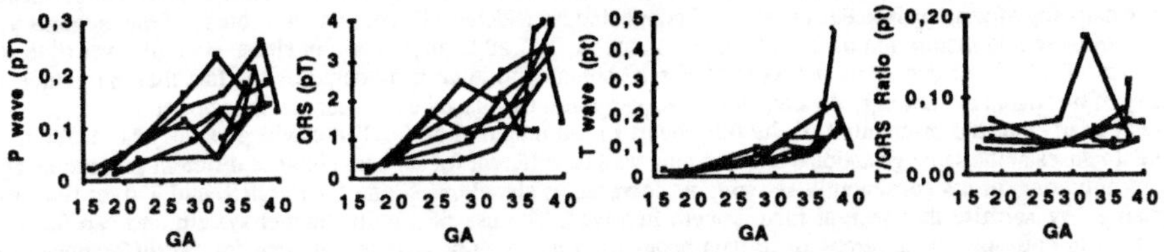

Fig. 2: Course of the amplitude of P, QRS, T and T/QRS during pregnancy for the individual fetuses.

Table 2: Slope and r^2 of the linear regression of amplitudes with respect to gestational age for each subject.

ID	P / r^2	QRS / r^2	T / r^2	(T / QRS) / r^2
Subj. 1	0.007 / 0.66	0.094 / 0.99	0.003 / 0.98	-0.001 / 0.87
Subj. 2	0.006 / 0.40	0.068 / 0.25	0.007 / 0.49	0.002 / 0.97
Subj. 3	0.008 / 0.75	0.168 / 0.61	0.017 / 0.48	0.002 / 0.37
Subj. 4	0.006 / 0.89	0.043 / 0.41	0.005 / 0.95	0.003 / 0.81
Subj. 5	0.005 / 0.54	0.091 / 0.61	0.005 / 0.78	0.001 / 0.04
Subj. 6	0.010 / 0.94	0.268 / 1.00	0.012 / 1.00	0.000 / 0.00
Subj. 7	0.008 / 0.60	0.155 / 0.87	0.006 / 0.76	-0.001 / 0.60
mean ± SD	0.007 ± 0.002	0.127 ± 0.077	0.008 ± 0.005	0.001 ± 0.007
Min / Max	0.005 / 0.010	0.043 / 0.268	0.003 / 0.017	-0.001 / 0.003

SD, standard deviation.

Discussion

Previous work examining CTI in the course of pregnancy has generally reported an increase in the duration of ventricular depolarisation. Using an averaged abdominal FECG, Brambati and Pardi[3] were able to identify the QRS complex and found a steady increase in its duration during normal gestation. They were able to establish a clear relationship between duration and birth weight, indicating that longer QRS duration may well reflect the increase in conduction time due to the larger fetal cardiac muscle mass. This is supported by an anatomical study comparing fetal heart weights [5], which found consistently longer mean QRS duration among male fetuses whose heart weight was greater than that of the female. On the basis of FMCG, Zhuravlev et al.[6] also examined various CTI and amplitude ratios and noted a significant increase in the duration of the QRS complex, from 36 to 50 ms, throughout the gestational period. They did not however report changes in other time intervals. Furthermore, Quinn et al.[2] reported increases in PR, QRS, and QT intervals with FMCG.

The results of our analysis confirm the increase in QRS duration but there remain some discrepancies with respect to the other CTI. In our study design we took several precautions in order to reduce interindividual variability and to enhance the information in the FMCG signal. By registering data from the same fetuses repeatedly at regular intervals as pregnancy progressed, we were able to trace the course of single developing cardiac systems and thereby exclude variability due to observation of different systems at different times. Furthermore, the use of a multichannel system with sensors as far as 19cm apart permitted an optimization of probe placement. Thus, efforts were made to maximize observed signal strengths of opposing polarity. In the evaluation of the averaged signal, the consideration of each individual channel as well as the integrated signal improved the accuracy in the identification of signal deflections. Moreover, the CTI chosen included the basic time components which can be derived from FMCG signal morphology. This ensured that observed changes (or lack of them) in intervals composed of more than one such component could not be a result of diverging changes in the constituent components. On that basis, our data suggests that only the atrial and ventricular depolarisation times, as reflected in increase the duration of the P wave and the QRS complex, change in the course of pregnancy. In spite of our efforts to enhance the identification of events, T wave begin and end were difficult to determine and therefore, conclusions on any interval dependent on these events must be viewed with caution. This may explain why, in contrast to Quinn, we were not able to find a clear increase in QT interval over time.

Other time intervals have been proposed as indicators of fetal well-being in labor. For instance, PR and RR correlate positively in healthy fetuses but negatively in acidotic fetuses [7]. Overall, the diagnostic values of CTI may be seen in their capability to reflect fetal functional disturbances and growth retardation as well as to identify specific pathology forms such as RH disease and congenital heart defects [1], early in the course of the pregnancy.

Growth and maturation of the fetal heart undoubtedly lead to an increase in signal strength over time. The estimation of this development on the basis of FMCG signal amplitude is hindered by the fact that magnitude and polarity of the signal are strongly dependent on two variables: the distance between the detector and the source (i.e. the fetal heart) and the orientation of the fetal heart in the maternal body. These will generally be different for different fetuses at the same gestational age and may well be different for the same fetus at different gestational ages. Accordingly, our results consistently showing an increase in signal amplitude for P, QRS, and T over time were surprising. We surmise that several factors were involved. The use of a multichannel system and careful probe placement to optimize signal strengths at data acquisition may compensate to a degree for the differences in the position of the fetal heart. In addition, by estimating amplitude on the basis of the integral signal, all available information on signal strength is utilized. The development of new sensors with a larger sensor array diameter will then augment the information about direction and intensity of the signal in the depolarisation and the repolarisation phases, allowing reasonable estimation of the orientation of the heart. Furthermore, by observing the development in individual fetuses, variability is reduced, in particular as gross changes in body position become increasingly unlikely as pregnancy progresses.

The results presented here suggest that the estimation of changes in signal strength in the course of pregnancy is possible using FMCG. Nonetheless uncertainty about the validity of absolute values remain. This may be countered by the use of relative values. Indeed, a number of ratios have been proposed in the literature: P/R, R/P, R/R, R/Q, R/S, R/T, T/QRS, and even PR/RR [4,6]. Among them, T/QRS has gained more and more popularity as a marker for fetal well-being under labor with typical normal values around 0.15 ± 0.05, whereas under acidemic conditions these rose to 0.26 ± 1.9 [4]. In our data, values well below those reported above were consistently found and these changed minimally over time. This finding agrees with that of Wang et al. [8] who, using FMCG, also reported an increase in T/QRS measured neonatally in the same subjects. This suggests that, because of its stability, this ratio may serve as a suitable parameter for fetal well-being.

As Deans [4] recommended, combining the information on interval duration, wave amplitude and shape may enhance clinical diagnostic information. In summary, the examination of fetal CTI and signal amplitudes using FMCG in the course of pregnancy seems to hold promise for use as a clinical tool in prenatal diagnostics of fetal development and well-being.

References

[1] Pardi, G., Ferrazi, E., Cetin, I. The clinical relevance of the abdominal fetal electrocardiogram, J Perinat Med, 1986, 14: 371-377.

[2] Quinn, A., Weir, A., Shahani, U., Bain, R., Maas, P., Donaldson, G. Antenatal fetal magnetocardiography: a new method for fetal surveillance? Br J Obstet Gynaecol, 1994, 101: 866-870.

[3] Brambati, B., Pardi, G. The intraventricular conduction time of fetal heart in uncomplicated pregnancies, Br J Obstet Gynaecol, 1980, 87: 941-948.

[4] Deans, A.C., Steer, P.J. The use of the fetal electrocardiogram in labour, Br J Obstet Gynaecol, 1994, 101: 9-17.

[5] Schulz, D.L., Giordano, D.A. Hearts of infants and children, Arch Pathol, 1962, 74: 464-471.

[6] Zhuravlev, Y.E., Rossi, D., Emery, S.J. Clinical assessment of fetal magnetocardiography, In: Baumgartner, C., Deecke, L., Stroink, G., Williamson, S.J. Biomagnetism: Fundamental Research and Clinical Applications, Amsterdam, Elsevier/IOS Press, 1995: 700-703.

[7] Murray, H.G. The fetal electrocardiogram: current clinical developments in Nottingham, J Perinat Med, 1986, 14: 399-403.

[8] Wang, M., Wakai, R.T., Henry, J., Forbes, E., Martin, C.B. Mapping of the fetal magnetocardiogram, In: Baumgartner, C., Deecke, L., Stroink, G., Williamson, S.J. Biomagnetism: Fundamental Research and Clinical Applications, Amsterdam, Elsevier/IOS Press, 1995: 696-699.

Fetal Magnetocardiography in Twin Pregnancy

Van Leeuwen, P.[1] and Schüßler, M.[2]

[1]*Department of Biomagnetism, Research and Development Center for Microtherapy (EFMT), Bochum, Germany;* [2]*Department of Obstetrics, Marienhospital, Witten, Germany*

Introduction

The frequency of multiple pregnancies, twins as well as higher grade, is increasing, among other things, due to fertilizing therapy. Such pregnancies are known to be of elevated risk for mother and child and, as a consequence, a more intensive surveillance is required. It has been shown that fetal magnetocardiography (FMCG) is a facile method of registering fetal cardiac activity from the second trimester onward [1,2]. We would like to present the possibility of using the fetal magnetocardiogram as an adjunct to standard diagnostic tools in twin pregnancy.

Methods

Subjects: Five recordings were acquired in three women with healthy twin pregnancies. Two cases were measured once, in the 32nd and 33rd week of gestation, respectively; in the third case, three data acquisitions, in the 29th, 33rd and 35th week, were performed.

Data Acquisition: FMCG recordings were performed with a 37 channel biomagnetometer (Siemens Krenikon). The diameter of the probe is 19 cm. Each sensor consists of a first order gradiometer with a 7 cm baseline. System noise is less than 30 fT/Hz^{-1} for frequencies > 2 Hz and 10 fT/Hz^{-1} for frequencies > 10 Hz. To reduce the influence of electromagnetic artifacts, the measurements were performed in a shielded room (Vakuumschmelze). The position of the probe was placed over the maternal abdomen such that the fetal signal of the twin fetuses could be clearly identified and differentiated (Fig.1). By recording a limb lead maternal ECG, the mother's QRS complexes could be distinguish from the fetal deflections. Data was recorded for five minutes with a bandpass of 1-200 Hz at a sampling rate of 1 kHz.

Data analysis: In order to extract the fetal signals, the maternal MCG signal was digitally subtracted on the basis of the maternal ECG. In all cases the QRS complexes of the fetuses differed in signal polarity and intensity in individual channels. An appropriate template of the QRS complex of the fetus with more easiliy identifiable signal morphology was used to identify all beats of this fetus. Complexes not automatically recognized were visually identified and added manually and similarly, falsely marked artifacts were removed. Then, on the basis of the identified complexes, a new data set was created by subtracting the signal of this twin in order to make identification of the second twin easier. The QRS complexes of the second twin were then marked as described above for the first twin. For each twin, the identified complexes were used to calculate an averaged beat (Fig. 2) and a time series with a beat period accuracy ±0.5 ms (Fig.3).

The averaged beat was used to calculate the cardiac time intervals and signal amplitudes of each twin. Changes in heart rate variability and complexity were determined by examining the time series using parameters from the time domain (standard deviation (SD), root mean square of successive differences (RMSSD)), the frequency domain (power in the low frequency (LF: 0.04-0.15 Hz) and high frequency (HF: 0.15.-0.40 Hz) bands) as well as the phase domain (approximate dimension (ApD1), approximate entropy (ApEn)).[3-5]

Results

Here, we present the data of the twin pair with acquisitions in the 29th, 33rd and 35th week of gestation. Both twins were born vaginally in the 37th week, the second with forceps. The birth weights were 2860g and 2790g, the 5 and 10 min. APGAR scores were 10,10 and 9,10, respectively. The raw FMCG data permitted identification of QRS complexes of both twins in all data (Fig.1). Averaging the data seperately for each fetus permitted estimation of cardiac time intervals and wave amplitudes (Fig.2 and Table 1).

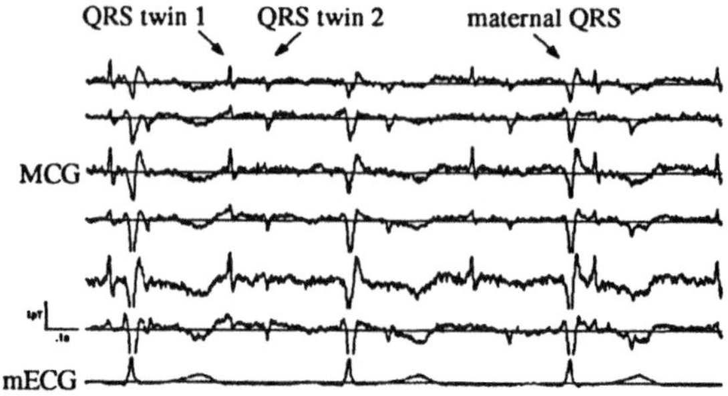

Fig. 1: Raw data in the 29th week of gestation including six magnetic (MCG) and one maternal electric (mECG) channel.

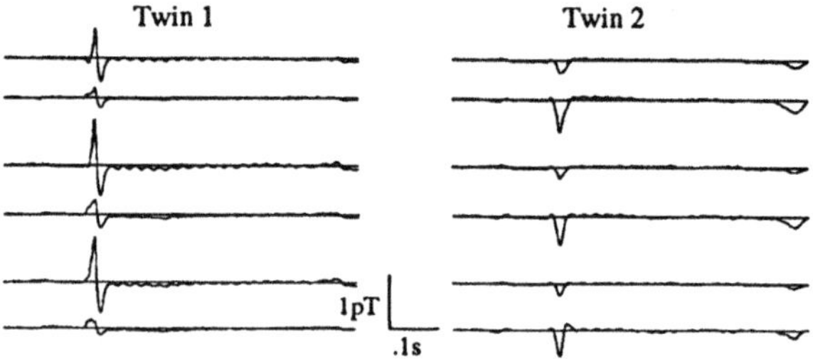

Fig. 2: Averaged beat for twin 1 and twin 2 (29th week, same channels as in Fig.1).

Table 1: Cardiac time intervals and wave amplitudes at different weeks of gestation.

		Twin 1			Twin 2		
GA (weeks)		29	33	35	29	33	35
Cardiac time intervals (ms)	P wave	52	60	72	51	60	62
	P_{end} - Q_{begin}	50	28	20	56	42	43
	QRS complex	44	42	42	39	47	50
	ST segment	42	56	107	102	108	91
	T wave	111	87	91	95	125	112
Amplitudes (pT)	P wave	.013	.014	.024	.021	.031	.031
	QRS complex	.374	.473	.592	.304	.400	.481
	T wave	.037	.052	.016	.012	.022	.018
	T/QRS	.098	.110	.028	.041	.055	.038

On the basis of the complete time series for each twin and week, changes in heart rate variability could be documented. Time domain values showed: lower heart rates in week 33, in particular for twin 1; lower SD earlier in pregnancy (week 29), in particular for twin 2; and greater variation in RMSSD in twin 1 (Fig 3. Table 2). In the frequency domain, the LF and HF bands tended to increase as pregnancy progressed, apart from a drop for twin 1 in the 35th week. LF/HF fell continuously for twin 1 and remained constant in twin 2. The phase domain values showed an overall tendency to increase apart from higher values in twin 1 at week 33.

Discussion

Fetal magnetocardiography opens up new diagnostic perspectives for multiple pregnancy presently not possible with other methods. The development of the individual twin fetus may be investigated by analysing the raw

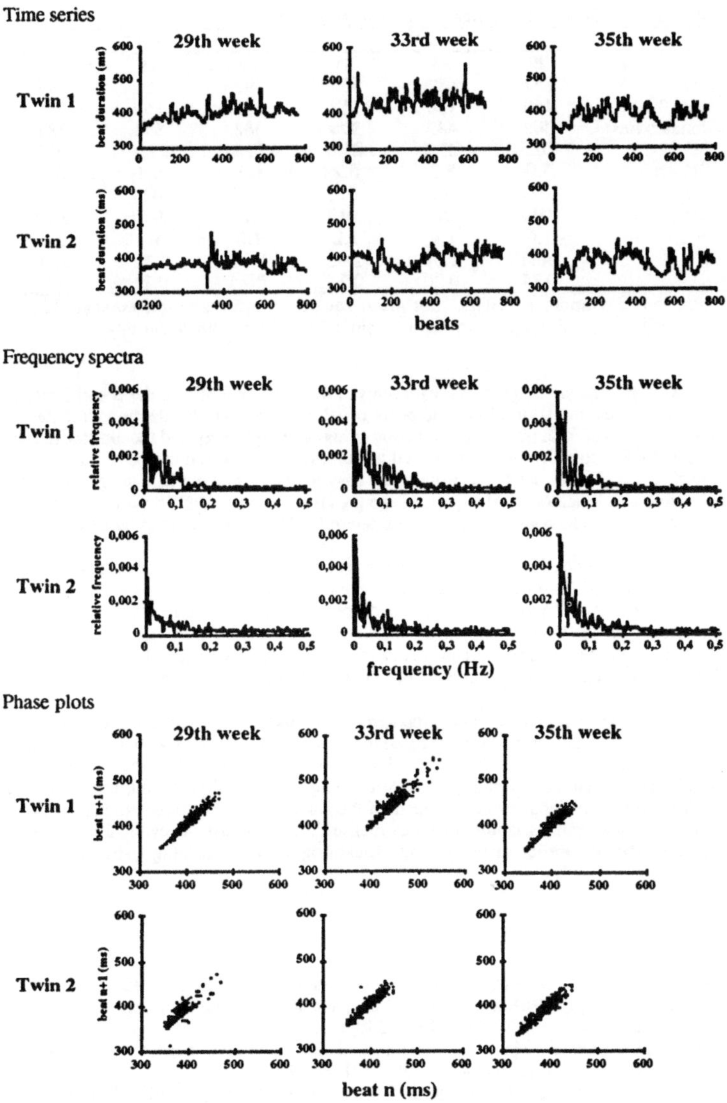

Fig. 3: Heart period time series, frequency spectra and phase plots of both twin fetuses in the 29th, 33rd and 35th week of gestation.

Table 2: Time, frequency and phase domain values at different weeks of gestation.

		Twin 1			Twin 2		
GA (weeks)		29	33	35	29	33	35
	n (beats)	757	682	759	795	756	785
	Heart rate (bpm)	150	136	152	157	151	158
Time Domain	Beat duration (ms)	402	442	396	382	398	383
	SD (ms)	20.5	23.3	23.8	13.9	22.3	26.1
	RMSSD (ms)	5.0	8.3	6.2	6.1	6.1	6.8
Frequency domain	LF (ms)	.027	.032	.023	.021	.023	.029
	HF (ms)	.017	.023	.019	.020	.020	.024
	LF/HF	1.6	1.4	1.2	1.0	1.1	1.2
Phase domain	ApD1	1.21	1.51	1.36	0.00	1.15	1.30
	ApEn	0.11	0.20	0.14	0.14	0.11	0.14

GA, gestational age; SD, standard deviation; RMSDD, root mean square of successive differences, LF, low frequency band; HF, high frequency band; ApD1, approximate dimension; ApEn, approximate entropy.

signal, the averaged signal as well as changes in heart rate variability. The raw signal allows the identification of deviations from sinus rhythm and the examination of ectopic beats which may indicate conduction irregularities. Furthermore, the calculation of an averaged beat permits examination of signal morphology and the exact determination of time intervals. P wave and QRS complex duration as well as amplitude tend to increase over time, corresponding to an increase of muscle mass [3]. The T wave is usually not easy to detect, making its evaluation difficult. One may also compare the relative amplitude of various parts of the signal such as the T/QRS ratio, where higher values may indicate fetal compromise [6]. The corresponding values determined here for both twin fetuses agree with such results.

The possibility to reconstruct a time series of consecutive heart beats with great temporal accuracy offers new possibilities for the evaluation of fetal heart rate variability as a marker for fetal well-being. Coinciding with a decrease in fetal heart rate during pregnancy, common statistical measures of variability such as the standard deviation tend to demonstrate an increase. Development of autonomic nervous activity can be investigated by performing spectral analysis of the time series and then examining changes in relevant frequency bands [4]. Furthermore, by applying methods of nonlinear dynamics, one may describe the degree of complexity qualitatively as well as quantitatively [5].

The results reported here show fetal heart rate development on the basis of these parameters in varying degrees. Tendencies towards lower heart rates, more heart rate variability and complexity as well as an increase in LF/HF are partially present. Conflicting results within the single domains may be explained by the fact that only a short span of six weeks was observed. Had data been acquired over a longer period of gestation, trends could be expected to be more visible. Furthermore, differences in the course of the parameters of the three domains may well reflect the complementary aspects of heart rate variability that they measure. In summary, the evaluation of the case data presented here demonstrates the possibility of using magnetocardiography in the diagnostics of multiple pregnancy.

References

[1] Dunajski, Z., Peters, M. Development of the magnetocardiograms from the 13th week of gestation onward. In: Baumgartner, C., Deecke, L., Stroink, G., Williamson, S.J. Biomagnetism: Fundamental Research and Clinical Applications, Amsterdam, Elsevier/IOS Press, 1995: 340-341.

[2] Quinn, A., Weir, A., Shahani, U., Bain, R., Maas, P., Donaldson, G. Antenatal fetal magnetocardiography: a new method for fetal surveillance? Br J Obstet Gynaecol, 1994, 101: 866-870.

[3] Van Leeuwen, P., Lange, S., Schüßler, M., Lajoie-Junge, L. Determination of Changes in Fetal Cardiac Time Intervals During Pregnancy using Magnetocardiography, this volume.

[4] Van Leeuwen, P., Hartmann, M., Schüßler, M., Lange, S. Fetal Magnetocardiography: Spectral Analysis of Fetal Heart Rate Variability in the Course of Pregnancy, this volume

[5] Van Leeuwen, P., Bettermann, H., Schüßler, M., Lange, S. Magnetocardiography in the determination of fetal heart rate complexity, this volume.

[6] Deans, A., Steer, P. The use of the fetal electrocardiogram in labour, Br J Obstet Gynaecol, 1994, 101: 9-17.

Magnetocardiographic Monitoring of the Fetal Autonomic Nervous System

Zhuravlev, Y.E.[1,3], Lewis, M.J.[1], Rassi, D.[1], and Emery, S.J.[2]

University of Wales[1], Swansea, UK; Singleton Hospital[2], Swansea, UK; Institute of Radioengineering and Electronics[3], Moscow, Russia

Introduction

At present there is no non-invasive method of obtaining functional information about the fetal nervous system. Fetal heart rate (FHR) variability is widely used, in an empirical way, as an indicator of fetal status. This does not provide a quantitative analysis with regard to a physiological understanding of the underlying cardiac control mechanisms. Studies on both term and preterm infants suggest that new information may be obtained from accurate beat-to-beat heart rate variability analysis [1,2].

Human heart rate is governed by a number of interacting physiological control systems that operate on widely different time scales [3,4]. To be able to assess this complex process for the fetus, the FHR needs to be measured with millisecond accuracy. Ultrasonography lacks the necessary temporal resolution, and abdominal fetal electrocardiography (FECG) has the disadvantage of severe signal distortion due to the maternal ECG. Fetal magnetocardiography (FMCG) using Superconducting Quantum Interference Device (SQUID) instrumentation has the required accuracy, is not affected by maternal signals, and is therefore more appropriate for such studies than ultrasound or FECG [5]. We have previously demonstrated that SQUID measurements can be reliably carried out in an unshielded hospital ward [6].

In the present study, FMCG recordings were used to obtain accurate beat-to-beat FHR variability curves. To assess the complex neural control of the fetal heart rate we used data processing methods that provide a quantitative evaluation of the dynamic interaction between short-term FHR variability (associated with parasympathetic or vagal control) and other, slower components (due to sympathetic neural activity). Such interaction is believed to reflect the sympatho-vagal balance and could provide a quantitative means of assessing the condition of the fetal nervous system.

Methods

FMCG measurements, using a single-channel SQUID system, were carried out on 24 healthy volunteers in the course of normal, uncomplicated pregnancy. For six of these subjects, regular two-weekly recordings were obtained, starting at around the 23rd week of gestation. Details of the apparatus and the experimental procedures have been given elsewhere [6]. The SQUID output was band-pass filtered from 0.1 to 200 Hz before being digitised at a sampling rate of 500 Hz. The FMCG data were digitised using a 16-bit analogue-to-digital converter and all data acquisition and processing was performed on a microcomputer. An algorithm based on autocorrelation R-wave detection was used to improve the accuracy of the R-R interval measurement, and an interpolation procedure was used to achieve a time resolution of 1 ms.

Our dynamic analysis of the FHR variability data consists of two stages. In the first stage we separated the different types of oscillations within the instantaneous FHR variability curve according to their characteristic time scales. Three distinct frequency bands of the FHR variations are considered: very low frequencies (VLF) below 0.04 Hz, low frequencies (LF) $0.04<f<0.2$ Hz and high frequencies (HF) $f>0.2$ Hz. Smooth interpolation of the raw heart-rate data is performed using a cubic spline routine to increase the number of points by a factor of five, and appropriate band pass filtering is applied to extract the three components.

The second stage consists of calculating quantitative indices for the activity of the two branches of the fetal autonomic nervous system. We took the average amplitude of the HF oscillations within a 4-second moving time window to represent vagal activity, and the amplitude of the LF oscillations to characterise the activity of the sympathetic nervous system. These two components were then plotted for the entire data set, typically recorded for 3 minutes. It is the dynamic correlation of these two curves, and their ratio during periods of high correlation, with which we are mainly concerned.

Results

Fig. 1(a) shows the interpolated raw FHR variability curve obtained from a 39 week old fetus. The VLF, LF and HF components extracted from this curve are shown in Fig. 1(b, c and d) respectively. The frequency bandwidths used here were based on the results of power spectral analysis of our HRV data, and were in close agreement with previously reported values [2,3,7]. According to current physiological understanding, the VLF component is associated with thermoregulation, the LF component is attributed mainly to the intrinsic oscillations of the baroreceptor reflex loop, and the HF component corresponds to respiratory activity.

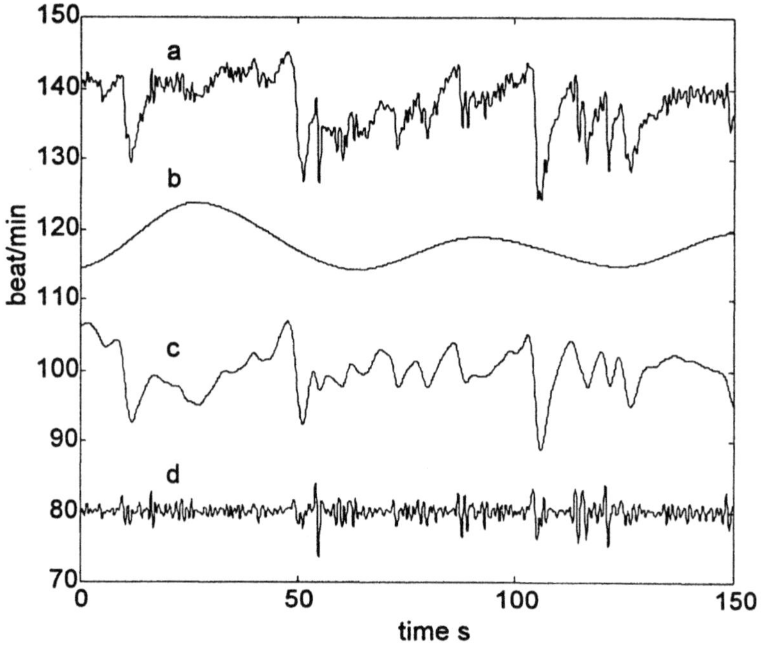

Fig. 1(a) Interpolated raw FHR variability curve obtained from a 39 week old fetus. The VLF, LF and HF components extracted from this curve are represented by the curves (b), (c) and (d), respectively

In Fig. 2 the 'average amplitude' HF (a) and LF (b) curves are plotted synchronously; the HF curve has been amplified by a factor of three in order to facilitate visual comparison. The reciprocal correlation, with a slight time shift, between these two curves is clearly visible. This correlation is quantitatively demonstrated in Fig. 3(a). The correlation coefficient was calculated, within a 20-second time window, between the LF and HF components. Moreover, optimum correlation was achieved by shifting the HF curve forward by 2 seconds along the (horizontal) time axis. The ratio of these two components, calculated in the same 20-second time window and with the same time shift, is shown in Fig. 3(b).

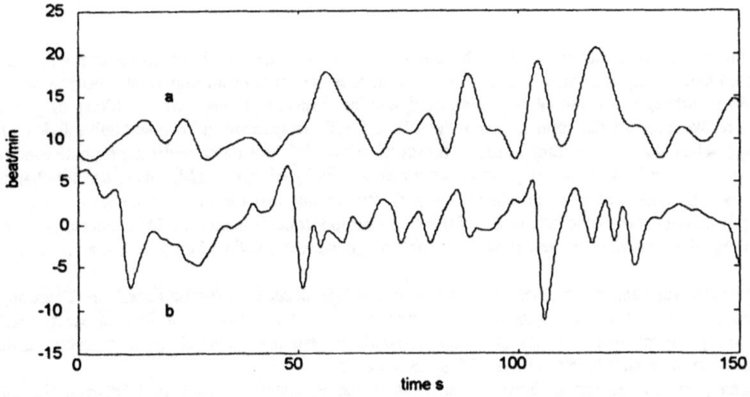

Fig. 2 The average amplitude HF (a) and LF (b) curves. The HF curve has been amplified by a factor of three in order to facilitate visual comparison

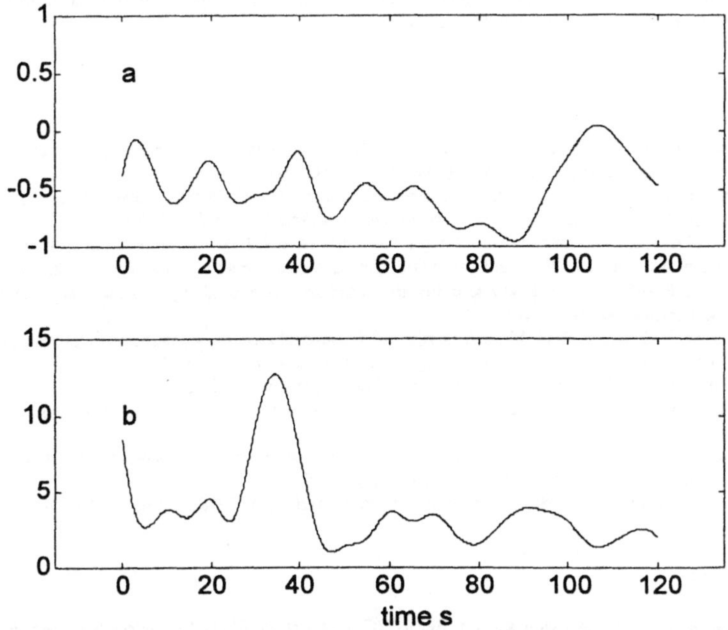

Fig. 3 (a) Correlation between the LF and average amplitude HF curves calculated within a 20-second time window and with the HF curve shifted forward along the horizontal axis for 2 seconds; (b) The ratio of these two components, calculated in the same 20-second time window and with the same time shift

Discussion

We believe that the observed correlation in the dynamic behaviour of the LF and HF components of fetal heart rate variability is related to the sympatho-vagal balance. Such a balance is known to be an important feature of the heart-rate control mechanisms although, because of the complex dynamics involved, it has proved difficult to provide a quantitative description. We propose the ratio between the LF and HF components as a quantitative index of the sympathetic-parasympathetic balance. Care must be taken, however, to calculate this index under appropriate conditions. The high correlation between the LF and HF components is periodically disrupted, presumably due to sudden changes in the systemic conditions (e.g. movement or sleep related changes). We assume that the LF/HF ratio is only meaningful during periods of highly correlated dynamic behaviour of these two components. This ratio could be used to quantify the gain of the fetal parasympathetic system, and may provide valuable diagnostic information during the entire third trimester and possibly earlier.

We investigated the variation of the LF/HF ratio as a function of gestational age for the same fetus. There appears to be a decrease in this ratio as the fetus develops, which presumably reflects the maturation of the fetal parasympathetic system. An increase with postnatal age in the gain of the parasympathetic system has been observed in preterm neonates [2] for whom it was possible to monitor breathing and ECG simultaneously.

An interesting parameter emerging from our analysis is the apparent temporal shift between the indices representing sympathetic and parasympathetic activity. A similar time shift in the reaction time of the adult heart-rate to sympathetic system control has been previously reported [8]. It is also worth noting that the very low-frequency component of the FHR variability curve appears to interact with both the LF and HF components in such a way that both these latter components are suppressed at the maxima of the VLF curve. This could indicate the involvement of the thermoregulation system together with other possible mechanisms in the control of the cardiovascular system. In order to investigate this effect more comprehensively, longer data sets need to be analysed.

References

[1] Divon M.Y., Winkle H., Yeh S-Y., Platt L.D., Langer O. and Merkatz I.R. Diminished respiratory sinus arrhythmia in asphyxiated term infants 1986 Am.J.Obstet.Gynecol. 155(6):1263-5.

[2] Van Ravenswaaij-Arts C., Hopman J., Kollée L., Stoelinga G. amd Van Geijn H. Spectral analysis of heart rate variability in spontaneously breathing very preterm infants 1994 Acta Pædiatr. 83:473-80.

[3] Akselrod S., Gordon D., Ubel F.A., Shannon D.C., Barger A.C. and Cohen R.J. Power Spectrum Analysis of Heart Rate Fluctuation: A Quantitative Probe of Beat-to-Beat Cardiovascular Control 1981 Science 213:220-222.

[4] Malliani A., Lombardi F. and Pagani M. Power spectrum analysis of heart rate variability: a tool to explore neural regulatory mechanisms 1994 Br.Heart.J. 71:1-2.

[5] Wakai R.T., Wang M., Pedron S.L., Reid D.L. and Martin Jr C.B. Spectral analysis of antepartum fetal heart rate variability from fetal magnetocardiogram recordings 1993 Early Hum Dev. 35:15-24.

[6] Zhuravlev, Y., Rassi, D., and Emery, S. Clinical assessment of fetal magnetocardiography 1995 In (eds. C.Baumgartner *et al.*) Biomagnetism: Fundamental Research and Clinical Applications, Amsterdam, Elsevier/IOS-press pp700-703.

[7] Karin J., Hirsh M., Sagiv C. and Akselrod S. Fetal autonomic nervous system activity monitoring by spectral analysis of heart rate variations 1992 Proc . Computers in Cardiol. 479-482

[8] Hill-Smith I. and Purves R.D. Synaptic delay in the heart: an ionophoretic study 1978 J.Physiol. 279:31-54

Acknowledgements

The authors wish to acknowledge the following support: EPSRC (Grant GR/J72110), the Royal Society (project grant), the University of Wales, Swansea (studentship MJL) and the School of Postgraduate Studies in Medical and Health Care, Swansea (provision of nursing support).

GPSR Compliance

The European Union's (EU) General Product Safety Regulation (GPSR) is a set of rules that requires consumer products to be safe and our obligations to ensure this.

If you have any concerns about our products, you can contact us on ProductSafety@springernature.com

In case Publisher is established outside the EU, the EU authorized representative is:

Springer Nature Customer Service Center GmbH
Europaplatz 3
69115 Heidelberg, Germany

Batch number: 10844661

Printed by Printforce, the Netherlands

Biomag 96
Volume II

Springer Science+Business Media, LLC

Cheryl J. Aine Yoshio Okada
Gerhard Stroink Stephen J. Swithenby
Charles C. Wood

Editors

Biomag 96

Proceedings of the Tenth International Conference on Biomagnetism

Volume II

Springer

Cheryl J. Aine
Radiology Service Albuquerque
VA Medical Center
Building 49, 114M
1501 San Pedro SE
Albuquerque, NM 87108
USA

Yoshio Okada
Department of New Mexico School
of Medicine
Albuquerque, NM 87131
USA

Gerhard Stroink
Department of Physics
Dalhousie University
Halifax, Nova Scotia, Canada
B3H 3J5

Stephen J. Swithenby
Department of Physics and Astronomy
The Open University
Milton Keynes MK7 6AA, UK

Charles C. Wood
Biophysics Group (P-21), MS D454
Los Alamos National Laboratory
P.O. Box 1663
Los Alamos, NM 87545
USA

Library of Congress Cataloging-in-Publication Data
International conference on Biomagnetism (10th : 1999 : Sante Fe, N.M.)
Biomag 96 : proceedings of the Tenth International Conference on Biomagnetism / editors, Cheryl J. Aine . . . [et al.].
p. cm.
Includes bibliographical references and index.
ISBN 978-1-4612-7066-9 (hardcover : alk. paper)
1. Biomagnetism Congresses. I. Aine, Cheryl J. II. Title.
QP345.I56 1999
571.4′7—dc21 99-42829

Printed on acid-free paper.

Production managed by Jenny Wolkowicki; manufacturing supervised by Joseph Quatela.
Camera-ready copy provided by the editors.

9 8 7 6 5 4 3 2 1 SPIN 10742866

ISBN 978-1-4612-7066-9 ISBN 978-1-4612-1260-7 (eBook)
DOI 10.1007/978-1-4612-1260-7

Contents
Volume II

IV Biomagnetic Measures of the Gastro-Intestinal System, Susceptometry and Other Biomagnetic Measures

1. **Biomagnetic Research in Gastroenterology** 594
Baffa, O., Basile, M., Bradshaw, A., Forsman, M., Nakaya, Y. and Richards,W.O.

Biomagnetic Measures of the G-I System

2. **Magnetic Field Measurement of Rabbit Colonic Electrical Activity** 608
Bradshaw, L.A., Allos, S.H., Wikswo, J.P., Jr. and Richards, W.O.

3. **Esophageal Transit Time evaluated by means of the Biomagnetic and Scintigraphic Methods** 612
Daghastanli, N.A., Braga, F.J.H.N., Oliveira, R.B. and Baffa, O.

4. **Measurements of Colonic Motility using Ferrimagnetic Powder** 616
Forsman, M. and Abrahamsson, H.

5. **Detection of Stomach Contractions using a Biosusceptometer** 620
Miranda, J.R.A., Baffa, O., Oliveira, R.B., Sousa, P.L. and Braga F.J.H.

6. **Biomagnetic Measurement of Gastric Mechanical Motility using a High-Temperature SQUID Imaging.** 624
Nomura, M., Saijyo, T., Haruta, Y., Itozaki, H., Toyoda, H., Nakaya, Y., Ito, S. and Kado, H.

7. **Single and Multichannel Magnetic Field Measurements of Gastrointestinal Activity in the Pre- and Post-Prandial States** 628
Petrie, R.J., van Leeuwen, P., Brandts, B., Turnbull G., Veldhuyzen van Zanten, S.J.O. and Stroink, G.

8. **Magnetogastrograms using 64-Channel SQUID System: Study on Clinical Usefulness** 631
Saijyo, T., Nomura, M., Haruta, Y., Itozaki, H., Toyoda, H., Nakaya, Y., Ito, S. and Kado, H.

9. **Noninvasive Measurement of the Vector Magnetic Field from Human Gastrointestinal Sources** 635
Staton, D.J., Allos, S.H., Henry, V.K., Bradshaw, L.A., Ladipo, J.K., Richards, W.O. and Wikswo, J.P., Jr.

Susceptometry And Other Biomagnetic Measures

10. **Human Alveolar Macrophage Activation in Healthy Subjects and in COPD-Patients Evaluated by Magnetopneumography** 639
Barth, W., Möller, W., Meier-Sydow, J., Pohlit, W. and Stahlhofen, W.

11. **Magnetic Markers linked to Monoclonal Antibodies allow *in vitro* Evaluation of Cell Population Size** 643
Del Gratta, C., Della Penna, S., Battista, P., Di Donato, L., Vitullo, P., Romani, G.L. and Di Luzio, S.

12. **Iron Quantification by Non-Invasive Biomagnetic Organ Susceptometry** 647
Engelhardt, R., Fischer, R., Nielsen, P. and Gabbe, E.E.

13. **The Use of Biomagnetic Liver Susceptometry in the Ferrara-Hamburg-Turin study on Thalassemia** 651
Fischer, R., Piga, A., Tricta, F., Nielsen, P., Engelhardt, R., Garofalo, F., Di Palma, A. and Vullo, C.

14. **Detection of Biodegradation of Ferrofluid by Relaxation Measurements** 655
Kötitz, R., Trahms, L., Pfefferer, D., Kresse, M., Semmler, W. and Weitschies, W.

15. **Distribution and Biotransformation Dynamics of Magnetic Contrast Agents in Organism on Evidence derived from SQUID-Magnetometry** 659
Mikhailik, O.M., Pankratov, Yu.V., Dudchenko, A.K., Sosnitsky, V.N., Budnik, N.N., Minov, Yu.D., Sutkovoj, P.I. and Bakai, E.A.

16. **Human Alveolar Long-Term Clearance in Healthy and Disease evaluated by Magnetopneumography** 663
Möller, W., Barth, W., Meier-Sydow, J., Pohlit, W., Siekmeier, R. and Stahlhofen, W.

17. A Model of Phagosome Motion within Cells incorporating Viscoelasticity based on Cytomagnetometric Measurements 667
Nemoto, I.

18. Magnetic Susceptibility Tomography with nonuniform Field 671
Parente Ribeiro, E., Wikswo, Jr., J.P., Costa Ribeiro, P. and Szczupak, J.

19. MCG studies in Rats using Ferrofluids 675
Saligram, U., Moran, J.E. and Tepley, N.

20. Magnetic Field and Electric Potential of Excited Plant Cell Chara Corallina: Calculation and Comparison with Experiment 679
Slibar, M., Trontelj, Z., Jazbinsek, V., Thiel, G. and Müller, W.

21. System for Magnetic Susceptibility Investigations of Human Blood and Liver 683
Sosnitsky, V.N., Budnik, N.N., Minov, Yu.D., Sutkovoj, P.I. and Vojtovich, I.D.

22. Registration of Iron Overload by Means of SQUID-Magnetometry when coupled with Low-Temperature ESR Spectroscopy 687
Sosnitsky, V.N., Mikhailik, O.M., Lubyanova, I.P., Gapelyuk, A.V., Budnik, N.N., Minov, Yu.D., Sutkovoj, P.I., Pankratov Yu.V. and Bakai, E.A.

V Magnetoencephalography (MEG): Basic Research

23. MEG Studies of Cognition 692
Salmelin, R.

Basic Human MEG

24. The Magnetic Field Components evoked by an Optical Illusion 697
Adachi, Y., Kohda, M., Morita, T. and Takebayashi, M.

25. Spatiotemporal Imaging of Human Cortical Areas sensitive to Visual Motion 701
Ahlfors, S.P., Aronen, H.J., Belliveau, J.W., Dale, A.M., Huotilainen, M., Ilmoniemi, R.J., Korvenoja, A., Liu, A.K., Simpson, G.V., Tootell, R.B.H., and Virtanen, J.

26. Task-Specific Magnetic Fields from the left Human Prefrontal Cortex 705
Basile, L.F.H., Simos, P.G., Tarkka, I.M. and Papanicolaou, A.C.

27. **Searching for the Oddball Response in the EEG and MEG** 709
Berg, P., Kakigi, R., Koyama, S., Dobel, C., Zobel, E. and Scherg, M.

28. **Magnetoencephalographic Measurements of a Developing Infant: A Pilot Study** 713
Bowyer, S.M., Moran, J. and Tepley, N.

29. **The Steady State Visual evoked Response and Estimates of Phase Velocity** 717
Burkitt, G.R., Silberstein, R.B. and Wood, A.W.

30. **Nonlinear Spatio-Temporal Interactions and Neural Connections in Human Vision using Transient and m-Sequence Stimuli** 721
Chen, H.-W., Aine, C.J., Flynn, E.R. and Wood, C.C.

31. **Spatial Frequency tuning Functions and Contrast Sensitivity at different Eccentricities in the Visual Field** 725
Chen, H.-W., Aine, C.J., Flynn, E.R. and Wood, C.C.

32. **Neuromagnetic Fields Evoked by Sine Tones, Sine Tone Complexes, Vowel Formants, and Two-Formant Vowels** 729
Diesch, E., Luce, T.A. and Eyferth, K.

33. **Source Location and Orientation, Dipole Moment, and Latency of Magnetic Mismatch Fields may mirror the Perceptual Salience of Phonetic Contrasts** 733
Diesch, E., Luce, T.A. and Eyferth, K.

34. **Auditory evoked Magnetic Fields after Stimulation with Test-Tones and Words-Studies on Experimentees with Good Hearing and with Long-Term Hearing Impediments** 737
Emmerich, E., Meyer, R., Nowak, H., Huonker, R., Gießler, F., Meißner, W. and Beleites, E.

35. **Analysis of Movement-Related Brain Activities elicited by External Instructions** 741
Endo, H., Takeda, T., Kizuka, T., Masuda, T. and Kumagai, T.

36. **Visual evoked Magnetic Fields: Bilateral Dipole Pattern for Full-Field Stimuli** 745
Fujita, S., Nakasato, N., Seki, K., Kanno, A., Kawamura, T., Ohtomo, S., Fujiwara, S. Tamura, I. and Yoshimoto, T.

37. **Evoked Magnetic Responses from Human V1: Evidence for a Chromatic-Sensitive Motion Mechanism** 749
Fylan, F., Singh, K.D., Holliday, I.H., Anderson, S.J. and Harding, G.F.A.

38. Evoked magnetic responses to Isoluminant Chromatic Stimuli from Human V1: Spatial Frequency Characteristics 753
Fylan, F., Singh, K.D., Holliday, I.H., Anderson, S.J. and Harding, G.F.A.

39. EEG and MEG Activity accompanying spontaneous Reversals of the Necker Cube 757
Gaetz, M., Weinberg, H., Jantzen, K.J., Rzempoluck, E. and Millbank, A.

40. MEG and EEG Responses to a Pitch Naming and Identification Task: Slow DC Shift predicts Performance 761
Gordon, R., Weinberg, H., Cheyne, D. and Jantzen, K.

41. Magnetic Detection of Activity of presumed Inhibitory Interneurons in the Human Somatosensory Cortex 765
Hashimoto, I., Mashiko, T. and Imada, T.

42. Effect of Color on Visual evoked Magnetic Fields with Pattern-Reversal Stimulation 769
Hatanaka, K., Seki, K., Nakasato, N., Kanno, A., Ohtomo, S., Fujiwara, S. and Yoshimoto, T.

43. Evoked Magnetic Fields at Semantic Judgment of Simple Japanese Sentences 773
Hirata, Y., Sakai, T. and Kuriki, S.

44. Neuromagnetic Correlates of Endogenous Auditory Sensations 777
Hoke, E.S., Hoke, M. and Ross, B.

45. Spatial and Temporal Properties of the slow Components of the Contingent Magnetic Variation in a warned Choice Reaction Time Task 781
Hultin, L., Högstedt, P., Pizzella, V., Romani, G.L., Rossini, P. and Tecchio, F.

46. Dependence of SEFs on the Interstimulus Interval: A Model for SEF Generation Mechanism at the Primary Sensorimotor Cortex 785
Huttunen, J., Wikström, H., Korvenoja, A., Aronen, H. and Ilmoniemi, R.J.

47. Somatosensory evoked Fields from SMI and SII During 'Interfering' Finger Movements and Tactile Stimulation 789
Huttunen, J., Wikström, H., Korvenoja, A., Aronen, H. and Ilmoniemi, R.J.

48. Time Course of Neuromagnetic Signals related to Reading Japanese Phonograms 793
Imada, T., Kawakatsu, M., Ikeya, A., Kadota, R., Miyamoto, J., Nakayama, S. and Kotani, M.

49. Electric and Magnetic Brain Responses evoked by Deviant Tones in Melodies 797
Jürgens, R., Lopez, L., Diekmann, V., Becker, W., Ried, S., Grözinger, B. and Erné, S.N.

50. Multidipole Estimation of the Neuromagnetic Fields related to Auditory Discrimination 801
Kikuchi, Y., Yoshizawa, S., Kita, M., Nishimura, C., Tanaka, M., Endo, H., Kumagai, T. and Takeda, T.

51. Measurement of Gustatory-evoked Magnetic Fields with Sharp Stimulation 805
Kobayakawa, T.K., Endo, H.E., Saito, S.S., Ayabe-Kanamura, S.A., Kikuchi, Y.K., Yamaguchi, Y.Y., Ogawa, H.O. and Takeda, T.T.

52. Source Location Outside the Auditory Cortex: Auditory Event related Field to Tone-Duration in a Discrimination Task 809
Kofoed, B., Bak, C.K., Csepe, V., Pantev, C. and Saermark, K.

53. Measurement of MEG evoked by Visual Discrimination Task 813
Kumagai, T., Takeda, T. and Endo, H.

54. Cerebral Activities in the Occipital and Temporal Regions when Subjects perform a matching Task of Visual Stimuli 817
Kuriki, S. and Hirata, Y.

55. Magnetoencephalographic Study on Cerebral Cortical Activities related to Speech 821
Kyuhou, S., Sasaki, K., Nambu, A., Matsuzaki, R., Tsujimoto, T. and Gemba, H.

56. Magnetic Recording of the Propagation of Motor Unit Action Potentials in the Human Leg Muscles 825
Masuda, T., Endo, H., Kumagai, T. and Takeda, T.

57. Effect of Luminosity Change on Visual evoked Field 829
Matsuhashi, M., Oyanagi, H., Kumagai, T., Endo, H. and Takeda, T.

58. Evoked Magnetic Fields for Gustatory Stimuli on the Tongue 833
Murayama, N., Nakasato, N., Hatanaka, K., Fujita, S., Igasaki, T., Kanno, A. and Yoshimoto, T.

59. Different Effects of Stimulus Rate on the Somatosensory evoked Magnetic Fields and Potentials 837
Nagamine, T., Mäkelä, J.P., Mikuni, N., Mima, T., Nishitani, N., Sato, T., Ikeda, A., Kimura, J. and Shibasaki, H.

60. Binaural Fusion of Virtual Pitch in the Human Auditory Cortex 841
Pantev, C., Elbert, T., Ross, B., Eulitz, C. and Terhardt, E.

61. Left and Right Supratemporal Auditory Cortices are differentially modulated by Linguistic-Attentional Task Demands 845
Poeppel, D., Yellin, E., Phillips, C., Roberts, T.P.L., Rowley, H., Wexler, K. and Marantz, A.

62. Extensive Somatosensory Stimulation alters Somatosensory evoked Fields 848
Rockstroh, B., Vanni, S., Elbert, T. and Hari, R.

63. Laterality of Hippocampal Responses to Infrequent and unpredictable Omissions of Visual Stimuli 852
Rogers, R.L., Taylor, S.A., Akhtari, M. and Sutherling, W.W.

64. MEG Measurement of Auditory Sensory Memory Persistence via the M100 in Children and Adults 856
Rojas, D.C., Sheeder, J.L., Teale, P., Walker, J.R., Robertson, B.A., Selig, S. and Reite, M.

65. The Relation between Gustatory-Evoked Magnetic Fields and Reaction Times to different Taste Qualities 860
Saito, S.S., Kobayakawa, T.K., Ayabe-Kanamura, S.A., Endo, H.E., Yamaguchi, Y.Y., Kikuchi, Y.K., Ogawa, H.O. and Takeda, T.T.

66. SQUIDs and Stereopsis 864
Shahani, U., Weir, A.I., Lang, G., Mansfield, D.C., Halliday, D.M., Hutson, D., Maas, P.M. and Donaldson, G.B.

67. Source Localization of a N400-like Negativity in a Sentence-Reading Paradigm using Evoked Magnetic Fields and Magnetic Resonance Imaging 868
Simos, P.G., Basile, L.F.H., Zouridakis, G. and Papanicolaou, A.C.

68. Information Theory Predicts the Magnitude of an MEG Response related to the Operation of the Auditory Sensory Memory 872
Sinkkonen, J., Tiitinen, H., May, P., Ilmoniemi, R.J. and Näätänen, R.

69. **Determination of the Nature of the Connection between the Auditory and Visual Cortices** 874
Stephen, J.M., Broadhurst, J.H., Knuth, K.H. and Schwartz, B.J.

70. **Visualization of Tonotopy of Auditory Cortex using a newly developed 129 Channel Vector Magnetoencephalography** 877
Takanashi, Y., Iwamoto, K., Yoshikawa, K., Makino, M., Tomita, T., Tomita, S., Kajihara, S., Kondo, Y., Okamura, S., Yoshida, Y., Ueda, M. and Nakajima, K.

71. **Measurement of Accommodation related MEG Responses** 881
Takeda, T., Hashimoto, K., Kumagai, T. and Endo, H.

72. **Electromagnetic source Localization of Cognitive Processes related to Attention to Single or Omitted tones** 885
Tarkka, I.M. and Stokić, D.S.

73. **A Cortical Substrate for Hand Skill: An MEG Evaluation** 889
Thoma, R.J., Lewine, J.D., Davis, J.T. and Orrison, W.W., Jr.

74. **Separation of Overlapping Sources induced by mixed Stimulations using 3-Dimensional MEG Mappings** 892
Uchikawa, Y., Kobayashi, K., Sakawa, K. and Kotani, M.

75. **Measurements of Visual evoked MEG Responses associated with Color Discrimination** 896
Ueno, K., Ueno, S. and Weinberg, H.

76. **Measurement and Source Analysis of MEG Activities associated with Short-Term Memory and Identification Processes** 900
Yoshida, H., Ueno, S. and Weinberg, H.

MEG Measures Of Brain Oscillatory Responses

77. **Oscillatory Brain Activities.** 904
Lopes da Silva, F., Basar, E., van Dijk, B., Narici, L., Pantev, C., Ribary, U. and Salenius, S.

78. **Time-Frequency Analysis of Oscillatory Gamma-Band Activity: Wavelet Approach and Phase-Locking Estimation** 919
Bertrand, O., Tallon-Baudry, C. and Pernier, J.

79. **Inverse Imaging of Distributed Oscillatory Activity** 923
Bradshaw, L.A. and Wikswo, J.P., Jr.

80. **MEG Measurements of 40Hz Auditory evoked Response in Human Brain** 927
Ciulla, C., Takeda, T., Morabito, M., Kumagai, T. and Endo, H.

81. **Oscillatory Neuromagnetic Activity induced by Verbal and non-Verbal Stimuli presented in Visual and Auditory Modalities** 931
Eulitz, C., Maess, B., Pantev, C., Friederici, A., Feige, B. and Elbert, T.

82. **Singular Spectrum Analysis applied to Brain Signals** 935
Gaeta, R., Benzi, R., Peresson, M. and Narici, L.

83. **Characterization of Photosensitive Discharges by Oscillatory responses to Driving Stimuli** 939
Karhu, J. and Tesche, C.D.

84. **Equivalent Current Sources of the Auditory evoked Gamma Band and Steady State Fields** 943
Knuth, K.H., Broadhurst, J.H. and Schwartz, B.J.

85. **Local Modulation of MEG Rhythmic Activities related to the alternation of Binocular Rivalry and Binocular Fusion** 947
Kobayashi, T., Kato, K. and Kuriki, S.

86. **Magnetic Oscillatory Components of Human Flash-VEP: Luminance and Eccentricity Functions and effect of Acute Scopolamine** 951
Lopez, L., Narici, L., Conforto, S., Romani, G.L. and Sannita, W.G.

87. **Increased Variability of the Period Duration for Waveforms in Epochs of MEG Data as a Function of Arousal State** 955
Moran, J.E. and Tepley, N.

88. **Suitability of FFT Spectra versus DPA Period Spectra for Identification of Arousal States** 959
Moran, J.E. and Tepley, N.

89. **Chaos in the Brain: A New Strategy to Discriminate Deterministic Low Dimensional Dynamics in the Spontaneous Activity of the Human Cortex** 963
Narici, L., Boccaletti, S., Giaquinta, A. and Arecchi, T.

90. **A New Procedure to discriminate and study different Rhythmical Cortical Activities on the basis of their responsiveness to Simple Sensory Stimulation** 967
Narici, L. and Peresson, M.

91. **Cognitive Temporal Binding and its Relation to 40Hz Activity in Humans: Alteration during Dyslexia** 971
Ribary, U., Joliot, M., Miller, S.L., Kronberg, E., Cappell, J., Tallal, P. and Llinás, R.

92. **A Technique for the Identification of Hippocampal Theta from MEG Data** 975
Tesche, C.D.

VI Magnetoencephalography (MEG): Clinical Studies

93. **Clinical Applications of MEG** 980
Mäkelä, J.P., Elbert, T., Kakigi, R., Lewine, J., Lopez, L., Nagamine, T., Nakasato, N., Reite, M. and Sannita, W.

Clinical MEG

94. **Changes of the Functional Organization of the Somatosensory Cortex in Chronic Pain Patients** 991
Braun, C., Flor, H., Birbaumer, N. and Elbert, T.

95. **Clinical MEG II: Development of a Normative Database** 995
Davis, J.T., Lewine, J.D., Edgar, J.C., Hoesing, J. and Orrison, W.W., Jr.

96. **Cortical Reorganization in Arm Amputees: Alterations of the Somatosensory Representation of the Intact Arm** 999
Elbert, T., Sterr, A., Rockstroh, B., Charbonnier, D., Flor, H., Pantev, C., Wienbruch, C., Knecht, S. and Taub, E.

97. **Comparison of Magnetoencephalographic and Intraoperative Electrocorticographic Localization of Sensory Response** 1003
Greenblatt, R., Bucholz, R., Sobel, D., Schwartz, B. and Hirschkoff, E.

98. **Origin of Slow Wave observed in Cerebrovascular Disease** 1007
Hatanaka, K., Nakasato, N., Kanno, A., Ohtomo, S., Seki, K., Fujiwara, S. and Yoshimoto, T.

99. **Magnetic Source Imaging in the Evaluation of Chronic Alcohol Abuse** 1011
Hill, D., Waldorf, V.A., Lewine, J.D., Provencal, S.L., Moyers, T. and Yeo, R.

100. **MEG and MCG in a Clinical Environment: BioMag Laboratory, Helsinki** 1014
Ilmoniemi, R.J., Ahonen, A., Alho, K., Aronen, H.J., Huttunen, J., Karp, P., Mäkijärvi, M., Näätänen, R., Paetau, R., Pekkonen, E., Karhu, J., Standertskjöld-Nordenstam, C.-G., Tiihonen, J., Toivonen, L. and Katila, T.

101. **MEG Traces of Painful Processes** 1018
Kakigi, R., Kitamura, Y., Koyama, S., Hoshiyama, M., Shimojo, M., Watanabe, S., Takeshima, Y. and Nagata, O.

102. Normalized N100m Latency of the Auditory evoked Fields after Surgical Removal of Temporal Lobe Gliomas 1022
Kanno, A., Nakasato, N., Ohtomo, S., Seki, K., Kawamura, T., Fujita, S., Kumabe, T., Kayama, T. and Yoshimoto,T.

103. Reversed Hemispheric Asymmetry of Auditory N100m in Schizophrenics 1026
Karhu, J., Tiihonen, J., Pekkonen, E., Katila, H., Huotilainen, M., Virtanen, J. and Ilmoniemi, R.

104. Neuromagnetic Identification of the Somatosensory Cortex in cases with Arteriovenous Malformation Adjacent to the Central Sulcus 1030
Kawamura, T., Nakasato, N., Ohtomo, S., Seki, K., Kanno, A., Fujita, S., Fujiwara, S. and Yoshimoto, T.

105. Time Course of Abnormal MEG Activity associated with Transient Ischemic Attacks 1033
Kober, H., Vieth, J.B., Stippich, C., Kassubek, J.R. and Hopfengaertner, R.

106. Preoperative Localization of Sensorimotor Cortex: MRI versus MEG 1037
Lewine, J.D., Baldwin, N.G., Bucholz, R.D., Sanders, J.A., Halliday, A.L., Anson, J.A., Shih, J., Stearley, J., Davis, J.T., Astur, R.S., Paulson, K. and Orrison, W.W., Jr.

107. Clinical MEG I: Towards a Standardized Examination 1041
Lewine, J.D., Davis, J.T., Davis, L.E., Canive, J., Roberts, B., Graeber, D., Shih, J., Edgar, J.C., Provencal, S.L., Paulson, K., Meyers, J., Christner, R., Silveri, J., Rawcliffe, N., Tessman, C., Espinosa, M., Depper, M., Sanders, J.A. and Orrison, W.W., Jr.

108. Two Components of Giant Somatosensory evoked Magnetic Fields in Patients with Cortical Myoclonus 1045
Mima, T., Mikuni, N., Nishitani, N., Nagamine, T., Ikeda, A., Fukuyama, H., Takigawa, T., Kimura, J. and Shibasaki, H.

109. Deviations from the Tonotopic Map are correlated with Tinnitus Strength 1049
Mühlnickel, W., Flor, H. and Elbert, T.

110. The Determinism of MEG in Tinnitus Patients 1053
Mühlnickel, W., Hoke, E.S., Hoke, M. and Flor, H.

111. Auditory Signal Processing is impaired in Alzheimer's Disease: A Whole-Head MEG Study 1057
Pekkonen, E., Huotilainen, M., Virtanen, J., Näätänen, R., Ilmoniemi, R.J. and Erkinjuntti, T.

112. SEFs to Median Nerve and Finger Stimulation: Analysis of Interhemispheric Asymmetries in Healthy Subjects and following Hemispheric Lesion 1060
Pizzella, V., Rossini, P.M., Rossi, S., Tecchio, F., Lopez, L. and Romani, G.L.

113. Effects of Acute, Oral Administration of Phenobarbital (1.6 mg/kg) to Healthy Subjects 1064
Sannita, W.G., Lewine, J.D., Maclin, E.L., Orrison, W.W., Jr. and Robinson, S.

114. Visual evoked Fields for Pattern Reversal Stimuli in Patients with Occipital Lobe Lesions 1067
Seki, K., Nakasato, N., Fujita, S., Kanno, A., Hatanaka, K., Kawamura, T., Ohtomo, S., Fujiwara, S., Takahashi, A. and Yoshimoto, T.

115. MEG and fMRI Localisation of Primary Sensory-Motor Cortex in Normals and a Patient with a Low-Grade Astrocytoma 1071
Singh, K.D., Furlong, P.L., Weale, P., Harding, G.F.A. and Walsh, A.R.

116. Comparison between the Processing of Static Images of Human Faces in High Functioning Autistic Subjects and Normal Controls 1075
Swithenby, S.J., Bräutigam, S., Bailey, A.J., Jousmaki, V. and Tesche, C.

117. A Comparison of MEG and EEG for the Assessment of Traumatic Brain Injury 1079
Weinberg, H., Gaetz, M., Jantzen, K., Cheyne, D. and Deecke, L.

118. Somatosensory evoked Magnetic Fields in Acute Cerebral Infarction 1083
Wikström, H., Huttunen, J., Roine, R.O., Salonen, O., Aronen, H.J. and Ilmoniemi, R.J.

Clinical MEG: Epilepsy

119. Magnetoencephalographic vs electrocorticographic Source Localization in Patients with Epilepsy 1087
Aung, M., Sobel, D., Schwartz, B., Squires, K. and Hirschkoff, E.

120. Propagation of Epileptic Activity during Single (unaveraged) Interictal Spikes as studied with Magnetic Field Tomography (MFT) 1090
Bamidis, P.D., Hellstrand, E. and Ioannides, A.A.

121. Presurgical Evaluation of Methohexital-Induced Epileptiform Activity: Comparison of EEG-, MEG-, and Simultaneous EEG/MEG Models 1094
Diekmann, V., Becker, W., Jürgens, R., Grözinger, B., Kleiser, B., Wollinski, K.H. and Richter, H.P.

122. Intracortical Spread of Epileptic Spikes-Simultaneous Measurement of Magnetic Field and Electric Potential in Rabbits 1098
Eiselt, M., Zwiener, U., Gießler, F., Wagner, H., Haueisen, J., Huonker, R., Nowak, H. and Schack, B.

123. Effects of Volume Current on Localization of Focal Epileptic Activity 1102
Ikeda, H., Takeuchi, R., Osozawa, S., Ishiyama, A., Kasai, N. and Minami, T.

124. Evaluation of Abnormal Low Frequency Magnetic Activity (ALFMA) in Epilepsy 1106
Lewine, J.D., Davis, J.T. and Orrison, W.W., Jr.

125. Event-Related Magnetic Fields Associated with Auditory Oddball Paradigm-Postoperative Study in Patients with Temporal Lobe Epilepsy 1110
Nishitani, N., Nagamine, T., Mikuni, N., Ikeda, A., Kimura, J. and Shibasaki, H.

126. Use of Magnetic Source Imaging in Surgical Planning for Epilepsy Patients who have Undergone Prior Resection 1114
Sobel, D., Aung, M., Schwartz, B., Squires, K. and Hirschkoff, E.

127. Methohexital-Induced Changes in Spectral Power of Neuromagnetic Signals: Reduced β-Band Enhancement over the Hemisphere Ipsilateral to the Epileptogenic Focus 1117
Wienbruch, C., Eulitz, C., Lehnertz, K., Brockhaus, A., Elger, C.E., Elbert, T. and Hoke, M.

VII Multimodality Comparisons And Integration

128. Evaluation of Source Localization Models for Somatosensory evoked MEG with PET Verification 1122
Akhtari, M., McNay, D., Levesque, M., Rogers, R.L., Brown, C., Mandelkern, M., Babai, A. and Sutherling, W.W.

129. Combination of Point and Surface Matching Techniques for Accurate Registration of MEG and MRI 1126
Bamidis, P.D. and Ioannides, A.A.

130. EEG and MEG Source Analysis of Somatosensory evoked Responses to Mechanical Stimulation of the Fingers 1130
Cheyne, D., Roberts, L.E., Gaetz, W., Bosnyak, D., Weinberg, H., Johnson, B., Nahmias, C. and Deecke, L.

131. Dynamic Neuroimaging by MEG, constrained by MRI and fMRI 1134
George, J.S., Schmidt, D.M., Mosher, J.C., Aine, C.J., Ranken, D.M., Wood, C.C., Lewine, J.D., Sanders, J.A. and Belliveau, J.W.

132. Combining Modalities to investigate Cortical Retinotopy 1138
Hurdal, M.K. and McElwain, D.L.S.

133. fMRI constrained Linear Estimation of Cortical Activity from MEG Measurements: A Model Study 1142
Liu, A.K., Sereno, M.I., Rosen, B.R., Belliveau, J.W. and Dale, A.M.

134. Multiple Modality Biomagnetic Analysis System 1146
Martin, K., Erne, S., Law, C., Conforto, S., Mallick, J. and Tatar, B.

135. Measurement of Visually evoked EEG and MEG associated with Verbal Cognitive Processes 1150
Nakagawa, S., Ueno, S., Iramina, K. and Weinberg, H.

136. Superadditive Information from Simultaneous MEG/EEG Data 1154
Pflieger, M.E., Simpson, G.V., Ahlfors, S.P. and Ilmoniemi, R.J.

137. Multimodal Functional Imaging of the Sensorimotor Cortex with fMRI and Magnetic Source Imaging (MSI) 1158
Roberts, T.P.L. and Rowley, H.A.

138. Quantitative Comparison of Motor Control Studied with MEG and fMRI 1162
Russell, D.P., Schwartz, B.J. and Hirschkoff, E.C.

139. Generating Cortical Constraints for MEG Inverse Procedures using MR Volume Data 1166
Sandor, S.R., Leahy, R.M. and Timsari, B.

140. Correlation between Depth and MEG/EEG Surface recordings in Auditory Cortex. Contribution of Numerical Simulations 1170
Schwartz, D., Badier, J.M., Scarabin, J.M. and Liégeoix-Chauvel, C.

141. Brain Activities evoked by Missing Auditory Stimuli: A Combined Study using MEG and fMRI Data 1174
Takino, R., Miyauchi, S., Sasaki, Y., Pütz, B. and Sekihara, K.

142. Integration of Functional MRI and MEG of Median Nerve Stimulation 1178
Verbiesen, M.J.R., Peters, A.R., den Boer, J.A., Knösche, T.R. and Peters, M.J.

143. Comparison of Functional Localization in Human Visual Cortices using MEG and fMRI: A Preliminary Report 1182
Yoshikawa, K., Takanashi, Y., Iwamoto, K., Yoshida, Y., Ueda, M., Tanaka, C., Umeda, M., Ebisu, T., Fukunaga, M., Naruse, S., Sato, H. and Nakajima, K.

Contents
Volume I

I Biomagnetic Instrumentation

1. **Biomagnetic Instrumentation** 3
   Koch, H.

2. **Reference Noise Cancellation: Optimizations and Caveats** 7
   Barnes, G.R., and Ioannides, A.A.

3. **Positioning of MCG Measurement Equipment** 11
   Borchardt, M., Krause, E., and Lindner, G.

4. **A Low-Noise, Integrated DC SQUID Magnetometer for Applications in Biomagnetism** 15
   Cantor, R., Vinetskiy, V. and Matlashov, A.

5. **Magnetocardiograms Recorded by First-Order Gradiometers using HTS rf SQUIDs** 20
   Copetti, C.A., Soltner, H., Zhang, Y., Schuett, H., Rongen, H., Brandenburg, G., Clemens, U., Zimmermann, E., Borgmann, J., Tavrin, Y., Otto, R., Bousack, H., and Halling, H.

6. **A Realtime, High-Bandwidth, Programmable Digital Signal-Processing System for Monitoring Acquisition, Filtering, and Archival Storage of Biomagnetic Signals** 24
   Corby, N.R., Hogle, R.A., Miller, P.D., and Pasquarelli, A.

7. **Magnetic Measurement Techniques for Locating Foreign Bodies in Humans** 28
   Costa Monteiro, E., Kelber, C., Andrade Lima, E., Costa Ribeiro, P., Bruno, A.C., Hall Barbosa, C., Parente Ribeiro, E., and Boechat, P.

8. **Optimisation of Co-Registration Procedures for Biomagnetic Measurements using a Thermoplastic Mask** 32
   Dammers, J., Barnes, G.R., Weise, F., Gross, J., Clewett, B.S., Ioannides, A.A., Müller-Gärtner, H.-W.

9. **Design of a Magnetically Shielded Room for Development of HTS SQUID Systems for Magnetocardiography** 35
Davidson, C.R., Griffin, R.H., Best, K.J., Maltin, L., Wolf, W., Zhang, Y., Bousack, H. and Soltner, H.

10. **A Digital-Signal-Processor (DSP) Flux-Locked-Loop (FLL) for LTC SQUIDs with Automatic Reset and Feedback Cancellation** 39
Flynn, E.R., Bracht, R., Kraus, R.H., Jr., Maas, P.M., Ruminer, P. and Stettler, M.

11. **A 32-Channel MCG System cooled by a GM/JT Cryocooler** 43
Fujimoto, S., Sata, K., Fukui, N., Haraguchi, E., Kido, T., Nishiguchi, K. and Kang, Y.M.

12. **A Cryocooled SQUID System with Minimal Clearance between the Pick-up Coil and the Specimen** 47
Fukui, N., Sata, K., Fujimoto, S., Kang, Y.M.

13. **A Comparison of Vector and Radial Magnetometer Arrays for Whole-Head Magnetoencephalography** 51
Hughett, P. and Miyauchi, S.

14. **32 Channel High Tc SQUID System for Biomagnetism** 55
Itozaki, H., Kugai, H., Nagaishi, T., Toyoda, H., Hirano, T., Haruta, Y., Tanaka, S. and Kado, H.

15. **Source Detectability for Multichannel Biomagnetic Sensors** 59
Johnson, R.T., Black, W.C. and Buchanan, D.S.

16. **A Digital Flux-Locked Loop for High Temperature SQUID Magnetometer and Gradiometer Systems with Field Cancellation** 63
Kraus, R.H., Jr., Bracht, R., Flynn, E.R., Jia, Q., Maas, P.M., Reagor, D. and Stettler, M.

17. **Characterisation of a Planar Second-Order Gradiometer in a Lightly Shielded Environment** 67
Lang, G., Hutson, D., Weston, R.G., Klein, U., Shahani, U., Weir, A.I., Maas, P.M., Pegrum C.M. and Donaldson, G.B.

18. **Magnetic Field Error of a Current Dipole in a Superconducting Shield** 71
Matsuba, H.

19. **High Tc Superconducting Shielded Biomagnetometer System** 75
Matsuba, H., Shintomi, K., Yahara, A., Imai, K., Irisawa, D. and Masuda, M.

20. Current Dipole Localization Errors as a Function of the System Noise and the Number of Sensors 79
Matsuba, H., Vrba, J. and Cheung, T.

21. Auditory Brainstem evoked Fields measured in a High Tc Superconducting Shield 83
Nogai, T., Yahara, A., Matsuba, H. and Kuriki, S.

22. High Performance Magnetically Shielded Room for Clinical Measurements 87
Paavola, M.A.J., Ilmoniemi, R.J., Sohlström, L., Meinander, T., Penttinen, A. and Katila, T.E.

23. An AC Susceptometer with a large Magnetised Volume for Biological Samples 91
Pasquarelli, A., Del Gratta, C., Della Penna, S., Di Luzio, S., Pizzella, V. and Romani, G.L.

24. A Six-Channel SQUID Biomagnetometer optimised for Fetal magnetocardiography 95
Rassi, D., Zhuravlev, Y.E., Josephs, O. and Swithenby, S.J.

25. High Temperature Superconducting Quantum Interference Devices in a Superconductor-Normal-Superconductor Geometry 99
Reagor, D., Jia, Q.X., Mombourquette, C., Foltyn, S., Houlton, R. and Wu, X.D.

26. A Digital SQUID Controller for Unshielded Biomagnetic Measurements 103
Robinson, S.E.

27. Investigation of the thin Film SQUIDs using Double Layer Model of the Secondary Sources 107
Romanovych, S.S.

28. DC-MEG Studies: Development of Techniques for resolving DC Fields 110
Saligram, U., Moran, J.E. and Tepley, N.

29. Development of a 61-Channel MEG System cooled by a GM/JT Cryocooler 114
Sata, K., Fujimoto, S., Fukui, N., Haraguchi, E., Kido, T., Nishiguchi, N. and Kang, Y.M.

30. Low-Noise, Single Layer $YBa_2Cu_3O_{7-\delta}$ DC SQUID Magnetometers for Shielded and Unshielded Operation 118
Schmidt, J.M., Lee, L.P., Matlashov, A., Teepe, M., Vinetskiy, V., Cantor, R. and DiIorio, M.S.

31. The Use of Software Magnetometers for Ambient Noise Cancellation in Unshielded Environment 122
Shabalin, A.V., Golyshev, N.V., Greenberg, Ya.S., Motorin, S.V., and Rogachevskij, B.M.

32. The use of a Current Feedback Loop for active Compensation of Magnetic Noise in Superconducting Magnetocardiograph 126
Shabalin, A.V., Golyshev, N.V., Motorin, S.V., and Rogachevskij, B.M.

33. A Skull-Based Multiple Dipole Phantom for EEG and MEG Studies 130
Spencer, M.E., Leahy, R.M. and Mosher, J.C.

34. High-Tc SQUID Gradiometers for Magnetocardiography in an Unshielded Environment 134
Tralshawala, N., Claycomb, J.R., Ying, Q., Miller, J.H., Jr., Nesteruk, K. and Jackson, D.R.

35. 143 Channel Whole-Cortex MEG System 138
Vrba, J., Angus, V., Betts, K., Burbank, M.B., Cheung, T., Fife, A.A., Haid, G., Kubik, P.R., Lee, S., Ludwig, W., McCubbin, J., McKay, J., McKenzie, D., Robinson, S.E., Smith, M., Spear, P., Taylor, B., Tillotson, M., Cheyne, D. and Weinberg, H.

36. SEFs from the Human Spinal Cord using a Novel Second Order Planar Gradiometer 142
Weir, A.I., Shahani, U., Hutson, D., Lang, G., Maas, P.M., Pegrum, C.M. and Donaldson, G.B.

37. High-Tc Superconducting rf Receiver Coils for High Resolution Magnetic Resonance Imaging of the Spine 146
Wosik, J., Nesteruk, K., L.M. Xie, Zhang, X.P., Jiao, C., Gierlowski, P. and Miller, J.H., Jr.

38. 3-D Magnetic Field Analysis of a Magnetically Shielded Room for Biomagnetic Measurements 150
Yamazaki, K., Uchikawa, Y. and Fujiwara, K.

39. A 129 channel Vector Neuromagnetic Imaging System 154
Yoshida, Y., Arakawa, A., Kondo, Y., Kajihara, S., Tomita, S., Tomita, T., Takanashi, Y. and Matsuda, N.

40. Magnetocardiography using SQUIDs cooled by Liquid Nitrogen 158
Zhang, Y., Schütt, H., Bousack, H., Wolf, W., Schirdewan, A., Burghoff, M. and Trahms, L.

II Models for the Biomagnetic Forward and Inverse Problems

41. Comparison of Different Approaches to the Biomagnetic inverse Problem—Workshop Report 163
Lutkenhöner, B., Greenblatt, R., Hämäläinen, M., Mosher, J., Scherg, M., Tesche, C., and Valdes-Sosa, P.

42. Distributed Source Reconstruction using a Non-Linear Spatiotemporal Regularization Method: An Alternative to LORETA 177
Baillet, S., Garnero, L. and Renault, B.

43. Localization of Extended Sources using a modified Linear Estimation Approach 181
Brauer, H., Ziolkowski, M., Tenner, U., Kosch, O., Wiechmann, H. and Arlt, A.

44. Inverse Magnetography for Reentry Currents 185
Brio, M., Marchesin, D., Costa Monteiro, E., Bruno, A.C. and Costa Ribeiro, P.

45. Frequency Domain Distributed Inverse Solutions 189
Casanova, R., Valdes-Sosa, P., Garcia, F.M., Aubert, E., Riera, J.J., Korin, W., and Lins, O.

46. Small Volume Resistivity Changes Observed using MEG 193
Clifton, Y.C., Moran, J.E., and Tepley, N.

47. A Weighting Matrix to remove Depth Bias in the Linear Biomagnetic Inverse Problem with Application to Cardiology 197
Crowley, T.A., Haupt C.D. and Kynor, D.B.

48. Imaging Properties of "Myocardial Current Imaging" 201
Dössel, O., Katscher, U., Köhler, Th., Fuchs, M., Wischmann, H.-A., Wagner, M., Drenckhahn, R. and Hoppe, W.

49. Practicability of Neuromagnetic Source Analysis using the Finite-Element Method 205
Drüen, K., Steinsträter, O. and Lütkenhöner, B.

50. A Numerically Stable Approximation for the Magnetic Field of the Conducting Spheroid close to the Symmetry Axis 209
Fieseler, T., Ioannides, A.A., Liu, M.J. and Nowak, H.

51. Source Reconstructions by Spatial Deviation Scans 213
Fuchs, M., Wischmann, H.-A., Wagner, M., Drenckhahn, R. and Köhler, Th.

52. **Generalized Inverse Solution for Bimodal Electro-Magnetic Source Images** 217
Gencer, N.G. and Williamson, S.J.

53. **Can Compact Currents be uniquely determined?** 221
Gorodnitsky, I.F.

54. **Estimation of Distributed Sources with adjustable Contrast** 225
Grunwald, A. and Lütkenhöner, B.

55. **Aspects of Non-Uniqueness: How one Source affects the Reconstruction of Another** 229
Hasson, R. and Swithenby, S.J.

56. **The Theoretical Basis of Iterative Distributed Solutions to the Biomagnetic Inverse Problem** 233
Hasson, R. and Swithenby, S.J.

57. **Anatomically constrained vs unconstrained Dipole Fit in the Combination of MEG and MRI** 237
Haueisen, J., Hajek, M., Huonker, R., Nowak, H. and Sauer, H.

58. **One Effect of a Maximum possible Current Strength** 241
Huang, M. and Heller, L.

59. **Multi-Source Localization in MEG using Simulated Annealing: Model Order Determination and Parameter Accuracy** 245
Huang, M., Supek, S., Aine, C.

60. **A Three-Dimensional Rule assisted Interpolation Algorithm for Iso-Field Mapping of Neuromagnetic Signals** 249
Hultin, L.

61. **The Influence of Glial Buffering Currents on Sustained Neuromagnetic Fields** 253
Hultin, L.

62. **Magnetic Field Tomography: Theoretical basis, Performance Measures and Limitations** 257
Ioannides, A.A., Bolton, J.P.R., Liu, M.J., Bamidis, P.D., Liu, L.C., Dammers, J. and Müller-Gärtner, H.W.

63. **Algorithm for Magnetocardiography Analysis based on Multi Channel SQUID and Thoracic MR Images** 261
Ishii, H., Niki, N., Nakaya, Y., Nishitani, H. and Kang, Y.M.

64. **Estimation of Distributed Sources of Auditory evoked Magnetic fields in the Human Brain** 265
Iwaki, S. and Ueno, S.

65. A New Approach to the Reconstructing Source Distribution 270
Kajihara, S., Tomita, T., Tomita, S., Kondo, Y., Arakawa, A., Yoshida, Y. and Matsuda, M.

66. A Comparative Study of Minimum Norm Inverse Methods for MEG Imaging 274
Leahy, R.M., Mosher, J.C. and Phillips, J.W.

67. Development, Implementation and Applications of Single Epoch Analysis of MEG Signals 278
Liu, L.C., Ioannides, A.A., Heer, H., Wagener, M., Delonge, F., Halling, H. and Müller-Gärtner, H.W.

68. Dipole Localization Accuracy as a Function of the Area Covered by the Magnetometer System 282
Lütkenhöner, B.

69. Towards a Standardized Representation of Neuromagnetic Data 286
Lütkenhöner, B., Grunwald, A. and Steinsträter, O.

70. Multiple Current-Dipole Distribution reconstructed by modified Selective Minimum-Norm Method 290
Matsuura, K. and Okabe, Y.

71. An Estimation Algorithm of Neuromagnetic Source using MRI Information 294
Mino, K., Niki, N., Nakasato, N. and Yoshimoto, T.

72. Realistic Modeling, Analysis and Animated 3D Visualization for MEG Studies 298
Morabito, M., Takeda, T., Kobayakawa, T. and Endo, H.

73. Subspace Angles: A Metric for Comparisons in EEG and MEG 302
Mosher, J.C.

74. Comparison of the Constant and Linear Boundary Element Method for EEG and MEG forward Modeling 306
Mosher, J.C., Chang, C.H. and Leahy, R.M.

75. MEG-based Imaging of Focal Neuronal Current Sources 310
Phillips, J.W., Leahy, R.M. and Mosher, J.C.

76. A Four-Sensor Skeleton Configuration for Single Dipole Localization 314
Prakash, N.M., Spelman, F.A. and Read, P.

77. The new Methods of Localization and Reconstruction of Biomagnetic Signal Sources 318
Primin, M.A. and Nedayvoda, I.V.

78. **Effects of Tissue Conductivity Variations on Cardiac Magnetic Fields** 322
Ramon, C., Czapski, P., Huntsman, L.L., Bardy, G.H. and Kim, Y.

79. **Discrete Spline Electric-Magnetic Tomography (DSPET) based on realistic Neuroanatomy** 326
Riera, J.J., Aubert, E., Valdes-Sosa, P., Casanova, R., and Lins, O.

80. **Strategy for objective Comparisons of Inverse Solutions** 330
Schlitt, H.A., Zhao, S.-R., Halling, H. and Müller-Gärtner, H.-W.

81. **A method for locating Regions containing Neural Activation at a given Confidence Level from MEG Data** 334
Schmidt, D.M. and George, J.S.

82. **Generalized Wiener Estimation of Three-Dimensional Current Distribution from Biomagnetic Measurements** 338
Sekihara, K. and Scholz, B.

83. **Localizing dipoles: The Correlation Coefficient and the Goodness of Fit in Source Localization of Filtered Noise** 342
Siedenberg, R., Goodin, D.S., Aminoff, M.J., Rowley, H.A. and Roberts, T.P.L.

84. **Monte-Carlo Analysis and Confidence Region Ellipsoids for Equivalent Current Dipole Solutions to EEG/MEG Data** 346
Singh, K.D. and Harding, G.F.A.

85. **Localization of Current Sources with Constraints of Solution Area** 350
Sugaya, H., Osozawa, S., Ishiyama, A. and Kasai, N.

86. **Source Localization by Genetic Algorithm** 354
Takeuchi, R., Ikeda, H., Ishiyama, A. and Kasai, N.

87. **Optimization of Conductivity and Permeability Parameters for Brain, CSF, Skull and Scalp using Implanted Sources in the Human Cranium** 358
Taylor, S.A., Rogers, R., McNay, D., Levesque, M., Akhtari, M. and Sutherling, W.

88. **Modeling of Spreading Cortical Depression using a realistic Head Model** 361
Tepley, N., Roth, B.J. and Wijesinghe, R.S.

89. **A Simulation Study on the Nature of Electrical Sources of the MCG** 365
Tilg, B., Wach, P., Lafer, G., Fleischmann, P. and Rucker, W.

90. **Global Optimization in the Localization of Brain Activity** 369
Uutela, K., Hämäläinen, M. and Salmelin, R.

91. **Variable Resolution Electric-Magnetic Tomography** 373
Valdes-Sosa, P., Marti, F., Garcia, F., and Casanova, R.

92. **Spatio Temporal Distributed Inverse Solutions** 377
Valdes-Sosa, P., Riera, J.J., and Casanova, R.

93. **Quantification of Deviations between Spherical and realistically Shaped Forward Models** 381
v. Rango, J., Schlitt, H.A., Halling, H., Terstegge, A. and Müller-Gärtner, H.-W.

94. **The Influence of Holes in a Layer with Low Conductivity on the Potential and Magnetic Field** 385
van den Broek, S.P., Reinders, K.F.J. and Peters, M.J.

95. **Estimation of Cardiac Sources on the Epicardium and Endocardium from Simulated and Measured MCG Data** 389
Wach, P., Tilg, B., Lafer, G., Kynor, D. and Rucker, W.

96. **Current Density Reconstructions using the L1 Norm** 393
Wagner, M., Wischmann, H.-A., Fuchs, M., Köhler, T. and Drenckhahn, R.

97. **Modelling Closed Surface Geometry for Skull and Torso using Snapping** 397
Yamrom, B.

98. **Locally Refined Head Geometries used to Evaluate Spherical Model Accuracy** 401
Yvert, B., Bertrand, O., Thévenet, M. and Pernier, J.

99. **Mesh Refinement of Boundary Element Head Models in MEG Source Reconstructions** 405
Zanow, F. and Peters, M.J.

III Magnetocardiography (MCG)

100. **New Trends in Clinical Magnetocardiography.** 410
Mäkijärvi, M., Brockmeier, K., Leder, U., Lesh, M., Moshage, W., Nakaya, Y., Nenonen, J., Nowak, H., Oeff, M., Stroink, G., Fenici, R.

101. **Revisiting the Fetal Magnetocardiogram and Fetal Magnetoencephalogram** 418
Wakai, R.T., Leuthold, A.C. and Martin, C.B.

102. **The Fetal Magnetocardiogram during Different Stages of Pregnancy** 422
Achenbach, S., Menéndez, T., Moshage, W., Flüg, M., Beinder, E., Bittl, A., Bachmann, K., and Lang, N.

103. Clinical Experience with the Combination of Multichannel Magnetocardiography and Magnetic Resonance Imaging for the Localization of Idiopathic and Secondary Ventricular Cardiac Arrhythmias 425
Achenbach, S., Moshage, W., and Bachmann, K.

104. Localization of the Origin of Ventricular Arrhythmias: Comparison of ECG Morphology and Localization based on ECG Algorithms to Multichannel Magnetocardiography 429
Achenbach, S., Moshage, W., and Bachmann, K.

105. Comparison of the QRS duration and RMS40 in Signal-Averaged Recordings of the Multichannel MCG to the Signal-Averaged ECG 433
Achenbach, S., Moshage, W., Flüg, M., Diem, B., Schibgilla, V., and Bachmann, K.

106. Investigation of Fetal Cardiac Arrhythmias by Magnetocardiography 437
Achenbach, S., Moshage, W., Menéndez, T., Beinder, E., Bittl, A., Hofbeck, M., Buheitl, G., Singer, H., and Bachmann, K.

107. Contraventions of Homogeneity of Ventricular Repolarization according to Data of Magnetocardiography and their Connection with Ventricular Tachycardia 440
Bobrov, V.A., Gapelyuk, A.V., Chaikovsky, I.A., Stadnyuk, L.A., Sosnitsky, V.N. and Kozlovsky, V.I.

108. Ventricular Repolarization Disturbances' Diagnostics in Chronic Ischemia Patients Evidence Derived from MCG 444
Chaikovsky, I.A., Lutay, M.I., Lomakovsky, A.N., Gapelyuk, A.V., Sosnitsky, V.N., Gushcha, B.G., and Minov, Yu.D.

109. Induced Atrial Flutter in Rabbit Hearts Tissue and Inverse Magnetocardiography 448
Costa Monteiro, E., Andrade Lima, E., Ornelas, P.H., Cruz, R.S., Picon, A.P., Kühner, G.S., Kelber, C., Parente Ribeiro, E., Bruno, A.C., Costa Ribeiro, P., Marchesin, D., Brio, M., and Leite, T.T.

110. Cellular Automata Simulation in the Magnetic Study of Rotating Propagation Models Associated to Cardiac Arrhythmias 452
Costa Monteiro, E., Correia, M.P., Hall Barbosa, C.R. and Costa Ribeiro, P.

111. Comparison of Measured and Simulated MCGs with a Realistic Heart-Torso Model 456
Czapski, P., Haueisen, J., Ramon, C., Nowak, H., Huntsman, L.L., Leder, U., Bardy, G.H., and Kim, Y.

112. High Resolution Finite Element Simulations of Magnetic Fields resulting from Myocardial Infarction 460
Czapski, P., Ramon, C., Huntsman, L.L., Bardy, G.H., and Kim, Y.

113. Invasive Evaluation of MCG-Localization Results Importance of Measured Electrophysiological Time Intervals 464
Gapelyuk, A., Copetti, C.A., Schirdewan, A., Schütt, H., Wiedemann, M., Meyerfeldt, U., Primin, M.A. and Maslenikov, Y.

114. Changes in Magnetic Field Maps at QRS-Onset after Myocardial Infarction 467
Hailer, B., Van Leeuwen, P., Donker, D., Rahn, N., Lange, S. and Wehr, M.

115. The Automatic Selection of Decisive Parameters in the Magnetocardiographic Localization of the Accessory Pathway in Patients Suffering from the WPW Syndrome 471
Jazbinsek, V., Trontelj, Z., Oeff, M. and Burghoff, M.

116. Source Localization of Initial QRS Vector in Patients with Idiopathic Cardiomyopathy: Comparison with Normal Subjects and Patients with Anterior Myocardial Infarction 475
Kishi, F., Nomura, M., Yukinaka, M., Saito, K., Nakaya, Y. and Ito, S.

117. Linear and Nonlinear Analysis in the Parallel HR-MCG/HR-ECG Experiment in Arrhythmogenic Right Ventricular Cardiomyopathy 479
Kowalik, Z.J., Mäkijärvi, M. Wichter, T., Fetsch, T., Elbert, T. Breithard, G. and Hoke, M.

118. His-Purkinje Conduction System Maps: Fact or Fiction? 483
Lamothe, M.J.R., Stroink, G. and Gardner, M.J.

119. Spectral Analysis of Body Surface Potential Maps and Magnetic Field Maps of Patients with and without Ventricular Tachycardia 487
Meeder, R.J.J., Stroink, G., Ritcey, S.P. and Gardner, M.J.

120. Magnetocardiography—A New Approach for early Detection of Arrhythmogenic Right Ventricular Dysplasia 491
Meyerfeldt, U., Schirdewan, A., Wiedemann, M., Schütt, H., Dietz, R., Thierfelder, L., Burghoff, M. and Trahms, L.

121. **Magnetocardiographic Functional Imaging Studies in the BioMag Laboratory** 494
Montonen, J., Ahonen, A., Hämäläinen, M., Ilmoniemi, R.J., Laine, P., Nenonen, J., Paavola, M., Simelius, K., Simola, J. and Katila, T.

122. **Spatial Aspects in Magnetocardiographic Time Domain and Spectrotemporal Post-Infarction Arrhythmia Risk Assessment** 498
Montonen, J., Korhonen, P., Paavola, M., Mäkijärvi, M., Toivonen, L. and Katila, T.

123. **Comparison of Different Imaging Modalities as Anatomical Reference Methods for Magnetocardiographic Localization** 502
Moshage, W., Achenbach, S. and Bachmann, K.

124. **Comparison of the Magnetocardiographic Localization Accuracy using Simultaneous and sequential recordings in a single and multiple planes** 506
Moshage, W., Achenbach, S., Flüg, M., Lohse, P. and Bachmann, K.

125. **Localization of Kent Bundles prior to HF Catheter Ablation: A Quantitative Comparison of the Localization Accuracy based on an Inverse Solution applied to Electrical and Magnetic Field Distributions** 509
Moshage, W., Achenbach, S., Killmann, R., Bruder, H. and Bachmann, K.

126. **Analysis of Electromotive Force by Magnetocardiogram—Is the Origin of Initial QRS Force (Septal Vector) of Electrocardiogram Intraventricular Septum?** 513
Nakaya, Y., Yukinaka, Y., Nomura, M., Kishi, F. and Saito, K.

127. **The Algorithm of Spatial-Temporary Analysis of Cardiomagnetic Signals** 517
Nedayvoda, I.V. and Primin, M.A.

128. **Measurement of Tangential Components of the Magnetic Field in the PR Segment in Man** 521
Noda, Y., Mitsui, T., Terada Y., Tsukada, K., Adachi, A., Yoshida, T. and Haruta, Y.

129. **Observation of a Difference Signal in MCG Measurements in Patients with Ventricular Extrasystoles** 525
Pedersen, J.R., Bloch Thomsen, P.E., Kofoed, B., Bak, C.K. and Saermark, K.

130. **The Analysis of Magnetocardiosignal at Switching of Heart from Rhythm to Rhythm: The Preliminary Researches** 529
Primin, M., Nedayvoda, I. and Gapelyuk, A.

131. Vector Fetal Magnetocardiography 533
Rassi, D., Zhuravlev, Y.E., Lewis, M.J. and Emery, S.J.

132. Biomagnetic Evidence for Extreme Anisotropy—a Mechanism Limiting Conduction Capability in Accessory Pathways 537
Schirdewan, A., Meyerfeldt, U., Wiedemann, M., Schütt, H., Burghoff, M., Gapelyuk, A., Copetti, C.A. and Dietz, R.

133. Magnetocardiogram in 3 Patients after Aborted Sudden Cardiac Death without overt underlying Heart Disease 540
Schütt, H., Stellbrink, C., Dammers, J., Boers, F., Wiedenau, G., Ioannides, A.A., Müller-Gärtner, H.W., Gapelyuk, A. and Schirdewan, A.

134. Magnetocardiography and other Methods of Investigation in Diagnostics of Heart Remodelling after Myocardial Infarction and in Hypertension 544
Shabalin, A.V., Nikitin, Y.P., Golyshev, N.V., Motorin, S.V., Rogachevsky, B.M., Kytmanov, A.V., Ermakova, E.N. and Tselmin, S.G.

135. Comparative Data on Magnetocardiography and Clinical-Instrumental Methods of Investigation when some Hereditary Cardiovascular Diseases 547
Shabalin, A.V., Nikitin, Y.P., Golyshev, N.V., Motorin, S.V., Rogachevsky, B.M., Lisithenko, O.V., Ermakova, E.N., Dolgich, M.M. and Kuroedov, A.Y.

136. Atrial Depolarization Homogeneity Diagnostics on Evidence derived from MCG 550
Stadnyuk, L.A., Chaikovsky, I.A., Sosnitsky, V.N., Gapelyuk, A.V., Minov, Yu.D. and Chervonopiskaya, E.M.

137. Magnetocardiographic Localization of Focal Arrhythmias in Conjunction with Catheter Ablation 554
Steiner, P.R., Kynor, D.B., Haupt, C.D., Rowley, H.A. and Lesh, M.D.

138. Magnetocardiographic Assessment of Left Ventricular Hypertrophy: Correlation with Echocardiography 558
Takala, P., Montonen, J., Aalto, T., Mäkijärvi, M., Nenonen, J., Yildirim, Y., Nieminen, M.S. and Katila, T.

139. Localization of Atrial Depolarization Activity from Non-averaged MCG's in Congenital Heart Disease 562
Thieme, R., Brockmeier, K., Brümmerhoff, A., Kissenbeck, S., Burghoff, M. and Bein, G.

140. Tangential (x- and y-) Component Measurement of Cardiac Magnetic Field and Comparison with Conventional z-Component Measurement 565
Tsukada, K., Kandori, A., Yoshida, T., Haruta, Y., Noda, Y., Horigome, H., Terada, Y. and Mitsui, T.

141. Visualization for Electric Activities in a Human Heart 569
Ueda, T., Kasai, N., Sagane, M., Chinone, K. and Awano, N.

142. Magnetocardiography in the Determination of Fetal Heart Rate Complexity 573
Van Leeuwen, P., Bettermann, H., Schüßler, M. and Lange, S.

143. Fetal Magnetocardiography: Spectral Analysis of Fetal Heart Rate Variability in the Course of Pregnancy 577
Van Leeuwen, P., Hartmann, M., Lange, S. and Schüßler, M.

144. Determination of Changes in Fetal Cardiac Time Intervals and T/QRS Ratio during Pregnancy using Magnetocardiography 581
Van Leeuwen, P., Lange, S., Schüßler, M. and Lajoie-Junge, L.

145. Fetal Magnetocardiography in Twin Pregnancy 585
Van Leeuwen, P. and Schüßler, M.

146. Magnetocardiographic Monitoring of the Fetal Autonomic Nervous System 589
Zhuravlev, Y.E., Lewis, M.J., Rassi, D. and Emery S.J.

IV. Biomagnetic Measures of the Gastro-Intestinal System, Susceptometry and Other Biomagnetic Measures

BIOMAGNETIC RESEARCH IN GASTROENTEROLOGY

Baffa, O.[1], Basile, M.[2], Bradshaw, A.[3], Forsman, M.[4], Nakaya, Y.[5] and Richards, W.O.[3,6*]

Departamento de Física e Matemática, FFCLRP-Universidade de São Paulo, 14040-901 Ribeirão Preto-SP, Brazil[1], Istituto di Tecnologie Avanzate Biomediche-Universitá "G. D'Annunzio", 66.100 Chieti, Italy[2], Living State Physics Group, Vanderbilt University, Nashville, TN, USA[3], Department of Applied Electronics, Chalmers University of Technology, S-412 96 Gothenburg, Sweden[4], Department of Nutrition, School of Medicine, The University of Tokushima, Japan[5] and Department of Surgery at the Veterans Affairs Medical Center and Department of Surgery School of Medicine, Vanderbilt University, Nashville, TN, USA[6]

Introduction:

The gastrointestinal system is responsible for providing the body with a constant supply of nutrients, water and electrolytes. The food must be digested - chemically and physically broken down to the molecular level - in order to be absorbed by the small intestine. Since both processes are rate dependent, the flow rate of the food (chime) must be compatible with the absorption rate of the small and large intestine.

Motility is an important attribute of the gastrointestinal tract and as with blood flow through the heart, it is controlled by electrical signals. In the case of the heart, electrical measurements are straightforward to perform and can yield useful clinical information. However, in the case of gastrointestinal motility, qualitative and semi-qualitative methods based mainly on the use of X-rays and radioisotopes have been employed. For quantitative evaluations the direct measurement of parameters, such as pressure, flow and force were the ones performed more routinely in clinical and basic research. Only recently have electrical measurements been investigated further to extract information about motility. Still the primary methods to investigate motility rely on X-rays, scintigraphy, and catheters inserted into the gastrointestinal tract.

The use of magnetic measurements to study the gastrointestinal tract can be divided in two categories, one concerned with magnetic fields produced intrinsically by the electrical currents of organs and the other with magnetic fields produced by magnetic materials that are ingested. In the first group, magnetic measurements are used to image electrical currents associated with the basic electric rhythms (BER) of the gastrointestinal tract, and two new words -magnetogastrogram and magnetoenterogram- were coined to describe these recordings. In the second group, the fields produced by magnetic markers (MM), or tracers (MT) can be followed by measurements at the surface of the torso. In this context, MM are localized magnetic particles, while MT are dispersed magnetic particles.

Information can be gathered regarding position, time course, quantity, state of order, among others parameters related to gastrointestinal motility. Biomagnetic techniques represent in some cases an alternative and in others a unique way to study the gastrointestinal tract. It is believed that this approach, combined with the already established methods, could contribute to new knowledge in this area. Reflecting this belief, the two last conferences of this series have seen growing interest in this topic, leading to the present chapter. In particular, during the 8th International Conference on Biomagnetism there was a special session on biomagnetic applications to gastroenterology, fully documented in the proceedings book [1]. In

* Authors are listed in alphabetical order, ordering of appearance has no meaning to their importance in this paper

the following sections, a survey of the work performed by some of the groups working in the area will be reviewed and updated.

Magnetic tracers to study GI motility

a-Studies Using Magnetic Markers

The initial studies using MM were methodologically simple. In 1957 a magnetometer was used to study the motility of the small bowel by having human subjects ingest a magnetic stirring bar (ALNICO V, 5 mm diameter and 12 mm long length) and detecting fluctuations of the magnetic fields produced by changes in its orientation of less than 1 degree relative to the detector [2]. A study was carried out involving 180 subjects, subjected to various physiological conditions, including cold pressor, hyperventilation, exercise and drugs.

A different marker was used by the biomagnetism group at the University of Chieti [3], where a magnetized steel sphere of 2 mm O.D. was inserted in a tube, yielding an apparent density D = 1.9 g/cm^3. The MM was ingested and was followed by magnetic measurements made with a SQUID biogradiometer. Fig. 1 shows the typical arrangement for the measurements. By fitting the isofield lines with the isofield of a magnetic dipole, it was possible to locate the MM and to overlay the location onto MRI images to determine its position in the intestine. Repeated mapping allowed the measurement of segmental transit time as shown in Fig. 2. Using this marker, mean oroanal transit time was (56 ± 5) hours, the mouth to caecum transit time was (13 ± 1.7) hours, and the total colonic transit time was (43.5 ± 5) hours [4].

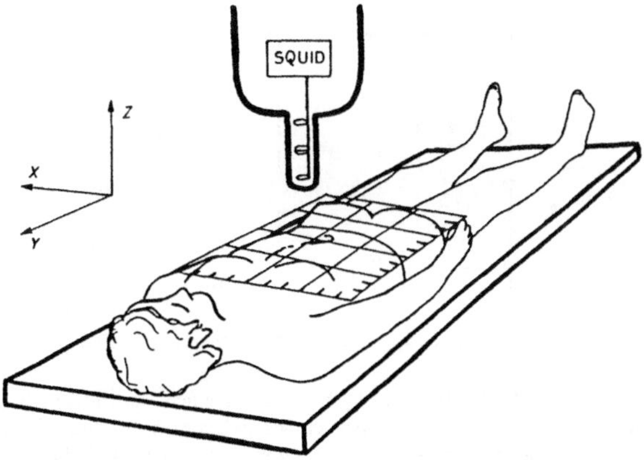

Fig.1-Typical arrangement for measurements of the biomagnetic fields produced by electrical currents of the GI tract and ingested magnetic particles. The SQUID-based biogradiometer can in some experiments be replaced by a fluxgate or active induction coils. Shielding can also be employed to improve signal-to-noise ratio.

The biomagnetism group at the Chalmers University of Technology in Sweden used a spherical permanent magnet made with epoxy and microspheres (D = 1.4 g/cm^3). A fluxgate gradiometer system, originally designed for magnetopneumography (MPG) research, was employed to study GI transit time. The test meal (cheese sandwich and milk) was ingested together with MM and other radiopaque markers. The investigation involved modeling and *in vivo* measurements, where four subjects took part. They ingested the magnetic marker together with radiopaque markers. In a one-day measurement, the subject ingested a magnetic marker in the early morning. The magnetic field was then sampled outside the subject at 15-minute intervals, 3 to 15 hours after ingestion. The MPG instrument was used to measure field maps in two planes. A nonlinear least-mean-square procedure was applied to the field maps to estimate the six free parameters, comprising the location and the moment of the magnetic dipole. Radiographs validated the estimated marker positions and showed that the magnetic marker behaved similarly to the radiopaque markers in the gut as shown in Fig. 3 [5]. The modeling calculations confirmed adequate precision of the MM localization.

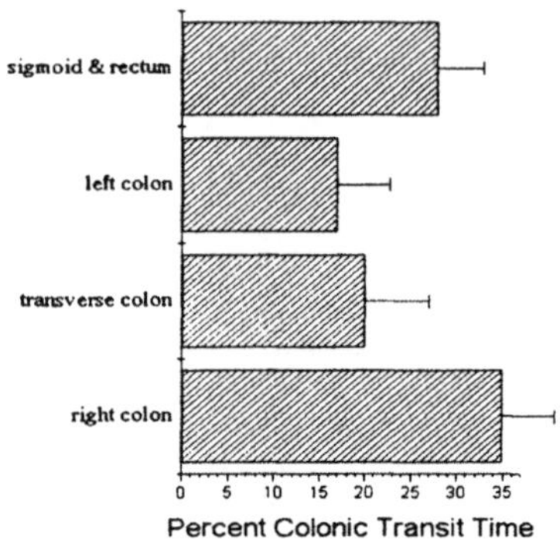

Fig.2-Proportion of time spent by the MM in different colonic regions determined by magnetic methods.

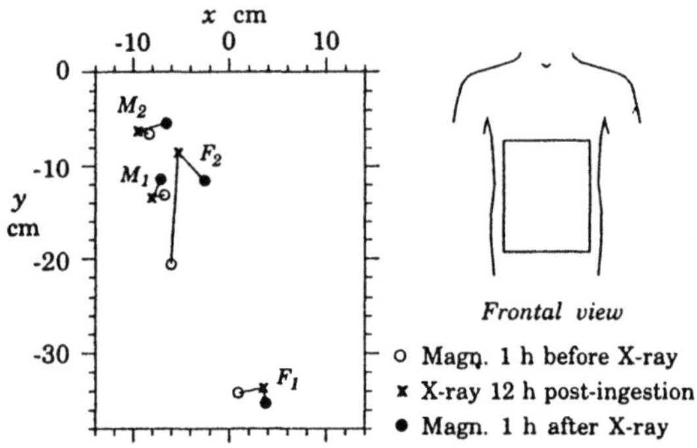

Fig.3-Location of the MM determined by X-ray and magnetic measurements. Crosses indicate the position determined by x-rays and open circles indicate the position found by fitting of the magnetic data. The radiographical determination was done one hour before and after the magnetic measurement.

MM has been also useful to study the pharmacodynamics of drug delivery. The group from PTB-Berlin prepared insoluble sucrose pellets coated with magnetite and pellets of epoxy mixed with magnetite (70 % w/w). These pills were magnetized in a 0.1 T magnetic field and they were followed by magnetic field mapping using a seven-channel system [6].

More recently the group at Tokushima University used a 16 channel high Tc SQUID magnetometer placed inside a small shielded enclosure to study stomach motility, by recording fields produced by a 1.3 mm diameter steel ball placed at the antrum. Studies were performed under different fasting conditions and are reported in more detail in these proceedings [7].

b-Studies Using Magnetic Tracers

Magnetic tracers can be ingested with a test meal, endowing the GI tract with a strong magnetic signal. Depending on the material used as a tracer either the magnetic permeability or the remanent magnetization can be measured. Thus, changes in magnetic signal due to magnetic permeability or remanent magnetization would be the correlates of specific processes under study in particular regions of the GI tract.

Frei et al [8] followed by Benmair et al [9] were the first to use ferromagnetic tracers of high magnetic permeability to study gastric emptying. They used an astatic pair of coils to detect the magnetic field produced by an excitation coil powered by 60 Hz voltage mains. Changes of magnetic reluctance of one pair of coils, due to the presence of the high permeability material produces a voltage signal. The tracer used was a magnesium ferrite, that proved to be innocuous in a previous test as an X-ray contrast agent. The test meal was made by mixing 50 grams of ferrite with 200 grams of chocolate pudding, giving a final ferrite concentration of 20 %. The subject laid in a wood bed with the detector positioned dorsally, and data was acquired as a function of time. A total of 53 normal subjects and patients were studied.

Manganese ferrite powder mixed with an yogurt was used also as a MT by the biomagnetism group at the University of São Paulo-Ribeirão Preto to study different functions and parameters of gastrointestinal tract. In a first study, gastric emptying was measured using an AC biosusceptometer (ACB) that, due to improvements in the instrumentation, was easy to operate and could detect test meals with a 3% ferrite concentration [10]. The instrument can be viewed as two air core transformers, one working as a reference and the other as a measuring transformer. To maximize the signal-to-noise ratio the detection coils are positioned as near the subject as possible, and phase sensitive detection is employed. It can be shown that the signal amplitude is very sensitive to the distance of the magnetic material from the detector ($\sim r^{-6}$) and this can be exploited to measure gastric contractions, as described below.

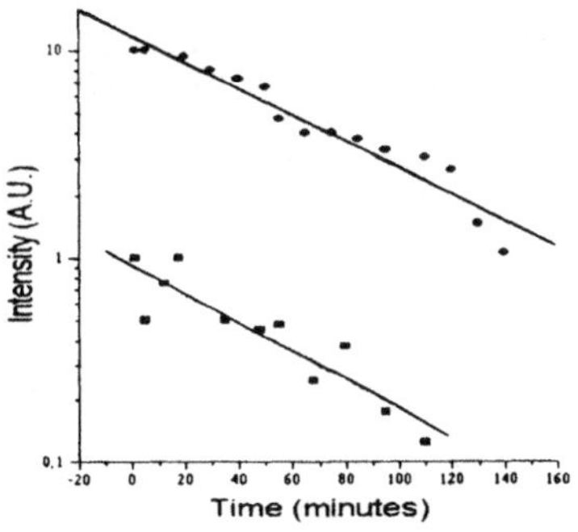

Fig.4-Gastric emptying measured by scintigaphy and the AC biosusceptometer. The test meal was labeled with technetium and ferrite.

In Fig. 4 a measurement of gastric emptying with the AC biosusceptometer is shown in comparison with a simultaneous measurement of the same function with a scintigraphic technique. The test meal was double labeled with ^{99m}Tc and ferrite.

This same instrument was used to measure the orocaecal transit time [11,12] with advantages over the method of the hydrogen expired breath, produced by a lactulose test meal. In Fig. 5 a typical curve showing the arrival of the food head at the caecum area is presented. It is important to notice that the biomagnetic method gave meaningful results in situations that the hydrogen test failed, either due to the absence of or a high basal H_2 level, thus being insensitive to interference of the colonic bacteria status and small gut infection [13].

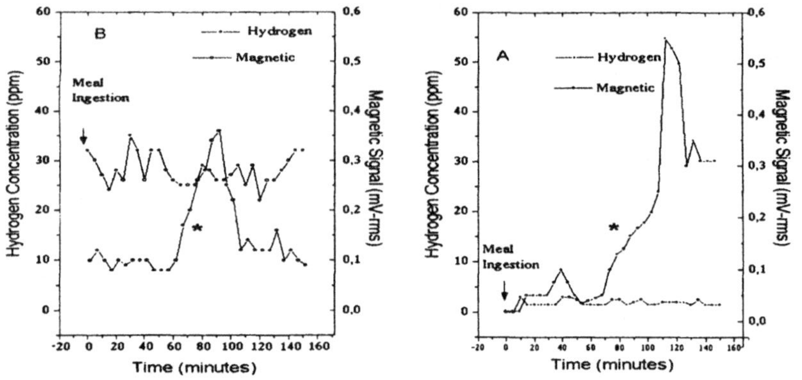

Fig.5-Orocaecal transit time determined by AC biosusceptometer compared to the hydrogen breath test in two cases where the hydrogen test fails.

Partial esophagus transit time (ETT) was studied with the same marker and a variation of the test meal. A bolus test meal, consisting of 5 ml of yogurt mixed with 5 g of ferrite powder, was followed by placing one gradiometer at the proximal esophagus (furcula) and the other on the mid distal esophagus (xiphoid process). The ETTs obtained by the biomagnetic method were compared in 6 subjects with ETTs obtained by the scintigraphy method, and were found to be in good agreement [14].

Another application of the AC biosusceptometer (ACB) was the study of the contractions of the antrum [15]. The instrument was placed near the umbilicus after the subject had ingested the magnetic test meal, and the mechanical contractions associated with the stomach slow waves (SSW) could be detected. The ACB could detect the SSW because these waves produce undulations of the stomach wall, varying the distance of the test meal to the detector and, consequently, the signal intensity. Simultaneous measurements were performed with electrogastrography (EGG) and scintigraphy, showing good agreement between the two methods. The effect of drugs known to affect the GI motility was also studied. It could be shown that the amplitude of the antral contraction were reduced immediately after the administration of Buscopan and recovered to 50% of its initial value after 40 minutes.

In a second category of experiments the test meal was magnetized and the remanent field was measured using a fluxgate magnetometer [16]. The decay of magnetization clearly shows the SSW and contains information relevant to the mixing function of the stomach, although some modeling assumptions are necessary to ascertain this relationship.

The group at Chalmers University used the same principle to measure gastric emptying. In these experiments the subjects ingested a solid test meal containing ferrimagnetic particles (γ-Fe_2O_3, 400 mg), The strength of the remanent field was measured with fluxgate magnetometers outside the stomach (fig. 6) and the amount of magnetic tracer was estimated [17].

The procedure was repeated 17 times up to two hours postprandially. The first measurement was taken about ten minutes after the beginning of ingestion. The amount of gastric content was then estimated every fifth minute during the first hour, and every tenth minute during the second hour.

In vivo measurements were carried out on 16 healthy male volunteers. The estimated retained magnetic tracer after the two-hour measurement time was (31 ± 2) % (mean ± SD) and the lag phase time was (28 ± 10) minutes. The corresponding scintigraphic curve from 16 males showed (40 ± 14) % retained isotope after two hours.

The magnetic detection of electrical activity of the gastrointestinal tract

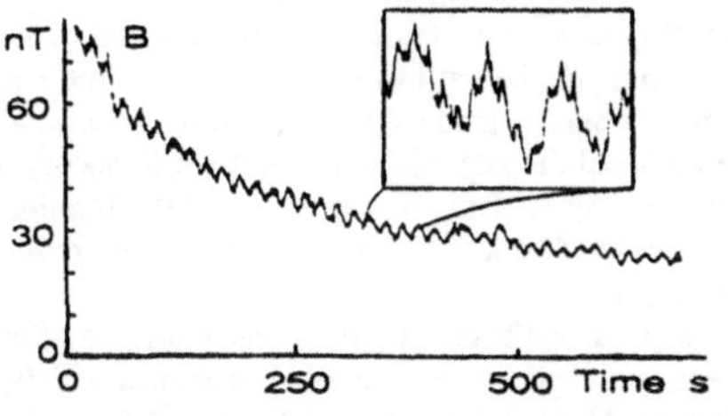

Fig.6-A typical decay curve of the remanent magnetization of a test meal in the stomach, the inset shows the characteristic 3 per minute SSW.

Although signals from magnetic markers in the stomach were detected as early as 1957, the detection of slow wave magnetic fields produced by the electrical currents occurring in the gastrointestinal tract had to wait until very recently when high sensitivity detectors (SQUIDs) were employed to detect these fields. Measurements were initially performed in humans in an unshielded environment [18] and more recently in shielded enclosures [19] in animals and humans. The gastric electrical activity known as Basic Electric Rhythm (BER) was first detected by the group at the University of Chieti using a second order biogradiometer in an unshielded environment. The characteristic BER of 3 oscillations per minute could be observed directly in the magnetic field recording and by Fourier analysis. Fig. 7 shows their first measurements.

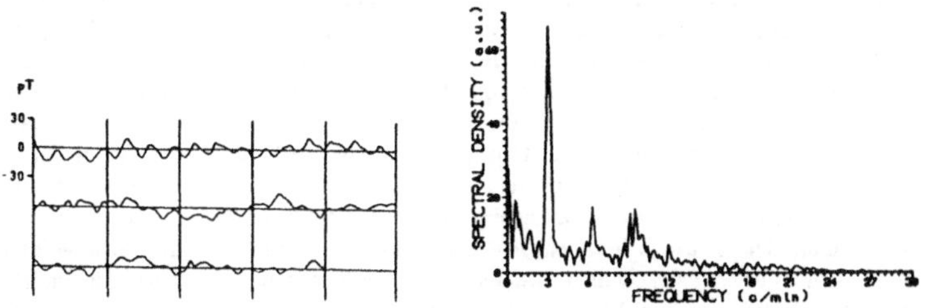

Fig.7- Magnetic signal (MGG) produced by the stomach and its Fourier transform.

The magnetogastrogram of normal volunteers has also been studied at University of São Paulo, before and after 100 ml of whole milk was ingested. Comparisons were made with the EGG, showing a strong correlation between both signals. Statistical analysis of the frequency of the MGG before and after the test meal showed a decrease in frequency ($p < 0.01$), confirming previous studies done with the EGG.

The biomagnetism group at the Vanderbilt University is also pursuing research in this subfield and measurements were performed in the small intestine, which are quite different from those recorded from the stomach. Electrically, the gastric signals are easier to detect because there is less internal fat near the stomach; however, small bowel signals are virtually impossible to see electrically without extensive adaptive filtering. Magnetically, the small bowel signals are quite clear. In initial studies, animals had their gut exposed by surgery and the magnetic signals associated with the small intestine electrical activity was measured. In a very interesting work it was shown that the magnetic measurements could diagnose the intestinal ischemia induced in rabbits by the administration of thrombin. This experiment simulated the clinically important case of mesenteric vessel thrombosis, which has no method of early detection and is associated with high mortality. More details will be given in one of the following sections.

Sources of Gastrointestinal Biomagnetic Fields: A Modeling Perspective

Two types of spontaneous gastrointestinal electrical activity may be recorded with external SQUID magnetometers. A low-frequency slow wave potential known as the Basic Electrical Rhythm (BER) mediates a higher-frequency spiking action potential that is associated with the contraction of the bowel.

The BER oscillates continuously with periods typically ranging from 3-25 cycles-per-minute (cpm) depending upon the location of the activity and the animal under investigation. In humans, the stomach BER is around 3 cpm, and the small bowel exhibits a natural gradient in its frequency. The proximal duodenum oscillates near 12 cpm and progressively more distal sites oscillate at lower frequencies. In the terminal ileum, the oscillation occurs at about 8 cpm. The normal frequency gradient is not linear; it is stepwise. The duodenal slow wave frequency of about 12 cpm persists until the mid-jejunum, where the frequency drops abruptly to another plateau. The frequency plateaus become successively shorter at more distal locations.

In diseased states, the BER changes. Experimental studies with rabbits have shown a decrease of as much as 40% to 50% in BER frequency during ischemia [20].

The small intestine consists of two layers of muscle fibers. The thick inner layer of fibers is oriented circumferentially. Contraction of these fibers is associated with the basic electrical activity and serves to mix intestinal contents. The thinner outer layer of fibers runs longitudinally along the bowel and is responsible for the longitudinal bowel contractions associated with peristalsis [21].

Electrical activity in the stomach arises at one location in the greater curve of the antrum known as the pacemaker site. Slow waves generated by the pacemaker are passively propagated via cell-to-cell coupling toward the pylorus at a rate of about 3 waves per minute. Originally, researchers believed that the small intestine also contained one pacemaker site in the upper duodenum with slow waves passively propagated down the entire length of intestine. However, the observed natural gradient in BER frequency along the intestine would suggest that certain numbers of slow waves would have to be occasionally dropped as they propagated down the bowel. Diamant and Bortoff [22] provided the first experimental evidence that multiple pacemakers are present in the small intestine. With electrodes spaced 2 cm apart down a length of intact intestine, they recorded the previously-established stepwise decrease in frequency with distance along the bowel. They subsequently cut the bowel in 2 cm segments and repeated the recording. With the bowel cut into segments, they observed a frequency decrease that was nearly linear. They also noticed that near electrodes in a transition region where the frequency changed in the intact bowel preparation, a "waxing and waning" characteristic was evident. Further investigation proved that the waxing and waning patterns resulted from the superposition of activity on either side of the electrode in the transition region. These results led them to suggest that the intestine consists of a series of electrically coupled oscillators where each oscillator has its own intrinsic "driving" frequency. The intrinsic frequencies of the oscillators decrease in a linear fashion with distance along the gut, but because of the electrical coupling, large regions of intestine may be "entrained" by the activity in a fairly localized region, thus creating a frequency plateau. When the electrical coupling is insufficient, or when the intrinsic frequency of oscillators in the chain becomes too low, the next oscillator in the series takes over as a new pacemaker oscillator. Thus, the small intestine consists of several pacemaker oscillators at different frequencies separated by regions of propagation of the electrical activity at the same frequency.

Modeling efforts for describing gastrointestinal electrical activity began in the late sixties with Nelsen and Becker's observation that the slow wave activity is consistent with coupled relaxation oscillators described by the Van der Pol differential equation [23]. By assigning intrinsic frequencies and coupling constants to a group of Van der Pol oscillators, Sarna performed a computer simulation of gastric electrical activity [24]. Smout et al. described the cutaneous electrogastrogram using a series of depolarization-repolarization dipole pairs spaced to result in an oscillatory behavior at the abdominal surface 5 cm away [25]. Recently, Familoni et al. combined the relaxation oscillator method and the dipole models to obtain a series of coupled dipoles describing the electrical activity underlying the cutaneous electrogastrogram [26].

We extended this type of model for forward and inverse calculations of the magnetic field from gastrointestinal electrical activity. A 2-dimensional "chain" model for the gastrointestinal system was developed using pacemaker and propagation oscillating dipoles. Sine waves were used to simulate the oscillatory activity since sinusoids can be described succinctly by their amplitudes, frequencies, and phases, and since the observed basic electrical activity is approximately sinusoidal. In the region representing the stomach, the pacemaker dipole was located along the greater curve of the pylorus oriented toward the pylorus in the direction of electrical propagation in the stomach. The propagation of the electrical activity was represented by four additional dipoles along a line from the antrum to the pylorus whose phases were shifted in a manner consistent with uniform propagation. For the small intestine, six dipoles act as pacemakers. The first is located in the upper duodenum with a frequency of 12 cpm, and the other five are located at progressively more aboral locations with frequencies of 11.5, 11, 10, 9, and the last pacemaker dipole in the terminal ileum is assigned a frequency of 8 cpm. Two dipoles are placed immediately following each of the pacemakers with the same frequency but with phases that shift in a manner consistent with uniform propagation of intestinal electrical activity. The arrangement of dipoles is illustrated in Fig. 8.

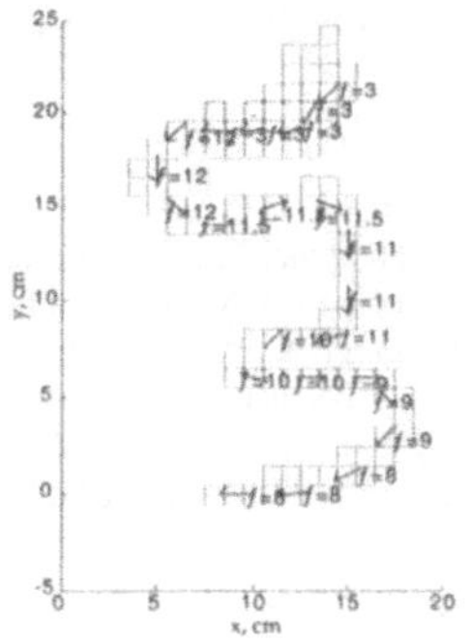

Fig. 8. Arrangement of dipoles in the "chain gut" model. Frequencies of the dipoles are indicates. Phases of same-frequency dipoles shift in a manner consistent with uniform propagation.

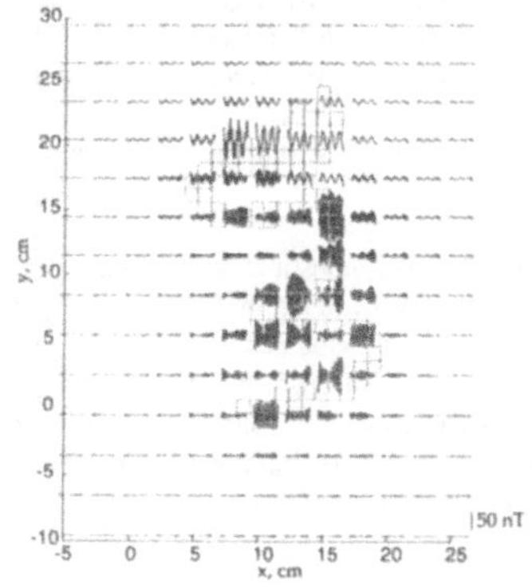

Fig. 9. Magnetic field distribution in a plane 1 cm above the source plane. Sixty seconds of data calculated for each channel are plotted. Variations in the frequency of the electrical activity are evident in the magnetic field data.

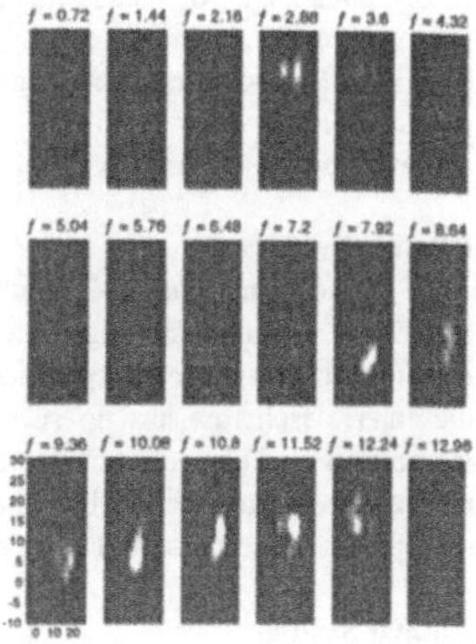

Fig.10. Spatial maps of the frequency content in the magnetic field of Fig. 2 for 18 different frequencies. Gray-scale intensity is proportional to strength of activity.

Sixty seconds of magnetic field data that would be recorded by a hypothetical 182-channel magnetometer array whose coils are arranged in the plane 1 cm above the source plane are plotted in Fig. 9. The magnetic field recorded in a single magnetometer channel contains contributions from a number of underlying sources. Above the gastric region, a 3 cpm characteristic is evident in the magnetic field corresponding to the frequency of the gastric dipoles. The frequency of the magnetic field along the intestinal tract also reflects the frequency of the underlying sources, and in several locations the magnetic field contains several frequencies from dipoles with different frequencies.

The amount of activity at a particular frequency in a specific region can be visualized by use of spectral analysis of the magnetic field data. Fig. 10 shows spatial maps of the magnitude of spectral decompositions of the magnetic field data in each channel for 18 frequencies from 0.72 cpm to 12.96 cpm. The grayscale intensity indicates the amount of activity. Gastric activity is clearly seen in the upper part of the map at f = 2.88 cpm where the stomach is located, and the small bowel activity is indicated in maps from 7.92 to 12.24 cpm near the locations of small bowel dipoles as indicated in Fig. 8. These type of maps could be useful for identifying diseased states of the bowel. For example, an ischemic episode with activity at a lower than normal frequency would result in bright spots in the spatial maps of frequency content at abnormal frequencies.

Future modeling efforts should incorporate more accurate source representations than the sine wave approximation used here. The picture presented by this 2-dimensional model was enlightening but potentially deceptive if it is compared with the actual situation. The 3-dimensional nature of the GI tract must be accounted for when attempting to compare models with actual data. In addition, models should begin to compare the external magnetic fields with external potentials to demonstrate the superiority of magnetic measurements because of the intervening layers of low-conductivity fat present in the abdomen. Clearly, these models must include accurate conductivity and geometrical considerations to precisely determine volume conductor effects.

Noninvasive measurement of gastrointestinal magnetic fields for early detection of ischemia

Previous sections have described in detail the electrical activity of the gastrointestinal tract and some of the advantages of measuring these signals noninvasively using SQUID magnetometers. From a clinical prospective, the foremost reason to be excited about the measurement of gastrointestinal magnetic fields is the high signal-to-noise ratio of magnetic field recordings. Cutaneous electrodes have many drawbacks including an extremely low signal-to-noise ratio and the inability to localize or discriminate dipole sources within the abdomen because of the layers of insulators present within the abdominal wall.

Clinical Problems

Acute mesenteric ischemia of the gastrointestinal tract is a serious condition associated with as much as a 90% mortality and significant morbidity in those that survive [27]. While the patient with myocardial ischemia is diagnosed expeditiously with the changes seen on the electrocardiogram, the patient with mesenteric ischemia has no such diagnostic tool, and thus the diagnosis can only be made at exploratory laparotomy. The ideal test for mesenteric ischemia diagnosis would be noninvasive and fast, would detect ischemia prior to irreversible changes, would have high sensitivity and specificity, would be available for multiple studies, and would be easily performed in critically ill patients.

Animal Studies

CHANGES IN BASIC ELECTRICAL RHYTHM (BER) RECORDED USING A SQUID MAGNETOMETER DURING ACUTE ISCHEMIA.
One of the first studies done at Vanderbilt University utilized a rabbit model in which the animal was anesthetized and the bowel exteriorized into a warm, plexiglas dish. Bipolar electrodes were placed on the bowel and the magnetic fields of the exteriorized small intestines were recorded using microSQUID. After stabilization for 15 minutes the mesenteric artery was thrombosed using intra arterial injection of thrombin to cause complete ischemia. The BER frequency determined by electrodes was identical to that determined by the noncontact SQUID magnetometer. After 5 minutes of injection, there was a 27% drop in frequency ($p < 0.05$) [28].

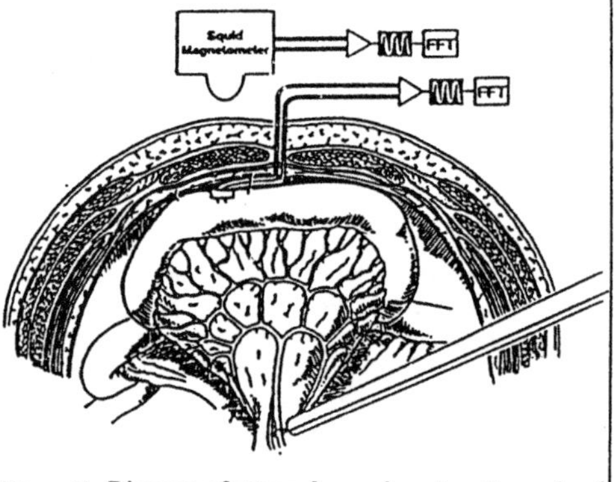

Figure 11: Diagram of set-up for noninvasive diagnosis of intestinal ischemia. Trough a midline laparotomy, a loop of ileum is isolated and the segmental mesenteric artery is dissected and encircled with a monofilament snare that is exteriorized. Bipolar electrodes are sutured onto the serosal surface of the bowel, and the bowel itself is sutured to the abdominal wall to ensure the same loop of intestine is measured by the SQUID magnetometer. After closing the abdomen, the SQUID is approached to the abdominal wall of the anesthetized animal. The eight rabbits were studied for a 15 minutes baseline period, with continuous recordings of BER from the electrodes and SQUID magnetometer. The segmental mesenteric artery was occluded for 15 minutes while continuous recordings of BER were taken by both methods.

CORRELATION BETWEEN PATHOLOGY AND ELECTRICAL ACTIVITY DURING ACUTE ISCHEMIA. The next study looked at the timing of BER changes during arterial ischemia and whether or not there are irreversible pathologic changes during these early abnormalities of BER frequency. Mesenteric ischemia was induced by ligating the superior mesenteric artery in anesthetized rabbits. There was a significant decrease in BER frequency 20 minutes after the onset of ischemia [21.5 ± 1.9 cycles per minute (cpm) to 14.2 ± 1.5 cpm ($p= 0.02$ Student's T-test)]. Microscopic evaluation revealed no pathologic changes until 60 minutes of complete arterial ischemia. Sixty minutes after the occlusion of the SMA the pathologic changes of sloughing of the mucosal villi and edema were minimal. These studies show that the electrical changes occur well before the onset of pathologic changes [29].

NONINVASIVE DIAGNOSIS OF MESENTERIC ISCHEMIA IN AN ANIMAL MODEL. Another study utilized an anesthetized rabbit shown in Fig. 11 [19]. The snare around the superior mesenteric artery was used to occlude the blood supply to the intestine for 30 minutes and then loosened to allow reperfusion of the bowel. There was an immediate and dramatic drop in BER frequency during occlusion which returned to baseline levels after 15 minutes of reperfusion (Fig. 12).

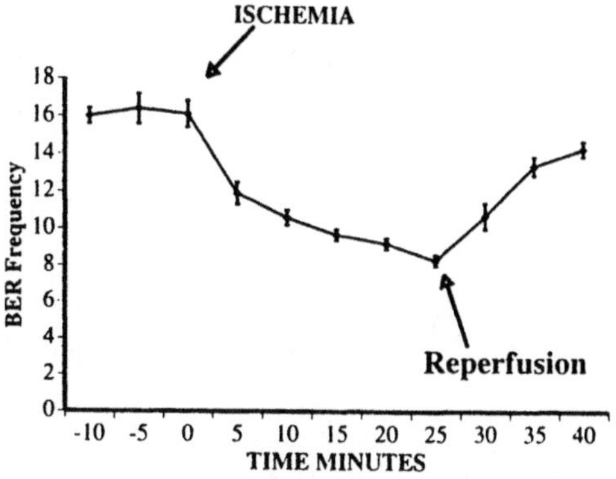

Fig.12-BER frequency determined with the SQUID magnetometer in ten animals before, during, and after snare occlusion of the main superior mesenteric artery. There was a significant decrease in frequency 5 minutes after occlusion of the superior mesenteric artery. On reperfusion, BER frequency rises to near baseline values 15 minutes after the start of reperfusion.

The frequency determined by the noninvasive SQUID magnetometer was highly correlated (r =.91) to those obtained with invasive serosal electrodes (Fig. 13). This study demonstrated that magnetic field measurements could be obtained noninvasively in the animal model and could detect changes in BER frequency during periods of ischemia. Moreover these changes in electrical activity appear to reflect the viability of smooth muscle within the intestine.

MAGNETIC FIELD MEASUREMENTS DURING MESENTERIC VENOUS THROMBOSIS. While arterial thrombosis or embolism represents most of the patients with mesenteric ischemia, there is a significant subset of patients who have thrombosis of the superior mesenteric vein. We performed studies to determine whether or not changes in BER frequency could be identified noninvasively using a SQUID magnetometer in an animal model of mesenteric vein thrombosis [30]. Anesthetized rabbits were prepared in a fashion similar to that described for the arterial occlusion studies. The superior mesenteric vein was isolated, and after 15 minutes of stable recordings thrombin was injected into the superior mesenteric vein to produce complete thrombosis of the venous system. The BER dropped from 16.42 ± 0.69 cpm to 8.80 ± 0.74 cpm at 30 minutes and to 6.82 ± 0.7 after 90 minutes ($p < 0.0001$ paired T-Test). Pathology at this time shows ischemic mucosa with coagulative necrosis and sloughing of the epithelium. These studies suggest that the changes in BER frequency correlate to the viability of the smooth muscle in the intestine and that these changes occur

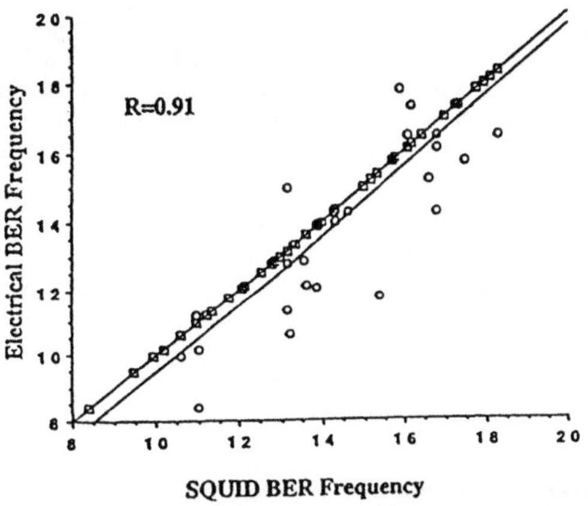

Figure 13: The frequency of the BER with the frequency determined by fast-Fourier transformation recorded by the electrode is plotted on the abscissa. The magnetic field BER frequencies obtained at the same time are plotted on the ordinate axis. There was a high level of correlation between the simultaneous recordings. (r=0.91).

before irreversible pathologic changes within the smooth muscle. Moreover, these changes could be identified noninvasively through the abdominal wall in our animal model.

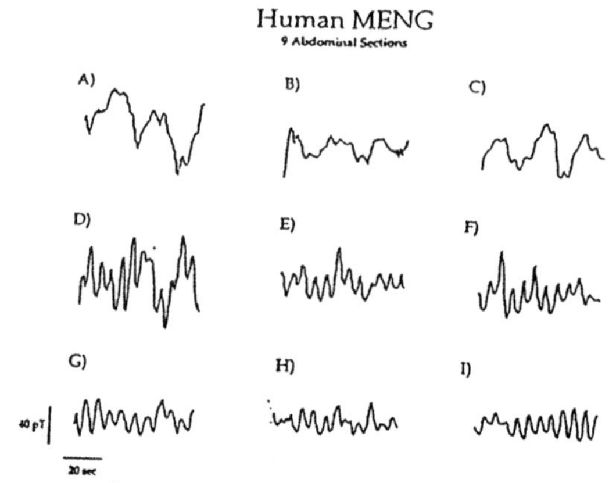

Figure 14: MENG recordings were taken from the recorder using the MAGNES™ 37 channel magnetometer from 9 locations in the same volunteer; A) Right upper quadrant, B) epigastrium, C) left upper quadrant, D) right mid quadrant, E) periumbilical, F) left mid quadrant, G) right lower quadrant, H) hypogastrium, I) left lowes quadrant. The recording were sampled at 1041 Hz and are shown after median filtering of 1040 point or 1 second wide. All recordings were obtaneid while the volunteers suspended respiration, recordings identified by A,B and C refers to gastric and D,E,F,G,H and I to small bowel in the same volunteer the small intestine.

Human Studies

A number of studies have been performed looking at the magnetic field measurements of the gastrointestinal tract in human volunteers [31]. Twenty-one human volunteers (7 male; 14 female) ages 26 - 55 (mean 33) have been studied to examine the magnetic field activity within the abdomen. Cyclical wave forms consistent with gastric and small bowel BER have been identified in every volunteer [32]. There has been great progress made in developing algorithms to exclude the magnetocardiogram contamination.

Although we were able to measure the magnetic field activity in every volunteer, in order to identify segments of ischemic bowel it will be necessary to identify the spatial distribution of electrical activity within the gastrointestinal tract. The stomach has a BER frequency of three cycles per minute that originates on the greater curvature and propagates towards the pylorus while the duodenum has a BER frequency of 12 cpm. There is a stepwise decrease in BER frequency known as the frequency gradient within the small intestine. The intestinal BER frequency gradually decreases as you move aborally in the GI tract (9 cpm in the distal ileum). A mapping study was performed in order to identify this frequency gradient within the small bowel. The volunteer was asked to lie quietly while the 37 channel magnetometer was positioned in 9 regions over the abdomen. While the volunteer suspended respirations, recordings were obtained and from each of these recordings we could identify BER activity. Recordings were subjected to autoregressive spectral analysis to identify the dominant frequencies within each of these sections. In the upper quadrants, the BER ranged between 2.5 and 2.7 cpm. In the right middle quadrant recordings, a peak of 12 cpm can be seen corresponding to the duodenal frequency. In the right lower quadrant, the BER frequency is 9.7 cpm (Fig. 14 and 15). This corresponds to the expected locations within the abdomen for gastric and small bowel electrical activity [33]. This demonstrates the potential usefulness of magnetic field recordings at multiple sites to identify the spatial variations in frequency within the gastrointestinal tract. This line of study has been extremely exciting and shows great promise for clinical applicability. There are however significant questions that remain to be answered prior to the deployment of these devices and techniques in clinical practice.

Questions and Aims

1) Do other conditions such as infection, peritonitis, or bowel obstruction affect BER frequency ?
2) Can small areas of ischemic bowel be detected within large amounts of normal activity ?
3) Can we detect the absence of signal in dead bowel if other normal signals exist around it ?
4) Since many of these patients are clinically ill and are on ventilators or in the emergency room, it will be necessary to perform measurements in an unshielded hospital environment. This is a very noisy environment and multiple technical factors will have to be overcome in order to make this a useful clinical instrument.

5) In all medical meetings today there is a great amount of talk about cost containment and the issues of cost effectiveness. Although on the surface, this appears to be a costly technology there is no competing technology capable of measuring small bowel electrical activity. Thus, new technology will have to show that it is cost effective in order for hospitals to purchase this type of equipment.

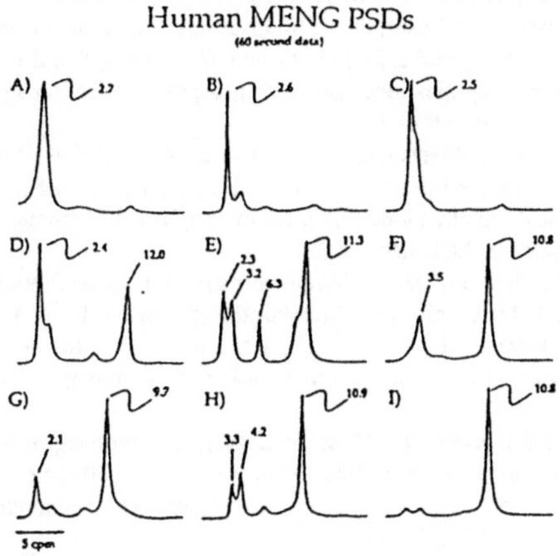

Fig. 15-Autoregressive spectral analysis was used on 60 second periods of data to determine the dominant frequency of the BER. The 9 graphs correspond to the locations described in fig. 14. Since none of the 60 second recordings were taken simultaneously, there is temporal variation between each of the 9 recordings. The Frequency of gastric BER (**A,B,C**) ranged between 2.5-2.7 cpm. In (**D**) the right upper quadrant recordings a peak at 12.o cpm can be seen that corresponds to the predicted BER frequency of the duodenum. In the middle and lower quadrants the BER frequency was found to range from (**E**) 11.3 cpm to (**F**) 10.8 cpm to (**G**) 9.7 cpm to (**H**) 10.9 cpm to (**I**) 10.8 cpm that corresponds to the predicted frequency gradient within the small intestine.

Summary

Animal studies of acute mesenteric ischemia demonstrate that changes in BER frequency occur prior to pathologic changes and reflect the viability of the bowel. Magnetic field measurements have a high signal-to-noise ratio, are highly correlated with invasive serosal electrode recordings, and can detect ischemic bowel noninvasively. Human studies have been started and show great promise. The BER of the stomach and small intestine can be recorded noninvasively. Further work to create an optimized instrument for noninvasive magnetic field diagnosis of acute mesenteric ischemia is ongoing.

Concluding Remarks:

The future of the biomagnetic methodology in gastroenterology is linked to numerous facts. First, it is necessary to find applications where biomagnetic measurements can provide unique information, or the information provided by existing methods are poor when compared to that obtained by biomagnetic measurements. Second, the instrumentation used should be affordable to hospitals, and if possible of simple operation and low cost. Third, other aspects of the G.I. tract should be investigated.

It is strong believed that there is an enormous space for improvements in the examples above mentioned that could lead to new applications and refinements in the present research and diagnose methods in gastroenterology.

References:

[1] Hoke, M., Erné, S.N., Okada, Y.C., and Romani, G.L. Biomagnetism: Clinical Aspects, Amsterdam, Excerpta Medica, 1992.

[2] Wenger, M. A., Henderson, E.B., and Dinning, J.S. Magnetic method for recording gastric motility, science, 1957, 990-991.

[3] Di Luzio, S., Comani, S., Romani, G.L., Basile, M., Del Gratta, C., and Pizzela, V. A biomagnetic method for studying gastro-intestinal activity, Il Nuovo Cimento, 1989, 11D(12): 1853-1859.

[4] Basili, M., Neri, M., Carriero, A., Casciardi, S., Comani, S., Del Gratta, C., Donato, L. G., Di Luzio, S., Macri, M.A., Pasquarelli, A., Pizzela, V., and Romani, G. L. Measurement of segmental transit time in man, Digestive Disease Sciences, 1992, 37(10):1537-1543.
[5] Forsman, M., Hultin, L., and Abrahamsson, H. Measurements of gastrointestinal transit using fluxgate magnetometers, In: Baumgartner, C., Deecke, L., Stroink, G. and Williamson, S.J. Biomagnetism: Fundamental Research and Clinical Applications. Studies in Applied Electromagnetics and Mechanics, Amsterdam, Elsevier Science Publisher, 1995.
[5] Forsman, M., Hultin, L., and Abrahamsson, H. Measurements of gastrointestinal transit using fluxgate magnetometers, In: Baumgartner, C., Deecke, L., Stroink, G. and Williamson, S.J. Biomagnetism: Fundamental Research and Clinical Applications. Studies in Applied Electromagnetics and Mechanics, Amsterdam, Elsevier Science Publisher, 1995.
[6]-Weitschies W., Wedemeyer J., Stehr R., and Trahms L. Magnetic markers as a noninvasive tool to monitor gastrointestinal transit. IEEE Transactions on Biomedical Engineering, 1994, 41:192-195.
[7] Nomura, M., Saijyo, T., Haruta, Y., Itozaki, H., Toyoda, H., Nakaya, Y., Ito, S., and Kado, H. Biomagnetic measurement of gastric mechanical motility using a high-temperature SQUID imaging. This volume.
[8] Frei E.H., Benmair Y., Yerashalmi S., and Dreyfuss F. Measurements of the emptying of the stomach with a magnetic tracer, IEEE Transactions on Magnetics, 1970, Vol MAG-6:348-349.
[9] Benmair, Y., Dreyfuss, F., Fischel, B., Frei, E.H., and Gilat, T. Study of gastric emptying using a ferromagnetic tracer. Gastroenterology, 1977,73:1041-1045.
[10] Miranda J.R.A., Baffa O., Oliveira R.B., and Mitiko M. N. An AC biosusceptometer to study gastric emptying, Medical Physics, 1992, 19(2): 445-449.
[11] Oliveira, R.B., Miranda, J.R.A., Baffa, O., Cambrea, C.R., and Troncon, L.E.A. A biomagnetic technique for orocaecal transit time measurement, In: Hoke, M., Erné, S.N., Okada, Y., Romani, G.L. Biomagnetism: Clinical Aspects, Excerpta Medica, Amsterdam,1992.
[12] Baffa O., Oliveira R. B., Arruda Miranda J. R., and Troncon L. E. A. Analysis and development of AC biosusceptometer for orocaecal transit time measurements, Medical & Biological Engineering & Computing, 1995, 33:353-357.
[13] Oliveira, R.B., Baffa, O., Troncon, L. E. A Miranda J. R. A., and Cambrea C. R.. Evaluation of a biomagnetic technique for measurement of orocaecal transit time, European Journal of Gastroenterology and Hepatology, 1996, in press.
[14] Daghastanli, N.A., Braga, F.J.H.N., Oliveira, R.B., and Baffa, O. Esophageal transit time evaluated by means of the biomagnetic method, This volume.
[15] Miranda, J.R.A., Baffa, O., Oliveira, R.B., Braga, F.J.H., and Sousa, P.L. Detection of Stomach Contractions Using a Biosusceptometer, This volume.
[16] Baffa, O., Oliveira, R.B., Miranda, J.R.A., and Sousa, P.L. Mixing power of food in the stomach evaluated by a biomagnetic technique, In: Baumgartner, C., Deecke, L., Stroink, G. and Williamson, S.J. Biomagnetism: Fundamental Research and Clinical Applications. Studies in Applied Electromagnetics and Mechanics, Amsterdam, Elsevier Science Publisher, 1995.
[17] Forsman M. Assessment of gastric movements by monitoring magnetic field decline of magnetized trace particles, Physics in Medicine and Biology, 1994,39a: p. 62.
[18] Comani, S., Basile, M., Casciardi, S., Del Gratta, C., Di Luzio, S., Erné, S.N., Macri, M., Neri, M., Peresson, M., and Romani, G. L. Extracorporeal direct magnetic measurement of gastric activity. In: Hoke, M., Erné, S.N., Okada, Y., Romani, G.L. Biomagnetism: Clinical Aspects, Excerpta Medica, Amsterdam,1992.
[19] Saijyo, T., Nomura, M., Haruta, Y., Itozaki, Y., Toyoda, H., Nakaya, Y., Ito, S., and Kado, H. Magnetogastrograms using 64-channel SQUID system: study on clinical usefulness, This volume.
[20] Richards W.O., Garrard C.L., Allos S.H., Bradshaw L.A., Staton D.J., and Wikswo J.P., Jr. Noninvasive diagnosis of mesenteric ischemia using a SQUID magnetometer, Annals of Surgery, 1995, 221(6): 696-705.
[21] Liu L.W.C., and Huizinga, J.D. Electrical coupling of circular muscle to longitudinal muscle and interstitial cells of Cajal in canine colon, Journal of Physiology (Lond.), 1993, 470: 445-461.
[22] Diamant N.E., and Bortoff A. Nature of the intestinal slow-wave frequency gradient, American Journal of Physiology, 1969, 216(2): 301-307.

[23] Nelsen T.S., and Becker J.C. Simulation of the electrical and mechanical gradient of the small intestine, American Journal of Physiology, 1968, 214: 749-757.
[24] Sarna S.K., Daniel E.E., and Kingma Y.J. Simulation of the electrical activity of the stomach by an array of relaxation oscillators, Digestive Diseases Science, 1972, 17: 299-310.
[25] Smout A.J.P.M., Van der Schee E.J., and Grashuis J.L. What is measured in electrogastrography ? Digestive Disease Sciences, 1980, 25: 179-187.
[26] Familoni B.O., Abell T.L., and Bowes K.L. A model of gastric electrical activity in health and disease, IEEE Transactions on Biomedical Engineering, 1995, 42(7): 647-657.
[27] Williams L. Mesenteric Ischemia. In: Sawyers J, Williams L, The Acute Abdomen. Philadelphia: W.B. Saunders, 1988:331-353. Surgical Clinics of North America; vol 68.
[28] Golzarian J., Staton D.J., Wikswo J.P., Jr., Friedman R.N., Richards W.O. Diagnosing intestinal ischemia using a noncontact Superconducting Quantum Interference Device, American Journal of Surgery, 1994, 167:586-592.
[29] Garrard C.L., Halter S., Richards W.O. Correlation between pathology and electrical activity during acute intestinal ischemia, Surgery Forum, 1994, 45:368-371.
[30] Allos S., Sorrell M., Slater S., Staton D., Wikswo J.P., Richards W. SQUID magnetometer diagnosis of mesenteric vein thrombosis, Gastroenterology, 1995, 108(4):A269.
[31] Richards W.O., Staton D., Golzarian J., Friedman R.N., Wikswo J.P., Jr. Non-invasive SQUID magnetometer measurement of human gastric and small bowel electrical activity. In: Baumgartner, C., Deecke, L., Stroink, G. and Williamson, S.J. Biomagnetism: Fundamental Research and Clinical Applications. Studies in Applied Electromagnetics and Mechanics, Amsterdam, Elsevier Science Publisher, 1995.
[32] Garrard C.L., Wikswo J.P., Jr., Staton D., Golzarian J., Gallen C., Richards W.O. Noninvasive measurement of small bowel electrical activity, Gastroenterology, 1994, 106(4):A502.
[33] Richards W.O., Bradshaw L..A., Staton, D.J., Garrard, C. L., Liu, F., Buchanan, S., and Wikswo J.P., Jr. Magnetoenterography (MENG): Noninvasive measurement of bioelectric activity in human small intestine, Submitted Digestive Diseases and Sciences, 1996.

Acknowledgments:

The work presented in this chapter is the result of the endeavor of several research groups, to whom many contributors have collaborated, we thank all of them listed in the above referenced papers. We are also grateful to the partial financial support of the agencies: Department of Veterans Affairs Research Service, FAPESP, CNPq, CAPES, TWAS, the Veterans Administration, the NIH, Conductus, Inc, National Research Council of Italy and Swedish National Board of Industrial and Technical Development (NUTEK).

Magnetic Field Measurement of Rabbit Colonic Electrical Activity

Bradshaw, L.A., Allos, S.H., Wikswo, J.P., Jr. and Richards, W.O.

Vanderbilt University, Nashville, TN USA

Introduction

The smooth muscle of the gastrointestinal (GI) system produces two types of electrical activity that may be detected electrically or magnetically. The slow-wave, or basic electrical rhythm (BER) is a sinusoidal type of activity that oscillates about 3 times per minute in the human stomach and 8-12 times per minute in the small intestine. The electrical response activity is a higher-frequency spiking potential that is strongly associated with segmental contractions of the bowel that propel intestinal contents along the GI tract. Our previous studies on intestinal electrical activity have shown that magnetic measurements taken outside the abdomen correlate strongly with recordings from electrodes placed on the serosal surface of the bowel, and that external magnetic measurements are superior to cutaneous electrode measurements because of the attenuation and smearing of the potential by low-conductivity layers in the abdomen [1, 2].

Electrode measurements of colonic electrical activity have found that, in contrast to the well-defined frequency patterns of the gastric and small bowel systems, colonic activity is highly variable in both frequency and amplitude [3, 4]. Two possible explanations for this observation have been suggested. First, gap junctional coupling between cells in the colon may be weaker gap junctions may be less prevalent than in other parts of the gastrointestinal system [5]. In that case the electrical activity generated in one section of the colon would be somewhat less likely to be correlated with activity generated in another section than would activity in the intestine, for example. Another possible reason for the large variability in colonic electrical activity is the fact that the network of interstitial cells of Cajal (ICC), which acts as a pacemaker for the electrical activity, is less dense in the colon than in other parts of the GI tract [6]. Thus, electrical activity generated by the ICC network does not propagate into colonic smooth muscle in the same way that it does in the case of the intestine. Therefore, one may expect to see a variety of frequencies in colonic electrical activity.

In this study, we want to demonstrate the usefulness of a biomagnetometer system for the detection of colonic electrical activity. We expect that the frequencies of colonic activity that we record with serosal electrodes and with a magnetometer outside the subject will vary in a manner similar to the previous electrode studies. Further, we expect that frequencies detected by the magnetometer will correlate well with frequencies detected with the internal electrode.

Methods

Five male New Zealand white rabbits were anesthetized with Ketamine (30 mg/kg) and Acepromazine (5 mg/kg). The rabbits were fed a nonmagnetic diet for at least 72 hours before the experiment. A midline laparotomy was performed and a segment of the colon was exteriorized. A serosal electrode platform was sutured to the colon segment, and then the colon was replaced in the abdomen. The electrode platform consisted of eight bipolar pairs of nonmagnetic silver electrodes separated by 1 cm embedded in Silastic® silicone adhesive (Dow Corning, cat. no. 891). The laparotomy incision was closed and the animal was placed with the section of the abdomen containing the electrode platform directly under the magnetometer.

The magnetometer was the four-channel high-resolution μSQUID designed for tissue studies. The pick-up coils in μSQUID are 3 mm in diameter arranged at the corners of 4.4 mm square. The coils may be lowered to within 1.5 mm of the room-temperature sample.

Simultaneous recordings of the external magnetic fields and internal serosal potentials were obtained for 10 minutes. Since colonic electrical activity is known to be highly variable, we divided the analysis of these signals into long-term analysis and short-term or instantaneous analysis. For the long-term analysis, the frequency content of both magnetic and electric signals were analyzed using the Fast Fourier Transform (FFT) based power spectral density (PSD) of the entire 10 minute data set, and the frequency content common to both electric and magnetic signals was determined using the cross spectral density (CSD). Welch's averaged periodogram method was used for computation of both the PSD and the CSD. Because it is not clear that the electrical activity necessarily occurs at a single frequency, we applied a peak detection algorithm to determine the peaks in the PSDs and CSDs. The cross spectrum

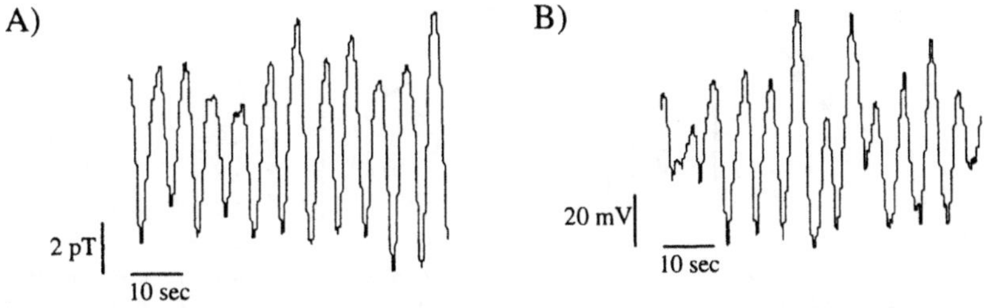

Figure 1. (A) Colonic electrical activity recorded by an external SQUID magnetometer. (B) Colonic activity recorded by serosal electrode. The two recordings were obtained simultaneously.

itself indicates how the electric and magnetic recordings are correlated at different frequencies, but to further test our hypothesis that electrodes and magnetometers record the same activity, we determined the degree of correlation between peaks in the magnetic PSD and the electric PSD, between peaks in the magnetic PSD and the CSD, and between peaks in the electric PSD and the CSD.

A high degree of variability is expected from colonic electrical activity, so we wished to further analyze the spectral content of short-duration data samples. Since the frequency resolution of the FFT is inversely proportional to the sample duration, short sample durations result in unacceptably poor frequency resolution. Thus, we applied an autoregressive (AR) method to the spectral analysis of successive 30 second time segments within the 10 minute recording period. The AR method can provide adequate frequency resolution for short-duration samples [7]. Again, the peak detection algorithm was applied to the AR PSDs, and the degree of correlation between peaks in the electric PSD and the magnetic PSD was determined. Thus, the short-term analysis will give us an idea of the instantaneous correlation of the electric and magnetic signals while the long-term analysis provides a picture of the overall correlation.

Results

Figure 1 shows 60 second tracings of typical simultaneously-recorded serosal potentials and external magnetic fields from one of the studies. A clear oscillatory rhythm is evident in both traces. Visual inspection suggests that the frequency of the oscillation is around 11.5 cpm, possibly modulated by a lower-frequency activity. The power spectral densities of the entire 10 minute electric and magnetic signals as well as the cross spectrum of the electric and magnetic signals are shown in Figure 2. The largest peaks in the power spectra are seen at 11.4 cpm in the mag-

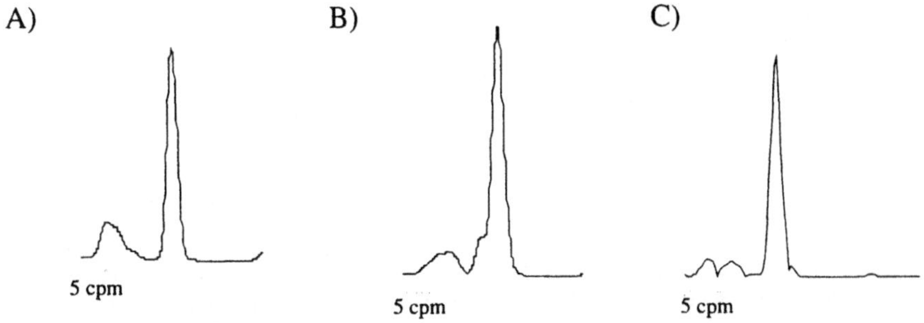

Figure 2. (A) PSD of the entire 10-minute external SQUID recording. (B) PSD of the entire 10-minute serosal electrode recording. (C) CSD of the electrode and SQUID recordings.

netic signal and 12.1 cpm in the serosal electric signal. Broad smaller peaks in the spectra appear in the 3-6 cpm range. The cross spectrum has its largest peak at 11.7 cpm, with smaller peaks at 3.0 and 5.7 cpm, indicating the these

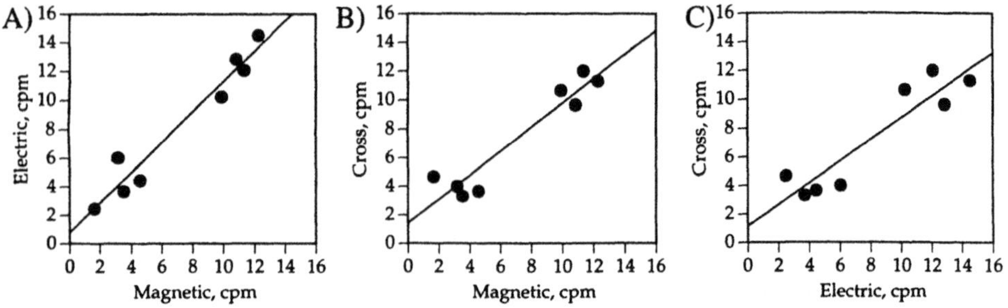

Figure 3. Peak frequencies in (A) electric vs. magnetic recordings, (B) cross spectrum vs. magnetic recordings, and (C) cross spectrum vs. electric recording. The peak frequencies determined by each method are strongly correlated (r = 0.97 for magnetic/electric; r = 0.95 for cross/magnetic; r = 0.92 for cross/electric).

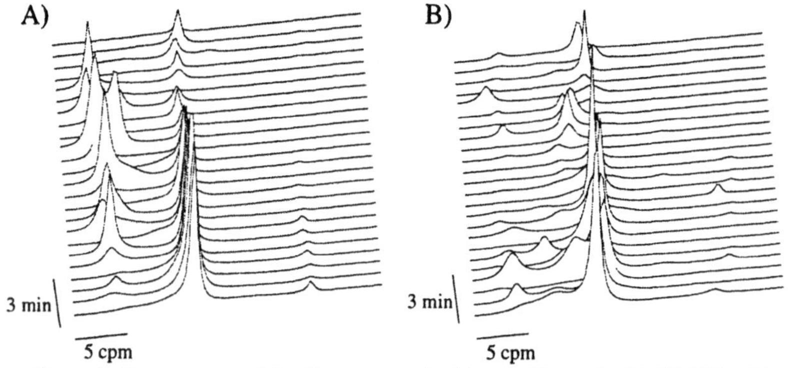

Figure 4. AR power spectral densities at successive 30 second intervals. (A) AR PSDs of the signals from the SQUID magnetometer; (B) AR PSDs of the serosal electric signal.

frequencies are common to both SQUID and electrode measurements.

Peaks such as these were seen in all experiments. During one of the studies, the magnetic signal was lost due to environmental disturbance, so that study was excluded from the correlation analysis. The other four experiments showed peaks in the 2-6 cpm range as well as the 10-14 cpm range. Peaks in the electric signal are plotted against peaks in the magnetic signal in Figure A, and peaks in the cross spectra are plotted against peaks in the magnetic and electric signals in Figure B and C respectively. The correlation of frequencies detected by each analysis method is quite strong (r = 0.97 for electric vs. magnetic, r = 0.95 for cross vs. magnetic, r = 0.92 for cross vs. electric).

Changes in the raw data during different time intervals were observed. Though activity was usually seen consistently in the 8-12 cpm range with only slight variation, lower frequency activity in the 2-6 cpm range was occasionally seen particularly in magnetic recordings. Figure 4 shows plots of the AR power spectra at successive 30 second intervals over the entire 10-minute duration of the same experiment represented in Figure 1. Although consistent peaks are seen in the 8-12 cpm range, lower frequency activity in the 2-6 cpm range is also apparent during some intervals, particularly in the magnetic recordings. From these traces, we see that the frequency peaks change with time, and that the magnetometer detects electrical activity that is not as clearly detected in the serosal electrodes.

The correlation of the short-term peaks in the electric and magnetic signals is somewhat smaller than correlation of the long-term peaks. Figure 5 shows the peaks in the short-term electric signals plotted against the peaks in the short-term magnetic signals. The instantaneous correlation coefficient (the correlation of the peaks in 30 second duration samples) is 0.82.

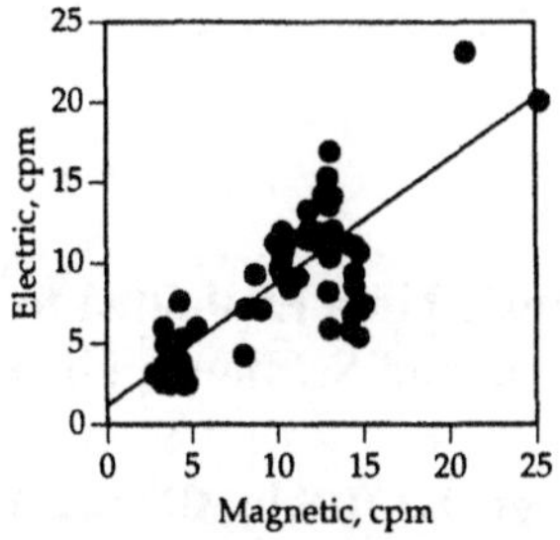

Figure 5. Instantaneous correlation of the serosal electric and magnetic signals. The peaks detected in successive 30 second intervals are shown with peaks in the electric signals plotted against peaks in the magnetic signals.

Discussion

This work represents the first report of the magnetic detection of colonic electrical activity. The electrical activity of the colon was detected with invasive serosal electrodes at a frequency of 8.3 ± 1.7 cpm, and by a noninvasive SQUID magnetometer at a frequency of 7.2 ± 1.5 cpm. Activity was typically seen in the 8-12 cpm range, and less dominant activity was observed in the 2-6 cpm range. As in previous electrode studies, the colonic electrical rhythm was seen to be highly variable in its frequency and amplitude.

SQUID recordings were very similar to serosal electrode data. The frequency of the electrical activity recorded in the SQUID was strongly correlated with that recorded in the electrodes ($r = 0.97$). Cross spectral peaks also correlated well with both electric and magnetic peaks ($r = 0.92$ and $r = 0.95$, respectively). Short-term correlation was somewhat smaller ($r = 0.82$) due to the fact that magnetometers record activity over a larger region than the serosal electrodes and so include contributions from other electrically active sources.

This work suggests that noninvasive SQUID magnetometers may be used to characterize colonic electrical activity that was previously possible only with invasive serosal electrodes.

References

[1] Bradshaw, L. A., Allos, S. H., Wikswo, J. P., Jr., and Richards, W. O. Transabdominal magnetic detection of small bowel electrical activity. In preparation.

[2] Bradshaw, L. A. *Measurement and Modeling of Gastrointestinal Electrical Activity.* Ph.D. Dissertation, Vanderbilt University, 1995.

[3] Riezzo, G., Pezzolla, F., Maselli, M. A., and Giorgio, I. Electrical activity recorded from abdominal surface before and after right hemicolectomy in man. *Digestion*, 1994, 44: 185-190.

[4] Chambers, M. M., Bowes, K. L., Kingma, Y. J., Bannister C., and Cote, K. R. In vitro electrical activity in human colon. *Gastroenterology*, 1981, 81: 502-508.

[5] Liu, L. W. C., Sperelakis, N., and Huizinga, J. D. Pacemaker activity and intercellular communication in the gastrointestinal musculature. In: Huizinga, J. D. *Pacemaker Activity and Intercellular Communication*, Boca Raton, CRC Press, 1995.

[6] Christensen, J., Rick, G. A., and Lowe, L. S. Distributions of interstitial cells of Cajal in stomach and colon of cat, dog, ferret, opossum, rat, guinea pig, and rabbit. *J. Autonomic Nervous System*, 1992, 37: 47-56.

[7] Marple, S. L. *Digital Spectral Analysis with Applications*, Englewood Cliffs, Prentice-Hall, 1987.

Acknowledgments

This work was supported in part by grants from the Veterans Administration and by an NIH SBIR grant to Conductus, Inc.

Esophageal Transit Time Evaluated by Means of the Biomagnetic and Scintigraphic Methods

Daghastanli, N.A.[1], Braga, F.J.H.N.[1,2], Oliveira, R.B.[2] and Baffa, O.[1]

Physics Department-FFCLRP, USP[1]-14040-901-Ribeirão Preto-SP, Brazil and Faculdade de Medicina de Ribeirão Preto[2], 14049-900 Ribeirão Preto-SP, Brazil

Introduction:

Esophageal motor and transit disorders are important causes of dysphagia and chest pain. Scintigraphy is a well-established method to assess esophageal transit in healthy persons as well as in patients with esophageal motility disorders [1,2]. Studies have been performed showing important aspects in health and pathologic situations, such as Chagas' disease and other pathological situations. However, scintigraphy has important limitations: the burden of ionizing radiation and the use of expensive equipment. On the other hand, the biomagnetic method has some advantages[3-6]. In such a way, pregnant women could be evaluated; additional costs with radiation protection could be eliminated and the biomagnetic method could be used in many institutions that are not well equiped as a consequence of the lack of money. Recently, a simple and easy to perform magnetic method employing an inexpensive biosusceptometer was developed for measurements of the gastric emptying and small bowel transit. We herein describe a method to measure the esophageal transit time (ETT) employing a susceptometric technique[7,8], and the initial results of its use. The aim of this study is to compare scintigraphic and biomagnetic thoracic ETT (between proximal and mid-distal esophagus).

Materials and Methods:

In vivo studies were perfomed after approval of the experimental protocols by the ethics committee of the University Hospital. We have studied 11 normal volunteers (8 males, 3 females; aged 20-31 years old). A semi-liquid meal test was used in both cases (10 ml yoghurt). For scintigraphic studies the test meal was labeled with 350 MBq of ^{99m}Tc-phytate; 100 frames (0.2 s/frame) were acquired in a Siemens-Orbiter Scintillation Camera linked to a computer. Two Regions-of-Interest (ROIs) were drawn to generate time/activity curves: one on the proximal esophagus (furcula) and the other on the mid-distal part of the organ (xiphoid process). For biomagnetic studies the test meal was labeled with 5 g of ferrite and swallowing was monitored by 2 gradiometers, each of them placed on the very same areas where ROIs were to be drawn (external marks were used). ROI size was determined after evaluation of the point spread function (PSF) and its width at the half maximum. Scintigraphic and biomagnetic measurements were performed in the same day. The gradiometers were coupled to an AC biosusceptometer previously described [7,8]. Acquisition of the biomagnetic signal was done by a 12 bits resolution A/D board at 10 Hz frequency. Data analysis and visualization were performed in a PC microcomputer. In both experiments the subjects stood in the upright position. The signal was filtered using the MatLab software.

Results and Discussion:

Fig. 1 shows both typical biomagnetic and scintigraphic curves. The radioactive counting at the two regions shows two clearly defined positive peaks. The first maximum was due to the passage of the meal by the furcula and the second by the xiphoid. The time difference between the two maxima was used to determine ETT. The voltage produced by the biosusceptometer due to the presence of the magnetic test meal is shown in B. The passage of the test meal in front of the detectors produced two negative peaks, that allowed the determination of ETT. The ETT is the elapsing time between the two peaks in these curves. Fig. 2 shows a histogram comparing ETT measurements in each volunteer. Statistical analysis shows that the ETT determined by scintigraphy (mean = 4.2; SD = 1.1) and susceptometry (mean = 4.6; SD = 1.0) are not different (paired t-Test, $p > 0.05$). In a general way ETT determined by the biomagnetic method is slightly larger than the one determined by scintigraphy. This could be the consequence of the a larger density of the meal when labeled with ferrite as compared to the one with liquid (^{99m}Tc-phytate).

One possibility to check the density effect of the test meal on the prolongation of ETT measured by susceptometry is to make measurements with the subject laying supine; this approach would eliminate gravity effects that are more prominent with the subject in the upright position. A technical difficulty with the present susceptometric set-up was the proximity of gradiometers: magnetic fields of one gradiometer were coupled to the other, depending on distance. The solution was to place the gradiometers as far as possible from each other at an appropriate angle.

The gradiometric system presently in use for ETT assessment was developed to evaluate stomach emptying and orocecal transit time using a magnetic test meal[9]. In these cases the meal is closer to the anterior abdominal wall and the best approach is to have magnetization and detection at this plane. However, the anatomy of the esophagus makes it to be closer to the anterior wall at the furcula but is becomes progressively posterior as it gets closer to the stomach. In this case, an alternative to the gradiometric system could be a simple transformer where the nucleus would be the esophagus. This approach is already being tested with very promising results.

Another point that is under investigation is how to turn the method observer independent. The extrema of the signal were determined by inspection and ETT calculated. A fitting procedure or a derivative of the signal can be employed to detect the extrema automatically. The second procedure has been used with good results, specially when the signal-to-noise ratio is high.

ETT determinations is an important parameter in many clinical situations. Diseases may causes several signs and symptoms due to esophageal disfunction and medical approach may be clinical or surgical. Thus, it is clear that a precise-and better, inexpensive- evaluation of the organ's function is highly desirable. In summary, these initial results suggest that ETT can be evaluated by means of the biomagnetic method and that it is possible to improve it.

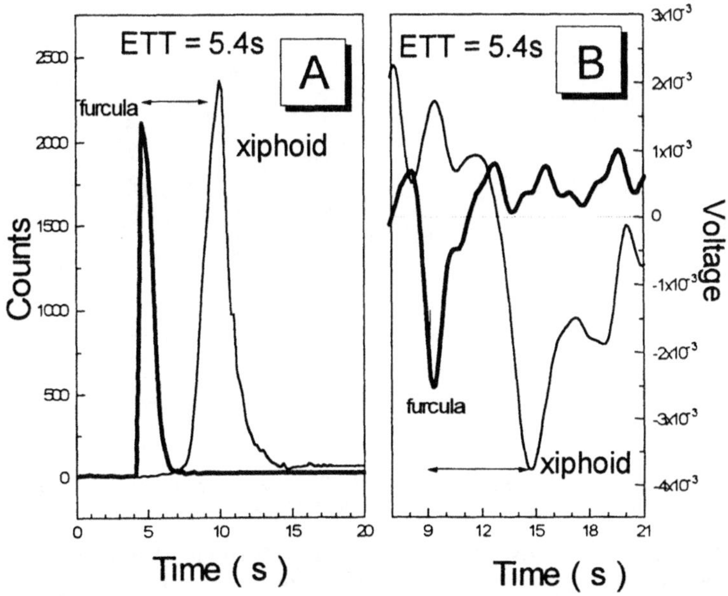

Fig. 1-Typical scintigraphic (A, volunteer 11) and biomagnetic (B, volunteer 10) recordings of the ingestion of a bolus of meal. Solid line shows the data collection at the furcula and narrow line at the xiphoid process.

volunteers	scintigraphic (s)	biomagnetic (s)
1	2.4	2.6
2	4.2	3.7
3	5.0	5.1
4	3.6	4.8
5	3.6	3.9
6	4.2	4.9
7	3.6	5.1
8	4.2	3.8
9	4.2	5.0
10	6.0	5.4
11	5.4	6.3

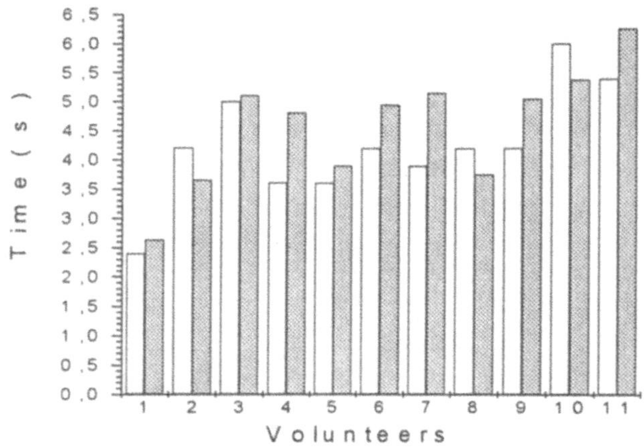

Fig. 2-Table and histogram comparing ETT measured by the biomagnetic (shaded) and scintigraphy methods.

References :

[1] Oliveira, R.B., Rezende, J., Dantas, R.O., Iazigi, N. The Spectrum of esophageal Motor Disorders in Chagas'Diseases. Am. J. Gastroenterology, 1995, 90: 1119-24.

[2] Russel C.O.H., Hill,L.D., Holmes, E.R. III, et al. Radionuclide Transit : A sensitive screening test for Esophageal dysfunction. Gastroenterology, 1981, 80:887-92.

[3] Forsman, M., Assessment of gastric movements by monitoring magnetic field decline of magnetized trace particles , Physics in Medicine and Biology, 1994, 39a: 62.

[4] Baffa, O., A brief account of biomagnetic research in gastroenterology, Physics in Medicine and Biology, 1994, 39a: 50.

[5] Forsman, M., Hultin, L. and Abrahamsson, H. Measurement of gastrointestinal transit time using fluxgate mangnetometers, In: Baumgartner C, Deecke L, Stroink G, Williamson SJ Biomagnetism: Fundamental research and clinical applications. Studies in applied electromagnetics and mechanics, Vol 7, Amsterdam, Elsevier Sci. Publ., 1995.

[6] Basile, M., Neri, M., Carriero, A., Casciardi, S., Comani, S., Del Gratta, C., Donato, L. G., Di Luzio, S., Macri, M.A., Pasquarelli, A., Pizzella V. and Romani, G.L. Measurement of segmental transit time trough the gut in man, Digestive Diseases Sciences, 1992, 37 (10):1537-1543.

[7] Miranda J.R.A., Baffa O., Oliveira R.B. and Matsuda N. M. An AC biosusceptometer to study gastric emptying, Medical Physics, 1992, 19(2): 445-449.

[8] Baffa O., Oliveira R. B., Arruda Miranda J. R. and Troncon L. E. A. Analysis and development of AC biosusceptomter for orocaecal transit time measurements, 1995, Medical & Biological Engineering & Computing, 1995, 33:353-357.

[9] Miranda, J.R.A.1,2,3, Baffa, O.1, Oliveira, R.B.3, Braga, F.J.H.N.3, and Sousa, P.L.1. Detection of Stomach Contraction Using a Biosusceptometer, 10th Int. Conf. on Biomag.: Biomag'96. Santa Fe-USA. Feb/16-21/96.

Acknowledgments:

The authors are grateful to J.G. Guasti Jr for discussions and his endeavor in development the AC biosusceptometer, Prof. N. Iazigi for his support and to E. de Paula, E. Navas, L. Aziani and M. Oliva for technical help. Partial financial support was received from Conselho Nacional de Desenvolvimento Científico e Tecnológico (CNPq), Coordenação de Aperfeiçoamento do Pessoal de Ensino Superior (CAPES), Fundação de Amparo à Pesquisa do Estado de São Paulo (FAPESP), Third World Academy of Sciences (TWAS) and FAEPA -HCFMRP-USP.

Measurements of Colonic Motility Using Ferrimagnetic Powder

Forsman, M.[1] and Abrahamsson, H.[2]

[1]Lindholmen Development, Box 8714, S-402 75 Gothenburg and Department of Applied Electronics, Chalmers University of Technology, S-412 96 Gothenburg; [2]Department of Internal Medicine, University of Gothenburg, Gothenburg, Sweden

Introduction

Biomagnetic methods for measurements of gastrointestinal motility may constitute a future clinical complement to invasive or radiological methods. The biomagnetic alternative is especially attractive for measurements in pregnant women, children and healthy persons. In previous studies, a swallowed magnetic marker was successfully localized inside the large intestine using a SQUID-magnetometer [1] and fluxgates [2]. A disadvantage, revealed when the transit of the magnetic marker was compared to that of ten radiopaque markers, was that a bulk of markers ingested simultaneously might disperse widely in the gut [2]. Hence, a single marker might poorly represent the mean progression. The aim of this study was to evaluate the usefulness of ferrimagnetic powder, mixed in a solid meal, to study gastrointestinal transit and intraluminal movements.

To locate the intraluminal tracer, the abdomen was first briefly exposed to a magnetic field, then the remanent field was measured, and the location of the particles was visually estimated from the field maps. Since the magnetization aligns the magnetic moments of the tracer particles, the movements of the particles can be assessed via the post-magnetization field strength decay. This decay is due to rotations of the particles, i.e., a consequence of the forces that mix and propel the chyme. The principle, originally proposed by Cohen [3], was recently implemented in measurements of intragastric movements [4,5]. Information about particle movements that can be assessed in this way can not be obtained from currently used clinical methods without using high radiation doses.

Methods

Ten male volunteers, aged 21 to 30 years, with no history of intestinal diseases, gave their informed consent to participate in the study. Maghemite powder (Bayferrox 8140, Bayer) was used as a tracer. This ferrimagnetic material is also known as γ iron oxide, γ-Fe_2O_3. The solubility of these particles was tested *in vitro*. The particles were added to hydrochloric acid, pH=1. No solubility could be observed in one week. During preparation of the test meal, 2 grams of the tracer were mixed with moderately heated pancake batter, which were then baked into pancakes. The mass of one portion was 130-140 g and the approximate energy value was 1340 kJ. The particles were sufficiently small so that none of the subjects could feel them while chewing.

The measurement took place 24 hours after the ingestion, and was divided into two parts. The location of the tracer particles was first sought for. During the measurement the subject rested on a pneumatically driven bench, allowing computer controlled horizontal movements [6]. The bed was initially positioned by the guidance of an optical beam, so that the horizontal origin was located at the xiphoid process. The intraluminal particles were then briefly magnetized by a homogeneous field pulse (40 mT, 2s), which was generated by two coils in a Helmholtz configuration. The vertical magnetic flux density, B_z, was then measured with first-order fluxgate gradiometers (baselength=140 mm, noise amplitude=45 pT/$\sqrt{Hz}$). B_z was measured in a grid in two planes, in six-twelve transversal scans, 40 mm apart. Each scan contained 40 points, with an inter-sampling distance of 15 mm. The anterior sensor was adjusted as close to the skin as possible. The distance between the proximal coil and the skin was about 35 mm. The posterior sensor was fixed with the proximal coil 62 mm beneath the surface of the bench.

After inspection of the anterior field map (an example is shown in Fig. 1, left) the bed was positioned to measure B_z at the point of maximum field strength. The abdomen was then magnetized a second time, after which a baseline was sampled as the subject was moved 0.5 m away from the sensors in the beginning and at the end of a 12-minute sampling period. The magnetic signals were digitized at a frequency of 3.3 Hz. To reject disturbances induced by respiration, a digital low pass filter with a cut off frequency of 0.2 Hz and no delay was applied to the data. The respiration was registered in order to facilitate a comparison with the magnetic signals, although these had been filtered. The respiration was registered by using a small air-filled cushion, placed under a rubber band, near the umbilicus. The cushion was coupled with a tube to a pressure sensor, the signal of which was amplified, lowpass filtered (1 Hz), digitized at 3.3 Hz and digitally highpass filtered (0.08 Hz and no delay). The subjects were asked to keep quiet and to avoid deep breaths during the measurement.

An *in vitro* experiment was carried out to control the remanent field stability. The test meal was placed on the bench in a plastic pot. A magnetic measurement was then performed as described above.

The obtained field maps were inspected visually. The decay of the sampled field strength was plotted as function of time. For each subject, mono- and bi-exponential models were fitted to the anterior and posterior field amplitude decays. The nonlinear regression was carried out by using a least mean square procedure. The rms residual was calculated as a percentage of the rms field and used to compare the results of the two models. Normalized mean curves were calculated for anterior and posterior respectively and modeled as described above.

Results

The field maps of 5 out of the 10 subjects showed two or more peaks, on both planes. One subject's field maps are shown in Fig. 1. The height, sharpness, and location of the major anterior peak show that a large part of the tracer is located in the transverse colon (tc). The maximums are 192 and 133 nT anterior and posterior, respectively. There is also an accumulation of particles in the recto-sigmoid region (r-s). The ratio between the anterior and posterior peak values for the second peak is 73 to 64, i.e. close to unity. The major part of the test meal in each person was classified into the four colonic regions (Table 1). In subject (S1) the contents were distributed equally in tc and r-s, therefore S1 contributed a "classification half" to each of the two regions in Table 1.

Table 1. Locations of the major part of the test meal, 24 hours post-ingestion, in the ten subjects.

Colonic region	# Classifications
ascending colon	1
transverse colon	5.5
descending colon	1
recto-sigmoid	2.5

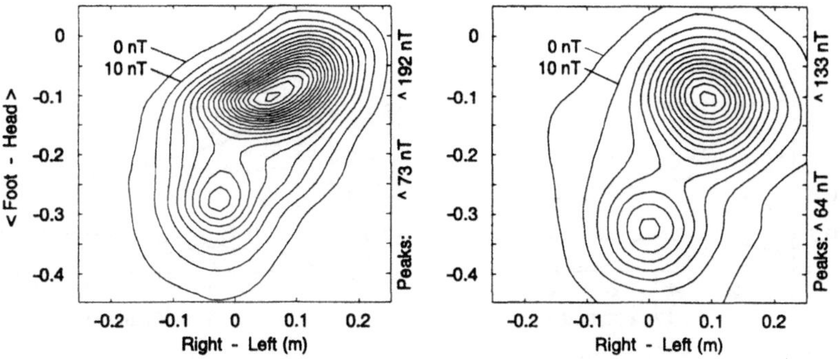

Fig. 1. Anterior (left) and posterior (right) field maps outside the colon with magnetized ferrimagnetic powder, 24 hours after ingestion. Origin was located at the xiphoid process.

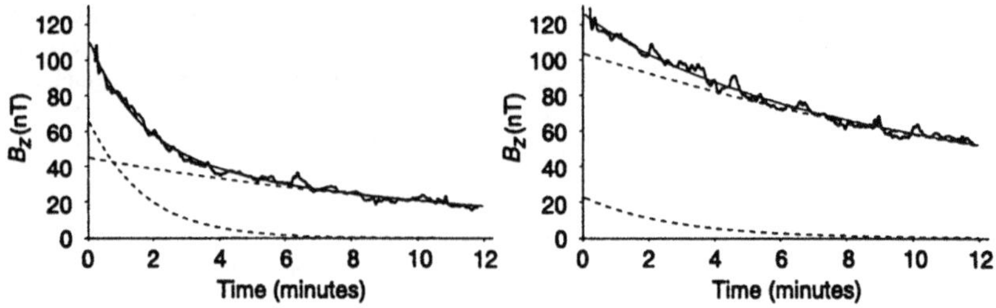

Fig. 2. Anterior field curves after magnetization from Subject S1. The left diagram illustrates the decay (thick line) outside the transverse colon, 120 mm caudally and 20 mm to the left of the xiphoid process. The thin line shows the result of an applied bi-exponential model, where the dashed lines are the two exponential components. The right diagram illustrates the decay (thick line) outside the recto-sigmoid region, 300 mm caudally and 30 mm to the right of the xiphoid process.

The *in vitro* experiment showed that the remanent field of the test meal was very stable. The remanent field of the magnetized meal decreased 0.5-1% during the 12-minute period *in vitro*. The lower and upper signals were similar, with B_z at approximately 380 nT. In six repeated measurements, the peak-to-peak noise amplitude never reached above 2 nT.

Two field decay measurements were performed in subject S1, for whom approximately equal accumulations were found in tc and in r-s. The decay rate above tc was twice that above r-s, see Fig. 2. The curves in Fig. 2 also represent the two different types that were seen among the subjects. The residual rms of the applied bi-exponential function was 5.1% of the signal rms (Fig 2. left). In Fig. 2, right, the residual/signal rms was 3.4%, which was almost the same as for the residual of a single-exponential model (3.9%).

The mean residual rms to signal rms ratios from the decay curves were 5.6% anterior and 4.1% posterior. All raw curves correlated with the measured respiration, i.e. higher/lower amplitudes anterior/posterior during inhalation. For a few subjects some respiration disturbances could be seen also after the decay curves were digitally filtered. The mean decay over twelve minutes from the ten subjects (n = 11, since two decay curves were measured for S1), was well approximated by a bi-exponential function (Fig. 3). The residuals were 1.7% anterior and 1.0% posterior. The anterior decay was faster for all subjects.

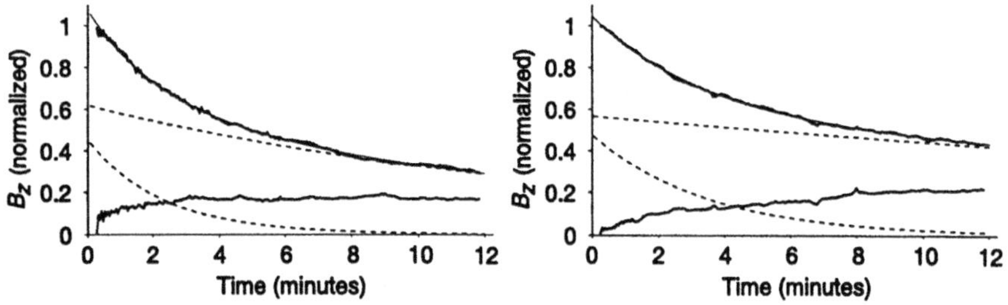

Fig. 3. Mean decay curves anterior (left) and posterior (right). Eleven curves were included in the two mean curves, the SD is shown as the lower thick line the diagrams. The SD is zero at start, since all curves were normalized prior to the averaging. The bi-exponential model is illustrated as in Fig. 2.

Discussion

The location of the test meal was easy to estimate from the field maps. Comparison of the anterior and posterior B_z facilitated estimation of the depth of the source. We did not use any inverse solution. However, it would be possible to calculate a compensated (for the source not being central between the measurement planes) source value at each (x,y) point in the maps. An integration of such values should hint whether some particles have been lost in the feces or not, which is hard to tell directly from the maps. The test meal main accumulations (Table 1) are in line with what could be expected from 24-hour post-ingestion radiological measurements on healthy males [7].

With powder as a tracer, we have overcome the problem that a single marker may not be representative for the mean progression in the colon after a certain time. Fig. 1. clearly shows that it is possible to localize a dispersed test meal. In this study the progression was represented by an "infinite" number of markers. Another difficulty remains: to relate the results from biomagnetic measurements to transit in specific structures. However, in the clinical situation, the most important question is if a transit delay involves the proximal colon, as in "slow transit costipation." This would be easy to answer with the present method.

The ratio of signal to instrumental and external noise was high in these measurements. This was confirmed by the *in vitro* measurements and by the low mean rms value of the individual regressions of the decay curves. The flux density is proportional to the inverse cube of the distance to the source-particles. Therefore the respiration disturbance was a significant source of error in the raw signal, especially for subjects who mixed normal breaths with deep ones. Most of this error was rejected by filtering.

The two mean curves (Fig. 3) were well described by the four parameters of the bi-exponential model. Also the individual curves approximated a bi-exponential decay, although examples where a single exponential fit was sufficient were seen as well (e.g. Fig. 2, right). A theory for the bi-exponential model is that the particles separate into two compartments: a semifluid compartment (chyme), where the particles are rotated quickly, and a more solid compartment where the particles rotate slowly. The factor parameters of the fitted bi-exponential function give a ratio between the amounts of particles in the two compartments.

We conclude that the present noninvasive method is valid in measuring colonic transit. The possibility to identify intraluminal movements has a potential to be useful for evaluation of motility in constipated patients.

References

[1] Basile, M., Neri, M., Carriero, A., Casciardi, S., Comani, S., Del Gratta, C., Di Donato, L., Di Luzio, S., Macri, M.A., Pasquarelli, A., Pizzella, V., and Romani, G.L. Measurements of segmental transit through the gut in man, Dig Dis Sci, 1992, 37;10:1537-1543.

[2] Forsman M., Hultin L., and Abrahamsson, H. Measurements of gastrointestinal transit using fluxgate magnetometers, In: Baumgartner, C., Decke, L., Stroink, G., and Williamson, S.J. Biomagnetism: Fundamental Research and Clinical Applications, Amsterdam, IOS Press,1995, 739-742.

[3] Frei, E.H., Benmayr, Y., Yerashalmi, S., Dreyfuss, F. Measurements of the emptying of the stomach with a magnetic tracer. *IEEE Trans. Mag.*, 1970, *MAG*-6(2): 348-349.

[4] Baffa, O., Oliveira, R. B., Miranda, J. R. A., and Sousa, P. L. Mixing power of food in the stomach evaluated by a biomagnetic technique, In: Baumgartner, C., Decke, L., Stroink, G., and Williamson, S.J. Biomagnetism: Fundamental Research and Clinical Applications, Amsterdam, IOS Press,1995, 753-756.

[5] Forsman, M. Assessment of gastric movements by monitoring magnetic field decline of magnetized trace particles, Physics in Medicine and Biology, 1994, 39a: 62.

[6] Kalliomäki, K., Aittoniemi, K., Kalliomäki, P.-L., Moilanen, M. Measurement of lung-retained contaminants in vivo among workers exposed to metal aerosols. *Am. Ind. Hyg. Assoc. J.* 1981, 42:234-238.

[7] Abrahamsson, H., Antov, S., and Bosaeus, Gastrointestinal and colonic segmental transit time evaluated by a single abdominal x-ray in healthy subjects and constipated patients, I. Scand J Gastroenterol, 1988, 23; suppl. 152: 72-80.

Detection of Stomach Contractions Using a Biosusceptometer

Miranda, J.R.A.[1,2,3], Baffa, O.[1], Oliveira, R.B.[4], Sousa, P.L.[1] and Braga F.J.H.[4]

[1]Departamento de Física e Matemática-FFCLR-USP-14040-901-Ribeirão Preto-SP, Brazil; [2]Instituto de Física de São Carlos-USP, 13560-970 São Carlos-SP, Brazil; [3]Departamento de Física e Biofísica, Instituto de Biociencias-UNESP, 18618-000 Botucatu-SP, Brazil; [4]Faculdade de Medicina de Ribeirão Preto, 14049-900 Ribeirão Preto-SP, Brazil

Introduction:

Symptoms presumably related to gastrointestinal (GI) motor disorders are among the most frequently encountered conditions in clinical practice [1]. Dearanged gastric antral contractility (GAC) is considered to play an important role in pathogenesis of several gastrointestinal motor disorders [2]. Intraluminal manometry, the standard method for assessment of GAC is both invasive and somewhat inaccurate [3]. More recently scintigraphic techniques to measure GAC were developed [4], but radiation burden and high cost of equipment limits its wide use. Magnetic Resonance Imaging (MRI) is also being investigated to measure GAC [5] with the advantage of no radiation exposure, but requiring even more expensive equipment than scintigraphic techniques. Biomagnetic methods based on the detection of magnetic flux variation produced by an ingested test meal labeled with magnetic particles can be an attractive alternative for GAC assessment.

Materials and Methods:

In vivo studies were initiated after approval of the experimental protocols by the ethics committee of the University Hospital (Medical School Ribeirão Preto). The subjects ingested a 900 kcal mixed meal followed, within a 1 hour period after, by 100 ml of a chocolate flavored custard (80 kcal) containing 3 g of manganese ferrite ($MnFe_2O_4$). The measurements were performed within 1 hour after the ingestion of the magnetic test meal and each recording was 2 to 10 minutes long. The subject stood in the upright position during the measurements, having the magnetic detector with a slight pressure against the abdominal wall over the epigastrium. In two studies, less than 370 MBq (10 mCi) of the radioisotope ^{99m}Tc-phytate was also added to the custard and GAC was simultaneously assessed by scintigraphy and biomagnetic methods. In these experiments 238 frames (1 s/frame) were acquired posteriorly in a Siemens-Orbiter Scintillation Camera linked to a computer and the magnetic signal acquired anteriorly. A region of interest (ROI) was drawn in the antrum area to generate time activity curves. Although the gamma camera has a great deal of iron, the spatial discrimination of the gradiometric system was good enough to allow simultaneous measurements. One hundred forty tracings of GAC were obtained by the magnetic method from 27 healthy volunteers (5 women and 22 men, median age 28 years, ranging from 20 to 40 years).

The magnetic detector employed in this study is an AC biosusceptometer (ACB), based on a differential (first order gradiometer) coil arrangement of two air core transformers, having one transformer for reference, and the other working as a measuring transformer on which the magnetic tracer having a high magnetic permeability acts like an external nucleus. The coils were mounted coaxily in a PVC tube and fixed with epoxy resin with a final diameter of 3.5 cm and 20 cm long. The primaries or excitation coils are fed by a power amplifier and the imbalance voltage produced by the test meal at the secondaries or detection coils, is detected by a lock-in amplifier. A 8 pole low pass filter (HP 5489A) with a cutoff frequency set at 3 Hz was used. The signal was digitized at 10 Hz frequency yielding data files up to 6 Kb. The ACB was previously used in studies of gastric emptying (6) and orocaecal transit time measurements [7].

Results and Discussion:

Fig. 1.a shows a typical raw tracing of the GAC. The 3 cycles/minute oscillation can be clearly recognized. In some measurements (fig. 1.b) the signal appeared superimposed to a higher frequency signal due to respiration; asking the subject to stop breathing for some time would remove this high frequency oscillation. Figs. 1c and 1.d show the Fourier transform of signals shown in fig. 1.a and 1.b., respectively, the 50 mHz power amplitude is clearly from the background signals. A typical oscillatory pattern of about 50 mHz (3 cycles/minute) was conspicuous in 120 of the magnetic measurements. In 17 of the remaining tracings the 3 cycles/minute was not recognizable in the crude tracings, but it was possible to distinguish 50 mHz signal from the background after Fourier transform. In only three measurements this signal was not detected by this method, indicating a rate of failure of only 2 %.

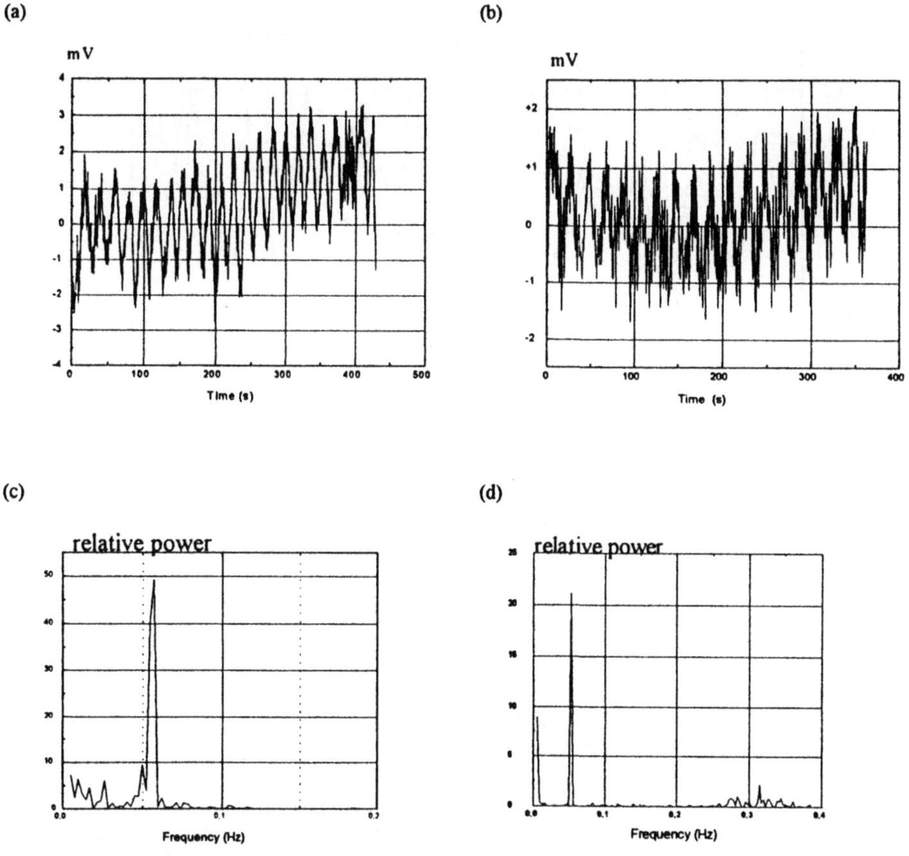

Fig. 1-Typical GAC tracing detected by the biomagnetic method, a clean signal (a) and in presence of high intensity breathing (b). The plot of the power spectrum obtained after Fourier transforms is shown in (c) and (d), respectivaly.

The question whether or not the magnetic signal was associated with the gastric contraction was further investigated by comparing GAC detected by the scintigraphic and biomagnetic techniques. Simultaneous recording of the radioactive counting of a region located at the distal antrum was made with the ACB. Fig. 2 shows one of these simultaneous recordings of the GAC. As the antrum contracts food is moved away and the radioactive counting varies proportionally to the cross section seen in the images. A cross correlation between the two tracings is given in Fig. 2.b. The coincidence between the two measurements allow us to infer that the magnetic measurements are actually reporting the mechanical activity of the stomach. The mean frequency value (mean= 53.71 mHz, SD= 3.55 mHz) measured by the biomagnetic methods is in very good agreement to the values that have been measured by EGG in a normal population of volunteers.

(a)

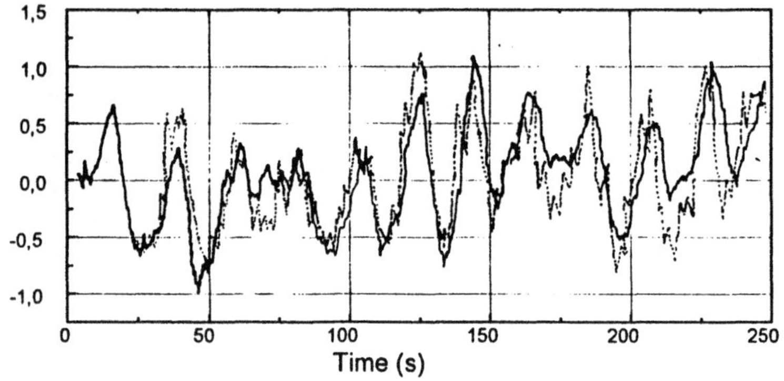

(b)

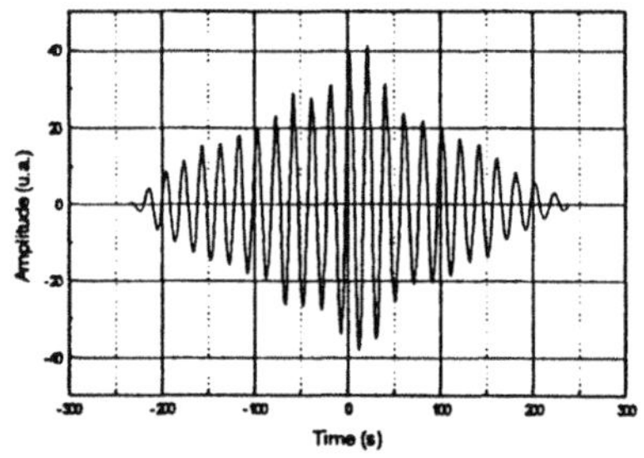

Fig. 2-Simultaneous recording of the GAC by the scintigraphic (dashed) and biomagnetic (continuous line) methods (a). The radioactive counting and ACB amplitude were normalizad to the initial amplitude for better comparison. In (b) the cross correlation function between the two signals is shown.

In the present study frequency is most easy quantity to be measured. It is possible that combining this method with another technique that could give the distance of the test food or antrum wall to the detector, the voltage amplitude detected could be translated in mechanical displacement of the antrum. These results show that a biomagnetic technique may be successfully employed to record GAC. The finding that the profiles of the variations of the magnetic activity were quite similar to those recorded by a scintigraphic method simultaneously applied, supports the notion that the magnetic signal detected is tied to the mechanical counterpart gastric antrum activity.

In summary, the biomagnetic method employed here is very straighforward and could detect the GAC very easily, having none of the drawbacks present in the EGG and other measuring methods.

References:

[1] Drossaman D.A. Clinical research in the functional digestive disorders, Gastroenterology, 1987, 92:1267-1269.

[2] Malagelada J.R. and Stanghellini V. Manometric evaluation of functional upper gut symptoms Gastroenterology, 1985, 88:1223-1231.

[3] Wingate, D.L. Small bowel manometry, The American Journal of Gastroenterology, 1995, 90(4):536-539.

[4] Urbain J-L.C., Vekemans, M.C., Bouillon, R., Bex, M. and Mayeur, S.M. Characterization of gastric antral motility disturbances in diabetes using a scintigraphic technique. Journal of Nuclear Medicine, 1993, 34:576-581.

[5] Fraser R., Schwizer, W., Borovicka, J., Asal, K. and Fried, M. Gastric motility measured by MRI, Digestive Disease Sciences, 1994, 39(12):20S-23S.

[6] Miranda J.R.A., Baffa O., Oliveira R.B. and Mitiko M. N. An AC biosusceptometer to study gastric emptying, Medical Physics, 1992, 19(2): 445-449.

[7] Baffa O., Oliveira R. B., Arruda Miranda J. R. and Troncon L. E. A. Analysis and development of AC biosusceptomter for orocaecal transit time measurements, Medical & Biological Engineering & Computing, 1995, 33:353-357.

Acknowledgments:

The authors are grateful to J.G. Guasti Jr for discussions and his endeavor in development the BAC, and to E. de Paula, E. Navas, L. Aziani and M. Oliva for technical help. Partial financial support was received from Conselho Nacional de Desenvolvimento Científico e Tecnológico (CNPq), Coordenação de Aperfeiçoamento do Pessoal de Ensino Superior (CAPES), Fundação de Amparo à Pesquisa do Estado de São Paulo (FAPESP), Third World Academy of Sciences (TWAS) and FAEPA -HCFMRP-USP

Biomagnetic Measurement of Gastric Mechanical Motility Using a High-Temperature SQUID Imaging

Nomura, M.[1], Saijyo, T.[1], Haruta, Y.[2], Itozaki, H.[2], Toyoda, H.[2], Nakaya, Y.[3], Ito, S.[1] and Kado, H.[2]

[1]Second Department of Internal Medicine and [3]Department of Nutrition, The University of Tokushima, Tokushima, 770, Japan; [2]Superconducting Sensor Laboratory, Chiba, 270-13, Japan

INTRODUCTION

Biomagnetic measurement using superconducting quantum interference device (SQUID) is an effective non-invasive tool for detecting gastric function. In the present study, the weak magnetic field originating from gastric mechanical motility was detected using a newly developed high temperature (-196°C) SQUID system by liquid nitrogen.

SUBJECTS AND METHODS

Two healthy volunteers were selected for the present study. The magnetic field was recorded in the magnetically shielded room using a 16-channel high temperature SQUID system developed by Superconducting Sensor Laboratory in Japan (Figure 1). A magnetized steel ball (OD, 1.3 mm) attached to the tip of a stomach tube (12 F), was inserted from nose, positioned fluoroscopically in the antrum. The magnetic fields were recordedin fasting or fed (600-ml liquid meal) states. A magnetic sensor was placed above the epigastric region and magnetic signals produced by gastric mechanical motility were measured. Three recording bandwidths (DC-50 Hz, DC-1 Hz and DC-0.1 Hz) were used, and fast Fourier transform analysis was performed using spectral analyzer (HP3562A DYNAMIC SIGNAL ANALYZER). After recording the magnetic signal, the stomach tube with steel ball was removed.

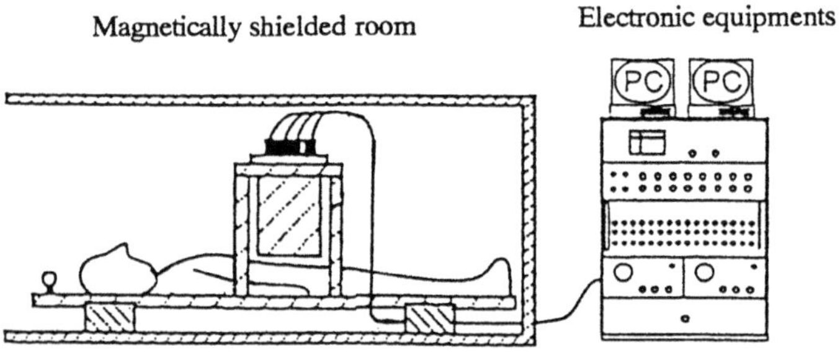

Figure 1 System diagram of 16-channel high temperature SQUID system

RESULTS

1. Back ground noise of a 16-channel high temperature SQUID system

Figure 2 shows the frequency spectra of the back ground noise of a 16-channel high- temperature SQUID system. The noise levels were 0.4 $pT/Hz^{1/2}$ and 4 $pT/Hz^{1/2}$ at 500 mHz and 50 mHz, respectively.

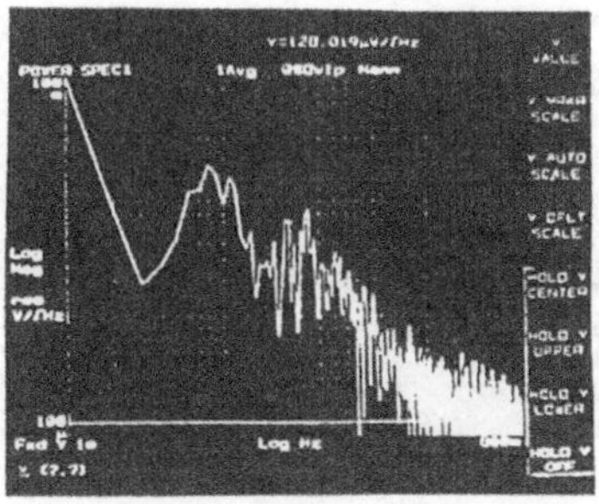

Figure 2 Frequency spectra of the back ground noise of 16-channel SQUID system

2. Magnetic fields originating from gastric mechanical motility

Figure 3 shows the magnetic signal from the steel ball in the stomach at fasting state. The recording bandwidth was DC-1 Hz. Magnetic signal around 13-18 cycle/min was detected. After the magnetic recording filtered with bandwidth of DC-0.1 Hz, the regular 3 cycle/min signal, which indicated gastric motility, was clearly visible. Figure 4 shows the magnetic signal at fasting and fed states using bandwidth of DC-0.1 Hz. Three cycle/min magnetic signal was visible at each state. The amplitude of the magnetic field at the fed state was stronger than that of fast state.

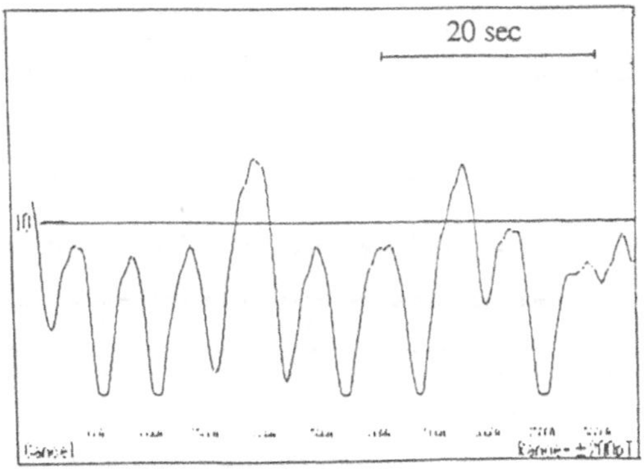

Figure 3 Magnetic signal at fasting state (bandwidth filter; DC-1 Hz)

3. Fast Fourier transform analysis in fasting and fed states

Figures 5 shows the frequency spectra in fasting and fed states, respectively. The two spectral peaks were visible at 0.05 Hz and 0.15 Hz, and amplitude of magnetic field at 0.05 Hz in fed state was stronger than in fasting state. Table 1 shows the magnetic intensity and peak frequency at fasting and fed states at 0.05 Hz. The magnetic field was greater after ingestion of food than in the fasting state ($p<0.01$). The peak frequency of gastric motility was not statistically different between the two states (n.s.).

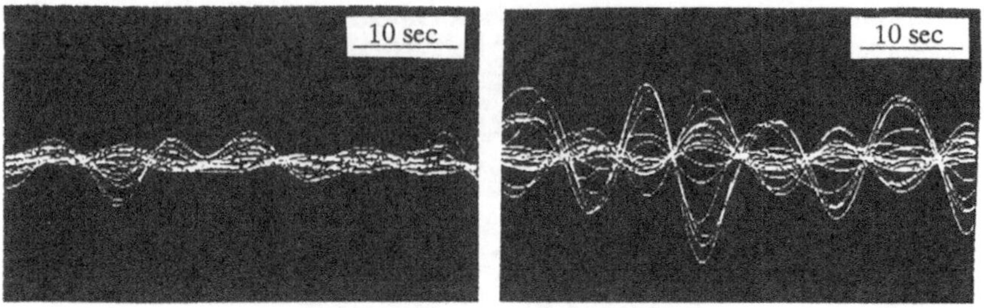

a b

Figure 4 Magnetic signal at fasting (a) and fed (b) states (bandwidth filter; DC-0.1 Hz)

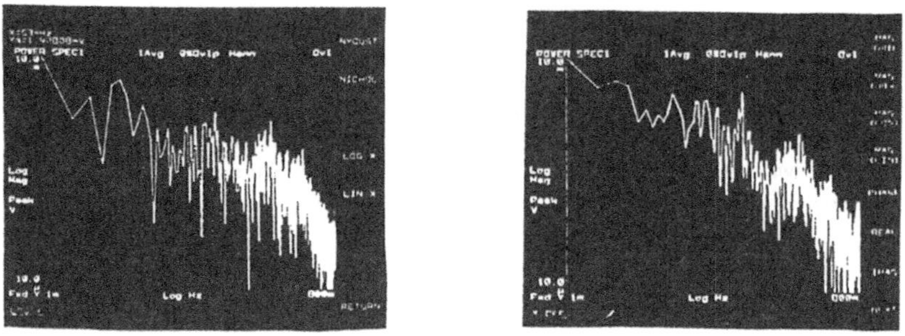

a b

Figure 5 Frequency spectra in fasting (a) and fed states (b).

Table 1 Magnetic intensity and peak frequency at fasting and fed states

	Fasting state	Fed state
Magnetic intensity (pT)	9.0±4.0	21.2±3.8
Peak frequency (cycle/min)	3.1±0.2	3.2±0.1

(Values are mean±SEM. pT; pico-tesla)

DISCUSSION

We succeededin recording gastric motility using a high-temperature SQUID system. This is the first multi-channel high-Tc SQUID system and its operation is much simpler than that of the conventional SQUID system that uses liquid helium.

The magnetogastrogram was recordedby Comani et al.[1] and Basile et al.[2] They detected the 3 cycle/min magnetic signals originated from the stomach without using magnetized tracer. In the present study, we recordedthe magnetic signal from the magnetized tracer positioned in the stomach in order to detect the mechanical gastric movement.

An electrogastrogram is the summation of gastro-intestinal electrical activity. On the other hand, the magnetogastrogram without using magnetized tracer reflect only the gastric electricai activity because the SQUID system is superior in positional resolution to the electrogastrogram. Moreover, the magnetic signals generated by a magnetized-steel ball in the stomach reflect the mechanical gastric movement.

Using fast Fourier transform analysis, two spectral peaks (3 cycle/min and 13-18 cycle/min peaks) were detected. Three cycle/min magnetic signals reflected the gastric mechanical movement and 13-18 cycle/min signals represent the respiratory movement. These results coincide well with the previous reports[1-3].

Gastric electrical activity might disagree with gastric mechanical movement in some diseases. This new method represents gastric motility more specifically, and it is expected to contribute to diagnosis and management of various mechanical gastric disorders[4,5].

REFERENCES

[1] Comani, S., Basile, M., Casciardi, S., Gratta, CD., Luizo, CD., Erne, SN., Macri, M., Neri, M., Peresson, M., Romani, GL. Extraporeal direct magnetic measurement of gastric activity. In: Hoke, M. et al. Biomagnetism: Clinical aspects. Amsterdam, London, New York, Tokyo, Elsevier Science Publishers, pp639, 1992.

[2] Basile, M. The biomagnetic approach to the study of gastrointestinal activity. In: Hoke, M. et al. Biomagnetism:Clinical aspects. Amsterdam, London, New York, Tokyo, Elsevier Science Publishers, pp613, 1992.

[3] Alvarez, W.C. The electrogastrogram and what it shows. JAMA, 1922, 78:1116.

[4] Nomura, M., Saijyo, T., Itozaki, H., Toyoda, H., Nakaya, Y., Ito, S., Kado, H. Magnetic measurement of gastric activity using a high-temperature superconducting quantum interference device (SQUID). Gastroenterol 1995, 108, 108:A658.

[5] Saijyo, T., Nomura, M., Haruta, Y., Itozaki, H., Toyoda, H., Nakaya, Y., Ito, S., Kada, H. Biomagnetic measurement of gastric electrical activity by 64-channel magnetogastrographic imaging. Gastroenterol 1995, 108:A680.

Single and Multichannel Magnetic Field Measurements of Gastrointestinal Activity in the Pre- and Post-Prandial States

Petrie, R.J.[1], van Leeuwen, P.[2], Brandts, B.[3], Turnbull G.[4], Veldhuyzen van Zanten, S.J.0.[4] and Stroink, G.[1]

[1]*Dalhousie University, Dept. of Physics, Halifax, Canada;* [2]*EFMT, Bochum, German;* [3]*Ruhr University, Dept. of Physiology, Bochum, Germany;* [4]*Dalhousie University, Dept. of Medicine, Division of Gastroenterology, Halifax, Canada*

Introduction

Current techniques for assessing gastrointestinal activity are highly invasive. These techniques such as radiology, scintigraphy and endoscopy yield results that are qualitative and non-specific. A method of evaluating GI function that is both non-invasive and quantitative is needed.

Magnetic signals corresponding to electrical depolarization of gastrointestinal smooth muscle can be measured with SQUID gradiometers [1]. The purpose of this research is to investigate GI motility by studying these magnetic signals. Motility is a term that encompasses electrical activity and innervation of smooth muscle, contraction of muscle and propulsion of GI contents [2]. This work specifically focuses on two aspects of motility: contractions resulting from action potentials and slow wave depolarization of smooth muscle.

The goal in studying contractile activity is to identify action potentials in the magnetic signal by correlating these signals with contractions measured simultaneously using a manometry catheter. Slow wave depolarization was studied by making measurements over the abdominal surface with a multichannel gradiometer system. We have previously shown that sequential, single channel measurements over the abdominal region are not sufficient to represent spatial information about gastrointestinal function. Both the pre- and post-prandial states were studied. These signals were then examined with respect to frequency content, spatial distribution and time dependence.

Methods

I. Contractile Activity: Two subjects participated in this research. Neither had a history of gastrointestinal dysfunction. Subjects were adult male (average age 42 years). Prior to recording, a manometry catheter was inserted through the nasal passages of the subject approximately 60 cm into the GI tract. Both the nostril and pharyngeal regions were anesthetized with Viscous 2% (2g/100ml) Xylocaine oral topical anesthetic and Xylocaine endotracheal spray (20mg measured dose) respectively. The catheter was designed to record intraluminal pressure at six different locations (10 cm apart) along the length of the tube. Once in place the subject was positioned under the single channel, second order SQUID gradiometer (noise=40 $fT/\sqrt{Hz}$, sensing coil diameter= 2.54cm, baseline=4cm) so that the probe was located approximately above tbe pyloric sphincter.

Measurements were made for thirty minutes in both the pre- and post- prandial states. The magnetic signal was low pass filtered at 16Hz. The pre-prandial state was assured by the subjects fasting for eight hours before recording while the fed motility state was induced by ingestion of a standard test meal: 250ml orange juice, 130ml yogurt, 1 bran muffin (7 g protein, 295 cal).

II. Slow Wave Depolarization of Smooth Muscle: Subjects participating in this study were healthy adults (three females and two males) without history of GI dysfunction. Mean age: 32 years. Subjects presented having fasted for eight hours. Both pre- and post- prandial motility states were studied.

GI slow wave depolarization was measured with a 37 channel, first order gradiometer system (Siemens Krenikon, noise=30 $fT/\sqrt{Hz}$, baseline=7cm, diameter of sensor array=19cm). The probe was positioned along the central axis of the body, half way between the sternum and line joining the iliac crests. Subjects were recorded for 60 minutes in each of the motility states.

Analysis: All multichannel data were first analyzed for frequency content using power spectral analysis. Bands exhibiting significant power were identified. For each of these bands isoharmonic maps were generated. Isoharmonic maps spatially display the average power distribution over the region corresponding to the sensor array for a 10 minute period of time. The power at the sites with maximum power was then plotted as a function of time over the 60 minute record. The result is information about the spectral, spatial and temporal dependence of magnetic GI signals.

Results and Discussion

1. Contractile Activity: Gastric and duodenal slow waves or basic electrical rhythm (BER) could be identified in the magnetic record in all subjects (Fig. 1) both visually and through spectral analysis. It is clear that the gastric signal (2.8 ± 0.1 cpm) dominated during the first 10 minutes after ingestion. The duodenal signal (12.2 ± 0.1 cpm) does not become of significant amplitude until 16 minutes after ingestion.

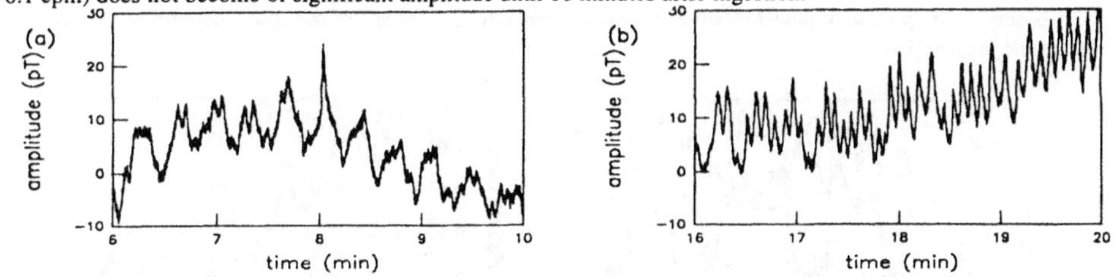

Figure 1: Example of basic electrical rhythm (post- prandial state) of the (a) stomach and (b) duodenum. The frequency of the gastric and duodenal slow waves is 2.8 ± 0.1 cpm and 12.2 ± 0.1 cpm respectively.

Correlation of the magnetic signals with the internal GI pressure data (Fig. 2) did not identify action potentials. Some difficulties have been encountered with this protocol. It is now evident that a radiological scan of the catheter location is necessary to guarantee its correct positioning prior to measurement. Also, it is clear that cardiac artifacts in the magnetic signal may partially obscure action potentials in the smooth muscle.

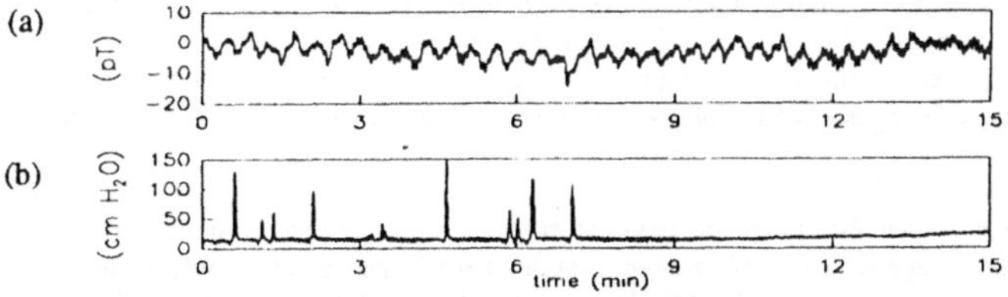

Figure 2: (a) Magnetic and (b) one (of six) internal pressure recordings (post-prandial) used in correlation study for identifying electrical action potentials that result in contraction of intestinal smooth muscle.

Thus, although action potentials have been observed magnetically in vivo in rabbit small intestine [3], we did not observe spiking activity in the magnetic measurement of human GI smooth muscle.

II. *Slow Wave Depolarization of Smooth Muscle:* From power spectral estimation several different frequency bands were identified for further investigation. There was some minor variance in these bands between individuals. The bands of relevance to this paper are: 45 - 55 mHz, (corresponding to the 3 cpm frequency of gastric smooth muscle depolarization), 190 - 205mHz (12 cpm frequency of duodenum), 140 - 160mHz (9 cpm frequency of distal jejunum) and 90 - 105mHz (6 cpm frequency of distal ileum). The integrated power in each band was calculated for every channel in the array. The spatial distribution plots of signal power in these bands over the abdominal region are shown in Fig 3.

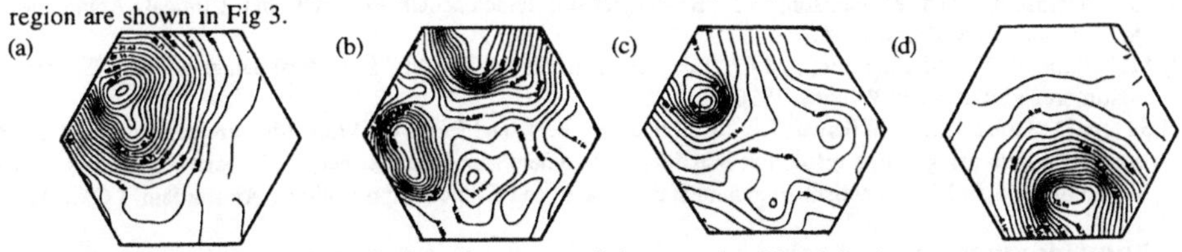

Figure 3: Typical spatial distribution of post - prandial, magnetic, GI activity in the (a) 45 - 55 mHz (b) 90 - 105 mHz (c) 140 - 160 mHz and (d) 190 - 205mHz bands in healthy subjects.

It is clear that the maxima of different frequencies are at different abdominal locations. Maximum power indicates strong GI activity underneath the probe at this location. This is what is expected from GI physiology which states that different segments of the gastrointestinal tract function at different frequencies. For brevity, only the post prandial - state is shown. However, individuals exhibit maps of similar morphology in both the pre-prandial states but the magnitude of the signals is less. These results suggest that magnetic field measurements can spatially distinguish gastric and small bowel activity.

It is important to investigate how this activity behaves as a function of time. Iso◊◊harmonic maps were generated over 10 minute intervals during the 60 minute recording period. No significant changes in morphology were observed. Magnitude of power relative to the maximum power at the site of maximum activity was observed to be time dependent (Fig. 4)

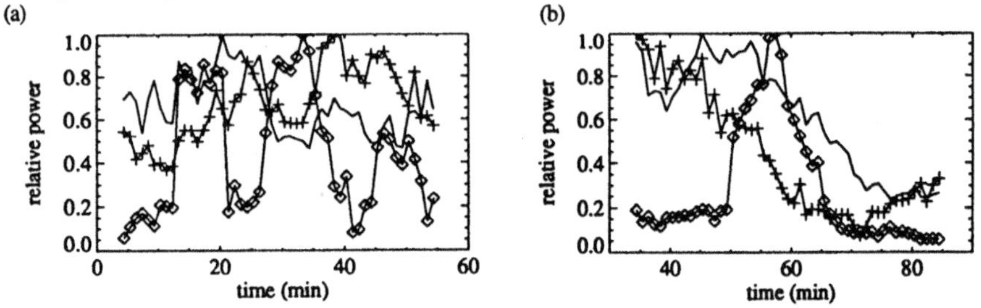

Figure 4: Pre- (a) and post-prandial (b) plots of relative power at sites of GI activity over the 60 minute recording period. Shown is 45 - 55 mHz (------), 190 - 205mHz (◊◊◊◊◊◊) and 140 - 150 mHz (++++++).

During fasting-the signals exhibit no well established pattern. These same frequency bands show that maximum power occurs at different times after ingestion for different segments of the GI tract. The gastric signal reaches maximum power 45 minutes after ingestion. The duodenal signal peaks 57 minutes after ingestion. From this we see that magnetic signals can be followed over a period of time that is typical of GI function.

Conclusion:

It has been shown that magnetic field measurements are capable of recording gastrointestinal activity generated by electric depolarization of smooth muscle. Gastric and small bowel slow wave activity can be distinguished through spectral analysis. Although slow waves have been recorded using this technique, simultaneous magnetic and internal pressure measurements have not yielded observation of action potentials.

Changes in gastrointestinal activity can be measured and quantified in terms of spatial and temporal distribution of power in the various GI frequency bands. Gastric and small bowel activity although not demonstrating a clear pattern in the pre-prandial state, exhibit peak activity at distinct times after ingestion.

We conclude that a magnetic multichannel system can be used to repetitively and non-invasively measure and analyze magnetic gastrointestinal signals.

References:

[1] Richards,W.O., Staton, D., Golzarian, J., Friedman, R.N. and Wikswo, J.P. Jr. Non-invasive SQUID magnetometer measurement of human gastric and small bowel electrical activity, In: Baumgartner, C., Deecke, L., Stroink, G. and Williamson, S.J. Biomagnetism: Fundamental Research and Clinical Applications, Amsterdam, I.O.S. Press, 1995.

[2] McCallum, R.W. and Champion, M.C., Eds., Gastrointestinal Motility Disorders: Diagnosis and Treatment. Baltimore, Williams and Wilkins, 1990.

[3] Staton, D., Golzarian, J., Wikswo, J.P. Jr., Friedman, R.N. and Richards, W.O. Measurement of small bowel activity in vivo using a high resolution SQUID magnetometer, In: Baumgartner, C., Deecke, L., Stroink G. and Williamson, S.J. Biomagnetism: Fundamental Research and Clinical Applications, Amsterdam, I.O.S. Press, 1995.

Acknowledgments:

This research has been funded by the Natural Sciences and Engineering Research Council and the Camp Hill Medical Centre Research Fund.

Magnetogastrograms Using 64-Channel SQUID System: Study on Clinical Usefulness

Saijyo, T.[1], Nomura, M.[1], Haruta, Y.[2], Itozaki, H.[2], Toyoda, H.[2], Nakaya, Y.[3], Ito, S.[1] and Kado, H.[2]

[1]Second Department of Internal Medicine and [3]Department of Nutrition, The University of Tokushima, Tokushima, 770, Japan; [2]Superconducting Sensor Laboratory, Chiba, 270-13, Japan

INTRODUCTION

Magnetogastrograms (MGGs) using a superconducting quantum interference device (SQUID) gradiometer provides excellent temporal and spatial resolution, and are effective non-invasive tool for detecting functional disorder. It is reported that slow motility waves originate high on the greater curvature of the stomach, an area that acts as a pacemaker, which coordinate contractions of the gastric smooth muscle [1, 2]. The gastric electrical activity produces a weak magnetic field. The purpose of the present study was to assess the magnetic field due to gastric electrical activity and to localize the gastric pacemaker site using a 64-channel SQUID system.

SUBJECTS AND METHODS

Two healthy volunteers were selected for the study. The MGG was recorded at 64 points over the epigastric region with a first-derivative dc-SQUID gradiometer at Superconducting Sensor Laboratory in Chiba, and the magnetic field tangentially to the abdominal wall was measured. The MGGs were recorded in a magnetically shielded room in three states (fasting, 60 min after oral medication with cisapride, and 10 min after ingestion of a 500-kcal meal). The recording bandwidths were 50 mHz - 30 Hz. Fast Fourier transform analysis was performed at each state and the peak frequency spectra and magnetic fields were compared among states using spectral analyzer (HP356A Dynamic Signal Analyzer). Magnetic isofield maps were constructed and the locations of gastric electrical activities were computed by the least squares method. These localized activities were projected onto the magnetic resonance image providing anatomical localization in the stomach.

RESULTS

1) MGGs at 64 recording positions in fasting and after fed state

Figures 1 show the MGGs in fasting (panel a) and fed (panel b) states, respectively. The magnetocardiograms were visible on the low frequency wave at each recording position. After the magnetic recording was filtered out with bandwidth at 50 mHz - 100 mHz, the regular 3-cycle/min signal was clearly visible (Figures 2 a, b). The magnitude of the magnetic field after fed state was stronger than that of fasting state.

2) Fast Fourier transform analysis in three states

Figures 3 shows an example of the frequency spectra in three states; fasting (panel a), 60 min after oral medication with cisapride (panel b), and 10 min after ingestion of a 500-kcal meal (panel c), respectively. Two frequency components were visible at each state. Peak frequency were at 52 mHz, 57 mHz and 50mHz, and the magnitude of magnetic field were 274 fT, 385 fT and 760fT at each state, respectively. This component corresponds to gastric electrical activity. Another component around 150 mHz represent the breathing frequency.

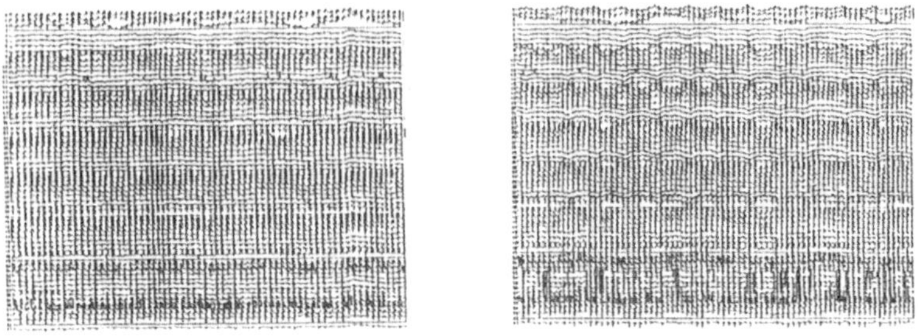

a b

Figure 1 MCGs in fasting (a) and fed (b) states

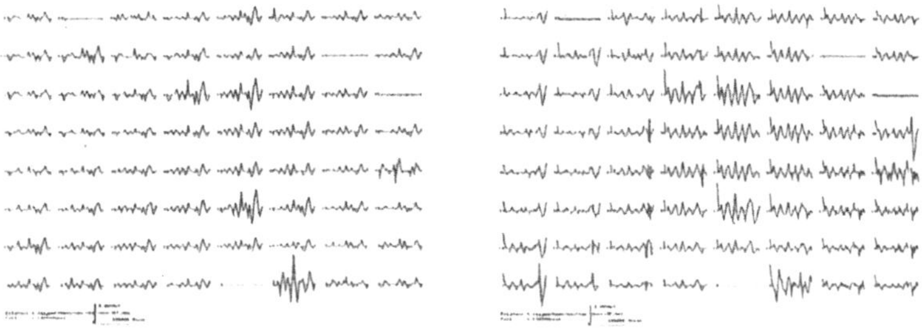

a b

Figure 2 Filtered magnetogastrographic signals at fasting (a) and fed states

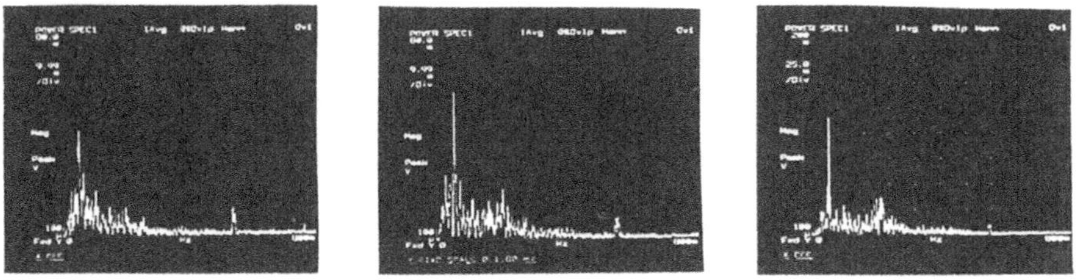

a b c

Figure 3 Fast Fourier transform analysis at each stage

(a): fasting state, (b):after medicating cisapride, (c):after taking meal. The 3 cycle/min signal indicated gastric motility, and around 15 cycle/min signal, which reflected respiratory movement, were detected

Figure 4 shows the magnitude of magnetic field (panel a) and peak frequency spectra (panel b) among three states at around 50 mHz. The magnetic field increased after ingestion of cisapride (2.7$\pm$0.1 vs. 4.4 $\pm$0.4

pico tesla; $p<0.05$). Moreover, the magnetic field after the meal was 1.25-fold greater than that after administration of cisapride (4.4 ±0.4 vs. 5.5±2.5 pico tesla; $p<0.05$). The frequency spectra of gastric electrical activity were 3.0-3.4 cycles/min but there were no statistically significant differences among the three states.

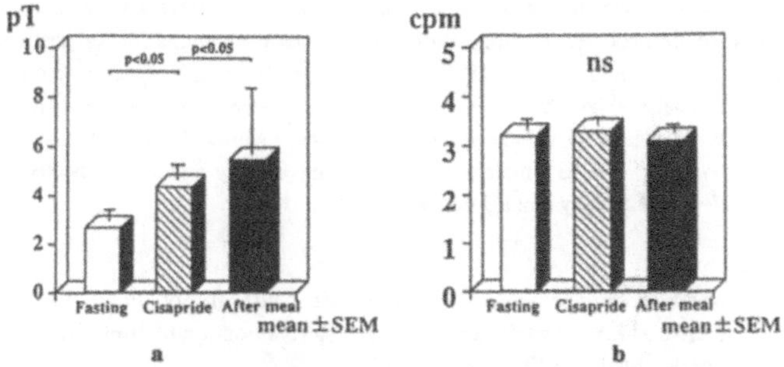

Figure 4 Magnitude of magnetic field (a) and peak frequency spectrum (b) among three states

3) Localized activities onto the MRI

Figure 5 shows the magnetic isofield maps after meal.The isofield maps were shown every 20msec. In Figure 6, the localized activities were projected onto the MRI in order to provide anatomical localization by the least squares method. The initial electrical activities were on the stomach.

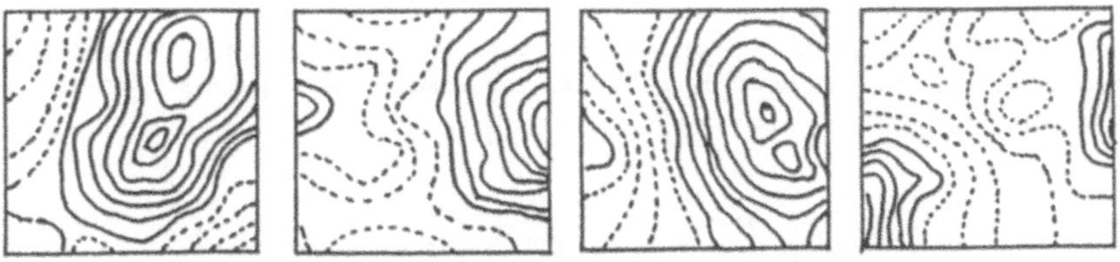

Figure 5 Magnetic isofield maps after meal

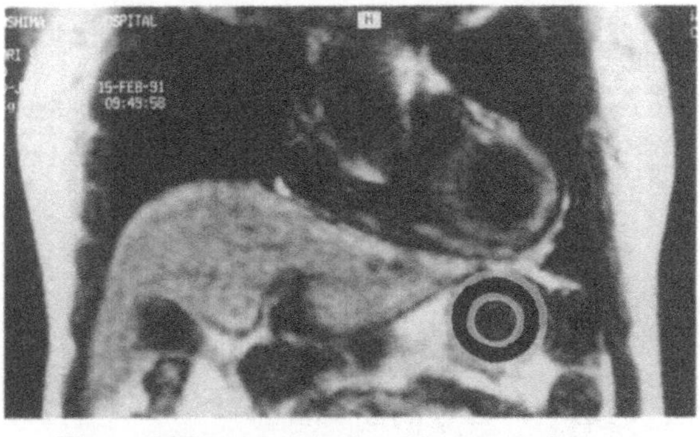

Figure 6 Magnetic resonance image of abdomen
(superimposed gastric electrical activities)

DISCUSSION

We succeeded in recording gastric motility using 64-channel SQUID gradiometer. The isofield map also showed gastric motility with 3 cycle/min in agreement with previous studies [3,4]. After ingestion of meals the amplitude of magnetic field was increased but activation sequences and frequency of slow wave motility was similar with those of control. An electrogastrogram indicates the summation of gastro-intestinal electrical activity but the magnetic field generated in the stomach reflects only gastric electrical activity [5, 6]. Multichannel recordings also showedthe location of pacemaker site and slow wave motility sequentially. In this study we only measured in healthy volunteer. The present study shows that the MGGs is useful for detecting gastric electrical activity in various clinical conditions noninvasively and it is expected to contribute to the detection of gastric electrical activity propagation and disorder.

REFERENCES

[1] Alvarez, W.C. New methods of studying gastric peristalsis. Am J Physiol, 1922, 201:287.

[2] Brown, B.H., Smallwood R.H., Duthie H.L., et al. Intestinal smooth muscle electrical potentials recorded from surface electrodes. Med Biol Eng, 1975, 13:97.

[3] Basile, M. The biomagnetic approach to the study of gastrointestinal activity. Biomagnetism:Clinical aspects. Amsterdam, London, New York, Tokyo, Elsevier Science Publishers, pp613, 1992.

[4] Comani, S., Basile, M., Casciardi, S., Gratta, CD., Luizo, CD., Erne, SN., Macri, M., Neri, M., Peresson, M., Romani, GL. Extraporeal direct magnetic measurement of gastric activity. In: Hoke, M. et al. Biomagnetism: Clinical aspects. Amsterdam, London, New York, Tokyo, Elsevier Science Publishers, pp639, 1992.

[5] Nomura, M., Saijyo, T., Itozaki, H., Toyoda, H., Nakaya, Y., Ito, S., Kado, H. Magnetic measurement of gastric activity using a high-temperature superconducting quantum interference device (SQUID). Gastroenterol 1995, 108, 108:A658.

[6] Saijyo, T., Nomura, M., Haruta, Y., Itozaki, H., Toyoda, H., Nakaya, Y., Ito, S., Kada, H. Biomagnetic measurement of gastric electrical activity by 64-channel magnetogastrographic imaging. Gastroenterol 1995, 108:A680.

Noninvasive Measurement of the Vector Magnetic Field from Human Gastrointestinal Sources

Staton, D.J.[1], Allos, S.H.[3], Henry, V.K.[3], Bradshaw, L.A.[1], Ladipo, J.K.[3], Richards, W.O.[2,3] and Wikswo, J.P., Jr.[1]

[1]Dept. of Physics and Astronomy, Vanderbilt University; [2]Depts. of Surgery, Dept. of Veterans Affairs Medical Ctr. and [3]Vanderbilt University Medical Ctr., Nashville, TN USA

INTRODUCTION

A vector SQUID gradiometer (Conductus Instruments, Inc., San Diego, CA) was used to measure magnetic fields associated with bioelectric sources in the stomach and small intestine of normal, healthy, human volunteers. These measurements are an extension to previous work [1,2,3] in which a single component of the magnetic field was measured. Magnetic recordings associated with gastrointestinal current sources, known as the Magneto-ENteroGram (MENG), contain frequency information about a specific type of activity known as the Basic Electrical Rhythm (BER). In normal subjects, the BER frequency of the stomach is approximately 3 cycles per minute (cpm), and varies in the small intestine from 12 cpm in the duodenum to 8 cpm in the terminal ileum. Certain disease conditions can change BER frequency [4], and it is therefore of interest to localize sources of particular frequencies. One of the advantages of using a vector instrument to measure these magnetic fields is that we are sensitive to current sources of all geometric orientations in the abdomen. One of the challenges of using a single channel vector instrument is to decouple the contributions from multiple sources in the magnetic recordings. Since the multiple current sources in the abdomen are of varying frequency, are spatially distributed, and are generally of different orientations, it follows that at a given location, the vector magnetic fields from each current source should point in different directions. The purpose of this preliminary study is to measure the vector MENG in order to examine the possibility that using all components of the magnetic field could be more suitable than using only a single component for localizing current sources in the gastrointestinal tract of humans.

METHODS

Ten normal, healthy volunteers (6 male, 4 female) were subjected to an overnight fast. Each volunteer was dressed in nonmagnetic clothing and placed in a supine position in a magnetically shielded room. After several deep breaths, respirations were suspended during recording to eliminate motion artifact. Recordings were taken at 10 positions over the abdominal surface using the following geometry: the gradiometer was positioned with the normal channel oriented along the axis normal to the frontal plane of the volunteer; for this channel and the two remaining tangential channels, the gradient was measured along the normal direction with a 10 cm long baseline. Data were collected using a DC to 5 Hz bandwidth and 250 Hz sampling frequency. The cardiac contributions in the vector magnetic recordings were digitally removed using a running boxcar median filter of width one second (250 data points). The frequency content of the magnetic recordings was analyzed by calculating the approximated power spectra, using a maximum entropy method, also referred to as autoregressive (AR) filtering.

To test the hypothesis that distributed sources in the abdomen produce magnetic fields of varying frequency and direction, we examined projections of the vector magnetic recordings in many directions. This was done by taking various linear combinations of the three simultaneous recordings from the vector gradiometers. Let the recorded magnetic field be represented by $\mathbf{B} = B_x\mathbf{i} + B_y\mathbf{j} + B_z\mathbf{k}$, and the arbitrary direction of interest be denoted by the unit vector $\hat{\mathbf{v}} = \sin\theta\cos\varphi\mathbf{i} + \sin\theta\sin\varphi\mathbf{j} + \cos\theta\mathbf{k}$, where θ is the polar angle and φ is the azimuth angle describing the orientation of $\hat{\mathbf{v}}$ in spherical coordinates. The "virtual" magnetometer recording in the direction $\hat{\mathbf{v}}$, denoted by B_v, is thus the dot product $B_v = \mathbf{B}\cdot\hat{\mathbf{v}}$. We calculated the projection B_v in 32 directions (corresponding to the vertices of a rhombic triacontahedron) which represented a uniform discretization of the possible combination of angles in spherical coordinates. We then visually examined the projected magnetic component to search for the minimal and maximal components for each gastric and intestinal frequency.

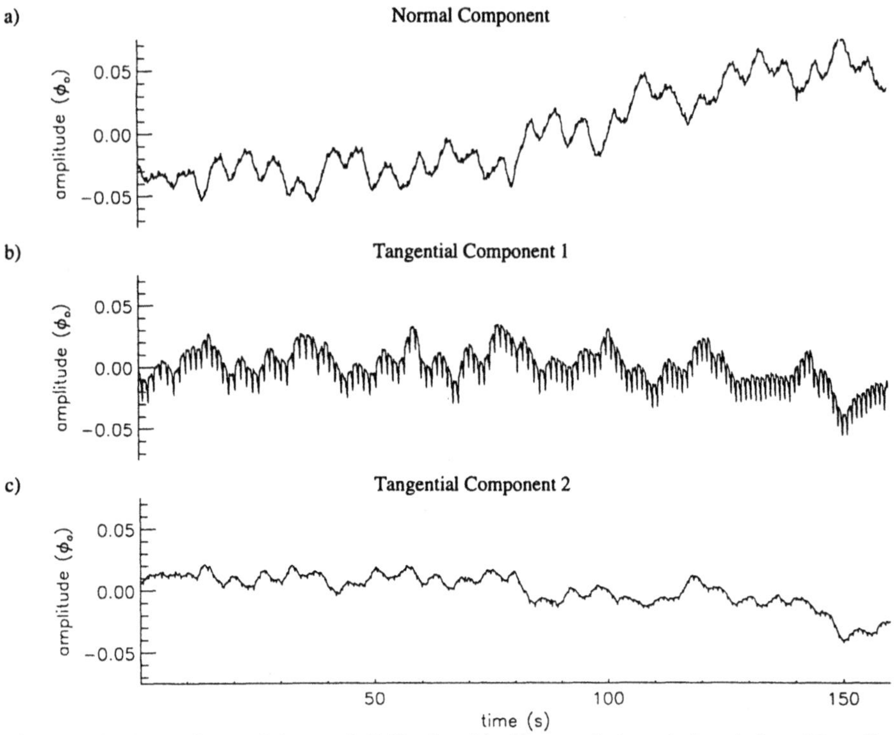

Figure 1. Simultaneously recorded magnetic fields of combined human a,b,c) gastric, intestinal , and b) cardiac electrical activity from three channels of a vector SQUID gradiometer (10 cm baseline). These raw traces were collected from the lower abdominal region (CLQ) using a 250 Hz sampling frequency and DC-5 Hz recording bandwidth. The orthogonal components of these magnetic recordings are oriented a) normally and b,c) tangentially to the frontal plane of the human volunteer.

RESULTS

Magnetic fields associated with gastrointestinal activity were recorded in all ten volunteers. A representative, unprocessed vector magnetic recording is shown in Figure 1. The three components contain contributions from multiple current sources located in the heart, stomach, and small bowel, and also environmental magnetic noise. The measurement was made in the lower abdominal region of a human volunteer. These recordings are typical in that at most locations over the abdomens of all the volunteers, the magnetometer was sensitive to multiple current sources of different BER frequency. The power spectra of the representative vector traces are shown in Figure 2. These spectra show contributions from two primary current sources of frequency 2.7 cycles per minute (cpm) and 9.9 cpm, corresponding to stomach and small bowel, respectively, as well as a weak component at 5 to 6 cpm that may be associated with the colon.

Shown in Figure 3 are projections of the magnetic field that were chosen to optimize the signals from the stomach, the small bowel, and noise. In this particular case, the unit vector projections corresponding to sources in the stomach and intestine appear to be approximately orthogonal The angles are as follows: stomach θ = 37.37°, φ = 108.0°; intestine θ = 37.37°, φ = 36.0°; noise θ = 100.8°, φ = 72.0°. The digitized angles are determined by the rhombic triacontahedron that was used, so even greater separation of the signals may be possible using more optimized unit projections. Note that this procedure is suited for systems such as the gut, where fixed-orientation current sources are turned on and off periodically, with sources at different locations having different frequencies, so

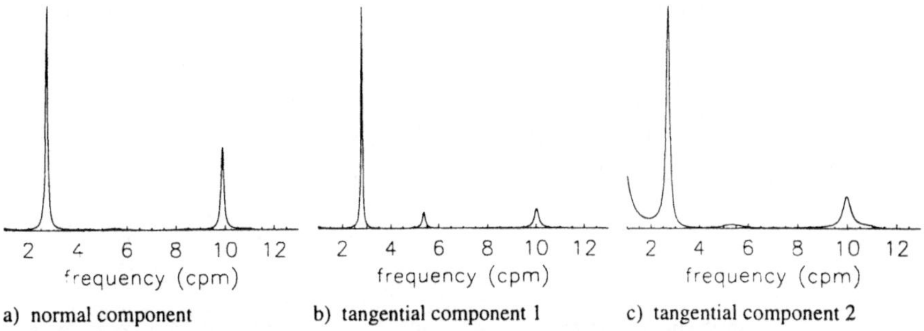

Figure 2. Normalized power spectra estimates (maximum entropy method) of the measured magnetic fields shown in Fig. 1. Frequency is given in cycles per minute (cpm). These spectra show contributions from two primary current sources of frequency 2.7 cycles per minute (cpm) and 9.9 cpm, corresponding to stomach and small bowel, respectively, as well as a weak component at 5 to 6 cpm that may be associated with the colon.

that a single direction is more applicable for a particular measurement location to separate the signals produced by different sources. In other situations, such as the magnetocardiogram, the orientation of the fields changes over a limited range during the cardiac cycle; it may be possible, however, to determine the field component for which the fetal R-wave was optimized over the maternal R-wave.

DISCUSSION

Although little is known about current sources in the gastrointestinal tract, it does not seem likely that the magnetic field vector from a given gastrointestinal source is statically directed in particular direction and varies only in field strength. It is more likely that the field vector corresponding to a particular source varies in direction and magnitude with time. We are presently investigating methods for modeling the recorded magnetic fields to determine the temporal variation in direction and magnitude of the fields corresponding to each source.

In most other biomagnetic measurements, the current sources are highly localized; for example, MEG measures the response of specific regions of the cerebral cortex responding to a particular stimulus, and the MCG measures electrical activity contained within the heart. In these situations, the challenge is to place enough magnetometer sensors in the immediate vicinity of the source. While there are contributions to the field from the unrelated activity of adjacent regions of the brain, the problem is not usually the separation of two competing signal sources. In the gastrointestinal measurements, the sources are distributed over the entire abdomen, and one challenge is to separate the signals that overlap in both space and time. Our power spectral analyses and the vector measurements can be combined into a powerful method to separate the contributions; Figure 3 shows that it is possible to do this without the need for extensive mapping of the magnetic fields. It would be potentially advantageous, in terms of both cost and convenience, if a small number of vector SQUID magnetometers could be used to record the MENG, rather than using a larger array of SQUIDs that measured only one component of the magnetic field. Our preliminary studies demonstrate that the vector MENG can be recorded with high quality, and that the components can contain differing information. The proposed research is directed toward exploring the potential advantages of the vector approach to provide the requisite clinical data at low cost.

REFERENCES

[1] Richards WO, Staton DJ, Golzarian J, Friedman RN, Wikswo JP, Jr.. (Ed) Deecke L. Baumgartner C., Stroink G., and Williamson S.J. Biomagnetism: Fundamental Research and Clinical Applications - Proceedings of the 9th International conference on Biomagnetism.

[2] Garrard CL, Wikswo JP, Jr., Staton DJ, Golzarian J, Gallen C, Richards WO. Gastroenterology, Vol. 106, No. 4, Part 2, A502 (1994).

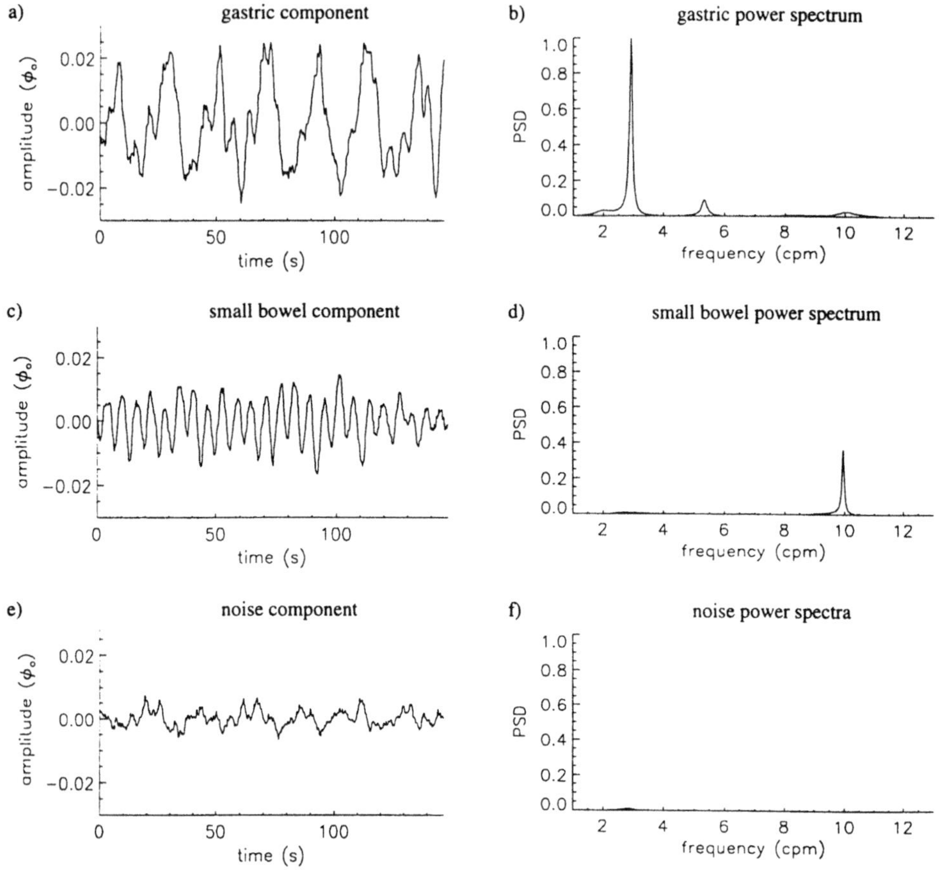

Figure 3. Components of the vector magnetic field and power spectra corresponding to a,b) gastric and c,d) small bowel electrical activity, and e,f) noise in the measurement. Power spectra densities are normalized relative to the gastric power spectra.

[3] Comani S, Basile M, Casciardi S, Del Gratta C, Di Luzio S, Erne SN, Macri M, Neri M, Peresson M, Romani GL. (Ed) Biomagnetism: Clinical Aspects, pp. 639-642, Amsterdam, Elsevier Sci. Publ., 1992.

[4] Richards WO, Garrard LC, Allos SH, Bradshaw LA, Staton DJ, Wikswo JP, Jr.. Annals of Surgery, Vol. 221, No. 6 (1995).

Human Alveolar Macrophage Activation in Healthy Subjects and in COPD-Patients Evaluated by Magnetopneumography

Barth, W.[1], Möller, W.[2], Meier-Sydow, J.[3], Pohlit, W.[1], Stahlhofen, W.[2]

[1]*Institut für Biophysik, J. W.-Goethe Universität, Paul-Ehrlich-Str. 20, D-60596 Frankfurt,*
[2]*GSF - Institut für Inhalationsbiologie, Ingolstädter Landstr. 1, D-85764 Oberschleissheim and*
[3]*Abteilung für Pneumologie, Medizinische Klinik II, D-60590 Frankfurt, Germany*

Introduction

Magnetopneumography (MPG) has been used to study long-term particle clearance of the human lungs in healthy and disease [1, 2]. Ferrimagnetic iron-oxide test particles are inhaled and used as a tracer. The particles have a sufficient stability in order to monitor alveolar clearance over a year without any burden to the subjects. Within hours after particle deposition they are phagocytized by alveolar macrophages (AM). This moves the particles to the intracellular environment, now being under the cellular motile activity. After magnetization the particles form small remanent magnets, all being oriented parallel to the external field. This allows to detect a small macroscopic lung field (LF). The LF decays within ≈20 minutes. Because of the remanent magnets the decay must be due to a stochastic rotation of the particles, finally leading to zero LF. It has been shown that this decay, called *relaxation*, reflects intracellular phagosome motion and thereby is a monitor for the motile activity of the AM-cells. In this study intracellular phagosome motion was investigated in healthy non-smokers (NS) and smokers (S) and in patients with chronic obstructive pulmonary disease (COPD), answering the question whether smoking might affect cellular motility. Further investigations have shown that relaxation can be influenced by inert particle exposure [3]. This study shows that an AM-activation was induced by the deposition of the magnetic test-particles, that depends on cigarette smoking. MPG provides a method to detect AM-activation *in vivo* without medical treatment, like a bronchoalveolar lavage (BAL).

Materials and methods

Preparation of magnetic tracer particles, inhalation and detection in the human lungs

Monodisperse ferrimagnetic iron oxide particles (Fe_3O_4) were used in this study as a tracer. The particles were produced by nebulizing a colloidal non-magnetic iron oxide solution (Fe_2O_3) with a spinning top aerosol generator (STAG) [4]. The particles were reduced into the magnetic form of iron oxide (Fe_3O_4) in a furnace at 800 °C. About 40 breaths under controlled breathing conditions (1 l tidal volume, 250 cm^3/s air flow) were used for the deposition of about 0.5 mg of particles. The particles are spherical with 1.3 µm geometric (2.9 µm aerodynamic) diameter and monodisperse (geometric standard deviation $\sigma_g < 1.1$). Under this standardized conditions the particles are mainly deposited in the alveolar region of the human lungs [2]. The MPG-system for detection of the particles in the human lungs consists of a 40 cm diameter magnet. A capacitor battery (1 mF, 1000 V) is discharged into the magnet, producing a 100 mT, 30 ms magnetic field pulse. This field is sufficient to magnetize the particles and to align the remanent magnetic dipoles parallel to the magnetizing field. All micro-magnets together produce a macroscopic LF (≈100 pT), being detected by a sensitive superconducting SQUID-device. The whole system is located within a magnetically shielded room, where a lower detection limit of 20 µg of Fe_3O_4-particles within the whole lung is achieved.

Relaxation

Relaxation is a monitor for intracellular phagosome motion. Relaxation describes a stochastic particle twisting process and was model as rotational Brownian movement [5], where thermal energy kT was replaced by a cellular energy E_Z. When using spherical particles, the decay should follow an exponential function according to:

$$\frac{B(t)}{B_0} = e^{-t/\tau_r} \qquad \tau_r = \frac{\kappa V \eta}{2E_Z}, \tag{1}$$

where V is the volume of the particles, η is the surrounding viscosity and κ is the rotational shape factor (κ = 6 for spheres). *In vivo* and *in vitro* investigations of relaxation with spherical particles yield deviations from a single exponential behaviour [6]. An initial fast decaying phase (over the first 30 s) was followed by a slower exponential phase lasting up to 20 min. Therefore the decay was fitted by a double exponential function:

$$\frac{B(t)}{B_0} = (1 - A_{r\_s})\exp(-\frac{t}{\tau_f}) + A_{r\_s}\exp(-\frac{t}{\tau_s}), \tag{2}$$

where τ_f and τ_s are the time-constants for the fast and for the slow relaxation phase, respectively, A_{r_s} describes the fraction of relaxation following the slow phase. Further studies have shown that both, the fast and the slow decaying fraction depend on the particle size [6]. For comparison with other studies the time constant for the initial phase, τ_a, was analyzed.

Measurement protocol

9 healthy non-smokers (NS, age: 53 ± 5 years) and 8 healthy smokers (S, 53 ± 9 years) were involved into the study. At the beginning of the study a detailed questionnaire was performed, allowing to exclude subjects with chronic diseases or persons who are under a permanent drug administration. All subjects gave their informed consent according to a protocol approved by the Ethical Committee of the University Hospital in Frankfurt. In detailed lung function tests hyperreactivity of the subjects was excluded. Cigarette smoke consumption was classified in pack-years according to the number of pack's of cigarettes smoked per day times the years of cigarette smoking. NS never smoked during their live span, S are active smokers with a mean smoke-consumption of 45 ± 23 pack-years (PY). Additionally 19 COPD-patients (age: 60 ± 8 years) were investigated, 2 were NS, 3 were S (32 ± 18 PY) and 14 were ex-smokers (ES, 36 ± 26 PY). MPG-measurements (including relaxation) were performed directly (30 min) after inhalation, as well as 1 day, 1 week, 1 month, 4 month and 9 month after inhalation.

Results and discussion

The particles are phagocytized within hours after deposition by AM-cells. Our studies have shown that non-phagocytized (free) particles do not contribute to relaxation nor to particle rotation [7]. The non-rotatable (fixed) fraction of particles was interpreted being non-phagocytized (free) and thereby subtracted from the relaxation curves. Relaxation in humans shows an influence on the time after particle deposition, as demonstrated in Fig. 1a. Directly after inhalation the decay is fastest while it slows down with increasing time after particle deposition. Earlier investigations have shown, that an inert particle deposition can induce an AM-activation with a faster relaxation [3]. From this we follow that the initially faster relaxation is a measure of a non-specific AM-activation due to the deposition of the iron-oxide test-particles. The corresponding change of the relaxation time-constants is shown in fig 1b, where the fast as well as the slow time-constant increase with time after particle deposition and thereby demonstrate the deceleration of the decay. From one week after particle deposition till the end of the study (9 month) relaxation was stable. Therefore the measurements obtained from the period between 1 week and 9 month were used to form mean values for the fast (τ_f), the initial (τ_a) and the slow (τ_s) decaying constant and for the fraction of relaxation showing the slow decay (A_{r_s}). The initially faster decay was described by a AM-activation factor, calculated from the ratio of the time-constant in the stable phase compared to the time-constant measured directly after particle exposure (τ_s/τ_{a_1}). An AM-activation detected by a faster decay of relaxation is reflected by a value > 1.

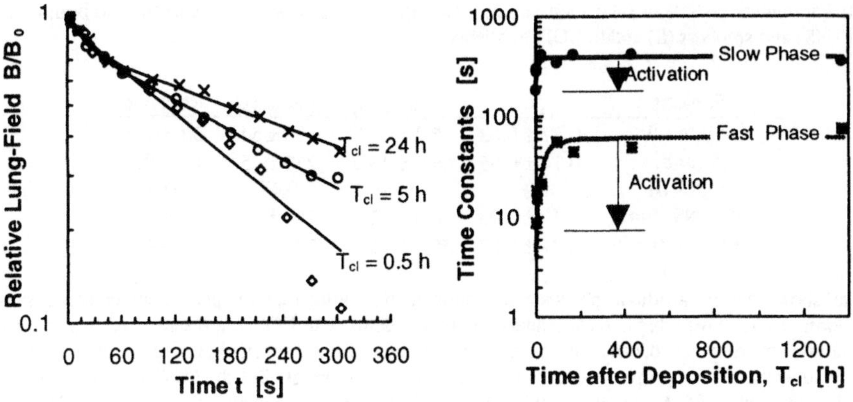

Figure 1: a) *In vivo* relaxation of the ingested magnetic microparticles within macrophage cells depending on the time after particle deposition, T_{cl}. b) Change of the fast and the slow relaxation time constant with time after particle deposition due to an AM-activation

In Table 1 the results for the mean values of relaxation in the stable phase (1 week - 9 month) are shown for healthy subjects and for patients with COPD. Relaxation shows no differences between NS and S, although there seems to be a tendency for a slower decay in smoking subjects. This results are surprising because we expected a strong influence of cigarette smoke on AM-motile activity and thereby on the relaxation process. Two in vitro studies on AM-relaxation in smokers and non-smokers are available. In both studies macrophages were harvested by broncho-alveolar lavage (BAL) from the human lungs and then incubated together with magnetic micro particles. After 24 h relaxation was measured in a cytomagnetometric system. In the study by Valberg et al. [8] relaxation was faster in smokers compared to non-smokers. In this study the cigarette consumption is not clearly reported, but the subjects seem to be younger persons. The faster relaxation was interpreted as a result of the permanent cigarette smoke exposure, causing a permanent AM-activation. In the second study by Gehr et al. [9] relaxation showed no differences between NS and S. The subjects in this study had a mean smoke consumption of 8 PY. In both studies no investigations on non-rotatable (free) particles were performed. These corrections significantly influence the results on relaxation. Our data suggest that even with a very high cigarette smoke consumption of 45 PY cell organelle motions are not influenced. The organism seem to respond to the permanent cigarette smoke exposure with an increased number of AM cells in the lungs. But it is known that several biochemical and immunological factors are raised in S compared to NS. This activation's seem to have no correlation to cytoskeletal activities. Beside the missing corrections on non-rotatable particles the differences found in one of the *in vitro*-studies additionally might be influenced by the BAL-procedure and by incubation. In COPD relaxation is faster only by tendency (τ_a: p=0.06, τ_s: p=0.27).

Cytoskeletal AM-activation measured by relaxation in healthy non-smokers and smokers and in patients with COPD is shown in Table 1. In healthy non-smokers AM-activation is very strong, directly after particle deposition the decay is accelerated twofold compared to the stable phase. A comparable AM-activation is not present in smokers. We must conclude that the missing activation in S either shows an adaptation of the AM-cells to the permanent cigarette smoke exposure or reflects a damage in cellular defense reaction due to cigarette exposure. In COPD-patients AM-activation is also present, but less than strong than in NS. Because most of the COPD-patients are ex-smokers this might reflect to a beginning back-formation of the smokers lung to non-smoking conditions.

Table 1: Parameters (± SD) of relaxation and cellular energy together with AM-activation in healthy non-smokers (NS) and smokers (S) and in COPD-patients

Subjects	A_{r_s}	initial τ_a [s]	slow τ_s [s]	activation
NS (n=9)	0.77 ± 0.03	65.2 ± 17.2	250 ± 54	2.1 ± 1.2
S (n=8)	0.74 ± 0.05	65.9 ± 15.7	272 ± 65	1.0 ± 0.4
p (NS-S)	0.17	0.93	0.44	0.05
COPD(NS) (n=2)	0.79 ± 0.03	56.9 ± 22.9	253 ± 41	1.6 ± 0.9
COPD(ES) (n=14)	0.74 ± 0.07	55.3 ± 15.1	239 ± 66	1.7 ± 0.9

Our data show that intracellular phagosome motion is not influenced by permanent cigarette smoke consumption *in vivo*. From this one can understand that patients with COPD, a disease that mostly results form cigarette smoking over decades of years, show only little differences in phagosome motion compared to healthy subjects. COPD primarily occurs in the airways of the lungs. But the MPG-method is sensitive to the alveolar region of the lungs. In this case the results are in agreement with expectations. The investigations show that MPG is not only a tool to study long-term particle clearance in the human lungs in healthy and disease, but also a tool to monitor motile functions of AM cells *in vivo* in the lungs. AM-motile activity is an important factor in the potency of the cellular defense reaction of the human lungs against foreign substances. Reduced AM-motility might enhance the possibility of lung infections and thereby might be a co-factor for certain lung diseases.

Acknowledgement: This work was supported by the Deutsche Forschungsgemeinschaft (Po 108/7-1) and by the CEC (FI4P-CT95-0026).

References

[1] Cohen, D., S.F. Arai, J.D. Brain (1979), Smoking Impairs Long-Term Dust Clearance from the Lung, *Science*, Vol. 294, 514-517.

[2] Stahlhofen W, Möller W (1993) Behaviour of magnetic micro-particles in the human lung, *Radiation and Environmental Biophysics*, Vol 32, 221-238

[3] Stahlhofen W, Möller W (1994) In vivo and in vitro studies of the cellular defense system of the human lung, *Toxicology Letters*, Vol. 72, 127-136.

[4] Möller W, Stahlhofen W, Roth C (1990) Improved spinning top aerosol generator for the production of high concentrated ferrimagnetic aerosols, *Journal of Aerosol Science*, Suppl. 1, Vol. 21, S435-S438

[5] Nemoto, I. (1982), A Model for Magnetization and Relaxation of Ferrimagnetic Particles in the Lung, *IEEE Transactions on Biomedical Engineering*, Vol. BME-29, 745-752.

[6] Möller W, Guzijan V, Pohlit W, Stahlhofen W, Wenisch T, Wiegand J (1992) Cytomagnetometry with ferrimagnetic micro-particles - influence of particle size and dispersity, *Journal of Aerosol Science*, Vol. 23, Suppl. 1, S519-S522

[7] Barth W, Möller W, Pohlit W, Stahlhofen W, Wiegand, W (1994) Magnetopneumographic estimation of particle phagocytosis in the human lungs, Journal of Aerosol Science, Vol. 25, Suppl. 1, S491-492

[8] Valberg. P.A., W.A. Jensen, R.M. Rose (1990), Cell Organelle Motions in Bronchoalveolar Lavage Macrophages from Smokers and Nonsmokers, *American Review of Respriatory Disease*, Vol. 141, 1272-1279.

[9] Gehr, P., M. Klauser, V. Im Hof (1989), Intracellular Motility of Pulmonary Macrophages from Smokers and Nonsmokers, *Advances in Biomagnetism*, Editors: S.J. Williamson, M. Hoke, G. Stroink, M. Kotani, Plenum Press New York and London, 473 - 476.

Magnetic Markers Linked to Monoclonal Antibodies Allow *in vitro* Evaluation of Cell Population Size

Del Gratta, C.[1,2], Della Penna, S.[2], Battista, P.[3], Di Donato, L.[1], Vitullo, P.[3], Romani, G.L.[1,2], and Di Luzio, S.[1,2]

[1]Institute of Medical Physics; [2]Institute of Advanced Biomedical Technologies; [3]Institute of Human Pathology and Social Medicine, Gabriele D'Annunzio University, Chieti, Italy

INTRODUCTION

Biomagnetic instrumentation has been used for the detection of magnetic particles in the human body. Recently, macroscopic markers were used for specific physiological measurements, like monitoring bowel movements. At a molecular level, liver biosusceptometry allows the measurement of iron concentration in the liver of patients affected by diseases such as hemocromatosis and thalassemia. The motivation of the work to be described hereafter is the assessment of a method utilising biomagnetic instrumentation and microscopic markers, smaller than cellular size, to be directed at specific cell populations. These markers may be linked to target cells via specific antibodies.

Monoclonal antibodies (Mabs) directed against human lymphocytes markers and tumors associated antigens (TAAs) have proven to be extremely useful in defining either subpopulations of lymphocytes or differences between human malignant and normal cell populations. While different diagnostic methods, utilising Mabs, are available in vitro, the use of Mabs for the immunodiagnosis *in vivo* requires their conjugation with a radioactive marker such as ^{125}I. The use of magnetic markers linked to Mabs would result in a relative non-invasiveness for patient studies, and a high sensitivity due to the high field sensitivity of biomagnetic instrumentation.

The assessment of our method requires an interdisciplinary research effort, with contributions from the biological as well as from the physical side; the former concerning cell targeting and bio-compatibility, and the latter concerning the measurement of very weak magnetic fields and their analysis, like for instance the solution of the magnetostatic inverse problem.

Our research programme on this topic ranges from feasibility studies [1] to the final goal of in-vivo experimentation. We report here on an in-vitro evaluation of the specificity of the method [2].

MATERIALS AND METHODS

The biomagnetic sensor is equiped with a rf-SQUID coupled to a second-order gradiometer with 1.5 cm pick-up coil and 7 cm baseline. The peak-to-peak noise in the recording bandwidth of DC-5Hz is 2 pT. The SQUID output is analog filtered with 48 db/oct slope and then digitally converted for digital off-line analysis and data storage at a sampling rate of 100Hz.

Application of an external magnetising field was necessary for signal enhancement. We used a pair of Helmoltz coils placed so that their axis coincided with the gradiometer axis. The radius of the coils was about 60 cm allowing a comparatively large region of homogeneous field for the measurements. The coils were driven by a stable DC power supply unit and it was possible to apply a field of up to 4.5×10^{-4} T in the region of measurement. The fluctuations of the field due to thermal noise in the copper wires raises the overall noise of the experimental set up, resulting in a peak-to-peak noise of 4 pT.

The markers consisted of small beads 4.5 μm in diameter with a ferrite polycristalline core surrounded by a polystyrene shell and are commercially available as a product for cell separation from Dynal Norway AS. The volume magnetic susceptibility was measured by means of our instrumentation and was found to be 1.0 SI units. The collective behaviour of these beads is referred to as superparamagnetic since they feature a high susceptibility without the characteristics of ferromagnetic materials; in particular these beads do not show any hysteresis or saturation, at least in the low field range of our experiments. Moreover, their field scales linearly with concentration, as it was experimentally checked [1].

The field generated by the markers linked to the cells in a preparation was measured by scanning with the sensor an area of about 14 cm × 22 cm over the preparation. The preparation was moved under the sensor on a wooden table monitored by a fiberoptic system allowing the determination of the position along a scan with a precision of 3 mm.

Details of the immunoreactions are described elsewhere [2]. We used cultures of normal human lymphocytes and colon carcinoma cells. Sample preparations were obtained by cytospin and consisted of two circular spots 7.2 mm in diameter per slide. Preparations were exposed to beads linked to appropriate markers or antibodies and afterwards copiously washed to remove unbound beads. Negative control preparations were also used to test the specificity of the reactions.

The model for the forward field calculation, consisted of two thin, uniformly magnetised disks. We considered the field contribution of an infinitesimal element of the disk, assuming that is consisted of a single layer of beads. The total field of a circular disk was obtained by approximating the integral by means of a seven point integration technique. The effect of finite coil dimensions was taken into account by means of the same approximation. This yielded theoretical maps of the mean magnetic field at the gradiometer location in the measurement plane. Before comparing the theoretical maps to the experimental ones, net field maps were obtained by subtracting from each experimental map a map of an analogous preparation, non exposed to the beads. This was done in order to eliminate the diamagnetic contribution of the glass and of the biological substance. The net maps were finally fitted, by means of the least squares method, with the theoretical maps, where the free parameter was the susceptibility of the distribution. The negative control maps were analysed in the same way.

Two further steps must be taken to obtain the number of cells from the estimated mean susceptibility. The comparison of the latter with the susceptibility of a bead yields the number of beads. The number of cells may then be obtained if the mean number of beads attached to a cells is known. We called this figure the targeting ratio (TR). In our case it had to be estimated independently and this was done by visual inspection of the preparations.

RESULTS AND DISCUSSION

In the following two ways of counting the cells in a preparation are compared: 1) the mean number of cells per unit area - obtained by visual inspection - times the area of the preparation; and 2) the total number of beads divided by the TR. The first figure is indicated in the text whereas the second is indicated in the last column of the tables.

Preparations containing lymphocytes:

The measurements were performed in an applied field of 4.63×10^{-4} T. The mean TR value was estimated to be 9.2. The mean number of cells for the three preparations was 68,000, which gives, considering the usual percentatges of T- and B-lymphocytes of 75% and 15%-19% respectively, 51,000 and 10,200-12,900 cells respectively.

Table 1: T- lymphocyte count

N.	Susceptibility	Beads	Target cells
1	0.152 ± 0.008	585,000 ± 88,000	64,000 ± 10,000
2	0.154 ± 0.008	595,000 ± 89,000	65,000 ± 10,000
3	0.117 ± 0.006	450.000 ± 67,000	49,000 ± 7,000

Table 2: B- lymphocyte count

N.	Susceptibility	Beads	Target cells
1	0.050 ± 0.002	191,000 ± 29,000	21,000 ± 3,000
2	0.047 ± 0.002	181,000 ± 27,000	20,000 ± 3,000
3	0.048 ± 0.002	185,000 ± 28,000	20,000 ± 3,000

Preparations containing tumor cells:
The measurements were performed in an applied field of 4.66×10^{-4} T. The mean number of cells for the three preparations was 61,000. The TR value was estimated to be 2.8.

Table 3: colon carcinoma cell count

N.	Susceptibility	Beads	Target cells
1	0.058 ± 0.003	180,000 ± 27,000	64,000 ± 8,000
2	0.041 ± 0.002	130,000 ± 20,000	46,000 ± 6,000
3	0.058 ± 0.003	180,000 ± 27,000	64,000 ± 8,000

In addition to these, a single preparation containing a greater number of the same cells was measured in the same conditions. The number of cells was 195,000, the TR value was estimated to be about 2.7.

Table 4: colon carcinoma cell count

N.	Susceptibility	Beads	Target cells
1	0.19 ± 0.01	610,000 ± 90,000	225,000 ± 35,000

The most important conclusion that can be drawn from this series of experiments is that the specificity requirement was met, both from a qualitative and from a quantitative point of view. From the qualitative point of view, the negative control preparations showed no magnetic signal above the noise level. From a quantitative point of view, when performed in triplicate, results were repeatable; the mismatch between calculated and estimated values of population was low in the preparations of tumor cells, being between 5% and 15% , but was higher in the case of lymphocyte preparations, although the population ratio value was adequate. In this later case, we believe that the direct counting procedure was not accurate. Overall, the population values obtained are acceptable, considering the large errors that are intrinsic to cell counting procedures. Finally, the peroxidase method, conducted in parallel proved to be less specific than the use of Mabs.

The direct counting of the cells was perhaps the weakest part of our procedure; a more precise counting method could be useful only for evaluating the precision of the biomagnetic method. With a more accurate counting the reported errors may be found to be lower. Also, the determination of the targeting ratio may seem awkward; however, it is likely that TR may be fixed once and for all in various experimental situations: this is suggested by the estimation of the TR regarding the tumor cell preparation. Moreover, the knowledge of TR is not mandatory when only qualitative results are needed. The objective of extending this method to in vivo applications, requires improved performances of the instrumentation as well as of the data analysis system. In this case, the marker distribution may be located at larger distances from the sensor and also have an irregular shape. Regarding the former problem, enhanced sensitivity may be obtained with a more sensitive magnetic sensor and a larger magnetising field. Regarding the latter problem, data analysis procedures must be used that do not require a precalculated field pattern. Finally a parallel reasearch should be carried on to assess the biological issues of in vivo marker administration and cell targeting.

The experiments described in this paper were performed with a rather simple and out of date instrumentation, being equipped with a rfSQUID coupled to a wire wound gradiometer. However, the use of a rfSQUID proved to be successful due to its noise properties in the low part of the spectrum. Present day technique instruments equipped with dcSQUIDs featuring a higher 1/f noise level would require the application of a slowly varying magnetising field. The use of multichannel instrumentation could speed up the measurements. Applied magnetising fields of larger intensity and appropriate spatial distribution may improve overall and local sensitivity of the instrumentation. From the point of view of data analysis, more elaborate susceptibility distributions may be retrieved from the data by solving the magnetostatic inverse problem, leading to the imaging of the target population.

ACKNOWLEDGEMENTS

Work partially supported by the National Research Council of Italy (CNR), under the Progetto Finalizzato "A.C.R.O". P.B. is partially supported by MURST-40% funding.

REFERENCES

[1] Macrì, M.-A., Del Gratta, C., Di Donato, L., Di Luzio, S., Romani, G.L., Della Penna, S., and Pasquarelli, A., 1994, Biomagnetic measurements utilising a superparamagnetic marker: a feasibility study, Nuovo Cimento, D16:425-432.

[2] Del Gratta, C., Della Penna, S., Battista, P., Di Donato, L., Vitullo, P., Romani, G.L., and Di Luzio, S., 1995, Detection and counting of specific cell populations by means of magnetic markers linked to monoclonal antibodies, Phys. Med. Biol., 40:671-681.

Iron Quantification by Non-Invasive Biomagnetic Organ Susceptometry

Engelhardt, R., Fischer, R., Nielsen, P. and Gabbe, E.E.

Abteilung Medizinische Biochemie, UKE, University Hamburg, Germany

Introduction:

In the last years (1989 - 1996) SQUID biosusceptometry has been used to assess iron stores in more than 600 patients suspected for iron overload. In the following we report on applications to 3 major clinical problems in iron overload diseases. From more than 1600 biomagnetic measurements in humans, we determined the iron concentration in liver and spleen tissue with respect to hereditary hemochromatosis (I), transfusional hemosiderosis (II) and screening for hereditary hemochromatosis in prospective blood donors (III).

The iron concentration in different organs is an important parameter in the diagnosis and staging of human iron overload diseases such as genetic hemochromatosis (GHC) or transfusional hemosiderosis.

In hemochromatosis, the inborn regulation defect of intestinal iron absorption leads to a progressive iron overload, mainly of the liver. The biochemical parameter such as serum iron (> 200 µg/dl), transferrin saturation (> 60 %), serum ferritin (> 300 µg/l) are typically increased in homozygous patients. Undiagnosed and therefore untreated patients have a high risk to develop severe organ damages such as liver cirrhosis, diabetes and cardiomyopathy, especially when higher liver iron concentrations above 2 mg Fe/g_{liver} are present. The standard therapy is the iron depletion by weekly phlebotomies of 500 ml blood, until normal liver iron concentration (0.1 - 0.5 mg Fe/g_{liver}) is achieved.

Patients with ß-thalassemia major or other transfusion dependent chronic anemias need regular blood transfusions. This therapy also results in a severe accumulation of iron in the body. Although the distribution of iron in non-parenchymal and parenchymal cells of the liver, spleen and other tissue is quite remarkably different from genetic hemochromatosis, the degree of iron-loading in this secondary hemosiderosis is also a factor which is correlated with the expectation of life. A severe siderosis can result in a cardiomyopathy, diabetes or liver cirrhosis. In order to prevent organ damages and prolong life, the daily use of subcutaneous infusions of the iron chelating compound *deferoxamine* (DFO) is recommended in patients under regular blood transfusions. Preliminary data indicate that the oral iron chelator *deferiprone* (DFP, former names L1, CP20) may have similar effectiveness in promoting iron excretion in iron-loaded patients.

Currently, the classical liver biopsy with histologic determination of Prussian blue stainable iron in liver slices and the chemical determination of iron in the liver biopsy specimens is still the standard method in the diagnosis of a clinical relevant iron overload. During the past decade, several methods for non-invasive liver iron quantification (CT, MRI, nuclear resonance scattering, SQUID biomagnetic susceptometry) have been studied mostly in smaller groups of patients. Among this, only MRI and biomagnetic susceptometry seemed to have the principle capability to measure liver iron concentration sensitive enough to be of clinical use.

In the present study, the benefit of non-invasive iron quantification by biomagnetic organ susceptometry (BOS) under clinical conditions was studied in a large group of patients suspected for iron overload diseases.

Methods:

Biomagnetic organ susceptometry was performed by lowering the patient in the known localized magnetic dc-field (B_{max}= 20 mT, gradiometer 1storder) of the Hamburg Biosusceptometer (BTi) with water as the reference medium. The fluxintegral contributions for organs and torso were calculated in advance for certain distances and radii (3 - 25 cm) as ellipsoids or hemispheres (liver and spleen) and as hemispheres or cylinders (torso tissue). The calculated fluxintegrals were fitted over distances and radii and the coefficients were tabulated and stored in the computer for the analyzing of measurements.

In BOS, the magnetic flux change (rf-SQUID voltage output) detected by two 2ndorder gradiometer pickup coils was fitted by tissue (ribs, muscle, fat; near field) and liver or spleen (far field) flux integral contributions simultaneously in a 2-layer model [2]. From bedside sonographic imaging, the actual distance from the respective organ to the torso surface and the organ volumes were determined. The total body iron stores were estimated by multiplying the biomagnetically measured liver and spleen iron concentrations (the major storage organs), with the respective organ volumes obtained by sonography.

Results and discussion:

I. Hereditary hemochromatosis

In hereditary hemochromatosis it is most important to prevent a marked iron overload in the affected individuals before the onset of irreversible clinical symptoms. The gene frequency for hemochromatosis (HFE-gen) in caucasoids is about 10 % and between 0.2 - 0.6 % of the general population are affected in a homozygous (clinical relevant) form. One approach for a screening study for hemochromatosis is to measure liver iron concentration in subjects with pre-known increased parameters of iron metabolism (serum iron, serum ferritin). Due to absolute non-invasive character of the BOS-method, the indication for such a magnetic biopsy can be made very widely. Another way is to check family members of subjects with known homozygous hemochromatosis.

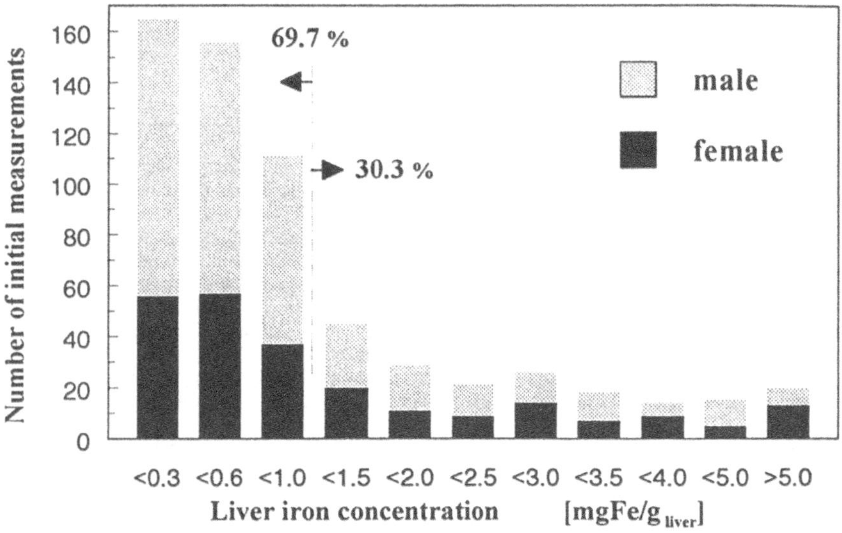

Fig. 1 Frequency distribution of liver iron concentrations measured by SQUID biosusceptometry in patients suspected for iron overload (n = 621).

Under the mentioned approach, more than 600 patients, suspected for hemochromatotic iron overload, were investigated with the biomagnetic iron quantification (Fig. 1). Pre-treated homozygotes or patients with secondary siderosis were excluded from this study. Out of these patients, 86 were diagnosed as homozygotes for hereditary hemochromatosis. As a consequence of this diagnosis, these patients underwent an iron depletion therapy by weekly phlebotomies, until the iron concentration in the liver reaches normal values (0.1 - 0.5 mg Fe/g_{liver}). The effect of phlebotomies (roughly 0.05 $mgFe/g_{liver}$ for a 500 ml venesection) was monitored during the therapy by BOS [3]. After the successful depletion of the excessive iron stores, a maintenance therapy with regular phlebotomies must be performed life-long. For this phase of long-term treatment of hemochromatosis patients, the BOS-method is a valuable tool to avoid reverted iron accumulation.

Following family members of 86 identified homozygous carriers, 73 relatives were found (parents, children) which are obligate heterozygote due to the autosomal recessive mode of inheritance in this disease. The results from BOS demonstrated the significant lower amount of liver siderosis in this group.

In the remaining group, in which no relevant liver siderosis was present, a homozygous idiopathic hemochromatosis was excluded although, one or two blood parameters of iron metabolism were occasionally or permanently increased (serum-iron, serum-ferritin, transferrin-iron saturation). In one part of this group, secondary factors were found (acute liver cell damage, oestrogen medication, vitamin-B12-deficiency etc) which are known to influence parameters of iron metabolism. In the absence of secondary factors, a subgroup of 151 subjects was classified as suspected heterozygous carriers.

Subjects, in which a conclusive diagnosis (homozygous/heterozygous for hemochromatosis) was not possible at the moment, should be followed within the next years in order to detect or exclude changes in the liver iron concentration.

II. Transfusional hemosiderosis

Monitoring of iron chelation therapy is the main purpose of SQUID biosusceptometry in patients with posttransfusional siderosis and/or iron loading anemias (ß-thalassemia major, - intermedia, sickle cell anemia) were monitored by BOS (n = 39, age: 4-33 y). Liver and spleen iron concentrations ranged from 0.4 to 13 $mgFe/g_{liver}$ and 0.3 to 3.7 $mgFe/g_{spleen}$ respectively. The correlation of serum ferritin with the liver iron concentration was very poor in patients under treatment with parenteral DFO-injections (n = 33) and/or oral DFP (19) as shown in Fig. 2. The organ volumes were measured by ultra sound scanning. The total body iron stores were estimated by assuming that 90 % of the iron stores are present in liver and spleen.

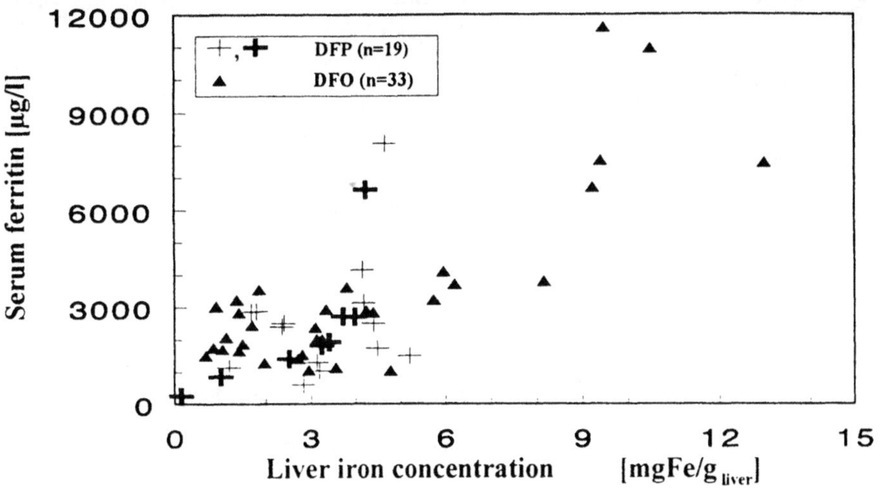

Fig. 2 Serum ferritin versus liver iron concentration measured by BOS in patients with secondary iron overload. Crosses represent patients under treatment with DFP (thin crosses for 0.2 - 2 years, bold crosses for 2- 5.5 years.

Repeated measurements of liver and spleen iron concentrations as well as sonographical determination of liver and spleen volume were available in six patients under DFP treatment for 3 - 15 month. In this group, detailed information was obtained on the whole body iron store (5 - 22 g) and the iron excretion rates (10 - 43 mg/d) for each patient. As indicated by decreasing liver iron concentrations, 4 of 6 subjects showed a negative iron balance with iron excretion rates of 70 - 84 mg/d, which could be assessed by a 2-compartment model, taken into account the blood transfusion rate (influx of iron) and the actual iron stores [4].

The non-invasive measurement of liver (spleen) iron offers some advantages in the therapy control of patients under chelation therapy. From BOS and sonographic imaging, a precise and reliable information on the iron balance is available and an individual iron chelation therapy can be assessed. This is of special value in the clinical evaluation of new oral iron chelators. For the individual patient, the reliable information on his actual iron overload may help to increase the acceptance of chelation therapy and thus may stabilize the compliance.

III. Screening in prospective blood donors

In a study on the prevalence of hereditary hemochromatosis in North-Germany, 2812 prospective blood donors (mean age: 26y, female = 1409, male = 1403) from our local blood transfusion department were screened. The screening was performed in a 3-step filter for elevated serum iron (> 180 µg/dl), serum ferritin (> 100 µg/l, female; > 200 µg/l, male), transferrin saturation (> 50 %) and erythrocyte ferritin (> 50 ag/RBC).

At the 3rd filter level, 63 subjects suspected for iron overload were measured by BOS. The 3rd filter was passed by 20 subjects with iron concentration > 0.5 $mgFe/g_{liver}$.

The final diagnosis was made by taking into account the increased liver iron determined by BOS and the finding of an increased intestinal iron absorption (> 50 %) from a 10 μmol $^{59}Fe(II)$ascorbate test dose. As a result of this study, 7 cases of homozygous hemochromatosis (without clinical symptoms) were found. This confirms the current frequency estimation (0.2 - 0.6 % in Western countries) also for North-Germany. The benefit from the screening by BOS is the improved health care in the presymptomatic state of hemochromatosis. Moreover, these persons suffering sofar only from a disregulation of iron absorption, may be engaged as "super blood donors".

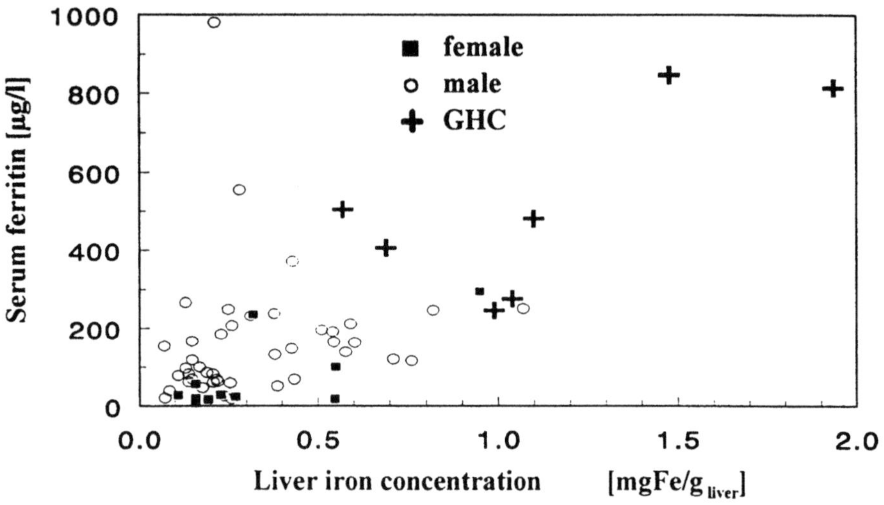

Fig. 3 Serum ferritin versus liver iron concentration in the third-step-filter. The crosses indicate subjects diagnosed as homozygous hemochromatosis.

In conclusion, the Hamburg Biosusceptometer is a dedicated system for non-invasive liver and spleen iron quantification over the whole range of possible iron stores (deficient to overload). For the diagnosis of genetic hemochromatosis, BOS substitutes the conventional liver needle biopsy completely. The method is precise, practicable and convenient for patients from 3 to 80 years and gives the clinicians relevant information for diagnoses, therapy and prognosis in iron overload diseases.

The results of our studies should have a strong impact on the design of the next generation biosusceptometer. It should be optimized for a better spatial resolution and for measurements in the low iron concentration range, because about 70 % of our patients had liver iron concentrations below 1 $mgFe/g_{liver}$ (Fig. 1).

References:

[1] Powell, L.W. In: Brock. J.H., Halliday, J.W., Pippard, M.J. and Powell, L.W. Iron metabolism in health and disease, London, Saunders, 1994.

[2] Fischer, R., Eich, E, Engelhardt, R., Heinrich, H.C., Kessler, M. and Nielsen, P. The calibration problem in liver iron susceptometry . In: Williamson, S.J. et al. (ed), Advances in Biomagnetism, Plenum Press, New York, pp 501-504, 1989.

[3] Fischer, R., Engelhardt, R., Nielsen, P., Gabbe, E.E., Heinrich, H.C., Schmiegel, W.H. and Wurbs, D. Liver iron quantification in the diagnosis and therapy control of iron overload patients. In: Hoke, M. et al. (ed), Biomagnetism: Clinical Aspects, Elsevier, Amsterdam, pp 585-588. 1992.

[4] Nielsen, P., Fischer, R., Engelhardt, R., Tondüry, P., Gabbe, E.E. and Janka, E.E. Liver iron stores in patients with secondary haemosiderosis under iron chelation therapy with deferoxamine or deferiprone. Brit.J. Haematol.1995, 91:827-833.

The Use of Biomagnetic Liver Susceptometry in the Ferrara-Hamburg-Turin Study on Thalassemia

Fischer, R.[1], Piga, A.[2], Tricta, F.[2], Nielsen, P.[1], Engelhardt, R.[1], Garofalo, F.[2], Di Palma, A.[3] and Vullo, C.[3]

[1]Medizinische Biochemie, Universität Hamburg, Germany; [2]Centro Microcitemie, Università di Torino, Italy; [3]Divisione Pediatrica, Arcispedale S. Anna, Ferrara, Italy

Introduction

Thalassemia is a genetic disease which requires lifelong blood transfusions resulting in progressive iron overload. This disease is rare in Northern Europe, but frequent in Mediterranian and Asian countries (e.g. 6000 patients in Italy). Current treatment for iron overload is by subcutaneous administration of the iron chelator deferoxamine (DFO) or in clinical trials, by the oral iron chelator deferiprone (DFP). In some patients where histocompatible donors are available also bone marrow transplantation (BMT) can be used. Most of these transfusion-dependent thalassemic patients have been chelated for more than 14 years at an age of 20. In these patients it has become particularly important to determine as precisely as possible the level of iron stores so as to take into account:

- the individual adjustment of chelating therapy;
- the prevention of the toxic effects of chelation attributed to DFO, mainly in young patients;
- the assessment of compliance with iron chelation therapy, as this is strictly correlated to survival.

Patients and Methods

In the present work we report on measurements of the iron stores by SQUID biomagnetic liver susceptometry (BLS) in 175 thalassemic patients from Northern Italy (Turin and Ferrara) (age: mean 18 y, range 4 - 38 y). Biomagnetic spleen susceptometry (BSS) was also possible in 50 % of the 73 nonsplectomized patients with enlarged organs. The main diagnoses were ß-thalassemia major (n=150), combined sickle cell anemia / ß-thalassemia (n=4), and ex-thalassemia after bone marrow transplantation (BMT) 1 to 7 years ago (n=18).

Measurements took place at 6 instants from December 94 to Oct 95 at the Hamburg Biosusceptometer. The system (Ferritometer: BTi, San Diego) with 2nd order gradiometers (2 channels) and rf-SQUIDS in the nonuniform field of 1st order field coil gradiometers (2 channels) has been described elsewhere [1]. Sonography (positioning, liver/spleen contour and volume estimation) in one patient and biomagnetometry in another were accomplished simultaneously at different places with 2 patients/hour. The patient's position was transfered from one place to the other by laser-cross alignment.

Full data analysis was done by forward modeling of the magnetic flux change from spherical organ/torso geometries as function of detector coil distance. Magnetic volume susceptibilities of liver/spleen (i.e. liver/spleen iron concentration) and body (torso) tissue were fitted to SQUID voltages as function of precalculated fluxintegrals (spherical torso-liver model) [2]. For online analysis the magnetic body susceptibility (χ_{tis}) was derived in 1st order approximation from the patient's body mass index (BMI = height/weight²) as: $\chi_{tis} = (-9.130 + 0.805/[1+226/(BMI-13.3)^2])\cdot 10^{-6}$ [SI] (unpublished data). The liver iron concentration was calculated from a linear fit to SQUID voltages as function of fluxintegrals from a hemispherical liver after subtraction of the magnetic tissue contribution from a cylindrical torso (range: 13 to 31 kg/m², corresponding with $\chi_{tis} = (-9.13 \text{ to } -8.66)\cdot 10^{-6}$ SI).

Clinical evaluation of biomagnetometry liver and spleen iron concentration was made by serum ferritin (SF), total transfused iron input, total infused DFO, mean annual compliance, total chelated iron extrapolated from periodical measurement of urinary iron excretion, and by liver biopsy with histological determination of the hepatic iron score (grading: 1 - 20). Mean annual compliance was calculated as mean annual number of daily DFO-infusions from the total number of DFO-infusions(~ 1.5g DFO/infusion) during the years of chelation therapy done by each patient. As each of these 7 parameters were not known in all patients at the same time a group of patients with ß-thalassemia major (n=36) was selected where a linear cross-correlation could be done.

Results and Discussion

Liver iron concentrations ranged from 170 (normal: 100-400) to 9600 μg/g_{liver} (severe iron overload) with a median value of 1390 μg/g_{liver} and an interquartile range (50%) from 910 to 1990 μg/g_{liver}. Spleen iron concentrations ranged from 280 to 2350 μg/g_{liver} with a meadian value of 1320 μg/g_{liver} and an interquartile range from 990 to 1530 μg/g_{liver}. The ratio of spleen/liver iron concentration was found to be 1.21 ± 0.76. Approximatively, one could apply liver iron concentration values to those spleens which were not measured. At least 75 % of our patients have iron stores well below the levels associated with complications typical in thalassemics (e. g. myocardiopathy, liver fibrosis and cirrhosis, diabetes mellitus, hypothyroidism). Online liver results differed by 5 ± 11 % from more precise results obtained by spherical torso-liver modeling. This was different in BSS where geometry mapping for small-sized spleen ellipsoids had to be done first by sonography in order to apply individual fluxintegrals.

Serum ferritin is the routine biochemical parameter used in the monitoring of individual therapy in thalassemic patients. Although this parameter is strongly influenced by inflammation and tissue damage processes, it is still used to estimate individual body iron stores and to monitor iron mobilization. The limited use of serum ferritin determinations for this purpose was already shown in biomagnetic measurements by Brittenham et al [3] and Nielsen et al. [4] and is demonstrated on a large scale in Fig. 1. Standard logarithmic transformation has been applied to the skewed distributions of serum ferritin and liver iron concentrations. The poor correlation coefficient of r=0.52 indicates that only 27% of serum ferritin values are within the ± 1 s.d. range of the respective regression line. Even in the interquartile range of liver siderosis (910 - 1990 μg/g_{liver}), serum ferritin values ranged from 81 to 5329 μg/l (normal: 35 - 235).

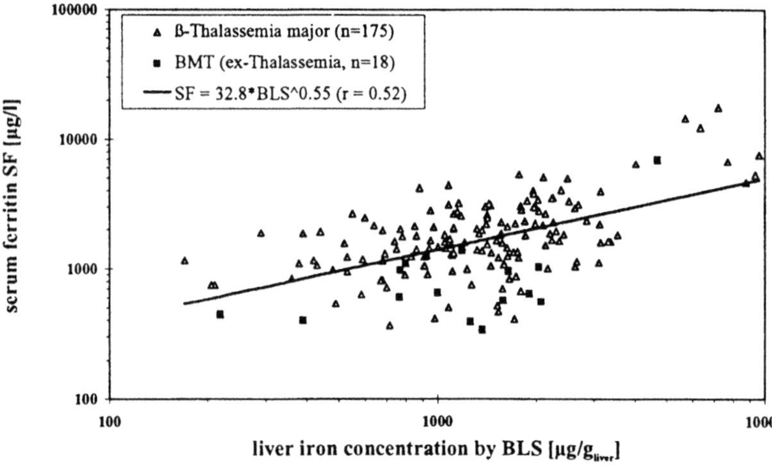

Fig. 1 Correlation between serum ferritin and liver iron concentration in patients with ß-thalassemia major (*triangles*) and in ex-thalassemic patients after bone marrow transplantation (BMT: *squares*). Liver iron was measured by SQUID biomagnetic susceptometry (BLS).

Patients after bone marrow transplantation have still elevated iron stores between 218 and 4700 μg/g_{liver}, whereas the corresponding serum ferritin values are relatively low (see Fig. 1). The reduction of iron stores after BMT needs further investigation. Biomagnetometry may be a very useful tool to study this problem in relation to the other influencing parameters as blood transfusions, DFO infusions, and time after BMT.

In 30 % of the patients (n=50) also liver biopsies were executed. Qualitative estimation of the degree of liver iron siderosis was performed by standard staining techniques in the corresponding histological sections. Hepatic iron scores from 3 to 18 were registered. The interquartile range was between 8 and 13 with a median value of 11. For the linear correlation with liver iron concentrations measured by BLS a correlation coefficient of r = 0.65

was achieved. The physico-chemical measurement of iron concentration in biopsy specimens was lacking from several problems: (i) the size of several biopsies was below 3-5 $mg_{w.w.}$, (ii) a variety of patients are affected by hepatitis with fibrotic and cirrhotic infiltration of liver tissue, (iii) the portion of the biopsy specimen assigned for reliable chemical analysis was not inspected by a pathologist with respect to cirrhosis, substantial scarring, tissue from the capsule, or iron-free foci [5]. Only a poor correlation was achieved with LIC-BLS, which was even worse than with hepatic iron scoring. However, the discrepancy between in-vitro physico-chemical quantitation of biopsies and in-vivo measurement of liver iron stores by SQUID biosusceptometry in thalassemic patients needs further exploration. As the calibration of the Hamburg Biosusceptometer is based on physical procedures only (air-water susceptibility difference of an object with known geometry), and as this is in contrast to findings in genetic hemochromatosis [6], the iron quantitation in the liver biopsy as the accepted reference standard may be question in this kind of patients. Moreover, a recent investigation in autopsy livers from two patients with ß-thalassemia has shown a striking variability of hepatic iron between samples [7].

In the selected group of 36 patients with ß-thalassemia major a linear cross-correlation was performed with all parameters typically used in therapy monitoring of these patients (see Tab. 1). After several years of therapy well chelated patients are in a steady-state with the total amount of transfused blood/iron (tot. trsf. Fe), the total amount of infused deferoxamine (total DFO), and the total amount of chelated iron (tot. chel. Fe). This is indicated by their high correlation coefficients between each two of the respective parameters in Tab. 1. The correlation of liver iron concentration measured by SQUID biosusceptometry (LIC-BLS) with all other parameters was surprising:

- after years of chelation therapy, the total amount of transfused blood (iron) does not determine the degree of iron overload any longer ($r = 0.26$);
- on the other side, the chelation therapy (i. e. total amount of DFO infusions) does influence the liver iron stores to a certain degree ($r = -0.64$);
- urinary iron excretion correlates only poorly with liver iron stores ($r = -0.24$), which can be explained by fecal iron as main excretion pathway in patients kept at high hemoglobin values [8];
- the relative close correlation coefficient with serum ferritin ($r = 0.76$) in contrast to the correlation test with all thalassemic patients ($r = 0.52$!) is explained by the skewed distribution of serum ferritin, where only few patients with high ferritin values determine the slope of the corresponding regression line;
- correlation with grading of liver iron overload by hepatic scoring ($r = 0.68$) was similar to the correlation test in the larger group of biopsied patients ($r = 0.65$!);
- surprisingly, the closest correlation was found for the mean daily number of DFO infusions during the years of individual chelation therapy, i. e. compliance. The relation for compliance as function of liver iron concentration was found as $COMP = 1 - 0.2 \cdot LIC$ [mg Fe/g_{liver}]. Values of compliance better than 0.9 were found only for patients with LIC-BLS < 1000 µg Fe/g_{liver}.

Tab. 1 Linear cross-correlation between clinical parameters used for therapy monitoring in selected patients with ß-thalassemia major (n=36) and iron concentrations measured by biomagnetic liver susceptometry [LIC-BLS](r^* significant at $p < 0.05$).

Parameter	tot. trsf. Fe	total DFO	compliance	tot. chel. Fe	ferritin	hep. score	LIC-BLS
tot. trsf. Fe	1	0.89*	0.14	0.90*	0.18	0.12	-0.26
total DFO	0.89*	1	0.55*	0.82*	0.51*	0.37*	-0.64*
compliance	0.14	0.55*	1	0.12	0.76*	0.66*	**-0.94***
tot. chel. Fe	0.90*	0.82*	0.12	1	0.16	0.17	-0.24
ferritin	0.18	0.51*	0.76*	0.16	1	0.50*	0.76*
hep. score	0.12	0.37*	0.66*	0.17	0.50*	1	0.68*
LIC-BLS	0.26	0.64*	**0.94***	0.24	0.76*	0.68*	1

The survival of patients is strongly correlated to compliance [9, 10], e. g. a mean annual compliance of at least 0.68 DFO infusions per day (~ 1.0 g DFO/d) is necessary to achieve a survival of 95 % beyond an age of 20 years. This corresponds with a liver iron concentration of less than 2000 $\mu g/g_{liver}$. As the accurate determination of compliance is difficult to achieve, annual measurement of liver iron concentration by SQUID biosusceptometry may become a simple and useful method for this purpose. Moreover, BLS may also be used as a reliable method for the cross-check of compliance and thus, may improve the patient's compliance. The closest correlation between compliance and biomagnetometry could be explained by the high specificity and sensitivity of the SQUID biomagnetometry method.

In conclusion, the study has shown that SQUID biomagnetic liver susceptometry can be applied routinely to all kinds of thalassemic patients, even in measurement sessions where a high daily throughput of patients is required. In contrast to standard current methods, more reliable information can be obtained from BLS. Annual assessment of liver iron by BLS could become the method of choice for evaluation of the iron burden in transfusion dependent thalassemia patients and for the assessment of chelating therapy. This study may help to enhance the clinical acceptance of this technology and may help to develop a next instrument generation at one of the thalassemia centers in Italy.

Acknowledgements

We thank our co-workers R. Kongi, I. Zimmermann, C. Albert, S. Hoppe, B. Probst, K. Reller, B. Dresow, E.E. Gabbe (Hamburg), M.R. Gamberini, R. Govoni, M. Sprocati (Ferrara), and U. Bertola, R. Pimazzoni, L. Sacchetti (Torino), who were involved in the measurements and in taking care of the patients with enthusiam. Moreover, we thank our patients from Ferrara and Torino for accepting the challenge to go for a SQUID measurement to Amburgo. We would also like to thank the patient's associations for thalassemia: Associazione del Bambino Talassemico and Red Cross of Torino, who supported this investigation.

References

[1] Paulson, D.N., Fagaly, R.L., Toussaint, R.M., and Fischer, R. Biomagnetic susceptometer with SQUID instrumentation, IEEE Transactions on Magnetics, 1990, MAG-27: 3249-3252.

[2] Fischer, R., Eich, E., Engelhardt, R., Heinrich, H.C., Kessler, M., and Nielsen, P. The calibration problem in liver iron susceptometry, In: Williamson, S.J., et al. Advances in Biomagnetism, New York, Plenum Press, 1990.

[3] Brittenham, G.M, Cohen, A.R., McLaren, C.E., Martin, M.B., Griffith, P.M., Nienhuis, A.W., Young, N.S., Allen, C.J., Farrell, D.E., and Harris, J.W. Hepatic iron stores and plasma ferritin concentration in patients with sickle cell anemia and thalassemia major. American Journal of Hematology, 1993, 42: 81-85.

[4] Nielsen, P., Fischer, R., Engelhardt, R., Tondüry, P. Gabbe, E.E., and Janka, G.E. Liver iron stores in patients with secondary haemochromatosis under iron chelation therapy with deferoxamine or deferiprone, British Journal of Haematology, 1995, 91: 827-833.

[5] Ludwig, J., Batts, K.P., Moyer, T.P., Baldus, W.P., Fairbanks, V.F. Liver biopsy diagnosis of homozygous hemochromatosis: a diagnostic algorithm. Mayo Clinic Proceedings 1993, 68: 263-267

[6] Fischer, R., Engelhardt, R., Nielsen, P., Gabbe, E.E., Heinrich, H.C., Schmiegel, W.H., and Wurbs, D. Liver iron quantification in the diagnosis and therapy control of iron overload patients, In: Hoke, M., et al. Advances in Biomagnetism '91, Amsterdam, Elsevier, 1992.

[7] Ambu, R., Crisponi, G., Sciot, R., VanEyken, P., Parodo, G., Iannelli, S., Marongiu, F., Silvagni, R., Nurchi, V., Costa, V., Faa, G., and Desmet, V.J. Uneven hepatic iron and phosphorous distribution in beta-thalassemia, Journal of Hepatology, 1995, 23: 544-549.

[8] Di Palma, A., Moratelli, A., Tolomelli, P., Giuberti, P., Tenan, R., Fagioli, F., Grandi, L., Toffoli, C., Atti, G., and Vullo, C. Pattern of iron excretion in relation to haemoglobin level and iron load in 8 Haematological patients following the administration of subcutaneous deferioxamine. European Journal of Haematology, 1994, 53: 197-200.

[9] Brittenham, G.M, Griffith, P.M., Nienhuis, A.W., McLaren, Young, N.S., Tucker, E.E., Allen, C.J., Farrell, D.E., and Harris, J.W. Efficacy of deferoxamine in preventing complications of iron overload in patients with thalassemia major. New England Journal of Medicine, 1994, 331: 567-573.

[10] Gabutti, V. and Piga, A. Results of long-term iron chelating therapy. In: Hershko, C. Chelators in Medicine, Basel, Karger, 1996 (in press).

Detection of Biodegradation of Ferrofluid by Relaxation Measurements

Kötitz, R.[1], Trahms, L.[1], Pfefferer, D.[2], Kresse, M.[2], Semmler, W.[2] and Weitschies, W.[2]

[1]*Physikalisch-Technische Bundesanstalt, Berlin, Germany;* [2]*Institut für Diagnostikforschung GmbH, Berlin, Germany*

Introduction

Ferrofluids are colloidal solutions of ferro- or ferrimagnetic particles of typically 10 nm size, coated with a surfactant and dispersed in a carrier liquid. Due to their size these particles are single domain and magnetized to saturation. Ferrofluids are superparamagnetic i.e. gain a very high magnetization in the presence of a magnetizing field. Thus the high sensitivity of SQUID sensors allows the detection of low concentrations of these particles. Fig.1 gives an overview of the components and some relevant parameters of ferrofluids.

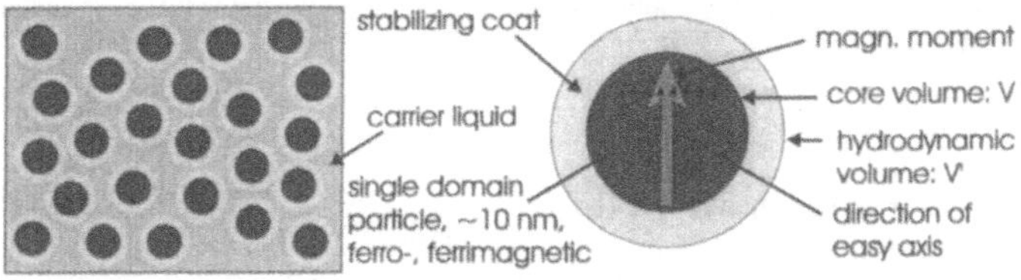

Fig.1 Some basics about ferrofluids

Ferrofluids were developed in the 1960's mainly for industrial applications and are now widely used in a variety of technical applications. During the past years their application in biomedicine has become increasingly important. When bound with specific substances they are commonly used as carriers for drug targeting and for magnetic cell separation. Furthermore ferrofluids are established as contrast agents in MRI investigations.
A new application is their use as magnetic markers for biomagnetic measurements [1,2]. After the injection of dilute ferrofluids into the body it is useful to determine the distribution of magnetic particles in the body tissue. This is usually done in analogy to susceptibility maps where the magnetization of the sample is measured in the presence of a magnetizing field. However biomagnetic susceptibility measurements are often complicated by the presence of diamagnetic tissue or paramagnetic contamination which especially at low ferrofluid concentrations mask the superparamagnetism of the injected particles [2]. Moreover when using SQUID sensors usually the signal hast to be modulated either by sample or sensor movement or variation of the magnetizing field, in order to generate a time dependent signal.
Here we present **M**agneto**R**ela**X**ometry (MRX) as a novel measurement technique which is based on the relaxation processes in ferrofluids. The principles of MRX are shown in Table 1.

Table 1: Principles of MRX

0. incorporation of fine magnetic particles into the object under study
1. magnetization of the object for a time of t_{mag} in a field of H_{mag}
2. switching off the magnetizing field, H=0
3. measurement of the relaxation of the sample magnetization, M=M(t)

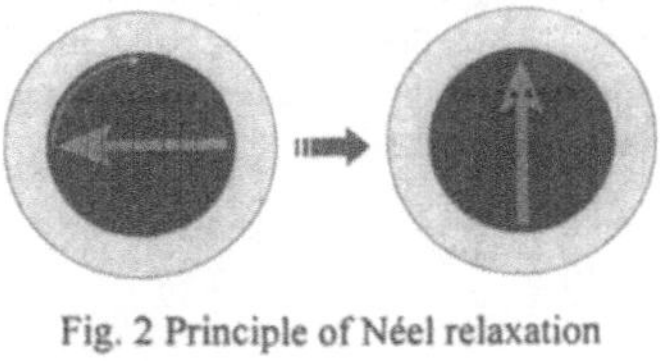

Fig. 2 Principle of Néel relaxation

Here we focus on the Néel relaxation mechanism which is characterized by the rotation of the internal magnetization vector inside the particle [3]. This is schematically shown in Fig. 2. The Néel relaxation process is characterized by a time constant of $\tau_N = \tau_0 \exp(KV/k_BT)$, where τ_0 is usually taken as 10^{-9} s, V is the core volume of the magnetic particle and K the anisotropy constant of the bulk material, k_B Boltzmann's constant and T temperature.

Due to the distribution of particle sizes, the Néel relaxation of magnetic systems after magnetization in a weak field for a finite time obeys a universal time dependence[5]:

$$M \sim \log(1+t_{mag}/t) \qquad (1)$$

So we face a measurement signal which is determined only by the magnetization parameters. Thus signal processing methods like triggering, averaging and curve fitting can easily be adapted to this universal time dependence to increase the signal to noise ratio.
Since the measurement signal has an inherent time dependence SQUID detection is possible without modulation of the sample signal. In contrast to susceptibility investigations MRX measurements are not affected by para- or diamagnetic components within the body [2] and therefore yield a signal specific only for the applied marker.
In a simple phantom experiment it was shown that the spatial distribution of ferrofluid can be reconstructed by measuring its Néel-relaxation [1]. In this paper we report the use of MRX for studying the accumulation and degradation of ferrofluids in biological tissue.

Methods

We used 2 water based ferrofluids consisting of magnetite cores of about 10 nm with different coatings (dextran and silan) resulting in hydrodynamic sizes of 40 - 60 nm. The measurements were performed with the PTB 37 channel dc SQUID Magnetometer [4] in the Berlin Magnetically Shielded Room (BMSR). The application of a magnetizing field inside the shielded chamber causes problems since (i) the shielded room must not be magnetized permanently and (ii) the ferromagnetic material of the chamber walls generates relaxation signals that interfere with the sample signal. Therefore a specially designed twin coil was used which consists of a magnetizing and a compensation coil connected in series opposition. The magnetic fields of the individual coils add at the location of the sample and cancel at a distance. The design is depicted in Fig. 3.

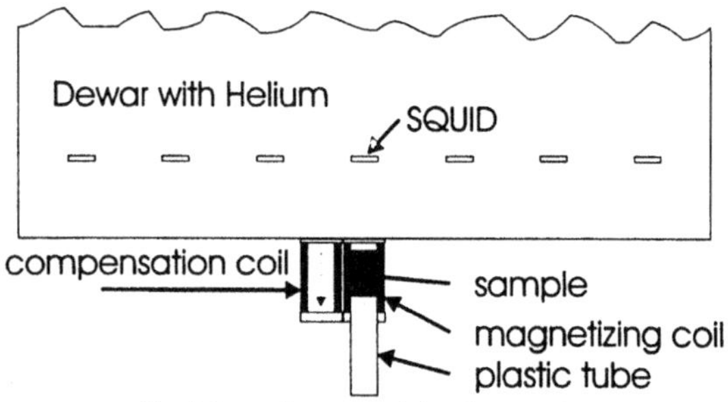

Fig. 3 Setup for magnetizing the samples

The current feeding the coil is software controlled. All samples were magnetized for 10 s in a field of 5 mT. Relaxation was measured between $t_0 = 200$ ms and $t_e = 90$ s after switching off the field. An electronic gradiometer cancels artifacts, that are still generated by the relaxation of the chamber wall.

Results

In order to estimate the sensitivity of the MRX technique different dosages of dextran coated magnetite (100, 50, 10 and 2 µmol Fe/kg body weight, and for control, 0 µmol Fe/kg body weight) were injected into rats (Han-Wistar SPF, 100 g, male). Samples of liver and spleen were harvested 4 h post injection (p.i.) and investigated by MRX. Fig. 4 shows the relaxation signal of a liver sample observed for the lowest applied dosage of 2 µmol Fe/kg. The amplitude of the relaxation signal is about 3 pT / 515 mg. Note that this is the result of a single shot, filtered by a low pass of 0.5 Hz.

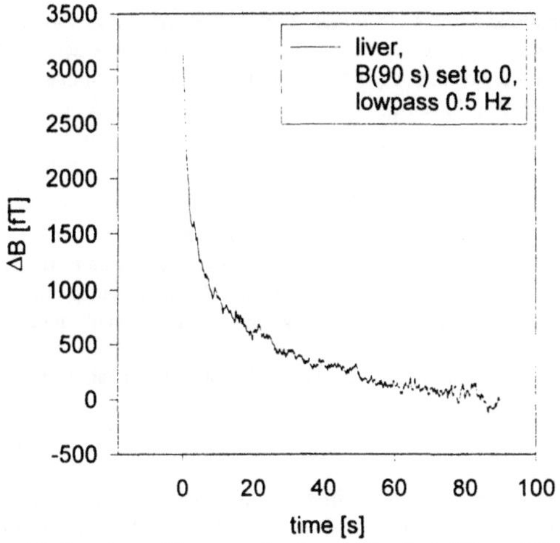

Fig. 4 Relaxation of liver sample after application of 2 μmol Fe/kg ferrofluid

There was no signal detectable for reference samples of untreated animals. This confirms the specificity of the relaxation signal which is independent of the natural iron background of the organs under investigation. For comparison the relaxation amplitude ΔB was reduced to the mass of the samples. As expected the time dependence of the relaxation was similar for all samples containing ferrofluid, according to eq. (1).

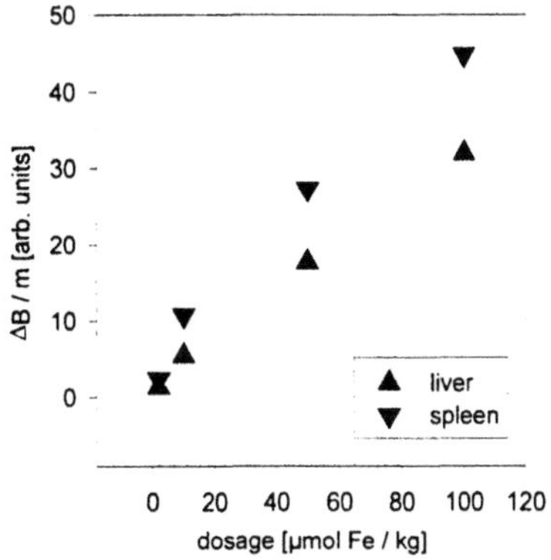

Fig.5 Dependence of reduced relaxation amplitude on applied dosage ferrofluid

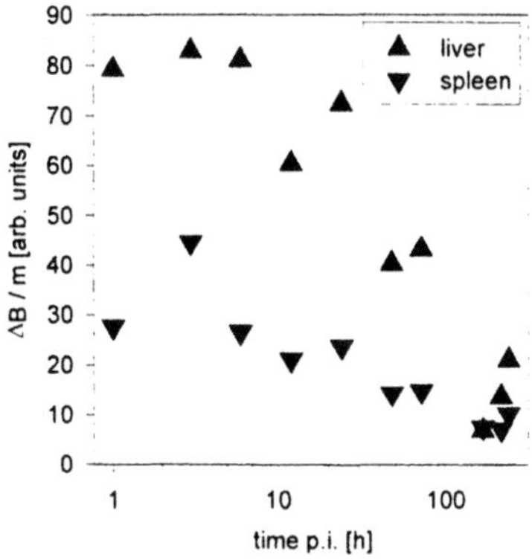

Fig. 6 Dependence of reduced relaxation amplitude on time after application of 20 μmol Fe / kg ferrofluid

Fig. 5 shows that the observed reduced relaxation amplitudes $\Delta B = B(t_0) - B(t_e)$ in both organs were proportional to the applied dosages. Since it can be assumed that the relaxation amplitude is proportional to the amount of particles in the sample it is a measure for the accumulation of ferrofluid in the particular organ.

In order to study the biodegradation of the ferrofluid particles, samples of rat liver and spleen from 1 h p.i. to 10 d p.i. were investigated by MRX. The injected ferrofluid consisted of a suspension of silan coated magnetite. The given dosage was 20 μmol Fe / kg body weight. The result is depicted in Fig. 6.

In both organs the maximum relaxation signal was found 3 h after injection. For later time points the relaxation signals decrease. This indicates a biodegradation with a time constant of about 3 days. It is similar for both organs.

However the relaxation amplitude and hence the amount of particles accumulated in the different organs varies for particles with different coatings. Silan coated particles accumulate to a higher extend in the liver than in the spleen whereas the spleen accumulates more dextran coated particles per gram of organ than the liver. This is possibly due to the different surface properties of the coating.

Discussion

The MRX method presented here is a novel tool for the highly sensitive determination of the accumulation, organ distribution and biodegradation of injected ferrofluids in biological tissue. Due to the specificity of the relaxation signal there is no interference with diamagnetic tissue susceptibilty or paramagnetic components. Consequently MRX is not influenced by the high natural iron background of the organs under investigation. The use of radioactive markers is avoided.
The application of the MRX technique for an *in vivo* study will be a further challenge.

References

[1] Kötitz, R., Trahms, L., Koch, H., and Weitschies, W. In: Baumgartner, C., Deecke, L., Stroink, G., and Williamson, S.J. Biomagnetism: Fundamental Research and Clinical Applications, Amsterdam, Elsevier Science, 1995, 785-788.
[2] Thomas, I.M., and Friedman, R.N. In: Baumgartner, C., Deecke, L., Stroink, G., and Williamson, S.J. Biomagnetism: Fundamental Research and Clinical Applications, Amsterdam, Elsevier Science, 1995, 809-813.
[3] Kötitz, R., Fannin, P.C., and Trahms, L., J. Mag. Magn. Mat., 1995, 149:42-46.
[4] Koch, H., Cantor, R., Drung, D., Erné, S.N., Matthies, K.P., Peters, M., Ryhänen, T., Scheer H.J., and Hahlbohm, H.D. IEEE Trans. Magn., 1991, MAG-27: 2793.
[5] Berkov, D.V., and Kötitz, R. Journal of Physics C, accepted for publication.

Distribution and Biotransformation Dynamics of Magnetic Contrast Agents in Organism on Evidence Derived from SQUID-Magnetometry

Mikhailik, O.M.[1], Pankratov, Yu.V.[1], Dudchenko, A.K.[1], Sosnitzky, V.N.[2], Budnik, N.N.[2], Minov, Yu.D.[2], Sutkovoj, P.I.[2] and Bakai, E.A.[1]

[1]Research SONAR Center for Biotechnical Systems, Ukrainian Acad. Sci., Kiev, Ukraine;
[2]Institute of Cybernetics, Ukrainian Acad. Sci., Kiev, Ukraine

Introduction

The development of magnetic contrast agents is the important and promising field of ferromagnetics use in medicine [1]. In order to outline contrast effect it is possible to apply different techniques, such as MRI and SQUID-magnetometry techniques. SQUID-magnetometry could permit accurate noninvasive localization of contrast agents in biological tissues (Magnetic Source Imaging) [2]. Storage of magnetic particles in tissue changes it magnetic susceptibility which becomes evident when measuring magnetic signal. In MRI the contrast effect is based on the effect of proton relaxation time variation by their interaction with magnetic field generated by particles. Thus the important advantage of the high sensitive magnetometry method as compared to that of nuclear magnetic resonance is the direct coupling of measured magnetic field with particle distribution in space. The lack of qualitative and the absence of quantitative data about the aspects associated with the finely dispersed ferromagnetics injection into organism hampers their use as contrast agents for magnetic tomography. The aim of the present study is the investigation of distribution and biotransformation dynamics of finely dispersed iron and magnetite as a base for magnetic contrast agents in organism of experimental animals by supersensitive magnetometry when coupled with low temperature ESR spectroscopy.

Metods

Finely dispersed iron (FDI) or magnetite suspensions in physiological solution were injected into the eye sine of mice at a dose of 200 mg/kg per animal body weight. The mean diameter of particles with different surface structure was about 80 mm. The surface of particles was stabilised via plasmocondevsation of γ aminopropiltriethoxysilane (sample I), graft polimerization of 2-hydroxyethyl methacrylate, acrylamide and N,N'-methylen *bis*-acrylamide (1:1:0.1) (Specimen II) and divinylbenzene (Specimen III). The magnetic field maps have been registered 15 mm over the mice. The measurements were made with single channel SQUID-magnetometer under the following parameters: sensitivity - 30 $fT/Hz^{1/2}$, frequency band 0.1-100 Hz. The ESR spectra have been obtained on a EPA-10 mini spectrometer (St. Petersburg Instruments, Russia) at temperatures varying from 100 to 200 K and a super high frequency of 9.30 GHz. Metallic iron concentration has been quantifed from the signals of magnetic particles aggregates (g-factors 5-6 and 2-3 and line width 2000-5000 Oe)[3], ferritin concentration - from the signals with g-factors 4.0-4.5 and 2.0-2.5 and line width 300-800 Oe. Low molecular iron concentration has been determined from iron nitrosyl complexes spectrum (g-factor 2.03) after tissue treatment with nitrite. Ferromagnetic signals have been calibrated against ferromagnetic suspensions while paramagnetic ones - with a CuEDTA standard solution and ferritin ESR signals - against Mossbauer data. The Mossbauer spectra have been registered on a conventional constant acceleration NP255 spectrometer with a 150 mCi ^{57}Co source in a chromium matrix.

Results and discussion

Experiments with animals proved the method of high-sensitive magnetometry to be able to register the picture of magnetic field formed by finely dispersed iron distributed in animal organisms (Fig.1A). Registered outside of the body magnetic field is determined mainly not only by concentrations of exogenic magnetic particles but also by their aggregative state. According to our model of ferromagnetic resonance in the aggregates of small

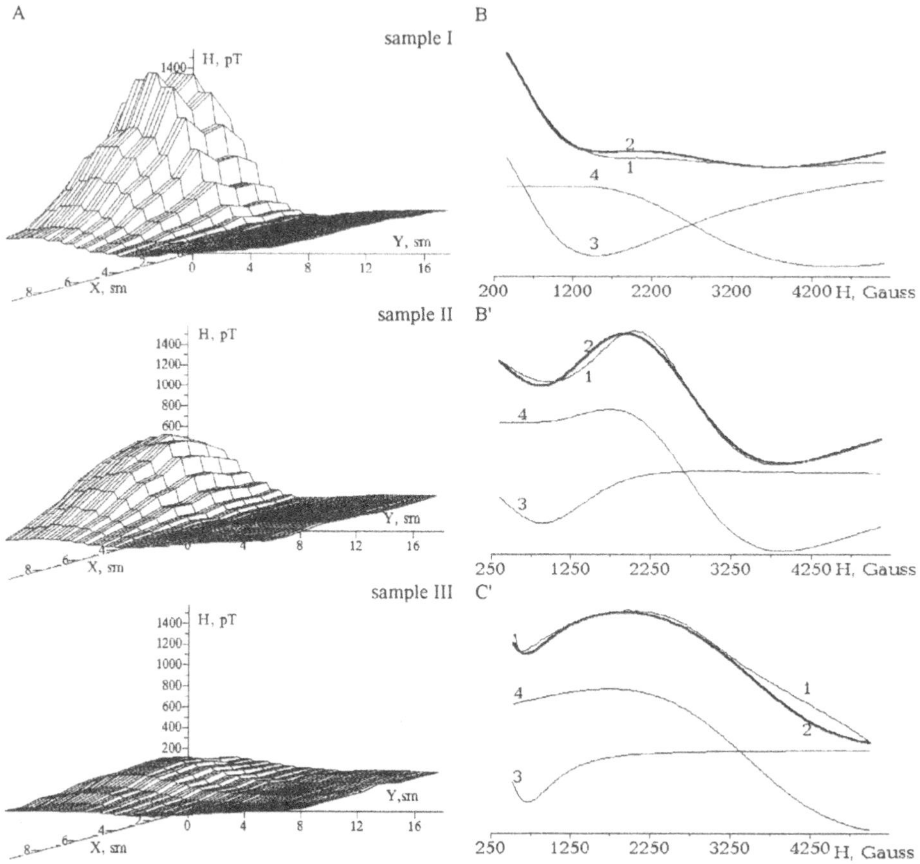

Fig. 1. Magnetic field maps (A) 15 mm over the mice bodies 5 min after intravenous injection of FDI and ESR spectra (B) of suspensions of FDI I,II,III (1 - experimental ESR data; 2 - the sum of curves corresponding to chain (3), spherical (4) aggregates, "flat" layers (5) of metallic iron particles, frequency 9.30 GHz, microwave power 5 mW, modulation amplitude 10 G, T=150K).

magnetic particles the ESR signals corresponding to finely dispersed magnetic particles have been represented (with an error of 2 - 10%) as the sum of the spectra for particle aggregates of three types: chain, spherical and "flat" layers of particles sorbed on the surface of tissues [3] (Fig.1B).

The type of aggregates is responsible for a g-factor value whereas their width is dependent on the thickness of a non magnetic layer on the particle surface. The field created by spherical aggregates is much less then the field generated by the chain ones because of random orientation of magnetic moments of particles in spherical aggregates. That is why the same dose injected FDI with greater relative amount chain aggregates generates the higher magnetic field outside the body (Fig. 1).

Using our model of ferromagnetic resonance in the aggregates of small magnetic particles one can obtain quantitative data on content of metallic phase in each form of aggregates. In our experiments, concentration of iron particles in animal organs was 0.1-0.3 mg/ml while magnetic field strength was in a range of 200-1500 pT without external magnetizing field. In experiments on MRI contrasting [4] accumulation of magnetite

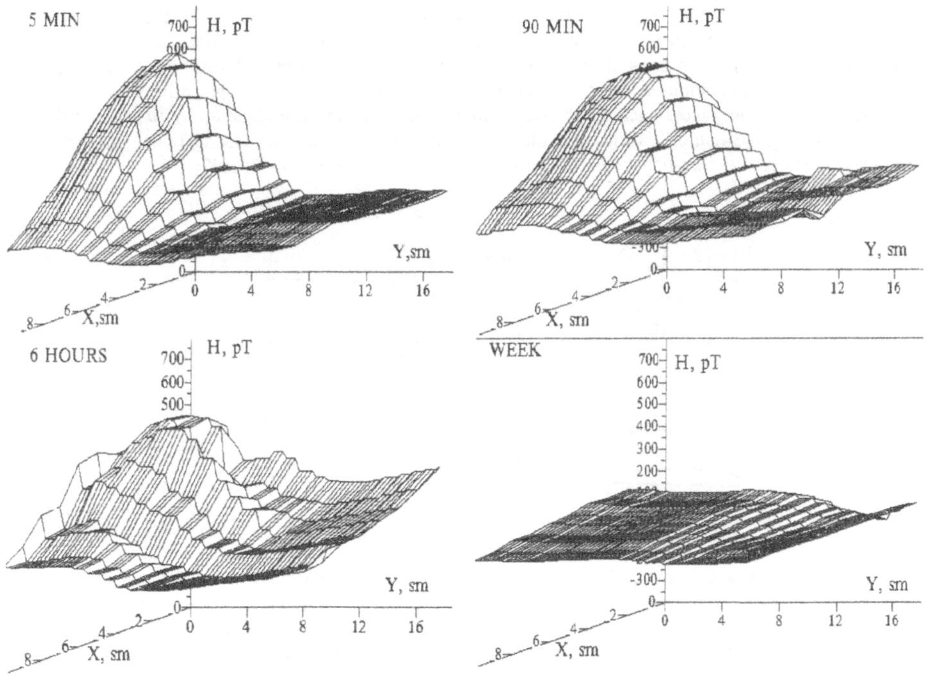

Fig. 2. Magnetic field maps 15 mm over the mice bodies in different times after intravenous injection of FDI suspension in a dose of 200 mg per kg body weight, sample II.

microspheres with concentration 0.13 mg/ml was shown to reduce the T2 relaxation time of liver tissue by 30%. One could believe that SQUID magnetometry could be a real competitor to NMR tomography in detecting of small amounts of ferromagnetic contrast agents.

Our experiments with animals proved the method of high-sensitive magnetometry to be able to register the picture of magnetic field formed by finely dispersed iron and magnetite distributed in animal organisms. It allows also to trace the dynamics of their redistribution even in organism of small animals (mice) (Fig.1). After a week magnetic field over the body decreases almost to the background picture thus supporting the biodegradation of ferromagnetic particles in the organism (Fig.2). ESR results also have verified the rather fast biodegradation of finely dispersed metal iron in organism. In all tissues under investigations a far greater exogenic iron fraction has been detected in ferritin and only a small one (not more than 10%) in 'free' iron pool and a still lesser fraction (not more than 1%) in transferrin. The iron content in ferritin increased approximately 2-10 times as compared to that of check test whereas the concentration of low molecular iron complexes only 1.5 - 3.5 times. Hence any detrimental effects that may be exerted by FDI on a living body due to intensification of free radical oxidation processes are limited by the time of biotransformation.

We have studied also the dependence of the ESR spectra of ferritin in tissues on orientation in a magnetic field after sample freezing in rather weak magnetic field (250 Gauss) (Fig. 3). The variations in the position and shape of lines in ferritin spectra upon rotation of samples about a magnetic field suggest an anisotropic distribution of magnetic moments in the sample volume. Consequently one could believe that ferritin molecules with high iron content can form ordered aggregate structures displaying strong magnetic anisotropy and therefore create rather strong local magnetic field. The local magnetic field of ferritin aggregates have been estimated from experimental ESR spectra using approaches of our model for ferromagnetic resonance in magnetic particles aggregates and was shown to exceed 1000 Oe. The matter is that the shift of low-field component (L) with respect to high-field component (H) may be represented as shift due to the local magnetic field while maximum position and line width of high-field component should be accounted for magnetic anisotropy. Ferritin molecules display magnetic anisotropy, they can be in magnetically ordered or superparamagnetic state depending on temperature and degree of saturation with iron [6]. So, one could suggest that magnetic field generated by ferritin, as in the case of any magnetic particles, is determined mainly not only by concentrations but also by aggregative state and magnetic ordering. These circumstances should be taken into account on iron quantification by tissue susceptometry. Evidently, it the aggregation of ferritin molecules that is the cause of nonlinear dependence of magnetic biopsy data versus liver iron concentration observed in patients with heavy liver iron overload [7].

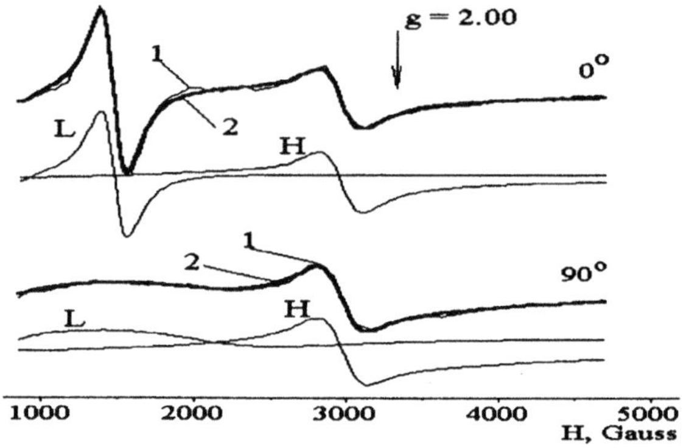

Fig.3. ESR spectra performed on mice liver sample rapidly frozen in a magnetic field of 250 Oe depending on the angle between magnetic field vectors upon sample freezing and spectrum registration moments: (1), thin lines - experimental data; (2), thick lines - the sum of curves corresponding to low-field (L) and high-field (H) components; frequency 9.30 GHz, microwave power 50 mW, T=150 K, modulation amplitude 10 G.

Acknowledgments

This work was supported by the Ukrainian Academy of Sciences as well as by Ukrainian State Committee for Sciences and Technology.

References

[1] Weissleder. R. . Elizondo, G., Wittenberg, J. et al. Ultrasmall superparamagnetic iron oxide: characterization of a new class of contrast agents for MR imagine, Radiology, 1990, 175: 489-93.

[2] Thomas, I.M., Ma, Y.P., Tan, S. et al. Spatial resolution and sensitivity of magnetic susceptibility imaging, IEEE Trans. App. Superconductivity, 1993, 3: 1937-1940.

[3] Mikhailik, O.M., Pankratov, Yu.V. and Bakai, E.A. Biotransformation of intravenously injected finely dispersed iron powders, J. Magn. Magn. Mater., 1993, 122: 379-382.

[4] Majumdar, S., Zoghbi, S.S., Gore, J.C. Pharmacokinetics of superparamagnetic iron-oxide MR contrast agents in the rat, Invest.Radiol.,1990, 25: 771-7.

[5] Smith, A.G., Carthew, P., Francis, J.E. et al. Characterization and accumulation of ferritin in hepatocyte nuclei of mice with iron overload, Hepatology, 1990, 12: 1399-1405.

[6] Kaufman K.S, Papaefthymiou G.C., Frankel R.B., Rosenthal A. Nature of iron deposits on the cardiac walls in β-thalassemia by mossbauer spectroscopy, Bioch.Biophys.Acta, 1980, 629:522-529.

[7] Fisher, R., Engelhardt, R., Nielsen, P. et al. Liver iron quantification in the diagnosis and therapy control of iron overload patients, In: Hoke, M. et al. Biomagnetism: clinical aspects, Amsterdam, Elsvier Science Publishers, 1992.

Human Alveolar Long-Term Clearance in Healthy and Disease Evaluated by Magnetopneumography

Möller, W.[1], Barth, W.[2], Meier-Sydow, J.[3], Pohlit, W.[2], Siekmeier, R.[3] and Stahlhofen, W.[1]

[1]GSF-Institut für Inhalationsbiologie, D-85758 Oberschleissheim; [2]J.-W.-Goethe Universität, Institut für Biophysik, D-60596 Frankfurt, Germany; [3]Medizinische Klinik II, Abt. für Pneumologie, Universität Frankfurt, D-60590 Frankfurt, Germany

Introduction

Magnetopneumography (MPG) was used to investigate alveolar long-term clearance of the human lungs in healthy and disease. Further investigations have shown that this method can be used to record alveolar clearance of the human lungs non-invasively over a period of more than one year [1, 2]. The method of magnetopneumography prevents ethical objections when doing the investigations on humans. Another method for the investigation of alveolar clearance is the radioactive labeling technique. But radionuclides with long half-lives (≈0.5 year) are necessary for alveolar clearance studies. Therefore the radioactive labeling technique is only applicable in animal studies. Further studies have shown that alveolar clearance might be impaired by cigarette smoking. Therefore the data were analyzed with respect to smoking conditions. Additionally patients with chronic obstructive lung disease (COPD), sarcoidosis (SARK) and idiopathic lung fibrosis (FIBR) were investigated. This study gives new information on disturbed alveolar defense mechanisms, that might be a co-factor in the formation of the disease.

Methods

0.5 mg of spherical monodisperse ferrimagnetic iron-oxide particles (Fe_3O_4, 2.9 μm aerodynamic, 1.35 μm geometric diameter, $\sigma_g < 1.1$) were deposited in the lungs by voluntary inhalation. The particles are produced by a Spinning Top Aerosol Generator (STAG) [3]. Particles with 3 μm aerodynamic diameter are deposited mainly in the alveolar region of the lungs. Therefore no or only little fast clearance (within one day) was expected. After deposition the particles were magnetized in an external 100 mT, 30 ms field pulse in homogeneous field technique [2]. All particles together produce a weak remanent magnetic field (rmf), being detected by a SQUID-device. The whole MPG-system is located within a magnetically shielded room, allowing a lower detection limit of 20 μg magnetite in the whole lungs. Before inhalation the natural ferromagnetic contamination of the lung was recorded for each subject and subtracted from all further measurements. For statistical analysis the groups were compared with a ttest, using SAS-software package.

Subjects

A group of younger (age 20 - 39 years) and a group of older (age 40 - 65 years) healthy subjects were involved in the study, both being subdivided into never-smokers (NS) and asymptomatic smokers (S). The mean age and the cigarette smoke consumption are reported in Table 1. Cigarette smoke consumption was classified in pack-years (PY) according to the number of pack's of cigarettes smoked per day times the years of cigarette smoking. At the beginning of the study lung function tests and a detailed questionnaire were performed, allowing to exclude subjects with chronic diseases or persons who are under a permanent drug administration. All subjects gave their informed consent according to a protocol approved by the Ethical Committee of the University Hospital in Frankfurt. Among COPD-patients more than 50 % are ex-smokers (ES). MPG-measurements were performed directly (30 min) as well as 2 days, 1 week, 1 month, 5 months and 9 months after particle inhalation.

Table 1: Age and cigarette smoke consumption (in pack-years, PY) of the subjects involved into the MPG-study on alveolar long-term clearance

Group	N	Age [Yrs]	+/-SD	PY	+/-SD
NS (20-39)	10	29	3	0.0	0.0
S (20-39)	9	31	4	11.7	10.9
NS (40-65)	9	54	5	0.0	0.0
S (40-65)	9	53	9	44.9	22.6
COPD (NS)	1	53		0.0	0.0
COPD (S)	3	61	11	32.0	18
COPD (ES)	6	61	8	42.0	25.6
FIBR (NS)	5	49	13	0.0	0.0
FIBR (S)	7	49	17	14.7	6.4
SARK (NS)	10	46	13	0.0	0.0
SARK (S)	5	43	8	18.3	12.9

Results and discussion

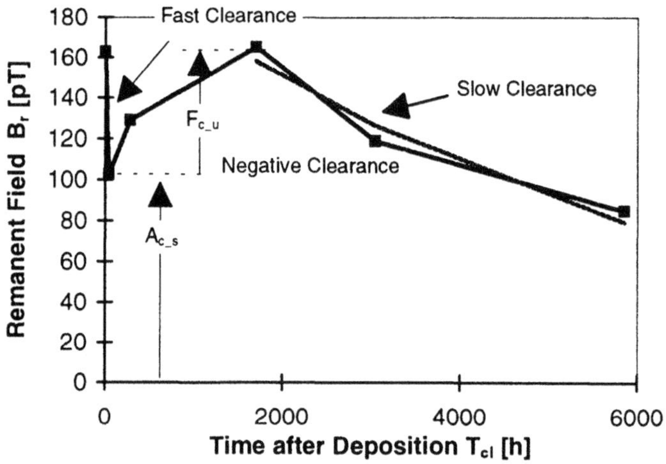

Figure 1: Typical course of the rmf of a healthy subject after deposition of 1.3 μm magnetic tracer particles.

A typical course of the rmf after particle deposition in the human lungs is shown in Fig. 1. We can clearly distinguish **three** phases:

i) a *fast clearance* phase occurring within the first day,

ii) a phase with an increase of the signal, called *'negative' clearance*, happening between day 2 and the first two months, and

iii) the *slow clearance* phase, beginning after 1-2 months.

The following parameters were estimated from this curve:

1) the *fraction of slow cleared material*, called $\mathbf{A_{c\_s}}$,

2) the *strength of 'negative' clearance*, $\mathbf{F_{c\_u}}$, describes the ratio between the highest rmf after 1-2 month and the rmf after 1 day (fast clearance completed), and

3) the *half-time of slow clearance*, $\mathbf{T_{c\_s}}$, using all rmf's from 1 month and later.

This data are shown for healthy subjects (NS and S) and for patients with sarcoidosis, fibrosis and with chronic obstructive pulmonary disease in Tab. 2.

Table 2: Parameters of clearance measurements in healthy subjects, subdivided into non-smokers (NS) and smokers (S) and in patients.

Group	N	A_{c_s} ± SD	F_{c_u} ± SD	T_{c_s} ± SD [days]
NS (20-39)	10	0.940±0.164	1.173±0.179	124.5±66.4
S (20-39)	9	0.864±0.143	1.139±0.175	208.4±82.4
NS (40-65)	11	0.865±0.199	1.584±0.743	161.9±119.5
S (40-65)	9	0.804±0.185	1.165±0.174	459.2±334.0
SARK (NS&S)	15	0.863±0.178	1.308±0.230	330.3±219.5
FIBR (NS&S)	12	0.677±0.334	1.278±0.455	602.3±435.5
COPD (NS&S)	15	0.494±0.200	1.800±0.615	259.7±70.7 (N=10)

Fast Clearance

Despite to the assumption, that all particles would penetrate to the alveolar region, several subjects showed a small fraction of particles being cleared within the first two days. Because the design of the study was concentrated to the slow clearance phase the time course of fast clearance could not be analyzed. The mean amount of slow cleared particles (A_{c_s}) is shown in Tab. 2. In younger healthy subjects more than 90 % of the particles remain in the lungs after 2 days and follow the slow clearance mechanism. There is a tendency to higher fraction of fast cleared particles ($1-A_{c\_s}$) in smokers and in older subjects (40-65). Fast cleared particles result from a deposition in the ciliated (bronchial) regions of the lung. An increasing amount of fast cleared particles demonstrates an increased fraction of bronchial deposition in this subjects. In patients with COPD 50 % of the particles were cleared within the first two days, none of the patients was without a fast clearance phase. This demonstrates a high bronchial deposition due to bronchial obstructions, that are known from an increased airway-resistance. The increased fast clearance phase in FIBR is also a measure for bronchial obstructions. The results in healthy smokers as well as in older subjects give indications for beginning airway obstructions.

Negative Clearance

The phenomenon of ***'negative'*** clearance, that describes an increase of the rmf of the lungs, is typical for MPG-studies using a magnetic tracer. In earlier studies [4] it was estimated that particle translocation might be the reason for this process. An analysis with clustered steel-spheres shows that the formation of aggregates results in an amplification of the rmf due to a mutual magnetization. A doublet has double the rmf compared to two separated spheres. From this analysis we follow that particle aggregation additionally can induce a negative clearance effect. Particle aggregates can be formed when macrophages die an release particles. Other particle-loaden macrophages can then phagocytize the particles. The results for negative clearance are reported in Tab. 2. There is no difference between NS and S. The effect is strongest in COPD and seem to correlate with bronchial obstructions and the high bronchial deposition.

Slow Clearance

Half-times for alveolar *long-term clearance* in younger and older subjects are shown in Tab. 2 and Fig. 2. In younger as well as in older smokers long-term clearance is significantly impaired ($p<0.05$). I two subjects of the group S(40-65) no clearance could be detected within the period of investigation. In this case half-time was set to 1000 days. Alveolar clearance capacity seems to decay with increasing PY. A correlation between clearance half-time T_{c_s} and pack year PY yields:

without NS T_{c_s} = 191±71 days + 5.0±2.1 (days/pack-year) * PY, p=0.026

with NS T_{c_s} = 159±31 days + 5.7±1.3 (days/pack-year) * PY, p<0.01.

An analysis without NS yields an overestimation for PY = 0, additionally including NS gives a higher significance. Finally we can conclude that cigarette smoke consumption reduces alveolar clearance by: **≈ 6 days/pack-year**.

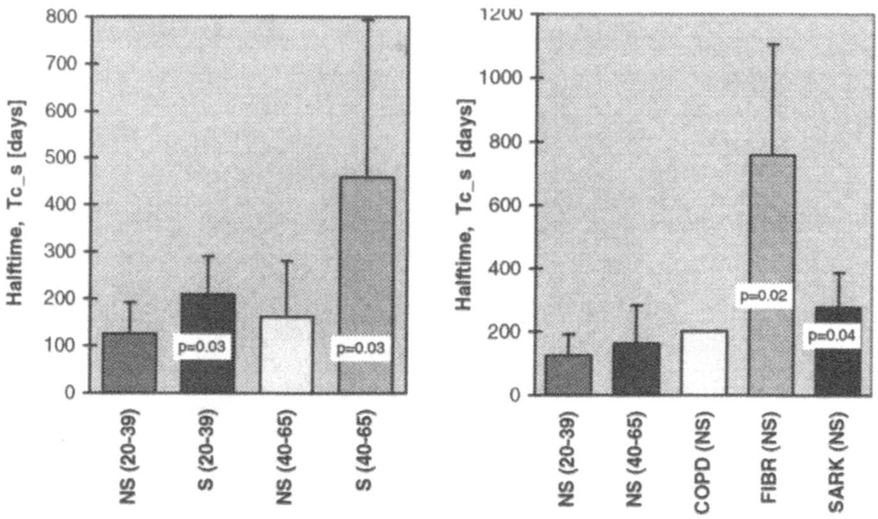

Figure 2: Half-times for alveolar clearance in healthy NS and S in different age groups (left) and in non-smoking patients (right).

Alveolar Clearance is highly impaired in FIBR-patients. Because of the significant influence of cigarette smoking in healthy subjects, only non-smoking patients were compared to older healthy subjects NS (40-65). Half of the FIBR(NS) showed no clearance within the period of observation. In FIBR alveolar clearance is strongly disturbed and thereby might be a co-factor for inducing this disease. SARK also show an influence on alveolar clearance, but the effect is less pronounced compared to healthy S. Despite the fact that COPD have high PY, alveolar clearance is faster compared to healthy S. This gives a hint that part of the influence of cigarette smoking on alveolar clearance might be reversible, after subjects stop smoking. This results show that MPG provides a sensitive and non-invasive method to study early influences on alveolar clearance and to monitor a disturbed lung defense in disease.

Acknowledgement: This study was supported by the DFG (Po 108/7-1) and by the CEC (FI4P-CT95-0026).

References

[1] Cohen, D., S.F. Arai, J.D. Brain (1979), Smoking Impairs Long-Term Dust Clearance from the Lung, *Science*, Vol. 294, 514-517.

[2] Stahlhofen W, Möller W (1993) Behaviour of magnetic micro-particles in the human lung, *Radiation and Environmental Biophysics*, Vol 32, 221-238

[3] Möller W, Stahlhofen W, Roth C (1990) Improved spinning top aerosol generator for the production of high concentrated ferrimagnetic aerosols, *Journal of Aerosol Science*, Suppl. 1, Vol. 21, S435-S438

[4] Freedman, A.P., S.E. Robinson, M.R. Street (1988), Magnetopneumographic Study of Human Alveolar Clearance in Health and Disease, *Annals of Occupational Hygiene*, Vol. 32, Supplement I, 809-820 (Inhaled Particles VI).

A Model of Phagosome Motion within Cells Incorporating Viscoelasticity Based on Cytomagnetometric Measurements

Nemoto, I.

Tokyo Denki University, Hatoyama, Saitama, Japan

Introduction

Cytomagnetometry was proposed more than ten years ago [1][2] but has not yet acquired the citizenship in biophysics. The main reason seems to be that the method is macroscopic and crude while there are advanced microscopic biophysical techniques available to handle the molecules themselves in the cells. Indeed, these new techniques have made it possible to measure even the force produced by one actin molecule. However, there are still quite limited possibilities for measuring the activities within the cells *in vivo*. We believe that cytomagnetometry using magnetic particles provides such possibilities; the movement of the intracellular structures can be detected non-invasively by measuring the magnetic field from the cells, and furthermore, magnetic particles within the cells can be manipulated by an external magnetic field easily.

Fig. 1 shows the basic idea of cytomagnetometry. Magnetic particles are introduced into the cells such as pulmonary macrophages by phagocytosis. The paticles then exist within the organelles called phagosomes. (They change names as the cellular process proceeds, but we simply call them 'phagosomes' for simplicity.) A strong field magnetizes the particles and the phagosomes act like small magnets. A weak field is measured from the cells which decays with a typical half time of 1-2 minutes. This magnetic relaxation is due to the randomization of the direction of the magnetic moments of phagosomes caused by their mechanical motions. Therefore, the relaxation reflects the mechanism of the intracellular motion and also the rheological properties of the cytoplasms. Manipulation of the phagosomes from outside can be done by applying a magnetic field. Experiments may be devised to estimate the values of the parameters involved in the intracellular motion (especially of the organelles), such as the energy (E_r), the viscosity and the elasticity. Phagosome motion is important not only in its own sake but also in its relationship to the cellular motion in general including muscle contraction[3].

Viscoelasticity and relaxation

Cytomagnetometry has shown that the movement of the phagosomes are impeded not only by the apparent viscosity but also by elasticity[2][4]. These studies suggested that the relaxation curve might be influenced by the elasticity as well. Indeed, if the particles are aligned by a strong field for a certain period of time (instead of by pulse

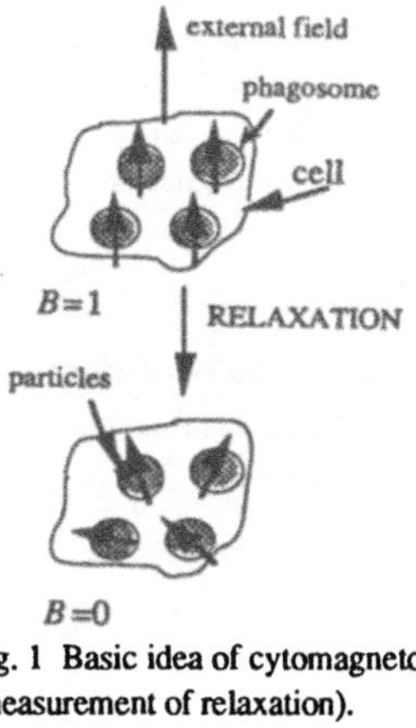

Fig. 1 Basic idea of cytomagnetometry (measurement of relaxation).

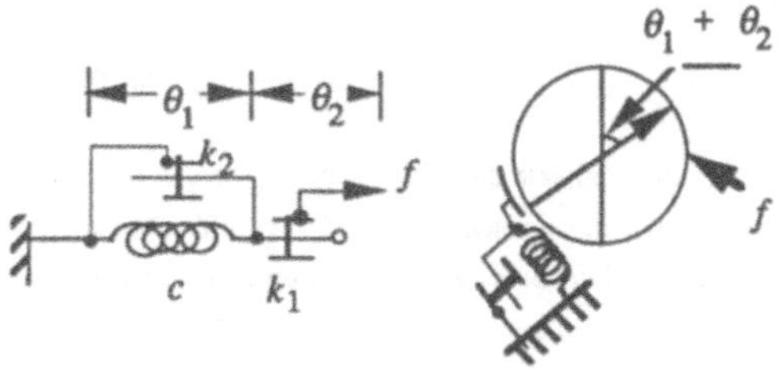

Fig. 2 The viscoelasticity model against the rotary motion of the phagosome.

field), the relaxation rate, especially in its beginning, may be considerably increased by the elastic recoil. Therefore, in relaxation measurements, pulse magnetization has been mostly used. However, we also have to consider possible effects of elasticity on the movement of the phagosomes produced by the (seemingly) random forces responsible for the relaxation. For this purpose, we have the 1-dimensional model of visco-elastic body pulled by an external force, be it the still undefinite random force responsible for the relaxation or the magnetic torque produced by an external field (Fig. 2). The equation of motion for this model is written

$$\frac{d\theta_1}{dt} = -\frac{c}{k_2}(\theta_1 - \overline{\theta_1}) + \frac{f}{k_2}, \quad \frac{d\theta_2}{dt} = \frac{f}{k_1} \tag{1}$$

where θ_1 and θ_2 are defined as shown in Fig. 1 such that $\theta = \theta_1 + \theta_2$ is the the direction of the magnetic moment of the phagosome measured from that of the measurement. $\overline{\theta_1}$ is the value of θ_1 for which the elastic body is in equilibrium. For the relaxation process, we take f to be the formal differentiation of the Wiener process $W(t)$ of which the mean is 0 and the variance is αt where α is some constant to which we will give a specific form later. Then the solution of the stochastic differential equation is formally written as

$$\theta_3(t) = \overline{\theta_1}\left(1 - e^{-\frac{c}{k_2}t}\right) + \int_0^t e^{-\frac{c}{k_2}(t-s)} dW(s), \quad \theta_2(t) = \int_0^t \frac{1}{k_1} dW(s) \tag{2}$$

where the integrals are the Wiener integrals. Note that the initial condition $\theta_1(0) = \theta_2(0) = 0$ is used. The mean and the variance of the solution of this equation can be calculated to yield

$$E[\theta] = 0, \quad V[\theta(t)] = E[\overline{\theta_1}^2](1 - e^{-\beta t})^2 + \alpha\left\{\frac{1 - e^{-2\beta t}}{2\beta k_2^2} + \frac{2(1 - e^{-\beta t})}{\beta k_1 k_2} + \frac{t}{k_1^2}\right\} \tag{3}$$

where $\beta = c/k_2$. Since θ is disbributed normally with mean 0 and variance $V[\theta(t)]$, the cell field is given by

$$B(t) = E[\cos\theta(t)] = \frac{2}{\sqrt{2\pi V[\theta(t)]}} \int_0^\infty \cos\xi \exp\left\{-\frac{\xi^2}{2V[\theta(t)]}\right\} d\xi = \exp\{-V[\theta(t)]/2\} \tag{4}$$

which shows that when $t \to \infty$, the relaxation follows an exponential decay but that for finite t, the relaxation curve is certainly influenced by elasticity.

We next want to relate α to E_r which we previously defined to be the energy (corresponding to kT in case of thermal Brownina motion) responsible for the relaxation[1]. In our model, the mean square distance during the infinitesimal time Δt was given by $E[\Theta^2] = 4\Delta t E_r / \nu$ where ν is the viscosity factor. Here we have

$$\lim_{\Delta t \to 0} \frac{V[\theta(t)]}{\Delta t} = \alpha\kappa^{-2}, \quad \kappa = \left(\frac{1}{k_1} + \frac{1}{k_2}\right)^{-1} \tag{5}$$

Comparing these two, we get $\alpha = 4E_r\kappa^2/\nu$. We assume that ν can be replaced by κ and thus $\alpha = 4E_r\kappa$.

Pulse magnetization and complete alignment

We employ two methods of magnetizing the particles; pulse magnetization and complete alignment. If the particles are homogeneous, there would be no change in the magnetization of the whole ensemble of the phagosomes. However, if there is visoelasticity, the relaxation curve would be different because the elastic body is in general, not in equilibrium at any time, that is to say $\overline{\theta_1}$ can be disbritubed. The variance of $\overline{\theta_1}$ should be equal to the variance of the solution of the equation $d\xi = -(c/k_2)\xi\, dt + (1/k_1)dW_t$ at the limit $t \to \infty$, which is

$$\lim_{t\to\infty} V[\xi(t)] = \lim_{t\to\infty}\left[V[\xi(0)]e^{-2\beta t} + \frac{\alpha(1 - e^{-2\beta t})}{2\beta k_2^2}\right] = \frac{\alpha}{2\beta k_2^2} \tag{6}$$

With the complete alignment scheme, we can put $E[\overline{\theta_1}^2] = 0$ because after a sufficient period of alignment, the elastic body will have nearly reached its equilibrium state. In summary we have two relaxation curves corresponding to two kinds of variance; $V[\theta^p(t)]$ for pulse magnetization and $V[\theta^a(t)]$ for complete alignment:

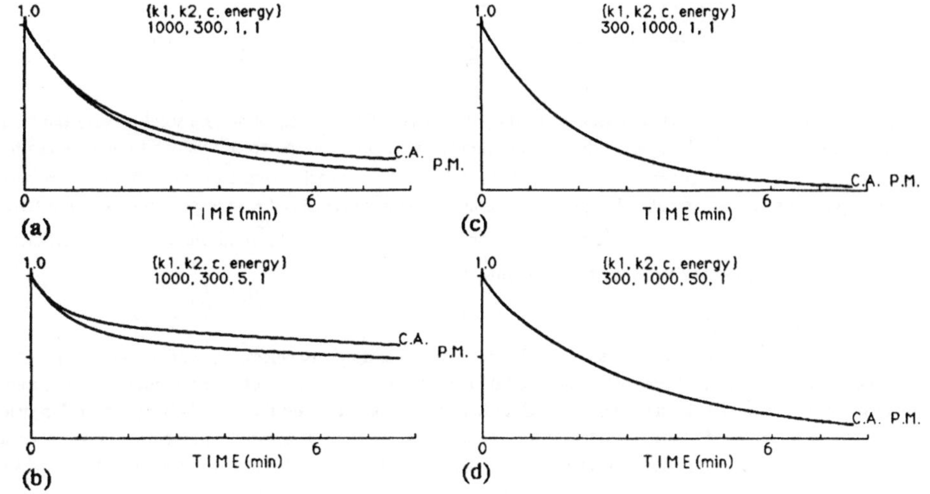

Fig. 3. Relaxation curves calculated from eq.(7). C.A. and P.M. stand for complete-alignment and pulse-magnetization, respectively. The parameter values are normalized by E_r.

$$V[\theta^p(t)] = 4E_r\kappa\left\{\frac{(1-e^{-\beta t})}{\beta k_2{}^2} + \frac{2(1-e^{-\beta t})}{\beta k_1 k_2} + \frac{t}{k_1{}^2}\right\} \quad V[\theta^a(t)] = 4E_r\kappa\left\{\frac{(1-e^{-2\beta t})}{2\beta k_2{}^2} + \frac{2(1-e^{-\beta t})}{\beta k_1 k_2} + \frac{t}{k_1{}^2}\right\}. \quad (7)$$

Fig. 3 shows a few examples of relaxation curves corresponding to the variances calculated by eq. (7). It is seen that when $k_1 > k_2$, the effect of elasticity and the difference between the two magneitzing schemes are apparent.

Viscoelastic recoil

Assume that we apply a medium field H (i.e., not sufficient to alter the magnetization of each particle but sufficient to overcome the randomizing force) perpendicularly to the primary-magnetization direction right after the primary magneti-zation for T_2 seconds and remove the field. Then the equations describing theese processes are:

$$\text{for } 0 \le t \le T_2, \quad \frac{d\theta_1}{dt} = -\frac{c}{k_2}\theta_1 + \frac{1}{k_2}\mu H\cos(\theta_1+\theta_2), \quad \frac{d\theta_2}{dt} = \frac{1}{k_1}\mu H\cos(\theta_1+\theta_2), \quad \text{and}$$

$$\text{for } t \ge T_2, \quad \frac{d\theta_3}{dt} = -\frac{c}{k_1}\theta_3 + \frac{1}{k_2}\frac{dW}{dt}, \quad \frac{d\theta_2}{dt} = \frac{1}{k_1}\frac{dW}{dt}. \quad (8)$$

Here we assume for simpicity that $E[\overline{\theta_1}^2] = 0$. μ is the magnetic moment of the phagosome. The first part is a nonlinear differential equation and can only be solved numerically. The second part is the stochastic differential equation already mentioned above. The initial condition for the firstp part is given by $\theta_1(0) = \theta_2(0) = 0$ and that of the second part is given by the solution of the first part at T_2. The numerical solution for the first part would be inconvenient for parameter estimation by fitting the measurement data. Therefore, we approximate the first part by

$$\frac{d\theta_1}{dt} = -\frac{c}{k_2}\theta_1 + \frac{1}{k_2}\mu H\left(1-0.5\cdot\theta_1{}^2(1+k_2/k_1)^2\right), \quad \frac{d\theta_2}{dt} = \frac{1}{k_1}\mu H\left(1-0.5\cdot\theta_2{}^2(1+k_1/k_2)^2\right) \quad (9)$$

The two variables are now separate, which makes the solution much easier. Its rigorous solution is given by

$$\theta_i(t) = -\frac{r_{1i}r_{2i}\left(\exp(r_{1i}t)-\exp(r_{2i}t)\right)}{a_i\left(r_{2i}\exp(r_{1i}t)-r_{1i}\exp(r_{2i}t)\right)}, \qquad i = 1,2 \quad (10)$$

where

$$r_{1i} = \frac{b_i + \sqrt{b_i^2 - 4a_i d_i}}{2}, \quad r_{2i} = \frac{b_i - \sqrt{b_i^2 - 4a_i d_i}}{2}, \quad a_1 = -\frac{\mu H}{2k_2}\left(1+\frac{k_2}{k_1}\right)^2, a_2 = -\frac{\mu H}{2k_1}\left(1+\frac{k_1}{k_2}\right)^2$$

$$b_1 = -\lambda, \quad b_2 = 0, \quad d_1 = \mu H / k_2, \quad d_2 = \mu H / k_1.$$

Fig. 4 shows the whole time courses described by the above differential equation (the solid curves) and the one using eq. (9) for the vertical magnetization process (the black circles). The parameter values are normalized by E_r taken to be the unit value. These curves (especially the form of the curves in (d)) are similar to those experienced in cytomagnetometric measurements. The parameter values shown are normalized with respect to E_r which is taken to be 1. A typical value $\sim 5\times10^{-18} J$ for E_r [1] and $k_i = 100, c = 10$ would yield the viscocity coefficient of $\sim 10^2$ Ns/m^2, and the rigidity moduous of ~1.5 N/m^2.

Concluding Remark

A one-dimensional model of cell field behavior incorporating viscoelasticity was proposed. Preliminary analysis of the experimental data with this model suggests the involvement of microtubules and intermediate filaments in producing the elasticity, but we need further verification for the estimation. The proposed 1-dimensional model should be extended to the spherical model formulated in [1], but it is not trivial to incorporate viscoelasticity in the spherical model. The model of interacation between filaments and phagosome membranes proposed in [3] should also be modified to take into account the viscoelasticity.

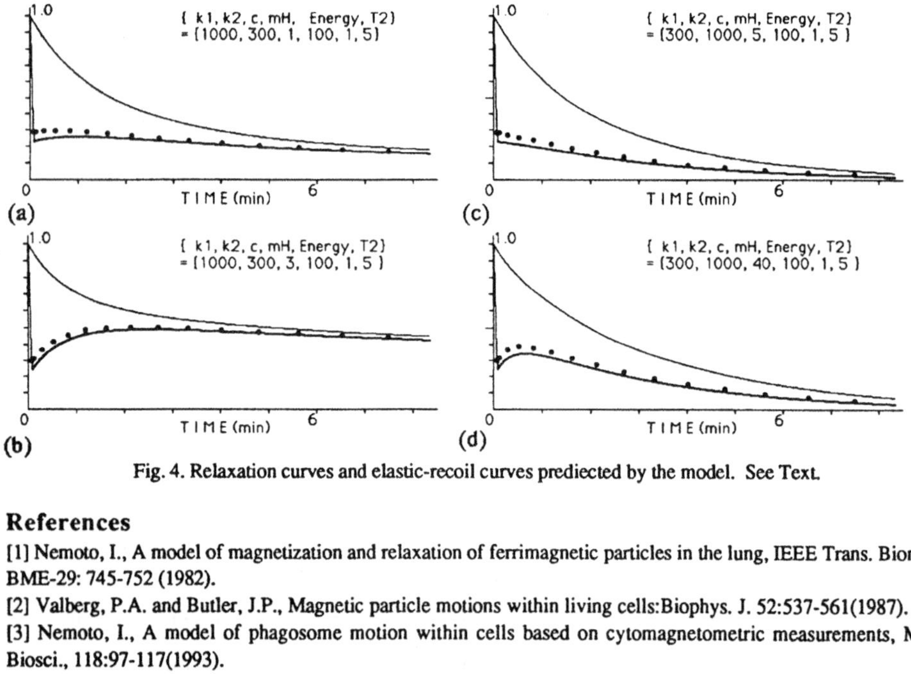

Fig. 4. Relaxation curves and elastic-recoil curves prediected by the model. See Text.

References

[1] Nemoto, I., A model of magnetization and relaxation of ferrimagnetic particles in the lung, IEEE Trans. Biomed. BME-29: 745-752 (1982).

[2] Valberg, P.A. and Butler, J.P., Magnetic particle motions within living cells:Biophys. J. 52:537-561(1987).

[3] Nemoto, I., A model of phagosome motion within cells based on cytomagnetometric measurements, Math Biosci., 118:97-117(1993).

[4] Wiegand, J., Moeller, W., Nemoto, I. and Stahlhofen,W., Magnetic microparticles can measure intracellular viscoelasticity in vivo, IN:Baumgartner, C., Deecke, L., Stroink, G. and Williamson, S.J. Biomagnetism, Amsterdam, Elsevier, 1995.

Magnetic Susceptibility Tomography with Nonuniform Field

Parente Ribeiro, E.[1,2], Wikswo, Jr., J.P.[2], Costa Ribeiro, P.[1] and Szczupak, J.[1]

[1]*PUC-Rio, Rio de Janeiro, Brazil;* [2]*Vanderbilt University, Nashville, TN, USA*

Abstract

Imaging systems that generate slice views of the body are called tomographic scanners. Several modalities of scanners that generate such cross sectional images have been developed for biomedical applications, such as X-ray tomography, PET and MRI. Magnetic susceptibility tomography (MST) may be a new tool for visualizing the interior of the body. Previous works have shown that it is possible to reconstruct a $4\times4\times4$ susceptibility distribution from the perturbation produced under a uniform field. In this paper the solution to the inverse problem is addressed in the framework of reconstruction from projections. A nonuniform field is considered together with uniform field.

Introduction

The number of tomographic scanners available in almost every hospital illustrates the importance of such tools for physicians. From the more popular x-ray computerized tomography (CT), magnetic resonance imaging (MRI), positron emission tomography (PET) to the emerging electrical impedance tomography (EIT), each modality employs a different method to provide 2D slice views of the body [1]. All of them share the same underlying mathematical problem of reconstructing the object data from external measurements made at different 'angles', also referred to as reconstruction from projections [2,3].

Magnetic susceptibility measurements on humans have been made for some time [4,5,6]. It has found useful clinical application, particularly in liver iron overload assessment [7] and it has been used in gastroenterology research to track paramagnetic tracers. Imaging of planar samples has already been demonstrated [8] but the possibility of 3D reconstruction without the assumption of a particular model was only recently established [9,10]. A $4\times4\times4$ voxel model representing a 3D susceptibility distribution could be successfully reconstructed using a uniform magnetizing field, in the presence of a small amount of noise. The problem is very ill-posed and an adequate geometry and applied field must be identified to enable convergence at higher resolutions and higher noise levels.

Methods

When an object is exposed to a magnetic field of strength H, it develops a magnetization M. The magnetic susceptibility χ (a dimensionless quantity in S.I.) is M/H. The magnetic flux density B is given by $B=\mu_o(H+M)$, where μ_o is the permeability of free space. It can also be expressed in term of the susceptibility: $B=\mu_o(1+\chi)H$. The permeability is B/H and the quantity $(1+\chi)$ is the relative permeability (μ_r). If the orientation of the magnetization is opposite to the applied field, the object has negative susceptibility and is said to be diamagnetic. Water, which is the major constituent of the body, has χ=-9x10^{-6}. If the magnetization has the same direction as the applied field and it is small ($0<\chi<10^{-3}$) the object is paramagnetic. Ferromagnetic materials have high permeability, in the range ~ $10^3 < \mu_r < 10^5$ and they may exhibit histeresis. For diamagnetic and paramagnetic materials (which includes biological tissue) the Born approximation holds, which means that the magnetization of any one region is too weak to affect an adjacent region.

The magnetic flux density $\boldsymbol{B}(\boldsymbol{r})$ at point $\boldsymbol{r}$ in space due to an object with susceptibility $\chi(\boldsymbol{r}')$ and volume v' is given by the dipole field equation

$$\boldsymbol{B}(\boldsymbol{r}) = \frac{\mu_o}{4\pi}\int_{v'} \frac{3\chi(\boldsymbol{r}')\boldsymbol{H}_a(\boldsymbol{r}')\cdot(\boldsymbol{r}-\boldsymbol{r}')}{|\boldsymbol{r}-\boldsymbol{r}'|^5}(\boldsymbol{r}-\boldsymbol{r}') - \frac{\chi(\boldsymbol{r}')\boldsymbol{H}_a(\boldsymbol{r}')}{|\boldsymbol{r}-\boldsymbol{r}'|^3} dv'^3 . \tag{1}$$

This can be rewritten as

$$\boldsymbol{B}(\boldsymbol{r}) = \int_{v'} G(\boldsymbol{r},\boldsymbol{r}',\boldsymbol{H}_a)\chi(\boldsymbol{r}')dv' , \tag{2}$$

where

$$G(\boldsymbol{r},\boldsymbol{r}',\boldsymbol{H}_a) = \frac{\mu_o}{4\pi}\left\{\frac{3\boldsymbol{H}_a(\boldsymbol{r}')\cdot(\boldsymbol{r}-\boldsymbol{r}')}{|\boldsymbol{r}-\boldsymbol{r}'|^5}(\boldsymbol{r}-\boldsymbol{r}') - \frac{\boldsymbol{H}_a(\boldsymbol{r}')}{|\boldsymbol{r}-\boldsymbol{r}'|^3}\right\} . \tag{3}$$

The measured magnetic perturbation is thus the projection of the susceptibility distribution on the Green's function. One can collect several projections by varying $\boldsymbol{r}$ and $\boldsymbol{H}_a$, i.e., taking measurements in several locations and with different applied fields, and attempt to estimate the original susceptibility distribution. The first question which arises is whether this mapping is one-to-one or, in other words, whether the inverse solution is unique. A necessary and sufficient condition for that to happen is that the null space of the mapping N be equal to $\{0\}$, i.e., there must exist no magnetically silent source configurations.

If only a uniform applied field is considered, there are only three linear independent options for $\boldsymbol{H}_a$ because $\boldsymbol{G}$ is linear with respect to $\boldsymbol{H}_a$. Furthermore, any concentric sphere with uniformly distributed susceptibility has the same multipole expansion coefficients (being all equal to zero but the dipole term). Hence, sources of different size will produce the same dipolar field pattern, creating an infinite set of magnetically silent configurations. This will not be the case if a nonuniform field is also applied.

To tackle the problem numerically, the source object is discretized into n elements of volume v_i', where $i = 1,2,\ldots,n$ and the measurements are made at m different points r_j, where $j = 1,2,\ldots,m$. Equation 2 becomes

$$\boldsymbol{B}(\boldsymbol{r}_j) = \sum_{i=1}^{m} G(\boldsymbol{r}_j, \boldsymbol{r}_i', \boldsymbol{H}_a)\chi(\boldsymbol{r}_i'). \tag{4}$$

In matrix notation, a $(n \times 1)$ column matrix **X** denotes the susceptibility distribution. A $(m \times n)$ matrix **G** represents the Green's function and the magnetic flux density is described by a $(m \times 1)$ column matrix **B**. Only the z component is considered. Equation (4) is then **B=GX.**

If $m=n$, the number of measurements equals the number of unknowns and the system of equations will be exactly determined. However due to the nature of the Green's functions, it will be very ill-conditioned. An exact solution could be found, but if noise is present in the measurement, the solution will have large errors. To improve matrix conditioning, more measurements are incorporated so that the system becomes over-determined ($m>n$) and a least squares solution can be attempted.

In the forward problem, the magnetic field matrix **B** is calculated using a sample-generated susceptibility distribution **X**, and then **B** is truncated to a fixed number of digits. This is equivalent to adding noise N: **B=GX+N.**

The generalized inverse $\mathbf{G}^{\#}$ is evaluated using singular value decomposition (SVD) and the inverse problem reduces to $\mathbf{X}^{e} = \mathbf{G}^{\#}\mathbf{B}$. The error is the difference between the estimated and actual distributions. The relative root mean square error *rrmse*, obtained by dividing the rms error by the mean value of the original distribution, is used to evaluate the reconstruction quality.

In Fig. 1a, the *rrmse* of a $4 \times 4 \times 4$ random susceptibility cube reconstructed from magnetic data with different truncation levels (3, 4 and 5 digits) is plotted together with the condition number of the matrix **G**, which is the ratio between the largest and smallest singular values. This is a number between unity (well conditioned) and infinity (singular). A large number indicates an ill-conditioned matrix. High errors are encountered when the problem becomes ill-posed. The curves correlate well and they should be even more similar if each error plot is averaged over an ensemble of different source distributions.

When resolution is increased, the matrix becomes more ill-conditioned. Even with a large number of measurements and a uniform field applied along the other two axes, the condition number does not seem to go below a certain level ($\approx 5 \times 10^6$), as can be seen on Fig. 1b.

To improve the conditioning of the problem, we analyzed the use of a nonuniform field. The field due to a single coil encircling the sample was considered. Contrary to the uniform field, such a field is not rejected by the gradiometric detecting device. This rejection is important since the susceptibility values are very small, making the field perturbation several orders of magnitude smaller than the applied field. Were such an instrument actually built, a cancellation scheme would have to be utilized to subtract the field coupled directly to the SQUID.

Results

Several field configurations were analyzed for $8 \times 8 \times 8$ susceptibility voxels by varying the position and direction of the sample and the field. Measurements were considered to be taken at 0.6 units above and below the sample. Two scenarios were investigated: In the first, the external uniform and nonuniform field was applied only on the same direction as the detection device. In the second, the field was applied in all three directions. They are summarized on Fig. 3. A reduction of the condition number to 5×10^5 was achieved.

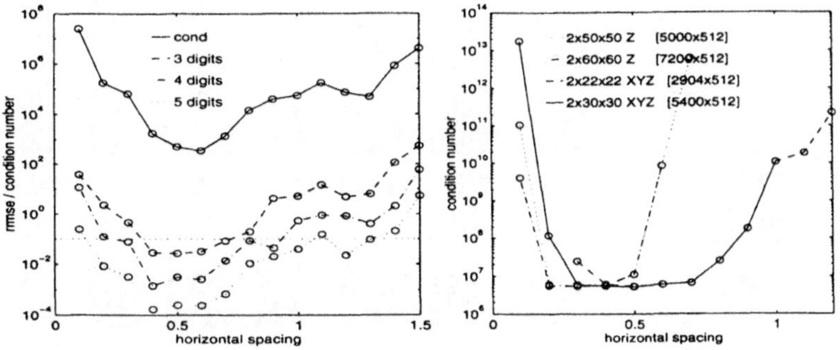

(a) (b)

Fig. 1 - (a) The matrix G describing the forward problem of a sample of $4 \times 4 \times 4$ susceptibility voxels exposed to uniform field (unitary spacing). An array of 8×8 measurement locations at 1 unit above and below the sample is considered. The calculated magnetic field is truncated with different numbers of significant digits (3,4,5) prior to inversion. The condition number is also plotted as a function of measurement point spacing. The horizontal line establishes a threshold of acceptable noise (subjectively) to be 10%. (b) The condition number for uniform field reconstruction of a $8 \times 8 \times 8$ susceptibility cube. Measurements are made on plane 0.6 units above and below. The legend shows the number of measurements, the direction of applied field and size of Green's matrix (inside brackets).

The condition number could be further reduced to 9.5×10^4 by also taking some measurements on the sides of the sample. This configuration was used to simulate the forward and the inverse problem for different digits of precision in the measured field. A susceptibility cube with values selected to be either 0 or 1 was used. Figure 4 shows the original planes and the reconstructed solutions for 5, 4 and 3 digits of precision.

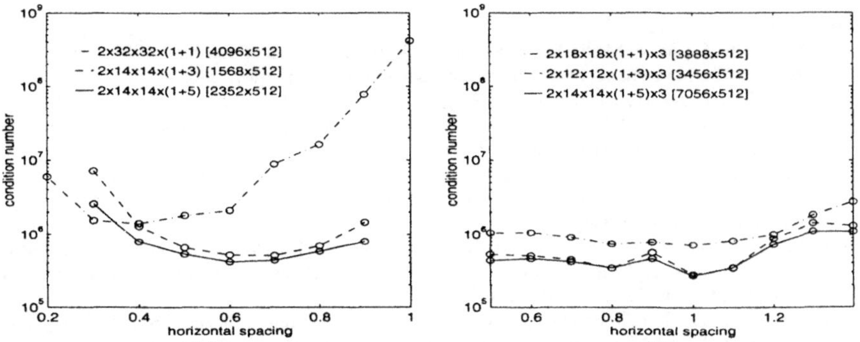

(a) (b)

Fig. 3 - The condition number for an $8 \times 8 \times 8$ voxel susceptibility distribution, with different geometries. A uniform field and the field due to a single coil with diameter larger than the sample are considered. (a) Fields are applied only on one direction. (b) Fields are applied on all three directions. Each plot shows three traces corresponding to: coil placed at the center, coil placed at three axial positions (-2,0,2) and coil placed at five axial positions (-2,-1,0,1,2). Inside the brackets is the matrix size for each measurement.

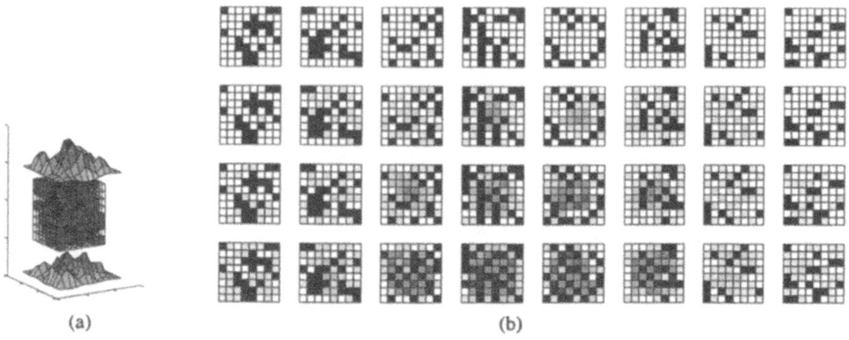

Fig. 4 - (a) Forward problem: 3D display of $8 \times 8 \times 8$ sample and simulated field. (b) Graylevel image of each sample plane. The first row is the original distribution. The following rows are reconstruction's from data with 5, 4 and 3 digits of precision, respectively. The *rrmse* for each reconstruction is 8%, 38% and 257%.

Discussion

We have established a mathematical framework to treat the three dimensional magnetic susceptibility inverse problem as a reconstruction from projections. Condition number helped explain why under certain configurations the inverse solution exhibited large error.

An improvement of about one order of magnitude on the condition number was verified by also employing together with the uniform field the nonuniform field generated by a single coil enclosing the sample. That additional information was confirmed to be linearly independent from the information obtained with the uniform field only. Reconstruction of the $8 \times 8 \times 8$ susceptibility distribution from the measured magnetic perturbation in the presence of noise was demonstrated to be possible from magnetic data with small noise (5 digits). Novel nonuniform configurations may provide further increase in resolution.

This three dimensional imaging technique could greatly improve the accuracy of already established applications of susceptometry such as quantitative liver iron measurements [7], and may also be able to image different tissues types, launching new applications with appealing imaging and non-invasive characteristics.

References:

[1] Webb, S. The Physics of Medical Imaging, Bristol, Institute of Physics Publishing, 1988.

[2] Natterer, F. The Mathematics of Computerized Tomography, John Wiley,1986.

[3] Herman, G.T. Image Reconstruction from Projections, Academic Press, 1980.

[4] Wikswo, J.P., Jr., Opfer, J.E., Fairbank, W.M. Noninvasive magnetic detection of cardiac mechanical activity: Experiments, Medical Physics, 1980, 7:307-313.

[5] Rassi, D., Hoare, A.,Samadian, V., Melville, D. SQUID measurements and computational modeling of a simple thorax phantom. In: Atsumi, K., Kotani, M., Ueno, S., Katila, T., Williamson, S.J. Biomagnetism'87. Tokio, Tokio Denki University Press, 1988.

[6] Di Luci, S., Obletter, G., Comani, S., Del Gratta, C., Romani, G.L. Magnetic mapping of dc fields related to tissue susceptibility in the human body. In: Williamson, S.J., Hoke, M., Stroink, G., Kotani, M. Advances in Biomagnetism. New York, Plenum Press, 55-508, 1989.

[7] Fischer, R., Heinrich, H.C. Biosusceptometry: Current status of clinical diagnostics and biomagnetic research, In: Hoke, M., Erne S.N., Okada, Y.C., Romani, G.L. Biomagnetism: Clinical Aspects, 1992.

[8] Roth, B.J., Sepulveda, N.G., Wikswo, J.P., Jr. Using a magnetometer to image a two-dimensional current distribution, J. Appl. Phys. 65, 361-372, 1989.

[9] Wikswo, J.P., Jr., Sepulveda, N.G., Thomas, I.M. Three-dimensional biomagnetic imaging with magnetic susceptibility tomography, In: Baumgartner, C., Deecke, L., Stroink, G., Williamson, S.J. Biomagnetism: Fundamental research and clinical applications. Amesterdam, IOS Press, 780-784, 1995.

[10] Sepulveda, N.G., Thomas, I.M., Wikswo, J.P., Jr. Magnetic susceptibility tomography for three-dimensional imaging of diamagnetic and paramagnetic objects, IEEE Transactions on Magnetics, 1994, 30:5062-5069.

MCG Studies in Rats Using Ferrofluids

Saligram, U.[1,2], Moran, J.E.[1] and Tepley, N.[1,2]

[1]*Henry Ford Hospital, Detroit, MI, USA;* [2]*Oakland University, Rochester, MI, USA*

INTRODUCTION

Ferrofluids are colloidal suspensions of single-domain ferrite particles in a liquid carrier. The liquid carrier used in biological applications is water. The particles are coated with a proprietary stabilizing dispersing agent which prevents particle agglomeration.

Ferrofluids have a broad range of biological applications. They have been used *in vivo* in animal studies to target chemotherapeutic agents to specific anatomical sites [1]. They have also been used in blood flow studies as a magnetic tracer [2,3]. In a previous study [2] with Wistar rats an amplification in Magnetocardiogram (MCG) was observed after introducing ferrofluid into the blood circulation. However this apparent amplification was observed in only one of the four rats injected with ferrofluid and this led to the speculation that this particular animal might have had damaged cardiac tissue, and the ferrofluid from the blood had leaked into the tissue. Based on this assumption additional studies were carried out. In these studies the various factors that could be responsible for this apparent increase in amplitude were examine. The ferrofluid used in this study EMG 507, is manufactured by FERROFLUIDICS CORPORATION, Nashau, NH. The liquid carrier is water; average particle size is 100°A; saturation magnetization is 100 Gauss; viscosity is ~2CP @ 27°C; particle concentration by volume is 1.8%; initial susceptibility is 0.38; density is 1.15 gm/ml; ionic nature of the surfactant is anionic and the pH is in the range 6-8. The manufacturer claims that the ferrofluids are non-toxic.

METHODS

Male wistar rats with mean weight of 250 gms were used. A seven channel BTi model 607 magnetometer was used to gather data (0.1 - 50 Hz). The sampling rate was 250 Hz Each rat was anesthetized using sodium pentobarbital (40 mg/kg). The magnetometer was positioned over the rat such that the channel at the center of the hexagonal array was directly above the heart. This magnetometer geometry was kept the same in all the studies. Thirty seconds of baseline MCG data were collected.

The right femoral artery was cannulated. The rat was then placed on a table inside the shielded room. 0.25 cc of ferrofluid diluted with saline (1 part ferrofluid in 20 parts saline) was introduced into the blood stream through the cannulated artery. Thirty seconds of data were then collected immediately after the introduction of ferrofluid. Ten second epochs of data were collected at intervals of 5 minutes. This study was a repeat of the previous study [2], referred to earlier.

A second study was then carried out, in which a fine needle was used to puncture the cardiac tissue. Thirty seconds of data were collected immediately after the puncture. Ten seconds of data were collected 5 minutes after the puncture, in order to determine if the act of puncturing displaced the heart from its normal position in the thoracic cavity, thereby causing changes in the amplitude of the MCG. 0.25 cc of ferrofluid diluted with saline (1 part ferrofluid in 20 parts saline) was then injected directly into the heart. The site of injection was similar in all the studies. Thirty seconds of data were collected immediately after the injection. Five minutes after the injection 10 second epochs of data were collected at 5 minute intervals. This five minute time interval was used to minimize the effect of the possible displacement of the heart due to the injection on the MCG signal amplitude. (If the heart is displaced due to the injection, its position with respect to each of the 7 magnetometer channels changes. Some of the channels would then show an increase in the amplitude of the signal that they record, and this increase would be due to the heart shifting closer to these channels and not due to the ferrofluid. Even though the heart may be displaced due to the injection, it is expected to return to its original position within a few seconds. The five minute interval after the injection allows eneough time for this to happen, thereby minimising the effect on the amplitude of the signal due to the shifting of the heart.) One hour's worth of data were collected in each study. The effect of an external magnetic field was studied using a circular magnet with its poles along the axis. The magnet (pole strength 1100 gauss-cm^2) was placed under the rat's heart at a vertical distance of 24 inches (field due to the magnet at this point was 7.8 mgauss), with its north pole facing up. Thirty seconds of data were collected at intervals of two minutes. This was repeated with the south pole facing up.

A third study using saline instead of ferrofluid was carried out. In this study 0.25 cc of saline was injected into the heart after the initial puncture. The protocol used to collect data in this study was the same as that used for ferrofluid study.

An average of 12 epochs of data were collected in each of the above studies.
For the time line of each of the studies refer to Fig . 1.

DATA ANALYSIS

The MCG signals were digitized using a HP 694A multiprogrammer. The primary comparison was the analysis of difference in signal amplitude before and after introduction of ferrofluid and saline. The baseline signal amplitude was compared to the following :

1) Signal amplitude after ferrofluid was introduced into the animals bloodstream through the cannulated femoral artery.
2) Signal amplitude after cardiac tissue puncture.
3) Signal amplitude after ferrofluid was injected into the heart.
4) Signal amplitude after saline was injected into the heart.

The mean amplitude of the MCG signal was calculated for each of the seven channels for each of the cases listed above. The mean amplitude of the baseline signal was then subtracted from the mean amplitude in each of the above cases. Standard deviation was also calculated for each of the channels for each run and for all the runs in each of the cases listed above. A paired t-test was done for each channel to determine if the amplitude difference from the baseline was significant. The testing level for each of these tests was set at .007. In the case where a magnet was used to create an external field, the data were filtered using a high pass filter of 1 Hz to remove the artifact due to breathing. This artifact was introduced into the signal by the presence of ferrofluid in the thoracic cavity.

RESULTS

No discernible change in the amplitude in case 1, i.e; ferrofluid introduced into the bloodstream through the cannulated artery without cardiac puncture was observed. Four rats were used in this study [Fig . 2].

No discernible change in the amplitude in case 2, i.e; after cardiac puncture was observed. Four rats were used in this study [Fig . 3].

A statistically significant ($p <= 0.007$) increase in the amplitude of the signal was found in case 3, i.e; after ferrofluid was injected into the heart. Seventeen rats were used in this study. Of the seventeen rats thirteen showed a significant increase in amplitude from the base line amplitude [Fig . 4]. Four rats did not show a significant increase in amplitude. One of the rats died due to an overdose of sodium pentobarbital.

No discernible increase in amplitude was observed in case 4, i.e; after saline was injected into the heart. Five rats were used in this study [Fig. 5].

DISCUSSION

The increase in the amplitude of the MCG signal only when ferrofluid is injected into the heart and not when it is introduced into the blood stream, justifies our assumption that the increase in amplitude is caused by ferrofluid leaking into the damaged cardiac tissue. The major difficulty with the procedure is the inability to induce uniform tissue damage in size, severity and location in every animal. This would make it difficult to study the effect of ferrofluid on tissue at the same location in every trial. This is an important consideration for this study because the site of tissue damage may have an effect on the magnitude of the increase in the signal amplitude.

This property of ferrofluids i.e; increasing the amplitude of electrophysiological signals, could probably be used to study weak signals from other electrophysiologically active tissues which would normally go undetected. In the case of cardiac tissue, the increase in amplitude is observed only after the tissue is damaged; this may or may not be the case for other tissue types. If ferrofluids leak into damaged tissue only, and if the tissue is electrophysiologically active, the disruption in the normal tissue function due

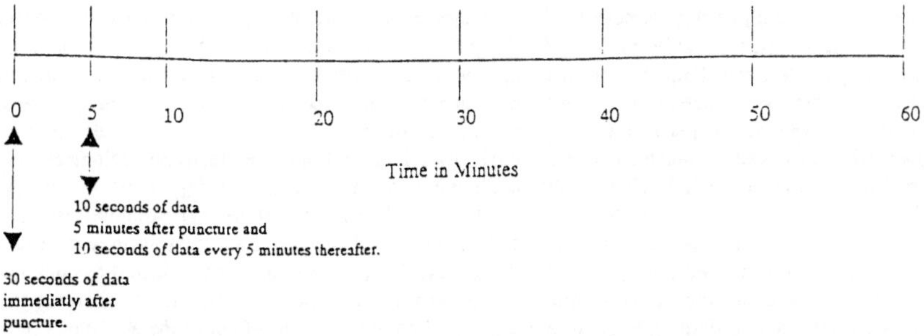

Fig. 1 Time line : 30 seconds of data were collected immediately following puncture of cardiac tissue or injection of ferrofluid or saline. 10 seconds of data were collected 5 minutes later and 10 seconds of data were collected every 5 minutes thereafter. Total time of each study was 60 minutes.

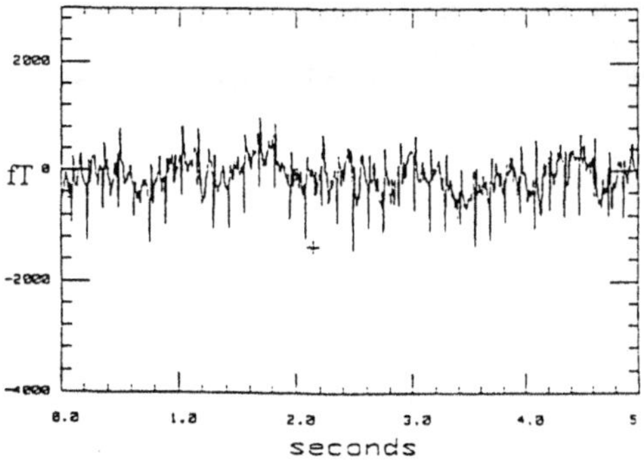

Fig. 2 MCG of rat after cardiac tissue was punctured.

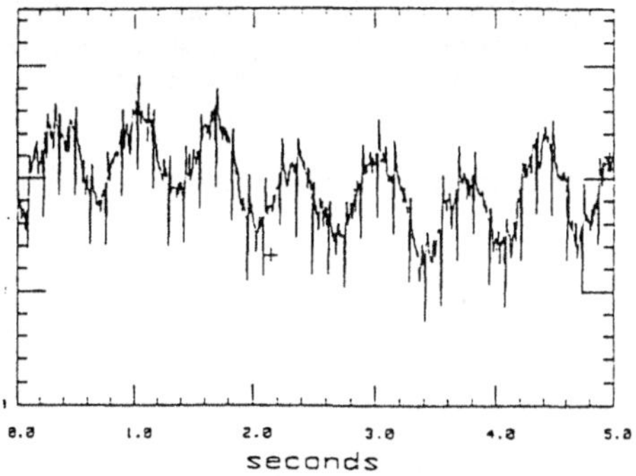

Fig. 3 MCG of rat after ferrofluid was introduced into the bloodstream through cannulated femoral artery.

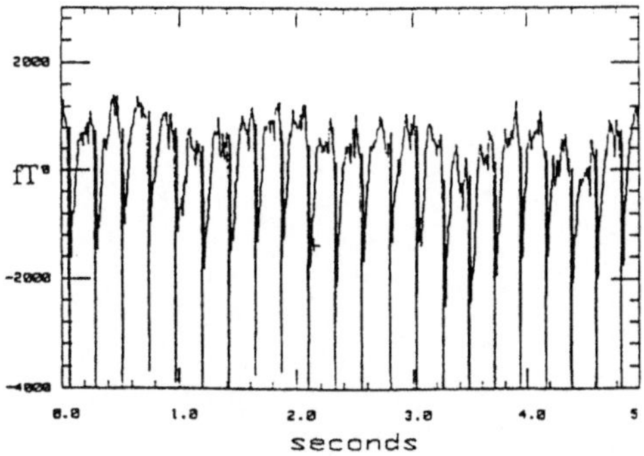

Fig. 4 MCG of rat after ferrofluid was injected directly into the heart.

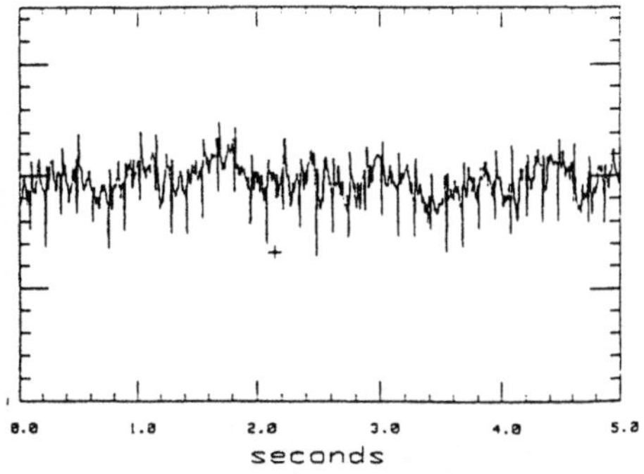

Fig. 5 MCG of rat after saline was injected directly into the heart.

to the damage could possibly be detected in the presence of ferrofluids. In addition repair and restoration of normal tissue function could be monitored. Using the spatial information of the relaxation processes of the magnetite particles, registered by the neuromagnetometer, a spatial image of the ferrofluid distribution in biological tissue can be determined. This could then be used to determine areas of damage. On the other hand if the amplitude increase due to the presence of ferrofluids is not restricted to electrophysiological signals from damaged tissue, then it could prove useful in studying the electrophysiological functions of normal tissue in greater detail. Ferrofluids could possibly be reconcentrated using an external magnetic field and used repeatedly. It could also be concentrated in specific regions to study that particular tissue.

The size of the magnetite particles and the charge of the surfactant coating may contribute to the actual mechanism responsible for the amplitude increase described in this study. Since the actual mechanism is not thoroughly understood at this time, it is not possible to draw any conclusions regarding the various factors that may be contributing to the increase in signal amplitude. The effect of the above two properties of ferrofluids may in turn be different for different tissue types.

REFERENCES:

[1] Senyei, A., Widder, K., Czeslinsky, G. Magnetic Guidance of Drug Carrying Microspheres, Journal of Applied Physics, 1978 49: 3578-3583.

[2] Chopp, M., Chen, Q., Moran, J.E., Tepley, N. Biomagnetic Measurements Utilizing Ferrofluids, Proceedings of the 8th International Conference on Biomagnetism.

[3] NewBower, Ronald.S. Magnetic Fluids in the Blood, IEEE Transactions on Magnetics, 1973, MAG-9, 3: 447-450.

This research supported by NIH/NINDS, grant no. 1R01 NS30914.

Magnetic Field and Electric Potential of Excited Plant Cell Chara Corallina: Calculation and Comparison with Experiment

Šlibar[1], M., Trontelj[1], Z., Jazbinšek[1], V., Thiel[2], G., Müller[3], W.

[1]*Institute of Mathematics, Physics and Mechanics, University of Ljubljana, SI–1111 Ljubljana, Slovenia;* [2]*Pflanzenphysiologisches Institut, Universität Göttingen, D–37073 Göttingen, Germany;* [3]*Physikalisch–Technische Bundesanstalt, Institut Berlin, D–10587 Berlin, Germany*

Introduction

It has been shown that the propagation of a single action potential in the internodal cell of green algae Chara corallina can be also detected magnetically [1, 2]. From this experiment some parameters, like the velocity of the depolarization wave along the excited cell, the length of the depolarized area and the peak value of the intracellular current were determined.

The relatively simple geometry and structure of the internodal cell of Chara corallina offer also a possibility to calculate the magnetic field distribution in the cellular vicinity if the transmembrane potential is known from measurements. In the inverse calculation the transmembrane potential can be obtained from the measured magnetic field. The results of these calculations will be shown.

Similarities in the electrophysiological behavior between these large single plant cells and large animal nerve axons are of considerable help here. Both, the cylindrical symmetry and quasi–stationary equations for electric and magnetic quantities can be applied in the Chara corralina case. So we can start with results obtained by Clark and Plonsey [3] and by Woosley, Roth and Wikswo [4]. They have formulated relations between the action potential, cellular currents and magnetic field for the single nerve axon.

Calculations

The potential calculation is based on the Laplace equation

$$\triangle\Phi = 0.$$

with the boundary conditions: The transmembrane potential Φ_m is equal to the difference of intracellular Φ_i and extracellular potential Φ_e

$$\Phi_m(z) = \Phi_i(a, z) - \Phi_e(a, z),$$

and the normal component of current density at the membrane surface is continuos

$$\mathbf{n} \cdot \mathbf{J}_i(a, z) = \mathbf{n} \cdot \mathbf{J}_e(a, z),$$

where $\mathbf{n}$ is the normal to the mebrane surface, $\mathbf{J}_i$ and $\mathbf{J}_e$ are intracellular and extracellular curent densities, respectively. The cylindrical symmetry and the meaning of all quantities are evident from Fig. 1.

Solution of Laplace equation for Φ_i and Φ_e was obtained by Clark and Plonsey [3]

$$\Phi_i(\rho, z) = \frac{1}{2\pi}\int_{-\infty}^{\infty} \frac{I_0(|k|\rho)}{\beta(|k|a)I_0(|k|a)}\phi_m(k)e^{-ikz}\,dk,$$

$$\Phi_e(\rho, z) = \frac{1}{2\pi}\int_{-\infty}^{\infty} \frac{K_0(|k|\rho)}{\alpha(|k|a)K_0(|k|a)}\phi_m(k)e^{-ikz}\,dk,$$

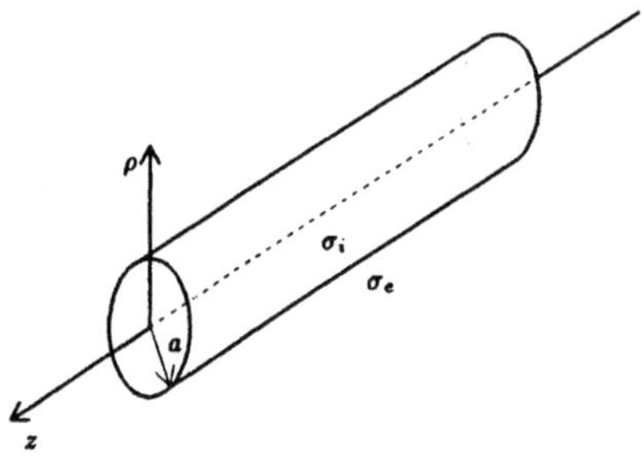

Figure 1: The cell model. $\sigma_i = 1.2\,\Omega^{-1}m^{-1}$, $\sigma_e = 0.025\,\Omega^{-1}m^{-1}$ and $a = 1\,mm$.

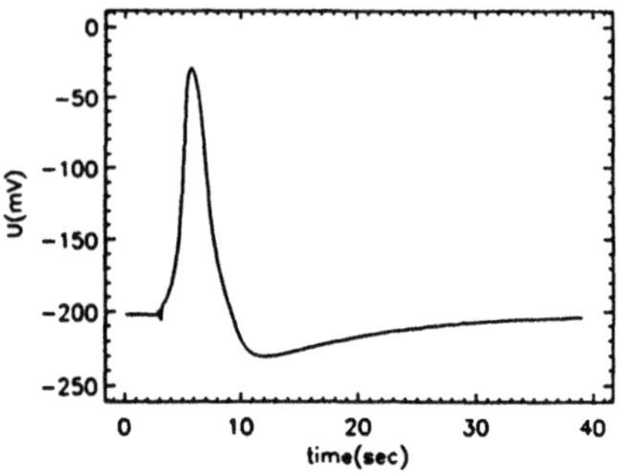

Figure 2: The whole sequence of measured action potential.

with

$$\alpha(|k|a) = -(1+\gamma(|k|a)), \quad \beta(|k|a) = 1 + \frac{1}{\gamma(|k|a)}, \quad \gamma(|k|a)) = \frac{\sigma_e K_1(|k|a) I_0(|k|a)}{\sigma_i K_0(|k|a) I_1(|k|a)},$$

where I_0, I_1, K_0 and K_1 are modified Bessel functions, and $\phi_m(k)$, the spatial Fourier transform of the transmembrane potential

$$\phi_m(k) = \int_{-\infty}^{\infty} \Phi_m(z) e^{+ikz}\, dz.$$

Further Fourier transformations give, as pointed out in [4]

$$\phi_i(\rho, k) = \frac{I_0(|k|\rho)}{\beta(|k|a) I_0(|k|a)} \phi_m(k),$$

and

$$\phi_e(\rho, k) = \frac{K_0(|k|\rho)}{\alpha(|k|a) K_0(|k|a)} \phi_m(k).$$

The magnetic field $\mathbf{B}_i$ due to the intracellular current density is obtained by the application of the law of Biot-Savart (or the Amper's law) and can be formulated, as it is shown in [4]

$$\mathbf{B}_i(\rho, z) = \int_{-\infty}^{\infty} \mathbf{G}(\rho, a, z - z') J_i^z(a, z')\, dz',$$

where $\mathbf{G}$ is the proper Green's function and J_i^z the intracellular current density component along the cell. More useful is the expression in the k space, since it is directly related to $\phi_m(k)$

$$\mathcal{B}_i(\rho, k) = i\mu_0 a \sigma_i k \frac{I_1(|k|a) K_1(|k|\rho)}{\beta(|k|a)} \phi_m(k).$$

Similarly, $\mathbf{B}_e(\rho, z)$ and $\mathcal{B}_e(\rho, k)$ due to the extracellular current density are obtained. The sum of the both $\mathcal{B}_i(\rho, k)$ and $\mathcal{B}_e(\rho, k)$ gives the $\mathcal{B}(\rho, k)$ which is connected via the Fourier transformation to the $\mathbf{B}(\rho, z)$. The whole transformation path in both directions from $\Phi_m(z)$ to $\mathbf{B}(\rho, z)$ is shown schematically on Fig. 3. The results of calculated and measured action potentials and magnetic field are shown on Figs. 4, 5, 6, 7.

$$\Phi_m(z) \underset{F^{-1}}{\overset{F}{\rightleftarrows}} \phi(k) \rightleftarrows \mathcal{B}(\rho,k) \underset{F}{\overset{F^{-1}}{\rightleftarrows}} \mathrm{B}(\rho,z)$$

Figure 3: Calculation steps from the transmembrane potential to the magnetic field. F denotes the Fourrier transformation, F^{-1} the inverse Fourier transformation.

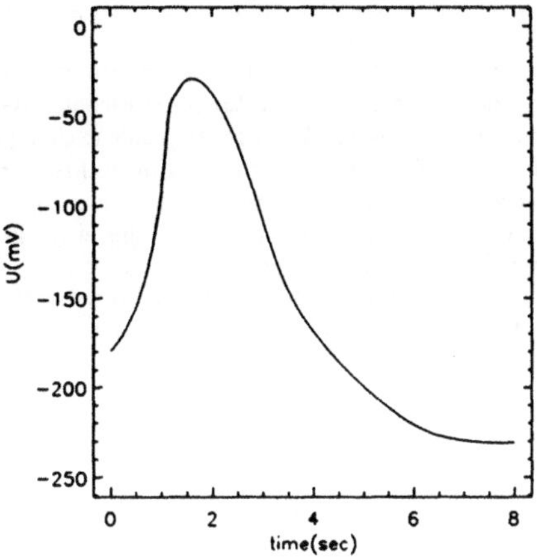

Figure 4: The first 8 seconds of measured action potential in the Chara corralina internodal cell.

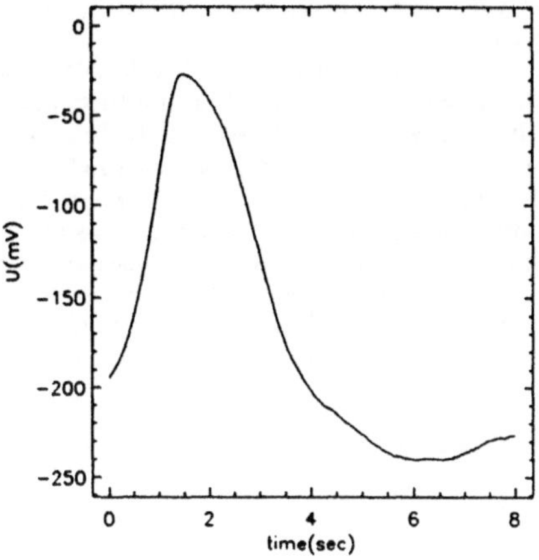

Figure 5: Action potential in the Chara corralina internodal cell calculated from the measured magnetic field shown in Fig. 6.

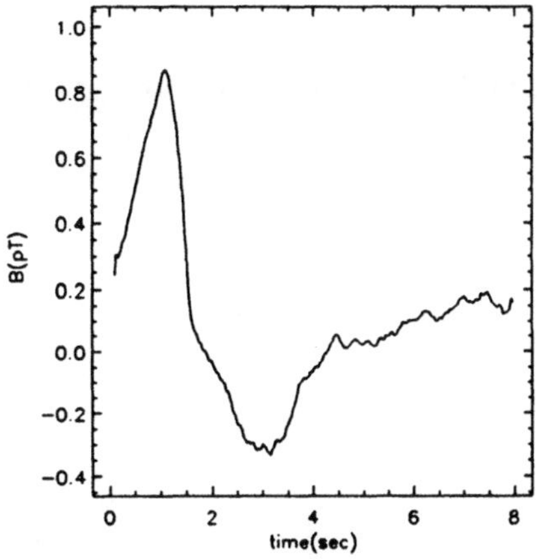

Figure 6: Measured magnetic field $6\,cm$ away from the Chara corallina internodal cell.

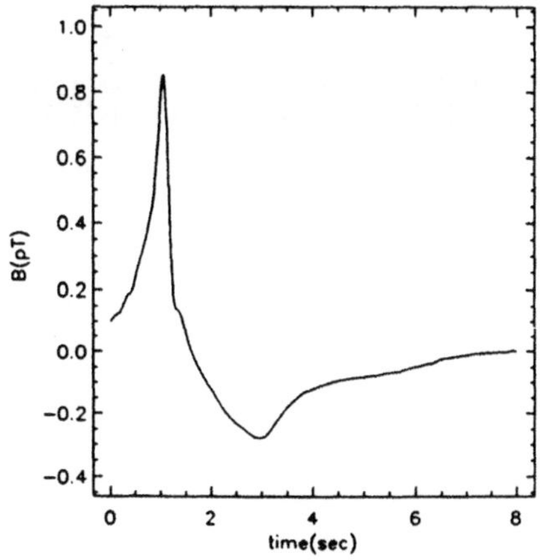

Figure 7: Magnetic field $6\,cm$ away from the Chara corallina internodal cell calculated from the measured action potential shown in Fig. 4.

Discussion and conclusions

Comparing the measured and calculated magnetic field and action potential for the internodal cell of green algae Chara corralina we notice a relatively good agreement between the calculated and measured signals, however, there are differences in the shape of signals.

We have to remember that mainly the intracellular current determines the measured magnetic signal in our experiment [1, 2]. The cylindrical symmetry, though valid within narrow area arround the Chara corralina internodal cell when considering the extracellular current density contribution, is correctly applied in our case.

It has to be mentioned that the internodal cell of Chara corralina consists actually of two membranes: The outer plasmalemma and the inner tonoplast which separates the cytoplasm from the inner vacuole. We believe our approximation where only the plasmalemma takes complete responsibility for the ionic exchange between extracellular and intracellular space is to big extent correct. That agrees with experimental results of plant electrophysiologists [5]. However, for the more precise calculation both membranes have to be considered separately and not only integrally as we did in our case. It is to expect that this way the improved agreement between experimental and calculated signal shape will be achieved.

We can conclude that expressions for electric potentials and magnetic fields of single large nerve axon, described in [3] and [4], can be applied also for the case of single internodal cell of green algae Chara corallina at least for the contributions of intracellular current density.

References

[1] Trontelj, Z., Zorec, R., Jazbinšek, V., and Erné, S.N. Magnetic detection of a single action potential in Chara corralina internodal cell, Biophysical Journal, 1994, 66:1694-1696.

[2] Jazbinšek, V., Trontelj, Z., Zorec, R., Erné, S.N., Lužnik, J., Pirnat, J., and Jagličić, Z. Magnetically detected intracellular current in single cell of the green algae Chara corralina, In: Baumgartner, C., Deecke, L., Stroink, G., and Williamson, S.J. Biomagnetism: Fundamental Research and Clinical Applications, Amsterdam, Elsevier, 1995.

[3] Clark, J., and Plonsey, R. A mathematical evaluation of the core conductor model, Biophysical Journal, 1966, 6: 95–112.

[4] Woosley, J.K., Roth, B.J., and Wikswo jr., J.P. The magnetic field of a single axon: A volume conductor model, Mathematical Biosciences, 1985, 76:1–36.

[5] Beilby, M.J., Electrophysiology of Giant Algal Cells, Methods in Enzymology, 1989, 174:403–442.

System for Magnetic Susceptibility Investigations of Human Blood and Liver

Sosnitsky, V.N., Budnik, N.N., Minov, Yu.D., Sutkovoj, P.I. and Vojtovich, I.D.

Biomagnetic Laboratory, Institute of Cybernetics, Kiev, Ukraine

Introduction

Recent years there is growing evidence of iron overloading involvement in the pathology of atherosclerosis, cancer, hepatitis, diabetes, diseases of joints. Reactions of free radical oxidation are involved in the pathology of most, if not all, human diseases. Free radical oxidation seems to be involved also in disease development in "iron overloaded" patients with high levels of the so-called "free iron" in tissues (the catalysis of peroxidation). This idea is supported, in particular, by severe cardiovascular disorders observed when excess ascorbic acid is given to iron-overloaded patients. This as well as our observations suggest the problem of exogenic iron role in steel and cast iron welders' pathology development.

We have created computer-aided systems in which these and some other problems are solved. One of the systems is used for recording of magnetic fields of the cardiovascular system, the other -- for investigation of magnetic susceptibility of various organs and blood. Signal processing and representation are performed by specially designed algorithms and programs.

Susceptibility Biomagnetic System

The biomagnetic system used to register the signals coming from blood and human body organs (induced by a test magnetic field) has the same basic components as the MCG system [1,2]. A movable patient's bed permits to move a patient in vertical and horizontal directions. The test magnetic field is created by the system of square Helmholtz coils. The case of each coil is made from the aluminum details when side equals 2 m. The distance between coils is 108.9 cm. The parameters of one coil are as follows: the numbers of turns, resistance and inductance are equal to 72, 14 Ω and 53 mH, respectively. The coils provide the alternating (14 Hz) test magnetic field of about 78 μT in the system center. The coils are driven by an alternating low-frequency signal from generator by the power amplifier. The equality of the currents in the coils is caused by them serial connection. It provides heterogeneity of magnetic field at least than 10^{-5} within 70×70×120 mm volume (in the center of the system). The patient's bed and the cryostat are situated in the middle of the coils' system so that the bottom coil of gradientometer was allocated exactly in the center of the system of the Helmholtz coils. A rubber bag is fastened to

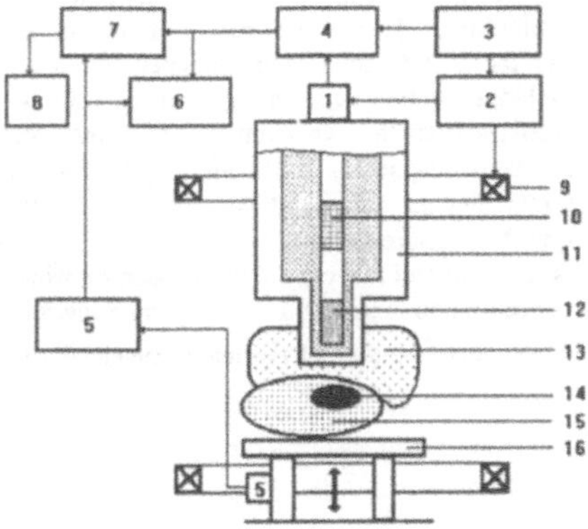

Fig.1. Cryoelectronic system for magnetic susceptibility measurement of human blood and liver: 1 - SQUID electronic; 2 - power amplifier; 3 - generator; 4 - lock-in amplifier; 5 - position transducer; 6- 2-channel chart recorder; 7 - ADC; 8 - PC; 9 - Helmholtz's coils; 10 - SQUID-detector; 11 - radiotransparent helium cryostat; 12 - gradientometer; 13 - water bag; 14 - liver; 15 - patient's body; 16 - pneumatic driven bed

the cryostat tail and this is one for a parasitic paramagnetic air signal to be removed. The upper part of the bag is a Plexiglas disk having Ø45 cm (with the circular region for the tail, where the thickness was reduced) which rigidly fastened to the cryostat. The bag is filled with the distilled water. Under the movement of a patient upwards the volume of the bag is decreased and the water is poured into expanding reservoir by the connective pipe and lets a patient be taken down at the speed of 0.5 cm/s. The movement of the bed is caused by the pneumatic driver. The bed distance is registered by a position transducer. Despite of the undertaking efforts to receive a high degree of balance of antenna and homogeneity of the magnetic field , the signal created by the current in the coil fill a full dynamic range of SQUID-magnetometer at the value of this current beginning from 0.2 A. To eliminate this effect the compensation of the signal created by the magnetic field is used. Here a part of the signal from the output of the power amplifier is fed in the feedback circuit of the magnetometer. The output signal of the magnetometer is fed by lock-in amplifier to chart recorder or by ADC on the PC. The design of the system is presented in Fig.1.

Magnetic biopsy of liver.

Method of measurement. A test magnetic field with the frequency of 14 Hz is used under the magnetic biopsy. Before measuring the amplitude and phase of a signal which enters from an output of a power amplifier in a circuit of feedback of the magnetometer, are fitted so that the value of a "parasitic" signal for the frequency of 14 Hz (and its harmonic) is minimized from the output of the magnetometer. A measuring procedure is as follows. A patient idles (on a bed) at an angle of 45°, where a liver is in an upper position. The bed is lifted in an upper position, then the bag fits tightly the body of the patient. Thus, a "parasitic" signal because of air is removed. Under measuring the bed is lowered to a distance of 5 cm for a time of 10 s. The output signal of the magnetometer, which is proportional to the magnetic susceptibility of the liver, is input of a lock-in amplifier. A signal, which is proportionate to a distance between the liver and a bottom coil of the gradientometer, is registered from an output of a transducer of bed displacement. These signals are input of a 2-channel chart recorder. Further, the mentioned signals are sent conveyed to PC through ADC. The measurements are performed a few times to improve the SNR. Strictly speaking, the signal from the output of the lock-in amplifier is proportionate to the difference of magnetic susceptibilities of the liver and distilled water. The desired concentration is found by the expression $C=K[dV(C)-dV(0)]$, where K, dV(C) and dV(0) are the calibrating coefficient, the change of the signal from the liver depending on the distance, and the change of signal for the case of iron absence, respectively.

The value of the change of the signal depending on the distance is determined as a difference of the signal between two points X_1 and X_2, where the dependence of the signal on the distance is about linear $dV=V(X_1)-V(X_2)$. Choice of these points is complied with by trials to obtain the above mentioned most linear dependence. For this case, points have corresponded to the distances of 2 cm and 5 cm between the bottom coil of the gradientometer and surface of patient body.

The "parasitic" signals, which are present in the process of the measurements, can be divided into two classes. The signal from the Helmholtz coils and the vortex currents, generated by test magnetic field, fall into first class. These signals are minimized by mechanical balancing of the gradientometer and, finally, by fitting the compensation of both amplitude and phase in the feedback circuit of the magnetometer. The above mentioned signals do not depend on the distance to the liver and, therefore, are removed by data processing since they give the same contribution in the output signal for both a neighboring and outlying position of the liver.

The signals from the bed and body of patient fall into second class of the "parasitic" signals. These signals depend on the distance to the liver and, therefore, give a contribution in the difference dV. However, the signal from bed is taken into account by the calibration. This is also true for almost all parts of body surrounding the liver. The magnetic susceptibility of these parts equals approximately the susceptibility of distilled water. Under calibrating the signals from both lungs and blood circulating in the liver are not taken into account. The signal from lungs (its susceptibility differs from the susceptibility of water because of the presence of air in lungs) is very small that is argued by the corresponding calculations [3]. The liver contains 15-18 ml of blood for 100 g of the tissue of the liver, that give the equivalent concentration of iron in 78-94 μg/g. If it is necessary, this value can be subtracted from the measured signal.

The calibration has been carried out by glass (250 ml) and rubber (500 ml) capacities, which have been filled up by a solution of $FeCl_3 6H_2 0$ with equivalent concentration of iron of 0.1; 0.5; 1 and 5 mg/ml. Comparing the received result with a theoretical dependence of the output signal of the magnetometer on the magnetic susceptibility, we obtain a calibration characteristic for arbitrary concentration of iron in the liver. Here it was enough to measure two values of the concentration, since both the theoretical calculations [3] and experimental results show a good linear relation between the output signal and the concentration of iron. It can be concluded analyzing the shape of this characteristic that the concentration of iron of 0.1 mg/g and higher can be measured by the described device. The calibration signal is accumulated for several measurements, then this signal is averaged and the value dV is calculated. The calibration coefficient is computed by $K = [C_1 - C_2]/[dV(C_1) - dV(C_2)]$, where C_1-C_2 and $dV(C_{1,2})$ mean the difference of the concentrations for two measurements and the corresponding difference of the output signal, respectively. Such calibration by a "chemical" standard should be implemented after each engaging of the device. However, it is inconvenient for clinic conditions. An "electric" calibration by a coil, which is fed by rectangular impulses of a current of 14 Hz, is more convenient. The amplitude of the impulses in the coil is calibrated so that the signal from the output of the magnetometer equals the signal from the capacity with the known concentration of iron.

The main systematic error changes of dependence on the concrete patient and relates to an uncertainty of a thickness of the skin-liver layer. According to the published results, the nominal value of the thickness is equal to 20 mm, a range is within 15 and 25 mm. This range leads to an uncertainty of a value of iron concentration of 2.5 mg/g; this value is completely improper. To decrease this error the ultrasonic technique should be used, which has a precision of determination of the mentioned thickness of 1mm, that allows to diminish the error by about 10.

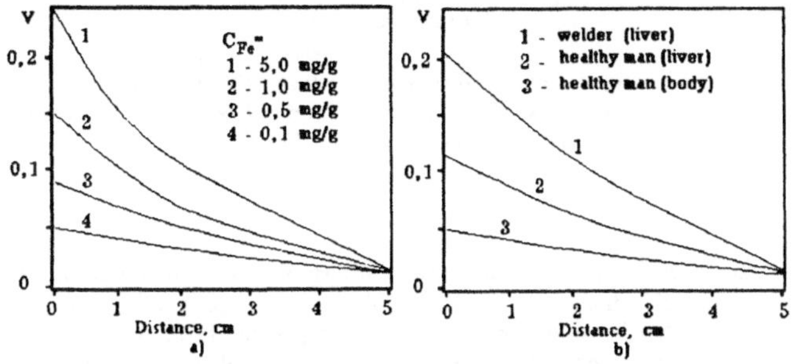

Fig. 2. The dependence of the output signal of biosusceptometer on a distance to the set of the rubber capacities with different calibration values of iron concentration (a), the liver of a healthy person and welder (b).

It is seen from Fig.2a that the iron concentration higher than 0.1 mg/g liver can be measured by the described measuring system. This means that a normal concentration (0.1-1 mg/g) and an overloading (more than 1 mg/g) can be recorded. Fig.2b presents the results of the measurements of the magnetic susceptibility of the liver of both a healthy man and a welder suffering from occupational diseases. A concentration of 0.8 mg/g liver has been derived for the healthy man. It corresponds to the normal value. The iron concentration for a group of healthy subjects varies within 0.5-0.8 mg/g. A high iron concentration about 3 mg/g has been registered for the welder. The distance to object is measured from its upper position. The signal from the body (curve 3) has been received for the position of the patient when the liver is situated in the lower position.

Magnetoplethysmography (MPG)

MPG means the measurement of the blood volume changing for a cardiocycle from human organs by a signal from blood in external test magnetic field. This signal can be elicited due to a difference between the magnetic susceptibilities of blood $\chi_b = 5 \cdot 10^{-6}$ and body (χ_{water}). The change of the blood volume in heart for cardiocycle leads to the change of magnetic field in the bottom coil of the gradientometer. The essence of the method of MPG developed in [4] is based on the application of a DC test magnetic field with a value of 135 A/m (170 μT). We use the AC field with a frequency of 114 Hz, then the noise of the magnetometer and various external noises are considerably decreased. Furthermore, a "parasitic" MCG signal has no influence for this frequency.

Along with obtaining the MPG the system is used for the magnetic biopsy of the liver, where technical changes are introduced as follows. The LFF with a cut-off frequency of 30 Hz is included after the lock-in detector. A capacitance of 22 μF is included serially to the Helmholtz coils for compensating its inductance. This makes it possible to increase a current in the coils up to 1,1 A. Here the ECG is used as an input signal for the reference channel instead of the signal of a bed displacement. Before measuring, the amplitude and phase of a signal of 114 Hz (which is sent from the power amplifier output to the NF circuit of the magnetometer) are fitted in an appropriate manner.

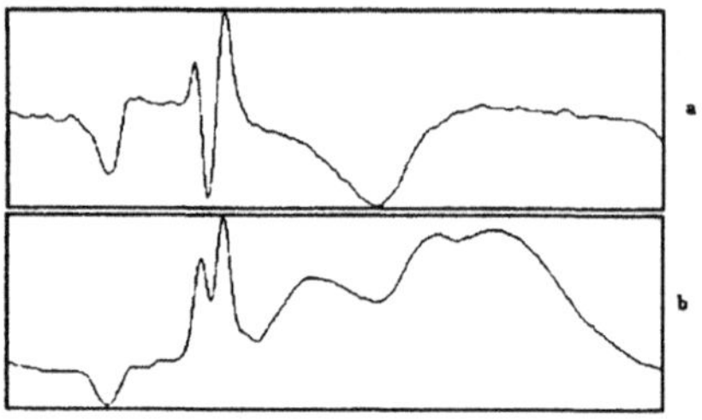

Fig.3. The magnetoplethysmogram of heart of a healthy subject (a) and a welder with iron overloading (b)

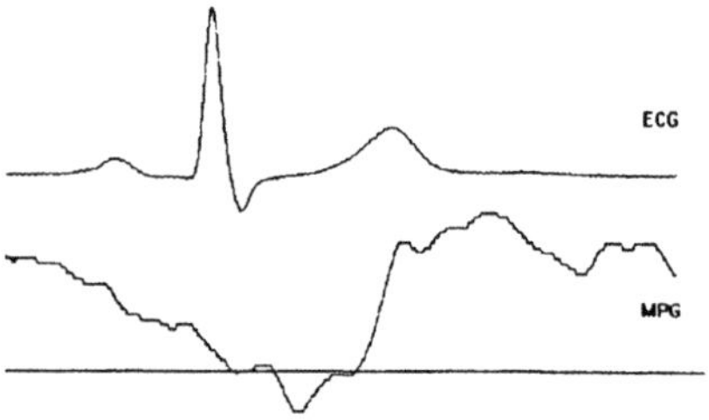

Fig.4. The magnetoplethysmogram of liver of a welder (ECG is presented for reference).

The MPG of healthy man and welder having an augmented content of iron in blood presented in Fig. 3. This iron overloading is registered by biochemical analysis too. The iron content level in blood is so high among some welders that it became possible for us to have registered liver MPG for the first time (Fig.4).

Conclusion

Computer-aided systems for recording, processing and representation of own or induced magnetic signals emitted by various human organs are created. Relaxation-oscillation (RO) SQUID-magnetometers are the basis of these systems [1]. When ROs are used in a SQUID-magnetometer, it becomes possible to refuse from the modulation-demodulation scheme applied in the conventional analog mode. The result is that we can pass from the microvolt signal registration to the millivolt pulse signals. There appears the ability to simplify the electronics, to enhance SR and noise-immunity, to widen the dynamic range, to speed up the searching process. This circumstance provides the reliable operation of the device in the presence of the high industrial magnetic noise level without the special means for magnetic vacuum creation (magnetic shielding room, active Helmholtz coil compensation).

Over the last several years a group of scientists from Institute for Occupational Health of the Ukrainian Academy of Medical Sciences, Biomagnetic Laboratory of Institute of Cybernetics and Laboratory for Finely Dispersed Ferromagnetics of the Research SONAR Center for Biotechnical Systems of the National Academy of Sciences of the Ukraine is engaged in research of studying the impairments in welders' health due to exposure to welding aerosols. The methods that were use: standard biochemical, immunological methods as well as biomagnetometry.

References

[1] Budnik, N.N., Minov, Yu.D., Sosnitsky, V.N., Sutkovoj, P.I., Vojtovich, I.D. Pulse-relaxation oscillation SQUID-magnetometer, Proc. XIII IMEKO World Congress, Torino(Italy), 1994, III, 2383-2387.

[2] Budnik, N.N., Gapelyuk, A.V., Minov, Yu.D., Primin, M.A., Sosnitsky, V.N., Vojtovich, I.D. Human biomagnetic signal measurement for diagnosing some diseases, Ibid., 2509-2514.

[3] Romanovich, S., Sosnitsky, V., Vojtovich, I. Investigation of biomagnetic field using two-dimensional model of secondary sources, Proc. 9th Int. Conf. Biomagnetism, Vienna (Austria), 1993, 455-456.

[4] Maniewski, R., Katila, T., Poutanen, T. et al., Magnetic measurement of cardiac mechanical activity, IEEET, 1988, BME-35, 662-670.

Registration of Iron Overload by Means of SQUID-Magnetometry When Coupled with Low-Temperature ESR Spectroscopy

Sosnitsky, V.N.[1], Mikhailik, O.M.[2], Lubyanova, I.P.[3], Gapelyuk, A.V.[1], Budnik, N.N.[1], Minov, Yu.D.[1], Sutkovoj, P.I.[1], Pankratov Yu.V.[2] and Bakai, E.A.[2]

[1]Institute of Cybernetics, Ukrainian Acad. Sci., Kiev, Ukraine; [2]Research SONAR Center for Biotechnical Systems, Ukrainian Acad. Sci., Kiev, Ukraine; [3]Institute for Occupational Health, Ukrainian Acad. Med. Sci., Kiev, Ukraine

Introduction

At present new noninvasive biomagnetic methods of tissue iron quantification for primary and secondary hemochromatosis diagnosis are gaining fast development mainly as liver susceptometry [1]. However, often iron accumulation in other organs (spleen, heart, pancreas etc.) is also of importance. Studies have been performed on ferromagnetic contaminants in the lung and estimation of lung clearance, it was determined that in a year after contamination of the lungs by magnetite ranged from 10% to 50% elimination of ferromagnetic materials occurred [2,3]. Whether these particles enter the blood flow, whether they accumulate in other organs have not yet been clear. Our examinations of more than 2000 of iron welders have revealed the possibility of iron overloading syndrome development during iron inhalation intake from welding aerosol. In hemosiderotic patients accumulation of iron in cardiac muscle results in impairment of cardiac activity, enlargement of myocardium mass of the left ventricle [4], this was found to be also true for welders with pneumoconiosis [5]. The purpose of the present work was to elucidate the peculiarities of iron state in patients with iron body overload resulting from inhalation exposure to iron containing welding aerosols by means of magnetocardiography (MCG), magnetic-susceptibility plethysmography (MSPG) and low temperature electron spin resonance spectroscopy (ESR) methods.

Methods

Men-welders with occupational diseases of bronchopulmonary system were examined. The diagnosis of iron overload was confirmed primarily by biochemical parameters (serum iron and serum iron binding capacity). In the process of MCG studies a component of the magnetic field perpendicular to the chest surface in 36 points of right-angle measuring grid with a step of 4 cm was registered. The measurements were made with single channel SQUID-magnetometer under the following parameters: sensitivity - 30 $fT/Hz^{1/2}$, frequency band 0.1-100 Hz. Recording duration for each point amounted to 30 sec, discretization frequency - 500 Hz. MSPG were registered over the end of left ventricular in the alternating magnetic field of about 0.1 mT with a frequency of 114 Hz to decrease the noise of magnetometer as well as external noises. The ESR spectra of blood and serum samples were registered at microwave radiation frequency ranged from 9.28 to 9.3O GHz, microwave power - 5-50 mW, magnetic field - 500-5500 Oe, temperature 100-150K. The levels of serum transferrin, iron in the form of serum ferritin as well as the degree of serum transferrin and ferritin saturation with iron were calculated on evidence derived from low temperature ESR data with calibration of ESR ferritin spectrum according to gamma-resonance spectroscopy data.

Results and discussion

Among welders with pneumoconiosis patients with iron serum concentration above 37 μmol/l were registered frequently than in patients with chronic bronchitis. Total serum iron binding capacity in welders amounts to 91±5 μM. The major differences in morphology of the MCG curves in the group of welders relative to the control group were observed at ST-T interval corresponding to the period of mechanical systole at the absence of heart electric activity (Fig.1), MCG curves showed the appreciable deviation from magnetic isoline (10-30 pT) in comparison with the control one (1-2 pT). This deviation was chosen as an index A_d (Fig.1b). The average integral coefficient for all 36 points $\langle A_d \rangle$ obtained when analyzing magnetograms of 10 men from the control group which were never exposed to iron inhalation and had no diseases of cardiovascular system did not exceed 0,5 relative units. This index in 18 examined welders was higher than 0,5 and amounted to 4.1 relative units.

Evidently, the welders MCG signal is superposition of the signal due to electric cardiac activity and magnetic susceptibility signal directly related to cardiac volume changes and movements only with the Earth magnetic field as magnetizing one. In the magnetic field of the Earth a magnetic susceptibility signal of about 1 pT at most can be expected [6]. The changes of the amplitude approximately by ±2 pT when changing the direction of the magnetized external field strength 135 A/m were also observed [7] and explained as occurring due to moving of blood with paramagnetic species as a result of changes in heart volume. When the heart is exposed to magnetic field 520 A/m the maximum amplitude of the observed signal is about 8 pT [8]. In our case the signal amplitude up to 30 pT was observed only at the Earth magnetic field.

The considerable increase of $\langle A_d \rangle$ and A_{dmax} was noticed in welders with pneumoconiosis and increased liver echogenicity. It is likely to suppose that the registered magnetic signal is associated with deposition of magnetic iron species in the lungs or liver. However, according to Maniewski's estimations [6] magnetic susceptibility plethysmography signal contributes 85 to 90% of the total magnetic susceptibility signal measured above the heart and the signals caused by displacement of the chest surface and by the motion of the chest wall-lung boundary contribute 15 to 20 and 5 to 10%, respectively. Even considerable increase of concentrations of paramagnetic blood constituents as transferrin and low-molecular iron complexes cannot elucidate observed ten-fold increase of the magnetic component amplitude. Four cases of patients with high level of serum iron and transferrin in blood but with normal signal amplitude at TP interval support the idea of the other source of magnetic component which is registered in a number of subjects. At the same time 1/3 of welders with high A_d showed absence of changes in the lung roentgenogram typical for pneumoconiosis in welders. In 1/3 of welders of this group echogenic liver properties were not changed, a high magnetic component was registered in two welders with normal echogenic liver properties. Thus, there is a reason to suppose that intensive signal in the area of quadrants C3-C5, D3-D5, E3-E5 according to the measuring grid can be associated with high concentration of ferritin in the blood and/or ferromagnetic inclusions entering the blood flow and/or cardiac muscle from the lungs of the examined welders.

The examination of blood and serum samples of patients with high MCG deviation from the magnetic isoline at T-P interval (Fig.3) and their comparison with the spectra of ferritin (Fig. 3f) has shown that in addition to the components which can be referred to ferritin [9] new ones are registered. Similar spectra were registered in tissues of experimental animals after biodegradation of intravenously administered finely dispersed metallic iron. It is important to note that the intensity of new registered components in the whole blood significantly exceeds that in serum samples. Thus, the sources of these signals may partially deposit with blood elements when receiving the serum and the index content of iron in the blood serum does not reflect the true content of non-heme iron in the whole blood. The changes in the position and width of the absorption lines under rotation of the blood samples frozen in the residual field of ESR-spectrometer (250 Oe) (Fig. 3a-3e) testify sufficiently high magnetic properties of the absorbing species. The clusters and small particles of oxidized iron with a few magnetic subgrids and their aggregates can be considered as these particles [10]. One could attribute these lines to ferritin molecules aggregates displaying strong magnetic anisotropy and/or to iron magnetic impurities that penetrate from the lungs into the blood flow.

MPCG curves for examined patients with iron overload are of rather complex character. It was found that parameters iron in the form of serum ferritin and degree of serum ferritin saturation with iron are in interrelation with MPCG curves. High (up to 30 pT) amplitudes of plethysmograms (Fig. 2c) suggest the idea about the presence of species with high magnetic properties in the blood of some patients and impulsive deviations toward positive magnetic fields in a phase of heart contraction seems to be an evidence of magnetic species accumulation in a heart muscle (Fig. 2b). It is noteworthy that for patients with such kind of deviations in MPCG increased organ echogenicity was registered and the total iron binding serum ferritin ability was low. The highest amplitude of MPCG curve (Fig.2b) was observed in a patient with maximum iron concentration in serum ferritin (5600 μg Fe/l) and highest degree of serum saturation with iron (47%). It seems that the low ability to apoferritin synthesis leads to accumulation of iron in organs whereas high capacity for ferritin production to some extent prevents from deposition of iron in tissues. As in the case of liver magnetic biopsy [6] common parameter as transferrin level as well as the degree of its saturation with iron have no diagnostic value for estimation of nonheme iron in organs whereas amplitudes of MPCG curves and the parameters A_d correlates with observed frequency of pneumoconiosis and chance of increased liver echogenicity (Fig.4).

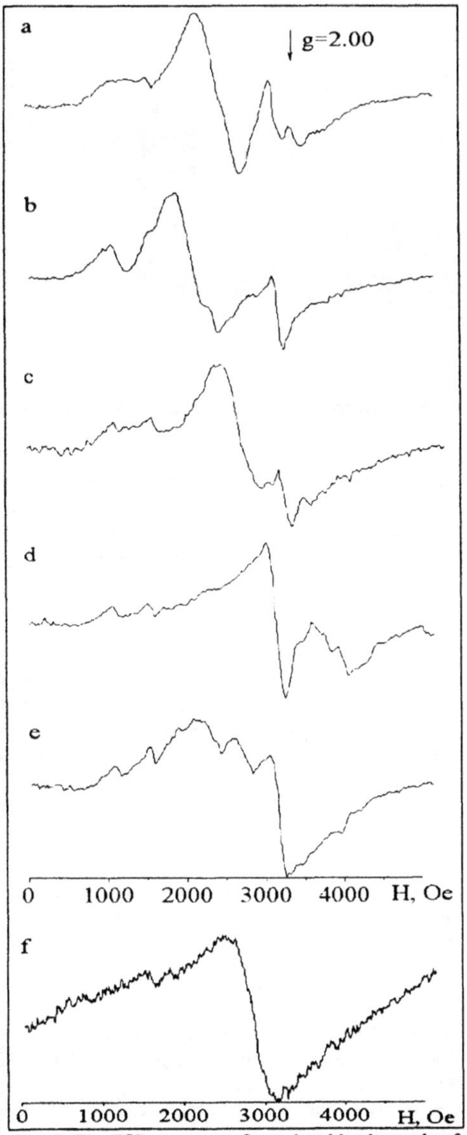

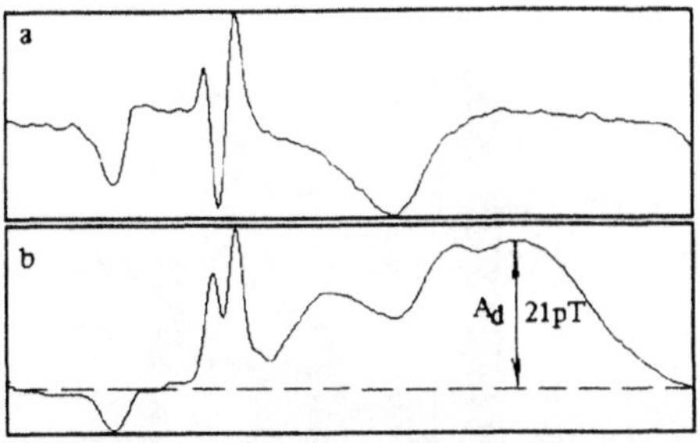

Fig.1. Magnetocardiograms of a healthy patient (a) and of a welder with iron overload (b) in C4 points of the spatial grid.

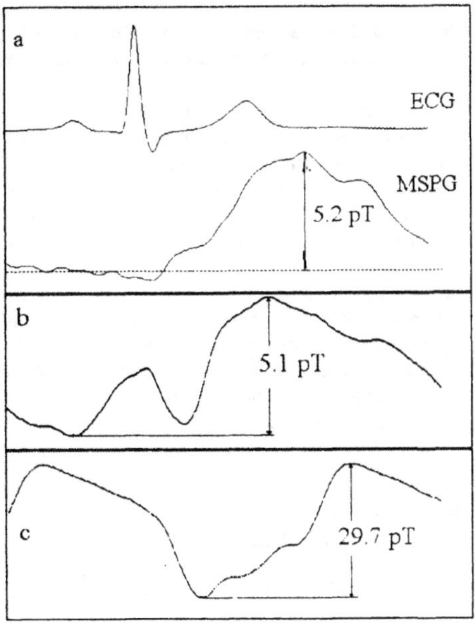

Fig. 2. Electrocardiogram (ECG) and magnetic-susceptibility plethysmograms (MSPG) for a healthy subject (a) and for patients with iron overload as a result of iron inhalation entry, iron in serum ferritin 220 and 5600 μg/l, respectively (b and c).

Fig. 3. The ESR spectra performed on blood sample of a welder with maximum deviation (A_d=31pT) from isoline in MCG (a) and after sample fast freezing in a magnetic field of 250 Oe at the angles between the magnetic field vectors during sample freezing and spectrum registration 0° (b), 30° (c), 60° (d), 90° (e), 180° (a) and ESR spectrum of highly saturated ferritin (f) performed on serum sample of a welder with heavy iron overload (5600μ g/l iron in ferritin); ν=9.3 GHz, microwave power 50 mW, HF-modulation amplitude 10 G, T=124K.

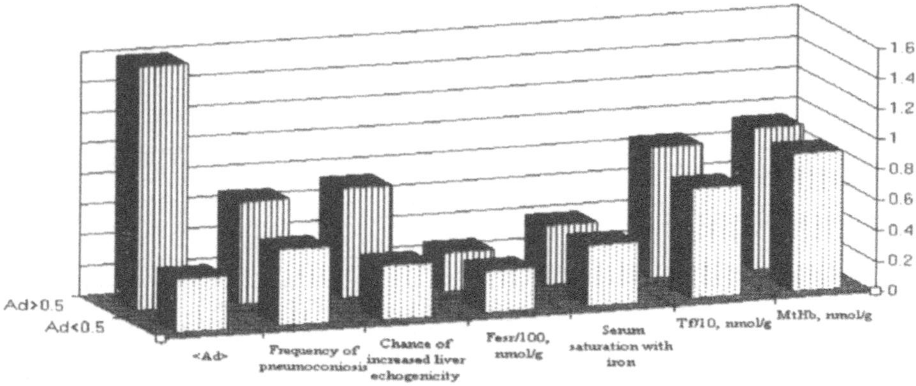

Fig. 4. Average deviation from isoline Ad (rel. un.) at ST-T interval of magnetocardiogrames and some characteristics of iron metabolism for two groups of patients.

Conclusion can be made that common calibration in biomagnetometry for quantitative iron determination using paramagnetic iron compounds and magnetic susceptibility determined for ferritin with low degree of saturation with iron is not appropriate for cases of significant iron overload in the presence of magnetic species. The index - content of iron in the blood serum - does not reflect the true content of non-heme iron in the whole blood under inhalation iron entry, the parameters iron in the form of serum ferritin and degree of serum ferritin saturation with iron are in interrelation with MSPG curves. To elucidate the nature of the revealed magnetic signals both in magnetograms and in the ESR spectra is of importance for solving calibration problem in the case of body iron overload. Not only the enhancement of 'free' iron concentration but also the local magnetic field of iron magnetic species could be considered as an important factor biological action (in particular, as promoter of lipid peroxidation) and one of the main mechanisms of pathology development. In this context the development of methods ferromagnetic species detection in tissues seems rather important. MCG and MSPG when coupled with ESR spectroscopy has much potential for yielding information about kinds and content of iron pools in organism under the iron overload.

References

[1] Fisher, R. and Heinrich. H.C. Biosusceptometry - current status of clinical diagnostics and biomagnetic research, In: Hoke, M. et al. Biomagnetism: clinical aspects, Elsvier Science Publishers, 1992.

[2] Cohen, D. Ferromagnetic Contamination in the Lungs and Other Organs of Human Body, Science, 1973, 180: 745-748.

[3] Kalliomaki, P.L., Aittoniemi, K. and Kalliomaki, K. In: Williamson, S.J. et al. (eds) Biomagnetism: An Interdisciplinary Approach, New York, Plenum Press, pp. 559-568.

[4] Yeliseev, O. M. Heart lesion in hemochromatosis, Ter. Arkhiv., 1988, No 8: 141-147 (in Russian).

[5] Krasnyuk, E. P., Lubyanova, I. P., Koshkina et al. The state of hemodynamics of chronic bronchitis and pneumoconiosis in welders, Vrachebnoye delo, 1988, No 3: 113-115 (in Russian).

[6] Maniewski, R. Magnetic studies on mechnical activity of the heart, Clin. Rev. Biomed. Eng., 1991, 19: 203-229.

[7] Katila, T., Maniewski, R., Tuomisto, T. et al. Magnetic measurement of cardiac volume changes, IEEET, 1982, BME-29: 18-25.

[8] Manievski R., Katila, T., Poutanen, T., Magnetic measurements of cardiac mechanical activity, IEEE Trans. Biomed. Eng., 1988, 35: 662.

[9] Weir I D, Peters T J, Gibson J F. Electron Spin Resonance studies of spleen ferritin and hemosiderin. Biochim. Biophys. Acta 1985; N 828: 298-305.

[10] Vonsovsky, S.V. Magnetism, Moskow, Nauka, 1971 (in Russian).

V. Magnetoencephalography (MEG): Basic Research

MEG Studies of Cognition

Salmelin, R.
Low Temperature Laboratory, Helsinki University of Technology, Espoo, Finland

Introduction

In this paper, the vast potential of MEG in revealing cortical processes of cognition will be discussed through experiments on picture naming, visual word recognition in normal and dyslexic subjects, and visual and motor imagery.

Methods

The measurements were done with a Neuromag-122™ whole-head magnetometer [1]. In this device, the 122 sensors are arranged in a helmet-shaped holder. In each block, there are two figure-of-eight planar pickup coils which are most sensitive to currents in the two orthogonal directions along the plane of the helmet. These gradiometers detect maximum signal just above a current source, where the field gradient is largest.

In addition to the on-line averaged evoked responses, also the on-going cortical activity was recorded and stored on a magneto-optical disk for later off-line analysis. Cortical rhythms often show systematic event-related changes. The oscillations are, however, not exactly time-locked to the stimulation or task and, thus, will not show in ordinary averaged responses. The modulation of cortical rhythms was quantified using the Temporal Spectral Evolution (TSE) method [2], where the signal is first filtered through a passband suggested by the frequency spectrum, typically around 10 and 20 Hz, and then its absolute value is taken. When this signal is averaged with respect to stimulus/task onset, one gets the average amplitude level of the narrow-band oscillations as a function of time. The cortical rhythms are modulated over periods of a few seconds whereas the time-locked evoked responses typically have a time scale of less than half a second.

Picture Naming

In this study the subjects were shown pictures of everyday objects. The measurement started with a passive viewing condition. In the overt naming task, the subjects were told to name the objects aloud as soon as possible. During the covert naming task, they were asked to name the pictures mentally, not preparing for vocalization [3].

Figure 1 shows the cortical dynamics of picture naming in a single subject. The occipital visual cortex was activated first. During the next 200 ms, bilateral signals occured over the temporo-parieto-occipital junctions. The early frontal source presumably indicates motor preparation for mouth movements. About 500 ms after picture onset, signals were observed in the inferior parts of the frontal lobes; the left-hemisphere region likely represents Broca's area. A site close to the vertex, slightly frontal to the foot area, then became active, apparently reflecting involvement of the supplementary motor area in coordination of vocalization. The early visual response and the signal in the right temporo-parieto-occipital junction were the same for all tasks. All other sources showed naming-specific responses, most strongly those in the left hemisphere. The traditional view maintains that the left hemisphere dominates in language tasks. In this experiment, however, signal processing during picture naming clearly advanced symmetrically in both hemispheres, from the visual to language- and eventually to vocalization-related areas.

Figure 2 illustrates the results of the TSE analysis of the 10-Hz activity during passive viewing and covert and overt picture naming, in the posterior, midline, and frontal areas. In fact, our original assumption was that very little would be seen in the phase-locked responses because of the large variation in latencies one might expect in this kind of task and that one would rather have to rely on the TSE analysis. During passive viewing, rhythmic activity was reduced only in the posterior region, that is, only the visual areas were strongly involved in this task. This effect was sharpened by covert naming. Overt naming led to a prominent suppression in the back of the head, lasting for about 3 s. It was followed by suppression in the frontal sites and, finally, along the central sulcus. The gradient isocontour maps of the overt naming condition illustrate how, immediately after stimulus onset, the parieto-occipital rhythms are strongly diminished. At about 500 ms, during the maximum suppression of the posterior rhythms, 10-Hz activity in the hand areas is briefly enhanced. About 900 ms after stimulus onset, suppression spreads to the inferior frontal cortex in both hemispheres. The signal processing thus seems to advance from visual to language and eventually to motor areas, in general agreement with the time-locked responses.

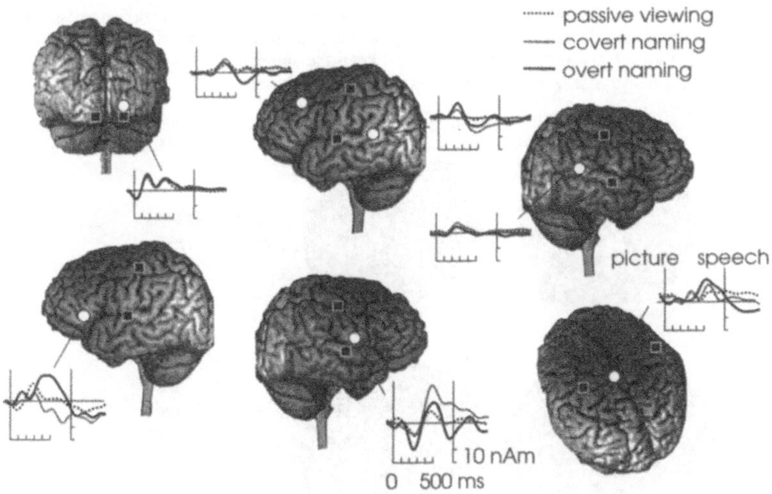

Fig. 1. Activation sequence in one subject during picture naming. The black squares indicate the visual, auditory, and hand somatosensory cortices for reference. The orange dots show the active cortical sites identified during the overt naming condition. The curves depict the temporal behaviour of the source areas over a period of about 1 s. The picture was shown at the first vertical line and vocalization started at the other vertical line.

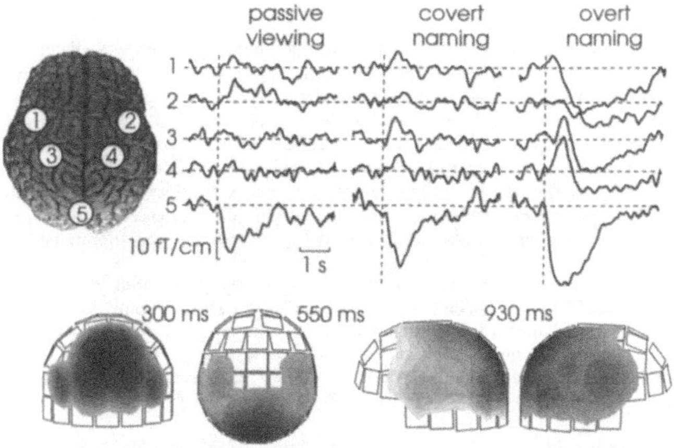

Fig. 2. TSE curves of 10-Hz oscillations during passive viewing and covert and overt picture naming in the numbered brain areas, plotted over a period of 5 s, starting 1 s before picture onset. The base levels are shown by the thin horizontal lines. Levels below and above zero indicate suppression and enhancement of rhythmic activity, respectively. The gradient isocontours below show the spatial extent of the maximum suppressions/enhancements at selected time instants during overt picture naming. Adapted from [3].

Visual Word Recognition in Normal and Dyslexic Subjects

Six subjects with developmental dyslexia and 8 control subjects participated in the experiment [4]. The dyslexic and control groups were matched for age and level of education and all subjects were also tested behaviourally. These dyslexics have eventually learned to read fluently but are still quite slow compared to normals.

The subjects were shown 7-8 letter Finnish words. Four categories alternated randomly. There were concrete words, abstract words, pronounceable pseudowords with no meaning, and meaningless random letter strings with the same amount of consonants and vowels as in the other classes. To keep the subjects alert, they were instructed to say aloud the rarely occurring word 'giraffe' which was not included in the analysis.

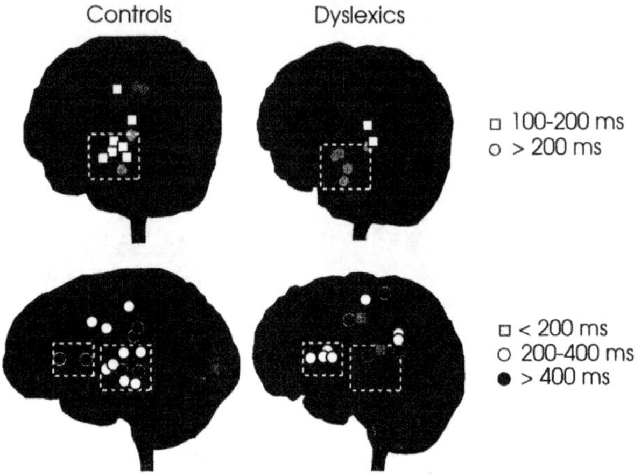

Fig. 3. Cortical areas activated by visually presented words in control *(left)* and dyslexic subjects *(right)*. Only the left posterior *(above)* and lateral views, summarizing areas with clear differences between the two groups (dashed rectangles), are displayed. The symbols indicate the different time windows.

The dominant cortical areas and the latencies of maximum activation were determined for all subjects. Figure 3 summarizes the active cortical areas for all control subjects on the left and for all dyslexics on the right. Activation within 200 ms after stimulus onset occurred predominantly in the posterior parts of the brain. After 200 ms, activity spread to the temporal and frontal areas. Controls and dyslexics showed clearly different early activation in the left inferior temporo-occipital cortex. In controls, strong, transient signals were recorded in this area within a very narrow time window about 180 ms after word presentation. In the dyslexics, however, the activation peaked only at about 500 ms, if at all. The inferior temporal sulcus or fusiform gyrus has recently been reported to show letter-string specific responses, in addition to the previously found face-specific signals [6, 7]. Thus, it appears that the coherent response to letter strings is missing or delayed in the language-dominant left hemisphere of the dyslexics.

After 200 ms, temporal and frontal cortices became involved. In controls, the left temporal lobe showed strong activation between 200 and 400 ms but, in dyslexics, this area was practically silent. This result supports the view that phonological processing in the left temporal lobe is impaired in dyslexics. Interestingly, left inferior frontal cortex, approximately agreeing with Broca's area, showed activation within 400 ms after word presentation in dyslexics but not in controls. These signals raise the possibility that the dyslexics, in order to compensate for the deficits in the automatic word recognition, in fact try to 'guess' the words, thus adopting rather a top-down than a bottom-up approach.

Our results suggest that, in addition to the phonological impairment, the visual word form perception is impaired in dyslexics, which could be the immediate reason for the manifest difficulties with written words. The obvious question is whether the visual and phonological problems are independent or whether the phonological deficit could in fact lead to the observed visual impairment. We first learn to listen to words and only much later make the connection between the symbolic written words and the original phonological code. Impaired phonological processing could thus impede the formation of a word recognition unit which would automatically set letters apart from other objects and facilitate fast reading.

Visual Imagery

There is an on-going debate of whether the same brain areas are involved both in actual sensory perception or motor performance and in imagery. Simple visual stimulation results in a brief suppression of the posterior cortical rhythms, also illustrated in Fig. 1. In fact, the posterior 10-Hz activity does not originate only in the calcarine fissure. Figure 4 shows how another source cluster is found along the midline, in the parieto-occipital sulcus, clearly distinct from the primary visual cortex. Rhythmic activity in both of these areas is suppressed by visual stimulation.

But what happens when, instead of an actual visual stimulus, one imagines a picture in the 'mind's eye'? In this study, the subjects had their eyes closed and they were given a name of a letter through earphones and asked to visualize its upper-case form [8]. After 3 s, they were asked a question about that letter, for example, whether the letter was symmetrical with respect to the vertical axis. To answer yes or no, they lifted the left or right index finger. Figure 4 shows the averaged modulation of the parieto-occipital 10-Hz rhythm as a function of time in two subjects. In Subject 1, rhythmic activity remains suppressed throughout the imagery task, whereas Subject 2 shows two distinct suppressions, one after the letter is given through the earphones and an even stronger one when he has to decide about its visual properties. The gray curves show the result of a control task which required auditory instead of visual imagery and, obviously, did not reduce the parieto-occipital 10-Hz rhythm. The same cortical areas are thus affected in a very similar way both by actual visual stimulation and by visual imagery.

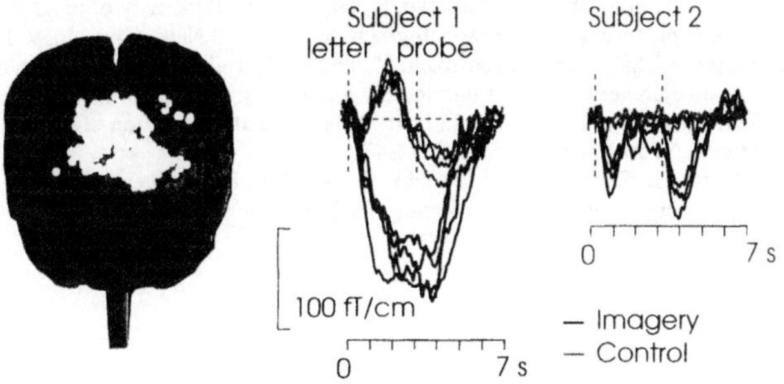

Fig. 4. Sources of posterior 10-Hz rhythm, with two distinct clusters. The curves show suppression of the 10-Hz rhythm during visual imagery and during a control task requiring auditory imagery in two subjects, with replications. Adapted from [8].

Motor Imagery

In this study, the right median nerve was stimulated at the wrist to induce a sharp thumb twitch. Just like voluntary movement, electric median nerve stimulation is also followed by bursts of 20-Hz activity [2, 9]. The 20-Hz oscillations originate predominantly in the motor cortex [2, 10]. When the stimulated right hand was at rest, there was a typical burst of 20-Hz activity in the left central sulcus, with the TSE curve peaking at around 400 ms after stimulus onset. When the subject continuously explored an object with the stimulated hand, the 20-Hz burst was completely suppressed. When he stretched his fingers, keeping them tense, the rhythms were also somewhat reduced. But, most interestingly, just imagining the exploration task resulted in a prominent suppression of the post-stimulus 20-Hz activity, without any appreciable signal in the simultaneously recorded electromyogram. Thus, the primary motor cortex is involved in motor imagery [11].

Another interesting feature is the phase-locked evoked response. The sharp 20-ms response is followed by a 30-ms deflection with opposite polarity and a strong 60-ms component. The 30-ms deflection is suppressed and even reversed by simultaneous exploring movements. But, neither stretching nor imagined exploration affect the early components in any appreciable way. Thus, we have an interesting situation where the somato*sensory* evoked response is not changed by motor imagery whereas the simultaneously recorded spontaneous rhythmic activity of the *motor* cortex is strongly affected.

Conclusion

Clearly, whole-head neuromagnetometers have made MEG studies of cognition truly feasible. Both phase-locked evoked responses and modulation of cortical rhythms, which is also less sensitive to slight variation between repetitions, are extremely useful in these studies. MEG provides the accurate timing which is quintessential for understanding the functional role of distinct cortical areas in cognition, even if the localization accuracy certainly suffers from the inverse problem. And finally, analysis of individual data sets, instead of group averages, allows one to fully appreciate – and also suffer from – differences between individual activation patterns which derive, at least partly, from distinct strategies in cognitive processing.

References

[1] Ahonen, A.I., Hämäläinen, M.S., Kajola, M.J., Knuutila, J.E.T., Laine, P.P., Lounasmaa, O.V., Parkkonen, L.T., Simola, J.T., and Tesche, C.D. 122-channel SQUID instrument for investigating the magnetic signals from the human brain, Physica Scripta, 1993, T49: 198-205.
[2] Salmelin, R., and Hari, R. Spatiotemporal characteristics of sensorimotor MEG rhythms related to thumb movement, Neuroscience, 1994, 60: 537-550.
[3] Salmelin, R., Hari, R., Lounasmaa, O., and Sams, M. Dynamics of brain activation during picture naming, Nature, 1994, 368: 463-465.
[4] Salmelin, R., Service, E., Kiesilä, P., and Uutela, K. Cortical dynamics of word recognition in normal and dyslexic subjects: a neuromagnetic study, Society for Neuroscience Abstracts, 1995, 21: 1763.
[6] Nobre, A.C., Allison, T., and McCarthy, G. Word recognition in the human inferior temporal lobe, Nature, 1994, 372: 260-263.
[7] Allison, T., McCarthy, G., Nobre, A., Puce, A., and Belger, A. Human extrastriate visual cortex and the perception of faces, words, numbers, and colors, Cerebral Cortex, 1994, 5: 544-554.
[8] Salenius, S., Kajola, M., Thompson, W.L., Kosslyn, S., and Hari, R. Reactivity of magnetic parieto-occipital alpha rhythm during visual imagery, Electroencephalography and clinical Neurophysiology, 1995, 95: 453-462.
[9] Salenius, S., Schnitzler, A., Salmelin, R., Jousmäki, V., and Hari, R. Reactivity of human cortical sensorimotor rhythms associated with median nerve stimulation, Society for Neuroscience Abstracts, 1995, 21: 115.
[10] Salmelin, R., Hämäläinen, M., Kajola, M., and Hari, R. Functional segregation of movement-related rhythmic activity in the human brain, NeuroImage, 1995, 2: 237-243.
[11] Schnitzler, A., Salenius, S., Salmelin, R., Jousmäki, V., and Hari, R. Involvement of primary sensorimotor cortex in motor imagery: a neuromagnetic study, Society for Neuroscience Abstracts, 1995, 21: 518.

The Magnetic Field Components Evoked by an Optical Illusion

Adachi, Y.[1], Kohda, M.[2], Morita, T.[2] and Takebayashi, M.[1]

[1]Electrotechnical Laboratory, Amagasaki, Hyogo, Japan; [2]Dept. of Engineering, Osada University, Suita, Osaka, Japan

Introduction

We often misperceive some physical phenomena whose spatio-temporal properties may be beyond the ability of our given sensory system. When two sensory stimuli are presented in a short stimulus onset asynchrony (SOA), the perception of the preceding stimulus is influenced by the following stimulus. It is well known as the backward masking that can evoke misperception[1]. This is observed in visual, auditory and tactile perception. Backward masking is thought of as the phenomenon that the responses to the following stimulus influence those of the preceding stimulus inversely in the signal processing system.

When two visual stimulus dots located at separate sites out of the central part of visual field are flashed serially in a short time, we often perceive the position of the preceding stimulus to be closer to the position of the following stimulus. This misperception is one kind of backward masking called optical saltation. The origin of this optical illusion should be in the information process of the central nervous system (CNS) because the preceding dot can be perceived on the blind spot on which we can usually perceive nothing under some spatial and temporal conditions.

There are two well-known facts that the optical illusion does not occur with a longer SOA or a larger spatial distance between two stimuli, and that the responses of specific adjacent retinal cells activate specific adjacent cells in both the laterale geniculate body and primary visual cortex[2]. These facts suggest that the neural responses caused by a single stimulus need some spatio-temporal extension for normal perception in the tissue of the brain, and that the illusion can occur when the extensions of each response overlap partly and interact with each other.

In this paper, we examined responses related to this optical illusion in the human brain by three-dimensional analysis of neuromagnetic signals that has high temporal resolution in attempts to discover the neurological origin of backward masking.

Methods

Two healthy men aged 23 and 25 years were examined as subjects. The subject sat in a magnetically shielded room without illumination. The flashes and fixation point were presented into the shielded room through optical fibers from red light-emitting diodes (λ=660 nm). The diameter of the stimulus dot was 1.2 mm and the duration of each visual stimulus was 10 msec.

Figure 1(a) shows the arrangement of the subject and the stimuli. Two stimulus dots L1 (upper dot) and L2 (lower dot) were arranged on a vertical line horizontally apart from the fixation point which was at the visual angle of 15 degrees. The visual angle between L1 and L2 was 5 degrees. All stimulus dots were 40 cm away from subject's eye. The subject was instructed to watch the fixation point at the center of his visual field with his dominant eye during the measuring so as not to allow eye movement. His non-dominant eye was covered with an eye mask. The stimuli were given on the opposite side of the stimulated eye.

The temporal arrangement of stimuli is illustrated in Fig. 1(b). One sequence consisted of five flashes. SOA between flashes *c* and *d* was 60 msec and an illusion was induced in this period. Flash *b* was the control stimulus so that the subject could make certain of the true location of L1 before the illusion occurred. Flash *a* was the sign of the biginning of the sequence. SOAs of *a-b* and *b-c* were 1000 msec. Flash *e* was given 1600 msec after the illusional stimulus (flash *c* and *d*), in order to inform the subject of the end of the sequence. After flash *e*, subjects were allowed to blink and relax until the next sequence. The interval between sequences was 2 sec. The illusional stimulus (flash *c* and *d*) were delivered every 5.6 sec in this experiment. The SOA between flashes *c* and *d*) and the spatial arrangement of L1 and L2 were decided according to reference[1], and it was confirmed in pre-psychophysical experiments that the subject was able to perceive the illusion under these conditions.

The neuromagnetic signals elicited from the brain were measured with a 122-channel whole-cortex magnetometer that consists of planar first-order SQUID gradiometers[3]. The head of the subject was positioned in the

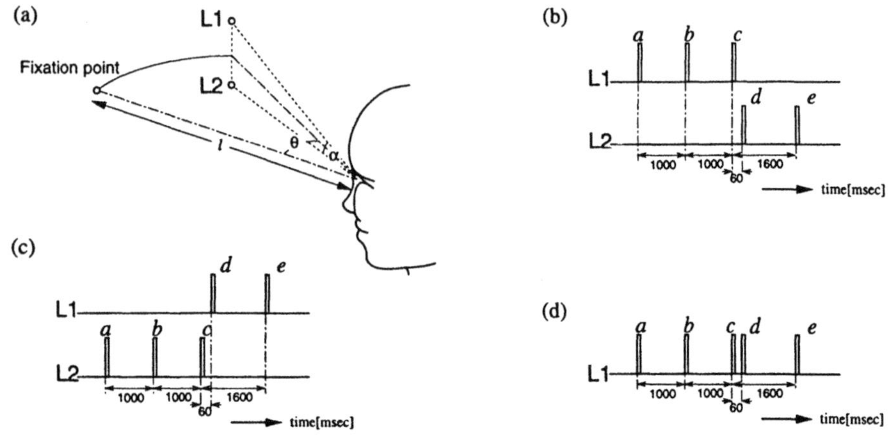

Fig. 1 (a) The spatial arrangement of the light stimuli. In this paper θ=15°, α=5° and l=40 cm. (b)(c)(d) Schematic illustrations of the temporal arrangements of stimulus sequences.

helmet-shape magnetometer array and supported with the head fixer, made of non-magnetic material, to prevent movement. The detected signals from the magnetometer were filtered with frequency limits of 0.03-100 Hz and digitized with sampling rate of 400 Hz and stored on a workstation for following off-line analysis. In the off-line analysis the signals were filtered with digital band-pass filter of 0.3-30 Hz. The responses to each stimulus were averaged in order to improve the signal-noise ratio.

In order to measure the response to the single L2 stimulus, we conducted a similar experiment as shown in Fig. 1(c). The subject could still perceive the illusion with this order. In addition we measured with the stimulus sequence that consisted of only L1 as shown in Fig. 1(d).

The electro-oculogram (EOG) was recorded to monitor ocular and muscle artifacts during the stimulating process. Any epochs which included such artifacts were to be rejected.

Results

Figure 2 shows the grand-average of subject 1's responses to flash *b* (single L1 stimulus; thin line) and flash *c* (illusional stimulus; bold line). The averaged responses were analyzed for 1000 msec, between 100 msec prior and 900 msec after the each stimulus. The responses to flash *c* and to flash *d* overlapped and merged, because SOA between them was 60 msec. It is obvious that the primary and/or secondary visual cortices in the contralateral occipital lobe were activated (in latency 100-250 msec) because the planar first-order gradients straight above the activated location of the cortex delivered the largest amplitude of the signal[3].

The MEG signals from the channel eliciting the largest peak amplitude (channel A) are shown in Fig. 2(b). The response to the single L2 stimulus at channel A is also shown in Fig. 2(b) (dashed line). The responses to the double L1 stimulus (Fig. 1(d)) at channel A are shown in Fig. 2(c). The illusional stimulus elicited the slow components with latency ca. 370 msec above the right temporal lobe (channel B) as shown in Fig. 2(d). The equivalent current dipoles (ECDs) for the three major peaks of the responses seen in Fig. 2(b), corresponding to the single L1 stimulus and the illusional (L1->L2) stimulus, were estimated. The locations of ECDs are shown in Fig. 3.

The contralateral occipital lobe components and the slow components above the right temporal lobe were obtained from subject 2 as well as from subject 1. However, the occipital components from subject 2 were not as clear as those from subject 1.

Discussion

The presence of the slow components whose latencies were 250 msec or more was reported in several odd ball paradigm experiments. These components are called P300 and P3. The origin of the P3 magnetic fields is under discussion. The deep source of the P3 was suggested to be around the hippocampus[4] or the temporal lobe[5]. It is plausible that the right temporal slow components evoked by the illusional stimulus might be associated with P3 components according to their locations and latencies.

It was difficult for the subject to perceive the 50 msec interval between flash *c* and flash *d* clearly in the double L1 sequence. More often the subject perceived the series of two L1 stimuli as mostly continuous. In Fig. 2(c), the observed responses to the double L1 stimulus were identical to ones to the preceding L1 only. In the case of the L1->L2 stimulus the subject could easily perceive the difference in location of the two flashes and judge which flash was preceding. Nevertheless the component for the following L2 stimulus was missing in the MEG signal corresponding to the L1->L2 stimulus as well as the double L1 stimulus. The incoming signal in the visual cortex induced by the two flashes was comparable to that caused by the single preceding flash. It is supposed from these results that the subject perceives two flashes by signal processing in the mid-brain, including the laterale geniculate body and superior colliculus, before cortical activation. If the two flashes were perceived by the visual cortex or higher cortical structure, the two components corresponding to the preceding L1 stimulus (flash *c*) and the following L2 stimulus (flash *d*)

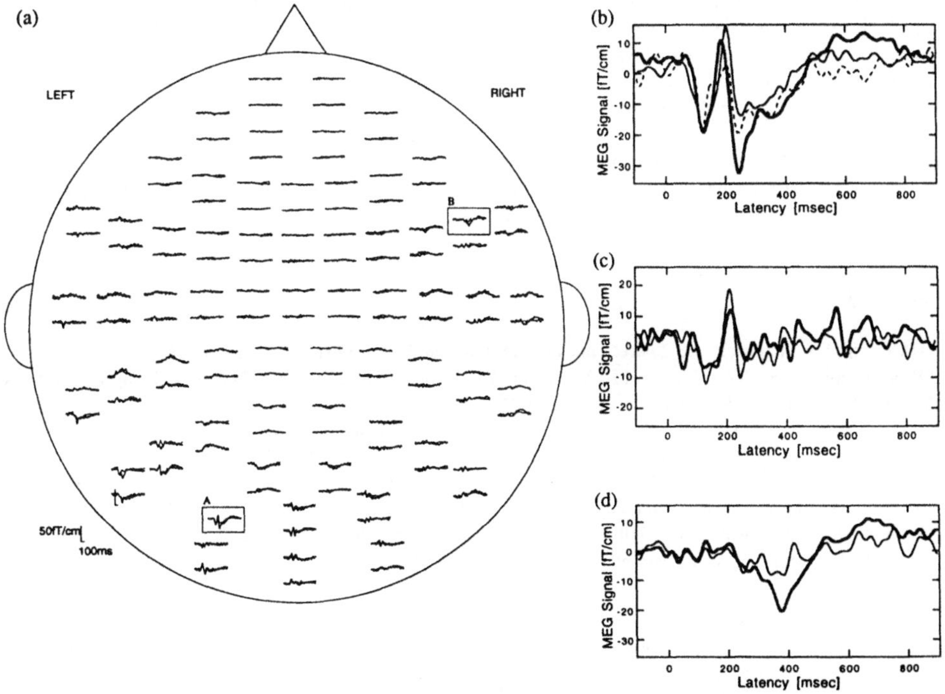

Fig. 2 (a) The MEG signal of subject 1 caused by the single L1 stimulus (thin line) and by the illusional stimulus (bold line). The stimuli were given on the right visual field of the left eye. Upper channels and lower channels at each measuring point are the latitudinal derivatives and the longitudinal derivatives respectively. (b) The thin line, bold line and dashed line correspond to the single L1 stimulus, the illusional stimulus and the single L2 stimulus respectively. (c) The response caused by the L1->L1 stimulus (bold line) at channel A. The thin line represents the response to the single L1 stimulus in the double L1 sequence. (d) The slow component at channel B.

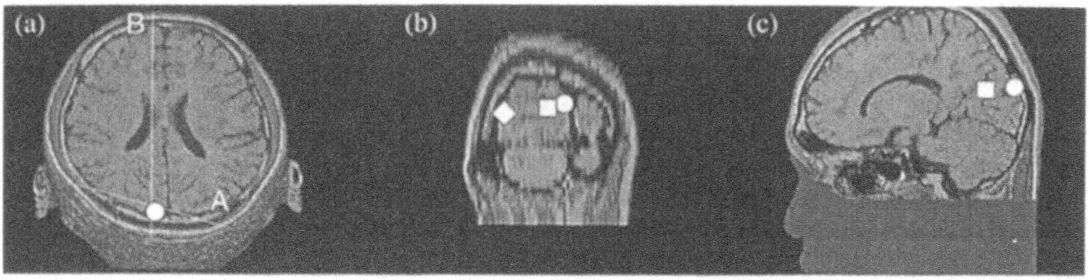

Fig. 3 The location of the ECDs for the three peaks of the response to the illusional stimulus (circle, 120 msec; square, 180 msec; diamond, 240 msec). We can see the shifts in location of the activation from the primary visual cortex to the secondary visual cortex in this period. (b) and (c) show the cross section cutting through plane A and B in (a) respectively. The locations of the sources of the response to the single L1 didn't differ from that of the L1->L2 significantly. The MR images were acquired with Yokogawa 0.5 T MRI system, MR Vectra at Fukai Hospital.

should be observed in the visual cortex. Consequently, it is probable that the optical illusion caused by the interaction of the two responses also occurs in the mid-brain.

The latencies of the three third major peaks of the signal from channel A, as caused by the single L1 and L2 stimuli and the illusional stimulus (Fig. 2 (b)), were nearly equivalent, all being approx. 250 msec. Furthermore, the amplitude of the peak caused by the illusional stimulus was equal to the linear sum of the amplitudes of the peaks caused by the single L1 and L2 stimulus. Accordingly it is suggested that the information processing of both the preceding and the following stimuli might be carried out at the same time in the secondary visual cortex, and that this could be associated with evoking the optical illusion. However, we need to confirm the reproducibility of these results not only by increasing the number of subjects, but also by measuring the changes in response under conditions with various SOAs and spatial distances between the two stimuli.

We are grateful to Dr. Akiba and Dr. Sato for advice and encouragement, Ms. Nakamura for fruitful discussion and Dr. Kakui for his continuous encouragement.

References:

[1] Geldard, F.A., Sherrick, C.E. Space, Time and touch, Scientific American, 1986, 255: 84-89

[2] Nieuwenhuys, R., Voogd, J., and Hujizen, Chr. The Human Central Nervous System-A Synopsis and Atlas, Springer-Verlag, Berlin, Heidelberg, New York, 1989

[3] Hämäläinen, M., Hari, R., Ilmoniemi R.J. et al Magnetoencephalography-theory, instrumentation, and applications to noninvasive studies of the working human brain, Rev. Mod. Phys., 1993, 65: 413-498

[4] Rogers, R.L., Papanicolaou, A.C., Baumann, S.B. et al Late magnetic fields and positive evoked potentials following infrequent and unpredictable omissions of visual stimuli, Electroenceph. Clin. Neurophysiol., 1992, 83: 146-152

[5] Okada, Y.C., Kaufman, L. and Williamson, S.J. The hippocampal formation as a source of the slow endogeneous potentials, Electroenceph. Clin.Neurophysiol., 1983, 55: 417-426

Spatiotemporal Imaging of Human Cortical Areas Sensitive to Visual Motion

Ahlfors, S.P.[1], Aronen, H.J.[2], Belliveau, J.W.[3], Dale, A.M.[4], Huotilainen, M.[5], Ilmoniemi, R.J.[6], Korvenoja, A.[2], Liu, A.K.[3], Simpson, G.V.[1], Tootell, R.B.H.[3] and Virtanen, J.[5]

[1]Albert Einstein College of Medicine, Bronx, NY; [2]Radiology, Helsinki University Central Hospital, Finland; [3]Massachusetts General Hospital, NMR Center, Charlestown, MA; [4]University of Oslo, Norway; [5]Cognitive Brain Research Unit, University of Helsinki, Finland; [6]BioMag Laboratory, Helsinki University Central Hospital, Finland

Introduction

Movement in the visual field is an important part of perceptual information. Previous imaging studies have identified a number of cortical areas sensitive to visual motion [1,2], including what is considered to be the human homologue of area MT/V5 [3,4]. Our study sought to reproduce and expand upon a previous functional magnetic resonance imaging (fMRI) investigation [4] and to examine the timing of activity in the network of regions participating in the processing of visual motion information with magnetoencephalography (MEG).

We measured brain activity in the same subjects using both MEG and fMRI, thereby combining the good spatial resolution of fMRI and the millisecond-scale temporal information obtained with MEG [5,6]. Special considerations, however, are necessary, when combining these brain imaging methods, because at present, the exact relationship between the hemodynamic (as measured with fMRI) and electrophysiological (MEG) responses is not known.

Methods

Four subjects (all male, aged 25-45 years) were recorded both with whole-head 122-channel MEG (Neuromag Ltd, dc-SQUIDs, planar first-order gradiometers, analog filtering 0.03-100 Hz, sampling at 397 Hz) and with functional MRI (GE 1.5 T with Advanced NMR echo planar, asymmetric spin echo sequence, head coil, bite bar, 64 sets of 25 slices were collected in each run, TR = 7 s, slice thickness 5 mm). The computer-generated visual stimuli consisted of expanding and contracting concentric low-contrast rings. About 400 epochs of MEG/EEG responses were averaged, triggered from the reversal of the direction of motion occuring every 3 s. The speed of motion was 2.4 deg/s. The analysis period was –200 to 800 ms, the averaged responses were lowpass filtered at 0-40 Hz using FFT, and the zero level was adjusted using the prestimulus baseline –100 to 0 ms. The fMRI measure was the Fourier component of the fMRI signals in response to the stimulus pattern being alternatingly moving and stationary in cycles of either 30 or 40 s.

Sources of the MEG responses were modeled with multiple equivalent current dipoles. As a first attempt, the locations of the dipoles were determined by a non-linear least-squares fit to the MEG data only. A single dipole was fitted at one instant of time (or a range of latencies around the peak signal), and the contribution from this source was projected out using the signal-space projection technique [7]. Then, another single dipole was fitted to the part of the data that could not be explained by the first one. These two dipoles were used as the initial guess for a two-dipole fit performed to the original data. The orientations of these MEG-derived dipoles were assumed not to change in time. This process signal-space projection followed by a multi-dipole fit was repeated, adding dipoles until a satisfactory explanation was obtained for the measured responses. To determine the locations and orientations, usually only two or three dipoles were fitted at a time to part of the data (in space and time, for example, to sensors over one hemisphere), and only in the end, the optimal amplitudes for all the dipoles were obtained using the whole set of data.

The multiple dipole model becomes prohibitively difficult to solve when the complexity of the source configuration increases, partly because the non-linear least-squares minimization algorithm tends to stop in a local, sub-optimal minimum, partly because a dipole may not be an adequate model for the source. Therefore, we examined solutions in which the locations of the dipoles were constrained to the loci of the fMRI data. The orientation of these dipoles was allowed to vary and was optimized, together with the dipole moment, for each time instant separately using the linear pseudo-inverse method. Some sets of fMRI locations had to be represented by a single dipole, because nearby dipoles often resulted in an unstable solution with a quadrupole-like combination of dipoles with large opposing amplitudes, the field patterns of which mostly cancel each other.

Locations of active regions were identified and visualized within each subject's high-resolution anatomical MRI, co-registered with MEG using the nasion and preauricular points as fiducial points.

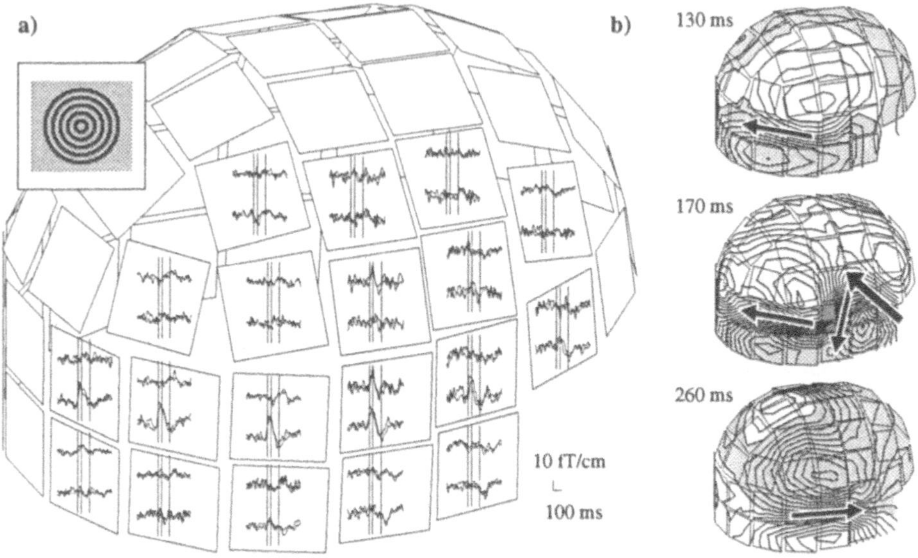

Fig. 1 a) Visual evoked magnetic fields as a function of time over the right hemisphere, in response to the reversal of the direction of the motion of concentric rings (inset at top left). The superimposed curves show responses to two types of reversal (expanding to contracting, and vice versa). The vertical markers indicate latencies 130, 170, and 260 ms. The rectangles depict the locations of sensor pairs. The upper and lower waveforms correspond to $\partial B_z/\partial x$ and $\partial B_z/\partial y$, where B_z is the normal component of the magnetic field; x, y, and z refer to the local coordinate system of the sensor units. b) Isocontour maps of the estimated B_z at three latencies. The arrows indicate the location and orientation of equivalent current dipoles determined by a least-squares fit to the MEG signals. Subject SA.

Results

Averaged MEG responses are shown in Fig. 1a. The largest deflections peak at 170 and 260 ms after the reversal of the direction of motion. The spatial distribution of the measured magnetic field at three different latencies is depicted in Fig. 1b. At 130 ms the field map is characteristic of a field that would be generated by a single source, but at 170 ms the field pattern is more complex, indicating multiple simultaneously active brain regions. A multi-dipole analysis of the case shown Fig. 1 suggests at least three sources at 170 ms (two of them located anteriorly to the one that is active at 130 ms) and one dipole at 260 ms which is located close to the middle dipole at 170 ms, but with an orthogonal orientation.

The fMRI data showed in all subjects a pattern of activity consisting of three regions: one in the lateral surface of the occipital lobe, another in the occipito-temporal junction, and a third one in the vicinity of the superior temporal sulcus near the posterior termination of the Sylvian fissure. In some subjects, fMRI activity was also found in inferior occipital lobe and parietal lobe.

Fig. 2 depicts the amplitudes of source activity when the locations of dipoles were constrained to fMRI loci. This combined MEG-fMRI analysis suggests that there are temporally overlapping activities in several cortical regions. Dipole #1 has the most uncertainty with respect to the cortical region it represents, since it is probably accounting for activity from multiple sources near the pole and inferior surface of the occipital lobe. Attempts to model these regions with multiple dipoles resulted in large interactions due to their proximity, and thus they are represented by a single source in Fig. 2. The first areas to be activated are those represented by dipole #1, and dipoles #4,5 which may correspond to the human MT/V5 regions bilaterally. These regions onset at about 140 ms, followed by activation of the posterior dipoles #2,3 and by the anterior dipoles #8,9 (not identified with fMRI, see below). The temporal data illustrate that, in addition to staggered onset times, all of these areas have transient responses that last for hundreds of milliseconds but do not continue for the entire duration of the motion stimulation. However, the fMRI-derived anterior motion area represented by dipoles #6,7 has a later onset (after 250 ms) which

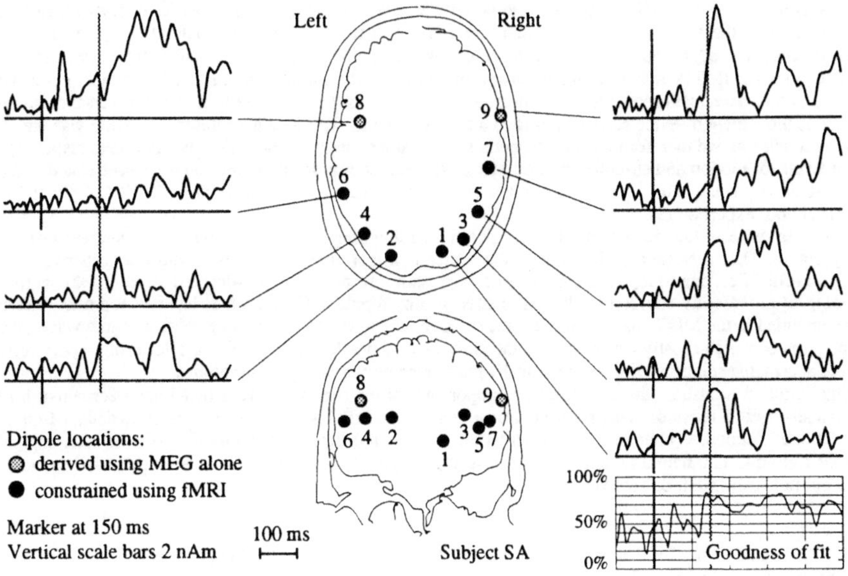

Fig. 2 Source activity for the MEG response to the reversal of motion. Locations of equivalent current dipoles derived from the MEG data and regions of hemodynamic activity as measured with fMRI are shown within the outline of the head. The amplitude waveforms of the dipole moments were scaled individually by a factor proportional to the inverse square of the distance between the dipole and the array of sensors, thereby approximately normalizing the noise level.

occurs after the other areas have reached maximal or near maximal amplitude, and then sustains its activity throughout the 800 ms analysis period.

While the fMRI based sources explained the majority of the MEG data, they did not adequately account for the most anterior regions of MEG activity. MEG derived sources in the region at or near the precentral sulcus were needed to explain this remaining MEG data (indicated by the gray dots #8 and #9 in Fig. 2). These have prominent peaks at 180 ms (right, #9) and at around 350 ms (left, #8; this dipole also has a sharp peak very early, at 40 ms, but that was not seen in other subjects).

Discussion

The combined MEG and fMRI measurements indicated a complex spatial and temporal pattern of activity, involving a number of cortical regions related to processing of visual motion information. The fMRI data consistently showed activity in three areas in the posterior lateral surface of each hemisphere. Previous PET and fMRI studies have given evidence that two of these regions might correspond to the human homologues of the monkey areas V3A and MT/V5 [3,4]; the third one is close to the regions which in previous PET studies have been reported to be activated by vestibular [8,9] and visual motion [1,2] stimuli. In addition, MEG indicated activity in a region near the precentral sulcus, which may correspond to the frontal eye field. No activity was detected there with fMRI, and further studies will be employed to examine this region. The fMRI-constrained dipole analysis of the MEG responses suggested that there is temporally overlapping transient activity in MT/V5 and regions posterior to it, as well as the in the region of the frontal eye field. These regions have early, yet slightly differing onset times and are not sustained throughout the recording epoch. Whereas the fMRI-derived anterior motion area showed later onset and sustained activity.

The exact relationship between hemodynamics and electrophysiology, as measured with fMRI and MEG, respectively, is not known at present. There are several possible causes for the differences in the locations of fMRI activity and MEG sources. Besides technical matters, like measurement noise and coregistration error, and changes in the state of the subject during the separate recording sessions, there are some fundamental problems in comparing the two measurements. First, fMRI is sensitive to activity all over the brain, whereas MEG is most sensitive to activity in sulci near the surface of the head and relatively insensitive to radially oriented, deep, or mutually cancelling sources. Second, because the time scales of the two physiological phenomena are different (milliseconds for MEG, seconds for fMRI), it is difficult to ensure that sensory stimuli and the task of the subject contribute equally in the two measurement modalities. In our study, MEG reflects the responses to transient reversal of motion and continuous movement over the period leading to the next reversal. The baseline referent for MEG was the 100 ms preceding a reversal and therefore includes responses to sustained motion. The fMRI also measures responses to transient reversals of motion and sustained motion. However, the referent for these motion responses was the same pattern in a stationary state. This may partly explain the difference between the locations of fMRI activity and the MEG sources in our experiments.

Our data indicate that an assumption of a one-to-one mapping of fMRI loci and MEG sources does not necessarily obtain. Thus a further exploration of the source configuration is warranted, and may otherwise give misleading results. For example, if a true source is missing from the fMRI-derived dipoles, the activity corresponding to the missing source may be assigned to nearby dipoles. On the other hand, if too many nearby sources are included in the MEG analysis, the inverse solution may have large, highly correlated source waveforms, the field patterns of which mostly cancel out. In the latter case, a smaller number of non-interfering dipoles often gives a more acceptable solution, with physiologically reasonable primary current amplitudes.

Our results demonstrate the usefulness and importance of the combined MEG-fMRI approach. First, fMRI can suggest a structure in the underlying source configuration for the MEG which may be difficult to deduce from the MEG data alone. Second, a change in the orientation of a source dipole as a function of time can separate contributions from different neural populations within a single region in the fMRI. Third, a source may be detected with MEG but not with fMRI, or vice versa.

References

[1] Dupont, P., Orban, G.A., De Bruyn, B., Verbruggen, A., and Mortelmans, L. Many areas in the human brain respond to visual motion, Journal of Neurophysiology, 1994, 72: 1420-1424

[2] Cheng, K., Fujita, H., Kanno, I., Miura, S., and Tanaka, K. Human cortical regions activated by wide-field visual motion: an $H_2{}^{15}O$ PET study, Journal of Neurophysiology, 1995, 74: 413-427

[3] Zeki, S., Watson, J.D.G., Lueck, C.J., Friston, K.J., Kennard, C., and Frackowiak, R.S.J. A direct demonstration of functional specialization in human visual cortex, Journal of Neuroscience, 1991, 11: 641-649.

[4] Tootell, R.B.H., Reppas, J.B., Kwong, K.K., Malach, R., Born, R.T., Brady, T.J., Rosen, B.R., and Belliveau, J.W. Functional analysis of human MT and related visual cortical areas using magnetic resonance imaging, Journal of Neuroscience, 1995, 15: 3215-3230.

[5] Simpson, G.V., Pflieger, M.E., Foxe, J.J., Ahlfors, S.P., Vaughan, Jr., H.G., Hrabe, J., Ilmoniemi, R.J., and Lantos, G. Dynamic neuroimaging of brain function, Journal of Clinical Neurophysiology, 1995, 12: 432-449

[6] George, J.S., Aine, C.J., Mosher, J.C., Schmidt, D.M., Ranken, D.M., Schlitt, H.A., Wood, C.C., Lewine, J.D., Sanders, J.A., and Belliveau, J.W. Mapping function in the human brain with magnetoencephalography, anatomical magnetic resonance imaging, and functional magnetic resonance imaging, Journal of Clinical Neurophysiology, 1995, 12: 406-431

[7] Tesche, C.D., Uusitalo, M.A., Ilmoniemi, R.J., Huotilainen, M., Kajola, M., and Salonen, O. Signal-space projections of MEG data characterize both distributed and well-localized neuronal sources, Electroencephalography and Clinical Neurophysiology, 1995, 95:189-200.

[8] Friberg, L., Olsen, T.S., Roland, P.E., Paulson, O.B., and Lassen, N.A. Focal increase of blood flow in the cerebral cortex of man during vestibular stimulation, Brain, 1985, 108:609-623.

[9] Bottini, G., Sterzi, R., Paulesu, E., Vallar, G., Cappa, S.F., Erminio, F., Passingham, R.E., Frith, C.D., and Frackowiak, R.S.J. Identification of the central vestibular projections in man: a positron emission tomography activation study, Experimental Brain Research, 1994, 99:164-169.

Acknowledgements

We thank Drs. R. Näätanen, B. Rosen, M. Sereno, and C.-G. Standertskjöld-Nordenstam. This study was supported by grants from the United States Public Health Service: NS27900, MH50054, EY07980, by grants form the Human Frontier Science Program, and by the Academy of Finland, and conducted during the tenure of an established investigatorship from the American Heart Association to JWB.

Task-Specific Magnetic Fields from the Left Human Prefrontal Cortex

Basile, L.F.H.[1], Simos, P.G.[1], Tarkka, I.M.[2] and Papanicolaou, A.C.[1]

[1]Deparment of Neurosurgery, University of Texas, Houston, USA; [2]Baylor College of Medicine, Houston, USA

Introduction

Recent studies have provided evidence for task-specific activity in distinct areas of the frontal cortex [1,2,3]. In a previous study, we used magnetoencephalography (MEG) to test the hypothesis of functional specialization of the frontal cortex [1] by asking subjects to perform two visual recognition tasks. A spatial recognition task was associated with sources of magnetic fields in the right frontal lobe, which were localized more anteriorly and inferiorly, than the sources of activity obtained during a visual pattern recognition task. Both tasks involved presentation of common visual stimuli and required common motor responses. These results were analogous to the findings from a multi-unit recording study in the monkey [3]. That study showed concentrations of neurons in dorsolateral and inferior convexity prefrontal areas, that responded preferentially during the performance of a visuo-spatial and a visual-pattern delayed response task, respectively. Another experiment from our laboratory (unpublished data) provided evidence for different sources of activity in the right frontal lobe, when a familiar-person recognition task was compared to a visual pattern recognition task. Sources in more medial areas of the right prefrontal cortex were obtained when subjects performed the familiar-person recognition task. A study by Petrides et al. [2], using positron emission tomography in humans, also reported regional prefrontal cortical activity corresponding to the performance of two different tasks.

The purpose of the present experiment was to extend our results on task-specific frontal cortical magnetic activity with the use of a verbal task. In this case, we expected to obtain consistent sources of magnetic fields from the left frontal cortex in right handed subjects. The expected hemispheric asymmetry in activity would be consistent with the view that certain portions of the left frontal cortex play an essential role in the performance of even simple verbal tasks. To enhance the possibility of observing this asymmetry in activity, we eliminated the requirement of an overt motor response, which in our previous experiments might have resulted in the bilateral activity seen in most subjects. In fact, results from a pilot study in which a button press response was added to the present task, confirmed our suspicion regarding bilateral frontal activity.

Methods

Six normal subjects, three males and three females, ranging in age between 24 and 54 years, volunteered to participate in two or three two-hour experimental sessions. All subjects had normal vision and no history of neurological disorders. They were required to sign consent forms and they were paid for their participation. Stimuli were presented on the screen of a computer that also sent trigger pulses to start the MEG data acquisition. The stimuli consisted of 216 three-to-ten letter words which were drawn white on a black background, and subtended a maximum of approximately 3 ° of visual angle. All words were concrete nouns, that belonged to nine well defined semantic categories (e.g., tools, fruits, vehicles, etc.). The exposure duration of each stimulus was 370 ms. Each trial consisted of the successive presentation of two words. The inter-stimulus interval was 2000 ms, whereas the inter-trial interval varied randomly between 2000 and 5000 ms. Each experimental run consisted of a series of 120 trials. Across the different runs, the percentage of word pairs matching semantically ranged from 45 to 55%. The order of trials containing matches or mismatches was randomized within each run. Subjects were instructed to fixate their gaze on a small white square of light, that was continuously present in the center of the screen during the interstimulus interval. They were asked to decide whether the first word in a trial matched the second one, i.e., whether the two words belonged to the same semantic category. Finally, they were required to count silently the number of matches and report their count after each experimental run. Specific instructions were given to the subjects to avoid blinking, swallowing, or moving their tongue, face or mandibles until shortly after the end of each trial.

Magnetic evoked fields were recorded with a 7-channel Neuromagnetometer (Biomagnetic Technologies, Inc., Model 607), using second-order gradiometers with a 4 cm baseline and 1.8 cm coil diameter. Recordings were done in a magnetically shielded room. Each average was computed from a minimum of 60 epochs, from the total of 120 recorded during each run. The MEG analog filtering was set at 0 to 50 Hz and the digitization rate was 200 Hz. A Nicolet Compact Four system and Ag/AgCl electrodes were used for acquisition of the electro-oculogram (EOG). The EOG was recorded for the purpose of subsequent exclusion of MEG trials containing eye movement or blink artifacts. Before averaging, trials on which the EOG signal exceeded a peak-to-peak amplitude of 70 μV were rejected. The magnetometer itself was also a source of some artifacts. Those artifacts were easily recognizable, and they typically consisted of transient high amplitude noise bursts usually confined to a single MEG channel. Visual inspection of all MEG epochs ensured that trials contaminated with this type of artifact were not used in subsequent analysis. When more than 60 epochs for a given condition and channel contained artifacts (leaving less than 60 epochs to be averaged), that channel was eliminated from further analysis. The position of the magnetometer relative to the subject's head was automatically recorded with a BTi probe position indicator, consisting of a radio-frequency transmitting device attached to the dewar and three sensors placed on the head. Error tolerance for the location of sensors was less than 3 mm and the accuracy 0.1 mm. Each subject's fiduciary points, the nasion and preauricular points, served as basis for the MEG Cartesian coordinate system. The Y axis was determined by the two preauricular points, the X axis by the nasion and the mid distance between the two preauricular points, and the Z axis was normal to the X-Y plane, passing through the X-Y intersection (the center of the head). Five to six placements of the magnetometer on each side of the head covered the frontal, centro-parietal and temporal areas. Following artifact elimination and averaging, the MEG data were baseline adjusted relative to the mean field amplitude during the 500 millisecond period preceding the onset of the first word. The averages were then digitally filtered (bandpass 0 to 5 Hz, Butterworth 4 pole filter, 24dB/octave roll-off) and used to create isofield contour maps. The isofield contour maps were constructed at 5 ms intervals, from 100 to 2700 ms after the onset of the first word. The intracranial sources of the observed fields were modeled as equivalent current dipoles (ECDs) using a finite difference version of the Levenberg-Marquardt algorithm incorporated into the BTi software. The dipole fitting method was applied at every 5 ms, from 100 to 2700 ms of the trial duration. The resulting dipole parameters (spatial coordinates, orientation and strength) were computed with reference to the best-fitting sphere, the curvature of which was defined by the surface of each subject's scalp at the region of the head containing the field extrema. The criterion we used for accepting solutions as satisfactory, was a correlation of at least .85 between the measured fields and the predicted fields, at each time point. The predicted fields were calculated as solutions of the forward problem, from a set of hypothetical dipole parameters. The hypothetical parameters were set by the BTi software as initial values in the iterative dipole-fitting process. Magnetic resonance images (MRIs) from five subjects were used to identify the brain locations of the calculated dipolar sources. The procedure for transforming the MEG coordinates into MRI coordinates involved the use of lipid markers in the nasion and preauricular points during the MRI session. The MRI coordinate system was then aligned to the MEG coordinate system using a computer algorithm.

Results

Clear dipole-like isofield maps were observed in all six subjects on the left fronto-temporal region of the scalp. On the right side, only the dipole-like maps corresponding to the event-related fields were observed. The sources of these fields were restricted to temporal lobe regions and will not be further discussed. The middle row in Figure 1 shows representative isofield contour maps from two subjects during the course of the experimental trial. For illustration purposes these maps were selected at the time of highest peak-to-peak (i.e., outward and inward flux) field amplitude. A common angle of view relative to the subjects' coordinate systems (65 degrees anteriorly in the X-Y plane, starting from the left preauricular point, and 70 degrees relative to the Z axis) was used in the figure. The shaded areas represent magnetic flux entering the head, and white areas represent flux exiting the head. The similarity in the maps between the subjects is remarkable, despite the fact that we observed a large degree of variability in the temporal pattern of their corresponding average fields (see waveform morphology on the top of the Figure). As it was also observed in our previous experiment [1], the spatial distribution of extrema for a given subject was approximately the same throughout the trial duration.

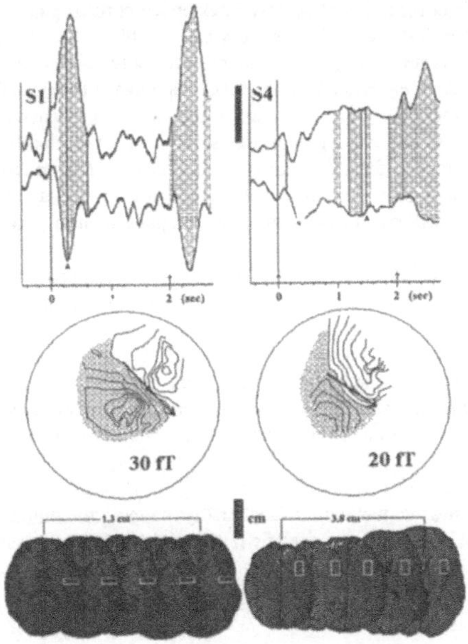

Figure 1 - Magnetic field extrema, isofield contour maps and range of dipole positions for subjects S1 and S4, who represent the extreme examples of phasic and tonic types of frontal activity.

The upper part of the Figure shows the MEG waveforms recorded at the positions of the magnetic field extrema, for the two representative subjects. One recording contains the highest amplitude average field exiting the head (upward deflection) whereas the other contains the highest amplitude average field entering the head (downward deflection). Subjects S1, S2 and S3 presented a more phasic pattern of activity, as opposed to subjects S4, S5 and S6, who showed more sustained activity throughout the trial. Since we had never observed those extreme cases of phasic activity in the previous studies, we repeated the recordings for two subjects, S1 and S3. The results were practically identical across the two replications, leading to the conclusion that the observed temporal pattern of field strength is a characteristic of each individual. Intracranial sources of the measured fields, in the form of satisfactory dipole solutions located in the frontal lobes, were obtained only for the left hemisphere in all six subjects. This was the case for several time points during the trial. The number of solutions and their distribution in time varied across subjects. We observed time windows with good solutions from as early as the onset of the first word (indicated by arrowheads in the Figure), to as late as after the offset of the second word. The durations of these time windows ranged from a few tens to many hundreds of milliseconds. The shaded areas in the waveforms of the Figure mark the time periods during which dipolar maps and corresponding satisfactory dipole solutions were found. In all subjects, there were relatively small time-dependent changes in dipole location. The bottom part of the figure shows the cubic regions containing the entire range of dipole locations for the two subjects. These regions were defined by the ranges of dipole coordinates in the x, y and z dimensions of our coordinate system, across all time windows with satisfactory solutions for a given subject. The actual positions of dipoles obtained during the task were points near or underneath the cortex corresponding to Brodmann's areas 6 and 44, in the left posterior frontal lobe (in the right side of MRI scans in the figure).

Discussion

The purpose of the present study was to use elicited magnetic fields as a means of detecting task-specific regional activity from the frontal cortex. This attempt was successful in that all six subjects displayed dipole-like sources restricted to posterior regions of the left frontal lobe (in the vicinity of Brodmann's areas 6 and 44), during a verbal task. We interpret the dipoles as unevenly weighed centers of tangential current distribution, therefore as indirect measures of activity in extended cortical patches. This interpretation is based on two assumptions. The first comes from studies of local blood flow changes in humans. They show that the metabolic changes occur in areas much larger than those corresponding to a few cortical columns [4]. This was indeed the case in experiments using tasks that engage the frontal cortex [2,4,5]. The second assumption is based on the fact that, given that there is activity in an extended cortical surface, a dipole representing it is expected to lie in any sub-region of this surface. That is, the dipole can lie in its center, borders, or underneath the cortical surface, depending on the particular geometry of the cortical currents in question [6]. Specifically, it is only the portion of cortex showing an asymmetry in current distribution in the total active surface, that will result in a dipole-like field pattern on the scalp [7].

An unexpected result in this study was the variability across subjects in the temporal pattern of elicited magnetic activity, and it is possible that this pattern reflects an underlying variability in the distribution of more phasically or tonically active cells ('delay cells') across subjects, and thus their contribution to the externally recorded fields. Alternatively, the variability that we noted may reflect a fundamental difference across subjects in the timing of the engagement of their cortical areas into larger functional circuits. This last possibility could, in principle, be related to different personal strategies in the performance of the task. To our knowledge, a combination of those explanations cannot be ruled out in the present.

MEG has proven to be useful as a non-invasive method for functionally mapping the human frontal cortex. In the present case, posterior left frontal cortex activity, specific to the performance of a verbal task was obtained in all subjects

References

[1] Basile, L.F.H.; Rogers, R.L.; Bourbon, W.T. and Papanicolaou, A.C.. Slow magnetic fields from human frontal cortex. Electroenceph. Clin. Neurophysiol., 1994, 90:157-165.

[2] Petrides, M.; Alivisatos, B.; Evans, A.C. and Meyer, E. Dissociation of human mid-dorsolateral from posterior dorsolateral frontal cortex in memory processing. Proc. Natl. Acad. Sci. USA, 1993a, 90:873-877.

[3] Wilson, F.A.W.; Scalaidhe, S.P.O. and Goldman-Rakic, P.S. Dissociations of object and spatial processing domains in primate prefrontal cortex. Science, 1993, 260:1995-1958.

[4] Roland, P.E. and Friberg, L. Localization of cortical areas activated by thinking. J. Neurophysiol., 1985, 53:1219-1243.

[5] Petrides, M.; Alivisatos, B.; Meyer, E. and Evans, A.C. Functional activation of the human frontal cortex during the performance of verbal working memory tasks. Proc. Natl. Acad. Sci. USA, 1993b, 90:878-882.

[6] Kaufman, L.; Schwartz, B.; Salustri, C. and Williamson, S.J. Modulation of spontaneous brain activity during mental imagery. Journal of Cognitive Neuroscience, 1990, 2:124-132.

[7] Kaufman, L.; Kaufman, J.H. and Wang, J. On cortical folds and neuromagnetic fields. Electroenceph. clin. Neurophysiol., 1991, 79:211-226.

Searching for the Oddball Response in the EEG and MEG

Berg, P.[1], Kakigi, R.[2], Koyama, S.[2], Dobel, C.[1], Zobel, E.[1], and Scherg, M.[3]

[1]University of Konstanz, Konstanz, Germany; [2]National Institute for Physiological Sciences, Okazaki, Japan; [3]University of Heidelberg, Heidelberg, Germany

Introduction

We start out from the observation that the P300 component is often visible in single trial EEG data. When the data are appropriately filtered (we used 1-8 Hz band-pass), the signal becomes clearer and the auditory N100/P200 is often also apparent. Using a new spatio-temporal pattern search method, integrated into the FOCUS program [1], an objective test was available to see if such patterns could be detected in the continuously recorded raw data on the basis of the average waveforms. By applying the search to both EEG and MEG data from the same subject, a comparison between the two measurement procedures could be made. In order to test reliability of the EEG oddball response over sessions, averages from one session were used to detect responses in other sessions.

Spatio-temporal pattern search method

Instead of searching for abstract characteristics of a pattern (e.g. a generic spike-wave, an eyeblink) on a single channel [2], pattern search is based on a template taken from a multiple channel data set. For example, it can be used to detect further occurrences of spikes in a data set after one occurrence is chosen as a template. The method is illustrated in Fig. 1 using a template pattern consisting of an average ERP. The template is characterized in terms of the first 5 PCA components (topographies) and their waveforms. Pattern search is performed by multiplying the topography matrix with a test window in the data, generating 5 test waveforms. For a pattern match these waveforms should be similar to the template waveforms. A pattern match is detected when the correlation between the 5 concatenated test and template waveforms exceeds a threshold. The test window is then adjusted in time to find the maximum correlation, determining the final match.

Methods

One male 49-year-old subject performed oddball tasks in multiple sessions for both EEG and MEG over a period of 3 months.

Pure tones were presented binaurally through headphones (EEG) or a loudspeaker (MEG). Tones were 50 ms in duration, 80 dB SPL, with 5 ms rise and fall times. Tone pitches were 800 Hz and 1200 Hz. Stimuli were presented in blocks of 200, with an ISI of 1.5 s. Each block consisted of a pseudorandom sequence containing 20% of one stimulus ('rare') and 80% of the other ('frequent'). In separate blocks the subject was required to respond with either a left or a right hand button press, to either the rare or the frequent tone.

In 8 sessions on separate days with four measurements per session, 64-channel EEG was recorded continuously during the oddball tasks. EEG electrodes were distributed as evenly as possible over the head, including infra-orbital, F9, F10, mastoid, and cerebellar placements, and referred to Cz. EEG data were converted to average reference. Similarly, 74-channel MEG was recorded during 3 sessions. Dual 37-channel MEG sensor arrays (BTI Magnes) were centered over T3 and T4. Data for EEG and MEG were each divided into 32 measurements containing responses to 200 stimuli. Averages over the time range 200 ms before to 1000 ms after the stimulus were computed for each stimulus type in each measurement, using an amplitude threshold for artifact rejection. Separately for each measurement, using the average as a template, a pattern search was made on the 1-8 Hz band-pass filtered data using a 50% correlation threshold. The proportions of correct detections, 'good' false alarms (where the response to the wrong stimulus was detected), and 'bad' false alarms (where the detected pattern was not synchronized to either stimulus type) were recorded (cf. Table 1).

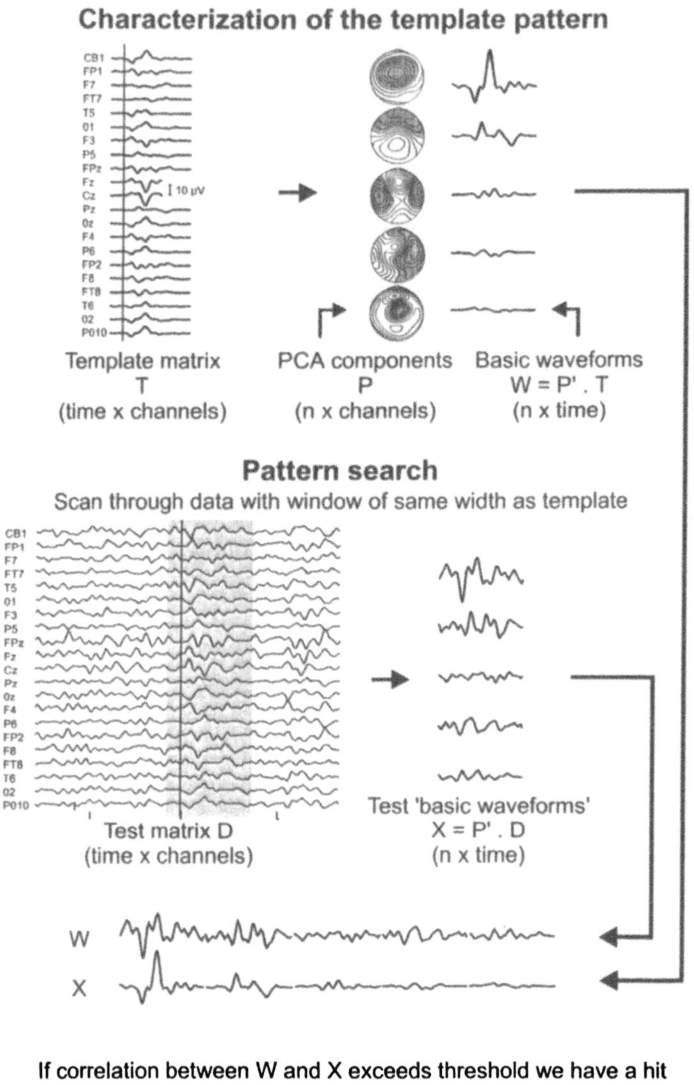

Fig. 1. Steps to the pattern search method. Top: The template matrix (in this case an EEG average to the rare stimulus, width 1200 ms) is characterized in terms of the first 5 PCA components (**P**), represented by the maps, and the basic waveforms (**W**) associated with each component. Bottom: In a scan through the data with a moving window, test waveforms (**X**) are generated. When the correlation between the concatenated waveforms exceeds a threshold, we have a pattern match. In the raw data illustrated, such a match is shown. Triggers are shown by vertical lines beneath the curves (4 triggers: frequent tone, button press, rare tone, rare tone). The oddball response is visible in the raw data, particularly at the Cz electrode, in the shaded region and in the response to the following trigger.

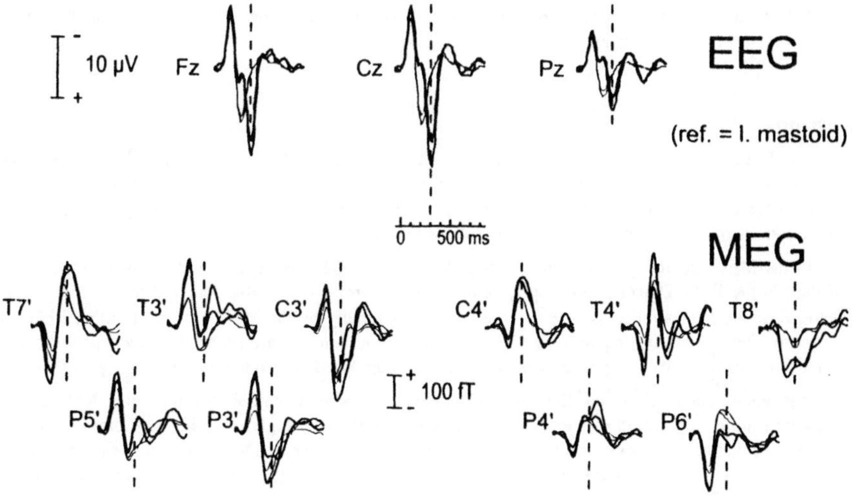

Fig. 2. Selected averages for the EEG (top) and MEG (bottom) to the stimulus that did not require a button press. Vertical dotted lines mark the latency of the peak of the EEG P300. Separate averages over odd (left hand) and even (right hand button press) days are overplotted for responses to rare (thick lines) and frequent (thin lines) tones. MEG labels show the 10-10 equivalent sensor locations.

For the EEG, averages from day 1 were used as templates for pattern search on days 3, 5, and 7. Similarly, averages from day 2 were used as templates for search on days 4, 6, and 8. Twenty-four independent pattern searches were made.

Results

Fig. 2 shows selected means for rare and frequent stimuli in the EEG and the MEG. In the EEG this subject had a large amplitude P300 (>20 µV between Cz and left mastoid) at latency 310 ms with a peak between Cz and Fz. In the MEG, peaks occur both before and after this latency, indicating a complex generator structure.

Pattern recognition results are shown in Table 1. Values are given in terms of percentage of the number of to-be-detected responses. Hit rates for both EEG and MEG were the same for the oddball response. False alarm rates were generally low, but significantly lower for the MEG. For responses to frequent stimuli, hit rates were higher for the MEG. MEG pattern search obtained **no** bad false alarms, i.e. if a match was obtained, it was always synchronized to the occurrence of a stimulus.

Table 1. Mean percentage correct detections and false alarms for pattern detection of EEG and MEG responses to rare and frequent stimuli. The third row shows the results of one-tailed t-test comparisons (df=31) between EEG and MEG.

	rare stimuli			frequent stimuli		
	correct	good FA	bad FA	correct	good FA	bad FA
EEG	71.0	10.5	2.5	22.0	1.7	0.7
MEG	72.0	7.2	0.0	32.0	1.7	0.0
t-test	N.S.	p < .05	p<.001	p<.001	N.S.	p<.001

Pattern search using a template from one day to detect oddball responses on data from another day resulted in a 61% hit rate (0.14 good false alarms, 0.04 bad false alarms).

Discussion

The fact that templates recorded on one day could be used to detect 61% of the oddball responses on data recorded on other days indicates that to a large extent the same brain processes were elicited in each session. Therefore, differences in results for EEG and MEG are probably due to the differences in signals picked up by the type of measurement rather than to differences in signals generated in the subject's brain.

In the MEG recordings, the pattern search detected no 'bad' false alarms and more responses to frequent stimuli than in the EEG. Detection of responses to frequent stimuli is probably better than the EEG because the main generators are tangential sources which are detected optimally by the MEG sensors. In other EEG data with visual stimuli (Berg et al., in preparation), in which large radial sources are probably activated in the secondary visual areas, the pattern detection rate, at 34%, was similar to the MEG detection rate for auditory stimuli.

It is remarkable that for both EEG and MEG, the pattern search could detect more than 70% of the oddball responses, indicating both a high efficiency of the search method, and a high replicability of brain responses to the rare stimuli. In other subjects with smaller P300 responses we have observed lower detection rates (mean 24%). In the same subjects' responses to novel auditory stimuli, which produced large amplitude responses, the hit rates were much higher (mean 70%). The presence of artifact signals such as alpha waves and eye movements were observed to reduce the hit rate. These results indicate that the success of pattern search depends on signal strength and signal-to-noise ratio.

Although the MEG did not display a 'P300' at the latency obtained in the EEG, several consistent peaks could be observed up to 600 ms after the stimulus, suggesting a degree of complexity of the oddball response often not recognized in the EEG literature.

References

[1] Scherg, M., and Ebersole, J.S. Neurophysiol. Clin., 1994, 24: 51-60.

[2] Gotman, J., and Wang, L. Electroenceph. clin. Neurophysiol., 1992, 83: 12-18.

Acknowledgements

We would like to acknowledge support for the first author from the Ministry of Education, Science, and Culture of Japan to carry out the MEG study.

Magnetoencephalographic Measurements of a Developing Infant: A Pilot Study

Bowyer, S.M.[1,2], Moran, J.[1] and Tepley, N.[1,2]

[1]*Henry Ford Hospital, Detroit, MI, USA;* [2]*Oakland University, Rochester, MI, USA*

Introduction

MEG time-series recording of spontaneous activity is being used in our laboratory to study the adult brain in association with the mechanisms for stroke [7] and Migraine [9, 1]. In the present study MEG is used to study the development of the Central Nervous System (CNS) in a developing infant. In the past electroencephalography (EEG) has been used for similar studies. EEG has recorded the organization of the frequencies associated with CNS development [2, 4, 6]. MEG studies are much more easily conducted on a sleeping infant. One feature of the data obtained, sleep spindles are bursts of activity in the 11-15 Hz range lasting from tenths of seconds to approximately four seconds [5]. Sleep spindles of early childhood first occur during the third month of life and are an indication of normal sleep patterns being established. This particular activity in the 11-15 Hz tends toward an adult 14 Hz rhythm between 18 and 24 months of age, when a more mature (shorter spindle train length with a less spike-like negative component) sleep spindle activity is developing [3]. Early childhood sleep spindles have been seen on EEG and are considered to be a good indication of a normal developing Central Nervous System [6, 8]. Sleep spindles are observed in the frontal or central cortex during stage 2, and stage 3 sleep [5]. There tends to be an asynchrony in the appearance of sleep spindles between hemispheres in infants under one year of age [6, 8]. The absence or decrease of early childhood sleep spindle activity indicates an abnormality of the CNS such as cerebral dysfunction [3, 6, 8]. A second feature of the data obtained is the low frequency activity in the 4-10 Hz range. Low frequencies tend to decrease over time in the maturing CNS [5]. Magnetoencephalographic measurements were performed on a developing infant during this longitudinal study.

Method

The subject was a healthy male, full-term, infant (birth weight: 6 lbs. 4 ozs.). There were no known abnormalities or defects of the Central Nervous System and no family history of neurological disease. No medication was administered at any time during the study. Informed consent was obtained from the parent.

The subject was studied once a week from 18 weeks of age to 65 weeks of age. During the first half of the study, from 18 to 43 weeks of age, mother's milk was used to nurse the infant to sleep. Formula was used thereafter to nurse the infant to sleep (44 to 65 weeks). After he fell asleep the infant was placed under a 7 channel Superconducting Neuromagnetometer (Bti model 607), inside of a magnetically shielded room. The probe was positioned over the left hemisphere of the head first then over the right hemisphere. In reference to the International 10-20 system of electrode placement, the central coil in the 7 channel array was centrally located over the frontal-central region in the T4 location on the right side and in the T3 location on the left side of the head. (Fig. 1) Six minutes of data were collected on one side of the infants head. He was then rotated 180° and repositioned under the probe while he slept. Three more minutes of data were collected on this side for a total of six minutes of data. Data was stored on optic disk for analysis later.

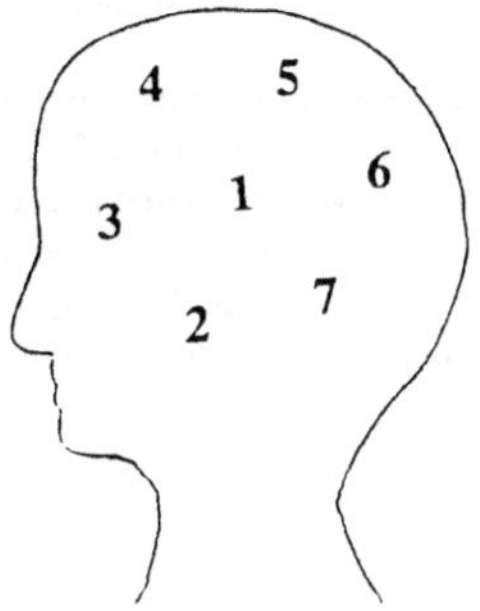

Fig. 1 Magnetometer channel locations

Results

Only the data for the channel over the frontocentral region of the infant's brain was used. This was where the sleep spindles were most prominent. Fig. 2 and 3 display the real time data showing typical sleep spindle bursts (underlined) as their length decreased over time.

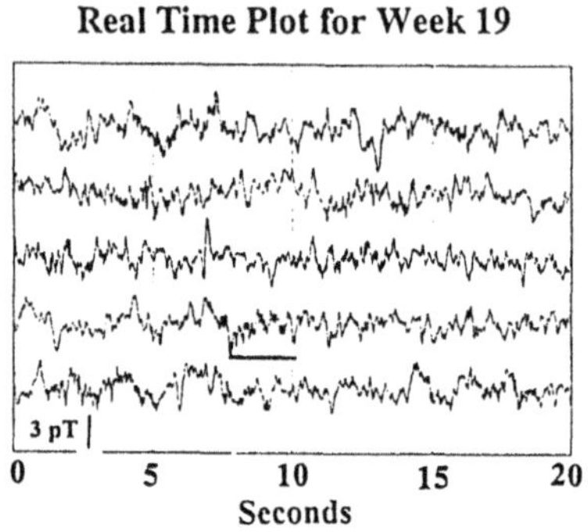

Fig. 2 Raw data collected from the right hemisphere during week 19. Each horizontal line represents 20 seconds of continuous data. The above plot represents 80 seconds of data. The sleep spindle burst is underlined.

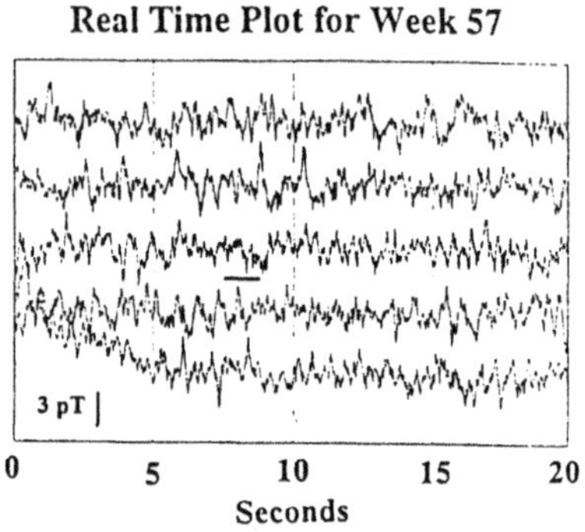

Fig. 3 Raw data collected from the right hemisphere during week 57. Each horizontal line represents 20 seconds of continuous data. The above plot represents 80 seconds of data. The sleep spindle burst is underlined.

Fig. 4 and 5 are FFTs performed on week 19 and 57 showing the peak frequency increased from 13.23 Hz in week 19 to 14.04 Hz in week 57.

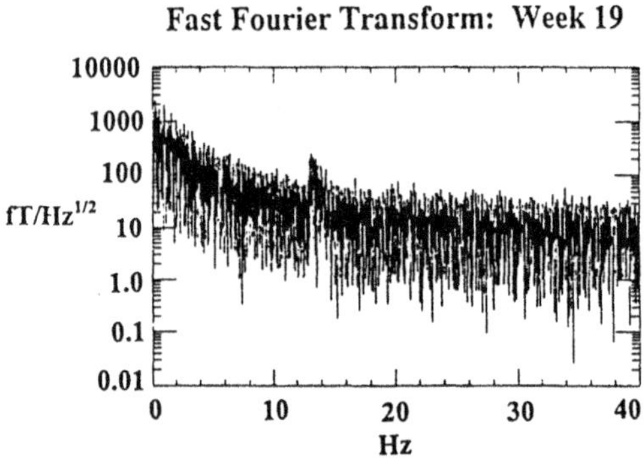

Fig. 4 FFT of data for week 19

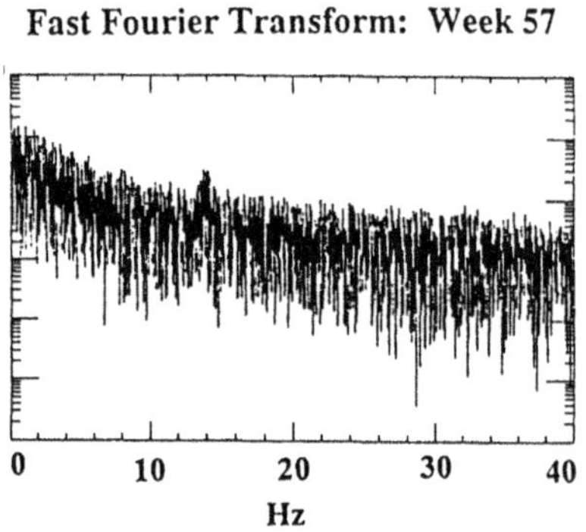

Fig. 5 FFT of data for week 57

The six minutes of data were broken up into 18 epochs of 20 seconds, each was spectrally analyzed by a Fast Fourier Transform (FFT). Analysis was performed on the frequency range from 3.5 Hz to 15.5 Hz. The data was converted into a percentage of the daily spectrum. An overall time averaged FFT was obtained. The average of the sequential FFTs was used to quantify the spectrum for each date. To quantify the changes in the study we calculated the percent changes in the relative spectral composition of the daily spectrum from this average spectra. The Frequencies were classified into two ranges (4-10

Hz) and 11-15 Hz) and these bands were grouped together and graphed over time. Linear regressions were plotted over the data spectra. A linear increase in the percent change over time was found for the frequency range 11-15 Hz and a decrease for the frequency range 4-10 Hz. These findings were significant in each hemisphere and for both frequency ranges: sleep spindle band (11-15 Hz), and low frequency (4-10 Hz). Table I displays the correlations and significant p-values.

TABLE I

		correlation	p-value
Sleep spindle 11-15 Hz:	Right hemisphere	0.504	0.002
	Left hemisphere	0.523	0.003
Low Frequency 4-10 Hz:	Right Hemisphere	0.578	0.0003
	Left Hemisphere	0.589	0.0005

Fig. 6 through 9 display graphically the percent change over time of the daily averaged FFT from the overall time averaged FFT.

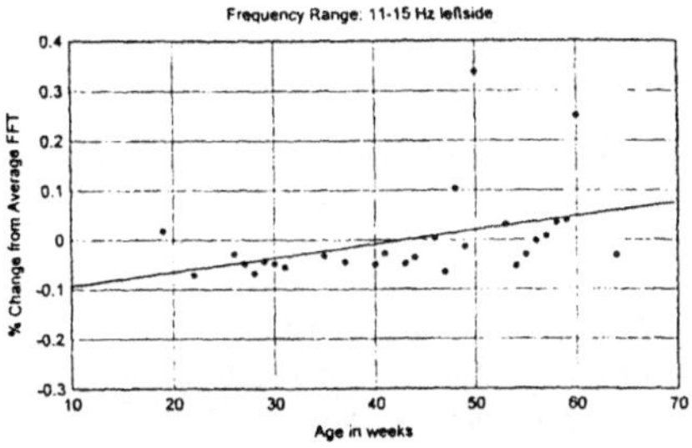

Fig. 6 Percent change in Sleep spindle activity of the left hemisphere.

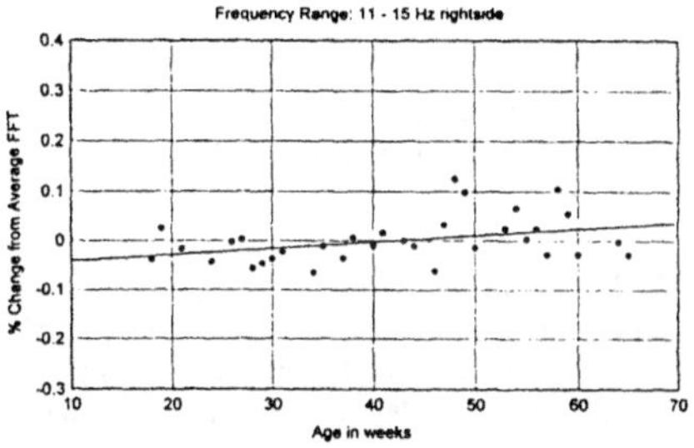

Fig. 7 Percent change in Sleep spindle activity of the right hemisphere.

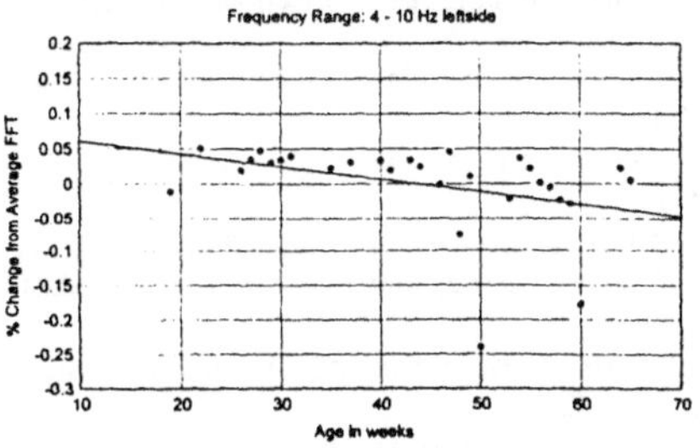

Fig. 8 Percent change in Low Frequency activity of the left hemisphere.

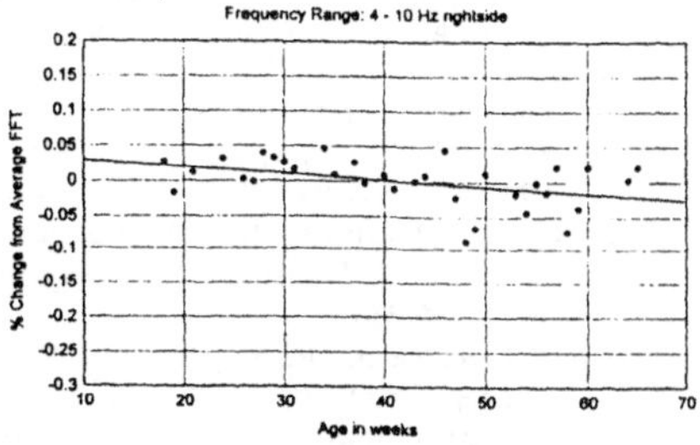

Fig. 9 Percent change in Low Frequency activity of the right hemisphere.

Discussion

Magnetoencephalography can be used to determine the significance of the CNS developments over time in a developing infant. This procedure has a number of advantages over

Electroencephalography. MEG is less time consuming; it takes approximately 20 minutes to place a full array of electrodes on an infant, while it takes less then 2 minutes to position the probe over the area of interest on the infants head. Electrodes are not required to collect data for MEG making it easier for the technician to prepare the infant for the study. Infants can peacefully fall asleep as opposed to being forcibly restrained in papooses used to facilitate the electrode placement. In papooses they tend to scream and cry placing undue stress on them prior to the studies.

The results showed a difference in the increase and decrease of the linear regressions from hemisphere to hemisphere. These results can be explained by observing that the infant was always studied on the left side first. During this time he was most likely entering into a state of arousal conducive to sleep spindle activity. (i.e. light sleep, stage 2) During the second half of the study he fell into a deeper sleep in which sleep spindle activity decreases or ceases. Since we cannot collect simultaneous data on both sides of the head we are unable to make a conclusion regarding the asyncronisity of sleep spindle appearance.

The data in this study was based on 6 minutes of light, daytime sleep and can not truly be compared to the over-night studies completed in EEG. In addition the stages of sleep were not scored. One of the purposes of this study was to see if the results from prior EEG studies could be replicated. We have shown that similar results can be obtained using MEG. These developmental changes need to be further studied to determine their significance over a larger sample. To the best of our knowledge this is the first study of this kind performed on MEG. This can be a clinically useful technique for obtaining information about the developing Central Nervous System.

Acknowledgment

Thank you to Michael for being a good subject, and to Uma Saligram and Yeaton Clifton for their help in collecting the data. This research supported by NIH/NINDS Grant No 1R01 NS30914

References

[1] Barkley, G., Tepley, N., Nagel-Leiby, S., Moran, J., Simkins, R., Welch, K.M.A. Magnetoencephalograhic Studies of Migraine. Headache Journal, 1990; 30 (7), 428-434.

[2] Ellingson, R. Development of Sleep Spindle Bursts During the First Year of Life. Sleep, 1982 5 (1): 39-46

[3] Jankel,W.R., Niedermeyer, E. Sleep Spindles. Journal of Clinical Neurophysiology, 1985, 2 (1): 1-35.

[4] Louis, J., Zhang, J.X., Reuol, M., Debilly, G., Challamel, M.J. Ontogenesis of nocturnal organization of sleep spindles: a longitudinal study during the first 6 months of life. Electroencephalography and Clinical Neurophysiology, 1992, 83: 289-296.

[5] Niedermeyer, E. Lopes Da Silva, F. Electroencephalography, Basic principles, clinical , applications and related fields, Maryland, Williams and Wilkins 1993.

[6] Shibagaki, M. Kiyono, s., Kazuyoshi, W. Spindle Evolution in Normal and Mentally Retarded Children: A Review. Sleep, 1982, 5 (1): 47-57.

[7] Tepley, N. Spreading Cortical Depression and Related DC Phenomena, In: Hoke, M., Ernd, S. N., Okada, Y. C., Romani, G. L. Biomagnetism: Clinical Aspects, Amsterdam, Excerpta Medica 1992, 329-335.

[8] Tyner, F. S., Knott, J.R., Mayer, Jr., W. B. Fundamentals of EEG Technology. Volume 1: Basic Concepts and Methods. New York, Raven Press p 242-245.

[9] Welch, K.M.A., Barkley, G., Tepley, N., Ramadan, N., Central Neurogenic Mechanisms of Migraine. Neurology, 1993; 43 (suppl 3): S21-S25.

The Steady State Visual Evoked Response and Estimates of Phase Velocity

Burkitt, G.R., Silberstein, R.B. and Wood, A.W.

Swinburne Centre for Applied Neurosciences, Swinburne University of Technology, Melbourne, Australia

Introduction

Several models of brain function predict the existence of wave like phenomena where electrical activity travels from one cortical region to another [1-3]. A model which can explain the rhythmogenesis of the MEG and EEG should approximate neural mass dynamics at a spatial scale consistent with that of scalp recordings. In this respect, Nunez's global resonance model is of particular relevance since it allows direct comparison between scalp recorded electrical activity and theory [3]. This model predicts waves mediated by the rapidly conducting cortico-cortical fibres with speeds between 6 to 9 m/sec. Hence, a specific test of the model can be made by measuring the velocity of MEG and EEG signals and comparing estimated values with the above range. A direct approach is to use the Steady State Visual Evoked Response (SSVER) and measure the spatial change in phase across the scalp. In this paper the SSVER phase is measured at regular intervals on the scalp and the resulting phase topography yields an estimate of phase velocity at the stimulus frequency. In addition, estimates of wavenumber are obtained from the same measurement.

Method

Concurrent MEG and EEG signals were recorded from midline scalp sites while 21 subjects (mean age = 26.2 yrs, SD = 7.1 yrs) with normal or corrected vision viewed a sinusoidally alternating checkerboard. Subjects lay face down on a cushioned wooden bench while the stimulus was observed through a 45° mirror positioned directly beneath. All MEG recordings (unshielded) were performed with a BTi biomagnetometer (model 601) consisting of a single channel DC SQUID and second-order gradiometer with a 19.8 mm diameter pick-up coil and 50.4 mm baseline. The distance from the bottom of the pick-up coil to the bottom outside surface of the dewar was 12 mm. Accurate alignment of the dewar tail over each recording position was assisted by grid markings on a customised electrocap which also served to locate the EEG electrodes. MEG sites were marked at 1.5 cm intervals anterior from the inion while 10 EEG electrodes were placed at 3 cm intervals. To reduce volume conduction distortion of the EEG signal, the midline EEG electrodes were arranged as a bipolar chain. Checkerboard stimuli were presented for approximately two minutes at each MEG scalp site. Between trials subjects were permitted to move and if necessary the individual was rested for a short time. Over the duration of a session, 8 to 9 MEG scalp sites were recorded from midline.

Maximum and minimum luminances of the checkerboard pattern were 8 and 0.7 cd/m^2 respectively, thus the spatial contrast for the checkerboard was 84%. The viewing distance was 1.08 metres yielding a screen size of 10.9° (horizontal) by 8.25° (vertical) and the check size was set to 11'.

It has been suggested that alpha rhythm frequencies contribute to global resonance and travelling waves [3], thus the checkerboard alternation rate was matched to the peak alpha frequency of each subject. Peak alpha frequencies were determined at the beginning of the recording session from 5 minutes of spontaneous EEG (eyes closed) recorded from a bipolar electrode at Oz.

The SQUID output was amplified to give a sensitivity of 3.65 pT/Volt. EEG signals were amplified by 50 000. All signals were band-pass filtered (MEG: 1 to 40 Hz, EEG: 0.1 to 40 Hz) and digitised to 12 bit accuracy (±5 Volts) at a rate of 160 Hz. MEG signals underwent additional filtering to reduce noise at 50 Hz using Twin-T notch filters centred at 50, 100 and 150 Hz and an inverse Chebyshev filter with a stop-band notch at 50 Hz. Checkerboard patterns were generated on a NEC 3FG multisync monitor with a vertical frame rate of 83 Hz. The alternation rate of the checkerboard was controlled with a sinusoidal frequency generator consisting of a phase-locked loop, ROM sine wave lookup table and an 8 bit digital-to-analog converter.

SSVER phase was determined at the alternation frequency of the stimulus by evaluating the cosine and sine Fourier coefficients during each stimulus cycle. High frequency resolution and noise resistance was then obtained by collapsing all the coefficients calculated over the recording period into a single average.

Results

Typical examples of the Steady State Visual Evoked Field (SSVEF) and Steady State Visual Evoked Potential (SSVEP) obtained from conventional averaging methods are illustrated in Fig. 1. MEG sites extended only half the distance of the midline due to the positioning limitations of the probe and the prolonged time required to record from separate sites.

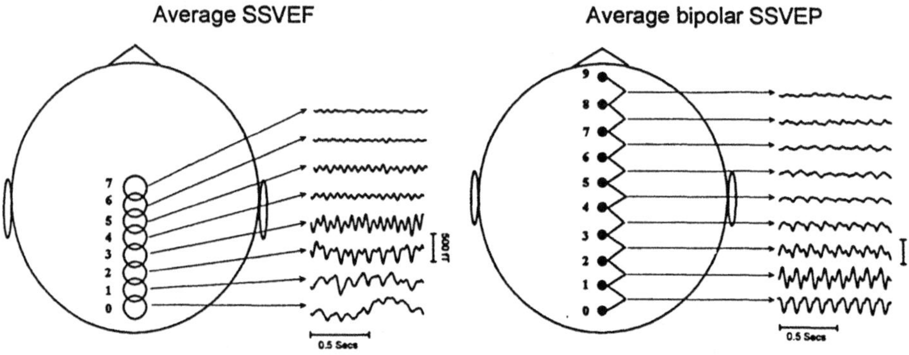

Fig. 1. The average SSVEF and SSVEP recorded from midline sites. In these illustrations, the average SSVEF (left) was obtained from 80 epochs (subject AW) while the average bipolar SSVEP (right) was obtained from 200 epochs (subject GH). The SSVEF demonstrates greater variability compared to the SSVEP since a smaller number of stimulus cycles were recorded during magnetic measurements.

In all subjects studied, the spatial variation of the SSVER was examined by plotting the phase at each midline recording site. Two distinctly different spatio-temporal features were observed. These comprised a smooth variation of phase with distance and a sharp phase discontinuity of approximately π radians along the midline. When present, the discontinuity occurred at parietal or central sites. Discontinuous variations may be due either tangential dipoles or standing wave nodes. As this paper addresses the issue of phase velocity estimation, it will be restricted to a consideration of smooth variations of phase with distance.

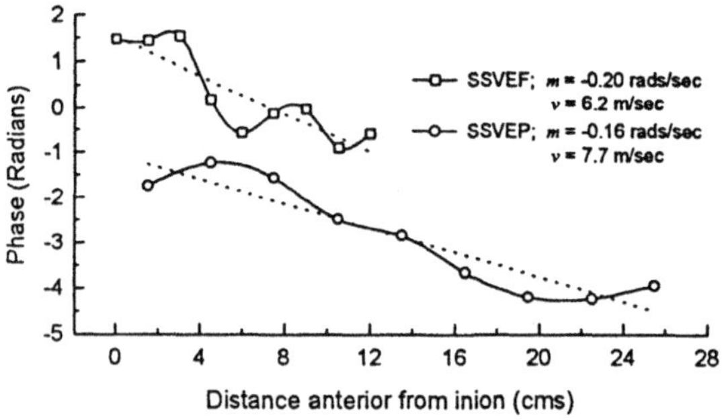

Fig. 2. Linear regression analysis (dotted line) of the phase topography (solid line) from one subject. MEG and EEG phase measurement sites along the midline are shown by open squares and circles respectively.

The spatial gradient of smooth phase variations was estimated from linear regression methods and only those gradients with a correlation coefficient exceeding an arbitrary level of 0.9 were included in the subsequent analysis. It was then possible to estimate phase velocity by treating the changes in phase as a time delay elapsed over a given scalp distance. Phase velocities were calculated from the best fit gradient of the spatial phase variation. Fig. 2. illustrates the linear regression technique with associated phase gradient and velocity estimates.

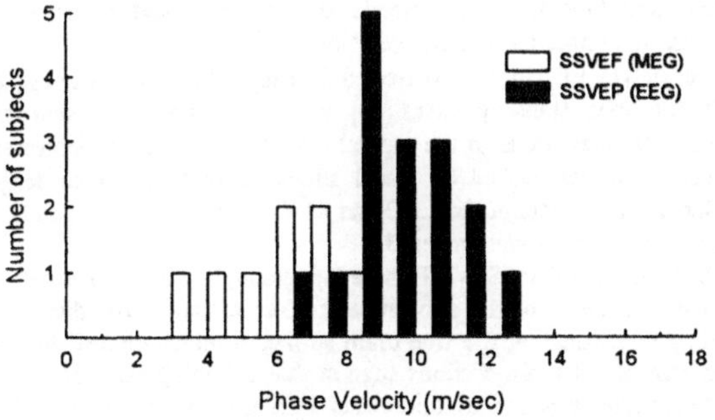

Fig. 3. Histogram of phase velocity estimates obtained from the SSVEF (open bars) and bipolar SSVEP (solid bars). Phase velocities from these two methods overlap within a range of 6 to 8 m/sec but are statistically different ($p = 2.4\times10^{-6}$, unpaired t-test). Note that all values were corrected for fissures with a constant of 2.2.

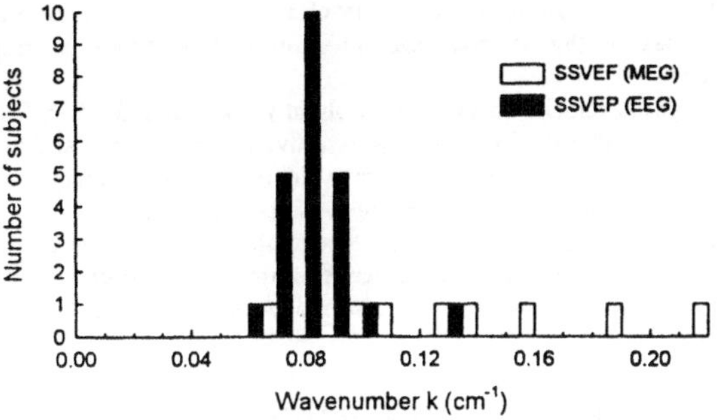

Fig. 4. Histogram of wavenumber estimates obtained from the SSVEF (open bars) and bipolar SSVEP (solid bars). Bipolar SSVEP wavenumbers are clustered between 0.06 and 0.10 cm^{-1}. No distinct wavenumber peak can be observed from SSVEF phase variations. Note that all values were corrected for fissures with a constant of 2.2

The phase velocity is given by;

$$v = \frac{2\pi fc}{m}$$

where v is the phase velocity (m/sec), f is the stimulus frequency (Hz), c is a folding ratio (2.2) to correct for the distances hidden within cortical fissures and m is the phase gradient (rads/m).

A histogram of all estimated phase velocity values is illustrated in Fig. 3. The SSVEF histogram (open bars) is characterised by a smaller number of subjects since fewer measurements met acceptable signal to noise criteria. Also note that SSVEF values show a tendency towards lower values compared to the SSVEP. The bipolar SSVEP (solid bars) generated the most sharply peaked distribution with all values bound between 6 and 13 m/sec. The mean SSVEF phase velocity was 5.1 ± 1.7 (SD) m/sec while mean SSVEP phase velocity was 8.7 ± 1.4 (SD) m/sec.

Gradients measured as radians per cm (cm^{-1}) can be treated as estimates of wavenumber (k). The distribution of wavenumber obtained from the SSVEF and bipolar SSVEP is illustrated in Fig. 4. The bipolar SSVEP is characterised by a sharp peak at 0.08 cm^{-1} while the SSVEF exhibits a broader distribution.

Discussion

The SSVER phase topography show features that are consistent with travelling waves. For instance, most estimates of phase velocity range from 5 to 10 m/sec, a range very similar to those estimated for cortico-cortical fibres (6 to 9 m/sec) from axon diameter studies of the corpus callosum [4, 5] and the frontal lobe [6, 7]. Previous data providing estimates of scalp velocity suggest wide variation ranging from as low as 3 m/sec to as high as 20 m/sec [8, 9]. This contrasts with the present finding of a narrow velocity range. The difference may be attributable to volume conduction distortion of the scalp topography since past estimates of velocity were calculated from monopolar EEG recordings. The monopolar EEG method is prone to volume conducted spread and reference electrode effects and this can reduce the apparent phase differences between electrode sites thereby resulting in larger estimates of phase velocity. In the current study, these difficulties were reduced by use of MEG and bipolar

EEG recording methods. In this respect, MEG and bipolar EEG methods appear well-suited to accurate measurement of the spatial changes in SSVER phase and estimates of phase velocity.

More speculatively, the mean wavenumber (0.075 ± 0.014 cm^{-1}, corrected for fissures) derived from Fig 4. is within the range anticipated (0.1 to 0.07 cm^{-1}) for global standing waves [10]. It is interesting to note that the corresponding wavelength (83 ± 15 cm) is of a similar magnitude to the sagittal circumference of the neocortex (~100 cm, corrected for fissures). Hence, travelling waves evoked by visual stimuli may extend over large distances. Such a finding is consistent with a global wave model of the EEG and MEG where periodic boundary conditions of the closed surface of the neocortex determine the wavenumber [3].

While the phase velocity and wavenumber estimates from SSVER phase topography are close to expected ranges, their accuracy is compromised by limited knowledge of the neocortical surface distance. All distances across the scalp were corrected for cortical folding by estimating the true brain surface distance divided by the exposed brain surface distance. Perimeter measurements of brain sections suggest that a folding ratio of 2.2 is appropriate [10] although it is probable considerable inter-subject variation exists. Furthermore, it is known that the ratio depends on the direction across the neocortex in which the perimeter is measured. Folding ratios that take into account such variation suggest a range of 2.0 to 2.5 [10]. At this stage, it is not clear whether a folding ratio of 2.2 is suitable for the midline surface of the neocortex but this potential inaccuracy can be avoided by measuring the sagittal neocortical perimeter with MRI methods.

In general, it was found that the SSVEF phase velocity estimates were slightly lower than SSVEP. The reasons for this finding are uncertain. One possibility is that the MEG is more sensitive to spatio-temporal noise arising from environmental magnetic fields. The difference may also be due to the volume conduction processes to which MEG is less sensitive. Further work is required to differentiate between these two possibilities.

In summary, the present findings indicate that features of the SSVER topography are compatible with a global model of EEG and MEG. It is not suggested that travelling wave mechanisms are the sole process which give rise to SSVER phase variations but merely that travelling waves may contribute to the observed data.

References

[1] Wilson, H.R., and Cowan, J.D. A mathematical theory of the functional dynamics of cortical and thalamic nervous tissue, Kybernetik, 1973, 13: 55-80.

[2] van Rotterdam, A., Lopes da Silva, F.H., van den Ende, J., Viergever, M.A., and Hermans, A.J. A model of the spatial-temporal characteristics of the alpha rhythm, Bull. Math. Biology., 1982, 44: 283-305.

[3] Nunez, P.L. Neocortical dynamics and human EEG rhythms, New York, Oxford University Press, 1995.

[4] Tomasch, J. Size distribution and number of fibres in the human corpus callosum, Anat. Rec., 1954, 119: 119-135.

[5] Aboitiz, F., Scheibel, A.B., Fisher, R.S., and Zaidel E. Fibre composition of the human corpus callosum, Brain Res., 1992, 598: 143-53.

[6] Bishop, G.H., and Smith, J.M. The size of nerve fibres supplying cerebral cortex, Exp. Neurol., 1964, 9: 483-501.

[7] Swadlow, H.A., Rosene, D.C., and Waxman, S.G. Characteristics of interhemispheric impulse conduction between prelunate gyri of the Rhesus monkey, Exp. Brain Res., 1978, 33: 445-467.

[8] Thatcher, R.W. Krause, P.J., and Hrybyk, M. Cortico-cortical associations and EEG coherence: A two-compartmental model, Electroenceph. Clin. Neurophysiol., 1986, 64: 123-143.

[9] Hughes, J.R., Kuruvilla, A., and Kino, J.J. Topographic analysis of visual evoked potentials from flash and pattern reversal stimuli: Evidence for "Travelling Waves", Brain Topography, 1992, 4: 215-228.

[10] Nunez, P.L. Electric Fields of the Brain, New York, Oxford University Press, 1981.

Nonlinear Spatio-temporal Interactions and Neural Connections in Human Vision Using Transient And M-Sequence Stimuli

Chen, H.-W., Aine, C.J., Flynn, E.R., and Wood, C.C.

Biophysics Group, MS D-454, Los Alamos National Laboratory
Los Alamos, NM 87545, U.S.A.

INTRODUCTION

There are two classes of connections among neurons in cat and monkey cortices [1]: (a) the traditional receptive field connections (feedforward connections), and (b) the dynamic association of neurons in different cortical areas which form functionally coherent assemblies (reciprocal connections). The feedforward connections have been well studied using traditional single-unit recording techniques. As pointed out by Singer [1], there are two strategies for characterizing the dynamic association of neurons in different cortical areas (reciprocal connections): 1) examining changes in discharge rate, and 2) examining temporal synchrony (the stimulus evoked coherent oscillations are in the gamma range 30-110 Hz). The first was evidenced by spatio-temporal interactions beyond the classical receptive fields (RFs) for V1 cortical neurons. The second was revealed recently using multiple-unit recording techniques and cross-correlation analysis, which have shown that cortical neurons can synchronize their discharges very precisely (e.g., [2]). The reciprocal connections can be categorized as tangential intra-areal, interhemispheric, and/or feedback (between cortex and LGN) connections, and may play an important role in integrating and binding widely distributed and parallel representations in different cortical areas.

Reciprocal connections, in essence, are the dynamic wiring (connections) of the neural network circuitry [1]. Given the high complexity of the neural circuitry in human brain, it is quite a challenge to study the dynamic wiring of highly parallel and widely distributed neural networks. The measurements of stimulus evoked coherent oscillations provide indirect evidence of dynamic wiring. In this study, in addition to the coherent oscillation measurements, we provide two more techniques for testing possible dynamic wiring: 1) measurements of spatio-temporal interactions beyond the classical receptive fields, and 2) neural structural testing using nonlinear systems analysis.

Spatio-temporal interaction and its relation to reciprocal connections. Animal studies have revealed the existence of facilitatory and/or inhibitory integration fields (IF) beyond the classical receptive fields (RFs) for retinal ganglion cells and striate cortical neurons. The extent of the integration fields is about 2-5 times the classical receptive field size [3]. Polat and Sagi [4] have recently found, in human vision, suppression and facilitation lateral interactions between spatial frequency (SF) channels revealed by lateral masking psychophysical experiments conducted at the fovea. The mechanism mediating these interactions at different eccentricities is, however, still an open question. One hypothesis for the interactions is based on the existence of multiple spatial frequency (SF) channels in the human visual system. Each channel may involve many neurons, and the receptive field of an SF channel may be defined as overlapping receptive fields of all the involved neurons (i.e., the RF of an SF channel is larger than that of a single neuron). Therefore, the spatial lateral interaction beyond the classical RFs of single neurons may be explained by the interactions between neighboring SF channels (feedforward connections) having spatially overlapping receptive fields. Another hypothesis is that the multiple SF channels are parallel and distributed in different cortical areas, and they are connected and bind via dynamic wiring. In this case, the distance of spatial lateral interactions would be large. Indeed, Ts'o and coworkers [5] have found excitatory horizontal intracortical connections from cat primary visual cortex, and suggested that these connections might contribute to properties beyond classical RF analysis. Therefore, measurements of spatio-temporal interactions may provide information on both traditional feedforward and reciprocal connections.

Nonlinear Systems analysis and Neural Circuitry testing. Several approaches to studying unknown biological systems consider the system initially as a "black box." In one approach, knowledge of the input-output (I/O) behaviors of the system are acquired by identifying the system's I/O transfer function. For example, the Laplace transfer function of a linear system is directly related to the I/O relationship of the system described by a linear differential equation. Unlike analyses of simple linear systems, analyses for nonlinear systems present a greater challenge since higher-order transfer functions are required for full characterization of a nonlinear system's I/O behaviors. One method used to characterize a nonlinear system (which is generally time-invariant or stationary, causal, and continuous), expresses the I/O relationship in terms of Volterra or Wiener functional expansions [6]. Such expansions are fully specified by a set of Volterra (or Wiener) kernels which can be estimated from the I/O measurements of the system under study. The system kernels provide a characterization of the system I/O relationship, and thus allow one to predict the response of a physical or physiological system to an arbitrary stimulus. Furthermore, the measured lower-order (the 1st- and 2nd-order) system kernels can provide information about the internal structure of the practical system under study [7]-[9], which suggests a number of potential applications in the areas of neural system modeling and identification. For example, we are interested in characterizing the I/O relationship of populations of neurons in cortical areas of human brain, obtained from noninvasive electromagnetic measurements. Neuronal responses can be modeled as circuitry, and nonlinear analysis may ultimately allow one to characterize the structure of neural system circuitry, using system kernels.

We have recently developed techniques using sparse-stimulation and short m-sequence stimuli [10], and have successfully measured the 1st- and 2nd-order kernels from neuromagnetic responses. The advantages for using this method are: 1) easier and fast kernel calculation; 2) short data acquisition times and less measurement errors; 3) short m-sequences with different lags (delays) are orthogonal to one another which permit their use in studying multiple input systems (i.e., the simultaneous presentation of m-sequences). Furthermore, the results from MEG experiments [10] indicate that the S/N ratio of responses evoked by m-sequences can be 2-3 times higher than responses evoked by transient target stimuli when stimulus presentation time is equated.

In this study, we have measured spatio-temporal interactions in human vision using transient and m-sequence stimuli. The estimated cross-correlations (system kernels) obtained using m-sequence stimuli were examined in hope of identifying possible neural connections (structures) and mechanisms underlying both feedforward and reciprocal connections. Measurements of coherent oscillations using transient stimuli will also be briefly discussed.

METHODS

Evoked neuromagnetic fields were recorded with a BTi 7-channel SQUID-coupled gradiometer system in a magnetically-shielded chamber. Four subjects with normal vision (HWC, CA, WEN, and JE) participated in this study. In the transient studies, we presented two pattern onset target stimuli (circular sinusoidal gratings, 266 ms duration and ~1 Hz rate of presentation) either simultaneously or sequentially with different SFs, to a specific eccentricity. These paired stimuli were presented at different eccentricities (0.8°, 1.6°, 3°, and 6°) in the visual field. The spatial separation between them was varied in order to quantify the nonlinear summation area. The nonlinear summation index was calculated as the peak response to the simultaneous presentation of stimuli divided by the sum of the peak responses to the two separate stimulus presentations. In the m-sequence studies, the paired m-sequence stimuli were presented at the same locations as the two target stimuli used for the transient study. Both m-sequence and transient target stimuli were presented for each placement of the dewar, for comparison. Three different m-sequence stimuli were used: 1) *Counter-phase stimuli*. Paired target patterns were counter-phase modulated by two orthogonal short binary m-sequences between two states (in-phase and off-phase). Subjects rested for ~5-10s after each of 6 binary m-sequence trials. 2) *In-phase stimuli*. Paired target stimuli were modulated between two different contrast levels (in-phase) of the same patterns by two orthogonal binary short m-sequences. The key feature of this stimulus is that it can be decomposed as a steady grating pattern component and a counter-phase modulated grating pattern component. 3) *Unpatterned stimuli*. Paired unpatterned stimuli were modulated above and below the background mean luminance by two orthogonal short binary m-sequences. MEG responses were digitized at 200 Hz.

The nonlinear dynamics and interactions between the spatial locations were revealed by the cross-correlations between the response and the input m-sequences.

RESULTS

The dashed curves in Fig. 1 are evoked neuromagnetic responses to the simultaneous presentation of the two stimuli, whereas the solid curves represent the sum of the responses to the individual presentations of the two stimuli. The dashed and solid curves should be approximately equal in peak amplitude if the spatial interactions between the stimuli were linear. On the other hand, a (large) difference between the two curves suggests a (strong) nonlinear interaction. Figs. 1a and 1c show responses of 2 sensors for stimuli 0.93° apart for subject WEN. Responses in Figs. 1b and 1d are responses to the stimulation with greater separation (1.6°). The results indicate stronger nonlinear interactions for stimuli which are closer together. Figs. 1e and 1f compare responses to paired stimuli (99% contrast) at different eccentricities for subject CA: in Fig. 1e the stimulus eccentricity and distance were 0.8° and 0.61°, respectively; whereas the stimulus eccentricity and distance were 3° and 2.3° for the responses in Fig. 1f, respectively. Figs. 1g and 1h are responses to the paired stimuli (4.5 c/d, 3° eccentricity) at two different

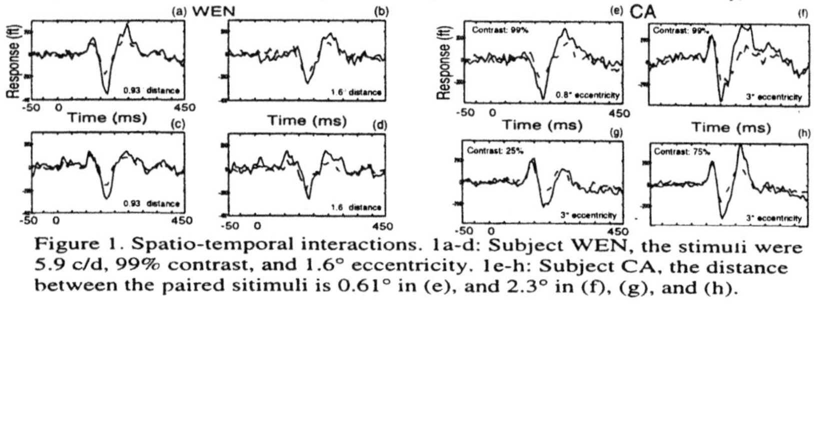

Figure 1. Spatio-temporal interactions. 1a-d: Subject WEN, the stimuli were 5.9 c/d, 99% contrast, and 1.6° eccentricity. 1e-h: Subject CA, the distance between the paired sitimuli is 0.61° in (e), and 2.3° in (f), (g), and (h).

contrast levels 25% and 75%, respectively for CA.

Fig. 2 shows measured 2nd-order cross-correlations for subjects WEN and HWC using counter-phase and in-phase stimuli. The contours of these cross-correlations provide information on neural circuitry structures. The kind of contour picture in Fig. 2a suggests simple feedforward (cascade) connections, whereas the more complex pictures in Figs. 2b and 2c suggest involvement of more complex neural circuitry. Fig. 3 shows the power spectra (using short-time FFT, duration: 115 ms, step: 20 ms) of waveforms (subject CA) stimulated by transient stimuli (50% contrast) at 3 different SFs (2.0, 4.5, and 8.5 c/d). The equipment and system noise (60 Hz) was suppressed using a threshold technique and low-pass filtering. Stronger 40 Hz coherent oscillations under stimulus conditions of SF=4.5 c/d (Fig. 3c) is evidenced, compared with other SF conditions (Figs. 3a and 3d).

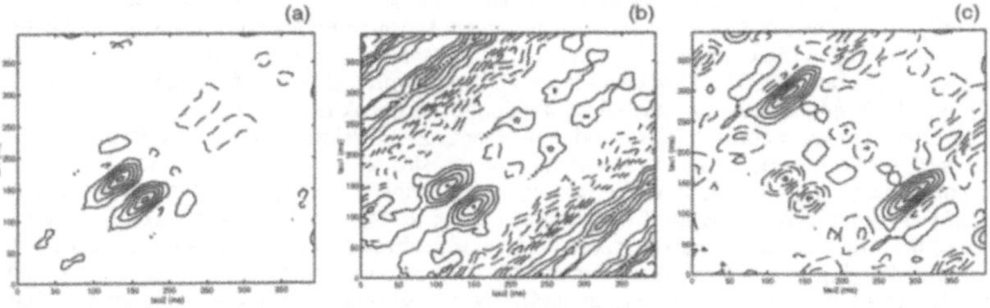

Figure 2. Measured 2nd-order cross-correlations (system kernels). Positive and negative contour values are shown as solid- and dashed-lines, respectively. (a): Subject HWC, counter-phase stimuli, 1.6° eccentricity, 25% contrast, 4.5 c/d. (b): Subject WEN, counter-phase stimuli, 1.6° eccentricity, 25% contrast, 2.0 c/d. (c): Subject HWC, in-phase stimuli, 3° eccentricity, contrast change from 20% to 80%, 4.5 c/d.

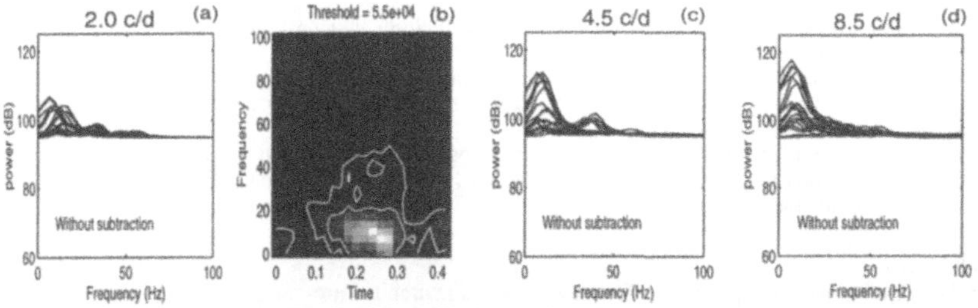

Figure 3. Measurements of coherent oscillations using short-time Fourier transform. Subject CA, 1.6° eccentricity, 50% contrast. (a): SF=2.0 c/d. (b) and (c): SF=4.5 c/d. (d): SF=8.5 c/d.

For a more quantitative analysis of nonlinear spatial summation, we introduce a summation Index (SI) by dividing the peak amplitude (e.g., the dashed curve in Fig. 1a) of the response to paired stimuli by the peak amplitude (e.g., the solid curve in Fig. 1a) of the response to the sum of the two individual stimulation conditions. An index value SI=1 indicates a linear summation, whereas SI < 1 indicates a nonlinear suppressive interaction. For example, an SI < 0.5 means that at least one of the peak amplitudes for single stimulus presentations was larger than that of the simultaneous stimulation. SIs were calculated for the strong second negative peak of the waveforms in Figs. 1a-d for subjects WEN and CA (not shown in Fig. 1), and are shown in Table 1.

Table1. Summation Index (SI) for two stimulus distances and their ratio

Subject	SI for Distance 0.93°	SI for Distance 1.6°	Ratio
CA (Sensor #1)	0.57	0.76	0.75
CA (Sensor #6)	0.44	0.58	0.76
WEN (Sensor #2)	0.52	0.71	0.73
WEN (Sensor #7)	0.55	0.72	0.76

DISCUSSION AND CONCLUSIONS

The results in Fig. 1 indicate that the response waveforms evoked by transient stimuli showed strong nonlinear summation for shorter separations between the two stimuli. Both subjects (in Table 1) showed strong nonlinear summations for shorter distances between the two stimuli. The interactions were weaker when the distance increased (as indicated by larger SI values). As discussed above, the distance of spatial interactions may be larger for reciprocal connections than for feedforward connections (SF channels). Therefore, it is possible that the interactions resulted from both feedforward connections and reciprocal connections when the two stimuli were closer. As the distance increases, the involvement of feedforward connections decreases, and thus the spatial interactions may be dominated by reciprocal connections. It was also found in Table 1 that the SI ratio between different separations was quite consistent between subjects. Figs. 1e and 1f indicate that the foveal responses have smaller spatial interaction areas than peripheral responses. The SI values for the 1st negative peak and the 2nd positive peak in Fig. 1g (25% contrast, 3° eccentricity) are 0.48 and 0.77, and are 0.50 and 0.46 in Fig. 1h (75% contrast, 3° eccentricity), respectively. It appears that the interaction for the 2nd positive peak is more sensitive to the contrast change. It is less likely that stimulus contrast would change the RF size of an SF channel, therefore, the interaction evidenced at the higher contrast level in the 2nd positive peak may primarily result from the reciprocal connection (Singer [1] and Eckhorn et al., [2] have shown that coherence oscillations were sensitive to visual stimulus parameters such as position, orientation, movement direction and velocity, etc. Here, we show that the reciprocal connection may also be sensitive to the stimulus contrast). The SI values for the 1st positive peak in Figs. 1f, 1g, and 1h are 1.06, 0.84, and 0.92 respectively, which indicates that the spatial summation for this peak is more linear.

From the m-sequence studies, the 1st- and 2nd-order cross-correlations were measured using unpatterned, counter-phase, and in-phase stimulation conditions. In general, strong 2nd- (nonlinear component) but weak 1st-order (linear component) cross-correlations for the unpatterned and counter-phase stimulus conditions resulted, whereas a strong 1st-order cross-correlation was apparent when in-phase stimuli were used. The estimated cross-correlations have been further used for testing possible neural feedforward and reciprocal connections (structures) and mechanisms underlying spatial nonlinear interactions. The contours of Fig. 2a is typical for the 2nd-order unpatterned kernels and some counter-phase kernels under specific stimulus conditions. This kind of kernel picture suggests simple feedforward (cascade) connections, as corroborated by computer simulations [10]. On the other hand, the more complex pictures of the 2nd-order counter-phase and in-phase kernels (under certain stimulus conditions) in Figs. 2b and 2c suggest involvement of more complex neural circuitry (parallel, and/or feedback, etc.). Ongoing efforts have been directed to find appropriate neural circuitry models (e.g., the recurrent thalamic-cortical-thalamic feedback connection and/or cortico-cortical reciprocal connection, etc.) which can generate these kernel pictures using computer simulations.

Fig. 3(b) and (c) indicate possible 40 Hz coherent oscillations under stimulus conditions of SF=4.5 c/d and 50% contrast. The 40 Hz component is notably weaker under other SFs (Figs. 3a and 3d) and contrast conditions (not shown), which suggests that the coherent oscillations (and thus reciprocal connections) may be sensitive to stimulus parameters such as SFs and contrast levels. These results agree with our results from studies of spatio-temporal interactions, as discussed above.

In summary, the three methods (measurements of spatio-temporal interactions, system kernels, and coherent oscillations) used in this study may provide reliable measurements for understanding complex neural circuitry structures with reciprocal connections.

REFERENCES

[1] Singer, W. Development and plasticity of cortical processing architectures. *Science*, 1995, 270:758-764.

[2] Eckhorn, R., Bauer, R., Jordan, W., Brosch, M., Kruse, W., Munk, M., and Reitboeck, H. Coherent oscillations: a mechanism of feature linking in the visual cortex? *Biol. Cybern.*, 1988, 60:121-130.

[3] Li, C.Y. and Li, W. Extensive integration field beyond the classical receptive field of cat's striate cortical neurons--classification and tuning properties. *Vision Res.*, 1994, 34:2337-2355.

[4] Polat, U. and Sagi, D. Lateral interactions between spatial channels: Suppression and facilitation revealed by lateral masking experiments. *Vision Res.*, 1993, 33:993-999.

[5] Ts'o, D., Gilbert, CD, and Wiesel, T. Relationships between horizontal interactions and functional architecture in cat striate cortex as revealed by cross-correlation analysis. *Neurosci.*, 1986, 6:1160-1170.

[6] Marmarelis, P.Z. and Marmarelis, V.Z. Analysis of physiological system: the white noise approach. New York: Plenum Press, 1978.

[7] Chen, H.-W., Ishii, N., and Suzumura, N. Structural classification of non-linear systems by input and output measurements. *Int J Systems Sci.*, 1986, 17:741-774.

[8] Chen, H.-W., Jacobson, L.D., and Gaska, J.P. Structural classification of multi-input nonlinear systems. *Biological Cybernetics*, 1990, 63:341-357.

[9] Chen, H.-W. Modeling and identification of parallel nonlinear systems: structural classification and parameter estimation methods. *Proceedings of The IEEE*, 1995, 83:39-66.

[10] Chen, H.-W., Aine, C.J., Best, E., Ranken, D., Harrison, R.R., Flynn, E.R., and Wood, C.C. Nonlinear analysis of biological systems using short m-sequences and sparse-stimulation techniques. *Annals of Biomedical Engineering*, 1996, 24:4 (In press).

ACKNOWLEDGMENT

This work was supported in part by a grant from NEI (EY08610) and by the United States Department of Energy (DOE) Contract W-7405-ENG-36.

Spatial Frequency tuning functions and Contrast Sensitivity at Different Eccentricities in the Visual Field

Chen, H.-W., Aine, C.J., Flynn, E.R., and Wood, C.C.

Biophysics Group, D454, Los Alamos National Laboratory
Los Alamos, NM 87545, U.S.A.

INTRODUCTION

Threshold and suprathreshold vision. The human luminance spatial frequency contrast sensitivity function (CSF) has been well studied using psychophysical measurements by detecting spatial frequency (SF) grating patterns at *threshold* [1], [2]. *Threshold* CSFs at different eccentricities have proven to be quite useful in both basic and clinical vision research. However, near threshold, the CSF is measured at a linear area of the saturating contrast-response curve. In contrast, most of our everyday vision may be at suprathreshold levels, and thus may function most of the time at the nonlinear area of the contrast-response curve. Furthermore, since the CSF is measured near threshold, it is quite possible that only the most sensitive retinal or LGN cells may contribute to the measured CSF. Since the M (magnocellular) cells in retina and LGN are about 10 times more sensitive than the P (parvocellular) cells at low to medium spatial frequencies [3], [4], the measured threshold CSF may reflect more M cell contributions. There have been recent attempts to measure suprathreshold contrast responses using contrast matching techniques [5], [6], [7]. These results show that the contrast response functions at suprathreshold levels are flatter than CSF curves measured at threshold. However, whether the results from contrast matching reflect the spatial frequency tuning functions at suprathreshold for human vision is still unclear since contrast matching is quite different from threshold measurements [4].

Contrast-response (CR) functions. In monkey studies, the contrast-response (CR) functions at different contrast levels have been measured for M and P cells in the retina and LGN as well as cortical neurons in V1 and MT [3], [8]. A Michaelis-Menten-like equation can be used to describe the contrast-response function well:

$$R(c) = R_{max}\, c^n/(c^n + c^n_{50}) + M; \qquad (1)$$

where R is the amplitude of the response, c is contrast, R_{max} the maximum response, c_{50} the contrast at which the response reaches half of its maximum value, n the steepness (gain) of the curve, and M the dc level (spontaneous discharge). Equation (1) provides a quantitative way to describe the contrast-response relationship. For example, the parameter c_{50} can be used as an index for contrast sensitivity. The lower the c_{50} value, the higher contrast sensitivity the CR function will be. For the M and P cells in retina, the average c_{50} values were 0.13 and 1.74 respectively [3], and for cortical neurons on average, the c_{50} value was 0.33 for V1 and 0.07 for MT [8].

In this study, in order to better characterize the CSF at normal contrast levels, we measured the SF tuning functions as well as the CR functions at different *suprathreshold* contrast levels and different eccentricities of the visual field using noninvasive MEG techniques. Traditionally, peak analysis has been used for the averaged transient evoked waveform [9]. However, sometimes, it is difficult to define a single peak when the signal-to-noise ratio of the waveforms is low (the noise will cause multiple peaks), or when the waveforms are complex due to overlapping components. In this study, in addition to peak analysis, we have also developed more reliable "averaged power" analysis methods using FFT: the averaged power can be calculated from the entire waveform (or portion of).

Although equation (1) can generally fit the measured data well, the fitting errors will be large if the CR functions are super-saturated (i.e., the response decreases when contrast increases, which may be caused by inhibition between mechanisms at high contrast levels). In order to reduce the fitting errors and to obtain a quantitative measure of the inhibition at high contrast, we introduce a fifth exponential parameter, in addition to the 4 parameters in equation (1):

$$R(c) = e^{-ac} R_{max}\, c^n/(c^n + c^n_{50}) + M; \qquad (2)$$

where, a is the exponential parameter. Equation (2) reduces to equation (1) when $a=0$, and the larger the a value, the stronger the super-saturation.

METHODS

Experiments. Evoked magnetic fields were recorded with a BTi 7-channel SQUID-coupled gradiometer system in a magnetically-shielded chamber. Four subjects with normal vision participated in this study. Transient target stimuli (circular sinusoid, 266 ms duration and ~1 Hz rate of presentation) were presented one at a time in a sequential fashion to 3 different eccentricities (1.6°, 3°, and 6°). Three SFs (2.0, 4.5 and 8.5 c/d) were presented at

1.6° eccentricity, three SFs (1.5, 4.5, 8.5 c/d) were presented at 3° eccentricity, and three SFs (1.0, 4.5, 8.5 c/d) were presented at 6° eccentricity (The low SFs differed slightly across eccentricities in order to have at least 1.5 cycles of grating for each target). For each specific SF and eccentricity, there were 5 different contrast levels (10%, 25%, 50%, 75%, and 99%). Each of the 5 contrast conditions in a block of trials contained 20 stimulus presentations, making a total of 100 presentations for each block of trials (less than 2 min for a block of trials). Subjects rested for about 10s after 5 blocks of trials which produced a total of 100 responses for each contrast level. The presentations were randomly interleaved for the 5 contrast conditions. The responses to the 3 different SFs at 3 different eccentricities were recorded from the same dewar location. Cortical magnification factor was also considered. The sizes of the stimuli at 1.6°, 3°, and 6° eccentricities were calculated to be 0.76°, 1°, and 1.5° respectively.

Data analysis. In peak analysis, the SF tuning curves and the c_{50} values were obtained from the peak amplitude of the 1st negative peak (with latency ~160-190 ms) of the response waveforms for different stimulus conditions. In averaged power analysis, the SF tuning curves and the c_{50} values were obtained from the averaged power of the whole waveform (0-500 ms) or from a fraction of the waveform (90 ms duration around the 1st negative peak (with latency ~180 ms). The averaged power was calculated by first taking the Fourier transform of the waveform, and then averaging the spectral power of different temporal frequencies.

For each eccentricity, there are 5 SF tuning curves at 5 different contrast levels, and each curve has 3 data points at 3 different SFs (i.e., 3x5=15 data points). To more quantitatively describe potential contributions of the M and P systems to the responses, we propose a two-stream model whose parameters were estimated by fitting the model to the data using a least-square-error program (Matlab, Signal processing toolbox). The model is expressed as:

$$R(sf, c) = M(sf)CR_m(c) + P(sf)CR_p(c) \quad (3)$$

where $R(sf, c)$ is the response at 3 different SFs (sf) and 5 different contrast levels (c); $M(sf)$ and $P(sf)$ are SF tuning functions for the M- and P-like systems, respectively; and $CR_m(c)$ and $CR_p(c)$ are the contrast-response (weighting) functions for the M- and P-like systems, respectively. There are 3 unknown parameters (at 3 SFs) for each $M(sf)$ and $P(sf)$, and 4 unknown parameters (at 4 contrast levels) for each $CR_m(c)$ and $CR_p(c)$. The values at contrast level 0.1 can be normalized as a unit for $CR_m(c)$ and $CR_p(c)$ (i.e., $CR_m(0.1) = Cr_p(0.1) = 1.0$). The total unknowns in equation (3) is 14, and the independent data values in the data set is 15. Therefore, we can uniquely estimate the 14 unknowns using a least-square-error fitting program. Equation (2), then, was used to estimate the 5 parameters (R_{max}, c_{50}, n, M, and a) from the obtained $CR_m(c)$ and $Cr_p(c)$ for the M- and P-like systems.

RESULTS

Both peak analysis and averaged power analysis were used to analyze the data measured from 4 subjects in order to construct SF tuning curves at 5 different contrast levels CR (contrast-response) functions for each specific SF. The CR functions can then be fitted by equation (2), resulting in estimations of the 5 parameters (R_{max}, c_{50}, n, M, and a). The results (SF tuning curves) from peak analysis for HWC were plotted in Figs. 1a-c, and those from averaged power analysis were plotted in Figs. 1d-f. The SF tuning curves in Figs. 1a-c are similar to those in Figs. 1d-f, which suggests that both peak analysis and averaged power analysis methods provided similar results. The CR functions and estimated 5 parameters at 6° eccentricity are shown in Figs. 1g-i.

The estimated SF tuning curves from averaged power analysis were further fitted by a two-stream (M & P) model (equation (3)) for all the 4 subjects at 3° eccentricity. The 1st-row in Fig. 2 shows the fitting accuracy: the minimum-square-error (MSE) is 9.3% for HWC (Fig. 2a), 5.6% for CA (Fig. 2d), 11.7% for WEN (Fig. 2g), and 6.9% for JE (Fig. 2j). The estimated SF tuning functions for the M and P systems ($M(sf)$ and $P(sf)$) were shown in the 2nd-row of Fig. 2. The estimated $CR_m(c)$ and $CR_p(c)$ (the contrast-response (weighting) functions for the M and P systems) were shown in the 3rd-row of Fig. 2 as the 'o' and '+' symbols, respectively. Equation (2), was then used to fit and estimate the 5 parameters from the obtained $CR_m(c)$ and $Cr_p(c)$. The fitting curves are shown in the 3rd-row of Fig. 2 as the dashed curves. The estimated parameters are listed in Tables 1 and 2. The fitting results in the 2nd- and 3rd-rows of Fig. 2 show similar features for all the 4 subjects.

Table 1. Parameters estimated from $CR_m(c)$.

Subjects	R_{max}	c_{50}	n	M	a	MSE
HWC	135	0.21	6.0	0.0	2.29	3.6%
CA	63	0.19	5.5	0.0	1.29	3.7%
WEN	2	0.11	10.0	1.0	2.74	8.6%
JE	983	0.33	7.0	2.0	10.24	9.1%

Table 2. Parameters estimated from $CR_p(c)$

Subjects	R_{max}	c_{50}	n	M	a	MSE
HWC	480	0.81	3.2	1.0	1.81	0.4%
CA	44	0.39	5.5	0.0	0.15	6.1%
WEN	8	0.47	10.0	1.0	1.48	4.4%
JE	664	0.66	7.0	0.0	3.93	1.5%

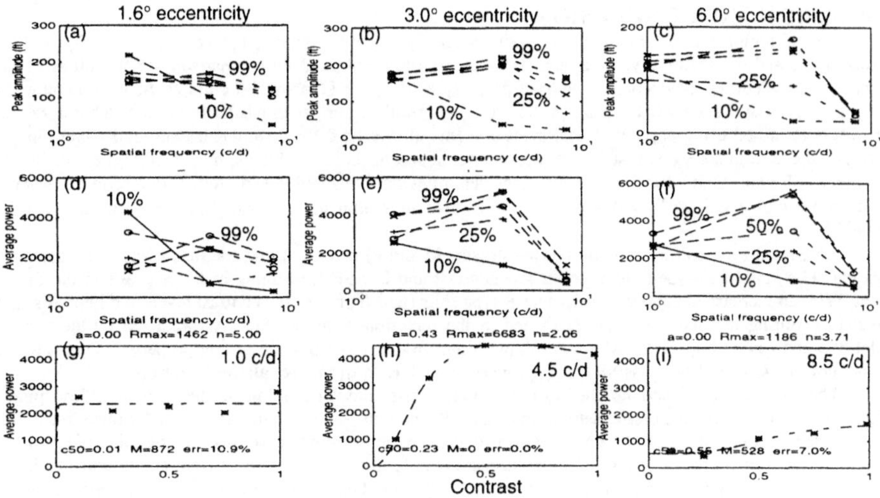

Figure 1. Spatial frequency tuning functions and contrast-response functions (Subject: HWC). 1a-c: Peak analysis (the 1st negative peak, latency ~160-190 ms); (a) 1.6 degree eccentricity; (b) 3 degree eccentricity; (c) 6 degree eccentricity. Figs. 1d-f: Averaged power analysis (waveform duration: 0-500 ms). 1g-i: Contrast-response functions fitted by equation (2): 6 degree eccentricity.

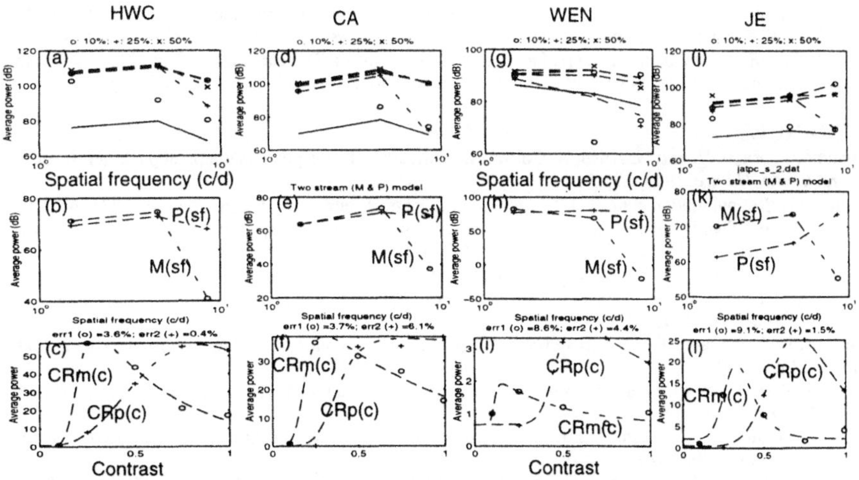

Figure 2. Data from averaged power anaysis (waveform duration: 90 ms; latency: ~180 ms) and fitted by a two-stream (M & P) model (equation (3)). Four subjects: HWC, CA, WEN, and JE. 1st column: HWC; 2nd column: CA; 3rd column: WEN; 4th column: JE. 1st row: Data were plotted as symbols; the curves were predictions from the model (equation (3)). 2nd row: Estimated M(sf) and P(sf) for the 4 subjects. 3rd row: estimated CRm(c) and CRp(c) for the 4 subjects.

DISCUSSION AND CONCLUSIONS

The SF tuning curves (obtained from both peak analysis and the averaged power analysis in Fig. 1) for high contrast levels at 1.6° eccentricity, were very flat with peak SFs > 8.5 c/d. In comparison, for stimuli at 3° and 6° eccentricities, the curves show a broad bandpass property with a peak SF about 3-5 c/d. For the lowest contrast level (10%), the curve shows a lower SF peak (< 1.5 c/d). This result is similar to results from other laboratories where CSF was determined using psychophysical measurements at threshold for a similar background mean luminance. Therefore, these results suggest that human eyes, under suprathreshold conditions, may have higher spatial resolution (higher SF gain) than what the CSF functions determined at threshold would predict. Furthermore, when comparing foveal with parafoveal conditions, our results also show that more foveal stimulation results in wider SF bandwidth.

These results are consistent with the two stream (M and P) theory: at low contrast levels, the majority of responding cells may be M-type (which prefer low contrast and low SFs), resulting in lower peaks in the SF tuning curves. With an increase in contrast level, more P-type cells (which prefer high contrast levels and high SFs) may be activated, resulting in curves with peaks shifted in the direction of higher SFs. As predicted by the two-stream model (the 2nd-row of Fig. 2), the M system shows a narrow-band low-pass SF tuning property, whereas the P system shows a broad-band band-pass SF tuning property, which is consistent for all the 4 subjects.

The measured CR functions (the 3rd-row of Fig. 1) also show that, at low SF the c_{50} value is low, indicating a high contrast sensitivity and earlier saturation of the CR function. As shown in Tables 1 and 2 (also in the 3rd-row of Fig. 2), the c_{50} value of $CR_m(c)$ (the M system) is lower (i.e., higher contrast sensitivity) than that of $CR_p(c)$ (the P system). These results are consistent with magno-parvocellular distinctions derived from single-unit monkey studies. The results in Tables 1 and 2 (and the 3rd-row of Fig. 2) indicate that the techniques developed in this study may provide useful information about the relative strengths of the M and P systems at different eccentricities. Therefore, based on these results, it appears that the c_{50} value and the estimated $CR_m(c)$ and $Cr_p(c)$ functions may provide information of the relative involvement of magno-like and parvo-like systems in humans. It is worth noting that (as shown in the 3rd-row of Fig. 2 and Tables 1 and 2) both M and P systems show super-saturation at high contrast levels as evidenced by the exponential a values in Tables, which may be caused by inhibition between mechanisms at high contrast levels. The super-saturation is generally stronger (higher a values) for the M-like system.

Single-unit studies in nonhuman primates indicate that different cortical areas may be differentially innervated by M or P streams. Based on our previous retinotopy studies using source localization strategies and our current results (i.e., SF tuning curves and CR functions), noninvasive MEG techniques may be able to shed more light on functions of neural groups in different cortical areas related to the two stream (M vs. P) theory.

REFERENCES

[1] Rovamo, J., Virsu, V. and Nasanen, R. Cortical magnification factor predicts the photopic contrast sensitivity of peripheral vision. *Nature*, 1978, 271:54-56.

[2] Kelly, D.H. Retinal inhomogeneity. I. Spatiotemporal contrast sensitivity. *J. Opt. Soc. Am.* 1984, A1:107-113.

[3] Kaplan, E. and Shapley, R.M. The primate retina contains two types of ganglion cells, with high and low contrast sensitivity. *Proc. Natl. Acad. Sci.*, 1986, 83:2755-2757.

[4] De Valois, R. and De Valois, K. Spatial Vision. Oxford Univ. Press, 1988.

[5] Georgeson, M.A., and Sullivan, G.D. Contrast constancy: deblurring in human vision by spatial frequency channels. *J. Physiol.*, 1975, 252:627-656.

[6] Bowker, D.O. Suprathreshold spatiotemporal response characteristics of the human visual system. *J. Opt. Soc. Am.* 1983, 73:436-440.

[7] Davis, E.T. Modeling shifts in perceived spatial frequency between the fovea and the periphery. *J. Opt. Soc. Am.* 1990, A7:286-296.

[8] Sclar, G., Maunsell, J.H.R. and Lennie, P. Coding of image contrast in central visual pathways of the macaque monkey. *Vision Res.*, 1990, 30:1-10.

[9] Regan, D. Human Brain Electrophysiology--Evoked Potential and Evoked Magnetic Fields in Science and Medicine. New York: Elsevier Science Publishing Co., 1989.

ACKNOWLEDGMENT

This work was supported in part by a grant from NEI (EY08610) and by the United States Department of Energy (DOE) Contract W-7405-ENG-36.

Neuromagnetic Fields Evoked by Sine Tones, Sine Tone Complexes, Vowel Formants, and Two-Formant Vowels

Diesch, E., Luce, T.A., and Eyferth, K.

Department of Psychology, Technical University of Berlin, Berlin, Germany

Introduction

In an earlier study [1], stimulus related differences were found in source location both among N1m sources and sustained field (SF) sources elicited by synthetic five-formant vowels which did not seem to correspond in any straightforward manner to the tonotopic functional organization of the auditory cortex established with sine tones [2]. This apparent discrepancy may be due to spectral complexity differences as well as the dimensionality difference between multivariate vowel sets and univariate sine tones. More generally, the question arises as to how the response of the auditory cortex to spectrally complex acoustic stimuli like vowels may be related to the responses to their spectral components. Spectral components of vowels may be considered at the level of harmonic structure and at the level of formant structure. In order to further investigate this issue, the magnetic field response to sine tones and vowels radically reduced in complexity and dimensionality was studied in two experiments.

Methods

In the first experiment, 14 subjects were presented with a 600 Hz and a 2100 Hz sine tone and the first (F1) and second formant (F2) of a synthetic two-formant instance of the vowel /æ/. F1 and F2 formant frequency values were 600 Hz and 2100 Hz. Additionally, the subjects received the two-tone complex and the two-formant vowel generated by additive superposition of the respective components. In the second experiment, a 200 Hz (F1a), a 400 Hz (F1b), a 800 Hz (F1c), a 2600 Hz vowel formant (F2), and three two-formant vowels generated by superimposing F1a, F1b, and F1c with F2 were presented. The two-formant vowels were perceived as /i/, /e/, and /æ/ and the single formants F1a and F1b as /u/ and /o/. 15 subjects were investigated. In both experiments, stimulus duration was 600 msec. Component stimuli were equated in subjective loudness, formants and vowels in fundamental frequency. Stimuli were delivered to the right ear in a random sequence with an ISI of 2 sec +/- 100 msec. There were 150 replications of each stimulus. In the first experiment, subjects were asked to count rare light flashes, in the second experiment, to count an additional rare vowel.

The magnetic field was recorded over the left hemisphere with a 37-channel magnetometer [3] installed in a magnetically shielded room. The bandwidth of the recording was 0.16-64 Hz. After artefact rejection, 1200 msec epochs including a 200 msec prestimulus baseline interval were selectively averaged and digitally lowpass filtered at 20 Hz for N1m analysis and 5 Hz for SF analysis. The model of a single spatio-temporal equivalent current dipole in a spherical volume conductor was used for source analysis. Source parameters were computed in a head-centered coordinate system. Spatio-temporal dipoles were fitted to a 20 msec interval around the N1m rms peak and to the SF in the 400-600 msec latency interval. Source solution were accepted only if polarity reversal was evident in the field pattern and the goodness of fit was 0.90 or better. In experiment 2, satisfactory single-dipole source solutions for the SF could be found only for a minority of subjects and stimulus conditions. Multi-dipole source solutions proved instable and variable across subjects. Therefore, in order to separate source contributions to the SF and to isolate the predominant contribution, SF waveforms in experiment 2 were submitted to principal components analysis (PCA) and the PCA waveform component accounted for by the first PCA factor was entered into the dipole source analysis.

The source parameters were submitted to repeated measures ANOVAs. Two types of analysis were performed. The first type of analysis sougt to determine the frequency dependence of source parameters. Thus, the component and composite stimulus sources, respectively, were compared among each other. The second type of analysis investigated whether the source fitted to the field elicited by a composite stimulus was equal to the model source fitted to the linear sum of the fields elicited by the stimulus components. With some kind of nonlinear addition, a systematic deviation of the source activated by composite stimuli from the the source predicted by the linear addition model should be observed. Thus, the composite stimulus sources were compared with the expected composite stimulus.

Results

A representative example of the evoked magnetic field waveforms and field distributions recorded in experiments 1 and 2 is presented in Fig. 1. The field average shows a pronounced N1m and a distinct SF component. The field distribution of both the N1m and the SF can be accounted for by a single equivalent current dipole. In experiment 1, the

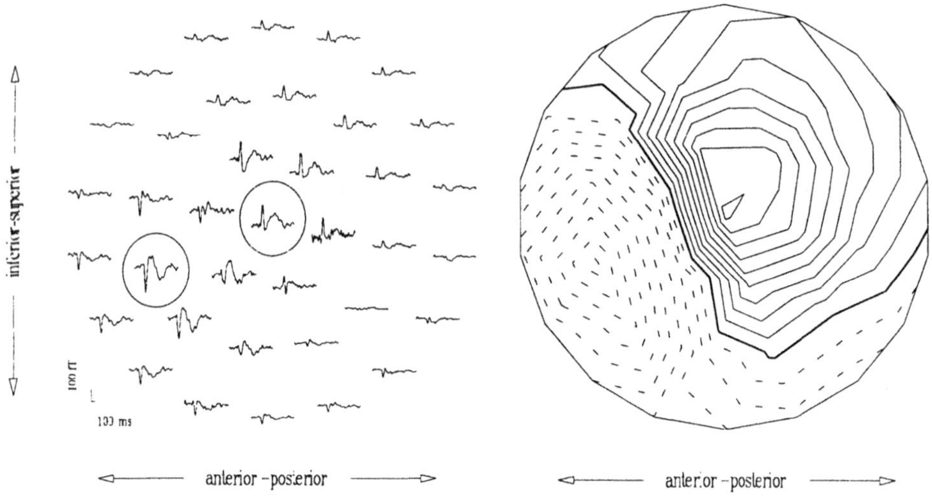

Fig. 1. Left: The 37 averaged evoked magnetic field waveforms recorded in response to the 600/ 2100 Hz compound tone in one of the subjects of experiment 1. The two waveforms with across-channels maximum and minimum N1m amplitude, respectively, are marked by circles. Right: The field distribution at the maximum N1m field power latency.

composite stimulus N1m sources were shorter in latency (F[1,11] = 58.78, P < 0.001), weaker in dipole moment amplitude (F[1,11] = 55.72, P < 0.001), and more lateral in source location (F[1,11] = 10.50, P < 0.01) than the expected composite stimulus N1m sources. 2100 Hz N1m sources were more medial in location (F[1,11] = 5.67, P < 0.05) than 600 Hz N1m sources. Tone stimulus N1m sources were more posterior (F[1,11] = 5.99, P < 0.05) and more medial in location (F[1,11] = 9.79, P < 0.01) and shorter in latency (F[] = 13.66, P < 0.01) than formant stimulus N1m sources. The composite stimulus SF sources were weaker in dipole moment amplitude (F[1,10] = 18.50, P < 0.001) than the expected composite stimulus SF sources. The tone stimulus SF dipole moments were weaker than the formant stimulus SF sources (F[1,10] = 16.59, P < 0.001). The SF source location pattern mirrored the N1m location pattern, but the SF source location differences fell short of statistical significance. Fig. 2 displays the N1m and SF source topographies.

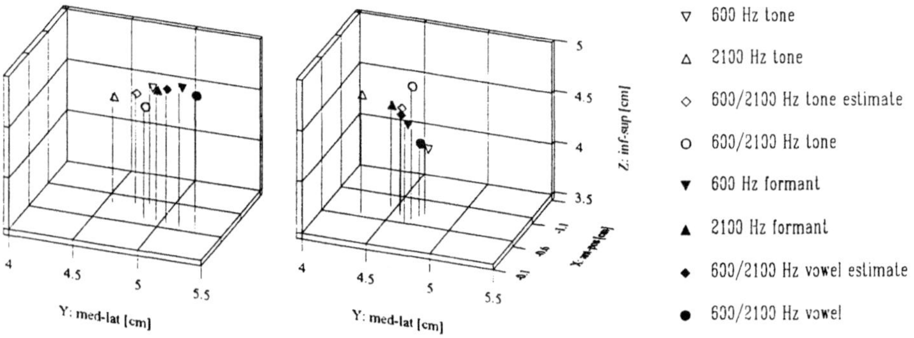

Fig. 2. Across-subjects average N1m (left) and SF (right) dipole source locations in experiment 1.

As in experiment 1, in experiment 2 the dipole moment of composite stimulus N1m sources was smaller than the dipole moment of expected composite stimulus N1m sources ($F[1,7] = 20.68$, $P < 0.01$). Apart from this finding, results were rather different for experiment 2. Composite stimulus N1m source location was somewhat more lateral than expected composite stimulus N1m source location, but the difference failed to attain statistical significance. The location of one-formant N1m sources was a function of formant frequency ($F[3,21] = 9.67$, $P < 0.002$) and the location of two-formant vowel N1m sources showed a similar first formant frequency dependence ($F[2,22] = 4.83$, $P < 0.05$). N1m sources associated with lower first-formant frequency were located more posterior than N1m sources associated with higher frequencies. Two-formant vowels were shorter in latency than one-formant vowels ($F[1,11] = 61.15$, $P < 0.001$). The latency of one-formant vowels was strongly related to formant frequency ($F[3,21] = 19.20$, $P < 0.001$). Analysis of the first PCA waveform component of the N1m analysis interval yielded comparable results. In the analysis of the first PCA waveform component of the SF the composite stimulus SF sources were weaker in dipole moment amplitude ($F[1,10] = 18.50$, $P < 0.001$) than the expected composite stimulus SF sources. As N1m source locations, SF source locations were related to first-formant frequency ($F[2,14] = 3.97$, $P < 0.05$). However, the SF frequency dependence was almost a mirror image of the N1m frequency dependence. Fig. 3 displays the average N1m and SF source locations. Fig. 4 shows individual subjects data for N1m source location on the anterior-posterior axis.

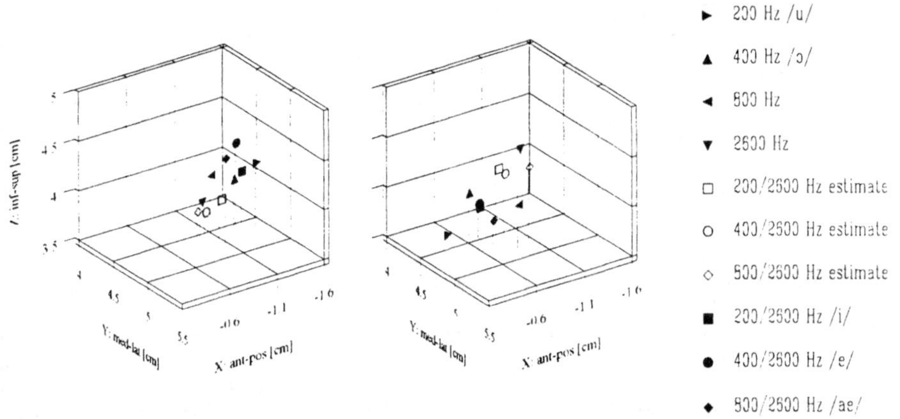

Fig. 3. Across-subjects average N1m (left) and SF (right) dipole source locations in experiment 2.

Comparison of the source location vs. frequency results suggests that different cortical fields may have contributed towards the generation of the evoked response in experiments 1 and 2. For ten subjects who participated in both experiments, this hypothesis could be tested by comparing centroid source locations. The centroid source locations were slightly more posterior and medial in experiment 2 than in experiment 1 both for N1m and SF. However, as none of the comparisons attained statistical significance, a gross anatomical hypothesis is rendered implausible.

Discussion

The latency of compound stimulus N1m sources was shorter than the latency of component stimulus N1m sources. The dipole moment of compound stimulus N1m sources was smaller than the sum of the dipole moments of stimulus component N1m sources. Both findings strongly suggest that the cortical response to two-tone stimuli and two-formant vowels cannot be accounted for in terms of the linear addition of the responses to sinusoidal components and single formants, respectively.

With regard to the tone activated sources, the finding in the first experiment that the 2100 Hz sources were more medial than the 600 Hz sources is consistent with the established tonotopic map of sound frequency [2]. Furthermore, that formant and vowel sources were more lateral than tone and compound-tone sources is compatible with the hypothesis of a cortical pitch map [3]. However, unless we allow for a timbre effect, the more medial location of the 2100 Hz formant source, compared with the 600 Hz formant source, seems incompatible with the pitch-map hypothesis.

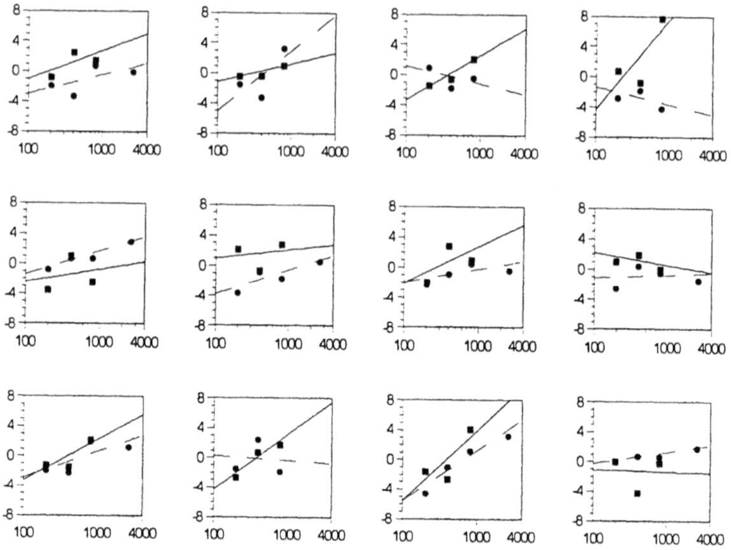

Fig. 4. Anterior-posterior N1m source location coordinate values as function of first-formant frequency. Circles: one-formant stimuli. Squares: two-formant vowels. Frequency is scaled in log Hz, source location in mm (ipsative values).

It seems likely that the tonotopic maps of formant frequency observed in the second experiment are spurious, i.e. not really maps of formant frequency. If they were, then, given their orientation, one would expect these maps to occupy an area of cortical space separate from the area occupied by the sources in the first experiment. However, no difference in source location was found between the first and the second experiment. A hint towards an account may be taken from the finding that, in cat auditory cortex, units are more narrowly tuned in the center than the periphery of the isofrequency strips oriented perpendicular to the frequency axis [5]. Vowel recognition depends on the extraction of the spectral envelope. One step of a variety of methods to do this is to use broadly tuned filters. Whereas frequency difference limens are correlated with frequency, the spacing of the harmonics of the fundamental frequency of a vowel is constant across the frequency range. Thus, the source topography in the second experiment may reflect the degree of frequency resolution diminution that is required to abstract from the harmonic structure of a formant peak.

References

[1] Diesch, E., Eulitz, C., Ross, B., Hampson, S. The neurotopography of vowels as mirrored by evoked magnetic field measurements. Brain and Lang. (in press)

[2] Pantev, C., Hoke, M., Lehnertz, K., Lütkenhöner, B., Anogianakis, G., Wittkowski, W. Tonotopic organization of the human auditory cortex revealed by transient auditory evoked magnetic fields. Electroencephal.Clin. Neurophysiol., 1988, 69: 160-170.

[3] Drung, D., Zimmermann, R., Cantor, R., Erné, S.N., Koch, H., Matthies, K.P., Peters, M., Scheer, H.J., Stollfuß, D. A 37-channel DC SQUID magnetometer system. Clin. Phys. Physiol. Meas., 1991, 12. suppl. B: 21-29.

[4] Pantev, C., Hoke, M., Lütkenhöner, B., Lehnertz, K. Tonotopic organization of the auditory cortex: Pitch versus frequency representation. Science, 1989, 246: 486-488.

[5] Heil, P., Rajan, R., Irvine, D.R.F. Sensitivity of neurons in cat primary auditory cortex to tones and frequency-modulated stimuli. II: Organization of response properties along the 'isofrequency' dimension. Hear. Res., 1992, 63: 135-156.

Acknowledgements

The research was supported by the *Deutsche Forschungsgemeinschaft*. The support provided by the staff of the Biomagnetism laboratory of the *Physikalisch-Technische Bundesanstalt* is gratefully acknowledged.

Source Location and Orientation, Dipole Moment, and Latency of Magnetic Mismatch Fields May Mirror the Perceptual Salience of Phonetic Contrasts

Diesch, E., Luce, T.A., and Eyferth, K.

Department of Psychology, Technical University of Berlin, Berlin, Germany

Introduction

Evoked magnetic field measurements have provided evidence for topographic differences between vowels and consonants with regard to the N1m component of the magnetically evoked response [1]. However, it has not yet been shown, whether the magnetic mismatch field (MMF) is sensitive to the distinction between vowels and consonants. The mismatch response reflects the detection of a difference between current stimulus input and a trace in sensory memory [2]. Mismatch responses are elicited by rare deviant stimuli presented in the context of frequent standard stimuli. For the electric counterpart of the MMF, the mismatch negativity, it has been shown that the amplitude of the mismatch response is correlated with the discriminability of the eliciting stimulus contrast [3,4]. It is relatively easy to discriminate perceptually among vowels even when they are presented in noise [5]. Under comparable conditions of ambient noise, it may be difficult to discriminate among plosive stop consonants differing in place of articulation [6]. Thus, it is to be expected that MMF sources activated by vowel and consonant contrasts, respectively, differ in dipole moment amplitude.

Methods

In two experiments, the MMF elicited by vowel and stop plosive consonant place-of-articulation contrasts was studied. In experiment 1, vowels occurred in isolation. In experiment 2, they were embedded within a syllabic context. In experiment 1, the vowels [a], [i], and [u] and the consonant-vowel syllables [ba], [da], and [ga] were presented. In vowel blocks, [a] was the standard (80%) and [i] and [u] were deviants (10% each). In consonant blocks, [da] was the standard and [ba] and [ga] were deviants (10% each). The sequence of blocks was counterbalanced and the within-block stimulus sequence was randomized. Half of the subjects received natural, the other half synthetic speech stimuli. Stimulus duration was 114 msec for the natural and 100 msec for the synthetic stimuli. In experiment 2, subjects were presented with one random stimulus sequence including [da] as standard (80%) and [di], [du], [ba], and [ga] as deviants (5% each). All stimuli were synthesized. Stimulus duration was 150 msec. The stimuli were equated for fundamental frequency and subjective loudness and delivered at an individually adjusted listening level trough a magnetically neutral transmission system to the subject's right ear. A constant offset-to-onset ISI of 400 msec was used. In both experiments, random intermittent light flashes were presented which the subjects were instructed to count. 13 subjects participated in the first and 12 subjects in the second experiment.

The magnetic field was recorded over the left hemisphere with a 37-channel magnetometer [7] inside a magnetically shielded room. The recording was continuous with a passband of 0.16-64 Hz. The ECG and EOG were monitored through separate electrical leads. The position and orientation of the dewar relative to the subject's head was obtained by measuring the magnetic field produced by five small induction coil triplets attached to the scalp. The data was segmented offline into 500 msec epochs (including a prestimulus baseline interval of 100 msec) and, after artefact rejection, selectively averaged. Only responses to standards immediately preceding a deviant were included in the standard averages. In order to improve the signal-to-noise ratio and to facilitate source analysis, the averages were digitally bandpass filtered with cut-offs at 0.5 Hz and 10 Hz. MMF waveforms were obtained by subtracting the standard from the deviant averages. Single spatio-temporal equivalent dipole sources were fitted for the 20 msec interval centered on the MMF rms peaks. On the basis of the induction coil measurements, an orthogonal right-handed headframe was defined with positive x projecting from the inion through the nasion and positive z projecting through the 10-20 location Cz. Source solutions were accepted if and only if the MMF field pattern showed polarity reversal, the distance of the source to the midsagittal xz-plane was larger than 25 mm, the distance to the horizontal xy-plane was larger than 10 mm, and the goodness of fit attained was 0.90 or better. The source parameters were submitted to a repeated measures ANOVA. The Greenhouse-Geisser epsilon correction was applied where appropriate.

Results

In both experiments, pronounced MMF deflections were registered in the vowel deviant and difference waveforms. Generally, the MMF deflections occurring in the consonant deviant and difference waveforms were distinctly smaller in amplitude. Figs. 1 and 2 present an example.

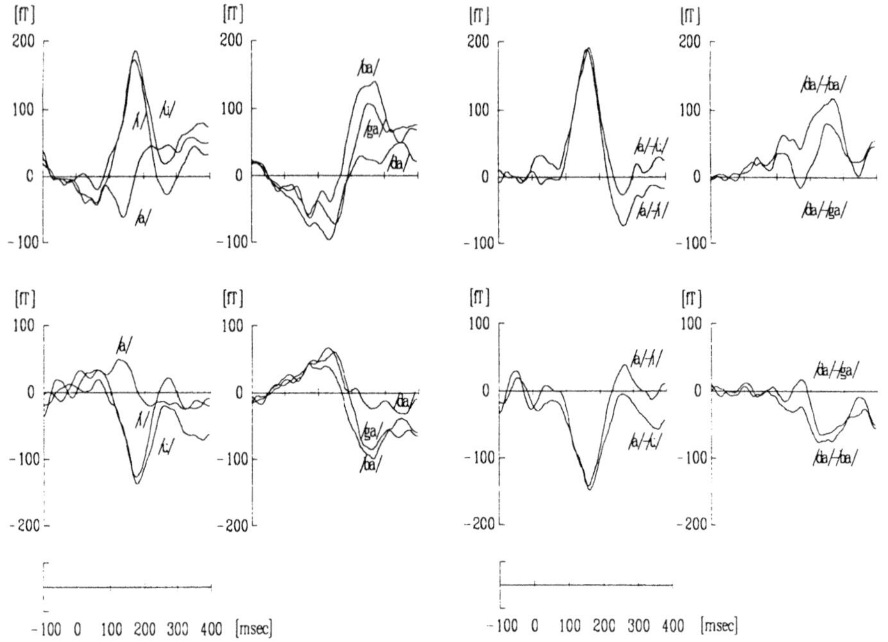

Fig. 1. Left part: Maximum and minimum averaged evoked magnetic field waveforms recorded in response to the standard (vowel condition: [a], consonant condition: [da]) and deviant stimuli (vowel condition: [i], [u], consonant condition: [ba], [ga]) in one subject in experiment 1. Right part: The difference waveforms computed by subtraction of the respective standard waveform from the deviant waveforms. Flux entering the skull is positive, flux leaving the skull negative.

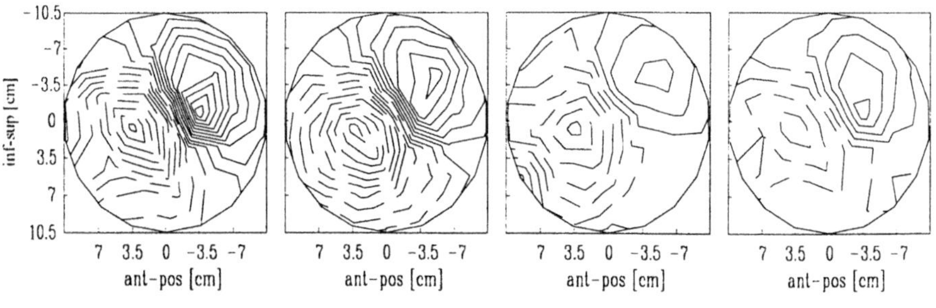

Fig. 2. The field distribution of the MMF difference waveforms presented in Fig. 1 at their respective rms peak latency. The spacing of the isocontour lines is 20 fT. Ant-pos: anterior-posterior, med-lat: medial-lateral, inf-sup: inferior-superior. From left to right: [a]-[i], [a]-[u], [da]-[ba], [da]-[ga]. Solid lines mark magnetic flux leaving the head, dashed lines magnetic flux entering the head.

Data of three subjects in the first experiment and one subject in the second experiment were discarded, because they failed to meet the acceptance criteria defined above. The remaining subjects exhibited mismatch field distributions which could be accounted for by single equivalent current dipoles.

In experiment 1, no difference was found between natural and synthetic speech stimuli. Therefore, the data from the two stimulus conditions were pooled. The vowel contrasts were found to elicit MMFs with shorter latency ($F[1,9] = 19.67$, $P < 0.002$) and stronger dipole moment ($F[1,9]=16.94$, $P < 0.003$) than the place-of-articulation consonant contrasts. MMF average peak latencies were 150 msec for [a]-[i], 157 msec for [a]-[u], 198 msec for [da]-[ga], and 189 msec for [da]-[ga]. The corresponding average dipole moment amplitudes were 26, 20, 13, and 9 nAm. Significant spatial separation was obtained between the source locations of vowel and the source locations of consonant contrast MMFs ($F[1,9] = 8.43$, $P < 0.02$). The consonant contrast MMF sources were located more anteriorly than the vowel contrast MMF sources. Post-hoc comparisons using Tamhane's T^2 showed that the effect was due mainly to the 10 mm distance, on the average, between the [da]-[ga] MMF sources and the vowel MMF sourcess along the anterior-posterior axis.

The results of experiment 2 were similar with respect to latency and dipole moment. Vowel contrast MMFs occurred with shorter latency ($F[1,10] = 12.9$, $P < 0.01$) and had stronger dipole moment amplitudes ($F[1,10] = 14.1$, $P < 0.01$) than consonant contrast MMFs. On the average, the MMF peaked at 142 msec for da]-[di] with a dipole moment amplitude of 16 nAm, at 175 msec for [da]-[du] with 15 nAm, at 232 msec for [da]-[ba] with 7 nAm, and at 200 msec for [da]-[ga] with 8 nAm. Vowel contrast MMF sources were located more laterally than consonant contrast MMF sources ($F[1,10] = 5.56$, $P < 0.05$). The average location difference between vowel and consonant sources was 9 mm. The MMF sources activated by the [da]-[ga] contrast were more anterior than the MMF source activated by the [da]-[ba] contrast ($F[2,20] = 5.12$, $P < 0.05$). On the average, this location difference amounted to 5 mm. Vowel and consonant contrast MMFs differed in their inclination towards the anterior-posterior axis of the head frame ($F[1,10] = 10.4$, $P < 0.01$). Again, post-hoc comparisons with Tamhane's T^2 showed that the effect was mainly due to the difference between [da]-[ga] and the vowel contrasts.

The spatial configuration of the average MMF sources was rather similar across the two experiments (Fig. 3). Vowel sources were spatially separated from consonant sources, but were not spatially separated from one another. The consonant sources were spatially separate, with the [da]-[ga] source in a more anterior location than the [da]-[ba] source.

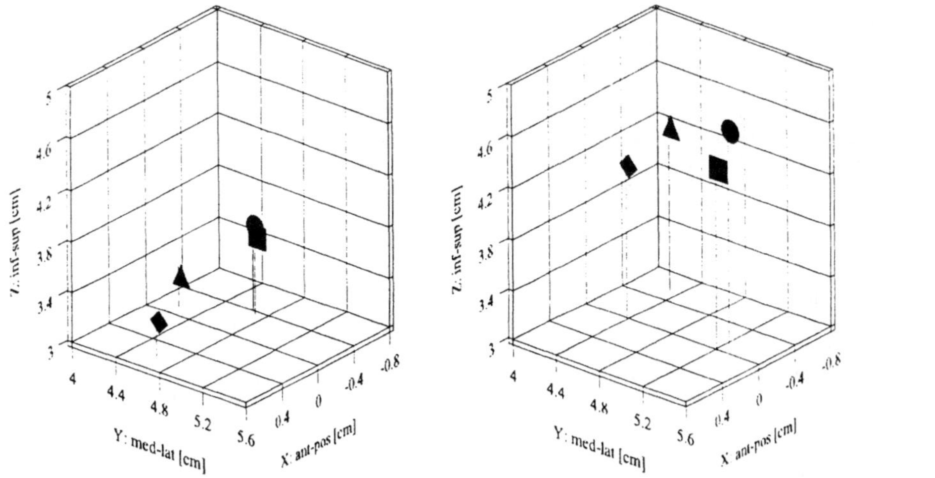

Fig.3. Across-subjects mean mismatch field source locations in head frame coordinates. Left: Experiment 1: circle: [a]-[i], square: [a]-[u], triangle: [da]-[ba], diamond: [da]-[ga]. Right: Experiment 2: circle: [da]-[di], square: [da]-[du], triangle: [da]-[ba], diamond: [da]-[ga]. Ant-pos: anterior-posterior, med-lat: medial-lateral, inf-sup: inferior-superior.

However, there were some differences. Vowel and consonant sources were spatially separated along the anterior-posterior axis in experiment 1, but along the medial-lateral axis in experiment 2. Furthermore, compared with the first experiment, the vowel sources seemed slightly more anterior and the consonant sources somewhat more posterior in the second experiment. These differences may be due either to stimulus factors, to subject factors, or both. Another difference between the two studies concerns the dipole moment amplitudes which were smaller for vowel and consonant contrasts in the second experiment. Most likely, this differences is due to the stimulus protocols used. Whereas vowel and consonant contrasts were presented in separate blocks in the first experiment, they were presented together in one random sequence in the second.

Discussion

Vowel contrasts and plosive stop consonant place-of-articulation contrasts elicited magnetic mismatch fields. Their field distributions were bipolar in structure and could be accounted for by single equivalent current dipoles. In both experiments, the consonant contrasts elicited MMFs with later latencies and smaller dipole moments than did the vowel contrasts. Vowel MMF sources were spatially seperated from consonant MMF sources. There was spatial separation between the consonant MMF sources, but not between the vowel sources. The pattern of these findings corresponds to the difficulty-of-discrimination pattern for vowels [5] and consonants [6]. The rate of confusion of [ba] and [ga] with [da] is much higher than the rate of confusion of [i] and [u] with [a]. Confusion of [da] with [ga] is more likely to occur than confusion of [da] with [ba]. It seems plausible to relate the latency and dipole amplitude findings to these perceptual data: The brain may take longer to generate a signal indicative of a weak stimulus contrast and the signal it generates may be smaller in amplitude. The source location findings may require another account. It has been shown that the topography of the electric mismatch negativity depends on the kind of stimulus contrast (frequency, intensity, duration) it is elicited by [8]. Accordingly, vowel and consonant MMF sources may be spatially separated because vowel and consonant contrasts differ in quality. Different consonant MMF sources may be spatially separated, because they are activated by different feature contrasts. However, this hypothesis leaves the lack of spatial separation between the vowel MMF sources without explanation. Thus, even the spatial separation effects may have to be accounted for in terms of the ease-of-discrimination hypothesis. A larger number of cells and synaptic connections may be involved in the perceptual extraction of the more subtle stimulus contrasts. As computation in neural nets involves activation and inhibition, this would not necessarily result in larger amplitude evoked magnetic fields. However, unless an increase of activity spreads in concentric circles across the cortical surface, there would be a shift in the location of an equivalent current dipole.

References

[1] Kuriki, S., Murase, M. Neuromagnetic study of the auditory responses in right and left hemispheres of the human brain evoked by pure tones and speech sounds, Exp. Brain Res., 1989, 77:127-134.

[2] Näätänen, R. Attention and brain function. Hillsdale, New Jersey, LEA, 1992.

[3] Sams, M., Paavilainen, P., Alho, K., Näätänen, R. Auditory frequency discrimination and event-related potentials. Electroencephal. Clin. Neurophysiol. 1985, 62: 437-448.

[4] Näätänen, R., Schröger, E., Karakas, S., Tervaniemi, M., Paavilainen, P. Development of a memory trace for a complex sound in the human brain. Neuroreport, 1993, 4: 503-506.

[5] Pickett, J.M. Perception of vowels heard in noises of various spectra. J. Acoust. Soc. Am., 1957, 29: 613-620.

[6] Miller, G.A., Nicely, P. An analysis of perceptual confusions among some English consonants. J. Acoust. Soc. Am., 1955, 27, 338-352.

[7] Drung, D., Zimmermann, R., Cantor, R., Erné, S.N., Koch, H., Matthies, K.P., Peters, M., Scheer, H.J., Stollfuß, D. A 37-channel DC SQUID magnetometer system. Clin. Phys. Physiol. Meas., 1991, 12, suppl. B: 21-29.

[8] Giard, M.H., Lavikainen, J., Reinikainen, K., Perrin, F., Bertrand, O., Pernier, J., Näätänen, R. Separate representation of stimulus frequency, intensity, and duration in auditory sensory memory: An event-related potential and dipole-source analysis.

Acknowledgements

The research was supported by the *Deutsche Forschungsgemeinschaft*. The support given by the staff of the Biomagnetism laboratory of the *Physikalisch-Technische Bundesanstalt* is gratefully acknowledged.

Auditory Evoked Magnetic Fields After Stimulation with Test-Tones and Words - Studies on Experimentees with Good Hearing and with Long-Term Hearing Impediments

Emmerich, E.[2], Meyer, R.[1], Nowak, H.[3], Huonker, R.[3], Gießler, F.[3], Meißner, W.[1] and Beleites, E.[1]

[1]Clinical Department of Audiology; [2]Institute of Physiology I and [3]Biomagnetic Center, Friedrich-Schiller-University Jena, Germany

Introduction:

The aim of the present study was to evaluate results of the distorted central auditory process. Localization of human cortical responses to acoustical stimulation with tones and speech signals began with magnetoencephalographic (MEG) studies on primary auditory cortex performed by R. HARI and her working group [1,2]. The MEG studies in connection with MRT-picture have demonstrated that the sources of the magnetic fields (peak-latencies 100 ms) are in the primary auditory cortex .

Our questions are partly based on clinical studies of U.S. IVARSSON et al. and N. KRAUS et al. about speech recognition after noise exposure [3,4]. The clear results of pure tone audiometry differ from those of speech audiometry in patients with noise induced hearing loss.

Method:

1. Clinical audiological tests on patients and on controls in the deparment of Audiology

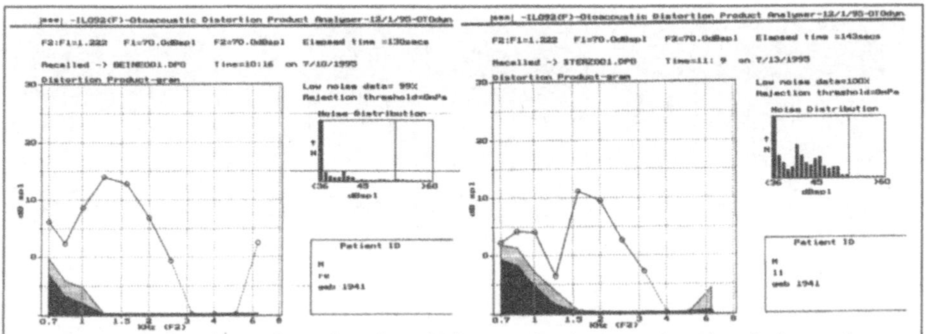

Fig. 1 Otoacoustic distortion products of a patient with long term hearing loss and problems in the speech recognition

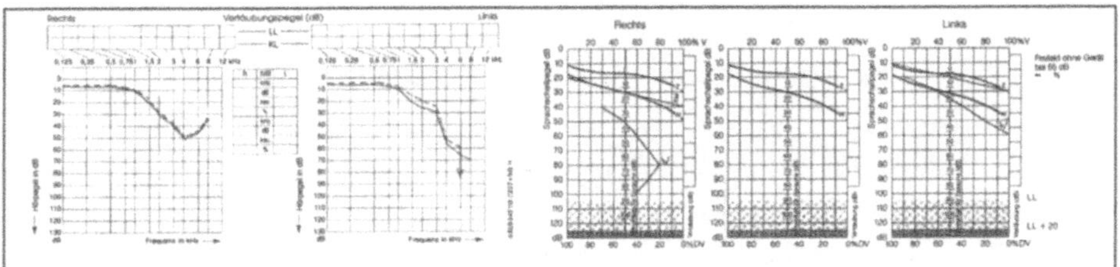

Fig.2 Mean hearing thresholds by stimulation with pure tone - with words („Freiburger Sprachtest")

2. Examinations of patients and controls in the Biomagnetic Center

The auditory evoked magnetic fields (AEF) and auditory evoked potentials (AEP) of 12 young healthy subjects with good hearing (average age 24, 10 male, 2 female) and patients with long term hearing loss (average age 45, 3 male) were derived using test-tones with *varying intensity and frequency* (60, 80, 100 dB SPL; 0,5, 1 and 2 kHz) with the help of the 50-channel Philips system [5]. Through the application of combinations of test-tone - test-word (sensical and non-sensical words) central auditory processes and word recognition were studied. The strength and position of a current dipole is reconstructed by the magnetic field distribution and in this way the source localization in the MRT- picture can be carried out. The highly time-sensitive localization results of the MEG combined with the MRT give us pictures for the functionally determined magnetic field source localization. Subsequently the AEF of hearing-impaired test subjects was derived using the test-tones and test-tone - test-word combinations. Using the AEF dipole localization and current density distribution were calculated and applied to the MRT-pictures. The stimuli were applied to subjects contralaterally to the 31-channel derivation of the AEF through a tested "Tube and Funnel System". The results of test from „Tube and Funnel Systems" are demonstrated in Fig.3a and b.

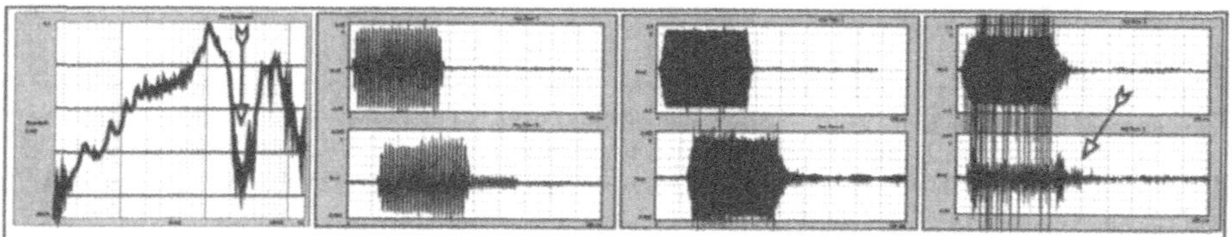

Fig. 3a Test tone 1-5 kHz: transfer function,
- upper trace: sound pressure at loudspeaker output outside MSR,
- lower trace: sound pressure at the ear inside MSR

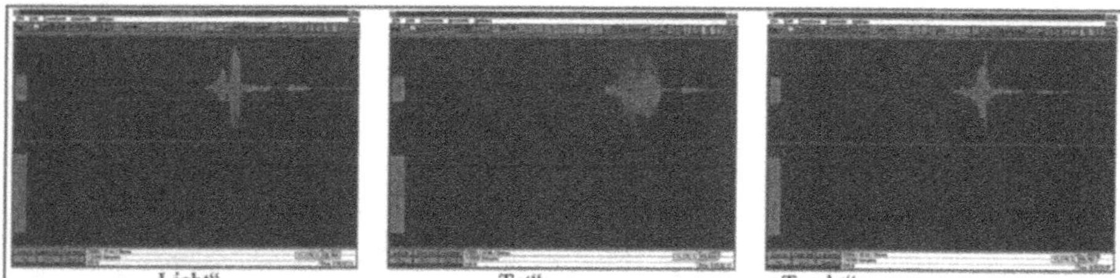

„Licht" „Tat" „Tracht"

Fig.3b Tone - word stimulation
- upper trace: sound pressure at loudspeaker output outside MSR,
- lower trace: sound pressure at the ear inside MSR

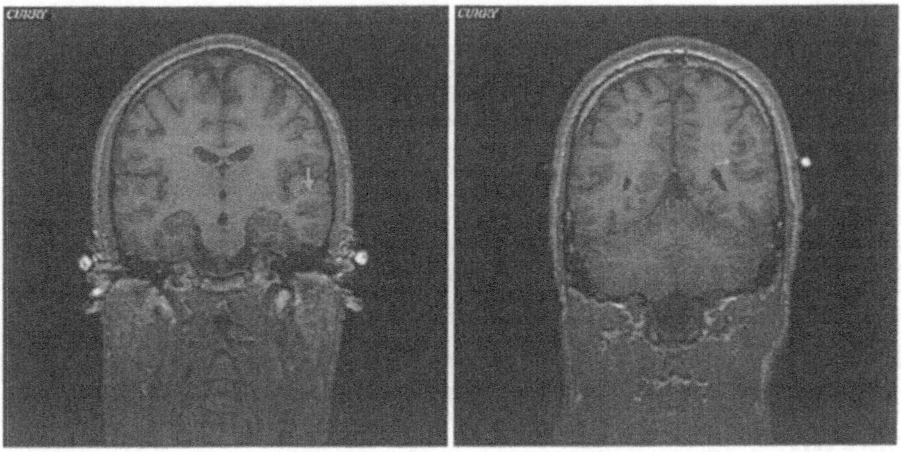

Fig.4 Source localisation pure tone stimulation and onset word - controls

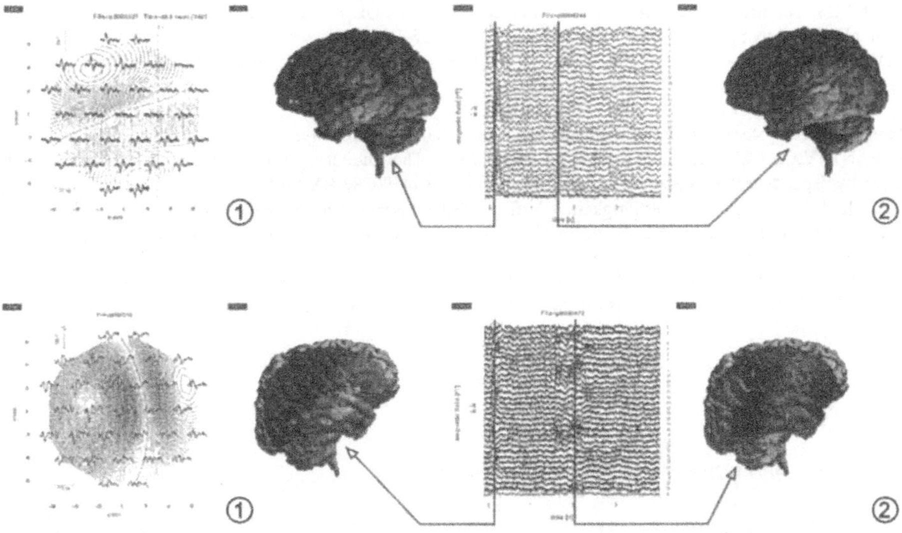

Fig.5 Current density reconstruction (1) test-tone 100 ms and (2) onset word: upper trace : - controls
lower trace : - patients

Scale: gray - maximum curent density

Results and discussion :

It is apparent that the amplitudes of AEF components$\leq$ 100 ms depend on stimulus intensity. It is also clear that the dipoles depiction of the components near 100 ms are reproducible with a large degree of stability, and that late components > 200 ms of these AEFs are displaced in the localization. Latent shifts of the AEF components could be found, as documented in the literature i.e. by CH. PANTEV et al. [5]. The results of current density reconstruction show that for more complex stimuli this localization of sources is a possible method.
We found different maximum neuronal activity by stimulation with pure tone and words. The results are demonstrated in Fig. 4 for controls.Changes in the source localization and current density distribution of patients are shown in Fig 5. Here we can see first results for patients. The clinical checkup of that patient shows in accordance with the literature H.G. DIEROFF, [7], that there is a difference in pure tone and speech audiometry. This phenomenon is generally known as late-onset auditory deprivation and discussed as process of inactivation in the brainstem and in the auditory cortex (G.M. GERKEN, [8] and R.J. SALVI, [9]).
Our results after word stimulation on patiens suggested that it is impossible to find a focus of neuronal activity in our patient in contrast to good hearing control persons. Identification of cortical areas after the different auditory stimulation may allow a specification of the role that these regions play in perception of pure tone stimuli and word stimuli.

References :

[1] HARI, R. "The Neuromagnetic Method in the Study of the Human Auditory Cortex" in F.Grandori, M.Hoke and G.L.Romani (Eds.), Auditory evoked magnetic Fields and Potentials Advances in Audiology, Vol.6, Karger, Basel, 222-282, 1990

[2] McEVOY, L.; MÄKELÄ, J.P.; HÄMÄLÄINEN, M.; HARI, R. "Effect of interaural time differences on middle-latency and late auditory evoked magnetic fields" Hear. Res. 78: 249-257, 1994

[3] KRAUS, N.; McGEE, T.; SHARA, A.; CARRELL,L.; NICOL,T. "Mismatch Negativity Event-Related Potential Elicited by Speech Stimuli" Ear and Hearing 13: 158-164, 1992

[4]IVARSSON, U. and ARLINGER, ST. „ Speech recognition in noise before and after work-days noise exposure „ Scand. Audiol. 23: 159- 163, 1995

[5] DÖSSEL,O.; DAVID, B:; FUCHS, M.; KRÜGER, J.; LÜDEKE, K.M'.; WISCHMANN, H.A. „ A 31 - channel SQUID systems for biomagnetic imaging." Appl. Superconductivity, vol.1, 10-12: 1813-

[6] PANTEV,C.; EULITZ,C.; ELBERT,T.; HOKE,M. " The auditory evoked sustained field: origin and frequency dependence" Elektroencephalogr. Clin. Neurophysiol. 90: 82-90, 1994

[7] DIEROFF, H.G, Lärmschwerhörigkeit, 3.Auflage Gustav Fischer, Jena -Stuttgart, 1994

[8] GERKEN, G.M., Alteration of central auditory processing of brief stimuli: A review and a neural model. J. Acoust. Soc. Am. 93, 2038-2049, 1993

[9] SALVI, R.J.; SAUNDERS, S.S.; GRATTON, M.A.; AREHOLE, S.; POWERS, N.: „Enhanced evoked response amplitudes in the inferior colliculus of the chinchilla following acoustic trauma." Hear. Res.50, 245-258, 1990

Acknowledgment :

This work was partly supported by BMBF (FKZ 01 ZZ 9104)

Analysis of Movement-Related Brain Activities Elicited by External Instructions

Endo, H., Takeda, T., Kizuka, T., Masuda, T., and Kumagai, T.
National Institute of Bioscience and Human-Technology, Tsukuba, Japan

Introduction

MEG recording has made it possible to identify activities in the sensorimotor cortex. A few studies have been done about a self-paced voluntary movement [1][2][3], an event-related activity with sensory guidance [4] and the functional organization [5][6]. However, it is still not clear how the temporal feature of each activity in the primary motor cortex is affected by the motivation that starts movement. Because sources of activity in the primary motor and the primary sensory cortex are located on the opposite wall in the central sulcus and are active during movement, the estimation of their activities has been difficult and no one has succeeded to clarify the temporal features of each. We have been studying movement-related brain activities and found that if the movement is triggered by an external Go-stimulus, a steeper change in magnetic fields is observed between the Go-stimulus and the EMG onset [7]. We named this activity as the motor field (MF) refered to the peak of the readiness field at the EMG onset [1]. In this study, we focused on the MF and identified the temporal aspects of the motor activities elicited by different conditions of the movement onset using LED visual stimuli.

Methods

Experimental procedure

The experiments were carried out on 4 right-handed subjects. The subjects were asked to move a right index finger briskly in the flexional-abductional direction. The subjects were asked to react to the Go-stimulus as quickly as possible. Fifty movements were performed in each experimental condition. Movement-related brain activities were investigated under the following 4 conditions of the movement onset:

(1) Self-Paced Movement (SPM): The subjects start their finger movement at free intervals without any stimuli.

(2) Visually Guided Movement with an LED Go-stimulus (VGM1): The subjects start after sensing an LED Go-stimulus presented at randomized intervals around 8 s.

(3) VGM1 with an LED warning stimulus (VGM2): The subjects start after sensing the LED Go-stimulus which is preceded by an LED warning stimulus.

(4) Go/NoGo task with an LED warning stimulus (VGM3): A Go or a NoGo-stimulus is presented after the warning stimulus (probability of the Go or NoGo-stimulus is 0.5).

The magnetic fields were recorded using a 64-channel whole-cortex SQUID system (CTF Systems Inc., Canada). Data were acquired at a rate of 250 samples/s. The EMG was recorded from the first dorsal interosseous muscle. A 3-dimensional head coordinate system was defined for each subject with 3 positioning coils put on the nasion and both tragions. A head shape was digitized on the each head coordinate system.

Data analysis

Averaging was done after off-line elimination of artefact-contaminated trials. The onset of the rectified EMG burst was used as the trigger to synchronize the averaging. The data were also averaged with respect to the Go-LED onset. The first 500 ms prior to any elicited activities was used as the baseline for the averaged data. The data were filtered with a zero phase shift, 20 Hz low-pass filter.

Equivalent current dipoles were used for activity estimation with the head modeled as a spherical volume conductor. The sphere model was adjusted to fit the shape of parietal and occipital areas of each digitized head shape. The strength of dipole's current moment was determined by fitting with the least square error method (LSEM), and the nonlinear parameters such as the dipole position and orientation, were searched by the Nelder-Meade simplex method (NMSM). The dipole was assumed to be fixed at the most probable position during the analysis period.

Results

Movement with self pace

Fig. 1 shows the results of the SPM. Fig. 1 (a) is the superimposed waveforms of 64 channels' data and Fig. 1 (b) is iso-contour maps of the MF at the EMG onset and the movement evoked field (MEF) at the peak latency of the

MEF. Fig. 1 (c) and (d) are the positions and the strength of three dominant activities: two motor activities and one sensory activity. The activity of contralateral MF (solid line) lasted after the movement onset, but on the other hand the activity of ipsilateral MF (dotted line) dropped off after the movement onset. Because an increasing motor activity prior to the EMG onset was not observed, the MF activity estimated under the assumption of one concentrated location could not represent the activity of neurons that activate the descending motor pathway. The activity of the MEF in the contralateral sensory cortex started after a few tens ms from the EMG onset. In contrast to the strength of magnetic fields at the peak latencies, the current moment of the MEF activity was estimated larger than that of the MF.

Movement with Go-stimulus

Fig. 2 shows the superimposed waveforms of the VGM1. A steeper change of magnetic fields instead of a slow shift of the readiness field was observed just before the movement onset (Fig. 2 (a)) The average triggered by Go-LED made it clear that this change started about 100 ms after the onset of the Go-stimulus (Fig. 2 (b)). This change was spatiotemporally distinguishable from the onset of the VEF. The steeper activity of the MF was observed clearly in three among four subjects.

Fig. 3 shows the relation between the reaction time and the onset latency of this change in three Go-stimulus conditions: VGM1, VGM2 and VGM3. The onset latency was determined from the waveforms when the magnetic fields started increasing. Soild lines indicate the reaction time and dotted lines indicate the onset latency of the MF. Different lines show the experiments were done on different days. In spite of variations of the reaction time, the onset latency does not change around 100 ms.

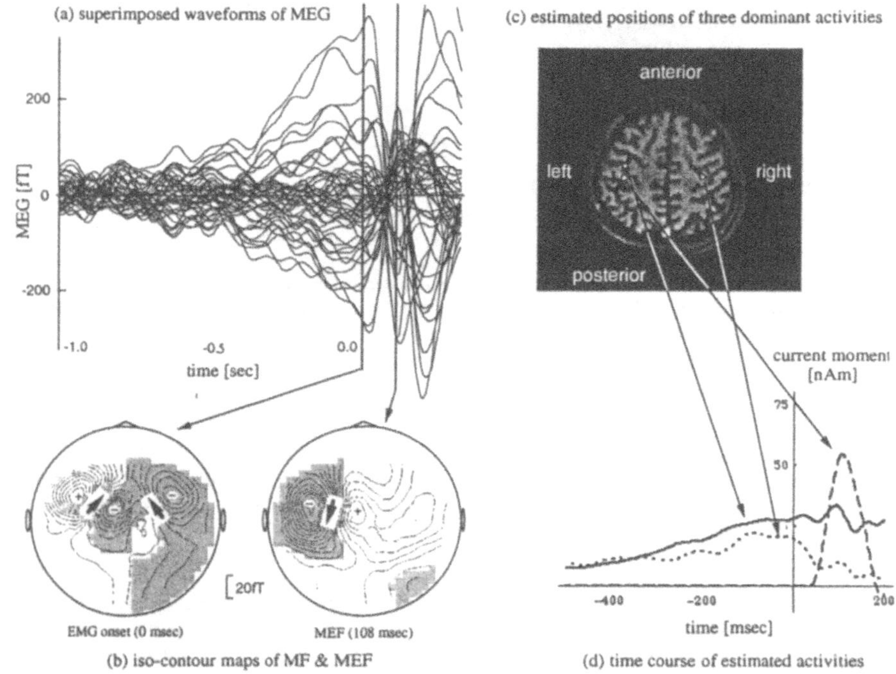

Fig. 1 Movement-related brain activites of a self-paced right finger movement. (a) Superimposed waveforms of 64 channels' data. (b) Iso-contour maps of the MF at the EMG onset and MEF at the peak latency of the MEF. (c) Estimated dipoles of three dominant activities. (d) The time course of estimated current moment by the LSEM.

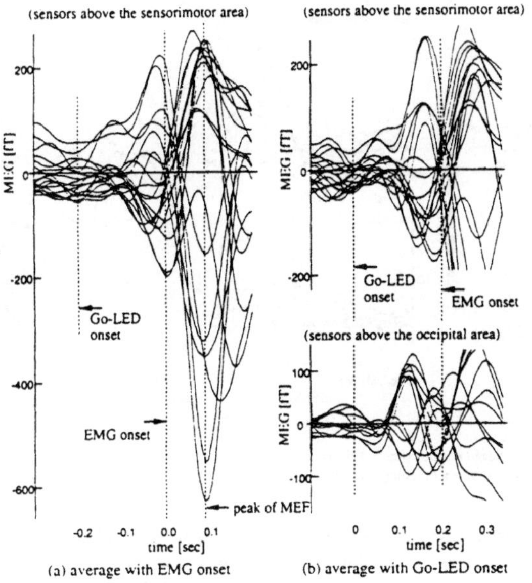

Fig. 2 MEG of the VGM1. (a) Data were averaged with respect to the movement. (b) Data were averaged with respect to the Go-stimulus. The MF shows a clear steeper change synchronize with the Go-stimulus.

In order to make each activity clear, the estimation of the time course of each activity was done in the VGM1 condition where there was no contamination of the CNV activity. Fig. 4 shows the time course of the activities before movement onset. The results of two subjects were superimposed. In the estimation, we assumed that there were 4 dipoles: 2 dipoles for the VEF and 2 for the MF. As for the sources of the VEF, the dipole positions were determined from the 1st large peak of the VEF by the warning LED stimulus. As for the motor activities, the dipole positions estimated in the SPM were used for the positions of the MF. The strength of the current moment was calculated with 4 fixed dipoles by the LSEM. The VEF started around 90 ms and the MF started around 120 ms. The MF reached its peak around 160~170 ms and dropped off. The time course of the motor activities was different from the activities estimated in the SPM. The onset and peak latency of the MF did not change according to the reaction time as observed from the waveforms and we could not get the consistent results of the relation between the activities of the MF and the reaction time.

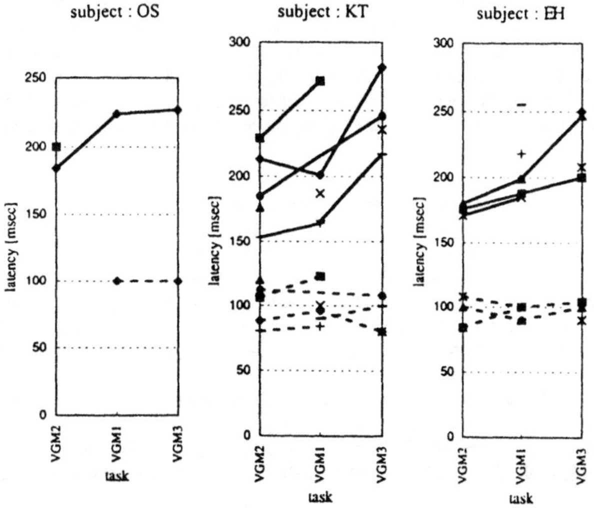

Fig. 3 The relation between the reaction time and the onset latency of the MF. The results of three subjects are plotted. Sold lines indicate the reaction time and dotted lines show the onset latency. Different lines show that the experiments were done on different days.

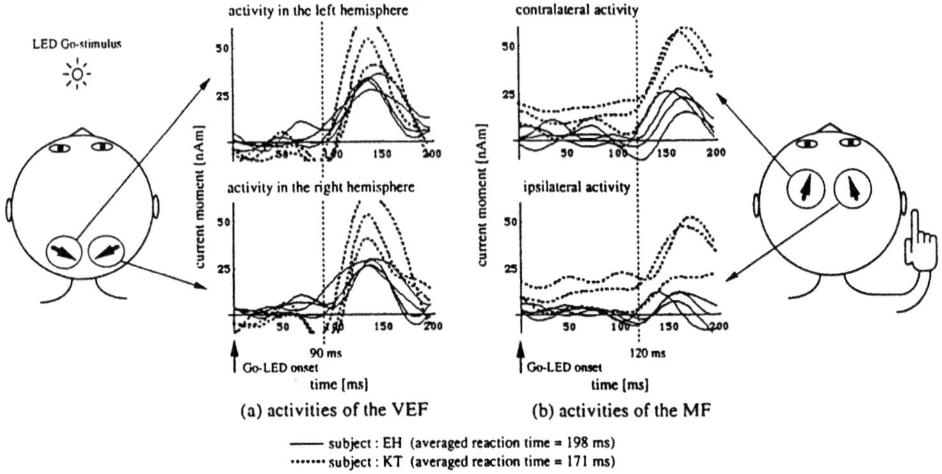

Fig. 4 The time course of estimated activities before the movement in the VGM1. (a) Activities of the VEF. (b) Activities of the MF. Upper trace: Activities in the left hemisphere. Lower trace: Activities in the right hemisphere. The activities estimated from several experiments of two subjects are plotted.

Discussion

In order to clarify how the motor activities change with the conditions of the movement onset, movement-related magnetic fields were examined in 4 kinds of movement onset conditions: one self-paced condition and three Go-stimulus conditions. In the conditions of the Go-stimulus, the steeper change of magnetic field was observed at 100 ms after the Go-stimulus and 120 ms from the results of the activity estimation. The change of the MF elicited by the Go-stimulus did not change with the onset conditions. Also, its latency did not coincide with the delay of the reaction time. As seen in the SPM, because the estimated MF activities do not synchronize with the EMG, they may not clearly represent the activity of neurons that activate the descending motor pathway. However, it is considered that there is a closely related pathway from the first stage of the stimulus processing system to the movement preparation.

References

[1] Kristeva, R., Cheyne, D., and Deecke, L. Neuromagnetic fields accompanying unilateral and bilateral voluntary movements: topography and analysis of cortical sources, Electroenceph. clin. Neurophysiol., 1991, 81: 284-298.

[2] Chiarenza, G.A., Hari, R.K., Karhu, J.J., and Tessore, S. Brain activity associated with skilled finger movements: multichannel magnetic recordings, Brain Topog., 1991, 3(4): 433-439.

[3] Nagamine, T., Toro, C., Balish, M., Deuschl, G., Wang, B., Sato, S., Shibasaki, H., and Hallett, M. Cortical magnetic and electric fields associated with voluntary finger movements, Brain Topog., 1994, 6(3): 175-183.

[4] Sasaki, K., Gemba, B., Nambu, A., and Matsuzaki, R. No-Go activity in the frontal association cortex of human subjects, Neurosci. Res., 1993, 18(3): 249-252.

[5] Cheyne, D., Kristeva, R., and Deecke, L. Homuncular organization of human motor cortex as indicated by neuromagnetic recording, Neurosci-Lett., 1991, 122: 17-20.

[6] Kristeva, R., Walter, H., Lutkenhoner, B., Hampson, S., Ross, B., Knorr, U., Steinmetz, H., and Cheyne, D. A neuromagnetic study of the functional organization of the sensorimotor cortex, Eur. J. Neurosci., 1994, 6(4): 632-639.

[7] Endo, H., Takeda, T., Kizuka, T., Kikuchi, Y., Masuda, T., and Kumagai, T. Estimation of movement-related brain activities with visual stimuli, Electroenceph. clin. Neurophysiol. Suppl., (in printing).

Visual Evoked Magnetic Fields: Bilateral Dipole Pattern for Full-Field Stimuli

Fujita, S.[1], Nakasato, N.[2,3], Seki, K.[2], Kanno, A.[3], Kawamura, T.[2], Ohtomo, S.[2], Fujiwara, S.[2,3], Tamura, I.[1] and Yoshimoto, T.[2]

[1]*R&D Center, Osaka Gas Co. Ltd., Osaka, Japan;* [2]*Department of Neurosurgery, Tohoku University School of Medicine, Sendai, Japan;* [3]*MEG Laboratory, Kohnan Hospital, Sendai, Japan*

Introduction

Pattern reversal stimulus is most frequently used for clinical application of visual evoked potentials (VEPs). Half-field stimulus has been employed to separate the left and right hemispheric responses. In the partial-field stimulus, however, subjects have to fix their eyes on a given point during the entire procedure; if the visual fixation is not strict, bilateral occipital responses may interfere with each other. In a clinical applications, therefore, the partial-field stimulus may not be suitable for inexperienced patients.

Recently, whole head magnetoencephalography (MEG) systems, which allow simultaneous recordings over the entire head, have been developed and used for clinical applications. In our recent studies of auditory evoked magnetic fields, a whole head MEG system revealed a clear dipole pattern of N100m responses, indicating two sources on bilateral superior temporal planes [1]. In the present study, we applied whole head MEG recordings to visual evoked fields (VEFs), in order to separate left and right hemispheric sources for full-field stimuli.

Materials and Methods

Four healthy males participated in this study. Green-black checkerboard pattern reversal stimuli were presented by a non-magnetic stimulator, which consisted of optical fiber bundles and LEDs on the full-field monoculary. The stimulus pattern was reversed once per second and individual checks, and full-field size, subtended visual angles of 1.4 and 16.8 degrees respectively at the subject's eye. The mean luminance level of the green checks was about 24 cd/m^2, with a contrast greater than 98% relative to the black checks. VEFs were measured using a whole head MEG system (CTF-Osaka Gas) linked to MRI. For source modeling purposes, three dimensional MRI head shape data were used to determine the best fit sphere for each subject. In consideration of the effect of the volume current, the equivalent current dipole was estimated using the Grynszpan-Geselowitz equation and superimposed on MRI slices for comparison with the anatomy directly.

Results

The major peaks appeared around 100msec after the pattern reversal in both hemispheres (Fig.1). The P100m peaks were identified visually as the maximum value. In all subjects, isofield maps of the P100m component indicated a two-dipole pattern over the bilateral occipital area. All the P100m dipoles estimated using a two-dipole model were localized at the lateral end of the calcarine fissures on each subject's MRI, although the individual map patterns were variable in shape (Fig.2).

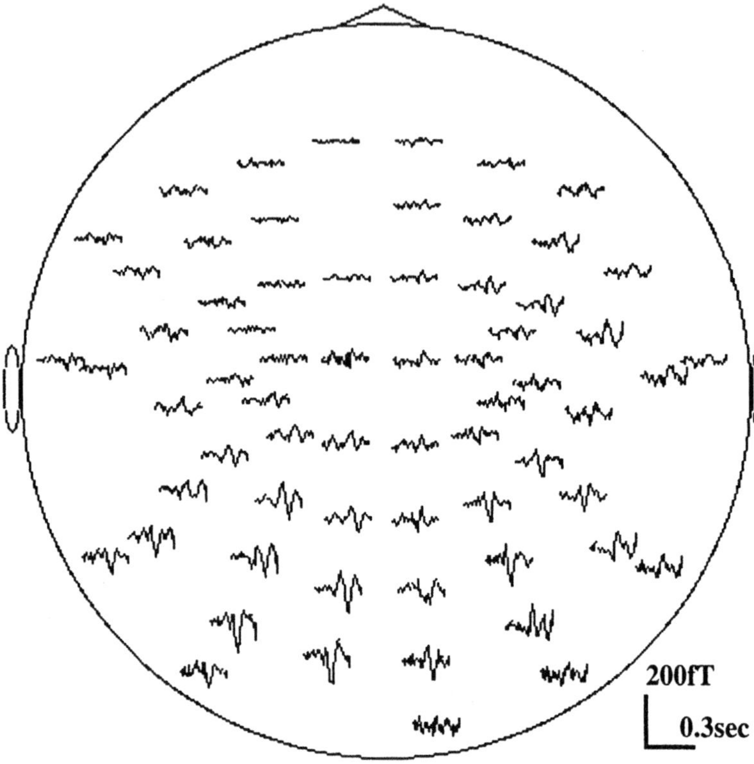

Fig.1 Waveforms of visual evoked field for the full-field stimulus. Major peaks appeared around 100msec after the pattern reversal. The positive and negative peaks were observed in both hemispheres.

Discussion

The anatomy of the occipital lobe is individually variable. Various map patterns of the P100m responses may be explained by the anatomical variations of the occipital lobes, such as depth and/or orientation of the calcarine fissures. In spite of the variation of the occipital anatomy, the estimated dipoles were always localized at the lateral end of the calcarine fissures, corresponding to our previous half-field studies [2]. For the stimulus pattern described under Methods, we believe the origin of P100m exists within the limited area of the striate cortex, because an extended origin should indicate the variation in the dipole position.

Our present study demonstrated the clinical utility of the full-field stimulus and a two-dipole model in visual evoked responses. Previous VEP studies required a partial-field stimulus; subjects have to fix their eyes on a given point. If the fixation is not strict, not only the contralateral but also the ipsilateral hemisphere generate P100m responses. In such cases, it would be necessary to use a two-dipole model instead of a single dipole model for accurate source localization. On the other hand, a two-dipole model requires a wide area measurement, because a two-dipole pattern spreads over not only occipital but also temporal and/or parietal areas. Due to the combination of the whole head measurement and a two-dipole model, bilateral P100m sources for full-field stimulus were separated

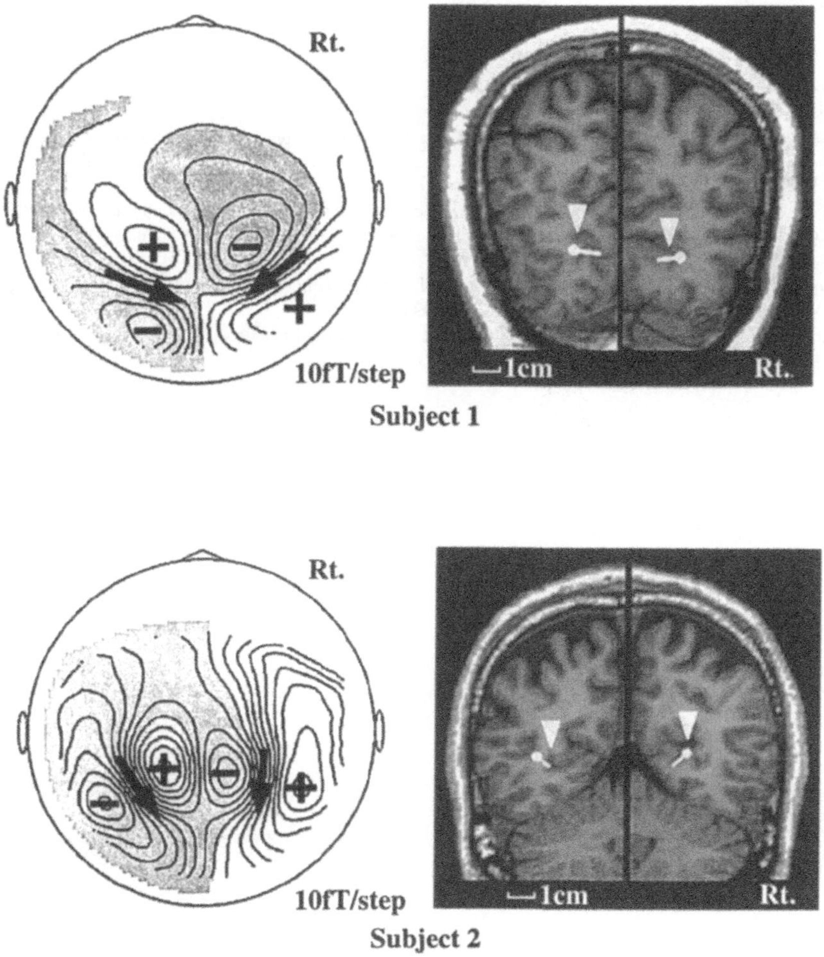

Fig.2 Visual evoked field for the full-field stimulus in two cases. The P100m isofield maps indicate a two-dipole patterns over the occipital area. The MRI coronal images indicate the P100m dipoles localized at the lateral end of the bilateral calcarine fissures(arrow heads). Circles and bars indicate the dipole positions and orientations respectively.

successfully. Our results accelerate further clinical application of VEFs to a wider range of patients [3,4], because the full-field VEF requires no strict fixation of the eyes.

References

[1] Nakasato, N., et al. Functional localization of bilateral auditory cortices using an MRI-linked whole head magnetoencephalography(MEG) system. Electroencephalography and Clinical Neurophysiology, 1995, 94: 183-190.

[2] Seki, K., et al. P100m of pattern reversal visual evoked magnetic field: source localization using an MRI-linked helmet-shaped magnetoencephalography system. Electroencephalography and Clinical Neurophysiology (Society proceedings), 1995, 95: 86-94.

[3] Nakasato, N., et al. Clinical application of Visual evoked fields using an MRI-linked whole head MEG system. Frontiers of Medical and Biological Engineering, 1995, in press

[4] Seki, K., et al. Visual Evoked Fields for Pattern Reversal Stimuli in Patients with Occipital Lobe Lesions. To be presented in this conference.

Evoked Magnetic Responses from Human V1: Evidence for a Chromatic-Sensitive Motion Mechanism

Fylan, F., Singh, K.D., Holliday, I.H., Anderson, S.J. and Harding, G.F.A.

Department of Vision Sciences, Aston University, Birmingham, B4 7ET, U.K.

Introduction

Within the human visual system various hierarchical processing models indicate a clear dichotomy of colour and motion processing [1] with predominantly luminance information being input to cortical motion centres via the magnocellular pathway. This separation of information is consistent with reports of poor responses from cortical motion centres to isoluminant chromatic stimuli. Recent evidence suggests that chromatic information may be used in motion analysis [2]. This study will explore the temporal response characteristics of chromatic-sensitive parvocellular units in human visual cortex by examining the variation in magnitude of the evoked magnetic response to isoluminant chromatic stimulation. Co-registration of MEG and MRI data will be examined to determine the cortical origins of the response.

Method

Five adult humans volunteered for the experiment (three females, two males, age range 27-39). The visual evoked magnetic response was recorded to isoluminant chromatic (red/green) gratings using the Aston 19-channel SQUID magnetometer [3] which was housed in a magnetically shielded room. A control stimulus comprised achromatic gratings of 80% contrast and was matched for the luminance of the chromatic gratings, at 12cd/m². The stimuli were generated using a Cambridge Research Systems VSG2/2 grating generator and displayed on an Eizo Flexscan T560i. The gratings were presented horizontally using an abrupt-onset temporal envelope of 1s duration. The interstimulus interval was 500ms, during which the gratings were replaced with a homogeneous background of the same mean hue and luminance. The spatial frequency was 2cpd and the temporal frequency was varied between 0Hz and 40Hz. The stimuli were viewed binocularly through a system of front-silvered mirrors, and subtended 4 degrees vertically by 6 degrees horizontally. To restrict input to a discrete region of cortex and to avoid activation of multiple regions of cortex due to dual representation of the horizontal and vertical meridians, the stimuli were positioned in the right lower quadrant of the visual field and displaced 0.5 degrees from the principal meridians.

The subjects were seated and stabilised using a bite bar and forehead rest and the dewar was centred above International 10-20 position Oz. Responses were sampled at 1KHz for 512ms and 100 trials were averaged for each stimulus condition. The resultant signals were bandpass filtered at 2-30Hz. For each experimental session the position of the subject's bite-bar, and therefore head, was digitised using a Polhemus Isotrak three-dimensional digitising system and converted to a standard co-ordinate system. Four bite-bar reference markers (oil filled drill holes) appeared as small, white spots on the MRI allowing accurate co-registration of MEG and MRI data. The MRIs were obtained using a 1.5T Siemens magnetic resonance scanner with 1mm x 1mm x 1.5mm voxel size. Localisation of the responses was performed using single and multiple current dipole models and Monte-Carlo analysis [4] provided 95% confidence regions which were co-registered with the subject's MRI.

Results

The response to isoluminant chromatic gratings was dominated by a single major component. For each of the average evoked responses, the global field power provided a measure of the total power in the recorded signal. The maximum global field power was plotted as a function of temporal frequency. This tuning function obtained was bimodal, peaking at 0Hz and 4Hz, shown in Fig. 1.

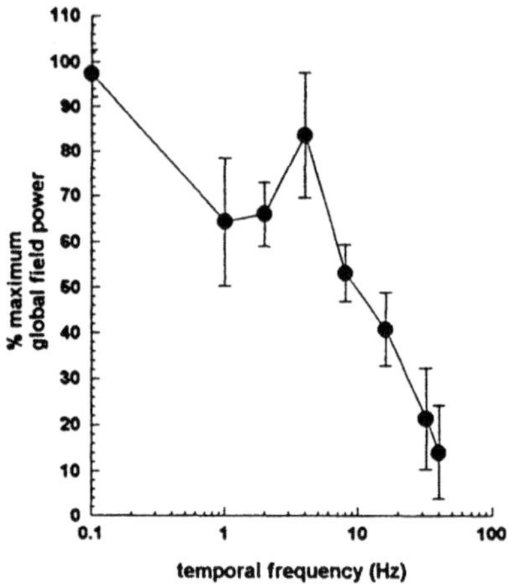

Fig. 1 Temporal frequency tuning characteristics of the chromatic evoked magnetic response (data points represent the mean of 5 subjectsand error bars show 2 standard deviations).

The latency of the major component increased gradually with temporal frequency up to approximately 16Hz and then declined rapidly to a level below that obtained for stationary patterns. To exclude the possibility that the chromatic responses arose from luminance artefacts, the experiment was repeated with an achromatic control stimulus for subject JD (Fig. 2). The control stimulus yielded low field strength signals and the latency of the response was independent of temporal frequency and at least 10ms earlier than the earliest chromatic response.

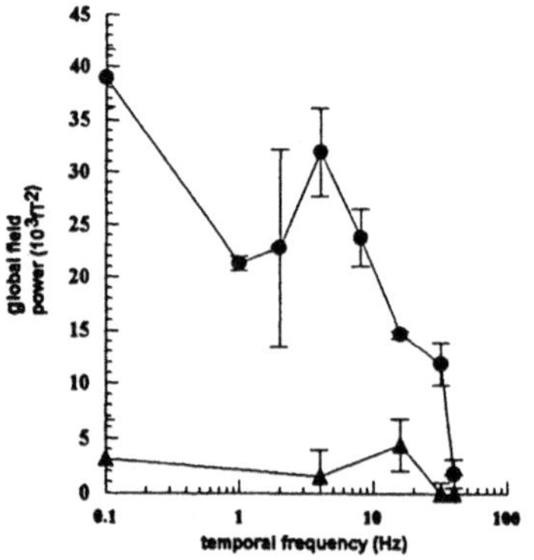

Fig. 2 Temporal frequency tuning characteristics of the response to chromatic (circles) and luminance (triangles) gratings (subject JD). Error bars show the anti-average noise data.

Each average evoked response was well modelled by a single equivalent current dipole with an orientation consistent with the anatomy of striate cortex [5]. To determine the cortical location of the underlying response generators the data was co-registered with MR images, shown in Fig. 3. This reveals the dipole confidence ellipsoids for each temporal frequency to lie along the upper surface of the calcarine fissure, consistent with a V1 origin.

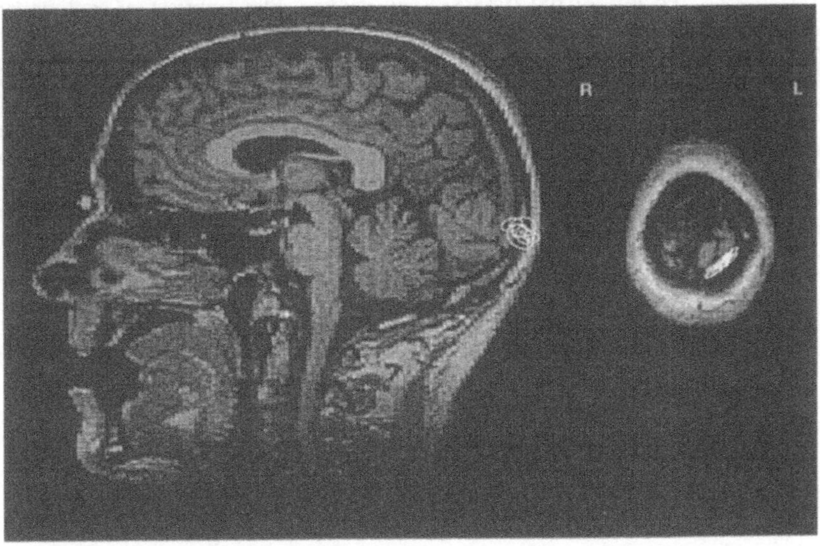

Fig. 3 95% confidence ellipsoids co-registered with sagittal and coronal MR images for the response to 2cpd red/green gratings of temporal frequency 0, 1, 2,4,8 and 16Hz (subject IH)

Discussion

The evoked magnetic response was recorded to 2cpd isoluminant chromatic gratings drifting at between 0Hz and 40Hz. The response was dominated by a major component whose latency and field power varied with the drift temporal frequency of the stimulus. The latency of the response increased with increasing temporal frequency up to 16Hz above which the latency decreased sharply. The temporal frequency tuning function was bimodal, peaking at 0Hz and at 4Hz. This consistent peak at 4Hz is surprising as psychophysical techniques do not indicate an increase in sensitivity to chromatic gratings at this temporal frequency. However, such an increase in amplitude of the VEP has also been noted previously for blue/yellow gratings [6]. The results may be indicative of two temporal frequency selective chromatic mechanisms, one sensitive to stationary targets and the other to moving targets.

To confirm the response did not arise from luminance artefacts in the stimulus the experiment was repeated with achromatic gratings. The response evoked by this luminance-modulated stimulus was of low field strength over the entire range of temporal frequencies tested. The latency of the luminance response was invariant with temporal frequency and was at least 10ms earlier than the earliest chromatic response. We therefore conclude that the responses obtained with chromatic stimuli cannot be attributed to luminance artefacts and represent the activity of chromatic-sensitive units.

Each average evoked response was well modelled by a single equivalent current dipole which was co-registered with MR images. This revealed the dipole confidence ellipsoids to lie within area V1 for each temporal frequency. This stability in the location of the response generator provides evidence to suggest that units in V1 encode the motion of chromatic targets over a wide range of temporal frequencies.

References

[1] Livingstone, M.S. & Hubel, D.H. (1988) Segregation of form, color, movement and depth: anatomy, physiology and perception. *Science* 240: 740-750.

[2] Cavanagh, P., Favreau, O.E. (1985) Color and luminance share a common motion pathway. *Vision Res.* 25: 1595-1601; Cavanagh, P., Anstis, S. (1991) The contribution of color to motion in normal and color-deficient observers. *Vision Res.* 31: 2109-2148.

[3] Matlashov, A.N., Slobodchikov, V., Bakharev, A, Zhuravlev, Y & Bodarenko, N. (1993) Biomagnetic multichannel system built with 19 cryogenic probes. In Deecke, L. et al. (eds) *Recent Advances in Biomagnetism: Proc. 9th International Conference on Biomagnetism* pp 284-285.

[4] Singh, K.D. & Harding, G.F.A. (1996) Monte-Carlo analysis and confidence region ellipsoids for equivalent current dipole solutions. This volume.

[5] Fylan F., Holliday I.E., Singh K.D., Anderson S.J. & Harding G.F.A. (1995) Cortical origins of visual evoked magnetic responses differ in humans for chromatic and luminance stimuli. *Human Brain Mapping (Suppl.)* 1:51.

[6] Rabin, J., Switkes, E., Crognale, M., Scheck, M.E. & Adams, A.J. (1994) Visual evoked potentials in three-dimensional color space: correlates of spatio-chromatic processing. *Vision Res.* 34: 2657-2671.

Acknowledgements

This research was supported by the Dr Hadwen Trust for humane research.

Evoked Magnetic Responses to Isoluminant Chromatic Stimuli from Human V1: Spatial Frequency Characteristics

Fylan, F., Singh, K.D., Holliday, I.H., Anderson, S.J. and Harding, G.F.A.

Department of Vision Sciences, Aston University, Birmingham, B4 7ET, U.K.

Introduction

In the human visual system, segregation of information on a functional basis commences as early as the retinal ganglion cells. Both chromatic and luminance information is conveyed within the parvocellular pathway to the primary visual area V1. Descriptions of the chromatic evoked response tend to concentrate on the amplitude and latency of the principle response peak rather than to characterise its distribution or to determine its cortical origins. A more intensive investigation of both the cortical origins of the response and the spatial tuning properties of these chromatic sensitive cells was therefore considered appropriate. This study will examine the magnitude of the evoked magnetic response to isoluminant chromatic gratings over a range of spatial frequencies. Co-registration of MEG and MRI data will be examined to determine the cortical origin of the response.

Method

Six adult humans volunteered for the experiment (four females, two males, age range 26-39). The visual evoked magnetic response was recorded to isoluminant chromatic (red/green) gratings using the Aston 19-channel SQUID magnetometer [1] which was housed in a magnetically shielded room. The stimuli were generated using a Cambridge Research Systems VSG2/2 grating generator and displayed on an Eizo Flexscan T560i. The mean luminance of the display was 12cd/m^2 and the contrast of the red and green component gratings was 100%. The gratings were presented horizontally using an abrupt-onset temporal envelope of 1s duration. The interstimulus interval was 500ms, during which the gratings were replaced with a homogeneous background of the same mean hue and luminance. The spatial frequency was varied between 0.25 and 8cpd and the temporal frequency was 0Hz. The stimuli were viewed binocularly through a system of front-silvered mirrors, and subtended 4 degrees vertically by 6 degrees horizontally. To restrict input to a discrete region of cortex and to avoid activation of multiple regions of cortex due to dual representation of the horizontal and vertical meridians, the stimuli were positioned in the right lower quadrant of the visual field and displaced 0.5 degrees from the principal meridians.

The subjects were seated and stabilised using a bite bar and forehead rest and the dewar was centred above International 10-20 position Oz. Responses were sampled at 1KHz for 512ms and 100 trials were averaged for each stimulus condition. The resultant signals were bandpass filtered at 2-30Hz. For each experimental session the position of the subject's bite-bar, and therefore head, was digitised using a Polhemus Isotrak three-dimensional digitising system and converted to a standard co-ordinate system. Four bite-bar reference markers (oil filled drill holes) appeared as small, white spots on the MRI allowing accurate co-registration of MEG and MRI data. The MRIs were obtained using a 1.5T Siemens magnetic resonance scanner with 1mm x 1mm x 1.5mm voxel size. Localisation of the responses was performed using single and multiple current dipole models and Monte-Carlo analysis [2] provided 95% confidence regions which were co-registered with the subject's MRI.

Results

The response to isoluminant chromatic gratings was dominated by a single major component whose latency increased with increasing spatial frequency, from 87ms at 0.25cpd (subject FF) to 163ms at 6cpd (subject KS). This resulted in an increase in the latency of the response of over 60ms when the spatial frequency was incresed from 0.25-6cpd, shown in Fig. 1. For each of the average evoked responses, the global field power provided a measure of the total power in the recorded signal. The maximum global field power was plotted as a function of spatial frequency, yielding a bandpass tuning curve, peaking at 1-2cpd and falling to the level of the noise at around 6cpd (Fig. 2).

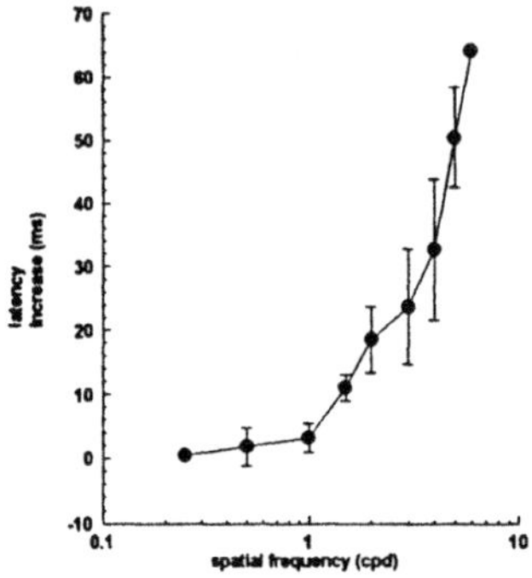

Fig. 1 Increase in latency of the major response component with increasing spatial frequency (data points are the mean of 6 subjects and error bars show 2 standard deviations).

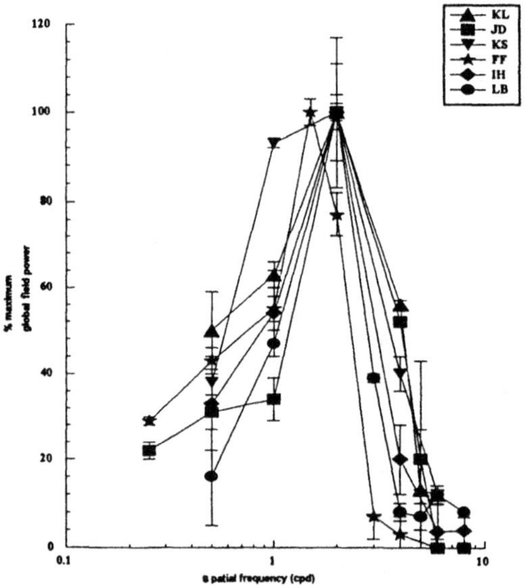

Fig. 2 Spatial frequency tuning characteristics of the chromatic evoked magnetic response. Data are shown for 6 subjects and error bars indicate the anti-average noise data.

Each average evoked response was well modelled by a single equivalent current dipole and the orientation of the dipole was consistent with the anatomy of striate cortex [3]. To determine the cortical location of the underlying response generators the data was co-registered with MR images, shown in Fig. 3. This reveals the dipole confidence ellipsoids to lie along the upper surface of the calcarine fissure, consistent with a V1 origin. The cortical location of the dipole generator was found to be invariant with spatial frequency, in agreement with the spatial frequency properties of area V1 [4].

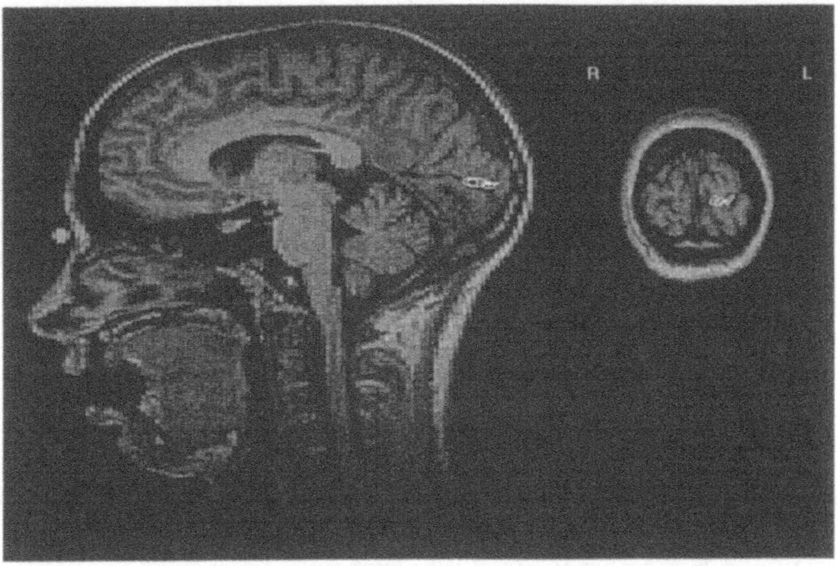

Fig. 3 95% confidence ellipsoids co-registered with sagittal and coronal MR images for dipoles to red/green gratings of spatial frequency 0.5, 1.0, 1.5 and 2cpd (subject FF).

Discussion

The evoked magnetic response was recorded to isoluminant chromatic gratings of 0.25-8cpd. The response was dominated by a single major component with a latency that increased with increasing spatial frequency, consistent with previous evoked potential studies [5]. The spatial frequency tuning response recorded here was bandpass, in contrast to the lowpass function determined psychophysically [6]. However, our experimental setup required a long viewing distance, precluding the use of more than one cycle of sinusoid for the lowest spatial frequency displayed. The sensitivity to gratings varies with the number of visible cycles, causing a reduction in field power at the lower spatial frequencies. This may also explain the reduction in visual evoked potential amplitudes at low spatial frequencies with chromatic stimuli [5].

The difference in our results from psychophysical observations can be explained by considering the neuronal activity necessary for detection by both techniques. To perceive a stimulus, relatively few neurones must signal the presence of the stimulus [7] but to record an evoked magnetic response that is sufficiently large to be distinguishable from the background noise many neurones must be simultaneously active. Hence although a reduction in field power at low spatial frequencies was caused by the low number of cycles in the display, it remains possible that the chromatic contrast sensitivity function is formed by an envelope of channels tuned to different spatial frequencies with more neurones tuned to 1-2 cycles/degree than to lower and higher spatial frequencies.

A further point of difference is that psychophysical measurements have indicated a high frequency cutoff of 11-12cpd (5) while the evoked magnetic response recorded here falls to the level of the noise at around 6cpd, a value comparable with previous evoked potential results (6). The difference between the evoked magnetic response data and psychophysical threshold measurements can again be explained by the requirement that many neurones must simultaneously respond before an evoked response can be recorded. As the sensitivity of the chromatic system decreases, fewer units will respond and while there may be sufficient numbers to allow detection of a stimulus, their may not be sufficient numbers to produce a clear evoked magnetic response. Hence MEG may provide a more accurate measure of the magnitude of cells tuned to a particular spatial frequency than detection thresholds traditionally used in psychophysics.

References

[1] Matlashov, A.N., Slobodchikov, V., Bakharev, A, Zhuravlev, Y & Bodarenko, N. (1993) Biomagnetic multichannel system built with 19 cryogenic probes. In Deecke, L. et al. (eds) *Recent Advances in Biomagnetism: Proc. 9th International Conference on Biomagnetism* pp 284-285.

[2] Singh, K.D. & Harding, G.F.A. (1996) Monte-Carlo analysis and confidence region ellipsoids for equivalent current dipole solutions. This volume.

[3] Fylan F., Holliday I.E., Singh K.D., Anderson S.J. & Harding G.F.A. (1995) Cortical origins of visual evoked magnetic responses differ in humans for chromatic and luminance stimuli. *Human Brain Mapping (Suppl.)* 1:51.

[4] Tootell, R.B.H., Silverman, R.S., Hamilton, S.L., Switkes, E. & DeValois, R. L. (1988e) Functional anatomy of macaque striate cortex. V. Spatial Frequency. *J. Neurosci.* 8: 1610-1624.

[5] Rabin, J., Switkes, E., Crognale, M., Scheck, M.E. & Adams, A.J. (1994) Visual evoked potentials in three-dimensional color space: correlates of spatio-chromatic processing. *Vision Res.* 34: 2657-2671.

[6] Mullen, K.T. (1985) The contrast sensitivity of human colour vision to red-green and blue-yellow chromatic gratings. J. *Physiol.* 359: 381-400.

[7] Newsome, W.T., Britten, K.H. & Movshon, J.A. (1989) Neuronal correlates of a perceptual decision. *Nature* 341: 52-54.

Acknowledgements

The research was supported by the Dr Hadwen Trust for humane research.

EEG and MEG Activity Accompanying Spontaneous Reversals of the Necker Cube

Gaetz, M., Weinberg, H., Jantzen, K.J., Rzempoluck, E. and Millbank, A.

Brain Behaviour Laboratory, School of Kinesiology, Simon Fraser University, Burnaby, Canada

Introduction

Synchronous oscillations between assemblies of cortical neurons have been proposed as a functional mechanism for linking or binding sensory information [1, 2]. Synchrony between groups of cells has been reported to occur at inter-electrode distances greater than one centimetre [3, 4]. Edelman's theory of neuronal group selection (TNGS) proposes that complex synchronous or *reentrant* activity serves as the basis for processing complex information [5]. Reentry is a process described as temporally continuous parallel interactions between distributed maps along ordered anatomical pathways. Reentry is not feedback, but parallel signalling in the time domain between spatially distributed maps, more similar to a process of dynamically changing correlations between distributed systems. Based on the TNGS three experimental questions were posed: 1) Can distributed cortical patterns present during perceptual reversals of a Necker cube be classified differently using a neural network ? 2) Does correlated activity increase during perception of Necker cube reversals? 3) Can distributed systems operating synchronously be identified ?

Methods

The subjects in the EEG phase were two males and two females (right handed) between 18 - 34 years of age (mean = 26.6 years). The EEG amplifiers were Nihon-Kohden model EEG-4217 interfaced with software designed to record single trials. The single trial epochs occurred between -1000 to 0 milliseconds, encapsulating a one second period prior to the trigger, and were digitised at 1024 points per second. Data collection for single trials was initiated by an electrical potential generated from the nasion electrode referenced to Cz. The EEG was filtered on-line 0.03 Hz to 70 Hz with a 60 Hz notch filter.

Five "naive" right handed adult subjects, three males and two females, participated in the second (MEG) phase of these experiments. Subjects were between 18 - 32 years of age, with a mean age of 26.4 years. A whole head CTF Systems MEG was used to record magnetic fields. The sensor array consisted of 64 first order gradiometers uniformly distributed over both hemispheres. The baseline of each 1st order gradiometer was 5 cm and each coil was 2 cm in diameter. Third gradient response was computed in software using a reference system of coils and squids. The single trials were one second epochs prior to the subject's response. The sampling rate was 250 points per second with low and high filter settings at DC and 100 Hz respectively. A 60 Hz filter and artifact rejection were used. Magnetic shielding was not necessary when third order spatial gradients were used.

Visual stimuli consisted of a square and a Necker cube, each presented on 8 1/2 x 11" sheets of white paper. Two indices of Necker cube reversal were defined: The perception of a *"shift-up"*, when the upper left-most corner of the cube face was in the upper left-most quadrant of the figure, or when perception of shift was in the opposite direction called a *"shift-down"*. The Necker cube and square were presented in random order. During perception of the control stimulus, the subject was instructed to close their eyes briefly, at approximately 3-5 second intervals, and to focus on the subjective centre of the control stimulus. While viewing the Necker cube, subjects were instructed to close their eyes for approximately 0.5 - 1.0 second immediately following perception of a reversal in the appropriate direction. Subjects were instructed to ignore reversals that occurred incompletely or in rapid succession. Consequently, the single trials included a subset of reversals that were of a subjective "high quality". Subjects were instructed to allow spontaneous perception of reversals without attempting to control them by modifying gaze. Data were collected until 20-30 EEG and over 30 MEG single trials were available for analysis. For experimental single trials, measures were taken to ensure eye closures signalled perceptual reversals (video monitoring and rapid blinking if an eye closure was not intended to mark a reversal).

EEG single trials were processed prior to neural network analysis. A subset of electrodes (T3, C3, C4, T4, T5, P3, P4, T6, O1 and 02) were selected for analysis. Correlation matrices were calculated for each single trial, resulting in a 10 x 10 matrix. Redundant information was removed from each matrix and the remaining data was transformed into a single vector (1 x 45). Vectors were subsequently organised into testing and training matrices. A subset of the MEG sensors was selected to closely approximate data analysed in the EEG phase (Fig. 1b). Correlation matrices were calculated for the subset resulting in a 32 x 32 matrix for each single trial. Redundant information was removed from each matrix and the remaining matrix values were transformed into vectors (1 x 496). The procedure for creating an MEG training matrix was similar to that described for EEG, with the exception that longer vectors were used.

Neural Network Analysis: The purpose of the neural network analysis was to discriminate EEG and MEG recorded one second preceding the Necker cube reversal from that recorded in the control condition. The network analysis used was a generalised regression neural network (GRNN) developed by Specht [35, 38]. Input to the GRNN consisted of individual exemplar row vectors from the training matrix. The GRNN estimated a number representing the probability that a single trial vector belonged to the experimental group from that training matrix. The probabilities were then grouped into control or experimental columns within subject for statistical analyses (i.e., for each subject, two statistical comparisons of probabilities were generated, *shift-up* vs. control and *shift-down* vs. control). One tailed equal variance t-tests were used to statistically test differences between GRNN classifications.

Correlation Subset Analysis: An *a priori* model was developed (see Fig. 1a and b) based on Edelman's description of reentrant pathways in visual cortex. The model started with the same electrode/sensor positions used for the GRNN analysis. Subsets of correlations chosen for analysis, (12 of 45 for EEG and 95 of 496 for MEG) were intended to spatially represent reentrant connections described by Edelman [8]. Averages were calculated across single trials, within subject and experimental condition. A sum of each average was then calculated, which resulted in a single value representing the magnitude of correlations within condition and experiment. The summed averages were analysed using the stepwise Bonferroni procedure for multiple comparisons.

Figure 1. Lines represent correlations between electrode/sensors that were included for EEG (a) and MEG (b).

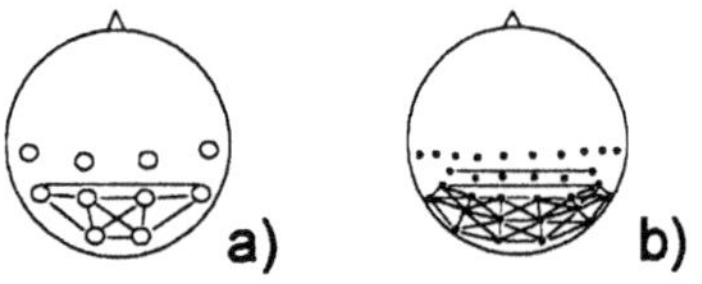

Overlapping Time Window Analysis (MEG): To determine whether distributed systems that function during perceived reversals could be detected, correlation matrices were calculated for each single trial using overlapping time intervals: 1-200, 100-300, 200-400, 300-500, 400-600, 500-700, 600-800, 700-900, 800-1000 (msec). The time window with the highest absolute sum of correlations was selected for each single trial and added to a matrix within subject and condition. Average correlation vectors were then calculated for each matrix. Average vectors from the control condition were subtracted from both experimental average vectors. Only correlations greater than -0.01 were analysed. This analysis was intended to be a conservative estimate of distributed systems with synchronous properties that produce reversals.

Results

Neural Network Analysis: EEG classifications were significant ($p < 0.05$) for each of the four subjects. The *shift-up* vs. control and *shift-down* vs. control were classified significantly by the GRNN for two of four subjects. For the remaining two subjects, only one of the two comparisons classified significantly (*shift-up* vs. control for one, *shift-down* vs. control for the other). MEG classifications were significant ($p < 0.05$) for both comparisons in all five subjects. The t-test p values did not indicate that *shift-up* could be classified as different than the control figure more easily than *shift-down* or vice versa for EEG or MEG.

Correlation Subset Analysis: For EEG, the summed averages were analysed using the stepwise Bonferroni t-test for multiple comparisons. The p value for the *shift-up* vs. control was $p = 0.019$, and *shift-down* vs. control was $p = 0.029$. The *shift-up* vs. control comparison was removed from the next iteration, and the remaining p value was significant at $p. < 0.05$.

For MEG, the summed averages were analysed using t-tests with the stepwise Bonferroni for multiple comparisons. The p value for the *shift-up* vs. control was $p = 0.039$. The p value for the *shift-down* vs. control was $p = 0.017$. The *shift-down* vs. control comparison was removed from the next iteration, and the remaining p value was significant at $p.< 0.05$. The correlation subsets were chosen *a priori*, therefore no further corrections were used.

Overlapping Time Window Analysis (MEG): Differences in correlation between Necker cube reversal and control average vectors are shown in Figure 2. Only correlation differences greater than .10 are shown. Differences in correlation between sensors are relatively consistent within subject and condition, but differ between individuals.

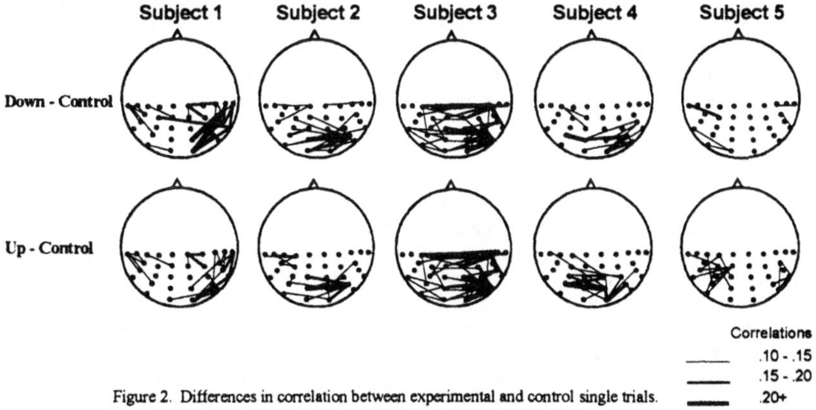

Figure 2. Differences in correlation between experimental and control single trials.

Discussion

Implicit in the TNGS is the concept that distributed groups of neurons operating within a system will display complex patterns of correlated activity. The GRNN was effective in detecting differences in correlation between experimental and control vectors, however, EEG and MEG did classify differently. The EEG classifications were less powerful than MEG; only six of eight experimental conditions were classified differently than control conditions. MEG data were classified significantly 10 out of 10 times. The magnitude of classifications was also considerably greater for the MEG.

There are *at least* three possible reasons for the difference in classifications of EEG and MEG. First, the MEG provides better spatial resolution of cortical sources. This may have increased the spatial sampling density for MEG and may have contributed to an increased signal to noise. Furthermore, the signals recorded with MEG are not degraded by bone and scalp and therefore could be expected to be differentiated to a greater extent than signals recorded at the same locations with EEG. Second, more MEG single trials were recorded. This is important because with enough information, the GRNN "is guaranteed to converge to a Bayesian classifier (the usual definition of optimality)" [7] (p.35). Finally, assuming a spherical model, the MEG is primarily sensitive to tangentially oriented sources, whereas the EEG may result from a summation of radial and tangential sources [9].

Gilbert and Wiesel [10] discuss tangentially oriented collaterals of extraordinary richness that extend over widespread areas of cortex, forming a number of distinct and repeating clusters. These horizontal cortical connections may extend beyond cortical columns and, in fact, may extend beyond hypercolumns [11]. Intercommunication of cortical clusters is an important attribute of information processing [12]. Tangential interconnections between distributed neuronal groups are presumed to be important components of cortico-cortical networks active during visual information processing, the function of which is probably non-linear. Therefore the evidence suggests that non-linear relationships between neuronal groups have an anatomical basis in the tangential network of collaterals and that MEG is more sensitive to these tangential sources.

According to the TNGS, reentry between neural groups is similar to a process of correlation, and therefore, one would expect correlations between proximal and distal recording sites to increase during increased information processing. Patterns of correlations recorded during Necker cube reversals were significantly higher than for the control stimulus. Using a moving time window analysis of MEG single trials, differences in correlation between Necker cube reversal and control average vectors were observed (Fig. 2). The correlations between sensors were relatively consistent within subject and condition, but differed between individuals. The results are considered to be consistent with reentrant properties in a distributed system that processes visual information.

GRNN classifications coupled with correlation matrix transformations were successful in demonstrating that patterns of CNS activity present during perception of Necker cube reversals were different from those preceding the perception of a two dimensional non-reversing figure. Increased correlations preceding Necker cube reversals were interpreted to be consistent with reentrant processes between distributed neuronal groups. The results provide evidence in agreement with Edelman's concept of reentry as a basis for information processing.

References

[1] Eckhorn, R., Bauer, R., Jordan, W., Brosch, M., Kruse, W., Munk, et al. Coherent oscillations: A mechanism of feature linking in the cat visual cortex?[1] Biological Cybernetics, 1988, 60:121-130.

[2] Gray, C. and Singer, W. Stimulus specific neuronal oscillations in orientation columns of cat visual cortex [2], Proceeding of the National Academy of Sciences USA, 1989, 86:1698-1702.

[3] Engel, A., Konig, P., Gray, C. and Singer, W. Stimulus-dependent neuronal oscillations in cat visual cortex: Inter-columnar interaction as determined by cross-correlation analysis [3], European Journal of Neuroscience, 1990, 2:588-606.

[4] Desmedt, J. and Tomberg, C. Transient phase-locking of 40 Hz electrical oscillations in prefrontal and parietal human cortex reflects the process of conscious somatic perception [4], Neuroscience Letters, 1994, 168:126-129.

[5] Edelman, G. Neural Darwinism: Selection and reentrant signalling in higher brain function [5], Neuron, 1993, 10:115-125.

[6] Specht, D. F. A general regression neural network [6], IEEE Transactions on Neural Networks, 1991, 568-576.

[7] Wasserman, P. Advanced Methods in Neural Computing, New York: Van Nostrand Reinhold, 1993.

[8] Edelman, G. The Remembered Present, New York, Basic Books, 1989.

[9] Cohen, D. and Cuffin, N. Demonstration of useful differences between magnetoencephalogram and electroencephalogram [9], Electroencephalography and Clinical Neurophysiology, 1983, 56:38-51.

[10] Gilbert, C. and Wiesel, T. Clustered intrinsic connections in cat visual cortex [10], Journal of Neuroscience, 1983, 3:1116-1133.

[11] Gilbert, C. Horizontal integration in the neocortex [11], Trends in Neuroscience, 1985, 8:160-165.

[12] Zeki, S., and Shipp, S. The functional logic of cortical connections [12], Nature, 1988, 335: 311-317.

MEG and EEG Responses to a Pitch Naming and Identification Task: Slow DC Shift Predicts Performance.

Gordon, R., Weinberg, H., Cheyne, D., and Jantzen, K.

Brain Behaviour Laboratory, Simon Fraser University, Burnaby, Canada

Introduction.

Damasio [1] has proposed that long-term memory storage relies on neural populations located in cortical regions responsible for perceptual processing. Memory occurs when primary and first-order sensory association areas are reactivated through conscious effort. Evidence in support of this position has recently been obtained in the visual modality [2]. In the auditory domain, there exits a population of individuals who, through extensive practice, have developed a unique long-term memory skill. Musicians who possess the skill of absolute pitch can name notes of specific frequency according to Western musical notation without reference. If Damasio's theory is correct, we would expect that individuals with absolute pitch would utilize neural generators normally reserved for sensory encoding for long-term memory storage of pitch information. The supratemporal cortex seems a likely cortical region for such processing. Also, in order to provide linguistic labels to specific frequency information, it is suspected that other neural generators may be involved. Recent MRI evidence suggests absolute pitch musicians have an enlarged left planum temporale [3]. Thus, both supratemporal cortex and left side planum temporale are hypothesized to play a role in absolute pitch processing.

Mechanisms of auditory memory storage have been indexed through the use of Mismatch Negativity, a negative potential occurring some 200 msec after stimulus onset. MMN and its magnetic counterpart, MMNm, are most easily elicited in oddball paradigms where an infrequent rare tone is presented in a series of standards. Averaged responses to the rare tone produce an enhanced negativity thought to represent neuronal populations responding to a change in stimulus characteristics [4]. The contents of the MMN storage are known to decay with time. An interstimulus interval of no more than 7-10 seconds is necessary to elicit the response [5]. If MMN generators play a role in memory storage of sensory information, it may in fact be these generators that are reactivated when discrete information such as pitch is recalled.

The current study examined MEG and EEG responses to a pitch identification task from subjects claiming to possess perfect pitch and from those who did not. It was hypothesized that neural generators responsible for absolute pitch ability would be localized to cortical areas normally reserved for processing perceptual information, i.e., MMN. As well, additional activation of left side parietal regions should also occur as an indication of linguistic processing.

Methods.

Collection parameters: Neuromagnetic activity was recorded using a 64-channel whole-cortex magnetoencephalographic system (CTF Systems Inc) without shielding. Electrical responses were recorded using a 21 channel Nihon Khoden system coupled with a 486 PC for digitization and collection of averages. Waveforms were sampled at a rate of 250 Hz for MEG data, 128 Hz for EEG. Signals were bandpassed DC to 50 Hz for MEG; .2 to 70 Hz for EEG. Linked ears were used as reference for EEG.

Subjects: Twelve subjects responded to an advertisement placed at local music schools requesting volunteers who possessed perfect pitch. Six of these subjects participated in the MEG sessions and 6 in the EEG. Subjects were given a pitch processing test consisting of a random presentation of tones spanning a chromatic two octave range. Results from this test were used to classify subjects either as good or moderate performers. Seven control subjects who did not have any formal musical training participated in both MEG and EEG conditions.

Procedure: For MEG recording, subjects were seated upright under the dewar and were instructed to view a computer monitor placed 5 meters away. Note names from Western musical notation were flashed onto the computer screen with a duration of 300 msec. Subjects were instructed to imagine the pitch of this note. Two seconds later, a binaural tone was presented via plastic earphones (to avoid magnetic artifact) that was either the same frequency of the visually presented note name (Match condition, 55% probability) or 1 semitone above or below (Mismatch condition, 45% probability). Eight tones and corresponding note names from an ascending chromatic scale (starting at Ab - 237 Hz) were used and presented in random order. Tones were sine waves of 100 msec duration with a 10 msec ramp. Once the tone was presented, a random silent interval of 1-2 seconds occurred after which the computer monitor prompted the subjects for an answer regarding the relationship between the visual prompt and tone (i.e., match or mismatch). Subjects responded verbally. The intertrial interval was variable from 7-8 seconds. 130 trials were collected. Recording epoch was 1.1 seconds triggered by the onset of the tone. Recordings were baselined 100 msec before tone onset. All stimulus parameters were identical for the EEG condition.

Results

Visual inspection revealed no differences between match and mismatch trials. Thus, data were combined and averaged across all trials. Figure 1 depicts superimposition of averaged MEG data for three subjects representative of the different skill levels. Figure 2 shows the corresponding topographic distributions. The most salient feature of these waveforms is a sustained potential occurring for the good and moderate task performers.

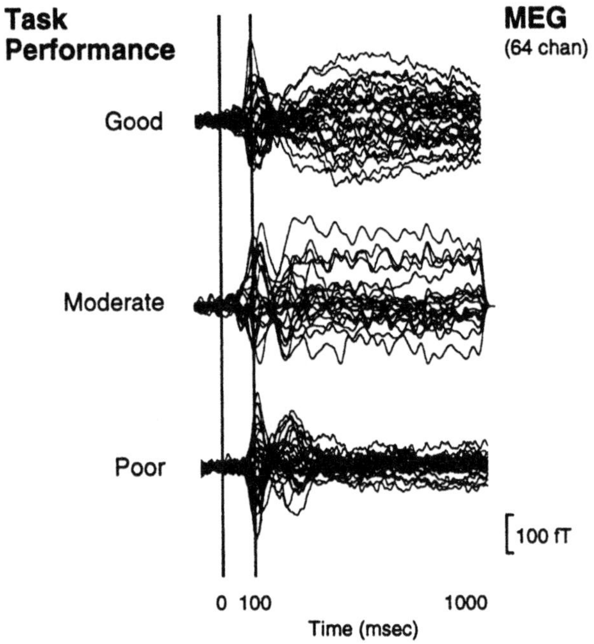

Figure 1. Superimposition of selected MEG channels. *Upper:* Averaged responses to tones for subject showing 100% accuracy on task. *Middle* Subject with 80% correct. *Lower:* Subject with chance performance. Averages contain approximately 100 artifact free trials.

This sustained potential begins to develop approximately 200 msec after the tone and, for good performers, has a broad flat peak occurring between 400-600 msec. In contrast, individuals with poor pitch processing skills produced no organized brain activity after 200 msec. For good performers, topographic distributions for the sustained potential resemble closely the N1m response although slightly more anterior (Figure 2). For moderate performers, the sustained potential resembles the N1m distribution but is missing the frontal fields seen in lateral and anterior sensors. Also noticed is that for both poor and moderate performers, a field pattern emerges at 200 msec that is not seen in good performers (see Figure 2). Given this pattern is peaking near 200 msec it is suspected to be part of the MMNm complex. EEG results (not shown) also reveal distributions consistent with MMN at 200 msec, but only for moderate and poor task performance. The topography of the sustained potential seen with EEG is characterized by a broad positivity at parietal and occipital electrodes and negativity at frontal locations. The negativity is more pronounced in the moderate performers than good performers. Noted in the MEG recordings (400 msec, Figure 2) are ingoing fields detected by left side parietal sensors for good and moderate task performance. Although there appears other asymmetries associated with sensor magnitude, these effects were not consistent between individuals nor related to task performance.

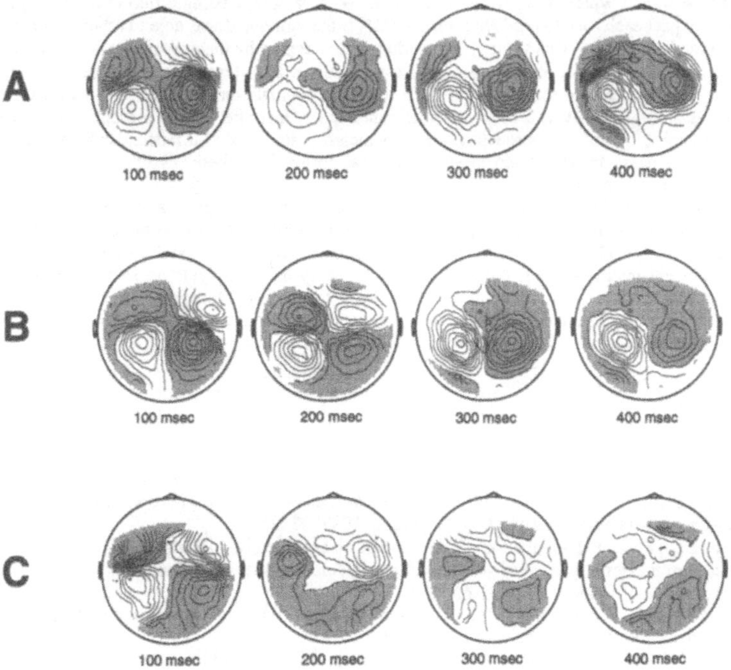

Figure 2. Isocontour plots for magnetic fields from averaged responses to tones. Time indicated is after tone onset. Gray indicates ingoing field, contours are in steps of 32 fT. **A**: Responses from subject showing good performance (100% correct) on pitch identification task. **B**: Responses from subject with moderate performance (80% correct). **C**: Responses from subject with poor task performance (chance).

Discussion

Its seems reasonable to suggest results are in agreement with Damasio's views on memory processes. Though analyses are only descriptive, topographic mapping indicates that individuals who possess tonal memory abilities produce a response that appears similar to the initial N1m topography. Although similar, it is also clear that the distribution is not identical to the N1m. Given the link of MMN generators to auditory memory [5], it is possible that it is access to these neuronal populations as opposed to N1 generators that gives rise to the absolute pitch phenomenon. What is unclear from the current study is if the sustained potential seen in good task performers is in fact similar to MMNm field generators. Dipole fits are not stable as both MEG and EEG data indicate multiple source activation. Nevertheless, the sustained potential appears to suggest an anterior movement of current sources which would be in agreement with at least one component of the MMNm [6]. Thus, one element of absolute pitch processing requires activation of supratemporal cortex.

A notable difference between good and moderate performance is the field distribution at 200 msec after stimulus onset. For moderate performance, a distinct dipolar field pattern emerges. This may in fact be a true MMNm component. Sams [5] reported MMNm for ISIs up to 12 seconds. For the current study, time between tones was variable from 11-12 seconds. It is conceivable that moderate performers utilized a strategy that relied on remembering the previous tone. Moderate performers, therefore, would be thought to possess the skill of relative pitch and not absolute pitch. Musicians possessing relative pitch skills are able to name notes but only when a known reference is provided. As the tones were presented randomly, only in rare cases did the same frequency occur twice in a row. With each tone being different from the previous one, we would expect each tone to produce a small MMN if the individual was able to retain in memory the frequency of tone heard prior. This strategy was perhaps also adopted by the poor performers. For absolute pitch subjects, however, this strategy is not necessary; with an long-term memory store of exact pitch information there is no need to remember previous tones for comparison purposes. In this manner, each tone is heard as a single entity and no MMNm is generated. Once the tone is presented, however, the absolute pitch subject is able to retroactively access memory stores of that particular frequency. This memory storage appears to be located in supratemporal cortex and likely involves similar generators initially involved in sensory encoding.

As a final note, left side, posterior sensors are active for both good and moderate performers beginning 300-400 msec after stimulus onset. Although tentative, this may be taken to reflect activity from the planum temporale region which has been implicated in absolute pitch processing [3]. Taken together, the current results support the retroactivation theory as individuals possessing absolute pitch produce a sustained endogenous potential that topographically closely resembles magnetic field patterns associated with the transient sensory response. As well, additional sources from left side parietal lobe and possibly frontal cortex appear to be involved in pitch processing ability.

References

[1] Damasio, A.R., Time-locked multiregional retroactivation: A system level proposal for the neural substrates of recall and recognition. Cognition, 1989, 33: 25-62.

[2] Rosler, F. Heil, M. and Hennighausen, E. Distinct cortical activation patterns during long-term memory retrieval of verbal, spatial and colour information. Journal of Cognitive Neuroscience, 7(1): 51-65.

[3] Schlaug, G., Jancke, L., Yanxiong, H., and Steinmetz, H. In vivo evidence of structural brain asymmetry in musicians, Science, 1995, 267: 699-700.

[4] Naatanen, R. Attention and Brain Function, New Jersey, Erlbaum, 1992.

[5] Sams, M., Hari, R. and Knuutila, J. The human auditory sensory memory trace persists for about 10 sec: neuromagnetic evidence. Journal of Cognitive Neuroscience, 5(3), 363-370.

[6] Tiitinen, H. Alho, K., Huotilainen, M., Ilmoniemi, R. J., Simola, J., and Naatanen, R. Tonotopic auditory cortex and the magnetoencephalographic (MEG) equivalent of the mismatch negativity, Psychophysiology, 30, 1993.

Magnetic Detection of Activity of Presumed Inhibitory Interneurons in the Human Somatosensory Cortex

Hashimoto, I.[1], Mashiko, T.[2] and Imada, T.[2]

[1]Department of Psychophysiology, Tokyo Institute of Psychiatry, Tokyo, Japan; [2]Information Science Research Laboratory, NTT Basic Research Laboratories, Atsugi-shi, Japan

Introduction

Previous studies of somatosensory evoked potentials (SEPs) and fields (SEFs) have shown that high-frequency signals in the range of 300-900 Hz concur with the primary response (N20 and N20m) of the somatosensory cortex [1,2]. However, the physiological mechanisms of the high-frequency oscillations remained undetermined. We addressed the issue by analyzing magnetic fields during wakefulness and sleep over the left hemisphere to right median nerve stimulation. Results obtained from this study have been described in detail elsewhere [3].

Methods

Eleven subjects (Ss) participated in the experiments; 10 Ss were studied during wakefulness and 7 Ss were tested during natural sleep. Brief electrical stimuli with a 0.2 msec duration were delivered to the right median nerve at the wrist. The stimulus intensity was about 3 times sensory threshold and elicited a mild twitch of the abductor pollicis brevis muscle. The stimuli were delivered at regular intervals with the repetition rate of 4 Hz. Magnetic recordings (bandpass 0.1-2,000 Hz) were sequentially taken over the left hemisphere with two 7-channel SQUID gradiometer systems (BTi Model 607).

An epoch of 50 msec duration (a 10 msec pre- and 40 msec post-stimulus) was digitized at a 5 kHz/channel sampling rate and 5,000 responses were averaged off-line. For isolation of the high-frequency oscillations from the underlying N20m, the wide-band (0.1-2,000 Hz) recorded responses were digitally high-pass (> 300 Hz) and low-pass (< 300 Hz) filtered. In order to define the same 3-dimensional head coordinate system for MEG and MRI measurements, the same 3 fiducial points, marked with vitamin pills, were used such that the estimated magnetic source locations could be projected on to the appropriate points on the MRI slices of the individual subjects.

Results

<u>SEFs during wakefulness</u>

The magnetic signals from the somatosensory cortex consisted of high amplitude N20m and P27m within 40 msec after median nerve stimulation. Superimposed upon the N20m, several small inflections could be recognized in the unfiltered original SEF records (Fig. 1A). FFT analysis of the wide-band records revealed two to four frequency peaks of signal energy. The main signal energy was distributed broadly between 20 and 300 Hz with a peak around 40-60 Hz, and a weaker signal energy with one to three peaks was found above 300 Hz. Among the weaker peaks above 300 Hz, the higher frequency peak from 580 Hz to 780 Hz carried the principal signal energy. After high-pass filtering of the original records, these small inflections buried in the predominant N20m were isolated and could be discriminated from the background noise (Fig. 1B).

High-frequency oscillations start approximately at or later than the onset of the N20m and end in the middle of the second slope. Similar to N20m, each peak of the high-frequency oscillations reversed polarity medio-laterally at exactly the same location. The field patterns at the peaks of the high-frequency oscillations and the low-pass filtered SEFs at the corresponding period showed a close similarity (Fig. 2). Estimated location of the dipole for each peak of the high-frequency oscillations agreed with activation of the 3b area of somatosensory cortex and was very close to that for N20m. The mean absolute dipole strength was 1.2nA.m for the high-frequency peaks and 17.0 nA.m for the low-pass filtered N20m at the corresponding points in time.

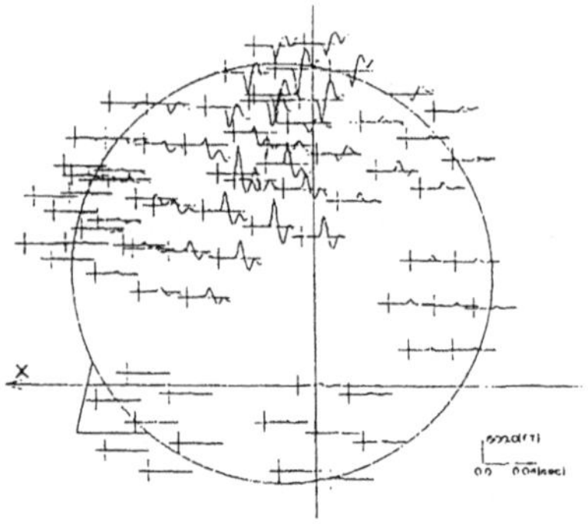

Fig. 1A: Wide-band recorded SEFs.

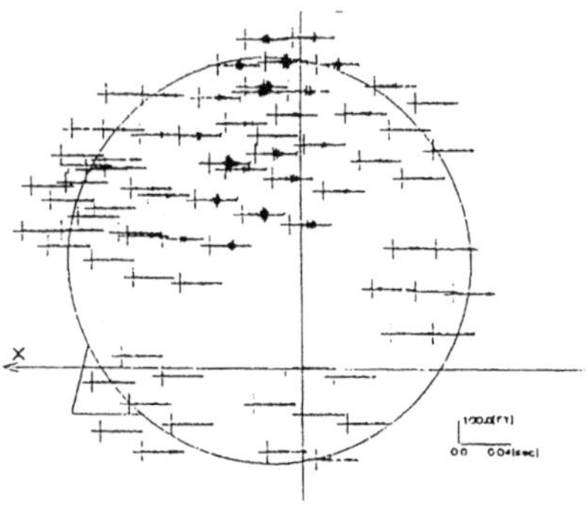

Fig. 1B: High-pass filtered SEFs.

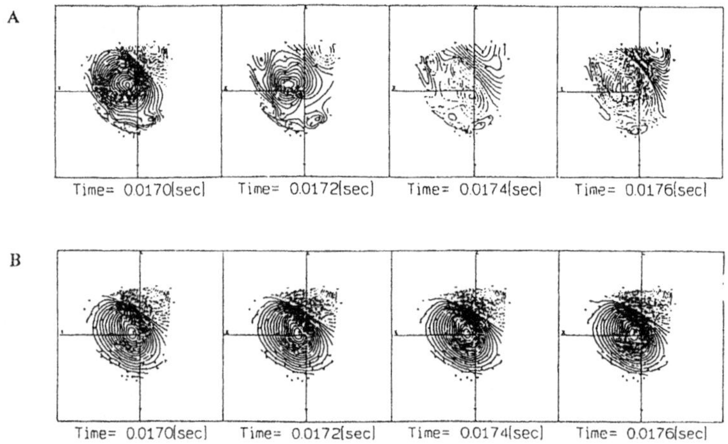

Fig. 2, A: Isofield maps of the high-frequency oscillations over the left hemisphere. The maps are shown at 0.2 msec intervals. Continuous lines indicate magnetic flux out of the head and dashed lines flux into the head. The isofield lines are separated by 2.5 fT. B: Isofield maps of the low-pass filtered SEFs during the same period as in A. The isofield lines are separated by 20 fT.

SEFs during sleep

Although the N20m showed a longer mean latency during sleep, this difference (mean, 0.6 msec) did not reach statistical significance. However, the mean N20m peak amplitude was significantly larger during sleep ($p < 0.005$). While strong high-frequency oscillations were present during waking (Fig. 3A), they were reduced in amplitude (Fig. 3B).

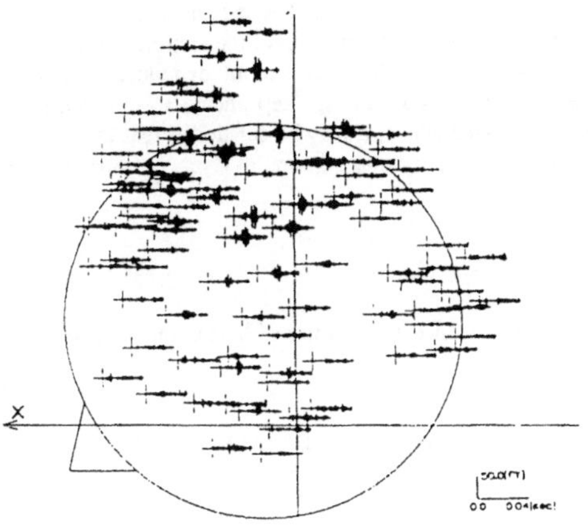

Fig. 3A: High-frequency oscillations during waking.

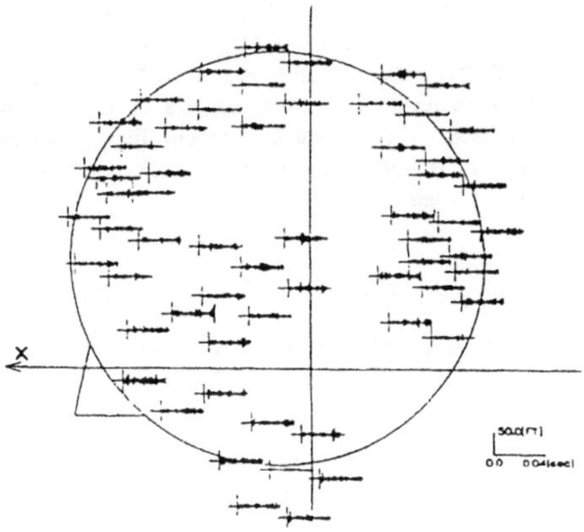

Fig. 3B: High-frequency oscillations during sleep. Note remarkable amplitude reduction.

Discussion

The physiological features of the high-frequency somatic signals can be summarized as follows. First, the equivalent dipoles for the high-frequency peaks and the underlying N20m are temporally overlapped and spatially colocalized within the 3b area. Second, the high-frequency oscillations present during wakefulness attenuate or disappear during sleep. Third, the underlying N20m in contrast increases its amplitude during sleep. Thus, there exists a reciprocal relation between the high-frequency oscillations and the N20m during a wake-sleep cycle. The reciprocal relation between the high-frequency oscillations and the N20m representing ensemble EPSPs of the pyramidal neurons in 3b area led us to assume that some inhibitory mechanisms are involved in the generation of high-frequency oscillations.

Morphological studies demonstrate that GABAergic inhibitory interneurons receive direct thalamic axon synapses and have their own axon terminations focused on the cell somata, proximal dendrites and initial segments of the pyramidal cells. The findings suggest that GABA mediated inhibition plays a powerful role in the shaping of the response profiles of pyramidal neurons. Furthermore, the patterns of connectivity of many GABAergic interneurons are bidirectional in the vertical direction between deep and superficial layers as well as in the horizontal direction [4]. This parallel arrangement of the interneurons provides the anatomical basis for spatial summation of activity of individual neurons. Thus, synchronous monosynaptic activation of a population of GABAergic interneurons, by summating individual intracellular currents spatiotemporally, may produce magnetic fields strong enough to be detectable outside the head.

Electrophysiological characteristics of cortical interneurons are high-frequency spike bursts (200-1,000 Hz) in response to synaptic volleys. In addition, the extracellular spike waveform originating from the interneurons is short in duration and about one-half that of the "regular spikes" produced by the pyramidal neurons, reflecting mainly different cell morphology [5].

The activity of cortical inhibitory interneurons studied in animal experiments bears striking similarities to the high-frequency oscillations in the somatosensory cortex in humans as described above. Furthermore, GABAergic interneurons are activated by acetylcholine and serotonin upon arousal, which causes enhancement of inhibition of down-stream glutamatergic pyramidal cell activity [6]. This GABA mediated feed-forward inhibition or disinbition of the pyramidal neurons may underlie the reciprocal relation of the amplitude between the high-frequency oscillations and the N20m during a wake-sleep cycle. Thus, an amplitude reduction of the high-frequency oscillations is associated with the increase in the N20m during sleep and vice versa during wakefulness. From these convergent lines of evidence as discussed above, it may be concluded that the high-frequency oscillations represent the activity of the GABAergic inhibitory interneurons in layer 4 of area 3b of the primary somatosensory cortex.

Acknowledgments

This research was supported by the Ministry of Education, Science and Culture of Japan grant 05404087.

Reference

[1] Yamada, T., Kameyama, S., Fuchigami, Y., Nakazumi, Y., Dickins, Q. S., and Kimura, J. Changes of short latency somatosensory evoked potential in sleep, Electroenceph. clin. Neurophysiol., 1988, 70:126-136.

[2] Curio, G., Mackert, B. M., Burghoff, M., Koetitz, R., Abraham-Fuchs, K., and Härer, W. Localization of evoked neuromagnetic 600 Hz activity in the cerebral somatosensory system, Electroenceph. clin. Neurophysiol., 1994, 91:483-487.

[3] Hashimoto, I., Mashiko, T., and Imada, T. Somatic evoked high-frequency magnetic oscillations reflect activity of inhibitory interneurons in the human somatosensory cortex, Electroenceph. clin. Neurophysiol. (in press)

[4] Jones, E. G. Varieties and distribution of non-pyramidal cells in the somatic sensory cortex of the squirrel monkey, J. Comp. Neurol., 1975, 160:205-268.

[5] McCormick, D. A., Connors, B. W., Lighthall, J. W., and Prince, D. A. Comparative electrophysiology of pyramidal and sparsely spiny stellate neurons of the neocortex, J. Neurphysiol., 1985, 54:782-806.

[6] Steriade, M., Jones, E. G., and Llinás, R. R. Thalamic oscillations and signaling. New York, John Wiley & Sons, 1990.

Effect of Color on Visual Evoked Magnetic Fields with Pattern-Reversal Stimulation

Hatanaka, K.[1], Seki, K.[2], Nakasato, N.[2], Kanno, A.[3], Ohtomo, S.[2], Fujiwara, S.[2] and Yoshimoto, T.[2]

[1]KRI, 17 Chudoji Minami-machi, Shimogyo-ku, Kyoto, Japan; [2]Department of Neurosurgery, Tohoku University School of Medicine, 1-1 Seiryo-cho, Aoba-ku, Sendai, Japan; [3]MEG Laboratory, Kohnan Hospital, 4-20-1 Nagamachiminami, Taihaku-ku, Sendai, Japan

Introduction

Visual image is supposed to be processed by parallel pathways that analyze different properties of vision such as form, depth, motion, and color. Positron emission tomography (PET) studies of functional specializations of the human visual system have demonstrated that the fusiform gyrus is playing a role in color processing[1,2]. Poor time resolution of PET, however, gave no information about the time course of the activity in the brain. Although visual evoked potentials (VEPs) studies with dipole source analysis [3,4] supplemented PET results, no direct comparison with the anatomy of the visual cortex of subjects was made. This is because spatial resolution of source localization in VEPs is low[4]. We have been investigating visual evoked magnetic fields (VEFs), which in principle have much better spatial resolution than VEPs, and found that the most prominent response (P100) evoked with pattern-reversal stimulation was located in the primary visual cortex[5].

The most important factor in visual evoked response is the type of stimulus. Earlier investigations for the effect of color on visual response [6,7] used simple colored disc shaped pattern. An abstract display of colored squares and rectangles (a 'Mondrian') was occasionally employed [1-4] because this type of stimulus is believed to excite only color component. Pattern-reversal stimulation, which consists of reversing white and black checkerboard pattern, is widely used for physiological studies as well as clinical application. Pattern-reversal stimulation modulates luminance, contrast, form, and motion properties of the vision. If we use colored checkerboard pattern, color component is also modulated. Since color of the pattern-reversal stimulation has effects on response[8,9], investigations of the difference in VEFs give much insight in the color processing mechanism. Moreover, with excellent spatial and temporal resolution of VEFs, we can naturally expect to trace the time course of the information processing in human visual cortex.

Methods

We recorded VEFs from five healthy subjects with a 64-channel whole-head SQUID magnetometer (CTF Systems Inc., Canada). Stimulus consisted of green-black or red-black checkerboard pattern-reversal stimulation. The brightness of each color was adjusted to 22 Cd/m^2 and the contrast was better than 99 %. Distance between subject's eye and the stimulator was about 1 m. Individual check size was 1° and the total stimulating field subtended rectangular visual field of 5° by 10°. Right-half visual field stimulation to right eye was carried out, and the fixation point was 1° lateral to the center of the left edge of the pattern. Stimulation rate was 2 Hz with pattern reversal period of 500 msec. An analysis time of 450 msec (150 msec pre- and 300 msec post-pattern-reversal) was used and 200 trials passed through 100 Hz low pass filter and 50 and 100 Hz notch filters were digitized at 625 Hz and averaged. For each color, two sessions were carried out to confirm reproducibility of VEFs.

Source locations of the most common components of N75, P100, and N145 were analyzed by single current dipole model. The effect of volume current was taken into account using Grynszpan and Geselowitz spherical head shape approximation. Position and radius of 'best sphere' (*i.e.* sphere which fit into subject's head in the least square sense) were determined by magnetic resonance imaging (MRI) of the subject. Conversion of VEFs coordinates system to MRI was carried out by reference to three fiducial points (two preauricular points and nasion), each marked by small coils for VEFs and oil-containing capsules for MRI. Positions and orientations of the estimated equivalent current dipoles (ECDs) were directly superposed on the MRI of the subjects.

Results

All five subjects showed well-defined P100 and smaller N75 and N145 responses for green-black pattern-reversal stimulation. As for red-black stimulation, responses differed between subjects, but generally N145 waves were most markedly observed and P100 and N75 waves were also noticed. Results for source analysis of a subject are shown in Table 1. Source locations of N145 for green-black stimulation and N75 component for red-black stimulation were not clearly defined. Note that for green-black stimulation, P100 wave has the largest magnitude and lowest fitting error while for red-black case N145 becomes major component.

Averaged parameters of N75, P100 and N145 waves are summarized in Table 2. Estimated locations and orientations of ECDs on the MRI of the subject are shown in Fig. 1. The P100 dipoles were localized in the bottom of the calcarine fissure contralateral to the stimulation. Distance between the locations of P100 dipole for green-black stimulation and red-black one is 0.58 cm. The position of N145 for red-black stimulation were also located in much the same place for P100 (distance is 0.46 cm). The N75 dipoles for green-black stimulation were located about 0.9 cm anterior and 0.5 cm lateral to P100 dipoles (distance is 1.11 cm). The localization error of the VEFs system is supposed to be about 0.3 cm. Therefore the locations of P100 are seemed to be unrelated to colors, and P100 and N145 waves originate from the same place.

Table 1. Anatomical locations of the ECDs evoked with color-black pattern-reversal stimulation of the right-half visual field (x, y, z: anterior, left lateral, and upward coordinates of the VEFs coordinate system; φ: azimuthal angle of dipole, 0 for+x direction; θ: declination angle of dipole, 0 for +z direction; Q: magnitude of current dipole)

Session	Color	Wave	Latency (msec)	x (cm)	y (cm)	z (cm)	φ (deg)	θ (deg)	Q (nAm)	Error (%)
1	Green	N 75	75.2	-3.57	2.21	5.16	62.1	67.2	11.5	22.3
		P100	107.2	-4.36	1.55	4.71	254.1	114.9	25.3	3.6
2	Green	N 75	75.2	-3.10	1.87	4.88	64.5	62.3	15.3	11.8
		P100	105.6	-4.19	1.46	4.78	254.2	113.7	26.5	4.4
3	Red	P100	94.4	-4.38	0.91	4.71	260.7	108.5	12.0	15.4
		N145	153.6	-4.68	1.22	4.67	78.2	69.2	19.9	7.7
4	Red	P100	102.4	-4.47	1.02	4.46	261.2	119.3	12.6	12.0
		N145	155.2	-4.59	1.27	4.58	77.9	70.3	21.5	5.0

Table 2. Averaged parameters of ECDs for the same subject shown in Table 1

Wave	Color	Latency (msec)	x (cm)	y (cm)	z (cm)	φ (deg)	θ (deg)	Q (nAm)	Error (%)
N 75	Green	75.2	-3.34	2.04	5.02	63.3	64.8	13.4	17.1
P100	Green	106.4	-4.28	1.51	4.75	254.2	114.3	25.9	4.0
P100	Red	98.4	-4.43	0.97	4.59	261.0	113.9	12.3	13.7
N145	Red	154.4	-4.64	1.25	4.63	78.1	69.8	20.7	6.4

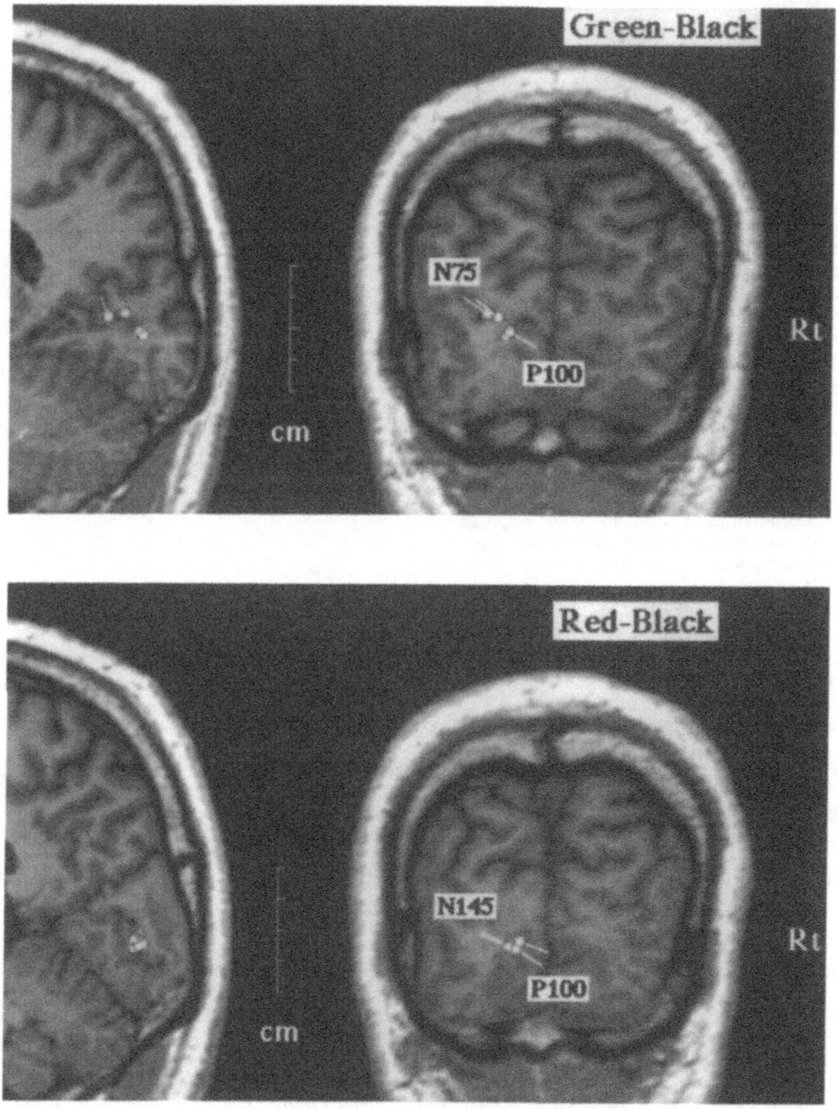

Fig. 1.
ECDs evoked by right-half visual field pattern-reversal stimulation to the right eye. Circles and bars indicate positions and orientations of the dipoles. Dipoles of successive two sessions are shown for each waves. Positions of the ECDs are along the bottom of the calcarine fissure. Note the inversions of the dipoles in time sequence of N75, P100 and N145.

Discussions

Although the precise locations of N75 and P100 ECDs might be different, three major responses of N75, P100, and N145 stayed along the bottom of the calcarine fissure (V1). They changed only directions, and during these time sequences, processing such as projection to the higher visual cortex such as V4 (color) and V5 (movement) or back-projection to V1 must occur.

We can express color-black pattern-reversal stimulation schematically as shown in Fig. 2. Although luminance is adjusted to the same intensity, chromaticity is apparently much stronger in red-black stimulation. Hence, red-black pattern-reversal stimulation has much stronger effect on color component; or we can assume red-black stimulation mainly as color (and brightness) stimulation. For green-black stimulation, the effect of color is small and the response is similar to the conventional white-black, or brightness, stimulation. Stronger response of N145 for red-black stimulation, therefore, suggests N145 is related to the color component. However, the locations of N145 dipoles are almost same as P100 component and localized in the bottom of the calcarine fissure (V1), and not in the fusiform gyrus (V4). One hypothesis we could propose is that N145 component is the result of back projection from V4 to V1. Buchner *et al.*[4] pointed out that activity peaked at 135 msec could be located in fusiform gyrus. Although visual stimulation is quite different, their findings may support our hypothesis since their method(VEPs) is sensitive to dipoles in gyrus (V4, 135 msec) whereas ours (VEFs) is sensitive to sources in sulcus (V1, 145 msec).

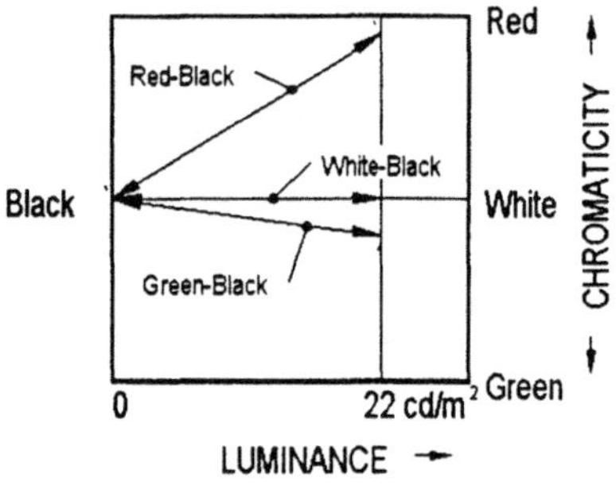

Fig. 2

Schematic representation of color-black pattern-reversal stimulation.

References

[1] Lueck, C. J., Zeki, S., Friston, K. J., Deiber, M. -P., Cope, P., Cunningham, V. J., Lammertsma, A. A., Kennard, C., and Frackowiak, R. S. J. The colour centre in the cerebral cortex of man, *Nature*, 1989, 340: 386-389.

[2] Zeki, S., Watson, J.D.G., Lueck, C.J., and Friston, K.J. A direct demonstration of functional specialization in human visual cortex, *J. Neurosci.*, 1991, 11: 641-649.

[3] Plendl, H., Paulus, W., Roberts, I. G., Bötzel, K., Towell, A., Pitman, J. R., Scherg, M., and Halliday, A. M. The time course and location of cerebral evoked activity associated with the processing of colour simuli in man, *Neuroscience Letters*, 1993, 150: 9-12.

[4] Buchner, H., Weyen, U., Frackowiak, R.S.J., Romaya, J., and Zeki. S. The timing of visual evoked potential activity in human area V4, *Proc. R. Soc. Lond.* B, 1994, 257: 99-104.

[5] Seki, K., Hatanaka, K., Nakasato, N., Otsuki, T., and Yoshimoto, T. The origin of P100 component in visual evoked response, *J. Funct. Neurophysiol.* 1990, 3:270.

[6] Paulus, W. M., Hömberg, V., Cunnigham, K., Halliday, A. M. and Rohde, N. Colour and brightness components of foveal evoked potentials in man, *Electroencephalogr. Clin. Neurophysiol.*, 1984, 58: 107-119.

[7] Krauskopf, J., Klemic, G., Lounasmaa, O. V., Travis, D., Kaufman, L. and Williamson, S. J. Neuromagnetic measurements of visual responses to chromaticity and luminance, In: Williamson, S. J. *et al. Advances in Biomagnetism*, 209-212, Plenum Press, New York, 1990.

[8] Taghavy, A., and Kügler, C. F. A. Colour-black pattern reversal visual evoked potentials (colour-black-PVEPs): neurobiological aspects and clinical applicability of a new method, *Intern. J. Neuroscience*, 1988, 43: 225-236.

[9] Hatanaka, K., Seki, K., Nakasato, N., and Yoshimoto, T. Visual evoked magnetic fields with red-black and green-black pattern-reversal stimulation, In: Hoke, M. *et al.*, *Biomagnetism: Clinical aspects*, 197-201, 1992.

Acknowledgments

This work has been supported in part by Grants-in-Aid for Scientific Research No. 03404042 and No. 064050 from the Ministry of Education, Science and Culture of Japan.

Evoked Magnetic Fields at Semantic Judgment of Simple Japanese Sentences

Hirata, Y.[1], Sakai, T.[1], and Kuriki, S.[1,2]

[1]*Research Institute for Electronic Science, Hokkaido University, Sapporo, Japan*

[2]*National Institute for Physiological Sciences, Okazaki, Japan*

Introduction

Most studies of auditory evoked magnetic fields of the human brain have been conducted by recording responses reflecting early processes to sound such as N1m or MMF. From lesion and PET studies, it has been hypothesized that spoken words first activate the primary auditory cortex, then the information is processed at Wernicke's area, and lastly a frontal cortex processes semantic association [1][2]. Magnetoencephalography is a suitable method for tracing such activities spatio-temporally. In this study, we used some Japanese sentences, in which each word was formed from syllables consisting of the combination of a consonant-vowel or a vowel. We measured the magnetic fields from the temporal area of subjects by using the word stimuli with an equalized duration.

Experiment I

Method Four Japanese subjects (three right-handed, one left-handed) participated in the first experiment. The combination of a noun followed by a verb was known as one of the most simplest Japanese sentences. Stimuli used in this experiment consisted of a Japanese noun (400 ms duration) followed by a 30 ms blank, a Japanese verb (600 ms), a 1.57 s blank, and a sound cue (500 Hz, 40 ms) with a repetition rate of 5.6-6.1 s (Fig. 1). In order to adjust the duration of the stimuli, two-syllable nouns and three-syllable verbs were selected. Spoken words digitized at a rate of 10 kHz were presented to both ears of the subjects through plastic ear tubes at sound intensity of 45 dB SPL. The subjects were instructed to listen to a simple Japanese sentence which consisted of a noun and verb and to judge quickly whether the sentence was congruous or not. After listening to the cue of a beep, they were required to push a left button of a mouse when the sentence was congruous, or a right button when the sentence was incongruous. One hundred responses to both congruous and incongruous stimuli were recorded in each session. This experiment was done for each subject several times by using different sentences of similar kinds. Magnetic fields evoked from the subjects' brain were recorded in a magnetically shielded room with a dual 37-channel SQUID magnetometer (BTi Inc.), which covered the right and left temporal areas and thus two hemispheres' activities were simultaneously recorded. The signals were digitized at 297.6 Hz. The period of 2.9 s from the onset of the noun including a 0.4 s pre-stimulus baseline was analyzed. The responses were averaged and digitally filtered with a bandpass of 1-40 Hz. In order to measure the reaction time of the same semantic task, an experiment using the same stimuli, except for a tone cue, was also carried out.

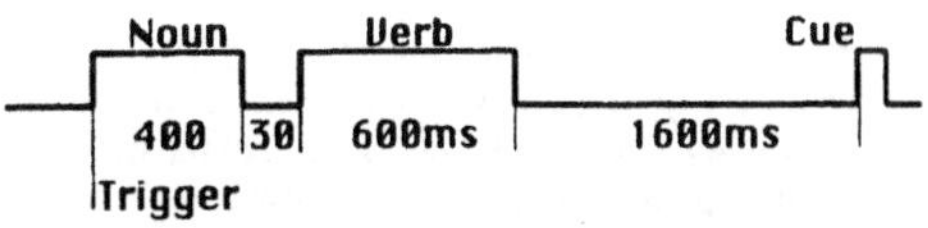

Fig. 1 . Stimulus sequence of experiment I.

Results Figure 2 shows superimposed 37 magnetic fields which were recorded from the right and left hemispheres of two subjects. The large signal at 0 ms latency is an artifact of electrical noise from a D/A converter. Although this kind of noise was also observed at other latencies of 400 ms, and 1.0 s, it was clear that the noise did not affect the baseline of the recorded fields. Two out of the four subjects showed a large amplitude of about 350 fTpp of the N1m elicited by the first syllable of the noun. However, no obvious N1m which was elicited by the syllables of the verb was observed in all the subjects. Comparing to the power (squared amplitude of the fields) of both hemispheres the left power was greater than that of the right one in several sessions. This observation is compatible with the PET study which reports the dominancy of the left hemisphere activation in lexico-semantic processes [3].

The averaged reaction time was 1170 ms from the onset of the noun to the button press time of the subjects. The simple reaction time to the movement of the index finger muscles by a magnetic transcranial stimulation of the motor cortex is about 120 ms [4]. Therefore, the semantic judgment is supposed to be completed at the latencies of around 1050 ms. In this respect, a slow wave of the fields was observed in several sessions in all subjects, where its

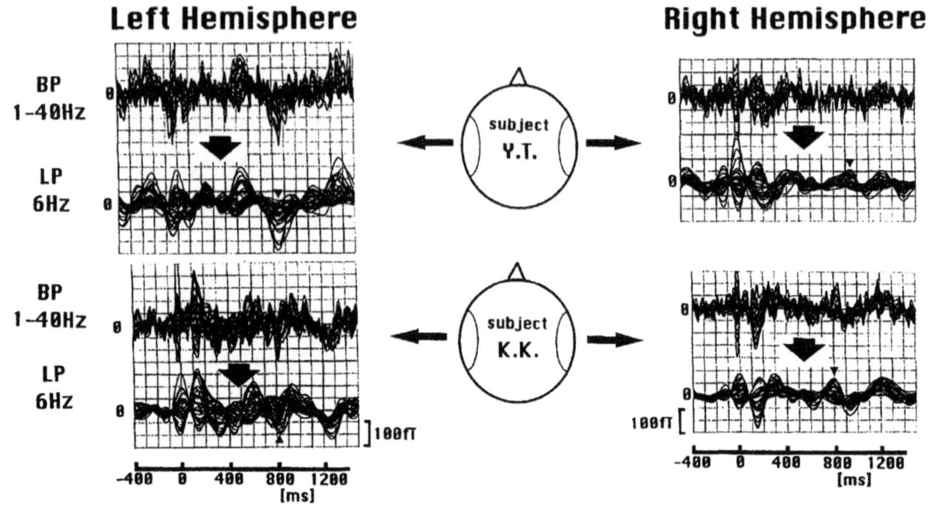

Fig. 2. Superimposed 37 waveforms (1-40 Hz) recorded from both hemispheres of two subjects. Low-pass filtered data below 6 Hz were drawn to examine the existence of the slow wave at latencies of 0.8-1 s. The wave observed at 0 ms was an electrical artifact, and it did not affect the baseline.

latency varied between 800 ms and 1.0 s (Fig. 2). The slow wave was positive in the right hemispheres and negative in the left hemispheres and was synchronized in both hemispheres. A digital bandpass filter from 1 to 6 Hz was applied to all data to confirm the existence of the wave. The wave was observed in 14 sessions of the grand total 25 sessions among all subjects. This slow wave might be related to the semantic judgment.

Experiment II

The combinations of a noun and a three-syllable verb used in the experiment I might be not appropriate to keep the subjects' judgment with equal latency. Because there was a possibility that the subjects could judge the meaning of the whole sentences when listening to the first or second syllable of the three-syllable verbs.

Method Three right-handed subjects participated in this experiment. The selected sentences, which were also a combination of a noun and a verb, could not be judged before hearing the second syllable of the verb. The nouns had two or three or four syllables. We made 184 pairs of congruous and incongruous sentences, which differed only in their second syllables of their verbs. The subjects were required to judge quickly whether the each verb was coherent to the noun when listening to the sentence. Each stimulus sentence consisted of a noun, followed by a 1.1-1.6s blank, a two-syllable verb (650 ms), a 1.5s blank, a tone cue (500 Hz, 40 ms), and a 1.2s blank (Fig. 3). All the nouns and the verbs were synthesized by a computer software. The original duration of the synthesized verbs was 400 ms. Two syllables of the verbs were separated from each other, and the second syllable was presented 400 ms after the onset of the first syllable. The trigger point was provided at the onset of the second syllable of the verbs. The recording duration was 3.6 s, including 800 ms for the pre-trigger period. The subjects were

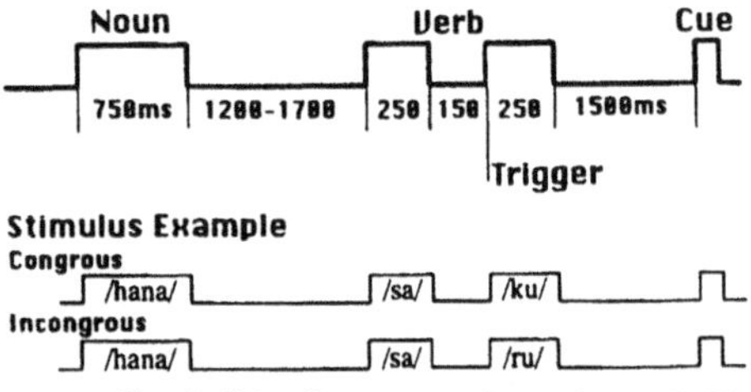

Fig. 3. Stimuli sequence of experiment II. The subjects could not be judged before hearing the second syllable of the verb.

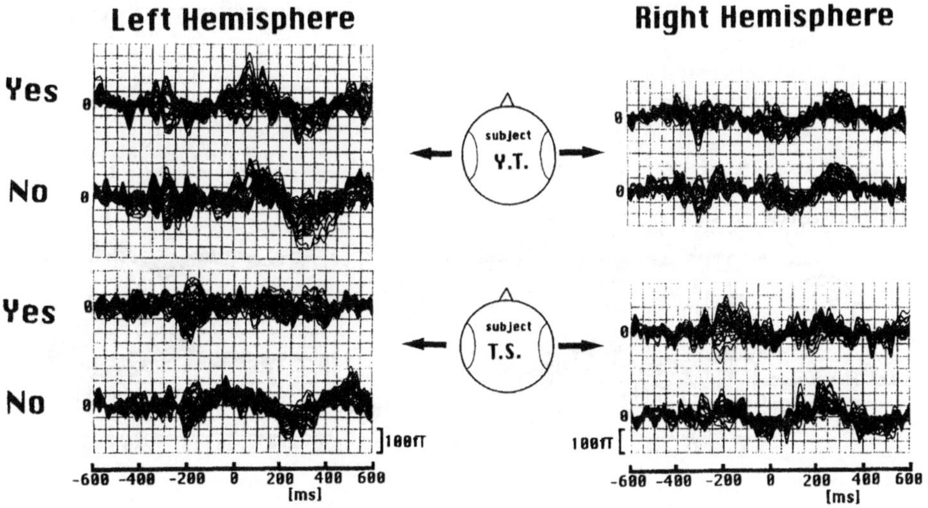

Fig. 4. Superimposed 37 waveforms (1-40 Hz) recorded from two hemispheres of the two subjects, where the responses were selectively averaged by the subjects' judgment: semantically congruous (Yes) and incongruous (No). Three evoked magnetic fields by the first syllable of the verbs were observed, and they were identified as P1m, N1m, and P2m based on the latencies from the onset of the verb. Occurrence of those which were evoked by the second syllable of the verbs was not clear. All the eight recorded responses of subject Y.T. and one of No (incongruous) magnetic fields of the subject T.S. showed slow waves at latencies between 250 and 300 ms.

instructed to follow the same procedures to Experiment I. They were also required not to press the button when they were not convinced whether the sentence was congruous or not.

Results Each subject participated in the experiment twice. Figure 4 shows the selectively averaged magnetic fields from the left and right hemispheres by the subject's judgment of congruous (Yes) and incongruous (No). Three evoked field responses to the first syllable of the verbs were observed , and they were identified as P1m, N1m, and P2m based on their latencies from the onset of the verb. However, the occurrence of those which were evoked by the second syllable of the verbs was not clear. All the eight recorded responses of the subject Y.T. showed slow waves at latencies between 250 and 300 ms. On the other hand, the subject F.T.'s magnetic fields did not show obvious slow waves. The subject T.S. showed the slow wave in one out of two incongruous responses. The slow wave was larger in the semantically incongruous responses than congruous ones in some sessions.

By high-pass filtering above 3 Hz and readjusting their baseline with 400 ms duration before the onset of the verbs, activities corresponding to the second syllables of the verbs were also clearly observed. Hence, the observed waveform of the magnetic fields could be interpreted as the superimposition of the responses of the P1m, the N1m, and the P2m from the auditory cortex and other event related fields of the slow wave. To isolate the slow wave, the fields data were low-pass filtered below 3 Hz (Fig. 5). The resulting isofield map at latency of around 300 ms showed the magnetic flux out of the skull in the right hemisphere, and magnetic flux into the skull in the left hemisphere. The maximum flux-in and -out points were located at the anterior temporal region of the subject.

DISCUSSION

In the Wernicke-Geschwind model for language, the Wernicke's area is located in the Broadmann's area 22 [5], but we could not find any obvious activity in the posterior part of the measured region. The activity might take place in the gyrus or may not be synchronized with the stimuli. The slow wave observed in the experiment I and II seem to be the same activity because of the same semantic task. From the latency (400-600 ms) from the onset of the verb and the task in the experimental I, the slow wave might be N400, which appears in semantically incongruous stimuli. On the other hand, from the latency (250-300 ms) from the second syllable of the verbs in the experiment II,

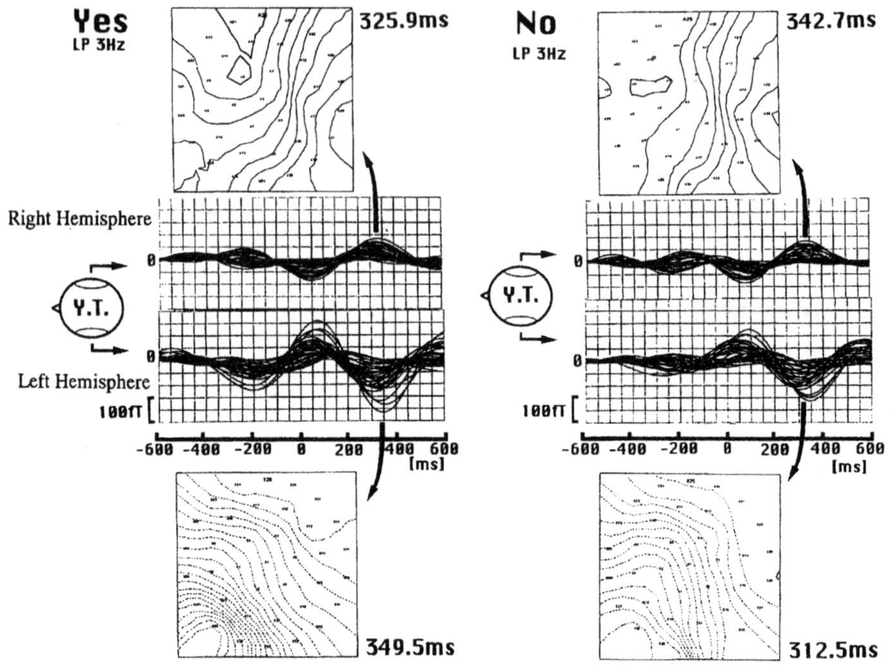

Fig. 5. The slow wave low-pass filtered (3 Hz) from the recorded magnetic fields of the subject Y.T.. The resulting isofield map at latency of around 300 ms showed the magnetic flux out of the skull in the right hemisphere, and the magnetic flux into the skull in the left hemisphere. The maximum flux-in and -out points were located at the anterior temporal region of the subject.

the slow wave might be P300, which reflects the perception and usually is observed in oddball tasks. The iso-field maps, however, indicate that the source of the slow wave is deep within the skull, and it may be located near in the hippocampus or the cingulate gyrus, whose function is related to memory.

CONCLUSION

A slow wave was observed in the two semantic experiments. It may be related with higher auditory process from its latency. However, the slow wave was observed in about half of the sessions of the experiments. Further measurements and the dipole source estimation are required to clarify the characteristic of the slow wave.

References:

[1] Geshwind, N., "Specializations of the human brain," Scientific American, 1979, **241**: 158-168.

[2] Petersen, S.E., Fox, P.T., Posner, M.I. Mintun, M., and Raichle, M.E., "Positron emission tomographic studies of the cortical anatomy of single-word processing," Nature, 1988, **331**: 585-589.

[3] Démonet, J.F., Chollet, F., Ramsay, S., Cardebat, D., Nespoulous, J.L., et al., "The anatomy of phonological and semantic processing in normal subjects," Brain, 1992, **115**: 1753-1768.

[4] Pascual-leone, A., Brasil-neto, J.O., Valls-solé, J., Cohen L.G., and Hallett M., "Simple reaction time to focal transcranial magnetic stimulation. comparison with reaction time to acoustic, visual and somatosensory," Brain, 1992, **115**:109-122.

[5] Mayeux, R., and Kandel, E.R. "Disorders of language: the aphasias," In: Kandel, ER., Schwartz, J.H., and Jessell, T.M. Principles of neural science, New York, Elsevier,1991.

Neuromagnetic Correlates of Endogenous Auditory Sensations

Hoke, E.S., Hoke, M. and Ross, B.

Institute of Experimental Audiology, University of Münster, Münster, Germany

Introduction

Endogenous auditory sensations occur without *simultaneous* reception of an external acoustic stimulus. They can occur under physiological conditions, but can also be associated with pathological processes. A commonly known sensation of the second kind is tinnitus, or ringing in the ear, a frequent symptom which is often associated with several hearing disorders. A sensation of the first kind is the auditory after-image. Contrary to the vision system in which after-images have been known for a long time, auditory stimulation was, in general, not assumed to lead to clear-cut after-effects. An exception is the exposure to acoustic overstimulation which has been known to lead to a temporary threshold shift which is sometimes accompanied by an auditory sensation, tinnitus.

It was in 1964 when Zwicker, for the first time, described a clear-cut and reproducible auditory after-effect that followed a specific acoustic stimulation [1]. This after-effect, which he termed "negative afterimage in hearing" and which was later coined "Zwicker tone" in honour of its discoverer [2], is an endogenous auditory sensation which occurs after the offset of a band-suppressed noise that was presented for some seconds. The sensation of this monaural phenomenon is most similar to that of a pure tone with a frequency corresponding to the centre frequency of the gap and with an equivalent level of 10-15 dB above the auditory threshold. The sensation decays gradually, and it may last as long as 10 s, depending on how long the evoking noise was presented.

While the physiological basis of the after-image in vision has been uncovered some decades ago, the search for a neurophysiological correlate of the auditory afterimage has not been successful so far.

This Zwicker tone study [3] was undertaken in the search for a human model of tinnitus, because there are many similarities between the two auditory sensations:

- Beats cannot be generated by interference with an external tone [4].
- Both sensations do not follow the masking laws which exist for real sound signals (e.g., contralateral masking).
- No binaural fusion of the Zwicker tones occurs if different exciting signals are presented to the two ears.
- The loudness of both sensations is only slightly above the subjective threshold.

These many similarities led us to look for a neurophysiological correlate of the Zwicker tone, hoping that this might be helpful in our search for neurophysiological correlates of tinnitus. Studies of the neuromagnetic activity of the auditory cortex were expected to be the most promising ones because of the power they have demonstrated in localizing neurophysiological correlates of other purely subjective auditory phenomena [5].

Methods

Four female and seven male subjects, all right-handed and at the age of 21 to 30 years (25 ± 3 years, median 25 years), participated in this study. All subjects had a normal audiological status and were able to perceive the Zwicker tone.

Three different stimulation paradigms were employed successively in separarte experimental blocks of 500 s duration each. In each paradigm, noise pulses were monaurally presented with a sensation level of 45 dB. The duration of the bursts was 3 – 5 s (randomly distributed), and pauses of 1 s were interposed between successive bursts. The series of three blocks was repeated four times for each subject.

- The first paradigm was the *notched noise condition (NN)* in which bursts of notched noise were presented. The cut-off frequencies were 3.2 and 5.2 kHz, and the minimum of the spectral level was at 4 kHz. The pauses of 1 s were silent. In this condition, a Zwicker tone was supposed to be heard.
- The second paradigm was the *white noise condition (WN)* in which bursts of white noise were presented. The pauses of 1 s were silent. In this condition, no Zwicker tone was supposed to be heard.
- The last paradigm was the *white noise with interposed tone condition (WNT)* in which also bursts of white noise were presented. During the pauses, however, a pure tone was presented with an intensity of 10 dB sensation level and a frequency of 4 kHz. In this condition, an auditory sensation similar to that of a Zwicker tone was supposed to be heard.

The MEG recordings were performed in a magnetically shielded room (Vacuumschmelze) with a 37-channel neuromagnetometer (MAGNES®, Biomagnetic Technologies). The sensor array was placed over the left temporal plane. During the experimental blocks of 500 s, magnetic data were continuously recorded with a bandwidth from DC to 100 Hz and a sampling frequency of 294 Hz.

For each stimulus condition, epochs of 2 s, beginning 500 ms before the offset of the noise pulses, were averaged after artifact rejection. The averaged data were digitally low-pass filtered at 20 Hz (second-order zero phase-shift Butterworth characteristics). In order to derive correlates of the Zwicker tone, difference waveforms of the averaged and filtered MEG signals were computed i) between the NN and the WN conditions, and, for comparison, ii) between the WNT and the WN conditions.

The next steps consisted in calculating rms waveforms for the original and for the difference waveforms, and generating isofield contour plots for the original and for the difference waveforms.

Finally, source analyses based on a single moving equivalent current dipole (ECD) model in a spherical volume conductor were applied for each sampling point of the averaged original and of the difference waveforms. Only dipole localizations with a goodness of fit > 0.95 were considered.

For the difference waveforms, also mean locations were calculated for that epoch during the pause in which a clear-cut dipolar field distribution was recognizable, the goodness of fit exceeded 0.95, and the dipole coordinates remained sufficiently stable.

Results

Fig. 1 shows in the upper row the grand means of the original averaged rms waveforms for the three conditions, and in the bottom row the grand means of the difference waveforms. It is obvious that the three grand means of the original waveforms look very similar.

The N_1m deflection is clearly expressed approximately 120 ms after the offset and, weaker, after the onset of the noise burst, with almost similar amplitude in the different conditions. During the pause, the niveau of the activity is elevated, with a greater elevation in the NN and the WNT conditions as compared to the WN condition. This activity during the pause is clearly expressed in the difference waveforms, in which identical contributions like the transient components have obviously been almost totally removed by the subtraction operation.

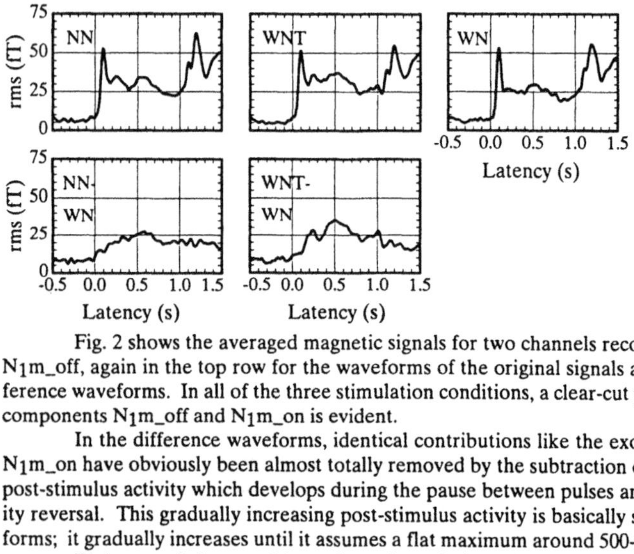

Fig. 1 Grand means of the original averaged rms waveforms for the three conditions, NN, WNT and WN, (upper row), and the grand means of the difference waveforms NN-WN and WNT-WN (bottom row). (After [3])

Fig. 2 shows the averaged magnetic signals for two channels recorded over the field extrema of the N_1m_off, again in the top row for the waveforms of the original signals and, in the bottom row, for the difference waveforms. In all of the three stimulation conditions, a clear-cut polarity reversal for the exogenous components N_1m_off and N_1m_on is evident.

In the difference waveforms, identical contributions like the exogenous components N_1m_off and N_1m_on have obviously been almost totally removed by the subtraction operation. However, there exists a post-stimulus activity which develops during the pause between pulses and which exhibits a clear-cut polarity reversal. This gradually increasing post-stimulus activity is basically similar in the two difference waveforms; it gradually increases until it assumes a flat maximum around 500-600 ms and then decreases.

Eight out of eleven subjects showed an obvious dipolar field distribution in the difference waveforms for a certain time during the pause. In the remaining three subjects, we have not been able to find a distinct dipolar field distribution.

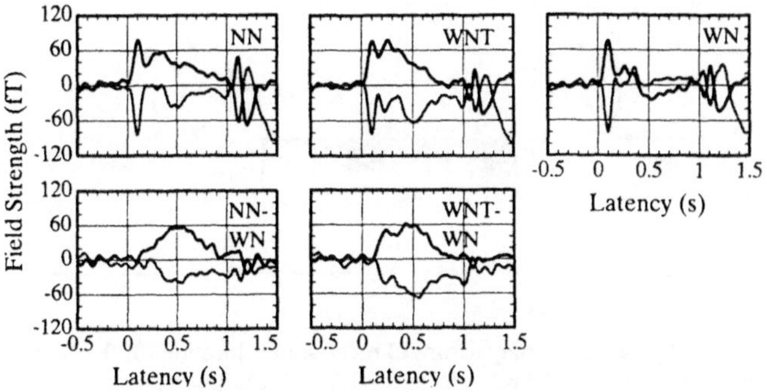

Fig. 2 Averaged magnetic signals for two channels recorded over the field extrema of the N1m_off, again in the top row for the waveforms of the original signals, NN, WNT and WN, and, in the bottom row, for the difference waveforms, NN-WN and WNT-WN. (After [3])

Fig. 3 represents, for three subjects (p3, S14, S15), a comparison of the magnetic field distribution patterns of both the original and the difference waveforms at selected instants of time.

The first column shows the field pattern 100 ms prior to the offset of the burst. No regular pattern can be observed. The second column shows the field pattern around the maximum of the N1m after the offset of the noise burst. On account of the similarity of this response component in the waveforms of the three stimulation conditions, averages were calculated for this component. The evoked field shows a distinct dipolar distribution.

The next two columns show the field patterns for the difference waveforms NN-WN (third column) and WNT-WN (fourth column), 500 ms after the offset of the noise. A distinct dipolar field distribution can be observed in both cases. The fifth column, finally, shows the field patterns for the white noise waveforms, again 500 ms after noise offset. In some cases a dipolar pattern of different location and different orientation is slightly expressed, whereas in other cases only a random pattern exists.

It becomes obvious that the distribution patterns of the post-stimulus activity resemble those of the transient responses; but the dipole orientation of the post-stimulus activity appears to be slightly more inclined than that of the transient responses.

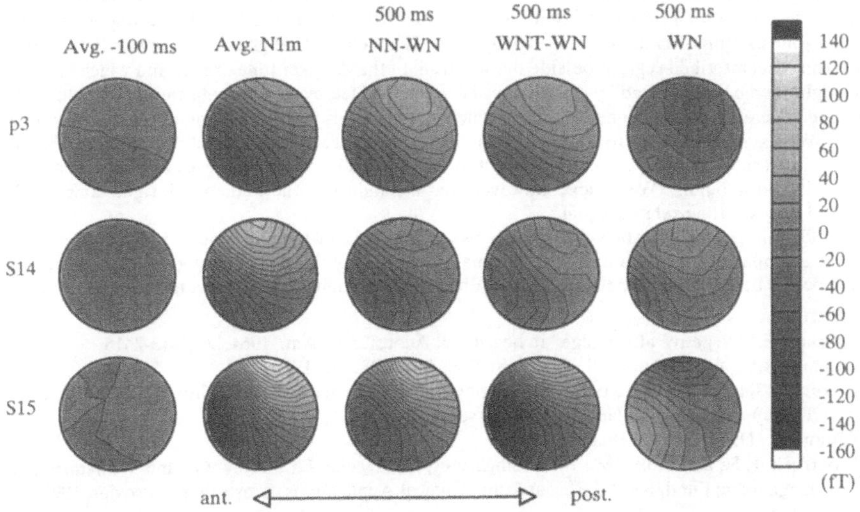

Fig. 3 Comparison of field distribution patterns of the original and the difference waveforms. (After [3])

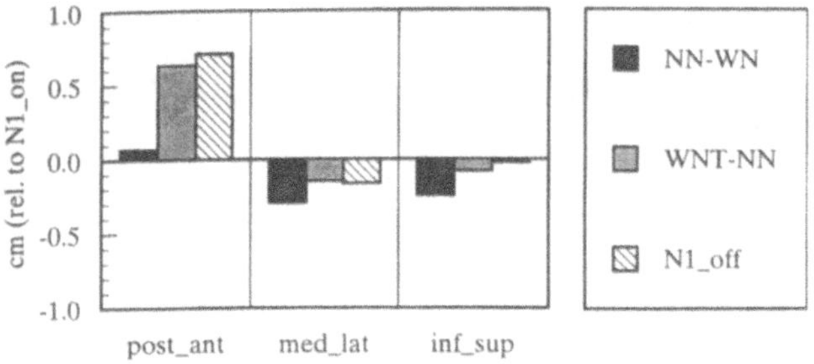

Fig. 4 Comparison of mean source locations of the N1m_off, of the poststimulus activity during the pause in the NN-WN and in the WNT-WN conditions. (After [3])

Fig. 4 shows comparisons of mean source locations of the N1m_off, and of the poststimulus activity during the pause in the NN-WN and WNT-WN conditions.

Source locations were normalized relative to that of the N1m_on, and the analysis was restricted to those eight out of eleven subjects in which the prerequisites as mentioned above were fulfilled. It is obvious that the source locations of the different activities do not differ significantly, especially in the medial-lateral and inferior-superior direction.

Discussion

A neuromagnetic study in normal-hearing subjects was performed in order to look for a neurophysiological correlate of the Zwicker tone in the auditory cortex. With a stimulation paradigm especially designed for this study we have been able to derive post-stimulus activity which appears to be related to the Zwicker tone and which originates in the supratemporal auditory cortex. It is a longer lasting neuromagnetic activity which is distinctly weaker than that of the onset and offset responses to the tone burst, and which shows a clear-cut dipolar field distribution.

It appears that the assumed Zwicker tone-related activity has certain similarities with that of the sustained response. The hypothesis is put forward that during the sensation of the Zwicker tone a process takes place in the auditory cortex which is similar to that underlying the sustained response, and which gives rise to the Zwicker tone sensation. However, while the sustained response is due to neural activity generated by a *simultaneously* presented external acoustic stimulus, the assumed sustained response is due to a temporary absolute or relative reduction of neural activity originating from those regions in which the Zwicker tone exciting stimulus caused an adaptation. The relatively augmented neural activity which originates from circumscribed regions outside the spectrum of the Zwicker tone exciter and which lasts for some time might then be interpreted by higher neural levels as elicited by an external acoustic stimulus. This may cause the generation of a sustained response which, in turn, leads to the sensation of the Zwicker tone.

We have failed so far to confirm for the Zwicker tone-related sustained response a source location which is, in all subjects, systematically different to those of the transient responses to the onset and offset of the evoking noise bursts. We believe, however, that this failure is due to the weak signal amplitude of the supposed Zwicker tone-related signal.

Certain psychoacoustic similarities between the Zwicker tone and tonal tinnitus rise the question whether a similar effect might exist also in tonal tinnitus. A preliminary study with a comparable experimental design has shown that neural activity is obviously recordable which might be related to tinnitus.

References

[1] Zwicker, E. "Negative afterimage" in hearing, J. Acoust. Soc. Am., 1964, 36: 2413-2415
[2] Lummis, R.C., and Guttman, N. J. Acoust. Soc. Am., 1972, 51: 1930-1944
[3] Hoke, E.S., Hoke, M., and Ross, B. Neurophysiological correlate of the auditory after-image ("Zwicker Tone"), Audiology & Neuro-Otology (submitted)
[4] Krump, G. Dissertation, München, 1993
[5] Proefrock, E.S., and Hoke, M. In: Baumgartner, C., Deecke, L., Stroink, G., and Williamson, S.J. Biomagnetism: Fundamental Research and Clinical Applications, Amsterdam, Elsevier, 1995, 234

Acknowledgements

This work has been supported by grants from the Deutsche Forschungsgemeinschaft (Ho 847/6) and is part of a Ph.D. thesis.

Spatial and Temporal Properties of the Slow Components of the Contingent Magnetic Variation in a Warned Choice Reaction Time Task

Hultin, L.[1], Högstedt, P.[1], Pizzella, V.[2], Romani, G.L.[3], Rossini, P.[4,5] and Tecchio, F.[2]

[1]*Department of Applied Electronics, Chalmers University of Technology, Gothenburg, Sweden;* [2]*Istituto di Elettronica Stato Solido - C.N.R., Roma;* [3]*Istituto di Fisica Medica, Universitá G.D'Annunzio, Chieti, Italy;* [4]*Dipartimento di Neurologia, Ospedale "Fatabenefratelli", Isola Tiberina, Roma;* [5]*I.R.C.C.S. "S. Lucia," Via Ardeatina, Roma, Italy*

Introduction

Electrically recorded contingent negative variation, eCNV, has been reported as a two-component wave [1]. The early component, the O-wave (orientation-wave), is related to the processing of the warning stimulus (S1) and is sensitive to information that S1 provides with respect to the imperative stimuli (S2). The late component, the E-wave (expectation-wave), is considered to be a manifestation of at least two major processes: motor preparation and stimulus anticipation. The electrical O-wave has a frontocentral distribution, with right hemisphere preponderance, while the E-wave is located precentrally, with a preponderance contralateral to the side of the movement. The CNV is a cognitive process where the neurological responses are constantly changing during an experimental session, due to alterations in the state of alertness of the subject, the process of learning and the reorganization of mental strategy to perform the task. During an initial learning period the amplitudes of the eCNV are decreasing and the potential distribution changes. Recordings of Contingent Magnetic Variation (CMV), the neuromagnetic counterpart of the eCNV, have been reported by several authors [e.g. 2, 3]. The scope of this study was to analyze the presence, distribution, and origin of slow components of CMV in a warned choice reaction time task after the initial learning period has been performed.

Methods

Neuromagnetic signals from 12 healthy subjects were recorded using the 28-channel MEG instrument installed at IESS, Rome [4]. The paradigm used was a modified version of the classical CNV paradigm, first proposed by Walter et al. [5], and has previously been reported by Hultin et al. [2, 3]. A low-intensity optical pulse of short duration was used as warning signal. A 70-ms tone of 400 or 1400 Hz, presented to the left ear, was used as S2 (imperative stimulus). S2 followed S1 by 1500 ms and the probability for the high-pitch tone was 0.66. The subject was asked to press a nonmagnetic dummy push button with the right index finger immediately on hearing the high-pitch tone, but not to push when the low-pitch tone was heard. The EEG was recorded through Ag-AgCl electrodes at Cz, Oz, T3, and T4. Linked earlobes were used as reference, left mastoid process as ground. Electro-oculogram, EOG, was recorded from medial supra-orbital and lateral infra-orbital electrodes at the left eye. EMG signals, from the first dorsal interosseous muscle on the right hand, were rectified and then recorded. All signals were bandpass-filtered (0.048 - 64Hz) and digitized with a sampling rate of 250 Hz. Six different probe positions were obtained for each subject and seventy epochs were recorded for each position. Epochs were rejected if they contained eye artifacts, magnetic or electric artifacts, EMG activity before S2, or if the reaction time exceeded mean reaction time plus 2 SD. The training session that preceded the recordings consisted of 45 epochs.

An ellipsoidal model of the head was used for field-maps and source localization [3]. Coordinates refer to a Cartesian system corresponding to the axes of the ellipsoidal model. The y-axis is parallel, and one centimeter superior, to a line between the inion and the nasion. The positive z-axis passes through the skull at the vertex, and the positive x-axis passes through the skull at the midpoint on a line from the inion to the nasion, approximately 2 - 3 cm superior of the right pre-auricular point. Sources were modeled as two separate equivalent current dipoles. Start values for the inverse solution were given manually under visual guidance of simulated forward solution. The strength and localization of the two current dipoles were determined as the best fit to the observed data using an iterative least-mean-squares fitting Marquardt algorithm. The extent to which the fitted dipoles provided an adequate model was expressed as the degree to which the calculated field, Bc, could account for the observed data, Bo, and is referred to here as the goodness of fit, $g = 100\ (1 - \sum (B_o - B_c)^2 / \sum B_o^2)$.

Results

Large interindividual differences are seen in the morphology of the CMV, as well as in the eCNV. According to the classification proposed by Tecce [6] the morphology of the CNV can be differentiated on the basis of the rise time of the ascending limb. Furthermore, large differences are seen in the early CNV complex that follows immediately on the warning stimuli. The large interindividual differences, the complex neurological process, and the limited number of subjects in this study thus reduce the possibilities of reaching general conclusions in terms of dipole localizations. The spatial and temporal morphology of the CMV obtained from the recorded subjects reveal, nevertheless, several common features. Based on the spatial and temporal properties of the recorded neuromagnetic signals, four slow components could be separated from the CMV (Fig. 1). The first three components are here considered to be subcomponents of the O-wave, and are thus labeled O1, O2 and O3. The late component is considered to be related to the expectation wave and is thus labeled E.

The first slow component, O1, followed immediately on the visual evoked response. The component peaked about 300 ms after S1 and lasted approximately 100 ms. The electric potentials showed a slight right-sided lateralization and the amplitudes of the occipital potentials were in all cases stronger than, or equal to, the amplitude at vertex (Fig. 1). The spatial morphology of magnetic flux and the obtained source localization suggested occipital activity as well as bilateral frontocentral activity (Fig 2, 3). Visual inspection and dipole localization indicated further that the activity in the sensorimotor cortex is, in most subjects, active as early as 300 ms after the warning stimuli. The second early subcomponent, O2, peaked at 460-520 ms after S1 and had a duration of up to 150 ms. During this time period a two equivalent current dipole model failed to achieve proper source localization, indicating a complex neural activity (Fig. 3). Visual inspection of the obtained field-maps indicates activity in the posterior part of the right parietal lobe, as well as bilateral frontocentral activity. The third early component, O3, peaked at 620-720 ms after S1 and was present up to 400 ms before S2. The corresponding electrical component was not clearly distinguishable in all subjects. The electrical potentials were strongest at vertex and a left-sided lateralization was seen in most subjects. Source localization suggests, in several subjects, activity in the sensorimotor areas.

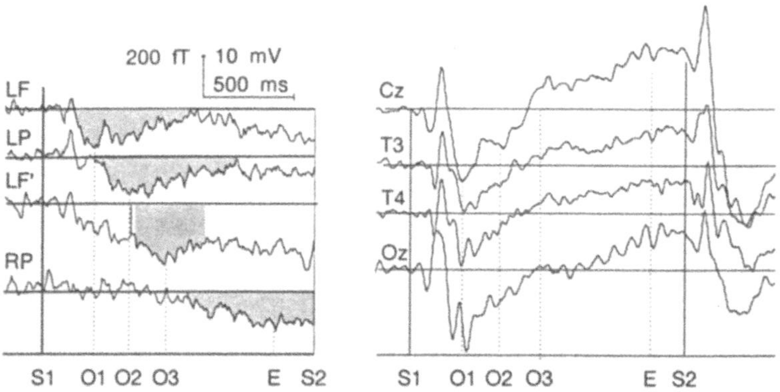

Fig. 1: One example of CMV (left) and eCNV (right), recorded in one subject (FT). The eCNV is the grand average of the electric signals recorded in six different probe positions. Negativity at the active electrode, and outgoing magnetic flux, are seen as an upward deflection. The magnetic signals are recorded over the left frontal (LF and LF'), left parietal (LP), and right parietal (RP) hemispheres. Based on the spatial and temporal properties of the recorded neuromagnetic signals four slow components (O1, O2 ,O3 , E) could be separated in the CMV.

The late component, related to the expectation wave and thus labeled E, started 500 - 700 ms before S2 and ended abruptly at the delivery of the imperative stimuli. Field-maps as well as the obtained dipole localization showed a close relation between O3 and the late component. The obtained results suggested that the dominant sources of the late component are symmetrically located in the left and right premotor areas [3].

Discussion

CNV reflects a complex cognitive process where the magnetic fields span large parts of the scalp. Thus CMV has to be recorded over the entire head simultaneously or at multiple probe positions. It is known that the amplitude of the eCNV varies during a sequence of recordings [7]. The ambiguity resulting from measuring the non-stationary CNV process at multiple probe positions must be carefully taken into consideration in the experimental setup, the analysis of the data and the interpretation of the results. The morphological structures of field-maps are only partly degraded due to this variability. The inverse solution, on the other hand, might be affected by a significant error. For future studies, recently developed MEG helmets with the ability to record magnetic fields from the entire scalp simultaneously will partly solve this problem.

As a rule of thumb, the time constant should be at least three times the interstimulus interval (1.5 seconds) to avoid distortion of the data [8]. Due to instrumental setup and noise in the very low frequency band, the measurements were performed using a time constant of 3.3 seconds. It cannot, therefore, be excluded that the time constant we used has influenced the obtained data.

The ability to separate the late component of the CNV from the early component is highly dependent on the interval between the warning and the imperative stimuli. At an inter-stimulus interval of 1.5 seconds, which has been used in this study, there might be a significant contribution from O3 at the time of the E-wave. Although the morphology of the CMV suggests that the magnetic components are more separated that their electric counterparts, and that an ISI of 1.5 seconds would be satisfactory in most cases, a longer ISI would be preferable.

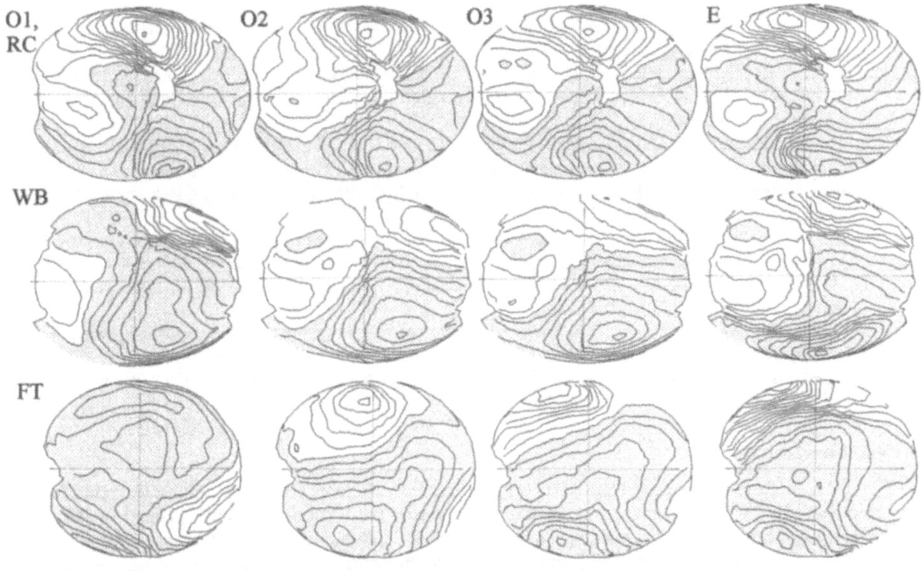

Fig. 2: Field-maps, based on the CMV, from three subjects (RC, WB and FT) at four time periods (O1, O2, O3 and E) corresponding to the slow components of the CMV. The field-maps are seen from the top with the nose directed to the left; shaded areas represent ingoing magnetic flux.

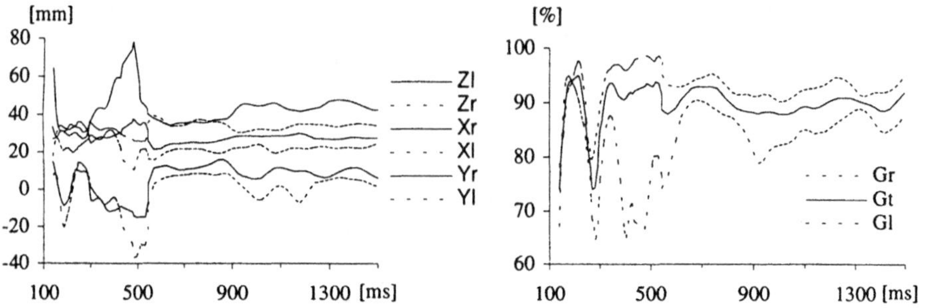

Fig. 3 Localization for least-mean-squares fitted dipoles for one subject, RC, as a function of time (left). The contralateral side and the ipsilateral side were fitted simultaneously. Position coordinates are referred to a Cartesian system corresponding to the ellipsoidal model used. Goodness of fit, G, refers to the degree to which the calculated field could account for the observed data, and is calculated for sensor positions on the contralateral side, Gl, and ipsilateral side, Gr, as well as for all sensor positions, Gt (right). The slow components of RC peaked at 320 ms (O1), 480 ms (O2), 680ms (O3) and 1400 ms (E).

References

[1] Rohrbaugh, J.W., Syndulko, K., and Lindsley, D.B. Brain wave components of the contingent negative variation in humans, Science, 1976, 191: 1055-1057.

[2] Hultin, L., Rossini, P., Romani, G.L., Högstedt, P., Stanzione, P., Tecchio, F., and Pizzella, V. A method for investigation of magnetic fields related to CNV in normal subjects and Parkinson patients, In: Deecke, L., Baumgartner, C., Stroink, G., Williamson, S.J. Biomagnetism: Fundamental research and clinical applications, Elsevier, Amsterdam. 1995.

[3] Hultin, L., P. Rossini, G. L. Romani, P. Högstedt, F. Tecchio and V. Pizzella. Neuromagnetic localization of the late component of the contingent negative variation, Electroencephalogr Clin Neurophysiol, Accepted for publication, 1996.

[4] Torrioli, G., Casciardi, S., Del Gratta, C., Foglietti, V., Gallagher, W.J., Ketchen, M.B., Kleinsasser, A.W., Pasquarelli, A., Pizzella, V., Romani, G.L., Sandstrom, R.L. (1992) Twenty-eight-channel hybrid neuromagnetometer. In: Hoke, M., Erné, S.N., Okada, Y.C., Romani, G.L (Eds.), Biomagnetism: Clinical Aspects. Elsevier, Amsterdam.

[5] Walter, G., Cooper, R., Aldrige, V. J., and McCallum, W. Contingent negative variation: an electric sign of sensorimotor association and expectancy in the human brain, Nature, 1964, 203: 380-384.

[6] Tecce, J.J. Contingent Negative variation and individual differences: a new approach in brain research, Arch Gen Psychiatry, 1971, 24: 1-16.

[7] Peters, J.F., Billinger, T.W., Knott, J.R. Event related potentials of brain (CNV and P300) in a paired associate learning paradigm, Psychophysiology, 1977, 14: 579-585.

[8] Cooper, R., Osselton, and J.W., Shaw, J.C. EEG Technology, London, Butterworth & Co., 1980.

Acknowledgments

This work was supported in part by the Swedish National Board of Industrial and Technical Development (NUTEK) under grant No. 623-90-0183, and by the National Research Council of Italy (CNR) under the Progetto Finalizzato "Superconductive and Cryogenic Technologies." The authors want to thank Dr. Alberto Pasquarelli for adapting the MEG system at CNR to the experimental setup. We also want to express our gratitude to Professor Roland Kadefors for valuable comments on the manuscript.

Dependence of SEFs on the Interstimulus Interval: A Model for SEF Generation Mechanism at the Primary Sensorimotor Cortex

Huttunen, J.[1], Wikström, H.[1], Korvenoja, A.[2], Aronen, H.[2] and Ilmoniemi, R.J.[1]

[1]BioMag Laboratory, Medical Engineering Centre; [2]Department of Radiology, Helsinki University Central Hospital, Helsinki, Finland

Introduction

Somatosensory evoked fields (SEFs) following median nerve stimulation can be recorded from the contralateral primary sensorimotor cortex (SMI), from the second somatosensory cortex (SII) bilaterally and from contralateral posterior parietal cortex (PPC) [1,2]. At SMI, the response can be divided into distinct N20m, P35m and P60m deflections, while SII responses are biphasic with peaks at 80–100 ms and about 150 ms. The latency of the PPC response is usually 80–100 ms.

It is generally agreed upon that the first cortical response, the N20m, represents summated excitatory postsynaptic potentials (EPSPs) in area 3b pyramidal neurones as these are excited at the basal dendrites by thalamocortical axons. However, the neurophysiological mechanisms underlying the later deflections are completely unknown. A better understanding of these mechanisms would obviously improve meaningful interpretations of SEF changes under various physiological or pathophysiological conditions.

The present study reports dependence of the SEFs on the interstimulus interval (ISI) in a population of healthy subjects. Based on remarkable similarities between the ISI dependence of SEFs and the ISI dependence of intracellularly recorded evoked potentials in sensorimotor cortical neurones, as reported in the literature, we present a tentative model for the generation mechanism of the SEFs from SMI.

Methods

A detailed description of the methods will be presented elsewhere [3].

SEFs were recorded in 9 healthy volunteers (20–48 years, 4 women) with a 122-channel planar gradiometer array (Neuromag Ltd.).

The right median nerve was stimulated at the wrist with 0.2-ms constant-current pulses above motor threshold. 100—200 responses were averaged on-line at least twice with each of the ISIs of 0.15, 0.3, 1, 3 and 5 s, which were presented in random order. The subjects watched a movie during the recording with no task related to stimulation.

The signals were band-pass filtered between 0.03 Hz and 310 Hz, digitized at 940 Hz and averaged on-line. Automatic rejection procedures were used to exclude blink and eye-movement artefacts.

A single equivalent current dipole (ECD) was used to model local cortical activity at the peaks of the identified deflections from SMI, SII and PPC. The spherical head model was obtained individually from each subject's magnetic resonance images. The head coordinate system was defined with a digitizer; the head and magnetometer coordinates were aligned with an automatic process defining the head position within the magnetometer dewar. ECDs explaining more than 80% of the field variance in the subgroup of channels (30—40) used for ECD fitting were accepted for analysis.

Results

With long ISIs, the response from the left (contralateral) SMI consisted of clear N20m, P35m and P60m peaks with an intervening 'negative'-going N45m, which usually did not cross the baseline at ISIs over 1 s. The field patterns at the peaks of these deflections were dipolar, the corresponding ECDs pointing in the anterior direction for N20m and in the posterior direction for P35m and P60m. The ECD locations agreed well with the hand area of SMI. The ECDs of the individual peaks were located close to each other (Table 1), P35m, however, being slightly more medially located than N20m in most subjects.

A PPC response could be identified in 7 of the 9 subjects; it is characterized by its more posterior location compared with the earlier peaks from the SMI region. A contralateral SII response was observed in all subjects and an ipsilateral in 7. The SII responses were small in 4 subjects, the corresponding dipole moments being below 10 nAm.

Table 1. Mean peak latencies and ECD coordinates for the different SEF deflections (standard deviations in parentheses). The ISI was 5 s with the exception of N45m, for which the ISI was 0.15 s. The *x*-axis passed from the left to the right preauricular point; the positive *y*-axis through the nasion; positive *z* was thus upwards. PPC = posterior parietal cortex, SIIc and SIIi = contra- and ipsilateral secondary somatosensory cortices, respectively.

Deflection	Latency/ms	*x*/mm	*y*/mm	*z*/mm
N20m	21 (1.6)	-37 (2.6)	12 (7.1)	94 (7.0)
P35m	35 (3.7)	-34 (6.2)	13 (7.3)	99 (9.1)
N45m	46 (8.2)	-37 (6.0)	16 (9.8)	95 (8.9)
P60m	69 (18.3)	-35 (5.4)	14 (7.4)	100 (8.6)
PPC	88 (19.0)	-35 (6.8)	1 (9.1)	106 (10.1)
SIIc	104 (9.5)	-29 (35.4)	19 (13.0)	68 (5.4)
SIIi	110 (7.2)	38 (11.6)	21 (20.0)	67 (6.8)

With shortening ISI, N20m remained stable from 5 to 0.3 s, but was slightly attenuated at 0.15 s (Fig.1, Table 2). P35m and P60m were largest at the 5-s ISI and showed steady declines toward shorter ISIs, being nearly undetectable at the ISI of 0.15 s. We should like to emphasize that both were diminished between 5-s and 1-s ISIs, *i.e.*, at relatively long ISIs. With the redution of P35m and P60m, the N45m showed a gradual increase with shortening ISI. At 0.15 s the N45m dominated the SMI responses.

Both the PPC and the SII responses could only be seen at ISIs of 1 s or longer. On the average, these were largest at the 5-s ISI, declining rapidly toward shorter ISIs. The P35m, P60m, PPC and SII deflections were not clearly saturated at the 5-s ISI. The ECD locations did not depend on ISI.

Table 2. The average ECD strengths in nAm (standard deviations) of the SEF deflections at different ISIs. The F statistics were obtained from an analysis of variance for repeated measures.

Defletion	5 s	3 s	1 s	3 s	0.15 s	F(4,28)	p
N20m	15 (1)	15 (2)	18 (2)	13 (2)	9 (2)	5.8	0.002
P35m	28 (2)	27 (2)	19 (2)	9 (2)	6 (2)	48.0	0.000
N45m	10 (5)	12 (5)	10 (4)	19 (5)	19 (3)	2.7	0.05
P60m	33 (1)	28 (2)	21 (2)	9 (3)	9 (3)	16.8	0.000
PPC	17 (4)	9 (3)	4 (2)	-	-	8.2	0.000
SIIc	20 (6)	12 (4)	7 (4)	6 (3)	1 (1)	5.2	0.003

Discussion

Intracortical microelectrode studies, the anatomy of thalamocortical connections, previous SEP and SEF studies all indicate that the N20m deflection represents summated excitatory postsynaptic potentials (EPSPs) from layer IV or IIIb pyramidal neurones of Brodmann's area 3b [cf. 3]. No explicit models have been presented, however, to account for the physiological mechanisms underlying any of the later deflections from SMI, or of the PPC and SII responses.

Evoked intracellular inhibitory postsynaptic potentials (IPSPs) are known to be very sensitive to repetitive stimulation with short ISIs [*e.g.*, 4–6]. This decline of IPSPs occurs already with ISIs in the range of seconds [5]; stimulation with ISIs shorter than 0.2 s can completely abolish phasic IPSPs [4]. This ISI behaviour of IPSPs is strikingly similar to the ISI behaviour of P35m and P60m. Furthermore, the latencies of P35m and P60m are consistent with latencies of fast and slow IPSPs, respectively, in intracellular recordings. The ECD orientations of P35m and P60m are consistent with current flow from cortical surface toward cortical depth at area 3b. Such direction of current flow can be caused by active current sources near the cell somata of pyramidal neurones within the middle layers of area 3b, i.e., by soma-near inhibitory currents.

In intracellular recordings from the neurones at SMI, shortening the ISI of afferent stimulation to 0.1—0.2 s results in a strengthening of a secondary EPSP, which follows the primary EPSP by about 20 ms [*e.g.*, 6]. This secondary EPSP coincides with a prominent increase of the field potentials at ISI of 0.1–0.2 s, known as the 'augmenting' response. Such behaviour is similar to the increase of N45m SEF deflection with shortening of the ISI. The nearly identical ECD location of N45m with that of N20m and the orientation of the N45m ECD are consistent with EPSPs arising in much the same neuronal population that is initially excited during N20m.

In summary, the ISI behaviour of the SEFs from SMI suggests the following tentative model: N20m reflects primary EPSP, P35m fast IPSPs, N45m secondary EPSPs and P60m slow IPSPs in the pyramidal neurones of Brodmann's area 3b. While the ISI behaviour alone obviously does not provide direct evidence to support the model, pharmacological interventions, e.g., with agonists and antagonists of γ-amino-butyric acid, could be used to test the hypothesis in humans or in an animal model.

On the basis of the ISI behaviour, similar hypotheses appear more difficult for the long-latency PPC and SII responses.

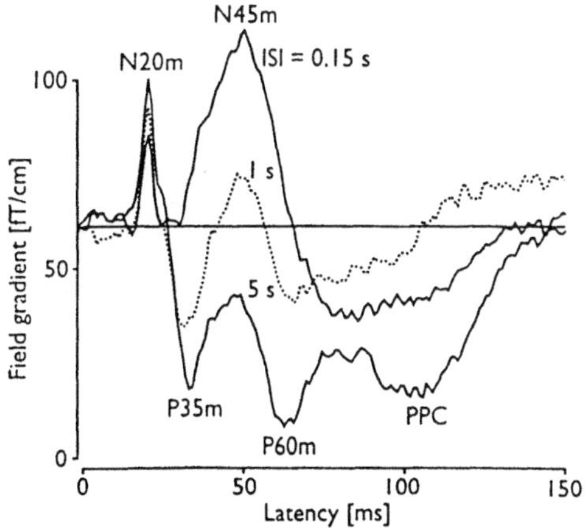

Figure 1. The longitudinal field gradients of SEFs from one channel overlying the left SMI at different ISIs. Note the marked reductions of P35m and P60m already between 5-s and 1-s ISI, and the enhancement of N45m toward the short ISIs.

References:

[1] Hari, R., Reinikainen, K., Kaukoranta, E., Hämäläinen, M., Ilmoniemi, R., Penttinen, A., Salminen, J., and Teszner, D. Somatosensory evoked cerebral magnetic fields from SI and SII in man. Electroenceph. clin. Neurophysiol., 1984, 57: 254—263.

[2] Forss, N., Hari, R., Salmelin, R., Ahonen, A., Hämäläinen, M., Kajola, M., Knuutila, M., and Simola, J. Activation of the human posterior parietal cortex by median nerve stimulation. Exp. Brain Res., 1994, 99: 309-315.
[3] Wikström, H., Huttunen , J., Korvenoja, A., Virtanen, J., Salonen, O., and Ilmoniemi, R.J. Effects of interstimulus interval on somatosensory evoked magnetic fields: a hypothesis concerning SEF generation at the primary sensorimotor cortex.. Electroenceph clin Neurophysiol., 1996, submitted for publication.
[4] Nacimiento, A.C., Lux, H.D., and Creutzfeldt, O.D. Postsynaptische Potentiale von Nervenzellen des motorischen Cortex nach elektrischer Reizung spezisfischer und unspezifischer Thalamuskerne. Pfluegers Archiv, 1964, 281: 152—169.
[5] Deisz, R.A., and Prince, D.A. Frequency-dependent depression of inhibition in guinea-pig neocortex *in vitro* by $GABA_B$ receptor feed-back on GABA release. J. Physiol., 1989, 412: 513—541.
[6] Klee, M. R. Different effects on the membrane potential of motor cortex units after thalamic and reticular stimulation, In: Purpura, D.P., and Yahr, M.D. The Thalamus, New York, Columbia University 1966, pp. 287—322.

Somatosensory Evoked Fields from SMI and SII During 'Interfering' Finger Movements and Tactile Stimulation

Huttunen, J.[1], Wikström, H.[1], Korvenoja, A.[2], Aronen, H.[2] and Ilmoniemi, R.J.[1]

[1]BioMag Laboratory, Medical Engineering Centre; [2]Department of Radiology, Helsinki University Central Hospital, Helsinki, Finland

Introdution

Ascending sensory impulses play an important role in providing feedback information necessary for the programming of movements. Therefore, it could be expected that sensory input to the cortex is facilitated from a moving limb, as demonstrated in monkeys [1].

During movements, however, the conscious perception of cutaneous stimuli is dampened. Furthermore, the early cortical somatosensory evoked potentials (SEPs) are reduced during various types of active or passive finger movements [e.g., 2]. The effect of passive movement indicates that the observed changes of SEPs could be due to occlusion or inhibition by reafferent impulses associated with the movements rather than due to the ongoing motor program. This view is supported by the fact that continuous cutaneous stimulation similarly suppresses the SEPs as does movement [3]. Hence, SEPs have shown no evidence for an enhancement of sensory input to the primary sensorimotor cortex (SMI), which would specifically correspond to enhanced input during active movements.

The somatosensory evoked fields (SEFs) bear the advantage over SEPs that the underlying generators can be better identified. Consequently, using SEFs it has been possible to realiably locate separate generators at SMI and at the second somatosensory cortex (SII) [4], and recently also at the posterior parietal cortex (PPC) [5].

The functional significance of SII has remained relatively poorly understood. However, some observations suggest that SII can play a role in performing active skilful movements requiring sensory feedback [6]. To date, the sensory responsiveness of SII neurones has been shown to be enhanced in monkeys when a behaviourally important motor response is expected [7].

In the present study, we wanted to determine whether tactile interfering stimulation and finger movements have differential effects on the sensory responsiveness of SMI and SII. To achieve this, we recorded SEFs to electrical median nerve stimulation in four conditions. In the first condition the subjects had no task related to the stimulation (resting condition). In the second condition the subjects performed active finger movements continuously during the nerve stimulation. In the third and fourth conditions interfering continuous tactile stimulation was applied either to the same or to another part of the body which was being stimulated electrically.

Methods

Somatosensory evoked fields (SEFs) to alternating bilateral median nerve stimulation at an interstimulus interval of 1.3 s (2.6 s per each nerve) were recorded with a 122-channel helmet-shaped planar gradiometer (Neuromag Ltd.) in 6 healthy volunteers. SEFs were obtained at least twice in a resting condition with the subject watching a film and being motionless, and in the following test conditions:

(1) Finger opposition: the subjects touched with the tip of the right thumb the tips of the other digits in a specified sequence,

(2) Tactile-palm: the right palm of the subject was brushed by the experimenter with a cotton pad,

(3) Tactile-face: the right lower face region of the subject was brushed in the same way.

Conditions 1–3 will be called interference conditions; also during these the subject watched a film. Because the interference condition was always applied to the right side, we could assess SEF modulations separately when the same part of the body (right median nerve) or the homologous site on the opposite side of the body (left median nerve) was stimulated electrically to elicit SEFs.

The recording pass-band was 0.03–310 Hz. Signals were digitized at 940 Hz, and averaged on-line after passing an automatic artefact rejection algorithm. The analysis time extended from 50 ms before stimulus to 350 ms beyond it.

The single-dipole model was used in the analysis. Dipole fitting to the SII responses was done after subtracting the influence of simultaneous activity at SMI from the responses with a signal-space projection method. The head was modelled with a sphere obtained from individual magnetic resonance images. The N20m, P35m and P60m SEFs from left and right SMI, a P100m generated near SMI but possibly getting contribution from the posterior parietal cortex [5], and the bilateral SII responses after stimulation on each side were analyzed. Statistical differences were evaluated with the Wilcoxon matched-pairs signed ranks test.

Results

As described in earlier studies, the SMI responses could be divided into distinct N20m, P30m and P60m peaks. The corresponding field patterns were dipolar with equivalent current dipoles (ECDs) located at the SMI hand area close to each other. The SII responses had 2 peaks at 65–100 ms and at 120–170 ms; the 1st of these was used in the analysis. P100m showed a dipolar field pattern with ECDs lying near SMI, which tended, however, to be 1–2 cm medial and posterior to the N20m ECD, thus possibly being generated in the posterior parietal cortex [5].

<u>1. Responses from the left SMI to right median nerve stimulation.</u>

In this case, both the electrical stimulation used to elicit SEFs and the interferences were applied to the contralateral (right) hand. Tactile interference at the palm and finger opposition had qualitatively similar effects (Fig. 1): both induced a marked reduction of P35m (mean ECD strengths 21 nAm at rest, 5 nAm during finger opposition and 8 nAm during tactile-palm, $p<0.05$ for both conditions), a smaller decrease of N20m (mean ECD strengths 14 nAm at rest, 11 nAm during finger opposition and 7 nAm during tactile-palm, $p<0.1$ for finger opposition and $p<0.05$ for tactile-palm), and no significant change in P60m. P100m was markedly reduced during finger opposition and tactile-palm (mean ECD strengths 18 nAm at rest, 9 nAm during finger opposition and 7 nAm during tactile-palm, $p<0.05$ for both). P100m was, however, difficult to define as a separate deflection in some subjects, especially during the interference conditions; the changes of P100m nevertheless reflect altered activity around 100 ms after stimulus.

Tactile interference to the right side of the face had no significant effect at the group level, although P35m was enhanced in 2 subjects. It is of interest to note that tactile interference to the palm appeared to reduce N20m slightly more than did finger opposition, although P35m was even more reduced during finger opposition than during tactile interference.

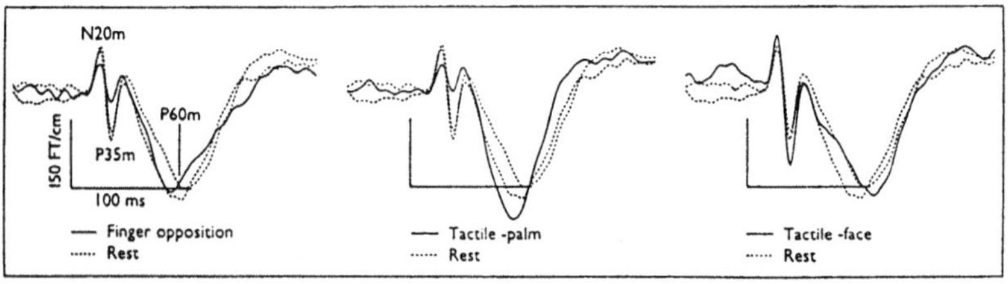

Figure 1. Longitudinal SEF gradients from the channel with maximum signals from the left SMI after right median nerve stimulation in one subject. Note reductions of N20m and especially P35m during finger opposition and tactile stimulation of the palm. In this subject, P35m was enhanced during tactile stimulation of the face.

2. Responses from the right SMI to left median nerve stimulation

In this situation, the interferences were applied to the ipsilateral hand, while SEFs were triggered by stimulation of the contralateral (left) median nerve. The SEFs were similarly affected by all 3 conditions: P35m was slightly increased (mean ECD strengths 19 nAm at rest, 27 nAm during finger opposition, 26 nAm during tactile-palm and 24 nAm during tactile-face, $p<0.05$ for all interferences), while N20m, P60m and P100m were not significantly affected.

3. Responses from SII after right median nerve stimulation

Here the interferences and electrical trigger stimuli were applied to the same hand. The SII responses from right and left SII regions (ipsi- and contralateral to median nerve stimulation) behaved in the same way in all interference conditions. During finger opposition, the SII responses were enhanced whereas both tactile interference of the palm and face markedly reduced them (Fig. 2; mean ECD strengths for the 80–100 ms deflection at the contralateral SII were 11 nAm at rest, 23 nAm during finger opposition, 7 nAm during tactile-palm and 5 nAm during tactile-face, $p<0.05$ for all interferences).

4. Responses from SII after left median nerve stimulation

When the left median nerve was stimulated, the interferences and median nerve stimulation were on the opposite sides. In this situation, tactile interference to the right palm and face again suppressed the SII responses bilaterally (mean ECD strengths at the contralateral SII were 15 nAm at rest, 1 nAm during tactile-palm and 2 nAm during tactile-face, $p<0.05$ for both interferences). However, right-sided finger opposition did not enhance the SII responses, but rather tended to reduce them (mean ECD strength 8 nAm, $p>0.1$). Responses from both left and right SII behaved again in the same way in all conditions.

In summary, at contralateral SMI, tactile interference applied to the palm and finger opposition movements markedly reduced P35m and P90m, caused a smaller reduction in N20m, and had no significant effect on P60m. Tactile interference to the right side of the face had no significant effects. At the ipsilateral SMI, tactile interference to the palm, to the face and finger opposition movements all caused a significant increase in P35m, while the other deflections were not affected. At SII, tactile interferences reduced the responses regardless of whether they were applied to the same or to the opposite side of the body compared with the electrical median nerve stimuli. However, finger opposition movements enhanced the SII responses, when performed with the same hand that was being stimulated to elicit SEFs, but not when performed with the opposite hand.

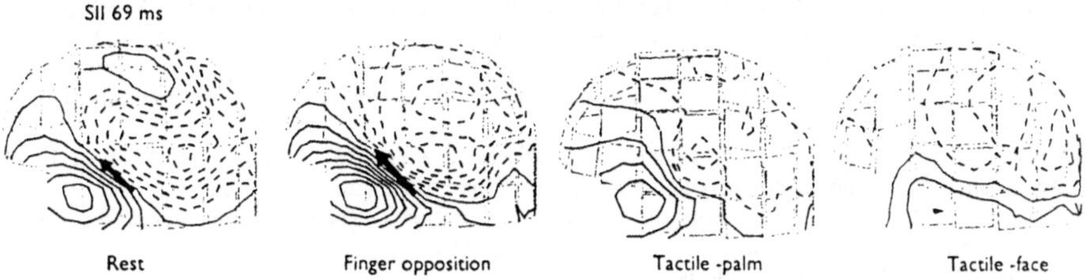

Figure 2. Isofield contour maps and surface projections of corresponding ECDs from the left SII region to right median nerve stimulation at rest, during right-hand finger opposition, tactile stimulation of the right palm and right face. Simultaneous activity at the left SMI has been subtracted. Note the increase of the response during finger opposition, but reductions during tactile interfering stimulations. In the tactile-face condition, no successful dipole fitting could be achieved in the SII region.

Discussion

Because of the similar modifications of SEFs from the contralateral SMI during finger opposition movements and tactile interference, the results do not disclose a modification of afferent input to SMI that would be specific to movement and could be interpreted as enhancement of sensory transmission from the moving limb to the SMI cortex [1]. This conclusion is in agreement with recent reports also describing similar SEF modulations at SMI during movements and during passive touching of objects [8,9]. One reason for a lack of movement-specific enhancement of SEFs at SMI could be that the responsiveness is increased only within the motor cortex, which does not necessarily participate in SEF generation.

The enhancement of P35m from the ipsilateral SMI during tactile interference was also reported recently [11]. In that study the effects in the ipsilateral hemisphere were interpreted as being due to interhemispheric interactions via callosal fibers. However, our finding that the enhancement of P35m was not caused only by tactile interference applied to the hand but also to the face, suggests that the effect could be more non-specific in nature. It is difficult to explain the effect of tactile interference to the face as being caused by transcallosal facilitation or inhibition, because the callosal fibers are known to interconnect mainly homologous parts of the body. Because all 3 interference conditions similarly enhanced the ipsilateral P35m, we suspect that this effect could result from an overall change in cortical excitability, e.g., in the average membrane potential level, possibly due to a change in the arousal of the subject.

At SII, the reductions of the responses during tactile interferences regardless of the side of median nerve stimulation and the site of the interference is consistent with the well-known property of large bilateral receptive fields of the SII neurones. During tactile interferences, the continuous afferent bombardment probably renders the neurones refractory to intervening impulses due to electrical nerve stimuli. The enhancement of the SII responses to right, but not to left, median nerve stimulation during right-sided finger movements suggests that the SII neurones are 'tuned' to respond more briskly to afferent impulses from the moving limb than from a non-moving limb. This feature indicates that SII plays an important role in integrating relevant sensory inputs to ongoing motor programs.

References:

[1] Evarts, E.V., and Fromm, C. The pyramidal neuron as summing point in a closed-loop control system in the monkey, In: Desmedt, J.E.. Cerebral Motor Control in Man: Long Loop Mechanisms, Basel, Karger, 1978, pp. 56-69.

[2] Huttunen, J., and Hömberg, V. Modification of cortical somatosensory evoked potentials during tactile exploration and simple active and passive movements, Electroenceph. clin. Neurophysiol., 1991, 81:216-223.

[3] Jones, S.J., and Power, C.N. Scalp topography of human somatosensory evoked potentials: the effect of interfering tactile stimulation applied to the hand, Electroenceph. clin. Neurophysiol., 1984, 58: 25-36.

[4] Hari, R., Reinikainen, K., Kaukoranta, E., Hämäläinen, M., Ilmoniemi, R., Penttinen, A., Salminen, J., and Teszner, D. Somatosensory evoked cerebral magnetic fields from SI and SII in man, Electroenceph. clin. Neurophysiol., 1984, 57: 254-263.

[5] Forss, N., Hari, R., Salmelin, R., Ahonen, A., Hämäläinen, M., Kajola, M., Knuutila, J., and Simola, J. Activation of human posterior parietal cortex by median nerve stimulation, Exp. Brain Res., 1994, 99: 309-315.

[6] Caselli, R.J. Ventrolateral and dorsomedial somatosensory association cortex damage produces distinct somesthetic syndromes in humans, Neurology, 1993, 43: 762-771.

[7] Poranen, A., and Hyvärinen, J. Effects of attention on multiunit responses to vibration in the somatosensory regions of the monkey's brain, Electroenceph. clin. Neurophysiol., 1982, 53:525-537.

[8] Schnitzler, A., Witte, O., Cheyne, D., Haid, G., Vrba, J., and Freund, H.-J. Modulation of somatosensory evoked magnetic fields by sensory and motor interferences, NeuroReport, 1995, 6: 1653-1658.

[10] Kakigi, R., Koyama, S., Hoshiyama, M., Watanabe, S., Shimojo, M., and Kitamura, Y. Gating of somatosensory evoked responses during active finger movements: magnetoencephalographic studies, J. Neurol. Sci., 1995, 128: 195-204.

[11] Schnitzler, A., Salmelin, R., Salenius, S., Jousmäki, V., and Hari, R. Tactile information from the human hand reaches the ipsilateral primary somatosensory cortex, Neurosci. lett., 1995, 200: 25-28.

Time Course of Neuromagnetic Signals Related to Reading Japanese Phonograms

Imada, T.[1,2], Kawakatsu, M.[3], Ikeya, A.[3], Kadota, R.[3], Miyamoto, J.[3], Nakayama, S.[3] and Kotani, M.[3]

[1]Basic Research Laboratories, Nippon Telegraph and Telephone Corporation, Atsugi, Japan; [2]Low Temperature Laboratory, Helsinki University of Technology, Espoo, Finland; [3]Dept. of Electronic Engng., Faculty of Engng., Tokyo Denki University, Tokyo, Japan

Introduction

The Japanese language has three writing systems: kanji script, or ideograms, hiragana script, or phonograms and katakana script, another type of phonograms. Modern Japanese has 71 hiragana and 71 katakana characters, which have one-to-one correspondence. A hiragana character has only one phonetic value, which represents a mora, or a syllable, in spoken Japanese. A recent positron emission tomography (PET) study [1] reported that overt reading of words written in hiragana activated bilaterally the medial and lateral occipital gyri, the posterior inferior temporal areas, Heschl's gyri, the posterior inferior frontal areas (including Broca's area), and the supplementary motor area. In addition, unilateral activation was seen in the left basal ganglia and the right thalamus. This PET study, however, did not reveal the temporal dynamics of the hiragana reading process. We thus attempted to follow the spatiotemporal activation sequence of cortical signals while subjects read hiragana.

Methods

Visually evoked cortical magnetic fields were recorded from five healthy right-handed Japanese males (22-46 years) with a whole-scalp Neuromag-122™ magnetometer [2] in a magnetically shielded room at the Helsinki University of Technology. We employed a visual attention paradigm, where the subject, looking at the fixation point in the middle of the screen, attended to the left visual field (LVF) in one block and to the right visual field (RVF) in the other block. Stimuli were presented, in random order within one block, either to the attended visual field or to the unattended visual field. To the attended visual field, two of 43 hiragana characters (Fig. 1a) were presented, and the subject was instructed to release a response key when these two hiragana had identical consonants in Japanese pronunciation, say /sa/ and /shi/, and not to release the key when they had different consonants. To the unattended visual field, two of 43 deformed hiragana characters (Fig. 1b) were presented; these could not be identified as hiragana and thus could not be read. The presentation probability was 0.4 for the hiragana with different consonants, 0.2 for the hiragana with the same consonant, and 0.4 for the deformed hiragana. Stimulus duration was 170 ms and the interstimulus interval varied randomly between 1.05 and 1.2 s.

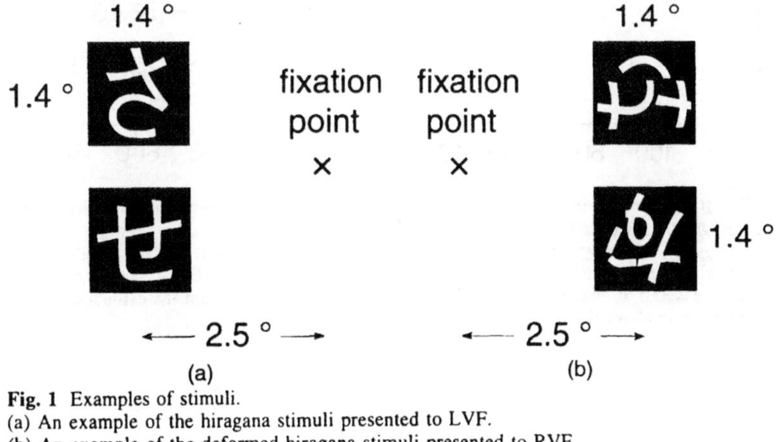

Fig. 1 Examples of stimuli.
(a) An example of the hiragana stimuli presented to LVF.
(b) An example of the deformed hiragana stimuli presented to RVF.

Responses (passband 0.03-100 Hz) were digitized at 0.5 kHz. Epochs with electro-oculograms exceeding 100 μV or with magnetic fields beyond 3000 fT/cm were rejected. More than 100 epochs were averaged separately according to the following conditions: (A) read hiragana in LVF, (B) read hiragana in RVF, (C) ignore deformed hiragana in LVF, and (D) ignore deformed hiragana in RVF. Responses to hiragana with the same consonant were not analyzed because they included motor reactions. Averaged responses were digitally lowpass-filtered at 40 Hz. The amplitude was measured with respect to a 100-ms prestimulus baseline. Equivalent current dipoles (ECDs), searched using a spherical head model, were selected according to the following four criteria [3]: *(i)* goodness-of-fit > 80%, *(ii)* 95%-confidence-volume < 268 mm^3, *(iii)* chi square probability* P_{obs} > 0.95, and *(iv)* continuous activation of the dipole for at least 10 ms. If the chi square probability is close to 1.0, the single dipole model is considered adequate.

Results

Figure 2 shows response waveforms of subject S2. In the occipital area, a strong deflection was observed between 100 and 200 ms, with larger peak amplitude in the "read" than in the "ignore" condition. In the left temporal area, from about 200 through 600 ms, the hiragana reading elicited slow activity, which was not seen in the "ignore" condition. All subjects showed the same tendency, and the maximum peak amplitude between 106 and 169 ms was significantly larger in the "read" than in the "ignore" condition for LVF stimuli (78 and 66 fT/cm; $p < 0.01$), but was not significant for RVF stimuli (98 and 91 fT/cm). At 80-180 ms, all subjects showed ECDs in one or more of the following locations: the contralateral occipital regions near the calcarine fissure, the bottom of the occipital lobe, the cuneus and the precuneus. From about 150 ms onwards, ECDs appeared in the ipsilateral occipital cortex. Figure 3 illustrates the time course of the activation in subject S2 in the "read" condition. At 88-138 ms, the activation moved from the area close to the right calcarine fissure to the right lateral cortex near the fusiform gyrus, where it stayed at 144-166 ms. At 156-166 ms, another activity was observed in the left lateral occipital cortex close to the left temporal cortex.

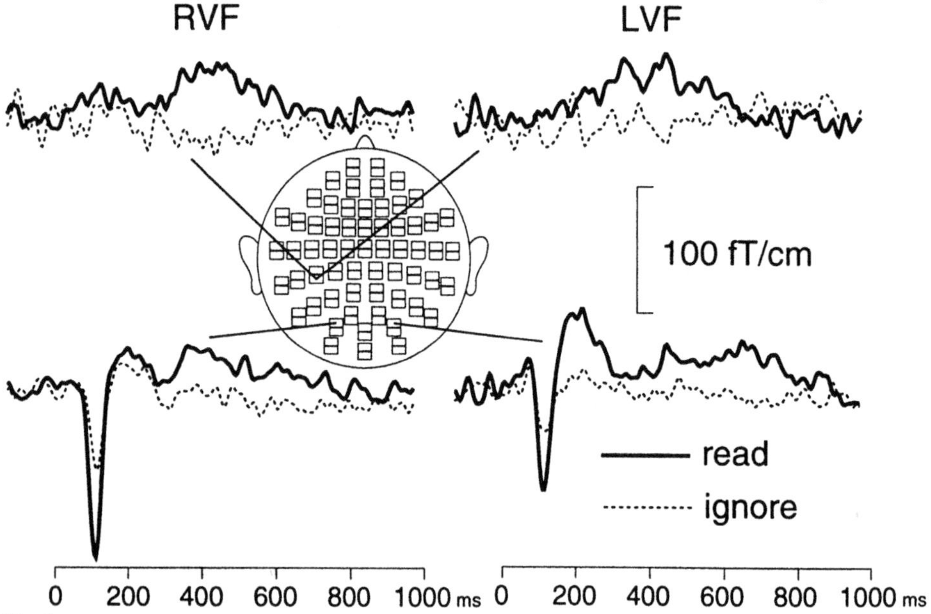

Fig. 2
Responses of subject S2 responding to attended hiragana (solid line) and ignored deformed hiragana (dashed line). The stimulus was presented to the RVF and LVF.

* $P_{obs} = P(\chi^2_{n-r} < \chi^2_{obs})$, where χ^2_{n-r} is the value of the chi square distribution with the freedom of n-r (n: the number of channels, r: the number of ECD parameters), and $\chi^2_{obs} = \Sigma^{n}_{i=1} (B_i - B'_i)^2 / \sigma_i^2$, where B_i is the magnetic signal measured at channel i, B'_i is the magnetic data calculated from the estimated ECD, and σ_i^2 is variance of the noise in channel i.

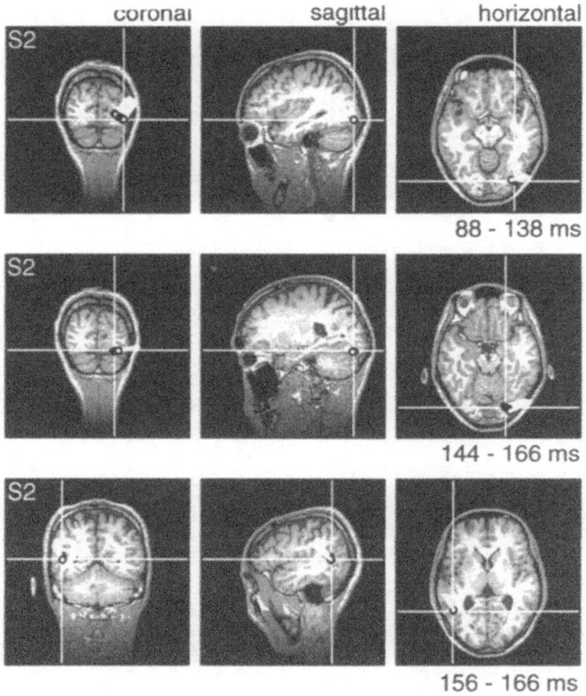

Fig. 3
ECD clusters from subject S2 until 200 ms after hiragana was presented to the LVF. The sources are projected on the subject's magnetic resonance images. The activation time period is indicated below the images. The ECD clusters are depicted by small white circles. The white lines starting at these circles indicate the direction of the ECD current .

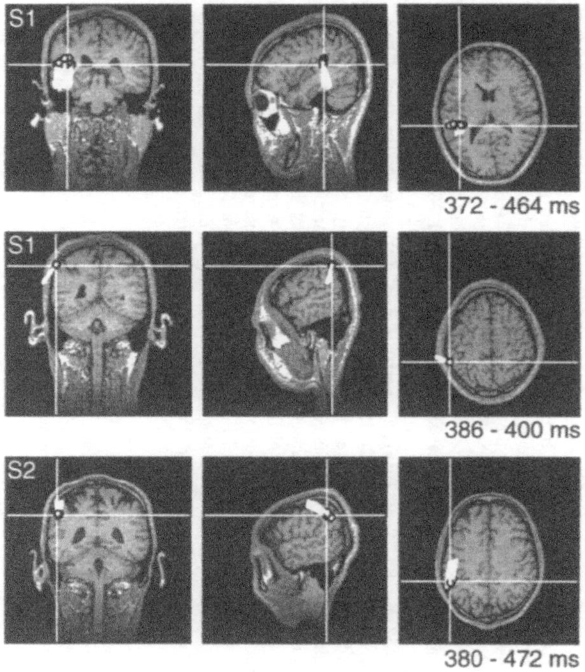

Fig. 4
ECD clusters from subjects S1and S2 200-500 ms after the hiragana presentation. The activation time period is indicated below the images.

Magnetic resonance images were taken at the Communications Research Laboratory, Ministry of Posts and Telecommunications.

In subject S2, at 200-500 ms, the left hemisphere was most clearly activated by the hiragana characters in the LVF and RVF. In fact, for LVF stimuli, there were 16 ECD clusters in the left hemisphere while there were two ECD clusters in the right. In the same subject, the left posterior inferior frontal cortex was activated at 276-328 ms by the LVF hiragana and at 292-310 ms by the RVF hiragana. Then, the left parietal cortex near the supramarginal gyrus and the angular gyrus became active 392-462 ms after the LVF hiragana presentation and 382-398 ms after the RVF hiragana presentation.

In all subjects, the common areas activated by hiragana in the LVF and RVF were mostly in the left hemisphere; the cortices around the supramarginal and the angular gyri (S1 386-400 ms, S2 380-472 ms, S5 214-338 ms), the posterior part of the floor of the Sylvian fissure (S1 374-394 ms, S5 350-414 ms), the superior temporal sulcus (S4 438-458 ms, S5 284-342 ms), the posterior part of the inferior temporal sulcus (S4 410-480 ms), and the lateral occipital cortex (S4 126-136 ms). Figure 4 illustrates these common areas from subjects S1 and S2. Two other subjects displayed right-hemisphere activity as well: in the primary visual cortex (S4 166-214 ms and 388-398 ms) and in the cortices around the supramarginal and the angular gyri (S5 198-212 ms and 286-346 ms). The subject-by-subject comparison showed that the left parietal cortex, the left superior temporal sulcus and the posterior floor of the left Sylvian fissure were active 200-500 ms after the hiragana presentation, but were not activated by deformed hiragana in the unattended visual field. The ignore condition produced fewer ECD clusters for both LVF stimuli (read: 22.6, ignore: 3.8, $p=0.03$) and RVF stimuli (read: 29.4, ignore: 11.8, $p=0.066$). In the left hemisphere in particular, the difference was significant for both LVF stimuli (read: 15.0, ignore: 2.2, $p=0.044$) and RVF stimuli (read: 17.8, ignore: 7.0, $p=0.04$).

Discussion

The results can be summarized as follows: Reading hiragana characters activates, until about 200 ms, mainly the occipital visual cortices. After 200 ms, the activation moves to the left hemisphere, including areas related to visual and/or language processing (for example, the cortices around the supramarginal and the angular gyri, the posterior part of the floor of the Sylvian fissure, and the superior temporal sulcus). Neuropsychological studies have shown that patients with alexia and/or aphasia often show damages in these cortical regions [4,5], suggesting their involvement in reading. Thus, after preprocessing in the occipital visual cortices, hiragana reading might continue in these regions for the duration of 200-500 ms after the stimulus onset. These claims are also supported by the results in which the "ignore" condition produced far fewer ECD clusters, especially in the left hemisphere.

In the PET study of Sakurai et al. [1], however, the activity in the left angular gyrus was significantly decreased by the task of overt hiragana reading. This discrepancy might be explained by the difference between their task and ours.

The activated areas varied depending on the subject, although all areas belonged to the visual and/or language processing areas. This variability can be explained by assuming that all these areas were more or less activated in all subjects but the dipoles producing weak magnetic fields outside the head were ignored by the applied strict screening criteria.

References

[1] Sakurai, Y., Momose, T., Iwata, M., Watanabe, T., Ishikawa, T. and Kanazawa, I. Semantic process in kana word reading: activation studies with positron emission tomography. NeuroReport, 1993, 4: 327 - 330.

[2] Ahonen, A., Hämäläinen, M.S., Kajola, M.J., Knuutila, J.E.T., Laine, P.P, Lounasmaa, O.V., Parkkonen, L.T., Simola, J.T. and Tesche, C.D. 122-Channel SQUID instrument for investigating the magnetic signals from the human brain. Physica Scripta, 1993, T49: 198-205.

[3] Hämäläinen, M.S., Hari, R., Ilmoniemi, R.J., Knuutila, J., and Lounasmaa, O.V. Magnetoencephalography -- theory, instrumentation, and applications to noninvasive studies of the working human brain. Reviews of Modern Physics, 1993, 65: 413 - 497.

[4] Sasanuma, S. Kana and kanji processing in Japanese aphasics. Brain and Language, 1975, 2: 369 - 383.

[5] Iwata, M. Kanji versus Kana; Neuropsychological correlates of the Japanese writing system. Trends in Neurosciences, 1984, 7: 290 - 293.

Acknowledgement

We are very grateful to Professor Riitta Hari (Low Temperature Laboratory, Helsinki University of Technology) for her helpful discussions and valuable comments on our draft.

Electric and Magnetic Brain Responses Evoked by Deviant Tones in Melodies

Jürgens, R.[1], Lopez, L.[2,3], Diekmann, V.[1], Becker, W.[1], Ried, S.[1], Grözinger, B.[1], and Erné, S.N.[2]

[1]*Sektion Neurophysiologie, Universität Ulm, 89081 Ulm, Germany;* [2]*Zentralinstitut für Biomedizinische Technik, Universität Ulm, 89081 Ulm, Germany;* [3]*Instituto Tecnologie Avanzate Biomediche, Universita "G.D'Annunzio", Chieti, Italy*

Introduction

The perception of deviant tones in an otherwise stereotyped sequence of auditory stimuli creates a characteristic mismatch activity (MMA) in the auditory cortex, which typically has a latency of about 150 ms and which can be recorded with EEG or MEG (mismatch negativity, MMN or mismatch field, MMF) [1]. MMA is generally viewed as reflecting the "hits" of an ever ongoing search for novel events based on the comparison of current events with a memory trace of the preceding ones, that is, on ***short term memory*** [2]. Alternatively, however, one could conceive of a more generalized mechanism based on the brain's ability to compare actual with ***predicted*** events. To pursue this question we investigated whether detection of "wrong" tones in pieces of classical music would also evoke MMA.

Methods

20 naive subjects (Ss) took part in the experiment; eleven rated themselves as musically ungifted and/or unknowledgeable while the others were performing amateurs. All Ss were given Bentley's test of musical talents (an established procedure in German schools, comprising pitch discrimination, resolution of chords, memorizing melodies and rhythms) [3]; its score differed significantly between the gifted and ungifted groups.

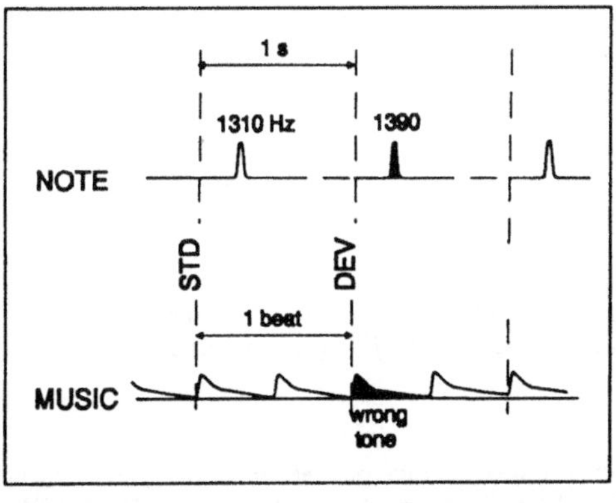

Figure 1. Principle of stimulus presentation. Traces represent sound amplitude as a function of time.

NOTE: typical odd-ball paradigm wit.. standard (STD, 300 repetitions) and deviant (DEV, 100 randomized repetitions) single frequency tones (40 ms duration, 10 ms rise/fall time). 1 s inter-stimulus interval.

MUSIC: continuously played pieces of classical music, synthetized piano sound, interspersed with wrong tones (DEV, 1/2 tone above correct, STD tones) see text for further details.

There were three NON-MUSIC paradigms in which frequent events (75% of cases; STD) alternated randomly with rare, deviant ones (25%; DEV). Three types of stimuli were used: TONE (STD=1310 Hz, DEV= 1390 Hz, cf. Fig. 1), CHORD (three simultaneous tones; STD=1040,1310,1560 Hz, DEV=1040,1390,1560 Hz) and ARPEGGIO (three consecutive tones, frequencies as in CHORD). The MUSIC paradigms comprised a simple and well known melody (MOZART, 40 repetitions of the same piece with a total of 478 beats, of which 120 started with a wrong tone, cf. Fig. 1) and a complex melody (BACH, 15 repetitions, 508 beats, 120 DEVs). All tones were generated by a sound blaster and presented binaurally with 60 dB HL. Ss were to listen to the stimuli and to count the number of deviations.

EEG (19 channels according to 10-20 system and 6 extra electrodes around C3/C4; 64 Hz / 1.0 s; averaged ears as reference) and MEG (DORNIER System, 22 software gradiometers operated in shielded room; DC - 64 Hz) were simultaneously recorded at 500 Hz sampling rate. Experiments were split into two sessions with the MEG aimed at the right and left hemispheres, respectively. STD and DEV events were selectively averaged off-line, based on their different trigger channel attribution.

Results

In all Ss, STD and DEV stimuli elicited typical auditory N100/N100m responses in EEG and MEG; however, there was considerable idiosyncratic variability with regard to amplitude and, for N100m, also to shape. The BACH melody evoked an almost sinusoidal response due to the high tone repetition frequency; its amplitudes ranged from 0.5 to 1.5 µV and were larger for Ss with high Bentley score (ANOVA, p=0.007).

MMA could be recognized in the majority of our Ss, even for the MOZART and BACH paradigms. As an example, the responses of one S to MOZART are depicted in Figure 2. There are three N100/N100m responses corresponding to the three tones occurring during the depicted 900-ms epoch. Following the first N100/N100m wave, a clear MMA elicited by the wrong tone at the sweep onset is recognized (shaded). The wrong tone evoked also a late P300 activity (typically over electrode Pz) and a P300-related MEG response (which was lacking in a number of other Ss, however).

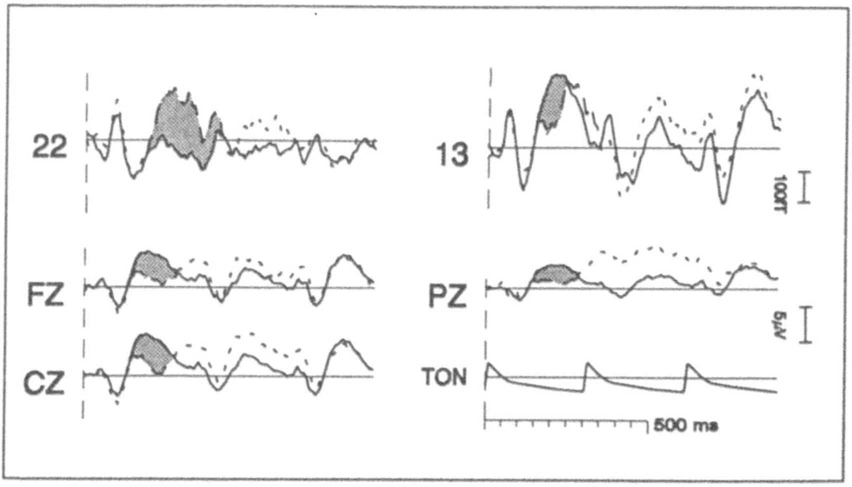

Figure 2. Averaged magnetic and electric responses to MOZART melody. One subject. EEG: positive=up. Recording position of MEG channel 13 posterior and 22 anterior to acoustic cortex. Solid lines, STD; broken lines, DEV.

MMN amplitudes exhibited a significant positive correlation with musical ability as expressed by the Bentley score (NOTE: p=0.048, CHORD: p=0.035, ARPEGGIO: p=0.0003, BACH: p=0.003, ANOVA, electrodes F3/F4).

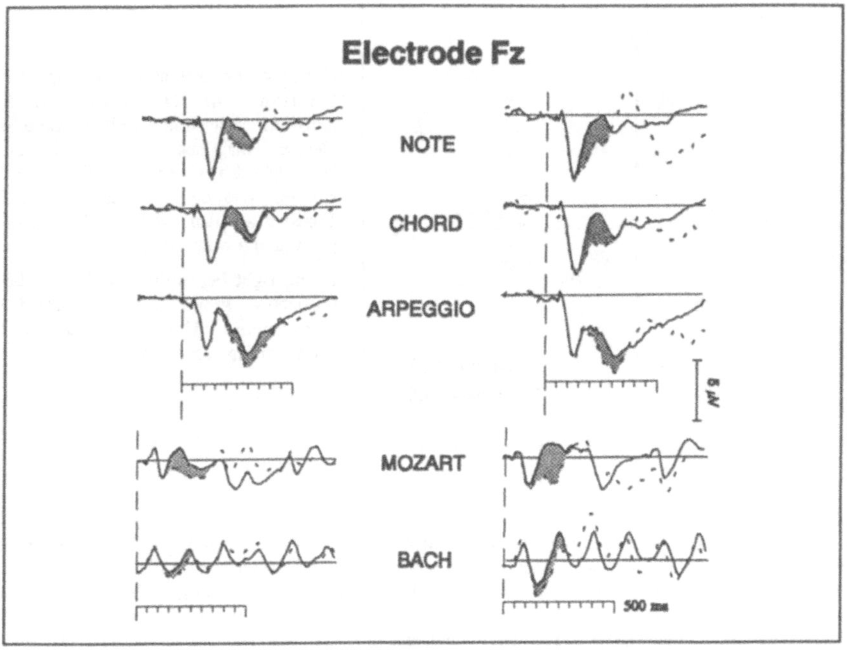

Figure 3. Grand averages of electrical responses from eight Ss with Bentley score < 37 (left column) and six Ss with a score > 40 (right column); positive=up; solid lines, STD; broken lines, DEV.

Further differences between Ss with low and high scores of musical ability were examined on the basis of grand averages from the two extreme groups of Ss. This step is illustrated in Figure 3. Whereas in the three NON-MUSIC paradigms the "low" and "high" groups showed quite similar N100 and MMN amplitudes, the musically gifted Ss produced clearly larger N100 and MMN in the MOZART and BACH conditions.

The amplitudes of the P300 potential as measured at Pz also differed between the two groups (generally larger in the gifted group as already observed by others [4], not shown). Interestingly however, these differences did not exactly mirror those of MMN; differences were largest for the perceptually difficult ARPEGGIO, whereas in MOZART, where deviants were easy to detect, the P300 amplitudes were almost equal. These results are compatible with the notion that the P300 reflects more conscious perceptual levels than MMA; conceivably, MMA corresponds to a precursor of conscious sensation [1].

Single equivalent dipole locations of N100/N100m in the NOTE paradigm calculated from a simultaneous MEG/EEG model [5] revealed a hemispheric asymmetry for musically ungifted Ss which was not observed in the gifted Ss (Figure 4). Similar differences were found also in "CHORD" and "ARPEGGIO". In "MOZART" and "BACH" no differences could be detected because of a larger scatter of localizations. Attempts to extend localization to MMA by means of the "simple" (one dipole/hemisphere, one instant) simultaneous electric/magnetic model used for N100, remained unsatisfactory, so far. MMA onset in EEG and MEG is often quite asynchronous (see Fig. 2); hence, appropriate modelling calls for spatio-temporal methods.

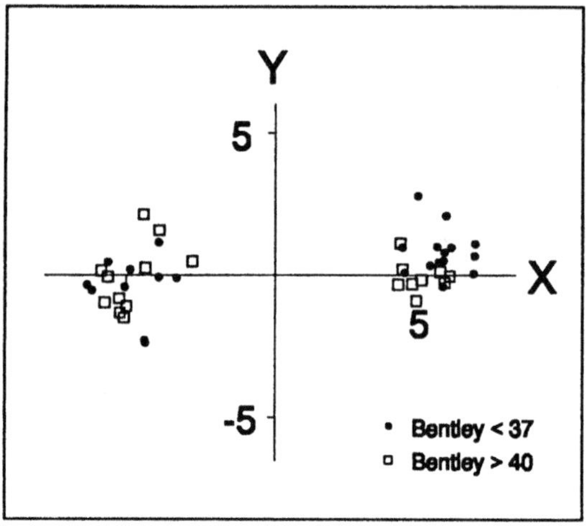

Figure 4. Equivalent current dipole (ECD) locations of N100/N100m in NOTE paradigm calculated with a simultaneous electric / magnetic model (one instant, four layer spherical model, one dipole per hemisphere; average goodness of fit, 95%). Transversal plane defined by nasion and preauricular point markers.

In the right hemisphere, ECDs of the "low musicality" group (dots) are located 6 mm more anterior (U-test, p=0.009) than in the "high" group (squares).

Discussion

The observation that wrong tones in melodies evoke MMA is difficult to explain on the basis of a comparison with a memory trace of the physical properties of the most recently perceived preceding tones. Continuously played music differs qualitatively from all NON-MUSIC paradigms including ARPEGGIO in that its "regularity" cannot be stored as a fixed reference pattern in short term memory. Therefore, we suggest that MMA does not only reflect short term memory mechanisms but also comparisons between ***expected*** and actual events in a more general sense. There are two conceivable processes whereby expectation could be built up in a MUSIC paradigm:

1) By storing a complete melody in ***long term memory***. Every forthcoming tone would be immediately available for comparison to actual tones; for the simple MOZART melody, such a strategy seems possible even for musically ungifted subjects.
2) ***Learned rules*** would provide a set of "allowed" future tones on the basis of the recently heard tones; the selection of these rules would be determined by the perceived musical context. If this idea applies to our BACH paradigm it is interesting to note that MMA latency was still comparable to that of other paradigms, although rule application would seem to involve highly complex processes of analysis.

References

[1] Näätänen, R. and Alho, K. Mismatch negativity - a unique measure of sensory processing in audition. Intern. J. Neuroscience, 80, 317-337, 1995
[2] Sams, M., Hari, R., Rif, J. and Knuutila, J. The human auditory sensory memory trace persists about 10 sec: Neuromagnetic evidence. Journal of Cognitive Neuroscience 5 No 3, 363-370, 1993
[3] Bentley, A. A study of some aspects of musical ability amongst young children, including those unable to sing in tune. Unpublished thesis, University of Reading, U.K., library, 1963
[4] Besson, M., Faïta, F. and Requin, J. Brain waves associated with musical incongruities differ for musicians and non-musicians. Neuroscience Letters 168, 101-105, 1994
[5] Diekmann, V., Becker, W., Grözinger, B., Jürgens, R., Westphal, K.P. and Kornhuber, H.H. Localisation of focal epileptic activity with a simultaneous EEG and MEG model. In: Baumgartner et al. (Eds.) Biomagnetism: Fundamental Research and Clinical Applications, pp 23-27, Elsevier, 1995

Multidipole Estimation of the Neuromagnetic Fields Related to Auditory Discrimination

Kikuchi, Y.[1], Yoshizawa, S.[2], Kita, M.[3], Nishimura, C[4], Tanaka, M.[2], Endo, H.[5], Kumagai, T.[5] and Takeda, T.[5]

[1]Tokyo Medical and Dental University, 1-5-45 Yushima, Bunkyo-ku, Tokyo 113, Japan; [2]University of Tokyo, 7-3-1 Hongo, Bunkyo-ku, Tokyo 113, Japan; [3]National Institute of Mental Health, National Center of Neurology and Psychiatry, 1-7-3 Kohnodai, Ichikawa City 272, Japan; [4]Toho University School of Medicine, 5-21-16 Ohmori-nishi, Ohta-ku, Tokyo 143, Japan; [5]National Institute of Bioscience and Human Technology, 1-1 Higashi, Tsukuba City, Ibaragi 305, Japan

Introduction

Auditory P3, which is the event-related potential (ERP) usually recorded during auditory discrimination tasks, is related to cognitive, memory and attentional process, and is also useful in clinical evaluation of the psychiatric disorders such as schizophrenia and so on. While the cognitive correlates of the P3 components, as well as their changes with respect to disease and development, have been extensively investigated, their synaptic generators remain controversial despite efforts using scalp EEG recordings in brain-lesioned subjects, animal models, and recording directly from the depth of the human brain.

MEG recordings are most useful in investigating the normal brain functions non-invasively. However, MEG measurements have been thus far made with a single- or a several-channel system ([6]), so a comprehensive magnetic field map covering a whole head had to be obtained over multiple and prolonged recording sessions. This may have been a limitation to the control of a subject's attention levels. Recently, according to the introduction of whole-head type SQUID systems, some of the problems of habituation and changing attention levels of subjects have been attenuated. We carried out MEG recordings by using a 64-channel whole head type SQUID system, and investigated the source locations of the components in the event-related field related to auditory discrimination.

Methods

Event-related fields (ERFs) were recorded from four normal-hearing adults with no neurological disorders in an auditory oddball paradigm. The 64-channel whole-head type SQUID system (CTF Inc., Canada) using first-order gradiometers with a coil diameter of 2.0 cm and a 5.0 cm baseline was used to measure external evoked magnetic fields over a whole head. The Cartesian coordinate system was defined based on three landmarks: the nasion, the left and the right preauricular points. The spatial parameters (Cartesian coordinates), as well as the strength and the orientation of the equivalent current dipoles (ECDs), were estimated from the 64 magnetic field measurements every 3.2 msec during the period between 280 msec and 450 msec post stimulus.

Pure tones of 2kHz and lkHz were applied as a target and as a non-target respectively. The presentation of sounds was randomized with targets occurring 20% of the time and non-targets 80%. Both sounds were delivered binaurally to the subjects at about 60 dB above threshold, via plastic tube terminating in high-quality insert earphones (Cabot Inc., USA). Subjects were required to mentally count the targets as correctly as possible with their eyes closed. MEGs were bandpassed from 0 to 40Hz and digitized at 625 Hz. Furthermore, off-line digital filtering reduced the bandpass to 0.01-30 Hz, and forty responses to target rare tones were averaged for the ERFs. The entire recording procedure for each subject lasted approximately 10 minutes. Eye movement artifacts were checked by monitoring the magnetic fields from the neuromagnetic sensors placed anteriorly.

Following the experimental session, magnetic resonance images (MRIs) were taken of each subject using a 0.5 Tesla system. All images were T1 weighted to emphasize anatomy and consisted of 2.0 mm x 2.0 mm pixels. 128 images were obtained, and all images were saved to a magneto-optical disk for off-line analysis. We carried out multi-dipole (2-4 dipoles) estimation, by considering anatomical symmetries and for minimizing error, based on Grynszpan-Geselowitz's equation. The coordinates of the dipoles' gravity centers were overlaid on individual MRI to show the corresponding locations in the brain.

Results

Clear evoked magnetic wave forms to both frequent and rare tones were obtained from all the subjects. In all the subjects, the averaged fields in response to targets showed prominent with latencies longer than 300 msec as compared with those to non-targets. Observations of the isomagnetic field maps during the time generally showed that the fields directed outward from the anterior region in the left hemisphere and the posterior region in the right hemisphere, while the direction was reverse over the posterior region in the left hemisphere and the anterior region in the right hemisphere, in each subject (Fig.1).

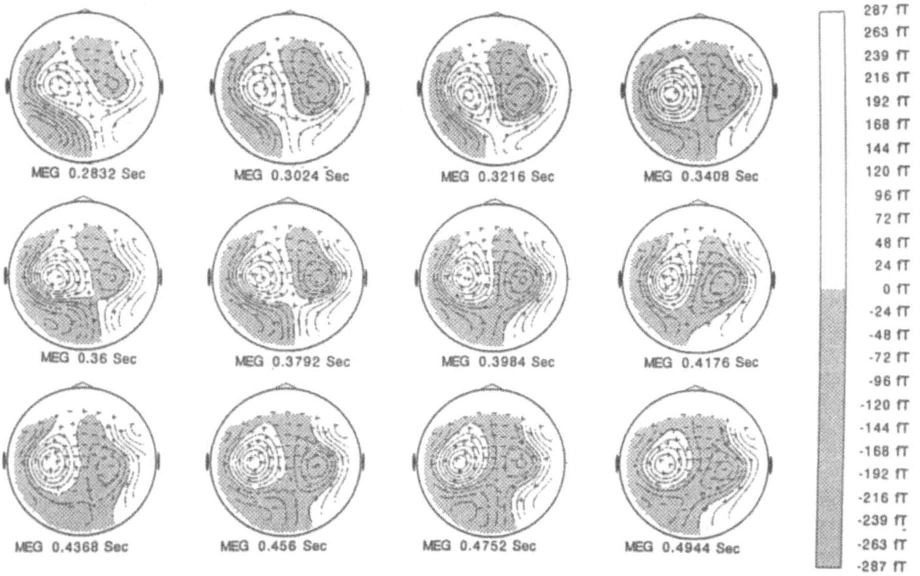

Fig. 1. Temporal change of the iso-magnetic field maps during the period between 283.2 msec and 494.4 msec. White area (+) shows the outward magnetic field, and hatched area shows the inward field.

In order to study the temporal changes of the sources responsible for the late components, dipole solutions were calculated at successive 3.2 msec intervals for the wave form between 280 msec and 450 msec post stimulus. Fig. 2 shows the averaged MRI images superimposed by the ECDs estimated in subject A, during the period between 355 msec and 432 msec post stimulus. In this case, a two dipole model was applied, and the gravity centers of ECDs were found around the hippocampus, the posterior parts of the parahippocampal gyrus, the collateral sulcus, the lingual gyrus and the fusiform gyrus. Every dipole directed upwards. In the case of subject B, ECDs were also estimated in the hippocampus, the posterior parts of the parahippocampal gyrus, the collateral sulcus, the lingual gyrus and the fusiform gyrus during the period between 280 msec and 350 msec. In addition, they were found in the superior temporal sulcus (STS) and the superior temporal plane (STP) between 350 msec and 450 msec. Their directions were identical to those of subject A. In subject C, their locations were around the STS, the STP, the collateral sulcus and the lingual gyrus between 293 msec and 304 msec, and were in the collateral sulcus, the lingual gyrus, the central sulcus, the precentral sulcus and the inferior frontal sulcus between 314 msec and 390 msec. The locations of ECDs estimated in subject D were around the central sulcus and the precentral sulcus during the period between 302 msec and 328 msec, and were around the parietal and the subparietal cortices between 338 msec and 380 msec.

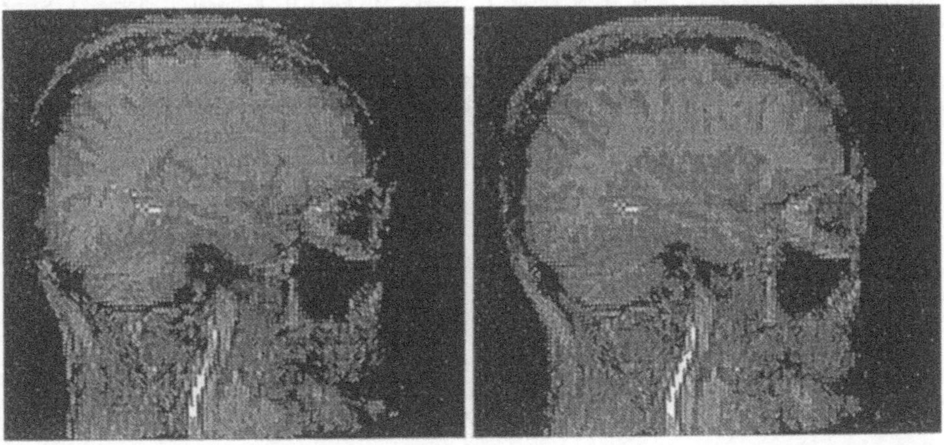

Subj.A/2kHz/355-432/L Subj.A/2kHz/355-432/R

Fig. 2 Averaged sagittal MRIs superimposed by the gravity centers of the ECDs estimated in subject A, between 355 msec and 432 msec. White pixels represent the gravity centers. (left:left, right:right)

Discussion

The P3 is subdivided into a fronto-central P3a relatively enhanced by non-target rare or novel stimuli, and a later parietal P3b enhanced by certainty of target detection. The studies by depth electrode recordings ([1], [2] [3]) showed that the responses are classified into two patterns: The first sharp triphasic short-latency sequence resembling the scalp N2/P3a/SW pattern is observed in all recorded neocortical areas and is most often focal or large in the anterior or the posterior middle temporal gyrus, the posterior parahippocampal gyrus and the fusiform gyrus. The second pattern is a broad, usually monophasic, relatively long-latency, waveform that resembled the scalp P3b. It is localized to the hippocampus, to a part of the superior temporal sulcus, and to rhinal cortex underlying the amygdala. The N2/P3a/SW pattern is also found in supramarginal and posterior cingulate gyri, as well as in multiple frontal sites, including the anterior cingulate gyrus, gyrus rectus and dorsolateral prefrontal cortex. While the temporal order of each active area in the brain has not been so clear in the present study, the source locations obtained in normal subjects by multi-dipole estimation, were in good agreement with parts of those found in epileptic patients by the depth electrode recordings.

References

[1] Halgren, E., Baudena, P., Clarke, J.M., Heit, G., Liegeois, C., Chauvel, P., Musolino, A. Intracerebral potentials to rare and distractor auditory and visual stimuli. I. Superior temporal plane and parietal lobe. Electroenceph. clin. Neurophysiol., 94:191-220, 1995.

[2] Halgren, E., Baudena, P., Clarke, J.M., Heit, G., Marinkovic, K., Devaux, B., Vignal, J-P., Biraben, A. Intracerebral potentials to target and distractor auditory and visual stimuli. II. Medial, lateral and posterior temporal lobe. Electroenceph. clin. Neurophysiol., 94:229-250, 1995.

[3] Baudena, P., Halgren, E., Heit, G., Clarke, J.M. Intracerebral potentials to target and distractor auditory and visual stimuli. III. Frontal cortex. Electroenceph. clin. Neurophysiol., 251-264, 1995.

[4] Kikuchi, Y., Yoshizawa, S., Kita, M., Nishimura, C., Tanaka, M., Endo, H., Kumagai, T., Takeda, T. Source estimation of the auditory event-related neuromagnetic fields (ERFs) using a whole-cortex type SQUID system. "Visualization of Information Processing in the Human Brain: Recent Adavances in MEG and Functional MRI", The 10th Tokyo Institute of Psychiatry International Symposium, Tokyo, Japan, Oct. 12-13, 1995.

[5] Kikuchi, Y., Yoshizawa, S., Kita, M., Nishimura, C., Tanaka, M., Endo, H., Kumagai, T., Takeda, T. Source localization of event-related neuromagnetic fields using a whole-cortex type SQUID system. Electroenceph. clin., 97:S124, 1995.

[6] Rogers, R.L., Baumann, S.B., Papanicolau, A.C., Bourbon, T.W., Alagarsamy, S., Eisenberg, H.M. Localization of the P3 sources using magnetoencephalography and magnetic resonance imaging. Electroenceph. clin. Neurophysiol., 79:308-321, 1991.

Measurement of Gustatory-Evoked Magnetic Fields with Sharp Stimulation

Kobayakawa, T.K.[1], Endo, H.E.[1], Saito, S.S[1], Ayabe-Kanamura, S.A.[2], Kikuchi, Y.K.[3], Yamaguchi, Y.Y.[1], Ogawa, H.O.[4], and Takeda, T.T.[1]

[1]National Institute of Bioscience and Human Technology, MITI, Tsukuba, Japan; [2]University of Tsukuba, [3]Tokyo Medical and Dental University; [4]Kumamoto University

Introduction

Non-invasive methods to measure brain function, particularly the methods of magnetoencephalography (MEG) and functional magnetic resonance imaging (fMRI), have improved rapidly and remarkably. With these methods, researchers have measured higher cortical functions of the living human brain, especially visual and auditory functions. MEG has the special advantage that it can trace the movement of cortical activity.

In the case of gustation, few experiments have addressed human gustatory-evoked potentials (GEPs) [1][2][3][4], because of difficulties involved in precise control and presentation of gustatory stimuli. In imaging studies performed using PET or fMRI, it has been reported that the insula, parietal cortex, cingulate gyrus, parahippocampal gyrus, and thalamus were activated during taste discrimination (PET)[5][6], and insula and frontal operculum were activated by simple taste stimulation (fMRI)[7]. These imaging methods, however, offer relatively low temporal resolution. MEG offers much better resolution and we have therefore employed it to trace the flow of gustatory information in the brain. As MEG requires strict stimulus control, we have developed a gustatory apparatus to present the stimulus with a short rise-time. First, we measured the gustatory-evoked magnetic fields (GEMs) and then, we sought to estimate the area of gustatory activity in the human brain using a dipole model. Next, before measuring GEMs, we presented the substance which surpress a sensitivity to a certain taste, and investigated the change of responses.

Methods

Taste Delivery System

The stimulating apparatus was composed of three sub-systems: (1) the system that supplied the tastants and deionized rinsing water, (2) the system for stimulus presentation, and (3) the system that sensed the timing of the stimulus. The supply system and the presentation system were connected by teflon tubing (Fig.1).

The deionized water and tastants were driven through the system by compressed air and separated from each other by a small amount of air injected between them. The water, tastants, and air were switched by solenoid valves controlled by a personal computer. Stimuli were presented through a hole (4 mm x 10 mm) in the wall of a teflon tube. The subject bit gently the tube, then covered the hole with his or her tongue. Liquid and air from inside the tube could then flow across the exposed position of the tongue. Since the flow of the liquid and air generated slight negative pressure, the tongue of the subject was sucked slightly into the hole and hence the liquid did not leak into the mouth.

A precise trigger was necessary for proper averaging of data. The tastant took about 27 s to go from the solenoid valve to the hole (stimulus presentation position). Arrival time would vary over too wide a range, if timing to control the solenoid valve served for triggering. To obtain a stricter trigger, an optical sensor was placed as close as possible to the subject's mouth. The tastants were colored red, whereas the air and the deionized water were uncolored. Light was generated outside the room and fed inside by a bundle of optical fibers. The light passed through the teflon tube and was modulated by the liquid or air in the vicinity of the subject's mouth. The signal was then brought out of the room via optical fibers and converted into a voltage by the photo-transistor (Fig. 2). The distance between the sensor and the hole in the tube equalled 6.5 cm. The liquid took 300 ms (sd=2.9 ms) to cover that distance. Rise-time to the 80% level of the signal equalled 16.5 ms (sd=1.49 ms). Such performance was accurate enough for strict triggering.

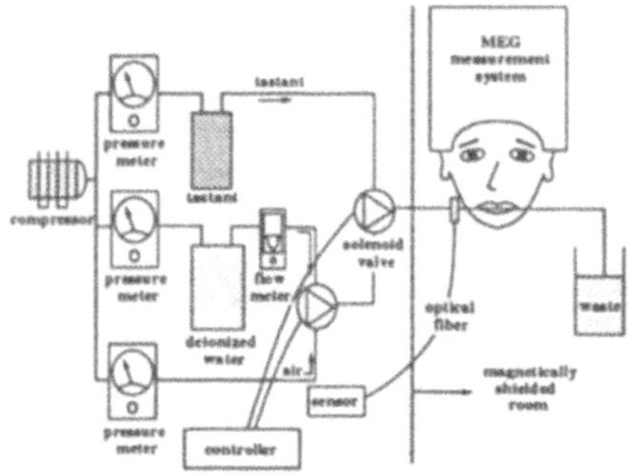

Fig. 1: The taste delivery system

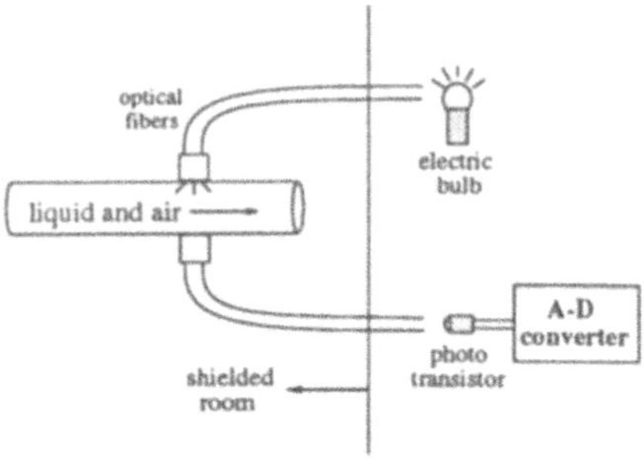

Fig. 2: The optical sensor

The small amount of air that separated the tastant and water was released into the tube for as short a duration as possible to minimize the likelihood that the subject could sense a difference of pressure between the liquid and air.

Evoked Magnetic Field Measurements

For measurement of magnetic fields, we used a 64-channel whole-head SQUID system (CTF Systems Inc., Canada). This system has sensors for measuring environmental magnetic noise. Subtraction of the outer noise from the measured data leads to extraction of clearer signals from the brain.

Since magnetic substances would interfere with the measurements in the magnetically shielded room, a special wooden chair with a pneumatically operated cushion was developed [8]. In addition, the subject's jaw was supported by a wooden chin rest. The tube through which the liquids flowed was brought to the subject's tongue and held in place by a frame attached to the chair. The cushioned chair and the chin rest tended to forestall feelings of fatigue during a session.

Solutions of 1 M NaCl and of 3 mM saccharin were used as tastants. Next, we measured GEMs from saccharin after with a subject chewed gymnemic acid gum for 15 min. The tastants were presented at the central, front edge of the subject's tongue, the part known to have high sensitivity to these tastes. After a tastant was presented, deionized water was presented as a rinse. The duration of the stimulus was about 400 ms and that of the rinse was approximately 30 s. The tastant and rinse were maintained at the temperature of the tongue. Thirty trials were presented for each tastant and accompanying rinse.

Six subjects, which included 3 females, participated in the experiment. They ranged from 21 to 32 years of age. Sampling rate for the MEG signal was 250 Hz and the low-pass filter was set at 40 Hz. During recording of the MEGs, the subject watched a fixation point. After the experiment, signals from the optical sensor were used as a trigger for off-line averaging of the data. Trials contaminated by eye movements were rejected. The number of trials used for averaging was about 35.

Results

Fig. 3 shows the magnetic isocontour map of the brain surface 400 ms after presentation of saccharin. The magnetic field indicated force flow from the left-front to the left-back of the head and from the right-back to the right-front.

In this investigation, dipoles were estimated under the assumption that the shape of the brain was a simple sphere, though of course the actual shape departs from a sphere. In the future, the dipoles need to be estimated for a real brain model calculated from MRI.

We estimated the dipoles on the assumption that the pattern of this map had two dipoles, which we superimposed on the MRI. In 86% experiments, the dipoles were estimated in the upper side of insula region (goodness of fit was from 80% to 96%) (Fig. 4). In five subjects, the directions of these dipoles were superior, posterior and medial.

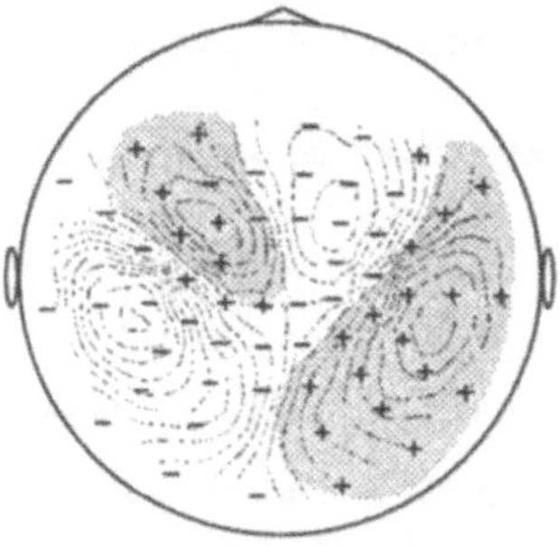

Fig. 3: Contour map of the surface

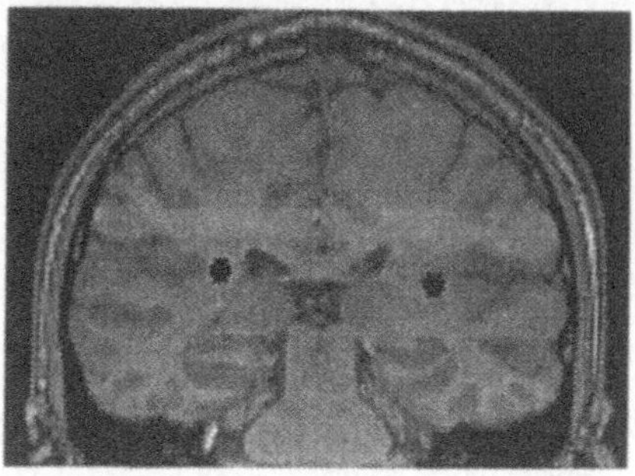

Fig. 4: Activated cortex by gustatory stimulation

In four of five subjects, the latencies at which the clear pattern appeared were much longer (from 150 ms to 200 ms) after subjects chewed gymnemic acid gum than when they did not chew the gum. Figs. 5 and 6 depict averaged GEMs of 64 channels aligned according to the stimulus onset and superimposed on the same graph in one subject. Fig. 5 shows the response of saccharin, and Fig. 6 shows that of saccharin after subject chewed gymnemic gum

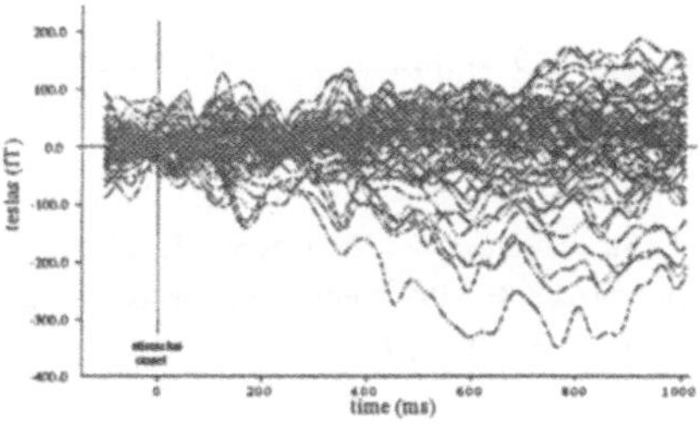

Fig. 5: response of saccharin

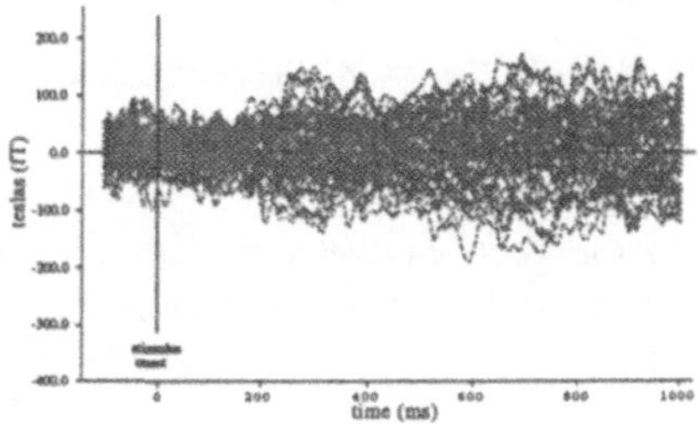

Fig. 6: response of saccharin after gymenmic acid

respectively. These figures show the amplitude of the response of saccharin after gymnemic acid presentation was much smaller than that of saccharin.

Discusstion

Previous studies using PET [5][6] reported that the insula, parietal cortex, cingulate cortex, parahippocampal gyrus, and thalamus were activated by gustatory stimulation. In those experiments, cortical activity was measured during a discrimination task. Several areas including the insula were activated by the task; some may have been related to recognition or motor activity.

In the case of the Macaque monkey, the primary gustatory cortex has been reported to be in the insula[9]. The present results from humans are consistent with that finding.

In an fMRI study[7] when a solution of NaCl was presented, both hemispheres were activated. In contrast, when a solution of sucrose was presented, just one hemisphere was predominantly activated. For NaCl, the stimulus in common in that experiment and the present one, we also found both hemispheres to be activated.

All subjects felt a sweet taste to saccharin, although some subjects also said they could feel sour, bitter or a salty taste to saccharin. After presentation of gymnemic acid, subjects reported a sour or salty taste; (of course, they did not report sweet taste). The sodium ion is included in saccharin solution and it seems this causes the sour, bitter, or salty taste. Therefore, in the experiment using gymnemic acid the response was not surpressed completely. In order to surpress the sweet taste, we would be better to use another substance (i.e. aspartame) as a sweet tastant instead of saccharin.

References

[1] Kobal, G. (1985) Gustatory evoked potentials in man. Electroenceph. clin. Neurophysiol., 62: 449-454.

[2] Plattig, K., Bekiaroglou, P. and Schell, S. (1994) Gustatory-evoked potentials in man (A long history and final results). In: K. Kurihara, N. Suzuki and H. Ogawa (Eds.), Olfaction and Taste XI, Springer-Verlag, Tokyo: 638-641.

[3] Prescott, J. (1994) Brain event-related potential to primary tastes. In: K. Kurihara, N. Suzuki and H. Ogawa (Eds.), Olfaction and Taste XI, Springer-Verlag, Tokyo: 642-645.

[4] Maetani, C., Notake, K., Takemoto, I., Hizuka, I., Hagino, J., Matsunaga, T., Yoshimura, S. and Tonoike, M. (1988) Measurement of EEG caused by gustatory stimulation. Proceedings of the 22nd. Japanese Symposium on Taste and Smell: 97-99.

[5] Fukuda, H., Yamada, K., Kinomura, S., Kawashima, R. and Cuo, S. (1991) High order gustatory projections in the human brain studies with positron emission tomography. The Salt Science Research Foundation - Annual Research Report: 337-347.

[6] Kinomura, S., Kawashima, R., Yamada, K., Ono, S., Itoh, M., Yoshioka, S., Yamaguchi, T., Matsui, H., Miyazawa, H., Itoh, H., Goto, R., Fujiwara, T., Satoh, K. and Fukuda, H. (1994) Functional anatomy of taste perception in the human brain studied with positron emission tomography. Brain Res., 659: 263-266

[7] Hirsch, J., DelaPaz, R., Relkin, N., Victor, J., Bartoshuk, L., Norgren, R. and Pritchard, T. (1994) Localization of human gustatory cortex using functional magnetic imaging. Chem. Sens., 19: 486.

[8] Takeda, T., Endo, H., Kumagai, T. and Maurizio, M (1995b) Evaluation of a whole-head type MEG system and its application. J. Jpn. Biomag. Bioelectromag. Soc., 8: 128-131.

[9] Ogawa, H. (1972) Taste response in the Macaque monkey chorda tympani. Physiol. Behav., 9: 325-331.

Source Location Outside the Auditory Cortex: Auditory Event Related Field to Tone-Duration in a Discrimination Task

Kofoed, B.[1], Bak, C.K.[1], Csepe, V.[2], Pantev, C.[3] and Saermark, K.[1]

[1]Department of Physics, Technical University, Lyngby, Denmark; [2]Department of Psychophysiology, Hungarian Academy of Sciences, Budapest, Hungary; [3]Institute of Experimental Audiology, University of Münster, Germany

Introduction

The purpose of the present experiment was to locate the sources of the "mismatch field" (MMF), when the discrimination was based on tone duration differences. In the oddball paradigm applied, standard tones of duration D1 are randomly replaced by deviant tones of duration D2<D1. It is well known, that deviants invoke a MMF approximately 200 ms after the deviation, in these experiments the termination of the the target tone. The subjects were required to attend to and silently count the number of deviants. By choosing D2 appropiately, the latency of the MMF signal can be moved away from the M200 complex, thus permitting a localization of the MMF sources. Source location of the MMF to pitch differences and to tone differences where the MMF signal appears in the M100-200 region has earlier been reported [1],[2]. In those experiments a single dipole fit has been used for the purpose of localization. In the present experiment the MUSIC-algorithm is used.

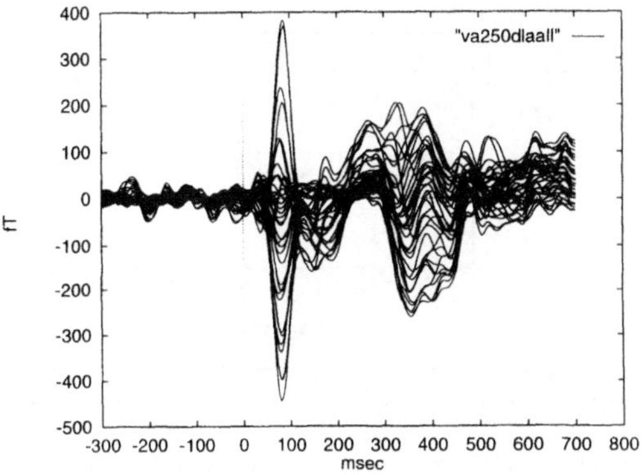

Figure 1: The figur shows the averages for the target tone (A-averages). All 37 coils are shown. The duration of the target tone is 250 ms., and the tone onset is at 0 s. The MMF-signal is seen in the interval 380-520 ms after onset of the stimulus tone. Probably more than one component is involved.

Methods

The measurements were performed by means of a BTI-37 channel SQUID system equipped with a probe position indicator. The sound was led to the right ear via a plastic tube, and the measurements were made on the left hemisphere. The duration of the frequent tone was D1=500ms, and it occurred with the probability $p = 0.67$. The duration of the target tone was D2=250 ms or D2=200 ms, and it occured with the probanility $p = 0.33$. Both stimuli were of 1kHz pitch and of intensity 70 db above the psychoacoustic threshold. The interstmulus interval [ISI] varied randomly in the range 1.25 - 2.75s or in the range 2.5 - 5.5 s.

The signals were highpass filtered, using a hardware filter with a cut-off frequency 1 Hz. The sampling rate was 298 Hz. The collected epoch was 1000 ms including 300 ms prestimulus interval. The average was performed across the epochs for the target tone (A-averages) and across the tones immediately preceeding a target epoch and across the remaining epochs. After averaging frequencies higher than 30 Hz were digitally filtered out. Inspecting the measured time-series visually (Fig.1), the MMF-signal does not look dipolar. More then one source is involved, and in time the sources are overlapping. Futhermore, the combination of highpass filtering and the size of the interstimulus interval used will affect the data in such a way, that a single dipole fit can not be applied. A method to separate more than one source, overlapping in time, is the MUsic SIgnal Classification (MUSIC) algorithm [3]. An inherent property of this method is, that source localization is not affected by filtering of the signals.

Results

Fig.1 shows the MMF signal appearing in the time interval 300 - 500ms and only in averages across the targets. The MUSIC algorithm is applied to 50 time slices around the MMF signal. Following the procedure given by Mosher J.C. et al.[3] the so called costfunction is calculated. In xy-planes, xz-planes or yz-planes contour plots for the reciprocal costfunction is made. x, y and z are the usual axes for the head. The points where the reciprocal costfunction has a maximum specify the locations for the sources.

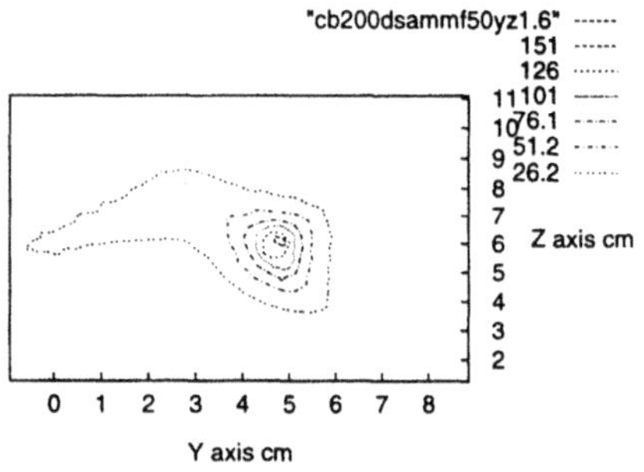

Figure 2: Contourplot for a two-dimensional slice (constant x) of the reciprocal costfunction. The yz-plane shown is the plane where the reciprocal costfunction has its maximum value for one of the sources.

In the latency range of the MMF we find two sources. Fig.2 and Fig.3 show the contourplots in the yz-planes where the reciprocal costfunction has its maxima for the two sources respectively. For comparison, Fig.4 shows the contourplot using 50 time slices around the M100 signal for the same A-averages. The source shown in Fig.3 is therefore located in the auditive cortex and according to its latency, it can not be excluded, that it is the off-response to the stimulus. The other source, however, is found outside the auditive cortex, as shown in Fig.2. Its location is in good agreement with our results reported earlier [4].

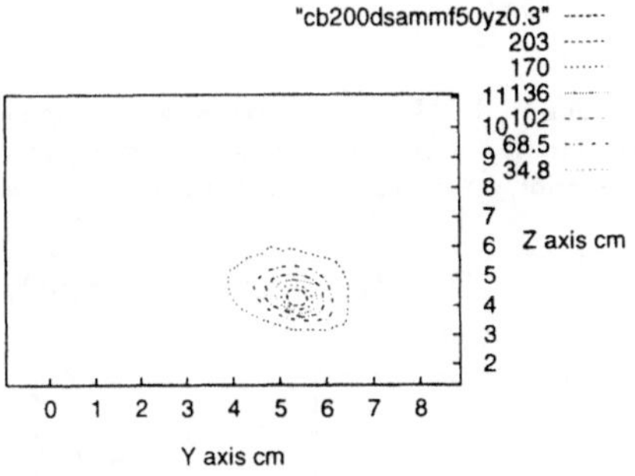

Figure 3: Contourplot for a two-dimensional slice (constant x) of the reciprocal costfunction. The yz-plane shown is the plane where the reciprocal costfunction has its maximum value for the other source.

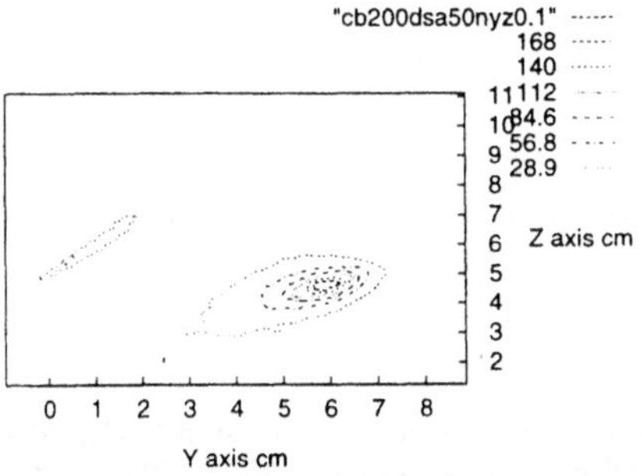

Figure 4: Contourplot for a two-dimensional slice (constant x) of the reciprocal costfunction. The yz-plane shown is the plane where the reciprocal costfunction has its maximum value the for M100 source.

Four subjects have participated in the experiment. For each subject D1 has been 500ms and D2 = 200ms or D2 = 250ms. The interstimulus interval has been 1.25 - 2.75s or 2.5 - 5.5s for each D2. For each subject we therefore have 4 runs. Making grand average, the total number of localizations for one source is 16.

In Fig.5 the grand averages over the 4 subjects is shown. Each source location is measured relative to the position of the M100 source as determined from the same A-average. In the figur the zero point corresponds to the location of the M100 source. Around the locations for the two sources the standard deviations are plotted.

Conclusion

Using the MUSIC algorithm to the MMF signal two souces are found to be involved, when the discrimination is based on tone duration. One of the sources is located very near the M100 source and the other is located about 1 cm anterior, 1.8 cm superior and 0.5cm inferior relative to the M100 source.

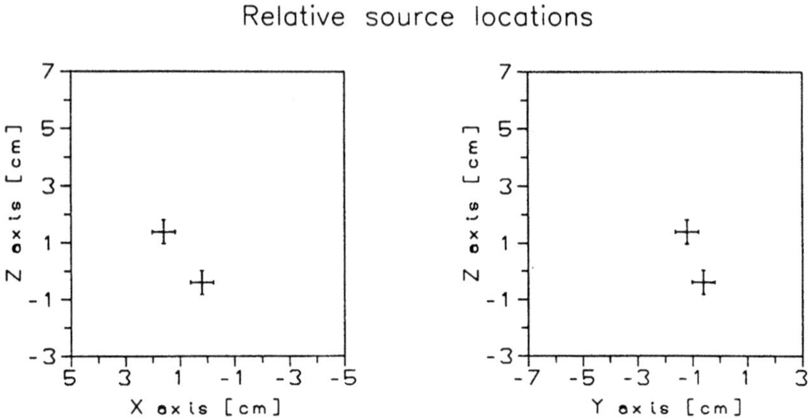

Figure 5: The xz- and yz-location of the two sources found. The locations are plotted relative to the position of the M100 source. The figure shows the grand averages for all subjects and all runs. The figur also shows the error bars for the standard deviation in the measurements

References

[1] Sams M., Kaukoranta E., Hamalainen M.. Naatanen R. Cotical activity elicited by changes in the auditory stimuli: different sources for the magnetic N100m and the mismatch responses. Psychophiology. 1991. 28. p21-29.

[2] Csepe V., Pantev C., Hoke M., Hampson S. and Ross B. Evoked magnetic responses of the human auditory cortex to minor pitch changes: localization of the mismatch field Electroencaph. Clin. Neurophys.1992, 84, p538-548.

[3] Mosher J.C., Liwis P.S. and Leahy R.M. Multiple dipole modeling and localization from spatio-temporal MEG data. IEEE Transactions on Biomedical Engineering, 1992, 39, p541-557.

[4] Kofoed B., Bak C.K., Rahn E. and Saermark K. Auditory event-related magnetic fields in a tone-duration discrimination task. Source localization for the mismatch field and for a new component M2". Acta Neurol. Scandinav. 1995. 91, 362-371.

Acknowledgements

We are very grateful to the Danish Natural Science Research Council for financial support.

Measurement of MEG Evoked by Visual Discrimination Task

Kumagai, T., Takeda, T. and Endo, H.

National Institute of Bioscience and Human-Technology, Tsukuba, Japan

Introduction

The oddball paradigm task, a simple discrimination task using infrequent and frequent stimuli, evokes late components after the latency of 300 msec in EEG measurements. It is known that they are composed of many components which are called P3a, P3b and slow wave. Responses to infrequent stimuli are larger than those to the frequent stimuli. They are more sensitive to the cognitive parameters than to the stimulus parameters of the task. Hence late components are defined as endogenous components. It is important to investigate the functions and their sources in time domain in order to understand the style of cerebral information processing.

The late components have been investigated by EEG, PET, implanted electrodes and MEG. However, sources and functions of those components have not been clarified. MEG has a high time resolution like EEG, and it also has a good spatial resolution if the activated areas are fairly concentrated. Although an increasing number of auditory P3 studies with MEG have been reported, there are only few reports on the visual evoked P3[1][2]. Those studies have been carried out using systems covering only a part of the cortex, hence the whole brain activity has not been reported yet.

In this study we examined the magnetoencephalographic (MEG) signals evoked by a visual discrimination task using a 64-channel whole-cortex MEG system. Since P3 may involve a widespread activity of the brain, it is important to measure the entire cortex simultaneously to understand the underlying neural activities. The activated area was estimated with multiple equivalent current dipoles (ECDs) using a simple spherical brain model.

Methods

Equipment:

The measurements were performed using a whole-cortex type MEG system (CTF, Inc.) in which 64 sensors cover the whole brain. The sensors are first order gradiometers whose diameters and base lines are 2 cm and 5 cm respectively. The system was set up in a magnetically shielded room with a reclining chair equipped with a pneumatic lifting mechanism(Fig.1). The visual stimuli were shown to the subjects using a liquid crystal projector system. A translucent screen was set in front of the subject seat. The projector, positioned in front of the magnetically shielded room, projected the stimuli on the screen through the doorway of the magnetic shielded room. In previous experiments we checked that the system can cancel the environmental noises from the doorway through the software noise reduction using reference sensors [3]. The distance between subject's eyes and the screen was 1.5 m.

Stimulus and Paradigm:

The oddball paradigm was composed of yellow and purple color onset stimuli. The frequent stimulus was either yellow or purple while the infrequent stimulus was another color. Frequent and infrequent stimuli were showed with a probability one of 0.8 and 0.2, respectively. The subject had to count silently the number of times in which the infrequent stimulus appeared and to report this number after each trial. The infrequent stimulus appeared between 40 and 45 times in each trial, while the frequent stimulus appeared about four times as often. Fig.2 shows the color onset stimuli used in this experiment. When the stimulus was turned off, a simple gray circle was shown to the subject. While the stimulus was turned on, a purple or a yellow concentric circle was presented on the gray circle. The visual angle of color circles and the gray circle were 3.5 degrees. The luminance of the gray circle, the purple annulus and the yellow annulus were almost the same. A small black fixation point was put at the center of the stimulus. The duration of the stimulus was 200 msec. The inter stimulus interval was chosen from 2 to 2.2 s at random.

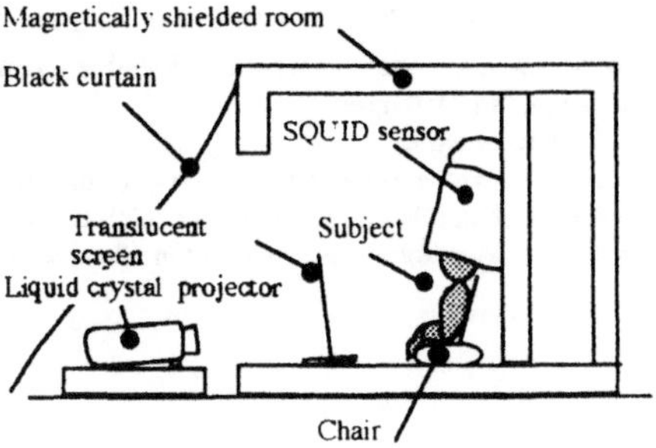

Fig.1 The experimental system

Measurment:and Analysis:

The subjects were four right handed normal volunteers ranging from 20 to 40 years old, indicated in the following as S1, S2, S3 and S4. The MEG signals were measured from 200 msec before to 500 msec after the onset of stimulus. The sampling rate was 325 Hz. The measurements that included a magnetic change larger than 900 fT were rejected, since it might be contaminated by artifacts. The responses to infrequent and frequent stimuli were averaged respectively in every trials.

Averaged data were analyzed as follows. DC offset was removed from each channel so that the integration evaluated from 200 msec before to the stimulus onset. After the software noise reduction, we filtered the signal of each sensors with a low-pass filter of 30 Hz, 20th order. Source estimations were performed by a simplex method using Equivalent Current Dipole (ECD) model. Estimation error (E) was defined as $\Sigma\{(B_{ti} - B_{mi})^2 / B_{mi}{}^2\}$. Where B_{ti} is a magnetic field that is calculated from estimated sources at channel i, B_{ti} is a magnetic field measured at channel i. B_{ti} were calculated by the Grynszpan-Gesolowitz equation using a sphere model. In the sphere model we used, the surface of the sphere touched internally the scalp according to the medial sagittal MRI slice of each subject.

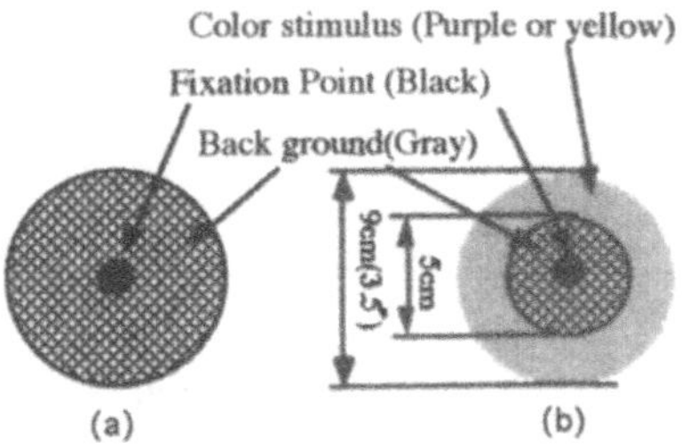

Fig.2 The visual stimulus. (a) The stimulus is turned on. (b) The stimulus is turned off.

Results

Superimpose waveforms:

The fig.3 shows the superimposed waveforms of responses to the purple frequent and the yellow infrequent stimuli. There were no remarkable responses to both waveforms until about the latency of 150 msec. Noise level of responses to infrequent responses was larger than noise level of responses to frequent responses, since the averaging number of infrequent responses was small. Visual evoked responses of the latency of almost 100 msec were not clear since stimuli had little luminance changes.

The responses to frequent stimuli had peaks around 220 and 380 msec. The component around 220 msec was not clear in S3. The component around 380 msec was not clear in S1 and late in S2. S4 had another component at 150 msec. The infrequent stimulus evoked large activities from 150 msec to the end of measurments in S1 and S2. In this response many waves inverted their polarity at the latency of 300 msec. In S3 and S4, the infrequent stimuli evoked activities from 300 msec to the end of measurements. According to above results responses to the infrequent stimulus were able to be classified into two components at least. The first component was from 150 to 300 msec and the second component was from 300 msec to the end of recordings.

Fig.4 Shows the responses of a typical sensor (the right lateral sensor) that recorded the activities to the infrequent stimulus after 300 msec. All subjects showed a large difference between responses to frequent and infrequent stimuli. The rise time of the responses after 300 msec changed slightly for each subject. Therefore it was supposed that the responses after 300 msec were the late complex responses which included P3a, P3b and slow waves, and the responses from 150 msec to 300 msec included N2 and P2 components which were in advance of P3.

Above mentioned characteristics were also shown when we exchanged infrequent and frequent stimuli, i.e. the yellow stimulus was frequent and the purple stimuli was infrequent. Hence, It is believable that the subjects performed the oddball paradigm well and the responses after 300 msec included the P3 components.

Morphology of isocontour maps:

Fig.5 Shows the isocontour maps of the responses to the infrequent stimulus at the latency of 350 and 450 msec of S1. The isocontour maps at 450 msec had sources of magnetic flux at the left frontal and right occipital areas, and sinks of magnetic flux at the right frontal and left occipital areas in all subjects. According to these maps, it seems that there was only one localized activity area in each hemisphere. The isocontour maps of the responses to the infrequent stimulus at 350 msec had more complex patterns than the maps at 450 msec. Therefore it reasonable that there were many active areas. The isocontour maps at 350 msec were similar to the maps at 450 msec except that the maps at 350 msec had a sink of magnetic flux at the left frontal area. The above characteristics of responses to the infrequent stimulus were also found in S3 and S4. However, in subject S2, the maps at 350 msec was similar to the maps at 450 msec.

Source Estimation:

Next we estimated the sources of the late components using two-ECD model. The sources of responses to the infrequent stimulus at the latency of 450 msec were estimated near the bilateral parahippocampal or occipitotemporal gyri. This estimation was consistent in all subjects. In each subject, the E(the estimation error) was lower than 10%.

The strengths of ECDs were from 200 to 400 nAm. The ECDs pointed to the medial direction. Fig.6 shows an example of the estimated source locations.

Sources of the responses to the infrequent stimulus at the latency of 350 msec could not be estimated through two-ECD model with a low estimation error. Contrary the sources of responses to the frequent stimulus at 350 msec were estimated around the deep part of occipital lobe with E less than 15 %. It might mean that the area activated by the frequent stimulus was more localized than the area activated by the infrequent stimulus. However estimated positions were not consistent in all subjects.

Discussion

The sources of the infrequent response at 450 msec were estimated near the bilateral parahippocampal or occipitotemporal gyri. These results are plausible since similar results were obtained using implanted electrodes by Halgren et al.[4]. The results of Halgren suggested that the sources were broad activities and the dipole strength was rather large in our results. Hence, our estimation may be the center of gravity of these broad active areas.

The sources of the responses to the infrequent stimulus at 350 msec could not be estimated by the two-ECD model with a low error. It suggests that these components contained many source generators. These responses probably contain some of P3a, P3b and slow wave. Rogers et al.[2] estimated sources of these late complex responses to the infrequent stimulus on the assumption that ECDs are spatially separated enough to have little effect on each other. They estimated each source separately using single-ECD model. They estimated sources in the medial temporal and occipital lobe. However, in some case it is thought that this assumption is not correct. Hence we tried to estimate the signal source from MEG whole cortex using a four-dipole model. In a few cases, the sources were estimated near the bilateral parahippocampal gyri and the insula with Es less than 10 %. Those results agree with the study of Halgren et al.[4] which reported the activities of the hippocampus and the superior temporal sulcus at 380 msec. Since estimation result with multi-ECD model could have many local solutions, we cannot accept this estimation results as it is. We believe that new evaluation standards through mathematical analysis and anatomical and physiological knowledge as a constraint of estimation should be developed to multiple source estimation.

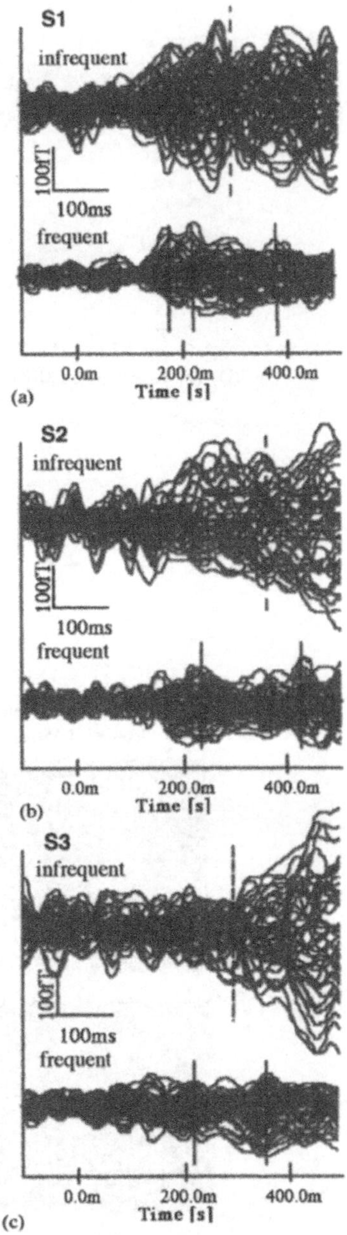

Fig.3 The superimposed MEG waveforms of S1, S2 and S3

References

[1]Okada T.C., Kaufman L. and Williamson S.J. The hippocampal formation as a source of the slow endogenous potentials. Electroenceph. clin. Neurophysiol., 1983, 55:417-426.

[2]Rogers R.L., Baumann S.B., Papanicolaou A.C., Bourbon T.W., Alagarsamy S. and Eisenberg, H.M. Localization of the P3 sources using magnetoencephalography and magnetic resonance imaging. Electroenceph. clin. Neurophysiol., 1991, 79:308-321.

[3]Takeda T., Morabito M., Kumagai T. and Endo H., use of CRT as a visual stimulator in MEG measurements, Proc.XI Int. Conf. on Event-related Potentials of the Brain Y.Koga, ed., Elsevier Int Congr Ser

[4]Halgren E., Baudena P., Clarke J.M., Heit G. Liegeois C., Chauvel P. and Musolino A., Intracerebral potentials to rare target and distractor auditory and visual stimuli. I. Superior temporal plane and parietal lobe, Electroencephalor. Clin. Neurophysiol., 1995, 94:191-220.

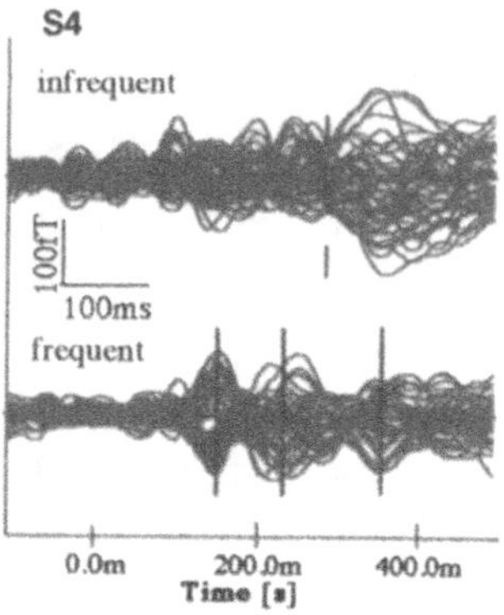

Fig.3(D) The superimposed MEG waveforms of S4

Fig.4 The MEG response of right lateral sensor

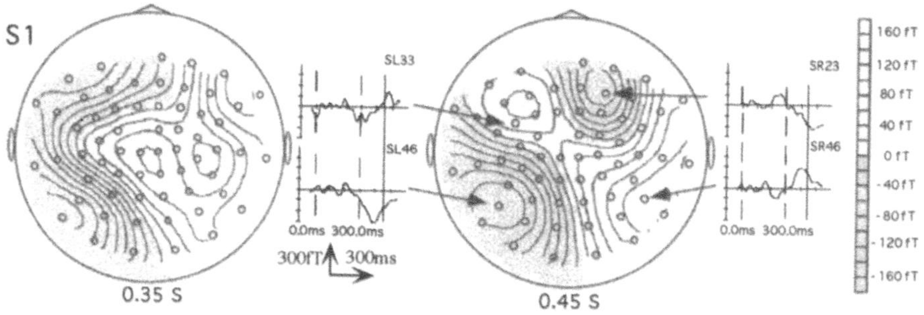

Fig.5 The isocontour maps. The reponses to the infrequent stimulus at the latency of 350 and 450 msec

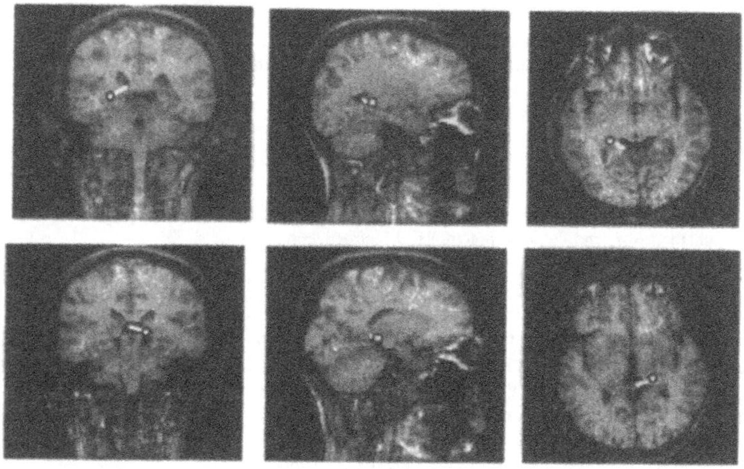

back view right view top view

Fig.6 Estimated generator sources at the latency of 460 ms for S2. E was 7 %. The source strength was 210 nAm for the left source and 230 nAm for the right source.

Cerebral Activities in the Occipital and Temporal Regions When Subjects Perform a Matching Task of Visual Stimuli

Kuriki, S.[1,2] and Hirata, Y.[1]
Hokkaido University[1,2], Sapporo, Japan
National Institute for Physiological Sciences[2], Okazaki, Japan

Introduction

One of the most exciting possibilities in the MEG study is that the time course of cerebral activation can be obtained within the brain anatomy of individual subjects during various information processing from sensory to higher functions. Previous MEG studies describe the brain activities associated with visual and higher processing such as object viewing [1] and naming [2]. The main objective of this study is to visualize the time course of the cerebral activation during the recognition of unreadable pseudo-characters in terms of graphical structure. This work would be contrasted with our previous studies on the visual and phonological processing of readable real-characters [3,4].

Method

Five right- and one left-handed male subjects (21-41 ys) participated in the experiments. In each of the 100 trials in a measurement session, the subject was presented with sequential stimuli of three characters (sample), a single character (probe), and a white circle (mark) (Fig.1). The visual stimuli were displayed on a screen in front of the subject's eyes with a visual angle of 0.9 deg of each character at a luminance of 4.3 cd/m^2 in the darkened magnetically shielded room. The three samples and the single probe were displayed symmetrically around the fixation point and at the left lower quadrant of 0.8 deg eccentricity, respectively. The stimulus characters were unreadable, which had been synthesized from Japanese Katakana scripts (readable characters) by shifting the line elements in the vertical or horizontal direction. These pseudo-characters were used as the sample and probe. The subject was instructed to fixate his eyes during the recording period after the probe onset before the mark, and to judge whether the shape of the probe agreed with that of the samples. He replied the match or mismatch by pressing one of two keys after the mark. After pilot measurements it was determined to use identical single pseudo-characters as the three samples to assure high performance of the task. The matching probability between the sample and the probe was adjusted to about 50 %.

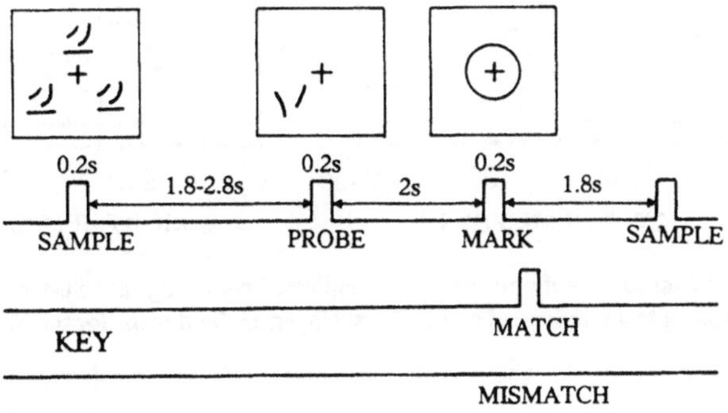

Fig. 1. Visual stimuli of pseudo-characters and the sequence of stimuli for the delayed matching task between the probe and sample characters.

MEG recordings were carried out in the magnetically-shielded room using a dual 37-channel SQUID with upper and lower 37-channel coil sets. The right and left sides of the head were simultaneously measured, varying slightly the recording area from anterior to posterior side in 6-8 sessions. We measured the responses to the single probe character in a 2.0 s recording period after the probe onset. The averaged responses over 100 trials were digitally filtered to 1-40 Hz, and the average in a 0.4 s prestimulus period served as the baseline. Match and mismatch responses were separately averaged. In the waveforms of these responses no significant difference was observed, and therefore we averaged the match and mismatch responses off-line.

Localization of single equivalent current dipoles (ECDs) was made, using a spherical conductor model, separately for the right and left cerebral hemisphere from the 37-channel fields recorded over each hemisphere. The localization was performed every 3.5 ms during the recording period, from which the ECDs having goodness-of-fit values more than 90 % in about 15 ms around the root-mean-squared (rms) peak were selected and averaged. Further averaging of the selected ECDs across recording sessions were made in individual subjects. In all subjects magnetic resonance (MR) tomography images, and surface images in some subjects, were obtained. The MR coordinates were transformed into the head coordinates, and the ECD location was registered in the MR images.

Results

The average of the reaction time for key press, which was measured separately for all subjects was 606 ms. The correct answer of the task recorded during the MEG measurements was above 90 % in most subjects, which indicates that the subjects correctly performed the task. Although the amplitude of the responses depended on the subject, three main peak components were observed in all subjects. The range of the rms peak latency of the 1st, 2nd, and 3rd component was 125-180, 180-250, and 250-460 ms, respectively. Fig. 2 illustrates the responses in duplicated recordings at selected coil locations of high-amplitude positive and negative fields for the three components. Isofield maps for the 2nd component, showing dipolar patterns, are also illustrated.

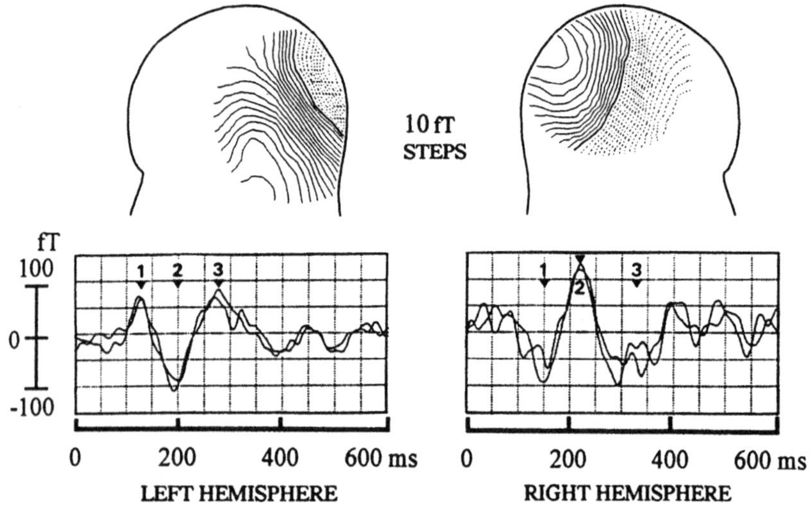

Fig. 2. Lower : waveforms of the responses in duplicated recordings at selected coil locations, where three main components are indicated by numbers. Upper: isofield maps for the 2nd component.

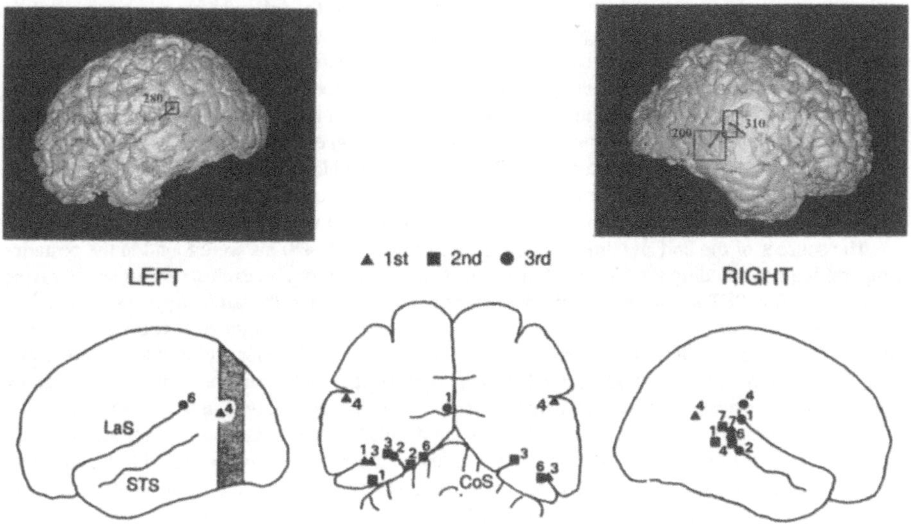

Fig. 3. ECD sources projected on the surface MR images and on the schematic views of the brain, where the sources were classified into three components and numbered to indicate subjects. The locus of the coronal sections is indicated by the hatched area shown on the left lateral surface.

The computed ECD sources were found at specific locations in, or close to, sulci, reflecting the fact that cortical currents tangential to the skull exclusively contribute to the neuromagnetic field. From the projection on the MR images the source location in the individual brain structure was determined in reference to the main sulci. The obtained brain sites of the three sources in all subjects are plotted on the schematic maps shown in Fig.3, together with representative source location on the brain surface MR images. These results indicate that the 1st and some of the 2nd component sources are located at the inferior part of the temporo-occipital region around the collateral sulcus (CoS). This area is included in the posterior part of the fusiform gyrus (FG) and the inferior temporal gyrus (ITG). In the lateral temporal region the 2nd and 3rd component sources are located at the posterior part of the superior temporal sulcus (STS) and the posterior end of the lateral sulcus which is circumscribed by the supramarginal gyrus (SmG). The right-hemisphere preponderance of these sources is a marked contrast to the left-hemisphere dominant activities observed in the readable character task [3,4].

Discussion

From the average reaction time of 606 ms and plausible transmission time from the motor cortex to the muscle for the key press, the judgment of the matching of the probe with the samples should be completed before about 500 ms. Therefore, the observed 1st to 3rd components before 460 ms would reflect the processes involved in the visual perception, recognition, and comparison.

The density of the 1st and 2nd component sources in the FG and ITG in the inferior temporo-occipital extrastriate cortex was left hemisphere dominant, suggesting stronger neural activity in the left side than the right. Sakurai et al. [5] reported the activation of the regional cerebral blood flow in the extrastriate cortex in the posterior inferior temporal region, in addition to that in the medial and lateral occipital region, during a reading task of Japanese characters (words and non-words). The activation in the inferior temporal region is significant only on the left side for non-words, and on both sides for real-words but stronger on the left than the right. On the other hand, in visual attention during a shape-matching task of squares and rectangles, significant activation is observed in the occipital inferior region in the lingual gyrus and in the occipitotemporal FG, with stronger amplitude on the left side than the right [6]. Enhancement of the activity in the FG by selective attention is also reported [7]. Considering these results of PET studies, it is suggested that the early responses in the FG and ITG in our pseudo-character task reflect the early graphical processing of the visual stimuli, but are not specific to the scripts or characters. The pseudo-characters are not familiar to the subjects, so that there is strong demand on the analysis of the structure and visual attention.

The sources of the 2nd and 3rd components occurring in 180-460 ms were found in the posterior temporal region including the STS and SmG. This temporal activity was exclusively unilateral on the right side. A few PET studies have delineated the cortical activities in the lateral temporal region. In a shape attention task the activity was significantly elevated along the right STS (Broadmann's area (BA) of 21/22) [6], while in an object discrimination task the left inferior and middle temporal gyri (BA 20, 21) were more strongly activated than those on the right, but the SmG was bilaterally activated [8]. In macaques, areas in the inferior temporal cortex are involved in the object recognition and discrimination [9]. In our previous study of the delayed matching task using Japanese readable characters [3,4] the posterior STS and SmG in the left temporal region are suggested to be related to the phonological process, which is required in the recognition of real characters. These active sites are homologous to the location of the 2nd and 3rd component sources in the right hemisphere in the present study. Thus it may be suggested that the left and right human temporal cortices are specialized into the phonological (more generally, linguistic) and graphical/structural processing. Further studies on the specific functional role of the activities in the right posterior STS and SmG are needed.

In conclusion, the MEG study combined with cognitive task of delayed matching of visually-presented unreadable pseudo-characters revealed the early neural activities in the inferior temporo-occipital extrastriate cortex and the later activities in the right posterior temporal cortex. These activities may be the neural substrate which subserves the visual and subsequent structural processing of the pseudo-characters.

References

[1] Lu ST, Hämäläinen MS, Hari R, Ilmoniemi RJ, Lounasmaa OV, Sams M, Vilkman V, Seeing faces activates three separate areas outside the occipital visual cortex in man. Neurosci. 1991, 43:287-290.

[2] Salmelin R, Hari R, Lounasmaa OV, Sams M, Dynamics of brain activation during picture naming. Nature, 1994, 368:463-465.

[3] Kuriki S, Hirata Y, Fujimaki N, Tsuchiya N, Kobayashi T, Neuromagnetic study of human cortical areas related to visual/auditory information processing. Human Brain Mapping, 1995, Suppl.1: 27.

[4] Kuriki S, Hirata Y, Fujimaki N, Kobayashi T. Cerebral activities of the human brain in a delayed matching task of visual characters. Electroenceph. clin. Neurophysiol. (to be published).

[5] Sakurai R, Momose T, Iwata M, Watanabe T, Ishikawa T, Kanazawa I, Semantic process in Kana word reading: activation studies with positron emission tomography. NeuroRep. 1993, 4:327-330.

[6] Corbetta M, Miezin F, Dobmeyer S, Shulman GL, Petersen SE, Attentional modulation of neural processing of shape, color, and velocity in humans. Science 1990, 248:1556-1559.

[7] Heintz HJ, Mangun GR, Burchert W, Hinrichs H, Scholz M, Münte TF, Gös A, Scherg M, Johannes S, Hundeshargen H, Gazzaniga MS, Hillyard SA, Combined spatial and temporal imaging of brain activity during visual selective attention in humans. Nature, 1994, 372:543-546.

[8] Sergent J, Ohta S, Macdonald B, Functional neuroanatomy of face and object processing. Brain, 1992, 115:15-36.

[9] Gross CG, Rocha-Miranda CE, Bender DB, Visual properties of neurons in inferior temporal cortex of the macaque. J. Neurophysiol. 1972, 35:96-111.

Magnetoencephalographic study on cerebral cortical activities related to speech

Kyuhou, S., Sasaki, K., Nambu, A., Matsuzaki, R., Tsujimoto, T. and Gemba, H.

National Institute for Physiological Sciences, Okazaki, Japan

Introduction

Human motor speech function has been investigated neuropsychologically [1]. Recently by using non-invasive techniques measuring regional cerebral blood flows such as the functional magnetic resonance imaging (fMRI) [3] and the positron emission tomography (PET) [5], it was demonstrated that several cortical areas not only in the dominant hemisphere but also in the non-dominant hemisphere were activated during speech tasks. As these methods, however, have poor time resolution, it is difficult to get temporal information of the cortical activities. The magnetoencephalography (MEG) is suitable for investigating the temporal sequence of the activation of cortical areas on speech, since MEG has good temporal and spatial resolution. The present study is a trial to identify functionally the motor speech center in the frontal lobe of human cerebral cortex by using multichannel SQUID gradiometers. Parts of the results have been reported shortly [8].

Methods

Subjects were six healthy adults of 29-64 years old, five males (four of right handedness, one of left handedness) and one female (right handedness). Handedness was assessed according to Edinburgh Handedness Inventory by Oldfield [4]. An examinee lay on a bed in a magnetically shielded room and two sets of 37-channel SQUID gradiometers (Magnes Model 700, BTi) were attached to both sides of the head for simultaneous magnetic recordings from both the left and right hemispheres. EEG electrodes were also placed over the frontal-parietal part of the head according to the international 10-20 method. Light stimuli of two different colors were delivered in front of a subject for 500 ms in irregular order and random time intervals of 6.0-9.0 s. The subject should respond to either of the stimuli by uttering a short word (noun), e.g., 'pan' [pan] ("bread" in Japanese), 'en' [en] ("round" in Japanese) or the other by vocalizing a short simple voice, e.g., 'pa' [pa], 'e' [e], respectively. The initial sounds of both voices are to be the same, i.e., 'pa' [pa], 'e' [e] in these examples. The subject was asked to recall the meaning of the word every time it was uttered, or to think nothing while uttering the simple voice. Magnetic fields were respectively averaged by the onset pulses of light stimuli 100 times. EMGs were recorded from perioral muscles with silver surface electrodes, rectified and averaged by the pulse. The onset of voice was also monitored with a microphone and was found to be considerably later than that of perioral EMGs.

Results

Evoked magnetic fields (EMFs) preceding utterance of a short word and vocalization of a simple voice are shown respectively in Fig. 1 A and B. The subject is right-handed. Two components of EMFs preceding the utterance were observed over the lateral frontal parts on both sides. The early component at a peak latency of 120-165 ms from the onset of light stimulus was prominent on utterance of the word but this component was hardly seen on that of the simple voice. The late component of magnetic changes at a peak latency of 160-190 ms was observed similarly on both the utterances. The late component appeared about 20-40 ms before the start of perioral EMGs. Perioral EMGs started about 170-220 ms after the onset pulse of light stimuli and differences were hardly observed between latencies of the EMGs on the word and simple voice utterances. These findings suggest that the early activity is important for speech of the word. On the other hand, the late component is thought to be related to the execution of the articulatory movements because latencies between this component and the onset of the articulatory movements were constant. By using the dual 37-channel system, magnetic fields were simultaneously recorded from both the left and right hemispheres (Fig. 2). The time courses of some of the EMFs were similar in both the left and right hemispheres as exemplified in Fig. 2 A and B. There were no detectable differences between respective latencies of the EMFs on both sides in these examples. Locations of the current dipoles responsible for

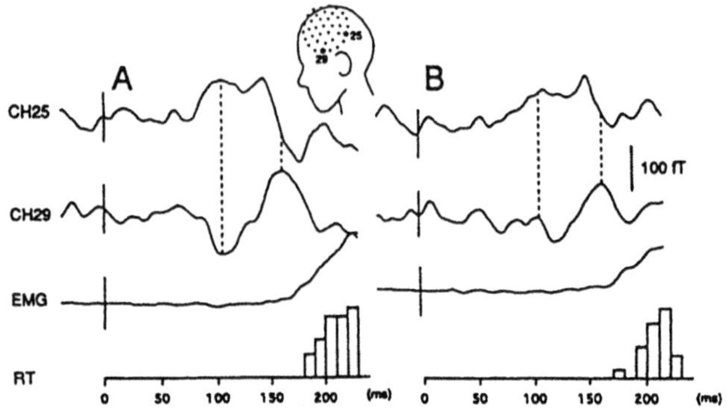

Fig. 1: Evoked magnetic fields on utterance of a short word (A) and on that of a simple voice (B) recorded from the left dorsolateral part of the head of a right-handed subject. Two sites of specimen records (CH25, 29) were indicated by asterisks among dots for 37-channel sites on the head. Reaction times (RT) were measured from the onset of the visual stimuli to that of the EMGs of perioral muscles.

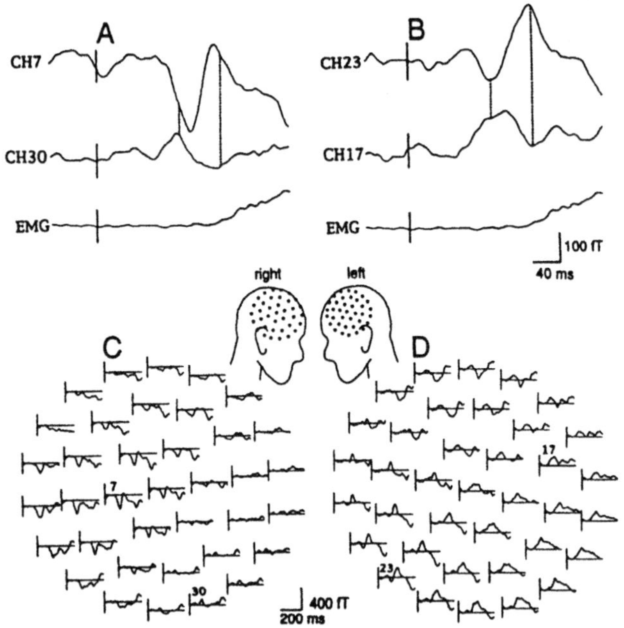

Fig. 2: Simultaneous recordings of magnetic fields from both the right (A, C) and left (B, D) hemispheres on utterance of a short word by using a dual 37-channel system. Records of the channels 7 and 30 in C and the channels 23 and 17 in D are respectively enlarged in A and B with EMGs of perioral muscles rectified. All records were aligned 100 times by the onset pulse of light stimuli. Positions of sensor coils are shown by dots on the right and left side view of the head shapes in the middle. The subject is the same as in Fig. 1.

the two components of EMFs were estimated and were plotted on MRI pictures of the subject (Fig. 3). The estimated current dipoles for the early component were located in the inferior lateral part of the prefrontal cortex (Fig. 3 A1, A2, B1, B2). Those for the late component were positioned on the lateral part of the frontal-parietal cortical areas around the central sulcus (Fig. 3 C1, C2, D1, D2). The estimated current dipoles appeared in the quite symmetrical places in the left and right hemispheres. These results indicate that the inferior lateral prefrontal and motor-sensory cortices are sequentially activated prior to speech in both the hemispheres.

Four other subjects including one of left handedness also showed findings similar to those described above. The remaining right-handed subject revealed current dipoles at latencies of 151-165 ms only in the right prefrontal cortex. Similar activities were sometimes noted also in the left hemisphere but not so conspicuous as in the right hemisphere. This subject showed the frontal mental theta wave on calculation more predominantly in the right frontal lobe than in the left one [7] and the dominant hemisphere would be presumed on the right side.

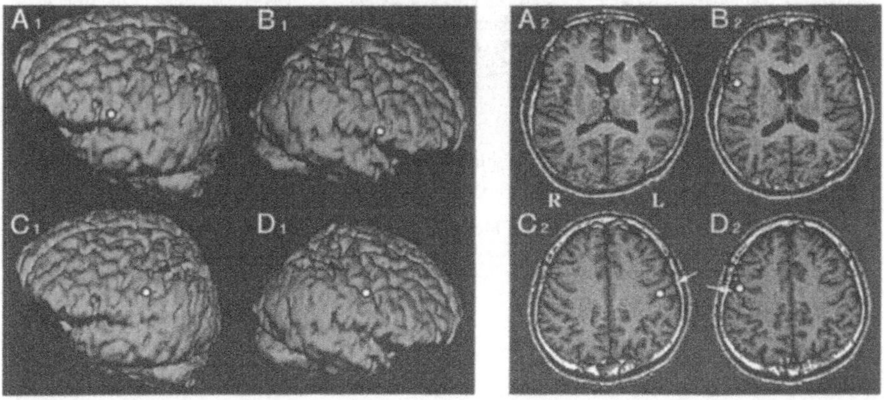

Fig. 3: Locations of the equivalent current dipoles (ECDs) were plotted on the 3-dimensionally reconstructed MRI pictures of the brain (A1-D1). The same results were also plotted on the horizontal sections of MRI pictures (A2-D2). Open circles in A and B indicate the locations of the ECDs responsible for the early component (138 ms after the onset of visual stimulus on both the left and right sides) while those in C and D show the ECDs responsible for the late component (167 and 186 ms after the visual stimuli respectively on the left and right side). The capital letters (R and L) in A2 respectively indicate the right and left side of the head. The arrows indicate the central sulcus. The same subject in Fig. 1 and Fig. 2.

Discussion

It was found in the present study that the inferior lateral prefrontal cortex and the motor-sensory cortex were sequentially activated prior to utterance of a word while only the motor-sensory cortex was activated in preceding vocalization of a simple voice. Pronounced activation of the inferior lateral prefrontal cortex before uttering a word suggests that this cortical area is the motor speech center related to organizing words. The area appears to correspond to the Broca's speech area. In fact, the equivalent current dipoles were located in the frontal opercular region (Broadmann's area 44-45) that is thought to be the Broca's speech area. The motor-sensory cortex activated both on utterance of a word and a simple voice probably corresponds to the face area because it was located in the cortex of the lateral part of the central sulcus. Activation of the face area may induce the articulatory movements for production of words and simple voices.

The present study demonstrated that the bilateral frontal cortices were activated prior to the utterance. Recent MEG [6] and EEG [2] studies also reported the activation of the bilateral frontal cortices on speech tasks such as the picture naming and self-paced utterance. These results indicate that not only the dominant hemisphere but also the non-dominant hemisphere plays important roles on utterance in the normal speech, although the speech center in the dominant hemisphere would be indispensable and crucial.

References:

[1] Broca, P. Remarques sur le siège de la faculté du langage articulé, suivies d'une observation d'aphémie, Bull. Soc. Anat., 1861, 36: 330-357.

[2] Deecke, L., Engel, M., Lang, W. and Kornhuber, W.W. Bereitschaftspotential preceding speech after holding breath, Exp. Brain Res., 1986, 65: 219-223.

[3] Hinke, R.M., Hu, X., Stillman, A.E., Kim S.G., Merkel, H., Salmi, R. and Ugurbil, K. Functional magnetic resonance imaging of Broca's area during internal speech, Neuroreport, 1993, 4: 675-678.

[4] Oldfield, R.C. The assessment and analysis of handedness: the Edinburgh Inventory., Neuropsychologia, 1971, 9: 97-113.

[5] Roland, P. Brain Activation, New York, Wiley-Liss, 1993.

[6] Salmelin, R., Hari, R., Lounasmaa, O.V. and Sams, M. Dynamics of brain activation during picture naming, Nature, 1994, 368: 463-465.

[7] Sasaki, K., Tsujimoto, T., Nambu, A., Matsuzaki, R. and Kyuhou, S. Dynamic activities of the frontal association cortex in calculating and thinking, Neurosci. Res., 1994, 19: 229-233.

[8] Sasaki, K., Kyuhou, S., Nambu, A., Matsuzaki, R., Tsujimoto, T. and Gemba, H. Motor speech centres in the frontal cortex, Neurosci. Res., 1995, 22: 245-248.

Acknowledgements

We thank Mr. O. Nagata and Mr. Y. Takeshima for their technical help.

Present Address

Sasaki, K. and Tsujimoto, T. : Faculty of Medicine, Kyoto University, Kyoto 606-01, Japan.
Gemba, H., Kyuhou, S. and Matsuzaki, R. : Kansai Medical University, Moriguchi 570, Japan.
Nambu, A. : Tokyo Metropolitan Institute for Neuroscience, Fuchu 183, Japan.

Magnetic Recording of the Propagation of Motor Unit Action Potentials in the Human Leg Muscles

Masuda, T., Endo, H., Kumagai, T., and Takeda, T.

National Institute of Bioscience and Human-Technology, Tsukuba, Japan

Introduction

We have developed multichannel electromyography (EMG) for detecting the propagation of motor unit action potentials [1]. With this non-invasive technique it becomes possible to define motor endplate regions based on the source of the propagation. The measurement of muscle fiber conduction velocity has also become possible (Fig. 1). The purpose of the present study was to clarify the possibility of magnetic recording technique for detecting the activity in human skeletal muscles and to develop another new non-invasive method for analyzing and diagnosing human motor functions. With the magnetic recording we can expect better localization of current sources compared with the electrical recording.

Methods

We used a 64-channel SQUID system developed by CTF Systems (Canada). Since this system was designed to detect the magnetic field originating from the human brain, the detecting coils were arranged on a surface of hemisphere. The inter-coil distance was about 40 mm.

The subject were three normal males. They put the knee in the hollow of the Dewar vessel with fully flexing

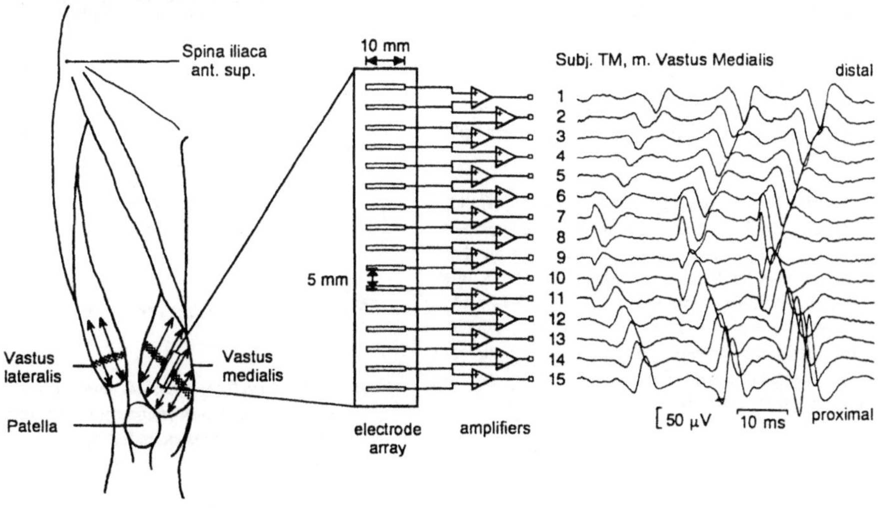

Fig. 1 Propagation of motor unit action potentials in the vastus medialis and the vastus lateralis detected with a linear array of surface electrodes. The shaded area indicates the motor endplate regions defined by the source of the propagation.

the knee joint (Fig. 2). The subject isometrically exerted the force of knee extension. We recorded the magnetic field for 10 s with a sampling frequency of 1.25 kHz. After the recording the sampled signals were high-pass filtered above 5 Hz with a Butterworth filter to eliminate motion artifact. In this setup the motor endplate regions of the vastus medialis and the vastus lateralis were located near the edge of the recording area and the proximal side of the muscles was outside of the SQUID sensors.

To detect discharges of single motor units we used a technique of EMG feedback. Bipolar electrodes were attached on the belly of the vastus medialis. One of the electrodes was placed on the motor endplate regions. The onset of the EMG therefore indicates the start of the depolarization on muscle fibers (Fig. 1). The EMG signal was amplified and presented to the subject as an amplitude modulated sound. The subject controlled the contraction force to produce spike trains that appear at regular intervals with the same intensity. After the recording we averaged the signals from SQUID sensors by using the zero-crossings of the EMG as triggers.

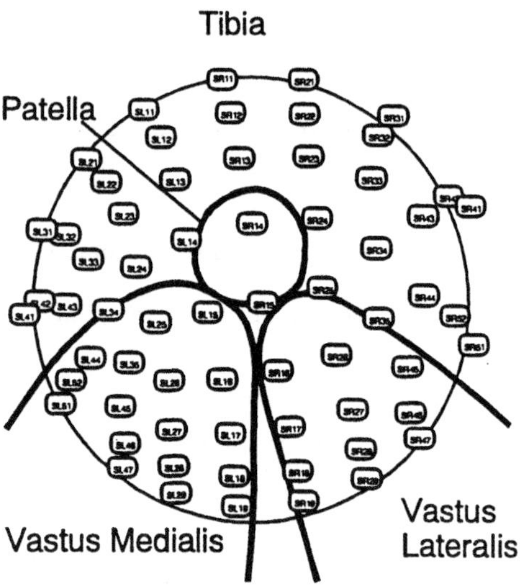

Fig. 2 Configuration of the SQUID sensors and the location of the knee extensor muscles in the recording area.

Results

The amplitude of the magnetomyographic (MMG) field recorded during moderate voluntary contraction was 10-20 pT peak-to-peak (Fig. 3), which was 10-100 times greater than that of magnetoencephalogram. There was no difficulty in recording MMG signals. MMG signals were sensitive to motion artifact, which was however removed by high-pass filtering above 5 Hz.

The isofield maps in Fig. 4 showed magnetic field spreading from the proximal side of the vastus medialis to the direction of the patella. The similar spreading was observed for some of the peaks and may correspond to the propagation of motor unit action potentials shown in Fig. 1. However most of the fields showed more complex distribution that was difficult to be explained by a simple moving dipole.

To clarify the components constituting MMG field, we analyzed MMG of a single motor unit. The averaged MMG and EMG signals are shown in Fig. 5. The number of averaging was 84 for this record. MMG of a single motor unit had smaller amplitude of 1-2 pT peak-to-peak. The isofield map of the averaged MMG still showed complex distribution preceding a clear and strong dipole pattern of reversed polarity (Fig. 6). The first map was at the onset of the depolarization.

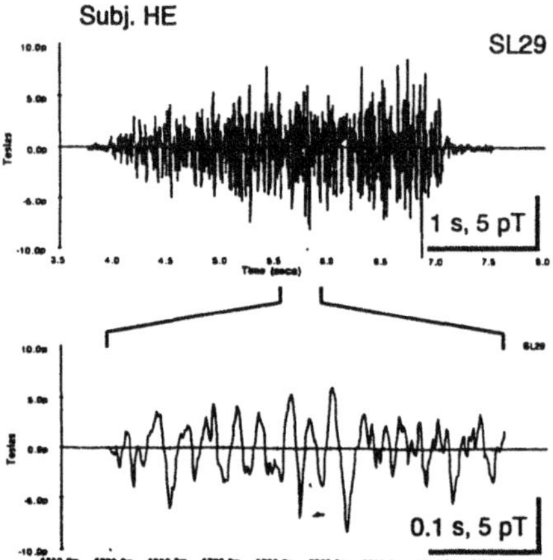

Fig. 3 Raw waveform of MMG detected under moderate voluntary contraction. A portion of the waveform is extended in time (below).

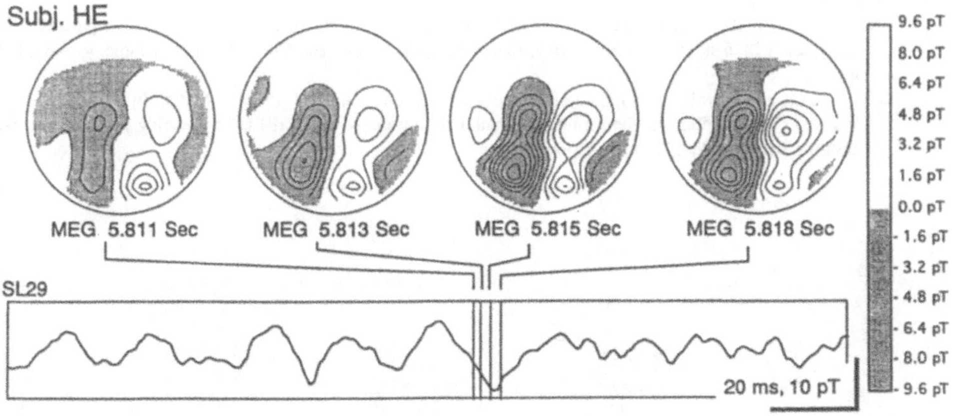

Fig. 4 Isofield maps of the MMG shown in Fig. 3.

Discussion

The magnetic field produced by discharges of single motor unit is explained by a pair of tripoles that arise at the motor endplate regions and propagate in opposite directions along the muscle fibers (Fig. 7) [2]. The currents in each tripole flow in opposite directions and therefore the magnetic field produced by the tripole is cancelled by each other and weak. The resulting magnetic field is equivalent to that produced by a dipole of a current flowing in the direction of the propagation. When the depolarization reaches the tendon, the first current of the tripole diminishes and the second current produces a strong magnetic field with an opposite direction. In the field shown in Fig. 6 the tripole travelling in the proximal direction reaches the tendon earlier than the distal tripole.

If the SQUID sensors are closer to the muscle fibers and have a higher resolution of, for example, 5 mm, the magnetic field of the tripole may become evident. With the resolution of 40 mm, only a subtracted current of the tripole is detected as a dipole. The low resolution causes the complex magnetic field for single motor unit action potentials and more complex field for interference MMG.

As a conclusion we must detect discharges of single motor unit to analyze the magnetic field originating from skeletal muscles or improve the spatial resolution of the SQUID system. If these conditions are satisfied, the magnetic recording technique could have an advantage in localizing the position and in estimating the absolute strength of current sources compared with the conventional EMG recordings.

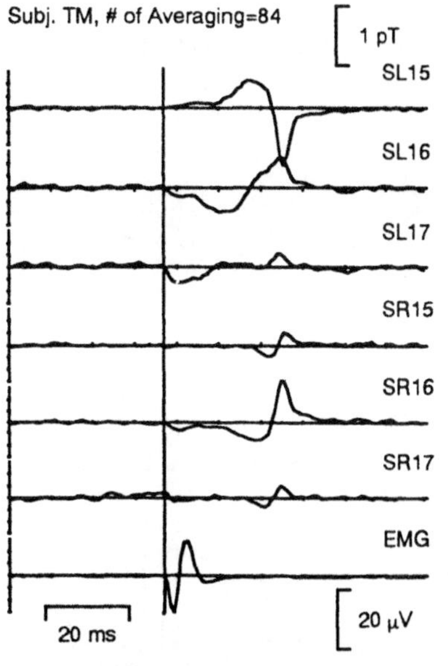

Fig. 5 MMG and EMG of a single motor unit after averaging. The positive-going zero-crossings of EMG were used as triggers.

References:

[1] Masuda, T., and Sadoyama, T. Skeletal muscles from which the propagation of motor unit action potentials is detectable with a surface electrode array, Electroencephalography and Clinical Neurophysiology, 1987, 67: 421-427.

[2] McGill, K. C., and Huynh, A. A model of the surface-recorded motor-unit action potential, In: Harris, G., and Walker, C. Proceedings of the Annual International Conference of the IEEE Engineering in Medicine and Biology Society, 1988, 10: 1697-1699.

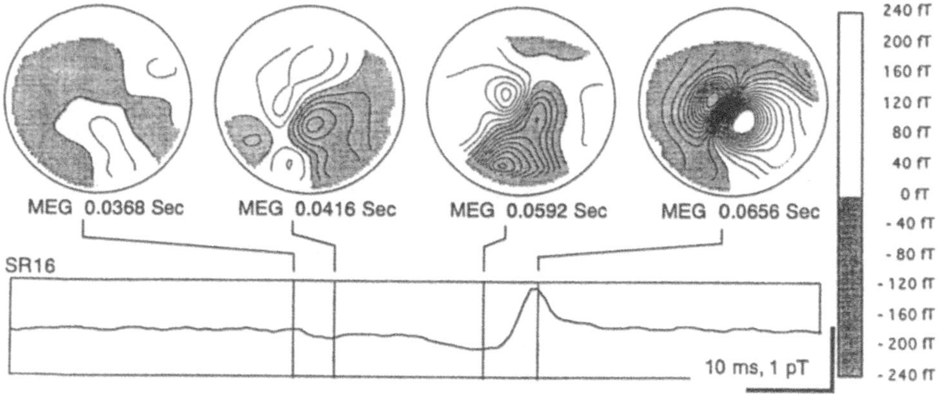

Fig. 6 Four typical isofield maps of MMG produced by a single motor unit shown in Fig. 5.

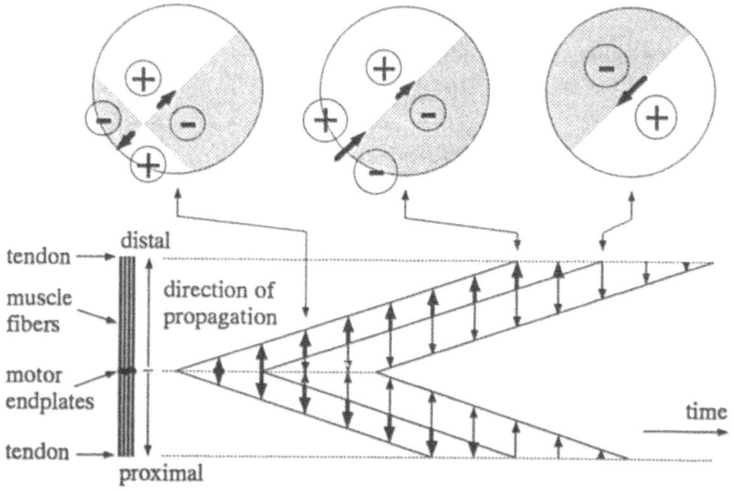

Fig. 7 Schematic diagram explaining the magnetic field produced by a single motor unit. The current in muscle fibers is modeled as a pair of tripoles that propagate in opposite directions to the tendons.

Effect of Luminosity Change on Visual Evoked Field

Matsuhashi, M.[1], Oyanagi, H.[1], Kumagai, T.[2], Endo, H.[2] and Takeda, T.[2]

[1]*Department of Ophthalmology, Toho University School of Medicine, Tokyo, Japan;*
[2]*National Institute of Bioscience and Human-Technology, Tsukuba, Japan*

Introduction

The study of vision in an objective manner is important for a better understanding of the factors which influence the normal visual system [1] [2]. By using the visual evoked potential (VEP), efforts have been made to measure the vision objectively [3]. Responses elicited by flash stimulation are called flash or luminance VEPs. Increase of luminosity in flash stimulation shows larger responses in the flash VEPs. This study is to show the effects of flash intensity on the flash visual evoked brain magnetic field (VEF).

Methods

VEFs were recorded by 64 channel whole head type magnetoencephalogram (MEG) systems by CTF Systems Inc. of Canada. The dewar was designed especially for Japanese people. Other than MEG system, reference SQUIDs for enviromental noise reduction, head positioning system were also included in the whole systems. Measurements were carried out in an AK3b magnetically shielded room by Vacuumschmelze GmbH of Germany.

Setup of hardware, recording and processing of data were made by the MEG software of CTF Systems. Butterworth low pass filter at 40 Hz and notch filters at 50, 100 and 150 Hz were used for recording. Artifact rejection was set on for noise produced by blinking.

Responses were recorded 100 msec for pretrigger and 600 msec for posttrigger. In this paper, however, results are shown only from -100 msec to 400 msec. Signals were averaged 32 times. For data processing, offsets were removed, digital filter of low pass at 40 Hz was applied and third gradient formation was done with the software for the noise reduction [4]. Head positioning data was recorded each time.

A red light emission diode (LED) was used for the flash stimulation. Drive unit for LED was hand made and it was controlled by a personal computer. Stimulus frequency was randomized and its mean was 0.7 Hz, stimulus duration was 10 msec. An LED for fixation was placed 50 cm in front of the eyes and one for stimulus was put 2.9 degrees to the left. The intensity of stimulus was changed by use of neutral density filters to reduce the relative intensity to 10^{-1}, 10^{-2}, and 10^{-3}.

Responses were recorded from three normal subjects. Responses by each stimulus intensity were compared and analyzed. All subjects showed major peaks at around 100 msec and 160 msec. They were called P100-N100*m* and P160-N160*m* respectively in this paper. Amplitude was measured peak-to-peak, from P100-N100*m* to P160-N160*m*. Peak time latency was measured from the onset of the stimulus to the each peak.

Results

With the increase of stimulus intensity, larger responses were recorded. Fig. 1 shows responses at sensor SR17 in subject 1. Amplitude of P160-N160*m* became larger with the increase of stimulus intensity. As for the peak time, they did not differ significantly both at P100-N100*m* and P160-N160*m* in these recordings. Maps at P100-N100*m* in each stimulus intensity are shown in Fig.2, and the location of sensor SR17 is indicated.

Responses at sensor SR27 in subject 2 are shown in Fig.3. Similar rersults in amplitude were recorded. The peak time of P160-N160*m* became shorter in stronger stimulus intensities in these recordings.

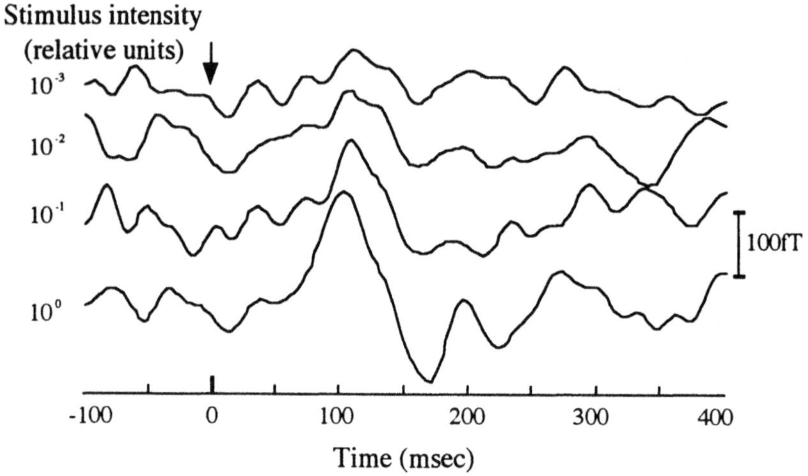

Fig. 1 Visual evoked magnetic fields recorded at sensor SR17 in subject 1. Note that the stronger the stimulus intensity was, the larger the responses became at P100-N100*m* and P160-N160*m*. This phenomenon has been recognized in all subjects. Peak time did not differ significantly both at P100-N100*m* and P160-N160*m* in these recordings. Arrow indicates stimulus onset in this and following figures.

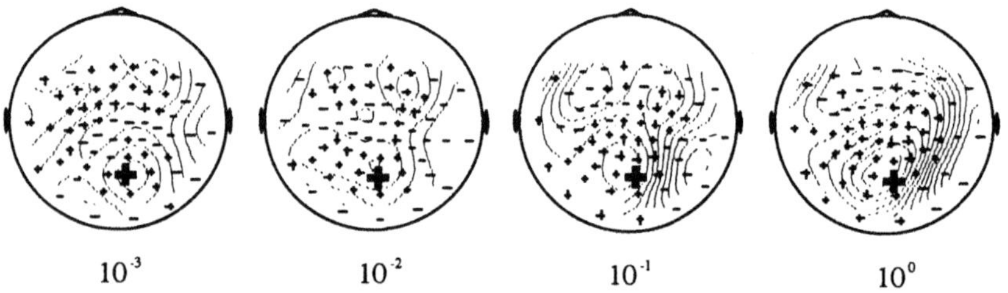

Fig. 2 Maps at P100-N100*m* in subject 1. These maps showed similar topography, suggesting that responses of P100-N100*m* by each stimulus intensity represented similar magnetic field changes. Large bold plus designates sensor SR17, from where responses are shown in Fig. 1.

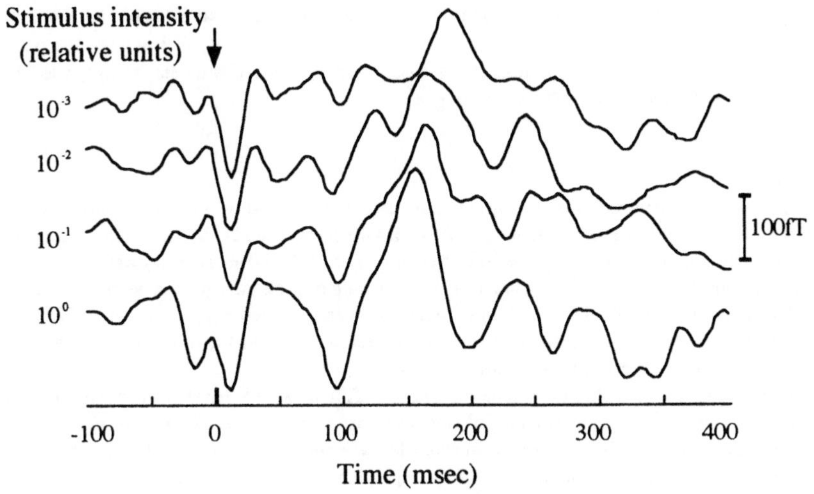

Fig. 3 Visual evoked magnetic fields recordedat sensor SR27 in subject 2. Note that the stronger the stimulus intensity was, the larger the responses became at P100-N100*m* and P160-N160*m* as found in Fig. 1. As for the peak time, peak time delay was recognizedat P160-N160*m* in accordance with the diminution of stimulus intensity in these recordings.

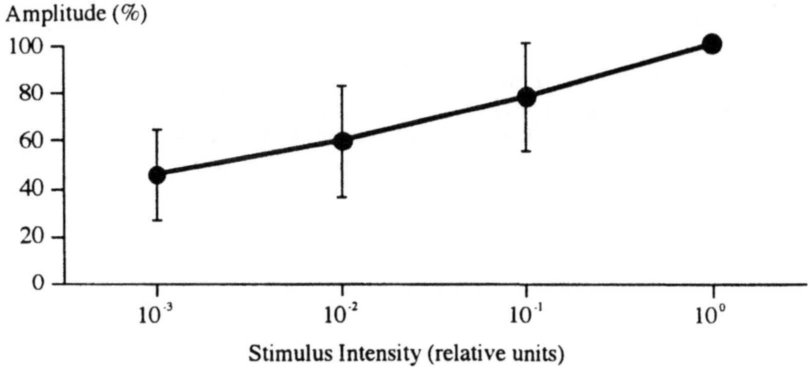

Fig. 4 The amplitude (peak-to-peak) of the averaged evoked magnetic field of P160-N160*m* as a function of the stimulus intensity. The mean amplitute got larger in accordancewith the increase of stimulus intensity. By the present stimulus intensity, the amplitude kept rising and there was no steady or decrease. The vertical lines indicate one standard deviation.

In all subjects, with the increase of stimulus intensity, larger responses were recorded. Amplitude of P160-N160*m* was compared in each subject. Amplitude with the stimulus intensity of 10^0 was made 100 per cent and relative amplitude was calculated. The mean and one standard deviation with each stimulus intensity was 77.9 ± 22.5 at 10^{-1}, 59.5 ± 23.5 at 10^{-2}, and 45.7 ± 19.1 at 10^{-3} (Fig. 4).

Discussion

This study showed the effect of stimulus intensity change on the flash VEFs. The stronger the luminosity was, the larger the responses became as found in VEPs. By the present condition, the amplitude kept rising and there was no steady or decrease. This suggested that the intensity of LED used in this study was not too strong. As for the latency, with the increase of the flash intensity, shortening of peak time was found in most recordings as shown in Fig. 2, however, peak time didnot differ significantly in some recordings as shown in Fig.1. Because peak time change was not always consistent, it was not analyzed and illustrated as in Fig.4 in this study. Further study would be needed for analyzing the characteristics in peak time.

Maps shown in Fig. 2 showed similar topography. This suggested responses compared represented the similar ones. However, topograms were not exactly the same configuration. And the source modeling was done in each record and the location of the equivalent current dipole was estimated. Also, they did not show exactly the same position. Therefore, comparing the size of the equivalent current dipole was not applied in this study. However, when those results could.show exactly the same configuration and position, it would be worth comparing the size of the equivalent current dipole. In this study, definite results were found by comparing the amplitude of responses.

References

[1] Matsuhashi, M., Kumagai, T., Endo, H., Nishimura, C., and Takeda, T. Visual evoked multichannel magnetoencephalography, Japanese Journal of Clinical Ophthalmology, 1995, 49:403-406.

[2] Matsuhashi, M.,Oyanagi, H., Kumagai, T., Endo, H., and Takeda, T. Binocular summation in visual evoked brain magnetic fields by pattern stimulation, The Journal of Japan Biomagnetism and Bioelectromagnetics Society, 1995, 8:80-83.

[3] Matsuhashi, M., Oguchi, Y., and Uemura, Y. Clinical ERG and VECP from unsedated children aged under one year, Japanese Journal of Ophthalmology, 1979, 23(supp.):279-285.

[4] Takeda, T., Endo, H., and Kumagai, T. On the measurement reliability of the 64-channel whole-cortex MEG system, The Journal of Japan Biomagnetism and Bioelectromagnetics Society, 1994, 7:224-227.

Evoked Magnetic Fields for Gustatory Stimuli on the Tongue

Murayama, N.[1], Nakasato, N.[2], Hatanaka, K.[3], Fujita, S.[4], Igasaki, T.[1], Kanno, A.[5] and Yoshimoto, T.[2]

[1]*Department of Electrical Engineering & Computer Science, Kumamoto University, Kumamoto, Japan;* [2]*Department of Neurosurgery, Tohoku University School of Medicine, Sendai, Japan;* [3]*KRI, Kyoto, Japan;* [4]*R&D Center, Osaka Gas Co. Ltd., Osaka Japan;* [5]*MEG Laboratory, Kohnan Hospital, Sendai, Japan*

Introduction

Animal studies have suggested that gustatory information primarily projects to the bilateral frontal opercula and dorsal insulae [1, 2, 3]. However, clinical studies showed that lesions at the postcentral gyrus [4, 5] or anterior insula [6] caused gustatory disturbances. Scalp topography of gustatory evoked potentials (GEPs) has identified the largest amplitude (about 450 ms in latency) at the central vertex [7]. However, it is difficult to localize the primary gustatory area by non-invasive scalp GEP recording, due to inhomogeneous head conductivity and possible overlaps of bilateral activities. Recently, magnetoencephalography (MEG) has been introduced to localize the sources of various evoked responses. MEG is known to be less influenced by the inhomogeneous head conductivity than electroencephalography. For example, the bilateral cortical sources of auditory evoked magnetic fields were successfully separated and localized precisely on the primary auditory cortex [8]. In the present study, we measured evoked magnetic fields for gustatory stimuli on the tongue in an attempt to identify the location of the primary gustatory area in the human cerebral cortex.

Methods

Five normal subjects participated in this experiment. The subject was seated in a magnetically shielded room with the chin resting on a frame. Taste stimuli were delivered through a tube to the anterior region of the tongue which was protruded slightly out of the mouth. The starting point of the stimulation was detected by an optical sensor mounted at a distance of 5 mm from the nozzle of the tube. As the taste stimuli, 10% glucose and 0.3 mol NaCl were used. Responses obtained by these taste stimuli were compared with that of distilled water (non-taste control). The duration of taste stimuli was about 10 sec. During the interstimulus periods, the tongue was rinsed with distilled water delivered for about 10 s. During the interstimulus period, the tongue was rinsed with distilled water for about 10 s. MEG recording used a helmet-shaped MEG system with 64 channels (CTF Systems - Osaka Gas). Evoked magnetic fields were measured over the entire head except lower part of the face. Signals were averaged over 40 stimulus presentations and two separate sessions were performed for each stimulus condition. A two-dipole model was employed for source estimation of evoked responses, using the best fit sphere for each subject's head surface based on MRI data. Estimated dipoles were superimposed on MRI. Somatosensory evoked fields (SEFs) for median nerve stimuli were measured as physiological landmarks of the central sulcus.

Results

Figure 1 shows the typical waveforms of evoked fields due to the stimulus of 10% glucose on the tongue. Bilateral responses were found over the scalp for glucose, NaCl and distilled water. The most prominent peak, M175, indicated a bilateral two-dipole pattern with a latency of about 130-210 ms (Fig. 2). The amplitude of M175 was larger for 10% glucose than for distilled water.

Sources of the bilateral M175 responses were estimated using a two-dipole model. When the signal/noise ratio was good enough, generators of the M175 were estimated on the most medial parts of the bilateral opercula near the insulae (Fig. 3). These sources were approximately on the border of the frontal and parietal operculum (Fig. 3). The dipole-fit error was lowest for 10% glucose, possibly due to the higher signal/noise ratio than for the other taste stimuli.

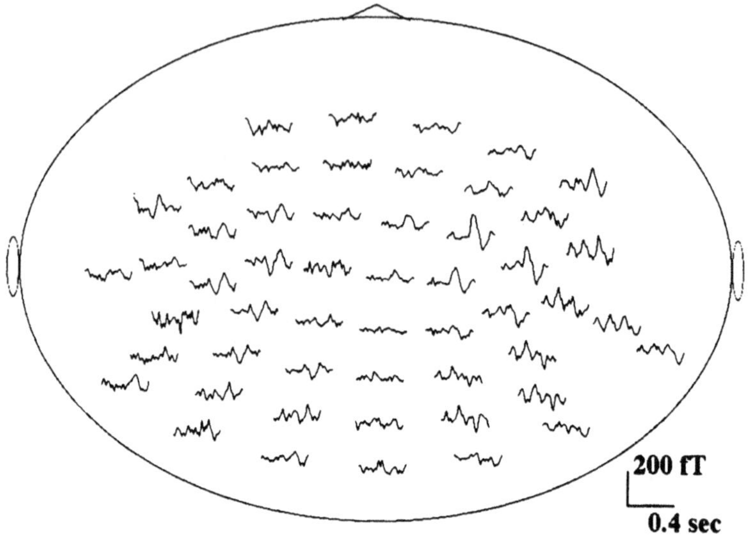

Fig.1 Wave forms of evoked magnetic fields for 10% glucose stimuli on the tongue in subject 1. All 64 MEG channels over the entire head are shown from above.

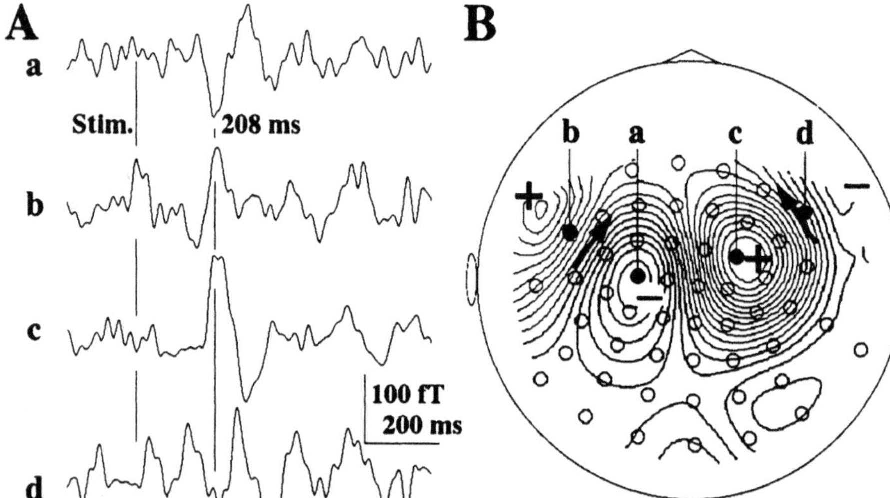

Fig.2 Evoked magnetic fields for 10% glucose stimuli on the tongue in subject 1. A: The first peak (M175) of the evoked fields appeared at a latency of about 200 ms. B: Isofield map indicates a bilateral two-dipole pattern at the M175 peak. Arrows indicate the M175 currents approximately.

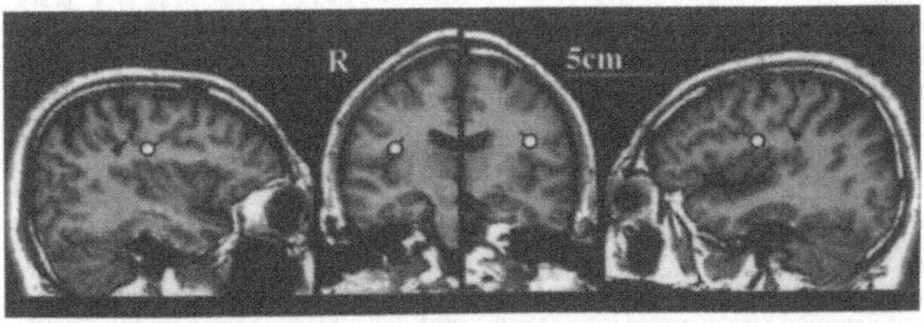

Fig.3 Source estimation of the evoked magnetic fields for 10% glucose stimuli on the tongue in subject 1. Circles and bars indicate the M175 dipole positions and orientations, respectively. Arrowheads indicate the central sulcus identified by the somatosensory evoked fields.

Table 1 M175 component in evoked magnetic fields for taste and non-taste stimuli on the tongue in 5 subjects.

Subjects Age/Sex	Stimulus on Tongue	M175 Response		
		Latency	Isofield Map Pattern	Dipole Estimation
1) 44/M	glucose	208 ms	two-dipole	bilateral opercula
	water	211 ms	two-dipole	n/a*
2) 36/M	glucose	147 ms	two-dipole	bilateral opercula
	water	152 ms	two-dipole	n/a*
3) 23/M	glucose	152 ms	two-dipole	left: operculum right: n/a**
	NaCl	126 ms	two-dipole	left: operculum right: n/a**
	water	131 ms	two-dipole	left: operculum right: n/a**
4) 40/M	glucose	163 ms	two-dipole	left: operculum right: n/a**
5) 32/M	glucose	160 ms	two-dipole	n/a*

glucose = 10% glucose; NaCl = 0.3 mol NaCl; water = distilled water; n/a* not available, dipole fit error > 20%; n/a** = not available, small dipole in the right hemisphere

Table 1 shows a summary of the M175 responses in all five subjects. Although the peak latencies varied , the sources for the gustatory stimuli were estimated on the bilateral opercula in two subjects and unilateral operculum in the two subjects. The latter two subjects indicated an asymmetrical two-dipole pattern with smaller signal amplitude on the right hemisphere. We also found two a dipole-pattern for the distilled water stimuli, but their sources were not localized. We observed some later wave peaks following M175, but the reproducibility of the signals was not good enough for dipole estimation.

Discussion

The M175 response observed in the present experiment might be the response to tactile stimulation of the tongue. However, we believe that gustatory information also contributes to M175 for the following three reasons. First, the two gustatory stimuli evoked larger responses than distilled water. Second, the M175 dipole positions were localized in the most medial part of the operculum near the insula, suggesting areas deeper than the tongue somatosensory areas. Third, the M175 latency (130-210 ms) seems longer than the response from the primary somatosensory area (SI). Karhu et al. [9] recorded SEFs to electrical stimulation of the tongue and observed two peaks (about 50 ms and 140 ms in latency). They speculated that the earlier source originates in the primary somatosensory area (SI) on the contralateral side to the tongue stimulation and that the later source originates in the secondary somatosensory area (SII) in the bilateral hemispheres. More late components were reported in a scalp GEP study. Funakoshi and Kawamura [10] obtained two potential components. They interpreted the earlier one (about 150 ms in latency) as the tactile response, and the later ones (500 ms -1500 ms in latency) as the gustatory response. However, the later potential might be related to higher brain functions such as gustatory recognition.

Although we successfully measured evoked magnetic fields for gustatory stimuli to the tongue, further study is necessary to confirm the location of the human gustatory cortex.

References:

[1] Benjamin, R.M., and Burton, H. Projection of tongue nerve afferents to anterior opercular-insular cortex in squirrel monkey (*Saimiri sciureus*), Brain Research, 1968, 7: 221-231.

[2] Ogawa, H. Gustatory cortex of primates: anatomy and physiology, Neuroscience Research, 1994, 20: 1-13.

[3] Scott, T.R., Yaxley, S., Siekiewicz, Z.J., and Rolls, E.T. Gustatory responses in the frontal opercular cortex of the alert cynomulgus monkey, Journal of Neurophysiology, 1986, 56: 676-890.

[4] Bornstein, W.S. Cortical representation of taste in man and monkey. I. Functional and anatomical relations of taste, olfaction and somatic sensibility. Yale J. Biol. Med., 1940a, 12: 719-736.

[5] Bornstein, W.S. Cortical representation of taste in man and monkey. II. The localization of the cortical taste area in man and a method of measuring impairment of taste in man. Yale J. Biol. Med., 1940b, 13: 133-156.

[6] Motta, G. I centri corticali del gusto. Boll. Sci. Med., 1959, 131: 480-493.

[7] Kobal, G. Gustatory evoked potentials in man. Electroencephalography and Clinical Neurophysiology, 1985, 62: 449-454.

[8] Nakasato, N., Fujita, S., Seki, K., Kawamura, T., Matani, A., Tamura, I., Fujiwara, S., and Yoshimoto, T. Functional localization of bilateral auditory cortices using an MRI-linked whole head magnetoencephalography (MEG) system, Electroencephalography and Clinical Neurophysiology, 1995, 94: 183-190

[9] Karhu, J., Hari, R., Lu, S.T., Raetau, R., and Rif, J. Cerebral magnetic fields to lingual stimulation, Electroencephalography and Clinical Neurophysiology, 1991, 80: 459-468.

[10] Funakoshi, M., and Kawamura, Y. Summated cerebral evoked responses to taste stimuli in man, Electroencephalography and Clinical Neurophysiology, 1971, 30: 205-209.

Acknowledgments

This work has been supported in part by Grants-in-Aid for Scientific Research No. 03404042 and No. 06404050 from the Ministry of Education, Science and Culture of Japan; and Magnetic Health Science Foundation.

Different Effects of Stimulus Rate on the Somatosensory Evoked Magnetic Fields and Potentials

Nagamine, T.[1], Mäkelä, J.P.[2], Mikuni, N.[1,3], Mima, T.[1], Nishitani, N.[1], Sato, T.[1], Ikeda, A.[1], Kimura, J.[4] and Shibasaki, H.[1]

[1]Departments of Brain Pathophysiology, [3]Neurosurgery, [4]Neurology, Kyoto University School of Medicine, Kyoto, Japan; [2]Central Military Hospital, Helsinki, Finland

Introduction

In order to execute the processing of continuing information without delay, sensory cortices require rapid recovery from the preceding stimuli. Depending on the received stimuli, every information is analyzed serially and conveyed parallel through several brain areas. Although the brain areas responsible for these processes have been roughly estimated, exact location remains still unclear. Both somatosensory evoked potentials (SEPs) [1-3] and magnetic fields (SEFs) [4,5] with different interstimulus intervals (ISIs) have been used to characterize the recovery function of the somatosensory cortex. Magnetoencephalography (MEG) is sensitive to intracellular currents flowing tangential to the head surface whereas electroencephalography (EEG) reveals both tangential and radial currents. We have employed simultaneous recordings of electroencephalographic and magnetoencephalographic somatosensory evoked responses (SERs) to elucidate the temporal behavior of the sensory cortex by utilizing recovery function.

Methods

Nine normal volunteers gave informed written consent for the study. The subject was sitting in a chair in a magnetically shielded room (Kyoto University School of Medicine), with the forearm supinated. The subjects were requested to keep their eyes open and to avoid frequent blinking. Unilateral median nerve was stimulated at the wrist with the constant current of electric shock lasting 0.2 ms. Five out of nine subjects were tested on both sides. The intensity of the electric shock was adjusted to invoke weak thumb twitch. Stimulus rate of once every 0.9 s was applied first and then that of 4 s was tested.

A 122-channel DC-SQUID neuromagnetometer, covering the whole scalp [6], was utilized to record the magnetic fields produced by the brain. 61 pairs of two orthogonally oriented figure-of-eight planar first-order gradiometers was designed to detect the two orthogonal derivatives of the radial magnetic field at each pair location. This configuration of the gradiometers allowed us to catch the source current just below the largest signal. The locations of three head position indicator (HPI) coils placed on the scalp were measured with respect to the three anatomical landmarks (bilateral preauricular points and nasion) using a 3-D digitizer (Isotrak 3S1002, Polhemus Navigation Sciences, Colchester, Vermont, U.S.A.).

EEG signals were collected from cup electrodes placed at locations C3, CP3, P3, C4, CP4, P4 and Oz (according to the modified International 10-20 System), referenced to the linked earlobe electrodes. Two electro-oculograms (EOGs) linking the right infraorbital with either the right earlobe or the right supraorbital position were adopted to monitor the eye movements. The impedance of all electrodes was kept below 5 kΩ.

Signals of all EEG, EOG channels and eight selected MEG channels from bilateral central and occipital areas were plotted continuously on the paper so that the vigilance of the subject could be checked.

The signals were digitized at 860 Hz after filtered with analog passband of 0.03–280 Hz for MEG and that of 0.07-280 Hz for EEG and EOG. On-line averaging was done with respect to the electric shock with the time window of 600 ms including 100-ms pretrigger interval. Averaging started after giving five electric shocks. Big artifacts exceeding the preset rejection level of 3 pT/cm for MEG and 150 μV for EEG and EOG were excluded from the averaging automatically.

Averages of around 200 sweeps were obtained for one session, and at least two sessions were employed for 0.9-s stimulus interval condition to confirm the reproducibility. For the responses to 4-s stimulus interval, the first subaverages of 50 responses and the following 50 subaverages were obtained and compared. Responses with reproducible wave forms were group-averaged across sessions.

Amplitude of the responses was measured by taking the time window from 100 to 5 ms prior to the electric shock as a baseline. Component configurations were determined at the contralateral somatomotor area.

Results

In all subjects, N20 as well as magnetic counterpart, N20m, was clearly demonstrated over the contralateral somatomotor area. Two typical subjects are shown in Fig. 1. SEF responses showed P30m, P40m and P60m, among which P60m demonstrated the biggest amplitude, and some intervening smaller peaks. Although some subjects had subsequent deflections around 80 - 200 ms, they were much smaller than those at 20 - 60 ms. SEP recordings showed subsequent deflections, labeled as P30, N35, P45, N75, P100, N150 and P250. Individual variation was evident in the amplitudes, which were clearly larger for stimulus interval at 4 s than that at 0.9 s (Table). Both SEF and SEP components later than 40 ms were readily suppressed by the 4-s stimulus interval condition (Fig. 2). Although SEF responses became almost stable later than 150 ms, SEP demonstrated bigger amplitude than those of early latencies. Especially P250 was strongly enhanced with 4-s stimulus interval (Fig. 2).

Table Latency and absolute amplitude of each component for 0.9-s and 4-s stimulus interval condition. Measurements were performed at the channels showing biggest N20m or N20 component. Responses following the right median nerve stimulation and those with the left median nerve are put together.

	stimulus interval									
	0.9 s					4 s				
	Latency (ms)		Amplitude (fT/cm)		n	Latency (ms)		Amplitude (fT/cm)		n
	mean	SD	mean	SD		mean	SD	mean	SD	
N20m	21.6	1.6	61.5	28.7	14	21.6	1.6	60.2	27.7	14
P30m	28.5	2.1	22.0	18.5	10	28.5	2.1	21.3	18.1	10
N35m	31.6	2.0	26.8	18.3	5	31.6	2.0	27.2	17.6	5
P40m	38.0	3.5	67.9	46.5	11	38.1	3.2	87.8	60.7	11
N50m	43.7	4.7	43.8	36.0	7	43.9	4.5	54.5	44.8	8
P60m	60.4	8.9	89.7	40.4	12	59.2	8.4	102.9	47.8	13
P100m	91.8	12.6	50.3	39.8	8	91.8	12.6	81.7	57.6	8
N150m	130.5	26.4	20.3	12.8	9	130.5	26.4	38.4	24.3	9
	Latency (ms)		Amplitude (μV)		n	Latency (ms)		Amplitude (μV)		n
	mean	SD	mean	SD		mean	SD	mean	SD	
N20	20.0	1.6	2.06	0.90	14	20.0	1.6	1.59	0.74	14
P30	26.1	1.4	1.11	0.52	8	26.1	1.4	0.87	0.40	8
N35	32.0	3.1	1.42	0.75	8	31.0	1.6	0.97	0.75	8
P45	40.7	3.4	2.11	1.37	13	42.2	5.1	3.43	2.12	6
N75	65.3	10.9	1.78	1.53	9	64.7	10.5	2.85	2.51	10
P100	97.7	7.9	2.23	1.59	5	96.6	6.8	4.25	2.85	7
N150	151.2	23.4	2.46	1.74	8	135.6	20.0	4.31	1.85	12
P250	276.4	16.7	4.04	2.24	4	291.3	19.4	11.59	4.75	12

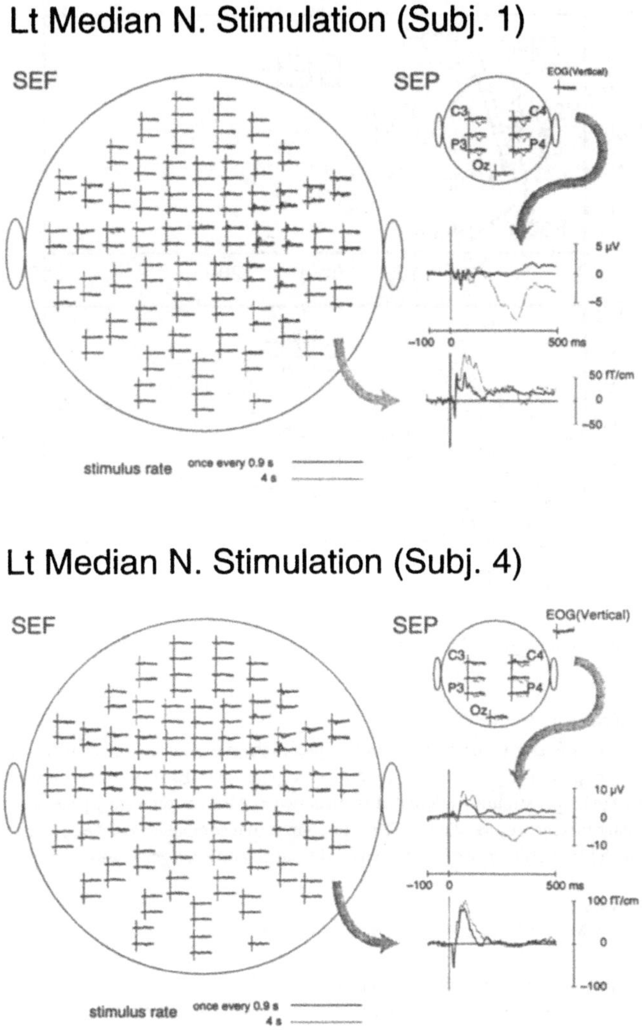

Fig. 1
Simultaneous recording of SEF and SEP in two subjects.
Responses with stimulus rate of once every 0.9 s and 4 s conditions are superimposed. The biggest responses marked by shading are magnified in the right lower corner. One of the MEG channels was not available because of the machine trouble.

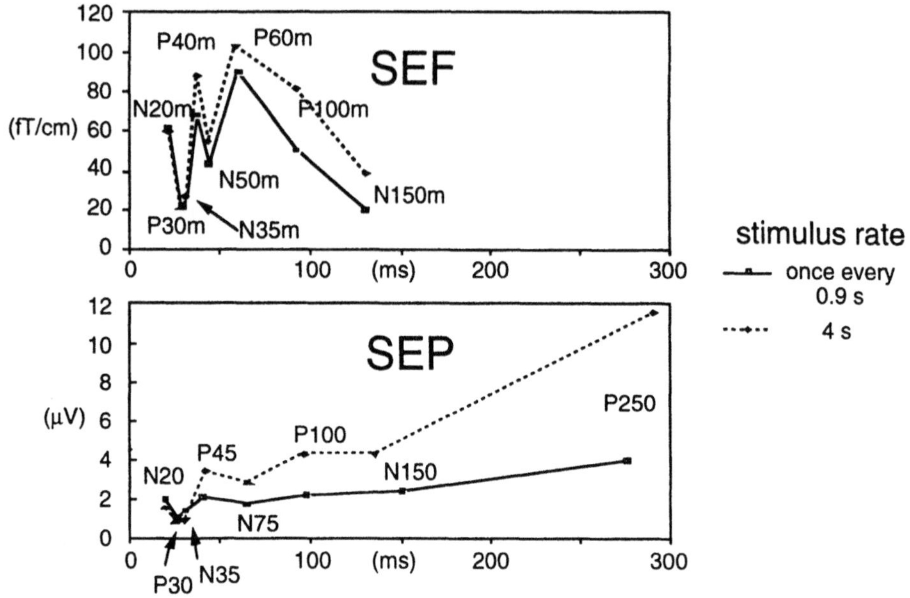

Fig. 2
Mean absolute amplitude of SEF and SEP as a function of latency for different stimulus rate.

Discussion

Clear difference in the behavior with respect to the stimulus rate between SEF and SEP gives us some hint for the sensory processing. Since the prominent responses in SEF have already ended within 100 ms and less affected by the longer stimulus interval, it is suggested that the corresponding generators are responsible mainly in the early stage of sensory processing. Therefore, tangential components of the sensory cortex seem to play a main role during the first 100 ms and be replaced by the radially oriented generators.

References

[1] Nakamura, M., Shibasaki, H., Nishida, S., Neshige, R. 1990, IEEE Trans. Biomed. Eng., 37: 738-740

[2] Emori, T., Yamada, T., Seki, Y., Yasuhara, A., Ando, K., Honda, Y., Leis, A.A., Vachatimanont, P. Electroencephalography and clinical Neurophysiology, 1991, 78:116-123.

[3] Romani, A., Bergamaschi, R., Versino, M., Callieco, R., Calabrese, G., Cosi, V. Electroencephalography and clinical Neurophysiology, 1995, 96:475-478.

[4] Huttunen, J., Ahlfors, S., Hari, R. Electroencephalography and clinical Neurophysiology, 1992, 82:176-181.

[5] Hari, R., Karhu, J., Hämäläinen, M., Knuutila, J., Salonen, O., Sams, M., Vilkman, V. 1993 European Journal of Neuroscience, 1993, 5: 724-34.

[6] Ahonen, A.I., Hämäläinen, M.S., Kajola, M.J., Knuutila, J.E.T., Laine, P.P., Lounasmaa, O.V., Parkkonen, L.T., Simola, J.T. and Tesche, C.D. 122- channel SQUID instrument for investigating the magnetic signals from the human brain. Physica Scripta, 1993, T49: 198-205.

Binaural Fusion of Virtual Pitch in the Human Auditory Cortex

Pantev, C.[1], Elbert, T.[2], Ross, B.[1], Eulitz, C.[1] and Terhardt, E.[3]

[1]Institut for Experimental Audiology, Münster, Germany; [2]Department of Psychology, Konstanz, Germany; [3]Institute of Electroacoustics, Munich, Germany

The vowels of human speech and the sounds produced by many musical instruments are harmonic complex tones. These are comprised of pure tones that are integer multiples of a fundamental frequency. The perceived pitch of complex tonal stimuli is qualitatively different from the one of pure tones. The pitch of pure tones is a basic auditory percept which is monotonically related to its frequency, a phenomenon known as spectral pitch. The pitch of complex tones is termed virtual pitch. It can be assessed by auditory matching to pure tones. The virtual pitch of harmonic complex tones corresponds to a frequency approximating the frequency difference between the components, i.e. to the fundamental frequency, even if the fundamental frequency is missing. The fact that the first six to eight harmonics of a complex tone can be perceived as separate spectral pitches led Terhardt [1] to conclude that the virtual pitch is extracted by a type of "Gestalt" recognition, i.e. virtual pitch is an auditory attribute that is established on a higher level of processing than spectral pitch. The distinction between spectral pitch and virtual pitch can be likened to the distinction between primary visual contours and virtual ("illusory") contours in vision [2]. Just as a visual virtual contour can be regarded as being "inferred" from a set of primary contours, virtual pitch is regarded as the result of a two-stage process which involves peripheral frequency analysis and virtual-pitch formation by a pertinent "pitch processor"[3]. It is believed that the pitch processor is located in more central regions of the auditory system. An important feature is binaural fusion, i.e. the ability to combine spectral components regardless of the ear to which they are presented. It was therefore hypothesized that neural activity ascending from both ears would fuse into one percept [4].

Here we provide first evidence from magnetic source imaging that a pitch processor fuses the information coming from separate ears to produce a virtual pitch represented in the same cortical area as the spectral pitch.

Methods

In the present investigation, the magnetic responses of pure-tones and complex-tones were measured in eight normal hearing, right handed subjects. Four different burst stimuli (duration 500 ms) were presented with a 10 ms cosine slope. The four stimuli differed in their spectral contents (Fig. 1b) as follows: 1) a 250 Hz pure tone; 2) 1500 Hz pure tone; 3) a complex tone composed of eight components, these being the 4th to 11th harmonics of the missing fundamental frequency of 250 Hz (presented monaurally); and 4) two complex tones composed of the even and odd harmonics of the eight-component complex tone (presented binaurally). Stimulus intensity was 60 dB relative to the individual hearing threshold. The interstimulus interval was randomized between 2.5 and 3.5 s. The eight-component complex tone shows a 250 Hz periodicity and although it does not include spectral power below 1000 Hz, 250 Hz periodicity was audible. All subjects were able to match this virtual pitch successfully with a spectral pitch of or close to 250 Hz, that was simultaneously presented to the opposite ear (Fig. 1c, top). In the same way the complex tones containing the even or odd harmonics of the eight-component complex tone (periodicity of 500 Hz) were matched with a spectral pitch of or close to 500 Hz (Fig. 1c, bottom).

Two magnetic sensor arrays (BTi) were used to record the neuromagnetic brain activity (bandwidth 0.1-100 Hz) simultaneously from both auditory cortices. Stimulation was presented in 16 blocks of 150 stimuli each and the block sequence was varied across subjects in a randomized order. Thus each AEF was an average of 600 stimulus related epochs. In order to reduce possible variations caused by changes in attention and vigilance, subjects watched cartoons during the session. Due to the head-sensor constellation the sensing coils of the bottom sensor were further away from the scalp as compared with the upper sensor. This fact causes a reduction of the AEF amplitudes recorded from the right hemisphere (c.f. Fig. 1a).

For each AEF distribution corresponding to the four stimulus conditions a single equivalent current dipole model located in a best fitting local sphere was used to estimate the moment and the space coordinates of a dipole, representing the cortical activity and the center of the activated cortical area.

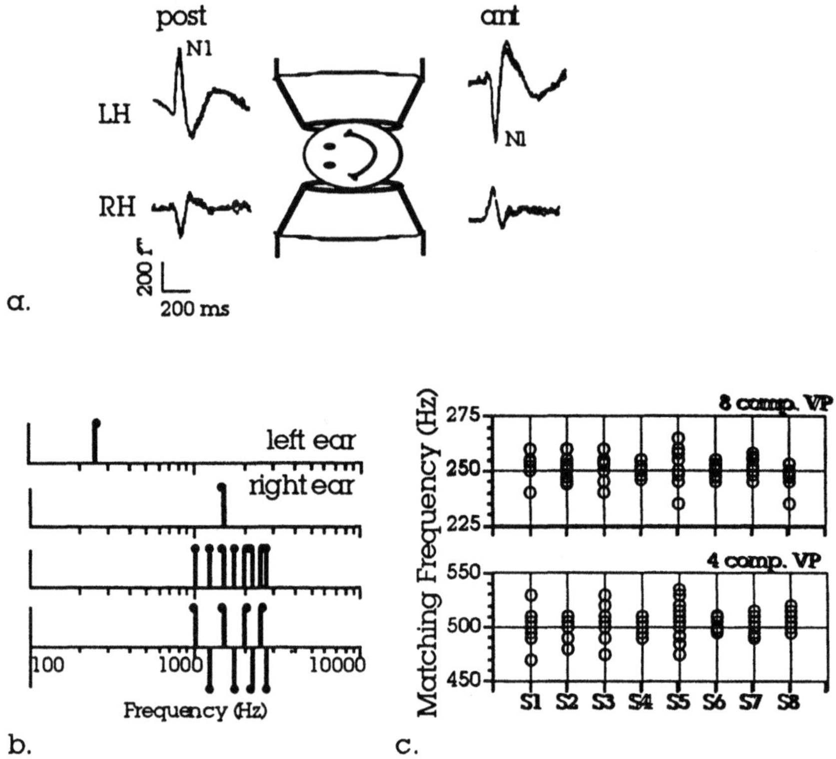

Fig. 1: **a)** Simultaneous magnetoencephalographic recording of auditory evoked activity from the left and the right hemispheres. The waveforms at the posterior and anterior field extremes are superimposed for four repeated blocks of identical stimulation. **b)** Scheme of the spectral content of the stimuli: **First** and **second rows** show spectral pitches of 250 and 1500 Hz presented to the left ear (upward pointing pin). **Third row**: virtual pitch consisting of the 4th to 11th harmonics of the missing fundamental frequency of 250 Hz presented to the left ear. **Fourth row**: two complex tones composed of the even and the odd harmonics of the eight components complex tone presented to the left and right ears, correspondingly. **c)** Ten repeated matching results for each subject. The eight components virtual pitch was matched with a spectral pitch of or close to 250 Hz, presented to the opposite ear (**top row**). The even or odd harmonics of the eight-component complex tone (periodicity of 500 Hz) were matched with a spectral pitch of or close to 500 Hz presented to the opposite ear (**bottom row**).

The averaged value of the source coordinates of the data points around the N1 peak, having the highest two percents of goodness of fit was taken as a measure of the source locations for each of the four different stimulus conditions.. Only data with a goodness of fit greater than 95% were accepted. The data qualified only four subjects suitable for the analyses of the right hemisphere. Therefore, one-way analyses of variance (ANOVA) were calculated separately for the left and the right hemispheres, using STIMULUS CLASS (the four different stimuli) as a repeated measures factor. Probabilities were Greenhouse-Geisser adjusted. Post-hoc Scheffé tests were used to resolve significant ANOVA effects.

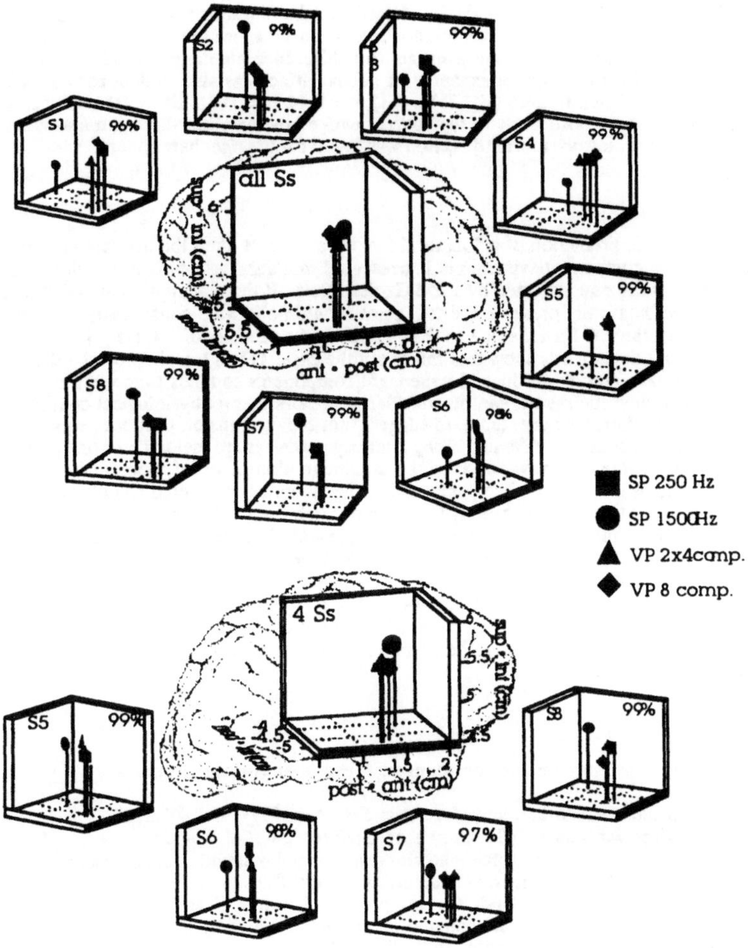

Fig. 2: **(top)** Source locations presented individually for each subject around a sketched brain (subject's number and goodness of fit are listed). The enlarged view at the center represents the spatial grand average across subjects. Squares denote spectral pitch of 250 Hz, circles spectral pitch of 1500 Hz, diamonds virtual pitch of the eight components complex tone (*all presented monaurally*), triangles virtual pitch of the even and odd harmonics of the eight components complex tone (*presented binaurally*). The anterior-posterior axis is denoted by a filled bar, the dimension of each cube is 1.5 x 1.5 x 1.5 cm.
(bottom) Source locations in the right hemisphere as obtained in four out of eight subjects.

Results

The estimated cortical source locations are presented separately for each subject and also as spatial grand averages across subjects in Fig. 2. Goodness of fit equalled 99% in the majority of the cases, confirming the adequacy of a single dipole model. The spatial source coordinates were separately tested for stimulus-related differences.. The separation of the representation of the 1500 Hz source from a location around which the other three sources cluster (Fig. 2) was corroborated by statistical analysis (ANOVA). In the medio-lateral direction the effect was significant for the *left* ($F(3, 21) = 16.39$; $\varepsilon = 0.45$; $p < 0.003$) and the *right* hemisphere ($F(3, 9) = 31.62$; $\varepsilon = 0.72$; $p < 0.001$). Post-hoc Scheffé testing revealed that this effect was exclusively due to the 1500 Hz spectral pitch condition which produced a location different from each of the other three conditions (all $p<0.001$). On the average, this difference amounted to 6.3 mm for the left hemisphere and to 5.5 mm for the right hemisphere. A similar effect was also evident for the posterior-anterior direction over the *left* hemisphere ($F(3,21) = 9.35$; $e = 0.44$; $p < 0.02$). The 1500 Hz spectral pitch source was located 2.9 mm more posteriorly to the sources evoked by the other three stimulus conditions. No significant difference in posterior-anterior direction was found for the *right* hemisphere.

Discussion

If the psychoacoustic prediction of a "binaural fusion" is correct, then the activated brain regions corresponding to the eight components virtual pitch, presented monaurally or binaurally, should activate the same cortical region as the one activated by a 250 Hz pure tone. If there were no binaural fusion, then the brain sources for the binaurally presented virtual pitch would be located consistently more medially than the representation of the 250 Hz-tone due to their matching frequencies of 500 Hz [5]. However, the statistical evaluation of our data clearly show that this is not the case. Given the absence of a difference in the location of the cortical excitation produced by the eight components complex tone when presented either monaurally or binaurally, the results can be considered the first neurophysiological confirmation of binaural fusion in hearing. Furthermore, the data suggest that binaural fusion produces equal results in both hemispheres. Neural excitations travelling along auditory pathways are melded into one pattern at or before the cortical level. At the cortex, we observe the nonlinear addition of information from both ears, having a quality that pertains to a virtual pitch composed of the binaurally presented complex tone. In conclusion, virtual pitch is not determined separately for the input to each ear, but is the result of binaural fusion in much the same way as the perception of monaurally presented complex tones. The finding that the same cortical region is activated by the 250 Hz spectral pitch and the 250 Hz virtual pitch (the complex tone composed of eight harmonics of the missing fundamental frequency of 250 Hz) confirms that the perceived pitch and not simply the spectral content of the auditory stimulus determines the representation within the tonotopic cortical map. This implies that the virtual pitch has been determined at a stage prior to the AEF N1-peak, i.e. the hypothesized pitch processor should be realized in a neuronal network of the primary auditory cortex or in subcortical structures.

References

1. Terhardt, E. Zur Tonhöhenwahrnehmung von Klängen. I. Psychoakustische Grundlagen. Akustica, 1972, 26: 173-186.
2. Terhardt, E. Pitch, consonance, and harmony. J. Acoust. Soc. Am., 1974, 55: 1061-1069.
3. Terhardt, E. Warum hören wir Sinustöne? Naturwissenschaften, 1989, 76: 496-504.
4. van den Brink, G. Monotic and dichotic pitch matchings with complex sounds. In: E. Zwicker and T. E. (Eds.), Facts and models in hearing. Springer, Heidelberg, 1974: 178-189.
5. Pantev, C., Bertrand, O., Eulitz, C., Verkindt, C., Hampson, S., Schuirer, G. and Elbert, T. Specific tonotopic organizations of different areas of the human auditory cortex revealed by simultaneous magnetic and electric recordings. EEG and clin. Neurophysiol., 1995, 94: 26-40.

Acknowledgement

This work has been supported by a grant from the Deutsche Forschungsgemeinschaft (Klinische Forschergruppe Biomagnetismus and Biosignalanalyse). The authors thank Drs. Jablonski, Buchanon and Lawson (BTi) for technical support.

Left and Right Supratemporal Auditory Cortices are Differentially Modulated by Linguistic-Attentional Task Demands

Poeppel, D.[1], Yellin, E.[2], Phillips, C.[2], Roberts, T.P.L.[1], Rowley, H.[1], Wexler, K.[2] and Marantz, A.[2]

[1]*University of California San Francisco, San Francisco, USA;* [2]*Massachusetts Institute of Technology, Cambridge, USA*

Introduction

An important goal in the study of the neural basis of speech perception is to understand the physiological basis of speech processing as a sensory input process and relate it to the well-known anatomical asymmetry for language representation. One advantage gained by focusing on speech perception as a subroutine of language comprehension is that one can use stimuli the acoustic properties of which are explicitly defined and easy to manipulate parametrically.

Beyond the interest in questions of speech and language lateralization, there is an extensive literature devoted to the study of auditory attention, both using event-related potentials and using MEG. Two classes of experimental protocols are commonly employed: tone oddball paradigms, in which one measures the mismatch response to an oddball stimulus in a sequence of standards [2], and 'alternating ear' or dichotic listening paradigms, in which one is asked to selectively attend to the input delivered to the left or right ears [1,2,3].

It is established that 'ear-specific' attention can modulate the M100 auditory evoked field [3]. When subjects were presented dichotically with tones of different frequencies, the M100 generated by a stimulus presented to the attended ear was larger than the M100 generated by the same stimulus presented to the unattended ear. The present study investigated whether 'content-specific, ear-independent' attention also modulates aspects of the M100, and whether there is a *differential* hemispheric effect. This study thus evaluates the possibility of an *attention-dependent gating* of activity that affects the two hemispheres differentially. The same speech stimuli are presented in two experimental conditions, passive listening and performing a simple discrimination task. The prediction was an increased M100 amplitude in the left temporal cortex in the attended condition. There was no prediction about the response of the right hemisphere.

Methods

We investigated the M100 auditory evoked response and its lateralization by recording from the left and right supratemporal planes of six subjects (3 women, 3 men). Four syllables were presented to the contralateral ear 100 times each in pseudorandom order. The four syllables (/bæ/, /pæ/, /dæ/, and /tæ/) were synthesized using a Klatt formant synthesizer. The syllables were 300ms in duration and the voice onset time was 20ms for the two voiced syllables (/bæ/ and /dæ/) and 80ms for the voiceless syllables, reflecting typical values used in natural speech. In one condition, subjects passively listened to the stimuli. In a second condition, subjects were required to make an overt phonetic classification by pressing a response key. Subjects had to group the voiced versus the voiceless syllables. The task was performed with ease, with 5 subjects performing at ceiling and one subject performing at 94%. Neuromagnetic fields were acquired in a magnetically shielded room using a 37-channel system (Magnes, BTi, San Diego, CA). Fields were recorded with a 1.0Hz high-pass cutoff and a sampling rate of 1041.7 Hz (bandwidth 400Hz). The evoked field data were signal-averaged and band-pass filtered (1-20Hz) for further analysis.

Single equivalent current dipole estimates of the M100 field were coregistered with MR images. For each subject, high resolution volumetric magnetic resonance images (SPGR sequence, 128x128x124 matrix, resolution ~ 1x1x1.5mm, TR=36ms, TE=8ms, flip=70deg) were acquired using a 1.5 Tesla SIGNA magnetic resonance scanner (GE Medical Systems, Milwaukee, WI).

Results

The latency of the M100 RMS peak ranged from 82ms to 129ms post stimulus onset. Analyses of variance of the latency data for both conditions revealed no significant main effects or interactions. In particular, there was no effect of the variable hemisphere. Figure 1 plots the M100 amplitude differences. The amplitude data showed the following patterns: (a) In the four right-handed subjects (2 women and 2 men) the task *increased* the mean RMS value in the left hemisphere on 14 of 16 measurements (4 subjects x 4 syllables), with the mean increases over the passive condition ranging from 1.4 femtoTesla (fT) to 54.8fT. (b) In the right-hemisphere measurements from the same four subjects, 15 of 16 measurements showed a *decrease* of RMS peak compared to the passive condition, with

Figure 1 M100 rms difference (task-passive) for each subject and each syllable

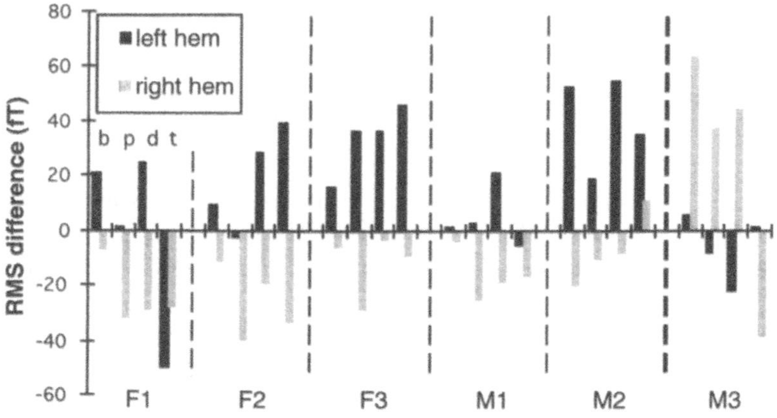

Figure 1 Task - passive M100 rms difference values. Each bar represents the task-passive difference of the M100 rms peak. Positive values indicate that the response amplitude in the task condition was larger than the corresponding measurements in the passive condition. Negative values indicate that the activation in the passive condition was greater than the discrimination condition for that hemisphere. Black bars represent the comparison for the left hemisphere, grey bars represent the comparison for the right hemisphere. Plotted are the mean amplitude changes for the four syllables for each subject. F1-F3, female subjects; M1-M3, male subjects.

Figure 2 Single dipole location for syllables (passive & discriminate) for subject M1

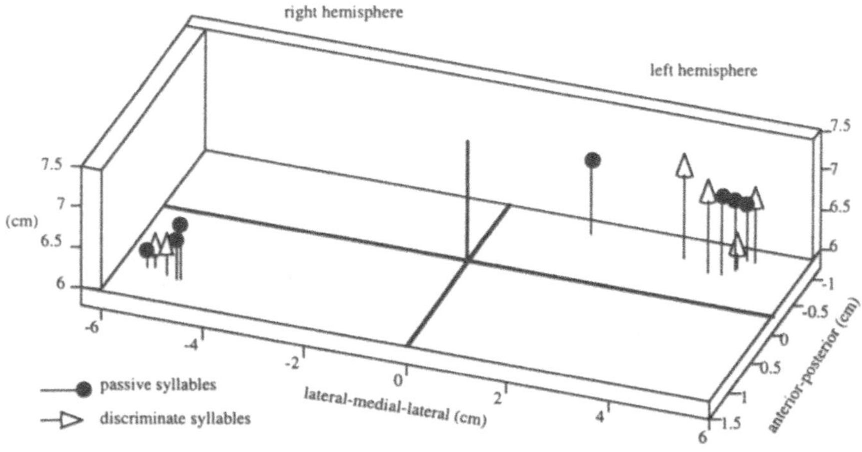

Figure 2 Dipoles elicited by syllable stimuli for one subject. The filled circles are dipoles derived for the passive condition, the triangles are dipoles for the VOT discrimination condition. MRI shows that all dipoles localize to supratemporal auditory cortex. The left-right asymmetry observed for the cluster of dipoles in this subject is observed by other investigators as well. This localization asymmetry is typically attributed to the size asymmetry between the plana, with the left planum temporale being larger in many subjects.

the mean decreases ranging from 3.1fT to 39.5fT. (c) The two left-handed subjects patterned in two ways: One subject's data showed the same pattern described for the right-handers. The other subject patterned in the opposite way, with the task decreasing the RMS peak in the left and increasing it in the right hemisphere recordings. With few exceptions, the task-related changes in RMS were very large when compared with the baseline (passive) RMS peak values. Typical changes were on the order of tens of percent. Analyses of variance of the RMS data show no main effect of hemisphere in the passive condition, $F(1, 5)=0.089$, $p=0.78$, but a significant main effect of hemisphere in the attention condition, $F(1, 5)=11.282$, $p=0.02$, despite the low power of the study. Also, the hemisphere (left/right) x task (passive/attended) interaction was highly significant, $F(1, 5)=55.30$, $p=0.001$. Together, these findings suggest that *the task induced the lateralization.* Single equivalent current dipole modeling (Figure 2) and coregistration of the dipoles with structural magnetic resonance images of the individual subjects showed that both conditions activated the same small area in supratemporal auditory cortex, thus excluding the explanation that the presence of the task simply led to the recruitment of different cortical areas.

Discussion

A number of recent results derived from hemodynamic functional brain imaging methods (PET, fMRI) suggest that speech perception is at least partially mediated bilaterally in the superior temporal gyri. However, when subjects execute certain task demands, typically 'meta-linguistic' tasks such as phoneme categorization or semantic classification, the observed neural activity is more highly lateralized. The findings reported here are similar. Passive listening to syllables revealed complex but bilaterally symmetric responses. In contrast, the execution of a categorization task with the same stimuli showed that the same region in left temporal cortex increased its response amplitude (large M100 amplitude increases) while the the region in right temporal cortex showed large M100 amplitude decreases. Attention to speech sounds is thus reflected in lateralized supratemporal cortical responses possibly concordant with hemispheric language dominance. Taken together, the data suggest that lateralization can be induced by the attentional demands of a given experiment rather than necessarily reflecting underlying hemispheric specialization.

The effect of attention needs to be clarified in future experiments to ascertain precisely what kind of attentional modulation is occurring. The issue that requires clarification is what kind of attentional manipulation leads to this response attenuation or enhancement. Several possibilities exist. (1) Only attending to the *particular stimulus set* changes the outcome. The effect is therefore due to subjects having to make overt linguistic discriminations on the presented material. (2) Attending to *any speech or language stimuli* during the task will result in this effect. It is not crucial to attend to the stimuli proper; attention to any linguistic material, even though it is not part of the task, will result in the change. (3) Finally, it is possible that *any* attentional task will lead to such a modulation, including, for example, the execution of a visual task.

References

[1] Arthur, D., Lewis, P., Medvick, P., and Flynn, E. A neuromagnetic study of selective auditory attention. Electroenceph. clin. Neurophysiol., 1991, 78: 348-360.
[2] Rif, J., Hari, R., Hämäläinen, M., and Sams, M. Auditory attention affects two different areas in the human supratemporal cortex. Electroenceph. clin. Neurophysiol., 1991, 79: 464-472.
[3] Woldorff, M.G., Gallen, C.C., Hampson, S.A., Hillyard, S.A., Pantev, C., Sobel, D., Bloom, F.E. Modulation of early sensory processing in human auditory cortex during auditory selective attention. Proc. Natl. Acad. Sci. USA, 1993, 90: 8722-8726.

Acknowledgments

We are grateful to Biomagnetic Technologies Inc. for equipment support and to Susanne Honma and Mary Mantle for their help in running experiments.

Extensive Somatosensory Stimulation Alters Somatosensory Evoked Fields

Rockstroh, B.[1], Vanni, S.[2], Elbert, T.[1], and Hari R.[2]

Department of Psychology[1], University of Konstanz, Germany;
Low Temperature Laboratory[2], Helsinki University of Technology, Espoo, Finland

Introduction

Changes in afferent input lead to functional cortical modifications, i.e., to cortical reorganization resulting in alterations in the cortical responses to stimuli. Animal studies by Jenkins, Merzenich and coworkers [1, 2] demonstrated that amputation of the digits result in an altered representation of the hand in area 3b. Such plastic changes can also be observed in humans using noninvasive measurement techniques. For instance, magnetoencephalographic studies [3,4] demonstrate that after amputation the sensory receptive fields of the face may invade areas which formerly had represented the hand and the digits. On the other hand, increased afferent input, e.g., by behavioral training may produce an enlargement of the cortical area that is activated by a particular task [1,2,5]. Prolonged increase in tactile stimulation to the distal pad of one or two phalanxes has been reported to result in a greatly increased cortical representation specific to that circumscribed area of the hand in owl monkeys [1]. In extreme cases, extensions spanning an area of several millimeters have been observed in the animal brains [6]. In patients suffering from syndactyly, Mogilner et al. [7] reported reorganization of the somatosensory cortex within weeks after surgical separation of the fingers. Training-induced somatosensory plasticity in humans has been demonstrated in blind persons suggesting increased cortical representation of the index fingers used in Braille reading [8]. With MEG-based source imaging Elbert et al. [9] found that in violinists who regularly played and practised their instrumentd the representation of the digits of the left hand in the somatosensory cortex was substantially expanded compared to the digits of the left hand in non-musician controls. The expansion was less pronounced for the left hand, which has the less active task of holding the neck of the instrument, compared to the digits engaged in fingering. The center for cortical responsivity for tactile stimulation of the digits for the left hand was shifted medially in the violinists as compared to age-matched control subjects. The amount of cortical reorganization for the digits´ receptive field was significantly correlated with age of onset of musical training and years of training.

On this background the present study was designed to assess the variability of activity in the primary somatosensory cortex SI (area 3b) across time and to determine to what extent intense somatosensory stimulation might change the cortical somatosensory representation. For this purpose repeated magnetoencephalographic measurements were used. The somatotopical organisation of SI is clearly reflected in the distribution of somatosensory evoked magnetic fields (SEFs), and sources of SEFs generally correspond with the somatosensory homunculus derived from intracranial stimulation studies. Separate representations have been described for thumb and little finger, both for early somatosensory evoked fields (assumed to be generated in SI) and for long-latency SEFs, possibly representing SII-activity [10,11], as also documented, for instance, for lingual, trigeminal, facial, tibial, and median nerve SEFs. Despite of the observed interindividual variability in locations of the equivalent current dipoles (ECDs) for early SEFs (SI activity), ECDs of each of the 10 subjects examined by Hari and colleagues [10,11] were concentrated to strips of less than 2 cm in length, 5 to 9 cm lateral to the midline.

Methods

One right-handed adult was trained to read Braille text with the three middle fingers of the left hand. Following the training the same three fingers were kept immobilized and deprived from sensory stimulation by applying local anesthetics (lidocain-prilocain creme) and bandaging. The experimental protocol included 8 baseline measurements, one measurement after 16 hours of Braille training and one after 18 hours, one measurement following 27 hours of sensory deprivation, and 3 additional measurements distributed across 3 weeks following sensory deprivation. The first three baseline sessions comprised stimulations of the left hand only (see Fig. 2).

Somatosensory evoked magnetic fields (SEFs) were measured with a Neuromag–122™ whole-scalp neuromagnetometer which contains 122 planar SQUID gradiometers in a helmet shaped array. Head position indicators (HPI) were attached to the subject's head to give exact information of the location and orientation of the sensor array with respect to the head. Preauricular points and the nasion were marked in order to allow replication of HPI positioning across measurements. During the recordings, the subject was sitting in a magnetically shielded room, with the head supported against the helmet-shaped magnetometer, keeping her head immobile and fixating a point in front of her. Cortical representations of adjacent fingers were determined by monitoring SEFs to electrical stimulation of the of the thumb (D1) and little finger (D5) of both hands. Two Medelec ST 10 stimulators were used for delivering 0.3-ms constant current pulse stimuli at constant intervals of 1 s at intensities of 7-8 mA. Within each session, two series of 500 stimuli were delivered, 250 to the left thumb and 250 to the right thumb in alternating order followed by 500 stimuli alternating between the left and the right little finger.

Signals were bandpass filtered (0.03-300 Hz), digitized at 997 Hz, and 500 responses were averaged on-line for each finger. Averages included a 50-ms prestimulus baseline and 150 ms following stimulation. Trials contaminated by ocular artefacts or large-scale MEG shifts were excluded on-line from the average. Data of two baseline sessions (one prior to and one following experimental manipulations) had to be discarded because of obvious artifacts. Source currents were identified by modelling the signal distribution by equivalent current dipoles (ECDs) during discrete time periods; for each time period the ECD best explaining the magnetic field variance was determined by a least-squares search. For source analysis, a subset of 20 channels over the respective cortex (left for right hand stimulation, right for left hand stimulation) was selected. The goodness-of-fit [12] of the model was calculated and only ECDs explaining more than 90% of the field variance were considered. After localizing estimated sources at different time points a time-varying multidipole model was computed using these sources to explain the entire time period. The sites and orientation of the dipoles were kept fixed but their amplitudes were allowed to change as a function of time. Source localization was further accomplished by superimposing ECDs to the individual magnetic resonance (MR) images. The MRI scan was obtained using a 1.5-T Siemens Magnetom system.

Results

SEFs consisted of prominent deflections at 25-27 ms (M27) and 41-43 ms (M42). The locations of the ECDs during these peaks demonstrated good replicability across baseline measurements (Fig. 1). For the *left D1*, the absolute positions varied in mediolateral, inferosuperior and anteroposterior directions by 4.3 mm, 7.6 mm, and 6.1 mm, respectively (standard deviations across 5 measurements). The corresponding values were 5.5/8.8/7.4 mm for the *left D5*, 3.8/3.2/8.6 mm for the *right D1*, and 3.9/2.9/2.7 mm for the *right D5*. The respective variability in the left D1-D5 distance was 4.3/4.4/4.2 mm. Sources of M27 agreed with somatotopic organization of the SI cortex (with a mean difference of 25.4 ± 4.1 mm for the Euclidean distance between D1- and D5-representations in the left hemisphere and 24.0 ± 4.2 for the right hemisphere/left hand stimulation). Locations of M42 did not systematically differ for the two digits.

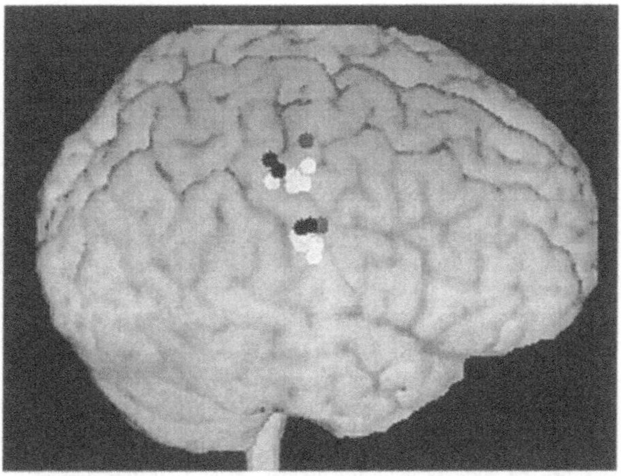

Fig.1. The cortical surface has been reconstructed from the individual MRI. The locations of the ECDs at 25 (D1) and 27 ms (D5) after stimulus onset were superimposed onto this cortical surface. The upper cluster of dots represents the locations for D5, the lower one those for D1 (left hand, right hemisphere). White dots: locations during pre- and post-baseline sessions, black dots: locations following Braille training, grey dots: locations following anesthesia.

Braille training altered representations of the stimulated left (non-dominant), but not of the right hand: The M27 dipole moments for D5, and after the 2nd training period also for D1, were stronger after the training than during any of the baseline or control sessions. (Left D5 baseline: 7.6 ± 1.7 nAm, Braille: 10.5 ± .3 nAm, $p < .05$; right D5 baseline: 11.4 ± 1.8nAm, Braille 8.9 ± .6 nAm; Fig. 2). After training, the distance of left D1-D5 along the curved surface of area 3b tended to be enlarged (Fig. 2). Superposition of M27-ECDs onto the individual magnetic resonance image suggested a shift of the left D5 in medial direction along the central sulcus (Fig. 1).

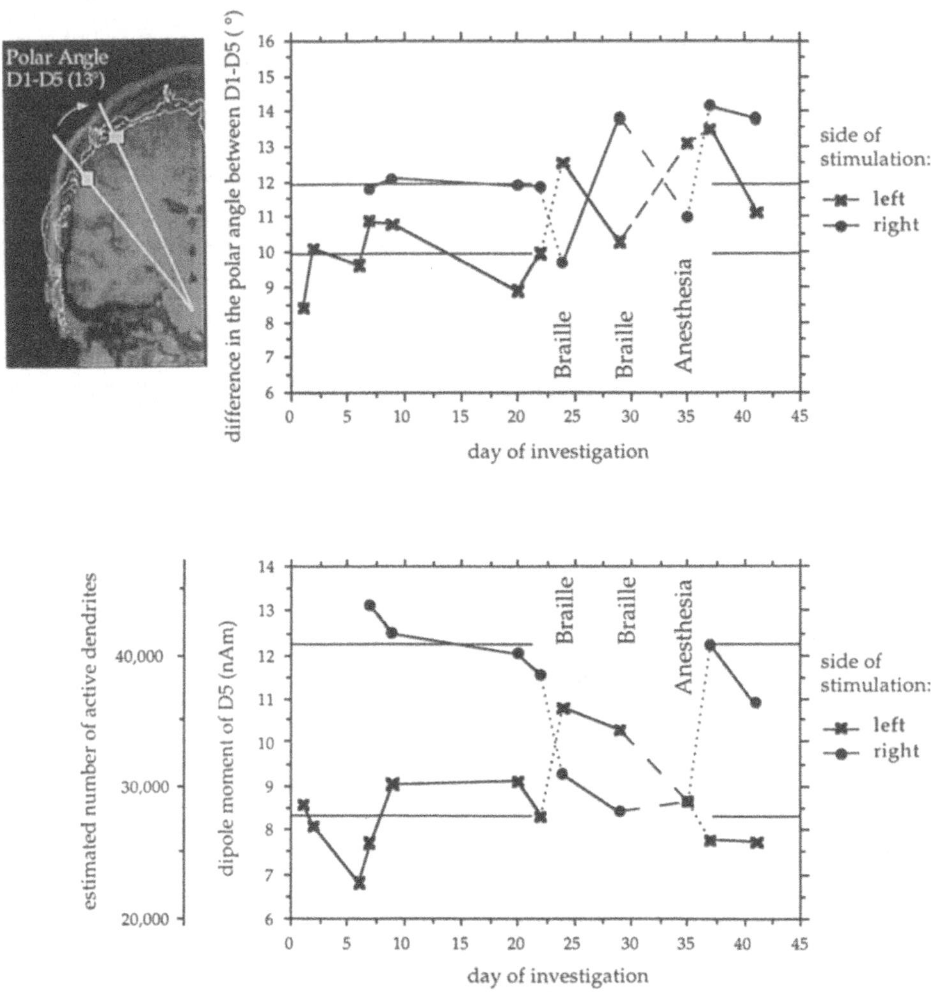

Fig.2. Change of dipole parameters across time; the abscissa indicates the days relative to the start of the investigation. The symbols mark days during which MEG measurements were obtained. The upper panel illustrates the change in the distance between the location of the ECDs for D1 and D5. The distance was measured as the declination (polar angle) between the locations of the represenations of the two digits as viewed from the center of the head, the latter being defined as the midpoint between the two meatuses. The overlay of D1 and D5 locations onto the corresponding coronal MRI section in the upper left corner illustrates this measure. During baseline measurements the polar angle is smaller for left than for right hand stimulation. This difference vanishes following Braille training and anesthesia. The lower panel shows the alteration of the dipole moment following Braille training in nAm. The ordinate on the left is a rough estimate for the number of neurons that may be involved during the cortical response [13]. Again, during baseline the right hand response is larger than the one for the left hand. Following Braille training, this "handedness" effect is reversed.

Discussion

Animal studies have consistently demonstrated that cortical plasticity can be observed during learning if a heavy schedule of training is designed to ensure continuous performance improvements. Thereby the training has to be conducted under conditions of a high motivational drive [1,2,5]. The results of the present study coincide with these considerations. Nevertheless, they must be considered with caution as the study was based on one subject only. To date, repeated measures with whole-head systems are rare. From the variation across baseline sessions it can be concluded that long-term within-subject replicability of MEG recordings should be further improved by advancing methods for head fixation, in order to allow the measurement of small variations in cortical organization. There is considerable evidence that intense stimulation of fingers during a meaningful behavioral tasks seems to be related to changes in cortical representation. Increased use of a limb leads to an expansion of the cortical representation and to a reduction of receptive fields. But alterations are more complex. The present results indicate that changes in representations are not directly related to the enhanced stimulation. Disuse is thought to cause invasion of the representational zone from nearby sites. If this was the case for the present study, then a reduced and not an enhanced neuronal response to stimulation of left D5 should have been seen. While the shift in cortical representation of D5 following training is weak compared to the variations between sessions, the dipole strenghts is clearly enhanced after training. Increased neuronal interconnectivity which is triggered by the trainining may be responsible for the observed enhancement. Similarly, the decrease in the dipole moment of right D5 cannot be explained by simplistic relationships. The laws of interhemispheric transfer and the those underlying cortical reorganization in general have to be studied further.

References

[1] Jenkins, W. M., Merzenich, M. M., Ochs, M. T., Allard, T., Gufc-Robles, E. Functional reorganisation of primary somatosensory cortex in adult owl monkeys after behaviorally controlled tactile stimulation, Journal of Neurophysiology, 1990, 63: 82-104

[2] Jenkins,W.M., Merzenich, M.M. Cortical representational plasticity: Some implications for the bases of recovery from brain damage. In: von Steinbüchel, N., von Cramon, D.Y.,, Pöppel, E. (Eds.) Neuropsychological Rehabilitaion. Heidelberg, Springer, 1992 (pp.20-35.)

[3] Yang, T.T., Gallen, C.C., Schwartz, B., Bloom, F.E., Ramachandran, V.S., Cobb., S. Sensory maps in the human brain. Nature, 1994, 368.

[4] Elbert T., Flor H., Birbaumer N., Knecht S., Hampson S.,Taub E. Evidence for Extensive Reorganization of the Somatosensory Cortex in Adult Humans after Nervous System Injury. Neuroreport, 1994, 5: 2593-2597.

[5] Recanzone, G. H., Merzenich, M. M., Jenkins, W. M., Grajski, A., Dinse, H. R., J. Plasticity in the frequency representation of primary auditory cortex following discrimination training in adult owl monkeys. Journal of Neuroscience, 1992, 67: 1031-1056.

[6] Pons, T., Garraghty, P. E., Ommaya, A. K., Kaas J. H., Taub, E., Mishkin M. Massive Cortical Reorganisation After Sensory Deafferation in Adult Macaques. Science, 1991, 252: 1857-1860.

[7] Mogilner, A., Grossmann, J.A.I., Ribary, U., Joliot, M., Volkmann, J., Rapaport, D., Beasley, R.W., Llinas, R.R. Somatosensory cortical plasticity in adult humans revealed by magnetoencephalography. Proceedings of the National Academy of Science USA, , 1993, 90: 3593-3597.

[8] Pascual-Leone A., Torres, F. Plasticity of the sensorimotor cortex representation of the reading finger in Braille readers. Brain, 1993, 116: 39-52.

[9] Elbert T, Wienbruch C, Rockstroh,B., Pantev C, Taub E. Increased use of the left hand in string players associated with increased cortical representation of the fingers. Science, 1995, 220: 305-307

[10] Hari, R., Reinikainen, K., et al. Somatosensory evoked cerebral magnetic fields from SI and SII in man. Electroencephalography and Clinical Neurophysiology , 1984, 57: 254-263.

[11] Hari, R., Karhu, J., Hämäläinen, M., Knuutila, J., Salonen, O., Sams, M., Vikman, V. Functional Organization of the human first and second somatosensory cortices: a neuromagnetic study. European Journal of Neuroscience, 1993, 5: 724-734.

[12] Kaukoranta, E., Hari, R., et al. Cerebral magnetic fields evoked by peroneal nerve stimulation. Somatosensory Research , 1986, 3: 309-321.

[13] Williamson, S.J., Kaufmann, L. Theory of neuroelectric and neuromagnetic fields. In: Grandori, F., Hoke, M., Romani,G.L. (Eds.) Auditory Evoked Magnetic Fields and Electric Potentials. Basel, Karger, 1990 (pp. 1-39).

Acknowledgement

Research was sponsored by the EU Human Capital and Mobility Programme through the BIRCH Installation in the Low Temperature Laboratory of Helsinki University of Technology and by the DFG.

Laterality of Hippocampal Responses to Infrequent and Unpredictable Omissions of Visual Stimuli

Rogers, R.L., Taylor, S.A., Akhtari, M., and Sutherling, W.W.

Neuromagnetism Laboratory, Epilepsy & Brain Mapping Center, Hospital of the Good Samaritan, Los Angeles, CA 90017 USA

Information obtained in the context of lesion studies and invasive electrical recordings as well as non-invasive event related potential (ERP) and MEG studies points to the possibility that several anatomically distinct sources contribute to the generation of the P3 component (Halgren et al., 1989; Smith et al., 1990; Yamaguchi and Knight, 1991; Rogers et al., 1992; 1993a; 1993b). In our previous work, we have identified two main clusters of sources of P3s irrespective of the stimulus modality: One in the sensory cortex specific to the particular modality of the stimulus and a second, subcortical cluster the precise location of which was not easy to discern. It appeared to be, however, in the general vicinity of the hippocampus and/or the lateral thalamus. In this study we attempted to identify the location of the subcortical cluster of sources, in the context of a visual P3 task. Moreover, in view of suggestions that P3s associated with spatial as opposed to auditory tasks may be generated mostly by right hemisphere sources (e.g. Johnson 1989a; 1989b), we recorded magnetic fields over both hemispheres during the same task. We identified sources bilaterally and subsequently we compared the relative strength of these two clusters of subcortical sources.

Materials and Methods

Ten neurologically normal, right-handed, adult subjects ranging in age from 23 to 53 years (mean = 29.1) were recruited from the university setting.

Magnetic fields and electrical potentials were recorded during a standard oddball paradigm in which a visual stimulus was presented for 50 msec. every 600 msec. and on 10% of the trials was omitted. Omissions were randomly dispersed with restrictions that no fewer than 3 and no more than 15 frequent stimuli were presented. Stimuli consisted of four points of light arranged as a 6 cm square. Light sources were presented by an IBM PC computer with an internal digital-to-analog card. The 5 volt analog output channels were used to power four light emitting diodes outside of the magnetically shielded chamber. The light from the diodes were then transmitted into the room via fiber optic cables that terminated on a panel in front of the subject. Subjects were instructed to fixate on a mark in the center of the square so that all four visual quadrants would be simultaneously stimulated. Subjects were also instructed to count the number of omissions so that their attention was directed toward the infrequent missing stimuli. One hundred such epochs were averaged for each trial, however, subjects were actually presented with between 100 and 120 missing stimulus trials. This was done so that they would not have prior knowledge of the number of omitted stimuli. Trials were repeated for each placement of the neuromagnetic probe.

Magnetic fields were measured using a seven channel probe system (Biomagnetic Technologies, Inc.) which utilized second-order gradiometers with a 4 cm baseline and 1.8 cm diameter of the superconductive sensor coils. A complete trial was performed separately for each probe position. The order of left-right trials was counterbalanced. The number of dewar placements required to cover the lateral aspects of each hemisphere ranged from 5 to 7 (35 to 49 sensor locations). The location of the magnetic sensors in relation to the subjects' heads was determined using three low frequency magnetic field sensors placed on the head with a transmitter mounted on the cryogenic container. Error tolerance was set so that maximum location error would not exceed 2 mm.

Averaged electrical potentials were obtained simultaneously with all MEG recordings. The active electrode sites were Oz, Pz, and Fz referenced to linked mastoids with the common ground on the forehead. Additionally, eye movements in the horizontal and vertical directions were also obtained. Both magnetic field and electric potential measurements were band passed on-line between .1 and 50 Hz.

To ensure the integrity of the average wave forms, each individual epoch was inspected for eye movements as well as variability and normalcy of the peaks. Single trials with obvious artifacts were removed from the averages, but these never exceeded 4% of the total number of sweeps. Variances of the magnetic wave forms were comparable to their electrical counterparts. Frequency histograms of field and potential values were normally distributed about the means.

Dipole parameters including location (Cartesian coordinates), orientation and strength were calculated for each peak using a finite difference version of the Levenberg-Marquardt algorithm for a spherical head model. The size and origin of the sphere was determined separately for each hemisphere by a least-squares fit from the digitization of the temporal areas. Since a single dipole source will produce a distinctive external field pattern with two extrema around which the field strength diminishes in all directions, isofield contour maps were created for each peak and inspected in order to determine that they were consistent with a dipole source. In addition, a minimum least-squares fit which accounted for at least 80% of the total variance between the measured fields and the calculated forward solution was required.

Magnetic resonance images were obtained for projecting the calculated sources on to the appropriate brain image. The procedures and details of coregistration of MEG and MRI data are described in detail elsewhere (Rogers, et al., 1991).

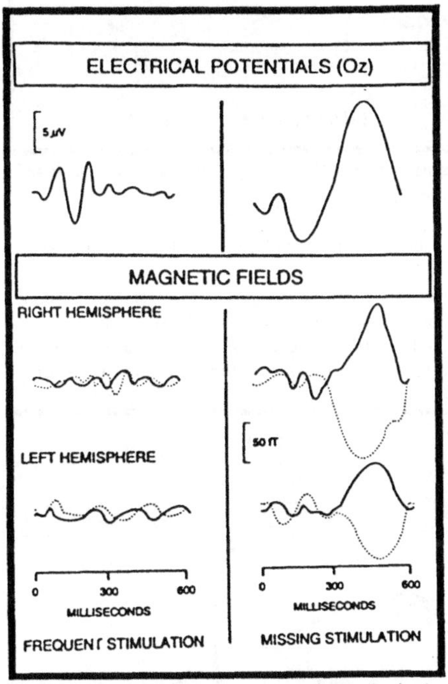

Figure 1: Shows grand average waveforms for both potentials (Cz) and field extrema for the frequent and missing stimulus conditions.

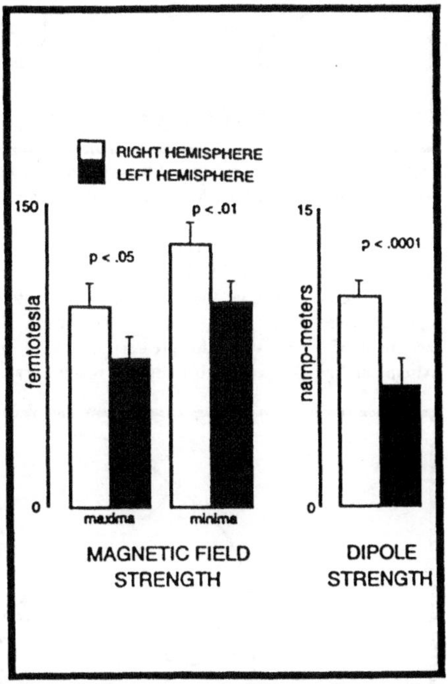

Figure 2: Demonstrates the difference between the two groups for both magnetic field extrema and dipole strength.

Results

Electrical potentials at Oz produced very consistent and predictable results with the frequent stimulation producing the common three peaks as seen in the top-left portion of Figure 1, which represents the average wave forms across all trials for all ten subjects. The omitted stimulus produced two peaks including a negative peak at 221.0 $\pm$19.2 msec. followed by a positive peak seen at 420.1 $\pm$23.1 msec.

Grand average magnetic flux wave forms over the right and left cerebral hemispheres are illustrated in the bottom portions of Figure 1. Magnetic flux to frequent stimulation (left portion of figure) did not show any discernible peaks above noise level. On the other hand, magnetic flux to infrequent omissions of visual

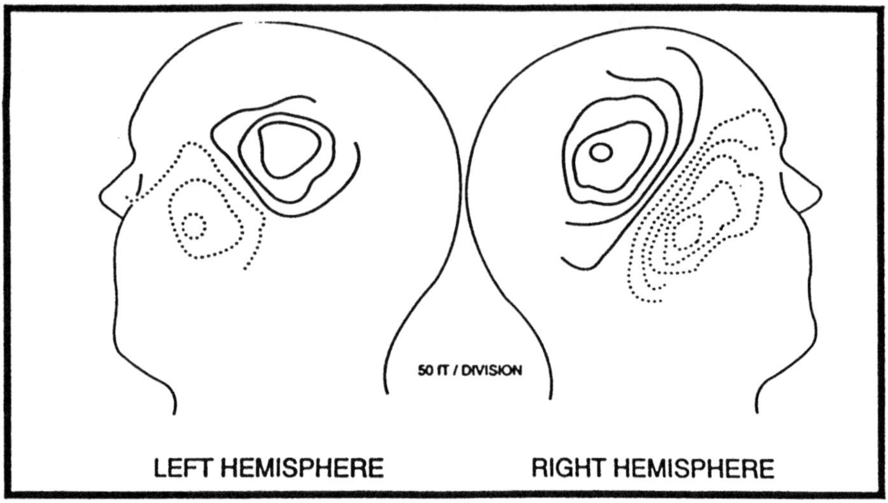

Figure 3: Isocontour maps over the right and left hemisphere during the missing stimulus condition at 455 msec following stimulation. Note the greater strength on the right side.

stimulation produced large distinctive dipole patterns over both the right and left hemispheres in all subjects at latency ranges from 380 to 450 msec.

Figure 2 shows the magnetic flux pattern over the right versus left hemisphere in a single representative subject at a latency of 435 msec. Each contour line in these isofield maps represents an increment of 50 fT, thus, the greater number of contour lines in the right hemisphere indicates that the extrema and gradients are substantially greater over the right hemisphere. Since the location and spacing of

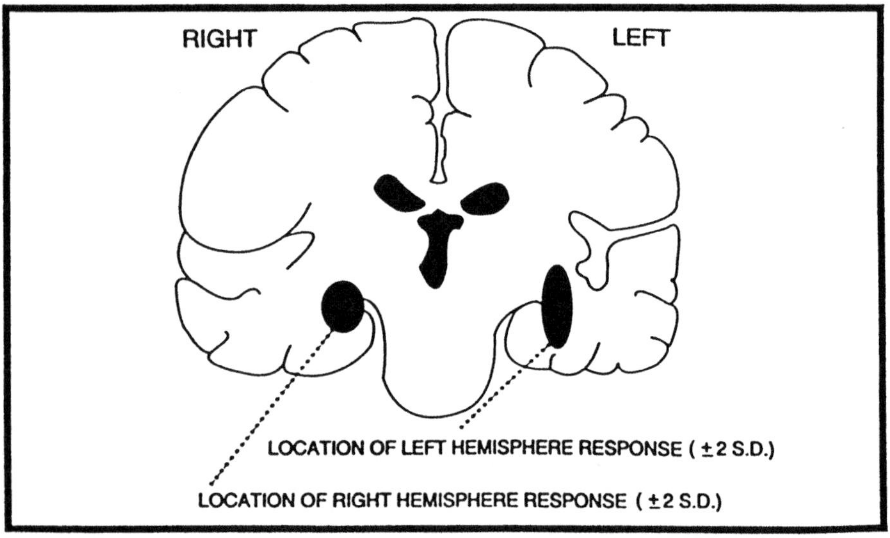

Figure 4 shows the distribution across subjects of the location of right and left hemisphere activity and their 95% confidence intervals.

the extrema are similar in both hemispheres, this would indicate that the generator source emanates from similar locations and orientations in both cases with the major difference being the strength of the sources.

Univarite F-test were performed on the differences between right and left hemisphere responses on the peak amplitudes of both extrema and on the calculated dipole strength. Figure 3 summarizes these results indicating that significantly greater responses occurred on all three of these measures (maxima: $p < .01$; minima: $p < .05$; dipole strength: $p < .0001$). Dipole strength was greater over the right hemisphere for all ten subjects without exception. Since physical laws dictate that magnetic field strength attenuates 50% per centimeter, it is not conceivable that the fields over the two hemispheres are from the same source. Figure 4 shows the distribution across subjects of the location of right and left hemisphere activity and their 95% confidence intervals.

Discussion

Previous studies have demonstrated that externally measured magnetic field patterns are indicative of activity in the vicinity of the right hippocampal formation during infrequent and unpredictable intrusions or omissions of visual stimuli in an oddball evoked response paradigm. These fields occur coincident with late endogenous evoked potential components that are consistently recorded in similar situations. In the present study, magnetic fields temporally corresponding to the late P300 component of simultaneously recorded evoked potentials were accounted for by sources in the vicinity of the left and right hippocampus in addition to previously reported sources in the vicinity of the primary visual cortex. Projection of these sources onto MRIs suggested that both hippocampal structures are simultaneously active and that there is an amplitude and strength-related dominance of the right hippocampal sources to visual stimulation.

References

Halgren, E., Squires, N.K., Wilson, C.L., Rohrbaugh, J.W., Babb, T.L. and Crandell, P.H. Endogenous potentials generated in the human hippocampal formation and amygdala by infrequent events. Science, 1980, 210: 803-805.

Johnson, Jr., R. Auditory and visual P300s in temporal lobectomy patients: evidence for modality-dependent generators. Psychophysiology, 1989a, 26: 633-650.

Johnson, Jr., R. Developmental evidence for modality-dependent P300 generators: a normative study. Psychophysiology, 1989b, 26: 651-667.

Rogers, R.L., Papanicolaou, A.C., Baumann, S.B. and Eisenberg, H.M. Late magnetic fields and positive evoked potentials following infrequent and unpredictable omissions of visual stimuli. Electroenceph. clin. Neurophysiol., 1992, 83: 146-152.

Rogers, R.L., Basile, L.F.H., Papanicolaou, A.C. and Eisenberg HM. Magnetoencephalography reveals two distinct sources associated with late positive evoked potentials during visual oddball task. Cerebral Cortex, 1993a, 3: 163-169.

Rogers, R.L., Basile, L.F.H., Papanicolaou, A.C., Bourbon, T.W. and Eisenberg, H.M. Visual evoked magnetic fields reveal activity in the superior temporal sulcus. Electroenceph. clin. Neurophysiol., 1993b, 85(5): 344-347.

Smith, M.E., Halgren, E., Sokolik, M., Baudena, P., Musolino, A., Liegeois-Chauvel, C. and Chauvel, P. The intracranial topography of the P3 event related potential elicited during auditory oddball. Electroenceph. clin. Neurophysiol., 1990, 76: 235-248.

Yamaguchi, S. and Knight, R.T. Anterior and posterior association cortex contributions to the somatosensory P300. J. Neurosci., 1991, 11: 2039-2054.

MEG Measurement of Auditory Sensory Memory Persistence via the M100 in Children and Adults

Rojas, D. C.,[1] Sheeder, J. L.,[1] Teale, P.,[1] Walker, J. R.,[1] Robertson, B. A.,[1] Selig, S.,[1] and Reite, M.[1,2]

[1]University of Colorado Health Sciences Center, Denver, USA; [2]Veterans Administration Medical Center, Denver, USA

Introduction:

There have been several papers linking the auditory evoked field (AEF) component M100 to the auditory sensory memory trace, or echoic memory [1,2,3]. Recently, a preliminary effort to describe the AEF in children has been reported [4]. It is apparent that children do not show the same component structure as adults when presented with the same stimuli. The most striking finding of this report was the apparent lack of M100 response in the younger children, a finding which we have also replicated (see below). One possible explanation, proposed by Paetau, et al. [4], is that the M100 refractory period is longer in children than adults. Thus, the field strength of the M100 would be much weaker than adults at similar interstimulus intervals. Also, if the refractory period is longer in children than adults, then the echoic memory trace duration would also be longer, assuming that M100 indexes the trace duration, as proposed by Lü, et al. [1]. As this possibility is theoretically important to both psychology and MEG research, two experiments were undertaken in children and adults to explore the development of the AEF M100 component.

Our first attempt to describe the M100 in children was a standard AEF mapping experiment to pure tones delivered at one interstimulus interval (ISI), similar to the procedure we typically employ when mapping the M100 in adults. The results of this experiment led to a second experiment in which we positioned the intrument over a single, putative M100 extremum, and measured the AEF while systematically varying the ISI between tones. Both experiments are described in this paper.

Methods:

A BTI Model 607 7-channel, second order gradiometer with 1.8 cm coil diameters and a 4.0 cm baseline was used for MEG recordings in both experiments. Subjects were recorded reclining in a custom non-magnetic bed. All recordings were made in a magnetically-shielded room constructed from 6.3 cm mu-metal and aluminum and lined with 0.76 mm thick mu-metal. The room provides 35 dB noise reduction at DC. MEG data were acquired via BTI amplifiers and anti-aliasing low pass filters, and digitized at 1 kHz with a 12-bit, 64-channel Scientific Solutions analog-to-digital converted. Neuroscan SCAN software running on an 80386 computer was used for signal acquisition and processing. MEG was written out on a Grass Model 78 ink-writer unit to facilitate visual inspection of the raw signal for magnetic artifact, eye movement, subject movement and other potential artifacts. Coil locations and orientations relative to the head were determined using a sonic digitizer, as described in Reite et al. [5]. Custom software permitted co-localization in space of the subject's head (referenced to nasion and the left and right pre-auricular points), each recording site, and each gradiometer coil (and orientation). The accuracy of the digitizer system is ± 1 mm.

In the first experiment, we recorded data from left and right hemispheres in 17 right-handed children (6 girls and 11 boys) ranging in age from 6-9 years old. To map the distribution of the AEF, we recorded auditory evoked responses using 25 msec (3 msec rise/fall), 1 kHz toneburst stimuli. Tones were delivered at a constant rate every 2 seconds, and were presented at 85 dB SPL over insert earphones. Auditory stimuli were generated using a SoundBlaster SB16, 16-bit sound card housed in an 80286 PC. Responses to 100 stimuli, recorded from 35 sites per hemisphere (5 dewar positions), were averaged together. Six of these children (4 males and 2 females) were recorded again one year after their first recording session to ascertain any changes over time.

Five children (4 males and 1 female ages 8-10 years old) and 12 adults (6 males and 6 females ages 22 to 47 years old) participated in the second, variable ISI experiment. Subjects listened to 1 kHz tones presented monaurally at 85 dB SPL over insert earphones. Interstimulus interval was varied in randomly presented sessions of 1, 1.5, 2, 3, 4, and 6 seconds. MEG recordings were made from the contralateral hemisphere with a BTI Model 607-21 7 channel gradiometer, positioned over the putative anterior M100 extremum. Recordings were made from both hemispheres at all 6 ISIs while subjects watched silent movies of their choice. MEG data were bandpass filtered between 0.1 and 300 Hz, digitized at 2 kHz, low pass filtered off-line at 40 Hz, and baseline corrected using a 200 msec pre-stimulus window for further analyses.

The time constant of decay of the echoic trace, or τ, was computed by curve fitting the measured M100 amplitude data from the 6 ISIs to the equation

$$F(t) = A\left(1 - e^{-(t-t_0)/\tau}\right)$$

where $F(t)$ is the field amplitude in fT at time t, A represents the asymptotic amplitude, t_0 is the time axis intercept, and τ is the time constant, or trace lifetime of echoic memory [1]. Curves were fit to the data using a non-linear regression procedure (Marquardt-Levenberg algorithm) to minimize the sums of squares as A, t_0, and τ were varied. The trace lifetime τ represents the time necessary for 2/3rds of the trace to decay. Two constraints were placed on the fits a priori: a) A was not allowed to be higher than 1000 fT which is reasonable given both the theoretical constraints upon source generator strength and our own practical experience with the M100, and b) t_0 was not allowed to be less than 0.0 seconds, reasonable in light of the fact that negative values of the time intercept would imply memory trace formation prior to the presentation of the stimulus.

Results:

In the first experiment, all children exhibited identifiable auditory evoked magnetic fields, with several dipolar-like components. However, the latency and polarity of the components varied considerably. Fig. 1 shows overlapping AEF data from both hemispheres of children and adults to the same stimulus paradigm. Several components are evident in both children and adults, but differences between them are also clearly identifiable. The children's AEF morphology consistently differed from that typically seen in adults both in the number of components within a 500 msec window and the variability between children. Table 1 lists the number of

Table 1. Number of Dipolar Components Evident in 35 Channel Recordings

			Year 1			Year 2		
Subj. ID	**Sex**	**Age**	**Left**	**Right**	**M100?**	**Left**	**Right**	**M100?**
1	male	9	5	4	R(2)	4	5	R(2)
2	female	8	5	4	R(2)	5	4	R(2)
3	male	9	6	3	-	-	-	-
4	male	8	4	3	-	3	4	-
5	male	6	4	2	-	3	5	L(1)
6	male	9	4	4	R(2)	-	-	R(2)
7	male	6	3	5	R(2)	-	-	R(2)
8	male	9	3	4	-	4	4	R(1)
9	female	8	6	5	R(2)	-	-	R(2)
10	female	9	5	5	R(2)	3	5	R(2)
11	female	9	5	3	-	-	-	-
12	female	7	5	5	R(2)	-	-	R(2)
13	male	9	4	4	-	-	-	-
14	male	7	6	4	-	-	-	-
15	male	9	5	7	R(2)	-	-	R(2)
16	female	7	5	5	R(2)	-	-	R(2)
17	male	9	4	4	R(2)	-	-	R(2)
mean	n/a	8.12	4.65	4.18	n/a	3.67	4.50	n/a
SD	n/a	1.11	0.93	1.13	n/a	0.82	0.55	n/a

Note to Table 1. Left and right refer to the two hemispheres. Numbers in right and left hemisphere columns reflect the number of dipolar components evident in the 35 channel AEF recordings. R = Right, and L= Left. Number in parentheses reflects serial position within AEF of component with similar polarity and latency to an adult M100.

components seen in both hemispheres for all children, including those recorded a second time 1 year apart. There were no significant correlations between age and number of components in either left or right hemisphere, $r = .10$ and $r = .08$, $p > .05$, respectively. Left and right hemisphere topographic maps constructed from a representative female child and adult are also illustrated in Fig. 1.

In the second, variable ISI experiment, the adults' mean τ was $1.56 \pm .30$ (SEM) for the left hemisphere, and $1.01 \pm .20$ for the right. The left hemisphere mean for the children was 3.44 ± 1.5 and the right hemisphere mean was $1.91 \pm .93$. Fig. 2 illustrates the curve fits for each group separately.

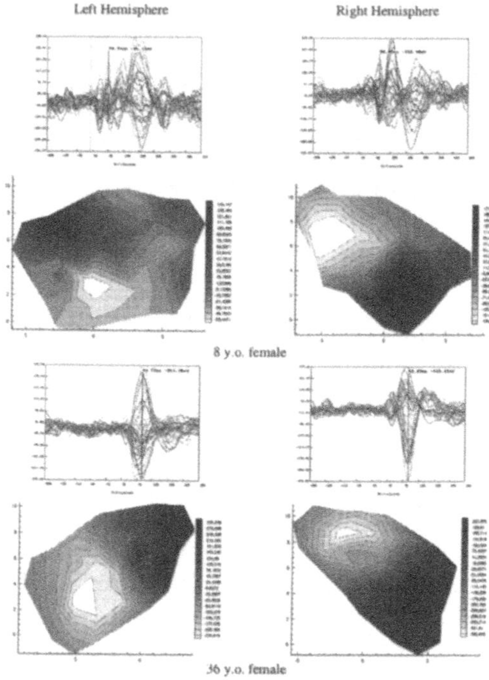

Fig. 1. Waveforms and topographic maps for 35 channel recordings. Scale for both in fT. Axes on maps are in cm, and reflect headframe coordinates. Positive y is superior, and positive x is anterior.

Discussion:

Most of the children did not show adult-like morphology and latency. Specifically, the M100 ingoing and outgoing extremum were not seen at comparable latencies to adults in almost all of the child recordings. At this point, we are not confident that any of the components seen are analogs to the adult M100, since in most of the children a large component of the correct polarity was not seen until 200 msec post-stimulus. Of the children who did show adult-like morphology and latency, it was only present in the right hemisphere recordings. We interpret this finding to mean that there is independent development of the right hemisphere from the left with respect to the auditory system. The preliminary data from the second experiment support the contention that children have longer M100 refractory periods than adults, and perhaps also longer echoic trace lifetimes. This would explain why the children in the first experiment did not exhibit M100-like responses, since at the same ISI an adult would have a much stronger response than a child. Interestingly, the left hemisphere appears to have slightly longer trace lifetimes than the right in the children. As with the mapping experiment, which revealed more adult-like responses in the right

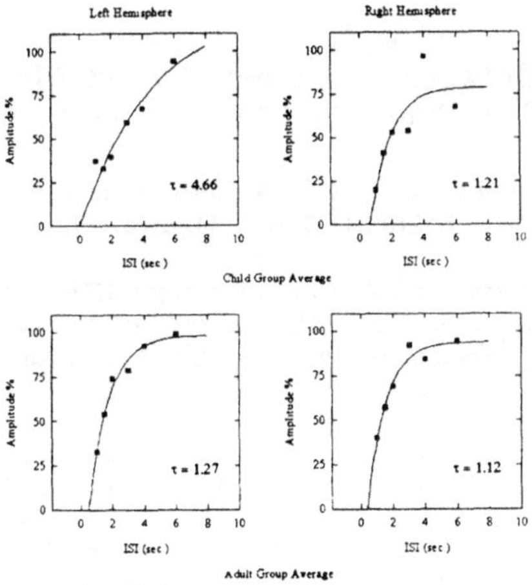

Fig. 2. Amplitude - ISI relationship for putative M100 component in adults and children. Amplitude is represented as normalized response, expressed as percentage of largest response in fT. Taus are curve-fit to mean responses.

hemisphere, this suggests that the auditory cortex of the two hemispheres may develop independently.

More data will need to be collected to further delineate these developmental changes. First, although there is clear evidence of a change between ages 6-9 and adulthood, the timecourse of that maturation will need to be explored in greater detail. There is also a pressing need to demonstrate that the children's putative M100 response is related to echoic memory, as has been done before in adults [1] with the use of psychoacoustic methods. If the relationship holds, these data suggest that echoic trace lifetime is longer in children than adults.

References:

[1] Lü, Z. -L., Williamson, S. J., and Kaufman, L. Science, 1992, 258: 1668-1670.
[2] Lü, Z. -L., Williamson, S. J., and Kaufman, L. Brain Research, 1992, 572: 236-241.
[3] Sams, M., Hari, R., Rif, J., and Knuutila, J. Journal of Cognitive Neuroscience, 1993, 5: 363-370.
[4] Paetau, R., Ahonen, A., Salonen, O., and Sams, M. Journal of Clinical Neurophysiology, 1995, 12: 177-185.
[5] Reite, M., Adams, M., Simon, J., Teale, P., Sheeder, J., Richardson, D. and Grabbe, R. Cognitive Brain Research, 1994, 2: 13-20.

Acknowledgments:

This work was supported by a grants from the Developmental Psychobiology Research Group at the UCHSC School of Medicine, #250, USPHS MH47476, and a grant from the Alzheimer's Association, PRG-93-143. Salary support for D. C. Rojas was provided by USPHS 5T32 MH15442 and M. Reite salary support was provided by USPHS grant 5K02 MH46335.

The Relation Between Gustatory-Evoked Magnetic Fields and Reaction Times to Different Taste Qualities

Saito, S.S.[1], Kobayakawa, T.K.[1], Ayabe-Kanamura, S.A.[2], Endo, H.E.[1], Yamaguchi, Y.Y.[1], Kikuchi, Y.K.[3], Ogawa, H.O.[4] and Takeda, T.T.[1]

[1]National Institute of Bioscience and Human Technology, MITI, Tsukuba, Japan; [2]University of Tsukuba, Tsukuba, Japan; [3]Tokyo Medical and Dental University, Tokyo, Japan; [4]Kumamoto University, Kumamoto, Japan

Introduction

Kobayakawa et al. reported the development of a new gustatory stimulation apparatus for which the rise-time was less than 20 ms for a measurement of EEG [1]. They also modified the apparatus for a measurement of MEG, and reported the estimation of the source of the gustatory-evoked magnetic fields (GEMs) in the human cerebral cortex [2][3]. In the present study we measured the reaction times (RTs) to the different taste qualities and compared them with the GEMs of different taste qualities to find the mutual relation between the reaction times and the onset of GEMs.

Methods

In separate experiments we measured the RTs and the GEMs, using the same sharp stimulation method. Six

Table 1. Results of RTs and GEMs onset

		RTs					GEMs		
Subj.	Stimulus	ave. (ms)	Diff.* (ms)	SD (ms)	n	subject's introspection	GEMs onset (ms)	Diff.*	subject's introspection
AS	NaCl	350	81	111	37	salty	40	60	at first sour and salty
	saccharine	431		83	40	sour > sweet	100		at first sour and sweet
YY	NaCl	354	117	76	40	salty	70	110	salty, moderately weak
	saccharine	471		150	36	moderate sweet	180		fairly sweet
KT	NaCl	372	66	107	37	salty	30	50***	salty
							210		salty
	saccharine	438		106	34	sour > sweet	340		sweet
							80		sour and sweet
TA	NaCl	629	247	110	39	salty	130	150	salty
	saccharine	876		244	36	sweet and a little sour	280		sweet and a little sour
MK	NaCl	730	81	184	40	salty	60	80	salty
	saccharine	811		186	37	sweet	140		sweet
EH	NaCl	369	-11	95	36	distinct salty	140	-20	clearly salty
	saccharine	350		68	37	strange sweet tooth paste with salt taste (not just salt)	120		strange sweet tooth paste with salt (not just salt)
Ave.	NaCl	467	96	114	38		93**	72	
	saccharine	563		140	37		172**		

* difference between NaCl and Saccharine

** the data were averaged for a subject having two GEMs onsets

*** 80ms (sour and sweet) - 30ms (first onset)

subjects (3 females) participated in this experiment; their ages ranged from 21 to 32 years old. A subject was seated on a wooden chair in a magnetically shielded room, and his or her jaw was supported by a chin rest attached to the chair. A Teflon tube through which a taste solution flowed was placed on the subject's tongue using a guide attached to the chair.

A solution of 1 M NaCl was used as a salty taste and a solution of 3 mM saccharin was used as a sweet taste. We used saccharin instead of sucrose because saccharin presents a sweet taste in low concentration, which would not contaminate the apparatus and not wound the membranes of taste cells. However, saccharin caused other taste qualities (sour, bitter, or salty) depending on the location of the placement of a hole of the tube on the tongue. We instructed subjects to position the hole of the Teflon tube on the part of their tongue where they perceived only or primarily a sweet taste. The duration of each stimulus was 400 ms. During the inter-stimulus interval, deionized water flowed to rinse the tongue for about 30 s. The temperature of both the solution and the deionized water was maintained at the temperature of the tongue's surface.

Results

Table 1 shows for each subject, RTs, the latency until the GEMs onset, and subjects' introspection for NaCl and saccharin.

Subject's introspection: No subject reported a tactile or temperature differences between the taste solution and deionized water. All subjects perceived NaCl as being salty. Two subjects reported an "only sweet taste" for saccharin. Subject EH reported a "strange sweet-like tooth paste with salt taste" for saccharin. Two other subjects reported "primarily sour with some sweet", and one subject reported "primarily sweet with some sour".

Reaction Time: For each subject, an averaged reaction time was calculated for two tastants. The saccharin showed significantly longer RTs than NaCl except for subject EH.

GEMs: When treating the GEMs data, the trials which contained eye movement were rejected and the remainder of the trials were averaged. The number of trials for averaging was about 35 per subject. The sampling rate was 250 Hz. Averaged GEMs of the 64 channels were superimposed on the same graph, aligned according to the time after the raising of the stimulus. We estimated the onset of GEMs of each tastant for each subject.

The onset of GEMs was significantly later for saccharin than NaCl, for five of the six subjects. Fig. 1 and 2 show the GEMs of NaCl or saccharin for subject YY; the pattern is typical of all subjects.

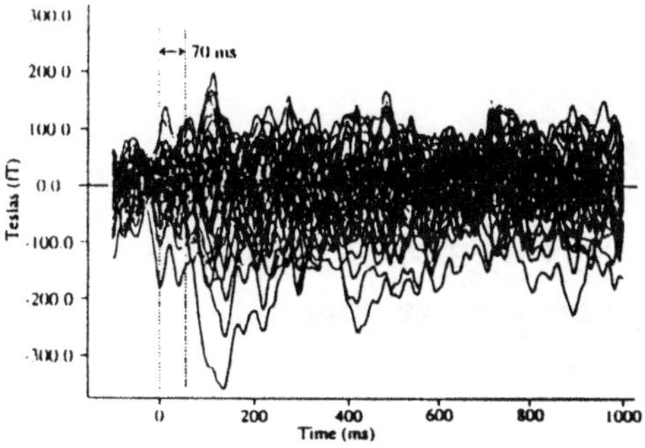

Fig. 1 GEMs onset of NaCl (subject YY)

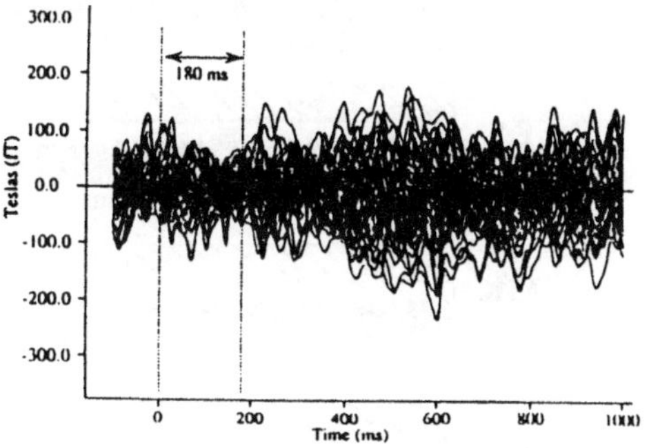

Fig. 2 GEMs onset to saccharin (subject YY)

In the case of subject KT, we measured two GEMs' experiments of 40 trials for each stimuli and estimated each onset for each GEMs' experiment; the GEMs onsets were very different in both taste qualities (Table 1). For saccharin solution the perception of "only sweet" for saccharin showed the later GEMs onset (Fig. 3) than the perception of "sour and sweet" (Fig. 4). For NaCl solutions perceived as salty in both experiments, one had a GEMs onset of 30 ms after stimulation, another had a GEMs onset of 210 ms after stimulation. The GEMs containing the faster onset of GEMs showed a second onset at the same time as the later GEM onset. In the cases of other subjects, subject AS, MK, and YY showed the GEMs onset at 40, 70 or 60 ms after the stimulation like to the faster onset of subject KT, and two of them had the second onset at 190 ms (subject AS) or 260 ms (subject YY) after the stimulation. Subject TA and EH showed the GEMs onset 130 or 140 ms after the onset of stimulation like to the later onset of subject KT.

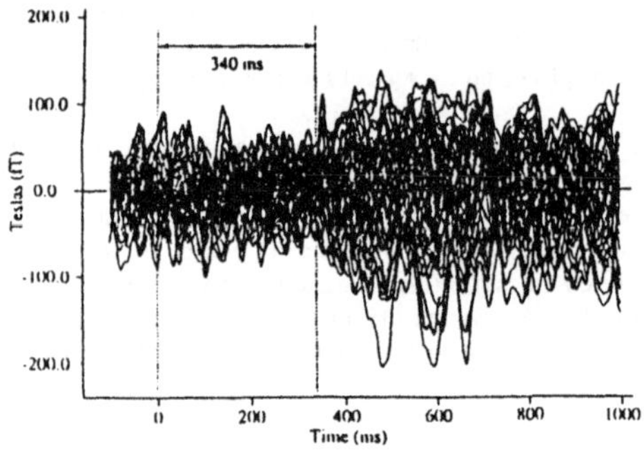

Fig. 3 GEMs onset to saccharine (perceived as only sweet, subject KT)

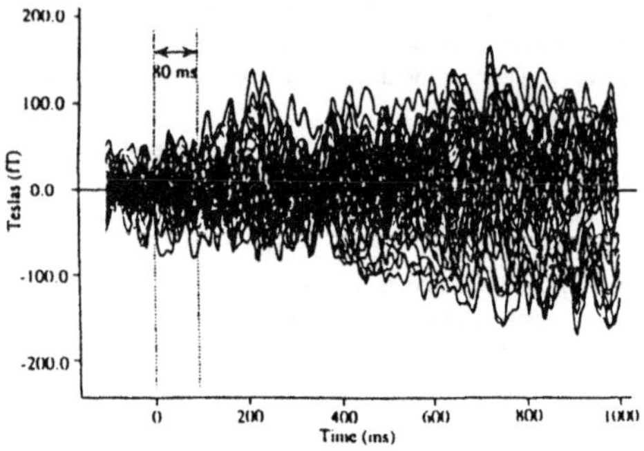

Fig. 4 GEMs onset to saccharine (perceived as sour and sweet)

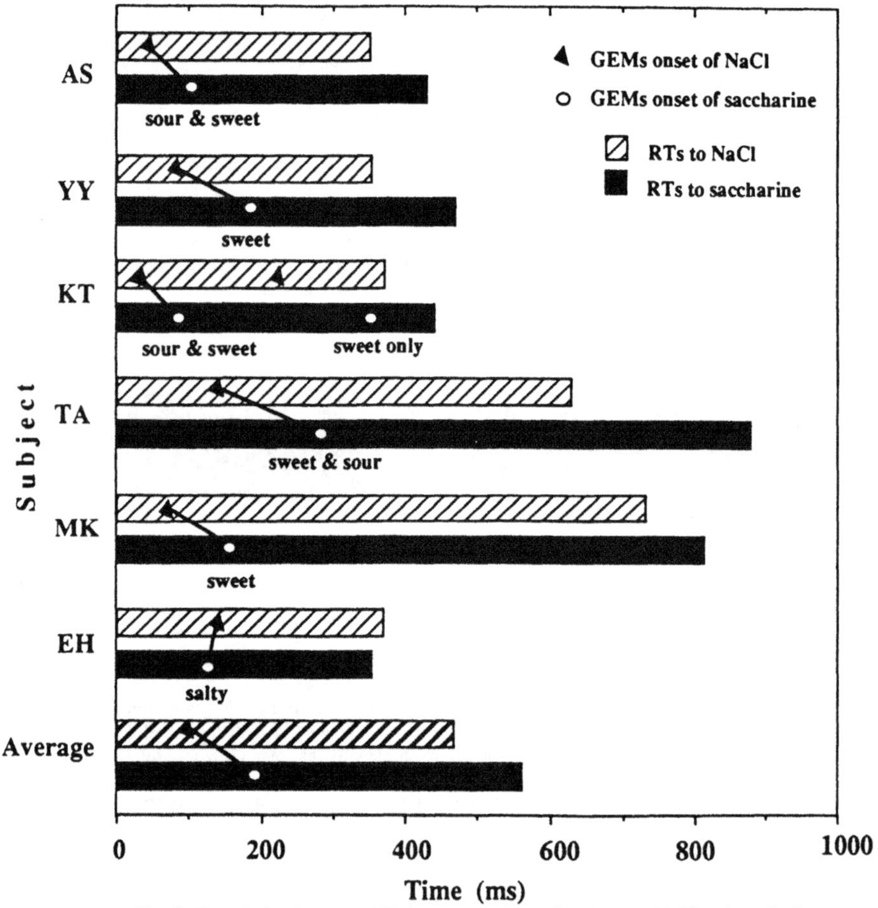

Fig. 5 The relation between GEMs onsets and reaction times to NaCl and saccharine

YY showed that the GEMs of NaCl began to respond 70 ms after the stimulation, whereas the GEMs of saccharin began to respond about 180 ms after the onset of the stimulus. The difference of RTs to the two tastants 117 ms, which was similar to the difference (110 ms) of the GEMs onsets for the two tastants. Similarly, subject AS showed that the GEMs of NaCl began to rise 40 ms after the stimulation, and the GEMs of saccharin began to rise about 100 ms after the onset of the stimulus. Also, the difference between the RTs to the two tastants was 81 ms, which was similar to the difference (60 ms) between the GEMs onsets for the two tastants.

Discussion

Reaction time to 1 M NaCl was reported as 433 ms that was shorter than RT to sweet taste of sucrose in Yamamoto & Kawamura [4]. The averaged RTs to 1 M NaCl measured in this study was 467 ms that was consistent with the finding of Yamamoto & Kawamura. Reaction times measured in this study showed longer RTs to saccharine than NaCl except for subject EH. In the case of subject EH, there was no difference between two tastants for the RTs and the latency until the GEMs onset. Subject EH reported a salty taste in saccharin, which may account for the same RTs to NaCl. This result confirms the GEMs findings by Kobayakawa [2][3].

Besides, in every subject, a difference between the RTs to NaCl and the RTs to saccharin was nearly identical with a difference between the latency until the GEMs onset of NACl and that of saccharin. It means the RT subtracted from the latency until the GEMs onset is constant in spite of tastant. It's supposed that this definite time consists of processes of cognition and reaction to button except for sensation and perception, and that the difference between tastants in RTs or GEMs onset is caused by the difference of receptor mechanisms in sensation and perception process. In fact, the receptor mechanism of a salt by ion activity is assumed to be simpler than that of a sweetener through receptor site of protein.

In the case of subject KT, the perception of "only sweet" for saccharin showed the longer GEMs onset than the perception of "sour and sweet". This suggests that the onset of GEMs is related to perceived phenomena. Otherwise in the case of KT's NaCl, the GEMs data containing the faster GEMs onset showed a second onset at the same time as the later onset. Besides, other subjects showed the fast GEMs onset, the late GEMs onset or both GEMs onsets. More experiments are needed to explain exactly these results, but one speculation is that there are two peaks within 1000 ms of the onset on NaCl stimulation.

Conclusion

For all subjects, the latency until the GEMs onset is correlated to the RTs; this result confirms the measurement of GEMs by Kobayakawa [3][4]. Then, the difference between RTs for two tastants and the difference between the onset times of GEMs for two tastants were virtually identical.

Acknowledgment

We are grateful to Dr. T. Yamamoto, Dr. B. P. Halpern and Dr. B. Challis for their valuable advice and comments.

References

[1] Kobayakawa, T., Saito, S., Ayabe-Kanamura, S., and Yoshimura, S. (1995) Gustatory-Evoked potentials (GEPs) varied according to the concentration, Proceedings of the 17th meeting of Association for Chemo-reception Sciences: 263.

[2] Kobayakawa, T., Endo, H., Saito, S., Ayabe-Kanamura, S., Kikuchi, Y., Yamaguchi, Y., Ogawa, H., and Takeda, T. (1996) Trial measurement of gustatory-evoked magnetic fields. Electroencephalography and clinical Neurophysiology (in press).

[3] Kobayakawa, T., Endo, H., Saito, S., Ayabe-Kanamura, S., Kikuchi, Y., Yamaguchi, Y., Ogawa, H., and Takeda, T. (1996) Measurement of gustatory-evoked magnetic fields with sharp stimulation. Biomag96.

[4] Yamamoto, T., and Kawamura, Y. (1981) Gustatory reaction time in human adults. Physiology and Behavior, 26: 715-719.

SQUIDs and Stereopsis

Shahani, U.[1], Weir, A.I.[1], Lang, G.[1], Mansfield, D.C.[2], Halliday, D.M.[3], Hutson, D.[1], Maas, P.M.[1] and Donaldson, G.B.[1]

[1]Wellcome Biomagnetism Unit & Department of Physics and Applied Physics, University of Strathclyde; [2]Department of Ophthalmology, Southern General Hospital; [3]Department of Physiology, University of Glasgow, Glasgow, Scotland

INTRODUCTION

Stereopsis implies the perception of solid form and is generally used to denote the perception of solid objects by the human visual system. Although monocular cues are important for depth perception at a distance, for objects less than a 100 feet away, this process is also mediated by stereoscopic vision. Fixation at a point with both eyes renders the image of that point falling on the centre of the retina in each eye. However, because the two eyes are about six centimetres apart, three-dimensional objects produce slightly different images on different parts of the 2 retinae. The degree of this binocular disparity depends on the distance of the object from the plane of fixation of the 2 eyes. Thus, points on a three-dimensional object just outside the fixation plane stimulate different points on each eye and multiple disparities provide cues for stereopsis - which results in the perception of depth. The hypothesis that binocular disparity is the effective stimulus for depth perception is attributed to Kepler, the astronomer but the demonstration of that disparity alone can produce depth sensation was first provided by Wheatstone in 1838 with his invention of the stereoscope [1]. Bela Julesz[2] discovered that apparent three-dimensionality of objects can be perceived by the human visual system in random-dot stereograms which contain hidden images. This fact suggests that the main clue required for the perception of stereopsis is retinal disparity which excites the "cyclopean" group of neurones in the brain. These neurones force fusion of the discrete retinal images to give the perception of apparent third dimension in an essentially two-dimensional image.

More than three decades of research has gone into the study of the human and primate visual system. Physiological evidence of the existence of binocular cells in the primate brain comes from several neurophysiological studies of disparity sensitive single neurones in the primary visual and visual association cortices [see 1 for review].

Human stereopsis was studied using techniques which primarily measured retinal disparity by presenting spatially disparate images to each eye and asking human subjects to either fuse these images or have displaced images in each eye as a result of diplopia. Several hypothesis were used to describe the nature of these disparity sensitive neurones in the cortex the discussion of which is beyond the scope of this paper [1].

EEG and EP studies of stereoscopic depth processing providing physiological correlates to the psychophysical studies in man have been few [1]. However, several recent studies have concentrated not so much on mapping depth processing as on other related areas of the brain involved in verbal tasks [3,4], visual imagery and visual representation using multichannel EEG and MEG as well as functional imaging using either PET or MRI.

In this study we report the response of the cortical neurones involved in stereopsis recorded from over the occipital and parietal cortices using a conventional single channel axial SQUID neuromagnetometer. We have used frequency analysis to study the resonance phenomena that have been described as a characteristic of higher order processing in the brain [5].

METHODS

In our eddy current shielded room, free running magnetoencephalograms were recorded from over the occipital and parietal cortices in humans in order to study the frequency responses of cortical neurones during stereoscopic vision. Points $O_{1/2}$ and $P_{3/4}$ of the 10-20 electrode system were used for the location of the SQUID neuromagnetometer. The measurement points cover areas V_1, V_2, V_3 and V_4 of the visual system all of which are involved in processing depth information.

Bandpass settings were 0.3 to 100 Hz and signals were amplified using a Grass Model 12 Neurodata Acquisition System. Sequential recordings were made from subjects with eyes closed, eyes open and when perceiving the cyclopean *wallpaper* pattern under conditions of uniform luminance. Recordings of at least one minute were made of the subjects' response to each of the experimental conditions. For the first minute of recording subjects were asked to keep their eyes closed. Another minute recording was made of the subjects' with their eyes open but perceiving the 2-d wallpaper pattern. Finally subjects were asked to focus on the pattern in an attempt to perceive the hidden 3-D object. Subjects indicated perception of the 3-D object and its loss. At least one minute of sustained 3-D perception was recorded over each point. The subjects were then asked to close their eyes and relax and a further resting recording was made.

The results were recorded on a DAT tape recorder and analysed off-line in the frequency domain with a Hewlett-Packard 3561A Dynamic Signal Analyzer using the exponential window function and rms averaging. The exponential window function uses a slow overlapping temporal slice to average the data so that the oldest sample is given the least weight. The segments of data were then subjected to statistical analysis by the discrete Fourier transform and parameters derived from it, all of which were estimated by dividing the records into a number of disjoint sections of equal duration and estimating spectra by averaging across these discrete sections. Following standard time series practice a log transformation was used to obtain confidence limits for frequency spectra allowing distinct periodicities in the MEG recordings to be characterised [6].

RESULTS

MEG signals from all of our subjects showed increased activity at frequencies between 8 and 12 Hz in the alpha range in the eyes closed condition which attenuated on eye opening. Examples of each of these taken from Subject DMcK from over the right parieto-occipital cortex at point P_4 of the 10-20 system are shown in Fig 1. These data are consistent with a recent report in the literature which suggests that MEG activity at 8-12 Hz attenuates on eye opening and visual imagery recorded from the parieto-occipital cortex [4]. On perception of the three-dimensional image, this alpha rhythm was further attenuated and there was a marked increase in the spectral component at several frequencies between 35 and 75 Hz. There was substantial inter-individual variation of peak frequency but for constant location intra-individual variation was small. Figure 2 shows the activity of Subject DMcK on perception of the hidden 3-D image in the wallpaper pattern. There is an increase in general background activity. This is most marked at frequencies of 42-43 Hz and 67-68 Hz. Recordings over point P_3 on the left parieto-occipital hemisphere showed increases in power at similar frequencies on perception of the hidden 3-D image. While recording over point P_4, the subject lost the image and then regained it. Spectral content at frequencies associated with stereopsis decreased dramatically on loss of perception of the 3-D image. A decrease was seen in the spectral content at all frequencies. Regaining perception of the hidden 3-D image resulted in an increase in the spectral content at 42-43 Hz and 65-70Hz . There was an increase in similar frequency ranges seen when he first perceived the 3-D image. All data presented in the figures was also subjected to statistical analysis by the discrete Fourier transform and parameters derived from it, all of which were estimated by dividing the records into a number of disjoint sections of equal duration and estimating spectra by averaging across these discrete sections and a log transformation was used to obtain 95% confidence limits for frequency spectra which were statistically significant.

1.

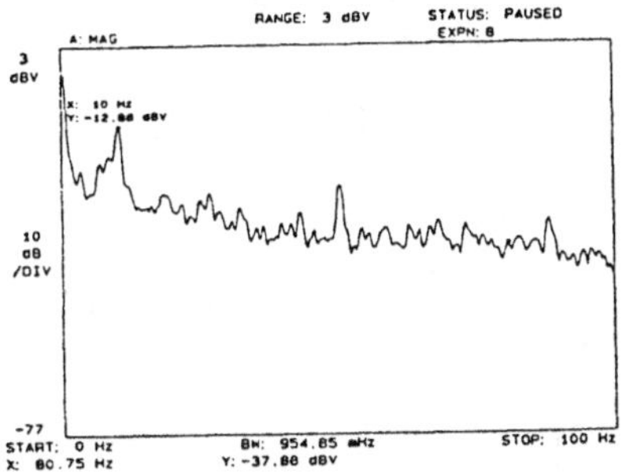

2.

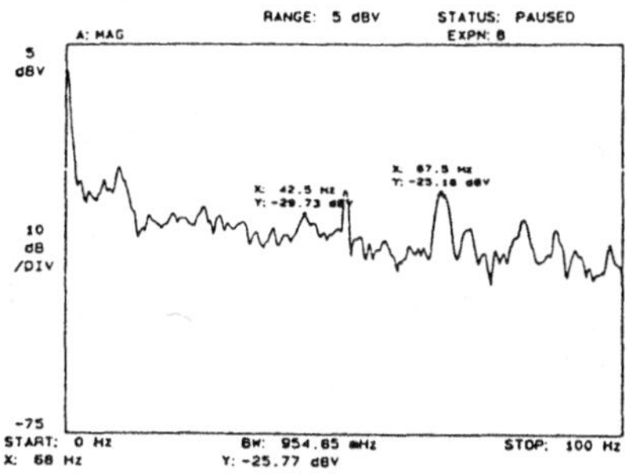

Fig.1 Eyes closed (point P_4. Subject DMcK). Increase in alpha rhythm. Fig.2 Perception of 3D (Subject DMcK). Increase in frequencies of 40-50 Hz and 60-70 Hz. Each figure is a RMS exponential weighting average of MEG. In addition, the data was found to be statistically significant.

In order to test whether the increase in spectral content of frequencies of 40-50 Hz and 60-70 Hz was directly associated with stereoscopic vision, we performed a series of control experiments.

The first set of control experiments were devised to test whether the increase in the spectral content of the frequency ranges associated with stereopsis was due to perception of 3-D per se and not the sudden recognition of an identifiable object. Subjects were presented with a non-cyclopean wallpaper pattern without a hidden image. The picture itself could be seen in 3-D. Fig. 3 shows results obtained from Subject GL in response to one such non-cyclopean wallpaper pattern. There was an increase in spectral content at frequencies of 40-50 Hz and at 69Hz. Fig 4 shows results from the same Subject GL in response to a cyclopean wallpaper pattern in which a hidden image is perceived in 3-D. There was an increase in spectral content at 58-59 Hz in addition to that seen in the previous Fig suggesting that object recognition could influence the activity of cortical neurones during stereopsis.

3.

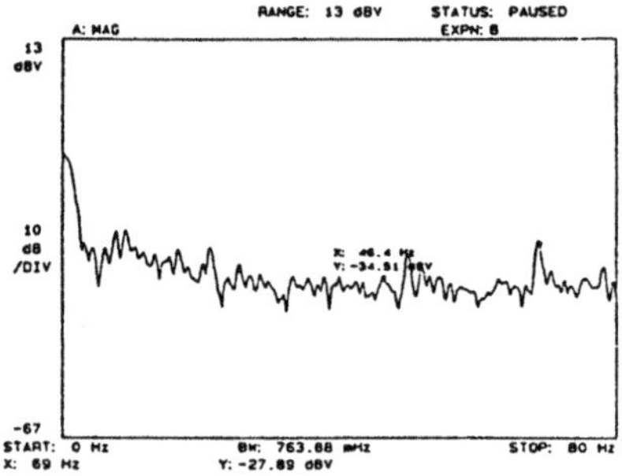

4.

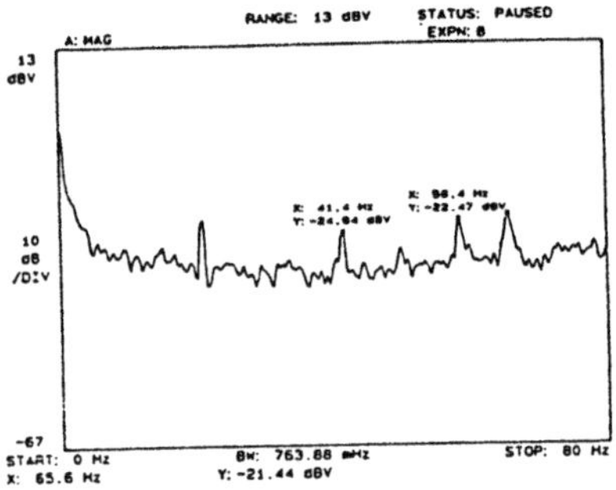

Fig.3 Perception of 3D. Perception of non-cyclopean wallpaper pattern. Fig.4 Perception of 3D (Subject GL). Each figure is a RMS exponential weighting average of MEG. In addition, the data was found to be statistically significant.

The second set of control experiments were devised in order to test whether the increase in activity at frequencies associated with the perception of depth was not a result of converging or diverging the eyes. Convergence or divergence of the eyes through the 2-D plane of focus is the technique most commonly used to focus on a wallpaper pattern or a random dot stereogram in order to perceive depth. To control for convergence or divergence we used a 6-dioptre prism in front of one eye. In addition, the wallpaper pattern was rotated through 90 degrees in order to prevent perception of depth. Results from subjects suggested that convergence or divergence alone without stereoscopic vision did not produce an increase in the spectral content at the frequencies associated with stereopsis. Figs.5 and 6 show the results of this experiment in Subject DCM.

5.

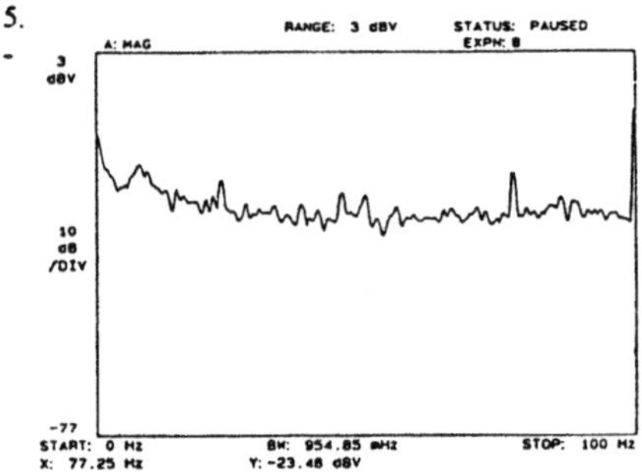

6.

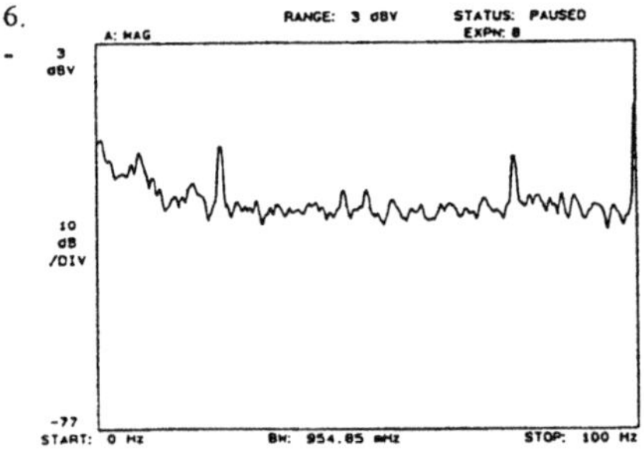

Fig.5 Convergence and Fig.6 Divergence (DCM). Each figure is a RMS exponential weighting average of MEG. In addition, the data was found to be statistically significant.

SUMMARY

We have found that the spectral content of frequencies in the range of the alpha rhythm recorded from the parieto-occipital cortex was reduced on eye opening and perception of a visual image. Perception of 3-D during stereoscopic vision resulted in a further attenuation of this alpha-rhythm. These results are in agreement with reports in the literature on the reduction of the alpha rhythm during tasks such as visual imagery and verbal association [4].

We have seen increased cortical activity at frequencies specific to stereopsis and the perception of depth. This increase at specific frequencies may be influenced by object recognition. The increased activity at frequencies when perceiving 3-d in a stereogram is not the same when convergence or divergence occurs without stereopsis.

Our results are consistent with reports in the literature which suggests that frequencies above 40 Hz are involved in focused attention, pattern recognition and higher order visual activity [see 5].

REFERENCES:

[1] Regan, D. Human Brain Electrophysiology. Evoked potentials and evoked magnetic fields in science and medicine. 1989 Elsevier.

[2] Julesz, B., Vision Research, 1986, 26: 1601-1612.

[3] Gevins, A., Leong, H, Smith, M.E., Le, J and Du, R. Mapping cognitive brain function with modern high-resolution electroencephalography. TINS, 1995, 18(10) : 429-436.

[4] Salenius, S., Kajola, M., Thompson, W.L., Kosslyn, S. and Hari, R. Reactivity of magnetic parieto-occipital rhythm during visual imagery. EEG Journal 1995, 95 : 453-462.

[5] Basar, E., Basar-Eroglu, Demiralp,T. and Schurmann, M. Time and frequency analysis of the Brain's Distributed Gamma-band system. IEEE 1995 July/August : 400-410.

[6] Conway, B.A., Halliday, D.M., Farmer, S.F., Shahani, U., Maas, P., Weir, A.I. and Rosenberg, J.R. Synchronization between motor cortex and spinal motonuronal pool during the performance of a maintained motor task in man. J.Physiol. 1995, 489 (3) :917-924.

Acknowledgements : This work was supported by the Wellcome Trust and EPSERC.

Source Localization of a N400-like Negativity in a Sentence-Reading Paradigm Using Evoked Magnetic Fields and Magnetic Resonance Imaging

Simos, P.G., Basile, L.F.H., Zouridakis, G., and Papanicolaou, A.C.

Department of Neurosurgery, University of Texas -Houston Health Science Center, Houston, Texas, USA

Introduction

The N400 component of the Event-Related Potential (ERP) corresponds to a negative-going deflection that peaks at approximately 400 ms after the onset of the final word in a sentence [1]. N400 amplitude is believed to be an index of the degree that a given concept is deemed "appropriate" within a particular linguistic context [2]. Little is known regarding the sources of neural activity that give rise to N400-like negativities measured from the scalp. Identifying the neural substrates of N400 activity could provide important information regarding the brain structures that participate in language comprehension. Intracranial recordings have revealed sources of electrical activity, that become active at the same latency as the scalp-recorded N400 response, in paleo- and archeocortical regions as well as in neocortical areas. The former included the hippocampus and the parahippocampal gyrus [3, 4]. Potential neocortical sources may include the inferior parietal cortex, temporal areas surrounding the collateral sulcus [3], and the middle temporal gyrus [5]. Magnetoencephalographic (MEG) measurements show excellent temporal resolution and combined with magnetic resonance imaging (MRI) have enabled researchers to localize the intracranial sources of several ERP components within neurologically plausible anatomical regions [6, 7, 8]. However, the only reported attempt to localize the sources of the magnetic equivalent of the N400 component (N400m) in the brain [9] has produced inconclusive results.

The present investigation employed a standard paradigm for eliciting the N400 response [1, 10], in which the subjects are asked to read sentences presented one word at a time in order to be able to recognize them after the testing session. Half of the sentences ended with a semantically inappropriate word (e.g., *He was stung by a towel*). Using standard MEG analysis techniques, we attempted to identify brain areas that give rise to the magnetic equivalent of the N400 response.

Methods

Eight right-handed adults, native speakers of English were tested in this study. The subjects were between 18 and 45 years old, had normal or corrected-to-normal vision, and no history of neurological disorders. They were required to sign consent forms and they were paid for their participation. Eight hundred five-to-eleven word sentences with highly predictable final words served as stimuli in the recording sessions. Sentence completion norms were available for all of the sentences. Half of these sentences were transformed into semantically meaningless sentences by substituting an inappropriate word for the original ending. Congruous and incongruous final words were matched for word frequency and length. Word endings in the incongruous sentences were always grammatically and syntactically appropriate. The stimuli were randomly assigned to seven lists of 80 sentences each. The sentences were presented one word at a time in the center of a computer screen. Each word was flashed for 200 ms with an inter-word interval (onset to onset) of 800 ms. The inter-sentence interval was randomly varied between 2 and 4 s. The stimuli were drawn white on a black background and the final words subtended between 0.8° and 2.0° of horizontal visual angle. To ensure that subjects were engaged in the task, they were given after each session a list of 16 sentence bodies and were asked to mark the ones that they had seen in the previous block (sentence recognition task).

The ERPs were recorded from Fz and Pz referenced to the right mastoid, using a Nicolet Compact Four system. The electro-oculogram (EOG) was monitored and recorded to identify epochs contaminated by eye movements. The EEG data were analog-filtered from 0.01 to 30 Hz. Further analysis of the electrical signals was done in common with the magnetic signals. MEG was simultaneously recorded with a 7-channel Neuromagnetometer (Biomagnetic Technologies, Inc., Model 607), using second-order gradiometers with a 4 cm baseline and 1.8 cm coil diameters. Recordings were done in a magnetically shielded room. The position of the magnetometer relative to the subject's head was automatically recorded with a BTi probe-position indicator which had an error tolerance of less than 3 mm and an accuracy of 0.1 mm. Each subject's fiduciary points -- i.e., the nasion and the two preauricular points -- served as the basis for the MEG Cartesian coordinate system. The Y axis in that system was defined by the two preauricular points, the X axis by the nasion and the center of the head, while the Z axis was normal to the X-Y plane and passing through the X-Y intersection (i.e., the center of the head). The MEG analog filters were set at 0 to 50 Hz. The signal was digitized at 200 Hz starting 100 ms before the onset of the final word and continuing for 1 s after the onset. All the MEG epochs were filtered off-line with a bandpass of 0.01 to 20Hz (Butterworth 4 pole filter, 24dB/octave rolloff) in order to improve signal quality. Before averaging, trials on which an eye movement or blink had occurred (as indicated by a peak-to-peak amplitude in the EOG channel in excess of 70 μV) were rejected. Artifacts that were produced by the magnetometer itself typically consisted of high amplitude noise bursts confined to a single MEG channel. Those artifacts were easily recognizable by visual inspection of the MEG epochs and excluded from subsequent analyses. The data were then baseline adjusted relative to the mean amplitude in the 100-ms prestimulus period, subjected to a computerized linear detrending procedure, and averaged separately for each condition (i.e., Congruous, Incongruous) and recording session. A minimum of 30 single-trial MEG records were used to calculate each averaged waveform. Neuromagnetic measurements were obtained from 49 locations on the left side of the scalp corresponding to seven consecutive placements of the recording device. The intracranial sources of the observed evoked magnetic fields (EMFs) were modeled as equivalent current dipoles, by using a finite-difference version of the Levenberg-Marquardt algorithm. The dipole fitting method was applied at every 5 ms, from 200 to 800 ms after stimulus onset. Dipole parameters (spatial coordinates, orientation [Psi] and strength [Q]) were computed with reference to the best-fitting sphere, the curvature of which was defined by the surface of each subject's scalp over the region of the head containing the field extrema. The criterion we used for accepting a dipole solution as satisfactory was a correlation of at least .85 between the measured and the predicted fields at each time point. MRI scans from three subjects were used to identify the brain regions that corresponded to the estimated locations of the equivalent current dipoles. For each of the remaining five subjects, spatial dipole coordinates were projected on MRI scans from three other neurologically normal individuals with closely matching head size. A computer algorithm was used to convert the MEG dipole coordinates into corresponding MRI coordinates [6].

Results

All subjects reached the 85% correct criterion in the sentence recognition task in every recording session. The amplitude of the N400 response, that was measured between 320 and 420 ms post-stimulus (240 and 320 ms in S#3), was more negative in the Incongruous than in the Congruous condition, $F(1,156)=20.88$, $p=.0001$. N400 amplitude varied in the same direction across conditions in every subject. No systematic pattern of amplitude or latency variation emerged between recording sessions. Figure 1A shows representative average ERPs from two subjects.

In the Incongruous condition EMF waveforms displayed a prominent deflection during the course of the electrical N400 wave. In every case the peak-to-peak field strength (i.e., the algebraic sum of outward and inward flux) was smaller in the Congruous condition. Figure 1B displays representative magnetic waveforms from the two subjects. Individual contour maps (Figure 1C) contained a single pair of extrema. Seven subjects produced satisfactory dipole solutions. The vertical lines drawn in Figures 1A and 1B indicate the range of time points associated with satisfactory dipole solutions during the N400 latency window. In every subject, for whom reliable dipole parameters were computed in both conditions (n = 5), the entire range of Q values computed during the N400 latency window was larger in response to incongruous words than in response to congruous words. These findings provide strong evidence that the N400 response and its magnetic equivalent (N400m) belonged to the family of negativities which are presumably sensitive to linguistic context [2]. After superimposing the dipole coordinates onto the appropriate horizontal MRI slice, we used a standard stereotaxic atlas of the human brain [11] to identify the cerebral structures (e.g., sulci, and gyri) and cytoarchitectonic areas where dipoles were localized. MRI slices from the three subjects whose MRI scans were available are presented in the upper row in Figure 1E. The bottom row in Figure 1E displays the estimated locations of the sources of the N400m response for the five subjects for whom MRIs were not available. Given that there was close agreement in the dipole regions across the three different MRIs used in each case we present only one slice per subject. Five subjects demonstrated dipole-like activity localized in posterior temporal neocortex near the temporo-parietal junction, typically in the middle temporal gyrus. These areas lie near the banks of the superior temporal sulcus. In one subject (#6) dipole regions extended more posteriorly into the temporo-occipital association cortex (area 37). In two subjects (#1 & #7) solutions fell in the medial temporal lobe, in the vicinity of the hippocampus and the parahippocampal gyrus.

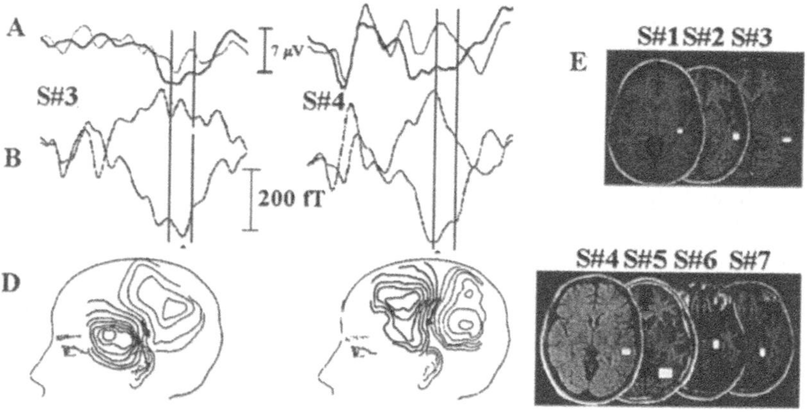

Discussion

The present study was successful at recording the magnetic equivalent of the N400 response elicited by the final words in sentences. In every case dipole solutions were localized in temporal lobe structures, including neocortical areas on the lateral surface of the left hemisphere as well as more medially located paleo- and archeocortical regions. The finding of dipole activity in posterior temporal regions near the temporo-parietal junction is in agreement with the results of a recent study that combined ERPs and metabolic imaging using Positron Emission Tomography (PET) [12], and with the outcome of intracranial recordings [5]. On the basis of our data it seems likely that the sources of the magnetic equivalent of the N400 response reflect focal activity in areas specialized for those aspects of language which are presumably indexed by that ERP component. These processes are involved in lexical access that occurs within a previously established semantic context and leads to word recognition. In agreement with this claim, several lines of converging evidence have established a firm link between temporal and temporo-parietal cortex and language processes such as word recognition [13] and semantic comprehension of written material, including lesion [14] and PET studies [15, 16].

In addition to evidence for superficial sources, our data showed dipole solutions in more medial aspects of the temporal lobe, in the vicinity of the hippocampus and the parahippocampal gyrus. This finding is consistent with reports of intracranially-recorded electrical potentials from medial temporal lobe structures which shared many of the characteristics of the scalp-recorded N400 response [3, 4]. The functional anatomy of the hippocampal formation makes this structure ideal for mediating some of the neural processes involved in recent memory [17]. Inasmuch as the N400 is a sign of certain operations involved in word recognition and it is affected by experimental manipulations that influence the course of this process (such as semantic context, subjective expectancy, etc.), it seems likely that the intracranial generation of this component is closely linked to neural events in the medial temporal lobe.

References

[1] Kutas, M., & Hillyard, S.A. (1980). *Biol. Psych.*, 11, 99-116.

[2] Halgren, E. (1990). Insights from evoked potentials into the neuropsychological mechanisms of reading. In A.B. Scheibel & A.F. Wechsler (Eds.), *Neurobiology of Higher Cognitive Function*. New York, NY: The Guilford Press.

[3] McCarthy, G., Nobre, A.C., Bentin, S., & Spencer, D.D. (1995). Language-related field potentials in the anterior temporal lobe: I. Intracranial distribution and neural generators. *J. Neurosci., 15,* 1080-1089

[4] Smith, M.E., Stapleton, J.M., & Halgren, E. (1986). *Electroencephal. Clin. Neurophys., 63,* 145-159.

[5] Guillem, F., N'Kaoua, B., Rougier, A., & Claverie, B. (1995). Intracranial topography of event-related potentials (N400/P600) elicited during a continuous recognition memory task. *Psychophysiol. 32,* 382-392

[6] Rogers, R.L., Baumann, S.B., Papanicolaou, A.C., Bourbon, T.W., Alagarsamy, S., & Eisenberg, H.M. (1991). Localization of the P3 sources using magnetoencephalography and magnetic resonance imaging. *Electroencephal. Clin. Neurophys., 79,* 308-321.

[7] Papanicolaou, A.C., Baumann, S.B., Rogers, R.L., Saydjari, C., Amparo, & E.G., Eisenberg, H.M. (1990). Localization of auditory response sources using MEG and MRI. *Arch. Neurol., 47,* 33-37.

[8] Basile, L.F.H., Rogers, R.L., Bourbon, W.T., & Papanicolaou, A.C. (1994). Slow magnetic fields from human frontal cortex. *Electroencephal. Clin. Neurophys., 90,* 157-165.

[9] Schmidt, A.L., Arthur, D.L., Kutas, M., & Flynn, E. (1989). Neuromagnetic responses during reading meaningful and nonmeaningful sentences. *Psychophys. Abstr., 26,* S6.

[10] Polich, J. (1985). Semantic categorization and event-related potentials. *Brain Lang., 26,* 304-321.

[11] Talairach, J. & Tournoux, P. (1988). *Co-planar stereotaxic Atlas of the human brain.* New York, NY: Thieme Medical Publishers.

[12] Nenov, V.I., Halgren, E., Smith, M.E., Badier, J-M., Ropchan, J., Blahd, W.H., et al. (1991). Localized brain metabolic response correlated with potentials evoked by words. *Behav. Brain Res.,*

[13] Damasio, A.R. & Damasio, H. (1983). The anatomic basis of pure alexia. *Neurol. (Cleveland), 33,* 1573-1583.

[14] Hart, J. & Gordon, B. (1990). Delineation of single-word semantic comprehension deficits in aphasia, with anatomical correlation. *Ann. Neurol., 27,* 226-231.

[15] Price, C.J., Wise, R.J.S., Watson, J.D.G., Patterson, K., Howard, D., & Frackowiak, R.S.J. (1994). Brain activity during reading: The effects of exposure duration and task. *Brain, 117,* 1255-1269.

[16] Howard, D., Patterson, K., Wise, R., Brown, W.D., Friston, K., Weiller, C., et al. (1992). The cortical localization of the lexicons. *Brain, 115,* 1769-1782.

[17] Squire, L.R. (1986). Mechanisms of memory. *Science, 232,* 1612-1619.

Information Theory Predicts the Magnitude of an MEG Response Related to the Operation of the Auditory Sensory Memory

Sinkkonen, J.[1,2], Tiitinen, H.[1], May, P.[3], Ilmoniemi, R. J.[4], and Näätänen, R.[1]

Cognitive Psychophysiology Research Unit[1], Department of Psychology, University of Helsinki, Finland; Department of Radiology[2], Helsinki University Central Hospital, Finland; Department of Mathematics[3], King's College, London, UK; BioMag Laboratory[4], Medical Engineering Centre, Helsinki University Central Hospital, Finland

Introduction

An electric field generated by the primate brain, MMN, and its magnetic counterpart, MMNm, have established their status as physiological correlates of a transient auditory memory[5].

Small changes in the energy consumption of a limited area of the cortex should be linearly proportional to amplitude changes of the scalp-recorded ElectroMagnetic Fields: EMFs are mainly due to the summation of post-synaptic potentials of cortical neurons which together with their cause, the action potentials, consume most of the energy available for the nervous system[1,3,4].

Information, as defined by Shannon[6], is proportional to $\log p$. In the absence of any *a priori* meaning of the stimuli to the organism under observation, Shannon information can be used as a measure of the importance of the stimuli. *Optimal resource sharing* should then express itself in stimulus-evoked changes of the EMF being directly proportional to the information of the stimulus, $-p \log p$, where p is the probability of the stimulus event: this way the same amount of processing capacity becomes allocated for each incoming unit of information.

Methods

We studied 8 subjects with a whole-scalp magnetometer[2] in an auditory environment containing an unpredictable sequence of two mutually distinguishable sinusoidal tones (stimulus duration 60 ms including 10-ms rise and fall times, ISI 200 ms, frequencies 1000 Hz and 1150 Hz; the subject received the stimuli to her left ear while reading self-selected texts). The probability of the 1 150-Hz tone was varied from 0.02 to 0.40 in eight steps; the presentation order was interleaved and counter-balanced. For illustration, see Fig. 1.

For each subject, one channel over the right auditory cortex with good S/N ratio was selected, and temporal average (100-200ms; shaded in the Fig. 2) of the signal from the selected channel was used as a measure of the resource usage of the cortex. The signal consisted almost exclusively of MMNm, as N1m and other EMF components are attennuated in nonattending subjects hearing fast stimulus sequences with small deviations[7]. See Fig. 2.

Results

Results, averaged over subjects, are presented in Fig. 3. They closely matched our prediction based on the principle of optimal resource sharing.

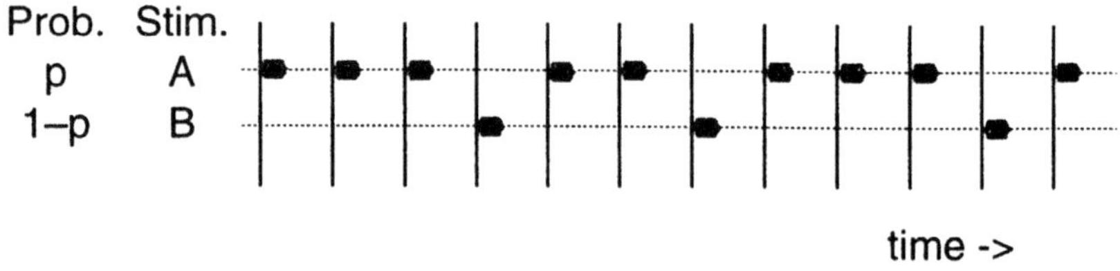

Figure 1: The stimulus paradigm consisted of independent but regularly occurring sinusoidal tones of two frequency. The probabilities of the tones were varied (see text for details).

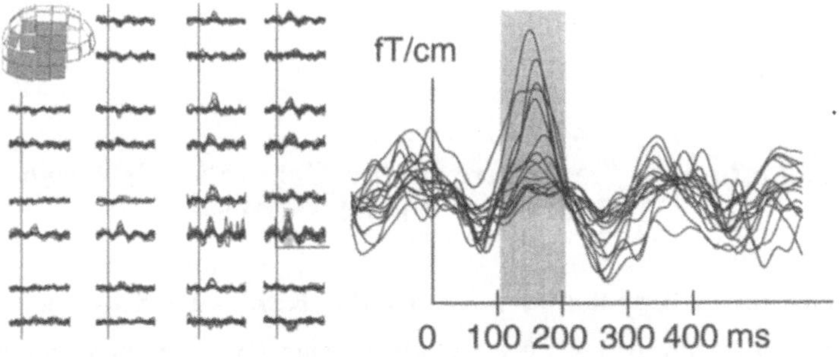

Figure 2: Data of a representative subject for each stimulus probability, measured over the right auditory cortex. Location of the illustrated sensors relative to the whole-head SQUID array is shown in the inset (left). One channel with good S/N ratio was selected for each subject, and the data was integrated from 100 ms to 200 ms after stimulus onset to get a quantification of the MMNm amplitude (right).

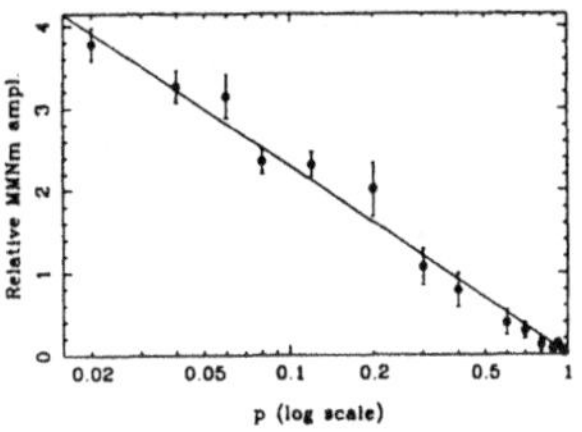

Figure 3: The predicted linearity of MMNm amplitude (line) and the averaged response amplitude of eight subjects (dots with bars) as a function of logarithmic stimulus probability. Because of subject-wise variation of the overall amplitude of the responses, normalization of the data over subjects was necessary before averaging. Bars denote the standard error of the mean. Fluctuations from linearity are within random variation.

Discussion

Energy invested for the processing of unpredictable stimuli seems to be proportional to their information content. The present result suggests that optimal resource allocation is one of the principles of information processing of the cortex. The characterization of the behaviour of the MMNm amplitude as a function of probability also allows optimization of the auditory odd-ball paradigm for applications where fast data collection is important.

References

[1] Creutzfeld, O. & Houchin, J. in *Handbook of Electroencephalography and Clinical Neurophysiology* (ed A. Remond) Vol. 2C, 5-55 (Elsevier, Amsterdam, 1974).

[2] Knuutila, J. *et al. IEEE Trans. Magn.* **29,** 3315-3320 (1993).

[3] Lorente de No, R. *A Study of Nerve Physiology* **132,** part 2 (Rockefeller Institute, New York, 1947).

[4] Martin, J. H. in *Principles of Neural Science* 3rd ed. (eds Kandel, E. R., Schwartz, J. H. & Jessell, T. M.) 779 (Elsevier, New York, 1991).

[5] Ritter, W. *et al. Ear and Hearing* **16,** 52-67 (1995).

[6] Shannon, C. E. *Bell Syst. Tech. J.* **27,** 379-423 (1948).

[7] Tiitinen, H. *et al. Nature* **372,** 90-92 (1994).

Acknowledgements

We thank M. Huotilainen, K. Kaila, S. Kaski, J. Lavikainen, and G. Nyman for their help in conducting the experiments and in preparation of this manuscript and the associated poster. This work was supported by the Academy of Finland.

Authors' contact information: e-mail: `janne@iki.fi`; fax: +358 0 471 5781; phone: +358 40 5464798.

Determination of the Nature of the Connection Between the Auditory and Visual Cortices

Stephen, J.M.[1], Broadhurst, J.H.[1], Knuth, K.H.[1*] and Schwartz, B.J.[2]

[1]Physics Department, University of Minnesota, Minneapolis, USA; [2]Scripps Research Institute, La Jolla, USA

Introduction

Studies confirming the influence of visual information on the response of the auditory cortex to speech sounds [1] lead us to question the nature of the connection between the visual and auditory cortices. The superior colliculus exhibits visual-sensitive neurons that can be modulated (both inhibited and excited) by auditory stimuli [2]. Since the superior colliculus is one of the nuclei that the optic nerve passes through as it travels to the visual cortex, this suggests that one might expect a modulation of the visual response with a simultaneous auditory stimuli.

However, the superior colliculus is not traversed by the main auditory nerve as it passes through the brain stem to the auditory cortex. Therefore, it is first important to consider if there is any influence of the auditory response with a corresponding visual stimulus and if there is to determine the possible region of the convergence of information. This led us to study the response of the auditory cortex to a click and unrelated light flash. In animal studies some interesting results that show that some neurons located in the auditory cortex are excited by a visual stimulus [3]. Amongst conflicting results in electroencephalographic studies, one study claimed that the influence of a visual stimulus to the auditory cortex was a linear superposition of the individual responses [4]. In the present study the results show that the response to the light flash and click is not a linear summation of the responses to the individual stimuli.

Methods

Seven normal subjects were presented with three different stimuli. The response of the auditory cortex to the three different stimuli was measured using a 37-SQUID biomagnetometer. First, the subject was presented with a red light point flash of duration 0.5 msec at a distance of approximately one meter. The second stimulus was a click of duration 0.5 msec at sound pressure level between 60-100 dB depending on the subject. Third, the subject was presented with the light flash and click simultaneously. Each stimulus was presented many times and the response was obtained by on-line signal averaging.

To determine if there is any interaction between the incoming visual and auditory information we determined if the response to the combined stimulus was a simple linear superposition of the individual responses. This was done by adding the response of the auditory cortex to the light flash (Fig. 1a) to the response of the auditory cortex to the click (Fig. 1b) and subtracting off the combined response (Fig. 1c). These responses were from one and the same channel, that was chosen based on the high signal-to-noise ratio in that channel. This difference response was then compared across subjects to determine if the pattern persisted among different individuals. The comparison was done by taking a correlation between two subjects' difference plots and determining if the correlation was significant (based on the t-test).

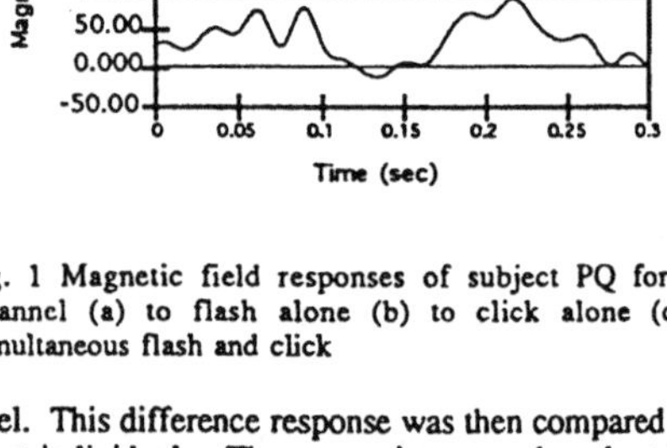

fig. 1 Magnetic field responses of subject PQ for one channel (a) to flash alone (b) to click alone (c) to simultaneous flash and click

* Now at Kresge Hearing Institute of the South, New Orleans, USA

Results

If there were no interaction between the two modalities we would expect the difference plot to be at the level of the random noise. For example, if the magnetic fields measured were the result of different areas of the brain being activated by the different stimuli without any merging of information, when the two stimuli are presented together the total magnetic field should be a simple summation of the magnetic fields to the individual responses. As can be seen in the difference plot of subject PQ (Fig. 2), the result is well above the expected noise level. The difference plots for all seven subjects were consistent with this result.

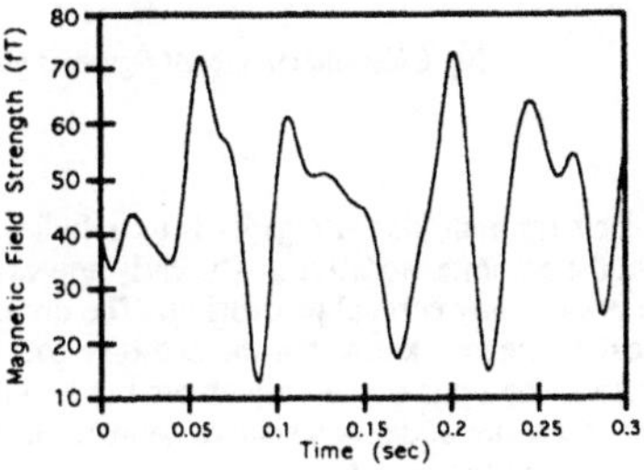

fig. 2 Difference plot for subject PQ

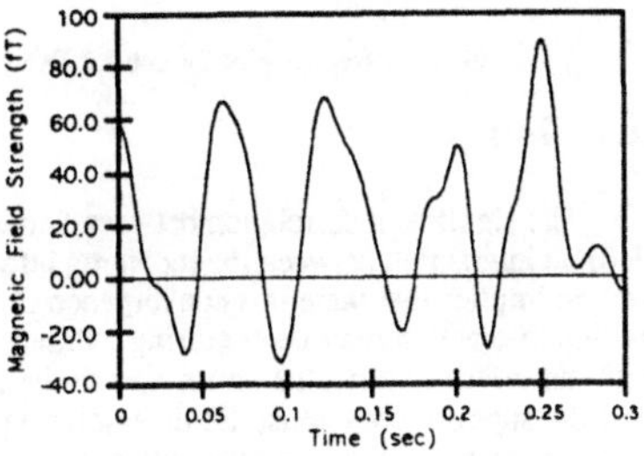

fig. 3 Difference plot for subject EH

By comparing the difference plots of Fig. 2 and Fig. 3, one can see that not only did all seven subjects have a nonzero result, but there is also a similar pattern across subjects. The apparent similarities were confirmed by the finding that there is a significant correlation between Fig. 2 and Fig. 3 with an offset of 2 msec of Fig. 2 with respect to Fig. 3 as can be seen in Fig. 4. The correlation at lag time 2 msec was 0.26±0.08. Similar results can be seen in Fig. 6, which is the correlation of two other difference plots, Fig. 3 and Fig. 5. Although there is a slightly different offset for the maximum correlation, this plot also has a significant correlation at lag time 12 msec of 0.30±0.08.

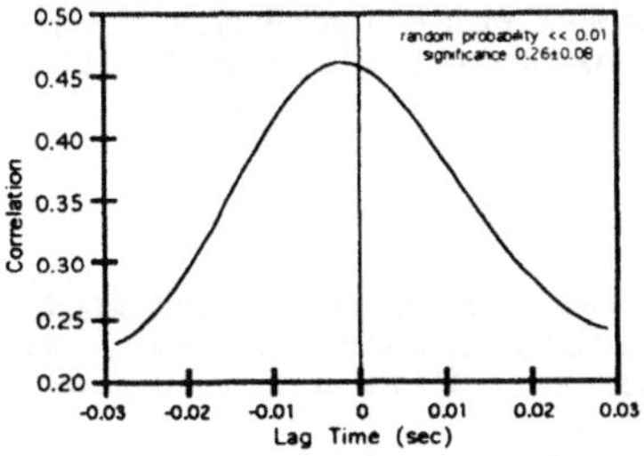

fig. 4 Correlation plot of figures 2 and 3

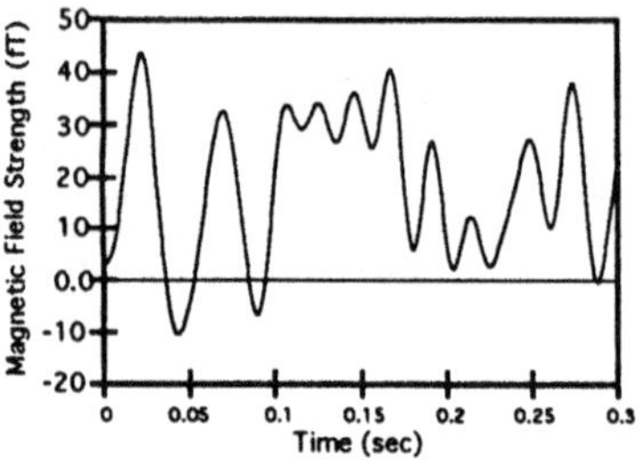

fig. 5 Difference plot for subject DO

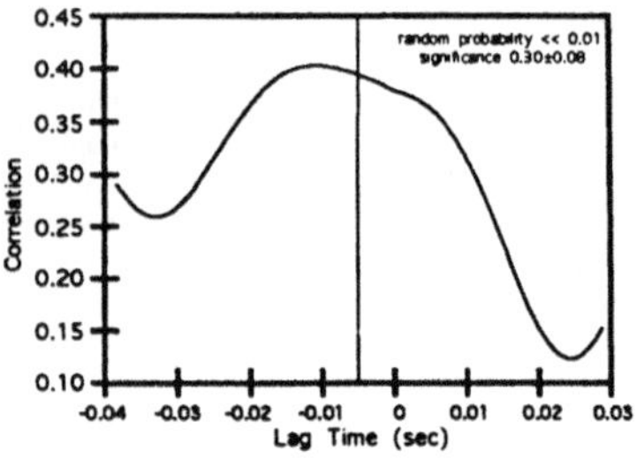

fig. 6 Correlation plot of figures 3 and 5

Discussion

The significant correlation between two different subjects for multiple pairings leads us to believe that there is a definite interaction between the incoming information from the separate modalities. The early appearance of this interaction implies that there is a convergence of information prior to any cortical processing. The question then is where this interaction may be occurring. Based on anatomical studies we know that the auditory pathway passes through the inferior colliculus on its way to the auditory cortex. The optic nerve only enters the brainstem at the level of the superior colliculus. Some studies have shown that there are auditory sensitive neurons in the superior colliculus, despite the fact that conventional knowledge does not point to the auditory pathway passing through this portion of the colliculus. This implies that there might be some connection between the inferior and superior colliculus that not only allows some auditory information to pass to the superior colliculus but also allows some information on the visual stimulus to pass to the inferior colliculus to affect the auditory signal.

To thoroughly cover the study of the connection between the auditory and visual cortices we plan on testing the influence of an auditory stimulus on the visual response.

References:

[1] Sams, M., Aulanko, R., Hämäläinen, M., Hari, R., Lounasmaa, O.V., Lu, S.-T., Simola, J. Seeing speech: visual information from lip movements modifies activity in the human auditory cortex, Neuroscience Letters, 1991, 127: 141-145.

[2] Meredith, M.A., and Stein, B.E. Visual, auditory, somatosensory convergence on cells in superior colliculus results in multisensory integration, Journal of Neurophysiology, 1986, 56: 640-662.

[3] Watanabe, J., Iwai, E. Neuronal activity in visual, auditory and polysensory areas in the monkey temporal cortex during visual fixation tasks, Brain Research Bulletin, 1991, 26: 583-592.

[4] Ciganek, L. Evoked potential in man: interaction of sound and light, Electroenceph. Clin. Neurophysiol., 1966, 21: 28-33.

Acknowledgments

This study was supported in part by the University of Minnesota, Biomagnetic Technologies Inc. and the Scripps Research Institute. We thank Lacy Kurelowich, Joslyne Foley, and Patti Quint, the technicians at Scripps, for helping collect the data, Steve Cobb at BTI for the technical support, and Dr. Benjamin Bayman for our discussions on the analysis of the data.

Visualization of Tonotopy of Auditory Cortex Using a Newly Developed 129 Channel Vector Magnetoencephalography

Takanashi, Y.[1], Iwamoto, K.[1], Yoshikawa, K.[1], Makino, M.[1], Tomita, T.[2], Tomita, S.[2], Kajihara, S.[2], Kondo, Y.[2], Okamura, S.[2], Yoshida, Y.[2], Ueda, M.[2,3] and Nakajima, K.[1]

[1]Department of Neurology, Research Institute for Neurological Diseases and Geriatrics, Kyoto Prefectural University of Medicine, Kyoto, Japan; [2]Technology Research Laboratory, Shimadzu Corporation, Kyoto, Japan; [3]Superconducting Sensor Laboratory, Chiba, Japan

Introduction

The existence of the human tonotopic organization has been revealed using the neuromagnetic measurement. These early studies on the tonotopic organization in the human were performed by measuring sequentially at many sites using the magnetoencephalography (MEG) with a few channels[1, 2, 3], so these studies needed a vast effort and time to fulfill a study. This meant that we could not use mangentoencephalography (MEG) to investigate the auditory function of patients in clinical setting. Now, many MEG systems, which are commercially available, consist of multi-channel first-order axial gradiometers or planar gradiometers. These have some advantages; specially, axial gradiometer is sensitive to the radial component of magnetic fields, and planar gradiometer is favorable to obtain their tangential component. Even using these two systems, however, we could not obtain three dimensional components about magnetic fields. Consequently, we have developed a new 129 channel vector magnetometer system[5].

The purpose of the present study is to evaluate the reliability of source localization in our new MEG system, and to visualize the human tonotopic organization on MR imaging.

Subjects and methods

Nine normal-hearing subjects (9 males, all right-handed), with no history of otological or audiological disorders and aged among 20 and 46 years, participated in this study after giving us their informed consent. Five subjects were investigated in both hemispheres; two subjects, the left one; remaining two subjects, the right one. Tone-burst with 100 ms duration (rising time, 10 ms; falling time, 10 ms) and an interstimulus interval 1.00-1.86 msec were at random delivered to subject's contralateral ear, and white noise with 65 dB strength was delivered to ipsilateral ear to the examined hemisphere through a plastic tube. The carrier frequencies were 500, 1000, 2000 Hz with 85 dB, and 4000 Hz with 95 dB strength.

Magnetic auditory evoked fields (AEFs) were recorded in a magnetically shielded room using a novel 129-channel vector neuromagnetic imaging system[5]. In brief, our MEG system consisted of 43 units of three orthogonally-oriented ellipse-shaped pick-up coils. Each sensor coil was positioned at the same center point and the three planes of coils intersected each other. The effective diameter of the recording area was 190 mm. The baseline was 50 mm. The distance between the centers of the sensor coils was 25 mm. Our system could cover half of the cerebral hemisphere. Our MEG system can measure changes of the magnetic fields at 43 sites, and can determine the three dimensional direction of magnetic fields at each site. Each subject calmly lay on his side on a bed in a magnetically shielded room. We determined the exact location of the subject's head in relation to each pick-up coil by measuring the magnetic signals of three positioning coils placed at known sites on the head. We ensured reliable positions by averaging the signals 50 times prior to the experiments. If two extrema were not included in the obtained magnetic fields, the position of sensors was changed to cover the two extrema. The signals were filtered with a band-pass filter (0.7-100 Hz), and were digitized at 1024 Hz. Two hundred single responses were averaged off-line, including signals for 50 ms before (pre stimulus baseline) and 300 ms after the stimulus. To identify sources of the equivalent current dipoles (ECDs) with high correlation values were found by a least-squares algorithm on the basis of a one-dipole model for 300-ms period after stimulation. Generator sites of the ECDs were superimposed on appropriate axial, coronal and sagittal MRIs. Magnetic resonance images were obtained for all subjects with 0.5T MAGNEX-a (Shimadzu, Japan) after MEG measurement. In order to create the MEG 3D coordinate system in the MR images three hollow disks filled with NiCl solution (0.5%) were affixed at sites where three positioning coils were placed in each subject before MRI measurement.

Results

Fig. 1 shows two sets of data; one is raw data, and the other is calculated data from one of our subjects obtained with 500 Hz tone-burst. Three dimensional components derived from the raw data were clearly shown, and enables us easily recognize the direction of the vector of magnetic field obtained at each recording site.

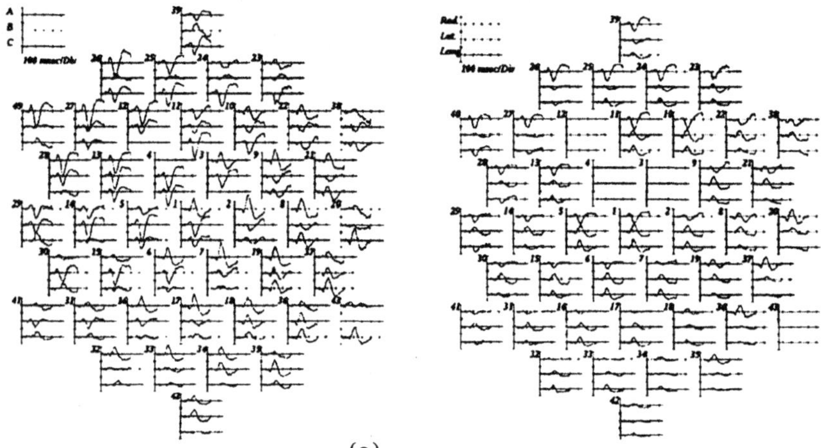

(a)

(b)

Fig. 1 Two sets of 129 averaged waves evoked by 500 Hz tone-burst stimulation. (a) show the raw data set obtained from the right hemisphere, (b) is the set of converted data to the radial, longitudinal, and latitudinal component on the scalp surface. Note that 4 sets of data were not calculated due to the lack of one of three data from A, B, and C channel.

Table 1 shows the latencies, the dipole moment (the strength of current), and the goodness of fit of N100m, which were obtained from both hemispheres with each carrier frequency in 5 subjects. We could not find any significant difference between the right and the left hemisphere on the latency of N100m, however, the dipole moment of ECD in the right hemisphere is stronger than that in the left one. The values of goodness of fit of N100m in the right hemisphere were generally larger than that in the left.

Table 1: The parameters of N100m in both hemisphere

	Latency (msec)		Dipole moment (nAm)		Goodness of fit (%)	
	Rt.	Lt.	Rt.	Lt.	Rt.	Lt.
500 Hz	103 ± 9.8	112.6 ± 7.0	42.8 ± 7.5	26.6 ± 11.2	97.8 ± 1.1	94.6 ± 3.0
1000 Hz	98.2 ± 10	107.1 ± 14	43.6 ± 9.7	27.9 ± 8.8	97.0 ± 2.0	94.0 ± 3.2
2000 Hz	99.0 ± 11.2	105.5 ± 6.0	32.2 ± 13.1	24.5 ± 8.3	96.6 ± 1.5	92.3 ± 8.6
4000 Hz	103.2 ±11	107.8 ± 4.8	36.0 ± 22.6	16.9 ± 7.8	95.8 ± 2.2	90.1 ± 6.1

n = 5

Fig. 2 shows the estimated distance of the ECDs of N100m from the center of a sphere model. These graphs indicate that the location of N100m ECDs is found to vary depending upon the tone-burst frequency. However, their locations at 4000 Hz tone-burst varied among subjects.

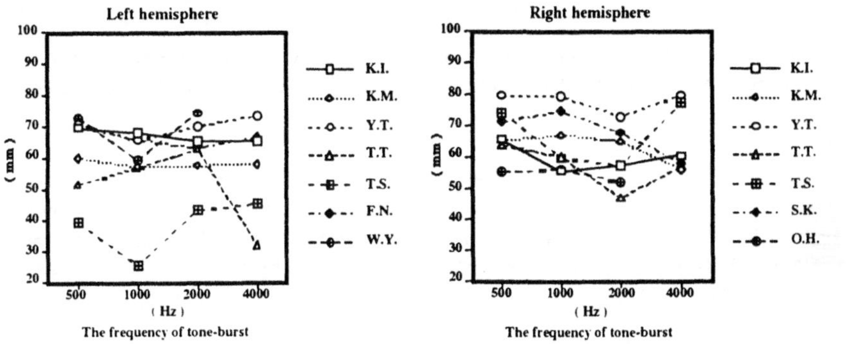

Fig. 2. The estimated distance of the equivalent current dipole of N100m from the center of a sphere model in each subject. The left column shows data obtained from the left hemisphere, the right column from the right hemisphere.

Fig. 3 shows the locations of the estimated N100m dipoles evoked by 4 different carrier frequency. Although the locations of estimated N100m dipoles with each career frequency were different each other, we superimposed each dipole of N100m on an axial and a sagittal slice of MRI obtained from subject Y.T. in order to compare four different dipoles each other on the same anatomical location.

Discussions and conclusions

We have developed a new 129 channel vector magnetometer system linked with MRI. The location of each ECD dipole was estimated using a single-dipole model in a spherical model determined individually according to the MRI images. There was found to be the tendency that the location of N100m ECD was dependent upon the tone-burst frequency. However, the locations at 4000 Hz tone-burst varied among subjects.

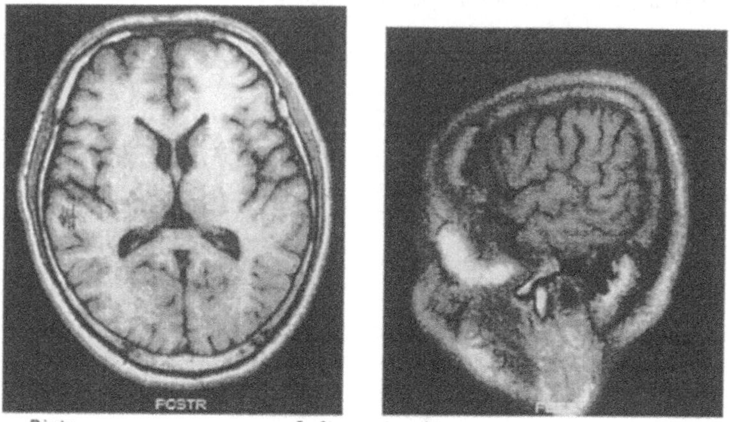

Fig. 3. The estimated dipoles of N100m on axial and sagittal MRIs obtained from subject Y.T. Four estimated dipoles were superimposed on each MRI. Note that all dipoles are located in the superior temporal gyrus

Previous studies on the visualization of the human tonotopy have always discussed based on the values of coordinate system[1, 2, 3]. Recently, Pantev et al. superimposed the estimated N100m source locations on individual MRI images[4]. Our results are in good agreement with their data. This study clearly demonstrates that a newly developed vector magnetometer system has the good reliability of source localization when visualizing the human tonotopy.

References

[1] Romani, G.L., Williamson, S.J. and Kaufman, L. Tonotopic organization of the human auditory cortex. Science, 1982, 216: 1339-1340.

[2] Elberling, C., Bak, C., Kofoed, B., Lebeck, J. and Saermark, K. Audiotorymagnetic fields. Source location and 'tonotopical organization' in the right hemisphere of the human brain. Scand. Audiol., 1982, 11:61-65.

[3] Pantev, C., Hoke, M., Lehnertz, K., Lutkenhoner, B., Anogianakis, G., and Wittkowski, W. Tonotopic organization of the human auditory cortex revealed by transient auditory evoked fields. Electroenceph. clin. Neurophysiol., 1988, 69: 160-170.

[4] Pantev, C., Bertrand, O., Eulitz, C., Verkindt, C., Hampson, S., Schuierer, G., and Elbert, T. Specific tonotopic organizations of different areas of the human auditory cortex reveald by simultaneous magnetic and electric recordings. Electroenceph. clin. Neurophysiol., 1995, 94: 26-40.

[5] Yoshida, Y., Arakawa, A., Kondo, Y., Kajihara, S., Tomita, S., Tomiata, T., Takanashi, Y., and Matsuda, N. A 129-channel vector neuromagnetic imaging system. in This volume.

Measurement of Accommodation Related MEG Responses

Takeda, T., Hashimoto, K., Kumagai, T. and Endo, H.

Human Informatics Department, National Institute of Bioscience and Human-Technology
1-1 Higashi, Tsukuba City, Ibaraki, 305, JAPAN

Introduction

We have been developing optometers to measure accommodation, eye movement and pupil diameter simultaneously under real working conditions[1,2,3]. By using them, we have found that the accommodation is sensibly controlled by apparent depth perception judged by brains[4]. Yet, there is much to be clarified on the control mechanism of the accommodation. Hence, the study of accommodation both from ocular responses and brain activities has high possibility to light a new insight into brain and visual functions.

A special relay lens system has been developed to impose visual stimuli and at the same time to measure accommodation responses with a dynamic refractometer[1]. As for the brain response, a 64-channel Whole-cortex MEG systems (CTF Inc., Canada) was employed. Using those systems, MEG and accommodation responses were measured simultaneously for the first time (Fig. 1).

Methods

Fig.1 Measurement System.

One of the most prominent characteristics of the MEG system is a software noise reduction (SNR) technique[5], which uses noise data sensed by reference coils and subtract them from the measured data. The efficiency of the software noise reduction technique was examined and confirmed by a phantom head examination and a method, which allowed measurement with opening the door of a magnetic shielded room, was established[6].

In order to measure the accommodation, we have made a special relay lens system (see Fig.1), which optically transfers the subject's eye image to the outside of the magnetically shielded room. It consists of 4 identical spherical lenses with 400 mm focal lengths, forming a focal system. Thereby it moves the eye by 3200 mm to form a real image just in front of the dynamic refractometer where the subject's eye is placed in normal measurement. The accommodation measurement is done with the real image of the eye.

The dynamic changes of the focus point of the eye were measured by a servo-controlled optical system, using the retinal reflection of injected infra-red beams. A target in the system was moved by a personal computer (NEC, PC-9800). The resolution of the measurement is about 0.1 D (Diopter; it is defined by the reciprocal of a focal point) and the speed of measurement is 6.4 Hz. Details of the optical system of the dynamic refractometer are explained elsewhere [2].

Subjects were 3 right-handed volunteers with adequate accommodative power (HK, 36, male with 4D of Accommodative Power (AP); EH, 32, male with 4D of AD; KC, 26, female with 6D of AP) and no ocular problems other than myopia.

The AP was measured with the dynamic refractometer by moving a target from a position located farther than the far point (the farthest point which the subject can focus) to another position nearer than the near point (analogously defined) [1]. The target was changed stepwise with a random time interval of 5 ± 0.5s. The amplitude of the stimulus was set to be about 60 % of each subject's AP to prevent excessive visual fatigue. Subjects were instructed to try to watch the target as clearly as possible.

The measurements were done on the right eye, occluding the left eye with an eye patch. The light in the magnetically shielded room was turned off to dilate the pupil diameter. The experiments consisted of more than 64 recordings; the collected data were averaged after discarding those contaminated by eye blinks. The sampling rate was set to 250 Hz. The recording time started 1s before the onset of the stimulus and lasted 2.5s. The evoked fields, the accommodation responses and the target positions were simultaneously recorded by a personal computer. The trigger signal for MEG measurement was made by a predetermined threshold of the target position. The MEG signals were removed its offset using 1s pre-recording and filtered with a zero phase shift, 1-40 Hz band-pass filter.

Results

Only the responses of subject HK are shown in this paper to save space. The responses of the other subjects were similar with the presented data.

Fig. 2 depicts a control response of HK when he gazed at the target in the refractometer which was positioned at the farthest (-5D) without moving. MEG recordings of all the 64 channels are superimposed. It shows noise-like response with intensity of about 50 fT all the time.

Then the target was moved between -5 and -8 D, because HK was -5 D myopic. Fig. 3 depicts whole channels' responses for the target movement of far to near, shown on a circular projection of the head. Clear responses are found on the sensors located over occipital and lateral lobes.

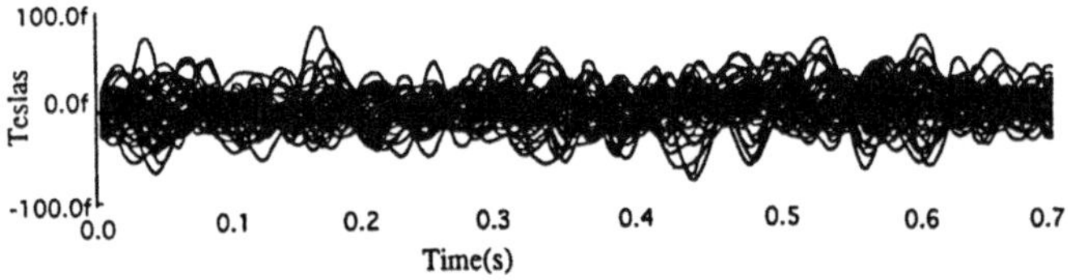

Fig.2 Superimposed MEG responses of HK when the target was not moved.

Fig. 4 shows the accommodation and MEG responses for the same stimulus together with the target movement. The time lag of the accommodative response was 288 $\pm$50 ms. It was calculated by averaging the time lags of the 64 records determined by inspection. The start of accommodative response is indicated by an arrow in the figure. The averaged time lag was stable within each subject. Though the amplitudes do not match with the stimuli, it is natural and is called as the accommodation lag[7]. The mean time lag and the shape of the accommodation responses were in good agreement with the literature[8]. Superimposed MEG responses indicated well synchronized activation of the brain around 100 and 200 ms. Both peaks showed the phase reversal phenomenon. On the other hand, the responses were rather quiet around 290 ms when the average accommodative response began to rise.

Fig. 5 shows the accommodation and MEG responses when the target was moved from

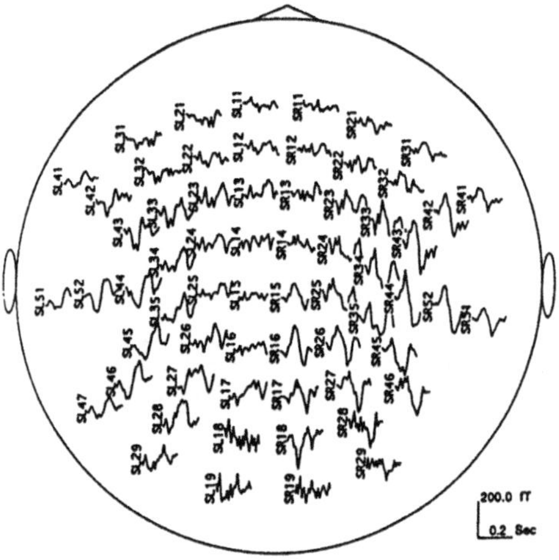

Fig.3 Responses of all the 64 channels which show large signals on the channels located over occipital and lateral lobes.

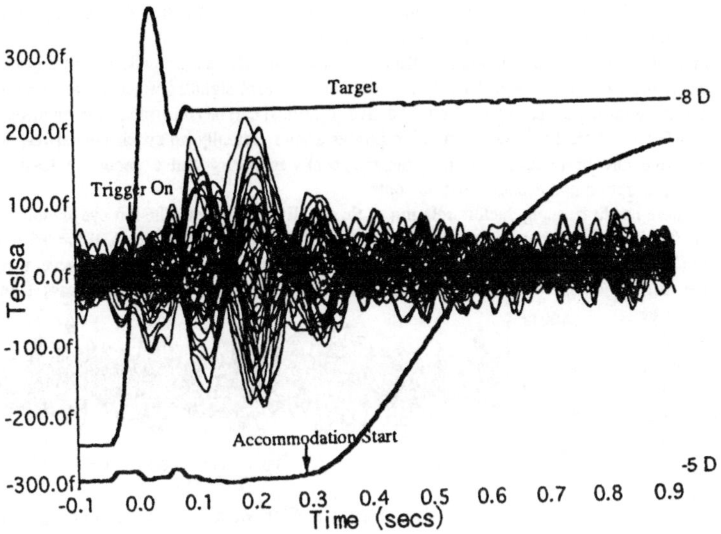

Fig.4 Superimposed MEG and accommodation responses together with target change which was abruptly moved from far (-5 D) to near (-8 D).

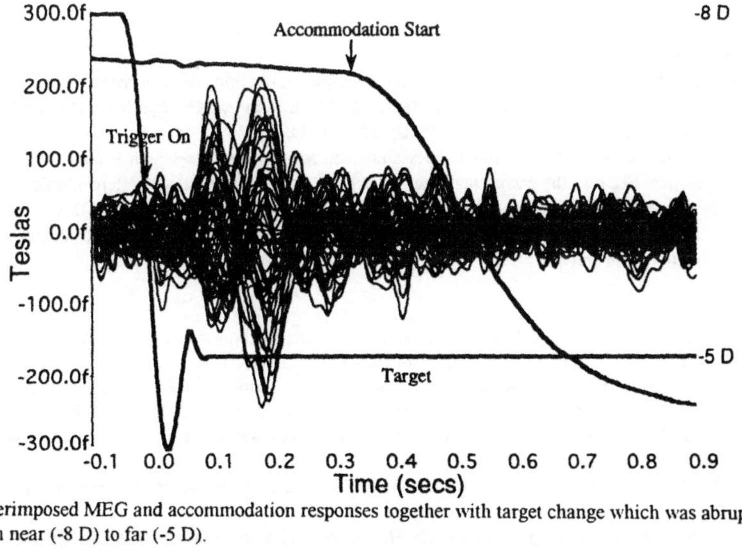

Fig.5 Superimposed MEG and accommodation responses together with target change which was abruptly moved from near (-8 D) to far (-5 D).

near to far, together with the target movement. The time lag of the accommodative response was 341 ± 56 ms. Superimposed MEG responses showed a tendency similar to the ones in Fig. 4. There were at least two phase reversal peaks around 100 and 190 ms and a relatively low activation when the accommodation started.

Discussion

There are no previous reports on the accommodation related MEG measurement. On the other hand, there are a few studies concerning the location of the control center of the accommodation, employing implanted electrodes in animals[9]. However no basic relationship between cerebral activity and accommodation response was found nor was identified the control center of the accommodation.

In the present study, the superimposed MEG responses and the accommodation responses (Fig. 4,5) clearly suggest that perception of blur and determination of some control signals for accommodation are made at around 100 and 200 ms after the onset of the accommodative stimuli. It can be confirmed by comparing the control responses shown in Fig.2, where the average of 64 responses shows virtually no synchronized responses except about 50 fT of external and brain noises. Hence, the two peaks in Fig. 4 and 5 should at least have strong correlation with blur perception and accommodation control.

However, there might be many factors relating to these MEG responses. The left eye should be moved up to 10° by accommodative vergence, because right eye position is fixed to the optical axis and the left eye should be directed from -5D (20 cm) to -8 D (12.5 cm). The sensors located near the left eye recorded several hundreds fT drifts. Though the slow components of the eye movement were removed by the band-pass filter, the fast components might be involved in Fig. 4 and 5

Fig.6 shows superimposed MEG responses when HK gazed at a spot presented on a CRT, which was moved up to 10° for the same period of the accommodative stimuli. The MEG data filtered with the same band showed much smaller responses compared with the responses shown in Fig. 4 and 5. Hence eye movement should not have major importance in those responses.

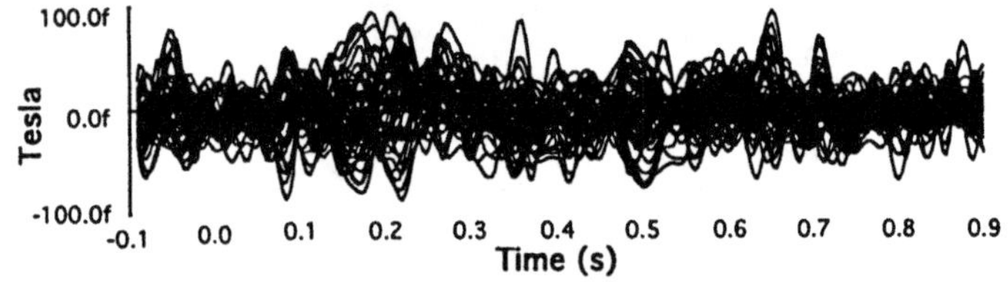

Fig.6 Superimposed MEG responses of HK when target's direction was moved slowly.

Near accommodation drives pupil constriction in general. As the change of pupil diameter is not so large judging from TV monitoring, it must have not great influence on the MEG response in those conditions. Anyway, it will be checked soon.

It is noteworthy that the MEG data in Fig.4 shows another activation after the onset of the accommodation and data in Fig. 5 dose not show. This might be related to the fact that the ciliary muscle is activated for the far to near accommodation but is relaxed for the near to far accommodation [7].

In concluding, the present experiment does show the MEG responses which should correlate with the accommodation control. Because the accommodation must have close correlation with stereoscopic perception[8], the present study on accommodation should have tremendous importance in developing 3-D TVs in the future. However, more detailed experiments are needed to clarify the timing, location and characteristics of the accommodation-related brain activities.

References

[1]Takeda, T., Ostberg O., Fukui Y. and Iida, T., Dynamic Accommodation Measurements for Objective Assessment of Eyestrain and Visual Fatigue, J. of Human Ergology, 17, 1, 21-30, 1988

[2] Takeda, T., Fukui, Y. and Iida, T., Three-dimensional Optometer, Applied Optics, 27, 12, 2595-2602, 1988.

[3] Takeda, T., Fukui, Y. and Iida, T., Three-dimensional Optometer Ⅲ, Applied Optics, 32, 22, 4155-4168, 1993.

[4] Takeda, T., Iida, T. and Fukui, Y., Dynamic eye accommodation evoked by apparent distances, Optometry and Vision Science, 67, 6, 450-455, 1990.

[5] Vrba J. *et al.*, 1982, Spatial discrimination in SQUID gradiometers and 3rd order gradiometer performance, Can. J. Phys., 60, 1060-1073.

[6] Takeda, T., Morabito, M., Kumagai, T. and Endo, H., Use of CRT as a visual stimulator in MEG measurements, Event-related potentials of the brain edited by Y. Koga, Elsevier, 1995 (in press).

[7] Stark, L., 1987, Presbyopia in light of accommodation, in Presbyopia edited by L. Stark and G. Obrecht, Fairchild Publications New York, 264-274.

[8] Campbell, F.W. and Westheimer, G., 1959, Factors influencing accommodation responses of the human eye, J. Opt . Soc. Am., 49, 568-571.

[9] Bando, T. and Toda, H., 1991, Cerebral cortical and brainstem areas related to the central control of lens accommodation in cat and monkey, Comp. Biochem. Physiol. 98C, 1, 229-237.

Electromagnetic source localization of cognitive processes related to attention to single or omitted tones

Tarkka, I. M., and Stokić, D. S.
Baylor College of Medicine, Division of Restorative Neurology, Houston, Tx, USA

Introduction

Magnetic and electric recordings have confirmed that the P300, a late positive component of the human event-related potentials (ERP), can be evoked not only by a target discrimination task but also by the omission of targets [1,2,3]. Recently it was shown that also single target stimuli, delivered without distractors, can evoke a P300 response [4]. In this case the task is simpler than the oddball paradigm which may render it more useful for clinical applications. However, the analogy of the single stimulus P300 to the classical P300 needs to be shown.

The purpose of the present study was to extend our previous work on the identification of the brain regions generating the P300 potential elicited by an auditory oddball paradigm. Our goal was to find out if the P300 responses elicited by rare or single tones or the omission of tone differ in their morphology and/or in the underlying cerebral structures responsible for generating the P300. Noninvasive electric source localization technique was employed by using the Brain Electrical Source Analysis (BESA) software in order to develop spatio-temporal equivalent electric dipole models for the scalp-recorded electric data. Individual data was modeled using individual head radius for the spherical head model.

Methods

Subjects
Seven healthy subjects (1 woman, 6 men, mean age 32 y, from 25 y to 39 y) volunteered for the study and gave an informed consent prior to the experiment. The experimental protocol was approved by the local committee for human subject research. None of the subjects had a history of neurological symptoms and they were not under any medical treatment.

Experimental protocols
Three different paradigms using auditory stimuli constituted the experiment. First, a classical oddball paradigm was used delivering tones at the rate of 1.1 Hz. The frequent tones were low pitched (1000 Hz) and the target tones, occurring randomly 18 % of the time, were high pitched (2000 Hz). The plateau duration of the tone was 50 ms with rise and fall times of 10 ms. Tones were delivered binaurally through ear-phones at an 85 dB intensity level with 60 dB masking with white noise. Subjects were asked to silently count the number of rare, high pitched tones. Three to four short sessions were recorded, each consisting of about 40 rare tones. Short breaks were given to the subjects between sessions.

Second, no frequent tones were delivered, however the target tones (2000 Hz) remained delivered as in the previous oddball condition. Thus, at irregular intervals single tones were presented and the subjects were asked to silently count them. Again each short session consisted of about 40 tones.

In the third condition, the frequent tones were delivered as in the rare tone condition but this time target tones were omitted. Therefore the irregular absence of the tone was to be counted. Data collection was performed in a same way for the rare tone, single tone, and omitted tone conditions. All subjects performed the three tasks with better than 96% target detection rate.

ERP recording

ERPs were recorded with 30 silver/silverchloride cup electrodes (diameter 10 mm) evenly spaced on the scalp. During the recording linked earlobes served as a reference, for further analysis average reference was used. Electrode impedances were kept below 3 kΩ. The EEG bandpass was from 0.07 Hz to 80 Hz. The 700 ms recording window consisted of a 100 ms pre-stimulus baseline and 600 ms for the post-stimulus epoch. During recording the frequent and rare wave forms corresponding to each condition were sorted into two memory buffers and averaged on-line. Automatic artefact rejection was used to control the quality of the data.

Source analysis

Prior to source analysis all the wave forms were base-line corrected and filtered at 30 Hz high cut-off frequency. The Brain Electric Source Analysis (BESA) program, version 2.0, [5,6,7] was used to develop equivalent electric dipole models to suggest locations of the active brain regions. Individual head radius was used for each subject's 4-shell spherical head model. In addition to descriptive measures and topographic mapping of the scalp recorded data, the program allowed iterative simultaneous or sequential fitting of the location and orientation of equivalent electric dipoles. The development of the models for the rare, single and omitted tone conditions was done independently from each other. The strategy for model development was to first develop models for grand average data of each condition. This was done by dividing the 600 ms window into two parts; one corresponding to the auditory evoked potential (0-180 ms) and one to the cognitive potential (180-600 ms). These two models were then combined and fitted in order to model the 600 ms window of the grand average data. The models explaining grand average wave forms were used as starting parameters for developing the individual models. The locations, orientations and strengths of the dipoles were compared between the three conditions for each of the subjects.

Results

A late positive component, P300, was evoked in all studied conditions. The peak latency of the P300 evoked by single stimuli was shorter than that evoked by the rare ($p<.05$) or the omission ($p<.01$) condition. The P300 amplitudes of the rare and single tone conditions were similar, in the order of 15 μV, and the P300 evoked by the omitted tone had a smooth and widespread morphology with a peak amplitude of about 10 μV. The spatio-temporal models for rare and single tone conditions both consisted of 6 dipoles and the model for omitted tones consisted of 3 dipoles. The grand average models for each of the conditions are shown in Fig. 1. The residual variances (RV %) of the models of the individual subjects, indicating the goodness of fit, ranged from 5.1 % to 11.9 % in the rare tone condition, from 5.4 % to 14.1 % in the single tone condition and from 3.1 % to 25.7 % in the omitted tone condition. The differences in dipole locations of the individual models between rare and single tone conditions were less than 15 mm for all three axis in 95.3 % of the data. As the equivalent electric dipole presumably accounts for neural activity of an area of 10 mm in diameter, it makes the differences in dipole locations between the rare and single tone conditions unremarkable.

When the parameters of the average dipole locations were transferred into a brain atlas [8], a fairly lateral region of the superior temporal gyrus bilaterally was implicated by the dipoles number 1 and 2. The source potentials of dipoles 1 and 2 indicated that their major contribution was at the time of the late auditory evoked potential in both rare and single tone conditions (see Fig. 1.). A deeper region, bilaterally, in the superior temporal gyrus was implicated by dipoles 3 and 4. The source potentials of these dipoles implied that these source were active both at the time of the late auditory evoked potential and the cognitive potential. The dipoles 5 and 6, bilaterally, located deeper in the medial temporal lobe, more than 10 mm posterior to dipoles 3 and 4, suggesting hippocampal and parahippocampal regions as active sources. Dipoles 5 and 6 had the majority of their strength during the P300 component. The 3-dipole model explaining the P300 evoked by omitted tones consisted of a right unilateral dipole deep in the superior temporal gyrus showing most of its strength at the time of the long latency auditory evoked potential as seen in the rare or single tone conditions. The P300 component appeared to be explained by two deep dipoles, one in temporal lobe region and one in the posterior frontal lobe.

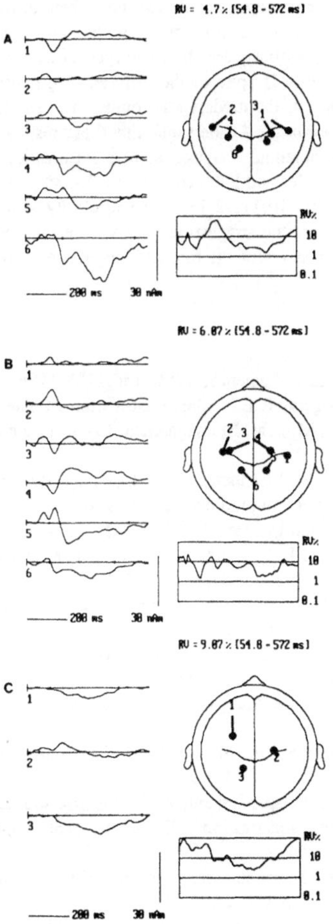

Fig. 1. Spatio-temporal multiple dipole source models developed for the grand average ERP data (7 subjects) of rare tone (A), single tone (B) and omitted tone (C) conditions. On the left are the source potentials for each of the dipoles and the RV % in the middle shows the residual variance throughout the analyzed window. The overall RV % is shown at the top for the entire window. Note, that the models for rare and single tone conditions are fairly similar.

Discussion

The target tones delivered in a classical oddball paradigm and the single tone paradigm produced similar P300 components in their peak amplitude and scalp topography. The peak latency of the P300 produced by single tone was shorter than that of the rare tone or the P300 evoked by the omission of tone. In this aspect the single tone evoked P300 resembles the novel P300. The comparison of the dipole locations of the individual models revealed close similarity of the models which implies that the rare and single tone tasks activated adjacent anatomical regions. These two models support the notion that bilateral superior temporal lobe areas, primary and secondary auditory cortices, are active especially at the time of the long latency auditory evoked potential and furthermore they suggest that the deeper medial temporal lobe sources, in part in the deep superior and transverse temporal gyri and in part in the hippocampal/parahippocampal area, contribute to the late cognitive component. The parameters of the models for the rare tone in the present study were also compared with the parameters of the multiple dipole models we presented previously for the same type of auditory evoked P300 response but recorded in a different group of subjects [9]. In spite of the difference in radiuses of the head models used in these two studies, a close correspondence of the models was found. The similarity of the two independent models adds credibility to the modeling approach. At the same time these results imply the same active neural generators for auditory evoked P300 potential. Also, source localization performed on magnetoencephalographic auditory evoked long latency fields and P300 components has suggested similar areas for the generators of the human P300 as the present study [10,11,12,13]. The similarity found in the models for rare and single tone evoked P300 appears to support the suggestion of Polich et al. 1994 [4] that the underlying neural and cognitive mechanisms may be the same for the rare and single tone conditions.

References

[1] Sutton, S., Braren, M., Zubin, J., and John, E. R. Science, 1965, 150:1187-1188.

[2] Papanicolaou, A.C., Baumann, S.B., and Rogers, R.L. Source estimation of late components of omitted tone evoked magnetic fields. In: Hoke, M. et al. Biomagnetism: Clinical aspects, Amsterdam, Elsevier, 1992, pp. 177-180.

[3] Rogers, R.L., Papanicolaou, A.C., Baumann, S.B., Eisenberg, H.M. Late magnetic fields and positive evoked potentials following infrequent and unpredictable omissions of visual stimuli. Electroencephalography and clinical Neurophysiology, 1992, 83:146-152.

[4] Polich, J., Eischen, S.E., and Collins, G.E. P300 from a single auditory stimulus. Electroencephalography and clinical Neurophysiology, 1994, 92:253-261.

[5] Scherg, M., and von Cramon, D. Evoked dipole source potentials of the human auditory cortex. Electroencephalography and clinical Neurophysiology, 1986, 65:344-360.

[6] Scherg, M., Vajsar, J., and Picton, T.W. A source analysis of the late human auditory evoked potentials. Journal of Cognitive Neuroscience, 1989, 1:336-355.

[7] Scherg, M., and Berg, P. Brain Electrical Source Analysis Handbook, 1994, V. 2.0., Herndon, Virginia.

[8] Talairach, J., and Tournoux, P. Co-Planar Stereotaxic Atlas of the Human Brain. Thieme Medical Publishers, Inc., New York, 1988.

[9] Tarkka, I.M., Stokić, D.S., Basile, L.F.M., and Papanicolaou, A.C. Electric source localization of the auditory P300 agrees with magnetic source localization. Electroencephalography and clinical Neurophysiology, 1995, 96:538-545.

[10] Hari, R., Rif, J., Tiihonen, J., and Sams, M. Neuromagnetic mismatch fields to single and paired tones. Electroencephalography and clinical Neurophysiology, 1992, 82:152-154.

[11] Okada, Y.C., Kaufman, L. and Williamson, S.J. The hippocampal formation as a source of the slow endogenous potentials. Electroencephalography and clinical Neurophysiology, 1983, 55:417-426.

[12] Rif, J., Hari, R., Hamalainen, M.S., Sams, M. Auditory attention affects two different areas in the human supratemporal cortex. Electroencephalography and clinical Neurophysiology, 1991, 79:464-472.

[13] Rogers R.L., Papanicolaou, A.C., Baumann, S.B., Bourbon, W.T., Alagarsamy, S., Eisenberg, H.M. Localization of P3 sources using magnetoencephalography and magnetic resonance imaging. Electroencephalography and clinical Neurophysiology, 1991, 79:308-321.

Acknowledgements

This work was supported by the Vivian L. Smith Foundation for Restorative Neurology, Houston, Tx, USA.

A Cortical Substrate For Hand Skill: An MEG Evaluation

Thoma, R. J., Lewine, J.D., Davis, J.T., and Orrison, W.W., Jr.

The New Mexico Institute of Neuroimaging, The New Mexico Regional Federal Medical Center, Albuquerque, New Mexico, USA.

INTRODUCTION

It is well established that, independent of culture, approximately 90% of the human population demonstrates a right hand preference on most tasks. This population asymmetry is especially noteworthy because of its magnitude and because it is uniquely human. The current study is an attempt to relate differential hand skill directly to a measure of cortical functional anatomy. Specifically, it was hypothesized that more sensorimotor cortex would be dedicated to the dominant versus non-dominant hand. To assess this, we utilized magnetic source imaging, a method that combines magnetoencephalographic (MEG) data on brain function with magnetic resonance (MR) data on brain structure. Current flow associated with neuronal activity produces a magnetic field, just as does the current flow in a wire. The magnetic field that can be measured outside the head selectively reflects current flow within the dendrites of pyramidal cells oriented parallel to the skull surface. Using signal averaging techniques it is possible to identify the magnetic signals time-locked to particular motor or sensory events.

Evoked fields associated with self-paced movements are characterized by a slow field shift beginning several hundred milliseconds prior to the movement, and a large magnetic signal during/after the movement. The field prior to movement is believed to reflect predominantly activities in motor cortex, whereas the sharp signal at the time of movement reflects mostly feedforward/feedback activation of somatosensory cortex. The somatosensory component is typically of larger amplitude and more easily localized. By combining simplifying assumptions about the conductivity properties of the brain, skull, and scalp with assumptions about the configuration and number of neural generators, it is often possible to infer the spatial location of those cells that generate specific magnetic field components. In particular, the head is assumed to be well modeled by a spherical volume conductor, and it is assumed that the magnetic field pattern at any particular instant in time can be modeled as though it were generated by an equivalent current dipole intended as a mathematical representation of the neuronal currents of a focal population of cells. Available data indicate that the dipole model is adequate for characterizing the primary motor field and the associated somatosensory field, with sources localizing to within 5 mm of the central sulcus.

To infer the amount of cortex devoted to sensorimotor processing for the right versus left hand, we took advantage of the somatotopic organization of sensorimotor cortex. Specifically we localized sources associated with movements of the thumb and little finger and took the distance between these sources to be indicative of the amount of cortex devoted to the hand.

METHODS

Subjects: Data were successfully collected from 9 volunteers drawn from a population comprised of students and faculty of the Psychology Department at the University of New Mexico and staff from the Department of Radiology at the Albuquerque VA Medical Center (VAMC). The sample consisted of 4 right-handed females, 3 right-handed males and 2 left-handed males. Subjects ranged in age from 18 to 41 years old. The MEG data collection often lasted several hours and subjects were compensated for their time with pizza.

MEG Data Collection: All MEG data were collected at the Albuquerque VAMC using a large array biomagnetometer (BTi) equipped with 37 first-order, axial gradiometers. Prior to the experiment, a head-centered coordinate frame was defined by touching the stylus of a 3D digitizer to right and left preauricular points and to the nasion. The stylus was also used to digitize the entire headshape in order to define a best-fitting sphere for the source modeling calculations. During each experimental test block, subjects were required to make 150 self-paced movements of a specified digit (right thumb, right little finger, left thumb, left little finger) at a rate of about 1 per 3 seconds. The movements caused the closing of a photo-optic switch that provided a time stamp to the data acquisition system. The magnetometer was initially positioned over the parietal/frontal region contralateral to the hand being tested. The unit was repositioned during the course of the experiment, as needed, to ensure that data were recorded over both the positive and negative field extrema (region of maximal exiting and entering flux, respectively). Fifteen hundred msec of data were acquired prior to each movement, and 500 msec were acquired after each movement. Data were filtered on-line with a bandpass of 0.1 - 50 Hz, and they were digitized at a rate of approximately 200 Hz. Each block of trials lasted 20-30 minutes.

MEG Data Analysis: Prior to signal averaging, trials where the magnetic field on any sensor exceeded 10 picoTesla were rejected. Typically, less than 5% of trials were contaminated. The data were signal averaged, baseline corrected, and subjected to a 20Hz low-pass digital filter. For this project, analyses focused on the large component coincident with the movement, because this was recorded more reliably in all conditions. Also, this component is better characterized by the dipole-in-a-sphere model. The dipole-in-a-sphere modeling algorithm provides the location, orientation and strength of that current dipole which best accounts for the magnetic field pattern. A dipole fit was considered valid if, and only if, it accounted for more than 90% of the variance in the field map and the confidence volume was less than one cubic centimeter. The spatial parameters of the dipole were taken as indicative of the coordinates of the relevant sensorimotor cortex. To assess the extent of sensorimotor cortex devoted to the hand, a linear distance between dipole sources for the first and fifth digit was calculated. Ideally, distances should be measured along the actual cortical surface, but prior experience indicated that this is a very time-consuming process, and that the linear distance correlates with the cortical distance.

Magnetic Resonance Imaging (MRI): A 3D sagittal volumetric data set (1.5 mm slice thickness, no gap) was obtained using a Seimens 1.5T whole-body imaging system. CEMAX VIP software running on a SUN workstation was used to create a volume rendering of the data. By identification of the MEG head-frame fiducials (preauricular points and nasion) on the MR data set, it was possible to plot MEG dipoles on the brain volume. The data were then inspected to confirm that dipole sources localized to the sensorimotor cortex.

Handedness Measures: During the MEG data collection phase of the project, experimenters were blind as to the handedness of subjects. After the MEG data were analyzed, all subjects were given the Waterloo Handedness Questionnaire as a measure of hand preference. The questionnaire allows subjects to circle answers (left always, left usually, equal, right always, right usually) to questions about which hand they use for certain tasks, for example, holding a toothbrush. Hand skill was measured with a pegboard task [1]. Each trial on the task requires subjects to move 10 pegs from one holder and fit them sequentially into holes in another holder approximately 3 inches away. Five trials with each hand are completed by each subject.

RESULTS

Signals were recorded successfully from the four test conditions for nine subjects. In all cases, dipole sources localized to the sensorimotor cortex of the contralateral hemisphere. In every case, the linear distance between localized source points for the first and fifth finger was larger contralateral to the subject's dominant hand than the corresponding measure in the opposite hemisphere [Fig 1].

For right-handers, the average left hemisphere distance was 14.5 mm, and average right hemisphere distance was 11.7 mm. In left-handlers, the averages were 10.3 mm and 14.5 mm, respectively. Differences between cortical distances for the two hemispheres (right-hemisphere thumb to fifth finger distance minus left-hemisphere thumb to fifth finger distance) were significantly correlated with differences in total times for right and left hands on the pegboard task (right-hand score minus left-hand score) $r= .81$, $p=.008$. That is, the more sensorimotor cortex devoted to a hand, the faster the performance of the contralateral hand. Additional analyses showed a significant correlation between Waterloo hand preference total scores and cortical distance difference scores ($r=.7153$, $p=.046$). The Waterloo questionnaire also correlated with pegboard difference scores ($r=-.762$, $p=.028$).

DISCUSSION

Through the use of MSI, we were able to demonstrate that the linear distance between finger representations on the post-central gyrus is greater contralateral to the dominant hand than the respective distance contralateral to the non-dominant hand. In addition, differences calculated by subtracting subjects' left-hemisphere from right-hemisphere cortical measurements were highly correlated with differences (R-L) in pegboard scores. Subjects' reported hand preference was also correlated with cortical distance scores.

Unfortunately, the present analysis does not conclusively define the causal relationship between handedness and cortical representation. There are two, not necessarily mutually exclusive, possible causal chains. Given that there is evidence for right thumb sucking preference by 10 weeks gestational age [2], one possibility is that a larger hand representation in sensory-motor cortex is established very early in life, and that this is the causal substrate of hand preference. Alternatively, increased hand representation in the dominant hemisphere may be an effect of preferential useage of the contralateral hand. Recent research with animals has shown changes in somato-sensory cortex occur as a result of dynamic interaction between cortical tissue and behavioral stimulation [3]. Differential hand use, early in life, may cause a cascade of environment/brain feedback events that results in the pattern of hand-use and cortical measures in the current dataset.

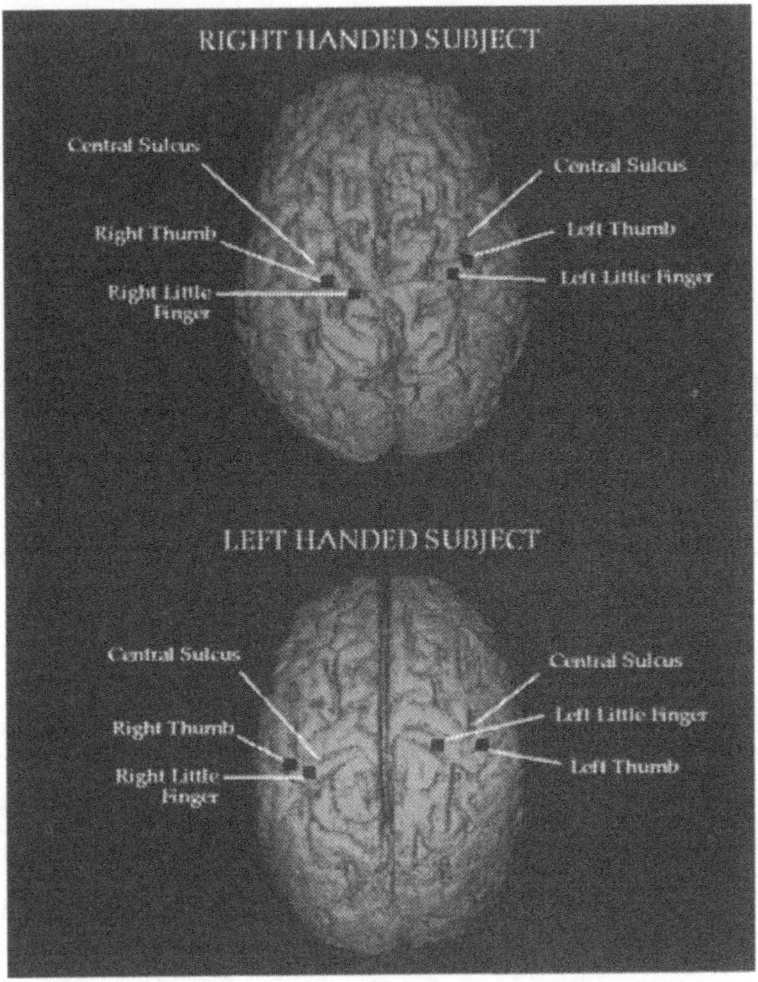

Fig 1: MEG results for a right versus left handed patient. For the right hander the distance between thumb and little finger responses is greater over the left hemisphere. For the left hander, the difference is greater over the right hemisphere.

REFERENCES

[1] Annett, M. Handedness in families, Annals of Human Genetics, 1973, 37: 93-105.
[2] Hepper, P.G., Shahidullah, S., and White, R., Handedness in the human fetus. Neuropsychologia, 1991, 28: 1107-1111.
[3] Rencanzone, G., Merzenich, M., & Schreiner, C., Changes in the distributed temporal response properties of SI cortical neurons reflect improvements in performance on a temporally based tactile discrimination task. Journal of Neurophysiology, 1992, 67.

Separation of Overlapping Sources Induced by Mixed Stimulations Using 3-Dimensional MEG Mappings

Uchikawa, Y.[1], Kobayashi, K.[1], Sakawa, K.[1] and Kotani, M.[2]

[1]*Faculty of Science and Engineering, Tokyo Denki University, Saitama 350-03 Japan;* [2]*Faculty of Engineering, Tokyo Denki University, Chiyodaku, Tokyo 101 Japan*

INTRODUCTION

There is a problem of separating multi-sources overlapping in time when many distinct areas of the cortex are active. Several methods for evaluating source localization from spatio-temporal MEG data have been reported [1,2,3,4]. Almost all SQUID magnetometers for MEG study are for measuring magnetic field perpendicular to the scalp. There is a limitation in spatial resolution, using flux-detection coils with finite radius in regular devices, to separate multi-sources [5]. To get more accurate information about source configuration, we have proposed a new 3-D measurement of MEG, namely vector magnetoencephalogram [6].

This paper describes the separation of two sources overlapping in time, located in the brain, by using a 3-D second-order gradiometer [6]. First, to discriminate and to separate multi-sources, we used 3-D MEG data from magnetic fields produced by two current dipoles implanted in a cranium model. Second, to apply and to confirm for human MEGs with the results of model study, we carried out 3-D MEG measurements for mixed AEFs and SEFs overlapping in time.

METHODS

The cranium model shown in Fig. 2 and 3 was based on a skull sample for medical education (The circumference of nasion-inion measured 32.3 cm, T3-T4 in the international 10-20 system measured 21.8 cm) and was made with colorless epoxy resin filled with salt water. The electrode for a current dipole (6 μAm) was made with an insulated copper wire (0.5 mm diameter, 6 mm length) and was set in the head model with a depth of 1 cm from the inner surface and parallel to the scalp. The 3-D MEG measurement was made by a 3-D second-order gradiometer. In the case of two current dipole sources, the magnetic fields were generated by two different frequencies of 30 and 70 Hz with constant current, and measured by separating the frequencies. As shown in Fig.1, a 3-D second-order gradiometer, which can simultaneously measure magnetic fields both of perpendicular to and tangential to the scalp, consists of three separate second-order gradiometers. Each coil is orthogonally wound with Nb-Ti wire on the rectangular solid of 3 x 3 x 6 cm. These three coils are connected to three rf-SQUIDs. The base line of the detecting coil for Br, which is the magnetic field perpendicular to the scalp, is 2.8 cm and that of the coils for Bθ and Bϕ, which are the magnetic field tangential to the scalp, are 1.4 cm

Fig.1 Coordinate system for measuring 3-D magnetoencephalogram using a 3-D second-order gradiometer.

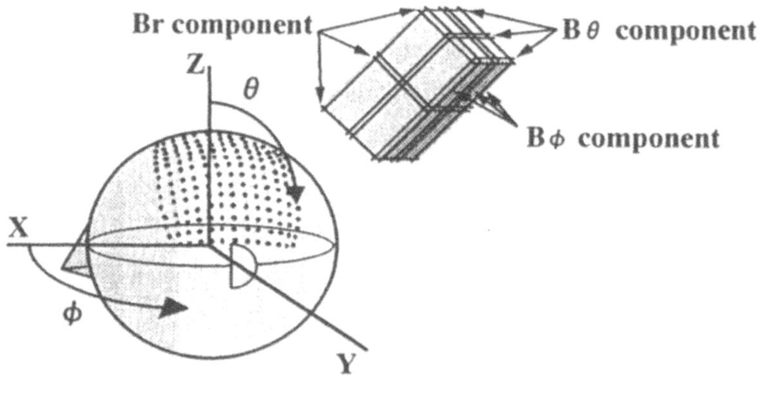

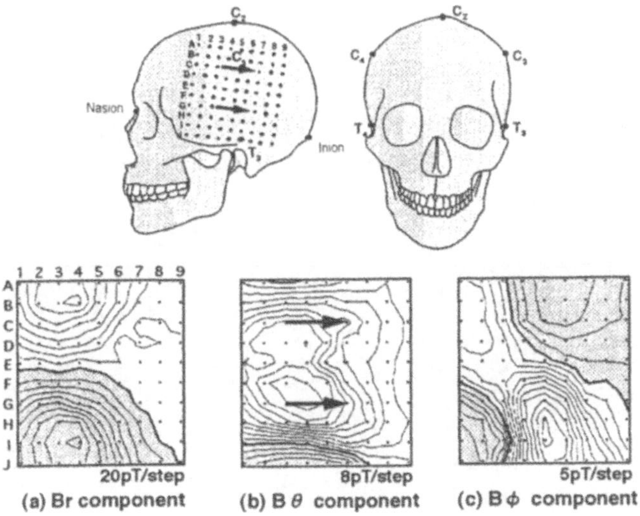

Fig. 2 Isofield contour maps for two cuurent dipoles, oriented in the same direction, implanted at a depth of 1 cm from the inner surface of a human cranium model. The separation of the dipoles is 4 cm. Arrows indicate the cuurent dipoles.

respectively. Coordinate system for measuring 3-D MEG is shown in Fig. 1. Gradiometers for Bθ and Bϕ, in particular, can provide the maximum and/or minimum field with a sign just above the location of a single current source [6,7]. This is a merit for separating multi-sources and is maybe the same thing with other coil configurations such as the 2-D coil [8].

RESULTS

Fig. 2 shows example of isofield contour maps for two current dipoles with the same direction, at C5 and G5, implanted at a depth of 1 cm from the inner surface of a cranium model. The separation of the dipoles is 4 cm.

In the Br component (Fig. 2(a)), a dipole distribution with the maximum of 140 pT (magnetic flux outflow) at B4 and the minimum of -126 pT (magnetic flux inflow) at I4 can be seen. It is possible to see that the magnetic field distribution is organized by a single current dipole located at E3, because two current dipoles go in the same direction. However, the Bθ component (Fig. 2(b)) shows two extremes of about 80 pT with same polarity at both of D4 and F4. In this case, by referring to the map of Bθ, we can clearly say that this map consists of "two sources with same polarity" underlying each extreme corresponded to the measurement points of D4 and F4. The direction and the location of two sources estimated by visual inspection were represented by arrows on the scalp as shown in Fig. 2(b).

Fig. 3 shows example of isofield contour maps for two current dipoles with the opposite direction, at C5 and G5, implanted at a depth of 1 cm from the inner surface. The separation of the dipoles is 4 cm. This source configuration is important for MEG measurement since electric potentials are attenuated at the scalp while magnetic fields are not. As shown in Fig. 3(a), the map of Br significantly misses the dipole pattern and shows one extreme of - 160 pT at E5. However, the map of Bθ clearly shows two extremes of -96 pT at C5 and 112 pT at G5 with opposite polarity. Positions of C5 and G5 correspond to the places implanted each current dipole. We can clearly say that this map consists of "two sources with opposite polarity" underlying each extreme corresponded to the measurement points of C5 and G5. The direction and the location of two sources estimated by visual inspection were represented by arrows on the scalp as shown in Fig. 3(b)

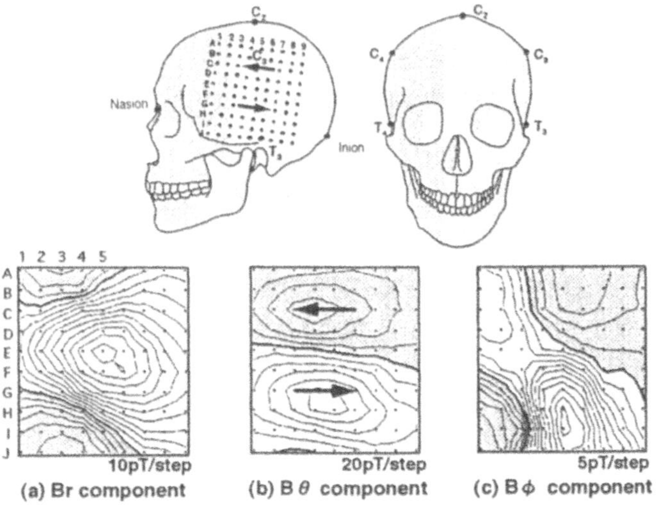

Fig. 3 Isofield contour maps for two current dipoles, oriented in the opposite directions, implanted at a depth of 1 cm from the inner surface of a human cranium model. The separation of the dipoles is 4 cm. Arrow indicate the current dipoles.

APPLICATION FOR MIXED AEF AND SEF

To separate multi-sources for human MEG with the results of model study, we carried out 3-D MEG measurement of mixed AEFs and SEFs overlapping in time. AEFs were elicited by 1 kHz tone bursts of 50 ms duration to the right ear and SEFs were elicited by electric pulses of 0.2 ms duration with 6 to 8 mA to the median nerve of the right wrist. The interval of stimulation was 500 ms. Stimulation to the median nerve was delivered 40 ms later than to that of the tone burst. There were two evoked field, namely mixed AEF and SEF, overlapping in time. All magnetic data were average by 300 to 400 measurements. The sampling interval was 0.5 ms. A band-pass filter was used in the range of 0.5 to 40 Hz.

Fig. 4 shows example of isofield contour maps on the left hemisphere at 120 ms latency of AEF and at 80 ms latency of SEF. As shown in Fig. 4(a), map of Br shows one extreme with about 950 fT at measurement position C3, but it is a quite similar to that of Fig. 3(a). Map of Br, in this case, was not helpful to estimate the location and the number of sources due to lack of a dipole pattern. By referring to maps of Bθ and Bϕ shown in Fig. 4(b) and Fig. 4(c), which are tangential magnetic field, we can clearly see two extremes with opposite polarity in both map. The locations and the directions of sources estimated by visual inspection from these maps were indicated by arrows on the scalp. This field pattern is almost the same as that of Bθ map in Fig. 3(b). In this case, we can clearly predict that these maps consist of the magnetic fields generated by "two sources with opposite polarity" underlying each extreme corresponded to the locations on the scalp indicated by arrows. These results from the maps of Bθ and Bϕ, relating to the estimation of source configuration, were confirmed by comparison with coronal MRI images of a subjet's head showing the estimated source locations as shown in Fig. 4(d).

CONCLUSION

We have developed a 3-D second-order gradiometer, which can simultaneously detect field components perpendicular to and tangential to the scalp. It was found that this gradiometer becomes a useful tool for

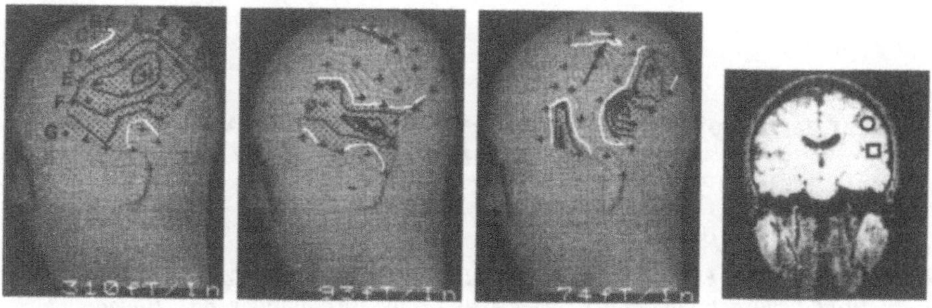

Fig. 4 Isofield contour maps consisting of mixed AEFs and SEFs. The latency of an AEF is 120 ms. The latency of a SEF for median nerve stimulation in the right wrist is 80 ms. Coronal MRI of a subject's head showing the source localization of AEFs (square) and SEFs (circle).

separating multi-sources overlapping in time, in the cortex. Two gradiometers for detecting tangential field component, in particular, work as spatial band-pass filter having the characteristic of the second derivative tangential to the scalp. Namely, we conclude that this 3-D second-order gradiometer can provide the information of the constraint conditions by visual inspection for calculating the inverse problem with multi-sources, which is difficult in so far as we have used the MEG data of magnetic field perpendicular to the scalp.

REFERENCES

[1]Smith, W. E., Estimation of the Spatio-Temporal Correlations of Biological Electrical Sources from Their Magnetic Fields, IEEE Trans. Biomed. Engng., 1992, 39 : 997-1004.

[2]Baumgartner, C., Sutherling, W. W., Shi Di and Barth, D.S., Spatiotemporal modeling of cerebral evoked magnetic fields to median nerve stimulation, Electroencephalography and clinical Neurophysiology, 1991, 79 : 27-35.

[3]Achim, A., Richer, F. and Saint-Hilaire, J.M., Methodological considerations for the evaluation of spatio-temporal source models, Electroencephalography and clinical Neurophysiology, 1991, 79 : 227-240.

[4]Achim, A., and Richer, F., Methods for Separating Temporally Overlapping Sources of Neuroelectric Data, Brain Topography, 1988, 1 : 22-28.

[5]Okada, Y., Discrimination of localized and distributed current dipole sources and localized single and multiple sources, Biomagnetism:Applications and Theory, ed. by H. Weinberg, G. Strolnk, T. Katila, 1985 : 266-272.

[6]Uchikawa, Y., Matsumura, F., Kobayashi, K. and Kotani, M., Application to Discrimination of Multi-Sources to Somatosensory Evoked field using 3-Dimensional second-order gradiometer, In : Deecke, L., Baumgartner, C., Stroink, G. and Williamson, S.J (eds.), Biomagnetism :Fundamental research and clinical application, 1993 : 219-220.

[7]Ahonen, A., Hamalainen, M., Kajola, M., Knuutila, J., Lounasmaa, O., Simola, J., Tesche, C., and Vilkman, V., Multichannel SQUID systems for brain research, IEEE Trans. Magn., 1991, 27 : 2786-2792.

[8]Cohen, D., Proc. 12th Intern. Conf. on Med. and Bio. Eng., Jerusalem, Israel, 1979.

ACKNOWLEDGMENT

This study has been supported by the grant from the Center for Research, Tokyo Denki University.

Measurements of Visual Evoked MEG Responses Associated with Color Discrimination

Ueno, K.[1], Ueno, S.[1] and Weinberg, H.[2]

[1]*Institute of Medical Electronics, Faculty of Medicine, University of Tokyo, Tokyo, Japan;*
[2]*Brain Behaviour Laboratory, Simon Fraser University, Burnaby, B.C., Canada*

Introduction

Recent rapid progress of SQUID techniques made it possible to measure MEG signals over the whole cortex of the brain simultaneously. It is extremely important for evoked MEG research in which the signals have to be averaged. With auditory, visual and somatosensory stimulation, the electrical activities evoked by stimuli can be recorded as a response of the brain. Recently, investigations on higher brain function such as memory and cognition have been progressing. For example, P300 has been introduced as a neuro-electro-magnetic phenomenon associated with the higher brain function. The P300 is an electric response of the brain to target stimuli around 300msec after the onset of stimulation. Investigations of P300 have mainly focused on auditory evoked MEG[1]-[4]. We studied the P300 evoked by visual stimulation using a discrimination task consisting of two different colours and estimated the source of the P300 response using a 2-dipole model.

Method

The experiments were carried out on 3 healthy adults (A: male, 32 years, B: male, 32 years, C: female, 28 years). The visual evoked MEG was recorded using a multi-channel 3rd-order gradiometer system (CTF Systems Inc.). The paradigm to measure the P300 component was an oddball paradigm. Visual stimuli were displayed in the center of a monitor screen. The distance from the subject's eyes to the center of the monitor screen was 3.64m. The stimulus was a rectangle (27.5x18.5cm) filled with a single colour. The colours used as stimulation were red and blue.

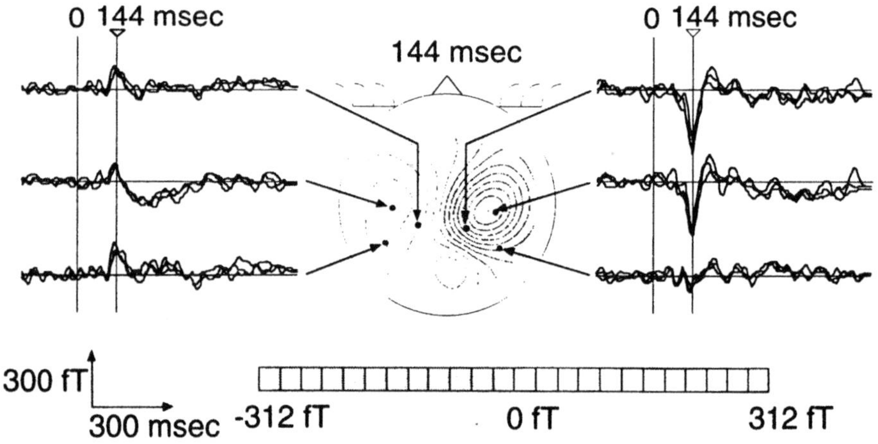

Fig.1 Signal traces and a topography of visual evoked MEG in control experiments for subject A. Each trace is the average of 100 response to repeated stimuli (red).

Each stimulus was randomly presented for 1 sec at irregular intervals of 1 to 2 sec. Low frequency colour, red, was the target. It was presented 100 times at a frequency of 25%. The subject's task was to count the number of targets. For comparison, we also stimulated 300 times using a single colour red. The data were collected at a digitizing rate of 250Hz and averaged after being filtered through a 30Hz lowpass filter. In the P300 experiment, we averaged only the data associated with target colour.

Results and Discussion

Fig. 1 shows the visual evoked MEGs with repeated single colours as the stimulus for subject A. Each trace is the average of 100 responses. Three waves were very similar and had clear peaks at latencies of about 150 and 200 msec. The responses are associated with the visual processing of colour in the brain. The sources at a latency of about 150 msec were estimated with a 2-dipole model for all three subjects. The results were shown in Fig.2. The two dipole were located in the occipital areas of each of the right and left hemisphere, corresponding to the visual cortex. The dipole moments were between about 15 to 45 nAm.

Fig. 3 shows the superposition of two MEG waveforms for subject C. One is the response to red stimulus in the control experiment, the other is the response to red, the target stimulus, in the P300 experiment. A clear difference could be found at the latency from 250msec where there is a reversal between posterior and anterior. and also between right and left. The peak of the difference was at a latency of about 400 msec. The difference appeared to be a P300 response.

Fig. 4 shows the topographies of the differences at a latency of 396 msec for all three subjects. The topographies have similar patterns which are reversed between anterior and posterior, and also between right and left hemispheres, so it is possible to assume the sources of these topographies are also the same. The sources with a 2-dipole model was estimated as shown in Fig. 5. The estimated sources for all three subjects was located near the center of the

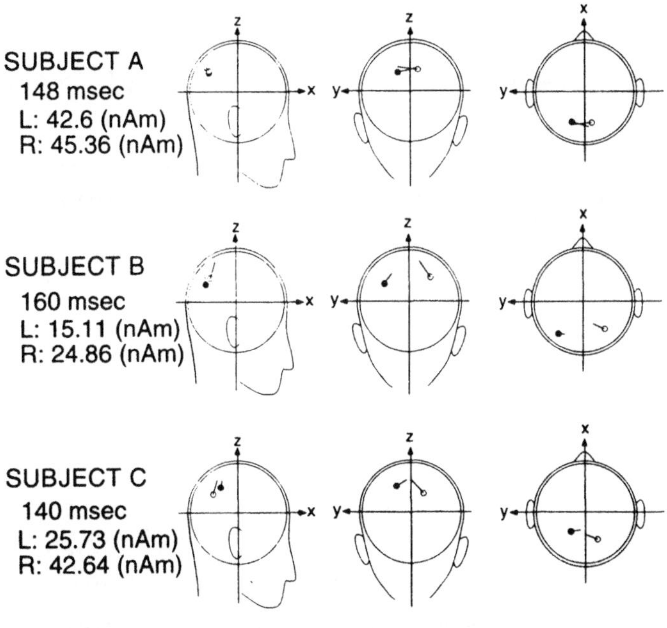

Fig.2 The estimated sources of the visual evoked MEG at latency about 150 msec in control experiment for three subjects.

head model. All the P300 responses recorded in these three subjects can be produced by similar sources in the brain. However, each dipole has a much larger moment than the estimated sources on control wave at about the latency of 150 msec in Fig.2. This means that the dipole model cannot be used in source estimation. Because a dipole doesn't have any spread, its moment should be adequately small. The sources of the P300 component might be a distributed system or might be composed of several partial sources. It is necessary to estimate the sources by taking into account anatomical and functional brain structures.

Acknowledgements

This research was supported in part by the Grants from the Ministry of Education, Science and Culture in Japan, and the Collaborative Research And Development Grant of the Natural Sciences and Engineering Council of Canada.

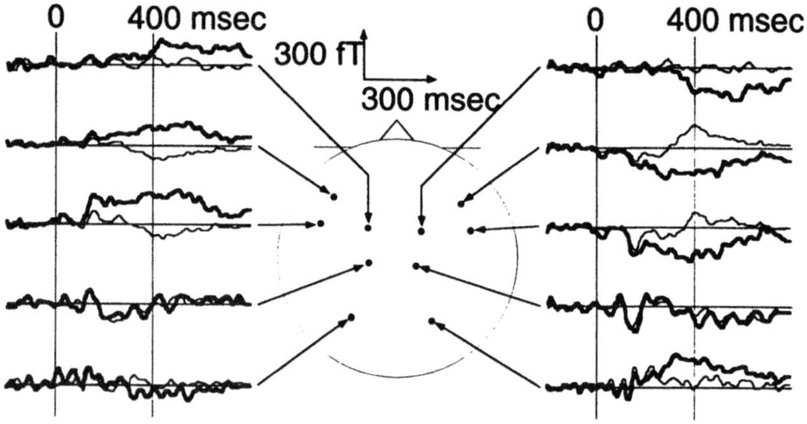

Fig.3 The superposition of two MEG waveforms for subject C. The bold trace is the response to red stimulus in the control experiment, the thin trace is the response to red, the target stimulus, in the P300 experiment.

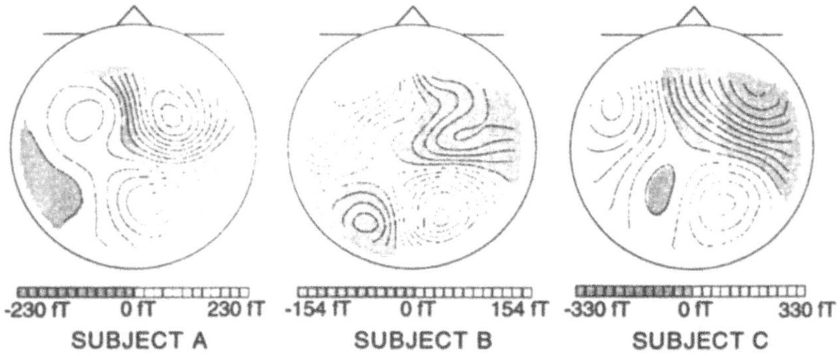

Fig.4 The topographies of the differences at a latency of 396 msec for three subjects. The topographies have similar patterns which are reversed between anterior and posterior, and also between right and left hemispheres.

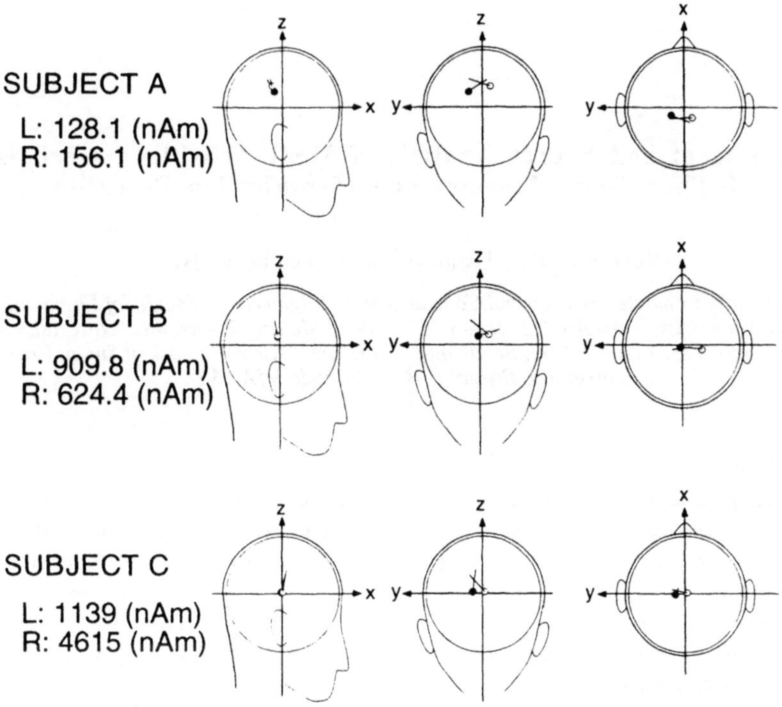

Fig.5. The estimated sources of the P300 component at latency of 396 msec for three subjects.

References

[1] E. Gordon et al., Magnetoencephalography: locating the source of P300 via magnetic field recording., Clinical & Experimental Neurology. 23:101-10. 1987.

[2] E. Gordon, C. Rennie and L. Collins. Magnetoencephalography and late component ERPs., Clinical & Experimental Neurology. 27:113-20. 1990.

[3] R. L. Rogers, A. C. Papanicolaou, S. B. Baumann and H. M. Eisenberg. Recording of event-related potentials (P300) from human cortex., Electroencephalography & Clinical Neurophysiology. 83(2):146-52, 1992 Aug.

[4] R. Neshige and H. Luders. Recording of event-related potentials (P300) from human cortex. , Journal of Clinical Neurophysiology. 9(2):294-8. 1992 Apr.

Measurement and Source Analysis of MEG Activities Associated with Short-Term Memory and Identification Processes

Yoshida, H.[1], Ueno, S.[2] and Weinberg, H.[3]

[1]Department of Computer Science and Communication Engineering, Faculty of Engineering, Kyushu University, Fukuoka 812, Japan; [2]Institute of Medical Electronics, University of Tokyo, Tokyo 113, Japan; [3]Brain Behaviour Laboratory, Discovery Park at Simon Fraser University, Burnaby, B.C., Canada V5A 1S6

Introduction

Sternberg's paradigm [1] has been very useful in examining the short-term memory process, and many EEG studies have been undertaken using this paradigm. Some studies reported slow wave [2] or DC potential shifts [3] in memory tasks. We thought that MEG was more applicable than EEG in measuring low frequeny signals. In this study, subjects memorized the order of four tones and compared it with that of another set of four tones. They had 1.6 seconds to memorize the order of the first set of tones, 1.4 sec as an inter-stimulus interval, and 1.6 sec to compare the order of the last set of tones. Extremely slow magnetic fields were observed by using this paradigm. We estimated the current dipole sources underlying short-term memory for tones.

Methods

The whole-cortex 64 channels of MEG data were digitized with a sampling frequency of 125 Hz, and filtered using a zero phase shift, low-pass digital filter with a cut-off at 20 Hz. Fig.1 shows the sequence of events in each trial. Two kinds of tones (two octaves apart: the 1046.5 Hz tone and the 261.6 Hz tone) were used. Three right-handed male subjects (age 26, 28, 50 years old) participated. The subjects memorized an order of four tones, which was chosen randomly. These four tones were called the memory set. After 1.4 sec, the other four tones called the probe set were presented. The subjects discriminated between the order of the first set of tones from that of the last set of tones. When another long (1 sec) tone was presented 2.4 sec later, the subjects were to push either a true or a false button. Pushing the true button meant that the order of the first four tones was the same as that of the last in the forward order. Pushing the false button meant that the order was not the same. In a control task, subjects had to hear the memory and probe set without treating. Data were averaged twenty times. The power spectra was calculated using the 2.56 sec data from the start of the trial. Additionally, it were also calculated using the data from 3.0 sec to 5.56 sec. In the estimation of the equivalent current dipolar sources, the head was assumed to be a homogeneous spherical volume conductor with a radius of 100mm and the Grynszpan-Geselowitz equation [4] was solved.

Results

Fig.2 shows the MEG recordings at C4 of the 10-20 international system. In comparison with the control magnetic fields (a), the magnetic fields during the task (b) clearly included very low frequency components. Fig.3 shows the power spectra at C4 of the 10-20 international system. Fig.3(a) is the control power spectra. Fig.3(b) is the power spectra during the memory task, and Fig.3(c) is the power spectra during the identification task. The intensity of the control power spectral components at 0.5 Hz was less than 15 fT(rms)/√Hz. However, the intensity of the power spectral component at 0.5 Hz during the memory task was 25 fT(rms)/√Hz, and that during the identification task was 45 fT(rms)/√Hz. Because the memory processes affected the very low frequency components of the MEG, the current dipole source models were made from the data filtered from below 0.5 Hz. Fig.4 shows the MEG topographies for three subjects. The control topographies showed very weak and complicated patterns. However, all the topographies during the task showed one dipole pattern in the frontal region of the head. The polarity of the MEG topographies was opposite for the time the subjects were listening to the tones (a), (c) and afterward (b), (d). Fig.5 shows the two-dipole models. Fig.5(a),(b) are the two-dipole models during the memory task, and Fig.5(c),(d) are the two-dipole models during the identification task. Three pairs of two-dipole sources were superimposed. The two-dipole sources were very close and their directions were contrary to each other. Compared with the source models during the memory task, the locations of the two-dipole sources during the identification task were slightly anterior to the head. We suggested that the location of the estimated sources should be in the cingulate

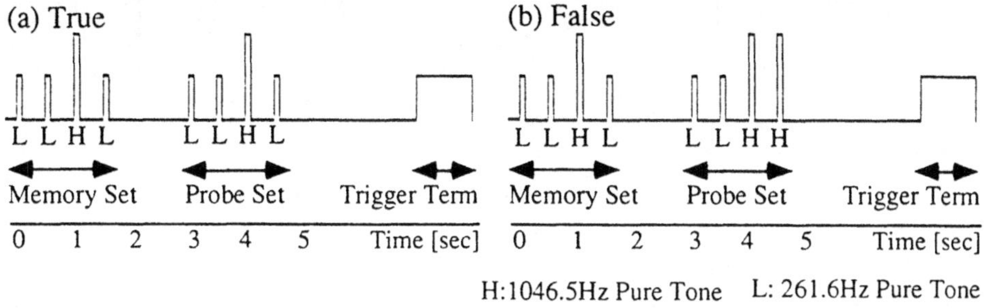

Fig.1 The time chart of the short-term memory and identification experiment.

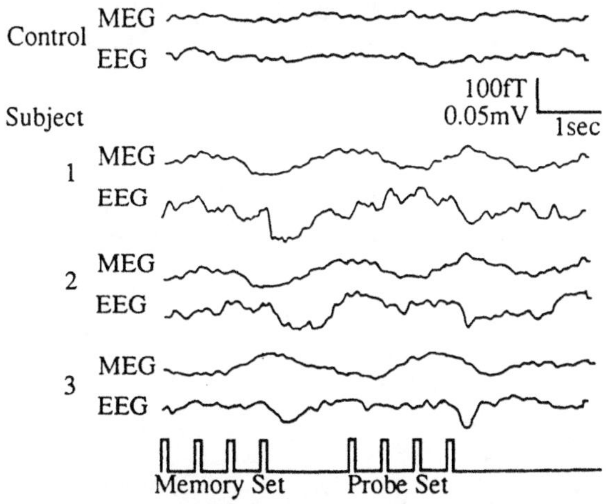

Fig.2 The recordings of MEG and EEG at C4 of the 10-20 international system (0-5.0Hz).

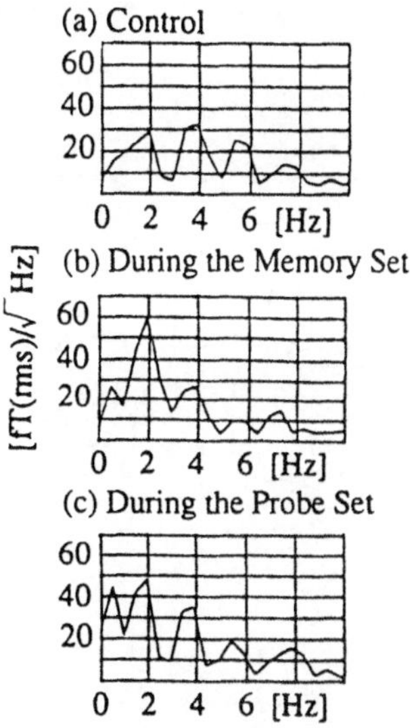

Fig.3 The power spectra at C4 of MEG data of the subject 1.

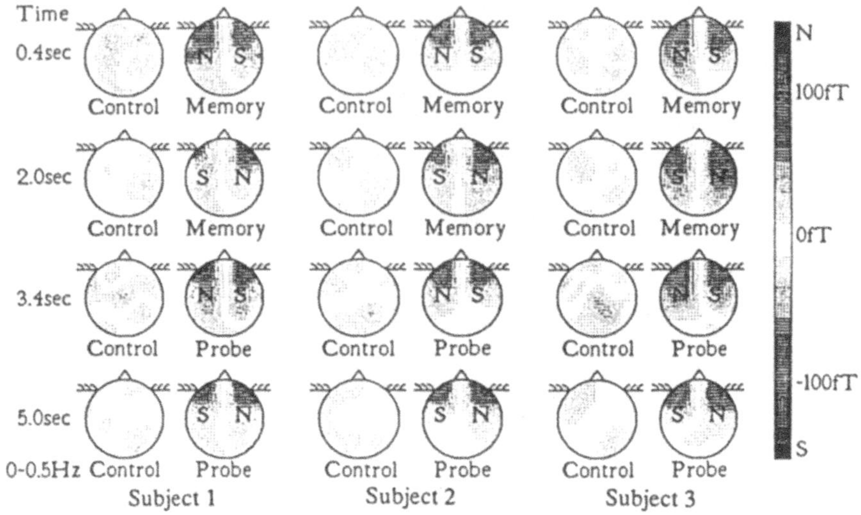

Fig.4 The MEG topographies (0-0.5 Hz) for three subjects.

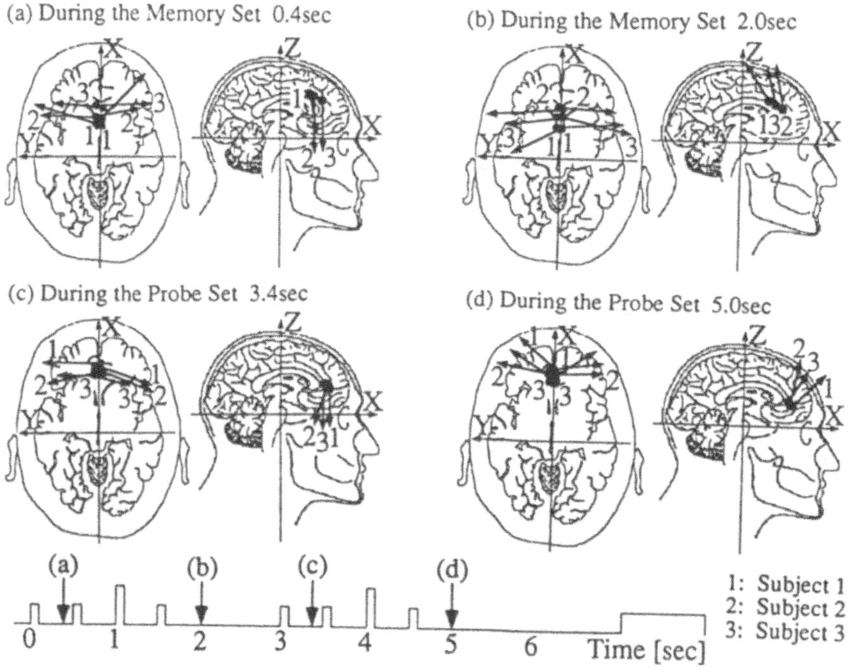

Fig.5 The two-dipole source models (0-0.5 Hz) for three subjects.

gyrus. Because many cases have been reported in which illness of the cingulate gyrus caused amnesia, we thought that the estimated locations as proper[5].

Discussion

The Sternberg paradigm employed many kinds of tones, and it was difficult to discriminate a certain tone from others. That was why we employed only two kinds of tones. The correction ratio was increased (more than 95%) in every trial. Information was given to subjects in a serial format, and the analysis term was prolonged. As a result, we could observe extremely slow magnetic fields.

We also tried using a single dipole estimation before estimating with two dipoles. However, the estimated location of the single dipole was in the longitudinal cerebral fissure. It wasn't proper. Based on the results, we supposed that there were two current dipole sources in the brain.

The polarity of the MEG topographies was opposite for the time during which the subject was listening to the tones and the time afterward. We suggested that the mass current in the neurons turned over because of the feedback during both the memorizing process and the identification process.

Conclusions

We investigated the influence of a short-term memory task using a whole-cortex MEG sensor. We observed that MEG activities during the task included extremely slow waves. Based on the results, we proposed a two-dipole source model. Two current dipole sources were estimated in the cingulate gyrus. They were very close and their directions were contrary to each other.

Acknowledgements

This study was supported in part by grants of the ministry of education, culture, and science on Japan, and collaborative research and development grant by natural sciences and engineering council of Canada.

References:

[1] Sternberg, S., High-speed scanning in human memory. Science, 1966, 153: 652-654.

[2] Ruchkin, D.S., Sutton, S., Kietzman, M.L. and Silver, K. Slow wave and P300 in signal detection. Electroenceph. Clin. Neurophysiol., 1980, 50: 35-47.

[3] Lang, M., Lang, W., Uhl, F., Kornhuber, A., Deecke, L. and Kornhuber, H.H. Slow negative potential shifts indicating verbal cognitive learning in a concept formation task. Hum.Neurobiol., 1987, 6: 183-190.

[4] Geselowitz, D.B. On the magnetic field generated outside an inhomogeneous volume conductor by internal sources. IEEE Trans. on Magn. MAG-6, 1970, No.2: 346-347.

[5] Iversen, S.D. Do hippocampal lesions produce amnesia in animal? Int. Rev. Neurobiol. 1976, 19: 1-49.

Oscillatory Brain Activities

Lopes da Silva, F. (Organizer)[1], Basar, E.[2], van Dijk, B.[3], Narici, L.[4], Pantev, C.[5], Ribary, U.[6], and Salenius, S.[7]

[1]Institute of Neurobiology, Graduate School Neurosciences Amsterdam, University of Amsterdam, The Netherlands; [2]Institute of Physiology, Medical University Lubeck, Germany; [3]Institute of Opthalmological Research, Graduate School Neurosciences Amsterdam, The Netherlands; [4]Dipartamento di Fisica, Università di Roma "Tor Vergata", Italy; [5]Center of Biomagnetism, Institute of Experimental Audiology, University of Münster, Germany; [6]Center for Neuromagnetism, Department of Physiology and Biophysics, New York University Medical Center, New York, USA; [7]Low Temperature Laboratory, Helsinki University of Technology, Espoo, Finland.

Introduction

The workshop was an attempt to present an overview of new lines of research with respect to the problems of how to detect, analyze and interpret signals generated by neuronal populations particularly with an oscillatory character. Indeed in the last decade there was a resurgence of interest for the recording and the analysis of oscillatory activities of the brain. This can be ascribed to three main developments; at the cellular level, the unraveling of the conditions under which neurons can display membrane oscillations even *in vitro* [18]; at the level of neuronal populations, the finding that cortical neurons can display enhanced correlated oscillations within a frequency range between 20 and 60 Hz (beta and gamma band) in response to specific stimuli [13][9] and, at the macroscopic level, the possibility of analyzing similar types of oscillations in the human brain during cognitive tasks using the new technology of Magnetic/Electric functional brain mapping. The most important issues related to these aspects that were discussed in the Workshop are summarized here.

Nature of Oscillatory Activities

Intrinsic membrane processes and synaptic circuit properties

The basic mechanisms responsible for the generation of oscillations in the neuronal networks of the brain reflect both intrinsic and synaptic, i.e. local circuit, properties, as it has been clearly shown in thalamocortical circuits [44]. Indeed in the intact brain, both synaptic interactions and input signals play an essential role in setting the conditions that are necessary for oscillations to occur. The frequency of the latter depends on the intrinsic (non-linear) properties of the neurons and on the strength and time constants of the synaptic interactions [21]. The existence of feedback circuits shapes the dynamics of the neuronal networks creating the conditions for the occurrence of different types of oscillatory behaviour. In this respect an essential property is the capacity of the neurons, within certain populations, of being active in synchrony, due to mutual synaptic interactions both excitatory and inhibitory.

Types of oscillations: linear vs. non-linear

Under certain conditions there are classes of brain oscillations that may be described as resonance phenomena of neuronal networks that behave in a linear regimen submitted to relatively weak (random) inputs. Nevertheless, in general, these networks exhibit nonlinear behaviour. This behaviour becomes evident when spectral analysis of the corresponding field potentials reveals the presence of significant relationships between harmonic components. This has been shown for some alpha rhythms of the visual cortex and beta/gamma activities of the olfactory bulb and cortex (review in [22]). The nonlinear character of the dynamic behaviour of, for example, the neuronal networks of the olfactory cortex is also evident from a number of physiological changes that occur during processing of olfactory information. One of these changes is that the median frequency of the gamma activity increases during odour sampling [11], while the subharmonic component is strongly enhanced [6]. Particularly interesting is the fact that this increase in subharmonics, revealing strong nonlinear dynamics, is most conspicuous when the odour being sampled has acquired a special meaning through conditioning (Fig 1). This highlights the fact that neuronal networks exhibit nonlinear dynamic behaviour while analyzing significant information.

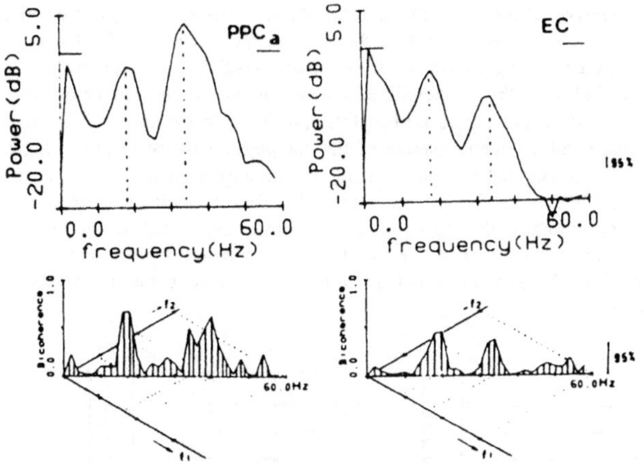

Fig. 1: Evidence for non-linear spectral properties of EEG signals recorded from the depth of two areas of the temporal cortex of the cat: PPCa = anterior prepyriform cortex, and EC = entorhinal cortex. Above: power spectra of signals recorded simultaneously from the two cortical areas. Below: the corresponding bi-coherence that shows the phase relationship between different frequency components of each signal. Note the existence of two main frequency peaks around 36 and 18 Hz, with the former dominating in the PPCa and the latter in the EC. In addition the bi-coherence shows two main peaks at the intersection between 18 and 18 Hz, indicating a harmonic relation betweenl8 Hz and 18+18 Hz, and another around the intersection 36 Hz and 36 Hz indicating a harmonic relation between 36 and 36+36 Hz, that cannot be distinguished in the power spectra.

Transfer Between Synaptic Activity and Neuronal Firing in Neuronal Populations (Correlation Images)

To estimate the functional connectivity of the brain, in vivo, requires recording of the activity at many sites in the cortex simultaneously and a model of the signal transfer. The first requirement has to be met because bivariate measurements of correlations suffer from the well known problems of a common source resulting in an overestimate of coherence, or of a missing source resulting in an underestimate of coherence. The second requirement is purely mathematical.

The data set used [43] consisted of pairs of radial cortical activity profiles. We recorded the low frequency (3-200Hz) content of the intra-cortical EEG at 14 sites at varying depths with respect to the visual cortical surface simultaneously. We called this profile the *synaptic activity profile.* To do this we applied spatial filtering with a 3 point 2nd-order derivative along the array to ensure that our recordings were from a restricted area. The other profile, called the *spiking activity profile,* was the envelope of the high frequency (1000-SOOOHz) content of the intra-cortical EEG at 16 sites at varying depths. These two profiles were recorded simultaneously from area 18 of the lightly anaesthetized cat. We used a linear non-parametric predictive model, and tested wether we could predict these profiles from each other. In these experiments we observed that a linear model quite accurately predicted even the individual features in each recording from the simultaneously recorded second data set. This held true in general. We found multiple coherencies in the range 0.7 to 0.9 using this straightforward linear model. The prediction of the synaptic activity profile from the spiking activity profile was a little more accurate than the reverse prediction. An obvious explanation would be that every spike results in some form of synaptic activity, while not every synaptic activity results in spikes.

The compound (multi-in/multi-out) transfer function of the complete predictive filter describes how the signals from one of the profiles influence the signals of the other profile. We therefore used these compound transfer functions to estimate the functional connectivity within the cortical column from which we recorded. This estimate is probably not precise, given that the cortex is certainly not a linear system, and given that we probably undersampled spatially. However, we wanted to test the *hypothesis* that *the functional connectivity in the visual cortex changes upon visual stimulation.* Thus we do not need a precise estimate as long as the estimate is consistent.

Figure 2 shows data from an experiment where we used three different stimuli: a homogeneous nonmodulated grey TV screen (no-stimulus), a homogeneous flashing TV screen (flash stimuli), and a sinusoidally moving rectangular black bar on a white TV screen (moving bar stimulus) in random alternation. In this figure we depict the impulse response functions of the filter model that predicts the synaptic activity profile from the spiking activity profile. Specifically in each panel we show the synaptic activity profile that would result if there would be a unit pulse change in the spiking activities at the electrodes nrs.8 through 11, corresponding to the input layers of the cortex (bottom half of layer IV). The predicted synaptic activity at the surface of the cortex is depicted at the top (electrode nr. 1) each subsequent electrode lies 150 mm deeper in the cortex. From the figure we see that both the response wave shape and the spatial pattern of the compound impulse response function, depend on the type of stimulus used. We also see that for each stimulus the impulse response functions show oscillatory components, typically in the 50-70Hz band, which are most outspoken for the moving bar stimulus. These oscillations can not be due to an oscillatory common driving signal, since these would be nulled out by the analysis method used (any signal that is common to more than 2 electrode sites is removed by the non-parametric predictive model). The conclusion drawn from this experiment is that indeed the functional connectivity of the visual cortex changes upon visual stimulation.

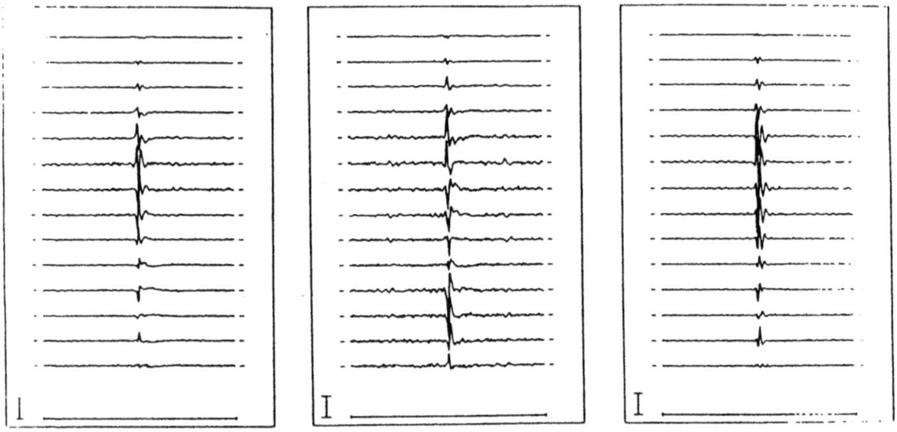

Fig. 2: Compound impulse response functions as calculated for three different stimulus conditions: Non-stimulated (left column), Flash-stimuli (middle column), and Moving-bar-stimulus (right column). The traces within each column depict the predicted synaptic activities that would result at the intra-cortical electrode array, if the spiking activity would be a unit pulse of I sample duration (1/1024 s) at electrodes close to the bottom of layer IVC and identically zero at all other electrode sites. The pulsing electrodes are marked at the right side of each column. The traces at the top of each column represent activity from superficial electrodes, the traces at the bottom from deep electrodes. The horizontal scale bars correspond to 1024ms, the vertical scale bar corresponds to 50 uV.

To better understand the interplay between changes in functional cortical connectivity and changes in the dynamics of cortical activity, we studied much simpler "toy systems of artificial neurons, 50% excitatory and 50% inhibitory, with spiking dynamics and synaptic plasticity resembling long term potentiation (LTP) and long term depression (LTD), that have been characterized in different brain areas notably in the hippocampus. A full account of this simulation studies can be found in Kalitzin and van Dijk [15].

For such a simple network, we can calculate the dependence of the average connectivity strength on the average activity strength from the plasticity dynamics, and we can also calculate the dependence of the average activity strength on the average connectivity strength from the spiking dynamics. Where these two relations cross the network is in equilibrium. We found that there is a parameter range where the system has multiple equilibrium states of which two (or more) can be stable. This implies that when the system is started at a random initial connectivity, the system goes to one of the stable equilibrium states. For instance the network could go to either a state of low average connectivity and low average activity or to a state of high average connectivity and high average activity. If the network is at a high connectivity equilibrium the neurons of the network are more synchronized.

If we supply the network with a diffuse external driving signal (affecting either all the neurons in the network or only a randomly drawn subset of the neurons) we can force the network to go into another equilibrium state. For example when the network is in a low connectivity state, then a sufficiently strong correlated input will drive the network to the high connectivity state. When the external input is switched off the network will stay in the new state, with more strongly coherent ongoing activity either forever, or until internal noise, or a noisy input, drives it out of that equilibrium again. Which of these two evolutions occur depends on the actual parameter settings; in this context the parameters that determine the rate of the plasticity changes are the most crucial. The response of the network to an external stimulus has two components: one that is time- and phase-locked (evoked), governed by the spiking dynamics of the network, and one that is time locked but not phase locked (induced), governed by the plasticity dynamics and driven by the synchronicity of the external stimulus. Anti-synchronous input can force the network to lower connectivity equilibria, and less coherent ongoing activity. The network functions as a *synchrony enhancer,* where the synchrony of the ongoing activity reflects the synchrony of the last external stimulus, i.e. as a *memory mechanism.* We postulate a similar role for synchrony and plasticity in the sensory cortex. Our experimental data have shown that the fast *plasticity* of the functional connectivity is present in the visual cortex.

These experimental results and model studies allow establishment of the dynamic relationships between the activity at the level of single neurons and that of neuronal populations.

Spatio-temporal Distributions of Oscillations in the Brain: The Role of MEG/MFT

Among all non-invasive functional brain imaging techniques available today, MEG/EEG have the unique potential of a high time resolution, offering the possibility to analyze oscillatory activities in the human brain. MEG offers, in addition, a better accuracy in localizing generator sources within the brain. The proper choice of MEG recordings permits to select small time segments of raw data and to extract several frequency components, or rhythms, in the human brain, which are present simultaneously at deeper and higher levels [35]. In addition, the high temporal resolution of the technique further enables to monitor the three-dimensional distribution of these individual oscillations in the human brain in time steps of less than 1 msec. In our laboratory a multi-channel MEG system was used to record the magnetic activity from human subjects, the raw signals were filtered to extract the oscillatory activity at around 40Hz [19]. By analyzing the *spatio-temporal distribution of this oscillatory activity* (Fig. 3), we could demonstrate a possible correlation to some sort of human cognitive brain functions [14].

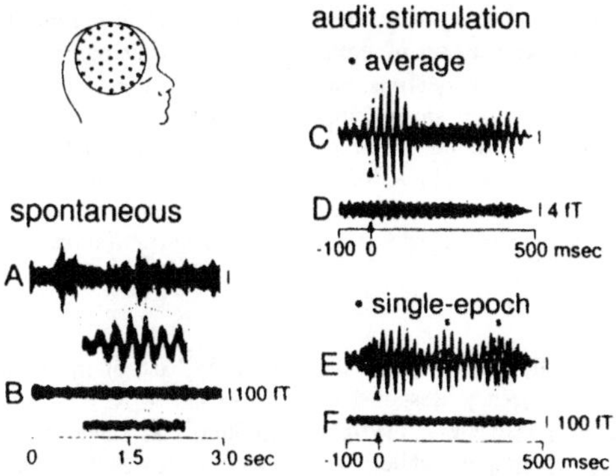

Fig. 3: Synchronized and coherent 40-Hz oscillations recorded form the right hemisphere of the human brain. A: spontaneous 40-Hz oscillations in the absence of any sensory stimuli, compared with the system noise (B). Reset of spontaneous 40-Hz activity by auditory stimulation (300 msec tones), and induced synchronized activity with respect to auditory onset, as seen in average data (C) and in single epochs (E), compared to the appropriate system noise (D,F). (modified from Llinas and Ribary 1993 [19]).

Earlier findings indicate that the oscillatory activity at around 40-Hz in humans is continuously generated by the central nervous system, and is altered during different states of consciousness [19]. We demonstrated large spontaneous 40-Hz coherent magnetic activity in the awake and in REM sleep states but very reduced such activity during delta sleep. Because spontaneous 40-Hz oscillation was seen in wakefulness and in dreaming, we proposed it as a correlate of cognition. In addition, spontaneous 40-Hz oscillation has been shown to be reset or modified by sensory stimuli in the awake state but this was not observed during REM or delta sleep [19]. These data indicate that the 40Hz oscillations could be involved in integrating outside sensory information with the ongoing cognitive activity of the individual, probably resulting from coherent 40-Hz resonance between thalamocortical specific and nonspecific loops [19].

Using Magnetic Field Tomography (MFT) on humans [35], we earlier demonstrated that such 40-Hz coherence displayed well defined cortico-subcortical correlations with a time shift that is consistent with thalamo-cortical conduction times. In particular, the MFT results indicated that the onset of activity at the thalamic level was followed by widespread activation of the thalamo-cortical system which resulted in large coherent thalamo-cottical 40-Hz oscillations organized in space and time, which was altered in pathological states such as Alzheimer's disease [34][35]. These studies provided, in conjunction with intracellular recordings from cortical interneurons [20] and thalamic neurons [45], further support of the hypothesis that recurrent thalamo-cortico-thalamic activity is involved in organizing and supporting coherent 40-Hz oscillations as a major synchronized event during sensory or cognitive processing [19]. As such, consciousness may arise by the resonant 40-Hz coactivation of at least two systems, which would temporarily conjoin cerebral cortical sites specifically activated at or around 40-Hz frequency. In this manner the specific system would provide the content, and the nonspecific system the temporal binding of such content into a single cognitive experience [19].

Spontaneous Oscillations vs. Event-Related Activities: Phase-Locked and Time-Locked (Evoked) vs. Time-Locked but not Phase-Locked (Induced) Activities

Basic mechanisms

As indicated above two types of changes in the electrical activity of the cortex may occur upon sensory stimulation: one change is *time-locked and phase-locked (evoked)* and can be extracted from the ongoing activity by simple linear methods such as *averaging;* the other is *time-locked but not phase-locked (induced)* and can only be extracted trough some non-linear method such as *envelope detection.* Which mechanisms underlie these types of responses? The time- and phase-locked response can easily be understood in terms of the response of a stationary system to the external stimulus, the result of the existing wiring of the cortex. The second response component can not be understood in such terms. We demonstrated that the non-phase locked component can be understood as a changed ongoing activity, resulting from changes in the functional connectivity within the cortex. We based this conclusion on the results using two data sets: intra-cortical recordings from lightly anaesthetized cat visual cortical area 18, and simulated responses from artificial neural systems with well defined plasticity.

The anatomical connectivity is the pattern of "wires" that interconnect neurons, or local ensembles of neurons, such as cortical areas or laminas, i.e. a spatial and directed set of possible signal transfer pathways; the functional connectivity reflects the ways in which the activity of neurons, or of local ensembles of neurons, influences the activity of other neurons or local ensembles, i.e. a spatio-temporal directed set of actual signal transfer. In artificial neural systems the anatomical connectivity would be the general layout of the network, the functional connectivity would be the complete set of synaptic weights. Both the anatomical and the functional connectivity of the cortex change continuously, the changes in functional connectivity occur probably faster, and as explained above (section 3) can account for the emergence of *time-locked but not phase-locked oscillations,* i.e. of *induced activity,* elicited by a sensory event.

Analysis methodologies: FRP

The reactivity of underlying brain rhythmicity to stimulation in the same frequency band of the rhythms themselves, was studied using the recently introduced Frequency Responsiveness Procedure (FRP) [23]. This procedure allows to discriminate on the sole basis of responsiveness, the brain activities that share the same frequency band and that often feature a widely overlapping topography, while it significantly increases the inter- and intra-individual reproducibility with respect to simple spontaneous activity measurements. The different ways rhythmical activities are affected by simple sensory stimulations can be compared to results of recent cognitive investigations that suggest similar discriminations [16][17]. The combined information relative to the same activity (cognitive correlates plus sensory responsiveness) may improve our ability of building functional models. Moreover FRP can be used as a simple rhythm classifier. In consideration of the good reproducibility of FRP responses, and of the quite short measurement time (<10 min), it may also be used to build functional/pathological indicators.

Alzheimer's disease, for instance, has been recently described to produce effects on the frequency responsiveness of the cortex [32]. The first FRP tests, performed with the 28 channel neuromagnetometer (in use at CNR-ESS - Rome) were able to discriminate activities in the lower and higher alpha band and to measure stimulation-rate-dependent alpha depression as well as a slowing down of the alpha rhythm following rhythmical stimulation [23]. Further experiments, performed with the Neuromag 122 channel neuromagnetometer (LTL laboratory - Helsinki University of Technology) permitted to thoroughly investigate the interaction between sensory stimulation and cortical spontaneous activities [24].

The FRP stimulation consists in delivering intermittent bursts of stimuli at a given rate intermingled by a pause of constant length. The entire stimulation epoch (burst plus pause) is repeated several times for each rate, in order to perform off-line averaging of the response spectra. The rates are scanned within the extended alpha band (from 6.0 Hz to 14.0 Hz at step of 0.5 Hz). Preliminary studies set the number of repetitions to 10 (but probably 5 could have been enough to achieve comparable results) to rule out possible influence of the scanning order of the stimulation. Two of the FRP findings were described in more detail: (i) the co-existence of two rhythms within the alpha band with strongly overlapping topography but different responsiveness to rhythmical stimuli; and (ii) the induction of more "Organized States" following FRP stimulation.

Co-existence of activities in the alpha band (Fig. 4):

The magnetic field associated with the bioelectrical activity evoked by the stimulation was measured with the 28 channel superconductive system in use at the CNR-ESS laboratory in Rome [10]. The data were acquired at a rate of 1.0 kHz (recording bandwidth 0.48 Hz - 250 Hz). The visual stimuli were short light pulses delivered by two red LEDS, mounted on a blackened scuba-mask (about 5 cm from the eyes, at an eccentricity of about 5 degrees) in the left hem)field. The subject was asked to maintain fixation on two green LEDS, constantly 'on', positioned in the mask with zero eccentricity.

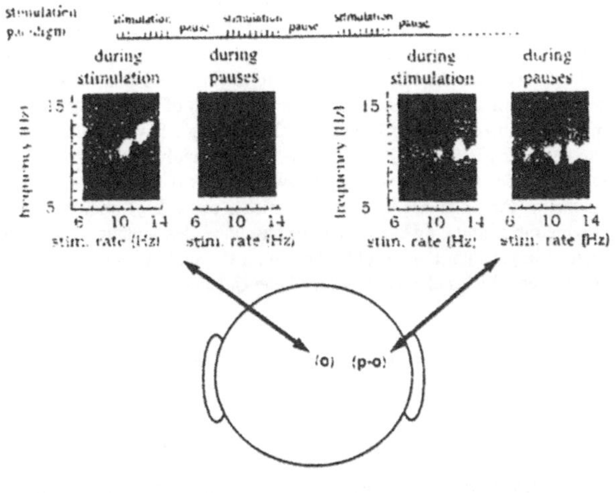

Fig. 4: Different activities are present in the same frequency band (10-12 Hz) with overlapping topography and can be discriminated by the FRP procedure. Measurements following left hemifield visual stimulation (led short flashes) from two sites (out of 28 neuromagnetic sensors) are shown. Each image is a collection of 17 vertical spectra, each relative to the responses to rhythmic stimulation at rates ranging from 6.0 Hz to 14.0 Hz, step 0.5 Hz. Spectral responses during the stimulation and in the pause immediately following it are shown. The spectral amplitude is gray coded, white being maximal. One activity, peaking at '10 Hz, present mostly in the parieto-occipital area (p-o), is not influenced by the stimulation train, even when "resonant" (i.e. at 10 Hz). This is also the most evident activity seen in spontaneous measurements. Another activity peaking at '10 Hz and .12 Hz, stronger in the occipital region (o), is enhanced by the stimulation and is neither present in the pauses nor in the spontaneous measurements (not shown). Further correlation studies confirm that the former is not locked by the stimulating train while the latter is. For further details see the original work [23].

The FRP investigation managed to discriminate two activities within the 9 Hz-12 Hz band featuring a strongly overlapped topography, on the sole basis of their responsiveness. One activity, mostly on the low end of the band (<10 Hz) was quite undisturbed by the stimuli: it was present indifferently during the stimulating burst and during the pauses, and did not show improvements in the correlation across channels due to the stimulating bursts, even when these were delivered at 10 Hz (the peak of this on-going, or spontaneous, activity). The second one was, instead, locked to the stimululi as shown by a high amplitude measured during the burst while this activity decreased during the pauses, as well as by the sign)ficant improvement of cross-channels ccrrelation during bursts.

This second activity featured a wider spectrum covering the band 10 Hz - 12 Hz, however peaking at about 12 Hz. The phase-locked activity was localized slightly more occipital than the latter that was stronger in the parieto-occipital area.

These findings support the hypothesis of a lower alpha band reflecting "attentional" processes, constantly "on" during the test, not affected by the stimulation, as opposed to a higher alpha band reflecting stimulus related (cognitive) processes [16][17]. An innovative, non-linear, analysis technique, able to extract the time behaviour of oscillating activities buried in noise, permitted also to confirm this discrimination on the basis of the different dynamics of the two types of oscillatory activities with respect to the stimulating bursts [12].

Induction of organized states

Preliminary analysis from measurements performed with the 122 channel Neuromag whole-head neuromagnetometer (LTL/ Helsinki University of Technology) shows the induction of activity depression following FRP stimulating protocol [24]. In this study FRP was performed using the stimulation of three sensory modalities (for a total of 21 measurement sessions): somatosensory, visual and auditory. The data were acquired at 0.5 kHz with a passband 0.03 Hz - 160 Hz. Somatosensory stimuli consisted of short current pulses delivered at the median nene of the left wrist, with an intensity aufficient to elicit a painless thumb twitch. Visual stimuli, 30 ms long, were delivered by a matrix of 16 red leds (4 x 4 array, 10 mm x 10 mm) at an eccentricity of about 7 degrees in the left hem)field. Auditory stimuli were noise bursts 20 ms long, generated with a 44 KHz sampling rate.

Regardless of stimulating modality and rate, a depressive effect on the cortical activity outside the major activity peaks, in the band between 5 and 25 Hz, was obsened following the bursts of stimuli. The same effect was not measured when stimulating with a single stimulus. The depression of MEG activities is a whole cortex effect (appearing as a sort of more "Organized" State of the cortex: OS), lasting a few tens of seconds. A number of stimuli larger than three appears to be needed to induce it. The intra- and inter-individual reproducibility of the induction of OS is remarkable. The OS induction can be described by a new indicator: the "Global Activity Factor" (GAF). GAF is calculated for each channel as the difference between percentages of "active" frequency bins measured during the pauses immediately following the stimulating bursts, and the baseline (spontaneous) condition. A channel is defined "active" when its amplitude is larger than a threshold (choosing a threshold between 2 and 10 fT produce the same results). GAF grand average over subjects clearly shows the independence of OS induction from stimulation modality and cortical region.

The depression of activities over all cortical sites strongly suggests thalamic involvement in inducing the OS. OS could be described as an "alert" short-term cortical condition following trains of stimuli. The FRP appears a useful discriminating tool to investigate overlapping (in time-frequency and space) rhythmical activity and to study their functional correlates.

Brain "Resonance" Responses

The group of Basar has carried out a long series of experiments both in cat and human, where the relationships between evoked and spontaneous activities, particularly of the auditory and visual systems, were analyzed in detail [3]. Evoked potentials were studied in the frequency domain by means of the Fourier transform and adaptive digital filtering was applied. They found resonances mainly at 5, 10, 20, and 40Hz frequency bands, depending on the sites of measurement and the stimulus modality [1]. They concluded that evoked activities have components that are superpositions of induced rhythms caused by resonance phenomena in neuronal networks. They used the expression "gamma responses" for the 40Hz potentials that were shown to be widely distributed in the cat and human brain, not only upon sensory stimulation, but also after cognitive stimulation [3][21]. Brain gamma oscillations appear to have a broad spectrum of correlates with functional states, since they are widespread in a large variety of species (from human to snails) and are observed in multiple brain areas. Therefore one can speak of distributed gamma systems. Altough the research related to rhythmic activities in the beta/gamma frequency band have been very popular in recent years, one should also realize that other induced rhythms, particularly in the alpha frequency range, manifest functional correlates, namely with primary sensory and cross-modality processing [2].

Psychophysiological Correlates: Evoked and Induced Activities

The finding that after transient sensory stimulation two types of activity in gamma-band range can be differentiated: *evoked (time and phase-locked) and induced (time but not phase-locked) is* illustrated in Fig. 5. The evoked gamma-band activity is stimulus time- and phase-locked and it can be extracted from the noise in time-domain using the averaging procedure. In contrast, the induced gamma-band activity is not phase-locked to the stimulus or it strongly jitters, and thus it can not be extracted in the time domain.

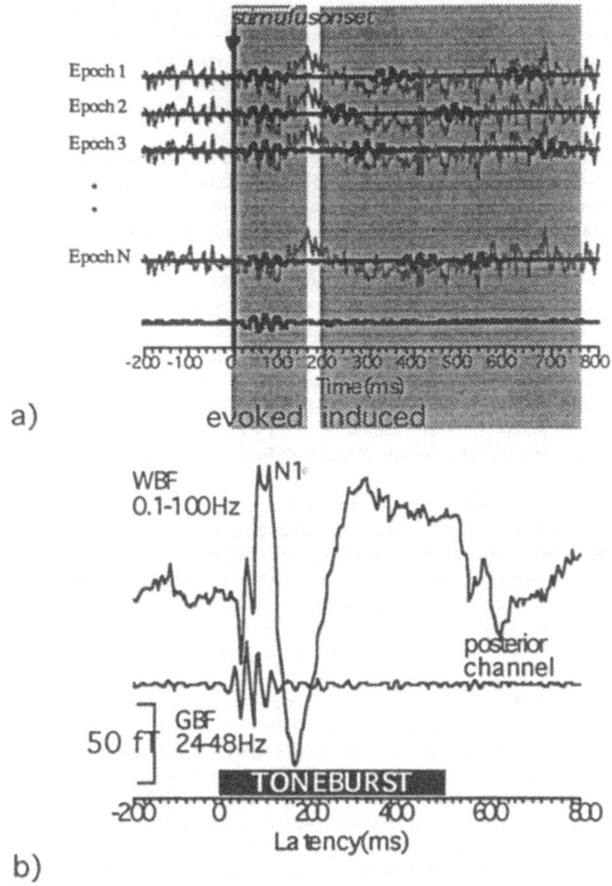

Fig. 5: (a) Schematic representation of evoked and induced gamma-band activity (black line) as superimposed on noise (grey line).
(b) Wide-band (top) and gamma-band evoked field due to tone-burst stimulation (posterior field maximum, N=384).

A transient auditory stimulus elicits not only the well known large slow wave components with the most prominent N1 peak, but also a noticeable gamma-band activity during the first 100 ms after the stimulus onset with spectral energy between 30 and 40 Hz (c.f. Fig. Sb). The equivalent source of the transient gammaband field (GBF) was estimated to be located on the supratemporal plane, few millimetres more medial and anterior as compared with the equivalent N1 source [29]. With series of functional studies we tried to increase our understanding about the transient gamma-band response as well as the function it may serve for sensory processing in the human auditory cortex. The functional behaviour of the gamma-band field was investigated with respect to changes in the stimulus parameters: 1) different kinds of auditory stimuli (in order to investigate the morphology of the gamma-band response); 2) different stimulus repetitions rates (in order to examine the recovery function of the gamma-band response); 3) different stimulus rise times (in order to differentiate between the middle latency and the gamma-band evoked responses as well as to test the gamma-band response originality and independence); 4) the dependence upon the carrier frequency of the stimulus (in order to test whether the gamma-band response represents the tonotopy of the underlying cortical equivalent sources) and 5) the dependence of the evoked gamma-band response upon attentional variations (in order to study how far the gamma-band response demonstrates an electrophysiological correlate of attention). In all these studies we compared the functional behaviour of the gamma band response in time-, frequency and "source space"-domain with the corresponding behaviour of the best investigated N1 peak of the slow evoked response.

We compared the morphology of the evoked transient GBFs elicited by different type of simuli, such as click, tone-pip and tone-bursts, that differed in duration (500 or 2000ms) or in quality (pure tone or complex tone burst). The different stimulus configurations did not produce sign)ficant changes in the gamma-band response morphology. For all different stimuli we were able to verify consistently an oscillatory response, configured by four to five waves with a global duration between 100 and 200ms.

Using tone pips with carrier frequency of lOOOHz and stimulus rates of 1/8, 1/4, 1/2 and lHz the recovery functions of the evoked GBF and the N1 component of the slow auditory evoked field (sAEF) were compared within four different subjects [27]. Both the GBF and N1 amplitudes decreased at high stimulus rates, however N1 amplitude decreased faster. The normalized rate-amplitude characteristics, i.e. the dependence of the root mean square field amplitude (RMS, calculated over the 37 recording channels) upon the stimulus rate, showed different slopes, indicating different recovery patterns. This finding was confirmed by the results in source space, demonstrating two separate groups of source locations in the supratemporal cortex with the GBF sources medial and anterior to N1. Thus, these results suggested that the GBF and the N1 have different origins and may arise from different processes in the auditory pathway.

A major problem when trying to separate the evoked GBF, is the possible involvement of the middle latency components, that are in the same frequency range. Therefore in ten subjects the waveforms of the evoked GBFs, elicited by 50 ms tone burst with short rise/decay time of 2 ms and long rise/decay time of 25 ms were compared. The first stimulus possesses the ability to elicit the middle latency reponse, due to its high degree of synchronization, but the second one does not. The GBF, when using a short stimulus, is accompanied by a characteristic time course in the initial part, that coincides with Na and Pa waves of the middle latency response. This effect could not be observed for the GBF elicited by the long stimulus. In contrast no obvious difference was observed in the final part of the GBFs, evoked by short and long stimuli.

The tonotopy is one of the general principles of functional organization of the auditory system. The depth of the N1 source of the sAEF indicates this tonotopic organization: the higher the stimulus frequency, the deeper the equivalent source. Moreover, in most subjects the depth of N1 increases linearly with the logarithm of the stimulus frequency [28][25]. Twelve subjects were investigated at 3 different stimulus frequencies: 500, 1000 and 4000 Hz (short-tone bursts of 50 ms duration, 3 ms rise/decay time, intensity of 60 dB nHL and interstimulus interval randomized between 600 and 800 ms) and EEG and MEG were recorded simultaneously [5]. The N1 scalp potential maps demonstrated the well known topographic changes with increasing stimulus frequency: an increase of the postero-lateral potential extremum was observed. In contrast, on the gamma-band scalp potential maps no clear effects were discernible. The magnetic data also showed obvious results. The estimated gamma-band and the N1 source locations, that were obtained in each single subject were averaged across subjects and presented as a function of the stimulus frequency. Significant dependence on frequency was observed only for the medial-lateral source coordinate of N1, but not for the transient GBF.

Functional Relations Between Brain Oscillations, Perception, Attention, and Other Cognitive Processes

Gamma-band MEG fields

The effects of attention and processing on the transient GBF were studied in six subjects by Pantev and Elbert [26]. In a classical odd-ball paradigm rare stimuli (low pitch of 667 Hz, probability of 33%) were randomly interspersed in a sequence of frequent stimuli (high pitch of 1500 Hz; 67% probability). Stimuli were Gaussian tone-pulses of the above mentioned frequencies with half-value time of 6 ms and an intensity of 60 dB nHL. The interstimulus interval varied randomly from 1.5 to 2.5 s. The stimuli were presented in twelve blocks, with one block having a random length between 150 and 250 stimuli. The subjects' task was either to attend to the oddball stimuli or to process a completely different task in order to distract from the oddball stimuli. During the "attend condition", the subject was instructed to concentrate on the appearance of rare stimuli and to count them. During the "distract condition", the subject had to perform a numbers-to-letter task involving mental arithmetic with considerable demand on memory. With the begin of the stimulus block, the subject started to calculate the sum of the numbers corresponding to the rank of the letters of a certain well known sentence. The RMS field values of the wide-band and gamma-band evoked responses were calculated separately for the rare and frequent stimuli of the "attend" and "distract" conditions and then averaged across subjects. A clear difference was obtained within the time-range of P300 for the "attend" and "distract" conditions, indicating that the subjects complied well with the experimental task. Higher amplitudes of Nl were obtained for the rare stimuli in both "attend" and "distract" conditions. This effect was explained by the well-known short term habituation effect in the auditory system. No substantial difference was obtained for the GBF, neither between the rare and frequent stimuli, nor between the "attend" and "distract" conditions. In addition, no sign)ficant results were observed for the "attend" and the "distract" condition in source space.

The results that we obtained in different studies of auditory evoked transient GBF, led us to conclude that the transient evoked gamma-band activity (GBF) represents an early distinctive event of the human auditory cortex. This effect reflects the synchronization of a neuronal network, involving primary auditory cortical areas and probably some part of the primary integration region, when an adequate stimulus enters the auditory system. This rather global cortical synchronization mirrors the auditory cortical perception and probably serves as a dynamical link between cortical subregions that are involved in the evaluation of the physical parameters of the auditory stimulus.

"40-Hz" coherent MEG activity

The specific "40Hz" activity was also shown to be related to auditory perception, i.e. coherent 40Hz activity appears to correlate to the temporal binding of incoming auditory stimuli [14]. Ribary's group tested, in particular, the hypothesis that the minimal interval required to identify separate auditory stimuli correlates with the time needed for the reset of the 40Hz MEG signal.

In control subjects, experimental and modeling results indicated a stimulus-interval-dependent response with a critical interval of 12-15 msec. At shorter intervals only one 40-Hz response, to the first stimulus, was observed. With longer intervals, a second 40-Hz response appeamd, which coincided with the subject's perception of a second distinct auditory stiumulus. These results indicated that oscillatory activity near 40-Hz, recorded during the first 100 msec post stimulus, represent a neurophysiological correlate to the early temporal processing of auditory stimuli underlying cognition [14]. It is important to note that this finding, the 12-15 msec interval required for perceptual separation, does not simply reflect the conduction time of the auditory system, which is in the order of 15-20 msec. Neuropsychological and psychophysical observations indicate that the brain is capable to distinguish two auditory stimuli (separated by more than 2-3 msec) and to identify two distinct stimuli (separated by more than 15-20 msec). In addition, it is well known that the auditory system is able to detect delays orders of magnitude smaller, in order to localize sound in space. Indeed, the auditory system is known for its excellent time resolution, as opposed to the visual system which has a delay that is deeply embedded in the integration time for the retinal system. Therefore, our data suggest that the 40-Hz activity does not only correlate with the primary processing of auditory information per se but with temporal processing of auditory events, which happens at a critical time period of 12-15 msec, in order to bind single elements as one cognitive event.

These findings are demonstrating one example, that we can use an MEG system, record the magnetic activity from a human brain, extract the oscillatory activity at around 40Hz, analyze its spatio-temporal distribution and demonstrate a coding-mechanism which starts at 25-30 msec post stimulus and relates to the perception or the awareness of auditory input [14], and more importantly, a coding-mechanism which is altered during cognitive impairment such as dyslexia (see in this book: Ribary et al.). This sign)ficant evidence indicates that the 40-Hz response could represent a major mechanism underlying quantum cognitive brain function. In addition, it also indicates that the MEG technology has many more potential applications than it is commonly used today, i.e., that the MEG technique can do better and be an important competitive tool to other functional brain imaging techniques.

Sensorimotor rhythms, mu and "40-Hz"

MEG easily detects the sensorimotor mu rhythm over the central sulcus in practically all human subjects. This rhythm typically consists of two distinct frequency components, one around 10- and one around 20-Hz, which may be transiently phase-locked giving rise to the classical comb-shaped mu rhythm. Detailed source analysis shows that the 10-Hz component arises mainly posterior to the central sulcus, whereas the 20Hz component arises slightly more anteriorly and medially [40]. Both components are attenuated during movement; however, after the movement, a transient increase above the pre-movement baseline, a rebound, typically occurs. The post-movement rebound for the 20-Hz component systematically occurs 300-500 ms earlier than the 10-Hz component [39]. These results, led to the hypothesis that the 20-Hz component is somatomotor whereas the 10-Hz component is mainly somatosensory, and that the components may have different functional roles. The hypothesis gains further support from a recent study showing that the 20-Hz rhythm, in contrast to the 10-Hz rhythm, displays what one could call "motorotopy": for the 20-Hz rhythm the attenuation during movement and the post-movement rebound follow the "motor homunculus", whereas for the 10-Hz rhythm the rebound centers in the somatosensory hand area regardless of which body part is moved [41]. Analysis and quantification of reactivity may be facilitated by focusing on rhythms modulated by peripheral stimulation; these reactions may be more robust and easier to replicate and quantify than the reactivity of ongoing or spontaneous rhythms possibly reflecting lower degrees of the induced rhythms. An example is provided by the transient mu rhythm enhancement occurring after median nerve stimulation (and after movements). This enhancement is suppressed by movements and sensory stimulation and the suppression is easily quantified [38]. Interestingly the enhancement is suppressed also by motor imagery, indicating involvement of primary sensorimotor cortex during imagery [42][38][36].

Although movement-related increase of rhythmic activity around 40-Hz has been reported repeatedly [8][30], most subjects do not show sign)ficant sensorimotor activity in the frequency bands above 20 Hz. Nonetheless, we have recently studied such activity in detail in a subject known from EEG recordings to show enhancement of 40-Hz activity in association with finger movements. The movement-related 40-Hz rhythm was found to increase before and during self-paced finger movements and to originate in the precentral motor cortex [37]. Furthermore, the cortical 40-Hz rhythm was found to be coherent with the rhythmicity of the electromyogram simultaneously recorded from the contracting muscle (Fig 6).

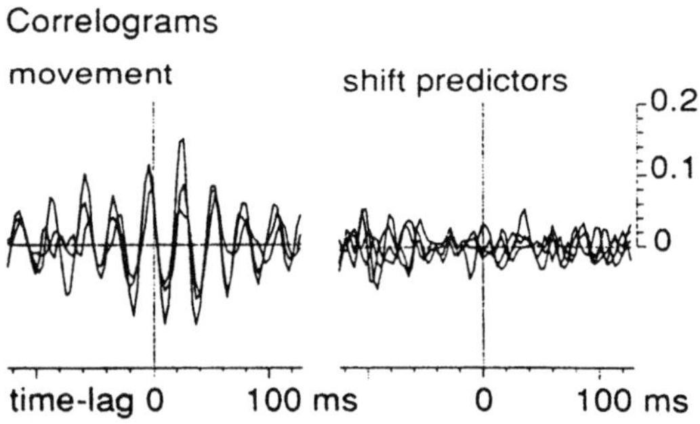

Fig. 6: Averaged cross-correlograms between rolandic MEG and extensor EMG, computed during contralateral index-finger movements, and compared with shift predictors, i.e. cross-correlograms computed using MEG and EMG signals from different movements. Adapted from Salenius et al. 1996 [37].

Conway et al. [7] were the first to report on coherence between MEG and EMG, although in association with sustained contraction and not movements. They found coherence between 15 and 30 Hz in different subjects. Based on the above findings one may speculate that the 40-Hz activity obsened in association with movement planning and execution in our subject could represent an exception, and that the functionally relevant enhancement of movement-related rhythmic activity in most subjects occurs at frequencies between 15 and 35 Hz. The coherence between MEG and EMG obviously reflects communication between the cortex and the periphery. The cortical rhythms may reflect "binding" in the motor cortex, that is, synchronization of control of different muscles, by means of a common frequency and phase, although it is still possible to argue that they are epiphenomenal [38].

The sensorimotor mu rhythm can thus be decomposed in partly independent frequency components. Further studies may confirm already suggested subdivisions of, for example, the parieto-occipital alpha rhythm, based on frequency or main source areas, with partly independent reactivity and possibly functional differences (see above 5.2). However, functionally corresponding frequency components may sometimes show sign)ficant interindividual variability, as suggested by the coherence between motor cortex and the motor units, which varies at least between 15 and 40 Hz in different subjects [7][37]. Furthermore, even though characterization of the reactivity of distinct frequency components may yield important information about the functional states of a neural network, a functional state can indeed be associated with oscillatory activity at several frequencies, or with aperiodic behavior. In this case an approach based on analysis of narrow frequency bands may yield a simplified or, in the worst case, misleading picture of the oscillatory dynamics of the network under study. Interpreting MEG results in terms of activity of underlying neuronal populations requires careful consideration of data from invasive measurements. Source analysis with multi-channel MEG enables a tight coupling of macroscopic rhythms to the underlying functional anatomy, and may thus, in combination with analysis of temporal behavior and coherence, help in elucidating the functional sign)ficance of cortical rhythms.

General Points of Discussion and "Take Home Messages"

(i) The question of which analysis methods are most suitable to extract and to quantify oscillatory activities from the MEG, was generally discussed in the Workshop. A general conclusion was that it is important to apply wide frequency band spectral analysis before any narrow-band pass filtering, in order to be able to determine whether different, mutually related, frequency components, i.e. different harmonics , contribute to the oscillatory phenomenon of interest. If this is not done one may be led to overemphasize some very specific frequency components. Of course, the latter may be prominently present such as an alpha rhythm between 10 -12Hz, or a beta/gamma rhythm of 38-42Hz, but it is always indicated to start the analysis of the correlations between MEG/EEG signals and cognitive processes using wide-band spectral analysis. After the relevant frequency bands, within a given context, are identified, the latter may be extracted for further analysis, for example, to obtain MEG/EEG brain maps of their spatio-temporal distributions.

(ii) Furthermore there is an increased awareness that neuronal networks have intrinsic nonlinear dynamic properties, that are reflected in the characteristics of the oscillations they generate, namely in their harmonic composition. Changes in the neuronal network activity may be reflected in several properties of an oscillation: - an increase/decrease in amplitude of a given oscillatory component, representing an increase/decrease in the proportion of synchronously active neurons within a population; - a change in frequency, and/or in the relative amplitude of different harmonic components, representing a change in the operating point of the underlying neuronal network that lies in its nonlinear range of operation. The latter can take place due to a change in the strength of the input to the network and/or of in that of modulating signals.

(iii) In this respect, but also in general terms, it is important to emphasize the need to make bridges between studies of oscillatory brain dynamics in experimental animals and in human subjects. An interesting research line into the sign)ficance of human MEG/EEG oscillations with respect to brain functional states, is to analyze the dynamics of these oscillations under well chosen behavioural and cognitive conditions as examplif~ed above (section 6.1 and 6.2). In this respect a particularly fruitful approach is to calculate brain maps and MFT distributions, of the changes in the MEG/EEG relevant parameters, representing transitions between dynamic states that correspond to different stages of cognitive/perceptual or motor tasks. This poses a number of technical problems since it involves analysing nonstationary processes that may have complex dynamics but this topic goes beyond the scope of this overview [31][33][46]. Nevertheless the use of *timevarying spectral analysis* offers a practical possibility for estimating spectral changes in MEG/EEG relative to well defined events.

Finally, the following "take home messages" were formulated:

- **Urs Ribary**: use temporal information of the MEG to study global brain networks, their interactions and the correlation of their activities with human brain cognitive functions.

- **Christo Pantev**: the transient auditory evoked gamma-band activity mirrors global cortical synchronization, reflects cortical perception and probably serves as a dynamic link between cortical subregions involved in the evaluation of stimulus parameters.

- **Erol Basar**: the brain's elementary oscillations, from delta to gamma frequencies, are distributed and have multiple functional correlates.

- **Stephan Salenius**: source localization, in combination with analysis of reactivity and coherence, reveals functional subdivisions of macroscopic cortical rhythms.

- **Livio Narici**: use the frequency responsivity procedure (FRP) as a rhythm discriminating method. In this way "lockable" and "unlockable" rhythms within the same frequency band can be readily distinguished.

- **Bob van Dijk**: sensory (visual) stimuli can change the functional connectivity of the cortex. These changes, in turn, lead to changes in the dynamics of ongoing activity. the latter are time-locked but not phase-locked response components.

- **Fernando Lopes da Silva**: MEG/EEG signals indicate the dynamical states of distributed neuronal networks of the brain. These networks have essential nonlinear properties. The analysis of the characteristics of the corresponding dynamical states using multi-channel recordings and both linear and nonlinear methods of quantification (e.g. time-varying (bi)spectra), can help to formulate specific hypothesis about the dynamics of functional states of brain systems with respect to cognitive functions.

References

[1] Basar, E., Basar-Eroglu, C., Rahn, E,. and Schdrmann, M. Sensory and cognitive components of bMin resonance responses. An analysis of responsiveness in human and cat brain upon visual and auditory stimulation. Acta Otolaryngol. Suppl., 1991, 491: 25-34.

[2] Basar, E., and Schurmann, M. Functional aspects of evoked alpha and theta responses in humans and cats. Occipital recordings in "cross modality" experiments. Biol. Cybern., 1994, 72: 175-183.

[3] Basar, E. and Bullock, Th. Induced rhythms in the brain, Boston, Birkhauser, 1992.

[4] Basar, E. and Schurmann, M. Alpha rhythms in the brain; functional correlates. News in Physiol. Sciences, 1996. In Press.

[5] Bertrand, O. and Pantev, C. Stimulus frequency dependence of the transient oscillatory auditory evoked responses (40 Hz) studied by electric and magnetic recordings in human. In: Pantev, C., Elbert, T. and Lutkenhoner, B. Oscillatory Event Related Brain Dynamics, New York, Plenum, 1994: 231-24.

[6] Boeijinga, P.H. and Lopes da Silva, F.H. Modulations of EEG activity in the entorhinal cortex and forebrain olfactory areas during odour sampling. Brain Res., 1989, 478: 257-268.

[7] Conway, B.A., Halliday, D.M., Farmer, S.F., Shahani, U., Maas, P., Weir, A.I., and Rosenberg, J.R. Synchronization between motor cortex and spinal motoneural pool during the performance of a maintained motor task in man. J. Physiol., 1995, 489: 917-924.

[8] DeFrance, J. and Sheer, D. Focused arousal, 40-Hz EEG, and motor programming. In: Giannitrapani, Murri The EEG of Mental Activities. (Karger, Basel) (1988) pp. 153-168.

[9] Eckhorn, R., Bauer, R., Jordan, W., Brosch, M., Kruse, W., Munk, M. Reitboeck. Coherent oscillations, a mechanism of feature linking in the visual cortex? Biol. Cybern., 1988, 60: 121-130.

[10] Foglietti, V., Del Gratta, C., Pasquarelli, A., Pizzella, V., Torrioli, G., Romani, G.L., Gallagher, W.J., Ketchen, M.B., Kleinsasser, A.W., Sandstrom, R.L. 28-channel hybrid system for neuromagnetic measurements. IEEE Trans-MAG, 1991, 27: 2959-2962.
[11] Freeman, W.J. and Viana di Prisco, G. Relation of olfactory EEG to behavior: time series analysis. Behav. Neurosci., 1986, 100: 753-763.
[12] Gaeta, R., Benzi, R., Peresson, M. and Narici, L. Singular Spectrum analysis applied to neuromagnetic signals. Biomag 1996, this volume.
[13] Gray, C.M., Koenig, P., Engel, K.A., Singer, W. Oscillatory responses in cat visual cortex exhibit intercolumnar synchronization which reflects global stimulus properties. Nature, 1988, 338: 334-337.
[14] Joliot, M., Ribary, U. and Llinas, R. Neuromagnetic coherent oscillatory activity in the vicinity of 40-Hz coexists with cognitive temporal binding in the human brain. Proc. Natl. Acad. Sci. USA, 1994, 91: 1174811751.
[15] Kalitzin, S., van Dijk, B.W., Spekreijse, H. and Van Leeuwen, W.A. Coherency and connectivity in oscillating neural networks: linear partialization analysis. Biological Cybernetics, in press.
[16] Klimesch, W., Schimke, H., Ladurner, G., and Pfurtscheller, G. J. Alpha frequency and memory performance, Psychophysiol., 1990, 4: 381-390
[17] Klimesch, W., Schimke, H. and Pfurtscheller, G. Alpha frequency cognitive load and memory performance, Brain Topography, 1993, 5: 241-251.
[18] Llinas, R.R. The intrinsic electrophysiological properties of mammalian neurons: insights into central nervous function. Science,1988, 242: 1654-1664.
[19] Llinas, R. and Ribary, U. Coherent 40-Hz oscillation characterizes dream state in humans. Proc. Natl. Acad. Sci. USA 1993, 90: 2078-2081.
[20] Llinas, R. Grace, A.A. and Yarom, Y. I vitro neurons inmammalian cortical layer 4 exhibit intrinsic activity in the 10 to 50Hz frequency range. Proc. Natl. Acad. Sci. USA, 1991, 88: 897-901.
[21] Lopes da Silva, F.H. Neural mechanisms underlying brain waves: from neural membranes to networks. Electroenceph. clin. Neurophysiol., 1991, 79: 81-93.
[22] Lopes da Silva, F.H. The generation of electric and magnetic signals of the brain by local networks. In: Greger, R. and Windhorst, U. Comprehensive Human Physiology, Berlin-Heidelberg, Springer-Verlag, 1996.
[23] Narici, L. and Peresson, M. Discrimination and study of rhythmical brain activities in the alpha band: a neuromagnetic frequency responsiveness test. Brain Res., 1995, 703: 3144.
[24] Narici, L., Portin, K., Salmelin, R. and Hari R. In preparation.
[25] Pantev, C., Bertrand, O., Eulitz, C., Verkindt, C., Hampson, S., Schuirer, G. and Elbert, T. Specific tonotopic organizations of different areas of the human auditory cortex revealed by simultaneous magnetic and electric recordings. EEG and clin. Neurophysiol., 1995, 94: 26-40.
[25] Pantev, C. and Elbert, T. The transient auditory evoked gamma-band field. In: Pantev, C., Elbert, T. and Lutkenhoner, B. Oscillatory event related brain dynamics, New York, Plenum, 1994: 219-230.
[27] Pantev, C., Elbert, T., Makeig, S., Hampson, S., Eulitz, C. and Hoke, M. Relationship of transient and steady-state auditory evoked fields. Electroenceph. clin. Neurophysiol.,1993, 88: 389-396.
[28] Pantev, C., Eulitz, C., Elbert, T. and Hoke, M. The auditory evoked sustained field: origin and frequency dependence. EEG and clin. Electrophysiology, 1994, 90: 82-90.
[29] Pantev, C., Makeig, S., Hoke, M., Galambos, R., Hampson, S. and Gallen, C. Human auditory evoked gamma-band magnetic fields. Proc Natl Acad Sci USA, 1991, 88: 8996-9000.
[30] Pfurtscheller, G. and Neuper, C. Simultaneous EEG 10 Hz desynchronization and 40 Hz synchronization during finger movements. NeuroReport, 1992, 3: 1057-1060.
[31] Pijn, J.P., van Neenen, J, Noest, A. and Lopes da Sillva, F.H. Chaos or noise in EEG signals: dependence on state and on brain site. Electroenceph. clin. Neurophysiol., 1991, 79: 371-381.
[32] Politoff, A.L., Monson, N., Hass, P., and StadLer, R. Decreased alpha bandwidth responsiveness to photic driving in Alzheimer disease. Electroencephalogr. Clin. Neurophysiol., 1992, 82: 45-52.
[33] Pritchard, W.S. and Duke, D.W. Measuring chaos in the bMin: a tutorial review of nonlinear dynamical analysis. Int. J. Neurosci., 1992, 67: 31-80.
[34] Ribary, U., Llinas. R., Kluger, A., Suk, J. and Ferris, S.H. Neurophathological dynamics of magnetic, auditory, steady-state responses in Alzheimer's disease. In: Williamson, S., Hoke, M., Stroink, G. and Kotani, M. Advances in Biomagnetism, New York, Plenum Press, 1989, pp. 311-314.

[35] Ribary, U., Ioannides, A.A., Singh, K.D., Hasson, R., Bolton, J.P.R., Lado, F., Mogilner, A. and Llinas, R. Magnetic field tomography (~;T) of coherent thalamo-cortical 40-Hz oscillation in humans. Proc. Natl. Acad. Sci. USA, 1991, 88: 11037-11041.
[36] Salenius, S., Kajola, M.,Thompson, W.L., Kosslyn, S. and Hari, R. Reactivity of magnetic parietooccipital alpha rhythm during visual imagery. Electroenceph clin Neurophysiol., 1995, 95: 453-462.
[37] Salenius, S., Salmelin, R., Neuper, C., Pfurtscheller, G. and Hari, R. Human cortical 40-Hz rhythm is closely related to EMG rhythmicity. Neurosci. Lett., 1996, In Press.
[38] Salenius, S., Schnitzler, A., Salmelin, R., Jousmaki, V., Hari, R. Modulation of human rolandic rhythms during natural sensorimotor tasks. Submitted.
[39] Salmelin, R. and Hari, R. Characterization of spontaneous MEG rhythms in healthy adults. Electroenceph clin Neurophysiol., 1994a, 91: 237-248.
[40] Salmelin, R. and Hari, R. Spatiotemporal chaMcteristics of sensorimotor MEG rhythms related to thumb movement. Neurosci., 1994b, 60: 537-550.
[41] Salmelin, R., Hamalainen, M., Kajola, M. and Hari, R. Functional segregation of movement-related rhythmic activity in the human brain. Neuroimage, 1996, 2: 237-243.
[42] Schnitzler, A., Salenius, S., Salmelin, R., Jousmaki, V. and Hari, R. Involvement of primary sensorimotor cortex in motor imagery: a neuromagnetic study. Soc. Neurosci. Abstr., 1995, 21: 558.
[43] Spekreijse, H., van Dijk, B.W., Kalitzin, S.N. and Vijn, P.C. Non-linear dynamics of columns of cat visual cortex revealed by simulation and experiment. In: Ciba Foundation Symposium, 1994, 184: 88-99.
[44] Steriade, M. and Llinas, R R. The functional states of the thalamus and the associated neuronal interplay. Physiol. Rev., 1988, 68: 649-742.
[45] Steriade, M., Curro Dossi, R., Pare, D. and Oakson, G., Fast oscillations (20-40Hz) in thalamocortical systems and their potentiation by mesopontine cholinergic nuclei in the cat. Proc. Natl. Acad. Sci. USA, 1991, 88: 4396-4400.
[46] Theiler, J. and Rapp, P.E. Re-examination of the evidence for low-dimensional, nonlinear srtucture in the human electroencephalogram. Electroenceph. clin. Neurophysiol., 1996, 98: 213-222.

Acknowledgements

We would like to acknowledge the excellent secretarial assistance of T. Ngo-Ha.

Time-Frequency Analysis of Oscillatory Gamma-Band Activity: Wavelet Approach and Phase-Locking Estimation

Bertrand, O., Tallon-Baudry, C., and Pernier, J.

Brain Signals and Processes Laboratory, INSERM U280, Lyon, France

Introduction

High-frequency oscillatory activity above 20 Hz (gamma-band activity) has been reported from both animal and human recordings (see [1, 2] for a review). Theoretical works have proposed a specific role for oscillatory synchronization of distributed neural assemblies in feature binding and sensory information processing [3]. Most human EEG and MEG studies have been considering the transient 40 Hz activity occurring in the first 100 ms following the stimulus onset in the auditory modality [4]. However, although non phase-locked 40 Hz activity was observed in animal data after visual [5], auditory [6, 7] or somatosensory [8] stimulations, only few human studies reported stimulus induced gamma-band response [9, 10], suggesting a role in focused arousal and/or cognitive information processing. In recent studies in the visual modality, we have shown that a 40 Hz non phase-locked activity occurs 200-300 ms after the stimulus onset and that it is specifically related to coherent visual stimuli (illusory triangles) compared to low-frequency evoked components [11, 12].

Detecting phase-locked and non phase-locked gamma-band activity in single trials and averaged responses requires some specific tools and related statistical tests. We present here time-frequency analysis methods allowing to detect and quantify both evoked and induced activities. The sensibility of the methods will be presented in simulation studies, and some experimental results will be briefly discussed in the visual and auditory modalities.

Methods

Our goal was to detect and characterize stimulus-related oscillatory activities in the gamma-band (ranging from 20 to 100 Hz). As both latency and frequency of oscillatory bursts are a priori not known, a time-frequency representation of the signal is required. For that purpose, we used a wavelet transform of the signals. This approach, compared to the more classical moving short-time Fourier transform, provides a better compromise between time and frequency resolution [13].

The signal to analyze (EEG or MEG, single-trial epoch or averaged response) is convoluted by complex Morlet's wavelets (sometimes called Gabor wavelets) $w(t,f_0)$, having a gaussian shape both in the time domain (standard deviation σ_t) and in the frequency domain (standard deviation σ_f) around its central frequency f_0: $w(t,f0) = A.\exp(-t^2/2\sigma_t^2).\exp(2i\pi f_0 t)$, with $\sigma_f = 1/2\pi\sigma_t$. Wavelets are normalized so that their total energy is equal to 1 ($A = 1/(\sigma_t\sqrt{\pi})^{1/2}$). A wavelet family, as opposed to classical windowed short-time Fourier transform, is defined by a constant ratio f_0/σ_f, which should be chosen in practice greater than about 5, i.e., the same number of cycles under the gaussian envelopes for any f_0. The time-frequency resolution, defined by $2\sigma_t$ and $2\sigma_f$, is thus varying with frequency (with $f_0/\sigma_f = 7$, this means 111 ms / 6 Hz in the 20 Hz band, and 22 ms / 29 Hz in the 100 Hz band). Below 20 Hz, wavelet duration is of hundreds of ms, thus leading to a poor time resolution. The time-varying energy $E(t,f_0)$ of the signal in a frequency band around f_0 is the squared norm of the result of the convolution of the complex wavelet $w(t,f_0)$ with the signal $s(t)$: $E(t,f_0) = |w(t,f_0)*s(t)|^2$. Repeating this calculation for a family of wavelets will provide a time-frequency representation of the energy of the signal (TF energy).

This analysis can be applied on the averaged evoked response to get the oscillatory activity phase-locked to the stimulus onset which remains after averaging. It can also be applied to each single trials and then all TF energies can be averaged. In this case, both phase-locked and non phase-locked activities are summed up, and bursts of high signal-to-noise ratio will emerge as long as their time-jitter does not exceed too much the wavelet time-duration. The TF energy of the post-stimulus epoch will be referred to a baseline level define as the mean value of the pre-stimulus TF energy in each frequency band.

To statistically test the phase-locking of a particular time-frequency component, we adapted the "phase-averaging" methods previously proposed in the frequency domain [14]. We consider the normalized complex time-varying energy of each single trial i associated to each wavelet $w(t,f_0)$: $P_i(t,f_0) = w(t,f_0)*s_i(t)/|w(t,f_0)*s_i(t)|$. For all single-trials, these normalized complex values are distributed on a unit circle in the complex plane. When a given time-frequency component is rather phase-locked to the stimulus, all complex points have roughly the same orientation on this circle. Conversely, when it is absolutely non phase-locked to the stimulus, the points are randomly distributed on the unit circle. The vector average of these points gives a representation of the phase distribution in a given time-frequency region: its length will further be called the phase-ordering factor. It is independent of the signal amplitude, and it is ranging from 0 (purely non phase-locked activity) to 1 (strictly phase locked activity). With this method, even very low amplitude signals can be identified provided that they are rather strictly phase-locked. Statistical approach (Rayleigh test) has been proposed [15] to test the uniformity of the angle distribution on the unit circle. Critical values for the Rayleigh test are known exactly and are widely tabulated [15]: for 200 single-trials, for instance, the 0.01 error level corresponds to phase-ordering factors greater than 0.152.

To statistically detect a non-phase-locked oscillatory burst as emerging from the baseline activity, we use a non-parametric sign test applied to each time-frequency region. It consists in comparing, for all single trials, each post-stimulus time-frequency energy to its corresponding pre-stimulus baseline value in the same frequency band. Due to the fact that TF energy distribution across trials are generally not gaussian, we choose the median value of the pre-stimulus period as a representative of the baseline energy. A Z value can be computed at each time-frequency point: $Z(t,f_0)=\left(N^+(t,f_0)-1/2N\right)/1/2\sqrt{N}$, where $N^+(t,f_0)$ denotes the number of post-stimulus TF energy values $E(t,f_0)$ above the baseline median value, and N is the total number of single trials. Tabulated probability values show that $p<0.01$ when $|Z(t,f_0)|$ is greater than 2.33. Positive and negative values can be obtained for Z, corresponding to increased and decreased activity compared to baseline.

Four types of time-frequency representation of the data is thus proposed : TF energy of the averaged response and TF phase-ordering factor for phase-locked activity, average of the single-trial TF energies and TF sign test Z values for non phase-locked activity.

Simulation study

To evaluate how these TF representations varies with the time-jitter and the amplitude of the oscillations, several simulations were performed with oscillatory bursts embedded in real EEG noise. Sets of data were made of 200 single-trial EEG epochs on which a wavelet-like signal was added. Gamma-band bursts of frequency ranging from 15 to 60 Hz were considered. Fig. 1 shows the four TF representations in case of 40 Hz signals peaking at 200 ms with different time-jitters and amplitudes (2ms/1μV, 20 ms/2 μV, and 40 ms/4 μV).

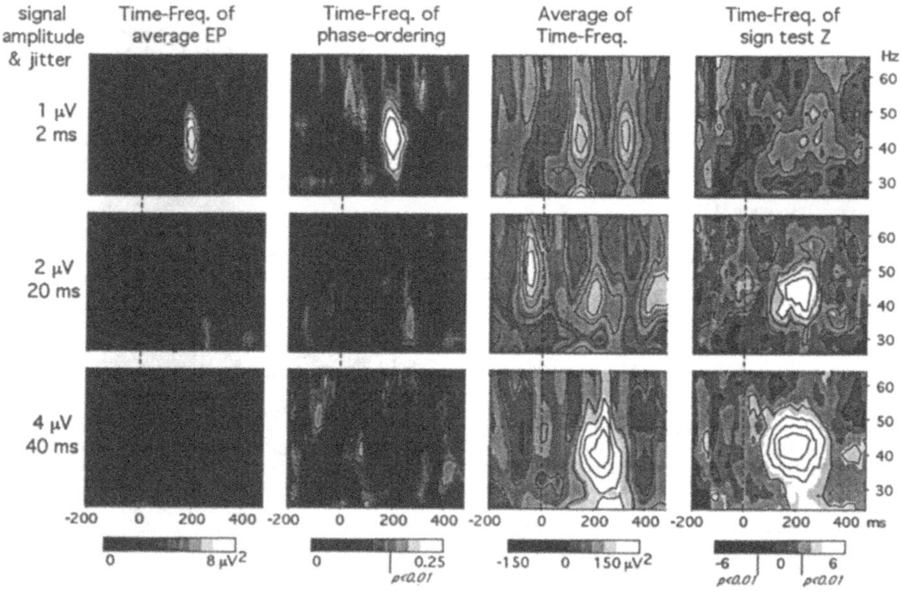

Fig.1. Time-frequency representations obtained on simulated data with different time-jitters and amplitudes combinations (see text for more details).

It appears that signal having a small time-jitter (2ms) are clearly discernible on the TF energy of the averaged response and the TF phase-ordering factor, even with amplitude as low as 1μV. However, it is not discernible on the average of the single-trial TF energies. Conversely, signal having an important time-jitter is no longer visible on phase-locked TF representations while it appears on the average of the single-trial TF energies as long as its signal-to-noise ratio is high enough (amplitude 4 μV). Nevertheless, for signal of lower amplitude (2 μV), while the 200ms/40Hz peak is not very clearly detectable on the average of the single-trial TF energies, the TF sign test Z value improves the detection of this non phase-locked activity, with a rather good elimination of noisy peaks.

A more systematic evaluation with time-jitters ranging from 0 to 150 ms, and signal amplitudes ranging from 0 to 5 μV, was conducted on wavelet-like oscillatory bursts (frequency ranging from 15 to 60 Hz). By means of the statistical tests based on the phase-ordering factor and the sign test Z value, limits (p=0.01) have been computed in the time-jitter/amplitude plane to characterize the detectability of phase-locked and non phase-locked signals (Fig. 2). Due to the low level of EEG background noise in the upper part of the gamma-range (40 Hz and above), phase-locked signals as low as 0.5-1 μV can be detected provided that their time-jitter does not exceed 5 ms. Detectability of non-phase locked signals is less sensitive to time-jitter in the lower part of the gamma-band (below 40 Hz), but requires higher amplitudes of about 2-3 μV.

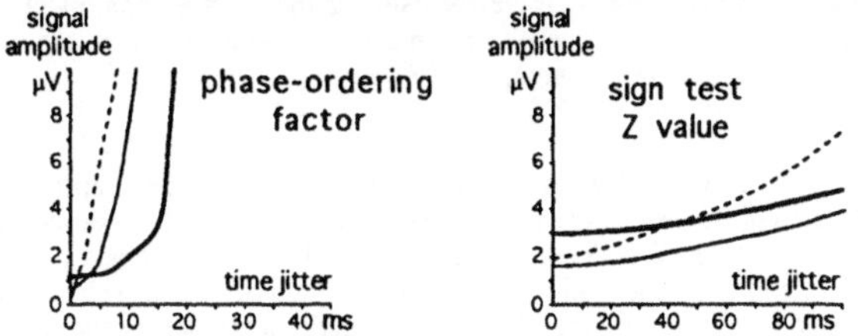

Fig. 2. Limits of statistical detection of oscillatory bursts (thick line: 15 Hz, thin line: 40 Hz, dotted line: 60 Hz). All signals having a combination of time-jitter and amplitude above those lines are detectable with the phase ordering factor (rather phase-locked activity) or the sign test Z value (induced non phase-locked activity).

Experimental studies

It has been suggested that feature binding could be achieved in the visual cortex by synchronization of spatially distributed neurons in the gamma-band. We have recently tested this hypothesis in human [18], using three types of visual stimuli: two coherent stimuli (a Kanizsa illusory and a real triangle) and a non-coherent one ("no-triangle"). The time-frequency analysis of averaged as well as single-trial EEG epochs (8 subjects) was performed to characterize phase-locked and non phase-locked high-frequency activities. An early phase-locked 40 Hz component was found maximal at electrodes Cz-C4, and does not vary with stimulation type. We described a second 40 Hz component, emerging from pre-stimulus baseline level around 280 ms, not phase-locked to stimulus onset. This component shows a higher energy in response to a coherent triangle, no matter whether real or illusory: it could thus reflect a mechanism of feature binding based on high-frequency synchronization. At the same latencies, the low-frequency evoked components behave differently, suggesting that low- and induced high-frequency activities have different functional roles.

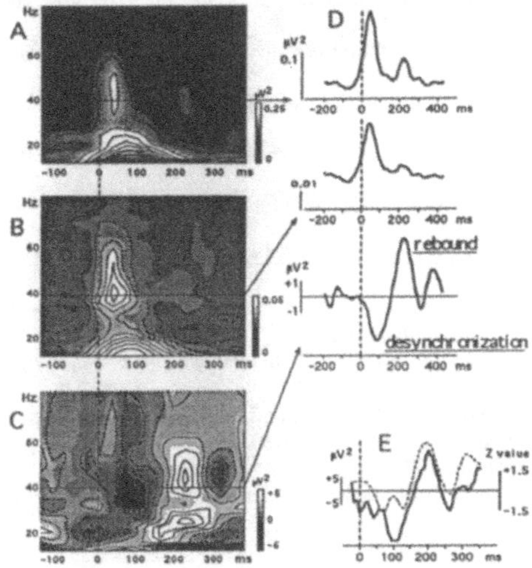

Fig.3. 40 Hz activity acoustic tone-burst presentation. Time-frequency representations (grand average over 10 subjects) of (A) the averaged evoked potential, (B) the phase-ordering factor, and (C) the average of the single-trial energies. (D) shows more precisely the time-varying energy in the 40 Hz band. (E) shows for one single subject the average of the single-trial energies (dotted line) and the related sign test Z value (thin line) around 40 Hz.

In the auditory modality, recent animal studies [7] have reported spontaneous 40 Hz activity recorded on an ECoG epipial electrode array placed over the auditory cortex of a rat. Single-trial spectral analysis of click responses indicated an inhibition of gamma-band oscillations during 150 ms, with subsequent enhanced gamma-activity between 250 and 350 ms. The epipial distributions of pre-stimulus and enhanced post-stimulus gamma-oscillations were the same. We present here preliminary time-frequency results in human EEG (grand-average of 10 subjects) of 40 Hz activity related to tone-burst presentation (1000 Hz, 50 ms duration) (fig. 3). A clear transient 40 Hz response peaking around 50 ms and phase-locked to the stimulus onset was observed (fig. 3A,B) as reported in previous papers [2]. The representations of the average of the single-trial TF energies and of the TF sign test Z values show a clear 40 Hz activity peaking around 200 ms, emerging from pre-stimulus baseline level (fig. 3C). Since it almost does not appear in the averaged data and phase-ordering factor, it most probably corresponds to a non phase-locked induced activity. This is consistent with previous findings in human recording reacting to click stimuli [9]. However, we also observed a significant reduction of 40 Hz activity immediately after the stimulus onset, with a minimum around 100 ms (fig. 3C,E). These 40 Hz de-synchronization and rebound phenomena are consistent with those observed on rat auditory cortex [6]. The transient 40 Hz response is clearly visible in the averaged phase-locked response even though it has a very small amplitude. However, it is very small in the single-trials compared to the spontaneous and induced activities, and thus does not show up in the average of TF energies.

Discussion

We have presented here methods to analyze EEG or MEG signals in the time-frequency domain for the detection and characterization of event-related gamma-band oscillatory activity. These methods have been evaluated with EEG simulations and applied in EEG experimental studies. Transient phase-locked and induced non phase-locked 40 Hz bursts have been observed in auditory and visual modalities. The visual data provides some evidence for a specific functional role of induced gamma-band in feature binding. Additional experimental studies should be performed to further clarify the respective functional roles of evoked and induced gamma-band, compared to low-frequency components.

References :

[1] Basar, E., and Bullock, T.H. Induced Rythms in the Brain, Boston, Birkhauser, 1992.

[2] Pantev, C., Elbert, T., and Lütkenhöner, B. Oscillatory Event-Related Brain Dynamics, New York, Plenum Press, 1994.

[3] von der Malsburg, C., and Schneider, W. A neural coktail-party processor, Biol. Cybern., 1986, 54: 29-40.

[4) Pantev, C., Makeig, S., Hoke, M., Galambos, R., Hampson, S. and Gallen, C. Human auditory evoked gamma-band magnetic fields, Proc. Nat. Acad. Sci. USA, 1991, 88: 8996-9000.

[5] Eckhorn, R., Schanze, T., Brosch, M., Salem, W., and Bauer, R. Stimulus-specific synchronization in cat visual cortex: multiple microelectrode and correlation studies from several cortical areas. In: Induced Rythms in the Brain, Basar, E., and Bullock, T.H. Boston, Birkhauser, 1992: 47-82.

[6] Basar-Eroglu, C. and Basar, E. A compound P300-40 Hz response of the cat hippocampus. Int. J. Neuroscience, 1991, 60: 227-237.

[7] Franowicz, M.N., and Barth, D.S. Comparison of evoked potentials and high-frequency (gamma-band) oscillating potentials in rat auditory cortex, J. Neurophysiol., 1995, 74, 1: 96-112.

[8] MacDonald, K.D., and Barth, D.S. High-frequency (gamma-band) oscillating potentials in rat somatosensory and auditory cortex, Brain Research, 1995, 694: 1-12.

[9] Jokeit, H. and Makeig, S. Different event-related patterns of gamma-band power in brain waves of fast- and slow-reacting subjects. Proc. Nat. Acad. Sci. USA, 1994, 91: 6339-6343.

[10] Lutzenberger, W., Pulvermüller, F., and Birbaumer, N. Words and pseudo-words elicit distinct patterns of 30 Hz EEG responses in humans. Neurosci. Letters, 1994, 176: 115-118.

[11] Tallon, C. Bertrand, O., Bouchet, P., and Pernier, J. Gamma-range activity evoked by coherent visual stimuli in humans, Eur. J. Neurosci., 1995, 7: 1285-1291.

[12] Tallon-Baudry, C., Bertrand, O., Bouchet, P., and Pernier, J. Variation of gamma-range oscillations correlates with stimulus coherency in human. Human Brain Mapping, 1995, suppl. 1: 41.

[13] Sinkkonen, J., Tiitinen, H. and Näätänen, R. Gabor filters: an informative way for analysing event-related brain activity, J. Neurosci. Methods, 1995, 56: 99-104.

[14] Jervis, B.W., Nichols, M.J., Johnoson, T.E., Allen, E., and Hudson, N.R. A fundamental investigation of the composition of auditory evoked potentials, IEEE Trans. Biomed. Eng., 1983, 30: 43-50.

[15] Mardia, K.V. Statistics of directional data, London, Academic, 1972.

[16] Tallon-Baudry, C., Bertrand, O., Delpuech, C., and Pernier, J. Stimulus specificity of phase-locked and non phase-locked 40 Hz visual responses in human. Submitted.

Inverse Imaging of Distributed Oscillatory Activity

Bradshaw, L.A. and Wikswo, J.P., Jr.

Vanderbilt University, Nashville, TN, USA

Introduction

Several types of bioelectric current source produce external electric and magnetic fields with oscillatory characteristics. Often, these sources are not well localized and may be distributed over a wide spatial range or may consist of several simultaneously-active localized sources distributed over a wide area. Additionally, the oscillation associated with such activity may not be synchronized over space -- activity in different regions may oscillate at different frequencies. The brain's alpha rhythm is one such example, with activity spread over a large region of the brain oscillating at a characteristic frequency between 8 and 12 Hz [1]. The basic electrical activity of the gastrointestinal system is another example. Frequencies of oscillation in gastrointestinal electrical activity include a 3 cycle-per-minute (cpm) slow-wave in the stomach and slow-waves that range from 8-12 cpm in the small intestine, with more distal segments of the small bowel typically having lower frequencies than more proximal segments [2,3]. We present a technique to address the problem of estimating the source current distribution from external measurements of the electrical activity.

Methods

Distributed oscillatory activity may be considered as collections of localized oscillators. We will represent these localized oscillators as current dipoles for purposes of evaluating the algorithm. In this formulation, the positions of the oscillators that comprise the source current distribution are considered fixed and the magnitudes of the dipole strengths vary according to the electrical activity. The current dipole strength oscillates according to a sine wave with a characteristic amplitude and phase. In this manner, we can treat bioelectric activity for which the oscillation is sinusoidal in nature. Other more complicated oscillatory patterns may also be treated by this technique if the oscillation can be characterized according to certain parameters.

Two situations arise with multiple oscillators characterized in this way. Each oscillator may have a different characteristic frequency, or two or more oscillators may have the same frequency with different phases. The former case arises with activity such as the gastrointestinal basic electrical rhythm (BER) where pacemaking oscillators appear along the bowel at progressively lower frequencies with more distal locations. The latter situation occurs when any type of propagating electrical activity is present.

The phenomenon of propagation can be characterized by phase shifts in the signal at different time instants. Uniform propagation implies that the phase shifts linearly with time for an oscillator at a given location. Since the frequency of the propagation oscillator remains the same as that of nearby oscillators, one may utilize the phase spectrum of the oscillation pattern to determine the nature of the propagation [4].

External multichannel measurements of the magnetic fields from bioelectric current sources record the magnetic fields adjacent to a region that typically includes the source activity. Several methods are available for estimating the source current distribution from external measurements. A review of some of these methods is provided in reference [5]. These methods all have varying degrees of usefulness for tracking the time variation in the signals, but none explicitly addresses oscillatory characteristics.

We describe a method for localizing sources of oscillatory activity by decomposing the frequency content of the magnetic field pattern. By exploiting both magnitude and phase information that results from a frequency-domain analysis, we effectively restrict the number of possible solutions for the inverse procedure.

Before the algorithm may be applied, the nature of the oscillatory activity and its characterizing parameters must be determined. The parameters define a basis function that represents the oscillatory activity. For the purposes of demonstrating the technique, we will use sinusoidal oscillations that are characterized by their amplitude and phase. The basis function in this case is thus a sine wave. More complex waveforms may be necessary for precise studies of activity in which the oscillation cannot be easily characterized as sinusoidal, but as long as oscillation parameters can be defined, the algorithm still applies with minor adjustments.

Once the oscillation has been characterized, it is necessary to determine the frequency or frequencies present in the recorded magnetic field pattern. Frequency decomposition techniques are applied to the entire magnetic field pattern and the results are averaged to obtain an idea of which frequencies are most prominently represented in the magnetic data. The magnitude spectrum of the Fourier transform is one such technique for identifying multiple frequencies of oscillation present in the signal.

The next step is to determine the parameters of the basis function that most closely describe the recorded data for each frequency present in the recording. Using the sine wave as a basis function, we find the amplitude and phase of the sine wave that best fits the observed data in each channel for each frequency present in the signal. This process is basically a complicated form of Fourier decomposition.

Let $B_i(t)$ represent the data recorded in channel i of an n-channel magnetometer array, and nf represent the total number of frequencies present in the signal as determined by the frequency decomposition. For each channel and for each frequency present in the recording, we wish to determine the amplitude and phase of the sine wave that best matches the recorded data. In other words, we wish to know $A'_{i,j}$ and $\phi_{i,j}$ given that

$$B_i(t) = A'_{i,j}\sin(\omega_j t + \phi_{i,j})\,;\ i = 1, \ldots, n\,;\ j = 1, \ldots, nf\,, \tag{1}$$

where ω_j is one of the signal frequencies. Once the amplitudes and phases are determined, the phase information in $\phi_{i,j}$ may be evaluated to determine the presence or absence of propagation in the magnetic field pattern. Linear increases in best-fit phase indicates uniform propagation. If propagation is not indicated by the phase spectrum, single dipole fits to the sine wave amplitude pattern $A'_{i,j}$ isolate oscillators at different frequencies. If propagation is indicated, multiple dipole fits determine the location of the several oscillators in the propagation pattern whose frequencies are the same, but whose phases differ.

To illustrate this algorithm, we apply it to two hypothetical situations, one in which the underlying current distribution results in a magnetic field pattern with different frequency components, and another in which all oscillators have the same frequency but different phases simulating uniform propagation in a "V" pattern.

Results

The first example uses six oscillators whose distinct frequencies are the odd harmonics of a square wave. The amplitude and phases of the oscillators are chosen with their locations to result in a square wave magnetic field in the middle channel of the magnetometer array. The magnetic fields in the array are calculated at points that would coincide with the coils in the 37 channel Biomagnetic Technologies, Inc. Magnes™ system. Figure 1 shows the dipole distribution and resulting magnetic field pattern used for evaluation of the Frequency-Domain Inverse Algorithm. Fast Fourier Transformation (FFT) of the data in each channel was performed and the results of the 37 FFTs were averaged to obtain the average magnitude spectrum. The six frequencies corresponding to each of the dipolar oscillators were apparent in the average spectrum. For each of the six frequencies present, amplitudes and phases of sine waves with the given frequency were determined. Single dipole fits to the six sine wave patterns resulted in the six derived dipoles indicated in Figure 2A. The actual dipole distribution is also plotted, although it is not apparent since the actual and derived dipole distributions are so nearly equivalent. For comparison, moving dipole solutions to

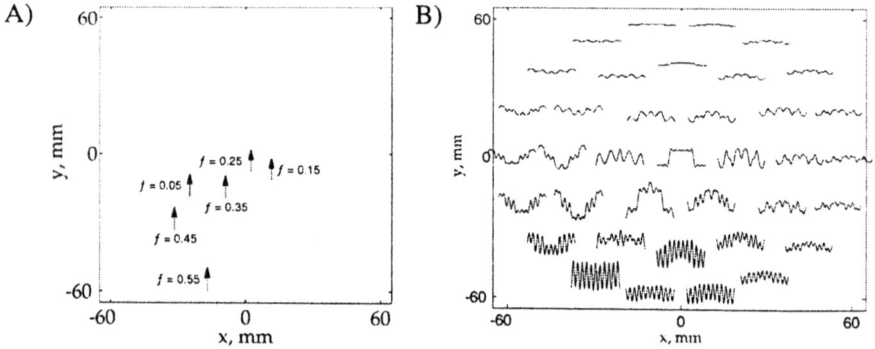

Figure 1. (A) Six dipoles with different oscillation frequencies indicated are used to evaluate the Frequency-Domain Inverse Algorithm. (B) The magnetic field pattern calculated for the 37 channel system from the six dipoles in (A).

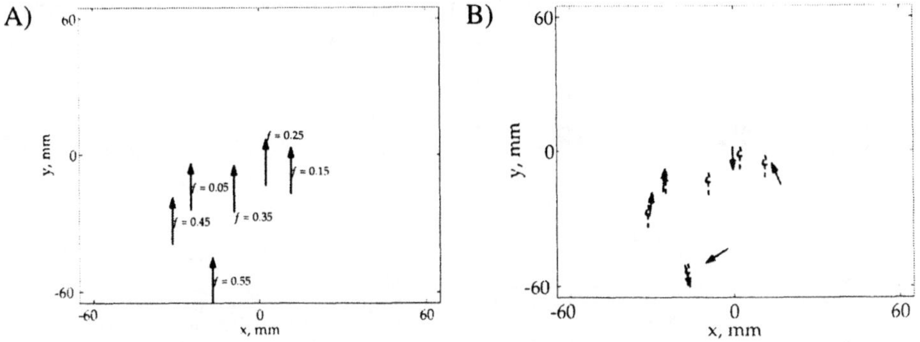

Figure 2. (A) Frequency-Domain Inverse Algorithm solutions to the magnetic field pattern in Figure 1B. (B) A six-dipole fit to a single time instant of the magnetic field data in Fig. 1B. In each case, the true dipole distribution is indicated by dashed arrows and the derived distribution by solid arrows.

the magnetic field patterns at successive time instants as well as a six-dipole fit to the magnetic data at single time instants were calculated. The results of the six-dipole fit are shown in Figure 2B.

The Frequency-Domain Inverse Algorithm requires minor modification when applied to data in which propagating oscillatory activity is present. Since propagation is characterized by several oscillators with a particular frequency and different phases, the single dipole fits to sine waves of one frequency that were used in the previous example will not be sufficient to fully describe patterns of propagation, and multiple dipole fits are necessary.

We simulate the propagation of electrical activity by assigning five dipoles arranged in a "V" pattern identical frequencies and phases that vary linearly according to the position of the dipoles, as shown in Figure 3. In this case, the same frequency decomposition that was applied to the square wave magnetic field example yields only one peak in the average magnitude spectrum at the frequency of the five dipoles. However, the phase spectrum of each dipole suggests the presence of propagating activity. The amplitude and phase of the best-fit sine wave are again calculated. We know that multiple dipole fits are necessary to adequately describe propagating activity, but the number of dipoles to fit is unclear; in fact, it is somewhat arbitrary. Since propagation is in essence a distributed activity, it follows that the more dipoles used in the fit, the better agreement with the actual data. However, including more dipoles in an inverse procedure makes the procedure more computationally intensive, and an optimal number of dipoles to fit

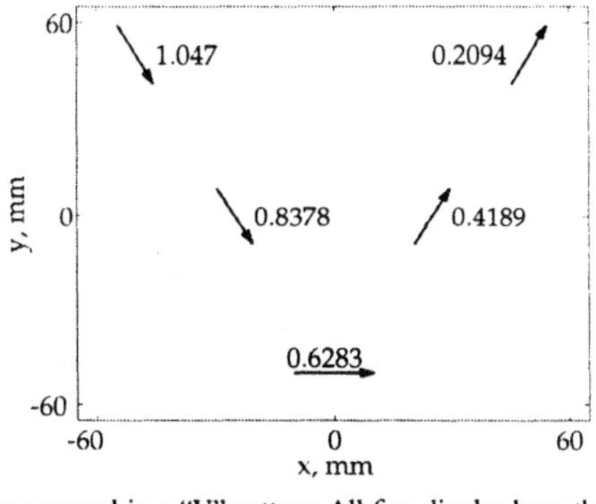

Figure 3. Five dipoles arranged in a "V" pattern. All five dipoles have the same frequency. The phases vary in a linear manner with dipole position to simulate uniform propagation of electrical activity. The phase values (in radians) are listed next to the arrows representing the dipoles.

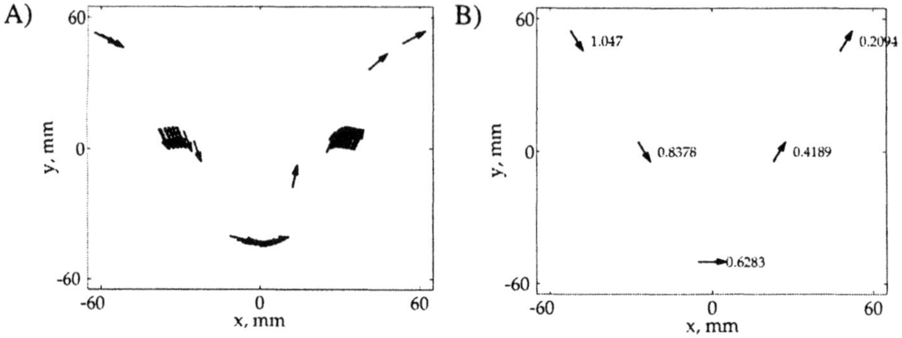

Figure 4. (A) Moving single dipole solutions to the magnetic field pattern from the propagating dipole distribution of Figure 3. (B) Five dipoles localized by the Frequency-Domain Inverse Algorithm. The actual dipole distribution is also displayed but cannot be seen since it corresponds so nearly to the inverse solution.

is desired. To this end, we perform moving single dipole fits to the magnetic field in successive time instants and cluster the results of the moving dipole analysis to determine the minimum number of dipoles required to properly describe the propagation pattern. Five clusters are apparent in the results of the moving dipole approach shown in Figure 4A. The geometric centers of these five clusters are used as starting parameters for the Levenberg-Marquardt fit of five dipoles to the magnetic field pattern. The results of the multiple dipole fit are shown in Figure 4B.

Discussion

The Frequency-Domain Inverse Algorithm was applied to two simulations of distributed oscillatory activity. In both cases, the dipole oscillators were correctly identified from distant magnetic field calculations. Using the goodness-of-fit parameter [5] commonly used in biomagnetism, it was found that the Frequency-Domain Inverse Algorithm solutions accounted for 99.8% of the square wave magnetic field pattern and 99.9% of the propagating "V" magnetic field pattern. In contrast, a multiple-dipole approach to the square wave magnetic field pattern provided a goodness-of-fit of only 65.0% while the average moving-dipole solution accounted for only 2.0% of the magnetic field pattern. This algorithm could be applied with little or no modification to many of the problems in which characterization of periodic or oscillatory activity is required from external electric or magnetic measurements.

References

[1] Nunez, P. L. *Electric Fields of the Brain*. New York, Oxford University Press, 1981.

[2] Christensen, J. D., Schedl, H. P., and Clifton, J. A. The small intestinal basic electrical rhythm (slow wave) frequency gradient in normal men and in patients with a variety of diseases, *Gastro*, 1966, 50: 309-315.

[3] Smout, A. J. P. M., Van Der Schee, E. J., and Grashuis, J. L. What is measured in electrogastrography? *Dig. Dis. and Sci.*, 1980, 25: 179-187.

[4] Bradshaw, L. A. *Measurement and Modeling of Gastrointestinal Bioelectric and Biomagnetic Fields*. Ph.D. Dissertation, Vanderbilt University, 1995.

[5] Hämäläinen, M., Hari. R., Ilmoniemi, R. J., Knuutila, J. and Lounasmaa, O. V. Magnetoencephalography - theory, instrumentation, and applications to noninvasive studies of the working human brain. *Rev. Mod. Phys, 1993,* 65(2): 413-497.

Acknowledgments

This work was supported in part by grants from the Veterans Administration and by an NIH SBIR grant to Conductus, Inc.

MEG Measurements of 40 Hz Auditory Evoked Response in Human Brain

Ciulla, C., Takeda, T., Morabito, M., Kumagai, T. and Endo, H.

Behaviour Control Laboratory, National Institute of Bioscience and Human-Technology, 1-1, Higashi, Tsukuba, Ibaraki 305 Japan

Introduction

Steady-state responses (SSRs) evoked by auditory stimulation in the brain were discovered by Galambos using EEG [1]. The amplitude of the response was found to be dependent on repetition rate, frequency and intensity of the stimulus, and it was maximum at repetition rates near 40 per second. Another EEG study conducted by Galambos [2] described intermodality cancellation of tactile and auditory 40 Hz potentials evoked by steady-state stimulations. A subcortical structure of the brain (possibly the rostral reticular formation) was assumed to be the generation site of the 40 Hz evoked activity. A model of the 40 Hz auditory-event related potentials was proposed and evaluated by EEG in Spydell et al., [3]. Comparison of the phase lag of the EEG response at the stimulation rate of 40 Hz between normal and thalamic and midbrain lesioned subjects, suggested again subcortical origin of the 40 Hz response.

More recently MEG was successfully employed in studies concerned with 40 Hz auditory evoked activity in the human brain. Makela et al., [4] reported that the presumed generators of the 40 Hz magnetic response were located at the primary auditory cortex deep within the sylvian fissure, close to the sources of the components of the transient response. The generation mechanism of the 40 Hz steady-state response was explored by Hari et al., [5]. SSRs were obtained in response to clicks presented between 10 and 70 Hz. Assuming that each click evokes a similar response at all repetition rates, it was proposed that the amplitude enhancement at 40 Hz was due to the superposition of the responses evoked by single clicks. The results of those studies were achieved by measurements carried out on only one hemisphere.

The present research aimed to study the origin in the human brain of the auditory SSR at 40 Hz. The use of a whole-cortex MEG system allowed simultaneous detection of the neuromagnetic responses from both hemispheres and estimation of the equivalent dipolar sources.

Methods

(a) Subjects, stimuli and neuromagnetic recordings

Five normal-hearing subjects (3 females and 2 males) aged between 27 and 48 years participated in the experiments. All of the subjects were right handed. Subjects were instructed to rest in relaxed wakefulness while eyes closed. Stimuli were 500 Hz tone bursts repeated at the rate of 40 per sec (6 ms duration, 1 ms rise and fall time). They were delivered through a 4.75 m plastic tube to the right ear at the intensity of 60 dB. Neuromagnetic fields were recorded by a whole-cortex 64-channels MEG system (CTF System Inc.) placed in a magnetically shielded room. Detection coils were 1st order axial gradiometers, 20 mm in diameter, and the distance between coils was 42 mm center to center. The pulse generator, the amplifier, and the transducer were placed outside the shielded room in order to avoid contamination of the neuromagnetic signal with the stimuli. Three coils placed at the preauricular points (tragions) and the nasion, defined the right-handed MEG head coordinate system with the origin located at the mid point between the tragions. The same fiducial points marked with pills having a 4 mm diameter were used in order to reproduce the head coordinate system on magnetic resonance images. MRI were determined individually for each of the subjects. The thickness of the MRI slices was 2 mm.

(b) Experimental session

Prior to the experiments, the estimation of the MEG system instrumental noise, the determination of the baseline spontaneous brain activity, and a control experiment were conducted. MEG measurements were lowpass filtered on-line (0-100 Hz) with a sampling rate of 1250 Hz. Six hundred epochs of 1500 ms analysis period were recorded and averaged. The use of on-line notch filters removed the powerline noise (50 Hz). Trials containing artifacts produced by eye blinks were eliminated automatically by software. Both baseline spontaneous brain activity and the control experiment were determined in one subject instructed to rest while eyes closed in relaxed wakefulness. The aim of the control experiment was to ascertain whether or not 40 Hz activity could be evoked when the stimulus was a 500 Hz continuous tone.

In order to check the replicability of source estimation, the experiment was repeated twice on different days (subjects KA and AY). In an additional experiment with two subjects, 40 Hz SSRs were obtained delivering the stimuli to the left ear (subjects KA and KI).

(c) Data analysis

The signal was extracted by off-line bandpass filtering (35-45 Hz). An equivalent current dipole (ECD) was used to represent the source of the brain activity, and its location was estimated in a spherical homogeneous volume conductor. The sphere gave the best fit to the curvature of the brain at the auditory cortices of both hemispheres, and it was determined for each of the subjects. The radius and the position of its center with respect to the origin of the MEG system were determined on the basis of magnetic resonance images.

An iterative least square minimization algorithm based on the Grynszpan-Geselowitz formulae provided the location and the orientation of the estimated dipolar sources in respect to the MEG head coordinate system. Only estimates with goodness of fit greater than 92% were considered. The depth of the sources was computed as radial distance from the surface of the spherical head model. The locations of the sources on MRI were determined applying the rotation/translation matrix of the MRI and the MEG head coordinate systems. Correction matrices were determined individually for each of the subjects.

Results

The instrumental noise of the MEG system (less than 20 fT for single epoch) was reduced to less than 1 fT (20 fT/$\sqrt{600}$) by averaging (n=600). The frequency spectra of the MEG sensors had an amplitude level lower than 1 fT rms/$\sqrt{Hz}$ (Fig. 1). The baseline of the averaged (n=600) spontaneous brain activity had a frequency spectra around 40 Hz comparable to that of the instrumental noise. The frequency spectra of the averaged brain activity had no peak around 40 Hz when the stimulus was a 500 Hz continuous tone (control experiment). A clear and distinct peak around 40 Hz was seen when the stimuli (500 Hz tone bursts) were delivered at the rate of 40 per sec (Fig. 1). Forty-Hertz magnetic SSRs were recorded with relatively high signal to noise ratio (about 10). Fig. 2 shows a time window of the raw MEG signals (0-100 Hz) recorded, the waves present a clear and consistent rhythm at 40 Hz. The neuromagnetic isocontours had for each hemisphere two extrema of opposite polarity located over the fronto-temporal and the temporo-parietal areas of the head (see Fig. 3). Considerable stability during the analysis period (1500 ms) was observed in the location of the fields extrema, suggesting the activation of a restricted area of the brain. Variability in the amplitude of the response between subjects was noticed. The maximum amplitude recorded was about 40 fT (subject KA). Comparisons between the ECDs estimated in the two hemispheres (left and right) was conducted on the basis of the dipole moment and the depth. With both right and left ear stimulation, the source located in the left hemisphere was estimated deeper (20-36 mm) and with slightly bigger dipolar moment (1.2-5.7 nAm) with respect to that of the right hemisphere (7-24 mm and 1.2-4 nAm). Table I summarizes for 5 subjects and for each hemisphere the maximum amplitude recorded by the MEG sensors, and the results of dipole estimation (location: x, y, and z; declination: d; dipolar moment: q; depth: D; and goodness of fit: g). The declination was computed with respect to the z axis of the MEG head coordinate system.

For all of the subjects 3-dimensional MRI locations of both contralateral and ipsilateral ECDs agreed with the activation of the supratemporal plane (SP) deep within the sylvian fissure. ECDs declination varied within 20-50 deg and 23-43 deg respectively for the left and the right hemisphere (ranges across subjects), compatibly with the anatomy of the supratemporal plane. Fig. 4 shows dipolar sources estimated for subject KA superimposed on MRI. Coronal and sagittal sections and top views are given for both left and right hemisphere.

(a)

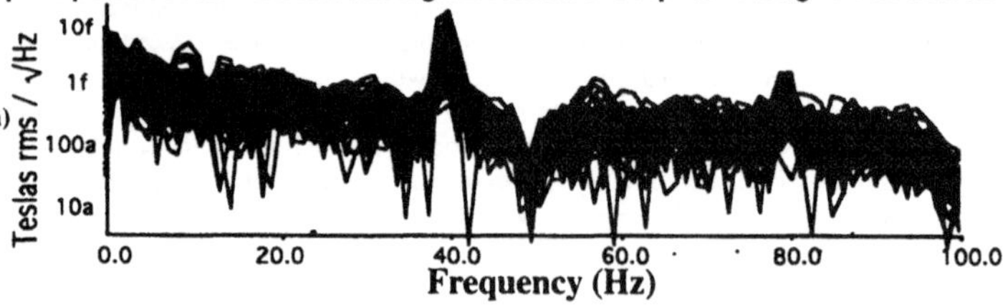

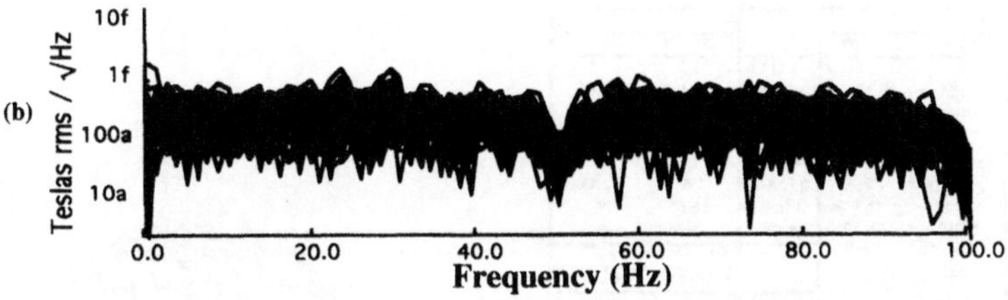

Fig. 1. Frequency spectra of the MEG signals. (a) Typical MEG response to an auditory steady-state stimulation at 40 Hz. Signals from all sensors of the MEG system are superimposed (0-100 Hz), units are in Tesla rms/$\sqrt{Hz}$. The brain response is peaked around 40 Hz. (b) The noise level of the MEG system. The instrumental noise was below 1 fT rms/$\sqrt{Hz}$ (n=600).

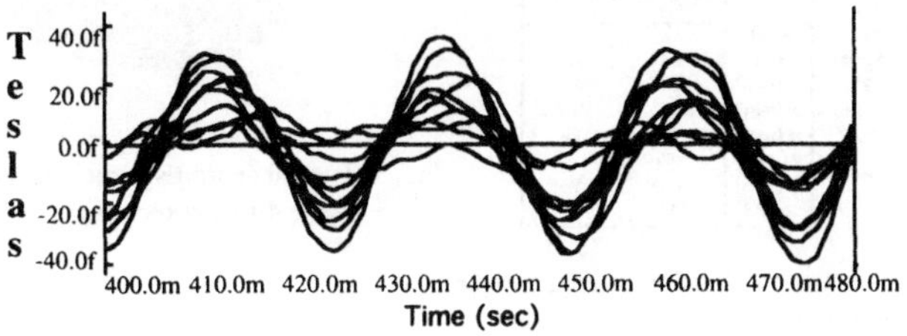

Fig. 2. Time window of the MEG signals recorded in response to auditory steady-state stimulation at 40 Hz.

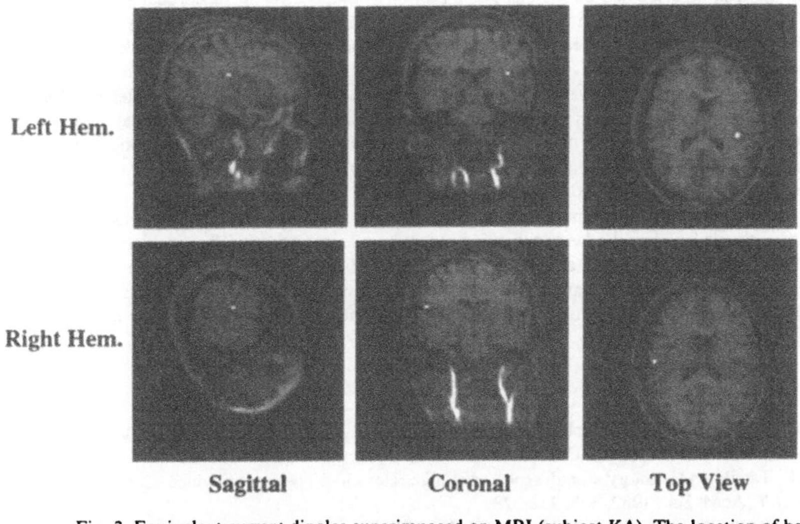

Fig. 3. Equivalent current dipoles superimposed on MRI (subject KA). The location of both contralateral and ipsilateral ECDs was in agreement with the activation of the supratemporal plane deep within the sylvian fissure. Sources of the left hemisphere were estimated deeper than those of the right hemisphere both with left and right ear stimulation. Sagittal section, coronal section and top view are shown.

	KA		AY	
	Right Hem	Left Hem	Right Hem	Left Hem
A (fT)	22	15	18	12
	20	20	20	12
x (mm)	1 05	-1 83	-5 09	-9 056
	-2 49	-8 3	5 69	0 66
y (mm)	-56 56	34 18	-48 36	31 73
	-52 44	38 2	-41 81	29 99
z (mm)	61 12	56 93	47 1	46 13
	61 79	63 37	56 12	59 32
d (mm)	42	50	26	23
	26	26	35	30
q (nAm)	2 85	4 06	3 7	4 9
	2 07	4 66	3 99	5 68
D (mm)	7 25	31 36	20 38	33 71
	11 63	28 77	24 6	36 02
g (%)	96 8		98 1	
	97		96 2	

	KA (left ear stim.)		KI (left ear stim.)	
	Right Hem	Left Hem	Right Hem	Left Hem
A (fT)	25	38	20	10
x (mm)	-2 356	-2 77	4 7	-2 11
y (mm)	-53 96	47 58	-58 82	49 76
z (mm)	57 01	58 13	49 79	52 96
d (mm)	30	20	43	28
q (nAm)	3 45	4 73	1 67	2 15
D (mm)	10 28	19 6	12	26 22
g (%)	93 4		96 2	

	KI	
	Right Hem	Left Hem
A (fT)	20	18
x (mm)	7 1	-3 86
y (mm)	-53 53	47 21
z (mm)	50 65	63 44
d (mm)	37	30
q (nAm)	1 87	2 83
D (mm)	16 89	28 2
g (%)	95 4	

	NH	
	Right Hem	Left Hem
A (fT)	8	12
x (mm)	1 11	-1 58
y (mm)	-53 2	52 83
z (mm)	45 78	45
d (mm)	23	20
q (nAm)	1 18	1 83
D (mm)	20 52	25 34
g (%)	95 9	

	TA	
	Right Hem	Left Hem
A (fT)	12	8
x (mm)	12 47	6 22
y (mm)	-50 52	46 63
z (mm)	62 03	59 66
d (mm)	40	29
q (nAm)	1 85	1 17
D (mm)	11 46	23 65
g (%)	96 2	

Table I
Maximum amplitude of the 40 Hz response, localization, strength, depth and goodness of fit of the equivalent current dipoles (5 subjects).

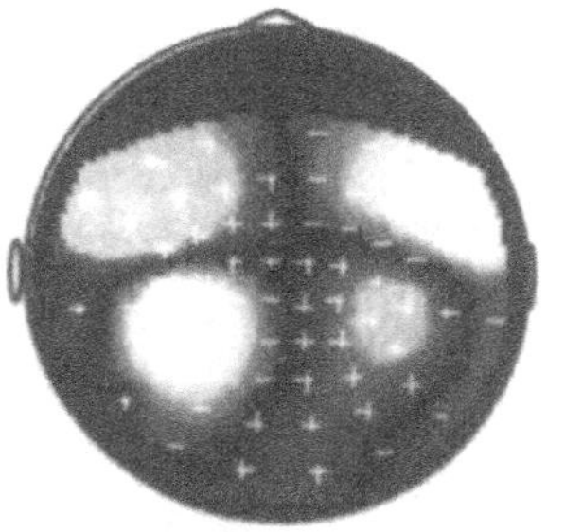

Fig. 3. Typical neuromagnetic isocontour of the 40 Hz response.

Discussion

The present research provides evidence for the anatomical origin of the 40 Hz steady-state response to ipsilateral stimuli and can be discussed on the basis of the former research. While subcortical generators were proposed [2], [3], evidence for at least one cortical generator (the contralateral) were presented by [4], rendering improbable the hypothesis of a single midbrain generator. The cases studied by this research extended the findings that were reported earlier by Makela et al., [4]. MRI confirmed that the ECDs of 40 Hz steady-state response to both contralateral and ipsilateral stimuli were located in the primary auditory cortex at the supratemporal plane. Sources of the right hemisphere were found to have depths remarkably shallower and smaller dipolar moment than those of the left hemisphere. To attribute this fact to a predominance of the contralateral subcortical activity as suggested by Hashimoto et al., [6], and Moller et al., [7], seems questionable since the difference in both depth and dipolar moment remained the same also with left ear stimulation.

On the basis of the results presented here, it is likely to assume that the 40 Hz magnetic steady-state response recorded at the scalp is largely due to activation of the supratemporal plane. Thus an exclusive midbrain origin seems at least doubtful. However as proposed earlier by Galambos [2], waves might originate in the reticular formation and then generate a similar response at cortical level by conduction of impulses. If this was true then would be likely to expect that subcortical structures might contribute to the generation of the neuromagnetic response recorded at the scalp.

References:

[1] Galambos, R., Makeig, S., and Talmachoff, P.T. A 40-Hz auditory potential recorded from the human scalp. Proc. Natl. Acad. Sci. USA., 1981, 78: 2643-2647.

[2] Galambos, R. Tactile and auditory stimuli repeated at high rates (30-50 per sec) produce similar event related potentials. Ann. N.Y. Acad. Sci., 1982, 338: 722-728.

[3] Spydell, J.D., Pattee, G., and Goldie, W.D. The 40 Hz auditory event-related potential: normal values and effects of lesions. Electroenceph. clin. Neurophysiol., 1985, 62: 193-202.

[4] Makela, J.P., and Hari, R. Evidence for cortical origin of the 40 Hz auditory evoked response in man. Electroenceph. clin. Neurophysiol., 1987, 66: 539-546.

[5] Hari, R., Hamalainen, M., and Joutsiniemi, S.L. Neuromagnetic steady-state responses to auditory stimuli. J. Acoust. Soc. Am., 1989, 86: 1033-1039.

[6] Hashimoto, I., Mashiko, T., Yoshikawa, K., Mizuta, T., Imada, T., and Hayashi, M. Neuromagnetic measurements of the human primary auditory response. Electroenceph. clin. Neurophysiol., 1995, 96: 348-356.

[7] Moller, A.R., and Jannetta, P.J. Evoked potentials from the inferior culliculus in man. Electroenceph. clin. Neurophysiol., 1982, 53: 612-620.

Oscillatory Neuromagnetic Activity Induced by Verbal and Non-Verbal Stimuli Presented in Visual and Auditory Modalities

Eulitz, C.[1,3], Maess, B.[2], Pantev, C.[3], Friederici, A.[2], Feige, B.[1] and Elbert, T.[1]

[1]*University of Konstanz, Germany;* [2]*Max-Plank-Institute for Cognitive Neuroscience, Leipzig, Germany;* [3]*University of Münster, Germany*

Introduction

The evaluation of brain rhythms has been used as an alternative method for quantifying functional changes in brain activity [1, 2, 3]. Recently it was applied successfully to the investigation of processes during speech perception and production [4, 5].

The neurophysiological plausibility regarding the evaluation of brain rhythms in different frequency bands is given by a number of experimental findings reviewed by Lopes da Silva [6]. As to the question of whether oscillations in neuronal networks, as reflected in brain waves, are merely epiphenomena, or reflect general functions, it has been suggested that higher frequency oscillations (above 20 Hz) may mediate the formation of assemblies of neurons that represent a given stimulus pattern. The activity of such an assembly of neurons is characterized by the coherent firing of large groups of neurons. Experimental evidence for coherent neuronal activity induced by changes in essential features of sensory stimuli comes from local slow wave field-, single- and multi-unit recordings from the olfactory system and the visual system of animals (for review see [7]). If these coherently activated neurons represent some part of a cortical cell assembly, then cell assemblies can be assumed to be the units by which elementary cortical functions are realized. Thus, one way of exploring these distinct cognitive processes in the human brain could be to measure coherent oscillations with non-invasive techniques such as EEG or MEG.

For the detectability of cell assembly activity using MEG recordings, the number of coherently activated neurons and their spatial distribution is critical. Previously, only simulation studies were available to estimate the number of neurons belonging to one cell assembly. Aertsen and coworkers [8] recently reported sizes of cell assemblies ranging from a few thousand to several ten thousand neurons, with a fraction of 1/10 to 1/5 of the assembly activated at any given moment (within a 2-4 ms time-window). This means in the ideal case, that about twenty thousand neurons of one cell assembly could be activated at a time. As pointed out by Wikswo [9], the combined activity of ten to one-hundred thousand neurons is necessary to generate a magnetic field detectable outside the skull. From these considerations it is clear that only large, not too widely distributed cell assemblies can be detected noninvasively. Given that the estimated number of neurons belonging to one cell assembly is correct, MEG recordings of cell assembly activation may operate near the boundary of detectability.

The first goal of this study was to quantify stimulus induced changes in oscillatory brain activity and to analyze whether or not these changes were different for the perception of language compared to non-verbal stimuli. Secondly, the study examined the dependence of oscillatory brain activity on the modality of stimulus presentation, and analyzed topographical aspects, in particular, hemispheric lateralization. A third question was whether or not the changes in oscillatory brain activity could be explained by an activity of word-specific cell assemblies, as suggested by the "neurological theory of language" [10]. This model states that the activation of word specific cortical cell assemblies can be measured as an enhancement of the normalized spectral power in frequency bands above 20 Hz across activated brain regions. If this is the case, there should be an enhancement of spectral power in the gamma band in response to the auditory and visual presentation of word stimuli across brain regions relevant for language processing, but one should detect topographically different alterations in gamma band activity when non-verbal stimuli are presented.

Methods

MEG (37-channel system, Magnes®, BTi) was recorded over the left and right hemispheres in twelve right-handed subjects. During the session, words two syllables in length and physically similar non-verbal stimuli were presented with equal probability in a randomized order in either the visual or auditory modality. Approximately 15% of these stimuli served as targets and each subject responded to them by deflecting a beam of light. The experiment consisted of 4 recording blocks (auditory and visual stimulation respectively, for the left and the right hemisphere). The order of blocks was balanced across subjects with the constraint that half of the subjects were measured over the left hemisphere first and that half of these subjects started with the auditory modality. For each subject, a total of 252 stimuli per experimental condition were delivered.

Power spectral estimates were obtained for subsequent 107 ms Welch-windowed segments (overlapping by one half) during a pre-stimulus baseline of 600 ms and 1200 ms post-stimulus onset. The geometric mean was calculated across trials. The mean time-dependent spectra were transformed with respect to baseline-related changes in two different ways. First, normalized mean power spectra were calculated by baseline dividing. A second parameter was the portion of single trial spectral power estimates above or below the mean single trial baseline spectral power (relative gain). Effects were considered only if both the normalized spectral power and the relative gain indicated statistically significant differences between experimental conditions.

Results

Fig. 1 illustrates changes of normalized spectral power values in time and frequency. These spectral power values represent the verbal condition in the auditory modality. Spectral power was grand averaged across all channels over the left hemisphere and then across all subjects. There are two major time/frequency regions where the spectral power is enhanced compared to the baseline and one time/frequency region showing a reduction of spectral power in most of the experimental conditions. Regions of enhanced spectral power occur around 129 ms in the 9.3–18.6 Hz bands, and around 883 ms in the 13.9–32.5 Hz bands. A reduction of normalized spectral power is seen around 345 ms in the 13.9–41.8 Hz range. When comparing these changes between experimental conditions, no modality independent and hemisphere specific reduction or enhancement of spectral power was found.

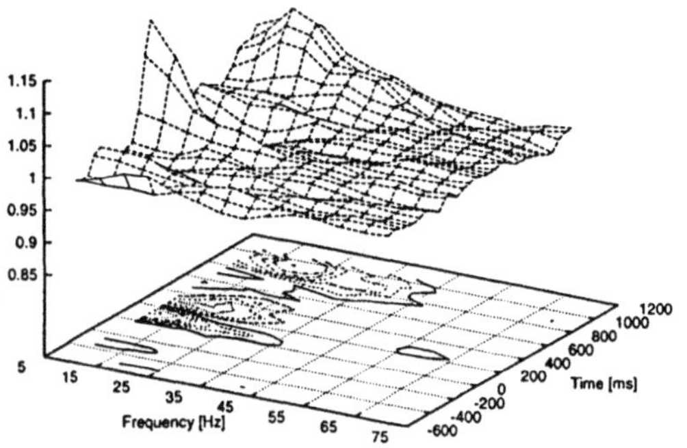

Fig. 1: Plot shows changes of normalized spectral power over the left hemisphere for the verbal condition in the auditory modality grand averaged across all subjects. The x-axis represents the frequency, the y-axis the latency, and the z-axis the normalized spectral power. The surface plot indicates the changes in normalized spectral power in the time-frequency plane. The same information is illustrated in the bottom half using an isocontour plot with a contour step of 1.5%. A normalized spectral power of 1 corresponds to the mean spectral power across nine pre-stimulus time windows (separately calculated for each frequency band). A normalized spectral power of 1.15 indicates an enhancement by 15% above baseline level.

In the higher frequency range, Fig. 1 shows a slight enhancement of the mean normalized spectral power in the time window centered at 237 ms in the 60.3 Hz and 65.0 Hz bands. A three-way univariate analyses of variance (ANOVA) for a 2 MODALITIES (auditory vs. visual) x 2 STIMULUS CLASSES (verbal vs. non-verbal) x 2 HEMISPHERES (left vs. right) design (repeated measures) delivered a statistically significant STIMULUS CLASS by HEMISPHERE interaction ($F(1/9)=7.42$; $p<0.03$; only results for the 60.3 Hz band will be reported in detail). There was no influence of MODALITY on this effect. For a more detailed topographical analysis of this effect, subaverages of the normalized spectral power from channels over the frontotemporal, the centroparietal and the temporooccipital brain regions were calculated. Statistical analyses using the BRAIN REGIONS (3 levels) as an additional repeated measures factor revealed on top of the STIMULUS CLASS by HEMISPHERE interaction ($F(1/9)=9.51$; $p<0.02$) a significant main effect of the BRAIN REGIONS ($F(2/14)=4.44$; $p<0.04$). The BRAIN REGIONS did not, however, significantly interact with other repeated measures factors. Post-hoc Scheffé comparisons suggested that the main effect was mostly caused by differences between centroparietal and frontotemporal brain regions ($p<0.04$). Mean values of the normalized spectral power showed a well expressed enhancement over the right hemispheric centroparietal and temporooccipital brain regions for the non-verbal condition, whereas for the verbal condition the enhancement of spectral power was largest over the corresponding left hemispheric brain regions for both the auditory and the visual modality. An extended view of the normalized spectral power changes for the 60.3 Hz frequency band as a function of time is presented in Fig. 2.

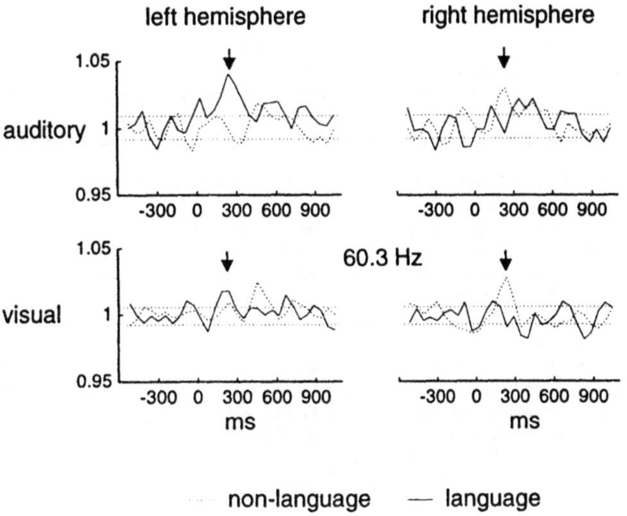

Fig. 2: The waveforms show the changes of normalized spectral power as a function of time for the frequency band around 60.3 Hz. They are averaged across channels over centroparietal brain regions for the auditory modality and across temporooccipital brain regions for the visual modality (and then across all subjects). The dotted line represents the non-verbal condition; the solid line indicates the verbal condition.

Discussion

The results of this study show that an enhancement of the normalized spectral power compared to a baseline level in the frequency band around 60 Hz and the latency range around 300 ms might be interpreted as a correlate of specific cognitive processing. In an experimental situation where simple lexical processing took place, the normalized spectral power was predominantly enhanced over temporooccipital and centroparietal brain regions of the language-dominant left hemisphere. In contrast, during the perception of visually and acoustically presented non-verbal patterns, the enhancement of the normalized spectral power was largest over centroparietal areas of the right hemisphere. This result might be taken as evidence for the activation of those cortical cell assemblies responsible for language processing located for the most part in the left hemisphere, thus confirming Braitenberg and Pulvermüller's [5, 10] cell assembly theory of language processing, i.e. word presentation should lead to a cell assembly ignition, because an adequate stimulus has been perceived, while presentation of non-verbal stimuli should fail to ignite specific assemblies in the same brain regions. Gamma-band responses (above 25 Hz) to words should be stronger than responses to non-verbal stimuli over brain regions processing the words. This hold true for the present results. As all subjects were right-handed one could expect that speech would be predominantly processed in this region [11]. The results for the non-verbal stimuli indicated a predominant activation of brain regions in the right hemisphere, as assumed for the processing of visual-spatial information [12] or complex acoustic pattern and prosodic information. This might be taken as evidence for the activation of other cortical cell assemblies situated for the most part in the right hemisphere thus accounting for the processing of non-verbal patterns. However, the percentage of change in the normalized spectral power was quantitatively small. Given that a). cell assemblies exist, b). specific cell assemblies were activated during the experimental procedure and c). the applied methods of data analyses were adequate, one can conclude that MEG recordings presumably of cell assembly activation were slightly above the threshold of detectability.

Changes of the normalized spectral power in other time/frequency bins indicated differences in the processing of stimulus classes not directly related to linguistic processing. The reduction in normalized spectral power (15-45 Hz) mirrored task specificity rather than stimulus specificity alone, thus not confirming previously published results [5]. In contrast, the left hemispheric enhancement of spectral power present around 240 ms in the 60-65 Hz band seems to reflect oscillatory patterns specific to the processing of words.

References

[1] Kaufman, L., Curtis, S., Wang, J.-Z., and Williamson, S. J. Changes in cortical activity when subjects scan memory for tones, Electroenceph clin Neurophysiol, 1991, 82: 266-284.

[2] Pfurtscheller, G. Event-related synchronization (ERS): an electrophysiological correlate of cortical areas at rest, Electroenceph. clin. Neurophysiol., 1992, 83: 62-69.

[3] Salmelin, R., and Hari, R. Characterization of spontaneous MEG rhythms in healthy adults, Electroenceph. clin. Neurophysiol., 1994, 91: 237-248.

[4] Klimesch, W., Pfurtscheller, G., Mohl, W., and Schimke, H. Event-related desynchronisation, ERD-mapping and hemispheric differences for words and numbers, Int J Psychophysiol., 1990, 8: 297-308.

[5] Pulvermüller, F., et al. Brain rhythms, cell assemblies and cognition: evidence from the processing of words and pseudowords, PSYCOLOQUY, 1994, 5(48): brain-rythms.1.Pulvermueller.

[6] Lopes da Silva, F.H. Neural mechanisms underlying brain waves: from neural membranes to networks, Electroenceph. clin. Neurophysiol., 1991, 79: 81-93.

[7] Singer, W. Synchronization of cortical activity and its putative role in information processing and learning, Annu. Rev. Physiol., 1993, 55, 349-374.

[8] Aertsen, A., Erb, M., Palm, G., and Schütz, A. Coherent assembly dynamics in the cortex: multi-neuron recordings, network simulations and anatomical considerations, In: Pantev, C., Elbert, T., and Lütkenhöner, B. Oscillatory Event Related Brain Dynamics, New York, London, Plenum Press, 1994, pp. 59-83.

[9] Wikswo, J.P. Biomagnetic sources and their models, In: Williamson, S.J., Hoke, M., Stroink, G., and Kotani, M. Advances in Biomagnetism, New York, Plenum Press, 1989, pp. 1-18.

[10] Braitenberg, V., and Pulvermüller, F. Entwurf einer neurologischen Theorie der Sprache, Naturwissenschaften, 1992, 79: 103-117.

[11] Rasmussen, T., and Milner, B. The role of early left-brain injury in determining lateralisation of cerebral speech functions, In: Dimond, S., and Blizzard, D., Evolution and lateralization of the brain, New York, New York Academy of Sciences, 1977.

[12] Rösler, F., Heil, M., and Henninghausen, E. Distinct cortical activation patterns during long-term memory retrieval of verbal, spatial and colour information, Journal of Cognitive Neuroscience, 1995, 7: 51-65.

Acknowledgement

Research was supported by grants from the Deutsche Forschungsgemeinschaft

Singular Spectrum Analysis applied to brain signals

Gaeta, R.[1], Benzi, R.[1], Peresson, M.[2] and Narici, L.[1]

[1]Dipartimento di Fisica - Università di Roma "Tor Vergata", Roma, Italy; [2] Divisione di Neurologia, Ospedale Fatebenefratelli - S. Pietro, Roma, Italy

Introduction

A large number of concurrently active processes are usually responsible for the spontaneous or evoked Electro/Magneto Encephalogram (EEG & MEG). Rhythmical activity, often referred to as "oscillating" activity, is most often measurable in many areas of the cortex. Interest in this kind of activities is increasing fast, due to the recent evidences of their correlation with functional, cognitive and perceptual activities. It has been shown that these are linked to rhythms within narrow frequency bands and that several times activities with different functional correlates share the same frequency window and often approximately the same source location.

The investigation of these "oscillations", their behavior during or following sensory stimulation or performance of cognitive tasks, is, therefore, a most interesting study in order to discriminate among similar activities with different correlates, as well as to approach the undestanding of processes such as learning and habituation. However most of the analysis tools to measure the dynamics of the many rhythmical activities sharing the same frequency band present important disadvantages. Time studies of narrow band filtered signals require prior knowledge of the band and risks to distort the real signal cutting important parts of it. Tools such as running FFT techniques [1] rely on same sort of time smoothing and therefore partially loose the time information, others (for example Single Sweep Analysis [2]) rely on pre-determined models of the oscillation and may force non-existent features in the dynamics.

In this paper we propose a novel technique (Singular Spectrum Analysis, SSA[3,4]) that is able to discriminate oscillatory activities in noisy signals, and to follow their dynamics. As a first test we apply this technique to neuromagnetic measurements of the activity (spontaneous as well as evoked by bursted rhythmical visual stimulation) over the occipital cortex of a normal subject.

The SSA technique has been recently developed and applied to studies of oscillatory components in the oceanographic dynamics or to the investigation of slow oscillation in the atmospheric dynamics. In these studies this novel technique proved to be quite useful. However in such cases the Signal to Noise Ratio (SNR) was favorable. The poor SNR proper of brain signals required a revision of the SSA procedure to identify oscillations in order to achieve meaningful results. SSA technique combined with our novel oscillation-retrieving procedure permitted to succesfully identify several oscillations in the extended α band (6.0 Hz - 14.0 Hz) on a relaxed subject, over the occipital cortex , and to compare them to the oscillation measured following repetitive rhythmic visual stimulation.

In the following we present a description of the SSA technique and of our oscillation-retrieving procedure. The reader interested in more technical details about SSA is addressed to the original papers [3,4]. The results from the first test of this technique, performed on neuromagnetic signals, are discussed and compared to those obtained. on the same set of data, with Fourier-based analyses [5,6]. We mention also first results of the dynamical studies relative to the found rhythmical activities.

Methods

The SSA is an application of the Karhunen-Loève expansion theorem. It is based on a Principal Component Analysis (PCA), performed in the vector space of delay coordinates for a single time series. It expands a process $X(t_i)$ as a sum of M components where M is an arbitrarily chosen 'embedding dimension', in contraddistinction with the fixed dimension of the data vectors in PCA. Vautard and Ghil (VG) [3,4] have shown that whenever in the $X(t_i)$ process a periodic oscillation is present, then a couple of "nearly equal" eigenvalues exists in the spectrum of the autocovariance matrix and the associated eigenvectors are orthogonal. In more detail, let be **X** $= (X_1, X_2, \dots, X_n)$ a stationary process with zero average. We construct the M x (N-M+1) matrix **Y** which has as i^{th} row the vector $(X_i, X_{i+1}, \dots, X_{i+N-M})$ $[1 \leq i \leq M]$. Let **T** be the autocovariance matrix of the process **X** (estimated by $\mathbf{Y} \cdot \mathbf{Y}^T/N$). We define **S** as the MxM matrix which has the autocovariance eigenvectors (Φ) as rows. We also define

$\Theta = S \cdot Y$ the M x (N-M+1) matrix featuring as rows the M 'principal components'. It has been shown that the process **X** can be succesfully reconstructed as a sum of M Reconstructed Components (RCs) and that the sum of the RCs relative to a couple of "nearly equal" eigenvalues λ of **T** describe the time behavior of the oscillation with no phase shift with respect to the original oscillation [3,4].

To maintain the maximum possible resolution when scanning for existent oscillations the highest possible M value should be used. External information (neurophysiological, models) should then be used to re-assemble those oscillations relative to the same physiological source.

Using the oscillation-retrieving procedure suggested by VG only oscillations that could account for a consistent part of the signal (SNR >> 0.5) were succesfully retrieved. Such procedure was infact tested with signals with quite low noise levels. We performed further tests (simulated and with the real neuromagnetic signals): we can show that this procedure does not work for brain signals, where the SNR may be quite lower than 0.5 for long time intervals [7]. The VG procedure was based on the comparison between couples of eigenvectors featuring similar eigenvalues. The spectra of these eigenvalues often featured more than one peak. The choice of the peak to be considered in the frequency comparison strongly depended on the used algorithm. This was, of course, an unwanted feature. To solve such problem, we propose to perform the comparison between the frequencies using the two RCs instead of the eigenvectors themselves. In this way it is possible to show that the results does not depend by the algorithm used and is quite robust, leading to a percentage of retrieving success several times better (more than 10 times) than with the VG procedure. Briefly: once all the Φ are calculated, as well as the RCs relative to similar λ and their respective Fourier spectra, we define "oscillation" of frequency

$$f = \frac{f_a+f_b}{2} \pm \frac{n}{2N}$$

the sum of the two "a" and "b" RCs ($\lambda_a = \lambda_b$), if the condition

$$|f_a-f_b| \leq \frac{n}{N}$$

is satisfied (n=acquisition rate; N=total number of points).

In most cases, when the frequency window where oscillations are looked for is centered on values $f_0 << 1/n$ it might be useful to reduce the number of points per second averaging K consecutive points. In this case the useful frequency window is ≈ [0, n/K] Hz

The neuromagnetic signals used to test the SSA technique and our novel retrieving procedure are measured with the 28 channel neuromagnetometer in use at the IESS-CNR in Rome [8] over the occipital cortex of a normal subject. We measured spontaneous activity and activity following intermittent bursted visual stimulation at rates r finely scanned ($\Delta r = 0.5$ Hz) within the extended a band (6 Hz - 14 Hz) [5,6]. Visual stimuli were delivered via two red leds positioned within a blackened mask, at an eccentricity of ≈ 5 degrees, in the left hemifield (fixation point was constantly indicated by two centered green leds). The magnetic field was measured over the right occipital region. Each stimulating burst contained 10 stimuli, the non stimulated periods between bursts was 1.3 s. Consequently the length of the entire epoch variated between 2.8 s (for r=6.0Hz) and 1.9 ms (for r=14.0 Hz). A complete SSA analysis was performed on the signal measured by the center channel of the neuromagnetometer. To check whether or not the found oscillation were localized on such channel we performed the same analysis, for a single stimulation rate (r=12.5 Hz), on the first Principal Component - PC - (relative to a PCA over all the channels), which explained 92% of the total variance.

Results

In Fig. 1 we show the retrieved oscillation from the center channel during the entire spontaneous and stimulated session. In the abscissa we present the stimulation rates and in the ordinate the frequency of the magnetic response. The dots correspond to the found oscillations. The diameters of the dots are linearly proportional to the rms amplitude of the reconstructed oscillations. The continuous horizontal line represent the largest oscillation found in the spontaneous measurement. The frequencies of the oscillations relative to the RCs of the first PC (r=12.5 Hz) are reported in the plot with asterisks.

The frequency values found for the existent oscillations in the signal appear to cluster around three regions indicated in the figure with dashed lines: i) on the f=r line, ii) on the f=2r line and iii) on the f≈9.5 Hz line. There is an appreciable slowing between the spontaneous main activity (≈ 9.8 Hz) and the oscillations apparently unaffected by the stimulation and clustered about 9.5 Hz [(iii) region]. This is more so farther from those r that elicit the strongest actvities on the (i) and (ii) regions (≈ 6 Hz, ≈10 Hz, ≈12 Hz).

Fig. 2 illustrates the behaviour of the sum of the two RCs with a frequency ≈ 12 Hz, as well as the sum of the RCs at ≈ 10 Hz (both relative to r=12.0 Hz). The dynamics of these two rhythms are compared with the the stimulation bursts, shown at the bottom. Zooms of the plots for a shorter time span are shown on the right.

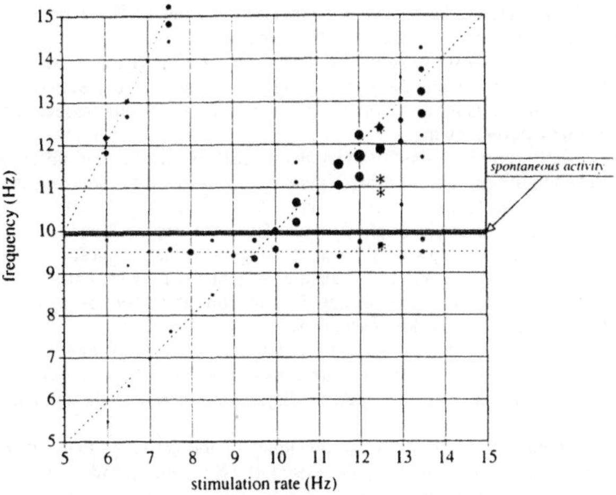

FIGURE 1 Plot of all the found oscillations in function of their frequency and rate of stimulation. Each RC is represented by a dot. The dot diameter is proportional to the rms amplitude of the RC. The horizontal solid line corresponds to the frequency of the RC found in the analysis of the spontaneous measurement. The asterisks correspond to the RCs relative to the first PC (obtained from all the channels) of the r=12.5 Hz measurement.

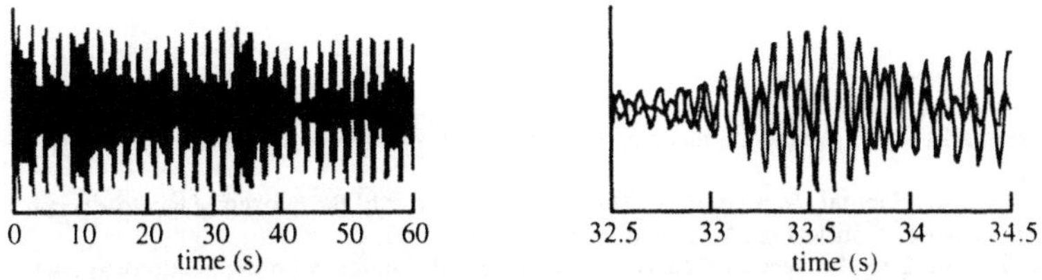

FIGURE 2 Time dynamics of the two oscillations (r=12 Hz) at ≈ 12 Hz (gray) and at ≈ 9.7 Hz (black) compared with the stimulating bursts (at the bottom, in gray). Note that only the former one is locked by the stimulation.

Discussion

The found oscillations, in the three 'regions', (i), (ii) and (iii) have been already evidenced with Fourier techniques [5,6]. The (i) region is the "driving region": cortical activity follows the driving of the stimulation. This driving effect is quite smaller at rates slower than the "spontaneous" a rhythm, that in this subject is at about 9.8 Hz. Also the (ii) region can be interpreted as "driving": it corresponds to the second harmonic of the stimuli, that, being sharp pulses, contains all the harmonics of the fundamental rate. On the basis of the inter channel cross correlations and of the topographical distribution of the activities [6], rhythms in the (iii) region have been interpreted as either unaffected by the stimulation rate, or locked by the stimulating bursts, as a component of the activity featuring its major peak at about 12 Hz.

It is interesting to note that most of the found oscillations appear in coupled frequencies with differences that are equal to the epoch rate (Figs. 1 and 2). This is due to the combined presence in the stimulation of the burst rate as well as of the epoch rate. This is not so in the case (iii) where the activities cluster around a constant ≈ 9.5 Hz frequency, indicating a lack of correlation with the stimulation rate. This is clearly evident in Fig. 2 where we present the results relative to r=12 Hz. At this rate the inter epoch interval is 2.05 s corresponding to 0.5 Hz. This corresponds to the difference between the center frequencies of the two added RCs (relative to the ≈12 Hz response).

A clear correlation with the stimulating bursts can be appreciated for the 12 Hz response while the ≈10 Hz one appears to be quite independent from the presentation of the stimuli.

These findings appear to confirm most of those obtained with Fourier analysis on the same data, while providing improvements in frequency resolution, and offering new insights. In this single channel test it is already possible to appreciate the presence of at least two distinct actvities within the same frequency band: one that follows the stimulating bursts, one undisturbed by them.

In this case the two discriminated activities feature different frequencies and different time dynamics. The first one (≈10 Hz), not linked to the stimuli, is probably generated by some of the sources also responsible for the α rhythm, as it is similar in frequency (and time behaviour) to the spontaneous α rhythm. The second (≈ 12 Hz) is strongly stimulus correlated. This response might generate from the cortical/thalamic pathways. Its dynamics (building up during the burst, and decaying after the end of the stimuli) may suggest that under visual stimulation the circuits between thalamo-cortical and reticular-thalamic cells first oscillates, under the stimulus driving, and then, at the end of the external driving, are dampened by the powerful inhibitory post synaptic potentials generated by neurons of the reticular nucleus in the thalamo-cortical pathways.

The SSA test on the first PC relative to r=12.5 Hz (asterisks in Fig. 1) shows clearly that the effect is not confined to the single channel studied but that the same activities are present over most of the channels. It also suggests a new activity at ≈ 11 Hz (also locked by the stimulation), not present in the studied channel, but evidently detected by most of the others.

The proposed technique allows to study the dynamics of the activities that will not show in averaged evoked-MEG measurements (such as our stimulus uncorrelated ≈10 Hz rhythm), and to investigate in more detail the effect of the length of the non stimulated period. However, to fully exploit it, it is mandatory to obtain separate information, possibly neurophysiological, to perform the correct reconstruction of the rhythms. As an example we had reasons to separate the ≈ 10 Hz activity from the ≈ 12 Hz activity in Fig. 2, as they responded differently to stimulation. But in general we may not have enough information to decide which of the RCs describe the same activity and therefore should be combined together.

In summary this quite powerful technique may be used to retrieve dynamical information on cortical oscillations in many situations and, when used in conjunction with other neurophysiological information may allow a complete reconstruction of rhythms dynamics.

References

[1] Jones, D.L. and Parks, T.W., A resolution comparison of several time-frequency representations, IEEE-Trans. Sign. Proc. 40(2).

[2] Liberati, D., Narici, L., Santoni, A., Cerutti, S., The dynamic behavior of the evoked Magneto-EncephaloGram detected through parametric identification, J. Biomed. Eng., 1992, 14:57-63

[3] Vautard, R., Ghil, M.,Singular spectrum analysis in nonlinear dynamics, with applications to paleoclimatic time series, Physica D, 1989, 35:395

[4] Vautard, R., Yiou, P., Ghil, M., Singular-spectrum analysis: a toolkit for short, noisy chaotic signals, Physica D, 1992, 58:95

[5] Narici, L, Peresson, M., A new procedure to discriminate and study different rhythmical cortical activities on the basis of their responsiveness to simple sensory stimulation, 1996, this volume.

[6] Narici, L, Peresson, M., Discrimination and study of rhythmical brain activities in the α band: a neuromagnetic frequency responsiveness test, Brain Res., 1995, 703:31-44.

[7] Gaeta R., Singular Spectrum Analysis applicata a segnali cerebrali, Thesis (in italian), 1995, Physics Department - University of Rome "Tor Vergata".

[8] Foglietti, V., Del Gratta, C., Pasquarelli, A., Pizzella, V., Torrioli, G., Romani, G.L., Gallagher, W.J., Ketchen, M.B., Kleinsasser, A.W., Sandstrom, R.L., 28-channel hybrid system for neuromagnetic measurements, IEEE Trans-MAG, 1991, 27:2959-2962.

Acknowledgements

We are grateful to V. Pizzella and F. Tecchio for the help in the measuring session.

Characterization of Photosensitive Discharges by Oscillatory Responses to Driving Stimuli

Karhu, J.[1] **and Tesche, C.D.**[2]

[1]*Department of Clinical Neurophysiology, Kuopio University Hospital, Kuopio, Finland;*
[2]*Low Temperature Laboratory, Helsinki University of Technology, Espoo, Finland*

Introduction

Photosensitivity has drawn an increasing amount of interest since the first report of epileptic fits induced by television [1], mainly because of the dramatically increased exposure of children and adolescents to videoscreens and videogames. Two stimulus-dependent mechanisms seem to be involved in photosensitive reactions. In some individuals electroencephalographic (EEG) discharges and/or seizures are induced by the flicker of the screen, which is scanned at the same frequency as that of the A.C. mains supply. In others, seizures may be induced by the raster pattern, which presents a grating pattern alternating at half the mains supply [2]. In addition to pure sensory mechanisms, there is growing evidence of seizure induction by cognitive effort during videogames.

The sign of photosensitivity in EEG is a widespread discharge following intermittent photic or flickering stimulus. This paroxysmal discharge is called a photoconvulsive response (PCR) and it consists of sharp waves and/or atypical (poly)spikes and waves. The PCR may be accompanied by jerking movements of the muscles, i.e., myoclonus, by absence, or by no visible clinical signs at all. The most common combination is a generalized spike and wave discharge with slow wave frequency at 3-3.5 Hz and an associated absence. Several determinants of experimental stimuli which trigger PCRs are well known (for a review, see [2]). The frequency of intermittent photic simulation (IPS) is optimally 15-20 Hz. Longer duration and higher stimulus intensities result in a higher incidence of photosensitive reactions. Adding patterns to intermittent visual stimulation markedly increases the probability of evoking abnormal responses as compared with diffuse flashes. Patterns of small squares formed by narrow lines (raster patterns) and closely spaced parallel lines seem to be most effective.

Although numerous studies have investigated the stimuli which trigger photosensitive seizures, the brain structures and neuronal networks which mediate the sensitizing input and elicit PCRs are poorly known. EEG methods have met with limited success because of the "generalized" (or distributed) nature of the PCR. Other functional imaging tools such as PET and fMRI do not have sufficient temporal resolution to evaluate the dynamics of rapidly spreading cerebral discharges. In addition, several pathophysiological and methodological problems may hamper the study of triggered epileptogenic brain activity. The epileptogenic activity may be triggered by a diffuse, widespread brain structure such as the reticular system, it may initiate in subcortical structures, or it may be multifocal. Finally, the experimentally triggered discharges may not represent the cerebral activity which causes clinical symptoms.

Magnetoencephalographic (MEG) arrays provide both sufficient temporal resolution and spatial information to enable the identification of components of very rapidly spreading brain activity. The millisecond-scale temporal resolution is particularly valuable for the characterization of changes of responses to the fast-frequency stimuli which are an essential component of photosensitivity. This study was directed towards the evaluation of distributed brain activity during PCRs utilizing novel signal analysis tools developed for the exploitation of topographic MEG information [3,4]. We first characterized neuronal populations in the frequency domain by their topographic field patterns and resonant behaviors during intermittent photic stimulation (IPS). The abnormal brain activity during PCRs was then spatially filtered with respect to these topographies to reveal waveforms and interactions identified with activation of each topography.

Methods

A 122-channel whole-head MEG array (Neuromag-122™) was used to record ongoing spontaneous activity and magnetic responses evoked by a bright flash (Grass stroboscope; distance from the eyes 2 m). Results are reported here for data recorded from a 9-year old non-medicated subject. The photosensitivity of the patient had been previously verified in several paper-EEGs and in a 24-hour videotelemetry recording. Photic stimuli were delivered in 4-s trains at 1, 3, 6, 10, 20, and 35 Hz. The subject was seated in a magnetically shielded room (Euroshield Ltd.) during the recordings (passband 0.03-100 Hz, sampling rate 397 Hz). The accurate position of the subject's head inside the helmet-shaped magnetometer was detected by measuring the magnetic field produced by three coils attached to the scalp.

In time-domain analysis of electromagnetic signals recorded on the scalp, a model with one or more equivalent current dipoles (ECDs) is frequently used to approximate localized sources of synchronized post-synaptic activity in the brain [5, 6, 7]. In MEG, magnetic field patterns are calculated as a function of the amplitude, location and orientation of the ECDs. These parameters are varied until a reasonable fit (usually above 90%) is obtained between the calculated and measured field values (for a review see [8]). Information about underlying sources may also be obtained from data which has been transformed from the time domain to the frequency domain by taking the fast Fourier transform (FFT) [3, 9]. In particular, any inversion algorithm which is applicable to time domain data can also be formally applied to FFT transformed data. ,

Time-domain ECD models are often not efficient in the case of distributed or complex brain sources such as those eliciting PCRs and other "generalized" epileptic discharges. Spatially distributed neuronal networks have oscillatory properties with characteristic physiologic frequencies for example at 6 Hz, 10 Hz, and 20 Hz [10]. These networks may be "driven" by external stimuli around these frequencies [11]. Since photosensitivity clearly depends on stimulation frequency and most patients are sensitive to stimulation within the physiological frequency range, we characterized the neuronal sources activated during photic stimulation by scanning the spectral behavior of the recorded brain signals. Topographies for prominent responses at the frequency of stimulation and its (sub)harmonics were selected from the FFT of each of the 122 channels of evoked data. Spectral "peaks" generated during ongoing spontaneous activity were also used to define MEG topographies. The FFT of a single epoch of a data generated a complex number characterized by an amplitude A_k and complex phase $Ø_k$. Each topography was characterized by the specific choice of frequency and complex phase.

Evoked and spontaneous FFT field patterns which were clearly dipolar (see Fig. 1) were characterized by ECDs. These frequency-domain ECDs represent activity of neuronal populations with specific oscillatory content (see Fig. 2) [3]. A different strategy was used for distributed or complex field patterns which were not adequately modeled by a limited number of frequency-domain ECDs. The most general description of activity in the brain from MEG or EEG data is provided by the values recorded by sensors. Signal-space projection (SSP) utilizes the pattern of responses in the detector array to characterize components of the signal which may be generated by either distributed or localized brain sources [4]. Although specific distributions of current flow in the brain determine unique vectors s_i in a 122-dimensional "signal-space", it is impossible to extract a unique current distribution from a set of sensor readings. However, waveforms for the ongoing activity of a set of SSP components may be extracted from the data [4]. In this study, we used the 122 numbers which generated the *i*th frequency-domain topography to define unit vectors s_i in signal-space. Thus the FFT field patterns generated during photic stimulation and spontaneous brain activity were used to construct signal-space "filters" which were applied subsequently to data recorded during PCRs. The resultant SSP waveforms described the evolution of discharge in terms of the responses of distributed oscillatory networks in brain.

Results

Fig. 1 shows topographic FFT field patterns in the photosensitive patient for spontaneous oscillatory activity during relaxed wakefulness with eyes closed, for 3-Hz activity during 3-Hz photic "driving" stimulation, and for 10-Hz activity during 10-Hz "driving" stimulation. The field pattern which was elicited by the about 10.5-Hz spontaneous activity was dipolar with an ECD location consistent with those of occipital MEG alpha rhythm. The oscillatory activity during the 10-Hz IPS produces a complex topographic field pattern dominated by responses over occipital and parieto-occipital areas. This response was modeled adequately by two ECDs (goodness-of-fit over 90%). The oscillatory activity driven by the 3-Hz photic stimulation was most complex with some dipolarity over frontal brain areas and could not be modeled adequately by dipoles.

The field patterns for the frequency-domain 10-Hz ECDs are shown in Fig. 2. Although these generators have similar oscillatory content, the topograpbies are consistent with distinct ECD source locations and orientations in parieto-occipital and in occipital areas (called parieto-occipital and occipital SSP components). The parieto-occipital field pattern closely resembles that elicited by the spontaneous 9-Hz activity.

We used the FFT field patterns shown in Figs. 1 and 2 to define signal-space components for the decomposition of PCRs. Fig. 3 shows the temporal waveforms in parieto-occipital and occipital SSP components during 3-Hz, 6-Hz, 10-Hz, and 20-Hz photic stimulation *(left).* The frontal SSP component was derived from the FFT topography produced by the 3-Hz IPS (see Fig. 1). Frontal, parieto-occipital and occipital SSP components were all utilized in the decomposition of PCRs *(right).* The amplitude of responses for each SSP waveform is scaled to the values derived from the field patterns [4]. The occipital SSP component shows the least activation, whereas prominent activation of frontal and parieto-occipital SSP components shows synchronized activity during the discharge at around 4-5 Hz.

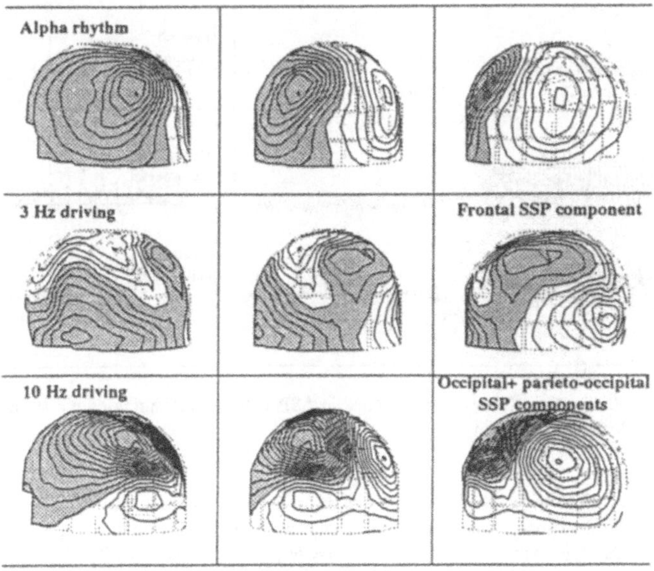

Figure 1 shows the FFT field patterns elicited by spontaneous activity and 3-Hz and 10-Hz photic stimulation.

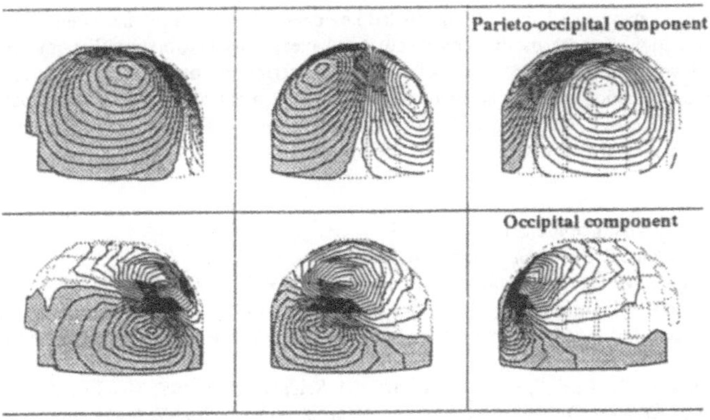

Figure 2 shows the simulated topographic field patterns for the two frequency-domain 10-Hz ECDs.

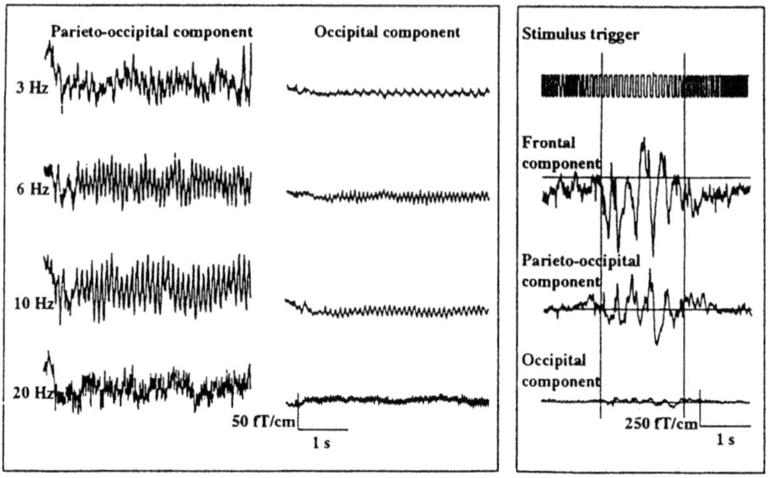

Figure 3 shows the spatially filtered temporal waveforms in SSP components during photic stimulation at various frequencies *(left)* and during a PCR (right).

Discussion

This case report from an ongoing study of photosensitivity demonstrates that complex oscillatory responses could be characterized by their topographies and resonant properties during driving utilizing FFT field patterns and frequency-domain ECDs. When a distributed photoconvulsive response was spatially filtered with respect to the SSP components defined in frequency-domain, activation in frontal and parieto-occipital components showed a time-locked, synchronized oscillatory behaviour during the discharge. The smaller amplitude of the response in the occipital SSP component suggests that either this component was not equally active during the PCR or that its activation was enhanced during driving stimuli relative to the activation during the PCR.

The oscillatory behaviour of spatially distributed neuronal networks in human brain is quite well defined, even if the overall function of these oscillations is not fully clear. The network oscillations are almost always present and can be manipulated by external stimuli. Characterization of these multiple networks in the frequency-domain and evaluation of their behavior during complex brain activation utilizing spatial filtering in signal-space seems to be a very promising and physiologically relevant technique for the study of "generalized" epileptic discharges and other distributed cerebral responses.

References

[1] Livingston, S. Am. J. Dis. Children, 1952, 83, 409.

[2] Hardmg, G. and Jeavons, P. Photosensitive epilepsy, 1994.

[3] Tesche, C. and Kajola, M. Electroenceph. Clin. Neurophysiol., 1993, 87:408-416.

[4] Ilmoniemi, R.J. and Williamson, S.J. Soc. Neurosci. Abstr., 1987, 13:46. Tesche, C.D., Uusitalo, M.A. Ilmoniemi, RJ., Huotilainen, M., Kajola, M. and Salonen, O. Electroenceph. Clm. Neurophysiol., 1995, 95: 189-200.

[5] Brenner, D., Lipton, J., Kaufman, L. and Williamson, S.J. Science, 1978, 199: 81-83.

[6] Scherg, M. and von Cramon, D. Electroenceph. Clin. Neurophysiol., 1986, 65: 344-360.

[7] Mosher, J.C., Lewis, P.S. and Leaby, R.M.. IEEE Trans. on Biomed. Engin., 1992, 39: 541-557.

[8] Hämäläinen, M., Hari, R, Ilmoniemi, R., Knuutila, J. and Lounasmaa, O.V. Rev. Mod. Phys., 1993, 65: 413-497.

[9] Lütkenhöner, B. In: 5th International Congress: International Society for Brain Electromagnetic Topography. Westfälische Wilhelms-Universität, Münster, Germany, 1994: pp. 54.

Equivalent Current Sources of the Auditory Evoked Gamma Band and Steady State Fields

Knuth, K.H.[1,2], Broadhurst, J.H.[2] and Schwartz, B.J.[3]

[1]Kresge Hearing Research Laboratory, New Orleans, USA; [2]University of Minnesota, Minneapolis, USA; [3]Scripps Clinic, La Jolla, USA

Introduction

Neural activity in the auditory cortex of the human brain can be evoked by presenting transient acoustic stimuli to the ear. Using the Magnes Biomagnetometer System at the Scripps Clinic in La Jolla CA, we measured the magnetic fields generated by this neural activity. The auditory steady-state response (SSR), first studied by Galambos et al.[1], is a sinusoidal response elicited by periodic trains of auditory click transients presented at rates ranging from 15-70 Hz. Galambos et al. proposed that the SSR was generated by a superposition of the individual transient middle-latency responses.

By presenting single and multiple auditory click transients at 25 ms intervals, we found that the response evoked by a long train of clicks (SSR) is generated by a nonlinear superposition of transient responses, and verified that several peaks in the transient response (middle-latency peaks also known as the auditory evoked gamma-band response, AEGBR), which occur at approximately 25 ms intervals, were responsible for the SSR. [2]

We used the sampled field data from the SQUID array to locate the positions of the source currents representing the activity. The response to a single click was found to be due to two sources, the mPa-mNb and the mNb'-mPb, named after their corresponding EEG peaks. One main source of the response to multiple click stimuli was found, the mPab-mNbc, which is probably the best fit current dipole to the field generated by the simultaneous activity of the mPa-mNb and the mNb'-mPb.

Methods

Three healthy right-handed subjects, two women and one man, with normal hearing and no history of otological or neurological disorders, aged between 25 and 40 years participated in the study. Magnetoencephalographic (MEG) measurements were made of the radial component of the evoked magnetic field using the Biomagnetic Technologies 37-channel biomagnetometer at the Scripps Research Institute. The subjects were placed in a magnetically shielded room to reduce background magnetic noise. To improve the signal to noise ratio, the responses from 256 to 1024 stimuli were averaged.

The sound stimuli were generated digitally by an Amiga 3000 computer and were delivered into the room via a plastic tube coupled to an earplug. The sound stimuli consisted of sets of single and multiple clicks, 0.5 ms square wave condensation clicks, numbering from 1-5 per set, with a separation of 24 ms between the individual clicks. This way, as the number of clicks in the stimulus are increased, the onset of the SSR could be studied. The interstimulus intervals between successive stimuli were randomly varied 0.8334 +- 0.1667 sec (the two-click stimulus had an ISI of 0.8584 +- 0.1667 s). The separate sets of stimuli (one click, two click, etc.) were not randomly mixed in their presentation, but each set was presented the designated number of times before proceeding to the next set.

The sounds were delivered to the right ear and the magnetic fields were measured over the contralateral hemisphere. Responses were digitized at a sampling rate of 1041.7 Hz and conditioned between 0.1 and 400 Hz using digital high and low pass filters. Further analysis was performed off-line. To perform source localizations of the 40 Hz activity, the magnetic field data was bandpass filtered about 40 Hz with a half-bandwidth of 12 Hz. Head shape data were taken to define the headframe coordinate system, and to locally fit a sphere to the measurement area. The system used the Levenberg-Marquardt nonlinear search algorithm[3] to solve the inverse problem. The results of the dipole fits were examined and only the dipoles with goodness of fit parameters greater than 0.95 were considered acceptable. Good responses were obtained in two of the three subjects, one man and one woman. Reproducibility of the dipole locations and magnitudes across experiments for a given subject allowed averages and standard deviations of the positions and moments to be calculated. Two dipoles were considered to be distinct as long as their two sigma volumes did not intersect. Due to the large variability in the dipole locations between subjects, the equivalent current dipoles were not averaged over subjects. In one subject the locations of these sources were superimposed on previously obtained MRI images.

Results

Acceptable dipole fits were obtained for two subjects using single and multiple click stimuli. The locations of these dipoles were extremely reproducible for both subjects across experiments allowing the averages and standard deviations of the dipole positions and magnitudes to be calculated. Fig. 1 shows the left anterior MEG recorded from Subject 1 in response to a single auditory click. The lower waveform represents the 40 Hz activity obtained by bandpass filtering the upper waveform. The vertical lines represent the times at which successful equivalent current dipole, ECD, fits were obtained. The dipoles were named according to the EEG peaks with which they coincide. Four distinct ECDs were found, mPa, mNb', mNb, and mPb. The ECD positions and magnitudes showed that these four sources represent two dipole pairs. Essentially two distinct true sources were detected, the mPa-mNb', and the mNb-mPb, for the single click stimulus.

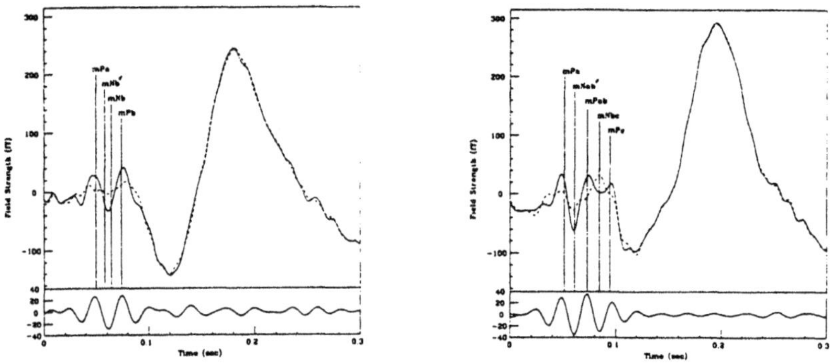

Fig. 1 The left anterior MEG in Subject 1 in response to a single auditory click (left) and two auditory clicks presented with an interval of 24 ms (right).

To demonstrate the effect of removing the DC component of the field, Fig. 2 shows the magnetic field contour plots at an interval of 12.4 ms, corresponding to the ECDs mPa (48.6 ms) and mNb (64.0 ms). The fields, after filtering, are of approximately the same form, but with opposite polarity. The arrows represent the approximate ECD location corresponding to the field. Most sources obtained were found to be a member of a source pair. This result lends credibility to the belief that the evoked 40 Hz activity is comprised of distinct sources.

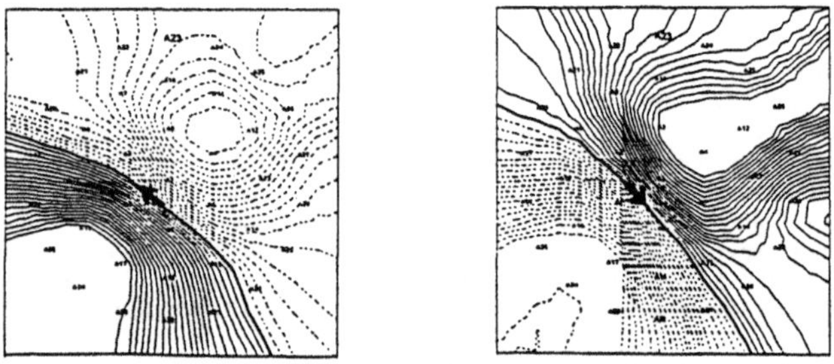

Fig. 2 A contour plot of the radial component of the magnetic field from Subject 1 found by interpolating the data sampled at the detector locations A1 through A37, at latencies 48.6 ms corresponding to mPa (left) and 64.0 ms corresponding to mNb (right). The approximate locations of the ECDs have been drawn in. Solid lines represent flux emerging from the head and dashed lines represent flux entering the head. The dark line represents zero field strength. The contour step is 1 fT.

Examining the responses of Subject 1 elicited by a two click stimulus, we found five distinct ECDs (Fig. 1 right). Similar ECDs were obtained for the other multiple (3-5) click responses. They are named according to the EEG peaks with which they coincide, but in this case the peaks from the two stimuli are overlapping. For example, mPab occurs at a time where the Pa and Pb peaks should be overlapping. The ECDs, mPab and mNbc, constitute a dipole pair. The source, mPab-mNbc, probably represents the best fit current dipole to the simultaneous activation of mPa-mNb and mNb'-mPb. We would have expected mPa to have its dipole pair mNb, but mNab' was found to be distinct from mNb at the two sigma level. This is interesting in that it suggests that the source mPa-mNb was modified by the simultaneous activity of mNb'-mPb during its activation. The final ECD, mPe, was found to be distinct from mPb and mPab. Similar sources were obtained for subject 2, but they were found to be more medial and anterior to those of subject 1.

The locations of the dipoles found above are displayed in Fig. 3 below. The left figure shows the ECDs due to a single click stimulus and the right figure shows the ECDs due to the two click stimulus. These figures show only the (x,y) coordinates of the ECDs in the headframe coordinate system, (i.e. the view is oriented so as to look down on the top of the head with the x-axis protruding from the nasion, and the y-axis protruding from the left ear). The boxes surrounding the ECDs represent two standard deviations in the positions of the ECDs.

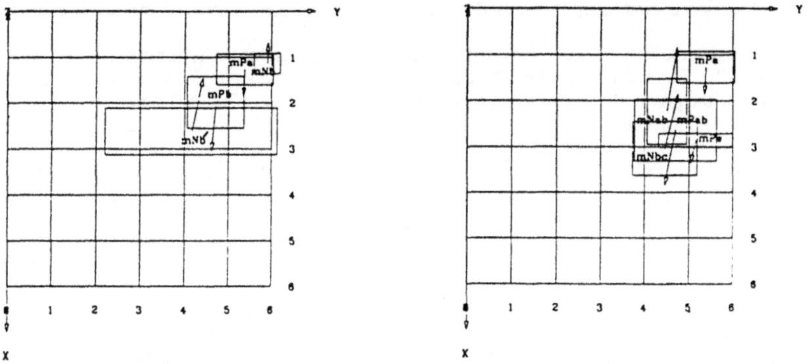

Fig. 3 The locations and the magnitudes of the ECDs obtained in Subject 1 elicited by a single click stimulus (left) and a two click stimulus (right). The view is looking down on the z-axis. The boxes surrounding the ECDs represent two standard deviations in their positions.

The ECD positions obtained for Subject 1 were superimposed on the subject's previously obtained MRI images, Fig. 4. These images show that the dipoles are within or fairly close to the Sylvian fissure.

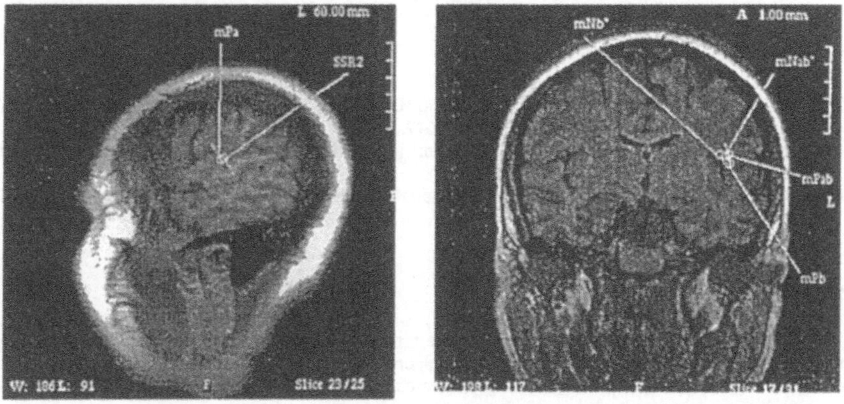

Fig. 4 The MRIs from Subject 1 superimposed with the locations of some of the obtained ECDs.

Discussion

By presenting stimuli consisting of small numbers of clicks (numbering from one to five), we were able to study both the AEGBR in response to single and multiple click stimuli. This allowed us to examine the relationship between the AEGBR and the onset of the SSR. By examining the amplitudes and durations of the gamma band activity we determined that the steady state response is generated by a nonlinear superposition of the AEGBR, also known as the 40 Hz early middle latency responses [2]. Using the magnetic field data we were able to locate the equivalent current dipoles for the sources responsible for the AEGBR, and the multiple click evoked GBR, which represents the onset of the SSR.

To locate the current sources responsible for the 40 Hz evoked activity, we bandpass filtered the data about 40 Hz with a 12 Hz half-bandwidth. This filtering removes the DC component of the field. This is significant in that a field that is simply turning on and off will appear as being generated by two alternating equivalent current dipoles at the same location, but with opposite dipole moments appearing at one half cycle intervals. Thus any current source will appear as a "dipole pair". The true current source, however, consists of the sum of these dipoles and the current source that generated the original field DC offset. This effect was observed with most of the sources we obtained which lends credibility to the finding that the 40 Hz evoked activity is the result of a few distinct sources.

We found that the single click evoked gamma band response was due to the activation of two sources, mPa-mNb and mNb'-mPb. The fact that the AEGBR consists of two independent sources which are not simultaneously active explains why the phase of the evoked gamma band field is observed to "sweep" across the cortex. Studies of the multiple click evoked gamma band field revealed a main source, the mPab-mNbc, which is probably the best fit current dipole to the two sources mPa-mNb and mNb'-mPb. In addition to this main source, the mPa was always observed; however, its "pair" which should have been mNab' was not in the same location. This implies that the true source mPa-mNb was affected by the second click during its activation. This may provide some insight into the nonlinear interactions occurring between mPa-mNb and mNb'-mPb during their simultaneous activation.

Similar results were obtained with Subject 2 for the multiple click stimuli (no acceptable fits were obtained for the single click stimulus in this subject). For the two click stimulus, five ECDs were obtained at similar latencies as Subject 1. In addition, these ECDs had the same relationships as the ECDs obtained for Subject 1. The relative locations of these ECDs also were comparable to those found for Subject 1; however, the absolute positions were found to be more anterior and medial than in Subject 1. Acceptable M100s were not obtained in this study. Other experiments performed examining the behavior of the SSR yielded too few acceptable dipoles for comparison, which is probably because the studies were concerned with amplitude and phase of the SSR and too few averages were taken to reduce noise.

These findings support our earlier results[2] that the multiple click GBR and the SSR are generated by a nonlinear superposition of 40 Hz middle latency responses (the AEGBR). The fact that the superposition was found to be nonlinear implies an interaction between the sources mPa-mNb and mNb'-mPb of the AEGBR, which was observed in the multiple click ECDs as the difference between mPa's pair in the single click (mNb) and multiple click (mNab) stimuli.

References

[1] Galambos, R., Makeig, S., and Talmachoff, P.J. A 40 Hz auditory potential recorded from the human scalp., Proc. Natl. Acad. Sci. USA, 1981, 78:2643-2647.

[2] Knuth, K.H., Broadhurst, J.H., Schwartz, B.J. Dynamical behavior of the generation of the auditory evoked 40 Hz steady state response. In preparation.

[3] Marquardt, D. An algorithm for least-squares estimation of nonlinear parameter. SIAM Journal of Applied Mathematics, 1963, 11:431-441.

Acknowledgements

We would like to thank the following people for their technical assistance: Steve Cobb, Joslyne Foley, Lacey Kurelowich, and Patti Quint. This work was supported in part by the University of Minnesota, Biomagnetic Technologies Inc., Scripps Clinic, Kresge Hearing Research Laboratory, and NIH NIDCD 5 T32 DC00007.

Local Modulation of MEG Rhythmic Activities Related to the Alternation of Binocular Rivalry and Binocular Fusion

Kobayashi, T., Kato, K. and Kuriki, S.

Research Inst. for Electronic Science, Hokkaido University, Sapporo, Japan

INTRODUCTION

Binocular rivalry (BR) is the process where dissimilar images presented to the left and right eyes at the same time are perceived alternately at any moment in time. This phenomenon has been studied primarily through psychophysical experiments [1, 2]. Recently, BR is considered to be a clue for understanding aspects of visual awareness [3]. Several physiological studies on cats and monkeys have been carried out in the past in an attempt to demonstrate a neural basis of BR [4, 5]. The next step in understanding BR is to elucidate the neural activities underlying this phenomenon in humans. However, experiments on human subjects require noninvasive techniques for observing the processes.

In previous papers [6, 7] we reported that altered neurophysiological processes with respect to the states of BR and binocular fusion (BF) may be reflected as variations in the attenuation of EEG power in the 8-13 Hz band in humans. In the present study, we measured and analyzed the spatio-temporal modulation of MEG power in the 8-13 Hz band related to the alternation from BF to BR and vice versa to study cortical activities involved in BR and/or BF.

METHOD

Five healthy subjects (males, 24-25 years old) with normal or corrected-to-normal visual acuity participated in the experiments. Two pairs of vertical and/or horizontal gratings used in the experiments are shown in Fig.1 with the time course of the stimuli's alternation and MEG data acquisition in the experiment. Each grating of the two pairs, 1° x 1° in visual angle, white, 10 cd/m^2, were projected on a screen (0.2 cd/m^2) in a shielded room, and presented to the left and right eyes separately through a mirror stereoscope. One of the two gratings was replaced with a grating which had an orientation orthogonal to the previous grating in intervals of 5 to 8 s. During the course of the experiment, subjects perceived BF and BR alternately by the replacement of the grating. We repeated this procedure until the selective average of the MEGs of 200 trials was completed. The subjects perceived a fused grating in BF when identical gratings were presented. On the other hand, when orthogonal gratings were presented, the subjects perceived one of the two gratings exclusively at any given time and then perception switched to the other grating in BR. This subjective switching occurred repeatedly during BR.

A SQUID magnetometer, with 19 second-order gradiometers covering a circular area of 146 mm in diameter on a spherical surface with a radius of 120 mm, was used for the measurements. The MEG data obtained were

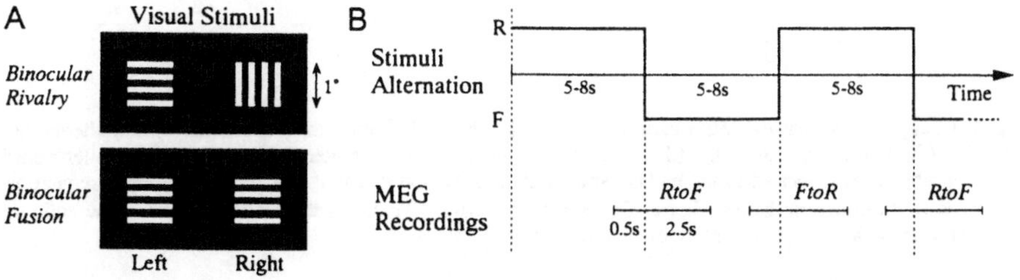

Fig. 1 A: Two pairs of gratings, 1° x 1° in a visual angle, white, 10 cd/m^2, on a black screen (0.2 cd/m^2), used as the visual stimuli of binocular rivalry (R) and binocular fusion (F). B: Time course of the stimuli alternation of R and F and MEG data acquisition.

digitally filtered in the 8-13 Hz band. To determine how the power in the 8-13 Hz band varies with time, the variance in 200 trials in the band was obtained with the time of alternation of the images as a reference trigger. Furthermore, we calculated the variance normalized by the averaged variance between -500 and 0 ms prior to the alternation from BF to BR. MEG data were obtained from two areas, where the center channel was located in the middle between P_Z and P_3, and P_Z and P_4 of the international 10-20 EEG electrode system. The measurement area covered the parietal, occipital and posterior-superior temporal regions in left and right hemispheres.

For two subjects (KK and TN), EEGs of 100 trials were measured and analyzed for the comparison with MEGs from 20 electrodes (F_{p1}, F_{p2}, F_7, F_3, F_z, F_4, F_8, T_3, C_3, C_z, C_4, T_4, T_5, P_3, T_6, P_z, P_4, O_1, O_z and O_2) placed on the scalp referenced to linked ears by the same procedure described above.

RESULTS

The sample plots of the normalized MEG variances in the 8-13 Hz band as a function of latency from the subject KK are shown in Fig. 2. A clear local modulation of the variances in the band was found after the alternation of visual stimuli. A similar modulation was observed in all subjects. The changes in normalized MEG variance in all subjects and their average at the channel located in the vicinity of P_3, where the largest variation was observed, are presented in Fig.3. The same modulation was observed at the corresponding locations in the right hemisphere. The variance was greatly attenuated until 700 ms after switching from BF to BR and then slightly recovered. On the other hand, no or only slight attenuation was observed in the variance until 700 ms and a subsequent rapid recovery to a stable level was observed in the alternation from BR to BF. The variance in BF, which averaged between -500 to 0 ms prior to the alternation of visual stimuli, was significantly larger ($n=5$, $p<0.005$; paired t-test) than that in BR.

The normalized variance in BR, which averaged between -500 to 0 ms, for both EEGs and MEGs in the 8-13 Hz band for two subjects, are shown in Fig.4 as topographies. The topographies were obtained by linear interpolation of the values at 20 electrode positions for EEGs and 19 sensor positions for MEGs. The top of each topography is the rostral direction. The normalized variance of EEGs was small in the occipital, parietal and posterior temporal regions of the two subjects. These spatial features are consistent with these of our previous results [6, 7]. On the other hand, large local variations were observed in the topographies of MEGs. Although there

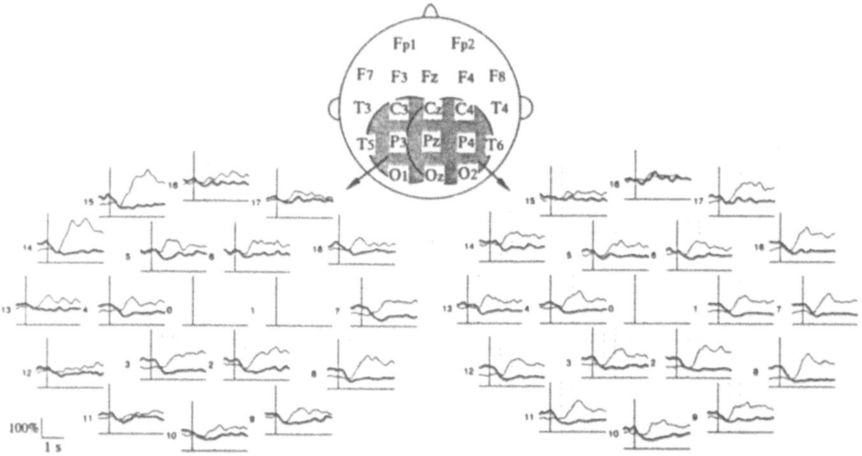

Fig.2 Changes in normalized MEG variance in the 8-13 Hz band (from Sub.KK) elicited by the alternation from BF to BR (thick line) and BR to BF (thin line) at two areas where the center channel #0 (unusable temporarily in these measurements) is located in the middle between P_Z and P_3, and P_Z and P_4. A part of two measurement areas is overlapped as illustrated in the top figure. The top of the channel layout is the rostral direction. The solid vertical line in each graph indicates the time of alternation of visual stimuli.

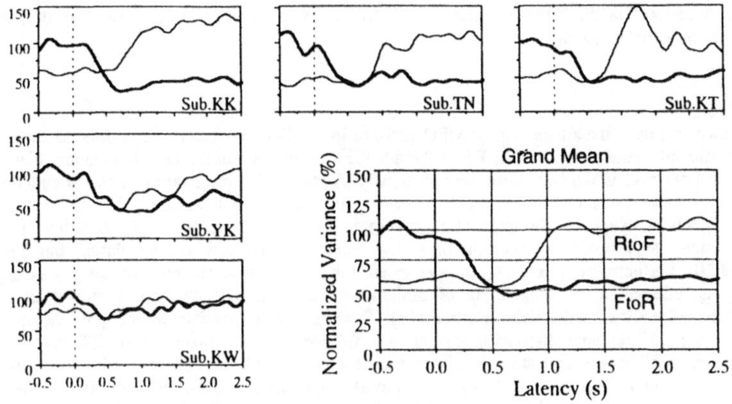

Fig.3 Changes in normalized MEG variance in the 8-13 Hz band elicited by the alternation from BF to BR (FtoR; thick line) and BR to BF (RtoF; thin line) in all subjects and their grand mean at the channel located in the vicinity of P_3, where the largest variation was observed in each subject.

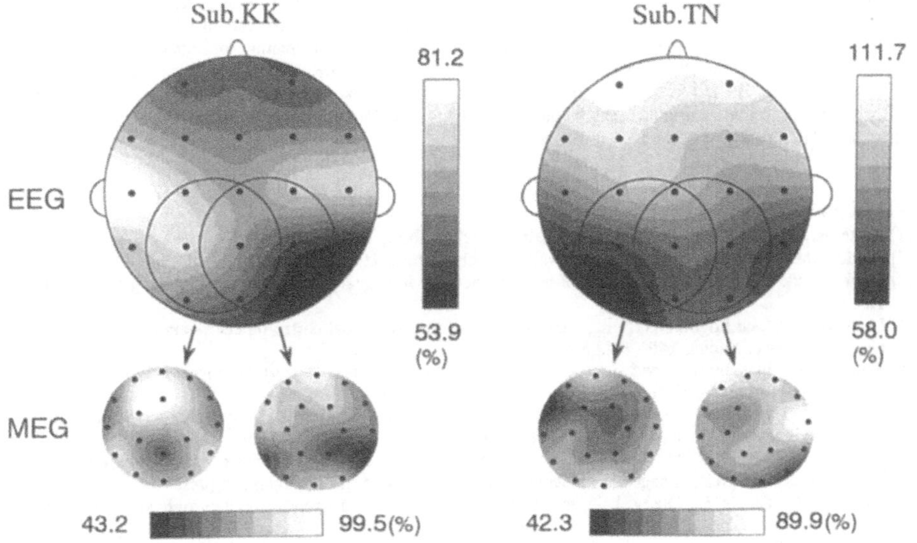

Fig.4 Topographical maps of the normalized EEG and MEG variance in BR, which averaged between -500 to 0 ms prior to the alternation of visual stimuli in the 8-13 Hz band, for two subjects (KK and TN). The top of the channel layout is the rostral direction. The EEG and MEG topographies were drawn, maintaining the relative size of their measurement areas. The 20 electrode positions for EEGs and the 19 sensor positions for MEGs are shown by dots in the figures. The scale bar was divided into 10 levels between the maximum and minimum values of variance in each map.

were individual differences in the MEG topographical distributions, a common feature was that they showed the smallest value in the parietal region.

DISCUSSION

As shown in Fig.3, the attenuation of MEG variance in the 8-13 Hz band in BR detected over the parietal region was significantly larger than that in BF, although the stimulus strength, i.e., luminance, size, shape and contrast, given to both eyes in the two states were physically identical. The mechanism of the attenuation of MEG power in the 8-13 Hz band, which is thought to be the same phenomenon as that of EEG, has not been clarified yet. However, it is widely accepted that the attenuation represents a change in the nature of excitation from basically synchronous to more asynchronous, occurring in a large number of cortical neurons during the period of the attenuation [8,9,10]. Furthermore, recent studies have revealed that cortical neurophysiological processing related to mental activity is reflected in the modulation of EEG power, and this phenomenon is known as event-related desynchronization and event-related synchronization [11]. Therefore, it is possible to interpret the modulation of MEG variance in the 8-13 Hz band observed in this study as follows. The attenuation until 700 ms and subsequent recovery after the alternation of visual stimuli reflects transient neural activities related to the alternation since the modulation was commonly observed in both cases of stimuli alternation between BF and BR. On the other hand, differences in the variance in the 8-13 Hz band at -500 to 0 ms prior to the stimuli alternation, where the variance seemed stable, reflects the altered cortical activities involved in BR and BF.

It is difficult to understand exactly what these altered activities are from only the results of the present experiments. Blake and Overton [12] tried to discover the site of binocular suppression by pshycophysical experiments, and proposed that the suppression occurs at a site in the human visual system after the locus of grating adaptation and, hence, after the striate cortex. Physiological and anatomical findings in the primate visual system suggest that different components of visual information processing are segregated into largely independent parallel pathways [13]. Livingstone and Hubel [14] have suggested, based on results of their psychophysical experiments, that BR results from activity in the magnocellular pathway. Although we are not certain whether the human visual system is similar to that of primates due to the lack of neuroanatomical information for humans, the idea of parallel pathways may explain the altered neurophysiological activities between BR and BF. The parvocellular pathway may be involved both in BR and BF because the form of visual stimuli may be processed commonly in the two state, whereas the magnocelullar pathway may be involved only in BR. Further study is needed to estimate where in the cortex and which pathway is involved in BR and BF.

REFERENCES:

[1] Blake, R., A neural theory of binocular rivalry, Psychol. Rev., 1989, 96:145-167.

[2] Wolfe, J.M. , Stereopsis and binocular rivalry, Psychol. Rev., 1986, 93:269-282.

[3] Crick, F. and Koch, C., Are we aware of neuronal activity in primary visual cortex?, Nature, 1995, 375:121-123.

[4] Leopold, D.A. and Logothetis, N.K., Activity changes in early visual cortex reflect monkeys' percepts during binocular rivalry, Nature, 1996, 379:549-553.

[5] Sengpiel, F., Blakemore C. and Harrad, R., Interocular suppression in the primary visual cortex: a possible neural basis of binocilar rivalry, Vision Res., 1995, 35:179-195.

[6] Kobayashi, T., Kato, K., Owada, T. and Kuriki, S., Difference of EEG spectral powers observed between binocular rivalry and binocular fusion, Fornt. Med. Biol. Engng., 1995, 7:11-19.

[7] Kobayashi, T., Kato, K. and Kuriki, S., Scalp distribution of the attenuated EEG rhythmic activities for dissimilar and/or moving images in binocular vision, Fornt. Med. Biol. Engng., in press.

[8] Kaufman, L., Curtis, S., Wang, J.-Z. and Williamson, S.J., Electroenceph. clin. Neurophysiol., 1991, 79:211-226.

[9] Markand, O.N., Alpha rhythm. Journal of Clinical Neurophysiology, 1990, 7:163-189.

[10] Steriade, M., Gloor, P., Llinás, R.R., Lopes da Silva, F.H. and Mesulam, M.-M., Basic mechanisms of cerebral rhythmic activities. Electroenceph. clin. Neurophysiol., 1990, 76:481-508.

[11] Pfurtscheller, G., Event-related synchronization (ERS): an electrophysiological correlate of cortical areas at rest. Electroenceph. clin. Neurophysiol., 1992, 83:62-69.

[12] Blake, R. and Overton, R., The site of binocular rivalry suppression. Perception, 1979, 8:143-152.

[13] Felleman DF and Van Essen DC. Distributed hierarchical processing in the primate cerebral cortex. Cerebral Cortex. 1991;1:1-47.

[14] Livingstone, M.S. and Hubel, D.H., Psychological evidence for separate channels for the perception of form, color, movement, and depth. The Journal of Neuroscience, 1987, 7:3416-3468.

Magnetic Oscillatory Components of Human Flash-VEP: Luminance and Eccentricity Functions and Effect of Acute Scopolamine.

Lopez, L.[1], Narici, L.[2], Conforto, S.[1], Romani, G.L.[1] and Sannita, W.G.[3,4]

[1]Institute of Advanced Biomedical Technologies, University of Chieti, Italy; [2]Department of Physics, University of Tor Vergata, Rome, Italy; [3]Center for Neuropsychoactive Drugs, University of Genova, Genova, Italy; [4]Department of Psychiatry, SUNY, Stony Brook, NY, USA

A full-field flash stimulation elicits a series of oscillatory potentials (100- 120 Hz) that are superimposed on retinal electroretinogram (ERG) [1] and are enhanced by highpass filtering at about 60-80 Hz. Comparable potentials were recorded from scalp locations [2,3] with longer latency (+25-30 ms). There is evidence [4] that a cortical/subcortical origin of these potentials is conceivable.

The functional significance of the oscillatory potentials has not been clarified yet, although there seem to be marked similarities between these potentials and the gamma band responses obtained with patterned stimulation [5], which would point to a stimulus-dependent synchronization that involves portions of the thalamo-cortical pathway and is instrumental in stimulus processing [6].

In the present work, a characterization of these potentials has been attempted, by means of different stimulus luminance and eccentricities and by cholinergic blockade.

Methods

Eight healthy subjects were recruited for a pharmacological study, and signed an informed consent. The study protocol involved a baseline recording with flash stimuli full field (at intensities ranging from 0.9 to 3.5 c.s.m^2) and in spots of varying diameter and eccentricity (1-3 cm; 5-30 deg from fixation indicated by a red LED point). In all cases, stimulus was monocular with duration of 20 μs and repetition rate of 0.8 Hz; white noise at 60 dB over threshold was delivered to mask strobe artefacts. The same sequence of stimuli was repeated at 30' from cholinergic muscarinic antagonist scopolamine in acute 0.5 mg i.m. doses).

Recordings at retinal level were obtained with conventional techniques (dermal electrodes on inferior orbital rim referred to linked mastoids); scalp recordings were made electrically (O1 and O2 vs Fpz) and magnetically (CNR-IESS 28-channel neuromagnetometer, 16 gradiometers, 9 magnetometers, 3 cancellation coils). MEG recordings were contralateral to the stimulated eye. After continuous acquisition, 65-75 artefact-free epochs were averaged and digitally filtered with 80-150 Hz bandpass.

Results

Oscillatory potentials (OPs) were recorded at retinal and scalp locations, in response to both full field and spot stimuli and were phase locked to stimuli and superimposed in latency to the ERG and the earlier waves of flash-VEP respectively. When the spot stimuli were used, there were no ERG, or retinal OPs, Scalp MEG recordings displayed definite oscillatory and broad band responses, as long as the stimuli were presented in the first 5° of eccentricity (Fig.1). An increase in latency of about 10-20 ms was observed as compared to full-field stimuli, but both latency and amplitude of the oscillatory responses were almost unaffected by the different spot diameters in the 5° position. When stimuli where presented in the 15° or 25° there was no oscillatory response, while the broad band VEP could still be recorded with graded increase in latency and decrease in amplitude.

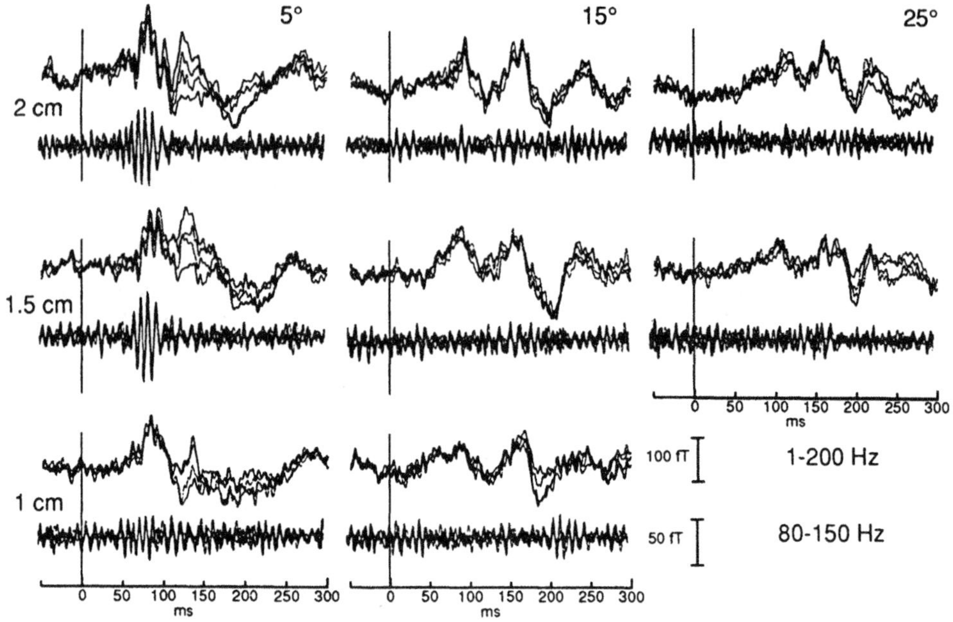

Fig. 1. MEG recordings in response to spot stimulation. Upper tracings Rows: increasing eccentricities; columns: increasing spot diameter. All spots were presented in the superior left half field

After injection of acute 0.5 mg scopolamine a marked reduction of the broad band late components was observed as described earlier [7], while the oscillatory responses were still recordable in all subjects. The oscillatory activity after scopolamine seemed to be enhanced. (Fig 2). These data were also confirmed by the analysis of the "locking indicator" L, which is a normalized version of a parameter recently introduced by other investigators [8,9] and describes the degree of locking of a narrow band response to an external stimulus. It is obtained by filtering of the raw data and calculation of i) the square of the averaged signal (where the average is performed with respect to the stimuli; ii) average of the squared signal. An integration is then performed and the square root of the ratio i/ii is taken. If the oscillating activity is locked by the stimulus L goes to 1. If the activity is mostly unlocked L goes to smaller values.

When we applied this analysis to our pre- and post-drug data, we observeed, in all subjects, an increase in locking of the 80-150 Hz activity in response to full field, after administration of scopolamine. The locking activity was also organized in two peaks of locked activity which occurred at about 50 ms and 80 ms from the stimulus, the second peak being generally more prominent and having longer duration (50 ms) (Fig. 3). When the same analysis was applied to the responses to spot stimuli we observed a reduction of locked activity after scopolamine.

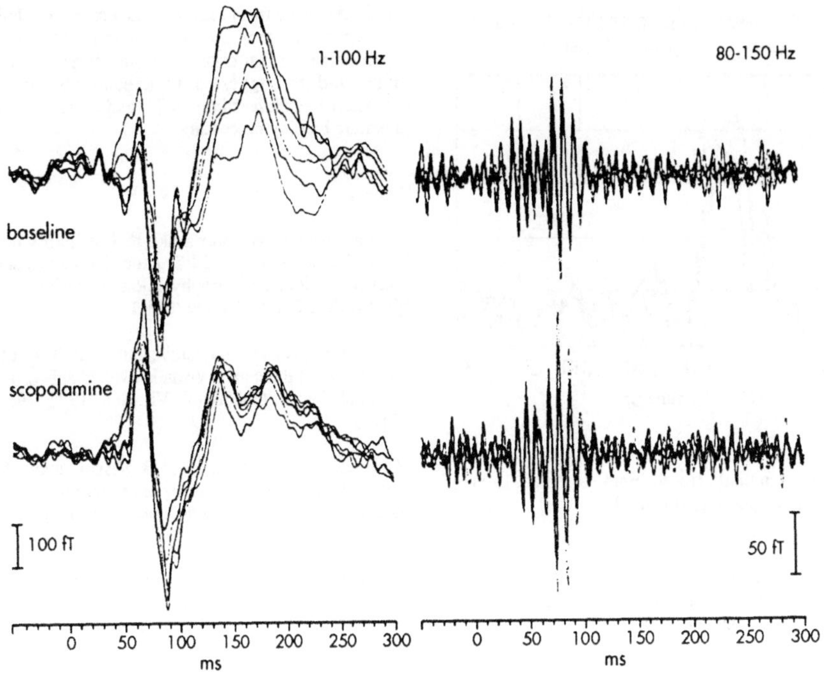

Fig. 2. MEG broad-band and oscillatory responses to full-field flash stimulation, before and after cholinergic blockade.

Discussion

These observations are consistent with experimental studies, in which *in vivo* recordings from latero-geniculate nucleus lamina 6 [10] indicate negligible contributions from volume-conducted pre-cortical phenomena to scalp-recorded oscillatory events.

In the present study, we observed the different behavior of the oscillatory activity in response to spot stimulus as compared to the broad band response. Namely, the oscillatory activity falls off as the stimulus is moved away from the central 5-10 degrees of visual filed, which would suggest independent cortical generators for broad-band VEP and oscillatory activity. The latter may reflect a predominance of events triggered by foveal/parafoveal stimulation in the generation of cortical. In this sense they would be the result of a parallel and partially independent stimulus processing with respect to the broad-band responses. The dichotomy between the effects of scopolamine on the

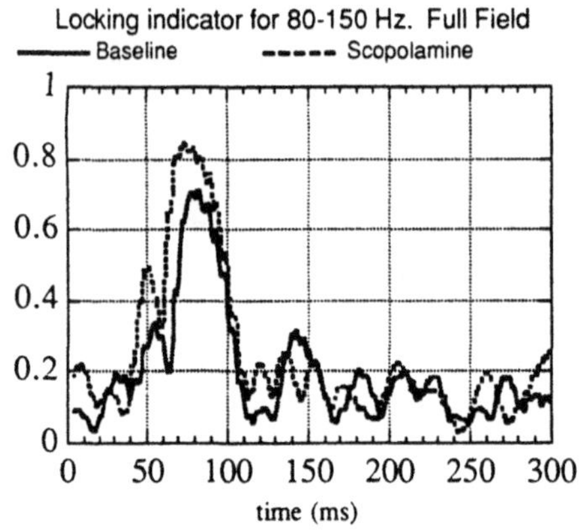

Fig. 3. Locking indicator diagram for the 80-150 Hz activity, in response to full-field

oscillatory responses to full-field and spot stimulation may also suggest a role of acetylcholine in the stimulus-related synchronization that occurs in visual cortex upon sensory input, and is suggested to depend on the physical characteristics of the stimulus [6] and is probably involved in feature binding processing.

References

[1] Karwoski, C. and Kawasaki, K. Oscillatory potentials. In: Heckenlively, J.R. and J.B. Arden Principles and practice of clinical electrophysiology of vision, St. Louis, Mosby-Year Book, 1991, pp 125-131.

[2] Harding, G.F.A. and Rubinstein, M.P. The scalp topography of the human visually evoked subcortical potential. Invest. Ophtalmol. Vis. Sci., 1980, 19: 318-321.

[3] Cracco, R.Q. and Cracco, J.B. Visual evoked potentials in man: early oscillatory potentials. Electroencephalography and clinical Neurophysiology, 1978, 45: 731-739.

[4] Lopez, L., Pasquarelli, A., Romani, G.L., Torrioli, G. and Sannita, W.G. Magnetic recording of cortical oscillatory potentials in response to flash stimulation in man. In: Baumgartner, C., L. Deecke, . Stroink and S.J. Williamson Biomagnetism: fundamental research and clinical applications, 7, Amsterdam, Elsevier, 1995, pp 170-173.

[5] Sannita, W.G., Lopez, L., Piras, C. and Bon, G.D. Scalp-recorded oscillatory potentials evoked by transient pattern-reversal visual stimulation in man. Electroenceph. clin. Neurophysiol., 1995, 96: 206-218.

[6] Gray, C.M., Engel, A.K., König, P. and Singer, W. Stimulus-Dependent Neuronal Oscillations in Cat Visual Cortex: Receptive Field Properties and Feature Dependence. European Journal of Neuroscience, 1990, 2: 607-619.

[7] Sannita, W.G., Balestra, V., Bon, G.D., Marotta, V. and Rosadini, G. Human flash-VEP and quantitative EEG are independently affected by acute scopolamine. Electroenceph. clin. Neurophysiol., 1993, 86: 275.

[8] Salmelin, R. and Hari, R. Spatiotemporal characteristics of sensorymotor MEG rhythms related to thumb movement. Journal of Neuroscience, 1994, 60: 537-550.

[9] Pfurtscheller, G. Event-related synchronization (ERS): an electrophysiological correlate of cortical areas at rest. Electroenceph. clin. Neurophysiol., 1992, 83: 62_69.

[10] Schroeder, C.E., Tenke, C.E. and Givre, S.J. Subcortical contributions to the surface recorded flash-VEP in the awake macaque. Electroenceph. clin. Neurophysiol., 1992, 84: 219-231.

Increased variability of the period duration for waveforms in Epochs of MEG data as a function of arousal state

Moran, J. E.[1], Tepley, N.[1,2]

Henry Ford Hospital[1], Detroit MI, USA; Oakland University[2], Rochester MI, USA

Introduction

In clinical evaluation of spontaneous magnetoencephalography (MEG) or EEG time series data, the state of arousal is assigned of one of 5 NREM scores (awake, stages 1-4) based on visual examination if the data for the presence of specific waveforms and the general amplitudes and frequencies of these waveforms[1]. However, this method of quantifying the state of arousal is not linearly related to changes in the spectrum of the data and totally insensitive to small shifts of arousal state. These deficiencies required us to develop a new system to quantify changes of arousal for our long duration MEG studies. Our arousal measurement parameters are based on a spectrum of period duration versus power, generated by our DPA (dynamic period analysis) technique. DPA spectra are not complete spectra, such as produced by FFT methods. We found FFT spectra to be ill suited for quantifying arousal state because of large subject to subject spectral differences dominated arousal spectral change. In our DPA spectra, spectral differences between individuals are relatively small compared to spectral changes associated with the state of arousal. Further, a DPA spectrum for any state of arousal can be divided into a linear combination of just two unique DPA spectra. One of these unique spectra is identical to a typical awake DPA spectrum and the other we have been able to identify as the time averaged power spectrum of a highly random component of the MEG data which is only present during states of arousal other than awake. We established these two unique DPA spectra by applying a principal component analysis of a large set of DPA period spectra and obtaining the two unique principal component spectra[2]. During shifts from awake to stages 2, 3, and 4 arousal states the fraction of the awake principal component DPA spectrum is significantly decreased while the proportion of the drowsy/sleep principal component DPA spectrum increases from the zero amplitude awake value. We have found the drowsy/sleep principal component DPA spectrum to exhibit changes in amplitude and amplitude variability which make it especially useful for distinguishing fluctuations of the awake state from a transition state of arousal that signifies the onset of stage 2 sleep. Also, we have established a relationship of our arousal parameters to the traditional classification system such that the results of our system can be related to this familiar and widely used scale of arousal state.

Methods

MEG recordings of spontaneous cortical activity obtained from six subjects were used to develop the two principal component DPA arousal state spectra. The MEG data were acquired with a 7 channel Neuromagnetometer (BTi model 607) equipped with second order gradiometers. The center channel was positioned at P3 of the international 10-20 EEG measurement system and 45 minutes of data was acquired from each subject lying on his/her side. For these subjects, the data was partitioned into 100 second epochs and DPA spectra generated for each epoch. For two additional subjects, included in this report, simultaneous EEG recordings were made with electrode placements, P3-A2 and Cz-F2. The 45 minutes of EEG and MEG data were partitioned into 25 second epochs. DPA spectra were produced for the center channel of the sensor array and the epochs of EEG data were scored for arousal state (awake, stages 1 - 4). For each DPA spectra, the amplitudes for the awake principal component spectrum and the drowsy/sleep principal component spectrum were generated.

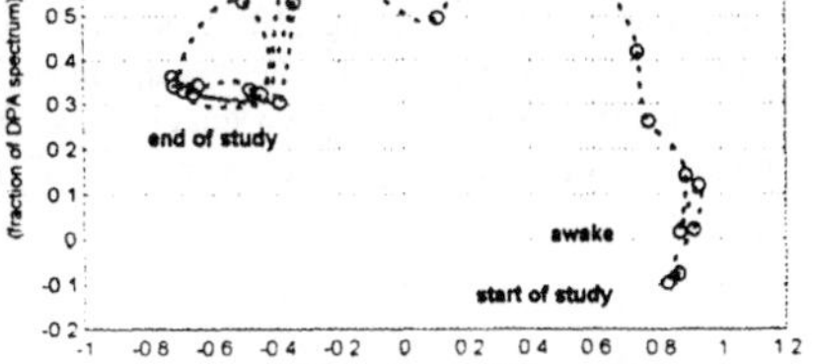

Figure 1 DPA spectral composition during 45 minute study, awake to sleep.

With these amplitudes plotted as X-Y coordinates, the angle from the horizontal (awake) axis was used to relate our two new parameters for quantifying arousal state to the existing system (see Fig. 1).

During subsequent analysis, these two arousal coordinate amplitudes were generated for DPA produced from epochs of durations 4, 8, 16, 32, 64, and 128 seconds. Since DPA spectra are time averaged spectra, these data were used to investigate changes in both the amplitude and amplitude variability of our principal component arousal parameters as a function of data epoch length and arousal state. The standard deviation of the amplitude of both the awake and drowsy/sleep components during extended data segments of a specific state of arousal were used to quantify the amplitude variability associated with the state of arousal and DPA epoch duration. The average amplitude for these same data segments was used to quantify the amplitude of the awake and drowsy/sleep components of the composite DPA spectra.

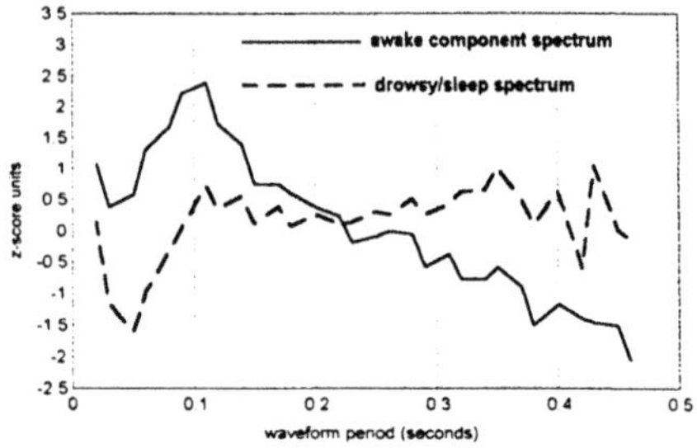

Figure 2 Awake and drowsy/sleep DPA spectra.

Results

The DPA spectra of the two principal components extracted from the DPA data using a principal component analysis are shown in Fig. 2. Our data indicates the onset of drowsiness gives rise to a shift in cortical function which introduces progressively greater durations of activity characterized by high variability in both frequency and amplitude. In Fig. 3, we demonstrate that the drowsy/sleep principal component is the time average spectrum of this random cortical process. For this Figure, 17 epochs (25 second epochs used to generate DPA spectra), recorded during stage 1 and 2 arousal states were averaged after subtracting the awake DPA spectral component. This averaged spectrum is nearly identical to the drowsy/awake spectrum in Fig. 3. The single epoch DPA spectrum plotted in Fig. 3 reflects the low rate of occurrence and random amplitude of these waveforms.

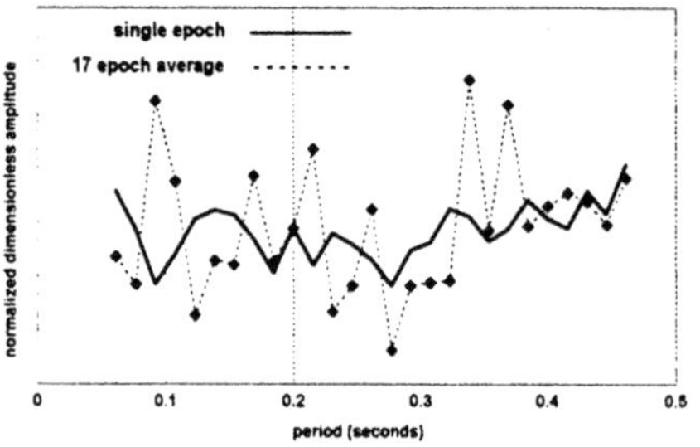

Figure 3 Single epoch DPA spectrum and 17 epoch average spectrum, after removal of awake spectral component.

In Fig. 4, the amplitude and amplitude variability of the fraction of the a DPA spectrum corresponding to the awake DPA principal component is graphed as a function of both arousal state and epoch duration used to generate the DPA spectrum. For both awake and stage 2 arousal states this fractional amplitude of the awake principal component requires a relatively short epoch of data before reaching a stable value with very low variability. In Fig. 4, both the amplitude and amplitude variability are nearly constant, especially for DPA epochs longer than 16 seconds. Relative to the awake state, the amplitude of the awake principal component of a stage 2 DPA spectrum is less and amplitude variability increased. These amplitude differences are nearly the same for all epoch durations, such that, the graphs of awake and stage 2 are almost parallel.

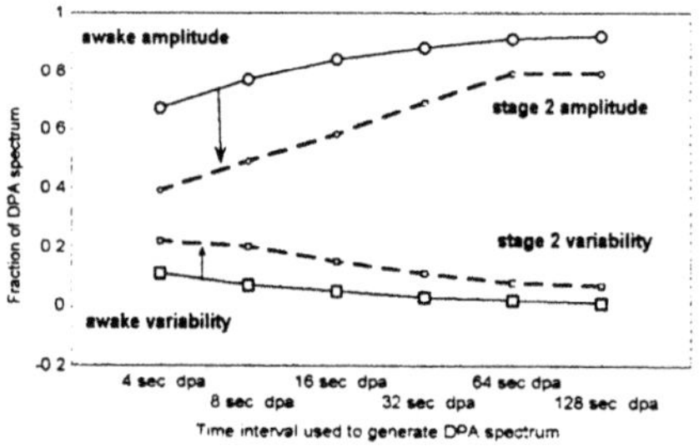

Figure 4 Change in the awake component of the DPA spectrum during transition from awake to stage 2 arousal state.

However, the amplitude and amplitude variability of the drowsy/sleep component content of awake and stage 2 DPA spectra demonstrate a more complex relationship to the state of arousal and the epoch duration used to generate the DPA spectra. As a result, this component is especially useful for recognizing the onset of sleep. This relationship allows the DPA epoch duration to be optimized for recognizing and quantifying changes of arousal. In Fig. 5, the drowsy/sleep content of the awake DPA spectra is nearly zero for all DPA epoch durations. However, for stage 2 DPA spectra, the amplitude of the drowsy/sleep fraction of the DPA spectrum is significantly increased when a longer epoch duration is used to generate these DPA spectra. Therefore, a larger amplitude difference of the drowsy/sleep component between awake and stage 2 arousal states is created when a longer DPA epoch duration is used. Even more interesting and potentially useful is the amplitude variability of the drowsy/sleep component as a function of the duration used to generate the DPA spectra. An adequate sample of the waveforms contributing to the awake DPA spectrum is obtained with a short epoch of data. Thus, random variability of the awake DPA spectrum is reduced when a longer epoch duration is used to generate DPA spectra. However, a very long epoch of data must be used to acquire all waveforms contributing to the stage 2 arousal state DPA spectrum. As a result, the variability of the drowsy/sleep component remains nearly constant until greater than 60 second epochs are used (see Fig. 6). Therefore, the difference in amplitude variability of the drowsy/sleep component between awake and stage 2 arousal states also increases when longer epoch durations are used. During sleep onset, the epoch duration effect on amplitude variability of the drowsy/sleep DPA component can be more dramatic than changes of the mean amplitude. In Fig. 7, a large increase in the variability of the drowsy/sleep component is easy to detect during the stage 1 sleep recorded just prior to the transition to stage 2 (1400 seconds to 1700 seconds). Notice, during this interval, the amplitude variability of the drowsy/sleep component is the nearly same as after the transition to stage 2 (1700 seconds to 2100 seconds). However, during this stage 1 epoch, the amplitude of the drowsy sleep component has an average amplitude identical to the awake state and abrupt increases with stage 2 onset. This sustained increase in amplitude variability is absent during other epochs of stage 1 sleep which alternated with epochs of awake. For this graph, a 32 second epoch duration optimized the compromise between maximizing the drowsy/sleep component amplitude difference between awake and sleep arousal states and the ability to resolve temporal changes of the arousal state. In actual application of this technique, the time series of drowsy/sleep amplitudes for more than one DPA epoch duration are examined for arousal state changes.

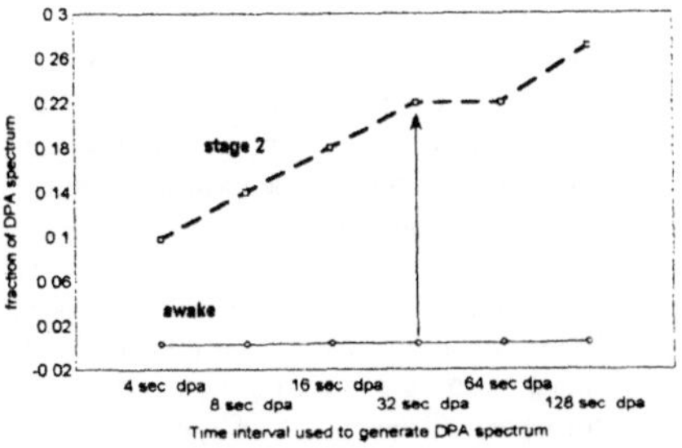

Figure 5 Amplitude of drowsy/sleep component of the DPA spectrum. Awake versus stage 2 arousal.

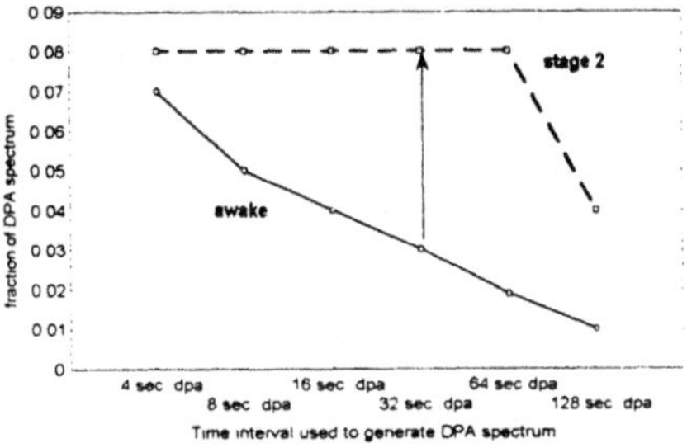

Figure 6 Variability of the drowsy/sleep component of the DPA spectrum. Awake versus stage 2 arousal.

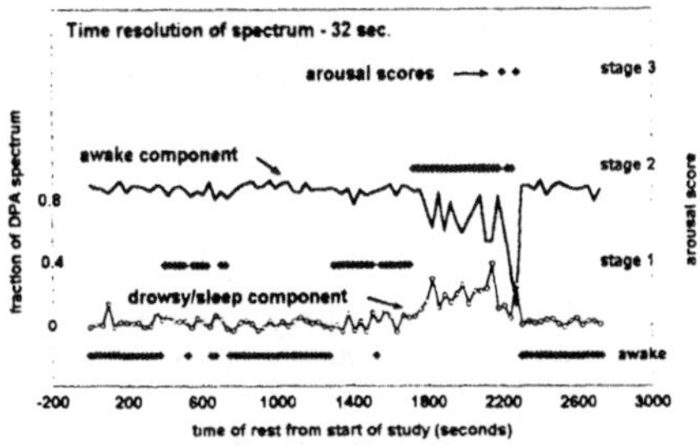

Figure 7 Time series of clinical arousal scores versus amplitudes of awake and drowsy/sleep fraction of the DPA spectra.

Discussion

Our results indicate that during the transition from awake to sleep some portion of the cortex begins to function in a mode distinctly different than the awake state cortex. The time average spectrum of the MEG waveforms arising from this drowsy/sleep mode of cortical function is identical to the DPA drowsy/sleep principal component we extracted. This spectrum has the property that the same time average power occurs at all period durations less than 0.46 seconds, included in our DPA spectra (see Fig. 2). DPA spectra are formed by counting the number of waveforms with each specific period duration and the corresponding waveform power. Therefore, during this mode of cortical function, for each period duration, the number of waveforms multiplied by the time average power of the waveform is equal to a constant for all period durations. If the power and duration of a waveform are on average proportional to the area of cortex generating the recorded activity, then this result implies that the number of waveforms arising from cortex of specific area is inversely proportional to the size of that area for this mode of cortical function.

Compared to awake state waveforms, drowsy/sleep waveforms occur infrequently and are characterized by large amplitude variability. Thus, a very long epoch duration is required to accumulate enough waveforms of all periods to produce a completely stable DPA spectra during sleep arousal levels. As a result, both the amplitude of the drowsy/sleep DPA component and the amplitude variability increase when longer epochs are used for DPA spectra generated from less than 60 seconds of data. Therefore, the amplitude variability of the drowsy/sleep fraction of DPA spectra can be used to detect the switch from awake to sleep cortical function (see Figs. 5, 6 and 7).

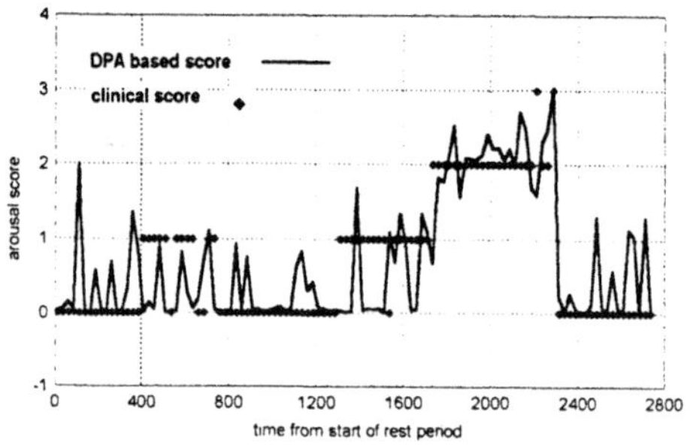

Fig. 8 The time series of DPA based and clinical arousal scores for 45 minute study of single subject.

Also, we have developed a mathematical formula which relates our new arousal parameters to the present clinical scores. This was done to facilitate the use of our new parameters which provide a continuous parameter of changes in the spectrum of MEG data associated with arousal state. As previously described, the amplitude of the awake and drowsy/sleep content of each DPA spectra can be used as X, Y coordinates (see Fig. 1). The angle from the awake coordinate axis was used to develop the relationship. We found the relationship is best described by;

log(DPA angle / 140) = 1.15 (clinical arousal score) - 3.6 angle in degrees

Using this formula on a time series of DPA spectra the same trend in arousal change and approximately the same arousal level is predicted as found in the actual time series of clinical scores (see Fig 8).

References

[1] Rechtchaffen, A., Kales, A., eds. A manual of standardized terminology, techniques and scoring system for sleep stages of human subjects. Public Health Service, U. S. Government Printing Office, Washington D.C., 1968.

[2] Norusis, M.J., Factor analysis in: SPSS-X advanced statistics guide., McGraw-Hill, 1985, pp. 125-163.

Acknowledgements

Research supported by NIH/NINDS Grant No. 1R01 NS30914

Dr. Timothy Roehrs, Sleep Disorders Center, Henry Ford Hospital, who provided arousal scores and useful suggestions.

Suitability of FFT Spectra versus DPA Period Spectra for Identification of Arousal States

Moran, J. E.[1], Tepley, N.[1,2]

Henry Ford Hospital[1], Detroit MI, USA; Oakland University[2], Rochester MI, USA

Introduction

To detect changes in arousal which often occurred during long duration DC MEG studies in our study of migraine, we found the present clinical method inadequate. Small fluctuations of arousal can not be quantified using the 5 stage scores available for NREM arousal states [1]. In addition, this scale is not linearly related to changes in the spectral activity observed in the MEG data. Therefore, we developed a zero crossing period measurement technique [2] with a built-in frequency selection filter (dynamic period analysis, DPA). DPA produces a period based power spectrum quantifying the dominant waveform activity recorded in a segment of data. For all subjects, changes of arousal level resulted in characteristic changes in the corresponding DPA spectra. This motivated us to apply a principal component analysis [3] to extract a set of principal component spectra for quantifying the small changes of arousal which we needed to detect. A set of DPA spectra were used in this analysis. Two principal spectral components were extracted and used to construct a single arousal parameter. Unlike clinical arousal scores (awake, stage 1, etc.), this arousal parameter is continuous and allows us to detect even small changes of arousal and quantify the associated DPA spectral changes. Because FFT methods of spectral analysis are more familiar, we attempted to construct a similar set of arousal parameters based on FFT spectra generated from sequential epochs of MEG data for these subjects. Unfortunately, the principal component analysis of these FFT spectra did not extract component spectra which described a common set of arousal changes shared by all individuals. Rather than corresponding to arousal states, the two FFT principal components primarily formed a basis for differentiating between individual subjects.

Methods

MEG recordings of spontaneous cortical activity obtained from six subjects were used to develop the two principal component DPA arousal state spectra. The MEG data were acquired with a 7 channel Neuromagnetometer (BTi model 607) equipped with second order gradiometers. The center channel was positioned at P3 of the international 10-20 EEG measurement system and 45 minutes of data was acquired from each subject lying on his/her side. This site was accessible to our MEG system and data recordings contained less muscle and eye movement artifacts. For these subjects, the data was partitioned into 100 second epochs and a DPA spectrum was generated for each epoch. For the principal component analysis, it was necessary to limit the frequency band to frequencies greater than 0.5 Hz. Characteristic amplitude changes of the DPA power spectra for these frequencies are sufficient to distinguish all states of arousal. FFT spectra were generated for these same epochs of data. These FFT spectra were constructed from the same 100 seconds data epochs by averaging seven FFT spectra obtained from data within 40 second time windows which overlapped by 75%. This is a commonly employed technique for constructing an average FFT spectrum for semi-stationary data when the FFT software incorporates a window function. Window functions are used to minimize significant leakage of spectral power from the true frequency component into other nearby frequency components [4, 5]. However, if the spectral properties of the data are not completely stationary throughout the data segment then window functions may weight the data in a manner in which the final FFT does not represent the average spectrum of the data. We used FFT software which incorporated a Hanning window to weight the data within each 40 second window prior to the FFT transformation [6]. The FFT frequency spectra were transformed to FFT period duration spectra such that they could be compared to the corresponding DPA spectra and also to consolidate the power of the high frequency components into fewer components such that shifts in power to lower frequencies occurring with drowsiness and sleep are given more weight in the mathematical analysis.

For two additional subjects, included in this study, simultaneous EEG recordings were made with electrode placements, P3-A2 and Cz-F2. The 45 minutes of EEG and MEG data were partitioned into 25 second epochs. DPA spectra were produced for the center channel of the sensor array and the epochs of EEG data were scored for

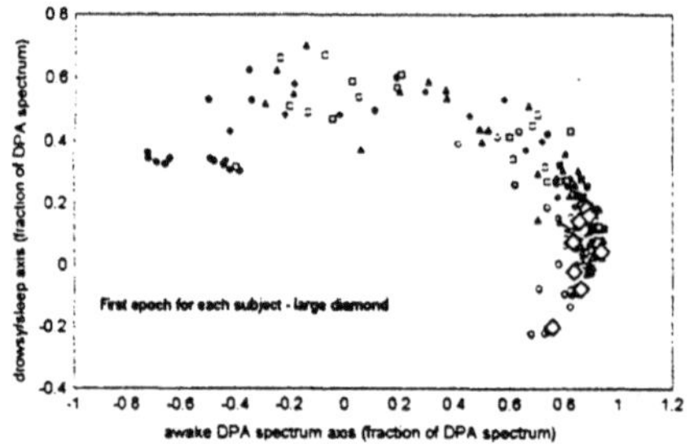

Fig. 1 Plot of β_{awake} and β_{drowsy} amplitudes for DPA spectra of eight subjects. Awake state nearly the same.

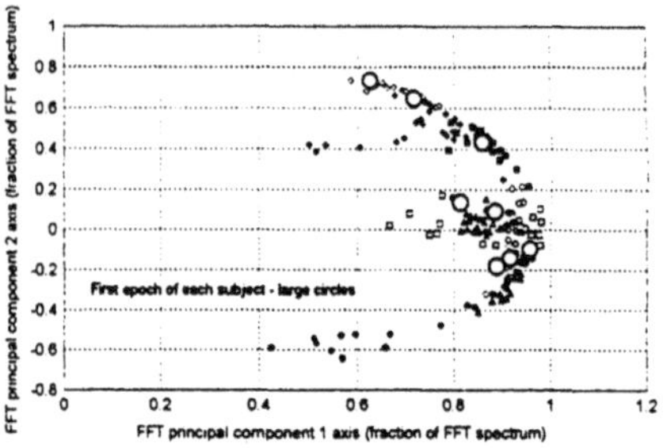

Fig. 2 FFT spectra principal component amplitudes for eight subjects. Awake state co-ordinates very different.

arousal state (awake, stages 1 - 4).

A principal component analysis was performed on the set of 162 DPA spectra from six individuals. Similarly, a principal component analysis was applied to the set of FFT spectra and two principal component spectra were generated.

Results

For the DPA spectra, the principal component analysis generated two principal component spectra. One spectrum was identical to the average awake spectrum of these subjects. The other component was found to be the time averaged power spectrum of waveforms which are present in the MEG data only in states of arousal other than awake. For any subject, a linear combination of the two principal component DPA spectra can be used to obtain a DPA spectra corresponding to the arousal state of the subject. Mathematically this relationship is:

$$DPA_{arousal\ state} = \beta_{awake}\ DPA_{awake} + \beta_{drowsy}\ DPA_{drowsy} + \text{Random}$$

with: $DPA_{arousal\ state}$ = DPA spectrum of subject during an epoch of the study
DPA_{awake} = principal component DPA spectrum corresponding to awake state
DPA_{drowsy} = principal component DPA spectrum of waveforms occurring if not awake
β_{awake} = amplitude of the fraction of $DPA_{arousal\ state}$ which is associated with DPA_{awake}
β_{drowsy} = amplitude of the fraction of $DPA_{arousal\ state}$ which is associated with DPA_{drowsy}
Random is the spectrum of random spectral fluctuations that remain in the DPA spectrum

Using β_{awake} as an X coordinate amplitude and β_{drowsy} as a Y coordinate amplitude, the set of 162 DPA spectra for these six subjects are plotted in Fig. 1. In Fig. 1, the awake DPA spectrum generated from the first 100 seconds of data for each subject is closely aligned with the horizontal positive awake axis. Sleep state DPA spectra from subjects that were found asleep at the end of the study were at the other end of the arc of data points which defines an arousal state path through this two component vector space.

The results of the principal component analysis on the FFT spectra can be described by an equation similar to the DPA relationship above, except the awake state spectra for each of the six subjects were located very far apart in the X,Y coordinate system plot in Fig. 2. The awake spectra from the first epoch can not be used to define a set of awake state coordinates. While arousal changes are easily seen for each subject, there is no common state of arousal path. These FFT principal component spectra are seen to primarily differentiate the spectra according to properties unique to the individual rather than change of arousal state.

During the first 100 second epoch of the study each subject was awake. Therefore, in order to quantify individual spectral differences, we cross-correlated each DPA spectrum from this first epoch with the first epoch

DPA spectra of the other subjects. We did the same cross-correlation analysis for the FFT spectra. For the first 100 second DPA spectra the average cross-correlation for all subjects was 0.81. The corresponding FFT cross-correlation average was only 0.74. To quantify changes of spectrum due to a shift in the state of arousal we cross correlated the DPA from the first 100 second epoch to the DPA spectra generated from the 26 subsequent 100 second epochs of data for each subject. The same cross-correlations were performed on the FFT spectra. The average arousal state change cross-correlation for the DPA spectra was 0.70 and 0.91 for the FFT spectra. Thus, FFT spectra exhibit relatively large subject to subject differences and relatively small spectral change with shifts of the state of arousal. The DPA spectra have the opposite characteristics. Changes in the DPA spectrum caused by a shift in the arousal state are greater than the differences in spectra between individuals. This was investigated in more detail using the two subjects for which simultaneously recorded EEG data was scored for arousal state. In Fig. 3, the sequence of cross-correlation amplitudes of the first DPA to the subsequent 26 DPA spectra is plotted with the corresponding FFT cross correlations and the time series of arousal scores. In Fig. 3, the DPA spectra are characterized by large spectral changes with a shift in arousal state and the corresponding FFT spectra change much less. These differences in sensitivity to arousal state are easily seen in plots of individual DPA and FFT spectra. In Fig. 4, the awake state DPA and FFT spectra for this subject are over-plotted. In Fig. 5, the stage 2 arousal state DPA spectrum and the corresponding FFT spectra are plotted. Comparing Fig. 4 and Fig. 5, the DPA spectra exhibits greater shifts of spectral power between arousal states than the corresponding FFT spectra.

In Fig. 1, the state of arousal is observed to be defined by the angular location away from the horizontal awake spectrum axis and the maximum angle is approximately 140 degrees. Therefore, we defined a new arousal parameter to be equal to the angle of DPA spectrum plotted in these coordinates divided by 140. With this scaling, zero corresponds to the awake state and 1.0 would correspond to the stage 4 arousal state. However, stage 1, 2 and 3 are not located at equal intervals along this scale. The relationship of our arousal state score (0.0 to 1.0) to the present clinical scores is not linear.

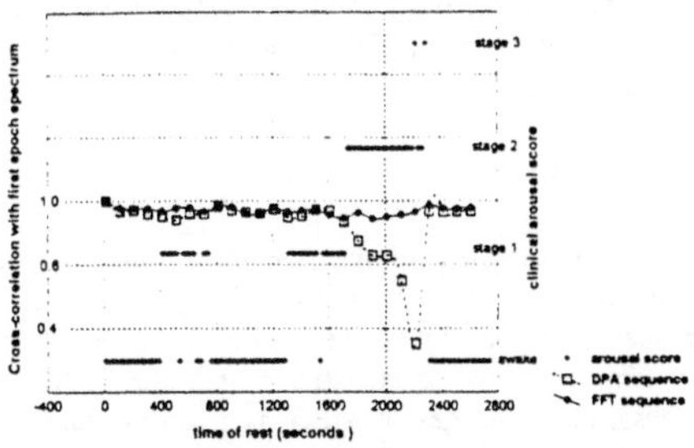

Fig. 3 The time series of DPA and FFT spectral change correlation amplitudes with the corresponding arousal scores.

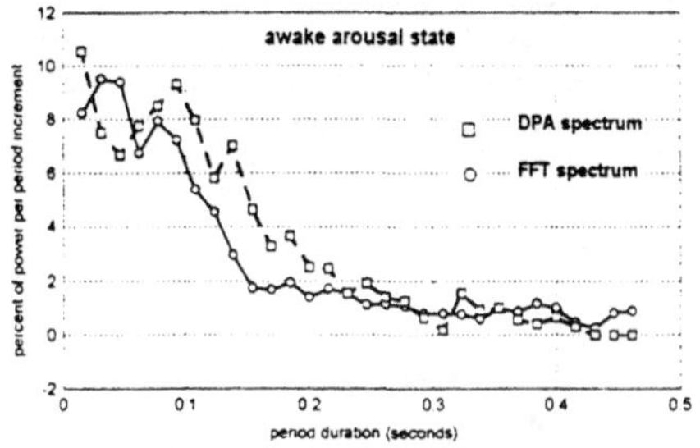

Fig. 4 The corresponding awake DPA and FFT period spectra.

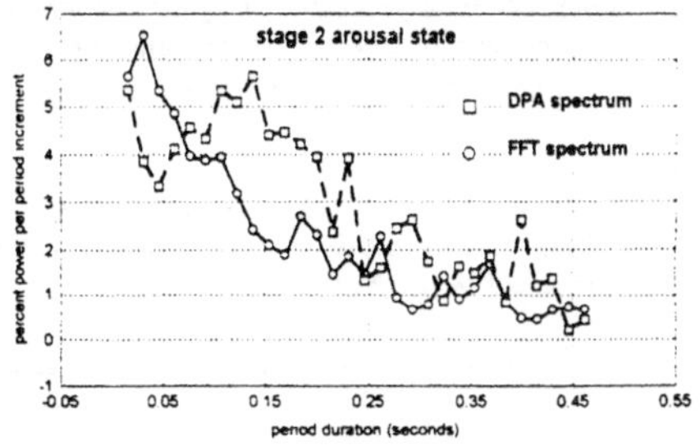

Fig. 5 The stage 2 DPA and FFT spectra.

For our data, mathematically,

$$\log(\text{DPA arousal score}) = 1.15\ (\text{clinical arousal score}) - 3.6$$

We used this relationship to predict the clinical arousal score from our DPA arousal score. In Fig. 6, the time series of predicted and actual scores are plotted. Both measures of arousal state display the same trends but do not completely agree at each epoch. While this plot is useful to those more familiar with the clinical scale, we find time series plots of the β_{awake} and β_{drowsy} amplitudes of the DPA spectra to be more useful for assessment of changes of arousal state.

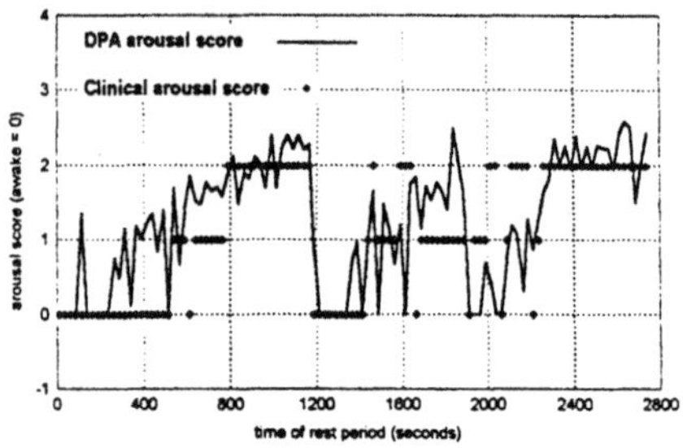

Fig. 6 Predicted arousal state score based on DPA parameters versus standard clinical scores.

Discussion

Assessment of the frequency content of data is primarily performed by either visual inspection or advanced mathematical techniques, such as FFT based power spectra. In this study, we demonstrated that the continuum of arousal state changes of MEG data can be quantified by applying a principal component analysis to DPA spectra and generating a set of unique arousal spectra. All DPA spectral changes occur in the corresponding FFT spectra. However, within the complete FFT spectrum the arousal related changes are small relative to subject to subject spectral differences. Thus FFT spectra are not suitable as a basis for development of arousal state parameters using a principal component analysis.

We use the arousal parameters, β_{awake} and β_{drowsy}, to identify and quantify amplitude and spectral changes of MEG data associated with shifts of the arousal state. However, these parameters may also be useful in assessing the physiological state of arousal based on changes observed in MEG or EEG data. The variability of the β_{drowsy} component may be especially useful for determining sleep onset.

References

[1] Rechtchaffen, A., Kales, A.. eds. A manual of standardized terminology, techniques and scoring system for sleep stages of human subjects. Public Health Service, U. S. Government Printing Office, Washington D.C., 1968.

[2] Frost J.D. Jr. Mimetic techniques. In: Gevins, A.S., Remond, A. eds., Methods of analysis of brain electrical and magnetic signals. EEG handbook, Elsevier, Amsterdam 1987 (revised series, Vol 1)

[3] Norusis, M.J., Factor analysis in: SPSS-X advanced statistics guide., McGraw-Hill, 1985, pp. 125-163.

[4] Lopes da Silva F., EEG analysis: theory and practice. In: Niedermeyer E., Lopes da Silva F., eds. Electroencephalography: basic principles, clinical applications, and related fields., 3rd. ed. Baltimore, USA: Williams & Wilkins 1993:1097-1123

[5] Dumermuth G., Molinari L., Spectral analysis of EEG background activity. In: Givens A.S., Remond A. eds., Methods of analysis of brain electrical and magnetic signals. EEG handbook, Elsevier, Amsterdam 1987 (revised series, Vol. 1)

[6] BTi Neuromagnetometer: Programmer's Reference Manual. Biomagnetic Technologies, Inc., 1987

Acknowledgements

Research supported by NIH/NINDS Grant No. 1R01 NS30914

Dr. Timothy Roehrs, Sleep Disorders Center, Henry Ford Hospital, who provided arousal scores and useful suggestions.

Chaos in the Brain: A New Strategy to Discriminate Deterministic Low Dimensional Dynamics in the Spontaneous Activity of the Human Cortex

Narici, L.[1], Boccaletti, S.[2], Giaquinta, A.[2] and Arecchi, T.[2]

[1]*Dipartimento di Fisica - Università di Roma "Tor Vergata", Roma, Italy;* [2]*Istituto Nazionale di Ottica, Arcetri, Firenze, Italy*

Introduction

Spontaneous activity of the human brain in normal subjects is not likely to be "low dimensional" being the result of superpositions of many correlated and uncorrelated activities. These activities are not constantly present, neither detectable uniformly over the whole scalp. It is likely that this difficulty in defining the "measured system" and the large time variability of the Signal to Noise Ratio (SNR) is one of the reasons for the wide spectrum of results that can be found in the literature on chaotic estimators especially when investigating spontaneous activity. As an example the Correlation Dimension (CD) for the waking stage (mostly α rhythm) ranges from 2.4 to 11 (see, for example, [1]). While some agreement can be found on the dimension decrease under epileptic seizures or other specific pathologies the attempts for a dimensional classification of α rhythm have been mostly unsuccessful.

The difficulties in obtaining reliable estimates of the correlation dimension and of the other dynamical indicators in these cases are often recognized (see for example [1-3]). However in several instances some authors suggest that even though absolute estimate were questionable, comparative results were significant. While the comparative use of these parameters appear to be 'safer', we object that the values of the chaotical evaluators given by the algorithms may be strongly influenced by the noise (including the brain concurrent activities that are, most often, outside the control of the experimenters). Thus, differences between two chaotic estimators, obtained in two different conditions may be due not only to the differences in the studied underlying dynamics, but also to the different way in which 'noise' and experimental conditions affect the calculation. Therefore we believe it is important to set precise limits for acceptance of the estimators and to propose 'standardized' procedures to calculate them, featuring suitable tests able to check their validity.

The α rhythm, largest cortical activity, appears in a narrow frequency band (α band $\approx$8 Hz to 12 Hz) and in *spindle* at irregular time intervals. Outside these *spindles* the SNR<<1. Passband filtering would sharply cut the amount of independent points. A very simple minded calculation shows this number to be $N \approx 2T\Delta f$, where T is the data length in time and Δf the filter bandwidth. As it may be plausible to define a sort of 'maximum stationarity time window' during sleep ($< 10^3$ s) and awake stages (α rhythm)($< 10^2$ s) this impose an upper bound to N ($\approx 10^3$ for α rhythm, passband filtered). This translates in an upper limit for the calculus of CD: ≤ 6 [4], or even ≤ 4 [5].

In this work we refer, for simplicity, to all activities present in the α band in the occipital region as 'α rhythm', even if it has been recently shown that different activities are sharing the same frequency window and often the same approximate source location (see, for example, [6,7]). We also assume that the reader is familiar with the most general procedures used to calculate chaotic estimators from experimental data (the interested reader is addressed to the many reviews available, for example: [1,8-10]).

Methods

Our contribution is two-fold: we propose a procedure to look for determinism and chaotic estimators (CD) in EEG/MEG data, using recent tests described in the literature and we propose a novel filtering technique [11], based on the local features of the wavelet tranformation. The procedure is briefly sketched in the following.

i) - We first calculate the autocorrelation function and Mutual Average Information [12] by which embedding delay time is chosen. ii) - The suitable embedding dimension m is then provided by the false nearest neighbor algorithm [10,13] that calculates the lowest dimension needed to eliminate all false self crossings of the orbit. The number of false crossing must go to zero for increasing m for a signal to be deterministic. This algorithm can therefore be applied to discriminate the original signal from a surrogate (a particular realization of a coloured noise which has the same distribution and autocorrelation function of the signal but all the phases of the Fourier component randomly chosen): if this discrimination is not possible or if the found embedding dimension is not sufficilently low the data must be filtered first. iii) - Test for determinism is carried out (β-test [14]) and chaotic estimators are calculated (Grassberger and Procaccia algorhithm, GP [15]), both for the original and for the surrogate data. iv) - If needed [see (ii)] the signal is filtered in wavelet space (see below); (ii) and (iii) are therefore repeated with the filtered signal.

The procedure is tested on spontaneous occipital neuromagnetic activity measured in a normal subject. The magnetic field over the left occipital cortex of a healthy subject is measured with a 28-channel superconductive system in use at the CNR-IESS laboratory in Rome [16]. The subject was lying prone in a bed wearing a blackened rubber mask, watching fixedly two centered green LEDs to maintain steady eyeball position. We measured 1 minute of occipital activity at an acquisition frequency f=1 KHz, with a band pass filter 0.4 Hz-250 Hz. The raw data were stored in an optical disc for the off-line analysis.

The advantage of filtering in the wavelet space, is that wavelets represent a localized transform. As a consequence, a signal *s* picked at a given frequency, that gives rise in the Fourier space to very few channels different than zero in which information on the signal is stored, in the *WT* space preserves many components as a consequence of the fact that a single frequency is delocalized in time. In this work we choose as the analysing wavelet the second derivative of the Gaussian function. To determine the best filtering thresholds for the *WT* we propose to use a strategy recently introduced [11,14] based on the use of an adaptive recognition of the dynamics of the signal [14]. This strategy, here thereafter called βWT, permits to fix these thresholds in such a way to achieve the extraction of the deterministic component of the original signal. The details of this procedure can be found in the original papers. Let us mention that for each embedding dimension m a coefficient β can be calculated that scales as $\exp(-k_1 m)$ for a deterministic signal whereas as $\exp(k_2 m)$ ($k_1,k_2>0$) for noise. In the case of the surrogate of a deterministic system, one half of the total number of degrees of freedom (the phases of each Fourier component) scales like noise, while the other half (the amplitudes of each Fourier component) scales like the deterministic signal. As a result, we expect an intermediate global scaling law. Other kinds of colored noise should also feature an intermediate result. An adaptive algorithm based on the minimization of the slope of $\beta(m)$ permits to implement a new type of filter for the noisy background in a temporal series. This criterion can, of course, be applied also to filters acting in other transform spaces (Fourier, Laplace, etc..), but, as mentioned, the localized feature of the WT may offer an important advantage when dealing with brain signals. The study of the $\log\beta(m)$ can thus be used for two purposes: as a criterion for fixing the optimal *WT* filtering thresholds, and as a test for the presence of determinism in the analysed signals. We will show that the βWT filter leaves mostly unaltered those frequencies featuring the largest interesting signal power (≈6 Hz to ≈ 40 Hz).

Results

The results presented here are relative to one channel of the neuromagnetometer, where the presence of activity in the α band was most prominent, in the perieto-occipital region. Studies on other channels also featuring strong α activity produced quite similar results.

Both mutual average information algorithm and evaluation of the the zeroes of the autocorrelation function provided the same estimate for the embedding delay time which permits to fully reconstruct the invariant set of points of the attractor: τ=27 ms. The false nearest neighbors algorithm showed no saturation, probably as a consequence of the high level of additive noise. Therefore, the signal was filtered using the βWT filter. The frequency window containing the most important part of the signal is unaffected by the βWT filter. This can be easily appreciated from Fig. 1 where we report the βWT filter results. We show for comparson the Fourier spectra, as well as a 0.1 s long period of the time behavior, of the raw and βWT filtered data. Very little differences can be appreciated by eyes.

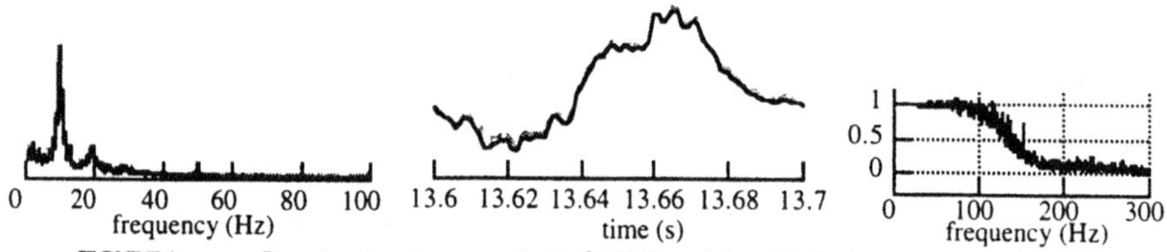

FIGURE 1 Raw data (gray) compared with βWT filtered data (black) in the frequency and time domain. On the right the resulting filter transfer function. Very little differences can be appreciated by eyes.

The study of log$\beta(m)$ is presented in Fig. 2. We show log$\beta(m)$ for the raw, for the βWT filtered signals and for the surrogate of the βWT filtered signal.

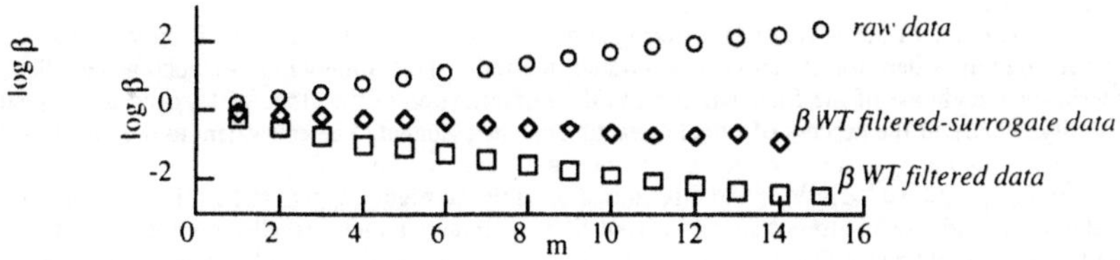

FIGURE 3 Plot of the log of the β parameter (see text) vs. embedding dimension m.

The βWT technique worked only partially on our data as it was not possible to determine the minimum slope of logβ(m). We found increasingly negative slopes when increasing the filtering action: a compromise had to be done in order to maintain a sufficiently large number of independent data. Therefore in choosing the filtering thresholds, a certain degree of subjectivity remained. However, for those choices of thresholds for which the filtered signal showed negative slopes of the logβ(m), the corresponding surrogate was clearly discriminated as shown in Fig. 3. The filtered signal keeps about 16% of the information originally contained in the original signal (about 10,000 independent points). Data has been filtered also in Fourier space, keeping the same amount of independent points (corresponding to $\Delta f \approx 70$ Hz), centered on the α band. In this case the slope of logβ(m) was again negative, but it was not possible to discriminate the surrogate.

The false nearest neighbors algorithm applied on the βWT filtered data provided $m \approx 7$ as embedding dimension while when applied on the surrogate we found no saturation.

Finally we calculated the correlation integral C_2 (GP algorhithm, [15]) after having properly reconstructed the invariant set in a space of dimension 7: no reliable plateau indicates that CD > 7. As a check we performed the GP algorhithm, reconstructing the data set in dimensions $2 \leq m \leq 12$, and produced the same results (Fig. 4).

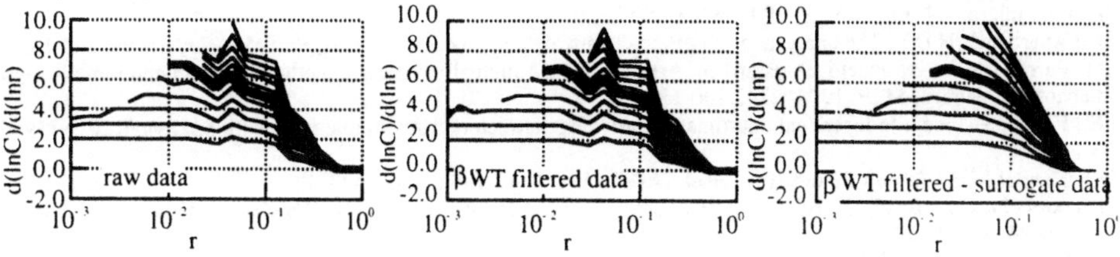

FIGURE 4 GP algorhithm results (local scaling exponents dlnC/dlnr vs logr for different embedding dimension ($2 \leq m \leq 12$); C: correlation integral; r: box size). The thicker line refer to m=7, the embedding dimension found with the false nearest neighbour algorhithm. No plateau is found.

Discussion

With the use of the βWT filter we increased N and thus the CD limits to 6 [5] or 8 [4]. Even with these larger limits CD could not be calculated. However our procedure allows to claim the deterministic nature of some of the waking cortical activity. The β-test did in fact show that our signal is mostly deterministic and that a colored noise, as a surrogate extracted from our signal, can be discriminated. This result opens a new perspective in studying the brain activity in other situations (as e.g. in stimulated cases) where the signal to noise ratio is more favorable and hence the influence of the filter is weaker. We will address such question elsewhere.

The unsuccess in the determination of the minimum of the logβ slope is probably linked to the high level of noise in the original signal as well as to the possibly many processes concurrently present. However, as shown in our results, a compromise can be reached extracting from the signal a mostly deterministic component, that can be discriminated from the stocastic component, and that is actually containing all the frequency band where are the most significant signals.

The localized action of the Wavelet Transform permits a time selective filtering that is most useful in those cases, as cortical rhythms, where the signal is concentrated in erratic, short, time windows, such as spindles. The proposed criteria for the choice of the filter parameters, the minimization of the slope of $\log\beta$ vs. m , appears of larger applicability, whenever the process of signal extraction can be assumed to be equivalent to the discrimination of the deterministic components in the raw data. Even when, as in our case, minimization cannot be performed, the combined requirements, negative $\log\beta$ slope and discrimination of the surrogate, appear to be sufficient for a reliable filtering. Finally we would like to stress that a passband filter maintains a much smaller number of independent points and fails the false nearest neighbors test. Even using the same number of independent points passband filtering produces results that cannot be discriminate from surrogate.

References

[1] Basar E., Chaotic dynamics and resonance phenomena in brain function: progress perspectives and thoughts, in: Basar, E. ed. Chaos in Brain Functions Berlin-Heidelberg-New York, Springer, 1990.

[2] Pijn J.P., Van Neerven J., Noest A. and Lopes da Silva F.H., Chaos or noise in EEG signals; dependence on state and brain site, Electroencephal. Clin. Neurophysiol., 1991, 79: 371-381.

[3] Achermann P., Hartmann R., Gunzinger A., Guggenbühl W and Borbély A.A., All-night sleep EEG and artificial stochastic control signals have similar correlation dimensions, Electroencephal. Clin. Neurophysiol., 1994, 90: 384-387.

[4] Ruelle, D., Deterministic chaos: the science and the fiction, Proc. R. Soc. Lond. A, 1990, 427:241.

[5] Kantz, H. and Schreiber, T., Dimension estimates and physiological data, Chaos, 1995, 5:143-154.

[6] Salmelin, R., Hari, R., Characterization of spontaneous MEG rhythms in healthy adults, Electroencephal. Clin. Neurophysiol., 1994, 91: 237-248.

[7] Narici, L., Peresson, M., Discrimination and study of rhythmical brain activities in the α band: a neuromagnetic frequency responsiveness test, Brain Res., 1995, 703:31-44.

[8] Rapp P.E., Bashore T.R., Martinerie J.M., Albano A.M., Zimmerman I.D., Mees A.I., Dynamics of brain electrical activity, Brain Topography, 1989, 2: 99-118.

[9] Pritchard W.S. and Duke D.W., Measuring chaos in the brain: a tutorial review of nonlinear dynamical EEG analysis, Intern. J. Neuroscience, 1992, 67: 31-80.

[10] Abarbanel, H.D.I., The analysis of observed chaotic data in physical systems, Rev.Mod.Phys., 1993, 65:1331.

[11] Perrone, A.L., Boccaletti, S., Basti, G., Arecchi, T.F., A mutually recursive method to detect and remove noise in chaotic dynamics, SPIE, 1994, 2242:130-139.

[12] Fraser, Swinney, Independent coordinates for strange attractors from mutual information, Phys. Rev. A. 1986, 33:1134.

[13] Kennel M.B., Brown R. and Abarbanel, H.D.I. Phys. Rev., 1992, A45:3403

[14] Arecchi, F.T., Basti, G., Boccaletti, S., Perrone A.L., Adaptive recognition of a chaotic dynamics. Europhys. Lett., 1994, 26:327-332.

[15] Grassberger, P. and Procaccia, I.,Measuring the strangeness of strange attractors, Physica D, 1983, 9:189

[16] Foglietti, V., Del Gratta, C., Pasquarelli, A., Pizzella, V., Torrioli, G., Romani, G.L., Gallagher, W.J., Ketchen, M.B., Kleinsasser, A.W., Sandstrom, R.L., 28-channel hybrid system for neuromagnetic measurements, IEEE Trans-MAG, 1991, 27:2959-2962

Acknowledgements

We are grateful to M. Peresson, V. Pizzella and F. Tecchio for the help in the measurement session, and to H. Kantz for providing the software for the CD calculation.

A New Procedure to Discriminate and Study Different Rhythmical Cortical Activities on the Basis of their Responsiveness to Simple Sensory Stimulation

Narici, L.[1] **and Peresson, M.**[2]

[1]*Dipartimento di Fisica - Università di Roma "Tor Vergata", Roma, Italy;* [2]*Divisione di Neurologia, Ospedale Fatebenefratelli -S. Pietro, Roma, Italy*

Introduction

The investigation of the spontaneous as well as stimulus related rhythmical brain activities has been receiving increasing attention in the recent years. Cognitive functions as well as arise of pathologies, have been correlated with time and frequency behaviour of cortical rhythms [1-9]. However the understanding of such correlations and of the genesis of the rhythms is still in a primitive stage. The erratic behaviour of the power of most of the spontaneous rhythms, appearing in spindles at random instants, and the quite large inter-individual variability caused difficulties when investigating such correlations or looking for functional and/or pathological indicators. Moving from the observation that those brain structures generating spontaneous oscillations have higher responsiveness to sensory stimulation in the same frequency band where such oscillations are detected [10], we propose to classify the activities on the basis of their degree of responsiveness to rhythmic sensory stimulation. Here we present a procedure able to discriminate these activities within the α band on such basis.

The measurement procedure is fast: less then 10 minutes. In the following we describe the stimulation paradigm, data analysis, data presentation that allow for studying frequency selective enhancements or suppressions of the cortical responses in selected frequency windows. This Frequency Responsiveness Procedure (FRP) combines, in one single test, multichannel detection, a finely rate scanned intermittent stimulation that permits to study its effect on the underlying activity and a compact presentation to appreciate the whole rate/frequency/topographical dependency of the spectral responses. This is a promising new tool for the study of cortical responsiveness to sensory stimuli. The classification of the underlying activity, on the basis of responsiveness indicators may become useful in the diagnosis of pathologies that have frequency selective effects on the reactivity of spontaneous rhythms to burst of stimuli such as the Alzheimer's disease [8].

The FRP is tested (under separate visual and somatosensory stimulation) on neuromagnetic measurements on healthy normal subjects: a first non-invasive study of the frequency responsiveness within the extended alpha band (6 Hz to 14 Hz) [11]. The degree of irregularity of the rhythms, as well as the intra- and inter-individual variability is quite reduced. Robustness of the test and stationarity of the responses is improved. Several of the results presented in this work suggest that we are eliciting similar responses than those obtained with complex cognitive experiment. This would help in the understanding of the linking mechanism between cognitive functions - sensory stimuli - rhythmical activity.

Methods

The stimulation consists in intermittent burst of stimuli (10 stimuli each burst) at a given rate. The pause between the burst is fixed to 1.3 s in this test. The entire stimulation epoch (burst plus non stimulated period) is repeated several times for each rate, in order to perform off-line averaging of the response spectra. The rates are scanned from 6.0 Hz to 14.0 Hz at step of 0.5 Hz. Preliminary studies are used to set the number of repetition to 10 (but probably 5 could have been enough to achieve comparable results), and to ruled out possible influence of the scanning order of the stimulation.

The FRP was tested on five normal subjects. The visual stimuli were short light pulses delivered by two red leds, mounted on a blackened scuba-mask (about 5 cm from the eyes, at an eccentricity of about 5 degrees) in the left hemifield. The subject was asked to maintain fixation on two green leds, constantly 'on', positioned in the mask with zero eccentricity. The somatosensory stimuli were current pulses ($\approx$0.1 ms long) delivered to the right median nerve at wrist with an amplitude just above motor threshold. In the somatosensory modality we tested two different subject conditions (open and closed eyes) to investigate possible cross-modality interactions between somatosensory stimulation and α activity. One minute of spontaneous activity was measured from the same cortical region at the beginning of each session to be used as baseline.

The magnetic field associated with the bioelectrical activity evoked by the stimulation was measured with the 28 channel superconductive system in use at the CNR-IESS laboratory in Rome[1] [12]. The data were acquired at a rate of 1.0 kHz (recording bandwidth 0.48 Hz - 250 Hz), and stored on optical disk for the off line analysis.

Analysis is performed separately in two segments in each epoch. The first during the stimulation, and the second following it, during the pauses (starting one inter-stimulus interval after the delivery of the last stimulus in the previous burst). Each segment is 1024 ms long. In each segment we calculate the spectrum, and then average them over the 10 repetitions relative to the same stimulation rate. The procedure is repeated for all stimulation rates. The spectra of the spontaneous activity is calculated during fifteen 4 s long periods.

For each channel seventeen averaged spectra for the stimulated and non stimulated periods (corresponding to the seventeen rates of stimulation), as well as fifteen spectra for the spontaneous activity (corresponding to the fifteen 4 s long time intervals) have been calculated. These spectra are collected in three interpolated maps, where the amplitude of the frequency response is gray mapped and the two independent variables are the frequency of the response and either the rates of stimulation (for the stimulated and non stimulated period maps) or the time (for the spontaneous maps). For each analysis (spontaneous, during stimulation, during pauses) we collected all the interpolated maps in a composite image, where the maps relative to each channel are positioned at the approximate scalp location of the channel.

Results

Complete results are shown for subject S1 in Fig. 1. In both modalities we show the composite images relative to the time behavior of the spontaneous activity (left) as well as to the rate behavior of the responses during stimulation and pauses (center and right). For the somatosensory test both open and closed eyes sets of results are presented. In Fig. 2 we present the results for three subjects (only somatosensory test, stimulated period, open eyes) to show the good inter-individual reproducibility.

visual test In the spontaneous activity we observe a single, broad, irregularly intermittent peak at about 10 Hz. Responses tho the FRP procedure show at least two active frequencies: $\approx$ 10 Hz and $\approx$ 12 Hz. Both of them present during stimulation ('driven'), closer to the calcarine fissure, while only the $\approx$10 Hz activity is present in the non stimulated period, confined to the parieto-occipital cortical region. When stimulating with $r<10$ Hz all the activities, during and following stimulation, are strongly attenuated. A hint for a negative shift in frequency of the major 10 Hz peak between the spontaneous measurement and the measurements during pauses is apparent after careful investigation. This is evidenced in Fig. 3 where we show the ratio between 'spontaneous' and 'pause' mean frequency of the 10 Hz peak for the 7 parieto-occipital channels (see also [11,13]).

somatosensory test In the somatosensory test it is strikingly evident the reduction of the parietal $\approx$ 10 Hz activity when opening the eyes. This is the known behavior of the α rhythm. The spontaneous activity is almost absent with the eyes open. With the eyes closed it is centered at about 10 Hz, and it is very strong in the parietal region, weaker and more irregular in the rolandic region. The driven activity during stimulation is present for most of the stimulation rates (slightly stronger at $\approx$10 Hz) and it is confined in the rolandic region, across the rolandic fissure. A much weaker $\approx$ 10 Hz activity can be found also during pauses, both in the rolandic and in the parietal regions, even in the open eyes condition, where very little spontaneous activity was found. Note the attenuation for $r < 10$ Hz similar to what found during visual test; this is not as evident, however, for all the studied subjects.

Discussion

Recent investigations have suggested that activities in the α band may have different functional correlates. Many authors prefer to divide the α band into a 'lower' sub-band (6-10 Hz) and a 'higher' sub-band (10-14 Hz), the former recently related to activity linked mainly to attentional processes, the latter to activity linked to stimulus correlated cognitive processes [5]. The results obtained with the FRP, show different 'reactivities' within the α band, and support this subdivision idea providing information about the interaction of these rhythms (and their functional correlates) with simple sensorial stimulation. For example, we can speculate that the apparent frequency shift (Fig. 3) could be explained referring to the recent work by Klimesh and coworkers [5] where it was demonstrated a decrease in α frequency when attentional demands and/or memory loading was increased. In our case we can speculate that the rate 'decoding' process (needed to allow for the rate-dependent response behavior) might be linked to more attentional

---

[1] At the time of the visual measurements, this system featured 16 vertical gradiometers plus 12 outer planar gradiometers. To simplify data interpretation, we are presenting only the results coming from the inner 16 channels. One of the inner 16 channels was not working properly during the visual measurements, therefore we present results from a total of 15 channels. The somatosensory measurements were performed with a modified version of the system, where the 12 outer planar gradiometers were substituted by 12 magnetometer, three of which used for electronic noise cancellation. These results are therefore presented from a total of 25 channels.

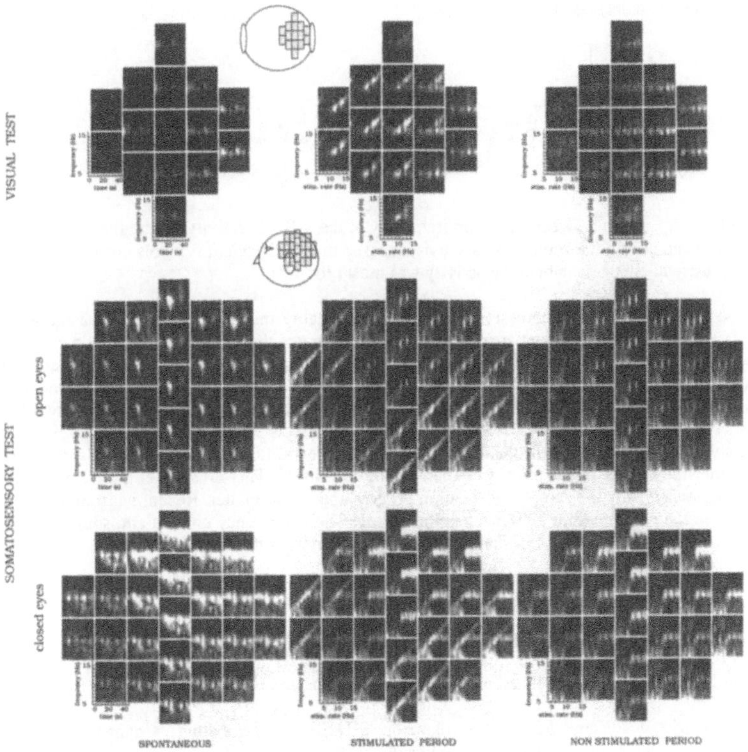

FIGURE 1 Composite images relative to subject S1. Each interpolated, gray scaled map (white: largest amplitude) is the collection of all the spectra measured by the same channel, and represent the amplitude of the responses in function of time and frequency (left, spontaneous measurements), or in function of the stimulation rate and frequency. The positions of the channels are indicated in the head silhouettes.

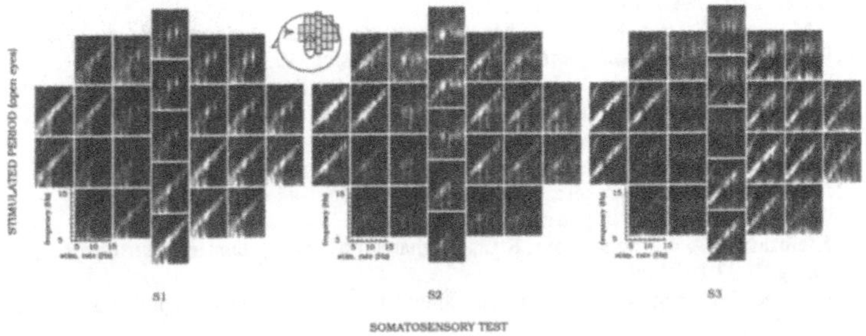

FIGURE 2 Composite images collecting all the spectra relative to three subjects (Somatosensory stimulation, open eyes, during the stimulating burst). The positions of the channels are indicated in the head silhouette.

demands and, therefore, to a decrease in α frequency.

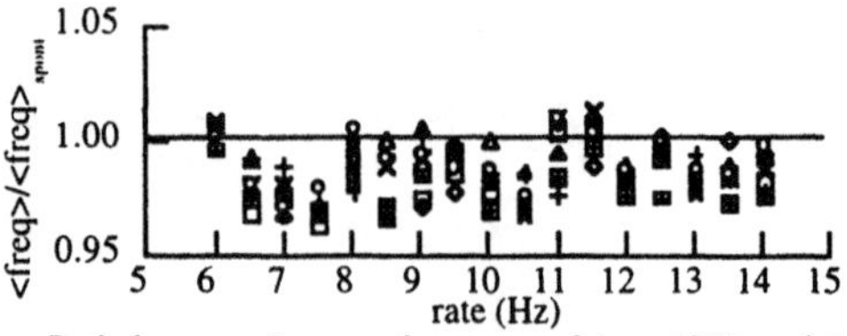

FIGURE 3 Ratio between the mean frequency of the ≈ 10 Hz activity during the pauses of the burst stimulation and during the spontaneous activity measurements. Different symbols correspond to seven different parietal channels, where the activity was most prominent.

The rate dependent alpha suppression appears to be modality independent. Studies on the activity in the cortical regions close to the hand and foot areas [6,7] have demonstrated the increase in amplitude of occipital α and concurrent decrease of μ rhythm, following voluntary movements. Our paradigm is somewhat different (the thumb twitch is not voluntary), but for $r<10$ Hz, the occipital α rhythm (closed eyes) appears strongly attenuated, however for $r>10$ Hz the rhythm seems unaffected by the stimulation. The same behavior can be observed to a lesser extent in the open eyes condition. Other studies described α suppression effects and related them to memory scanning [3]. In our case we suggest that memory is probably involved for rate decodification, and speculate a different effect on the rhythm whether the rate is faster or slower than the rhythm itself. This effect seems not to be linked to the specific sensory pathway: only the rate decodification process appears to matter. Recent neuromagnetic studies [9] showed activities in the occipital and parieto-occipital cortex. Their frequency content and source location agreed with what found here. The present work evidences that these two rhythms have also distinct sensorial 'reactivity'.

References

[1] Basar, E., Lopes, F., Hari, R., Schürmann, M. Eds. Alpha Processes: Sensory and Cognitive Behavior, Boston, Basel, Berlin, Birkäuser *in press*

[2] Basar, E. and Schürmann, M., Functional aspects of evoked alpha and theta responses in humans and cats, *Biol. Cybern. in press.*

[3] Kaufman, L., Curtis, S., Wang, J.Z., and Williamson, S.J. Electroencephalogr., Changes in cortical activity when subjects scan memory for tones, Clin. Neurophysiol., 1991, 82: 266-284

[4] Klimesch, W., Schimke, H., Ladurner, G., and Pfurtscheller, G. J., Alpha frequency and memory performance, Psychophysiol., 1990, 4: 381-390

[5] Klimesch, W., Schimke, H. and Pfurtscheller, G., Alpha frequency cognitive load and memory performance, *Brain Topography*, 1993, 5:241-251.

[6] Pfurtscheller, G., Event-related synchronization (ERS): an electrophysiological correlate of cortical areas at rest, *Electroencephalogr. Clin. Neurophysiol.*, 1992, 83:62-69.

[7] Pfurtscheller, G., Neuper, C., Event-related synchronization of mu rhythm in the EEG over the cortical hand area in man, *Neurosci. Lett.*, 1994, 174:93-96.

[8] Politoff, A.L., Monson, N., Hass, P., and Stadter, R., Decreased alpha bandwidth responsiveness to photic driving in Alzheimer disease, *Electroencephalogr. Clin. Neurophysiol.*, 1992, 82:45-52.

[9] Salenius, S., Kajola, M., Thompson, W.L., Kosslyn, S., Hari, R., Reactivity of magnetic parieto-occipital alpha rhythm during visual imagery, Electroencephal. Clin., Neurophysiol., 1995 in press.

[10] Basar, E., EEG-Brain Dynamic, Amsterdam, Elsevier, 1980

[11] Narici, L., Peresson, M., Discrimination and study of rhythmical brain activities in the α band: a neuromagnetic frequency responsiveness test, Brain Res., 1995, 703:31-44

[12] Foglietti, V., Del Gratta, C., Pasquarelli, A., Pizzella, V., Torrioli, G., Romani, G.L., Gallagher, W.J., Ketchen, M.B., Kleinsasser, A.W., Sandstrom, R.L., 28-channel hybrid system for neuromagnetic measurements, IEEE Trans-MAG, 1991, 27:2959-2962..

[13] Gaeta, R., Benzi, R., Peresson, M. and Narici, L., Singular Spectrum Analysis applied to neuromagnetic signals, 1996, this volume.

Acknowledgements

The authors express gratitude to V. Pizzella and F. Tecchio for the help during the measurement sessions.

Cognitive Temporal Binding and its Relation to 40Hz Activity in Humans: Alteration During Dyslexia

Ribary, U.[1], Joliot, M.[2], Miller, S.L.[3], Kronberg, E.[1], Cappell, J.[1], Tallal, P.[3] and Llinás, R.[1]

[1]Center for Neuromagnetism, Dept. of Physiology and Neuroscience, New York University Medical Center, New York, USA; [2]SHFJ, DRIPP-CEA, Orsay, France; [3]Center for Molecular and Behavioral Neuroscience, Rutgers University, Newark, USA

Introduction

Spontaneous oscillatory electrical activity in the human brain at a frequency near 40-Hz and its reset within thalamo-cortical systems by sensory stimulation has been proposed earlier [1-3] and has been suggested to be related to cognitive processing and to the temporal binding of sensory stimuli [3,4]. In particular, we demonstrated spontaneous 40-Hz coherent magnetic activity in the awake and in REM sleep states that is reduced during delta sleep [3]. This 40-Hz magnetic oscillation has been shown to be reset by sensory stimuli only in the awake state but not during REM or delta sleep. Because spontaneous 40-Hz oscillation was seen in wakefulness and in dreaming, we proposed it as a correlate of cognition, probably resultant from coherent 40-Hz resonance between thalamocortical specific and nonspecific loops [3].

Using Magnetic Field Tomography (MFT) on humans, we earlier demonstrated that such 40-Hz coherence displayed well defined cortico-subcortical correlations with a time shift that is consistent with thalamo-cortical conduction times [2]. In particular, the MFT results indicated that the onset of activity at the thalamic level was followed by widespread activation of the thalamo-cortical system which resulted in large coherent thalamo-cortical 40-Hz oscillation, organized in space and time, which is altered in pathological states such as Alzheimer's disease [2].

40-Hz oscillatory activity has been suggested to subserve cognition by conforming relevant sensory information into specific time segments, that implement cognitive binding [3]. It was also proposed that 40-Hz binding abnormalities—a dyschronia [5]—could be one of the neurophysiological correlates of the temporal deficits recorded in reading disabilities [6]. In particular, we were interested 1) in testing this hypothesis and in determining specifically whether the minimal interval required to identify separate auditory stimuli correlates with the reset of the 40-Hz magnetic signal, and 2) in analyzing any alterations in the 40Hz magnetic activity in subjects with language-based learning disabilities (dyslexia).

Methods

We used a 37-channel MEG system (BTi), in order to record magnetic activity from nine awake human subjects in response to one or a pair of auditory clicks [4], presented at interstimulus intervals of 3, 6, 9, 12, 15, 18, 24, and 30 msec. The subjects were asked to lie on a bed with their eyes closed and to stay awake and attentive. The magnetic-sensor array was positioned over the auditory area of the right hemisphere. Stimulus pairs were presented with interstimulus-pair intervals of 130 ± 15 msec. MEG activity was recorded from 10 msec before to 100 msec after the onset of the first stimulus (bandpass 1-400 Hz, sample rate 1041 Hz). Transient responses were averaged using the onset of the first stimulus as a trigger.

In order to analyze differences in magnetic recordings from subjects with language-based learning disabilities (diagnosed at Rutgers University), we further modified the technical approach within the paradigm, and implemented a new set of recordings on 9 healthy controls and on 6 dyslexic subjects. In particular, we increased the recording epoch to 140 msec (-10, 0, 130 msec), and changed the presentation of interstimulus intervals to 6, 12, 15, 18, 21, 24, 30 and 36 msec. The Stimulus pairs were presented with interstimulus-pair intervals of 160 ± 15 msec. Our laboratory is presently equipped with a dual 37-channel MEG system (BTi) which allowed us to record, for the first time, bilateral magnetic activity from 4 control subjects and 2 dyslexic subjects, taking into account the responsiveness of both brain hemispheres.

Two distinct models were utilized in the analysis of MEG data. Data-simulation-procedures were implemented in this study in order to account for the MEG data [4]. Following magnetic recordings, a perceptual task was performed on each subject [4], using 6 sets of 2 click presentations with the same interstimulus intervals, and the perceptual threshold was computed.

Results and Discussion

A power spectral analysis of the raw data revealed a significant component near 40 Hz, indicating the presence of a synchronized 40-Hz event in the human brain [4]. The data was filtered at 20-50 Hz for further analysis to remove a 10-Hz component, which modified the baseline in the raw data in each of the 37 channels differently, depending on the 10-Hz component in that particular channel. The high frequency rhythm near 40 Hz in the raw data was well correlated with the 40-Hz response in the filtered data.

Data showed a clear reset of 40-Hz oscillatory activity in control subjects, followed by a stable synchronized and coherent time-locked response with respect to one auditory click [4]. We define the 40-Hz response produced by a single stimulus as a 2.5 oscillatory cycle, demonstrating two and a half oscillations at 40-Hz. Using the dual probe MEG system in control subjects, preliminary data indicate a similar response in both hemispheres. In dyslexics, preliminary data suggest an unstable reset and a prolonged period in synchronized activity in response to auditory stimuli.

Our evidence supports the hypothesis that the minimal interval required to identify separate auditory stimuli correlates with the reset of the 40-Hz magnetic signal (Fig.1). In particular, perceptual tests in control subjects indicated that 15-20 msec are required for the perceptual separation of two stimuli, as reported earlier in the literature. Neuropsychological and psychophysical observations indicated that the auditory system is capable of tonality discrimination of two stimuli that are separated by only 1-2 msec [7], while 15-20 msec are required for the perceptual separation of two stimuli [8]. Experimental and modeling results, based on MEG data, further indicated a stimulus-interval-dependent response with a critical interval of 12-15 msec. At shorter intervals only one 40-Hz response, to the first stimulus, was observed. With longer intervals, a second 40-Hz wave appeared, which coincided with the subject's perception of a second distinct auditory stimulus [4]. These results indicate that oscillatory activity near 40-Hz, recorded during the first 100 msec post stimulus, represents a neurophysiological correlate to the early temporal processing of auditory stimuli underlying cognition.

Using the modified paradigm, perceptual tests in control subjects confirmed our previous findings, that 15-20 msec are required for the perceptual separation of two stimuli. In subjects with language-based learning disabilities, preliminary results indicated a delayed perceptual threshold of 20-30 msec for the identification of a second auditory stimulus (Fig.2), in agreement with earlier findings [9]. In addition, our findings indicate a different time interval necessary for the appearance of a second 40-Hz wave (Fig.3). In particular, data may suggest an increased time period for the processing of incoming stimuli, resetting to the first stimulus, or resetting to the second stimulus and deleting, at least parts, of the first stimulus. This would indicate that, by

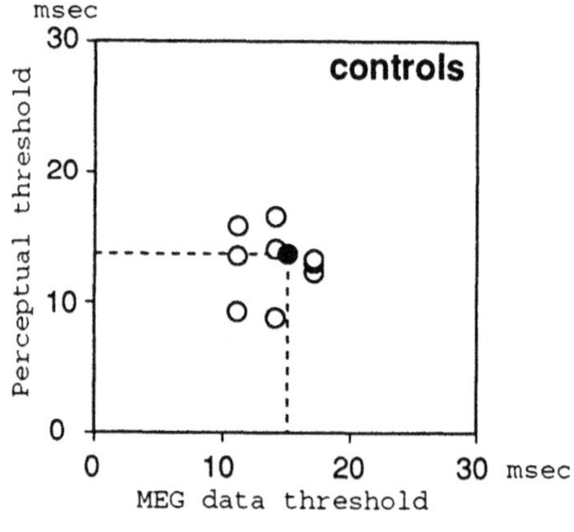

Fig.1: Relationship between perceptual-threshold (average: 13.7 ± 1.32 msec) and MEG-threshold (average: 15.0 ± 2.6 msec) for identifying two clicks with a minimum interstimulus interval for control subjects. The graph plots the thresholds for each of the 9 subjects (hollow circles) and the mean across all subjects (filled circle). (Modified from [3]).

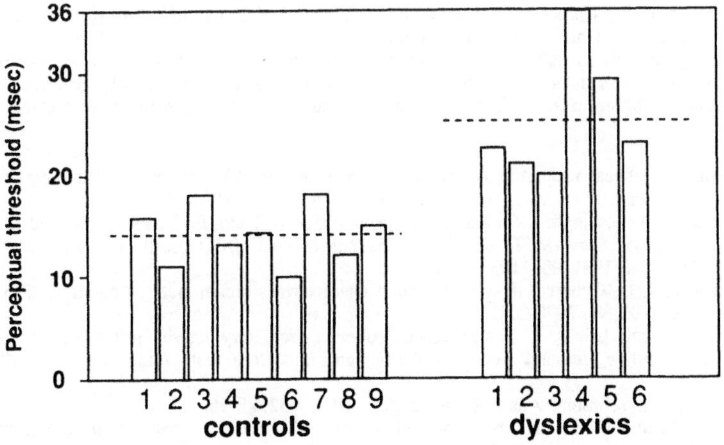

Fig.2: Perceptual-thresholds (minimum interstimulus intervals) required for identifying two separate clicks for 9 individual control subjects and for 6 subjects with language-based learning disabilities. Dashed lines indicate the average values for controls (14.06 ± 2.95 msec) and dyslexics (25.29 ± 6.16 msec).

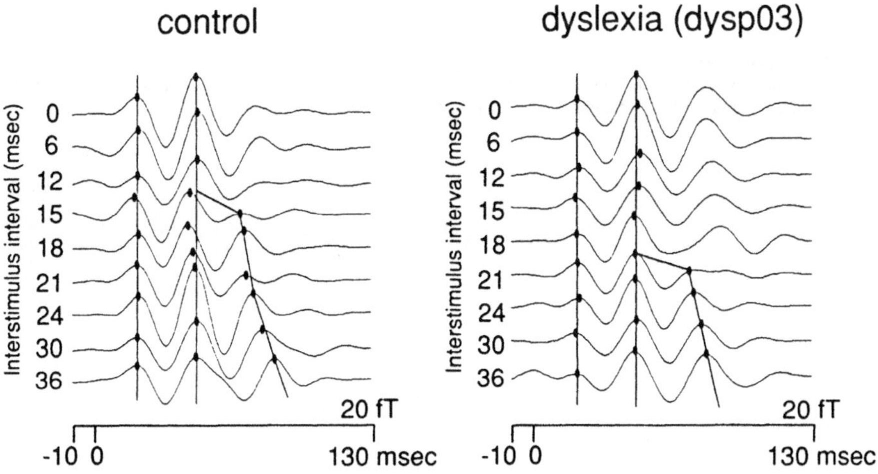

Fig.3: Effect of increasing the interstimulus interval on MEG activity in one control and in one dyslexic subject. The largest of the 37 responses, obtained by the single-click experiment and filtered at 20-50 Hz, was selected and that same channel recording is the actual one illustrated for different interstimulus intervals (6-36 msec). The peak latencies of the responses are marked and the solid lines indicate the reset to one or two clicks. Note that the minimal interval required for a second reset in 40Hz activity is increased in the dyslexic subject (around 21 msec) compared to the control (around 15 msec).

using our recording and analysis procedures, it may be possible to distinguish different levels of severity or different subgroups of dyslexia and to find a neurophysiological marker underlying sensory information processing of auditory stimuli in subjects with language-based learning disabilities, as a powerful new methodology to understand and detect dyslexia for further therapeutical procedures [10]. However, a large population of subjects will be necessary before this phenomena can be characterized more completely.

References

[1] Llinás, R. Intrinsic electrical properties of mammalian neurons and CNS function, In: Fidia Research Foundation Neuroscience Award Lectures, Raven Press, New York, 1990, 4: 175-194.

[2] Ribary, U., Ioannides, A.A., Singh, K.D., Hasson, R., Bolton, J.P.R., Lado, F., Mogilner, A., and Llinás, R. Magnetic field tomography (MFT) of coherent thalamo-cortical 40-Hz oscillation in humans, Proc. Natl. Acad. Sci. USA, 1991, 88: 11037-11041.

[3] Llinás, R., and Ribary, U. Coherent 40-Hz oscillation characterizes dream state in humans, Proc. Natl. Acad. Sci. USA, 1993, 90: 2078-2081.

[4] Joliot, M., Ribary, U., and Llinás, R. Neuromagnetic coherent oscillatory activity in the vicinity of 40-Hz coexists with cognitive temporal binding in the human brain, Proc. Natl. Acad. Sci. USA, 1994, 91: 11748-11751.

[5] Llinás, R. Is dyslexia a dyschronia? Annal NY Acad. Sci., 1993, 682: 48-56.

[6] Tallal, P., Miller, S. and Fitch, R.H. Neurobiological basis of speech: A case for the preeminence of temporal processing, Annal NY Acad. Sci., 1993, 682: 27-47.

[7] Miller, G.A., and Taylor, W.G. The perception of repeated bursts of noise, J. Acoust. Soc. Am., 1948, 20: 171-182.

[8] Hirsh, I.J. Auditory perception of temporal order, J. Acoust. Soc. Am., 1959, 31: 759-767.

[9] Tallal, P. Auditory temporal perception, phonics, and reading disabilities in children, Brain and Language, 1980, 9: 182-198.

[10] Tallal, P., Miller, S.L., Bedi, G., Byma, G., Wang, X., Nagarajan, S.S., Schreiner, C., Jenkins, W.M., and Merzenich, M.M. Language comprehension in language-learning impaired children improved with acoustically modified speech, Science, 1996, 271: 81-84.

Acknowledgements

The support of BTi and the excellent technical assistance by Richard Jagow is greatly acknowledged. This study was supported by The Charles A. Dana Foundation.

A Technique for the Identification of Hippocampal Theta from MEG Data

Tesche, C. D.

Low Temperature Laboratory, Helsinki University of Technology, 02150 Espoo, Finland

Introduction

Rhythmic macroscopic field potentials in hippocampus between 4–12 Hz (theta) are prominent during attention and exploratory behavior in rats and other small mammals [1]. Theta oscillations modulate the firing of individual hippocampal and entorhinal cortical neurons and are believed to be essential for normal memory trace formation [2]. Observations of hippocampal theta in primates are scarce. There are only two reports of human hippocampal theta, both from depth electrode recordings in patients [3], and no reports of the observation of ongoing spontaneous activity in normal human hippocampus. Population responses in human hippocampal formation have been inferred from averaged MEG responses evoked by attention to oddball targets embedded in a sequence of standard stimuli [4,5]. This paper describes a method utilized to determine ongoing spontaneous activity in normal human hippocampus from MEG data recorded with a whole-head array [6].

Methods

MEG data was recorded with a 122-channel array of planar gradiometer detectors covering the whole scalp (Neuromag-122^{TM}). Seven healthy right-handed adults performed two tasks after providing informed consent. In a mental calculation task, subjects silently added and subtracted numbers corresponding to the position in the alphabet of printed letters. For example, c a m g ... was the calculation 3 + 1 - 13 + 7 .... The second task was passively viewing a picture. Data was recorded in 60 s epochs (bandpass filter 0.03–330 Hz, sampling rate 1 kHz). A vertical electro-oculogram was used to identify eye movements and blinks.

Current flow at a set of source locations in the neighborhood of the hippocampus was modelled as a vector sum of three orthogonal equivalent current dipole (ECD) sources. Parietal-occipital alpha and rolandic mu, if prominently represented in the data, were modelled by equivalent current dipole (ECD) sources determined from a least-squares fit between the FFT of the data at specific frequencies and simulated magnetic field patterns [7]. Signal-space projection (SSP) [8,9] was used to derive waveforms for cortical and hippocampal sources from the raw data. Magnetic field patterns were calculated for a set of N ECDs using a spherical head model for superficial sources and a single-compartment boundary-element model for the conducting volume of the head for deeper sources [10]. The 122 sensor readings computed for the ith ECD defined a unit vector $\mathbf{s}_i$ in a 122–dimensional space. The measured MEG signal M(t) defined a time-dependent vector in the same space, where

$$\mathbf{M}_T(t) = \sum_{i=1}^{N} a_i(t)\mathbf{s}_i + \mathbf{n}(t). \qquad (1)$$

The signal-space waveform $a_i(t)$ describes the strength of the component $\mathbf{s}_i$ as a function of time. The time-dependent component $\mathbf{n}(t)$ includes both intrinsic system noise and current flow within the brain which are not explicitly modelled by the vectors $\mathbf{s}_i$. A pseudoinverse of the source matrix $\mathbf{S} = (\mathbf{s}_1, \mathbf{s}_2, \ldots, \mathbf{s}_N)$ determines SSP waveforms for the N sources $\mathbf{a}(t) = \{a_1(t), a_2(t), \ldots, a_N(t)\}^T$,

$$\mathbf{a}(t) = \mathbf{S}^+ \mathbf{M}_T(t). \qquad (2)$$

where $\mathbf{S}^+ = [\mathbf{S}^T\mathbf{S}]^{-1}\mathbf{S}^T$. This approximation yields an unbiased estimate for $\mathbf{a}(t)$ when $\mathbf{n}(t)$ is spatially uncorrelated, normally distributed noise [9].

Results

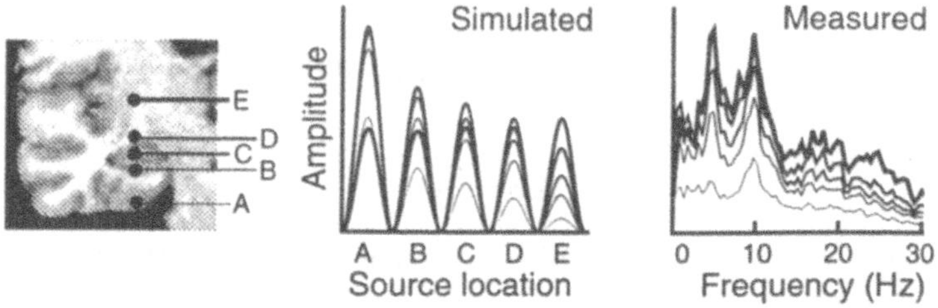

Fig. 1. SSP waveforms computed for a family of "probe" ECD sources in the neighborhood of left anterior hippocampus (left). The orientations of the ECDs are perpendicular to the plane of the MR image. The simulated data contains an isolated source at one of the locations A–E. An example of SSP spectra computed from the data for the same set of "probe" ECDs is shown on the right. The task was mental calculation.

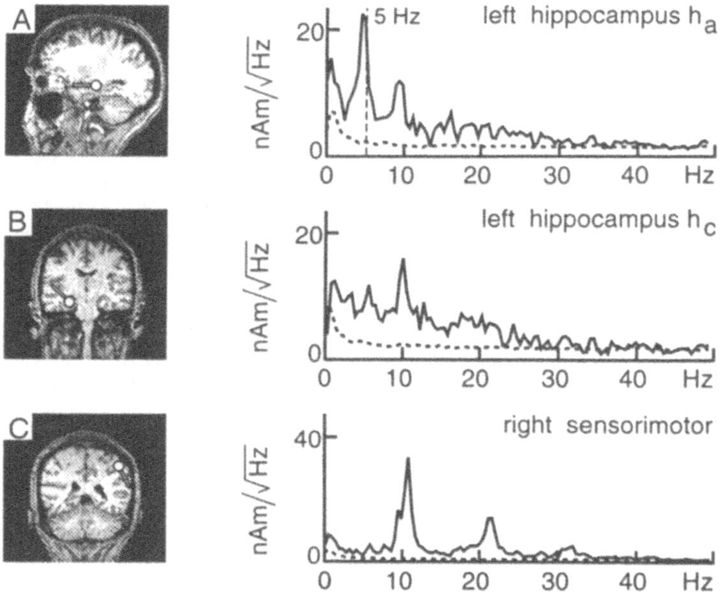

Fig. 2. MR images of the locations and orientations of ECDs in (A–B) left anterior hippocampus and (C) right sensorimotor cortex. Spectral densities as a function of frequency for 5.6 s epochs of mental calculation for each source are shown on the right. The dashed lines indicate the system noise for each source.

There is no unique solution to the inverse problem in MEG. However, waveform scaling has been applied to the study of evoked hippocampal responses to attended auditory oddballs to check the consistency between the SSP results and the MEG data [5]. In this study, waveform scaling was extended to the frequency domain. Averaged spectra were computed from the SSP waveforms for a set of "probe" ECD components with locations in the vicinity of the hippocampus. Results were obtained for the same set of "probe" ECDs for data generated by sources located in the neighborhood of the hippocampus. Fig. 1 shows that the dependence of the spectral components on the location of the "probes" is consistent with that observed from simulations for results generated by an ECD source at location D.

Fig. 2 shows results for subject #4. Both spectral peaks at 5 and 9 Hz and considerable broadband spectral components are visible for activity in hippocampus. In contrast, in sensorimotor cortex the spectral peaks at 11 and 22 Hz (rolandic mu) are much larger than the broadband spectral components.

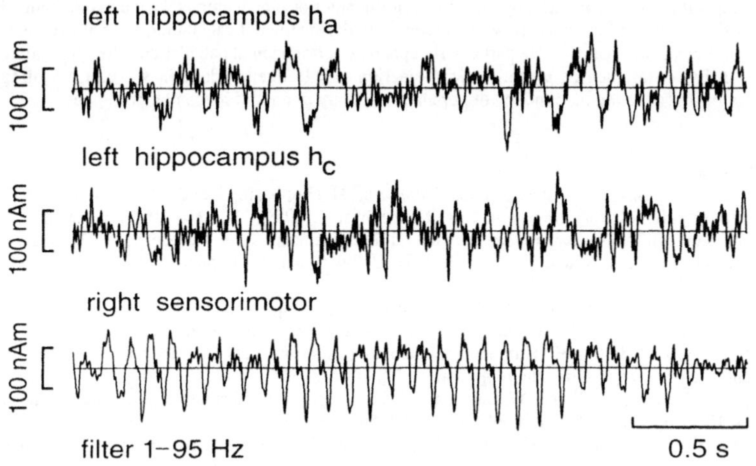

Fig. 3. Waveforms for hippocampal and sensorimotor activity for a 3 s interval of the epoch used to generate the spectra shown in Fig. 2.

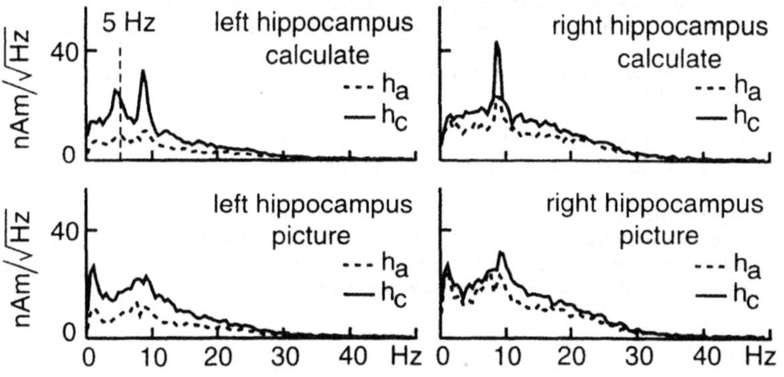

Fig. 4. Spectra for hippocampal activity during the performance of two different tasks; mental calculation (calculate) and passive viewing of a picture (picture). Additive system noise has been subtracted from these spectra.

Short bursts of oscillatory activity were only occasionally seen in hippocampal waveforms (Fig. 3) In contrast, the relatively smaller broadband component and the harmonic structure of the sensorimotor spectra is reflected in the well-defined comb-shaped waveform typical of rolandic mu.

Task-dependent features of hippocampal activity are difficult to extract directly from the time-domain data. As a result, the ability to compute detailed averaged spectra for specific components of the signal [9] is an important tool for the study of human hippocampal population activity. Fig. 4 shows task-dependent features in spectra averaged over 60 s epochs (subject #1). A spectral peak at 5 Hz is present in component h_c during mental calculation, but not during passive viewing of a picture. Spectral peaks at 9–10 Hz also show task dependent features.

Discussion

Spectra consistent with population activity in anterior hippocampus were derived from whole-head MEG data. Broadband spectral components were present both during mental calculation and passive viewing of a picture, with spectral peaks between 4–7 Hz particularly apparent during mental calculation. The application of new analysis tools for MEG data, such as signal-space projection and frequency-domain waveform scaling, opens up the possibility of investigating the function of population activity in normal human hippocampus.

References

[1] Grastyan, E., Lissak, K., Madarasz, I. and Donhoffer, H. Hippocampal electrical activity during the development of conditioned reflexes. Electroenceph. Clin. Neurophysiol., 1959, 11: 409–430. Vanderwolf, C.H. Hippocampal electrical activity and voluntary movement in the rat. Electroenceph. Clin. Neurophysiol., 1969, 26: 407–418.

[2]Buzsáki, G., Leung, L.S. and Vanderwolf, C.H. Cellular basis of hippocampal EEG in the behaving rat. Brain Res., 1983, 6: 139–171.

[3] Arnolds, D.E.A.T., Lopes da Silva, F.H., Aitink, J.W., Kamp, A. and Boeijinga, P. The spectral properties of hippocampal EEG related to behaviour in man. Electroenceph. Clin. Neurophysiol., 1980, 50: 321–328. Halgren, E., Babb, T.L. and Crandall, P.H. Human hippocampal formation EEG desynchronizes during attentiveness and movement. Electroenceph. Clin. Neurophysiol., 1978, 44: 778–781.

[4] Okada, Y.C., Kaufman, L. and Williamson, S.J. The hippocampal formation as a source of the slow endogenous potentials. Electroenceph. Clin. Neurophysiol., 1983, 55: 417–426.

[5] Tesche, C., Karhu, J. and Tissari, S. Non-invasive detection of neuronal population activity in human hippocampus. Cog. Brain Res., in press.

[6] Tesche, C.D., Non-invasive detection of theta activity in human hippocampal formation with a whole-head magnetoencephalographic array. Soc. Neurosci. Abstr., 1996, 22 (1): 276.

[7] Tesche, C.D. and Kajola, M. A comparison of the localization of spontaneous neuromagnetic activity in the frequency and time domains. Electroenceph. Clin. Neurophysiol., 1993, 87: 408–416.

[8] Ilmoniemi, R.J. and Williamson, S.J. Analysis for the magnetic alpha rhythm in signal space. Soc. Neurosci. Abstr., 1987, 13 (2): 46.

[9] Tesche, C.D., Uusitalo, M.A., Ilmoniemi, R.J., Huotilainen, M., Kajola, M. and Salonen, O. Signal-space projections of MEG data characterize both distributed and well-localized neuronal sources. Electroenceph. Clin. Neurophysiol., 1995, 95: 189–200.

[10] Hämäläinen, M. and Sarvas, J. Realistic conductivity geometry model of the human head for interpretation of neuromagnetic data. IEEE Trans. Biomed Eng., 1989, 36: 165–171.

Acknowledgements

I thank J. Karhu for useful discussions and R. Hari for comments on the manuscript. This work was supported by NIH National Research Service Award 1 F33 NS09623-01H from the NINDS. The MRI scans were obtained at the Radiology Department of the Helsinki University Central Hospital.

VI. Magnetoencephalography (MEG): Clinical Studies

MEG: Clinical Applications

[1]Mäkelä, J.P., [2]Elbert, T., [3]Kakigi, R., [4]Lewine, J., [5]Lopez, L., [6]Nagamine, T., [7]Nakasato, N., [8]Reite, M., and [9]Sannita, W.

[1]Low Temperature Laboratory, Helsinki University of Technology, Espoo; Central Military Hospital, Helsinki, Finland, [2]University of Konstanz, Konstanz, Germany, [3]Department of Integrative Physiology, National Institute for Physiological Sciences, Okazaki, Japan, [4]The New Mexico Institute of Neuroimaging, Albuquerque, New Mexico, U.S.A., [5]Institute for Advanced Biomedical Technologies, Chieti, Italy, [6]Department of Brain Pathophysiology, Kyoto University School of Medicine, Kyoto, Japan, [7]Tohoku University, Department of Neurosurgery, Sendai, Japan, [8]University of Colorado, Health Sciences Center, Denver, Colorado, U.S.A., [9]Center of Neuropsychoactive Drugs, University of Genova, Genova, Italy; State University of New York, New York, U.S.A.

During the last two decades, magnetoencephalography (MEG) has become available for monitoring human cortical activity. Although impressive results have been obtained with instruments having smaller coverage, new whole-scalp neuromagnetometers [1;2] began a new era in the neuromagnetic patient studies. The simultaneous activity of different widely spaced cortical areas can now be observed. This is an advantage in studying propagation of activity, *e.g.* in epilepsy, and also in studying functional alterations after lesions, which may occur far from the site of the lesion. Furthermore, individual variation is remarkable even in healthy subjects' MEG responses to simple stimuli. For instance, some control subjects may show gross hemispheric differences in auditory evoked fields (AEFs) to simple tones [3;4]. Increased recording speed allows collection of large normative data bases, necessary for clinical applications.

Non-invasive MEG measurements may help in clinical decision making in individual patients. There are examples of epileptic surgery guided, in part, by MEG information. MEG source localization may also aid in locating functionally irretrievable areas before neurosurgery. In addition, MEG may give information of mechanisms underlying diseases and recovery. Furthermore, one can also view lesions in neurologic diseases as "experiments of nature" and record MEG in patients to obtain indirect information about normal brain functions.

Search for MEG clinical applications should be directed to areas where no good objective tools exist at present, or where the nowadays applied methods are inaccurate, expensive or risky for the patients. The non-invasiveness and excellent time resolution of MEG should be taken into account in search of such alternatives.The following chapters describe approaches which are potentially interesting in this respect.

Functional localization before neurosurgery

MEG source localization may aid in pinpointing functionally irretrievable areas before neurosurgery. Tumours or vascular malformations may distort brain anatomy so as to render localization of. *e.g.*, motor areas impossible on the basis of anatomic landmarks. Functional localization of the eloquent cortical areas may prevent postoperative hemiparesis, or to encourage neurosurgeon to operate in an otherwise unclear situation.

MEG sources are now routinely superimposed on MR images using co-registration techniques. An MRI-linked whole head MEG system [2] has been applied for more than 400 subjects including normal volunteers and patients with various types of neurological diseases. The sources of the somatosensory evoked fields (SEFs) to median nerve stimuli are localized on the posterior wall of the central sulcus [5;6]; this localization can be applied especially usefully in neurosurgery since primary motor cortex, to be protected during operations, lies on its anterior wall. Although MRI scans may identify "anatomically" the central sulcus, it is controversial as to whether the "functional" central sulcus is shifted due to structural lesions such as arteriovenous malformations (AVMs). AVMs are congenital and can lead to central nervous system (CNS) reorganzation; anatomical localization does not take into account the plasticity of brain function. In patients with AVM adjacent to the central sulcus, the N20m sources for the median nerve SEF in the affected hemispheres were localized on the anatomical central sulcus with a 4-mm standard deviation [7]. This was so even in two cases with a large and diffuse AVM occupying almost the entire frontal lobe, suggesting that AVM may not shift the localization of the functional central sulcus.

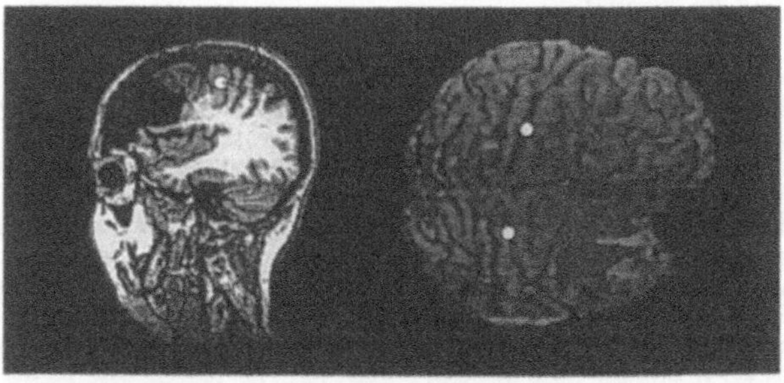

Fig. 1. Sources of SEF P30m deflection overlied on 3-D MR image of a patient after removal of right frontal glioma. Two years after operation, the patient's symptoms reappeared and reoperation was planned. Sagittal section shows that tumor infiltrates the primary motor cortex, immediately anterior to the SEF sources. Because of high probability of postoperative hemiparesis, reoperation was cancelled. Unpublished data, courtesy to Dr. Forss.

The N100m sources of the auditory evoked magnetic fields have been localized in the vicinity of the primary auditory cortex bilaterally on superior temporal surfaces [3;8;9]. In AEFs measured with the MRI-linked whole head MEG system [4], averaged position of the estimated N100m dipoles was precisely on the upper surface of temporal lobes with a vertical 2-mm standard deviation. Disappearance of N100m response [10;11] and delay of the N100m latency [12] have been reported in patients with temporal lobe lesions. Current MEG techniques cannot localize speech-related cortical areas directly. However, functional localization of the auditory cortex is useful for surgery especially on the left temporal lobe, because the left auditory cortex is surrounded by the language-related cortex.

Visual evoked fields (VEFs) are equally applicable in the functional localization of eloquent cortical regions. In recordings of pattern reversal VEF in a patient with homonymous left hemianopsia due to right occipital AVM, a two-dipole pattern of P100m appeared over the occipital area bilaterally. In this patient, the medial part of the calcarine fissure was occupied by the AVM whereas the lateral part was spared [13]. The results suggest good correspondence between functional localization and cortical anatomy. Further MEG studies, especially in patients with brain lesions, are warranted to provide more information about cortical sources of evoked responses.

The parameters of function used in presurgical evaluation need to be stable across populations of patients and age-matched controls. SEF N20m stability has been evaluated in detail [14].The median nerve, thumb and little finger of the two hands were stimulated, and N20m localization, distance between homologous equivalent current dipoles (ECDs) in the two hemispheres, distance between thumb and little finger ECDs, and the difference in this distance in the two hemispheres were calculated. In patients 2-7 mo after hemispheric stroke, at least one of these parameters was aberrant, implying functional reorganization [15] In the study of plasticity (see below) it is also important to have predictive parameters that describe the intra-individual relationship between homologous areas in the two hemispheres.

Functional MR imaging (FMRI) has recently been applied to location of sensory, motor, visual, and auditory activities. Possibilities of making functional measurements with 1.5 T instruments, already available in hospitals, are specially alluring. At present, no clear understanding exists of the spatial relations of neural activity and altered blood flow and changes of oxygenation. Comparison of FMRI and MEG localization of unilateral motor activity resulted in 17 mm difference between the centers of activity estimated with these methods [16]. Methodological improvements of FMRI are progressing rapidly. MEG measurements may give one reference point for this development.

AVMs, containig blood vessels, probably provide difficulties in FMRI localization because of aberrant blood flow and vessel permeability. In addition to activations in the regions expected for the paradigms employed in a FMRI study of AVM patients, qualitatively reproducible activations were observed in unexpected regions. This was attributed to plasticity due to AVM [17]. Comparison with MEG measurements could be beneficial in selecting functionally important activations from those due to AVM.

Detection of abnormal slow-wave activity

Brain lesions are often accompanied by increased EEG activity in the theta and delta bands. Whereas this increase in slowing has been reported in the EEG literature for years, the MEG counterpart of the phenomenon has been characterized only recently. In EEG, slowing is typically diffuse, often encompassing all of the electrodes over one or both hemispheres. In contrast, slowing in the MEG is often focal, especially in patients with circumscribed lesions. MEG field patterns associated with slow waves are typically complex, implying contributions from multiple, simultaneously active brain regions. Nevertheless, for some slow waves, an equivalent current dipole model provides an excellent account of the field pattern, presumably because a relatively focal area of dysfunctional tissue is responsible for its generation. Whereas it would be of interest to disentangle the multiple sources contributing to complex slow waves, presently available analysis algorithms preclude this on a routine basis. It is thus fortunate that clinical information can be ascertained by focusing analysis on the slow waves with a dipolar field configuration.

A challenge in the evaluation of slow waves is distinguishing pathological slowing from normal one, such as that associated with sleep. This is particularly problematic in patients, because they often become drowsy during an examination. To deal with this problem, several investigators have developed specific criteria for identification of pathological slowing [19-22]. One of the most critical of these criteria relates to spatial clustering of sources. In healthy subjects, it is rare to find co-localization of the dipolar sources from separate slow wave events within a 1cm^3 volume. In patient groups, clustering of sources is common. For example, when clustering analyses are used [19], clusters of slow wave sources are found in less than 10% of normal subjects, invariably at occipital or high parietal locations, but in more than 70% of patients with gross brain lesions, *e.g.*, neoplasms or encephalomalacia associated with cerebrovascular accidents. Fig. 2 shows data from a patient with a clear MRI lesion. Delta activity is obvious within the raw data, and as shown by the isofield contour map, the indicated slow wave has a highly dipolar field pattern. In this case, sources localize all along the lateral margin of the lesion, as shown on the MR section.

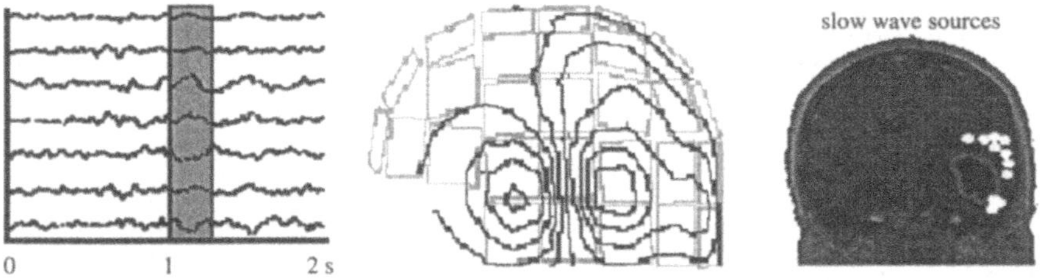

Fig. 2. Left: slow wave data from a patient with a left temporal lesion with cystic cavity, ring enhancement and surrounding edema. Middle: Field pattern of a slow wave, overlied on the sensor helmet of a 122-channel instrument, seen from the left. Right: slow wave sources, overlied on the patient's MRI, localize to the lateral margins of the lesion.

In stroke patients, clustering of slow wave sources is common along the margins of the lesion, in the ischemic penumbra. The data from patients with neoplasms and strokes lends "face validity" to the notion that cluster analysis of slow wave activity provides for good identification of areas of cortical dysfunction. However, a less intense mirror focus of slowing is found in the healthy hemisphere in 20-30% of patients. Also, it is not uncommon to find slow wave sources localizing deep into the white matter. This probably reflects erroneous modelling of an extended cortical sheet with a single dipole. Given this situation, slow wave activity must be interpreted within the entire clinical context, with acknowledgment that source localizations of slow waves are probably accurate only to within 1-2 cm, especially in the depth direction.

One interesting clinical development in the assessment of slow waves has been the identification of significant dipolar slowing in patients that have neurobiological or psychiatric dysfunction, but no gross anatomical lesions. For example, slowing is found in 80% of patients with epilepsy, including those without obvious anatomical changes. In patients with both interictal spikes and slowing, spikes and slow wave sources co-localize. Dipolar slowing is also found in 65% of patients with post-concussive syndromes subsequent to mild head trauma [20]. This is true even in cases with normal MRI, normal CT, and even normal clinical EEG. Other conditions that show slowing include transient ischemia [22], schizophrenia, dementia, and chronic alcohol abuse.

Although the presence of dipolar slow wave clusters is believed to be a strong indication of underlying pathology, the converse is not necessarily true; an absence of slow wave clusters does not always indicate that all is well. Some patients have clearly abnormal delta and theta activity in EEG and MEG power spectra, but no dipolar slow wave clusters. This is believed to reflect the fact that the relevant pathology is not focal. The precise relationship between the location of slow waves and underlying primary pathology must be considered carefully. For example, thalamic lesions generate cortical slow waves [18]. That is, the slowing localizes cortically, even though the critical lesion is subcortical.

Despite the care that must be given to the clinical interpretation of slowing, available data suggest that MEG characterization of slow waves is of clinical utility. MEG slowing provides objective evidence of brain dysfunction that is lacking through other neuroradiological and neurophysiological procedures.

Pain representation in CNS

Pain is a common clinical problem. To date, no objective pain measurement system is available; subjective psychophysical pain assessment is the only alternative. Objective pain indexes would be of considerable interest. To develop them it would be useful to understand CNS pain representation.

Painful electric or CO2-laser stimulation elicits a complex pattern of activation in MEG [23]. In addition to primary somatosensory activation, bilateral second somatosensory areas (SII) [24;25] and cingulate gyrus [26] are activated. However, the signals about 200 ms onwards are complex, show large intraindividual variation and cannot easily be attributed to dipolar sources [26]. Probably, attentional networks are strongly activated by painful stimuli. Therefore, more experimentation is needed to subtract pain-specific components from those related to general attention.

Cortical plasticity in adult humans

Structure and function of various brain areas are modified following injury or as a result of experience. The follow-up of this mutability by non-invasive methods may be interesting in evaluation of rcovery and results of rehabilitation. One can, in some cases, see objective signs of recovery and reorganization after the lesion in the CNS by MEG. The cortical representation of the sensory perception relates in an orderly way to the spatial arrangement of peripheral receptors. Changing the relative weighting of connections by increased stimulation or by inactivating a pathway may alter these maps.

Somatosensory plasticity

In animal studies, increased use of a limb expands its cortical representation and reduces receptive field size, whereas deafferentation or disuse generates an invasion in the representational zone from nearby sites. Synchronous, behaviorally relevant stimulation at nearby sites (*e.g.*, different fingers) results in the fusion of representations, whereas asynchronous stimulation to two different sites segregates the representations [27;28]. Such a reorganization in adult humans can be detected by MEG. The somatotopic map that spans across a relatively large distance along the posterior wall of the central sulcus can be studied particularly well with this technique.

Patients with unilateral upper extremity amputation have altered topographic representation of the cortical face area and the cortical upper arm (stump) areas which lie adjacent to the former receptive field of the amputated limb. Sources of SEFs elicited by facial stimulation were shifted 1-2 cm towards the receptive field which would normally receive input from the now amputated hand and fingers [29;30]. A similar tendency was observed for the receptive field for the stump of the upper arm. These alterations imply extensive plastic cortical reorganization in adult humans following injury. Furthermore, in human upper extremity amputees a strong correlation exists between extent of cortical reorganization and the amount of experienced phantom limb pain [31] as well as with the amount of body surface from which painful stimuli evoked sensations that were perceived to be emanating from the the phantom limb [32]. Although phantom pain is a disadvantageous consequence of injury, it nevertheless suggests that cortical reorganization can have an outcome of perceptual significance for the individual. These studies also demonstrated fluctuation of cortical reorganization, phantom pain and referred sensations [33]. Furthermore, a pilot investigation, using MEG, suggests that the persistent, incoming stimulation of chronic pain may alter the central representation of body parts related to the pain process. Patients who experience chronic noxious stimulation over extended periods of time seem to produce more extensive activation of neural assemblies than controls, when phasic somatosensory stimuli are applied to body sites related to their chronic pain [34]. It is of interest to try to analyze why alterations related to SI area relate to pain; as seen above, pain-specific responses ocur primarily in the SII-cortices.

The phenomena leading to phantom pain may be aberrant derivatives of the mechanism permitting the processes associated with growth, learning, and adjustment to environmental demands. For instance, the cortical representation of the left hand digits of string players, which carry out the demanding task of fingering the strings, is larger than that in controls [35]. No such differences were observed for the representations of the right hand digits. Furthermore, extensive training of Braille reading enlarges the representation of the reading fingers within days [36]. In patients with syndactyly, where the fingers are congenitally fused, the SEF sources of finger stimulation are not organized somatotopically. When the fingers were surgically separated, they became individually represented in the cortex within weeks [37]. Obviously, the representation of different parts of the body in the human primary somatosensory cortex changes in a use-dependent manner. In amputees, the presumed increase of afferent input from the intact hand, based on the requirement that the hand should be used more to compensate for the loss of the contralateral one, should consequently result in enlargements of the intact finger representatios. This was shown to be so in a study comparing the cortical organisation of the representation of the intact arm and the amputation zone in upper extremity amputees with the one in controls [38].

On cellular level, receptive fields are shaped, in part, by lateral inhibition. A small denervation causes disinhibition of viable, but normally suppressed inputs from areas adjacent to the affected region and leads to the expression of new or expanded receptive fields. These processes may be related to phantom pain. Cortical alterations have, in part, been explained by arborization of thalamocortical neurons that had retained their normal sources of input and that were adjacent to the affected cortical areas that had lost their afferents: however, this mechanism can account for only a small fraction of the massive reorganization covering up to a few cm, observed after the arm amputation in man [29;30]. In addition to cortical alterations, plasticity in the lower levels of somatosensory pathways probably contribute to the effects observed in the cortex. Small changes in the local circuits of subcortical maps can be relayed to cortex and appear as major reorganizations [39].

Auditory plasticity

In a similar vein, MEG can be used to detect alterations of tonotopic organization in the auditory cortex. Several tonotopic maps have been described and their plasticity is currently under investigation: The sources of N100m for higher frequencies lie in progressively deeper locations, whereas the sources of the middle-latency P30m for higher frequencies are located progressively more superficial as frequency increases [40] . Tinnitus, a ringing in the ears, may be an auditory phantom perception. Patients with tinnitus have deviance in their tonotopic map; the amount of deviation is correlated with the subjective tinnitus strength [41]. However, tonotopy of N100m is not always apparent in controls (for references, see *e.g.*, [42]). In animal experiments, frequency specificity of single cells in AI is affected by stimulus intensity [43]. Obviously, carefully delineated studies of *e.g.*, stimulus intensity vs. frequency are needed before reliable clinical studies can be performed.

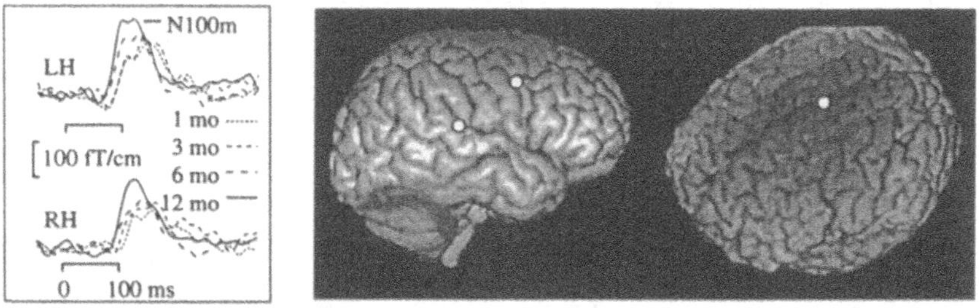

Fig. 3. Left: Development of AEFs to healthy ear stimulation in a patient with left-sided deafness after operation of acoustic neuroma. LH indicates left and RH right hemisphere. Adapted from [44]. Right: Sources of AEFs to non-attended healthy-ear stimulation in a patient after sudden unilateral hearing loss. Adapted from [45].

Other types of auditory reorganization have been observed as well. Figure 3 shows AEFs in a patient deafened in the left ear after an acoustic neuroma operation; the stimuli were delivered to the healthy, right ear. On the basis of plastic somatosensory reorganization, one could expect that the N100m ipsilateral to the stimulated healthy ear would increase in amplitude during the follow-up, because there would be reorganization due to loss of input from the contralateral, operated nerve. This turned out not to be so: the responses were initially dampened in both hemispheres and recovered towards normal values during one year after the operation, even exceeding them in some cases [44]. The activation of both auditory cortices appeared to become more brisk and concise during the recovery. This may be due to enhanced coherence of postsynaptic potentials underlying N100m. No systematic changes were seen in the AEF source locations.

Three out of 8 patients with sudden unilateral hearing loss of unknown origin showed atypical AEFs over head midline. Furthermore, over the right hemisphere, the same three patients had two active cortical areas, an early posterior and a late anterior area. The later, anterior source reflected activation in the frontal association areas ([45]; see Fig.3). It is possible that the increased difficulty in the analysis of auditory stimuli due to unilateral hearing defects has generated a compensatory increase in the processing of stimuli by auditory polysensory pathways and enhanced their cortical source areas so as to make them visible in AEFs.

In animal experiments, cortical and subcortical maps reorganize after restricted damage of the cochlear receptors. The cortical area normally representing frequencies of the lesioned cochlear portion is occupied by neighbouring frequencies. However, this does not mean central compensation of peripheral hearing loss; sensitivity to those frequencies that would normally produce activation in the damaged region of the cochlea is not restored [46]. On the other hand, enhanced mean responses of single cells, and increase of the coherence between activities of different neurons participating in the same processing assembly would lead also to behavioural improvement by means of increased reliability and shortening of required processing time [47]. Little attention has been paid to changes in temporal properties of neural signalling during development of lesion-induced changes, although MEG's superior temporal resolution is optimal for such studies. The results of [44] show that indeed, shortening of the response latencies and increased coherence of neural activity, reflected as larger signals, occur in the human central nervous system during recovery after peripheral nerve lesions.

Diagnostics of motor disorders

Currently available MEG techniques for investigating motor disorders are extensions of approaches previously used in EEG. One possibility to detect abnormality of voluntary movement is to measure movement-related magnetic fields. They are produced, in part, in the primary motor cortex: its location may be shifted by lesions around somatomotor area. Onset of spontaneous movements, detected by electromyography, may be used as a fiduciary point for backward averaging. Jerk-locked back-averaging is an useful tool to explore cortical hyperexcitability in patients with involuntary movements: it can be shown that some myoclonic jerks in patients with progressive myoclonus epilepsy are cortically initiated.

Analysis of magnetic spontaneous activity has been applied to elucidate CNS mechanisms of tremor in parkinsonian patients. If tremor is produced by cortical functions subserving normal movements, somatomotor mu rhythm disruption might occur during tremor as well as during other movements. In patients with hemi-parkinsonism, the spectra of spontaneous activity showed a clear dampening of 10-Hz rhythm during tremor [48]. Apparently, parkinsonian tremor activates cortical mechanisms normally producing rapid, alternating movements. Recently, synchronized and tremor-related MEG activity over the frontoparietal cortical regions was observed in parkinsonian patients. This activity could be attributed to source areas that were previously ascribed for voluntary movements, and was suggested to be generated by rhythmic activity in the diencephalon, generating central motor loop oscillations [49].

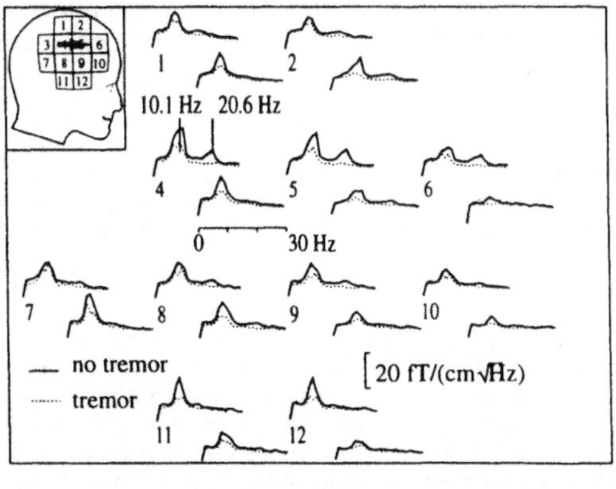

Fig. 4. Dampening of 10-Hz rhythm during parkinsonian tremor. The inset shows the measurement location; the arrow depicts the N20m source for median nerve stimulation. Modified from [48].

MEG may reveal abnormalities in primary sensorimotor cortices and, in special cases, also in premotor and supplementary motor cortices. Activities in basal ganglia can only be estimated indirectly by current MEG techniques. Responses from primary motor areas show extensive variation in normal subjects, exceeding clearly the variability of *e.g.*, SEFs; this should be kept in mind when evaluating signals from patients. However, careful observation of clinical symptoms and carefully planned experimental setups, strenghtened by advances in MEG technique, will undoubtedly clarify the mechanisms of motor disorders.

Pharmaco-MEG

Analysis of alteration of CNS activity by different pharmacological manipulations by MEG is still in its infancy. Complex and interacting subcortical structures modulate cortical activity. It is not possible on the basis of MEG measurements only to conclude at which level the drug effects on cortical/subcortical function take place. Preliminary MEG data, however, suggest that in humans this approach can provide information on drug action that is complementary to that obtained by conventional quantitative EEG methods [50;51]. Chemically specified subsystems exist in sensory systems. A functional dissection of neural circuits by pharmacological manipulation has allowed segregation of evoked response components [52;53], and showed local drug interference at synaptic and non-synaptic level [54]. Cortical and precortical evoked phenomena depend upon characteristics of the information processing in the CNS that *e.g.*, in visual system, can be activated independently and are selectively affected by drugs. Hypotheses exist about the functional roles of dopaminergic, GABAergic and cholinergic transmission in the CNS, and the effects on brain electrophysiology of compounds interfering with these systems have been described [53;55;56]. Interaction between cortical-subcortical acetylcholine modulation and the carrier systems mediating in the transfer of sensory information *e.g.*, visually evoked oscillatory potentials [57;58], is also conceivable. Preliminary human data [59] indicate that scopolamine increases the synchronization of cortical oscillatory response after full-field luminance stimulation while reducing it when stimulation is in spots.

By MEG, it is feasible to study activity within a limited cortical region, to use *e.g.*, dipole modelling to create a "spatial filter" representing activity of a cortical region. Thereafter, it is possible to observe modulations of the activated region when changing different stimulus parameters, or by changing the state of the subject with different pharmacological manipulations.The best example at present is modulation of AEF sources by intravenous lidocaine in patients with tinnitus. In three out of four patients with tinnitus, transient amelioration of the symptoms during lidocaine injection was reflected in shortened and sharpened activation of the N100m source [60].

Anesthesia generates strong effects on brain electric activity. Monitoring these changes is feasible by MEG: This was demonstrated with an experiment, showing burst-suppression pattern of MEG signals generated by enflurane anesthesia in a dog [61]. The result itself is by no means unexpected: however, the experiment demonstrates that it is perfectly feasible to record MEG despite apparatus providing inhalation anesthesia and monitoring the subject state. Selection of epileptic drugs is at present time-consuming. It is possible to monitor epileptic cortical activity by MEG and study *e.g.*, with intravenous injections of how the selected drug alters the activity of the epileptic focus (Knutsson, pers. comm.). The optimal selection of antiepileptic agent may avoid several outpatient visits and additional epileptic seizures during tests of ineffective medication.

Several pharmacologic agents mediate their effects through receptors affecting ion channels in cell membranes. The effects of direct modulation of these ionic channels on MEG signals has been analyzed in vitro in slabs of turtle cerebellum [62]. This type of basic research will undoubtedly increase potential of pharmaco-MEG, depending on the extent to which animal and human data can be comprehensively compared.

In some cases, the topography of spontaneous activity or evoked responses recorded at scalp level matches the available evidence of cortical generators. However, localization may be less important in human neuropharmacology than the definition of relative activation of discrete brain portions and of the dynamics of drug effect, connected to characteristics of neurotransmission and pharmacological mechanisms of action. "Transparency" of low-pass filtering skull, and independence from reference favour MEG application in neuropharmacology.

MEG in psychiatry

The importance of MEG's time resolution, compared with other imaging techniques, may become useful in psychiatry. The combination of MEG to other techniques should enable clinicians to find diagnostic tools for psychiatric diseases, in which the understanding of the brain information processing is crucial. The earliest applications of MEG in psychiatry have studied AEF N100m deflections. In controls, the sources of N100m are more anterior over the right than the left hemisphere. In schizophrenics, this difference is smaller than in controls [63]. Probably this reflects, at least in part, structural differences of temporal lobes evident in CT and MRI as well.

Transient auditory hallucinations delay N100m as does real speech in controls, suggesting that auditory cortex is activated during hallucinations by endogeneous neural firing [64], in good agreement with recent FMRI data [65].

The anomalous asymmetry demonstrated by relative antero-posterior left-right N100m source localization in schizophrenia appears to be sex-specific. In actively psychotic males with schizophrenia, there is less asymmetry of N100m source localization than in control males; schizophrenic males have sources relatively further anterior on the left superior temporal gyrus (STG) and relatively more posterior on the right STG (see Fig. 5). On the other hand, actively psychotic females with schizophrenia demonstrate increased asymmetry, with N100m sources further anterior on the right STG compared with control females [66]. The same group of schizophrenic males also demonstrated smaller left STGs than controls, whereas in females with schizophrenia the STG volume was normal [67].

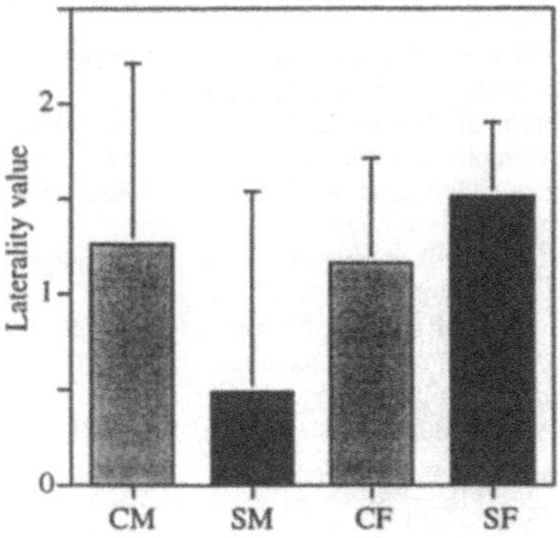

Fig. 5. Laterality value ± S.D. of control males (CM), schizophrenic males (SM), control females (CF) and schizophrenic females (SF). The laterality value was calculated by subtracting the left N100m Y coordinate value from the right Y coordinate value for each group. The antero-posterior Y axis runs through the nasion, perpendicular to the line connecting preauricular points. Modified from [66].

Spontaneous 10-Hz activity in the vicinity of the N100m sources is suppressed when subjects scan memory to recall melodies [68]. This suppression is clear over both left and right STG in controls. In schizophrenic males, however there is a deficient suppression over the left but not over the right STG during such task [69]. In a pilot study with schizophrenic females, normal suppression was also observed over the right hemisphere but not over the left one. Thus, only males with schizophrenia had anatomic abnormalities in the left STG whereas both males and females showed functional deficit over the left hemisphere. This suggests relative independence of of functional and structural abnormalities in schizophrenia.

Electroconvulsive therapy in depression increases slow EEG rhythms. Similar slowing is evident in MEG spontaneous activity: the frequencies below 8 Hz increase in a dose-dependent manner during treatment and disappear about 1 month after the treatment. The alterations appear to occur in the precentral and central areas, in the posterior temporal lobes, and around temporo-pariteo-occipital junctions [70]. The relation of the changes to the effect of the therapy are unknown at present. Obviously, increase in understanding of the basic processes involved in attention, learning, and memory will further the study of psychiatric disorders by MEG.

Conclusions

MEG is useful at the level of single patients in defining the location of functionally irretrievable areas before surgery. MEG studies may reveal the mechanisms of diseases and recovery and also allow conclusions to be drawn regarding how the normal brain functions, on the basis of findings in neurologic patients. The temporal resolution of MEG is useful in studying how signal coherence and response speed improve after CNS lesions. The future will show if more complex methods of signal analysis, *e.g.*, coherence calculations, provide additional benefit from this time resolution in the study of patients. In general, most MEG measurements have been performed in research units distant from hospitals; the studied patient groups are small and selected. Recently, whole-head MEG systems have been installed in medical centers as well, and the near future will clarify the role of MEG in clinical use.

References

[1] Ahonen, A, Hämäläinen, M, Kajola, M, et al. (1993) 122-channel SQUID instrument for investigating the magnetic signals from human brain. Physica Scripta, T40: 198-205.

[2] Vrba, J, Betts, K, Burbank, M, et al. (1993) Whole cortex 64 channel SQUID biomagnetometer system. IEEE Transactions of Applied Superconductors, 3: 1878-1882.

[3] Mäkelä, JP, Ahonen, A, Hämäläinen, M, et al. (1993) Functional differences between auditory cortices of the two hemispheres revealed by whole-head neuromagnetic recordings. Human Brain Mapping, 1: 48-56.

[4] Nakasato, N, Fujita, S, Seki, K, et al. (1995) Functional localization of bilateral auditory cortices using an MRI-linked whole head magnetoencephalography (MEG) system. Electroencephalography and clinical Neurophysiology, 94: 183-190.

[5] Wood, C C, Cohen, D, Cuffin, B N, et al. (1985) Electric sources in the human somatosensory cortex: identification by combined magnetic and potentials field recordings. Science, 227: 1051-1053.

[6] Kawamura, T, Nakasato, N, Seki, K, et al. (1996) Neuromagnetic evidence of pre-and post-central cortical sources of somatosensory evoked responses. Electroencephalography and clinical Neurophysiology, 100: 44-50.

[7] Kawamura, T, Nakasato, N, Ohtomo, S, et al. Neuromagnetic identification of the somatosensory cortex in cases with arteriovenous malformation adjacent to the central sulcus. In this volume.

[8] Hari, R, Aittoniemi, K, Järvinen, M L, et al. (1980) Auditory evoked transient and sustained magnetic fields of the human brain. Localization of neural generators. Experimental Brain Research, 40: 237-240.

[9] Pantev, C, Hoke, M and Lehnertz, K (1990) Identification of source of brain neuronal activity with high spatiotemporal resolution through combination of neuromagnetic source localization (NMSL) and magnetic resonance imaging (MRI). Electroencephalography and clinical Neurophysiology, 75: 173-184.

[10] Mäkelä, JP, Hari, R, Valanne, L, et al. (1991) Auditory evoked magnetic fields after cerebral vascular lesion. Annals of Neurology, 30: 76-82.

[11] Nakasato, N, Fujita, S and Matani, A Clinical application of the whole-head MEG: Auditory evoked response in patients with intracranial structural lesions. In: Baumgartner, C, Deecke, L, Stroink, G, Williamson, SJ (Eds.), Biomagnetism: Fundamental research and clinical applications. Studies in applied electromagnetics and mechanics. Elsevier/IOS press, Amsterdam, 1995: 186-190.

[12] Kanno, A, Nakasato, N, Ohtomo, S, et al. Normalized N100m latency in the auditory evoked fields after surgical removal of temporal lobe gliomas. In this volume.

[13] Seki, K, Nakasato, N, Fujita, S, et al. (1996) Neuromagnetic evidence that the P100m component of the pattern reversal visual evoked response originates in the bottom oc calcarine fissure. Electroencephalography and clinical Neurophysiology, in press.

[14] Rossini, PM, Narici, L, Martino, G, et al. (1994) Analysis of interhemispheric asymmetries of somatosensory evoked magnetic fields to right and left median nerve stimulation. Electroencephalography and clinical Neurophysiology, 91: 476-482.

[15] Rossini, P.M, Pizzella, V, Rossi, S, et al. SEFs to median nerve and finger stimulation: analysis of interhemispheric asymmetries in healthy subjects and following hemispheric lesions. In this volume.

[16] Beisteiner, R, Gomiscek, G, Erdler, M, et al. (1995) Comparing localization of conventional functional magnetic resonance imaging and magnetoencephalography. European Journal of Neuroscience, 7: 1121-1124.

[17] Maldjian, J, Atlas, S, Howard, R, et al. (1996) Functional magnetic resonance imaging of regional brain activity in patients with intracerebral artriovenous malformations before surgical or endovascular therapy. Journal of Neurosurgery, 84: 477-483.

[18] Mäkelä, JP, Salmelin, R, Kotila, M, et al. (1994) Neuromagnetic correlates of memory disturbances caused by infarction in anterior thalamus. Society for Neuroscience Abstracts, 20(1): 810.

[19] Lewine, JD, Orrison, WW, Astur, RS, et al. Explorations of pathophysical spontaneous activity by macnetic source imaging. In: Baumgartner, C, Deecke, L, Stroink, G, Williamson, SJ (Eds.), Biomagnetism:Fundamental research and Clinical applications. Studies in applied electromagnetics and mechanics. Elsevier/IOS Press, Amsterdam, 1995: 55-59.

[20] Lewine, JD, Orrison, WW, Sloan, JH, et al. Neuromagnetic assesment of pathophysiological brain activity induced by minor head trauma. In this volume.

[21] Vieth, J, Kober, H and Grummich, P (1996) Sources of spontaneous slow waves associated with brain lesions, localised by using the MEG. Brain Topography, 8: 215-221.

[22] Vieth, JB, Kober, H, Stippich, C, et al. Time course of abnormal MEG activity associated with transient ischemic attacks. In this volume.

[23] Kakigi, R, Koyama, S, Hoshiyama, M, et al. (1995) Pain-related magnetic fields following painful CO2 laser stimulation in man. Neuroscience Letters, 192: 45-48.

[24] Hari, R, Kaukoranta, E, Reinikainen, K, et al. (1983) Neuromagnetic localization of cortical activity evoked by painful dental stimulation in man. Neuroscience Letters, 42: 77-82.
[25] Huttunen, J, Kobal, G, Kaukoranta, E, et al. (1986) Cortical responses to painful CO2 stimulation of the nasal mucosa. Electroencephalography and clinical Neurophysiology, 64: 347-349.
[26] Kitamura, Y, Kakigi, R, Hoshiyama, M, et al. (1995) Pain-related somatosensory evoked magnetic fields. Electroencephalography and clinical Neurophysiology, 95: 463-474.
[27] Jenkins, WM, Merzenich, MM and Recanzone, G (1990) Neocortical representational dynamics in adult primates: implications for neuropsychology. Neuropsychologia, 28: 573-584.
[28] Wang, X, Merzenich, MM, Sameshima, K, et al. (1995) Remodeling of hand representation in adult cortex determined by timing of tactile stimulation. Nature, 378: 71-75.
[29] Elbert, T, Flor, H, Birbaumer, N, et al. (1994) Evidence for extensive reorganization of the somatosensory cortex in adult humans after nervous system injury. NeuroReport, 5: 2593-2597.
[30] Yang, TT, Gallen, C, Ramachandran, VS, et al. (1994) Noninvasive detection of cerebral plasticity in adult human somatosensory cortex. NeuroReport, 5: 701-704.
[31] Flor, H, Elbert, T, Knecht, S, et al. (1995) Phantom-limb pain as a perceptual correlate of cortical reorganization following arm amputation. Nature, 375: 482-484.
[32] Knecht, S, Henningsen, H, Elbert, T, et al. (1995) Cortical reorganization in human amputees and mislocalization of painful stimuli to the phantom limb. Neuroscience Letters, 201: 262-264.
[33] Knecht, S, Henningsen, H, Elbert, T, et al. (1996) Reorganizational and perceptual changes after amputation. Brain, In press.
[34] Flor, H, Braun, Ch, Birbaumer, N, et al. Chronic pain enhances the magnitude of the magnetic field evoked at the site of pain. In: Baumgartner, C, Deecke, G, Stroink, G, Williamson, SJ (Eds.), Biomagnetism:Fundamental research and Clinical applications. Studies in applied electromagnetics and mechanics. Elsevier/IOS Press, Amsterdam, 1995: 107-111.
[35] Elbert, T, Pantev, C, Wienbruch, C, et al. (1995) Increased cortical representation of the fingers of the left hand in string players. Science, 270: 305-307.
[36] Rockstroh, B, Vanni, S, Elbert, T, et al. Extensive somatosensory stimulation alters somatosensory evoked fields. In this volume.
[37] Mogilner, A, Grossman, A, Ribary, U, et al. (1993) Somatosensory cortical plasticity in adult humans revealed by magnetoencephalography. Proceedings of the National Academy of Sciences of USA, 90: 3593-3597.
[38] Elbert, T, Sterr, A, Rockstroh, B, et al. Cortical reorganization in amputees: alterations of somatosensory representation in the hemisphere contralateral to the intact side. In this volume.
[39] Florence, SL, Wall, JT and Kaas, JH (1991) Cortical projections from the skin of the hand in squirrel monkeys. Journal of Comparative Neurology, 311: 563-578.
[40] Pantev, C, Bertrand, O, Eulitz, C, et al. (1995) Mirror-image tonotopy of different areas of human auditory cortex revealed by simultaneous magnetic and electric recordings. Electroencephalography and clinical Neurophysiology, 94: 26-60.
[41] Mühlnickel, W, Flor, H, Elbert, T, et al. Deviations from the tonotopic map are correlated with tinnitus strength. In this volume.
[42] Mäkelä, JP and Hari, R Long-latency auditory evoked magnetic fields. In: Sato, S. (Eds.), Magnetoencephalography: Clinical Applications and Comparison of EEG. Raven Press, 1990: 177-191.
[43] Phillips, DP, Semple, MN, Calford, MB, et al. (1994) Level-dependent representation of stimulus frequency in the cat primary auditory cortex. Experimental Brain Research, 102: 210-226.
[44] Vasama, J-P, Mäkelä, JP, Pyykkö, I, et al. (1995) Abrupt unilateral deafness modifies function of human auditory pathways. NeuroReport, 6: 961-964.
[45] Vasama, J-P and Mäkelä, JP (1995) Auditory pathway plasticity in adult humans after idiopathic sudden unilateral sensorineural hearing loss. Hearing Research, 87: 132-140.
[46] Irvine, DRF and Rajan, RR (1993) Plasticity in the frequency organization of auditory cortex of adult mammals with restricted cortical lesions. Biomedical Research, 14: 55-59.
[47] Ahissar, E and Ahissar, M (1994) Plasticity in auditory cortical circuitry. Current Opinion in Neurobiology, 4: 580-587.
[48] Mäkelä, JP, Hari, R, Karhu, J, et al. (1993) Suppression of magnetic mu rhythm during parkinsonian tremor. Brain Research, 617: 189-193.
[49] Volkmann, J, Joliot, M, Mogilner, A, et al. (1996) Central motor loop oscillations in parkinsonian resting tremor revealed by magnetoencephalography. Neurology, 46: 1359-1370.
[50] Sannita, WG, Lewine, JD, Maclin, EL, et al. Quantitative EEG and MEG effects in healthy subjects of acute oral phenobarbital (100 mg). In this volume.

[51] Wiebruch, C, Eulitz, C, Lehnertz, K, et al. Methohexital-induced changes in spectral power of neuromagnetic signals: reduced enhancement of beta-band power over the hemisphere ipsilateral to the epileptogenic focus. In this volume.
[52] Kraut, MA, Arezzo, JC and Vaughan, H (1990) Inhibitory processes in the flash evoked potential of the monkey. Electroencephalography and clinical Neurophysiology, 76: 440-452.
[53] Arakawa, K, Peachey, NS, Celesia, GG, et al. (1993) Component-specific effects of physostigmine on the cat visual evoked potential. Experimental Brain Research, 95: 271-276.
[54] Mavroudakis, N, Brunko, E, Nogueira, MC, et al. (1991) Acute effects of diphenylhydantoin on peripheral and central somatosensory conduction. Electroencephalography and clinical Neurophysiology, 78: 263-266.
[55] Bodis-Wollner, I (1990) Visual deficits related to dopamine deficiency in experimental animals and Parkinson's disease patients. Trends in Neurosciences, 13: 296-302.
[56] Sannita, WG, Balestra, V, Di Bon, G, et al. (1993) Human Flash-VEP and quantitative EEG are independently affected by acute scopolamine. Electroencephalography and clinical Neurophysiology, 86: 275-282.
[57] Singer, W (1993) Synchronization of cortical activity and its putative role in information processing and learning. Annual Review of Physiology, 55: 349-374.
[58] Sannita, WG, Lopez, L, Piras, C, et al. (1995) Scalp-recorded oscillatory potentials evoked by transient pattern-reversal stimulation in man. Electroencephalography and clinical Neurophysiology, 96: 206-218.
[59] Lopez, L, Narici, L, Conforto, S, et al. Oscillatory retinal and cortical responses to luminance stimulation: eccentricity function and effect of acute scopolamine. In this volume.
[60] Shiomi, Y, Fujiki, N, Hirano, S, et al. (1996) Tinnitus remission by lidocaine demonstrated by auditory evoked magnetoencephalogram: preliminary report. Association for Resarch in Otolaryngology, book of abstracts, 19: 13.
[61] Jäntti, V, Baer, G, Yli-Hankala, A, et al. (1995) MEG burst suppression in an anaesthetized dog. Acta Anesthesiologica Scandinavica, 39: 126-128.
[62] Lopez, L, Chan, CY, Okada, YC, et al. (1991) Multimodal characterization of population responses evoked by applied electric field in vitro: extracellular potential, magnetic evoked field, transmembrane potential and current source density analysis. Journal of Neuroscience, 11: 1998-2010.
[63] Reite, M, Teale, P, Goldstein, L, et al. (1989) Late auditory evoked magnetic sources may differ in the left hemisphere of schizophrenic patients. Archives of General Psychiatry, 46: 565-572.
[64] Tiihonen, J, Hari, R, Naukkarinen, H, et al. (1992) Modified activity of the human auditory cortex during auditory hallucinations. American Journal of Psychiatry, 149: 255-257.
[65] David, AS, Woodruff, PWR, Howard, R, et al. (1996) Auditory hallucinations inhibit exogeneous activation of auditory association cortex. NeuroReport, 7: 932-936.
[66] Reite, M, Sheeder, J, Teale, P, et al. (1996) Magnetic source imaging evidence of sex differences in cerebral lateralization in schizophrenia. Archives of General Psychiatry, in press.
[67] Reite, M, Teale, P, Sheeder, J, et al. (1996) Neuropsychiatric applications of MEG. Electroencephalography and clinical Neurophysiology, Suppl. Visualization of Information processing in the human brain; Recent advances in MEG and Functional MRI: in press.
[68] Kaufman, L, Curtis, S, Wang, J-Z, et al. (1991) Changes in cortical activity when subjects scan memory for tones. Electroencephalography and clinical Neurophysiology, 82: 266-284.
[69] Reite, M, Teale, P, Sheeder, J, et al. (1996) MEG evidence of abnormal early auditory function in schizophrenia. Biological Psychiatry, in press.
[70] Salmelin, R, Mäkelä, JP, Heikman, P, et al. (1996) Human brain rhythms and electroconvulsive therapy. Society for Neuroscience Abstracts.

Changes of the Functional Organization of the Somatosensory Cortex in Chronic Pain Patients

Braun, C.[1], Flor, H.[1,2], Birbaumer, N.[1] and Elbert, T.[3]

[1]Institute of Medical Psychology, University of Tübingen, Tübingen, Germany; [2]Department of Psychology, Humbolt University, Berlin, Germany; [3]Department of Psychology, University of Konstanz, Konstanz, Germany

Introduction

Does the persistent noxious stimulation experienced by patients suffering from chronic back pain lead to cortical reorganization? Previous research has shown that changes in sensory inputs lead to changes in the cortical representation of the relevant somatic structures. Cessation of peripheral input leads to the reduction in size of the cortical representation of the corresponding area, while adjacent cortical areas expand [1]. If a body part is used extensively, however, the corresponding representation in somatosensory cortex expands [2]. Chronic pain patients have experienced years of persistent noxious stimulation in the lumbar region. Cortical reogranization should thus be present in an area of the somatosensory cortex that represents the back, but not in unrelated areas of somatosensory cortex, such as those representing the finger.

We investigated the cortical responses of chronic back pain patients and control subjects to electrical stimulation of different magnitudes to either the back or the finger. If the cortical representation of the painful body region is increased, this should be reflected in the magnetic field generated in response to the stimulation. The magnetic fields should exhibit higher amplitudes and a broader topographical distribution when evoked from the affected body part of the chronic pain patients. Furthermore, dipole analysis should reveal a shift in the cortical locations of the back in chronic pain patients.

Methods

Three groups of subjects participated in the experiment. The patients suffering from continuous, long term lower back pain were divided into two groups: Chronic patients (n=5), suffering for an average duration of 18 years from back pain, and Subchronic patients (n=5), suffering for an average of 7 years from back pain. Age- and gender-matched healthy subjects (n=9) comprised the control group.

Electrical bipolar pulses with a duration of 10 ms. and a preexperimentally determined current intensity (max. 5 mA) served as pain stimuli. Stimuli were applied through an intracutaneous gold electrode to two different sites: (1) the left back, in the region of the most intense pain, and (2) the left index finger. Tactile perception and pain threshold were determined in two ascending and descending series using the method of limits. Three stimulus intensities were presented: (1) sub-threshold, which was determined for each subject to be the point midway between the perception and pain thresholds; (2) painful, which was determined to be 50% above the pain threshold, and (3) a standard stimulus of 500 μA. Each stimulus was presented to the finger and back electrodes in a blocked design. After each block of 24 trials, subjects were asked to rate (on a verbal 10 point scale) the stimulus intensity and the current level of clinical pain.

Magnetic fields were recorded from 37 locations centered over the right C4, contralateral to the side of stimulation. In the BTi Neuromagnetometer system, the coils were arranged in a circular array (diameter 14.4 cm) on a spherical surface (with 12.2 radius). The coil diameter was 2.0 cm and the distance between the center of two adjacent coils was 2.2 cm. Data were sampled at a rate of 297 pts/s.

Trials were excluded from analysis if the difference between the maximum and the minimum (computed across a 1 second epoch) exceeded 2.5 pT in any of the MEG channels. This criterion also served to exclude larger eye movements or blinks which would have led to excessive amplitudes in most anterior channels. An average of 230 artifact free trials per subject were analyzed. Responses were averaged for each subject separately for the different experimental conditions. Parameters were extracted from the time course of the RMS at 80 to 125 ms across all MEG channels.

The center of the activated source was determined for all stimulation types using a single moving dipole model. A sphere was fit locally to the digitized head shape in the region proximal to the sensor array for each subject. The location (x,y,z positions), orientation, and amplitude of a best fitting equivalent current dipole were estimated for each point in time. The anterior-posterior, medial-lateral, and inferior-superior coordinates of the dipole location with their confidence volumes were calculated in the head-frame based coordinate system. Only estimates with a correlation between the measured and the expected field of >.95 and a confidence volume of less than 1 cm^3 were used.

Results

Amplitude of the evoked magnetic field at the back ($r=5$, $p<.05$), but not at the finger ($r=0$, n.s.) correlated with the chronicity of the pain problem. With increasing duration of the illness, a higher response of the magnetic field is found.

An ANOVA of the M100 RMS activity revealed a significant difference between the chronic pain patients and the combined subchronic and control subjects. The 100 ms component was significantly enhanced during back stimulation, but not finger stimulation, in the chronic patients compared to subchronic patients and the control subjects ($F(1,16)=5.34$, $p<0.35$) (see Figure 1).

The latency window between 70 and 80 msec yielded the most pronounced peak and was chosen for the analysis of the location of the cortical activity in response to painful stimulation. Neither the inferior-superior nor the anterior-posterior dipole coordinates were significantly different between the groups, although they differed between the back and finger (see Figure 2). In the medial-lateral direction of the back stimulation response site, however, the dipole analysis revealed a source location which was shifted 2.5 cm medially for the chronic patients compared to the other groups.

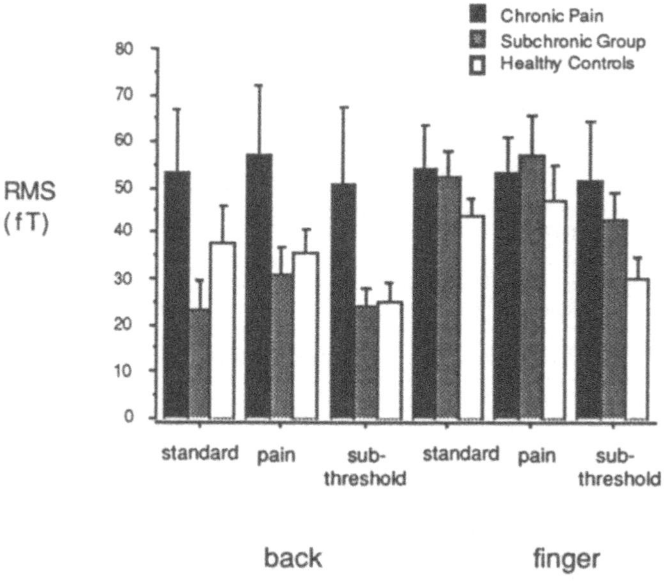

Figure 1. Peak amplitudes of the root mean square (RMS) across all 37 MEG-channels, computed for the range 80-125 ms.

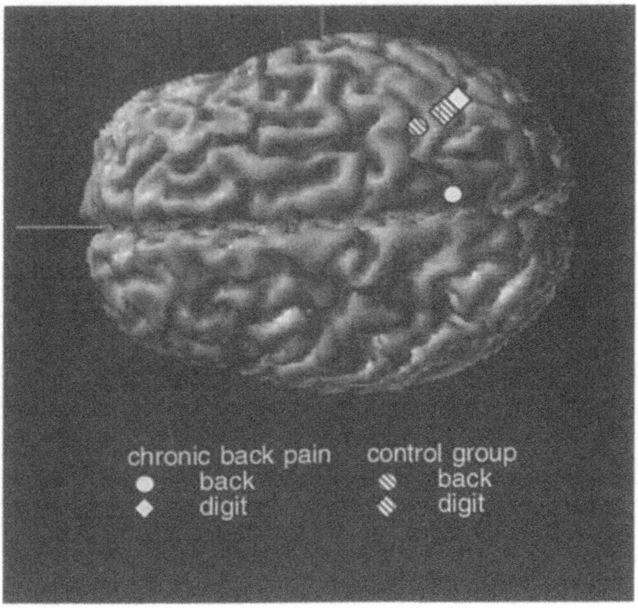

Figure 2. Localization of the dipoles for control subjects and chronic pain patients when stimulated at the back or finger.

Discussion

The results of this study support the hypothesis that chronic pain leads to reorganization of the cortical representation of the affected body region. Those patients who had suffered from chronic noxious stimulation of the lower back region for the longest amount of time were those with the most enhanced MEG response to stimulation of the back. Finger stimulation revealed no differences between patient and control groups: only when the back was stimulated did the differences between patients and controls emerge [3].

The amplitude of the magnetic field evoked at the finger was positively correlated with chronicity of the pain problem, suggesting that long-lasting proprioceptive input led to enhanced cortical responses to tactile stimulation. This result is in accordance with our assumption about stimulation-produced plasticity of the somatosensory cortex. Furthermore, dipole analysis revealed differences in dipole location between the groups for the back, but not the finger stimulation. Taken together, these results suggest that the cortical representation of the chronic pain patients' affected body region is increased. Years of painful stimulation may have led to cortical reorganization that may now maintain the pain without peripheral input.

References

[1] Merzenich, M.M., Nelson, R.J., Stryker, M.P., Cynader, M.S., Schoppman, A., Zook, J.M. Somatosensory cortical map changes following digit amputation in adult monkeys. Journal of Comparative Neurology, 1984, 224: 591-605.

[2] Recanzone, G.H., Merzenich, M.M., Jenkins, W.M., Grajsi, K.A., Dinse, H.R. Topographic reorganization of the hand representation in cortical area 3b of owl monkeys trained in a frequency discrimination task. Journal of Neurophysiology, 1992, 65: 12031-1056.

[3] Flor, H., Braun, C., Birbaumer, N., Elbert, T., Ross, B., Hoke, M. Chronic pain enhances the magnitude of the magnetic field evoked at the site fo pain, In: Baumgartner, C. et al. Biomagnetism: Fundamental Research and Clinical Applications, Elsevier Science, 1995.

Acknowledgement

We thank Laura Helmuth for helpful comments and for checking this manuscript. The study was supported by a grant of the Deutsche Forschungsgemeinschaft (Bi 195/24).

Clinical MEG II: Development Of A Normative Database

Davis, J.T., Lewine, J.D., Edgar, J.C., Hoesing, J., and Orrison, W.W., Jr.

The New Mexico Institute of Neuroimaging, The New Mexico Regional Federal Medical Center, Albuquerque, New Mexico, USA.

INTRODUCTION

As clinical MEG matures, the need for normative data on both evoked responses and spontaneous brain activity is becoming acute. Towards this end, The New Mexico Institute of Neuroimaging has developed a comprehensive series of protocols for examining brain activity in both patient and normal control populations [1]. This protocol, which can be completed in about two hours, evaluates spontaneous brain activity and visual, auditory, and somatosensory, functioning. A major objective of this program is to develop techniques for quantitative MEG that will be similar to those previously developed for quantitative EEG, but hopefully, of superior clinical utility, owing to MEG's exceptional spatial resolution. We are in the process of developing a database of information on 100+ neurologically normal individuals between the ages of 20 and 60.

A wide range of parameters are being included in the database, including information on the latency, amplitude, and source locations for multiple evoked potential components in auditory, visual and somatosensory domains. Spontaneous data are evaluated for spectral composition in both eyes open and eyes closed conditions. We are also beginning to apply algorithms that characterize interhemispheric, interlobar, and intralobar coherence in delta, theta, alpha, and beta frequency bands. Once a normative database is completed, it will be possible to use multivariate statistical procedures to identify abnormalities in patient data sets. Ultimately, it should be possible to develop discriminant functions that allow for rapid patient diagnosis.

METHODS

Subjects: Prior to study, subjects complete a comprehensive questionnaire that includes questions about developmental milestones, head trauma, and psychiatric and neurologic dysfunction. Only subjects with normal developmental histories and no evidence of substance abuse, head trauma, or neurologic or psychiatric dysfunction are included in the normative data base. Subjects must also demonstrate a normal magnetic resonance imaging examination prior to inclusion in the normative data base and they must have a normal EEG.

Evoked Data: Information on evoked potential paradigms are presented in reference [1]. For auditory activity, we are evaluating responses to 500 Hz, 2000 Hz, and 4000 Hz tone pips presented binaurally (duration 50 msecs), and also monaural 2000 Hz tones. Each subject's response is ultimately characterized by the latency and amplitude of M50, M100, and M200 components over right and left hemispheres, plus dipole parameters. For somatosensory activity, responses to electrical stimulation of the median and tibial nerves are being characterized. The latency and amplitude of M20 and M30 components are being assessed, as is information on associated sources. Visual responses to stimuli in right and left upper and lower visual fields are being evaluated with emphasis of the latency and amplitude of M100 and M 130 components.

Spontaneous Data: The details of data collection are described in reference [1]. Spontaneous data are visually inspected for gross abnormalities including focal slowing and epileptiform transients. It should be noted that the presence of these signals per se is not considered an exclusionary criteria from the data base. Exclusion is based on other than MEG factors, as the use of MEG as an exclusionary parameter would be circular to the development of a normative database. Following visual inspection, average power spectra are calculated from artifact free data segments. A Fast Fourier Transform is applied to successive windows of approximately 3.4 seconds in length, which are slid by 1.7 second steps. Epochs with signals in excess of 3 picotesla are excluded from analysis. At each non artifact step, a transform is calculated and included in the average, until a total of one hundred averages is obtained. Despite attempts at automatic artifact rejection, processed data are sometimes contaminated with various forms of low amplitude artifact that adds power to the spectra. Typically, the artifactual contribution can be recognized in the spectral data itself and a subspace projection method is used to remove the artifact signal (e.g. heartbeat artifact). In order to obtain regional measures of spectral values, the 122 sensors are grouped so that the data from each group approximates a traditional head region. The number of sensors in each region is kept similar, except for the midline occipital group, which has only 2 sensors, and the midline frontal group which is excluded from further analyses. Within each regional grouping, a grand-average power spectra is calculated across all encompassed sensors. Figure 1 shows the sensor groupings. Several parameters are entered into the database for each grouping, including regional power in each band.

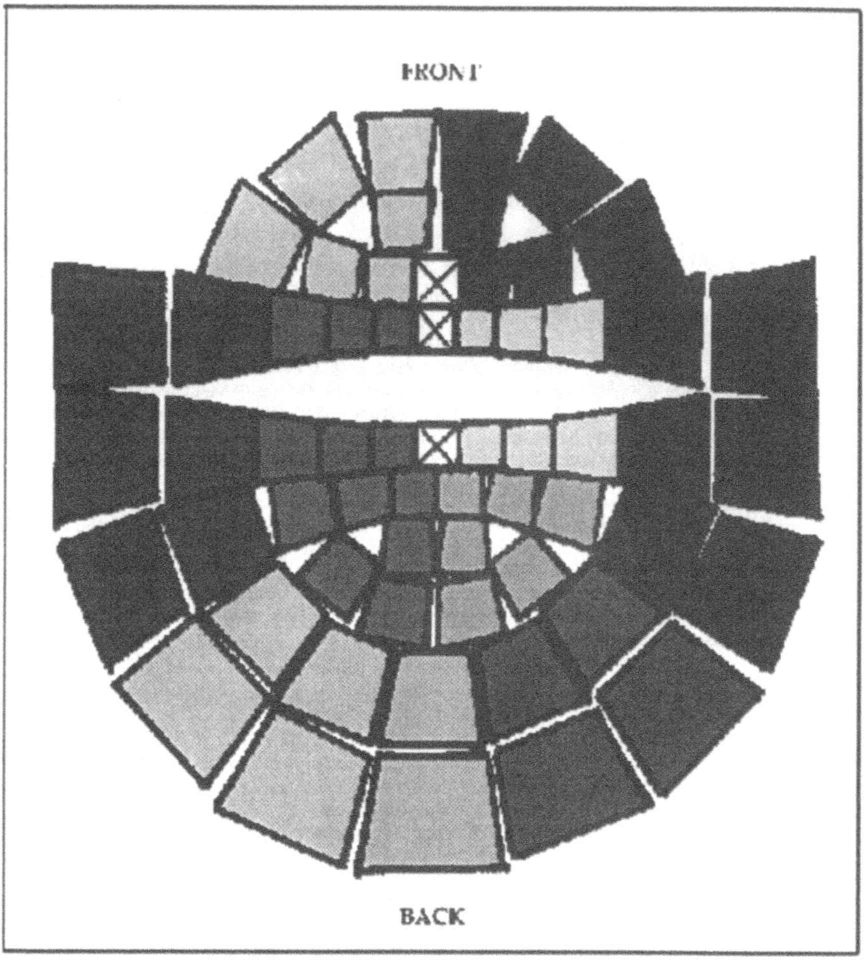

Fig 1: Sensor groupings for spectral analyses. Each box represents two of the 122 channels in our biomagnetometer.

RESULTS

Full exposition of all of the contents of the database is well beyond the scope of this brief report, so only a subset of the data is described here.

Subset 1: At present, auditory data have been completely analyzed for 12 right handed males between the ages of 25 and 40 years old. The mean latency of the M100 response to bilateral 2000 Hz tones is 107 msec with a standard deviation of 10 msec. The magnitude of the associated current dipoles range from 23 to 30 nA, with no significant difference between right and left hemispheres in either latency or dipole magnitude. The orientation of the M100 dipole response invariably has a negative z component (ventrally directed) which is usually the largest component. The y component is usually negative (posteriorly directed), and the x component (right-left) is near zero. The relative location of the right and left hemispheric dipoles is such that the right is on average 10 mm anterior to the left.

This admittedly small database has already proven useful in the identification of abnormalities in several patient populations. For example, figure 2 shows a comparison of auditory evoked data (2000 Hz, binaural tone) from a typical control subject and an individual with a dual diagnosis of bipolar disorder and Parkinson's disease. Clearly the M100 component of the auditory response is delayed for the patient.

Subset 2: Spectral data have also been analyzed from 10 right handed females between the ages of 20 and 40. Figure 3A shows delta and theta regional power for a representative subject. The overall amount of delta and theta power is low, and there are no gross regional asymmetries. The situation for a female patient with a left temporal-parietal tumor is markedly different. Excessive delta and theta is seen, with marked regional asymmetries consistent with the patients pathology.

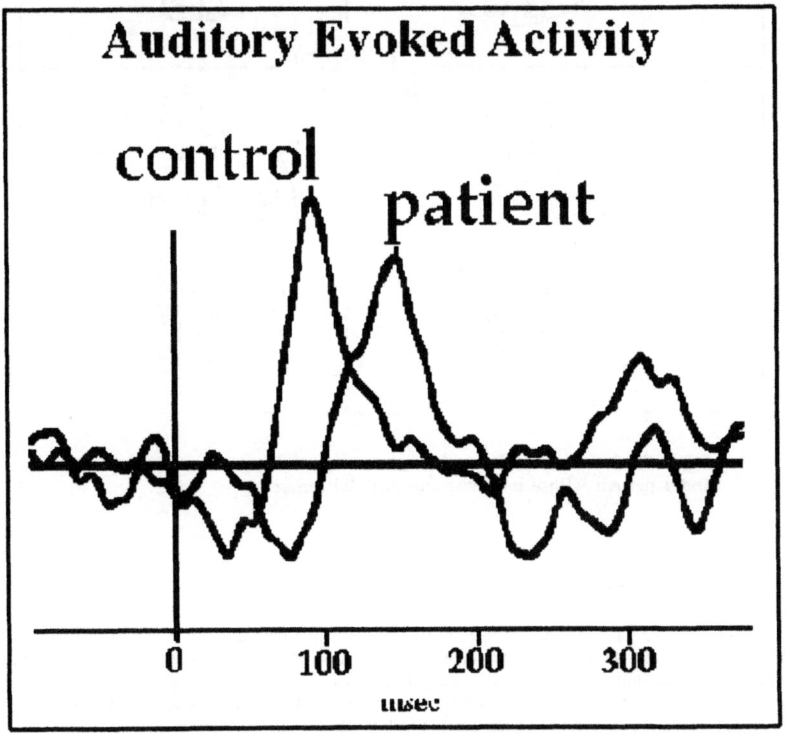

Fig 2: Auditory evoked activity from a control subject and a patient with bipolar disorder and Parkinson's disease. The patient's M100 response is delayed significantly.

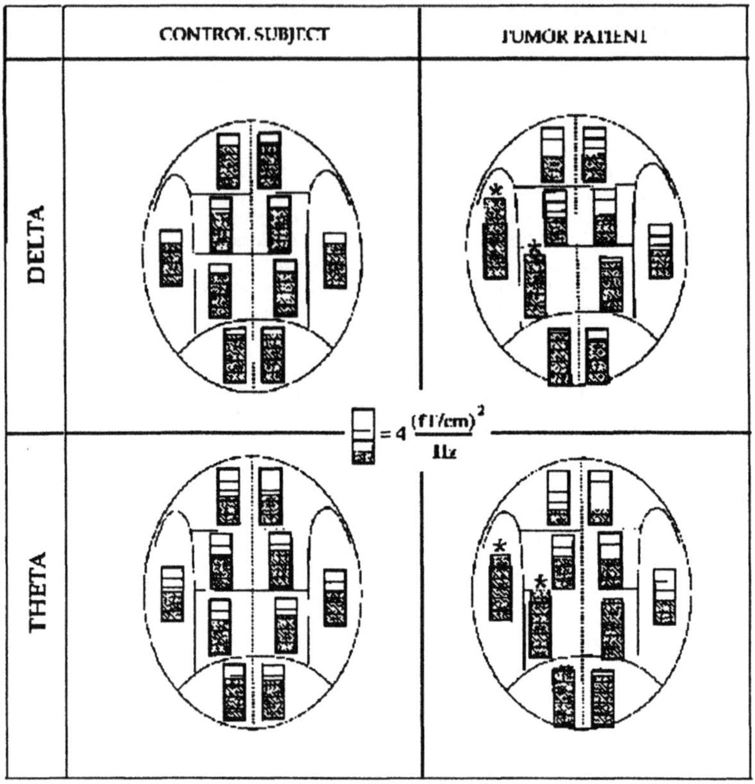

Fig 3: Regional spectral data in delta and theta bands for a control subject and a patient with a left temporal-parietal tumor.

DISCUSSION & CONCLUSIONS

The normative database that we are starting to compile has already proven useful in the objective identification of abnormal data sets. As the database grows, we anticipate greater sensitivity to subtle pathologies. As has been done for quantitative EEG, it is anticipated that the normative profile can be used in a multivariate approach, in which disorders, such as schizophrenia, can be recognized by the deviance from the overall normative profile. The current task is to continue to accumulate the normative data.

REFERENCES

[1] Lewine, J.D., Davis, J.T., Davis, L.E., et al., Clinical MEG I: Towards a standardized examination, Proceedings, Biomag'96, this volume.

Cortical Reorganization in Arm Amputees: Alterations of the Somatosensory Representation of the Intact Arm

Elbert, T.[1], Sterr, A.[1], Rockstroh, B.[1], Charbonnier, D.[1], Flor, H.[2], Pantev, C.[3], Wienbruch, C.[3], Knecht, S.[3], and Taub, E.[4]

University of Konstanz[1], Germany; Humboldt University[2], Berlin, Germany; University of Münster[3], Germany; University of Alabama[4], Birmingham, U.S.A..

Introduction

Modulation in the flow of somatosensory input is followed by changes in the cortical representation of the body in the somatosensory cortex. Intense, behaviorally relevant stimulation of a set of receptors leads to an expansion in the cortical representation of the stimulated parts of the body [5,7]. Magnetic source imaging (MSI) has provided evidence that use-dependent reorganization also occurs in humans [3,8]. String players, for example, provide a model for the effects of differential afferent input to the two hemispheres. It was found that the representation of the digits of the left hand proved to be larger than that in controls [3]. Disuse of limbs also brings about changes in the representation of the body in the somatosensory cortex. Such phenomenona have been observed after total loss of input. In owl monkeys, Pons and coworkers [6] reported massive reorganization of primary somatosensory cortex subsequent to sensory deafferentation of an entire arm 12 years earlier. These investigators also demonstrated that the adjacent representations from the face had expanded into the hand areas, so that stimulation of the face evoked responses in the area formerly occupied by the hand and digits. Extensive cortical reorganization following somatosensory deafferentation also occurs and persists in humans following upper extremity amputation, as demonstrated by MSI [2,4,9]. It has also been shown that the magnitude of cortical reorganization is strongly associated with the amount of phantom limb pain experienced by the amputees [5].

A loss of sensory input from the amputated limb causes the alterations in the hemisphere contralateral to the amputated side. The question arises if there is a measurable change also in the hemisphere contralateral to the intact side. Two different mechanisms may be involved in this alteration: First, homotopic regions of somatosensory cortex are linked such that plasticity induced in one hemisphere is mirrored in the other hemisphere [1]. Second, the obvious increase of afferent input from the intact hand may result in enlargements of the representational zones of the fingers.

The present study was designed to investigate the nature of the change of somatotopic representation in upper arm amputees contralateral to the intact hand. Using magnetic source imaging, the representation of digits D1 and D5 was compared with that of controls.

Methods

Six upper arm amputees (2 with left, 4 with right arm amputation) and six control subjects participated in the study. The amputation had occurred 2 to 49 years (mean 21 years) prior to the current investigation.

Somatosensory stimulation was delivered separately to the first and fifth digit of the hand (both hands in controls) and to the lower lip at the corners of the mouth. The stimulation consisted of a brief and light, non-painful superficial pressure delivered through a pneumatic device. One thousand stimuli were delivered to each of the stimulation sites (first and fifth digit, left and right corners of the mouth) at an average rate of 0.5 Hz. The interval between stimulus onsets varied between 50 and 500 ms. The sequence of stimulation was balanced across subjects. The sensor array of magnetic detectors was positioned over the hemisphere (centered over C3 or C4) contralateral to the stimulated limb and lip site.

Within the range of 30-75 msec a first major peak was identified in each of the evoked waveforms (Fig.1). For this peak a single equivalent current dipole (ECD) model (best fitting local sphere) was fitted and the medians of the dipole moment and the dipole location were computed from a selection of time points within a 20 msec time segment (eleven sampling points) around the maximal root mean square (RMS) across the 37 channels. Points were selected if they met the following criteria: (1) RMS indicating a signal-to-noise ratio > 3, (2) goodness of fit of the ECD-model to the measured field > .95 and (3) a minimal confidence volume of the ECD location < 300 mm^3.

The representation of hand and face in area 3b of controls was compared to the representation of hand and face in the deafferented hemisphere (contralateral to the amputation) and contralateral to the intact side of the amputees. Euclidean distances between lip- and D1-, lip- and D5-, and D1- and D5-locations, were calculated to uncover changes in the cortical representation of the intact hand. In order to evaluate the changes in lip representation, if any, due to amputation ECD-values of both hemispheres were averaged for the controls and compared with (a) parameters of deafferented hemispheres and (b) locations in hemipsheres contralateral to the intact side. Anovas and t-tests were calculated to confirm differences.

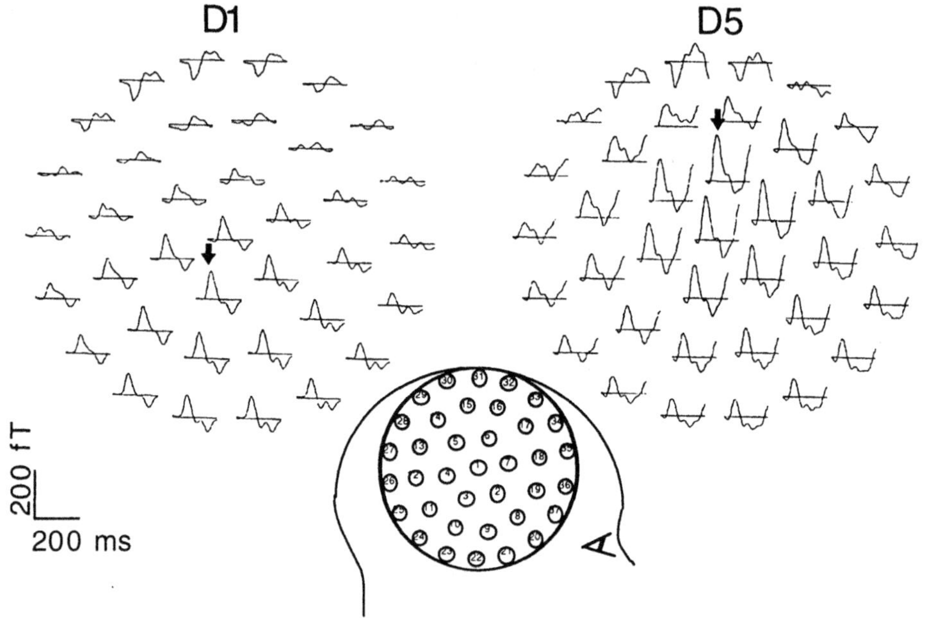

Fig. 1: The lower insert represents the sensor array placement. The set of waveforms indicates the magnetically evoked responses averaged for D1 (left) and D5 (right) respectfully. The black arrow marks the first major peak chosen for dipole modeling . The field maxima (black arrows) suggest that D1 is located more inferiorly than D5. This homuncular organization is confirmed by the dipole analysis.

Results

A typical example for the averaged evoked fields is presented in Fig.1. There was no obvious difference in waveshape between amputees and controls. In all subjects the peak for D1 was inferior to the peak for D5.

In amputees the cortical representations of D1 and D5 of the intact hand encompassed a greater area (Euclidean distance between digits: 12.4 ± 5.5 mm) than in controls (6.9 ±1.6 mm), $t = 2.4$, $p<.05$. The average difference on the inferior-superior plane was 5.5 mm larger in amputees than in controls ($t=3.2$, $p<.01$). When compared to the cortical representations in control subjects, only the D1-representation in the amputees had shifted along the central sulcus. The shift was limited to the lateral direction (7.7 mm, $t=2.1$, $p<.1$) and not significant in the inferior direction (2.6 mm). The distance between the representations of lip and D5 was 5.6 mm greater in amputees than in control subjects (t=2.6 $p<.05$). The distance between D1 and lip did not differentiate groups. There was also no group difference in the magnitude of the dipole moments of the digits.

In the amputees' deafferented hemisphere the cortical representation of the lip had shifted in a superior direction (13.9 mm, $t=2.9$, $p<.01$) compared to controls. The location of the lip contralateral to the intact side corresponded to the lip location in control subjects. The locations of the ECDs obtained in this study are schematically summarized in Fig. 2.

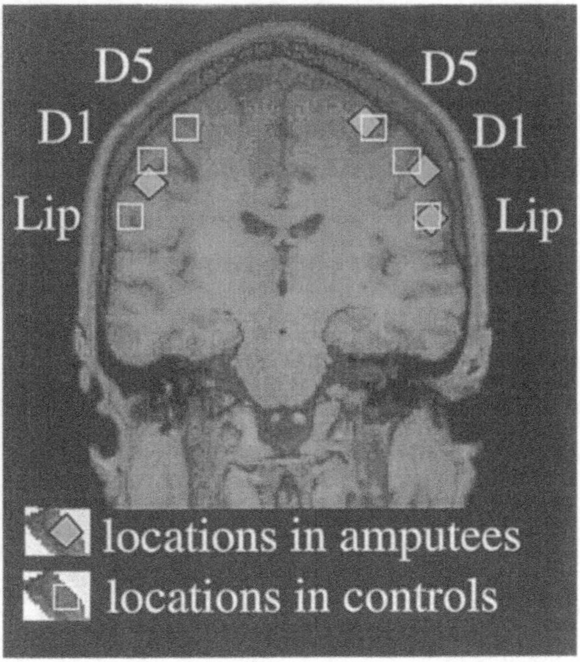

Fig.2. Schematic locations of the ECD for D1-, D5- and Lip-stimulation are seen here as projections onto a coronal MR-slice (of one control subject). Open squares indicate the average locations for controls, filled diamonds represent average locations for arm amputees. For amputees, the left half illustrates representations in the deafferented hemisphere. Compared to controls, the lip location in amputees has shifted contra- but not ipsilateral to the amputated side. Contrallateral to the intact side, the finger area of amputees has expanded, mainly as a result of the shift in D1-location.

Discussion

The results of the present study indicate that cortical reorganization after amputation is not restricted to the hemisphere contralateral to the amputated side. After an upper arm amputation, the representation of the intact fingers become enlarged. The shaping of the receptive field features of a cortical neuron by the input flow constitutes one basic mechanism underlying cortical reorganization [5]. Therefore, the extension of the hand area in amputees could be explained by the increased use of the remaining extremity and the subsequent increase of somatosensory input. The shift in the intact side was most pronounced for D1 and smaller for D5. Differences in the frequency of use of the separate digits may account for this result.

An invasion of the lip representation into the area formerly occupied by the amputated hand has been previously reported. Our present results suggest that such an invasion is not transferred to the hemipshere contralateral to the intact hand, the plastic changes in one hemisphere are not mirrored in the other.

The plastic alterations of the somatosensory cortex in amputees result from the interplay of two principles which govern cortical reorganization: (1) As the organization of the adult human brain seems to be use-depended and continuously "adapts" to the current needs and experiences of the individual the representational zone of the fingers of the remaining hand becomes enlarged [3]. (2) In the hemisphere deprived of input from the fingers, the face and lip representations expand into the now deafferented region [2]. This expansion may be responsible for the asymmetry of the location of the lip-response, observed in amputees. It could be speculated that the expansion encompassed by the digits may not alter the lip represention in the hemisphere contralateral to the intact side. Flor et al. [4] reported a strong relationship between hemispheric asymmetries in the lip representation and phantom limb pain. We can now conclude, that this relationship arises from alterations in the deafferented hemisphere and not from changes contrallateral to the intact side. Future investigations should study the temporal stability and the time course of the observed alterations.

References

[1] Calford, M., and Tweedale, R. Interhemispherictransfer of plasticity in the cerebral cortex. Science, 249: 805-807, 1990.

[2] Elbert, T., Flor, H., Birbaumer, N., Knecht, S., Hampson, S., and Taub, E. Evidence for extensive reorganization of the somatosensory cortex in adult humans after nervous system injury. Neuroreport 5, 2593-2597, 1994.

[3] Elbert, T., Pantev, C., Wienbruch, C., Rockstroh, B., and Taub, E. Increased use of the left hand in string players associated with increased cortical representation of the fingers. Science, 220: 305-307, 1995.

[4] Flor, H., Elbert, T., Knecht, S., Wienbruch, C., Pantev, C., Birbaumer, N., Larbig, W., and Taub, E. Phantom limb pain as a perceptual correlate of cortical reorganization following arm amputation. Nature, 375: 482-484, 1995.

[5] Jenkins, W. M., Merzenich, M. M., Ochs, M. T., Allard, T., and Gufc-Robles, E. Functional reorganisation of primary somatosensory cortex in adult owl monkeys after behaviorally controlled tactile stimulation, Journal of Neurophysiology, 63: 82-104, 1990.

[6] Pons, T., Garraghty, P. E., Ommaya, A. K., Kaas, J. H., Taub, E., and Mishkin, M. Massive cortical reorganisation after sensory deafferation in adult macaques. Science, 252: 1857-1860, 1991.

[7] Recanzone, G. H., Merzenich, M. M., Jenkins, W. M., Grajski, A., Dinse, H.R. Plasticity in the frequency representation of primary auditory cortex following discrimination training in adult owl monkeys. Journal of Neuroscience, 67: 1031-1056, 1992,

[8] Rockstroh, B., Vanni, S., Elbert, T., and Hari, R., Extensive somatosensory stimulation alters somatosensory evoked fields. This volume.

[9] Yang, T.T., Gallen, C., Schwartz, B., Bloom, F.E., Ramachandran, V.S., and Cobb., S. Sensory maps in the human brain. Nature, 368: 592-593, 1994.

Acknowledgement

This work was supported by grants from the Deutsche Forschungsgemeinschaft.

Comparison of Magnetoencephalographic and Intraoperative Electrocorticographic Localization of Sensory Response

Greenblatt, R.[1], Bucholz, R.[2], Sobel, D.[3], Schwartz, B.[3,4] and Hirschkoff, E.[3,4]

[1]Source/Signal Imaging, San Diego, CA, USA; [2]Saint Louis University, St. Louis, CA, USA; [3]The Scripps Clinic and Research Foundation, La Jolla, CA, USA; [4]Biomagnetic Technologies, Inc., San Diego, CA, USA

Introduction

Magnetoencephalography (MEG) has been applied to identify regions of eloquent cortex potentially endangered by a planned surgical procedure. The resulting presurgical functional maps serve as an aide for the neurosurgeon during presurgical evaluation and planning. Prior work examined the reliability of this method [1] and assessed its accuracy by comparison with a variety of intraoperative results [2, 3]. Comparison of the three dimensional location of a dipolar source of somatosensory response as determined from MEG data with the corresponding location determined intraoperatively using electrocorticography (ECoG) has been limited by two primary factors: 1) the localization precision of routine intraoperative electric measurements, and 2) the limitation of routine intraoperative measurements to the cortical surface. Efforts at comparison have been further complicated by the effect of the relative orientation of the neurons responding to the stimulus; routine intraoperative electric recording usually makes the simple assumption that the source lies directly below the point of voltage phase reversal as measured on the surface of the exposed cortex[4], an assumption that can lead to substantial error [5]. Prior work attempted to account for this orientation effect in an approximate manner [6].

The present study was undertaken to take into account the fact that the source of sensory response generally lies below the surface of the cortex and may have any spatial orientation to provide a better test of the accuracy of non-invasive MEG somatosensory localization. Four patients who were scheduled to undergo open craniotomy were subject to presurgical mapping of somatosensory cortex using a 37 channel biomagnetometer (Biomagnetic Technologies Magnes). Subsequently, electrocorticographic localization of the response to median nerve stimulation was undertaken during surgery using a grid of EEG electrodes mounted on the exposed surface of the cortex. Single and multiple dipole modeling [7] of the somatosensory response was then undertaken for both the magnetic and electric data and the results compared.

Methods

Patients. 4 patients with tumors near eloquent cortex scheduled to undergo open craniometry were studied presurgically with MEG and intraoperatively with ECoG. Informed consent was obtained from all patients.

MEG. A 37 channel Magnes biomagnetometer (Biomagnetic Technologies, Inc.) in a magnetically shielded room was used to record evoked field data. Fields were evoked by tactile mechanical stimuli to individual fingers or lips.

ECoG. A 4x5 rectangular electrode grid (1 cm spacing) was placed over the exposed cortex. Responses were evoked by electrical stimulation of individual fingers or median nerve, and recorded with a Nicolet Viking system (10-13.5mA, 0.2 msec). 120 to 630 sweeps were used to generate on-line averages.

MRI. A T1-weighted volumetric image set was obtained from each subject using a 1.5T MRI system (General Electric).

Coregistration. 12 indelible scalp marks were placed on each patient, (including nasion and left and right preauricular points). The locations of these fiducial marks were measured prior to MEG acquisition using a Fastrak digitizer (Polhemus). The same marks were measured in the operating room with an LED digitizer (Picsys), as were the locations of the individual electrodes, permitting coregistration of the MEG and ECoG data. The nasion and preauricular points were marked with vitamin E capsules prior to magnetic resonance imaging, and these 3 fiducials were used for coregistration of the ECoG, MEG, and MRI data.

Source localization. Middle latency MEG data were fit with a single equivalent current dipole, using the Magnes software. The dipole at the latency with the largest correlation with the data was used for comparison with the ECoG source location. ECoG data were analyzed with EMSE software (Source/Signal Imaging) using the PRoMS algorithm [7], and a 1 shell sphere model with origin fit to the electrode location curvature. The location of the peak of the resulting source map was selected manually for comparison with the MEG dipole.

Results

Fig. 1 shows representative waveforms for ECoG (top) and MEG (bottom) for one subject. Several points are worth noting. Clearly the ECoG signal is more complex than the MEG signal, simply by inspection. This is consistent with the results of single dipole fitting. Good single dipole fits could always be obtained from the MEG data, but not from the ECoG data, which generally required 2 or more dipoles to account for greater than 95% residual variance. In addition, there are significant differences in time course between the MEG and ECoG signals, which are probably due to the difference in stimulus method (tactile for MEG vs. electrical for ECoG). Lastly, we have observed considerable intersubject differences in the ECoG data, which in part may be accounted for by the parietal lesions in these patients.

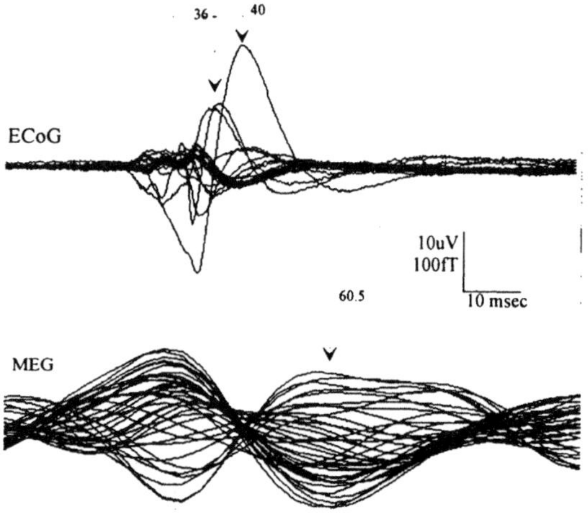

Fig. 1. (A) Averaged electrocorticogram for patient 1489, referenced to contralateral ear. The stimulus to the contralateral thumb was presented at t=0, indicated by the vertical mark. The 20 individual traces have been superimposed to a common origin to emphasize the temporal patterns. The arrows indicated the latencies for which ECoG-derived source maps are displayed in figure 2. (B) Averaged magnetoencephalogram for the same patient. The stimulus was presented at t=0, indicated by the vertical mark. The 37 individual traces have been superimposed to a common origin to emphasize the temporal patterns. The arrow indicates the latency for which the MEG-derived source map is displayed in figure 2.

Source maps from ECoG data for this patient are shown in fig. 2. Similar source maps could be obtained from 3 of the 4 patients who had adequate signal/noise in the ECoG. In 1 patient, data was used from the independent stimulation of 2 digits. For these 4 datasets, the location of the peak of the deeper, more tangential dipole from the ECoG source map was compared with the MEG single equivalent dipole location. The mean discrepancy between the 2 localization methods was 8±3 mm (mean±standard deviation).

The locations of each of the 4 pairs of sources were transferred to the corresponding MRI images, which were read by a trained observer. The location of each of the sources was consistent with a location in the postcentral gyrus.

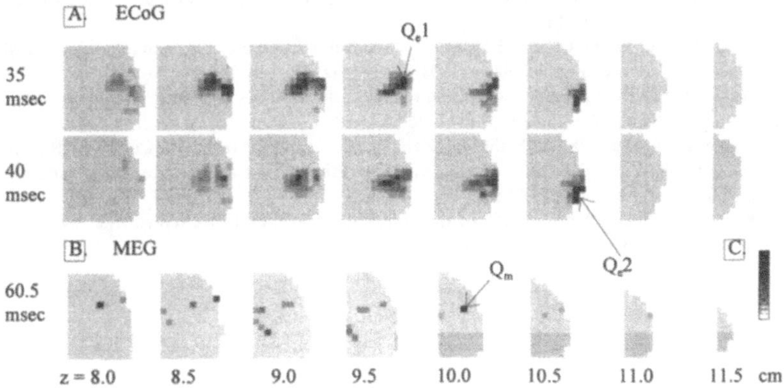

Figure 2. **A**. Source maps from ECoG at 36 (top) and 40 (middle) msec. poststimulus latencies following stimulation of the contralateral thumb. Source maps are displayed as axial slices, 5 mm resolution, with the nose (+X) towards the top of the figure, the right ear (-Y) towards the right. The cortical surface was determined by extrapolation from the measured electrode locations. The 36 msec slice is dominated by the activity of a primarily tangential source (Q_e1). The 40 msec slice is dominated by the activity of a primarily radial source (Q_e2). **B**. Source maps from MEG at 60 msec. poststimulus latencies. Source maps are displayed as in A, but span slightly different cortical extents, the cortical surface was approximated from scalp measurements in B.. The slice is dominated by the activity of a source (Q_m) at the approximate location of Q_e1. C. Source map scale maximum is 3×10^{-10} ampere meters for A, 3.5×10^{-8} ampere meters for B. The difference in source strength may be attributed to the more distributed source pattern in A.

Conclusion

Middle latency MEG data was adequately modeled as a single equivalent current dipole. Middle latency ECoG data required more complex modeling, usually involving at least two sources. Both MEG and ECoG-derived sources could be located reliably to postcentral gyrus. The location of the ECoG-derived tangential source agreed with that of the MEG-derived dipolar source to within about 8 mm on average. This discrepancy may be a limitation imposed by the combined imprecision of the MEG method, the ECoG method, and the coregistration process. These data suggest that the use of appropriate modeling techniques may improve the utility of ECoG data, and provide additional support for MEG non-invasive functional mapping of somatosensory cortex.

Literature cited

[1] Gallen, C.C., Schwartz, B.J., Rieke, K., Pantev, C., Sobel, D., Hirschkoff, E., and Bloom, F.E., Intrasubject reliability and validity of somatosensory source localization unsing a large array biomagnetometer. *Electroenceph and Clin Neurophys.*, **1994**, 90:145-156.

[2] Gallen, C.C., Sobel, D.F., Waltz, T., Aung, M., Copeland, B., Schwartz, B.J., Hirschkoff, E.C., and Bloom, F.E., Noninvasive presurgical neuromagnetic mapping of somatosensory cortex. *Neurosurgery*, 1993, **33**:260-268.

[3] Morioka, T., Yamamoto, T., Katsuta, T., Fujii, K., and Fukui, M., Presurgical three-dimensional magnetic source imaging of the somatosensory cortex in a patient with peri-rolandic lesion: tecnical note. *Neurosurgery*, **1994**, 34:930-934.

[4] Allison, T., McCarthy, G., Wood, C.C., and Jones, S.J. Potentials evoked in human and monkey cerebral cortex by stimulation of the median nerve. *Brain*, 1991 **114**:2465-2503.

[5] Suzuki, A., Yasui, N., J. Interoperaticve localization of the central sulcus by cortical somatosensroy evoked potentials in brain tumor. *Neurosurgery*, 1992, **76:867-870**.

[6] Gallen, C., Schwartz, B., Bucholz, R., Malik, G., Barkley, G., Smith, J., Tung, H., Copeland, B., Bruno, L., Assam, S., Hirschkoff, E., Bloom, F., Presurgical localization of functional cofrtex using magnetic soruce imaging. *J Neurosurgery*, 1995, **82**:988-994.

[7] Greenblatt, RE., Probabilistic reconstruction of multiple soruces in the bioelectromagnetic inverse problem. *Inverse Problems*, 1993, **9**:271-284.

Origin of Slow Wave Observed in Cerebrovascular Disease

Hatanaka, K.[1], Nakasato, N.[2], Kanno, A.[3], Ohtomo, S.[2], Seki, K.[2], Fujiwara, S.[2] and Yoshimoto, T.[2]

[1]*KRI, 17 Chudoji Minami-machi, Shimogyo-ku, Kyoto, Japan;* [2]*Department of Neurosurgery, Tohoku University School of Medicine, 1-1 Seiryo-cho, Aoba-ku, Sendai, Japan;* [3]*MEG Laboratory, Kohnan Hospital, 4-20-1 Nagamachiminami, Taihaku-ku, Sendai, Japan*

Introduction

Due to excellent spatio temporal resolution and noninvasive nature of the measurements, magnetoencephalography (MEG) becomes a powerful tool for clinical application. This is especially the case in evoked MEG applied for presurgical diagnosis of the patients who suffered brain diseases. However, in spontaneous activity such as epileptic discharges, applicability is not so simple. This is because, for spontaneous waves, several brain activities or noises were superimposed on the detected signal, and separation for target activity is generally difficult. Small array of sensing coils also restricted the spatial resolution and sometimes brought erroneous results simply because they could not cover the entire brain at once. The appearance of whole-head MEG directly linked to magnetic resonance imaging (MRI) reduces data acquisition time, patient fatigue and localization error. It also provides objective information about patients which is essential for correct diagnosis. Spontaneous MEG studies used to be concentrated to epileptic spike activity. However, slow wave activity, typically observed by electroencephalography (EEG), is quite common in stroke, tumor, epilepsy, cerebrovascular disease, or head injury. Earlier investigations of slow wave activity with MEG [1-3] were restricted to small (1 to 37) channel system. Here we present the results of whole head MEG and source analysis of a slow wave activity.

Methods

A 13-year-old male subject, who had brain infarctions due to a cerebrovascular disease, was seated on a chair in a magnetically shielded room. We employed a 64 channel MEG (CTF Systems Inc., Canada) and a standard international 10-20 system EEG (Nihonkohden Inc., Japan) for measurements and recording. Subject was instructed to stay at rest, and eventually he fell asleep during the recording. By monitoring the EEG chart, we could pick up twenty 10 seconds epochs of prominent slow wave activities from the session lasting for about an hour. Sampling clock of MEG was 250 Hz and the signal filtered by 0.5 to 40 Hz band pass and 50 and 100 Hz notches was stored on a disk for further analysis. Seven channels of EEG, FP1, CZ, F3, F4, P3, O1 and O2, were analog to digital converted and passed through the same filters and stored simultaneously with MEG data for comparison.

Source analysis of MEG data was carried out after measurement using single current dipole model. The effect of volume current was considered using Grynszpan and Geselowitz spherical head shape approximation. Position and radius of ' best sphere' (*i.e.* sphere which fit into subject's head in the least square sense) were determined by magnetic resonance imaging (MRI) of the subject. Conversion of MEG coordinates system to MRI was carried out by reference to three fiducial points (two preauricular points and nasion), each marked by small coils for MEG and oil-containing capsules for MRI. Positions and orientations of the estimated equivalent current dipoles (ECDs) were directly superposed on the MRI of the subject.

Results

We found that slow wave activities included delta wave, theta wave, and sharp (spike) and slow wave complex. These observations naturally contained normal (*e.g.* slow wave sleep) as well as pathological activity. We rejected artifacts such as eye movement or blink by carefully reading the EEG chart. Although magnetic noises from moving vehicles were mostly reduced by the low cut filter (0.5 Hz) of the band pass, we lost dc or lower frequency components [4,5] of slow wave activities at the same time. It is essential to find out pathological activities peculiar to the brain disease in the recording. Examples of a pathological slow wave are shown in Fig. 1. These activities were observed in light sleep (stage 2) epochs.

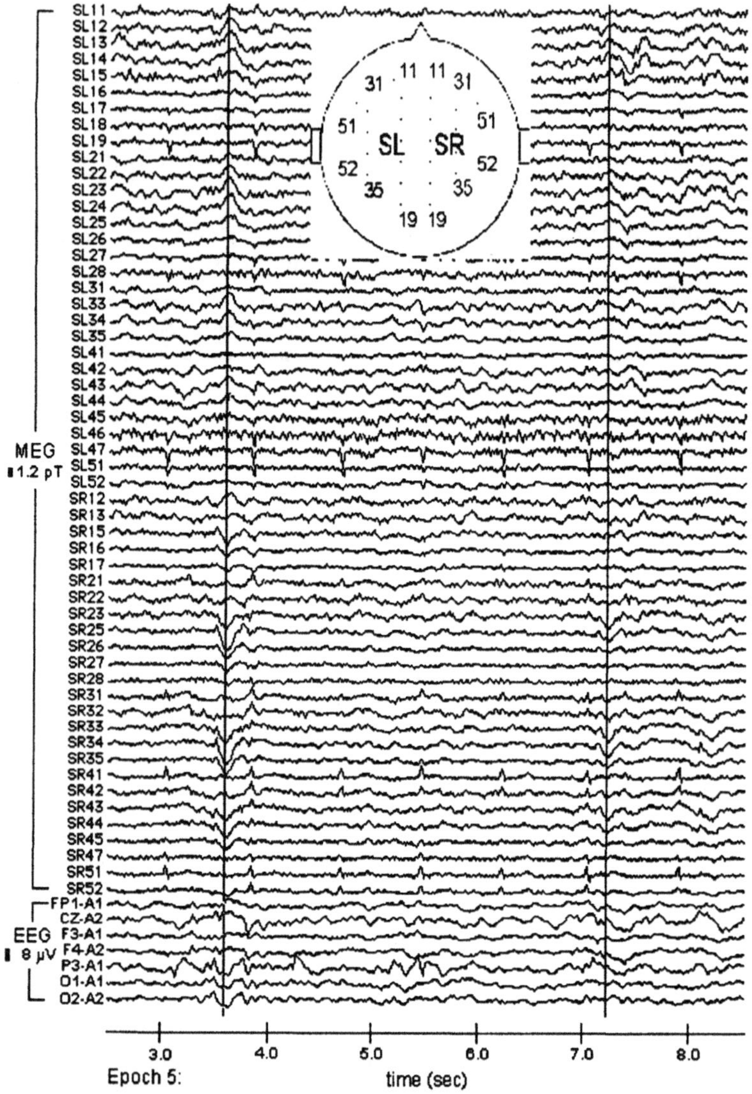

Fig. 1

An example of the simultaneous recording of MEG and EEG during light sleep stage. Bandpass filter was 0.5 to 40 Hz. The positions of the pickup coils of MEG are indicated on schematic head. Some noisy MEG channels are deleted in the above chart, and seven EEG recording stored in the MEG system are shown. Two pathological slow wave activities are observed: one peaked at 3.612 sec and the other at 7.216 sec. Note that slow wave peak at 3.612 sec is observed both in MEG and EEG, while peak at 7.216 sec is only detected by MEG. This is because magnitude of the dipole is about half for the later wave (see Table 1). Spikes of about 0.8 sec period observed in SL19 etc. are magnetocardiogram detected by MEG pick-up coils.

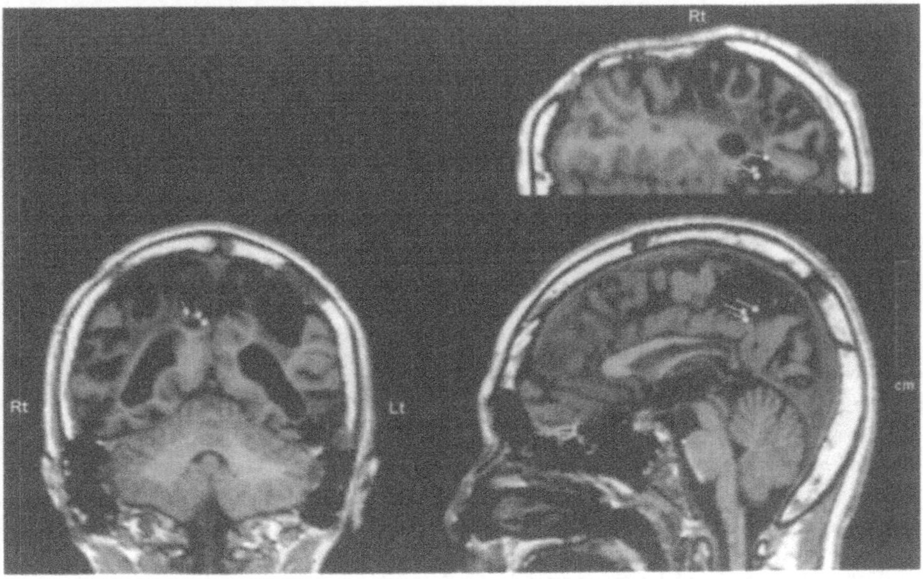

Fig. 2

Positions (circles) and orientations (bars) of the ECDs estimated from the slow waves in a case of cerebral infarction.

Table 1. The ECD parameters of the slow waves in a case of cerebral infarction (x, y, z: anterior, left lateral, and upward coordinates of the MEG system; **φ**: azimuthal angle of dipole, 0 for+x direction; **θ**: declination angle of dipole, 0 for +z direction; Q: magnitude of current dipole)

Dipole	Epoch:	time (sec)	x (cm)	y (cm)	z (cm)	φ (deg)	θ (deg)	Q (nAm)	Error (%)
1	5:	3.612	-0.98	-0.32	8.16	332	82.9	650	7.8
2	5:	7.216	-0.78	-0.57	8.47	358	83.7	333	15.2
3	6:	5.816	-1.24	-1.16	8.72	354	79.1	443	5.5

Results of the source analysis of the slow wave activities are shown in Fig. 2, and the parameters of ECDs are summarized in Table. 1. The ECDs were localized at the lower bottom surface of a large infarct on right parietal lobe, and the orientations were anterior-posterior direction.

Discussion

Completed infarct foci are replaced by cerebrovascular fluid or connective tissues and thus electrically silent. Therefore, abnormal slow waves are believed to be originated in the surrounding brain tissues. ECDs might be distributed or extended in space. Surprisingly, our present study demonstrated that some of abnormal slow waves indicated clear single dipole patterns with relatively low fitting errors even for the whole head data. These findings suggest that electrically active tissue exist at least a part of the boundary of infarct foci.

Infarction grows from ischemia, and cerebrovascular diseases generally accompany focal neurological symptoms and signs resulting from blood vessels. MEG studies in animal models have demonstrated that focal cerebral anoxia [4] or ischemia [5] brought slow or dc (direct current) variations of magnetic fields. As Suthering pointed out [6], it is very important to confirm these results because there could be potential diagnostic value of detecting slow wave in human cerebrovascular dieseases. Our results, though not dc-MEG, support this comment.

EEG may detect the peculiar slow wave activity we found in a cerebrovascular disease patient. MEG can, however, detect and localize sources of the activity accurately. Further studies of whole head MEG in cerebrovascular disease patients will reveal the occurrence and development mechanism of these diseases in future.

References

[1] Vieth, J.B. Magntoencephalography in the study of stroke (cerebrovascular accident), In: Sato, S. *Advances in Neurology* 54: 261-269, New York, Ravan Press, 1990.

[2] Vieth, J. B., Kober, H., Weise, E., Daun, A., Moeger, A., Friedrich, S. Functional 3D localization of cerebrovascular accidents by magnetoencephalography (MEG), *Neurological Research* ,1992, 14: S132-S134.

[3] Gallen, C.C., Sobel, D.F., Schwartz, B., Copeland, B., Waltz, T., and Aung, M. Magnetic source imaging Present and future, *Investigative Radiology*, 1993, 28, Supplement 3: S153-S157.

[4] Takanashi, Y., Chopp, M., Levine, S.R., Kim, J., Moran, J.E., Tepley, N., Chen, Q., Barkley, G.L., and Welch, K.M.A. Magnetic fields associated with anoxic depolarization in anesthetized rats, *Brain Research*, 1991, 562:13-16.

[5] Chen, Q., Chop, M., Chen, H., and Tepley, M. Magnetoencephalography of focal cerebral ischemia in rats, *Stroke*, 1992, 23:1299-1303.

[6] Suthering, W.W. Editorial comment, *Stroke*, 1992, 23:1303.

Acknowledgments

This work has been supported in part by Grants-in-Aid for Scientific Research No. 03404042 and No. 064050 from the Ministry of Education, Science and Culture of Japan.

Magnetic Source Imaging in the Evaluation of Chronic Alcohol Abuse

Hill, D., Waldorf, V.A., Lewine, J.D., Provencal, S.L., Moyers, T., Yeo, R.

The New Mexico Institute of Neuroimaging, The New Mexico Regional Federal Medical Center, Albuquerque, New Mexico, USA

INTRODUCTION

Chronic alcohol abuse is a significant problem for a large number of individuals, especially in the population of U.S. Veterans. Research to date has documented some of the long-term deleterious effects associated with chronic abuse. Cognitive difficulties may include impairments in short-term memory, paired associate learning, visual-spatial skills, complex problem-solving, and psychomotor functioning. Neuroradiographic findings show that cortical atrophy and white matter changes are prominent at late stages of alcoholism, but intermediate effects are less well documented. From a brain functional perspective, evoked potential studies demonstrate that subjects in the late stages of alcohol abuse show changes in P3 amplitude for both the visual and auditory stimuli.

A limitation of previous research has been the focus on patients with severe, chronic conditions such as Wernicke-Korsakoff syndrome. The present study was focused on specification of neurobiological and neuropsychological deficits that may manifest at intermediate stages of alcohol abuse. In particular, we used magnetoencephalography, neuropsychology, and magnetic resonance imaging to determine if patients with a history of substance abuse demonstrate focal abnormalities.

METHODS

Subjects: Four volunteer subjects enrolled in the out-patient alcohol treatment program at the Albuquerque Veterans Administration Medical Center (VAMC) participated in this project. The subjects were all right-handed. All four subjects had a diagnosis of alcohol dependence, based on the Diagnostic and Statistical Manual of Mental Disorders-Third Edition (APA, 1987). Average daily alcohol intake was 106 ounces (range= 32-226), with an average of 112.3 days elapsed between last drink and testing (range= 26-180). No history of head trauma or psychoactive substance dependence was reported. Case histories were as follows: Case 1 was a 49 year old Caucasian male with 12 years of education and a 27 year drinking history. Case 2 was a 39 year old Native American male with 12 years of education and a 23 year drinking history. Case 3 was a 42 year old Hispanic male with 16 years of education and a 20 year drinking history. Case 4 was a 50 year old Caucasian male with 15 years of education and a 26 year drinking history.

Neuropsychological Testing: Each subject was administered a comprehensive battery of neuropsychological tests assessing the cognitive domains of general intelligence, perceptual-motor skills, motor skills, memory skills, executive planning skills, and language skills. Standard norms were used to evaluate performances of each subject, and age and education appropriate norms were used whenever possible.

Magnetic Resonance Imaging (MRI): Magnetic resonance imaging was performed with a 1.5 Tesla Magnetom system (Siemens). Images were obtained using modified gradient and spin echo sequences with the following parameters: Sagittal - 1.5 mm contiguous slices, TR=40, TE=6; Coronal - 5.0 mm slices with 1 mm gaps, TR=800, TE=15; Axial - 5.0 mm slices with 1 mm gaps, TR=2500, TE=90. The images were examined for pathology by a neuroradiologist who was blind to the clinical status of the subjects. The sagittal volumetric data set was re-formatted and re-sliced into 1.0mm thick coronal slices for the purpose of whole-brain volumetric measurement. Volume analyses were performed using the NIH Image 1.52 program.

Magnetoencephalography (MEG): The studies were performed in a magnetically and radio-frequency shielded room equipped with a large array 37-channel biomagnetometer or a recently installed whole-head system with 122-channels. For each subject, 3 minutes of continuous data was recorded in an eyes closed state from each of multiple positions to provide for a whole-head examination. MEG examinations typically required 2 hours of subject time. MEG data were examined for epileptiform transients, abnormal power spectra, and abnormal low frequency magnetic activity (ALFMA). For the evaluation of ALFMA, data were digitally filtered with a bandpass of 1-4 Hz in order to isolate activity in the delta range. After identification of the latencies of local amplitude peaks where the magnetic field strength was greater than 200 fT, single dipole modelling was performed. Only highly dipolar slow waves were considered in further analyses. A spatial clustering algorithm was used to determine if a particular cortical region was responsible for generation of multiple dipolar low frequency events. This is a critical aspect of the analysis procedure because available data suggest that it is the repetitive generation of dipolar low frequency activity that provides the most reliable sign that a region of tissue is dysfunctional. A 1-cm radius volume element was considered pathological only if it was responsible for generation of five distinct low frequency events for each minute of recorded data. Previous work indicates that these particular parameter values provide for good separation of control and patient populations.

RESULTS

As summarized below, and in table 1, abnormalities were found on one or more tests for three of the four patients.

Case 1: This subject's general intelligence was in the above average range. The skill domains of perceptual-motor, memory, executive planning and language were all found to be within the normal limits. However, this individual showed significant impairment on the motor task. This subject's MRI was read as normal. Whole brain volume was $1298 cm^3$, a value within normal limits. MEG examination was abnormal, with extensive ALFMA over left temporal regions. In addition, occasional epileptiform transients were identified bi-parietally, R>L.

Case 2: Although this subject showed a similar drinking history to the other subjects, his performance on the neuropsychological battery was within or above normal limits. MRI was normal. Whole brain volume was $1297 cm^3$. MEG examination was normal.

Case 3: Given this individual's educational level, general intelligence scores were lower than would be expected. Relative strengths were noted in perceptual-motor and language skills. However, tests of higher cognitive abilities showed mild deficits, although the scores fell within the low normal range. The motor skills test scores suggested a lateralized deficit, as dominant hand performance was significantly below average, whereas non-dominant fell within the normal range. MRI results indicated diffuse white matter changes. Whole brain volume was $1170 cm^3$. This is almost two standard deviations below normal. MEG results indicated abnormal power spectra (excessive delta) over right and left temporal regions, but dipolar sources of ALFMA were not identified. In addition, some left parietal epileptiform transients were identified.

Case 4: This subject's general intelligence, memory, language, and motor skills were in the average range. However, this individual showed impairment on tests of executive functioning and perceptual-motor skills. MRI revealed diffuse white matter changes, and note was made of cerebral volume loss, especially in the frontal lobes. Whole brain volume was $1162 cm^3$, a value almost two standard deviations below normal. MEG revealed abnormal power spectra (excessive delta) over left and right temporal lobes and left parietal areas. Much of the left parietal slowing had dipolar fields with sources of ALFMA localizing around the central sulcus. It is noteworthy that this patient was re-evaluated with MEG 14 months after cessation of alcohol abuse. During this follow-up, MEG power spectra were normal and no ALFMA was identified.

TABLE 1. Assessment Summary: Neuropsychological, MRI, and MEG Findings.

	Neuropsych	MRI	Brain Volume	MEG
Case 1	Motor Abnormalities	Normal	$1298 cm^3$	LT-A,P-E
Case 2	Normal	Normal	$1297 cm^3$	Normal
Case 3	Motor Abnormalities	White Matter Changes	$1170 cm^3$	R/LT-S, LP-E
Case 4	Executive & Perceptual-Motor Abnormalities	White Matter Changes & Atrophy	$1162 cm^3$	R/LT-S, LP-S, LP-A

MEG Abnormalities

location	type
L: Left	A: ALFMA
R: Right	S: Spectral (high delta)
T: Temporal	E: Epileptiform transients
P: Parietal	

DISCUSSION

In this study, a novel multimodal approach was utilized to further understand the long-term effects of chronic alcohol use. Techniques for assessing cognitive function (neuropsychological evaluation), brain structure (magnetic resonance imaging) and electrophysiology (magnetoencephalography) were all used with a common patient population to explore the extent to which alcohol abuse induces focal abnormalities.

Neuropsychological findings indicated that 3 of the 4 cases demonstrated clinically significant deficits. Findings included motor and perceptual-motor impairments, as well as problem-solving difficulties. These deficits are consistent with those found in other research with this population.

MRI findings revealed white matter changes in two of the patients, a finding consistent with other radiological studies of alcoholics. Two cases had abnormal MRI examinations indicating white matter changes. For both, atrophy was also present, whole brain volumes being almost two standard deviations below normal (1313 cm^3, sd= 89 cm^3). Both of these patients demonstrated neuropsychological deficits and MEG abnormalities.

Overall, MEG identified abnormalities in 3 of the 4 patients. Abnormalities included abnormal low frequency magnetic activity, excessive delta in power spectra, and in some cases, epileptiform transients (Fig 1). The patient without MEG abnormalities was also found to be normal on MRI and all neuropsychological tests. It is also noteworthy that Case 3 had a normal followup examination after 14 months of sobriety. This suggests that MEG assessed brain dysfunction may be partly reversible.

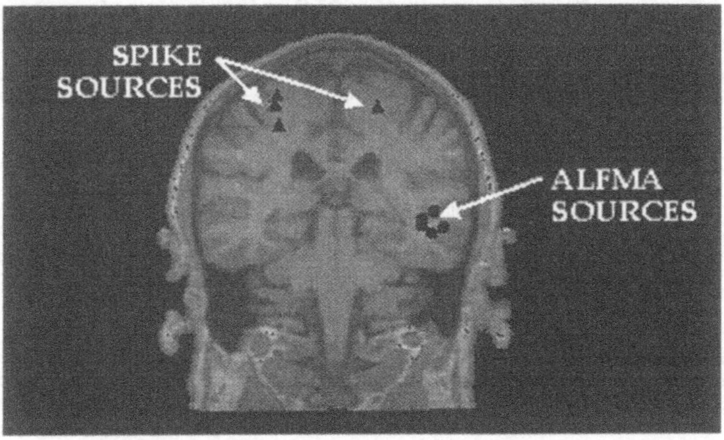

Fig 1: Case 1, MEG shows left temporal ALFMA and bi-parietal epileptiform transients. MRI was normal, as was whole brain volume. Neuropsychological testing revealed multiple deficits including memory and perceptual-motor problems.

CONCLUSIONS

The preliminary results presented here suggest that MEG may be useful for providing objective evidence of brain dysfunction in patients with a history of chronic alcohol abuse. An ability to demonstrate to patients the negative impact of alcohol abuse on their brain function may be beneficial in convincing patients to stay in therapy programs. Also, a better understanding of how alcohol effects brain physiology may ultimately lead to better interventional strategies.

MEG and MCG in a Clinical Environment: BioMag Laboratory, Helsinki

Ilmoniemi, R.J.[1], Ahonen, A.[4], Alho, K.[5], Aronen, H.J.[2], Huttunen, J.[1], Karhu, J.[6], Karp, P.[1], Mäkijärvi, M.[1,3], Näätänen, R.[5], Paetau, R.[7], Pekkonen, E.[5], Standertskjöld-Nordenstam, C.-G.[2], Tiihonen, J.[8], Toivonen, L.[1,3] and Katila, T.[9]

[1]BioMag Laboratory/Medical Engineering Centre; [2]Department of Radiology; [3]1st Department of Medicine, Helsinki University Central Hospital; [4]Neuromag Ltd, Helsinki; [5]Cognitive Brain Research Unit, University of Helsinki; [6]Department of Clinical Neurophysiology, Kuopio University Hospital, Kuopio; [7]Children's Castle Hospital, Helsinki; [8]Department of Forensic Psychiatry, Niuvanniemi Hospital, Kuopio; [9]Laboratory of Biomedical Engineering, Helsinki University of Technology, Espoo, Finland

Introduction

The success of biomagnetism will critically depend on how useful the technique is as a clinical tool and what kinds of applications exist. We have started work at the new BioMag Laboratory of the Helsinki University Central Hospital to bring this technology clinically available and to evaluate its possibilities. The project follows extensive development and pioneering work for over two decades at the Helsinki University of Technology in neuromagnetism [1] and cardiomagnetism [2] done in collaboration with the Helsinki University Central Hospital. While some clinical applications already have been established [1–4] and we are actively working on them, much of our research is conducted on various patient groups and healthy subjects in order to evaluate and assess this new technology and to extend our knowledge of the functioning of the brain and the heart.

Here, we first describe some goals, design criteria, and technical solutions in building our new biomagnetism laboratory. Then, ongoing clinical and technical projects will be summarized.

Methods

The large-array magnetometers made available by Neuromag Ltd. were the starting point for the design of the laboratory. In addition to the whole-scalp 122-channel neuromagnetometer, a 68-channel cardiomagnetometer was developed and built for this project by Neuromag in collaboration with the Medical Engineering Laboratory of the Helsinki University of Technology. Although in an ideal case brain and heart research would have separate shielded rooms, we decided to place both systems in the same room (see Fig. 2), requiring that the switch from brain to heart measurements, and vice versa, is easy and quick and that as much as possible of the electronics and computing facilities can be shared between the systems. Although the magnetometers, together with gantries and counterweights, weigh about 270 kg each, they can be interchanged easily on air cushions by one person.

The site for the laboratory was selected on the basis of careful measurements of the external magnetic noise level and the requirement that the distance from cardiology and neurology departments is not very long. The level of mechanical vibrations at the site is low (Typical rms levels for acceleration, speed, and displacement of the underlying floor in the bandwidth 5–100 Hz were 0.35 mm/s^2, 1.2 µm/s, and 14 nm, respectively; transient levels could be 2 orders of magnitude higher.). Otherwise, mechanical damping for the shielded room would have been considered. The recorded environmental noise spectrum in the selected location is shown in Fig. 1. It was decided that a shielding factor of 30 (at 0.1 Hz) to 80 dB (100 Hz) would be sufficient at this site. Actually, this minimum shielding requirement was exceeded by 10–20 dB (Euroshield Ltd., see [5]).

A ramp mechanism was built to allow easy wheeling of the patient bed into the shielded room [5]. Lighting was arranged via glass fibers to exclude magnetic interference from the lamps. A rectangular hole was built in the wall to allow the subject to look at a computer screen or a video display (up to 15"). Another opening enables cardiac monitoring inside the room during intervention studies. In many experiments, in particular with children, elderly people or demented patients, movies shown to the subjects help them to sit still during the measurement.

Proper use of the biomagnetic technique requires the utilization of other imaging modalities to help solve the inverse problem (MRI, fMRI, EEG/ECG) and to display the results (MRI). We have built in the facility a 64-channel EEG/ECG system [6] to be used simultaneously with magnetic measurements. The laboratory's local area network is connected to the hospital's fast image network to allow easy transfer of MRI images from the Radiology Department to be used in conjunction with data analysis. In addition, we have realized a fast connection to the Helsinki University of Technology so that data can be analyzed directly at the site where MCG software is being developed and where much of the data are analyzed. Software maintenance by Neuromag Ltd. is performed telematically via the networks as well. A special consideration in the setup of these connections is the strict requirement of patient data security. Unauthorized access to information about patients is prevented in two ways. First, patient names are separated from the data. Second, connecting to the computers from the outside is possible by authorized users only.

To maximize effective recording time, the laboratory (about 200 m^2) was designed so that a patient can be prepared (change of clothes, attachment of electrodes and marker coils, digitization of head or chest, etc.) while the previous patient is still being studied. Space was assigned in the laboratory also for data analysis, technical development work, and for cardiac surface-potential measurements. A registered nurse accompanies most patients in the shielded room, making sure that the patient stays still and performs the assigned task. MCG measurements usually take less than 5 minutes; MEG paradigms for patients are designed to last at most 40–60 minutes.

The 122-channel helmet-shaped MEG magnetometer array [7] consists of 61 identical plug-in units, each containing two orthogonal planar gradiometer flux transformers and dc SQUIDs. Although the 16.5-mm baselength of the flux transformers is relatively short and the sensitivity pattern of each sensor is quite narrow, the array as a whole is able to see as deep in the brain as an array of axial sensors. The advantages of the planar design include very good balance and precise calibration; in addition, the raw signals are easy to interpret, because the measured signal is at maximum right above the source. Furthermore, because of the tighter sensitivity pattern collecting brain noise, the signal-to-noise ratio in the SQUID output from planar gradiometers is typically higher than from axial gradiometers [7].

The 68-channel MCG device [8] has 62 planar and 6 axial gradiometers over a slightly curved dewar bottom (diameter 300 mm). Since the advantage of the planar gradiometer geometry in MCG may not be as clear as in MEG, the axial gradiometers were built in order to evaluate the gradiometer types experimentally. The high intrinsic balance and ease of calibration were lost, but the signal level was increased in these channels as expected.

Data from the two magnetometer arrays are collected by a real-time computer, which also controls the head-position indicator electronics and provides stimulus triggers when needed. On-line and off-line data analysis is performed by HP 725/100-series PA-RISC workstations interconnected by a local area network. An X-terminal in the patient preparation room is used for entering patient data, for 3D digitizing, and for data analysis. Additional terminals and workstations can be connected to the local area network directly or via ATM-level networks.

Liquid helium transfer is carried out 3 times per 2 weeks. For quality assurance purposes, an empty room measurement and phantom measurement should be performed at least once a week. A preventive maintenance service is scheduled for four days per system in a year.

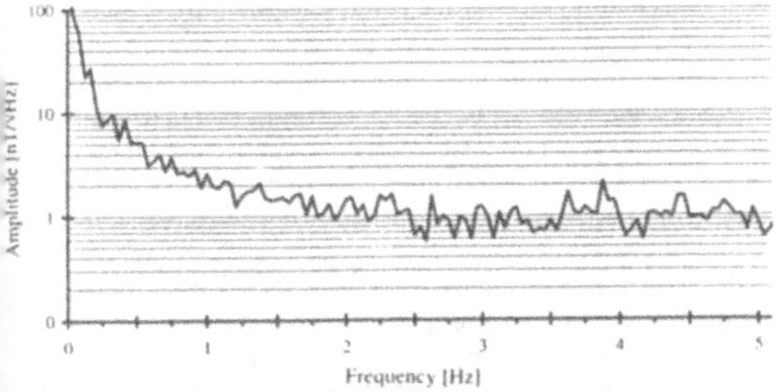

Figure 1. Magnetic field noise level at the site of the laboratory prior to construction of the shielded room. At some other candidate sites at the hospital, 2–5 times higher noise levels were measured.

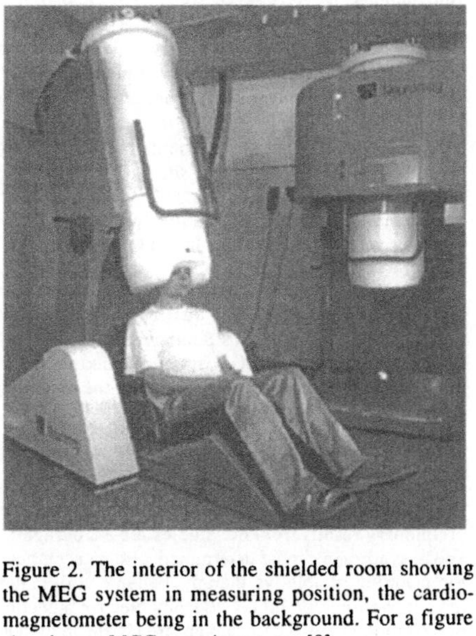

Figure 2. The interior of the shielded room showing the MEG system in measuring position, the cardiomagnetometer being in the background. For a figure showing an MCG experiment, see [8].

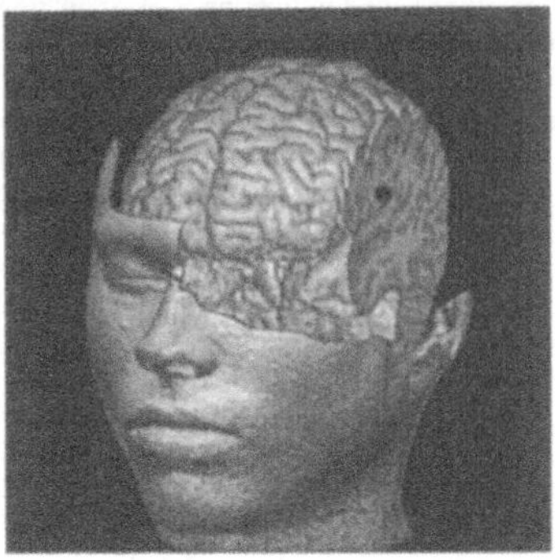

Figure 3. 3D-rendering of a subject's head showing the location of an auditory response.

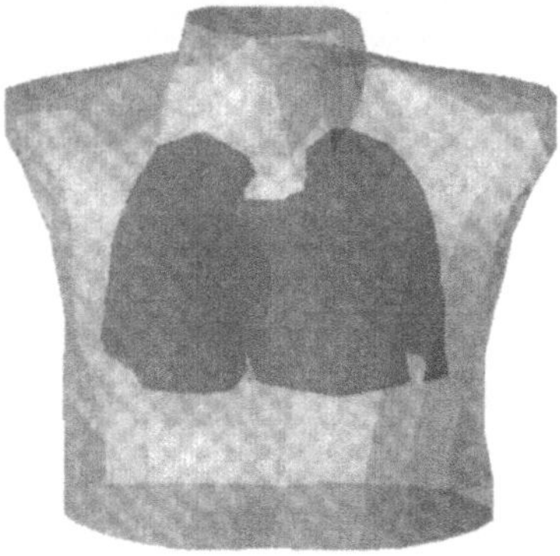

Figure 4. Torso model used in realistic-geometry computations of cardiac magnetic fields.

Results

Our major clinical projects with MEG include the localization of epileptic foci and the determination of the location of the central sulcus in patients going to brain surgery. Extensive work has been initiated to study patients with Alzheimer's and Parkinson's diseases, schizophrenia [9,10], alcoholism, or developmental disorders [11]. We have also used MEG in studies of neuronal changes associated with disorders of central and peripheral auditory processing and after brain infarction. Heart studies [8] include the determination of myocardiac arrhythmogenic foci, the estimation of the risk of sudden cardiac death, repolarization disorders and left-ventricle hypertrophy. During the first year of operation, we completed a total of 767 measurement sessions, of which 288 were on patients.

We have compared MEG and fMRI localization of somatosensory activation [12] in normal subjects and in patients scheduled for surgery. On the basis of these studies, the method of locating the central sulcus with MEG looks very promising, as previously reported by Lewine *et al.* [4]. In some cases, however, artefacts may be a problem. To eliminate the effect of external noise, eye movements, or magnetic particles left inside the head during previous surgery, we have successfully used the Signal-Space Projection method [13].

We have studied somatosensory and motor functions in 20 patients with acute brain infarction in order to detect plastic changes during recovery. While it has become clear that patients with even a moderate paresis can be studied up to 60 minutes, the current plans of manufacturers to allow recording also from supine position is more than welcome. Equally important are efforts to help maintain the head position during data collection.

Our experience with the mismatch paradigm [14,15] shows that MEG studies lasting up to 45 minutes can be performed even in severely demented subjects (Mini-Mental State score down to 14/30). During these experiments, a motion picture is often shown in order to keep the patient vigilant and comfortable.

In patients being evaluated for surgical treatment of epilepsy, low-amplitude fast cortical activity not seen in scalp EEG has been detected in ictal whole-head MEG. In such patients, MEG seems to reduce the need for invasive preoperative recordings. In patients with different clinical epilepsy syndromes, the determination of cortical generators of the underlying activity seems to help in understanding the pathogenesis and in the planning of therapy.

Very young children may fall asleep or are unable to sustain conscious attention to head position even during a brief measurement. The quality of the data was greatly improved in such children by using a special examination seat, allowing exact positioning of the head. For MEG examination of infants, the MCG equipment is used.

Preliminary MEG results obtained in schizophrenics showed a reversed interhemispheric asymmetry of temporal auditory areas [9,10], although the MRIs in these patients were normal. It thus appears that MEG might be capable of revealing subtle changes in the functional anatomy in major psychiatric disorders. This kind of application may be very useful in the screening of high-risk populations.

MCG has been performed on over 100 patients who have the Wolff–Parkinson–White (WPW) syndrome, life-threatening ventricular tachyarrhythmias after myocardial infarction, cardiomyopathy, or congenital long-QT syndrome [8]. A routine measurement takes just 5 minutes, while patient preparation takes about 10 minutes. The preliminary results from our studies are encouraging. The 3-dimensional location of the overt accessory pathway in WPW syndrome can be determined with an accuracy of 1–1.5 cm. The preoperative information has been helpful in the planning of ablative therapy. Patients with malignant ventricular tachyarrhythmias after myocardial infarction have low-amplitude high-frequency electromagnetic activity, so called late fields, at the end of the ventricular depolarization phase. These signals are absent in patients without arrhythmic events after myocardial infarction. Using spectral turbulence analysis, the patients with tachycardias after infarction can be separated from those without.

Similar abnormal ventricular activity can also be found in patients with cardiomyopathy and ventricular tachyarrhythmias. Long-QT patients have an inhomogeneous pattern of repolarization in MCG, clearly distinguishable from the normal pattern of their healthy relatives. In summary, multichannel MCG seems a powerful new clinical tool for identifying risk factors and for locating the foci in patients with life-threatening arrhythmias.

Discussion

In clinical work, an essential requirement is reliability. The hardware and software must work without interruption; in addition, one has to be able to rely on results from the inverse problem algorithms used; operator errors must be minimized and wrong judgements avoided. To reach high reliability with a new complicated technology, effective training of the personnel as well as of the clinicians is necessary. Reliability and efficiency from the human point of view is best achieved by developing standardized experimental and analysis procedures. We have started systematic work in this direction. Standardized procedures will help in collecting normative data banks, which can be used as reference in patient studies. These data banks will facilitate studies with new patient groups and in pharmacological evaluations and support assessment studies for developing clinical practices.

To support the clinical studies and to push the technologies further, work is done at the BioMag Laboratory in modeling, signal and image processing, multimodal functional imaging and in transcranial magnetic stimulation.

References

[1] Hämäläinen, M., Hari, R., Ilmoniemi, R.J., Knuutila, J., and Lounasmaa, O.V. Magnetoencephalography—theory, instrumentation, and applications to noninvasive studies of the working human brain, Rev. Mod. Phys., 1993, 65: 413–497.

[2] Nenonen, J. and Katila, T. Noninvasive functional localization by biomagnetic methods, J. Clin. Eng., 1991, 16: 423–434, 495–503.

[3] Hari, R. Magnetoencephalography as a tool of clinical neurophysiology, In: Niedermeyer, E., Lopes da Silva, F. Electroencephalography: Basic Principles, Clinical Applications, and Related Fields, Baltimore, Williams & Wilkins, 1993.

[4] Lewine, J., Baldwin, N., Sanders, J., Anson, J., Halliday, A., and Orrison, W. Noninvasive localization of somatosensory cortex in neurosurgical patients—A comparison of structural (MRI) and functional (MEG) approaches with validation by intraoperative monitoring, Soc. Neurosci. Abstr., 1995, 21, Part 2: 1419.

[5] Paavola, M., Ilmoniemi, R.J., Sohlström, L., Meinander, T., Penttinen, A., and Katila, T. High performance magnetically shielded room for clinical measurements, these proceedings.

[6] Virtanen, J., Parkkonen, L., Huotilainen, M., Teder, W., Pekkonen, E., Ilmoniemi, R., and Näätänen, R. Facilities for simultaneous EEG and MEG recording, Human Brain Mapping, 1993, Suppl. 1: 111.

[7] Knuutila, J.E.T., Ahonen, A.I., Hämäläinen, M.S., Kajola, M.J., Laine, P.P., Lounasmaa, O.V., Parkkonen, L.T., Simola, J.T., Tesche, C.D. A 122-channel whole-cortex SQUID system for measuring the brain's magnetic fields, IEEE Transactions on Magnetism, 1993, 29: 3315–3320.

[8] Montonen, J., Ahonen, A., Hämäläinen, M., Ilmoniemi, R.J., Laine, P., Nenonen, J., Paavola, M., Simelius, K., Simola, J., and Katila, T. Magnetocardiographic functional imaging studies in BioMag Laboratory, these proceedings.

[9] Tiihonen, J., Katila, H., Pekkonen, E., Aronen, H., Ilmoniemi, R.J., Huotilainen, M., Räsänen, P., Virtanen, J., and Karhu, J. Cerebral asymmetry in schizophrenia, submitted for publication.

[10] Karhu, J., Tiihonen, J., Pekkonen, E., Katila, H., Huotilainen, M., Virtanen, J., and Ilmoniemi, R.J. Reversed hemispheric asymmetry of auditory N100m in schizophrenics, these proceedings.

[11] Sipilä, L., Heikkilä, E., Autti, T., Sainio, K., Huttunen, J., Aronen, H.J., Korvenoja, A., Ilmoniemi, R.J., and Santavuori, P., Somatosensory evoked magnetic fields from primary sensorimotor cortex in juvenile neuronal ceroid lipofuscinosis, submitted for publication.

[12] Korvenoja, A., Wikström, H., Huttunen, J., Virtanen, J., Laine, P., Aronen, H.J., Seppäläinen, A.–M., and Ilmoniemi, R.J. Activation of ipsilateral primary sensorimotor cortex by median nerve stimulation, NeuroReport, 1995, 6: 2589–2593.

[13] Tesche, C.D., Uusitalo, M.A., Ilmoniemi, R.J., Huotilainen, M., Kajola, M., and Salonen, O. Signal-space projections of MEG data characterize both distributed and well-localized neuronal sources, Electroenceph. Clin. Neurophysiol., 1995, 95: 189–200.

[14] Huotilainen, M., Tiitinen, H., Lavikainen, J., Ilmoniemi, R.J., Pekkonen, E., Sinkkonen, J., Laine, P., and Näätänen, R. Sustained fields of tones and glides reflect tonotopy of the auditory cortex, NeuroReport, 1995, 6: 841–844.

[15] Pekkonen, E., Huotilainen, M., Virtanen, J., Sinkkonen, J., Rinne, T., Ilmoniemi, R., and Näätänen, R. Age-related functional differences between auditory cortices: a whole-head MEG-study, NeuroReport, 1995, 6: 1803–1806.

The support of TEKES, SITRA, the Academy of Finland, and the Ministry of Education is gratefully acknowledged.

MEG Traces of Painful Processes

Kakigi, R, Kitamura, Y., Koyama, S., Hoshiyama, M., Shimojo, M., Watanabe, S., Takeshima, Y. and Nagata, O.

Department of Integrative Physiology, National Institute for Physiological Sciences, Okazaki, 444, Japan

Introduction

The cerebral representation of pain perception in man is poorly understood as compared with other modalities of sensations such as touch or vibration. This is mainly because of a lack of the appropriate instrumentation for stimulation and recording. Therefore, we recently studied magnetoencephalographies (MEGs) following painful electricai stimulation [1,2] and CO_2 laser stimulation [3,4] to detect the cerebral areas which are responsible for pain perception in man, and reviewed them.

Methods

Five normal volunteers (1 female and 4 males; mean age 32 years, range 28 to 41 years) were studied. Informed consent was obtained from all participants prior to the study.

The electric stimulus was a constant voltage square-wave pulse delivered transcutaneously to the right index finger. We adopted the three levels of intensities, "weak", "moderately painful" and "very painful", depending on the subjective feeling of each subject. Actually, 2-4 mA, 5-7 mA and 10-13 mA were used for weak, moderately painful and very painful session, respectively. Stimuli was applied at random. The interstimulus interval was between 1500 and 5000 msec and the stimulus duration was 1 msec. Responses were filtered with a 0.1-100 Hz bandpass filter, and digitized at a sampling rate of 1041.7 Hz. The analysis time was 100 msec before and 900 msec after the stimuli, and DC offset was achieved using a pre-stimulus period as the baseline. Each session was comprised of an average of two hundred trials, and at least two averages were obtained to ensure reproducibility. The order of each session was at random.

A special CO_2 laser stimulator (Nippon Infrared Industries, Kawasaki, Japan) was used to record MEGs.. The laser wavelength was 10.6 μm, and the diameter of the irradiation beam was about 2 mm. The stimulus duration was 20 msec. The intensity was approximately 18 mJ/mm^2, which elicited a sharp pain that all subjects described as tolerable, "like a pin- prick". To avoid causing burns or habituation, the irradiated points were moved slightly for each stimulus. The interstimulus interval of the laser beam was at random, between 3 to 10 seconds. To avoid magnetic noises, the stimulator was settled outside a shielded room, and the laser beam was carried through the optical fibers approximately 3.5 m in length, which penetrated the wall of the shielded room. The beam was applied to the bilateral arms and legs in separate sessions. Responses were recorded with a 0.1-50 Hz bandpass filter, followed by that at 1-30 Hz. Responses were digitized at a sampling rate of 512 Hz. Fifty trials were averaged in one session.

MEGs were measured with dual 37-channel biomagnetometers (Magnes, Biomagnetic Technologies Inc., San Diego, CA). The detection coils of the biomagnetometer were arranged in a uniformly distributed array in concentric circles over a spherically concave surface. The device was 144 mm in diameter and its radius was 122 mm. The outer coils were 72.5^0 apart. Each coil was 20 mm in diameter and the distance between the centers of each coil was 22 mm. Each coil was connected to a superconducting quantum interference device (SQUID). We recently reported several papers of MEGs using this magnetometer [5-11]. The measurement matrix was centered at the five different positions; around the C3, C4, Fz, Cz and Pz of the International 10-20 System in each subject. The whole head was mostly covered by these positions of magnetometer placement. A spherical model was fitted to the digitized shape of the head of each subject, and the location, orientation and amplitude of a best-fitted single equivalent current dipoles (ECD) were estimated at each time point.

Magnetic resonance imaging (MRI) were obtained using a GE Signa 1.0 T system. The T1-weighted coronal and axial images with contiguous 3 mm slice thickness were adopted for overlays with ECD sources detected by MEG.

Results

(1) MEGs following painful electrical stimulation

Four main deflections and their counterparts were clearly identified in all subjects in all "weak", "moderately painful" and "very painful" sessions less than 100 msec in latency (Fig .1). The position of ECD of each deflection was located around the hand sensory area of the SI in the left hemisphere which was contralateral to the stimulated finger.

Two main middle-latency deflections, N100m-P100m and P250m-N250m, were identified only in the "moderately painful" and "very painful" sessions, and they were much larger in the latter session (Fig. 1). Their ECDs were located around the superior bank of the Sylvian fissure, probably in the second sensory cortex (SII) in bilateral hemispheres. The following large deflection, P250m-N250m, was clearly identified in 4 of 5 subjects. However, the position of their ECDs was unexpected from the early deflections. When the probe was centered at the C3 or C4, ECDs were located in bilateral SII (Fig. 2) like the N100m-P100m. When the probe was centered at the Cz position, clear polarity-reversal deflections were also identified in 4 subjects and their ECDs were located along the interhemispheric fissure, probably in the cingulate cortex (Fig. 2).

When the probe was centered at the Fz and Pz positions which covered the mid-frontal and mid-parietal areas, no clear polarity-reversed deflections were identified. No consistent ECD considered to be generated in the mid-frontal and mid-parietal areas were identified.

When the lower limb was stimulated, the results were similar to the results following upper limb stimulation in terms of the waveform and ECD location, although the latency was different.

(2) MEGs following painful CO_2 laser stimulation

Consistent, clear magnetic fields were identified in bilateral cerebral hemispheres following the stimulation of each hand and foot in 4 of 5 subjects. Their duration was relatively long, being about 80-100 msec. The peak latencies varied in each subject, being about 180- 210 and 220-250 msec for arm and foot stimulation, respectively. No definite difference in latency was identified in either hemisphere. For example, when the probe was centered around C4, the latencies following the left arm and the right arm stimulation were similar. The amplitude following arm stimulation was 2 to 3 times larger than that following leg stimulation in each subject. The position of ECDs was located along the superior bank of the Sylvian fissure, around the SII in bilateral hemispheres. Their positions were very similar to those of the N100m-P100m and N250m-P250m following painful electrical stimulation (see Fig. 2). However, consistent ECD was identified in neither SI corresponding to the stimulated area nor the cingulate cortex, even when the probe was centered at the Fz, Cz and Pz.

Discussion

To our knowledge, only a few systematic studies of MEG following painful stimulation have been reported. Hari et al. [12] studied the painful electrical stimulation of tooth pulp, Huttunen et al. [13] used CO_2 stimulation of the nasal mucosa, Hawland et al. [14] applied high-intensity painful electric finger stimulation, and recently Laudahn et al. [15] reported MEGs following painful CO_2 laser stimulation applied to the temple. In the present studies, we used 2 sets of a large array of the magnetometer with a small intra-coil distance, and used two different ways of painful stimulation to elucidate the activated areas in the brain following painful stimulation in man in more detail, and found the ECDs of the pain-related components in SII in bilateral hemispheres and in the cingualate cortex. These findings were generally compatible with the previous studies [12-15], in which ECDs were found in SII or the surrounding areas, but activation in the cingulate cortex has not been reported previously. It may be due to the difference of method, because they did not focus on the activities in the cingulate cortex.

Although the functions of SII are not clear, SII probably provides for a very fundamental somatosensory capacity because this area has been found in nearly every mammalian species, including animals lacking extensive manipulative skills [16]. SII is the first cortical stage of the somatosensory system in which neurons process information from both sides of the body [16]. Therefore, it is reasonable to assume that the bilateral SII are activated following painful stimulation.

Cerebral activities following the initial excitation of neurons in SII to painful stimulation should be very complex. Whereas the periods of the large deflections, P250m-N250m, were definitive, we did not consider that they were generated by a single ECD. We considered that multiple areas including bilateral SII and cingulate cortices were activated simultaneously and independently. Therefore, a large inter-individual difference in location of the ECD was recognized, because magnetic fields generated by multiple areas should

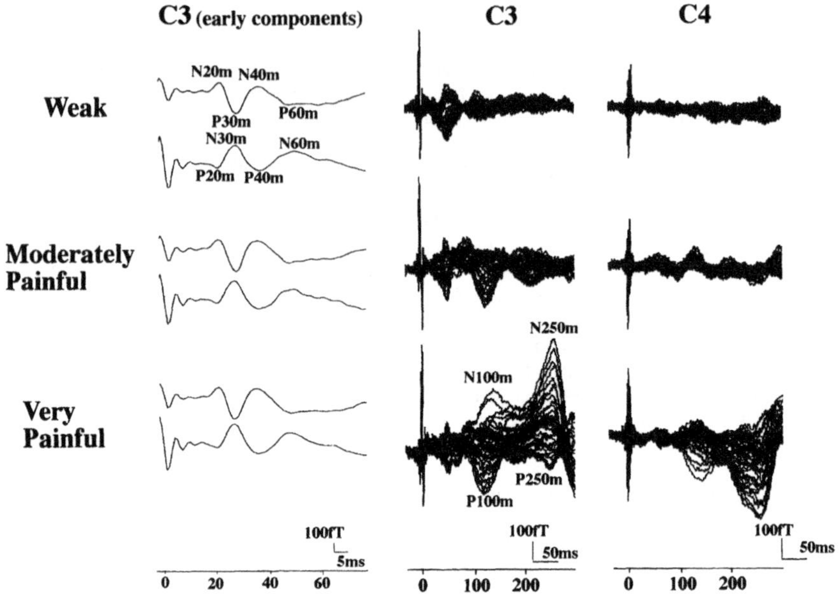

Fig. 1: MEGs waveforms recorded in C3 and C4 positions following "weak", "moderately painful" and "very painful" electrical stimulation applied to the right finger in a subject. The left figure shows the early components recorded at 2 representative channels at the C3 position. No significant difference was found in each condition. The middle and the right figure shows the superimposition of all 37 channels in each condition. The N100m-P100m and M250m-P250m were identified only in "moderately painful" and "very painful" sessions, being much larger in the latter.

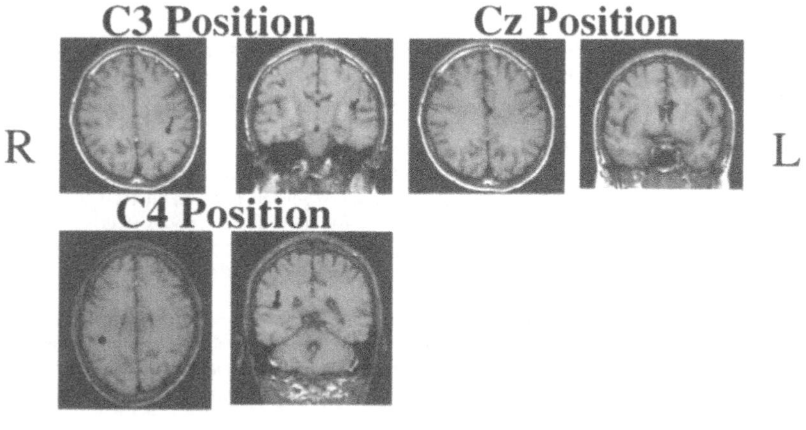

Fig. 2: Localization of representative ECDS of the N250m-P250m following very painful electrical stimulation applied to the right finger recorded at the C3, C4 and Cz positions overlapped on axial and coronal views of MRI. They were located in bilateral SII and the cingulate cortex.

interfere with each other. Positron emission tomography (PET) has demonstrated that painful heat stimuli causes significant activation of SI in the hemisphere contralateral to the stimulated hand, bilateral SII and the cingulate cortex [17], fundamentally compatible with the present study. The reason why the activation in the cingulate cortex was not found following CO_2 laser stimulation was unknown, but the small S/N ratio may be one reason.

In conclusion, when we receive painful stimulation, bilateral SII are initially activated. It is now not clear whether SII are activated by signals receiving from SI or by signals reaching directly from the thalamus. Then, multiple areas including SII and cingulate cortex are activated simultaneously and independently.

References

[1] Kitamura, Y., Kakigi, R., Hoshiyama, M., Koyama, S., Shimojo, M. and Watanabe, S. Pain-related somatosensory evoked magnetic fields. Electroencephalogr. Clin. Neurophysiol., 1995; 95: 463-474..

[2] Kitamura, Y., Kakigi, R., Hoshiyama, M., Koyama, S., Shimojo, M. and Watanabe, S. Pain-related somatosensory evoked magnetic fields following lower limb stimulation. 1996, submitted.

[3] Kakigi, R., Koyama, S., Hoshiyama, M., Kitamura, Y., Watanabe, S. and Shimojo, M. Pain-related magnetic fields following painful CO_2 laser stimulation in man. Neurosci. Lett., 1995, 192: 45-48.

[4] Kakigi, R., Koyama, S., Hoshiyama, M., Kitamura, Y., Watanabe, S. and Shimojo, M. Pain-related brain responses following CO_2 laser stimulation: magnetoencephalographic studies. In: Hashimoto, I., Visualization of Information Processing in the Human Brain, Amsterdam, Elsevier, 1996, in press.

[5] Kakigi, R. Somatosensory evoked magnetic fields following median nerve stimulation. Neurosci. Res., 1994, 20: 165-174.

[6] Kakigi, R., Koyama, S., Hoshiyama, M., Watanabe, S., Shimojo, M. and Kitamura, Y. Gating of somatosensory evoked responses during active finger movements: magnetoencephalographic studies. J. Neurol. Sci., 1995, 128: 195-204.

[7] Kakigi, R., Hoshiyama, M., Koyama, S., Shimojo, M., Kitamura, Y. and Watanabe, S. Topography of somatosensory evoked magnetic fields following posterior tibilal nerve stimulation. Electroencephalogr., Clin. Neurophysiol., 1995; 95: 463-474..

[8] Hoshiyama, M., Kakigi, R., Koyama, S., Kitamura, Y., Shimojo, M. and Watanabe, S. Somatosensory evoked magnetic fields after mechanical stimulation of the scalp in humans. Neurosci. Lett., 1995, 195: 29-32.

[9] Hoshiyama, M., Kakigi, R., Koyama, S., Kitamura, Y., Shimojo, M. and Watanabe, S. Somatosensory evoked magnetic fields following stimulation of the lip in humans. Electroencephalogr. Clin. Neurophysiol., 1996, in press.

[10] Shimojo, M., Kakigi, R., Hoshiyama, M., Kitamura, Y., Koyama, S. and Watanabe, S. Intracerebral interactions caused by bilateral median nerve stimulation in man: a magnetoencephalographic study. Neurosci. Res., 1996; 24: 175-181.

[11] Shimojo, M., Kakigi, R., Hoshiyama, M., Kitamura, Y., Koyama, S. and Watanabe, S. Differentiation of receptive fields in the sensory cortex following stimulation of various nerves of the lower limb in man. J. Neurosurg., 1996, in press.

[12] Hari, R., Kaukoranta, E., Reinikainen., K, Huopaniemie, T. and Mauno, J. Neuromagnetic localization of cortical activity evoked by painful dental stimulation in man. Neurosci. Lett., 1983; 42: 77-82.

[13] Huttunen, J., Kobal, G,. Kaukoranta, E. and Hari, R. Cortical responses to painful CO_2 stimulation of nasal mucosa; A magnetoencephalographic study in man. Electroencephalogr. Clin. Neurophysiol., 1986, 64: 347-349.

[14] Howland, E.W., Wakai, R.T., Mjaanes, B.A., Balog, J.P. and Cleeland, C.S. Whole head mapping of magnetic fields following painful electric finger shock. Cogn. Brain Res., 1995, 2: 165-172.

[15] Laudahn, R., Kohlhoff, H. and Bromm, B. Magnetoencephalography in the investigation of the cortical pain processing system. In: Bromm, B., Desmedt, J.E., Pain and the Brain. From nociception to Cognition. Advances in Pain Research and Therapy, Vol. 22, New York, Raven Press, 1995, 267-282.

[16] Burton, H. Second sensory cortex and related areas. In: Jones, E.G., Peters, A.,. The Cerebral Cortex, Vol. 5, New York: Plenum Press, 1986: 31-98.

[17] Tarbot, J.D., Marrett, S., Evans, A.C., Meyer, E., Bushnell, M.C., Duncan, G.H. Multiple representations of pain in human cerebral cortex. Science, 1991; 251: 1355-1358.

Normalized N100m Latency of the Auditory Evoked Fields After Surgical Removal of Temporal Lobe Gliomas

Kanno, A.[1], Nakasato, N.[1,2], Ohtomo, S.[2], Seki, K.[2], Kawamura, T.[2], Fujita, S.[4], Kumabe, T.[2], Kayama, T.[3], and Yoshimoto, T.[2]

MEG Laboratory, Kohnan Hospital[1], Sendai, Japan., Department of Neurosurgery, Tohoku University School of Medicine[2], Sendai, Japan; Department of Neurosurgery, Yamagata University School of Medicine[3], Yamagata, Japan; and R&D Center, Osaka Gas Co. Ltd.[4] Osaka, Japan

Introduction

Auditory evoked potentials have been measured in patients with several temporal lobe diseases. The N100 responses disappear in patients with lesions on bilateral superior temporal gyrus[1]. However, separation of a unilateral abnormality in evoked potentials, is difficult due to the spearing effect by tissue layers with inhomogeneous electric conductivities. Activity on the normal hemisphere may interfere with abnormal activity on the diseased hemisphere. Magnetoencephalography (MEG) is known to be less influenced by the inhomogeneous head conductivity. We have found that the whole head MEG is especially suitable to identify differences in bilateral cerebral function [2-4].

MEG detects N100m as the most prominent peak of auditory evoked fields (AEFs). Both left and right monaural stimuli cause a two-dipole pattern of N100m over the bilateral hemispheres with a latency of about 100 ms. In normal subjects, the contralateral N100m is known to appear earlier than the ipsilateral response to the stimulated ear. N 100m dipoles are localized on the upper surface of temporal lobes [2-5].

There have been only few reports on clinical application of the AEFs. Disappearance of the unilateral N100m response was reported in patients with temporal lobe infarctions[5]. In this study. we measured AEFs to evaluate auditory function in two patients with temporal lobe gliomas.

Materials and Methods

Two patients with left temporal gliomas participated in the present study.

Case 1 is a 27-year-old man with mild aphasia. MRI revealed a cystic tumor in the left temporal lobe. After the first surgery and radio-chemotherapy, the patient returned to work with no language or hearing disturbances. Histological examination indicated anaplastic astrocytoma. Six month after the surgery, recurrence of the tumor was discovered. Left anterior temporal lobectomy was performed for total removal of the tumor. The patient had no hearing or language disturbances afterwards.

Case 2 is a 45-year-old man with mild sensory aphasia. MRI revealed a cystic tumor in the left temporal lobe. The tumor was totally removed and histological examination indicated anaplastic astrocytoma. After radio-chemotherapy, the patient returned home with persistent mild aphasia.

AEFs were measured before and after each surgery, using a helmet-shaped 66-channel MEG system (CTF Systems-Osaka Gas) linked to MRI[2,3].The MEG signals were averaged based on 50 monaural stimuli of 2000 Hz tones. We measured peak latency of the N100m response in each hemisphere. All N100m latencies were compared to our normal range obtained from 20 male volunteers: i) for the left hemispheric N100m, the contralateral response was 86.9±9.4 ms (mean±s.d.) and the ipsilateral response was 100.1±7.3 ms; and ii) for the right hemispheric N100m, the contralateral response was 84.2±9.0 ms and the ipsilateral response was 98.2±9.3 ms [2]. Finally, source locations of the N100m was estimated using single- or double-dipole models. Dipole positions were superimposed on the MRI for each subject.

Results

All AEF measurements detected clear, bilateral N100m responses in both patients. Dipole estimation indicated the bilateral temporal planes on MRI (Figs.1 and 3). Before the first surgery, both patients had delayed N100m latencies in their left, lesion side, hemispheres. After the first surgery, left N100m latencies shortened into the normal range (Figs. 2 and 4). The contralateral N100m latency indicated more clear-cut recovery than the ipsilateral response.

Discussion

The present study demonstrated the unilateral abnormality of the auditory function in patients with temporal lobe gliomas. Although several pathways have been proposed for N100m responses, we believe the cortical lesion introduced delay in the N100m latencies. At least in these two patients, delayed N100m was due to mass effect rather than invasion of the tumors, because the N100m latencies normalized after removal of the tumors. Our results further suggest that AEF can be applied to evaluation of auditory function in patients with other types of brain lesions. Our helmet-shaped MEG system is especially useful for clinical application, because of the simultaneous coverage of two cerebral hemispheres and short measurement time.

References

[1] Woods, D.L., Clayworth, C.C., Knight, R.T., Simpson, G.V., and Naeser, M.A. Generators of middle and long-latency auditory evoked potentials: implications from studies of patients with bitemporal lesions. Electroencephalography and Clinical Neurophysiology, 1987, 68: 132-148.

[2] Nakasato, N., Fujita, S., Seki K., Kawamura T., Matani, A., Tamura, I., Fujiwara, S., and Yoshimoto, T. Functional localization of bilateral auditory cortices using an MRI-linked whole head magnetoencephalography (MEG) system. Electroencephalography and Clinical Neurophysiology, 1995, 94: 183-190.

[3] Nakasato, N., Fujita, S., Matani, A., Tamura, I., Fujiwara, S., and Yoshimoto, T. Clinical application of the whole head MEG: Auditory evoked response in patients with intracranial structural lesions. In: Baumgartner, C., Deecke, L., Stroink, G., and Williamson, S.J. Biomagnetism: Fundamental Research and Clinical Applications, Amsterdam, Elsevier/IOS Press, 1995, pp 186-190.

[4] Kanno, A., Nakasato, N., Fujita, S., Seki, K., Kawamura, T., Ohtomo, S., Fujiwara, S., and Yoshimoto, T. Right hemispheric dominance in the auditory evoked magnetic fields for pure-tone stimuli. Electroencephalography and Clinical Neurophysiology, in press.

[5] Mäkelä, J.P. Auditory evoked magnetic fields in stroke. Physiological Measurement, 1993, 14: 51-54.

Acknowledgements

This work has been supported in part by Grants-in-Aid for Scientific Research No.03404042 and No.06404050 from the Ministry of Education, Science and Culture of Japan; and Magnetic Health Science Foundation.

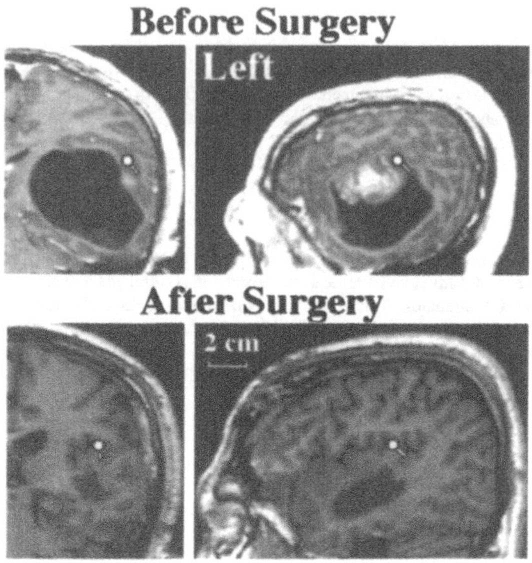

Fig. 1 N100m dipole positions before and after the first surgery in Case 1. Circles and bars indicate dipole positions and orientations of the left hemispheric N100m after contralateral ear stimuli. Note that stable dipole localizations on the upper surface of the temporal lobes, both with and without cystic tumor, and before and after craniectomy.

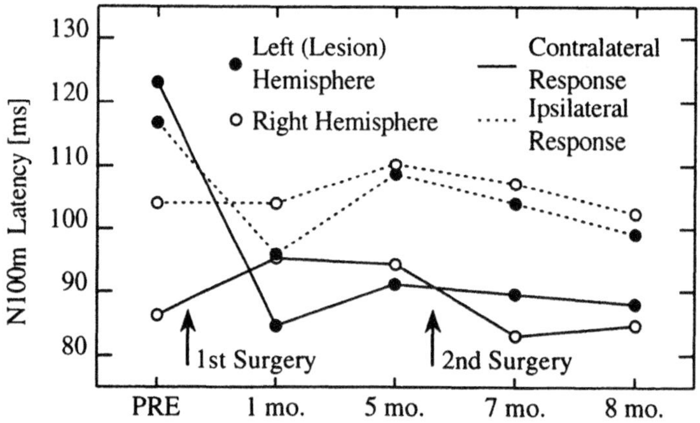

Fig. 2 N100m latency of AEFs in Case 1. Latency of the left hemispheric N100m due to contralateral ear stimuli (123.2 ms) was longer before the first surgery but normalized (88.0 ms) afterwards.

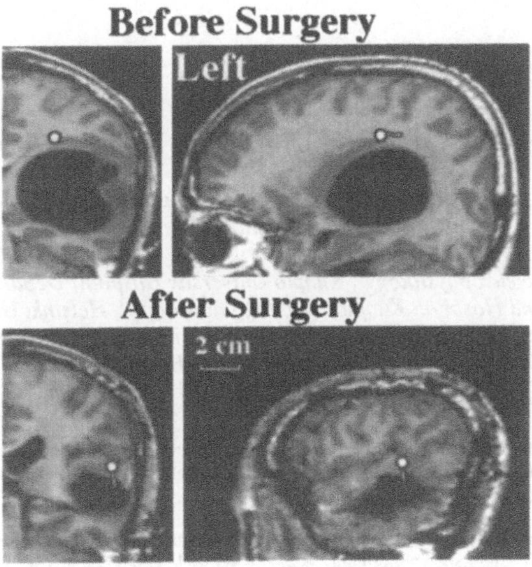

Fig. 3 N100m dipole positions before and after the surgery in Case 2. Circles and bars indicate dipole positions and orientations of the left hemispheric N100m due to contralateral ear stimuli. Note that stable dipole localizations on the upper surface of the temporal lobes, both with and without cystic tumor, and before and after craniectomy.

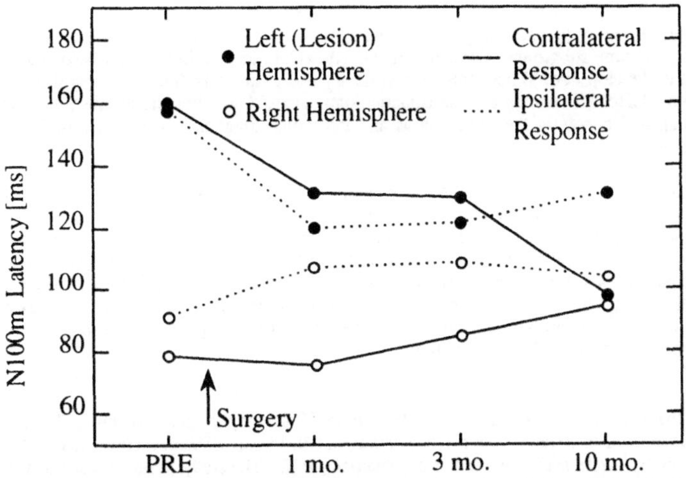

Fig. 4 N100m latency of AEFs in Case 2. Latency of the left hemispheric N100m due to contralateral ear stimuli (161.6 ms) was longer before the surgery but normalized 10 months after the surgery (99.2 ms).

Reversed Hemispheric Asymmetry of Auditory N100m in Schizophrenics

Karhu, J.[1], Tiihonen, J.[2], Pekkonen, E.[3,4], Katila, H.[5], Huotilainen, M.[3,4], Virtanen, J.[3,4], and Ilmoniemi, R.[3]

Department of Clinical Neurophysiology[1], Kuopio University Hospital; Department of Forensic Psychiatry[2], Niuvanniemi Hospital, Kuopio; BioMag Laboratory[3], Helsinki University Central Hospital; Cognitive Psychophysiology Research Unit[4], University of Helsinki; Department of Psychiatry[5], Helsinki University Central Hospital, Finland

Introduction

Human brain hemispheres are asymmetric. Individual anatomical differences involve widespread cortical and subcortical areas and are more prominent among males [1,2]. The interhemispheric asymmetry is particularly clear in temporal lobe structures and speech areas [3,4]).

Schizophrenia has been recently associated with a large number of anatomical abnormalities. In most structural imaging studies, the only well established abnormality is lateral ventricle enlargement with very limited diagnostic usefulness. Practically all imaging studies have related a single symptom or category of symptoms to a single brain region, with the exception of a magnetic resonance imaging (MRI) study showing thalamic disorganization in schizophrenia and attempting to explain multiple symptoms by dysfunction of this main relay station in brain [5]. So far, the most clearly defined *in vivo* findings show alterations in temporal lobe and limbic structures and have been associated with positive symptomatology of schizophrenia, which often modulates perception at early stages. For example, schizophrenics have decreased volumes of superior temporal gyri (STG) [6], and auditory N100m, which is generated mainly in STG, is altered by auditory hallucinations [7].

Post mortem [8], structural imaging [9], and magnetoencephalographic (MEG) studies [10] have shown a loss of temporal area asymmetry in some subgroups of schizophrenia. To explain this finding, it has been suggested that schizophrenics may fail to develop normal hemispheric asymmetry due to an early failure in neural development, possibly because of a gene defect influencing the hemispheric dominance [8]. Several structural studies have given contradictory results [11], and currently the implications and individual diagnostic value of possible abnormal lateralization remain elusive.

We evaluated the lateralization of primary auditory areas in temporal cortex by comparing the source locations of magnetic auditory N100m in schizophrenics and in normal subjects. MEG and electrocorticography recordings have shown that auditory evoked responses at 18–50 ms are generated in the primary auditory cortex in Heschl's gyrus and that magnetic 100 ms-responses have major source areas in or near primary auditory cortex [12]. Since N100m reflects the coherent activation of a functional neuronal network, it may in principle be more sensitive to subtle disorganization of functional anatomy than other structural and metabolic *in vivo* measures are. In addition to the hypothesis of abnormal hemispheric lateralization, we also wanted to test the diagnostic usefulness of simple and rapid MEG recording and analysis methods in scizophrenia.

Methods

We compared the source locations of auditory N100m in 10 schizophrenics (age range 19–34 years; 6 males; recordings were made within 10 days from hospitalization) and in 14 normal subjects (age range 19–50 years; 8 males). Psychiatric diagnoses were made by a senior psychiatrist (H. K.). The study protocol was accepted by the ethical committees in Kuopio and Helsinki University Hospitals and volunteers provided their written informed consent after a description of the study.

Each auditory stimulus block contained standard (80%) tone bursts (duration 50 ms; 0.8 kHz) and randomly embedded deviant tones (duration 25 ms; 0.8 kHz), which were delivered monaurally wih an interstimulus interval of 2.5 s. The subjective hearing threshold was measured before the recordings and the stimulus intensity was adjusted to 60 dB over the threshold for each ear. The subjects watched silent video presentations during the experiments and were instructed to ignore the stimuli.

Auditory evoked fields (AEFs) were measured with a 122-channel whole-scalp covering magnetometer (Neuromag Ltd.) in a magnetically shielded room (Euroshield Ltd.). The recording passband was 0.03–100 Hz with a sampling rate of 397 Hz. The accurate position of the subject's head inside the helmet-shaped magnetometer was detected by measuring the magnetic field produced by three marker coils attached to the scalp. Source locations were modelled by single equivalent current dipoles (ECDs) [13], which were determined with a least-squares fit at 1-ms-intervals from 80–130 ms for the responses elicited by contralateral standard tone stimulation. The calculation was done for each hemisphere separately utilizing a spherical head model and a subset of 40 channels over temporal brain areas (*cf.* Fig. 1).

Only ECDs which explained more than 85% of the recorded dipolar magnetic field and had a stable field pattern for over 10 ms around N100m deflection were accepted. The analysis was done blindly to the entire studied group and the subjects were divided to "patient" and "control" groups for further analysis only after completing the source localization routine.

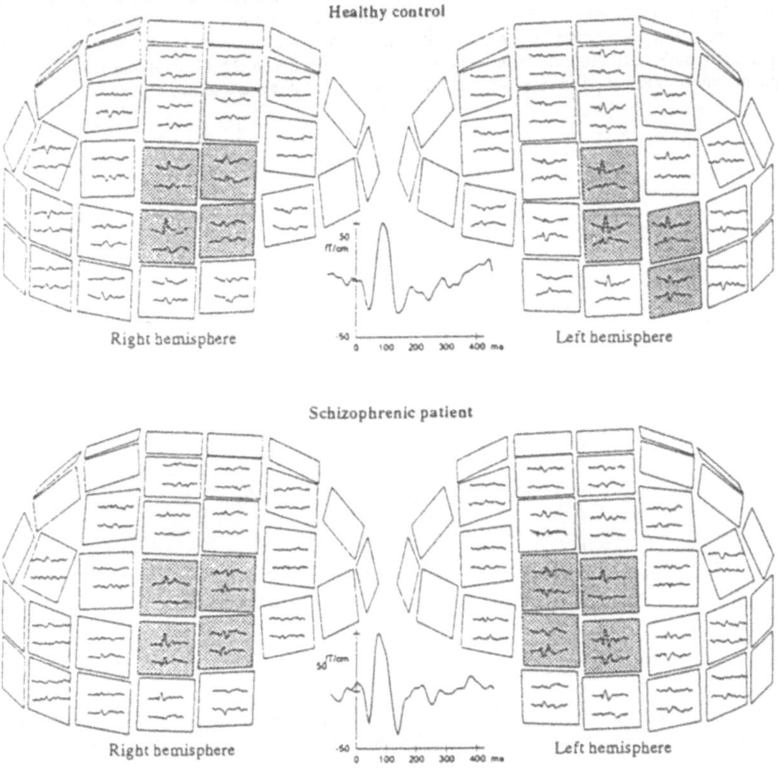

Figure 1. The averaged AEFs elicited by contralateral stimulation in a normal subject (above) and in a schizophrenic (below). The number of averages is about 100 and the responses have been digitally low-pass filtered at 30 Hz. The channel with the largest N100m is shown in the insert. The subsets of channels which show AEF waveforms in the figure were utilized to calculate ECDs for N100m deflections. Four channel pairs with the largest total signal in each hemisphere have been shaded.

Results

Fig. 1 shows the AEFs elicited by contralateral standard tone stimulation over both hemispheres in a normal subject (above) and in a schizophrenic (below). Since the maximum total signal ($\sqrt{(\partial B_r/\partial x)^2 + (\partial B_r/\partial y)^2}$) of planar gradiometers is observed directly above a cerebral source area, the location of the channels with largest N100m gives a rough estimate of its cortical source location. In the normal subject, these channel pairs are slightly posterior and inferior in the left hemiphere when compared with the right hemisphere, which suggests more posterior source(s) for N100m in the left hemisphere. In the schizophrenic, the same channel pairs show the largest N100m in both hemispheres.

N100m sources could be localized by a single ECD reliably in both hemispheres in 8 schizophrenics and in 12 normal subjects (Goodness-of fit values were in all subjects over 78% for N100m sources. Two normal subjects and two patients were rejected from further analysis because of the preset 85% limit for the acceptance of ECD.) All normal subjects showed more posterior N100m sources to contralateral ear stimulation in the left temporal lobe (mean antero-posterior difference 6.5 mm; range 3–15 mm; antero-posterior was defined as a line passing the nasion and perpendicular to the line connecting preauricular points), whereas 5 patients showed a reversed interhemispheric asymmetry between the temporal lobes (mean antero-posterior difference 6.0 mm; range 1–8 mm). Both schizophrenic females had more anterior location of N100m sources in the left temporal lobe. The individual inter-hemispheric antero-posterior differences in the N100m source locations are shown in Fig. 2.

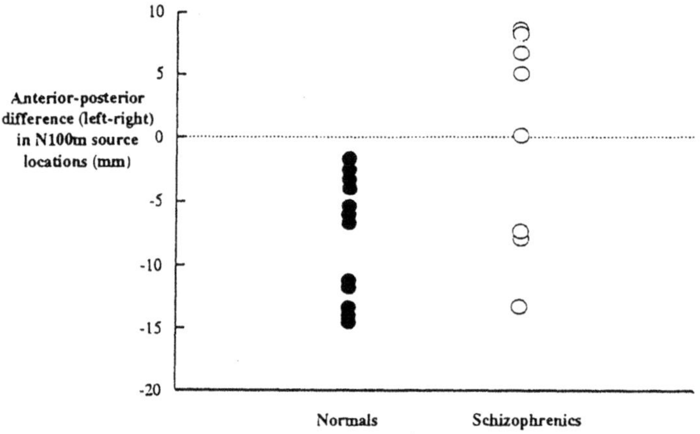

Figure 2. Interhemispheric differences (left minus right, mm) in the anterior-posterior location of the N100m sources. Normal subjects are depicted by filled circles and schizophrenics with open circles.

Discussion

These preliminary results suggest strongly that abnormalities in the individual functional anatomy of temporal structures can be detected by MEG in non-chronic schizophrenics. The interhemispheric asymmetry in the anterior-posterior location of auditory N100m equivalent sources was reversed in 50% of the studied schizophrenics. Since these sources are activated by any auditory input and probably reflect the activation of primary auditory cortices, the finding may reflect a fine-grained anatomical anomaly in the auditory areas of temporal lobes, caused for example by a neural migration disorder.

Studies of larger patient groups and an evaluation of schizophrenics with different spectra of clinical symptoms are obviously essential. We modelled the N100m deflection as a single point-like source, and thus we cannot rule out the possibility of diminished or increased activity in one of several underlying sources which could in principle lead to a source location shift in schizophrenia. Since this would require changes in one hemisphere only to cause the observed interhemispheric source location reversal, the explanation does not seem probable. If the specificity of our finding remains high in the future studies (at present 100 %), this simple auditory test may prove to be very helpful in the diagnostics and screening of high-risk populations.

References

[1] Galaburda, A.M., LeMay, M., Kemper, T., and Gershwind, N. Right-left asymmetries in the brain, Science, 1978, 199: 852–856.

[2] Weinberger, D.R., Luchins, D.J.,Morihisa, J., and Wyatt, R.J. Asymmetrical volumes of the right and left frontal and occipital regions of the human brain, Annals of Neurology, 1982: 97–100.

[3] Gerschwind, N. and Levitsky, W. Human brain: left-right asymmetries in the temporal speech region, Science, 1968, 161: 186–187.

[4] Witelson, S.F. and Kigar, D.L.Sylvian fissure morphology and asymmetry in men and women: bilateral differences in relation to handedness in men, Journal of Comparative Neurology, 1992, 323: 363–340.

[5] Andreasen, N.C., Arndt, S., Swayze II, V., Cizadio, T., Flaum, M., O'Leary, D., Ehrhardt, J.C., and Yuh, W.T.C. Thalamic abnormalities in schizophrenia visualized through magnetic resonance image averaging, Science, 1994, 266: 294–298.

[6] Shenton, M.E., Kikinis, R., Jolez, F.A., et al. Abnormalities of the left temporal lobe and thought disorder in schizophrenia. A quantitative magnetic resonance imaging study, New England Journal of Medicine, 1992, 327: 604–612.

[7] Tiihonen, J., Hari, R., Naukkarinen, H., Rimon, R., Jousmäki, V., and Kajola, M. Modified activity of the human auditory cortex during auditory hallucinations, 1992, American Journal of Psychiatry, 149: 255–257.

[8] Crow, T.J., Ball, J., Bloom, S.R., et al. Schizophrenia as an anomaly of development of cerebral asymmetry, Archives of General Psychiatry, 1989, 46: 1145–1150.

[9] Luchins, D.J., Weinberger, D.R., and Wyatt, R.J. Schizophrenia and cerebral asymmetry detected by computed tomography, American Journal of Psychiatry, 1982, 139: 753–757.

[10] Reite, M., Teale, P., Godstein, L., Whalen, J., and Linnville, S. Late auditory magnetic sources may differ in the left hemisphere of schizophrenic patients, Archives of General Psychiatry, 1989, 46: 565–572.

[11] Kuluych, J.J., Vladar, K., Fantie, B., Jones, D., and Weinberger, D.R. Normal asymmetry of planum temporale in patients with schizophrenia, British Journal of Psychiatry, 1995, 166: 742–749.

[12] Hari, R., Aittoniemi, K., Järvinen, M.-L., Katila, T. and Varpula, T. Auditory evoked transient and sustained magnetic fields of the human brain. Experimental Brain Research, 1980, 40:23724.

[13] Hämäläinen, M., Hari, R., Ilmoniemi, R.J., Knuutila, J., and Lounasmaa, O.V. Magnetoencephalography - theory, instrumentation, and applications to noninvasive studies of the working human brain. Reviews of Modern Physics, 1993, 65: 413–498.

Neuromagnetic Identification of the Somatosensory Cortex in Cases with Arteriovenous Malformation Adjacent to the Central Sulcus

Kawamura, T.[1], Nakasato, N.[1,2], Ohtomo, S.[1], Seki, K.[1], Kanno, A.[2], Fujita, S.[3], Fujiwara, S.[1,2] and Yoshimoto, T.[1]

[1]Department of Neurosurgery, Tohoku University School of Medicine, Sendai, Japan; [2]MEG Laboratory, Kohnan Hospital, Sendai, Japan; [3]R&D Center, Osaka Gas Co. Ltd., Osaka, Japan

Introduction

It is critical to preserve eloquent cortices during surgery of cerebral arteriovenous malformations (AVMs). Although MRI scans may identify "anatomical" central suclus, it is controversial whether the "functional" central sulcus can be shifted due to AVMs. Cortical recording of somatosensory evoked potentials (SEPs) can be used to recognize the central sulcus during open surgery. However, the cortical SEPs are not available during surgery for large AVMs, intravascular surgery, or stereotaxic radiosurgery. In the present study, somatosensory evoked fields (SEFs) were measured to localize the "functional" central sulcus non-invasively in cases with AVM adjacent to the central sulcus.

Method

Fourteen patients, 7 males and 7 females, aged 9 to 67 year-old (mean 39.5), participated in the study (Table.1). All cases had AVM adjacent to the central sulcus. All patients underwent three-dimensional MRI scans. Anatomical central sulcus was defined as the one-anterior-sulcus to the marginal branch of the cingulate sulcus on a paramedian sagittal MRI image. For median nerve stimuli, SEFs were measured using a helmet-shaped 66-channel MEG system (CTF Systems - Osaka Gas) linked to MRI [1]. At the peak latency of the first component (N20m), signal source was estimated using a single-dipole model. Location of the N20m dipole was superimposed on MRI. Finally, the dipole deviation, in the anterior-posterior direction, was measured from the posterior bank of the anatomical central sulcus.

Results

In 13 cases, the N20m dipoles in the lesion hemispheres were localized within one gyrus to the anatomical central sulcus (Fig.1). In only one case, case 3 with fronto-parietal subcortical hemorrhage, the N20m dipole disappeared in the lesion hemisphere. Even in two cases with a large AVM which occupied almost entire frontal lobe, N20m dipole was localized on the anatomical central sulcus (Fig.2).

Case	Sex	Age	Lesion	Initial Symptom
1	F	30	Rt. F	Headache and lt. facial palsy due to ICH
2	F	53	Lt. P	Rt heminumbness
3	F	47	Lt. F	Headache due to ICH
4	M	17	Lt. F	Attack of consciousness loss
5	M	50	Rt. P	Lt heminumbness due to ICH
6	F	50	Rt. F(Large)	Lt hand weakness and numbness
7	F	49	Lt. F(Large)	Rt hemiparesis and hemihypesthesia
8	F	9	Rt. F	Seizure
9	M	50	Lt. F	Headache due to ICH
10	M	23	Rt. P	Incidental
11	F	52	Lt. FP	Memory disturbance, Dysarthria
12	M	13	Rt. F	Partial seizure
13	M	67	Rt. P	Attack of consciousness loss
14	M	25	Lt. P	Incidental

Table.1 Summary of the cases with AVM adjacent to the central sulcus. F: frontal lobe, P: parietal lobe, ICH: intracerebral hemorrhage

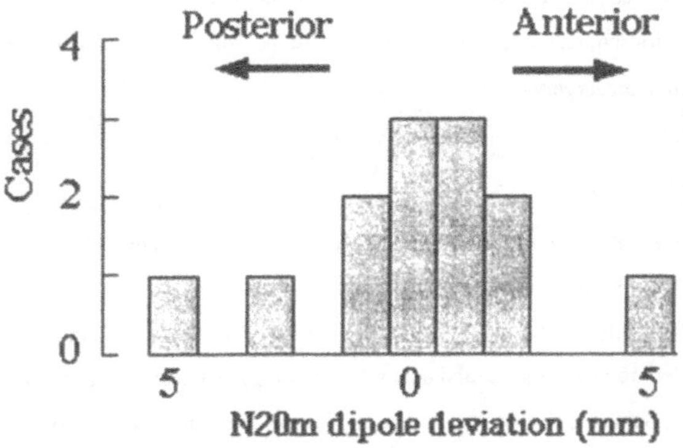

Fig.1 Anterior or posterior deviation of the N20m dipole from the surface of the posterior bank of the "anatomical " central sulcus.

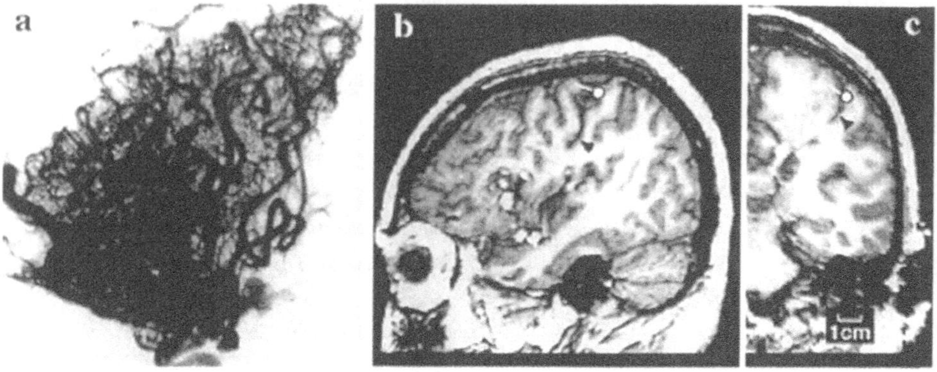

Fig.2 N20m source localization in a 49 year-old female patient (case 7) with a large AVM in the left frontal lobe. a: Lateral view of the left carotid angiogram revealing a large AVM in the left frontal lobe. b and c: N20m dipole position estimated on the "anatomical" central sulcus (arrow heads).

Discussion

Previous SEF studies indicated that the origin of the N20m is in the posterior wall of the central sulcus "area 3b" [2,3]. It is well known that mass lesions, such as brain tumors, may shift the central suclus. However, our results suggest that AVM does not shift the localization of somatosensory function. We believe cortical functions may be preserved in parenchyma adjacent to the nidus or among abnormal vessels of AVM. This non-invasive method can be applied for clinical decision making before not only conventional surgery but also intravascular surgery or stereotaxic radiosurgery.

References

[1] Kawamura, T., Nakasato, N., Seki, K., Shimizu H., Fujiwara, S., and Yoshimoto, T. Locating the central sulcus using three dimensional image fusion system of helmet shaped whole head MEG and MRI. No Shinkei Geka 22: 737-741, 1994

[2] Wood, C.C., Cohen, D., Cuffin, B.N., Yarita, M. and Allison, T. Electrical sources in human somatosensory cortex: Identification by combined magnetic and potential recordings. Science 227: 1051-1053, 1985

[3] Kawamura, T., Nakasato, N., Seki, K., Kanno, A., Fujita, S., Fujiwara, S., and Yoshimoto, T. Neuromagnetic evidence of pre - and post - central cortical sources of somatosensory evoked responses. Electroencephalography and Clinical Neurophysiology, in press.

Acknowledgements

This research was supported by Grants-in-Aid for Scientific Research No.03404042 and No.06404050 from the Ministry of Education. Science and Culture of Japan.

Time Course of Abnormal MEG Activity Associated with Transient Ischemic Attacks

Kober, H., Vieth, J. B., Stippich, C., Kassubek, J.R., Hopfengaertner, R.
Dept. of exp. Neuropsychiatry, University of Erlangen-Nürnberg, Germany

Introduction

Cerebrovascular emboli may cause irreversible deficits (brain infarcts) or reversible ones (transient ischemic attacks (TIA)). If the reason is a high grade stenosis (>70%) of the cervical internal carotid artery, this symptomatic stenosis should be treated surgically according to large North-American and European joint studies [3]. According to the location in the brain clinical symptoms may or may not occur. But not only clinically "silent"

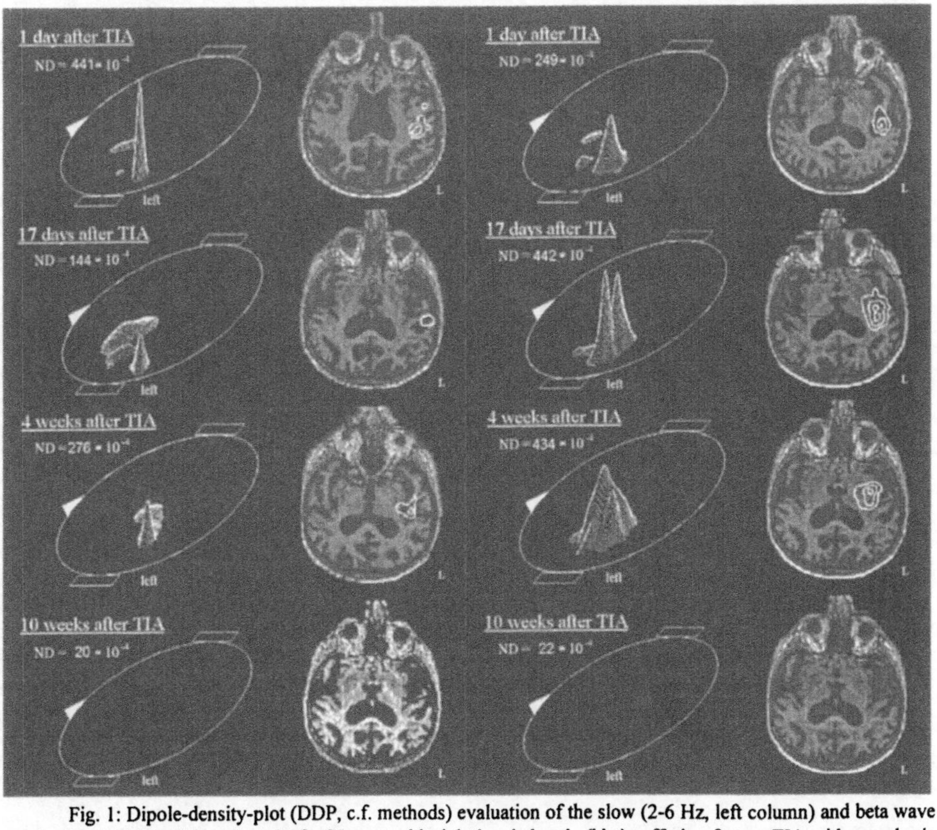

Fig. 1: Dipole-density-plot (DDP, c.f. methods) evaluation of the slow (2-6 Hz, left column) and beta wave activity (12.5-30 Hz, right column) of a 85 years old, right handed male (hkc) suffering from a TIA with an aphasia which lasted for half an hour. Upper row to lower row: time course of the normalized dipole density (ND, c.f. methods) in dipoles per ccm showing abnormal activity still four weeks after the TIA for both slow and beta wave activity. In the last measurement of the follow up study (lowest row, 10 weeks after the TIA) slow and beta wave activity have become normal.

infarcts [1] - shown by imaging techniques - but also reversible functional impairments ("silent TIA") may occur. So a "silent TIA" is another cause that a carotid stenosis is symptomatic.

It is well known that cerebrovascular lesions of the brain are the source of neuronal dysfunctions, which can produce focal slow wave activity (2-6 Hz). Recently, we could show that also focal beta wave activity (12.5-30 Hz) can be associated with these lesions [4,9]. Slow wave activity has not only been observed with *structural lesions* (infarcts), but also with TIAs [7,8]. Both, slow and beta wave activity, can be localized by magnetoencephalography (MEG) [2,7,8]. Therefore MEG is a suitable mean to investigate the time course of neuronal dysfunction due to cerebrovascular lesions.

In order to have a kind of screening test for "silent TIAs", it is of great clinical interest to know whether a "silent TIA" has occurred and where the corresponding abnormal activity is located. In order to use the MEG for this purpose we have to know how reliable the activity can be shown of a structural lesion (brain infarct) and of the non structural lesion, the TIA with clinical symptoms, in comparison to a base line group and how long the reversible effect of the TIA lasts.

Methods

Spontaneous measurements of 19 cases with TIAs (no structural lesions in the CT or MRI) and of 14 normal subjects (baseline group) were recorded with the biomagnetic 37-channel system KRENIKON[R] (Siemens). The recordings (10 min) were done sequentially over both hemispheres. In 2 cases with a TIA follow up studies were performed: 1 day, 1 week, 2.5 weeks, 4 weeks and 10 weeks after the cerebrovascular event.

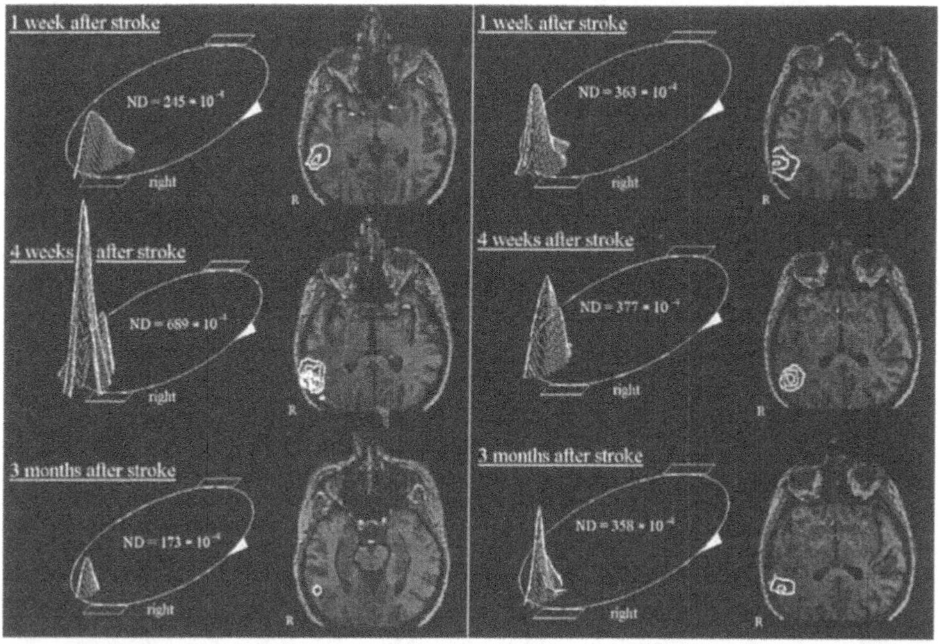

Fig. 2: Dipole-density-plot (DDP, c.f. methods) evaluation of the slow (2-6 Hz, left column) and beta wave activity (12.5-30 Hz, right column) of a 54 years old female (msf) suffering from a right temporo occipital brain infarct followed by a left side hemiparesis. Upper row to lower row: time course of the normalized dipole density (ND, c.f. methods) in dipoles per ccm. In contrast to the TIA patient (c.f. Fig. 1) abnormal activity is still found three month after the brain infarction.

After digitally filtering the whole data set from 2 to 6 Hz (delta/theta activity) and from 12.5 to 30 Hz (beta activity) the multisource problem of localizing spontaneous activity was handled by the Dipole-Density-Plot (DDP) [6], a temporal integration of localization results showing concentrations of dipoles over time in a standardized form. To use the single dipole model adequately, a Principal Component Analysis (PCA) first selects the time epochs with one dominant component (typically: 5 seconds selected out of 10 minutes).

The standardization is achieved by normalizing the dipole density to the number of dipoles with a minimal correlation of measured and calculated field distribution, used for the calculation of the 3-dimensional dipole-density distribution. Thus yielding a measure for the predominance of the focal abnormal activity over the background activity. The DDP results were inserted in form of isocontour lines of the activity's distribution into MRI slices using a transformation obtained from a contour fit algorithm [5]. The accuracy of this very precise coregistration method is better than 2 mm.

Results

Figures 1 and 2 show location and time course of the abnormal activity in the case of a TIA and a brain infarction respectively. Fig. 3 shows that the stroke activity is at least 4 times higher and the TIA activity at least two times higher than the activity of the base line group. The DDP yielded abnormal activity still 4 weeks after the event for the TIA and 3 months for the infarction. The DDP localizations of the slow and the beta wave activity are well in accordance with the clinical symptoms in both cases presented in Fig. 1 and Fig. 2. For the infarct patient suffering from a left side hemiparesis the localizations were found close to the right temporo occipital infarction area in all measurements. For the right handed TIA patient suffering from an aphasia the localizations were found in the speech dominant (left) hemisphere of the brain.

In all TIA cases we found either the center of the slow and the fast wave activity in the correct hemisphere with a different time course and with a different extent and shape of the DDP. But in all cases the TIA activity reached normal values between 4 and 10 weeks.

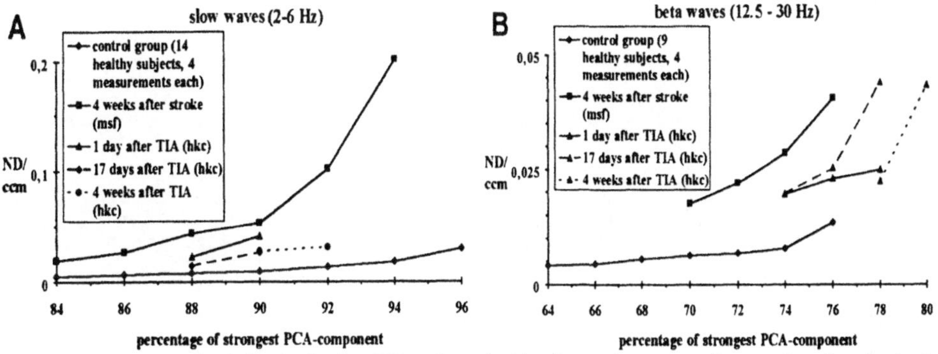

Fig. 3: Normalized dipole density (ND, c.f. methods) of a stroke patient four weeks after the brain infarction (c.f. Fig. 2) and a TIA patient one day, 17 days and four weeks after the incident (c.f. Fig. 1) compared to the average dipole density of a healthy baseline group for different percentages of predominace of the strongest principal component. A: Theta/delta activity. B: Beta activity. The pathologic brain activity induced by the brain infarct has the highest intensity. For the TIA patient the intensity of the dipole density is above the baseline group still four weeks after the occurence of the cerebrovascular lesion. For all cases an increase of ND with the PCA-percentage of the strongest signal component (dominance of a single source) can be observed. This is a consequence of the more adequately applied dipole model to the PCA-selected time slices. Thus reducing the scattering ("biological noise") of dipole locations.

Discussion

Our study shows abnormal activity for at least 4 weeks after a TIA. This would be sufficient for a screening of "silent TIAs" using the MEG in combination with the DDP. Thus this method could be used clinically as an additional indication for the surgical treatment of patients with a stenosis of the carotid artery. But further studies must show when the normal values are really reached by applying a more dense pattern of measurements between 4 and 10 weeks. In addition the different behavior of the slow and the fast wave band should be studied as well as the correlation to PET results.

References

[1] *Caplan L.R.* Silent brain infarcts, Cerebrovasc. Dis., 1994, 4 (Suppl 1):32-40.

[2] *Gallen C.C., Schwartz B.J., Pantev C., Hampson S., Sobel D., Hirschkoff E., Rieke K., Otis S., Bloom F.* Detection and localization of delta frequency activity in human strokes, In: *Hoke M et al.* Biomagnetism: Clinical Aspects, Amsterdam, Elsevier, 1992.

[3] Guidelines for Carotid Endarterectomy. A multidisciplinary consensus statement from the ad hoc committee, American Heart Association, Stroke, 1995, 26:188-201.

[4] *Grummich P., Vieth J.B., Kober H., Pongratz H. and Ulbricht D.* Focal spontaneous beta wave activity localized close to structural lesions, Biomed. Engineer., 1993, 38 (Suppl):197-198.

[5] *Kober H., Grummich P., and Vieth J.B.* Fit of the digitized head surface with the surface, reconstructed from MRI-tomography, Biomagnetism: Fundament. Res., Amsterdam, Elsevier, 1995.

[6] *Kober H., Vieth J.B., Grummich P., Daun A., Weise E. and Pongratz H.* The factor analysis used to improve the dipole-density-plot (DDP) to localize focal concentrations of spontaneous magnetic brain activity, Biomed. Eng. (Berlin), 1992, 37 (Suppl 2):164-165.

[7] *Vieth J.B.* Magnetoencephalography in the study of stroke (cerebrovascular accidents), In: *Sato S (Ed)* Magnetoencephalography, Adv. in Neurol. Vol. 54, New York, Raven Press, 1990.

[8] *Vieth J.B., Kober H., Weise E., Daun A., Moger A., Friedrich S. and Pongratz H.* Functional 3D localization of cerebrovascular accidents by magnetoencephalography (MEG), Neurol. Res., 1992, 14:132-134.

[9] *Vieth J.B., Kober H., Grummich P., Pongratz H. and Ulbricht D.* Slow wave and beta wave activity associated with white matter structural brain lesions, localized by the Dipole-Density-Plot (DDP), Biomagnetism: Fundament. Res., 1995, Amsterdam, Elsevier.

Preoperative Localization Of Sensorimotor Cortex: MRI versus MEG

Lewine, J.D., Baldwin N.G., Bucholz R.D., Sanders J.A., Halliday, A.L., Anson J.A., Shih, J., Stearley J., Davis, J.T., Astur, R.S., Paulson, K., and Orrison, W.W., Jr.

The New Mexico Institute of Neuroimaging The New Mexico Regional Federal Medical Center, Albuquerque, New Mexico, USA.

INTRODUCTION

Accurate knowledge of cortical functional organization is important in the surgical treatment of brain disorders, because the relevant pathology is often within, or adjacent to, cortical regions that are critical for essential sensory, motor, or cognitive functioning. Avoidance of sensorimotor cortex is especially important because damage to this area can cause a hemiparalysis that reduces the quality of a patient's post-operative life. It is well established that the central sulcus normally demarcates the boundary between primary motor and somatosensory areas, but presurgical identification of the central sulcus by magnetic resonance imaging (MRI) can be difficult, especially in neurosurgical patients where lesions can cause significant distortion of the local neuroanatomy. The present study compared central sulcus identifications using MRI versus MEG in patients that underwent subsequent intraoperative monitoring for validation of preoperative inferences.

METHODS

Subjects: Presurgical MSI mapping of sensorimotor cortex was performed with 52 neurosurgical candidates. Nine patients had been previously determined to have inoperable lesions because MR data were initially thought to indicate that the lesion could not be resected without induction of a significant hemiparalysis.

MR: One hundred twenty-eight, 1.5 mm thick, contiguous T1-weighted sagittal images were obtained using a 1.5 Tesla whole-body imager (Magnetom, Siemens, Islen, NJ). A neuroscientist (JDL) and a neuroradiologist (WWO) reviewed the full series of 2D images in an attempt to identify the central sulcus by anatomical methods. Three different 2D strategies were used, as described by Sobel and colleagues [1]. In addition, 3D surface rendered images were examined.

MEG: The MEG examination was performed using a 37-channel Magnes biomagnetometer system (Biomagnetic Technologies Incorporated, San Diego, CA) or a 122-channel whole head biomagnetometer (Neuromag Ltd., Helsinki, FI). Median nerve stimulation was accomplished using a constant current stimulator (model S88, Grass Electronics, Quincy MA). The stimulator delivered current pulses that were 200 usec in duration. The magnitude of each stimulus pulse was set to be just above the threshold needed to elicit a noticeable thumb twitch (typically 5-8 mA).

A minimum of 100 stimuli were delivered at a rate of 3/sec. MEG data epochs (50 msec prestimulus, 200 msec poststimulus) were recorded with a bandpass of 1-800 hz, at a digitization rate of more than 2Khz. After signal averaging of the data epochs, baseline correction, and additional digital filtering (10-150 hz), a dipole-in-a-sphere model was used to characterize the magnetic field pattern at the time of the 20 msec (M20) neuromagnetic response. Common fiducials were identifed for MEG and MRI data and magnetic source localization images showing the anatomical location of the M20 dipole were generated.

ECoG: Intraoperative electrocorticographic monitoring of the median nerve somatosensory evoked potential (SEP) was performed using an electrode strip or grid. Data were examined for the line of "phase-reversal" for the N20 component, and the point along the line at the intersection of a line connecting regions of maximal positive and negative potential was taken as the position of the hand representation of the somatosensory cortex. In 10 patients the spatial coordinates of this point were determined in MEG space, using a frameless stereotaxy system.

Comparison of MEG and ECoG: For all surgical patients, MEG and ECoG data were compared qualitatively to determine if the methods agreed on identification of the central sulcus. For the 10 patients where stereotaxic coordinates were obtained for the ECoG data, a quantitative comparison was made between the surface projected M20 dipole position and the surface position for the ECoG determined location of the somatosensory hand area.

RESULTS

Complete agreement in the central sulcus identifications of all 2D and 3D MR anatomical methods was found for 18 patients. For the other 34 patients, one or more of the anatomical methods failed to provide useful data for identification of the central sulcus, because the lesion obscured critical landmarks (e.g., the marginal ramus and central sulcus notch for the mid-sagittal method). Nevertheless, in 14 of these cases, there was sufficient congruence in the results from the non-obscured methods that there was confidence in the overall neuroanatomical identification of the central sulcus. The correctness of these identifications was validated by corticography in 13 cases, the 14th case not going to surgery because the lesion was located in primary motor cortex. For the remaining 20 patients (40% of the total population), the lesions caused so much distortion of the anatomy at the parietal-frontal border that identification of the central sulcus by anatomical methods was impossible, or uncertain, occasionally with disagreements between various methods (Fig. 1).

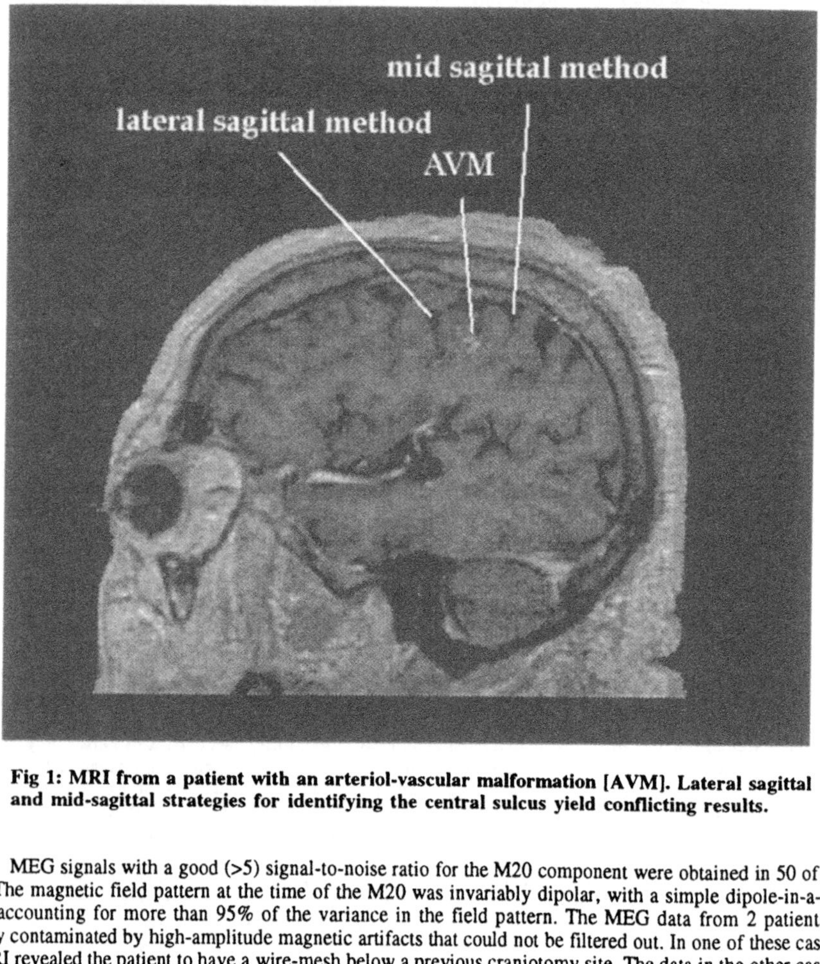

Fig 1: MRI from a patient with an arteriol-vascular malformation [AVM]. Lateral sagittal and mid-sagittal strategies for identifying the central sulcus yield conflicting results.

MEG signals with a good (>5) signal-to-noise ratio for the M20 component were obtained in 50 of the 52 cases. The magnetic field pattern at the time of the M20 was invariably dipolar, with a simple dipole-in-a-sphere model accounting for more than 95% of the variance in the field pattern. The MEG data from 2 patients were severely contaminated by high-amplitude magnetic artifacts that could not be filtered out. In one of these cases, CT and MRI revealed the patient to have a wire-mesh below a previous craniotomy site. The data in the other case were contaminated by magnetic artifacts believed to have originated from the patient's dental work which included a reconstructed tooth thought to have a metal post inside. Fortunately, in both cases, neuroanatomical methods provided for correct identification of the central sulcus as confirmed by corticography.

Forty-eight of the 50 patients with good MEG data underwent resective surgeries with electro-corticographic mapping of somatosensory cortex. In each case, ECoG validated MEG based identification of the post-central gyrus. Two patients did not have invasive surgery, because MEG indicated the patient's lesion to invade primary motor cortex. Ten patients were initially thought to have been inoperable because radiological data suggested that their lesion encroached upon primary motor cortex. MEG "confirmed" this suspicion in one case, but in the other nine, MEG indicated that the the pathological tissue could be resected without inducing significant motor deficits, because the lesion had displaced (rather than invaded) the motor cortex.

In 10 patients, precise stereotaxic coordinates were obtained for the ECoG determined location of the hand representation of somatosensory cortex. Surface projected MEG dipole locations showed good agreement with the corticography determined location of hand somatosensory cortex (mean 4.5 mm, range 2-8 mm, see Fig 2)

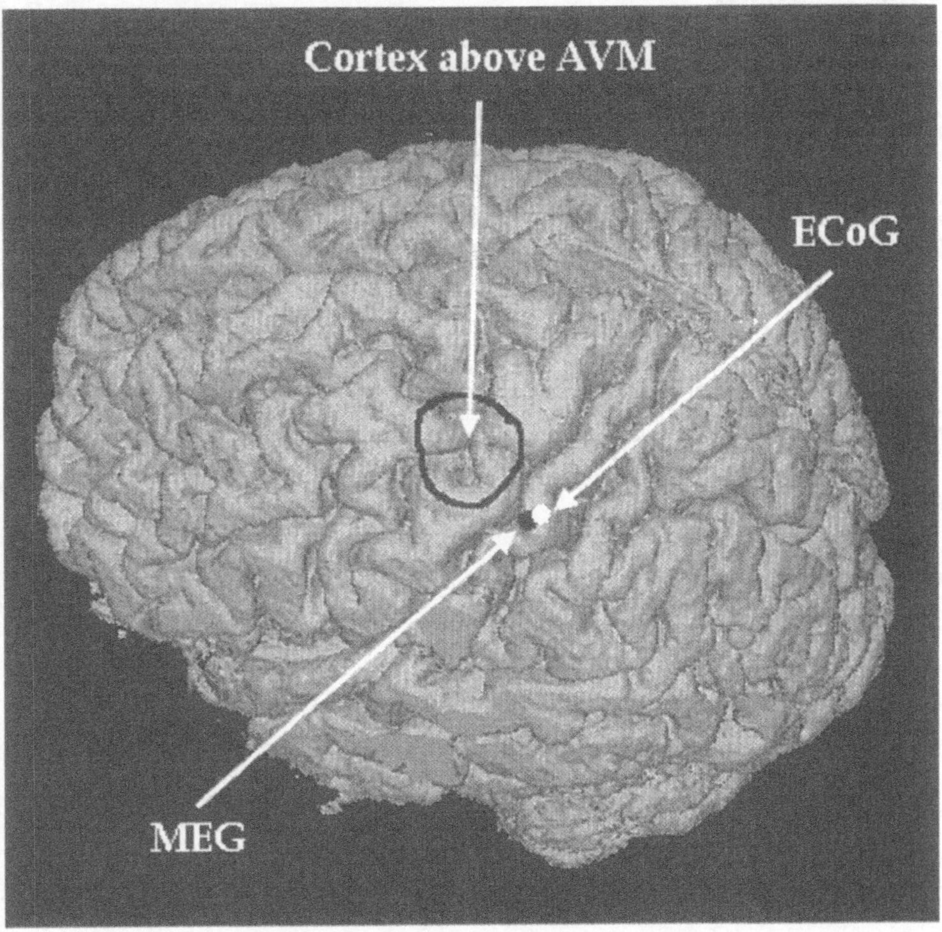

Fig 2: Preoperative MEG localization of the somatosensory cortex (black circle) shows excellent agreement with intraoperative ECoG findings (white circle). The data are from the same patient as shown in Fig 1, and they indicate that the AVM is anterior to the central sulcus and that the mid-sagittal anatomical method was correct.

DISCUSSION

MRI anatomical methods are often assumed to be useful in the noninvasive identification of the central sulcus, but the situation is complex in neurosurgical patients with lesions because of distortions of the local neuroanatomy. In this study, neuroanatomical methods were inadequate for 20 patients (38%). This rather poor ability of MRI to localize the central sulcus in patients with frontal and parietal mass lesions was markedly contrasted by the excellent MEG results in which the central sulcus was identified in 50 of 52 patients (97%) -- the two failures being caused by obvious artifacts in the data.

The clinical utility of MEG was clear in the 20 patients for whom anatomical methods had failed. Ten of these patients were originally judged to be inoperable, but in nine cases MEG indicated that the pathological tissue could be excised without induction of significant motor deficits. That the tenth patient should not have surgery was an equally important finding.

MEG is not the only functional imaging method that may be capable of providing for accurate presurgical localization of the sensorimotor areas of the brain. Individual case reports are now available for positron emission tomography (PET), functional magnetic resonance imaging (fMRI), and high-density scalp recordings of the somatosensory evoked potential. However, there have yet to be any large-scale validation studies of these methods, and each method has its limitations. For example, fMRI is very sensitive to movement artifacts, and success rate in patients with vascular malformations has been low. Dipole modeling of high density recordings of the scalp somatosensory evoked potential is often suggested as a cost-effective alternative to MEG evaluation of sensorimotor function, and in some cases this is probably true. However, neuroelectric evaluations are likely to be more severely compromised by electrical conductivity inhomogeneities at a lesion boundary, and also by skull defects from prior surgeries. Skull defects cause very severe distortions of scalp electrical potential patterns, but minimal distortions of the MEG field pattern. Of the 50 patients in the present study, 10 had significant skull defects.

CONCLUSIONS

MEG mapping of the median nerve evoked field is more effective than structural MRI for identification of the central sulcus in neurosurgical patients with lesions near the fronto-parietal junction. Preoperatively obtained MEG identification of the hand somatosensory area shows good agreement with intraoperative identification of this region and it serves as a valuable presurgical planning tool. The MEG portion of the examination takes less than 30 minutes to execute and magnetic source localization images can be made available to surgeons within minutes of completion of MEG and MRI examinations.

REFERENCES

[1] Sobel, D.F., Gallen C.C., Schwartz, B.J., et al., Central sulcus localization in humans: comparison of MRI-anatomic and magnetoencephalographic functional methods, AJNR, 1993, 14: 915-925.

Clinical MEG I: Towards a Standardized Examination

Lewine, J.D., Davis, J.T., Davis, L.E. Canive, J., Roberts, B., Graeber, D., Shih, J., Edgar, J.C., Provencal S.L., Paulson, K., Meyers, J., Christner, R., Silveri, J., Rawcliffe, N., Tessman, C., Espinosa, M., Depper, M., Sanders, J.A., and Orrison, W.W., Jr.

The New Mexico Institute of Neuroimaging, The New Mexico Regional Federal Medical Center, Albuquerque, New Mexico, USA

INTRODUCTION

Recent availability of large-array and whole-head biomagnetometer systems has finally allowed magnetoencephalography to begin to develop into the clinical tool foreshadowed by initial single-channel SQUID systems. Whereas clinical experiments used to take several hours or even days to complete, the newest generation of sensors allows for a basic neurodiagnostic examination to be completed in less than two hours.

The New Mexico Institute of Neuroimaging (NMIN), initially (1992-1994) equipped with a 37-channel biomagnetometer (BTi), and now equipped with a 122-channel whole-head biomagnetometer (Neuromag Ltd.), employs magnetoencephalography as part of the standard of diagnostic care for a wide range of patients with neurologic and psychiatric disorders. Dominant areas of direct clinical utility are in (1) presurgical mapping of sensorimotor functions, (2) characterization of epileptiform activity, and (3) characterization of the neurobiological effects of stroke and minor head trauma. Clinical research activities include investigations of (1) psychiatric conditions, (2) progressive dementias, (3) CNS effects of pharmacological agents, and (4) fetal cardiac and brain activities. Since 1992, the facility has performed examinations with more than 1500 patients.

Recently, we have developed a standardized neurodiagnostic examination that is used with most patients. This examination is divided into several sub-tests that provide for assessment of each of the three major sensory systems as well as spontaneous brain activity. In some cases, this standard examination is supplemented by additional specialized testing (e.g., addition of a P300 odd-ball paradigm in the testing of schizophrenic patients). To date, approximately 150 subjects have received the complete battery of sub-tests, with another 350 subjects having received most of the tests.

STRATEGY

The standardized examination has been developed for implementation of the 122-channel whole-head biomagnetometer system developed by Neuromag Ltd.

Staff: At NMIN, the dedicated staff is comprised of 4 technologists, 1 electronics technician, and 2 senior scientists. In addition, this staff is supported by the part-time efforts of 2 neuroradiologists, and 5% effort by each of two electroencephalographers.

Subject Preparation (time 10-30 minutes): Following completion of an informed consent document (approved by the Human Research Review Committee of the University of New Mexico) and a brief clinical interview, the patient is fitted with a flexible cap that has holders for 3 small coils that can be energized during the experiment. The positions of the coils, relative to a head coordinate system defined by preauricular points and the naision, are defined using a 3D-digitizer (polhemus). For patients with a seizure history, a specially designed electrode cap is used. This cap has 62 electrodes and can be applied in about 20 minutes. When applicable, electrode positions are also digitized. For all patients, cardiac and EOG electrodes are also applied. EEG data are collected using SynAmplifiers (NeuroScan) interfaced directly with the Neuromag unit and also with the SCAN software and hardware package of NeuroScan Inc. Following completed preparation, the subject is led into a magnetically shielded chamber and positioned under the sensor unit. Because small children often have difficulty staying still in the sensor unit, we routinely sedate with chloral hydrate or elavil.

Experimental Protocols (time 60-90 minutes): As specified below, each subject participates in 7 sub-tests. These are: (1) spontaneous - eyes open, (2) spontaneous - eyes closed, (3) spontaneous - reactivity, (4) visual, (5) contrast, (6) auditory, and (7) somatosensory. Depending upon a patient's clinical history, additional testing may be performed. All subjects also receive an MRI examination following the MEG examination. When appropriate, neuropsychological testing and SPECT examination may be performed.

Stimuli: Stimuli for evoked response studies are delivered using the STIM software package of NeuroScan Inc. For somatosensory experiments, electrodes are attached to right and left median and tibial nerves. Electrical pulses are delivered using Grass Stimulators, controlled through the STIM GenTask program. For auditory stimulation, the GenTask program is used with signals relayed through etymotic speaker units within the shielded room. The STIM GenTask program controls visual stimulation which is effected by a monitor positioned outside of a 12x12 inch hole in the shielded room.

Data Analysis: Data analysis is a two step procedure. The initial step is performed by a group of technologists (which is also responsible for patient handling and data collection). The technologists perform pre-specified spectral analyses of the continuous data. Dipole analyses for pre-specified signal components in each sensory domain is performed also.

Reporting: The technologists relay a data packet containing spectral maps, contour maps, dipole fits, activity profiles, and magnetic source localization images to senior scientists who (1) evaluate the accuracy and clinical significance of the initial analyses, (2) visually inspect continuous data for epileptiform discharges and delta activity (with multiple dipole spatio-temporal modeling and MRI mapping as needed), and (3) generates special images and a clinical report for clinicians. For presurgical cases, a complete analysis of all data (including 3D surface rendered MRI images) is made available to physicians within 24 hours. For routine clinical cases (e.g., stroke, trauma, psychiatric), a complete analysis is available within 48 hours. Evaluation of data sets with epileptiform activity is especially complex because spikes must be visually identified and then characterized using multiple dipole spatio-temporal models in order to specify propagation pathways. These cases may take as much as 1 week for complete processing.

SUBTESTS 1 & 2:
FIVE MINUTES OF CONTINUOUS DATA IN EACH OF EYES OPEN AND EYES CLOSED CONDITIONS

Objective: Identification of abnormal resting physiology, with an emphasis on epileptiform transients and abnormal low frequency magnetic activity in the delta band.

Acquisition Parameters

Length of Record:	5 minutes	(additional recording is common for patients with epilepsy)
Digitization Rate:	300 hz	
On-line Filters:	0.1-100 hz	

Data Analysis Procedures

1. Spectral analysis of data: Average of one hundred, 3.4 sec-long artifact free epochs with 50% shift between successive epochs. Spectra are assessed for abnormal dominance by low frequency activity, and unusual topographies or asymmetries. Significant delta activity in the eyes open state is a definite sign of pathology.

2. Visual inspection of data: Data are visually inspected for epileptiform transients (spikes and sharpwaves) and bursts of delta activity. Visual inspection of data is imperative because spectral analyses involving averaging of epochs recorded over long time periods may fail to reveal intermittent abnormalities. Utilization of compressed spectral arrays rather than average spectra may help here.)

3. When abnormal activity is identified, spatio-temporal dipole modeling is applied.

SUBTEST 3:
Reactivity Protocol

Objective: To assess reactivity of spontaneous rhythms to changes in subject state.

Acquisition Parameters

Length of Record:	4 minutes -	30s eyes open; 30s bilateral fists; 30s eyes open;30s eyes closed; 30s eyes open; 30s eyes closed; 30s bilateral fists; 30s eyes closed.
Digitization Rate:	300 hz	
On-line Filters:	0.1-100 hz	

Data Analysis Procedures

1. Spectral analysis of data in 30s blocks: Check for reactivity of alpha rhythm to eye opening and closing and reactivity of mu rhythm to hand movements.

2. Visual inspection of data: Data are also visually inspected for epileptiform transients and bursts of delta activity. The reactivity paradigm distinguishes mu rhythm from fronto-parietal sharpwaves.

SUBTEST 4:
Visual Mapping

Objective: To assess the integrity of the visual system with special reference to field cuts caused by lesions and delayed responses associated with optic neuritis or MS

Stimuli : 100 trials each, randomized, with random ISI of 200-300 msec. Stimulus duration is 50 msec. Subjects count the number of stimuli presented at fixation.

location	type	size	eccentricity	spatial frequency
fixation	blue/yellow	0.5 degree	0 degree	10 cycles/degree
URVF	red/green	1.0 degrees	1 degree	05 cycles/degree
ULVF	red/green	1.0 degrees	1 degree	05 cycles/degree
LRVF	red/green	1.0 degrees	1 degree	05 cycles/degree
LLVF	red/green	1.0 degrees	1 degree	05 cycles/degree
URVF	black/white	2.0 degrees	3 degrees	2.5 cycles/degree
ULVF	black/white	2.0 degrees	3 degrees	2.5 cycles/degree
LRVF	black/white	2.0 degrees	3 degrees	2.5 cycles/degree
LLVF	black/white	2.0 degrees	3 degrees	2.5 cycles/degree

Acquisition Parameters

Digitization Rate: 150 hz
On-line Filters: 1-50 hz
Epoch: -100 to 400 msec

Data Analysis Procedures

1. Digital filter 3-30 hz, baseline correct -100 to 0
2. Single dipole analysis of 90-110 msec component for periphery and 110-140 msec component for parafoveal stimuli. Rapid stimulus presentation biases system to dipolar responses from primary visual cortex with clear retinotopic organization.

SUBTEST 5:
Contrast Following

Objective: To assess the ability of the visual system to follow contrast reversing stimuli. Previous EEG data suggest that following ability may be a good predictor to response to medication in some populations.

Stimuli: 90 seconds of contrast reversal at each of 4, 8, 12, and 16 hz, for an 8x10 degree fixated black and white checkerboard pattern (1 cycle/degree).

Acquisition Parameters

Digitization Rate: 300 hz
On-line Filters: .1-100 hz

Data Analysis Procedures

1. Spectral analysis at each frequency.
2. Examination of amplitude and topography of spectra with assessment of harmonic responses and asymmetries.

SUBTEST 6:
Auditory Mapping

Objective: To assess the integrity of the auditory system with special reference to asymmetries in response topography, latency, amplitude or source location.

Stimuli: 100 trials each, randomized, with random ISI of 700-800 msec. Stimulus duration is 50 msec. Subjects count the number of low intensity tones.

type	frequency	intensity
binaural	500 hz	85 dB
binaural	2000 hz	85 dB
binaural	4000 hz	85 dB
right ear only	2000 hz	85 db
left ear only	2000 hz	85 db
binaural	2000 hz	70 dB

Acquisition Parameters

Digitization Rate:	150 hz
On-line Filters:	1-50 hz
Epoch:	-100 to 400 msec

Data Analysis Procedures

1. Digital filter 1-30 hz, baseline correct -100 to 0
2. Dipole analysis of M100 response using bilateral dipole model. Modeling of other components is possible but these show high variability of response in normals. Data are assessed for abnormal latencies, amplitudes, asymmetries and topographies. A special interest is placed on abnormal topographies for particular tonal frequencies and monaural versus binaural conditions.

SUBTEST 7: Somatosensory Mapping

Objective: To assess the integrity of the somatosensory system with special reference to asymmetries in response latency, amplitude or source location. Additivity in bilateral condition is of special interest in patients with simultaneous extinction. Also of interest is the spatial location of central sulcus relative to a lesion.

Stimuli: 100 trials each, randomized, with random ISI of 450-550 msec. Stimulus duration is 200 usec and adjusted to just above twitch threshold. In patients with poor cortical responses because of lesions, an increased ISI is recommended.

type	location
unilateral pulse	right median nerve
unilateral pulse	right tibial nerve
unilateral pulse	left median nerve
unilateral pulse	left tibial nerve
bilateral pulse	right and left median nerves

Acquisition Parameters

Digitization Rate:	1250 hz
On-line Filters:	1-400 hz
Epoch:	-100 to 400 msec

Data Analysis Procedures

1. Digital filter 10-150 hz (for SI), baseline correct -100 to 0. For SII, digital filter 1 - 50 hz.
2. Dipole analysis of M20 response for localization of primary somatosensory cortex.
3. Multiple dipole analysis required for assessment of later responses because of recurrent activation of SI.

Two components of giant somatosensory evoked magnetic fields in patients with cortical myoclonus

Mima, T.[1], Mikuni, N[2], Nishitani, N[1], Nagamine, T.[1], Ikeda, A.[1], Fukuyama, H.[1], Takigawa, T.[3], Kimura, J.[4], and Shibasaki, H.[1]

Department of Brain Pathophysiology[1], Neurosurgery[2] and Neurology[4], Kyoto University, School of Medicine, Kyoto, Japan; National Utano Hospital, Kyoto, Japan[3]

Introduction

Pathologically enlarged cortical somatosensory evoked potentials (giant SEPs) are one of the important features of cortical myoclonus [1-4]. Magnetoencephalographic studies have revealed that the somatosensory evoked magnetic fields (SEFs) to median nerve stimulation are also enlarged, and that their generator sources can be localized at the somatosensory cortex [5; 6]. However, based on the SEP scalp topography using EEG, we have reported that the tangential (N30-P30) and "radial" (P25-N35) components can be distinguishable in giant SEPs [7]. Since giant somatosensory evoked responses (SERs) in cortical myoclonus are regarded as the indicator of the hyperexcitability of sensorimotor cortex in those patients, studies of the generator mechanisms of giant SERs will be of great help to understand the pathogenesis of cortical myoclonus. The purpose of this study is to identify MEG correlate of each component of giant SEPs, and to locate the generator sources on individual MRI.

Methods

We investigated 3 patients with cortical myoclonus (Patient 1; Lafora disease, Patients 2 and 3; familial cortical myoclonic tremor). Their clinical features are summarized in Table 1. All subjects gave informed consent before the experiments according to the approval by the Ethical Committee of Kyoto University School of Medicine.

SEFs were recorded by a helmet-shaped magnetometer array with 122 first-order planar gradiometers. The planar gradiometer detects the largest signal above the local current source. MEG signals were recorded by the sampling rate of 0.9 kHz (bandpass; 0.03 to 300 Hz). SEPs were simultaneously recorded from at least 4 sites (C3, P3, C4 and P4). All electrodes were referenced to linked ear lobe electrodes. The electrooculogram (EOG) at the right inferior lateral canthus referred to the right earlobe was simultaneously recorded. The median nerve trunk at wrist was stimulated with electric pulse of 0.2 ms duration. MEG and EEG signals were averaged with respect to the stimulus onset. Any response with amplitude exceeding 150 μV in the EOG channel or 3000 fT/cm in any of MEG channels was rejected from the average. The analysis time window was 600 ms including the prestimulus 100 ms. In each session, 2 trials of 150 responses each were averaged separately to confirm the reproducibility of the responses. Amplitude of each recognizable peak of SEPs was measured from the baseline determined by averaging the prestimulus segment. N20 was identified as an initial local negative potential around 20 ms at the contralateral parietal area (P3/P4). P30 was identified as a positive peak at the same site. P25 and N35 were recognized at the contralateral central area (C3/C4). The terminology of the peaks followed the studies of normal subjects [8].

The MEG channel with the local extreme signal in each record was chosen for the measurement of the latency. The magnetic fields contour patterns were constructed by using a minimum-norm estimate. Parameters of an equivalent current dipole (ECD) that best explain the most dominant source at each peak were determined by a least-squares search, using subsets of channels over the response areas. ECDs explaining more than 80 % of the fields over the selected period of time were used for further analysis. MEG coordinates were matched with MRI by using the 3 head position indicators. ECDs were superimposed on individual brain MRI.

Results

EEG recordings demonstrated four sets of early components of SEPs (N20, P25, P30 and N35). Peak latency and amplitude of each component are summarized in Table 2.

In MEG recording, the initial cortical component of SEFs was N20m at the contralateral sensorimotor area, which was followed by a peak of opposite polarity at the same MEG channel (P30m). Another MEG peak between N20m and P30m could be recognized at the different channel at the same or nearby site (referred to as P25m). P25m was followed by the next peak of opposite polarity (N35m) (Fig. 1).

Contour maps of magnetic fields and the parameters of ECDs were calculated for each recognizable SEF component. Magnetic field pattern of N20m showed clear single dipolar pattern around the contralateral sensorimotor area in all 3 patients. Magnetic field contours of P30m were similar to those of N20m, though the flux direction was inverted. ECDs of N20m and P30m had the same location posterior to the central sulcus, and their directions were inverted each other (anterior-posterior orientation). Magnetic field contours of P25m were clearly different from those of N20m or P30m. Magnetic fields of N35m were similar to those of P25m in 5 nerves

TABLE 1 Clinical features of 3 patients

Patient	Diagnosis	Sex	Age	Clinical Symptoms	Medication (per day)
1	PME (Lafora disease)	M	20	seizure, mental deterioration	VPA: 1000 mg, CZP: 3 mg
2	Cortical Tremor	M	68	none	none
3	Cortical Tremor	F	58	none	VPA: 600 mg, CZP: 4 mg, zonisamide: 400 mg

F: female, M: male, PME: progressive myoclonus epilepsy, VPA: valproic acid, CZP: clonazepam

TABLE 2 Parameters of ECD for each SEF and SEP component

Patient	SEF							SEP	
		source location			dipole	g-value	confidence		
(side)	latency (ms)	x axis (mm)	y axis (mm)	z axis (mm)	moment (nAm)	(%)	volume (mm^3)	latency (ms)	amplitude (μV)
N20m								**N20**	
1(L)	22	39.3	19.5	100.8	38.2	95.3	0.9	20	−1.2
(R)	23	−47.1	17.6	98.6	19.3	93.1	8.2	20	−0.1
2(L)	22	43.8	−2.2	92.7	39.7	98.3	0.6	21	−4.8
(R)	22	−45.5	−3.3	97.0	23.1	96.0	3.8	21	−1.2
3(L)	20	41.0	17.9	94.9	22.6	95.9	10.5	20	−2.0
(R)	21	−38.4	6.0	91.1	35.5	98.5	2.7	20	−2.0
P25m								**P25**	
1(L)	27	33.4	17.0	115.2	18.0	95.1	2.5	25	11.3
(R)	28	−38.7	11.4	92.4	43.1	87.2	5.4	25	7.2
2(L)	25	39.4	−6.4	98.3	26.4	92.8	1.6	24	27.3
(R)	25	−34.8	−7.8	103.4	35.5	95.1	1.2	25	21.1
3(L)	24	34.0	22.7	97.0	55.4	96.1	1.2	23	20.2
(R)	23	−31.0	13.1	101.8	29.6	93.4	2.7	23	18.6
P30m								**P30**	
1(L)	31	37.4	17.8	106.1	46.0	97.7	0.3	28	9.1
(R)	40	−43.4	11.3	93.4	61.9	96.9	0.8	40	4.2
2(L)	27	43.0	−5.9	94.3	113.7	98.2	0.0	26	19.2
(R)	28	−44.0	−8.2	95.0	131.2	98.4	0.0	27	18.3
3(L)	26	37.1	19.1	91.5	72.9	96.0	0.8	25	11.1
(R)	26	−38.4	2.1	94.9	55.4	98.9	0.4	25	14.2
N35m								**N35**	
1(L)	50	30.7	17.1	99.4	37.9	94.3	3.7	49	−3.5
(R)	52	−46.3	12.8	98.4	17.7	85.0	16.5	52	−8.3
2(L)	40	32.1	−12.5	99.1	52.0	90.2	0.4	39	−60.8
(R)	39	−38.8	−7.1	103.0	57.7	91.8	0.2	39	−35.6
3(L)	30	33.5	25.4	99.1	36.1	96.7	3.1	29	−16.9
(R)	29	−30.6	15.3	92.5	53.6	94.4	1.8	29	−11.1

In this table, EEG amplitudes were measured from the baseline determined by the prestimulus segment. The 3-dimensional locations (x, y and z) of ECDs were measured from the midpoint of the two preauricular points. The x axis runs from the left to the right preauricular point. The y axis runs from back to the front through nasion. The z axis runs from the base to the top of the head through the center of the two preauricular point.

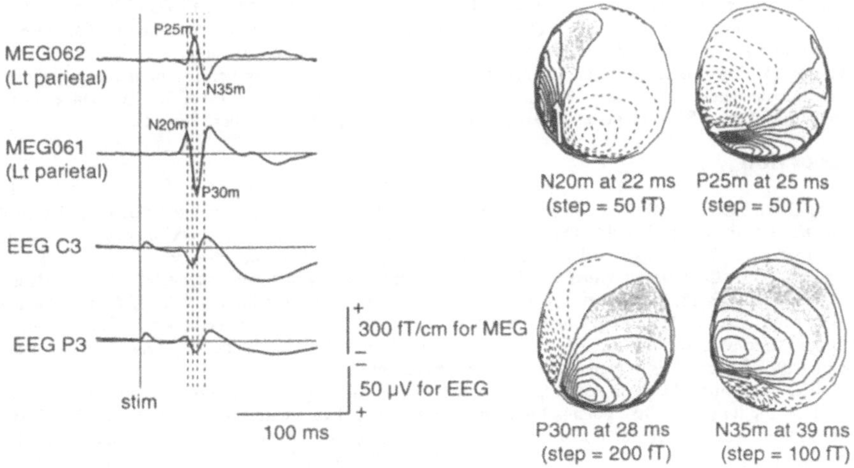

Fig. 1

Left figure: Giant SEFs to the right median nerve stimulation in Patient 3. Time window is 100 ms including the prestimulus segment of 20 ms. The vertical line indicate the stimulus onset. Representative MEG channels and EEG electrodes are demonstrated. Negative deflexion of the EEG signals are shown upward. Each vertical dashed line indicates each SEF component (N20m, P25, P30m and N35m, respectively).

Right figure: Magnetic field patterns in the same patient. The shadowed areas indicate magnetic flux out of, and the white area flux into the head. Each step of isocontour and the latency are shown below the contourmap. White arrow shows the locations and orientations of the ECDs.

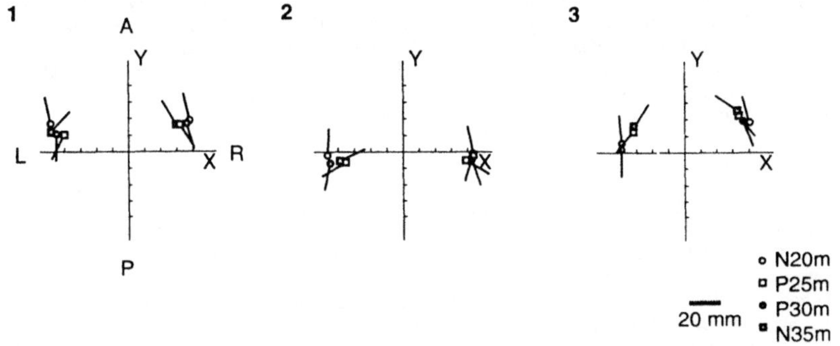

Fig. 2

Source location and orientation of each SEF component projected on the topview. N20m (open circle), P25m (open square), P30m (filled circle) and N35m (filled square) are presented. The direction fo the ECDs are shown as a bar.

except for the left median nerve stimulation of Patient 2. In Patients 1 and 2, ECDs of P25m and N35m were located posteromedial to that of N20m (postcentral gyri), whereas they were located anteromedial to that of N20m (precentral gyri) in Patient 3. ECDs of P25m had the posterolateral direction, and those of N35m showed anteromedial direction in the other subjects except for that to left median nerve stimulation in Patient 2 which showed posteromedial direction. Parameters of ECDs are summarized in Table 2. The source locations and orientations of these SEF components were projected onto the topview of the head (Fig. 2).

Discussion

We could discern MEG counterparts of four EEG components of giant SEPs (N20m, P25m, P30m and N35m). Both EEG and MEG reflect the same electromagnetic activities in human brain, but MEG detects the radial components of the magnetic fields which are generated by tangentially oriented intracellular currents. The magnetic counterpart of P25 and N35 in SEPs were not always recognized clearly, possibly because these SEP components are generated by the electrical current radial to the brain surface, which were suggested by the analysis of scalp topography of SEPs [8; 9]. However, even in normal subjects, MEG correlates of P25 and N35 (P25m and N35m, respectively) could be identified, which are thought to reflect the tangential components of the so-called "radial" electrical dipoles described in EEG scalp topography [10; 11]. This is the first report of P25m and N35m components of giant SEFs in patients with cortical myoclonus.

In MEG recording, the strongest response was P30m in all subjects, whereas P25 was the largest in SEPs. Therefore, the largest positive potential in giant SEP does not always correspond to the largest MEG peaк, which suggests that "radial" electrical dipole has some contribution to the generation of the exaggerated cortical responses to somatosensory stimuli. ECDs of N20m were localized at the posterior wall of the central sulcus (Brodmann area 3b). Those of enlarged P30m were located at the same site as those of N20m. ECDs of P25m and N35m were localized at the postcentral sensory cortices in Patients 1 and 2, whereas in Patient 3 they were at the precentral motor cortices. Recent monkey studies suggested that the monkey counterpart of P25 was caused by the simultaneous activation of both motor and sensory cortices [12]. It is most likely that P25 and N35 on EEG are also the summation of the potentials generated at both the precentral and postcentral gyri, and that the pathological enhancement of the cortical responses may occur separately at each site.

References

[1] Rothwell, J.C., Obeso, J.A. and Marsden, C.D. On the significance of giant somatosensory evoked potentials in cortical myoclonus, Journal of Neurology, Neurosurgery, and Psychiatry, 1984, 47: 33-42.

[2] Shibasaki, H., Kakigi, R. and Ikeda, A. Scalp topography of giant SEP and pre-myoclonus spike in cortical reflex myoclonus, Electroencephalography and Clinical Neurophysiology, 1991, 81: 31-37.

[3] Kakigi, R. and Shibasaki, H. Generator mechanisms of giant somatosensory evoked potentials in cortical reflex myoclonus, Brain, 1987, 110: 1359-1373.

[4] Hallett, M., Chadwick, D. and Marsden, C.D. Cortical reflex myoclonus, Neurology, 1979, 29: 1107-1125.

[5] Kahru, J., Hari, R., Pateau, R., Kajola, M. and Mervaala, E. Cortical reactivity in progressive myoclonus epilepsy, Electroencephalography and Clinical Neurophysiology, 1994, 90: 93-102.

[6] Uesaka, Y., Ugawa, Y., Yumoto, M., Sakuta, M. and Kanazawa, I. Giant somatosensory evoked magnetic field in patients with myoclonus epilepsy, Electroencephalography and Clinical Neurophysiology, 1993, 87: 300-305.

[7] Ikeda, A., Shibasaki, H., Nagamine, T., Xu, X., Terada, K., Mima, T., Kaji, R., Kawai, I., Tsatsuoka, Y. and Kimura, J. Peri-rolandic and fronto-parietal components of scalp-recorded giant SEPs in cortical myoclonus, Electroencephalography and Clinical Neurophysiology, 1995, 96: 300-309.

[8] Allison, T., McCarthy, G., Wood, C.C. and Jones, S.J. Potentials evoked in human and monkey cerebral cortex by stimulation of the median nerve, Brain, 1991, 114: 2465-2503.

[9] Desmedt, J.E., Nguyen, T.H. and Bourguet, M. Bit-mapped color imaging of human evoked potentials with reference to the N20, P22, P27 and N30 somatosensory responses, Electroencephalography and Clinical Neurophysiology, 1987, 68: 1-19.

[10] Tiihonen, J., Hari, R. and Hämäläinen, M. Early deflexions of cerebral magnetic responses to median nerve stimulation, Electroencephalography and Clinical Neurophysiology, 1989, 74: 290-296.

[11] Kakigi, R. Somatosensory evoked magnetic fields following median nerve stimulation, Neuroscience Research, 1994, 20: 165-174.

[12] Peterson, N.N., Schroeder, C.E. and Arezzo, J.C. Neural generators of somatosensory evoked potentials in the awake monkey, Electroencephalography and Clinical Neurophysiology, 1995, 96: 248-260.

Acknowledgements

This study was partly supported by Grant-in-Aid for International Scientific Research 07044258 from the Japan Ministry of Education, Science and Culture.

Deviations from the Tonotopic Map are Correlated with Tinnitus Strength

Mühlnickel, W.[1], Flor, H.[1], and Elbert, T.[2]

[1]Department of Psychology, Humboldt-University of Berlin, Germany; [2]Department of Psychology, University of Konstanz, Germany

INTRODUCTION

Tinnitus is characterized by phantom perception of acoustic signals without any external source of sound. Based on previous studies that reported reorganization of the somatosensory cortex in amputees with phantom sensation [1], we hypothesized that tinnitus might be based on plastic changes in the tonotopic map in primary auditory cortex. In accordance with the finding that phantom limb pain is positively related to reorganization in primary somatosensory cortex, it was assumed that tinnitus strength should be positively correlated with amount of cortical reorganization.

METHOD

Seventeen right-handed tinnitus patients (tonal tinnitus, maximal hearing loss 25 dB HL) and seventeen matched (age, tender) right-handed normal-hearing controls without tinnitus participated in the study. See table 1 for clinical aspects of the patients in the tinnitus group.

Table 1: Clinical aspects of the tinnitus group.

Pat-No.	Age	Sex	Strength	T.ear	T. freq.
2	31	m	1.0	left	4256
3	38	m	-	right	4000
4	35	w	4.3	left	8000
5	32	m	-	both	6000
6	35	m	4.0	both	6000
7	18	w	3.3	right	4000
8	16	m	-	right	200
9	19	w	2.6	both	noisiform
10	27	m	3.0	both	250
11	36	w	2.0	right	4000
12	44	w	3.0	left	1000
13	31	w	4.6	both	6000
14	33	m	2.6	right	8000
15	29	m	3.6	right	3000
16	31	w	-	both	1000
17	32	m	4.6	both	4000
18	20	m	4.3	left	3680
$\bar{x}$	30.3±2.4		3.4±1.05		4024±2485

Pure tone pulses (500 ms duration, 70 dB HL, four different carrier frequencies in random order) were sequentially presented to both ears, with an interstimulus interval (ISI) of 2 s. In the control group, carrier frequencies of 1000, 2000, 4000 or 8000 Hz were employed. In the tinnitus group three of these standard frequencies were employed. The tinnitus frequency was introduced as the fourth frequency and the standard frequency that was closest to the tinnitus frequency was deleted from the list of stimuli.

Magnetoencephalographic recordings were carried out in a magnetically shielded room using a 37 channel neuromagnetometer (Magnes^{TM}, Biomagnetic Technologies). 200 stimulus-related epochs of the magnetic field were recorded over the contralateral hemispheres. The magnetic signals were digitally filtered using a 20 Hz low pass filter and selectively averaged for each stimulus frequency. The location of wave N1m of the stimulus-related field was estimated using a equivalent current dipole model (ECD) in a homogeneous spherical volume conductor [2].

The ECD locations of the healthy controls that corresponded to the four different frequencies formed a trajectory in three-dimensional space representing the tonotopic map. The ECD locations of the standard frequencies are a logarithmic function of the stimulus frequencies with a main orientation in medial-lateral direction [3]. In the tinnitus group, the dipole locations of the three standard frequencies were used to reconstruct the map.

Next, the Euclidian distance of the location of the tinnitus frequency and the trajectory of the standard frequencies was computed. Reorganization of primary auditory cortex related to tinnitus might result in a shift of the tinnitus frequency into the upper or lower frequency range or in a shift into adjacent cortical aereas. Since tinnitus frequencies were not uniform it was impossible to test the shift into the higher or lower frequency range. It was, however, possible to compute the deviation of the tinnitus frequency from the trajectory of standard frequencies as a measure of invasion of the tinnitus frequency into neighbouring regions. This shift was modeled by computing the Euclidian distance of the location of the tinnitus frequency from the trajectory of standard frequencies. Figure 1 shows an enlarged twodimensional projection of the standard frequencies in the auditory cortex in one tinnitus patient. The filled black circles represent the ECD locations of the magnetic field distribution to the three standard stimuli. The open circle represents the ECD location of the tinnitus frequency. The deviation of the tinnitus frequency from the trajectory of the standard frequencies was determined by computing the Euclidian distance of the tinnitus frequency to the trajectory. The distancies were separately computed for both hemispheres. This procedure computes the minimum deviation of the tinnitus frequency from the normal tonotopic map. It is conceivable that the true pre-tinnitus frequency might has been further removed on this trajectory, the Euclidian distance would in this case underestimate the shift and is thus a conservative measure of reorganization. In a subject with chronic tinnitus the ECD location of the magnetic response of the tinnitus frequency is shifted outside of the area of the putative tonotopic map.

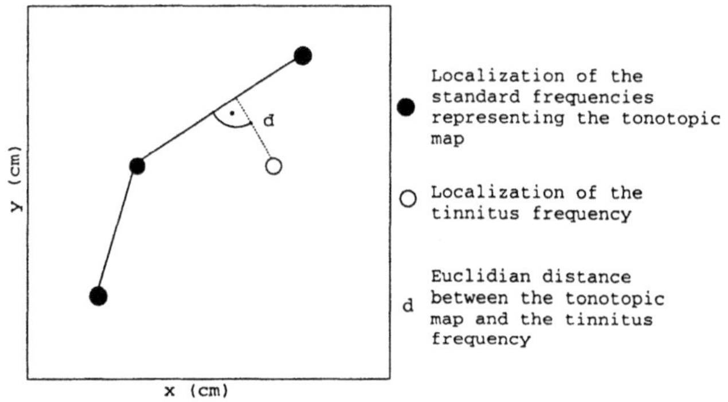

Figure 1: A model of computing the Euclidian distance d.

RESULTS

The deviation from the tonotopic map of the tinnitus patients was compared with the shift of the tonotopic map of the controls. Based on the typical tinnitus frequency range (see table 1) of the patients, tinnitus frequencies of 2,000 Hz and 4,000 Hz were "created" for the controls. The deviation of the 2 kHz or 4 kHz tone from the trajectory formed by the standard tones was calculated by using the Euclidian distance measure. It was expected that the Euclidian distance of the controls should be significantly smaller than those of the patients, thus indicating a lack of cortical reorganization. An ANOVA showed a significant difference between the Euclidian distance measures of the patients and the controls ($F(1,31) = 6.36$, $p < .05$). When the deviations of 2 kHz and 4 kHz tones were summarized, there was no significant difference between them. Whereas the Euclidian distance of the tinnitus patients was on the average of 4.15 ± 2.2mm, the Euclidian distance of the controls amounted to 2.8 ± 1.0mm. This confirms the hypothesis of a shift of the cortical representation of the tinnitus frequency outside the area of the tonotopic map.

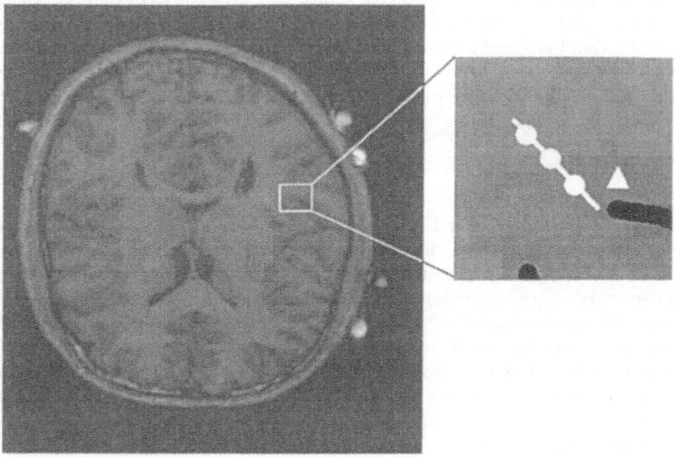

Figure 2: A horizontal MRI slice of Brodman area 41 of the right hemisphere. The results of one subject are shown. The line represents the trajectory of the dipole locations of three tones. The triangle represents the location of the corresponding tinnitus frequency.

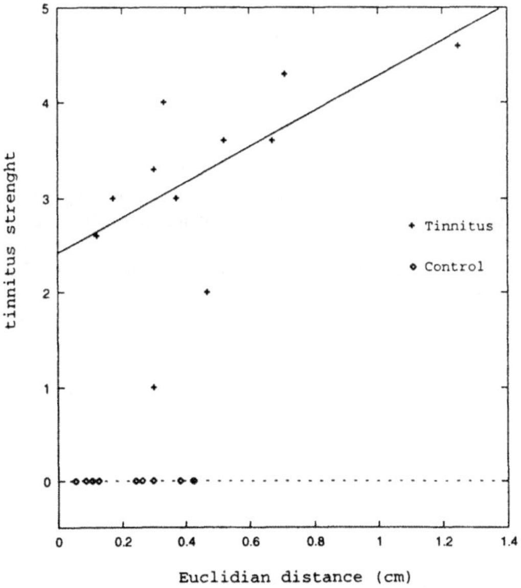

Figure 3: Scatterplot of the tinnitus strenght vs. the Euclidian distance. The distances of the controls were the distances representing the 4 kHz tone with respect to the average tinnitus frequency of the tinnitus group.

In order to determine the relationship of the cortical map and the subjective tinnitus strength a correlation coefficient was computed between the cortical distance measure and a tinnitus inventory. Tinnitus strength was assesed by a tinnitus-specific version of the West Haven-Yale Multidimensional Pain Inventory (WHYMPI) [4], where the expression *pain* was replaced by *earsound*. The correlation of the subjective tinnitus strength and the Euclidian distance was significant at $r(10) = 0.817$, $p < .05$, for the stimulation of the left ear and recordings of the right hemisphere. Right ear tinnitus was to infrequent ($n = 3$, see table 1) to compute correlation coefficients.

DISCUSSION

This study yielded two important results. First, tinnitus seems to be related to reorganization of primary auditory cortex as indicated in the shifts. The shifts have no preferred direction, e.g. the medial-lateral direction, suggesting an distortion rather than a mere shift of the tinnitus frequency into upper or lower frequency ranges of the auditory cortex. Second, in striking concordance with the results reported on phantom limb pain, tinnitus strenght and amount of cortical reorganization seem to be positively related.

These preliminary results need to be further substantiated. It would have been desirable to use the same frequencies in the patient and control group (e.g. by employing yoked control design). Moreover although the plastic changes were observed in auditory cortex, in the origin of the reorganization might be on the thalamic or lower levels.

The following hypothesis might be derived: The abnormal activity which is generated at the site of lesion (for instance, in the auditory nerve or in the cochlea) results in a loss of lateral inhibition in more centrally locted neural structures, which then in turn would generate the neural activity that underlies the sensation of tinnitus. After a period of extended abnormal input to these central structures, neural connections become strengthened, so that the abnormal activity continues, even after cessation of the abnormal input that originally caused the problem. This zone would be no longer be available for the representation of the tones at or near the vicinity of the tinnitus frequency. The representational zone for this frequency should consequently be displaced. Alternatively, one might argue that a distortion of the map has occured in the first place and tinnitus, as an auditory phantom perception, might be a consequence of such alterations.

This research was supported by the Deutsche Forschungsgemeinschaft (Fl 156/16; Ho 847/6) and is part of W.M.'s PhD thesis.

REFERENCES

[1] Flor, H., Elbert, T., Knecht, S., Wienbruch, C., Pantev, C., Birbaumer, N., Larbig, W. and Taub, E. Phantom-limb pain as a perceptual correlate of cortical reorganization following arm amputation, Nature, 1995, 375: 482-484.

[2] Hämäläinen, M.S. and Ilmoniemi, R.J. Interpreting measured magnetic fields of the brain. Estimates of current distributions, Report TKK-F-A559, 1984.

[3] Pantev. C, Eulitz, C., Elbert, T. and Hoke, M. The auditory evoked sustained field: origin and frequency dependence, Electroencephalography and clinical Neurophysiology, 1994, 90: 82-90.

[4] Flor, H., Rudy, T.E., Birbaumer, N., Streit, B. und Schugens, M.M. Zur Anwendbarkeit des West Haven-Yale Multidimensional Pain Inventory im deutschen Sprachraum, Der Schmerz, 1990, 4: 82-87.

The Determinism of MEG in Tinnitus Patients

Mühlnickel, W.[1], Hoke, E.S.[2], Hoke, M.[2], and Flor, H.[1]

[1]*Department of Psychology, Humboldt-University of Berlin, Germany;* [2]*Institute of Experimental Audiology, University of Münster, Germany*

INTRODUCTION

Research during the last decade has shown that specific aspects of brain dynamics can be determined by using tools from nonlinear system's theory [1]. Based on a suggestion of Kaplan and Glass [2] it is possible to obtain information about the determinism of a time series even for short measurement intervals. The present study compares the determinism of the MEG in tinnitus patients with that of controls. Given that the auditory cortex is the final link in a chain producing tinnitus, then different patterns of brain dynamics may appear in the auditory cortex of patients. The MEG is ideally suited to measure the fluctuations in the tangentially oriented current flow of cortical areas. Therefore, we can expect to see differences in the determinism of the MEG between tinnitus sufferers and controls, when recorded over temporal regions.

THE DETERMINISTIC TEST OF A TIME SERIES

The goal of the presently introduced method for EEG analyses is to obtain information about the determinism of a time series, even for a short time interval. Kaplan & Glass [2] developed a direct test for determinism in a time series. This method is based on the fact that the tangent of the trajectory of a deterministic system is a function of the region in the phase space. The estimation is briefly described in the following steps:

- If the attractor is embedded in a d-dimensional phase space, then the phase space is divided into n^d boxes (where n is an integer)
- If a trajectory passes through a box j, then a vector v_{kj} (k pass) is constructed. This vector is defined by the coordinates of the starting-point and the end-point of that part of the trajectory inside the box j.
- The sum of all v_{kj} yields a vector $\mathbf{V}_j = \frac{1}{n_j}\sum_k \mathbf{v}_{k,j}$. The V_j characterizes the attractor.
- The average $\bar{L}_n^d = \langle \|\mathbf{V}_j\| \rangle_{n_j=n}$ over each box containing n passes, shows a statistic over all V_j with n passes (to gain deterministic pattern in a time series this last step was modified in so far, as $\bar{L}_n^d$ was calculated each time the trajectory passes a **given** box).

$\bar{L}_n^d = 1$ is valid only for deterministic systems with infinite time series and a box edge lenght $\rightarrow 0$. White noise falls off with $\bar{L}_n^d = n^{-\frac{1}{2}}$.

To gain deterministic patterns in a system the statistic term $L(d,n) = \bar{L}_n^d$ was replaced by a dynamic version $LD(d,n) = \bar{LD}_n^d$ [3]. The static version estimates $L(d,n) = \bar{L}_n^d$ after the complete reconstruction of the attractor. If, for example, the number of passes is $n = 5$, then in $L(d,5)$ there would be boxes with $n = 5$ passes. Another possibility would be to consider boxes that could contain more than $n = 5$ passes, meaning that the sum vector V_j was calculated subsequently for each pass. If one given box was passed $n = 20$ times, then this box was considered for each pass up to $n = 20$. The result for $LD(d,n)$ shows a gain of determistic patterns. Figure 1 shows the statistic term $\bar{LD}_n^d$ for a threedimensional embedding for different time series. The three time series have the same autocorrelation function and consited of 68,000 points. The delay time was $\tau = 80$ data points and the phase space was divided into $n = 16^3$ boxes.

As can be seen, the deterministic Glass-Mackey-System and the noise are identical. They differ if the embedding dimension increases, see Fig. 2.

Table 1 compares the statistic terms $\bar{L}_n^d$ and $\bar{LD}_n^d$ for an embedding dimension of $d = 7$ and a delay time of $\tau = 5$. The presented values were averages of the interval $u = [11, 20]$ passes. Each time series contains 3,000 data points and their autocorrelation functions are identical. Table 1 includes results from

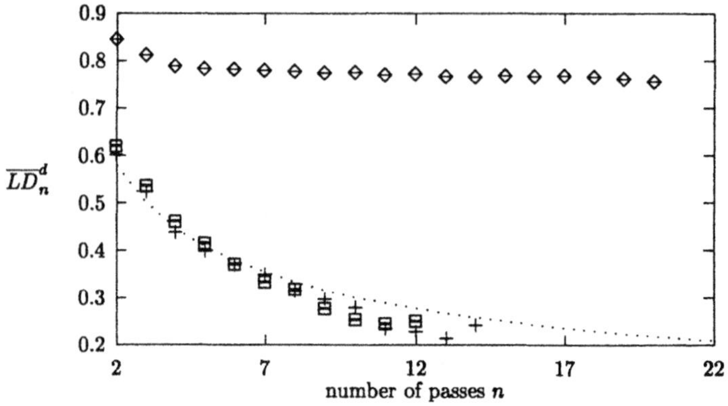

Figure 1:
Statistic term $\bar{LD}_n^d$ for a threedimensional embedding of: (a) ◇ - the Lorenz-, (b) □ - the Glass-Mackey- and (c) + - a noisiform trajectory. The doted line represents the theoretically decrease of white noise - $1/\sqrt{n}$.

Table 1: Comparisons for the static and the dynamic version of the test for determinism.

Nonlinear system	Static $\bar{L}_n^d$	Dynamic $\bar{LD}_n^d$
Lorenz	.943 ±.001	.977 ±.001
van der Pol-type	.601 ±.01	.779 ±.007
Glass-Mackey	.542 ±.002	.545 ±.001
MEG (epileptic p.)	.338 ±.021	.389 ±.001
EEG (epileptic p.)	.402 ±.030	.378 ±.001
MEG (control)	.365 ±.002	.381 ±.001
filtered noise	.300 ±.005	.303 ±.002

time series of corresponding length generated by (a) the Lorenz attractor, (b) the van-der-Pol-like oscillator, (c) the Glass-Mackey equations with the result for (d) the noise and (e) the EEG and MEG-recordings measured over central locations (C_4). Two important outcomes are obvious: (1) a higher deterministic value for the MEG and for the EEG than for the white noise and (2) an increase in the dynamic patterns in a system in the dynamic version $\bar{LD}_n^d$. Furthermore, the results for the epileptic patient were even higher than those for a control subject, an observation that is in agreement with the literature [1]. MEG-analyses tend to produce higher values for determinism than EEG-based computations, a result that confirms the higher specificity of MEG.

METHOD

The data were taken from an investigation especially designed for the study of the contigent negative variation (CNV) in tinnitus patients [4]. In the present study dynamic aspects of these data were investigated. The estimation of the determinism of a given time series was chosen because this analysis requires only a short time series. 23 patients (tonal and noisiform tinnitus together) and 9 controls participated in the study. MEG signals were recorded from both temporal regions using a Magnes 37-channel system (sampling frequency 122 Hz). The test for determinism, adapted for the analysis of biosignals [3], was applied to separately examine the MEG for the three levels of the imperative stimulus (1,3 and 5 dB, resp.). The selected time interval of 10-13 s started 1 s before the onset of the warning stimulus (S1) and ended 1 s prior to the next S1. The embedding dimension was set to d = 7 and the delay time to t = 5 points. The values

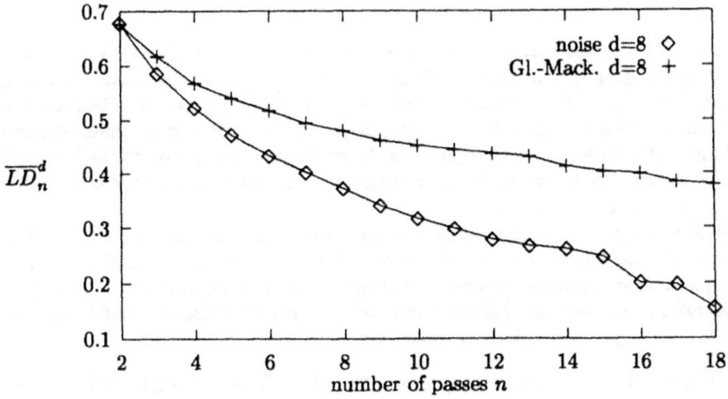

Figure 2:
Statistic term $\bar{LD}_n^d$ for a Glass-Mackey- and a noisiform trajectory. The embedding dimension was $d = 8$.

of $\bar{LD}_n^d$ were averaged within the interval of $u = [15, 25]$ passes.

RESULTS

Results averaged over five runs are presented in fig. 3. A greater value corresponds to a higher determinism of the system. The group difference of the averages is not significant (F(1,27)=2.4). As can be seen, however, the patient group displays a considerably larger variability than controls.

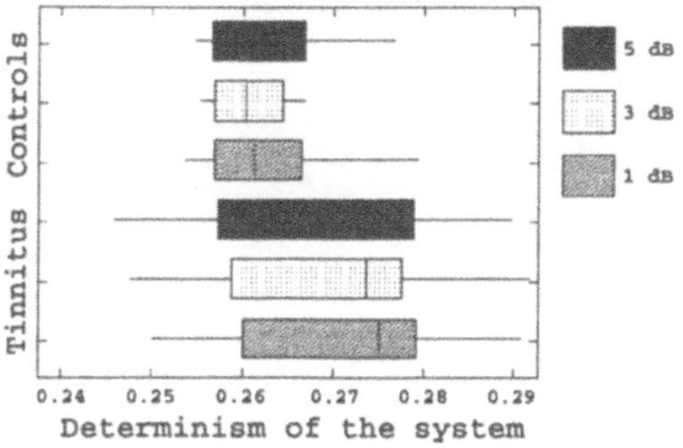

Figure 3: Boxplot of the averaged values of determinism.

But a trend can be seen in an ANOVA comparing the different levels of imperative stimulus (F(2,27)=3.16, p=.053). The lowest level (1 dB) of the imperative stimulus generates a more deterministic MEG than the higher level tones. This is consistent for both groups. Further analysis considering left or right ear tinnitus, the tinnitus frequency and the type of tinnitus were not significant.

DISCUSSION

One important result of the present study is the increased determinism of the MEG signal when the loudness of the imperative stimulus was decreased. This may indicate an increased attention of the subjects to the lower level tones. This differentiation between levels of loudness was not obtained when the spectral power was computed. This suggests that tools developed in the theory of nonlinear dynamics may yield important additional information on brain dynamics. Since the motor reaction to the imperative stimulus is an identical process to any level, the trend is an endogenous response reflecting the more difficult decision-making process in perceiving near-threshold tones.

In general, a larger variability of the deterministic measure of the MEG recorded over the auditory cortex in tinnitus patients than in the control group may indicate that different types of pathological processes are involved in the symptom of tinnitus. Therefore it could be of interest to see whether subjects with an atypical dynamical pattern can also be differentiated on the basis of traditional audiological and psychophysical methods. This investigation is currently under way.

This research was supported by the Deutsche Forschungsgemeinschaft (Ho 847/6) and is part of W.M.'s PhD thesis.

REFERENCES

[1] Elbert T., Ray W.J., Kowalik Z.J., Skinner J.E., Graf K.E., Birbaumer N. Chaos and physiology - Deterministic Chaos in excitable cell assemblies, Phys. Rev., 1994, 74: 1-47.

[2] Kaplan, D.T. and Glass, L. Direct Test for determinism in a time series, Phys. Rev. Lett., 1992, 68: 427-430.

[3] Mühlnickel, W, Rendtorff, N., Kowalik, Z.J., Rockstroh, B., Miltner, W. and Elbert, T. Testing the Determinism of EEG and MEG, Integrative Physiological and Behavioral Science, 1994, 29: 262-269.

[4] Proefrock, E.S. and Hoke, M. Contingent Magnetic Variation (CMV) studied with stimuli close to the hearing Threshold in normal subjects and tinnitus patients, In: Baumgartner, C., Deecke, L., Stroink, G. and Williamson, S.J. Biomagnetism: Fundamental Research and Clinical Applications, Amsterdam, Elsevier, 1995, pp. 234-239.

Auditory Signal Processing is Impaired in Alzheimer's Disease: A Whole-Head MEG Study

Pekkonen, E.[1,2], Huotilainen, M.[1], Virtanen, J.[1], Näätänen, R.[1], Ilmoniemi, R.J.[2] and Erkinjuntti, T.[3]

[1]*Cognitive Brain Research Unit, Department of Psychology, University of Helsinki, Finland;* [2]*BioMag Laboratory, Medical Engineering Centre, Helsinki University Central Hospital, Finland;* [3]*Department of Neurology, Helsinki University Central Hospital, Finland*

Introduction

Auditory stimuli elicit in or near the primary auditory cortex a prominent magnetic response called N100m that is often preceded by the smaller P50m response [1, 2]. The N100m, which together the P50m reflects automatic stimulus processing, is larger and earlier over the contralateral than ipsilateral auditory cortex with respect to the ear stimulated [3, 4]. In addition, aging seems to delay ipsilateral, but not contralateral N100m in the auditory cortex [5].

The aim of this study was to investigate whether Alzheimer's disease (AD) impairs parallel auditory processing between the hemispheres. The 122-channel whole-head magnetometer offers a unique facility to measure activity over both hemispheres simultaneously without interference of vigilance changes.

Methods

Eleven patients with AD and 11 age-matched healthy subjects were studied with approval of the local Ethical Committee. The Mini-Mental State Examination (MMSE) [6] was performed to all AD patients and to control subjects over 60 years of age. The control subjects had no history of neurologic, psychiatric, or any other severe diseases. Subjects in each group had normal hearing and none of them used drugs affecting the central nervous system. In all patients, the head-MRI had been performed to exclude ischemic changes.

Recordings were performed in a magnetically shielded room (Euroshield Ltd.) and the subject sat under the helmet-shaped dewar with his/her head against the bottom of the instrument. The subject watched a silent video and was instructed not to attend to the 700-Hz tones presented monaurally. The stimulus intensity was adjusted to 60 dB over the subjective hearing threshold. Each stimulus block consisted of standard (80%) and randomly embedded deviant tones (20%) differing in duration, 50 and 25 ms, respectively. Stimulus blocks with 0.5- and 2.5-s interstimulus intervals (ISIs) were separately presented to the left ear. For 18 subjects, an additional stimulus block with 0.5-s ISI was presented to the right ear.

Auditory evoked fields (AEF) were recorded using a 122-channel whole-head magnetometer (Neuromag Ltd.). The analysis period was 750 ms, including a 150-ms prestimulus period. The recording passband was 0.03-100 Hz and the sampling rate 397 Hz. The electro-oculograms were recorded to exclude eye artefacts. Over 300 responses were averaged for the standard tone in each condition. Peak latency and amplitude values of the standard responses were measured from the channel showing the largest response. AEF sources were modelled as equivalent current dipoles (ECDs) found by a least-squares fit to the 2.5-s ISI data from a subset of 34 channels over each auditory cortex, separately. A spherical head model was used for the dipole fitting [7].

The MEG data were analyzed with the multivariate analysis of variance for repeated measures (MANOVA). Additional unpaired and paired *t*-tests were employed when appropriate.

Results

Figure 1 shows the evoked magnetic fields (AEFs) of one AD patient and one control subject to the standard tones in the 2.5-s condition. In each subject, the standard tone elicited a clear N100m response over both hemispheres, the responses being larger in the control subject. The ipsilateral, but not the contralateral, N100m is clearly delayed in the AD patient.

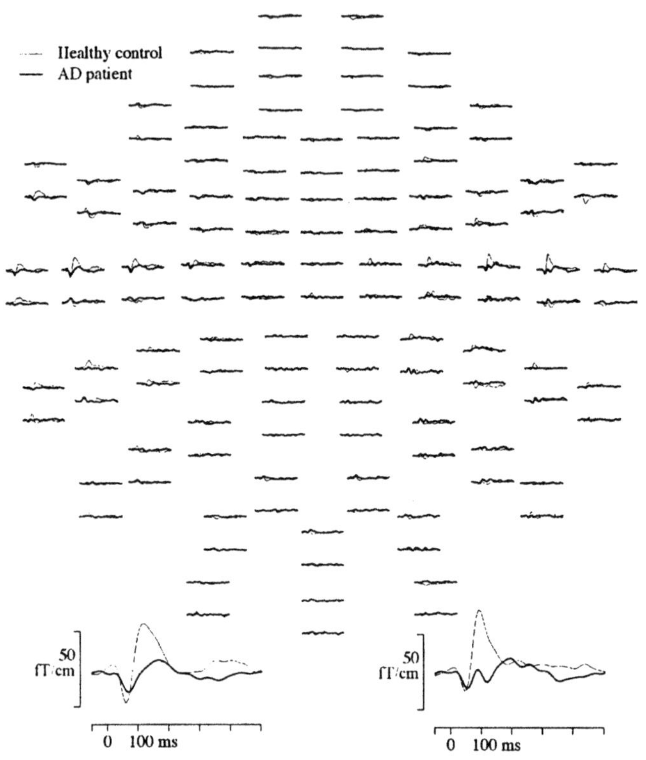

Fig. 1. Auditory evoked magnetic fields to the standard tones presented to the left ear in one healthy control subject and one AD patient with 2.5-s ISI. AEFs were recorded from 61 measurement sites, and in each pair the upper and lower AEFs show the latitudinal and longitudinal derivatives of the normal field component, respectively. Maximum responses over each auditory cortex of the hemisphere are shown enlarged. Note the delayed ipsilateral (left) N100m in the AD patient.

The ipsilateral P50m was significantly delayed in the AD group compared with the control group (MANOVA, F (1, 20) = 5.43, $p < 0.05$). Patients had significantly delayed ipsilateral P50m with 0.5-s ISI (unpaired t-test, $p < 0.05$), whereas the contralateral P50m latency was about the same in both groups.

AD patients had significantly delayed ipsilateral N100m to the standard tone (MANOVA; F (1, 20) = 14.56, $p < 0.01$). In contrast, the contralateral N100m latency was about the same in the two groups. In the 2.5-s condition, each response was significantly earlier contralaterally in both groups (P50m and N100m in each group, $p < 0.001$, paired t-test). No significant differences were observed in the amplitudes and the source locations of the P50m and N100m between the groups.

Discussion

This is the first MEG study demonstrating that ipsilateral, but not contralateral, auditory processing is impaired in patients with AD. It is proposed that impaired sound localization in AD reflects the dysfunction of the primary auditory cortex [8]. P50m and N100m, as previously stated, seem to be mainly generated in the primary auditory cortex [1]. Therefore, impaired parallel auditory processing between hemispheres might contribute to attenuated sound localization in AD.

In AD, there is both cortical and subcortical loss of neurons and accumulation of amyloid plaques. These degenerative changes also appear in the primary auditory cortices and in the subcortical auditory structures [9]. In the auditory system, a majority of the ascending fibers reach the contralateral auditory cortex, whereas fewer fibers reach the ipsilateral auditory cortex. If one assumes that degenerative changes in AD are rather symmetrical, then the smaller, ipsilateral auditory pathway will be first impaired, leaving the larger auditory pathway more intact.

In conclusion, MEG seems to be a useful tool to objectively study functional changes of the auditory system in AD.

Acknowledgements

This study was supported by the Academy of Finland and the Finnish Medical Foundation. We wish to thank Ms. Suvi Heikkilä for her important contribution in the measurements.

References

[1] Hari, R., Aittoniemi, K., Järvinen, M.-L., Katila, T. and Varpula, T. Auditory evoked transient and sustained magnetic fields of the human brain, Experimental Brain Research, 1980, 40: 237–240.

[2] Näätänen, R., Ilmoniemi, R.J. and Alho, K. Magnetoencephalography in studies of human cognitive brain function, Trends in Neuroscience, 1994, 17: 389–395.

[3] Reite, M., Zimmermann, T. and Zimmermann, E. Magnetic auditory evoked fields: interhemispheric asymmetry, Electroencephalography and Clinical Neurophysiology, 1981, 51: 388–392.

[4] Mäkelä, J.P., Ahonen, A., Hämäläinen, M., Hari, R., Ilmoniemi, R., Kajola, M., Knuutila, J., Lounasmaa, O.V., McEvou, L., Salmelin, R., Salonen, O., Sams, M., Simola, J., Tesche, C. and Vasama, J.-P. Functional differences between auditory cortices of the two hemispheres revealed by whole-head neuromagnetic recordings. Human Brain Mapping, 1993, 1: 8–56.

[5] Pekkonen, E., Huotilainen, M,. Virtanen, J., Sinkkonen, J., Rinne, T., Ilmoniemi, R.J. and Näätänen, R. Age-related functional differences between auditory cortices: A whole-head magnetic study, NeuroReport, 1995, 6: 1803–1806.

[6] Folstein, M., Folstein, S. and McHugh, P.R. "Mini-mental status." A practical method for grading the cognitive state of patients for the clinician, Journal Psychiatry Research, 1975, 12: 189–198.

[7] Hämäläinen, M., Hari, R., Ilmoniemi, R.J., Knuutila, J. and Lounasmaa, O.V. Magnetoencephalography - theory, instrumentation, and applications to noninvasive studies of the working human brain, Rev Modern Physics, 1993, 65: 413–497.

[8] Kurylo, D.D., Corkin, S., Allard, T., Zatorre, R.J. and Growdon, J.H. Auditory function in Alzheimer's disease, Neurology, 1993, 43: 1893–1899.

[9] Terry, R.D., Katzman, R. and Bick, K.L. Alzheimer disease, New York, Raven Press, 1994.

SEFs to Median Nerve and Finger Stimulation: Analysis of Interhemispheric Asymmetries in Healthy Subjects and Following Hemispheric Lesion

Pizzella, V.[1], Rossini, P.M.[2,5], Rossi, S.[2], Tecchio, F.[2], Lopez, L.[4] and Romani, G.L.[3,4]

[1]*Institute of Solid State Electronics, CNR, Rome, Italy;* [2]*Ospedale "Fatebenefratelli", Rome, Italy;* [3]*Institute of Medical Physics and* [4]*Institute of Advanced Biomedical Technologies, Gabriele D'Annunzio University, Chieti, Italy;* [5]*IRCCS "S. Lucia", Rome, Italy*

Introduction

A systematic investigation of the asymmetries between Equivalent Current Dipoles (ECDs) activated in the two hemispheres after left and right median nerve, thumb, and little finger electric stimulation has been performed on 20 healthy subjects and 10 patients with unilateral hemispheric lesion. We studied the location and strength of the ECDs activated in the primary somatosensory cortex contralateral to the stimulated side corresponding to the first cortical component of the Somatosensory Evoked magnetic Field (SEF), namely N20m. Aim of this work is twofold: i) define normal limits for the parameters of the ECD generating N20m component on normal population; ii) use these limits to look for sign of brain plasticity in patients following an hemispheric lesion. On this line a previous paper by ours was devoted to the analysis of interhemispheric asymmetries between ECDs elicited by median nerve stimulation [1]. In the present study we enlarged the sample of our previous control population and, in order to make the test more accurate, we added the stimualtion of single digits (thumb and little finger). In this way the "hand extension", that is the dimension of the parietal region receiving the sensory input from the contralateral hand, was also calulated. In fact, we calculated the "hand extension" as the distance - expressed in millimeters - of the generators responsible for the N20m cortical responses to separate stimulation of thumb and little finger in the contralateral hand.

Materials and Methods

The magnetic field distribution over the scalp following electrical stimulation of the hand was recorded in 20 healthy volunteers (9 males, 11 females, age range 27-60 years) and in 10 patients with unilateral hemispheric lesion. Among the 10 patients, 7 had a striatocapsular lesion, and 3 had a cortical lesion. All examined patients were in the chronic stage (between 2 and 7 months after the stroke), 7 had a complete recovery, while 2 evidenced residual hypoesthesia and the last one had a complete loss of hand sensation.

Measurements were performed using the 28-channel system [2], available at IESS-CNR in Rome. This system features 16 first-order axial gradiometers (1.8 cm coil diameter and 8 cm baseline) and 9 magnetometers (pick-up area 81 mm^2) plus 3 balancing magnetometers for noise cancellation coupled to low noise dc-SQUIDs, with an overall sensitivity of about 5-7 fT /Hz. The 25 measuring sites are regularly distributed on a spherical surface covering an area of about 180 cm^2. All the measurements were performed inside a magnetically shielded room (Vacuumschmelze GMBH).

Left and right little finger, thumb (via ring electrodes) and median nerve at wrist (via surface disks) were separately stimulated by 0.2 ms electric pulses with an interstimulus interval of 641 ms. Stimulus intensities were set at about twice the subjective threshold of perception for fingers and just above motor threshold for the median nerve. About 300 artefact-free trials were acquired (0.48 - 250 Hz bandwidth, 1 KHz of sampling rate) and averaged for each stimulation modality. The entire recording procedure lasted about 1 hour. The amplitude of SEFs was calculated for each channel with respect to a baseline level chosen as the mean value of the 5-15 ms post-stimulus epoch.

We analysed the initial deflection (N20m) of the SEFs, which is quite stable and repeatable, independent of subject attention and therefore more suitable for follow-up studies both in normals and in patients with hemispheric lesions. Since the brain activity at these post-stimulus latencies are exclusively generated by contralateral sources located close to the central sulcus, proper dewar positioning on the hemisphere contralateral to the stimulated side allowed to record the response with a single position of the recording system. The center of the dewar tail was placed over a scalp site approximately corresponding to the C3 and C4 positions of the 10-20 International EEG System. Exact dewar position with respect to the subject head was detected by using five coils positioned on nasion, the two preauricular points, vertex and inion, whose 3D-positions were digitized (Polhemus Isotrak) at the beginning and the end of the recording session.

ECD characteristics (spatial coordinates and strength) were calculated at 1 ms intervals in the 15-50 ms post-stimulus epoch using a moving dipole model inside an homogeneously conducting sphere. The localisation result was accepted only if the explained variance was above 90%. The ECD locations are expressed using a reference system built for each subject/patient according to the following steps (Fig. 1):

- XZ plane is defined by the vertex and the two preauricular points.
- Y axis passes through the nasion and hits normally the XZ plane defining in this way the origin. The nasion y value is assumed to be positive.
- Z axis goes from the origin to the vertex (positive toward the vertex).
- X axis is identified from the previous two using the right hand rule (positive toward right ear).

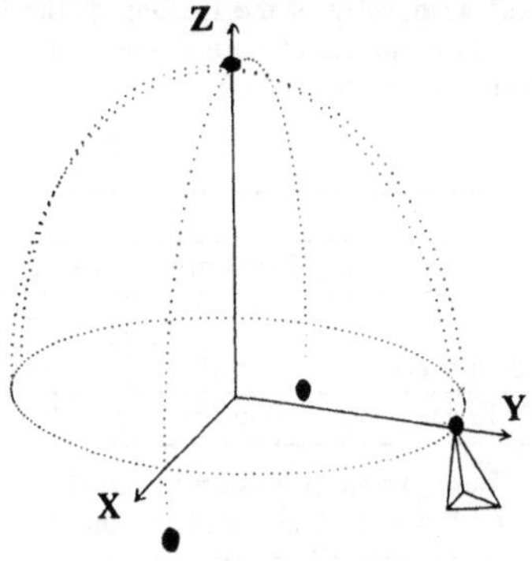

Figure 1 Used reference system. Anatomical landmarks are indicated with black a dot.

Results

For each normal subject the ECD locations relative to median nerve, thumb and little finger were found to be in agreement with the sensory homunculus mapping. The average ECD parameters, together with their standard deviation, relative to all stimulation modalities for the first cortical component (N20m) are reported in Table I.

Table I Average ECD location for N20m component in normal subjects

LEFT hemisphere					RIGHT hemisphere			
x (mm)	y (mm)	z (mm)	s (nA·m)		x (mm)	y (mm)	z (mm)	s (nA·m)
-44 ± 7	-6 ± 9	66 ± 7	6 ± 3	thumb	46 ± 7	-2 ± 6	70 ± 6	6 ± 4
-39 ± 6	-13 ± 6	69 ± 7	15 ± 5	median nerve	41 ± 6	-8 ± 6	72 ± 4	14 ± 7
-37 ± 7	-12 ± 9	77 ± 8	6 ± 5	little finger	36 ± 9	-9± 7	77 ± 8	5± 3

Four parameters were identified to define a normative data set:

loc: distance between the individual ECD and the average of the same ECD across all subjects for each stimulated district (Fig. 2a).

loc_asy: distance between homologous ECDs (i.e. thumb with thumb) in left and right hemispheres after mirroring left hemisphere into the right one, i.e. taking the same positive sign for the x coordinate (Fig. 2b).

hand: the "hand extension" calculated as the distance - expressed in millimeters - between the ECDs relative to thumb and little finger of the same hand (Fig. 2c).

hand_asy: the difference between left and right hand extension.

For each of the above parameters we identified the parameter interval including at least 98% of the normal population. In order to do so, the specific distribution of each parameter was considered. For instance the *loc* parameter is distributed as χ^2 with three degrees of freedom (one for each coordinate x, y, z). The *loc* values which can be considered normal are therefore included between 0 and 2.79×σ, where σ is the maximum value of the standard deviation among the three coordinates. Similarly for the *loc_asy* and *hand* parameters. The *hand_asy* parameter is normally distributed with zero mean (no difference has been found between left and right hand extension) and the normal range is simply $0.\pm 1.96\times\sigma$.

The normal parameters calculated from the normative database are summarized in Table II. This table is to be used as follows: if, for instance, the *loc_asy* parameter for the thumb is greater than 27 mm, there is an "abnormal" asymmetry in the location of the left and right thumb ECDs. This is a sign of a possible brain plasticity. Only one set of normal intervals were identified for *loc* and *hand* parameters since left and right hemispheres were merged together.

Table II Normal parameter intervals

	normal parameter intervals			
	loc (mm)	loc_asy (mm)	hand (mm)	hand_asy (mm)
thumb	0 to 24	0 to 27	0 to 27	-13 to 13
median nerve	0 to 18	0 to 26		
little finger	0 to 26	0 to 32		

These parameter limits were used for the identification of possible brain plasticity in patients following an hemispheric lesion. In 8 out of the 10 studied patients we found that SEFs preserved their morphology, while in two patient no clear N20m response was found from the lesioned hemisphere. These patients were therefore excluded from the analysis. Moreover other four ECD location could not be succesfully identified (the explained variance was below 90%) leaving out a total of 44 successfully identified ECDs. In summary, in 5 cases out of the 10 studied, an abnormal value of one or more different parameter was found. In detail, the first two parameters evidenced the following results:

- subcortical lesions (5 cases): in 2 cases all the three districts (thumb, median nerve, little finger) showed "abnormal" values both for *loc* and *loc_asy* parameters, in one case only loc parametsr for median nerve was out of normality, while in the last two cases all values were within normal limits.
- cortical lesions (3 cases): 2 cases exhibit abnormal parameter values: one case had displacement of little finger ECD and the other of median nerve ECD, with a marked asymmetry respect to the healthy hemisphere.

The last two parameters evidenced the following results:

- subcortical lesions (5 cases): in 1 case the extension of the region reserved to the hand in the lesioned hemisphere was larger than normal. All other cases had parameter values within normal limits.
- cortical lesions (3 cases): in one case the hand extension was enlarged in both hemispheres.

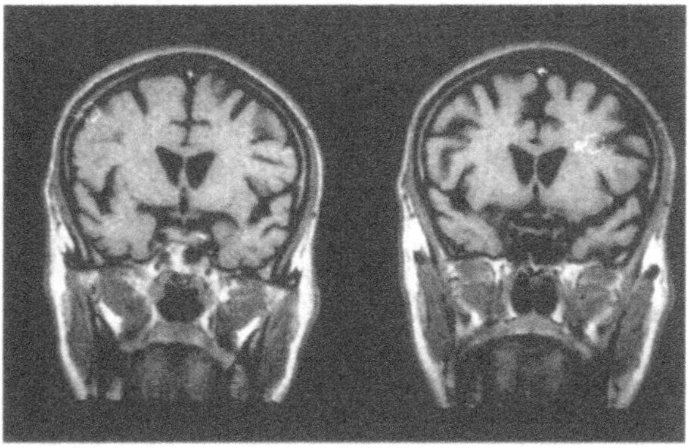

Figure 2 - Example of *loc_asy* "abnormal" value for thumb on a patient with subcortical left lesion. left) ECD location in the right hemisphere relative to left thumb stimulation; right) ECD location in the left hemisphere relative to right thumb stimulation.

Discussion

As it is known from pioneering Penfield's studies the extension of the somatotopical representation in SI of a given body district is proportional to the role it plays for sensory perception and is based upon the "density of cortical neurones directly connected with that body district" [3]. Additionally the short-latency generator sources strictly depict transient or stable reorganization of sensorimotor cortex following transient or stable sensory and motor deficits [4,5].

N20m ECD localisation for thumb and little finger showed the expected "homunculus" distribution [6]; this finding is consistent and repeatable at different times intra-subjectively, and it emphasises the reliability of the spatial ECDs coordinates as a parameter characterising the individual somatotopical organisation of the sensorimotor cortex.

Our findings show that the examined parameters are differently affected by subject's characteristics and instrumental errors. For what concerns the inter-subject variability of loc parameter, it may be ascribed not only to the variability of the sensory cortex somatotopy, but also to the differences in head dimensions as well as of the two central sulci position with respect to the anatomical landmarks. The instrumental error includes sensor noise, sensor position evaluation, limited spatial extension of the biomagnetic sensor and model inadequacy. Some of these elements are removed when absolute ECD coordinates are subtracted to each other. In fact, the loc_asy parameter is not affected by the left and right central sulci positions respect to the anatomical landmarks and the hand parameter is unaffected the sensor position. This justifies the use of several parameter when screening patients looking for brain plasticity.

At the present time, we only evaluated the cortical representation of the hand in the two hemispheres, while the analysis of possible relationships between the parameter "hand extension" and hemispheric "dominance" remained unexplored and beyond the scope of this investigation. Also work is in progress to include following components - such as P30m - and ECD strenght in the analysis to further improve the accuracy of the method.

In conclusion, we have defined the spatial properties of the neural generators embodied in the region of the primary sensory cortex devoted to hand in normals. Additionally, we showed that this non-invasive and safe technique permits to evaluate the dimension of the cortical region receiving sensory information from the hand. The properties of these generators have been used to screen patients following hemispheric lesion looking for sign of brain plasticity. This, in our opinion, will provide a unique opportunity also to follow-up reorganising processes in the sensory cortex somatotopy after unilateral hemispheric lesion and may help in the assessment of the subsequent rehabilitative therapy.

References

[1] Rossini, P.M., Narici, L., Martino, G., Pasquarelli, A., Peresson, M., Pizzella, V., Tecchio, F., and Romani G.L. Analysis of interhemispheric asymmetries of somatosensory evoked magnetic fields to right and left median nerve stimulation. Electroencephalogr. clin Neurophysiol., 1994, 91:476-482.

[2] Foglietti V, Pasquarelli A, Pizzella V, Torrioli A, Romani GL, Casciardi S, Gallagher WJ, Ketchen MB, Kleinsassers AW, Sandstrom RL. Operation of a hybrid 28-channels neuromagnetometer. IEEE Trans. Appl. Sup, 1993, 3: 1890-1893.

[4] Rossini, P.M., Martino, G., Narici, L., Pasquarelli, A., Peresson, M., Pizzella, V., Tecchio, F., Torrioli, G., Romani, G.L. Short-term brain plasticity in humans: transient finger representation changes in sensory cortex somatotopy following ischemic anesthesia, Brain Research, 1994, 642: 169-177.

[5] Flor, H., Elbert, T., Knecht, S., Wienbruch, C., Pantev, C., Birbaumer, N., Larbig, W., Taub, E. Phantom-limb pain as a perceptual corelate of cortical reorganization following arm amputation, Nature, 1995, 375:482-484.

[6] Narici, L., Modena, I., Opsomer, R.J., Pizzella, V., Romani, G.L., Torrioli, G., Traversa, R., Rossini, P.M. Neuromagnetic somatosensory homunculus: a non-invasive approach in humans, Neuroscience Letters, 1991, 121: 51-54.

Acknowledgments

We would like to thank Alberto Pasquarelli for helpful discussions and management, Iole Indovina for technical assistance during data acquisition, Patrizio Pasqualetti, Simone Rossi and Roberto Carluccio for statistical evaluations and preparation of the manuscript.

Effects of Acute, Oral Administration of Phenobarbital (1.6 mg/kg) to Healthy Subjects

Sannita, W.G.[2,3], Lewine, J.D.[1], Maclin, E.L., Orrison, W.W., Jr.[1] and Robinson, S.

[1]New Mexico Institute of Neuroimaging and Department of Radiology, UNM Medical School, Albuquerque, NM, USA; [2]Department of Psychiatry, State University of New York, Stony Brook, NY, USA; [3]Center of Neuropsychoactive Drugs, DISM-Neurophysiopathology, University of Genova, Genova, Italy

INTRODUCTION

The effects of barbiturates on the background EEG signal depend on the administered dose, with increments of fast frequency (peculiarly in the ~15.0-25.0 Hz range) or low frequency EEG activities following administration at small to moderate or at large doses respectively. The increment of fast frequency EEG activity is referred to as distinctive of barbiturates [1-3] and has been documented in healthy volunteers 30-120 min after single, oral doses Phenobarbital (PB) as low as of 50 to 100 mg [4,5], in spite of low drug plasma concentrations (below 6.9 μg/ml) and with pharmacokinetics and brain dynamics that were consistent with available evidence of early CNS bioavailability [6]. The ~15.0-25.0 Hz frequency interval appears to be a spectral parameter with peculiar sensitivity to PB, with substantial correlation between EEG effect and drug plasma concentration [4-5]. The power increase in this frequency segment occurs earlier and at lower doses on central electrodes, therefore allowing inference about possible generators

METHOD

Seven healthy volunteers (5 males and 2 females; age: 30-47 yr. [mean 38.7±6.8 yr.]; body weight: 59-80 kg) were recruited among faculty and staff after excluding any history or clinical evidence of relevant pathology, allergy, or long-term drug use/abuse. Screening EEGs were normal, with the "alpha" rhythm being evident at the visual inspection for at least 60% of recording time in standard resting conditions. Subjects were aware of the study rationale and acquainted with equipment and data acquisition procedures; they agreed to avoid neuroactive compounds in the 72 h before experimental session. MEG signals (BTi System, 38 sensors; 3 antero-lateral locations of the probe) were obtained in baseline conditions and at regular intervals during the 2-h period following oral administration of PB, 1.6 mg/kg. Data acquisition was on the 1.0-100.0 Hz bandwidth, with sampling rate of 297.6 Hz. The raw signal was band-pass filtered at 13.0-23.0 Hz, and the best fitting dipole was determined on every one of 8-ms consecutive data segments by a sequential dipole fitting algorithm. For data analysis, the constrain was artificially set that the root mean square (RMS) signal strength recorded across all sensors at relevant time points be at least 100 fT to focus analysis on signals of relevant strength; in addition, 97% variance criteria were applied to isolate dipolar events under data conditions suggesting that a focal cortical area dominate the recorded signal. A temporal constrain allowed only one valid dipole fit each 40 ms of data, i.e. approximately one dipole per cycle of the beta activity waveforms.

RESULTS

An increase of current density and beta bursting was evident at anterior locations after PB (with drug plasma concentrations of 3.3±0.8 μg/ml), though with limited pre- vs. postdrug differences in absolute power. Most MEG activity in the 13.0-23.0 Hz frequency interval could be modelled by dipoles with approximately 80-90% of variance being accounted for by the model. However, at several instances in time (especially in baseline conditions) the dipole model accounted for less than 80% of the variance, conceivably depending on erroneous

TABLE: number of dipole fits in baseline condition prior to and after administration of Phenobarbital (1.6 mg/kg)

subject	baseline (predrug)	postdrug (time: 30 min)
1	9	64
2	58	161
3	91	175
4	356	371
5	104	427
6	7	98
7	35	69
mean (sd)	94.3 (121)	195.0 (164.4) paired t-test: $p<0.01$

source localization with a single dipole model or implying multiple generators. A reasonable dipole fit was obtained under the specific constraints involving temporal selection, RMS signal strength and variance criteria. The number of valid dipole fits increased in all subjects after PB (predrug: 94.3±121; postdrug: 195.0±164.4; paired t-test: $p<0.01$). In the subjects whose 13.0-23.0 Hz activity in baseline was accounted for by a sufficient number of dipole fits, these were distributed in superior temporal and inferior frontal-parietal locations. After drug administration, the increase in the number of frontal-parietal dipole fits was preponderant compared to superior temporal locations.

DISCUSSION

These preliminary observations suggest that PB action results in increased number and stronger activation of multiple cortical generators of fast activity, particularly those that in baseline are at superior temporal and frontal-parietal locations. At therapeutic doses and with moderate plasma levels of barbiturate: i) the transmembrane resistance and threshold to direct stimulation are increased in in vitro models, while the resting membrane potential is virtually unchanged; these effects are conceivably due to inactivation of Na carrier mechanism, with reduction of Na conductance change, and are common to other compounds active on the CNS; ii) GABAergic inhibition is enhanced, with characteristics (i.e. enhancement of postsynaptic GABA-receptor current with binding at sites different from those of benzodiazepines and by means of regulation of different properties of the GABA-receptor channel) that may in part be peculiar of PB; and iii) barbiturates reduce presynaptic neurotransmitters release and affect Ca channels, including Ca-regulated T-currents involved in the generation of pacemaker currents in the thalamus. A dual role of $GABA_B$ receptors in excitation and inhibition of thalamocortical cells (conceivably mediated by Ca potentials) is also known and is suggested to play a role in the generation of 'burst' firing excitation [8-10]. The experimental conditions of the study and the preliminary results do not allow inference favouring any of these mechanisms of Phenobarbital action on structures/functions generating/modulating the spontaneous electrophysiological activities of the brain. Further studies with selective pharmacological manipulation of these mechanisms are needed for this purpose. Consistent with theory and experimental evidence, the peculiarities of MEG techniques potentially favour application in (human) neuropharmacology.

REFERENCES

1. Fink, M EEG and human psychopharmacology. Ann. Rev. Pharmacol. 1969, 9: 241-258,
2. Sato, H. Relationship between serum levels and fast EEG activities in rats by a single administration of Phenobarbital. Electroenceph. clin. Neurophysiol. 1980, 50: 509-514,
3. Schwartz, J., Feldstein, S., Fink, M., Shapiro, D.M. and Itil, T.M. Evidence for a characteristic EEG frequency response to thiopental. Electroenceph. clin. Neurophysiol. 1971, 31: 149-153

4. Sannita, W.G., Rapallino, M.V., Rodriguez, G. and Rosadini, G. EEG effects and plasma concentrations of Phenobarbital in volunteers. Neuropharmacology 1980, 19: 927-930
5. Sannita, W.G., Balbi, A., Giacchino, F. and Rosadini, G. Quantitative EEG effects and plasma concentration of Phenobarbital, 50 mg and 100 mg single-dose oral administration to healthy volunteers: evidence of early CNS bioavailability. Neuropsychobiol. 1990, 205-212
6. Viswanathan, C.T., Booker, H.E. and Welling, P.G. Pharmacokinetics of Phenobarbital following single or repeated doses. J. clin. Pharmacol. 1979, 3: 382-389
8. MacDonald, R. Antiepileptic drug actions. Epilepsia 1989, 30:19-28
9. Miller, J.W. and Ferrendelli, J.A. Characterization of GABAergic seizure regulation in the midline thalamus. Neuropharmacology 1990, 29: 649-655
10. Crunelli, V. and Leresche, N. A role for $GABA_B$ receptors in excitation and inhibition of thalamocortical cells. Trends Neurosci. 1991, 14: 16-21

Visual Evoked Fields for Pattern Reversal Stimuli in Patients with Occipital Lobe Lesions

Seki, K.[1], Nakasato, N.[1,2], Fujita, S.[3], Kanno, A.[2], Hatanaka, K.[4], Kawamura, T.[1], Ohtomo, S.[1], Fujiwara, S.[1,2], Takahashi, A.[2] and Yoshimoto, T.[1]

[1]*Department of Neurosurgery, Tohoku University School of Medicine, Sendai, Japan;* [2]*MEG Laboratory, Kohnan Hospital, Sendai, Japan;* [3]*R&D Center, Osaka Gas Co. Ltd, Osaka, Japan;* [4]*KRI, Kyoto, Japan*

Introduction

The "cruciform model" of the visual cortex suggests that the P100 generators of the visual evoked responses to pattern reversal (PR) stimuli are located throughout the entire striate cortex, including the interhemisphere surface and calcarine fissures. Our recent studies of visual evoked fields (VEFs) to PR stimuli [1-3] have localized the P100m sources near the lateral end of the calcarine fissures. We suggested that a smaller part of the striate cortex contributes to the P100m response than in the cruciform model. In the present study, we measured PR-VEFs in patients with homonymous hemianopsia due to unilateral occipital lobe lesions, to demonstrate the correlation between P100m patterns and occipital lesions.

Methods

Five patients, 4 males and a female, aged 17 to 68 (mean 42.6) years old, with occipital lesions (four arteriovenous malformations [AVMs] and one brain tumor) participated in this study. All patients had homonymous hemianopsia due to the lesions. Informed consent was obtained after the experimental procedures were explained.

Green-black checkerboard PR stimuli, reversed once per second, were sequentially delivered through a fiber optics system to the full-visual fields. Patients were instructed to watch the fixation point on the screen monocularly. Individual checks and full-field size subtended visual angles of 1.4 and 17 degrees at the patient's eye, respectively. Mean luminance level of the green checks was about 20 cd/m^2, with a contrast greater than 98% to the black checks.

VEFs were measured using a helmet-shaped MEG system (CTF - Osaka Gas) [1-3]. The MEG signals were filtered with 50 and 100 Hz notch filters and direct current to 100 Hz low pass filters and digitized at 625 Hz. The resulting data were averaged based on 200 PR presentations, including a 200 ms pre-trigger baseline and 300 ms after the PR. The P100m peak in the averaged signal was identified within a latency window of 90 to 140 ms after the PR. When a clear two-dipole pattern appeared, the source location was estimated using a two-dipole model. When a one-dipole pattern appeared, the source location was estimated using a one-dipole model. All subjects also underwent three-dimensional MR image. Three fiduciary points, nasion and preauricular positions were marked with small coils for MEG and small oil-containing capsules for MRI. These fiduciary points were used to coordinate the localized dipole position and the MR imaging anatomy.

Results

The P100m peaks of PR-VEFs were observed in all patients (Table 1). In four patients, a one-dipole pattern of P100m appeared over the normal occipital area, corresponding to their homonymous hemianopsia. The P100m dipole positions were estimated at the lateral end of the calcarine fissures in the normal hemisphere (Fig. 1).

In the other patient (*Case 4*), a two-dipole pattern of P100m appeared over the bilateral occipital areas, contradictory to his homonymous left hemianopsia. In this patient, the lateral part of the calcarine fissure was spared where the P100m dipole was localized; only the medial part of the calcarine fissure was occupied by the AVM (Fig. 2).

Table 1

Visual evoked fields in patients with occipital lobe lesions.

Case No.	Age /Sex	Diagnosis /Side	Side of homonymous hemianopsia	P100m latency	Map pattern /Side
1	31/F	AVM /Rt	Lt	105.6 ms	One-dipole /Lt
2	48/M	AVM /Lt	Rt	91.2 ms	One-dipole /Rt
3	17/M	AVM /Lt	Rt	112.0 ms	One-dipole /Rt
4	49/M	AVM /Rt	Lt	116.8 ms	Two-dipole /Bilateral
5	68/M	Brain tumor /Rt	Lt	128.0 ms	One-dipole /Lt

AVM: brain arteriovenous malformation

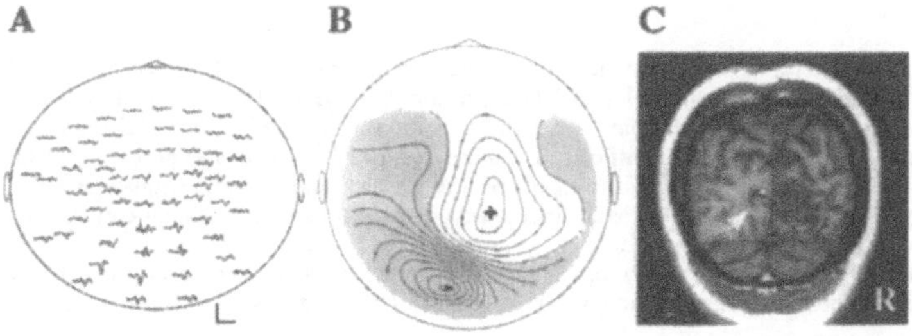

Fig. 1: Visual evoked fields for the pattern reversal full-field stimuli in *Case 1*.
A: All 64 channel data are shown as a view from above. Calibrations are 400 fT and 200 ms.
B: Isofield map of the P100m indicates a one-dipole pattern over the normal left occipital area (15 fT/step).
C: The P100m dipole is estimated near the lateral end of the left calcarine fissure (arrow) on the coronal slice of MRI. Note that the large AVM occupies the right calcarine fissure. Calibration is 1 cm.

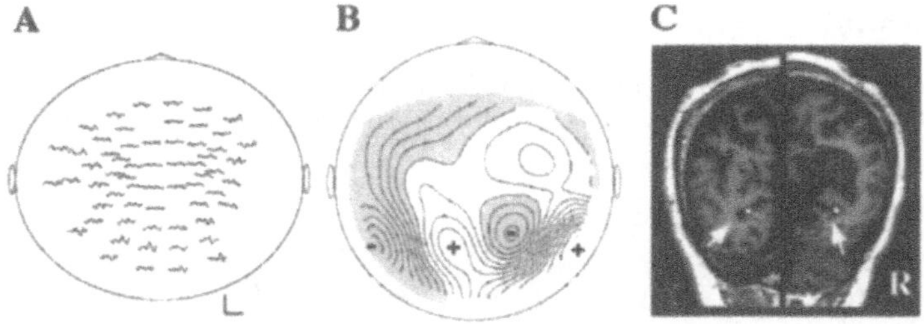

Fig. 2: Visual evoked fields for the pattern reversal full-field stimuli in *Case 4*.
A: All 64 channel data are shown as a view from above. Calibrations are 400 fT and 200 ms.
B: Isofield map of the P100m indicates a two-dipole pattern over the bilateral occipital areas (10 fT/step).
C: The P100m dipoles are estimated near the lateral ends of the bilateral calcarine fissures (arrows) on the coronal slice of MRI. Note that the AVM occupies medial part of the right calcarine fissure, and that the lateral part is spared where the P100m dipole is localized. Calibration is 1 cm.

Discussion

Our recent studies [2] have demonstrated that a two-dipole pattern of the P100m appears over the bilateral occipital areas due to PR stimuli of the full visual fields in normal subjects. The present study found a unilateral one-dipole pattern in four patients with homonymous hemianopsia. Although we showed the lesion-side P100m disappeared due to occipital lesions, it is difficult to confirm the specific cortical area generating P100 responses. However, the other patient had the normal two-dipole pattern, despite homonymous hemianopsia due to a large AVM occupying the medial part of the calcarine fissure and the interhemispheric occipital cortex. We believe this finding supports our hypothesis that the P100m is generated by the lateral part of the calcarine fissure only. PR-VEF may not completely represent the retinotopy of the visual cortex, because of the discrepancy between bilateral P100m responses and homonymous hemianopsia. Further studies are necessary to confirm the cortical origin of PR-VEFs.

References:

[1] Nakasato, N., Seki, K., Fujita, S., Kanno, A., Kawamura, T., Ohotomo, S., Fujiwara, S., and Yoshimoto, T. Functional localization of visual cortex using an MRI-linked whole head MEG system, Human Brain Mapping, 1995, Suppl 1: 63.

[2] Nakasato, N., Seki, K., Fujita, S., Hatanaka, K., Kawamura, T., Ohtomo, S., Kanno, A., Ikeda, H., and Yoshimoto, T. Clinical application of visual evoked fields using an MRI-linked whole head MEG system, Frontiers of Medical and Biological Engineering, in press.

[3] Seki, K., Nakasato, N., Kawamura, T., Fujita, S., and Yoshimoto. T. Source localization of P100m in pattern reversal visual evoked magnetic response. Japanese Journal of Electroencephalography and Electromyography, 1994, 22: 369-374.

Acknowledgements

This work has been supported in part by Grants-in-Aid for Scientific Research No. 03404042 and No. 06404050 form the Ministry of Education, Science and Culture of Japan; and Magnetic Science Foundation.

MEG and fMRI Localisation of Primary Sensory-Motor Cortex in Normals and a Patient with a Low-Grade Astrocytoma

Singh, K.D.[1], Furlong, P.L.[1], Weale, P.[2], Harding, G.F.A.[1] and Walsh, A.R.[2]

[1]*Clinical Neurophysiology Group, Dept. of Vision Sciences, Aston University, Birmingham, UK;* [2]*Queen Elizabeth Medical Centre, Birmingham, UK*

Introduction

The aim of this study was to compare MEG and fMRI localisations of primary sensory-motor cortex, in normal subjects and patients scheduled for neurosurgery. The study is a precursor to the development of a pre-surgical mapping program. The results indicate that in both normals and patients, fMRI and MEG localisations are confirmatory and match well with localisations obtained during surgery. However, in one subject, the MEG localisations were displaced significantly from the central sulcus. We discuss the implications and possible generating mechanisms of this displacement.

Methods

MEG Acquisition

The subjects were six healthy control volunteers and one surgical candidate. The patient was a 44 year old right handed male who presented with a history of tonic-clonic seizures and partial seizures involving myoclonic jerks of the right hand and arm. An MRI scan revealed what appeared to be a large cystic tumour in close proximity to the sensory-motor cortex. The seizures were partially controlled with anti-convulsants but the seizure frequency remained unacceptable. A surgical option was offered.

The proximity of the lesion to the sensory-motor cortex was a cause for concern and in order to investigate the precise juxtaposition of the functional anatomy, Magnetoencephalography (MEG) and Functional Magnetic Resonance Imaging (fMRI) were performed. The same procedures were also performed in the control volunteers for comparison.

MEG evoked signals were measured using a 19 channel magnetometer in response to median nerve and posterior tibial stimulation using surface mounted electrodes. The stimulus duration was 0.1ms with a 600ms inter-stimulus interval. In each measurement, 250 epochs were collected with the dewar optimally positioned over the relevant portion of the head. The data collection rate was 2kHz for an epoch length of 64ms - 128ms. Analogue filters were used with a lowpass range of 0-424Hz. In addition, noise rejection magnetometers were used for spatial filtering. After collection and averaging, digital filtering was used with a bandpass of 2-250Hz.

Precise head location with respect to the magnetometer was achieved using bitebar and head based fiducial markers, which contained oil filled reservoirs for MRI co-registration. MEG source localisations were performed using a single equivalent current dipole model. Once a valid localisation had been obtained at a particular latency, Monte-Carlo analyses [1] were used to assess the estimated cortical region of activation as a 3D ellipsoid [2]. These ellipsoids were then displayed on appropriate MRI views.

Functional MRI

fMRI data were collected using a standard Siemens 1.5T clinical scanner. A FLASH imaging protocol was used (TE = 60 msec, TR = 75 msec, Flip Angle = 40 degrees). The acquisitions were made in a similar manner to those of Jack et al. [3].

The activated state for the subject was finger-thumb opposition, which contains both sensory and motor stimulation. In each run, 10 rest images and 10 activated images were acquired, each image consisting of seven seconds of rest or stimulation. The runs were interleaved in blocks of 5.

Functional data was collected from 4 slices, each 5mm thick, and positioned to encompass the region of interest. The functional data was analysed by subtraction of the averaged rest image from the averaged activated image, and thresholding to leave only high intensity pixels in the functional image. This final image was then superimposed on a high resolution (1mm x 1mm x 1.5 mm) anatomical MRI. Co-registration of function and anatomy was automatic, as the anatomical image was acquired in the same run as the functional image.

Results

MEG Evoked Responses

In the case of the patient, stimulation of the right median nerve at the wrist yielded a recognisable evoked response (Fig. 1a) which was normal in latency and amplitude in comparison with the control group. Two clear peaks of activation were noted at 21ms and 40ms and single equivalent dipole source localisations were calculated at these latencies. Ellipsoid confidence volumes for the source localisations were superimposed onto the MRI (Fig. 1b). The white grid lines were derived from the co-ordinate system of Talairach and Tournoux [4] with the grid origins passing through the anterior and posterior commissures.

Data from stimulation of the right posterior tibial nerve was computed in the same manner at two latencies (43 and 78ms) and yielded similar findings (Fig. 2a and Fig. 2b). Both datasets implicated a location on the posterior bank of the lesion for the site of primary sensory-cortex. This is confirmed by the source localisations being between the two commissural axes of the Talairach and Tournoux co-ordinate system.

Functional Magnetic Resonance Imaging

Analysis of pixel image intensity in relation to the activation sequence revealed areas of significant activation in the sensory-motor cortex region in three control subjects and the patient.

In the control subjects, the fMRI locus was either adjacent to the MEG localisation or anterior to it, as seen in Fig. 3a. In the patient data (Fig. 3b) the MEG localisation is clearly anterior to the fMRI localisation.

Surgical validation

A large craniotomy was performed over the left fronto-tempero-parietal regions to expose both the lesion and the expected area encompassing the sensory-motor cortex. Once the cortex was exposed, the patient was aroused from general anaesthesia to respond to direct electrical stimulation of the cortex. A bipolar probe delivering an alternating current at 50Hz was used. Verbal feedback to sensory stimulation and observation of motor signs enabled mapping of the cortex which was labelled with numeric indicators. For validation purposes, the surgical area was photographed and the gyral patterns compared with 3 dimensional surface rendered images of the cortex derived from the MRI (Fig. 4b). Using this technique, the Rolandic fissure was determined as the first sulcus posterior to the lesion and the appropriate landmark is indicated by the bold black lines in Fig. 4a and Fig. 4b.

Histological section of the lesion determined it to be a low grade Astrocytoma. This was successfully excised and the patient made a complete recovery with no permanent neurological deficit.

Discussion

In this work we have found that in the control subjects, 95% confidence ellipsoids for the MEG dipole fits for median and posterior tibial stimulation were visually identified to be close to the expected location of the central sulcus. fMRI data yielded more widespread activation, but the primary activity was again close to the central sulcus and was therefore consistent with the MEG localisations.

In a single patient, the MEG localisation was displaced anteriorly from the fMRI localisation. This was confirmed by intra-operative cortical mapping which suggested that the MEG data was displaced anteriorly by 8mm from the sensory cortex, for both median and posterior tibial nerve data. In this individual case, therefore, fMRI appeared to be a more accurate predictor for the location of primary sensory-motor cortex.

A possible explanation for this displacement is that the Astrocytoma was a large conductivity perturbation and so the true conductivity profile may not be the homogeneous sphere model used in the source localisation algorithm. If this is so, this has important implications for the use of MEG/EEG pre-surgical mapping in subjects who have large tumours (for example) close to the region of interest. It may be that in these circumstances, realistic conductivity models of the head and any lesions may be required. Alternatively, the displacement might reflect cortical re-organisation of primary sensory-motor cortex which has occured during the development of the Astrocytoma.

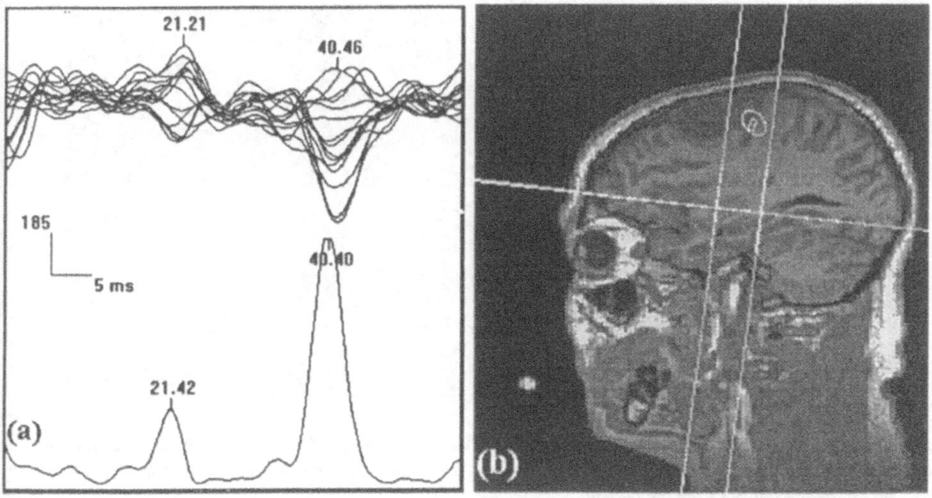

***Fig. 1a** (left upper) shows the MEG evoked responses, from the patient, following Right Median nerve stimulation. The lower trace shows the integrated Global Field Power of the same dataset, indicating two latencies of maximum activation. **Fig. 1b** (right) shows the 95% confidence ellipses for source localisations at the two latencies indicated. The white grid lines show the axes of the co-ordinate system of Talairach and Tournoux.*

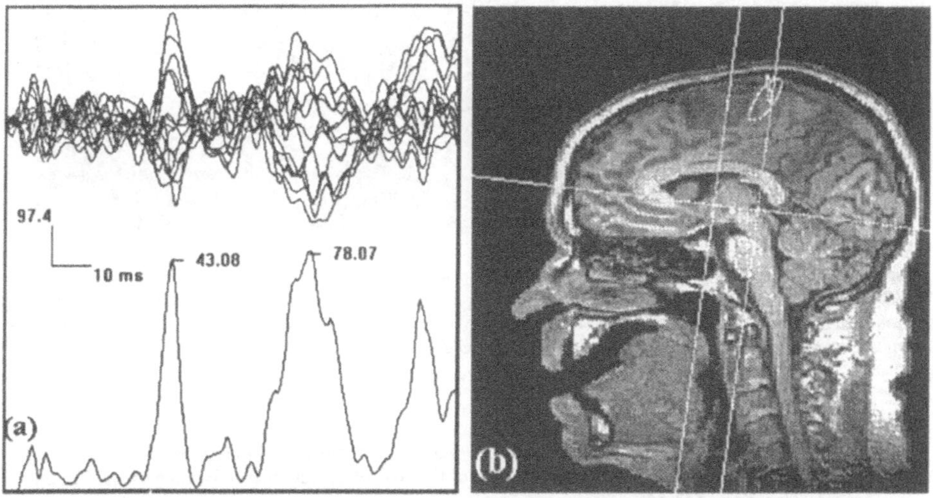

***Fig. 2a** (left upper) shows the MEG evoked responses, from the patient, following Right Posterior Tibial nerve stimulation. The lower trace shows the integrated Global Field Power of the same dataset indicating two latencies of maximum activation. **Fig. 2b** (right) shows the 95% confidence ellipses for source localisations at the two latencies indicated. The white grid lines show the axes of the co-ordinate system of Talairach and Tournoux.*

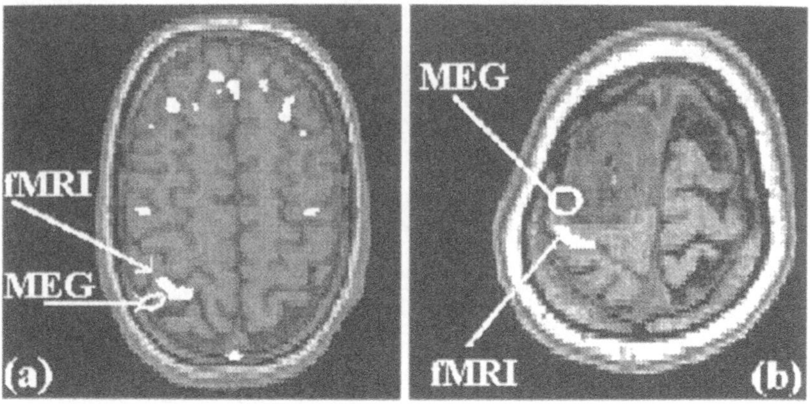

Fig. 3a (left) indicates the location of maximal fMRI activation following a finger-thumb opposition sequence, together with the MEG localisation following Median nerve stimulation, in a control subject. ***Fig. 3b*** *(right) shows the corresponding dataset from the patient.*

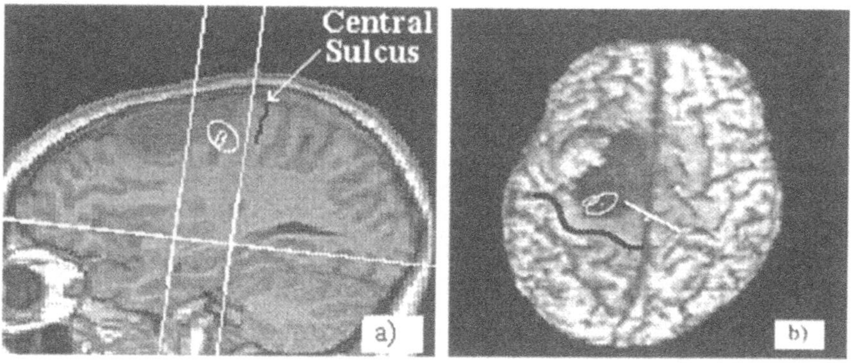

Fig. 4a (left) shows the position of the central sulcus as determined by peroperative cortical stimulation (bold black line) together with the MEG localisation following Median nerve stimulation (white ellipses). ***Fig. 4b*** *(right) shows both Median nerve and Posterior Tibial nerve MEG localisations superimposed onto a 3D surface rendered cortical image with the central sulcus indicated by the bold black line.*

References

[1] Medvick, P., Lewis, P.S., Aine, C.J., and Flynn, E.R. Monte Carlo analysis of localisation errors in magnetoencephalography. In: Williamson, S.J. et al (Eds), 1990, Advances in Biomagnetism. Plenum Press, New York, 543-546.

[2] Singh, K.D., Harding, G.F.A. This Volume.

[3] Jack, C.R., Thompson, R.M., Butts, R.K., Sharbrough, F.W., Kelly, P.J., Hanson, D.P., Riederer, S.J., Ehman, R.L., Hangiandreou, N.J., and Cascino G.D. Sensory motor cortex: correlation of presurgical mapping with Functional MR imaging and invasive cortical mapping. Radiology, 1994, 190:85-92.

[4] Talairach, J., and Tournoux, P., Co-planar stereotaxic atlas of the human brain. Stuttgart, Thieme, 1988.

Comparison Between the Processing of Static Images of Human Faces in High Functioning Autistic Subjects and Normal Controls

Swithenby, S.J.[1], Bräutigam, S.[1], Bailey, A.J.[2], Jousmaki, V.[3] and Tesche, C.[3]

[1]The Open University, Department of Physics, Milton Keynes, MK7 6AA, UK; [2]MRC Child Psychiatry Unit, 16 De Crespigny Park, London SE5 8AF, UK; [3]Low Temperature Laboratory, Helsinki University of Technology, 02150 Espoo, Finland

1 Introduction

We have studied neural processing of photographs of faces by normal and autistic individuals using the Neuromag 122^{TM} channel helmet system.

Autism is a severe neurodevelopmental disorder that involves a variety of fundamental cognitive, psychological and neurobiological abnormalities and is not associated with a characteristic brain lesion. The abnormalities which tend to occur in an autistic individual are;

- deficits in social interactions, specifically a lack of social reciprocity
- a delay in language development and continuing communication problems
- a tendency to stereotyped and repetitive patterns of behaviour.

Studies on autistic subjects using other techniques have revealed deficits in face processing abilities including impairments in recognition of emotional expression and identity (for recent reviews [1], [2]).

Previous studies on normal subjects with both MEG and EEG have revealed face specific activation in the extrastriate cortex in association with presentation of images showing faces and non-face objects[4,5]. Guided by these observations we developed four experimental tasks to probe face specific responses in normal subjects and possible abnormalities in autistic individuals, specifically in the early stages of neural processing of images. The subjects had to;

1. identify faces (or control images) in a sequence of randomly presented images (adult faces, motor bikes, abstract and dot patterns, familiar objects and animals).
2. identify identical faces (or control images) amongst a series of image pairs (boys' faces, wallpaper and dot patterns, motor bikes and mugs).
3. identify neutral (smiling) faces in a series of faces.
4. identify the same individual amongst pairs of faces expressing different emotions.

In each task, stimuli were followed by a visual cue requesting the subject to respond by pressing an appropriate key on a non-magnetic keypad. The combination of visual cue and keypad response was used to control attention conditions, to record behavioural data and to separate motor activity from visually evoked responses.

2 Method

Subjects
10 normal (6 males and 4 females, ages between 24 and 50) and 4 autistic (3 males and 1 female, ages between 24 and 35) subjects were tested. The autistic subjects were classified as high functioning and recruited from Great Britain. No significant attempt has been made to match these two groups with respect to age, gender and psychological measures, such as verbal or non-verbal IQ.

MEG
All experiments were carried out at the Low Temperature Laboratory in Helsinki using a 122 whole-head system. Data was sampled at a rate between 370 and 400 Hz, with an anti-aliasing filter set at 130 Hz. One-line averaging (after artefact rejection) was performed for the different stimulus categories within a given task. Prior to further analysis, the data sets were clipped in time range, baseline corrected and digitally filtered (0.3 - 45 Hz).

Task Protocols
Image presentation, trigger signals for the data acquisition system, and driving of a non-magnetic keypad to record subject responses were controlled by a laptop computer and custom software. All images were presented as grey-scale 8-bit bitmaps via a projection monitor. The images were controlled for luminance, contrast and size. The diagram below shows one cycle in the procedure for Task 1 (similar timings and sequences were used for the other tasks)

Image →	Pause →	Prompt Cue →	Response
100ms	1s +/- 200ms	200ms	max 1.8s

Analysis
MEG data were analysed in several steps comprising 1) visual inspection of channel output $S_i(t)$, 2) tabulation of the statistics of latencies and amplitudes of easily recognisable peaks, 3) visual inspection of signal topography via rotated gradient plots, and 4) examination of the statistics of integrated power summed over groups of channels. Behavioural data (correctness of key-presses and reaction times) were analysed by standard statistical methods. Integrated power ratios for a given interval are defined as

$$IPR = \frac{\Sigma_{i=1}^{N} \int_{t_1}^{t_2} S_i^2(t)dt}{\Sigma_{i=1}^{N} \int_{-b}^{0} S_i^2(t)dt}$$

where N denotes the number of channels in a given channel group, and b determines the stretch of data used for baseline correction. All integrals were numerically estimated by a Gill-Miller algorithm. The channel groups were carefully selected to sharpen the power content over a given brain region without blurring the analysis due to overlapping areas of activation.

3 Results

Behavioural Data
Task performance, as measured by correct responses, was essentially perfect for normal subjects in all tasks. The four autistic subjects completed tasks 1 and 2 equally well. However, their performance on tasks 3 and 4 (tasks involving classification of emotion) was significantly impaired. There are no simple and clear trends in the reaction times

Qualitative MEG Observations
In normal subjects, there is a clear and characteristic face specific response in the right occipito-temporal region at 130 to 140 ms. This response appears to be independent of the specific nature of a task, as shown in Fig. 1a. This observation holds true in 8 out of 10 normal subjects. One subject showed an equivalent signal pattern but in the left occipito-temporal region. The signal topography, given by rotated gradient plots, between 130 and 140 ms is similar across this subject group but has varying degree of complexity. In most subjects, the dominant current contribution is consistent with activation in the fusiform gyrus. The averaged signals did not reveal any significant evidence for face specific activity at other latencies.

In our 4 autistic subjects, qualitative observations failed to show any clear evidence for face specific responses. However, evoked responses, clearly distinguished from noise, were observed in the right occipito-temporal region in all cases (see Fig. 1b).

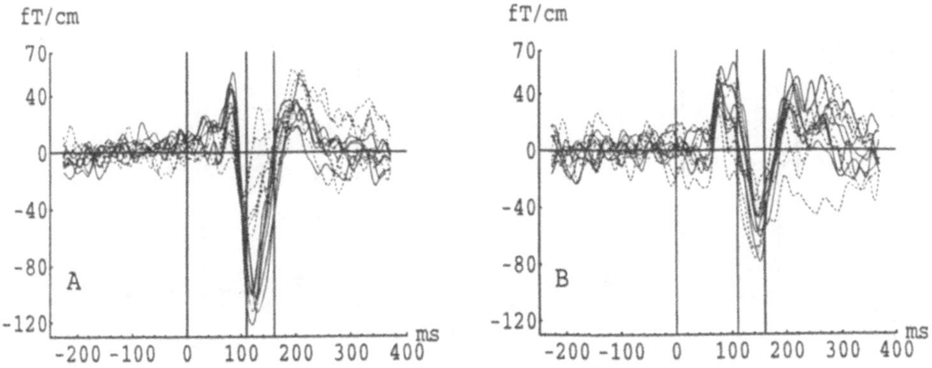

Figure 1: Face (solid) versus non-face (dashed) responses across all tasks for a normal, A, and an autistic, B, subject.

Quantitative MEG Observations

Summing signal power of a channel group sharpens specific features in the evoked responses (Fig 2). In the right channel group alone, there is a clear peak at approximately 130 ms, denoting the face specificity. Integrating over this peak (between 110 and 160 ms) provides a numerical measure which we have computed for all subjects. Fig. 3 shows relevant cohort means. There is a clear difference between face and non-face responses in normal but not autistic subjects.

An analysis of variance[1] for the data shown in Fig. 3 confirmed the clear distinction between face and non-face stimuli ($F_{3,36} = 4.3$, $p = 0.01$) in normal subjects. This pattern was further confirmed by the task 2 data, but with slightly less significance ($F_{2,18} = 3.3$, $p < 0.07$). This distinction could not be established for either task in the autistic subjects ($F_{3,12} = 0.77$, $p > 0.5$ and $F_{2,9} = 0.16$, $p > 0.8$). In both subject groups, in task 1, there is an hemispherical asymmetry in the response to faces (normals : $F_{1,14} = 4.5$, $p \approx 0.05$; autistics : $F_{1,6} = 3.8$, $p < 0.1$).

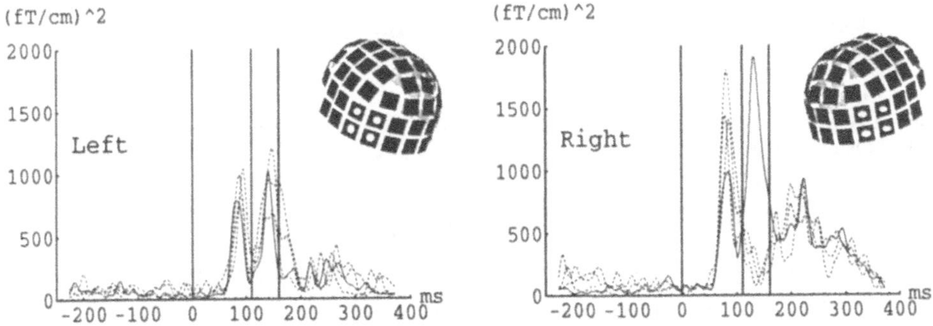

Figure 2: Channel-summed signal power in a normal subject (face:solid, non-face:dashed). The insets show the channel groups used for summation.

[1] Skewness was acceptable in most cases. In cases with higher skewness, logarithmic transformation did not alter the statistics significantly and analysis was performed on the raw ratios

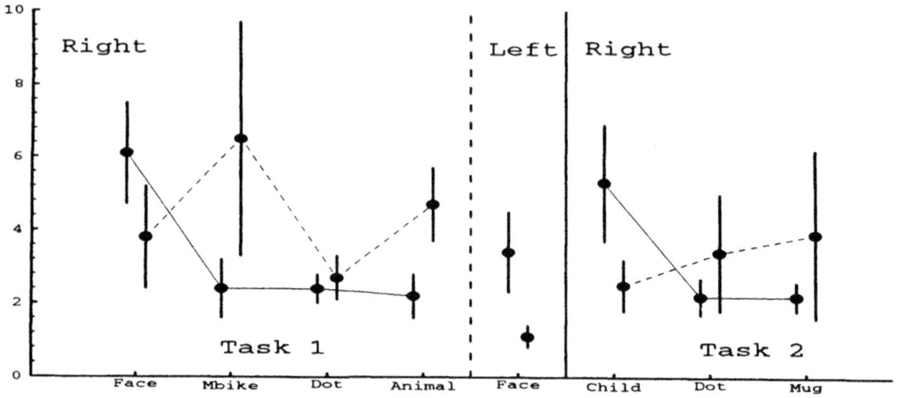

Figure 3: Cohort mean values of power ratios for normal (solid) and autistic (dashed) subjects.

4 Concluding Remarks

In this study, we have departed from common practice in not attempting any detailed source identification. The complexity of the signal topography and the results of preliminary modelling studies suggested that localised generator models would not be appropriate. Instead we have used a measure based on integrated regional power to compare responses of different subject groups and tasks. In normal subjects, our results are broadly consistent with face specific processing predominantly in the right hemisphere, as shown in previous studies. However, the associated latencies are generally earlier than reported. The underlying activity may reflect the earliest task independent stages in Bruce and Young's face processing model[3].

The autistic group was able to carry out those face processing task not involving classification of emotion, but there was no significant evidence for face specific activity. These observation suggest they use an abnormal strategy. It is unclear how this relates to current suggestions that autistic people analyse in terms of components parts, rather than using whole face representations.

Further work involving a larger group will emphasise possible differences between normal and autistic individuals at later latencies, specifically in pair discrimination tasks. It will also include source distribution analysis.

This study was supported by the EC HCM program BIRCH at the Low Temperature Laboratory, Helsinki, and by the MRC and EPSRC, UK.

[1] Bailey A, Phillips W and Rutter M J. Child Psychol. and Psychiatry, 1996, Vol 36 No 1

[2] Boucher J and Lewis V J. Child Psychol. and Psychiatry, 1992, Vol 33 No 5, pp 843-859

[3] Young A W Phil. Trans. R. Soc. Lond. B, 1992, 335, pp 47-54

[4] Lu et al. Neuroscience, 1991, Vol 43 No 2/3, pp 287-290

[5] K. Bötzel, S. Schulze and S. Stodieck Experimental Brain Research, 1995, Vol 104, pp 135-143

A Comparison of MEG and EEG for the Assessment of Traumatic Brain Injury

Weinberg, H.[1], Gaetz, M.[1], Jantzen, K.[1] Cheyne, D.[1] and Deecke, L.[2]

[1]Brain Behaviour Laboratory, School of Kinesiology at Simon Fraser University; [2]Neurological University Clinic, Vienna, Austria

Introduction

The current series of experiments are designed to establish methods for using MEG to assess information processing deficits after mild head injury and for the identification of which functional neurophysiological systems are impaired. The eventual goal of these experiments is to facilitate rehabilitation by accurately targeting the functional anatomy of the impaired systems through analysis of MEG and EEG. One important aspect of these procedures is to determine the plastic capacity of impaired systems. What are reported here are initial results with respect to the utilisation of MEG in EEG paradigms we have been using for the assessment of head injury.

Methods

EEG data was collected with 21 electrodes placed according to the international 10/20 electrode placement system. Third gradient MEG data was collected using a CTF 143 channel whole head MEG system in an unshielded environment.

Stimulation parameters: Stimulation parameters were kept as similar as possible in the EEG and MEG paradigms. However, due to the magnetic properties of the stimulus transducers, the auditory stimuli for MEG recordings were delivered through plastic tubes, resulting in a stimulus delay of 23 ms. Visual stimuli were presented at a distance of 1 meter for EEG recordings whereas in the MEG paradigm, a computer monitor had to be positioned at a distance of two meters from the MEG sensors. Adjustments in stimulus size were made to maintain a constant visual angle for both EEG and MEG recordings. Auditory mid latency responses were evoked with alternating clicks presented at 70 dB every 100 ms. Clicks were presented to the left and right ears separately; a masking noise (-20 dB) was presented in the non stimulated ear. Stimulation for the pattern visual evoked responses was monocular full visual field presented seperately to the left and right visual field of each eye. The stimulus consisted of a black and white checkerboard pattern which reversed at 3 Hz. Each check subtended a visual angle of approximately 35.93 minutes of arc.

Visual and auditory p300 event related responses were collected using the standard oddball paradigm. Three types of visual stimuli were used, shapes, numbers and words. The auditory stimuli were two sine wave tones. The stimulus duration was 100 ms with an ISI of 1000 +- 100 ms. The infrequent stimulus was presented according to a pseudo-random schedule in 20 percent of the trials.

For CNV's, as with the P300's, one auditory and three visual conditions were collected. The stimuli for the visual conditions consisted of shapes, numbers and words; the auditory stimuli were sine wave tones. The stimulus duration was 100 ms., The interval between S1 and S2 was 2000 ms and the ISI was 2000 +- 100 ms. The "go" stimulus, in a standard go-no-go CNV paradigm was presented on 50 percent of trials according to a pseudo-random schedule.

Data Collection Parameters: Mid-latency Auditory Evoked Potentials (MAEP) were digitized at a sample rate of 4000 Hz for a period of 64 ms and filtered online between 10 and 1000 Hz. The MAEPs were digitized at a sample rate of 1250 Hz for 100 ms. and were filtered online between 0 and 500 Hz. Averages consisted of 700 to 1000 artifact free trials. Pattern Visual Evoked Potentials (PVEP) were sampled at 1000 Hz and filtered online between 1 and 100 Hz. The PVEP were sampled at 1250 hz for 200 ms. and filtered online between 0 and 300 Hz. Each average consisted of 150 to 200 artifact free single trials. EEG P300 data was sampled at 250 Hz for 1024 ms and filtered online between 1 and 30 Hz. MEG P300's were sampled at 250 for 1000 ms and lowpass filtered below 35 Hz. Each average consisted of 15 to 30 artifact free single trials. MEG CNV's were sampled at 62.5 Hz for 4096 ms. and lowpassed online below 30 Hz. The Magnetic counterpart of the CNV was sampled at 125 Hz for 3504 ms and lowpassed online below 25 Hz. Each average consisted of 15 to 30 single trials.

Results

Auditory and Visual P300 MEG and EEG: Fig. 1 (auditory) and 2 (visual) are P3 responses that were significantly delayed in both EEG and MEG when compared to the normative EEG data base (greater than 2.5 SD). Although the EEG shows a vertex maximum distribution suggesting a single source, the MEG shows two dipolar like distributions, one in each hemisphere, for both auditory and visual modalities. Further, the MEG data suggest a different distribution for visual and auditory P300's whereas the EEG distributions are the same for both modalities.

Figure 1. Auditory P300 Responses

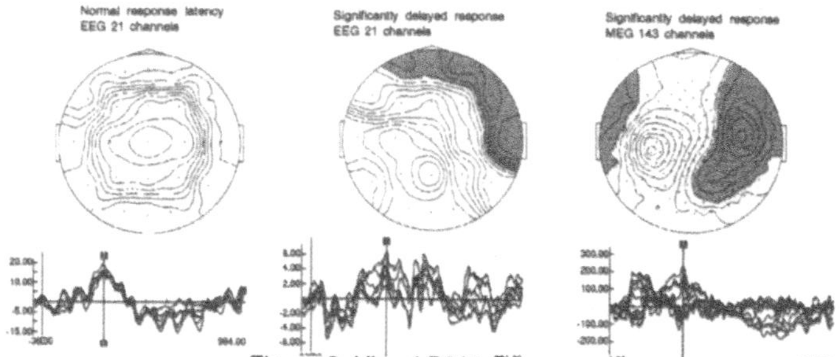

Figure 2. Visual P300 Responses

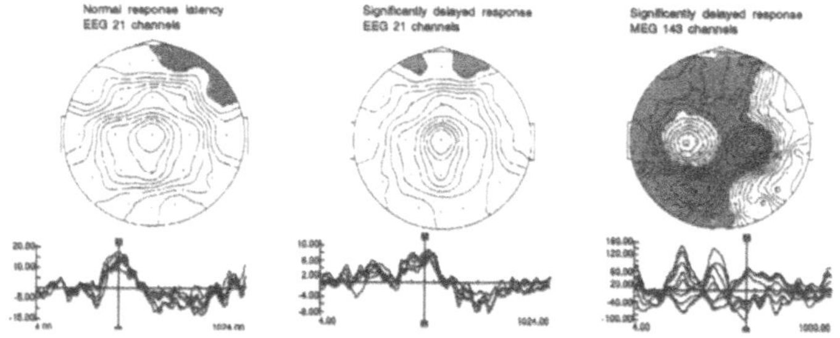

Checkerboard Reversal MEG and EEG: Fig. 3 shows a checkerboard response for right monocular full field stimulation. Both EEG and MEG showed a delayed response compared to the EEG normal data base. The EEG shows the usual paradoxical localisation [1] resulting from the orientation of sources in calcarine fissure. However, the MEG checkerboard response, shows the correct lateralisation. A comparison of PVEPs using EEG and a whole head MEG in the same subject has not been previously reported as far as we know although PVEPs using a whole head system has been reported [2].

Middle-Latency Responses: Figure 4 shows one of the first reports of MEG mid-latency response using a whole head gradiometer system. Kakela et al., 1994 [3] reported a midlatency response using a planar gradiometer. In the four head inured patients used there were no abnormalities in latencies in either EEG or MEG. EEG mid-latency delays are seen in head injured patients when there is clear thalamic involvement in sensory deficits. However, in these patients there were no such delays. The EEG distributions shows no localized activity whereas the MEG reveals a clear bilateral dipolar response in posterior temporal lobe.

Contingent Negative Variation (CNV): The EEG CNV maps (figure 5) shows a typical vertex maximal distribution which is not uncommon in both normal and head injured subjects, unless there is significant frontal

Figure 3. PVEP and PVEF responses to checkerboard reversals.

Right Monocular Full Field

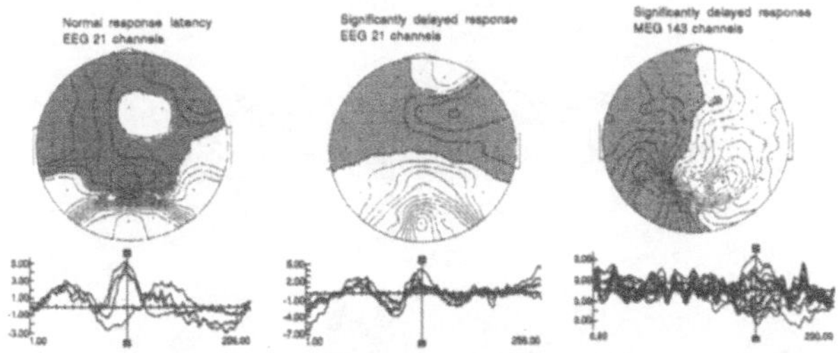

Right Monocular Right Visual Field

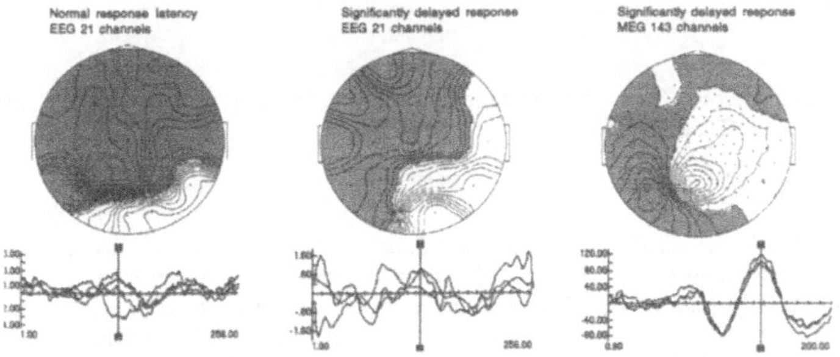

lobe damage. However, the MEG, unlike the EEG, shows clearly defined bilateral source fields for visual stimuli in S1 and S2. Compared to EEG, the whole head MEG recorded, appears to provide similar information regarding response latencies, but a much greater definition of localised activity. The MEG adds additional information about the distribution and strength of various components of evoked and event related potentials; although the EEG generally shows a similar distribution for different components of an evoked potential. This is particularly important in mid-latency responses where the MEG results in dipolar-like bilateral fields whereas the EEG does not. The checkerboard MEG response is similar to EEG for full field stimulation, however monocular half-field stimulation resulted in a clearly defined activity in the correct hemisphere and did not display paradoxical localisation - a result which will have important advantages for the analysis of abnormalities resulting from the thalamo-cortical systems responsible for this response. For example, the N2-P2 latencies observed in the EEG and MEG were similar whereas the EEG resulted in a distribution with a vertex-parietal maximum. The distribution of

Figure4. MidLatencyAuditoryEvokedResponses

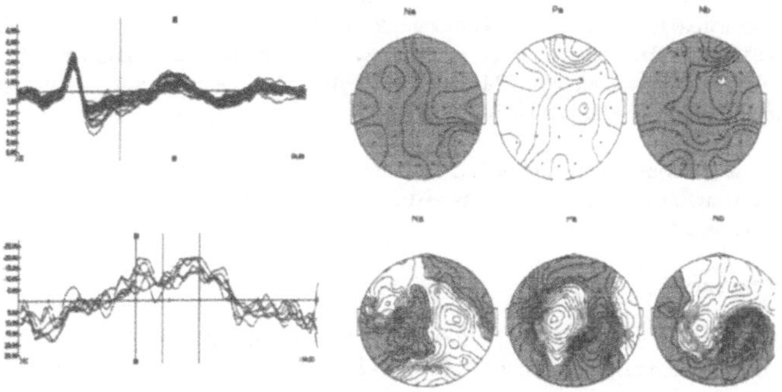

Figure5. ContingentNegativeVariation

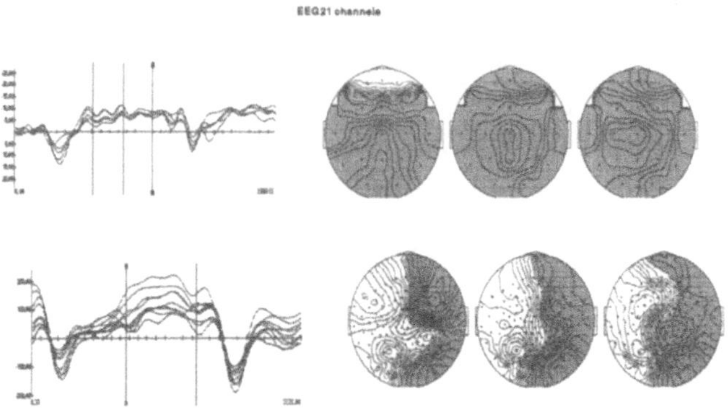

the MEG clearly indicated bilateral sources. This increased definition of the complexity of distributions is also demonstrated in the measurement of MEG and EEG, CNVs

Discussion

Compared to EEG, the whole head MEG recorded, appears to provide similar information regarding response latencies, but a much reater defintion of localised activity. The MEG adds additional information about the distribution and strength of various components of evoked and event related potentials; although the EEG generally shows a similar distribution for different components of an evoked potential. This is particularly important in mid-latency responses where the MEG results in dipolar-like bilateral fields whereas the EEG does not. The checkerboard MEG response is similar to EEG for full field stimulation, however monocular half-field stimulation resulted in a clearly defined activity in the correct hemisphere and did not display paradoxical localisation - a result which will have important advantages for the analysis of abnormalities resulting from the thalamo-cortical systems responsible for this response. For example, the N2-P2 latencies observed in the EEG and MEG were similar whereas the EEG resulted in a distribution with a vertex-parietal maximum. The distribution of the MEG clearly indicated bilateral sources. This increased definition of the complexity of distributions is also demonstrated in the measurement of MEG and EEG, CNVs.

These experiments are reported as initial results in an ongoing program of research intended to use the MEG for the assessment of processing deficits associated with mild head injury and the definition of which functional source systems are impaired.

References

1] Barret G., Blumhardt L., Halliday AM., Halliday E and Kriss A. (1979). A paradox in the lateralization of the visual evoked response. Natrue, 261: 253-255.

[2] Seki, K., et al. P100m of pattern reversal visual evoked magnetic field: source localization using an MRI - linked helmet-shaped magnetoencephalography system. EEG Clin. Neurophysiol. (Society Proceedings), 1995, 95: 86-94

[3] Makela J.P., Hamalainen M., Hari R., and McEvoy L. (1994). Whole-head mapping of middle-latency auditory evoked magnetic fields. Electroencephalography and Clinical Neurophysiology, 92, 414 - 421.

Somatosensory Evoked Magnetic Fields in Acute Cerebral Infarction

Wikström, H.[1,2], Huttunen, J.[2], Roine, R.O.[1], Salonen, O.[3], Aronen, H.J.[3] and Ilmoniemi, R.J.[2]

[1]*Department of Neurology;* [2]*BioMag Laboratory, Medical Engineering Centre;* [3]*Department of Radiology, Helsinki University Central Hospital, Helsinki, Finland*

Introduction

Cerebral infarction interfering with sensorimotor functions causes abnormalities in the somatosensory evoked potentials (SEPs) [1,2]. Somatosensory evoked fields (SEFs) [3] reflect partially the same phenomena as SEPs, but allow more accurate spatial discrimination of the underlying neural generators. We present preliminary results of SEFs from the primary sensorimotor cortex (SMI) in the acute stage of cerebral infarction with an emphasis on correlating SEF abnormalities with clinical impairments.

Methods

Sixteen patients with their first unilateral cerebral infarction were studied within 1 week from the onset of symptoms. The patients were categorized according to their clinical picture to those with purely motor paresis, those with selectively sensory symptoms and those with sensorimotor paresis. On the basis of clinical testing of touch, vibration, position, pain and temperature senses, sensory dysfunction was categorized as mild, moderate, or severe. Motor paresis was considered mild if almost normal function of the affected limb was possible, only skilled movements being impossible. A moderate paresis was characterized with preserved ability to lift the affected limb and in severe paresis the affected limb could not oppose the force of gravitation.

Left and right median nerves were stimulated alternately at the wrist with 0.2-ms constant-current pulses causing a thumb twitch. The interstimulus interval at each side was 3 s. The patient sat motionless in a magnetically shielded room watching a movie while SEFs were recorded with a 122-channel helmet-shaped first-order planar gradiometer array (Neuromag Ltd.) [4].

The signals were band-pass filtered at 0.03–310 Hz and digitized at 940 Hz. 100–200 responses were averaged for each run. Vertical and horizontal electro-oculograms (EOG) were monitored and sweeps with EOG signals exceeding 150 μV were automatically rejected.

An individual cartesian coordinate system was defined for each patient using a digitizer so that the x-axis passed through the preauricular points with the positive direction to the right; the positive y-axis passed through the nasion; thus, the positive z-axis pointed upwards. Three position indicator coils were attached to the head; their coordinates were also determined with the digitizer. Before each SEF trial, when the subject's head was positioned under the magnetometer dewar, current was passed through the coils; the resulting magnetic fields were recorded to locate the coils, allowing for a transformation between head and magnetometer coordinate systems.

The amplitudes of N20m, P35m, and P60m were expressed as dipole moments of equivalent current dipoles (ECDs). The strengths and locations of the ECDs were determined with least-squares searches using data from 18–22 channels covering both field extrema of the corresponding SEF deflection. The head was modelled with a sphere best fitting the inner surface of the skull in each subject's magnetic resonance (MR) images. The transformation between head and MR coordinate systems was done on the basis of the three anatomical landmarks easily identified from the MR images: the nasion and the two preauricular points. All ECDs obtained in this way explained over 80% of the field variance and were accepted as models of local cortical activity.

Results

There were no significant asymmetries in the ECD locations between the hemispheres. Neither did we observe changes in SEF laterality, i.e., no abnormally enhanced SEFs from the ipsilateral hemisphere after stimulation of the affected hand were seen. The data from individual patients are presented in the Table 1.

SEF amplitude asymmetries and clinical picture

Four patients had purely sensory symptoms. Three of them had symmetric SEFs, in the fourth patient they were missing on the affected side.

Three patients suffered from selectively motor paresis. Two of them had severe and one mild paresis. All shared mild SEF asymmetries.

Nine patients suffered from sensorimotor paresis. Four had severe paresis; three of them had also severe asymmetry of SEFs. However, one patient with severe paresis had no SEF asymmetry at all. Three patients had moderate symptoms. One of these had severe SEF asymmetry and the two remaining had moderate asymmetry. Two subjects with mild paresis had either mild or moderate SEF asymmetry.

Four patients, one with sensory, one with motor, and two with sensorimotor paresis had an enlarged P35m or P60m deflection in the non-affected hemisphere. Their pareses were classified as moderate or severe. Three of the patients had large cortical-subcortical infarcts and one had a small pontine lesion. The data from patient EW, featured by enlarged P60m deflection on the healthy side, is shown in the Fig. 1.

Table 1. Patient characteristics and SEFs in the infarcted hemisphere compared with the unaffected side. C = cortical, SC = subcortical, 0 = no asymmetry, – = moderately reduced, — = severely reduced, + = enhanced, a.b = absent bilaterally. In the symptoms column, the size of the letters indicate magnitude of involvement of the corresponding system, for example, Sm indicates dominance of sensory symptoms over motor symptoms.

Patient	Age	Sex	Side	Infarction	Symptoms	Severity	N20m	P35m	P60m	SEF abnormality
KL	65	M	L	C SC	S	moderate	0	0	0	none
JP	51	M	R	SC	S	moderate	—	—	—	severe
HG	48	M	L	C SC	S	severe	0	0	0	none
AP	59	F	L	C SC	S	mild	0	0	0	none
ML	56	M	L	C SC	M	mild	0	–	+	mild
SS	49	M	L	SC	M	severe	–	–	0	mild
PE	59	M	R	SC	M	severe	+	0	0	mild
VR	77	M	L	C SC	SM	moderate	—	—	a.b	severe
SH	26	F	L	SC	sM	severe	+	0	0	mild
UH	53	M	L	SC	SM	mild	0	+	0	mild
AW	62	M	R	SC	SM	mild	—	–	a.b	mild
JR	48	M	L	C SC	SM	severe	—	—	—	severe
HV	63	M	R	C SC	Sm	moderate	—	–	—	mild
RL	66	M	R	C SC	Sm	moderate	0	—	a.b	mild
AV	56	M	R	C SC	SM	severe	—	a.b	—	severe
EW	64	M	L	C SC	SM	severe	—	—	—	severe

Discussion

In the present study we had 16 patients with their first unilateral cerebral infarction. The data of four of them with only sensory dysfunction suggests that in purely sensory paresis the SEF amplitudes can be symmetric. Purely motor paresis, observed in three patients, was associated with mild amplitude asymmetries. Sensorimotor paresis had more profound effects on the patients' SEFs. Of our nine patients with sensorimotor dysfunction, all had asymmetric SEFs and in four the asymmetry was severe. The severity of SEF amplitude asymmetry correlated poorly, however, with the severity of sensory symptoms; even severe sensory paresis could be associated with symmetric SEFs. An interesting finding is the enhancement of P35m or P60m in the non-affected hemisphere of four patients with moderate to severe paresis. Three of them had large cortical infarctions and one small pontine lesion. This enhancement can be in association with impaired tonic interhemispheric interactions in patients with extensive infarctions (2).

The present results did not show any changes in the laterality of sensory functions in the acute stage of infarction; i.e. SEFs after stimulation of the affected limb, if present, had source locations at SMI on the affected side, and no major activity was seen in the unaffected hemisphere. A follow-up study will reveal, whether changes in the laterality occur in the long run, especially in those patients who recover from their clinical symptoms.

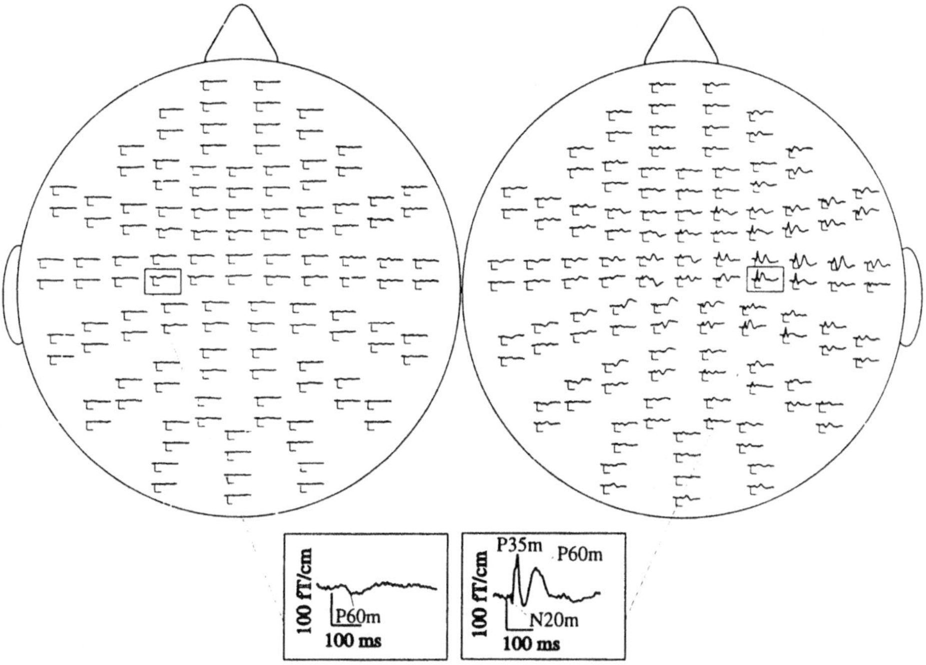

Figure 1. 122-channel SEFs in patient EW. Large left-sided cortical-subcortical infarction with severe sensorimotor paresis. Left: right median nerve stimulation. Right: left median nerve stimulation. On the infarcted side, only a small P60m could be detected. Note the enlarged P60m deflection on the healthy side.

References:

[1] Mauguière, F., Desmedt, J.E., and Courjon, J. Astereognosis and dissociated loss of frontal or parietal components of somatosensory evoked potentials in hemispheric lesions, Brain, 1983, 106: 271–311.

[2] Nakashima, K., Kanba, M., Fujimoto, K., Sato, T., and Takahashi, K. Somatosensory evoked potentials over the non-affected hemisphere in patients with unilateral cerebrovascular lesions, Journal of the Neurological Sciences, 1985, 70: 117–127.

[3] Hari, R., and Kaukoranta, E. Neuromagnetic studies of somatosensory system: principles and examples, Progress in Neurobiology, 1985, 24: 233–256.

[4] Ahonen, A.I., Hämäläinen, M.S., Kajola, M.J., Knuutila, J.E., Laine, P.P., Lounasmaa, O.V., Parkkonen, L.T., Simola, J.T., and Teche, C.D. 122-channel SQUID instrument for investigating the magnetic signals from the human brain, Physica Scripta, 1993, 49: 198–205.

Magnetoencephalographic vs electrocorticographic source localization in patients with epilepsy

Aung, M. [1], Sobel, D.[1], Schwartz, B.[2], Squires, K.[2], and Hirschkoff, E.[2]

1.The Scripps Clinic and Research Foundation, La Jolla, CA , USA
2. Biomagnetic Technologies, Inc, San Diego, CA , USA

Introduction

Surgical resection of cortical tissue responsible for seizure activity has proven to be an effective therapy for patients with intractable epilepsy. Its success, however, depends on the patient having one or a few focal seizure sources located in areas for which excision can be carried out without creating undue deficit. Conventional non-invasive methods for localizing sources of interictal or ictal epileptogenic activity generally provide only crude spatial information, and many surgery candidates without localized structural lesions undergo invasive electroencephalographic (iEEG) monitoring with implanted grid or depth electrodes to localize epileptogenic sources. In some cases, even after extended observation with iEEG, intraoperative monitoring with depth electrodes preparatory to surgery will indicate that resection would be too much of a risk to eloquent cortex. (One such case is included in our study). Magnetic source imaging (MSI) offers the potential to localize these sources non-invasively in a short out-patient procedure.

In this study, 100 epilepsy patients judged by their physicians to be candidates for surgery, and who showed frequent interictal epileptogenic activity on routine surface EEG studies, and who were likely to undergo invasive EEG monitoring were recruited to add a MSI to their evaluation program. Patients were referred from eleven epilepsy centers throughout the U.S. and Canada. Non-invasive tests at those centers prior to MSI typically included EEG, MRI, neurological examination, neuropsychological testing, and phase I video EEG monitoring. In many cases, PET and/or SPECT test results were also available.

Methods

Subsequent to the MEG measurement, each of the patients underwent invasive EEG "phase II" monitoring with implanted electrode grids and/or strips. For the MEG phase of the study, each patient underwent a concurrent MEG and EEG examination utilizing a 74 channel biomagnetometer (Biomagnetic Technologies Magnes II®) and a 21 channel EEG recording system (Nihon Khoden) to record spontaneous incidents of interictal epileptogenic activity. A radio frequency 3-D digitizer was used to determine the position of the individual gradiometers relative to a head-centered 3-D orthogonal coordinate system defined by the nasion and by markers in the right and left outer ear canal. Each patient underwent an MR scan with fiducial markers attached, providing a common coordinate system to overlay MEG source localizations [1].

Patient anti-seizure medication levels were not reduced for the MEG study. With the patient lying on one side and the head voluntarily stabilized, the two 37-channel probes were positioned over homotopic areas of the two hemispheres. A sequence of four to five dual probe placements and measurements assured coverage of the whole head, and afforded the ability to distinguish between synchronous and asynchronous bilateral sources. Recording sessions typically lasted from two to four hours, depending on the patient's level of interictal activity. EEG and MEG waveforms were monitored in real-time by observers trained to recognize and distinguish the morphology of epileptogenic events as against artifacts. (EEG data were later reviewed and interpreted separately by an experienced epileptologist.) Twenty 5-6 second epochs of data were obtained at each recording site. Raw data were digitized at 860 Hz and band-pass filtered at 0.1-200 Hz. Data epochs were then digitally filtered off line with a bandpass of 3-70 Hz. Off line data analysis proceeded with visual identification of MEG and EEG interictal spikes or sharp waves. Temporal windows including these events were then analyzed at each time point in the data sample with a single equivalent current dipole (ECD) model. Each event for which the dipole model correlated with the recorded data to at least 98% and for which the equivalent dipole strength was physiologically realistic was represented by a single focal source plotted on an MRI image of the patient's brain. Images showing all ECD modeled epileptiform events (Fig 1) were presented along with waveform information and the EEG tracings for review by an experienced epileptologist. The epileptologist indicated the likely source of seizure activity and target for surgical resection, if such an inference could reasonably be drawn from the MSI data. The diagnostic conclusion drawn from the MSI assessment was then compared with that from the consensus of all other measures including invasive EEG monitoring undertaken during the patient's normal course of care [2].

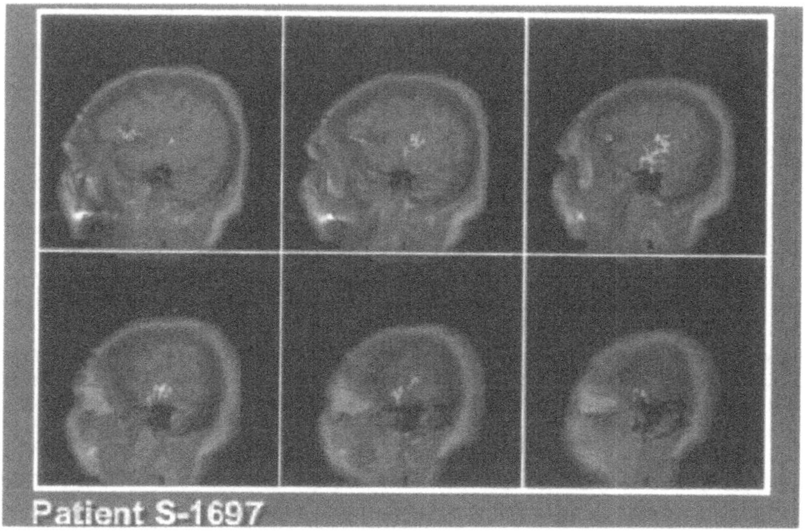

Fig. 1. MEG dipole overlays on sagittal images for patient #S1697.(Spike sources are indicated as yellow triangles, with lines indicating orientation.) The patient had suffered complex partial seizures with secondary generalization from age 12. Invasive EEG monitoring had shown left mesiotemporal sources. Subsequently, the patient underwent a left temporal lobectomy, and was seizure free in a follow up 11 months later.

Results

Of the total of twenty patients studied, MSI detected and localized interictal spikes in 16 and slow waves in another two cases. In eight of the cases MSI and invasive EEG ("Phase II") results indicated the same focal area. In five other cases, MSI and invasive EEG were in agreement in indicating multi-focal or bilateral epileptogenic zones. In two cases, MEG spikes were detected over a broader area which included that indicated by the interpretations based on invasive EEG. One case produced a discordant result, with MEG indicating a region inconsistent with that based on invasive EEG. Data recorded from the remaining four patients produced insufficient MEG spike signals for source localization.

Eighteen of the twenty cases went on to have surgery, of whom follow up information has been obtained for seventeen. Twelve patients have been seizure free for the 3-12 months following the surgery. One additional patient ha been seizure free but has suffered visual field loss . Three patients have shown a 75-90% reduction om seizure occurance, and one patient has suffered an increase in seizure frequency since surgery.

Discussion

MSI is often consistent with results from a consensus of current methods including invasive EEG. Given its concordance with invasive EEG in particular, MSI has the potential to aid in guiding invasive monitoring, especially in patients with no structural lesions, to ensure adequate cortical coverage. MSI may even be able in some cases to obviate the need for invasive EEG prior to surgery. Such a change would save not only considerable cost to the patient and the insurer, but would provide a much less intrusive procedure for the patients, many of whom are quite young.

References

[1] Gallen, C.C., Hirschkoff E.C., Buchanan, D.S. Magnetoencephalography and Magnetic Source Imaging. Neuroimaging Clinics of North America, 5(2), May, 1995.

[2] Smith, J.R., Schwartz, B.J., Gallen C.C., Orrison, W., Lewine, J., Murro A.M., King, D.W. Park, Y.D. (1995) Utilization of Multichannel Magnetoencephalography in the Guidance of Ablative Surgery. Journal of Epilepsy 8:119-130.

Acknowledgements

The authors would like to thank the following collaborators for their assistance: Dr. Izak Fried and Dr. Sachar Eliashiv (UCLA Medical Center), Dr. Vicente Iragui and Dr Evelyn Tecoma (UCSD Medical Center, La Jolla, CA), Dr. Brien Smith (Henry Ford Hospital, Detroit, MI), Dr. Joseph Smith and Dr. Yong Park(Medical College of Georgia, Augusta, GA), Dr. Thomas Waltz (Scripps Clinic Research and Foundation, La Jolla, CA, Dr. James Wheless and Dr. Valerie Curtis (Hermann Hospital, Houston, TX)

Propagation of epileptic activity during single (unaveraged) interictal spikes as studied with Magnetic Field Tomography (MFT)

Bamidis, P.D.[1], Hellstrand, E.[2], Ioannides, A.A.[1,3]

[1]Dept. of Physics, The Open University, Milton Keynes, U.K.; [2]Dept. of Clin. Neurophysiology, Karolinska Hospital, Stockholm, Sweden; [3]IME, Research Centre Jülich, Germany

Introduction

The potential in multichannel MEG systems to analyse interictal epileptic data directly, without signal averaging, has been underlined by many MEG investigators [1]. Since spikes of similar shapes may arise from different cortical areas, analysis of individual, non-averaged spikes would be preferable to that of averaged spike signals [2]. An averaging technique might give good localisation results at the maximum of the spike (in consensus with some unaveraged spikes), but one cannot guarantee that the propagation of epileptic activity as appearing in the averaged record would represent the actual epileptic propagation. The study of unaveraged epileptic events, however, necessitates a robust analysis method that: (i) does not include any preconceptions on the number and position of sources; (ii) allows for distributed sources, since extended and/or multiple brain areas may be involved in the generation of epileptiform activity [3,4]. Magnetic Field Tomography (MFT) meets these requirements [5]. Earlier studies exploited the feasibility of analysing unaveraged interictal multichannel MEG data (i.e. spikes, sharp waves) with MFT [6]. More recently, MFT was used to study the spatio-temporal evolution of interictal activity during single spike events [7]. In this work, emphasis is drawn on the interactions between neocortex and deep temporal structures, as well as the spatio-temporal characteristics of interictal activity in the depth of the temporal lobe of two epileptic patients.

Materials and Methods

Patient #1 was an 18-year old girl with a history of pharmaco-resistant complex partial epilepsy (CPE) since the age of 14, with daily seizures of dizziness and impaired consciousness. Patient #2 was a 42-year-old male, suffering from CPE since the age of 13 without known aetiology. His seizures started with visual symptoms, with more acoustic sensations, like "noise", dominating later on. Generalisations of seizures occurred frequently.

The MEG recordings were made with the 37-channel KRENIKON®, Siemens system placed over the temporal area. The data were band-pass filtered on-line through 0.5-70.0 Hz, digitised at 0.4 kHz (Patient #1) or 0.5 kHz (Patient #2). The present analysis has been limited to those datasets off-line filtered through 1-20 Hz.

Three dimensional MFT estimates of the primary current density, $\mathbf{J}^p$, were obtained within a partial hemispherical volume encompassing the left (Patient #1) or right (Patient #2) hemisphere. A succinct description of MFT has been recently given in [5]; more details can be found in other reports of this volume as well as in [8]. In short, MFT uses continuous probabilistic estimates of $\mathbf{J}^p$, in order to construct tomographic displays of brain activity. In our work, the instantaneous MFT estimates were studied, as well as the temporal integrals of the intensity, $|\mathbf{J}^p|^2$. We use integrals of intensity in space (activation curves) to study the temporal evolution of activity within a region of interest (ROI), as well as displays of temporal integrals of intensity to obtain summaries of activations within a given time window. Whenever activation curves within specific ROIs or levels of the source space (e.g. at the depth of the temporal lobe - hippocampal level) indicate strong activation, the spatial changes in the maximum activity are studied; such changes within a given image/level are displayed by connecting areas of strong activity with arrows. This provides an image of the various anatomical regions of the brain which are directly or indirectly functionally connected, allowing for propagations and/or sequential shifts of activity to be studied.

Results

In earlier publications [6,7] we have observed that certain cortical areas seemed to be activated both before and after deep temporal activation. The interplays between deep and superficial activation occurred with various time constants. For the two patients involved in this study, the time differences of the activations from cortical to deep or deep to cortical areas varied between 10 to 60 ms. An example from each patient is shown in Fig. 1. During the initial, rising phase of the spike, the activity appeared in superficial, cortical areas of the temporal lobes (a); around the peak of the spike (b) in the amygdala-hippocampal complex, and then back to the cortex (c).

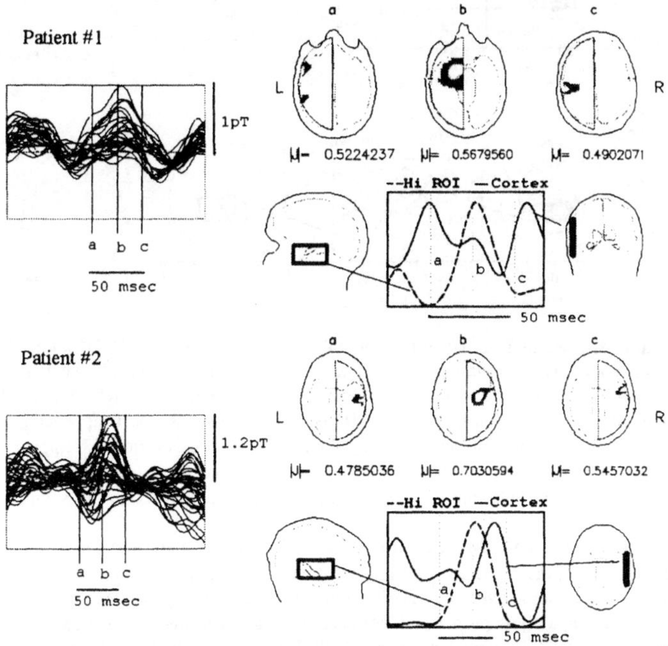

Fig 1: Top half; Patient #1; Left: left-hemisphere interictal spike with superimposed signals of all channels (note time bars a-c). Right; *top row:* instantaneous MFT estimates of intensity on axial MR images (note time instants a-c and |J^P| values in arbitrary units); *bottom row; middle:* graph of temporal evolution of activity in the left hippocampal area (dotted line) and temporal cortex (full line) (note time bars a-c); *bottom row; left and right:* MR images of ROIs at the level of hippocampus (sagittal view) and temporal cortex (coronal view). Bottom half: similar, but for the right hemisphere of patient #2. See text.

However, as different cortical areas are involved in the pre- and post-deep activation, one could easily assume that the observed cortical activations could just be random variations of noise. To exclude this possibility, we back-averaged a number of single epoch events, time-locked to the peaks of strong hippocampal activations. This procedure selectively amplifies activations from areas which maintain a consistent time relation with the activation in the hippocampus. The results are summarised in Fig. 2, where 18 single event solutions were back-averaged for patient #1, while 19 for patient #2.

The involved cortical ROIs remain "silent" during the interval of hippocampal activation, but, they do fire before and after this interval. This is clearly shown in the cortical activation curves, which reveal distinct peaks above the noise level before and after the hippocampal firing. Similar patterns occur for both patients. Note also that the cortical areas appear extended rather than "focal". Furthermore, as the schematics at the far right of Fig. 2 demonstrate, there seems to be an interplay of activation between the different regions of the cortex. In both patients, the hippocampal areas appear to be interacting mostly with the anterior cortical regions.

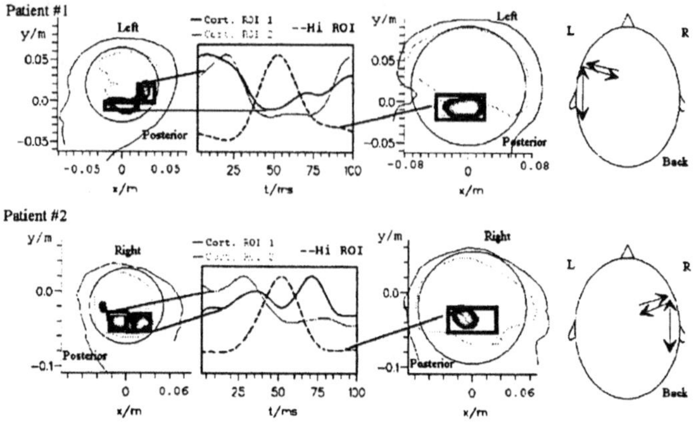

Fig 2: Top row: (Patient #1): *Left*: superficial integrals of intensity over 100 ms, after back-averaging 18 single event solutions of strong hippocampal activation. Two main ROIs are revealed in the sagittal view: one fronto-infero-temporal and one more posterior in the temporo-parietal region. *Middle*: graph of activations of the cortical and hippocampal ROIs. Note the pre- and post-hippocampal activation of the cortex. *Right*: sagittal view of deep intensity integrals after back-averaging. Hippocampal activation is clearly shown. *Extreme Right*: schematic axial view of the neocortico-mesiobasal interactions emerging from the back-averaged solutions. Bottom: Similar, but after back-averaging 19 solutions for Patient #2.

For both patients, the MFT estimates at the peaks of the spikes indicated activation at the depth of the temporal lobe. The shapes of the solutions were consistent with activation of the hippocampal areas. Studying the spatial changes in the maximum activity in a sagittal section through the amygdala and hippocampus complex, we noticed that for each patient, activity propagated mostly in the anterior-posterior direction, suggesting spread from amygdala to hippocampus (12/26 events for patient #1, 7/15 for patient #2). The effect was more prominent for patient #1, for whom the shapes of the integrated intensity during such intervals appeared usually along an arch, indicating propagation from structures anterior to the hippocampus along its main axis to its posterior end (cf. Fig. 3a and [7]). However, shifts of activity (Fig 3b) from anterior to posterior and then back to anterior and vice versa were also observed (8/26 for patient #1, 4/15 for patient #2). Propagations from posterior to anterior were less frequent for both patients (4/26 for patient #1, 4/15 for patient #2).

Discussion and Conclusions

There has been an increasing interest in the study of unaveraged epileptic data [9]. The ECD model is mostly used to explain small segments of the unaveraged record, usually around the peaks of spikes or sharp waves. This leaves part of the event unexplained. We are using a more general method (MFT) to obtain estimates of activity throughout the whole time period of the various events, that is from the rising phase of a spike to its peak, and then the repolarisation. In doing so, we have observed the interactions between neocortical areas and deep temporal structures during a number of unaveraged interictal spikes in 2 epileptic patients. Superficial activity was seen to precede and follow deep activity. The time differences of the activations from cortical to deep or deep to cortical areas varied in the range between 10-60 ms. Back-averaging of the individual solutions was performed to enhance any orderly series of activations. Two rather extended cortical areas in each patient emerged after back-averaging. Interplays of activity were consistently found between these two cortical sites, and between the most anterior cortical area and the hippocampal complex. The image revealed from the back-averaged solution (cf. schematics in Fig. 2) may not necessarily represent the actual sequence of activations in each individual event, but it does mean that it occurs frequently (more than half of the events studied for each patient), and certainly above chance.

The persistent shifts in the maximum of the MFT estimates over time during an interictal spike occur consistently over different epochs. However, the evolution of activity in different areas of the brain such as superficial cortical structures and deep situated amygdala-hippocampus areas may not necessarily indicate an established anatomical pathway - intermediate steps may be missing (e.g. activity on the contralateral side [7]), partly due to the set-up of the sensors. MFT simulations and tests of the validity of this analysis could not be included here due to space limitations, but are provided elsewhere [8].

In conclusion, MEG measurements and MFT analysis offer the opportunity to non-invasively recover the functional connectivities of the various anatomical structures from single epoch data, avoiding the dangers associated with averaging.

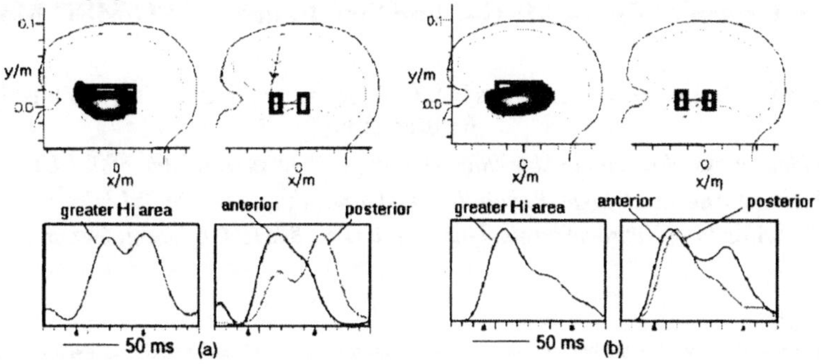

Fig 3: Patient #1. (a) Sagittal view of integrated intensity over 60 ms in a sagittal level through hippocampus (left). Instantaneous shifts of maximum activity computed every 2.5 ms (right). Propagation from anterior to posterior hippocampus is observed. Note the shape of the integrated intensity and the activation curves for the two indicated ROIs (bottom right). Activation of the greater hippocampal area is given in bottom left. (b) Similar but with anterior-posterior-anterior sequential shifts.

References:

[1] Sato, S. Current status of biomagnetic research in epileptology. In: Hoke, M., Erne, S.N., et al. Biomagnetism: Clinical Aspects, Amsterdam: Elsevier, 1992.

[2] Witte, O.W., Dorn, T., and Uhlig, S. Contribution of different areas of epileptic foci to the generation of interictal epileptic discharges. In: Hoke, M., Erne, S.N., et al. Biomagnetism: Clinical Aspects, Amsterdam: Elsevier, 1992.

[3] Barth, D.S., Sutherling, W.W., Engel, J., et al. Neuromagnetic evidence of spatially distributed sources underlying epileptiform spikes in the human brain. Science 1984, 223:293-296.

[4] Graf, M., Niedermeyer, J., Schiemann, J., et al. Electrocorticography: information derived from intra-operative recordings during seizure surgery. Clin. Electroenceph., 1984, 15(2):83-91.

[5] Ioannides, A.A. Estimates of 3d brain activity ms by ms from biomagnetic signals: method (MFT), results and their significance, In: Eiselt, E., Zwiener, U., et al. Quantitative and Topological EEG and MEG analysis, Jena, Universitätsverlag Druckhaus-Maayer Gmbh, 1995.

[6] Ioannides, A.A., Hellstrand, E., Bamidis, P.D., et al. Estimates of brain activity from unaveraged interictal multichannel MEG signals, In: Baumgartner, C., Deecke, L., et al. Biomagnetism: Fundamental research and Clinical Applications, Amsterdam, Elsevier, 1995.

[7] Bamidis, P.D., Hellstrand, E., Lidholm, H., et al. MFT in complex partial epilepsy: spatio-temporal estimates of interictal activity, Neuroreport, 1995, 7(1):17-23.

[8] Bamidis, P.D., PhD Thesis, The Open University, Milton Keynes, UK, 1996.

[9] Lewine J.D., Orrison, W.W. Jr, Astur, R.S., et al. Explorations of pathophysiological spontaneous activity by magnetic source imaging. In: Deecke L, Baumgartner C, et al. Biomagnetism: Fundamental research and Clinical Applications, Amsterdam, Elsevier, 1995.

Presurgical Evaluation of Methohexital–Induced Epileptiform Activity: Comparison of EEG–, MEG–, and Simultaneous EEG/MEG Models

Diekmann, V.[1], Becker, W.[1], Jürgens R.[1], Grözinger, B.[1], Kleiser, B.[2], Wollinski, K.H.[3], Richter, H.P.[4]

[1] Sektion Neurophysiologie and [2]Abteilung Neurologie, Universität Ulm, 89081 Ulm, Germany;
[3] Abteilung Anästhesiologie, Rehabilitationskrankenhaus Ulm, 89081 Ulm, Germany;
[4] Abteilung Neurochirurgie, Universität Ulm, 89312 Günzburg, Germany

Introduction

We have developed a method of *simultaneous* recording of EEG and MEG and localizing the origin of epileptic spikes by calculating single equivalent current dipoles (ECDs) on the basis of electric, magnetic, and *simultaneous* electric–magnetic models. We here compare the results of these 3 methods in view of the question whether one of these models presents specific advantages in the search for foci of simple or complex partial epilepsy (SCPE).

Methods

Our report draws on a series of 32 SCPE patients (Ps) referred to the Dept. of Neurology for presurgical diagnosis and a total of 37 sessions (some Ps were repeatedly examined) in which EEG (25 channels) and MEG (Dornier; 22 signal/6 compensation channels software gradiometer [1, 2]) were simultaneously recorded inside a shielding room (Vacuumschmelze Type AK3a). In 2 of these Ps, ECoG (from intra–cranial strip electrodes), instead of EEG, was recorded in conjunction with MEG. Also, in some Ps, EEG was complemented by foramen ovale or sphenoidal recordings. EEG electrodes were arranged according to the 10–20–system, and 6 additional electrodes were placed over the suspected epileptogenic area where electrode distance was thus reduced to less than 2 cm. Similarly, the two cryostats of the MEG system were conjointly centered over the suspected area.

Patients' head was immobilized by vacuum pads. Since the frequency of spontaneous epileptiform activity can be low in SCPE patients, it is often not possible to collect a sufficient number of spikes within the limited time a patient can be immobilized. In 22 of the above 32 patients we therefore resorted to provocation by i.v. injection of the barbiturate methohexital. Methohexital induces a brief anesthesia accompanied by seizurelike brain activity with foci matching those of spontaneous seizures [3]. During anesthesia, patients were given oxygen while respiration, heart rate, blood–pressure , blood–O_2–saturation, and expiratory CO_2 were monitored. On average, during the 3–min epoch preceding methohexital injection, less than 1 spike (range 0–5) could be clearly recognized in surface recordings (EEG, MEG), whereas there were 15 (0–58) such events in the 5–min period following injection.

Spikes were identified off–line by visual inspection and isolated by subtracting a linear interpolation of pre– and post–spike background activity. Using a 4–layer sphere fitted to the patient's head on the basis of MR–images, single electric and magnetic ECDs were first calculated separately from electric (EM) and magnetic (MM) models; EEG and MEG then were weighted with the goodness of fit of EM and MM, respectively, by taking EEG/σ_e and MEG/σ_m (σ_e^2 and σ_m^2 , residual variance of EM and MM). Finally, the weighted data were fitted with a simultaneous electric–magnetic model (SM) [4]. ECD positions and moments were transferred into the patients' MRI for visualization (Fig. 1). Observed spike activity was summarized by calculating the spatial dipole density distribution [5]. To this end, each ECD was associated with a 3D–Gaussian distribution centered at its locus with variances reflecting the Cramer–Rao lower bounds [6] of the locus; iso–density lines then were derived from the sum of all Gauss–distributions and transferred into the patient's MRI (see Figs. 2 – 4).

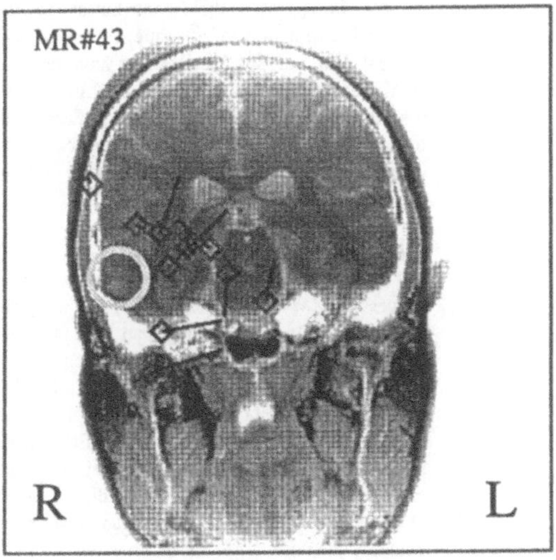

Fig. 1. ECDs obtained by SM in a patient with temporobasal tumor (circled). Locations (◊) and moments (bars) projected into coronal MRI–slice containing centre of tumor (R right, L left).

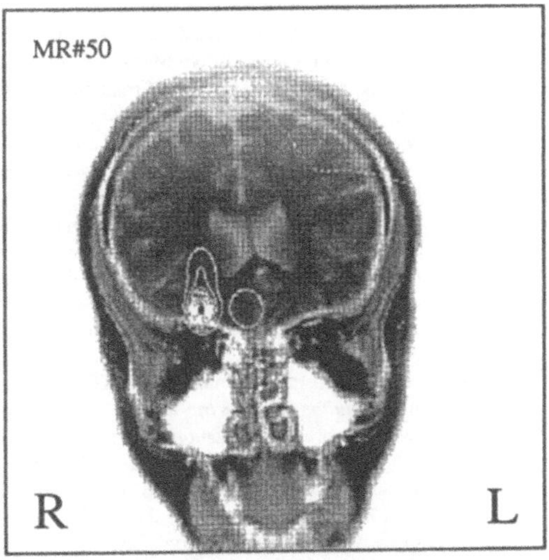

Fig. 2. Same P as in Fig. 1, dipole density obtained from EM. Coronal MRI–slice cutting through maximum of EM dipole density with isodensity curves (density Δ = 0.5 $cm^{-3)}$

Results

Among our 32 patients we selected candidates for a verification of ECD localization by successively applying the following criteria:
(1) More than 5 localizable spikes. Spikes were considered localizable if SM explained more than 75% of the EEG+MEG variance (although this condition may appear weak by the standards for EM and MM fits, it is a strong one for the combined data). Criterion (1) reduced the number of cases to 16.
(2) "Classical" clinical examinations (video-monitoring, intra-cranial electrodes, semiology) suggest either a single epileptogenic focus or two clearly distinct foci (one in each hemisphere); criterion (2) left ten cases.
(3) Clinical reasoning and postoperative outcome associates foci with a clearly recognizable structural change in MRI (neoplasm, infarction, etc.); criterion (3) left over six cases, which on average yielded 27 localizations.

In these six Ps, the locus of peak dipole density was established for each of the three models (EM, MM, SM) and its 3D-distance to the nearest recognizable lesion border in MRI was measured (Fig. 2). The distances obtained in this way from EM and MM averaged 3.0 and 3.3 cm respectively, whereas the average distance indicated by SM was clearly smaller (2.1 cm). At the individual level, SM distance was closest to the lesion border in four of the six Ps. Interestingly, in each of these four patients MM was worse, in terms of distance from lesion border, than EM - an observation which underscores the possible pit-falls of relying solely on MEG. The patient shown in Figs. 1-4 constitutes a typical example for this type of behaviour. He suffered from a ganglioglioma in his right temporobasal cortex (Fig. 1). Peak dipole density derived from EEG occurred at a distance of 4.0 cm from the tumor border (Fig. 2; note that consecutively numbered MRI slices have 6 mm distance; hence, the coronal plane shown in Fig. 2 is 4.2 cm anterior to the centre of the tumor shown in Fig. 1). The P's MM dipole density peaked at 5.9 cm from the nearest tumor border (Fig. 3), whereas the maximum of SM density was at 2.4 cm (Fig.4).

In the remaining two Ps, SM-distance was intermediate to those of EM and MM with one case of EM and one of MM yielding the closest localization.

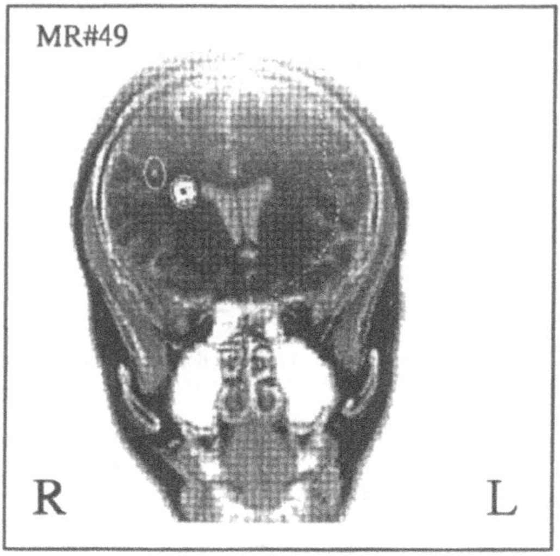

Fig. 3. Same P as in Fig. 1, dipole density obtained from MM. Coronal MRI–slice cutting through maximum of MM dipole density with isodensity curves (density Δ = 1 cm^{-3})

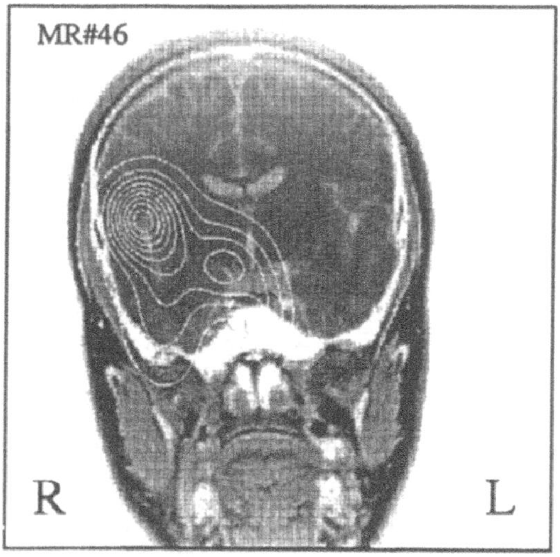

Fig. 4. Same P as in Fig. 1, dipole density obtained from SM. Coronal MRI–slice cutting through maximum of SM dipole density with isodensity curves (density Δ = 0.05 cm^{-3})

In terms of their contribution to the over all evaluation of a patient's presurgical clinical data, our SM localizations have led to a significant reappraisal in two cases; in the remaining four Ps, including the P of Figs. 1-4, the outcome of SM localization confirmed the conclusions reached independently on the basis of standard clinical examinations.

Discussion

Although several recent studies report successful MM localization of interictal and ictal spike activity to regions congruent with those indicated by other diagnostic procedures [e.g. 7, 8], an appraisal of the clinical importance of MEG as compared to EEG is still difficult.

What is obvious is that MEG cannot simply replace EEG since it would miss activity emanating from predominantly radial sources. One of our patients provided an instructive example hereof: A focus in her extra-mesial temporal cortex radiated large EEG spikes accompanied by very small MEG activity which, when viewed in isolation, would have been difficult to interpret; EM indicated an almost radial source. Thus, EEG remains a mandatory method in the search for epileptic activity.

On the other hand, in view of its cost, few institutions will afford MEG if it does not provide more than only a confirmation of EEG. The basic question therefore is whether *adding* MEG will *improve* the quality of localization.

In order to try a first answer to this question, we have selected six SCPE patients (out of a population of 32) by applying strict criteria as to both the mathematical quality of spike localization (S/N-ratio >4) and the verification of the epileptogenic region so that we could quantitatively compare the results of EM, MM, and SM. The results of this comparison suggest that the combination of EEG and MEG in a *simultaneous electric magnetic model* may indeed improve localization in that it tends to point to a locus closer to the border of the lesion in cases of lesion-related SCPE than either EEG or MEG alone. However, this conclusion is expressed with several reservations:

(1) The number of well documented cases is still very small. (2) By combining EEG and MEG we have increased the spatial sampling density in comparison to the situation where either only EEG or MEG is considered. One could argue that

with double the number of EEG-channels, EM might yield results comparable to SM. Yet, we feel that the resulting √2-fold increase of sampling density cannot engender shifts in localization by up to 3.5 cm (maximum in our group of Ps), and this all the more as we had spaced electrodes overlying the presumed spiking area more closely (cf. Methods) than called for by the Nyquist frequency of superficial sources (0.5 cm^{-1}). (3) We have arbitrarily considered the border of the lesions (as discernible in MRI) as the most likely area of origin of the spike activity (unless there was clear clinical evidence to the contrary). Clearly, a better delineation on the basis of intra-cranial grid recordings is needed which, ideally, would be accompanied by simul-taneous surface EEG and MEG measurements. On the other hand, the MEG machine used in the present experiments had only a limited angle of view on the Ps' heads. It may well turn out that with a whole head system, SM will give even better results.

Finally, independent of the questions regarding the specific contribution of MEG, provocation by methohexital proved to be extremely useful for simultaneous MEG/EEG measurements in SCPE patients. Methohexital induces quasi-ictal activity [3] (which is more relevant than interictal activity for the identification of the epileptogenic area), and often reduces the time required to collect a sufficient number of spikes from hours to minutes.

References:

[1] Becker, W., Diekmann, V., Jürgens, R. First experiences with a multichannel software gradiometer recording normal and tangential components of MEG. *Physiol. Meas.*, 1993, 14: A45-A50

[2] Diekmann, V., Jürgens, R., Becker, W., Elias, H., Ludwig, W., and Vodel, W. RF- to DC-SQUID upgrade of a 28-channel magnetoencephalography (MEG) system. *Measurement Science and Technology*, in press

[3] Hufnagel, A., Burr, W., Elger, C.E., Nadstawek, J., Hefner, G. Localization of the epileptic focus during methohexital-induced anesthesia, Epilepsia, 1992,33: 271-284

[4] Diekmann, V., Becker, W., Grözinger, B., Jürgens, R., Westphal, K.-P., Kornhuber, H.H. Localization of focal epileptic activity with a simultaneous EEG and MEG model, In: Baumgartner, C., Deecke, L., Stroink, G., and Williamson, S.J., Biomagnetism: Fundamental Research and Clinical Applications, Amsterdam, IOS-Press, 1995, pp 23-27

[5] Vieth, J., Kober, H., Weise, E., Daun, A., Moegner, A., Friedrich, S. and Pongratz, H. Functional 3D localization of cerebrovascular accidents by magnetoencephalography, *Neurological Research*, 1992, 14, Suppl.: 132-134

[6] Mosher, J.C., Spencer, M.E., Leahy, R.M., Lewis, P.S. Error bounds for EEG and MEG dipole source localization, *Electroenceph. clin. Neurophysiol.*, 1993, 86: 303-321

[7] Smith, J.R., Gallen, C., Orrison, W., Lewine, J., Murro, M., King. D.W., Gallagher, B.B. Role of multichannel magnetoencephalography in the evaluation of ablative seizure, *Stereotact. Funct. Neurosurg.*, 1994,62: 238-244

[8] Stefan, H., Schneider, S., Feistel, H., Pawlik, G., Schüler, P., Abraham-Fuchs, K., Schlegel, T., Neubauer, U., Huk, W.J. Ictal and interictal activity in partial epilepsy recorded with multichannel magnetoencephalography: correlation of electroencephalography/electrocorticography, magnetic resonance imaging, single photon emission computed tomography, and positron emission tomography findings, *Epilepsia*, 1992, 33(5): 874-887

Intracortical Spread of Epileptic Spikes - Simultaneous Measurement of Magnetic Field and Electric Potential in Rabbits

[1]Eiselt, M., [1]Zwiener, U., [2]Gießler, F., [1]Wagner, H., [2]Haueisen, J., [2]Huonker, R., [2]Nowak, H., and [3]Schack, B.

[1]Inst. Pathophysiology; [2]Biomagnetic Centre; [3]Inst. of Medical Statistics, Informatics and Documentation, Clinic of the Friedrich Schiller University, 07740 Jena, Germany

Introduction

In many experiments we observed differences between simultaneously measured ECoG and MEG signals concerning the temporal relationship of characteristic features like maxima or minima. This was not surprising despite MEG and extracellular electric potential are functionally related. They are caused by different bioelectric processes. The MEG is mainly caused by intracellular currents of neuronal cell populations. The extracellular electric potential is caused by currents through the cellular membrane. These currents can be elucidated by current source density analysis (CSD).

The following questions should be answered: How do different cortical depths of spike foci influence the temporal relationship between ECoG and MEG ? Which part of spike generation processes is detected by the MEG and which part may influence the electric potential at the cortical surface ?

Methods

Initiation of epileptic spikes

Interictal spikes were investigated by CSD in 5 and the MEG in 12 adult rabbits. Additionally, in all animals the ECoG was recorded. Spikes were initiated by stereotactic, intracortical application (600 - 750 µm below the cortical surface) of small amounts of penicillin (200 I.E.). For CSD investigations penicillin was injected in superficial brain surface (Regio retrosplenialis). For MEG investigations two locations of penicillin application were used: superficial cortical area (Regio retrosplenialis; 3 animals; main current dipole with radial direction) and Gyrus cinguli of the left or right hemisphere 2000 µm below the superficial brain surface (9 animals; main current dipole tangential direction to the cortical surface; fig. 1). Both regions had similar architecture of histology.

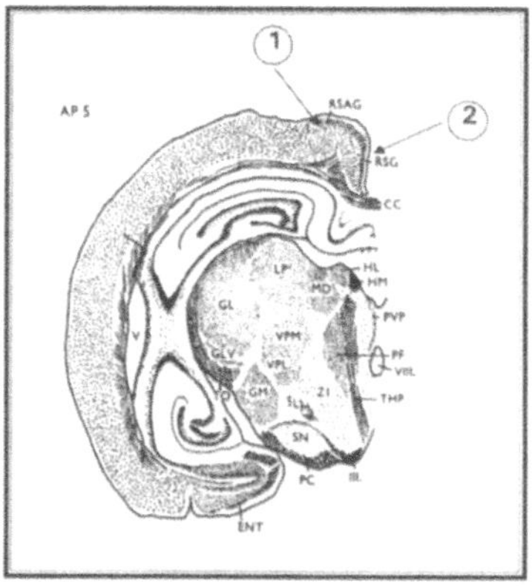

Fig. 1. Rabbit brain with cortical regions of penicillin application: 1 = Regio retrosplenialis - current dipole in radial direction; 2 = Gyrus cinguli - current dipole in tangential direction.

Measurements

Intracortical electric potential distribution perpendicular to the cortical surface was detected by 16 electrodes spaced 150 µm apart. ECoG was measured on the brain surface immediately above the region of penicillin application. The magnetic field was detected by a 7 channel 1st order gradiometer system (sensitivity 30 fT/ √Hz, diametre 9.5 mm). The measurement was repeated five times and thus, 35 channel MEG measurement could be performed.

MEG and ECoG data were sampled at a rate of 1000 Hz, intracortical potential at a rate of 512 Hz. They were filtered with band pass 3 - 30 Hz. This has changed the moment of maximum and minimum appearance in ECoG and MEG in the same way. In comparison to raw data the time shift caused by filtering was smaller than 2 msec. To improve signal to noise ratio (> 30 : 1) 200 spikes detected by MEG were averaged.

Testing the recording systems with an artificial signal the frequency dependent phase shift between magnetic signal and electric signal was at 20 Hz 1.5° which corresponds to 0.2 msec. The electric signal preceded the magnetic signal independent from the frequency band investigated.

Calculations

One dimensional CSD analysis was based on the 2nd spatial derivative of intracortical electric potential. Calculation of CSD and fitting of data sets were done by algorithms published by Rappelsberger et al. [5] and the Neville algorithm [7].

For CSD and the magnetic field hilliness (which corresponds to global field power [4]) and global similarity (a parameter similar to the parameter published by Desmedt and Chalkin [1]) were calculated millisecond by millisecond (fig. 2). With the help of these parameters the occurrence of greatest amplitude of the whole signal at each instant and the strongest changes over time could be detected.

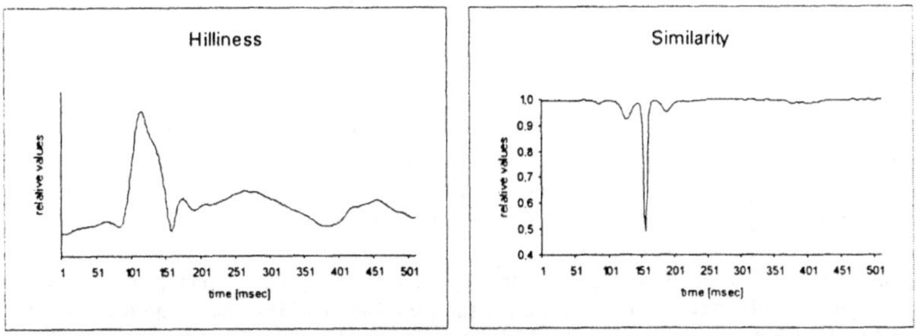

Fig. 2. Hilliness and similarity calculated from the magnetic field. In the left diagram a peak of hilliness can be observed at about 120 msec followed by three further peaks with lower amplitude. In the right diagram the reduction of similarity indicates strong change within the magnetic field. Both may indicate that there are multiple processes and the first is different from the following.

For CSD and MEG the temporal relationship between peaks of hilliness were compared with the occurrence of the first negative maximum and the following positive maximum of ECoG spikes. Events occurring earlier were marked by negative time.

Results

There were no differences between spikes detected by ECoG in the three groups (CSD, MEG with radial directed dipole and MEG with tangential directed dipole) concerning the temporal relationship of the negative and positive maximum occurrence.

The maxima of CSD occurred simultaneously with the negative (1 ± 2.6 msec; not significant - n.s.) or positive maximum (1 ± 2.7 msec; n.s.) of the ECoG spike (fig. 3).

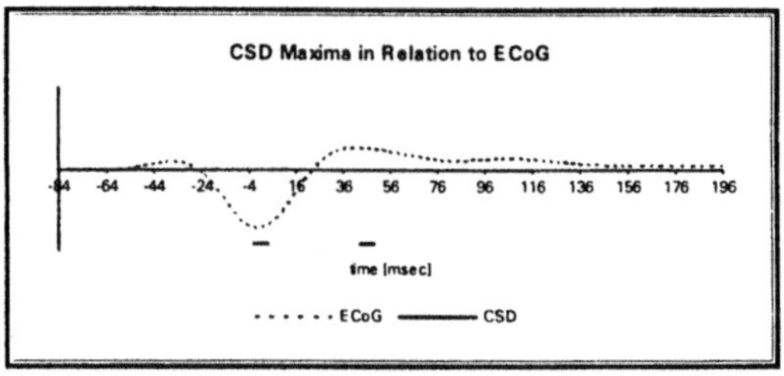

Fig. 3. The averaged moments of occurrence of CSD maxima in relation to ECoG spike (n = 5; lines = mean ± SD of the moments of occurrence of CSD maxima; dashed line = electric potential of ECoG).

Similar observations were made between spikes detected by MEG and ECoG when penicillin was injected in the superficial cortical area. There were no significant temporal differences between ECoG and MEG (negative ECoG maximum - 1st maximum MEG: -2.1 ± 5.3 msec, n.s.; positive ECoG maximum - 2nd MEG maximum: -2.9 ± 8.6 msec, n.s.; fig. 4).

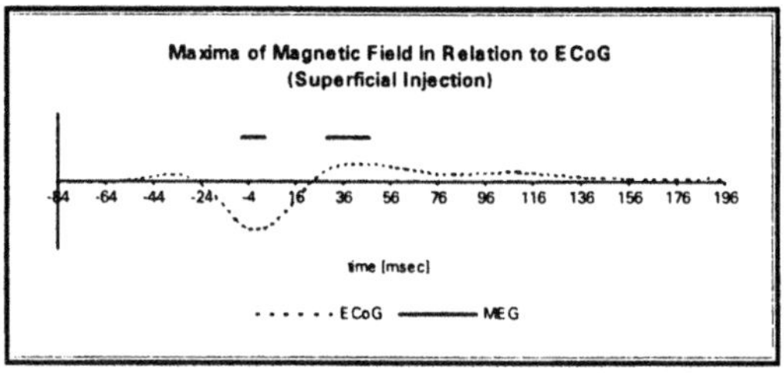

Fig. 4. The averaged moments of occurrence of magnetic field maxima in relation to ECoG spike when penicillin was injected in superficial brain surface (n = 3; lines = mean ± SD of the moments of occurrence of magnetic field maxima; dashed line = electric potential of ECoG).

Spikes detected by MEG appeared earlier than spikes detected by ECoG when penicillin was injected in the region of Gyrus cinguli (2000 μm below the brain surface). The first MEG maximum occurred -19.3 ± 14.1 msec ($p < 0.001$) before the negative maximum of the spike detected by ECoG and the second MEG maximum -10.0 ± 21.7 msec (n.s.) before the positive maximum (fig. 5).

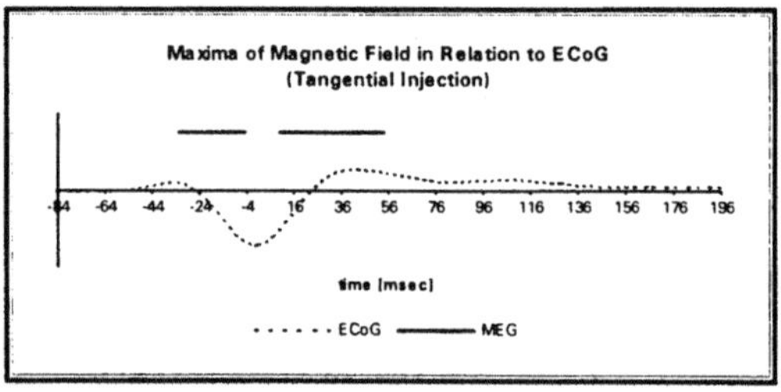

Fig. 5. The averaged moments of occurrence of magnetic field maxima in relation to ECoG spike when penicillin was injected in Gyrus cinguli 2000 μm below the brain surface (n = 9; lines = mean ± SD of the moments of occurrence of magnetic field maxima; dashed line = electric potential of ECoG).

The mean difference between areas of penicillin application and the location of the current dipoles by the method of spatial filtering [6, 8] was between 0.7 - 1 mm.

Discussion

In this study neuronal processes initiated by focal penicillin application were investigated by three different methods. The one dimensional CSD perpendicular to the cortical surface describes neuronal processes immediately influenced by penicillin application. These are synchronised excitation of great amounts of neurones followed by inhibitory processes responsible for avoidance of repetitive excitations observable during seizures. During these periods intracellular currents are directed towards cortical depth or cortical surface. Simultaneously to this process neuronal activity propagates towards the surrounding tissue involving other neuronal populations and initiate there inhibition which limits the further spread [2]. If the location of the spike generation is near the cortical surface the ECoG is mainly influenced by the first excitatory processes. If spike generation occurs in deeper cortical areas the ECoG is mainly influence by the events caused by the spread of neuronal activity within the surrounding tissue. The MEG will be influenced by the whole neuronal activity but mainly by the synchronised excitation involving many neurones in the same way. This will cause a signal strong enough to be detected outside the brain.

There are different time shifts between ECoG and MEG during the initial process describing the excitation and the following process. We suppose that neuronal activity during the negative spike occurs in restricted regions and propagates towards the cortical surface and thus causing a time delay between ECoG spike and MEG. Neuronal activity is more distributed during the positive wave [3]. Thus, it reaches earlier the cortical surface and the time delay between ECoG and MEG is smaller.

References

[1] Desmedt, J.E., and Chalkin, V. New method for titrating differences in scalp topographic patterns in brain evoked potential mapping, Electroencephalography and clinical Neurophysiology, 1989, 74: 359-366.

[2] Eiselt, M., Zwiener, U. , Wagner, H. Conditions for a precise analysis of penicillin-induced spikes by CSD-analysis, In: Eiselt, M., Zwiener, U., and Witte, H. Quantitative and topological EEG and MEG analysis, Jena, Universitätsverlag Druckhaus Mayer, (1995) 231 - 236.

[3] Goldensohn, E.S. and Salazar, A.M., Temporal and spatial distribution of intracellular during generation and spread of epileptogenic discharges, In: Delgado-Escueta, A.V., Ward, A.A., Woodbury D.M. and Porter, R.J., Advances in Neurology, New York, Raven Press, 1986, 44: 559-589.

[4] Lehmann, D., Ozaki, H. and Pal, I. EEG alpha map series: brain micro-states by space-oriented adaptive segmentation, Electroencephalography and clinical Neurophysiology, 1987, 67: 271 - 288.

[5] Rappelsberger, P., Pockberger, H. and Petsche, H. Current Source Density analysis: Methods and application to simultaneously recorded field potentials of the rabbit's visual cortex, Pflügers Archiv, 1981, 389: 159 - 170.

[6] Robinson, S.E. and Rose, D.F. Current source image estimation by spatially filtered MEG. In: Hoke, M., Erné, S.N., Okada Y.C. and Romani, G.L., Biomagnetism: Clinical aspects, Amsterdam: Excerpta medica, 1992, 337 - 338.

[7] Stoer, G. Einführung in die numerische Mathematik I., Berlin, Heidelberg, New York, Springer, 1976.

[8] Wagner, H., Eiselt, M. Zwiener, U. Investigation of the exactness of the source analysis from biomagnetic signals by the method of spatial filtering with particular consideration of the analysis of interictal spike activity - a computer simulation. Medical & Biologica Engineering & Computing 1995 (submitted)

supported by BMBF P 01 ZZ9104

Effects of Volume Current on Localization of Focal Epileptic Activity

Ikeda, H.[1], Takeuchi, R.[1], Osozawa, S.[1], Ishiyama, A.[1], Kasai, N.[2], and Minami, T.[3]

Department of Electrical Engineering, Waseda University[1], Tokyo, Japan
Electrotechnical Laboratory[2], Tsukuba, Japan
Faculty of Medicine, Kyushu University[3], Fukuoka, Japan

Introduction

In the localization of focal epileptic activity, there are some cases that the estimated positions scatter. In these cases, the improvement on the localization of focal epileptic activity is expected. Usually the foci of epilepsy are localized with the spherical model assuming a single current dipole for a focus from MEG (Magnetoencephalogram). However, skull and palate are in a head, then a head is not spherical and is inhomogenous. We previously investigated the effects of volume current on the source localization by experiments and computer simulation with a single-layer hemispherical phantom. From the results, it was found that the contribution of volume current to the MEG increases when the current dipole exists on the region near the bottom and in the deep place of the head [1][2][3]. In this paper, we investigate the possibility of improvement on the localization of focal epileptic activity with the MEGs of three subjects of focal epilepsy by considering the effects of the volume current.

Methods

The MEGs of subject A (1 year old, female) and subject B (20 years old, male), and the MEG of subject C (10 years old, male) were recorded at Kyushu University (at rate of 520.83 samples/sec) and Tokyo University (at rate of 1041.7 samples/sec) respectively. All MEGs were recorded with a Magnes™ 37-channel Biomagnetometer (Biomagnetic Technologies Inc.). After recording, the MEGs were filtered to retain only the frequency band 1-30 Hz for subject A and 1-60 Hz for subject B, and 8-48 Hz for subject C to select for focal epileptic waves. An example of the focal epileptic waves for subject B is shown in Fig.1. It is difficult to know the zero-level in MEG. We take the output of each magnetometer at the beginning of the wave, t1 as the zero-level. The time t1 is shown by an arrow in Fig.1. We localized the focus of epilepsy using the MEG at the time near the peak of the waves. The isofield map at t2 of the waves shown in Fig.1 is shown in Fig.2.

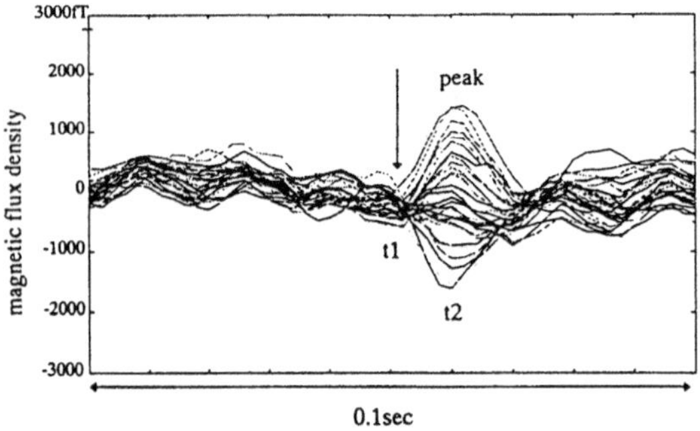

Fig.1 Focal epileptic waves for subject B
The all waves measured are shown together.

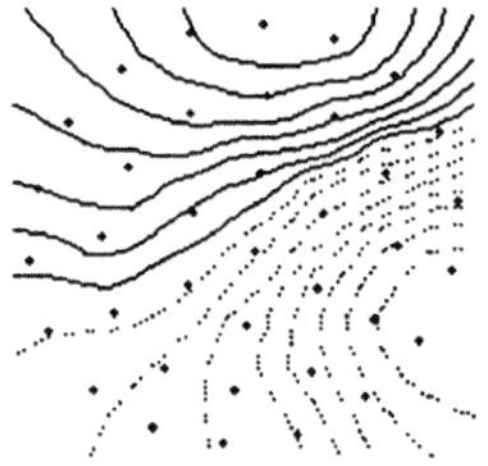

max 1295.09 fT
min -1422.47 fT
step 171.60 fT

Fig.2 Isofield map at t2
The solid lines show plus values and the dotted lines show minus values, and the circles show the position of pick up coils.

We assumed the focus of epilepsy as a single current dipole. In the source localization, the volume current was considered by the equation (1) [4].

$$B(r_P) = \frac{\mu_0}{4\pi}\int_v J_i \times \nabla \frac{1}{|r - r_P|} dv - \frac{\mu_0}{4\pi} \sum_j \left(\sigma_j^{in} - \sigma_j^{out}\right) \int_{S_j} V n_j \times \nabla \frac{1}{|r - r_P|} dS \quad (1)$$

where σ_j^{in} and σ_j^{out} are the conductivities of two different adjacent regions separated by the surfaces S_j, J_i is the current sources at r, n_j is the outer unit normal to the surface S_j, and V is the electrical potential. In above equation, the first term is the field due to the current sources and the second term denotes the field due to the volume current. The models for boundary elements were constructed by extracting the outline of the brain from the MRI or X-ray CT of each subjects to calculate the second term in equation (1). To compare our model with the spherical model, the localization of the focus is also carried out by the spherical model. The equation (2) calculated by Sarvas was used in the localization for the spherical model [5].

$$B(r_p) = \frac{\mu_0}{4\pi F^2}\left(FQ \times r - Q \times r \cdot r_p \nabla F\right) \quad (2)$$

where $F = a\left(r_p a + r_p{}^2 - r \cdot r_p\right), a = r_p - r, a = |a|, r_p = |r_p|, Q = \int_v J_i(r) dv$

The models for boundary elements are shown in Fig.3 with the position of pick up coils that are shown as black dots. Subject A has a chipped region in her head. The models which consist of two parts, brain and cerebrospinal fluid were used as shown in Figs.3 (a) and (b). The conductivities of cerebrospinal fluid and the brain ,and the outside of the brain were assumed to be 1.0 S/m, 0.3 S/m, and 0.0 S/m, respectively. (c) is the model for subject B, and (d) is the model for subject C. The CT of subject C is the only upper side of his brain, therefore the model for subject C looks like a hemisphere as shown in (d). All models are looked from the left in Fig.3.

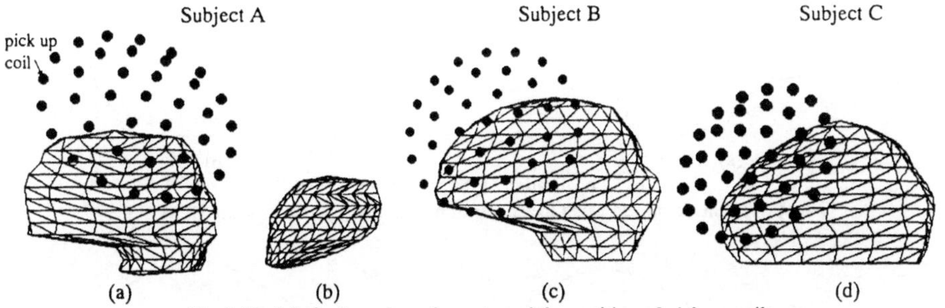

Fig.3 Models for boundary elements and the position of pick up coils
(a): the boundary elements of the brain with subject A, (b): the boundary elements of the chipped region with subject A , (c): the boundary elements of the brain with subject B , and (d): the boundary elements of brain with subject C

Results

The localizations of the foci were carried out with 35 waves for subject A and 79 waves for subject B, and 28 waves for subject C. The results of the localization of the focus are shown on the outline of the brain below. Only the results which have the dipolarity higher than 98% are shown. We applied the equation (3) as the dipolarity (DP).

$$DP = \sqrt{1 - \frac{\sum_i \left(B_i^c - B_i^m\right)^2}{\sum_i \left(B_i^m\right)^2}} \times 100 \quad [\%] \quad (3)$$

where B_i^c is the computed flux density value at the point of the pick up coil i, B_i^m is the measured flux density value at the point of the pick up coil i.

The results for subject A are shown in Fig.4. (a) is the result localized using the spherical model and (b) is the result localized using boundary element method (BEM). The results are looked from the left. The dashed line shows the outline of chipped region.

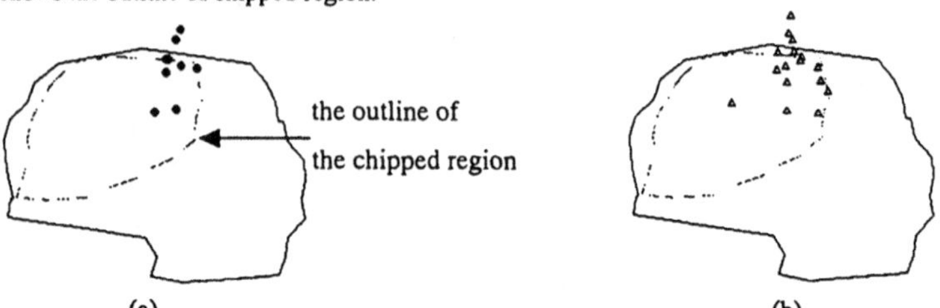

(a) (b)

Fig.4 Localization of the focus with subject A

(a): results localized using the spherical model and (b): results using BEM

Both of results seems to be similar. It shows that the inhomogenity slightly influences the localization of the focus.

The results for subject B and subject C are shown in Fig.5 and in Fig.6 respectively. (a) is the result localized using the spherical model and (b) is the result localized using BEM in Fig.5 and in Fig.6. These results are looked from the left.

(a) (b)

Fig.5 Localization of the focus with subject B

(a): results localized using the spherical model and (b): results using BEM

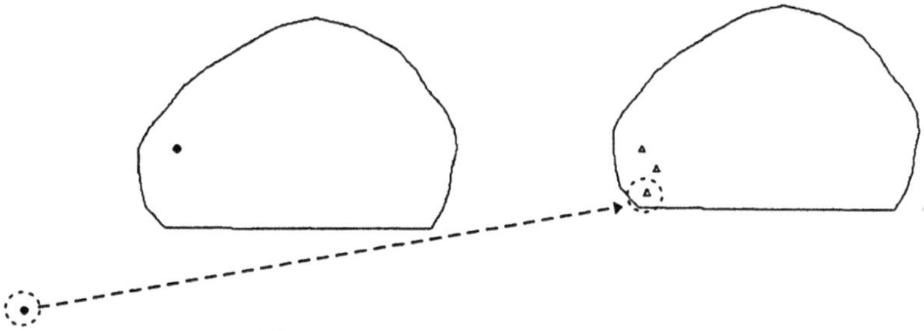

(a) (b)

Fig.6 Localization of the focus with subject C

(a): results localized using the spherical model and (b): results using BEM

In the case of subject B, the estimated positions using BEM are almost same as the estimated positions using the spherical model.

Though, in the case of subject C, one of the estimated positions using the spherical model is out of the brain as shown by the dotted circle in Fig.6, however the estimated positions using BEM are in the brain. It suggests that there is a possibility of improvement on the localization of the focus using BEM.

Discussion

We investigated the contribution of volume current to the MEG depending the dipole position by a computer simulation with a single-layer hemispherical phantom model. The model we used is shown in Fig.7. We simulated the contribution of volume current to the MEG by changing positions of the current dipole. d is the depth from the surface of hemispherical model. The result of the simulation is shown in Fig.8 [2][3]. To estimate the contribution of volume current to the MEG, the equation (4) is used.

$$f_t = \frac{\sum_i (B_i^{vol} - B_i^{source})^2}{\sum_i (B_i^{source})^2} \qquad (4)$$

where B_i^{vol} is the field due to the current sources and volume (return) current, B_i^{source} is the field due to the current sources. From the result, it is found that the contribution of volume current to the sum of magnetic fields at all measurement points increases when the current dipole exists on the region near the bottom and in a deep place of the model. It suggests that the localization of the focus considering the volume current using BEM will improve the accuracy when the current dipole exists in the frontal lobe or temporal lobe. The results of 3 subjects are in agreement with the result of the simulation.

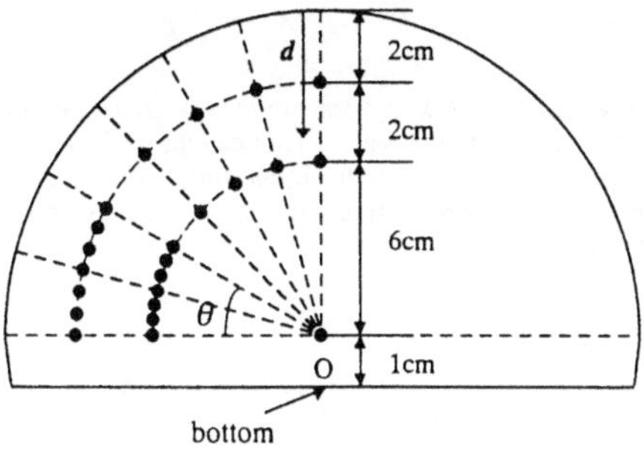

Fig.7 Cross section of a single-layer hemispherical phantom

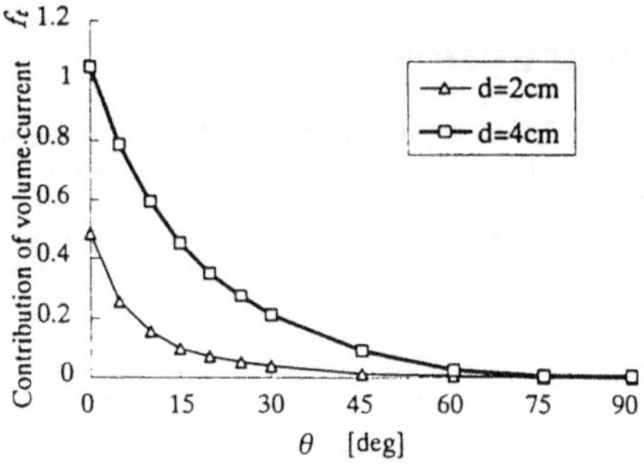

Fig.8 Contribution of volume current to the MEG depending the dipole position

Summary

We have investigated the possibility of improvement on the localization of focal epileptic activity with the MEGs of three subjects of focal epilepsy by considering the effects of volume current. From the result, it suggests that the method of the localization considering the volume current has a potential to improve the accuracy when the current dipole exists in the frontal lobe or temporal lobe.

Acknowledgements

The authors would like to thank Dr. Kaneko for his helpful suggestions and discussions.

References:

[1] I. Kanai, Y. Nagasaka, A. Ishiyama, Source Estimation by a Method Combines MEG and EEG, The Journal of Japan Biomagnetism and Bioelectromagnetics Society (in Japanese), 1992, 5: 204-207.

[2] M. Kobayashi, I. Tonegawa, A. Ishiyama, N. Kasai, Influence of Volume Current on MEG Using a Semi-spherical Phantom Model, The Journal of Japan Biomagnetism and Bioelectromagnetics Society (in Japanese),1993,6: 102-105.

[3] R. Takeuchi, S. Osozawa, T. Miyashita, A. Ishiyama, N. Kasai, Influence of Volume Current and Dipole Position on MEG Source Localization, The Journal of Japan Biomagnetism and Bioelectromagnetics Society (in Japanese), 1994, 7: 154-157.

[4] F. Grynspan, D.B.Geselowitz, Model studies of the magnetocardiogram, Biophys.J., 1973, 13: 911-925.

[5] J.Sarvas, Basic mathematical and electromagnetic concepts of the biomagnetic inverse problem, Phys. Mad. Biol, 1987, 32, 11-22.

Evaluation Of Abnormal Low Frequency Magnetic Activity (ALFMA) In Epilepsy

Lewine, J.D., Davis, J.T., and Orrison, W.W. Jr.

The New Mexico Institute of Neuroimaging, The New Mexico Regional Federal Medical Center, Albuquerque, New Mexico, USA.

INTRODUCTION

MEG attempts to define regions of epileptic pathology in patients with a seizure history are compromised by several factors. First, most studies are interictal in nature, an issue of concern to many epileptologists. Also, in some cases epileptic magnetic field patterns are highly complex and not amenable to simple single or even multiple dipole modeling. A final concern is the observation that most patients have seizures only very intermittently, with their interictal EEG often devoid of epileptic spikes and sharpwaves.

We have therefore sought to determine if aspects of MEG signals beyond spikes and sharpwaves might provide insights into epileptic pathology. Of particular interest were preliminary observations that many patients with a history of seizures, including those with interictal MEG and EEG devoid of spikes and sharpwaves, show abnormal low frequency magnetic activity (ALFMA) that can provide clinically useful information on the location of an epileptic focus [1,2]. Presented herein, are preliminary results of a large scale assessment of ALFMA in epileptic patients.

METHODS

Subjects: Sixty sequential patients with a history of seizures but without mass lesions (tumors of vascular malformations) or other gross neuroanatomical pathology (e.g., cortical dysplasia) were included in this study. The ages of the patient population ranged from 14 to 66 years.

MEG Examination: Neuromagnetic data were collected using a 37-channel, axial-gradiometer system (Magnes, Biomagnetic Technologies Inc., San Diego, CA). The study was conducted in a magnetically and radio-frequency shielded room. For the characterization of ALFMA, three minutes of continuous data (collected as ninety sequential two-second long epochs) were collected at each of eight placements of the array. The placements were over the following cortical regions: (1) midline frontal, (2) left parietal, (3) left temporal-occipital-parietal junction, (4) left temporal/inferior frontal, (5) right parietal, (6) right temporal-occipital-parietal junction, (7) right temporal/ inferior frontal, and (8) midline occipital. Additional data were collected over regions of interest for evaluation of spike and sharpwave, but the data from these recordings were not included in the analyses reported below.

Data were collected at a digitization rate of 300 Hz, with on-line filtering from 1-100 Hz. The data set from each probe position was analyzed separately. The dipole analysis of ALFMA involved four steps: (1) filtering and peak selection; (2) dipole modeling; (3) cluster analysis; and (4) generation of magnetic source localization images. First, MEG data were digitally filtered with a bandpass of 1-4 Hz in order to isolate ALFMA in the delta band. Independent analyses were performed on the data collected at each dewar position. After identification of the latencies of local amplitude peaks where the magnetic field strength was greater than 200 femtoTesla, single dipole modeling was performed every 8 msec during a 40-msec long window spanning each amplitude peak. In most cases, more than 5000 dipole fits were calculated for the data stream collected at each dewar position. Only the most dipolar of these fits ($r > 0.97$) were considered in further analyses. As a consequence of this strict criteria for dipolarity, more than 99% of the total amount of data collected is rejected from further consideration. A spatial clustering algorithm was used to determine if a particular cortical region was responsible for generation of multiple dipolar low frequency events. This is a critical aspect of the analysis procedure because available data suggest that it is the repetitive generation of dipolar low frequency activity that provides the most reliable sign that a region of tissue is dysfunctional. A 1-cm radius volume element was considered pathological only if it was responsible for generation of fifteen distinct low frequency events. This analysis strategy has been used to assess several clinical populations and more than 40 neurologically normal subjects. It yields a low false positive rate of only 8%.

MRI Examination: Magnetic resonance imaging was performed with a 1.5 Tesla whole body imaging system (Magnetom, Siemens Medical Systems, Islen, NJ). Images were obtained using modified gradient and spin echo sequences.By identifying the MEG coordinate fiducials on corresponding MR images it was possible to reconcile MEG and MR spatial frames to generate magnetic source localization images where the locations of the dipole sources of ALFMA are marked on appropriate MR sections

RESULTS

The MEG data from ten patients were uninterpretable because of low frequency magnetic artifacts from dental work or excessive eye movements. For the remaining fifty patients, interictal spikes and sharpwaves were found for twenty (40%) whereas ALFMA was found for forty-one (82%). ALFMA activity tended to be aperiodic, as shown in the below neuromagnetic waveforms recorded over the left temporal cortex of a sixteen-year-old female patient [Fig 1]. The early portion of epoch 30 show a relatively normal pattern of neuromagnetic activity. In contrast, the later portion of epoch 30 and all of 31 are punctated by low frequency delta and theta bursts. Whereas MEG power spectra for this recording position are normally dominated by 8-13 hz activity with little delta, the average power spectrum (across epochs and sensors) for this patient showed abnormal dominance of delta activity. Most slow waves have complex magnetic field patterns, but some have highly dipolar patterns. This suggests relatively focal generation, and single dipole modeling strategies can provide a first order localization of the relevant neuronal sources. In many cases the generators of ALFMA are found to localize to a single brain lobe, a result strongly suggestive of focal dysfunction [Fig 1D].

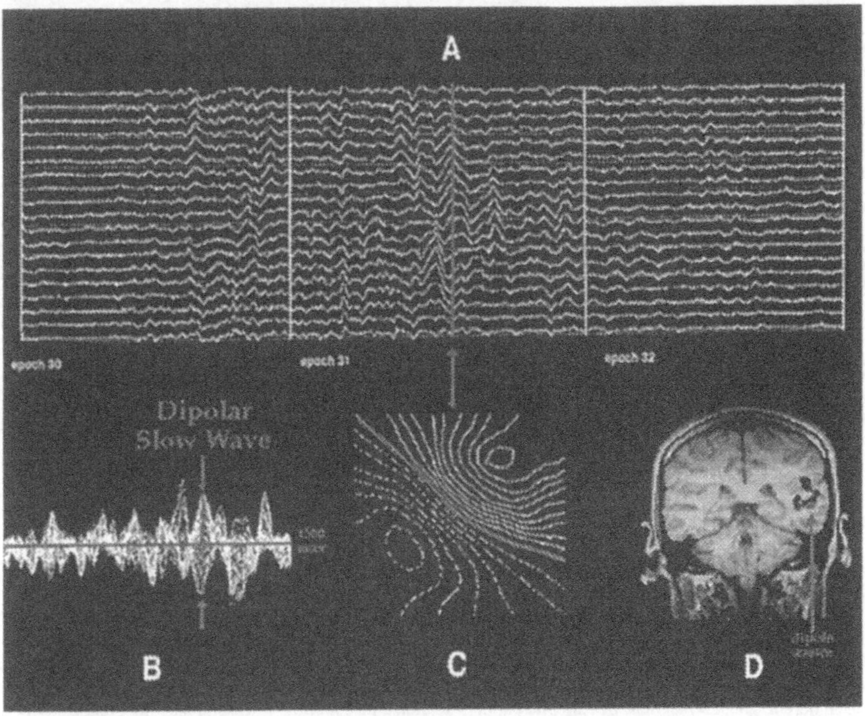

Fig 1: (A) Continuous data showing aperiodic slow waves. (B) Data overlay of 37-channels. Some slow waves have a dipolar field pattern, as shown in (C). (D) The sources of dipolar abnormal low frequency magnetic activity often cluster to a common area, a result that suggests the region to be dysfunctional. [reproduced from Lewine and Orrison, Neuroimaging Clinics of North America, 1995, 5(4), page 593]

Individual ALFMA source sometimes localize to white matter, a result suggestive of mis-modelling of activity generated by an extended cortical sheet. Nevertheless, even a mis-localized deep source provides information on the side and lobe of dysfunction. The distribution of ALFMA sources within a lobe may be scattered, and each source position is probably accurate to within only 1-2 centimeters. Nevertheless, in most cases ALFMA analyses clearly implicate a particular hemisphere and lobe in the epileptic pathology of the patient, and the result is consistent with clinical history and data from other neuroimaging procedures.

Several observations support the ALFMA approach to characterization of epilepsy. ALFMA is present for about 90% of those patients that do show interictal MEG spikes or sharpwaves, and when present, there has always been co-localization of spikes and some ALFMA. Admittedly, in many cases, several sites of ALFMA may be identified, although the most intense focus invariably correlates with spike locations [Fig 2].

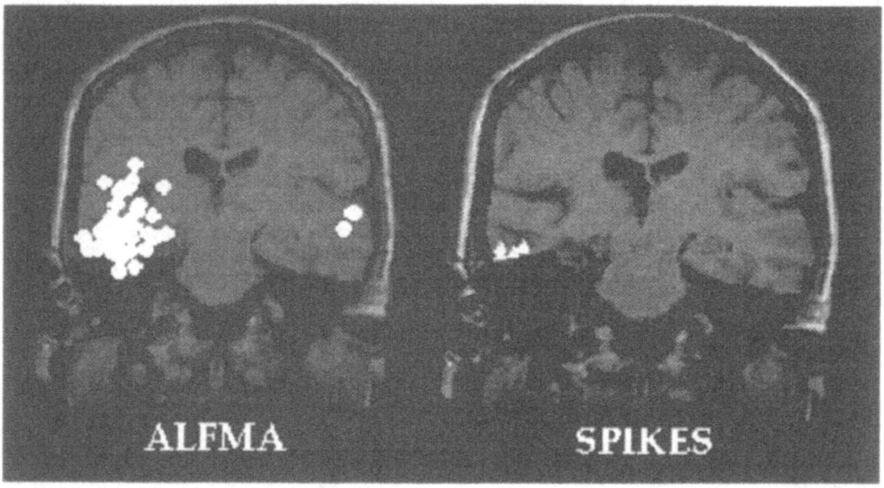

Fig 2: Data are from a 36 year-old female with seizures that proved resistance to partial right anterior temporal lobectomy. Dipolar sources of ALFMA are identified bitemporally. More activity is identified on the right. Several epileptic spikes were recorded also, with sources localizing to the right temporal lobe, just posterior to the original resection zone.

DISCUSSION

Biomagnetic characterization of patients with epilepsy is a challenging problem because examinations are mostly limited to the interictal period. At present, there continues to be tremendous debate over the utility of interictal data relative to ictal data. It is generally accepted that ictal data is preferred, but this is not always available. Within the interictal domain, MEG is faced with the problem that many patients fail to show sufficient interictal spiking to allow for confident identification of a focus. In fact, when non-selected patients are considered, only about 40% of patients demonstrate spikes and sharpwaves during a typical 2 hour MEG examination. Fortunately, as presented herein, greater than 80% of epileptic patients demonstrate abnormal low frequency magnetic activity during the interictal period. These patients also demonstrate slow waves in EEG recordings, but the activity tends to be relatively diffuse, even in cases suspected to have focal epilepsy. In contrast, slow waves in MEG are often focal and sources associated with separate slow waves cluster into circumscribed brain regions.

On the one hand, it is surprising that the ALFMA analysis technique, which is unable to adequately characterize more than 99% of the collected data, nevertheless provides clinically useful information. Single dipole modeling of spontaneous brain activity is considered by many to be inappropriate, but the present results indicate that this approach can be useful in the clinical identification of pathophysiology.

The fundamental rationale for this strategy is that the abnormal signals from focal dysfunctional regions are so large that they dominate the recorded magnetic field pattern, with the simple dipole model being only minimally perturbed by the lower amplitude activity from nearby normal tissue. In evoked response studies where MEG is used to identify the location of primary somatosensory cortex, source locations are thought accurate to within a few millimeters. This is almost certainly not the case in ALFMA studies where there is no signal averaging and the magnitude of the extraneous background brain activity is high. Dipole sources of ALFMA often localized in the white matter, rather than at the cortical mantle as expected on the basis of biophysical considerations. This most likely indicates that an extended region of the overlying cortex was actually responsible for ALFMA generation, with dipole modeling of this extended activity giving an erroneously deep source location (the magnetic field pattern of an extended cortical sheet and a deeper dipole source are often indistinguishable). Nevertheless, ALFMA source locations are probably an accurate reflection of the physiology to within 1-2 cm. Whereas this could be inadequate resolution for some neurosurgical planning, it is more than sufficient for identifying which gross brain regions are compromised by epilepsy.

At present the nature of the pathological processes that cause ALFMA are not fully understood. ALFMA is found in many conditions including mild head trauma, ischemic disease, and dementia. Multiple physiological mechanisms may all have similar biomagnetic manifestations.

In considering the clinical impact of the examination of ALFMA it is clear that the presence of ALFMA alone is inadequate for directing surgical intervention. However, when ALFMA data agree with data from other noninvasive methods, a strong case can be made for surgical intervention without the need for invasive monitoring. In those cases where invasive monitoring is still indicated, data on ALFMA can be used to guide the placement of subcortical grids and depth probes.

CONCLUSIONS

The ALFMA strategy outlined here has promise for augmenting the clinical utility of MEG evaluations of patients with epilepsy. ALFMA is present in a high percentage of patients, even in the absence of interictal spikes and sharp waves. Preliminary results suggest that the clinical effectiveness of the ALFMA investigation is superior to that for interictal SPECT or PET.

REFERENCES

[1] Lewine, J.D., and Orrison, W.W., Spike and slow wave localization by magnetoencephalography, Neuroimaging Clinics of North America - Epilepsy, 1995, 5(4): 575-596.

[2] Vieth, J., Grummich, P., Kober H., et al. Localization of slow and beta MEG waves associated with epileptogenic lesions. Epilepsia, 1993, 34 (supplement 6): 143.

Event-Related Magnetic Fields Associated with Auditory Oddball Paradigm - Postoperative Study in Patients with Temporal Lobe Epilepsy

Nishitani, N.[1], Nagamine, T.[1], Mikuni, N.[1,2], Ikeda, A.[1], Kimura, J.[3] and Shibasaki, H.[1]

[1]Departments of Brain Pathophysiology; [2]Neurosurgery; [3]Neurology, Kyoto University School of Medicine, Kyoto, Japan

Introduction

Regarding the late positive component of event-related potentials (ERPs) associated with auditory oddball paradigm which is now well known as P300, many studies have been done about its characteristics and neural generators. Since P300 is considered to reflect higher brain functions, some investigators have studied the topographical distribution of the scalp-recorded P300 as well as invasive recording using intracranial depth electrodes or chronically implanted subdural electrodes to identify its neural generators and functional significance [1-4]. In addition, there are several studies using magnetoencephalography (MEG), which has been developed as one of the non-invasive technology, to elucidate the brain function. Two regions have been suggested regarding the sources of the event-related magnetic fields corresponding to P300; the mesial region of the temporal lobe, possibly the hippocampus [5,6], and the superficial temporal lobe [7]. Rogers et al. [8] also suggested that the responses to the target stimuli arose in the deep regions of brain initially and were transferred to the auditory cortex. However, since these studies recorded the magnetic responses from the restricted area with one or several channels of magnetometer and analyzed them with one dipole model, it might not be enough to disclose the cognitive processing which activates many regions simultaneously and parallel. Therefore, in order to elucidate the generator sources of P300, it would be better to acquire the responses of all the areas in the brain simultaneously and to analyze them by adopting the multidipole model. We studied the event-related magnetic fields (ERFs) associated with the auditory oddball paradigm in normal volunteers and in patients after the operation for the temporal lobe epilepsy by using a whole head magnetometer with 122-channel planar first-order gradiometers to clarify the possible brain areas involved in the cognitive processing.

Methods

Six healthy volunteers (4 males, 2 females, 20-38 years, rt-handed) and seven patients after the operation for the intractable temporal lobe epilepsy (1 male, 6 females, 18-38 years, rt-handed) participated in this study. The lesion of the epileptic foci lay on the unilateral temporal lobe. Out of seven patients, five were operated with standard anterior temporal lobectomy, one with lateral temporal resection and one with selective amygdalohippocampectomy. An informed consent was obtained from the subjects after the full explanation of this study.

The subjects sat in a glassfiber chair in a quiet and magnetically shielded room. They were requested to keep their eyes open. A standard auditory oddball paradigm was used as a stimulus sequence (50 ms duration, 10 ms rise/fall and 30 ms plateau). The intensity of both tones was adjusted to 85 dBSPL at the exit of plastic tube. The order of tone presentation was randomized with a varying rate of once per 2.2 s on the average (2.0-2.4 s). The subject was requested to count the number of the presented target stimuli silently. All the measurement of magnetic field was done by using the whole head coverage magnetometer with 122-channel planar first-order gradiometers at 61 measurement sites. The simultaneous electroencephalographic (EEG) recording in all subjects was obtained from 4 electrodes according to the International 10-20 System (Fz, Cz, Pz and Oz) referenced to linked earlobe electrodes. To monitor eye movements and blinks, the electrooculogram (EOG) was recorded from an electrode placed below the right outer canthus and referenced to linked earlobe electrodes. The recording passband was 0.03-100 Hz for MEG, and 0.07-120 Hz for EEG, and sampling rate was 404 Hz. The responses were averaged on-line with respect to the onset of target and non-target stimuli independently after excluding epochs containing big artifacts such as eye blinks. Analysis window for the on-line averaging was from 200 ms before to 800 ms after the stimulus. One session consisted of 100 epochs. At least 3 sessions were performed to confirm the reproducibility. Amplitude measurement was done from the baseline, which was determined by averaging the 100 ms segment before the stimulus onset for each channel of EEG and MEG. The peak latency of N100 component was visually determined at Cz, and that of N1m component was also visually determined at the channel showing maximum magnetic deflection with the peak latency around 100 ms after the onset of the stimulus, on each hemisphere. P300 component was determined as the positive deflection of the response seen only following the target stimuli at Cz or Pz at the latency between 250 and 600 ms. The magnetic deflections corresponding to P300 were detected in the response to the target stimuli at the latency between 250 and 600 ms. Only the magnetic responses to the target stimuli were analyzed for the following dipole analysis of the late magnetic component corresponding to the P300 component. To identify the current source underlying the measured signals, a single spherical head model, which best fitted to each subject's brain surface determined on one's own MRI, was adopted for the calculation of the equivalent current dipoles (ECDs). ECDs were assumed from the sub-region consisting of 20-30 channels showing significant difference between deflections to the target and those to the non-target stimuli. ECDs indicated the most dominant source acquired by the least-squares search during several segmental time periods by using a subset of channels over the interested areas. The goodness of fit of the model was also calculated, to demonstrate the ratio of correspondence between the estimate magnetic field distribution by ECD and the actual magnetic field distribution. Only ECDs with the goodness of fit more than 60 % at selected periods of time were adopted for the further analysis. Then the analysis time and channel were extended to the whole recording time period and all channels. At first ECD for N1m was determined on each hemisphere independently. The estimated waveforms by introducing two dipole model for N1m were superimposed on the measured responses. Channels showing the significant discrepancy between the simulated responses and the actually measured ones were, if any, selected visually for the next subset channels to estimate ECDs. Every time after obtaining a new ECD, simulation was performed with various combinations of the estimated ECDs. These analysis processes were repeated until the discrepancy between the simulated and measured responses became small and the goodness of fit could not get significantly higher value. Estimated dipoles by these processes were superimposed on the subject's own MRI.

Results

All subjects performed well the task of counting the number of target stimuli silently. N1m was recognized clearly in response to both target and non-target stimuli at about 100 ms after the onset of stimuli, which was almost the same as the latency of N100. In all patients and all healthy subjects, P300 was recognized with the maximum amplitude at Pz at the mean latency of 331 ms and 343 ms, respectively, between which there was no statistically significant difference (t-test; $p>0.05$)(Fig. 1). In healthy subjects, responses to the target stimuli showed an additional magnetic deflection at around 400 ms on the anterior temporal, middle temporal and parietal regions on each hemisphere, but not in the responses to the non-target stimuli (M400). Their latency was longer than that of P300 by 20-100 ms. Dipoles of M400 lay in the bilateral mesial temporal, superior temporal and parietal regions. The simulated signals assumed by 6 dipole model for M400 could explain the measured signals. Further attempt to detect an additional dipole was failed. On the other hand, in 4 out of 5 patients after standard temporal lobectomy, M400 could not be detected over the anterior temporal area of the operated side, whereas M400 over other areas could be estimated (Fig. 2). In the patient after lateral temporal resection, M400 could be observed over the operated temporal area as well as other areas. The latency of M400 in patients was longer than that of P300 by 100-150 ms. In the remaining two patients, M400 was not detected due to large artifacts around the region of craniotomy.

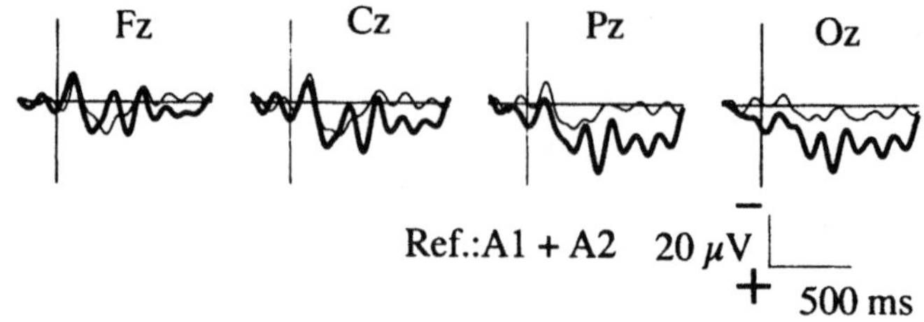

Fig. 1

Group average responses of auditory ERP after the left standard temporal lobectomy in a patient with temporal lobe epilepsy. Thick and thin lines show wave forms in response to target and non-target stimuli, respectively. P300 is clearly seen 340 ms after the target stimuli.

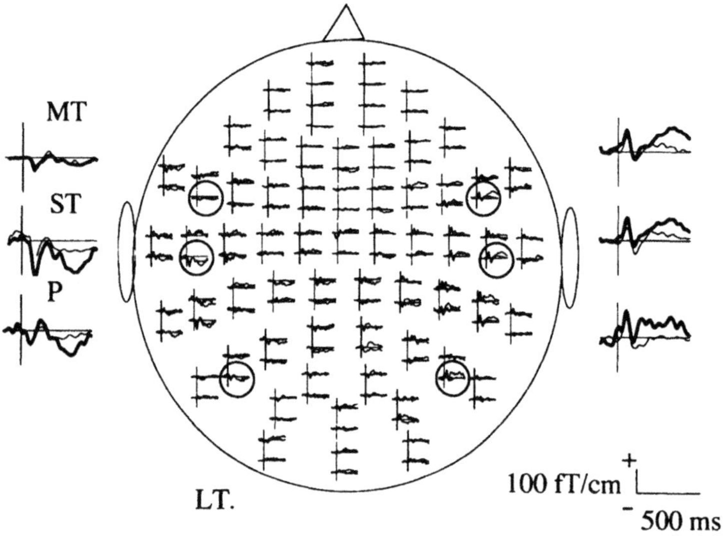

Fig. 2

Group average responses of auditory ERF in the same patient as shown in Fig.1. Selected channels from the mesial temporal (MT), superior temporal (ST) and parietal (P) regions on each hemisphere are magnified. The early component was deteced in the bilateral temporal areas. During the later period between 400 and 500 ms after the onset of the target stimuli, there are significant differences between deflections to the target and to the non-target stimuli in the channels except for the one on the left MT area.

Discussion

In the present study M400 recorded by the whole head type MEG is considered to reflect the cognitive processing like P300, because it was recognized only in the responses to the target stimuli. The estimated dipoles of M400 were located in the mesial temporal, superior temporal and parietal regions on each hemisphere. These areas are compatible with the previous EEG studies showing the possible sources of P300 in the mesial temporal, superior temporal and temporoparietal regions [1,3 and 4]. In the previous MEG studies, only the hippocampus and the superior temporal region were detected as the source areas of ERFs [5 - 8], possibly because they were estimated by one dipole model based on the recordings with one or several channels magnetometer.

Although P300 could be detected in all patients, M400 was not observed over the anterior temporal region of the operated side in the patient after the standard temporal lobectomy. In the patient after the lateral temporal lobe resection, M400 was detected in all six areas including the ipsilateral temporal area. Polich and Squire [2], who studied P300 in amnesic patients with bilateral hippocampal lesions, concluded that the hippocampal lesion did not contribute significantly to the scalp-recorded P300. These results indicate the minimal contribution of the mesial temporal region to the scalp recorded P300, even if it might still be one of the generator sources of P300. Therefore, it can be concluded that the cognitive processing in the mesial temporal region would be detected more clearly by MEG.

Conclusions

We studied the ERFs in normal subjects and in patients after the operation for the intractable temporal lobe epilepsy, and clarified the possible generators of M400. These findings suggested the followings. 1) The possible sites involved in of the cognitive processing associated with the auditory oddball paradigm lie in the bilateral mesial temporal, superior temporal and parietal regions. 2) Cognitive function may be carried out not only by the mesial temporal region but by the combination of these multiple areas.

References

[1] Knight, R.T., Scabini, D., Woods, D.L. and Clayworth, C.C. Contributions of temporal-parietal junction to the human auditory P3, Brain Res., 1989, 502: 109-116.

[2] Polich, J. and Squire, L.R. P300 from amnesic patients with bilateral hippocampal lesions. Electroenceph. clin. Neurophysiol., 1993, 86: 408-417.

[3] McCarthy, G. and Wood, C.C. Intracranial recordings of endogeneous ERPs in humans. Electroenceph. clin. Neurophysiol., 1987, 39: 331-337.

[4] Halgren, E., Baudena, P., Clarke, J.M., Heit, G., Liegeois, C., Chauvel, P. and Musolino, A. Intracerebral potentials to rare target and distractor auditory and visual stimuli. I Superior temporal plane and parietal lobe, Electroenceph. clin. Neurophysiol., 1995, 94: 191-220.

[5] Okada, Y.C., Kaufman, L. and Williamson, S.J. The hippocampal formation as a source of the slow endogenous potentials, Electroenceph. clin. Neurophysiol., 1983, 55: 417-426.

[6] Lewine, J.D., Roeder, S.B., Oakey, M.T., Arthur, D.L., Aine, C.J., George, J.S. and Flynn, E.R., Localization of the generators of the magnetic P3, In: Williamson, S., Hoke, M., Stroink, G. and Kotani, M. Advances in Biomagnetism. Plenum Press, New York, 229.

[7] Gordon, E., Slogget, G., Harvay, I., Kraiuhin, C., Pennie, C., Yiannikas, C. and Meares, R. Magnetoencephalography: locating the source of P300 via magnetic field recording, Clin. Exp. Neurol., 1987, 23: 101-110.

[8] Rogers, R.L., Baumann, S.B., Papanicolaou, A.C., Bourbon, T.W., Alagaarsamy, S. and Eisenberg, H.M. Localization of the P3 sources using magnetoencephalography and magnetic resonance imaging. Electroenceph. clin. Neurophysiol., 1991, 79: 308-321.

Use of Magnetic Source Imaging in Surgical Planning for Epilepsy Patients who have Undergone Prior Resection

Sobel, D.[1], Aung, M. [1], Schwartz, B.[1,2] Squires, K.[2], and Hirschkoff, E.[1,2]

1. The Scripps Clinic and Research Foundation, La Jolla, CA , USA
2. Biomagnetic Technologies, Inc, San Diego, CA , USA

Introduction

While surgical intervention often substantially reduces seizure activity, a number of cases arise in which the seizure activity following resection is either not reduced as much as desirable or is initially reduced but increases in frequency and severity over time. In these cases, further surgery may be indicated. A quick, non-invasive means to assess the location of remaining epileptogenic activity would be a valuable tool for this purpose. A key issue is whether the continuing seizures originate from zones adjacent to prior surgery, or whether they represent multifocal regions or areas not adjacent to prior surgery which were previously undetected which had developed since the surgery.

The present study was undertaken to determine the utility of magnetic source imaging as a tool to evaluate candidates for surgery in cases where initial resection has not reduced seizure activity to an acceptable level or where seizure activity has resumed. The cases evaluated in this study were part of a larger study evaluating 100 candidates for epilepsy surgery referred from eleven epilepsy centers throughout the U.S. and Canada. Evaluations at those centers prior to MSI typically included EEG, MRI, neurological examination, neuropsychological testing, and video EEG (Phase I) monitoring, and in some cases invasive EEG (Phase II) monitoring.. In many cases, PET and/or SPECT test results were also available. Earlier work had demonstrated the capacity of Magnetic Source Imaging (MSI) to localize focal sources of interictal epileptogenic activity, in agreement with more invasive methods and with electrocorticographic monitoring during and after surgery.[1]

Methods

Twenty-four patients who had had resective surgery but who now were experiencing uncontrollable seizure activity underwent MEG recording during which interictal epileptiform events were recorded concurrently on a 74-channel biomagnetometer *(Biomagnetic Technologies Magnes II®)* and a 21 channel EEG recording system (Nihon Khoden) to record spontaneous incidents of interictal epileptogenic activity. A radio frequency 3-D digitizer determined the position of the individual gradiometers relative to a head-centered 3-D orthogonal coordinate system defined by the nasion and by markers in the right and left outer ear canal. Each patient underwent an MR scan with fiducial markers attached, providing a common coordinate system to overlay MEG source localizations [2]. Normal seizure control medications were maintained.

With the patient lying on one side and the head voluntarily stabilized, the two 37-channel probes were positioned over homotopic areas of the two hemispheres. A sequence of four to five dual probe placements and measurements assured coverage of the whole head, and afforded the ability to distinguish between synchronous and asynchronous bilateral sources. Recording sessions typically lasted from two to four hours, depending on the patient's level of interictal activity. EEG and MEG waveforms were monitored in real-time by observers trained to recognize and distinguish the morphology of epileptogenic events as against artifacts. (EEG data were later reviewed and interpreted separately by an experienced epileptologist.) Twenty 5-6 second epochs of data were obtained at each recording site. Raw data were digitized at 860 Hz and band-pass filtered at 0.1-200 Hz. Data epochs were then digitally filtered off line with a bandpass of 3-70 Hz. Off line data analysis proceeded with visual identification of MEG and EEG interictal spikes or sharp waves. Temporal windows including these events were then analyzed at each time point in the data sample with a single equivalent current dipole (ECD) model. Each event for which the dipole model correlated with the recorded data to at least 98% and for which the equivalent dipole strength was physiologically realistic was represented by a single focal source plotted on an MRI image of the patient's brain.

Images showing all ECD modeled epileptiform events (Fig 1) were presented along with waveform information and the EEG tracings for review by an experienced epileptologist. The epileptologist indicated the likely source of seizure activity and target for surgical resection, if such an inference could reasonably be drawn from the MSI data. The diagnostic conclusion drawn from the MSI assessment was then compared with that from the consensus of all other measures including invasive EEG monitoring undertaken during the patient's normal course of care [2].

Five of these patients later underwent invasive EEG Phase II monitoring with implanted grid or depth electrodes. The source locations suggested by the MEG data were then compared with the results of conventional diagnostic testing (MRI, Phase I or Phase II monitoring).

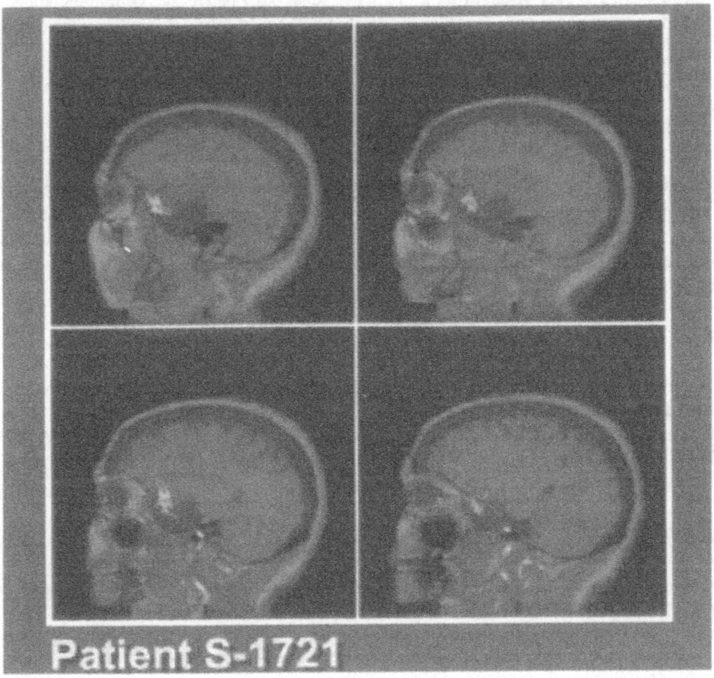

Fig. 1. MEG dipole overlays on sagittal images for patient #S1721.(Spike sources are indicated as yellow triangles, with lines indicating orientation.) The patient had suffered complex partial since age 12. Prior surgery was a right anterior temporal lobectomy, evident in sagittal slices. Spike sources are seen at the superior and anterior margins of prior surgery.

Results

MSI showed abnormal spike activity in 21 of the 24 patients. In three cases there was insufficient data for MSI to localize spike activity. MSI localizations agreed with the consensus of Phase I and Phase II EEG indications in 14 cases, in the sources indicated were adjacent to the region of prior surgery in 12 patients. In an additional 4 cases, MSI and conventional measures were concordant in indicating multi-focal or bilateral sites of activity. An additional two cases were only in partial concordance, with MSI indicating a wider and more diffuse area of abnormal activity in relation to Phase I and Phase II EEG measures. There was one case with an inconsistent localization between MSI and EEG measures.

Of the 24 patients measured, 10 have gone on to a second surgery. Of these cases, 6 are seizure free after 3-6 months, with 2 cases experiencing visual or motor deficits. One other patient has continuing auras 6 months post-surgery, and two more cases show only partial or minimal improvement over 3-6 months. One case had an increase in seizure frequency as of 3 months after surgery.

Discussion

MEG localization has the potential to indicate quickly and non-invasively whether or not epileptogenic source localizations are consistent with regions adjacent to prior surgery. Where bilateral or contralateral MEG signals are found, MEG may help indicate or even guide further invasive monitoring. Where MEG indicates abnormalities adjacent to prior resection, it may help increase confidence or reduce the time needed to determine whether further surgery is feasible or advisable.

References

[1] Gallen, C.C., Hirschkoff E.C., Buchanan, D.S. Magnetoencephalography and Magnetic Source Imaging. Neuroimaging Clinics of North America, 5(2), May, 1995.

[2] Smith, J.R., Schwartz, B.J., Gallen C.C., Orrison, W., Lewine, J., Murro A.M., King, D.W. Park, Y.D. (1995) Utilization of Multichannel Magnetoencephalography in the Guidance of Ablative Surgery. Journal of Epilepsy 8:119-130.

Acknowledgements

The authors would like to thank the following collaborators for their assistance: Dr. Izak Fried and Dr. Sachar Eliashiv (UCLA Medical Center), Dr. Vicente Iragui and Dr Evelyn Tecoma (UCSD Medical Center, La Jolla, CA), Dr. Brien Smith (Henry Ford Hospital, Detroit, MI), Dr. Joseph Smith and Dr. Yong Park(Medical College of Georgia, Augusta, GA), Dr. Thomas Waltz (Scripps Clinic Research and Foundation, La Jolla, CA, Dr. James Wheless and Dr. Valerie Curtis (Hermann Hospital, Houston, TX)

Methohexital-Induced Changes in Spectral Power of Neuromagnetic Signals: Reduced β-Band Enhancement Over the Hemisphere Ipsilateral to the Epileptogenic Focus

Wienbruch, C.[1], Eulitz, C.[3], Lehnertz, K.[2], Brockhaus, A.[2], Elger, C.E.[2], Elbert, T.[3] and Hoke, M.[1]

[1]Institute of Experimental Audiology, University of Münster, Germany; [2]Clinic of Epileptology, University of Bonn, Germany; [3]Department of Psychology, University of Konstanz, Germany

Introduction

Effects of anesthetics on electroencephalographic (EEG) activity are used to localize brain lesions as well as epileptogenic areas. Small doses of barbiturates (up to 200 mg) result in an increase of fast activity in the β-band of the EEG. This increase of β-band activity is known to be larger over normal brain areas as compared to areas with cerebral lesions (e.g. [1]). Furthermore, it has been reported that some barbiturates as amobarbital, thiopental, or methohexital or other narcotics as propofol may provoke epileptiform activity resulting in spike activity and spike-burst-suppression patterns. As an additional feature, various authors ([1] [2]) described qualitatively a loss of methohexital-induced β-band activity in the electrocorticogramm (ECoG) over the epileptogenic area. A quantitative analysis of the spectral contents of the ECoG is mandatory before the clinical use of methohexital induced changes in β-band activity can be rated.

Drug-induced activation of the epileptogenic focus provides a rapid and safe method to obtain a sufficient amount of information relevant for the lateralization and localization of the primary epileptogenic area. We investigated the value of methohexital induced activity for the presurgical lateralization and localization of the epileptogenic zone using MEG. We measured simultaneously the MEG and EEG during methohexital induced narcosis in 13 patients all suffering from temporal lobe epilepsy (TLE) of temporomesial origin. The aim of this study was to quantify anesthesia-induced changes in the spectral power characteristics of MEG and EEG signals and to analyze the topographical aspects of these changes with respect to the side of the primary epileptogenic area in patients suffering from temporal lobe epilepsy (cf. [3]).

Method

Thirteen patients suffering from pharmacoresistant unilateral temporal lobe epilepsy were investigated. Before performing this study, all patients underwent invasive presurgical evaluation (cf. Bonn protocol of presurgical evaluation, [4]) using electrocorticographic and stereo-EEG recordings for the localization and delineation of the primary epileptogenic area. As confirmed by the post-operative outcome a left temporal seizure origin was found in 5 patients and a right temporal seizure origin in 8 patients. Selective amygdalohippocampectomy (sAHE) was performed in 8 cases, AHE with partial temporal resection in 4 cases and a resection of a ganglioglioma in one case. After surgery all but one patient are completely free of seizures for at least 6 month.

Neuromagnetic data were recorded using a 37-channel neuromagnetometer (Magnes^{TM}; Biomagnetic Technologies, Inc.). Measurements were carried out in a magnetically shielded room. The sensor array was centered sequentially over the left and the right supra-temporal cortex about 1.5 cm superior to T3 or T4 electrode position of the international 10-20 system. The head position relative to sensor pickup coils was measured by a sensor position indicator. EEG data were simultaneously recorded (SYN-AMPS^{TM}; Neuro Scan, Inc.) from temporal, frontal, central and parietal electrode positions according to the international 10-20 system with the nose used as a reference. Subjects were lying on their side with their head and body fixated by a vacuum cushion. They were instructed to avoid eye blinks and head movements. Continuous data were recorded from left and right hemisphere in 10 minute blocks using a sampling rate of 297.6 Hz and a passband of 0.03 to 100 Hz.

After a 2 min baseline recording a dose of 100 mg methohexital was applied intravenously (i.v.) in a bolus injection according to the conditions used during the invasive presurgical evaluation [2]. After a break of approximately 20 min exactly the same procedure was repeated on the opposite hemisphere. This break is sufficient for the recovery of the patient with respect to consciousness and recovery of MEG and EEG back to the baseline due to the short half life time of approximately 2 min. Anesthetic monitoring and standby was provided, and precautions were taken to counteract anesthetic complications. During the measurement an anesthesist was within the magnetically shielded room. Additional video monitoring of the patient and the anesthesist was performed. All patients were informed about the procedure before the investigation and declared their informed consent.

Off-line the data were digitally bandpass filtered using a 4-48 Hz recursive Butterworth-filter (2nd order). The spectral power was estimated for each magnetic and electric channel. Data sets of 300 seconds length starting 20 seconds before i.v. injection of methohexital were analyzed using an overlapping segmentation technique. The length of the Parzen-windowed segments was 512 points which corresponds to 1.720 s. According to [5] the segments were overlapped by one half of their length to obtain the smallest spectral variance per data point. The power spectra of these segments were averaged within 19.785 seconds time windows and normalized by dividing each mean spectrum by the spectrum of the time before the injection of methohexital separately for each channel. The spectral band power in the θ-, α-, β- and γ-band represents the integrated power within the border frequencies (θ (4.07-7.55 Hz), α (8.13-12.79 Hz), β (13.37-23.83 Hz) and γ (24.41-47.66 Hz)). To further reduce the data, spectral band power was averaged across neighboring channels. MEG channels were collapsed over frontocentral, temporoparietal and frontotemporal brain regions, the EEG-channels were averaged across frontocentral and temporal electrodes separately for the left and the right hemisphere.

The resulting normalized spectral band power values of each frequency band were statistically tested for differences between hemispheres and brain regions as dependent variables. For MEG data we used two-way univariate analyses of variance (ANOVA) for a 3 *brain-regions* (frontocentral vs. frontotemporal vs. temporoparietal) x 2 *side-of-focus* (side of epileptogenic focus vs. contralateral to the side of epileptogenic focus) design. For the statistical analyses of EEG data the factor *brain-regions* had only two steps (frontocentral vs. temporal). Wherever appropriate, a Greenhouse-Geisser correction of the degrees of freedom was applied.

Results

No adverse effects during or after anesthesia were noticed by either the patients or the investigators. Patients recovered quick and had no clinical excitatory effects or signs like vomiting and nausea. No seizure was elicited.

In contrast to the baseline recording, a high amplitude β-band activity is obtained around 2 minutes after administration of methohexital. A well pronounced difference of the MEG β-band spectral power between the hemispheres ipsi- and contralateral to the epileptogenic focus can be observed approximately two minutes after the injection of methohexital for a period of 1-2 minutes, while the EEG β-band spectral power on the ipsilateral side is even more enhanced compared to the contralateral side. The normalized spectral power estimates derived from MEG signals of the third minute after methohexital activation measured over frontocentral, frontotemporal and temporoparietal brain regions of both hemispheres were used as dependent variables for a two-way ANOVA. Only main effects of the *side-of-focus* or *side-of-focus* by *brain-regions* interactions will indicate a usefulness of changes in spectral power for the lateralization of the side of the primary epileptogenic focus. In the β-band changes in normalized spectral power were influenced by the factor *side-of-focus* (ipsilateral vs. contralateral side) which revealed a statistically significant main effect ($F(1/12) = 10.56; p < 0.007$). The *side-of-focus* effect did not differ between the three separated brain regions.

As illustrated in Fig. 1, during the third minute after methohexital injection, normalized β-band spectral power was more pronounced over the hemisphere contralateral to the epileptogenic focus as compared to the ipsilateral side in ten of thirteen patients, which holds true for both the left and right TLE patients. Only one patient (R16) showed a clear reversal of that typical lateralization pattern. In this particular patient methohexital induced high-amplitude spike activity followed by short intervals of isoelectric line or very slow δ-activity with a rate of approximately one per second was seen in MEG and EEG recorded over the hemisphere ipsilateral to the epileptogenic focus. The *side-of-focus* effect was evident over all brain regions with a slight predominance over frontocentral regions, without preferring one of them.

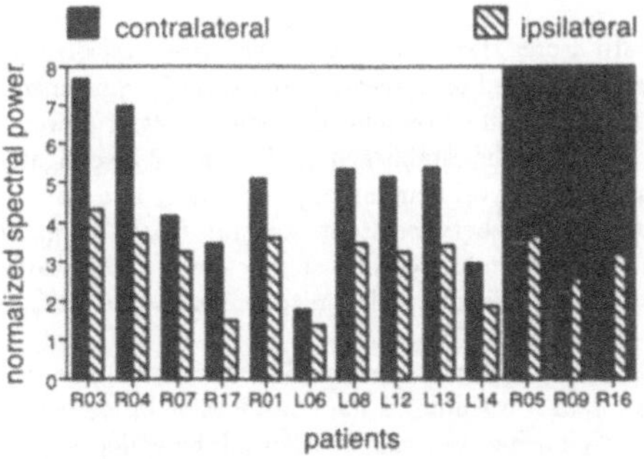

Figure 1: Methohexital induced changes of the normalized spectral power in the β-band of the MEG data for each individual patient. The figure shows the mean normalized β-band spectral power averaged across all brain regions for the third minute after the injection of methohexital. The first letter of the patient identification indicates whether the hemisphere with the primary epileptogenic focus was the right (R) or the left (L) one. The gray shaded area marks those patients showing ipsilateral to the epileptogenic focus more pronounced enhancement of the mean normalized β-band spectral power.

For the θ-, α- and γ-band exclusively effects of ***brain-regions*** were found while *side-of-focus* effects were not evident in these frequency bands. The fine structure of differences between brain regions was different for the various frequency bands as illustrated in Fig. 2.

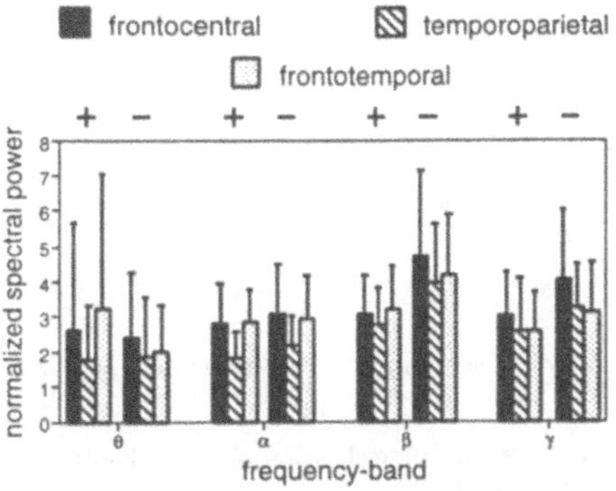

Figure 2: Normalized spectral power of the MEG data over three separated brain regions for the θ-, α-, β- and γ-band. Each column represents the mean $\pm$ standard deviation of the normalized spectral power across all patients. The hemisphere ipsilateral to the side of the epileptogenic focus is marked with a (+), the contralateral side with a (−).

For the θ-band a main effect of ***brain-regions*** was found ($F(2/24) = 4.55; \epsilon = 0.82; p < 0.03$). Post-hoc Scheffé comparisons revealed that the effect of ***brain-regions*** was mainly caused by the difference between frontotemporal areas where the normalized θ-band spectral power exceeded 2.6 times the baseline level, and the temporoparietal regions where the spectral power was enhanced only 1.8 times ($p < 0.04$). A similar pattern was found for the α-band (main effect of ***brain-regions***: $F(2/24) = 14.60; \epsilon = 0.94; p < 0.0001$). Post-hoc Scheffé comparisons confirmed that in this frequency band the normalized spectral power over the temporoparietal brain region (2.0 times) was remarkable lower than over both the frontotemporal (2.9 times: $p < 0.0005$) and the frontocentral area (2.9 times: $p < 0.0005$). The ***brain-regions*** effect in the γ-band ($F(2/24) = 6.22; \epsilon = 0.86; p < 0.008$) was mainly originated by the marked enhancement of normalized spectral power over the frontocentral regions (3.5 times) as compared to the temporoparietal (2.9 times: $p < 0.03$) as well as the frontotemporal brain area (2.8 times: $p < 0.02$).

The EEG data did not show lateralized effects in the β-band. Brain region effects occurred only in the θ-band ($F(1/12) = 13.14; p < 0.004$) and the α-band ($F(1/12) = 19.17; p < 0.002$). The pattern of differences between brain areas was similar to that obtained for the MEG data.

Based on the differences in normalized β-band spectral powers of MEG recordings between the hemispheres ipsi- and contralateral to the epileptogenic area during the third minute after methohexital injection we were able to correctly lateralize the primary epileptogenic focus in 77% (10) of our patients.

Nevertheless, false positive decisions would have been made in 23% (3) of our patients.

Discussion

The aim of the presurgical evaluation of patients with pharmacoresistant epilepsies is the correct localization and delineation of the primary epileptogenic area. Our results, based on a spectral analysis of simultaneous MEG and EEG recordings, confirm a loss of methohexital-induced enhancement of β-band spectral power in the MEG over the side of the primary epileptogenic area. This could be established in 10 of the 13 investigated patients. The effect was clearly expressed during a 1-2 minute interval starting approximately two minutes after the administration of methohexital. No significant differences between fronto-central, fronto-temporal or temporo-parietal brain regions were found. Although a general induction of fast β-activity could also be observed in our EEG recordings, a reduction of this activity over the ipsilateral hemisphere is not statistically significant.

Our results indicate that the loss of methohexital-induced enhancement of β-band spectral power in the MEG over the side of the primary epileptogenic area might contribute to the non-invasive lateralization of the primary epileptogenic area in patients suffering from pharmacoresistant temporal lobe epilepsy. This holds true particularly in those cases not exhibiting an increase of drug-induced spike activity or spike-burst-suppression (SBS) patterns in the MEG or EEG recordings, the latter pattern being of high significance for the lateralization and localization of epileptogenicity to the ictogenic temporal lobe based on invasive ECoG recordings via subdural strip electrodes [2]. Although the number of patients investigated so far is still too small in order to determine the value of the presented method for non-invasive presurgical evaluation, we would be able to correctly lateralize the primary epileptogenic area in 77 % of the cases. Nevertheless, false positive decisions about the side of the ictogenic temporal lobe would have been made in 23 % of our patients.

In conclusion, the reduced enhancement of β-band spectral power of MEG recordings can be used as an additional non-invasive tool for the presurgical lateralization of the primary epileptogenic area. In order to determine the value of this method for non-invasive presurgical evaluation, additional investigations on a larger number of cases, especially with single or multiple epileptogenic foci in brain regions other than temporo-mesial structures are required.

References

[1] G. Pampiglione. Induced fast activity in the EEG as an aid in the location of cerebral lesions. *Electroenceph clin Neurophysiol*, 4:79–82, 1952.

[2] A. Hufnagel, W. Burr, C. E. Elger, J. Nadstawek, and G. Hefner. Localization of the epileptic focus during methohexital-induced anesthesia. *Epilepsia*, 33:271–284, 1992.

[3] C. Wienbruch, C. Eulitz, K. Lehnertz, A. Brockhaus, C. E. Elger, T. Elbert, and M. Hoke. Methohexital-induced changes in spectral power of neuromagnetic signals: β-augmentation is smaller over the hemisphere containing the epileptogenic focus. *submitted to: Electroenceph clin Neurophysiol.*

[4] J. Engel Jr., editor. *Surgical Treatment of the Epilepsies*, pages 740–742. Raven Press, 2 edition, 1993.

[5] W. H. Press, B. P. Flannery, S. A. Teukolsky, and W. T. Vetterling. *Numerical Recipes in C.* Cambridge University Press, Cambridge, 1988.

Acknowledgment

This research was supported by grants from the Deutsche Forschungsgemeinschaft (Klinische Forschergruppe Biomagnetismus & Biosignalanalyse, Ho 847/6) and is part of a PhD thesis.

The authors wish to thank A. Kowalik MD, A. Baborie MD, M. Möllmann MD, N. Mertes MD, L. Spinne, G. Westermeier, C. Goeters MD, M. Reuss MD, W. Frebel MD, U. Ruta MD, T. Jakob, and S. Lorenz for their technical assistance.

VII. Multimodality Comparisons And Integration

Evaluation of Source Localization Models for Somatosensory Evoked MEG With PET Verification

Akhtari, M. [1], McNay, D. [1], Levesque, M. [1], Rogers, R.L. [1], Brown, C. [2], Mandelkern, M. [2,3], Babai, A. [4] and Sutherling, W.W. [1]

[1] Neuromagnetism Laboratory, Epilepsy & Brain Mapping Center, Hospital of the Good Samaritan, Los Angeles, CA; [2] Department of Nuclear Medicine, West Los Angeles VA Medical Center, Los Angeles, CA; [3] Department of Physics, University of California at Irvine, Irvine, CA; [4] UCSD School of Medicine, San Diego, CA

Introduction

We used MRI to segment the gray/white parenchymal surface and applied these results to constrain the cortical surface to solve for the location of the current sources generating externally measured magnetic field patterns evoked by somatosensory stimuli [1]. This method was applied to four normal subjects . We obtained somatosensory evoked PET data from one subject under identical stimulus conditions that were used to obtain the MEG data. Using prominent anatomical landmarks, the results of the MRI overlaid on the PET study show good agreement with the cortically constrained MEG localization. Furthermore, we compared this MRI constrained model to random search sphere model analyses and obtained similar localization results. The results from cortically constrained boundary element method (BEM) showed agreement with all the above methods of localization.

Method

Four neurologically normal subjects (1 female and 3 males, ages 22 to 31 years old) were studied. Somatosensory evoked responses were measured during non-simultaneous stimulation of the right median and the posterior tibial nerves using a Grass S88 stimulator and SIU7 DC current stimulus unit. 7-15 mA stimuli were delivered at 3.1 Hz, with 0.3 ms duration. Data were recorded and digitized at a rate of 4096 Hz. The first 256 time points following each stimulation were saved and two independent sets of 512 recordings each were averaged and stored separately (Figure 1). Magnetic fields were measured at 63 locations using a 7 channel SQUID-coupled second order gradiometer (Biomagnetic Technologies. Inc., Model 607). Eighteen overlapping channels were deleted and data from 45 channels were used for analysis. The measurements were recorded over the left hemisphere, and over the calvarium, for median nerve, and posterior tibial nerve stimulations, respectively.

Sixty-four sequential 3-mm cranial images were acquired on a 1.5 T G. E. Signa 5 MRI machine using a double spin echo sequence (TR=4000-6000 μs; TE= 60-80 μs; FoV=220-280 mm). We used custom molded bite plates and/or vitamin E capsules to mark the common fiducials between MRI and MEG coordinate systems. The MRI segmentation was performed using the General Electric workstation software package. The gray/white interface was digitized and the information was used as coordinates and orientations of 2-3x105 possible current sources. We used the conventional sphere model [2] with Nelder Mead simplex algorithm to randomly search for the location of the best fitting dipole. One subject's segmentation data was used to triangulate the brain/skull, skull/scalp, and scalp/air interfaces.(Fig. 2) We input the segmentation and triangulation data along with the best fitting source element from the cortically constrained primary model to a BEM code developed by the General Electric company. Several cortically constrained surfaces of various extents, all centered on the best fitting element were also evaluated as potential active extended sources using BEM.

The PET data were obtained after bolus injections of four nonconsecutive doses of H215O during electrical stimulation of the right median nerve. The PET data from left posterior tibial stimulation were obtained in the same manner. The PET data were overlaid on standard MRI available on PET analysis software. The overlays were performed using transformations utilizing the Talairach atlas.

Results

The location of the MRI constrained sources are displayed on the centroid of the highlighted area in Figure 1 for sagittal, axial and coronal views for all four subjects. Note that the locations of these sources are posterior and in close proximity to the central sulcus. Location of central sulcus was determined by visually tracking the cingulate sulcus on the sagittal images.

Table 1 compares the MRI constrained free space model with the spherical model using 1024 trials per averaged waveform. The goodness of fit values, (i.e. % variance accounted for), ranged from 87.1% to 90.7% for the MRI constrained model, and 86.4% to 95.4% for the spherical model. Intra-subject Euclidean difference in localization between the two models ranged from 2.0 to 15.5 mm. The coordinate system is right handed with the x-axis toward the left ear and the z-axis directed superiorly; the center of the best sphere was located 4-5 cm above the nasion-pa plane in all four subjects.

Figure 2 represents the results of MRI constrained MEG localization of right and left posterior tibial stimulation of subject #2. The active element is located at the cross-hair.

Figure 3 represents the locations of the cortical representation of right median, and left tibial nerves of subject #1 as obtained by PET scan.

The different boundaries of the head as obtained from reconstructed MRI of subject #1, and the corresponding triangulations are represented in Figure 4. The goodness of fit for MRI constrained best fitting element improved by approximately 4%. The goodness-of-fit values for the extended sources reached a peak value of 94% for an extended source that had an area of approximately 140 mm2. This area is displayed in Figure 1), subject #1. The extended sources of lesser or higher area had lower goodness of fit values.

Discussion

Results obtained using the MRI constrained model are in good agreement with the PET results. These response locations are also consistent with the known functional neuroanatomy, based on human [3] studies. In the present results, the goodness-of-fit values were similar for the three single element models. The discrepancy between the results of the sphere model and MRI constrained model (Table 1) for subjects 2 and 4, suggests that the results of the sphere model may be in error. The sphere model localized the sources in the two subjects by more than ten mm posterior to the central sulcus. Somatosensory localization of the median nerve has never been observed to be more than ten mm posterior to the central sulcus in nonpathological subjects. Moreover, with cortical constraints, extended active sources with realistic conformity to the cortical shape can be readily studied. For MEG localization of these rather superficial sources, the results of single element BEM analysis indicate only a slight improvement of goodness of fit compared to the cortically constrained primary model. However, the comparison of cortically constrained free space extended sources, and that of BEM remains to be carried out.

References

[1] Akhtari M, McNay D, Mandelkern M, Teeter B, Cline HE, Mallick J, Clark G, Tatar R, Lufkin R, Chan K, Law C, Sutherling WW. Somatosensory evoked response source localization using actual cortical surface as the spacial constraint. Brain Topography, 1994 Fall, 7 (1) : 63-9.

[2] Stok CJ. The influence of model parameters on EEG/MEG single dipole source estimation. IEEE Trans. Biomed. Eng. 1987;BME-34:289-296.

[3] Sutherling WW, Crandall PH, Darcey TM, Becker DP, Levesque MF, Barth DS. The magnetic and electric fields agree with intracranial localizations of somatosensory cortex. Neurology 1988;38:1705-1714.

Acknowledgements

Supported by grant NS20806 from the National Institute of Neurological, Communicative Disorders and Stroke, Epilepsy Branch, the Visions of the Brain foundation, and the Mathers charitable foundation. We are greatly indebted to Ms. Winifred L. Monaco B.A. for her invaluable contribution to this project.

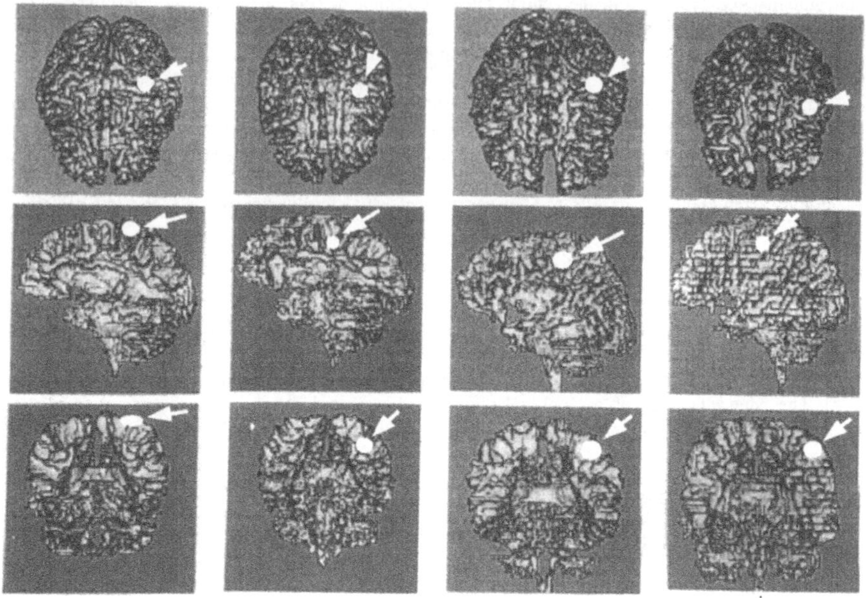

Fig. 1) Somatosensory evoked MEG localization of the right median nerve of four subjects (centroid of the highlighted area, arrow)

	MRI CONSTRAINED MODEL					SPHERICAL MODEL					MRI vs. SPHERE
SUBJECT	LAT	X	Y	Z	%V	LAT	X	Y	Z	%V	
1	19.6	54	-7	8	87.1	20.3	52	-7	8	93.4	2.0
2	26.4	74	5	-6	90.7	26.4	63	6	-3	86.4	11.4
3	23.4	53	-1	4	89.7	23.2	52	1	1	95.4	3.7
4	23.9	82	-2	0	89.8	23.9	68	4	-3	91.1	15.5

Latency in ms.
All distances are measured in mm.
%V: percent variance accounted for.

Table 1) The locations and the goodness-of-fit values of the best fitting MRI constrained current elements, and the best fitting single dipoles using the sphere model.

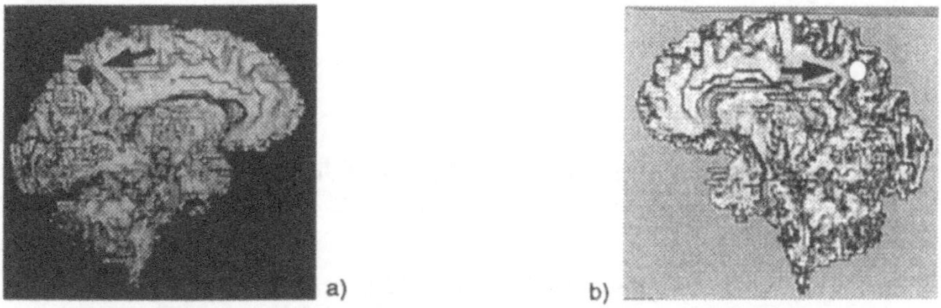

Fig. 2) a) Somatosensory evoked MEG localization of the right posterior tibial nerve (centroid of the highlighted area, arrow); and b) the corresponding localization of the left posterior tibial nerve (centroid of the highlighted area, arrow).

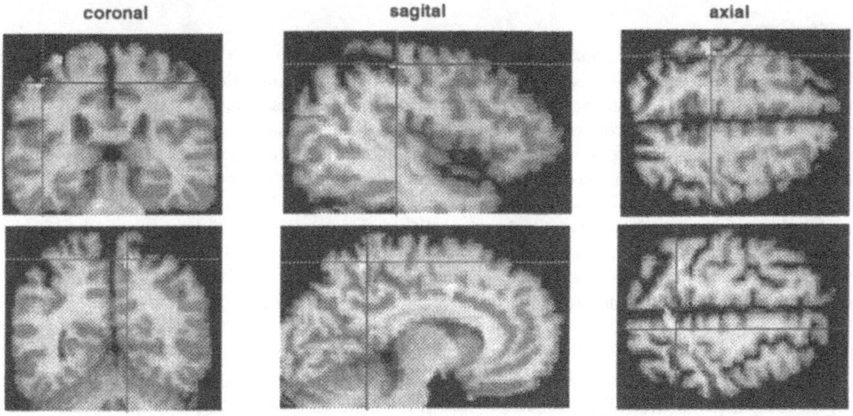

Fig. 3) Somatosensory evoked PET localization of the right median nerve (cross-hair, top pannel), and the left posterior tibial nerve (cross-hair, bottom pannel).

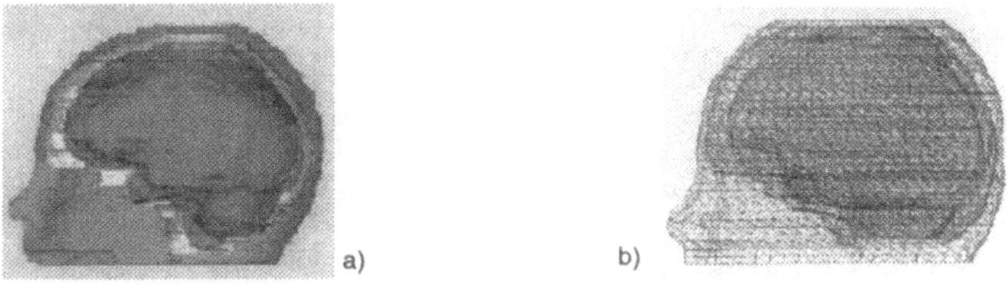

Fig. 4) a) MRI reconstruction of the three cranial surfaces; b) direct tessellation of the corresponding surfaces from a).

Combination of point and surface matching techniques for accurate registration of MEG and MRI

Bamidis, P.D.[1]**, Ioannides, A.A.**[1,2]

[1]*Physics Department, The Open University, Milton Keynes, U.K.;*
[2]*Institute of Medicine, Research Centre Jülich, Germany*

Introduction

The easiest and most commonly used method to register the MRI and MEG datasets is point matching, where a few 'fiducial' points, i.e. nasion, 2 preauriculars, are marked during the acquisition of either technique. Another methodology involves the use of a stereotactic frame. In either case, the registration of the two modalities requires the definition of the correct transformation matrix with respect to scaling, translation and rotation parameters. We have exploited both the above strategies and, in general, root mean square (RMS) mismatches per point of less than 1 mm (simulations) or 6 mm (general practice with real data) were achieved.

However, we faced situations where a nominal accuracy of millimetres was obtained for the fiducial points, but the surface of the head as outlined in the MEG (with a 3D digitizer) did not fit with the MRI obtained head surface, as a result of inconsistent definitions of the fiducials. In this work, we show how we resolved such problems, and how accuracy can be improved by combining point matching with a more general surface matching approach, thereby practically demonstrating the need to move to a more robust registration methodology.

Motivation

Fig. 1 displays the situation where the fit of the pre-auricular and the nasion points (marked with arrows in (a)) is excellent (crosses in (b)) but the description of the digitised scalp in the MEG system (dots in (b)) does not fit with the outlines as extracted from the presumably aligned MRIs (light contours in (b)). This was probably due to the inconsistent definition of the fiduciary skull landmarks in each acquisition (MEG and MRI). Note that similar discrepancies occurred in two individual cases (subjects). In such cases, an inaccuracy on the order of 1-2 cm ruined the possibility to relate the MEG solutions to precise anatomical sites. The need to move to a more robust and less subjective alignment procedure is, therefore, obvious.

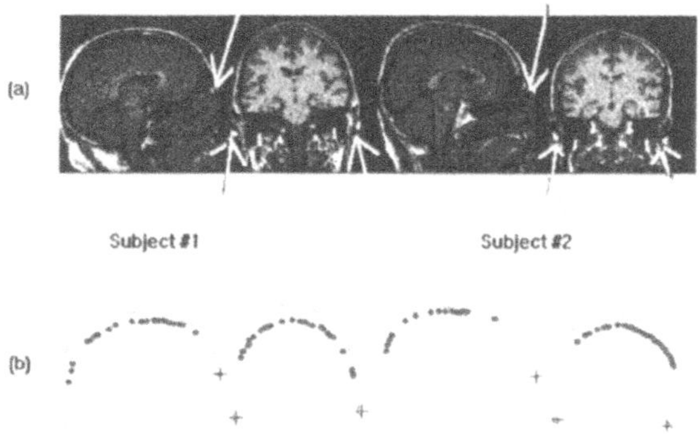

Fig. 1: (a): the vitamin pills used to perform the point-matching (arrows). (b): The fit of the MEG (crosses) and MRI fiducials (circles) is perfect. However, the head shape outlines in the MEG system (dots) do not match the MRI scalp outlines (light contours).

Methods

Motivated from such problems, we developed a system that combines the point matching with a surface matching approach. The head-hat analogy [1] is followed for a rigid body translation and rotation, in a way similar to that used by others [2,3]. A cost function, F, is formed by the sum of the squared distances of the MEG points from the MRI head surface, and Powell's method is employed to minimise F. We have adopted the following strategy in an attempt to optimise the registration results.

In the initial stage, a point matching is performed (3-5 points) and the accuracy is evaluated not in terms of the RMS mismatch per point, but by visualisation of the fitting of MEG head outlines on the MRI scalp. Sometimes satisfactory results are obtained and the process stops there. Usually, however, a stepwise surface matching is attempted: the program is given the parameters found in the first stage as initial approximations. Then the translation is roughly fixed by making the centroids of MEG (hat) and MRI (head) surfaces to coincide, and a 3D minimisation of F with respect to the three angles of rotation is tried (scaling is assumed to be known [3]). Finally, F is minimised with respect to all six parameters together, to fine-tune the fitting.

Results

Simulations have shown an overall accuracy in the range of 1-3 mm, that is, less than 1 mm translation and less than 2 degrees rotation errors (see Table 1). Given the intrinsic noise in real data, 3-5 mm RMS mismatches per point have been considered quite satisfactory. Whenever point registration was considered as unreliable, surface matching improved the accuracy by 1-2 cm, while in the general case, an improvement of 1-4 mm was achieved. The number of MEG points used, varied from a few hundred up to a few thousand. The fit was not significantly dependent on the number of points as such, but often points obtained in perpendicular directions (e.g. midsagittal, midcoronal lines) accelerated the minimisation and improved the overall performance.

Table 1: Results from three simulated MEG/MRI registrations. The starting positions (initial errors) are varied, and so is the outcome. In all cases, however, the final misfit per point is less than 3 mm.

	Translation (mm)			*Rotation (degrees)*			
Errors	b_x	b_y	b_z	θ_x	θ_y	θ_z	*RMS misfit/point (mm)*
initial	+10.0	-10.0	+10.0	-5.0	+2.0	+6.0	14.18
final	-0.14	-0.78	-0.10	-1.4	-0.1	-0.4	2.46
initial	-15.0	+10.0	+15.0	+8.0	+4.0	-10.0	12.7
final	-1.23	+0.08	+1.76	+1.02	+2.1	-0.18	3.0
initial	+5.0	+5.0	+5.0	-2.0	-2.0	-2.0	6.40
final	-0.1	+0.66	-0.1	-1.3	-0.05	+1.0	1.40

In Fig. 2 we show the final registration product from a real experimental set-up where the final mismatch per point was some 3 mm, having started from some 23 mm per point. Different visualisation angles reveal that the fit between MEG head shape points and MRI scalp outlines is excellent throughout. Finally, Fig. 3 provides a summary of the whole registration procedure in another real case. On the most left are the initial datasets (unregistered), approximation by point matching in the middle, and final result after surface matching on the right.

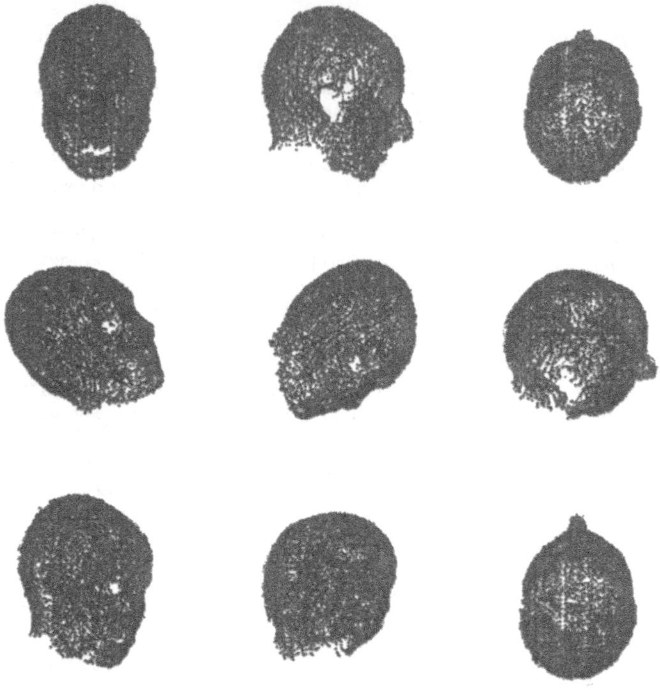

Fig. 2: Visualisation of registration results in different perspective angles. MEG headshape points (dark grey) are superimposed on MRI outlines (light grey). The fit is excellent throughout. The RMS mismatch/point is some 3 mm.

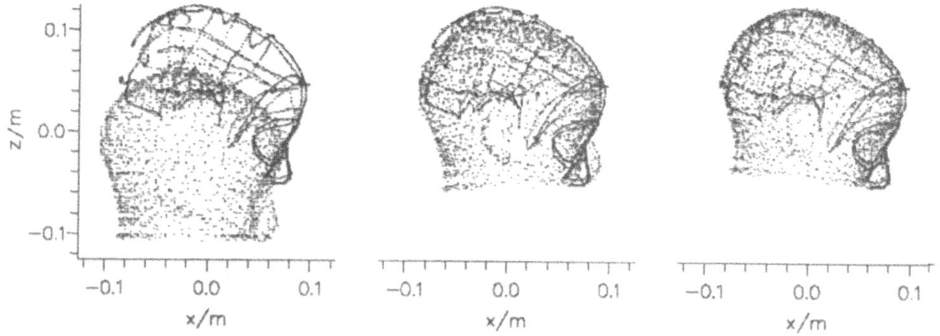

Fig. 3: Skull outlines in MRI (light grey) and MEG (dark grey) co-ordinate systems: before registration (left), after point matching (centre), and after surface matching (right).

Discussion and Outlook

This is not the first time that the surface matching registration technique is applied on MEG/MRI datasets. However, it is the first time the need to move from the most common and easy point matching strategy to the more robust and less subjective methodology of surface matching is shown through the use of practical examples. The improvement in accuracy obtained from the combination of the prior and the latter technique was not shown before through visual examples.

The accuracy we have achieved is comparable with that achieved by others [3]. Our simulations gave better results than the applications to actual data. However, the real data we have been using were always part of an actual MEG experiment (e.g. auditory odd-ball, CNV etc). In other words, we have not attempted any model experiments as such, in which all the individual steps would have been carefully followed to allow for a proper registration accuracy evaluation.

Furthermore, the software we have developed does not include a proper Graphics User Interface (GUI) yet. Efforts to combine it with a proper GUI as well as to optimise the code for faster minimisations are currently being undertaken.

High registration accuracy is the first necessary step for any MEG modelling. For Magnetic Field Tomography (MFT) [4], our method of choice, accuracy is mandatory. First, more complex source spaces [5] can be designed to follow the anatomy closer. Second, the source space itself can also be changed iteratively [6,7] to model deep generators. These are intermediate steps before constraining the source space to the cortical mantle or other important "deeper" structures so that physiologically more interesting questions can be tackled.

However, before the latter can occur, one should aim for optimisation of all the individual steps taken during a MEG recording. These are, avoidance of subject/patient movements, more accurate acquisition of MEG headshape points through the use of better defined MEG/Dewar systems, and automation of the procedure.

As the MEG research is indulging onto its new era of whole head cover systems and multimodality imaging, accuracy in registration becomes more important. We believe, that our experiences will provide a useful reference for the MEG researcher, and more importantly will help solving problems by combining the simple (point matching) with the more difficult and general (surface matching) techniques.

References:

[1] Pelizzari, C.A., Chen, G.T.Y., Spelbring, D.R., et al. Accurate 3d correlation of CT, PET, and/or MR images of the brain, J. Comp. Assist. Tomography, 1989, 13: 20-26

[2] Abraham-Fuchs K., Lindner T., Wegener P. Fusion of biomagnetism with MR or CT images by contour-fitting, Biomed. Eng., 1991, 36(Suppl): 88-89

[3] Kober, H., Grummich, P., Vieth, J. In: Baumgartner, C., Deecke, L., et al. Biomagnetism: Fundamental research and Clinical Applications, Amsterdam, Elsevier, 1995.

[4] Ioannides, A.A. In: Ho, M.W., Popp, F.A. and Warnke, U., Bioelectrodynamics and Biocommunication, Singapore, World Scientific, 1994.

[5] Liu, M.J., and Ioannides, A.A. In Deecke L, Baumgartner C et al. Biomagnetism: Fundamental Research and Clinical Applications, Elsevier, Amsterdam, 1995.

[6] Bamidis, P.D., Hellstrand, E., Lidholm, H., et al. MFT in complex partial epilepsy: spatio-temporal estimates of interictal activity, Neuroreport, 1995, 7(1): 17-23.

[7] Bamidis, P.D., PhD Thesis, The Open University, Milton Keynes, UK, 1996.

EEG and MEG Source Analysis of Somatosensory Evoked Responses to Mechanical Stimulation of the Fingers

Cheyne, D.[1], Roberts, L.E.[2], Gaetz, W.[2], Bosnyak, D.[2], Weinberg, H.[1], Johnson, B.[1], Nahmias, C.[3] and Deecke, L.[4]

[1]Brain Behaviour Laboratory, Simon Fraser University, Burnaby, Canada; Departments of [2]Psychology and [3]Nuclear Medicine, McMaster University, Hamilton, Canada; [4]Neurological Clinic, University of Vienna, Vienna, Austria.

Introduction

Previous MEG and EEG studies have successfully utilized electrical stimulation of the digits in order to study generators of early responses in the human primary somatosensory cortex [1-3]. Natural stimulation of the somatosensory system using vibratory pulses to the tips of the digits has been found to elicit qualitatively different responses in the EEG [4, 5] presumably due to the activation of different neuronal pathways. Similar responses have been observed in the MEG to transient mechanical stimuli producing large responses at latencies of 50 msec, corresponding to the electrical P50 response [6, 7]. A more recent study [8] also noted MEG responses at 70 msec latencies in some subjects resembling the N70 component described by Hämäläinen and co-workers [4]. The earlier, P50 response is presumed to reflect activation of primary sensory cortex (SI). A study of mechanically evoked epidural and single unit responses in waking monkeys [5] found that this component was associated with a period of inhibitory input to neurons in areas 3b and 1, whereas the later, slow component (N70) was not associated with activity in SI and appeared to arise from other cortical areas, such as SII. The magnetically recorded P50 (P50m) appears to be the largest and most consistent MEG event recorded during transient tactile stimulation in humans and its potential application for somatotopic mapping studies [6, 7] warrants further investigation of its neural generation. Moreover, the marked orthogonality of the EEG and MEG topographies of the P50 response makes it an ideal candidate for the comparison and/or combination of EEG and MEG localization methods. The current study compared separate high-density, 32 channel EEG and 143 channel MEG recordings in two subjects using identical stimulation paradigms in order to compare dipole source locations in somatosensory cortex obtained separately for each method. In addition, 3-dimensional MRI was obtained for both subjects in order to constrain source model information as well as aid in the integration of coordinate systems for the MEG and EEG source models.

Methods

EEG recordings were made during mechanical stimulation of the finger tips of the right hand in two subjects using a high density 32 channel EEG system (NeuroScan, Herndon, VA) consisting of a 19 channel grid overlying the left Rolandic fissure and 13 distributed 10-20 locations. MEG recordings were obtained separately using a whole cortex MEG system (CTF Systems, Inc) consisting of 143 1st-order gradiometers covering the whole head. Recordings were performed without magnetic shielding with the aid of software 3rd-order gradient formation with an overall white noise level of about 6 $fT/\sqrt{Hz}$ [9]. Stimuli consisted of 5 msec duration pulses produced by an electromechanical stimulator with an inter-stimulus interval of 750 msec. Acoustic artifacts from the stimulator were masked with white noise and for MEG recordings, the electromechanical stimulator was driven from a distance of several meters to reduce magnetic artifact which was noticeable in some posterior channels. Trials of 256 msec duration were averaged using a 64 msec pre-stimulus baseline and off-line bandpass filtered between 1 and 100 Hz. For EEG data two blocks of 800 trials (1000 samples/sec) were collected and for MEG one block of 600 trials (1250 samples/sec). Comparison of responses for the first and second blocks of EEG data indicated good replicability and dipole fits were performed on the first block of data for comparison with the MEG. For both subjects, three-dimensional proton density magnetic resonance images were obtained with vitamin E markers placed at fiduciary landmarks used to create the head-based MEG coordinate system and selected 10-20 locations. An off-line procedure allowed the identification of these markers in the MRI for the co-registration the MRI and MEG coordinate systems.

Results

MEG responses showed a clear field reversal peaking at 50-60 msec and a slower component peaking at latencies of 70-90 msec, similar to the P50m and N70m responses observed previously [7]. The P50 component

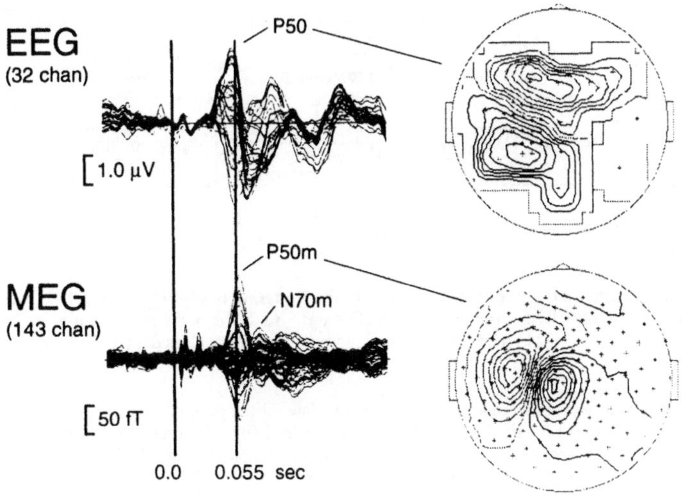

Fig. 1. Left: Superposition of 32 EEG channels (upper) and 143 MEG channels (lower) for averaged response to mechanical stimulation of the right index finger in one subject. Right: Isocontour maps of the electrical and magnetic field patterns at the peak of the P50 response. Small crosses indicate gradiometer/electrode location and darker contours represent outgoing flux and positive potential in steps of 23 fT and 0.37 μV, respectively. Small deflections visible in early phase of MEG response are due to electromechanical stimulator artifact.

was characterised by orthogonal dipolar patterns of magnetic flux and electrical potential at similar latencies suggestive of a tangential dipole source with opposite orientation of intracellular and extracellular current measured by MEG and EEG, respectively (Fig. 1). The EEG also showed monopolar distributions of peak responses at latencies of 60 - 70 msec in some conditions, indicative of radially oriented sources not observed in the MEG.

Localization of single dipole sources was obtained for the MEG using a least-squares fitting algorithm with corrections for non-radial field contributions [9]. A best fit sphere was chosen for each subject based on the structural MRI data, aided by the MRI markers of fiduciary and 10-20 locations. EEG source analysis was achieved using a 4-shell spherical model after Stok [10], based on digitized locations of the electrode montage projected onto a sphere with the same radius. For comparison with MEG results, EEG dipole locations in the spherical model were transformed to a common head-based coordinate system by appropriate translation by the MEG sphere origin (no rotation was assumed). The overall approach for the integration of these three modalities is summarized in Figure 2. Since MEG source localization which includes non-radial measurements (due to helmet geometry) is dependent on the sphere origin (i.e., definition of radial direction) confirmation of the choice of best-fit sphere was attempted by a trial and error procedure, comparing changes in residual error with slight deviations in the sphere location (indicated by "adjust X,Y,Z" step in Fig. 2). Most interestingly, we found that when the sphere origin was iteratively adjusted in this manner, the initial sphere location based on the MRI produced the lowest error in almost all cases and was therefore accepted as the default origin in subsequent analysis. The sphere origin also determines the location of the EEG sources relative to the MEG sources, since EEG source modelling was performed relative to a sphere with the same radius onto which digitized locations of the electrodes had been projected. Appropriate translation of the EEG source locations by this origin, therefore allows direct comparison of EEG and MEG source locations. As also indicated by Fig. 2, when comparing final source locations based on the two modalities, MEG sources are assumed to provide more accurate information regarding dipole moment (for cases where radial contributions are negligible) and might therefore be used to adjust the EEG spherical model conductivity values.

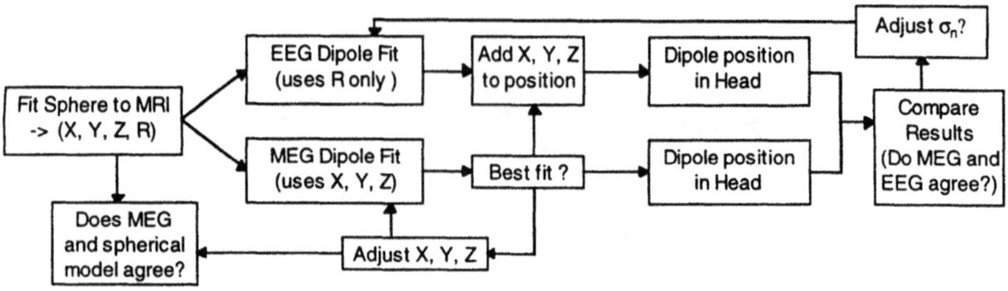

Fig. 2. Flow chart of iterative steps used in the integration of EEG and MEG source modelling procedures with 3-dimensional MRI data. X, Y, Z = sphere location in head coordinate system. R = sphere radius. σ_n = conductivity for spherical model (shells 1.. n).

Tangential dipole sources were localized from MEG data for the peak of the P50m component with residual errors ranging from 2.1% to 6.1%. EEG source localization yielded fits of sources for the P50 with errors ranging from 1.5% to 5.5%. Comparisons of EEG locations to those obtained with MEG were made in the common coordinate system after translation by the sphere origin as described above. Table 1 shows dipole locations obtained for three digits (1 = thumb, 3 = middle, 5 = pinky) for both subjects obtained from EEG and MEG data and the resulting differences in source location and strength indicating generally good agreement between source locations for the P50 component. One notable difference was the degree of lateral dispersion of the sources which tended to be greater for EEG fits. Additionally, there is some indication of systematic vertical and lateral shifts between EEG and MEG sources in subject 1. Other differences did not show any systematic trend and therefore do not likely reflect incorrect specification of the sphere origin. Constraining the EEG dipoles to a tangential orientation resulted in only minor changes in location and error indicating that this component is generated by a largely tangential source in SI, most likely area 3b. Dipole moment for EEG fits tended to be lower than that of the MEG by approximately 25 to 50%. Since higher moments would be expected for any contributions of a radial source component, this suggests a possible correction to the spherical model conductivities for the EEG based on the MEG dipole moments.

Fits attempted for the peak of the N70 component in both MEG and EEG data tended to be less consistent and higher in error, possibly due to the lower signal to noise ratio of these responses. These sources also tended to locate lateral and deeper to those for the P50, consistent with an SII origin but were not analyszed further in the current study. Dipole locations for the P50 projected onto the appropriate MRI slice indicated location in the region

Table 1. Comparison of dipole source parameters for EEG and MEG dipole solutions in 2 subjects.

Subject 1 Digit	Latency (ms)	Position X (cm)	Y (cm)	Z (cm)	Moment (nA-m)	Error (%)
(EEG)				(+5.5)		
1	59.0	0.52	3.21	9.07	6.3	1.7
3	58.0	0.59	3.27	9.39	7.1	2.8
5	59.0	0.23	3.16	9.63	6.0	1.5
(MEG)						
1	59.2	0.45	4.12	9.96	10.3	6.1
3	60.0	0.93	3.76	9.84	15.4	4.9
5	61.6	0.91	4.17	10.28	8.9	3.5
Difference	(EEG -	MEG)				
1		0.08	-0.91	-0.89	-4.0	
3		-0.35	-0.49	-0.45	-8.3	
5		-0.68	-1.01	-0.65	-2.9	
Mean		-0.32	-0.80	-0.66	-5.0	

Subject 2 Digit	Latency (ms)	Position X (cm)	Y (cm)	Z (cm)	Moment (nA-m)	Error (%)
(EEG)		(-0.5)		(+5.2)		
1	56.0	0.69	4.52	8.88	13.9	5.5
3	54.0	0.02	3.21	9.09	19.1	2.8
5	54.0	-0.03	3.13	8.27	17.5	3.0
(MEG)						
1	57.6	-0.32	4.06	8.83	17.5	4.5
3	57.6	-0.04	3.98	8.84	23.7	2.3
5	60.0	-0.09	4.16	9.01	17.5	2.1
Difference	(EEG -	MEG)				
1		1.01	0.46	0.05	-3.6	
3		0.06	-0.78	0.25	-4.6	
5		0.06	-1.03	-0.74	0.0	
Mean		0.38	-0.45	-0.15	-2.7	

of the postcentral gyrus with a tangential orientation (anterior directed for EEG, posterior directed for MEG) consistent with surface negativity at the posterior bank of the central sulcus, due to either deep inhibitory, or superficial excitatory input to neuronal populations in area 3b. Sources for different fingers tended to be distributed along a line in agreement with the known somatotopy of the hand although these differed in their extent for the MEG and EEG solutions. Also, sources tended to move downward and medial in one subject, and in the other subject upward and medial, suggesting differences in gyral folding which was also evident in the individual MR images.

Discussion

Comparisons of early components of electrically evoked MEG and EEG somatosensory responses have suggested that MEG sensitivity to only tangential sources may provide additional constraints for EEG source analysis [3]. In the current study, somatosensory responses to tactile stimulation of the digits indicated that EEG and MEG responses produce highly orthogonal field patterns at latencies of 50 msec, most likely due to tangential generators in SI, with little contribution from radial sources, allowing for direct comparison of dipole modelling results from EEG and MEG data independently. This result is similar to that found by Buchner et al., [3] for the 30 msec (N30/P30) response evoked by electrical stimulation. In addition, we tested strategies for combination of the MEG and EEG coordinate systems, which confirmed structural MRI as a useful tool for estimating the best fit sphere, both for determining contributions of non-radial magnetic fields for MEG dipole solutions, as well as for the translation of EEG dipole solutions to the MEG coordinate system. This approach yielded similar results for absolute EEG and MEG source locations with overall deviations of about half a centimeter, although some systematic discrepancies remained. Combined solutions for EEG and MEG sources using simultaneously acquired data may provide more accurate source location estimates, although the correct integration of the two coordinate systems will be an important factor for the successful combination of these two methods.

Acknowledgments

This project was supported in part, through NSERC operating grants (Roberts-OGP0000132), (Weinberg-OGP0008322), an NSERC CRD grant (Weinberg-661044492), the NATO Division of Scientific Affairs (Roberts/Elbert-940648), and a Deutsche Forschungsgemeinschaft Fellowship (DFG BE1538) to L. Roberts.

References

[1] Hari, R. and Kaukoranta E. Neuromagnetic studies of somatosensory system: principles and examples. Prog. Neurobiol., 1985, 24: 233-256.

[2] Buchner, H., Adams, L., Müller, A., Ludwig, I., Knepper, A., Thron, A., Niemann, K. and Scherg, M. Somatotopy of human hand somatosensory cortex revealed by dipole source analysis of early somatosensory evoked potentials and 3D-NMR tomography. Electroenceph. clin. Neurophysiol., 1995, 96: 121-134.

[3] Buchner, H., Fuchs, M., Wischmann, H-A., Dössel, O., Ludwig, I., Knepper, A. and Berg., P. Source analysis of median nerve and finger stimulated somatosensory evoked potentials: Multichannel simultaneous recording of electric and magnetic fields combined with 3D-MR tomography. Brain Topog., 1994: 299-310

[4] Hämäläinen, H. Kekoni, J., Sams, M., Reinikainen, K. and Näätänen R. Human somatosensory evoked potentials to mechanical pulses and vibration: contributions of SI and SII somatosensory cortices to P50 and P100 components. Electroenceph. clin. Neurophysiol., 1990, 75:13-21.

[5] Gardner, E.P., Hämäläinen, H.A., Warren, S., Davis, J. and Young W. Somatosensory evoked potentials (SEPs) and cortical single unit responses elicited by mechanical tactile stimuli in awake monkeys. Electroenceph. clin. Neurophysiol., 1984, 58: 537-552.

[6] Suk, J., Ribary U., Cappell, J., Yamamoto T. and Llinás R. Anatomical localization revealed by MEG recordings of the human somatosensory system. Electroenceph. clin. Neurophysiol., 1991, 78:185-196.

[7] Yang, T.T., Gallen C.C., Schwartz, B.J., and Bloom, F.E. Noninvasive somatosensory homunculus mapping in humans by using a large-array biomagnetometer. Proc. Natl. Acad. Sci., 1993, 90: 3098-3102.

[8] Cheyne, D., Weinberg H., Hattori, H., Gordon, R., Vrba, J. and Burbank M. In: N. Tepley and G. L. Barkley (Eds.) Characterisation of human sensorimotor cortex using whole-cortex MEG: implications for clinical use. Proceedings of the North American Biomagnetism Action Group Annual Meeting. Detroit, 1994.

[9] Vrba J., Angus, V., Betts K., Burbank M., Cheung T., Fife A., Haid G., Kubik P., Lee S., McCubbin J., McKay J., McKenzie, Robinson, S., D., Spear P., Taylor B., Tillotson M., Cheyne D. and Weinberg H. A 143 channel, whole cortex MEG system (these proceedings).

[10] Stok C. J. The inverse problem in EEG and MEG with application to visual evoked responses. Ph.D. Thesis, 1986, Krips Repro Meppel.

Dynamic Neuroimaging by MEG, Constrained by MRI and fMRI

George, J.S.[1], Schmidt, D.M.[1], Mosher, J.C.[1], Aine, C.J.[1], Ranken, D.M.[1], Wood, C.C.[1], Lewine, J.D.[2], Sanders, J.A.[2] and Belliveau, J.W.[3]

[1]Los Alamos National Laboratory, Los Alamos, New Mexico, USA; [2]New Mexico Regional Federal Medical Center, Albuquerque, New Mexico, USA; [3]Massachusetts General Hospital NMR Center, Charlestown, Massachusetts, USA

Introduction

MEG is a direct measure of the electrical activity of populations of neurons, with excellent temporal resolution. However, the spatial characterization of sources depends on the solution of an ill-posed inverse problem. MRI provides high resolution volumetric data defining the anatomy of the head and brain; alternative MRI data acquisition strategies allow mapping of hemodynamic correlates of neural function. However, functional MRI (fMRI) and related techniques are limited by the slow timecourse of the hemodynamic response and the ill-defined relationship between neural activation and associated hemodynamic changes. Given such complementary strengths and weaknesses, the integration of multiple imaging technologies should provide dynamic functional neuroimaging capabilities with optimal spatial and temporal resolution. We are exploring a range of strategies for the integrated analysis of data from MRI, fMRI and MEG.

Anatomical and functional MRI provide constraints on the location and configuration of putative neural sources. We assume that the neural sources measured with MEG are confined to the thin sheet of gray matter comprising neocortex and similar structures in the human brain. For some analyses we further assume that the net current moment in any patch of cortex is oriented normal to the local surface. The geometry of cortex is defined from MRI using software tools developed in our laboratory. Analysis of fMRI data depends on the identification of clusters of significant activation in statistical parametric images. Our general stratgey is to use MEG to define the detailed temporal dynamics of sources localized (at least in part) on the basis of MRI data.

Methods

MRI data described here were obtained on a 1.5 T Siemens Magnetom at the New Mexico Regional Federal Medical Center (NMRFMC). Anatomical data were acquired using a 3-D volume acquisition technique, MP-RAGE. In-plane resolution was typically 1mm (256^2 voxels) with slice resolution of 1.5 or 2 mm (128 slices). The volume was typically resampled to an isotropic resolution of 2 mm for segmentation and other analyses. Multislice functional MRI was acquired with the same instrument, using a T2* weighted FLASH aquisition. Typically, 6 slices were acquired with 2 mm in-plane resolution, and with 5 or 6 mm between contiguous slices. Interleaved blocks of 4 control or stimulated image sets (2 minutes per block) were acquired. Volume data segmentation, analysis and visualization was performed with MRIVIEW [1] and other codes developed in our laboratory. These systems are written in IDL, a high level programming and analysis system developed by Research Systems Inc. Figure 1 illustrates several features of our analysis of anatomical and functional MRI data.

MEG data were collected on a 37 channel neuromagnetometer (BTi, Magnes) at the NMRFMC employing 2 to 4 replications of the experiment at different sensor array placements. Because of differences in the physics and underlying physiology of fMRI and MEG measurements, we initially compared these alternative measures of neural activation. Although correspondence between identified sources was not perfect, and optimal experimental conditions for MEG and fMRI may vary considerably, the results encouraged us to pursue integrated analyses [2,3].

Statistically significant voxels in fMRI (95% confidence level), were identified based on t-statistic difference values, taking into account the (large) number of independent measurements. Analyses were repeated for 3-D clusters of contiguous voxels of increasing size, which lowered the threshold value for statistical significance. The average value of a voxel or region of interest in a stimulated trial was required to exceed the average value of adjacent control periods, across all stimulus/control cycles. This method is similar to the widely used temporal correlation techniques but is applicable even given the limited temporal sampling of our conventional MRI acquisition scheme. This method is particularly useful for eliminating large artifacts that may arise through transient movement of the subject.

Results

Our first approach for integrated analysis employed simplified equivalent sources to account for observed MEG responses. Given a cluster of active voxels derived from fMRI, current orientations defined by MRI geometry, and an estimate of the relative activation of voxels within the cluster based on functional MRI, we found an equivalent current dipole that best accounted for the fields associated with the composite source. Dipole parameters

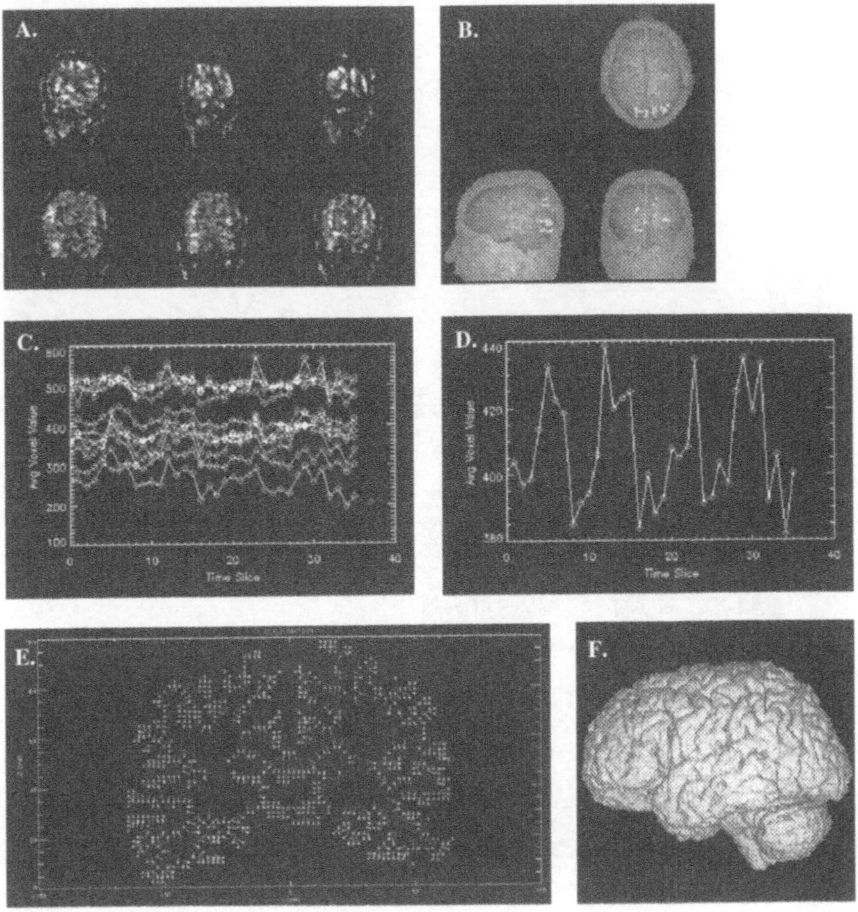

Figure 1. Structural and functional MRI data
A. A visual fMRI experiment with six coronal slices through occipital and adjacent cortex.
B. Projection images illustrating the location and extent of each of the fMRI clusters
C. fMRI timecourses for each of 11 identified clusters.
D. The average timecourse across all 11 clusters showing ~7% signal over most stimulus periods.
E. Locations and normals of cortical surface voxels from segmentation of 3-D data with MRIVIEW.
F. Reconstruction of the cortical surface from data as in E.

were determined by fitting the predicted field distributions, or locations were set equal to the center of mass of the cluster. A set of such sources was fit to MEG data to establish activation timecourses. An example of this sort of analysis is illustrated in figure 2, A and B. Dipole orientations were either estimated and fixed based on MRI data, or were allowed to rotate during the fitting procedure. While rotating dipoles produced a better fit (with many more parameters), we did not see striking differences between optimized and center of mass equivalent dipoles. However, many studies suggest that the best equivalent source may not correspond to the actual region of activation when the underlying current distribution is complex or extended. The early peaks in the timecourse data (0-50 ms latency) reflect patterns in the field maps arising from "wrap-around" due to the fixed and inadequate interstimulus interval used in these MEG experiments for compatibility with fMRI protocols.

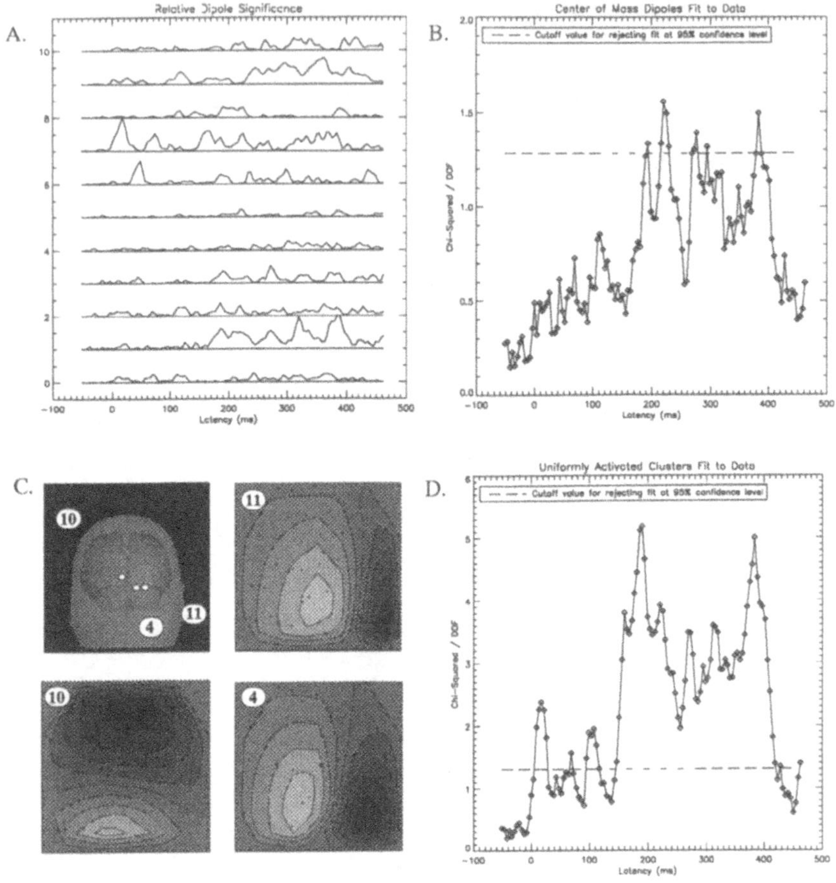

Figure 2. Parametric estimates of MEG source activation constrained by MRI
A. Significant currents for rotating dipoles at the centers of mass of the fMRI clusters.
B. Reduced Chi-square error across time for the rotating dipoles.
C. Basis fields associated with three of the 11 uniformly activated clusters identified from fMRI.
D. Reduced Chi-square error across time in composite model reconstruction.

The second approach similarly employed anatomical and functional MRI to define the location and geometry of the source, but did not employ a simplified equivalent source. Instead, a detailed forward calculation employing all active voxels was used to define a single basis vector for the entire cluster. A set of such sources was scaled to produce a best fit to the observed data. This approach explicitly accounts for distributed sources and is physically interpretable. In the analysis illustrated, our automatic cluster-finding procedure was used to identify 11 significant clusters in the fMRI data, and field distributions corresponding to composite sources were constructed assuming uniform activation within each cluster. Time courses were estimated by linear regression on the MEG spatial temporal data using the normalized forward calculations (such as illustrated in figure 2c) as basis vectors. However, basis vectors derived in this manner did not form an orthogonal set, so that the decomposition of the MEG data into a set of timecourses was ambiguous, consistent with the observation that similar temporal peaks appeared in several clusters. Further, in this study, uniformly activated clusters did not adequately fit the data. This might reflect physiological heterogeneity in activation levels within an area, or missing sources. The limited extent (in depth) of the fMRI data in this study insured that some visual areas were missed.

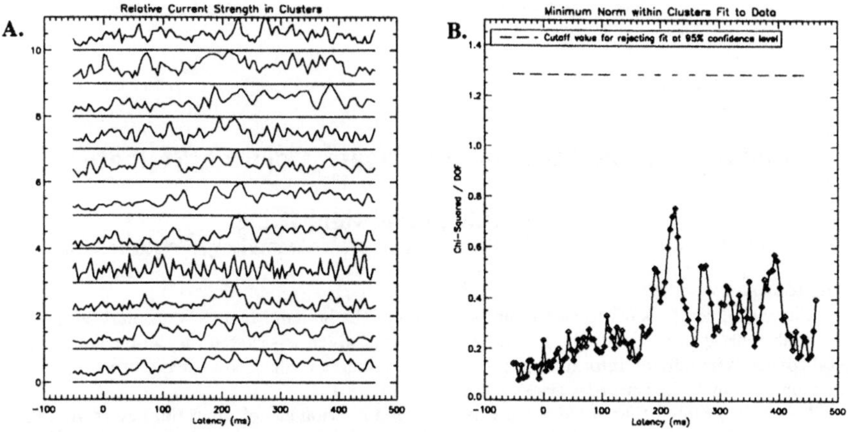

Figure 3. Constrained Minimum Norm Reconstruction
A. Reconstructed current timecourses from MEG data for each of the clusters constrained by MRI.
B. Reduced Chi-square error in the constrained MEG reconstruction as a function of time.

Our third approach was an extension of methods we have developed for distributed source reconstruction from MEG data incorporating anatomical constraints. Linear estimator procedures based on the Moore-Penrose pseudoinverse allow reconstructions of distributed current distributions of minimum Euclidean norm that account for the observed data. MRI derived constraints on source location and orientation may provide more accurate estimates of source location and extent. In the analysis illustrated in figure 3, "active" voxels identified by statistical analysis of fMRI data were used as a mask to select a small subset of the cortical voxels identified from anatomical segmentation. Currents were reconstructed using a weighted pseudoinverse procedure. The resulting inverse problem remained underdetermined (requiring more parameters than MEG sensors) yet might be overconstrained by MRI data. A more flexible paradigm would allow fMRI and MEG data to each contribute to the estimation of source spatial configuration. An alternative approach employs fMRI data used as a soft constraint, increasing the a priori estimate of the probability of current within a region. This prior probability together with experimental data can be used to compute a spatial probability metric based on Bayesian decision theory (Schmidt and George, this volume).

Discussion

The complementary strengths of multiple imaging techniques are required for mapping the structure and function of the human brain. Neural electromagnetic techniques provide excellent temporal resolution of neural population dynamics, but the requirement to solve an ill-posed inverse problem limits the resolution of source location and extent, particularly when multiple sources may be active simultaneously. MRI provides excellent spatial resolution. However, the physiology of the underlying hemodynamic changes limit the temporal resolution of activation measurements to a few seconds, and several problems can produce artifactual regions of apparent activation. A knowledge of individual anatomy can improve neuromagnetic current reconstructions by providing constraints on parameters. Sophisticated computational tools now exist to allow these prospects to be realized. We have provided an initial demonstration of several techniques for the integrated analysis of human MRI, fMRI, and MEG data. Future work will be necessary to establish the validity and scientific utility of these methods.

References

[1] Ranken, D., George, J. MRIVIEW: An interactive computational tool for investigation of brain structure and function, In: Nielson, G.M., Bergeron, D. IEEE Visualization '93, IEEE Computer Society Press, 1993

[2] George, J.S., Sanders, J.S., Lewine, J.D., Caprihan, A., and Aine, C.J. Comparative Studies of brain activation with MEG and Functional MRI. In: Deecke, L., Baumgartner, C., Stroink,G. and Williamson, S.J., Biomagnetism: fundamental research and clinical applications. Elsevier/IOS, Amsterdam, 1995

[3] George, J.S.,Aine, C.J., Mosher, J.C., Ranken, D.M. Schlitt, H.A., Wood, C.C., Lewine, J.D., Sanders, J.A., Belliveau, J.W. Mapping function in the human brain with MEG, anatomical MRI and functional MRI, J. Clin.Neurophysiol., 1995, 12(5):406-431

This work was supported by NIH grants EY08610 and DA/MH09972 and DOE contract KP0601000.

Combining Modalities to Investigate Cortical Retinotopy

Hurdal, M.K., McElwain, D.L.S.
Queensland University of Technology, School of Mathematics, Brisbane, Australia

Introduction

The visual cortex is often used in investigations of the human brain. The primary visual cortex, or V1, is mainly buried in the calcarine fissure, a deep sulcus which is situated on the medial surface of the occipital cortex. Virtually all information in the visual system is recognized as being processed by V1 first, and then passed out to higher order systems [1].

It is well accepted that there is a precise map of how segments of the retina are related to areas of V1. V1 has a retinotopic map in that one spot in the visual field maps directly to a spot on V1 [2]. The overall details of this map are well known from animal studies and the study of lesions in humans [3, 4]. It is agreed that more area of V1 is devoted to representing central vision rather than peripheral vision. Thus, the scale of this mapping changes as a function of retinal location. The term 'cortical magnification' [5], denoted by M, is used to describe this mapping and refers to an area of cortex that is devoted to representing a small area of visual field.

In addition, upper visual cortex receives signals from the lower visual field and similarly, lower visual cortex processes information from the upper visual field. The right visual cortex process the left field of view and vice versa. Central vision is represented at the pole of occipital cortex [3]. This general representation is referred to as the cruciform model [6].

Beyond this general agreement of this mapping, there is disagreement as to the amount of cortex that is allocated to representing central vision or other portions of the visual field [7]. There is much controversy regarding the precise details of cortical magnification and this retinotopic map. Furthermore, individual anatomical variations in the calcarine fissure seem to indicate that the cruciform model in not adequate to describe cortical retinotopy [8, 9].

Combining modalities is an ideal approach to investigate cortical retinotopy. Dipole source localization [10, 11] is one method of non-invasively determining which region of cortex is active. A neural source of magnetic and electrical activity can be modeled by a single dipole. The location of a source can then be determined by estimating a dipole that explains observed MEG or EEG data. Cortical anatomy also needs to be considered [12] and magnetic resonance (MR) images are well suited to this purpose. In order to be able to use these different modalities together effectively, a model of the visual cortex is required.

Methods

The Theoretical Model

Let the visual field be represented in polar co-ordinates (r, θ) and let the surface of the cortex be described in Cartesian co-ordinates (x, y). Cortical magnification is a function of retinal location and this relates a small area of cortex to a small area of visual field. Mathematically, this can be represented by derivatives. A small area $dxdy$ in the cortical plane corresponds to a small area of visual field $rdrd\theta$ adjusted by the cortical magnification parameter $M(r, \theta)$ [13]:

$$dxdy = M(r,\theta)rdrd\theta. \tag{1}$$

In addition, cortical magnification is a point to point mapping which transforms the visual field into its neuronal representation [14]. This means that the number of points in the retina is equivalent to the number of points in the cortex. Let $C(x, y)$ represent the density of the points in the cortex and let $\rho(r, \theta)$ represent the density in the retina. Then

$$C(x,y)dxdy = \rho(r,\theta)rdrd\theta \tag{2}$$

Equating (1) and (2) yields

$$M(r,\theta) = \frac{\rho(r,\theta)}{C(x,y)}, \tag{3}$$

giving an expression for cortical magnification in terms of physiological properties.

The Practical Model

In order to incorporate cortical anatomy into the investigation of cortical retinotopy, a practical and realistic model of V1 is required. Cortical features were obtained from T1 weighted MR scans taken in a 1.5 Tesla field. Scans were taken 1.5 mm apart with contiguous spacings of 1.5 mm. Points located along the calcarine fissure were systematically extracted from the MR scans (see Fig. 1a). These points formed the basis of the model of V1.

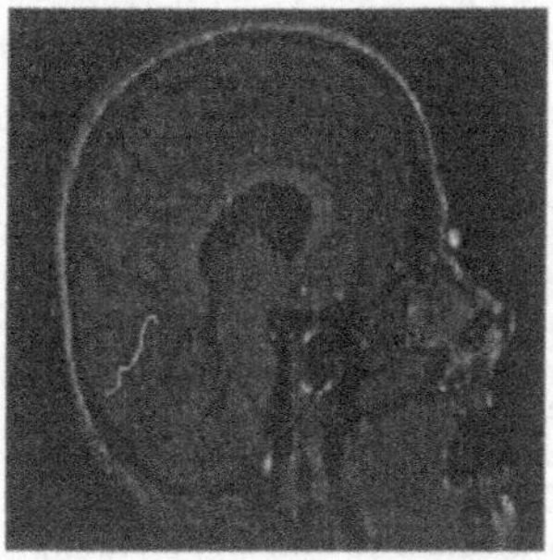

(a) MRI with Calcarine Fissure Highlighted

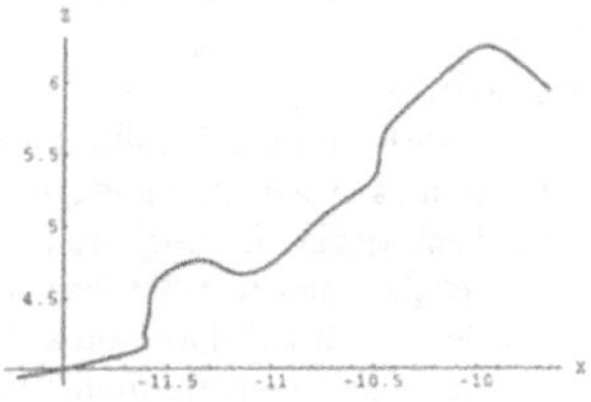

(b) Spline Curve which Models Calcarine Fissure

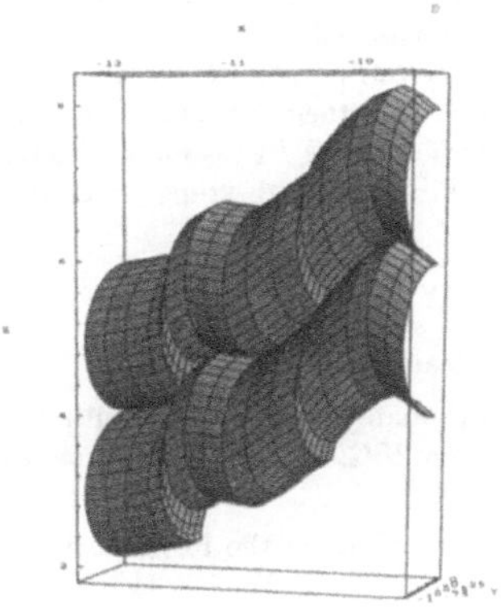

(c) Cylinder Model of V1

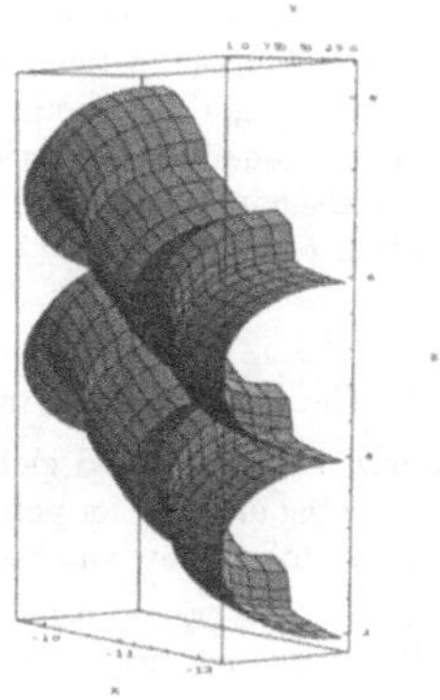

(d) Different Orientation of Cylinder Model

Fig. 1: Points representing the calcarine fissure were extracted from MRI data (Fig. 1a). A cubic spline was then fitted to the data (Fig. 1b). V1 is modeled by two cylinders representing the upper and lower visual cortex in one hemisphere of the brain. The cylinders lie beside each other and abut along the calcarine fissure. Since the calcarine fissure is modeled by the spline curve, the cylinders follow the realistic shape of the calcarine fissure (Fig. 1c and Fig. 1d).

Results

Cylindrical Model of Primary Visual Cortex

The points that were extracted from the MR scans were fitted to a cubic spline curve, resulting in the curve shown in Fig. 1b. As is apparent from comparing Fig. 1a and Fig. 1b, this curve approximates the shape of the calcarine fissure extremely well.

The visual cortex in one hemisphere of the brain was then modeled as two adjacent cylinders. The upper visual cortex is represented by one cylinder, as is the lower visual cortex. The two cylinders lie along side each other and abut along the length of the calcarine fissure. Since the fissure is modeled by the spline curve, the cylinders also follow the shape of the spline curve (see Fig. 1c and Fig. 1d). The result is a three dimensional cylindrical model of V1 which incorporates the realistic shape of the calcarine fissure. Fig. 1c and Fig. 1d show different orientations of this model.

Discussion

A mathematical definition for cortical magnification and a procedure for estimating human cortical magnification have been discussed in this paper. A theoretical model has been presented which is a formal and novel mathematical definition and has been obtained through new equations. By incorporating physiological properties of the point to point mapping between the visual field and the visual cortex, the concept of cortical magnification has been placed on a firmer mathematical footing.

A new and realistic model of the calcarine fissure has also been developed. The upper and lower visual cortex are modeled as two cylinders which follow the shape of the calcarine fissure. This cylinder model can now be used in conjunction with dipole source localization to estimate cortical magnification. Visual evoked potential data has been collected from a checkerboard pattern reversal experiment and the neural sources have been estimated using a dipole source localization algorithm [9]. The cylindrical model can now be used to relate the position of the estimated sources to the positions of the visual stimuli. In this manner, a truly human estimate for cortical magnification can be obtained.

The advantage of this approach is that this methodology can be applied to other areas of brain research. For example, the motor cortex and auditory cortex both have their own unique mapping [15]. By combining the modalities and techniques of dipole source localization, MEG / EEG data collection, MR images and mathematical modeling, similar approaches can be used to define and estimate a cortical mapping in these other regions.

References

[1] Zeki, S. The visual image in mind and brain, Scientific American, 1992, 267: 42–50.

[2] Schwartz, E.L. Local and global functional architecture in primate striate cortex: outline of a spatial mapping doctrine for perception, In: Rose, D., and Dobson, V.G. Models of the Visual Cortex, New York, John Wiley and Sons, Ltd., 1985.

[3] Holmes, G. The organization of the visual cortex in man, Proceedings of the Royal Society of London, Series B. Biological Sciences, 1945, 132: 348–361.

[4] Zeki, S. A Vision of the Brain, Oxford, Blackwell Scientific Publications, 1993.

[5] Daniel, P.M., and Whitteridge, D. The representation of the visual field on the cerebral cortex in monkeys, Journal of Physiology (London), 1961, 159: 203–221.

[6] Ahlfors, S.P., Ilmoniemi, R.J., and Hämäläinen, M.S. Estimates of visually evoked cortical currents, Electroencephalography and Clinical Neurophysiology, 1992, 82: 225–236.

[7] Horton, J.C., and Hoyt, W.F. The representation of the visual field in human striate cortex: a revision of the classic Holmes map, Arch. Opthalmol., 1991, 109: 816–824.

[8] Chorlton, M.C., Hurdal, M.K., Fulham, W.R., Finlay, D.C., and McElwain, D.L.S. Visual evoked potentials to small stimuli presented along a vertical meridian: Individual differences and dipole modelling, Australian Journal of Psychology, 1994, 46: 87–94.

[9] Hurdal, M.K. Dipole modelling for the localization of human visual evoked scalp potential sources, Master of Science thesis, University of Newcastle, Australia, 1994.

[10] Rush, S., and Driscoll, D.A. Current distribution in the brain from surface electrodes, Anesthesia and Analgesia, 1968, 47: 717–723.

[11] Ary, J.P., Klein, S.A., and Fender, D.H. Location of sources of evoked scalp potentials: correction for skull and scalp thickness, IEEE Transactions on Biomedical Engineering, 1981, 28: 447–452.

[12] Aine, C., Supek, S., George, J., Ranken, D., Best, E., Tiee, W., Vigil, V., Flynn, E., and Wood, C. MEG studies of human vision: retinotopic organization of V1. In: Baumgartner, C. et al. Biomagnetism: Fundamental Research and Clinical Applications, Amsterdam, Elsevier Science, 1995.

[14] Murray, J.D. Mathematical Biology, Heidelberg, Springer-Verlag, 1989.

[13] Fischer, B. Overlap of receptive field centers and representation of the visual field in the cat's optic tract, Vision Research, 1973, 13: 2113–2120.

[15] Schwartz, E.L. Spatial mapping in the primate sensory projection: analytic structure and relevance to perception, Biological Cybernetics, 1977, 25: 181–194.

Acknowledgements

The authors wish to acknowledge Dr. Krish Singh of the Department of Psychology, Royal Holloway, University of London for the use of his software in displaying and presenting MR data.

fMRI Constrained Linear Estimation of Cortical Activity from MEG Measurements: A Model Study

Liu, A.K.[1,2], Sereno, M.I.[3], Rosen, B.R.[1], Belliveau, J.W.[1] and Dale, A.M.[1]

[1]*Massachusetts General Hospital NMR Center, Charlestown, MA*
[2] *Harvard-MIT Division of Health Sciences and Technology, Cambridge, MA*
[3]*University of California, San Diego Department of Neurosciences, La Jolla CA*

Introduction

The localization of activity within the brain based on magnetoencephalography (MEG) measurements is inherently ill-posed. It has been suggested that information from functional magnetic resonance images (fMRI) can be used to constrain the ill-posedness [1,2,3,4]. Using the MEG temporal data and a linear estimation approach [1] or weighted minimum norm, we wish to compute the time courses of activated areas observed by fMRI.

One can foresee situations where the hemodynamic and magnetic measurements differ, although the actual region of cortical activation is the same. In these modeling studies, we have examined the effects of several "mis-specifications" on various types of weightings of the linear approach.

The three mis-specifications examined were:
1) mis-registration - the number of areas of activation on fMRI correspond to the number of generators of magnetic fields, but the two are offset by some distance
2) extra fMRI source - an area of fMRI activation does not generate magnetic fields
3) missing fMRI source - a generator of magnetic fields is not measured as an area of fMRI activation.

The spatial weightings were:
1) minimum norm - weights all locations equally
2) depth-weighted minimum norm - weights locations approximately inversely proportional to depth
3) fMRI constrained - spatial information from fMRI is used.

Methods

The derivation of the linear inverse operator [1] begins with a set of measurements

$$\mathbf{x} = \mathbf{As} + \mathbf{n}$$

where **x** are the measurements, **A** is the gain matrix, **s** is the strength of each dipole component, and **n** is the noise vector. We would like to calculate a linear inverse operator **W** that minimizes the expected difference between the estimated and the correct source solution. The expected error can be defined as

$$Err_W = \left\langle |\mathbf{Wx} - \mathbf{s}|^2 \right\rangle$$

Assuming both **n** and **s** are normally distributed with zero mean and covariance matrices **C** and **R**, respectively, we can rewrite the expected error as

$$\begin{aligned} Err_W &= \left\langle |\mathbf{W}(\mathbf{As}+\mathbf{n}) - \mathbf{s}|^2 \right\rangle \\ &= \left\langle |(\mathbf{WA}-\mathbf{I})\mathbf{s} + \mathbf{Wn})|^2 \right\rangle \\ &= \left\langle |\mathbf{Ms} + \mathbf{Wn})|^2 \right\rangle \quad \text{where } \mathbf{M} = \mathbf{WA} - \mathbf{I} \\ &= \left\langle |\mathbf{Ms}|^2 \right\rangle + \left\langle |\mathbf{Wn}|^2 \right\rangle \\ &= tr(\mathbf{MRM}^T) + tr(\mathbf{WCW}^T) \end{aligned}$$

This last expression can be explicitly minimized by taking the gradient, setting it to zero and solving for **W**. This gives a linear inverse operator

$$\mathbf{W} = \mathbf{R}\mathbf{A}^{T}(\mathbf{A}\mathbf{R}\mathbf{A}^{T}+\mathbf{C})^{-1}$$

Information from fMRI, or any other spatial activation information, is incorporated into the **R** matrix. Noise characteristics of the sensors may be included in the **C** matrix.

For these model studies, we used a spherical head model with a radius of 10 cm and a 122 channel MEG sensor description (Neuromag, Inc.). Noise was assumed to be spatially uncorrelated and white with a SNR of 10.

The fMRI weighting was systematically varied from no weighting (i.e. minimum norm) to complete fMRI weighting (all locations not specified by fMRI were set to 0). A depth-weighted minimum norm was also computed for comparison [3]. The depth-weighted minimum norm weighted each location by the inverse of the power at the detectors given a unit activation at that location. Only the optimum fMRI weighting case, chosen to minimize the effect of the missing dipole, is presented below.

Either 5 or 10 sources were located at random within a single hemisphere or a single octant of the head model. Then the average percentage of incorrectly specified power per dipole pair was computed for each type of spatial weighting (minimum norm, depth-weighted minimum norm, and fMRI constrained) and mis-specification between fMRI and actual electromagnetic source. This random placement was repeated 10 times in the hemisphere and 100 time in the octant, and the average ± standard error of the mean of the percentage of incorrectly specified power was computed.

Results

Shown below are the percentages of incorrectly specified power for each of the spatial weightings (Minimum Norm, Depth-Weighted Minimum Norm, and fMRI Constrained) due to the following mis-specifications: no mis-specification between fMRI and the actual sources, an extra fMRI source that is not an electromagnetic generator, a missing fMRI source that is generating electromagnetic fields, and a spatial mis-registration between fMRI and the actual sources. The percentages of incorrectly specified power ± the standard error of the mean for 5 dipoles and 10 dipoles placed in a single hemisphere are shown separately in Tables 1 and 2. The percentages of incorrectly specified power ± the standard error of the mean for 5 and 10 dipoles placed in a single octant are shown in Tables 3 and 4.

Each column within the table describes slightly different values:
None: percentage of power that is mis-localized among locations that are correctly specified by fMRI.
Extra fMRI source: percentage of power from the correct locations that is placed on the extra fMRI source.
Missing fMRI source: percentage of power from the missing fMRI source that is placed on the locations correctly specified by fMRI.
Spatial: percentage of power that is mis-localized when there is a 1cm spatial discrepancy between fMRI and the actual electromagnetic generators

Spatial Weightings	Mis-specifications			
	None	Extra fMRI source	Missing fMRI source	Spatial (1cm)
Minimum Norm	25±2.5	28±5.0	28±5.3	26±2.5
Depth-Weighted	20±2.3	14±1.4	21±5.7	21±2.3
fMRI Constrained	6±1.6	6±1.3	28±4.8	15±1.6

Table 1: 5 Dipoles in Hemisphere. Percentage of incorrectly specified power (mean ± standard error of the mean). n=10

Spatial Weightings	Mis-specifications			
	None	Extra fMRI source	Missing fMRI source	Spatial (1cm)
Minimum Norm	28±1.5	28±2.9	31±1.8	28±1.5
Depth-Weighted	19±1.7	17±3.8	21±1.9	20±1.7
fMRI Constrained	9±1.4	9±1.4	27±2.1	16±1.2

Table 2: 10 Dipoles in Hemisphere. Percentage of incorrectly specified power (mean ± standard error of the mean). n=10

Spatial Weightings	Mis-specifications			
	None	Extra fMRI source	Missing fMRI source	Spatial (1cm)
Minimum Norm	30±0.5	30±1.3	29±1.3	30±0.5
Depth-Weighted	26±0.6	19±1.0	26±1.5	27±0.6
fMRI Constrained	9±0.5	8±0.5	29±1.4	19±0.4

Table 3: 5 Dipoles in an Octant. Percentage of incorrectly specified power (mean ± standard error of the mean). n=100

Spatial Weightings	Mis-specifications			
	None	Extra fMRI source	Missing fMRI source	Spatial (1cm)
Minimum Norm	31±0.3	30±1.1	30±1.1	31±0.3
Depth-Weighted	27±0.4	21±0.9	27±1.1	28±0.4
fMRI Constrained	11±0.3	11±0.5	27±1.1	20±0.2

Table 4: 10 Dipoles in an Octant. Percentage of incorrectly specified power (mean ± standard error of the mean). n=100

Conclusions

There is only a slight difference between 5 and 10 sources in the amount of incorrectly specified power, suggesting that the number of sources will not be problematic. Also, when the sources are placed in closer proximity (i.e. placement in one octant versus placement in a hemisphere), there is a small increase in the amount of incorrectly specified power. More sources and sources that are closer together will both increase the error in power specification. However, the increase in incorrectly specified power due to the number of sources and the proximity of sources would, on average, be small.

Regarding the various spatial constraints, minimum norm was the least successful at dealing with the various types of mis-specifications we examined. The addition of a depth-dependent weighting on the minimum norm produces very little improvement (< 5% for the various metrics). However, the addition of the fMRI weighting to the minimum norm correctly localizes more of the activation, produces less crosstalk, and also places less activation on an extra fMRI source. Unfortunately, none of these various weighted minimum norm schemes handles a source that is missing from fMRI. Although the inclusion of fMRI weighting does not worsen the situation, in even the best cases a hemodynamically undetected source would still perturb the other sources by approximately 28%, on average.

In conclusion, with only spatial mis-specifications or extra fMRI areas of activation, these simulations demonstrate that the use of a priori fMRI information allows us to correctly predict spatiotemporal brain activity. Additional refinements will be required to adequately describe the spatiotemporal record in the presence of hemodynamically inactive, but magnetically active, sources.

References

[1] Dale, A.M., Sereno, M.I. Improved Localization of Cortical Activity by Combining EEG and MEG with MRI Cortical Surface Reconstruction: A Linear Approach. Journal of Cognitive Neuroscience, 1993, 5: 162-176.
[2] Fox, P.T., Woldorff M.G. Current Opinion in Neurobiology, 1994, 4:151-6.
[3] Hamalainen, M., Hari, R., Ilmoniemi, R.J., Knuutila, J., Lounasmaa, O.V. Magnetoencephalography - theory, instrumentation, and applications to noninvasive studies of the working human brain. Review of Modern Physics, 1993, 65: 413-497.
[4] Belliveau J.W. In: Uemura K., Lassen N.A., Jones T., and Kanno I., eds. MRI techniques for functional mapping of the human brain: integration with PET, EEG/MEG and infrared spectroscopy. Quantification of Brain Function. Amsterdam: Elsevier Science Publishers, 1993: 639-667.

Acknowledgements

This study was supported by grants from the United States Public Health Service: MH50054, EY07980, by grants from the Human Frontier Science Program, and conducted during the tenure of an established investigatorship from the American Heart Association to JWB.

Multiple Modality Biomagnetic Analysis System

Martin, K.[1], Erne, S.[2], Law, C.[1], Conforto, S.[3], Mallick, J.[1] and Tatar, B.[1]

[1]GE Corporate R&D, Schenectady, New York, USA; [2]Institute for Biomedical Engineering, University Ulm, Ulm, Germany; [3]Institute of Advanced Biomedical Technologies, Chieti, Italy

Introduction

As an emerging new modality, biomagnetic analysis presents exciting opportunities while raising some difficult issues. During the past four years researchers at General Electric, The University of Ulm and The University of Chieti have created a state of the art biomagnetic analysis system, and have addressed many of these difficulties. One of the critical aspects of a biomagnetic analysis system is how it handles different types of data. Many of the clinical requirements for a successful system require fast, efficient and flexible data management. This paper describes what those requirements are and how we addressed them. We start by providing a high level overview of our system. We then list some of the pertinent clinical requirements, present our solutions and describe how well they have worked. Finally we discuss future directions.

Before going into detail on the clinical requirements, we will provide a high level overview of our system. The five main components of our system are; the sensor head, the BSS (biomagnetic sensor system), the DAPS (digital acquisition processing system), the BOC (biomagnetic operators console) and the BAC (biomagnetic analysis console). The sensor head measures the signals, while the BSS performs analog to digital conversion and a variety of control functions. The BSS and DAPS communicate through a high speed fiber optic connection. The DAPS performs additional data processing, and records the data to disk. The BOC communicates to the DAPS through ethernet and shared memory. The BOC coordinates the data acquisition and is the control center for the sensor head, BSS and DAPS. The end product of the BOC is a runfile, and the process is described in Corby [1]. A runfile contains all the configuration information and data from an acquisition. The BAC is responsible for analyzing the runfile once it has been acquired. The BAC can obtain a runfile from a variety of sources ranging from mounting disks off of the BOC to inserting a magnetic tape. These five components are designed to support generating a runfile with 256 channels of 16 bit data being acquired at 10,000 Hertz.

Clinical Requirements
1) Handle large amounts of data.
2) Handle high data rates.
3) Display raw and filtered data in real time.
4) Support multiple modalities.
5) Quick turnaround time.
6) Multiple platform support.
7) Handle different sensor heads.

The first requirement we will address is handling large amounts of data. From the above discussion, you can see that a runfile can consist of a gigabyte of data. At this point in time, handling a gigabyte of data is a significant issue for most PCs and workstations. Similarly, displaying 256 channels of 10KHz data in real time is a challenge. A clinician wants to see a high quality stripchart of the acquired data, displayed in real time and properly subsampled. As the third item in our list indicates, most clinicians want to perform band pass filtering or base line correction in real time as well.

The fourth requirement demands that we support MR, CT, and 3D surface data in addition to the temporal data from the sensor head. This is important for displaying dipole positions within image data, or relative to a 3D model. The fifth requirement stems from a desire to perform a quick analysis before the patient has left the facility. The sixth and seventh requirements are important due to the variety of hardware available. We do not want to lock the clinician into a specific computer platform or a specific sensor head.

Methods

To meet the needs of a biomagnetic analysis system we had to design a specialized data architecture. Traditional techniques that load in the data and then overwrite it when filtering would not work. Additionally, constraints that the system handle multiple runfiles being open at once, required us to use no global variables for data storage. Our solution consists of three data types, two data pipelines, a complete runfile and a hardware independent abstraction. We will first discuss the three data types: display, image and signal.

Display data can be thought of as 3D geometric data, typically consisting of polygons, lines, triangle strips and points. Display data is used to store and represent torso and skull models, sensor arrays, dipole and min-norm results, fiducial points and iso-surfaces. Display data can have attributes such as normals, vectors, scalars, texture coordinates and time. Image data is used to store a series of 2D scalar slices such as MR or CT. Image data is assumed to be static. Signal data is used to store temporal signals such as those from an EEG or MEG. There may be any number of channels of data and any sample rate. The patient position, electric and magnetic data are stored as signal data.

There are two data pipelines: the display data pipeline and the signal data pipeline. There are a number of qualities that they have in common. Both pipelines are object oriented and represent filtering operations as objects (or nodes) in the pipeline. These pipelines have multiple inputs and outputs, are demand driven and rely on implicit control and conditional execution. Schroeder [2] provides a good introduction to data pipelines. Fig. 1 shows a simple

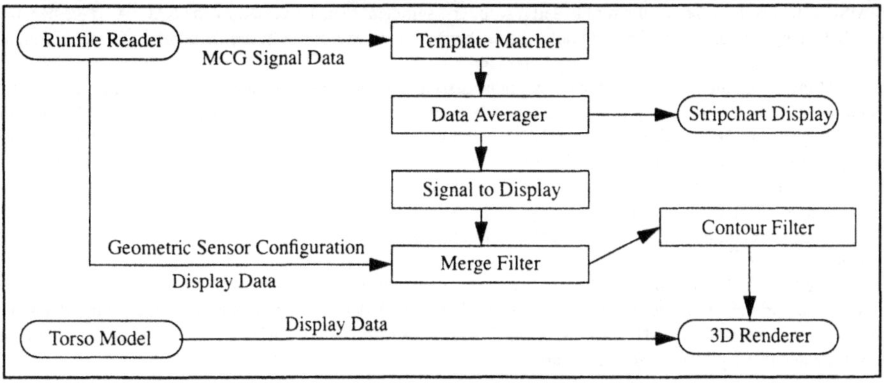

Fig. 1, Sample example data pipeline.

example of a data pipeline. The runfile reader and torso model are data sources for the pipeline. The runfile reader generates both display data and signal data. The signal data the goes through a template matcher followed by a data averager. This will average a number of heartbeats into a composite heartbeat with an improved signal to noise ratio. This result is then displayed on a stripchart and also sent to a signal to display filter. This converts signal data into scalar values for display data. These scalar values are then merged with the geometric configuration of the sensor array. This then gets passed through a contour filter to generate contour lines which are then displayed by a 3D renderer. The torso model is simple read in and then displayed in the 3D renderer along with the contour lines.

The display data pipeline provides little beyond what was discussed above. In contrast the signal data pipeline supports multiple ring buffer output caches. The caching allows us to handle very large amounts of data since only a small portion resides in memory. Given the demand driven nature of the pipeline, only the portions of data required for a given operation are read in. Likewise, filtering is also a demand driven process. The following example will serve to illustrate the caching operation. The clinician is looking at a stripchart showing two seconds of raw MEG data. As the stripchart scrolls forward in real time, it requests the additional data it needs to display from its input. The input checks to see if all or part of the data is already cached. It then requests whatever data it needs, if any, from its input. Eventually the request may get back to the runfile reader where the required data will be read from the disk.

Now the clinician turns on band pass filtering, this results in a filter being inserted between the stripchart and the runfile reader. The stripchart will refresh its entire display, requesting two seconds of data. The band pass filter then checks its cache to determine if any of the data is already present. Nothing will be there, so it will request the full two seconds from the runfile reader which should already have that data cached. The band pass filter operates on the data and stores its result in its cache. The stripchart then displays the result and continues to scroll forward. To perform its scrolling the stipchart may need about a tenth of a seconds worth of data. The actual amount will depend on the load on the machine and the desired rate of scrolling. The band pass filter then needs to generate this additional tenth of a seconds worth of data. Due to the limitations of band pass filtering, the filter may require a minimum of a half second worth of input data to properly work. To handle this the filter requests the minimum amount of data it needs to work, performs the filtering, and then writes the entire result into the cache. This means that the band pass filter will already have cached the next few updates worth of data for the stripchart. If the clinician were to change the frequencies of the band pass filter, its entire output cache would be considered invalid and cleared. But the band pass filter's input would still have the required data cached for quick recomputation.

The multiple output caching is used to support comparative analysis. A clinician may want to perform two different operations of the same data and then compare the results. By supporting multiple output caches, a data pipeline can be split off into two paths at any point without a loss in performance. The multiple output caches effectively prevent thrashing of the data. The same approach can be used within a filter that requires frequent access to different parts of the raw data, such as analyzing evoked response data for a particular stimulus.

Now that we have discussed the three types of data and the two data pipelines, we will consider the complete runfile and hardware abstraction. The runfile is designed to contain all the information necessary to analyze its data. Any MSI system could generate the runfile and it would be sufficient to be processed by the BAC. A fringe benefit of this is that changes in the sensor head do not require changes in the analysis software. In addition to the information contained in this file, its organization is very important.

Many of our analysis routines start by preprocessing the raw data over time to generate an averaged result. This preprocessing step frequently relies on a few reference channels of data for selecting templates and performing temporal correlation. These reference channels are stored in a contiguous space in the runfile. This enables the initial scanning of the reference channels, selection of templates, template matching and correlation tree generation to be done without having to scan through the entire runfile. Then the averaging step can only read from disk the portions that are required. This can create a huge savings in time. We go from having to scan the entire runfile twice, to only having to read select parts of it. Given that scanning through a gigabyte file takes about five minutes this is an important issue.

The hardware abstraction is what we use to maintain platform independence. Since we have restricted ourselves to three primary data types, we have created a hardware independent abstraction for displaying them. This allows us to run under WindowsNT displaying our 3D graphics (display data) using OpenGL, or run on a HP UNIX workstation and display it using Starbase graphics.

Results

We have built a biomagnetic analysis console (BAC) on top of the data architecture described above. The resulting system supports a cornucopia of features and is flexible enough so that trying out new algorithms does not require significant changes to the software. As the field continues to develop this is an important feature. The BAC has been run on HP, Sun, SGI and WindowsNT platforms. With the current hardware we can properly display and filter a few stripcharts with a sample rate of about 10,000 Hz. For a larger number of channels the display must be slowed down or the display can be approximated by subsampling the data to approximately 0.1 pixel resolution.

The BAC can work with multiple CT/MR, runfile, and 3D data sets at once and different data sets can be registered so that dipoles appear as markers on the MR/CT slices. Initial 3D surface data can be obtained through Marching Cubes iso-surface extraction [3] or laser range scanners [4]. To generate closed surfaces required for some forms of analysis we use the surface construction algorithm of Yamrom [5]. The system supports an electronic notebook for recording pictures and results. Most anything that the clinician sees on the screen can be stored into this electronic notebook. This is made possible by the restriction of our data to the three primary data types. There is support for both neuro and cardiac analysis, continuous and evoked recording. Pipeline filters have been created for: baseline correction, band pass filtering, template matching, temporal averaging, correlation tree calculation, artifact rejection, geometric channel selection, power spectrum, cross correlation, minimum norm, multiple dipole fit, rigid body fiducial registration, contour line extraction and a selection of traditional MR/CT operations.

We have found that the resulting system can perform a quick analysis of a runfile within a few minutes after the acquisition has been completed. Further analysis of the runfile can occur while a new data set is being acquired by the BOC, avoiding a potential bottleneck.

Discussion

Throughout the medical community, the software demands of data acquisition systems have become considerable. The BAC consists of over 200,000 lines of code with tough performance demands. Without a good software foundation such a system quickly degenerates into an unmanageable mess. Within the MSI community this is compounded by a lack of clearly defined and accepted clinical procedures. While much progress has been made on this front, a successful MSI system must still provide a very high degree of flexibility.

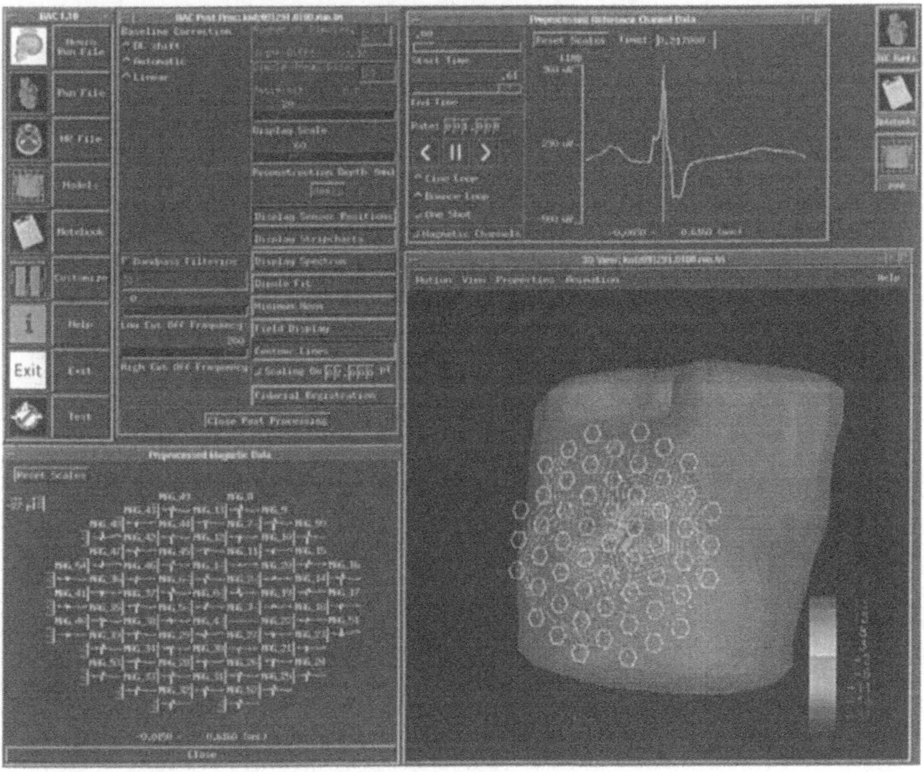

Fig. 2, Sample screen from the BAC.

References

[1] Corby, N.R., Jr., Miller, P.D., Hogle, R.A., "A realtime, high-bandwidth, programmable digital signal-processing system for monitoring, acquisition, filtering, and archival storage of biomagnetic signals" Biomag '96, Santa Fe NM, February 16-21, 1996.

[2] Schroeder, W., Martin, K., Lorensen, W., "The Visualization Toolkit: An Object Oriented Approach to 3D Graphics" Prentice Hall 1996.

[3] Lorensen, W.E., Cline, H.E., "Marching cubes: a high resolution 3D surface construciton algorithm," Computer Graphics, 1987, 21 (3): 163-169.

[4] Cyberware, 2110 Del Monte Avenue, Monterey, CA 93940.

[5] Yamrom, B. "Modelling closed surface geometry for skull and torso using snapping" Biomag '96, Santa Fe NM, February 16-21, 1996.

Measurement of Visually Evoked EEG and MEG Associated with Verbal Cognitive Processes

Nakagawa, S.[1], Ueno, S.[1], Iramina, K.[1] and Weinberg, H.[2]

[1]Institute of Medical Electronics, Faculty of Medicine, University of Tokyo, Tokyo 113, Japan;
[2]Brain Behaviour Laboratory, Simon Fraser University, B.C., Burnaby, Canada

Introduction

Magnetoencephalography (MEG) is a useful method for the diagnosis of functional brain diseases as well as for the study of higher brain function, such as cognition, memory and blending of brain functions. In recent years, substantial advances in understanding the functional organization of the human brain have been made through the use of MEG. Most of these investigations were concerned with evoked magnetic fields. Only a few neuromagnetic studies on higher brain function have been reported [1],[2].

In this study, we focused on the measurement of verbal activities in the human brain. Kutas et al. reported a late negative slow wave elicited by words, which is attenuated if the word is repeated or primed by a related word or by sentence context [3],[4]. We observed verbal cognitive activities by measuring electroencephalograms (EEGs) and MEGs evoked by visually cognitive tasks. We analyzed the EEG and MEG data based on topographic maps [5], and discussed the source estimation using MEG data.

Methods

Seven male volunteers with no history of neurological disease took part in this experiment. Their ages ranged between 21 - 30 years. All the subjects were right handed. Visual evoked potentials (VEPs) and magnetic fields were recorded simultaneously. Visual stimulation was projected on the screen. (Fig.1) Four letter English words, meaningless sets of four letters (nonsense words) and random dots were presented for subjects as visual stimuli. (Fig. 2) The visual angle of each letter was 1.3 degrees. The intensity of the stimulation of random dots was adjusted to be identical to the stimuli of letters. Each stimulus was randomly presented for 1.4 sec at an interval of 2.0 ~ 3.0 sec. Subjects were asked to maintain fixation of their eyes at center of screen.

The EEG data were also recorded from 14 locations over the surface of the head, corresponding to the points defined by the international 10-20 system. The magnetic field components perpendicular to the surface of the head were measured using a 7-channel DC-SQUID first order gradiometer and a 64-channel third order gradiometer system. After being filtered with a bandpass filter (0.5 - 30 Hz), the EEG and MEG data were averaged. 110 - 300 responses were averaged at each position. The analysis time was 1.0 sec including 0.2 sec prestimulus baseline.

Results and Discussion

Figure 3 shows simultaneous measurements of EEG and MEG. The bold solid lines denote responses evoked by English words, the bold dotted lines denote responses evoked by nonsense words, and the fine lines denote responses evoked by random dots. All waves have a large peak at about 150 msec . The differences among the three stimulus conditions are observed between latencies of 300 and 600 msec.

Figure 4 shows EEG topographies of each stimulation. Topography at 152 msec shows that the activity

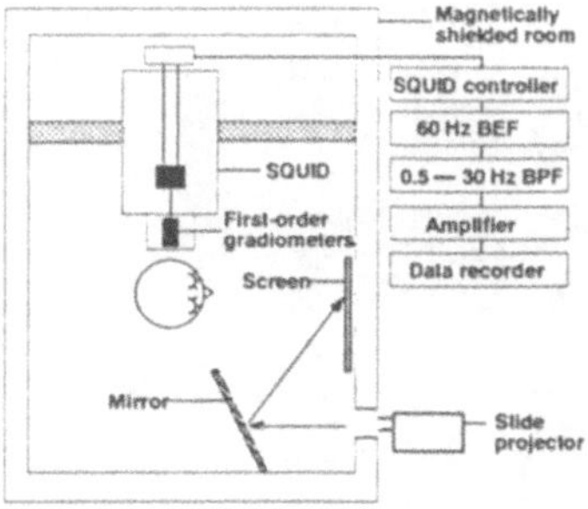

Fig.1. System of measurement. Visual stimulation was projected onto a screen in a magnetically shielded room.

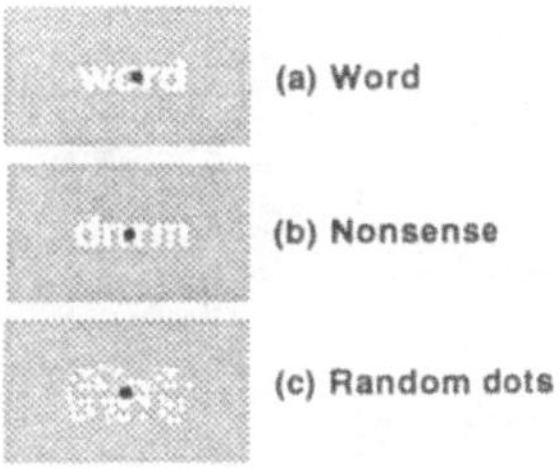

Fig.2. Patterns of visual stimulation: (a) English word, (b) Nonsense word, (c) Random dots. Subjects were asked to maintain fixation of their eyes at center of screen.

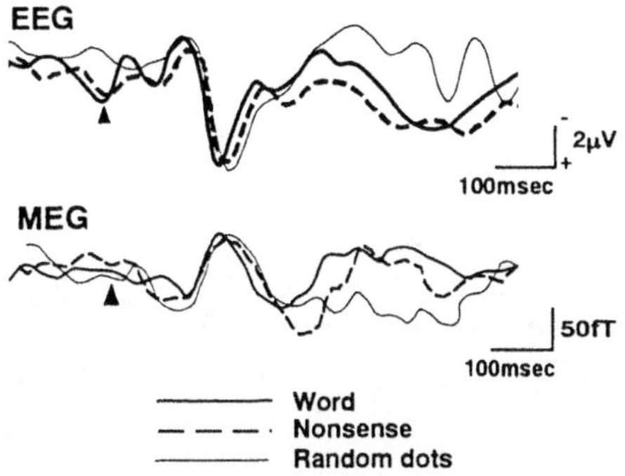

Fig.3. Averaged wave forms of visual evoked potentials and magnetic fields recorded at C3 (subject A). The bold solid lines denote responses evoked by English words, the bold dotted lines denote responses evoked by nonsense words, and the thin solid lines denote responses evoked by random dots.

is dominant in the occipital region. The sources at a latency of about 160 msec were estimated with a 2-dipole model using MEG data. The results are shown in Figure 5. Two equivalent current dipoles are localized in or near the primary visual cortex.

We estimated the sources for recognition of English words from the MEG topographic data. The MEG topographic patterns obtained here were complicated, and it was difficult to estimate the source with a single-dipole model. A 2-dipole model was used to estimate sources as shown in Figure 6. In subject A, an asym-

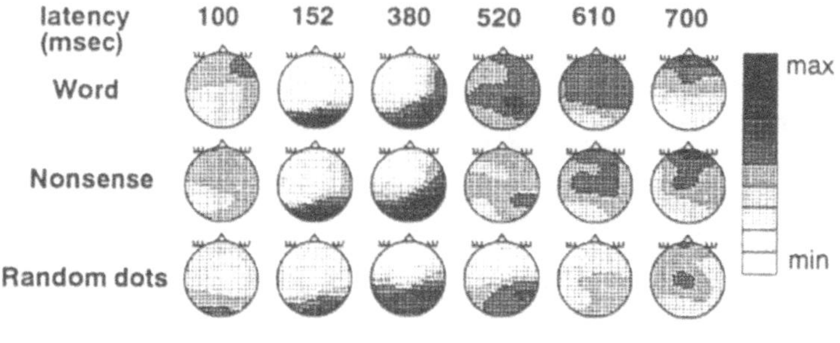

Fig. 4. The EEG topographies of VEPs evoked by English words, nonsense words, and random dots (subject A). The potentials are quantified into 10 levels. The topographies of English words and random dots show dominant activity in the frontal and central regions between latencies of 520 and 700 msec.

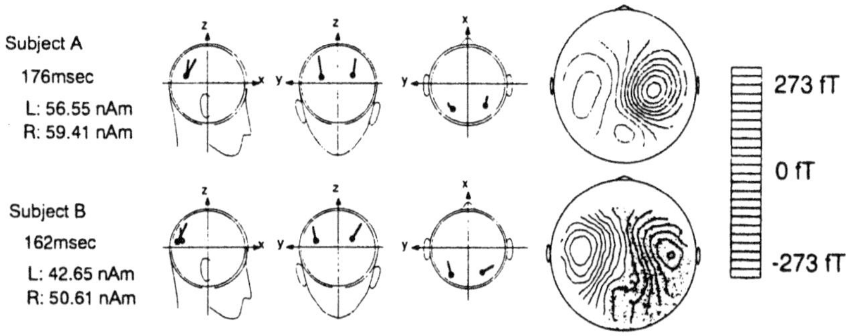

Fig. 5. The estimated sources of visual evoked MEG at a latency about 160 msec. Two equivalent current dipoles are localized in or near the primary visual cortex.

metrical distribution of dipoles was estimated. In subject B, the estimated source was located near the center in the head. In three of the four subjects for which we estimated sources, the dipole in the left-hemisphere was larger than the dipole in the right hemisphere.

In estimation of source with a latency of 160-170 msec, the sources with 50 nAm seems rather intense. These intense sources can be caused by undermodeling the data with too few sources. Aine et al. estimated the sources of visually evoked magnetic fields with 5-dipole model [6]. It is essential to estimate sources by taking into account anatomical and functional organization of the brain.

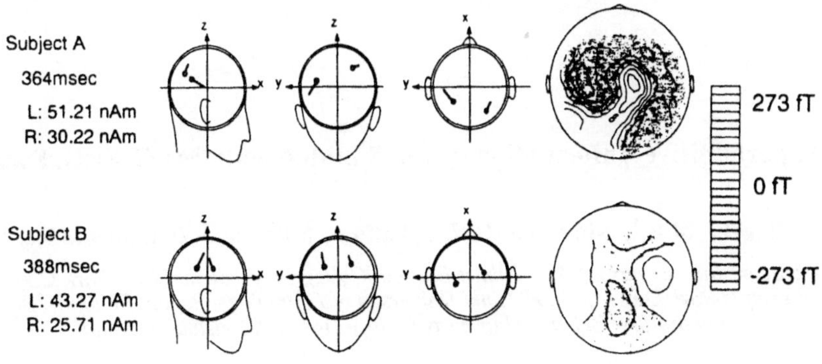

Fig.6. The estimated sources of two subjects for recognition of English words.

References :

[1] Lu, S.T., Hämäläinen, M., Hari, R., Ilmoniemi, R., Lounasmaa, O., Sams, M., and Vilkin, V., Seeing faces activates three separate areas outside the occipital visual cortex in man, Neuroscience, 1991, 43: 287-290.

[2] Gordon, E., Rennie, C., and Collins, L., Magnetoencephalography and late component ERPs, Clinical & Experimental Neurology, 1990, 27 : 113-20.

[3] Kutas, M., Hillyard, S. A., Reading senseless sentences : Brain potentials reflect semantic incongruity, Science, 1980, 207: 203-204.

[4] Kutas, M., Hillyard, S. A., Brain potentials during reading reflect word expectancy and semantic association, Nature, 1984, 11: 161-163.

[5] Ueno, S., Matsuoka, S., Mizogochi, T., Nagashima, M., and Chen, C., Topographic computer display of abnormal EEG activities in patients with CNS disease, Memoirs of the Faculty of Engineering, Kyushu Univ., 1975, 34 : 165-209.

[6] Aine, C., Supek, S., George, J., Ranken, D., Best, E., Tiee, W., Vigil, V., Flynn, E., and Wood, C., MEG studies of human vision: Retinotopic organization of V1, In: Baumgartner, C., Deecke, L., Stroink, G., Williamson, S.J., Biomagnetism; Fundamental Research and Clinical Applications, Amsterdam, IOS Press, 1995.

Superadditive Information from Simultaneous MEG/EEG Data

Pflieger, M.E.[1], Simpson, G.V.[2], Ahlfors, S.P[2,3] and Ilmoniemi, R.J.[4]

[1]*Neuro Scan, Inc., El Paso, USA;* [2]*Albert Einstein College of Medicine, New York City, USA;* [3]*Low Temperature Laboratory, Helsinki University of Technology, Espoo, Finland;* [4]*BioMag Laboratory, Helsinki University Central Hospital, Helsinki, Finland*

Introduction

Although MEG and EEG measurement modalities are empirically distinct, there is a theoretically unified statistical-biophysical framework for posing and (in a limited sense) solving the electromagnetic inverse problem. In addition, biomagnetic and bioelectric measurements are differentially sensitive to the same type of intracranial signals, i.e., macroscopic ionic source currents. Since MEG and EEG share a common foundation and probably provide complementary information about the same kind of source signals, a truly integrated electromagnetic source imaging (EMSI) modality appears theoretically immanent. In practice, EMSI requires combined analysis of MEG and EEG acquired (in the ideal case) simultaneously.

Combined MEG/EEG analysis must first surmount the *measurement scaling problem* which arises because the physical units for MEG (femtotesla) and EEG (microvolts) are not directly comparable. Consequently, the direct sum of an *N*-channel MEG measurement space with an *M*-channel EEG measurement space does not yield an (*N*+*M*)-channel MEG/EEG measurement space with a usable metric. Greenblatt [1] showed how to resolve the scaling problem by using background noise variance-covariance statistics to normalize the combined data. Rescaling all data channels with respect to the inverse symmetric square root of the background noise matrix (see below) produces a common frame of reference in which magnetic field and electric potential units are converted to *unitless* signal-to-noise ratios (SNRs). This multichannel normalization operation is called here an *SNR transform* of the data.

Simultaneous acquisition of MEG and EEG is clearly advantageous from the standpoint of experimental design since between-session variables (e.g., subject state and response variability) are eliminated by constructing average event-related field (ERF) and potential (ERP) waveforms from exactly the same set of trials. Moreover, from the standpoint of the SNR transform, note that *between-modality* background noise covariance statistics cannot be computed unless acquisition is simultaneous. In this paper we show that there are substantial benefits to be gained by incorporating these inter-modal background noise covariances in the SNR transform.

To quantify such benefits, we compute a measure of *empirical information available for source estimation* (based on the SNR transform) that meets several criteria. First, for the sake of generality, the information measure does not use specific modeling assumptions that are otherwise required to actually estimate sources. In particular, the computation does not use a head model (compare [2]), nor does it make specific assumptions about the nature or configuration of the current sources. Second, the information measure is logarithmic following Shannon's rationale [3]. Third, the computed quantity of information is invariant with respect to any nonsingular linear reformatting of the multichannel data; i.e., all such reversible transformations are "information preserving" (in a linear sense). As a corollary, it follows that information is independent of the scale utilized for each channel; i.e., the units of physical measurement need not be uniform across channels. Fourth, information computed for a noise-only data sample is close to zero. Fifth, information is additive for independent sources of signal variance when the background noise is uncorrelated across sensors. Sixth, redundant sampling of a signal at the sensors increases information just to the extent that uncertainty due to noise is reduced. And seventh, from the perspective of ERFs and ERPs, information ultimately measures our ability to *spatially distinguish* event-related signals from the residual background noise. A procedure for computing an information measure that satisfies these criteria is described below.

The phenomenon of *superadditive* information is theoretically possible for a channel partition $A+B$ (i.e., the information of $A+B$ can exceed the information of A plus the information of B) if the event-related A signal is sufficiently uncorrelated with the event-related B signal while at the same time the background noise is sufficiently correlated between A and B. Superadditivity cannot occur if the noise covariances between A and B are neglected. Thus, superadditivity for an MEG+EEG partition is *a priori* possible only if acquisition is simultaneous.

Methods

Data acquisition. The analyzed data were derived from 122 Neuromag MEG channels (retaining 120 good channels) with 29 simultaneous EEG channels (using silver electrodes and copper leads) sampled at 319 Hz (0.03 to 80 Hz bandpass) from one subject (GVS) in 1993 at the Low Temperature Laboratory of the Helsinki University of Technology. Event-related averages were replicated 6 times (~70 epochs each) for 4 visual stimuli consisting of wedge-shaped checkerboard pattern onsets (50 ms duration) in four quadrants (upper right, upper left, lower left, and lower right). The event-related averages were baseline adjusted and 40 Hz lowpass filtered offline.

Computation of event-related information: 8 special cases. The composite ERF/ERP signal for each of the 4 stimulus conditions was estimated by the grand average across the 6 replications. The grand average $\vec{y}_i(t)$ for each condition i consisted of 149 channels by 118 time points in a window of 30 ms to 400 ms post-stimulus. The associated residual background MEG/EEG noise covariance $\mathbf{C}_i$ for each condition i was estimated by computing spatial covariance (149×149) statistics for each time sample across the 6 replications, dividing by 6 to rescale single replication covariances to the level of residual grand average covariances, and averaging over the 118 points in the time interval. Time averaging is necessary in this case since each point sample covariance matrix is singular (5:149) whereas the lumped interval covariance matrix is not (590:149). The partitioned MEG/EEG grand average signal $\vec{y}_i(t)$ and noise covariance $\mathbf{C}_i$ statistics were restricted and constrained to derive 8 special cases:

- M0: MEG only with off-diagonal noise covariances forced to zero;
- M1: MEG only utilizing all estimated between-channel noise covariances;
- E0: EEG only with off-diagonal noise covariances forced to zero;
- E1: EEG only utilizing all estimated between-channel noise covariances;
- ME0: MEG+EEG with off-diagonal covariances forced to zero;
- ME1: MEG+EEG utilizing all estimated between-channel covariances;
- ME01: MEG+EEG with zero inter-modal covariances, utilizing intra-modal covariances; and
- ME10: MEG+EEG utilizing inter-modal covariances, with zero intra-modal covariances.

For each special case and stimulus condition, the inverse symmetric square root $\mathbf{C}_i^{-0.5}$ of the noise spatial covariance matrix was obtained via singular value decomposition (SVD) of $\mathbf{C}_i$ by taking the reciprocal square root of the singular values (which are all positive since $\mathbf{C}_i$ is positive definite). There are other senses of taking the "square root" of a positive definite matrix, such as Cholesky decomposition, but only the symmetric square root is invariant with respect to channel number reorderings. The grand average ERF/ERP spatial vectors $\vec{y}_i(t)$ were then normalized by the SNR transform $\vec{z}_i(t) = \mathbf{C}_i^{-0.5}\vec{y}_i(t)$ which yielded unitless signal-to-noise ratio vectors. For each special case, spatial SVD was performed on 472 $\vec{z}$ vectors (4 stimulus conditions $\times$ 118 time points) in order to span the signal space for the experimental conditions and time window of interest. The resulting singular values w_j are signal-to-noise ratios for their respective spatial components. Components with SNRs less than 1.0 were discarded. Finally, total information was computed in units of "bits above noise" [4] for all R retained components as $H = \sum_{j=1}^{R} \log_2 w_j$.

Random channel permutations: How typical is the MEG/EEG partition? From the standpoint of computing information, is there anything special about the distinction between MEG and EEG channels? What information quantities would be obtained for arbitrary partitions of 120+29 channels made at random without regard for modality? Note that of the 4 special cases that include all 149 channels, the computations for cases ME0 and ME1 are unaffected by the distinction between MEG and EEG. However, cases ME01 and ME10 depend on an explicit modal partition of 120 MEG and 29 EEG channels: case ME01 forces 6960 inter-modal covariances to zero, and case ME10 forces 15,092 intra-modal off-diagonal covariances to zero. We guessed that MEG and EEG channels "specially complement" each other from an information perspective. Specifically, we hypothesized that (a) the information computed for ME01 (obtained by *deleting* the MEG/EEG noise interactions from case ME1) should be *less* than would be observed for a typical partition, and (b) the information computed for ME10 (obtained by *adding* the MEG/EEG noise interactions to case ME0) should be *greater* than would be observed for a typical partition. To test this hypothesized relation, 750 arbitrary partitions were generated by randomly permuting the channels but retaining (incorrectly) the original MEG/EEG channel designations. ME01 and ME10 information computations were performed for each random partition. Each 120+29 partition X (either the actual MEG/EEG partition or a random partition) was assigned a score equal to the number of partitions for which the hypothesized relation was true for X. The typicality of the MEG/EEG partition was estimated by the fraction of random partitions with scores greater than or equal to its score.

Results

Computation of empirical information in units of bits above noise for the 8 special cases yielded: 46.3 for M0, 132.7 for M1, 14.7 for E0, 37.0 for E1, 53.3 for ME0, 194.3 for ME1, 142.2 for ME01, and 69.5 for ME10. These results are graphed in Fig. 1.

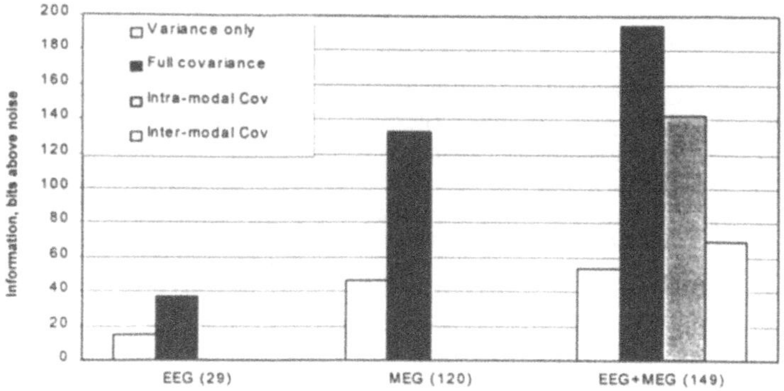

Fig. 1. Empirical event-related information computed for the 8 special cases.

From these numbers, the benefit of using the full noise covariance matrix in the SNR transform versus variances only can be expressed as information ratios. Thus, the full covariance versus variance only benefit ratios are: 2.52 for EEG only, 2.87 for MEG only, and 3.65 for MEG+EEG.

The *additivity ratio* α for a channel partition $A+B$ is defined as $\alpha \equiv H(A+B)/(H(A)+H(B))$, where $H(A+B)$ is the information computed for all channels, $H(A)$ is the information computed for the A channels only, and $H(B)$ is the information computed for the B channels only. Information is said to be *additive* if α=1.0 ("the whole equals the sum of the parts"), *subadditive* if α<1.0 ("the whole is less than the sum of the parts"), and *superadditive* if α>1.0 ("the whole is greater than the sum of the parts"). The results above give rise to the following additivity ratios: for the variance only cases, ME0/(M0+E0) = 0.87 (subadditive); for the full covariance cases, ME1/(M1+E1) = 1.14 (superadditive); for the mixed case where inter-modal noise covariances are deleted from the full covariance matrix, ME01/(M1+E1) = 0.84 (subadditive); and for the mixed case where inter-modal noise covariances are added to the variance only matrix, ME10/(M0+E0) = 1.14 (superadditive).

The estimated typicality of the MEG/EEG partition was 0.052 -- 39 out of 750 random partitions had scores that equaled or exceeded the MEG/EEG partition score (=527).

Further computations. A few exploratory computations were made to further investigate the conditions for superadditivity. The 120 MEG channels were 60 pairs of tangential gradiometers at each location. The odd-numbered channels covered all 60 locations with one tangential orientation, and the even-numbered channels had the other orientation. Full covariance information for a 60+60 odd-even partition of the 120 MEG channels was superadditive (α=1.15=132.7/(57.3+58.5)). A 60+60 (odd-even)-(even-odd) partition with equal representation of the two tangential orientations in both parts was also superadditive (α=1.12=132.7/(58.9+59.3)). Further subdivisions were made. A 30+30 partition of the 60 odd-numbered MEG channels was subadditive (α=0.95=57.4/(29.7+30.6)); likewise, a 30+30 partition of the 60 even-numbered channels was subadditive (α=0.93=58.5/(31.1+32.1)). Four 30MEG+29EEG partitions were all closer to being additive: α=1.00 (66.5/(29.7+37.0)), α=1.00 (67.3/(30.6+37.0)), α=0.98 (67.0/(31.1+37.0)), and α=0.97 (67.0/(32.1+37.0)).

Discussion

Without using specific modeling assumptions that are required to actually estimate sources, the foregoing analysis has been made to study the potential benefits of combining MEG and EEG for the purpose of source estimation. The method has been to compare the raw information empirically available at the sensors for 8 special cases derived from an actual simultaneous 120-channel MEG plus 29-channel EEG visual event-related recording.

An SNR transform of the bimodal event-related data converts the MEG and EEG data to a common frame of reference derived from the covariance matrix of the residual background noise. Although the noise matrix can be simplified by considering within-channel variances only, this assumes that the background MEG/EEG is statistically independent at the sensors. As a consequence, the total noise in the multichannel system is overestimated since the background activities picked up at two channels are actually correlated (to some extent) due to common neural origins. Overestimation of the total noise results in underestimation of SNRs which directly reduces the information estimates. For a single modality, incorporation of between-channel covariances resulted in information yields 2.5-2.9 times the variance only yields. The improvement in yield was even greater (over 3.6 fold) for the combined MEG+EEG data. Therefore, there is considerable *common* background noise picked up at the sensors.

Superadditive information was observed for the MEG/EEG partition, and there was a 27% information loss when inter-modal covariances were neglected. Superadditivity was also observed for 60+60 within-MEG partitions, but not for 30+30 partitions. In a 256-channel ERP study that used the same stimuli, we did not observe superadditivity for 128+128 EEG partitions [5]. Thus, superadditive event-related information has been shown to be *empirically possible* for MEG/EEG partitions and for within-MEG partitions. It of course remains to be seen how typical these results are.

As for the question of whether MEG and EEG channels complement each other in an atypical way, there is a hint (but no conclusive evidence) that they do. Only about 5% of random partitions were more complementary than the actual MEG/EEG partition; however, further discernment is needed to tease apart the effects of modality from location (e.g., occipital versus frontal sensors). In this regard, it is certainly possible that location could take precedence over modality. Another bit of evidence in favor of a special relationship between MEG and EEG is the observation that the 30+29 MEG+EEG partitions had higher additivity ratios than the 30+30 within-MEG partitions.

Finally, there are grounds for expecting that the empirical information gain obtained by considering between-channel noise covariances will translate to improved source estimation. For example, a quasi-realistic simulation study using 64-channel EEG data demonstrated improved accuracy (2×) and stability (20×) of estimated ERP source waveforms for the full covariance versus the variance only cases [6]. However, the actual benefits of combined MEG/EEG source estimation remain to be determined in the context of specific modeling assumptions. Moreover, the issue of different accuracies for magnetic versus electric head models needs to be considered.

References

[1] Greenblatt, R.E. Combined EEG/MEG source estimation methods. In: Baumgartner, C., Deecke, L., Stroink, G., and Williamson, S.J. *Biomagnetism: Fundamental Research and Clinical Applications*, Amsterdam, Elsevier/IOS-Press, 1995, pp. 402-405.

[2] Kemppainen, P.K., and Ilmoniemi, R.J. Channel capacity of multichannel magnetometers. In: Williamson, S.J., Hoke, M., Stroink, G., and Kotani, M. *Advances in Biomagnetism*, New York, Plenum Press, 1989, pp. 635-638.

[3] Shannon, C.E., and Weaver, W. *The Mathematical Theory of Communication*. Urbana, University of Illinois Press, 1949.

[4] Pflieger, M.E., and Sands, S.F. Information growth in multichannel EEG. *Human Brain Mapping*, 1995, Supp. 1: 99.

[5] Pflieger, M.E., and Sands, S.F. 256-channel ERP information growth. *NeuroImage* 3(3): S10, 1996.

[6] Pflieger, M.E., Simpson, G.V., and Vaughan, H.G., Jr. Improved estimation of ERP source activities in the presence of realistic background EEG. *Human Brain Mapping*, 1995, Supp. 1: 101.

Multimodal Functional Imaging of the Sensorimotor Cortex with fMRI and Magnetic Source Imaging (MSI)

Roberts, T.P.L. and Rowley, H.A.

University of California San Francisco, San Francisco, USA

Introduction

The clinical advent of echo planar imaging (EPI) combined with the appreciation of the blood oxygen level dependent (BOLD) contrast mechanism has given rise to rapid development in 'functional magnetic resonance imaging' (fMRI) of the cortical response to peripheral stimulation and cognitive task performance [1,2]. Studies tend to involve stimulation paradigms lasting from seconds to minutes, with high speed MR images being acquired at a rate of the order of 1/sec. The BOLD contrast mechanism relies on the fact that a regional cortical blood flow increase occurs in response to stimulation but that this is not accompanied by a concomitant increase in local tissue oxygen extraction [3]. This leads to a relative increase in the proportion of oxygenated to deoxygenated blood in the capillary bed. Since deoxyhemoglobin is paramagnetic (and thus accelerates transverse magnetization dephasing) such a change will lead to an attenuation of this dephasing effect and somewhat increased signal intensity on imaging sequences such as gradient-recalled-echo (GRE) and echo-planar imaging (EPI); small signal increases of up to 4% have been reported at clinical field strengths [1,2,4]. This hemodynamic response to stimulation is not instantaneous but exhibits a gradual approach towards a plateau level, with a time constant of the order of a few seconds. So hemodynamic rise-time essentially limits the meaningful temporal resolution achievable in such studies.

However, electrophysiological studies demonstrate that significant neural events occur within milliseconds of stimulus presentation. Particularly if the goal of the study is to track the propagation of signal from one brain area to others, significantly higher temporal resolution is required. Electroencephalography (EEG) studies with scalp-placed electrodes, while providing such temporal resolution, may not be adequate for the localization of the neuronal current source. Magnetoencephalography (MEG) shares the temporal resolution of EEG, limited only by analog-to-digital conversion rates (typically 1-4 kHz). However, since MEG detects the magnetic rather than the electric component of the extracranial field, it can form the basis of a more robust neuronal source localization method, provided an array of magnetic field detectors can cover sufficient spatial extent to allow adequate modeling of the current source. When MEG source models are combined with high-resolution MRI, magnetic source images (MSI) display functional information in anatomic context [5,6].

An important clinical role for functional brain imaging is presurgical mapping to allow definition of eloquent cortex in relation to mass lesions which may be treated by resection or alternative non-surgical approaches, such as focused irradiation (gamma knife) or chemotherapy/radiotherapy, according to the functional nature of nearby brain tissue [7].

This study evaluated two functional imaging techniques, fMRI and MSI, for mapping of the sensorimotor cortex. The methods were compared in normal volunteers for rate of achieving localization with comparable stimuli. Assessing the comparability and relative merits of each technique in the spatial and temporal domains leads to the tantalizing opportunity to combine information derived from each examination to obtain a more accurate or more complete description of the underlying brain functional activity.

Methods

Studies were performed with the approval of the UCSF human studies committee. Eight normal volunteers were studied using both MRI and MEG (3F, 5M, mean age 30 yrs, range 25-36 yrs). Painless tactile stimulation was performed in both MRI and MSI environments using a balloon diaphragm, clipped to the fingertip. The diaphragm was driven with bursts of compressed air (15-30 p.s.i.) lasting approximately 30ms and repeated at an inter-stimulus-interval of 0.5-1s. Stimulation was applied sequentially to the left and right index fingers. Additionally, fMRI was used to observe the cortical activity associated with the performance of a simple motor task, involving the flexion of the index finger.

fMRI was performed using a GE SIGNA 1.5T system (GE Medical Systems, Milwaukee, WI), equipped with gradient coils that can produce ±20 mT/m with a rise time of 230 mT/m/ms. For fMRI studies, periods of activation were interleaved with periods of rest. Each period was of 20s duration. Throughout the protocol, multislice echo planar images (128x128 matrix, 40cmx20cm field-of-view) were acquired with the interval between successive images of the same slice being 2.5s. Thus 8 multislice image sets were acquired during each 20 second period. In each multislice set, 5 axial slices of 7mm thickness and 3mm interslice interval were collected, covering the extent from the corpus callosum to the vertex. In addition to gradient-recalled echo EPI (TE=60ms), spin-echo (TE=100ms) EPI sequences were used in three subjects.

The fMRI time-series of images were analyzed as follows: the first four images were discarded to eliminate signal intensity variations arising from progressive saturation. Subsequently, on a pixel-by-pixel basis, signal intensity variations over the time-course of the image series were correlated with the stimulus function and both *r*, the correlation coefficient and the corresponding *t*-statistic were determined. A t-test was then performed to determine the correlation significance (threshold: $p<0.01$). Further a spatial constraint was imposed: to eliminate spurious activation-noise, it was required that at least 5 contiguous pixels (with no geometric constraints) be similarly correlated.

MEG was performed using a 37-channel biomagnetometer (MAGNES, BTi., San Diego, CA). Data were collected in 300ms epochs, centered on the stimulation trigger, with a sampling rate of 297.8 Hz per channel. 256-512 epochs, collected with identical stimuli and pseudo-random inter-stimulus intervals in the range 500ms-1s were averaged to improve signal-to-noise ratio. Latencies within the range 30ms to 70ms post stimulus were examined. Extracranial magnetic fields were modeled using the single equivalent dipole method to obtain the spatial coordinates of the neuronal current source. Co-registration of MEG localizations with high-resolution 3D SPGR MRI was performed (using skin landmarks) to depict the anatomic origin of the activity.

Results

In eight subjects with a total of sixteen stimulation sites, fMRI achieved true positive localization in eleven cases; of the stimulations which failed to elicit a detectable fMRI response, one case was a bilateral failure and the remaining three were unilateral. MSI achieved satisfactory localization in all cases (Fig. 1).

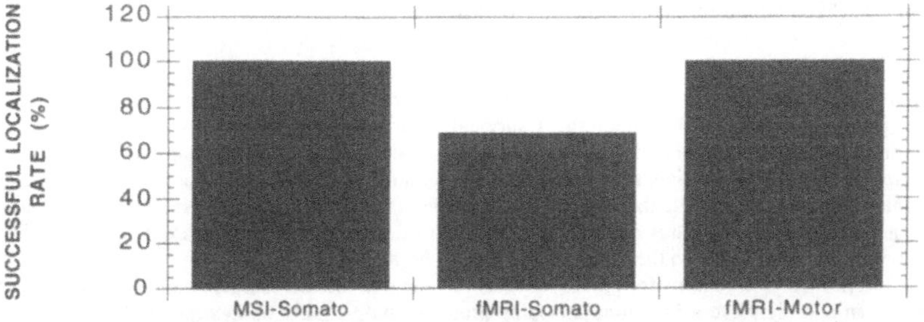

Figure 1 Localization Rate : fMRI vs. MSI with Somatosensory Stimulation

Ipsilateral and frontal/pre-motor activation was detected with fMRI in three cases; this was inherently avoided using MSI by the spatial position of the detector probe and its restricted field-of-view. In the majority of cases, fMRI and MSI localizations appear in accordant gyri and at similar axial levels (Fig.2).

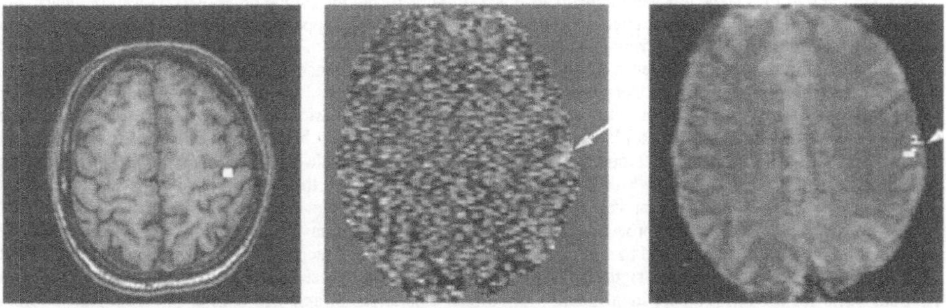

Figure 2 fMRI and MSI localization in good anatomic agreement.
Right index finger stimulation: (left) MSI; (middle) fMRI correlation coefficient map; (right) fMRI overlay

There was an observed tendency for the fMRI localization to lie more superficially than the MSI localization. A typical time course of signal intensity changes from pixels, identified on fMRI overlay maps is illustrated in figure 3. fMRI also has the potential for false negatives; some subjects demonstrate no clearly identifiable activation (5/16); however, MSI provides a reliable post-central localization with a similar stimulus. In the three subjects in whom spin-echo echo planar imaging was performed with a similar stimulus protocol, no significant activation was detected using the above statistical approach.

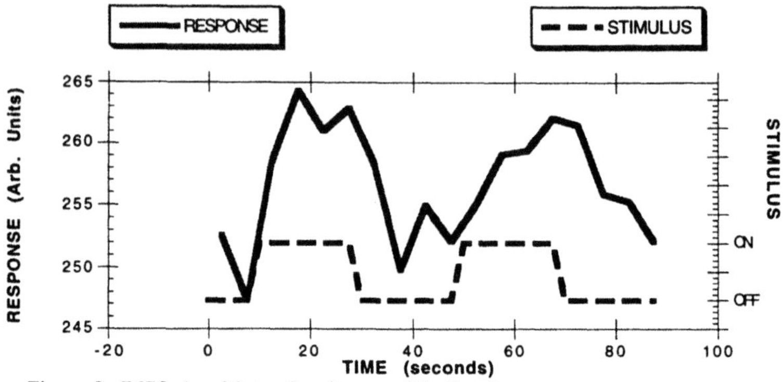

Figure 3 fMRI signal intensity change with stimulus

Discussion

Mapping the sensorimotor cortex provides the neurosurgeon with vital information regarding the functional organization of cortical tissue, in the planning of surgical approach route for the resection of mass lesions, epileptogenic tissues and other intracranial anomalies [4,7]. Currently intraoperative mapping by direct cortical stimulation is required; this is expensive, time-consuming and requires a craniotomy. A non-invasive, pre-operative mapping alternative could help indicate the inherent risk of surgical intervention and suggest alternative therapeutic strategies. Both fMRI and MSI offer this potential; the purpose of this study was to compare the two modalities for the reliability with which they depicted the sensorimotor cortical organization.

fMRI offers attractive benefits for clinical cortical mapping. Most neurosurgery candidates undergo a pre-operative MRI scan and therefore a functional mapping protocol could simply be incorporated into the standard, avoiding the need for a separate scan and associated errors of image co-registration (as well as penalties associated with time and cost). Furthermore the sensitivity of fMRI does not vary with the depth or geometric extent or orientation of the activated source and so it is amenable for use with a wide variety of stimuli to map a range of primary and associated functional activities. Some regions (e.g. brainstem and frontal pole) may be less successfully imaged with fMRI since the technique is inherently sensitive to magnetic susceptibility artifacts commonly present in these anatomic regions. To achieve clinical utility, it is necessary to have fMRI coverage of more than a single anatomic plane or slice. This is particularly important in cases where functional areas might be displaced from their expected anatomical site, either by mass lesions or adaptation. To this end, multislice or 3D approaches should be advocated. In order to achieve this and maintain adequate temporal resolution (of the order of 1-2s), EPI sequences are required. Conventional gradient-recalled-echo imaging, while providing appropriate contrast, cannot satisfy such multislice capabilities without incurring a loss of temporal resolution.

In this study, we employed a gradient-recalled-echo EPI sequence with a TE of 60ms to give strong magnetic susceptibility sensitivity. As discussed by Weisskoff and others, this sequence tends to preferentially highlight larger vessels, such as veins and venules rather than simply capillaries. In 3 of the subjects we additionally therefore employed a spin-echo echo-planar imaging sequence, which might be predicted to be rather insensitive to larger vessels and therefore reflect only changes in oxygenation present at the capillary level. However, no statistically significant correlation with the stimulus function was found in these subjects with this sequence. This was true even when regions of interest were targeted using the gradient-recalled-echo activation response.

A number of statistical approaches for the analysis of fMRI data have been proposed. All seek to identify pixels within the image that respond to stimulation or task performance. Since the observed response is only of the order of a few percent and image signal to noise itself may be poor, these methods must attempt to provide a rigorous basis for the unequivocal identification of cortical activation. Simple subtraction of "resting-state" from "activated-state" images (similar to the method used in PET studies of stimulation and task performance) generally suffers from inadequate signal intensity difference compared to random image noise. To counter this, several acquisitions in each state may be averaged. However, this necessarily reduces temporal resolution. We performed a correlation analysis to allow identification of pixels whose signal intensity variations correlated with the stimulus presentation. However, a simple temporal correlation is insufficient (with a threshold of $p<0.01$, an image of matrix 128x128 may contain many "random" or false positive correlations). We invoked a requirement for spatial connectivity, reducing the incidence of such correlation "noise".

Neuronal localizations based on MEG data were derived using the single equivalent dipole (SED) approach. This simplification of the neuronal environment, although not a faithful description of human brain activation, has been widely used to provide an estimate of a single current source that might give rise to the measured extracranial fields. Several studies have indicated that in the case of simple cortical processes, the model provides adequate accuracy of localization, compared to alternative standards such as invasive electrocorticography. The modeling process involves the fitting of dipole strength, localization and orientation, to minimize the difference between observed and predicted extracranial fields. The fitting process provides a measure of data-model correlation. Following previous investigators, we accepted "dipole localizations" that satisfied the observed data with a confidence volume of less than $1cm^3$ and with $r>0.98$. Changing such selection criteria influences the rate of localization "success".

Somatosensory stimulation with a painless pneumatic tapping to the fingertips proved inadequate for routine use with fMRI (69% statistically acceptable localizations), although in the 11/16 cases where localizations were achieved, good correspondence with MSI and anatomical expectation was found. Thus it seems that failed localizations result from poor signal-to-noise and statistical power rather than an inherent inability of the technique. Simple motor task performance seems more reliable, providing 16/16 successful localizations with fMRI. MSI was not performed with motor task performance, due to the complexity of the motor evoked neuromagnetic field. Also, although attempts were made to provide identical somatosensory stimuli for fMRI and MSI studies, the pressure and duration of the stimulus pulse, as well as the inter-stimulus-interval may have varied in the different environments (the MSI stimulus was computer generated and regulated; to operate in the fringe fields of the magnet during fMRI, the stimulus was manually-driven).

The general agreement between the two distinctly different modalities is encouraging, particularly when considering the rather different physiological aspects they probe (hemodynamic versus electrical). The lack of precise colocalization might be anticipated especially since to achieve detectable fMRI localizations we adopted the GRE-EPI sequence with its dominant venous contribution. Furthermore the simplistic MSI modeling adopted limits the confidence of MSI source localization. Improved MEG modeling strategies with more complex current source descriptions might become feasible if limited in range by some spatial prior knowledge, provided for example by fMRI. However, clearly strict spatial constraints imposed by fMRI should be avoided due to underlying physiological differences. Although not specifically required in the application of presurgical mapping of the sensorimotor strip, it appears that the temporal resolution of MSI might allow the elucidation of temporal sequences of multiple sites of activation identified by fMRI (and inherently time-averaged and thus unresolvable in time). Thus the combination of techniques, even without the requirement for strict spatial colocalization, might provide valuable information about neural signal propagation, unobtainable from either method alone: the combination might provide information which is "more than the sum of the parts".

References

1. Kwong KK. Functional MRI with echo planar imaging. Magnetic Resonance Quarterly 1995; 11(1):1-20.
2. Kwong KK, Belliveau JW, Chesler DA, et al. Dynamic magnetic resonance imaging of human brain activity during primary sensory stimulation. Proceedings of the National Academy of Sciences, USA 1992; 89:5675-5679.
3. Fox P, Raichle M. Focal physiological uncoupling of cerebral blood flow and oxidative metabolism during somatosensory stimulation in human subjects. Proceedings of the National Academy of Sciences, USA 1986; 83:1140-1144.
4. Hammeke TA, Yetkin FZ, Mueller WM, et al. Functional magnetic resonance imaging of somatosensory stimulation. Neurosurgery 1994; 35(4):677-681.
5. Baumgartner C, Doppelbauer A, Deecke L, et al. Neuromagnetic investigation of somatotopy of human hand somatosensory cortex. Exp Brain Res 1991; 87(3):641-8.
6. Gallen CC, Sobel DF, Schwartz B, Copeland B, Waltz T, Aung M. Magnetic source imaging. Present and future. Invest Radiol 1993; 28 Suppl 3:S153-7.
7. Roberts TPL, Rowley HA, Kucharczyk J. Applications of magnetic source imaging to presurgical brain mapping. In: Kucharczyk J, Roberts TPL, Moseley ME, Orrison W, ed. Neuroimaging Clinics of North America "Functional Neuroimaging" Saunders, 1995; 5(2): 251-266.

Acknowledgments

We are grateful to BTi for equipment support and to Susanne Honma, R.T. for technical help.

Quantitative Comparison of Motor Control Studied with MEG and fMRI

Russell, D.P.[1], Schwartz, B.J.[2,3] and Hirschkoff, E.C.[2,3]

[1]The Neurosciences Institute, La Jolla, CA, USA; [2]Biomagnetic Technologies, La Jolla, CA, USA; [3]The Scripps Research Institute, La Jolla, CA, USA

Introduction

We have compared the data resulting from magnetoencephalography (MEG) and functional MRI (fMRI) mapping of the same subjects under the same stimulus conditions in an evoked motor response paradigm. The goals of this study were threefold: 1) to compare the activation foci revealed by each modality as a cross-validation check; 2) to compare the quantitative aspects of the signal yielded by each modality, in this case parametric changes in signal strength as a function of stimulus intensity; and 3) to lay a foundation for future work in which results from both modalities will be combined in such a way as to enhance the information content of both.

Methods

The motor task studied was movement of the dominant-hand index finger. The subject's hand was placed in a holder which helped maintain a constant position of the other fingers, and which provided a trigger bar against which the index finger was placed. During MEG acquisition movement initiation/conclusion was indicated by the leading/trailing edges of the trigger signal, and movement onset provided the reference point for around which MEG epochs were averaged. The same arrangement sans trigger signal was used in the MRI suite to provide a uniform mechanical environment across imaging modalities. The finger was flexed and relaxed against an elastic load. Loads of four different magnitudes (three in MEG) were employed, the lowest (force level 1) being approximately 1.4 Newtons at 1.5 cm displacement, and ranging to level 4, each unit adding another 1.4 N at 1.5 cm displacement. Level 1 generally felt quite modest to subjects, while level 4 (5.6 N) was moderately strenuous. The amplitude of finger movement was mechanically limited. Subjects were instructed to concentrate on applying only as much force as was needed to reach the stop point, and to keep the movement time as constant as possible.

Seven subjects (two female, five male; six right-handed, one left) were each submitted to both MEG and fMRI within a two-month period. Additional subjects were studied using only one modality or the other.

MEG was performed on a Biomagnetic Technologies Magnes system with 37 SQUID detector channels. Finger movement was initiated by a visual cue (LED) with a single flexion/relaxation repeated at an average interval of two seconds which was randomly varied within a half-second window. Each force-level trial consisted of 256 repetitions, which were averaged, bandpass filtered at 1-20 Hz, and submitted to equivalent current dipole (ECD) source analysis [1] and other quantitative and statistical measures as described in the results. Whenever possible, all three force levels (1, 2, and 4) were studied, and selected levels repeated for increased within-subject statistics. A separate trial in which the index finger was pneumatically stimulated was also performed in order to unambiguously localize somatosensory response.

Functional MRI was performed on a General Electric Signa 1.5-tesla scanner retrofitted for echo-planar imaging (EPI) with the Advanced NMR Instascan system. A set of eight spatially consecutive gradient-echo EPI axial images through the relevant region enclosing areas M1/S1 was collected every two seconds (the TR, or repetition time) for 4 minutes, alternating between the stimulated and resting conditions every 32 seconds. A blood-oxygenation-level-dependent (BOLD) contrast was achieved by using an echo time (TE) of 50 msec. Each of the four force levels was applied for a 4-minute scan session. Finger movement was gradual and at a rate of one flexion cycle per second as timed by the EPI scan sounds. These images were submitted to a time-course analysis to detect pixels with statistically significant activation and their activation amplitude. MR images for anatomical localization of MEG-derived sources were also collected in the same session so that the coordinate transformation placing the MEG results in the MRI reference frame would apply to the fMRI maps as well.

Results

Good accordance was found between ECD locations as determined by MEG and the activation regions found in fMRI, with both motor (M1) and sensory (S1) components identified in each. An example from one subject is shown in Fig. 1. In Fig. 1(a) three separate dipoles are shown localized on MRI anatomical views of the same subject, with the orientation indicated by the line segment. The black open square indicates the dipole at a typical latency of 30 msec prior to movement onset, consistent with a location within the precentral gyrus. The white open square is at a latency of 48 msec post-movement-onset. Note the more posterior location and opposite orientation, consistent with the postcentral gyrus, as is typically the case. The white filled square is taken from the somatosensory-stimulation run, again indicating a postcentral-gyrus location quite similar to the so-called somatosensory component of the motor task. All dipole fits showed high correlation ($r > 0.95$) with the point-dipole model.

Fig. 1(b) shows the two motor-task dipoles, now localized on the fMRI activation maps from the same subject and force level. The significantly activated pixels (dark tones) are shown on a sample lower-resolution echo-planar image from the time series (brighter gray tones). Significance is defined on an individual-pixel basis as Student's $t > 4$. Note the close correspondence between the MEG dipole locations and the fMRI activity. Activity is seen primarily in the contralateral S1/M1 region of this right-handed subject.

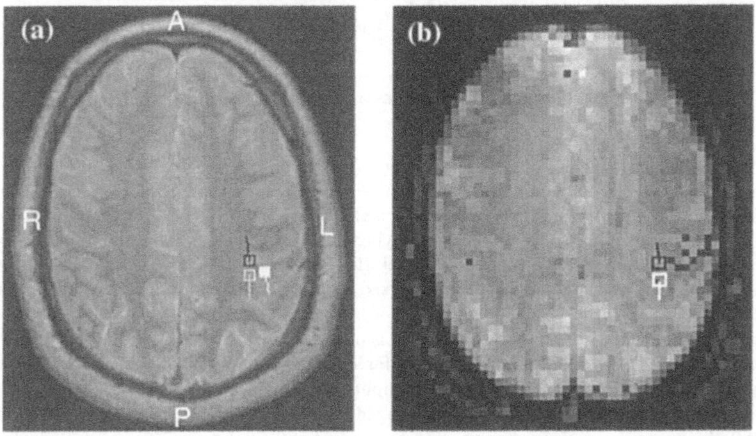

Figure 1: Sample MEG and fMRI localizations of motor activity in one subject

The amplitude of both MEG and the fMRI activation amplitude varied consistently with applied force level and with movement rate. The data for all nine subjects studied with MEG are shown in Fig. 2. Shown is the magnitude $|Q|$ of the dipole identified as the somatosensory component, found in a latency range of 20-100 msec post-trigger in all subjects, the particular latency identified as that characterized by the highest correlation coefficient to the dipole model. This typically coincided closely with the peak RMS power across all detectors of the averaged waveform. The data have been normalized by subtracting from each subject's data the mean across all force levels for that subject in order to remove intersubject variance and emphasize force-dependent behavior.

The mean across subjects of the (unnormalized) MEG data at each force level is plotted in Fig. 3(a). The corresponding fMRI data is shown in Fig. 3(b). Error bars are the standard error on the mean, which now includes the variance due to intersubject differences in average activation level. The fMRI index of activity is taken here as the sum of activation amplitude from all pixels within a region of interest (ROI) surrounding the S1/M1 region and which includes all pixels significantly activated at *any* force level. Note the striking similarity in the quantitative behavior of these two activity indices, taken from two imaging modalities derived from distinct physiological substrates.

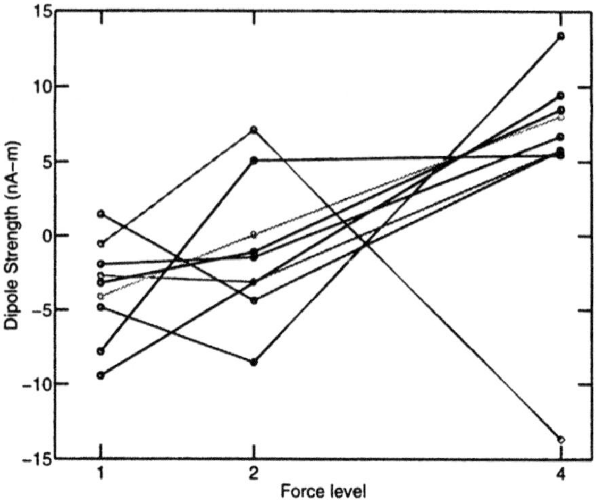

Figure 2: MEG dipole magnitude as a function of force load in all subjects

This similarity is further emphasized in Fig. 4(a). Here the average MEG dipole magnitude at each force is plotted against the corresponding average fMRI amplitude for the three forces common to both. The plotted line is simply drawn from the origin to the highest data point (force level 4) to make the point that the data are consistent with a zero intercept, implying that the two indices are functionally analogous as a measure of differential activation specific to the task (force control). If, however the data from each subject are averaged over all force loads and similarly analyzed for correlations on a subject-to-subject basis, no such result emerges.

Another index of MEG activation strength is proposed as a more likely candidate of subject-specific activation. In Fig. 4(b) the data are pooled across force loads within each subject, and each subject's MEG activation amplitude is plotted against the fMRI amplitude. Here, however, the MEG activation is measured as the RMS power (across all detectors) of the averaged (across epochs) MEG waveforms, integrated over the interval of 0-300 msec post-movement onset. The correlation of these data is 0.52, not statistically significant. Nevertheless, the correspondence, which is quite good with the exception of two subjects, is suggestive and worth further investigation.

Discussion

The MEG and fMRI signals obviously originate from rather distinct substrates of neural physiology. Hence it is important to characterize the degree to which these and other modalities offer common as well as complementary information about the underlying neural processes. Beyond simply cross-validating the results of each modality and utilizing the obvious spatiotemporal synergy that exists between them, it is hoped that combining data from such distinct techniques will provide information about an expanded subset of the rich ensemble of neural activity subserving such mental processes as motor control.

The most obvious validation consists of comparing the spatial information obtained from each, as is done here and in other studies. In addition, it is interesting to compare the quantitative (i.e., activation amplitude) aspects of the data for several reasons, one of which is the usefulness such knowledge will be in planning and carrying out any further studies in which the quantitative aspects of the signal are relevant. Such might be the case in attempts to observe more subtle mental state changes than the simple "on/off" pattern employed in blood-flow-mapping techniques such as fMRI, or in observing transient brain states that cannot be robustly averaged over many trials as is the case in evoked electromagnetic-field methods such as MEG.

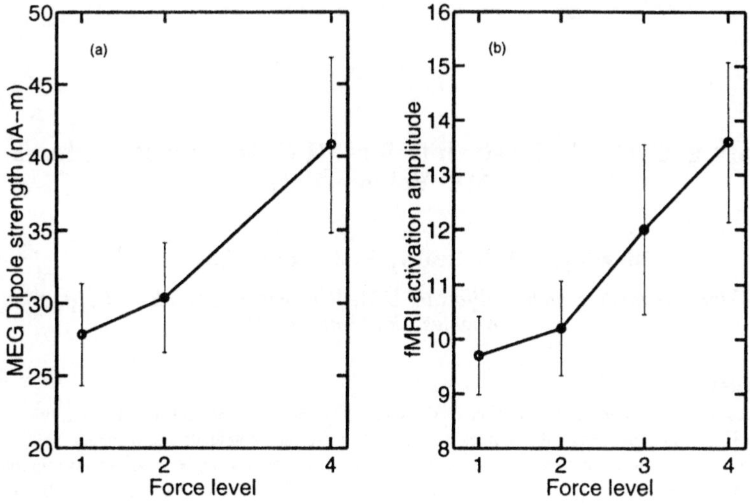

Figure 3: Average MEG and fMRI activation amplitudes as a function of force load, averaged over all subjects

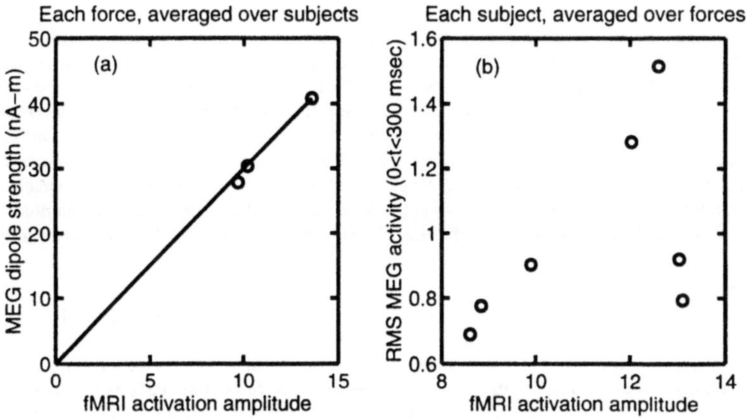

Figure 4: Quantitative correlations between MEG and fMRI activation amplitudes

References

[1] Sarvas, J. Phys. Med. Biol., 1987, 32: 11–22.

Generating Cortical Constraints for MEG Inverse Procedures Using MR Volume Data

Sandor, S.R.[1], Leahy, R.M.[2] and Timsari, B.[2]

[1]*TRW Inc., Redondo Beach, California, USA;* [2]*University of Southern California, Los Angeles, California, USA*

Introduction

Accurate knowledge of the location of the cortical surface is important in solving forward and inverse problems in EEG and MEG. Since the sources of event related surface potentials and magnetic fields are generally assumed to be confined to the cortex, knowledge of the shape of the cortex can be used to constrain the location of sources when solving the inverse problem [3]. Accurate knowledge of the cortical surface can also be used in solving the forward problem. By assuming that the head consists of a set of homogeneous regions of constant conductivity (e.g. brain, skull, scalp) the forward problem relating the measured EEG or MEG signal and the primary current sources can be solved using a boundary element method (BEM). The brain region for the BEM can be found using the method described here.

Our primary objective is to develop a technique that is automatic and accurate in extracting the cortical surface and which can track the major sulci. The MR images we use are T_1-weighted where cerebrospinal fluid (CSF) appears dark compared with white and gray matter. In these images the cortex is a ribbon with an outer boundary defined by the surface of the brain and an inner boundary by gray-white matter transitions. In general the outer cortical surface is clearly defined by a gray-CSF border, but due to noise, finite resolution and partial volume effects, surface extraction based on thresholding or simple edge detection typically results in links between the brain and other structures. Also, cortical structure may not be apparent from any one 2-D slice through the brain. Therefore, our approach processes these images in 3-D and separates the cortex from surrounding structures. We use the following pre-processing steps: 1) anisotropic filtering, an optional step which is used to sharpen image boundaries and 2) 3-D Marr-Hildreth filtering which locates surfaces within the head volume. The filtered image is then processed using a sequence of 3-D morphological operations which extract a binary brain volume from the skull and connective tissues.

Edge Detection

In our work, anatomical boundaries in MR brain images are located by edge detection. In order to select an edge detector capable of extracting accurate detailed information of brain anatomy from the gray-scale MR images we performed a subjective comparison between detectors which have been reported in the literature as giving the best results when applied to medical images - the Canny[2], Deriche[6], and Marr-Hildreth[5] methods. An ideal operator for our purpose is one which has a low computational cost and accurately detects closed boundaries of important anatomical structures. Our morphological algorithm requires as input a binary image of the closed boundaries of tissues inside the head. The algorithm can deal with some degree of noise in the form of disconnected edge elements and scattered false contours. However, if too many artifacts of this type occur, then preprocessing as described in the following section is required.

The Canny and Deriche edge finding methods can both produce disconnected edge segments and require costly edge linking procedures to produce a set of closed boundaries. In contrast, the Marr-Hildreth operator always produces a set of closed contours. While the Canny and Deriche methods gave better localization of edge segments in some regions of the image, these segments are often too dense to produce accurate closed boundaries using a simple edge linking procedure, consequently we decided to use the Marr-Hildreth operator for edge finding. While it is easy to show that this operator can produce spurious boundaries, we have found that these are removed when used in conjunction with the morphological processing step. Our experimental experience indicates that the spatial resolution of a Marr-Hildreth detector can be chosen so that it is effective in finding the outlines of different tissues. We have found that at a resolution appropriate for extracting the brain surface, this processor detects sulci boundaries corresponding to white-gray matter transitions rather than gray-CSF transitions [1]. Thus the final cortical surface that we extract has a tendency to move between the outer cortical boundary on the gyri and the inner cortical surface in the sulcal folds.

Anisotropic Filtering

The accuracy of anatomical boundaries obtained by edge detection is a direct consequence of the quality of the input image. An edge enhancement preprocessor appears to be essential when dealing with low contrast MR images with a low signal to noise ratio. Conventional linear spatial filtering methods for image enhancement reduce noise at the cost of reduced resolution. This results in edge blurring and is not a good preprocessor for our purpose. Instead we have found nonlinear anisotropic diffusion filtering to be especially appropriate for enhancing MR image data because of its edge preserving properties. In this method a filtered image is viewed as the solution to the anisotropic diffusion equation [7]. In this equation, the diffusion coefficient adaptively controls the diffusion strength to smooth the image within a region and stop smoothing across a boundary. This coefficient is a monotonically decreasing function of some measure of edge strength, usually the magnitude of the gradient of the image intensity. By selectively smoothing the image while preserving edges, we generate a result which gives improved edge definition with fewer spurious edges when processed with a conventional edge detector.

To demonstrate the efficiency of anisotropic filtering in reducing noise and enhancing region boundaries in MR images, in figure (1-a) and (1-c) we show a transaxial slice of a volume MR image before and after applying 3D anisotropic filtering with an exponential diffusion coefficient. Homogenous structures in the filtered image appear smoother and their boundaries are more clearly defined. The improved differentiation between adjacent regions of similar intensity characteristics facilitates the performance of the edge detector [figure (1-b,d)].

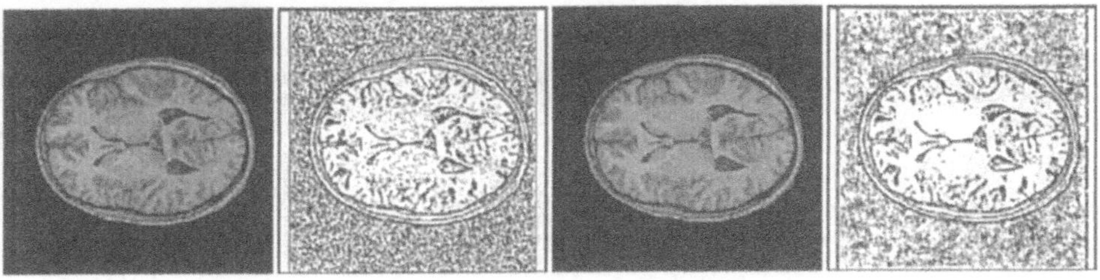

(a) (b) (c) (d)

Fig. 1. (a) Transaxial slice of an MR volume (b) Edges detected for that slice by the 3-D Marr-Hildreth operator (c) The same slice after applying anisotropic filtering (d) Edges detected for the filtered slice by the 3-D Marr-Hildreth operator

A Morphological Algorithm for Cortex Extraction

Our morphological algorithm requires as input a binary image volume, which we generate by applying the Marr-Hildreth edge detector to the MR brain data. In the resulting edge image voxels are either black or white depending on whether or not they represent region boundaries. We represent a binary head image by a 3-D set, denoted *X*, which is made up of two types of voxels: $X = X_{Int} \cup X_{Bound}$ where X_{Int} represents the object interior voxels which are connected only to other object interior voxels, and X_{Bound} represents object voxels that are connected to the background. In our work, "connected" is defined as 6-connectedness, i.e. voxels are connected only if they share a face, but not an edge or a corner. Set *X* represents other anatomical regions, not just the brain. The brain is a connected 3-D subset of *X*, where any two points in this subset can be joined by a 3-D path entirely contained within the brain volume. Because of noise, partial volume effects, or true anatomical connections, *X* may contain voxels that link brain regions to extraneous surrounding structures, such as dura matter or skin.

To sever unwanted connections and extract only the brain, we first perform an erosion, which shrinks the brain volume and eliminates all regions smaller than the structuring element. We have chosen a 3-D rhombus structuring element of discrete size one, which is a 3-D cross three voxels wide in the x, y, and z directions. This element deletes narrow connections without globally damaging or distorting an image. In particular, erosion by a rhombus will eliminate all boundary voxels X_{Bound} from X and leave X_{Int} unchanged. For definitions of morphological operations and their sizes see [4]. For an isotropically sampled MR volume with $1mm^3$ voxels, transforming by R1 eliminates regions which have a size of 2 mm or less in any direction and shrink a majority of the brain surface by 1 mm. After erosion we carry out a 3-D flood filling operation. This step is necessary because although erosion eliminates undesirable connections, non-brain regions, such as eyeballs or skull, still remain. We want to select only brain voxels. A 3-D flood filling routine finds all voxels connected to a seed point in the brain (figure (2-a)). We will denote this eroded and flood filled set of brain voxels as X_{EBrain}, where $X_{EBrain} \subseteq X_{Int}$. Since erosion shrinks the brain surface and widens holes in the volume, X_{EBrain} is not a true representation of an entire brain. To expand the brain from the eroded image we first dilate it by R1 to restore some voxels eliminated by erosion and denote the result X_{DBrain} (figure (2-b)). Then we fill holes in X_{DBrain} by closing the set with an octagon structuring element of size two, denoted O2. This isotropic element, which is a digital approximation to a sphere in Euclidean space, has a width of 9 voxels. We choose size two because an octagon of size one is only 5 voxels wide and may not be large enough to close all holes that were widened by erosion. Besides filling image holes, a closing operation replaces portions of an object's boundary with the structuring element's boundary. Therefore, this operation will result in a brain volume in which boundaries are smooth and all holes less than 9 voxels wide are closed (figure (2-c)). We will denote the result of this operation X_{CBrain}. Set X_{CBrain} is a smooth image that serves as a template on which we can reintroduce fine details from the original binary volume, X. When we take the set difference $X_{CBrain} \backslash X$ we obtain a set which represents all gaps and holes in the brain which include those corresponding to the internal cavities of the brain as well those resulting from the foldings of the brain cortex. We denote this set X_{Holes}. A difference operation between the closed brain and its set of holes yields the brain volume X_{VBrain} (figure (2-d)). The current X_{VBrain} contains some holes, not connected to the outer brain surface, which clearly do not contribute to the cortex. To remove these hole, we apply the 3-D flood filling process to $X^c{}_{VBrain}$, which corresponds to the background of X_{VBrain}. The output, X_{Back} can be considered as a mold for brain, and its complement gives us the solid brain volume, X_{Brain}. The outer contour of X_{Brain} is the cortical surface. Figure (2-e) shows a slice from a brain extracted by this morphological processing.

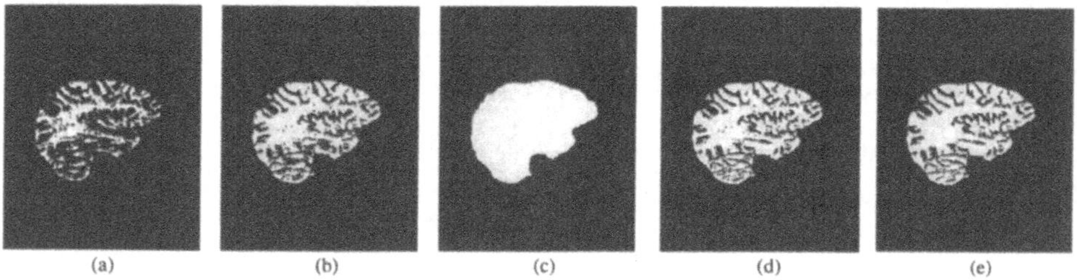

Fig. 2. Example: (a) a slice through the volume after erosion by a 3-D rhombus structuring element and applying flood-filling (X_{EBrain}), (b) the slice after the set of eroded brain voxels has been dilated by a 3-D rhombus structuring element (X_{DBrain}), (c) the slice after 3-D closing of the dilated image by an octagon of size 2 (X_{CBrain}), (d) the result of taking a difference operation between the set of brain holes and the closed image (X_{VBrain}), (e) the final result of the morphologic extraction algorithm(X_{Brain}).

Below, we summarize the steps of our algorithm for extracting a brain from an MR head image. The symbols $\ominus, \oplus, \bullet$ represent morphological erosion, dilation, and closing respectively.

1. $X_{Int} = X \ominus R1$
2. Flood-fill X_{Int} to arrive at X_{EBrain}
3. $X_{DBrain} = X_{EBrain} \oplus R1$
4. $X_{CBrain} = X_{DBrain} \bullet O2$
5. $X_{Holes} = X_{CBrain} \backslash X$
6. $X_{VBrain} = X_{CBrain} \backslash X_{Holes}$
7. Flood-fill $X^c{}_{VBrain}$ to arrive at X_{Back}
8. $X_{Brain} = X^c{}_{Back}$
9. Find the outer boundary of X_{Brain} to get X_{Cortex}

Experimental Results

We have implemented the brain extraction method described above, including edge detection and anisotropic filtering. The entire procedure is written in C and takes approximately 10 minutes on a 60MHz SPARC20 workstation to extract the brain from a typical stack of 120 images. An additional 10 minutes are required for 3 iterations of the anisotropic pre-filtering. Our implementation gives the user the option to: 1) select whether or not anisotropic filtering is desired before edge detection and 2) chose the size of the Gaussian operator used in Marr-Hildreth edge detection, where a larger operator smooths the image and detects fewer fine details. With this software we have thus far processed 30 different MR data sets, and in each instance, a visual inspection of the extracted brain surface indicated that our algorithm was successful in automatically extracting the cortex from these head images. These results show that the major sulci are found using this method. We have noted in a few cases that the cerebellum is removed from the brain volume.

Shown in figure (3) is an example of a cortical surface that was automatically extracted from a T_1-weighted volume acquisition (0.5T Picker Outlook) with a 2.5mm slice thickness. In figure (3-b) a typical slice through the brain volume is shown with its corresponding cortical boundaries superimposed on the slice. A rendering of the cortical surface is displayed in figure (4-a). Figure (4-b) shows the smoothed brain surface obtained by applying a morphological closing operation to the high resolution cortical surface.

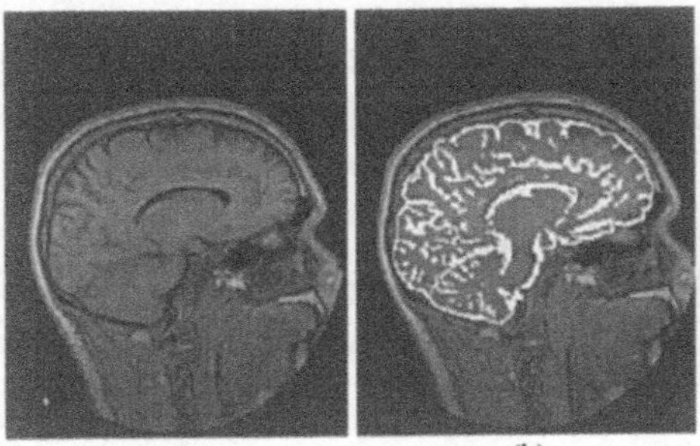

(a) (b)

Fig. 3. (a) a slice through the original binary MR volume, (b) outer boundaries of the same slice superimposed on it

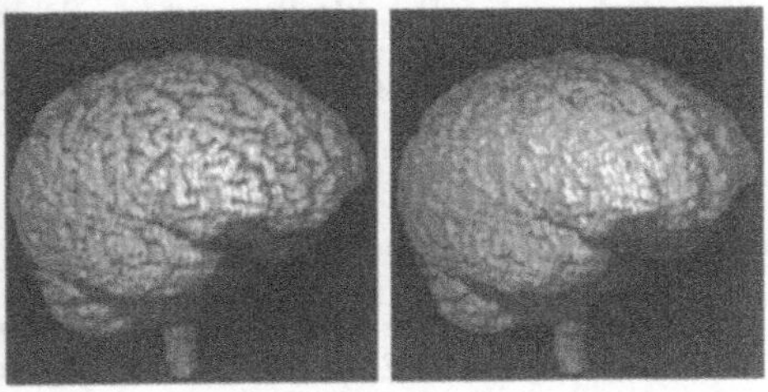

(a) (b)

Fig. 4. (a) a rendering of the cortical surface extracted with morphological processing, (b) a rendering of smoothed cortex

Conclusion

We have described an image processing technique which will automatically find the cortical surface from T_1-weighted volume magnetic resonance (MR) data. This method uses a combination of non-linear smoothing, edge finding and binary morphology procedures to find the surface. It requires no user interaction other than specification of a few input parameters. We have tested the method on approximately 30 data sets with a range of slice thicknesses (~1mm-2.5mm) and scan parameters. The method successfully extracts the cortical surface and tracks the major sulci.

REFERENCES

[1] M. Bomans, K-H Hohne, U. Tiede and M. Riemer, "3-D segmentation of MR images of the head for 3-D display," IEEE Trans. Medical Imaging, vol. MI-9, pp 177-183, 1990.

[2] J. Canny, "A Computational Approach to Edge Detection," IEEE Transactions on Pattern Analysis and Machine Intelligence, vol. PAMI-8, pp. 679-698, November, 1986.

[3] A. M. Dale, M. I. Sereno, "Improved Localization of Cortical Activity by Combining EEG and MEG with MRI Cortical Surface Reconstruction: A Linear Approach", Journal of Cognitive Neuroscience, vol. 5, pp. 162-176, 1993.

[4] P. Maragos, "Pattern Spectrum and Multiscale Shape Representation," IEEE Transactions on Pattern Analysis and Machine Intelligence, vol. 11, pp. 701-716, July, 1989.

[5] D. Marr, E. Hildreth, "Theory of Edge Detection,"Proceedings of Royal Society of London, vol. 207, pp. 187--217, 1980.

[6] O. Monga, R. Deriche, J. M. Rocchisani, "3-D Edge Detection Using Recursive Filtering: Application to Scanner Images," CVGIP: Image Understanding, vol. 53, no. 1, pp. 76-87, Jan 1991.

[7] P. Perona and J. Malik, "Scale-Space and Edge Detection Using Anisotropic Diffusion," IEEE Transactions on Pattern Analysis and Machine Intelligence, vol. 12, pp. 629-639, July, 1990.

Acknowledgements

This work supported by NIMH, Grant No. R01-MH53213, and by the TRW Inc. Doctoral Fellowship Program.

Correlation Between Depth and MEG/EEG Surface Recordings in Auditory Cortex. Contribution of Numerical Simulations

Schwartz, D.[1], Badier, J.M.[2], Scarabin, J.M.[1] and Liégeois-Chauvel, C.[2]

[1]*Laboratoire Signaux et Images en Médecine, Rennes, France;* [2]*Clinique Neurologique, CHR Pontchaillou, Rennes, France*

Introduction

The main purpose of our study is to combine three different methods of brain activity recordings: depth recordings (stereotactic EEG: SEEG) and surface recordings (EEG/MEG) in order to improve localizations of the sources of the brain activity. We present here the first step of this study. We built a model of a region of interest (ROI) within the brain based on MRI information. Our goal was to obtain an accurate description of the surface of the ROI. According to this model, we defined realistic dipole layers in respect to the anatomy (orientation, shape, extension). By activating those dipole layers we produced SEEG, EEG and MEG theoretical signals. In order to validate this concept on a physiological point of view, we chose the auditory cortex as the ROI. We compared the simulations with in vivo signals (Auditory evoked potential (AEP)) for which Liégeois-Chauvel et al [1] studied the localization of the middle latency components using SEEG. We performed the validation by taking into account the auditory cortex anatomy and built the simulations for each patient studied. We then compared the theoretical results with the real SEEG data. Furthermore we used those simulations to test the reliability of several MEG sources modeling algorithms.

Materials

We studied patients with intractable temporal lobe epilepsy undergoing SEEG for localization of seizure foci [2]. Informed consents were obtained for all patients. Those patients underwent the following examinations:

□ A MRI examination after the withdrawal of SEEG electrodes. The original examination was reconstructed in 3D with a trilinear interpolation (isotropic volume, voxel size=0.9375mm^3).

□ A simultaneous MEG/EEG examination with the 37 channels BTi Magnes system and a standard 10-20 EEG electrical arrangement. The AEP and auditory evoked field (AEF) were recorded with pure tones: tone burst, Frequency=(500HZ, 1kHz, 2kHz), duration=40ms, rise and decay=5ms, 500 stimulations (1/sec), stimulus level=70 dB HL (hearing level). The averaged signals were filtered with a band-pass filter (1-120 Hz).

□ A SEEG examination (monopolar recording). Same stimulation as the MEG examination except duration=24ms and 100 stimulations. The averaged signals were filtered with a band-pass filter (1-350Hz).

Methods

ROI Segmentation

The auditory cortex and the surrounding areas were manually segmented on MRI by an anatomist. The segmentation was performed in a 3D environment allowing a constant checking of the accuracy in sagittal, coronal and axial directions. The description contained the Heschl gyrus, the superior-temporal gyrus, the planum temporale and the the Sylvius fissure.

Model description and simulations

The list of voxels obtained above was not suitable in our context. Therefore we chose a B-spline model to match the data. The local characteristics of the ROI such as normal to the actual surface were easier to

compute with this mathematical description.
We defined an area of activity (AoA) as a dipole layer constrained on the segmented cortical surface, i.e. defined by a given part of the cortical surface (location and spatial extent) and a uniform density of single dipoles located on the surface with a direction normal to this surface. We added to this geometrical description three temporal parameters: the initial time of activation, a window of activation and the shape of the activation (mono or bipolar). Each dipole contribution was computed (accordingly to the SEEG leads position, and to the MEG/EEG sensors positions) by the following methods:

□ EEG/SEEG: The potentials were computed assuming the surface lying in a three shells sphere. An integration of potential was performed along each lead in order to take their size into account.

□ MEG: The magnetic field was computed assuming a surface lying in an homogeneous sphere.

The MEG, EEG and SEEG signals generated by a AoA were the sum of the contributions of each single dipole. In the same way we combined AoAs in order to generate the complete MEG, EEG and SEEG simulated signals. **i)** We simulated standing sources (standing AoA) with three of four AoA overlapping in time (parameters of those AoA: 100mm^2, 600 to 900 dipoles, time of activation: n°1: 30ms, n°2: 50ms, n°3: 60ms, n°4: 80ms, window of activation: ±30ms) (fig. 1). **ii)** We simulated activity propagating along the gyrus (moving AoA) with a AoA (50mm^2, 150 dipoles) moved along a trajectory (from the medial part to the lateral part of Heschl gyrus, Fig. 2) with a given speed (0.1mm/ms). In the comparisons with real signals, the SEEG leads and the MEG/EEG sensors were localized in their actual positions for a given patient. In this case an accurate MEG, EEG, SEEG to MRI registration were used.

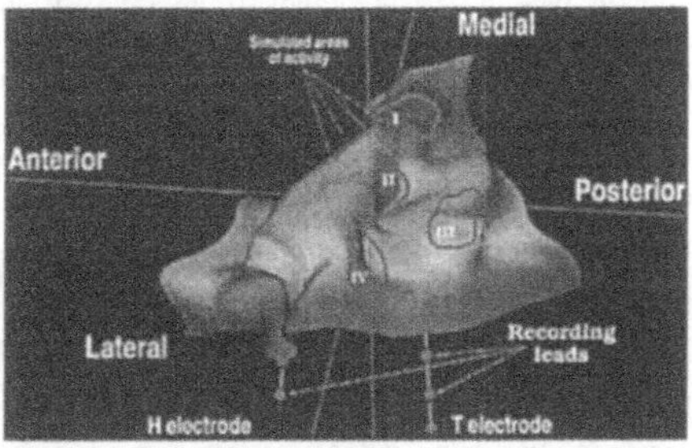

Fig. 1: 3D view of the model from the patient DR with two actual SEEG electrodes. The location on medial to lateral axe of AoAs n°1, n°2 and n°4 were chosen according to the Liégeois-Chauvel's results [1]. The AoA n°3 was chosen to test the effect of a source interfering with the others.

Sources modeling algorithms

We computed localizations on MEG simulated signals with the following algorithms: a monodipolar moving dipole fit (MDF) and a monodipolar spatio-temporal algorithm (SPT). Both algorithms searched for an equivalent dipole to match the MEG signals. The MDF performs a localization for a given latency. The SPT performs a localization by analyzing the signals on a time window around a given latency [3].

Results

Validation of simulations

We verified the capabilities of the model to reproduce SEEG like signals from simulations by analyzing simulated signals obtained on theoretical SEEG electrodes. Virtual electrodes were defined to explore the entire surface of the ROI. From moving AoA or standing AoA we calculated the shapes of simulated signals and we compared them with typical shapes of real signals recorded in AEP. We found typical shapes, curves presenting polarity inversions between two recordings leads of the same electrode and polarity inversions between the recording leads of two electrodes as well as time shift between peaks of activity. Those characteristics are used to localize the activities with depth recordings. Those inversions show the effects of an AoA lying between two leads of the same electrode or lying between two electrodes (Fig. 2). The polarity inversions between two leads of the same electrodes showed a gradient of large amplitude. On the contrary for a polarity inversion between two electrodes the gradient of amplitude was smaller (2 to 10 times inferior to the gradient between two leads of an electrode). Each inversion was studied in respect to the localization of the recording site (Fig. 2). The simulation showed that a lead is sensitive to only the small volume that surrounds it. The relative amplitude decreased very fast when the distance increases between the recording site and the sources. In real signals, it implied that with a small number of averages only the closest sources to a lead were recorded the others disappearing in physiological noise. All those properties can be used to establish the precise location of the AEP activities when recorded from the depth.

Fig. 2: Polarity inversions on simulated signals according to the location of the recordings sites (between two leads of an electrode (left) and between two electrodes (right)).

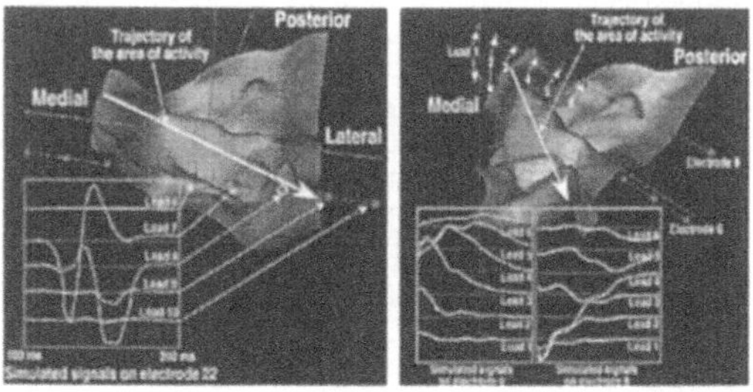

At this stage, we performed the simulations with the real position of two depth electrodes surrounding the Heschl gyrus of patient DR. The data recorded from the patient showed a inversion between the electrode T and the electrode H (Fig. 3). According to the latency of the inversion we defined three standing AoAs (the first in the medial part of Heschl (time of activation=55ms), the second 9 mm more laterally (time of activation 80ms) and the third on the lateral part of Heschl (time of activation=95ms). With those AoAs we retrieved the shape of the real signal especially the inversion (explained by the orientation of the second AoA on posterior to anterior direction). This simulation shows that we can reproduce useful part of the real signals. It implies that the assumptions for the model were correct in our context of studies on auditory cortex.

Fig. 3: Comparisons between simulation and real signals on SEEG electrodes H&T of patient DR. The locations of the AoAs used in the simulations are shown on left figure and right we present the resulting simulated signals and the real signals.

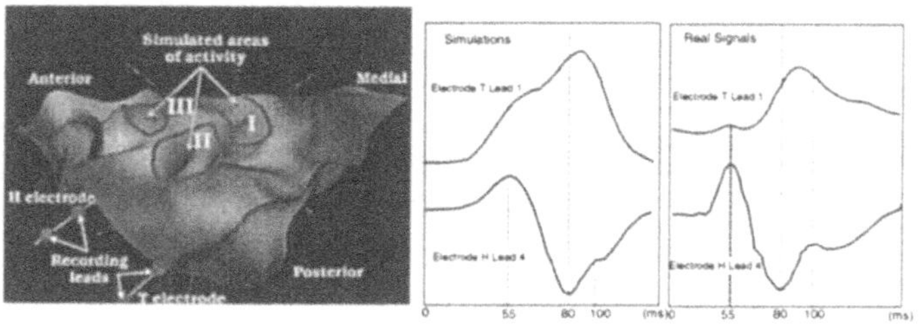

Evaluation of source modeling algorithms

Preliminary results of the evaluation of a moving dipole fit algorithm and a spatio-temporal algorithm in MEG context are presented here.

Tests with a moving AOA: We tested the effects of the shape of the ROI with a moving AoA. On both algorithms, we noticed an increase of the localization errors and a decrease of the stability of the localization when the shape of the structure supporting the dipole layer became more complex. When the AoAs had a part of their single dipoles almost radially oriented (on the medial to lateral direction) or when the contributions of the single dipoles of the AoAs cancelled each others the errors of localization increased.

Tests with standing AoAs: We tested the effects of AoAs overlapping in time. We combined three or four standing AoAs. With three AoAs (n°1, n°2 and n°4) (Fig. 1) and an optimal position of the sensor (i.e. magnetic field map showing a symmetrical inversion) the SPT algorithm showed a better stability in time and in space of the solution than the MDF. The separation between the activities was better and the time of stability for each AoA was greater i.e. each AoA was localized by the SPT algorithm during a larger amount of time and during this time the error of localization was stable (Fig. 4). Moreover the MDF algorithm produced equivalent dipoles localized between the AoAs. These localizations were consistent with the anatomy (i.e. they were located at 1 to 3 mm under or above the surface) but they were not relevant

of the actual AoAs. We tested the effect of a shift of the sensor in the Cz direction. It increased the errors especially for the localization of the AoA n°2 which overlapped in time with the AoA n°1 and n°3.
With four AoAs (Fig. 1) the MDF algorithm didn't find a reliable localization of the AoA n°2 and n°3 because of the overlap in time between those AoAs. On the contrary the SPT algorithm succeeded in localize the four AoAs during a large amount of time with a good accuracy (Fig. 4).

Latency errors: We noticed that the time overlapping produced an important error in term of latency of the localization of the AoA (for both algorithms and especially the MDF, Fig. 4). A AoA was rarely localized at its peak of activity but before or after this peak according to the overlapping of the others AoAs. In conclusion, overlapping sources may produce spatial and temporal localizations errors. This result is of main interest in the study of the processes involved in AEP and others evoked potentials.

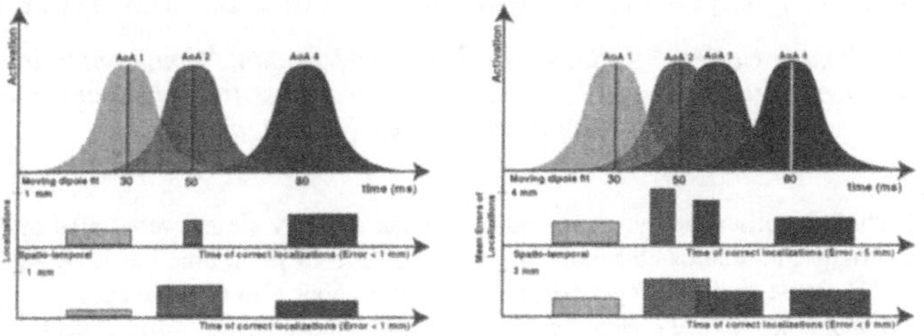

Fig. 4: Results of the localization for three standing AoAs (right) and four standing AoAs (left). The curves give the time of activation and the overlap of the AoAs. The graphics represent the results of the localization: The horizontal axis represents the time where the algorithms found a localization for the AoA with an error inferior to the following thresholds: 1 mm (3 AoAs) and 5 mm (4 AoAs). The vertical axis represents the mean errors localization for each AoA. The errors of localization calculated were the distance between the localization and the center of gravity of a given AoA.

Discussion and conclusion

Those simulations provide useful information for the analysis of AEP and AEF. The use of the anatomical model and the actual locations of depth electrodes for a given patient allowed the study of the different combinations of AoAs which reproduced the main characteristics of the real signal (especially inversions which may be explained accordingly to the ROI). In a following stage, it will be of main interest to integrate this information in localizations process from MEG or EEG signals. With the results of the evaluation of the source modeling algorithms we have the possibility to analyze our MEG localizations from real AEF in term of time and space stability to identify the sources of the main components of the AEF. In this evaluation we used only monodipolar algorithms. The SPT gave better results than the MDF in term of accuracy of localization on the spatial and temporal point of view. The SPT succeeded in localizing AoAs despite the overlap of the sources and the use of a single dipole model. The use of multi-dipolar algorithms may improve the localization results when the overlap in time between sources becomes more complicated. In conclusion, this method give the possibility to refine source spatial and temporal localization by the use of combined EEG/MEG as well as SEEG based on the individual anatomy.

References

1. Liégeois-Chauvel, C., Mussolino, A., Badier, J.M., Marquis, P. and Chauvel, P. Evoked potentials recorded from the auditory cortex in man: evaluation and topography of the middle latency components, Electroenceph. and clin. Neurophysiol., 1994, 7(4):618-625.
2. Bancaud, J., Talairach, J., Bonis, A., Schaub, C, Szikla, G., Morel, P. and Bordas-Ferrer, M. La Stéréo-encéphalographie dans l'épilepsie: Informations neurophysiopathologiques apportées par l'investigation fonctionnelle stéréotaxique. Masson, Paris, 1965.
3. Bouliou, A., Bihoué, P., Toulouse, P., Scarabin, J.M. Modelisation of the cerebral activity by a probability distribution upon the cortical surface from spatio-temporal MEG data. Human Brain Mapping, 1995, (Sup.1):105.

Brain Activities Evoked by Missing Auditory Stimuli: A Combined Study Using MEG and fMRI Data

Takino, R.[1], Miyauchi, S.[2], Sasaki, Y.[2], Pütz, B.[2] and Sekihara, K.[3]

[1]Shiraume Gakuen College, Kodaira-shi, Tokyo 187, Japan; [2]Communications Research Laboratory, Koganei-shi, Tokyo 184, Japan; [3]Central Research Laboratory, Hitachi Ltd., Kokubunji-shi, Tokyo 185, Japan

Introduction

In MEG studies, cortical activities around the human auditory cortex were found to be evoked by randomly occurring infrequent stimulus changes, e.g., in tone frequency or in tone intensity, among a monotonous sequence of tones [1]. Some studies using random stimulus omissions also found evoked 'endogenous' responses by the 'missing' stimulus itself [2, 3]. However, the location of the activities differed between the two types of studies. In this study, we employed the same 'missing' tone sequence in MEG and fMRI measurements and investigated the activities around the superior temporal auditory cortex.

Methods

Subjects and Stimulus

Seven (six males and one female, age 28–47y) and five (four males and one female, age 30–39y) healthy right-handed subjects participated in the fMRI and MEG experiments, respectively. They were instructed to just listen to the tones with their eyes open without exerting any mental activities, like counting tones. In both experiments, a simple repetition of 1kHz tones (ISI 300ms, duration 20ms) were presented dichotically with either a headset through a plastic tube (fMRI experiment) or a speaker at the outside of the magnetically shielded room (MEG experiment). Stimulus intensity was adjusted at a level that the subjects could clearly hear the tones, usually the audio-amplifier output level being set at 90dB. We employed three types of tone sequences: (1) *standard,* consisting of only 1kHz tones, (2) *missing,* consisting of 1kHz tones (85%) and randomly occurring stimulus omissions (15%) following more than three consecutive tones, and (3) *no tone,* in which no tone stimuli were presented.

In the MEG experiment, only the *missing* tone sequence was presented, and selective averaging for the missing tones and for the standard tones being presented just before the missing ones were performed. 150 epochs were averaged respectively over 350ms beginning 100ms before stimulus onset.

In the fMRI experiment, the subjects underwent two kinds of experiment. In order to define the location of the primary auditory cortex, the *standard* tone sequence was presented in the test condition and the *no tone* sequence was presented in the control condition. In the second experiment, we studied missing tone effects. The subjects listened to the *missing* tone sequence during the test condition while the *standard* tone sequence was presented during the control condition. The computed t-values that exceeded the statistical significance level ($p < 0.01$) were superimposed onto the corresponding anatomical images, and areas of activation were thus determined.

Data acquisition and analysis

MEG experiments

Magnetic fields were measured over both the left and right auditory cortices using a dual 37-channel SQUID system (second order gradiometer, BTi). The MEG signals were bandpass filtered 1–60Hz and digitized at a 500Hz sampling rate. Dipole fits with a correlation with the measured field of 0.95 or better were accepted. For one subject, we employed a Neuromag-122 whole-scalp magnetometer system and applied the spatio-temporal analysis of the Multiple-Signal Classification (MUSIC) algorithm [4] to the data with latencies between 0 and 250ms.

fMRI experiments

fMRI studies were carried out on a 1.5T Siemens scanner equipped with an EPI booster. We collected 6 to 8 EPI functional images (transverse, 5mm thick, FoV=200–250mm) capturing the sylvian fissure. The test and the control conditions were alternately repeated 20–30 times. EPI scans were triggered after 15 seconds of sequence presentation in either condition. Data were analyzed by pixel to pixel Student's paired t-test. For one subject, we employed one additional tone sequence in which, instead of omitting tones as in the *missing* tone sequence, deviant tones that only differ in frequency (2 kHz) from the other tones that were presented (*deviant* tone sequence). Each tone sequence was presented 24s and EPI images were collected every 8s. A set of four tone sequences was repeated 10 times. Data were analyzed via cross correlation calculation for each pixel [5].

Results

MEG experiments

All five subjects showed a distinctive peak following the standard stimuli with a peak latency of 60–83ms. The electric current dipoles (ECDs) were located in the superior temporal gyrus, pointing upwards, almost orthogonal to the sylvian fissure in both hemispheres. Four of the five subjects showed one stable peak after the presentation of missing stimuli with a peak latency of 105–148ms. The ECDs were found near the standard stimuli's ECDs locations, but pointing downwards in both hemispheres.

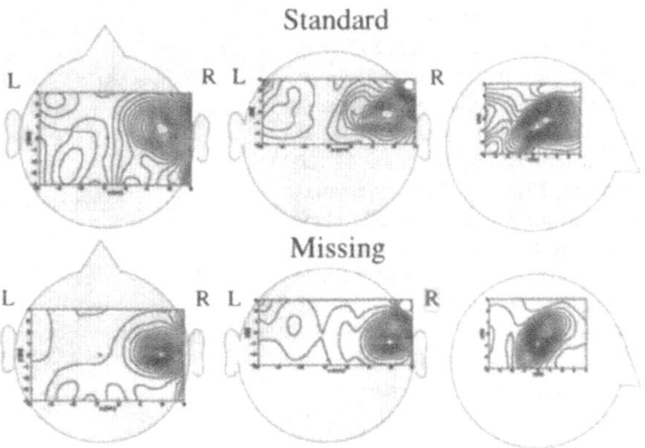

Fig. 1 MUSIC source localization results

Figure 1 shows the results of calculating the MUSIC-Mosher localizer in a three-dimensional volume. The results are projected onto the transverse, coronal, and sagittal planes. The contours in Fig. 1 show the relative value of the Mosher MUSIC localizer; and the area where the localizer forms a peak is considered to be the location of the dipole source. These results revealed that a single source dominates in the right hemisphere for both the standard stimuli and the missing stimuli. However, the source for the missing stimuli is located nearly 1cm posterior and 0.5cm lateral to the source for the standard one.

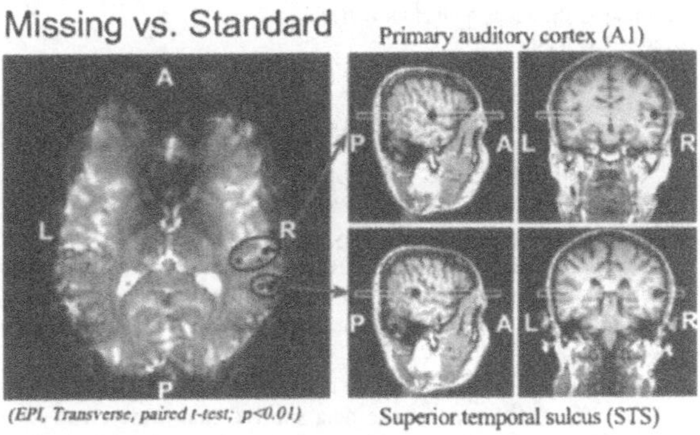

Fig. 2 Areas activated by missing auditory stimuli

fMRI experiments

During the first experiment defining the location of the primary auditory cortex, all subjects showed statistically significant activation in the superior temporal regions of both hemispheres, especially in the Heschl's gyri. In the second experiment, although the *missing* tone sequence has 15% fewer stimuli in number, it activated the area around the Heschl's gyri more than the *standard* tone sequence in all subjects. We found another activation in the auditory association cortex in the middle to posterior part of the superior temporal sulcus (STS; Fig. 2). The activation patterns were consistent when the same experiments were repeated using the same subjects. In five subjects, the activated area in the STS was found in the right hemisphere. The other two subjects showed the activation in the both hemispheres.

Cross correlation analysis shows that the small areas near the primary auditory cortex were also activated by the missing and the deviant tones compared to the standard tones, but did not clearly overlap with the activation areas determined by the comparison between the *standard* tone sequence and the *no tone* sequence. The most prominent activation was found in the right STS in the presentation of the *missing* tone sequence. The location of activation in the STS was almost the same as the one determined by the MEG MUSIC analysis (Fig. 3). The time course analysis for the ROI in the STS shows the activation by the missing tones. The deviant tones activated the same areas, but less effectively than the missing tones. The standard tones did not activate these areas. Rather, MR signals obtained during the presentation of the standard tones were the same level or less than the ones for the no tone sequence (Fig. 4). The cross correlation coefficient for the averaged data and the reference function that is shown as a shadow pattern in Fig. 3 is 0.893.

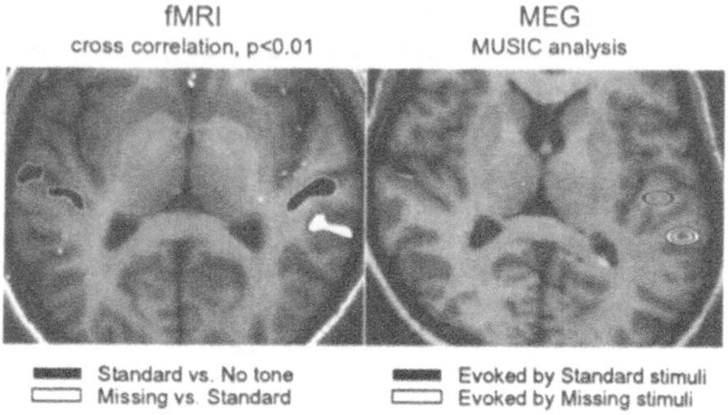

Fig.3 Comparison of fMRI and MEG results

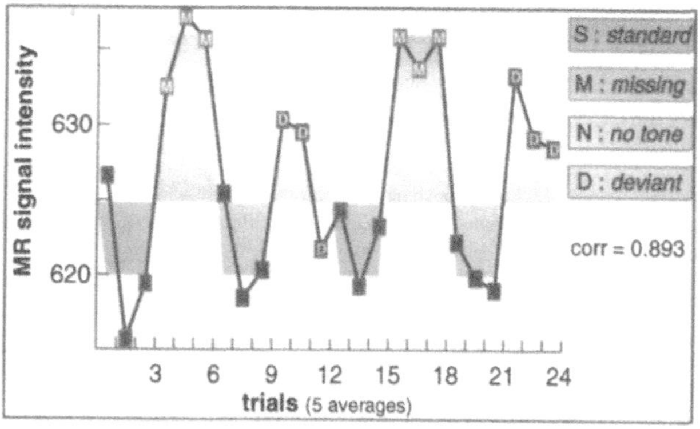

Conclusion

Both MEG and fMRI experiments results showed that the areas around the human auditory cortex were more activated by the missing (or deviant) stimuli than by monotonous tone stimuli. Our fMRI experiments showed that additional areas in the superior temporal sulcus are also involved in the detection of auditory stimulus changes. Although we had examined only one subject with the MEG MUSIC analysis, this method showed exactly the same areas activated during the presentation of the *missing* tone sequence.

The fMRI time course data showed that those areas were activated when there were some unexpected stimulus changes in the tone sequences (*missing* and *deviant*). The *standard* tone sequence had no effect on these areas. Activation levels were equal to or less than the ones by the *no tone* sequence. This would imply that these areas are not activated by a real stimulus, e.g., a tone presentation. They rather seem to be involved in the detection processes of global changes based on some endogenous neuronal activities.

References:

[1] Hari, R., Joutsiniemi, S.L., Hämäläinen, M., and Vilkman, V. (1989) Neuroscience. Letter, 99:164–168

[2] Joutsiniemi, S.L. and Hari, R. Omissions of auditory stimuli may activate frontal cortex, European J. of Neuroscience, 1989, 1:524–528

[3] Miyauchi, S., Takino, R., Sasaki, Y., Pütz, B., and Okamura, H. Missing auditory stimuli activate the primary and associate auditory cortices: a combined study MEG with fMRI, The 10th Tokyo Institute of Psychiatry International Symposium Abstract., 1995, pp.72–73

[4] Mosher, J.C., Lewis, P.S., Leahy, R. Multiple Dipole Modeling and Localization from Spatio-Temporal MEG Data, IEEE Trans. Biomed. Eng., 1992, 39:541–557

[5] Bandettini, P.A., Jesmanowicz, A., Wong, E.C., and Hyde, J.S. Processing strategies for time-course data sets in functional MRI of the human brain, Magn. Reson. Med., 1993, 30:161–173

Integration of Functional MRI and MEG of Median Nerve Stimulation

Verbiesen, M.J.R.[1], Peters, A.R.[2], den Boer, J.A.[2], Knösche, T.R.[1] and Peters, M.J.[1]

[1]Biomagnetic Centre Twente, University of Twente, Enschede, The Netherlands; [2]MR Research Medisch Spectrum Twente, Enschede, The Netherlands

Introduction

In this study we combined two non-invasive methods of brain imaging, functional magnetic resonance imaging (fMRI) and magnetoencephalography (MEG), in order to obtain a more detailed and integrated view of stimulus processing by the brain. MRI has conventionally been used to obtain an anatomical frame for interpreting EEG and MEG localisation results. In recent years, fMRI has been developed. It is based on the blood oxygenation level dependency (BOLD) effect of active brain regions, which is linked to neuronal activity. The effect of BOLD mostly reflects the microvasculature for venous blood [1]. It has been proven that there is a linear relation between the level of oxygenation and the measured T2 values [2]. fMRI has a good spatial resolution (1 mm) but poor temporal resolution (5 s). MEG reflects the currents within excitated pyramidal neurons and has a high temporal (around 1 ms) but a rather poor spatial (several cm) resolution.

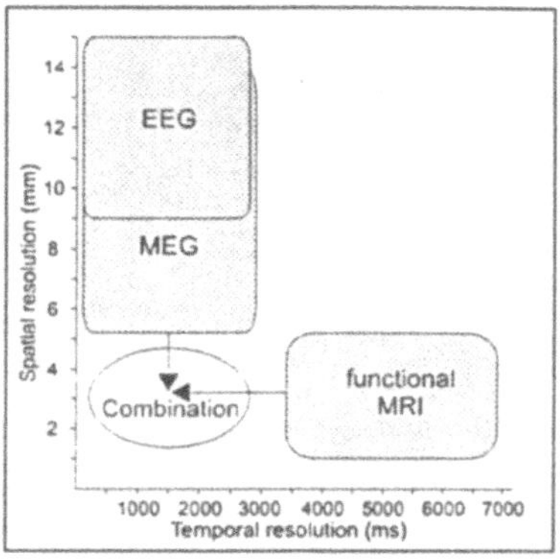

Fig. 1: Combining EEG, MEG, fMRI to improve both temporal and spatial resolution

In order to explore the combination of these techniques, we choose to use a well-known paradigm. Such a paradigm is the stimulation of one median nerve using an electric wrist stimulator. We used this stimulus for several reasons. First, it was feasible for both modalities. Second, much is known about the cortical responses to this stimulus from EEG, ECoG, and MEG studies.

Methods

In our study we performed repeated measurements of three right-handed healthy young volunteers (2M, 1F, 24-25 years). For fMRI we held several measurement sessions with a long time interval in between, in order to test the measurements' reproducibility. For MEG we did several measurements within one session.

We stimulated the median nerve (MN) of the right hand.. The stimulation sequences for fMRI and MEG had to differ, because of safety precaution during fMRI scanning. For this modality we used a special MR compatible stimulator (Digilog). PC-based software controls both the stimulation pattern and the timing of the MR scanner, to make sure that no stimulation is given when gradient fields are present in the MR scanner. The stimulation sequence consisted of 5 electric pulses with a duration of 500 μs and an interval of 50 ms during every scan cycle of 470 ms. The stimulation sequences for MEG consisted of a single electric pulse with a duration of 500 μs every 500 or 1000 ms. The intensity of the pulse was in all cases the maximum that could be tolerated by the volunteer.

For the fMRI scans we used a 1.5T ACS-II system (Philips M.S.) without any special hardware or software. The fMRI sequence is a Turbo Field Echo, this sequence is sensitive to the BOLD effect, but also sensitive to changes in blood flow, especially in the draining veins. With this sequence we covered a large volume of the brain (192x144x35 mm^3). The scan could be acquired in 30 s and consisted of 7 transversal slices (thickness=5 mm, scan matrix: 64^2, reconstruction matrix: 256^2). Consequently, we had a voxel size of 3x3x5 mm. With this scan, we obtained 30 dynamical frames in a regime of 3 without and 3 with stimulation (total duration 15 min). The dynamical scans during stimulation are averaged and their variance computed. So are the ones during rest periods. We identified the significant differences between the two resulting images with a two-sided t-test (pooled variance) for each pixel. A mask is made by threshholding the t-image at 95% significance. This way we defined the activated areas. We determined the actual effect of the stimulus from the intensity of the ratio between stimulation and rest periods. The masked ratio image was projected onto an anatomical T1-weighted (IR) MR image of the corresponding area, resulting in a colour image containing all the information. We reproduced position and orientation of the scans over several sessions with high accuracy (better than 2 mm) by computer matching of survey images. For each of our volunteers, we masked the two data sets with the highest response at $P<0.05$. These two masks were multiplied, resulting in the activated areas that could be reproduced in both sessions ($P<0.0025$).

The MEG measurements were performed with a 19-channel MEG system (University of Twente) at two locations: superior and at the left lateral side of the head, so at a total of 2x19=38 positions. Several measurements during one session were carried out applying two different interstimulus intervals: 500 ms (300 trials) and 1000 ms (200 trials). We made use of a 3D space tracker device and a special coil-cap on the head of the subject to determine the position of the head with respect to the gradiometer pickup coils. The positions of 3 marker points (left/right ear and nasion) with respect to the MRI co-ordinate system (visible in the MRI) were used to couple the co-ordinate systems of the two modalities. After averaging and low pass filtering (<100 Hz) the MEG data, we performed non-linear dipole fits using the Marquardt algorithm. We fitted rotating dipoles in several time intervals between 12 and 120 ms after the stimulus. The head model we used, was a realistic brain model extracted from the MRI of a different person. It was adjusted to match with the individual shape and position of the volunteer brains. We performed both single and double dipole fits. The program ASA (University of Twente) was used to perform these fits.

In order to combine both sorts of measurements we applied different strategies. First, we superimposed the positions of the current dipoles fitted to the MEG data, to the fMRI results. Second, we fixed dipoles at the locations found by fMRI and subsequently fitted their directions and strength to the MEG data. The residual variances gave an indication, how well and at which latencies these dipole positions match in the MEG data. Third, we used the fMRI results to estimate the number of active dipoles and use this information to obtain better dipole fits from MEG.

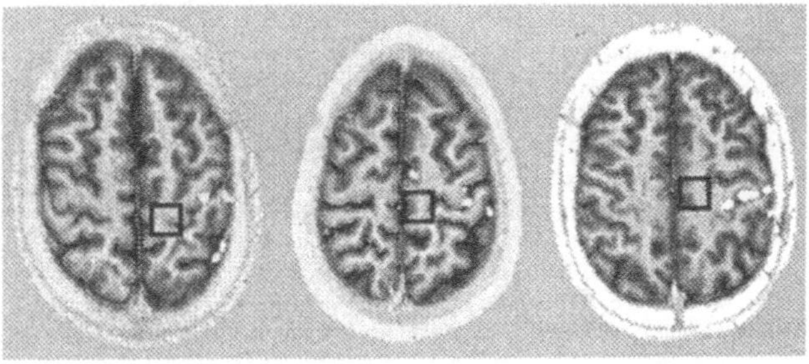

Fig. 2: One slice of reproducibly activated fMRI areas (white) for the three volunteers and projected position of one fitted rotating dipole (box) in time window 12-25 ms after the stimulus; 3D distances between closest fMRI area and dipole position ranges between 17-32 mm.

Results

The results from the fMRI data show that even with a stimulus as simple as the one employed in this study, differences in position and extent of the reproducibly activated areas for the different volunteers do occur (see Fig. 2). Also the intensities of the activated areas differ between two session with the same volunteer. The significant areas exhibit a change in signal amplitude between 1% and 8%. It seems that the larger changes are mostly due to the vein draining the cerebral tissue adjacent to the central sulcus. The cortex shows intensity changes up to 6%. With the MN nerve stimulation, there appears activity in both walls of the contralateral central sulcus. We observed also areas with significant signal decrease (1-2%) in both the ipsilateral and contralateral hemisphere of the brain. These locations are scattered and not reproducible among different session with the same volunteer. One of the volunteers consistently showed a large area of signal increase in the contralateral medial grey matter, presumably the supplementary motor area (SMA), which is remarkable for a sensory stimulus. Also the other volunteers had several small reproducible areas of signal increase outside the contralateral central sulcus.

The results of the analysis of the MEG data also show differences between the three volunteers. The time courses of the MEG data are consistent between the different measurements and volunteers. The location for the fitted dipoles vary among the volunteers (see Fig. 2). The qualities of the fits vary as well. For example, a fit of a rotating dipole in the time interval 12-25 ms yielded residual variances (RV) of 7-40 %. In the case of the high RV, the data could be explained much better with 2 dipoles (RV 11%). In all cases, the RV was the lowest in the time interval 45-70 ms for a single dipole fit. For other time intervals, a better fit could be achieved with more than one dipole.

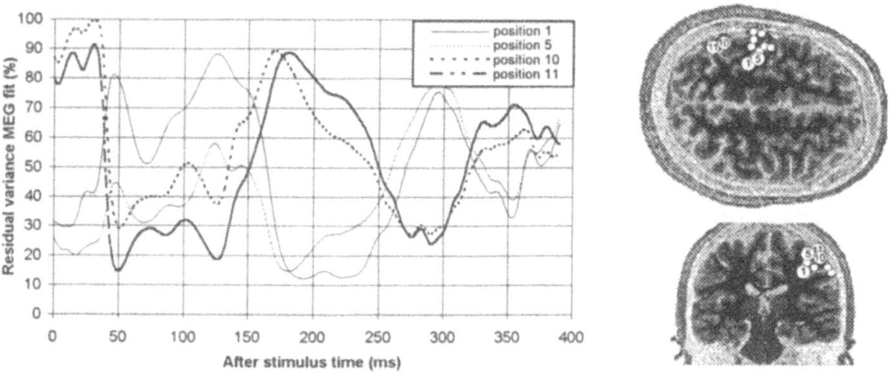

Fig. 3: Residual variance for a single rotating dipole fit in MEG data with a fixed position in four reproducible fMRI areas for different latencies. Left: anatomical positions of the fMRI areas

Results independently obtained from fMRI and MEG (dipole fit) analysis are superimposed in Fig. 2. The distance between the position of the fitted rotating dipole in the time interval 12-25 s and the nearest fMRI area is ranging between 17 and 32 mm.

Our next step was to use the non venous fMRI active areas as *a priori* known positions for a single rotating dipole, which is fitted to the MEG data. In Fig. 3 can be seen that in this volunteer two groups of positions (1+5 and 10+11) give alternately a low RV, so these two groups can tentatively be assigned to different areas which have a different sequence of activation. The first region is clearly in the primary somatosensory cortex, which is activated at short latencies. About the second region we do not wish to speculate.

Finally, we performed different multiple dipole fits with the following conditions.

1. Three rotating dipoles with their position fixed at the three fMRI positions with the largest distance between each other.
2. Three rotating dipole with their position fitted to the MEG data, while the initial guess (which is always necessary for a dipole fit) was extracted from the fMRI positions.
3. Three rotating dipole with their position fitted to the MEG data and random initial guess.

The results can be seen in Fig. 4-6. The fit with the overall lowest RV is found with the second condition. The change in position of the three dipoles when fitted (condition 2) can be seen in Fig. 5. This fit appears not to change very much the mutual position of the dipoles. This shift is ranging from 15-18 mm. In Fig. 6 one can see that the magnitude of the three dipoles have a different course in time. It seems that there are at least two major areas (represented by dipoles 1 and 2). Dipole 3 exhibits a much smaller contribution to the MEG data, which could be due to a radial orientation of the dipole.

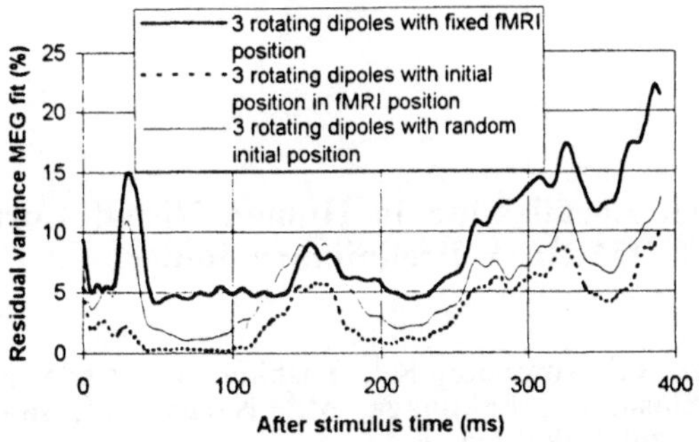

Fig. 4: RV for MEG fit with three rotating dipoles with fixed and fitted position and with random initial guess

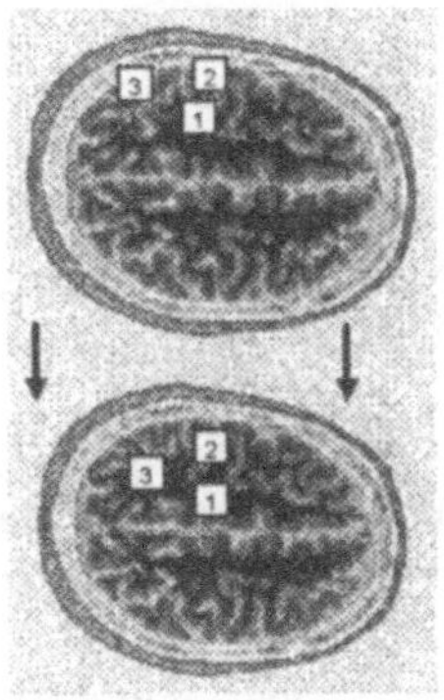

Fig. 5 Change of projected position of dipole before (fMRI position; top) and after (bottom) position fit

Discussion

First of all we have to consider that the location of the active areas in fMRI, which is linked with the oxygen consumption by the nerve cells, and the location of activity from MEG data, which is linked to electric activity of the nerve cells, can differ. Keeping this in mind we tried to combine the two modalities. Our study showed that the differences in location between the fMRI areas and the dipole positions of the MEG range between 17 and 32 mm , which is rather large. In a similar study, John A. Sanders et al [3] compared the location for motor tasks. They found differences of 7-14 mm.

There are three possible reasons for the differences in location we found. First, the underlying brain processes detected by the two modalities may indeed have different locations. Second, matching the two co-ordinate systems may introduce an error up to 4 mm. Third, there may be a localisation error in one of the modalities. Looking at both techniques, MEG has presumably the largest spatial error (volume conductor model, positioning gradiometers, etc.). In our multiple dipole fits, we fitted three dipoles to the MEG data with an initial guess at the fMRI locations. In this fit we saw that the mutual position of the locations as extracted from the fMRI results was largely preserved after fitting to the MEG data (Fig. 5). Apparently we are seeing the same sources in both modalities. Considering that with MEG the depths of the sources is difficult to estimate, we might attribute the observed position differences to inaccuracy of the MEG based dipole fit. This leads us to the conclusion that, when using the spatial information of the fMRI for the position and number of active areas and the temporal information of the MEG for the temporal behaviour of these areas, we may obtain better results than with each of the modalities alone. Further research is of course necessary to substantiate these preliminary results.

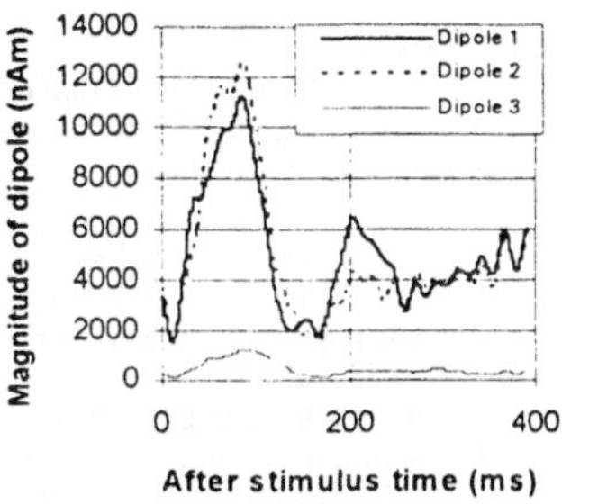

Fig. 6: Magnitude of the three position fitted rotating dipoles of Fig. 5

Another interesting phenomenon appeared in this study. Even with a simple stimulus such as MN, there are a lot of differences in responses among the volunteers in spite of similar conditions. These differences were present in both fMRI and MEG

References

[1] Orrison, W.W., Lewine J.D., Sanders, J.A., Hartshorne, M.F.: Functional Magnetic Resonance Imaging, Functional Brain Imaging, 1995: 145-181

[2] Brooks, R.A., DiChiro, G.: Magnetic resonance imaging of stationary blood: a review, Med Phys 14, 1987: 903-913

[3] Sanders, J.A., Lewine, J.D., George, J.S. et al: Correlation of fMRI with MEG, Proceedings of 12th SMRM Annual Meeting, New York, 1993: 1418.

Comparison of Functional Localization in Human Visual Cortices Using MEG and fMRI: A Preliminary Report

Yoshikawa, K.[1], Takanashi, Y.[1], Iwamoto, K.[1], Yoshida, Y.[3], Ueda, M.[3,4], Tanaka, C.[5], Umeda, M.[5], Ebisu, T.[5], Fukunaga, M.[5], Naruse, S.[2], Sato, H.[6] and Nakajima, K.[1]

[1]Department of Neurology, Research Institute for Neurological Diseases and Geriatrics, Kyoto, Japan; [2]Department of Neurosurgery, Kyoto Prefectural University of Medicine, Kyoto, Japan; [3]Keihanna Research Laboratory, Shimadzu Corporation, Kyoto, Japan; [4]Superconducting Sensor Laboratory, Chiba, Japan; [5]Department of Neurosurgery, Meiji College of Oriental Medicine, Kyoto, Japan; [6]GE-Yokogawa Medical System, Tokyo, Japan

Introduction

The application of computer technology to functional magnetic resonance imaging (fMRI) and magnetoencephalography (MEG) has enabled visualization of human brain function in vivo and has revolutionized clinical neurology [1, 2, 3, 4, 5].

Although MEG can identify neuronal activity directly with higher temporal resolution in the millisecond range, this technique is limited because it requires an estimate of current distribution and an assumption of the number of electrical current dipoles associated with neuronal activities. Moreover, we have to superimpose the estimated data appropriately on computer tomography or magnetic resonance imaging of the human brain to decide the anatomical localization.

On the other hand, fMRI has high spatial resolution and can estimate multiple activated areas with ease. But it's temporal resolution is not better than 1second.

We had reported the comparison of information from MEG and fMRI in response to the same visual stimuli.[10]. In the present study, we attempted to make use of the comparison in order to overcome the weaknesses of the electromagnetic approach.

Methods

Three normal right-handed subjects aged between 23 to 24 years with normal visual acuity participated in this study following informed consent. Each subject was presented with a checkerboard pattern at 1 Hz during MEG testing and at 2 Hz during fMRI. The visual stimulation sets were composed of left hemifield, right hemifield, and whole field patterns. The subject's task was to gaze at a fixation point in the center of the screen on which visual stimuli were presented by a rear projecting computer that generated images using a multimodal stimulator (ST10, Medlec, England). During MEG, the screen was positioned about 1 m from the subject who was sitting on a non-magnetic chair. During fMRI, the screen was reflected, and the images were presented by a mirror.

Magnetic visual evoked fields (VEF) were recorded in a magnetic-shielded room using a novel 129 channel vector neuromagnetic imaging system, which has been developed recently by Shimadzu Corporation, Japan. In the present study, the signals were band pass filtered (0.7-100 Hz); digitized at 1024 Hz; and averaged off-line. Signals were also recorded 50 msec pre-stimulus and 300 msec following the stimulus to determine the baseline. To identify the source of evoked responses, the equivalent current dipole (ECD) with the highest correlation value was found by a least-squares algorithm based on a one- or two-dipole model. Generator sites and vectors of ECDs were superimposed on appropriate axial, sagittal and coronal MRIs.

FMRI was obtained by a conventional 1.5 T MRI scanner (Sigma, GE) in the same subjects. FMR images were obtained by using a one-shot gradient echo planar imaging sequence (TR/TE/FA=4000 ms/50 ms/90 degrees) at rest and with stimulation alternately. Axial plane images of 5mm thickness were acquired. One examination consisted of 10 trials (off-on-off-on-off-on-off-on-off-on) with 400 total slices. The statistical significance of differences was evaluated by a cross correlation thresholding method. The activated areas were indicated by colored dots according to their correlation values.

Results

Typical VEFs were obtained at a latency of 100 to 170 ms from all subjects following three stimulation sets. The best correlation of ECD estimation was derived between 150 and 165 ms. Generator sites and vectors of the ECDs from a subject (O.M., 23-year-old, man) were superimposed on appropriate axial, sagittal, and coronal

MRI scans (right half of Fig. 1, 2 and 3). The activated areas in the functional MRI study of the same subject were visualized in the left half of Figs. 1, 2 and 3.

On hemifield stimulation (Figs. 1 and 2), the ECD generator site was localized in the calcarine sulcus contralateral to the stimulated field, which corresponded anatomically to the activated area in the fMRI study, respectively.

On whole field stimulation, a single generator site was localized in the right occipital lobe with a correlation of 83.7% based on a one-dipole model, while two generator sites were localized in subcortical areas of the occipital lobe bilaterally with a correlation of 94.4% based on a two-dipole model (Fig. 3). In fMRI, activated areas with a high cross correlation coefficient were distributed along the medial sides of the occipital lobes bilaterally as seen in the left half of Fig. 3. However, these areas did not correspond to the ECD sites determined by MEG.

Discussion

Our results revealed that, in each hemifield visual stimulation, the generator site of the ECD based on a one-dipole model corresponded well to the activated visual cortical areas in our fMRI study [6, 7, 8]. However, with whole field stimulation, ECD generator site using a one- or two- dipole model did not correspond to the activated areas determined by fMRI. The fMRI data appeared to represent a simple summation of those obtained with right- and left- hemifield stimulation.

In analyzing the neuromagnetic data, we did not have any criterion to objectively evaluate the validity of the predicted ECD generator site except for the correlation value. Our data, especially that obtained with hemifield stimulation, indicate that the combination of MEG and fMRI may confirm the accuracy of ECD localization.

Furthermore, the MEG data are limited by the need to predict the number of ECDs associated with neuronal events. However, in the present study, we assumed that two dipoles, rather than one, should be estimated prior to ECD analysis in whole field stimulation because at least two areas were activated during fMRI with the same stimulation. FMRI may predict the number of ECDs to be estimated in MEG data analysis. This is the second advantage of combination of MEG and fMRI.

Although a relatively high correlation was derived in a two-dipole model, the estimated generator site of the ECD was found to be different from that activated in fMRI. This discrepancy may be explained by the mutual interaction between the two dipoles which were evoked simultaneously and located very closely to each other [9, 11]. We do not have an alternative explanation for this discrepancy between estimated generator sites in MEG and activated areas in fMRI at present. Future studies are required to resolve these localization problems.

Based on this preliminary report, we hypothesize that localization of neuronal events in MEG, properly confirmed by fMRI, will contribute to the research of higher human brain function.

References:

[1] Kwong, K.K., Belliveau, J.W., et al. Dynamic magnetic resonance imaging of human brain activity during primary sensory stimulation, Proc. Natl. Acad. Sci. USA, 1992, 89: 5675-5679.

[2] Ogawa, S., Tank, D.W., et al. Intrinsic signal changes accompanying sensory stimulation : Functional brain mapping with magnetic resonance imaging, Proc. Natl. Acad. Sci. USA, 1992, 89: 5951-5955.

[3] Cohen, D. Magnetoencephalography: evidence of magnetic fields produced by alpha rhythm currents, Science, 1968, 161: 784-786.

[4] Tayler, T.J., Cuffin, B.N., et al. The visual evoked magnetoencephalogram, Life Sci., 1975, 17: 683-692.

[5] Aine, C.J., Bodis-Wollner, I., George, J.S. Generators of visual evoked neuromagnetic responses, Advances in Neurology Vol.54: Magnetoencephalography, New York, Raven Press, 1990: 141-155.

[6] Holms, G. Disturbances of vision from cerebral lesions, with special references to the cortical representation of the macula, Brain, 1916, 39: 34-73.

[7] Phelps, M.E., Kuhl, D.E., Mazziotta, J.C. Metabolic mapping of the brain's response to visual stimulations: Studies in humans, Science, 1981, 211: 1445-1448.

[8] Sereno, M.I., Dale, A.M., et al. Borders of multiple visual areas in humans revealed by functional magnetic resonance imaging, Science, 1995, 268: 889-893.

[9] Tan, S., Roth, J., et al. The magnetic field of cortical current sources: the application of a spatial filtering model to the forward and inverse problems, Electroencephalogr. Clin. Neurophysiol., 1990, 76: 73-85.

[10] Takanashi, Y., Yoshikawa, K., et al. Comparisons of Two Different Functional Localizations in Human Visual cortices using MEG and fMRI, In: Book of Abstracts, Visualization of information Processing in the Human Brain: Recent Advances in MEG and Functional MRI, Tokyo, Tokyo Institute of Psychiatry, 1995.

[11] Lutkenhoener, B. A simulation study of the resolving power of the biomagnetic inverse procedure, Clin. Phys. Physiol. Meas., 1991, 12 (Suppl.A): 73-78.

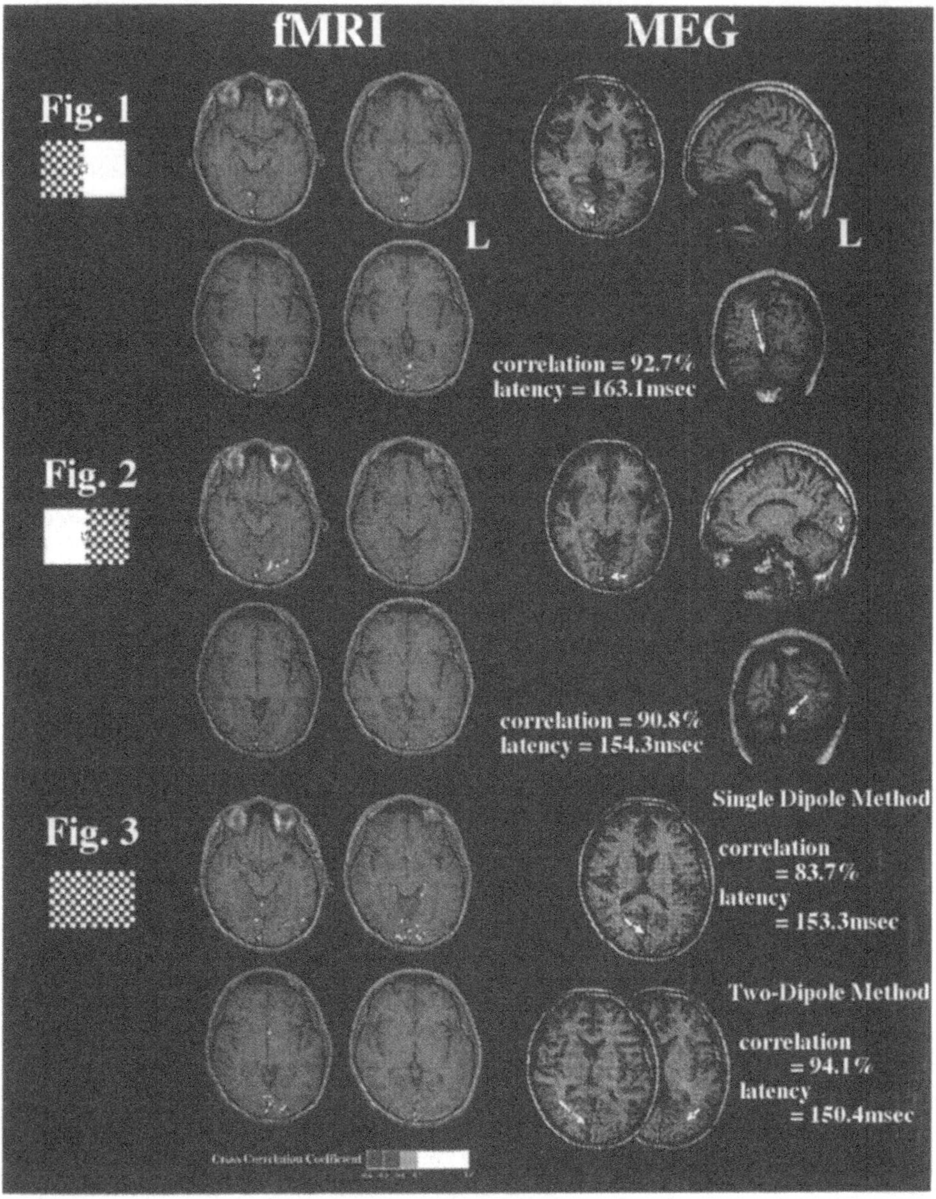

Figure Legends

Comparison of MEG and fMRI localization following visual stimulation. See details in the text.
Fig. 1: In left hemifield visual stimulation.
Fig. 2: In right hemifield visual stimulation.
Fig. 3: In whole filed visual stimulation.

GPSR Compliance

The European Union's (EU) General Product Safety Regulation (GPSR) is a set of rules that requires consumer products to be safe and our obligations to ensure this.

If you have any concerns about our products, you can contact us on ProductSafety@springernature.com

In case Publisher is established outside the EU, the EU authorized representative is:

Springer Nature Customer Service Center GmbH
Europaplatz 3
69115 Heidelberg, Germany

Batch number: 10844661

Printed by Printforce, the Netherlands